Perspectives
in
Mathematical Logic

Ω-Group:

R. O. Gandy H. Hermes A. Levy G. H. Müller
G. E. Sacks D. S. Scott

Ω-Bibliography of Mathematical Logic

Edited by Gert H. Müller

In Collaboration with Wolfgang Lenski

Volume IV

Recursion Theory

Peter G. Hinman (Editor)

Springer-Verlag Berlin Heidelberg New York
London Paris Tokyo

Gert H. Müller
Wolfgang Lenski
Mathematisches Institut, Universität Heidelberg
Im Neuenheimer Feld 288, D-6900 Heidelberg

Peter G. Hinman
Department of Mathematics
University of Michigan
Ann Arbor, MI 48109
U.S.A.

The series *Perspectives in Mathematical Logic* is edited by the Ω-Group of the Heidelberger Akademie der Wissenschaften. The Group initially received a generous grant (1970–1973) from the Stiftung Volkswagenwerk and since 1974 its work has been incorporated into the general scientific program of the Heidelberger Akademie der Wissenschaften (Math. Naturwiss. Klasse).

ISBN 3-540-15523-6 Springer-Verlag Berlin Heidelberg New York
ISBN 0-387-15523-6 Springer-Verlag New York Berlin Heidelberg

Library of Congress Cataloging in Publication Data
[Omega]-bibliography of mathematical logic.
(Perspectives in mathematical logic)
Includes indexes.
Contents: v. 1. Classical logic / Wolfgang Rautenberg, ed. - v. 2. Non-classical logics / Wolfgang Rautenberg, ed. - v. 3. Model theory / Heinz-Dieter Ebbinghaus, ed. [etc.]
1. Logic, Symbolic and mathematical - Bibliography. I. Müller, G. H. (Gert Heinz), 1923 - II. Lenski, Wolfgang, 1952 -. III. Title: Bibliography of mathematical logic. IV. Series.
Z6654.M26047 1987 [QA9] 016.5113 86-31426

Media conversion, printing and bookbinding: Appl, Wemding
2141/3140-543210

Dedicated
to
Alonzo Church

whose bibliographic work for the
Journal of Symbolic Logic
was a milestone in the
development of modern logic.

Table of Contents

Preface . . . IX

Introduction . . . XV

User's Guide . . . XXV

Ω-Classification Scheme . . . XXXIII

Subject Index . . . 1

- Automata, formal systems, and grammars . . . D03, D05 . . . 3
- Turing machines and related notions . . . D10 . . . 43
- Complexity of computation . . . D15 . . . 60
- Recursive functions and relations, subrecursive hierarchies D20 . . . 86
- Recursively enumerable sets and degrees . . . D25 . . . 113
- Other degrees; reducibilities . . . D30 . . . 129
- Undecidability and degrees of sets of sentences . . . D35 . . . 142
- Word problems, etc. . . . D40 . . . 157
- Effectiveness in mathematical structures . . . D45, C57 . . . 169
- Recursive equivalence types of sets and structures, isols . . D50 . . . 182
- Hierarchies . . . D55 . . . 187
- Recursion theory on ordinals, admissible sets, etc. . . . D60 . . . 205
- Higher-type and set recursion . . . D65 . . . 210
- Inductive definability . . . D70 . . . 215
- Abstract and axiomatic recursion theory . . . D75 . . . 219
- Applications . . . D80 . . . 225
- Proceedings, textbooks, surveys, and miscellaneous . . . D97, D98, D99 234

Author Index . . . 241

Source Index . . . 599

- Journals . . . 601
- Series . . . 621
- Proceedings . . . 626
- Collection volumes . . . 651
- Publishers . . . 659

Miscellaneous Indexes . . . 667

- External classifications . . . 669
- Alphabetization and alternative spellings of author names . . . 693
- International vehicle codes . . . 695
- Transliteration scheme for Cyrillic . . . 697

Preface

Gert H. Müller

The growth of the number of publications in almost all scientific areas, as in the area of (mathematical) logic, is taken as a sign of our scientifically minded culture, but it also has a terrifying aspect. In addition, given the rapidly growing sophistication, specialization and hence subdivision of logic, researchers, students and teachers may have a hard time getting an overview of the existing literature, particularly if they do not have an extensive library available in their neighbourhood: they simply do not even know what to ask for! More specifically, if someone vaguely knows that something vaguely connected with his interests exists somewhere in the literature, he may not be able to find it even by searching through the publications scattered in the review journals. Answering this challenge was and is the central motivation for compiling this Bibliography.

The Bibliography comprises (presently) the following six volumes (listed with the corresponding Editors):

I.	Classical Logic	W. Rautenberg
II.	Non-classical Logics	W. Rautenberg
III.	Model Theory	H.-D. Ebbinghaus
IV.	Recursion Theory	P. G. Hinman
V.	Set Theory	A. R. Blass
VI.	Proof Theory; Constructive Mathematics	J. E. Kister; D. van Dalen & A. S. Troelstra.

Each volume is divided into four main parts:

1) The *Subject Index* is arranged in sections by topics, usually corresponding to sections in the classification scheme; each section is ordered chronologically by year, and within a given year the items are listed alphabetically by author with the titles of the publications and their full classifications added.

2) The *Author Index* is ordered alphabetically by author, and contains the full bibliographical data of each publication together with its review numbers in Mathematical Reviews (MR), Zentralblatt für Mathematik und ihre Grenzgebiete (Zbl), Journal of Symbolic Logic (JSL), and Jahrbuch über die Fortschritte der Mathematik (FdM). We much regret that we were not able to include reviews from Referativnyj Zhurnal Matematika in this edition.

3) The *Source Index* gives the full bibliographical data of each source (journals and books) for which only abbreviated forms are used in the Author Index.

4) The *Miscellaneous Indexes* contain various further indexes and tables to aid the reader in using the Bibliography.

For a more detailed technical description of the Bibliography see the *Table of Contents* and the *User's Guide*.

The uniform classification of all entries is a central feature of the Bibliography. The basic framework is the 03 section of the (1985 version of the) 1980 classification scheme of Mathematical Reviews and Zentralblatt für Mathematik und ihre Grenzgebiete. However, this has been modified in a number of ways. Indeed, the 1980 scheme was designed for the classification of works written after 1980, whereas the majority of entries in the Bibliography come before this date. In some areas this has made the classification of older works difficult, and we have tried to cope with this by adding a few new sections and altering slightly the interpretation of others. We have not designated the classifications assigned to a work as primary and secondary, because of the difficulty in doing so in many cases. Each volume contains the full annotated classification scheme together with a description of its general features. In their *introductions* the Editors discuss specifically their interpretations of the classification sections falling in their respective volumes.

The Subject Index is another central feature of the Bibliography. Reading through this Index gives a *historical perspective* for each classification section and provides a rather quick overview of the literature in it. By browsing through the entries of a specific area the reader may be rewarded by finding things (literature, subjects, questions) he was not aware of or had forgotten.

An obvious question now is the extent to which one can rely on the *completeness* and *correctness* of the Bibliography and on the *accuracy of the classifications*. We comment on each of these aspects separately.

In an effort to be as complete as possible, we consulted all sources available to us and decided in favour of inclusion in doubtful cases (so that certainly some papers with little bearing on mathematical logic are listed here and there). As the historical starting point for the Bibliography we chose the appearance of Frege's *Begriffsschrift* (1879). A certain restriction on scope stems from our decision to concentrate on mathematical logic and in particular on those areas defined by the titles of the six volumes. A major source of material was provided by the review journals mentioned above; we used them both to identify publications in the less known journals and to find review numbers and other bibliographical data of items found in other sources. We also made use of various lists of literature contained in books, survey articles, mimeographed notes, etc. Some especially valuable newer sources were:

W. Hodges: A Thousand Papers in Model Theory and Algebra
M. A. McRobbie, A. Barcan and P. B. Thistlewaite: Interpolation Theorems: A Bibliography
D. S. Scott and J. M. B. Moss †: A Bibliography of Books on Symbolic Logic, Foundations of Mathematics and Related Subjects
C. A. B. Peacocke and D. S. Scott: A Selective Bibliography of Philosophical Logic.

Various strategies and crosschecks were used to ensure the completeness of the bibliographical data and in particular of the reviews mentioned above. For each item listed in the Bibliography we tried to include any translations, reprintings in alternative sources and errata, and to give cross references for a work appearing in several parts.

On the whole this Bibliography was compiled and organized for use by the practising mathematician; there is no claim that the most rigorous standards of librarianship are met.

It is hard to say how successful our striving for completeness was. This is especially true for the most recent literature. No 1986 items were included. We checked all the main journals in logic, the reviews in MR, Zbl and JSL and Current Mathematical Publications for literature published up to the end of 1985, but undoubtedly some gaps remain.

As for correctness, in any ordinary book we can tolerate a number of printing errors because of our knowledge of the language and the context, but, when one organizes data connected by (abstract) pointers in a computer program, almost every typing error has far-reaching consequences. Various consistency tests were used to check the program and the input data. There are, however, many other sources for mistakes and errors.

For some items our references contained incomplete or ambiguous information. Although we tried to complete the bibliographical data, this was often difficult, particularly in cases where, for example, the source was obscure or the publisher was given only by location. Another source of errors lies in the identification of author names. An author may publish using abbreviations of his first, his second or both of his given names. This is generally not a problem for authors with uncommon surnames, but if the surname is, e.g., Smith or Brown the possibility of misidentification arises. We may have identified two different authors or failed to identify two or more different forms of an author name.

It is unavoidable in a project of this scope that there will be errors, particularly in the classification, so perhaps it is worthwhile to explain briefly the process by which the classification was done. Items entered before 1981 were originally classified according to a scheme unrelated to the current one. To begin the conversion to the 1980 scheme we used the computer to change old categories to their new versions wherever there was a well-defined correspondence. Then every entry was checked and if necessary reclassified by hand. From 1981 each new entry was classified shortly after being entered in the database. For the most part this was done on the basis of titles, reviews, and other information, but without consulting the works themselves. This was necessary to preserve the finiteness of the enterprise, but it has inevitably led to errors, certainly in some cases egregious ones. These were constantly being corrected during the final editing process, but many will remain.

Although the Editors have to some extent used different strategies in classifying the entries falling into their respective volumes, finally a reasonable degree of uniformity has arisen. The user is referred to the Editors' introductions for further details on the classifying procedure.

A special apology goes to the native speakers of languages with diacritical marks. Our central difficulty was to get the right spelling of names used in different forms in such a variety of sources. In addition, entering diacritical marks in a computer introduces yet another source of errors; so they have almost all been ignored (the User's Guide and the Miscellaneous Indexes contain details of those that have been transliterated). We appreciate that, although the absence of, for example, accents in the text of a French title may not create undue problems, the lack of diacritical marks in author names is particularly unfortunate. We hope that this omission will not be too misleading.

The future

By its nature a bibliography has lasting value to the extent it succeeds in "completing the past". But it should also serve for some years as an aid to current research. We have various plans to extend the scope of the present Bibliography by including new areas such as universal algebra, sheaves, philosophical logic (subdividing the present volumes I and II appropriately), and philosophy of mathematics. The present six volumes cover only approximately 80% of the data on our computer files.

The possibility of extending the classification scheme by developing a so-called thesaurus system was discussed on several occasions. Certainly this would be desirable; to some extent *Alonzo Church* tried to create such a system in connection with his bibliography in the Journal of Symbolic Logic. However there are difficult scientific problems connected with the creation of such a system and their solution requires much time and expertise.

Another way to extend the Bibliography which would perhaps better serve the purpose of providing an overview of certain special areas would be to commission a series of survey papers to appear from time to time as, say, an additional issue of the Journal of Symbolic Logic; each paper would include an annotated listing of the literature taken from the Bibliography.

There are plans to establish a bibliographical centre for Mathematical Logic and adjacent areas. A central function of such a centre would be to collect information on all new publications (including mimeographed notes, theses, etc.) as well as to correct errors and omissions in the current data. It is hoped that all logicians would provide information concerning their own publications as they appear. A continuation of the Bibliography together with supplements (to appear periodically) would be prepared at the centre. We also hope to make available an on-line system. From these activities and the flow of information from the individual logician to the centre and vice versa a "living Bibliography" would emerge. This would provide a way to determine the main trends in the progress (or decline) of specified directions of work. So a centre would exist at which it would be possible to gain some oversight of the rapidly developing field of mathematical logic.

Acknowledgements

Work on the Bibliography started at the same time as the Ω-Group came into being, early in 1969. To begin with, index cards were used for storing bibliographical data; it was *Horst Zeitler* and *Diana Schmidt* who convinced me that we are living in the 20th century and that the data should be computerized. They, together with *Ann Singleterry-Ferebee,* first brought the Bibliography to a workable computerized form at the end of the seventies. In this period I also had the help of *Ulrich Felgner* and *Klaus Gloede,* in particular in classifying the literature. At about this time, others contributed in many useful ways. In particular, important problems of principle were highlighted by a long list of intriguing questions from *Dana Scott:* "How do you classify this or that item . . .?" *Robert Harrison* worked faithfully collecting data for the Source Index.

The second period, beginning in the early eighties, was characterized by the programming necessary to manage the data. This was carried out by *Ulrich Burkhardt* (†) and *Werner Wolf* and finally by the *outstanding work* of *Rolf Bogus.* In this period we also changed the classification system and for this I had the continuous and intensive help of *Andreas Blass* and *Peter Hinman.* In addition, both of them, together with *Heinz-Dieter Ebbinghaus,* gave me much advice about organization and technical arrangements. Over the last four years the work of large groups of students has been essential for collecting reviews, entering corrections and new items into the computer, etc. Again and again I have been overwhelmed by their idealism and energy. Among them I wish to mention particularly the continuous help of *Elisabeth Wette* and *Ulrike Wieland.*

The Bibliography would not have reached publishable form without the work of my collaborator *Wolfgang Lenski* (in the second period). It would have been unthinkable for me to interfere anywhere in the process of the growth of the Bibliography without discussing the matter with him beforehand. He has accumulated a detailed knowledge of every aspect of the project and has devoted his talents for many years to the common enterprise.

My secretary, *Elfriede Ihrig,* has willingly assisted in the work of the Ω-Group and the Bibliography from the beginning, over many years, filled with ups and downs and with all kinds of tasks. She has always maintained her warmhearted balance.

To all I express my personal warm thanks!

The *Journal of Symbolic Logic* sent information concerning papers for which reviews were never published. We also acknowledge permission to use computer tapes with lists of literature covering certain periods of time from *Mathematical Reviews* and from *Zentralblatt für Mathematik und ihre Grenzgebiete.*

Yuzuru Kakuda and *Tosiyuki Tugue* collected and prepared the Japanese literature for us. *Petr Hajek* and *Gerd Wechsung* helped us with updating the bibliographical references of so many sources not available to us. *Mo Shaokui* corrected data on the Chinese literature and added items of which we were not aware.

The Editors filled many gaps, corrected all mistakes which came to their attention and undertook the burden of checking - and changing if necessary - the classification of the entries in their special areas. Here again I would like to mention *Rolf Bogus* and *Wolfgang Lenski* who organized the enormous exchange service for the transfer of literature among the Editors and the inputting of the many changes and corrections. *Andreas Blass* and *Peter Hinman* were also instrumental in this exchange; their preliminary classification of each item added to the Bibliography during the final editing process meant that the Editors had mainly to look at items inside their own areas. *Jane Kister* read through the whole Source Index correcting mistakes and suggesting valuable changes in it.

In collecting and organizing the data for the Bibliography we have received much help from various sources, and especially in letters from colleagues all over the world, containing information and suggestions. I apologize for being unable to answer them all individually, but all were read carefully.

We thank all those concerned.

As everybody can guess, the whole enterprise was indeed expensive. *Financial support was provided by the Heidelberger Akademie der Wissenschaften in the framework of the Ω-Group project.*

Special thanks go to the firm APPL, who transformed our computer tapes to the present printed form, and to the editorial and production staff of SPRINGER-Verlag for their continuous help, notably in the traditionally fine realization of the six volumes.

Finally, through working so many years on this project I have come to understand and appreciate more and more the immense work of Alonzo Church in building his Bibliography of Logic and its adjacent areas, together with a detailed classification, that is contained in so many volumes of the Journal of Symbolic Logic. Understanding comes from doing.

Introduction

Peter G. Hinman

As is evident from the size of this volume, the literature of Recursion Theory is large; that we have found it expedient to divide it into twenty-one subfields indicates that it is diverse. This introduction is intended as a brief guide to these subdivisions with some indication of how they fit into their historical and mathematical contexts.

Lest the reader have inflated expectations, let me issue an early disclaimer: what follows is neither a history of Recursion Theory nor a survey of the highlights of the field. The former remains to be written while several of the latter will be found under [D98] in the Subject Index. It is rather an eclectic mixture of mathematics, history, and taxonomy whose goal is to describe the subareas of Recursion Theory as described by the Classification Schema. This is best accomplished in some cases via a chronological development, in others by a discussion of the current situation. Since the Classification Schema is the key to the organization of the book, we shall refer freely to its categories, e.g. [D25], and expect the reader to consult the schema for their definitions.

The proper starting point for Recursion Theory is probably Dedekind [1888] in which he proved that (what is now called) the schema of primitive recursion does define a unique function. Over the next forty years there was some interest in generalizing this schema to produce more complex recursions, but it was not until Ackermann [1928] and Sudan [1927] that a recursion schema was exhibited whose solution is not primitive recursive. As so often happens, once this first step was taken, others followed more easily, and there was an active development in the first half of the 1930's of hierarchies of complexity of recursions. A good reference is Péter [1951 (46603)].

Interest in recursive functions was heightened by their use in Gödel [1931 (15052)] as a key element in the proofs of the incompleteness theorems. Gödel used the German *rekursiv* to describe his class of functions, but they are exactly the primitive recursive functions. The next main development was Kleene's discovery that all of these functions, and indeed all functions definable by any known sort of recursion, are definable in Church's λ-calculus. This led Church in 1934 (published in Church [1936 (02127)]) to state his Thesis : the λ-definable functions are exactly those which are effectively (algorithmically) computable.

Because λ-definability is not a particularly intuitive notion, Church's Thesis needed sorely the evidence provided in 1934–1936. This consisted of two additional characterizations of the class of λ-definable functions which seemed to capture better the flavor of recursiveness and effective computability. The first was Gödel's adaptation of a suggestion of Herbrand that any system of functional equations of a certain kind which has a unique solution be considered a definition of a "recursive" function. Gödel added the restriction that equations giving values of the defined function be formally deducible from the defining system. The second was Turing's [1936 (13723)] notion of function computable by an abstract machine following a program of very elementary steps. The discovery by Kleene and Turing that these three classes of functions coincide – λ-definable, Herbrand-Gödel recursive, and Turing computable – marks the real beginning of Recursion Theory.

It is perhaps somewhat curious that such a cornerstone of the field as Church's Thesis is not a mathematical statement and is not subject to mathematical proof.

The Thesis asserts that a mathematically well-defined class of functions coincides with an intuitively understood and, from the mathematical point of view, poorly specified class of functions. Half of the Thesis, that every recursive function is algorithmically computable, is provable from a few intuitively natural hypotheses about algorithms. The converse inclusion, however, rests on assumptions about the totality of algorithms and has been the object of a good deal of philosophical debate. When an author states a theorem to the effect that a problem is algorithmically solvable and proves it by producing a recursive function which does the job, only the first and less problematic half of Church's Thesis is invoked. As we shall see below, however, the most interesting results of Recursion Theory have been of the form that a problem *cannot* be solved by a recursive function. The passage from this precise result to the much more interesting assertion that the problem is not algorithmically solvable makes essential use of the problematic half of Church's Thesis.

Over the years, however, evidence in favor of Church's Thesis has accumulated to the point that it is now almost universally accepted. An important component of this evidence is the existence of a large number of quite diverse characterizations of the class of recursive functions. This evidence together with other studies of the structure of this class form much of the subject matter of [D20]. Interestingly, the original goal of classifying all recursive functions according to a measure of the complexity of the recursions needed to define them remains unmet. [D15] is also concerned with subclasses of recursive functions but from quite a different perspective (see below).

Over the next few years - the late 1930's and early 1940's-several extensions of the notion of recursiveness began to be studied: partial recursive functions [D20], recursive enumerability [D25], relative recursion and degrees [D30], and the arithmetical hierarchy [D55]. Partial recursiveness arises naturally in all three of the basic formulations by relaxing the requirement that derivations in the λ- or equation calculus or Turing machine computations exist for all arguments. The relationship of recursive functions to formal systems via Gödel numbering, begun in Gödel [1931 (15052)] and continued in Church [1936 (02127)], led to the notion of recursive enumerability as an abstraction from the generative process of (formal) deduction. Turing [1939] introduced the idea of a Turing machine with an "oracle" and the corresponding notion of one recursively unsolvable problem being reducible to another.

Post [1944] was the first study devoted to reducibilities among recursively enumerable (r.e.) sets. This work introduced the notion of degree of unsolvability and posed what became known as Post's Problem: does there exist more than one degree among non-recursive r.e. sets. All such sets then known were of the highest possible degree and Post and others tried several techniques for constructing r.e. sets which would be "simpler" and therefore of lower degree, but still not recursive. Although this approach was not powerful enough to solve the problem, it initiated what is probably the most intricate theory among those spanned by this volume.

Post's Problem inspired much effort, but it turned out that its solution required a major new idea, the priority method. This was introduced independently by Friedberg [1957 (04635)] and Muchnik [1956 (09612)], and it soon proved to be a tool of extraordinary complexity and power. Indeed, to this day some elaboration of the priority principle is one of the main ingredients of nearly every proof of a theorem about r.e. sets. The solution was that there are many non-recursive r.e. degrees and, as it developed over the next few years, that they form a partially ordered set with many interesting properties. A landmark result was that of Sacks [1964] that this ordering is dense.

Degrees of arbitrary (not necessarily r.e.) sets [D30] were first studied in detail in Kleene-Post [1954]; this subject has also enjoyed a continuous history of activity since its birth. The properties of arbitrary degrees are quite different from those of r.e. degrees. For example, they are far from densely ordered and one of the major problems has been to discover which partial orderings or lattices can be embedded as initial segments.

The arithmetical hierarchy [D55], discovered independently by Kleene [1943] and Mostowski [1947], provided another way to compare the complexity of non-recursive sets and relations: by the quantifier prefix (applied to a recursive relation) needed to define them. Both authors were motivated by the fact that r.e. sets are exactly those definable using only existential quantifiers, but Mostowski was additionally inspired by the analogy with the projective hierarchy of Descriptive Set Theory, which had been developed in the 1920's and 1930's. He knew that there was one flaw in the analogy: the class of projective sets definable using only existential quantifiers (over real numbers - the analytic sets) has the separation property while the class of r.e. sets does not. It remained for Addison [1954] and [1958 (00171)] to show that the correct analogy is with the Borel hierarchy where the fundamental operation of countable union corresponds more closely to the existential number quantifier. This discovery led to a very fruitful interaction between Recursion Theory and Descriptive Set Theory which continues to this day and accounts for a large overlap between the categories [D55] and [E15].

The developments discussed so far are to a large extent elaborations of the definability notions inherent in two of the three original characterizations of recursiveness. The characterzation via Turing machines is based on different intuitions and these, too, were elaborated in the following years. Indeed, Turing machines were only the first of a large family of abstract computing models called automata. The variety of types which have been considered is staggering, but from today's perspective, most can be seen as idealizations of computers with differing sorts of input and output devices, processing units, and storage. Input and output are usually expressed as a finite sequence of symbols from an alphabet. The processing unit "reads" the input and, under the control of a "program", passes through a finite sequence of states, possibly writing to and reading from its storage along the way, and produces its output.

A class of machines is defined by specifying these parameters in more detail: how are input and output represented, how many states are available, what sorts of programs are admitted, how large is the storage device and how is it addressed, how many steps is the computation allowed, etc. A basic distinction is made between machines with bounded [D05] and unbounded [D10] storage capacity. With each class of machines is associated the class of functions computed, although this too is described in a variety of ways. A common variant allows only two outputs and describes the machine as recognizing the "language" consisting of those input sequences which lead to a designated output.

This notion of language is distinct from the general logical notion of formal language in that only a syntax is specified; there is (usually) no semantics or interpretation. Indeed, some varieties of automata are really best described as "grammars" for such a language; the strings of the language are those which result from applying given rules to members of an initially given set. Some of the earliest undecidability results were associated with simple systems of this sort [D03], and the line between these and explicitly grammatical systems [D05] is a fine one. At the other end of the scale, [D05] overlaps [B65] when the more formal aspects of grammars for natural languages are studied. Because sets of strings with the concatenation operator often form a semigroup, these sections have also entries in common with [D40].

With the growing importance of computers in the late 1950's and 1960's, the role of abstract automata as models for practical computing devices came to the fore. Although derivations in the λ- and equation calculi and computations of Turing machines and other automata are normally required to be finite, there are no *a*

Recursively enumerable sets and degrees [D25]

This classification is reserved for results about r.e. sets and degrees in the line descending principally from Post [1944]. The degrees may be Turing, many-one, bounded truth-table, or whatever, using recursive functions, but not polynomial [D15] or primitive recursive [D20]. Special properties of co-r.e. sets are also usually here, but r.e. sets seen merely as the first step in the arithmetical hierarchy are more often in [D55]. Generalizations of recursive enumerability to ordinals [D60], higher types or sets [D65], or general structures [D75] are generally not classified [D25] unless they are closely tied to traditional r.e. results or methods. Similarly, recursive enumerability of an undecidable problem does not earn it a [D25] classification unless some real use is made of the special properties of the class, for example to show that a class of r.e. degrees is represented by instances of the problem.

Other degrees; reducibilities [D30]

A somewhat arbitrary decision was made to include here degrees and reducibilities coarser than the traditional ones - (hyper)arithmetical, Δ^1_n, α-recursive, or constructible - but not the finer ones of [D15]-[D25]. The central topic is Turing degrees of arbitrary sets; other degrees are usually identified by additional classification numbers [D55]-[D65] or [E45].

Undecidability and degrees of sets of sentences [D35]

The central theme here is undecidability as it applies to formal languages, most often first- or second-order predicate languages, possibly enriched with other quantifiers or operators. This is meant to exclude "languages" which are merely formal strings of symbols without semantics [D05] and word problems for algebras [D40]. Other undecidable problems are classified under the type of problem, if it is logical, otherwise in [D80]. Papers which contain both decidability and undecidability results have both [B25] and [D35]. For uniformity, all works relating to Hilbert's tenth problem are classified here unless they are almost entirely concerned with representations of r.e. sets [D25]. A companion [B20] usually indicates treatment of a reduction class of the predicate calculus.

Word problems, etc. [D40]

The results here, although most could be formulated as being about sets of sentences, have a different flavor from those of [B25] and [D35]. All are about algebraic theories, and most involve a specific model or set of models. In contrast with the [B25]/[D35] split, we have here not segregated decidability and undecidability results.

Effectiveness in mathematical structures [D45] and [C57]

The variety of topics included here has been discussed above. Studies where the focus is on (Gödel) numbering of syntactic objects [F30] are not included unless the semantics is also brought into play, as, for example, with recursive element types. The line between the two classifications is delicate and is often one of tone, with [C57] designating a more model-theoretic orientation and [D45] indicating greater attention to the recursion-theoretic aspects of the problem.

Recursive equivalence types of sets and structures, isols [D50]

This is a narrowly conceived category, and there is generally little doubt when an item belongs here.

Hierarchies [D55]

The hierarchies admitted here all extend beyond the recursive; lower level ones are covered by [D15] or [D20]. There is a large overlap with [E15] (Descriptive Set Theory). We have attempted to include only works which in some fashion use the recursion-theoretic nature of the hierarchy, either in the statements or proofs of results. Since the hierarchies of Descriptive Set Theory can usually be formulated in a more or less recursion-theoretic way, this has involved a good deal of arbitrary choice. Hierarchies of definability in abstract set theory are excluded [E47].

Recursion theory on ordinals, admissible sets, etc. [D60]

Most papers in this category involve some notion of generalized finiteness; whether the underlying objects are ordinals or sets is less important. Use of ordinals just to measure levels of a hierarchy does not justify the inclusion of an item here.

Higher-type and set recursion [D65]

The cutoff between [D65] and [D20] is determined by whether the arguments of functions are at least as complex as type-2 functionals, which entails that computations are infinite. In most instances there is a clear distinction of methods between this category and [F10].

Inductive definability [D70]

Inductive definitions have been a part of mathematics for centuries and are a central technique in logic, but only since the 1950's have they been an explicit object of study. This category is intended for works which are related to this latter trend, either by results or methods. In general, this means that only inductions involving an infinitary rule, and thus having closure ordinal greater than ω, qualify. However, a number of historical precursors have been included. Of course this is still Recursion Theory, so formal theories of inductive definitions are banished to [F35].

Abstract and axiomatic recursion theory [D75]

"Abstract" here means "over structures not considered in preceding categories". The flavor is usually one of getting at the underlying assumptions of computability in a context where our intuitions are less influenced by historical practice. For want of a better place we have put here also algebras of recursive functions (also [D20]).

Applications [D80]

We group here all applications of recursion-theoretic notions to areas of mathematics not specifically covered above: topology, differential equations, graph theory, etc. This includes decidability questions but also effectiveness of constructions and operations.

Proceedings, textbooks, surveys, and miscellaneous [D97]-[D99]

Collections and proceedings are included here only when Recursion Theory is the main subject matter; otherwise they appear only in the Source Index. Textbooks and surveys, on the other hand, are listed in this section if there is substantial recursion-theoretic content regardless of the other topics covered. In all cases, items which concern mainly a specific subarea should have also that designation. Bibliographies of parts of Recursion Theory have been put in [D99]; other items in this class either have only a tangential connection to Recursion Theory or for lack of information could not be more accurately classified.

Finally, I want to thank the people who have made essential contributions to the preparation of this volume. I will not repeat the acknowledgements made in the General Preface except in a few special cases, but most of the people mentioned there were directly of great help to me and I support enthusiastically the thanks given them.

First must come Gert Müller without whom there would be no Bibliography. His tireless efforts over more than ten years have been an inspiration for us all. His genius for organization and unerring scientific vision have been the determining factors in the success of the enterprise.

Wolfgang Lenski played a tremendous role in the latter years in organizing masses of data into forms we could understand and process and was always ready to answer questions and provide special information at short notice. My fellow editors were a constant source of encouragement and helpful criticism. A generous grant from the Bell Telephone Laboratories provided computing equipment and the University of Michigan provided computer time; both greatly facilitated the later stages of the editing. Finally, special thanks go to Elizabeth and Mira who put up (almost) uncomplainingly with my frequent disappearances both to Heidelberg and into my study during the last few years.

User's Guide

Wolfgang Lenski & Gert H. Müller

§1. Introduction

After some opening remarks, the organization of this Guide follows the main division of the volume: *Subject Index, Author Index, Source Index, Miscellaneous Indexes*. For each part we give first a general explanation followed by a more detailed description of typical entries in the index in question. The reader will probably find the User's Guide most helpful when he comes across an unclear entry in the Bibliography: he can then turn directly to the corresponding section in this Guide for an explanation of the abbreviations and conventions used.

§2. General remarks

The main languages of the Bibliography are English, French, German, Italian, and Spanish. For other languages translations (of titles, names of sources, etc.) are used - with some few exceptions in cases for which we had no translation. These translations were taken from various available sources or made by the Editors.

For practical reasons, all entries are in the Roman alphabet and diacritical marks have not been used. Thus, for languages other than English certain conventions have been adopted.

The *transliteration of Cyrillic* names and titles, the treatment of *diacritical marks* and the *alphabetization and alternative spelling* of author names are explained in detail in the Miscellaneous Indexes.

The *abbreviations* of *sources* were either taken from one of the various reviewing journals or invented by us. Although we had to abbreviate long titles, we hope that in most cases the abbreviation will suggest the full title in a sufficiently understandable way. How successful we were is left to the user to decide.

The *review numbers* given with the entries in the Author Index are from *Mathematical Reviews* (MR), *Zentralblatt für Mathematik und ihre Grenzgebiete* (Zbl), *Journal of Symbolic Logic* (JSL), and *Jahrbuch über die Fortschritte der Mathematik* (FdM). We made a serious attempt to include all reviews of any given item but we have doubts concerning our success. We also tried to avoid listing two reviews for a given item in those cases in which the second "review" simply points to the original review and does not give any additional information.

In case of *multipart publications* pointers are given to the other parts, as far as they are known, in the *Remarks* to the publication in question. It is not always the case that the different parts of a publication all have the same classifications. Thus it may happen that, for example, part I has a classification in this volume and part II does not. In this case the Remarks for part I indicate the author(s) and year of publication of part II. The user will need to consult the other volumes for further bibliographic information on part II.

The general way to search through the Bibliography is to use certain *pointers:* From the Subject Index to the Author Index the pointer is [Author, Year, Title]; from the Author Index to the Source Index the pointers are 5-digit codes; e.g. (**J** 1234) is a code which sends the user to the J-section of the Source Index.

A *word* of *caution*: In order to use the Bibliography for quotations in future

§6. Miscellaneous Indexes

This part contains the following indexes:

1. External classifications
2. Alphabetization and alternative spellings of author names
3. International vehicle codes
4. Transliteration scheme for Cyrillic

In each case a description of the contents and use are given in the corresponding introductory texts.

Ω-Classification Scheme

Andreas R. Blass
Peter G. Hinman

The classification scheme used for the Ω-Bibliography is a modified version of the section "03: Mathematical Logic and Foundations" of the 1985 Mathematics Subject Classification of *Mathematical Reviews* and *Zentralblatt für Mathematik und ihre Grenzgebiete*. For the sake of uniformity we have labeled all sections with a letter followed by a two-digit number; the prefix 03 is superfluous and therefore omitted. This decision has led to the creation of new sections to replace 03-01 through 03-06 (cf. X96-X98 and A10) and several sections with prefix other than 03 which have substantial logical content. Examples of the latter sort are B70 (to replace 94C10) and B75 (to replace the "logical part" of 68B10) (68Q55 and 68Q60 since 1985).

An important category of differences between the two schemes arises from the fact that whereas the MR/Zbl system is intended to classify works written after 1980, the majority of entries in the Ω-Bibliography were written before 1980. The subject matter of Mathematical Logic has, of course, changed immensely over the years, and today's categories are not always sufficient to distinguish properly important lines of earlier research. To deal with this problem we have added a few new sections (e.g. B22, B28, B65, C07, E07, and E47), renamed others (e.g. B35, C35, and E10), and altered slightly the interpretation of others (e.g. B25 and D65). To aid the reader in learning our conventions we have added descriptors to the section names. Topics preceded by a + (−) sign are specifically included (excluded) from a section. When this is in conflict with current MR/Zbl practice, this fact is also noted.

A

A05 Philosophical and critical

A10 History, Biography, Bibliography
MR uses 03-03 and 01A for history and biography
MR puts bibliography under specific fields.

B GENERAL LOGIC

B03 Syntax of logical languages

B05 Classical propositional logic and boolean functions
+ Axiomatizations of classical propositional logic
+ Boolean functions (machine manipulation is also in B35); MR puts these in G05 and in 06E30 and 94C10.
− Fragments of propositional logic: see B20
− Switching circuits: see B70; MR also uses 94C10

B10 Classical first-order logic
+ Many-sorted logic
+ Syntax and semantics up to the Completeness Theorem
− Model theory: see Cn, particularly C07
− Proof theory: see Fn

B15 Higher-order logic and type theory
+ Higher-order algebraic and other theories
– Higher-order model theory: see C85
– Set theory with classes: see E30 and E70
– Intuitionistic theory of types: see F35

B20 Fragments of classical logic
+ Fragments of propositional and of first-order logic
+ Fragments used in model theory, set theory, etc.
+ Syllogistic
– Classical propositional logic: see B05
– Weak axiomatizations without restrictions on formulas: see B55, B60, F50 ("Fragment" refers to reduced expressive power, not reduced deductive power; MR heading "Subsystems of classical logic" includes both)

B22 Abstract deductive systems
+ Consequence relations
MR uses B99

B25 Decidability of theories and sets of sentences
+ Decidability of satisfiability
+ Decidable Diophantine problems
– Decidable word problems: see D40
– Other decidability results: see subject of problem, e.g. D05, or D80; MR includes these results here.
– Undecidability results: see D35, D40, D80, etc.

B28 Classical foundations of number systems
+ Natural numbers, real numbers, ordinal numbers
+ Axiomatic foundations and set-theoretic foundations
MR uses B30

B30 Logical foundations of other classical theories; axiomatics
+ Axiomatic method
+ Geometry, probability, physics, etc.
+ Models for non-mathematical theories
– Foundations of parts of logic: see that part.
MR heading: "Foundations and axiomatics of classical theories" includes also B28

B35 Mechanization of proofs and logical operations
+ Theorem proving, proof checking by machine
+ Minimization algorithms for Boolean functions
+ Optimization of logical operations
MR sometimes uses 03-04 or 68G15 (68T15 since 1985)

B40 Combinatory logic and lambda-calculus
+ Models of lambda-calculus

B45 Modal and tense logic
+ Intensional logic; see also A05
+ Normative and deontic logic
+ Other non-truth-functional systems

B46 Relevance and entailment
+ Fragments
– Primarily modal logic
MR uses B45

B48 Probability and inductive logic
See also A05 and C90
\+ Confirmation theory
– Foundations of probability: see B30; MR uses B48

B50 Many-valued logic
\+ Matrix interpretations of propositional connectives unless used only as a tool for investigating classical propositional logic.
– Boolean valued set theory: see E40
– Probability logic: see B48 or C90

B51 Quantum logic
– Algebraic study of Quantum logic: see G12
MR uses only G12

B52 Fuzzy logic
\+ Vagueness logic
– Papers demonstrating the fuzziness of the author's thought processes

B53 Paraconsistent logic
\+ Discussive and dialectical logic
MR uses B60

B55 Intermediate and related logics
\+ (Fragments of) propositional and predicate logics between intuitionistic or minimal and classical

B60 Other logics
– Intuitionistic logic: see F50 (MR uses B20)

B65 Logic of natural languages
– Computer languages: see B75
– Formal grammars unless applied to natural languages: see D05
– Natural language as a tool for the study of thought, reality, etc.: see A05
MR uses B65, B99, and 68Fn (68Sn since 1985)

B70 Logic in computer design; switching circuits
\+ Hardware related to logic
MR uses 94Cn

B75 Logic of algorithmic and programming languages
\+ Algorithmic and dynamic logic; MR uses B70 (formerly B45)
\+ Logical analysis of programs
\+ Logical aspects of database query languages and information retrieval
\+ Semantics of programming languages related to logic
\+ Software related to logic
– Specific algorithms: see subject of algorithm
MR uses B60, B70, 68Bn, 68Fn, and 68H05 (68Pn, 68Qn, and 68Tn since 1985)

B80 Other applications of logic
MR uses B99

B96 Collected works
\+ Selected works
– Collections (almost) entirely in one subfield: see that subfield
MR uses 01A75, 03-03, and 03-06

D60 Recursion theory on ordinals, admissible sets, etc.
+ Beta-recursion on inadmissible ordinals
– Classification of ordinary recursive functions using ordinals: see D20
– Ordinal notations: see D45 and F15
– Other aspects of admissibility: see C62, C70, or E45

D65 Higher-type and set recursion
+ Primitive recursive set functions
– Functionals in Proof Theory: see F10
– Recursion on the hereditarily finite sets: see D20
– Recursion with all arguments and parameters of type $\leqslant 1$: see D20; MR includes this in D65 as long as there are type 1 arguments

D70 Inductive definability
+ Constructions equivalent to inductive definitions, e.g. set derivatives, game sentences, etc.
+ Recursion theory of inductive definitions and their duals
– Inductive definitions in proof theory: see F35 and F50
– Mechanics of inductive definitions: see B28, E20, or E30

D75 Abstract and axiomatic recursion theory
+ Algebras of (partial) recursive functions; MR uses D20
+ Recursion over general structures

D80 Applications
+ Decidability or undecidability results in areas outside logic and algebra
+ Effective versions of problems outside logic and algebra

D96 Collected works
+ Selected works
– Collections (almost) entirely in one subfield: see that subfield
MR uses 01A75, 03-03, and 03-06

D97 Proceedings
+ Collections of papers by various authors, even if they do not derive from any actual conference
– Proceedings (almost) entirely in one subfield: see that subfield
– Proceedings not concentrated in this field: see Source Index
MR uses 03-06

D98 Textbooks, surveys
MR uses 03-01 and 03-02

D99 None of the above or uncertain, but in this section

E SET THEORY

E05 Combinatorial set theory
+ Partition relations, ideals, ultrafilters, trees named after people; MR uses also 04A20
– Finite combinatorics (excluded from the Ω-Bibliography); MR uses 05Xn

E07 Relations and orderings
+ Relation algebras: see also G15; MR uses G15
– Theories about ordering: see C65
MR uses E20, 04A05, 04A20, or 06An

E10 Ordinal and cardinal numbers
+ Cardinal algebras, ordinal algebras
+ Dedekind finite cardinals
– Cardinal exponentiation and the (generalized) continuum hypothesis: see E50; MR sometimes uses 04A10
– Combinatorial aspects of cardinals and ordinals: see E05
– Large cardinals: see E55

E15 Descriptive set theory
+ Definability properties of sets (in the real line or similar spaces)
+ Effective descriptive set theory
– General topology, measure theory, etc.: see E75
MR sometimes uses 04A15
See also D55

E20 Other classical set theory
+ Set algebra

E25 Axiom of choice and related propositions
+ Weak axioms of choice and their negations
MR sometimes uses 04A25

E30 Axiomatics of classical set theory and its fragments
+ Zermelo-Fraenkel set theory and minor variants
+ Gödel-Bernays set theory (also in E70)
– Morse-Kelley set theory (a second order theory: see E70)
– New Foundations, etc.: see E70

E35 Consistency and independence results
+ Forcing used to prove consistency

E40 Other aspects of forcing and Boolean-valued models
+ Forcing in generalized recursion theory: see also D60 and D65
– Model theoretic forcing: see C25

E45 Constructibility, ordinal definability and related notions
+ Other inner models, e.g. the core model

E47 Other notions of set-theoretic definability
+ Lévy hierarchy, indescribability
– Formalization of branches of mathematics within set theory

E50 Continuum hypothesis and Martin's axiom
+ Cardinal exponentiation
+ Variants of Martin's axiom
MR sometimes uses 04A30

E55 Large cardinals
+ Effective (denumerable) analogues of large cardinals
+ Weakly inaccessible and larger cardinals
– Axioms of infinity provable in ZFC
– Large proof-theoretic ordinals: see F15

E60 Determinacy and related principles which contradict the axiom of choice
+ Infinite exponent partition relations
+ Projective determinacy, definable determinacy
+ Other uses of infinite games in set theory and logic
– Applications of games outside set theory and logic
– Weak axioms that merely contradict choice

E65 Other hypotheses and axioms
+ Reflection principles
+ Combinatorial principles

E70 Nonclassical and second-order set theories
+ Leśniewski's Ontology and Mereology; MR uses B60
+ Nonstandard theories, e.g. New Foundations, Ackermann
+ Set theories formulated in non-classical logic
+ Theory of real classes (Morse-Kelley, and Gödel-Bernays set theory); MR uses E30

E72 Fuzzy sets

E75 Applications
+ Independence from set theory of mathematical propositions (also in E35)
+ Results in other branches of mathematics obtained by set theoretic methods
– Set-theoretical foundations of mathematics: see B28 and B30

E96 Collected works
+ Selected works
– Collections (almost) entirely in one subfield: see that subfield

MR uses 01A75, 03-03, and 03-06

E97 Proceedings
+ Collections of papers by various authors, even if they do not derive from any actual conference
– Proceedings (almost) entirely in one subfield: see that subfield
– Proceedings not concentrated in this field: see Source Index

MR uses 03-06

E98 Textbooks, surveys

MR uses 03-01 and 03-02

E99 None of the above or uncertain, but in this section

F PROOF THEORY AND CONSTRUCTIVE MATHEMATICS

F05 Cut elimination and normal form theorems
+ Hilbert's epsilon symbol
– Cut elimination and normal form theorems for modal systems: see B45

F07 Structure of proofs
– Proof schemas used rather than studied: see B10, C07, etc.

F10 Functionals in proof theory
– Typed lambda-calculus: see B40

F15 Recursive ordinals and ordinal notations
+ Ordinal notations even if not proof theory
+ Transfinite progressions of theories (Turing, Feferman; also in F30)

F20 Complexity of proofs
– Complexity of non-proof-theoretic procedures: see D15
– Purely qualitative (rather than quantitative) properties of proofs: see F07

F25 Relative consistency and interpretations
- – Consistency of systems of arithmetic: see F30 and F35
- – Set theoretic consistency results: see E35

F30 First-order arithmetic and fragments
- + Gödel incompleteness theorems
- + Metamathematics of intuitionistic arithmetic
- + Provability logic; MR uses also B45 and F40
- + Provably recursive functions; MR uses also D20
- + Recursive arithmetic
- – Model theory of arithmetic: see C62 and H15

F35 Second- and higher-order arithmetic and fragments
- + Metamathematics of intuitionistic analysis
- + Proof theory of systems of type theory
- + Proof theory of generalized inductive definitions
- – Model theory : see C62

F40 Gödel numberings in proof theory
- + Any use of Gödel numbering of syntax
- – Gödel numberings in recursion theory: see D20 and D45

F50 Metamathematics of constructive systems
- + Intuitionistic logic and subsystems; MR uses also B20
- + Model theoretic methods applied to constructive systems
- + Realizability
- – Metamathematics of predicative systems: see F65

F55 Constructive and intuitionistic mathematics
- + Bishop school of constructivism
- – Metamathematics: see F50

F60 Constructive recursive analysis
- + Classical recursive analysis
- + Soviet school of constructivism
- – Metamathematics: see F50

F65 Other constructive mathematics
- + Constructive trends not covered by F55 or F60
- + Predicative mathematics
- + Metamathematics of predicative systems
- – Other metamathematics: see F50

F96 Collected works
- + Selected works
- – Collections (almost) entirely in one subfield: see that subfield

MR uses 01A75, 03-03, and 03-06

F97 Proceedings
- + Collections of papers by various authors, even if they do not derive from any actual conference
- – Proceedings (almost) entirely in one subfield: see that subfield
- – Proceedings not concentrated in this field: see Source Index

MR uses 03-06

F98 Textbooks, surveys

MR uses 03-01 and 03-02

F99 None of the above or uncertain, but in this section

G ALGEBRAIC LOGIC

G05 Boolean algebras
+ Boolean rings, etc.
– Boolean functions : see B05; MR puts Boolean functions in G05, 06E30, and sometimes 94C10
– Pseudo-Boolean algebras : see G10

G10 Lattices and related structures
+ Heyting algebras; MR uses also 06D20
+ Semilattices, continuous lattices; MR uses 06B35
– Studies of "The lattice of..." where the lattice structure is not the main point

G12 Quantum logic
See also B51

G15 Cylindric and polyadic algebras, relation algebras

G20 Łukasiewicz and Post algebras
+ Lattices (or weaker structures) corresponding to many-valued logic

G25 Other algebras related to logic
+ Boolean algebras with provability and other operators
+ Implicative algebras, BCK algebras, etc.

G30 Categorical logic, topoi
+ Almost any connection between categories and logic, e.g. categories of models, logical foundations of category theory
– Pure category theory (Excluded from the Ω-Bibliography); MR uses 18Xn

G96 Collected works
+ Selected works
– Collections (almost) entirely in one subfield: see that subfield
MR uses 01A75, 03-03, and 03-06

G97 Proceedings
+ Collections of papers by various authors, even if they do not derive from any actual conference
– Proceedings (almost) entirely in one subfield: see that subfield
– Proceedings not concentrated in this field: see Source Index
MR uses 03-06

G98 Textbooks, surveys
MR uses 03-01 and 03-02

G99 None of the above or uncertain, but in this section

H NONSTANDARD MODELS

H05 Infinitesimal analysis in pure mathematics

H10 Other applications of infinitesimal analysis
+ Economics, physics, etc.

H15 Nonstandard models of arithmetic
+ Work emphasizing nonstandard methods
– Work emphasizing model theory : see C62

H20 Other nonstandard models

H96 Collected works
+ Selected works
– Collections (almost) entirely in one subfield: see that subfield
MR uses 01A75, 03-03, and 03-06

H97 Proceedings
+ Collections of papers by various authors, even if they do not derive from any actual conference
– Proceedings (almost) entirely in one subfield: see that subfield
– Proceedings not concentrated in this field: see Source Index
MR uses 03-06

H98 Textbooks, surveys
MR uses 03-01 and 03-02

H99 None of the above or uncertain, but in this section

Subject Index

D03 ∪ D05 Automata, formal systems, and grammars

1913

THUE, A. *Ueber die gegenseitige Lage gleicher Teile gewisser Zeichenreihen* ◇ D03 D40 ◇

1914

THUE, A. *Probleme ueber Veraenderungen von Zeichenreihen nach gegebenen Regeln* ◇ D03 D40 ◇

1936

POST, E.L. *Finite combinatory processes - formulation I* ◇ D03 D20 D40 ◇

TURING, A.M. *On computable numbers, with an application to the "Entscheidungsproblem"* ◇ D05 D10 D20 D35 F60 ◇

1937

HERMES, H. *Definite Begriffe und berechenbare Zahlen* ◇ D05 D10 D20 F60 ◇

TURING, A.M. *Computability and λ-definability* ◇ A05 B40 D05 D10 D20 F99 ◇

1943

MCCULLOCH, W.S. & PITTS, W. *A logical calculus of the ideas immanent in nervous activity* ◇ B60 D05 ◇

POST, E.L. *Formal reductions of the general combinatorial decision problem* ◇ D03 D40 ◇

1944

POST, E.L. *Recursively enumerable sets of positive integers and their decision problems* ◇ D03 D25 D40 ◇

1946

POST, E.L. *A variant of a recursively unsolvable problem* ◇ D03 D05 D40 ◇

1947

MARKOV, A.A. *On the impossibility of certain algorithms in the theory of associative systems I (Russian)* ◇ D03 D40 ◇

MARKOV, A.A. *The impossibility of certain algorithms in the theory of associative systems II (Russian)* ◇ D03 D40 ◇

POST, E.L. *Recursive unsolvability of a problem of Thue* ◇ D03 D10 D40 ◇

1948

POST, E.L. *Degrees of recursive unsolvability, preliminary report* ◇ D03 D30 D55 ◇

1951

MARKOV, A.A. *The impossibility of certain algorithms in the theory of associative systems (Russian)* ◇ D03 D35 D40 ◇

MARKOV, A.A. *The impossibility of algorithms for the recognition of certain properties of associative systems (Russian)* ◇ D03 D40 ◇

MARKOV, A.A. *The theory of algorithms (Russian)* ◇ D03 D20 ◇

1952

KALMAR, L. *Another proof to the Markov-Post theorem* ◇ D03 D20 D40 ◇

MARKOV, A.A. *On unsolvable algorithmic problems (Russian)* ◇ D03 D20 D25 D35 D40 ◇

MARKOV, A.A. *Theory of algorithms (Russian) (Hungarian summary)* ◇ D03 D20 ◇

1953

DETLOVS, V.K. *Normal algorithms and recursive functions (Russian)* ◇ D03 D20 ◇

FREUDENTHAL, H. *Machines pensantes* ◇ A05 D05 ◇

NAGORNYJ, N.M. *On strengthening the reduction theorem of the theory of algorithm (Russian)* ◇ D03 ◇

RIGUET, J. *Sur les rapports entre les concepts de machine de multipole et de structure algebrique* ◇ D03 D05 ◇

SHANNON, C.E. *Computers and automata* ◇ B70 D05 ◇

1954

BURKS, A.W. & WARREN, D.W. & WRIGHT, J.B. *An analysis of a logical machine using parenthesis-free notation* ◇ B03 D05 ◇

BURKS, A.W. & MCNAUGHTON, R. & POLLMAR, C. & WARREN, D.W. & WRIGHT, J.B. *Complete decoding nets: general theory and minimality* ◇ B70 D05 ◇

MARKOV, A.A. *Theory of algorithms (Russian)* ◇ D03 D20 D40 D98 ◇

SKOLEM, T.A. *Results in investigations in the foundations (Norwegian)* ◇ A05 B98 D03 F55 ◇

1955

MOISIL, G.C. *Contribution a l'etude algebrique des mecanismes automatiques (Romanian) (Russian and French summaries)* ◇ B70 D05 ◇

1956

KLEENE, S.C. *Representations of events in nerve nets and finite automata* ◇ D05 ◇

LEEUW DE, K. & MOORE, E.F. & SHANNON, C.E. & SHAPIRO, N.Z. *Computability by probabilistic machines* ◇ D05 ◇

MCCARTHY, J. & SHANNON, C.E. (EDS.) *Automata studies* ◇ D05 D97 ◇

MEDVEDEV, YU.T. *On the class of events able to be represented in a finite automation (Russian)* ◇ D05 ◇

MINSKY, M.L. *Some universal elements for finite automata* ◇ D05 ◇

MOORE, E.F. *Gedanken-experiments on sequential machines* ◇ D05 ◇

RIGUET, J. *Algorithmes de Markov et theorie des machines* ◇ D03 D05 ◇
TSEJTIN, G.S. *Associative calculus with unsolvable equivalence problem (Russian)* ◇ D03 D40 ◇
TSEJTIN, G.S. *On the problem of recognition of properties of associative calculi (Russian)* ◇ D03 D40 ◇

1957

BURKS, A.W. & WANG, HAO *The logic of automata I,II* ◇ D05 ◇
DUDA, W.L. *Post canonical language* ◇ D03 D20 ◇
DUNHAM, B. *Symbolic logic and computing machines. A survey and summary* ◇ D05 D10 ◇
RABIN, M.O. & SCOTT, D.S. *Remarks on finite automata* ◇ D05 ◇
RABIN, M.O. *Two-way finite automata* ◇ D05 ◇
TRAKHTENBROT, B.A. *On operators realizable in logical nets (Russian)* ◇ D05 ◇
WANG, HAO *Symbolic representations of calculating machines* ◇ B70 D05 D10 F30 ◇

1958

CHOMSKY, N. & MILLER, GEORGE A. *Finite state languages* ◇ B65 D05 ◇
DETLOVS, V.K. *The equivalence of normal algorithms and recursive functions (Russian)* ◇ D03 D20 ◇
ERSHOV, A.P. *On operator algorithms (Russian)* ◇ B75 D03 D20 ◇
GINSBURG, S. *On the length of the smallest uniform experiment which distinguishes the terminal states of a machine* ◇ D05 ◇
KULAGINA, O.S. *A method of determining grammatical concepts on the basis of set theory (Russian)* ◇ B65 D05 E75 ◇
MAYS, W. *Cybernetic models and thought processes* ◇ D05 ◇
NAGORNYJ, N.M. *On a minimal alphabet of algorithms over a given alphabet (Russian)* ◇ B75 D05 ◇
NAGORNYJ, N.M. *Some generalisations of the concept of normal algorithm (Russian)* ◇ D03 D20 ◇
NASLIN, P. *Circuits a relais et automatismes a sequences* ◇ B70 D05 ◇
NERODE, A. *Linear automaton transformations* ◇ D05 ◇
PORTE, J. *Systemes de Post, algorithmes de Markov* ◇ D03 ◇
RANEY, G.N. *Sequential functions* ◇ D05 ◇
ROSE, A. *Applications of logical computers to the construction of electrical control tables for signalling frames* ◇ B70 D05 ◇
TRAKHTENBROT, B.A. *Synthesis of logical nets whose operators are described by means of the calculus of one-place predicates (Russian)* ◇ B70 D05 ◇
TSEJTIN, G.S. *Associative calculus with unsolvable equivalence problem (Russian)* ◇ D03 D40 ◇

1959

ASSER, G. *Normierte Postsche Algorithmen* ◇ D03 ◇
ASSER, G. *Turing-Maschinen und Markowsche Algorithmen* ◇ D03 D10 ◇
ASSER, G. *Zwei Rekursionstheoreme fuer normale Algorithmen (Russian, English and French summaries)* ◇ D03 ◇
CHERNYAVSKIJ, V.S. *On a certain simplification of normal algorithms (Russian)* ◇ D03 ◇
CHERNYAVSKIJ, V.S. *On a class of normal Markov algorithms (Russian)* ◇ D03 ◇
GINSBURG, S. *A synthesis technique for minimal state sequential machines* ◇ D05 ◇
GINSBURG, S. *A technique for the reduction of a given machine to a minimal state machine* ◇ D05 ◇
HUZINO, S. *On the existence of Sheffer stroke class in the sequential machines* ◇ B05 D05 ◇
HUZINO, S. *Some properties of convolution machines and* σ*-composite machines* ◇ D05 ◇
KOBRINSKIJ, N.E. & TRAKHTENBROT, B.A. *Ueber den Aufbau einer allgemeinen Theorie der logischen Netze (Russian)* ◇ D05 ◇
NARASIMHAN, R. & SRINIVASAN, C.V. *On the synthesis of finite sequential machines* ◇ D05 ◇
RABIN, M.O. & SCOTT, D.S. *Finite automata and their decision problems* ◇ D05 ◇
SHEPHERDSON, J.C. *The reduction of two-way automata to one-way automata* ◇ D05 D10 ◇
TSEJTIN, G.S. *A simple example of an associative calculus with an unsolvable equivalence problem (Russian)* ◇ D03 D40 ◇
TSEJTIN, G.S. *On the problem of recognizing properties of associative calculi (Russian)* ◇ D03 D40 ◇

1960

AKUSHSKY, I.Y. & BAZILEVSKIJ, YU.YA. & SHREJDER, YU.A. *Logical, recursive and operator methods for the analysis and synthesis of automata* ◇ D05, D10 ◇
BUECHI, J.R. *Weak second-order arithmetic and finite automata* ◇ B15 B25 C85 D05 F35 ◇
CHERNYAVSKIJ, V.S. *On the reversibility of algorithms (Russian)* ◇ D03 D10 ◇
ERSHOV, A.P. *Operator algorithms. I. Basic concepts (Russian)* ◇ B75 D03 D20 ◇
GINSBURG, S. *Connective properties preserved in minimal state machines* ◇ D05 ◇
GINSBURG, S. *Some remarks on abstract machines* ◇ D05 ◇
GINSBURG, S. *Synthesis of minimal state machines* ◇ D05 ◇
GLUSHKOV, V.M. *On an algorithm of synthesis of abstract automata (Russian) (English summary)* ◇ D05 ◇
GLUSHKOV, V.M. *Ueber eine Methode der Analyse abstrakter Automaten (Ukrainisch)* ◇ D05 ◇
HUZINO, S. *On some sequential equations* ◇ D05 ◇
KARACUBA, A.A. *Solution to a problem in the theory of finite automata (Russian)* ◇ D05 ◇
LEE, C.Y. *Automata and finite automata* ◇ D05 D10 ◇
MCNAUGHTON, R. & YAMADA, H. *Regular expressions and state graphs for automata* ◇ D05 ◇
NAGORNYJ, N.M. *On the investigation of the isomorphisms of associative calculi (Russian) (German summary)* ◇ D05 D20 ◇

1961

AJZERMAN, M.A. & GUSEV, L.A. & ROZONOEHR, L.I. & SMIRNOVA, I.M. & TAL', A.A. *The algorithmic insolubility of the problem of recognizing the*

representability of recursive events in finite automata (Russian) ◇ D05 ◇
ARBIB, M.A. *Turing machines, finite automata and neural nets* ◇ D05 D10 ◇
ARDEN, D.N. *Delayed-logic and finite-state machines* ◇ D05 ◇
ASSER, G. & VUCKOVIC, V. *Funktionen-Algorithmen* ◇ D03 D20 ◇
BALAKRISHNAN, A.V. *Prediction theory for Markoff processes* ◇ D03 ◇
COURT, L.M. & HIZ, H. & LAMBEK, J. *Comments* ◇ B65 D05 ◇
CULIK, K. *Some notes on finite state languages and events represented by finite automata using labelled graphs (Czech and Russian summaries)* ◇ D05 ◇
ELGOT, C.C. *Decision problems of finite automata design and related arithmetics* ◇ B25 D03 D05 D35 F30 ◇
FLOYD, R.W. *A note on mathematical induction on phrase structure grammars* ◇ B65 D05 ◇
GINSBURG, S. *Compatibility of states in input-independent machines* ◇ D05 ◇
GINSBURG, S. *Distinguishability of a semigroup by a machine* ◇ D05 ◇
GINSBURG, S. *Sets of tapes accepted by different types of automata* ◇ D05 ◇
GLUSHKOV, V.M. *Abstract automata and the decomposition of free semigroups (Russian)* ◇ D05 ◇
GLUSHKOV, V.M. *The abstract theory of automata (Russian)* ◇ D05 ◇
KUZNETSOV, O.P. *Asynchronous logical networks and the representation of non-regular events* ◇ D05 ◇
LEE, C.Y. *Categorizing automata by W-machine programs* ◇ D05 ◇
MCNAUGHTON, R. *The theory of automata, a survey* ◇ D05 D98 ◇
MINSKY, M.L. *Recursive unsolvability of Post's problem of "tag" and other topics in the theory of Turing machines* ◇ D03 D10 ◇
NAGORNYJ, N.M. *On the realization of functions in alphabets by algorithms of certain classes (Russian)* ◇ D05 ◇
PARIKH, R. *Language generating devices* ◇ D05 ◇
ROHLEDER, H. *Zum Ausmultiplizieren der Klammern beim Verfahren von Nelson* ◇ B35 D05 ◇
SCHUETZENBERGER, M.-P. *A remark on finite transducers* ◇ D05 ◇
SCHUETZENBERGER, M.-P. *On the definition of a family of automata* ◇ D05 ◇
SENDOV, B. & SKORDEV, D.G. *On equations in words (Russian) (German summary)* ◇ D05 ◇
SMULLYAN, R.M. *Elementary formal systems* ◇ D03 D25 ◇
SMULLYAN, R.M. *Extended canonical systems* ◇ B45 D03 D25 ◇
SMULLYAN, R.M. *Monadic elementary formal systems* ◇ D03 D25 ◇
SMULLYAN, R.M. *Theory of formal systems* ◇ B98 D03 D05 D20 D25 D98 ◇
TRAKHTENBROT, B.A. *Finite automata and the logic of one-place predicates (Russian)* ◇ B20 D05 F35 ◇
TSETLIN, M.L. *Certain problems in the behavior of finite automata* ◇ D05 ◇

1962

BAR-HILLEL, Y. *Some recent results in theoretical linguistics* ◇ B65 D05 ◇
BUECHI, J.R. *Mathematische Theorie des Verhaltens endlicher Automaten* ◇ D05 ◇
BUECHI, J.R. *On a decision method in restricted second order arithmetic* ◇ B25 C85 D05 F35 ◇
BURKS, A.W. & WRIGHT, J.B. *Sequence generators, graphs and formal languages* ◇ B20 D05 ◇
BURKS, A.W. & WRIGHT, J.B. *Sequence generators and digital computers* ◇ B20 D05 ◇
CANTOR, DAVID G. *On the ambiguity problem of Backus systems* ◇ D05 ◇
CHERNYAVSKIJ, V.S. & KOZMIDIADI, V.A. *Ueber einige Begriffe der Theorie der mathematischen Maschinen (Russisch)* ◇ D05 D10 ◇
FRIEDMAN, JOYCE *A decision procedure for computations of finite automata* ◇ D05 ◇
GHIRON, H. *Rules to manipulate regular expressions of finite automata* ◇ D05 ◇
GILL, A. *Introduction to the theory of finite-state machines* ◇ D05 D10 D98 ◇
GINSBURG, S. & ROSE, G.F. *A comparison of the work done by generalized sequential machines and Turing machines* ◇ D05 D10 ◇
GINSBURG, S. *An introduction to mathematical machine theory* ◇ D05 D10 ◇
GINSBURG, S. *Examples of abstract machines* ◇ D05 ◇
HARTMANIS, J. & STEARNS, R.E. *Some dangers in state reduction of sequential machines* ◇ D05 ◇
HUZINO, S. *Finite automata and Asser's function algorithms* ◇ D03 D05 D20 ◇
HUZINO, S. & YONEYAMA, N. *On a proof of Shepherdson's theorem* ◇ D05 D10 ◇
KOBRINSKIJ, N.E. & TRAKHTENBROT, B.A. *Introduction to the theory of finite automata (Russian)* ◇ D05 D98 ◇
KOZMIDIADI, V.A. *Sets enumerable and solvable by automata (Russian)* ◇ D05 ◇
KUBINSKI, T. *On extension of the theory of syntactic categories (Polish summary)* ◇ D05 ◇
KUDRYAVTSEV, V.B. *A completeness theorem for a class of automata without feedback (Russian)* ◇ D05 ◇
LEVENSHTEJN, V.I. *The inversion of finite automata (Russian)* ◇ D05 ◇
MARKOV, A.A. *Computable invariants (Russian)* ◇ D05 D20 ◇
MASLOV, S.YU. *Transformation of arbitrary canonical calculi into canonical calculi of special types (Russian)* ◇ D03 D05 ◇
MINSKY, M.L. *Size and structure of universal Turing machines using tag systems* ◇ D03 D10 ◇
POGORZELSKI, H.A. *A note on an arithmetization of a word system in a denumerable alphabet* ◇ D05 D20 ◇
POGORZELSKI, H.A. *Word arithmetic: theory of primitive words* ◇ D05 D20 ◇
ROOTSELAAR VAN, B. *Algebraische Kennzeichnung freier Wortarithmetiken* ◇ D05 ◇

ROOTSELAAR VAN, B. *Die Struktur der rekursiven Wortarithmetik des Herrn v. Vucovic* ◇ D05 ◇
ROSE, G.F. *Output completeness in sequential machines* ◇ D05 ◇
SCHUETZENBERGER, M.-P. *Finite counting automata* ◇ D05 ◇
SINGLETON, R.C. *A test for linear separability as applied to self-organizing machines* ◇ D05 ◇
SMULLYAN, R.M. *On Post's canonical systems* ◇ D03 D25 ◇
TRAKHTENBROT, B.A. *Finite automata and the logic of one-place predicates (Russian)* ◇ B20 D05 F35 ◇
VUCKOVIC, V. *On some possibilities in the foundations of recursive arithmetics of words (English) (Serbo-Croatian summary)* ◇ D05 ◇
VUCKOVIC, V. *Recursive arithmetic of words and finite automata (Serbo-Croatian) (English summary)* ◇ D05 ◇
WINETT, J.M. *An α-state finite automaton for multiplication by α* ◇ D05 ◇
YANOV, YU.I. *On identical transformations of regular expressions (Russian)* ◇ D05 ◇
ZELEZNIKAR, A. *Some arithmetic normal algorithms (Serbo-Croatian summary)* ◇ D03 ◇

1963

AJZERMAN, M.A. & GUSEV, L.A. & ROZONOEHR, L.I. & SMIRNOVA, I.M. & TAL', A.A. *Logic, automata and algorithms (Russian)* ◇ B98 D05 D10 D98 ◇
ARBIB, M.A. *Monogenic normal systems are universal* ◇ D03 D10 ◇
ASSER, G. *Ueber die algebraische Theorie der Automaten* ◇ D05 ◇
BECVAR, J. *Finite and combinatorial automata. Turing automata with a programming tape* ◇ D05 D10 ◇
BODNARCHUK, V.G. *Systems of equations in the algebra of events (Russian)* ◇ D05 ◇
BOEHLING, K.H. *Netzwerke -Schaltwerke - Automaten. Ein Ueberblick ueber die synchrone Theorie* ◇ B70 D05 ◇
BRAFFORT, P. & HIRSCHBERG, D. (EDS.) *Computer programming and formal systems* ◇ B97 D05 D97 ◇
BRZOZOWSKI, J.A. *Canonical regular expressions and minimal state graphs for definite events* ◇ D05 ◇
BRZOZOWSKI, J.A. & POAGE, J.F. *On the construction of sequential machines from regular expressions* ◇ D05 ◇
CHURCH, A. *Logic, arithmetic, and automata* ◇ A10 B25 D05 D98 ◇
CULIK, K. *Some axiomatic systems for formal grammars and languages* ◇ D05 ◇
DAVIS, MARTIN D. *Unsolvable problems: a review* ◇ D03 D10 D35 D40 ◇
FLOYD, R.W. *Syntactic analysis and operator precedence* ◇ B75 D05 ◇
GINSBURG, S. & ROSE, G.F. *Operations which preserve definability in languages* ◇ D05 ◇
GINSBURG, S. & SPANIER, E.H. *Quotients of context-free languages* ◇ D05 ◇
GLADKIJ, A.V. *Configuration characteristics of languages (Russian)* ◇ D05 ◇
GLADKIJ, A.V. *On the recognition of replaceability in recursive languages (Russian)* ◇ D05 ◇
GLUSHKOV, V.M. *Theorie der abstrakten Automaten* ◇ D05 D98 ◇
GORN, S. *Processors for infinite codes of the Shannon-Fano type* ◇ B75 D05 ◇
GREIBACH, S.A. *The undecidability of the ambiguity problem for minimal linear grammars* ◇ D05 ◇
HARTMANIS, J. *Regularity preserving modifications of regular expressions* ◇ D05 ◇
HILTON, A.M. *Logic, computing machines, and automation* ◇ A05 B98 D05 ◇
HIROSE, K. & SEKI, S. *A remark on the theory of automata* ◇ D05 ◇
ION, I.D. *On the connection between the theory of algorithms and the abstract theory of automata* ◇ D05 D20 ◇
KRATKO, M.I. *On the reducibility of the combinatorial problem of Post to certain mass problems in the theory of finite automata (Russian)* ◇ D03 D05 ◇
KUZNETSOV, O.P. *On a class of regular events (Russian)* ◇ D05 ◇
LAEMMEL, A.E. *Application of lattice-ordered semigroups to codes and finite-state transducers* ◇ B70 D05 ◇
MAKAROV, S.V. *On the realization of stochastic matrices by finite automata (Russian)* ◇ D05 ◇
MARCUS, S. *Typologie des langues et modeles logiques* ◇ D05 ◇
MARKOV, A.A. *Indistinguishability by invariants in the theory of associative calculi (Russian)* ◇ D05 D20 ◇
MARKOV, A.A. *On certain algorithms related to systems of words (Russian)* ◇ D05 D40 ◇
MASLOV, S.YU. *Some methods for the definition of sets in generating bases (Russian)* ◇ D05 ◇
MASLOV, S.YU. *Strong representability of sets by calculi (Russian)* ◇ D03 D20 ◇
MCANDREW, M.H. *A partition problem* ◇ D05 ◇
MEZEI, J.E. *Structure of monoids with applications to automata* ◇ D05 ◇
NEANDER, J. *Aequivalenzkalkuel und orientierte Graphen* ◇ B70 D05 ◇
PATTON, T.E. *On n-adic representation of numbers* ◇ D03 D25 ◇
PERLES, M.A. & RABIN, M.O. & SHAMIR, E. *The theory of definite automata* ◇ D05 ◇
PETER, R. *Ueber die Rekursivitaet der Begriffe der mathematischen Grammatiken (Russian summary)* ◇ D05 D20 ◇
POENARU, V. *Expose sommaire de la theorie des algorithmes (d'apres A.A. Markov)* ◇ D03 D20 ◇
RABIN, M.O. *Probabilistic automata* ◇ D05 ◇
RABIN, M.O. *Real time computation* ◇ D05 D10 D15 ◇
RED'KO, V.N. *On commutative closing of events (Ukrainian) (Russian and English summaries)* ◇ D05 ◇
SCHUETZENBERGER, M.-P. *Certain elementary families of automata* ◇ D05 ◇

SCHUETZENBERGER, M.-P. *On context-free languages and push-down automata* ◇ D05 D10 ◇
SCOGNAMIGLIO, G. *Un metodo di calcolo dei prodotti delle matrici booleane elementari* ◇ B05 D03 ◇
SHAMIR, E. *On sequential languages and two classes of regular events. Introduction* ◇ D05 ◇
STARKE, P.H. *Ueber die Darstellbarkeit von Ereignissen in nicht-initialen Automaten* ◇ D05 ◇
URBANO, R.H. *Boolean matrices and the stability of neural nets* ◇ B05 D05 ◇
WANG, HAO *Dominoes and the AEA case of the decision problem* ◇ B20 B25 D05 D35 ◇
WANG, HAO *Tag systems and lag systems* ◇ D03 ◇
WATANABE, S. *Periodicity of Post's normal process of tag* ◇ D03 ◇
ZELEZNIKAR, A. *Some algorithm theory and its applicability (Serbo-Croatian summary)* ◇ D03 D05 ◇

1964

ARBIB, M.A. *Brains, machines, and mathematics* ◇ D05 D10 D80 F30 ◇
BAR-HILLEL, Y. & SHAMIR, E. *Finite state languages: formal representations and adequacy problems* ◇ B65 D05 ◇
BAR-HILLEL, Y. & GAIFMAN, C. & SHAMIR, E. *On categorial and phrase structure grammars* ◇ B65 D05 ◇
BAR-HILLEL, Y. & GAIFMAN, C. & SHAMIR, E. *On formal properties of simple phrase structure grammars* ◇ B65 D05 ◇
BRZOZOWSKI, J.A. *Derivatives of regular expressions* ◇ D05 ◇
BUECHI, J.R. *Regular canonical systems* ◇ D05 D35 ◇
COCKE, J. & MINSKY, M.L. *Universality of tag systems with $P = 2$* ◇ D03 ◇
CRESSWELL, M.J. *Propositional arithmetic* ◇ B05 D03 ◇
CULIK, K. *Applications of graph theory to mathematical logic and linguistics* ◇ B65 D03 D05 ◇
ELGOT, C.C. & RUTLEDGE, J.D. *RS-machines with almost blank tapes* ◇ D05 D10 F30 ◇
ELGOT, C.C. & ROBINSON, A. *Random-access stored-program machines, an approach to programming languages* ◇ B75 D05 D10 ◇
EVEN, S. *Rational numbers and regular events* ◇ D05 ◇
FREY, T. *Ueber die Konstruktion nichtvollstaendiger Automaten* ◇ D05 ◇
FREY, T. *Ueber die Konstruktion endlicher Automaten* ◇ D05 ◇
GINSBURG, S. & GREIBACH, S.A. & HIBBARD, T.N. *Solvability of machine mappings of regular sets to regular sets* ◇ D05 ◇
IGARASHI, S. *A formalization of the descriptions of languages and the related problems in a Gentzen-type formal system* ◇ B75 D05 ◇
KRATKO, M.I. *Algorithmic unsolvability of the problem of recognition of completeness for finite automata (Russian)* ◇ D05 ◇
KRATKO, M.I. *On the existence of non-recursive bases of finite automata (Russian)* ◇ D03 D05 ◇
KRATKO, M.I. *The algorithmic unsolvability of a problem in the theory of finite automata (Russian)* ◇ D05 ◇
KURODA, S.-Y. *Classes of languages and linear-bounded automata* ◇ D05 ◇
LANDWEBER, P.S. *Decision problems of phrase-structure grammars* ◇ B65 D05 ◇
LYUBICH, YU.I. *Estimates of the number of states that arise in the determinization of a nondeterministic autonomous automaton (Russian)* ◇ D05 ◇
MAKAROV, S.V. *Turing machines and finite automata (Russian)* ◇ D05 D10 ◇
MARCUS, S. *Grammar and finite automata (Romanian)* ◇ B65 D05 ◇
MARCUS, S. *Langues completement adequates et langues regulieres* ◇ D05 ◇
MARCUS, S. *Sur un modele de H.B. Curry pour le langage mathematique* ◇ D05 ◇
MARKOV, A.A. *Normal algorithms which compute Boolean functions (Russian)* ◇ B05 D03 ◇
MASLOV, S.YU. *On the "tag problem" of E.L.Post (Russian)* ◇ D03 ◇
MASLOV, S.YU. *Some properties of E.L.Post's apparatus of canonical systems (Russian)* ◇ D03 D05 D10 ◇
MCNAUGHTON, R. & YAMADA, H. *Regular expressions and state graphs for automata* ◇ D05 ◇
MOORE, E.F. (ED.) *Sequential machines: selected papers* ◇ D05 D97 ◇
PEAK, I. *Automaten und Halbgruppen I* ◇ D05 ◇
POGORZELSKI, H.A. *Commutative recursive word arithmetic in the alphabet of prime numbers* ◇ D05 D20 ◇
POGORZELSKI, H.A. *Primitive words in an infinite abstract alphabet* ◇ D05 D20 ◇
POGORZELSKI, H.A. *Skolem arithmetics on certain concrete word systems* ◇ D05 ◇
RED'KO, V.N. *On the determinate aggregate of relationships of the algebra of regular events (Ukrainian)* ◇ D05 ◇
SALOMAA, A. *Axiom systems for regular expressions of finite automata* ◇ D05 ◇
SALOMAA, A. *On the reducibility of events represented in automata* ◇ D05 ◇
SEDOL, YA.YA. *The free product of associative calculi with common subalphabet, and some related questions (Russian)* ◇ D05 D40 ◇
SINGLETARY, W.E. *A complex of problems proposed by Post* ◇ B20 D03 D25 D30 D35 ◇
STARKE, P.H. *Einige Probleme der Automatentheorie I, II* ◇ D05 ◇
VUCKOVIC, V. *On a class of regular sets* ◇ D05 ◇
YANOV, YU.I. *Invariant operations over events (Russian)* ◇ D05 ◇
ZAROVNYJ, V.P. *The group of automatic one-to-one mappings (Russian)* ◇ D05 D10 D75 ◇
ZASLAVSKIJ, I.D. *Graph schemes with memory (Russian)* ◇ B70 B75 D05 D10 ◇

1965

ARBIB, M.A. *A common framework for automata theory and control theory* ◇ D05 D10 ◇

Subject Index

BABIKOV, G.V. *Normale Algorithmen und Turingmaschinen (Russian)* ◇ D03 D10 ◇
BARZDINS, J. *Capacity of a medium and the behavior of automata (Russian)* ◇ D05 ◇
BECVAR, J. *Real-time and complexity problems in automata theory* ◇ D05 D10 D15 ◇
BOONE, W.W. *Finitely presented group whose word problem has the same degree as that of an arbitrarily given Thue system (application of methods of Britton)* ◇ D03 D30 D40 ◇
BUECHI, J.R. *Decision methods in the theory of ordinals* ◇ B15 B25 C85 D05 E10 ◇
BUECHI, J.R. *Transfinite automata recursions and weak second order theory of ordinals* ◇ B15 B25 C85 D05 E10 ◇
CUDIA, D.F. & SINGLETARY, W.E. *Post's correspondence problem and degrees of unsolvability; degrees of unsolvability in automata and grammars* ◇ D03 D05 D25 ◇
CURRY, H.B. *Two examples of algorithms* ◇ D03 ◇
DONG, YUNMEI & LI, KAIDE *P grammar and its decision and analysis problems (Chinese)* ◇ D05 ◇
DUDICH, A.M. & GLEBSKIJ, YU.V. & KOGAN, D.I. & LIOGON'KIJ, M.I. & MARKOV, A.A. *Algorithmen, die durch wiederholungsfreie Anwendung endlicher Automaten realisiert werden (Russisch)* ◇ D05 ◇
ELGOT, C.C. & MEZEI, J.E. *On relations defined by generalized finite automata* ◇ D05 ◇
EVEN, S. *Comments on the minimization of stochastic machines* ◇ D05 ◇
FISCHER, P.C. *Generation of primes by a one-dimensional real-time iterative array* ◇ D05 ◇
FISCHER, P.C. & RUBY, S.S. *Multi-tape and infinite-state automata, a survey* ◇ D05 ◇
FRISH, I. *On stop-conditions in the definitions of constructive languages* ◇ D05 ◇
GECSEG, F. & PEAK, I. *Automata with isomorphic semi-groups* ◇ D05 ◇
GINSBURG, S. & HIBBARD, T.N. & ULLIAN, J.S. *Sequences in context free languages* ◇ D05 ◇
GLADKIJ, A.V. *Algorithmic undecidability of inherent ambiguity in context-free languages (Russian)* ◇ D05 D10 ◇
GLADKIJ, A.V. *Certain decision problems for context-free grammars (Russian)* ◇ D05 ◇
GREIBACH, S.A. *A new normal-form theorem for context-free phrase structure grammars* ◇ D05 ◇
GUAN, JIWEN *Theory of linear inner automata (Chinese)* ◇ D05 ◇
HARRISON, M.A. *Introduction to switching and automata theory* ◇ B70 D05 D98 ◇
HARTMANIS, J. & LEWIS II, P.M. & STEARNS, R.E. *Memory bounds for recognition of context-free and context-sensitive languages* ◇ B65 D05 D15 ◇
KALMAR, L. (ED.) *Colloque sur les fondements des mathematiques, les machines mathematiques, et leurs applications* ◇ B97 D05 D10 D97 ◇
KRATKO, M.I. *A class of Post calculi (Russian)* ◇ D03 ◇
KUZ'MIN, V.A. *Realization of functions of the algebra of logic by means of automata, normal algorithms and Turing machines (Russian)* ◇ B05 D03 D05 D10 ◇
KUZNETSOV, O.P. *Representation of regular events in asynchronous automata (Russian)* ◇ D05 ◇
KUZNETSOV, O.P. *Ueber eine Klasse von regulaeren Ereignissen (Russisch)* ◇ D05 ◇
MARCUS, S. *Sur la notion de projectivite* ◇ D05 ◇
MEZEI, J.E. & WRIGHT, J.B. *Generalized ALGOL-like languages* ◇ B75 D05 ◇
NEPOMNYASHCHIJ, V.A. *On algorithms realized by re-application of finite automata (Russian)* ◇ D05 ◇
NOWAKOWSKI, R. *A structural theory of matrix-defined finite automata (Polish) (Russian and English summaries)* ◇ D05 ◇
PEAK, I. *Automata and semi-groups II (Russian)* ◇ D05 ◇
PEAK, I. *Certain extensions of semi-simple automata (Russian)* ◇ D05 ◇
POGORZELSKI, H.A. *Nonconcatenative abstract Skolem arithmetics I,II,III* ◇ D05 ◇
RICHALET, J. *Calcul operationnel dans un anneau fini* ◇ D05 ◇
SALOMAA, A. *On probabilistic automata with one input letter* ◇ D05 ◇
SHAMIR, E. *On sequential languages and two classes of regular events* ◇ D05 ◇
SHEPHERDSON, J.C. *Machine configuration and word problems of given degree of unsolvability* ◇ D05 D30 D35 D40 ◇
SPIVAK, M.A. *Algorithm for abstract synthesis of automata for an expanded language of regular expressions (Russian)* ◇ D05 ◇
STANCIULESCU, F.S. *Einige Bemerkungen zur Sequentiallogik der endlichen Automaten* ◇ D05 ◇
STARKE, P.H. *Die Imitation endlicher Medwedjew-Automaten durch Nervennetze* ◇ D05 ◇
STARKE, P.H. *Einfuehrung in die Theorie der Nervennetze* ◇ D05 ◇
STARKE, P.H. *Theorie stochastischer Automaten I,II (English and Russian summaries)* ◇ D05 ◇
TANG, CHISUNG *A recursive theory of computer instructions (Chinese)* ◇ B75 D05 D80 ◇
TAO, RENJI *Reduction of automata, universal automata and some problems regarding the behaviour of automata (Chinese)* ◇ D05 ◇
TITGEMEYER, D. *Einfuehrung in die Theorie der Kalkuele und Algorithmen* ◇ D03 D05 D10 ◇
TONOYAN, R.N. *Logical schemes for algorithms and their equivalent transforms (Russian)* ◇ B75 D05 ◇
WANG, HAO *Logic and computers* ◇ A05 B35 D03 D10 ◇
WANG, HAO *Remarks on machines, sets, and the decision problem* ◇ B20 B25 D03 D05 D10 D35 E30 ◇
WEEG, G.P. *The automorphism group of the direct product of strongly related automata* ◇ D05 G20 ◇
YOELI, M. *Lattice-ordered semigroups, graphs, and automata* ◇ D05 ◇

1966

AMAR, V. & PUTZOLU, G. *Generalizations of regular events* ◇ D05 ◇
ASSER, G. *Ueber eine Darstellung der rekursiven Wortfunktionen in endlichen Automaten* ◇ D05 D20 ◇

BERGER, R. *The undecidability of the domino problem* ◇ D05 D10 ◇
BOEHM, C. & JACOPINI, G. *Flow diagrams, Turing machines and languages with only two formation rules* ◇ B75 D05 D10 ◇
BOONE, W.W. *Word problems and recursively enumerable degrees of unsolvability. A first paper on Thue systems* ◇ D03 D25 D40 ◇
BUECHI, J.R. *Algebraic theory of feedback in discrete systems, part I* ◇ C05 D05 ◇
CAIANIELLO, E.R. (ED.) *Automata theory* ◇ D05 D97 ◇
CARACCIOLO, A. *Generalized Markov algorithms and automata* ◇ D03 D05 D10 ◇
CHOMSKY, N. *Topics in the theory of generative grammar* ◇ B65 D05 ◇
DEUEL, D.R. & GILL, A. *Some decision problems associated with weighted directed graphs* ◇ D05 D80 ◇
ELGOT, C.C. & RABIN, M.O. *Decidability and undecidability of extensions of second (first) order theory of (generalized) successor* ◇ B15 B25 C85 D05 D35 F30 F35 ◇
GINSBURG, S. & ROSE, G.F. *A characterization of machine mappings* ◇ D05 ◇
GINSBURG, S. & ULLIAN, J.S. *Ambiguity in context free languages* ◇ D05 ◇
GINSBURG, S. & SPANIER, E.H. *Bounded regular sets* ◇ D05 ◇
GINSBURG, S. & GREIBACH, S.A. *Deterministic context free languages* ◇ D05 ◇
GINSBURG, S. & GREIBACH, S.A. *Mappings which preserve context sensitive languages* ◇ D05 ◇
GINSBURG, S. & ULLIAN, J.S. *Preservation of unambiguity and inherent ambiguity in context-free languages* ◇ D05 ◇
GINSBURG, S. & ROSE, G.F. *Preservation of languages by transducers* ◇ D05 ◇
GINSBURG, S. & SPANIER, E.H. *Semigroups, Presburger formulas, and languages* ◇ D05 F30 ◇
GINSBURG, S. *The mathematical theory of context free languages* ◇ B65 D05 D98 ◇
GREIBACH, S.A. *The unsolvability of the recognition of linear context-free languages* ◇ D05 ◇
GROSS, M. *Sur certains procedes de definition de langages formels* ◇ D05 ◇
HIBBARD, T.N. & ULLIAN, J.S. *The independence of inherent ambiguity from complementedness among context-free languages* ◇ D05 ◇
HOOPER, P.K. *Monogenic Post normal systems of arbitrary degree* ◇ D03 ◇
HOOPER, P.K. *The immortality problem for Post normal systems* ◇ D03 ◇
HOTZ, G. *Eindeutigkeit und Mehrdeutigkeit formaler Sprachen* ◇ D05 ◇
HUZINO, S. *On the simulation of real-time Turing machines by a modified Shepherdson-Sturgis' machine* ◇ D05 D10 D15 ◇
HUZINO, S. *Simulatability of finite automata by Shepherdson and Sturgis' machines* ◇ D05 D10 ◇
KRATKO, M.I. *Formal Post calculi and finite automata (Russian)* ◇ D03 D05 ◇
KREIDER, D.L. & RITCHIE, R.W. *A universal two-way automaton* ◇ D05 ◇
LOEFGREN, L. *Explicability of sets and transfinite automata* ◇ D05 E25 E50 E55 ◇
LUCHKIN, V.D. *On the ranks of configurations of contex-free languages (Russian)* ◇ D05 ◇
MACHOVER, M. *Contextual determinacy in Lesniewski's grammar (Polish and Russian sumaries)* ◇ D05 ◇
MCNAUGHTON, R. *Testing and generating infinite sequences by a finite automaton* ◇ D05 ◇
MEDINA, A. *Logical analysis of organization in finite automata (Spanish summary)* ◇ D05 ◇
MIRKIN, B.G. *An algorithm for constructing a base in a language of regular expressions (Russian)* ◇ D05 ◇
MIRKIN, B.G. *On a language of pseudo-regular expressions (Russian) (English summary)* ◇ D05 ◇
MIRKIN, B.G. *On the theory of multitape automata (Russian) (English summary)* ◇ D05 ◇
NADIU, G.S. *On some theorems of recursion of normal algorithms* ◇ D03 ◇
NEPOMNYASHCHIJ, V.A. *On certain automata capable of computing a basis for recursively enumerable sets (Russian)* ◇ D05 D25 ◇
NOVOTNY, M. *Ueber endlich charakterisierbare Sprachen* ◇ D05 ◇
PARIKH, R. *On context-free languages* ◇ D05 ◇
RABIN, M.O. *Lectures on classical and probabilistic automata* ◇ D05 ◇
SALOMAA, A. *Axiomatization of an algebra of events realizable by logical network (Russian)* ◇ D05 ◇
SALOMAA, A. *Two complete axiom systems for the algebra of regular events* ◇ D05 G05 ◇
SCHREIBER, P. *Ueber die Entbehrlichkeit von Hilfsbuchstaben bei der Berechnung mehrstelliger Wortfunktionen durch Markowsche Algorithmen* ◇ D03 D20 ◇
SCHUETZENBERGER, M.-P. *On a family of sets related to McNaughton's L-language* ◇ D05 ◇
SEPER, K. *A note on normalizability of Ter-Zaharjan's quasi-normal algorithms and of Markov's branching, iteration and union operations* ◇ D03 ◇
STARKE, P.H. *Eine Bemerkung ueber homogene Experimente* ◇ D05 ◇
STARKE, P.H. *Einige Bemerkungen ueber nicht-deterministische Automaten* ◇ D05 ◇
STARKE, P.H. *Stochastische Ereignisse und Wortmengen* ◇ D05 ◇
STARKE, P.H. *Stochastische Ereignisse und stochastische Operatoren* ◇ D05 ◇
STARKE, P.H. *Theory of stochastic automata* ◇ D05 ◇
STEEL JR., T.B. (ED.) *Formal language description languages for computer programming* ◇ B75 D05 ◇
SWANSON, J.W. *A reduction theorem for normal algorithms* ◇ D03 ◇
ULLIAN, J.S. *Failure of a conjecture about context free languages* ◇ D05 ◇
YANOV, YU.I. *Certain subalgebras of events having no finite complete systems of identities (Russian)* ◇ C05 D05 ◇
ZELINKA, B. *Sur le PS-isomorphisme des langues* ◇ D05 ◇

1967

AANDERAA, S.O. & FISCHER, P.C. *The solvability of the halting problem for 2-state Post machines* ◇ D03 D05 ◇
BAER, ROBERT M. *Certain directed Post systems and automata* ◇ D03 D05 ◇
BEDNAREK, A.R. & WALLACE, A.D. *Finite approximants of compact totally disconnected machines* ◇ D05 ◇
BOEHLING, K.H. *Ueber eine Darstellungstheorie sequentieller Automaten* ◇ D05 ◇
BOOTH, T.L. *Sequential machines and automata theory* ◇ D05 ◇
COHEN, JOEL M. *The equivalence of two concepts of categorical grammar* ◇ D05 ◇
CULIK, K. & HAVEL, I.M. *On multiple finite automata* ◇ D05 ◇
DEUSSEN, P. *Some results on the set of congruence relations in a finite strongly connected automaton (German summary)* ◇ D05 ◇
EICKEL, J. *Loesung der Analyse-und Mehrdeutigkeitsprobleme durch Ueberfuehrung formaler Sprachen in sackgassenfreie formale Sprachen* ◇ D05 ◇
EILENBERG, S. & WRIGHT, J.B. *Automata in general algebras* ◇ D05 ◇
ELGOT, C.C. & ROBINSON, A. & RUTLEDGE, J.D. *Multiple control computer models* ◇ B70 D05 D10 ◇
FREJDZON, R.I. *The representability of algorithmically decidable predicates by Rabin machines (Russian)* ◇ D03 D10 D20 ◇
GINSBURG, S. & HARRISON, M.A. *Bracketed context free languages* ◇ D05 ◇
GINSBURG, S. & GREIBACH, S.A. & HARRISON, M.A. *One-way stack automata* ◇ D05 D10 ◇
GINSBURG, S. & GREIBACH, S.A. & HARRISON, M.A. *Stack automata and compiling* ◇ B75 D05 D10 ◇
GROSS, M. & LENTIN, A. *Notions sur les grammaires formelles* ◇ D05 ◇
HARTMANIS, J. *Context free languages and Turing machine computation* ◇ D05 D10 ◇
HARTMANIS, J. *On memory requirements for context free language recognition* ◇ D05 ◇
HAVEL, I.M. *On the modification of normal algorithms of Markov* ◇ D03 ◇
HOPCROFT, J.E. & ULLMAN, J.D. *An approach to a unified theory of automata* ◇ D05 D10 ◇
HOPCROFT, J.E. & WEINER, P. *Modular decomposition of synchronous sequential machines* ◇ D05 ◇
HOPCROFT, J.E. & ULLMAN, J.D. *Nonerasing stack automata* ◇ D05 D10 ◇
HOREJS, J. *Automata without internal memory (Czech) (Russian and English summaries)* ◇ D05 ◇
HOTZ, G. *Homomorphie und Aequivalenz formaler Sprachen* ◇ D05 ◇
KIMBALL, J. *Predicates definable over transformational derivations by intersection with regular languages* ◇ B65 D05 ◇
KUNZE, J. *Selektive Graphenschemata* ◇ D05 ◇
LORENTS, A.A. *Certain problems in the theory of finite probabilistic automata (Russian)* ◇ D05 F50 ◇
MARKOV, A.A. *An example of an independent system of words which cannot be included in a finite complete system (Russian)* ◇ D05 ◇
MARKOV, A.A. *Normal algorithms connected with computation of boolean functions (Russian)* ◇ B05 D03 ◇
MASLOV, S.YU. *Representation of enumerable sets by means of local calculi (Russian)* ◇ D05 D25 ◇
MASLOV, S.YU. *The concept of strict representability in the general theory of calculi (Russian)* ◇ D05 D20 ◇
MATIYASEVICH, YU.V. *Simple examples of undecidable associative calculi (Russian)* ◇ D03 ◇
MATIYASEVICH, YU.V. *Simple examples of unsolvable canonical calculi (Russian)* ◇ D03 ◇
MEZEI, J.E. & WRIGHT, J.B. *Algebraic automata and context-free sets* ◇ D05 ◇
MINSKY, M.L. *Computation: finite and infinite machines* ◇ D05 D10 D20 ◇
MOISIL, G.C. *Theorie structurelle des automates finis* ◇ D05 ◇
MOSZNER, Z. *On the theory of relations (Polish)* ◇ D03 E07 E98 ◇
NEUBER, S. & STARKE, P.H. *Ueber Homomorphie und Reduktion bei nicht-deterministischen Automaten* ◇ D05 ◇
PETRI, C.A. *Grundsaetzliches zur Beschreibung diskreter Prozesse* ◇ B70 D05 ◇
RABIN, M.O. *Mathematical theory of automata* ◇ D05 D98 ◇
ROSENBERG, ARNOLD L. *A machine realization of the linear context-free languages* ◇ D05 ◇
ROSENBERG, ARNOLD L. *Multitape finite automata with rewind instructions* ◇ D05 ◇
ROSENKRANTZ, D.J. *Matrix equations and normal forms for context-free grammars* ◇ D05 ◇
ROWICKI, A. *On the notion of occupancy of the tree* ◇ D05 D15 ◇
ROZENBERG, G. *Decision problems for quasi-uniform events (Russian summary)* ◇ D05 ◇
RUS, T. *Algebraic treatment of formalized languages (Romanian)* ◇ D05 G15 ◇
RUS, T. *The algebra of formalized languages (Romanian)* ◇ D05 G15 ◇
SALOMAA, A. *On m-adic probabilistic automata* ◇ D05 ◇
SALOVAARA, S. *On set theoretical foundations of system theory. A study of the state concept* ◇ D05 E75 ◇
SCHREIBER, P. *Normale Algorithmen ohne abbrechende Regeln* ◇ D03 ◇
SCOTT, D.S. *Some definitional suggestions for automata theory* ◇ D05 ◇
SHURYGIN, V.A. *Constructive sets with equality and their mappings (Russian)* ◇ D03 F60 ◇
SINGLETARY, W.E. *Recursive unsolvability of a complex of problems proposed by Post* ◇ D03 D25 D35 ◇
SINGLETARY, W.E. *The equivalence of some general combinatorial decision problems* ◇ D03 D25 ◇
SINTZOFF, M. *Existence of a van Wijngaarden syntax for every recursively enumerable set* ◇ D03 D05 D25 ◇
STARKE, P.H. *Huellenoperationen fuer nicht-deterministische Automaten* ◇ D05 ◇

STARKE, P.H. *Ueber Experimente an Automaten* ◇ D05 ◇
STARKE, P.H. & THIELE, H. *Zufaellige Zustaende in stochastischen Automaten* ◇ D05 ◇
SYRKIN, G.I. *A test for the validity of the translation theorem in the theory of normal algorithms (Russian)* ◇ D03 ◇
THATCHER, J.W. *Characterizing derivation trees of context free grammars through a generalization of finite automata theory* ◇ D05 ◇
TOMESCU, I. *On some combinatorical problems in the theory of classification (Romanian)* ◇ D05 ◇
ULLIAN, J.S. *Partial algorithm problems for context free languages* ◇ D05 ◇
UMEZAWA, T. *On uniform transition of states in a finite automaton* ◇ D05 ◇
WEISS, M. *Axiomatische Untersuchungen zur elementaren Theorie der freien Halbgruppen mit Substitution als undefiniertem Grundbegriff* ◇ B03 D05 ◇
YASUHARA, A. *A remark on Post normal systems* ◇ D03 ◇
YNTEMA, M.K. *Inclusion relations among families of context-free languages* ◇ D05 ◇
YOUNGER, D.H. *Recognition and parsing of context-free languages in time n^3 (Russian)* ◇ D05 D15 ◇
ZELEZNIKAR, A. *Overlapping algorithms* ◇ B75 D03 ◇

1968

AHO, A.V. & ULLMAN, J.D. *The theory of languages* ◇ D05 D10 D15 ◇
ARBIB, M.A. & GIVE'ON, Y. *Algebra automata I: Parallel programming as a prolegomena for the categorical approach. II:The categorical framework for dynamic analyisis* ◇ B75 C05 D05 G30 ◇
ARBIB, M.A. *The algebraic theory of machines, languages, and semigroups* ◇ C05 D05 D98 G30 ◇
ARBIB, M.A. *The automata theory of semigroup embeddings* ◇ D05 ◇
BRAUER, W. & INDERMARK, K. *Algorithmen, rekursive Funktionen und formale Sprachen* ◇ D05 D10 D20 D98 ◇
CAZANESCU, V.E. *Les notions fondamentales de la theorie des categories pour la categorie des automates sequentiels abstraits et la categorie des quasimachines* ◇ D05 G30 ◇
CODD, E.F. *Cellular automata* ◇ D05 ◇
CUDIA, D.F. & SINGLETARY, W.E. *Degrees of unsolvability in formal grammars* ◇ D03 D25 ◇
CUDIA, D.F. & SINGLETARY, W.E. *The Post correspondence problem* ◇ D03 D25 ◇
DALEN VAN, D. *Fans generated by nondeterministic automata* ◇ D05 F55 ◇
DAVIS, A.S. *Half-ring morphologies* ◇ D05 G25 ◇
DEMPSTER, J.R.H. *On the multiplying ability of two-way automata* ◇ D05 D10 ◇
DRAGALIN, A.G. *Lexicographical operator algorithms (Russian)* ◇ B75 D03 ◇
ENGELER, E. *Formal languages: automata and structures* ◇ D05 ◇
FISCHER, P.C. & MEYER, A.R. & ROSENBERG, ARNOLD L. *Counter machines and counter languages* ◇ D05 D10 D15 D20 ◇
FISCHER, P.C. & ROSENBERG, ARNOLD L. *Multitape one-way nonwriting automata* ◇ D05 ◇
FISCHER, P.C. & MEYER, A.R. & ROSENBERG, ARNOLD L. *Real-time counter machines* ◇ D05 ◇
FRISH, I. *Grammars with partial orderings of the rules* ◇ D05 ◇
GINSBURG, S. & ROSE, G.F. *A note on preservation of languages by transducers* ◇ D05 ◇
GINSBURG, S. & SPANIER, E.H. *Control sets on grammars* ◇ D05 ◇
GINSBURG, S. & SPANIER, E.H. *Derivation-bounded languages* ◇ D05 ◇
GINSBURG, S. & HARRISON, M.A. *On the elimination of endmarkers* ◇ D05 ◇
GINSBURG, S. & HARRISON, M.A. *One-way nondeterministic real-time list-storage languages* ◇ D05 ◇
GINZBURG, A. *Algebraic theory of automata* ◇ D05 ◇
GREIBACH, S.A. *A note on undecidable properties of formal languages* ◇ D05 ◇
GRIFFITHS, T.V. *Some remarks on derivations in general rewriting system* ◇ D03 ◇
GRODZKI, Z. *On the equivalence of Markov normal algorithms* ◇ D03 ◇
HAVEL, I.M. *Regular expressions over generalized alphabet and design of logical nets* ◇ D05 ◇
HOPCROFT, J.E. & ULLMAN, J.D. *Decidable and undecidable questions about automata* ◇ D05 D10 ◇
HOPCROFT, J.E. & ULLMAN, J.D. *Relations between time and tape complexities* ◇ D05 D10 D15 ◇
HOTZ, G. *Automatentheorie und formale Sprachen I. Turingmaschinen und rekursive Funktionen* ◇ D05 D10 D20 D98 ◇
HU, S.-T. *Mathematical theory of switching circuits and automata* ◇ B70 D05 D98 ◇
ION, I.D. *Sur l'axiomatisation de l'algebre des evenements reguliers* ◇ C05 D05 ◇
KAMEDA, T. *Generalized transition matrix of a sequential machine and its applications* ◇ D05 ◇
LORENTS, A.A. *Some problems in the constructive theory of finite probabilistic automata (Russian) (German summary)* ◇ D05 F65 ◇
NELSON, RAYMOND J. *Introduction to automata* ◇ D05 D10 D98 ◇
PAIR, C. & QUERE, A. *Definition et etude des bilangages reguliers* ◇ D05 ◇
POHL, H.-J. *Ueber die Reduzierung der Anzahl von Eingabesignalen von Automaten* ◇ D05 ◇
RABIN, M.O. *Decidability of second-order theories and automata on infinite trees* ◇ B15 B25 C85 D05 ◇
ROZENBERG, G. *Some remarks on Rabin and Scott's notion of multitape automaton (Russian summary)* ◇ D05 ◇
SALOMAA, A. *On events represented by probabilistic automata of different types* ◇ D05 ◇
SALOMAA, A. *On finite automata with a time-variant structure* ◇ D05 ◇
SALOMAA, A. *On finite time-variant automata with monitors of different types* ◇ D05 ◇

SALOMAA, A. *On languages accepted by probabilistic and time-variant automata* ◇ D05 ◇
SALOMAA, A. *On regular expressions and regular canonical systems* ◇ D05 ◇
SALOMAA, A. & TIXIER, V. *Two complete axiom systems for the extended language of regular expressions* ◇ D05 ◇
SHURYGIN, V.A. *Complete constructive sets with equality, and some of their properties (Russian)* ◇ D03 F60 ◇
STARKE, P.H. *Aequivalenz und Reduktion bei stochastischen Automaten* ◇ D05 ◇
STARKE, P.H. *Die Reduktion von stochastischen Automaten* ◇ D05 ◇
THATCHER, J.W. & WRIGHT, J.B. *Generalized finite automata theory with an application to a decision problem of second order logic* ◇ B15 B25 D05 ◇
TURQUETTE, A.R. *Dualizable quasi-strokes for m-state automata* ◇ B50 D05 ◇
VARTAPETOV, E.A. *Coding of abstract automata by normal codes I (Russian) (English summary)* ◇ D05 ◇
WAGNER, E.G. *Bounded action machines: toward an abstract theory of computer structure* ◇ D05 ◇

1969

AGASANDYAN, G.A. *Some problems of variable structure automaton theory (Russian)* ◇ D05 ◇
AHO, A.V. *Nested stack automata* ◇ D05 ◇
ANDERSON, MICHAEL *Note on the mortality problem for shift state trees* ◇ D05 ◇
ARBIB, M.A. *Theories of abstract automata* ◇ D05 D10 D98 ◇
ARIKAWA, S. *On the length functions of languages recognizable by linear bounded automata* ◇ D05 ◇
ASSER, G. *Malzew-Raeume* ◇ D05 ◇
AXT, P. & SINGLETARY, W.E. *Decision problems associated with complete deterministic normal systems* ◇ D03 ◇
AXT, P. & SINGLETARY, W.E. *On deterministic normal systems* ◇ D03 D25 ◇
BAER, ROBERT M. *A reduction theorem for normal algorithms* ◇ D03 ◇
BAER, ROBERT M. *Computability by normal algorithms* ◇ D03 D20 ◇
BAER, ROBERT M. & SPANIER, E.H. *Referenced automata and metaregular families* ◇ D05 ◇
BELETSKIJ, M.I. *The relationship between categorical and domination grammars I,II (Russian. English summary)* ◇ D05 ◇
BERSHTEJN, L.S. & KARELIN, V.P. & MELIKHOV, A.N. & PASHKEVICH, A.P. *Graphs of regular expressions, and abstract synthesis of automata (Russian)* ◇ D05 ◇
BOOK, R.V. & EVEN, S. & GREIBACH, S.A. & OTT, G. *Ambiguity in graphs and expressions* ◇ D05 ◇
BREJTBART, YU.YA. & KOZMIDIADI, V.A. *Two subclasses of Turing machines reducible to finite automata (Russian)* ◇ D05 D10 ◇
BUECHI, J.R. & LANDWEBER, L.H. *Definability in the monadic second-order theory of successor* ◇ B15 B25 C40 C85 D05 F35 ◇
BUECHI, J.R. & LANDWEBER, L.H. *Solving sequential conditions by finite-state strategies* ◇ D05 ◇
BURKHARD, HANS-DIETER *Ueber Experimente an nicht-deterministischen Automaten I* ◇ D05 ◇
COOPER, D.C. *Program scheme equivalences and second-order logic* ◇ B15 D05 ◇
EILENBERG, S. & ELGOT, C.C. & SHEPHERDSON, J.C. *Sets recognized by n-tape automata* ◇ D05 ◇
FREY, T. *Ueber die Reduktion und Konstruktion nicht vollstaendiger /partieller/ Automaten* ◇ D05 ◇
GALLAIRE, H. *Recognition time for context-free languages by on-line Turing machines* ◇ D05 D10 D15 ◇
GINSBURG, S. & GREIBACH, S.A. *Abstract families of languages* ◇ D05 ◇
GINSBURG, S. & GREIBACH, S.A. & HOPCROFT, J.E. *Pre-AFL* ◇ D05 ◇
GINSBURG, S. & GREIBACH, S.A. & HOPCROFT, J.E. *Studies in abstract families of languages* ◇ D05 ◇
HARRISON, M.A. *Lectures on linear sequential machines* ◇ D05 D98 ◇
HARTMANIS, J. *On the complexity of undecidable problems in automata theory* ◇ D05 D15 ◇
HONDA, N. & NASU, M. *Mappings induced by PGSM mappings and some recursively unsolvable problems of finite probabilistic automata* ◇ D05 E72 ◇
HOPCROFT, J.E. & ULLMAN, J.D. *Formal languages and their relation to automata* ◇ D05 D98 ◇
HOTZ, G. & WALTER, H. *Automatentheorie und formale Sprachen II. Endliche Automaten* ◇ D05 D98 ◇
JONES, N.D. *Context-free languages and rudimentary attributes* ◇ D05 ◇
KANOVICH, M.I. & PETRI, N.V. *Some theorems on the complexity of normal algorithms and computations (Russian)* ◇ D03 D15 ◇
KOZMIDIADI, V.A. & MARCHENKOV, S.S. *On multihead automata (Russian)* ◇ D05 D10 ◇
KWASOWIEC, W. *Some properties of machines (Russian)* ◇ B70 D05 ◇
LANDWEBER, L.H. *Decision problems for ω-automata* ◇ D05 D10 ◇
LANDWEBER, L.H. *Synthesis algorithms for sequential machines* ◇ B70 D05 ◇
LETICHEVS'KIJ, O.A. *Functional equivalence of discrete transformers I (Russian) (English summary)* ◇ D05 ◇
LORENTS, A.A. *Questions of the reducibility of finite probabilistic automata (Russian)* ◇ D05 ◇
LORENTS, A.A. *Synthesis of stable finite probabilistic automata (Russian)* ◇ D05 ◇
LORENTS, A.A. *The economy of states of finite probabilistic automata (Russian)* ◇ D05 ◇
MAKAREVSKIJ, A.YA. & STOTSKAYA, E.D. *Representability in deterministic multi-tape automata (Russian) (English summary)* ◇ D05 D10 ◇
MARCUS, S. *Contextual grammars* ◇ D05 ◇
MARUSCIAC, I. & NADIU, G.S. *A universal Turing machine for normal algorithms (Romanian) (French summary)* ◇ D03 D10 ◇
NOWAK, J. *A theory of finite many-output automata defined by matrices (Polish and Russian summaries)* ◇ D05 ◇
OSTROUKHOV, D.A. *Estimation of the complexity of normal algorithms (Russian)* ◇ D03 ◇

POHL, H.-J. *Darstellbarkeit von Ereignissen in Z-endlichen Automaten* ◇ D05 ◇
RABIN, M.O. *Decidability of second order theories and automata on infinite trees* ◇ B15 B25 C85 D05 ◇
REUSCH, B. *Lineare Automaten* ◇ D05 ◇
REVESZ, G. *On certain formal grammars and syntatic analysis without blind path* ◇ D05 ◇
REYNOLDS, J.C. *A generalized resolution principle based upon context-free grammars* ◇ B35 D05 ◇
ROVAN, B. *Bounded push down automata (Czech summary)* ◇ D05 ◇
ROWICKI, A. *On minimal and maximal orders of simple processes* ◇ D05 ◇
ROWICKI, A. *Ueber ein Kompliziertheitsmass einfacher Prozesse* ◇ D05 ◇
ROZENBERG, G. *p-automata and p-events* ◇ D05 ◇
ROZENBERG, G. *Finite memory address machines are universal* ◇ D05 ◇
ROZENBERG, G. *On the introduction of orderings into the grammars of Chomsky's hierarchy* ◇ B65 D05 ◇
SALOMAA, A. *On grammars with restricted use of productions* ◇ D05 ◇
SALOMAA, A. *On the index of a context-free grammar and language* ◇ D05 ◇
SALOMAA, A. *Probabilistic and weighted grammars* ◇ D05 ◇
SALOMAA, A. *Theory of automata* ◇ D05 D10 D98 ◇
SAMOJLENKO, L.G. *On a class of grammars of direct components (Russian)* ◇ D05 ◇
SAMOJLENKO, L.G. *Some subclasses of immediate-constitutent grammars which are weakly equivalent to context-free grammars (Russian)* ◇ D05 ◇
SCHEIN, B.M. *On some problems in the theory of partial automata (Czech summary)* ◇ D05 ◇
SCHMITT, A.A. *Zur Theorie der nichtdeterministischen und unvollstaendigen Automaten* ◇ D05 ◇
SCHNORR, C.-P. & WALTER, H. *Pullbackkonstruktionen bei Semi-Thuesystemen (English and Russian summaries)* ◇ D03 ◇
SCHWEIZER, B. & SKLAR, A. *A grammar of functions I* ◇ D05 ◇
SCHWEIZER, B. & SKLAR, A. *A grammar of functions II* ◇ D05 ◇
STARKE, P.H. *Abstrakte Automaten* ◇ D05 D75 D98 ◇
STARKE, P.H. *Schwache Homomorphismen fuer stochastische Automaten* ◇ D05 ◇
STARKE, P.H. *Ueber die Minimalisierung von stochastischen Rabin-Automaten* ◇ D05 ◇
TAJTSLIN, M.A. *Equivalence of automata with respect to a commutative semigroup (Russian)* ◇ D05 ◇
VARTAPETOV, E.A. *Coding of abstract automata by normal codes II (Russian) (English summary)* ◇ D05 ◇
WALIGORSKI, S. *Algebraic theory of automata (Polish)* ◇ D05 ◇
ZASLAVSKIJ, I.D. *On Shannon pseudofunctions (Russian)* ◇ D05 D15 F60 ◇

1970

AMOROSO, S. & COOPER, GERALD *The Garden-of-Eden theorem for finite configurations* ◇ D05 ◇
ARIKAWA, S. *Elementary formal systems and formal languages – simple formal systems* ◇ D05 ◇
BARANSKIJ, V.A. & TRAKHTMAN, A.N. *Subsemigroup graphs (Russian) (English summary)* ◇ C65 D05 ◇
BARTHOLOMES, F. & HOTZ, G. *Homomorphismen und Reduktionen linearer Sprachen* ◇ D05 ◇
BARZDINS, J. *Deciphering of sequential networks in the absence of an upper limit on the number of states (Russian)* ◇ D05 ◇
BARZDINS, J. & TRAKHTENBROT, B.A. *Finite automata. Behavior and synthesis (Russian)* ◇ D05 D98 ◇
BAVEL, Z. & MULLER, D.E. *Connectivity and reversibility in automata* ◇ D05 ◇
BEDNAREK, A.R. & NORRIS, E.M. *Congruences and ideals in machines I* ◇ C05 D05 ◇
BIRKHOFF, GARRETT & LIPSON, J.D. *Heterogeneous algebras* ◇ C05 D05 ◇
BOEHLING, K.H. & INDERMARK, K. *Endliche Automaten I* ◇ D05 ◇
BOEHLING, K.H. & SCHUETT, D. *Endliche Automaten II* ◇ D05 ◇
BRAUER, W. *Automates topologiques et ensembles reconnaissables* ◇ D05 ◇
BRAUER, W. *Ein elementarer Beweis des Kaskadezerlegungssatzes fuer endliche Automaten* ◇ D05 ◇
BUECHI, J.R. *Algorithmisches Konstruieren von Automaten und die Herstellung von Gewinnstrategien nach Cantor-Bendixson* ◇ D05 E60 ◇
BUECHI, J.R. & HOSKEN, W.H. *Canonical systems which produce periodic sets* ◇ D03 ◇
BURKHARD, HANS-DIETER *Ueber Experimente an nicht-deterministischen Automaten II* ◇ D05 ◇
BURKS, A.W. *Programming and the theory of automata* ◇ B75 D05 ◇
CERNY, J. *Two particular types of finite state languages* ◇ D05 ◇
CHEN, I-NGO & SHENG, C.L. *The decision problems of definite stochastic automata* ◇ D05 ◇
CLAUS, V. *Zerlegungen von Semi-Thue-Systemen* ◇ D03 ◇
COLLINS, D.J. *A universal semigroup (Russian)* ◇ C05 C57 D05 D40 ◇
CUDIA, D.F. *General Problems of formal grammars* ◇ D05 ◇
CULIK, K. *On conditional context-free grammars for programming and natural languages* ◇ B75 D05 ◇
CULIK II, K. *N-ary grammars and the description of mapping of languages* ◇ D05 ◇
DINCA, A. *Distribution classes, total non-ambiguity, and grammars with a finite number of states (Romanian) (French summary)* ◇ D05 ◇
DOERR, J. & HOTZ, G. (EDS.) *Automaten-Theorie und formale Sprachen. Bericht aus dem mathematischen Forschungsinstitut Oberwolfach 3* ◇ D05 ◇
DONER, J.E. *Tree acceptors and some of their applications* ◇ B25 C85 D05 ◇

EBBINGHAUS, H.-D. *Aufzaehlbarkeit* ◇ D03 D10 D20 D25 F30 ◇
EICKEL, J. *Vereinfachung der Struktur formaler Sprachen* ◇ D05 ◇
ELGOT, C.C. *The common algebraic structure of exit-automata and machines* ◇ D05 D10 ◇
GINSBURG, S. & HARRISON, M.A. *On the closure of AFL under reversal* ◇ D05 ◇
GINSBURG, S. & ROSE, G.F. *On the existence of generators for certain AFL* ◇ D05 ◇
GINSBURG, S. & GREIBACH, S.A. *Principal AFL* ◇ D05 ◇
GINSBURG, S. & SPANIER, E.H. *Substitution in families of languages* ◇ D05 ◇
GINSBURG, S. & HOPCROFT, J.E. *Two-way balloon automata and AFL* ◇ D05 ◇
GLEBSKIJ, YU.V. *A certain class of word sets (Russian) (English summary)* ◇ D05 ◇
GORAJ, A. & MIRKOWSKA, M. & PALUSZKIEWICZ, A. *On the notion of the description of the programm (Russian summary)* ◇ B75 D05 ◇
GRINBLAT, V.A. *On the applicability problem for normal algorithms (Russian)* ◇ D03 ◇
GRODZKI, Z. *The k-machines* ◇ D03 D05 ◇
GROSS, M. & LENTIN, A. *Introduction to formal grammars* ◇ B65 D05 D98 ◇
GUREVICH, Y. *Minsky machines and the case* $\forall\exists\forall$ & $\exists^{\infty}$ *of the decision problem (Russian)* ◇ B20 D05 D35 ◇
HARTMANIS, J. & HOPCROFT, J.E. *What makes some language theory problems undecidable* ◇ D05 D30 ◇
HERMES, H. *Entscheidungsproblem und Dominospiele* ◇ D05 D10 D35 ◇
JAERVI, T. *On control sets induced by grammars* ◇ D05 ◇
KAMEDA, T. & VOLLMAR, R. *Zur Umkehrkomplexitaet von Sprachen* ◇ D05 D10 D15 ◇
KORSHUNOV, A.D. *The invariant properties of finite automata (Russian)* ◇ D05 ◇
KOZMIDIADI, V.A. *A certain generalization of finite automata that produces a hierarchy analogous to A.Grzegorczyk's classification of primitively recursive functions (Russian)* ◇ D05 D10 D20 ◇
KRAL, J. *A modification of substitution theorem with some applications* ◇ D05 ◇
KROM, MELVEN R. *The decision problem for formulas in prenex conjunctive normal form with binary disjunctions* ◇ B20 D03 D35 ◇
LANGMAACK, H. & SCHMIDT, GUNTHER *Klassen unwesentlich verschiedener Ableitungen als Verbaende* ◇ D03 D40 G10 ◇
LETICHEVS'KIJ, O.A. *Functional equivalence of discrete transformers. II (Russian)* ◇ D05 ◇
LEWIS, F.D. *The classification of unsolvable problems in automata theory* ◇ D05 D25 ◇
LINNA, M. *Set of schemata of c-valid equations between regular expressions is independent of the basic alphabet* ◇ D05 ◇
MAGIDOR, M. & MORAN, G. *Probabilistic tree automata and context free languages* ◇ D05 ◇
MCCAWLEY, J.D. *English is a VSO language* ◇ B65 D05 ◇
MONTAGUE, R. *Universal grammar* ◇ A05 B65 D05 ◇
NOWAK, J. *A simplification of the basic theorem on automata (Polish and Russian summaries)* ◇ D05 ◇
NOWAKOWSKI, R. *Two definitions of the equivalence of automata (Polish and Russian summaries)* ◇ D05 ◇
OSTROUKHOV, D.A. *On the coding of natural numbers with the aid of schemes of normal algorithms (Russian)* ◇ D03 D20 ◇
PAGER, D. *The categorization of tag systems in terms of decidability* ◇ D03 ◇
PODLOVCHENKO, R.I. *Schemes of algorithms that are defined on situations (Russian)* ◇ D05 D20 ◇
RABIN, M.O. *Weakly definable relations and special automata* ◇ B15 B25 C40 C85 D05 ◇
REISCHER, C. & SIMOVICI, D.A. *On the reduced form of the linear boolean automata* ◇ B70 D05 ◇
SALOMAA, A. *On some families of formal languages obtained by regulated derivations* ◇ D05 ◇
SALOMAA, A. *Periodically time-variant context-free grammars* ◇ D05 ◇
SHURYGIN, V.A. *Constructive sets with equality and their mappings (Russian)* ◇ D03 F60 ◇
SIEFKES, D. *Decidable theories I. Buechi's monadic second order successor arithmetic* ◇ B15 B25 C85 D05 F35 F98 ◇
SIEFKES, D. *Decidable extensions of monadic second order successor arithmetic* ◇ B15 B25 C85 D05 F35 ◇
STARKE, P.H. & THIELE, H. *On asynchronous stochastic automata* ◇ D05 ◇
STARKE, P.H. & THIELE, H. *Ueber asynchrone stochastische Automaten* ◇ D05 ◇
STARKE, P.H. *Ueber regulaere nicht-deterministische Operatoren* ◇ D05 ◇
STARKE, P.H. *Ueber Minima von nicht-deterministischen Automaten* ◇ D05 ◇
THATCHER, J.W. *Self-describing Turing machines and self-reproducing cellular automata* ◇ D05 D10 ◇
THATCHER, J.W. *Universality in the von Neumann cellular model* ◇ D05 D10 ◇
T"RKALANOV, K.D. *A certain class of partially ordered semigroups with a solvable inequality problem (Russian)* ◇ D05 D40 ◇
TUZOV, V.A. *Graph-schemes with associative calculus (Russian)* ◇ B70 D05 ◇
TUZOV, V.A. *The equivalence of logical schemes with permutation operators (Russian) (English summary)* ◇ B75 D05 ◇

1971

AANDERAA, S.O. & BELSNES, D. *Decision problems for tag systems* ◇ D03 D25 D30 ◇
ALAD'EV, V.Z. *Certain estimates for Neumann-Moore structures (Russian) (English and Estonian summaries)* ◇ D05 ◇
ANISIMOV, A.V. *The group languages (Russian) (English summary)* ◇ D05 D40 ◇
BARNES, B.H. *A two-way automaton with fewer states than any equivalent one-way automaton* ◇ D05 ◇

BARTOL, W. *On the existence of machine homomorphisms* ◇ D05 D10 ◇
BASCA, O. *La synthese des automates finis par la methode de A.Church* ◇ D05 ◇
BLIKLE, A.J. *Algorithmically definable functions: a contribution towards the semantics of programming languages* ◇ B75 D05 D20 ◇
BOOK, R.V. & EVEN, S. & GREIBACH, S.A. & OTT, G. *Ambiguity in graphs and expressions* ◇ D05 ◇
BOONE, W.W. & COLLINS, D.J. & MATIYASEVICH, YU.V. *Embeddings into semigroups with only a few defining relations* ◇ D03 D40 ◇
BOUCHER, C. *Lecons sur la theorie des automates mathematiques* ◇ D05 D10 D15 D98 ◇
BREJTBART, YU.YA. *On the automaton and "zone" complexity of the predicate "to be a k-th power of an integer" (Russian)* ◇ D05 D15 ◇
BRZOZOWSKI, J.A. & CULIK, K. & GABRIELIAN, A. *Classification of noncounting events* ◇ D05 D25 ◇
BURKHARD, W.A. & VARAIYA, P.P. *Complexity problems in real time languages* ◇ D05 D15 ◇
CHANG, SHIKUO *Fuzzy programs – theory and applications* ◇ B75 D05 ◇
CONWAY, J.H. *Regular algebra and finite machines* ◇ D05 ◇
DALEN VAN, D. *A note on some systems of Lindenmayer* ◇ D03 D05 D10 D25 ◇
DEUSSEN, P. *Halbgruppen und Automaten* ◇ D05 ◇
DOEPP, K. *Automaten in Labyrinthen I,II (Russian summary)* ◇ D05 ◇
D'YACHENKO, V.F. *On transformation of schemes of algorithms (Russian)* ◇ D05 D20 ◇
EJSMUND, J. *On a certain class of m-address machines (Polish and Russian summaries)* ◇ D05 ◇
ENGELER, E. *Algorithmic approximation* ◇ B75 D05 D20 ◇
ENGELER, E. (ED.) *Symposium on semantics of algorithmic languages* ◇ B75 B97 D05 ◇
FEJNBERG, V.Z. & TYSHKEVICH, R.I. *The permutability of transformations I (Russian)* ◇ B75 C05 D05 ◇
GECSEG, F. & PEAK, I. *Algebraic theory of automata* ◇ D05 D98 ◇
GINSBURG, S. & SPANIER, E.H. *AFL with the semilinear property* ◇ D05 ◇
GINSBURG, S. & HOPCROFT, J.E. *Images of AFL under certain families of homomorphisms* ◇ D05 ◇
GINSBURG, S. & GOLDSTINE, J. *Intersection-closed full AFL and the recursively enumerable languages* ◇ D05 D25 ◇
GLEBSKIJ, YU.V. & KOGAN, D.I. *Additively controllable systems and languages: certain algorithmic problems (Russian) (English summary)* ◇ D05 ◇
GOLD, E.M. *Universal goal-seekers* ◇ D05 D10 D20 ◇
GROSCHE, G. *Syntaktische Untersuchungen zu mathematischen Sprachen I,II,III* ◇ D05 ◇
GROSS, M. & LENTIN, A. *Mathematische Linguistik* ◇ B65 D05 D98 ◇
GRZYMALA-BUSSE, J.W. *Operation-preserving functions and autonomous factors of finite automata* ◇ D05 ◇
HARTMANIS, J. & LEWIS II, P.M. *The use of lists in the study of undecidable problems in automata theory* ◇ D05 ◇
HERMAN, G.T. *When is a sequential machine the realization of another* ◇ D05 ◇
HOARE, C.A.R. *Procedures and parameters: an axiomatic approach* ◇ D05 ◇
HUGHES, C.E. & OVERBEEK, R.A. & SINGLETARY, W.E. *The many-one equivalence of some general combinatorial decision problems* ◇ D03 D05 D10 D25 D30 ◇
HUMMITZSCH, P. *Die Entscheidbarkeit der endlichen Ununterscheidbarkeit endlicher Automaten* ◇ D05 ◇
HUMMITZSCH, P. *Zur schwachen Aequivalenz von endlichen Automaten (Russian and English summaries)* ◇ D05 ◇
KASAMI, T. & TANIGUCHI, K. *Some decision problems for two-dimensional nonwriting automata (Japanese)* ◇ D05 ◇
KNUTH, D.E. *Examples of formal semantics* ◇ D05 ◇
KOBAYASHI, K. *Structural complexity of context-free languages* ◇ D05 ◇
KOHAVI, Z. & PAZ, A. (EDS.) *Theory of machines and computations* ◇ D05 D97 ◇
KURKI-SUONIO, R. *A programmer's introduction to computability and formal languages* ◇ B75 D05 D20 D98 ◇
LEHMANN, D.J. *Lr(k) grammars and deterministic languages* ◇ D05 ◇
LEPISTOE, T. *On commutative languages* ◇ D05 ◇
MCNAUGHTON, R. *A decision procedure for generalized sequential mapability-onto of regular sets* ◇ D05 ◇
METRA, I. *Comparison of the number of states of probabilistic and deterministic automata that represent given events (Russian)* ◇ D05 ◇
MUELLER, HORST *Endliche Automaten und Labyrinthe (English and Russian summaries)* ◇ D05 ◇
NIVAT, M. *Congruence de Thue et t-languages* ◇ D03 D05 D40 ◇
PAZ, A. *Introduction to probabilistic automata* ◇ D05 D98 ◇
PHAN DINH DIEU *On a class of stochastic languages* ◇ D05 ◇
PHAN DINH DIEU *On the languages representable by finite probabilistic automata* ◇ D05 D10 ◇
PRATHER, R.E. *An algebraic proof of the Paull-Unger theorem* ◇ D05 ◇
PRIESE, L. *Normalformen von Markov'schen und Post'schen Algorithmen. Eine Einfuehrung in die Theorie der normierten Algorithmen* ◇ D03 D40 D98 ◇
RABIN, M.O. *Decidability and definability in second-order theories* ◇ B15 B25 C40 C85 D05 ◇
RAJLICH, V. *Absolutely parallel grammars and two-way deterministic finite-state transducers* ◇ D05 ◇
RAS, Z. *Deductive systems of computing machines* ◇ B05 D05 ◇
RAS, Z. *On a relationship between the propositional calculus and a grammar (Russian summary)* ◇ B05 D05 ◇

RAS, Z. *Semi-Thue systems as deductive systems of certain computing machines (Russian summary)* ◇ D03 ◇
ROMOV, B.A. *The uniprimitive foundations of the maximal subalgebras of Post algebras (Russian) (English summary)* ◇ B50 D03 G20 ◇
ROZENBERG, G. *The unsolvability of the isomorphism problem for address machines* ◇ D05 ◇
SALOMAA, A. *Grammars with control languages* ◇ D05 ◇
SALOMAA, A. *The generative capacity of transformational grammars of Ginsburg and Partee* ◇ D05 ◇
SCHMITT, A.A. *Automaten. Algorithmen. Gehirne* ◇ D05 D10 ◇
SCHURMANN, A. *Functions computable by a computer (Polish and Russian summaries)* ◇ D05 D10 D20 ◇
STANULOV, N. *On the information structure of a class of finite automata* ◇ D05 ◇
STARKE, P.H. *Einige Bemerkungen ueber asynchrone stochastische Automaten* ◇ D05 ◇
STARKE, P.H. *Ueber die Transformation zweiseitig unendlicher Folgen durch determinierte Automaten* ◇ D05 ◇
TSEJTIN, G.S. *Lower estimate of the number of steps for an inverting normal algorithm and other similar algorithms (Russian) (English summary)* ◇ D03 D15 ◇
TSEJTIN, G.S. *Reduced form of normal algorithms and a linear acceleration theorem (Russian) (English summary)* ◇ D03 D15 ◇
ULLIAN, J.S. *Three theorems concerning principal AFLs* ◇ D05 ◇
URPONEN, T. *On axiom systems for regular expressions and on equations involving languages* ◇ D05 ◇
WALAT, A. *Some properties of k-machines (Russian summary)* ◇ D03 D05 ◇
YEH, R.T. *Some structural properties of generalized automata and algebras* ◇ C05 D05 ◇

1972

ALAD'EV, V.Z. *Computability in homogeneous structures (Estonian and Russian summaries)* ◇ D03 D05 D10 ◇
ALESHIN, S.V. *Finite automata and the Burnside problem for periodic groups (Russian)* ◇ D05 ◇
ANIKEEV, A.S. *Classification of derivable propositional formulas (Russian)* ◇ B15 D05 F20 F50 ◇
ANISIMOV, A.V. *Certain algorithmic questions for groups and context-free languages (Russian) (English summary)* ◇ D05 D40 ◇
AROLD, D. & BERNHARDT, L. & STAIGER, L. *Ueber die Entscheidung und Aufzaehlung regulaerer Wortmengen durch Markowsche Algorithmen* ◇ D03 D05 ◇
BAKER, J.L. *Grammars with structured vocabulary: A model for the ALGOL-68 definition* ◇ B75 D05 D25 ◇
BAKKER DE, J.W. *Recursion, induction and symbol manipulation* ◇ D05 D20 ◇
BARZDINS, J. *Prognostication of automata and functions* ◇ D05 D20 ◇
BIEN, Z. *Some properties of p-events* ◇ D05 ◇
BOOK, R.V. & GINSBURG, S. *Multi-stack-counter languages* ◇ D05 ◇
BRAINERD, W.S. & KNODE, R.B. *Some criteria for determining recognizability of a set* ◇ D05 ◇
BREJTBART, YU.YA. & KOZMIDIADI, V.A. *Certain subclasses of Turing machines that are reducible to finite automata (Russian) (English summary)* ◇ D05 D10 ◇
BUDACH, L. & HOEHNKE, H.-J. *Homologie und Automaten* ◇ D05 ◇
CHOMSKY, N. *Studies on semantics in generative grammars* ◇ B65 B98 D05 ◇
CIAPALA, E. *Produkte von n-Maschinen (Polish) (Russian and English summaries)* ◇ D05 ◇
CLAUS, V. & HOTZ, G. *Automatentheorie und formale Sprache III. Formale Sprachen* ◇ D05 ◇
COBHAM, A. *Uniform tag sequences* ◇ D03 ◇
COHEN, A.M. *Two algebraic proofs for regular languages* ◇ D05 ◇
COHEN, R.S. *Rank-non-increasing transformations on transition graphs* ◇ D05 ◇
COSTICH, O.L. *A Medvedev characterization of sets recognized by generalized finite automata* ◇ D05 ◇
FRIEDRICH, U. *Entscheidbarkeit der monadischen Nachfolgerarithmetik mit endlichen Automaten ohne Analysetheorem (Russian, English and French summaries)* ◇ B25 C85 D05 F35 ◇
GERMANO, G. & MAGGIOLO-SCHETTINI, A. *Equivalence of partial recursivity and computability by algorithms without concluding formulas* ◇ D03 D20 ◇
GINSBURG, S. & GREIBACH, S.A. *Multitape AFA* ◇ D05 ◇
GINSBURG, S. & GOLDSTINE, J. *On the largest full sub-AFL of an AFL* ◇ D05 ◇
GIULIANO, J.A. *Writing stack acceptors* ◇ D05 D10 ◇
GOLDSTINE, J. *Substitution and bounded languages* ◇ D05 ◇
GREIBACH, S.A. *Syntactic operators on full semi-AFL's* ◇ D05 ◇
HERINGER, H.J. *Formale Logik und Grammatik* ◇ B65 B98 D05 ◇
HOSKEN, W.H. *An asymmetric regular set* ◇ D05 ◇
HOSKEN, W.H. *Some Post canonical systems in one letter* ◇ D03 ◇
HUGHES, C.E. *Degrees of unsolvability associated with Markov algorithms* ◇ D03 D25 D55 ◇
ILYUSHKIN, V.A. *On the complexity of a grammatical description of context-free languages (Russian)* ◇ D05 ◇
ITKIN, V.E. *Term-logical equivalence of schemes of programs (Russian)* ◇ B75 D05 ◇
JARZABEK, S. *Product k-machines and their classification (Polish) (Russian and English summaries)* ◇ D05 ◇
JOSHI, A.K. & KOSARAJU, S.RAO & YAMADA, H. *String adjunct grammars I: Local and distributed adjunction. II: Equational representation, null symbols, and linguistic relevance* ◇ D05 ◇
KAIN, R.Y. *Automata theory: machines and languages* ◇ D05 D10 D98 ◇
KAMBAYASHI, Y. & YAJIMA, S. *Finite memory machines satisfying the lower bound of memory* ◇ D05 ◇
KLOETZER, G. & RAUTENBERG, W. *Im Grenzbereich Algebra, Logik, Maschinen. 10 Jahre Forschungsarbeit*

der Nowosibirsker Schule A.I. Malcev's (eine Studie) ◇ A10 B98 C05 D05 D98 ◇
KURMIT, A.A. *On the question of the enumerability and representability of sets of words by means of finite automata (Russian)* ◇ D05 ◇
KUZNETSOV, O.P. *Ueber die Kompliziertheit der Berechnungen in eindimensionalen iterativen Strukturen (Russian)* ◇ D05 D10 ◇
LEDGARD, H.F. *Embedding Markov normal algorithms within the λ-calculus* ◇ B40 D03 D15 ◇
LETICHEVS'KIJ, O.A. *Functional equivalence of discrete transformers. III (Russian)* ◇ D05 ◇
LETICHEVS'KIJ, O.A. *Optimization of the strategy of a control processor (Russian) (English summary)* ◇ D05 ◇
LEVIN, V.A. *Infinite-valued logic and transient processes in finite automata (Russian)* ◇ B50 D05 ◇
LEVIN, V.I. (ED.) *Fragen der Synthese endlicher Automaten (Russian)* ◇ D05 D98 ◇
LORENTS, A.A. *Elemente der konstruktiven Theorie stochastischer Automaten (Russian)* ◇ D05 D10 D98 F60 F98 ◇
MACHTEY, M. *Augmented loop languages and classes of computable functions* ◇ D05 D20 ◇
MAL'TSEV, I.A. *Congruences and automorphisms in cells of Post algebras (Russian)* ◇ C05 D03 G20 ◇
MAL'TSEV, I.A. *Some properties of cellular subalgebras of a Post algebra and their basic cells (Russian)* ◇ D03 G20 ◇
MASLOV, S.YU. *Deduction search in calculi of general type (Russian) (English summary)* ◇ B35 D03 F07 ◇
MASLOV, S.YU. & RUSAKOV, E.D. *Probabilistic canonical calculi (Russian) (English summary)* ◇ B48 D03 ◇
MAZURKIEWICZ, A.W. *Recursive algorithms and formal languages* ◇ D05 D20 ◇
MEZNIK, I. *G-machines and generable sets* ◇ D05 ◇
MILGRAM, D.L. & ROSENFELD, A. *Array automata and array grammars* ◇ D05 D10 ◇
MIZUMOTO, M. & TANAKA, K. & TOYODA, J. *General formulation of formal grammars* ◇ B65 D05 ◇
MOISEEVA, G.N. *Ueber die Synthese von Automaten auf der Basis von Algorithmengraphen mit parallelen Zweigen (Russisch)* ◇ D05 ◇
MUELLER, HORST *Endliche Automaten* ◇ D05 ◇
MUENTEFERING, P. & STARKE, P.H. *Ueber reduzible Ereignisse* ◇ D05 D10 ◇
NEKVINDA, M. *On a certain event recognizable in real time (Czech summary)* ◇ D05 D15 ◇
NEPOMNYASHCHIJ, V.A. *Conditions for the algorithmic completeness of systems of operations* ◇ D05 ◇
PAVLIDIS, T. *Linear and context-free graph grammars* ◇ D05 ◇
PIRICKA-KELEMENOUA, A. *Directable automata and directly subdefinite events (Czech summary)* ◇ D05 ◇
RABIN, M.O. *Automata on infinite objects and Church's problem* ◇ B15 B25 C85 D05 ◇
RAJLICH, V. *Absolutely parallel grammars and two-way finite-state transducers* ◇ D05 ◇
RAS, Z. *On a relationship between certain grammars and enumerable first order predicate calculi (Russian summary)* ◇ B20 D05 ◇
RITCHIE, R.W. & SPRINGSTEEL, F. *Language recognition by marking automata* ◇ D05 D10 ◇
ROZENBERG, G. *The equivalence problem for deterministic TOL-systems is undecidable* ◇ D03 ◇
SALOMAA, A. *Matrix grammars with a leftmost restriction* ◇ D05 ◇
SCHMIDTKE, K. *Classification of abstract computers with respect to the rho-inclusion (Polish) (Russian and English summaries)* ◇ D05 ◇
SMITH, L.W. & YAU, S.S. *Generation of regular expressions for automata by the integral of regular expressions* ◇ D05 ◇
STARKE, P.H. *Allgemeine Probleme und Methoden in der Automatentheorie* ◇ D05 ◇
STARKE, P.H. *Das Analyse-Synthese-Problem in der Automatentheorie* ◇ D05 ◇
STARKE, P.H. *Ueber die Minimisierung von Automaten mit halbgeordnetem Ausgabealphabet (Russian, English and French summaries)* ◇ D05 ◇
STARKE, P.H. *Ueber die Experimentmengen determinierter Automaten* ◇ D05 ◇
STARKE, P.H. *Ueber Finalexperimente an determinierten Automaten* ◇ D05 ◇
URPONEN, T. *On regular expressions possessing the empty word property* ◇ D05 ◇
URPONEN, T. *On regular expressions over one letter and on commutative languages* ◇ D05 ◇
WAGNER, E.G. *Bounded action machines II. The basic structure of tapeless computers (German summary)* ◇ D05 ◇
WAGNER, K. *Zur Axiomatisierung eines sequentiellen Aussagenkalkuels (Russian, English and French summaries)* ◇ B22 D05 ◇
WALL, R.E. *Introduction to mathematical linguistics* ◇ B65 D03 D05 D98 ◇
WECHSUNG, G. *Quasisequentielle Wortfunktionen* ◇ D05 ◇
WECHSUNG, G. *Ueber die Gruppe der eineindeutigen laengentreuen sequentiellen Funktionen* ◇ D05 ◇
WINKOWSKI, J. *Composed abstract machines (Russian summary)* ◇ D05 ◇
ZHAROV, V.G. *The complexity of the terms of constructive sequences of normal algorithms (Russian)* ◇ D03 D15 ◇

1973

ARBIB, M.A. *Coproducts and decomposable machines* ◇ D05 G30 ◇
ARBIB, M.A. & CULIK, K. *Sequential and jumping machines and their relation to computers* ◇ B70 D05 D10 ◇
BIRKHOFF, GARRETT *Current trends in algebra* ◇ A10 D05 D15 ◇
BLATTNER, M. *The unsolvability of the equality problem for sentential forms of context-free grammars* ◇ D05 ◇
BOEHLING, K.H. & INDERMARK, K. (EDS.) *Erste Fachtagung der Gesellschaft fuer Informatik ueber Automatentheorie und formale Sprachen (Bonn, 9. - 12. Juli 1973)* ◇ D05 D97 ◇
BRZOZOWSKI, J.A. & SIMON, IMRE *Characterizations of locally testable events* ◇ D05 ◇

BUTZBACH, P. *Une famille de congruences de Thue pour lesquelles le probleme de l'equivalence est decidable. Application a l'equivalence des grammaires separees* ◇ D03 D05 ◇

DAUSCHA, W. & NUERNBERG, G. & STARKE, P.H. & WINKLER, K.-D. *Theorie der determinierten zeitvariablen Automaten* ◇ D05 ◇

DOEPP, K. *Ein Dualitaetstheorem der Automatentheorie* ◇ D05 ◇

DOERFLER, W. *Halbgruppen und Automaten* ◇ D05 ◇

FLAJOLET, P. & STEYAERT, J.-M. *Complexite des problemes de decision relatifs aux algorithmes de tri* ◇ D05 D15 ◇

FLAJOLET, P. & STEYAERT, J.-M. *Decision problems for multihead finite automata* ◇ D05 ◇

GABRIELIAN, A. & GINSBURG, S. *Structured storage AFA* ◇ B75 D05 ◇

GERMANO, G. & MAGGIOLO-SCHETTINI, A. *A flow diagram composition of Markov's normal algorithms without concluding formulas* ◇ B75 D03 D20 ◇

GERMANO, G. & MAGGIOLO-SCHETTINI, A. *Markov's normal algorithms without concluding formulas* ◇ D03 ◇

GINSBURG, S. & GOLDSTINE, J. *Intersection-closed full AFL and the recursively enumerable languages* ◇ D05 D25 ◇

GINSBURG, S. & GREIBACH, S.A. *On AFL generators for finitely encoded AFA* ◇ D05 ◇

GLADKIJ, A.V. *Formale Grammatiken und Sprachen (Russisch)* ◇ D05 D10 D15 ◇

GLUSHKOV, V.M. & LETICHEVS'KIJ, O.A. *The theory of discrete transformators (Russian)* ◇ D05 ◇

GODLEVSKIJ, A.B. *Certain special cases of the stopping problem and of the functional equivalence of automata (Russian) (English summary)* ◇ D05 ◇

GORLOV, V.V. *Congruences on closed Post classes (Russian)* ◇ B50 D03 ◇

HAMBLIN, C.L. *Language types and logical theorems* ◇ B05 D05 ◇

HENKE VON, F.W. & INDERMARK, K. & WEIHRAUCH, K. *Hierarchies of primitive recursive wordfunctions and transductions defined by automata* ◇ D05 D20 ◇

HENKE VON, F.W. & WEIHRAUCH, K. *Klassifizierung von primitiv-rekursiven Transformationen und Automatentransduktionen* ◇ D05 D20 ◇

HOSCH, F.A. & LANDWEBER, L.H. *Finite delay solutions for sequential conditions* ◇ D05 ◇

HSIA, PEI & YEH, R.T. *Finite automata with markers* ◇ D05 ◇

HUGHES, C.E. & SINGLETARY, W.E. *Combinatorial systems with axiom* ◇ D03 D05 D10 D25 D30 ◇

HUGHES, C.E. *Many-one degrees associated with problems of tag* ◇ D03 D10 D25 D40 ◇

HUGHES, C.E. *Many-one degrees associated with semi-Thue systems* ◇ D03 D25 ◇

KANOVICH, M.I. *Komplexitaet der beschraenkten Entscheidbarkeit von Algorithmen (Russian)* ◇ D05 D15 D20 ◇

KARPINSKI, M. *Free structure tree automata I: Equivalence. II: Non-deterministic and deterministic regularity. III: Normalized climbing automata* ◇ D05 ◇

KUDRYAVTSEV, V.B. *The functional properties of logical nets* ◇ D05 ◇

KUDRYAVTSEV, V.B. *The function system $\mathscr{P}_{\Sigma}$ (Russian)* ◇ D05 ◇

LAING, R. *The capabilities of some species of artificial organism* ◇ D05 D80 ◇

LALLEMENT, G. *The role of regular and inverse semigroups in the theory of finite state machines and languages* ◇ D05 ◇

LETICHEVS'KIJ, O.A. *Practical methods for recognizing the equivalence of discrete transformers and program schemes (Russian) (English summary)* ◇ B75 D05 ◇

LETICHEVS'KIJ, O.A. *The equivalence of automata with respect to cancellative semigroups (Russian)* ◇ D05 D40 ◇

LINNA, M. *Finite power property of regular languages* ◇ D05 ◇

MAYER, O. *Die Darstellung indizierter Sprachen durch Ausdruecke* ◇ D05 ◇

MILGRAM, D.L. & ROSENFELD, A. *Parallel/sequential array automata* ◇ D05 ◇

MOISIL, G.C. *Sur la possibilite de modeler le fini par l'infini* ◇ D05 D10 ◇

MOON, B.A. *A Markov algorithm interpreter* ◇ D03 D15 ◇

MOSZNER, Z. *Structure de l'automate plein, reduit et inversible* ◇ D05 ◇

NAGORNYJ, N.M. *Separability with respect to invariants (Russian)* ◇ D03 D25 ◇

NASH, B. *Reachability problems in vector addition systems* ◇ D05 ◇

NEKVINDA, M. *On the complexity of events recognizable in real time* ◇ D05 D15 ◇

NICOLESCU, R. *Some remarks on the theory of normal algorithms* ◇ D03 ◇

NIVAT, M. (ED.) *Automata, languages and programming* ◇ B75 D05 D97 ◇

NIVAT, M. *Congruences parfaites et quasi-parfaites* ◇ D03 ◇

NOWAK, J. *On a certain congruence of automata with respect to connecting and coupling with retardation (Polish and Russian summaries)* ◇ D05 ◇

NOWAK, J. *Remarks on the notion of equivalence of automata (Polish and Russian summaries)* ◇ D05 ◇

NOWAKOWSKI, R. *On the homomorphisms of graphs of automata* ◇ D05 ◇

NOWAKOWSKI, R. *On the products of graphs of automata* ◇ D05 ◇

NOWAKOWSKI, R. *Some properties of connection of graphs of automata* ◇ D05 ◇

ORLOV, V.A. *A simple proof of the algorithmic undecidability of certain problems on the completeness of automaton bases (Russian) (English summary)* ◇ D05 ◇

ORLOV, V.A. *Ueber die Kompliziertheit der Realisierung beschraenkt-determinierter Operatoren durch Schemata in Automatenbasen (Russisch)* ◇ D03 D05 ◇

OSTROUKHOV, D.A. *Linearization of constructive sequences of normal algorithms (Russian)* ◇ D03 F60 ◇

OSTROUKHOV, D.A. *The complexity of terms of sequences of normal algorithms (Russian)* ◊ D03 D20 ◊
OTTMANN, T. *Ketten und arithmetische Praedikate von endlichen Automaten* ◊ D05 D55 D75 ◊
OTTMANN, T. *Ueber Moeglichkeiten zur Simulation endlicher Automaten durch eine Art sequentieller Netzwerke aus einfachen Bausteinen* ◊ D05 ◊
OVERBEEK, R.A. *The representation of many-one degrees by the word problem for Thue systems* ◊ D03 D30 D40 ◊
PATERSON, M.S. & VALIANT, L.G. *Deterministic one-counter automata* ◊ D05 D15 ◊
PAZ, A. & SALOMAA, A. *Integral sequential word functions and growth equivalence of Lindenmayer systems* ◊ D05 ◊
PETERS JR., P.S. & RITCHIE, R.W. *Nonfiltering and local-filtering transformational grammars* ◊ D05 ◊
PILIPOSYAN, A.G. *Certain transformations of logical schemes (Russian) (Armenian summary)* ◊ D05 ◊
PODLOVCHENKO, R.I. *R-schemes and equivalence relations between them (Russian)* ◊ D05 ◊
PODLOVCHENKO, R.I. *A complete system of similar transformations of R-schemes (Russian)* ◊ D05 ◊
RINE, D.C. *A correspondence between control logic of associative memories and Markov algorithms* ◊ B70 D03 ◊
ROSE, G.F. & WEIHRAUCH, K. *A characterization of the classes L_1 and R_1 of primitive recursive word functions* ◊ D05 D20 ◊
SALOMAA, A. *Developmental languages: a new type of formal languages* ◊ D05 ◊
SALOMAA, A. *Formal languages* ◊ D05 D15 D98 ◊
SALOMAA, A. *L-systems: a device in biologically motivated automata theory* ◊ D05 ◊
SALOMAA, A. *On exponential growth in Lindenmayer systems* ◊ D05 D15 ◊
SALOMAA, A. *On sentential forms of context-free grammars* ◊ D05 ◊
SALOMAA, A. *On some recent problems concerning developmental languages* ◊ B75 D05 ◊
SAVITCH, W.J. *Nondeterministic finite automata revisited* ◊ D05 D15 ◊
SCHMIDTKE, K. *Classification of computing machines with finite computations with respect to ρ-inclusion* ◊ D05 ◊
SCHUETZENBERGER, M.-P. *A propos du relation rationelles fonctionelles* ◊ D05 ◊
SEMENOV, A.L. *Algorithmic problems for power series and for context-free grammars (Russian)* ◊ D05 D80 ◊
SKOWRON, A. *Machines with input and output* ◊ D05 D10 ◊
SKRZYPKOWSKI, T. *On the inclusion of relational machines (Polish) (Russian and English summaries)* ◊ D05 ◊
STARKE, P.H. *Grammatiken und Sprachen* ◊ D05 ◊
STARKE, P.H. *On the sequential relations of time-variant automata* ◊ D05 ◊
STARKE, P.H. *Ueber diagnostische Strategien an determinierten Automaten* ◊ D05 ◊
STARKE, P.H. & THIELE, H. *Ueber die Experimentmengen nicht-deterministischer Automaten (Russian)* ◊ D05 ◊
STARKE, P.H. *Ueber die Experimentmengen schwach-initialer Automaten* ◊ D05 ◊
TAKAHASHI, MASAKO *Primitive transformations of regular sets and recognizable sets* ◊ D05 ◊
TAKAOKA, T. *Two measures over language space (Japanese)* ◊ D05 ◊
TITARENKO, L.N. *The complexity of recognizing functional properties of normal algorithms (Russian)* ◊ D03 D15 ◊
VASILEVSKIJ, M.P. *The detection of the disrepair of automata (Russian) (English summary)* ◊ D05 ◊
VIDAL-NAQUET, G. *Quelques applications des automates a arbres infinis* ◊ B25 D05 D35 ◊
WAJS, R. *On certain properties of relational machines* ◊ D05 ◊
WAKULICZ-DEJA, ALICJA *Eigenschaften von Maschinen, die durch partielle Uebergangsfunktionen charakterisiert werden* ◊ D05 ◊
WAND, M. *A concrete approach to abstract recursive definitions* ◊ D05 D75 G10 ◊
WANG, JUENTIN *On the representation of generative grammars as first-order theories* ◊ B10 D05 ◊
WECHSUNG, G. *Isomorphe Darstellungen der Kleeneschen Algebra der regulaeren Mengen* ◊ D05 ◊
WECHSUNG, G. *Quasisequentielle Funktionen* ◊ D05 E20 ◊
YAKU, T. *The constructibility of a configuration in a cellular automaton* ◊ D05 ◊
YANOV, YU.I. *Ueber das Problem aequivalenter Umformungen (Russian)* ◊ D05 ◊
ZEIGLER, B.P. *Every discrete input machine is linearly simulatable* ◊ D05 ◊

1974

AANDERAA, S.O. & LEWIS, H.R. *Linear sampling and the ∀∃∀ case of the decision problem* ◊ B20 D03 D05 D35 D80 ◊
ADAMEK, J. *Free algebras and automata realizations in the language of categories* ◊ D05 G30 ◊
AJTKHOZHAEVA, E.ZH. & BARASHENKOV, V.V. *The definition of some operator scheme relations (Russian)* ◊ D05 ◊
ALAD'EV, V.Z. *τ_n-grammars and languages generated by them (Russian) (English summary)* ◊ D05 ◊
ANDO, S. & ITO, T. *A complete axiom system of super-regular expressions* ◊ D05 ◊
ARBIB, M.A. & MANES, E.G. *Machines in a category: an expository introduction* ◊ C05 D05 D98 G30 ◊
BARZDINS, J. *On synthesizing programs given by examples* ◊ D03 ◊
BARZDINS, J. *Prediction and identification in the limit of finite automata (Russian)* ◊ D05 ◊
BIRKHOFF, GARRETT & LIPSON, J.D. *Universal algebra and automata* ◊ C05 D05 ◊
BOOTH, T.L. *Design of minimal expected processing time finite-state transducers* ◊ D05 ◊
BURKHARD, HANS-DIETER *Diagnose und Einstellung nicht-deterministischer Automaten bei regulaeren Unterscheidungsformen* ◊ D05 ◊
CASE, J. *Periodicity in generations of automata* ◊ D05 ◊

CHOUEKA, Y. *Theories of automata on ω-tapes: a simplified approach* ◇ D05 ◇

COURCELLE, B. & VUILLEMIN, J. *Semantics and axiomatics of a simple recursive language (French summary)* ◇ D05 ◇

DANG HUY RUAN & TSEJTIN, G.S. *An upper estimate of the finite-state complexity for a class of generating schemes containing complements and intersections (Russian)* ◇ D05 ◇

DASSOW, J. *Kleene-Mengen und Kleene-Vollstaendigkeit (English and Russian summaries)* ◇ D05 ◇

DASSOW, J. *Kleene-Vollstaendigkeit bei stark-definiten Ereignissen* ◇ D05 ◇

DASSOW, J. *Ueber zwei abgeschlossene Mengen von Automatenoperatoren* ◇ D05 ◇

DAVIS, CHARLES C. *Some semantically closed languages* ◇ A05 D05 ◇

DESCHAMPS, J.-P. *Asynchronous automata and asynchronous languages* ◇ D05 ◇

DIMITROV, V. & WECHLER, W. *R-fuzzy automata* ◇ D05 E72 ◇

EILENBERG, S. *Automata, languages, and machines. Vol. A* ◇ C05 D05 D98 ◇

FISCHER, P.C. & WARKENTIN, J.C. *Predecessor machines* ◇ D05 ◇

FRIEDMAN, E.P. *Relationships between monadic recursion schemes and deterministic context-free languages* ◇ D05 D20 ◇

GABRIELIAN, A. & GINSBURG, S. *Grammar schemata* ◇ D05 ◇

GALIL, Z. *Two way deterministic pushdown automaton languages and some open problems in the theory of computation* ◇ D05 D10 D20 ◇

GINSBURG, S. & SPANIER, E.H. *On incomparable AFL* ◇ D05 ◇

GINSBURG, S. & ROVAN, B. *On the periodicity of word-length in DOL languages* ◇ D05 ◇

GINSBURG, S. & ROSE, G.F. *The equivalence of stack-counter acceptors and quasi-realtime stack-counter acceptors* ◇ D05 D10 D15 ◇

GLEBSKIJ, YU.V. & GORDON, E.I. *Asynchronous automata with delays, and logical languages (Russian) (English summary)* ◇ D05 ◇

GODLEVSKIJ, A.B. & LETICHEVS'KIJ, O.A. (EDS.) *Automata and algorithm theory, and mathematical logic (Russian)* ◇ B35 B97 D05 D20 D97 ◇

GODLEVSKIJ, A.B. *Ueber einen Spezialfall des Problems der Funktionalaequivalenz diskreter Transformer (Russisch)* ◇ D05 ◇

GREIBACH, S.A. *Some restrictions on w-grammars* ◇ D05 ◇

GRODZKI, Z. *The Kolmogorov's complexity of the computation of the k-machines* ◇ D03 D15 D20 ◇

GUITART, R. *Remarques sur les machines et les structures* ◇ D05 G30 ◇

HANF, W.P. *Nonrecursive tilings of the plane I* ◇ D05 D80 ◇

HARDOUIN-DUPARC, J. *Paradis terrestre dans l'automate cellulaire de Conway* ◇ D05 ◇

HARTMANIS, J. & HUNT III, H.B. *The LBA problem and its importance in the theory of computing* ◇ D05 D15 ◇

HAVRANEK, T. *An application of logical-probabilistic expressions to the realization of stochastic automata* ◇ B70 B80 D05 ◇

HERSCHEL, R. *Einfuehrung in die Theorie der Automaten, Sprachen und Algorithmen* ◇ D05 D10 D98 ◇

HUGHES, C.E. *Single premise Post canonical forms defined over one-letter alphabets* ◇ D03 ◇

KARPINSKI, M. *Free structure tree automata IV: sequential representation* ◇ D05 ◇

KARPINSKI, M. *Probabilistic climbing and sinking languages* ◇ D05 ◇

KOERBER, P. & OTTMANN, T. *Simulation endlicher Automaten durch Ketten aus einfachen Bausteinautomaten* ◇ D05 ◇

KOLODIJ, A.N. *Two-sided nondeterministic automata (Russian) (English summary)* ◇ D05 ◇

KORSHUNOV, A.D. *A survey of certain trends in automata theory (Russian)* ◇ D05 ◇

KOSARAJU, S.RAO *Regularity preserving functions* ◇ D05 ◇

KUDRYAVTSEV, V.B. *On the functional system* $\mathscr{P}_{\Sigma}$ *(Russian)* ◇ D05 ◇

LEVIN, V.A. *Equations in infinite-valued logic, and transition processes in finite automata (Russian)* ◇ B50 D05 ◇

LINDNER, R. & WAGNER, K. *The axiomatisation of a certain sequential propositional calculus (Russian)* ◇ B60 B70 D05 D10 ◇

MARKOV, A.A. *On the language* A_1 *(Russian)* ◇ D03 D35 F50 ◇

MASLOV, A.N. *The hierarchy of indexed languages of an arbitrary level (Russian)* ◇ D05 ◇

MCNAUGHTON, R. *Algebraic decision procedures for local testability* ◇ D05 ◇

MYERS, D.L. *Nonrecursive tiling of the plane II* ◇ D05 D80 ◇

NAGORNYJ, N.M. *Ueber einige Verfahren zur Beschreibung der Arbeit von Systemen wechselwirkender Rechenmaschinen (Russian)* ◇ D05 ◇

NARUSHIMA, H. & NOJIMA, S. & OOHARA, S. *A tag type automaton with a double reading head* ◇ D03 D05 ◇

NOVOTNY, M. *On some operators reducing generalized grammars* ◇ D05 ◇

NOVOTNY, M. *Sets constructed by acceptors* ◇ D05 D25 ◇

OTTMANN, T. *Arithmetische Praedikate ueber einem Bereich endlicher Automaten* ◇ D05 D55 D75 ◇

PASCU, A. *On the recursiveness of context-sensitive languages* ◇ D05 ◇

PETROSYAN, G.N. *A certain basis of operators and predicates with an undecidable emptiness problem (Russian) (English summary)* ◇ B75 D05 D80 ◇

PLATEK, M. *Questions of graphs and automata in a generative description of language* ◇ D05 ◇

PODLOVCHENKO, R.I. *Non-determined algorithm schemata or R-schemata* ◇ D05 ◇

PRIESE, L. & ROEDDING, D. *A combinatorial approach to selfcorrection* ◇ D05 D10 ◇

RANGEL, J.L. *The equivalence problem for regular expressions over one letter is elementary* ◇ D05 D20 ◇

RASIOWA, H. *On ω^+-valued algorithmic logic and related problems* ◇ B75 D05 ◇
RAYNA, G. *Degrees of finite-state transformability* ◇ D05 D30 ◇
SALOMAA, A. *L systems* ◇ D05 ◇
SALOMAA, A. *Recent results on L-systems* ◇ D05 ◇
SALOMAA, A. *Solution of a decision problem concerning unary Lindenmayer systems* ◇ D05 ◇
SMIKUN, L.B. *The problem of the equivalence of automata with respect to the direct product of groups (Russian) (English summary)* ◇ D05 ◇
SPEHNER, J.-C. *Deux algorithmes relatifs aux sous-monoides d'un monoide libre* ◇ D05 ◇
STAIGER, L. & WAGNER, K. *Automatentheoretische und automatenfreie Charakterisierungen topologischer Klassen regulaerer Folgenmengen* ◇ D05 ◇
STARKE, P.H. *Das allgemeine Diagnose-Problem* ◇ D05 ◇
STARKE, P.H. *On diagnosing experiments with nondeterministic automata with final states* ◇ D05 ◇
STARKE, P.H. *Theorie der sequentiellen Automaten* ◇ D05 ◇
SUDBOROUGH, I.H. *Bounded-reversal multihead finite automata languages* ◇ D05 ◇
TOWNSEND, R. *A decidability result in algebraic language theory* ◇ D05 ◇
TSEJTLIN, G.E. & YUSHCHENKO, E.L. *Ueber die Darstellung von Sprachen in Bobsleigh-Automaten (Russisch)* ◇ D05 ◇
WANG, HAO *Notes on a class of tiling problems* ◇ D05 D80 ◇
WIJNGAARDEN VAN, A. *The generative power of two-level grammars* ◇ D05 ◇
YASUHARA, A. *Some non-recursive classes of Thue systems with solvable word problem* ◇ D03 D40 ◇
ZEMBRZUSKI, K. *On D-machines (Russian summary)* ◇ D05 ◇
ZEMBRZUSKI, K. *On D-machine representation of continuous functions (Russian summary)* ◇ D05 ◇
ZHAROV, V.G. *The complexity of the universal algorithm (Russian)* ◇ D03 D15 ◇

1975

ALAD'EV, V.Z. & OSIPOV, O. *About the complexity of parallel algorithms defined by homogeneous structures (Russian) (Estonian and English summaries)* ◇ D03 D05 D10 D15 ◇
AMOROSO, S. & COOPER, GERALD & PATT, Y.N. *Some clarifications of the concept of a Garden-of-Eden configuration* ◇ D05 ◇
ANDO, S. & ITO, T. *On applying Scott's logic to termination problems* ◇ B40 D05 ◇
ANISIMOV, A.V. & SEIFERT, F.D. *Zur algebraischen Charakteristik der durch kontext-freie Sprachen definierten Gruppen (English and Russian summaries)* ◇ D05 D40 ◇
ARBIB, M.A. & MANES, E.G. *A category-theoretic approach to systems in a fuzzy world* ◇ C05 D05 E72 G30 ◇
ARBIB, M.A. & MANES, E.G. *Adjoint machines, state behavior machines, and duality* ◇ D05 G30 ◇
ARBIB, M.A. & MANES, E.G. *Fuzzy machines in a category* ◇ B52 D05 G30 ◇
BAKER, J.L. *Factorization of Scott-style automata* ◇ D05 ◇
BARALT-TORRIJOS, J. & CHIARAVIGLIO, L. & GROSKY, W. *The programmatic semantics of binary predicator calculi* ◇ B40 D05 ◇
BARZDINS, J. *Inductive derivation of automata, function and programs* ◇ D05 D20 ◇
BIEN, Z. *The star measure of regular events (Russian summary)* ◇ D05 ◇
BOERGER, E. *Recursively unsolvable algorithmic problems and related questions reexamined* ◇ D03 D10 D20 ◇
BOFFA, M. *Introduction aux automates finis et aux langages reguliers* ◇ D05 ◇
BRAKHAGE, H. (ED.) *Automata theory and formal languages* ◇ D05 D97 ◇
BUDACH, L. & HOEHNKE, H.-J. *Automaten und Funktoren* ◇ D05 ◇
BUDACH, L. *On the solution of the labyrinth problem for finite automata* ◇ D05 ◇
BURKHARD, HANS-DIETER *Zustandsidentifizierung an asynchronen Automaten* ◇ D05 ◇
BURKS, A.W. *Logic, biology and automata - some historical reflections* ◇ A05 A10 D05 D10 D80 ◇
CALUDE, C. *Sur une classe de distances dans un demi-groupe libre* ◇ D05 ◇
COUSINEAU, G. & RIFFLET, J.M. *Langages d'interpretation des schemas recursifs* ◇ B75 D05 ◇
CREMERS, A.B. & GINSBURG, S. *Context-free grammar forms* ◇ D05 ◇
CURRY, H.B. *Representation of Markov algorithms by combinators* ◇ B40 D03 ◇
DASSOW, J. *Ueber metrische Kleene-Vollstaendigkeit (English and Russian summaries)* ◇ D05 ◇
FEAUTRIER, P. *Post grammars as a programming language description tool* ◇ B75 D03 ◇
FREJDZON, R.I. *Finite approximate approach to the study of the complexity of recursive predicates (Russian)* ◇ D05 D10 D15 ◇
GAINES, B.R. & KOHOUT, L.J. *Possible automata* ◇ B50 D05 ◇
GALIL, Z. *Functional schemas with nested predicates* ◇ B75 D05 ◇
GALIL, Z. *Two fast simulations which imply some fast string matching and palindrome-recognition algorithms* ◇ D05 D10 D15 ◇
GENRICH, H.J. *Extended simple regular expressions* ◇ B75 D05 ◇
GINSBURG, S. *Algebraic and automata-theoretic properties of formal languages* ◇ D05 D98 ◇
GINSBURG, S. & SPANIER, E.H. *Substitution on grammar forms* ◇ D05 ◇
GINSBURG, S. & ROZENBERG, G. *Tol systems and control sets* ◇ D05 ◇
GINSBURG, S. & GOLDSTINE, J. & GREIBACH, S.A. *Uniformly erasable AFL* ◇ D05 ◇
GLUSHKOV, V.M. & TSEJTLIN, G.E. & YUSHCHENKO, E.L. *Automata theory, and some questions of the synthesis of language processor structures (Russian) (English summary)* ◇ D05 ◇

GOGUEN, J.A. & THATCHER, J.W. & WAGNER, E.G. & WRIGHT, J.B. *Factorisations, congruences, and the decomposition of automata and systemi* ⋄ D05 ⋄
GRODZKI, Z. *Generalized Markov normal algorithms* ⋄ D03 ⋄
HAMILTON, W.L. & MERTENS, J.R. *Reproduction in tessellation structures* ⋄ D05 ⋄
HAVEL, I.M. *Finite branching automata: automata theory motivated by problem solving* ⋄ D05 ⋄
HOLLERER, W.O. & VOLLMAR, R. *On "forgetful" cellular automata* ⋄ D05 ⋄
HOTZ, G. *Strukturelle Verwandtschaften von Semi-Thue-Systemen* ⋄ D03 ⋄
HSIA, PEI & YEH, R.T. *Marker automata* ⋄ D05 ⋄
HUGHES, C.E. & SINGLETARY, W.E. *Combinatorial systems defined over one- and two-letter alphabets* ⋄ D03 ⋄
HUGHES, C.E. *Sets derived by deterministic systems with axiom* ⋄ D03 D25 ⋄
HUGHES, C.E. *The general decision problem for Markov algorithms with axiom* ⋄ D03 ⋄
JIRKU, P. *Towards an integrated theory of formal and natural languages* ⋄ B65 D05 ⋄
KANOVICH, M.I. *A hierachical semantic system with set variables (Russian)* ⋄ B20 B65 C40 D05 F50 ⋄
KARPINSKI, M. *Almost deterministic ω-automata with existential output condition* ⋄ D05 ⋄
KARPINSKI, M. *Decision algorithm for Havel's branching automata* ⋄ D05 ⋄
KRAMOSIL, I. *A probabilistic approach to automaton-environment systems* ⋄ D05 ⋄
LAGANA, M.R. & LEONI, G. & PINZANI, R. & SPRUGNOLI, R. *Improvements in the execution of Markov algorithms (Italian summary)* ⋄ D03 ⋄
LEONI, G. & SPRUGNOLI, R. *The compilation of Pointer Markov Algorithms* ⋄ D03 D15 ⋄
LEWIS, H.R. *Description of restricted automata by first-order formulae* ⋄ D05 ⋄
LINNA, M. *On ω-words and ω-computations* ⋄ D05 ⋄
LISOVIK, L.P. *Generalized finite operators, and the word problem of the definitor algebra of regular events (Russian)* ⋄ D05 ⋄
MANOHAR, R.P. & TREMBLAY, J.P. *Discrete mathematical structures with applications to computer science* ⋄ D05 D20 G05 G10 ⋄
MASLOV, S.YU. *Mutation calculi (Russian) (English summary)* ⋄ D03 D25 D80 ⋄
MIKULECKY, P. *On configurations in cellular automata* ⋄ D05 ⋄
NAKAMURA, A. *On causal ω^2-systems* ⋄ D05 ⋄
NAZARYAN, G.A. *On the realization of boolean functions in algorithmic languages (Russian)* ⋄ B05 B75 D03 ⋄
OTTMANN, T. *Mit regulaeren Grundbegriffen definierbare Praedikate* ⋄ D05 ⋄
OTTMANN, T. *Some classes of nets of finite automata* ⋄ D05 ⋄
OWINGS JR., J.C. *Splitting a context-sensitive set* ⋄ D05 D25 ⋄
PATERSON, M.S. & VALIANT, L.G. *Deterministic one-counter automata* ⋄ D05 ⋄
PERROT, J.F. *Une theorie algebrique des automates finis monogenes* ⋄ D05 ⋄
RAKHMATULIN, N.A. *Automata-theoretic characteristics for the spectra of formulas of finite levels (Russian)* ⋄ C13 D05 D10 ⋄
REEDY, A. & SAVITCH, W.J. *The Turing degree of the inherent ambiguity problem for context-free languages* ⋄ D05 D30 ⋄
REVESZ, G. *Phrase-structure grammars and dual pushdown automata (Hungarian) (English summary)* ⋄ D05 D10 ⋄
REYNOLDS, J.C. *On the interpretation of Scott's domains* ⋄ B75 D05 G10 ⋄
RINE, D.C. *Representation and design of production systems* ⋄ B70 D03 G20 ⋄
RUSIECKI, A. *Computability of recursive-defined functions by executing the program, equivalent to their definitions (Polish) (English and Russian summaries)* ⋄ B75 D05 ⋄
SALOMAA, A. *Formal power series and growth functions of Lindenmayer systems* ⋄ D05 ⋄
SALOMAA, A. *On some decidability problems concerning developmental languages* ⋄ B75 D05 ⋄
SALOMON, K.B. *The decidability of a mapping problem for generalized sequential machines with final states* ⋄ D05 ⋄
SCHULER, P.F. *A note on degrees of context-sensitivity* ⋄ D05 ⋄
SCHULER, P.F. *WCS-analysis of the context-sensitive* ⋄ D05 D25 ⋄
SKYUM, S. *Confusion in the Garden of Eden* ⋄ D05 ⋄
SONTAG, E.D. *On some questions of rationality and decidability* ⋄ D05 ⋄
STAIGER, L. & WAGNER, K. *Finite automata acceptation of infinite sequences* ⋄ D05 E15 ⋄
STARKE, P.H. *Application of two-tape automata to analysis and synthesis of nondeterministic generalized sequential machines* ⋄ B70 D05 ⋄
STARKE, P.H. *On the representation of relations by deterministic and nondeterministic multitape automata* ⋄ D05 ⋄
STARKE, P.H. *Ueber die Darstellbarkeit von Relationen in Mehrbandautomaten* ⋄ D05 ⋄
SUDBOROUGH, I.H. *On tape-bounded complexity classes and multihead finite automata* ⋄ D05 D10 D15 ⋄
TER-AKOPOV, A.K. *Fast response optimization of discrete transformers with respect to automata composition (Russian) (English summary)* ⋄ D05 ⋄
VAISER, A. *Complexity of computation and stability of separation of languages by finite probabilistic automata* ⋄ D05 D15 ⋄
WAGNER, K. *A hierarchy of regular sequence sets* ⋄ D05 ⋄
WAGNER, K. *Akzeptierbarkeitsgrade regulaerer Folgenmengen* ⋄ D05 ⋄
WAND, M. *An algebraic formulation of the Chomsky hierarchy* ⋄ D05 ⋄
WECHLER, W. *R-fuzzy automata with a time-variant structure* ⋄ D05 E72 ⋄

WECHLER, W. *R-fuzzy grammars* ◇ D05 ◇
WECHSUNG, G. *Eine algebraische Charakterisierung der linearen Sprachen* ◇ D05 ◇
WECHSUNG, G. *The axiomatization problem of a theory of linear languages* ◇ D05 ◇
WIEHAGEN, R. *Dechiffrierung von ND-Automaten* ◇ D05 ◇

1976

ADLER, S. *Optimale K-bilineare Realisierung endlicher Automaten* ◇ D05 ◇
ADLER, S. *Zur Existenz optimaler K-bilinearer Realisierungen von endlichen transitiven Automaten* ◇ D05 ◇
ALT, H. & MEHLHORN, K. *Lower bounds for the space complexity of context-free recognition* ◇ D05 D15 ◇
BENDA, V. & BENDOVA, K. *On specific features of recognizable families of languages* ◇ D05 ◇
BENDA, V. & BENDOVA, K. *Recognizable filters and ideals* ◇ D05 ◇
BENEJAM, J.-P. & PAGET, M. *La complexite des algorithmes de Markov (English summary)* ◇ D03 D10 D15 ◇
BREJTBART, YU.YA. *Some bounds on the complexity of predicate recognition by finite automata* ◇ D05 D15 ◇
BRUNNER, J. & WECHLER, W. *On the behaviour of R-fuzzy automata* ◇ B75 D05 ◇
BURGINA, E.S. *Classes of binary grammars, binary relations and their projections (Russian) (English summary)* ◇ D05 ◇
BURGINA, E.S. *Properties of universal languages on some classes (Russian) (English summary)* ◇ D05 ◇
BURKHARD, HANS-DIETER *Identifizierungsexperimente an asynchronen Automaten* ◇ D05 ◇
BURKHARD, HANS-DIETER *Zum Laengenproblem homogener Experimente an determinierten und nicht-deterministischen Automaten* ◇ D05 ◇
CANNON, R.L. *An algebraic technique for context-sensitive parsing* ◇ D05 ◇
CHAPENKO, V.P. & GOBZEMIS, A.YU. & KURMIT, A.A. & LORENTS, A.A. & PETRENKO, A.F. & VASYUKEVICH, V.O. & YAKUBAJTIS, EH.A. & ZAZNOVA, N.E. *Automata theory (Russian)* ◇ D05 D98 ◇
CHVALINA, J. *Autonomous automata and closures with the same endomorphism monoids* ◇ D05 ◇
DAKOVSKI, L.G. & VOJKOV, G.K. *Possibilities of realizing stable natural extensions (Russian)* ◇ D05 ◇
DASSOW, J. *Kleene-Mengen und trennende Mengen* ◇ B50 D05 ◇
DASSOW, J. & HARNAU, W. *Spezielle Gleichungen in freien Halbgruppen und eine Anwendung auf Normalformen regulaerer Ereignisse* ◇ D05 ◇
DOEMOESI, P. *On superpositions of automata* ◇ D05 ◇
DUSKE, J. *Zur Theorie kommutativer Automaten (English summary)* ◇ D05 ◇
EHRIG, H. & ROSEN, B.K. *Commutativity of independent transformations on complex objects* ◇ D03 ◇
FISCHER, P.C. & ROBERTSON, E.L. & SAXTON, L.V. *On the sequential nature of functions* ◇ A05 D05 ◇
FRIZEN, D.G. & STARKOV, M.A. & TROFIMOV, O.E. *An automaton for calculation of the number of objects (Russian)* ◇ D05 D15 ◇
FROEHLICH, R. *Kontextisolierte Regelgrammatiken* ◇ D05 ◇
GABBAY, D.M. *On Kreisel's notion of validity in Post systems* ◇ B50 D03 F50 ◇
GAINES, B.R. & KOHOUT, L.J. *The logic of automata* ◇ D05 ◇
GECSEG, F. *On products of abstract automata* ◇ D05 ◇
GERMANO, G. & MAGGIOLO-SCHETTINI, A. *A language for Markov's algorithms composition* ◇ D03 ◇
GINSBURG, S. & MAURER, H.A. *On strongly equivalent context-free grammar forms* ◇ D05 ◇
GINSBURG, S. & LYNCH, N.A. *Size complexity in context-free grammatical families* ◇ D05 D15 ◇
GINSBURG, S. & GOLDSTINE, J. & GREIBACH, S.A. *Some uniformly erasable families of languages* ◇ D05 ◇
GORDON, H.G. *Complete degrees of finite-state transformability* ◇ D05 D20 ◇
GRODZKI, Z. & ZURAWIECKI, J. *The (k,m)-computation sets* ◇ D03 D05 ◇
HANSON, DAVID R. *A simple variant of the boundary-tag algorithm for the allocation of coroutine environments* ◇ D03 ◇
HATCHER, W.S. & RUS, T. *Context-free algebras* ◇ D05 ◇
HEIDLER, K. *Die Unentscheidbarkeit des Eck-Dominoproblems* ◇ D05 ◇
HERMES, H. *Domino games* ◇ D05 D35 ◇
HONDA, N. & NASU, M. *A completeness property of one-dimensional tessellation automata* ◇ D05 ◇
HOPKIN, D.R. & MOSS, B.P. *Automata* ◇ D05 D10 D98 ◇
HULE, H. & MAURER, H.A. & OTTMANN, T. *OL forms* ◇ D05 ◇
HUNT III, H.B. & SZYMANSKI, T.G. *Dichotomization, reachability, and the forbidden subgraph problem* ◇ B75 D05 D15 ◇
IBARRA, O.H. & KIM, C.E. *A useful device for showing the solvability of some decision problems* ◇ D03 D10 D80 ◇
KARASEK, J. *Nondeterministic and deterministic k-machines* ◇ D03 ◇
KARPINSKI, M. *Multiplicity functions on ω-automata* ◇ D05 ◇
KINBER, E.B. *Frequenzberechnungen auf endlichen Automaten (Russian) (English summary)* ◇ D05 D20 ◇
LAING, R. *Automaton introspection* ◇ D05 ◇
LETICHEVS'KIJ, O.A. & SMIKUN, L.B. *On a class of groups with solvable problem of automata equivalence (Russian)* ◇ D05 D40 ◇
LEWIS, H.R. *Krom formulas with one dyadic predicate letter* ◇ B20 D05 D35 ◇
LITMAN, A. *On the monadic theory of ω_1 without AC* ◇ B25 C85 D05 E10 E25 ◇
LOBODA, M. *Generalized canonical systems for generation of regular sets* ◇ D05 ◇

LOECKX, J. *Algorithmentheorie* ◇ D03 D10 D98 ◇
MAN, V.D. *Die Aequivalenzeigenschaft der Konfigurationen der n-dimensionalen iterativen Automaten* ◇ D05 ◇
MASLOV, S.YU. *Absorption relation on regular sets (Russian) (English summary)* ◇ D05 ◇
MCNAUGHTON, R. & SEIFERAS, J.I. *Regularity preserving relations* ◇ D05 ◇
MEISSNER, H.G. *Limes-Erkennung von regulaeren und linearen Sprachen mit Optimalitaetseigenschaften* ◇ D05 ◇
MEISSNER, H.G. *Ueber die Fortsetzbarkeit von sequentiellen Baum-Operatoren mit endlichem Gewicht* ◇ B75 D05 ◇
MELZI, G. *I supporti fisici dell'inferenza formale. Vol. I* ◇ B10 B75 D05 ◇
MICHAELSON, S. & MILNER, R. (EDS.) *Automata, languages and programming. III* ◇ D05 D97 ◇
MILGRAM, D.L. *A region crossing problem for array-bounded automata* ◇ D05 D10 ◇
MIR, H. *Two families of context-sensitive languages generated by controlled operator grammars* ◇ D05 ◇
MORGAN, C.G. *Some undecidability results for construction problems in tessellation automata* ◇ D05 ◇
NABEBIN, A.A. *Multitape automata in a unary alphabet (Russian)* ◇ D05 D10 ◇
NAGORNYJ, N.M. *Normal algorithms and first order languages (Russian)* ◇ B10 D03 ◇
PHAN DINH DIEU *A note on iterative arrays of finite automata over the one-letter alphabet* ◇ D05 ◇
PRIESE, L. *On a simple combinatorial structure sufficient for sub-lying nontrivial self-reproduction* ◇ D03 D10 ◇
PRIESE, L. *Reversible Automaten und einfache universelle 2-dimensionale Thue-Systeme* ◇ D03 D05 D10 ◇
RED'KO, V.N. *Definitor-theoretic aspects of languages (Russian)* ◇ D03 D05 D25 ◇
REVESZ, G. *A note on the relation of Turing machines to phrase structure grammars* ◇ D05 D10 ◇
ROSENBLOOM, P.C. *Structural models for use in psychological research* ◇ D03 D80 ◇
ROZENBERG, G. & SALOMAA, A. *Context-free grammars with graph-controlled tables* ◇ D05 ◇
ROZENBERG, G. & RUOHONEN, K. & SALOMAA, A. *Developmental systems with fragmentation* ◇ D05 ◇
ROZENBERG, G. & SALOMAA, A. *The mathematical theory of L-systems* ◇ D05 ◇
SALOMAA, A. *L systems: a parallel way of looking at formal languages. New ideas and recent developments* ◇ D05 ◇
SALOMAA, A. *Recent results on L systems* ◇ D05 ◇
SALOMAA, A. *Sequential and parallel rewriting* ◇ D05 ◇
SALOMAA, A. *Undecidable problems concerning growth in informationless Lindenmayer systems (German and English summaries)* ◇ D05 D80 ◇
SEIFERT, F.D. *Eine Klassifizierung endlich erzeugbarer Gruppen durch formale Sprachen* ◇ D05 ◇
SHUKURYAN, S.K. *Solvability of the equivalence problem in a class of multitape multihead automata and flow charts over memory (Russian) (English summary insert)* ◇ D05 ◇
SHUKURYAN, S.K. *Some decidable cases of the special problem of functional equivalence for x-y-automata (Russian)* ◇ D05 ◇
SOLNTSEV, S.V. *Two remarks on the stopping problem for cellular growth models (Russian)* ◇ D05 ◇
SOLS, I. *Logic, foundations, geometry and automata in the realm of topoi* ◇ D05 E70 G30 ◇
STAIGER, L. *Regulaere Nullmengen* ◇ D05 E75 ◇
STAIGER, L. & WAGNER, K. *Zur Theorie der abstrakten Familien von ω-Sprachen (ω-AFL)* ◇ D05 ◇
STARKE, P.H. *Closedness properties and decision problems for finite multi-tape automata* ◇ D05 ◇
STARKE, P.H. *Decision problems for multi-tape automata* ◇ D05 ◇
STARKE, P.H. *Entscheidungsprobleme fuer autonome Mehrbandautomaten* ◇ D05 ◇
STARKE, P.H. *Mehrbandakzeptoren* ◇ D05 ◇
STARKE, P.H. *On the diagonals of n-regular relations* ◇ D05 ◇
STARKE, P.H. *Ueber die Darstellbarkeit von Relationen in Mehrbandautomaten* ◇ D05 ◇
SUDBOROUGH, I.H. *On deterministic context-free languages, multihead automata, and the power of an auxiliary pushdown store* ◇ D05 D10 D15 ◇
SUDBOROUGH, I.H. *One-way multihead writing finite automata* ◇ D05 ◇
VANDERVEKEN, D.R. *The Lesniewski-Curry theory of syntactical categories and the categorially open functors* ◇ B15 D05 G30 ◇
WAGNER, K. *Eine Axiomatisierung der Theorie der regulaeren Folgenmengen (English and Russian summaries)* ◇ D05 ◇
WECHSUNG, G. *Kompliziertheitstheoretische Charakterisierung der kontextfreien und linearen Sprachen* ◇ D05 D15 ◇
WOOD, D. *Iterated a-NGSM maps and Γ systems* ◇ D05 ◇
YAKU, T. *Surjectivity of nondeterministic parallel maps induced by nondeterministic cellular automata* ◇ D05 ◇
ZHAROV, V.G. *On lower and upper estimates for the complexity of the terms of constructive sequences of normal algorithms (Russian)* ◇ D03 D15 ◇

1977

ADLER, S. *On structured extensions of finite automata* ◇ D05 ◇
ALESHIN, S.V. *Ueber ein Vollstaendigkeitskriterium fuer Superpositionen von Automatenabbildungen* ◇ D05 ◇
AMBARTSUMYAN, A.A. & POTEKHIN, A.I. *Standard realization of an asynchronous automaton (Russian)* ◇ D05 ◇
ASSER, G. *Bemerkungen zum Labyrinth-Problem* ◇ D05 ◇
AVALISHVILI, I. & BERISHVILI, G. *Extension of automata (Russian)* ◇ D05 ◇
BABICHEV, A.V. & PRONINA, V.A. & TRAKHTENGERTS, EH.A. *The conversion of engendering grammars into a form observing precedence functions (Russian)* ◇ D05 ◇

BARTOL, W. & RAS, Z. & SKOWRON, A. *Theory of computing systems* ◇ B75 D05 ◇
BENDA, V. & BENDOVA, K. *Characterization of recognizable families by means of regular languages* ◇ D05 ◇
BENDA, V. & BENDOVA, K. *On families recognizable by finite branching automata* ◇ D05 ◇
BERTONI, A. & MAURI, G. & TORELLI, M. *Some recursively unsolvable problems relating to isolated cutpoints in probabilistic automata* ◇ D05 ◇
BLATTNER, M. & HEAD, T. *Single-valued a-transducers* ◇ D05 ◇
BLUM, M. & SAKODA, W.J. *On the capability of finite automata in 2 and 3 dimensional space* ◇ D05 ◇
BOASSON, L. & COURCELLE, B. & NIVAT, M. *A new complexity measure for languages* ◇ D05 D15 ◇
BOOK, R.V. & DOBKIN, D.P. & SELMAN, A.L. & WRATHALL, C. *Inclusion complete tally languages and the Hartmanis-Berman conjecture* ◇ D05 ◇
BREMER, H. *Ein vollstaendiges Problem auf der Baummaschine* ◇ D05 D10 D15 ◇
BRUNNER, J. & WECHLER, W. *Zur Theorie der R-fuzzy-Automaten I* ◇ D05 E72 ◇
BUDACH, L. *Environments, labyrinths and automata* ◇ D05 ◇
BUTNARIU, D. *L-fuzzy automata. Description of a neural model* ◇ D05 D80 E72 ◇
CENZER, D. *Non-generable RE sets* ◇ D03 D25 ◇
CHIRKOV, M.K. & EGLITIS, L.V. *Synthesis of partial automata by partial regular events (Russian)* ◇ D05 ◇
CLEAVELAND, J.C. & UZGALIS, R.C. *Grammars for programming languages* ◇ B75 D05 ◇
COHORS-FRESENBORG, E. *Mathematik mit Kalkuelen und Maschinen* ◇ D03 D10 D98 ◇
COY, W. *Automata in labyrinths* ◇ D05 ◇
CREMERS, A.B. & GINSBURG, S. & SPANIER, E.H. *The structure of context-free grammatical families* ◇ D05 ◇
CRESSWELL, M.J. *Categorial languages* ◇ B65 D05 ◇
CSUHAJ VARJU, E. *Generative grammar forms and operators* ◇ D05 ◇
DALIK, J. *Verbandstheoretische Eigenschaften von Sprachen II* ◇ D05 ◇
DASSOW, J. *Einige Bemerkungen zu einem modifizierten Vollstaendigkeitsbegriff fuer Automatenabbildungen* ◇ D05 ◇
DASSOW, J. *Some remarks on the algebra of automaton mappings* ◇ D05 ◇
DASSOW, J. *Ueber Abschlusseigenschaften von biologisch motivierten Sprachen* ◇ D05 D80 ◇
DAUCHET, M. *Grammaires transformationnelles et bimorphismes de magmoides* ◇ D05 ◇
DEUTSCH, M. *Zum Begriff der Wortmischung als Basis fuer die Arithmetik* ◇ B28 D05 D75 ◇
DIETERICH, E. *Grobstrukturen fuer kontextfreie Grammatiken* ◇ D05 ◇
DUMITRESCU, S. & STEFAN, G.M. *The modelling of digital logical structures as finite state systems* ◇ D05 ◇
EHRENFEUCHT, A. & ROZENBERG, G. *On simplifications of PDOL systems* ◇ D05 ◇
ELGOT, C.C. *Finite automaton from a flowchart scheme point of view* ◇ D05 ◇
EMDEN VAN, M.H. *Relational equations, grammars, and programs* ◇ B75 D05 E07 ◇
FREJVALD, R.V. *On sets recognizable by a probabilistic machine with isolated cut-point and by no deterministic machine (Russian)* ◇ D05 D10 ◇
GINSBURG, S. & LYNCH, N.A. *Derivation complexity in context-free grammar forms* ◇ D05 ◇
GINSBURG, S. & MAURER, H.A. *On quasi-interpretations of grammar forms* ◇ D05 ◇
GRANT, P.W. *Recognition of EOL languages in less than quartic time* ◇ D05 D15 ◇
HEMMERLING, A. *Allgemeine Untersuchungen zur Entscheidbarkeit und Trennbarkeit bezueglich Kodierungen* ◇ D05 ◇
HEMMERLING, A. *Vergleich einiger Kodierungen der natuerlichen Zahlen bezueglich der regulaeren Entscheidbarkeit von Zahlenmengen* ◇ D05 D20 ◇
HOTZ, G. *Automatentheorie und formale Sprachen* ◇ B75 D05 D97 ◇
HUZINO, S. & SHIBATA, R. *An elementary proof of R. W. Ritchie's theorem on the set of squares* ◇ D05 ◇
INOUE, K. & NAKAMURA, A. *Some properties of two-dimensional on-line tessellation acceptors* ◇ D05 D15 ◇
ISARD, S.D. *A finitely axiomatizable undecidable extension of K* ◇ B45 D05 D35 ◇
KANOVICH, M.I. *Complex properties of context-sensitive languages (Russian)* ◇ D03 D05 D15 ◇
KAPHENGST, H. & REICHEL, HORST *Initial algebraic semantics for non context-free languages* ◇ B75 D05 G30 ◇
KOREC, I. *Decidability (undecidability) of equivalence of Minsky machines with components consisting of at most seven (eight) instructions* ◇ D05 D10 ◇
LADNER, R.E. *Application of model theoretic games to discrete linear orders and finite automata* ◇ B15 B25 C07 C13 C65 C85 D05 E07 E60 ◇
LEEUWEN VAN, J. *Recursively enumerable languages and van Wijngaarden grammars* ◇ D05 D25 ◇
LEONI, G. & SPRUGNOLI, R. *Some relations between Markov algorithms and formal languages* ◇ D03 D05 ◇
LEVIN, V.I. *Logical determinants and automata with continuous time. I: Theory of logical determinants. II: Automata without memory. III: Automata with memory (Russian)* ◇ D05 D10 ◇
LIPTON, R.J. & MILLER, R.E. & SNYDER, LAWRENCE *Synchronization and computing capabilities of linear asynchronous structures* ◇ B40 B70 D05 ◇
LISOVIK, L.P. *The word problem for an algebra of regular events over semigroups (Russian) (English summary)* ◇ D05 ◇
MARINKOVIC, I. *The unsolvability of the emptiness problem of the two deterministic finite transducer mappings* ◇ D05 ◇
MAURER, H.A. & SALOMAA, A. & WOOD, D. *EOL forms* ◇ B75 D03 D05 ◇

MAURER, H.A. & OTTMANN, T. & SALOMAA, A. *On the form equivalence of L-forms* ◇ D05 ◇
MEISSNER, H.G. *Zu einigen Begriffen und Resultaten aus der Theorie der Baumautomaten* ◇ D05 ◇
MONIEN, B. *A recursive and a grammatical characterization of the exponential-time languages* ◇ D05 D10 D15 ◇
MONIEN, B. *The LBA-problem and the transformability of the class* $\mathfrak{E}^2$ ◇ D05 D15 ◇
MUELLER, HORST *A one-symbol printing automaton escaping from every labyrinth* ◇ D05 ◇
NABEBIN, A.A. *Expressibility in restricted second-order arithmetic (Russian)* ◇ B15 C40 D05 F35 ◇
NAKAMURA, A. & ONO, H. *Two-dimensional finite automata and their application to the decision problem of monadic first-order arithmetic* $A[P, F(x), G(x)]$ ◇ D05 D35 F30 F35 ◇
O'DONNELL, M.J. *Subtree replacement systems: a unifying theory for recursive equations, LISP, lucid and combinatory logic* ◇ B40 B75 D05 D20 ◇
PLOTKIN, G.D. *LCF considered as a programming language* ◇ B75 D05 ◇
RASIOWA, H. *Many-valued algorithmic logic as a tool to investigate programs* ◇ B75 D03 G20 ◇
ROJZEN, S.I. *A counterexample in the theory of automaton equivalence types (Russian)* ◇ D05 D50 ◇
ROJZEN, S.I. *An analogue of the additive system of cardinals that is connected with finite automata (Russian) (English summary)* ◇ D05 D50 ◇
ROZENBERG, G. & SALOMAA, A. *New squeezing mechanisms for L systems* ◇ A05 B75 D05 ◇
SAKAROVITCH, JACQUES *Sur les groupes infinis, consideres comme monoides syntaxiques de langages formels* ◇ D05 D40 ◇
SALOMAA, A. *Formal power series and language theory* ◇ D05 ◇
SOBOLEV, S.K. *On finite-dimensional superintuitionistic logics (Russian)* ◇ B55 D05 F50 G10 G25 ◇
SOIL, A. *Remarks on uniform tag sequences* ◇ D03 ◇
STAIGER, L. *Empty-storage-acceptance of* ω*-languages* ◇ D05 D10 ◇
STAIGER, L. & WAGNER, K. *Recursive* ω*-languages* ◇ D05 D10 D55 ◇
STARKE, P.H. *Analyse und Synthese von asynchronen ND-Automaten* ◇ D05 ◇
STARKE, P.H. *Closedness properties of multihead languages* ◇ D05 ◇
STARKE, P.H. *Multitape automata and languages* ◇ D05 ◇
TSEJTLIN, G.E. *The system of algorithmic algebras and some control schemes in homogeneous structures (Russian)* ◇ D05 G25 ◇
VASYUKEVICH, V.O. *Ueber die Formalisierung der Loesung von Problemen der statischen Analyse diskreter Automaten mit moeglichen Funktionsstoerungen (Russian)* ◇ D05 ◇
VELOSO, P.A.S. *Some bounds on quasi-initialised finite automata* ◇ D05 ◇
VERBEEK, R. & WEIHRAUCH, K. *Data presentation and computational complexity* ◇ D05 D15 ◇
WAGNER, K. *Eine topologische Charakterisierung einiger Klassen regulaerer Folgenmengen* ◇ D05 ◇
WIEHAGEN, R. *Identification of formal languages* ◇ B75 D05 ◇

1978

AGUILO, J. & FABREGAT, F. & VALDERRAMA, E. & VAQUERO, A. *State assignment for incomplete sequential machines (Spanish) (English summary)* ◇ D05 ◇
ALBERT, J. & MAURER, H.A. & ROZENBERG, G. *Simple EOL forms under uniform interpretation generating of languages* ◇ D05 ◇
AL'PIN, YU.A. & KUZNETSOV, S.E. *Estimates for the number of completely indecomposable and cohesive matrices and corollaries for automata theory (Russian)* ◇ D05 ◇
ANGLUIN, D. *On the complexity of minimum inference of regular sets* ◇ D05 D15 ◇
ANISIMOV, A.V. & LISOVIK, L.P. *Equivalence problems for finite automata mappings into a free and a commutative semigroup (Russian) (English summary)* ◇ D05 ◇
ANKUDINOV, G.I. *Questions of the theory of plex-languages (Russian)* ◇ D05 ◇
ARNOLD, A. & DAUCHET, M. *Caracterisation algebrique des ensembles recursivement enumerables d'arbres* ◇ D05 ◇
ARNOLD, A. & LATTEUX, M. *Recursivite et cones rationnels fermes par intersection* ◇ D05 D80 ◇
ARSLANOV, M.M. (ED.) *Algorithms and automata. Collection of scientific works (Russian)* ◇ D05 D20 D97 ◇
BABICHEV, A.V. *Automata that accept languages generated by precedence grammars (Russian)* ◇ D05 ◇
BACIU, A. & PASCU, A. *Leontief type languages* ◇ D05 ◇
BAKER, B.S. *Generalized syntax directed translation tree transducers, and linear space* ◇ D05 D15 ◇
BAKER, B.S. *Tree transducers and tree languages* ◇ D05 ◇
BARTH, G. *Grammars with dynamic control sets* ◇ B75 D05 ◇
BEAUQUIER, J. *Ambiguite forte* ◇ D05 ◇
BEAUQUIER, J. *Un generateur inheremment ambigu du cone des langages algebriques* ◇ D05 ◇
BERSTEL, J. *Ensembles reconnaissables de nombres* ◇ D05 ◇
BERSTEL, J. *Memento sur les transductions rationnelles* ◇ D05 ◇
BOASSON, L. & HORVATH, S. *On languages satisfying Ogden's lemma* ◇ D05 ◇
BOASSON, L. *Un critere de rationalite des langages algebriques* ◇ D05 ◇
BOERGER, E. *Ein einfacher Beweis fuer die Unentscheidbarkeit der klassischen Praedikatenlogik* ◇ B10 D03 D10 D35 ◇
BOGOMOLOV, A.M. & TVERDOKHLEBOV, V.A. *Synchronization of a deterministic finite automaton (Russian)* ◇ D05 ◇
BOOTH, K.S. *Isomorphism testing for graphs, semigroups, and finite automata are polynomially equivalent problems* ◇ D05 D15 ◇

BRANDENBURG, F.-J. *Die Zusammenhangskomplexitaet von nichtkontextfreien Grammatiken* ◇ D05 D15 ◇
BREIDBART, S. *On splitting recursive sets* ◇ D05 D15 D20 D25 ◇
BRUNNER, J. *Zur Theorie der R-fuzzy-Automaten II* ◇ D05 E72 ◇
BUCHBERGER, B. & ROIDER, B. *Input/output codings and transition functions in effective systems* ◇ D05 D20 D45 ◇
BUDACH, L. *Automata and labyrinths* ◇ D05 ◇
BUDACH, L. *Automaten in Labyrinthen (Russisch)* ◇ D05 ◇
BUKHARAEVA, Z.K. & GOLUNKOV, YU.V. *Complexity and running time of normal algorithms computing boolean functions (Russian)* ◇ B05 B35 D03 D15 ◇
BULITKO, V.K. *On computations with bounded write storage (Russian)* ◇ D05 ◇
BUNIMOVA, E.O. & KAUFMAN, V.SH. & LEVIN, V.A. *Description of language projections (Russian)* ◇ D05 ◇
BUTZBACH, P. *Sur l'equivalence des grammaires simples* ◇ D05 ◇
CAZANESCU, V.E. *Parties algebriques d'un monoide* ◇ D05 ◇
CHIRKOV, M.K. *On the abstract analysis of partial automata (Russian)* ◇ D05 ◇
CHOTTIN, L. & KIERSZENBAUM, F. *Sur les langages d'insertions* ◇ D05 ◇
CHOUEKA, Y. *Finite automata, definable sets, and regular expressions over* ω^n*-tapes* ◇ D05 ◇
CIOBOTARU, S. *The parallel matrix grammars and the languages of Thue* ◇ D03 D05 ◇
COHEN, R.S. & GOLD, A.Y. ω*-computations on deterministic pushdown machines* ◇ D05 ◇
COOK, C.R. *Context-free grammars with graph control* ◇ D05 ◇
COURCELLE, B. *A representation of trees by languages I* ◇ D05 D10 ◇
COURCELLE, B. *A representation of trees by languages II* ◇ D05 D10 ◇
CRESPI-REGHIZZI, S. & GUIDA, G. & MANDRIOLI, D. *Noncounting context-free languages* ◇ D05 ◇
CRESTIN, J. *Langages quasirationnels* ◇ D05 ◇
CULIK II, K. & MAURER, H.A. & OTTMANN, T. & RUOHONEN, K. & SALOMAA, A. *Isomorphism, form equivalence and sequence equivalence of PDOL-forms* ◇ D05 ◇
CULIK II, K. & SALOMAA, A. *On the decidability of homomorphism equivalence for languages* ◇ D05 ◇
CULIK II, K. & MAURER, H.A. & OTTMANN, T. *On two-symboled complete EOL-forms* ◇ D05 ◇
CULIK II, K. & WOOD, D. *Speed-varying OL systems* ◇ D05 ◇
CULIK II, K. *The ultimate equivalence problem for DOL systems* ◇ D05 ◇
CULL, P. *A matrix algebra for neural nets* ◇ D05 ◇
DANG HUU DAO *On the semigroups of directable automata (Preprint)* ◇ D05 ◇
DENNING, P.J. & DENNIS, J.B. & QUALITZ, J.E. *Machines, languages, and computation* ◇ B98 D05 D10 D98 ◇
DIKOVSKIJ, A.YA. & MODINA, L. *On three unambiguousness types of context-free languages (Russian)* ◇ D05 ◇
DOERFLER, W. *The Cartesian composition of automata* ◇ D05 ◇
EDMUNDSON, H.P. & TUNG, I.I. *A characterization of the recursiveness of an m-adic probabilistic language* ◇ D05 D20 ◇
EHRENFEUCHT, A. & ROZENBERG, G. *Elementary homomorphisms and a solution of the DOL sequence equivalence problem* ◇ D05 ◇
EHRENFEUCHT, A. & ROZENBERG, G. *On the structure of derivations in deterministic ETOL systems* ◇ D05 ◇
EHRENFEUCHT, A. & ROZENBERG, G. *Simplifications of homomorphisms* ◇ D05 ◇
EICHNER, L. *The semigroups of linearly realizable finite automata I,II* ◇ D05 ◇
EMDE BOAS VAN, P. & VITANYI, P.M.B. *A note on the recursive enumerability of some classes of recursive languages* ◇ D05 D25 ◇
ENGELFRIET, J. *A hierarchy of tree transducers* ◇ D05 ◇
ESIK, Z. *On decidability of injectivity of tree transformations* ◇ D05 ◇
FREJVALD, R.V. *Recognition of languages with high probability on different classes of automata (Russian)* ◇ D05 D10 ◇
FRIEDE, D. *Ueber deterministische kontextfreie Sprachen und rekursiven Abstieg* ◇ D05 ◇
FRIEDMAN, E.P. & GREIBACH, S.A. *On equivalence and subclass containment problems for deterministic context-free languages* ◇ D05 ◇
FRIEDMAN, JOYCE & WARREN, D.S. *A parsing method for Montague grammars* ◇ B65 D05 ◇
GABBASOV, N.Z. & MURTAZINA, T. *An improvement of the estimate in Rabin's reduction theorem (Russian)* ◇ D05 ◇
GALLEGO DE, M.S. *The abstract family of languages of categorical type* ◇ D05 ◇
GALYUSHKOV, B.S. *Compound grammars (Russian)* ◇ D05 ◇
GOESSEL, M. *A generalized principle of automata superposition relative to a pair of boolean functions (Russian) (English summary)* ◇ B05 D05 ◇
GRABOWSKI, J. & STARKE, P.H. *Petri-Netze* ◇ D05 D80 ◇
GRABOWSKI, M. *The algebraic properties of nets from a program-theoretical point of view* ◇ D05 D80 ◇
HALL PARTEE, B. *Fundamentals of mathematics for linguistics* ◇ B65 D05 D10 ◇
HARRISON, M.A. *Introduction to formal language theory* ◇ D05 D98 ◇
HUGHES, C.E. *The equivalence of vector addition systems to a subclass of Post canonical systems* ◇ D03 D80 ◇
HULE, H. & MAURER, H.A. & OTTMANN, T. *Good OL forms* ◇ D05 ◇
ISTRAIL, S. *Tag systems generating Thue irreducible sequences* ◇ D03 ◇
KALANDARISHVILI, N.G. *An estimate of the algorithm of analysis in the language of equivalent transformations (Russian) (Georgian and English summaries)* ◇ D05 ◇

KANOVICH, M.I. *Complexity of recognition of invariant properties (Russian)* ◇ D03 D15 ◇
KLOPOTOWSKI, J. *Transition operators for (Z,Q)-machines* ◇ D05 ◇
KORSHUNOV, A.D. *Enumeration of finite automata (Russian)* ◇ D05 ◇
LATTEUX, M. *Mots infinis et langages commutatifs* ◇ D05 ◇
LAURINOLLI, T. *Bounded quantification and relations recognizable by finite automata* ◇ D05 ◇
LEITSCH, A. *Unsolvability in systems of constructing automata* ◇ D05 ◇
LEVIN, A.G. *Generating an input sequence which brings an asynchronous circuit to a given state (Russian)* ◇ B70 D05 ◇
LISOVIK, L.P. *The comparison of finitely many-valued transformations of finite automata (Russian)* ◇ D05 ◇
MACHTEY, M. & WINKLMANN, K. & YOUNG, P. *Simple Goedel numberings, isomorphisms, and programming properties* ◇ D05 D10 D15 D20 D45 F40 ◇
MAIBAUM, T.S.E. *Pumping lemmas for term languages* ◇ D05 ◇
MAN, V.D. *Berechnung von Wortfunktionen auf deterministischen n-dimensionalen iterativen Automaten* ◇ D05 D20 ◇
MARINESCU, D. *A universal Turing machine with three tapes for normal algorithms* ◇ D03 D10 ◇
MASLOV, S.YU. *Macroevolution as deduction process* ◇ D03 ◇
MAURER, H.A. & ROZENBERG, G. *Increasing the similarity of EOL form interpretations* ◇ D05 ◇
MAURER, H.A. & SALOMAA, A. & WOOD, D. *On good EOL forms* ◇ D03 D05 ◇
MAURER, H.A. & SALOMAA, A. & WOOD, D. *Relative goodness of EOL forms* ◇ B75 D05 ◇
MAURER, H.A. & SALOMAA, A. & WOOD, D. *Uniform interpretations of L-forms* ◇ D03 D05 ◇
MOSTOWSKI, A.WLODZIMIERZ *Recursively enumerable degrees of programming problems* ◇ B75 D03 D25 ◇
NEPOMNYASHCHIJ, V.A. *Examples of predicates inexpressible by s-rudimentary formulas (Russian) (English summary)* ◇ D05 D20 ◇
NOWACZYK, A. *Categorial languages and variable-binding operators* ◇ A05 D05 ◇
ORMAN, G. *An automata realization of the functions associated with a derivation* ◇ D05 D10 ◇
OTTMANN, T. *Eine einfache universelle Menge endlicher Automaten* ◇ D05 ◇
PAGET, M. *Proprietes de complexite pour une famille d'algorithmes de Markov* ◇ B75 D03 D10 D15 ◇
PAUN, G. *An infinite hierarchy of contextual languages with choice* ◇ D05 ◇
PECKEL, J. *A deterministic subclass of context-free languages* ◇ D05 D15 ◇
PENTTONEN, M. & ROZENBERG, G. & SALOMAA, A. *Bibliography of L systems* ◇ A10 D05 D99 ◇
PODKOLZIN, A.S. *On universal homogeneous structures (Russian)* ◇ D05 ◇
POIGNE, A. *Context-free rewriting* ◇ D03 ◇
PRIESE, L. *A note on asynchronous cellular automata* ◇ D03 ◇
PRINOTH, R. *Starke Faerbbarkeit in Petri-Netzen* ◇ D05 D15 ◇
ROJZEN, S.I. *Unsolvability of the additive theory of automata isols over a unary alphabet (Russian)* ◇ D05 D50 ◇
ROMANOVS'KIJ, V.YU. *Undecidable properties of unary recursive schemes over a free group (Ukrainian) (English and Russian summaries)* ◇ D05 ◇
SALOMAA, A. & SOITTOLA, M. *Automata-theoretic aspects of formal power series* ◇ D05 D80 ◇
SCHWARTZ, DIETRICH *Kanonische Abbildungen und Eilenberg-Maschinen* ◇ D05 ◇
SEESE, D.G. *Decidability of ω-trees with bounded sets (German and Russian summaries)* ◇ B25 C65 C85 D05 ◇
SIEFKES, D. *An axiom system for the weak monadic second order theory of two successors* ◇ B15 B25 C85 D05 ◇
SMIKUN, L.B. *Degrees of unsolvability of algorithmic problems connected with automata operations on groups (Russian)* ◇ D05 D30 ◇
STANKIEWICZ, E. & ZAKOWSKI, W. *The (Z, Q)-systems* ◇ D05 ◇
STARKE, P.H. *Free Petri net languages* ◇ D05 ◇
STARKE, P.H. *Free Petri net languages* ◇ D05 D80 ◇
SUDBOROUGH, I.H. *On the tape complexity of deterministic context-free languages* ◇ D05 D10 D15 ◇
TAO, RENJI *Automata and its reduction (Chinese)* ◇ D05 ◇
TAO, RENJI *Some problems regarding the behaviour of automata (Chinese)* ◇ D05 ◇
TINHOFER, G. *On the simultaneous isomorphism of special relations (isomorphism of automata)* ◇ C13 D05 ◇
TSEJTLIN, G.E. & YUSHCHENKO, E.L. *Theory of parametric models of languages and networks of parallel automata. Program transformations* ◇ D05 ◇
TURAKAINEN, P. *On characterization of recursively enumerable languages in terms of linear languages and VW-grammars* ◇ D05 D25 ◇
WECHLER, W. *The concept of fuzziness in automata and language theory* ◇ B52 D05 E72 ◇

1979

AGGARWAL, S.K. & HEINEN, J.A. *A general class of noncontext free grammars generating context free languages* ◇ D05 ◇
AINHIRN, W. & MAURER, H.A. *On ε productions for terminals in EOL forms* ◇ D05 ◇
ALT, H. *Lower bounds on space complexity for contextfree recognition* ◇ D05 D15 ◇
ANASIMOV, P.A. *A choice of a sequence of the input alphabet symbols maximizing the efficiency of the work of an automaton (Russian)* ◇ D05 ◇
ANISIMOV, A.V. *Church-Rosser marking transducers and decidable properties in tree processing (Russian)* ◇ D05 ◇
ANISIMOV, A.V. *Discrete backtracking converters (Russian)* ◇ D05 ◇

ANTELMANN, H. & BUDACH, L. & ROLLIK, H.-A. *On universal traps* ◇ D05 ◇
ARBIB, M.A. & MANES, E.G. *Intertwined recursion, tree transformations, and linear systems* ◇ D05 D20 D80 G30 ◇
ASVELD, P.R.J. & ENGELFRIET, J. *Extended linear macro grammars, iteration grammars, and register programs* ◇ D05 D10 ◇
AUTEBERT, J.-M. *Operations de cylindre et applications sequentielles gauches inverses* ◇ D05 ◇
AUTEBERT, J.-M. & BEAUQUIER, J. & BOASSON, L. & NIVAT, M. *Quelques problemes ouverts en theorie des langages algebriques* ◇ D05 ◇
AUTEBERT, J.-M. *Relationships between AFDL's and cylinders* ◇ D05 ◇
AUTEBERT, J.-M. & BEAUQUIER, J. & BOASSON, L. & LATTEUX, M. *Sur la decidabilite de certaines questions liees a la linearite des langages (English summary)* ◇ D05 ◇
AUTEBERT, J.-M. *Une note sur le cylindre des langages deterministes* ◇ D05 ◇
BABINOV, YU.P. *Class of generalized context-sensitive precedence languages (Russian)* ◇ D05 ◇
BACIU, A. & PASCU, A. *Leontief type grammars and languages* ◇ D05 ◇
BAKER, B.S. *Composition of top-down and bottom-up tree transductions* ◇ D05 ◇
BALCER, M. & MIKA, A. & PIZON, T. *A theorem on the equivalence of the definitions of admissible words of a certain alphabet and arithmetic term expressions* ◇ D05 ◇
BALCER, M. & MIKA, A. & PIZON, T. *An algorithm for deciding the truth of an equation between two arbitrary arithmetic expressions* ◇ D05 ◇
BALCER, M. & MIKA, A. & PIZON, T. *The construction of admissible words of a certain alphabet as equivalent concepts by means of arithmetic term expressions* ◇ D05 ◇
BARTH, G. *Fast recognition of context-sensitive structures* ◇ D05 D15 ◇
BASARAB, I.A. *Equivalent transformations of metalinear LA(1) grammars and unary recursive schemes (Russian)* ◇ D05 ◇
BEAUQUIER, J. *Deux familles de langages incomparables* ◇ D05 ◇
BEAUQUIER, J. *Generateurs algebriques et systemes de paires iterantes* ◇ D05 ◇
BEAUQUIER, J. *Independence of linear and one-counter generators* ◇ D05 ◇
BEAUQUIER, J. *Strong non-deterministic context-free languages* ◇ D05 ◇
BERSTEL, J. *Sur la construction de mots sans carre* ◇ D05 ◇
BERSTEL, J. *Sur les mots sans carre definis par un morphisme* ◇ D05 ◇
BERSTEL, J. *Transductions and context-free languages* ◇ D05 ◇
BIANCO, E. *Informatique fondamentale. De la machine de Turing aux ordinateurs modernes* ◇ D03 D10 D98 ◇
BISWAS, S. & VAISHNAVI, V.K. *Tree adjunct languages as integer set recognizers* ◇ D05 ◇
BLATTNER, M. *Inherent ambiguities in families of grammars* ◇ D05 ◇
BLATTNER, M. & HEAD, T. *The decidability of equivalence for deterministic finite transducers* ◇ D05 ◇
BOASSON, L. *Context-free sets of infinite words* ◇ D05 ◇
BOASSON, L. *Un langage algebrique particulier* ◇ D05 ◇
BOCHMANN, D. & SIMON, F.U. *Automatengraphen und ihre analytische Behandlung* ◇ D05 ◇
BOCHMANN, G.V. *Semantic equivalence of covering attribute grammars* ◇ D05 ◇
BOERGER, E. *A new general approach to the theory of the many-one equivalence of decision problems for algorithmic systems* ◇ B25 D03 D05 D10 D25 D30 D40 ◇
BOOK, R.V. *A remark on tally languages and complexity classes* ◇ D05 D15 ◇
BOOK, R.V. *Complexity classes of formal languages (preliminary report)* ◇ D05 D15 ◇
BRAUNMUEHL VON, B. & VERBEEK, R. *Finite-change automata* ◇ D05 D10 ◇
BUCURESCU, I. *On the state minimization of incompletely specified sequential machines* ◇ D05 ◇
BUDA, A. *Generalized sequential machine maps* ◇ D05 ◇
BUDA, A. *The languages of sequential program machines are coregular (Russian) (English summary)* ◇ D05 ◇
CALUDE, C. & MARCUS, S. & PAUN, G. *The universal grammar as a hypothetical brain* ◇ B65 D05 D80 ◇
CAMERINI, P.M. & GALBIATI, G. & MAFFIOLI, F. *On the complexity of Steiner-like problems* ◇ D05 ◇
CAZACU, M. *A general mathematical notion of the algorithm* ◇ D03 D10 D20 ◇
CHANG, NINGSAN & FU, K.S. *Parallel parsing of tree languages for syntactic pattern recognition* ◇ D05 ◇
CHIRKOV, M.K. *On some types of incompletely specified automata* ◇ D05 ◇
CHOTTIN, L. *Strict deterministic languages and controlled rewriting systems* ◇ D05 ◇
CHOU, S.M. & FU, K.S. *Inference for transition network grammars* ◇ D05 ◇
CHRISTOL, G. *Ensembles presque periodiques k-reconnaissables* ◇ D05 ◇
COARDOS, V. *Equations associated with a context-free grammar* ◇ D05 ◇
COCHET, Y. *Church-Rosser congruences on free semigroups* ◇ D03 ◇
COOK, S.A. *Deterministic CEL's are accepted simultaneously in polynomial time and log squared* ◇ D05 D15 ◇
CULIK, K. & KARHUMAEKI, J. *Interactive L systems with almost interactionless behaviour* ◇ D05 ◇
CULIK II, K. & WOOD, D. *A mathematical investigation of propagating graph OL systems* ◇ D05 ◇
CULIK II, K. *A purely homomorphic characterization of recursively enumerable sets* ◇ D03 D25 ◇
CULIK II, K. & WOOD, D. *Doubly deterministic tabled OL systems* ◇ D05 ◇
CULIK II, K. & RICHIER, J. *Homomorphism equivalence on ETOL languages* ◇ D05 ◇

CULIK II, K. & KARHUMAEKI, J. *Interactive L systems with almost interactionless behaviour* ◇ D05 ◇
CULIK II, K. & MAURER, H.A. *On simple representations of language families* ◇ D05 ◇
CULIK II, K. *On the homomorphic characterizations of families of languages* ◇ D05 ◇
DAMM, W. *An algebraic extension of the Chomsky-hierarchy* ◇ B75 D05 ◇
DANECKI, R. & KARPINSKI, M. *Decidability results on plane automata searching mazes* ◇ D05 ◇
DANG HUU DAO *On the characteristic semigroups of Mealy-automata* ◇ D05 ◇
DANKO, W. *Definability in algorithmic logic* ◇ B75 D05 D20 ◇
DASSOW, J. *Concentration dependent OL systems* ◇ D05 ◇
DASSOW, J. *ETOL systems and compound grammars* ◇ D05 ◇
DASSOW, J. *On the circular closure of languages* ◇ D05 ◇
DAUCHET, M. & MONGY, J. *Transformations de noyaux reconnaissables. Capacite generative des bimorphismes des forets* ◇ D05 ◇
DEUSSEN, P. *One abstract accepting algorithm for all kinds of parsers* ◇ B75 D05 ◇
DUSKE, J. & PARCHMANN, R. & SPECHT, J. *Szilard languages of IO-grammars* ◇ D05 ◇
EHRENFEUCHT, A. & ROZENBERG, G. *A result on the structure of ETOL languages* ◇ D05 ◇
EHRENFEUCHT, A. & ROZENBERG, G. *An observation on scattered grammars* ◇ D05 ◇
EHRENFEUCHT, A. & ROZENBERG, G. *Finding a homomorphism between two words is NP-complete* ◇ D05 D15 ◇
EHRENFEUCHT, A. & ROZENBERG, G. *On arithmetic substitutions of EDTOL languages* ◇ D05 ◇
EHRENFEUCHT, A. & ROZENBERG, G. *On the structure of polynomially bounded DOL systems* ◇ D05 ◇
EHRENFEUCHT, A. & ROZENBERG, G. & VERMEIR, D. *On ETOL systems with rank* ◇ D05 ◇
EICHNER, L. *The semigroups of totally linearly realizable automata* ◇ D05 ◇
ENGELFRIET, J. & SLUTZKI, G. *Bounded nesting in macro grammars* ◇ D05 ◇
ENGELFRIET, J. & ROZENBERG, G. *Equality languages and fixed point languages* ◇ D05 ◇
ENGELFRIET, J. *Two-way automata and checking automata* ◇ D05 ◇
ESIK, Z. *On functional tree transducers* ◇ D05 ◇
ETTL, W. & LEITSCH, A. *Aufzaehlung subrekursiver Klassen und konstruierende Automaten* ◇ D05 D20 ◇
FISCHER, MICHAEL J. & PIPPENGER, N.J. *Relations among complexity measures* ◇ D05 D15 ◇
FRANZEN, H. & HOFFMANN, B. *Automatic determination of data flow in extended affix grammars* ◇ B75 D05 ◇
FREEMAN, M. *Aspects of the upper bounds of finite input-memory and finite output-memory sequential machines* ◇ D05 ◇
FREJVALD, R.V. *Language recognition using finite probabilistic multitape and multihead automata (Russian)* ◇ D05 ◇
FREJVALD, R.V. *On shortening the recognition time of some word sets by using a random number generator (Russian)* ◇ D05 D15 ◇
FRIEDE, D. *Partitioned LL(k) grammars* ◇ D05 ◇
FRIEDE, D. *Transition diagrams and strict deterministic grammars* ◇ D05 ◇
FRIEDMAN, E.P. & GREIBACH, S.A. *Superdeterministic DPDAs: The method of accepting does affect decision problems* ◇ D05 D10 D15 ◇
FROLOV, E.B. *Connectivity of the graph of a finite automaton with a large set of states (Russian)* ◇ D05 ◇
FROUGNY, C. *Langages tres simples generateurs* ◇ D05 ◇
GALLEGO DE, M.S. *A characterization of context-free languages* ◇ D05 ◇
GINSBURG, S. & LEONG, B. & MAYER, O. & WOTSCHKE, D. *On strict interpretations of grammar forms* ◇ D05 ◇
GOLUNKOV, YU.V. & KNYAZEV, E.A. *Completeness of systems of commands with a finite set of memory states (Russian)* ◇ D05 ◇
GRABOWSKI, J. *The unsolvability of some Petri net language problems* ◇ D05 D80 ◇
GREIBACH, S.A. *Formal languages: origins and directions* ◇ B70 D05 D98 ◇
GURARI, E.M. & IBARRA, O.H. *Simple counter machines and number-theoretic problems* ◇ D05 D10 ◇
GURARI, E.M. & IBARRA, O.H. *Some decision problems concerning sequential transducers and checking automata* ◇ D05 D10 ◇
GURARI, E.M. & IBARRA, O.H. *The complexity of the equivalence problem for counter machines, semilinear sets, and simple programs* ◇ D05 D15 ◇
HASHIGUCHI, K. *A decision procedure for the order of regular events* ◇ D05 ◇
HEAD, T. *Codeterministic Lindenmayer schemes and systems* ◇ D05 ◇
HEMMERLING, A. *Concentration of multidimensional tape-bounded systems of Turing automata and cellular spaces* ◇ D05 D10 ◇
HEMMERLING, A. *Zur Raumkompliziertheit mehrdimensionaler Zellularraeume und Turing-Automaten (Russian and English summaries)* ◇ D05 D10 ◇
HESSE, W. *A correspondence between W-grammars and formal systems of logic and its application to formal language description* ◇ D03 D05 ◇
HOPCROFT, J.E. & ULLMAN, J.D. *Introduction to automata theory, languages, and computation* ◇ D05 D10 D98 ◇
HOTZ, G. *Ueber die Darstellbarkeit des syntaktischen Monoides kontextfreier Sprachen* ◇ D05 ◇
HUNT III, H.B. *Observations on the complexity of regular expression problems* ◇ D05 D15 ◇
HUNYADVARI, L. *The L-fuzzy Kleene theorem* ◇ D05 ◇
ISTRAIL, S. *On complements of some bounded context-sensitive languages* ◇ D05 ◇

JOHANSEN, P. *The generating function of the number of subpatterns of a DOL sequence* ◇ D05 ◇
KLEINE BUENING, H. *Generalized vector addition systems with finite exception sets* ◇ B75 D05 D10 D80 ◇
KLETTE, R. & LINDNER, R. *Zweidimensional arbeitende Vektormaschinen und ihr Leistungsvermoegen bei der Loesung von Entscheidungsproblemen der Aussagenlogik* ◇ B35 D05 D15 ◇
KONDO, M. *Syntactic structure of mathematical words (Japanese)* ◇ B65 D05 ◇
KUDLEK, M. *Context free normal systems* ◇ D03 ◇
KURMIT, A.A. *Generation of classes of information - lossless automata by means of substitutions of a set of states (Russian)* ◇ D05 ◇
LANGE, E.E. *On the solution of a class of logical-combinatorial problems by a permutational method (Russian)* ◇ D05 ◇
LANGMAACK, H. *On a theory of decision problems in programming languages* ◇ B75 D05 D80 ◇
LEISS, E. *A note on infinite graphs with actions* ◇ D05 E20 ◇
LEISS, E. *On tractable unrestricted regular expressions* ◇ D05 ◇
LINNA, M. *Two decidability results for deterministic pushdown automata* ◇ D05 D10 ◇
LISOVIK, L.P. *Equivalence problem for finitely ambiguous finite automata over semigroups (Russian)* ◇ D05 ◇
LISOVIK, L.P. *The identity problem for regular events over the direct product of a free and a cyclic semigroup (Russian) (English Summary)* ◇ D05 ◇
MASLOV, S.YU. *Calculi with monotone deductions and their economic interpretation (Russian) (English summary)* ◇ D05 D80 ◇
MAURER, H.A. & SALOMAA, A. & WOOD, D. *Context-dependent L forms* ◇ B75 D05 ◇
MAURER, H.A. & PENTTONEN, M. & SALOMAA, A. & WOOD, D. *On non context-free grammar forms* ◇ B75 D05 ◇
MAURER, H.A. & ROZENBERG, G. & SALOMAA, A. & WOOD, D. *Pure interpretations of EOL forms* ◇ B75 D05 ◇
MEERSMAN, R. & ROZENBERG, G. & VERMEIR, D. *Persistent ETOL systems* ◇ D05 ◇
NAKAMURA, A. & ONO, H. *Undecidability of the first-order arithmetic* $A[P(x),2x,x+1]$ ◇ D05 D35 F30 F35 ◇
NIVAT, M. *Infinite words, infinite trees, infinite computations* ◇ B75 D05 D75 ◇
PAUN, G. *On the family of finite index matrix languages* ◇ D05 ◇
PRIESE, L. *Ueber ein 2-dimensionales Thue-System mit zwei Regeln und unentscheidbarem Wortproblem* ◇ D03 ◇
RIEDEMANN, E.H. *The control of parallel computations by labeled Petri nets: a study in terms of multiple-firing automata and parallel program schemata (Dissertation)* ◇ D05 D80 ◇
ROEDDING, D. & ROEDDING, W. *Network of finite automata* ◇ D05 ◇
RUOHONEN, K. *On some decidability problems for HDOL systems with nonsingular Parikh matrices* ◇ D05 ◇
RUOHONEN, K. *On the decidability of the OL-DOL equivalence problem* ◇ D05 ◇
SAGER, H. *The nature of reduction in Turing machines and its relation to reduction in automata* ◇ D05 D10 ◇
SARKISYAN, A.D. *Realization of arithmetic functions on iterative networks (Russian) (Armenian summary)* ◇ D05 D15 ◇
SELMAN, A.L. *P-selective sets, tally languages, and the behaviour of polynomial time reducibilities on NP (preliminary report)* ◇ D05 D15 ◇
SELMAN, A.L. *P-selective sets, tally languages, and the behaviour of polynomial time reducibilities on NP* ◇ D05 D15 ◇
SEPER, K. *Algorithmic constructions inspired by Caporaso* ◇ D05 D20 ◇
SHURYGIN, V.A. *Bounds on the complexity of normal algorithms (Russian)* ◇ D03 D15 ◇
SIMOVICI, D.A. *Polylocal languages* ◇ D05 ◇
SKORDEV, D.G. *Algebraic generalization of a result of Bohm and Jacopini* ◇ D05 ◇
STARKE, P.H. *On the languages of bounded Petri nets* ◇ D05 D80 ◇
STARKE, P.H. *Semilinearity and Petri nets* ◇ D05 D80 ◇
TAO, RENJI *The reversibility of finite automata (Chinese)* ◇ D05 ◇
TAO, RENJI *Universal automata (Chinese)* ◇ D05 ◇
THOMAS, WOLFGANG *Star-free regular sets of* ω*-sequences* ◇ C07 D05 ◇
TOCA, A. Ω*-automata* ◇ D05 ◇
UESU, T. *A complete system of grammars for plane graphs* ◇ D05 ◇
VOELKEL, L. *Language recognition by linear bounded and copy programs* ◇ D05 ◇
WAGNER, K. *On* ω*-regular sets* ◇ D05 ◇
WECHLER, W. *Fuzzy sets and languages* ◇ D05 E72 ◇
WECHSUNG, G. *The oscillation complexity and a hierarchy of context-free languages* ◇ D05 ◇
WEGNER, L. *Bracketed two-level grammars -- a decidable and practical approach to language definitions* ◇ D05 ◇
WIRSING, M. *Small universal Post systems* ◇ D03 ◇
WISNIEWSKI, K. *A notion of the acceptance of infinite sequences by finite automata* ◇ D05 E15 ◇
YAMASAKI, H. *On multitape automata* ◇ D05 ◇
ZIELONKA, W. *On the equivalence of Lambek's syntactic calculus and categorial calculi* ◇ D05 ◇

1980

ABLAEV, F.M. *The relation of automatic measures of the complexity of languages to Loveland's complexity of word resolution (Russian)* ◇ D05 D15 ◇
ABRAMSON, F.G. & BREJTBART, YU.YA. & LEWIS, F.D. *Complex properties of grammars* ◇ D05 D10 D15 ◇
ACKROYD, M.H. & CHIRATHAMJAREE, C. *A method for the inference of non-recursive context-free grammars* ◇ D05 ◇
AINHIRN, W. *How to get rid of pseudoterminals* ◇ D05 ◇
ALBERT, J. & WEGNER, L. *Languages with homomorphic replacements* ◇ D05 ◇

ALBERT, J. & MAURER, H.A. & ROZENBERG, G. *Simple EOL forms under uniform interpretation generating CF languages* ◇ D05 ◇
ALBERT, J. & CULIK II, K. *Test sets for homomorphism equivalence on context-free languages* ◇ D05 ◇
ANGLUIN, D. *Inductive inference of formal languages from positive data* ◇ D05 ◇
ASVELD, P.R.J. *On controlled iterated gsm mappings and related operations* ◇ D05 ◇
AUTEBERT, J.-M. & BEAUQUIER, J. & BOASSON, L. *Langages sur des alphabets infinis* ◇ D05 ◇
AUTEBERT, J.-M. & BEAUQUIER, J. & BOASSON, L. *Limites de langages algebriques (English summary)* ◇ D05 ◇
AUTEBERT, J.-M. & BEAUQUIER, J. & BOASSON, L. & LATTEUX, M. *Very small families of algebraic nonrational languages* ◇ D05 ◇
BABCSANYI, I. *Semigroup-automata* ◇ D05 ◇
BACHMANN, P. *Theoretische Untersuchungen zur Behandlung von Kontext- Sensitivitaeten* ◇ D05 ◇
BACIU, A. & PASCU, A. *A Bar Hillel type property of Leontief languages* ◇ D05 ◇
BEATTY, J.C. *Two iteration theorems for the LL(k) languages* ◇ D05 ◇
BECKHOFF, G. *A novel approach to the state reduction problem* ◇ D05 ◇
BENEY, J. & FRECON, L. *Langage et systeme d'ecriture de transducteurs* ◇ D05 ◇
BERMAN, P. *A note on sweeping automata* ◇ D05 D15 ◇
BEYNON, W.M. *On the structure of free finite state machines* ◇ D05 ◇
BOASSON, L. & NIVAT, M. *Adherences of languages* ◇ D05 ◇
BOASSON, L. *Derivations et reductions dans les grammaires algebriques* ◇ D05 ◇
BOCHMANN, D. *Zur Dynamik und Steuerung hierarchischer Automatennetze* ◇ D05 ◇
BOERGER, E. & KLEINE BUENING, H. *The r.e. complexity of decision problems for commutative semi-Thue systems with recursive rule set* ◇ D03 D25 ◇
BRANDENBURG, F.-J. *Multiple equality sets and Post machines* ◇ D05 D15 ◇
BRZOZOWSKI, J.A. *Developments in the theory of regular languages* ◇ D05 ◇
BRZOZOWSKI, J.A. & FICH, F.E. *Languages of R-trivial monoids* ◇ D05 ◇
BRZOZOWSKI, J.A. & LEISS, E. *On equations for regular languages, finite automata, and sequential networks* ◇ D05 ◇
BRZOZOWSKI, J.A. *Open problems about regular languages* ◇ D05 ◇
BUEVICH, V.A. *On τ-completeness in the class of automaton mappings (Russian)* ◇ D05 ◇
BUSZKOWSKI, W. *Logical complexity of some classes of tree languages generated by multiple-tree-automata* ◇ D05 D55 ◇
CAZACU, M. *The simulation relations between the classes of algorithms* ◇ D03 D10 D20 ◇
CAZANESCU, V.E. *On the uniqueness of the solution of a system of context-free equations* ◇ D05 ◇
CENZER, D. *Non-generable formal languages* ◇ D03 D25 D30 D70 ◇
CHAPUT, D. & SABIDUSSI, G. *Une generalization du theoreme de Ginsburg-Rose* ◇ D05 ◇
CHIEN, T.Y. *Decompositions of a free monoid into disjunctive languages* ◇ D05 ◇
CHU, TANIEN & SHI, HUIJAN *Some properties of quasi-disjunctive languages on a free monoid* ◇ D05 ◇
CONSTABLE, R.L. *The role of finite automata in the development of modern computing theory* ◇ D05 ◇
COOK, S.A. & RACKOFF, C.W. *Space lower bounds for maze threadability on restricted machines* ◇ D05 D15 ◇
CSIRMAZ, L. *Iterated grammars* ◇ D05 ◇
CULIK, K. *Parallel permit grammars and some graph representations of Petri nets* ◇ D05 D80 ◇
CULIK II, K. & DIAMOND, N.D. *A homomorphic characterization of time and space complexity classes of languages* ◇ D05 D10 D15 ◇
CULIK II, K. *Homomorphisms: decidability, equality and test sets* ◇ D05 ◇
CULIK II, K. & KARHUMAEKI, J. *On the equality sets for homomorphisms on free monoids with two generators* ◇ D05 D40 ◇
CULIK II, K. & SALOMAA, A. *Test sets and checking words for homomorphism equivalence* ◇ D05 ◇
CZAJA, L. *Deadlock and fairness in parallel schemas: A set-theoretic characterization and decision problems* ◇ B75 D05 E75 ◇
DANG HUU DAO *Further results on characteristic semigroups of Mealy-automata* ◇ D05 ◇
DANG HUU DAO *Quasi-Moore-semigroups and Moore-semigroups for some classes of Mealy-automata* ◇ D05 ◇
DANILOV, V.V. & PODKOPAEV, B.P. *Use of the spectral and autocorrelation properties of logic functions for synthesizing combination schemes on Maitra's cascades (Russian)* ◇ D05 ◇
DASSOW, J. *On some extensions of Indian parallel context free grammars* ◇ D05 ◇
DASSOW, J. *On Parikh-languages of L systems without interaction* ◇ D05 ◇
DAVID, R. & MITRANI, E. & TELLEZ-GIRON, R. *Emploi de cellules universelles pour la synthese de systemes asynchrones decrits par reseaux de Petri* ◇ D05 D80 ◇
DEKKING, F.M. *Regularity and irregularity of sequences generated by automata* ◇ D05 ◇
DUBITZKI, T. & WU, A. *Cellular d-graph automata with augmented memory* ◇ D05 ◇
DYER, C.R. *One-way bounded cellular automata* ◇ D05 ◇
EHRENFEUCHT, A. & MAURER, H.A. & ROZENBERG, G. *Continuous grammars* ◇ D05 ◇
EHRENFEUCHT, A. & ROZENBERG, G. *DOS systems and languages* ◇ D05 ◇
EHRENFEUCHT, A. & ROZENBERG, G. *Every two equivalent DOL systems have a regular true envelope* ◇ D05 ◇
EHRENFEUCHT, A. & ROZENBERG, G. & VERRAEDT, R. *Many-to-one simulation in EOL forms is decidable* ◇ D05 ◇

EHRENFEUCHT, A. & ROZENBERG, G. *On basic properties of DOS systems and languages* ◇ D05 ◇
EHRENFEUCHT, A. & ROZENBERG, G. *On the emptiness of the intersection of two DOS languages problem* ◇ D05 ◇
EHRENFEUCHT, A. & ROZENBERG, G. *The sequence equivalence problem is decidable for OS systems* ◇ D05 ◇
ENGELFRIET, J. & ROZENBERG, G. *Fixed point languages, equality languages, and representation of recursively enumerable languages* ◇ D05 D25 ◇
ENGELFRIET, J. & FILE, G. *Formal properties of one-visit and multi-pass attribute grammars* ◇ D05 ◇
ENGELFRIET, J. *Some open questions and recent results on tree transducers and tree languages* ◇ D05 ◇
ENGELFRIET, J. & LEEUWEN VAN, J. & SCHMIDT, ERIK MEINECHE *Stack machines and classes of nonnested macro languages* ◇ D05 D10 ◇
ENGELFRIET, J. & ROZENBERG, G. & SLUTZKI, G. *Tree transducers, L systems, and two-way machines* ◇ D05 ◇
ESIK, Z. *Decidability results concerning tree transducers I* ◇ D05 ◇
EVLADENKO, V.N. *A recognition algorithm for one class of quasirecursive parametric grammars (Russian)* ◇ D05 ◇
EZAWA, Y. *Structured control sets on grammars* ◇ D05 ◇
FALKENBERG, B. *Halteprobleme von Fang-Systemen (tag systems)* ◇ D03 D25 ◇
FAL'KOVICH, M.A. & GALIN, A.B. *Diagnosis of memoryless digital automata using a modified technique of natural logical deduction (Russian)* ◇ B70 D05 F99 ◇
FARADZHEV, R.G. *Design equations for linear sequential machines* ◇ D05 ◇
FELEA, V. *Relational VT-algebras over free theories* ◇ D05 ◇
FOERSTER, M. & PIRLING, C. & VOELKEL, L. *Zweiregister-Akzeptoren mit unterschiedlichen Druckbefehlen* ◇ D05 ◇
FUERER, M. *The complexity of the inequivalence problem for regular expressions with intersection* ◇ D05 D15 ◇
GAIFMAN, H. & SHAMIR, E. *Roots of the hardest context free language and other constructs* ◇ D05 ◇
GANDY, R.O. *Church's thesis and principles for mechanisms* ◇ A05 D05 D10 D20 F99 ◇
GENRICH, H.J. & LAUTENBACH, K. & THIAGARAJAN, P.S. *Elements of general net theory* ◇ B70 B75 D05 ◇
GINSBURG, S. *Methods for specifying families of formal languages - past-present-future* ◇ D05 ◇
GOLDSTINE, J. *Formal languages and their relation to automata: what Hopcroft & Ullman didn't tell us* ◇ D05 ◇
GURARI, E.M. & IBARRA, O.H. *Path systems: constructions, solutions and applications* ◇ D05 D10 ◇
HAUSSLER, D. & ZEIGER, H.P. *Very special languages and representations of recursively enumerable languages via computation histories* ◇ D05 ◇
HEILBRUNNER, S. *An algorithm for the solution of fixed-point equations for infinite words* ◇ D05 ◇
HONDA, N. & INAGAKI, Y. & OYAMAGUCHI, M. *The equivalence problem for real-time strict deterministic languages* ◇ D05 D15 ◇
HUET, G. *Confluent reductions: abstract properties and applications to term rewriting systems* ◇ D03 D05 ◇
HUET, G. & OPPEN, D.C. *Equations and rewrite rules: a survey* ◇ D03 D05 D98 ◇
HUGHES, C.E. *Derivatives and quotients of prefix-free context-free languages* ◇ D05 D25 ◇
HUGHES, C.E. & STRAIGHT, D.W. *Word problems for bidirectional single-premise Post systems* ◇ D03 D25 ◇
INOUE, K. & TAKANAMI, I. *A note on decision problems for three-way two-dimensional finite automata* ◇ D05 ◇
JANICKI, R. *Some remarks on deterministic Mazurkiewicz algorithms and languages associated with them* ◇ D05 ◇
JEANROND, H. *Deciding unique termination of permutative rewriting systems: choose your term algebra carefully* ◇ D05 ◇
JOSHI, A.K. & LEVY, L.S. & YUEH, KANG *Local constraints in programming languages I. Syntax* ◇ B75 D05 ◇
KALISZ, E. *The conversion of PRIM-URM conflicts in context-free grammars into PRIM-PRIM conflicts (Romanian, English summery)* ◇ D05 ◇
KARPINSKI, M. *The ultimate equivalence of iterated homomorphisms is recursively unsolvable* ◇ D05 ◇
KLESHCHEV, A.S. *Relational model of computations (Russian)* ◇ D05 ◇
KUDRYAVTSEV, V.B. *Functional systems (Russian)* ◇ B50 C05 D05 ◇
KURDUMOV, G.I. *An algorithm-theoretic method for the study of uniform random networks* ◇ D05 D10 ◇
KUZENKO, V.F. & SHKIL'NYAK, S.S. & ZUBENKO, V.V. *Syntactical definitions and the problem of reorientation (Russian)* ◇ D05 ◇
LADNER, R.E. *Complexity theory with emphasis on the complexity of logical theories* ◇ B25 C98 D05 D10 D15 ◇
LEVIN, V.A. *Some new data on logical determinants and their applications to dynamics of automata (Russian)* ◇ D05 ◇
LISOVIK, L.P. & RED'KO, V.N. *Regular events in semigroups (Russian)* ◇ D05 D40 ◇
MAURER, H.A. & SALOMAA, A. & WOOD, D. *On generators and generative capacity of EOL forms* ◇ D05 ◇
MAURER, H.A. & SALOMAA, A. & WOOD, D. *Pure grammars* ◇ D05 ◇
MAYR, E.W. *Ein Algorithmus fuer das allgemeine Erreichbarkeitsproblem bei Petrinetzen und damit zusammenhaengende Probleme (Dissertation)* ◇ D05 D80 ◇
METZ, J. *Abstrakte Rechnermodelle* ◇ D05 D98 ◇
MIKOLAJCZAK, B. *On the simulation problem of a set of automata* ◇ D05 ◇
MOLL, K.R. *Left context precedence grammars* ◇ D05 ◇

MONIEN, B. & SUDBOROUGH, I.H. *The interface between language theory and complexity theory* ◇ D05 D15 ◇
MONIEN, B. *Two-way multihead automata over a one-letter alphabet* ◇ D05 ◇
MOSTOWSKI, A.WLODZIMIERZ *Finite automata on infinite trees and subtheories of SkS* ◇ B25 C85 D05 ◇
MOSTOWSKI, A.WLODZIMIERZ *Nearly deterministic automata acceptation of infinite trees and a complexity of weak theory of SkS* ◇ D05 ◇
MUNDICI, D. *Natural limitations of algorithmic procedures in logic (Italian summary)* ◇ C40 D05 D80 ◇
NELSON, EVELYN *Categorical and topological aspects of formal languages* ◇ D05 G30 ◇
NGO THE KHANH *Simple deterministic machines and their languages* ◇ D05 ◇
NGUYEN XUAN MY *Some classes of semi-Thue systems (German and Russian summaries)* ◇ D03 ◇
NIGIYAN, S.A. *Reducibility to free schemes (Russian)* ◇ D05 ◇
NIJHOLT, A. *Context-free grammars: covers, normal forms, and parsing* ◇ D05 ◇
ORMAN, G. *An automatic realization of $\xi(h)$-function (Romanian summary)* ◇ D05 D10 ◇
PALTANEA, R. *Example of a grammar that generates the formulas of propositional calculus* ◇ B05 D05 ◇
ROZENBERG, G. *A survey of results and open problems in the mathematical theory of L systems* ◇ D05 ◇
RUS, T. *HAS-Hierarchy; a natural tool for language specification* ◇ D05 ◇
RYTTER, W. *Automata theory and complexity* ◇ D05 D15 D98 ◇
SALOMAA, A. *Morphisms on free monoids and language theory* ◇ D05 ◇
SILBERGER, D.M. *Universal terms of complexity three* ◇ D05 ◇
SIMMONS, H. *The word and torsion problems for commutative Thue systems* ◇ D03 D15 D40 ◇
SIMOVICI, D.A. *Computing of graphs of relations using generative grammars* ◇ D05 ◇
SIMOVICI, D.A. *Context-sensitive languages and Ackermann's function* ◇ D05 D20 ◇
SOLOV'EV, S.YU. *On a constructive procedure of grammatical inference (Russian)* ◇ D05 ◇
STARKE, P.H. *Beitraege zur Theorie der Mehrbandautomaten* ◇ D05 ◇
STARKE, P.H. *Petri-Netze* ◇ D05 D98 ◇
STARKE, P.H. *Remarks on Reusch's nondeterminism problem* ◇ D05 ◇
TRNKOVA, V. *General theory of relational automata* ◇ D05 ◇
VIRAGH, J. *Deterministic ascending tree automata. I* ◇ D05 ◇
WASILEWSKA, A. *On the Gentzen-type formalizations* ◇ B25 D05 F07 ◇
WINKLER, P.M. *Computational characterization of abelian groups* ◇ D05 D80 ◇
YEHUDAI, A. *The decidability of equivalence for a family of linear grammars* ◇ D05 ◇
YOKOMORI, T. *Stochastic characterizations of EOL languages* ◇ D05 ◇

1981

ADAM, ANDRAS *On the complexity of codes and pre-codes assigned to finite Moore automata* ◇ D05 ◇
AGUZZI, G. *The theory of invertible algorithms (French summary)* ◇ D03 ◇
ALBERT, J. & WEGNER, L. *Languages with homomorphic replacements* ◇ D05 ◇
ALBERT, J. & MAURER, H.A. & OTTMANN, T. *On sub regular OL forms* ◇ D05 ◇
ALBERT, J. & CULIK II, K. & KARHUMAEKI, J. *Test sets for context free languages and algebraic systems of equations over a free monoid* ◇ D05 ◇
AMBARTSUMYAN, A.A. & ZAPOL'SKIKH, E.N. *An approach to time decomposition of automata II (Russian)* ◇ D05 ◇
AMBARTSUMYAN, A.A. & ZAPOL'SKIKH, E.N. *Time decomposition of automata I (Russian)* ◇ D05 ◇
ANISIMOV, A.V. *Church-Rosser transformers and decidable properties of tree processing* ◇ D05 ◇
ANISIMOV, A.V. *Finite automaton semigroup mappings (Russian)* ◇ D05 ◇
ARAKI, T. & TOKURA, N. *Flow languages equal recursively enumerable languages* ◇ D05 D25 ◇
ASVELD, P.R.J. & ENGELFRIET, J. *A note on non-generators of full AFL's* ◇ D05 ◇
AUTEBERT, J.-M. & BEAUQUIER, J. & BOASSON, L. & LATTEUX, M. *Langages algebriques domines par des langages unaires* ◇ D05 ◇
AUTEBERT, J.-M. *Un resultat de discontinuite dans les familles de langages* ◇ D05 ◇
BABCSANYI, I. *Rees automata (Hungarian) (English summary)* ◇ D05 ◇
BAGYINSZKINE OROSZ, A. *Petri nets and their languages* ◇ D05 D80 ◇
BARON, G. & KUICH, W. *The characterization of nonexpansive grammars by rational power series* ◇ D05 ◇
BARRERO, A. & GONZALEZ, R.C. & THOMASON, M.G. *Equivalence and reduction of expansive tree grammars* ◇ D05 ◇
BARTHA, M. *An algebraic definition of attributed transformations* ◇ D05 ◇
BEAUQUIER, J. *A remark about a substitution property* ◇ D05 ◇
BEAUQUIER, J. *Substitution of semi-AFL's* ◇ D05 ◇
BLATTNER, M. & LATTEUX, M. *Parikh-bounded languages* ◇ D05 ◇
BOASSON, L. & NIVAT, M. *Centers of languages* ◇ D05 ◇
BOASSON, L. *Some applications of CFL's over infinite alphabets* ◇ D05 ◇
BOASSON, L. & COURCELLE, B. & NIVAT, M. *The rational index: a complexity measure for languages* ◇ D05 D15 ◇
BOERGER, E. *Logical description of computation processes* ◇ B25 D05 D15 D35 ◇
BOOK, R.V. *NTS grammars and Church-Rosser systems* ◇ D03 D05 ◇
BOOK, R.V. & O'DUNLAING, C. *Testing for the Church-Rosser property* ◇ D03 ◇

BOOK, R.V. *The undecidability of a word problem: on a conjecture of Strong, Maggiolo-Schettini and Rosen* ◇ D03 D40 ◇
BOOK, R.V. & O'DUNLAING, C. *Thue congruences and the Church-Rosser property* ◇ D03 ◇
BRANDENBURG, F.-J. *Three write heads are as good as k* ◇ D05 ◇
BRANDSTAEDT, A. *Closure properties of certain families of formal languages with respect to a generalization of cyclic closure* ◇ D05 ◇
BREAZU, V. & STANASILA, O. *A generalization of a theorem of Kleene and applications to structured programming (Romanian) (English summary)* ◇ B75 D05 ◇
BUCHER, W. *A note on a problem in the theory of grammatical complexity* ◇ D05 ◇
BUCHER, W. & CULIK II, K. & MAURER, H.A. & WOTSCHKE, D. *Concise description of finite languages* ◇ D05 ◇
BUCURESCU, I. & PASCU, A. *Fuzzy pushdown automata* ◇ D05 E72 ◇
BUCURESCU, I. & PASCU, A. *On the languages generated by context-sensitive fuzzy grammars* ◇ D05 E72 ◇
BUDACH, L. *Two pebbles don't suffice* ◇ D05 D15 ◇
BURKHARD, HANS-DIETER *Ordered firing in Petri nets* ◇ D05 ◇
BURKHARD, HANS-DIETER *Two pumping lemmata for Petri nets* ◇ D05 ◇
BUTZ, A.R. *Functions realized by consistent sequential machines* ◇ D05 ◇
CHEBOTAR', K.S. *Some modifications of Knuth's algorithm for verifying cyclicity of attribute grammars (Russian)* ◇ D05 ◇
CHOFFRUT, C. *A closure property of deterministic context-free languages* ◇ D05 ◇
CHOFFRUT, C. *Prefix-preservation for rational partial functions is decidable* ◇ D05 ◇
CLAUS, V. *The (n,k)-bounded emptiness-problem for probabilistic acceptors and related problems* ◇ D05 D10 ◇
COURCELLE, B. *An axiomatic approach to the Korenjak - Hopcroft algorithms* ◇ D05 ◇
COURCELLE, B. *Attribute grammars: Theory and applications* ◇ D05 ◇
CRESPI-REGHIZZI, S. & GUIDA, G. & MANDRIOLI, D. *Operator precedence grammars and the noncounting property* ◇ D05 ◇
CUMMINGS, L.J. *Overlapping substrings and Thue's problem* ◇ D03 ◇
DAMM, W. & GUESSARIAN, I. *Combining T and level-N* ◇ D05 ◇
DANECKI, R. *Multiple regularity and binary ETOL-systems* ◇ D05 ◇
DANG HUU DAO *Mealy-automata and their characteristic semigroups* ◇ D05 ◇
DASSOW, J. *Completeness problems in the structural theory of automata* ◇ D05 ◇
DASSOW, J. *Ein modifizierter Vollstaendigkeitsbegriff in einer Algebra von Automatenabbildungen* ◇ D05 ◇
DASSOW, J. *Equality languages and language families* ◇ D05 ◇
DASSOW, J. *On the congruence lattice of algebras of automaton mappings* ◇ D05 ◇
DAUCHET, M. & LATTEUX, M. *A propos d'un langage presque rationnel* ◇ D05 ◇
DERSHOWITZ, N. *Termination of linear rewriting systems* ◇ B75 D05 ◇
DIKOVSKIJ, A.YA. *Classifications of context-free languages by complexity of derivation (Russian)* ◇ D05 ◇
DIKOVSKIJ, A.YA. *Complexity of derivation of unambiguous context-free languages and grammars (Russian)* ◇ D05 ◇
DYER, C.R. *Relation of one-way parallel/sequential automata to 2-D finite-state automata* ◇ D05 ◇
DYER, C.R. & ROSENFELD, A. *Triangle cellular automata* ◇ D05 ◇
EHRENFEUCHT, A. & ROZENBERG, G. & RUOHONEN, K. *A morphic representation of complements of recursively enumerable sets* ◇ D05 D10 ◇
EHRENFEUCHT, A. & ROZENBERG, G. *FPOL systems generating counting languages* ◇ D05 ◇
EHRENFEUCHT, A. & ROZENBERG, G. *On the (generalized) Post correspondence problem with lists of length 2* ◇ D03 ◇
EHRENFEUCHT, A. & ROZENBERG, G. *On the subword complexity of square-free DOL languages* ◇ D05 ◇
EHRENFEUCHT, A. & ROZENBERG, G. & VERMEIR, D. *On ETOL systems with finite tree-rank* ◇ D05 ◇
EHRENFEUCHT, A. & PARIKH, R. & ROZENBERG, G. *Pumping lemmas for regular sets* ◇ D05 ◇
ENGELFRIET, J. & FILE, G. *Passes and paths of attribute grammars* ◇ D05 ◇
ENGELFRIET, J. & FILE, G. *Passes, sweeps and visits* ◇ D05 D15 ◇
ENGELFRIET, J. & FILE, G. *The formal power of one-visit attribute grammars* ◇ D05 ◇
FORYS, M. *On some application of F-grammars to pattern description (Polish and Russian summaries)* ◇ D05 ◇
FORYS, M. *On F-languages and their grammars* ◇ D05 ◇
FORYS, W. *F-languages and F-automata* ◇ D05 ◇
FREJVALD, R.V. *Probabilistic two-way machines* ◇ D05 D10 D15 ◇
FREJVALD, R.V. *Two-way finite probabilistic automata and tape-bounded Turing machines (Russian)* ◇ D05 D10 ◇
GABRIELIAN, A. *Pure grammars and pure languages* ◇ B05 D03 D05 D10 ◇
GALYUSHKOV, B.S. *Semi-context-free grammars (Russian)* ◇ D05 ◇
GEORGEFF, M.P. *Interdependent translation schemes* ◇ D05 ◇
GERGELY, T. *Algebraic representation of language hierarchies* ◇ D05 G15 ◇
GOSTEV, YU.G. *Atomary languages and grammars. On the theory of data structure families (Russian)* ◇ B05 D05 ◇
GREIBACH, S.A. *Formal languages: origins and directions* ◇ A10 D05 ◇
GURARI, E.M. & IBARRA, O.H. *The complexity of the equivalence problem for two characterizations of Presburger sets* ◇ B25 D05 D15 F30 ◇

GURARI, E.M. & IBARRA, O.H. *Two-way counter machines and Diophantine equations* ◇ D05 D35 ◇
HARRISON, M.A. & YEHUDAI, A. *Eliminating null rules in linear time* ◇ D05 D15 ◇
HARTMANIS, J. & MAHANEY, S.R. *Languages simultaneously complete for one-way and two-way log-tape automata* ◇ D05 D15 ◇
HARTMANIS, J. *Observations about the development of theoretical computer science* ◇ A10 D05 D15 ◇
HAUSSLER, D. *Model completeness of an algebra of languages* ◇ B25 C35 D05 ◇
HOEHNKE, H.-J. *On K-recursive definitions and bicategories* ◇ D05 G30 ◇
HONDA, N. & INAGAKI, Y. & OYAMAGUCHI, M. *The equivalence problem for two dpda's, one of which is a finite-turn or one-counter machine* ◇ D05 D10 ◇
HROMKOVIC, J. *Closure properties of the family of languages recognized by one-way two-head deterministic finite state automata* ◇ D05 ◇
HUNT III, H.B. & STEARNS, R.E. *On the equivalence and containment problems for unambiguous regular expressions, grammars, and automata* ◇ D05 ◇
IBARRA, O.H. & ROSIER, L.E. *On restricted one-counter machines* ◇ D05 ◇
JANIGA, L. *Another hierarchy defined by multihead finite automata* ◇ D05 ◇
JANSSENS, D. & ROZENBERG, G. *Decision problems for node label controlled graph grammars* ◇ D05 ◇
JANTZEN, M. *The power of synchronizing operations on strings* ◇ D05 D25 ◇
JAZAYERI, M. *A simpler construction for showing the intrinsically exponential complexity of the circularity problem for attribute grammars* ◇ D05 D15 ◇
KARHUMAEKI, J. *Generalized Parikh mappings and homomorphisms* ◇ D05 ◇
KARHUMAEKI, J. *On strongly cube-free ω-words generated by binary morphisms* ◇ D05 ◇
KEKLIKOGLOU, I. *Verallgemeinerte kommutative Sprachen (English summary)* ◇ D05 ◇
KLEINE BUENING, H. *Classes of functions over binary trees* ◇ B75 D05 D15 D20 ◇
KUR'EROV, YU.N. *Equiprobable canonical calculi (Russian)* ◇ B35 D03 ◇
LAZDINYA, G.K. *Probabilistic supercounter machines (Russian)* ◇ D05 ◇
LEISS, E. *On generalized language equations* ◇ D05 ◇
LEISS, E. *Succinct representation of regular languages by boolean automata* ◇ D05 ◇
LEWIS, F.D. *A note on context free languages, complexity classes, and diagonalization* ◇ D05 D15 ◇
LISOVIK, L.P. *Program schemata over cyclic interpretations (Russian)* ◇ D05 ◇
LISOVIK, L.P. *Recursive-deterministic abstract automata (Russian)* ◇ D05 ◇
MALENGE, J.P. & RIX, H. *Indecidable...pas sur! (English summary)* ◇ D05 ◇
MAURER, H.A. & SALOMAA, A. & WOOD, D. *Completeness of context-free grammar forms* ◇ B75 D05 ◇
MAURER, H.A. & ROZENBERG, G. *Sub context-free L forms* ◇ B75 D05 ◇
MICALI, S. *Two-way deterministic finite automata are exponentially more succinct than sweeping automata* ◇ D05 D15 ◇
MIKOLAJCZAK, B. *On the time computational complexity of several problems of simulation of a set of finite automata* ◇ D05 D15 ◇
MONIEN, B. *On the LBA problem* ◇ D05 ◇
MOSTOWSKI, A.WLODZIMIERZ *The complexity of automata and subtheories of monadic second order arithmetics* ◇ B15 B25 D05 D15 F35 ◇
MULLER, D.E. & SCHUPP, P.E. *Context-free languages, groups, the theory of ends, second-order logic, tiling problems, cellular automata, and vector addition systems* ◇ B15 B25 D05 D80 ◇
NAKAMURA, A. & ONO, H. *Undecidability of extensions of the monadic first-order theory of successor and two-dimensional finite automata* ◇ B15 B25 D05 D35 ◇
OTTMANN, T. & SALOMAA, A. & WOOD, D. *Sub-regular grammar forms* ◇ D05 ◇
PANSIOT, J.-J. *A note on Post's correspondence problem* ◇ D03 ◇
PARK, D. *Concurrency and automata on infinite sequences* ◇ D05 ◇
PAUN, G. *The languages of propositional calculus and the Chomsky hierarchy (Romanian) (English summary)* ◇ B65 D05 ◇
PAUN, G. *Thue languages and matrix grammars* ◇ D03 D05 ◇
PAVLENKO, V.A. *The combinatorial problem of Post with two pairs of words (Russian) (English summary)* ◇ D03 ◇
PNUELI, A. & SLUTZKI, G. *Automatic programming of finite state linear programs* ◇ D05 ◇
POIGNE, A. *Context-free languages of infinite words as least fixpoints* ◇ D05 ◇
RAOULT, J.-C. *Finiteness results on rewriting systems (French summary)* ◇ D05 ◇
ROVAN, B. *A framework for studying grammars* ◇ D05 ◇
ROZENBERG, G. *On subwords of formal languages* ◇ D05 ◇
RUOHONEN, K. *The decidability of the DOL-DTOL equivalence problem* ◇ B75 D05 ◇
SAKAROVITCH, JACQUES *Sur une propriete d'iteration des langages algebriques deterministes* ◇ D05 ◇
SALOMAA, A. *Jewels of formal language theory* ◇ D05 D98 ◇
SHU, YONGCHANG & WANG, YIZHI *Fuzzy languages and fuzzy grammars (Chinese)* ◇ B52 D05 ◇
STARKE, P.H. *A note on conflicts in Petri nets* ◇ D05 ◇
STARKE, P.H. *Processes in Petri nets* ◇ D05 ◇
STOKLOSA, J. *Properties of (α, k)-computations* ◇ D05 ◇
THOMAS, WOLFGANG *A combinatorial approach to the theory of ω-automata* ◇ B25 D05 ◇
TROFIMOV, V.I. *Growth functions of some classes of languages* ◇ D05 ◇
TURAKAINEN, P. *On nonstochastic languages and homomorphic images of stochastic languages* ◇ D05 ◇

TURAN, G. *On cellular graph-automata and second-order definable graph-properties* ◇ C85 D05 D80 ◇
VAZHENIN, YU.M. *Sur la liaison entre problemes combinatoires et algorithmiques* ◇ B25 D05 D15 D35 ◇
VOGEL, J. & WAGNER, K. *On a class of automata accepting exactly the languages which are elementary in the sense of KALMAR* ◇ D05 D20 ◇
YOSHIMOTO, Y. *The proof of sufficiency of McNaughton's condition on generalized sequentialmachine mappability-onto of regular sets* ◇ D05 ◇
ZHU, SHANGYONG *Fuzzy tree grammars and fuzzy forest grammars (Chinese)* ◇ D05 ◇
ZIELONKA, W. *Axiomatizability of Ajdukiewicz-Lambek calculus by means of cancellation schemes* ◇ B65 D05 ◇

1982

ADAMEK, J. & TRNKOVA, V. *Analyses of languages accepted by varietor machines in category* ◇ D05 ◇
ALBERT, J. *A note on the undecidability of contextfreeness (French summary)* ◇ D05 ◇
AUTEBERT, J.-M. & BEAUQUIER, J. & BOASSON, L. *Formes de langages et de grammaires* ◇ D05 ◇
AUTEBERT, J.-M. & BEAUQUIER, J. & BOASSON, L. & LATTEUX, M. *Indecidabilite de la condition IRS (English summary)* ◇ D05 ◇
BADER, C. & MOURA, A. *A generalization of Ogden's lemma* ◇ D05 ◇
BAETEN, J.C.M. & BERGSTRA, J.A. & KLOP, J.W. *Priority rewrite systems* ◇ B75 D05 D20 ◇
BAJRASHEVA, V.R. *Degrees of automated transformations (Russian)* ◇ D05 D30 ◇
BLATTNER, M. & GINSBURG, S. *Position-restricted grammar forms and grammars* ◇ D05 ◇
BLUM, N. *On the power of chain rules in context free grammars* ◇ D05 ◇
BOOK, R.V. *Confluent and other types of Thue systems* ◇ D03 ◇
BOOK, R.V. & JANTZEN, M. & WRATHALL, C. *Monadic Thue systems* ◇ D03 D05 ◇
BOOK, R.V. *When is a monoid a group? The Church-Rosser case is tractable* ◇ D03 D15 D40 ◇
BRANDENBURG, F.-J. *Extended Chomsky-Schuetzenberger theorems* ◇ B65 D05 ◇
BRANDSTAEDT, A. *Reversal-bounded realtime acceptors with only two tapes* ◇ D05 ◇
BRENT, R.P. & GOLDSCHLAGER, L.M. *Some area-time tradeoffs for VLSI* ◇ B70 D05 D15 ◇
BUDACH, L. & MEINEL, C. *Environments and Automata I., II (German and Russian summaries)* ◇ D05 D10 D80 ◇
BUDACH, L. & WAACK, S. *On the halting problem for automata in cones* ◇ D05 ◇
BURKHARD, HANS-DIETER *Generalized identification experiments for finite deterministic automata* ◇ D05 ◇
BUSZKOWSKI, W. *Some decision problems in the theory of syntactic categories* ◇ B25 D05 ◇
CARSTENS, H.G. & GOLZE, U. *Recursive paths in cross-connected trees and an application to cell spaces* ◇ D05 ◇
CASE, J. & LYNES, C. *Machine inductive inference and language identification* ◇ D05 D20 ◇
COHEN, JACQUES & HICKEY, T. & KATCOFF, J. *Upper bounds for speedup in parallel parsing* ◇ D05 D15 ◇
COMER, S.D. *Inductive domains and algebraic semantics of CF languages* ◇ D05 ◇
COURCELLE, B. *Fundamental properties of infinite trees* ◇ D05 D98 E75 ◇
DASSOW, J. *Weitere Bemerkungen zu einem modifizierten Vollstaendigkeitsbegriff in einer Algebra von Automatenabbildungen* ◇ D05 ◇
DOBERKAT, E.-E. *Stochastic behavior of initial nondeterministic automata* ◇ D05 ◇
EBBINGHAUS, H.-D. *Undecidability of some domino connectability problems* ◇ D05 ◇
EHRENFEUCHT, A. & KARHUMAEKI, J. & ROZENBERG, G. *The (generalized) Post correspondence problem with lists consisting of two words is decidable* ◇ D03 ◇
ETTL, W. & LEITSCH, A. *Enumerations of subrecursive classes and generations of automata* ◇ D05 ◇
FANG, ZHIXI *The closure property of certain classes of languages under bi-language forms (Chinese) (English summary)* ◇ D05 D15 D25 ◇
FARMER, F.D. *Homotopy spheres in formal language* ◇ D05 ◇
FISCHER, R. & UHLIG, D. *Fehlerkorrigierende Realisierungen zu Booleschen Funktionen und Automatenfunktionen mit linearer Kompliziertheit* ◇ B05 D05 ◇
FREJVALD, P.V. *Increase in number of states when finite probabilistic automata are made deterministic (Russian)* ◇ D05 ◇
GOLDMAN, J. & HOMER, S. *Quadratic automata* ◇ B70 D05 D45 ◇
GURARI, E.M. *The equivalence problem for deterministic two-way sequential transducers is decidable* ◇ D05 ◇
GURARI, E.M. & IBARRA, O.H. *Two-way counter machines and Diophantine equations* ◇ D05 D35 ◇
GUREVICH, Y. & HARRINGTON, L.A. *Automata, trees, and games* ◇ B25 C85 D05 ◇
HAGAUER, J. *On form-equivalence of deterministic pure grammar forms* ◇ D05 ◇
HASHIGUCHI, K. *Limitedness theorem on finite automata with distance functions* ◇ D05 ◇
HAY, L. *On the recursion-theoretic complexity of relative succinctness of representations of languages* ◇ D05 D30 ◇
HEMMERLING, A. *On the computational equivalence of synchronous and asynchronous cellular spaces (German and Russian summaries)* ◇ D05 ◇
HEMMERLING, A. *Zur Raumkompliziertheit von Absuchprozessen auf endlichen Graphen* ◇ D05 D15 ◇
HODGSON, B.R. *On direct products of automaton decidable theories* ◇ B25 D05 ◇
HUNYADVARI, L. *A characteristic of L-fuzzy languages generated by L-fuzzy grammars of type 3 (Russian)* ◇ B65 D05 ◇
HUNYADVARI, L. *Some properties of L-regular languages (Russian)* ◇ D05 ◇

IBARRA, O.H. *2DST mappings on languages and related problems* ◇ D05 ◇
ISTRAIL, S. *Generalization of the Ginsburg-Rice Schuetzenberger fixed-point theorem for context-sensitive and recursive-enumerable languages* ◇ D05 ◇
JANTZEN, M. & KUDLEK, M. *Homomorphic images of sentential form languages defined by semi-Thue systems* ◇ D03 D05 ◇
KARPINSKI, M. *Decidability of "Skolem matrix emptiness problem" entails constructability of exact regular expression* ◇ D05 ◇
KORTELAINEN, J. *On language families generated by commutative languages* ◇ D05 ◇
LIBUS, M. & OSTROWSKA, M. *Remarks on L.J.Stockmeyer's note "The complexity of decision problems in automata theory and logic" (Polish) (English summary)* ◇ D05 D10 D15 ◇
MAIDA, A. *Una revisitazione della logica classica in termini di grammatiche generative (English summary)* ◇ D05 ◇
MAURER, H.A. & SALOMAA, A. & WOOD, D. *Dense hierarchies of grammatical families* ◇ D05 ◇
MEINHARDT, D. & WAGNER, K. *Eine Bemerkung zu einer Arbeit von Monien ueber Kopfzahlhierarchien fuer Zweiwegautomaten (English and Russian summaries)* ◇ D05 ◇
MORITA, K. & SUGATA, K. & UMEO, H. *Deterministic one-way simulation of two-way real-time cellular automata and its related problems* ◇ D05 D15 ◇
MORIYA, T. *Pebble machines and tree walking machines* ◇ D05 ◇
NASYROV, I.R. *A reduction theorem for automata over trees and context-free grammars (Russian)* ◇ D05 ◇
NASYROV, I.R. *Representability of languages in deterministic and nondeterministic automata with a countable number of states (Russian)* ◇ D05 D10 ◇
NIJHOLT, A. *The equivalence problem for LL- and LR-regular grammars* ◇ D05 ◇
PAVLENKO, V.A. *The Post combinatorial problem for two pairs of words (Russian)* ◇ D03 ◇
RUZICKA, M. *Input-output systems, their types and applications for the automata theory* ◇ B75 D05 ◇
SANDRING, S. & STARKE, P.H. *A note on liveness in generalized Petri nets* ◇ D05 ◇
SLISENKO, A.O. *Context-free grammars as a tool for describing polynomial-time subclasses of hard problems* ◇ D05 ◇
STARKE, P.H. *A CE-system and its reachability graph* ◇ D05 ◇
STARKE, P.H. *Graph grammars and Petri net processes* ◇ D05 ◇
STARKE, P.H. *Praedikat-Transitions-Netze* ◇ D05 ◇
TAO, RENJI *A lower bound of kn^2 on time-complexity for one-tape automata on computations* ◇ D05 D15 ◇
THOMAS, WOLFGANG *Classifying regular events in symbolic logic* ◇ D05 ◇
TORBASOVA, V.P. *Machines over models (Russian)* ◇ D05 D75 ◇
TURAKAINEN, P. *A homomorphic characterisation of principal semiAFLs without using intersection with regular sets* ◇ D05 ◇
TURAKAINEN, P. *Rational stochastic automata in formal language theory* ◇ D05 ◇
UKKONEN, E. *Structure preserving elimination of null productions from context-free grammars* ◇ D05 ◇

1983

ABLAEV, F.M. *On the question of automaton complexity of languages and complexity of pattern recognition in the sense of Loveland (Russian)* ◇ D05 ◇
ALBERT, J. & OTTMANN, T. *Automaten, Sprachen und Maschinen fuer Anwender* ◇ D05 D10 D98 ◇
ANDZHAN, A.V. *Possibilities of automata for searching one-dimensional domains (Russian)* ◇ D05 ◇
ARBIB, M.A. & MANES, E.G. & STEENSTRUP, M. *Port automata and the algebra of concurrent processes* ◇ D05 ◇
BOERGER, E. *Undecidability versus degree complexity of decision problems for formal grammars* ◇ D05 D10 D25 ◇
BOOK, R.V. *A note on special Thue systems with a single defining relation* ◇ D03 ◇
BOOK, R.V. *Thue systems and the Church-Rosser property: replacement systems, specification of formal languages, and presentations of monoids* ◇ D03 ◇
BORGIDA, A.T. *Some formal results about stratificational grammars and their relevance to linguistics* ◇ B65 D05 ◇
BRANDSTAEDT, A. *Space classes, intersections of languages and bounded erasing homomorphisms* ◇ D05 D15 ◇
BUCHER, W. & HAGAUER, J. *It is decidable whether a regular language is pure context-free* ◇ B65 D05 ◇
BUECHI, J.R. & ZAIONTZ, C. *Deterministic automata and the monadic theory of ordinals $< \omega_2$* ◇ B15 B25 C85 D05 E10 ◇
BUECHI, J.R. *State-strategies for games in $F_{\sigma\delta} \cap G_{\delta\sigma}$* ◇ D05 D55 E15 E60 ◇
BUECHI, J.R. & SIEFKES, D. *The complete extensions of the monadic second order theory of countable ordinals* ◇ B15 C35 C85 D05 E10 ◇
BURKHARD, HANS-DIETER *Control of Petri nets by finite automata* ◇ D05 D80 ◇
BURKHARD, HANS-DIETER *On priorities of parallelism: Petri nets under the maximum firing strategy* ◇ D05 ◇
CULIK II, K. & WELZL, E. *Two way finite state generators* ◇ D05 ◇
DARE, V.R. & SIROMONEY, R. & SUBRAMANIAN, K.G. *Infinite arrays and infinite computations* ◇ B75 D05 ◇
DAVIS, MARTIN D. & WEYUKER, E.J. *Computability, complexity, and languages. Fundamentals of theoretical computer science* ◇ D05 D10 D20 D98 ◇
FASS, L.F. *Learning context-free languages from their structured sentences* ◇ D05 ◇
FREJVALD, R.V. *Undecidability of the emptiness problem for probabilistic finite multitape machines (Russian)* ◇ D05 ◇

GINSBERG, S. & GOLDSTINE, J. & SPANIER, E.H. *On the equality of grammatical families* ◇ B65 D05 ◇
HAUSSLER, D. *Insertion languages* ◇ D05 ◇
HEILBRUNNER, S. *A metatheorem for undecidable properties of formal languages and its application to LRR and LLR grammars and languages* ◇ D05 ◇
HODGSON, B.R. *Decidabilite par automate fini* ◇ B25 D05 D35 ◇
HUYNH, D.T. *Commutative grammars: The complexity of uniform word problems* ◇ D05 D15 ◇
JOUANNAUD, J.-P. *Confluent and coherent equational term rewriting systems application to proofs in abstract data types* ◇ B75 D03 ◇
KAPUR, D. & NARENDRAN, P. *The Knuth-Bendix completion procedure and Thue systems* ◇ D03 ◇
KARHUMAEKI, J. & LINNA, M. *A note on morphic characterisation of languages* ◇ B65 D05 ◇
KINBER, E.B. *A class of multitape automata with a decidable equivalence problem (Russian)* ◇ D05 ◇
KINBER, E.B. *The inclusion problem for some classes of deterministic multitape automata* ◇ D05 ◇
LATTEUX, M. & THIERRIN, G. *Semidiscrete context-free languages* ◇ D05 ◇
LEE, E.T. *Algorithms for finding Chomsky and Greibach normal forms for a fuzzy context-free grammar using an algebraic approach* ◇ D05 ◇
LISOVIK, L.P. *Finite transformers on marked trees and quasi-identities in a free semigroup (Russian)* ◇ D05 ◇
LISOVIK, L.P. *Minimal undecidable identity problem for finite-automaton mappings* ◇ D05 ◇
MO, SHAOKUI *A note on a regular language (Chinese) (English summary)* ◇ D05 ◇
MUCHNIK, A.A. *Supplement of the translator to the paper "On alternation. I, II" (Russian)* ◇ B45 D05 D10 D15 F50 ◇
MULLER, D.E. & SCHUPP, P.E. *Groups, the theory of ends, and context-free languages* ◇ D05 D40 ◇
NELSON, EVELYN *Iterative algebras* ◇ D05 ◇
NGO THE KHANH *Prefix-free languages and simple deterministic machines (Hungarian)* ◇ D05 ◇
NIWINSKI, D. *Fixed-point semantics for algebraic (tree) grammars* ◇ D05 ◇
O'DUNLAING, C. *Infinite regular Thue systems* ◇ D03 ◇
O'DUNLAING, C. *Undecidable questions related to Church-Rosser Thue systems* ◇ D03 ◇
PANSIOT, J.-J. *Hierarchie et fermeture de certaines classes de tag-systemes (English summary)* ◇ D03 ◇
PERRIN, D. *Varietes de semigroupes et mots infinis* ◇ C05 D05 ◇
RADENSKY, A.A. *Transformations in context-free languages* ◇ D05 ◇
ROEDDING, D. *Modular decomposition of automata* ◇ D05 ◇
RUOHONEN, K. *On some variants of Post's correspondence problem* ◇ D03 D05 ◇
RYSTSOV, I.K. *Polynomial complete problems in automata theory* ◇ D05 D15 ◇
RYTTER, W. *Remarks on trace languages* ◇ D05 ◇
STAIGER, L. *Finite state ω-languages* ◇ D05 D55 ◇
STAPLES, J. *Two-level expression representation for faster evaluation* ◇ B35 B40 D03 ◇
STARKE, P.H. *Graph grammars for Petri net processes* ◇ D05 ◇
STARKE, P.H. *Monogeneous fifo-nets and Petri nets are equivalent* ◇ D05 D80 ◇
STARKE, P.H. *On the concurrency of distributed systems* ◇ D05 D80 ◇
TCHUENTE, M. *Computation of Boolean functions on networks of binary automata* ◇ D05 ◇
THIELE, H. *A classification of propositional process logics on the basis of theory of automata* ◇ B75 D05 ◇
VALK, R. *Infinite behaviour of Petri nets* ◇ D05 ◇
ZHU, SHANGYONG *Error-correcting fuzzy tree grammar. I (Chinese) (English summary)* ◇ D05 ◇
ZHU, SHANGYONG *Error-correcting fuzzy tree grammars. II (Chinese) (English summary)* ◇ D05 ◇

1984

ABLAEV, F.M. & GABBASOV, N.Z. *Remarks on generalized reduction theorems for infinite automata (Russian)* ◇ D05 ◇
ALBERT, J. *On the Ehrenfeucht conjecture on test sets and its dual version* ◇ D05 ◇
ANGLUIN, D. & HOOVER, D.N. *Regular prefix relations* ◇ D05 ◇
AVENHAUS, J. & BOOK, R.V. & SQUIER, C.C. *On expressing commutativity by finite Church-Rosser presentations: a note on commutative monoids (French summary)* ◇ D05 ◇
BAUER, G. & OTTO, F. *Finite complete rewriting systems and the complexity of word problem* ◇ D05 D15 ◇
BAVEL, Z. & GRZYMALA-BUSSE, J.W. & SOO HONG KWANG *On the connectivity of the product of automata* ◇ D05 ◇
BERGSTRA, J.A. & KLOP, J.W. *The algebra of recursively defined processes and the algebra of regular processes* ◇ D05 D75 ◇
BERSTEL, J. *Some recent results on squarefree words* ◇ D05 ◇
BOOK, R.V. & SQUIER, C.C. *Almost all one-rule Thue systems have decidable word problems* ◇ D03 ◇
BOOK, R.V. *Homogeneous Thue systems and the Church-Rosser property* ◇ D03 ◇
BRUEGGEMANN, A. & PRIESE, L. & ROEDDING, D. & SCHAETZ, R. *Modular decomposition of automata* ◇ D05 ◇
BUDA, A. *A simple characterization of coregular tree-languages (Russian)* ◇ D05 ◇
BURKHARD, HANS-DIETER & STARKE, P.H. *A note on the impact of conflict resolution to liveness and deadlock in Petri nets* ◇ D05 D80 ◇
BURKHARD, HANS-DIETER *An investigation of controls for concurrent systems by abstract control languages* ◇ D05 D80 ◇
CERNY, A. *On generalized words of Thue-Morse* ◇ D03 ◇
COMPTON, K.J. *On rich words* ◇ C85 D05 D45 ◇
CULIK II, K. & YU, SHENG *Iterative tree automata* ◇ D05 ◇
DARONDEAU, P. & KOTT, L. *Towards a formal proof system for ω-rational expressions* ◇ D05 ◇

DURIS, P. & GALIL, Z. *A time-space tradeoff for language recognition* ◇ D05 D15 ◇
EHRENFEUCHT, A. & ROZENBERG, G. *On regularity of languages generated by copying systems* ◇ D05 ◇
FERMENT, D. *Principality results about some matrix languages families* ◇ D05 ◇
GECSEG, F. & STEINBY, M. *Tree automata* ◇ A05 B25 D05 D98 ◇
GIRAULT-BEAUQUIER, D. *Bilimites de langages reconaissables* ◇ D05 ◇
GIRE, F. & NIVAT, M. *Relations rationnelles infinitaires* ◇ D05 ◇
HAYASHI, T. & MIYANO, S. *Alternating finite automata on ω-words* ◇ B75 D05 ◇
HEAD, T. *Adherence equivalence is decidable for DOL languages* ◇ B75 D05 ◇
HEMMERLING, A. & KRIEGEL, K. *On searching of special classes of mazes and finite embedded graphs* ◇ D05 ◇
HONDA, N. & INAGAKI, Y. & IZUMI, H. *A complete axiom system for algebra of closed-regular expression* ◇ D05 ◇
HUYNH, D.T. *Deciding the inequivalence of context-free grammars with 1-letter alphabet is* Σ_2^p*-complete* ◇ D05 D15 ◇
IBARRA, O.H. & KIM, S.M. *A characterization of systolic binary tree automata and applications* ◇ B75 D05 ◇
IDT, J. *Automates a pile sur des alphabets infinis* ◇ D05 ◇
IVANOV, N.N. & MIKHAJLOV, G.I. & RUDNEV, V.V. & TAL', A.A. *Finite automata: equivalence and behaviour (Russian)* ◇ D05 ◇
JANTKE, K.P. *Polynomial time inference of general pattern languages* ◇ D05 D15 D80 ◇
JANTZEN, M. & KUDLEK, M. *Homomorphic images of sentential form languages defined by semi-Thue systems* ◇ D03 ◇
JANTZEN, M. *Thue systems and the Church-Rosser property* ◇ D03 ◇
KLEINE BUENING, H. *Complexity of loop-problems in normed networks* ◇ D05 D15 ◇
KOBAYASHI, K. & TAKAHASHI, MASAKO & YAMASAKI, H. *Characterization of ω-regular languages by first-order formulas* ◇ D05 ◇
KOLYADA, K.V. *Completeness of regular mappings (Russian)* ◇ D05 ◇
LATTEUX, M. & ROZENBERG, G. *Commutative one-counter languages are regular* ◇ D05 ◇
LEITSCH, A. *Enumerations of subrecursive classes and self-reproducing automata. I: Effective translations and decidable indexsets. II: Index-complexity and translation-complexity* ◇ D05 D20 D45 ◇
LIN, YUCAI *The decision problems about the periodic solutions of the domino problems* ◇ D03 ◇
LISOVIK, L.P. *Construction of decidable singular theories of two successor functions with an extra predicate (Russian)* ◇ B25 D05 ◇
LISOVIK, L.P. *Monadic second-order theories of two successor functions with an additional predicate (Russian)* ◇ B15 B25 C10 C85 D05 ◇
MARGOLIS, S.W. & PIN, J.-E. *Languages and inverse semigroups* ◇ D05 ◇
MATIYASEVICH, YU.V. *Studies in certain algorithmic problems of algebra and number theory (Russian)* ◇ D03 D05 D35 ◇
MCALOON, K. *Petri nets and large finite sets* ◇ D05 D20 D80 ◇
MCNAUGHTON, R. & NARENDRAN, P. *The undecidability of the preperfectness of Thue systems* ◇ D03 ◇
MICHEL, M. *Algebre de machines et logique temporelle* ◇ B45 D05 ◇
MUIR, A. & WARNER, M.W. *Lattice valued relations and automata* ◇ D05 E07 G10 ◇
NIVAT, M. & PERRIN, D. (EDS.) *Automata on infinite words* ◇ D05 ◇
NIWINSKI, D. *Fixed-point characterization of context free ∞-languages* ◇ D05 ◇
OTTO, F. *Conjugacy in monoids with a special Church-Rosser presentation is decidable* ◇ C05 D03 D05 D40 ◇
OTTO, F. *Some undecidability results for nonmonadic Church-Rosser Thue systems* ◇ D03 ◇
PANSIOT, J.-J. *Bornes inferieures sur la complexite des facteurs des mots infinis engendres par morphismes iteres* ◇ D05 ◇
PANSIOT, J.-J. *Complexite des facteurs des mots infinis engendres par morphismes iteres* ◇ D05 ◇
PECUCHET, J.-P. *Automates Boustrophedon, semi-groupe de Birget et monoide inversif libre* ◇ D05 ◇
PENTTONEN, M. *The reachability of vector addition systems and equivalent problems* ◇ D03 D05 D80 ◇
PERI, C. *Algorithms on sets of words that are isomorphic to Markov algorithms (Italian) (English summary)* ◇ D03 D20 ◇
PERRIN, D. *Recent results on automata and infinite words* ◇ D05 ◇
RESTIVO, A. & REUTENAUER, C. *Cancellation, pumping and permutation in formal languages* ◇ D05 ◇
RHODES, J. *Algebraic and topological theory of languages and computation I: Theorems for arbitrary languages generalizing the theorems of Eilenberg, Kleene, Schuetzenberger, and Straubing* ◇ D05 ◇
ROEDDING, D. *Some logical problems connected with a modular decomposition theory of automata* ◇ D05 ◇
SAJO, A. *On subword complexity functions* ◇ D05 ◇
SAOUDI, A. *Infinitary tree languages recognized by ω-automata* ◇ D05 ◇
SAVITCH, W.J. & VITANYI, P.M.B. *On the power of real-time two-way multihead finite automata with jumps* ◇ D05 ◇
SKOBELEV, V.G. *Enumeration of automata possessing homing sequences (Russian) (English summary)* ◇ D05 ◇
STARKE, P.H. *An uninvited address to "serializers" and "nontransitivists"* ◇ D05 ◇
STARKE, P.H. *Multiprocessor systems and their concurrency* ◇ B75 D05 ◇
STARKE, P.H. *Multiprocessor systems and their concurrency (German and Russian summaries)* ◇ B75 D05 ◇

Starke, P.H. *Ueber die Nebenlaeufigkeit von Multiprozessorsystemen* ◇ D05 ◇

Stefan, T. *A logical approach to regular languages* ◇ D05 ◇

Steinby, M. *Some decidable properties of Σ-rational and Σ-algebraic tree transformations* ◇ D05 ◇

Thomas, Wolfgang *An application of the Ehrenfeucht-Fraisse game in formal language theory* ◇ C07 D05 ◇

Welzl, E. *Encoding graphs by derivations and implications for the theory of graph grammars* ◇ D05 ◇

Winskel, G. *A new definition of morphism on Petri nets* ◇ D05 ◇

Wojciechowski, J. *Classes of transfinite sequences accepted by nondeterministic finite automata* ◇ D05 ◇

1985

Ablaev, F.M. *Possibilities of probalistic machines with respect to representation of languages in real time (Russian)* ◇ D05 D10 ◇

Arnold, A. *Deterministic and nonambigous rational ω-languages* ◇ D05 ◇

Beauquier, D. *Ensembles reconnaissables de mots biinfinis. Limite et determinisme* ◇ D05 ◇

Beauquier, D. *Mueller automata and bi-infinite words* ◇ D05 ◇

Blanchard, F. & Hansel, G. *Languages and subshifts* ◇ D05 ◇

Bloom, S.L. *Frontiers of one-letter languages* ◇ D05 ◇

Book, R.V. & Otto, F. *Cancellation rules and extended word problems* ◇ D03 D15 ◇

Book, R.V. *Thue systems as rewriting systems* ◇ D03 ◇

Borm, A.E. & Rosier, L.E. *A note on Parikh maps, abstract languages, and decision problems* ◇ D05 ◇

Braquelaire, J.P. & Courcelle, B. *The solution of two star-height problems for regular trees* ◇ D05 ◇

Bucher, W. & Haussler, D. & Main, M.G. *Applications of an infinite squarefree co-CFL* ◇ B75 D05 ◇

Bukharaev, R.G. *Foundations of the theory of probabilistic automata (Russian)* ◇ D05 ◇

Buszkowski, W. *Algebraic models of categorial grammars* ◇ B65 D05 ◇

Buszkowski, W. *The equivalence of unidirectional Lambek categorial grammars and context-free grammars* ◇ D05 ◇

Chandra, A.K. & Halpern, J.Y. & Meyer, A.R. & Parikh, R. *Equations between regular terms and application to process logic* ◇ B45 D05 ◇

Chrobak, M. *Hierarchies of one-way multihead automata languages* ◇ D05 ◇

Dare, V.R. & Siromoney, R. *A generalization of the Parikh vector for finite and infinite words* ◇ D05 ◇

Dare, V.R. & Siromoney, R. *On infinite words obtained by selective substitution grammars* ◇ D05 ◇

Darondeau, P. & Kott, L. *A formal proof system for infinitary rational expressions* ◇ D05 ◇

Dassow, J. & Wolter, U. *Remarks on equality languages* ◇ A05 B20 D05 ◇

Dauchet, M. & Tison, S. *Decidability of confluence for ground term rewriting systems* ◇ D05 ◇

Dauchet, M. *Decidability of yield's equality for infinite regular trees* ◇ D05 ◇

Dosen, K. *A completeness theorem for the Lambek calculus of syntactic categories* ◇ B65 C90 D05 ◇

Ehrenfeucht, A. & Kleijn, H.C.M. & Rozenberg, G. *Adding global forbidding context to context-free grammars* ◇ D05 ◇

Ehrenfeucht, A. & Hoogeboom, H.J. & Rozenberg, G. *On coordinated rewriting* ◇ D05 ◇

Emerson, E.A. *Automata, tableaux, and temporal logics* ◇ B45 D05 ◇

Fachini, E. & Napoli, M. *Synchronized bottom-up tree automata and L-systems* ◇ D05 ◇

File, G. *Tree automata and logic programs* ◇ B75 D05 ◇

Flajolet, P. *Ambiguity and transcendence* ◇ D05 ◇

Frougny, C. *Context-free grammars with cancellation properties* ◇ D05 ◇

Gonczarowski, J. & Warmuth, M.K. *Applications of scheduling theory to formal language theory* ◇ D05 D15 ◇

Gurevich, Y. & Shelah, S. *The decision problem for branching time logic* ◇ B25 B45 C65 D05 D25 ◇

Harel, D. *Recurring dominoes: Making the highly undecidable highly understandable* ◇ B25 B75 D05 D15 D55 ◇

Hayashi, T. & Miyano, S. *Finite tree automata of infinte trees* ◇ D05 ◇

Head, T. *The adherence of languages as topological spaces* ◇ D05 ◇

Hirose, S. & Yoneda, M. *On the Chomsky and Stanley's homomorphic characterization of context-free languages* ◇ D05 ◇

Hromkovic, J. *Fooling a two-way nondeterministic multihead automaton with reversal number restriction* ◇ D05 ◇

Hromkovic, J. *On the power of alternation in automata theory* ◇ D05 D10 ◇

Hunt III, H.B. & Rosenkrantz, D.J. *Testing for grammatical coverings* ◇ D05 ◇

Inoue, K. & Matsuno, H. & Takanami, I. & Taniguchi, H. *Alernating simple multihead finite automata* ◇ D05 ◇

Inoue, K. & Matsuno, H. & Takanami, I. & Taniguchi, H. *Alternating one-way multihead finite automata with only universal states* ◇ D05 ◇

Jantzen, M. *A note on a special one-rule semi-Thue system* ◇ D03 ◇

Jantzen, M. *Extending regular expressions with iterated shuffle* ◇ D05 ◇

Kapur, D. & Krishnamoorthy, M.S. & McNaughton, R. & Narendran, P. *An $O(|T|^3)$ algorithm for testing the Church-Rosser property of Thue systems* ◇ D03 D15 ◇

Kapur, D. & Krishnamoorthy, M.S. & McNaughton, R. & Narendran, P. *The Church-Rosser property and special Thue systems* ◇ D03 ◇

Kapur, D. & Narendran, P. *The Knuth-Bendix completion procedure and Thue systems* ◇ D03 ◇

KIERCZAK, J. *Rough grammars* ◇ D05 ◇
KOBAYASHI, K. & TAKAHASHI, M. & YAMASAKI, H. *Logical formulas and four subclasses of ω-regular languages* ◇ D05 ◇
KOLETSOS, G. *Church-Rosser theorem for typed functional systems* ◇ D05 ◇
KUR'EROV, YU.N. *On the embedding of an Ackermann class in a graph grammar (Russian) (English and Lithuanian summaries)* ◇ D05 D20 ◇
LANGE, K.-J. & WELZL, E. *String grammars with disconnecting* ◇ D05 ◇
LATTEUX, M. & LEGUY, B. & RATOANDROMANANA, B. *The family of one-counter languages is closed under quotient* ◇ D05 ◇
LECONTE, M. *Kth power free-codes* ◇ D05 ◇
LEISS, E. *Succinct representation of regular languages by Boolean automata.II* ◇ D05 ◇
METIVIER, Y. *Calcul de longueurs de chaines de reecriture dans le monoide libre* ◇ D05 ◇
MOSTOWSKI, A.WLODZIMIERZ *Regular expressions for infinite trees and a standard form of automata* ◇ D05 ◇
MUCHNIK, A.A. *Games on infinite trees and automata with dead ends. A new proof of decidability for the monadic theory with two successor functions (Russian)* ◇ B15 B25 C85 D05 ◇
MUELLER, HORST *Weak Petri net computers for Ackermann functions* ◇ D05 D20 ◇
MULLER, D.E. & SCHUPP, P.E. *Alternating automata on infinite objects, determinacy and Rabin's Theorem* ◇ B25 C85 D05 E60 ◇
OTTO, F. & WRATHALL, C. *A note on Thue systems with a single defining relation* ◇ D03 ◇
OTTO, F. *Deciding algebraic properties of monoids presented by finite Church-Rosser Thue systems* ◇ D03 ◇
OTTO, F. *Elements of finite order for finite monadic Church-Rosser Thue systems* ◇ D03 ◇
PAN, LUQUAN *On reduced Thue systems* ◇ D03 ◇
PANSIOT, J.-J. *On various classes of infinite words obtained by iterated mappings* ◇ D05 ◇
PAUN, G. & TATARAM, M. *Classes of mappings having context sensitive graphs* ◇ D05 ◇
PECUCHET, J.-P. *Automates boustrophedon sur des mots infinis* ◇ D05 ◇
PECUCHET, J.-P. *Automates boustrophedon et mots infinis (French) (English summary)* ◇ D05 ◇
PERRIN, D. *An introduction to finite automata on infinite words* ◇ D05 ◇
PERRIN, D. *Words over a partially commutative alphabet* ◇ D05 ◇
PIN, J.-E. *Star-free ω-languages and first order logic* ◇ D05 ◇
PLAISTED, D.A. *The undecidability of self-embedding for term rewriting systems* ◇ D05 ◇
RESTIVO, A. & SALEMI, S. *Overlap free words on two symbols* ◇ D05 ◇
RESTIVO, A. & SALEMI, S. *Some decision results on nonrepetitive words* ◇ D05 ◇
RUOHONEN, K. *Reversible machines and Post's correspondence problem for biprefix morphisms (German and Russian summaries)* ◇ D03 D05 ◇
RYTTER, W. *On the recognition of context-free languages* ◇ D05 ◇
SALOMAA, A. *Computation and automata* ◇ D05 ◇
SEEBOLD, P. *Generalized Thue-Morse sequences* ◇ D03 ◇
SEEBOLD, P. *Overlap-free sequences* ◇ D05 ◇
SEIDL, H. *A quadratic regularity test for nondeleting macro S grammars* ◇ D05 ◇
SHIELDS, M.W. *Deterministic asynchronous automata* ◇ D05 ◇
SISTLA, A.P. & VARDI, M.Y. & WOLPER, P. *The complementation problem for Buechi automata with applications to temporal logic (extended abstract)* ◇ B45 D05 ◇
STERN, J. *Complexty of some problems from the theory of automata* ◇ D05 D15 ◇
STOEVA, S.P. & TOPENCHAROV, V.V. *Fuzzy-topological automata* ◇ D05 ◇
VOGEL, J. & WAGNER, K. *Two-way automata with more than one storage medium* ◇ D05 ◇
WASILEWSKA, A. *Monadic second-order definability as a common characterization of finite automata, certain classes of programs and logics* ◇ B15 B75 C40 D05 ◇
WOJCIECHOWSKI, J. *Finite automata on transfinite sequences and regular expressions* ◇ D05 ◇
YAKU, T. *Wiring a Turing machine in the cellular automaton and the surjectivity problem for a parallel map* ◇ D05 D10 ◇

D10 Turing machines and related notions

1936

TURING, A.M. *On computable numbers, with an application to the "Entscheidungsproblem"* ◇ D05 D10 D20 D35 F60 ◇

1937

HERMES, H. *Definite Begriffe und berechenbare Zahlen* ◇ D05 D10 D20 F60 ◇

TURING, A.M. *Computability and λ-definability* ◇ A05 B40 D05 D10 D20 F99 ◇

1947

POST, E.L. *Recursive unsolvability of a problem of Thue* ◇ D03 D10 D40 ◇

1952

MOORE, E.F. *A simplified universal Turing Machine* ◇ D10 ◇

1954

JANICZAK, A. *On the reducibility of decision problems* ◇ D10 D20 D25 D30 ◇

TAMARI, D. *Une contribution aux theories de communication: machines de Turing et problemes de mot* ◇ D10 D40 ◇

TURING, A.M. *Solvable and unsolvable problems* ◇ B25 D10 D35 ◇

1955

HERMES, H. *Vorlesung ueber Entscheidungsprobleme in Mathematik und Logik* ◇ B25 B40 B65 D10 D20 D35 D98 ◇

1956

DAVIS, MARTIN D. *A note on universal Turing machines* ◇ D10 ◇

MCCARTHY, J. *The inversion of functions defined by Turing machines* ◇ D10 D20 ◇

SHANNON, C.E. *A universal Turing machine with two internal states* ◇ D10 ◇

1957

DAVIS, MARTIN D. *The definition of universal Turing machine* ◇ D10 ◇

DUNHAM, B. *Symbolic logic and computing machines. A survey and summary* ◇ D05 D10 ◇

ROSSER, J.B. *The relation between Turing machines and actual computing machines* ◇ D10 ◇

WANG, HAO *A variant to Turing's theory of computing machines* ◇ D10 ◇

WANG, HAO *Symbolic representations of calculating machines* ◇ B70 D05 D10 F30 ◇

WANG, HAO *Universal Turing machines: an exercise in coding* ◇ D10 ◇

1958

OBERSCHELP, W. *Varianten von Turingmaschinen* ◇ D10 ◇

PORTE, J. *Une simplification de la theorie de Turing* ◇ D10 D20 ◇

1959

ASSER, G. *Turing-Maschinen und Markowsche Algorithmen* ◇ D03 D10 ◇

CAMPEAU, J.O. *Simple Turing type computers* ◇ D10 ◇

HUZINO, S. *Turing transformation and strong computability of Turing computers* ◇ D10 ◇

KAPHENGST, H. *Eine abstrakte programmgesteuerte Rechenmaschine* ◇ D10 ◇

ROGERS JR., H. *The present theory of Turing machine computability* ◇ D10 ◇

SHEPHERDSON, J.C. *The reduction of two-way automata to one-way automata* ◇ D05 D10 ◇

TRAKHTENBROT, B.A. *Wieso koennen Automaten rechnen?* ◇ D10 D98 ◇

1960

AKUSHSKY, I.Y. & BAZILEVSKIJ, YU.YA. & SHREJDER, YU.A. *Logical, recursive and operator methods for the analysis and synthesis of automata* ◇ D05 D10 ◇

CHERNYAVSKIJ, V.S. *On the reversibility of algorithms (Russian)* ◇ D03 D10 ◇

HU, SHIHUA & LOH, CHUNGWAN *Kernel functions. Theory of recursive algorithms II (Chinese) (English summary)* ◇ D10 D20 ◇

HU, SHIHUA *Recursive algorithms. Theory of recursive algorithms I (Chinese) (English summary)* ◇ D10 D20 ◇

HUZINO, S. *On the pseudo universal Turing computer, its structure and programming* ◇ D10 ◇

LACOMBE, D. *La theorie des fonctions recursives et ses applications (expose d'une information generale)* ◇ D10 D20 D80 D98 ◇

LEE, C.Y. *Automata and finite automata* ◇ D05 D10 ◇

MONTAGUE, R. *Towards a general theory of computability* ◇ D10 D75 ◇

STARKE, P.H. *Bemerkungen zu der von Asser entwickelten Version der Turing-Maschine* ◇ D10 ◇

1961

ARBIB, M.A. *Turing machines, finite automata and neural nets* ◇ D05 D10 ◇

BELYAKIN, N.V. *The universality of a computing machine with potentially infinite exterior memory (Russian)* ◇ D10 ◇

KALMAR, L. *A practical infinitistic computer* ◇ B70 D10 ◇

LAMBEK, J. *How to program an infinite abacus* ◇ D10 D20 ◇
MELZAK, Z.A. *An informal arithmetical approach to computability and computation* ◇ D10 ◇
MINSKY, M.L. *Recursive unsolvability of Post's problem of "tag" and other topics in the theory of Turing machines* ◇ D03 D10 ◇
SLYAKHOVA, N.I. & TYUPA, V.G. *On algorithms of Turing type (Ukrainian) (Russian summary)* ◇ D10 ◇
VUCKOVIC, V. *Basic theorems on Turing algorithms* ◇ D10 ◇
VUCKOVIC, V. *Turing algorithms* ◇ D10 ◇

1962

BARZDINS, J. *On a class of Turing machines (Minsky machines) (Russian)* ◇ D10 ◇
BOEHM, C. *Macchine a indirizzi, dotate di un numero minimo di instruzioni* ◇ D10 ◇
BOEHM, C. & JACOPINI, G. *Nuovo tecniche di programmazione semplificanti la sintesi di macchine universali di Turing* ◇ B75 D10 ◇
BUECHI, J.R. *Turing machines and the "Entscheidungsproblem"* ◇ B20 D10 D35 ◇
CANNONITO, F.B. *The Goedel incompleteness theorem and intelligent machines* ◇ B35 D10 D25 F30 ◇
CHERNYAVSKIJ, V.S. & KOZMIDIADI, V.A. *Ueber einige Begriffe der Theorie der mathematischen Maschinen (Russisch)* ◇ D05 D10 ◇
GILL, A. *Introduction to the theory of finite-state machines* ◇ D05 D10 D98 ◇
GINSBURG, S. & ROSE, G.F. *A comparison of the work done by generalized sequential machines and Turing machines* ◇ D05 D10 ◇
GINSBURG, S. *An introduction to mathematical machine theory* ◇ D05 D10 ◇
HUZINO, S. & YONEYAMA, N. *On a proof of Shepherdson's theorem* ◇ D05 D10 ◇
KLEENE, S.C. *Turing-machine computable functionals of finite types I* ◇ D10 D30 D55 D65 ◇
KLEENE, S.C. *Turing-machine computable functionals of finite types II* ◇ D10 D30 D55 D65 ◇
MINSKY, M.L. *Size and structure of universal Turing machines using tag systems* ◇ D03 D10 ◇
RADO, T. *On non-computable functions* ◇ D10 D20 ◇
ROSE, A. *Remarque sur la machine universelle de Turing* ◇ D10 ◇
ROSE, A. *Sur les applications de la logique polyvalente a la construction des machines Turing* ◇ B50 D10 ◇
ZEMANEK, H. *Automaten und Denkprozesse* ◇ A05 D10 ◇

1963

AJZERMAN, M.A. & GUSEV, L.A. & ROZONOEHR, L.I. & SMIRNOVA, I.M. & TAL', A.A. *Logic, automata and algorithms (Russian)* ◇ B98 D05 D10 D98 ◇
ARBIB, M.A. *Monogenic normal systems are universal* ◇ D03 D10 ◇
BECVAR, J. *Finite and combinatorial automata. Turing automata with a programming tape* ◇ D05 D10 ◇
BELYAKIN, N.V. *Computation of effective operators on Turing machines with restricted erasure (Russian)* ◇ D10 ◇
BELYAKIN, N.V. *Distribution of intermediate information when computing on non-erasing Turing machines (Russian)* ◇ D10 ◇
BELYAKIN, N.V. *On a class of Turing machines (Russian)* ◇ D10 ◇
BELYAKIN, N.V. *Simulation of Turing machines on nets (Russian)* ◇ D10 ◇
CAFFIN, R.W. & GOHEEN, H.E. & STAHL, W.R. *Simulation of a Turing machine on a digital computer* ◇ D10 ◇
DAVIS, MARTIN D. *Unsolvable problems: a review* ◇ D03 D10 D35 D40 ◇
LECERF, Y. *Machines de Turing reversibles. Recursive insolublite en $n \in N$ de l'equation $u=\theta^n u$, ou θ est un "isomorphisme de codes"* ◇ D10 D20 ◇
LECERF, Y. *Recursive insolubilite de l'equation generale de diagonalisation de deux monomorphismes de monoides libres $\varphi x = \psi x$* ◇ D10 D20 ◇
LEE, C.Y. *A Turing machine which prints its own code script* ◇ D10 ◇
LEVIEN, R.E. *Set-theoretic formalizations of computational algorithms, computable functions, and general-purpose computers* ◇ B75 D10 D20 E75 ◇
LOEFGREN, L. *Self-repair as a computability concept in the theory of automata* ◇ D10 ◇
RABIN, M.O. *Real time computation* ◇ D05 D10 D15 ◇
RABIN, M.O. & WANG, HAO *Words in the history of a Turing machine with a fixed input* ◇ D10 ◇
RADO, T. *On a simple source for non-computable functions* ◇ D10 D20 ◇
SCHUETZENBERGER, M.-P. *On context-free languages and push-down automata* ◇ D05 D10 ◇
SHEPHERDSON, J.C. & STURGIS, H.E. *Computability of recursive functions* ◇ D10 D20 ◇
THATCHER, J.W. *The construction of a self-describing Turing machine* ◇ D10 ◇
TRAKHTENBROT, B.A. *Algorithms and automatic computing machines* ◇ D10 D20 D98 ◇
ZYKIN, G.P. *Remark on a theorem of Hao Wang (Russian)* ◇ D10 ◇

1964

ARBIB, M.A. *Brains, machines, and mathematics* ◇ D05 D10 D80 F30 ◇
BOEHM, C. *On a family of Turing machines and the related programming language* ◇ B75 D10 ◇
ELGOT, C.C. & RUTLEDGE, J.D. *RS-machines with almost blank tapes* ◇ D05 D10 F30 ◇
ELGOT, C.C. & ROBINSON, A. *Random-access stored-program machines, an approach to programming languages* ◇ B75 D05 D10 ◇
GREEN, M.W. *A lower bound on Rado's sigma function for binary Turing machines* ◇ D10 ◇
HO, C. *On completely reversible Turing machines (Russian)* ◇ D10 ◇
KREIDER, D.L. & RITCHIE, R.W. *Predictably computable functionals and definition by recursion* ◇ D10 D20 ◇
MAKAROV, S.V. *Turing machines and finite automata (Russian)* ◇ D05 D10 ◇

MASLOV, S.YU. *Some properties of E.L.Post's apparatus of canonical systems (Russian)* ◇ D03 D05 D10 ◇
MELZAK, Z.A. *An informal arithmetical approach to computability and computation II* ◇ D10 D15 ◇
PUTNAM, H. *Minds and machines* ◇ A05 D10 ◇
TRAKHTENBROT, B.A. *Turing computations with logarithmic delay (Russian)* ◇ D10 D15 ◇
TURING, A.M. *Computing machinery and intelligence* ◇ A05 B75 D10 ◇
ZAROVNYJ, V.P. *The group of automatic one-to-one mappings (Russian)* ◇ D05 D10 D75 ◇
ZASLAVSKIJ, I.D. *Graph schemes with memory (Russian)* ◇ B70 B75 D05 D10 ◇

1965

ARBIB, M.A. *A common framework for automata theory and control theory* ◇ D05 D10 ◇
BABIKOV, G.V. *Normale Algorithmen und Turingmaschinen (Russian)* ◇ D03 D10 ◇
BARZDINS, J. *Complexity of recognition of symmetry on Turing machines (Russian)* ◇ D10 D15 ◇
BECVAR, J. *A universal Turing machine with a programming tape* ◇ D10 ◇
BECVAR, J. *Real-time and complexity problems in automata theory* ◇ D05 D10 D15 ◇
BLUM, E.K. *Enumeration of recursive sets by Turing machine* ◇ D10 D20 ◇
FENG, JINGDONG *The inverse machine of Turing machines (Chinese)* ◇ D10 ◇
FISCHER, P.C. *On formalisms for Turing machines* ◇ D10 ◇
FREJVALD, R.V. *Complexity of recognition of symmetry on Turing machines with input (Russian)* ◇ D10 D15 ◇
GLADKIJ, A.V. *Algorithmic undecidability of inherent ambiguity in context-free languages (Russian)* ◇ D05 D10 ◇
HARTMANIS, J. & LEWIS II, P.M. & STEARNS, R.E. *Hierarchies of memory limited computations* ◇ D10 D15 ◇
HENNIE III, F.C. *One-tape, off-line Turing machine computations* ◇ D10 ◇
KALMAR, L. (ED.) *Colloque sur les fondements des mathematiques, les machines mathematiques, et leurs applications* ◇ B97 D05 D10 D97 ◇
KUZ'MIN, V.A. *Realization of functions of the algebra of logic by means of automata, normal algorithms and Turing machines (Russian)* ◇ B05 D03 D05 D10 ◇
LIN, S. & RADO, T. *Computer studies of Turing machine problems* ◇ D10 ◇
PETER, R. *Ueber die sequenzielle Berechenbarkeit von rekursiven Wortfunktionen durch Kellerspeicher* ◇ D10 D20 ◇
ROWICKI, A. *Remarks on order of simple processes* ◇ D10 D15 ◇
TITGEMEYER, D. *Einfuehrung in die Theorie der Kalkuele und Algorithmen* ◇ D03 D05 D10 ◇
WANG, HAO *Logic and computers* ◇ A05 B35 D03 D10 ◇
WANG, HAO *Remarks on machines, sets, and the decision problem* ◇ B20 B25 D03 D05 D10 D35 E30 ◇

1966

ARBIB, M.A. *Simple self-reproducing universal automata* ◇ D10 ◇
BELYAKIN, N.V. *Turing machines operating on a plane (Russian)* ◇ D10 ◇
BERGER, R. *The undecidability of the domino problem* ◇ D05 D10 ◇
BLIKLE, A.J. *Investigations in the theory of addressless computers* ◇ B75 D10 ◇
BOEHM, C. & JACOPINI, G. *Flow diagrams, Turing machines and languages with only two formation rules* ◇ B75 D05 D10 ◇
BRADY, A.H. *The conjectured highest scoring machines for Rado's $\Sigma(k)$ for the value $k=4$* ◇ D10 ◇
CARACCIOLO, A. *Generalized Markov algorithms and automata* ◇ D03 D05 D10 ◇
COOK, S.A. *The solvability of the derivability problem for one-normal systems* ◇ D10 ◇
FISCHER, P.C. *Turing machines with restricted memory access* ◇ D10 ◇
FREJVALD, R.V. *On the order of growth of exact time signalizers for Turing calculations (Russian)* ◇ D10 D15 ◇
GINSBURG, S. & SPANIER, E.H. *Finite-turn pushdown automata* ◇ D10 ◇
HANAK, J. *On real-time Turing machines* ◇ D10 D15 ◇
HENNIE III, F.C. *Online Turing machine computations* ◇ D10 ◇
HOOPER, P.K. *The undecidability of the Turing machine immortality problem* ◇ D10 ◇
HUZINO, S. *On the simulation of real-time Turing machines by a modified Shepherdson-Sturgis' machine* ◇ D05 D10 D15 ◇
HUZINO, S. *Simulatability of finite automata by Shepherdson and Sturgis' machines* ◇ D05 D10 ◇
KREIDER, D.L. & RITCHIE, R.W. *A basis theorem for a class of two-way automata* ◇ D10 D25 ◇
MELZAK, Z.A. *An informal arithmetical approach to computability and computation III* ◇ D10 D80 ◇
NEUMANN VON, J. *Theory of self-reproducing automata* ◇ D10 D98 ◇
NGUEN HYU NGI *Estimates for sequences of states of Turing machines (Russian)* ◇ D10 ◇
NOLIN, L. *Organigrammes et machines de Turing* ◇ A05 D10 D20 ◇
ROEDDING, D. *Der Entscheidbarkeitsbegriff in der mathematischen Logik* ◇ B25 D10 D20 D25 D35 ◇
SZEKELY, D.L. *On general purpose unifying automata* ◇ D10 ◇
TRAKHTENBROT, B.A. *Normed signalizers for Turing computations (Russian)* ◇ D10 D15 ◇
URBAN, J. *Die Minimalisierung der zur sequentiellen Berechnung der partiell-rekursiven Wortfunktionen notwendigen Kellerspeicher* ◇ D10 D20 ◇
ZAMA, N. *A generalization of algorithms and one of its applications I* ◇ D10 ◇

1967

ANDERSON, MICHAEL *Note on an inequality of Tibor Rado* ◇ D10 ◇
ARBIB, M.A. *Automaton automorphisms* ◇ D10 ◇
ARBIB, M.A. *Some comments on self-reproducing automata* ◇ D10 ◇
BECVAR, J. *Probleme der Komplexitaet in der Theorie der Algorithmen und Automaten* ◇ A05 B75 D10 ◇
ELGOT, C.C. & ROBINSON, A. & RUTLEDGE, J.D. *Multiple control computer models* ◇ B70 D05 D10 ◇
ENGELER, E. *Algorithmic properties of structures* ◇ B75 C75 D10 D75 ◇
FISCHER, P.C. *Turing machines with a schedule to keep* ◇ D10 ◇
FREJDZON, R.I. *The representability of algorithmically decidable predicates by Rabin machines (Russian)* ◇ D03 D10 D20 ◇
GINSBURG, S. & GREIBACH, S.A. & HARRISON, M.A. *One-way stack automata* ◇ D05 D10 ◇
GINSBURG, S. & GREIBACH, S.A. & HARRISON, M.A. *Stack automata and compiling* ◇ B75 D05 D10 ◇
GREIBACH, S.A. *A note on pushdown store automata and regular systems* ◇ D10 ◇
HARTMANIS, J. *Context free languages and Turing machine computation* ◇ D05 D10 ◇
HOPCROFT, J.E. & ULLMAN, J.D. *An approach to a unified theory of automata* ◇ D05 D10 ◇
HOPCROFT, J.E. & ULLMAN, J.D. *Nonerasing stack automata* ◇ D05 D10 ◇
KHODZHAEV, D. *On the complexity of calculations on Turing machines with an oracle (Russian)* ◇ D10 D15 ◇
MINSKY, M.L. *Computation: finite and infinite machines* ◇ D05 D10 D20 ◇
SCHWARTZ, J.T. (ED.) *Mathematical aspects of computer science* ◇ B35 B65 B75 B97 D10 D80 D97 ◇
SWANSON, J.W. *A variant of Turing machines requiring print instructions only* ◇ D10 ◇
TRAKHTENBROT, B.A. *Complexity of algorithms and computations (Russian)* ◇ D10 D15 D20 D98 ◇
WEINBERG, G.M. *Computing machines* ◇ B35 D10 ◇

1968

AHO, A.V. & ULLMAN, J.D. *The theory of languages* ◇ D05 D10 D15 ◇
ANDERSON, MICHAEL *Approximation to a decision procedure for the halting problem* ◇ D10 ◇
BLUM, M. & FISCHER, P.C. & HARTMANIS, J. *Tape reversal complexity hierarchies* ◇ D10 D15 ◇
BRAUER, W. & INDERMARK, K. *Algorithmen, rekursive Funktionen und formale Sprachen* ◇ D05 D10 D20 D98 ◇
DEMPSTER, J.R.H. *On the multiplying ability of two-way automata* ◇ D05 D10 ◇
DUNHAM, C. *An uncompletable function* ◇ D10 D20 ◇
FISCHER, P.C. & MEYER, A.R. & ROSENBERG, ARNOLD L. *Counter machines and counter languages* ◇ D05 D10 D15 D20 ◇
FISCHER, P.C. *Reduction of tape reversals for off-line one-type Turing machines* ◇ D10 ◇
FREJDZON, R.I. *An index of the complexity of recursive predicates which do not depend on a standardization of the concept of algorithm (Russian)* ◇ D10 D15 ◇
HARTMANIS, J. *Computational complexity of one-tape Turing machine computations* ◇ D10 D15 D20 ◇
HARTMANIS, J. *Tape-reversal bounded Turing machine computations* ◇ D10 D15 ◇
HERMAN, G.T. *Simulation of one abstract computing machine by another* ◇ D10 D25 D30 D55 ◇
HERMAN, G.T. *The halting problem of one state Turing machine with n-dimensional tape* ◇ D10 ◇
HOPCROFT, J.E. & ULLMAN, J.D. *Decidable and undecidable questions about automata* ◇ D05 D10 ◇
HOPCROFT, J.E. & ULLMAN, J.D. *Relations between time and tape complexities* ◇ D05 D10 D15 ◇
HOTZ, G. *Automatentheorie und formale Sprachen I. Turingmaschinen und rekursive Funktionen* ◇ D05 D10 D20 D98 ◇
JONES, N.D. *Classes of automata and transitive closure* ◇ D10 D15 ◇
NELSON, RAYMOND J. *Introduction to automata* ◇ D05 D10 D98 ◇
OCHRANOVA-DOLEZELOVA, R. *Real-time decidability, computability, countability and generability* ◇ D10 D15 ◇
PAWLAK, Z. *On the notion of a computer* ◇ D10 ◇
STRNAD, P. *On-line Turing machine recognition* ◇ D10 D15 ◇
ZASTAVKA, Z. *On algorithms with memory elements (Russian) (Czech summary)* ◇ D10 ◇

1969

AHO, A.V. & HOPCROFT, J.E. & ULLMAN, J.D. *On the computational power of pushdown automata* ◇ D10 D15 ◇
ARBIB, M.A. *Theories of abstract automata* ◇ D05 D10 D98 ◇
BAER, ROBERT M. *Definability by Turing machines* ◇ D10 D35 ◇
BARZDINS, J. *On computability by probabilistic machines (Russian)* ◇ D10 ◇
BREJTBART, YU.YA. & KOZMIDIADI, V.A. *Two subclasses of Turing machines reducible to finite automata (Russian)* ◇ D05 D10 ◇
CAZACU, C. *Machines Turing et T-schemas* ◇ D10 ◇
DZHANSEJTOV, K.K. *Cancellable and self-reproducing configurations (Russian) (Kazakh summary)* ◇ D10 ◇
FISCHER, P.C. *Quantificational variants on the halting problem for Turing machines* ◇ D10 ◇
FITCH, F.B. & ORGASS, R.J. *A theory of computing machines* ◇ B40 D10 ◇
GALLAIRE, H. *Recognition time for context-free languages by on-line Turing machines* ◇ D05 D10 D15 ◇
HERMAN, G.T. *A simple solution of the uniform halting problem* ◇ D10 ◇
HERMAN, G.T. *The uniform halting problem for generalized one-state Turing-machines* ◇ D10 ◇
HERMAN, G.T. *The unsolvability of the uniform halting problem for two state Turing-machines* ◇ D10 ◇

HERMES, H. *Basic notions and applications of the theory of decidability* ◇ D10 D20 D25 D30 D35 D50 D98 ◇
KOREC, I. *A complexity valuation of the partial recursive functions following the expectation of the length of their computations on Minsky machines* ◇ D10 D15 ◇
KOZMIDIADI, V.A. & MARCHENKOV, S.S. *On multihead automata (Russian)* ◇ D05 D10 ◇
LANDWEBER, L.H. *Decision problems for ω-automata* ◇ D05 D10 ◇
MAKAREVSKIJ, A.YA. & STOTSKAYA, E.D. *Representability in deterministic multi-tape automata (Russian) (English summary)* ◇ D05 D10 ◇
MARUSCIAC, I. & NADIU, G.S. *A universal Turing machine for normal algorithms (Romanian) (French summary)* ◇ D03 D10 ◇
MILNER, R. *Program schemes and recursive function theory* ◇ B75 D10 D20 ◇
MOSHCHENSKIJ, V.A. *On the question of the complexity of Turing computations (Russian)* ◇ D10 D15 D20 ◇
NOLIN, L. *Formalisation des notions de machine et de programme* ◇ B75 D10 D20 ◇
NOZAKI, A. *On the notion of universality of Turing machine (Czech summary)* ◇ D10 ◇
PAWLAK, Z. *Programmierte Maschinen* ◇ D10 ◇
SALOMAA, A. *Theory of automata* ◇ D05 D10 D98 ◇
SANTOS, E.S. *Probabilistic Turing machines and computability* ◇ D10 ◇
STRNAD, P. *On optimum time bounds for recognition of some sets of words by on-line Turing machines* ◇ D10 D15 ◇
TICHY, P. *Intension in terms of Turing machines (Polish and Russian summaries)* ◇ A05 D10 ◇
ULLMAN, J.D. *Halting stack automata* ◇ D10 ◇
YOUNG, P. *Toward a theory of enumerations* ◇ D10 D15 D20 D25 D45 ◇

1970

BELYAKIN, N.V. *Generalized computations and second order arithmetic (Russian)* ◇ C62 D10 D45 D65 D70 F35 ◇
BIELECKI, J. *The effect of the geometry of the tape of the Turing machine on the number of instructions in the program controlling it (Polish) (Russian and English summaries)* ◇ D10 D15 ◇
BURKS, A.W. *Von Neumann's self-reproducing automata* ◇ D10 ◇
EBBINGHAUS, H.-D. *Aufzaehlbarkeit* ◇ D03 D10 D20 D25 F30 ◇
EBBINGHAUS, H.-D. *Turing-Maschinen und berechenbare Funktionen I,III* ◇ D10 D20 ◇
ELGOT, C.C. *The common algebraic structure of exit-automata and machines* ◇ D05 D10 ◇
FISCHER, P.C. & MEYER, A.R. & ROSENBERG, ARNOLD L. *Time-restricted sequence generation* ◇ D10 D15 ◇
HERMES, H. *Entscheidungsproblem und Dominospiele* ◇ D05 D10 D35 ◇
JACOBS, K. *Turing-Maschinen und zufaellige 0-1-Folgen* ◇ D10 D80 ◇
KAMEDA, T. & VOLLMAR, R. *Note on tape reversal complexity of languages* ◇ D10 ◇
KAMEDA, T. & VOLLMAR, R. *Zur Umkehrkomplexitaet von Sprachen* ◇ D05 D10 D15 ◇
KATERINOCHKINA, N.N. *The equivalence of certain computational equipment (Russian) (English summary)* ◇ D10 D15 D20 ◇
KAZANOVICH, YA.B. *A classification of primitive recursive functions using Turing machines (Russian)* ◇ D10 D20 ◇
KOZMIDIADI, V.A. *A certain generalization of finite automata that produces a hierarchy analogous to A.Grzegorczyk's classification of primitively recursive functions (Russian)* ◇ D05 D10 D20 ◇
MAHN, F.-K. *Turing-Maschinen und berechenbare Funktionen II* ◇ D10 D20 ◇
MUELLER, HORST *Ueber die mit Stackautomaten berechenbaren Funktionen* ◇ D10 D20 ◇
NEPOMNYASHCHIJ, V.A. *Rudimentary predicates and Turing calculations (Russian)* ◇ D10 D20 ◇
NEPOMNYASHCHIJ, V.A. *The rudimentary interpretation of two-tape Turing computations (Russian) (English summary)* ◇ D10 D20 ◇
PRATHER, R.E. *On categories of infinite automata* ◇ D10 ◇
SAVITCH, W.J. *Relationship between nondeterministic and deterministic tape complexities* ◇ D10 D15 ◇
SCHMITT, A.A. *Die Zustands-Komplexitaetsklassen von Turingmaschinen* ◇ D10 D15 ◇
SCHOENHAGE, A. *Universelle Turing-Speicherung* ◇ D10 ◇
SOZANSKA-BIEN, Z. *Some properties of m-address machines (Polish and Russian summaries)* ◇ D10 ◇
STOSS, H.-J. *k-Band-Simulation von k-Kopf-Turing-Maschinen* ◇ D10 ◇
THATCHER, J.W. *Self-describing Turing machines and self-reproducing cellular automata* ◇ D05 D10 ◇
THATCHER, J.W. *Universality in the von Neumann cellular model* ◇ D05 D10 ◇
VALIEV, M.K. *Certain estimates of the computing time on Turing machines with input (Russian) (English summary)* ◇ D10 D15 ◇
WILLIS, D.G. *Computational complexity and probability constructions* ◇ D10 D15 ◇

1971

AMOROSO, S. & BLOOM, S.L. *A model of the digital computer* ◇ D10 D20 ◇
BARTOL, W. *On the existence of machine homomorphisms* ◇ D05 D10 ◇
BELSNES, D. *The immortality problem for non-erasing Turing machines* ◇ D10 ◇
BIELECKI, J. *Selfreproduction of modular Turing machines (Polish) (Russian and English summaries)* ◇ D10 ◇
BLOOM, S.L. *Some remarks on uniform halting problems* ◇ D10 D30 ◇
BOUCHER, C. *Lecons sur la theorie des automates mathematiques* ◇ D05 D10 D15 D98 ◇
CASE, J. *A note on degrees of self-describing Turing machines* ◇ D10 D25 ◇
COOK, S.A. *Characterizations of pushdown machines in terms of time-bounded computers* ◇ D10 D15 ◇

Subject Index

DALEN VAN, D. *A note on some systems of Lindenmayer* ◇ D03 D05 D10 D25 ◇
FELDMANN, H. & OBERQUELLE, H. & ORTLIEB, C. *Eine einfache universelle Turingmaschine in ALGOL 60 Simulation* ◇ B75 D10 ◇
FRIEDMAN, H.M. *Algorithmic procedures, generalized Turing algorithms, and elementary recursion theory* ◇ A05 D10 D20 D75 F99 ◇
GOLD, E.M. *Universal goal-seekers* ◇ D05 D10 D20 ◇
HARRISON, M.A. & SCHKOLNICK, M. *A grammatical characterization of one-way nondeterministic stack languages* ◇ D10 ◇
HERMAN, G.T. *Strong computability and variants of the uniform halting problem* ◇ D10 ◇
HUGHES, C.E. & OVERBEEK, R.A. & SINGLETARY, W.E. *The many-one equivalence of some general combinatorial decision problems* ◇ D03 D05 D10 D25 D30 ◇
IBARRA, O.H. *Characterizations of some tape and time complexity classes of Turing machines in terms of multihead and auxiliary stack automata* ◇ D10 D15 ◇
KONIKOWSKA, B. *Continuous machines, τ-computations and τ-computable sets* ◇ D10 D20 ◇
KONIKOWSKA, B. *Continuous machines* ◇ D10 ◇
KONIKOWSKA, B. *On some properties of continuous machines* ◇ D10 D20 ◇
KONIKOWSKA, B. *Properties of continuous machines* ◇ D10 D20 ◇
KRYUKOV, YU.A. *Turing machines with three states and two symbols and with one state and n symbols (Russian) (English summary)* ◇ D10 ◇
LEVY, S. *Computational equivalence* ◇ D10 D30 D55 ◇
MATIYASEVICH, YU.V. *Real-time recognition of the inclusion relation (Russian) (English summary)* ◇ D10 D15 ◇
MOSHCHENSKIJ, V.A. *A certain estimate of the state sequences of Turing machines (Russian) (English summary)* ◇ D10 ◇
MOSHCHENSKIJ, V.A. *The estimation of certain functions that characterize the performance of Turing machines (Russian) (English summary)* ◇ D10 D15 ◇
MUELLER, HORST *Stackautomaten in Labyrinthen* ◇ D10 ◇
PHAN DINH DIEU *On the languages representable by finite probabilistic automata* ◇ D05 D10 ◇
SANTOS, E.S. *Computability by probabilistic Turing machines* ◇ D10 D20 ◇
SCHMITT, A.A. *Automaten. Algorithmen. Gehirne* ◇ D05 D10 ◇
SCHREIBER, P. *Theseus im Labyrinth als Turingmaschine* ◇ D10 ◇
SCHURMANN, A. *Functions computable by a computer (Polish and Russian summaries)* ◇ D05 D10 D20 ◇
SHANK, H.S. *Records of Turing machines* ◇ D10 ◇
SHREJDER, YU.A. *Equality, resemblance, and order (Russian)* ◇ D10 E07 ◇
STOSS, H.-J. *Zwei-Band Simulation von Turingmaschinen* ◇ D10 ◇
STOTSKAYA, E.D. *Multitape deterministic automata without end markers (Russian)* ◇ D10 ◇
WEICKER, R. *Tabulator-Turingmaschinen und Komplexitaet (English summary)* ◇ D10 D15 ◇
YASUHARA, A. *Recursive function theory and logic* ◇ B98 D10 D20 D98 ◇

1972

ALAD'EV, V.Z. *Computability in homogeneous structures (Estonian and Russian summaries)* ◇ D03 D05 D10 ◇
BAXTER, RODNEY J. *On unlimited register machines* ◇ D10 ◇
BREJTBART, YU.YA. & KOZMIDIADI, V.A. *Certain subclasses of Turing machines that are reducible to finite automata (Russian) (English summary)* ◇ D05 D10 ◇
BULNES, J. *On the speed of addition and multiplication on one-tape, off-line Turing machines* ◇ D10 D15 ◇
CLARKE, J.J. *Turing machines and the mind-body problem* ◇ A05 D10 ◇
COOK, S.A. & RECKHOW, R.A. *Time-bounded random access machines* ◇ D10 D15 ◇
DEMBINSKI, P. *The machine realization of programs (Russian summary)* ◇ B75 D10 D20 ◇
EBBINGHAUS, H.-D. & HERMES, H. & JACOBS, K. & MAHN, F.-K. *Turing-Maschinen und rekursive Funktionen (Russian)* ◇ D10 D20 ◇
EBBINGHAUS, H.-D. & HERMES, H. *Universelle Turing-Maschinen* ◇ D10 ◇
FISCHER, P.C. & MEYER, A.R. & ROSENBERG, ARNOLD L. *Real-time simulation of multihead tape units* ◇ D10 D15 ◇
GIULIANO, J.A. *Writing stack acceptors* ◇ D05 D10 ◇
GUREVICH, Y. & KORYAKOV, I.O. *Remarks on R.Berger's paper on the domino problem (Russian)* ◇ D10 ◇
HAMLET, R.G. *A patent problem for abstract programming languages; machine-independent computations* ◇ B75 D10 D20 ◇
IBARRA, O.H. *A note concerning nondeterministic tape complexities* ◇ D10 D15 ◇
JONES, N.D. & SELMAN, A.L. *Turing machines and the spectra of first-order formulas with equality* ◇ C13 D10 D15 ◇
KAIN, R.Y. *Automata theory: machines and languages* ◇ D05 D10 D98 ◇
KAMEDA, T. *Pushdown automata with counters* ◇ D10 ◇
KOREC, I. *The mathematical expectation of the time for computations of partial recursive functions on Minsky machines (Russian) (German, English, and French summaries)* ◇ D10 D15 ◇
KUBINETS, M.V. *Recognition of a self-crossing of a plane trajectory by a Kolmogorov algorithm (Russian)* ◇ D10 D15 ◇
KUZNETSOV, O.P. *Ueber die Kompliziertheit der Berechnungen in eindimensionalen iterativen Strukturen (Russian)* ◇ D05 D10 ◇
LORENTS, A.A. *Elemente der konstruktiven Theorie stochastischer Automaten (Russian)* ◇ D05 D10 D98 F60 F98 ◇

LYNN, D.S. *New results for Rado's sigma function for binary Turing machines* ◇ D10 D20 ◇
MASLOV, A.N. *Probabilistic Turing machines and recursive functions (Russian)* ◇ D10 D20 ◇
MILGRAM, D.L. & ROSENFELD, A. *Array automata and array grammars* ◇ D05 D10 ◇
MOSHCHENSKIJ, V.A. *Zur Analyse von Turing-berechnungen (Russian)* ◇ D10 D15 ◇
MUELLER, HORST *Jede mit Stackautomaten berechenbare Funktion ist elementar* ◇ D10 D15 D20 ◇
MUENTEFERING, P. & STARKE, P.H. *Ueber reduzible Ereignisse* ◇ D05 D10 ◇
PATERSON, M.S. *Tape bounds for time-bounded Turing machines* ◇ D10 D15 ◇
RANDELL, B. *On Alan Turing and the origins of digital computers* ◇ A10 D10 ◇
RITCHIE, R.W. & SPRINGSTEEL, F. *Language recognition by marking automata* ◇ D05 D10 ◇
ROEDDING, D. & SCHWICHTENBERG, H. *Bemerkungen zum Spektralproblem* ◇ B15 C13 D10 D15 ◇
ROEDDING, D. *Registermaschinen* ◇ D10 ◇
SAVITCH, W.J. *Maze recognizing automata* ◇ D10 D15 ◇
SKARBEK, W. & ZEMBRZUSKI, K. *Some properties of continuous machines* ◇ D10 D20 ◇
SKOWRON, A. *Durch festprogrammierte Maschinen akzeptierte Mengen (Polish) (Russian and English summaries)* ◇ D10 ◇
SKOWRON, A. *Equivalence of generalized machines* ◇ D10 ◇

1973

ALFEROVA, Z. *Algorithmentheorie (Russian)* ◇ D10 D20 D98 ◇
ARBIB, M.A. & CULIK, K. *Sequential and jumping machines and their relation to computers* ◇ B70 D05 D10 ◇
CASPAR, K. & FENZL, W. & WEIMANN, B. *Untersuchungen ueber haltende Programme fuer Turing-Maschinen mit 2 Zeichen und bis zu 5 Befehlen* ◇ D10 ◇
CLEAVE, J.P. *Combinatorial systems III. Degrees of combinatorial problems of computing machines* ◇ D10 D25 D30 ◇
COOK, S.A. & RECKHOW, R.A. *Time bounded random access machines* ◇ D10 D15 ◇
FELDMAN, E.D. & OWINGS JR., J.C. *A class of universal linear bounded automata* ◇ D10 ◇
GLADKIJ, A.V. *Formale Grammatiken und Sprachen (Russisch)* ◇ D05 D10 D15 ◇
HECKER, H.-D. *Kompliziertheitsmasse bei Turingmaschinen* ◇ D10 D15 ◇
HERMAN, G.T. & JACKOWSKI, J.A. *A decision procedure using discrete geometry* ◇ D10 ◇
HUGHES, C.E. & SINGLETARY, W.E. *Combinatorial systems with axiom* ◇ D03 D05 D10 D25 D30 ◇
HUGHES, C.E. *Many-one degrees associated with problems of tag* ◇ D03 D10 D25 D40 ◇
KASHEF, R.S. & ROWAN, J.H. *A universal programmable cellular array* ◇ D10 ◇
LAING, R. *A note on Maxwell's demon and universal computation* ◇ D10 ◇
MEY VANDER, J.E. & PIERCE, J.C. & SINGLETARY, W.E. *Tutor – a Turing machine simulator* ◇ D10 ◇
MOISIL, G.C. *Sur la possibilite de modeler le fini par l'infini* ◇ D05 D10 ◇
MONIEN, B. *Relationships between pushdown automata and tape-bounded Turing machines* ◇ D10 D15 ◇
MOSHCHENSKIJ, V.A. *Lectures on mathematical logic (Russian)* ◇ B98 D10 D20 ◇
NEPOMNYASHCHIJ, V.A. *Rudimentary simulation of nondeterministic Turing calculations (Russian) (English summary)* ◇ D10 ◇
OVERBEEK, R.A. *The representation of many-one degrees by decision problems of Turing machines* ◇ D10 D25 ◇
PAVLOTSKAYA, L.M. *Solvability of the halting problem for certain classes of Turing machines (Russian)* ◇ D10 ◇
PENNER, V. *Ueber eine Hierarchie von push-down-entscheidbaren Mengen* ◇ D10 D20 ◇
POBEDIN, L.N. *Certain questions on generalized computability (Russian)* ◇ D10 D55 D75 ◇
RIEKSTINS, J. *The solvability of the halting problem for two-cycle Turing machines (Russian) (Latvian and English summaries)* ◇ D10 ◇
SAVITCH, W.J. *Maze recognizing automata and nondeterministic tape complexity* ◇ D10 D15 ◇
SKOWRON, A. *Functions computable by machines (Polish) (Russian and English summaries)* ◇ D10 ◇
SKOWRON, A. *Machines with input and output* ◇ D05 D10 ◇
SKOWRON, A. *Stored program machines* ◇ D10 D20 ◇
SLISENKO, A.O. *Identification of the symmetry predicate by means of multihead Turing machines with input (Russian)* ◇ D10 D15 ◇
TETRUASHVILI, M.R. *A realization of the Magnus algorithm on a Turing machine, and an upper bound of the complexity of the computations (Russian) (Georgian and English summaries)* ◇ D10 D15 D40 ◇
THOMAS, WILLIAM J. *Doubts about some standard arguments for Church's Thesis* ◇ A05 D10 D20 F99 ◇
TRAKHTENBROT, B.A. *Frequency computations (Russian)* ◇ D10 D15 D20 ◇
VOLLMAR, R. *Ueber Turingmaschinen mit variablem Speicher* ◇ D10 ◇
WAGNER, K. *Die Modellierung der Arbeit von Turingmaschinen mit n-dimensionalem Band durch Turingmaschinen mit eindimensionalem Band (English and Russian summaries)* ◇ D10 ◇
WAGNER, K. *Universelle Turingmaschinen mit n-dimensionalem Band (English and Russian summaries)* ◇ D10 ◇
WECHSUNG, G. *Funktionen, die von pushdown-Automaten berechnet werden* ◇ D10 ◇

1974

AANDERAA, S.O. *On k-tape versus (k – 1)-tape real time computation* ◇ D10 D15 ◇
AGAFONOV, V.N. & BARZDINS, J. *On sets associated with stochastic machines (Russian)* ◇ D10 D25 ◇
AITCHISON, P.W. & HERMAN, G.T. *A decision procedure using the geometry of convex sets* ◇ D10 D80 ◇

BAKER, B.S. & BOOK, R.V. *Reversal-bounded multi-pushdown machines* ◇ D10 D25 ◇
BARASHKO, A.S. *Tape complexity of languages that can be recognized by two-way finite-term pushdown automata (Russian) (English summary)* ◇ D10 D15 ◇
BARASHKO, A.S. *The number of push-down tape reversals of a two-way push-down store automaton (Russian)* ◇ D10 D15 ◇
BARTOL, W. *Algebraic complexity of machines* ◇ D10 D15 ◇
BARZDINS, J. *Bemerkung zur Programmsynthese anhand von rechnerischen Vorgeschichten (Russian)* ◇ D10 ◇
BOEHLING, K.H. & BRAUNMUEHL VON, B. *Komplexitaet bei Turingsmaschinen* ◇ D10 D15 D20 ◇
BOOK, R.V. *Comparing complexity classes* ◇ D10 D15 ◇
BOOK, R.V. *On the structure of complexity classes* ◇ D10 D15 ◇
BOOK, R.V. *Tally languages and complexity classes* ◇ D10 D15 ◇
DEIMEL JR., L.E. *Remark on the computational power of a Turing machine variant* ◇ D10 ◇
FISCHER, P.C. *Ueber Formalismen fuer Turingmaschinen* ◇ D10 ◇
FREJVALD, R.V. *Limitaere Berechnungen auf stochastischen Automaten (Russisch)* ◇ D10 D20 ◇
FREJVALD, R.V. & PODNIEKS, K.M. *Ueber limitaere Berechnungen auf nichtdeterministischen Turingmaschinen (Russian)* ◇ D10 D55 ◇
GALIL, Z. *Two way deterministic pushdown automaton languages and some open problems in the theory of computation* ◇ D05 D10 D20 ◇
GANOV, V.A. *Generalized computability and the descriptive theory of sets (Russian)* ◇ B75 D10 D55 E15 ◇
GILL III, J.T. *Computational complexity of probabilistic Turing machines* ◇ D10 D15 ◇
GINSBURG, S. & ROSE, G.F. *The equivalence of stack-counter acceptors and quasi-realtime stack-counter acceptors* ◇ D05 D10 D15 ◇
HAMLET, R.G. *Introduction to computation theory* ◇ D10 D15 D25 D98 ◇
HERSCHEL, R. *Einfuehrung in die Theorie der Automaten, Sprachen und Algorithmen* ◇ D05 D10 D98 ◇
HONDA, N. & IGARASHI, Y. *Deterministic multitape automata computations* ◇ D10 D15 ◇
HOPCROFT, J.E. & ULLMAN, J.D. *Some results on tape-bounded Turing machines (Russian)* ◇ D10 D15 ◇
IBARRA, O.H. *A hierarchy theorem for polynomial-space recognition* ◇ D10 D15 ◇
IBARRA, O.H. & MELSON, R.T. *Some results concerning automata on two-dimensional tapes* ◇ D10 ◇
JONES, JAMES P. *Recursive undecidability - an exposition* ◇ D10 D20 D35 D40 D98 ◇
JONES, N.D. & SELMAN, A.L. *Turing machines and the spectra of first-order formulas* ◇ C13 D10 D15 ◇
KREISEL, G. *A notion of mechanistic theory* ◇ A05 B30 D10 ◇
LAKINA, N.I. *On the question of programming of Turing machines (Russian)* ◇ D10 ◇
LEVIN, V.A. *Transients in simple asynchronous automata with memory (Russian)* ◇ D10 ◇
LINDNER, R. & WAGNER, K. *The axiomatisation of a certain sequential propositional calculus (Russian)* ◇ B60 B70 D05 D10 ◇
LUPANOV, O.B. *Methods to obtain complexity and calculability estimations for individual functions (Russian)* ◇ D10 D15 D20 ◇
LYNCH, N.A. *Approximations to the halting problem* ◇ D10 D20 D25 D45 ◇
MANNA, Z. *Mathematical theory of computation* ◇ B75 D10 D98 ◇
MONIEN, B. *Beschreibung von Zeitkomplexitaetsklassen bei Turingmaschinen durch andere Automatenmodelle* ◇ D10 D15 ◇
PRIESE, L. & ROEDDING, D. *A combinatorial approach to selfcorrection* ◇ D05 D10 ◇
SCHMIDTKE, K. *Classification of computing machines with infinite computations with respect to ρ-inclusion (Russian summary)* ◇ D10 ◇
SCHNORR, C.-P. *Rekursive Funktionen und ihre Komplexitaet* ◇ D10 D15 D20 D98 ◇
SEIFERAS, J.I. *Observations on nondeterministic multidimensional iterative arrays* ◇ D10 D15 ◇
SELMAN, A.L. *Relativized halting problems* ◇ D10 D25 D30 ◇
STOSS, H.-J. *k-Band-Simulation von k-Kopf-Turing-Maschinen (Russian)* ◇ D10 ◇
TRAKHTENBROT, B.A. *Bemerkungen ueber die Kompliziertheit von Berechnungen auf stochastischen Automaten (Russisch)* ◇ D10 D15 ◇
VALIANT, L.G. *The decidability of equivalence for deterministic finite-turn pushdown automata JA JC* ◇ D10 ◇
VALIANT, L.G. *The equivalence problem for deterministic finite-turn pushdown automata* ◇ D10 ◇
WAGNER, K. *Zellulare Berechenbarkeit von Funktionen ueber n-dimensionalen Zeichensystemen* ◇ D10 ◇
WEICKER, R. *Turing machines with associative memory access* ◇ D10 D15 ◇
WETTE, E. *The refutation of number theory. I* ◇ D10 F30 ◇

1975

ALAD'EV, V.Z. & OSIPOV, O. *About the complexity of parallel algorithms defined by homogeneous structures (Russian) (Estonian and English summaries)* ◇ D03 D05 D10 D15 ◇
AUGUSTON, M. *Universal Buchberger automata and real-time computations (Russian) (English summary)* ◇ D10 D15 D20 ◇
BAER, ROBERT M. *Self-assembly and differentiation as models of computability* ◇ D10 D80 ◇
BENESOVA, M. & KOREC, I. *Non-linear speed-up theorem for two register Minsky machines* ◇ D10 D15 ◇
BLUM, L. & BLUM, M. *Toward a mathematical theory of inductive inference* ◇ B48 D10 D15 ◇
BOERGER, E. *Recursively unsolvable algorithmic problems and related questions reexamined* ◇ D03 D10 D20 ◇

BURKS, A.W. *Logic, biology and automata - some historical reflections* ◇ A05 A10 D05 D10 D80 ◇
CHYTIL, M.P. *On complexity of nondeterministic Turing machines computations* ◇ D10 D15 ◇
CULIK, K. *A note on comparison of Turing machines with computers* ◇ B70 D10 ◇
ELGOT, C.C. *Monadic computation and iterative algebraic theories* ◇ C05 D10 D20 ◇
FREEDMAN, A.R. & LADNER, R.E. *Space bounds for processing contentless inputs* ◇ D10 D15 ◇
FREJDZON, R.I. *Finite approximate approach to the study of the complexity of recursive predicates (Russian)* ◇ D05 D10 D15 ◇
FREJVALD, R.V. *Fast computation by probabilistic Turing machines (Russian) (English summary)* ◇ D10 D15 ◇
FREJVALD, R.V. *Functions computable in the limit by probabilistic machines* ◇ D10 D20 ◇
GALIL, Z. *Two fast simulations which imply some fast string matching and palindrome-recognition algorithms* ◇ D05 D10 D15 ◇
HAUCK, J. *Turingmaschinen und berechenbare reelle Funktionen* ◇ D10 D20 F60 ◇
HAY, L. *Spectra and halting problems* ◇ C13 D10 D25 D30 ◇
HECKER, H.-D. *Eine Bemerkung zur Berechnungskompliziertheit auf Registermaschinen* ◇ D10 D15 ◇
IBARRA, O.H. & SAHNI, S.K. *Hierarchies of Turing machines with restricted tape alphabet size* ◇ D10 D15 ◇
KINBER, E.B. *On frequency real-time computations (Russian) (English summary)* ◇ D10 D15 D20 ◇
MEYER, A.R. *Weak monadic second order theory of successor is not elementary recursive* ◇ B25 D10 D15 D20 ◇
MONIEN, B. *About the deterministic simulation of nondeterministic* (log n)*-tape bounded Turing machines* ◇ D10 D15 ◇
OTTMANN, T. *Eine universelle Turingmaschine mit zweidimensionalem Band, 7 Buchstaben und 2 Zustaenden (English and Russian summaries)* ◇ D10 ◇
PAVLOTSKAYA, L.M. *The minimal number of different vertex codes in the graph of the universal Turing machine (Russian)* ◇ D10 ◇
POBEDIN, L.N. *The halt problem and theory of hierarchies (Russian)* ◇ D10 D55 ◇
RAKHMATULIN, N.A. *Automata-theoretic characteristics for the spectra of formulas of finite levels (Russian)* ◇ C13 D05 D10 ◇
REISIG, W. *Eine Verallgemeinerung des Berechenbarkeitsbegriffs durch Gleichungssysteme* ◇ D10 D75 ◇
REVESZ, G. *Phrase-structure grammars and dual pushdown automata (Hungarian) (English summary)* ◇ D05 D10 ◇
ROGOZHIN, YU.V. *The immortality problem for Post machines (Russian)* ◇ D10 ◇
ROGOZHIN, YU.V. *Unsolvability of the immortality problem for Turing machines with three states (Russian) (English summary insert)* ◇ D10 ◇
SAXTON, L.V. *An extension of the class of on-line machines* ◇ D10 D15 ◇
SCHNORR, C.-P. & STUMPF, G. *A characterization of complexity sequences* ◇ D10 D20 ◇
SUDBOROUGH, I.H. *On tape-bounded complexity classes and multihead finite automata* ◇ D05 D10 D15 ◇
VAISER, A. *Remarks on time and space functions of probabilistic machines* ◇ D10 D15 ◇
VOLLMAR, R. *On Turing machines with variable structure* ◇ D10 ◇
WAGNER, K. *Turing-Berechenbarkeit in linearer Zeit* ◇ D10 D15 ◇
WILKS, Y. *Putnam and Clarke and mind and body* ◇ A05 D10 ◇
YANOV, YU.I. *On some semantic characteristics of Turing machines (Russian)* ◇ D10 ◇

1976

ADLEMAN, L.M. & MANDERS, K.L. *Diophantine complexity* ◇ D10 D15 ◇
BAER, ROBERT M. & LEEUWEN VAN, J. *The halting problem for linear Turing assemblers* ◇ B75 D10 ◇
BAUR, W. *Zeitlich beschraenkte Turingmaschinen und polynomiale Reduktion* ◇ D10 D15 ◇
BECK, H. *Die mit Nestedstackautomaten berechenbaren Funktionen sind elementar* ◇ D10 D20 ◇
BEL'TYUKOV, A.P. *An iterative description of the class* $\mathfrak{E}^1$ *of Grzegorczyk's hierarchy (Russian)* ◇ D10 D20 ◇
BENEJAM, J.-P. & PAGET, M. *La complexite des algorithmes de Markov (English summary)* ◇ D03 D10 D15 ◇
BOERGER, E. *Ueber einige Interpretationen von Registermaschinen mit Anwendungen auf Entscheidungsprobleme in der Logik, der Algorithmentheorie und der Theorie formaler Sprachen* ◇ B25 D10 ◇
BRADY, J.M. *A programming approach to some concepts and results in the theory of computation* ◇ B75 D10 D20 ◇
CELONI, J.R. & PAUL, W.J. & TARJAN, R.E. *Space bounds for a game on graphs* ◇ D10 D15 ◇
CHANDRA, A.K. & STOCKMEYER, L.J. *Alternation* ◇ D10 D15 ◇
CHRISTEN, C. *Spektralproblem und Komplexitaetstheorie* ◇ B10 C13 D10 D15 D25 D35 ◇
CHYTIL, M.P. *Crossing-bounded computations and their relation to the LBA-problem* ◇ D10 D15 ◇
CHYTIL, M.P. *Return complexity of language generation* ◇ D10 D15 ◇
DATEL, V. *Anmerkungen zum Verhalten nichtdeterministischer Automaten (Tschechisch) (Russische Zusammenfassung)* ◇ D10 ◇
FUERER, M. *Simulation von Turingmaschinen mit logischen Netzen* ◇ B70 D10 D15 ◇
GILL III, J.T. & SIMON, I. *Ink, dirty-tape Turing machines, and quasicomplexity measures* ◇ D10 D15 ◇
GRIGOR'EV, D.YU. *Kolmogorov algorithms are stronger than Turing machines (Russian)* ◇ D10 D15 ◇
HAEUSSLER, A.F. *Polynomial beschraenkte nichtdeterministische Turingmaschinen und die Vollstaendigkeit des aussagelogischen Erfuellungsproblems* ◇ B05 D10 D15 ◇

HEINTZ, JOOS *Untere Schranken fuer die Komplexitaet logischer Entscheidungsprobleme* ◇ D10 D15 ◇
HEMMERLING, A. *Turing-Maschinen und Zellularraeume in der Ebene (English and Russian summaries)* ◇ D10 ◇
HOPKIN, D.R. & MOSS, B.P. *Automata* ◇ D05 D10 D98 ◇
HUWIG, H. *A machine independent description of complexity classes, definable by nondeterministic as well as deterministic Turing machines with primitive recursive tape or time bounds* ◇ D10 D15 ◇
IBARRA, O.H. & KIM, C.E. *A useful device for showing the solvability of some decision problems* ◇ D03 D10 D80 ◇
JEDRZEJOWICZ, J. *One-one degrees of Turing machines decision problems* ◇ D10 D25 ◇
KARMAZIN, V.N. *Mitteilung ueber minimale universelle Turingmaschinen (Russisch)* ◇ D10 ◇
KOZEN, D. *On parallelism in Turing machines* ◇ D10 D15 ◇
LADNER, R.E. & LYNCH, N.A. *Relativization of questions about log space computability* ◇ D10 D15 ◇
LOECKX, J. *Algorithmentheorie* ◇ D03 D10 D98 ◇
LYNCH, N.A. *Complexity-class-encoding sets* ◇ D10 D15 D25 ◇
MILGRAM, D.L. *A region crossing problem for array-bounded automata* ◇ D05 D10 ◇
MONIEN, B. *Transformational methods and their application to complexity problems* ◇ D10 D15 ◇
NABEBIN, A.A. *Multitape automata in a unary alphabet (Russian)* ◇ D05 D10 ◇
NURMEEV, N.N. *Computation of boolean functions by Turing machines (Russian)* ◇ B05 D10 D15 ◇
PRATT, V.R. & STOCKMEYER, L.J. *A characterization of the power of vector machines* ◇ D10 D15 ◇
PRIESE, L. *On a simple combinatorial structure sufficient for sub-lying nontrivial self-reproduction* ◇ D03 D10 ◇
PRIESE, L. *Reversible Automaten und einfache universelle 2-dimensionale Thue-Systeme* ◇ D03 D05 D10 ◇
REVESZ, G. *A note on the relation of Turing machines to phrase structure grammars* ◇ D05 D10 ◇
REVESZ, G. *Multicontrol Turing machines* ◇ D10 ◇
RICHARDSON, D.B. *Continuous self-reproduction* ◇ D10 ◇
SANTOS, E.S. *Fuzzy and probabilistic programs* ◇ B52 B75 D10 ◇
SAVAGE, J.E. *The complexity of computing* ◇ B70 D10 D15 D98 ◇
SCHNORR, C.-P. *The network complexity and the Turing machine complexity of finite functions* ◇ D10 D15 ◇
SHAKUOV, S.N. *Linear acceleration of the operating time of single-tape Turing machines (Russian)* ◇ D10 D15 ◇
SUDBOROUGH, I.H. *On deterministic context-free languages, multihead automata, and the power of an auxiliary pushdown store* ◇ D05 D10 D15 ◇
VAISER, A. *Stochastic languages and the complexity of computations on probabilistic Turing machines (Russian) (English summary)* ◇ D10 D15 ◇
VOELKEL, L. *Staerke-Relationen fuer URM-Befehlssysteme* ◇ D10 D15 D20 ◇
WAGNER, K. *Arithmetische Operatoren* ◇ D10 D55 ◇

1977

ABIAN, A. *Unsolvability of the halting problem for Turing machines* ◇ D10 D25 ◇
AE, T. *Direct or cascade product of pushdown automata* ◇ D10 ◇
BENEJAM, J.-P. *Algebraic characterizations of the satisfiability of first-order logical formulas and the halting of programs* ◇ C52 D10 G05 ◇
BORODIN, A. *On relating time and space to size and depth* ◇ D10 D15 ◇
BRANDSTAEDT, A. & SAALFELD, D. *Eine Hierarchie beschraenkter Rueckkehrberechnungen auf on-line Turingmaschinen* ◇ D10 D15 D20 ◇
BRAUNMUEHL VON, B. *Zwei-Zaehler-Automaten mit gekoppelten Bewegungen* ◇ D10 ◇
BREMER, H. *Ein vollstaendiges Problem auf der Baummaschine* ◇ D05 D10 D15 ◇
CHYTIL, M.P. *Comparison of the active visiting and the crossing complexities* ◇ D10 D15 ◇
COHORS-FRESENBORG, E. *Mathematik mit Kalkuelen und Maschinen* ◇ D03 D10 D98 ◇
COURCELLE, B. *On jump-deterministic pushdown automata* ◇ D10 ◇
DAVIS, MARTIN D. *Unsolvable problems* ◇ D10 D35 D40 ◇
FREJVALD, R.V. *On sets recognizable by a probabilistic machine with isolated cut-point and by no deterministic machine (Russian)* ◇ D05 D10 ◇
FREJVALD, R.V. *Probabilistic machines can use less running time* ◇ D10 D15 ◇
FREJVALD, R.V. & IKAUNIEKS, E. *Some advantages of nondeterministic machines over probabilistic ones (Russian)* ◇ D10 D15 ◇
GALIL, Z. & SEIFERAS, J.I. *Real-time recognition of substring repetition and reversal* ◇ D10 D15 ◇
GILL III, J.T. *Computational complexity of probabilistic Turing machines* ◇ D10 D15 ◇
GOLDSCHLAGER, L.M. *The monotone and planar circuit value problems are log space complete for P* ◇ B70 D10 D15 ◇
GRIGOR'EV, D.YU. *Imbedding theorems for Turing machines of different dimensions and Kolmogorov's algorithms (Russian)* ◇ D10 ◇
HEMMERLING, A. *Zur verallgemeinerten Realisierbarkeit von Pseudomusterfunktionen durch Zellularraeume* ◇ D10 D20 ◇
HOEHNKE, H.-J. *On partial recursive definitions and programs* ◇ D10 D75 ◇
IBARRA, O.H. *Counter machines and number-theoretic problems* ◇ D10 D35 ◇
KASAI, T. *Computational complexity of multitape Turing machines and random access machines* ◇ D10 D15 ◇
KLEINE BUENING, H. & OTTMANN, T. *Kleine universelle mehrdimensionale Turingmaschinen* ◇ D10 ◇
KLUPP, H. & SCHNORR, C.-P. *A universally hard set of formulae with respect to non-deterministic Turing acceptors* ◇ B25 D10 D15 ◇

KOREC, I. *Decidability (undecidability) of equivalence of Minsky machines with components consisting of at most seven (eight) instructions* ◇ D05 D10 ◇
KRATKO, M.I. & REVIN, O. *Turing machines operating on the plane, and stable iterative systems (Russian) (English summaries)* ◇ D10 D80 ◇
LEONG, B. & SEIFERAS, J.I. *New real-time simulations of multihead tape units* ◇ D10 D15 ◇
LEVIN, V.I. *Logical determinants and automata with continuous time. I: Theory of logical determinants. II: Automata without memory. III: Automata with memory (Russian)* ◇ D05 D10 ◇
LEWIS, H.R. *A measure of complexity for combinatorial decision problems of the tiling variety* ◇ B25 D10 D15 ◇
MONIEN, B. *A recursive and a grammatical characterization of the exponential-time languages* ◇ D05 D10 D15 ◇
MORITA, K. & SUGATA, K. & UMEO, H. *Computational complexity of L(m, n) tape-bounded two-dimensional Turing machines (Japanese)* ◇ D10 D15 ◇
MOSHCHENSKIJ, V.A. *A method of analysis for Turing computations. I (Russian) (English summary)* ◇ D10 D15 ◇
OTTMANN, T. *Lokale Simulierbarkeit zweidimensionaler Turingmaschinen* ◇ D10 ◇
PRATHER, R.E. *Structured Turing machines* ◇ D10 ◇
SAVITCH, W.J. & VITANYI, P.M.B. *Linear time simulation of multihead Turing machines with head-to-head jumps* ◇ D10 D15 ◇
SAVITCH, W.J. *Recursive Turing machines* ◇ D10 D15 ◇
SEIFERAS, J.I. *Iterative arrays with direct central control* ◇ D10 D15 ◇
SEIFERAS, J.I. *Linear-time computation by nondeterministic multidimensional iterative arrays* ◇ D10 D15 ◇
SETH, S.C. & STECKELBERG, J.M. *On a relation between algebraic programs and Turing machines* ◇ B75 D10 ◇
SHAKUOV, S.N. *Fast Turing computations and their linear speedup (Russian)* ◇ D10 D15 ◇
SIMON, J. *On the difference between one and many (Preliminary version)* ◇ D10 D15 ◇
SIMON, J. *Polynomially bounded quantification over higher types and a new hierarchy of the elementary sets* ◇ B15 D10 D15 D20 ◇
SLISENKO, A.O. *A simplified proof of the real-time recognizability of parlindromes on Turing machines (Russian) (English summary)* ◇ D10 D15 ◇
STAIGER, L. *Empty-storage-acceptance of ω-languages* ◇ D05 D10 ◇
STAIGER, L. & WAGNER, K. *Recursive ω-languages* ◇ D05 D10 D55 ◇
STILLWELL, J.C. *Concise survey of mathematical logic* ◇ B98 C98 D10 D35 ◇
STOCKMEYER, L.J. *The polynomial-time hierarchy* ◇ D10 D15 D40 D55 ◇
SUDBOROUGH, I.H. *Some remarks on multihead automata* ◇ D10 D15 ◇
TAKAHASHI, H. *Information transmission in one-dimensional cellular space and the maximum invariant set* ◇ D10 ◇
TAKAHASHI, H. *Undecidable equations about the maximum invariant set* ◇ D10 ◇
TANAKA, E. *The Turing machine constructed by trainable threshold elements* ◇ B70 D10 ◇
TRAKHTENBROT, B.A. *Algorithmen und Rechenautomaten* ◇ D10 D15 D40 D98 ◇
VALIEV, M.K. *Real time computations with restrictions on tape alphabet* ◇ D10 D15 ◇
VOELKEL, L. *Spracherkennung durch Registermaschinen* ◇ D10 ◇
YANOV, YU.I. *A convolution method for the solution of properties of formal systems (Russian)* ◇ D10 D20 ◇
ZAKOWSKI, W. *A generalization of the notions of a machine and computability* ◇ D10 D20 ◇

1978

APRILE, G. & RAIMONDI, T. *Homogeneous neuron networks* ◇ B70 D10 ◇
BISKUP, J. *Path measures of Turing machine computations* ◇ D10 D15 ◇
BISKUP, J. *The time measure of one-tape Turing machines does not have the parallel computation property* ◇ D10 D15 ◇
BOERGER, E. *Bemerkung zu Gurevich's Arbeit ueber das Entscheidungsproblem fuer Standardklassen* ◇ B20 D10 D35 ◇
BOERGER, E. *Ein einfacher Beweis fuer die Unentscheidbarkeit der klassischen Praedikatenlogik* ◇ B10 D03 D10 D35 ◇
BOOK, R.V. & GREIBACH, S.A. & WRATHALL, C. *Comparisons and reset machines* ◇ D10 D15 ◇
BRANDSTAEDT, A. *On a family of complexity measures on Turing machines, defined by predicates* ◇ D10 D15 ◇
BRANDSTAEDT, A. *Ueber den Einfluss eines zusaetzlichen Einweg-Eingabebandes bei Entscheidungen und Aufzaehlungen mit beschraenktem Regime sowie beschraenkter Rueckkehr bzw. beschraenkter dualer Rueckkehr auf einbaendrigen Turing-Maschinen* ◇ D10 D15 ◇
COHEN, R.S. & GOLD, A.Y. *ω-computations on Turing machines* ◇ D10 ◇
COURCELLE, B. *A representation of trees by languages I* ◇ D05 D10 ◇
COURCELLE, B. *A representation of trees by languages II* ◇ D05 D10 ◇
DALEY, R.P. *On the simplicity of busy beaver sets* ◇ D10 D15 D25 D30 ◇
DEGTEV, A.N. *Nonerasing Turing machines (Russian)* ◇ D10 ◇
DENNING, P.J. & DENNIS, J.B. & QUALITZ, J.E. *Machines, languages, and computation* ◇ B98 D05 D10 D98 ◇
FISCHER, MICHAEL J. & MEYER, A.R. & SEIFERAS, J.I. *Separating nondeterministic time complexity classes* ◇ D10 D15 ◇
FORTUNE, S. & WYLLIE, J. *Parallelism in random access machines* ◇ D10 ◇

Subject Index

FREJVALD, R.V. *Recognition of languages with high probability on different classes of automata (Russian)* ◇ D05 D10 ◇
FRIEDMAN, E.P. *A note on non-singular deterministic pushdown automata* ◇ D10 ◇
GALIL, Z. & SEIFERAS, J.I. *A linear-time on-line recognition algorithm for "palstar"* ◇ D10 D15 ◇
GALIL, Z. *Palindrome recognition in real time by a multitape Turing machine* ◇ D10 D15 ◇
GREIBACH, S.A. *One way finite visit automata* ◇ D10 ◇
GREIBACH, S.A. *Visits, crosses, and reversals for nondeterministic off-line machines* ◇ D10 D15 ◇
HALL PARTEE, B. *Fundamentals of mathematics for linguistics* ◇ B65 D05 D10 ◇
HARTMANIS, J. & IMMERMAN, N. & MAHANEY, S.R. *One-way log-tape reductions* ◇ D10 D15 ◇
HAVRANEK, T. *Enumeration calculi and rank methods* ◇ B50 D10 ◇
HEFFERNAN, J.D. *Discussion: some doubts about "Turing machine arguments"* ◇ A05 D10 ◇
HEIDLER, K. *Die Unentscheidbarkeit des Halteproblems auf der Grundlage der Programmiersprache Basic* ◇ B75 D10 ◇
HONDA, N. & OYAMAGUCHI, M. *The decidability of equivalence for deterministic stateless pushdown automata* ◇ D10 ◇
HUNT III, H.B. & ROSENKRANTZ, D.J. *Polynomial algorithms for deterministic pushdown automata* ◇ D10 D15 ◇
IBARRA, O.H. *Reversal-bounded multicounter machines and their decision problems* ◇ D10 ◇
IGARASHI, Y. *Tape bounds for some subclasses of deterministic context-free languages* ◇ D10 D15 ◇
INOUE, K. & TAKANAMI, I. *A note on closure properties of the classes of sets accepted by tape-bounded two-dimensional Turing machines* ◇ D10 ◇
INOUE, K. & TAKANAMI, I. *Closure properties of the classes of sets accepted by automata on a two-dimensional tape under two operations* ◇ D10 ◇
INOUE, K. & TAKANAMI, I. *Cyclic closure properties of automata on a two-dimensional tape* ◇ D10 ◇
LAING, R. *Anomalies of self-description* ◇ A05 D10 ◇
LILLO DE, N.J. *A note on Turing machine regularity and primitive recursion* ◇ D10 D20 ◇
LITOW, B. & SUDBOROUGH, I.H. *On non-erasing oracle tapes in space bounded reducibility* ◇ D10 D15 ◇
LYNCH, N.A. *Log space machines with multiple oracle tapes* ◇ D10 D15 ◇
MACHTEY, M. & WINKLMANN, K. & YOUNG, P. *Simple Goedel numberings, isomorphisms, and programming properties* ◇ D05 D10 D15 D20 D45 F40 ◇
MARINESCU, D. *A universal Turing machine with three tapes for normal algorithms* ◇ D03 D10 ◇
ORMAN, G. *An automata realization of the functions associated with a derivation* ◇ D05 D10 ◇
PAGET, M. *Proprietes de complexite pour une famille d'algorithmes de Markov* ◇ B75 D03 D10 D15 ◇
PAVLOTSKAYA, L.M. *Sufficient conditions for the solvability of the halting problem for Turing machines(Russian)* ◇ D10 ◇
PISANSKI, T. *Computability and solvability (Slovanian) (English summary)* ◇ D10 ◇
SCHNORR, C.-P. *Satisfiability is quasilinear complete in NQL* ◇ D10 D15 ◇
SIPSER, M. *Halting space-bounded computations* ◇ D10 D15 ◇
SLISENKO, A.O. *Models of computation based on address organisation of memory (Russian)* ◇ D10 ◇
SOLOMONOFF, R.J. *Complexity-based induction systems: Comparisons and convergence theorems* ◇ B30 D10 ◇
STAIGER, L. & WAGNER, K. *Rekursive Folgenmengen. I* ◇ D10 D55 D75 ◇
SUDBOROUGH, I.H. *On the tape complexity of deterministic context-free languages* ◇ D05 D10 D15 ◇
TLYUSTEN, V.SH. *Solvability of a uniform halting problem in the class of Post machines with two states (Russian) (English summary)* ◇ D10 ◇
WEBER VON, S. *Zur Definition eines speziellen algorithmischen Fehlers mittels Turingmaschinen* ◇ D10 ◇
WRATHALL, C. *Rudimentary predicates and relative computation* ◇ D10 D15 ◇
YANOV, YU.I. *Several theorems on convolutions (Russian) (English Summary)* ◇ D10 ◇
ZAKOWSKI, W. *On some mathematical models of computing machines (Polish)* ◇ D10 ◇

1979

ASVELD, P.R.J. & ENGELFRIET, J. *Extended linear macro grammars, iteration grammars, and register programs* ◇ D05 D10 ◇
BEL'TYUKOV, A.P. *A machine description and a hierarchy of initial Grzegorczyk classes (Russian) (English summary)* ◇ D10 D20 ◇
BERGSTRA, J.A. *Recursion theory on processes* ◇ D10 D30 D75 ◇
BIANCO, E. *Informatique fondamentale. De la machine de Turing aux ordinateurs modernes* ◇ D03 D10 D98 ◇
BOERGER, E. *A new general approach to the theory of the many-one equivalence of decision problems for algorithmic systems* ◇ B25 D03 D05 D10 D25 D30 D40 ◇
BOOK, R.V. & GREIBACH, S.A. & WRATHALL, C. *Reset machines* ◇ D10 D15 ◇
BRAUNMUEHL VON, B. & VERBEEK, R. *Finite-change automata* ◇ D05 D10 ◇
BRAUNMUEHL VON, B. & HOTZEL, E. *Supercounter machines* ◇ D10 D15 ◇
BRUCKNER, L.K. *On the Garden-of-Eden problem for one-dimensional cellular automata* ◇ D10 ◇
CAPORASO, S. *Consistency proof without transfinite induction for a formal system for Turing machines* ◇ D10 F05 ◇
CAZACU, M. *A general mathematical notion of the algorithm* ◇ D03 D10 D20 ◇
CHLEBUS, B.S. *On decidability of propositional algorithmic logic (Polish) (English summary)* ◇ B25 B75 D10 D15 ◇

DALEY, R.P. *On the simplification of constructions in degrees of unsolvability via computational complexity* ◇ D10 D15 D25 ◇
DEKHTYAR', M.I. *Complexity spectra of recursive sets of approximability of initial segments of complete problems* ◇ D10 D15 D30 F20 F30 ◇
FERRANTE, J. & RACKOFF, C.W. *The computational complexity of logical theories* ◇ B25 C98 D10 D15 ◇
FREJVALD, R.V. *Fast probabilistic verification of number multiplication (Russian)* ◇ D10 D15 ◇
FREJVALD, R.V. *Fast probabilistic algorithms* ◇ D10 D15 ◇
FREJVALD, R.V. *Finite identification of general recursive functions by probabilistic strategies* ◇ D10 D20 ◇
FREJVALD, R.V. *Running time for deterministic and nondeterministic Turing machines (Russian)* ◇ D10 D15 ◇
FRIEDMAN, E.P. & GREIBACH, S.A. *Superdeterministic DPDAs: The method of accepting does affect decision problems* ◇ D05 D10 D15 ◇
GREGUSOVA, L. & KOREC, I. *Small universal Minsky machines* ◇ D10 ◇
GREIBACH, S.A. *Linearity is polynomially decidable for realtime pushdown store automata* ◇ D10 D15 ◇
GRIGOR'EV, D.YU. *Time bounds of multidimensional Turing machines (Russian)* ◇ D10 D15 ◇
GURARI, E.M. & IBARRA, O.H. *An NP-complete number-theoretic problem* ◇ B25 D10 D15 D25 D80 ◇
GURARI, E.M. & IBARRA, O.H. *Simple counter machines and number-theoretic problems* ◇ D05 D10 ◇
GURARI, E.M. & IBARRA, O.H. *Some decision problems concerning sequential transducers and checking automata* ◇ D05 D10 ◇
HAJEK, P. *Arithmetical hierarchy and complexity of computation* ◇ D10 D15 D55 ◇
HARROW, K. *Theoretical and applied computer science: antagonism or symbiosis?* ◇ B75 D10 D80 ◇
HEINTZ, JOOS *Definability bounds of first order theories of algebraically closed fields* ◇ B25 C40 C60 D10 D15 ◇
HEMMERLING, A. *Concentration of multidimensional tape-bounded systems of Turing automata and cellular spaces* ◇ D05 D10 ◇
HEMMERLING, A. *Systeme von Turing-Automaten und Zellularraeume auf rahmbaren Pseudomustermengen* ◇ D10 ◇
HEMMERLING, A. *Zur Raumkompliziertheit mehrdimensionaler Zellularraeume und Turing-Automaten (Russian and English summaries)* ◇ D05 D10 ◇
HOPCROFT, J.E. & ULLMAN, J.D. *Introduction to automata theory, languages, and computation* ◇ D05 D10 D98 ◇
IBARRA, O.H. *Restricted one-counter machines with undecidable universe problems* ◇ D10 ◇
INOUE, K. & TAKANAMI, I. *Closure properties of three-way and four-way tape-bounded two-dimensional Turing machines* ◇ D10 ◇
INOUE, K. & TAKANAMI, I. *Three-way tape-bounded two-dimensional Turing machines* ◇ D10 ◇
IVANOV, A.G. *Theorems on the time hierarchy for random access machines (Russian) (English summary)* ◇ D10 D15 ◇
KLEINE BUENING, H. *Generalized vector addition systems with finite exception sets* ◇ B75 D05 D10 D80 ◇
KOBUCHI, Y. & SEKI, S. *Decision promblems of locally catenative property for DIL systems* ◇ D10 D80 ◇
LINNA, M. *Two decidability results for deterministic pushdown automata* ◇ D05 D10 ◇
LUGOWSKI, H. *Ueber Normalformen von Elementen freier Turing-Algebren* ◇ D10 ◇
MARCHENKOV, S.S. & MATROSOV, V.L. *Complexity of algorithms and computations (Russian)* ◇ D10 D15 D25 ◇
MEYER, A.R. & WINKLMANN, K. *The fundamental theorem of complexity theory (Preliminary version)* ◇ B75 D10 D15 ◇
ORMAN, G. *Control Turing machines for some derivational functions* ◇ D10 ◇
PAKHOMOV, S.V. *Machine-independent description of some machine complexity classes (Russian) (English summary)* ◇ D10 D15 ◇
PETRI, N.V. *Unsolvability of the problem of recognition of cancelling iterative nets (Russian)* ◇ D10 ◇
PRIESE, L. *Towards a precise characterization of the complexity of universal and nonuniversal Turing machines* ◇ D10 ◇
PRIESE, L. *Ueber eine minimale universelle Turing-Maschine* ◇ D10 ◇
SAGER, H. *The nature of reduction in Turing machines and its relation to reduction in automata* ◇ D05 D10 ◇
SAVITCH, W.J. & STIMSON, M.J. *Hierarchies of recursive computations* ◇ D10 D15 ◇
STRAIGHT, D.W. *Domino f-sets* ◇ D10 D20 ◇
THOMAS, WILLIAM J. *A simple generalization of Turing computability* ◇ D10 D75 ◇
WECHSUNG, G. *A crossing measure for 2-tape Turing machines* ◇ D10 D15 ◇
ZAK, S. *A Turing machine oracle hierarchy* ◇ D10 D15 ◇
ZAK, S. *A Turing machine space hierarchy* ◇ D10 D15 ◇
ZASLAVSKIJ, I.D. *The realization of three-valued logical functions by means of recursive and Turing operators (Russian)* ◇ B50 D10 D20 ◇
ZHAROV, V.G. *On the complexity of the terms of the constructive sequences of Turing machines (Russian)* ◇ D10 D45 F60 ◇

1980

AANDERAA, S.O. & COHEN, D.E. *Modular machines, the word problem for finitely presented groups and Collins' theorem* ◇ D10 D30 D40 ◇
AANDERAA, S.O. & COHEN, D.E. *Modular machines and the Higman-Clapham-Valiev embedding theorem* ◇ D10 D30 D40 ◇
ABRAMSON, F.G. & BREJTBART, YU.YA. & LEWIS, F.D. *Complex properties of grammars* ◇ D05 D10 D15 ◇

BENIOFF, P.A. *The computer as a physical system: a microscopic quantum mechanical Hamiltonian model of computers as represented by Turing machines* ◇ D10 ◇
BERMAN, L. *The complexity of logical theories* ◇ B25 D10 D15 F30 ◇
CAZACU, M. *The simulation relations between the classes of algorithms* ◇ D03 D10 D20 ◇
CHRISTODOULAKIS, D. *Eine einfache Basis fuer die Berechenbarkeit. Die Konstruktion der universellen Turingmaschine mit einem Zustand, zwei Symbolen und drei Baendern* ◇ D10 ◇
COHEN, D.E. *Degree problems for modular machines* ◇ D10 D25 D40 ◇
COHEN, R.S. & GOLD, A.Y. *On the complexity of ω-type Turing acceptors* ◇ D10 ◇
CULIK II, K. & DIAMOND, N.D. *A homomorphic characterization of time and space complexity classes of languages* ◇ D05 D10 D15 ◇
DALEY, R.P. *The busy beaver method* ◇ D10 D15 D25 ◇
EBBINGHAUS, H.-D. *Mathematische Logik und Informatik* ◇ C13 D10 D15 D98 E70 H15 ◇
ENGELFRIET, J. & LEEUWEN VAN, J. & SCHMIDT, ERIK MEINECHE *Stack machines and classes of nonnested macro languages* ◇ D05 D10 ◇
FRIEDMAN, E.P. & GREIBACH, S.A. *Superdeterministic PDAs: a subcase with a decidable inclusion problem* ◇ D10 ◇
GANDY, R.O. *Church's thesis and principles for mechanisms* ◇ A05 D05 D10 D20 F99 ◇
GURARI, E.M. & IBARRA, O.H. *Path systems: constructions, solutions and applications* ◇ D05 D10 ◇
HONDA, N. & INAGAKI, Y. & OYAMAGUCHI, M. *On the equivalence problem for two DPDA's, one of which is real-time* ◇ D10 D15 ◇
INOUE, K. & TAKANAMI, I. *A note on deterministic three-way tape-bounded two-dimensional Turing machines* ◇ D10 ◇
KHACHIYAN, L.G. & KOZLOV, M.K. & TARASOV, S.P. *The polynomial solvability of convex quadratic programming (Russian)* ◇ D10 D15 ◇
KINBER, E.B. *On inclusion problem for deterministic multitape automata* ◇ D10 ◇
KURDUMOV, G.I. *An algorithm-theoretic method for the study of uniform random networks* ◇ D05 D10 ◇
LADNER, R.E. *Complexity theory with emphasis on the complexity of logical theories* ◇ B25 C98 D05 D10 D15 ◇
LEWIS, H.R. *Complexity results for classes of quantificational formulas* ◇ B20 B25 D10 D15 ◇
MOLDESTAD, J. & STOLTENBERG-HANSEN, V. & TUCKER, J.V. *Finite algorithmic procedures and inductive definability* ◇ D10 D70 D75 ◇
MOLDESTAD, J. & STOLTENBERG-HANSEN, V. & TUCKER, J.V. *Finite algorithmic procedures and computation theories* ◇ D10 D75 ◇
MOSHCHENSKIJ, V.A. *Turing a-computations and essential complexity of a-inversions of binary words (Russian)* ◇ D10 ◇
ORMAN, G. *An automatic realization of $\xi(h)$-function (Romanian summary)* ◇ D05 D10 ◇
PASSY, S. *Structured programs for Turing machines* ◇ D10 ◇
PAUL, W.J. & PRAUSS, E.J. & REISCHUK, R. *On alternation* ◇ D10 D15 ◇
PAUL, W.J. & REISCHUK, R. *On alternation II. A graph-theoretic approach to determinism versus nondeterminism* ◇ D10 D15 ◇
ROGOZHIN, YU.V. *Universal one-place partial recursive functions (Russian)* ◇ D10 D20 ◇
SCHOENHAGE, A. *Storage modification machines* ◇ D10 ◇
SIPSER, M. *Halting space-bounded computations* ◇ D10 D15 ◇
TUCKER, J.V. *Computing in algebraic systems* ◇ D10 D35 D75 D80 ◇
VITANYI, P.M.B. *On the power of real-time Turing machines under varying specifications* ◇ D10 D15 ◇
VITANYI, P.M.B. *Real-time Turing machines under varying specifications* ◇ D10 D15 ◇
VITANYI, P.M.B. *Real-time Turing machines under varying specifications II* ◇ D10 D15 ◇
WEBER VON, S. *Zur Definition eines algorithmischen Fehlers 2. Art mittels Turingmaschinen* ◇ D10 ◇
WIEDMER, E. *Computing with infinite objects* ◇ D10 D20 ◇
ZAK, S. *A Turing machine oracle hierarchy I,II* ◇ D10 D15 ◇

1981

ADLEMAN, L.M. & LOUI, M.C. *Space-bounded simulation of multitape Turing machines* ◇ D10 D15 ◇
BRANDSTAEDT, A. *Pushdown automata with restricted use of storage symbols* ◇ D10 D15 ◇
CHANDRA, A.K. & KOZEN, D. & STOCKMEYER, L.J. *Alternation* ◇ D10 D15 ◇
CHARLESWORTH, A. *A proof of Goedel's theorem in terms of computer programs* ◇ D10 F30 ◇
CLAUS, V. *The (n,k)-bounded emptiness-problem for probabilistic acceptors and related problems* ◇ D05 D10 ◇
DALEY, R.P. *Busy beaver sets and the degree of unsolvability* ◇ D10 D15 D25 ◇
DALEY, R.P. *On the error correcting power of pluralism in inductive inference (Preliminary report)* ◇ D10 D20 ◇
DEUTSCH, M. *Registermaschinen ueber Quotiententermmengen* ◇ D10 D20 ◇
EHRENFEUCHT, A. & ROZENBERG, G. & RUOHONEN, K. *A morphic representation of complements of recursively enumerable sets* ◇ D05 D10 ◇
FREJVALD, R.V. *Probabilistic two-way machines* ◇ D05 D10 D15 ◇
FREJVALD, R.V. *Two-way finite probabilistic automata and tape-bounded Turing machines (Russian)* ◇ D05 D10 ◇
GABRIELIAN, A. *Pure grammars and pure languages* ◇ B05 D03 D05 D10 ◇
GINSBURG, S. & KINTALA, C.M.R. *Strict interpretations of deterministic pushdown acceptors* ◇ D10 ◇

GURARI, E.M. & IBARRA, O.H. *The complexity of decision problems for finite-turn multicounter machines* ◇ D10 D15 ◇
HONDA, N. & INAGAKI, Y. & OYAMAGUCHI, M. *The equivalence problem for two dpda's, one of which is a finite-turn or one-counter machine* ◇ D05 D10 ◇
IBARRA, O.H. & ROSIER, L.E. *On the decidability of equivalence for deterministic pushdown transducers* ◇ D10 ◇
KLOSINSKI, L.F. & SMOLARSKI, D.C. *Recognition algorithms for Fibonacci numbers* ◇ D10 D80 ◇
LEWIS, F.D. *Stateless Turing machines and fixed points* ◇ D10 D15 D20 ◇
LOUI, M.C. *A space bound for one-tape multidimensional Turing machines* ◇ D10 D15 ◇
LUGOWSKI, H. *Ueber die Identitaeten der Turing-Algebra I (English and Russian summaries)* ◇ C05 D10 D75 ◇
MELVILLE, R. *An improved simulation result for ink-bounded Turing machines* ◇ D10 ◇
MERCANTI, F. & PIAZZESE, F. *Neuromachines as Turing machines* ◇ D10 ◇
MERCANTI, F. *On a neuromachine equivalent to the universal Turing machine* ◇ D10 ◇
MEYER AUF DER HEIDE, F. & ROLLIK, A. *Random access machines and straight-line programs* ◇ D10 ◇
MOITRA, A. *Relation between algebraic specification and Turing machines* ◇ B75 D10 ◇
PAUL, W.J. *On heads versus tapes* ◇ D10 D15 ◇
PODNIEKS, K.M. *Prediction of the following value of a function (Russian)* ◇ D10 D20 ◇
RUZZO, W.L. *On uniform circuit complexity* ◇ D10 D15 ◇
SAVITCH, W.J. & VERMEIR, D. *On the amount of nondeterminism in pushdown automata* ◇ D10 ◇
SIMON, J. *Division in idealized unit cost RAMs* ◇ D10 D15 ◇
SIMON, J. *On tape-bounded probabilistic Turing machine acceptors* ◇ D10 D15 ◇
TAO, RENJI *On the computational power of automata with time or space bounded by Ackermann's or superexponential functions* ◇ D10 D15 D20 ◇
VOGEL, J. & WAGNER, K. *Two-way automata with more than one storage medium* ◇ D10 D15 ◇

1982

BACHEM, A. *Concepts of algorithmic computation* ◇ D10 D15 ◇
BENIOFF, P.A. *Quantum mechanical Hamiltonian models of Turing machines* ◇ D10 D80 ◇
BENIOFF, P.A. *Quantum mechanical models of Turing machines that dissipate no energy* ◇ D10 D80 ◇
BRANDSTAEDT, A. & VOGEL, J. *Kompliziertheitsbeschraenkte Checking-Stack-Baender und Raum-Zeit-Probleme (English and Russian summaries)* ◇ D10 D15 ◇
BUDACH, L. & MEINEL, C. *Environments and Automata I., II (German and Russian summaries)* ◇ D05 D10 D80 ◇
DALEY, R.P. *Busy beaver sets: characterizations and applications* ◇ D10 D25 D50 ◇
IBARRA, O.H. & MORAN, S. *On some decision problems for RAM programs* ◇ D10 ◇
INOUE, K. & ITO, A. & TAKANAMI, I. & TANIGUCHI, H. *Two-dimensional alternating Turing machines with only universal states* ◇ D10 ◇
KRYLOV, S.M. *Models of universal discrete-analog computers on the basis of Turing machines (Russian) (English summary)* ◇ D10 ◇
LIBUS, M. & OSTROWSKA, M. *Remarks on L.J.Stockmeyer's note "The complexity of decision problems in automata theory and logic" (Polish) (English summary)* ◇ D05 D10 D15 ◇
LOSEV, G.F. *Local algorithms for computing information with nonfixed memory (Russian)* ◇ D10 ◇
LOUI, M.C. *Simulations among multidimensional Turing machines* ◇ D10 ◇
MARCHENKOV, S.S. *Undecidability of the positive* $\forall\exists$*-theory of a free semigroup (Russian)* ◇ B20 D10 D35 ◇
MEINEL, C. *Embedding of the poset of Turing degrees in the poset* $[\mathfrak{U}, \leqslant]$ *(German and Russian summaries)* ◇ D10 D30 ◇
MONIEN, B. & SUDBOROUGH, I.H. *On eliminating nondeterminism from Turing machines which use less than logarithm worktape space* ◇ D10 D15 ◇
NASYROV, I.R. *Representability of languages in deterministic and nondeterministic automata with a countable number of states (Russian)* ◇ D05 D10 ◇
PAUL, W.J. *On-line simulation of* $k+1$ *tapes by* k *tapes requires nonlinear time* ◇ D10 D15 ◇
ROGOZHIN, YU.V. *Seven universal Turing machines (Russian)* ◇ D10 ◇
RYTTER, W. *A note on two-way nondeterministic pushdown automata* ◇ D10 ◇
SARKISYAN, A.D. *Classes of functions defined by their calculation time on iterative networks (Russian) (Armenian summary)* ◇ D10 D15 ◇
SIMOVICI, D.A. *Several remarks on the complexity of graph computations* ◇ D10 D15 ◇
UKKONEN, E. *The equivalence problem for some non-real-time deterministic pushdown automata* ◇ D10 ◇
WALSH, T.R.S. *The busy beaver on a one-way infinite tape* ◇ D10 ◇

1983

ALBERT, J. & OTTMANN, T. *Automaten, Sprachen und Maschinen fuer Anwender* ◇ D05 D10 D98 ◇
BANACHOWSKI, L. *On the Tarjan model of pointer machines* ◇ D10 ◇
BOERGER, E. *Undecidability versus degree complexity of decision problems for formal grammars* ◇ D05 D10 D25 ◇
BRADY, A.H. *The determination of the value of Rado's noncomputable function* $\Sigma(k)$ *for four-state Turing machines* ◇ D10 ◇
BRANDSTAEDT, A. & WAGNER, K. *Reversal-bounded and visit-bounded realtime computations* ◇ D10 D15 ◇
BURGIN, M.S. *Inductive Turing machines (Russian)* ◇ D10 D25 D55 ◇

Subject Index

CHAZELLE, B. & MONIER, L. *Unbounded hardware is equivalent to deterministic Turing machines* ◇ D10 ◇
DAVIS, MARTIN D. & WEYUKER, E.J. *Computability, complexity, and languages. Fundamentals of theoretical computer science* ◇ D05 D10 D20 D98 ◇
FREJVALD, R.V. *Space and reversal complexity of probabilistic one-way Turing machines* ◇ D10 D15 ◇
FREUND, R. *Real functions and numbers defined by Turing machines* ◇ D10 D20 D80 ◇
FURST, M. & LIPTON, R.J. & STOCKMEYER, L.J. *Pseudorandom number generation and space complexity* ◇ D10 D15 ◇
GUESSARIAN, I. *Pushdown tree automata* ◇ D10 ◇
HASENJAEGER, G. *Exponential diophantine description of a small universal Turing machine (UTM) using Matiyasevich's 1975/80 "masking"* ◇ D10 D15 D35 ◇
HEINTZ, JOOS *Definability and fast quantifier elimination in algebraically closed fields* ◇ B25 C10 C40 C60 D10 D15 ◇
HEMMERLING, A. *D ≠ ND fuer mehrdimensionale Turing-Automaten mit sublogarithmischer Raumschranke* ◇ D10 D15 ◇
HODGES, A. *Alan Turing: The Enigma* ◇ A10 D10 ◇
IBARRA, O.H. & ROSIER, L.E. *Simple programming languages and restricted classes of Turing machines* ◇ B75 D10 ◇
IBARRA, O.H. & MORAN, S. *Some time-space tradeoff results concerning single-tape and offline TMs* ◇ D10 D15 ◇
INOUE, K. & TAKANAMI, I. & TANIGUCHI, H. *Two-dimensional alternating Turing machines* ◇ D10 D15 ◇
ITZHAIK, Y. & YEHUDAI, A. *On containment problems for finite-turn languages* ◇ D10 ◇
LUGOWSKI, H. *Ueber die Identitaeten der Turing-Algebra II (English and Russian summaries)* ◇ C05 D10 ◇
MIYANO, S. *Remarks on multihead pushdown automata and multihead stack automata* ◇ D10 ◇
MUCHNIK, A.A. *Supplement of the translator to the paper "On alternation. I, II" (Russian)* ◇ B45 D05 D10 D15 F50 ◇
MUNDICI, D. *Natural limitations of decision procedures for arithmetic with bounded quantifiers* ◇ B25 D10 D15 F30 ◇
NAKAMURA, A. & ONO, H. *Pictures of functions and their acceptability by automata* ◇ D10 D20 ◇
PLOTNIKOVA, N.A. *Calculation with oracles* ◇ D10 D30 ◇
VOLGER, H. *Turing machines with linear alternation, theories of bounded concatenation and the decision problem of first order theories* ◇ B25 D10 D15 ◇
WATANABE, O. *The time-precision tradeoff problem on on-line probabilistic Turing machines* ◇ D10 D15 ◇
WINKLER, P.M. *Polynomial hyperforms* ◇ C05 D10 ◇
ZAK, S. *A Turing machine time hierarchy* ◇ D10 D15 ◇

1984

BERTONI, A. & MAURI, G. & SABADINI, N. *Non deterministic machines and their generalizations* ◇ D10 D15 ◇
BOYER, R.S. & MOORE, J.S. *A mechanical proof of the unsolvability of the halting problem* ◇ D10 ◇
BURGIN, M.S. *Inductive Turing machines with multihead, and Kolmogorov's algorithm (Russian)* ◇ D10 ◇
CHLEBUS, B.S. & CHROBAK, M. *Probabilistic Turing machines and recursively enumerable Dedekind cuts* ◇ D10 D25 D80 ◇
GABARRO, J. *Pushdown space complexity and related full-A.F.L.s* ◇ D10 D15 ◇
GIRAULT-BEAUQUIER, D. *Some results about finite and infinite behaviours of a pushdown automation* ◇ D10 ◇
HASENJAEGER, G. *Universal Turing machines (UTM) and Jones-Matiyasevich-masking* ◇ D10 D35 ◇
HEMMERLING, A. & MURAWSKI, G. *Zur Raumkompliziertheit mehrdimensionaler Turing-Automaten* ◇ D10 D15 ◇
HONG, JIAWEI *A tradeoff theorem for space and reversal* ◇ D10 ◇
HUNT III, H.B. *Terminating Turing machine computations and the complexity and/or decidability of correspondence problems, grammars, and program schemes* ◇ D10 D15 D20 ◇
IBARRA, O.H. & KIM, S.M. & ROSIER, L.E. *Space and time efficient simulations and characterizations of some restricted classes of PDAS* ◇ D10 D15 ◇
JONES, JAMES P. & MATIYASEVICH, YU.V. *Register machine proof of the theorem on exponential diophantine representation of enumerable sets* ◇ D10 D25 D35 ◇
LADNER, R.E. & LIPTON, R.J. & STOCKMEYER, L.J. *Alternation bounded auxiliary pushdown automata* ◇ D10 D15 ◇
MEINHARDT, D. *Tape reversal bounded Turing machines with an auxiliary pushdown or an auxiliary counter* ◇ D10 ◇
MONIEN, B. *Deterministic two-way one-head pushdown automata are very powerful* ◇ D10 ◇
NASYROV, I.R. *Representation of languages in Turing machines and automata with a countable number of states (Russian)* ◇ D10 ◇
OYAMAGUCHI, M. *Some remarks on subclass containment problems for several classes of dpda's* ◇ D10 ◇
PAUL, W.J. *On heads versus tapes* ◇ D10 D15 ◇
PLATEK, M. *Recognizing of languages by composition of deterministic pushdown transducers* ◇ D10 ◇
ROUHONEN, K. *On machine characterization of nonrecursive hierarchies* ◇ D10 D55 ◇
RUOHONEN, K. *A note on off-line machines with "Brownian" input heads* ◇ D10 ◇
SHAPIRO, E.Y. *Alternation and the computational complexity of logic programs* ◇ D10 D15 ◇
VITANYI, P.M.B. *On the simulation of many storage heads by one* ◇ D10 ◇
VOLGER, H. *Rudimentary relations and Turing machines with linear alternation* ◇ D10 D15 D20 ◇
VOLGER, H. *The role of rudimentary relations in complexity theory* ◇ D10 D15 ◇

1985

ABLAEV, F.M. *Possibilities of probalistic machines with respect to representation of languages in real time (Russian)* ◇ D05 D10 ◇

ALBERTS, M. *Complexity of computations on nondeterministic Turing machines (Russian)* ◇ D10 D15 ◇

ALBERTS, M. *Space complexity of alternating Turing machines* ◇ D10 D15 ◇

ALBERTS, M. *Tape complexity of nondeterministic Turing machines (Russian)* ◇ D10 D15 ◇

BUD'KO, A.E. *Two classes of Turing machines (Russian) (English summary)* ◇ D10 ◇

FREJVALD, R.V. *Space and reversal complexity of probabilistic one-way Turing machines* ◇ D10 ◇

GRANDJEAN, E. *Universal quantifiers and time complexity of random access machines* ◇ C13 D10 D15 ◇

HROMKOVIC, J. *On the power of alternation in automata theory* ◇ D05 D10 ◇

INOUE, K. & ITO, A. & TAKANAMI, I. & TANIGUCHI, H. *A space-hierarchy result on two-dimensional alternating Turing machines with only universal states* ◇ D10 D15 ◇

LADNER, R.E. & NORMAN, J.K. *Solitaire automata* ◇ D10 ◇

LITOW, B.E. *On efficient deterministic simulation of Turing machine computations below log-space* ◇ D10 D15 ◇

MAASS, W. *Combinatorial lower bound arguments for deterministic and nondeterministic Turing machines* ◇ D10 D15 ◇

MEYER AUF DER HEIDE, F. *Lower bounds for solving linear Diophantine equations on random access machines* ◇ D10 D15 ◇

MULLER, D.E. & SCHUPP, P.E. *The theory of ends, pushdown automata, and second-order logic* ◇ B15 B25 D10 D80 ◇

RYTTER, W. *The complexity of two-way pushdown automata and recursive programs* ◇ D10 D15 ◇

SCARPELLINI, B. *Lower bound results on lengths of second-order formulas* ◇ B15 C13 D10 F20 F35 ◇

VITANYI, P.M.B. *An* $n^1.618$ *lower bound on the time to simulate one queue or two pushdown stores by one tape* ◇ D10 D15 ◇

VOGLER, H. *Iterated linear control and iterated one-turn pushdowns* ◇ D10 ◇

YAKU, T. *Wiring a Turing machine in the cellular automaton and the surjectivity problem for a parallel map* ◇ D05 D10 ◇

D15 Complexity of computation

1956

MULLER, D.E. *Complexity in electronic switching circuits* ◇ B70 D15 ◇

1960

RABIN, M.O. *Degree of difficulty of computing a function and a partial ordering of recursive sets* ◇ D15 D20 ◇

1962

YAMADA, H. *Real-time computation and recursive functions not real-time computable* ◇ D15 ◇

1963

RABIN, M.O. *Real time computation* ◇ D05 D10 D15 ◇

RITCHIE, R.W. *Classes of predictably computable functions* ◇ D15 D20 ◇

1964

HARTMANIS, J. & STEARNS, R.E. *Computational complexity of recursive sequences* ◇ D15 ◇

KLOSS, B.M. *The definition of complexity of algorithms (Russian)* ◇ D15 D20 ◇

MELZAK, Z.A. *An informal arithmetical approach to computability and computation II* ◇ D10 D15 ◇

TRAKHTENBROT, B.A. *Turing computations with logarithmic delay (Russian)* ◇ D10 D15 ◇

ZHURAVLEV, YU.I. *Estimate of complexity of local algorithms for some extremum problems on finite sets (Russian)* ◇ D15 ◇

1965

ARBIB, M.A. & BLUM, M. *Machine dependence of degree of difficulty* ◇ D15 ◇

BARZDINS, J. *Complexity of recognition of symmetry on Turing machines (Russian)* ◇ D10 D15 ◇

BECVAR, J. *Real-time and complexity problems in automata theory* ◇ D05 D10 D15 ◇

COBHAM, A. *The intrinsic computational difficulty of functions* ◇ D15 ◇

FISCHER, P.C. & RUBY, S.S. *Translational methods and computational complexity* ◇ B70 D15 D20 ◇

FREJVALD, R.V. *Complexity of recognition of symmetry on Turing machines with input (Russian)* ◇ D10 D15 ◇

HARTMANIS, J. & LEWIS, P.M. & STEARNS, R.E. *Classification of computation by time and memory requirements* ◇ D15 ◇

HARTMANIS, J. & LEWIS II, P.M. & STEARNS, R.E. *Hierarchies of memory limited computations* ◇ D10 D15 ◇

HARTMANIS, J. & LEWIS II, P.M. & STEARNS, R.E. *Memory bounds for recognition of context-free and context-sensitive languages* ◇ B65 D05 D15 ◇

HARTMANIS, J. & STEARNS, R.E. *On the computational complexity of algorithms* ◇ D15 ◇

KOLMOGOROV, A.N. *Three approaches to the definition of the concept "quantity of information" (Russian)* ◇ D15 D20 D80 ◇

MATVEEVA, G.S. *On a theorem of Rabin concerning the complexity of computable functions (Russian)* ◇ D15 ◇

ROWICKI, A. *Remarks on order of simple processes* ◇ D10 D15 ◇

TRAKHTENBROT, B.A. *Optimal computations and the frequency occurrence of Ablonskij (Russian)* ◇ D15 ◇

1966

ARBIB, M.A. *Speed-up theorems and incompleteness theorems* ◇ D15 F30 ◇

BLUM, M. *Recursive function theory and speed of computation* ◇ D15 D20 ◇

FREJVALD, R.V. *On the order of growth of exact time signalizers for Turing calculations (Russian)* ◇ D10 D15 ◇

HANAK, J. *On real-time Turing machines* ◇ D10 D15 ◇

HUZINO, S. *On the simulation of real-time Turing machines by a modified Shepherdson-Sturgis' machine* ◇ D05 D10 D15 ◇

JAULIN, B. *Sur un aspect du calcul* ◇ B75 D15 D80 ◇

TRAKHTENBROT, B.A. *Normed signalizers for Turing computations (Russian)* ◇ D10 D15 ◇

1967

BLUM, M. *A machine-independent theory of the complexity of recursive functions* ◇ D15 ◇

BLUM, M. *On the size of machines* ◇ D15 ◇

KHODZHAEV, D. *On the complexity of calculations on Turing machines with an oracle (Russian)* ◇ D10 D15 ◇

MEYER, A.R. & RITCHIE, R.W. *Computational complexity and program structure* ◇ B75 D15 ◇

MEYER, A.R. & RITCHIE, D.M. *The complexity of loop programs* ◇ B75 D15 ◇

ROWICKI, A. *On the notion of occupancy of the tree* ◇ D05 D15 ◇

TRAKHTENBROT, B.A. *Complexity of algorithms and computations (Russian)* ◇ D10 D15 D20 D98 ◇

YOUNGER, D.H. *Recognition and parsing of context-free languages in time n^3 (Russian)* ◇ D05 D15 ◇

1968

AGAFONOV, V.N. *Complexity of computation of pseudorandom sequences (Russian)* ◇ D15 D80 ◇

AHO, A.V. & ULLMAN, J.D. *The theory of languages* ◇ D05 D10 D15 ◇

BARZDINS, J. *Complexity of programs to determine whether natural numbers not greater than n belong to a recursively enumerable set (Russian)* ◇ B75 D15 D25 ◇
BLUM, M. & FISCHER, P.C. & HARTMANIS, J. *Tape reversal complexity hierarchies* ◇ D10 D15 ◇
FISCHER, P.C. & MEYER, A.R. & ROSENBERG, ARNOLD L. *Counter machines and counter languages* ◇ D05 D10 D15 D20 ◇
FISCHER, P.C. & MEYER, A.R. *On computational speed-up* ◇ D15 ◇
FREJDZON, R.I. *An index of the complexity of recursive predicates which do not depend on a standardization of the concept of algorithm (Russian)* ◇ D10 D15 ◇
FREJVALD, R.V. *Completeness up to coding of systems of functions of k-valued logic and the complexity of its determination (Russian)* ◇ B50 D15 ◇
HARTMANIS, J. *Computational complexity of one-tape Turing machine computations* ◇ D10 D15 D20 ◇
HARTMANIS, J. *Tape-reversal bounded Turing machine computations* ◇ D10 D15 ◇
HOPCROFT, J.E. & ULLMAN, J.D. *Relations between time and tape complexities* ◇ D05 D10 D15 ◇
JONES, N.D. *Classes of automata and transitive closure* ◇ D10 D15 ◇
MEYER, A.R. & RITCHIE, D.M. *Classification of functions by computational complexity* ◇ D15 D20 ◇
OCHRANOVA-DOLEZELOVA, R. *Real-time decidability, computability, countability and generability* ◇ D10 D15 ◇
RENNIE, M.K. *A function which bounds truth-tabular calculations in S5* ◇ B25 B45 D15 ◇
STRNAD, P. *On-line Turing machine recognition* ◇ D10 D15 ◇

1969

AHO, A.V. & HOPCROFT, J.E. & ULLMAN, J.D. *On the computational power of pushdown automata* ◇ D10 D15 ◇
BENESOVA, M. *Real-time computable functions and almost primitive recursive functions* ◇ D15 D20 ◇
BORODIN, A. *Complexity classes of recursive functions and the existence of complexity gaps* ◇ D15 ◇
BORODIN, A. & CONSTABLE, R.L. & HOPCROFT, J.E. *Dense and non-dense families of complexity classes* ◇ D15 ◇
CHAITIN, G.J. *On the length of programs for computing finite binary sequences: statistical considerations* ◇ D15 D80 ◇
CHAITIN, G.J. *On the simplicity and speed of programs for computing infinite sets of natural numbers* ◇ D15 D80 ◇
GALLAIRE, H. *Recognition time for context-free languages by on-line Turing machines* ◇ D05 D10 D15 ◇
HARTMANIS, J. *On the complexity of undecidable problems in automata theory* ◇ D05 D15 ◇
HOREJS, J. *Recursive functions computable within $C\bar{f}\log\bar{f}$ (Czech summary)* ◇ D15 ◇
KANOVICH, M.I. *Estimates of the reduction complexity of algorithms (Russian)* ◇ B75 D15 ◇
KANOVICH, M.I. & KUSHNER, B.A. *Estimating the complexity of certain algorithmic problems of analysis (Russian)* ◇ D15 D20 D80 F60 ◇
KANOVICH, M.I. *On the decision complexity of algorithms (Russian)* ◇ B75 D15 ◇
KANOVICH, M.I. & PETRI, N.V. *Some theorems on the complexity of normal algorithms and computations (Russian)* ◇ D03 D15 ◇
KOLMOGOROV, A.N. *Logical basis of information theory and probability theory (Russian)* ◇ B30 D15 D20 D80 ◇
KOREC, I. *A complexity valuation of the partial recursive functions following the expectation of the length of their computations on Minsky machines* ◇ D10 D15 ◇
MARANDZHYAN, G.B. *Certain properties of asymptotically optimal recursive functions (Russian) (Armenian and English summaries)* ◇ D15 D20 ◇
MARANDZHYAN, G.B. *Hierarchies of recursive functions and asymptotic optimality (Russian) (Armenian summary)* ◇ D15 D20 ◇
MARANDZHYAN, G.B. *On complexity scales of natural numbers (Russian)* ◇ D15 D20 ◇
MCCREIGHT, E.M. & MEYER, A.R. *Classes of computable functions defined by bounds on computation: preliminary report* ◇ D15 D20 ◇
MCCREIGHT, E.M. & MEYER, A.R. *Properties of bounds on computation* ◇ D15 D20 ◇
MOSHCHENSKIJ, V.A. *On the question of the complexity of Turing computations (Russian)* ◇ D10 D15 D20 ◇
PETRI, N.V. *Algorithms connected with predicates and Boolean functions (Russian)* ◇ B35 D15 ◇
PETRI, N.V. *The complexity of algorithms and their operating time (Russian)* ◇ B35 D15 D20 ◇
PETRI, N.V. *Two theorems on the complexity of algorithms and computations (Russian)* ◇ D15 D25 ◇
STRNAD, P. *On optimum time bounds for recognition of some sets of words by on-line Turing machines* ◇ D10 D15 ◇
TRAKHTENBROT, B.A. *On the complexity of reduction algorithms in Novikov-Boone constructions (Russian)* ◇ D15 D40 ◇
YOUNG, P. *Toward a theory of enumerations* ◇ D10 D15 D20 D25 D45 ◇
ZASLAVSKIJ, I.D. *On Shannon pseudofunctions (Russian)* ◇ D05 D15 F60 ◇

1970

BASS, L. & YOUNG, P. *Hierarchies based on computational complexity and irregularities of class determined measured sets* ◇ D15 D20 ◇
BECVAR, J. *Programmkomplexitaet von Funktionen und Mengen* ◇ D15 ◇
BECVAR, J. *Programmkomplexitaet von berechenbaren Funktionen* ◇ D15 ◇
BIELECKI, J. *The effect of the geometry of the tape of the Turing machine on the number of instructions in the program controlling it (Polish) (Russian and English summaries)* ◇ D10 D15 ◇
FISCHER, P.C. & MEYER, A.R. & ROSENBERG, ARNOLD L. *Time-restricted sequence generation* ◇ D10 D15 ◇

KAMEDA, T. & VOLLMAR, R. *Zur Umkehrkomplexitaet von Sprachen* ◇ D05 D10 D15 ◇
KANOVICH, M.I. *On the complexity of enumeration and decision of predicates (Russian)* ◇ D15 ◇
KANOVICH, M.I. *On the decision complexity of recursively enumerable sets (Russian)* ◇ D15 D25 ◇
KATERINOCHKINA, N.N. *The equivalence of certain computational equipment (Russian) (English summary)* ◇ D10 D15 D20 ◇
KHOMICH, V.I. *The complexity of the algorithms that are connected with the realization of logico-arithmetical and propositional formulae (Russian)* ◇ D15 ◇
LITVINTSEVA, Z.K. *On the complexity of individual identity problems in semigroups (Russian)* ◇ D15 D40 ◇
LITVINTSEVA, Z.K. *On the complexity of some problems for groups and semi-groups (Russian)* ◇ D15 D40 ◇
PAGER, D. *On the efficiency of algorithms* ◇ D15 ◇
SAVITCH, W.J. *Relationship between nondeterministic and deterministic tape complexities* ◇ D10 D15 ◇
SCHMITT, A.A. *Die Zustands-Komplexitaetsklassen von Turingmaschinen* ◇ D10 D15 ◇
SCHNORR, C.-P. *Minimale Programmkomplexitaet und Zufaelligkeit* ◇ D15 D80 ◇
STRONG, H.R. *Depth-bounded computation* ◇ D15 D20 ◇
VALIEV, M.K. *Certain estimates of the computing time on Turing machines with input (Russian) (English summary)* ◇ D10 D15 ◇
WERNER, GEORGES *Quelques remarques sur la complexite des algorithmes (English summary)* ◇ D15 D20 ◇
WILLIS, D.G. *Computational complexity and probability constructions* ◇ D10 D15 ◇

1971

AUSIELLO, G. *Abstract computational complexity and cycling computations* ◇ D15 D75 ◇
BLUM, M. *On effective procedures for speeding up algorithms* ◇ D15 ◇
BOUCHER, C. *Lecons sur la theorie des automates mathematiques* ◇ D05 D10 D15 D98 ◇
BREJTBART, YU.YA. *On the automaton and "zone" complexity of the predicate "to be a k-th power of an integer" (Russian)* ◇ D05 D15 ◇
BURKHARD, W.A. & VARAIYA, P.P. *Complexity problems in real time languages* ◇ D05 D15 ◇
CONSTABLE, R.L. *Subrecursive-programming languages III: the multiple-recursive functions* ◇ D15 D20 ◇
COOK, S.A. *Characterizations of pushdown machines in terms of time-bounded computers* ◇ D10 D15 ◇
COOK, S.A. *The complexity of theorem proving procedures* ◇ D15 ◇
DALEY, R.P. *Recursive and pseudo-recursive sequence and the time bounded minimal-program complexity hierarchies* ◇ D15 ◇
EMDE BOAS VAN, P. *A note on the Meyer-McCreight naming theorem in the theory of computational complexity* ◇ D15 ◇
HARTMANIS, J. & HOPCROFT, J.E. *An overview of the theory of computational complexity* ◇ D15 D98 ◇
HARTMANIS, J. *Size arguments in the study of computation speeds* ◇ D15 ◇
HAVEL, I.M. *Weak complexity measures* ◇ D15 ◇
HELM, J.P. & YOUNG, P. *On size vs. efficiency for programs admitting speed-ups* ◇ D15 ◇
IBARRA, O.H. *Characterizations of some tape and time complexity classes of Turing machines in terms of multihead and auxiliary stack automata* ◇ D10 D15 ◇
KANOVICH, M.I. *On complexity of Boolean function minimzation (Russian)* ◇ B35 D15 ◇
KANOVICH, M.I. *On domains of definition of optimal algorithms (Russian)* ◇ B75 D15 ◇
LEWIS, F.D. *The enumerability and invariance of complexity classes* ◇ D15 D25 ◇
MACHTEY, M. *Classification of computable functions by primitive recursive classes* ◇ D15 D20 ◇
MARANDZHYAN, G.B. *Lattices of blocks of recursive functions (Russian) (Armenian summary)* ◇ D15 D20 ◇
MATIYASEVICH, YU.V. *Real-time recognition of the inclusion relation (Russian) (English summary)* ◇ D10 D15 ◇
MCCREIGHT, E.M. & MEYER, A.R. *Computationally complex and pseudo-random zero-one valued functions* ◇ D15 ◇
MOSHCHENSKIJ, V.A. *The estimation of certain functions that characterize the performance of Turing machines (Russian) (English summary)* ◇ D10 D15 ◇
PARIKH, R. *Existence and feasibility in arithmetic* ◇ A05 D15 F30 F65 H15 ◇
ROBERTSON, E.L. *Complexity classes of partial recursive functions* ◇ D15 D20 ◇
SCHNORR, C.-P. *Komplexitaet von Algorithmen mit Anwendung auf die Analysis* ◇ D15 D80 ◇
TOGER, A.V. *On the complexity of some functional classes (Russian)* ◇ D15 ◇
TSEJTIN, G.S. *Lower estimate of the number of steps for an inverting normal algorithm and other similar algorithms (Russian) (English summary)* ◇ D03 D15 ◇
TSEJTIN, G.S. *Reduced form of normal algorithms and a linear acceleration theorem (Russian) (English summary)* ◇ D03 D15 ◇
TUZOV, V.A. *Optimal schemes of algorithms (Russian)* ◇ D15 D20 ◇
WEICKER, R. *Tabulator-Turingmaschinen und Komplexitaet (English summary)* ◇ D10 D15 ◇
WERNER, GEORGES *Propriete d'invariance des classes de fonctions de complexite bornee* ◇ D15 D20 ◇
YOUNG, P. *A note on dense and nondense families of complexity classes* ◇ D15 ◇
YOUNG, P. *A note on "axioms" for computational complexity and computation of finite functions* ◇ D15 ◇
YOUNG, P. *Speed-up changing the order in which sets are enumerated* ◇ D15 D20 D25 ◇

1972

BAGCHI, A. & MEYER, A.R. *Program size and economy of descriptions: Preliminary report* ◇ D15 D25 ◇
BORODIN, A. *Computational complexity and the existence of complexity gaps* ◇ D15 ◇

BULNES, J. *On the speed of addition and multiplication on one-tape, off-line Turing machines* ◇ D10 D15 ◇
CONSTABLE, R.L. *Constructive mathematics and automatic program writers* ◇ B75 D15 F55 ◇
CONSTABLE, R.L. *Subrecursive programming languages I: Efficiency and program structure* ◇ B75 D15 ◇
CONSTABLE, R.L. *The operator gap* ◇ D15 ◇
COOK, S.A. & RECKHOW, R.A. *Time-bounded random access machines* ◇ D10 D15 ◇
DEMBINSKI, P. *Equivalence of programs and their complexity* ◇ B75 D15 D20 ◇
EMDE BOAS VAN, P. *Gap and operator gap* ◇ D15 ◇
ENDERTON, H.B. *Degrees of computational complexity* ◇ D15 ◇
FILOTTI, I. *On effectively levelable sets* ◇ D15 ◇
FISCHER, P.C. & MEYER, A.R. *Computational speed-up by effective operators* ◇ D15 ◇
FISCHER, P.C. & MEYER, A.R. & ROSENBERG, ARNOLD L. *Real-time simulation of multihead tape units* ◇ D10 D15 ◇
IBARRA, O.H. *A note concerning nondeterministic tape complexities* ◇ D10 D15 ◇
JONES, N.D. & SELMAN, A.L. *Turing machines and the spectra of first-order formulas with equality* ◇ C13 D10 D15 ◇
KANOVICH, M.I. *Complexity of bounded decidability of semirecursively enumerable sets (Russian)* ◇ D15 D25 ◇
KANOVICH, M.I. *On the universality of strongly undecidable sets (Russian)* ◇ D15 D20 D25 D30 ◇
KARP, R.M. *Reducibility among combinatorial problems* ◇ D15 ◇
KONIKOWSKA, B. *Formalisierung des Begriffs der stetigen Maschine (Polish) (Russian and English summaries)* ◇ D15 D20 ◇
KOREC, I. *The mathematical expectation of the time for computations of partial recursive functions on Minsky machines (Russian) (German, English, and French summaries)* ◇ D10 D15 ◇
KUBINETS, M.V. *Recognition of a self-crossing of a plane trajectory by a Kolmogorov algorithm (Russian)* ◇ D10 D15 ◇
LANDWEBER, L.H. & ROBERTSON, E.L. *Recursive properties of abstract complexity classes* ◇ D15 ◇
LEDGARD, H.F. *Embedding Markov normal algorithms within the λ-calculus* ◇ B40 D03 D15 ◇
MARANDZHYAN, G.B. *The strongly effective immunity of the pivots of additively optimal recursive functions (Russian) (Armenian and English summaries)* ◇ D15 D50 ◇
MAZURKIEWICZ, A.W. *Iteratively computable relations* ◇ D15 D20 ◇
MEYER, A.R. & MOLL, R. *Honest bounds for complexity classes of recursive functions* ◇ D15 D20 ◇
MOSHCHENSKIJ, V.A. *Zur Analyse von Turing-berechnungen (Russian)* ◇ D10 D15 ◇
MUELLER, HORST *Jede mit Stackautomaten berechenbare Funktion ist elementar* ◇ D10 D15 D20 ◇
NEKVINDA, M. *On a certain event recognizable in real time (Czech summary)* ◇ D05 D15 ◇
PATERSON, M.S. *Tape bounds for time-bounded Turing machines* ◇ D10 D15 ◇
ROEDDING, D. & SCHWICHTENBERG, H. *Bemerkungen zum Spektralproblem* ◇ B15 C13 D10 D15 ◇
SAVITCH, W.J. *Maze recognizing automata* ◇ D10 D15 ◇
STRNAD, P. *A hierarchy of optimal recognitions (Russian) (German, English and French summaries)* ◇ D15 D20 ◇
SYMES, D. *The computation of finite functions* ◇ D15 D20 ◇
ZHAROV, V.G. *The complexity of the terms of constructive sequences of normal algorithms (Russian)* ◇ D03 D15 ◇

1973

BASS, L. & YOUNG, P. *Ordinal hierarchies and naming complexity classes* ◇ D15 D20 ◇
BICHEVSKIJ, YA.YA. *An estimate of the complexity of bounded mass problems (Russian) (Latvian and English summaries)* ◇ D15 ◇
BIRKHOFF, GARRETT *Current trends in algebra* ◇ A10 D05 D15 ◇
BLUM, L. & BLUM, M. *Inductive inference: a recursion theoretic approach* ◇ B48 D15 D80 ◇
BLUM, M. & MARQUES, I. *On complexity properties of recursively enumerable sets* ◇ D15 D25 ◇
BLUM, M. & GILL III, J.T. *Some fruitful areas for research into complexity theory* ◇ D15 ◇
BORODIN, A. *Computational complexity: theory and practice* ◇ D15 ◇
CHERNIAVSKY, J.C. *The complexity of some non-classical logics* ◇ B45 D15 F20 F50 ◇
CHYTIL, M.P. *On changes of input/output coding I* ◇ D15 D20 ◇
COHORS-FRESENBORG, E. *Elementare und subelementare Funktionenklassen ueber binaeren Baeumen* ◇ D15 D20 ◇
CONSTABLE, R.L. *Two types of hierarchy theorem on axiomatic complexity classes* ◇ D15 ◇
COOK, S.A. *A hierarchy for nondeterministic time complexity* ◇ D15 ◇
COOK, S.A. & RECKHOW, R.A. *Time bounded random access machines* ◇ D10 D15 ◇
CULIK, K. *Main degrees of complexity of computer programs and computable functions* ◇ B75 D15 D20 ◇
DALEY, R.P. *An example of information and computation resource trade-off* ◇ D15 ◇
DALEY, R.P. *Complexity and randomness* ◇ D15 ◇
DALEY, R.P. *Minimal-program complexity of sequences with restricted resources* ◇ D15 ◇
DALEY, R.P. *On the learning of non-recursive sequences* ◇ D15 ◇
DALEY, R.P. *The process complexity in the understanding of sequences* ◇ D15 ◇
EMDE BOAS VAN, P. *A comparison of the properties of complexity classes and honesty classes* ◇ D15 ◇
FISCHER, P.C. & IRLAND, M.I. *A bibliography on computational complexity* ◇ A10 D15 D99 ◇

FISCHER, P.C. *Trends in computational complexity theory* ◇ D15 ◇
FLAJOLET, P. & STEYAERT, J.-M. *Complexite des problemes de decision relatifs aux algorithmes de tri* ◇ D05 D15 ◇
GLADKIJ, A.V. *Formale Grammatiken und Sprachen (Russisch)* ◇ D05 D10 D15 ◇
HECKER, H.-D. *Kompliziertheitsmasse bei Turingmaschinen* ◇ D10 D15 ◇
HELM, J.P. & MEYER, A.R. & YOUNG, P. *On orders of translations and enumerations* ◇ D15 D20 D25 D45 F40 ◇
KAMAE, T. *On Kolmogorov's complexity and information* ◇ D15 D20 ◇
KANOVICH, M.I. *Irreducibility of languages of the stepped semantic system (Russian)* ◇ D15 D30 F50 ◇
KANOVICH, M.I. *Komplexitaet der beschraenkten Entscheidbarkeit von Algorithmen (Russian)* ◇ D05 D15 D20 ◇
KHRAPCHENKO, V.M. *A quadratic estimate from below for complexity that is based on the continuity of the second derivative (Russian)* ◇ D15 ◇
LEVIN, L.A. *On the notion of a random sequence (Russian)* ◇ D15 D80 ◇
LEVIN, L.A. *Universal problems of full search (Russian)* ◇ D15 ◇
MACHTEY, M. *A notion of helping and pseudo-complementation in lattices of honest subrecursive classes* ◇ D15 D20 ◇
MARANDZHYAN, G.B. *On algorithms of minimal complexity (Russian)* ◇ D15 ◇
MARANDZHYAN, G.B. *The complexities of the representation of natural numbers by means of recursive functions (Russian)* ◇ D15 D20 ◇
MEHLHORN, K. *On the size of sets of computable functions* ◇ D15 D20 ◇
MEYER, A.R. & STOCKMEYER, L.J. *Word problems requiring exponential time: preliminary report* ◇ D15 D40 ◇
MONIEN, B. *Relationships between pushdown automata and tape-bounded Turing machines* ◇ D10 D15 ◇
MOON, B.A. *A Markov algorithm interpreter* ◇ D03 D15 ◇
MORAVEK, J. *Computational optimality of a dynamic programming method* ◇ B40 D15 ◇
MORGENSTERN, J. *Algorithmes lineaires tangents et complexite* ◇ D15 ◇
NAGASAKA, K. *On minimal-program complexity measure* ◇ D15 ◇
NEKVINDA, M. *On the complexity of events recognizable in real time* ◇ D05 D15 ◇
OPPEN, D.C. *Elementary bounds for Presburger arithmetic* ◇ B25 C10 D15 F30 ◇
PATERSON, M.S. & VALIANT, L.G. *Deterministic one-counter automata* ◇ D05 D15 ◇
RASIOWA, H. *Formalized ω^+ valued algorithmic systems (Russian summary)* ◇ B50 B75 D15 ◇
RUSTIN, R. (ED.) *Computational complexity. Courant computer science symposium* ◇ D15 D97 ◇
SALOMAA, A. *Formal languages* ◇ D05 D15 D98 ◇
SALOMAA, A. *On exponential growth in Lindenmayer systems* ◇ D05 D15 ◇
SAVITCH, W.J. *Maze recognizing automata and nondeterministic tape complexity* ◇ D10 D15 ◇
SAVITCH, W.J. *Nondeterministic finite automata revisited* ◇ D05 D15 ◇
SCHNORR, C.-P. *Process complexity and effective random tests* ◇ D15 ◇
SLISENKO, A.O. *Identification of the symmetry predicate by means of multihead Turing machines with input (Russian)* ◇ D10 D15 ◇
TETRUASHVILI, M.R. *A realization of the Magnus algorithm on a Turing machine, and an upper bound of the complexity of the computations (Russian) (Georgian and English summaries)* ◇ D10 D15 D40 ◇
TITARENKO, L.N. *The complexity of recognizing functional properties of normal algorithms (Russian)* ◇ D03 D15 ◇
TRAKHTENBROT, B.A. *A formalization of certain concepts in terms of complexity of computations* ◇ D15 ◇
TRAKHTENBROT, B.A. *Frequency computations (Russian)* ◇ D10 D15 D20 ◇
VERBEEK, R. *Erweiterungen subrekursiver Programmiersprachen* ◇ B75 D15 D20 ◇
VOELKEL, L. *Untersuchungen ueber die Kompliziertheit von Berechnungs- und Tabellenprogrammen* ◇ D15 D20 ◇
YOUNG, P. *Easy constructions in complexity theory: Gap and speed-up theorems* ◇ D15 D20 ◇

1974

AANDERAA, S.O. *On k-tape versus (k - 1)-tape real time computation* ◇ D10 D15 ◇
ADRIANOPOLI, F. & LUCA DE, A. *Closure operations on measures of computational complexity* ◇ D15 ◇
ALTON, D.A. & LOWTHER, J.L. *Non-existence of program optimizers in an abstract setting* ◇ B75 D15 ◇
AUSIELLO, G. *Relations between semantics and complexity of recursive programs* ◇ B40 B75 D15 D20 ◇
BARASHKO, A.S. *Tape complexity of languages that can be recognized by two-way finite-term pushdown automata (Russian) (English summary)* ◇ D10 D15 ◇
BARASHKO, A.S. *The number of push-down tape reversals of a two-way push-down store automaton (Russian)* ◇ D10 D15 ◇
BARTOL, W. *Algebraic complexity of machines* ◇ D10 D15 ◇
BARZDINS, J. *A property of limiting computable functionals (Russian)* ◇ D15 D20 ◇
BARZDINS, J. *Limitare Synthese von τ-Zahlen (Russian)* ◇ D15 D20 ◇
BARZDINS, J. & KINBER, E.B. & PODNIEKS, K.M. *On speed-up of synthesis and prognostication of functions (Russian)* ◇ D15 D20 ◇
BARZDINS, J. & FREJVALD, R.V. *Prognostizierung und limitaere Synthese effektiv aufzaehlbarer Klassen von Funktionen (Russian)* ◇ D15 D20 D45 ◇
BLOKH, A.SH. *Ueber die Kompliziertheit von Graph-Schemata (Russisch)* ◇ B75 D15 ◇

BLUM, M. & GILL III, J.T. *On almost everywhere complex recursive functions* ◇ D15 D20 ◇
BOEHLING, K.H. & BRAUNMUEHL VON, B. *Komplexitaet bei Turingsmaschinen* ◇ D10 D15 D20 ◇
BOOK, R.V. *Comparing complexity classes* ◇ D10 D15 ◇
BOOK, R.V. *On the structure of complexity classes* ◇ D10 D15 ◇
BOOK, R.V. *Tally languages and complexity classes* ◇ D10 D15 ◇
BRUDNO, A.A. *Topologische Entropie und Kompliziertheit nach A. N. Kolmogorov (Russisch)* ◇ D15 D80 ◇
BUCHBERGER, B. *On certain decompositions of Goedel numberings* ◇ D15 D20 D45 ◇
CHAITIN, G.J. *Information-theoretic limitations of formal systems* ◇ D15 D80 F20 ◇
CHAITIN, G.J. *Information-theoretic computational complexity* ◇ D15 D80 ◇
CHYTIL, M.P. *On changes of input/output coding II* ◇ D15 D20 ◇
COOK, S.A. & RECKHOW, R.A. *On the length of proofs in the propositional calculus: preliminary version* ◇ B05 D15 F20 ◇
COOK, S.A. & SETHI, R. *Storage requirements for deterministic polynominal time recognizable languages* ◇ D15 ◇
DALEY, R.P. *Non-complex sequences: characterizations and examples* ◇ D15 ◇
DALEY, R.P. *The extent and density of sequences within the minimal-program complexity hierarchies* ◇ D15 ◇
DEKHTYAR', M.I. *On the complexity of relative computations (Russian)* ◇ D15 D20 ◇
FAGIN, R. *Generalized first-order spectra and polynomial-time recognizable sets* ◇ C13 C85 D15 ◇
FISCHER, MICHAEL J. & RABIN, M.O. *Super-exponential complexity of Presburger arithmetic* ◇ B25 D15 F20 F30 ◇
FISCHER, P.C. *Further schemes for combining matrix algorithms* ◇ D15 ◇
FLAJOLET, P. & STEYAERT, J.-M. *On sets having only hard subsets* ◇ D15 D25 ◇
FROLOV, G.D. & KRINITSKIJ, N.A. & MIRONOV, G.A. *Certain criteria for the complexity of systems (Russian)* ◇ D15 ◇
GILL III, J.T. *Computational complexity of probabilistic Turing machines* ◇ D10 D15 ◇
GINSBURG, S. & ROSE, G.F. *The equivalence of stack-counter acceptors and quasi-realtime stack-counter acceptors* ◇ D05 D10 D15 ◇
GRODZKI, Z. *The Kolmogorov's complexity of the computation of the k-machines* ◇ D03 D15 D20 ◇
HAMLET, R.G. *Introduction to computation theory* ◇ D10 D15 D25 D98 ◇
HARTMANIS, J. & HUNT III, H.B. *The LBA problem and its importance in the theory of computing* ◇ D05 D15 ◇
HONDA, N. & IGARASHI, Y. *Deterministic multitape automata computations* ◇ D10 D15 ◇
HOPCROFT, J.E. & ULLMAN, J.D. *Some results on tape-bounded Turing machines (Russian)* ◇ D10 D15 ◇
HOTZ, G. *Komplexitaetsmasse fuer Ausdruecke (English summary)* ◇ D15 ◇
IBARRA, O.H. *A hierarchy theorem for polynomial-space recognition* ◇ D10 D15 ◇
JONES, N.D. & SELMAN, A.L. *Turing machines and the spectra of first-order formulas* ◇ C13 D10 D15 ◇
KANOVICH, M.I. *A certain extension of the step-by-step semantic system of A. A. Markov (Russian)* ◇ D15 F50 ◇
KANOVICH, M.I. *A theorem on speeding up in formal systems (Russian)* ◇ B35 D15 ◇
KANOVICH, M.I. *Complexity of the limit of Specker sequences (Russian)* ◇ D15 D20 F60 ◇
KARP, R.M. (ED.) *Complexity of computation* ◇ D15 D97 ◇
KOZMIDIADI, V.A. & MASLOV, A.N. & PETRI, N.V. (EDS.) *Complexity of computations and algorithms (Russian)* ◇ D15 D20 D98 ◇
LADNER, R.E. & LYNCH, N.A. & SELMAN, A.L. *Comparison of polynominal-time reducibilities* ◇ D15 D20 ◇
LEVIN, L.A. *Bezeichnung rekursiver Funktionen (Russisch)* ◇ D15 ◇
LONGO, G. *I problemi di decisione e la loro complessita* ◇ B25 B35 D15 F20 ◇
LUPANOV, O.B. *Methods to obtain complexity and calculability estimations for individual functions (Russian)* ◇ D10 D15 D20 ◇
MACHTEY, M. *The honest subrecursive classes are a lattice* ◇ D15 D20 ◇
MEHLHORN, K. *Polynomial and abstract subrecursive classes* ◇ D15 D20 ◇
METROPOLIS, N. & ROTA, G.-C. *Significance arithmetic on the algebra of binary strings* ◇ D15 F65 G05 ◇
MEYER, A.R. & MOLL, R. *Honest bounds for complexity classes of recursive functions* ◇ D15 D20 ◇
MONIEN, B. *Beschreibung von Zeitkomplexitaetsklassen bei Turingmaschinen durch andere Automatenmodelle* ◇ D10 D15 ◇
NOLIN, L. *Algorithmes universels* ◇ B75 D15 ◇
POGOSYAN, EH.M. *1-inductors and some of their properties (Russian) (Armenian summary)* ◇ D15 ◇
RABIN, M.O. *Theoretical impediments to artificial intelligence* ◇ B75 D15 D80 ◇
RACKOFF, C.W. *On the complexity of the theories of weak direct products: a preliminary report* ◇ B25 C13 C30 C60 D15 ◇
ROBERTSON, E.L. *Complexity classes of partial recursive functions* ◇ D15 D20 ◇
ROBERTSON, E.L. *Structure of complexity in the weak monadic second-order theories of the natural numbers* ◇ D15 D55 F35 ◇
SCHNORR, C.-P. *Optimal enumerations and optimal Goedel numberings* ◇ D15 D20 D25 D45 ◇
SCHNORR, C.-P. *Rekursive Funktionen und ihre Komplexitaet* ◇ D10 D15 D20 D98 ◇
SCHUBERT, L.K. *Iterated limiting recursion and the program minimization problem* ◇ B75 D15 D20 ◇
SCHUBERT, L.K. *Representative samples of programmable functions* ◇ B75 D15 D20 ◇

SEIFERAS, J.I. *Observations on nondeterministic multidimensional iterative arrays* ◇ D10 D15 ◇
SHURYGIN, V.A. *Einige Eigenschaften der Kompliziertheit konstruktiver reeller Zahlen (Russian)* ◇ D15 F60 ◇
TRAKHTENBROT, B.A. *Bemerkungen ueber die Kompliziertheit von Berechnungen auf stochastischen Automaten (Russisch)* ◇ D10 D15 ◇
WEICKER, R. *Turing machines with associative memory access* ◇ D10 D15 ◇
WERNER, GEORGES *Sous-classes recursivement enumerables d'une classe de complexite* ◇ D15 D20 ◇
ZHAROV, V.G. *The complexity of the universal algorithm (Russian)* ◇ D03 D15 ◇

1975

AGAFONOV, V.N. *Complexity of algorithms and computations. Part II (Russian)* ◇ B75 D15 D98 ◇
ALAD'EV, V.Z. & OSIPOV, O. *About the complexity of parallel algorithms defined by homogeneous structures (Russian) (Estonian and English summaries)* ◇ D03 D05 D10 D15 ◇
AUGUSTON, M. *Universal Buchberger automata and real-time computations (Russian) (English summary)* ◇ D10 D15 D20 ◇
BARASHKO, A.S. & ROJZEN, S.I. *On redundance of the points of partial recursive functions naming the computational complexity classes (Ukrainian, English summery)* ◇ D15 D20 ◇
BENESOVA, M. & KOREC, I. *Non-linear speed-up theorem for two register Minsky machines* ◇ D10 D15 ◇
BLUM, L. & BLUM, M. *Toward a mathematical theory of inductive inference* ◇ B48 D10 D15 ◇
CHYTIL, M.P. *On complexity of nondeterministic Turing machines computations* ◇ D10 D15 ◇
DALEY, R.P. *A note on a result of Kamae* ◇ D15 ◇
DALEY, R.P. *Minimal program complexity of pseudo-recursive and pseudo-random sequences* ◇ D15 D80 ◇
DEKHTYAR', M.I. *Reducibility with limited complexity (Russian)* ◇ D15 ◇
DEKHTYAR', M.I. *Relative computations of predicates, and their complexity (Russian)* ◇ D15 ◇
EHRENFEUCHT, A. *Practical decidability* ◇ B25 D15 ◇
EMDE BOAS VAN, P. *Ten years of speedup* ◇ A10 D15 D20 F20 G05 ◇
EMDE BOAS VAN, P. *The non-renameability of honesty classes (German summary)* ◇ D15 D20 ◇
ENGELER, E. *Algorithmic logic* ◇ B25 B75 D15 ◇
FARKAS, E. *On the evaluation of nested partial functions* ◇ D15 D20 ◇
FERRANTE, J. & RACKOFF, C.W. *A decision procedure for the first order theory of real addition with order* ◇ B25 C10 C60 D15 F35 ◇
FREEDMAN, A.R. & LADNER, R.E. *Space bounds for processing contentless inputs* ◇ D10 D15 ◇
FREJDZON, R.I. *Finite approximate approach to the study of the complexity of recursive predicates (Russian)* ◇ D05 D10 D15 ◇
FREJVALD, R.V. *Fast computation by probabilistic Turing machines (Russian) (English summary)* ◇ D10 D15 ◇
FREJVALD, R.V. *On complexity and optimality of the computation in the limit (Russian) (English summary)* ◇ D15 D20 ◇
GALIL, Z. *Two fast simulations which imply some fast string matching and palindrome-recognition algorithms* ◇ D05 D10 D15 ◇
HARPER, L.H. *A note on some classes of boolean functions* ◇ B05 D15 ◇
HASKELL, R. *Efficient implementation of a class of recursively defined functions* ◇ D15 ◇
HECKER, H.-D. *Eine Bemerkung zur Berechnungskompliziertheit auf Registermaschinen* ◇ D10 D15 ◇
IBARRA, O.H. & SAHNI, S.K. *Hierarchies of Turing machines with restricted tape alphabet size* ◇ D10 D15 ◇
IVSHIN, V.YU. *Complete complexity classes of computable functions (Russian)* ◇ D15 ◇
JONES, N.D. *Space-bounded reducibility among combinatorial problems* ◇ D15 ◇
KANOVICH, M.I. *On sets of complex-programmed numbers* ◇ D15 ◇
KHOMICH, V.I. *Weakly and strongly nonrealizable propositional formulae (Russian)* ◇ B20 D15 D20 F50 ◇
KINBER, E.B. *On frequency real-time computations (Russian) (English summary)* ◇ D10 D15 D20 ◇
KOSOVSKIJ, N.K. & VINOGRADOV, A.K. *Hierarchy of Diophantine representations of primitive recursive predicates (Russian)* ◇ D15 D20 ◇
LADNER, R.E. & LYNCH, N.A. & SELMAN, A.L. *A comparison of polynominal time reducibilities* ◇ D15 ◇
LAWLER, E.L. *Complexity of combinatorial computations* ◇ D15 ◇
LEONI, G. & SPRUGNOLI, R. *The compilation of Pointer Markov Algorithms* ◇ D03 D15 ◇
LISCHKE, G. *Flussbildmasse – Ein Versuch zur Definition natuerlicher Kompliziertheitsmasse* ◇ D15 ◇
LISCHKE, G. *Kompliziertheitsmasse, die gewisse Funktionalgleichungen erfuellen* ◇ D15 ◇
LISCHKE, G. *Ueber die Erfuellung gewisser Erhaltungssaetze durch Kompliziertheitsmasse* ◇ D15 D20 ◇
LUCA DE, A. *Complexity and information theory* ◇ D15 ◇
LYNCH, N.A. *"Helping": several formalizations* ◇ D15 D20 ◇
LYNCH, N.A. *On reducibility to complex or sparse sets* ◇ D15 ◇
MARQUES, I. *On degrees of unsolvability and complexity properties* ◇ D15 D20 D30 ◇
MARQUES, I. *On speedability of recursively enumerable sets* ◇ D15 D25 ◇
MATROSOV, V.L. *An analytic description of the complexity classes of computable functions (Russian)* ◇ D15 D20 ◇
MATROSOV, V.L. *The closure of certain complexity classes with respect to recursive operations (Russian)* ◇ D15 D20 ◇

MEERSMAN, R. *A survey of techniques in applied computational complexity* ◇ D15 D80 ◇

MEYER, A.R. *The inherent computational complexity of theories of ordered sets* ◇ B25 D15 ◇

MEYER, A.R. *Weak monadic second order theory of successor is not elementary recursive* ◇ B25 D10 D15 D20 ◇

MONIEN, B. *About the deterministic simulation of nondeterministic* (log *n*)*-tape bounded Turing machines* ◇ D10 D15 ◇

NABIALEK, I. & ZAKOWSKI, W. *On Z-computability of functions and sets of functions* ◇ D15 D20 ◇

OCHAKOVSKAYA, O.N. *Complexity of the derivation of equivalent Janov operator schemes (Russian) (English summary)* ◇ B75 D15 ◇

PATERSON, M.S. *Complexity of monotone networks for boolean matrix product* ◇ B70 D15 ◇

PKHAKADZE, SH.S. *A certain class of abbreviating symbols I (Russian) (Georgian summary)* ◇ B75 D15 ◇

POGOSYAN, EH.M. *Inductive inference with feedback (Russian) (Armenian summary)* ◇ B48 D15 ◇

PUDLAK, P. *Polynomially complete problems in the logic of automated discovery* ◇ B35 D15 ◇

PUDLAK, P. *The observational predicate calculus and complexity of computations* ◇ B10 D15 ◇

SAXTON, L.V. *An extension of the class of on-line machines* ◇ D10 D15 ◇

SUDBOROUGH, I.H. *On tape-bounded complexity classes and multihead finite automata* ◇ D05 D10 D15 ◇

SZYMANSKI, T.G. & ULLMAN, J.D. *Evaluating relational expressions with dense and sparse arguments* ◇ D15 D80 ◇

TETRUASHVILI, M.R. *Estimation of the complexity of a certain reduction (Russian)* ◇ D15 D40 ◇

TETRUASHVILI, M.R. *The complexity of Turing computations in terms of Magnus's algorithm (Russian) (Georgian summary)* ◇ D15 D40 ◇

TRAKHTENBROT, B.A. *On problems solvable by successive trials* ◇ D15 D50 ◇

ULLMAN, J.D. *NP-complete scheduling problems* ◇ D15 ◇

VAISER, A. *Complexity of computation and stability of separation of languages by finite probabilistic automata* ◇ D05 D15 ◇

VAISER, A. *Remarks on time and space functions of probabilistic machines* ◇ D10 D15 ◇

VALIEV, M.K. *On polynomial reducibility of the word problem under embedding of recursively presented groups in finitely presented groups* ◇ D15 D40 ◇

WAGNER, K. *Turing-Berechenbarkeit in linearer Zeit* ◇ D10 D15 ◇

WECHSUNG, G. *Minimale and optimale "Blumsche Masse" (English and Russian summaries)* ◇ D15 D20 ◇

WEIHRAUCH, K. *Program schemata with polynomial bounded counters* ◇ B75 D15 ◇

1976

ADLEMAN, L.M. & MANDERS, K.L. *Diophantine complexity* ◇ D10 D15 ◇

ADLEMAN, L.M. & MANDERS, K.L. *NP-complete decision problems for quadratic polynomials* ◇ D15 ◇

ALT, H. & MEHLHORN, K. *Lower bounds for the space complexity of context-free recognition* ◇ D05 D15 ◇

ALTON, D.A. *Diversity of speed-ups and embeddability in computational complexity* ◇ D15 ◇

AUSIELLO, G. *Difficult logical theories and their computer approximations* ◇ B25 B35 D15 ◇

BARASHKO, A.S. & ROJZEN, S.I. *Kernels of partially recursive functions that name classes of computational complexity (Russian)* ◇ D15 D20 ◇

BAUR, W. *Zeitlich beschraenkte Turingmaschinen und polynomiale Reduktion* ◇ D10 D15 ◇

BENEJAM, J.-P. & PAGET, M. *La complexite des algorithmes de Markov (English summary)* ◇ D03 D10 D15 ◇

BERMAN, L. & HARTMANIS, J. *On isomorphisms and density of NP and other complete sets* ◇ D15 ◇

BERMAN, L. *On the structure of complete sets: almost everywhere complexity and infinitely often speedup* ◇ D15 ◇

BREJTBART, YU.YA. *Some bounds on the complexity of predicate recognition by finite automata* ◇ D05 D15 ◇

CARDOZA, E. & LIPTON, R.J. & MEYER, A.R. *Exponential space complete problems for Petri nets and commutative semigroups* ◇ D15 D40 D80 ◇

CELONI, J.R. & PAUL, W.J. & TARJAN, R.E. *Space bounds for a game on graphs* ◇ D10 D15 ◇

CHAITIN, G.J. *Algorithmic entropy of sets* ◇ D15 D25 D80 ◇

CHAITIN, G.J. *Information-theoretic characterization of recursive infinite strings* ◇ D15 D20 D80 ◇

CHANDRA, A.K. & STOCKMEYER, L.J. *Alternation* ◇ D10 D15 ◇

CHERNIAVSKY, J.C. *Simple programs realize exactly Presburger formulas* ◇ B75 D15 D20 F30 ◇

CHOW, T.S. *Lower bounds on the complexity of 0-1-valued recursive functions* ◇ D15 D20 ◇

CHRISTEN, C. *Spektralproblem und Komplexitaetstheorie* ◇ B10 C13 D10 D15 D25 D35 ◇

CHYTIL, M.P. *Crossing-bounded computations and their relation to the LBA-problem* ◇ D10 D15 ◇

CHYTIL, M.P. *Return complexity of language generation* ◇ D10 D15 ◇

DEKHTYAR', M.I. *On the relativization of deterministic and nondeterministic complexity classes* ◇ D15 ◇

DEKHTYAR', M.I. *The complexity of the approximation of recursive sets (Russian) (German and English summaries)* ◇ D15 D20 ◇

DO MINH THAI *Programmkomplexitaet rekursiv aufzaehlbarer Mengen* ◇ D15 D25 D30 ◇

DUNHAM, B. & WANG, HAO *Towards feasible solutions of the tautology problem* ◇ B35 D15 ◇

EVEN, S. & TARJAN, R.E. *A combinatorial problem which is complete in polynomial space* ◇ D15 D40 ◇

FINE, T.L. *A computational complexity viewpoint on the stability of relative frequency and on the stochastic independence* ◇ D15 ◇

FISCHER, MICHAEL J. & LYNCH, N.A. & MEYER, A.R. *Relativization of the theory of computational complexity* ◇ D15 D20 ◇

FLAJOLET, P. & STEYAERT, J.-M. *Hierarchies de complexite et reductions entre problemes* ◇ D15 ◇
FLEISCHMANN, K. & MAHR, B. & SIEFKES, D. *Complexity of decision problems* ◇ D15 ◇
FRIZEN, D.G. & STARKOV, M.A. & TROFIMOV, O.E. *An automaton for calculation of the number of objects (Russian)* ◇ D05 D15 ◇
FUERER, M. *Polynomiale Transformationen und Auswahlaxiom* ◇ D15 E25 ◇
FUERER, M. *Simulation von Turingmaschinen mit logischen Netzen* ◇ B70 D10 D15 ◇
GALIL, Z. *Real-time algorithms for string-matching and palindrome recognition* ◇ D15 ◇
GAREY, M.R. & JOHNSON, D.S. & TARJAN, R.E. *The planar Hamiltonian circuit problem is NP-complete* ◇ D15 ◇
GENRICH, H.J. & THIELER-MEVISSEN, G. *The calculus of facts* ◇ B45 D15 ◇
GERMANO, G. & MAGGIOLO-SCHETTINI, A. *Recursivity, sequence recursivity, stack recursivity and semantics of programs* ◇ B75 D15 D20 ◇
GILL III, J.T. & SIMON, I. *Ink, dirty-tâpe Turing machines, and quasicomplexity measures* ◇ D10 D15 ◇
GINSBURG, S. & LYNCH, N.A. *Size complexity in context-free grammatical families* ◇ D05 D15 ◇
GRIGOR'EV, D.YU. *An application of separability and independence notions for proving lower bounds of circuit complexity (Russian) (English summary)* ◇ D15 ◇
GRIGOR'EV, D.YU. *Kolmogorov algorithms are stronger than Turing machines (Russian)* ◇ D10 D15 ◇
HAEUSSLER, A.F. *Polynomial beschraenkte nichtdeterministische Turingmaschinen und die Vollstaendigkeit des aussagelogischen Erfuellungsproblems* ◇ B05 D10 D15 ◇
HARTMANIS, J. *On effective speed-up and long proofs of trivial theorems in formal theories (French summary)* ◇ B25 B35 D15 D20 F20 ◇
HECKER, H.-D. *Zur Programmkomplexitaet rekursiv aufzaehlbarer Mengen* ◇ D15 D25 ◇
HEINTZ, JOOS *Untere Schranken fuer die Komplexitaet logischer Entscheidungsprobleme* ◇ D10 D15 ◇
HUNT III, H.B. & SZYMANSKI, T.G. *Dichotomization, reachability, and the forbidden subgraph problem* ◇ B75 D05 D15 ◇
HUWIG, H. *A machine independent description of complexity classes, definable by nondeterministic as well as deterministic Turing machines with primitive recursive tape or time bounds* ◇ D10 D15 ◇
JANTKE, K.P. *Zur Kompliziertheit der konsistenten Erkennung von Klassen allgemein-rekursiver Funktionen* ◇ D15 D20 ◇
KADANE, J.B. & SIMON, H.A. *Problems of computational complexity in artificial intelligence* ◇ B75 D15 ◇
KANOVICH, M.I. & KUSHNER, B.A. *Complexity of algorithms, and Specker sequences (Russian)* ◇ D15 D20 ◇
KANOVICH, M.I. *The relation between upper and lower bounds for the reduction complexity of enumerable sets (Russian)* ◇ D15 D25 ◇
KONDO, M. *Peut-on prolonger decimalement tout nombre naturel?* ◇ A05 D15 ◇
KOZEN, D. *On parallelism in Turing machines* ◇ D10 D15 ◇
LADNER, R.E. & LYNCH, N.A. *Relativization of questions about log space computability* ◇ D10 D15 ◇
LEMPEL, A. & ZIV, J. *On the complexity of finite sequences* ◇ D15 D20 ◇
LETICHEVS'KIJ, O.A. *Acceleration of the iteration of monotone operators (Russian)* ◇ D15 ◇
LEVIN, L.A. *On the principle of conservation of information in intuitionistic mathematics (Russian)* ◇ D15 D80 F50 F55 ◇
LEWIS, F.D. *On computational reducibility* ◇ D15 ◇
LEWIS, F.D. *Subrecursive reducibilities and completeness (Preliminary version)* ◇ D15 D20 ◇
LISCHKE, G. *Natuerliche Kompliziertheitsmasse und Erhaltungssaetze I* ◇ D15 ◇
LYNCH, N.A. *Complexity-class-encoding sets* ◇ D10 D15 D25 ◇
MACHTEY, M. *Minimal pairs of polynomial degrees with subexponential complexity* ◇ D15 ◇
MACHTEY, M. & YOUNG, P. *Simple Goedel numberings, translations, and the P-hierarchy: Preliminary report* ◇ D15 D20 D45 ◇
MATROSOV, V.L. *Complexity classes and classes of computable signaling functions (Russian)* ◇ D15 D20 ◇
MATROSOV, V.L. *Signalling computable functions for certain improvement of the complexity measure (Russian)* ◇ D15 ◇
MEHLHORN, K. *Polynomial and abstract subrecursive classes* ◇ D15 D20 ◇
MOLL, R. *An operator embedding theorem for complexity classes of recursive functions* ◇ D15 ◇
MONIEN, B. *Transformational methods and their application to complexity problems* ◇ D10 D15 ◇
MUCHNICK, S.S. *Computational complexity of multiple recursive schemata* ◇ D15 D20 ◇
NAZARYAN, G.A. *Complexity classes of sets of boolean functions (Russian) (Armenian summary)* ◇ B05 D15 ◇
NURMEEV, N.N. *Computation of boolean functions by Turing machines (Russian)* ◇ B05 D10 D15 ◇
PAPADIMITRIOU, C.H. *On the algebraic complexity of sets of functions* ◇ D15 ◇
PRATT, V.R. & STOCKMEYER, L.J. *A characterization of the power of vector machines* ◇ D10 D15 ◇
RACKOFF, C.W. *On the complexity of the theories of weak direct powers* ◇ B25 C13 C30 C60 D15 ◇
SAVAGE, J.E. *The complexity of computing* ◇ B70 D10 D15 D98 ◇
SAVITCH, W.J. *Three hardest problems* ◇ D15 ◇
SCHNORR, C.-P. *The combinational complexity of equivalence* ◇ B05 D15 ◇
SCHNORR, C.-P. *The network complexity and the Turing machine complexity of finite functions* ◇ D10 D15 ◇
SCHUSTER, P. *Probleme, die zum Erfuellungsproblem der Aussagenlogik polynomial aequivalent sind* ◇ B05 D15 ◇
SHAKUOV, S.N. *Linear acceleration of the operating time of single-tape Turing machines (Russian)* ◇ D10 D15 ◇

SKORDEV, D.G. *Certain combinatory spaces that are connected with the complexity of data processing (Russian)* ◇ B40 B75 D15 D75 ◇
SOLOVAY, R.M. *On sets Cook-reducible to sparse sets* ◇ D15 ◇
SPECKER, E. & WICK, G. *Laengen von Formeln* ◇ D15 ◇
SUDBOROUGH, I.H. *On deterministic context-free languages, multihead automata, and the power of an auxiliary pushdown store* ◇ D05 D10 D15 ◇
TOGER, A.V. *Asymptotic behavior of the complexity of the approximate computation of functions that satisfy a Lipschitz condition (Russian)* ◇ D15 ◇
TRAKHTENBROT, B.A. *Recursive program schemes and computable functionals* ◇ B75 D15 D55 D65 D75 ◇
TRAKHTENBROT, M.B. *Relationship between classes of monotonic functions* ◇ B75 D15 D20 ◇
VAISER, A. *Stochastic languages and the complexity of computations on probabilistic Turing machines (Russian) (English summary)* ◇ D10 D15 ◇
VOELKEL, L. *Staerke-Relationen fuer URM-Befehlssysteme* ◇ D10 D15 D20 ◇
WECHSUNG, G. *Kompliziertheitstheoretische Charakterisierung der kontextfreien und linearen Sprachen* ◇ D05 D15 ◇
WICHMANN, B. *Ackermann's function: A study in the efficiency of calling procedures* ◇ D15 D20 ◇
WUETHRICH, H.R. *Ein Entscheidungsverfahren fuer die Theorie der reell abgeschlossenen Koerper* ◇ B25 C60 D15 ◇
ZHAROV, V.G. *Codes of words that belong to an enumerable set (Russian)* ◇ D15 D25 ◇
ZHAROV, V.G. *On lower and upper estimates for the complexity of the terms of constructive sequences of normal algorithms (Russian)* ◇ D03 D15 ◇

1977

ALTON, D.A. *"Natural" complexity measures and a subrecursive speed-up theorem* ◇ D15 ◇
AUSIELLO, G. *On the structure and properties of NP-complete problems and their associated optimization problems* ◇ D15 ◇
AVENHAUS, J. & MADLENER, K. *Komplexitaet bei Gruppen: Der Einbettungssatz von Higman* ◇ D15 D40 ◇
AVENHAUS, J. & MADLENER, K. *Subrekursive Komplexitaet bei Gruppen I: Gruppen mit vorgeschriebener Komplexitaet* ◇ D15 D40 ◇
BARASHKO, A.S. *Bikernels and hierarchies of computational complexity classes (Ukrainian) (English summary)* ◇ D15 ◇
BENNISON, V.L. *On the problem of determining measure-independence* ◇ D15 D25 ◇
BENNISON, V.L. & SOARE, R.I. *Recursion theoretic characterizations of complexity theoretic properties* ◇ D15 D20 ◇
BERG, E.P. & LISCHKE, G. *Zwei Saetze fuer schwache Erhaltungsmasse* ◇ D15 D20 ◇
BERMAN, L. *Precise bounds for Presburger arithmetic and the reals with addition: preliminary report* ◇ B25 D15 F30 ◇
BISKUP, J. *Ueber Projektionsmengen von Komplexitatsmassen* ◇ D15 ◇
BOASSON, L. & COURCELLE, B. & NIVAT, M. *A new complexity measure for languages* ◇ D05 D15 ◇
BORODIN, A. *On relating time and space to size and depth* ◇ D10 D15 ◇
BRANDSTAEDT, A. & SAALFELD, D. *Eine Hierarchie beschraenkter Rueckkehrberechnungen auf on-line Turingmaschinen* ◇ D10 D15 D20 ◇
BREMER, H. *Ein vollstaendiges Problem auf der Baummaschine* ◇ D05 D10 D15 ◇
CHAITIN, G.J. *Program size, oracles, and the jump operation* ◇ D15 D25 D30 D80 ◇
CHYTIL, M.P. *Comparison of the active visiting and the crossing complexities* ◇ D10 D15 ◇
DALEY, R.P. *On the inference of optimal descriptions* ◇ B35 D15 ◇
ENGELER, E. *Structural relations between programs and problems* ◇ B75 D15 ◇
EVE, J. & KURKI-SUONIO, R. *On computing the transitive closure of a relation* ◇ D15 E07 ◇
FERRANTE, J. & GEISER, J.R. *An efficient decision procedure for the theory of rational order* ◇ B25 D15 ◇
FISCHER, P.C. & KINTALA, C.M.R. *Computations with a restricted number of nondeterministic steps* ◇ D15 ◇
FLEISCHMANN, K. & MAHR, B. & SIEFKES, D. *Bounded concatenation theory as a uniform method for proving lower complexity bounds* ◇ D15 D20 ◇
FREJVALD, R.V. *Probabilistic machines can use less running time* ◇ D10 D15 ◇
FREJVALD, R.V. & IKAUNIEKS, E. *Some advantages of nondeterministic machines over probabilistic ones (Russian)* ◇ D10 D15 ◇
GACS, P. & LOVASZ, L. *Some remarks on generalized spectra* ◇ B15 C13 D15 ◇
GALIL, Z. *On resolution with clauses of bounded size* ◇ B35 D15 ◇
GALIL, Z. *On the complexity of regular resolution and the Davis-Putnam procedure* ◇ B35 D15 ◇
GALIL, Z. & SEIFERAS, J.I. *Real-time recognition of substring repetition and reversal* ◇ D10 D15 ◇
GILL III, J.T. *Computational complexity of probabilistic Turing machines* ◇ D10 D15 ◇
GOLDSCHLAGER, L.M. *The monotone and planar circuit value problems are log space complete for P* ◇ B70 D10 D15 ◇
GRANT, P.W. *Recognition of EOL languages in less than quartic time* ◇ D05 D15 ◇
HAJEK, P. *Arithmetical complexity of some problems in computer science* ◇ D15 D55 ◇
HECKER, H.-D. *Ein Zusammenhang zwischen Universalitaet und Komplexitaet* ◇ D15 ◇
HENNIE III, F.C. *Introduction to computability* ◇ D15 D98 ◇
HOPCROFT, J.E. & PAUL, W.J. & VALIANT, L.G. *On time versus space* ◇ D15 ◇
HORVATH, S. *Complexity of sequence encodings* ◇ D15 D20 ◇
INOUE, K. & NAKAMURA, A. *Some properties of two-dimensional on-line tessellation acceptors* ◇ D05 D15 ◇

JACOBS, B.E. *On generalized computational complexity* ◇ D15 D60 ◇
JANSSEN, D. & MONIEN, B. *Ueber die Komplexitaet der Fehlerdiagnose bei Systemen* ◇ B70 D15 ◇
KANOVICH, M.I. *Complex properties of context-sensitive languages (Russian)* ◇ D03 D05 D15 ◇
KANOVICH, M.I. *On computability of Kolmogorov complexity* ◇ D15 ◇
KANOVICH, M.I. *On the precision of a complexity criterion for nonrecursiveness and universality (Russian)* ◇ D15 D20 D25 D30 ◇
KASAI, T. *Computational complexity of multitape Turing machines and random access machines* ◇ D10 D15 ◇
KINBER, E.B. *On speed-up of limiting identification of recursive functions by changing the order of questions (Russian) (English and German summary)* ◇ D15 D20 ◇
KLUPP, H. & SCHNORR, C.-P. *A universally hard set of formulae with respect to non-deterministic Turing acceptors* ◇ B25 D10 D15 ◇
KOREC, I. & PROCHAZKA, J. *Real-time computability of* $[\ |\alpha|X\ ]$ *and* $[X^{\alpha}]$ ◇ D15 ◇
KOZEN, D. *Lower bounds for natural proof systems* ◇ D15 F20 ◇
KRATKO, M.I. *On the axiomatic definition of the concept of computational complexity* ◇ D15 D75 ◇
LADNER, R.E. *The computational complexity of provability in systems of modal propositional logic* ◇ B45 D15 ◇
LEONG, B. & SEIFERAS, J.I. *New real-time simulations of multihead tape units* ◇ D10 D15 ◇
LEVIN, L.A. *On a concrete method of assigning complexity measures (Russian)* ◇ D15 D80 ◇
LEWIS, H.R. *A measure of complexity for combinatorial decision problems of the tiling variety* ◇ B25 D10 D15 ◇
LIPTON, R.J. & ZALCSTEIN, Y. *Word problems solvable in logspace* ◇ D15 D40 ◇
LISCHKE, G. *Natuerliche Kompliziertheitsmasse und Erhaltungssaetze II* ◇ D15 ◇
MOKATSYAN, A.A. *Some properties of relativizated Kolmogorov complexities (Russian)* ◇ D15 D30 ◇
MONIEN, B. *A recursive and a grammatical characterization of the exponential-time languages* ◇ D05 D10 D15 ◇
MONIEN, B. *The LBA-problem and the transformability of the class* $\mathfrak{E}^2$ ◇ D05 D15 ◇
MORITA, K. & SUGATA, K. & UMEO, H. *Computational complexity of L(m, n) tape-bounded two-dimensional Turing machines (Japanese)* ◇ D10 D15 ◇
MOSHCHENSKIJ, V.A. *A method of analysis for Turing computations. I (Russian) (English summary)* ◇ D10 D15 ◇
NABIALEK, I. *Bases of aggregable sets* ◇ D15 D20 ◇
NELSON, GREG & OPPEN, D.C. *Fast decision algorithms based on union and find* ◇ D15 D80 ◇
NIGMATULLIN, R.G. *The complexity of languages of type* $\cup M$ *(Russian)* ◇ D15 ◇
PLAISTED, D.A. *Sparse complex polynomials and polynomial reducibility* ◇ D15 ◇
PODNIEKS, K.M. *Forecasting strategies of limited complexity (Russian) (English Summary)* ◇ D15 D20 ◇
REYNVAAN, C. & SCHNORR, C.-P. *Ueber Netzwerkgroessen hoeherer Ordnung und die mittlere Anzahl der in Netzwerken benutzten Operationen* ◇ B70 D15 ◇
SAVITCH, W.J. & VITANYI, P.M.B. *Linear time simulation of multihead Turing machines with head-to-head jumps* ◇ D10 D15 ◇
SAVITCH, W.J. *Recursive Turing machines* ◇ D10 D15 ◇
SCHLIPF, J.S. *Ordinal spectra of first-order theories* ◇ C35 C50 C55 C62 C70 D15 D30 D70 ◇
SCHNORR, C.-P. *The network complexity and the breadth of Boolean functions* ◇ B05 B70 D15 ◇
SEIFERAS, J.I. *Iterative arrays with direct central control* ◇ D10 D15 ◇
SEIFERAS, J.I. *Linear-time computation by nondeterministic multidimensional iterative arrays* ◇ D10 D15 ◇
SHAKUOV, S.N. *Fast Turing computations and their linear speedup (Russian)* ◇ D10 D15 ◇
SHURYGIN, V.A. *On estimates of the complexity of algorithmic problems in constructive analysis (Russian)* ◇ D15 D20 F60 ◇
SIEFKES, D. *Degrees of circuit complexity* ◇ D15 ◇
SIMON, J. *On the difference between one and many (Preliminary version)* ◇ D10 D15 ◇
SIMON, J. *Polynomially bounded quantification over higher types and a new hierarchy of the elementary sets* ◇ B15 D10 D15 D20 ◇
SLISENKO, A.O. *A simplified proof of the real-time recognizability of parlindromes on Turing machines (Russian) (English summary)* ◇ D10 D15 ◇
SOARE, R.I. *Computational complexity, speedable and levelable sets* ◇ D15 ◇
SOLOVAY, R.M. *On random r.e. sets* ◇ D15 D25 ◇
STATMAN, R. *Herbrand's theorem and Gentzen's notion of a direct proof* ◇ D15 F05 F07 F20 ◇
STATMAN, R. *The typed* λ*-calculus is not elementary recursive* ◇ B40 D15 D20 ◇
STOCKMEYER, L.J. *The polynomial-time hierarchy* ◇ D10 D15 D40 D55 ◇
SUDBOROUGH, I.H. *Some remarks on multihead automata* ◇ D10 D15 ◇
TKACHEV, G.A. *Complexity of realization of one sequence of functions of k-valued logic (Russian)* ◇ B50 B70 D15 ◇
TRAKHTENBROT, B.A. *Algorithmen und Rechenautomaten* ◇ D10 D15 D40 D98 ◇
TRAKHTENBROT, B.A. *Frequency algorithms and computations* ◇ D15 D20 ◇
VALIEV, M.K. *Real time computations with restrictions on tape alphabet* ◇ D10 D15 ◇
VERBEEK, R. & WEIHRAUCH, K. *Data presentation and computational complexity* ◇ D05 D15 ◇
VINOGRADOV, A.P. *On the existence of a majorizing local algorithm in effectively describable problems of computing information (Russian)* ◇ D15 D20 ◇

WAGNER, K. & WECHSUNG, G. *Complexity hierarchies of oracles* ◇ D15 ◇
WECHSUNG, G. *Properties of complexity classes -- a short survey* ◇ D15 D98 ◇
WRATHALL, C. *Complete sets and the polynomial-time hierarchy* ◇ D15 ◇
YOUNG, P. *Optimization among provably equivalent programs* ◇ D15 ◇
ZHUKOV, S.A. *Some asymptotic estimates of the complexity of partial recursive functions (Russian)* ◇ D15 D20 ◇

1978

ADLEMAN, L.M. & BOOTH, K.S. & PREPARATA, F.P. & RUZZO, W.L. *Improved time and space bounds for boolean matrix multiplication* ◇ B05 D15 ◇
ADLEMAN, L.M. & MANDERS, K.L. *NP-complete decision problems for binary quadratics* ◇ D15 D35 ◇
AJZENSHTEJN, M.KH. *Ordering of complexity measures according to the effectiveness (Russian)* ◇ D15 D30 ◇
ANGLUIN, D. *On the complexity of minimum inference of regular sets* ◇ D05 D15 ◇
AVENHAUS, J. & MADLENER, K. *Subrekursive Komplexitaet bei Gruppen II: Der Einbettungssatz von Higman fuer entscheidbare Gruppen* ◇ D15 D40 ◇
BAKER, B.S. *Generalized syntax directed translation tree transducers, and linear space* ◇ D05 D15 ◇
BAKER, T.P. *"Natural" properties of flowchart step-counting measures* ◇ D15 ◇
BENNISON, V.L. & SOARE, R.I. *Some lowness properties and computational complexity sequences* ◇ D15 D25 D30 ◇
BERMAN, P. *Relationship between density and deterministic complexity of NP- complete languages* ◇ D15 ◇
BISKUP, J. *Path measures of Turing machine computations* ◇ D10 D15 ◇
BISKUP, J. *The time measure of one-tape Turing machines does not have the parallel computation property* ◇ D10 D15 ◇
BOOK, R.V. & GREIBACH, S.A. & WRATHALL, C. *Comparisons and reset machines* ◇ D10 D15 ◇
BOOK, R.V. & WRATHALL, C. *On languages accepted by time-bounded oracle machines* ◇ D15 ◇
BOOK, R.V. *Simple representations of certain classes of languages* ◇ D15 D25 D55 ◇
BOOTH, K.S. *Isomorphism testing for graphs, semigroups, and finite automata are polynomially equivalent problems* ◇ D05 D15 ◇
BRANDENBURG, F.-J. *Die Zusammenhangskomplexitaet von nichtkontextfreien Grammatiken* ◇ D05 D15 ◇
BRANDSTAEDT, A. *On a family of complexity measures on Turing machines, defined by predicates* ◇ D10 D15 ◇
BRANDSTAEDT, A. *Ueber den Einfluss eines zusaetzlichen Einweg-Eingabebandes bei Entscheidungen und Aufzaehlungen mit beschraenktem Regime sowie beschraenkter Rueckkehr bzw. beschraenkter dualer Rueckkehr auf einbaendrigen Turing-Maschinen* ◇ D10 D15 ◇
BREIDBART, S. *On splitting recursive sets* ◇ D05 D15 D20 D25 ◇
BRUSS, A.R. & MEYER, A.R. *On time-space classes and their relation to the theory of real addition* ◇ B25 D15 ◇
BUKHARAEVA, Z.K. & GOLUNKOV, YU.V. *Complexity and running time of normal algorithms computing boolean functions (Russian)* ◇ B05 B35 D03 D15 ◇
CALUDE, C. *Categorical methods in computability I,II (I: Recursive, non-primitive recursive functions) (Romanian) (English summary)* ◇ D15 D20 ◇
CASE, J. & MOORE, D.J. *The complexity of total order structures* ◇ D15 D20 D45 ◇
CHAITIN, G.J. & SCHWARTZ, J.T. *A note on Monte Carlo primality tests and algorithmic information theory* ◇ D15 D80 ◇
DALEY, R.P. *On the inference of optimal descriptions* ◇ D15 ◇
DALEY, R.P. *On the simplicity of busy beaver sets* ◇ D10 D15 D25 D30 ◇
DALEY, R.P. *Quantitative and qualitative information in computations* ◇ D15 D30 ◇
EMDE BOAS VAN, P. *Some applications of the McCreight-Meyer algorithm in abstract complexity theory* ◇ D15 ◇
FISCHER, MICHAEL J. & MEYER, A.R. & SEIFERAS, J.I. *Separating nondeterministic time complexity classes* ◇ D10 D15 ◇
FISCHER, P.C. *Computational complexity* ◇ D15 ◇
FORTUNE, S. & HOPCROFT, J.E. & SCHMIDT, ERIK MEINECHE *The complexity of equivalence and containment for free single variable program schemes* ◇ B75 D15 ◇
GALIL, Z. & SEIFERAS, J.I. *A linear-time on-line recognition algorithm for "palstar"* ◇ D10 D15 ◇
GALIL, Z. *Palindrome recognition in real time by a multitape Turing machine* ◇ D10 D15 ◇
GREIBACH, S.A. *Visits, crosses, and reversals for nondeterministic off-line machines* ◇ D10 D15 ◇
GRIGOR'EV, D.YU. *Multiplicative complexity of a pair of bilinear forms and of the polynomial multiplication* ◇ D15 D80 ◇
GURARI, E.M. & IBARRA, O.H. *An NP-complete number-theoretic problem* ◇ B25 D15 D80 ◇
HARROW, K. *The bounded arithmetic hierarchy* ◇ D15 D55 ◇
HARTMANIS, J. *Feasible computations and provable complexity properties* ◇ D15 ◇
HARTMANIS, J. & IMMERMAN, N. & MAHANEY, S.R. *One-way log-tape reductions* ◇ D10 D15 ◇
HECKER, H.-D. *Bemerkungen zu Enumerationssystemen und Zeitmassen* ◇ D15 D25 D45 ◇
HUNT III, H.B. & ROSENKRANTZ, D.J. *Polynomial algorithms for deterministic pushdown automata* ◇ D10 D15 ◇
IGARASHI, Y. *Tape bounds for some subclasses of deterministic context-free languages* ◇ D10 D15 ◇
JACOBS, B.E. *The α-union theorem and generalized primitive recursion* ◇ D15 D55 D60 ◇
KANOVICH, M.I. *An estimate of the complexity of arithmetic incompleteness (Russian)* ◇ D15 F20 F30 ◇

KANOVICH, M.I. *Complexity of recognition of invariant properties (Russian)* ◇ D03 D15 ◇
KAROL', A.M. *Complexity of the set of constructive real numbers (Russian)* ◇ D15 F60 ◇
KATS, B.E. *Regular Buechi systems and time-space trade-off in language recognition (Russian)* ◇ D15 ◇
KOSOVSKIJ, N.K. *On Kolmogorov's subrecursive algorithmic complexity (Russian)* ◇ D15 D20 ◇
KOSOVSKIJ, N.K. *The complexity of the solvability of boolean functional equations (Russian)* ◇ B05 D15 G05 ◇
KOZEN, D. *Indexing of subrecursive classes* ◇ D15 D20 ◇
KRICHEVSKIJ, R.E. *Digital enumeration of binary dictionaries (Russian)* ◇ D15 ◇
LEVITZ, H. *An ordinal bound for the set of polynomial functions with exponentiation* ◇ D15 D20 E07 E10 F15 F30 ◇
LEWIS, H.R. *Complexity of solvable cases of the decision problem for the predicate calculus* ◇ B20 B25 D15 ◇
LEWIS, H.R. *Renaming a set of clauses as a Horn set* ◇ B20 D15 ◇
LIPTON, R.J. *Model theoretic aspects of computational complexity* ◇ C62 D15 F30 F65 H15 ◇
LISCHKE, G. *Complexity measures defined by Mazurkiewicz-algorithms* ◇ D15 ◇
LITOW, B. & SUDBOROUGH, I.H. *On non-erasing oracle tapes in space bounded reducibility* ◇ D10 D15 ◇
LOVELAND, D.W. & REDDY, C.R. *Presburger arithmetic with bounded quantifier alternation* ◇ B25 C10 D15 F20 F30 ◇
LYNCH, N.A. *Log space machines with multiple oracle tapes* ◇ D10 D15 ◇
LYNCH, N.A. *Straight-line program length as a parameter for complexity measures* ◇ D15 D45 ◇
MACHTEY, M. & YOUNG, P. *An introduction to the general theory of algorithms* ◇ D15 D20 D98 ◇
MACHTEY, M. & WINKLMANN, K. & YOUNG, P. *Simple Goedel numberings, isomorphisms, and programming properties* ◇ D05 D10 D15 D20 D45 F40 ◇
MCLAUGHLIN, T.G. *On the relations between some rate-of-growth conditions* ◇ D15 ◇
MORAGA, C. *Comments on a method of Karpovsky* ◇ B70 D15 ◇
NAZARYAN, G.A. *Ueber eine Synthese von Algorithmen approximativ berechenbarer boolescher Funktionen (Russian)* ◇ B35 D15 D20 ◇
OPPEN, D.C. *A* $2^{2^{2^{pn}}}$ *upper bound on the complexity of Presburger arithmetic* ◇ B25 C10 D15 F30 ◇
PAGET, M. *Proprietes de complexite pour une famille d'algorithmes de Markov* ◇ B75 D03 D10 D15 ◇
PAOLA DI, R.A. *The operator gap theorem in* α*-recursion theory* ◇ D15 D60 ◇
PARIKH, R. *Effectiveness* ◇ D15 ◇
PAUL, W.J. *Complexity theory* ◇ D15 D98 ◇
PECKEL, J. *A deterministic subclass of context-free languages* ◇ D05 D15 ◇
PERL, J. *Entropie von Problemen* ◇ D15 ◇
PRINOTH, R. *Starke Faerbbarkeit in Petri-Netzen* ◇ D05 D15 ◇
RADZISZOWSKI, S. *Programmability and* $P = NP$ *conjecture* ◇ D15 ◇
SARKISYAN, G.Z. *Effective computability of arithmetic predicates and functions on the basis of schemes of functional elements (Russian) (English and Armenian summaries)* ◇ D15 ◇
SARKISYAN, G.Z. *Ueber eine Klasse arithmetischer Funktionen, die mit Hilfe von Schemata aus Funktionalelementen berechenbar sind (Russian) (Armenian summary)* ◇ D15 D20 ◇
SAVITCH, W.J. *Parallel and nondeterministic time complexity classes(Preliminary report)* ◇ D15 ◇
SAVITCH, W.J. *The influence of the machine model on computational complexity* ◇ D15 ◇
SCHAEFER, T.J. *The complexity of satisfiability problems* ◇ B05 D15 ◇
SCHNORR, C.-P. *Satisfiability is quasilinear complete in* NQL ◇ D10 D15 ◇
SELMAN, A.L. *Polynomial time enumeration reducibility* ◇ D15 ◇
SIPSER, M. *Halting space-bounded computations* ◇ D10 D15 ◇
SOLOMON, M.K. *Some results on measure independent Goedel speed-ups* ◇ D15 F20 ◇
SUDBOROUGH, I.H. *On the tape complexity of deterministic context-free languages* ◇ D05 D10 D15 ◇
TARJAN, R.E. *Complexity of monotone networks for computing conjunctions* ◇ B05 B70 D15 ◇
TETRUASHVILI, M.R. *The problem of conjugacy for one class of groups and the computational complexity (Russian)* ◇ D15 D40 ◇
TRAUB, J. *Recent results and open problems in analytic computational complexity* ◇ D15 ◇
VERBEEK, R. & WEIHRAUCH, K. *Data representation and computational complexity* ◇ D15 D45 ◇
WIEHAGEN, R. *Characterization problems in the theory of inductive inference* ◇ D15 D20 D45 ◇
WRATHALL, C. *Rudimentary predicates and relative computation* ◇ D10 D15 ◇

1979

AANDERAA, S.O. & BOERGER, E. *The Horn complexity of Boolean functions and Cook's problem* ◇ B05 D15 ◇
ADAMSON, A. & GILES, R. *A game-based formal system for* L_{∞} ◇ B25 B50 D15 ◇
ADLEMAN, L.M. & MANDERS, K.L. *Reductions that lie* ◇ D15 ◇
ALT, H. *Lower bounds on space complexity for contextfree recognition* ◇ D05 D15 ◇
ASPVALL, B. & PLASS, M. & TARJAN, R.E. *A linear-time algorithm for testing the truth of certain quantified boolean formulas* ◇ B35 D15 ◇
AVENHAUS, J. & MADLENER, K. *Complexity and independence of algorithmic problems in groups* ◇ D15 D40 ◇
BABORSKI, A. *The complexity of the original and the language of description* ◇ D15 ◇
BAKER, T.P. & SELMAN, A.L. *A second step toward the polynomial hierarchy* ◇ D15 ◇
BAKER, T.P. *On "provable" analogs of* $\mathscr{P}$ *and* $\mathscr{NP}$ ◇ D15 F30 ◇

BARTH, G. *Fast recognition of context-sensitive structures* ◇ D05 D15 ◇
BENNISON, V.L. *Information content characterizations of complexity theoretic properties* ◇ D15 D25 ◇
BERMAN, P. *Complexity of the theory of atomless boolean algebras* ◇ B25 D15 ◇
BIBEL, W. *Tautology testing with a generalized matrix reduction method* ◇ B35 D15 F05 ◇
BLUM, E.K. & LYNCH, N.A. *Relative complexity of operations on numeric and bit-string algebras* ◇ D15 D20 ◇
BOERGER, E. & KLEINE BUENING, H. *The reachability problem for Petri nets and decision problems for Skolem arithmetic* ◇ B25 D15 D35 D80 ◇
BOOK, R.V. *A remark on tally languages and complexity classes* ◇ D05 D15 ◇
BOOK, R.V. *Complexity classes of formal languages (preliminary report)* ◇ D05 D15 ◇
BOOK, R.V. *On languages accepted by space-bounded oracle machines* ◇ D15 ◇
BOOK, R.V. *Polynomial space and transitive closure* ◇ D15 ◇
BOOK, R.V. & BRANDENBURG, F.-J. *Representing complexity classes by equality sets (preliminary report)* ◇ D15 ◇
BOOK, R.V. & GREIBACH, S.A. & WRATHALL, C. *Reset machines* ◇ D10 D15 ◇
BRANDSTAEDT, A. & WECHSUNG, G. *A relation between space, return and dual return complexities* ◇ D15 ◇
BRAUNMUEHL VON, B. & HOTZEL, E. *Supercounter machines* ◇ D10 D15 ◇
BREJTBART, YU.YA. & LEWIS, F.D. *Combined complexity classes for finite functions* ◇ D15 ◇
CHANDRA, A.K. & STOCKMEYER, L.J. *Provably difficult combinatorial games* ◇ D15 ◇
CHLEBUS, B.S. *On decidability of propositional algorithmic logic (Polish) (English summary)* ◇ B25 B75 D10 D15 ◇
COMMENTZ-WALTER, B. *Size-depth tradeoff in monotone boolean formulae* ◇ B05 D15 ◇
COOK, S.A. *Deterministic CEL's are accepted simultaneously in polynomial time and log squared* ◇ D05 D15 ◇
COOK, S.A. & RECKHOW, R.A. *The relative efficiency of propositional proof systems* ◇ B05 D15 F20 ◇
DALEY, R.P. *On the simplification of constructions in degrees of unsolvability via computational complexity* ◇ D10 D15 D25 ◇
DANTSIN, E.YA. *Parameters defining the time of tautology recognition by the splitting method (Russian)* ◇ B35 D15 ◇
DEKHTYAR', M.I. *Bounds on computational complexity and approximability of initial segments of recursive sets* ◇ D15 D20 ◇
DEKHTYAR', M.I. *Complexity spectra of recursive sets of approximability of initial segments of complete problems* ◇ D10 D15 D30 F20 F30 ◇
DEMILLO, R.A. & LIPTON, R.J. *Some connections between mathematical logic and complexity theory* ◇ C57 D15 F30 H15 ◇
DEZANI-CIANCAGLINI, M. & RONCHI DELLA ROCCA, S. & SAITTA, L. *Complexity of λ-term reductions* ◇ B40 D15 ◇
DYMOND, P.W. *Indirect addressing and the time relationships of some models of sequential computation* ◇ D15 ◇
EHRENFEUCHT, A. & ROZENBERG, G. *Finding a homomorphism between two words is NP-complete* ◇ D05 D15 ◇
FARAT, V.M. & KOZHEVNIKOVA, G.P. *Algorithms for translation of algebraic expressions into an improved parenthesis-free notation and an analysis of their effectiveness(Russian)* ◇ B03 D15 ◇
FERRANTE, J. & RACKOFF, C.W. *The computational complexity of logical theories* ◇ B25 C98 D10 D15 ◇
FISCHER, MICHAEL J. & LADNER, R.E. *Propositional dynamic logic of regular programs* ◇ B75 D15 ◇
FISCHER, MICHAEL J. & PIPPENGER, N.J. *Relations among complexity measures* ◇ D05 D15 ◇
FORTUNE, S. *A note on sparse complete sets* ◇ D15 ◇
FREJVALD, R.V. *Fast probabilistic verification of number multiplication (Russian)* ◇ D10 D15 ◇
FREJVALD, R.V. *Fast probabilistic algorithms* ◇ D10 D15 ◇
FREJVALD, R.V. *On shortening the recognition time of some word sets by using a random number generator (Russian)* ◇ D05 D15 ◇
FREJVALD, R.V. *Running time for deterministic and nondeterministic Turing machines (Russian)* ◇ D10 D15 ◇
FRIEDMAN, E.P. & GREIBACH, S.A. *Superdeterministic DPDAs: The method of accepting does affect decision problems* ◇ D05 D10 D15 ◇
GAREY, M.R. & JOHNSON, D.S. *Computers and intractability.A guide to the theory of NP-completeness* ◇ D15 D98 ◇
GORDON, D. *Complexity classes of provable recursive functions* ◇ D15 D20 F30 ◇
GREIBACH, S.A. *Linearity is polynomially decidable for realtime pushdown store automata* ◇ D10 D15 ◇
GRIGOR'EV, D.YU. *Time bounds of multidimensional Turing machines (Russian)* ◇ D10 D15 ◇
GURARI, E.M. & IBARRA, O.H. *An NP-complete number-theoretic problem* ◇ B25 D10 D15 D25 D80 ◇
GURARI, E.M. & IBARRA, O.H. *The complexity of the equivalence problem for counter machines, semilinear sets, and simple programs* ◇ D05 D15 ◇
HAJEK, P. *Arithmetical hierarchy and complexity of computation* ◇ D10 D15 D55 ◇
HARTMANIS, J. *Relations between diagonalization, proof systems, and complexity gaps* ◇ D15 D20 ◇
HEINTZ, JOOS *Definability bounds of first order theories of algebraically closed fields* ◇ B25 C40 C60 D10 D15 ◇
HUNT III, H.B. *Observations on the complexity of regular expression problems* ◇ D05 D15 ◇
HUYNH, D.T. *On complexity measures which are induced by probability distributions* ◇ D15 D20 ◇

IMMERMAN, N. *Length of predicate calculus formulas as a new complexity measure* ◇ D15 ◇
IVANOV, A.G. *Theorems on the time hierarchy for random access machines (Russian) (English summary)* ◇ D10 D15 ◇
JACOBS, B.E. *α-naming and α-speedup theorems* ◇ D15 D60 ◇
JACOBS, B.E. *α-speedable and non α-speedable sets* ◇ D15 D60 ◇
JA'JA', J. *On the complexity of bilinear forms with commutativity* ◇ D15 ◇
KANOVICH, M.I. *A complexity version of Goedel's incompleteness theorem* ◇ D15 F20 F30 ◇
KANOVICH, M.I. *The complexity of search for short programs (Russian)* ◇ D15 D20 ◇
KHODZHAYANTS, M.YU. *Some characteristics of enumeration operators (Russian)* ◇ D15 D25 ◇
KLETTE, R. & LINDNER, R. *Zweidimensional arbeitende Vektormaschinen und ihr Leistungsvermoegen bei der Loesung von Entscheidungsproblemen der Aussagenlogik* ◇ B35 D05 D15 ◇
KOSOVSKIJ, N.K. *On decision procedures for invariant properties of short algorithms (Russian) (English summary)* ◇ D15 D20 D25 D45 ◇
KRAMOSIL, I. *A note on computational complexity of a statistical deducibility testing procedure* ◇ B35 D15 ◇
LANDWEBER, L.H. & LIPTON, R.J. & ROBERTSON, E.L. *On the stucture of sets in NP and other complexity classes* ◇ D15 ◇
LEWIS, F.D. *On unsolvability in subrecursive classes of predicates* ◇ D15 D20 D30 D55 ◇
LEWIS, H.R. *Satisfiability problems for propositional calculi* ◇ B05 D15 ◇
LI, XIANG *The productiveness of Blum's measure of effective computational complexity (Chinese)* ◇ D15 D25 ◇
LIEBERHERR, K.J. & SPECKER, E. *Complexity of partial satisfaction* ◇ B05 D15 ◇
LIPTON, R.J. *On the consistency of P= NP and fragments of arithmetic* ◇ D15 F30 ◇
LISCHKE, G. *Some considerations about complexity for Mazurkiewicz-algorithms* ◇ D15 ◇
MAKAROV, I.T. *A theorem on nonprovability of lower time bounds for a class of functions (Russian)* ◇ D15 ◇
MARCHENKOV, S.S. & MATROSOV, V.L. *Complexity of algorithms and computations (Russian)* ◇ D10 D15 D25 ◇
MEYER, A.R. & WINKLMANN, K. *The fundamental theorem of complexity theory (Preliminary version)* ◇ B75 D10 D15 ◇
MILLER, GARY L. *Graph isomorphism, general remarks* ◇ D15 ◇
MOSTOWSKI, A.WLODZIMIERZ *A note concerning the complexity of a decision problem for positive formulas in SkS* ◇ B15 B25 D15 ◇
NAZARYAN, G.A. *Ueber disjunkte Zerlegungen von Mengen boolescher Funktionen (Russian)* ◇ B05 D15 ◇
O'DONNELL, M.J. *A programming language theorem which is independent of Peano arithmetic* ◇ B75 D15 F30 ◇
PAKHOMOV, S.V. *Machine-independent description of some machine complexity classes (Russian) (English summary)* ◇ D10 D15 ◇
PAUL, W.J. *Kolmogorov complexity and lower bounds* ◇ D15 ◇
PAUL, W.J. *On time hierarchies* ◇ D15 ◇
PETROV, B.N. & UL'YANOV, S.V. & ULANOV, G.M. *Complexity of finite objects and informational control theory (Russian)* ◇ D15 D98 ◇
PRATT, V.R. *Axioms or algorithms* ◇ B25 B75 D15 ◇
PUDLAK, P. & SPRINGSTEEL, F. *Complexity in mechanized hypothesis formation* ◇ B35 D15 ◇
RADZISZOWSKI, S. *Logic and complexity of synchronous parallel computations* ◇ D15 ◇
SABEL'FEL'D, V.K. *Polynomial estimate of the complexity of the recognition of logic-term equivalence (Russian)* ◇ B75 D15 D20 ◇
SARKISYAN, A.D. *Realization of arithmetic functions on iterative networks (Russian) (Armenian summary)* ◇ D05 D15 ◇
SAVITCH, W.J. & STIMSON, M.J. *Hierarchies of recursive computations* ◇ D10 D15 ◇
SELMAN, A.L. *P-selective sets, tally languages, and the behaviour of polynomial time reducibilities on NP (preliminary report)* ◇ D05 D15 ◇
SELMAN, A.L. *P-selective sets, tally languages, and the behaviour of polynomial time reducibilities on NP* ◇ D05 D15 ◇
SHURYGIN, V.A. *Bounds on the complexity of normal algorithms (Russian)* ◇ D03 D15 ◇
SIMON, H.-U. *Word problems for groups and contextfree recognition* ◇ D15 D40 ◇
SLISENKO, A.O. *Complexity problems in the theory of computations (Russian) (English summary)* ◇ D15 ◇
STATMAN, R. *Intuitionistic propositional logic is polynomial-space complete* ◇ B40 D15 F20 F50 ◇
STATMAN, R. *The typed λ-calculus is not elementary recursive* ◇ B40 D15 D20 ◇
STORK, H.-G. *Remarks on the satisfiability problem of propositional logic* ◇ B05 D15 ◇
TETRUASHVILI, M.R. *The conjugacy problem for groups with one defining relation and the complexity of Turing calculations* ◇ D15 D40 ◇
VALIANT, L.G. *Completeness classes in algebra* ◇ D15 D75 ◇
VALIANT, L.G. *Negative results on counting* ◇ D15 ◇
VALIANT, L.G. *The complexity of enumeration and reliability problems* ◇ D15 ◇
VERBEEK, R. *On the naturalness of the Grzegorczyk hierarchy* ◇ D15 D20 ◇
WAGNER, K. *Bounded recursion and complexity classes* ◇ D15 D20 ◇
WECHSUNG, G. *A crossing measure for 2-tape Turing machines* ◇ D10 D15 ◇
WOEHL, K. *Zur Komplexitaet der Presburger Arithmetik und des Aequivalenz-Problems einfacher Programme* ◇ B25 B75 D15 F20 F30 ◇
WU, YUNZENG *Two problems in contemporary mathematical logic - CH and P= ?NP (Chinese)* ◇ D15 E50 ◇

YANG, DONGPING *α-operator gap theorem (Chinese)* ◇ D15 D60 ◇
ZAK, S. *A Turing machine oracle hierarchy* ◇ D10 D15 ◇
ZAK, S. *A Turing machine space hierarchy* ◇ D10 D15 ◇

1980

ABLAEV, F.M. *The relation of automatic measures of the complexity of languages to Loveland's complexity of word resolution (Russian)* ◇ D05 D15 ◇
ABRAMSON, F.G. & BREJTBART, YU.YA. & LEWIS, F.D. *Complex properties of grammars* ◇ D05 D10 D15 ◇
ADACHI, A. & KASAI, T. *A characterization of time complexity by simple loop programs* ◇ D15 D20 ◇
ADRIANOPOLI, F. *Conjugated measures of computational complexity* ◇ D15 D20 ◇
ALTON, D.A. *"Natural" programming languages and complexity measures for subrecursive programming languages: an abstract approach* ◇ B75 D15 D20 ◇
ANGLUIN, D. *On counting problems and the polynomial-time hierachy* ◇ D15 ◇
ASPVALL, B. *Recognizing disguised NR(1) instances of the satisfiability problem* ◇ B35 D15 ◇
ASVELD, P.R.J. *Space-bounded complexity classes and iterated deterministic substitution* ◇ D15 ◇
AUSIELLO, G. & MARCHETTI-SPACCAMELA, A. & PROTASI, M. *Toward a unified approach for the classification of NP-complete optimization problems* ◇ D15 ◇
AUSLANDER, L. & WINOGRAD, S. *The multiplicative complexity of certain semilinear systems defined by polynomials* ◇ D15 ◇
AVENHAUS, J. & MADLENER, K. *String matching and algorithmic problems in free groups* ◇ D15 D40 D80 ◇
BARASHKO, A.S. *On boundary sets in the theory of enumerations (Ukrainian) (English and Russian summaries)* ◇ D15 D25 D45 ◇
BAUR, W. & RABIN, M.O. *Linear disjointness and algebraic complexity* ◇ D15 D80 ◇
BECK, H. *Entscheidbarkeit und Komplexitaet der Funktionsgleichheit von Programmen (Dissertation)* ◇ B75 D15 ◇
BEL'TYUKOV, A.P. *Hierarchies of complexity of computation of partial functions with values 0 and 1 (Russian)* ◇ D15 ◇
BEN-ARI, M. *A simplified proof that regular resolution is exponential* ◇ B35 D15 F20 ◇
BENNISON, V.L. *Recursively enumerable complexity sequences and measure independence* ◇ D15 D25 D30 ◇
BERMAN, L. *The complexity of logical theories* ◇ B25 D10 D15 F30 ◇
BERMAN, P. *A note on sweeping automata* ◇ D05 D15 ◇
BERTONI, A. & MAURI, G. & TORELLI, M. *Sulla complessita di alcuni problemi di conteggio* ◇ D15 ◇
BLUM, M. & CHANDRA, A.K. & WEGMAN, M.N. *Equivalence of free boolean graphs can be decided probabilistically* ◇ D15 ◇
BOERGER, E. & KLEINE BUENING, H. *The reachability problem for Petri nets and decision problems for Skolem arithmetic* ◇ B25 D15 D35 D80 ◇
BOOK, R.V. & BRANDENBURG, F.-J. *Equality sets and complexity classes* ◇ D15 ◇
BRANDENBURG, F.-J. *Multiple equality sets and Post machines* ◇ D05 D15 ◇
BRUSS, A.R. & MEYER, A.R. *On time-space classes and their relation to the theory of real addition* ◇ B25 D15 ◇
COLBOURN, C.J. *Isomorphism complete problems on matrices* ◇ D15 ◇
CONSTABLE, R.L. & HUNT III, H.B. & SAHNI, S.K. *On the computational complexity of program scheme equivalence* ◇ D15 D20 ◇
COOK, S.A. & RACKOFF, C.W. *Space lower bounds for maze threadability on restricted machines* ◇ D05 D15 ◇
CULIK II, K. & DIAMOND, N.D. *A homomorphic characterization of time and space complexity classes of languages* ◇ D05 D10 D15 ◇
DALEY, R.P. *Quantitative and qualitative information in computations* ◇ D15 D30 ◇
DALEY, R.P. *The busy beaver method* ◇ D10 D15 D25 ◇
DEMILLO, R.A. & EISENSTAT, S.C. & LIPTON, R.J. *Space-time trade-offs in structured programming: An improved combinatorial embedding theorem* ◇ D15 ◇
DOSHITA, S. & ISHIBASHI, T. & YAMASAKI, S. *Unit resolution for a subclass of the Ackermann class* ◇ B35 D15 ◇
EBBINGHAUS, H.-D. *Mathematische Logik und Informatik* ◇ C13 D10 D15 D98 E70 H15 ◇
EMDE BOAS VAN, P. *Machine models and computational complexity* ◇ D15 ◇
EVEN, S. & YACOBI, Y. *Cryptocomplexity and NP-completeness* ◇ D15 D80 ◇
FOELDES, S. *On the complexity of the lattice embeddability problem* ◇ D15 D40 G10 ◇
FUERER, M. *The complexity of the inequivalence problem for regular expressions with intersection* ◇ D05 D15 ◇
GOETZE, B.G. & NEHRLICH, W. *The structure of LOOP programs and subrecursive hierarchies* ◇ D15 D20 ◇
GORALCIK, P. & GORALCIKOVA, A. & KOUBEK, V. *Testing of properties of finite algebras* ◇ C13 D15 D40 ◇
GRANT, P.W. *Some more independence results in complexity theory* ◇ D15 ◇
GRIGORAS, G. *On the completeness in abstract complexity classes of languages* ◇ D15 ◇
HAUSMANN, D. & KORTE, B. *The relative strength of oracles for independence systems* ◇ D15 ◇
HONDA, N. & INAGAKI, Y. & OYAMAGUCHI, M. *On the equivalence problem for two DPDA's, one of which is real-time* ◇ D10 D15 ◇
HONDA, N. & INAGAKI, Y. & OYAMAGUCHI, M. *The equivalence problem for real-time strict deterministic languages* ◇ D05 D15 ◇
KANELLAKIS, P.C. *On the computational complexity of cardinality constraints in relational databases* ◇ B75 D15 ◇

KHACHIYAN, L.G. & TARASOV, S.P. *Bounds of solutions and algorithmic complexity of systems of convex diophantine inequalities (Russian)* ◇ D15 ◇
KHACHIYAN, L.G. & KOZLOV, M.K. & TARASOV, S.P. *The polynomial solvability of convex quadratic programming (Russian)* ◇ D10 D15 ◇
KOSOVSKIJ, N.K. *An example of a sequence of symbols with large subrecursive Kolmogorov's complexity of initial segments (Russian)* ◇ D15 ◇
KOZEN, D. *Complexity of boolean algebras* ◇ B25 D15 G05 ◇
KOZEN, D. *Indexings of subrecursive classes* ◇ D15 D20 ◇
KRAMOSIL, I. *Computational complexity of a statistical verification procedure for propositional calculus* ◇ B35 D15 ◇
LADNER, R.E. *Complexity theory with emphasis on the complexity of logical theories* ◇ B25 C98 D05 D10 D15 ◇
LEITSCH, A. *Complexity of index sets and translating functions* ◇ D15 D20 ◇
LEWIS, H.R. *Complexity results for classes of quantificational formulas* ◇ B20 B25 D10 D15 ◇
LOVASZ, L. *Efficient algorithms: an approach by formal logic* ◇ B35 C13 D15 ◇
LUE, YIZHONG *Discussion on a computing complex problem* ◇ D15 ◇
LYNCH, N.A. *Straight-line program length as a parameter for complexity analysis* ◇ D15 D45 ◇
MAHANEY, S.R. *Sparse complete sets for NP: solution of a conjecture of Berman and Hartmanis* ◇ D15 ◇
MANDERS, K.L. *Computational complexity of decision problems* ◇ D15 ◇
MONIEN, B. & SUDBOROUGH, I.H. *The interface between language theory and complexity theory* ◇ D05 D15 ◇
NAKAMURA, A. & ONO, H. *On the size of refutation Kripke models for some linear modal and tense logics* ◇ B45 D15 ◇
NAZARYAN, G.A. *The relations of some complexity characteristics of sets of boolean functions (Russian) (Armenian summary)* ◇ B05 D15 ◇
OPPEN, D.C. *Complexity, convexity and combinations of theories* ◇ B25 B35 D15 ◇
OPPEN, D.C. *Reasoning about recursively defined data structures* ◇ B25 D15 ◇
PAUL, W.J. & PRAUSS, E.J. & REISCHUK, R. *On alternation* ◇ D10 D15 ◇
PAUL, W.J. & REISCHUK, R. *On alternation II. A graph-theoretic approach to determinism versus nondeterminism* ◇ D10 D15 ◇
PIPPENGER, N.J. *On another boolean matrix* ◇ B70 D15 ◇
PLAISTED, D.A. *The application of multivariate polynomials to inference rules and partial tests for unsatisfiability* ◇ B35 D15 F20 ◇
RYTTER, W. *Automata theory and complexity* ◇ D05 D15 D98 ◇
SABEL'FEL'D, V.K. *The logic-termal equivalence is polynomial-time decidable* ◇ B75 D15 ◇
SAZONOV, V.YU. *A logical approach to the problem "P=NP?"* ◇ D15 ◇
SAZONOV, V.YU. *Polynomial computability and recursivity in finite domains* ◇ D15 ◇
SCHINZEL, B. *Ueber die Kategorie der Programmbuendel (Hab.-Schr.)* ◇ D15 D20 D45 ◇
SCHNORR, C.-P. *A 3n-lower bound on the network complexity of boolean functions* ◇ B70 D15 ◇
SHAMIR, E. & SNIR, M. *On the depth complexity of formulas* ◇ D15 ◇
SIMMONS, H. *The word and torsion problems for commutative Thue systems* ◇ D03 D15 D40 ◇
SIPSER, M. *Halting space-bounded computations* ◇ D10 D15 ◇
SMITH, C.H. *Applications of classical recursion theory to computer science* ◇ B75 D15 D20 D80 ◇
STOCKMEYER, L.J. *Difficult computational problems* ◇ D15 ◇
STOLTENBERG-HANSEN, V. *On computational complexity in weakly admissible structures* ◇ D15 D60 D75 ◇
SU, ZHE *The significance of the "street vendor's load" problem (Chinese)* ◇ D15 ◇
VALIEV, M.K. *Decision complexity of variants of propositional dynamic logic* ◇ B25 B75 D15 ◇
VITANYI, P.M.B. *On the power of real-time Turing machines under varying specifications* ◇ D10 D15 ◇
VITANYI, P.M.B. *Real-time Turing machines under varying specifications* ◇ D10 D15 ◇
VITANYI, P.M.B. *Real-time Turing machines under varying specifications II* ◇ D10 D15 ◇
WAGNER, K. & WECHSUNG, G. *Kompliziertheitshierarchien* ◇ D15 ◇
WECHSUNG, G. *A note on the return complexity (German and Russian summaries)* ◇ D15 ◇
WILKIE, A.J. *Applications of complexity theory to Σ_0-definability problems in arithmetic* ◇ C62 D15 F30 ◇
WINKLER, P.M. *Classification of algebraic structures by work space* ◇ C05 D15 ◇
ZAK, S. *A Turing machine oracle hierarchy I,II* ◇ D10 D15 ◇

1981

AANDERAA, S.O. & BOERGER, E. *The equivalence of Horn and network complexity for boolean functions* ◇ B05 B70 D15 ◇
ADACHI, A. & KASAI, T. *A problem which requires $O(n^k)$ time* ◇ D15 ◇
ADLEMAN, L.M. & LOUI, M.C. *Space-bounded simulation of multitape Turing machines* ◇ D10 D15 ◇
AJZENSHTEJN, M.KH. *Subrecursive reducibility and accelerability (Russian)* ◇ D15 D20 ◇
AMIHUD, A. & CHOUEKA, Y. *Loop-programs and polynomially computable functions* ◇ D15 D20 ◇
ASVELD, P.R.J. *Time and space complexity of inside-out macro languages* ◇ D15 ◇
AUSIELLO, G. & D'ATRI, A. & PROTASI, M. *A characterization of reductions among combinatorial problems* ◇ D15 ◇
AUSIELLO, G. & D'ATRI, A. & PROTASI, M. *Lattice-theoretic ordering properties for NP-complete optimization problems* ◇ D15 ◇

AVENHAUS, J. & MADLENER, K. *P-complete problems in free groups* ◇ D15 D40 ◇
BANJEVIC, D. *On some basic properties of the Kolmogorov complexity* ◇ D15 ◇
BARASHKO, A.S. *Exact lower bounds for indicators of the enumeration of sets (Russian)* ◇ D15 D45 ◇
BENNETT, C.H. & GILL III, J.T. *Relative to a random oracle A,* $P^A \neq NP^A \neq co-NP^A$ *with probability 1* ◇ D15 ◇
BERTONI, A. & MAURI, G. *On efficient computation of the coefficients of some polynomials with applications to some enumeration problems* ◇ D15 D80 ◇
BISKUP, J. *A formal approach to null values in database relations* ◇ D15 ◇
BLOOM, S.L. & PATTERSON, D.B. *Easy solutions are hard to find* ◇ D15 ◇
BOASSON, L. & COURCELLE, B. & NIVAT, M. *The rational index: a complexity measure for languages* ◇ D05 D15 ◇
BOERGER, E. *Logical description of computation processes* ◇ B25 D05 D15 D35 ◇
BOOK, R.V. *Bounded query machines: on NP and PSPACE* ◇ D15 ◇
BOOK, R.V. & WRATHALL, C. *Bounded query machines: on NP( ) and NPQUERY( )* ◇ D15 ◇
BRANDSTAEDT, A. *Pushdown automata with restricted use of storage symbols* ◇ D10 D15 ◇
BUDACH, L. *Two pebbles don't suffice* ◇ D05 D15 ◇
CAZACU, C. *A depth-space analysis of algorithms* ◇ D15 D20 ◇
CHANDRA, A.K. & KOZEN, D. & STOCKMEYER, L.J. *Alternation* ◇ D10 D15 ◇
CHEW, P. & MACHTEY, M. *A note on structure and looking back applied to the relative complexity of computable functions* ◇ D15 ◇
CHLEBUS, B.S. *On the computational complexity of satisfiability in propositional logic programs* ◇ B75 D15 ◇
COOK, S.A. *Towards a complexity theory of synchronous parallel computation* ◇ B75 D15 ◇
DALEY, R.P. *Busy beaver sets and the degree of unsolvability* ◇ D10 D15 D25 ◇
DALEY, R.P. *Retraceability, repleteness and busy beaver sets* ◇ D15 D20 D50 ◇
DANTSIN, E.YA. *Two systems for proving tautologies based on the splitting method (Russian) (English summary)* ◇ B35 D15 F20 ◇
DEKHTYAR', M.I. *Structure of T-reducibilities with bounded complexity (Russian)* ◇ D15 D30 ◇
DIKOVSKIJ, A.YA. *A theory of complexity of monadic recursion schemes* ◇ D15 D20 ◇
DO MINH THAI *On minimal-program complexity of complex sets* ◇ D15 D25 ◇
DYMENT, E.Z. *On some properties of the Q-complexity of finite objects (Russian)* ◇ D15 ◇
EHRIG, H. & MAHR, B. *Complexity of algebraic implementations for abstract data types* ◇ D15 ◇
ENGELER, E. *Generalized Galois theory and its application to complexity* ◇ C60 D15 ◇
ENGELFRIET, J. & FILE, G. *Passes, sweeps and visits* ◇ D05 D15 ◇
ERNI, W. & LAPSIEN, R. *On the time and tape complexity of weak unification* ◇ B35 D15 ◇
ERSHOV, A.P. & KNUTH, D.E. (EDS.) *Algorithms in modern mathematics and computer science. Proceedings, Urgench, Uzbek SSR, September 16-22, 1979* ◇ D15 D20 D97 ◇
EVEN, S. & YACOBI, Y. *An observation concerning the complexity of problems with few solutions and its application to cryptography* ◇ D15 D80 ◇
FREJVALD, R.V. *Probabilistic two-way machines* ◇ D05 D10 D15 ◇
FUERER, M. *Alternation and the Ackermann case of the decision problem* ◇ B20 B25 D15 ◇
FURST, M. & SAXE, J.B. & SIPSER, M. *Parity, circuits, and the polynomial-time hierarchy* ◇ D15 ◇
GACS, P. *On the relation between descriptional complexity and algorithmic probability* ◇ D15 ◇
GERMANO, G. & MAGGIOLO-SCHETTINI, A. *Sequence recursiveness without cylindrification and limited register machines* ◇ B75 D15 D20 ◇
GOETZE, B.G. & NEHRLICH, W. *The number of loops necessary and sufficient for computing functions (German and Russian summaries)* ◇ D15 ◇
GRIGORAS, G. *On a generalization of gamma-reducibility* ◇ D15 ◇
GRIGOR'EV, D.YU. *Complexity of "wild" matrix problems and of the isomorphism of algebras and graphs (Russian) (English summary)* ◇ D15 D40 ◇
GURARI, E.M. & IBARRA, O.H. *The complexity of the equivalence problem for two characterizations of Presburger sets* ◇ B25 D05 D15 F30 ◇
GURARI, E.M. & IBARRA, O.H. *The complexity of decision problems for finite-turn multicounter machines* ◇ D10 D15 ◇
GURARI, E.M. & IBARRA, O.H. *The complexity of the equivalence problem for simple programs* ◇ D15 D20 ◇
HARRISON, M.A. & YEHUDAI, A. *Eliminating null rules in linear time* ◇ D05 D15 ◇
HARTMANIS, J. & MAHANEY, S.R. *Languages simultaneously complete for one-way and two-way log-tape automata* ◇ D05 D15 ◇
HARTMANIS, J. *Observations about the development of theoretical computer science* ◇ A10 D05 D15 ◇
HECKER, H.-D. *Zur Kompliziertheit der Beschreibung von Enumerationen rekursiv aufzaehlbarer Mengen* ◇ D15 D25 D45 ◇
HELLER, H. *Relativized polynomial hierarchies extending two levels* ◇ D15 ◇
HOMER, S. & JACOBS, B.E. *Degrees of non* α*-speedable sets* ◇ D15 D60 ◇
HONG, JIAWEI *On similarity and duality of computation (Chinese)* ◇ D15 ◇
HUWIG, H. *A definition of the P=NP-problem in categories* ◇ D15 D75 G30 ◇
IMMERMAN, N. *Number of quantifiers is better than number of tape cells* ◇ C07 D15 ◇
JAZAYERI, M. *A simpler construction for showing the intrinsically exponential complexity of the circularity problem for attribute grammars* ◇ D05 D15 ◇

JOHNSON, D.S. *The NP-completeness column: an ongoing guide* ◇ D15 ◇
JOSEPH, D. & YOUNG, P. *A survey of some recent results on computational complexity in weak theories of arithmetic* ◇ D15 F30 ◇
JUNG, H. *Relationships between probabilistic and deterministic tape complexity* ◇ D15 ◇
KANNAN, R. *A circuit-size lower bound* ◇ D15 ◇
KANNAN, R. *Towards separating nondeterministic time from deterministic time* ◇ D15 ◇
KANOVICH, M.I. *Complexity of a fixed point and of a shift algorithm (Russian)* ◇ D15 D20 ◇
KATSEFF, H.P. & SIPSER, M. *Several results in program size complexity* ◇ D15 ◇
KLEINE BUENING, H. *Classes of functions over binary trees* ◇ B75 D05 D15 D20 ◇
KO, KER-I & MOORE, D.J. *Completeness, approximation and density* ◇ D15 ◇
KORTE, B. & SCHRADER, R. *A survey on oracle techniques* ◇ D15 ◇
KOZEN, D. *Positive first-order logic is NP-complete* ◇ B25 D15 ◇
KRUPSKIJ, V.N. *On the complexity of a description of computable approximations for points of a metric space (Russian) (English summary)* ◇ D15 D80 ◇
LANDWEBER, L.H. & LIPTON, R.J. & ROBERTSON, E.L. *On the structure of sets in NP and other complexity classes* ◇ D15 ◇
LEGGETT JR., E.W. & MOORE, D.J. *Optimization problems and the polynomial hierarchy* ◇ D15 ◇
LEWIS, F.D. *A note on context free languages, complexity classes, and diagonalization* ◇ D05 D15 ◇
LEWIS, F.D. *Stateless Turing machines and fixed points* ◇ D10 D15 D20 ◇
LIEBERHERR, K.J. & SPECKER, E. *Complexity of partial satisfaction* ◇ B05 D15 ◇
LIPSHITZ, L. *Some remarks on the diophantine problem for addition and divisibility* ◇ B25 D15 ◇
LISCHKE, G. *Two types of properties for complexity measures* ◇ D15 ◇
LONG, T.J. *On γ-reducibility versus polynomial time many-one reducibility* ◇ D15 ◇
LOUI, M.C. *A space bound for one-tape multidimensional Turing machines* ◇ D10 D15 ◇
LUBIW, A. *Some NP-complete problems similar to graph isomorphism* ◇ D15 ◇
MACHTEY, M. & YOUNG, P. *Remarks on recursion versus diagonalization and exponentially difficult problems* ◇ D15 D20 ◇
MAHANEY, S.R. *On the number of P-isomorphism classes of NP-complete sets* ◇ D15 ◇
MAHR, B. & SIEFKES, D. *Relating uniform and nonuniform models of computation* ◇ D15 ◇
MAIER, W. *Graphen total rekursiver Funktionen* ◇ D15 D20 ◇
MATROSOV, V.L. *Complexity of computable functions for a generalized storage measure (Russian)* ◇ D15 D20 ◇
MAYR, E.W. & MEYER, A.R. *The complexity of the finite containment problem for Petri nets* ◇ D15 D20 D80 ◇
MICALI, S. *Two-way deterministic finite automata are exponentially more succinct than sweeping automata* ◇ D05 D15 ◇
MICHEL, P. *Borne superieure de la complexite de la theorie de N muni de la relation de divisibilite* ◇ B25 D15 ◇
MIKOLAJCZAK, B. *On the time computational complexity of several problems of simulation of a set of finite automata* ◇ D05 D15 ◇
MONIEN, B. & SUDBOROUGH, I.H. *Bounding the bandwidth of NP-complete problems* ◇ D15 ◇
MONIEN, B. & SUDBOROUGH, I.H. *Time and space bounded complexity classes and bandwidth constrained problems* ◇ D15 ◇
MORAN, S. & PAZ, A. *Non deterministic polynomial optimization problems and their approximations* ◇ D15 ◇
MORAN, S. & PAZ, A. *Nondeterministic polynomial optimization problems and their approximation* ◇ D15 ◇
MORAN, S. *Some results on relativized deterministic and nondeterministic time hierarchies* ◇ D15 ◇
MOSTOWSKI, A.WLODZIMIERZ *The complexity of automata and subtheories of monadic second order arithmetics* ◇ B15 B25 D05 D15 F35 ◇
MUNDICI, D. *Complexity of Craig's interpolation* ◇ C40 D15 ◇
MUNDICI, D. *Craig's interpolation theorem in computation theory* ◇ C40 D15 ◇
MUNDICI, D. *Irreversibility, uncertainty, relativity and computer limitations (Italian and Russian summaries)* ◇ A05 D15 ◇
NIGMATULLIN, R.G. *The problem of lower bounds of complexity and the theory of NP-completeness (Russian)* ◇ D15 ◇
ODIFREDDI, P. *Insiemi subcreativi (English summary)* ◇ D15 D25 ◇
PAUL, W.J. *On heads versus tapes* ◇ D10 D15 ◇
PAUL, W.J. & REISCHUK, R. *On time versus space II* ◇ D15 ◇
PELED, U.N. & SIMEONE, B. *A polynomial-time algorithm for recognizing threshold functions* ◇ D15 D80 ◇
PIPPENGER, N.J. *Computational complexity of algebraic functions* ◇ D15 ◇
RACKOFF, C.W. & SEIFERAS, J.I. *Limitations on separating nondeterministic complexity classes* ◇ D15 ◇
RADZISZOWSKI, S. *Logic and complexity of synchronous parallel computations* ◇ B75 D15 ◇
RUZZO, W.L. *On uniform circuit complexity* ◇ D10 D15 ◇
SABEL'FEL'D, V.K. *Tree equivalence of linear recursive schemata is polynomial-time decidable* ◇ B75 D15 ◇
SAVAGE, J.A. *Space-time tradeoffs - a survey* ◇ D15 D98 ◇
SCHMIDT, DIANA *An algebraic characterisation of P* ◇ D15 ◇
SCHNORR, C.-P. *On self-transformable combinatorial problems* ◇ D15 ◇
SCHOENING, U. *A note on complete sets for the polynomial-time hierarchy* ◇ D15 ◇
SCHOENING, U. *Untersuchungen zur Struktur von NP und verwandten Komplexitaetsklassen mit Hilfe verschiedener polynomieller Reduktionen* ◇ D15 ◇

SCHROEPPEL, R. & SHAMIR, A. *A $T=O(2^{n/2})$, $S=O(2^{n/4})$ algorithm for certain NP-complete problems* ◇ D15 ◇
SELMAN, A.L. *Some observations on NP real numbers and P-selective sets* ◇ D15 ◇
SHATROVA, N.P. *Upper bound of the degree of complexity of an algorithm for solving the conjugacy problem for a class of groups (Russian)* ◇ D15 D40 ◇
SIKDAR, K. *On the complexity classes and optimal algorithms for parallel evaluation of polynomials* ◇ D15 ◇
SIMON, J. *Division in idealized unit cost RAMs* ◇ D10 D15 ◇
SIMON, J. *On tape-bounded probabilistic Turing machine acceptors* ◇ D10 D15 ◇
SKYUM, S. & VALIANT, L.G. *A complexity theory based on Boolean algebra* ◇ D15 G05 ◇
SPRINGSTEEL, F. *Complexity of hypothesis formation problems* ◇ B35 D15 ◇
STAIGER, L. *Complexity and entropy* ◇ D15 ◇
TAO, RENJI *On the computational power of automata with time or space bounded by Ackermann's or superexponential functions* ◇ D10 D15 D20 ◇
TOMPA, M. *An extension of Savitch's theorem to small space bounds* ◇ D15 ◇
VAZHENIN, YU.M. *Sur la liaison entre problemes combinatoires et algorithmiques* ◇ B25 D05 D15 D35 ◇
VERBEEK, R. *Time-space trade-offs for general recursion* ◇ D15 ◇
VOGEL, J. & WAGNER, K. *Two-way automata with more than one storage medium* ◇ D10 D15 ◇
V'YUGIN, V.V. *Algorithmic entropy (complexity) of finite objects and its application to the definition of randomness and quantity of information (Russian)* ◇ D15 ◇
WAACK, S. *Tape complexity of word problems* ◇ D15 D40 ◇
WEGENER, I. *An improved complexity hierarchy on the depth of boolean functions* ◇ B05 B70 D15 ◇
WEGENER, I. *Boolean functions whose monotone complexity is of size $n^2 \log n$* ◇ B05 D15 ◇
WEIHRAUCH, K. *Recursion and complexity theory on CPO's* ◇ D15 D45 D75 ◇
WIETLISBACH, M.N. *Zur Komplexitaet von Entscheidungsalgorithmen, die auf dem Herbrand'schen Satz und regulaerer Resolution beruhen* ◇ B35 D15 ◇

1982

APOLLONI, B. & GREGORIO DI, S. *A probabilistic analysis of a new satisfiability algorithm (French summary)* ◇ B35 D15 ◇
BACHEM, A. *Concepts of algorithmic computation* ◇ D10 D15 ◇
BALCAZAR, J.L. & DIAZ, J. *A note on a theorem by Ladner* ◇ D15 ◇
BEL'TYUKOV, A.P. *Lower bounds of complexity for machine models of computations (Russian) (English summary)* ◇ D15 ◇
BLASS, A.R. & GUREVICH, Y. *On the unique satisfiability problem* ◇ D15 ◇
BOOK, R.V. & WRATHALL, C. *A note on complete sets and transitive closure* ◇ D15 ◇
BOOK, R.V. & WILSON, CHRISTOPHER B. & XU, MEIRUI *Relativizing time, space, and time-space* ◇ D15 ◇
BOOK, R.V. *When is a monoid a group? The Church-Rosser case is tractable* ◇ D03 D15 D40 ◇
BORODIN, A. *Structured vs. general models in computational complexity* ◇ B75 D15 ◇
BRANDSTAEDT, A. & VOGEL, J. *Kompliziertheitsbeschraenkte Checking-Stack-Baender und Raum-Zeit-Probleme (English and Russian summaries)* ◇ D10 D15 ◇
BRENT, R.P. & GOLDSCHLAGER, L.M. *Some area-time tradeoffs for VLSI* ◇ B70 D05 D15 ◇
BROWN, CYNTHIA A. & GOLDBERG, ALLEN & PURDOM JR., P.W. *Average time analyses of simplified Davis-Putnam procedures* ◇ B35 D15 ◇
BURGIN, M.S. *Complexity of parallel algorithms and computations (Russian)* ◇ D15 ◇
BURGIN, M.S. *Generalized Kolmogorov complexity and duality in computational theory (Russian)* ◇ D15 ◇
CALUDE, C. *Topological size of sets of partial recursive functions* ◇ D15 D20 E75 ◇
CHLEBUS, B.S. *On the computational complexity of satisfiability in propositional logics of programs* ◇ B75 D15 ◇
COHEN, JACQUES & HICKEY, T. & KATCOFF, J. *Upper bounds for speedup in parallel parsing* ◇ D05 D15 ◇
CORAY, G. *Complexite d'algorithmes modulaires* ◇ D15 ◇
DALEY, R.P. & MANDERS, K.L. *The complexity of the validity problem for dynamic logic* ◇ B75 D15 ◇
DAWES, A.M. *Splitting theorems for speed-up related to order of enumerations* ◇ D15 D25 ◇
DEMEL, J. & DEMLOVA, M. & KOUBEK, V. *Simplicity of algebras requires to investigate almost all operations* ◇ C05 D15 ◇
DIAZ, J. & VERGES, M. *A conjecture in the theory of complexity (Catalan)* ◇ D15 ◇
DOMANSKI, B. *The complexity of two decision problems for free groups* ◇ D15 D40 ◇
EBBINGHAUS, H.-D. *Some logical aspects of complexity theory* ◇ D15 ◇
EMDE BOAS VAN, P. *Machine models, computational complexity and number theory* ◇ D15 D98 ◇
EVANGELIST, M. *Non-standard propositional logics and their application to complexity theory* ◇ B80 D15 F20 ◇
EVEN, S. & LONG, T.J. & YACOBI, Y. *A note on deterministic and nondeterministic time complexity* ◇ D15 ◇
FANG, ZHIXI *The closure property of certain classes of languages under bi-language forms (Chinese) (English summary)* ◇ D05 D15 D25 ◇
FISCHER, MICHAEL J. & MEYER, A.R. & PATERSON, M.S. *$\Omega(n \log n)$ lower bounds on length of Boolean formulas* ◇ B05 D15 ◇
FRIEDMAN, H.M. & KO, KER-I *Computational complexity of real functions* ◇ D15 D80 ◇

FUERER, M. *The complexity of Presburger arithmetic with bounded quantifier alternation depth* ◇ B25 D15 ◇
HARTMANIS, J. *A note on natural complete sets and Goedel numberings* ◇ D15 D25 D45 ◇
HEMMERLING, A. *Zur Raumkompliziertheit von Absuchprozessen auf endlichen Graphen* ◇ D05 D15 ◇
HODGSON, B.R. & KENT, C.F. *An arithmetical characterization of NP* ◇ D15 ◇
HUWIG, H. *Ein Modell des P= NP-Problems mit einer positiven Loesung (English summary)* ◇ D15 G30 ◇
IMMERMAN, N. *Upper and lower bounds for first order expressibility* ◇ D15 ◇
JOHNSON, D.S. *The NP-completeness column: an ongoing guide* ◇ D15 D98 ◇
JUKNA, S. *A principle for obtaining lower bounds of arithmetical complexity (Russian) (English and Lithuanian summaries)* ◇ D15 ◇
JUKNA, S. *Arithmetical representations of machine complexity classes (Russian) (English and Lithuanian summaries)* ◇ D15 D20 ◇
KANNAN, R. *Circuit-size lower bounds and nonreducibility to sparse sets* ◇ D15 ◇
KARP, R.M. & LIPTON, R.J. *Turing machines that take advice* ◇ D15 ◇
KO, KER-I *Some negative results on the computational complexity of total variation and differentiation* ◇ D15 D80 ◇
KO, KER-I *The maximum value problem and NP real numbers* ◇ D15 D80 ◇
KOWALCZYK, W. *A sufficient condition for the consistency of P= NP with Peano arithmetic* ◇ D15 F25 F30 ◇
KREITZ, C. & WEIHRAUCH, K. *Complexity theory on real numbers and functions* ◇ D15 ◇
KRUPSKIJ, V.N. *Simultaneous approximability of real numbers (Russian)* ◇ D15 ◇
LAWLER, E.L. & LENSTRA, J.K. *Machine scheduling with precedence constraints* ◇ D15 ◇
LEIVANT, D. *Unprovability of theorems of complexity theory in weak number theories* ◇ D15 F30 ◇
LEWIS, H.R. & PAPADIMITRIOU, C.H. *Symmetric space-bounded computation* ◇ D15 ◇
LIBUS, M. & OSTROWSKA, M. *Remarks on L.J.Stockmeyer's note "The complexity of decision problems in automata theory and logic" (Polish) (English summary)* ◇ D05 D10 D15 ◇
LICHTENSTEIN, D. *Planar formulae and their uses* ◇ D15 ◇
LIEBERHERR, K.J. & VAVASIS, S.A. *Analysis of polynomial approximation algorithms for constant expressions* ◇ B05 D15 ◇
LONG, T.J. *A note on sparse-oracles for NP* ◇ D15 ◇
LONG, T.J. *Strong nondeterministic polynomial-time reducibilities* ◇ D15 ◇
LYNCH, J.F. *Complexity classes and theories of finite models* ◇ C13 D15 ◇
LYNCH, N.A. *Accessibility of values as a determinant of relative complexity in algebras* ◇ D15 ◇
MAHANEY, S.R. *Sparse complete sets for NP: solution of a conjecture of Berman and Hartmanis* ◇ D15 ◇
MARCHENKOV, S.S. *On the complexity of exponent calculation (Russian)* ◇ D15 ◇
MAYR, E.W. & MEYER, A.R. *The complexity of the word problems for commutative semigroups and polynomial ideals* ◇ D15 D40 ◇
MONIEN, B. & SUDBOROUGH, I.H. *On eliminating nondeterminism from Turing machines which use less than logarithm worktape space* ◇ D10 D15 ◇
MORITA, K. & SUGATA, K. & UMEO, H. *Deterministic one-way simulation of two-way real-time cellular automata and its related problems* ◇ D05 D15 ◇
MUELLER, HEINRICH *The complexity of the vertex coloring problem for some classes of graphs restricted by cycle properties* ◇ D15 D80 ◇
MUNDICI, D. *Complexity of Craig's interpolation* ◇ C40 D15 ◇
NAZARYAN, G.A. *Realization of Boolean functions in algorithmic languages under constraints on the running time of the algorithms (Russian) (Armenian summary)* ◇ B05 D15 ◇
PAPADIMITRIOU, C.H. & ZACHOS, S. *Two remarks on the power of counting* ◇ D15 ◇
PAUL, W.J. *On-line simulation of $k+1$ tapes by k tapes requires nonlinear time* ◇ D10 D15 ◇
PLOTKIN, J.M. & ROSENTHAL, J.W. *The expected complexity of analytic tableaux analysis in propositional calculus* ◇ B05 B35 D15 F20 ◇
POLJAK, S. & TURZIK, D. *A polynomial algorithm for constructing a large bipartite subgraph, with an application to a satisfiability problem* ◇ B05 D15 ◇
RACKOFF, C.W. *Relativized questions involving probabilistic algorithms* ◇ D15 ◇
SARKISYAN, A.D. *Classes of functions defined by their calculation time on iterative networks (Russian) (Armenian summary)* ◇ D10 D15 ◇
SCHOENHAGE, A. *Random access machines and Presburger arithmetic* ◇ B25 D15 ◇
SCHOENING, U. *A uniform approach to obtain diagonal sets in complexity classes* ◇ D15 ◇
SCHOENING, U. *On NP-decomposable sets* ◇ D15 ◇
SELMAN, A.L. *Analogues of semirecursive sets and effective reducibilities to the study of NP complexity* ◇ D15 ◇
SELMAN, A.L. *Reductions on NP and p-selective sets* ◇ D15 ◇
SIMOVICI, D.A. *Several remarks on the complexity of graph computations* ◇ D10 D15 ◇
SIPSER, M. *On relativization and the existence of complete sets* ◇ D15 ◇
SOARE, R.I. *Computational complexity of recursively enumerable sets* ◇ D15 D25 ◇
STATMAN, R. *Completeness, invariance and λ-definability* ◇ B40 D15 D20 ◇
STERN, J. *Quelques aspects du probleme P= NP* ◇ D15 ◇
TAO, RENJI *A lower bound of kn^2 on time-complexity for one-tape automata on computations* ◇ D05 D15 ◇
TETRUASHVILI, M.R. *On the conjugacy problem for finitely presented subgroups of finite index of a finitely presented group with unsolvable conjugacy problem (Russian) (English and Georgian summaries)* ◇ D15 D40 ◇

VALIANT, L.G. *Reducibility by algebraic projections* ◇ D15 ◇
WELSH, D.J.A. *Problems in computational complexity* ◇ D15 ◇
ZACHOS, S. *Robustness of probabilistic computational complexity classes under definitional perturbations* ◇ D15 ◇

1983

AJTAI, M. *Σ^1_1-formulae on finite structures* ◇ C13 C62 D15 H15 ◇
AUSTIN, A.K. *An elementary approach to NP-completeness* ◇ D15 ◇
BANACHOWSKI, L. & OKTABA, H. *A presentation of the scientific activities of the theory of computation group in the period from 1978–82* ◇ D15 D20 ◇
BLUM, N. *A note on the "parallel computation thesis"* ◇ D15 ◇
BOERGER, E. *From decision problems to problems of complexity* ◇ B25 D15 ◇
BOOK, R.V. & SELMAN, A.L. & XU, MEIRUI *Positive relativizations of complexity classes* ◇ D15 ◇
BOOK, R.V. & DONER, J.E. & XU, MEIRUI *Refining nondeterminism in relativizations of complexity classes* ◇ D15 ◇
BRANDSTAEDT, A. & WAGNER, K. *Reversal-bounded and visit-bounded realtime computations* ◇ D10 D15 ◇
BRANDSTAEDT, A. *Space classes, intersections of languages and bounded erasing homomorphisms* ◇ D05 D15 ◇
BRAUNMUEHL VON, B. & VERBEEK, R. *Input-driven languages are recognized in* $\log n$ *space* ◇ D15 ◇
BURGIN, M.S. *Multiple computations and the Kolmogorov complexity for such processes (Russian)* ◇ D15 ◇
CHAN, TATHUNG & IBARRA, O.H. *On the space and time complexity of functions computable by simple programs* ◇ D15 ◇
COOK, S.A. *The classification of problems which have fast parallel algorithm* ◇ D15 ◇
DOSHITA, S. & YAMASAKI, S. *The satisfiability problem for a class consisting of Horn sentences and some non-Horn sentences in propositional logic* ◇ B20 B25 D15 ◇
FRANCO, J. & PAULL, M. *Probabilistic analysis of the Davis-Putnam procedure for solving the satisfiability problem* ◇ B35 D15 ◇
FREJVALD, R.V. *Space and reversal complexity of probabilistic one-way Turing machines* ◇ D10 D15 ◇
FURST, M. & LIPTON, R.J. & STOCKMEYER, L.J. *Pseudorandom number generation and space complexity* ◇ D10 D15 ◇
GACS, P. *On the relation between descriptional complexity and algorithmic probability* ◇ D15 ◇
GASARCH, W.I. & HOMER, S. *Relativizations comparing NP and exponential time* ◇ D15 ◇
GRANDJEAN, E. *Complexity of the first-order theory of almost all finite structures* ◇ B25 C13 D15 ◇
GUREVICH, Y. *Algebras of feasible functions* ◇ D15 D20 D75 ◇
HAREL, D. *Recurring dominoes: making the highly undecidable highly understandable* ◇ B25 B75 D15 D55 ◇
HARTMANIS, J. & YESHA, Y. *Computation times of NP sets of different densities* ◇ D15 ◇
HARTMANIS, J. *On sparse sets in NP-P* ◇ D15 ◇
HARTMANIS, J. *On Goedel speed-up and succinctness of language representation* ◇ D15 F20 ◇
HASENJAEGER, G. *Exponential diophantine description of a small universal Turing machine (UTM) using Matiyasevich's 1975/80 "masking"* ◇ D10 D15 D35 ◇
HEINTZ, JOOS *Definability and fast quantifier elimination in algebraically closed fields* ◇ B25 C10 C40 C60 D10 D15 ◇
HEMMERLING, A. *$D \neq ND$ fuer mehrdimensionale Turing-Automaten mit sublogarithmischer Raumschranke* ◇ D10 D15 ◇
HODGSON, B.R. & KENT, C.F. *A normal form for arithmetical representation of NP-sets* ◇ D15 ◇
HOMER, S. & MAASS, W. *Oracle dependent properties of the lattice of NP sets* ◇ D15 ◇
HUNT III, H.B. & ROSENKRANTZ, D.J. *The complexity of monadic recursion schemes: executability problems, nesting depth, and applications* ◇ D15 D20 ◇
HUYNH, D.T. *Commutative grammars: The complexity of uniform word problems* ◇ D05 D15 ◇
IBARRA, O.H. & MORAN, S. *Some time-space tradeoff results concerning single-tape and offline TMs* ◇ D10 D15 ◇
IMMERMAN, N. *Languages which capture complexity classes* ◇ D15 ◇
INOUE, K. & TAKANAMI, I. & TANIGUCHI, H. *Two-dimensional alternating Turing machines* ◇ D10 D15 ◇
JOHNSON, D.S. *The NP-completeness column: an ongoing guide* ◇ D15 ◇
JOSEPH, D. *Polynomial time computations in models of ET* ◇ D15 F30 ◇
KANOVICH, M.I. *Complexity and convergence of algorithmic mass problems (Russian)* ◇ D15 D30 ◇
KO, KER-I *On self-reducibility and weak P-selectivity* ◇ D15 ◇
KO, KER-I *On the computational complexity of ordinary differential equations* ◇ D15 D80 ◇
KO, KER-I *On the definitions of some complexity classes of real numbers* ◇ D15 ◇
KOSOVSKIJ, N.K. *Polynomial lower bounds for the complexity of establishing the solvability of logical-arithmetical equations (Russian) (English summary)* ◇ D15 ◇
KRAMOSIL, I. *On extremum-searching approximate probabilistic algorithms* ◇ D15 D20 ◇
KREMPA, J. & PETTOROSSI, A. & SKOWRON, A. *$\mathfrak{F}$-computable numbers* ◇ D15 ◇
KURTZ, S.A. *On the random oracle hypthesis* ◇ D15 ◇
LAUTEMANN, C. *BPP and the polynomial hierarchy* ◇ D15 ◇
LIU, ZHENHONG *NP-complete problems and some approximate algorithms I,II (Chinese)* ◇ D15 ◇
LOLLI, G. *Complessita delle teorie* ◇ B25 D15 ◇
MANSFIELD, A. *On the computational complexity of a merge recognition problem* ◇ D15 ◇

MUCHNIK, A.A. *Supplement of the translator to the paper "On alternation. I, II" (Russian)* ◇ B45 D05 D10 D15 F50 ◇
MUNDICI, D. *A lower bound for the complexity of Craig's interpolants in sentential logic* ◇ B20 C40 D15 ◇
MUNDICI, D. *Natural limitations of decision procedures for arithmetic with bounded quantifiers* ◇ B25 D10 D15 F30 ◇
MYCIELSKI, J. *The meaning of the conjecture $P \neq NP$ for mathematical logic* ◇ B30 D15 ◇
ORPONEN, P. *Complexity classes of alternating machines with oracles* ◇ D15 ◇
REMESLENNIKOV, V.N. & ROMAN'KOV, V.A. *Model-theoretic and algorithmic questions of group theory (Russian)* ◇ B25 C60 C98 D15 D30 D40 ◇
RUMSAS, A. *On the complexity of computations over the ring of integers (Russian) (English summary)* ◇ D15 ◇
RYSTSOV, I.K. *Polynomial complete problems in automata theory* ◇ D05 D15 ◇
SANTHA, M. *Constructions d'oracles pour la hierarchie polynomiale relativisee (English summary)* ◇ D15 ◇
SAVAGE, J.E. & SWAMY, S. *Space-time tradeoffs for linear recursion* ◇ D15 ◇
SAVITCH, W.J. *A note on relativized log space* ◇ D15 ◇
SCHMIDT, DIANA *An alternative definition of NP* ◇ D15 ◇
SCHOENING, U. *A low and a high hierarchy within NP* ◇ D15 ◇
SCHOENING, U. *On the structure of Δ_2^P* ◇ D15 ◇
STILLWELL, J.C. *Efficient computation in groups and simplicial complexes* ◇ D15 D40 ◇
SUDBOROUGH, I.H. *Bandwidth constraints on problems complete for polynomial time* ◇ D15 ◇
TALJA, J. *On the complexity-relativized strong reducibilities* ◇ D15 D30 ◇
UKKONEN, E. *Two results on polynomial time truth-table reductions to sparse sets* ◇ D15 ◇
VAZIRANI, U.V. & VAZIRANI, V.V. *A natural encoding scheme proved probabilistic polynomial complete* ◇ D15 ◇
VOLGER, H. *A new hierarchy of elementary recursive decision problems* ◇ B25 D15 D20 ◇
VOLGER, H. *Turing machines with linear alternation, theories of bounded concatenation and the decision problem of first order theories* ◇ B25 D10 D15 ◇
WATANABE, O. *The time-precision tradeoff problem on on-line probabilistic Turing machines* ◇ D10 D15 ◇
WELSH, D.J.A. *Randomised algorithms* ◇ D15 ◇
YAP, C.K. *Some consequences of non-uniform conditions on uniform classes* ◇ D15 ◇
YESHA, Y. *On certain polynomial-time truth-table reducibilities of complete sets to sparse sets* ◇ D15 ◇
ZAK, S. *A Turing machine time hierarchy* ◇ D10 D15 ◇
ZEUGMANN, T. *A-posteriori characterizations in inductive inference of recursive functions (French, German and Russian summaries)* ◇ D15 ◇
ZHANG, GUOQIANG *NP-completeness and restricted partitions (Chinese) (English summary)* ◇ D15 ◇
ZIMAND, M. *Complexity of probabilistic algorithms* ◇ D15 ◇

1984

ALBRECHT, A. & BAUERNOEPPEL, F. & JUNG, H. & SAMMLER, O. *Einfuehrende Kapitel der Kompliziertheitstheorie. Seminarberichte* ◇ D15 ◇
AMBOS-SPIES, K. & FLEISCHHACK, H. & HAGEN, H. *p-generic sets* ◇ D15 E40 ◇
AMBOS-SPIES, K. *On the structure of polynomial time degrees* ◇ D15 ◇
AMBOS-SPIES, K. *P-mitotic sets* ◇ D15 ◇
BALCAZAR, J.L. *Separating, strongly separating, and collapsing relativized complexity classes* ◇ D15 ◇
BALCAZAR, J.L. & BOOK, R.V. & SCHOENING, U. *Sparse oracles, lowness, and highness* ◇ D15 ◇
BANJEVIC, D. *Note on the number of sequences of given complexity* ◇ D15 ◇
BAUER, G. & OTTO, F. *Finite complete rewriting systems and the complexity of word problem* ◇ D05 D15 ◇
BELAGA, EH.G. *Locally synchronous complexity in the light of the trans-box method* ◇ D15 ◇
BERTONI, A. & MAURI, G. & SABADINI, N. *Non deterministic machines and their generalizations* ◇ D10 D15 ◇
BLASS, A.R. & GUREVICH, Y. *Equivalence relations, invariants, and normal forms* ◇ D15 ◇
BLASS, A.R. & GUREVICH, Y. *Equivalence relations, invariants, and normal forms II* ◇ D15 E15 ◇
BLASS, A.R. & GUREVICH, Y. *Henkin quantifiers and complete problems* ◇ C80 D15 ◇
BOERGER, E. *Spektralproblem and completeness of logical decision problems* ◇ B25 C13 D15 ◇
BOOK, R.V. & SCHOENING, U. *Immunity, relativizations, and nondeterminism* ◇ D15 ◇
BOOK, R.V. & LONG, T.J. & SELMAN, A.L. *Quantitative relativizations of complexity classes* ◇ D15 ◇
BOOK, R.V. *Relativizations of complexity classes* ◇ D15 ◇
BUECHI, J.R. & MAHR, B. & SIEFKES, D. *Manual on REC - A language for use and cost analysis of recursion over arbitrary data structures* ◇ B75 D15 D80 ◇
BURGIN, M.S. *Complexity measures on systems of parallel algorithms (Russian)* ◇ D15 ◇
BYERLY, R.E. *Definability of recursively enumerable sets in abstract computational complexity theory* ◇ D15 D25 ◇
CENZER, D. *Monotone reducibility and the family of infinite sets* ◇ D15 D55 ◇
CHAZELLE, B. & OTTMANN, T. & SOISALON-SOININEN, E. & WOOD, D. *The complexity and decidability of SEPARATION* ◇ D15 ◇
CHEN, JIYUAN *The satisfiability problem for simple boolean expressions belongs to P (Chinese) (English summary)* ◇ B05 B25 D15 ◇
CHEN, ZHONGYUEH *On the fixpoints of nondeterministic recursive definitions* ◇ D15 ◇
CHERNIAK, C. *Computational complexity and the universal acceptance of logic* ◇ D15 ◇
CHESNOKOV, S.V. *Syllogisms in deterministic analysis (Russian)* ◇ D15 ◇

CHISTOV, A.L. & GRIGOR'EV, D.YU. *Complexity of quantifier elimination in theory of algebraically closed fields* ◇ C10 C60 D15 ◇

COLBOURN, C.J. & COLBOURN, M.J. & STINSON, D.R. *The computational complexity of recognizing critical sets* ◇ D15 ◇

DAHLHAUS, E. *Reduction to NP-complete problems by interpretations* ◇ C13 D15 ◇

DAHN, B.I. *The limit behaviour of exponential terms* ◇ C60 C65 D15 ◇

DALEY, R.P. & SMITH, C.H. *On the complexity of inductive inference* ◇ B75 D15 D20 ◇

DENENBERG, L. & LEWIS, H.R. *Logical syntax and computational complexity* ◇ B25 D15 ◇

DENENBERG, L. & LEWIS, H.R. *The complexity of the satisfiability problem for Krom formulas* ◇ B20 B25 D15 ◇

DURIS, P. & GALIL, Z. *A time-space tradeoff for language recognition* ◇ D05 D15 ◇

DURIS, P. & GALIL, Z. & PAUL, W.J. & REISCHUK, R. *Two nonlinear lower bounds for on-line computations* ◇ D15 ◇

EMDE BOAS VAN, P. & SAVELSBERGH, M.W.P. *Bounded tiling, an alternative to satisfiability?* ◇ D15 ◇

EMERSON, E.A. & SISTLA, A.P. *Deciding branching time logic: a triple exponential decision procedure for CTL* ◇ B75 D15 ◇

EMERSON, E.A. & SISTLA, A.P. *Deciding full branching time logic* ◇ B25 B45 D15 ◇

FRIEDMAN, H.M. *On the spectra of universal relational sentences* ◇ C13 D15 ◇

FRIEDMAN, H.M. *The computational complexity of maximization and integration* ◇ D15 D80 ◇

FUERER, M. *The computational complexity of the unconstrained limited domino problem (with implications for logical decision problems)* ◇ B20 D15 D80 ◇

FURST, M. & SAXE, J.B. & SIPSER, M. *Parity, circuits, and the polynomial-time hierarchy* ◇ D15 ◇

GABARRO, J. *Pushdown space complexity and related full-A.F.L.s* ◇ D10 D15 ◇

GRANDJEAN, E. *Spectre des formules du premier order et complexite algorithmique* ◇ C13 C98 D15 ◇

GRANDJEAN, E. *The spectra of first-order sentences and computational complexity* ◇ C13 D15 ◇

GRANDJEAN, E. *Universal quantifiers and time complexity of random access machines* ◇ C13 D15 ◇

GUREVICH, Y. & LEWIS, H.R. *A logic for constant-depth circuits* ◇ B75 D15 ◇

GUREVICH, Y. & STOCKMEYER, L.J. & VISHKIN, U. *Solving NP-hard problems on graphs that are almost trees and an application to facility location problems* ◇ D15 ◇

GUREVICH, Y. *Toward logic tailored for computational complexity* ◇ B75 C13 D15 ◇

HAREL, D. & PELEG, D. *On static logics, dynamic logics, and complexity classes* ◇ B75 D15 ◇

HARTMANIS, J. & YESHA, Y. *Computation times of NP sets of different densities* ◇ D15 ◇

HAUSSLER, D. & WARMUTH, M.K. *On the complexity of iterated shuffle* ◇ B75 D15 ◇

HECKER, H.-D. *Abstrakte Tempomasse und speed-up-Theoreme fuer Enumerationen rekursiv-aufzaehlbarer Mengen* ◇ D15 D25 ◇

HECKER, H.-D. *Das Kompressionstheorem fuer Tempomasse* ◇ D15 D25 ◇

HELLER, H. & ZACHOS, S. *A new characterization of BPP* ◇ D15 ◇

HELLER, H. *On relativized polynomial and exponential computations* ◇ D15 ◇

HELLER, H. *Relativized polynomial hierachies extending two levels* ◇ D15 ◇

HEMMERLING, A. & MURAWSKI, G. *Zur Raumkompliziertheit mehrdimensionaler Turing-Automaten* ◇ D10 D15 ◇

HERRE, H. *Bemerkungen zum Begriff der praktischen Berechenbarkeit* ◇ D15 D20 ◇

HUNT III, H.B. *Terminating Turing machine computations and the complexity and/or decidability of correspondence problems, grammars, and program schemes* ◇ D10 D15 D20 ◇

HUNT III, H.B. & ROSENKRANTZ, D.J. *The complexity of monadic recursion schemes: Exponential time bounds* ◇ D15 D20 ◇

HUYNH, D.T. *Deciding the inequivalence of context-free grammars with 1-letter alphabet is Σ_2^p-complete* ◇ D05 D15 ◇

IBARRA, O.H. & KIM, S.M. & ROSIER, L.E. *Space and time efficient simulations and characterizations of some restricted classes of PDAS* ◇ D10 D15 ◇

IZUMI, M. & TAKAHASHI, N. & TANAKA, H. *Some fine hierarchies on relativized time-bounded complexity classes* ◇ D15 ◇

JANTKE, K.P. *Polynomial time inference of general pattern languages* ◇ D05 D15 D80 ◇

JOHNSON, D.S. *The NP-completeness column: an ongoing guide* ◇ D15 ◇

KLEINE BUENING, H. *Complexity of loop-problems in normed networks* ◇ D05 D15 ◇

KO, KER-I *Reducibilities on real numbers* ◇ D15 D30 ◇

KORTAS, K. & KUBIAK, W. *The quasilinear complete number problems in NQL* ◇ D15 ◇

KOWALCZYK, W. *Some connections between presentability of complexity classes and the power of formal systems of reasoning* ◇ B75 D15 ◇

KRAMOSIL, I. *Recursive classification of pseudo-random sequences* ◇ D15 D80 ◇

LADNER, R.E. & LIPTON, R.J. & STOCKMEYER, L.J. *Alternation bounded auxiliary pushdown automata* ◇ D10 D15 ◇

LANGE, K.-J. *Nondeterministic logspace reductions* ◇ D15 ◇

LUCKHARDT, H. *Obere Komplexitaetsschranken fuer TAUT- Entscheidungen* ◇ B05 D15 ◇

MUNDICI, D. *Δ-tautologies, uniform and nonuniform upper bounds in computation theory (Italian summary)* ◇ B75 C40 D15 ◇

MUNDICI, D. *NP and Craig's interpolation theorem* ◇ C40 D15 ◇

MUNDICI, D. *Tautologies with a unique Craig interpolant, uniform vs. nonuniform complexity* ⋄ C40 D15 ⋄
ORPONEN, P. & SCHOENING, U. *The structure of polynomial complexity cores* ⋄ D15 ⋄
PAUL, W.J. *On heads versus tapes* ⋄ D10 D15 ⋄
PLAISTED, D.A. *Complete problems in the first-order predicate calculus* ⋄ B10 D15 ⋄
PUDLAK, P. *Bounds for Hodes-Specker theorem* ⋄ B05 D15 ⋄
REGAN, K.W. *Arithmetical degrees of index sets for complexity cases* ⋄ D15 D55 ⋄
REIF, J.H. *The complexity of two-player games of incomplete information* ⋄ D15 E60 ⋄
SANTHA, M. *La hierarchie polynomiale avec oracle* ⋄ D15 ⋄
SCARPELLINI, B. *Complexity of subcases of Presburger arithmetic* ⋄ B25 D15 F20 F30 ⋄
SCHMIDT, DIANA *Limitations on separating nondeterministic and deterministic complexity classes* ⋄ D15 ⋄
SCHMIDT, DIANA *The complement of one complexity class in another* ⋄ D15 ⋄
SCHOENING, U. *Minimal pairs for P* ⋄ D15 ⋄
SCHOENING, U. *Robust algorithms: a different approach to oracles* ⋄ D15 ⋄
SHAPIRO, E.Y. *Alternation and the computational complexity of logic programs* ⋄ D10 D15 ⋄
SIPSER, M. *A topological view of some problems in complexity theory* ⋄ D15 ⋄
STENGER, H.-J. *Algebraic characterisations of NTIME(F) and NTIME(F,A)* ⋄ B20 B35 D15 D20 ⋄
STRASSEN, V. *Algebraische Berechnungskomplexitaet* ⋄ D15 ⋄
TETRUASHVILI, M.R. *Computational complexity of recognizing word equality in semigroups of a certain class (Russian) (English and Georgian summaries)* ⋄ D15 D40 ⋄
TETRUASHVILI, M.R. *The computational complexity of the theory of abelian groups with a given number of generators* ⋄ B25 D15 ⋄
TOURLAKIS, G. *An inductive number-theoretic characterization of NP* ⋄ D15 ⋄
TOVEY, C.A. *A simplified NP-complete satisfiability problem* ⋄ D15 ⋄
URSIC, S. *A linear characterization of NP-complete problems* ⋄ D15 ⋄
VALIANT, L.G. *An algebraic approach to computational complexity* ⋄ D15 ⋄
VITANYI, P.M.B. *The simple roots of real-time computation hierarchies* ⋄ D15 ⋄
VOLGER, H. *Rudimentary relations and Turing machines with linear alternation* ⋄ D10 D15 D20 ⋄
VOLGER, H. *The role of rudimentary relations in complexity theory* ⋄ D10 D15 ⋄
WAGNER, K. *Compact descriptions and the counting polynomial time hierarchy* ⋄ D15 ⋄
WAGNER, K. *The complexity of problems concerning graphs with regularities* ⋄ D15 ⋄
WAGNER, K. *The complexity of combinatorial problems with compactly described instances* ⋄ D15 ⋄
YUKAMI, T. *Some results on speed-up* ⋄ D15 F20 F30 ⋄
ZEUGMANN, T. *Recursive operators versus recursive functions with respect to the generation of classes of functions having a fastest program* ⋄ D15 D20 ⋄

1985

AJZENSHTEJN, M.KH. *Strengthened theorems on nomenclature and pseudoacceleration for the capacity measure of complexity (Russian)* ⋄ D15 ⋄
ALBERTS, M. *Complexity of computations on nondeterministic Turing machines (Russian)* ⋄ D10 D15 ⋄
ALBERTS, M. *Space complexity of alternating Turing machines* ⋄ D10 D15 ⋄
ALBERTS, M. *Tape complexity of nondeterministic Turing machines (Russian)* ⋄ D10 D15 ⋄
AMBOS-SPIES, K. *On the relative complexity of subproblems of intractable problems* ⋄ D15 ⋄
AMBOS-SPIES, K. *Sublattices of the polynomial degrees* ⋄ D15 ⋄
ANSHEL, M. & DOMANSKI, B. *The complexity of Dehn's algorithm for word problems in groups* ⋄ D15 D40 ⋄
BAJAJ, C. *Geometric optimization and the polynomial hierarchy* ⋄ D15 ⋄
BALCAZAR, J.L. & SCHOENING, U. *Bi-immune sets for complexity classes* ⋄ D15 ⋄
BALCAZAR, J.L. & DIAZ, J. & GABARRO, J. *On some "nonuniform" complexity measures* ⋄ D15 ⋄
BALCAZAR, J.L. *Simplicity, relativizations and nondeterminism* ⋄ D15 ⋄
BOERGER, E. *Current trends in theoretical computer science* ⋄ B75 D15 ⋄
BOOK, R.V. & OTTO, F. *Cancellation rules and extended word problems* ⋄ D03 D15 ⋄
BOOK, R.V. & LONG, T.J. & SELMAN, A.L. *Qualitative relativizations of complexity classes* ⋄ D15 ⋄
BROY, M. *On the Herbrand-Kleene universe for nondeterministic computations* ⋄ D15 ⋄
BUDACH, L. & GRAW, B. *Nonuniform complexity classes, decision graphs and homological properties of posets* ⋄ D15 ⋄
CELLUCCI, C. *Proof theory and complexity* ⋄ D15 F05 F20 F98 ⋄
CHUNG, MOONJUNG & EVANGELIST, W.M. & SUDBOROUGH, I.H. *Complete problems for space bounded subclasses of NP* ⋄ D15 ⋄
DAHN, B.I. *Fine structure of the integral exponential functions below* 2^{2^z} ⋄ D15 ⋄
DIMITRACOPOULOS, C. *A generalization of a theorem of H.Friedman* ⋄ C62 D15 F30 ⋄
DYMOND, P.W. & TOMPA, M. *Speedups of deterministic machines by synchronous parallel machines* ⋄ D15 ⋄
EJSMONT, M. *On the log-space reducibility among array languages* ⋄ D15 ⋄
FAGIN, R. & KLAWE, M.M. & PIPPENGER, N.J. & STOCKMEYER, L.J. *Bounded depth, polynomial-size circuits for symmetric functions* ⋄ D15 ⋄
GONCZAROWSKI, J. & WARMUTH, M.K. *Applications of scheduling theory to formal language theory* ⋄ D05 D15 ⋄

GRANDJEAN, E. *Universal quantifiers and time complexity of random access machines* ◇ C13 D10 D15 ◇
GRIGORCHUK, R.I. *A relationship between algorithmic problems and entropy characteristics of groups (Russian)* ◇ D15 D40 ◇
HAKEN, A. *The intractability of resolution* ◇ B35 D15 D35 ◇
HAREL, D. *Recurring dominoes: Making the highly undecidable highly understandable* ◇ B25 B75 D05 D15 D55 ◇
HARTMANIS, J. & IMMERMAN, N. *On complete problems for NP ∩ CoNP* ◇ D15 ◇
HARTMANIS, J. & IMMERMAN, N. & SEWELSON, V. *Sparse sets in NP - P: EXPTIME versus NEXPTIME* ◇ D15 ◇
HINMAN, P.G. & ZACHOS, S. *Probabilistic machines, oracles, and quantifiers* ◇ C80 D15 ◇
HOMER, S. *Minimal polynomial degrees of nonrecursive sets* ◇ D15 ◇
HORVATH, S. *On defining the Blum complexity of partial recursive sequence functions* ◇ D15 D20 ◇
INOUE, K. & ITO, A. & TAKANAMI, I. & TANIGUCHI, H. *A space-hierarchy result on two-dimensional alternating Turing machines with only universal states* ◇ D10 D15 ◇
IWATA, S. & KASAI, T. *Gradually intractable problems and non-deterministic log-space lower bounds* ◇ D15 ◇
JOHNSON, D.S. *The NP-completeness column: an ongoing guide* ◇ D15 ◇
JOSEPH, D. & YOUNG, P. *Some remarks on witness functions for polynomial and noncomplete sets in NP* ◇ D15 ◇
JUNG, H. *On probabilistic time and space* ◇ D15 ◇
KAPUR, D. & KRISHNAMOORTHY, M.S. & MCNAUGHTON, R. & NARENDRAN, P. *An $O(|T|^3)$ algorithm for testing the Church-Rosser property of Thue systems* ◇ D03 D15 ◇
KELMANS, A.K. *The existence of a "most complex" problem in the class of problems verifiable in nonpolynomial time (Russian)* ◇ D15 ◇
KO, KER-I *Nonlevelable sets and immune sets in the accepting density hierachy in NP* ◇ D15 ◇
KO, KER-I & SCHOENING, U. *On circuit-size complexity and the low hierarchy in NP* ◇ D15 ◇
KO, KER-I *On some natural complete operators* ◇ D15 ◇
KOBAYASHI, K. *On proving time constructibility of functions* ◇ D15 ◇
KURTZ, S.A. *Sparse sets in NP – P: relativizations* ◇ D15 ◇
LI, MING *Lower bounds by Kolmogorov-complexity (extended abstract)* ◇ D15 ◇
LI, XIANG *On relativized nondeterministic polynomial-time bounded combinations* ◇ D15 ◇
LITOW, B.E. *On efficient deterministic simulation of Turing machine computations below log-space* ◇ D10 D15 ◇
LIVCHAK, A.B. *On polynomial computability (Russian)* ◇ D15 ◇
LONG, T.J. *On restricting the size of oracles compared with restricting access to oracles* ◇ D15 ◇
MAASS, W. *Combinatorial lower bound arguments for deterministic and nondeterministic Turing machines* ◇ D10 D15 ◇
MAHANEY, S.R. & YOUNG, P. *Reductions among polynomial isomorphism types* ◇ D15 ◇
MEYER AUF DER HEIDE, F. *Lower bounds for solving linear Diophantine equations on random access machines* ◇ D10 D15 ◇
ORPONEN, P. & RUSSO, D.A. & SCHOENING, U. *Polynomial levelability and maximal complexity cores* ◇ D15 ◇
RYTTER, W. *The complexity of two-way pushdown automata and recursive programs* ◇ D10 D15 ◇
SCHMIDT, DIANA *The recursion-theoretic structure of complexity classes* ◇ D15 ◇
SONTAG, E.D. *Real addition and the polynomial hierarchy* ◇ D15 ◇
STERN, J. *Complexty of some problems from the theory of automata* ◇ D05 D15 ◇
SZWAST, W. *On some properties of Horn's spectra (Polish) (English summary)* ◇ C13 D15 ◇
TULIPANI, S. *An algorithm to determine for any prime p, a polynomial-sized Horn sentence which expresses "the cardinality is not p"* ◇ B35 D15 ◇
VITANYI, P.M.B. *An n^1.618 lower bound on the time to simulate one queue or two pushdown stores by one tape* ◇ D10 D15 ◇
WANG, JIE *A necessary and sufficient condition for the existence for a given B of an A such that $P^A = NP^B$ (Chinese)* ◇ D15 ◇
WATANABE, O. *On one-one polynomial time equivalence relations* ◇ D15 ◇
WECHSUNG, G. *On sparse complete sets* ◇ D15 ◇
WECHSUNG, G. *On the Boolean closure of NP* ◇ D15 ◇
WEISPFENNING, V. *The complexity of elementary problems in archimedian ordered groups* ◇ B25 C10 D15 D40 ◇
YOUNG, P. *Goedel theorems, exponential difficulty and undecidability of arithmetic theories: an exposition* ◇ D15 D35 F20 F30 ◇

D20 Recursive functions and relations, subrecursive hierarchies

1888

DEDEKIND, R. *Was sind und was sollen die Zahlen?* ◇ B28 D20 E75 ◇

1923

SKOLEM, T.A. *Begruendung der elementaren Arithmetik durch die rekurrierende Denkweise ohne Anwendung scheinbarer Veraenderlichen mit unendlichem Ausdehnungsbereich* ◇ A05 B28 D20 F30 ◇

1926

HILBERT, D. *Ueber das Unendliche* ◇ A05 B28 D20 E10 E30 ◇

1927

SUDAN, G. *Sur le nombre transfini* ω^{ω} ◇ D20 E10 ◇

1928

ACKERMANN, W. *Zum Hilbertschen Aufbau der reellen Zahlen* ◇ B28 D20 F15 ◇

1931

GOEDEL, K. *Ueber formal unentscheidbare Saetze der "Principia Mathematica" und verwandter Systeme I* ◇ B28 B30 D20 D35 F25 F30 F35 ◇

1932

CHURCH, A. *A set of postulates for the foundation of logic* ◇ B10 B40 D20 ◇

POLITZER, R. *Rekursive Funktionen* ◇ D20 D98 ◇

1933

CHURCH, A. *A set of postulates for the foundation of logic (second paper)* ◇ B10 B40 D20 ◇

1934

PETER, R. *Ueber den Zusammenhang der verschiedenen Begriffe der rekursiven Funktionen* ◇ D20 ◇

1935

CHURCH, A. *A proof of freedom from contradiction* ◇ B40 D20 ◇

CHURCH, A. *An unsolvable problem of elementary number theory* ◇ B40 D20 D35 ◇

PETER, R. *Konstruktion nichtrekursiver Funktionen* ◇ D20 ◇

PETER, R. *Zur Theorie der rekursiven Funktionen (Hungarian) (German summary)* ◇ D20 ◇

1936

CHURCH, A. *A note on the "Entscheidungsproblem"* ◇ B20 D20 D35 ◇

CHURCH, A. *An unsolvable problem of elementary number theory* ◇ D20 D35 ◇

CHURCH, A. *Mathematical logic (mimeographed notes)* ◇ B40 B98 D20 ◇

CHURCH, A. & ROSSER, J.B. *Some properties of conversion* ◇ B40 D20 ◇

KEMPNER, A.J. *Remarks on "unsolvable" problems* ◇ D20 D35 ◇

KLEENE, S.C. *A note on recursive functions* ◇ D20 ◇

KLEENE, S.C. *General recursive functions of natural numbers* ◇ D20 D25 ◇

PETER, R. *Ueber die mehrfache Rekursion* ◇ D20 ◇

POST, E.L. *Finite combinatory processes - formulation I* ◇ D03 D20 D40 ◇

TURING, A.M. *On computable numbers, with an application to the "Entscheidungsproblem"* ◇ D05 D10 D20 D35 F60 ◇

1937

HERMES, H. *Definite Begriffe und berechenbare Zahlen* ◇ D05 D10 D20 F60 ◇

PETER, R. *Ueber rekursive Funktionen der zweiten Stufe* ◇ D20 ◇

TURING, A.M. *Computability and* λ*-definability* ◇ A05 B40 D05 D10 D20 F99 ◇

1940

CHURCH, A. *On the concept of a random sequence* ◇ D20 D80 E75 ◇

PETER, R. *Contribution to recursive number theory* ◇ D20 F30 ◇

1941

CHURCH, A. *The calculi of* λ *conversion* ◇ B40 D20 ◇

CURRY, H.B. *A formalization of recursive arithmetic* ◇ B28 D20 F30 ◇

1943

KALMAR, L. *Ein einfaches Beispiel fuer ein unentscheidbares arithmetisches Problem (Hungarian) (German summary)* ◇ D20 F30 ◇

1944

SKOLEM, T.A. *A note on recursive arithmetic* ◇ D20 F30 ◇

SKOLEM, T.A. *Remarks on recursive functions and relations* ◇ D20 ◇

SKOLEM, T.A. *Some remarks on recursive arithmetic* ◇ D20 F30 ◇

SKOLEM, T.A. *Some remarks on the comparison between recursive functions* ◇ D20 ◇

1946

POST, E.L. *Note on a conjecture of Skolem* ◇ D20 ◇

1947

CSILLAG, P. *Eine Bemerkung zur Aufloesung der eingeschachtelten Rekursion* ◇ D20 ◇

MARKOV, A.A. *On the representation of recursive functions (Russian)* ◇ D20 ◇
ROBINSON, R.M. *Primitive recursive functions* ◇ D20 ◇

1948

ROBINSON, R.M. *Recursion and double recursion* ◇ D20 ◇

1949

MARKOV, A.A. *On the representation of recursive functions* ◇ D20 ◇
MARTIN, R.M. *A note on nominalism and recursive functions* ◇ A05 D20 ◇

1950

KUZNETSOV, A.V. *On primitive recursive functions of large oscillation (Russian)* ◇ D20 ◇
MYHILL, J.R. *A system which can define its own truth* ◇ B30 D20 F25 F30 ◇
PETER, R. *Zusammenhang der mehrfachen und transfiniten Rekursionen* ◇ D20 ◇
ROBINSON, JULIA *General recursive functions* ◇ D20 ◇

1951

MARKOV, A.A. *The theory of algorithms (Russian)* ◇ D03 D20 ◇
PETER, R. *Probleme der Hilbertschen Theorie der hoeheren Stufen von rekursiven Funktionen* ◇ D20 ◇
PETER, R. *Rekursive Funktionen* ◇ D20 D98 ◇
SCHUETTE, K. *Eine Bemerkung ueber quasirekursive Funktionen* ◇ D20 ◇

1952

BERECZKI, I. *Existenz einer nichtelementaren rekursiven Funktion (Ungarisch) (Russische und deutsche Zusammenfassung)* ◇ D20 ◇
BERECZKI, I. *Loesung eines Markovschen Problems betreffs einer Ausdehnung des Begriffes der elementaren Funktion* ◇ D20 ◇
KALMAR, L. *Another proof to the Markov-Post theorem* ◇ D03 D20 D40 ◇
MARKOV, A.A. *On unsolvable algorithmic problems (Russian)* ◇ D03 D20 D25 D35 D40 ◇
MARKOV, A.A. *Theory of algorithms (Russian) (Hungarian summary)* ◇ D03 D20 ◇
PETER, R. *Transfinite Rekursionen (Grundlagenforschung und rekursive Funktionen)* ◇ D20 ◇

1953

CRAIG, W. *On axiomatizability within a system* ◇ C07 D20 ◇
DETLOVS, V.K. *Normal algorithms and recursive functions (Russian)* ◇ D03 D20 ◇
GRZEGORCZYK, A. *Some classes of recursive functions* ◇ D20 ◇
KOLMOGOROV, A.N. *On the concept of algorithm (Russian)* ◇ B25 D20 ◇
MOSTOWSKI, ANDRZEJ *A lemma concerning recursive functions and its applications* ◇ D20 ◇
PETER, R. *Rekursive Definitionen, wobei fruehere Funktionswerte von variabler Anzahl verwendet werden* ◇ D20 ◇
ROUTLEDGE, N.A. *Ordinal recursion* ◇ D20 F15 ◇
SKOLEM, T.A. *Some considerations concerning recursive functions* ◇ D20 ◇
USPENSKIJ, V.A. *The Goedel theorem and the theory of algorithms (Russian)* ◇ D20 D35 F30 ◇
USPENSKIJ, V.A. *Theorem of Goedel and theory of algorithms (Russian)* ◇ D20 D25 F30 ◇

1954

ADDISON, J.W. *On some points of the theory of recursive functions* ◇ D20 D55 ◇
GOODSTEIN, R.L. *The relatively exponential, logarithmic and circular functions in recursive function theory* ◇ D20 F30 F60 ◇
JANICZAK, A. *On the reducibility of decision problems* ◇ D10 D20 D25 D30 ◇
JANICZAK, A. *Some remarks on partially recursive functions* ◇ D20 ◇
MARKOV, A.A. *Theory of algorithms (Russian)* ◇ D03 D20 D40 D98 ◇
SCHUETTE, K. *Kennzeichnung von Ordnungszahlen durch rekursiv erklaerte Funktionen* ◇ D20 F15 ◇
SKOLEM, T.A. *Remarks on "elementary" arithmetic functions* ◇ D20 ◇

1955

GRZEGORCZYK, A. *Computable functionals* ◇ D20 ◇
GRZEGORCZYK, A. *On the definition of computable functionals* ◇ D20 F60 ◇
HERMES, H. *Vorlesung ueber Entscheidungsprobleme in Mathematik und Logik* ◇ B25 B40 B65 D10 D20 D35 D98 ◇
KALMAR, L. *The solution of a problem of K.Schroeter, concerning the definition of general recursive functions* ◇ D20 ◇
KALMAR, L. *Ueber ein Problem, betreffend die Definition des Begriffes der allgemein-rekursiven Funktion* ◇ D20 ◇
KUZNETSOV, A.V. & TRAKHTENBROT, B.A. *Investigation of partial recursive operators by means of the theory of Baire's space (Russian)* ◇ D20 D25 D30 ◇
LACOMBE, D. *Classes recursivement fermes et fonctions majorantes* ◇ D20 D55 ◇
MARKWALD, W. *Zur Eigenschaft primitiv-rekursiver Funktionen, unendlich viele Werte anzunehmen* ◇ D20 ◇
MO, SHAOKUI *On the definition of primitive recursive functions (Chinese) (English summary)* ◇ D20 ◇
MYHILL, J.R. & SHEPHERDSON, J.C. *Effective operations on partial recursive functions* ◇ D20 ◇
PETER, R. *Ein neuer Beweis fuer die Tatsache, dass die Klasse der primitiv-rekursiven Funktionen umfassender als die Klasse der elementaren Funktionen ist* ◇ D20 ◇
ROBINSON, JULIA *A note on primitive recursive functions* ◇ D20 ◇
ROBINSON, R.M. *Primitive recursive functions II* ◇ D20 ◇
ROSSER, J.B. *Deux esquisses de logique* ◇ B10 B40 D20 E30 ◇
ROSSER, J.B. *Logique combinatoire et λ-conversion* ◇ B40 D20 ◇

SHANIN, N.A. *On some logical problems of arithmetic (Russian)* ◇ D20 F30 F50 ◇
TRAKHTENBROT, B.A. *Tabular representation of recursive operators (Russian)* ◇ D20 ◇
USPENSKIJ, V.A. *On calculable operations (Russian)* ◇ D20 ◇

1956

HU, SHIHUA *The primitive recursiveness of a kind of recursion (Chinese) (English summary)* ◇ D20 ◇
KLEENE, S.C. *A note on computable functionals* ◇ D20 ◇
MCCARTHY, J. *The inversion of functions defined by Turing machines* ◇ D10 D20 ◇
MO, SHAOKUI *On the explicit form of general recursive functions (Chinese) (English summary)* ◇ D20 ◇
MOSTOWSKI, ANDRZEJ *Concerning a problem of H.Scholz* ◇ C13 D20 ◇
PETER, R. *Die beschraenkt-rekursiven Funktionen und die Ackermannsche Majorisierungsmethode* ◇ D20 ◇
RICE, H.G. *Recursive and recursively enumerable orders* ◇ D20 D25 D45 ◇
TRAKHTENBROT, B.A. *Signalizing functions and matrix operators (Russian)* ◇ B70 D20 ◇
USPENSKIJ, V.A. *Calculable operations and the notion of a program (Russian)* ◇ D20 D45 ◇

1957

AXT, P. *A subrecursive hierarchy and primitive recursive degrees* ◇ D20 ◇
CHURCH, A. *Binary recursive arithmetic* ◇ D20 F30 ◇
DUDA, W.L. *Post canonical language* ◇ D03 D20 ◇
GRZEGORCZYK, A. *On the definitions of computable real continuous functions* ◇ D20 F60 ◇
KALMAR, L. *On Church's hypothesis as foundation for studies related to so-called unsolvable mathematical problems (Hungarian)* ◇ D20 D35 ◇
KALMAR, L. *Ueber arithmetische Funktionen von unendlich vielen Variablen, welche an jeder Stelle bloss von einer endlichen Anzahl von Variablen abhaengig sind* ◇ D20 E20 ◇
KREISEL, G. & LACOMBE, D. & SHOENFIELD, J.R. *Effective operations and partial recursive functionals* ◇ D20 F60 ◇
KREISEL, G. & LACOMBE, D. & SHOENFIELD, J.R. *Fonctionnelles recursivement definissables et fonctionnelles recursives* ◇ D20 F60 ◇
KREISEL, G. *Sums of squares* ◇ C57 C60 D20 F07 F99 ◇
MAEHARA, S. *General recursive functions in the number-theoretic formal system* ◇ D20 F30 ◇
MOSTOWSKI, ANDRZEJ *On computable sequences* ◇ D20 F60 ◇
MROWKA, S. *Recursive families of sets* ◇ D20 ◇
NERODE, A. *General topology and partial recursive functionals* ◇ D20 D80 ◇
PETER, R. *The boundedly recursive functions of Grzegorczyk and the majorisation method of Ackermann (Hungarian)* ◇ D20 ◇
RABIN, M.O. *Effective computability of winning strategies* ◇ D20 D25 ◇
RICE, H.G. *On the relative density of sets of integers* ◇ D20 D30 ◇
SPECTOR, C. *Recursive ordinals and predicative set theory* ◇ D20 D55 E70 F15 F65 ◇

1958

DETLOVS, V.K. *The equivalence of normal algorithms and recursive functions (Russian)* ◇ D03 D20 ◇
ERSHOV, A.P. *On operator algorithms (Russian)* ◇ B75 D03 D20 ◇
FRIEDBERG, R.M. *Un contre-example relatif aux fonctionelles recursives* ◇ D20 ◇
FRIEDBERG, R.M. *4-quantifier completeness: A Banach-Mazur functional not uniformly partial recursive* ◇ D20 ◇
HERMES, H. *Zum Einfachheitsprinzip in der Wahrscheinlichkeitsrechnung* ◇ A05 B48 D20 D80 ◇
HU, SHIHUA & LOH, CHUNGWAN *Normal forms of general recursive functions (Chinese)* ◇ D20 ◇
KLEENE, S.C. *Extension of an effectively generated class of functions by enumeration* ◇ D20 ◇
KOLMOGOROV, A.N. & USPENSKIJ, V.A. *On the definition of an algorithm (Russian)* ◇ D20 ◇
MO, SHAOKUI *On the construction of number-theoretic functions (Chinese)* ◇ D20 ◇
MYHILL, J.R. *The foundations of mathematics, II. The theory of recursive functions* ◇ D20 ◇
NAGORNYJ, N.M. *Some generalisations of the concept of normal algorithm (Russian)* ◇ D03 D20 ◇
ORLOVSKIJ, E.S. *Some questions in the theory of algorithms (Russian)* ◇ D20 ◇
PETER, R. *Graphschemata und rekursive Funktionen* ◇ D20 ◇
PORTE, J. *Une simplification de la theorie de Turing* ◇ D10 D20 ◇
ROGERS JR., H. *Goedel numberings of partial recursive functions* ◇ D20 D45 F40 ◇
SHOENFIELD, J.R. *The class of recursive functions* ◇ D20 D55 ◇
YANOV, YU.I. *On logical schemata of algorithms (Russian)* ◇ B75 D20 ◇

1959

AXT, P. *On a subrecursive hierarchy and primitive recursive degrees* ◇ D20 ◇
KALMAR, L. *An argument against the plausibility of Church's thesis* ◇ A05 D20 F99 ◇
KALUZHNIN, L.A. *On the algorithmization of mathematical problems (Russian)* ◇ B75 D20 ◇
KREISEL, G. & LACOMBE, D. & SHOENFIELD, J.R. *Partial recursive functionals and effective operations* ◇ D20 F60 ◇
MO, SHAOKUI *The application of recursive functions on programming (Chinese)* ◇ D20 ◇
ORLOVSKIJ, E.S. *Algorithmic operators in the narrow sense (Russian)* ◇ D20 ◇
PETER, R. *Rekursivitaet und Konstruktivitaet* ◇ A05 D20 F60 ◇
PETER, R. *Ueber die Partiell-Rekursivitaet der durch Graphschemata definierten zahlentheoretischen Funktionen* ◇ B75 D20 ◇
ROGERS JR., H. *Recursive functions over well-ordered partial orderings* ◇ D20 D45 ◇

ROSE, A. *A note on the representation of general recursive functions and the μ-quantifier* ◇ D20 ◇
SHAPIRO, H.S. *Numbers and functions computable by means of rational recurrence formulae* ◇ D20 ◇
SKOLEM, T.A. *Some remarks on the constructions of functions by substitution* ◇ D20 ◇
SUZUKI, Y. *Enumeration of recursive sets* ◇ D20 ◇
TRAKHTENBROT, B.A. *Descriptive classifications in recursive arithmetics (Russian)* ◇ D20 F30 ◇
TRAKHTENBROT, B.A. *On effective operators and properties related to their continuousness (Russian)* ◇ D20 ◇
TSEJTIN, G.S. *Uniform recursiveness of algorithmic operators on general recursive functions and a canonical representation for constructive functions of a real argument (Russian)* ◇ D20 F60 ◇
USPENSKIJ, V.A. *Computable operations, computable operators and constructively continuous functions (Russian)* ◇ D20 F60 ◇
USPENSKIJ, V.A. *The concept of program and computable operators (Russian)* ◇ D20 D45 ◇
VUCKOVIC, V. *Partially ordered recursive arithmetics* ◇ D20 F30 ◇

1960

ASSER, G. *Rekursive Wortfunktionen* ◇ D20 ◇
ERSHOV, A.P. *Operator algorithms. I. Basic concepts (Russian)* ◇ B75 D03 D20 ◇
HU, SHIHUA *Kernel functions and normal forms of recursive functions (Chinese)* ◇ D20 ◇
HU, SHIHUA & LOH, CHUNGWAN *Kernel functions. Theory of recursive algorithms II (Chinese) (English summary)* ◇ D10 D20 ◇
HU, SHIHUA *Normal forms of recursive functions. Theory of recursive algorithms III (Chinese) (English summary)* ◇ D20 ◇
HU, SHIHUA *Recursive algorithms. Theory of recursive algorithms I (Chinese) (English summary)* ◇ D10 D20 ◇
HU, SHIHUA *Theory of recursive algorithms* ◇ D20 ◇
LACOMBE, D. *La theorie des fonctions recursives et ses applications (expose d'une information generale)* ◇ D10 D20 D80 D98 ◇
LIU, SHICHAO *A theorem on general recursive functions* ◇ D20 ◇
LIU, SHICHAO *An enumeration of the primitive recursive functions without repetition* ◇ D20 ◇
LIU, SHICHAO *An example of general recursive well-ordering which is not primitive recursive* ◇ D20 D45 ◇
LIU, SHICHAO *Proof of a conjecture of Routledge* ◇ D20 ◇
MCCARTHY, J. *Recursive functions of symbolic expressions and their computation by machine. Part I* ◇ B75 D20 ◇
NAGORNYJ, N.M. *On the investigation of the isomorphisms of associative calculi (Russian) (German summary)* ◇ D05 D20 ◇
PAVLOVA, E.A. *The lattice of denseness of sets of natural numbers (Russian) (Moldavian summary)* ◇ D20 D50 ◇
PETER, R. *Zu einem Rekursionsschema von Hu Shih-Hua* ◇ D20 ◇
PORTE, J. *Quelques pseudo-paradoxes de la "calculabilite effective"* ◇ A05 D20 ◇
POUR-EL, M.B. *A comparison of five "computable" operators* ◇ D20 ◇
RABIN, M.O. *Degree of difficulty of computing a function and a partial ordering of recursive sets* ◇ D15 D20 ◇
TRAKHTENBROT, B.A. *Algorithms and machine solutions of problems (Russian)* ◇ B25 B35 D20 ◇
USPENSKIJ, V.A. *Lectures on computable functions (Russian)* ◇ D20 D98 ◇
VUCKOVIC, V. *Rekursive Wortarithmetik* ◇ D20 ◇

1961

ASSER, G. & VUCKOVIC, V. *Funktionen-Algorithmen* ◇ D03 D20 ◇
ASSER, G. *Funktionen-Algorithmen und Graphschemata* ◇ D20 ◇
AXT, P. *Note on 3-recursive functions* ◇ D20 ◇
FRAISSE, R. *Une notion de recursivite relative* ◇ D20 D30 D75 ◇
GRZEGORCZYK, A. *Fonctions recursives* ◇ D20 D98 ◇
HANSON, N.R. *The Goedel theorem* ◇ D20 D35 F30 ◇
HARROP, R. *On the recursivity of finite sets* ◇ D20 ◇
KREISEL, G. & TAIT, W.W. *Finite definability of number-theoretic functions and parametric completeness of equational calculi* ◇ B20 D20 ◇
KREISEL, G. *Set theoretic problems suggested by the notion of potential totality* ◇ A05 C62 D20 D55 D65 ◇
LADRIERE, J. *Expression de la recursion primitive dans le calcul-λK* ◇ B40 D20 ◇
LAMBEK, J. *How to program an infinite abacus* ◇ D10 D20 ◇
PETER, R. *Primitiv-rekursive Wortbeziehungen in der Programmierungssprache "Algol 60" (Russian summary)* ◇ B75 D20 ◇
PETER, R. *Ueber die Verallgemeinerung der Rekursionsbegriffe fuer abstrakte Mengen als Definitionsbereiche* ◇ D20 D75 ◇
PETER, R. *Ueber die Verallgemeinerung der Theorie der rekursiven Funktionen fuer abstrakte Mengen geeigneter Struktur als Definitionsbereiche (Russian summary)* ◇ D20 D75 ◇
RIEGER, L. *On a critique of Church's thesis concerning general recursive functions in arithmetic (Czech)* ◇ A05 D20 F99 ◇
SMULLYAN, R.M. *Theory of formal systems* ◇ B98 D03 D05 D20 D25 D98 ◇
TAIT, W.W. *Nested recursion* ◇ D20 ◇
TEIXEIRA, M.T. *Recursive functions and the foundations of mathematics (Portuguese)* ◇ D20 ◇

1962

ERSHOV, A.P. *Operator algorithms II. A description of basic programming constructions (Russian)* ◇ B75 D20 ◇
FEFERMAN, S. *Classifications of recursive functions by means of hierarchies* ◇ D20 ◇
HUZINO, S. *Finite automata and Asser's function algorithms* ◇ D03 D05 D20 ◇

LACHLAN, A.H. *Multiple recursion* ◇ D20 ◇
LIU, SHICHAO *A generalized concept of primitive recursion and its application to deriving general recursive functions* ◇ D20 ◇
LORENZEN, P. *Metamathematik* ◇ A05 B98 C60 D20 D35 D98 F98 ◇
MARKOV, A.A. *Computable invariants (Russian)* ◇ D05 D20 ◇
MCLAUGHLIN, T.G. *A note on contraproduction domains* ◇ D20 D25 ◇
PETER, R. *Ueber die Rekursivitaet einiger Uebersetzungs-Transformationen I,II (II: Verwendung einer Linearisierungsweise des Kantorowitsch-schen Ausdrucks-graphen) (Russian summaries)* ◇ D20 ◇
PETER, R. *Ueber die Verallgemeinerung der Theorie der rekursiven Funktionen fuer abstrakte Mengen geeigneter Struktur als Definitionsbereiche (Fortsetzung)* ◇ D20 D75 ◇
POGORZELSKI, H.A. *A note on an arithmetization of a word system in a denumerable alphabet* ◇ D05 D20 ◇
POGORZELSKI, H.A. *Exponential chains of natural numbers and Vuckovic's recursion functions I (Polish) (Russian and English summaries)* ◇ D20 ◇
POGORZELSKI, H.A. *Word arithmetic: theory of primitive words* ◇ D05 D20 ◇
RADO, T. *On non-computable functions* ◇ D10 D20 ◇
SKOLEM, T.A. *Recursive enumeration of some classes of primitive recursive functions and a majorisation theorem* ◇ D20 D25 ◇
SKORDEV, D.G. *One more example of a recursively complete arithmetic operation (Bulgarian) (German summary)* ◇ D20 ◇
VUCKOVIC, V. *Einfuehrung von $\Sigma_f(x)$ und $\Pi_f(x)$ in der rekursiven Gitterpunktarithmetik* ◇ D20 F30 ◇

1963

AXT, P. *Enumeration and the Grzegorczyk hierarchy* ◇ D20 ◇
AXT, P. *Relativization of a primitive recursive hierarchy* ◇ D20 ◇
BUCK, R.C. *Mathematical induction and recursive definitions* ◇ B28 D20 ◇
CLEAVE, J.P. *A hierarchy of primitive recursive functions* ◇ D20 ◇
HU, SHIHUA & HUANG, ZULIANG *Addition and multiplication (Chinese)* ◇ D20 F30 ◇
ION, I.D. *On the connection between the theory of algorithms and the abstract theory of automata* ◇ D05 D20 ◇
LECERF, Y. *Machines de Turing reversibles. Recursive insolublite en $n\in N$ de l'equation $u=\theta^n u$, ou θ est un "isomorphisme de codes"* ◇ D10 D20 ◇
LECERF, Y. *Recursive insolubilite de l'equation generale de diagonalisation de deux monomorphismes de monoides libres $\varphi x=\psi x$* ◇ D10 D20 ◇
LEVIEN, R.E. *Set-theoretic formalizations of computational algorithms, computable functions, and general-purpose computers* ◇ B75 D10 D20 E75 ◇
MARKOV, A.A. *Indistinguishability by invariants in the theory of associative calculi (Russian)* ◇ D05 D20 ◇
MASLOV, S.YU. *Strong representability of sets by calculi (Russian)* ◇ D03 D20 ◇
MENDELSON, E. *On some recent criticism of Church's thesis* ◇ A05 D20 F99 ◇
MO, SHAOKUI *On general recursive functions (Chinese)* ◇ D20 ◇
PARTIS, M.T. *Commutative partially ordered recursive arithmetics* ◇ D20 F60 ◇
PETER, R. *Programmierung und partiell-rekursive Funktionen* ◇ B75 D20 ◇
PETER, R. *Ueber die Rekursivitaet der Begriffe der mathematischen Grammatiken (Russian summary)* ◇ D05 D20 ◇
PETER, R. *Ueber die Primitiv-Rekursivitaet einiger den Aufbau von Formeln charakterisierenden Wortfunktionen* ◇ D20 ◇
POENARU, V. *Expose sommaire de la theorie des algorithmes (d'apres A.A. Markov)* ◇ D03 D20 ◇
POLYAKOV, E.A. *Certain properties of algebras of recursive functions (Russian)* ◇ D20 D75 ◇
RADO, T. *On a simple source for non-computable functions* ◇ D10 D20 ◇
RIEGER, L. *Kleene's normal form for computable functions (Czech) (Russian and English summaries)* ◇ D20 ◇
RITCHIE, R.W. *Classes of predictably computable functions* ◇ D15 D20 ◇
ROSE, G.F. & ULLIAN, J.S. *Approximation of functions on the integers* ◇ D20 D25 ◇
SHAPIRO, N.Z. *Functions which remain partial recursive under all similarity transformations* ◇ D20 ◇
SHEPHERDSON, J.C. & STURGIS, H.E. *Computability of recursive functions* ◇ D10 D20 ◇
SKORDEV, D.G. *Computable and μ-recursive operators (Bulgarian) (Russian and German summaries)* ◇ D20 ◇
SMULLYAN, R.M. *Pseudo-uniform reducibility* ◇ D20 D25 D30 D35 ◇
TRAKHTENBROT, B.A. *Algorithms and automatic computing machines* ◇ D10 D20 D98 ◇
TRAKHTENBROT, B.A. *On the frequency computability of functions (Russian)* ◇ D20 D55 ◇
WANG, HAO *Computation* ◇ D20 D25 D30 D35 D60 ◇

1964

AXT, P. *On a problem of Skolem* ◇ D20 ◇
CELLUCCI, C. *Categorie ricorsive* ◇ C57 D20 G30 ◇
FENSTAD, J.E. *Algorithms in mathematics: an introduction to recursion theory and its applications I,II (Norwegian) (English summary)* ◇ D20 D98 ◇
HOREJS, J. *Note on definition of recursiveness* ◇ D20 ◇
HU, SHIHUA & YANG, DONGPING *On primitive recursiveness (Chinese)* ◇ D20 ◇
KLOSS, B.M. *The definition of complexity of algorithms (Russian)* ◇ D15 D20 ◇
KREIDER, D.L. & RITCHIE, R.W. *Predictably computable functionals and definition by recursion* ◇ D10 D20 ◇
LACHLAN, A.H. *Effective operations in a general setting* ◇ D20 D25 D75 ◇

MIKENBERG, A.M. *Complete systems of nondecreasing general recursive functions (Russian)* ◇ D20 ◇
MO, SHAOKUI *Studies on number-theoretic operators (Chinese)* ◇ D20 ◇
POGORZELSKI, H.A. *Commutative recursive word arithmetic in the alphabet of prime numbers* ◇ D05 D20 ◇
POGORZELSKI, H.A. *Primitive words in an infinite abstract alphabet* ◇ D05 D20 ◇
POLYAKOV, E.A. *Algebras of recursive functions (Russian)* ◇ D20 D75 ◇
POLYAKOV, E.A. *On some properties of algebras of recursive functions (Russian)* ◇ D20 D75 ◇
ROEDDING, D. *Ueber die Eliminierbarkeit von Definitionsschemata in der Theorie der recursiven Funktionen* ◇ D20 ◇
RUS, T. *Ueber ein formales System I,II (Rumaenisch)* ◇ B10 D20 ◇
SKORDEV, D.G. *On the concept of a recursively complete arithmetic operation (Bulgarian) (Russian summary)* ◇ D20 ◇
TSEJTIN, G.S. *A method of presenting the theory of algorithms and enumerable sets (Russian)* ◇ D20 D25 ◇
ZYKOV, A.A. *Recursively calculable functions of graphs* ◇ D20 D75 D80 ◇

1965

AXT, P. *Iteration of primitive recursion* ◇ D20 ◇
BLUM, E.K. *Enumeration of recursive sets by Turing machine* ◇ D10 D20 ◇
CROSSLEY, J.N. & DUMMETT, M. (EDS.) *Formal systems and recursive functions* ◇ B97 D20 D97 ◇
DAVIS, MARTIN D. *Introduction to Goedel's paper: On formally undecidable propositions of the Principia Mathematica and related systems I* ◇ A10 D20 F30 ◇
DAVIS, MARTIN D. (ED.) *The undecidable. Basic papers on undecidable propositions, unsolvable problems and computable functions* ◇ D20 D25 D35 D97 F30 ◇
FENSTAD, J.E. *A universal function for the class of all primitive recursive functions* ◇ D20 ◇
FISCHER, P.C. *Theory of provably recursive functions* ◇ D20 D30 F30 ◇
FISCHER, P.C. & RUBY, S.S. *Translational methods and computational complexity* ◇ B70 D15 D20 ◇
GOEDEL, K. *On undecidable propositions of formal mathematical systems* ◇ D20 D35 F30 ◇
GOLD, E.M. *Limiting recursion* ◇ D20 D55 ◇
HANF, W.P. *Model-theoretic methods in the study of elementary logic* ◇ B25 C07 D20 D25 D35 F25 ◇
HAY, L. *On creative sets and indices of partial recursive functions* ◇ D20 D25 ◇
HOREJS, J. *On generalizations of a theorem on recursive sets* ◇ D20 D25 ◇
HU, SHIHUA & YANG, DONGPING *About primitive recursivity (Chinese)* ◇ D20 ◇
ITO, MAKOTO *Boolean recursive functions and closure algebra* ◇ B25 B45 D20 ◇
KHUTORETSKIJ, A.B. *On recursive isomorphisms (Russian)* ◇ D20 D45 ◇
KOLMOGOROV, A.N. *Three approaches to the definition of the concept "quantity of information" (Russian)* ◇ D15 D20 D80 ◇
MAL'TSEV, A.I. *Algorithms and recursive functions (Russian)* ◇ D20 D98 ◇
MAL'TSEV, A.I. *Positive and negative numerations (Russian)* ◇ D20 D45 ◇
MAYOH, B.H. *Unsolvable problems in the theory of computable numbers* ◇ D20 D35 F60 ◇
MO, SHAOKUI & SHEN, BAIYING *Systems of primitive recursive arithmetic (Chinese)* ◇ D20 F30 ◇
MUELLER, GERT H. *Charakterisierung einer Klasse von rekursiven Funktionen* ◇ D20 ◇
NEGRESCU, I. & PAVALOIU, I. *A property of matrix schemes of algorithms (Romanian)* ◇ D20 ◇
PAVLOVA, E.A. *Certain arithmetic and algebraic properties of a system of densities (Russian)* ◇ D20 ◇
PETER, R. *Programmierung und partiell-rekursive Funktionen* ◇ B75 D20 ◇
PETER, R. *Ueber die sequenzielle Berechenbarkeit von rekursiven Wortfunktionen durch Kellerspeicher* ◇ D10 D20 ◇
PETER, R. *Zum Beitrag von F. Schwenkel "Rekursive Wortfunktionen ueber unendlichen Alphabeten"* ◇ D20 ◇
PUTNAM, H. *Trial and error predicates and the solution to a problem of Mostowski* ◇ C57 D20 D55 ◇
RICE, H.G. *Recursion and iteration* ◇ D20 ◇
RITCHIE, R.W. *Classes of recursive functions based on Ackermann's function* ◇ D20 ◇
ROEDDING, D. *Darstellungen der (im Kalmar-Csillagschen Sinne) elementaren Funktionen* ◇ D20 ◇
ROEDDING, D. *Einige aequivalente Praezisierungen des intuitiven Berechenbarkeitsbegriffs* ◇ D20 ◇
ROGERS JR., H. *On universal functions* ◇ D20 ◇
SCHWENKEL, F. *Rekursive Wortfunktionen ueber unendlichen Alphabeten* ◇ D20 D75 ◇
SKORDEV, D.G. *A class of primitive recursive functions (Russian) (English summary)* ◇ D20 ◇
TANG, ZHISONG *Recursiveness of systems of computer instructions (Chinese)* ◇ D20 ◇
ZAKREVSKIJ, A.D. *Algorithms of minimalization of weakly defined Boolean functions (Russian) (English summary)* ◇ B35 D20 ◇

1966

ASSER, G. *Ueber eine Darstellung der rekursiven Wortfunktionen in endlichen Automaten* ◇ D05 D20 ◇
AXT, P. *Iteration of relative primitive recursion* ◇ D20 ◇
BLUM, M. *Recursive function theory and speed of computation* ◇ D15 D20 ◇
CROSSLEY, J.N. *Constructive order types II* ◇ D20 D45 F15 ◇
DAVIS, MARTIN D. *Recursive functions - an introduction* ◇ D20 D35 ◇
DOEPP, K. *Ueber die Bestimmbarkeit des Verhaltens von Algorithmen* ◇ D20 D45 D55 ◇

HAY, L. *Isomorphism types of index sets of partial recursive functions* ◇ D20 D25 D30 ◇
LAVROV, I.A. & POLYAKOV, E.A. *Bases of algebras of recursive functions (Russian)* ◇ D20 D75 ◇
LOVELAND, D.W. *The Kleene hierarchy classification of recursively random sequences* ◇ D20 D55 D80 ◇
MCLAUGHLIN, T.G. *Two remarks on indecomposable number sets* ◇ D20 D25 ◇
NOLIN, L. *Organigrammes et machines de Turing* ◇ A05 D10 D20 ◇
OBERSCHELP, A. *Berechenbarkeit* ◇ D20 D98 ◇
OBERSCHELP, A. *Rechenmaschinen* ◇ D20 D98 ◇
PARIKH, R. *Some generalisations of the notion of well ordering* ◇ D20 D45 E07 F35 ◇
POLYAKOV, E.A. *Some properties of algebras of recursive functions (Russian)* ◇ D20 D75 ◇
RITTER, W.E. *Notation systems and an effective fixed point property* ◇ D20 D30 F15 ◇
ROEDDING, D. *Der Entscheidbarkeitsbegriff in der mathematischen Logik* ◇ B25 D10 D20 D25 D35 ◇
ROEDDING, D. *Ueber Darstellungen der elementaren Funktionen II* ◇ D20 ◇
SCHREIBER, P. *Ueber die Entbehrlichkeit von Hilfsbuchstaben bei der Berechnung mehrstelliger Wortfunktionen durch Markowsche Algorithmen* ◇ D03 D20 ◇
SHURYGIN, V.A. *Nontrivial constructive mappings of certain sets (Russian)* ◇ D20 D45 F60 ◇
URBAN, J. *Die Minimalisierung der zur sequentiellen Berechnung der partiell-rekursiven Wortfunktionen notwendigen Kellerspeicher* ◇ D10 D20 ◇
YANG, DONGPING *Primitive recursive simple sets and their hierarchy (Chinese)* ◇ D20 D25 ◇

1967

CLEAVE, J.P. & ROSE, H.E. *$\mathscr{E}_n$-arithmetic* ◇ D20 F30 ◇
DEVIDE, V. *A remark concerning the foundation of the theory of recursive functions (Serbo-Croatian summary)* ◇ D20 ◇
DOEPP, K. *Die Unentscheidbarkeit der Grzegorczyk-Klassen* ◇ D20 ◇
ERSHOV, YU.L. *Numerations of families of general recursive functions (Russian)* ◇ D20 D45 ◇
FREJDZON, R.I. *The representability of algorithmically decidable predicates by Rabin machines (Russian)* ◇ D03 D10 D20 ◇
GLADSTONE, M.D. *A reduction of the recursion scheme* ◇ D20 ◇
KLEENE, S.C. *Mathematical logic* ◇ B98 D20 D25 D55 D98 F30 F98 ◇
LAVROV, I.A. *The use of k-order arithmetic progressions in constructing the bases of the algebra of primitive recursive functions (Russian)* ◇ D20 D75 ◇
MAHN, F.-K. *Zu den primitiv-rekursiven Funktionen ueber einem Bereich endlicher Mengen* ◇ D20 D65 ◇
MASLOV, S.YU. *The concept of strict representability in the general theory of calculi (Russian)* ◇ D05 D20 ◇
MCCLEARY, S.H. *Primitive recursive computations* ◇ D20 ◇
MINSKY, M.L. *Computation: finite and infinite machines* ◇ D05 D10 D20 ◇
NASIBULLOV, KH.KH. *Some bases of the Robinson algebra (Russian)* ◇ D20 ◇
PARTIS, M.T. *Limited universal and existential quantifiers in commutative partially ordered recursive arithmetics* ◇ D20 F60 ◇
POLYAKOV, E.A. *Several properties of the algebra of general recursive functions (Russian)* ◇ D20 D75 ◇
RITTER, W.E. *Representability of partial recursive functions in formal theories* ◇ D20 F30 ◇
ROEDDING, D. *Primitiv-rekursive Funktionen ueber einen Bereich endlicher Mengen* ◇ D20 ◇
ROGERS JR., H. *Some problems of definability in recursive function theory* ◇ D20 D25 D98 ◇
SANCHIS, L.E. *Functionals defined by recursion* ◇ D20 F10 ◇
SHCHEGLOV, A.I. *The algebra of partially recursive functions (Russian)* ◇ D20 D75 ◇
SOKOLOV, V.A. *Isomorphisms of maximal subalgebras of R.Robinson's algebra (Russian) (English summary)* ◇ D20 D75 ◇
TRAKHTENBROT, B.A. *Complexity of algorithms and computations (Russian)* ◇ D10 D15 D20 D98 ◇

1968

BRAUER, W. & INDERMARK, K. *Algorithmen, rekursive Funktionen und formale Sprachen* ◇ D05 D10 D20 D98 ◇
CONSTABLE, R.L. *Extending and refining hierarchies of computable functions* ◇ D20 ◇
DUNHAM, C. *An uncompletable function* ◇ D10 D20 ◇
EILENBERG, S. & ELGOT, C.C. *Iteration and recursion* ◇ D20 D25 ◇
ENDERTON, H.B. *On provable recursive functions* ◇ D20 F30 ◇
FISCHER, P.C. & MEYER, A.R. & ROSENBERG, ARNOLD L. *Counter machines and counter languages* ◇ D05 D10 D15 D20 ◇
FITCH, F.B. *A note on recursive relations* ◇ B40 D20 ◇
HARRISON, J. *Recursive pseudo-well-orderings* ◇ D20 D45 ◇
HARTMANIS, J. *Computational complexity of one-tape Turing machine computations* ◇ D10 D15 D20 ◇
HERMES, H. *Praedikatenlogik und Theorie der rekursiven Funktionen* ◇ B10 D20 ◇
HOTZ, G. *Automatentheorie und formale Sprachen I. Turingmaschinen und rekursive Funktionen* ◇ D05 D10 D20 D98 ◇
KALMAR, L. *On the problem of full utilization of the technical possibilities of computers in devising appropriate approximation methods for the solution of numerical problems* ◇ D20 D80 ◇
KINO, A. *On provably recursive functions and ordinal recursive functions* ◇ D20 F05 F30 F35 ◇
KOZ'MINYKH, V.V. *One-place primitively recursive functions (Russian)* ◇ D20 G15 ◇
MALINOVSKIJ, V.I. *An equivalence problem in a certain class of address algorithms (Russian) (English summary)* ◇ D20 ◇

MEYER, A.R. & RITCHIE, D.M. *Classification of functions by computational complexity* ◇ D15 D20 ◇
NAGASHIMA, T. *On elementary functions of natural numbers* ◇ D20 ◇
NASIBULLOV, KH.KH. *Bases in algebras of partially recursive functions (Russian)* ◇ D20 D75 ◇
PARSONS, C. *Hierarchies of primitive recursive functions* ◇ D20 ◇
POLYAKOV, E.A. *Recursive subsets of sets of recursive functions (Russian)* ◇ D20 ◇
POLYAKOV, E.A. *Some problems in the theory of recursive functions (Russian)* ◇ D20 ◇
POLYAKOV, E.A. *Some properties of algebras of recursive functions (Russian)* ◇ D20 D75 ◇
RITCHIE, R.W. & YOUNG, P. *Strong representability of partial functions in arithmetic theories* ◇ D20 F30 ◇
ROBINSON, JULIA *Recursive functions of one variable* ◇ D20 D40 ◇
ROEDDING, D. *Klassen rekursiver Funktionen* ◇ D20 ◇
SHCHEGLOV, A.I. *Power of the set of maximal sub-algebras of an algebra of partial-recursive functions (Russian)* ◇ D20 D75 ◇
TAKAHASHI, MOTO-O *Ackermann's model and recursive predicates* ◇ D20 ◇
YANOV, YU.I. *The logical transformations of schemes of algorithms (Russian)* ◇ D20 ◇
YOUNG, P. *An effective operator, continuous but not partial recursive* ◇ D20 ◇
ZAMA, N. *On some algebraic formulation of algorithms* ◇ D20 ◇

1969

AANDERAA, S.O. & COOK, S.A. *On the minimum computation time of functions* ◇ D20 ◇
BAER, ROBERT M. *Computability by normal algorithms* ◇ D03 D20 ◇
BASU, S.K. *On classes of computable functions* ◇ D20 ◇
BENESOVA, M. *Real-time computable functions and almost primitive recursive functions* ◇ D15 D20 ◇
CANNONITO, F.B. & FINKELSTEIN, M. *On primitive recursive permutations and their inverses* ◇ D20 D45 ◇
CLEAVE, J.P. *The primitive recursive analysis of ordinary differential equations and the complexity of their solutions* ◇ D20 D80 F60 ◇
CONSTABLE, R.L. *Upward and downward diagonalizations over axiomatic complexity classes* ◇ D20 ◇
DEKHTYAR', M.I. *Impossibility of eliminating complete enumeration in computing functions from their diagrams (Russian)* ◇ D20 ◇
DUSHSKIJ, V.A. *Extension of partial recursive functions and functions with a recursive graph (Russian)* ◇ D20 ◇
FABIAN, R.J. & KENT, C.F. *Recursive functions defined by ordinal recursion* ◇ D20 ◇
GYURIS, L. *On the connection of Glushkovian microprogram-algebras and logical schemas of the algorithms* ◇ B75 D20 ◇
HART, J. *$\mathscr{E}^0$-arithmetic* ◇ C62 D20 ◇
HECKER, H.-D. *Definierbarkeit und Unentscheidbarkeit* ◇ D20 D35 ◇
HERMAN, G.T. *A new hierarchy of elementary functions* ◇ D20 ◇
HERMAN, G.T. *On hierarchies of elementary functions* ◇ D20 ◇
HERMES, H. *Basic notions and applications of the theory of decidability* ◇ D10 D20 D25 D30 D35 D50 D98 ◇
HERMES, H. *On the notion of constructivity* ◇ D20 F65 ◇
HEYTING, A. *What is computable ? (Dutch)* ◇ D20 D35 D55 ◇
HIRSCHELMANN, A. *Primitiv-rekursive Funktionen in Peano-Algebren* ◇ D20 D75 G25 ◇
KALMAR, L. *R.Peter's work in the theory of recursive functions* ◇ A10 D20 ◇
KANOVICH, M.I. & KUSHNER, B.A. *Estimating the complexity of certain algorithmic problems of analysis (Russian)* ◇ D15 D20 D80 F60 ◇
KAPHENGST, H. *Malzew-Raeume, ein allgemeiner Begriff der rekursiven Abbildung* ◇ D20 D75 ◇
KENT, C.F. *Reducing ordinal recursion* ◇ D20 ◇
KHASIN, L.S. *Complexity bounds for the realization of monotonic symmetrical functions by means of formulas in the basis* ∨ , ∧ , ¬ *(Russian)* ◇ B05 D20 ◇
KHUTORETSKIJ, A.B. *The reducibility of computable enumerations (Russian)* ◇ D20 D25 D45 ◇
KLEENE, S.C. *On the normal form theorem* ◇ D20 ◇
KOLMOGOROV, A.N. *Logical basis of information theory and probability theory (Russian)* ◇ B30 D15 D20 D80 ◇
KORSHUNOV, A.D. *Comparison of the complexity of the largest and shortest disjunctive normal forms and a lower estimate of the number of irredundant disjunctive normal forms for almost all Boolean functions (Russian)* ◇ B05 D20 ◇
MAHN, F.-K. *Primitiv-rekursive Funktionen auf Termmengen* ◇ D20 ◇
MARANDZHYAN, G.B. *Certain properties of asymptotically optimal recursive functions (Russian) (Armenian and English summaries)* ◇ D15 D20 ◇
MARANDZHYAN, G.B. *Hierarchies of recursive functions and asymptotic optimality (Russian) (Armenian summary)* ◇ D15 D20 ◇
MARANDZHYAN, G.B. *On complexity scales of natural numbers (Russian)* ◇ D15 D20 ◇
MARCHENKOV, S.S. *Elimination of recursion schemes in Grzegorczyk's class $\mathscr{E}^2$ (Russian)* ◇ D20 ◇
MCCREIGHT, E.M. & MEYER, A.R. *Classes of computable functions defined by bounds on computation: preliminary report* ◇ D15 D20 ◇
MCCREIGHT, E.M. & MEYER, A.R. *Properties of bounds on computation* ◇ D15 D20 ◇
MILNER, R. *Program schemes and recursive function theory* ◇ B75 D10 D20 ◇
MOSHCHENSKIJ, V.A. *On the question of the complexity of Turing computations (Russian)* ◇ D10 D15 D20 ◇
NASIBULLOV, KH.KH. & POLYAKOV, E.A. *Certain recursive schemes (Russian)* ◇ D20 ◇
NASIBULLOV, KH.KH. *Recursive functions of large scope (Russian)* ◇ D20 ◇

NOLIN, L. *Formalisation des notions de machine et de programme* ◇ B75 D10 D20 ◇
PAOLA DI, R.A. *Random sets in subrecursive hierarchies* ◇ D20 ◇
PETER, R. *Automatische Programmierung zur Berechnung der partiell-rekursiven Funktionen* ◇ B75 D20 ◇
PETER, R. *Ueber die Pair-schen freien Binoiden als Spezialfaelle der angeordneten freien holomorphen Mengen (Russian summary)* ◇ D20 D75 ◇
PETER, R. *Ueber zweistufig definierte Sprachen* ◇ B15 D20 ◇
PETRI, N.V. *The complexity of algorithms and their operating time (Russian)* ◇ B35 D15 D20 ◇
POGORZELSKI, H.A. *Goldbach sentences in abstract arithmetics* $\mathscr{A}^k(A)$ *I* ◇ D20 F30 ◇
POLYAKOV, E.A. *Formal definability in algebras of recursive functions (Russian)* ◇ D20 D75 ◇
POLYAKOV, E.A. *Precomplete classes of primitive recursive functions (Russian)* ◇ D20 ◇
ROBINSON, JULIA *Finitely generated classes of sets of natural numbers* ◇ D20 D55 E20 ◇
SCHNORR, C.-P. *Eine Bemerkung zum Begriff der zufaelligen Folge (English summary)* ◇ D20 D80 F65 ◇
SCHWICHTENBERG, H. *Rekursionsformeln und die Grzegorczyk-Hierarchie* ◇ D20 ◇
SHCHEGLOV, A.I. *A certain algebra of one-place primitive recursive functions (Russian)* ◇ D20 D75 ◇
VALIEV, M.K. *The complexity of the word problem for finitely presented groups (Russian)* ◇ D20 D40 ◇
YOUNG, P. *Toward a theory of enumerations* ◇ D10 D15 D20 D25 D45 ◇
ZASLAVSKIJ, I.D. *The axiomatic determination of constructive objects and operations (Russian) (Armenian and English summaries)* ◇ D20 F50 ◇

1970

BARZDINS, J. *On the relative frequency of solution of algorithmically unsolvable mass problems (Russian)* ◇ D20 D25 ◇
BASS, L. & YOUNG, P. *Hierarchies based on computational complexity and irregularities of class determined measured sets* ◇ D15 D20 ◇
BASU, S.K. *On the structure of subrecursive degrees* ◇ D20 ◇
BLIKLE, A.J. *Functions definable by means of programs* ◇ B75 D20 ◇
EBBINGHAUS, H.-D. *Aufzaehlbarkeit* ◇ D03 D10 D20 D25 F30 ◇
EBBINGHAUS, H.-D. *Turing-Maschinen und berechenbare Funktionen I,III* ◇ D10 D20 ◇
EILENBERG, S. & ELGOT, C.C. *Recursiveness* ◇ D20 D25 D98 ◇
FREJDZON, R.I. *A characterization of the complexity of recursive predicates (Russian)* ◇ D20 ◇
GRODZKI, Z. *The degree of complexity of the sets of computations I* ◇ D20 ◇
KATERINOCHKINA, N.N. *The equivalence of certain computational equipment (Russian) (English summary)* ◇ D10 D15 D20 ◇
KAZANOVICH, YA.B. *A classification of primitive recursive functions using Turing machines (Russian)* ◇ D10 D20 ◇
KOZMIDIADI, V.A. *A certain generalization of finite automata that produces a hierarchy analogous to A.Grzegorczyk's classification of primitively recursive functions (Russian)* ◇ D05 D10 D20 ◇
KOZ'MINYKH, V.V. *On the subalgebras of R. Robinson's algebras (Russian)* ◇ D20 D75 ◇
LEVIN, L.A. & ZVONKIN, A.K. *The complexity of finite objects and the basing of the concepts of information and randomness on the theory of algorithms (Russian)* ◇ D20 D80 ◇
LOEB, M.H. *A model theoretic characterization of effective operations* ◇ C62 D20 ◇
LOEB, M.H. & WAINER, S.S. *Hierarchies of number-theoretic functions I,II* ◇ D20 ◇
MAHN, F.-K. *Turing-Maschinen und berechenbare Funktionen II* ◇ D10 D20 ◇
MARCHENKOV, S.S. *Multiple recursions that are limited in the class of primitively recursive functions (Russian) (English summary)* ◇ D20 ◇
MAYOH, B.H. *The relation between an object and its name: notation systems and their fixed point theorems* ◇ A05 B03 D20 D35 D45 F30 F60 ◇
MILLER, W. *Recursive function theory and numerical analysis* ◇ D20 D80 ◇
MUCHNIK, A.A. *Ueber zwei Zugaenge zur Klassifikation rekursiver Funktionen (Russisch)* ◇ D20 ◇
MUELLER, GERT H. *Rekursionsformen in der Zahlentheorie* ◇ D20 ◇
MUELLER, HORST *Ueber die mit Stackautomaten berechenbaren Funktionen* ◇ D10 D20 ◇
NEPOMNYASHCHIJ, V.A. *Rudimentary predicates and Turing calculations (Russian)* ◇ D10 D20 ◇
NEPOMNYASHCHIJ, V.A. *The rudimentary interpretation of two-tape Turing computations (Russian) (English summary)* ◇ D10 D20 ◇
NISHIZAWA, T.W. *A proof of the equivalence of computability to recursiveness with no arithmetization* ◇ D20 ◇
OSTROUKHOV, D.A. *On the coding of natural numbers with the aid of schemes of normal algorithms (Russian)* ◇ D03 D20 ◇
OWINGS JR., J.C. *Commutativity and common fixed points in recursion theory* ◇ D20 ◇
PAIR, C. & QUERE, A. *Sur les fonctions recursives primitives de ramifications* ◇ D20 ◇
PLATEK, R.A. *A note on the failure of the relativized enumeration theorem in recursive function theory* ◇ D20 ◇
PODLOVCHENKO, R.I. *Schemes of algorithms that are defined on situations (Russian)* ◇ D05 D20 ◇
SCHNORR, C.-P. *Ueber die Zufaelligkeit und den Zufallsgrad von Folgen* ◇ D20 D80 ◇
SCHROEDER, M.E. *Hierarchien primitiv-rekursiver Funktionen im Transfiniten* ◇ D20 E10 E47 ◇
SKORDEV, D.G. *Some simple examples of universal functions (Russian)* ◇ D20 ◇
STENLUND, S. *Combinators as effectively calculable functions* ◇ B40 D20 ◇

STRONG, H.R. *Depth-bounded computation* ◇ D15 D20 ◇
VUCKOVIC, V. *Effective enumerability of some families of partially recursive functions connected with computable functionals* ◇ D20 D25 ◇
VUCKOVIC, V. *Recursive word-functions over infinite alphabets* ◇ D20 ◇
WAINER, S.S. *A classification of the ordinal recursive functions* ◇ D20 D25 ◇
WERNER, GEORGES *Quelques remarques sur la complexite des algorithmes (English summary)* ◇ D15 D20 ◇
ZAKHAROV, D.A. *Recursive functions (Russian)* ◇ D20 D98 ◇

1971

ALOISIO, P. *Una nota sulle funzioni ricorsive* ◇ D20 ◇
AMOROSO, S. & BLOOM, S.L. *A model of the digital computer* ◇ D10 D20 ◇
BAKKER DE, J.W. *Recursive procedures* ◇ B40 B75 D20 ◇
BLIKLE, A.J. *Algorithmically definable functions: a contribution towards the semantics of programming languages* ◇ B75 D05 D20 ◇
BORODIN, A. & CONSTABLE, R.L. *Subrecursive programming languages II. On program size* ◇ D20 ◇
CANNONITO, F.B. *A note on inverses of elementary permutations* ◇ D20 D45 F30 ◇
CONSTABLE, R.L. *Subrecursive-programming languages III: the multiple-recursive functions* ◇ D15 D20 ◇
D'YACHENKO, V.F. *On transformation of schemes of algorithms (Russian)* ◇ D05 D20 ◇
ENGELER, E. *Algorithmic approximation* ◇ B75 D05 D20 ◇
FINKELSTEIN, M. *Two conjugate primitive permutations not conjugate by a primitive recursive permutation* ◇ D20 ◇
FREJDZON, R.I. *Regular approximation of recursive predicates (Russian) (English summary)* ◇ D20 ◇
FRIEDMAN, H.M. *Algorithmic procedures, generalized Turing algorithms, and elementary recursion theory* ◇ A05 D10 D20 D75 F99 ◇
GLADSTONE, M.D. *Simplification of the recursion scheme* ◇ D20 ◇
GLINERT, E.P. *On restricted Turing computability* ◇ D20 ◇
GOLD, E.M. *Universal goal-seekers* ◇ D05 D10 D20 ◇
GRODZKI, Z. *The degree of complexity of the sets of computations II* ◇ D20 ◇
HERMAN, G.T. *The equivalence of different hierarchies of elementary functions* ◇ D20 ◇
JACOPINI, G. *Una molto semplice caratterizzazione delle funzioni calcolabili operanti su simboli* ◇ D20 ◇
KHUDYAKOV, V.V. & SOLON, B.YA. *On certain algebras of primitive recursive functions (Russian)* ◇ D20 ◇
KONIKOWSKA, B. *Continuous machines, τ-computations and τ-computable sets* ◇ D10 D20 ◇
KONIKOWSKA, B. *On some properties of continuous machines* ◇ D10 D20 ◇
KONIKOWSKA, B. *Properties of continuous machines* ◇ D10 D20 ◇
KOSOVSKIJ, N.K. *Algorithmic sequences from the initial class of the Grzegorczyk hierarchy (Russian) (English summary)* ◇ D20 ◇
KURKI-SUONIO, R. *A programmer's introduction to computability and formal languages* ◇ B75 D05 D20 D98 ◇
LACHLAN, A.H. *Solution to a problem of Spector* ◇ D20 D30 ◇
LEWIS, F.D. *Classes of recursive functions and their index sets* ◇ D20 ◇
MACHTEY, M. *Classification of computable functions by primitive recursive classes* ◇ D15 D20 ◇
MARANDZHYAN, G.B. *Lattices of blocks of recursive functions (Russian) (Armenian summary)* ◇ D15 D20 ◇
MINTS, G.E. *Quantifier-free and one-quantifier systems (Russian) (English summary)* ◇ B20 D20 F30 ◇
MORRIS JR., J.H. *Another recursion induction principle* ◇ B75 D20 ◇
NEPOMNYASHCHIJ, V.A. *The completeness of operations in operator algorithms (Russian)* ◇ D20 ◇
PATRIKEEV, V.L. *The computability of the translation functions of Janov's operator schemes (Russian)* ◇ B75 D20 ◇
PAYNE, T.H. *Effectively minimizing effective fixed-points* ◇ D20 ◇
PERETYAT'KIN, M.G. *Strongly constructive models and numerations of the boolean algebra of recursive sets (Russian)* ◇ C57 D20 D45 G05 ◇
ROBERTSON, E.L. *Complexity classes of partial recursive functions* ◇ D15 D20 ◇
SABBAGH, G. *Logique mathematique V. Decidabilite et fonctions recursives* ◇ B25 D20 D35 ◇
SANTOS, E.S. *Computability by probabilistic Turing machines* ◇ D10 D20 ◇
SCHURMANN, A. *Functions computable by a computer (Polish and Russian summaries)* ◇ D05 D10 D20 ◇
SCHWICHTENBERG, H. *Eine Klassifikation der ε_0-rekursiven Funktionen* ◇ D20 ◇
SUNDBLAD, Y. *The Ackermann functions. A theoretical, computational and formula manipulative study* ◇ D20 ◇
TSICHRITZIS, D.C. *A note on comparison of subrecursive hierarchies* ◇ D20 ◇
TSINMAN, L.L. *Certain examples and theorems from the theory of recursive functions (Russian)* ◇ D20 D25 ◇
TUZOV, V.A. *Optimal schemes of algorithms (Russian)* ◇ D15 D20 ◇
VIDAL-NAQUET, G. *Programmes formels et logique du second ordre* ◇ D20 ◇
WERNER, GEORGES *Propriete d'invariance des classes de fonctions de complexite bornee* ◇ D15 D20 ◇
YASUHARA, A. *Recursive function theory and logic* ◇ B98 D10 D20 D98 ◇
YOUNG, P. *Speed-up changing the order in which sets are enumerated* ◇ D15 D20 D25 ◇

1972

ALLISON, D.C.S. & HOARE, C.A.R. *Incomputability* ◇ B75 D20 ◇

ALOISIO, P. *Una nota sulle funzioni parziali ricorsive* ◇ D20 ◇
BAKKER DE, J.W. *Recursion, induction and symbol manipulation* ◇ D05 D20 ◇
BARZDINS, J. & FREJVALD, R.V. *On the prediction of general recursive functions (Russian)* ◇ D20 ◇
BARZDINS, J. *Prognostication of automata and functions* ◇ D05 D20 ◇
BEREZIN, S.A. *The generation of two-place primitive recursive functions (Russian)* ◇ D20 ◇
CHERNOV, V.P. *Topological variants of the continuity theorem for mappings and related theorems theorems (Russian) (English summary)* ◇ D20 F60 ◇
CHIKVASHVILI, R.I. *On the theory of effective functionals (Russian) (Georgian and English summaries)* ◇ D20 ◇
CHIKVASHVILI, R.I. *The index sets that are connected with effective functionals (Russian) (Georgian and English summaries)* ◇ D20 D55 ◇
CONSTABLE, R.L. & MUCHNICK, S.S. *Subrecursive program schemata. I. Undecidable equivalence problems. II. Decidable equivalence problems* ◇ B75 D20 ◇
CULIK, K. *Algorithmization of algebras and relational structures* ◇ D20 D45 ◇
DANTSIN, E.YA. *Increase of the complexity of functions by an application of multiple recursion (Russian) (English summary)* ◇ D20 ◇
DEMBINSKI, P. *Equivalence of programs and their complexity* ◇ B75 D15 D20 ◇
DEMBINSKI, P. *The machine realization of programs (Russian summary)* ◇ B75 D10 D20 ◇
DEMBINSKI, P. *The structure of program and its semantics* ◇ B75 D20 ◇
EBBINGHAUS, H.-D. & HERMES, H. & JACOBS, K. & MAHN, F.-K. *Turing-Maschinen und rekursive Funktionen (Russian)* ◇ D10 D20 ◇
ERSHOV, A.P. *Theory of program schemata* ◇ B75 D20 D98 ◇
FAHMY, M.H. *Equational calculus for primitive recursive rational-valued functions (Russian)* ◇ D20 D75 ◇
FISCHER, P.C. & WARKENTIN, J.C. *Predecessor machines and regressing functions* ◇ B70 D20 ◇
FREJDZON, R.I. *Families of recursive predicates of measure zero (Russian) (English summary)* ◇ D20 F60 ◇
GERMANO, G. & MAGGIOLO-SCHETTINI, A. *Equivalence of partial recursivity and computability by algorithms without concluding formulas* ◇ D03 D20 ◇
GONSHOR, H. *The category of recursive functions* ◇ D20 D75 G30 ◇
HAMLET, R.G. *A patent problem for abstract programming languages; machine-independent computations* ◇ B75 D10 D20 ◇
HAUCK, J. *Funktionale Rekursion* ◇ D20 D65 ◇
HAUCK, J. *Zur berechenbaren gleichmaessigen Finitheit rekursiver Funktionale* ◇ D20 F60 ◇
HOEHNKE, H.-J. *Superposition partieller Funktionen* ◇ C05 D20 ◇
KANOVICH, M.I. *On the universality of strongly undecidable sets (Russian)* ◇ D15 D20 D25 D30 ◇
KHACHATRYAN, V.E. & PETROSYAN, G.N. & PODLOVCHENKO, R.I. *Interpretations of schemes of algorithms, and various types of equivalence relations between schemes (Russian) (Armenian and English summaries)* ◇ B75 D20 ◇
KINBER, E.B. *Frequency calculations of general recursive predicates and frequency enumeration of sets (Russian)* ◇ D20 ◇
KONIKOWSKA, B. *Formalisierung des Begriffs der stetigen Maschine (Polish) (Russian and English summaries)* ◇ D15 D20 ◇
KOSOVSKIJ, N.K. *Properties of the solutions of equations in a free semigroup (Russian) (English summary)* ◇ D20 D25 D40 ◇
KOSOVSKIJ, N.K. *Recognition of invariant properties of algorithms (Russian) (English summary)* ◇ D20 ◇
KOZ'MINYKH, V.V. *Representation of partial recursive functions in the form of superpositions (Russian)* ◇ D20 ◇
KREISEL, G. *Which number theoretic problems can be solved in recursive progressions on Π^1_1-paths through $\mathcal{O}$?* ◇ A05 D20 D55 F15 F30 F99 ◇
KURDYUMOV, G.L. *Ueber Abbildungen mit rekursiven Graphen (Russisch)* ◇ D20 ◇
LYNN, D.S. *New results for Rado's sigma function for binary Turing machines* ◇ D10 D20 ◇
MACHTEY, M. *Augmented loop languages and classes of computable functions* ◇ D05 D20 ◇
MARCHENKOV, S.S. *Bounded recursions (Russian)* ◇ D20 ◇
MARCHENKOV, S.S. *The computable enumerations of families of recursive functions (Russian)* ◇ D20 D45 ◇
MASLOV, A.N. *Probabilistic Turing machines and recursive functions (Russian)* ◇ D10 D20 ◇
MAZURKIEWICZ, A.W. *Iteratively computable relations* ◇ D15 D20 ◇
MAZURKIEWICZ, A.W. *Recursive algorithms and formal languages* ◇ D05 D20 ◇
MEDVEDEV, YU.T. *Locally finitary algorithmic problems (Russian)* ◇ D20 F30 F50 ◇
MEYER, A.R. & RITCHIE, D.M. *A classification of the recursive functions* ◇ D20 ◇
MEYER, A.R. & MOLL, R. *Honest bounds for complexity classes of recursive functions* ◇ D15 D20 ◇
MUELLER, HORST *Jede mit Stackautomaten berechenbare Funktion ist elementar* ◇ D10 D15 D20 ◇
PAKHOMOV, S.V. *Properties of graphs of functions in the Grzegorczyk hierarchy (Russian) (English summary)* ◇ D20 ◇
PAYNE, T.H. *Sequences having an effectve fixed-point property* ◇ D20 ◇
PETER, R. *Zur Frage der Rekursivitaet der im "Algol 68" verwendeten zweistufigen Grammatik* ◇ B75 D20 ◇
ROSE, H.E. *$\mathcal{E}^{\alpha}$-arithmetic and transfinite induction* ◇ D20 F30 ◇
ROZINAS, M.G. *Algebra of multiple primitive recursive functions (Russian)* ◇ D20 ◇
SCHNORR, C.-P. *Optimal Goedelnumberings* ◇ D20 D45 ◇

SCHWICHTENBERG, H. *Beweistheoretische Charakterisierung einer Erweiterung der Grzegorczyk-Hierarchie* ◇ D20 F15 F30 ◇
SKARBEK, W. & ZEMBRZUSKI, K. *Computable real functions and their relation to the analog computer* ◇ D20 F60 ◇
SKARBEK, W. & ZEMBRZUSKI, K. *Some properties of continuous machines* ◇ D10 D20 ◇
SOKOLOV, V.A. *Certain properties of the algebra of all partially recursive functions (Russian)* ◇ D20 D75 ◇
SOKOLOV, V.A. *The maximal subalgebras of the algebra of all partially recursive functions (Russian) (English summary)* ◇ D20 D75 ◇
STRNAD, P. *A hierarchy of optimal recognitions (Russian) (German, English and French summaries)* ◇ D15 D20 ◇
SYMES, D. *The computation of finite functions* ◇ D15 D20 ◇
TURCHIN, V.F. *Equivalent transformations of recursive functions that are described in the language REFAL (Russian)* ◇ B75 D20 ◇
VIER, L.C. *Church's thesis in northern Dutch constructivism* ◇ D20 F50 ◇
WAINER, S.S. *Ordinal recursion, and a refinement of the extended Grzegorczyk hierarchy* ◇ D20 ◇

1973

ALFEROVA, Z. *Algorithmentheorie (Russian)* ◇ D10 D20 D98 ◇
BASS, L. & YOUNG, P. *Ordinal hierarchies and naming complexity classes* ◇ D15 D20 ◇
BELYAKIN, N.V. *On a class of recursive hierarchies (Russian)* ◇ D20 ◇
BEREZYUK, N.T. & FURMANOV, K.K. *Foundations of the theory of algorithms (Russian)* ◇ D20 D98 ◇
BIRD, R. *A note on definition by cases* ◇ D20 ◇
BOWIE, G.L. *An argument against Church's thesis* ◇ A05 D20 F99 ◇
BUNDY, A. *A note on omitting the replacement schema* ◇ D20 F30 ◇
CANNONITO, F.B. *The algebraic invariance of the word problem in groups* ◇ D20 D40 ◇
CANNONITO, F.B. & GATTERDAM, R.W. *The computability of group constructions. Part I* ◇ D20 D40 D45 ◇
CHYTIL, M.P. *On changes of input/output coding I* ◇ D15 D20 ◇
COHORS-FRESENBORG, E. *Elementare und subelementare Funktionenklassen ueber binaeren Baeumen* ◇ D15 D20 ◇
CULIK, K. *Main degrees of complexity of computer programs and computable functions* ◇ B75 D15 D20 ◇
ERSHOV, A.P. *The present state of the theory of program schemes (Russian)* ◇ B75 D20 D98 ◇
FRIEDMAN, E.P. *Equivalence problems in monadic recursion schemes* ◇ D20 ◇
GATTERDAM, R.W. *The computability of group constructions part II* ◇ D20 D40 D45 ◇
GATTERDAM, R.W. *The Higman theorem for primitive-recursive groups - a preliminary report* ◇ D20 D40 ◇
GATTERDAM, R.W. *The Higman theorem for $E^n(A)$ computable groups* ◇ D20 D40 ◇
GERMANO, G. & MAGGIOLO-SCHETTINI, A. *A flow diagram composition of Markov's normal algorithms without concluding formulas* ◇ B75 D03 D20 ◇
GERMANO, G. & MAGGIOLO-SCHETTINI, A. *Quelques characterisations des fonctions recursives partielles* ◇ D20 ◇
GOODSTEIN, R.L. *On limiting the applications of the uniqueness rules in the equation calculus* ◇ D20 ◇
HELM, J.P. & MEYER, A.R. & YOUNG, P. *On orders of translations and enumerations* ◇ D15 D20 D25 D45 F40 ◇
HENKE VON, F.W. & INDERMARK, K. & WEIHRAUCH, K. *Hierarchies of primitive recursive wordfunctions and transductions defined by automata* ◇ D05 D20 ◇
HENKE VON, F.W. & WEIHRAUCH, K. *Klassifizierung von primitiv-rekursiven Transformationen und Automatentransduktionen* ◇ D05 D20 ◇
HIRAI, T. *Total undecidability on the class of general recursive functions with a given property* ◇ D20 ◇
HU, SHIHUA *The description problem of algorithmic languages (Chinese)* ◇ B75 D20 ◇
KAMAE, T. *On Kolmogorov's complexity and information* ◇ D15 D20 ◇
KANOVICH, M.I. *Komplexitaet der beschraenkten Entscheidbarkeit von Algorithmen (Russian)* ◇ D05 D15 D20 ◇
KOBZEV, G.N. *Pointwise decomposable sets (Russian)* ◇ D20 D25 D30 ◇
KUSHNER, B.A. *A certain problem of Mostowski (Russian)* ◇ D20 F60 ◇
KUSHNER, B.A. *Computationally complex real numbers (Russian)* ◇ D20 F60 ◇
LEVIN, L.A. *On storage capacity for algorithms (Russian)* ◇ D20 D80 ◇
MACHTEY, M. *A notion of helping and pseudo-complementation in lattices of honest subrecursive classes* ◇ D15 D20 ◇
MANN, I. *Probabilistic recursive functions* ◇ D20 D75 ◇
MARANDZHYAN, G.B. *The complexities of the representation of natural numbers by means of recursive functions (Russian)* ◇ D15 D20 ◇
MARCHENKOV, S.S. *The existence of families without positive numerations (Russian)* ◇ D20 D30 D45 ◇
MEHLHORN, K. *On the size of sets of computable functions* ◇ D15 D20 ◇
MIKHEEV, V.L. *Classes of algebras of primitive recursive functions (Russian)* ◇ D20 D75 ◇
MOSHCHENSKIJ, V.A. *Lectures on mathematical logic (Russian)* ◇ B98 D10 D20 ◇
NABIALEK, I. *Some properties of τ-computable functions* ◇ D20 ◇
OREVKOV, V.P. *On the complexity of expansion of algebraic irrationalities in continued fractions (Russian)* ◇ D20 D80 ◇
OSTROUKHOV, D.A. *The complexity of terms of sequences of normal algorithms (Russian)* ◇ D03 D20 ◇

OWINGS JR., J.C. *Diagonalization and the recursion theorem* ◇ B40 D20 F30 ◇
PAYNE, T.H. *Effective extendability and fixed-points* ◇ D20 ◇
PENNER, V. *Ueber eine Hierarchie von push-down-entscheidbaren Mengen* ◇ D10 D20 ◇
PETER, R. *Veranschaulichung einer sequenziellen Berechnung der rekursiven Wortfunktionen durch "eisenbahnrangierende Graphen"* ◇ D20 ◇
PETER, R. *Zur Rekursivitaet der mathematischen Grammatiken* ◇ D20 ◇
POUR-EL, M.B. *Analog computers, digital computers, mathematical logic, differential equations - interrelations* ◇ D20 D75 D80 F65 ◇
PUTNAM, H. *Recursive functions and hierarchies* ◇ D20 D30 D35 D55 ◇
ROBINSON, JULIA *Axioms for number theoretic functions (Russian)* ◇ C62 D20 D75 F30 ◇
ROSE, G.F. & WEIHRAUCH, K. *A characterization of the classes L_1 and R_1 of primitive recursive word functions* ◇ D05 D20 ◇
SHKIRA, V.V. *Universal functions for certain classes of recursive functions and sets (Russian)* ◇ D20 D25 ◇
SKORDEV, D.G. *Some examples of universal functions that are recursively definable by means of small systems of inequalities (Russian)* ◇ D20 ◇
SKOWRON, A. *Stored program machines* ◇ D10 D20 ◇
STRIGIN, YU.D. *Hierarchies of general recursive operators and general recursive types of degrees of noncomputability (Russian)* ◇ D20 D30 ◇
STRIGIN, YU.D. *The hierarchy of general recursive functionals (Russian)* ◇ D20 D30 ◇
STRONG, H.R. & WALKER, S.A. *Characterizations of flowchartable recursions* ◇ D20 D75 ◇
TER-ZAKHARYAN, N.P. *On the language of multi-place recursive functions (Russian)* ◇ D20 ◇
TER-ZAKHARYAN, N.P. *Some entropy properties of algorithmic languages (Russian)* ◇ D20 ◇
THOMAS, WILLIAM J. *Doubts about some standard arguments for Church's Thesis* ◇ A05 D10 D20 F99 ◇
TRAKHTENBROT, B.A. *Frequency computations (Russian)* ◇ D10 D15 D20 ◇
TSICHRITZIS, D.C. *A model for iterative computation* ◇ D20 ◇
VERBEEK, R. *Erweiterungen subrekursiver Programmiersprachen* ◇ B75 D15 D20 ◇
VOELKEL, L. *Untersuchungen ueber die Kompliziertheit von Berechnungs- und Tabellenprogrammen* ◇ D15 D20 ◇
VUCKOVIC, V. *Local recursive theory* ◇ D20 D25 D75 D80 ◇
YOUNG, P. *Easy constructions in complexity theory: Gap and speed-up theorems* ◇ D15 D20 ◇
ZAKOWSKI, W. *Multidimensional continuous machines* ◇ D20 ◇
ZAKOWSKI, W. *On some properties of N-dimensional simple continuous machines* ◇ D20 ◇

1974

ALTAEV, A.V. *Operators defined by γ-enumeration and related reducibilities (Russian)* ◇ D20 D30 ◇
ANDZHAN, A.V. *Reduzierbarkeit von Funktionenklassen nach der Prognostizierung (Russisch)* ◇ D20 D30 ◇
ASSER, G. *Berechenbare Graphenabbildungen II* ◇ D20 D45 ◇
AUSIELLO, G. *Relations between semantics and complexity of recursive programs* ◇ B40 B75 D15 D20 ◇
AUSIELLO, G. *Un linguaggio di programmazione per le funzioni subricorsive* ◇ B75 D20 ◇
BAKER, T.P. & HARTMANIS, J. *On simple Goedel numberings and translations* ◇ B75 D20 D45 ◇
BARZDINS, J. *A property of limiting computable functionals (Russian)* ◇ D15 D20 ◇
BARZDINS, J. *Limitare Synthese von τ-Zahlen (Russian)* ◇ D15 D20 ◇
BARZDINS, J. & KINBER, E.B. & PODNIEKS, K.M. *On speed-up of synthesis and prognostication of functions (Russian)* ◇ D15 D20 ◇
BARZDINS, J. & FREJVALD, R.V. *Prognostizierung und limitaere Synthese effektiv aufzaehlbarer Klassen von Funktionen (Russian)* ◇ D15 D20 D45 ◇
BARZDINS, J. *Two theorems on the identification in the limit of τ-numbers (Russian)* ◇ D20 D45 ◇
BERRY, J.W. *N-ary almost recursive functions* ◇ D20 ◇
BLUM, M. & GILL III, J.T. *On almost everywhere complex recursive functions* ◇ D15 D20 ◇
BOEHLING, K.H. & BRAUNMUEHL VON, B. *Komplexitaet bei Turingsmaschinen* ◇ D10 D15 D20 ◇
BORZOV, YU.V. *Holographische Mengen (Russisch)* ◇ D20 D30 ◇
BUCHBERGER, B. *Certain decompositions of Goedel numbering and the semantics of programming languages* ◇ B75 D20 D45 ◇
BUCHBERGER, B. *On certain decompositions of Goedel numberings* ◇ D15 D20 D45 ◇
CHYTIL, M.P. *On changes of input/output coding II* ◇ D15 D20 ◇
DANKO, W. *Not programmable function derived by a procedure (Russian summary)* ◇ D20 ◇
DEKHTYAR', M.I. *On the complexity of relative computations (Russian)* ◇ D15 D20 ◇
FLAJOLET, P. & STEYAERT, J.-M. *Une generalisation de la notion d'ensemble immune* ◇ B75 D20 D25 D50 ◇
FREI-IMFELD, G. *Ueber eine Erweiterung der algebraischen Operationen* ◇ D20 G15 ◇
FREJDZON, R.I. *Table approximations to recursive predicates (Russian) (English summary)* ◇ D20 D25 ◇
FREJVALD, R.V. *Gleichmaessige und ungleichmaessige Prognostizierbarkeit (Russian)* ◇ D20 ◇
FREJVALD, R.V. *Limitaere berechenbare Funktionen und Funktionale (Russian)* ◇ D20 ◇
FREJVALD, R.V. *Limitaere Berechnungen auf stochastischen Automaten (Russisch)* ◇ D10 D20 ◇
FREJVALD, R.V. *On the limit synthesis of numbers of general recursive functions in various computable numerations (Russian)* ◇ D20 D45 ◇

FRIEDMAN, E.P. *Relationships between monadic recursion schemes and deterministic context-free languages* ◇ D05 D20 ◇

GACS, P. *On the symmetry of algorithmic information (Russian)* ◇ D20 ◇

GALIL, Z. *Two way deterministic pushdown automaton languages and some open problems in the theory of computation* ◇ D05 D10 D20 ◇

GANDY, R.O. *Set-theoretic functions for elementary syntax* ◇ B03 D20 D65 E47 ◇

GERMANO, G. & MAGGIOLO-SCHETTINI, A. *Sequence-to-sequence functions. A systematic study* ◇ B75 D20 ◇

GODLEVSKIJ, A.B. & LETICHEVS'KIJ, O.A. (EDS.) *Automata and algorithm theory, and mathematical logic (Russian)* ◇ B35 B97 D05 D20 D97 ◇

GOETZE, B.G. *Die Struktur des Halbverbandes der effektiven Numerierungen* ◇ D20 D45 ◇

GOETZE, B.G. *Unentscheidbarkeitsgrade rekursiver Funktionen* ◇ D20 ◇

GRODZKI, Z. *The Kolmogorov's complexity of the computation of the k-machines* ◇ D03 D15 D20 ◇

GUCCIONE, S. & LO SARDO, P. *Effective procedures and randomness* ◇ B75 D20 ◇

GUREVICH, I.B. *The noncomputability within a class of local algorithms of certain predicates connected with the minimization of Boolean functions (Russian)* ◇ B35 D20 ◇

HENKE VON, F.W. *Primitiv-rekursive Transformationen* ◇ D20 ◇

HIRSCHFELD, J. *Models of arithmetic and the semiring of recursive functions* ◇ C62 D20 H15 ◇

JAMIESON, D.W. & TEUTSCH, R.J. *Hockett on effective computability* ◇ B65 D20 ◇

JONES, JAMES P. *Recursive undecidability - an exposition* ◇ D10 D20 D35 D40 D98 ◇

KANOVICH, M.I. *"Complex" and "simple" numbers (Russian)* ◇ D20 ◇

KANOVICH, M.I. *Complexity of the limit of Specker sequences (Russian)* ◇ D15 D20 F60 ◇

KINBER, E.B. *Frequency calculations with small number of mistakes (Russian)* ◇ D20 ◇

KINBER, E.B. *On limiting synthesis of quasiminimal Goedel numbers (Russian)* ◇ D20 D45 ◇

KINBER, E.B. *Ueber Frequenzberechnungen auf unendlichen Auswahlmengen (Russisch)* ◇ D20 D25 ◇

KOBAYASHI, K. *A note on extending equivalence theories of algorithms* ◇ B75 D20 D25 ◇

KOSOVSKIJ, N.K. *Solutions of systems consisting of word equations and inequalities in lengths of words (Russian) (English summary)* ◇ D20 D25 ◇

KOZMIDIADI, V.A. & MASLOV, A.N. & PETRI, N.V. (EDS.) *Complexity of computations and algorithms (Russian)* ◇ D15 D20 D98 ◇

KOZ'MINYKH, V.V. *Representation of partial recursive functions with certain conditions in the form of superpositions (Russian)* ◇ D20 ◇

LADNER, R.E. & LYNCH, N.A. & SELMAN, A.L. *Comparison of polynominal-time reducibilities* ◇ D15 D20 ◇

LISCHKE, G. *Eine Charakterisierung der rekursiven Funktionen mit endlichen Niveaumengen ($\mathfrak{K}$-Funktionen)* ◇ D20 ◇

LOEB, M.H. & WAINER, S.S. *Hierarchies of number-theoretic functions (Russian)* ◇ D20 F15 ◇

LUPANOV, O.B. *Methods to obtain complexity and calculability estimations for individual functions (Russian)* ◇ D10 D15 D20 ◇

LYNCH, N.A. *Approximations to the halting problem* ◇ D10 D20 D25 D45 ◇

MACHTEY, M. *The honest subrecursive classes are a lattice* ◇ D15 D20 ◇

MASHURYAN, A.S. *Recursive definitions on induction models (Russian) (Armenian summary)* ◇ C13 D20 ◇

MEHLHORN, K. *Polynomial and abstract subrecursive classes* ◇ D15 D20 ◇

MEHLHORN, K. *The "almost all" theory of subrecursive degrees is decidable* ◇ B25 C80 D20 ◇

MEYER, A.R. & MOLL, R. *Honest bounds for complexity classes of recursive functions* ◇ D15 D20 ◇

MICHELIS DE, G. *Recursive functions not dependent on the computational rules* ◇ D20 ◇

MICHELIS DE, G. *Sulla computazione di funzioni ricorsive (English summary)* ◇ D20 ◇

MINTS, G.E. *Note on equation systems (Russian)* ◇ D20 ◇

NABIALEK, I. *Rational τ-computable functions* ◇ D20 ◇

NEPOMNYASHCHIJ, V.A. *Criteria for the algorithmic completeness of the systems of operators* ◇ D20 ◇

ONO, H. *A formal system of partial recursive functions* ◇ D20 ◇

PAKHOMOV, S.V. *A simple syntactic definition of all classes of the Grzegorczyk hierarchy (Russian)* ◇ D20 ◇

PATRIZIA, M. *Funzioni elementari* ◇ D20 ◇

PODNIEKS, K.M. *Comparison of the different types of identification in the limit and prediction of functions I (Russian)* ◇ D20 D45 ◇

POUR-EL, M.B. *Abstract computability and its relation to the general purpose analog computer (some connections between logic, differential equations and analog computers)* ◇ D20 D75 D80 F65 ◇

PROCHNOW, D. *Einfach-rekursive Programmschemata (English and Russian summaries)* ◇ B75 D20 ◇

RANGEL, J.L. *The equivalence problem for regular expressions over one letter is elementary* ◇ D05 D20 ◇

ROBERTSON, E.L. *Complexity classes of partial recursive functions* ◇ D15 D20 ◇

SCHNORR, C.-P. *Optimal enumerations and optimal Goedel numberings* ◇ D15 D20 D25 D45 ◇

SCHNORR, C.-P. *Rekursive Funktionen und ihre Komplexitaet* ◇ D10 D15 D20 D98 ◇

SCHUBERT, L.K. *Iterated limiting recursion and the program minimization problem* ◇ B75 D15 D20 ◇

SCHUBERT, L.K. *Representative samples of programmable functions* ◇ B75 D15 D20 ◇

SHMAIN, I.KH. *Extended calculus of recursive functions I (Russian)* ◇ D20 D75 F50 ◇

SKORDEV, D.G. *Recursively complete operations on words (Russian)* ◇ D20 ◇

VELIKANOV, K.M. *Specialization of the form of deduction in the extended calculus of recursive functions (Russian)* ◇ D20 ◇

WEIHRAUCH, K. *Teilklassen primitiv-rekursiver Wortfunktionen* ◇ D20 ◇

WERNER, GEORGES *Sous-classes recursivement enumerables d'une classe de complexite* ◇ D15 D20 ◇

ZAKOWSKI, W. *On τ-computability of functions and sets of functions of n real variables* ◇ D20 ◇

ZAKOWSKI, W. *On some properties of N-dimensional simple continuous machines* ◇ D20 ◇

ZASLAVSKIJ, I.D. *Recursive extrapolators (Russian)* ◇ D20 ◇

ZHAROV, V.G. *On an analog of a theorem of Specker (Russian)* ◇ D20 D80 F60 ◇

1975

ARON, E. & BARAK, A. *On the efficiency of ternary algorithms for multiplication and division* ◇ B50 B75 D20 ◇

ASSER, G. *Rekursive Graphenabbildungen. I* ◇ D20 D45 D80 ◇

AUGUSTON, M. *Universal Buchberger automata and real-time computations (Russian) (English summary)* ◇ D10 D15 D20 ◇

AUSIELLO, G. & PROTASI, M. *On the comparison of notions of approximation* ◇ D20 D80 ◇

BAKER, T.P. & HARTMANIS, J. *On simple Goedel numberings and translations* ◇ B75 D20 D45 ◇

BARASHKO, A.S. & ROJZEN, S.I. *On redundance of the points of partial recursive functions naming the computational complexity classes (Ukrainian, English summery)* ◇ D15 D20 ◇

BARZDINS, J. *Inductive derivation of automata, function and programs* ◇ D05 D20 ◇

BARZDINS, J. & FREJVALD, R.V. *The relation between predictability and identifiability in the limit (Russian) (English summary)* ◇ D20 ◇

BECK, H. *Zur Entscheidbarkeit der funktionalen Aequivalenz (English summary)* ◇ D20 D80 ◇

BEESON, M.J. *The nonderivability in intuitionistic formal systems of theorems on the continuity of effective operations* ◇ D20 F50 ◇

BERG, J. & CHIHARA, C.S. *Church's thesis misconstrued* ◇ A05 D20 F99 ◇

BERNHARDT, L. *Algorithmisches Konstruieren von Algorithmen* ◇ B75 D20 ◇

BIRD, R. *Non recursive functionals* ◇ D20 ◇

BOERGER, E. *Recursively unsolvable algorithmic problems and related questions reexamined* ◇ D03 D10 D20 ◇

CALDWELL, J. & POUR-EL, M.B. *On a simple definition of computable function of a real variable - with applications to functions of a complex variable* ◇ D20 D80 ◇

CANNONITO, F.B. & GATTERDAM, R.W. *The word problem and power problem in 1-relator groups are primitive recursive* ◇ D20 D40 ◇

CONSTABLE, R.L. & EGLI, H. *Computability concepts for programming language semantics* ◇ B75 D20 ◇

CORNELIUS, B.J. & KIRBY, G. *Depth of recursion and the Ackermann function* ◇ D20 ◇

CRISCUOLO, G. & MINICOZZI, E. & TRAUTTEUR, G. *Limiting recursion and the arithmetic hierarchy (French summary)* ◇ D20 D55 ◇

DEGTEV, A.N. *Reducibility of partial recursive functions (Russian)* ◇ D20 D25 ◇

EHRIG, H. & KUEHNEL, W. & PFENDER, M. *Diagram characterization of recursion* ◇ D20 ◇

ELGOT, C.C. *Monadic computation and iterative algebraic theories* ◇ C05 D10 D20 ◇

EMDE BOAS VAN, P. *Ten years of speedup* ◇ A10 D15 D20 F20 G05 ◇

EMDE BOAS VAN, P. *The non-renameability of honesty classes (German summary)* ◇ D15 D20 ◇

FARKAS, E. *On the evaluation of nested partial functions* ◇ D15 D20 ◇

FREJVALD, R.V. *Functions computable in the limit by probabilistic machines* ◇ D10 D20 ◇

FREJVALD, R.V. *Identifiability in the limit of the numbers of general recursive functions in various computable numerations (Russian) (English summary)* ◇ D20 D25 D45 ◇

FREJVALD, R.V. *Minimal Goedel numbers and their identification in the limit* ◇ D20 D45 ◇

FREJVALD, R.V. *On complexity and optimality of the computation in the limit (Russian) (English summary)* ◇ D15 D20 ◇

FUCHS, P.H. *Statistical characterization of learnable sequences* ◇ D20 ◇

GEORGIEVA, N.V. *A certain reflexive recursive scheme (Russian)* ◇ D20 ◇

GERMANO, G. & MAGGIOLO-SCHETTINI, A. *Sequence-to-sequence recursiveness* ◇ B75 D20 G30 ◇

GOETZE, B.G. & KLETTE, R. *Limes-rekursive Funktionen* ◇ D20 ◇

GOETZE, B.G. & KLETTE, R. *Some properties of limit recursive functions* ◇ D20 ◇

HARROW, K. *Small Grzegorczyk classes and limited minimum* ◇ D20 ◇

HAUCK, J. *Turingmaschinen und berechenbare reelle Funktionen* ◇ D10 D20 F60 ◇

HENKE VON, F.W. & INDERMARK, K. & ROSE, G.F. & WEIHRAUCH, K. *On primitive recursive wordfunctions (German summary)* ◇ D20 ◇

HIRSCHFELD, J. *Models of arithmetic and recursive functions* ◇ C62 D20 H15 ◇

JACOPINI, G. & MENTRASTI, P. *Funzioni ricorsive di sequenza* ◇ D20 ◇

JOCKUSCH JR., C.G. *Recursiveness of initial segments of Kleene's O* ◇ D20 F15 ◇

KANOVICH, M.I. *Dekker's construction and effective nonrecursiveness (Russian)* ◇ D20 D25 ◇

KHOMICH, V.I. *Weakly and strongly nonrealizable propositional formulae (Russian)* ◇ B20 D15 D20 F50 ◇

KINBER, E.B. *On comparison of limit identification and standardization of general recursive functions (Russian) (English summary)* ◇ D20 ◇

KINBER, E.B. *On frequency real-time computations (Russian) (English summary)* ◇ D10 D15 D20 ◇
KOSOVSKIJ, N.K. & VINOGRADOV, A.K. *Hierarchy of Diophantine representations of primitive recursive predicates (Russian)* ◇ D15 D20 ◇
KOSOVSKIJ, N.K. *Possibilities for the operations of one-place summation and one-place restricted multiplication (Russian) (English summary)* ◇ D20 ◇
KREJNIN, YA.L. *The basis and classification of recursive functions (Russian)* ◇ D20 ◇
KUEHNEL, W. & MESEGUER, J. & PFENDER, M. & SOLS, I. *Primitive recursive algebraic theories with applications to program schemes* ◇ B75 D20 G30 ◇
LISCHKE, G. *Ueber die Erfuellung gewisser Erhaltungssaetze durch Kompliziertheitsmasse* ◇ D15 D20 ◇
LYNCH, N.A. *"Helping": several formalizations* ◇ D15 D20 ◇
MACHTEY, M. *Helping and the meet of pairs of honest subrecursive classes* ◇ D20 ◇
MACHTEY, M. *On the density of honest sub-recursive classes* ◇ D20 ◇
MANOHAR, R.P. & TREMBLAY, J.P. *Discrete mathematical structures with applications to computer science* ◇ D05 D20 G05 G10 ◇
MARANDZHYAN, G.B. *Algorithmic languages that do not admit a mutual optimal translation (Russian) (Armenian summary)* ◇ D20 D45 ◇
MARQUES, I. *On degrees of unsolvability and complexity properties* ◇ D15 D20 D30 ◇
MATROSOV, V.L. *An analytic description of the complexity classes of computable functions (Russian)* ◇ D15 D20 ◇
MATROSOV, V.L. *The closure of certain complexity classes with respect to recursive operations (Russian)* ◇ D15 D20 ◇
MEYER, A.R. *Weak monadic second order theory of successor is not elementary recursive* ◇ B25 D10 D15 D20 ◇
MONK, L.G. *Elementary-recursive decision procedures* ◇ B25 C60 D20 ◇
NABIALEK, I. & ZAKOWSKI, W. *On Z-computability of functions and sets of functions* ◇ D15 D20 ◇
NAGASHIMA, T. *On a certain class of recursive functions* ◇ D20 ◇
NOLIN, L. *Algorithmes seriels, algorithmes paralleles* ◇ B40 B75 D20 ◇
NOLIN, L. *Theorie des algorithmes et semantique des langages de programmation* ◇ B40 B75 D20 ◇
PAYNE, T.H. *Computability of finite linear configurations* ◇ D20 ◇
PETER, R. *Die Rekursivitaet der Programmierungssprache "Lisp 1.5" in Spezialfaellen der angeordneten freien homomorphen Mengen (Russian summary)* ◇ B75 D20 ◇
PODNIEKS, K.M. *Comparison of the different types of identification in the limit and prediction of functions II (Russian)* ◇ D20 D45 ◇
PODNIEKS, K.M. *Probabilistic synthesis of enumerated classes of functions (Russian)* ◇ D20 ◇
PODNIEKS, K.M. *Probabilistic prediction of function values (Russian) (English summary)* ◇ D20 ◇
POGOSYAN, EH.M. & SARKISYAN, O.A. *On a formalization of inductive generalization (Russian) (Armenian summary)* ◇ D20 ◇
PRATHER, R.E. *A convenient cryptomorphic version of recursive function theory* ◇ D20 ◇
RAYMOND, F.H. *Note sur l'algebre des fonctions (English summary)* ◇ D20 D75 ◇
SALWICKI, A. *Procedures, formal computations and models* ◇ B75 D20 ◇
SCHNORR, C.-P. & STUMPF, G. *A characterization of complexity sequences* ◇ D10 D20 ◇
SIEFKES, D. *The recursive sets in certain monadic second order fragments of arithmetic* ◇ D20 F35 ◇
SOKOLOV, V.A. *A problem in the class of computable functions with the superposition operation (Russian)* ◇ D20 ◇
TER-ZAKHARYAN, N.P. *Quantitative characteristics of the language of many-argument recursive functions (Russian)* ◇ D20 ◇
THOMASON, S.K. *The logical consequence relation of propositional tense logic* ◇ B45 D20 D55 ◇
WECHSUNG, G. *Minimale and optimale "Blumsche Masse" (English and Russian summaries)* ◇ D15 D20 ◇
WERNER, GUENTER *Prognose von Folgen* ◇ D20 ◇
WIEHAGEN, R. *Inductive inference of recursive functions* ◇ D20 ◇
ZAKOWSKI, W. *On τ-computability and almost τ-computability of functions of n real variables* ◇ D20 ◇
ZAKOWSKI, W. *The (Z,Q)-machines* ◇ D20 ◇

1976

ASATRYAN, O.S. *Ueber Retrakte einer indizierten Kleeneschen Menge (Russian) (Armenian summary)* ◇ D20 D45 ◇
BARASHKO, A.S. & ROJZEN, S.I. *Kernels of partially recursive functions that name classes of computational complexity (Russian)* ◇ D15 D20 ◇
BECK, H. *Die mit Nestedstackautomaten berechenbaren Funktionen sind elementar* ◇ D10 D20 ◇
BEESON, M.J. *Derived rules of inference related to the continuity of effective operations* ◇ D20 F50 ◇
BEL'TYUKOV, A.P. *An iterative description of the class $\mathfrak{E}^1$ of Grzegorczyk's hierarchy (Russian)* ◇ D10 D20 ◇
BEREZIN, S.A. *An algebra of one-place primitive recursive functions with an iteration operation of general type (Russian) (English summary)* ◇ D20 D75 ◇
BERRY, GERARD *Bottom-up computation of recursive programs (French summary)* ◇ D20 D70 ◇
BRADY, J.M. *A programming approach to some concepts and results in the theory of computation* ◇ B75 D10 D20 ◇
CHAITIN, G.J. *Information-theoretic characterization of recursive infinite strings* ◇ D15 D20 D80 ◇
CHERNIAVSKY, J.C. *Simple programs realize exactly Presburger formulas* ◇ B75 D15 D20 F30 ◇
CHOW, T.S. *Lower bounds on the complexity of 0-1-valued recursive functions* ◇ D15 D20 ◇

COURCELLE, B. & VUILLEMIN, J. *Completeness results for the equivalence of recursive schemas* ◇ B75 D20 ◇

COY, W. *The logical meaning of programs of a subrecursive language* ◇ B75 D20 ◇

DALEY, R.P. *Noncomplex sequences: characterizations and examples* ◇ D20 ◇

DEKHTYAR', M.I. *The complexity of the approximation of recursive sets (Russian) (German and English summaries)* ◇ D15 D20 ◇

DEMUTH, O. *The domains of definition of effective operators over general recursive functions and of constructive functions of a real variable (Russian)* ◇ D20 F60 ◇

DEUTSCH, M. *Eine mengentheoretische Grundlegung der Theorie der Berechenbarkeit I* ◇ D20 D25 E47 ◇

FISCHER, MICHAEL J. & LYNCH, N.A. & MEYER, A.R. *Relativization of the theory of computational complexity* ◇ D15 D20 ◇

FUKSMAN, A.L. *Diagnostic and other simplified forms of push-down automata (Russian)* ◇ D20 ◇

GEORGIEVA, N.V. *Another simplification of the recursion scheme* ◇ D20 ◇

GEORGIEVA, N.V. *Classes of one-argument recursive functions* ◇ D20 ◇

GEORGIEVA, N.V. *Hierarchies of one argument recursive functions I* ◇ D20 ◇

GERMANO, G. & MAGGIOLO-SCHETTINI, A. *Recursivity, sequence recursivity, stack recursivity and semantics of programs* ◇ B75 D15 D20 ◇

GOETZE, B.G. *The structure of the lattice of recursive sets* ◇ D20 ◇

GORDON, H.G. *Complete degrees of finite-state transformability* ◇ D05 D20 ◇

GRABOWSKI, J. *Unpredictable sets* ◇ D20 ◇

GRANDJEAN, E. *Sur les classes de relations recursives definies par la methode de Trahtenbrot (English summary)* ◇ D20 ◇

GRANDJEAN, E. *Une notion de recursivite relative qui correspond a la recursivite au sens de Trahtenbrot (English summary)* ◇ D20 D30 ◇

HAFNER, I. *Regressive recursion* ◇ D20 ◇

HARTMANIS, J. *On effective speed-up and long proofs of trivial theorems in formal theories (French summary)* ◇ B25 B35 D15 D20 F20 ◇

HAVRANEK, T. *Statistics and computability* ◇ B75 D20 D80 ◇

HEIDLER, K. *Berechenbarkeit und Entscheidbarkeit auf der Grundlage der Programmiersprache Basic* ◇ B75 D20 ◇

HERZOG, T. & OWINGS JR., J.C. *The inequivalence of two well-known notions of randomness for binary sequence* ◇ D20 D80 ◇

IBRAGIMOV, M.YA. *Common fixed points (Russian)* ◇ D20 ◇

INDERMARK, K. & REISIG, W. *On recursively definable relations* ◇ D20 D25 ◇

JANTKE, K.P. *Zur Kompliziertheit der konsistenten Erkennung von Klassen allgemein-rekursiver Funktionen* ◇ D15 D20 ◇

KANOVICH, M.I. & KUSHNER, B.A. *Complexity of algorithms, and Specker sequences (Russian)* ◇ D15 D20 ◇

KINBER, E.B. *Frequenzberechnungen auf endlichen Automaten (Russian) (English summary)* ◇ D05 D20 ◇

KLETTE, R. *Erkennung allgemein rekursiver Funktionen (English and Russian summaries)* ◇ D20 ◇

KLETTE, R. *Indexmengen und Erkennung rekursiver Funktionen* ◇ D20 D25 D30 D55 ◇

KUSHNER, B.A. *On Grzegorczyk's theorem on the computability of an isolated extremum (Russian)* ◇ D20 D80 F60 ◇

LEMPEL, A. & ZIV, J. *On the complexity of finite sequences* ◇ D15 D20 ◇

LEWIS, F.D. *Subrecursive reducibilities and completeness (Preliminary version)* ◇ D15 D20 ◇

LIEPE, W. & WIEHAGEN, R. *Charakteristische Eigenschaften von erkennbaren Klassen rekursiver Funktionen* ◇ D20 ◇

LINDNER, R. & WERNER, GUENTER *Eine vergleichende Analyse von Stopstrategien fuer allgemeinrekursive Prognosen (English and Russian summaries)* ◇ D20 ◇

MAASS, W. *Eine Funktionalinterpretation der praedikativen Analysis* ◇ D20 F10 F35 F65 ◇

MACHTEY, M. & YOUNG, P. *Simple Goedel numberings, translations, and the P-hierarchy: Preliminary report* ◇ D15 D20 D45 ◇

MANNA, Z. & SHAMIR, A. *The theoretical aspects of the optimal fixedpoint* ◇ D20 ◇

MARANDZHYAN, G.B. *On weakly positive degrees of the sets of minimal indices of algorithms (Russian)* ◇ D20 D30 ◇

MATROSOV, V.L. *Complexity classes and classes of computable signaling functions (Russian)* ◇ D15 D20 ◇

MEHLHORN, K. *Polynomial and abstract subrecursive classes* ◇ D15 D20 ◇

MENTRASTI, P. *Sulle classi di funzioni elementari inferiori in una variabile (English summary)* ◇ D20 ◇

MILNER, R. *Models of LCF* ◇ B40 D20 ◇

MORRIS, P.H. *A reducibility condition for recursiveness* ◇ D20 D30 ◇

MUCHNICK, S.S. *Computational complexity of multiple recursive schemata* ◇ D15 D20 ◇

MUCHNICK, S.S. *The vectorized Grzegorczyk hierarchy* ◇ D20 ◇

OTTMANN, T. *Rekursive Prozeduren und partiell rekursive Funktionen* ◇ D20 ◇

PETER, R. *Rekursive Funktionen in der Computer-Theorie* ◇ B75 D20 D98 ◇

POLYAKOV, E.A. & ROZINAS, M.G. *Theory of algorithms (Russian)* ◇ D20 D30 D98 ◇

ROMANOVS'KIJ, V.YU. *On the solvability of the equivalence problem for a subclass of unary recursion schemes (Russian) (English summary)* ◇ D20 ◇

SAZONOV, V.YU. *Expressibility of functions in D.Scott's LFC language (Russian)* ◇ B40 D20 ◇

SAZONOV, V.YU. *Functionals computable in series and in parallel (Russian)* ◇ B40 D20 ◇

SCHMIDT, DIANA *Built-up systems of fundamental sequences and hierarchies of number theoretic functions* ◇ D20 E10 F15 ◇

SELIVANOV, V.L. *Enumerations of families of general recursive function (Russian)* ◇ D20 D25 ◇
SHAMIR, A. *The fixedpoints of recursive definitions* ◇ D20 ◇
SHEPHERDSON, J.C. *On the definition of computable function of a real variable* ◇ D20 F60 ◇
SKORDEV, D.G. *On Turing computable operators* ◇ D20 ◇
SKORDEV, D.G. *Some methods for building up the theory of recursive functions* ◇ D20 D25 ◇
TER-ZAKHARYAN, N.P. *Some nonoptimal languages of recursive functions (Russian) (English and Armenian summaries)* ◇ D20 ◇
TRAKHTENBROT, M.B. *Relationship between classes of monotonic functions* ◇ B75 D15 D20 ◇
TURCAT, C. & VERDILLON, A. *Recursion and testing of combinational circuits* ◇ B70 D20 ◇
VOELKEL, L. *Staerke-Relationen fuer URM-Befehlssysteme* ◇ D10 D15 D20 ◇
WETTE, E. *On the formalization of productive logic* ◇ B60 D20 ◇
WICHMANN, B. *Ackermann's function: A study in the efficiency of calling procedures* ◇ D15 D20 ◇
WIEHAGEN, R. *Limes-Erkennung rekursiver Funktionen durch spezielle Strategien* ◇ D20 ◇
WIEHAGEN, R. *Primitiv-rekursive Erkennung* ◇ D20 ◇
ZAKOWSKI, W. *Controlled (Z,Q)-machines and generalized (Z,Q)-computable sets (Russian summary)* ◇ D20 ◇

1977

APT, K.R. *Recursive embeddings of partial orderings* ◇ C57 D20 G05 ◇
ARSLANOV, M.M. & NADYROV, R.F. & SOLOV'EV, V.D. *A criterion for the completeness of recursively enumerable sets, and some generalizations of a fixed point theorem (Russian)* ◇ D20 D25 D55 ◇
BADAEV, S.A. *Computable enumerations of families of general recursive functions (Russian)* ◇ D20 D45 ◇
BEL'TYUKOV, A.P. *Maximum sequence of classes transformable by primitive recursion in a given class (Russian)* ◇ D20 ◇
BENNISON, V.L. & SOARE, R.I. *Recursion theoretic characterizations of complexity theoretic properties* ◇ D15 D20 ◇
BERG, E.P. & LISCHKE, G. *Zwei Saetze fuer schwache Erhaltungsmasse* ◇ D15 D20 ◇
BRADY, J.M. *Hints on proofs by recursion induction* ◇ B35 D20 ◇
BRANDSTAEDT, A. & SAALFELD, D. *Eine Hierarchie beschraenkter Rueckkehrberechnungen auf on-line Turingmaschinen* ◇ D10 D15 D20 ◇
DEGTEV, A.N. *A family of maximal subalgebras of R. Robinson's algebra (Russian)* ◇ D20 G25 ◇
DEGTEV, A.N. *Reducibility of partial recursive functions II (Russian)* ◇ D20 ◇
DEUTSCH, M. *Eine mengentheoretische Grundlegung der Theorie der Berechenbarkeit II* ◇ B15 D20 D35 E47 ◇
DIES, J.E. *Sur l'independance algorithmique des suites infinies de Bernoulli (English summary)* ◇ D20 D80 ◇
ENDERTON, H.B. *Elements of recursion theory* ◇ D20 D25 D98 ◇
FAHMY, M.H. *Equation calculi for Grzegorczyk's classes $\mathfrak{E}^n$ (Russian)* ◇ D20 ◇
FLEISCHMANN, K. & MAHR, B. & SIEFKES, D. *Bounded concatenation theory as a uniform method for proving lower complexity bounds* ◇ D15 D20 ◇
FRANCEZ, N. & KLEBANSKY, B. & PNUELI, A. *Backtracking in recursive computations* ◇ B75 D20 ◇
FREJVALD, R.V. & KINBER, E.B. *Identification in the limit of minimal Goedel numbers (Russian)* ◇ D20 D45 ◇
FUCHS, P.H. & SCHNORR, C.-P. *General random sequences and learnable sequences* ◇ D20 D25 D80 ◇
GERMANO, G. & MAGGIOLO-SCHETTINI, A. *Sequence-to-sequence partial recursive functions* ◇ B75 D20 ◇
GILLO, D. & GOETZE, B.G. & KLETTE, R. *Der iterierte Limes rekursiver Funktionen und die arithmetische Hierarchie* ◇ D20 D55 ◇
HAY, L. & MANASTER, A.B. & ROSENSTEIN, J.G. *Concerning partial recursive similarity transformations of linearly ordered sets* ◇ C57 C65 D20 D25 D30 D45 E07 ◇
HEIDLER, K. & HERMES, H. & MAHN, F.-K. *Rekursive Funktionen* ◇ D20 D98 ◇
HEMMERLING, A. *Vergleich einiger Kodierungen der natuerlichen Zahlen bezueglich der regulaeren Entscheidbarkeit von Zahlenmengen* ◇ D05 D20 ◇
HEMMERLING, A. *Zur verallgemeinerten Realisierbarkeit von Pseudomusterfunktionen durch Zellularraeume* ◇ D10 D20 ◇
HORVATH, S. *Complexity of sequence encodings* ◇ D15 D20 ◇
IWAMOTO, S. *A class of inverse theorems on recursive programming with monotonicity* ◇ D20 ◇
JUNG, H. & WIEHAGEN, R. *Rekursionstheoretische Charakterisierung von erkennbaren Klassen rekursiver Funktionen (English and Russian summaries)* ◇ D20 ◇
KANOVICH, M.I. *On the precision of a complexity criterion for nonrecursiveness and universality (Russian)* ◇ D15 D20 D25 D30 ◇
KINBER, E.B. *On a theory of inductive inference* ◇ D20 D45 ◇
KINBER, E.B. *On btt-degrees of sets of minimal numbers in Goedel numberings* ◇ D20 D30 D45 ◇
KINBER, E.B. *On identification in the limit of the minimal numbers for functions of effectively enumerable classes (Russian)* ◇ D20 D45 ◇
KINBER, E.B. *On speed-up of limiting identification of recursive functions by changing the order of questions (Russian) (English and German summary)* ◇ D15 D20 ◇
KOSOVSKIJ, N.K. *On classes of functions defined by addition (Russian)* ◇ D20 ◇
KUEHNEL, W. & MESEGUER, J. & PFENDER, M. & SOLS, I. *Primitive recursive algebraic theories and program schemes* ◇ B75 D20 G30 ◇

MARANDZHYAN, G.B. *On c-degrees of sets of minimal indices of algorithms (Russian) (Armenian and English summaries)* ◇ D20 D45 ◇
NABIALEK, I. *Bases of aggregable sets* ◇ D15 D20 ◇
O'DONNELL, M.J. *Subtree replacement systems: a unifying theory for recursive equations, LISP, lucid and combinatory logic* ◇ B40 B75 D05 D20 ◇
PAKHOMOV, S.V. *How to prove that some classes of simple primitive recursive functions are distinct (Russian)* ◇ D20 ◇
PODNIEKS, K.M. *Forecasting strategies of limited complexity (Russian) (English Summary)* ◇ D15 D20 ◇
RABIN, M.O. *Decidable theories* ◇ B25 C98 D20 ◇
SCHINZEL, B. *Decomposition of Goedelnumberings into Friedbergnumberings* ◇ D20 D45 ◇
SCHINZEL, B. *Struktur von Programmbuendeln* ◇ D20 G30 ◇
SEMENOV, A.L. *Presburgerness of predicates regular in two number systems (Russian)* ◇ B10 D20 F30 ◇
SHAPIRO, S. *On Church's thesis* ◇ A05 D20 F99 ◇
SHURYGIN, V.A. *On estimates of the complexity of algorithmic problems in constructive analysis (Russian)* ◇ D15 D20 F60 ◇
SIMON, J. *Polynomially bounded quantification over higher types and a new hierarchy of the elementary sets* ◇ B15 D10 D15 D20 ◇
STATMAN, R. *The typed λ-calculus is not elementary recursive* ◇ B40 D15 D20 ◇
TAERNLUND, S.-A. *Horn clause computability* ◇ B75 D20 ◇
TRAKHTENBROT, B.A. *Frequency algorithms and computations* ◇ D15 D20 ◇
VINOGRADOV, A.P. *On the existence of a majorizing local algorithm in effectively describable problems of computing information (Russian)* ◇ D15 D20 ◇
VOGEL, HELMUT *Ausgezeichnete Folgen fuer praedikative Ordinalzahlen und praedikativ-rekursive Funktionen* ◇ D20 F15 ◇
YANOV, YU.I. *A convolution method for the solution of properties of formal systems (Russian)* ◇ D10 D20 ◇
YANOV, YU.I. *Computations in a class of programs (Russian)* ◇ B75 D20 F30 ◇
YANOV, YU.I. *Equivalent transformations of computational trees (Russian) (English Summary)* ◇ B75 D20 ◇
ZAKOWSKI, W. *A generalization of the notions of a machine and computability* ◇ D10 D20 ◇
ZEMKE, F. *P.R.-regulated systems of notation and the subrecursive hierarchy equivalence property* ◇ D20 F15 ◇
ZHUKOV, S.A. *Some asymptotic estimates of the complexity of partial recursive functions (Russian)* ◇ D15 D20 ◇

1978

ARSLANOV, M.M. (ED.) *Algorithms and automata. Collection of scientific works (Russian)* ◇ D05 D20 D97 ◇
AVENHAUS, J. & MADLENER, K. *Algorithmische Probleme bei Einrelatorgruppen und ihre Komplexitaet* ◇ D20 D40 ◇
BEREZIN, S.A. *Maximal subalgebras of recursive function algebras (Russian)* ◇ D20 D75 ◇
BREIDBART, S. *On splitting recursive sets* ◇ D05 D15 D20 D25 ◇
BUCHBERGER, B. & ROIDER, B. *Input/output codings and transition functions in effective systems* ◇ D05 D20 D45 ◇
CALUDE, C. *Categorical methods in computability I,II (I: Recursive, non-primitive recursive functions) (Romanian) (English summary)* ◇ D15 D20 ◇
CALUDE, C. & FANTANEANU, B. *On recursive, non-primitive recursive functions* ◇ D20 ◇
CASE, J. & SMITH, C.H. *Anomaly hierarchies of mechanized inductive inference* ◇ B35 D20 ◇
CASE, J. & MOORE, D.J. *The complexity of total order structures* ◇ D15 D20 D45 ◇
CAZACU, C. & CAZACU, M. *An attempt to a general theory of algorithms* ◇ B75 D20 ◇
CHEN, ZHONGYUEH *Formalization of equivalence of recursively defined functions* ◇ D20 ◇
COLLINS, W.J. *Provably recursive real numbers* ◇ D20 F60 ◇
DAMM, W. & FEHR, E. & INDERMARK, K. *Higher type recursion and self-application as control structures* ◇ B40 B75 D20 ◇
EDMUNDSON, H.P. & TUNG, I.I. *A characterization of the recursiveness of an m-adic probabilistic language* ◇ D05 D20 ◇
EMDE BOAS VAN, P. *Least fixed points and the recursion theorem* ◇ B75 D20 ◇
ERSHOV, A.P. *Mixed computation in the class of recursive program schemata* ◇ B75 D20 ◇
FREJVALD, R.V. *Effektive Operationen und limitar berechenbare Funktionale (Russisch)* ◇ D20 ◇
FRIEDMAN, H.M. *Classically and intuitionistically provably recursive functions* ◇ B15 D20 E70 F30 F50 ◇
GEORGIEVA, N.V. *An extension of the decidable class of equations considered by Goodstein and Lee* ◇ D20 ◇
GOETZE, B.G. & KLETTE, R. *Ein Normalformentheorem fuer Σ_2-Funktionen* ◇ D20 D55 ◇
GOETZE, B.G. & NEHRLICH, W. *Loop programs and classes of primitive recursive functions* ◇ D20 ◇
GOLUNKOV, YU.V. *Precomplete classes of algorithms that preserve the membership in a set (Russian)* ◇ D20 D50 ◇
GOODMAN, NICOLAS D. *The nonconstructive content of sentences of arithmetic* ◇ D20 D30 D55 F30 F50 ◇
HEMMERLING, A. *Realisierung der Verbandsoperationen fuer Regularitaetsklassen von Kodierungen* ◇ D20 ◇
HOROWITZ, B.M. *Sets completely creative via recursive permutations* ◇ D20 D25 ◇
HUPBACH, U.L. *Rekursive Funktionen in mehrsortigen Peano-Algebren* ◇ D20 D75 ◇
JANTKE, K.P. *Universal methods for identification of total recursive functions* ◇ D20 ◇
KOLGANOV, N.A. *Identities in an algebra of recursive functions (Russian)* ◇ D20 D75 ◇
KOROL'KOV, YU.D. *Families of general recursive functions with a finite number of limit points (Russian)* ◇ D20 D45 ◇

KOSOVSKIJ, N.K. *On Kolmogorov's subrecursive algorithmic complexity (Russian)* ◇ D15 D20 ◇
KOZEN, D. *Indexing of subrecursive classes* ◇ D15 D20 ◇
LEVITZ, H. *An ordinal bound for the set of polynomial functions with exponentiation* ◇ D15 D20 E07 E10 F15 F30 ◇
LILLO DE, N.J. *A note on Turing machine regularity and primitive recursion* ◇ D10 D20 ◇
LORENTS, P.P. *The Lob-Wainer hierarchy and general recursive functions (Russian)* ◇ D20 ◇
MACHTEY, M. & YOUNG, P. *An introduction to the general theory of algorithms* ◇ D15 D20 D98 ◇
MACHTEY, M. & WINKLMANN, K. & YOUNG, P. *Simple Goedel numberings, isomorphisms, and programming properties* ◇ D05 D10 D15 D20 D45 F40 ◇
MAN, V.D. *Berechnung von Wortfunktionen auf deterministischen n-dimensionalen iterativen Automaten* ◇ D05 D20 ◇
MANNA, Z. & SHAMIR, A. *The convergence of functions to fixed points of recursive definitions* ◇ D20 ◇
MARCHENKOV, S.S. *A method of constructing maximal subalgebras in algebras of general recursive functions (Russian)* ◇ D20 D75 ◇
MELLISH, M. *Some prediction algorithms for nonrecursive sequences* ◇ D20 ◇
MIGNOTTE, M. *Some effective results about linear recursive sequences* ◇ D20 ◇
NAZARYAN, G.A. *Ueber eine Synthese von Algorithmen approximativ berechenbarer boolescher Funktionen (Russian)* ◇ B35 D15 D20 ◇
NEPOMNYASHCHIJ, V.A. *Examples of predicates inexpressible by s-rudimentary formulas (Russian) (English summary)* ◇ D05 D20 ◇
NEPOMNYASHCHIJ, V.A. & VOELKEL, L. *Zur Vollstaendigkeit von Befehlssystemen* ◇ D20 ◇
POGOSYAN, EH.M. *Training as a variety of induction inference (Russian)* ◇ D20 ◇
POLYAKOV, E.A. & ROZINAS, M.G. *Correlations between different forms of relative computability of functions (Russian)* ◇ D20 D30 ◇
POLYAKOV, E.A. (ED.) *Recursive functions (Russian)* ◇ D20 D30 D97 ◇
POUR-EL, M.B. & RICHARDS, I. *Differentiability properties of computable functions - a summary* ◇ D20 D80 F60 ◇
RYAN, W.J. *Goedel's second incompleteness theorem for general recursive arithmetic* ◇ D20 F30 ◇
SARKISYAN, G.Z. *Ueber eine Klasse arithmetischer Funktionen, die mit Hilfe von Schemata aus Funktionalelementen berechenbar sind (Russian) (Armenian summary)* ◇ D15 D20 ◇
SHAY, M. & YOUNG, P. *Characterizing the orders changed by program translators* ◇ B75 D20 D25 ◇
SOKOLOV, V.A. *"Zero" identities in the Robinson algebra (Russian)* ◇ D20 D75 ◇
VALIDOV, F.I. *Minimal numerations of families of general recursive functions (Russian)* ◇ D20 D45 ◇
VERBEEK, R. *Primitiv-rekursive Grzegorczyk-Hierarchien* ◇ D20 ◇
WIEHAGEN, R. *Characterization problems in the theory of inductive inference* ◇ D15 D20 D45 ◇

1979

ARBIB, M.A. & MANES, E.G. *Intertwined recursion, tree transformations, and linear systems* ◇ D05 D20 D80 G30 ◇
BEL'TYUKOV, A.P. *A machine description and a hierarchy of initial Grzegorczyk classes (Russian) (English summary)* ◇ D10 D20 ◇
BEL'TYUKOV, A.P. *Small classes based on bounded recursion (Russian)* ◇ D20 ◇
BLUM, E.K. & LYNCH, N.A. *Relative complexity of operations on numeric and bit-string algebras* ◇ D15 D20 ◇
BOUDOL, G. *A new recursion induction principle* ◇ D20 ◇
BROY, M. & FINANCE, J.P. & QUERE, A. & REMY, J.-L. & WIRSING, M. *Methodical solution of the problem of ascending subsequences of maximum length within a given sequence* ◇ B75 D20 ◇
CALUDE, C. & MARCUS, S. & TEVY, I. *The first example of a recursive function which is not primitive recursive* ◇ A10 D20 ◇
CARTWRIGHT, ROBERT & MCCARTHY, J. *Recursive programs as functions in a first order theory* ◇ B10 B75 D20 ◇
CAZACU, M. *A general mathematical notion of the algorithm* ◇ D03 D10 D20 ◇
COUSOT, P. & COUSOT, R. *Constructive versions of Tarski's fixed point theorems* ◇ D20 G10 ◇
DANKO, W. *Definability in algorithmic logic* ◇ B75 D05 D20 ◇
DEGANO, P. & SIROVICH, F. *Inductive generalization and proofs of function properties* ◇ B35 B75 D20 D80 ◇
DEKHTYAR', M.I. *Bounds on computational complexity and approximability of initial segments of recursive sets* ◇ D15 D20 ◇
ETTL, W. & LEITSCH, A. *Aufzaehlung subrekursiver Klassen und konstruierende Automaten* ◇ D05 D20 ◇
FACHINI, E. & MAGGIOLO-SCHETTINI, A. *A hierarchy of primitive recursive sequence functions* ◇ D20 ◇
FAL'K, V.N. & KUTEPOV, V.P. *Functional systems: theoretical and practical aspects (Russian)* ◇ D20 ◇
FERRO, A. & MICALE, B. *Sulla teoria della computabilita* ◇ D20 D25 ◇
FREJVALD, R.V. *Finite identification of general recursive functions by probabilistic strategies* ◇ D10 D20 ◇
FREJVALD, R.V. & WIEHAGEN, R. *Inductive inference with additional information* ◇ D20 ◇
FRIEDMAN, E.P. & GREIBACH, S.A. *Monadic recursion schemes: The effect of constants* ◇ B75 D20 ◇
GALLIER, J.H. *Recursion schemes and generalized interpretations* ◇ B75 D20 D75 ◇
GERMANO, G. & MAGGIOLO-SCHETTINI, A. *Computable stack functions for semantics of stack programs* ◇ B75 D20 ◇
GORDON, D. *Complexity classes of provable recursive functions* ◇ D15 D20 F30 ◇

HARAGUCHI, M. *The theoretical foundation of programming by examples* ◇ B75 D20 ◇
HARROW, K. *Equivalence of some hierarchies of primitive recursive functions* ◇ D20 ◇
HARTMANIS, J. *Relations between diagonalization, proof systems, and complexity gaps* ◇ D15 D20 ◇
HUYNH, D.T. *On complexity measures which are induced by probability distributions* ◇ D15 D20 ◇
HUZINO, S. & IIMORI, S. *On LISP-realization of partial recursive functions* ◇ B75 D20 ◇
JANTKE, K.P. *Automatic synthesis of programs and inductive inference of functions* ◇ D20 ◇
JANTKE, K.P. *Natural properties of strategies identifying recursive functions* ◇ B75 D20 D45 ◇
JUKNA, S. *On decidability of equivalence problem in some algebras of recursive functions (Russian)* ◇ B25 D20 D75 ◇
KANOVICH, M.I. *The complexity of search for short programs (Russian)* ◇ D15 D20 ◇
KLEENE, S.C. *Origins of recursive function theory* ◇ A10 D20 D98 ◇
KOROL'KOV, YU.D. *Families of general recursive functions without isolated points (Russian)* ◇ D20 D45 ◇
KOSOVSKIJ, N.K. *On decision procedures for invariant properties of short algorithms (Russian) (English summary)* ◇ D15 D20 D25 D45 ◇
LAVROV, I.A. *Computability of partial functions and enumerability of sets in Peano's arithmitic (Russian)* ◇ D20 D25 F30 ◇
LEWIS, F.D. *On unsolvability in subrecursive classes of predicates* ◇ D15 D20 D30 D55 ◇
LUCKHARDT, H. *A limit for higher recursion theory* ◇ D20 ◇
MARANDZHYAN, G.B. *On the sets of minimal indices of partial recursive functions* ◇ D20 D30 D45 D55 ◇
MARCHENKOV, S.S. *On the quasi-Peano property of recursive functions (Russian)* ◇ D20 ◇
MASHURYAN, A.S. *On a class of primitive recursive functions (Russian)* ◇ C13 D20 ◇
MO, SHAOKUI & YE, DAXING *On recursion schemas defining operators (Chinese) (English summary)* ◇ D20 ◇
MOORE, J.S. *A mechanical proof of the termination of Takeuchi's function* ◇ B35 D20 ◇
O'DONNELL, M.J. *A practical programming theorem which is independent of Peano arithmetic* ◇ B40 B75 D20 F05 F30 ◇
PETTOROSSI, A. *On the definition of hierarchies of infinite sequential computations* ◇ B40 D20 ◇
PRANK, R.K. *Expressibility in the elementary theory of recursive sets with realizability logic (Russian) (English summary)* ◇ B60 D20 F50 ◇
PROSKURIN, A.V. *Positive rudimentarity of the graphs of Ackermann and Grzegorczyk functions (Russian) (English summary)* ◇ D20 ◇
ROZINAS, M.G. & SOLON, B.YA. *Weakly semirecursive sets (Russian)* ◇ D20 D25 D30 ◇
SABEL'FEL'D, V.K. *Polynomial estimate of the complexity of the recognition of logic-term equivalence (Russian)* ◇ B75 D15 D20 ◇
SCHINZEL, B. *Classes of decompositions of a Goedelnumbering* ◇ D20 D45 ◇
SCHINZEL, B. *Uebersetzer zwischen Goedelnumerierungen* ◇ D20 D45 ◇
SEPER, K. *Algorithmic constructions inspired by Caporaso* ◇ D05 D20 ◇
SHANIN, N.A. *On canonical recursive functions and operations (Russian) (English summary)* ◇ D20 ◇
SMORYNSKI, C.A. *Some rapidly growing functions* ◇ D20 D80 F30 ◇
STATMAN, R. *The typed λ-calculus is not elementary recursive* ◇ B40 D15 D20 ◇
STRAIGHT, D.W. *Domino f-sets* ◇ D10 D20 ◇
SVENONIUS, L. *Two kinds of extensions of primitive recursive arithmetic* ◇ D20 F30 F35 ◇
VERBEEK, R. *On the naturalness of the Grzegorczyk hierarchy* ◇ D15 D20 ◇
WAGNER, K. *Bounded recursion and complexity classes* ◇ D15 D20 ◇
ZASLAVSKIJ, I.D. *The realization of three-valued logical functions by means of recursive and Turing operators (Russian)* ◇ B50 D10 D20 ◇
ZDEBSKAYA, G.V. *On some properties of computation in the limit (Russian)* ◇ D20 D30 ◇
ZHOU, CHAOCHEN *Program schemes and predicate calculus (Chinese)* ◇ B75 D20 ◇

1980

ADACHI, A. & KASAI, T. *A characterization of time complexity by simple loop programs* ◇ D15 D20 ◇
ADRIANOPOLI, F. *Conjugated measures of computational complexity* ◇ D15 D20 ◇
ALTON, D.A. *"Natural" programming languages and complexity measures for subrecursive programming languages: an abstract approach* ◇ B75 D15 D20 ◇
ARNOLD, A. & NIVAT, M. *Formal computations of nondeterministic recursive program schemes* ◇ B75 D20 ◇
BORZOV, YU.V. *Four theorems on holographic sets (Russian)* ◇ D20 D30 ◇
CALUDE, C. & MARCUS, S. & TEVY, I. *Recursive properties of Sudan function* ◇ D20 ◇
CAZACU, M. *The simulation relations between the classes of algorithms* ◇ D03 D10 D20 ◇
CHEN, HUOWANG *The decision problem of families of general recursive functions (Chinese)* ◇ D20 ◇
CONSTABLE, R.L. & HUNT III, H.B. & SAHNI, S.K. *On the computational complexity of program scheme equivalence* ◇ D15 D20 ◇
COSTE-ROY, M.-F. & COSTE, M. & MAHE, L. *Contribution to the study of the natural number object in elementary topoi* ◇ D20 F35 F50 G30 ◇
DEGTEV, A.N. *A category of partial recursive functions (Russian)* ◇ D20 D45 D75 G30 ◇
DEUTSCH, M. *Zur goedelisierungsfreien Darstellung der rekursiven und primitiv-rekursiven Funktionen und rekursiv aufzaehlbaren Praedikate ueber Quotiententermmengen* ◇ D20 D25 ◇
FALKINGER, J. *Reduzierbarkeit von berechenbaren Numerierungen von P_1* ◇ D20 D30 D45 ◇

FALKINGER, J. *Universalitaet von berechenbaren Numerierungen von partiell rekursiven Funktionen* ◇ D20 D30 D45 ◇
GANDY, R.O. *Church's thesis and principles for mechanisms* ◇ A05 D05 D10 D20 F99 ◇
GOETZE, B.G. & NEHRLICH, W. *The structure of LOOP programs and subrecursive hierarchies* ◇ D15 D20 ◇
GOLUNKOV, YU.V. *Precomplete classes of algorithms that preserve membership in a set I (Russian)* ◇ D20 D75 ◇
GONCHAROV, S.S. *Computable single-valued numerations (Russian)* ◇ D20 D45 ◇
HAUCK, J. *Stetigkeitseigenschaften berechenbarer reeller Funktionen* ◇ D20 F60 ◇
HOROWITZ, B.M. *Constructively nonpartial recursive functions* ◇ D20 D25 ◇
KHAKHANYAN, V.KH. *Comparative strength of variants of Church's thesis at the level of set theory (Russian)* ◇ D20 E70 F50 ◇
KHAKHANYAN, V.KH. *The consistency of intuitionistic set theory with Church's principle and the uniformization principle (Russian)* ◇ D20 E35 E70 F25 F50 ◇
KHAKHANYAN, V.KH. *The consistency of intuitionistic set theory with formal mathematical analysis (Russian)* ◇ D20 E35 E70 F25 F35 F50 ◇
KOZEN, D. *Indexings of subrecursive classes* ◇ D15 D20 ◇
LEITSCH, A. *Complexity of index sets and translating functions* ◇ D15 D20 ◇
MADATYAN, K.A. *On correction of the set of algorithms of pattern recognition by the schemes of functional elements (Russian)* ◇ D20 ◇
MANIN, YU.I. *The computable and the non-computable (Russian)* ◇ D20 D25 D98 ◇
MARCHENKOV, S.S. *A basis with respect to superposition in the class of functions that are elementary in the sense of Kalmar (Russian)* ◇ D20 ◇
MARCHENKOV, S.S. *Existence of superposition bases in countable primitive-recursively closed classes of one-place functions (Russian)* ◇ D20 ◇
MASHURYAN, A.S. *On the class of primitive recursive functions definable on finite models (Russian) (Armenian summary)* ◇ C13 D20 ◇
MATOS, A.B. & PORTO, A.G. *Ackermann and the superpowers* ◇ D20 ◇
MCALOON, K. *Progressions transfinies de theories axiomatiques, formes combinatoires du theoreme d'incompletude et fonctions recursives a croissance rapide* ◇ C62 D20 F15 F30 ◇
MCBETH, R. *A generalization of Ackermann's function* ◇ D20 ◇
MCBETH, R. *Exponential polynomials of linear height* ◇ D20 E07 ◇
POLYAKOV, E.A. *The theory of recursive functionals and effective operations (Russian)* ◇ D20 ◇
ROGOZHIN, YU.V. *Universal one-place partial recursive functions (Russian)* ◇ D10 D20 ◇
ROMANOVS'KIJ, V.YU. *Solvability of the equivalence of linear unary recursive schemes with individual constants (Ukrainian) (English and Russian summaries)* ◇ D20 ◇
SCHINZEL, B. *Ueber die Kategorie der Programmbuendel (Hab.-Schr.)* ◇ D15 D20 D45 ◇
SCHINZEL, B. *Zerlegung mit Vergleichsbedingungen einer Goedelnumerierung* ◇ D20 D45 ◇
SEMENOV, A.L. *An interpretation of free algebras in free groups (Russian)* ◇ D20 D35 ◇
SHEN', A.KH. *Axiomatic approach to the theory of algorithms and relativized computability (Russian)* ◇ D20 D75 ◇
SIMOVICI, D.A. *Context-sensitive languages and Ackermann's function* ◇ D05 D20 ◇
SMITH, C.H. *Applications of classical recursion theory to computer science* ◇ B75 D15 D20 D80 ◇
STEFANI, S. *Recursive functions with measurable oracles: An approach to a probabilistic recursion theory* ◇ D20 D30 ◇
WEBB, J.C. *Mechanism, mentalism, and metamathematics. An essay on finitism* ◇ A05 D20 F99 ◇
WIEDMER, E. *Computing with infinite objects* ◇ D10 D20 ◇
YABLONSKIJ, S.V. *On some results in the theory of function systems (Russian)* ◇ B50 D20 G25 ◇
YANOV, YU.I. *The complete limited system of equivalent transformation rules for programs computing total functions (Russian)* ◇ B75 D20 ◇

1981

AJZENSHTEJN, M.KH. *Subrecursive reducibility and accelerability (Russian)* ◇ D15 D20 ◇
AMIHUD, A. & CHOUEKA, Y. *Loop-programs and polynomially computable functions* ◇ D15 D20 ◇
AVENHAUS, J. & MADLENER, K. *An algorithm for the word problem in HNN extensions and the dependence of its complexity on the group representation* ◇ D20 D40 ◇
BAKKER DE, J.W. & ZUCKER, J.I. *Derivatives of programs* ◇ D20 ◇
BEICK, H.-R. & JANTKE, K.P. *Combining postulates of naturalness in inductive inference (German and Russian summaries)* ◇ D20 ◇
BRUIN DE, A. *On the existence of Cook semantics* ◇ B75 D20 D80 ◇
CALUDE, C. & VIERU, V. *A note on Knuth's iterated powers* ◇ D20 ◇
CALUDE, C. & VIERU, V. *An iterative normal form for partial recursive functions* ◇ D20 ◇
CALUDE, C. *Darboux property and primitive recursive functions* ◇ D20 ◇
CAZACU, C. *A depth-space analysis of algorithms* ◇ D15 D20 ◇
CHEN, HUOWANG *Effectively continuous functionals (Chinese)* ◇ D20 ◇
CROCIANI, C. *Su alcune proprieta ricorsive legate al concetto di densita (English summary)* ◇ D20 ◇
DALEY, R.P. *On the error correcting power of pluralism in inductive inference (Preliminary report)* ◇ D10 D20 ◇
DALEY, R.P. *Retraceability, repleteness and busy beaver sets* ◇ D15 D20 D50 ◇
DARE, V.R. & SIROMONEY, R. & SUBRAMANIAN, K.G. *Infinite arrays and infinite computations* ◇ D20 ◇

DEGTEV, A.N. *On (m,n)-computable sets (Russian)* ◇ D20 ◇
DEUTSCH, M. *Registermaschinen ueber Quotiententermmengen* ◇ D10 D20 ◇
DIKOVSKIJ, A.YA. *A theory of complexity of monadic recursion schemes* ◇ D15 D20 ◇
DIMA, N. *Sudan function is universal for the class of primitive recursive functions (Romanian) (English summary)* ◇ D20 ◇
DIMITROV, I. *Systematic transformation of recursive invariants in iterative algorithms (Russian)* ◇ D20 ◇
ENGELER, E. *An algorithmic model of strict finitism* ◇ D20 F65 ◇
ERSHOV, A.P. & KNUTH, D.E. (EDS.) *Algorithms in modern mathematics and computer science. Proceedings, Urgench, Uzbek SSR, September 16-22, 1979* ◇ D15 D20 D97 ◇
GERMANO, G. & MAGGIOLO-SCHETTINI, A. *Sequence recursiveness without cylindrification and limited register machines* ◇ B75 D15 D20 ◇
GURARI, E.M. & IBARRA, O.H. *The complexity of the equivalence problem for simple programs* ◇ D15 D20 ◇
HOWARD, W.A. *Computability of ordinal recursion of type level two* ◇ D20 F10 F15 F35 F50 ◇
IBARRA, O.H. & LEININGER, B.S. *Characterizations of Presburger functions* ◇ D20 F30 ◇
JEDRZEJOWICZ, J. *Undecidable problems associated with combinatorial systems and their one-one degrees of unsolvability* ◇ D20 D25 ◇
KANOVICH, M.I. *Complexity of a fixed point and of a shift algorithm (Russian)* ◇ D15 D20 ◇
KHAKHANYAN, V.KH. *The consistency of some intuitionistic and constructive principles with a set theory* ◇ D20 E35 E70 F50 ◇
KLEENE, S.C. *The theory of recursive functions, approaching its centennial* ◇ A10 D20 D98 ◇
KLEINE BUENING, H. *Classes of functions over binary trees* ◇ B75 D05 D15 D20 ◇
KOSOVSKIJ, N.K. *Elements of mathematical logic and its applications to the theory of subrecursive algorithms (Russian)* ◇ B98 D20 D98 F30 F60 F98 ◇
KUZ'MINA, T.M. *Structure of the m-degrees of the index sets of families of partial recursive functions (Russian)* ◇ D20 D30 ◇
LEWIS, F.D. *Stateless Turing machines and fixed points* ◇ D10 D15 D20 ◇
LI, XIANG *Study on the theory of computability and computational complexity of real numbers, I: Unsolvable problems in the constructive continuum* ◇ D20 ◇
LORENTS, P.P. *Generation of hierarchies of recursive functions and the solution of problems "A" and "B" of Loeb and Wainer using the method of correcting fundamental sequences (Russian) (Estonian and English summaries)* ◇ D20 ◇
MACHTEY, M. & YOUNG, P. *Remarks on recursion versus diagonalization and exponentially difficult problems* ◇ D15 D20 ◇
MAIER, W. *Graphen total rekursiver Funktionen* ◇ D15 D20 ◇
MARTIC, B. *Iterative systems and diagram algorithms* ◇ D20 ◇
MATROSOV, V.L. *Complexity of computable functions for a generalized storage measure (Russian)* ◇ D15 D20 ◇
MAYR, E.W. & MEYER, A.R. *The complexity of the finite containment problem for Petri nets* ◇ D15 D20 D80 ◇
MCBETH, R. *A note on exponential polynomials and prime factors* ◇ D20 F15 ◇
NASIBULLOV, KH.KH. *Some problems connected with lower classes of the Grzegorczyk hierarchy (Russian)* ◇ D20 ◇
NOLTEMEIER, H. *Informatik I: Einfuehrung in Algorithmen und Berechenbarkeit* ◇ D20 D98 ◇
PARIS, J.B. & WILKIE, A.J. *Models of arithmetic and the rudimentary sets* ◇ C62 D20 ◇
PODNIEKS, K.M. *Prediction of the following value of a function (Russian)* ◇ D10 D20 ◇
POLYAKOV, E.A. *Partial recursive functions with a recursive graph (Russian)* ◇ D20 ◇
SEMENOV, A.L. & USPENSKIJ, V.A. *What are the gains of the theory of algorithms: basic developments connected with the concept of algorithm and with its application in mathematics* ◇ D20 D98 ◇
SHAPIRO, S. *Understanding Church's thesis* ◇ A05 D20 F99 ◇
SHEN', A.KH. *Some remarks on numerations that are not natural (Russian)* ◇ D20 D45 ◇
SHEN, BAIYING *Axiom systems for primitive recursive word arithmetic WA (Chinese)* ◇ D20 D75 ◇
STAHL, G. *Character and acceptability of Church's thesis* ◇ D20 ◇
STATMAN, R. *Number theoretic functions computable by polymorphic programs* ◇ D20 ◇
TANG, TONGGAO *A note on the relative recursion theorem with a functional index (Chinese)* ◇ D20 ◇
TAO, RENJI *On the computational power of automata with time or space bounded by Ackermann's or superexponential functions* ◇ D10 D15 D20 ◇
VOGEL, J. & WAGNER, K. *On a class of automata accepting exactly the languages which are elementary in the sense of KALMAR* ◇ D05 D20 ◇
WILLIAMS, J.H. *Formal representations for recursively defined functional programs* ◇ B75 D20 ◇

1982

ARBIB, M.A. & KFOURY, A.J. & MOLL, R. *A programming approach to computability* ◇ B75 D20 D98 ◇
BAETEN, J.C.M. & BERGSTRA, J.A. & KLOP, J.W. *Priority rewrite systems* ◇ B75 D05 D20 ◇
BEICK, H.-R. *Zur Konvergenzgeschwindigkeit von Strategien der induktiven Inferenz (English and Russian summaries)* ◇ D20 ◇
BOERGER, E. *Note on bounded diophantine representation of subrecursive sets* ◇ D20 ◇
BYERLY, R.E. *An invariance notion in recursion theory* ◇ D20 D45 ◇
BYERLY, R.E. *Recursion theory and the lambda-calculus* ◇ B40 D20 ◇
CALUDE, C. *Note on a hierarchy of primitive recursive functions* ◇ D20 ◇

CALUDE, C. & CHITESCU, I. *Strong noncomputability of random strings* ◇ D20 ◇
CALUDE, C. *Topological size of sets of partial recursive functions* ◇ D15 D20 E75 ◇
CASE, J. & LYNES, C. *Machine inductive inference and language identification* ◇ D05 D20 ◇
CHEN, KEJIAN *Tradeoffs in the inductive inference of nearly minimal size programs* ◇ D20 ◇
CSIRMAZ, L. *Determinateness of program equivalence over Peano axioms* ◇ B75 D20 ◇
DEGTEV, A.N. *Comparison of linear reducibility with other reducibilities of tabular type (Russian)* ◇ D20 ◇
DEUTSCH, M. *Zur Komplexitaetsmessung primitiv-rekursiver Funktionen ueber Quotiententermmengen* ◇ D20 ◇
DYMENT, E.Z. *Existence of a degree of extendability of a partially recursive function, which is not a degree of separability (Russian)* ◇ D20 D30 ◇
ERSHOV, A.P. *Mixed computation: potential applications and problems for study* ◇ B75 D20 ◇
FACHINI, E. & MAGGIOLO-SCHETTINI, A. *Comparing hierarchies of primitive recursive sequence functions* ◇ D20 ◇
FAHMY, M.H. *A formal proof of Ackermann's theorem (Arabic summary)* ◇ D20 F30 ◇
FISHER, A. *Formal number theory and computability* ◇ B98 D20 F30 F98 ◇
FITTING, M. *A generalization of elementary formal systems* ◇ D20 D70 D75 ◇
FRAISSE, R. *Recursivite* ◇ D20 ◇
FREJVALD, R.V. & KINBER, E.B. & WIEHAGEN, R. *Inductive inference and computable one-one numberings* ◇ B48 D20 D45 ◇
HAY, L. & MILLER, DOUGLAS E. *A topological analog to the Rice-Shapiro index theorem* ◇ D20 E15 ◇
HOROWITZ, B.M. *Elementary formal systems as a framework for relative recursion theory* ◇ D20 D25 D30 ◇
JONES, JAMES P. & MATIYASEVICH, YU.V. *Exponential diophantine representation of recursively enumerable sets (French summary)* ◇ D20 D25 D35 ◇
JUKNA, S. *Arithmetical representations of machine complexity classes (Russian) (English and Lithuanian summaries)* ◇ D15 D20 ◇
KANOVICH, M.I. *Truth-table reducibilities of problems of the extension of partial recursive functions (Russian)* ◇ D20 D30 ◇
KLEINE BUENING, H. *Note on the* $E_1^* - E_2^*$ *problem* ◇ D20 ◇
KLETTE, R. *A few results on the complexity of classes of identifiable recursive function sets* ◇ D20 D55 ◇
KOROL'KOV, YU.D. *On the reducibility of index sets of families of general recursive functions (Russian)* ◇ D20 D45 ◇
KUCERA, A. *On recursive measure of classes of recursive sets* ◇ D20 D30 ◇
MAKAROV, V.P. *On theory of abstract algorithms* ◇ D20 ◇
MENTRASTI, P. & PROTASI, M. *Extended primitive recursive functions (French summary)* ◇ D20 ◇
MENZEL, W. & SPERSCHNEIDER, V. *Universal automata with uniform bounds on simulation time* ◇ D20 ◇
MINTS, G.E. & TYUGU, E.KH. *The completeness of structural synthesis rules (Russian)* ◇ B20 B75 D20 F50 ◇
MO, SHAOKUI *Theory of algorithms (Chinese)* ◇ D20 D98 ◇
MULRY, P.S. *Generalized Banach-Mazur functionals in the topos of recursive sets* ◇ D20 D45 G30 ◇
PLA I CARRERA, J. *On the R-representability of primitive recursive functions (Catalan)* ◇ D20 F30 ◇
PRIDA, J.F. *A non-standard study of the theory of relative recursivity (Spanish)* ◇ D20 D30 H05 ◇
RICCARDI, G.A. *The independence of control structures in programmable numberings of the partial recursive functions* ◇ D20 D45 ◇
SCHINZEL, B. *Complexity of decomposotions of Goedel numberings* ◇ D20 D45 ◇
SCHINZEL, B. *On decomposition of Goedel numberings into Friedberg numberings* ◇ D20 D45 ◇
SCHMERL, U.R. *Ueber die schwach und die stark wachsende Hierarchie zahlentheoretischer Funktionen* ◇ D20 ◇
SEBELIK, J. *Horn clause programs and recursive functions defined by systems of equations* ◇ D20 ◇
SHAPIRO, S. *Acceptable notation* ◇ D20 ◇
SMORYNSKI, C.A. *The varieties of arboreal experience* ◇ D20 F30 ◇
SOLOV'EV, V.D. *The structure of closed classes of computable functions and predicates (Russian)* ◇ D20 D75 ◇
SORBI, A. *Numerazioni positive, r.e. classi e formule (English summary)* ◇ D20 F30 ◇
STATMAN, R. *Completeness, invariance and* λ*-definability* ◇ B40 D15 D20 ◇
STEFANI, S. *Recursive functions with measurable oracles: the semilattice of the effectively splitting distributions (Italian summary)* ◇ D20 D30 ◇
SZABO, Z. *On the ability of some inductive inferential strategies* ◇ D20 ◇
TER-ZAKHARYAN, N.P. *The possibilities of coding messages in various languages of formal arithmetic (Russian) (Armenian summary)* ◇ D20 ◇
TERLOUW, J. *On definition trees of ordinal recursive functionals: reduction of the recursion orders by means of type level raising* ◇ D20 ◇
THIBAULT, M.-F. *Prerecursive categories* ◇ D20 D65 G30 ◇
VELINOV, YU.P. *Polycategories and recursiveness* ◇ D20 ◇
VUCKOVIC, V. *Relativized cylindrification* ◇ D20 D30 ◇
ZAKHAROV, S.D. *The algebra of enumeration operators (Russian) (English summary)* ◇ D20 D45 ◇
ZDEBSKAYA, G.V. *Recursively enumerable sets and limit computability (Russian)* ◇ D20 D25 ◇

1983

ANSHEL, M. & MCALOON, K. *Reducibilities among decision problems for HNN groups, vector addition systems and subsystems of Peano arithmetic* ◇ D20 D40 D80 F30 ◇

BANACHOWSKI, L. & OKTABA, H. *A presentation of the scientific activities of the theory of computation group in the period from 1978-82* ◇ D15 D20 ◇

BARENDREGT, H.P. & LONGO, G. *Recursion theoretic operators and morphisms of numbered sets* ◇ D20 D30 D45 ◇

BERTONI, A. & MAURI, G. & MIGLIOLI, P.A. *On the power of model theory in specifying abstract data types and in capturing their recursiveness* ◇ B75 C50 C57 D20 ◇

BOTUSHAROV, O.I. *A recursion theory characterization of inductive inference with additional information classes* ◇ B48 D20 ◇

BRAMHOFF, H. & JANTZEN, M. *Notions of computability by Petri nets* ◇ B75 D20 D80 ◇

CALUDE, C. & PAUN, G. *Independent instances for some undecidable problems (French summary)* ◇ D20 F30 ◇

CALUDE, C. *On the class of independent problems related to Rice theorem* ◇ D20 D25 ◇

CALUDE, C. & TATARAM, M. *Universal sequences of primitive recursive functions* ◇ D20 ◇

CASE, J. & SMITH, C.H. *Comparison of identification criteria for machine inductive inference* ◇ B35 B48 D20 ◇

CASE, J. *Pseudo-extensions of computable functions* ◇ D20 ◇

CICHON, E.A. & WAINER, S.S. *The slow-growing and the Grzegorczyk hierarchies* ◇ D20 F15 ◇

CLARES RODRIGUES, B. *An introduction to W-calculability: basic operations (Spanish) (English summary)* ◇ D20 D75 ◇

COLLINS, W.J. & YOUNG, P. *Discontinuities of provably correct operators on the provably recursive real numbers* ◇ D20 F60 ◇

DAVIS, MARTIN D. & WEYUKER, E.J. *Computability, complexity, and languages. Fundamentals of theoretical computer science* ◇ D05 D10 D20 D98 ◇

FREUND, R. *Real functions and numbers defined by Turing machines* ◇ D10 D20 D80 ◇

GASANOV, R.G. *On a mathematical model of algorithms (Russian) (English and Azerbaijani summaries)* ◇ D20 ◇

GLADKIJ, A.V. *Theory of algorithms (Russian)* ◇ D20 D98 ◇

GORDON, D. & SHAMIR, E. *Computation of recursive functionals using minimal initial segments* ◇ D20 ◇

GUREVICH, Y. *Algebras of feasible functions* ◇ D15 D20 D75 ◇

HAWRUSIK, F.M. & VENKATARAMAN, K.N. & YASUHARA, A. *A view of computability on term algebras* ◇ B75 D20 ◇

HUNT III, H.B. & ROSENKRANTZ, D.J. *The complexity of monadic recursion schemes: executability problems, nesting depth, and applications* ◇ D15 D20 ◇

JANTKE, K.P. *The recursive power of algebraic semantics* ◇ B75 D20 ◇

KANOVICH, M.I. *Effectiveness of the translation theorem and the recursion theorem (Russian)* ◇ D20 ◇

KHAKHANYAN, V.KH. *Set theory and Church's thesis (Russian)* ◇ D20 E35 E70 F30 F35 F50 ◇

KINBER, E.B. *A note on limit identification of c-minimal indices (German and Russian summaries)* ◇ D20 D45 ◇

KLEINE BUENING, H. *A classification of an iterative hierarchy* ◇ D20 ◇

KLEINE BUENING, H. *Durch syntaktische Rekursion definierte Klassen* ◇ D20 ◇

KRAMOSIL, I. *On extremum-searching approximate probabilistic algorithms* ◇ D15 D20 ◇

KUZ'MINA, T.M. *Reducibility by morphisms (Russian)* ◇ D20 D45 ◇

LONGO, G. *Recursiveness and continuity: an introduction (Italian)* ◇ D20 D65 ◇

MOHRHERR, J. *Kleene index sets and functional m-degrees* ◇ D20 D25 D30 D45 ◇

NAKAMURA, A. & ONO, H. *Pictures of functions and their acceptability by automata* ◇ D10 D20 ◇

NAUMOVIC, J. *A classification of the one-argument primitive recursive functions* ◇ D20 ◇

NIGIYAN, S.A. *Solvability of algorithmic problems* ◇ D20 ◇

ORLICKI, A. *On effective numberings of effective definitional schemes* ◇ D20 D45 ◇

PKHAKADZE, SH.S. *An example of an intuitively computable everywhere defined function, and Church's thesis (Russian) (English and Georgian summaries)* ◇ D20 ◇

RICHMAN, F. *Church's thesis without tears* ◇ D20 D75 F55 F60 ◇

SEIBT, H. *On isomorphic partial recursive definitions (German and Russian summaries)* ◇ D20 ◇

SKANDALIS, K. *Programmability in the set of real numbers and second-order recursion* ◇ B75 D20 D80 ◇

SPENCER, J.H. *Large numbers and unprovable theorems* ◇ D20 F30 ◇

SZALKAI, I. *The algebraic structure of primitive recursive functions (Russian and Hungarian summaries)* ◇ D20 D75 ◇

TAUTS, A. *Parallelizing of recursive computations (Russian) (German and Estonian summaries)* ◇ D20 ◇

TERZILER, M. *Une recherche sur les fonctions primitives recursives* ◇ D20 ◇

THURAISINGHAM, M.B. *Cylindrical decision problems for system functions* ◇ D20 D25 ◇

THURAISINGHAM, M.B. *Some elementary closure properties of n-cylinders* ◇ D20 D25 ◇

USPENSKIJ, V.A. *Post's machine (Russian)* ◇ D20 ◇

VINCENZI, A. *Experiments as abstract machines (Italian)* ◇ D20 ◇

VOLGER, H. *A new hierarchy of elementary recursive decision problems* ◇ B25 D15 D20 ◇

WEBB, J.C. *Goedel's theorems and Church's thesis: A prologue to mechanism* ◇ A05 D20 F99 ◇

1984

BEESON, M.J. & SCEDROV, A. *Church's thesis, continuity, and set theory* ◇ D20 E70 F50 ◇

BEKBAEV, U.D. & KHADZHIEV, D. *Extension of partially defined operations II (Russian)* ◇ D20 ◇

BOYER, R.S. & MOORE, J.S. *A mechanical proof of the Turing completeness of pure LISP* ◇ B35 D20 ◇
BUJ, D.B. & RED'KO, V.N. *Primitive program algebras I (Russian)* ◇ D20 D75 D80 ◇
BURRIS, S. *Model companions for finitely generated universal Horn classes* ◇ B25 C05 C10 C35 D20 ◇
BUZETEANU, S. & DIMA, N. *Arithmetization of the computation of Sudan's and Knuth's functions (Romanian) (English summary)* ◇ D20 ◇
CHIMEV, K.N. *Separable pairs of some classes of functions (Bulgarian) (English and Russian summary)* ◇ D20 E20 ◇
CHIMEV, K.N. *Separation of a union of sets of arguments of functions (Bulgarian) (English and Russian summary)* ◇ D20 E20 ◇
COHEN, D.E. *Modular machines, undecidability and incompleteness* ◇ D20 D25 D35 F30 ◇
DALEY, R.P. & SMITH, C.H. *On the complexity of inductive inference* ◇ B75 D15 D20 ◇
DENNIS-JONES, E.C. & WAINER, S.S. *Subrecursive hierarchies via direct limits* ◇ D20 F15 G30 ◇
DIMA, N. *Topological properties of recurrent sets* ◇ D20 E07 ◇
DYMENT, E.Z. *Rice numerations (Russian)* ◇ D20 D45 ◇
FEKLICHEV, A.V. *Program schemes with storage and markers and their decision properties (Russian)* ◇ D20 ◇
FREJVALD, R.V. & KINBER, E.B. & WIEHAGEN, R. *Connections between identifying functionals, standardizing operations, and computable numberings* ◇ D20 D45 ◇
FREJVALD, R.V. & KINBER, E.B. & WIEHAGEN, R. *On the power of probabilistic strategies in inductive inference* ◇ B48 D20 ◇
GANDY, R.O. *Some relations between classes of low computational complexity* ◇ D20 ◇
GEORGIEVA, N.V. *A theorem on representability of general recursive functions* ◇ D20 ◇
GERMANO, G. & MAZZANTI, S. *Partial closures and semantics of while: toward an iteration-based theory of data types* ◇ B75 D20 D75 D80 ◇
GOLUNKOV, YU.V. *On some families of maximal subsystems of algebras in a system of algorithmic algebras of partial recursive functions and predicates. II (Russian)* ◇ D20 D75 ◇
HERRE, H. *Bemerkungen zum Begriff der praktischen Berechenbarkeit* ◇ D15 D20 ◇
HONG, JIAWEI *On similarity and duality of computation I* ◇ D20 ◇
HORVATH, S. *A generating system for partial recursive functions on N∗* ◇ D20 ◇
HUNT III, H.B. *Terminating Turing machine computations and the complexity and/or decidability of correspondence problems, grammars, and program schemes* ◇ D10 D15 D20 ◇
HUNT III, H.B. & ROSENKRANTZ, D.J. *The complexity of monadic recursion schemes: Exponential time bounds* ◇ D15 D20 ◇
IVANOV, L.L. *Kleene-recursiveness and iterative operative spaces* ◇ D20 D75 ◇
KANOVICH, M.I. *On the independence of invariant propositions (Russian)* ◇ B30 D20 ◇
KANOVICH, M.I. *The solution of Roger's problem on the relation between the strong and weak recursion theorem (Russian)* ◇ D20 ◇
KASHAPOVA, F.R. *Constructive set theory with types, and consistency with Church's thesis (Russian)* ◇ D20 E70 F35 F50 ◇
KOMJATH, P. & SZABO, Z. *Orientation problems on sequences by recursive functions* ◇ D20 ◇
KRINITSKIJ, N.A. *Algorithms around us (Russian)* ◇ D20 ◇
LEITSCH, A. *Enumerations of subrecursive classes and self-reproducing automata. I: Effective translations and decidable indexsets. II: Index-complexity and translation-complexity* ◇ D05 D20 D45 ◇
MCALOON, K. *Petri nets and large finite sets* ◇ D05 D20 D80 ◇
MENZEL, W. & SPERSCHNEIDER, V. *Recursively enumerable extensions of R_1 by finite functions* ◇ D20 D25 ◇
MOSCHOVAKIS, Y.N. *Abstract recursion as a foundation for the theory of algorithms* ◇ D20 D75 ◇
NIKITCHENKO, N.S. *Computable compositions and universal imperative program logics (Russian)* ◇ B75 D20 ◇
NORMANN, D. *The infinite – a mathematical necessity* ◇ A05 D20 F30 ◇
PERI, C. *Algorithms on sets of words that are isomorphic to Markov algorithms (Italian) (English summary)* ◇ D03 D20 ◇
PKHAKADZE, SH.S. *An example of an intuitively computable everywhere defined function and Church's thesis (Russian)* ◇ D20 ◇
ROSE, H.E. *Subrecursion. Functions and hierarchies* ◇ D20 ◇
SELIVANOV, V.L. *On the hierarchy of limit computations (Russian)* ◇ D20 D25 ◇
SHEN, BAIYING *Explicit representations for inverse functions of number-theoretic functions (Chinese) (English summary)* ◇ D20 ◇
SHEN, BAIYING *Inverse functions of number-theoretic functions III (Chinese)* ◇ D20 ◇
SHEN, BAIYING *Primitive recursive arithmetic in the second class $A^0(D)$ (Chinese)* ◇ D20 F30 ◇
SKANDALIS, K. *Programmable real numbers and functions* ◇ D20 ◇
SLISENKO, A.O. *Linguistic considerations in devising effective algorithms* ◇ B35 D20 ◇
SORENSEN, R.A. *Unique alternative guessing* ◇ B45 D20 ◇
SOSKOV, I.N. *Simply calculable functions on a base set (Russian)* ◇ D20 ◇
SPERSCHNEIDER, V. *The length-problem* ◇ D20 D25 ◇
SPREEN, D. & YOUNG, P. *Effective operators in a topological setting* ◇ D20 D75 D80 ◇
STENGER, H.-J. *Algebraic characterisations of NTIME(F) and NTIME(F,A)* ◇ B20 B35 D15 D20 ◇

VOLGER, H. *Rudimentary relations and Turing machines with linear alternation* ◇ D10 D15 D20 ◇

ZASHEV, J.A. *Basic recursion theory in partially ordered models of some fragments of the combinatory logic* ◇ B40 D20 ◇

ZEUGMANN, T. *On the nonboundability of total effective operators* ◇ D20 ◇

ZEUGMANN, T. *Recursive operators versus recursive functions with respect to the generation of classes of functions having a fastest program* ◇ D15 D20 ◇

1985

BERARDUCCI, A. *A generalization of Goedel's recursive functions (Italian)* ◇ D20 ◇

BEZEM, M. *Isomorphism between HEO and HRO^E, ECF and ICF^E* ◇ D20 D75 ◇

BOTUSHAROV, O.I. *On inductive inference with additional information (Bulgarian summary)* ◇ B48 D20 ◇

BURTMAN, M.I. & GASANOVA, N.P. & GUSEJNBEKOVA, A.M. *Partially recursive functions on the rationals (Russian) (Aserbaid. and English summaries)* ◇ D20 ◇

CANTINI, A. *Majorizing provably recursive functions in fragments of PA* ◇ D20 F15 F30 ◇

CHLEBUS, B.S. *From domino tilings to a new model of computation* ◇ D20 ◇

CLOTE, P. *Applications of the low-basis theorem in arithmetic* ◇ C62 D20 D30 D55 F30 ◇

DEUTSCH, D. *Quantum theory, the Church-Turing principle and the universal quantum computer* ◇ D20 D80 ◇

GONCZAROWSKI, J. *Decidable properties of monadic recursive schemas with a depth parameter* ◇ D20 ◇

HELLER, A. *Dominical categories and recursion theory* ◇ D20 ◇

HORVATH, S. *On defining the Blum complexity of partial recursive sequence functions* ◇ D15 D20 ◇

HUANG, WENGI & NERODE, A. *Application of pure recursion theory in recursive analysis (Chinese)* ◇ D20 F60 ◇

JACOBSSON, C. & STOLTENBERG-HANSEN, V. *Poincare-Betti series are primitive recursive* ◇ D20 D80 ◇

JENSEN, F.V. & LARSEN, K.G. *Recursively defined domains and their induction principles* ◇ B75 D20 ◇

KOSTENKO, K.I. *Classes of algorithms and computations (Russian)* ◇ D20 D75 ◇

KOWALCZYK, W. *On the effectiveness of some operations on algorithms* ◇ D20 ◇

KUR'EROV, YU.N. *On the embedding of an Ackermann class in a graph grammar (Russian) (English and Lithuanian summaries)* ◇ D05 D20 ◇

LEIVANT, D. *Syntactic translations and provably recursive functions* ◇ D20 F30 F50 ◇

LEWIS, A.A. *On effectively computable realizations of choice functions* ◇ D20 ◇

LEWIS, A.A. *The minimum degree of recursively representable choice functions* ◇ D20 D30 ◇

LOLLI, G. *Foundational problems from computation theory* ◇ D20 ◇

MUELLER, HORST *Weak Petri net computers for Ackermann functions* ◇ D05 D20 ◇

NIWINSKI, D. *Equational μ-calculus* ◇ D20 ◇

ODIFREDDI, P. *Rapidly growing recursive functions and associated ordinals (Italian)* ◇ D20 ◇

RICHTER, M.M. & SZABO, M.E. *Nonstandard computation theory* ◇ D20 D75 H10 ◇

SCHAEFER, G. *A note on conjectures of Calude about the topological size of sets of partial recursive functions* ◇ D20 D75 ◇

SHEN, BAIYING *Primitive recursive arithmetic in the first class A^o. I (Chinese)* ◇ D20 F30 ◇

SZALKAI, I. *On the algebraic structure of primitive recursive functions* ◇ D20 D75 ◇

WAINER, S.S. *Subrecursive ordinals* ◇ D20 F15 ◇

WAINER, S.S. *The "slow-growing" Π^1_2 approach to hierarchies* ◇ D20 F15 ◇

WEIHRAUCH, K. *Type 2 recursion theory* ◇ D20 ◇

D25 Recursively enumerable sets and degrees

1936

KLEENE, S.C. *General recursive functions of natural numbers* ◇ D20 D25 ◇

ROSSER, J.B. *Extensions of some theorems of Goedel and Church* ◇ D25 D35 F30 ◇

1939

KONDO, M. *On the enumerable sets* ◇ D25 ◇

TURING, A.M. *Systems of logic based on ordinals* ◇ B40 D25 D30 F15 F30 ◇

1944

POST, E.L. *Recursively enumerable sets of positive integers and their decision problems* ◇ D03 D25 D40 ◇

1950

KLEENE, S.C. *A symmetric form of Goedel's theorem* ◇ D25 ◇

1951

HERMES, H. *Zum Begriff der Axiomatisierbarkeit* ◇ A05 B30 D25 ◇

1952

MARKOV, A.A. *On unsolvable algorithmic problems (Russian)* ◇ D03 D20 D25 D35 D40 ◇

ROBINSON, JULIA *Existential definability in arithmetic* ◇ D25 ◇

1953

DAVIS, MARTIN D. *Arithmetical problems and recursively enumerable predicates* ◇ D25 D35 D55 D80 ◇

DEKKER, J.C.E. *The constructivity of maximal dual ideals in certain Boolean algebras* ◇ D25 D45 G05 ◇

DEKKER, J.C.E. *Two notes on recursively enumerable sets* ◇ D25 ◇

MYHILL, J.R. *Three contributions to recursive function theory* ◇ D25 ◇

RICE, H.G. *Classes of recursively enumerable sets and their decision problems* ◇ D25 ◇

USPENSKIJ, V.A. *Theorem of Goedel and theory of algorithms (Russian)* ◇ D20 D25 F30 ◇

1954

DEKKER, J.C.E. *A theorem on hypersimple sets* ◇ D25 ◇

JANICZAK, A. *On the reducibility of decision problems* ◇ D10 D20 D25 D30 ◇

1955

DEKKER, J.C.E. *Productive sets* ◇ D25 ◇

KUZNETSOV, A.V. & TRAKHTENBROT, B.A. *Investigation of partial recursive operators by means of the theory of Baire's space (Russian)* ◇ D20 D25 D30 ◇

MEDVEDEV, YU.T. *On nonisomorphic recursively enumerable sets (Russian)* ◇ D25 D50 ◇

MYHILL, J.R. *Creative sets* ◇ D25 ◇

USPENSKIJ, V.A. *Systems of enumerable sets and their enumeration (Russian)* ◇ D25 D45 ◇

1956

MARKWALD, W. *Ein Satz ueber die elementararithmetischen Definierbarkeitsklassen* ◇ D25 D55 ◇

MUCHNIK, A.A. *On separability of recursively enumerable sets (Russian)* ◇ D25 ◇

MUCHNIK, A.A. *On the unsolvability of the problem of reducibility in the theory of algorithms (Russian)* ◇ D25 D30 D35 ◇

MYHILL, J.R. *A problem on recursively enumerable sets (problem 8)* ◇ D25 ◇

MYHILL, J.R. *A problem on recursively enumerable supersets (problem 9)* ◇ D25 ◇

RICE, H.G. *On completely recursively enumerable classes and their key arrays* ◇ D25 ◇

RICE, H.G. *Recursive and recursively enumerable orders* ◇ D20 D25 D45 ◇

ROBINSON, R.M. *Arithmetical representation of recursively enumerable sets* ◇ D25 F30 ◇

SHAPIRO, N.Z. *Degrees of computability* ◇ D25 D30 ◇

1957

DAVIS, MARTIN D. & PUTNAM, H. *Reductions of Hilbert's 10th problem* ◇ D25 D35 ◇

FEFERMAN, S. *Degrees of unsolvability associated with classes of formalized theories* ◇ D25 D35 ◇

FRIEDBERG, R.M. *The existence of a maximal set* ◇ D25 ◇

FRIEDBERG, R.M. *The fine structure of degrees of unsolvability of recursively enumerable sets* ◇ D25 ◇

FRIEDBERG, R.M. *Two recursively enumerable sets of incomparable degrees of unsolvability* ◇ D25 ◇

MONTAGUE, R. & TARSKI, A. *Independent recursive axiomatizability* ◇ B30 D25 ◇

MUCHNIK, A.A. *Negative solution of the reducibility problem of Post (Russian)* ◇ D25 ◇

RABIN, M.O. *Effective computability of winning strategies* ◇ D20 D25 ◇

SHOENFIELD, J.R. *Quasicreative sets* ◇ D25 ◇

USPENSKIJ, V.A. *Some notes on recursively enumerable sets (Russian) (English summary)* ◇ D25 ◇

1958

DAVIS, MARTIN D. & PUTNAM, H. *Reductions of Hilbert's tenth problem* ◇ D25 D35 ◇

DEKKER, J.C.E. & MYHILL, J.R. *Retraceable sets* ◇ D25 D30 D50 ◇

DEKKER, J.C.E. & MYHILL, J.R. *Some theorems on classes of recursively enumerable sets* ◇ D25 ◇

Friedberg, R.M. *Three theorems on recursive enumeration. I. Decomposition II. Maximal set. III. Enumeration without duplication* ◇ D25 ◇
Muchnik, A.A. *Isomorphism of systems of recursively enumerable sets with effective properties (Russian)* ◇ D25 D30 ◇
Muchnik, A.A. *Solution of Post's reduction problem and of certain other problems in the theory of algorithms I (Russian)* ◇ D25 D30 ◇
Shoenfield, J.R. *Degrees of formal systems* ◇ D25 D30 D35 D55 ◇
Smullyan, R.M. *Undecidability and recursive inseparability* ◇ D25 D35 ◇

1959

Ehrenfeucht, A. & Feferman, S. *Representability of recursively enumerable sets in formal theories* ◇ D25 F30 ◇
Muchnik, A.A. *Solution of the problem of reducibility of Post (Russian)* ◇ D25 ◇
Myhill, J.R. *Recursive digraphs, splinters, and cylinders* ◇ D25 ◇

1960

Craig, W. *Bases for first-order theories and subtheories* ◇ B10 D25 F05 ◇
Oberschelp, A. *Ein Satz ueber die Unloesbarkeitsgrade der Mengen von natuerlichen Zahlen* ◇ D25 ◇
Putnam, H. & Smullyan, R.M. *Exact separation of recursively enumerable sets within theories* ◇ D25 D35 ◇
Smullyan, R.M. *Theories with effectively inseparable nuclei* ◇ D25 D35 ◇
Ullian, J.S. *Splinters of recursive functions* ◇ D25 D30 ◇

1961

Cleave, J.P. *Creative functions* ◇ D25 ◇
Davis, Martin D. & Putnam, H. & Robinson, Julia *The decision problem for exponential diophantine equations* ◇ D25 D35 ◇
Ehrenfeucht, A. *Separable theories* ◇ D25 D35 ◇
Higman, G. *Subgroups of finitely presented groups* ◇ C60 D25 D45 ◇
Shepherdson, J.C. *Representability of recursively enumerable sets in formal theories* ◇ D25 F30 ◇
Shoenfield, J.R. *Undecidable and creative theories* ◇ D25 D35 ◇
Smullyan, R.M. *Elementary formal systems* ◇ D03 D25 ◇
Smullyan, R.M. *Extended canonical systems* ◇ B45 D03 D25 ◇
Smullyan, R.M. *Monadic elementary formal systems* ◇ D03 D25 ◇
Smullyan, R.M. *Theory of formal systems* ◇ B98 D03 D05 D20 D25 D98 ◇
Ullian, J.S. *A theorem on maximal sets* ◇ D25 ◇

1962

Boone, W.W. *Partial results regarding word problems and recursively enumerable degrees of unsolvability* ◇ D25 D40 ◇
Cannonito, F.B. *The Goedel incompleteness theorem and intelligent machines* ◇ B35 D10 D25 F30 ◇
Fridman, A.A. *Degrees of unsolvability of the word problem for finitely presented groups (Russian)* ◇ D25 D40 ◇
McLaughlin, T.G. *A note on contraproduction domains* ◇ D20 D25 ◇
McLaughlin, T.G. *On an extension of a theorem of Friedberg* ◇ D25 ◇
Rabin, M.O. *Diophantine equations and non-standard models of arithmetic* ◇ C62 D25 H15 ◇
Shoenfield, J.R. *Some applications of degrees* ◇ D25 D30 D35 D55 ◇
Skolem, T.A. *Proof of some theorems on recursively enumerable sets* ◇ D25 ◇
Skolem, T.A. *Recursive enumeration of some classes of primitive recursive functions and a majorisation theorem* ◇ D20 D25 ◇
Smullyan, R.M. *On Post's canonical systems* ◇ D03 D25 ◇
Tajtslin, M.A. *Effective inseparability of the set of identically true and the set of finitely refutable formulas of the elementary theory of lattices (Russian)* ◇ C13 D25 D35 G10 ◇
Vaught, R.L. *On a theorem of Cobham concerning undecidable theories* ◇ D25 D35 ◇
Yates, C.E.M. *Recursively enumerable sets and retracing functions* ◇ D25 D50 ◇

1963

Cleave, J.P. *Infinite sequence of pairwise recursively inseparable, recursively enumerable sets* ◇ D25 ◇
Cobham, A. *Some remarks concerning theories with recursively enumerable complements* ◇ D25 D35 ◇
Davis, Martin D. & Putnam, H. *Diophantine sets over polynomial rings* ◇ D25 D75 ◇
Davis, Martin D. *Extensions and corollaries of recent work on Hilbert's tenth problem* ◇ D25 D35 ◇
Lavrov, I.A. *Effective inseparability of the sets of identically true formulae and finitely refutable formulae for certain elementary theories (Russian)* ◇ C13 D25 D35 ◇
Martin, D.A. *A theorem on hyperhypersimple sets* ◇ D25 ◇
McLaughlin, T.G. *A remark on semiproductive sets* ◇ D25 D30 ◇
McLaughlin, T.G. *A theorem on productive functions* ◇ D25 ◇
Mullin, A.A. *On a theorem equivalent to Post's fundamental theorem of recursive function theory* ◇ D25 ◇
Patton, T.E. *On n-adic representation of numbers* ◇ D03 D25 ◇
Rose, G.F. & Ullian, J.S. *Approximation of functions on the integers* ◇ D20 D25 ◇
Sacks, G.E. *Degrees of unsolvability* ◇ D25 D30 D98 E75 ◇
Sacks, G.E. *On the degrees less than 0′* ◇ D25 D30 ◇
Sacks, G.E. *Recursive enumerability and the jump operator* ◇ D25 D30 ◇

SCARPELLINI, B. *Zwei unentscheidbare Probleme der Analysis* ◇ D25 D35 ◇
SMULLYAN, R.M. *Creativity and effective inseparability* ◇ D25 ◇
SMULLYAN, R.M. *Pseudo-uniform reducibility* ◇ D20 D25 D30 D35 ◇
TENNENBAUM, S. *Degree of unsolvability and the rate of growth of functions* ◇ D25 ◇
WANG, HAO *Computation* ◇ D20 D25 D30 D35 D60 ◇

1964

CLAPHAM, C.R.J. *Finitely presented groups with word problems of arbitrary degrees of insolubility* ◇ D25 D40 ◇
HIROSE, K. *A theorem on incomparable degrees of recursive unsolvability* ◇ D25 D30 ◇
HOWARD, W.A. & POUR-EL, M.B. *A structural criterion for recursive enumeration without repetition* ◇ D25 ◇
LACHLAN, A.H. *Effective operations in a general setting* ◇ D20 D25 D75 ◇
LACHLAN, A.H. *Standard classes of recursively enumerable sets* ◇ D25 ◇
MCLAUGHLIN, T.G. *A note on pseudo doubly creative pairs* ◇ D25 ◇
MCLAUGHLIN, T.G. *On contraproductive sets which are not productive* ◇ D25 ◇
MCLAUGHLIN, T.G. *Some observations on quasicohesive sets* ◇ D25 ◇
OHASHI, K. *A stronger form of a theorem of Friedberg* ◇ D25 ◇
OHASHI, K. *Enumeration of some classes of recursively enumerable sets* ◇ D25 ◇
POUR-EL, M.B. *Goedel numberings versus Friedberg numberings* ◇ D25 D45 ◇
PUTNAM, H. *On families of sets represented in theories* ◇ C40 D25 F30 ◇
SACKS, G.E. *A maximal set which is not complete* ◇ D25 ◇
SACKS, G.E. *A simple set which is not effectively simple* ◇ D25 ◇
SACKS, G.E. *The recursively enumerable degrees are dense* ◇ D25 ◇
SINGLETARY, W.E. *A complex of problems proposed by Post* ◇ B20 D03 D25 D30 D35 ◇
SMULLYAN, R.M. *Effectively simple sets* ◇ D25 ◇
TSEJTIN, G.S. *A method of presenting the theory of algorithms and enumerable sets (Russian)* ◇ D20 D25 ◇
YOUNG, P. *A note on pseudo-creative sets and cylinders* ◇ D25 D30 ◇

1965

CUDIA, D.F. & SINGLETARY, W.E. *Post's correspondence problem and degrees of unsolvability; degrees of unsolvability in automata and grammars* ◇ D03 D05 D25 ◇
DAVIS, MARTIN D. (ED.) *The undecidable. Basic papers on undecidable propositions, unsolvable problems and computable functions* ◇ D20 D25 D35 D97 F30 ◇
GLADSTONE, M.D. *Some ways of constructing a propositional calculus of any required degree of unsolvability* ◇ B20 D25 D35 ◇
HANF, W.P. *Model-theoretic methods in the study of elementary logic* ◇ B25 C07 D20 D25 D35 F25 ◇
HAY, L. *On creative sets and indices of partial recursive functions* ◇ D20 D25 ◇
HOREJS, J. *On generalizations of a theorem on recursive sets* ◇ D20 D25 ◇
IHRIG, A.H. *The Post-Lineal theorems for arbitrary recursively enumerable degrees of unsolvability* ◇ D25 ◇
LACHLAN, A.H. *Effective inseparability for sequences of sets* ◇ D25 ◇
LACHLAN, A.H. *On a problem of G.E. Sacks* ◇ D25 D30 ◇
LACHLAN, A.H. *On recursive enumeration without repetition* ◇ D25 ◇
LACHLAN, A.H. *Some notions of reducibility and productiveness* ◇ D25 D30 D55 ◇
MCLAUGHLIN, T.G. *On a class of complete simple sets* ◇ D25 ◇
MCLAUGHLIN, T.G. *Strong reducibility on hypersimple sets* ◇ D25 ◇
MUCHNIK, A.A. *On the reducibility of problems of the solvability of enumerable sets to problems of separability (Russian)* ◇ D25 D30 ◇
POUR-EL, M.B. & PUTNAM, H. *Recursively enumerable classes and their application to recursive sequences of formal theories* ◇ D25 ◇
SHOENFIELD, J.R. *Applications of model theory to degrees of unsolvability* ◇ C50 D25 D30 ◇
YANG, DONGPING *On creative pair and productive pair (Chinese)* ◇ D25 ◇
YATES, C.E.M. *Three theorems on the degrees of recursively enumerable sets* ◇ D25 ◇
YOUNG, P. *On semi-cylinders, splinters, and bounded-truth-table reducibility* ◇ D25 D30 ◇

1966

BOONE, W.W. & ROGERS JR., H. *On a problem of J.H.C.Whitehead and a problem of Alonzo Church* ◇ D25 D30 D40 ◇
BOONE, W.W. *Word problems and recursively enumerable degrees of unsolvability. A first paper on Thue systems* ◇ D03 D25 D40 ◇
BOONE, W.W. *Word problems and recursively enumerable degrees of unsolvability. A sequel on finitely presented groups* ◇ D25 D40 ◇
DAVIS, MARTIN D. *Diophantine equations and recursively enumerable sets* ◇ D25 ◇
HAY, L. *Isomorphism types of index sets of partial recursive functions* ◇ D20 D25 D30 ◇
KREIDER, D.L. & RITCHIE, R.W. *A basis theorem for a class of two-way automata* ◇ D10 D25 ◇
LACHLAN, A.H. *A note on universal sets* ◇ D25 ◇
LACHLAN, A.H. *Lower bounds for pairs of r.e.degrees* ◇ D25 ◇
LACHLAN, A.H. *On the indexing of classes of recursively enumerable sets* ◇ D25 ◇

LACHLAN, A.H. *The impossibility of finding relative complements for recursively enumerable degrees* ◇ D25 ◇
MARTIN, D.A. *Classes of recursively enumerable sets and degrees of unsolvability* ◇ D25 D30 ◇
MARTIN, D.A. *Completeness, the recursion theorem, and effectively simple sets* ◇ D25 ◇
MARTIN, D.A. *On a question of G.E.Sacks* ◇ D25 D30 ◇
MCLAUGHLIN, T.G. *Two remarks on indecomposable number sets* ◇ D20 D25 ◇
MITCHELL, R. *A generalization of productive sets* ◇ D25 ◇
NEPOMNYASHCHIJ, V.A. *On a basis for recursively enumerable sets (Russian)* ◇ D25 ◇
NEPOMNYASHCHIJ, V.A. *On certain automata capable of computing a basis for recursively enumerable sets (Russian)* ◇ D05 D25 ◇
PAOLA DI, R.A. *On sets represented by the same formula in distinct consistent axiomatizable Rosser theories* ◇ D25 F30 ◇
PAOLA DI, R.A. *Pseudo-complements and ordinal logics based on consistency statements* ◇ D25 F15 F30 ◇
PAOLA DI, R.A. *Some properties of pseudo-complements of recursively enumerable sets* ◇ D25 D30 F30 ◇
ROEDDING, D. *Der Entscheidbarkeitsbegriff in der mathematischen Logik* ◇ B25 D10 D20 D25 D35 ◇
YANG, DONGPING *Primitive recursive simple sets and their hierarchy (Chinese)* ◇ D20 D25 ◇
YATES, C.E.M. *A minimal pair of recursively enumerable degrees* ◇ D25 ◇
YATES, C.E.M. *On the degrees of index sets* ◇ D25 D30 D55 ◇
YOUNG, P. *A theorem on recursively enumerable classes and splinters* ◇ D25 ◇

1967

APPEL, K.I. *No recursively enumerable set is the union of finitely many immune retraceable sets* ◇ D25 D50 ◇
FLORENCE, J.B. *Infinite subclasses of recursively enumerable classes* ◇ D25 ◇
KLEENE, S.C. *Mathematical logic* ◇ B98 D20 D25 D55 D98 F30 F98 ◇
KRIPKE, S.A. & POUR-EL, M.B. *Deduction-preserving "recursive isomorphisms" between theories* ◇ B30 D25 D35 F30 ◇
LACHLAN, A.H. *The priority method I* ◇ D25 ◇
MASLOV, S.YU. *Representation of enumerable sets by means of local calculi (Russian)* ◇ D05 D25 ◇
MCLAUGHLIN, T.G. *Some remarks on extensibility, confluence of paths, branching properties, and index sets, for certain recursively enumerable graphs* ◇ D25 D80 ◇
MCLAUGHLIN, T.G. *Splitting and decomposition by regressive sets II* ◇ D25 D50 ◇
OWINGS JR., J.C. *Recursion, metarecursion, and inclusion* ◇ D25 D55 D60 ◇
PAOLA DI, R.A. *Some theorems on extensions of arithmetic* ◇ D25 D35 F15 F30 ◇
ROBINSON, R.W. *Simplicity of recursively enumerable sets* ◇ D25 ◇
ROBINSON, R.W. *Two theorems on hyperhypersimple sets* ◇ D25 ◇
ROGERS JR., H. *Some problems of definability in recursive function theory* ◇ D20 D25 D98 ◇
ROGERS JR., H. *Theory of recursive functions and effective computability* ◇ D25 D30 D55 D98 F15 ◇
SACKS, G.E. *On a theorem of Lachlan and Martin* ◇ D25 D30 ◇
SINGLETARY, W.E. *Recursive unsolvability of a complex of problems proposed by Post* ◇ D03 D25 D35 ◇
SINGLETARY, W.E. *The equivalence of some general combinatorial decision problems* ◇ D03 D25 ◇
SINTZOFF, M. *Existence of a van Wijngaarden syntax for every recursively enumerable set* ◇ D03 D05 D25 ◇
VUCKOVIC, V. *Creative and weakly creative sequences of r.e. sets* ◇ D25 ◇
YATES, C.E.M. *Recursively enumerable degrees and the degrees less than* $0^{(1)}$ ◇ D25 D30 ◇
YOUNG, P. *On pseudo-creative sets, splinters, and bounded truth-table reducibility* ◇ D25 D30 ◇

1968

ARSLANOV, M.M. *Two theorems on recursively enumerable sets (Russian)* ◇ D25 ◇
BARZDINS, J. *Complexity of programs to determine whether natural numbers not greater than n belong to a recursively enumerable set (Russian)* ◇ B75 D15 D25 ◇
BOONE, W.W. *Decision problems about algebraic and logical systems as a whole and recursively enumerable degrees of unsolvability* ◇ D25 D35 D40 ◇
CARPENTIER, M.A. *Creative sequences and double sequences* ◇ D25 ◇
CUDIA, D.F. & SINGLETARY, W.E. *Degrees of unsolvability in formal grammars* ◇ D03 D25 ◇
CUDIA, D.F. & SINGLETARY, W.E. *The Post correspondence problem* ◇ D03 D25 ◇
DAVIS, MARTIN D. *One equation to rule them all* ◇ D25 D35 ◇
EILENBERG, S. & ELGOT, C.C. *Iteration and recursion* ◇ D20 D25 ◇
ERSHOV, YU.L. *On computable enumerations (Russian)* ◇ D25 D45 ◇
FUKUYAMA, M. *A note on incomparable degrees* ◇ D25 D30 ◇
GLADSTONE, M.D. *A single-axiom implicational calculus of given unsolvability* ◇ B20 D25 D30 D35 ◇
GRINDLINGER, M.D. *Strengthening of a theorem of J.S.Smetanich (Russian)* ◇ D25 D40 ◇
HERMAN, G.T. *Simulation of one abstract computing machine by another* ◇ D10 D25 D30 D55 ◇
JOCKUSCH JR., C.G. *Semirecursive sets and positive reducibility* ◇ D25 D30 ◇
JOCKUSCH JR., C.G. *Supplement to Boone's "Algebraic systems"* ◇ D25 D40 ◇
JOCKUSCH JR., C.G. *Uniformly introreducible sets* ◇ D25 D30 ◇
LACHLAN, A.H. *Complete recursively enumerable sets* ◇ D25 ◇

LACHLAN, A.H. *Degrees of recursively enumerable sets which have no maximal supersets* ◇ D25 ◇
LACHLAN, A.H. *On the lattice of recursively enumerable sets* ◇ B25 C10 D25 ◇
LACHLAN, A.H. *The elementary theory of recursively enumerable sets* ◇ B25 D25 ◇
LAVROV, I.A. *Answer to a question of P.R.Young (Russian)* ◇ D25 ◇
MARTIN, D.A. & MILLER, W. *The degrees of hyperimmune sets* ◇ D25 D50 ◇
MATIYASEVICH, YU.V. *Arithmetic representations of powers (Russian)* ◇ D25 D35 ◇
MATIYASEVICH, YU.V. *The connection between Hilbert's tenth problem and systems of equations between words and lengths (Russian)* ◇ D25 D35 D40 ◇
MATIYASEVICH, YU.V. *Two reductions of Hilbert's tenth problem (Russian)* ◇ D25 D35 ◇
MAYOH, B.H. *Semi-effective numberings and definitions of the computable numbers* ◇ D25 D30 F60 ◇
PAOLA DI, R.A. *A note on diminishing the undecidable region of a recursively enumerable set* ◇ D25 ◇
POUR-EL, M.B. *Effectively extensible theories* ◇ B30 D25 ◇
POUR-EL, M.B. *Independent axiomatization and its relation to the hypersimple set* ◇ C07 D25 ◇
ROBINSON, JULIA *Finite generation of recursively enumerable sets* ◇ D25 ◇
ROBINSON, R.W. *A dichotomy of the recursively enumerable sets* ◇ D25 ◇

1969

ADLER, A. *Extensions of non-standard models of number theory* ◇ C62 D25 D35 H15 ◇
ARSLANOV, M.M. *Effectively hypersimple sets (Russian)* ◇ D25 ◇
ARSLANOV, M.M. *The embedding of recursively enumerable sets (Russian)* ◇ D25 ◇
AXT, P. & SINGLETARY, W.E. *On deterministic normal systems* ◇ D03 D25 ◇
CARPENTIER, M.A. *Complete enumerations and double sequences* ◇ D25 D45 ◇
COLLINS, D.J. *Recursively enumerable degrees and the conjugacy problem* ◇ D25 D40 ◇
ERSHOV, YU.L. *Hyperhypersimple m-degrees (Russian)* ◇ D25 ◇
ERSHOV, YU.L. & LAVROV, I.A. *On computable enumerations II (Russian)* ◇ D25 D45 ◇
ERSHOV, YU.L. *The theory of numerations. Part I: General theory of numeration (Russian)* ◇ D25 D30 D45 G30 ◇
FLORENCE, J.B. *Partially ordered sets representable by recursively enumerable classes* ◇ D25 ◇
FLORENCE, J.B. *Strong enumeration properties of recursively enumerable classes* ◇ D25 ◇
HAY, L. *Index sets of finite classes of recursively enumerable sets* ◇ D25 D30 ◇
HECKER, H.-D. *Riegeralgebren und rekursive Untrennbarkeit* ◇ D25 D35 ◇
HERMES, H. *Basic notions and applications of the theory of decidability* ◇ D10 D20 D25 D30 D35 D50 D98 ◇
HINMAN, P.G. *Some applications of forcing to hierarchy problems in arithmetic* ◇ D25 D30 D55 E40 ◇
JOCKUSCH JR., C.G. *Relationships between reducibilities* ◇ D25 D30 ◇
JONES, JAMES P. *Effectively retractable theories and degrees of undecidability* ◇ D25 ◇
JONES, JAMES P. *Independent recursive axiomatizability in arithmetic* ◇ D25 F30 ◇
KHUTORETSKIJ, A.B. *On nonprincipal enumeration (Russian)* ◇ D25 D45 ◇
KHUTORETSKIJ, A.B. *The reducibility of computable enumerations (Russian)* ◇ D20 D25 D45 ◇
KHUTORETSKIJ, A.B. *Two existence theorems for computable numerations (Russian)* ◇ D25 D45 ◇
PETRI, N.V. *Two theorems on the complexity of algorithms and computations (Russian)* ◇ D15 D25 ◇
POUR-EL, M.B. *A recursion-theoretic view of axiomatizable theories* ◇ C07 D25 D35 ◇
POUR-EL, M.B. *Independent axiomatization and its relation to the hypersimple set* ◇ C07 D25 D35 ◇
ROBINSON, JULIA *Diophantine decision problems* ◇ D25 D35 ◇
ROBINSON, JULIA *Unsolvable diophantine problems* ◇ D25 D35 ◇
SOARE, R.I. *Cohesive sets and recursively enumerable Dedekind cuts* ◇ D25 D30 ◇
SOARE, R.I. *Recursion theory and Dedekind cuts* ◇ D25 D30 F60 ◇
VUCKOVIC, V. *Almost recursive sets* ◇ D25 D50 ◇
YATES, C.E.M. *On the degrees of index sets II* ◇ D25 D30 D55 ◇
YOUNG, P. *Toward a theory of enumerations* ◇ D10 D15 D20 D25 D45 ◇

1970

ARSLANOV, M.M. *Complete hypersimple sets (Russian)* ◇ D25 ◇
BARZDINS, J. *On the relative frequency of solution of algorithmically unsolvable mass problems (Russian)* ◇ D20 D25 ◇
CASE, J. *Enumeration reducibility and partial degrees* ◇ D25 D30 ◇
CHUDNOVSKY, G.V. *Diophantine predicates (Russian)* ◇ D25 D35 D40 ◇
CLEAVE, J.P. *Some properties of recursively inseparable sets* ◇ D25 ◇
DEGTEV, A.N. *Remarks on retraceable, regressive and pointwise-decomposable sets (Russian)* ◇ D25 D50 ◇
DENISOV, S.D. & LAVROV, I.A. *Complete numerations with an infinite number of special elements (Russian)* ◇ D25 D45 ◇
DENISOV, S.D. *The m-degrees of recursively enumerable sets (Russian)* ◇ D25 ◇
EBBINGHAUS, H.-D. *Aufzaehlbarkeit* ◇ D03 D10 D20 D25 F30 ◇
EILENBERG, S. & ELGOT, C.C. *Recursiveness* ◇ D20 D25 D98 ◇
ERSHOV, YU.L. *Index sets (Russian)* ◇ D25 D30 D55 ◇
ERSHOV, YU.L. *On inseparable pairs (Russian)* ◇ D25 ◇

KANOVICH, M.I. *Complexity of resolution of a recursively enumerable set as a criterion of its universality (Russian)* ◇ D25 ◇
KANOVICH, M.I. *On the decision complexity of recursively enumerable sets (Russian)* ◇ D15 D25 ◇
KIPNIS, M.M. *Invariant properties of systems of formulae of elementary axiomatic theories (Russian)* ◇ D25 D35 ◇
KLEINBERG, E.M. *Recursion theory and formal deducibility* ◇ D25 F07 F40 ◇
LACHLAN, A.H. *On some games which are relevant to the theory of recursively enumerable sets* ◇ D25 E60 ◇
LERMAN, M. *Recursive functions modulo co-r-maximal sets* ◇ D25 D30 ◇
LERMAN, M. *Turing degrees and many-one degrees of maximal sets* ◇ D25 D30 ◇
LEWIS, F.D. *The classification of unsolvable problems in automata theory* ◇ D05 D25 ◇
MARTIN, D.A. & POUR-EL, M.B. *Axiomatizable theories with few axiomatizable extensions* ◇ D25 ◇
MATIYASEVICH, YU.V. *Enumerable sets are diophantine (Russian)* ◇ D25 D35 ◇
MATIYASEVICH, YU.V. *Solution of the tenth problem of Hilbert* ◇ B28 D25 D35 ◇
POUR-EL, M.B. *A recursion-theoretic view of axiomatizable theories* ◇ C07 D25 D35 ◇
SCHUPP, P.E. *A note on recursively enumerable predicates in groups* ◇ B25 D25 D40 ◇
TSEJTIN, G.S. *On upper bounds of recursively enumerable sets of constructive real numbers (Russian)* ◇ D25 F60 ◇
VUCKOVIC, V. *Effective enumerability of some families of partially recursive functions connected with computable functionals* ◇ D20 D25 ◇
WAINER, S.S. *A classification of the ordinal recursive functions* ◇ D20 D25 ◇

1971

AANDERAA, S.O. & BELSNES, D. *Decision problems for tag systems* ◇ D03 D25 D30 ◇
ALTON, D.A. *Recursively enumerable sets which are uniform for finite extensions* ◇ D25 ◇
AZRA, J.-P. *Relations diophantiennes et la solution negative du 10e probleme de Hilbert* ◇ D25 D35 ◇
BARZDINS, J. *Complexity and accuracy of the solution of initial segments of the problem of occurrence in a recursively enumerable set (Russian)* ◇ D25 ◇
BOONE, W.W. *Word problems and recursively enumerable degrees of unsolvability. An emendation* ◇ D25 D40 ◇
BRZOZOWSKI, J.A. & CULIK, K. & GABRIELIAN, A. *Classification of noncounting events* ◇ D05 D25 ◇
CASE, J. *A note on degrees of self-describing Turing machines* ◇ D10 D25 ◇
CAVINESS, B.F. & POLLACK, P.L. & RUBALD, C.M. *An existence lemma for canonical forms in symbolic mathematics* ◇ D25 F05 ◇
DALEN VAN, D. *A note on some systems of Lindenmayer* ◇ D03 D05 D10 D25 ◇
DEGTEV, A.N. *Hypersimple sets with retraceable complements (Russian)* ◇ D25 ◇
ELLENTUCK, E. *Incompleteness via simple sets* ◇ D25 F30 ◇
ERSHOV, YU.L. *Computable enumerations of morphisms (Russian)* ◇ D25 D45 ◇
ERSHOV, YU.L. *La theorie des enumerations* ◇ D25 D45 ◇
FENSTAD, J.E. *Hilbert's 10th problem (Norwegian) (English summary)* ◇ D25 D35 ◇
GINSBURG, S. & GOLDSTINE, J. *Intersection-closed full AFL and the recursively enumerable languages* ◇ D05 D25 ◇
HECKER, H.-D. *Allgemeine Trennbarkeitsbegriffe bei meta-mathematischen Untersuchungen* ◇ D25 D35 D75 ◇
HECKER, H.-D. *Rekursive Untrennbarkeit bei elementaren Theorien* ◇ D25 D35 ◇
HUGHES, C.E. & OVERBEEK, R.A. & SINGLETARY, W.E. *The many-one equivalence of some general combinatorial decision problems* ◇ D03 D05 D10 D25 D30 ◇
JOCKUSCH JR., C.G. & SOARE, R.I. *A minimal pair of Π_1^0 classes* ◇ D25 D30 ◇
KALLIBEKOV, S. *Index sets of degrees of unsolvability (Russian)* ◇ D25 D55 ◇
KALLIBEKOV, S. *The index sets of m-degrees (Russian)* ◇ D25 D55 ◇
KHUTORETSKIJ, A.B. *On the cardinality of the upper semilattice of computable enumerations (Russian)* ◇ D25 D30 D45 ◇
LERMAN, M. *Some theorems on r-maximal sets and major subsets of recursively enumerable sets* ◇ D25 D30 ◇
LEWIS, F.D. *The enumerability and invariance of complexity classes* ◇ D15 D25 ◇
MARCHENKOV, S.S. *On minimal numerations of systems of recursively enumerable sets (Russian)* ◇ D25 D45 ◇
MARCHENKOV, S.S. *On semilattices of computable numerations (Russian)* ◇ D25 D45 ◇
MATIYASEVICH, YU.V. *Diophantine representation of the set of prime numbers (Russian)* ◇ D25 D35 ◇
MATIYASEVICH, YU.V. *Diophantine representation of enumerable predicates (Russian)* ◇ D25 D35 ◇
MATIYASEVICH, YU.V. *Diophantine representation of recursively enumerable predicates* ◇ D25 D35 ◇
MCLAUGHLIN, T.G. *The family of all recursively enumerable classes of finite sets* ◇ D25 ◇
MILLER III, C.F. *On group theoretic decision problems and their classification* ◇ D25 D40 ◇
ROBINSON, R.W. *Interpolation and embedding in the recursively enumerable degrees* ◇ D25 ◇
ROBINSON, R.W. *Jump restricted interpolation in the recursively enumerable degrees* ◇ D25 ◇
SHOENFIELD, J.R. *Degrees of unsolvability* ◇ D25 D30 D98 ◇
SLISENKO, A.O. *A property of enumerable sets containing "complexly deducible" formulas (Russian) (English summary)* ◇ D25 F20 ◇
SPECKER, E. *Ramsey's theorem does not hold in recursive set theory* ◇ D25 D80 E05 F60 ◇
SUSSMANN, H.J. *Hilbert's tenth problem* ◇ D25 D35 ◇

THOMASON, S.K. *Sublattices of the recursively enumerable degrees* ◇ D25 ◇
TSINMAN, L.L. *Certain examples and theorems from the theory of recursive functions (Russian)* ◇ D20 D25 ◇
YOUNG, P. *Speed-up changing the order in which sets are enumerated* ◇ D15 D20 D25 ◇

1972

BAGCHI, A. & MEYER, A.R. *Program size and economy of descriptions: Preliminary report* ◇ D15 D25 ◇
BAKER, J.L. *Grammars with structured vocabulary: A model for the ALGOL-68 definition* ◇ B75 D05 D25 ◇
BERRY, J.W. *Almost recursively enumerable sets* ◇ D25 D55 ◇
COOPER, S.B. *Degrees of unsolvability complementary between recursively enumerable degrees I* ◇ D25 D30 ◇
COOPER, S.B. *Jump equivalence of the Δ_2^0 hyperimmune sets* ◇ D25 D30 D55 ◇
COOPER, S.B. *Minimal upper bounds for sequences of recursively enumerable degrees* ◇ D25 D30 ◇
DAVIS, MARTIN D. *On the number of the solutions of diophantine equations* ◇ D25 D35 ◇
DEGTEV, A.N. *The m-degrees of simple sets (Russian)* ◇ D25 ◇
ERSHOV, YU.L. *Relationship between sheaf spaces and numbered sets with the C_2^* property (Russian) (English summary)* ◇ C90 D25 D45 D65 ◇
GRASSIN, J. *Une generalisation du theoreme de Myhill-Sheperdson aux combinaisons boolennes d'ouverts dont une cle est recursivement enumerable* ◇ D25 ◇
HAY, L. *A discrete chain of degrees of index sets* ◇ D25 D30 ◇
HERMES, H. *Die Unloesbarkeit des zehnten Hilbertschen Problems* ◇ D25 D35 ◇
HUGHES, C.E. *Degrees of unsolvability associated with Markov algorithms* ◇ D03 D25 D55 ◇
JOCKUSCH JR., C.G. & SOARE, R.I. *Π_1^0 classes and degrees of theories* ◇ D25 D30 ◇
JOCKUSCH JR., C.G. *A reducibility arising from the Boone groups* ◇ D25 D30 D40 ◇
JOCKUSCH JR., C.G. *Degrees in which the recursive sets are uniformly recursive* ◇ D25 D30 ◇
JOCKUSCH JR., C.G. & SOARE, R.I. *Degrees of members of Π_1^0 classes* ◇ D25 D30 F30 ◇
KANOVICH, M.I. *Complexity of bounded decidability of semirecursively enumerable sets (Russian)* ◇ D15 D25 ◇
KANOVICH, M.I. *On the universality of strongly undecidable sets (Russian)* ◇ D15 D20 D25 D30 ◇
KOSOVSKIJ, N.K. *Properties of the solutions of equations in a free semigroup (Russian) (English summary)* ◇ D20 D25 D40 ◇
LACHLAN, A.H. *Embedding nondistributive lattices in the recursively enumerable degrees* ◇ D25 ◇
LACHLAN, A.H. *Recursively enumerable many-one degrees* ◇ D25 ◇
LACHLAN, A.H. *Two theorems of many-one degrees of recursively enumerable sets* ◇ D25 ◇
MATIYASEVICH, YU.V. *Arithmetic representation of enumerable sets with a small number of quantifiers (Russian) (English summary)* ◇ D25 ◇
MATIYASEVICH, YU.V. *Diophantine sets (Russian)* ◇ D25 D35 ◇
MATIYASEVICH, YU.V. *Diophantine representation of enumerable predicates (Russian)* ◇ D25 D35 ◇
MCLAUGHLIN, T.G. *Complete index sets of recursively enumerable families* ◇ D25 ◇
MCLAUGHLIN, T.G. *Supersimple sets and the problem of extending a retracing function* ◇ D25 D50 ◇
ROBINSON, R.M. *Some representations of diophantine sets* ◇ D25 D35 ◇
RUOHONEN, K. *Hilbert's tenth problem (Finnish) (English summary)* ◇ D25 D35 ◇
SALOMAA, A. *On a homomorphic characterization of recursively enumerable languages* ◇ D25 ◇
SOARE, R.I. *The Friedberg-Muchnik theorem re-examined* ◇ D25 ◇
V'YUGIN, V.V. *On discrete families of recursively enumerable sets (Russian)* ◇ D25 ◇
WRIGHT, J.B. *Characterization of recursively enumerable sets* ◇ D25 ◇

1973

ARTIGUE, M. *Une generalisation des ensembles cohesifs* ◇ D25 ◇
BARZDINS, J. *The frequency solution of the occurrence problem for a recursively enumerable set (Russian)* ◇ D25 ◇
BLUM, M. & MARQUES, I. *On complexity properties of recursively enumerable sets* ◇ D15 D25 ◇
BULLOCK, A.M. & SCHNEIDER, H.H. *On generating the finitely satisfiable formulas* ◇ B10 C07 C13 D25 ◇
CLEAVE, J.P. *Combinatorial systems III. Degrees of combinatorial problems of computing machines* ◇ D10 D25 D30 ◇
COLLINS, D.J. *The word, power and order problems in finitely presented groups* ◇ D25 D40 ◇
DAVIS, MARTIN D. *Hilbert's tenth problem is unsolvable* ◇ D25 D35 ◇
DEGTEV, A.N. *tt and m degrees (Russian)* ◇ D25 D30 ◇
ERSHOV, YU.L. & LAVROV, I.A. *The upper semilattice $L(\gamma)$ (Russian)* ◇ D25 D45 ◇
GINSBURG, S. & GOLDSTINE, J. *Intersection-closed full AFL and the recursively enumerable languages* ◇ D05 D25 ◇
HAY, L. *Discrete ω-sequences of index sets (1)* ◇ D25 D30 ◇
HAY, L. *Index sets in $\underline{0}'$ (Russian)* ◇ D25 D30 D55 ◇
HAY, L. *The class of recursively enumerable subsets of a recursively enumerable set* ◇ D25 D30 ◇
HAY, L. *The halting problem relativized to complements* ◇ D25 D30 ◇
HELM, J.P. & MEYER, A.R. & YOUNG, P. *On orders of translations and enumerations* ◇ D15 D20 D25 D45 F40 ◇

HIROSE, K. & IIDA, S. *A proof of negative answer to Hilbert's 10th problem* ◇ D25 D35 ◇
HIROSE, K. *On Hilbert's tenth problem (negative solution)* ◇ D25 D35 ◇
HUGHES, C.E. & SINGLETARY, W.E. *Combinatorial systems with axiom* ◇ D03 D05 D10 D25 D30 ◇
HUGHES, C.E. *Many-one degrees associated with problems of tag* ◇ D03 D10 D25 D40 ◇
HUGHES, C.E. *Many-one degrees associated with semi-Thue systems* ◇ D03 D25 ◇
JOCKUSCH JR., C.G. & SOARE, R.I. *Post's problem and his hypersimple set* ◇ D25 ◇
KALLIBEKOV, S. *On degrees of recursively enumerable sets (Russian)* ◇ D25 ◇
KALLIBEKOV, S. *Truth tabular degrees of recursively enumerable sets (Russian)* ◇ D25 ◇
KOBZEV, G.N. *btt-reducibility I,II (Russian)* ◇ D25 D30 ◇
KOBZEV, G.N. *Pointwise decomposable sets (Russian)* ◇ D20 D25 D30 ◇
KOREC, I. *Creative sets with prescribed density properties* ◇ D25 ◇
KUSABA, T. *On Hilbert's tenth problem (affirmative cases)* ◇ D25 D35 ◇
LACHLAN, A.H. *The priority method for the construction of recursively enumerable sets* ◇ D25 ◇
LADNER, R.E. *A completely mitotic nonrecursive r.e. degree* ◇ D25 ◇
LADNER, R.E. *Mitotic recursively enumerable sets* ◇ D25 ◇
MANIN, YU.I. *Hilbert's tenth problem (Russian)* ◇ D25 D35 D80 ◇
MATIYASEVICH, YU.V. *On recursive unsolvability of Hilbert's tenth problem* ◇ D25 D35 ◇
MCLAUGHLIN, T.G. *A non-enumerability theorem for infinite classes of finite structures* ◇ C13 D25 D55 ◇
MCLAUGHLIN, T.G. *On retraceable sets with rapid growth* ◇ D25 D50 ◇
NAGORNYJ, N.M. *Separability with respect to invariants (Russian)* ◇ D03 D25 ◇
OVERBEEK, R.A. *The representation of many-one degrees by decision problems of Turing machines* ◇ D10 D25 ◇
SHKIRA, V.V. *Universal functions for certain classes of recursive functions and sets (Russian)* ◇ D20 D25 ◇
TRAKHTENBROT, B.A. *Autoreducible and nonautoreducible predicates and sets (Russian)* ◇ D25 D30 ◇
VUCKOVIC, V. *Local recursive theory* ◇ D20 D25 D75 D80 ◇
V'YUGIN, V.V. *On minimal numerations of computable classes of recursively enumerable sets (Russian)* ◇ D25 D45 ◇
V'YUGIN, V.V. *On some examples of upper semilattices of computable enumerations (Russian)* ◇ D25 D45 ◇

1974

AGAFONOV, V.N. & BARZDINS, J. *On sets associated with stochastic machines (Russian)* ◇ D10 D25 ◇
ALTON, D.A. *Iterated quotients of the lattice of recursively enumerable sets* ◇ D25 ◇
ARTIGUE, M. *Etude d'un group de permutations des entiers naturels Lie au groupe des automorphismes du treillis des ensembles recursivement enumerables* ◇ D25 ◇
BADAEV, S.A. *On incomparable enumerations (Russian)* ◇ D25 D45 ◇
BAKER, B.S. & BOOK, R.V. *Reversal-bounded multi-pushdown machines* ◇ D10 D25 ◇
COOPER, S.B. *An annotated bibliography for the structure of the degrees below 0' with special reference to that of the recursively enumerable degrees* ◇ A10 D25 D30 ◇
COOPER, S.B. *Minimal pairs and high recursively enumerable degrees* ◇ D25 ◇
DEUTSCH, M. *Reduzierungen unentscheidbarer formaler Sprachen ueber induktiv definierten Bereichen* ◇ D25 D75 ◇
DUSHSKIJ, V.A. *Enumerational reducibility of recursively enumerable classes of sets (Russian)* ◇ D25 D30 D45 ◇
FLAJOLET, P. & STEYAERT, J.-M. *On sets having only hard subsets* ◇ D15 D25 ◇
FLAJOLET, P. & STEYAERT, J.-M. *Une generalisation de la notion d'ensemble immune* ◇ B75 D20 D25 D50 ◇
FREJDZON, R.I. *Table approximations to recursive predicates (Russian) (English summary)* ◇ D20 D25 ◇
GILL III, J.T. & MORRIS, P.H. *On subcreative sets and s-reducibility* ◇ D25 D30 ◇
GNANI, G. *Insiemi dialettici generalizzati* ◇ B60 D25 D55 ◇
HAMLET, R.G. *Introduction to computation theory* ◇ D10 D15 D25 D98 ◇
HAY, L. *A noninitial segment of index sets* ◇ D25 D30 ◇
HAY, L. *Index sets universal for differences of arithmetic sets* ◇ D25 D30 D55 ◇
JOCKUSCH JR., C.G. *Π_1^0 classes and boolean combinations of recursively enumerable sets* ◇ D25 D30 ◇
KINBER, E.B. *Frequency enumeration of sets (Russian)* ◇ D25 ◇
KINBER, E.B. *Ueber Frequenzberechnungen auf unendlichen Auswahlmengen (Russisch)* ◇ D20 D25 ◇
KOBAYASHI, K. *A note on extending equivalence theories of algorithms* ◇ B75 D20 D25 ◇
KOBZEV, G.N. *On complete Btt-degrees* ◇ D25 D30 ◇
KOSOVSKIJ, N.K. *Solutions of systems consisting of word equations and inequalities in lengths of words (Russian) (English summary)* ◇ D20 D25 ◇
LAVROV, I.A. *Certain properties of Post enumeration retracts (Russian)* ◇ D25 D45 ◇
LYNCH, N.A. *Approximations to the halting problem* ◇ D10 D20 D25 D45 ◇
MATIYASEVICH, YU.V. *Existence of noneffectivizable estimates in the theory of exponentially diophantine equations (Russian) (English summary)* ◇ D25 D35 ◇
MATIYASEVICH, YU.V. & ROBINSON, JULIA *Two universal three-quantifier representations of enumerable sets (Russian)* ◇ D25 D35 ◇

McLaughlin, T.G. *Closed basic retracing functions and hyperimmune sets* ⋄ D25 D50 ⋄
Novotny, M. *Sets constructed by acceptors* ⋄ D05 D25 ⋄
Sasso Jr., L.P. *Deficiency sets and bounded information reducibilities* ⋄ D25 D30 ⋄
Schnorr, C.-P. *Optimal enumerations and optimal Goedel numberings* ⋄ D15 D20 D25 D45 ⋄
Selman, A.L. *Relativized halting problems* ⋄ D10 D25 D30 ⋄
Singletary, W.E. *Many-one degrees associated with partial propositional calculi* ⋄ B20 D25 ⋄
Soare, R.I. *Automorphisms of the lattice of recursively enumerable sets* ⋄ D25 ⋄
Soare, R.I. *Automorphisms of the lattice of recursively enumerable sets I: Maximal sets* ⋄ D25 ⋄
Solov'ev, V.D. *Q-reducibility and hypersimple sets (Russian)* ⋄ D25 D30 ⋄
Vuckovic, V. *Almost recursivity and partial degrees* ⋄ D25 D30 ⋄
V'yugin, V.V. *On upper semilattices of numerations (Russian)* ⋄ D25 D45 ⋄
V'yugin, V.V. *Segments of recursively enumerable M-degrees (Russian)* ⋄ D25 ⋄
Yates, C.E.M. *Prioric games and minimal degrees below* $\underline{0}^{(1)}$ ⋄ D25 D30 E60 ⋄

1975

Alton, D.A. *A characterization of R-maximal sets* ⋄ D25 D35 ⋄
Alton, D.A. *Embedding relations in the lattice of recursively enumerable sets* ⋄ B25 D25 ⋄
Cohen, P.E. & Jockusch Jr., C.G. *A lattice property of Post's simple set* ⋄ D25 ⋄
Dawes, A.M. & Florence, J.B. *Independent Goedel sentences and independent sets* ⋄ D25 F30 ⋄
Degtev, A.N. *Reducibility of partial recursive functions (Russian)* ⋄ D20 D25 ⋄
Denisov, S.D. *Two-element separable enumeration of sets* γ *with undecidable problem* $P(\gamma,\gamma)$ *(Russian)* ⋄ D25 D45 ⋄
Deutsch, M. *Zur Darstellung koaufzaehlbarer Praedikate bei Verwendung eines einzigen unbeschraenkten Quantors* ⋄ D25 F30 ⋄
Deutsch, M. *Zur Theorie der spektralen Darstellung von Praedikaten durch Ausdruecke der Praedikatenlogik 1.Stufe* ⋄ B10 C13 D25 D35 ⋄
Florence, J.B. *On splitting an infinite recursively enumerable class* ⋄ D25 ⋄
Fraisse, R. *Cours de logique mathematique. Tome 3: Recursivite et constructibilite* ⋄ D25 D98 E35 E45 E98 ⋄
Frejvald, R.V. *Identifiability in the limit of the numbers of general recursive functions in various computable numerations (Russian) (English summary)* ⋄ D20 D25 D45 ⋄
Goetze, B.G. *Die Inseparabilitaetseigenschaften und die Ricesche Unentscheidbarkeitseigenschaft bei effektiven Numerierungen* ⋄ D25 D45 ⋄
Goodstein, R.L. *Hilbert's tenth problem and the independence of recursive difference* ⋄ D25 D35 ⋄
Hay, L. *Rice theorems for d.r.e. sets* ⋄ D25 D30 D55 ⋄
Hay, L. & Manaster, A.B. & Rosenstein, J.G. *Small recursive ordinals, many-one degrees, and the arithmetical difference hierarchy* ⋄ D25 D30 D55 ⋄
Hay, L. *Spectra and halting problems* ⋄ C13 D10 D25 D30 ⋄
Hughes, C.E. *Sets derived by deterministic systems with axiom* ⋄ D03 D25 ⋄
Kanovich, M.I. *Dekker's construction and effective nonrecursiveness (Russian)* ⋄ D20 D25 ⋄
Kobzev, G.N. *r-separated sets (Russian) (Georgian summary)* ⋄ D25 ⋄
Lachlan, A.H. *Uniform enumeration operations* ⋄ D25 D30 ⋄
Lachlan, A.H. *Wtt-complete sets are not necessarily tt-complete* ⋄ D25 D30 ⋄
Ladner, R.E. & Sasso Jr., L.P. *The weak truth table degrees of recursively enumerable sets* ⋄ D25 ⋄
Macintyre, A. & Simmons, H. *Algebraic properties of number theories* ⋄ C25 C40 C52 C62 D25 H15 ⋄
Marchenkov, S.S. *The existence of recursively enumerable minimal truth-table degrees (Russian)* ⋄ D25 ⋄
Marques, I. *On speedability of recursively enumerable sets* ⋄ D15 D25 ⋄
Maslov, S.Yu. *Mutation calculi (Russian) (English summary)* ⋄ D03 D25 D80 ⋄
Matiyasevich, Yu.V. & Robinson, Julia *Reduction of an arbitrary diophantine equation to one in 13 unknowns* ⋄ D25 D35 ⋄
McLaughlin, T.G. *A note concerning the* $\overset{*}{\vee}$ *relation on* Λ_R ⋄ D25 D30 D50 ⋄
Metakides, G. & Nerode, A. *Recursion theory and algebra* ⋄ C57 C60 D25 D45 ⋄
Morley, M.D. & Soare, R.I. *Boolean algebras, splitting theorems and* Δ^0_2 *sets* ⋄ D25 D55 G05 ⋄
Owings Jr., J.C. *Splitting a context-sensitive set* ⋄ D05 D25 ⋄
Schuler, P.F. *WCS-analysis of the context-sensitive* ⋄ D05 D25 ⋄
Scott, D.S. *Lambda calculus and recursion theory* ⋄ B40 B75 D25 D75 ⋄
Shoenfield, J.R. *The decision problem for recursively enumerable degrees* ⋄ B25 D25 ⋄
Stephan, B.J. *Compactness and recursive enumerability in intensional logic* ⋄ B45 D25 ⋄
Vershinin, V.A. *On the question of superposability of constructive sets with equality (Russian)* ⋄ D25 F60 ⋄
Vol'vachev, R.T. *Undecidability of the isomorphism and the conjugacy problem of commutative matrix groups and algebras (Russian)* ⋄ D25 D40 ⋄
Yates, C.E.M. *A general framework for simple* Δ^0_2 *and* Σ^0_1 *priority arguments* ⋄ D25 D55 ⋄

1976

ARSLANOV, M.M. *A theorem on density of recursively enumerable sets with respect to Turing reducibility (Russian)* ◇ D25 ◇
BELEGRADEK, O.V. *On m-degrees of word problems (Russian)* ◇ D25 D40 ◇
BOERGER, E. & HEIDLER, K. *Die m-Grade logischer Entscheidungsprobleme* ◇ D25 D35 D55 ◇
BOERGER, E. *On the construction of simple first-order formulae without recursive models* ◇ C57 D25 ◇
BUBIS, M.I. *On complete indexed sets (Russian)* ◇ D25 D55 ◇
CARSTENS, H.G. *Δ_2^0-Mengen* ◇ D25 D30 D55 ◇
CASE, J. *Sortability and extensibility of the graphs of recursively enumerable partial and total orders* ◇ D25 D45 ◇
CHAITIN, G.J. *Algorithmic entropy of sets* ◇ D15 D25 D80 ◇
CHRISTEN, C. *Spektralproblem und Komplexitaetstheorie* ◇ B10 C13 D10 D15 D25 D35 ◇
DAVIS, MARTIN D. & MATIYASEVICH, YU.V. & ROBINSON, JULIA *Hilbert's tenth problem: Diophantine equations: Positive aspects of a negative solution* ◇ D25 D35 D80 ◇
DEGTEV, A.N. *Minimal 1-degrees, and truth-table reducibility (Russian)* ◇ D25 D30 ◇
DEGTEV, A.N. *Partially ordered sets of 1-degrees, contained in recursively enumerable m-degrees (Russian)* ◇ D25 ◇
DEUTSCH, M. *Eine mengentheoretische Grundlegung der Theorie der Berechenbarkeit I* ◇ D20 D25 E47 ◇
DEUTSCH, M. *Zur Praefixoptimalitaet gewisser $\exists\forall\exists\ldots\exists$-Darstellungen aufzaehlbarer Praedikate* ◇ D25 ◇
DO MINH THAI *Programmkomplexitaet rekursiv aufzaehlbarer Mengen* ◇ D15 D25 D30 ◇
ERSHOV, YU.L. *Hereditarily effective operations (Russian)* ◇ D25 D65 ◇
FERNANDEZ-PRIDA, J. *Creative and productive recursive theories (Spanish)* ◇ D25 ◇
GONSHOR, H. *Effective density types* ◇ D25 D50 ◇
GUREVICH, Y. *Semi-conservative reduction* ◇ B20 D25 D35 ◇
HAY, L. *Boolean combinations of r.e. open sets* ◇ D25 D55 E15 ◇
HECKER, H.-D. *Zur Programmkomplexitaet rekursiv aufzaehlbarer Mengen* ◇ D15 D25 ◇
HERMES, H. *Hilbert's tenth problem* ◇ A10 D25 D35 ◇
HUGHES, C.E. *Two variable implicational calculi of prescribed many-one degrees of unsolvability* ◇ B20 D25 ◇
INDERMARK, K. & REISIG, W. *On recursively definable relations* ◇ D20 D25 ◇
JEDRZEJOWICZ, J. *One-one degrees of Turing machines decision problems* ◇ D10 D25 ◇
JOCKUSCH JR., C.G. & PATERSON, M.S. *Completely autoreducible degrees* ◇ D25 D30 ◇
JONES, JAMES P. & SATO, D. & WADA, H. & WIENS, D. *Diophantine representation of the set of prime numbers* ◇ D25 ◇
KANOVICH, M.I. *The relation between upper and lower bounds for the reduction complexity of enumerable sets (Russian)* ◇ D15 D25 ◇
KLETTE, R. *Indexmengen und Erkennung rekursiver Funktionen* ◇ D20 D25 D30 D55 ◇
KOBZEV, G.N. *Relationship between recursively enumerable tt- and w-degrees (Russian) (English summary)* ◇ D25 ◇
LACHLAN, A.H. *A recursively enumerable degree which will not split over all lesser ones* ◇ D25 ◇
LYNCH, N.A. *Complexity-class-encoding sets* ◇ D10 D15 D25 ◇
MARCHENKOV, S.S. *On the congruence of the upper semilattices of recursively enumerable m-powers and tabular powers (Russian)* ◇ D25 D30 ◇
MARCHENKOV, S.S. *One class of partial sets (Russian)* ◇ D25 ◇
MARCHENKOV, S.S. *Tabular powers of maximal sets (Russian)* ◇ D25 ◇
MATIYASEVICH, YU.V. *A new proof of the theorem on exponential diophantine representation of enumerable sets (Russian) (English summary)* ◇ D25 ◇
OMANADZE, R.SH. *The completeness of recursively enumerable sets (Russian) (Georgian and English summaries)* ◇ D25 ◇
OREJAS, F. *The priority method in degree theory (Spanish)* ◇ D25 ◇
RED'KO, V.N. *Definitor-theoretic aspects of languages (Russian)* ◇ D03 D05 D25 ◇
REMMEL, J.B. *Co-hypersimple structures* ◇ D25 D45 ◇
REMMEL, J.B. *Combinatorial functors on co-r.e. structures* ◇ D25 D30 D45 D50 ◇
SELIVANOV, V.L. *Enumerations of families of general recursive function (Russian)* ◇ D20 D25 ◇
SELIVANOV, V.L. *On computability of some classes of numerations (Russian)* ◇ D25 D45 ◇
SHOENFIELD, J.R. *Degrees of classes of RE sets* ◇ D25 D30 ◇
SKORDEV, D.G. *Some methods for building up the theory of recursive functions* ◇ D20 D25 ◇
SOARE, R.I. *The infinite injury priority method* ◇ D25 ◇
SOLOV'EV, V.D. *Some generalizations of the notions of reducibility and creativity (Russian)* ◇ D25 D30 ◇
SOLOV'EV, V.D. *Superhypersimple sets (Russian)* ◇ D25 ◇
ZHAROV, V.G. *Codes of words that belong to an enumerable set (Russian)* ◇ D15 D25 ◇
ZIEGLER, M. *Ein rekursiv aufzaehlbarer btt-Grad, der nicht zum Wortproblem einer Gruppe gehoert* ◇ D25 D40 ◇

1977

ABIAN, A. *Unsolvability of the halting problem for Turing machines* ◇ D10 D25 ◇
ARSLANOV, M.M. & NADYROV, R.F. & SOLOV'EV, V.D. *A criterion for the completeness of recursively enumerable sets, and some generalizations of a fixed point theorem (Russian)* ◇ D20 D25 D55 ◇
BADAEV, S.A. *Positive numerations (Russian)* ◇ D25 D45 ◇

BENNISON, V.L. *On the problem of determining measure-independence* ◇ D15 D25 ◇
CARSTENS, H.G. *The theorem of Matijasevic is provable in Peano's arithmetic by finitely many axioms* ◇ C52 D25 D35 F30 H15 ◇
CENZER, D. *Non-generable RE sets* ◇ D03 D25 ◇
CHAITIN, G.J. *Program size, oracles, and the jump operation* ◇ D15 D25 D30 D80 ◇
DYMENT, E.Z. *Supersets of recursively enumerable sets (Russian)* ◇ D25 ◇
ENDERTON, H.B. *Elements of recursion theory* ◇ D20 D25 D98 ◇
FUCHS, P.H. & SCHNORR, C.-P. *General random sequences and learnable sequences* ◇ D20 D25 D80 ◇
HAY, L. & MANASTER, A.B. & ROSENSTEIN, J.G. *Concerning partial recursive similarity transformations of linearly ordered sets* ◇ C57 C65 D20 D25 D30 D45 E07 ◇
HERRMANN, E. *A note on no hh-simple recursive enumerable sets (Russian summary)* ◇ D25 ◇
JOHNSON, N. *Rice theorems for* Σ_n^{-1} *sets* ◇ D25 D30 D55 ◇
KANOVICH, M.I. *On the precision of a complexity criterion for nonrecursiveness and universality (Russian)* ◇ D15 D20 D25 D30 ◇
KINBER, E.B. *On Turing degrees of hypersimple Post sets (Russian)* ◇ D25 D30 ◇
KOBZEV, G.N. *Maximal m-degrees (Russian) (Georgian and English summaries)* ◇ D25 ◇
KOBZEV, G.N. *Recursively enumerable bw-degrees (Russian)* ◇ D25 ◇
LAVROV, I.A. *Computable numberings* ◇ D25 D45 ◇
LEEUWEN VAN, J. *Recursively enumerable languages and van Wijngaarden grammars* ◇ D05 D25 ◇
MARCHENKOV, S.S. *On recursively enumerable minimal btt-degrees (Russian)* ◇ D25 ◇
MATIYASEVICH, YU.V. *Primes are non-negative values of a polynomial in 10 variables (Russian) (English summery)* ◇ D25 ◇
MATIYASEVICH, YU.V. *Some purely mathematical results inspired by mathematical logic* ◇ B30 D25 D35 D80 ◇
MCLAUGHLIN, T.G. *Degrees of unsolvability and strong forms of* $\Lambda_R + \Lambda_R \not\subseteq \Lambda_R$ ◇ D25 D30 D50 ◇
REMMEL, J.B. *Maximal and cohesive vector spaces* ◇ C57 D25 D45 ◇
SACKS, G.E. *RE sets higher up* ◇ D25 D60 D65 ◇
SANCHIS, L.E. *Data types as lattices: Retractions, closures and projections* ◇ B40 B75 D25 G10 ◇
SHORE, R.A. *Determining automorphisms of the recursively enumerable sets* ◇ D25 ◇
SIMPSON, S.G. *Degrees of unsolvability: A survey of results* ◇ D25 D30 D35 F35 ◇
SMORYNSKI, C.A. *A note on the number of zeros of polynomials and exponential polynomials* ◇ D25 D35 D80 ◇
SOLOVAY, R.M. *On random r.e. sets* ◇ D15 D25 ◇
STOLTENBERG-HANSEN, V. *A regular set theorem for infinite computation theories* ◇ D25 D75 ◇

1978

ARSLANOV, M.M. & SOLOV'EV, V.D. *Effectivizations of the definitions of classes of simple sets (Russian)* ◇ D25 ◇
BELEGRADEK, O.V. *On m-degrees of the word problem (Russian)* ◇ D25 D40 ◇
BENNISON, V.L. & SOARE, R.I. *Some lowness properties and computational complexity sequences* ◇ D15 D25 D30 ◇
BOOK, R.V. *Simple representations of certain classes of languages* ◇ D15 D25 D55 ◇
BREIDBART, S. *On splitting recursive sets* ◇ D05 D15 D20 D25 ◇
DALEY, R.P. *On the simplicity of busy beaver sets* ◇ D10 D15 D25 D30 ◇
DEGTEV, A.N. *m-degrees of supersets of simple sets (Russian)* ◇ D25 ◇
DEGTEV, A.N. *Solvability of the* ∀∃*-theory of a certain factor-lattice of recursively enumerable sets (Russian)* ◇ B25 D25 D50 ◇
DEGTEV, A.N. *Three theorems on tt-degrees (Russian)* ◇ D25 D30 ◇
DENEF, J. *Diophantine sets over* $Z[T]$ ◇ C60 D25 D35 ◇
DENISOV, S.D. *Structures of the upper semilattice of recursively enumerable m-degrees and related questions I (Russian)* ◇ D25 ◇
EMDE BOAS VAN, P. & VITANYI, P.M.B. *A note on the recursive enumerability of some classes of recursive languages* ◇ D05 D25 ◇
HECKER, H.-D. *Bemerkungen zu Enumerationssystemen und Zeitmassen* ◇ D15 D25 D45 ◇
HERMES, H. *Die Unloesbarkeit des zehnten Hilbertschen Problems* ◇ A10 D25 D35 ◇
HERRMANN, E. *Der Verband der rekursiv aufzaehlbaren Mengen (Entscheidungs-Problem) (Russian and English summaries)* ◇ B25 D25 D98 ◇
HOROWITZ, B.M. *Sets completely creative via recursive permutations* ◇ D20 D25 ◇
JOHNSON, N. *Classifications of generalized index sets of open classes* ◇ D25 D30 ◇
JONES, JAMES P. *Three universal representations of recursively enumerable sets* ◇ D25 F20 ◇
KALANTARI, I. *Major subspaces of recursively enumerable vector spaces* ◇ C57 C60 D25 D45 ◇
KOBZEV, G.N. *On the semi-lattice of tt-degrees (Russian) (English summary)* ◇ D25 D30 ◇
KOBZEV, G.N. *On tt-degrees of recursively enumerable Turing degrees (Russian)* ◇ D25 ◇
LERMAN, M. & SHORE, R.A. & SOARE, R.I. *r-maximal major subsets* ◇ D25 ◇
MAASS, W. *Contributions to* α*- and* β*-recursion theory* ◇ D25 D30 D60 E45 ◇
MOSTOWSKI, A.WLODZIMIERZ *Recursively enumerable degrees of programming problems* ◇ B75 D03 D25 ◇
OMANADZE, R.SH. *On some generalizations of the notion of set productivity (Russian)* ◇ D25 ◇
OMANADZE, R.SH. *On the reducibility on the class of recursive enumerable sets (Russian) (English summary)* ◇ D25 D30 ◇

PRANK, R.K. *On congruence relations in the lattice of recursively enumerable sets (Russian) (English summary)* ◇ D25 ◇
REMMEL, J.B. *A r-maximal vector space not contained in any maximal vector space* ◇ C57 D25 D45 ◇
REMMEL, J.B. *Realizing partial orderings by classes of co-simple sets* ◇ D25 D45 D50 ◇
REMMEL, J.B. *Recursively enumerable boolean algebras* ◇ C57 D25 D45 G05 ◇
RETZLAFF, A.T. *Simple and hyperhypersimple vector spaces* ◇ C57 D25 D45 ◇
SELIVANOV, V.L. *On index sets of computable classes of finite sets (Russian)* ◇ D25 D30 D55 ◇
SELIVANOV, V.L. *Some remarks about classes of recursively enumerable sets (Russian)* ◇ D25 ◇
SHAY, M. & YOUNG, P. *Characterizing the orders changed by program translators* ◇ B75 D20 D25 ◇
SHORE, R.A. *Controlling the dependence degree of a recursively enumerable vector space* ◇ D25 D45 D50 ◇
SHORE, R.A. *Nowhere simple sets and the lattice of recursively enumerable sets* ◇ D25 ◇
SMORYNSKI, C.A. *Avoiding self-referential statements* ◇ D25 F30 ◇
SOARE, R.I. *Recursively enumerable sets and degrees* ◇ D25 ◇
SOLON, B.YA. *e-powers of hyperimmune retraceable sets (Russian)* ◇ D25 D30 D50 ◇
TANAKA, H. *Recursion theory in analytical hierarchy* ◇ D25 D55 D60 E60 ◇
TURAKAINEN, P. *On characterization of recursively enumerable languages in terms of linear languages and VW-grammars* ◇ D05 D25 ◇
UESU, T. *A system of graph grammars which generates all recursively enumerable sets of labelled graphs* ◇ D25 D80 ◇

1979

ARSLANOV, M.M. *Weak recursively enumerable degrees and computability in the limit (Russian)* ◇ D25 D55 ◇
BENNISON, V.L. *Information content characterizations of complexity theoretic properties* ◇ D15 D25 ◇
BOERGER, E. *A new general approach to the theory of the many-one equivalence of decision problems for algorithmic systems* ◇ B25 D03 D05 D10 D25 D30 D40 ◇
BUKHARAEV, N.R. *On limit enumeration properties of complements of recursively enumerable sets (Russian)* ◇ D25 ◇
CULIK II, K. *A purely homomorphic characterization of recursively enumerable sets* ◇ D03 D25 ◇
DALEY, R.P. *On the simplification of constructions in degrees of unsolvability via computational complexity* ◇ D10 D15 D25 ◇
DEGTEV, A.N. & ZAKHAROV, D.A. *Recursively enumerable sets (Russian)* ◇ D25 ◇
DEGTEV, A.N. *Reducibilities of tabular type in the theory of algorithms (Russian)* ◇ D25 D30 ◇
DEGTEV, A.N. *Several results on upper semilattices and m-degrees (Russian)* ◇ D25 D30 ◇
EPSTEIN, R.L. *Degrees of unsolvability: structure and theory* ◇ D25 D30 D35 D98 ◇
FERRO, A. & MICALE, B. *Sulla teoria della computabilita* ◇ D20 D25 ◇
GURARI, E.M. & IBARRA, O.H. *An NP-complete number-theoretic problem* ◇ B25 D10 D15 D25 D80 ◇
HAY, L. & JOHNSON, N. *Extensional characterization of index sets* ◇ D25 ◇
HERRMANN, E. *The lattice structures of the recursively enumerable sets* ◇ D25 ◇
JEDRZEJOWICZ, J. *Inverse derivability and confluence problems of deterministic combinatorial systems* ◇ D25 ◇
JEDRZEJOWICZ, J. *One-one degrees of combinatorial systems decision problems* ◇ D25 ◇
JONES, JAMES P. *Diophantine representation of Mersenne and Fermat primes* ◇ D25 D35 ◇
KHODZHAYANTS, M.YU. *Some characteristics of enumeration operators (Russian)* ◇ D15 D25 ◇
KHODZHAYANTS, M.YU. *Structure of e-degrees (Russian)* ◇ D25 D30 ◇
KOBZEV, G.N. *tt-degrees of recursively enumerable Turing degrees II (Russian)* ◇ D25 D30 ◇
KOSOVSKIJ, N.K. *On decision procedures for invariant properties of short algorithms (Russian) (English summary)* ◇ D15 D20 D25 D45 ◇
KRUPSKIJ, V.N. *On completely enumerable sets in effectively metric spaces (Russian) (English summary)* ◇ D25 D45 ◇
LACHLAN, A.H. *Bounding minimal pairs* ◇ D25 ◇
LAVROV, I.A. *Computability of partial functions and enumerability of sets in Peano's arithmitic (Russian)* ◇ D20 D25 F30 ◇
LAVROV, I.A. *Retracts of Post's numbering and effectivization of quantifiers* ◇ D25 D45 ◇
LI, XIANG *A recursively unsolvable problem concerning the mu operator (Chinese) (English summary)* ◇ D25 ◇
LI, XIANG *The productiveness of Blum's measure of effective computational complexity (Chinese)* ◇ D15 D25 ◇
MARCHENKOV, S.S. & MATROSOV, V.L. *Complexity of algorithms and computations (Russian)* ◇ D10 D15 D25 ◇
MASLOVA, T.M. *Bounded m-reducibilities (Russian)* ◇ D25 D30 ◇
MATIYASEVICH, YU.V. *Algorithmic unsolvability of exponential Diophantine equations with three indeterminates (Russian)* ◇ D25 D35 ◇
OMANADZE, R.SH. *On some properties of Q-reducibility (Russian)* ◇ D25 D30 ◇
OMANADZE, R.SH. *On Q-reducibility (Russian)* ◇ D25 D30 ◇
PLYASUNOV, A.V. *Splinters and Turing degrees (Russian)* ◇ D25 ◇
ROZINAS, M.G. & SOLON, B.YA. *Weakly semirecursive sets (Russian)* ◇ D20 D25 D30 ◇
SHEN', A.KH. *The priority method and separation problems (Russian)* ◇ D25 D30 ◇
SOLON, B.YA. *e-degrees of productive sets (Russian)* ◇ D25 ◇

STOLTENBERG-HANSEN, V. *Finite injury arguments in infinite computation theories* ◇ D25 D30 D60 D75 ◇

VAEAENAENEN, J. *A new incompleteness in arithmetic (Finnish) (English summary)* ◇ A10 C62 D25 E05 F30 ◇

1980

ASATRYAN, O.S. *On creative equivalences (Russian) (Armenian and English summaries)* ◇ D25 D45 ◇

BARASHKO, A.S. *On boundary sets in the theory of enumerations (Ukrainian) (English and Russian summaries)* ◇ D15 D25 D45 ◇

BENNISON, V.L. *Recursively enumerable complexity sequences and measure independence* ◇ D15 D25 D30 ◇

BERNARDI, C. *Alcune osservazioni sugli insiemi produttivi e creativi (English summary)* ◇ D25 ◇

BERNARDI, C. & CROCIANI, C. *Insiemi semplici e massimali in ricorsivita (English summary)* ◇ D25 ◇

BOERGER, E. & KLEINE BUENING, H. *The r.e. complexity of decision problems for commutative semi-Thue systems with recursive rule set* ◇ D03 D25 ◇

BULITKO, V.K. *Reducibility via Zhegalkin's linear tables (Russian)* ◇ D25 D30 ◇

CENZER, D. *Non-generable formal languages* ◇ D03 D25 D30 D70 ◇

COHEN, D.E. *Degree problems for modular machines* ◇ D10 D25 D40 ◇

DALEY, R.P. *The busy beaver method* ◇ D10 D15 D25 ◇

DEGTEV, A.N. *On reducibility of numerations (Russian)* ◇ D25 D30 D45 ◇

DEUTSCH, M. *Zur goedelisierungsfreien Darstellung der rekursiven und primitiv-rekursiven Funktionen und rekursiv aufzaehlbaren Praedikate ueber Quotiententermmengen* ◇ D20 D25 ◇

ENGELFRIET, J. & ROZENBERG, G. *Fixed point languages, equality languages, and representation of recursively enumerable languages* ◇ D05 D25 ◇

FALKENBERG, B. *Halteprobleme von Fang-Systemen (tag systems)* ◇ D03 D25 ◇

GOLUNKOV, YU.V. *Precomplete classes of algorithms that preserve membership in a set II (Russian)* ◇ D25 D50 ◇

HECKER, H.-D. *Enumerationen in speziellen Standardklassen rekursiv-aufzaehlbarer Mengen* ◇ D25 ◇

HOROWITZ, B.M. *Constructively nonpartial recursive functions* ◇ D20 D25 ◇

HUGHES, C.E. *Derivatives and quotients of prefix-free context-free languages* ◇ D05 D25 ◇

HUGHES, C.E. & STRAIGHT, D.W. *Word problems for bidirectional single-premise Post systems* ◇ D03 D25 ◇

JOCKUSCH JR., C.G. *Fine degrees of word problems of cancellation semigroups* ◇ D25 D40 ◇

JONES, JAMES P. *Undecidable Diophantine equations* ◇ D25 D35 ◇

KHODZHAYANTS, M.YU. *Structure of e-degrees (Russian) (Armenian and English summaries)* ◇ D25 D30 ◇

KOREC, I. *Densities of first order theories* ◇ D25 ◇

LACHLAN, A.H. *Decomposition of recursively enumerable degrees* ◇ D25 ◇

LACHLAN, A.H. & SOARE, R.I. *Not every finite lattice is embeddable in the recursively enumerable degrees* ◇ D25 ◇

LERMAN, M. & SOARE, R.I. *d-simple sets, small sets, and degree classes* ◇ D25 ◇

LERMAN, M. & SOARE, R.I. *A decidable fragment of the elementary theory of the lattice of recursively enumerable sets* ◇ B25 C10 D25 ◇

MANIN, YU.I. *The computable and the non-computable (Russian)* ◇ D20 D25 D98 ◇

OMANADZE, R.SH. *On bounded Q-reducibility (Russian) (Georgian and English summaries)* ◇ D25 D30 ◇

PRANK, R.K. *Semantics of realizability for a language with variables for recursively enumerable sets (Russian)* ◇ B60 D25 F35 F50 ◇

RUOHONEN, K. *Hilbert's tenth problem (Swedish) (English summary)* ◇ D25 D35 ◇

SHORE, R.A. $\mathscr{L}^*(K)$ *and other lattices of recursively enumerable sets* ◇ D25 ◇

SOARE, R.I. *Constructions in the recursively enumerable degrees* ◇ D25 ◇

SOARE, R.I. *Fundamental methods for constructing recursively enumerable degrees* ◇ D25 ◇

SOARE, R.I. *Recursive enumerability* ◇ D25 ◇

STOLTENBERG-HANSEN, V. & TUCKER, J.V. *Computing roots of unity in fields* ◇ D25 D45 D80 ◇

ZDEBSKAYA, G.V. *On weakly hypersimple sets (Russian)* ◇ D25 ◇

1981

ARAKI, T. & TOKURA, N. *Flow languages equal recursively enumerable languages* ◇ D05 D25 ◇

ARSLANOV, M.M. *Some generalizations of a fixed-point theorem (Russian)* ◇ D25 D30 ◇

ASATRYAN, O.S. *Two theorems on bT reducibility (Russian) (Armenian summary)* ◇ D25 ◇

BERNARDI, C. *On the relation provable equivalence and on partitions in effectively inseparable sets* ◇ D25 D35 F30 ◇

BUKHARAEV, N.R. *T-degrees of differences of recursively enumerable sets (Russian)* ◇ D25 D30 D55 ◇

BUKHARAEV, N.R. *r-g-simple sets, ΦM sets and their degrees of unsolvability (Russian)* ◇ D25 ◇

CALUDE, C. & PAUN, G. *Global syntax and semantics for recursively enumerable languages* ◇ B03 D25 ◇

DALEY, R.P. *Busy beaver sets and the degree of unsolvability* ◇ D10 D15 D25 ◇

DEGTEV, A.N. *Relations between complete sets (Russian)* ◇ D25 D30 ◇

DO MINH THAI *On minimal-program complexity of complex sets* ◇ D15 D25 ◇

EPSTEIN, R.L. *Initial segments of degrees below 0'* ◇ D25 D30 D35 D98 ◇

FEJER, P.A. & SOARE, R.I. *The plus-cupping theorem for the recursively enumerable degrees* ◇ D25 ◇

HATCHER, W.S. & HODGSON, B.R. *Complexity bounds on proofs* ◇ D25 F20 ◇

HECKER, H.-D. *Zur Kompliziertheit der Beschreibung von Enumerationen rekursiv aufzaehlbarer Mengen* ◇ D15 D25 D45 ◇
HERRMANN, E. *Die Verbandseigenschaften der rekursiv aufzaehlbaren Mengen* ◇ D25 D98 ◇
HICKIN, K.K. *Bounded HNN presentations* ◇ D25 D40 ◇
JANTZEN, M. *The power of synchronizing operations on strings* ◇ D05 D25 ◇
JEDRZEJOWICZ, J. *Undecidable problems associated with combinatorial systems and their one-one degrees of unsolvability* ◇ D20 D25 ◇
JOCKUSCH JR., C.G. *Three easy constructions of recursively enumerable sets* ◇ D25 ◇
JONES, JAMES P. *Classification of quantifier prefixes over Diophantine equations* ◇ D25 D35 ◇
MAASS, W. & SHORE, R.A. & STOB, M. *Splitting properties and jump classes* ◇ D25 ◇
MARGENSTERN, M. *Le theoreme de Matiyassevitch et resultats connexes* ◇ D25 D35 ◇
MEYER, A.R. & MIRKOWSKA, G. & STREETT, R.S. *The deducibility problem in propositional dynamic logic* ◇ B75 D25 D35 ◇
MILLER, DAVID P. *High recursively enumerable degrees and the anticupping property* ◇ D25 ◇
ODIFREDDI, P. *Insiemi subcreativi (English summary)* ◇ D15 D25 ◇
ODIFREDDI, P. *Strong reducibilities* ◇ D25 D30 ◇
POSNER, D.B. & ROBINSON, R.W. *Degrees joining to 0'* ◇ D25 D30 ◇
POSNER, D.B. *The upper semilattice of degrees $\leqslant 0'$ is complemented* ◇ D25 D30 ◇
PRANK, R.K. *Expressibility in the elementary theory of recursively enumerable sets with realizability logic (Russian)* ◇ B60 D25 F50 ◇
PRANK, R.K. *On the quotient lattice of the lattice of recursively enumerable sets by the immunity congruence (Russian) (Estonian and English summaries)* ◇ D25 D50 ◇
SHORE, R.A. *The theory of the degrees below O'* ◇ D25 D30 D35 F30 ◇
SOLON, B.YA. *PC-degrees inside an e-degree of a hyperimmune retraceable set (Russian)* ◇ D25 D30 D50 H05 ◇

1982

BALDWIN, JOHN T. *Recursion theory and abstract dependence* ◇ C57 C60 D25 D45 ◇
CAI, MAOHUA & ZHANG, JINWEN *Recursively enumerable fuzzy sets I* ◇ D25 E72 ◇
CAI, MAOHUA *Recursively enumerable fuzzy sets II* ◇ D25 E72 ◇
DALEY, R.P. *Busy beaver sets: characterizations and applications* ◇ D10 D25 D50 ◇
DAWES, A.M. *Splitting theorems for speed-up related to order of enumerations* ◇ D15 D25 ◇
FANG, ZHIXI *The closure property of certain classes of languages under bi-language forms (Chinese) (English summary)* ◇ D05 D15 D25 ◇
FEJER, P.A. *Branching degrees above low degrees* ◇ D25 ◇
HARRINGTON, L.A. & SHELAH, S. *The undecidability of the recursively enumerable degrees* ◇ D25 D35 ◇
HARTMANIS, J. *A note on natural complete sets and Goedel numberings* ◇ D15 D25 D45 ◇
HOROWITZ, B.M. *An isomorphism type of arithmetically productive sets* ◇ D25 D55 ◇
HOROWITZ, B.M. *Arithmetical analogues of productive and universal sets* ◇ D25 D55 ◇
HOROWITZ, B.M. *Elementary formal systems as a framework for relative recursion theory* ◇ D20 D25 D30 ◇
ISHMUKHAMETOV, SH.T. *Families of recursively enumerable sets (Russian)* ◇ D25 D55 ◇
JONES, JAMES P. & MATIYASEVICH, YU.V. *A new representation for the symmetric binomial coefficient and its applications (French summary)* ◇ D25 ◇
JONES, JAMES P. & MATIYASEVICH, YU.V. *Exponential diophantine representation of recursively enumerable sets (French summary)* ◇ D20 D25 D35 ◇
JONES, JAMES P. *Universal diophantine equation* ◇ D25 ◇
KALANTARI, I. *Major subsets in effective topology* ◇ C57 C65 D25 D45 ◇
KALANTARI, I. & LEGGETT, A. *Simplicity in effective topology* ◇ C57 C65 D25 ◇
KALORKOTI, K.A. *Decision problems in group theory* ◇ D25 D40 ◇
KANOVICH, M.I. *On the complexity of the problem of separating recursively enumerable sets (Russian)* ◇ D25 D30 ◇
LERMAN, M. & REMMEL, J.B. *The universal splitting property I* ◇ D25 D45 ◇
MAASS, W. *Recursively enumerable generic sets* ◇ D25 E40 ◇
MILLER, DOUGLAS E. *Index sets and Boolean operations* ◇ D25 D45 D55 ◇
MOKATSYAN, A.A. *ω-mitotic but not btt-mitotic enumerable sets (Russian) (Armenian summary)* ◇ D25 ◇
NERODE, A. & REMMEL, J.B. *Recursion theory on matroids* ◇ C57 D25 D45 ◇
ROSENBERG, R. *Recursively enumerable images of arithmetic sets* ◇ D25 D55 F30 ◇
SELIVANOV, V.L. *On the index sets in the Kleene-Mostowski hierarchy (Russian)* ◇ D25 D45 D55 ◇
SHI, NIANDONG *Creative pairs of subalgebras of recursively enumerable boolean algebras (Chinese)* ◇ C57 D25 D45 G05 ◇
SIMPSON, S.G. *Four test problems in generalized recursion theory* ◇ D25 D55 D60 E15 ◇
SOARE, R.I. *Automorphisms of the lattice of recursively enumerable sets Part II:Low sets* ◇ D25 ◇
SOARE, R.I. *Computational complexity of recursively enumerable sets* ◇ D15 D25 ◇
SOARE, R.I. & STOB, M. *Relative recursive enumerability* ◇ D25 ◇
STOB, M. *Invariance of properties under automorphisms of the lattice of recursively enumerable sets* ◇ D25 ◇
THURAISINGHAM, M.B. *Representation of one-one degrees by decision problems for system functions* ◇ D25 ◇

ZDEBSKAYA, G.V. *Recursively enumerable sets and limit computability (Russian)* ◇ D20 D25 ◇

1983

BOERGER, E. *Undecidability versus degree complexity of decision problems for formal grammars* ◇ D05 D10 D25 ◇

BURGIN, M.S. *Inductive Turing machines (Russian)* ◇ D10 D25 D55 ◇

BYERLY, R.E. *Definability of r.e. sets in a class of recursion theoretic structures* ◇ D25 D45 ◇

CAI, MAOHUA *A projection theorem for recursively enumerable fuzzy sets (Chinese)* ◇ D25 E72 ◇

CALUDE, C. *On the class of independent problems related to Rice theorem* ◇ D20 D25 ◇

DEGTEV, A.N. *Relations between table-type degrees (Russian)* ◇ D25 D30 ◇

DEGTEV, A.N. *Relations between reducibilities of table type (Russian)* ◇ D25 ◇

DICHEV, A.V. *On the m-equivalence of Cartesian powers of sets of natural numbers (Russian)* ◇ D25 D30 ◇

DOWNEY, R.G. *Abstract dependence, recursion theory, and the lattice of recursively enumerable filters* ◇ D25 D45 ◇

DOWNEY, R.G. *Nowhere simplicity in matroids* ◇ D25 D45 ◇

DOWNEY, R.G. *On a question of A.Retzlaff* ◇ C57 C60 D25 D45 ◇

FEJER, P.A. *The density of the nonbranching degrees* ◇ D25 ◇

HERRMANN, E. *Definable Boolean pairs in the lattice of recursively enumerable sets* ◇ D25 D35 ◇

HERRMANN, E. *Orbits of hyperhypersimple sets and the lattice of Σ_3^0 sets* ◇ D25 D55 ◇

JOCKUSCH JR., C.G. & SHORE, R.A. *Pseudo jump operators I: The r.e. case* ◇ D25 D30 ◇

KALANTARI, I. & REMMEL, J.B. *Degrees of recursively enumarable topological spaces* ◇ C57 C65 D25 D45 ◇

KALANTARI, I. & LEGGETT, A. *Maximality in effective topology* ◇ C57 C65 D25 D45 ◇

KALLIBEKOV, S. *Index sets and degrees of some classes of recursively enumerable sets (Russian)* ◇ D25 D30 ◇

LERMAN, M. *The structures of recursion theory* ◇ D25 D30 D98 ◇

LI, XIANG *Effective immune sets, program index sets and effectively simple sets- generalizations and applications of the recursion theorem* ◇ D25 D50 ◇

MAASS, W. *Characterization of recursively enumerable sets with supersets effectively isomorphic to all recursively enumerable sets* ◇ D25 ◇

MAASS, W. & STOB, M. *The intervals of the lattice of recursively enumerable sets determined by major subsets* ◇ D25 ◇

MOHRHERR, J. *Kleene index sets and functional m-degrees* ◇ D20 D25 D30 D45 ◇

NERODE, A. & REMMEL, J.B. *Recursion theory on matroids II* ◇ D25 D45 ◇

SELIVANOV, V.L. *Effective analogues of A-, B- and C-sets and their application to index sets (Russian)* ◇ D25 D55 E15 ◇

STOB, M. *WTT-degrees and T-degrees of recursively enumerable sets* ◇ D25 ◇

THURAISINGHAM, M.B. *Cylindrical decision problems for system functions* ◇ D20 D25 ◇

THURAISINGHAM, M.B. *Some elementary closure properties of n-cylinders* ◇ D20 D25 ◇

THURAISINGHAM, M.B. *The concept of n-cylinder and its relationship to simple sets* ◇ D25 ◇

WANG, JIE *A note on a theorem about generating sets of recursively enumerable sets (Chinese)* ◇ D25 ◇

1984

AMBOS-SPIES, K. & JOCKUSCH JR., C.G. & SHORE, R.A. & SOARE, R.I. *An algebraic decomposition of the recursively enumerable degrees and the coincidence of several degree classes with the promptly simple degrees* ◇ D25 ◇

AMBOS-SPIES, K. *An extension of the non-diamond theorem in classical and α-recursion theory* ◇ D25 D60 ◇

AMBOS-SPIES, K. *Contiguous r.e. degrees* ◇ D25 ◇

AMBOS-SPIES, K. *On pairs of recursively enumerable degrees* ◇ D25 ◇

BALDWIN, JOHN T. *First-order theories of abstract dependence relations* ◇ C35 C45 C57 C60 D25 D45 ◇

BRANDENBURG, F.-J. *A truly morphic characterization of recursive enumerable sets* ◇ D25 ◇

BYERLY, R.E. *Definability of recursively enumerable sets in abstract computational complexity theory* ◇ D15 D25 ◇

BYERLY, R.E. *Some properties of invariant sets* ◇ D25 D30 ◇

CHLEBUS, B.S. & CHROBAK, M. *Probabilistic Turing machines and recursively enumerable Dedekind cuts* ◇ D10 D25 D80 ◇

COHEN, D.E. *Modular machines, undecidability and incompleteness* ◇ D20 D25 D35 F30 ◇

DING, DECHENG *The minimal pair of r.e. recursively inseparable sets (Chinese) (English summary)* ◇ D25 ◇

DOWNEY, R.G. *A note on decompositions of recursively enumerable subspaces* ◇ C57 C60 D25 D45 ◇

DOWNEY, R.G. *Bases of supermaximal subspaces and Steinitz systems. I* ◇ C57 C60 D25 D45 ◇

DOWNEY, R.G. & REMMEL, J.B. *The universal complementation property* ◇ C57 C60 D25 D45 ◇

ERSHOV, YU.L. *Strong nonseparability and k-heredity (Russian)* ◇ D25 ◇

HECKER, H.-D. *Abstrakte Tempomasse und speed-up-Theoreme fuer Enumerationen rekursiv-aufzaehlbarer Mengen* ◇ D15 D25 ◇

HECKER, H.-D. *Das Kompressionstheorem fuer Tempomasse* ◇ D15 D25 ◇

HERRMANN, E. *Classes of simple sets, filter properties and their mutual position* ◇ D25 ◇

HERRMANN, E. *Definable structures in the lattice of recursively enumerable sets* ◇ D25 D55 ◇

HERRMANN, E. *The undecidability of the elementary theory of the lattice of recursively enumerable sets* ◇ D25 D35 ◇

JOCKUSCH JR., C.G. & SHORE, R.A. *Pseudo jump operators II: Transfinite iterations, hierarchies and minimal covers* ◇ D25 D30 D55 ◇
JOCKUSCH JR., C.G. & KALANTARI, I. *Recursively enumerable sets and van der Waerden's theorem on arithmetic progressions* ◇ C57 D25 D80 ◇
JONES, JAMES P. & MATIYASEVICH, YU.V. *Register machine proof of the theorem on exponential diophantine representation of enumerable sets* ◇ D10 D25 D35 ◇
KHISAMIEV, Z.G. *Multiple m-reducibility of index sets (Russian)* ◇ D25 D30 ◇
LERMAN, M. & SHORE, R.A. & SOARE, R.I. *The elementary theory of the recursively enumerable degrees is not* $\aleph_0$ *categorical* ◇ C35 D25 ◇
LERMAN, M. & REMMEL, J.B. *The universal splitting property. II* ◇ D25 ◇
MAASS, W. *On the orbits of hyperhypersimple sets* ◇ D25 ◇
MENZEL, W. & SPERSCHNEIDER, V. *Recursively enumerable extensions of* R_1 *by finite functions* ◇ D20 D25 ◇
MILLER, DOUGLAS E. & REMMEL, J.B. *Effectively nowhere simple sets* ◇ D25 ◇
OMANADZE, R.SH. *Upper semilattice of recursively enumerable Q-degrees (Russian)* ◇ D25 ◇
PELZ, E. *Les debuts de la methode de priorite et les theorems de Friedberg-Muchnick* ◇ D25 ◇
SCHWARZ, S. *The quotient semilattice of the recursively enumerable degrees modulo the cappable degrees* ◇ D25 ◇
SELIVANOV, V.L. *On the hierarchy of limit computations (Russian)* ◇ D20 D25 ◇
SPERSCHNEIDER, V. *The length-problem* ◇ D20 D25 ◇
SPREEN, D. *On r.e. inseparability of cpo index sets* ◇ D25 D45 ◇
THURAISINGHAM, M.B. *System functions and their decision problems* ◇ D25 ◇
VALIDOV, F.I. *Recursively enumerable sets and discrete families of general recursive functions (Russian)* ◇ D25 ◇
WATNICK, R. *A generalization of Tennenbaum's theorem on effectively finite recursive linear orderings* ◇ C57 C65 D25 D45 ◇
WELCH, L.V. *A hierarchy of families of recursively enumerable degrees* ◇ D25 ◇

1985

AMBOS-SPIES, K. *Anti-mitotic recursively enumerable sets* ◇ D25 ◇
AMBOS-SPIES, K. *Generators of the recursively enumerable degrees* ◇ D25 ◇
ARSLANOV, M.M. *A class of hypersimple incomplete sets (Russian)* ◇ D25 ◇
ARSLANOV, M.M. *Effectively hyperimmune sets and majorants (Russian)* ◇ D25 ◇
ARSLANOV, M.M. *Families of recursively enumerable sets and their degrees of undecidability (Russian)* ◇ D25 ◇
COOPER, S.B. & MCEVOY, K. *On minimal pairs of enumeration degrees* ◇ D25 D30 ◇
CRESSWELL, M.J. *The decidable normal modal logics are not recursively enumerable* ◇ B25 B45 D25 D35 D80 ◇
DING, DECHENG *R.e. degrees of recursively inseparable sets (Chinese)* ◇ D25 ◇
DOWNEY, R.G. & HIRD, G.R. *Automorphisms of supermaximal subspaces* ◇ C57 C60 D25 D45 ◇
DOWNEY, R.G. *The degrees of r.e. sets without the universal splitting property* ◇ D25 ◇
FEJER, P.A. & SHORE, R.A. *Embeddings and extensions of embeddings in the r.e. tt and wtt - degrees* ◇ D25 ◇
GOLUNKOV, YU.V. & SAVEL'EV, A.A. *Systems of algorithmic algebras preserving ideals in a lattice of recursively enumerable sets (Russian) (English summary)* ◇ D25 ◇
GUREVICH, Y. & SHELAH, S. *The decision problem for branching time logic* ◇ B25 B45 C65 D05 D25 ◇
ISHMUKHAMETOV, SH.T. *Differences of recursively enumerable sets (Russian)* ◇ D25 ◇
JOCKUSCH JR., C.G. & MOHRHERR, J. *Embedding the diamond lattice in the recursively enumerable truth-table degrees* ◇ D25 ◇
JOCKUSCH JR., C.G. *Genericity for recursively enumerable sets* ◇ D25 E40 ◇
LERMAN, M. *On the ordering of classes in high/low hierarchies* ◇ D25 ◇
LERMAN, M. *The embedding problem for the recursively enumerable degrees* ◇ D25 ◇
LI, XIANG *Everywhere nonrecursive r.e. sets in recursively presented topological space (Chinese)* ◇ D25 D45 ◇
LI, XIANG *Turing degrees on pointwise r.e. open sets in effective Hausdorff spaces (Chinese)* ◇ D25 D30 D45 ◇
MAASS, W. *Major subsets and automorphisms of recursively enumerable sets* ◇ D25 ◇
MAASS, W. *Variations on promptly simple sets* ◇ D25 ◇
NERODE, A. & REMMEL, J.B. *Generic objects in recursion theory* ◇ D25 E40 ◇
PAOLA DI, R.A. *Creativity and effective inseparability in dominical categories* ◇ D25 G30 ◇
SOARE, R.I. *Tree arguments in recursion theory and the 0'''-priority method* ◇ D25 ◇
STOB, M. *Major subsets and the lattice of recursively enumerable sets* ◇ D25 ◇
TITOV, N.N. *On the finitely generatedness of some classes of recursively enumerable sets (Russian)* ◇ D25 ◇

D30 Other degrees; reducibilities

1939

TURING, A.M. *Systems of logic based on ordinals* ◇ B40 D25 D30 F15 F30 ◇

1948

POST, E.L. *Degrees of recursive unsolvability, preliminary report* ◇ D03 D30 D55 ◇

1953

KREISEL, G. *A variant to Hilbert's theory of the foundations of arithmetic* ◇ A05 B28 D30 D55 ◇

1954

JANICZAK, A. *On the reducibility of decision problems* ◇ D10 D20 D25 D30 ◇

KLEENE, S.C. & POST, E.L. *The upper semi-lattice of degrees of recursive unsolvability* ◇ D30 ◇

LACOMBE, D. *Sur le semi-reseau constitue par les degres d'indecidabilite recursive* ◇ D30 ◇

1955

KUZNETSOV, A.V. & TRAKHTENBROT, B.A. *Investigation of partial recursive operators by means of the theory of Baire's space (Russian)* ◇ D20 D25 D30 ◇

MEDVEDEV, YU.T. *Degrees of difficulty of the mass problem (Russian)* ◇ D30 D80 ◇

ROUTLEDGE, N.A. *Concerning definable sets* ◇ D30 D55 ◇

SPECTOR, C. *Recursive well-orderings* ◇ D30 D45 D55 F15 ◇

1956

MEDVEDEV, YU.T. *On the concept of the mass problem (Russian)* ◇ D30 D80 ◇

MUCHNIK, A.A. *On the unsolvability of the problem of reducibility in the theory of algorithms (Russian)* ◇ D25 D30 D35 ◇

SHAPIRO, N.Z. *Degrees of computability* ◇ D25 D30 ◇

SPECTOR, C. *On degrees of recursive unsolvability* ◇ D30 ◇

1957

FRIEDBERG, R.M. *A criterion for completeness of degrees of unsolvability* ◇ D30 ◇

RICE, H.G. *On the relative density of sets of integers* ◇ D20 D30 ◇

ROGERS JR., H. *Computing degrees of unsolvability* ◇ D30 D55 ◇

SHOENFIELD, J.R. *The non-enumerability of degrees of unsolvability* ◇ D30 ◇

SPECTOR, C. *Measure theory and higher order incomparability* ◇ D30 D55 ◇

1958

DEKKER, J.C.E. & MYHILL, J.R. *Retraceable sets* ◇ D25 D30 D50 ◇

MUCHNIK, A.A. *Isomorphism of systems of recursively enumerable sets with effective properties (Russian)* ◇ D25 D30 ◇

MUCHNIK, A.A. *Solution of Post's reduction problem and of certain other problems in the theory of algorithms I (Russian)* ◇ D25 D30 ◇

SHOENFIELD, J.R. *Degrees of formal systems* ◇ D25 D30 D35 D55 ◇

SPECTOR, C. *Measure-theoretic construction of imcomparable hyperdegrees* ◇ D30 D55 ◇

1959

FRIEDBERG, R.M. & ROGERS JR., H. *Reducibility and completeness for sets of integers* ◇ D30 ◇

KLEENE, S.C. *Recursive functionals and quantifiers of finite types I* ◇ D30 D55 D65 ◇

MEDVEDEV, YU.T. *On the concept of mass problem and its applications in the theory of recursive functions and in mathematical logic (Russian)* ◇ D30 D80 ◇

ROGERS JR., H. *Computing degrees of unsolvability* ◇ D30 D55 ◇

SHOENFIELD, J.R. *On degrees of unsolvability* ◇ D30 D55 ◇

USPENSKIJ, V.A. *On algorithmic reducibility (Russian)* ◇ D30 ◇

1960

GANDY, R.O. *On a problem of Kleene's* ◇ C62 D30 D55 ◇

SHOENFIELD, J.R. *An uncountable set of incomparable degrees* ◇ D30 ◇

SHOENFIELD, J.R. *Degrees of models* ◇ C57 D30 D35 ◇

ULLIAN, J.S. *Splinters of recursive functions* ◇ D25 D30 ◇

1961

FRAISSE, R. *Une notion de recursivite relative* ◇ D20 D30 D75 ◇

MYHILL, J.R. *Category methods in recursion theory* ◇ D30 ◇

MYHILL, J.R. *Note on degrees of partial functions* ◇ D30 ◇

SACKS, G.E. *A minimal degree less than O'* ◇ D30 ◇

SACKS, G.E. *On suborderings of degrees of recursive unsolvability* ◇ D30 ◇

1962

KLEENE, S.C. *λ-definable functionals of finite types* ◇ B40 D30 D55 D65 ◇

KLEENE, S.C. *Turing-machine computable functionals of finite types I* ◇ D10 D30 D55 D65 ◇
KLEENE, S.C. *Turing-machine computable functionals of finite types II* ◇ D10 D30 D55 D65 ◇
LIU, SHICHAO *Four types of general recursive well-orderings* ◇ D30 D45 D55 ◇
LIU, SHICHAO *Recursive linear orderings and hyperarithmetical functions* ◇ D30 D45 D55 ◇
SHOENFIELD, J.R. *Some applications of degrees* ◇ D25 D30 D35 D55 ◇

1963

ELLENTUCK, E. *Solution of a problem of R.Friedberg* ◇ D30 D50 ◇
FISCHER, P.C. *A note on bounded-truth-table reducibility* ◇ D30 ◇
KLEENE, S.C. *Recursive functionals and quantifiers of finite types II* ◇ D30 D55 D65 ◇
LIU, SHICHAO *A note an many-one reducibility* ◇ D30 ◇
LIU, SHICHAO *On many-one degrees* ◇ D30 ◇
MCLAUGHLIN, T.G. *A remark on semiproductive sets* ◇ D25 D30 ◇
MUCHNIK, A.A. *On strong and weak reducibility of algorithmic problems (Russian)* ◇ D30 ◇
SACKS, G.E. *Degrees of unsolvability* ◇ D25 D30 D98 E75 ◇
SACKS, G.E. *On the degrees less than 0′* ◇ D25 D30 ◇
SACKS, G.E. *Recursive enumerability and the jump operator* ◇ D25 D30 ◇
SMULLYAN, R.M. *Pseudo-uniform reducibility* ◇ D20 D25 D30 D35 ◇
WANG, HAO *Computation* ◇ D20 D25 D30 D35 D60 ◇

1964

ENDERTON, H.B. *Hierarchies in recursive function theory* ◇ D30 D55 ◇
ENDERTON, H.B. & LUCKHAM, D.C. *Hierarchies over recursive well-orderings* ◇ D30 D45 D55 ◇
HIROSE, K. *A theorem on incomparable degrees of recursive unsolvability* ◇ D25 D30 ◇
LACOMBE, D. *Deux generalisations de la notion de recursivite relative* ◇ D30 ◇
SINGLETARY, W.E. *A complex of problems proposed by Post* ◇ B20 D03 D25 D30 D35 ◇
SUZUKI, Y. *A complete classification of the Δ^1_2-functions* ◇ D30 D55 ◇
YOUNG, P. *A note on pseudo-creative sets and cylinders* ◇ D25 D30 ◇
YOUNG, P. *On reducibility by recursive functions* ◇ D30 ◇

1965

BOONE, W.W. *Finitely presented group whose word problem has the same degree as that of an arbitrarily given Thue system (application of methods of Britton)* ◇ D03 D30 D40 ◇
FISCHER, P.C. *Theory of provably recursive functions* ◇ D20 D30 F30 ◇
HENSEL, G. & PUTNAM, H. *On the notational independence of various hierarchies of degrees of unsolvability* ◇ D30 D55 F15 ◇
HIROSE, K. *On complete degrees* ◇ D30 ◇
HIROSE, K. *On the degree of unsolvability* ◇ D30 ◇
KREISEL, G. & SACKS, G.E. *Metarecursive sets* ◇ D30 D55 D60 ◇
LACHLAN, A.H. *On a problem of G.E. Sacks* ◇ D25 D30 ◇
LACHLAN, A.H. *Some notions of reducibility and productiveness* ◇ D25 D30 D55 ◇
MUCHNIK, A.A. *On the reducibility of problems of the solvability of enumerable sets to problems of separability (Russian)* ◇ D25 D30 ◇
SHEPHERDSON, J.C. *Machine configuration and word problems of given degree of unsolvability* ◇ D05 D30 D35 D40 ◇
SHOENFIELD, J.R. *Applications of model theory to degrees of unsolvability* ◇ C50 D25 D30 ◇
TITGEMEYER, D. *Untersuchungen ueber die Struktur des Kleene-Post'schen Halbverbandes der Grade der rekursiven Unloesbarkeit* ◇ D30 ◇
YOUNG, P. *On semi-cylinders, splinters, and bounded-truth-table reducibility* ◇ D25 D30 ◇

1966

BOONE, W.W. & ROGERS JR., H. *On a problem of J.H.C.Whitehead and a problem of Alonzo Church* ◇ D25 D30 D40 ◇
HAY, L. *Isomorphism types of index sets of partial recursive functions* ◇ D20 D25 D30 ◇
MARTIN, D.A. *Classes of recursively enumerable sets and degrees of unsolvability* ◇ D25 D30 ◇
MARTIN, D.A. *On a question of G.E.Sacks* ◇ D25 D30 ◇
MOSCHOVAKIS, Y.N. *Many-one degrees of the predicates $H_a(x)$* ◇ D30 D55 ◇
PAOLA DI, R.A. *Some properties of pseudo-complements of recursively enumerable sets* ◇ D25 D30 F30 ◇
RITTER, W.E. *Notation systems and an effective fixed point property* ◇ D20 D30 F15 ◇
SACKS, G.E. *Metarecursively enumerable sets and admissible ordinals* ◇ D30 D55 D60 ◇
SACKS, G.E. *Post's problem, admissible ordinals, and regularity* ◇ D30 D55 D60 ◇
SHOENFIELD, J.R. *A theorem on minimal degrees* ◇ D30 ◇
YATES, C.E.M. *On the degrees of index sets* ◇ D25 D30 D55 ◇
YOUNG, P. *Linear orderings under one-one reducibility* ◇ D30 ◇

1967

FRIDMAN, A.A. *Unloesbarkeitsgrade des Wortproblems fuer endlich-definierte Gruppen (Russisch)* ◇ D30 D40 ◇
GANDY, R.O. & SACKS, G.E. *A minimal hyperdegree* ◇ D30 D55 ◇
LIU, SHICHAO *Application of a general method for dealing with many-one degrees* ◇ D30 ◇
ROGERS JR., H. *Theory of recursive functions and effective computability* ◇ D25 D30 D55 D98 F15 ◇
SACKS, G.E. *Measure-theoretic uniformity in recursion theory and set theory (summary of results)* ◇ D30 D55 D60 E25 E35 E40 E50 ◇

SACKS, G.E. *Metarecursion theory* ◇ D30 D55 D60 ◇
SACKS, G.E. *On a theorem of Lachlan and Martin* ◇ D25 D30 ◇
THOMASON, S.K. *The forcing method and the upper semilattice of hyperdegrees* ◇ D30 E40 ◇
YATES, C.E.M. *Arithmetical sets and retracing functions* ◇ D30 D50 D55 ◇
YATES, C.E.M. *Recursively enumerable degrees and the degrees less than* $0^{(1)}$ ◇ D25 D30 ◇
YOUNG, P. *On pseudo-creative sets, splinters, and bounded truth-table reducibility* ◇ D25 D30 ◇

1968

BOKUT', L.A. *Degrees of unsolvability of the conjugacy problem for finitely-presented groups (Russian)* ◇ D30 D40 ◇
BOOLOS, G. & PUTNAM, H. *Degrees of unsolvability of constructible sets of integers* ◇ D30 D55 E45 ◇
DRISCOLL JR., G.C. *Metarecursively enumerable sets and their metadegrees* ◇ D30 D60 ◇
ERSHOV, YU.L. *On a hierarchy of sets I,II (Russian)* ◇ D30 D55 ◇
FUKUYAMA, M. *A note on incomparable degrees* ◇ D25 D30 ◇
FUKUYAMA, M. *A remark on minimal degrees* ◇ D30 ◇
GLADSTONE, M.D. *A single-axiom implicational calculus of given unsolvability* ◇ B20 D25 D30 D35 ◇
HERMAN, G.T. *Simulation of one abstract computing machine by another* ◇ D10 D25 D30 D55 ◇
HIROSE, K. *A conjecture on Hilbert's 10th problem* ◇ D30 D35 ◇
HIROSE, K. *An investigation on degrees of unsolvability* ◇ D30 ◇
JOCKUSCH JR., C.G. *Semirecursive sets and positive reducibility* ◇ D25 D30 ◇
JOCKUSCH JR., C.G. *Uniformly introreducible sets* ◇ D25 D30 ◇
LACHLAN, A.H. *Distributive initial segments of the degrees of unsolvability* ◇ D30 ◇
MAYOH, B.H. *Semi-effective numberings and definitions of the computable numbers* ◇ D25 D30 F60 ◇
MCLAUGHLIN, T.G. *A theorem on intermediate reducibilities* ◇ D30 ◇
ROSENSTEIN, J.G. *Initial segments of degrees* ◇ D30 ◇

1969

BOYD, R. & HENSEL, G. & PUTNAM, H. *A recursion-theoretic characterization of the ramified analytical hierarchy* ◇ D30 D55 ◇
ERSHOV, YU.L. *The theory of numerations. Part I: General theory of numeration (Russian)* ◇ D25 D30 D45 G30 ◇
HAY, L. *Index sets of finite classes of recursively enumerable sets* ◇ D25 D30 ◇
HERMES, H. *Basic notions and applications of the theory of decidability* ◇ D10 D20 D25 D30 D35 D50 D98 ◇
HINMAN, P.G. *Some applications of forcing to hierarchy problems in arithmetic* ◇ D25 D30 D55 E40 ◇
HUGILL, D.F. *Initial segments of Turing degrees* ◇ D30 ◇
JOCKUSCH JR., C.G. & MCLAUGHLIN, T.G. *Countable retracing functions and* Π_2^0 *predicates* ◇ D30 D55 ◇
JOCKUSCH JR., C.G. *Relationships between reducibilities* ◇ D25 D30 ◇
JOCKUSCH JR., C.G. *The degrees of bi-immune sets* ◇ D30 ◇
JOCKUSCH JR., C.G. *The degrees of hyperhyperimmune sets* ◇ D30 D50 ◇
LACHLAN, A.H. *Initial segments of one-one degrees* ◇ D30 ◇
LERMAN, M. *Some nondistributive lattices as initial segments of the degrees of unsolvability* ◇ D30 ◇
SACKS, G.E. *Measure-theoretic uniformity* ◇ D30 D55 E15 E25 E35 E40 E50 ◇
SACKS, G.E. *Measure-theoretic uniformity in recursion theory and set theory* ◇ D30 D55 D60 E15 E25 E35 E40 E50 ◇
SOARE, R.I. *A note on degrees of subsets* ◇ D30 ◇
SOARE, R.I. *Cohesive sets and recursively enumerable Dedekind cuts* ◇ D25 D30 ◇
SOARE, R.I. *Constructive order types on cuts* ◇ D30 D45 ◇
SOARE, R.I. *Recursion theory and Dedekind cuts* ◇ D25 D30 F60 ◇
SOARE, R.I. *Sets with no subset of higher degree* ◇ D30 ◇
THOMASON, S.K. *A note on non-distributive sublattices of degrees and hyperdegrees* ◇ D30 E40 ◇
YATES, C.E.M. *On the degrees of index sets II* ◇ D25 D30 D55 ◇

1970

CASE, J. *Enumeration reducibility and partial degrees* ◇ D25 D30 ◇
ERSHOV, YU.L. *Index sets (Russian)* ◇ D25 D30 D55 ◇
ERSHOV, YU.L. *On a hierarchy of sets III (Russian)* ◇ D30 D55 F15 ◇
FEINER, L. *The strong homogeneity conjecture* ◇ D30 ◇
GANDY, R.O. & SOARE, R.I. *A problem in the theory of constructive order types* ◇ D30 D45 ◇
HARTMANIS, J. & HOPCROFT, J.E. *What makes some language theory problems undecidable* ◇ D05 D30 ◇
JENSEN, R.B. *Definable sets of minimal degree* ◇ D30 D55 E35 E45 E47 ◇
JOCKUSCH JR., C.G. & SOARE, R.I. *Minimal covers and arithmetical sets* ◇ D30 D55 ◇
KUSHNER, B.A. *Some mass problems connected with the integration of constructive functions (Russian)* ◇ D30 F60 ◇
LACHLAN, A.H. *Initial segments of many-one degrees* ◇ D30 ◇
LERMAN, M. *Recursive functions modulo co-r-maximal sets* ◇ D25 D30 ◇
LERMAN, M. *Turing degrees and many-one degrees of maximal sets* ◇ D25 D30 ◇
PLATEK, R.A. *A note on the cardinality of the Medvedev lattice* ◇ D30 ◇
POULSEN, B.T. *The Medvedev lattice of degrees of difficulty* ◇ D30 D80 ◇

Sasso Jr., L.P. *A cornucopia of minimal degrees* ◇ D30 ◇
Thomason, S.K. *A theorem on initial segments of degrees* ◇ D30 ◇
Thomason, S.K. *On initial segments of hyperdegrees* ◇ D30 D35 E40 ◇
Thomason, S.K. *Sublattices and initial segments of the degrees of unsolvability* ◇ D30 ◇
Trakhtenbrot, B.A. *On autoreducibility (Russian)* ◇ D30 ◇
Yates, C.E.M. *Initial segments of the degrees of unsolvability I: a survey* ◇ D30 ◇
Yates, C.E.M. *Initial segments of the degrees of unsolvability - part II; minimal degrees* ◇ D30 ◇

1971

Aanderaa, S.O. & Belsnes, D. *Decision problems for tag systems* ◇ D03 D25 D30 ◇
Bloom, S.L. *Some remarks on uniform halting problems* ◇ D10 D30 ◇
Collins, D.J. *Truth-tables degrees and the Boone group* ◇ D30 D40 ◇
Hughes, C.E. & Overbeek, R.A. & Singletary, W.E. *The many-one equivalence of some general combinatorial decision problems* ◇ D03 D05 D10 D25 D30 ◇
Jockusch Jr., C.G. & Soare, R.I. *A minimal pair of Π^0_1 classes* ◇ D25 D30 ◇
Khutoretskij, A.B. *On the cardinality of the upper semilattice of computable enumerations (Russian)* ◇ D25 D30 D45 ◇
Kreisel, G. *Some reasons for generalizing recursion theory* ◇ A05 D30 D60 D65 D70 D75 ◇
Lachlan, A.H. *Solution to a problem of Spector* ◇ D20 D30 ◇
Lerman, M. *Initial segments of the degrees of unsolvability* ◇ D30 ◇
Lerman, M. *Some theorems on r-maximal sets and major subsets of recursively enumerable sets* ◇ D25 D30 ◇
Levy, S. *Computational equivalence* ◇ D10 D30 D55 ◇
Manaster, A.B. *Some contrasts between degrees and the arithmetical hierarchy* ◇ D30 D55 ◇
Owings Jr., J.C. *A splitting theorem for simple Π^1_1 sets* ◇ D30 D55 D60 ◇
Sacks, G.E. *F-recursiveness* ◇ D30 D55 D60 D75 E40 E45 ◇
Sacks, G.E. *On the reducibility of Π^1_1 sets* ◇ D30 D55 D60 E40 ◇
Selman, A.L. *Arithmetical reducibilities I* ◇ D30 D55 ◇
Shoenfield, J.R. *Degrees of unsolvability* ◇ D25 D30 D98 ◇

1972

Apt, K.R. *ω-models in analytical hierarchy (Russian summary)* ◇ C62 D30 D55 ◇
Berry, J.W. *A note on immune sets* ◇ D30 D50 ◇
Cleave, J.P. *Combinatorial systems I. Cylindrical problems* ◇ D30 ◇
Collins, D.J. *Representation of Turing reducibility by word and conjugacy problems in finitely presented groups* ◇ D30 D40 ◇
Cooper, S.B. *Degrees of unsolvability complementary between recursively enumerable degrees I* ◇ D25 D30 ◇
Cooper, S.B. *Jump equivalence of the Δ^0_2 hyperimmune sets* ◇ D25 D30 D55 ◇
Cooper, S.B. *Minimal upper bounds for sequences of recursively enumerable degrees* ◇ D25 D30 ◇
Degtev, A.N. *Hereditary sets and truth-table reducibility (Russian)* ◇ D30 ◇
Grilliot, T.J. *Omitting types; applications to recursion theory* ◇ C07 C62 D30 D55 D60 D75 ◇
Hay, L. *A discrete chain of degrees of index sets* ◇ D25 D30 ◇
Jockusch Jr., C.G. & Soare, R.I. *Π^0_1 classes and degrees of theories* ◇ D25 D30 ◇
Jockusch Jr., C.G. *A reducibility arising from the Boone groups* ◇ D25 D30 D40 ◇
Jockusch Jr., C.G. *Degrees in which the recursive sets are uniformly recursive* ◇ D25 D30 ◇
Jockusch Jr., C.G. & Soare, R.I. *Degrees of members of Π^0_1 classes* ◇ D25 D30 F30 ◇
Jockusch Jr., C.G. *Ramsey's theorem and recursion theory* ◇ D30 D55 D80 ◇
Jockusch Jr., C.G. *Upward closure of bi-immune degrees* ◇ D30 D50 ◇
Kanovich, M.I. *On the universality of strongly undecidable sets (Russian)* ◇ D15 D20 D25 D30 ◇
Kreczmar, A. *Degree of recursive unsolvability of algorithmic logic (Russian summary)* ◇ B75 D30 D35 ◇
Lerman, M. *On suborderings of the α-recursively enumerable α-degrees* ◇ D30 D60 ◇
Lerman, M. & Sacks, G.E. *Some minimal pairs of α-recursively enumerable degrees* ◇ D30 D60 ◇
Macintyre, A. *Omitting quantifier-free types in generic structures* ◇ C25 C57 C60 C75 D30 D40 ◇
Manaster, A.B. & Rosenstein, J.G. *Effective matchmaking (recursion theoretic aspects of a theorem of Philip Hall)* ◇ D30 D80 E05 ◇
Metakides, G. *α-degrees of α-theories* ◇ D30 D60 ◇
Mostowski, Andrzej *A transfinite sequence of ω-models* ◇ C62 D30 ◇
Polyakov, E.A. *On relative recursiveness and computability (Russian)* ◇ D30 D65 ◇
Sacerdote, G.S. *On a problem of Boone* ◇ C60 D30 D35 D40 ◇
Sacks, G.E. & Simpson, S.G. *The α-finite injury method* ◇ D30 D60 ◇
Selman, A.L. *Applications of forcing to the degree-theory of the arithmetical hierarchy* ◇ D30 D55 E40 ◇
Selman, A.L. *Arithmetical reducibilities II* ◇ D30 D55 ◇
Shore, R.A. *Minimal α-degrees* ◇ D30 D60 ◇
Stillwell, J.C. *Decidability of the "almost all" theory of degrees* ◇ B25 C80 D30 ◇

TANAKA, H. *A note on hyperdegrees* ◇ D30 D55 ◇
YATES, C.E.M. *Degrees of unsolvability* ◇ D30 ◇
YATES, C.E.M. *Initial segments and implications for the structure of degrees* ◇ D30 ◇

1973

BERNARDI, C. & GNANI, G. *Un'osservazione sulle classi (English summary)* ◇ D30 ◇
CLEAVE, J.P. *Combinatorial systems III. Degrees of combinatorial problems of computing machines* ◇ D10 D25 D30 ◇
COOPER, S.B. *Minimal degrees and the jump operator* ◇ D30 ◇
DEGTEV, A.N. *tt and m degrees (Russian)* ◇ D25 D30 ◇
ELLENTUCK, E. *Degrees of isolic theories* ◇ D30 D50 ◇
ELLENTUCK, E. *On the degrees of universal regressive isols* ◇ D30 D50 ◇
ERSHOV, YU.L. *Hierarchy of the sets of class* Δ_2^0 *(Russian) (English summary)* ◇ D30 D55 ◇
FEINER, L. *Degrees of nonrecursive presentability* ◇ C57 D30 D45 ◇
FRIEDMAN, H.M. Σ_1^1 *degree determinateness fails in all boolean extensions* ◇ D30 D55 E15 E40 E60 ◇
FRIEDMAN, H.M. *Borel sets and hyperdegrees* ◇ D30 D55 E15 ◇
HAY, L. *Discrete* ω *-sequences of index sets (1)* ◇ D25 D30 ◇
HAY, L. *Index sets in* $\underline{0}'$ *(Russian)* ◇ D25 D30 D55 ◇
HAY, L. *The class of recursively enumerable subsets of a recursively enumerable set* ◇ D25 D30 ◇
HAY, L. *The halting problem relativized to complements* ◇ D25 D30 ◇
HINMAN, P.G. *Degrees of continuous functionals* ◇ D30 D65 ◇
HUGHES, C.E. & SINGLETARY, W.E. *Combinatorial systems with axiom* ◇ D03 D05 D10 D25 D30 ◇
JOCKUSCH JR., C.G. *An application of* Σ_4^0 *determinacy to the degrees of unsolvability* ◇ D30 ◇
JOCKUSCH JR., C.G. & SOARE, R.I. *Encodability of Kleene's O* ◇ D30 D55 D65 ◇
JOCKUSCH JR., C.G. *Upward closure and cohesive degrees* ◇ D30 ◇
KANOVICH, M.I. *Irreducibility of languages of the stepped semantic system (Russian)* ◇ D15 D30 F50 ◇
KLEINBERG, E.M. *A characterization of determinacy for Turing degree games* ◇ D30 E60 ◇
KOBZEV, G.N. *btt-reducibility I,II (Russian)* ◇ D25 D30 ◇
KOBZEV, G.N. *Pointwise decomposable sets (Russian)* ◇ D20 D25 D30 ◇
MACINTYRE, J.M. *Minimal* α *-recursion theoretic degrees* ◇ D30 D60 ◇
MACINTYRE, J.M. *Noninitial segments of the* α *-degrees* ◇ D30 D60 ◇
MARCHENKOV, S.S. *The existence of families without positive numerations (Russian)* ◇ D20 D30 D45 ◇
MOSTOWSKI, A.WLODZIMIERZ *Uniform algorithm for deciding group-theoretic problems* ◇ D30 D40 ◇
MUZYUKINA, G.I. *mu -degrees of unsolvability of functions (Russian)* ◇ D30 ◇
OVERBEEK, R.A. *The representation of many-one degrees by the word problem for Thue systems* ◇ D03 D30 D40 ◇
PUTNAM, H. *Recursive functions and hierarchies* ◇ D20 D30 D35 D55 ◇
SASSO JR., L.P. *A minimal partial degree* $\leqslant 0'$ ◇ D30 ◇
SELMAN, A.L. *Sets of formulas valid in finite structures* ◇ C13 D30 ◇
STRIGIN, YU.D. *Hierarchies of general recursive operators and general recursive types of degrees of noncomputability (Russian)* ◇ D20 D30 ◇
STRIGIN, YU.D. *The hierarchy of general recursive functionals (Russian)* ◇ D20 D30 ◇
TANAKA, H. Π_1^1 *sets of sets, hyperdegrees and related problems* ◇ D30 D55 ◇
TRAKHTENBROT, B.A. *Autoreducible and nonautoreducible predicates and sets (Russian)* ◇ D25 D30 ◇

1974

ALTAEV, A.V. *Operators defined by* γ *-enumeration and related reducibilities (Russian)* ◇ D20 D30 ◇
ANDZHAN, A.V. *Reduzierbarkeit von Funktionenklassen nach der Prognostizierung (Russisch)* ◇ D20 D30 ◇
BERNARDI, C. *Aspetti ricorsivi degli insiemi dialettici (English summary)* ◇ D30 D35 G10 ◇
BORZOV, YU.V. *Holographische Mengen (Russisch)* ◇ D20 D30 ◇
BOZOVIC, I.B. & BOZOVIC, N.B. *On some unrecognizable relations among groups* ◇ D30 D40 ◇
CASE, J. *Maximal arithmetical reducibilities* ◇ D30 D55 ◇
CHONG, C.T. *Almost local non-* α *-recursiveness* ◇ D30 D60 ◇
COOPER, S.B. *An annotated bibliography for the structure of the degrees below 0' with special reference to that of the recursively enumerable degrees* ◇ A10 D25 D30 ◇
DEKKER, J.C.E. & ELLENTUCK, E. *Recursion relative to regressive functions* ◇ D30 D50 ◇
DENISOV, S.D. *Three theorems on elementary theories and tt-reducibility (Russian)* ◇ C65 D30 ◇
DUSHSKIJ, V.A. *Enumerational reducibility of recursively enumerable classes of sets (Russian)* ◇ D25 D30 D45 ◇
FRIEDMAN, H.M. *Minimality in the* Δ_2^1*-degrees* ◇ D30 D55 E45 ◇
GILL III, J.T. & MORRIS, P.H. *On subcreative sets and s-reducibility* ◇ D25 D30 ◇
GUASPARI, D. *Characterizing the constructible reals* ◇ D30 D55 E15 E45 ◇
HAJEK, P. *Degrees of dependence in the theory of semisets* ◇ D30 E40 E45 E70 ◇
HAY, L. *A noninitial segment of index sets* ◇ D25 D30 ◇
HAY, L. *Index sets universal for differences of arithmetic sets* ◇ D25 D30 D55 ◇
JOCKUSCH JR., C.G. Π_1^0 *classes and boolean combinations of recursively enumerable sets* ◇ D25 D30 ◇

KANOVEJ, V.G. *On degrees of constructibility and descriptive properties of the set of real numbers in an initial model and in its extensions (Russian)* ◇ C62 D30 E40 E45 ◇
KOBZEV, G.N. *On complete Btt-degrees* ◇ D25 D30 ◇
LERMAN, M. *Least upper bounds for minimal pairs of α-r.e. α-degrees* ◇ D30 D60 ◇
LUKAS, J.D. & PUTNAM, H. *Systems of notations and the ramified analytical hierarchy* ◇ D30 D55 E45 F15 F35 ◇
MACHTEY, M. *Minimal degrees in generalized recursion theory* ◇ D30 D60 ◇
MCLAUGHLIN, T.G. *Degrees of unsolvability within a regressive isol* ◇ D30 D50 ◇
NELSON, GEORGE C. *Many-one reducibility within the Turing degrees of the hyperarithmetic sets* $H_a(x)$ ◇ D30 D55 ◇
POLYAKOV, E.A. & ROZINAS, M.G. *Some algebraic problems in the theory of reducibilities of sets (Russian)* ◇ D30 ◇
RAYNA, G. *Degrees of finite-state transformability* ◇ D05 D30 ◇
ROGUSKI, S. *Degrees of nonconstructibility in Cohen's model* ◇ D30 E40 E45 ◇
ROZINAS, M.G. *Partial degrees and r-degrees (Russian)* ◇ D30 ◇
SASSO JR., L.P. *A minimal degree not realizing least possible jump* ◇ D30 ◇
SASSO JR., L.P. *Deficiency sets and bounded information reducibilities* ◇ D25 D30 ◇
SELMAN, A.L. *Relativized halting problems* ◇ D10 D25 D30 ◇
SHORE, R.A. Σ_n *sets which are* Δ_n*-incomparable (uniformly)* ◇ D30 D55 D60 E40 E47 ◇
SIMPSON, S.G. *Degree theory on admissible ordinals* ◇ D30 D60 ◇
SIMPSON, S.G. *Post's problem for admissible sets* ◇ D30 D60 ◇
SOLOV'EV, V.D. *Q-reducibility and hypersimple sets (Russian)* ◇ D25 D30 ◇
STEFANI, S. *Gradi di insolubilita e limiti* ◇ D30 ◇
THARP, L.H. *Continuity and elementary logic* ◇ C07 C80 C95 D30 D55 ◇
VUCKOVIC, V. *Almost recursivity and partial degrees* ◇ D25 D30 ◇
YATES, C.E.M. *Prioric games and minimal degrees below* $\underline{0}^{(1)}$ ◇ D25 D30 E60 ◇

1975

ADAMOWICZ, Z. *An observation on the product of Silver's forcing* ◇ D30 E40 E45 ◇
CLEAVE, J.P. *Combinatorial systems II. Non-cylindrical problems* ◇ D30 ◇
DUJOLS, R. *Une methode de construction directe de couples d'ensembles 1-reductibles ayant des types d'equivalence recursive incomparables (English summary)* ◇ D30 D50 ◇
EPSTEIN, R.L. *Minimal degrees of unsolvability and the full approximation construction* ◇ D30 ◇
GRIGORIEFF, S. *Minimalite des reels definis par "forcing" sur certaines familles d'arbres de suites finies d'entiers (English summary)* ◇ D30 E40 E45 ◇
HARRINGTON, L.A. & KECHRIS, A.S. *A basis result for* Σ^0_3 *sets of reals with an application to minimal covers* ◇ D30 D55 E15 E60 ◇
HAY, L. *Rice theorems for d.r.e. sets* ◇ D25 D30 D55 ◇
HAY, L. & MANASTER, A.B. & ROSENSTEIN, J.G. *Small recursive ordinals, many-one degrees, and the arithmetical difference hierarchy* ◇ D25 D30 D55 ◇
HAY, L. *Spectra and halting problems* ◇ C13 D10 D25 D30 ◇
KALLIBEKOV, S. *Hierarchy of sets for truth-table enumerability (Russian)* ◇ D30 D55 ◇
KANOVEJ, V.G. *The majorization of initial segments of degrees of constructivity (Russian)* ◇ D30 E40 E45 ◇
LACHLAN, A.H. *Uniform enumeration operations* ◇ D25 D30 ◇
LACHLAN, A.H. *Wtt-complete sets are not necessarily tt-complete* ◇ D25 D30 ◇
MARQUES, I. *On degrees of unsolvability and complexity properties* ◇ D15 D20 D30 ◇
MCLAUGHLIN, T.G. *A note concerning the* $\overset{*}{\vee}$ *relation on* Λ_R ◇ D25 D30 D50 ◇
NORMANN, D. *Forcing arguments and some degree-theoretic problems in higher type recursion theory* ◇ D30 D65 E40 ◇
REEDY, A. & SAVITCH, W.J. *The Turing degree of the inherent ambiguity problem for context-free languages* ◇ D05 D30 ◇
SASSO JR., L.P. *A survey of partial degrees* ◇ D30 ◇
SHORE, R.A. *Splitting an α-recursively enumerable set* ◇ D30 D60 ◇
SHORE, R.A. *The irregular and non-hyperregular α-r.e. degrees* ◇ D30 D60 ◇
SIMPSON, S.G. *Minimal covers and hyperdegrees* ◇ D30 D55 ◇
STEEL, J.R. *Descending sequences of degrees* ◇ C62 D30 ◇
STERN, J. *Some measure theoretic results in effective descriptive set theory* ◇ D30 D55 E15 E60 ◇

1976

ADAMOWICZ, Z. *On finite lattices of degrees of constructibility of reals* ◇ D30 E35 E45 ◇
BUSCH, D.R. *On the number of Solovay r-degrees* ◇ D30 D65 E15 ◇
CARSTENS, H.G. Δ^0_2*-Mengen* ◇ D25 D30 D55 ◇
CHONG, C.T. *An α-finite injury method of the unbounded type* ◇ D30 D60 ◇
CHONG, C.T. *Minimal upper bounds for ascending sequences of α-recursively enumerable degrees* ◇ D30 D60 ◇
CRISCUOLO, G. & GERLA, G. *tt-riducibilita e limiti ricorsivi* ◇ D30 ◇
DEGTEV, A.N. *Minimal 1-degrees, and truth-table reducibility (Russian)* ◇ D25 D30 ◇

DO MINH THAI *Programmkomplexitaet rekursiv aufzaehlbarer Mengen* ◇ D15 D25 D30 ◇
DYMENT, E.Z. *On some properties of the Medvedev lattice (Russian)* ◇ D30 ◇
ELLENTUCK, E. *Decomposable isols and their degrees* ◇ D30 D50 E10 ◇
FRIEDMAN, H.M. *Recursiveness in Π^1_1 paths through* $\mathcal{O}$ ◇ D30 D55 F15 ◇
FRIEDMAN, H.M. *Uniformly defined descending sequences of degrees* ◇ D30 D55 F30 F35 ◇
GRANDJEAN, E. *Une notion de recursivite relative qui correspond a la recursivite au sens de Trahtenbrot (English summary)* ◇ D20 D30 ◇
JOCKUSCH JR., C.G. & SIMPSON, S.G. *A degree theoretic definition of the ramified analytical hierarchy* ◇ D30 D55 E40 E45 H05 ◇
JOCKUSCH JR., C.G. & PATERSON, M.S. *Completely autoreducible degrees* ◇ D25 D30 ◇
KLETTE, R. *Indexmengen und Erkennung rekursiver Funktionen* ◇ D20 D25 D30 D55 ◇
KOBZEV, G.N. *About 1-degrees* ◇ D30 ◇
LACHLAN, A.H. & LEBEUF, R. *Countable initial segments of the degrees of unsolvability* ◇ D30 ◇
MARANDZHYAN, G.B. *On weakly positive degrees of the sets of minimal indices of algorithms (Russian)* ◇ D20 D30 ◇
MARCHENKOV, S.S. *On the congruence of the upper semilattices of recursively enumerable m-powers and tabular powers (Russian)* ◇ D25 D30 ◇
MARTIN, D.A. *Proof of a conjecture of Friedman* ◇ D30 D55 ◇
MORRIS, P.H. *A reducibility condition for recursiveness* ◇ D20 D30 ◇
OMANADZE, R.SH. *On one kind of reducibility (Russian) (English summary)* ◇ D30 ◇
POLYAKOV, E.A. & ROZINAS, M.G. *Functional degrees (Russian)* ◇ D30 ◇
POLYAKOV, E.A. & ROZINAS, M.G. *Theory of algorithms (Russian)* ◇ D20 D30 D98 ◇
REMMEL, J.B. *Combinatorial functors on co-r.e. structures* ◇ D25 D30 D45 D50 ◇
ROZINAS, M.G. *Jump operation for certain kinds of reducibilities (Russian)* ◇ D30 ◇
SACKS, G.E. *Countable admissible ordinals and hyperdegrees* ◇ D30 D55 D60 ◇
SAZONOV, V.YU. *Degrees of parallelism in computations* ◇ B40 B75 D30 D65 ◇
SHOENFIELD, J.R. *Degrees of classes of RE sets* ◇ D25 D30 ◇
SHORE, R.A. *On the jump of an α-recursively enumerable set* ◇ D30 D60 ◇
SHORE, R.A. *The recursively enumerable α-degrees are dense* ◇ D30 D60 ◇
SOLON, B.YA. *On non-minimal pm-degrees and pc-degrees (Russian)* ◇ D30 ◇
SOLOV'EV, V.D. *Some generalizations of the notions of reducibility and creativity (Russian)* ◇ D25 D30 ◇
V'YUGIN, V.V. *On Turing invariant sets (Russian)* ◇ D30 ◇
YATES, C.E.M. *Banach-Mazur games, comeager sets and degrees of unsolvability* ◇ D30 E60 E75 ◇

1977

ADAMOWICZ, Z. *Constructible semi-lattices of degrees of constructibility* ◇ D30 E35 E45 ◇
ADAMOWICZ, Z. *On finite lattices of degrees of constructibility* ◇ D30 E35 E45 ◇
ANDZHAN, A.V. *On a reducability of sets (Russian)* ◇ D30 ◇
BERRY, J.W. *Enumeration degrees and strong reducibilities* ◇ D30 ◇
BIENENSTOCK, E. *Sets of degrees of computable fields* ◇ C57 C60 D30 D45 ◇
CHAITIN, G.J. *Program size, oracles, and the jump operation* ◇ D15 D25 D30 D80 ◇
CHONG, C.T. *The minimal α-degree problem* ◇ D30 D60 ◇
HARRINGTON, L.A. & KECHRIS, A.S. *Π^1_2 singletons and $0^{\#}$* ◇ D30 D55 E45 E55 ◇
HAY, L. & MANASTER, A.B. & ROSENSTEIN, J.G. *Concerning partial recursive similarity transformations of linearly ordered sets* ◇ C57 C65 D20 D25 D30 D45 E07 ◇
HUGHES, C.E. & SINGLETARY, W.E. *The one-one equivalence of some general combinatorial decision problems* ◇ D30 D40 ◇
JOCKUSCH JR., C.G. & SOLOVAY, R.M. *Fixed points of jump preserving automorphisms of degrees* ◇ D30 ◇
JOCKUSCH JR., C.G. *Simple proofs of some theorems on high degrees of unsolvability* ◇ D30 ◇
JOHNSON, N. *Rice theorems for Σ^{-1}_n sets* ◇ D25 D30 D55 ◇
KANOVICH, M.I. *On the precision of a complexity criterion for nonrecursiveness and universality (Russian)* ◇ D15 D20 D25 D30 ◇
KINBER, E.B. *On btt-degrees of sets of minimal numbers in Goedel numberings* ◇ D20 D30 D45 ◇
KINBER, E.B. *On Turing degrees of hypersimple Post sets (Russian)* ◇ D25 D30 ◇
LIH, KUOWEI *Continuous degrees* ◇ D30 D65 ◇
MAASS, W. *On minimal pairs and minimal degrees in higher recursion theory* ◇ D30 D60 D65 ◇
MACINTYRE, J.M. *Transfinite extensions of Friedberg's completeness criterion* ◇ D30 D55 D60 ◇
MCLAUGHLIN, T.G. *Degrees of unsolvability and strong forms of $\Lambda_R + \Lambda_R \nsubseteq \Lambda_R$* ◇ D25 D30 D50 ◇
MOKATSYAN, A.A. *Some properties of relativizated Kolmogorov complexities (Russian)* ◇ D15 D30 ◇
ODIFREDDI, P. *A note on Suzuki's chain of hyperdegrees* ◇ D30 D55 E15 ◇
PARIS, J.B. *Measure and minimal degrees* ◇ D30 E75 ◇
POLYAKOV, E.A. & ROZINAS, M.G. *Enumeration reducibilities (Russian)* ◇ D30 ◇
ROZINAS, M.G. *Minimal non-recursively enumerable pm-degrees (Russian)* ◇ D30 ◇
SCHLIPF, J.S. *Ordinal spectra of first-order theories* ◇ C35 C50 C55 C62 C70 D15 D30 D70 ◇

SHORE, R.A. *α-recursion theory* ◇ D30 D60 E45 ◇
SIMPSON, S.G. *Degrees of unsolvability: A survey of results* ◇ D25 D30 D35 F35 ◇
SIMPSON, S.G. *First order theory of the degrees of recursive unsolvability* ◇ D30 D35 F35 ◇
SKANDALIS, K. *Finite lattices of degrees of definability* ◇ D30 E35 E45 ◇
SMOLSKA-ADAMOWICZ, Z. *On finite lattices of the degrees of constructibility of reals* ◇ D30 E35 E45 ◇
SOLON, B.YA. *Reducibility according to computability and e-interreducible sets (Russian)* ◇ D30 ◇

1978

AJZENSHTEJN, M.KH. *Ordering of complexity measures according to the effectiveness (Russian)* ◇ D15 D30 ◇
BALCAR, B. & HAJEK, P. *On sequences of degrees of constructibility (solution of Friedman's Problem 75)* ◇ D30 E45 ◇
BARENDREGT, H.P. & BERGSTRA, J.A. & KLOP, J.W. & VOLKEN, H. *Degrees of sensible lambda theories* ◇ B40 D30 ◇
BENNISON, V.L. & SOARE, R.I. *Some lowness properties and computational complexity sequences* ◇ D15 D25 D30 ◇
BERGSTRA, J.A. *Degrees of partial functions* ◇ D30 ◇
DALEY, R.P. *On the simplicity of busy beaver sets* ◇ D10 D15 D25 D30 ◇
DALEY, R.P. *Quantitative and qualitative information in computations* ◇ D15 D30 ◇
DEGTEV, A.N. *Three theorems on tt-degrees (Russian)* ◇ D25 D30 ◇
DEMUTH, O. & KRYL, R. & KUCERA, A. *An application of the theory of functionals partial recursive relative to number sets in constructive mathematics (Russian)* ◇ D30 D55 F60 ◇
EPSTEIN, R.L. & POSNER, D.B. *Diagonalization in degree constructions* ◇ D30 ◇
FRIEDMAN, S.D. *Negative solutions to Post's problem I* ◇ D30 D60 ◇
GAVRILOV, S. *Relations between different forms of relative computability of everywhere defined functions (Russian)* ◇ D30 ◇
GOODMAN, NICOLAS D. *The nonconstructive content of sentences of arithmetic* ◇ D20 D30 D55 F30 F50 ◇
HAJEK, P. *Some results on degrees of constructibility* ◇ D30 D55 E45 E55 ◇
HAY, L. *Convex subsets of 2^n and bounded truth-table reducibility* ◇ B05 D30 ◇
HEBEISEN, F. *Charakterisierung der Aufzaehlungsreduzierbarkeit* ◇ D30 ◇
HODES, H.T. *Uniform upper bounds on ideals of Turing degrees* ◇ D30 ◇
HRBACEK, K. *On the complexity of analytic sets* ◇ D30 D55 D65 E15 E45 ◇
JOCKUSCH JR., C.G. & POSNER, D.B. *Double jumps of minimal degrees* ◇ D30 ◇
JOHNSON, N. *Classifications of generalized index sets of open classes* ◇ D25 D30 ◇
KECHRIS, A.S. *Minimal upper bounds for sequences of Δ^1_{2n}-degrees* ◇ D30 D55 E60 ◇
KOBZEV, G.N. *On the semi-lattice of tt-degrees (Russian) (English summary)* ◇ D25 D30 ◇
LEGGETT, A. *α-degrees of maximal α-r.e. sets* ◇ D30 D60 ◇
LIH, KUOWEI *Type two partial degrees* ◇ D30 D65 ◇
MAASS, W. *Contributions to α- and β-recursion theory* ◇ D25 D30 D60 E45 ◇
MAASS, W. *High α-recursively enumerable degrees* ◇ D30 D60 ◇
MAASS, W. *The uniform regular set theorem in α-recursion theory* ◇ D30 D60 E45 ◇
MANSFIELD, R. *A footnote to a theorem of Solovay on recursive encodability* ◇ D30 D55 ◇
NELSON, GEORGE C. *Isomorphism types of the hyperarithmetic sets H_a* ◇ D30 D50 D55 ◇
NORMANN, D. & STOLTENBERG-HANSEN, V. *A non-adequate admissible set with a good degree-structure* ◇ D30 D60 ◇
OMANADZE, R.SH. *On the reducibility on the class of recursive enumerable sets (Russian) (English summary)* ◇ D25 D30 ◇
POLYAKOV, E.A. & ROZINAS, M.G. *Correlations between different forms of relative computability of functions (Russian)* ◇ D20 D30 ◇
POLYAKOV, E.A. (ED.) *Recursive functions (Russian)* ◇ D20 D30 D97 ◇
POLYAKOV, E.A. *Some remarks on the relative recursiveness (Russian)* ◇ D30 ◇
ROZINAS, M.G. *On the semilattice of e-degrees (Russian)* ◇ D30 ◇
ROZINAS, M.G. *Partial degrees of immune and hyperimmune sets (Russian)* ◇ D30 D50 ◇
SANCHIS, L.E. *Hyperenumeration reducibility* ◇ D30 D55 ◇
SCHMERL, J.H. *A decidable $\aleph_0$-categorical theory with a non-recursive Ryll-Nardzewski function* ◇ B25 C35 C57 D30 ◇
SELIVANOV, V.L. *On index sets of computable classes of finite sets (Russian)* ◇ D25 D30 D55 ◇
SHORE, R.A. *On the $\forall\exists$-sentences of α-recursion theory* ◇ D30 D60 ◇
SHORE, R.A. *Some more minimal pairs of α-recursively enumerable degrees* ◇ D30 D60 ◇
SIMPSON, S.G. *Sets which do not have subsets of every higher degree* ◇ D30 D55 ◇
SIMPSON, S.G. *Short course on admissible recursion theory* ◇ D30 D60 ◇
SMIKUN, L.B. *Degrees of unsolvability of algorithmic problems connected with automata operations on groups (Russian)* ◇ D05 D30 ◇
SOLON, B.YA. *e-powers of hyperimmune retraceable sets (Russian)* ◇ D25 D30 D50 ◇
SOLON, B.YA. *Quasiminimal pF-degrees (Russian)* ◇ D30 ◇
STEEL, J.R. *Forcing with tagged trees* ◇ C62 D30 D55 E40 F35 ◇

TRUSS, J.K. *A note on increasing sequences of constructibility degrees* ◇ D30 E45 ◇
WESEP VAN, R.A. *Wadge degrees and descriptive set theory* ◇ D30 E15 E60 ◇

1979

ABRAMSON, F.G. *Sacks forcing does not always produce a minimal upper bound* ◇ D30 D60 E40 ◇
BERGSTRA, J.A. *Recursion theory on processes* ◇ D10 D30 D75 ◇
BOERGER, E. *A new general approach to the theory of the many-one equivalence of decision problems for algorithmic systems* ◇ B25 D03 D05 D10 D25 D30 D40 ◇
BUDINAS, B.L. *Three linearly ordered degrees of constructibility of Δ^1_3 numbers (Russian)* ◇ D30 D55 E35 E45 ◇
CHONG, C.T. *Cones of degrees* ◇ D30 D60 E45 ◇
CHONG, C.T. *Generic sets and minimal α-degrees* ◇ D30 D60 E40 ◇
DEGTEV, A.N. *Reducibilities of tabular type in the theory of algorithms (Russian)* ◇ D25 D30 ◇
DEGTEV, A.N. *Several results on upper semilattices and m-degrees (Russian)* ◇ D25 D30 ◇
DEKHTYAR', M.I. *Complexity spectra of recursive sets of approximability of initial segments of complete problems* ◇ D10 D15 D30 F20 F30 ◇
EPSTEIN, R.L. *Degrees of unsolvability: structure and theory* ◇ D25 D30 D35 D98 ◇
GERLA, G. *Una generalizzazione della gerarchia di Ershov* ◇ D30 D55 ◇
HEBEISEN, F. *Masstheoretische Ergebnisse fuer WT-Grade* ◇ D30 ◇
HEBEISEN, F. *Ueber Halbordnungen von WT-Graden in e-Graden* ◇ D30 ◇
KHODZHAYANTS, M.YU. *Structure of e-degrees (Russian)* ◇ D25 D30 ◇
KOBZEV, G.N. *tt-degrees of recursively enumerable Turing degrees II (Russian)* ◇ D25 D30 ◇
LEWIS, F.D. *On unsolvability in subrecursive classes of predicates* ◇ D15 D20 D30 D55 ◇
MAASS, W. *On α- and β-recursively enumerable degrees* ◇ D30 D60 ◇
MARANDZHYAN, G.B. *On the sets of minimal indices of partial recursive functions* ◇ D20 D30 D45 D55 ◇
MASLOVA, T.M. *Bounded m-reducibilities (Russian)* ◇ D25 D30 ◇
NORMANN, D. *A jump operator in set recursion* ◇ D30 D65 E47 ◇
NORMANN, D. *Degrees of functionals* ◇ D30 D65 E45 E50 ◇
NORMANN, D. *Recursion in ${}^3E$ and a splitting theorem* ◇ D30 D65 ◇
OMANADZE, R.SH. *On some properties of Q-reducibility (Russian)* ◇ D25 D30 ◇
OMANADZE, R.SH. *On Q-reducibility (Russian)* ◇ D25 D30 ◇
RICHTER, L.J. *On automorphisms of the degrees that preserve jumps* ◇ D30 ◇
ROZINAS, M.G. *On the representation of e-degrees in the form of greatest lower bound of two incomparable e-degrees (Russian)* ◇ D30 ◇
ROZINAS, M.G. & SOLON, B.YA. *Weakly semirecursive sets (Russian)* ◇ D20 D25 D30 ◇
SANCHIS, L.E. *Reducibilities in two models for combinatory logic* ◇ B40 D30 ◇
SCHADE, W. *Indexmengen rekursiver reeller Zahlen* ◇ D30 D45 ◇
SELIVANOV, V.L. *Structures of the degrees of unsolvability of index sets (Russian)* ◇ D30 ◇
SHEN', A.KH. *The priority method and separation problems (Russian)* ◇ D25 D30 ◇
SHORE, R.A. *The homogeneity conjecture* ◇ D30 ◇
STOLTENBERG-HANSEN, V. *Finite injury arguments in infinite computation theories* ◇ D25 D30 D60 D75 ◇
ZDEBSKAYA, G.V. *On some properties of computation in the limit (Russian)* ◇ D20 D30 ◇

1980

AANDERAA, S.O. & COHEN, D.E. *Modular machines, the word problem for finitely presented groups and Collins' theorem* ◇ D10 D30 D40 ◇
AANDERAA, S.O. & COHEN, D.E. *Modular machines and the Higman-Clapham-Valiev embedding theorem* ◇ D10 D30 D40 ◇
AUTEBERT, J.-M. & BOASSON, L. *Generators of cones and cylinders* ◇ D30 ◇
BENNISON, V.L. *Recursively enumerable complexity sequences and measure independence* ◇ D15 D25 D30 ◇
BORZOV, YU.V. *Four theorems on holographic sets (Russian)* ◇ D20 D30 ◇
BUDINAS, B.L. *Partial ordering of Δ^1_n-degrees of subsets of natural numbers (Russian) (English summary)* ◇ D30 D55 E35 ◇
BULITKO, V.K. *Reducibility via Zhegalkin's linear tables (Russian)* ◇ D25 D30 ◇
CENZER, D. *Non-generable formal languages* ◇ D03 D25 D30 D70 ◇
CHONG, C.T. *Degree theory: from ω to singular cardinals* ◇ D30 D60 E45 ◇
CHONG, C.T. *Rich sets* ◇ D30 D60 ◇
DALEY, R.P. *Quantitative and qualitative information in computations* ◇ D15 D30 ◇
DEGTEV, A.N. *On reducibility of numerations (Russian)* ◇ D25 D30 D45 ◇
DYMENT, E.Z. *Exact bounds of denumerable collections of degrees of difficulty (Russian)* ◇ D30 ◇
FALKINGER, J. *Reduzierbarkeit von berechenbaren Numerierungen von P_1* ◇ D20 D30 D45 ◇
FALKINGER, J. *Universalitaet von berechenbaren Numerierungen von partiell rekursiven Funktionen* ◇ D20 D30 D45 ◇
FRIEDMAN, S.D. *Post's problem without admissibility* ◇ D30 D60 E05 E65 ◇
GANOV, V.A. *Generalized computability and jump operation (Russian)* ◇ D30 ◇

HODES, H.T. *Jumping through the transfinite: The master code hierarchy of Turing degrees* ◇ D30 E40 E45 ◇
HRBACEK, K. & SIMPSON, S.G. *On Kleene degrees of analytic sets* ◇ D30 D55 D65 E15 E35 E45 ◇
JOCKUSCH JR., C.G. *Degrees of generic sets* ◇ D30 E40 ◇
KHODZHAYANTS, M.YU. *Structure of e-degrees (Russian) (Armenian and English summaries)* ◇ D25 D30 ◇
LERMAN, M. *The degrees of unsolvability: Some recent results* ◇ D30 D98 ◇
NERODE, A. & SHORE, R.A. *Reducibility orderings: theories, definability and automorphisms* ◇ C62 D30 D35 ◇
NERODE, A. & SHORE, R.A. *Second order logic and first order theories of reducibility orderings* ◇ B10 B15 D30 D35 F35 F40 ◇
NORMANN, D. *Recursion on the countable functionals* ◇ D30 D55 D65 D98 ◇
OMANADZE, R.SH. *On bounded Q-reducibility (Russian) (Georgian and English summaries)* ◇ D25 D30 ◇
POSNER, D.B. *A survey of non-r.e. degrees ‹=0'* ◇ D30 ◇
SACKS, G.E. *Post's problem, absoluteness and recursion in finite types* ◇ D30 D65 ◇
SACKS, G.E. *Three aspects of recursive enumerability in higher types* ◇ D30 D65 ◇
SHINODA, J. *On the upper semi-lattice of J^s_a-degrees* ◇ D30 D55 D65 ◇
SHORE, R.A. *Some constructions in α-recursion theory* ◇ D30 D60 ◇
SIMPSON, S.G. *The hierarchy based on the jump operator* ◇ D30 D55 ◇
STEFANI, S. *Recursive functions with measurable oracles: An approach to a probabilistic recursion theory* ◇ D20 D30 ◇
ZIEGLER, M. *Algebraisch abgeschlossene Gruppen (English summary)* ◇ C25 C60 D30 D40 ◇

1981

ARSLANOV, M.M. *Some generalizations of a fixed-point theorem (Russian)* ◇ D25 D30 ◇
ASATRYAN, O.S. *Boolean subalgebras of m-degrees of quasimaximal sets (Russian) (Armenian summary)* ◇ D30 ◇
BERGSTRA, J.A. & TIURYN, J. *Algorithmic degrees of algebraic structures* ◇ D30 D75 ◇
BUDINAS, B.L. *Construction of definable degrees of constructibility (Russian)* ◇ D30 D55 E35 E45 E47 ◇
BUDINAS, B.L. *The selector principle and analytic definability of real numbers in extensions of the constructible universe (Russian)* ◇ D30 D55 E15 E35 E45 E47 ◇
BUKHARAEV, N.R. *T-degrees of differences of recursively enumerable sets (Russian)* ◇ D25 D30 D55 ◇
CASALEGNO, P. *Sui gradi delle funzioni parziali* ◇ D30 ◇
CLOTE, P. *A note on the leftmost infinite branch of a recursive tree* ◇ D30 D55 ◇
CLOTE, P. *A note on decidable model theory* ◇ B25 C15 C50 C57 D30 ◇
DEGTEV, A.N. *Relations between complete sets (Russian)* ◇ D25 D30 ◇
DEKHTYAR', M.I. *Structure of T-reducibilities with bounded complexity (Russian)* ◇ D15 D30 ◇
DICHEV, A.V. *An example of a set M of positive integers which is not m-equivalent to M^2* ◇ D30 ◇
EPSTEIN, R.L. & HAAS, R. & KRAMER, R.L. *Hierarchies of sets and degrees below O'* ◇ D30 D55 ◇
EPSTEIN, R.L. *Initial segments of degrees below 0'* ◇ D25 D30 D35 D98 ◇
FRIEDMAN, S.D. *Natural α-re degrees* ◇ D30 D60 ◇
FRIEDMAN, S.D. *Negative solutions to Post's problem II* ◇ D30 D60 ◇
HARRINGTON, L.A. & SHORE, R.A. *Definable degrees and automorphisms of $\mathscr{D}$* ◇ D30 D55 ◇
HODES, H.T. *Upper bounds on locally countable admissible initial segments of a Turing degree hierarchy* ◇ D30 E40 E45 ◇
JOCKUSCH JR., C.G. & POSNER, D.B. *Automorphism bases for degrees of unsolvability* ◇ D30 ◇
KECHRIS, A.S. *A note on Wadge degrees* ◇ D30 D55 E15 E60 ◇
KECHRIS, A.S. *Forcing with Δ perfect trees and minimal Δ-degrees* ◇ D30 D55 D75 E15 E40 E60 ◇
KECHRIS, A.S. & SOLOVAY, R.M. & STEEL, J.R. *The axiom of determinacy and the prewellordering property* ◇ D30 E15 E45 E60 ◇
KHODZHAYANTS, M.YU. *e-degrees, T-degrees and axiomatic theories (Russian) (Armenian summary)* ◇ D30 ◇
KUZ'MINA, T.M. *Structure of the m-degrees of the index sets of families of partial recursive functions (Russian)* ◇ D20 D30 ◇
LERMAN, M. *On recursive linear orderings* ◇ C57 C65 D30 D45 D55 E07 ◇
MAASS, W. *Recursively invariant β-recursion theory* ◇ D30 D60 ◇
MAL'TSEV, A.A. *The structure of an m-jump (Russian)* ◇ D30 ◇
NORMANN, D. *The continuous functionals; computations,recursions and degrees* ◇ D30 D65 ◇
ODIFREDDI, P. *Strong reducibilities* ◇ D25 D30 ◇
ODIFREDDI, P. *Trees and degrees* ◇ D30 ◇
POIZAT, B. *Degres de definissabilite arithmetique des generiques (English summary)* ◇ C25 C40 D30 D55 ◇
POSNER, D.B. & ROBINSON, R.W. *Degrees joining to 0'* ◇ D25 D30 ◇
POSNER, D.B. *The upper semilattice of degrees $\leqslant 0'$ is complemented* ◇ D25 D30 ◇
RICHTER, L.J. *Degrees of structures* ◇ C57 D30 D45 ◇
ROZINAS, M.G. *Expansion of partial degrees into r-degrees (Russian)* ◇ D30 ◇
SHORE, R.A. *The degrees of unsolvability: global results* ◇ C40 D30 D35 F35 ◇
SHORE, R.A. *The theory of the degrees below O'* ◇ D25 D30 D35 F30 ◇
SOLON, B.YA. *PC-degrees inside an e-degree of a hyperimmune retraceable set (Russian)* ◇ D25 D30 D50 H05 ◇

TANG, A. *Wadge reducibility and Hausdorff difference hierarchy in Pω* ◇ D30 D55 E15 ◇

1982

ARSLANOV, M.M. *A hierarchy of degrees of unsolvability (Russian)* ◇ D30 D55 ◇

BAJRASHEVA, V.R. *Degrees of automated transformations (Russian)* ◇ D05 D30 ◇

BUDINAS, B.L. *On the selector principle and the analytic definability of constructible sets (Russian)* ◇ D30 D55 E15 E35 E45 ◇

CHONG, C.T. *Double jumps of minimal degrees over cardinals* ◇ D30 D60 E45 ◇

COOPER, S.B. *Partial degrees and the density problem* ◇ D30 ◇

DEGTEV, A.N. *Small degrees in ordinary recursion theory* ◇ D30 ◇

DYMENT, E.Z. *Existence of a degree of extendability of a partially recursive function, which is not a degree of separability (Russian)* ◇ D20 D30 ◇

FARRINGTON, PATRICK *Constructible lattices of c-degrees* ◇ D30 E40 E45 ◇

FARRINGTON, PATRICK *The first-order theory of the c-degrees with the # -operation* ◇ D30 E45 E55 ◇

FRIEDMAN, S.D. *The Turing degrees and the metadegrees have isomorphic cones* ◇ D30 D60 ◇

HAY, L. *On the recursion-theoretic complexity of relative succinctness of representations of languages* ◇ D05 D30 ◇

HODES, H.T. *Jumping to a uniform upper bound* ◇ D30 ◇

HOROWITZ, B.M. *Elementary formal systems as a framework for relative recursion theory* ◇ D20 D25 D30 ◇

HU, ZHAOGUANG *The proof to Mizumoto and Tanaka's problem* ◇ D30 E72 ◇

KANOVICH, M.I. *On the complexity of the problem of separating recursively enumerable sets (Russian)* ◇ D25 D30 ◇

KANOVICH, M.I. *Truth-table reducibilities of problems of the extension of partial recursive functions (Russian)* ◇ D20 D30 ◇

KNIGHT, J.F. & NADEL, M.E. *Expansions of models and Turing degrees* ◇ C50 C57 D30 ◇

KNIGHT, J.F. & NADEL, M.E. *Models of arithmetic and closed ideals* ◇ C62 D30 ◇

KUCERA, A. *On recursive measure of classes of recursive sets* ◇ D20 D30 ◇

LOUVEAU, A. *La classification de Wadge des ensembles boreliens* ◇ D30 D55 E15 E60 ◇

MARKER, D. *Degrees of models of true arithmetic* ◇ C57 C62 D30 H15 ◇

MEINEL, C. *Embedding of the poset of Turing degrees in the poset* $[\mathfrak{U}, \leqslant]$ *(German and Russian summaries)* ◇ D10 D30 ◇

MELIKYAN, S.M. *Specker numbers and degrees of undecidability (Russian)* ◇ D30 D80 ◇

PRIDA, J.F. *A non-standard study of the theory of relative recursivity (Spanish)* ◇ D20 D30 H05 ◇

SELIVANOV, V.L. *A class of reducibilities in the theory of recursive functions (Russian)* ◇ D30 ◇

SELIVANOV, V.L. *The structure of degrees of generalized index sets (Russian)* ◇ D30 D45 ◇

SHORE, R.A. *Finitely generated codings and the degrees r.e. in a degree d* ◇ D30 D55 E40 ◇

SHORE, R.A. *On homogeneity and definability in the first order theory of the Turing degrees* ◇ D30 ◇

SHORE, R.A. *The Turing and truth-table degrees are not elementarily equivalent* ◇ D30 ◇

STEEL, J.R. *A classification of jump operators* ◇ D30 E45 E60 ◇

STEFANI, S. *Recursive functions with measurable oracles: the semilattice of the effectively splitting distributions (Italian summary)* ◇ D20 D30 ◇

STOB, M. *Index sets and degrees of unsolvability* ◇ D30 ◇

VUCKOVIC, V. *Relativized cylindrification* ◇ D20 D30 ◇

WEITKAMP, G. *Analytic sets having incomparable Kleene degrees* ◇ D30 D55 D65 E15 ◇

1983

BARENDREGT, H.P. & LONGO, G. *Recursion theoretic operators and morphisms of numbered sets* ◇ D20 D30 D45 ◇

CHONG, C.T. & FRIEDMAN, S.D. *Degree theory on* $\aleph_\omega$ ◇ D30 D60 ◇

CHONG, C.T. *Global and local admissibility. II. Major subsets and automorphisms* ◇ D30 D60 ◇

DEGTEV, A.N. *Relations between table-type degrees (Russian)* ◇ D25 D30 ◇

DICHEV, A.V. *On the m-equivalence of Cartesian powers of sets of natural numbers (Russian)* ◇ D25 D30 ◇

FARRINGTON, PATRICK *Hinges and automorphisms of the degrees of non-constructibility* ◇ D30 E45 ◇

GROSZEK, M.J. & SLAMAN, T.A. *Independence results on the global structure of the Turing degrees* ◇ D30 E35 ◇

HODES, H.T. *A minimal upper bound on a sequence of Turing degrees which represents that sequence* ◇ D30 ◇

HODES, H.T. *More about uniform upper bounds on ideals of Turing degrees* ◇ D30 ◇

HOMER, S. *Intermediate* β*-r.e. degrees and the half-jump* ◇ D30 D60 ◇

HOMER, S. & SACKS, G.E. *Inverting the half-jump* ◇ D30 D60 ◇

HRBACEK, K. *Degrees of analytic sets* ◇ D30 D65 E15 E45 ◇

JOCKUSCH JR., C.G. & SHORE, R.A. *Pseudo jump operators I: The r.e. case* ◇ D25 D30 ◇

KALLIBEKOV, S. *Index sets and degrees of some classes of recursively enumerable sets (Russian)* ◇ D25 D30 ◇

KANOVICH, M.I. *Complexity and convergence of algorithmic mass problems (Russian)* ◇ D15 D30 ◇

KANOVICH, M.I. *On the implicativity of the lattice of truth-table degrees of algorithmic problems (Russian)* ◇ D30 ◇

KANOVICH, M.I. *Reducibility via general recursive operators (Russian)* ◇ D30 ◇

KNIGHT, J.F. *Additive structure in uncountable models for a fixed completion of P* ◇ C50 C57 C62 C75 D30 ◇

KNIGHT, J.F. *Degrees of types and independent sequences* ◇ C35 C45 D30 ◇
KURTZ, S.A. *Notions of weak genericity* ◇ D30 E40 ◇
LERMAN, M. *Degrees of unsolvability. Local and global theory* ◇ D30 D98 ◇
LERMAN, M. *The structures of recursion theory* ◇ D25 D30 D98 ◇
MILLAR, T.S. *Persistently finite theories with hyperarithmetic models* ◇ C15 C50 C57 D30 ◇
MOHRHERR, J. *Kleene index sets and functional m-degrees* ◇ D20 D25 D30 D45 ◇
NORMANN, D. *R.e. degrees of continuous functionals* ◇ D30 D65 ◇
ODIFREDDI, P. *Forcing and reducibilities* ◇ D30 D55 E40 F30 ◇
ODIFREDDI, P. *Forcing and reducibilities. II: Forcing in fragments of analysis. III: Forcing in fragments of set theory* ◇ D30 D55 E40 ◇
ODIFREDDI, P. *On the first order theory of the arithmetical degrees* ◇ C65 D30 D55 ◇
PLOTNIKOVA, N.A. *Calculation with oracles* ◇ D10 D30 ◇
REMESLENNIKOV, V.N. & ROMAN'KOV, V.A. *Model-theoretic and algorithmic questions of group theory (Russian)* ◇ B25 C60 C98 D15 D30 D40 ◇
SHATROVA, N.A. *Degree of complexity of an algorithm for a class of groups (Russian)* ◇ D30 D40 ◇
SIMPSON, S.G. & WEITKAMP, G. *High and low Kleene degrees of coanalytic sets* ◇ D30 D55 D65 ◇
TALJA, J. *On the complexity-relativized strong reducibilities* ◇ D15 D30 ◇

1984

ABRAHAM, U. *A minimal model for ¬ CH: iteration of Jensen's reals* ◇ C62 D30 E35 E40 E45 E50 E65 ◇
ADAMOWICZ, Z. *Continuous relations and generalized* G_δ *sets* ◇ D30 E15 E40 ◇
BUJLINA, E.M. *Degrees of unsolvability of limit-enumerable sets (Russian)* ◇ D30 ◇
BYERLY, R.E. *Some properties of invariant sets* ◇ D25 D30 ◇
CHONG, C.T. & JOCKUSCH JR., C.G. *Minimal degrees and 1-generic sets below 0'* ◇ D30 ◇
COOPER, S.B. *Partial degrees and the density problem. Part 2: The enumeration degrees of the* Σ_2 *sets are dense* ◇ D30 ◇
FARRINGTON, PADDY *The first-order theory of the c-degrees* ◇ D30 D35 E45 E55 F35 ◇
GENG, SUYUN & ZHANG, LIANG *A note on the formal definition of "Turing reduction" (Chinese) (English summary)* ◇ D30 ◇
HODES, H.T. *Finite level Borel games and a problem concerning the jump hierarchy* ◇ D30 E45 E60 ◇
HOEVEN VAN DER, G.F. & MOERDIJK, I. *On choice sequences determined by spreads* ◇ D30 D55 F35 F50 G30 ◇
JOCKUSCH JR., C.G. & SHORE, R.A. *Pseudo jump operators II: Transfinite iterations, hierarchies and minimal covers* ◇ D25 D30 D55 ◇
KASTANAS, I.G. *The jump inversion theorem for* Q_{2n+1}*-degrees* ◇ D30 E15 E60 ◇
KHISAMIEV, Z.G. *Multiple m-reducibility of index sets (Russian)* ◇ D25 D30 ◇
KNIGHT, J.F. & LACHLAN, A.H. & SOARE, R.I. *Two theorems on degrees of models of true arithmetic* ◇ C57 C62 D30 ◇
KO, KER-I *Reducibilities on real numbers* ◇ D15 D30 ◇
MACINTYRE, A. & MARKER, D. *Degrees of recursively saturated models* ◇ C50 C57 C62 D30 D45 ◇
MOHRHERR, J. *Density of a final segment of the truth-table degrees* ◇ D30 ◇
OMAROV, A.I. *Elementary theory of D-degrees (Russian)* ◇ C20 C30 C50 D30 G10 ◇
SHORE, R.A. *The arithmetic and Turing degrees are not elementarily equivalent* ◇ D30 ◇
SHORE, R.A. *The degrees of unsolvability: the ordering of functions by relative computability* ◇ D30 ◇
YANG, DONGPING *On the embedding of* α*-recursive presentable lattices into the* α*-recursive degrees below* $0'$ ◇ D30 D60 ◇
ZAKHAROV, S.D. *e- and s-degrees (Russian)* ◇ D30 ◇

1985

ARSLANOV, M.M. *Lattice properties of the degrees below 0' (Russian)* ◇ D30 ◇
BULITKO, V.K. *Boolean classes of Turing reducibilities (Russian)* ◇ D30 ◇
CASALEGNO, P. *On the T-degrees of partial functions* ◇ D30 ◇
CLOTE, P. *Applications of the low-basis theorem in arithmetic* ◇ C62 D20 D30 D55 F30 ◇
CLOTE, P. *On recursive trees with a unique infinite branch* ◇ D30 D55 ◇
COOPER, S.B. & MCEVOY, K. *On minimal pairs of enumeration degrees* ◇ D25 D30 ◇
DEGTEV, A.N. *Semilattices of disjunctive and linear degrees (Russian)* ◇ D30 ◇
DIETZFELBINGER, M. & MAASS, W. *Strong reducibilities in* α*- and* β*- recursion theory* ◇ D30 D60 ◇
JOCKUSCH JR., C.G. & SHORE, R.A. *REA operators, R.E. degrees and minimal covers* ◇ D30 ◇
KIERSTEAD, H.A. & REMMEL, J.B. *Degrees of indiscernibles in decidable models* ◇ C30 C35 C57 D30 ◇
KUCERA, A. *Measure,* Π^0_1*-classes and complete extensions of PA* ◇ D30 E15 E75 F30 ◇
LERMAN, M. *Upper bounds for the arithmetical degrees* ◇ D30 D55 ◇
LEWIS, A.A. *The minimum degree of recursively representable choice functions* ◇ D20 D30 ◇
LI, XIANG *Turing degrees on pointwise r.e. open sets in effective Hausdorff spaces (Chinese)* ◇ D25 D30 D45 ◇
MAL'TSEV, A.A. *Structure of the semilattice of tt 1-degrees (Russian)* ◇ D30 ◇
MCEVOY, K. *Jumps of quasi-minimal enumeration degrees* ◇ D30 ◇

MILLAR, T.S. *Decidable Ehrenfeucht theories* ◇ B25 C15 C57 C98 D30 D35 D55 ◇

ODIFREDDI, P. *Global properties (automorphisms and definability) of m-degrees (Italian) (English summary)* ◇ D30 ◇

ODIFREDDI, P. *The structure of m-degrees* ◇ D30 ◇

SHORE, R.A. *The structure of the degrees of unsolvability* ◇ D30 ◇

SLAMAN, T.A. *Reflection and the priority method in E-recursion theory* ◇ D30 D65 E45 ◇

SLAMAN, T.A. *The E-recursively enumerable degrees are dense* ◇ D30 D65 E45 ◇

D35 Undecidability and degrees of sets of sentences

1900

HILBERT, D. *Mathematische Probleme* ◇ A05 D35 E30 E50 F25 ◇

1919

SKOLEM, T.A. *Untersuchungen ueber die Axiome des Klassenkalkuels und ueber Produktions- und Summationsprobleme, welche gewisse Klassen von Aussagen betreffen* ◇ B20 B25 C10 D35 G05 ◇

1922

BEHMANN, H. *Beitraege zur Algebra der Logik, insbesondere zum Entscheidungsproblem* ◇ B20 B25 D35 ◇

1923

BEHMANN, H. *Algebra der Logik und Entscheidungsproblem* ◇ B20 B25 D35 ◇

1926

FINSLER, P. *Formale Beweise und die Entscheidbarkeit* ◇ A05 D35 F25 ◇

1927

BEHMANN, H. *Entscheidungsproblem und Logik der Beziehungen* ◇ B25 D35 ◇

1928

ACKERMANN, W. *Ueber die Erfuellbarkeit gewisser Zaehlausdruecke* ◇ B25 D35 ◇

BERNAYS, P. & SCHOENFINKEL, M. *Zum Entscheidungsproblem der mathematischen Logik* ◇ B20 B25 D35 ◇

HILBERT, D. & ACKERMANN, W. *Grundzuege der theoretischen Logik* ◇ B25 B98 D35 ◇

1930

KALMAR, L. *Ein Beitrag zum Entscheidungsproblem* ◇ B25 D35 ◇

1931

GOEDEL, K. *Ueber formal unentscheidbare Saetze der "Principia Mathematica" und verwandter Systeme I* ◇ B28 B30 D20 D35 F25 F30 F35 ◇

ZERMELO, E. *Ueber Stufen der Quantifikation und die Logik des Unendlichen* ◇ A05 C75 D35 E07 E30 ◇

1932

KALMAR, L. *Zum Entscheidungsproblem der mathematischen Logik* ◇ B20 B25 D35 ◇

1933

GOEDEL, K. *Zum Entscheidungsproblem des logischen Funktionenkalkuels* ◇ B10 B20 B25 C13 D35 ◇

SCHUETTE, K. *Untersuchungen zum Entscheidungsproblem der mathematischen Logik* ◇ B20 B25 D35 ◇

1934

KIREEVSKIJ, N.N. *On the problem of the solvability of the decision problem (Russian)* ◇ B25 D35 ◇

1935

CHURCH, A. *An unsolvable problem of elementary number theory* ◇ B40 D20 D35 ◇

1936

ACKERMANN, W. *Beitraege zum Entscheidungsproblem der mathematischen Logik* ◇ B25 D35 ◇

CHURCH, A. *A note on the "Entscheidungsproblem"* ◇ B20 D20 D35 ◇

CHURCH, A. *An unsolvable problem of elementary number theory* ◇ D20 D35 ◇

KEMPNER, A.J. *Remarks on "unsolvable" problems* ◇ D20 D35 ◇

PEPIS, J. *Beitraege zur Reduktionstheorie des logischen Entscheidungsproblemes* ◇ B20 D35 ◇

ROSSER, J.B. *Extensions of some theorems of Goedel and Church* ◇ D25 D35 F30 ◇

SKOLEM, T.A. *Einige Reduktionen des Entscheidungsproblems* ◇ B20 D35 ◇

TURING, A.M. *On computable numbers, with an application to the "Entscheidungsproblem"* ◇ D05 D10 D20 D35 F60 ◇

1937

KALMAR, L. *Zur Reduktion des Entscheidungsproblems* ◇ B20 D35 ◇

KALMAR, L. *Zurueckfuehrung des Entscheidungsproblems auf den Fall von Formeln mit einer einzigen, binaeren Funktionsvariablen* ◇ B20 D35 ◇

PEPIS, J. *Ueber das Entscheidungsproblem des engeren logischen Funktionskalkuels (Polish) (German summary)* ◇ B20 B25 D35 ◇

SKOLEM, T.A. *Eine Bemerkung zum Entscheidungsproblem* ◇ B20 B25 D35 ◇

1938

PEPIS, J. *Ein Verfahren der mathematischen Logik* ◇ B20 D35 ◇

PEPIS, J. *Untersuchungen ueber das Entscheidungsproblem der mathematischen Logik* ◇ B20 D35 ◇

1939

KALMAR, L. *On the reduction of the decision problem I: Ackermann prefix, a single binary predicate* ◇ B20 D35 ◇

ROSSER, J.B. *An informal exposition of proofs of Goedel's theorems and Church's theorem* ◇ A05 B20 D35 F30 ◇

ZHEGALKIN, I.I. *Sur l'Entscheidungsproblem (Russian) (French summary)* ◇ B20 B25 D35 ◇

1940

SKOLEM, T.A. *Einfacher Beweis der Unmoeglichkeit eines allgemeinen Loesungsverfahren fuer arithmetische Probleme* ◇ D35 F30 ◇

1943

SURANYI, J. *Zur Reduktion des Entscheidungsproblems des logischen Funktionskalkuels (Hungarian) (German summary)* ◇ B20 D35 ◇

1947

KALMAR, L. & SURANYI, J. *On the reduction of the decision problem II: Goedel prefix, a single binary predicate* ◇ B20 D35 ◇

1948

JASKOWSKI, S. *Sur le probleme de decision de la topologie et de la theorie des groups* ◇ D35 D40 ◇

1949

KALMAR, L. *On unsolvable mathematical problems* ◇ A05 D35 ◇

NOVIKOV, P.S. *On the axiom of complete induction (Russian)* ◇ B25 B28 D35 F30 ◇

QUINE, W.V.O. *On decidability and completeness* ◇ B25 D35 F30 ◇

ROBINSON, JULIA *Definability and decision problems in arithmetic* ◇ D35 F30 ◇

SURANYI, J. *Reduction of the decision problem to formulas containing a bounded number of quantifiers only* ◇ B20 D35 ◇

1950

BETH, E.W. *Decision problems of logic and mathematics* ◇ B20 B25 C10 D35 ◇

DAVIS, MARTIN D. *On the theory of recursive unsolvability* ◇ D35 D55 ◇

KALMAR, L. *Contribution to the reduction theory of the decision problem I. Prefix* (x_1) (x_2) $(Ex_3)\ldots(Ex_{n-1})$ (x_n)*, a single binary predicate* ◇ B20 D35 ◇

KALMAR, L. & SURANYI, J. *On the reduction of the decision problem III: Pepis prefix, a single binary predicate* ◇ B20 D35 ◇

SURANYI, J. *Contributions to the reduction theory of the decision problem II: Three universal, one existential quantifier* ◇ B20 D35 ◇

TRAKHTENBROT, B.A. *The impossibility of an algorithm for the decision problem in finite domains (Russian)* ◇ C13 D35 ◇

TURQUETTE, A.R. *Goedel and the synthetic a priori* ◇ A05 D35 ◇

1951

GRZEGORCZYK, A. *Undecidability of some topological theories* ◇ D35 ◇

KALMAR, L. *Contributions to the reduction theory of the decision problem. III Prefix* $(x_1)(Ex_2)\ldots(Ex_{n-2})(x_{n-1})(x_n)$ *, a single binary predicate* ◇ B10 D35 ◇

KALMAR, L. *Contribution to the reduction theory of the decision problem IV. Reduction to the case of a finite set of individuals* ◇ D35 ◇

MARKOV, A.A. *The impossibility of certain algorithms in the theory of associative systems (Russian)* ◇ D03 D35 D40 ◇

ROBINSON, R.M. *Undecidable rings* ◇ C60 D35 ◇

SURANYI, J. *Contributions to the reduction theory of the decision problem V: Ackermann prefix with three universal quantifiers* ◇ B20 D35 ◇

1952

CHURCH, A. & QUINE, W.V.O. *Some theorems on definability and decidability* ◇ B15 D35 ◇

CRAIG, W. *Incompletability, with respect to validity in every finite nonempty domain of first order functional calculus* ◇ C13 D35 ◇

KALMAR, L. *Reduction of the decision problem to the satisfiability question of logical formulae on a finite set (Hungarian)* ◇ D35 ◇

MARKOV, A.A. *On unsolvable algorithmic problems (Russian)* ◇ D03 D20 D25 D35 D40 ◇

ROBINSON, R.M. *An essentially undecidable axiom system* ◇ D35 F30 ◇

SZMIELEW, W. & TARSKI, A. *Mutual interpretability of some essentially undecidable theories* ◇ D35 F25 ◇

1953

DAVIS, MARTIN D. *Arithmetical problems and recursively enumerable predicates* ◇ D25 D35 D55 D80 ◇

JANICZAK, A. *Undecidability of some simple formalized theories* ◇ B25 D35 ◇

MOSTOWSKI, ANDRZEJ & ROBINSON, R.M. & TARSKI, A. *Undecidability and essential undecidability in arithmetic* ◇ D35 F30 ◇

MOSTOWSKI, ANDRZEJ & ROBINSON, R.M. & TARSKI, A. *Undecidable theories* ◇ D35 F25 F30 ◇

TARSKI, A. *A general method in proofs of undecidability* ◇ D35 F25 ◇

TARSKI, A. *Undecidability of the elementary theory of groups* ◇ D35 ◇

TRAKHTENBROT, B.A. *On recursively separability (Russian)* ◇ B10 C13 D35 ◇

USPENSKIJ, V.A. *The Goedel theorem and the theory of algorithms (Russian)* ◇ D20 D35 F30 ◇

1954

KALICKI, J. *An undecidable problem in the algebra of truth-tables* ◇ B05 D35 ◇

TURING, A.M. *Solvable and unsolvable problems* ◇ B25 D10 D35 ◇

1955

HERMES, H. *Vorlesung ueber Entscheidungsprobleme in Mathematik und Logik* ◇ B25 B40 B65 D10 D20 D35 D98 ◇

SURANYI, J. *On the reduction theory of the decision problem of symbolic logic (Hungarian) (Russian and English summaries)* ◇ B20 D35 ◇

1956

COBHAM, A. *Reduction to a symmetric predicate* ◇ B10 D35 ◇

GRZEGORCZYK, A. *Some proofs of undecidability of arithmetic* ◇ D35 F30 ◇

JASKOWSKI, S. *Indecidability of first order sentences in the theory of free groupoids* ◇ D35 D40 ◇
KALMAR, L. *A direct proof of the unsolvability of the decision problem by means of a general recursive algorithm (Hungarian)* ◇ B20 D35 ◇
KALMAR, L. *Ein direkter Beweis fuer die allgemein-rekursive Unloesbarkeit des Entscheidungsproblems des Praedikatenkalkuels der ersten Stufe mit Identitaet* ◇ B10 D35 ◇
MUCHNIK, A.A. *On the unsolvability of the problem of reducibility in the theory of algorithms (Russian)* ◇ D25 D30 D35 ◇
MYHILL, J.R. *Solution of a problem of Tarski* ◇ D35 ◇
ROGERS JR., H. *Certain logical reduction and decision problems* ◇ B20 D35 ◇
TRAKHTENBROT, B.A. *The definition of a finite set and the deductive incompleteness of the theory of sets (Russian)* ◇ B28 D35 E30 ◇

1957

DAVIS, MARTIN D. & PUTNAM, H. *Reductions of Hilbert's 10th problem* ◇ D25 D35 ◇
DREBEN, B. *Systematic treatment of the decision problem* ◇ B20 B25 D35 ◇
EHRENFEUCHT, A. *Two theories with axioms built by means of pleonasmus* ◇ B25 C35 D35 ◇
FEFERMAN, S. *Degrees of unsolvability associated with classes of formalized theories* ◇ D25 D35 ◇
GRZEGORCZYK, A. *Decision problems* ◇ B25 C10 D35 D98 F30 ◇
KALMAR, L. *On Church's hypothesis as foundation for studies related to so-called unsolvable mathematical problems (Hungarian)* ◇ D20 D35 ◇
PUTNAM, H. *Decidability and essential undecidability* ◇ B25 D35 ◇
VAUGHT, R.L. *Sentences true in all constructive models* ◇ C13 C57 D35 ◇

1958

BERNAYS, P. *Remarques sur le probleme de la decision en logique elementaire* ◇ B10 D35 ◇
DAVIS, MARTIN D. & PUTNAM, H. *Reductions of Hilbert's tenth problem* ◇ D25 D35 ◇
GAL, L.N. *A note on direct products* ◇ B25 C30 D35 ◇
HASENJAEGER, G. *Formales und produktives Schliessen* ◇ A05 D35 ◇
NAGEL, E. & NEWMAN, J.R. *Goedel's proof* ◇ B28 D35 F30 F98 ◇
NOVIKOV, P.S. *Ueber einige algorithmische Probleme der Gruppentheorie* ◇ D35 D40 ◇
SHOENFIELD, J.R. *Degrees of formal systems* ◇ D25 D30 D35 D55 ◇
SMULLYAN, R.M. *Undecidability and recursive inseparability* ◇ D25 D35 ◇
TAJMANOV, A.D. *On classes of models, closed under direct product (Russian)* ◇ C30 C52 D35 ◇

1959

BAUMSLAG, G. & BOONE, W.W. & NEUMANN, B.H. *Some unsolvable problems about elements and subgroups of groups* ◇ D35 D40 ◇
HENKIN, L. & SUPPES, P. & TARSKI, A. (EDS.) *The axiomatic method, with special reference to geometry and physics* ◇ B30 B97 C65 D35 ◇
KALMAR, L. *On a hypothesis used in investigations of so-called unsolvable arithmetical problems (Russian)* ◇ A05 D35 ◇
KUZNETSOV, A.V. *On the equivalence and functional completeness problems (Russian)* ◇ B25 D35 ◇
NISHIMURA, T. *On Goedel's theorem (Japanese)* ◇ D35 F30 ◇
NOVIKOV, P.S. *On the unsolvability of some problems in algebra I (Russian)* ◇ D35 D40 ◇
ROBINSON, JULIA *The undecidability of algebraic rings and fields* ◇ C60 D35 ◇
STEGMUELLER, W. *Unvollstaendigkeit und Unentscheidbarkeit. Die metamathematischen Resultate von Goedel, Church, Kleene, Rosser und ihre erkenntnistheoretische Bedeutung* ◇ A05 D35 F30 F98 ◇
SURANYI, J. *Reduktionstheorie des Entscheidungsproblems im Praedikatenkalkuel der ersten Stufe* ◇ B20 D35 ◇
ZYKOV, A.A. *Remarks in connection with the reduction theorem for logical calculi (Russian)* ◇ B20 D35 ◇

1960

MAL'TSEV, A.I. *On free soluble groups (Russian)* ◇ C60 D35 ◇
MAL'TSEV, A.I. *On the undecidability of the elementary theory of certain fields (Russian)* ◇ D35 ◇
MAL'TSEV, A.I. *Some correspondences between rings and groups (Russian)* ◇ C60 D35 ◇
PUTNAM, H. *An unsolvable problem in number theory* ◇ D35 ◇
PUTNAM, H. & SMULLYAN, R.M. *Exact separation of recursively enumerable sets within theories* ◇ D25 D35 ◇
SCOTT, D.S. *On a theorem of Rabin* ◇ C35 C57 C62 D35 ◇
SHOENFIELD, J.R. *Degrees of models* ◇ C57 D30 D35 ◇
SMULLYAN, R.M. *Theories with effectively inseparable nuclei* ◇ D25 D35 ◇
VAUGHT, R.L. *Sentences true in all constructive models* ◇ C13 C57 D35 ◇

1961

DAVIS, MARTIN D. & PUTNAM, H. & ROBINSON, JULIA *The decision problem for exponential diophantine equations* ◇ D25 D35 ◇
EHRENFEUCHT, A. *Separable theories* ◇ D25 D35 ◇
ELGOT, C.C. *Decision problems of finite automata design and related arithmetics* ◇ B25 D03 D05 D35 F30 ◇
HANSON, N.R. *The Goedel theorem* ◇ D20 D35 F30 ◇
MAL'TSEV, A.I. *Effective inseparability of the set of identically true from the set of finitely refutable formulae of certain elementary theories (Russian)* ◇ D35 ◇
MAL'TSEV, A.I. *Remark on the paper "on the undecidability of the elementary theories of certain fields" (Russian)* ◇ D35 ◇

MAL'TSEV, A.I. *Undecidability of the elementary theory of finite groups (Russian)* ◇ C13 D35 ◇
OBERSCHELP, A. *Ueber die Unentscheidbarkeit gewisser Axiommengen* ◇ D35 ◇
RAUTENBERG, W. *Unentscheidbarkeit der euklidischen Inzidenzgeometrie* ◇ D35 ◇
ROSE, H.E. *On the consistency and undecidability of recursive arithmetic* ◇ D35 F30 ◇
SHOENFIELD, J.R. *Undecidable and creative theories* ◇ D25 D35 ◇
SMART, J.J.C. *Goedel's theorem, Church's theorem and mechanism* ◇ A05 B10 D35 F30 ◇

1962

BUECHI, J.R. *Turing machines and the "Entscheidungsproblem"* ◇ B20 D10 D35 ◇
DAVIS, MARTIN D. *Applications of recursive function theory to number theory* ◇ D35 ◇
DREBEN, B. & KAHR, A.S. & WANG, HAO *Classification of AEA formulas by letter atoms* ◇ B20 B25 D35 ◇
GODDARD, L. *Proof-making* ◇ A05 D35 ◇
GRZEGORCZYK, A. *An example of two weak essentially undecidable theories F and F** ◇ D35 ◇
KAHR, A.S. & MOORE, E.F. & WANG, HAO *Entscheidungsproblem reduced to the ∀∃∀ case* ◇ B20 D35 ◇
KRIPKE, S.A. *The undecidability of monadic modal quantification theory* ◇ B45 D35 ◇
LAVROV, I.A. *Undecidability of elementary theories of certain rings (Russian)* ◇ C60 D35 ◇
LORENZEN, P. *Metamathematik* ◇ A05 B98 C60 D20 D35 D98 F98 ◇
MARKOV, A.A. *Sur les invariants calculables* ◇ D35 D40 D45 ◇
RAUTENBERG, W. *Ueber metatheoretische Eigenschaften einiger geometrischer Theorien* ◇ B30 C65 D35 ◇
ROBINSON, JULIA *On the decision problem for algebraic rings* ◇ C60 D35 ◇
ROBINSON, JULIA *The undecidability of exponential diophantine equations* ◇ D35 ◇
SCARPELLINI, B. *Die Nichtaxiomatisierbarkeit des unendlichwertigen Praedikatenkalkuels von Lukasiewicz* ◇ B50 D35 ◇
SHOENFIELD, J.R. *Some applications of degrees* ◇ D25 D30 D35 D55 ◇
TAJTSLIN, M.A. *Effective inseparability of the set of identically true and the set of finitely refutable formulas of the elementary theory of lattices (Russian)* ◇ C13 D25 D35 G10 ◇
TAJTSLIN, M.A. *Undecidability of the elementary theory of commutative semigroups with cancellation (Russian)* ◇ D35 ◇
VAUGHT, R.L. *On a theorem of Cobham concerning undecidable theories* ◇ D25 D35 ◇

1963

COBHAM, A. *Some remarks concerning theories with recursively enumerable complements* ◇ D25 D35 ◇
DAVIS, MARTIN D. *Extensions and corollaries of recent work on Hilbert's tenth problem* ◇ D25 D35 ◇
DAVIS, MARTIN D. *Unsolvable problems: a review* ◇ D03 D10 D35 D40 ◇
ERSHOV, YU.L. & TAJTSLIN, M.A. *Undecidability of certain theories (Russian)* ◇ D35 ◇
GUMIN, H. *Digital computers, mathematical logic and principal limitations of computability* ◇ B75 D35 ◇
KAHR, A.S. *Improved reductions of the "Entscheidungsproblem" to subclasses of ∀∃∀ formulas* ◇ D35 ◇
KUZNETSOV, A.V. *Undecidability of the general problems of completeness solvability and equivalence for propositional calculi (Russian)* ◇ B22 D35 ◇
LAVROV, I.A. *Effective inseparability of the sets of identically true formulae and finitely refutable formulae for certain elementary theories (Russian)* ◇ C13 D25 D35 ◇
ROBINSON, R.M. *Undecidability of the elementary theory of the field of rational functions of one variable with rational coefficients (Russian)* ◇ D35 ◇
SCARPELLINI, B. *Zwei unentscheidbare Probleme der Analysis* ◇ D25 D35 ◇
SMULLYAN, R.M. *Pseudo-uniform reducibility* ◇ D20 D25 D30 D35 ◇
TAJTSLIN, M.A. *Undecidability of elementary theories of certain classes of finite commutative associative rings (Russian)* ◇ C13 D35 ◇
WANG, HAO *Computation* ◇ D20 D25 D30 D35 D60 ◇
WANG, HAO *Dominoes and the AEA case of the decision problem* ◇ B20 B25 D05 D35 ◇

1964

BUECHI, J.R. *Regular canonical systems* ◇ D05 D35 ◇
ERSHOV, YU.L. *Unsolvability of theories of symmetric and simple finite groups (Russian)* ◇ C13 C60 D35 ◇
GENENZ, J. *Reduktionstheorie nach der Methode von Kahr-Moore-Wang* ◇ B20 D35 D98 ◇
HARROP, R. *A relativization procedure for propositional calculi, with an application to a generalized form of Post's theorem* ◇ B22 D35 ◇
HERMES, H. *Unentscheidbarkeit der Arithmetik* ◇ D35 F30 ◇
HUBER-DYSON, V. *On the decision problem for theories of finite models* ◇ B25 C13 D35 ◇
KOSTYRKO, V.F. *The reduction class* $\forall\exists^n\forall$ *(Russian)* ◇ B20 D35 ◇
LACOMBE, D. *Theoremes de non-decidabilite* ◇ D35 D45 D98 ◇
ROBINSON, R.M. *The undecidability of pure transcendental extension of real fields* ◇ D35 ◇
SINGLETARY, W.E. *A complex of problems proposed by Post* ◇ B20 D03 D25 D30 D35 ◇
TAJTSLIN, M.A. *On elementary theories of free nilpotent algebras (Russian)* ◇ B25 C60 D35 ◇
YNTEMA, M.K. *A detailed argument for the Post-Linial theorems* ◇ B20 D35 ◇

1965

ALMAGAMBETOV, ZH.A. *On classes of axioms closed with respect to the given reduced products and powers (Russian)* ◇ B25 C20 C52 D35 ◇
AX, J. *The undecidability of power series fields* ◇ D35 ◇

CHOWLA, S. *The Riemann hypothesis and Hilbert's tenth problem* ◇ D35 ◇

CHOWLA, S. *The Riemann hypothesis and Hilbert's tenth problem. Mathematics and its applications Vol. 4* ◇ D35 D98 ◇

DAVIS, MARTIN D. (ED.) *The undecidable. Basic papers on undecidable propositions, unsolvable problems and computable functions* ◇ D20 D25 D35 D97 F30 ◇

ERSHOV, YU.L. & LAVROV, I.A. & TAJMANOV, A.D. & TAJTSLIN, M.A. *Elementary theories (Russian)* ◇ B25 C98 D35 D98 ◇

ERSHOV, YU.L. *Undecidability of certain fields (Russian)* ◇ D35 ◇

GLADSTONE, M.D. *Some ways of constructing a propositional calculus of any required degree of unsolvability* ◇ B20 D25 D35 ◇

GODDARD, L. *An augmented modal logic* ◇ B45 D35 ◇

GOEDEL, K. *On undecidable propositions of formal mathematical systems* ◇ D20 D35 F30 ◇

GUREVICH, Y. *Existential interpretation (Russian)* ◇ B20 B25 C60 D35 F25 ◇

HANF, W.P. *Model-theoretic methods in the study of elementary logic* ◇ B25 C07 D20 D25 D35 F25 ◇

HARROP, R. *Some generalizations and applications of a relativization procedure for propositional calculi* ◇ B22 B25 D35 ◇

HERMES, H. & ROEDDING, D. *A method for producing reduction types in the restricted lower predicate calculus* ◇ D35 ◇

MASLOV, S.YU. & MINTS, G.E. & OREVKOV, V.P. *Unsolvability in the constructive predicate calculus of certain classes of formulas containing only monadic predicate variables (Russian)* ◇ B20 D35 F50 ◇

MAYOH, B.H. *Unsolvable problems in the theory of computable numbers* ◇ D20 D35 F60 ◇

PATTON, T.E. *Church's theorem on the decision problem* ◇ D35 ◇

POST, E.L. *Absolutely unsolvable problems and relatively undecidable propositions* ◇ D35 ◇

RABIN, M.O. *A simple method for undecidability proofs and some applications* ◇ D35 F25 ◇

REICHBACH, J. *On characterization and undecidability of the first-order functional calculus* ◇ D35 ◇

ROBINSON, JULIA *The decision problem for fields* ◇ B25 C60 D35 ◇

SHEPHERDSON, J.C. *Machine configuration and word problems of given degree of unsolvability* ◇ D05 D30 D35 D40 ◇

SZCZERBA, L.W. & TARSKI, A. *Metamathematical properties of some affine geometries* ◇ B25 B30 C52 C65 D35 ◇

TAJTSLIN, M.A. *On the elementary theory of classical Lie algebras (Russian)* ◇ C60 D35 ◇

TAJTSLIN, M.A. *On the theory of finite rings with division (Russian)* ◇ C13 D35 ◇

WANG, HAO *Remarks on machines, sets, and the decision problem* ◇ B20 B25 D03 D05 D10 D35 E30 ◇

1966

DAVIS, MARTIN D. *Recursive functions - an introduction* ◇ D20 D35 ◇

DISHKANT, G.P. *Zur Frage der Entscheidbarkeit mechanischer Theorien (Russisch)* ◇ D35 ◇

ELGOT, C.C. & RABIN, M.O. *Decidability and undecidability of extensions of second (first) order theory of (generalized) successor* ◇ B15 B25 C85 D05 D35 F30 F35 ◇

ERSHOV, YU.L. *New examples of undecidable theories (Russian)* ◇ D35 ◇

GODDARD, L. & ROUTLEY, R. *Use, mention and quotation* ◇ A05 D35 ◇

GRANDY, R.E. *A note on the recursive unsolvability of primitive recursive arithmetic* ◇ D35 F30 ◇

GUREVICH, Y. *Certain algorithmic questions of the theory of classes of algebraic systems (Russian)* ◇ B25 D35 ◇

GUREVICH, Y. *Effective recognition of satisfiability of formulae of the restricted predicate calculus (Russian)* ◇ B20 B25 D35 ◇

GUREVICH, Y. *On the decision problem for pure restricted predicate logic (Russian)* ◇ B20 C13 D35 ◇

GUREVICH, Y. *The decision problem for the restricted predicate calculus (Russian)* ◇ B20 C13 D35 ◇

HAUSCHILD, K. *Ueber die Unentscheidbarkeit nichtassoziativer Schiefringe (English and French summaries)* ◇ D35 ◇

KENZHEBAEV, S. *Certain undecidable rings (Russian) (Kazakh summary)* ◇ C60 D35 ◇

KENZHEBAEV, S. *Undecidability of the elementary theory of a certain ring (Russian)* ◇ D35 ◇

KOGALOVSKIJ, S.R. *On the properties preserved under algebraic constructions (Russian)* ◇ B25 C30 C52 C60 D35 D45 ◇

KOSTYRKO, V.F. *On the AEA reduction class (Russian) (English summary)* ◇ B20 D35 ◇

QUINE, W.V.O. *Church's theorem on the decision problem* ◇ D35 ◇

ROEDDING, D. *Der Entscheidbarkeitsbegriff in der mathematischen Logik* ◇ B25 D10 D20 D25 D35 ◇

STEINER, H.-G. *Verschiedene Aspekte der axiomatischen Methode im Unterricht* ◇ D35 D98 ◇

TAJMANOV, A.D. *On formulae of Horn-type (Russian)* ◇ B20 D35 ◇

THATCHER, J.W. *Decision problems for multiple successor arithmetics* ◇ D35 ◇

1967

BOLLMAN, D.A. *Formal nonassociative number theory* ◇ B28 D35 ◇

BONDI, I.L. *The decidability or undecidability of certain formal logical calculi (Russian)* ◇ B25 D35 ◇

CLAPHAM, C.R.J. *An embbedding theorem for finitely generated groups* ◇ D35 D40 ◇

GALVIN, F. *Reduced products, Horn sentences, and decision problems* ◇ B20 B25 C20 C30 C40 D35 ◇

GUREVICH, Y. *A contribution to the elementary theory of lattice ordered abelian groups and K-lineals (Russian)* ◇ B25 C60 D35 ◇

GUREVICH, Y. *Hereditary undecidability of a class of lattice-ordered abelian groups (Russian)(English summary)* ◇ D35 ◇
KHISAMIEV, N.G. *Unsolvability of the elementary theory of a free lattice (Russian)* ◇ D35 G10 ◇
KILMISTER, C.W. *Language, logic and mathematics* ◇ A05 B98 D35 ◇
KRIPKE, S.A. & POUR-EL, M.B. *Deduction-preserving "recursive isomorphisms" between theories* ◇ B30 D25 D35 F30 ◇
KROM, MELVEN R. *The decision problem for segregated formulas in first-order logic* ◇ B20 B25 D35 ◇
LIFSCHITZ, V. *Deductive general validity and reduction classes (Russian)* ◇ B20 D35 ◇
LIFSCHITZ, V. *Some reduction classes and undecidable theories (Russian)* ◇ D35 ◇
LIFSCHITZ, V. *The decision problem for some constructive theories of equality (Russian)* ◇ B25 D35 F50 ◇
MIHAILESCU, E.G. *Decision problem in the classical logic* ◇ B25 D35 ◇
NAGEL, E. & NEWMAN, J.R. *Goedel's proof* ◇ D35 F30 ◇
OREVKOV, V.P. *The undecidability of a class of formulas containing just one single place predicate variable in modal calculus (Russian)* ◇ B45 D35 ◇
PAOLA DI, R.A. *A survey of Soviet work in the theory of computer programming* ◇ B25 B75 D35 G30 ◇
PAOLA DI, R.A. *Some theorems on extensions of arithmetic* ◇ D25 D35 F15 F30 ◇
PERKINS, P. *Unsolvable problems for equational theories* ◇ C05 D35 ◇
RAUTENBERG, W. *Elementare Schemata nichtelementarer Axiome* ◇ B25 C52 C60 C65 C85 D35 ◇
SINGLETARY, W.E. *Recursive unsolvability of a complex of problems proposed by Post* ◇ D03 D25 D35 ◇
TAJTSLIN, M.A. *Some further examples of undecidable theories (Russian) (English summary)* ◇ D35 ◇
VIDAL ABASCAL, E. *Undecidability in mathematics* ◇ D35 ◇
VUCKOVIC, V. *Mathematics of incompleteness and undecidability* ◇ D35 F30 ◇

1968

BOONE, W.W. *Decision problems about algebraic and logical systems as a whole and recursively enumerable degrees of unsolvability* ◇ D25 D35 D40 ◇
BOONE, W.W. & HAKEN, W. & POENARU, V. *On recursively unsolvable problems in topology and their classification* ◇ D35 D80 ◇
DAVIS, MARTIN D. *One equation to rule them all* ◇ D25 D35 ◇
ERSHOV, YU.L. *Restricted theories of totally ordered sets (Russian)* ◇ B25 C65 D35 E07 ◇
GLADSTONE, M.D. *A single-axiom implicational calculus of given unsolvability* ◇ B20 D25 D30 D35 ◇
HIROSE, K. *A conjecture on Hilbert's 10th problem* ◇ D30 D35 ◇
MATIYASEVICH, YU.V. *Arithmetic representations of powers (Russian)* ◇ D25 D35 ◇
MATIYASEVICH, YU.V. *The connection between Hilbert's tenth problem and systems of equations between words and lengths (Russian)* ◇ D25 D35 D40 ◇
MATIYASEVICH, YU.V. *Two reductions of Hilbert's tenth problem (Russian)* ◇ D25 D35 ◇
MEYER, R.K. *An undecidability result in the theory of relevant implication* ◇ B46 D35 ◇
OREVKOV, V.P. *Two undecidable classes of formulas in classical predicate calculus (Russian)* ◇ B20 D35 ◇
PIECZKOWSKI, A. *Undecidability of the homogeneous formulas of degree 3 of the predicate calculus (Polish and Russian summaries)* ◇ B20 D35 ◇
RAUTENBERG, W. *Unterscheidbarkeit endlicher geordneter Mengen mit gegebener Anzahl von Quantoren* ◇ B20 C07 C13 C65 D35 ◇
SHAYAKHMETOV, T.K. *Undecidability of certain theories with a supplemental predicate (Russian) (Kazakh summary)* ◇ D35 ◇
TAJTSLIN, M.A. *Elementary lattice theories for ideals in polynomial rings (Russian)* ◇ C60 D35 G10 ◇
TAJTSLIN, M.A. *On simple ideals in polynomial rings (Russian)* ◇ D35 ◇
TOURNEAU LE, J.J. *Decision problems related to the concept of operation* ◇ C05 D35 ◇
VALIEV, M.K. *A theorem of G.Higman (Russian)* ◇ C60 D35 D40 ◇

1969

ADLER, A. *Extensions of non-standard models of number theory* ◇ C62 D25 D35 H15 ◇
ADLER, A. *Some recursively unsolvable problems in analysis* ◇ D35 D80 ◇
BAER, ROBERT M. *Definability by Turing machines* ◇ D10 D35 ◇
EVANS, T. *An unsolvable problem concerning identities* ◇ D35 ◇
GINSBURG, S. & PARTEE, B. *A mathematical model of transformational grammars* ◇ B65 D35 ◇
GUREVICH, Y. *A decision problem for decision problems (Russian)* ◇ B25 D35 ◇
GUREVICH, Y. *The decision problem for the logic of predicates and operations (Russian)* ◇ B25 D35 ◇
HECKER, H.-D. *Definierbarkeit und Unentscheidbarkeit* ◇ D20 D35 ◇
HECKER, H.-D. *Riegeralgebren und rekursive Untrennbarkeit* ◇ D25 D35 ◇
HERMES, H. *Basic notions and applications of the theory of decidability* ◇ D10 D20 D25 D30 D35 D50 D98 ◇
HEYTING, A. *What is computable ? (Dutch)* ◇ D20 D35 D55 ◇
HUBER-DYSON, V. *On the decision problem for extensions of a decidable theory* ◇ B25 C13 C60 D35 ◇
KHMELEVSKIJ, YU.I. *On Hilbert's tenth problem (Russian)* ◇ D35 ◇
MEDVEDEV, YU.T. *A method for proving the unsolvability of algorithmic problems (Russian)* ◇ D35 F50 ◇
PAOLA DI, R.A. *The recursive unsolvability of the decision problem for the class of definite formulas* ◇ D35 ◇
POUR-EL, M.B. *A recursion-theoretic view of axiomatizable theories* ◇ C07 D25 D35 ◇
POUR-EL, M.B. *Independent axiomatization and its relation to the hypersimple set* ◇ C07 D25 D35 ◇

QUINE, W.V.O. *The limits of decision* ⋄ D35 ⋄
ROBINSON, JULIA *Diophantine decision problems* ⋄ D25 D35 ⋄
ROBINSON, JULIA *Unsolvable diophantine problems* ⋄ D25 D35 ⋄
SCHWARTZ, T. *A simple treatment of Church's theorem on the decision problem* ⋄ D35 ⋄
SLOMSON, A. *An undecidable two sorted predicate calculus* ⋄ B10 D35 ⋄

1970

APPLEBEE, R.C. & PAHI, B. *An unsolvable problem concerning implicational calculi* ⋄ B22 D35 ⋄
BOKUT', L.A. *Unsolvability of certain algorithmic problems in the class of associative rings (Russian)* ⋄ D35 D40 ⋄
CAVINESS, B.F. *On canonical forms and simplification* ⋄ B35 D35 ⋄
CHUDNOVSKY, G.V. *Diophantine predicates (Russian)* ⋄ D25 D35 D40 ⋄
COLLINS, G.E. & HALPERN, J.D. *On the interpretability of arithmetic in set theory* ⋄ B28 D35 E30 F25 F30 ⋄
GALVIN, F. *Horn sentences* ⋄ B20 C20 C30 C40 D35 ⋄
GUREVICH, Y. *Minsky machines and the case* $\forall\exists\forall$ & $\exists^{\infty}$ *of the decision problem (Russian)* ⋄ B20 D05 D35 ⋄
HAUSCHILD, K. & RAUTENBERG, W. *Universelle Interpretierbarkeit in Verbaenden (Russian, English and French summaries)* ⋄ D35 F25 G10 ⋄
HERMES, H. *Entscheidungsproblem und Dominospiele* ⋄ D05 D10 D35 ⋄
KIPNIS, M.M. *Invariant properties of systems of formulae of elementary axiomatic theories (Russian)* ⋄ D25 D35 ⋄
KOZLOV, G.T. *Unsolvability of the elementary theory of lattices of subgroups of finite abelian p-groups (Russian)* ⋄ C60 D35 ⋄
KROM, MELVEN R. *The decision problem for formulas in prenex conjunctive normal form with binary disjunctions* ⋄ B20 D03 D35 ⋄
MATIYASEVICH, YU.V. *Enumerable sets are diophantine (Russian)* ⋄ D25 D35 ⋄
MATIYASEVICH, YU.V. *Solution of the tenth problem of Hilbert* ⋄ B28 D25 D35 ⋄
MAYOH, B.H. *The relation between an object and its name: notation systems and their fixed point theorems* ⋄ A05 B03 D20 D35 D45 F30 F60 ⋄
NAKAMURA, A. *On a propositional calculus whose decision problem is recursively unsolvable* ⋄ B22 D35 ⋄
NAKAMURA, A. *On the undecidability of monadic modal predicate logic* ⋄ B45 D35 ⋄
POUR-EL, M.B. *A recursion-theoretic view of axiomatizable theories* ⋄ C07 D25 D35 ⋄
ROEDDING, D. *Reduktionstypen der Praedikatenlogik. Nach einer einstuendigen Vorlesung ausgearbeitet von Egon Boerger* ⋄ D35 ⋄
SHARONOV, V.I. & ZAMOV, N.K. *Amplifications of formulae of predicate calculus (Russian)* ⋄ B10 B20 D35 ⋄
TAJTSLIN, M.A. *On elementary theories of lattices of subgroups (Russian)* ⋄ B25 C60 D35 G10 ⋄
THOMASON, S.K. *On initial segments of hyperdegrees* ⋄ D30 D35 E40 ⋄

1971

AANDERAA, S.O. *On the decision problem for formulas in which all disjunctions are binary* ⋄ B20 D35 ⋄
AZRA, J.-P. *Relations diophantiennes et la solution negative du 10e probleme de Hilbert* ⋄ D25 D35 ⋄
BONDI, I.L. *The solvability problem for certain formal-logical calculi (Russian)* ⋄ B25 D35 ⋄
BRADFORD, R. *Cardinal addition and the axiom of choice* ⋄ D35 E10 E25 ⋄
DAVIS, MARTIN D. *An explicit diophantine definition of the exponential function* ⋄ D35 ⋄
DULAC, M.-H. *Decidabilite et operations entre theories* ⋄ B10 B25 D35 ⋄
EHRENFEUCHT, A. & MYCIELSKI, J. *Abbreviating proofs by adding new axioms* ⋄ D35 F20 ⋄
FENSTAD, J.E. *Hilbert's 10th problem (Norwegian) (English summary)* ⋄ D25 D35 ⋄
GARFUNKEL, S. & SHANK, H.S. *On the undecidability of finite planar graphs* ⋄ C13 D35 ⋄
GERMANO, G. *Incompleteness and truth definitions* ⋄ C40 D35 F30 ⋄
HAUSCHILD, K. & RAUTENBERG, W. *Interpretierbarkeit in der Gruppentheorie* ⋄ C60 D35 F25 ⋄
HAUSCHILD, K. & RAUTENBERG, W. *Interpretierbarkeit und Entscheidbarkeit in der Graphentheorie I* ⋄ B25 C65 D35 F25 ⋄
HECKER, H.-D. *Allgemeine Trennbarkeitsbegriffe bei meta-mathematischen Untersuchungen* ⋄ D25 D35 D75 ⋄
HECKER, H.-D. *Rekursive Untrennbarkeit bei elementaren Theorien* ⋄ D25 D35 ⋄
HERMES, H. *A simplified proof for the unsolvability of the decision problem in the case* $\forall\exists\forall$ ⋄ B20 D35 ⋄
KOLMOGOROV, A.N. & VARPAKHOVSKIJ, F.I. *The solution of Hilbert's tenth problem (Bulgarian)* ⋄ D35 ⋄
KOSOVSKIJ, N.K. *Diophantine representations of the sequence of solutions of the Pell equation (Russian) (English summary)* ⋄ D35 ⋄
KOSTYRKO, V.F. *The reduction class* $\forall x \forall y \exists z F(x,y,z) \wedge \forall^m \mathfrak{A}(F)$ *(Russian) (English summary)* ⋄ B20 C13 D35 ⋄
KOSTYRKO, V.F. *Undecidability of the elementary* $\exists$*-theory of groupoids (Russian)* ⋄ D35 ⋄
KRECZMAR, A. *The set of all tautologies of algorithmic logic is hyperarithmetical* ⋄ B75 D35 D55 ⋄
LINDNER, C.C. *Finite partial cyclic triple systems can be finitely embedded* ⋄ C05 C13 D35 ⋄
MACINTYRE, A. *On the elementary theory of Banach algebras* ⋄ C60 C65 D35 ⋄
MATIYASEVICH, YU.V. *Diophantine representation of the set of prime numbers (Russian)* ⋄ D25 D35 ⋄
MATIYASEVICH, YU.V. *Diophantine representation of enumerable predicates (Russian)* ⋄ D25 D35 ⋄
MATIYASEVICH, YU.V. *Diophantine representation of recursively enumerable predicates* ⋄ D25 D35 ⋄

McKenzie, R. *Negative solution of the decision problem for sentences true in every subalgebra of* $\langle N, + \rangle$ ◇ D35 F30 ◇
Murskij, V.L. *Nondiscernible properties of finite systems of identity relations (Russian)* ◇ C05 D35 D40 ◇
Orevkov, V.P. *On biconjunctive reduction classes (Russian) (English summary)* ◇ B20 D35 ◇
Palyutin, E.A. *Boolean algebras with a categorical theory in a weak second order logic (Russian)* ◇ B15 C35 C85 D35 G05 ◇
Paola di, R.A. *The relational data file and the decision problem for classes of proper formulas* ◇ B20 B75 D35 ◇
Pinus, A.G. *A remark on the paper by Ju.M.Vazhenin: "Elementary properties of transformation semigroups of ordered sets" (Russian)* ◇ B25 C65 D35 E07 ◇
Robinson, Julia *Hilbert's tenth problem* ◇ D35 ◇
Sabbagh, G. *Logique mathematique V. Decidabilite et fonctions recursives* ◇ B25 D20 D35 ◇
Siefkes, D. *Undecidable extensions of monadic second order successor arithmetic* ◇ B15 B25 C85 D35 F35 ◇
Suranyi, J. *Reduction of the decision problem of the first order predicate calculus to reflexive and symmetrical binary predicates* ◇ B20 D35 ◇
Sussmann, H.J. *Hilbert's tenth problem* ◇ D25 D35 ◇
Szczerba, L.W. *Undecidability of elementary Pasch-Free geometry (Russian summary)* ◇ D35 ◇
Turan, P. *On the work of Alan Baker* ◇ B25 D35 ◇

1972

Bollman, D.A. & Tapia, M. *On the recursive unsolvability of the provability of the deduction theorem in partial propositional calculi* ◇ B20 D35 ◇
Davis, Martin D. *On the number of the solutions of diophantine equations* ◇ D25 D35 ◇
Ershov, Yu.L. *Elementary group theories (Russian)* ◇ B25 C05 C13 C60 D35 ◇
Fridman, Eh.I. & Penzin, Yu.G. *The elementary and the universal theory of the ordered group of integers with maximal subgroups (Russian)* ◇ B20 B25 D35 ◇
Fridman, Eh.I. *Undecidability of the elementary theory of abelian torsion-free groups with a finite set of serving subgroups (Russian)* ◇ C60 D35 ◇
Gabbay, D.M. *Sufficient conditions for the undecidability of intuitionistic theories with applications* ◇ D35 F50 ◇
Garfunkel, S. & Shank, H.S. *On the indecidability of finite planar cubic graphs* ◇ C13 D35 ◇
Hauschild, K. & Herre, H. & Rautenberg, W. *Interpretierbarkeit und Entscheidbarkeit in der Graphentheorie II* ◇ B25 C65 D35 F25 ◇
Hauschild, K. *Universalitaet in der Ringtheorie* ◇ C60 D35 F25 ◇
Henson, C.W. *Countable homogeneous relational structures and* $\aleph_0$*-categorical theories* ◇ C10 C15 C35 C50 D35 ◇
Hermes, H. *Die Unloesbarkeit des zehnten Hilbertschen Problems* ◇ D25 D35 ◇
Kozlov, G.T. *Undecidability of the theory of abelian groups with a chain of subgroups (Russian)* ◇ D35 ◇
Kreczmar, A. *Degree of recursive unsolvability of algorithmic logic (Russian summary)* ◇ B75 D30 D35 ◇
Matiyasevich, Yu.V. *Diophantine sets (Russian)* ◇ D25 D35 ◇
Matiyasevich, Yu.V. *Diophantine representation of enumerable predicates (Russian)* ◇ D25 D35 ◇
Orevkov, V.P. *Undecidable classes of formulas for the constructive predicate calculus I (Russian)* ◇ D35 F50 ◇
Perkins, P. *An unsolvable provability problem for one variable groupoid equations* ◇ C05 D35 ◇
Pinus, A.G. *On the theory of convex subsets (Russian)* ◇ B25 C65 C85 D35 ◇
Robinson, R.M. *Some representations of diophantine sets* ◇ D25 D35 ◇
Ruohonen, K. *Hilbert's tenth problem (Finnish) (English summary)* ◇ D25 D35 ◇
Sacerdote, G.S. *On a problem of Boone* ◇ C60 D30 D35 D40 ◇
Seese, D.G. *Entscheidbarkeits- und Definierbarkeitsfragen der Theorie "netzartiger" Graphen I (Russian) (English and French summaries)* ◇ C80 C85 D35 ◇
Smith, D.D. *Non-recursiveness of the set of finite sets of equations whose theories are one-based* ◇ C05 D35 ◇

1973

Aanderaa, S.O. & Lewis, H.R. *Prefix classes of Krom formulas* ◇ B20 B25 D35 ◇
Adyan, S.I. *The works of P.S.Novikov and his students on algorithmic questions of algebra (Russian)* ◇ D35 D40 ◇
Anonymous *Problems* ◇ D35 D40 ◇
Bondi, I.L. *The decision problem for metric spaces (Russian)* ◇ D35 ◇
Bondi, I.L. *The decision problem for normed linear spaces and for Hilbert spaces (Russian)* ◇ B25 C65 D35 ◇
Davis, Martin D. *Hilbert's tenth problem is unsolvable* ◇ D25 D35 ◇
Fridman, Eh.I. *Positive element-subgroup theories (Russian)* ◇ B25 D35 ◇
Gabbay, D.M. *The undecidability of intuitionistic theories of algebraically closed fields and real closed fields* ◇ B25 C60 C90 D35 F50 ◇
Gao, Hengshan *The decision problem of modal predicate calculi. I: On Slomson's reductions (Chinese)* ◇ B45 D35 ◇
Gao, Hengshan *Two remarks on the decision problem (Chinese)* ◇ D35 ◇
Goldfarb, W.D. & Lewis, H.R. *The decision problem for formulas with a small number of atomic subformulas* ◇ B20 D35 ◇
Gurevich, Y. & Turashvili, T.V. *A strengthening of a certain result of J. Suranyi (Russian) (Georgian and English summaries)* ◇ B20 D35 ◇
Haken, W. *Connections between topological and group theoretical decision problems* ◇ B25 C60 C65 D35 D40 D80 ◇

HAVEL, I.M. *Ueber das zehnte Hilbertsche Problem (Tschechisch)* ⋄ D35 ⋄
HERRE, H. *Unentscheidbarkeit in der Graphentheorie* ⋄ D35 ⋄
HIROSE, K. & IIDA, S. *A proof of negative answer to Hilbert's 10th problem* ⋄ D25 D35 ⋄
HIROSE, K. *On Hilbert's tenth problem (negative solution)* ⋄ D25 D35 ⋄
KOKORIN, A.I. & MART'YANOV, V.I. *Universal extended theories (Russian)* ⋄ B25 C52 D35 ⋄
KUSABA, T. *On Hilbert's tenth problem (affirmative cases)* ⋄ D25 D35 ⋄
MANIN, YU.I. *Hilbert's tenth problem (Russian)* ⋄ D25 D35 D80 ⋄
MATIYASEVICH, YU.V. *On recursive unsolvability of Hilbert's tenth problem* ⋄ D25 D35 ⋄
MEYER, R.K. & ROUTLEY, R. *An undecidable relevant logic* ⋄ B46 D35 D40 ⋄
MILLER III, C.F. *Some connections between Hilbert's 10th problem and the theory of groups* ⋄ D35 D40 ⋄
PENZIN, YU.G. *The undecidability of fields of rational functions over fields of characteristic 2 (Russian)* ⋄ D35 ⋄
PINUS, A.G. *A weak second order theory of fixed sets (Russian)* ⋄ B15 D35 ⋄
PUTNAM, H. *Recursive functions and hierarchies* ⋄ D20 D30 D35 D55 ⋄
ROBINSON, JULIA *Solving diophantine equations* ⋄ B25 D35 D80 H15 ⋄
SHABUNIN, L.V. *The undecidability of certain formal systems of combinatory logic (Russian)* ⋄ B40 D35 ⋄
SMORYNSKI, C.A. *Elementary intuitionistic theories* ⋄ B25 C90 D35 F50 ⋄
TAKAHASHI, S. *A non-standard treatment of infinitely near points* ⋄ D35 H15 H20 ⋄
VASIL'EV, EH.S. *The elementary theories of complete torsion-free abelian groups with p-adic topology (Russian)* ⋄ B25 C20 C60 D35 ⋄
VERKHOZINA, M.I. *The undecidability of the separation problem for positive fragments of logical calculi (Russian)* ⋄ D35 F50 ⋄
VIDAL-NAQUET, G. *Quelques applications des automates a arbres infinis* ⋄ B25 D05 D35 ⋄

1974

AANDERAA, S.O. & LEWIS, H.R. *Linear sampling and the* $\forall\exists\forall$ *case of the decision problem* ⋄ B20 D03 D05 D35 D80 ⋄
ANDREWS, P.B. *Provability in elementary type theory* ⋄ B15 B25 D35 F35 ⋄
BAUDISCH, A. *Theorien abelscher Gruppen mit einem einstelligen Praedikat* ⋄ C60 D35 ⋄
BERNARDI, C. *Aspetti ricorsivi degli insiemi dialettici (English summary)* ⋄ D30 D35 G10 ⋄
BOERGER, E. Σ_3*-completude de l'ensemble des types de reduction* ⋄ B20 D35 D55 ⋄
BOERGER, E. *Beitrag zur Reduktion des Entscheidungsproblems auf Klassen von Hornformeln mit kurzen Alternationen* ⋄ B20 B25 D35 ⋄
BOERGER, E. *Die rekursive Unloesbarkeit des zehnten Hilbertschen Problems* ⋄ D35 ⋄
BOFFA, M. *Hierarchie analytique et ensembles constructibles* ⋄ D35 D55 E15 E45 ⋄
DURNEV, V.G. *Equations on free semigroups and groups (Russian)* ⋄ D35 ⋄
DURNEV, V.G. *Positive formulas in free semigroups (Russian)* ⋄ D35 D40 ⋄
ERSHOV, YU.L. *Theories of nonabelian varieties of groups* ⋄ C05 D35 ⋄
GARFUNKEL, S. & SCHMERL, J.H. *The undecidability of theories of groupoids with an extra predicate* ⋄ D35 ⋄
HAUSCHILD, K. *Rekursive Unentscheidbarkeit der Theorie der pythagoraeischen Koerper* ⋄ D35 ⋄
JONES, JAMES P. *Recursive undecidability - an exposition* ⋄ D10 D20 D35 D40 D98 ⋄
KESEL'MAN, D.YA. *The elementary theories of graphs and abelian loops (Russian)* ⋄ C65 D35 ⋄
MARKOV, A.A. *On the language* $Я_1$ *(Russian)* ⋄ D03 D35 F50 ⋄
MATIYASEVICH, YU.V. *Existence of noneffectivizable estimates in the theory of exponentially diophantine equations (Russian) (English summary)* ⋄ D25 D35 ⋄
MATIYASEVICH, YU.V. & ROBINSON, JULIA *Two universal three-quantifier representations of enumerable sets (Russian)* ⋄ D25 D35 ⋄
PIGOZZI, D. *The join of equational theories* ⋄ B25 C05 D35 ⋄
SHABUNIN, L.V. *Some algorithmic problems of calculi of combinatory logic (Russian) (English summary)* ⋄ B40 D35 ⋄
SOPRUNOV, S.F. *On the power of a real-closed field (Russian) (English summary)* ⋄ C20 C60 D35 ⋄
TURASHVILI, T.V. *A reduction of the decidability problem of first order predicate logic to a class with asymetric and irreflexive two-place predicates (Russian) (Georgian and English summaries)* ⋄ D35 ⋄
VAZHENIN, YU.M. *On the elementary theory of free inverse semigroups* ⋄ B25 C60 D35 ⋄
VAZHENIN, YU.M. *The elementary theories of symmetric groups and semigroups (Russian)* ⋄ D35 ⋄

1975

ALTON, D.A. *A characterization of R-maximal sets* ⋄ D25 D35 ⋄
BAUDISCH, A. *Elementare Theorien von Halbgruppen mit Kuerzungsregeln mit einem einstelligen Praedikat* ⋄ C05 C60 D35 ⋄
BAUR, W. *Decidability and undecidability of theories of abelian groups with predicates for subgroups* ⋄ B25 C60 D35 ⋄
BURRIS, S. & SANKAPPANAVAR, H.P. *Lattice-theoretic decision problems in universal algebra* ⋄ B25 C05 D35 ⋄
DENEF, J. *Hilbert's tenth problem for quadratic rings* ⋄ C60 D35 ⋄
DEUTSCH, M. *Die Unentscheidbarkeit und Nichtkalkuelisierbarkeit der Mengenlehre* ⋄ D35 E30 ⋄
DEUTSCH, M. *Zur Theorie der spektralen Darstellung von Praedikaten durch Ausdruecke der Praedikatenlogik 1.Stufe* ⋄ B10 C13 D25 D35 ⋄

FRIDMAN, EH.I. & SLOBODSKOJ, A.M. *Theories of Abelian groups with predicates specifying a subgroup (Russian)* ◇ C60 D35 ◇
GOLDFARB, W.D. & LEWIS, H.R. *Skolem reduction classes* ◇ B20 D35 ◇
GOODSTEIN, R.L. *Hilbert's tenth problem and the independence of recursive difference* ◇ D25 D35 ◇
GRASSELLI, J. *On the tenth Hilbert problem (Slovene) (English summary)* ◇ D35 ◇
HERRE, H. & WOLTER, H. *Entscheidbarkeit von Theorien in Logiken mit verallgemeinerten Quantoren* ◇ B25 C10 C55 C80 D35 ◇
JONSSON, B. & MCNULTY, G.F. & QUACKENBUSH, R.W. *The ascending and descending varietal chains of a variety* ◇ B25 C05 D35 ◇
KOPIEKI, R. & SARALSKI, B. & WALIGORA, G. *The research of Jaskowski on decidability theory of first order sentences I* ◇ A10 B20 B25 D35 ◇
MANASTER, A.B. *Completeness, compactness, and undecidability: an introduction to mathematical logic* ◇ B98 C07 D35 D98 ◇
MATIYASEVICH, YU.V. & ROBINSON, JULIA *Reduction of an arbitrary diophantine equation to one in 13 unknowns* ◇ D25 D35 ◇
MCKENZIE, R. *On spectra, and the negative solution of the decision problem for identities having a finite nontrivial model* ◇ B20 C05 C13 D35 ◇
OKEE, J. *A semantical proof of the undecidability of the monadic intuitionistic predicate calculus of first order* ◇ D35 F50 ◇
PODNIEKS, K.M. *The double-incompleteness theorem (Russian) (English summary)* ◇ D35 F30 ◇
PRESTEL, A. & ZIEGLER, M. *Erblich euklidische Koerper* ◇ C60 D35 ◇
RAUTENBERG, W. & ZIEGLER, M. *Recursive inseparability in graph theory* ◇ C13 D35 ◇
SEESE, D.G. *Ein Unentscheidbarkeitskriterium* ◇ D35 ◇
SHABUNIN, L.V. *Combinatory calculi I,II (Russian)* ◇ B40 D35 ◇
SHELAH, S. *The monadic theory of order* ◇ B25 C65 C85 D35 E50 ◇
THOMAS, WOLFGANG *A note on undecidable extensions of monadic second order successor arithmetic* ◇ D35 F35 ◇
TURASHVILI, T.V. *The decidability problem of first order predicate logic (Russian)* ◇ B20 D35 ◇
YABLON, P. *A generalized propositional calculus* ◇ B60 D35 ◇

1976

BARENDREGT, H.P. *The incompleteness theorems* ◇ D35 F30 ◇
BAUR, W. *Undecidability of the theory of abelian groups with a subgroup* ◇ D35 ◇
BOERGER, E. & HEIDLER, K. *Die m-Grade logischer Entscheidungsprobleme* ◇ D25 D35 D55 ◇
CAICEDO, X. *Algorithmen, Fibonaccizahlen und das zehnte Hilbertsche Problem (Spanish)* ◇ D35 D98 ◇
CHRISTEN, C. *Spektralproblem und Komplexitaetstheorie* ◇ B10 C13 D10 D15 D25 D35 ◇
DAVIS, MARTIN D. & MATIYASEVICH, YU.V. & ROBINSON, JULIA *Hilbert's tenth problem: Diophantine equations: Positive aspects of a negative solution* ◇ D25 D35 D80 ◇
FRIDMAN, EH.I. & SLOBODSKOJ, A.M. *Undecidable universal theories of lattices of subgroups of abelian groups (Russian)* ◇ D35 ◇
FRIED, M. & SACERDOTE, G.S. *Solving diophantine problems over all residue class fields of a number field and all finite fields* ◇ B25 C10 C13 C60 D35 ◇
FRIEDMAN, H.M. *On decidability of equational theories* ◇ B25 C05 D35 ◇
GAO, HENGSHAN *The decision problem of modal predicate calculi. II: On Slomson's reductions (Chinese)* ◇ B45 C13 D35 ◇
GONCHAROV, S.S. *Restricted theories of constructive boolean algebras (Russian)* ◇ B25 C57 D35 F60 G05 ◇
GUREVICH, Y. *Semi-conservative reduction* ◇ B20 D25 D35 ◇
GUREVICH, Y. *The decision problem for standard classes* ◇ B20 B25 D35 ◇
HACK, M. *The equality problem for vector addition systems is undecidable* ◇ D35 D80 ◇
HERMES, H. *Domino games* ◇ D05 D35 ◇
HERMES, H. *Hilbert's tenth problem* ◇ A10 D25 D35 ◇
HUGHES, C.E. *A reduction class containing formulas with one monadic predicate and one binary function symbol* ◇ B20 D35 ◇
LEWIS, H.R. *Krom formulas with one dyadic predicate letter* ◇ B20 D05 D35 ◇
LOEB, M.H. *Embedding first order predicate logic in fragments of intuitionistic logic* ◇ B10 D35 F50 ◇
MCNULTY, G.F. *The decision problem for equational bases of algebras* ◇ C05 D35 ◇
MCNULTY, G.F. *Undecidable properties of finite sets of equations* ◇ C05 D35 ◇
NORGELA, S.A. *On approximating reduction classes of CPC by decidable classes (Russian) (English summary)* ◇ B20 B25 D35 ◇
PENZIN, YU.G. *Undecidability of a theory of the integers with addition and predicate mutually disjoint (Russian)* ◇ D35 F30 ◇
PIGOZZI, D. *The universality of the variety of quasigroups* ◇ B25 C05 C52 D35 ◇
RUBIN, M. *The theory of boolean algebras with a distinguished subalgebra is undecidable* ◇ D35 G05 G25 ◇
TAJMANOV, A.D. *An algorithmic problem of number theory (Russian)* ◇ D35 D80 ◇
THIELER-MEVISSEN, G. *Zur Beschreibbarkeit der hyperarithmetischen reellen Zahlen mit analysiskonformen Mitteln* ◇ D35 D55 ◇
WEESE, M. *The universality of boolean algebras with the Haertig quantifier* ◇ C55 C80 D35 F25 G05 ◇
ZIEGLER, M. *A language for topological structures which satisfies a Lindstroem-theorem* ◇ B25 C40 C50 C65 C80 C90 C95 D35 ◇

1977

CARSTENS, H.G. *The theorem of Matijasevic is provable in Peano's arithmetic by finitely many axioms* ◇ C52 D25 D35 F30 H15 ◇

DAVIS, MARTIN D. & HERSH, R. *Hilbert's tenth problem* ◇ D35 ◇

DAVIS, MARTIN D. *Unsolvable problems* ◇ D10 D35 D40 ◇

DEUTSCH, M. *Eine mengentheoretische Grundlegung der Theorie der Berechenbarkeit II* ◇ B15 D20 D35 E47 ◇

GABBAY, D.M. *Undecidability of intuitionistic theories formulated with the apartness relation* ◇ D35 F50 ◇

HENSON, C.W. & JOCKUSCH JR., C.G. & RUBEL, L.A. & TAKEUTI, G. *First-order topology* ◇ B25 C65 C75 D35 H05 ◇

IBARRA, O.H. *Counter machines and number-theoretic problems* ◇ D10 D35 ◇

ISARD, S.D. *A finitely axiomatizable undecidable extension of K* ◇ B45 D05 D35 ◇

KRYNICKI, M. *Henkin quantifier and decidability* ◇ B25 C10 C80 D35 ◇

LIPSHITZ, L. *Undecidable existential problems for addition and divisibility in algebraic number rings II* ◇ D35 ◇

MART'YANOV, V.I. *Extended universal theories of the integers (Russian)* ◇ B25 D35 F30 ◇

MATIYASEVICH, YU.V. *Some purely mathematical results inspired by mathematical logic* ◇ B30 D25 D35 D80 ◇

NAKAMURA, A. & ONO, H. *Two-dimensional finite automata and their application to the decision problem of monadic first-order arithmetic* $A[P, F(x), G(x)]$ ◇ D05 D35 F30 F35 ◇

NORGELA, S.A. *On recursive nonseparability of the strategies of deduction-search in the classical predicate calculus (Russian)* ◇ B35 D35 ◇

PLISKO, V.E. *The nonarithmeticity of the class of realizable predicate formulas (Russian)* ◇ D35 D55 F50 ◇

POPOV, S.V. *Undecidable interval arithmetic (Russian) (English Summary)* ◇ B55 D35 ◇

SANKAPPANAVAR, H.P. *On the decision problem of the congruence lattices of pseudocomplemented semilattices* ◇ D35 G10 ◇

SEESE, D.G. *Second order logic, generalized quantifiers and decidability* ◇ B15 B25 C55 C65 C80 C85 D35 ◇

SIMPSON, S.G. *Degrees of unsolvability: A survey of results* ◇ D25 D30 D35 F35 ◇

SIMPSON, S.G. *First order theory of the degrees of recursive unsolvability* ◇ D30 D35 F35 ◇

SMORYNSKI, C.A. *A note on the number of zeros of polynomials and exponential polynomials* ◇ D25 D35 D80 ◇

SOBOLEV, S.K. *The intuitionistic propositional calculus with quantifiers (Russian)* ◇ B55 C80 C90 D35 F50 ◇

STILLWELL, J.C. *Concise survey of mathematical logic* ◇ B98 C98 D10 D35 ◇

TURASHVILI, T.V. *On the undecidable minimal classes of first order predicate logic (Russian) (English summary)* ◇ B20 D35 ◇

WEESE, M. *The undecidability of well-ordering with the Haertig quantifier (Russian summary)* ◇ C55 C65 C80 D35 E07 ◇

WIRSING, M. *Das Entscheidungsproblem der Klasse von Formeln, die hoechstens zwei Primformeln enthalten* ◇ B20 B25 D35 ◇

1978

ADLEMAN, L.M. & MANDERS, K.L. *NP-complete decision problems for binary quadratics* ◇ D15 D35 ◇

BOERGER, E. *Bemerkung zu Gurevich's Arbeit ueber das Entscheidungsproblem fuer Standardklassen* ◇ B20 D10 D35 ◇

BOERGER, E. *Ein einfacher Beweis fuer die Unentscheidbarkeit der klassischen Praedikatenlogik* ◇ B10 D03 D10 D35 ◇

BUSZKOWSKI, W. *Undecidability of some logical extensions of Ajdukiewicz-Lambek calculus* ◇ B65 D35 ◇

DELON, F. *Definition de l'arithmetique dans la theorie des anneaux de series formelles (English summary)* ◇ C60 D35 F25 F35 ◇

DENEF, J. *Diophantine sets over* $Z[T]$ ◇ C60 D25 D35 ◇

DENEF, J. & LIPSHITZ, L. *Diophantine sets over some rings of algebraic integers* ◇ C60 D35 ◇

DENEF, J. *The Diophantine problem for polynomial rings and fields of rational functions* ◇ C60 D35 ◇

HERMES, H. *Die Unloesbarkeit des zehnten Hilbertschen Problems* ◇ A10 D25 D35 ◇

HERRE, H. & PINUS, A.G. *Zum Entscheidungsproblem fuer Theorien in Logiken mit monadischen verallgemeinerten Quantoren* ◇ B25 C10 C55 C80 D35 ◇

KOKORIN, A.I. & PINUS, A.G. *Decidability problems of extended theories (Russian)* ◇ B25 C85 C98 D35 ◇

LIPSHITZ, L. *The diophantine problem for addition and divisibility* ◇ B25 D35 ◇

LIPSHITZ, L. *Undecidable existential problems for addition and divisibility in algebraic number rings* ◇ D35 ◇

MART'YANOV, V.I. *Undecidability of the theory of abelian groups with an automorphism (Russian)* ◇ D35 ◇

NORGELA, S.A. *Herbrand strategies of deduction-search in predicate calculus I (Russian) (English and Lithuanian summaries)* ◇ B25 B35 D35 F07 ◇

PENZIN, YU.G. *Unsolvable theories of a ring of continuous functions (Russian)* ◇ D35 ◇

SANKAPPANAVAR, H.P. *Decision problems: History and methods* ◇ A10 B25 D35 D98 ◇

SEESE, D.G. *A remark to the undecidability of well-orderings with the Haertig quantifier* ◇ C55 C65 C80 D35 ◇

SEESE, D.G. *Ueber unentscheidbare Erweiterungen von SC* ◇ D35 ◇

SHEKHTMAN, V.B. *An undecidable superintuitionistic propositional calculus (Russian)* ◇ B55 D35 ◇

SZABO, P. *The undecidability of the* D_A*-unification problem* ◇ B35 D35 ◇

THOMAS, WOLFGANG *The theory of successor with an extra predicate* ◇ B25 D35 ◇

WIRSING, M. *Kleine unentscheidbare Klassen der Praedikatenlogik mit Identitaet und Funktionszeichen* ◇ B20 D35 ◇
ZAMYATIN, A.P. *A non-abelian variety of groups has an undecidable elementary theory(Russian)* ◇ C05 C60 D35 ◇

1979

BOERGER, E. & KLEINE BUENING, H. *The reachability problem for Petri nets and decision problems for Skolem arithmetic* ◇ B25 D15 D35 D80 ◇
BOZOVIC, N.B. *A note on a generalization of some undecidability results in group theory* ◇ D35 D40 ◇
BOZOVIC, N.B. *On some classes of unrecognizable properties of groups* ◇ D35 D40 ◇
BRITTON, J.L. *Integer solutions of systems of quadratic equations* ◇ B25 D35 D80 ◇
BUSZKOWSKI, W. *Undecidability of the theory of lattice-orderable groups* ◇ C60 D35 ◇
DENEF, J. *The Diophantine problem for polynomial rings of positive characteristic* ◇ C60 D35 ◇
EPSTEIN, R.L. *Degrees of unsolvability: structure and theory* ◇ D25 D30 D35 D98 ◇
GUREVICH, Y. *Modest theory of short chains I* ◇ B25 C65 C85 D35 E07 ◇
GUREVICH, Y. & SHELAH, S. *Modest theory of short chains II* ◇ B25 C65 C85 D35 E07 E50 ◇
JONES, JAMES P. *Diophantine representation of Mersenne and Fermat primes* ◇ D25 D35 ◇
KRYAUCHYUKAS, V.YU. *Diophantine representation of perfect numbers (Russian)* ◇ D35 ◇
LEWIS, H.R. *Unsolvable classes of quantificational formulas* ◇ B20 B25 D35 D98 ◇
MATIYASEVICH, YU.V. *Algorithmic unsolvability of exponential Diophantine equations with three indeterminates (Russian)* ◇ D25 D35 ◇
NAKAMURA, A. & ONO, H. *Undecidability of the first-order arithmetic* $A[P(x),2x,x+1]$ ◇ D05 D35 F30 F35 ◇
PRESTEL, A. *Entscheidbarkeit mathematischer Theorien* ◇ B25 C60 C98 D35 ◇
ROMAN'KOV, V.A. *Universal theory of nilpotent groups (Russian)* ◇ B25 C60 D35 ◇
ROZENBLAT, B.V. *Positive theories of free inverse semigroups (Russian)* ◇ B20 C05 D35 D40 ◇
SCHOENFELD, W. *An undecidability result for relation algebras* ◇ D35 G15 ◇
SZCZERBA, L.W. & TARSKI, A. *Metamathematical discussion of some affine geometries* ◇ B25 B30 C35 C52 C65 D35 ◇

1980

ASH, C.J. & ROSENTHAL, J.W. *Some theories associated with algebraically closed fields* ◇ C60 D35 ◇
BECKER, J.A. & HENSON, C.W. & RUBEL, L.A. *First-order conformal invariants* ◇ C35 C65 D35 E35 E75 ◇
BECKER, J.A. & DENEF, J. & LIPSHITZ, L. *Further remarks on the elementary theory of formal power series rings* ◇ C60 D35 ◇
BOERGER, E. & KLEINE BUENING, H. *The reachability problem for Petri nets and decision problems for Skolem arithmetic* ◇ B25 D15 D35 D80 ◇
CHERLIN, G.L. *Rings of continuous functions: decision problems* ◇ B25 C60 C65 D35 ◇
CHLEBUS, B.S. *Decidability and definability results concerning well-orderings and some extensions of first order logic* ◇ B15 B25 C40 C65 C80 D35 E07 ◇
DANKO, W. *A criterion of undecidability of algorithmic theories* ◇ D35 ◇
ERSHOV, YU.L. *Regularly closed fields (Russian)* ◇ B25 C35 C60 D35 ◇
FELGNER, U. *The model theory of FC-groups* ◇ C35 C45 C60 D35 ◇
JAMBU-GIRAUDET, M. *Theorie des modeles de groupes d'automorphismes d'ensembles totalement ordonnes 2-homogenes (English summary)* ◇ C07 C60 C62 C65 D35 E07 F25 F35 ◇
JONES, JAMES P. *Undecidable Diophantine equations* ◇ D25 D35 ◇
KUDAJBERGENOV, K.ZH. *On constructive models of undecidable theories (Russian)* ◇ C15 C35 C57 D35 ◇
MADDUX, R. *The equational theory of* CA_3 *is undecidable* ◇ C05 D35 G15 ◇
MANASTER, A.B. & ROSENSTEIN, J.G. *Two-dimensional partial orderings: Undecidability* ◇ C57 C65 D35 E07 G10 ◇
MONTAGNA, F. *The undecidability of the first-order theory of diagonalizable algebras* ◇ D35 G25 ◇
NERODE, A. & SHORE, R.A. *Reducibility orderings: theories, definability and automorphisms* ◇ C62 D30 D35 ◇
NERODE, A. & SHORE, R.A. *Second order logic and first order theories of reducibility orderings* ◇ B10 B15 D30 D35 F35 F40 ◇
NERODE, A. & SMITH, RICK L. *The undecidability of the lattice of recursively enumerable subspaces* ◇ D35 D45 ◇
REICHBACH, J. *Decidability of mathematical sciences and their undecidability* ◇ B25 D35 ◇
ROMANOVSKIJ, N.S. *The elementary theory of an almost polycyclic group (Russian)* ◇ C60 D35 ◇
RUMELY, R. *Undecidability and definability for the theory of global fields* ◇ C40 C60 D35 ◇
RUOHONEN, K. *Hilbert's tenth problem (Swedish) (English summary)* ◇ D25 D35 ◇
SCHMERL, J.H. *Decidability and* $\aleph_0$*-categoricity of theories of partially ordered sets* ◇ B25 C15 C35 C65 D35 ◇
SEMENOV, A.L. *An interpretation of free algebras in free groups (Russian)* ◇ D20 D35 ◇
TUCKER, J.V. *Computing in algebraic systems* ◇ D10 D35 D75 D80 ◇

1981

BERNARDI, C. *On the relation provable equivalence and on partitions in effectively inseparable sets* ◇ D25 D35 F30 ◇
BOERGER, E. *Logical description of computation processes* ◇ B25 D05 D15 D35 ◇

CHERLIN, G.L. & DRIES VAN DEN, L. & MACINTYRE, A. *Decidability and undecidability theorems for PAC-fields* ◇ B25 C60 D35 ◇

CHERLIN, G.L. & SCHMITT, P.H. *Undecidable L^t theories of topological abelian groups* ◇ C60 C90 D35 ◇

CHONG, C.T. *Hilbert's tenth problem* ◇ D35 ◇

CHRISTIAN, C.C. *Das rekursive Inaccessibilitaetstheorem und der Goedelsche Unvollstaendigkeitssatz in ihrer Bedeutung fuer die Informatik* ◇ A05 D35 D80 F30 ◇

COMER, S.D. *The decision problem for certain nilpotent closed varieties* ◇ B25 C05 D35 ◇

DANKO, W. *A criterion of undecidability of algorithmic theories* ◇ B75 D35 ◇

DELON, F. *Indecidabilite de la theorie des anneaux de series formelles a plusieurs indeterminees (English summary)* ◇ C60 D35 ◇

DEUTSCH, M. *Zur Reduktionstheorie des Entscheidungsproblems* ◇ B20 D35 ◇

EPSTEIN, R.L. *Initial segments of degrees below 0'* ◇ D25 D30 D35 D98 ◇

ERSHOV, YU.L. *Undecidability of regularly closed fields (Russian)* ◇ C60 D35 ◇

FOELDES, S. & SABIDUSSI, G. *Recursive undecidability of the binding property for finitely presented equational classes* ◇ C05 D35 ◇

GOLDFARB, W.D. *On the Goedel class with identity* ◇ B20 B25 D35 ◇

GOLDFARB, W.D. *The undecidability of the second-order unification problem* ◇ B15 D35 F20 ◇

GURARI, E.M. & IBARRA, O.H. *Two-way counter machines and Diophantine equations* ◇ D05 D35 ◇

HAUSCHILD, K. *Zum Vergleich von Haertigquantor und Rescherquantor* ◇ B25 C10 C55 C80 D35 ◇

HERRE, H. *Miscellaneous results and problems in extended model theory* ◇ B25 C65 C80 C98 D35 ◇

JAMBU-GIRAUDET, M. *Interpretations d'arithmetiques dans des groupes et des treillis* ◇ C07 C62 C65 D35 E07 F25 F30 G10 ◇

JONES, JAMES P. *Classification of quantifier prefixes over Diophantine equations* ◇ D25 D35 ◇

KLEINE BUENING, H. *Some undecidable theories with monadic predicates and without equality* ◇ B20 D35 ◇

MANASTER, A.B. & REMMEL, J.B. *Partial orderings of fixed finite dimension: Model companions and density* ◇ C25 C65 D35 E07 ◇

MANASTER, A.B. & REMMEL, J.B. *Some decision problems for subtheories of two-dimensional partial orderings* ◇ B25 C10 C65 D35 ◇

MARGENSTERN, M. *Le theoreme de Matiyassevitch et resultats connexes* ◇ D25 D35 ◇

MATIYASEVICH, YU.V. *What should we do having proved a decision problem to be unsovable?* ◇ D35 ◇

MEYER, A.R. & MIRKOWSKA, G. & STREETT, R.S. *The deducibility problem in propositional dynamic logic* ◇ B75 D25 D35 ◇

NAKAMURA, A. & ONO, H. *Undecidability of extensions of the monadic first-order theory of successor and two-dimensional finite automata* ◇ B15 B25 D05 D35 ◇

PIZZI, C. *"Since", "even if", "as if"* ◇ B65 D35 ◇

POPOV, S.V. *Nondecidable intermediate calculus (Russian)* ◇ B55 D35 ◇

ROZENBLAT, B.V. & VAZHENIN, YU.M. *Decidability of the positive theory of a free countably generated semigroup (Russian)* ◇ B25 C05 D35 ◇

SHORE, R.A. *The degrees of unsolvability: global results* ◇ C40 D30 D35 F35 ◇

SHORE, R.A. *The theory of the degrees below O'* ◇ D25 D30 D35 F30 ◇

SLOBODSKOJ, A.M. *Unsolvability of the universal theory of finite groups (Russian)* ◇ C13 C60 D35 D40 ◇

URQUHART, A.I.F. *Decidability and the finite model property* ◇ B22 B25 D35 ◇

URQUHART, A.I.F. *The decision problem for equational theories* ◇ B25 C05 D35 ◇

VAZHENIN, YU.M. *Sur la liaison entre problemes combinatoires et algorithmiques* ◇ B25 D05 D15 D35 ◇

WEESE, M. *Decidability with respect to Haertig quantifier and Rescher quantifier* ◇ B25 C55 C80 D35 ◇

1982

AANDERAA, S.O. & BOERGER, E. & LEWIS, H.R. *Conservative reduction classes of Krom formulas* ◇ B20 C13 D35 ◇

AANDERAA, S.O. & BOERGER, E. & GUREVICH, Y. *Prefix classes of Krom formulae with identity (German summary)* ◇ B20 B25 D35 ◇

BLAIR, H.A. *The recursion-theoretic complexity of the semantics of predicate logic as a programming language* ◇ B75 D35 D55 ◇

BURRIS, S. & LAWRENCE, J. *Two undecidability results using modified boolean powers* ◇ C05 C60 D35 G05 ◇

GURARI, E.M. & IBARRA, O.H. *Two-way counter machines and Diophantine equations* ◇ D05 D35 ◇

GUREVICH, Y. *Existential interpretation II* ◇ B20 D35 F25 ◇

GUREVICH, Y. & SHELAH, S. *Monadic theory of order and topology in ZFC* ◇ B25 C65 C85 D35 E07 ◇

HARRINGTON, L.A. & SHELAH, S. *The undecidability of the recursively enumerable degrees* ◇ D25 D35 ◇

HAUSCHILD, K. *Model-theoretic properties of cause-and-effect structures* ◇ B45 C52 D35 ◇

HUBER-DYSON, V. & JONES, JAMES P. & SHEPHERDSON, J.C. *Some diophantine forms of Goedel's theorem* ◇ D35 F30 ◇

HUBER-DYSON, V. *Symmetric groups and the open sentence problem* ◇ C13 C60 D35 ◇

JENSEN, C.U. *Sur une classe de corps indecidables (English summary)* ◇ C60 D35 ◇

JONES, JAMES P. & MATIYASEVICH, YU.V. *Exponential diophantine representation of recursively enumerable sets (French summary)* ◇ D20 D25 D35 ◇

KESEL'MAN, D.YA. *Decidability of theories of certain classes of elimination graphs* ◇ B25 C65 D35 ◇

MARCHENKOV, S.S. *Undecidability of the positive $\forall\exists$-theory of a free semigroup (Russian)* ◇ B20 D10 D35 ◇

MART'YANOV, V.I. *Undecidability of the theory of Boolean algebras with automorphism (Russian)* ◇ D35 G05 ◇
MOROZOV, A.S. *Decidability of theories of Boolean algebras with a distinguished ideal (Russian)* ◇ D35 G05 ◇
RICHARD, D. *La theorie sans egalite du successeur et de la coprimarite des entiers naturels est indecidable. Le predicat de primarite est definissable dans le langage de cette theorie (English summary)* ◇ D35 F30 ◇
SEESE, D.G. & TUSCHIK, H.-P. & WEESE, M. *Undecidable theories in stationary logic* ◇ C55 C65 C80 D35 E05 E75 ◇
SHEKHTMAN, V.B. *Undecidable propositional calculi (Russian)* ◇ B22 B55 D35 ◇
SMORYNSKI, C.A. *The finite inseparability of the first-order theory of diagonalisable algebras* ◇ B45 C13 D35 F30 ◇
STREETT, R.S. *Global process logic is undecidable* ◇ B75 D35 ◇
TULIPANI, S. *A use of the method of interpretations for decidability or undecidability of measure spaces* ◇ B25 C65 D35 ◇
ZIEGLER, M. *Einige unentscheidbare Koerpertheorien* ◇ C60 D35 ◇

1983

BURRIS, S. *Boolean constructions* ◇ A10 B25 C05 C30 C98 D35 G05 ◇
CROSSLEY, J.N. & REMMEL, J.B. *Undecidability and recursive equivalence I* ◇ D35 D45 D50 ◇
GUREVICH, Y. *Decision problem for separated distributive lattices* ◇ D35 G10 ◇
GUREVICH, Y. & SHELAH, S. *Interpreting second-order logic in the monadic theory of order* ◇ B15 C65 C85 D35 E07 E50 F25 ◇
GUREVICH, Y. & SHELAH, S. *Random models and the Goedel case of the decision problem* ◇ B20 B25 C13 D35 ◇
GUREVICH, Y. & MAGIDOR, M. & SHELAH, S. *The monadic theory of ω_2* ◇ B15 B25 C65 C85 D35 E10 E35 ◇
HASENJAEGER, G. *Exponential diophantine description of a small universal Turing machine (UTM) using Matiyasevich's 1975/80 "masking"* ◇ D10 D15 D35 ◇
HERRMANN, E. *Definable Boolean pairs in the lattice of recursively enumerable sets* ◇ D25 D35 ◇
HODGSON, B.R. *Decidabilite par automate fini* ◇ B25 D05 D35 ◇
IVANOV, A.A. *Decidability of theories in a certain calculus (Russian)* ◇ B25 C10 C55 C60 C80 D35 ◇
JENSEN, C.U. *L'indecidabilite d'une classe de corps des fonctions meromorphes* ◇ D35 D80 ◇
KHARLAMPOVICH, O.G. *The universal theory of the class of finite nilpotent groups is undecidable (Russian)* ◇ C60 D35 ◇
LUO, LIBO *The τ-theory for free groups is undecidable* ◇ C35 C60 D35 ◇
NOSKOV, G.A. *Elementary theory of a finitely generated commutative ring (Russian)* ◇ C60 D35 ◇
NOSKOV, G.A. *The elementary theory of a finitely generated almost solvable group (Russian)* ◇ B25 C60 D35 ◇
PINUS, A.G. *Calculus with the quantifier of elementary equivalence (Russian)* ◇ C40 C55 C80 D35 ◇
PINUS, A.G. *The operation of Cartesian product* ◇ B25 C05 D35 ◇
RAGAZ, M. *Die Unentscheidbarkeit der einstelligen unendlichwertigen Praedikatenlogik* ◇ B50 D35 D55 ◇
RUBIN, M. *A Boolean algebra with few subalgebras, interval algebras and retractiveness* ◇ C55 C80 D35 E50 E65 G05 ◇
SCHWABHAEUSER, W. & SZMIELEW, W. & TARSKI, A. *Metamathematische Methoden in der Geometrie* ◇ B30 C65 C98 D35 ◇

1984

ASH, C.J. & DOWNEY, R.G. *Decidable subspaces and recursively enumerable subspaces* ◇ C57 C60 D35 D45 ◇
BAZHANOV, V.A. *Logic of quantum mechanics and the problem of its decidability (Russian)* ◇ B51 D35 ◇
BLAIR, H.A. *The intractability of validity in logic programming and dynamic logic* ◇ B75 C75 D35 D55 ◇
BOERGER, E. *Decision problems in predicate logic* ◇ B20 B25 D35 ◇
CANTOR, DAVID G. & ROQUETTE, P. *On diophantine equations over the ring of all algebraic integers* ◇ B25 C60 D35 ◇
CHERLIN, G.L. *Definability in power series rings of nonzero characteristic* ◇ C40 C60 D35 ◇
CHERLIN, G.L. *Undecidability of rational function fields in nonzero characteristic* ◇ C60 C85 D35 ◇
COHEN, D.E. *Modular machines, undecidability and incompleteness* ◇ D20 D25 D35 F30 ◇
CROSSLEY, J.N. & REMMEL, J.B. *Undecidability and recursive equivalence II* ◇ D35 D50 D75 ◇
DAHN, B.I. & WOLTER, H. *Ordered fields with several exponential functions* ◇ C25 C35 C65 D35 ◇
DENEF, J. & LIPSHITZ, L. *Power series solutions of algebraic differential equations* ◇ B25 C60 C65 D35 D80 ◇
DEUTSCH, M. *Reductions for the satisfiability with a simple interpretation of the predicate variable* ◇ B20 D35 ◇
ERSHOV, YU.L. *Regularly r-closed fields with weakly universal Galois group (Russian)* ◇ B25 C60 D35 ◇
FARRINGTON, PADDY *The first-order theory of the c-degrees* ◇ D30 D35 E45 E55 F35 ◇
GOLDFARB, W.D. *The unsolvability of the Goedel class with identity* ◇ B20 D35 ◇
GOLDFARB, W.D. *The Goedel class with identity is unsolvable* ◇ B20 D35 ◇
HARAN, D. *The undecidability of pseudo-real-closed fields* ◇ B25 C60 D35 ◇
HAREL, D. & PATERSON, M.S. *Undecidability of PDL with $L=\{a^{2^i} \mid i \geqslant 0\}$* ◇ B75 D35 ◇
HASENJAEGER, G. *Universal Turing machines (UTM) and Jones-Matiyasevich-masking* ◇ D10 D35 ◇

HEINDORF, L. *Regular ideals and boolean pairs* ◇ C30 C85 D35 G05 ◇

HERRMANN, E. *The undecidability of the elementary theory of the lattice of recursively enumerable sets* ◇ D25 D35 ◇

HODES, H.T. *The modal theory of pure identity and some related decision problems* ◇ B25 B45 D35 ◇

HODGES, W. *Finite extensions of finite groups* ◇ C13 C50 C60 D35 ◇

JENSEN, C.U. *Theorie des modeles pour des anneaux de fonctions entieres et des corps de fonctions meromorphes* ◇ C60 C65 D35 F35 ◇

JONES, JAMES P. & MATIYASEVICH, YU.V. *Register machine proof of the theorem on exponential diophantine representation of enumerable sets* ◇ D10 D25 D35 ◇

MATIYASEVICH, YU.V. *Studies in certain algorithmic problems of algebra and number theory (Russian)* ◇ D03 D05 D35 ◇

MURAWSKI, R. *Mathematical incompleteness of arithmetic (Polish)* ◇ D35 F30 ◇

RICHARD, D. *The arithmetics as theories of two orders (English and French summaries)* ◇ B28 C62 D35 F30 ◇

ROBINSON, D.J. *Decision problems for infinite soluble groups* ◇ D35 D40 ◇

SCHMITT, P.H. *Undecidable theories of valued Abelian groups* ◇ C60 D35 ◇

URQUHART, A.I.F. *The undecidability of entailment and relevant implication* ◇ B46 D35 G10 ◇

WEESE, M. *The theory of Boolean algebras extended by a group of automorphisms* ◇ C07 D35 G05 ◇

WEESE, M. *Undecidable extensions of the theory of Boolean algebras* ◇ D35 G05 ◇

WOLTER, H. *Some remarks on exponential functions in ordered fields* ◇ C25 C65 D35 ◇

1985

ARTEMOV, S.N. *Nonarithmeticity of truth predicate logics of provability (Russian)* ◇ B45 D35 D55 F30 ◇

BURRIS, S. *A simple proof of the hereditary undecidability of the theory of lattice ordered abelian groups* ◇ C05 C60 D35 G10 ◇

CRESSWELL, M.J. *The decidable normal modal logics are not recursively enumerable* ◇ B25 B45 D25 D35 D80 ◇

GUREVICH, Y. *Monadic second-order theories* ◇ B25 C65 C85 C98 D35 D98 ◇

HAKEN, A. *The intractability of resolution* ◇ B35 D15 D35 ◇

LACAVA, F. *Undecidability of the theory of L-algebras (Italian) (English summary)* ◇ D35 ◇

MAKANIN, G.S. *On the decidability of the theory of a free group (Russian)* ◇ B25 C60 C98 D35 ◇

MILLAR, T.S. *Decidable Ehrenfeucht theories* ◇ B25 C15 C57 C98 D30 D35 D55 ◇

PAPPAS, P. *A Diophantine problem for Laurent polynomial rings* ◇ C60 D35 D80 ◇

PINUS, A.G. *Applications of boolean powers of algebraic systems (Russian)* ◇ C05 C30 C55 C57 C80 C85 D35 E50 G05 ◇

URSINI, A. *Decision problems for classes of diagonalizable algebras* ◇ B25 B45 C05 D35 ◇

YOUNG, P. *Goedel theorems, exponential difficulty and undecidability of arithmetic theories: an exposition* ◇ D15 D35 F20 F30 ◇

D40 Word problems, etc.

1910

DEHN, M. *Ueber die Topologie des dreidimensionalen Raumes* ◇ D40 ◇

1912

DEHN, M. *Transformationen der Kurven auf zweiseitigen Flaechen* ◇ D40 ◇

1913

THUE, A. *Ueber die gegenseitige Lage gleicher Teile gewisser Zeichenreihen* ◇ D03 D40 ◇

1914

THUE, A. *Probleme ueber Veraenderungen von Zeichenreihen nach gegebenen Regeln* ◇ D03 D40 ◇

1931

MAGNUS, W. *Untersuchungen ueber einige unendliche diskontinuierliche Gruppen* ◇ D40 ◇

1932

MAGNUS, W. *Das Identitaetsproblem fuer Gruppen mit einer definierenden Relation* ◇ D40 ◇

1936

POST, E.L. *Finite combinatory processes - formulation I* ◇ D03 D20 D40 ◇

1943

POST, E.L. *Formal reductions of the general combinatorial decision problem* ◇ D03 D40 ◇

1944

POST, E.L. *Recursively enumerable sets of positive integers and their decision problems* ◇ D03 D25 D40 ◇

1946

POST, E.L. *A variant of a recursively unsolvable problem* ◇ D03 D05 D40 ◇

1947

MARKOV, A.A. *On the impossibility of certain algorithms in the theory of associative systems I (Russian)* ◇ D03 D40 ◇

MARKOV, A.A. *The impossibility of certain algorithms in the theory of associative systems II (Russian)* ◇ D03 D40 ◇

POST, E.L. *Recursive unsolvability of a problem of Thue* ◇ D03 D10 D40 ◇

TARTAKOVSKIJ, V.A. *On the process of extinction (Russian)* ◇ D40 ◇

TARTAKOVSKIJ, V.A. *On the problem of equivalence for certain types of groups (Russian)* ◇ D40 ◇

1948

JASKOWSKI, S. *Sur le probleme de decision de la topologie et de la theorie des groups* ◇ D35 D40 ◇

1949

HALL JR., M. *The word problem for semigroups with two generators* ◇ D40 ◇

HIGMAN, G. & NEUMANN, B.H. & NEUMANN, H. *Embedding theorems for groups* ◇ C60 D40 ◇

TARTAKOVSKIJ, V.A. *Application of the sieve method to the solution of the word problem for certain types of groups (Russian)* ◇ D40 ◇

TARTAKOVSKIJ, V.A. *Solution of the word problem for groups with a k-reduced basis for $k>6$ (Russian)* ◇ D40 ◇

TARTAKOVSKIJ, V.A. *The sieve method in group theory (Russian)* ◇ D40 ◇

1950

TURING, A.M. *The word problem in semi-groups with cancellation* ◇ D40 ◇

1951

EVANS, T. *The word problem for abstract algebras* ◇ D40 ◇

MARKOV, A.A. *The impossibility of certain algorithms in the theory of associative systems (Russian)* ◇ D03 D35 D40 ◇

MARKOV, A.A. *The impossibility of algorithms for the recognition of certain properties of associative systems (Russian)* ◇ D03 D40 ◇

SHEPHERDSON, J.C. *Inverses and zero divisors in matrix rings* ◇ B25 C60 D40 ◇

1952

KALMAR, L. *Another proof to the Markov-Post theorem* ◇ D03 D20 D40 ◇

MARKOV, A.A. *On unsolvable algorithmic problems (Russian)* ◇ D03 D20 D25 D35 D40 ◇

NOVIKOV, P.S. *Algorithmic unsolvability of the word problem in group theory (Russian)* ◇ D40 ◇

NOVIKOV, P.S. *On algorithmic unsolvability of the word problem (Russian)* ◇ D40 ◇

TARTAKOVSKIJ, V.A. *On primitive composition (Russian)* ◇ D40 ◇

1954

BOONE, W.W. *Certain simple unsolvable problems of group theory I,II* ◇ D40 ◇

EVANS, T. *An embedding theorem for semigroups with cancellation* ◇ D40 ◇

FEENEY, W.J. *Certain unsolvable problems in the theory of cancellation semi-groups* ◇ D40 ◇

KIMURA, NAOKI & TAMURA, T. *On decompositions of a commutative semigroup* ◇ D40 ◇

MARKOV, A.A. *Theory of algorithms (Russian)* ◇ D03 D20 D40 D98 ◇

NOVIKOV, P.S. *Unsolvability of the conjugacy problem in group theory (Russian)* ◇ D40 ◇

TAMARI, D. *Une contribution aux theories de communication: machines de Turing et problemes de mot* ◇ D10 D40 ◇

1955

ADYAN, S.I. *Algorithmic unsolvability of the problems of recognition of certain properties of groups (Russian)* ◇ D40 ◇

BOONE, W.W. *Certain simple unsolvable problems of group theory III,IV* ◇ D40 ◇

NOVIKOV, P.S. *On the algorithmic insolvability of the word problem in group theory (Russian)* ◇ D40 ◇

1956

BRITTON, J.L. *Solution of the word problem for certain types of groups I, II* ◇ D40 ◇

JASKOWSKI, S. *Indecidability of first order sentences in the theory of free groupoids* ◇ D35 D40 ◇

NOVIKOV, P.S. *On the unsolvability of the word problem for groups and some other problems of algebra (Russian) (English summary)* ◇ D40 ◇

TAKEUCHI, K. *The word problem of free algebras* ◇ D40 ◇

TSEJTIN, G.S. *Associative calculus with unsolvable equivalence problem (Russian)* ◇ D03 D40 ◇

TSEJTIN, G.S. *On the problem of recognition of properties of associative calculi (Russian)* ◇ D03 D40 ◇

1957

ADYAN, S.I. *Finitely generated groups and algorithms (Russian)* ◇ D40 ◇

ADYAN, S.I. *Finitely presented groups and algorithms (Russian)* ◇ B25 C60 D40 ◇

ADYAN, S.I. *Unsolvability of certain algorithmic problems in the theory of groups (Russian)* ◇ D40 ◇

BOONE, W.W. *Certain simple unsolvable problems of group theory V,VI* ◇ D40 ◇

RABIN, M.O. *Computable algebraic systems* ◇ C57 C60 D40 D45 D80 ◇

1958

ADYAN, S.I. & NOVIKOV, P.S. *Das Wortproblem fuer Halbgruppen mit einseitiger Kuerzungsregel (Russian) (German summary)* ◇ D40 ◇

ADYAN, S.I. *On algorithmic problems in effectively complete classes of groups (Russian)* ◇ C60 D40 ◇

BOONE, W.W. *An analysis of Turing's "The word problem in semigroups with cancellation"* ◇ D40 ◇

BOONE, W.W. *The word problem* ◇ D40 ◇

BRITTON, J.L. *The word problem for groups* ◇ D40 ◇

HALL, P. *Some word-problems* ◇ D40 ◇

MARKOV, A.A. *Zum Problem der Darstellbarkeit von Matrizen (Russisch)* ◇ D40 ◇

MIKHAJLOVA, K.A. *The occurence problem for direct products of groups (Russian)* ◇ D40 ◇

NOVIKOV, P.S. *Ueber einige algorithmische Probleme der Gruppentheorie* ◇ D35 D40 ◇

RABIN, M.O. *Recursive unsolvability of group theoretic problems* ◇ D40 ◇

TSEJTIN, G.S. *Associative calculus with unsolvable equivalence problem (Russian)* ◇ D03 D40 ◇

1959

ADYAN, S.I. *Undecidability of some algorithmic problems in group theory (Russian)* ◇ D40 ◇

BAUMSLAG, G. & BOONE, W.W. & NEUMANN, B.H. *Some unsolvable problems about elements and subgroups of groups* ◇ D35 D40 ◇

BOONE, W.W. *The word problem* ◇ D40 ◇

MIKHAJLOVA, K.A. *The occurence problem for free products of groups (Russian)* ◇ D40 ◇

NOVIKOV, P.S. *On the unsolvability of some problems in algebra I (Russian)* ◇ D35 D40 ◇

TSEJTIN, G.S. *A simple example of an associative calculus with an unsolvable equivalence problem (Russian)* ◇ D03 D40 ◇

TSEJTIN, G.S. *On the problem of recognizing properties of associative calculi (Russian)* ◇ D03 D40 ◇

1960

FRIDMAN, A.A. *On the relation between the word problem and the conjugacy problem in finitely defined groups (Russian)* ◇ D40 ◇

GRINDLINGER, M.D. *Dehn's algorithm for the word problem* ◇ D40 ◇

1961

GLADKIJ, A.V. *On simple Dyck words (Russian)* ◇ D40 ◇

SORKIN, YU.I. *Algorithmic solvability of isomorphism problems (Russian)* ◇ D40 ◇

WHITMAN, P.M. *Status of word problems for lattices* ◇ D40 G10 ◇

1962

BOONE, W.W. *Partial results regarding word problems and recursively enumerable degrees of unsolvability* ◇ D25 D40 ◇

EMELICHEV, V.A. *Solution of some algorithmic problems for commutative semigroups (Russian)* ◇ D40 ◇

FRIDMAN, A.A. *Degrees of unsolvability of the word problem for finitely presented groups (Russian)* ◇ D25 D40 ◇

MARKOV, A.A. *Sur les invariants calculables* ◇ D35 D40 D45 ◇

PETRESCO, J. *Algorithmes de decision et de construction dans les groupes libres* ◇ B25 D40 ◇

SHIRSHOV, A.I. *Some algorithmic problems for ε-algebras (Russian)* ◇ D40 ◇

SHIRSHOV, A.I. *Some algorithm problems for Lie algebras (Russian)* ◇ D40 ◇

1963

BRITTON, J.L. *The word problem* ◇ D40 ◇

DAVIS, MARTIN D. *Unsolvable problems: a review* ◇ D03 D10 D35 D40 ◇

FRENKEL', V.I. *Algorithmic problems in partially ordered groups (Russian)* ◇ C57 C60 D40 ◇

GRINDLINGER, M.D. *Solution of the isomorphism problem for a certain class of groups (Russian)* ◇ D40 ◇

MARKOV, A.A. *On certain algorithms related to systems of words (Russian)* ◇ D05 D40 ◇

1964

CLAPHAM, C.R.J. *Finitely presented groups with word problems of arbitrary degrees of insolubility* ◇ D25 D40 ◇

GLUKHOV, M.M. *Algorithimic solvability of the word problem for completely free modular lattices (Russian)* ◇ D40 G10 ◇

GRINDLINGER, E.I. *Solution of the word problem for a class of groups by Dehn's algorithm, and of the conjugacy problem by means of a generalization of Dehn's algorithm (Russian)* ◇ D40 ◇

GRINDLINGER, E.I. *Solution of the isomorphism problem for a class of semi-groups (Russian)* ◇ D40 ◇

GRINDLINGER, E.I. *The word problem for a class of semigroups with a finite number of defining relations(Russian)* ◇ D40 ◇

GRINDLINGER, M.D. *On Magnus's generalized word problem (Russian)* ◇ D40 ◇

GRINDLINGER, M.D. *Solution of the conjugacy problem for a class of groups, coinciding with their anti-centers, by means of the generalized Dehn algorithm (Russian)* ◇ D40 ◇

KHMELEVSKIJ, YU.I. *The solution of certain systems of word equations (Russian)* ◇ D40 ◇

LIPSCHUTZ, S. *An extension of Greendliger's results on the word problem* ◇ D40 ◇

SEDOL, YA.YA. *The free product of associative calculi with common subalphabet, and some related questions (Russian)* ◇ D05 D40 ◇

1965

BOONE, W.W. *Finitely presented group whose word problem has the same degree as that of an arbitrarily given Thue system (application of methods of Britton)* ◇ D03 D30 D40 ◇

EMELICHEV, V.A. *An algorithm for discerning regularity of a finitely defined commutative semigroup (Russian)* ◇ D40 ◇

FRENKEL', V.I. *The unsolvability of some algorithmic problems in groups, given by a system of generators and defining inequalities (Russian)* ◇ D40 ◇

GRINDLINGER, M.D. *On the word problem and the conjugacy problem (Russian)* ◇ D40 ◇

MATSUḾOTO, K. *Word problem for free lattice (Japanese)* ◇ D40 G10 ◇

ROTMAN, J.R. *The theory of groups. An introduction* ◇ D40 ◇

SHCHEPIN, G.G. *On the imbedding problem for the nilpotent product of finitely presented groups (Russian)* ◇ D40 ◇

SHEPHERDSON, J.C. *Machine configuration and word problems of given degree of unsolvability* ◇ D05 D30 D35 D40 ◇

1966

ADYAN, S.I. *Defining relations and algorithmic problems for groups and semigroups (Russian)* ◇ D40 ◇

BENSON, C. & MENDELSOHN, N.S. *A calculus for a certain class of word problems in groups* ◇ D40 ◇

BIRYUKOV, A.P. *Solution of certain algorithmic problems for finitely determined commutative semigroups (Russian)* ◇ D40 ◇

BIRYUKOV, A.P. *Solvability of the problem of isomorphism for finitely defined commutative semigroups with two generators (Russian)* ◇ D40 ◇

BOKUT', L.A. *On a property of the groups of Boone (Russian)* ◇ D40 ◇

BOONE, W.W. & ROGERS JR., H. *On a problem of J.H.C.Whitehead and a problem of Alonzo Church* ◇ D25 D30 D40 ◇

BOONE, W.W. *Word problems and recursively enumerable degrees of unsolvability. A first paper on Thue systems* ◇ D03 D25 D40 ◇

BOONE, W.W. *Word problems and recursively enumerable degrees of unsolvability. A sequel on finitely presented groups* ◇ D25 D40 ◇

CANNONITO, F.B. *Hierarchies of computable groups and the word problem* ◇ D40 ◇

EMELICHEV, V.A. *Regularity of a finitely determinated commutative semigroup (Russian)* ◇ D40 ◇

GRINDLINGER, E.I. *On the unsolvability of the word problem for a class of semigroups with a solvable isomorphism problem (Russian)* ◇ D40 ◇

GUREVICH, Y. *The word problem for certain classes of semigroups (Russian)* ◇ D40 ◇

KHMELEVSKIJ, YU.I. *Word equations without coefficients (Russian)* ◇ D40 ◇

MAL'TSEV, A.I. *Identical relations on varieties of quasigroups (Russian)* ◇ C05 D40 ◇

MATTHEWS, J. *The conjugacy problem in wreath products and free metaabelian groups* ◇ D40 ◇

MIKHAJLOVA, K.A. *The occurence problem for direct products of groups (Russian)* ◇ D40 ◇

MOSTOWSKI, A.WLODZIMIERZ *Computational algorithms for deciding some problems for nilpotent groups* ◇ D40 ◇

MOSTOWSKI, A.WLODZIMIERZ *On the decidability of some problems in special classes of groups* ◇ D40 ◇

TAJTSLIN, M.A. *On elementary theories of commutative semigroups with cancellation (Russian)* ◇ B25 C60 D40 ◇

WEINBAUM, C.M. *Visualizing the word problem, with an application to sixth groups* ◇ D40 ◇

1967

BOKUT', L.A. *On the Novikov groups (Russian) (English summary)* ◇ D40 ◇

CHEBOTAREVA, L.K. *On the belonging problem for free and almost free lattices (Russian)* ◇ D40 G10 ◇

CLAPHAM, C.R.J. *An embbedding theorem for finitely generated groups* ◇ D35 D40 ◇

FRIDMAN, A.A. *Unloesbarkeitsgrade des Wortproblems fuer endlich-definierte Gruppen (Russisch)* ◇ D30 D40 ◇

KHMELEVSKIJ, YU.I. *Solution of word equations in three unknowns (Russian)* ◇ D40 ◇

MAYOH, B.H. *Groups and semigroups with solvable word problems* ◇ C57 C60 D40 ◇

1968

Adyan, S.I. & Novikov, P.S. *Commutative subgroups and the conjugacy problem in free periodic groups of odd order (Russian)* ◇ D40 ◇

Adyan, S.I. & Novikov, P.S. *Defining relations and the word problem for free periodic groups of odd order (Russian)* ◇ D40 ◇

Bokut', L.A. *Degrees of unsolvability of the conjugacy problem for finitely-presented groups (Russian)* ◇ D30 D40 ◇

Boone, W.W. *Decision problems about algebraic and logical systems as a whole and recursively enumerable degrees of unsolvability* ◇ D25 D35 D40 ◇

Buskirk van, J. & Gillette, P. *The word problem and consequences for the braid groups and mapping class groups of the 2-sphere* ◇ D40 D80 ◇

Grindlinger, E.I. & Grindlinger, M.D. *The word problem for a class of semigroups (Russian)* ◇ D40 ◇

Grindlinger, M.D. *Strengthening of a theorem of J.S.Smetanich (Russian)* ◇ D25 D40 ◇

Jockusch Jr., C.G. *Supplement to Boone's "Algebraic systems"* ◇ D25 D40 ◇

Matiyasevich, Yu.V. *The connection between Hilbert's tenth problem and systems of equations between words and lengths (Russian)* ◇ D25 D35 D40 ◇

McCool, J. *Elements of finite order in free product sixth-groups* ◇ D40 ◇

Miller III, C.F. *On Britton's theorem A* ◇ D40 ◇

Neumann, B.H. *Lectures on topics in the theory of infinite groups* ◇ C05 C60 C98 D40 ◇

Osipova, V.A. *On the word problem for finitely presented semigroups (Russian)* ◇ D40 ◇

Richardson, D.B. *Some undecidable problems involving elementary functions of a real variable* ◇ D40 D80 ◇

Robinson, Julia *Recursive functions of one variable* ◇ D20 D40 ◇

Schupp, P.E. *On Dehn's algorithm and the conjugacy problem* ◇ D40 ◇

Stender, P.V. & Tartakovskij, V.A. *On the word problem in semigroups (Russian)* ◇ D40 ◇

Tajtslin, M.A. *On the isomorphism problem for commutative semigroups (Russian)* ◇ C05 C07 C60 D40 ◇

Truffault, B. *Sur le probleme des mots pour les groupes de Greendlinger* ◇ D40 ◇

Valiev, M.K. *A theorem of G.Higman (Russian)* ◇ C60 D35 D40 ◇

1969

Borisov, V.V. *Simple examples of groups with unsolvable word problem (Russian)* ◇ D40 ◇

Churkin, V.A. & Kargapolov, M.I. & Remeslennikov, V.N. & Roman'kov, V.A. & Romanovskij, N.S. *Algorithmic problems for σ-power groups (Russian)* ◇ D40 ◇

Collins, D.J. *On recognising Hopf groups* ◇ D40 ◇

Collins, D.J. *Recursively enumerable degrees and the conjugacy problem* ◇ D25 D40 ◇

Collins, D.J. *Word and conjugacy problems in groups with only a few defining relations* ◇ D40 ◇

Evans, T. *Some connections between residual finiteness, finite embeddability and the word problem* ◇ C05 D40 ◇

McCool, J. *The order problem and the power problem for free product sixth-groups* ◇ D40 ◇

Petresco, J. *Pregroupes de mots et probleme des mots* ◇ D40 ◇

Soldatova, V.V. *Solution of the word problem for a certain class of groups (Russian)* ◇ D40 ◇

Takeuchi, K. *The word problem for free distributive lattices* ◇ D40 G10 ◇

Tits, J. *Le probleme des mots dans les groupes de Coxeter* ◇ D40 ◇

Trakhtenbrot, B.A. *On the complexity of reduction algorithms in Novikov-Boone constructions (Russian)* ◇ D15 D40 ◇

Valiev, M.K. *The complexity of the word problem for finitely presented groups (Russian)* ◇ D20 D40 ◇

1970

Bendix, P.B. & Knuth, D.E. *Simple word problems in universal algebras* ◇ C05 D40 ◇

Bokut', L.A. *Unsolvability of certain algorithmic problems in the class of associative rings (Russian)* ◇ D35 D40 ◇

Chudnovsky, G.V. *Diophantine predicates (Russian)* ◇ D25 D35 D40 ◇

Collins, D.J. *A universal semigroup (Russian)* ◇ C05 C57 D05 D40 ◇

Collins, D.J. *On recognizing properties of groups which have solvable word problem* ◇ C60 D40 ◇

Day, A. *A simple solution of the word problem for lattices* ◇ D40 G10 ◇

Glukhov, M.M. *On free products and algorithm problems in R-varieties of universal algebras (Russian)* ◇ C05 D40 ◇

Grindlinger, E.I. & Grindlinger, M.D. *An algorithm for the solution of the word problem for certain semigroups (Russian)* ◇ D40 ◇

Grindlinger, E.I. *The word, divisibility and occurence problem for quotient semigroups of free products (Russian)* ◇ D40 ◇

Iskander, A.A. *Word problem for ringoids of numerical functions* ◇ D40 ◇

Kashintsev, E.V. *Graphs and the word problem for finitely presented semigroups (Russian)* ◇ D40 ◇

Kashintsev, E.V. *On the word problem (Russian)* ◇ D40 ◇

Klassen, V.P. *Inclusion problem for a certain class of groups (Russian)* ◇ D40 ◇

Langmaack, H. & Schmidt, Gunther *Klassen unwesentlich verschiedener Ableitungen als Verbaende* ◇ D03 D40 G10 ◇

Litvintseva, Z.K. *On the complexity of individual identity problems in semigroups (Russian)* ◇ D15 D40 ◇

Litvintseva, Z.K. *On the complexity of some problems for groups and semi-groups (Russian)* ◇ D15 D40 ◇

Litvintseva, Z.K. *The conjugacy problem for finitely presented groups (Russian)* ◇ D40 ◇

Lyapin, E.S. *Intersections of independent subsemigroups of a semigroup (Russian)* ◇ D40 ◇

McCool, J. *Unsolvable problems in groups with solvable word problem* ◇ D40 ◇
Remeslennikov, V.N. & Sokolov, V.G. *Some properties of a Magnus embedding (Russian)* ◇ C60 D40 ◇
Schupp, P.E. *A note on recursively enumerable predicates in groups* ◇ B25 D25 D40 ◇
Schupp, P.E. *On the conjugacy problem for certain quotient groups of free products* ◇ D40 ◇
Tamari, D. *The equivalence of associativity and word problems* ◇ C05 D40 ◇
Tetruashvili, M.R. *On the word problem for a certain class of finitely determined semigroups (Russian) (Georgian and English summaries)* ◇ D40 ◇
T"rkalanov, K.D. *A certain class of partially ordered semigroups with a solvable inequality problem (Russian)* ◇ D05 D40 ◇
Ustyan, A.E. *Examples of semigroups with an unsolvable word problem (Russian)* ◇ D40 ◇
Ustyan, A.E. *On the isomorphism problem for finitely presented semigroups (Russian)* ◇ D40 ◇
Yasuhara, A. *The solvability of the word problem for certain semigroups* ◇ D40 ◇

1971

Adam, Andras *A description of the finite right-congruences of finitely generated free semigroups* ◇ D40 ◇
Anisimov, A.V. *The group languages (Russian) (English summary)* ◇ D05 D40 ◇
Bakhturin, Yu.A. *A certain identity algorithm (Russian)* ◇ D40 ◇
Boone, W.W. & Collins, D.J. & Matiyasevich, Yu.V. *Embeddings into semigroups with only a few defining relations* ◇ D03 D40 ◇
Boone, W.W. *Word problems and recursively enumerable degrees of unsolvability. An emendation* ◇ D25 D40 ◇
Calugareanu, G. *Invariants de contraction dans les groupes (Romanian and Russian summaries)* ◇ D40 ◇
Clapham, C.R.J. *The conjugacy problem for a free product with amalgamation* ◇ D40 ◇
Cochet, Y. & Nivat, M. *Une generalisation des ensembles de Dyck* ◇ D40 ◇
Collins, D.J. *Truth-tables degrees and the Boone group* ◇ D30 D40 ◇
Glukhov, M.M. *Free expansions and algorithmic problems in R-varieties of universal algebras (Russian)* ◇ C05 D40 ◇
Iskander, A.A. *Word problem for ringoids of numerical functions* ◇ D40 ◇
Khmelevskij, Yu.I. *Equations in free semigroups (Russian)* ◇ D40 ◇
Kopytov, V.M. *Solvability of the membership problem in finitely generated solvable matrix groups over numbered fields (Russian)* ◇ D40 D45 ◇
Lipschutz, S. *Note on independent equation problem in groups* ◇ D40 ◇
McCool, J. *The power problem for groups with one defining relator* ◇ D40 ◇
Miller III, C.F. *On group theoretic decision problems and their classification* ◇ D25 D40 ◇
Murskij, V.L. *Nondiscernible properties of finite systems of identity relations (Russian)* ◇ C05 D35 D40 ◇
Nivat, M. *Congruence de Thue et t-languages* ◇ D03 D05 D40 ◇
Pavlov, R.D. *Nonsolvability of certain algorithmic problems of group theory in minimal alphabets (Russian)* ◇ D40 ◇
Pavlov, R.D. *On the problem of recognition of group properties (Russian)* ◇ D40 ◇
Pavlov, R.D. *The impossibility of certain algorithms for the recognition of group properties in minimal alphabets* ◇ D40 ◇
Priese, L. *Normalformen von Markov'schen und Post'schen Algorithmen. Eine Einfuehrung in die Theorie der normierten Algorithmen* ◇ D03 D40 D98 ◇
Sokolov, V.G. *An algorithm for the solution of the word problem for a certain class of solvable groups (Russian)* ◇ D40 ◇
T"rkalanov, K.D. *A solution of the inequality in a certain class of partially ordered semigroups by the method of Osipova (Russian and French summmaries)* ◇ B25 D40 ◇
T"rkalanov, K.D. & Zheleva, S. *On the problem of inequality for partially ordered semigroups (Bulgarian) (Russian and German summaries)* ◇ B25 D40 ◇
Weinbaum, C.M. *The word and conjugacy problems for the knot group of any tame, prime, alternating knot* ◇ D40 ◇

1972

Anisimov, A.V. *Certain algorithmic questions for groups and context-free languages (Russian) (English summary)* ◇ D05 D40 ◇
Appel, K.I. & Schupp, P.E. *The conjugacy problem for the group of any tame alternating knot is solvable* ◇ D40 D80 ◇
Bezverkhnij, V.N. & Rollov, Eh.V. *Solution of the conjugacy problem for subsemigroups of a free semigroup (Russian)* ◇ D40 ◇
Bokut', L.A. *Unsolvability of the equality problem, and subalgebras of finitely presented Lie algebras (Russian)* ◇ D40 ◇
Bokut', L.A. *Unsolvability of the word problem for Lie algebras (Russian)* ◇ D40 ◇
Collins, D.J. *Representation of Turing reducibility by word and conjugacy problems in finitely presented groups* ◇ D30 D40 ◇
Day, A. & Herrmann, C. & Wille, R. *On modular lattices with four generators* ◇ D40 G10 ◇
Gurevich, G.A. *On the conjugacy problem for groups with one defining relation (Russian)* ◇ D40 ◇
Jockusch Jr., C.G. *A reducibility arising from the Boone groups* ◇ D25 D30 D40 ◇
Kosovskij, N.K. *Properties of the solutions of equations in a free semigroup (Russian) (English summary)* ◇ D20 D25 D40 ◇
Levinson, H. *On the genera of graphs of group presentations. II* ◇ D40 ◇

Lifschitz, V. *Sufficient conditions for the solvability of the word problem in microprogram semigroups (Russian)* ◇ D40 ◇

Lindner, C.C. *Finite embedding theorems for partial Latin squares, quasi-groups and loops* ◇ D40 ◇

Macintyre, A. *Omitting quantifier-free types in generic structures* ◇ C25 C57 C60 C75 D30 D40 ◇

Macintyre, A. *On algebraically closed groups* ◇ C25 C57 C60 D40 ◇

Osipova, V.A. *On equations with one unknown in semigroups with a bounded measure of overlap of the defining words (Russian)* ◇ D40 ◇

Sacerdote, G.S. *On a problem of Boone* ◇ C60 D30 D35 D40 ◇

Sacerdote, G.S. *Some undecidable problems in group theory* ◇ D40 ◇

Ustyan, A.E. *On the word problem for finitely-generated semigroups (Russian)* ◇ D40 ◇

Ustyan, A.E. *The connection between the occurence problem and the word problem in semigroups (Russian)* ◇ D40 ◇

1973

Aanderaa, S.O. *A proof of Higman's embedding theorem using Britton extensions of groups* ◇ C60 D40 D80 ◇

Adyan, S.I. *Burnside groups of odd exponent and irreducible systems of group identities* ◇ D40 ◇

Adyan, S.I. *The works of P.S.Novikov and his students on algorithmic questions of algebra (Russian)* ◇ D35 D40 ◇

Anonymous *Problems* ◇ D35 D40 ◇

Balbes, R. *On free pseudo-complemented and relatively pseudo-complemented semi-lattices* ◇ D40 G10 ◇

Baur, W. *Eine rekursiv praesentierte Gruppe mit unentscheidbarem Wortproblem* ◇ D40 ◇

Boone, W.W. & Cannonito, F.B. & Lyndon, R.C. (eds.) *Word problems* ◇ D40 D97 ◇

Britton, J.L. *The existence of infinite Burnside groups* ◇ C60 D40 ◇

Cannonito, F.B. *The algebraic invariance of the word problem in groups* ◇ D20 D40 ◇

Cannonito, F.B. & Gatterdam, R.W. *The computability of group constructions. Part I* ◇ D20 D40 D45 ◇

Cannonito, F.B. & Gatterdam, R.W. *The word problem in polycyclic groups is elementary* ◇ D40 ◇

Cohn, P.M. *The word problem for free fields* ◇ D40 ◇

Collins, D.J. *The word, power and order problems in finitely presented groups* ◇ D25 D40 ◇

Fridman, A.A. *A solution of the conjugacy problem in a class of groups (Russian)* ◇ D40 ◇

Gatterdam, R.W. *The computability of group constructions part II* ◇ D20 D40 D45 ◇

Gatterdam, R.W. *The Higman theorem for primitive-recursive groups - a preliminary report* ◇ D20 D40 ◇

Gatterdam, R.W. *The Higman theorem for $E^n(A)$ computable groups* ◇ D20 D40 ◇

Gurevich, G.A. *On the conjugacy problem for groups with a single defining relation (Russian)* ◇ D40 ◇

Haken, W. *Connections between topological and group theoretical decision problems* ◇ B25 C60 C65 D35 D40 D80 ◇

Herrmann, C. *On the equational theory of submodule lattices* ◇ C05 D40 G10 ◇

Hughes, C.E. *Many-one degrees associated with problems of tag* ◇ D03 D10 D25 D40 ◇

Hutchinson, G. *Recursively unsolvable word problems of modular lattices and diagram chasing* ◇ D40 G10 ◇

Lentin, A. *Equations in free monoids* ◇ D40 ◇

Letichevs'kij, O.A. *The equivalence of automata with respect to cancellative semigroups (Russian)* ◇ D05 D40 ◇

Lipschutz, S. *On the word problem and T-fourth-groups* ◇ D40 ◇

Macintyre, A. *The word problem for division rings* ◇ D40 ◇

McKenzie, R. & Thompson, R.J. *An elementary construction of unsolvable word problems in group theory* ◇ D40 ◇

Meyer, A.R. & Stockmeyer, L.J. *Word problems requiring exponential time: preliminary report* ◇ D15 D40 ◇

Meyer, R.K. & Routley, R. *An undecidable relevant logic* ◇ B46 D35 D40 ◇

Miller III, C.F. *Decision problems in algebraic classes of groups (a survey)* ◇ D40 ◇

Miller III, C.F. *Some connections between Hilbert's 10th problem and the theory of groups* ◇ D35 D40 ◇

Mostowski, A.Wlodzimierz *Uniform algorithm for deciding group-theoretic problems* ◇ D30 D40 ◇

Mylopoulos, J. & Tourlakis, G. *Some results in computational topology* ◇ D40 D80 ◇

Neumann, B.H. *The isomorphism problem for algebraically closed groups* ◇ C25 C60 D40 ◇

Osipova, V.A. *An algorithm for recognizing the solvability of equations with one unknown in semigroups with a measure of overlap of the defining words that is less than 1/3 (Russian)* ◇ C05 D40 ◇

Osipova, V.A. *On the conjugacy problem in semigroups (Russian)* ◇ D40 ◇

Overbeek, R.A. *The representation of many-one degrees by the word problem for Thue systems* ◇ D03 D30 D40 ◇

Remeslennikov, V.N. *Example of a finitely presented group in the variety $\mathfrak{A}^5$ with the unsolvable word problem (Russian)* ◇ D40 ◇

Schiek, H. *Equations over groups* ◇ D40 ◇

Schupp, P.E. *A survey of small cancellation theory* ◇ C60 D40 ◇

Simmons, H. *The word problem for absolute presentations* ◇ D40 ◇

Tamari, D. *The associativity problem for monoids and the word problem for semigroups and groups* ◇ D40 ◇

Tetruashvili, M.R. *A realization of the Magnus algorithm on a Turing machine, and an upper bound of the complexity of the computations (Russian) (Georgian and English summaries)* ◇ D10 D15 D40 ◇

Timoshenko, E.I. *Algorithmic problems for metabelian group (Russian)* ◇ D40 ◇

ZHUK, I.K. *The word problem for a certain class of groups (Russian)* ◇ D40 ◇

1974

ANSHEL, M. & STEBE, P. *The solvability of the conjugacy problem for certain HNN groups* ◇ D40 ◇
APPEL, K.I. *On the conjugacy problem for knot groups* ◇ D40 D80 ◇
BELEGRADEK, O.V. *Algebraically closed groups (Russian)* ◇ C25 C60 D40 ◇
BEZVERKHNIJ, V.N. & ROLLOV, EH.V. *On subgroups of a free product of groups (Russian)* ◇ C05 D40 ◇
BEZVERKHNIJ, V.N. & ROLLOV, EH.V. *On subsemigroups of free semigroups (Russian)* ◇ C05 D40 ◇
BOKUT', L.A. *Undecidability of certain algorithmic problems for Lie algebras (Russian)* ◇ D40 D80 ◇
BOONE, W.W. & HIGMAN, G. *An algebraic characterization of groups with soluble word problem* ◇ D40 ◇
BOONE, W.W. *Between logic and group theory* ◇ C57 C60 D40 ◇
BOONE, W.W. & COLLINS, D.J. *Embeddings into groups with only a few defining relations* ◇ D40 ◇
BOZOVIC, I.B. & BOZOVIC, N.B. *On some unrecognizable relations among groups* ◇ D30 D40 ◇
DURNEV, V.G. *Positive formulas in free semigroups (Russian)* ◇ D35 D40 ◇
FINE, B. *The structure of $PSL_2(R)$; R, the ring of integers in a Euclidian quadratic imaginary number field* ◇ D40 ◇
GLASS, A.M.W. *The word problem for lattice ordered groups* ◇ D40 ◇
GLUKHOV, M.M. *Some algorithmic problems and free products in R-varieties of linear Ω-algebras (Russian)* ◇ C05 D40 ◇
HUBER-DYSON, V. *A family of groups with nice word problems* ◇ D40 ◇
JONES, JAMES P. *Recursive undecidability - an exposition* ◇ D10 D20 D35 D40 D98 ◇
LADZIANSKA, Z. *Poproduct of lattices and Sorkin's theorem* ◇ C05 D40 G10 ◇
LIPSHITZ, L. *The undecidability of the word problems for projective geometries and modular lattices* ◇ D40 D80 G10 ◇
MESKIN, S. *A finitely generated residually finite group with an unsolvable word problem* ◇ D40 ◇
ROMANOVSKIJ, N.S. *Some algorithmic problems for solvable groups (Russian)* ◇ D40 ◇
TAJTSLIN, M.A. *On the isomorphism problem for commutative semigroups (Russian)* ◇ D40 ◇
T"RKALANOV, K.D. *On the recognition of the applicability of a certain algorithm for divisibility (Bulgarian) (Russian and French summaries)* ◇ D40 ◇
YASUHARA, A. *Some non-recursive classes of Thue systems with solvable word problem* ◇ D03 D40 ◇

1975

ANISIMOV, A.V. & SEIFERT, F.D. *Zur algebraischen Charakteristik der durch kontext-freie Sprachen definierten Gruppen (English and Russian summaries)* ◇ D05 D40 ◇
AVENHAUS, J. & MADLENER, K. *$E_n - E_{n-1}$-entscheidbare Gruppen* ◇ D40 ◇
BELKIN, V.P. & GORBUNOV, V.A. *Filters in lattices of quasivarieties of algebraic systems (Russian)* ◇ C05 C60 D40 G10 ◇
BOONE, W.W. & HIGMAN, G. *An algebraic characterization of groups with soluble order problem* ◇ D40 ◇
CANNONITO, F.B. & GATTERDAM, R.W. *The word problem and power problem in 1-relator groups are primitive recursive* ◇ D20 D40 ◇
COHN, P.M. *Equations dans les corps gauches* ◇ D40 ◇
EDMUNDS, C.C. *On the endomorphisms problem for free groups* ◇ D40 ◇
EVANS, T. & MANDELBERG, K.I. & NEFF, M.F. *Embedding algebras with solvable word problems in simple algebras - some Boone-Higman type theorems* ◇ D40 ◇
FOX, C.D. *Commutators in orderable groups* ◇ D40 ◇
GORYAGA, A.V. & KIRKINSKIJ, A.S. *Decidability of the conjugacy problem does not carry over to finite extensions of groups (Russian)* ◇ D40 ◇
HERRMANN, C. *Concerning M.M.Gluhov's paper on the word problem for free modular lattices* ◇ D40 G10 ◇
HERRMANN, C. & HUHN, A.P. *Zum Wortproblem fuer freie Untermodulverbaende* ◇ D40 ◇
KIRKINSKIJ, A.S. & REMESLENNIKOV, V.N. *The isomorphism problem for solvable groups (Russian)* ◇ D40 ◇
KREISEL, G. *Was hat die Logik in den letzten 25 Jahren fur die Mathematik geleistet?* ◇ A05 B98 C75 D40 ◇
MACINTYRE, A. *Dense embeddings. I. A theorem of Robinson in a general setting* ◇ B25 C10 C35 C60 D40 ◇
MUZALEWSKI, M. *On the decidability of the identities problem in some classes of algebras* ◇ D40 ◇
ROMOV, B.A. *Algorithmically decidable problems that are connected with expressibility in Post algebras of finite degree (Russian)* ◇ D40 G20 ◇
TETRUASHVILI, M.R. *Estimation of the complexity of a certain reduction (Russian)* ◇ D15 D40 ◇
TETRUASHVILI, M.R. *The complexity of Turing computations in terms of Magnus's algorithm (Russian) (Georgian summary)* ◇ D15 D40 ◇
VALIEV, M.K. *On polynomial reducibility of the word problem under embedding of recursively presented groups in finitely presented groups* ◇ D15 D40 ◇
VOL'VACHEV, R.T. *Undecidability of the isomorphism and the conjugacy problem of commutative matrix groups and algebras (Russian)* ◇ D25 D40 ◇
ZIEGLER, M. *Gruppen mit vorgeschriebenem Wortproblem* ◇ D40 ◇

1976

ADYAN, S.I. *Transformations of words in a semigroup presented by a system of defining relations (Russian)* ◇ D40 ◇
ANSHEL, M. *Conjugate powers in HNN groups* ◇ D40 ◇
ANSHEL, M. & STEBE, P. *Conjugate powers in free products with amalgamation* ◇ B25 D40 ◇

ANSHEL, M. *Decision problems for HNN groups and vector addition systems* ◇ D40 D80 ◇
ANSHEL, M. *The conjugacy problem for HNN groups and the word problem for commutative semigroups* ◇ D40 ◇
BELEGRADEK, O.V. *On m-degrees of word problems (Russian)* ◇ D25 D40 ◇
BRUNS, G. *Free ortholattices* ◇ D40 G10 ◇
CARDOZA, E. & LIPTON, R.J. & MEYER, A.R. *Exponential space complete problems for Petri nets and commutative semigroups* ◇ D15 D40 D80 ◇
COMERFORD JR., L.P. & TRUFFAULT, B. *The conjugacy problem for free products of sixth-groups with cyclic amalgamation* ◇ D40 ◇
EVEN, S. & TARJAN, R.E. *A combinatorial problem which is complete in polynomial space* ◇ D15 D40 ◇
FLUM, J. *El problema de las palabras en la teoria de grupos* ◇ C25 C60 D40 ◇
GERASIMOV, V.N. *Distributive lattices of subspaces and the equality problem for algebras with a single relation (Russian)* ◇ D40 ◇
HURWITZ, R.D. *On the conjugacy problem in a free product with commuting subgroups* ◇ D40 ◇
LETICHEVS'KIJ, O.A. & SMIKUN, L.B. *On a class of groups with solvable problem of automata equivalence (Russian)* ◇ D05 D40 ◇
PAVLOV, R.D. *Group theoretic algorithmic problems in minimal alphabets* ◇ D40 ◇
SABBAGH, G. *Caracterisation algebrique des groupes de type fini ayant un probleme de mots resoluble (Theoreme de Boone-Higman, travaux de B. H. Neumann et Macintyre)* ◇ D40 ◇
SACERDOTE, G.S. *A characterization of the subgroups of finitely presented groups* ◇ C60 D40 ◇
SARKISYAN, O.A. *On the connection between algorithmic problems in groups and semigroups (Russian)* ◇ D40 ◇
T"RKALANOV, K.D. *On the o-isomorphism and broken-line problems, and the applicability of the algorithm for divisibility (Russian) (English summary)* ◇ D40 ◇
WEISPFENNING, V. *Negative-existentially complete structures and definability in free extensions* ◇ C25 C40 C60 C75 D40 G05 ◇
ZIEGLER, M. *Ein rekursiv aufzaehlbarer btt-Grad, der nicht zum Wortproblem einer Gruppe gehoert* ◇ D25 D40 ◇

1977

AVENHAUS, J. & MADLENER, K. *Komplexitaetsuntersuchungen fuer Einrelatorgruppen* ◇ D40 ◇
AVENHAUS, J. & MADLENER, K. *Komplexitaet bei Gruppen: Der Einbettungssatz von Higman* ◇ D15 D40 ◇
AVENHAUS, J. & MADLENER, K. *Subrekursive Komplexitaet bei Gruppen I: Gruppen mit vorgeschriebener Komplexitaet* ◇ D15 D40 ◇
BEZVERKHNIJ, V.N. *Loesung des Problems der Konjugiertheit von Untergruppen fuer eine Klasse von Gruppen I,II (Russian)* ◇ D40 ◇
BYRNES, C. & GAUGER, M.A. *Decidability criteria for the similarity problem, with applications to the moduli of linear dynamical systems* ◇ D40 D80 ◇
DAVIS, MARTIN D. *Unsolvable problems* ◇ D10 D35 D40 ◇
DURNEV, V.G. *On the question of equations of free semigroups (Russian)* ◇ D40 ◇
HUBER-DYSON, V. *Talking about free groups in naturally enriched languages* ◇ C60 D40 ◇
HUGHES, C.E. & SINGLETARY, W.E. *The one-one equivalence of some general combinatorial decision problems* ◇ D30 D40 ◇
HUTCHINSON, G. *Embedding and unsolvability theorems for modular lattices* ◇ D40 G10 ◇
KUKIN, G.P. *Problem of equality for Lie algebras (Russian)* ◇ D40 ◇
LALLEMENT, G. *Presentations de monoides et problemes algorithmiques* ◇ D40 D45 ◇
LIPTON, R.J. & ZALCSTEIN, Y. *Word problems solvable in logspace* ◇ D15 D40 ◇
MAKANIN, G.S. *The problem of the solvability of equations in a free semigroup (Russian)* ◇ B03 B25 D40 ◇
MAKANIN, G.S. *The problem of solvability of equations in a free semigroup (Russian)* ◇ B03 B25 D40 ◇
PERRAUD, J. *Sur les conditions de petite simplification et l'algorithme de Dehn (English summary)* ◇ D40 ◇
PRIDE, S.J. *The isomorphism problem for two-generator one-relator groups with torsion is solvable* ◇ C60 D40 ◇
ROMAN'KOV, V.A. *Unsolvability of the endomorphic reducibility problem in free nilpotent groups and in free rings (Russian)* ◇ D40 ◇
SACERDOTE, G.S. *Subgroups of finitely presented groups* ◇ D40 ◇
SACERDOTE, G.S. *The Boone-Higman theorem and the conjugacy problem* ◇ D40 ◇
SAKAROVITCH, JACQUES *Sur les groupes infinis, consideres comme monoides syntaxiques de langages formels* ◇ D05 D40 ◇
STOCKMEYER, L.J. *The polynomial-time hierarchy* ◇ D10 D15 D40 D55 ◇
TAYLOR, W. *Equational logic* ◇ C05 C98 D40 ◇
TRAKHTENBROT, B.A. *Algorithmen und Rechenautomaten* ◇ D10 D15 D40 D98 ◇

1978

ANSHEL, M. *Decision problems for HNN groups and commutative semigroups* ◇ B25 D40 ◇
ANSHEL, M. *Vector groups and the equality problem for vector addition systems* ◇ D40 D80 ◇
AVENHAUS, J. & MADLENER, K. *Algorithmische Probleme bei Einrelatorgruppen und ihre Komplexitaet* ◇ D20 D40 ◇
AVENHAUS, J. & MADLENER, K. *Subrekursive Komplexitaet bei Gruppen II: Der Einbettungssatz von Higman fuer entscheidbare Gruppen* ◇ D15 D40 ◇
BELEGRADEK, O.V. *Elementary properties of algebraically closed groups (Russian) (English summary)* ◇ C25 C60 D40 ◇
BELEGRADEK, O.V. *On m-degrees of the word problem (Russian)* ◇ D25 D40 ◇

BELYAEV, V.YA. *Subrings of finitely presented associative rings (Russian)* ◇ D40 ◇

BERGMAN, GEORGE M. *Terms and cyclic permutations* ◇ B03 D40 ◇

BRYLL, G. & MIKLOS, S. *The theory of concrete and abstract words (Polish) (English summary)* ◇ D40 ◇

BUNTING, P.W. & LEEUWEN VAN, J. & TAMARI, D. *Deciding associativity for partial multiplication tables of order 3* ◇ D40 ◇

EVANS, T. *An algebra has a solvable word problem if and only if it is embeddable in a finitely generated simple algebra* ◇ D40 ◇

EVANS, T. *Word problems* ◇ D40 ◇

KASHINTSEV, E.V. *An algorithm for the solution of the conjugacy problem for certain semigroups (Russian)* ◇ D40 ◇

KASHINTSEV, E.V. *On the word problem for special semigroups (Russian)* ◇ D40 ◇

KUKIN, G.P. *Algorithmic problems for solvable Lie algebras (Russian)* ◇ D40 ◇

OGANESYAN, G.U. *A class of semigroups with a decidable word problem (Russian)* ◇ D40 ◇

TETRUASHVILI, M.R. *The problem of conjugacy for one class of groups and the computational complexity (Russian)* ◇ D15 D40 ◇

1979

AUSTIN, A.K. *A note on decision procedures for identities* ◇ D40 ◇

AVENHAUS, J. & MADLENER, K. *Complexity and independence of algorithmic problems in groups* ◇ D15 D40 ◇

BELEGRADEK, O.V. *Algebraic equivalents of solvability of group-theoretic algorithmic problems (Russian)* ◇ D40 ◇

BOERGER, E. *A new general approach to the theory of the many-one equivalence of decision problems for algorithmic systems* ◇ B25 D03 D05 D10 D25 D30 D40 ◇

BOZOVIC, N.B. *A note on a generalization of some undecidability results in group theory* ◇ D35 D40 ◇

BOZOVIC, N.B. *On some classes of unrecognizable properties of groups* ◇ D35 D40 ◇

FREESE, R. & NATION, J.B. *Finitely presented lattices* ◇ D40 G10 ◇

GRUNEWALD, F. & SEGAL, D. *The solubility of certain decision problems in arithmetic and algebra* ◇ D40 D80 ◇

HOLLAND, W.C. & MCCLEARY, S.H. *Solvability of the word problem in free lattice-ordered groups* ◇ D40 ◇

KLEIMAN, J.G. *Identities and some algorithmic problems in groups (Russian)* ◇ D40 ◇

MACINTYRE, A. *Combinatorial problems for skew fields I. Analogue of Britton's lemma, and results of Adjan-Rabin type* ◇ D40 ◇

MAKANIN, G.S. *Identification of the rank of equations in a free semigroup (Russian)* ◇ D40 ◇

PAVLOV, R.D. *On the problem of recognizing group properties in bounded alphabets* ◇ D40 ◇

PAVLOV, R.D. *On the problem of recognizing homomorphisms of finitely presented groups* ◇ D40 ◇

PRESIC, S.B. *On quasi-algebras and the word problem* ◇ C05 D40 ◇

RASSIAS, G.M. *Stallings homomorphisms and the simply connectedness problem* ◇ D40 D80 ◇

REMESLENNIKOV, V.N. *An algorithmic problem for nilpotent groups and rings (Russian)* ◇ D40 ◇

ROMAN'KOV, V.A. *Equations in free metabelian groups (Russian)* ◇ D40 ◇

ROZENBLAT, B.V. *Positive theories of free inverse semigroups (Russian)* ◇ B20 C05 D35 D40 ◇

SARKISYAN, O.A. *Beziehungen zwischen Identitaets- und Teilbarkeitsproblemen in Gruppen und Halbgruppen (Russisch)* ◇ D40 ◇

SIMON, H.-U. *Word problems for groups and contextfree recognition* ◇ D15 D40 ◇

TETRUASHVILI, M.R. *The conjugacy problem for groups with one defining relation and the complexity of Turing calculations* ◇ D15 D40 ◇

1980

AANDERAA, S.O. & COHEN, D.E. *Modular machines, the word problem for finitely presented groups and Collins' theorem* ◇ D10 D30 D40 ◇

AANDERAA, S.O. & COHEN, D.E. *Modular machines and the Higman-Clapham-Valiev embedding theorem* ◇ D10 D30 D40 ◇

ADYAN, S.I. *Classifications of periodic words and their application in group theory* ◇ D40 ◇

ADYAN, S.I. & BOONE, W.W. & HIGMAN, G. (EDS.) *Word problems II. The Oxford book* ◇ D40 D97 ◇

AVENHAUS, J. & MADLENER, K. *String matching and algorithmic problems in free groups* ◇ D15 D40 D80 ◇

BAUMSLAG, G. *Problem areas in infinite group theory for finite group theorists* ◇ D40 ◇

BELEGRADEK, O.V. *Decidable fragments of universal theories and existentially closed models (Russian)* ◇ B25 C25 D40 ◇

BOKUT', L.A. *Decision problems for ring theory* ◇ D40 ◇

BOKUT', L.A. *Mal'cev's problem and groups with a normal form* ◇ D40 ◇

CANNONITO, F.B. *Two decidable Markov properties over a class of solvable groups (Russian)* ◇ D40 ◇

COHEN, D.E. *Degree problems for modular machines* ◇ D10 D25 D40 ◇

CULIK II, K. & KARHUMAEKI, J. *On the equality sets for homomorphisms on free monoids with two generators* ◇ D05 D40 ◇

EHRENFEUCHT, A. & FAJTLOWICZ, S. & MALITZ, J. & MYCIELSKI, J. *Some problems on the universality of words in groups* ◇ D40 ◇

EVANS, T. *Some solvable word problems* ◇ D40 ◇

FOELDES, S. *On the complexity of the lattice embeddability problem* ◇ D15 D40 G10 ◇

FREESE, R. *Free modular lattices* ◇ D40 G10 ◇

GORALCIK, P. & GORALCIKOVA, A. & KOUBEK, V. *Testing of properties of finite algebras* ◇ C13 D15 D40 ◇

GRUNEWALD, F. & SEGAL, D. *Some general algorithms I: Arithmetic groups. II: Nilpotent groups* ◇ D40 ◇

HAIMO, F. & SINGER, M.F. & TRETKOFF, M. *Remarks on analytic continuation* ◇ D40 ◇

HICKIN, K.K. & MACINTYRE, A. *Algebraically closed groups: embeddings and centralizers* ◇ C25 C60 D40 ◇

HURLEY, B. *A note on the word problem for groups* ◇ D40 ◇

HURWITZ, R.D. *On cyclic subgroups and the conjugacy problem* ◇ D40 ◇

JOCKUSCH JR., C.G. *Fine degrees of word problems of cancellation semigroups* ◇ D25 D40 ◇

JURA, A. *Some remarks on nonexistence of an algorithm for finding all ideals of a given finite index in a finitely presented semigroup* ◇ D40 ◇

KRAJNEV, V.A. *Words not containing sequential subwords are equal with respect to the frequency structure (Russian)* ◇ D40 ◇

KRYAZHOVSKIKH, G.V. *Approximability of finitely presented algebras (Russian)* ◇ C60 D40 ◇

LIPSCHUTZ, S. *Groups with solvable conjugacy problems* ◇ D40 ◇

LISOVIK, L.P. & RED'KO, V.N. *Regular events in semigroups (Russian)* ◇ D05 D40 ◇

LISOVIK, L.P. *Strict sets and finite semigroup coverings (Russian)* ◇ D40 ◇

MAKANIN, G.S. *Equations in a free semigroup (Russian)* ◇ B03 B25 D40 ◇

MARKSHAJTIS, G.N. *Solvability of the word problem for certain groups (Russian) (Lithuanian and English summaries)* ◇ D40 ◇

MCCLEARY, S.H. *A solution of the word problem in free normal-valued lattice-ordered groups* ◇ D40 ◇

PERRAUD, J. *Sur la condition de petite simplification C'(1/6) dans un produit libre amalgame (English summary)* ◇ D40 ◇

PERRAUD, J. *Sur le probleme des mots des quotients de groupes et produits libres* ◇ D40 ◇

REMESLENNIKOV, V.N. & ROMANOVSKIJ, N.S. *Algorithmic problems for solvable groups* ◇ D40 ◇

ROLLOV, EH.V. *Subsemigroups of a class of semigroups (Russian)* ◇ D40 ◇

ROMANOVSKIJ, N.S. *The embedding problem for abelian-by-nilpotent groups (Russian)* ◇ D40 ◇

SARKISYAN, R.A. *Algorithmic questions for linear algebraic groups I,II (Russian)* ◇ D40 ◇

SHIRVANYAN, V.L. *The word problem for groups with a recursive set of defining relations of the form* $A^n=1$ *(Russian) (Armenian summary)* ◇ D40 ◇

SIMMONS, H. *The word and torsion problems for commutative Thue systems* ◇ D03 D15 D40 ◇

TAJTSLIN, M.A. *The isomorphism problem for commutative semigroups solved positively (Russian)* ◇ C05 D40 ◇

TAMURA, S. *On the word problem for lo-semigroups* ◇ D40 ◇

THOMPSON, R.J. *Embeddings into finitely generated simple groups which preserve the word problem* ◇ D40 ◇

YAGZHEV, A.V. *Algorithmic problem of recognizing automorphisms among endomorphisms of free associative algebras of finite rank (Russian)* ◇ D40 ◇

ZIEGLER, M. *Algebraisch abgeschlossene Gruppen (English summary)* ◇ C25 C60 D30 D40 ◇

1981

AVENHAUS, J. & MADLENER, K. *An algorithm for the word problem in HNN extensions and the dependence of its complexity on the group representation* ◇ D20 D40 ◇

AVENHAUS, J. & MADLENER, K. *P-complete problems in free groups* ◇ D15 D40 ◇

BALLANTYNE, A.M. & LANKFORD, D.S. *New decision algorithms for fintely presented commutative semigroups* ◇ D40 ◇

BAUMSLAG, G. & CANNONITO, F.B. & MILLER III, C.F. *Computable algebra and group embeddings* ◇ D40 D45 ◇

BAUMSLAG, G. & DYER, E. & MILLER III, C.F. *On the integral homology of finitely presented groups* ◇ D40 D80 ◇

BAUMSLAG, G. & CANNONITO, F.B. & MILLER III, C.F. *Some recognizable properties of solvable groups* ◇ D40 ◇

BEZVERKHNIJ, V.N. *Solvability of the inclusion problem in a class of HNN-groups (Russian)* ◇ B25 C60 D40 ◇

BEZVERKHNIJ, V.N. & GRINBLAT, V.A. *The root problem in Artin Groups (Russian)* ◇ B25 C60 D40 ◇

BEZVERKHNYAYA, I.S. *Conjugacy of finite sets of subgroups in a free product of groups (Russian)* ◇ D40 ◇

BOOK, R.V. *The undecidability of a word problem: on a conjecture of Strong, Maggiolo-Schettini and Rosen* ◇ D03 D40 ◇

COMERFORD JR., L.P. & EDMUNDS, C.C. *Quadratic equations over free groups and free products* ◇ D40 ◇

COMERFORD JR., L.P. *Quadratic equations over small cancellation groups* ◇ D40 ◇

DO LONG VAN *Problemes des mots et de conjugaison pour une classe de groupes de presentation finie (English summary)* ◇ D40 ◇

DOBRITSA, V.P. *On constructivizable abelian groups (Russian)* ◇ C57 C60 D40 D45 ◇

GERHARD, J.A. & PETRICH, M. *The word problem for orthogroups* ◇ D40 ◇

GRIGOR'EV, D.YU. *Complexity of "wild" matrix problems and of the isomorphism of algebras and graphs (Russian) (English summary)* ◇ D15 D40 ◇

HICKIN, K.K. *Bounded HNN presentations* ◇ D25 D40 ◇

HORADAM, K.J. *A quick test for nonisomorphism of one-relator groups* ◇ D40 ◇

HORADAM, K.J. *The word problem and related results for graph product groups* ◇ D40 ◇

HUBER-DYSON, V. *A reduction of the open sentence problem for finite groups* ◇ C13 C60 D40 ◇

KHARLAMPOVICH, O.G. *A finitely presented solvable group with undecidable word problem (Russian)* ◇ D40 ◇

LOCKHART, J. *Decision problems in classes of group presentations with uniformly solvable word problem* ◇ D40 ◇

MILLER III, C.F. *The word problem in quotients of a group* ◇ D40 ◇

SARKISYAN, O.A. *Word and divisibility problems in semigroups and groups without cycles (Russian)* ◇ C05 D40 ◇

Shatrova, N.P. *Upper bound of the degree of complexity of an algorithm for solving the conjugacy problem for a class of groups (Russian)* ◇ D15 D40 ◇
Slobodskoj, A.M. *Unsolvability of the universal theory of finite groups (Russian)* ◇ C13 C60 D35 D40 ◇
Waack, S. *Tape complexity of word problems* ◇ D15 D40 ◇
Yabanzhi, G.G. *The word problem for some groups of the variety N_2A (Russian)* ◇ D40 ◇
Zeilberger, D. *Enumeration of words by their number of mistakes* ◇ D40 ◇

1982

Book, R.V. *When is a monoid a group? The Church-Rosser case is tractable* ◇ D03 D15 D40 ◇
Chandler, B. & Magnus, W. *The history of combinatorial group theory* ◇ A10 D40 ◇
Domanski, B. *The complexity of two decision problems for free groups* ◇ D15 D40 ◇
Huber-Dyson, V. *Decision problems in group theory* ◇ B25 D40 ◇
Huber-Dyson, V. *Finiteness conditions and the word problem* ◇ D40 ◇
Kalorkoti, K.A. *Decision problems in group theory* ◇ D25 D40 ◇
Kleiman, J.G. *On identities in groups (Russian)* ◇ C52 C60 D40 ◇
Lockhart, J. *Markov-type properties* ◇ D40 ◇
Mann, A. *A note on recursively presented and co-recursively presented groups* ◇ C60 D40 D45 ◇
Mayr, E.W. & Meyer, A.R. *The complexity of the word problems for commutative semigroups and polynomial ideals* ◇ D15 D40 ◇
McCleary, S.H. *The word problem in free normal valued lattice-ordered groups: A solution and practical shortcuts* ◇ D40 ◇
Myasnikov, A.G. & Remeslennikov, V.N. *Classification of nilpotent powered groups according to elementary properties (Russian)* ◇ C60 D40 ◇
Palasinski, M. *On the word problem for BCK-algebras* ◇ D40 G25 ◇
Rips, E. *Another characterization of finitely generated groups with a solvable word problem* ◇ D40 ◇
Romanovskij, N.S. *The word problem for centrally metabelian groups (Russian)* ◇ D40 ◇
Stillwell, J.C. *The word problem and the isomorphism problem for groups* ◇ D40 ◇
Tetruashvili, M.R. *On the conjugacy problem for finitely presented subgroups of finite index of a finitely presented group with unsolvable conjugacy problem (Russian) (English and Georgian summaries)* ◇ D15 D40 ◇

1983

Anshel, M. & McAloon, K. *Reducibilities among decision problems for HNN groups, vector addition systems and subsystems of Peano arithmetic* ◇ D20 D40 D80 F30 ◇
Glass, A.M.W. & Gurevich, Y. *The word problem for lattice ordered groups* ◇ D40 ◇
Herrmann, C. *On the word problem for the modular lattice with four free generators* ◇ C05 D40 G10 ◇
Hickin, K.K. & Phillips, R.E. *Isomorphism types in wreath products and effective embeddings of periodic groups* ◇ C57 C60 D40 ◇
Kukin, G.P. *The equality problem and free products of Lie algebras and of associative algebras (Russian)* ◇ D40 ◇
Muller, D.E. & Schupp, P.E. *Groups, the theory of ends, and context-free languages* ◇ D05 D40 ◇
Remeslennikov, V.N. & Roman'kov, V.A. *Model-theoretic and algorithmic questions of group theory (Russian)* ◇ B25 C60 C98 D15 D30 D40 ◇
Shatrova, N.A. *Degree of complexity of an algorithm for a class of groups (Russian)* ◇ D30 D40 ◇
Stillwell, J.C. *Efficient computation in groups and simplicial complexes* ◇ D15 D40 ◇

1984

Adyan, S.I. & Makanin, G.S. *Studies in algorithmic questions of algebra (Russian)* ◇ D40 D98 ◇
Avenhaus, J. & Madlener, K. *On the complexity of intersection and conjugacy problems in free groups* ◇ D40 ◇
Ballantyne, A.M. & Butler, G.A. & Lankford, D.S. *A progress report on new decision algorithms for finitely presented abelian groups* ◇ D40 ◇
Baumslag, G. *Algorithmically insoluble problems about finitely presented solvable, Lie, and associative algebras* ◇ D40 ◇
Billington, N. *Growth of groups and graded algebras* ◇ C60 D40 ◇
Cannonito, F.B. *On some algorithmic problems for finitely presented groups and Lie algebras* ◇ D40 ◇
Cannonito, F.B. & Robinson, D.J. *The word problem for finitely generated soluble groups of finite rank* ◇ D40 ◇
Gallier, J.H. & Pelin, A. *Solving word problems in free algebras using complexity functions* ◇ D40 ◇
Glass, A.M.W. *The isomorphism problem and undecidable properties for finitely presented lattice-ordered groups* ◇ C60 D40 G10 ◇
Glass, A.M.W. & Madden, J.J. *The word problem versus the isomorphism problem* ◇ D40 ◇
Gurevich, Y. & Lewis, H.R. *The word problem for cancellation semigroups with zero* ◇ D40 ◇
Hurwitz, R.D. *A survey of the conjugacy problem* ◇ D40 ◇
Juergensen, H. *Komplexitaet von Erzeugen in Algebren* ◇ D40 ◇
Mel'nichuk, I.L. *Unsolvability of problems of equality and divisibility in certain varieties of semigroups (Russian)* ◇ C05 D40 ◇
Oganesyan, G.U. *The isomorphism problem for semigroups with one defining relation (Russian)* ◇ D40 ◇
Otto, F. *Conjugacy in monoids with a special Church-Rosser presentation is decidable* ◇ C05 D03 D05 D40 ◇
Robinson, D.J. *Decision problems for infinite soluble groups* ◇ D35 D40 ◇

Scott, E.A. *A finitely presented simple group with unsolvable conjugacy problem* ◇ D40 ◇
Tetruashvili, M.R. *Computational complexity of recognizing word equality in semigroups of a certain class (Russian) (English and Georgian summaries)* ◇ D15 D40 ◇
Umirbaev, U.U. *Equality problem for center-by-metabelian Lie-algebras (Russian)* ◇ B25 C60 D40 ◇

1985

Anshel, M. & Domanski, B. *The complexity of Dehn's algorithm for word problems in groups* ◇ D15 D40 ◇
Baumslag, G. & Gildenhuys, D. & Strebel, R. *Algorithmically insoluble problems about finitely presented solvable groups, Lie and associative algebras II* ◇ D40 ◇
Bokut', L.A. *A remark on the Borisov-Boone group (Russian)* ◇ D40 ◇
Glass, A.M.W. *Effective extensions of lattices-ordered groups that preserve the degree of the conjugacy and the word problems* ◇ C57 C60 D40 ◇
Glass, A.M.W. *The word and isomorphism problems in universal algebra* ◇ C05 C57 D40 ◇
Grigorchuk, R.I. *A relationship between algorithmic problems and entropy characteristics of groups (Russian)* ◇ D15 D40 ◇
Grunewald, F. & Segal, D. *Decision problems concerning S-arithmetic groups* ◇ D40 ◇
Huynh, D.T. *Complexity of the word problem for commutative semigroups of fixed dimension* ◇ D40 ◇
Kandri-Rody, A. & Kapur, D. & Narendran, P. *An ideal-theoretic approach to word problems and unification problems over finitely presented commutative algebras* ◇ C05 D40 ◇
Weispfenning, V. *The complexity of elementary problems in archimedean ordered groups* ◇ B25 C10 D15 D40 ◇

D45 ∪ C57 Effectiveness in mathematical structures

1926

HERMANN, G. *Die Frage der endlich vielen Schritte in der Theorie der Polynomideale* ◇ C57 C60 D45 F55 ◇

1930

WAERDEN VAN DER, B.L. *Eine Bemerkung ueber die Unzerlegbarkeit von Polynomen* ◇ B25 C60 D45 F55 ◇

1950

KREISEL, G. *Note on arithmetic models for consistent formulae of the predicate calculus I* ◇ B10 C57 D45 D55 F30 ◇

1953

DEKKER, J.C.E. *The constructivity of maximal dual ideals in certain Boolean algebras* ◇ D25 D45 G05 ◇

HASENJAEGER, G. *Eine Bemerkung zu Henkin's Beweis fuer die Vollstaendigkeit des Praedikatenkalkuels der ersten Stufe* ◇ B10 C07 C57 D55 ◇

KREISEL, G. *Note on arithmetic models for consistent formulae of the predicate calculus II* ◇ B10 C57 D45 D55 F30 ◇

MOSTOWSKI, ANDRZEJ *On a system of axioms which has no recursively enumerable arithmetic model* ◇ C57 D45 E30 E70 ◇

1955

FROEHLICH, A. & SHEPHERDSON, J.C. *Effective procedures in field theory* ◇ C57 C60 D45 ◇

FROEHLICH, A. & SHEPHERDSON, J.C. *On the factorisation of polynomials in a finite number of steps* ◇ D45 D80 ◇

MOSTOWSKI, ANDRZEJ *A formula with no recursively enumerable model* ◇ C57 D45 ◇

SPECTOR, C. *Recursive well-orderings* ◇ D30 D45 D55 F15 ◇

USPENSKIJ, V.A. *Systems of enumerable sets and their enumeration (Russian)* ◇ D25 D45 ◇

1956

MOSTOWSKI, ANDRZEJ *Development and applications of the "projective" classification of sets of integers* ◇ C57 C80 D55 E15 E45 ◇

RICE, H.G. *Recursive and recursively enumerable orders* ◇ D20 D25 D45 ◇

USPENSKIJ, V.A. *Calculable operations and the notion of a program (Russian)* ◇ D20 D45 ◇

1957

HENKIN, L. *Sums of squares* ◇ C57 C60 ◇

KREISEL, G. *Sums of squares* ◇ C57 C60 D20 F07 F99 ◇

MOSTOWSKI, ANDRZEJ *On recursive models of formalised arithmetic* ◇ C57 C62 D45 ◇

RABIN, M.O. *Computable algebraic systems* ◇ C57 C60 D40 D45 D80 ◇

VAUGHT, R.L. *Sentences true in all constructive models* ◇ C13 C57 D35 ◇

1958

KREISEL, G. *Mathematical significance of consistency proofs* ◇ C57 C60 F05 F25 F50 ◇

NAGORNYJ, N.M. *Beispiel einer Gruppe mit nicht rekursivem Zentrum (Russisch)* ◇ D45 ◇

RABIN, M.O. *On recursively enumerable and arithmetic models of set theory* ◇ C57 C62 D45 ◇

ROGERS JR., H. *Goedel numberings of partial recursive functions* ◇ D20 D45 F40 ◇

1959

ROGERS JR., H. *Recursive functions over well-ordered partial orderings* ◇ D20 D45 ◇

USPENSKIJ, V.A. *The concept of program and computable operators (Russian)* ◇ D20 D45 ◇

1960

KRUSE, A.H. *Some developments in the theory of numerations* ◇ D45 E10 E25 E50 ◇

LIU, SHICHAO *An example of general recursive well-ordering which is not primitive recursive* ◇ D20 D45 ◇

RABIN, M.O. *Computable algebra, general theory and theory of computable fields* ◇ C57 C60 D45 ◇

SCOTT, D.S. *On a theorem of Rabin* ◇ C35 C57 C62 D35 ◇

SHOENFIELD, J.R. *Degrees of models* ◇ C57 D30 D35 ◇

VAUGHT, R.L. *Sentences true in all constructive models* ◇ C13 C57 D35 ◇

1961

HIGMAN, G. *Subgroups of finitely presented groups* ◇ C60 D25 D45 ◇

MAL'TSEV, A.I. *Constructive algebra I (Russian)* ◇ C57 C98 F60 F98 ◇

ROBINSON, A. *Model theory and non-standard arithmetic* ◇ C35 C50 C57 C60 C62 H15 ◇

SCOTT, D.S. *On constructing models for arithmetic* ◇ C20 C57 C62 H15 ◇

1962

GRZEGORCZYK, A. *A theory without recursive models* ◇ C57 C62 D55 ◇

KENT, C.F. *Constructive analogues of the group of permutations of the natural numbers* ◇ C57 D45 ◇

KRUSE, A.H. *Constructive methods of numeration* ◇ D45 E10 E25 ◇

Liu, Shichao *Four types of general recursive well-orderings* ◇ D30 D45 D55 ◇
Liu, Shichao *Recursive linear orderings and hyperarithmetical functions* ◇ D30 D45 D55 ◇
Mal'tsev, A.I. *On recursive abelian groups (Russian)* ◇ C57 C60 D45 ◇
Mal'tsev, A.I. *Strongly related models and recursively complete algebras (Russian)* ◇ C57 C60 C62 D45 ◇
Markov, A.A. *Sur les invariants calculables* ◇ D35 D40 D45 ◇
Mostowski, Andrzej *L'espace des modeles d'une theorie formalisee et quelques-unes de ses applications* ◇ B50 C07 C40 C57 C80 C85 C90 D45 ◇

1963

Frenkel', V.I. *Algorithmic problems in partially ordered groups (Russian)* ◇ C57 C60 D40 ◇
Mal'tsev, A.I. *Complete enumeration of a set (Russian)* ◇ D45 ◇

1964

Cellucci, C. *Categorie ricorsive* ◇ C57 D20 G30 ◇
Enderton, H.B. & Luckham, D.C. *Hierarchies over recursive well-orderings* ◇ D30 D45 D55 ◇
Frenkel', V.I. *On the effective partial ordering of finitely defined groups (Russian)* ◇ C57 C60 ◇
Lacombe, D. *Deux generalisations de la notion de recursivite* ◇ D45 D75 ◇
Lacombe, D. *Theoremes de non-decidabilite* ◇ D35 D45 D98 ◇
Mal'tsev, A.I. *On the theory of computable families of objects (Russian)* ◇ D45 ◇
Moschovakis, Y.N. *Recursive metric spaces* ◇ C57 C65 D45 F60 ◇
Pour-El, M.B. *Goedel numberings versus Friedberg numberings* ◇ D25 D45 ◇

1965

Crossley, J.N. *Constructive order types I* ◇ D45 ◇
Khutoretskij, A.B. *On recursive isomorphisms (Russian)* ◇ D20 D45 ◇
Mal'tsev, A.I. *Positive and negative numerations (Russian)* ◇ D20 D45 ◇
Putnam, H. *Trial and error predicates and the solution to a problem of Mostowski* ◇ C57 D20 D55 ◇

1966

Aczel, P. & Crossley, J.N. *Constructive order types III* ◇ D45 F15 ◇
Crossley, J.N. *Constructive order types II* ◇ D20 D45 F15 ◇
Crossley, J.N. & Schuette, K. *Non-uniqueness at ω^2 in Kleene's O* ◇ D45 F15 ◇
Doepp, K. *Ueber die Bestimmbarkeit des Verhaltens von Algorithmen* ◇ D20 D45 D55 ◇
Ehrenfeucht, A. & Kreisel, G. *Strong models of arithmetic* ◇ C57 C62 ◇
Kogalovskij, S.R. *On the properties preserved under algebraic constructions (Russian)* ◇ B25 C30 C52 C60 D35 D45 ◇
Moschovakis, Y.N. *Notation systems and recursive ordered fields* ◇ C57 C60 D45 F60 ◇
Nogina, E.Yu. *On effective topological spaces* ◇ C57 C65 ◇
Parikh, R. *Some generalisations of the notion of well ordering* ◇ D20 D45 E07 F35 ◇
Shurygin, V.A. *Nontrivial constructive mappings of certain sets (Russian)* ◇ D20 D45 F60 ◇

1967

Ershov, Yu.L. *Numerations of families of general recursive functions (Russian)* ◇ D20 D45 ◇
Lifschitz, V. *Constructive groups (Russian)* ◇ D45 F60 ◇
Mayoh, B.H. *Groups and semigroups with solvable word problems* ◇ C57 C60 D40 ◇
Vuckovic, V. *Recursive models for three-valued propositional calculi with classical implication* ◇ B50 C57 C90 F30 ◇

1968

Cleave, J.P. *Hyperarithmetic ultrafilters* ◇ C57 D55 E05 ◇
Crossley, J.N. *Recursive equivalence: a survey* ◇ D45 D50 ◇
Ershov, Yu.L. *Numbered fields* ◇ C57 C60 D45 ◇
Ershov, Yu.L. *On computable enumerations (Russian)* ◇ D25 D45 ◇
Hamilton, A.G. *An unsolved problem in the theory of constructive order types* ◇ D45 D50 ◇
Harrison, J. *Recursive pseudo-well-orderings* ◇ D20 D45 ◇
Madison, E.W. *Computable algebraic structures and nonstandard arithmetic* ◇ C57 C60 C62 D45 ◇
Madison, E.W. *Structures elementarily closed relative to a model for arithmetic* ◇ C57 C60 C62 ◇

1969

Cannonito, F.B. & Finkelstein, M. *On primitive recursive permutations and their inverses* ◇ D20 D45 ◇
Carpentier, M.A. *Complete enumerations and double sequences* ◇ D25 D45 ◇
Crossley, J.N. *Constructive order types* ◇ D45 ◇
Dekker, J.C.E. *Countable vector spaces with recursive operations. Part I* ◇ C57 C60 D45 D50 ◇
Ershov, Yu.L. *Completely enumerated sets (Russian)* ◇ D45 ◇
Ershov, Yu.L. & Lavrov, I.A. *On computable enumerations II (Russian)* ◇ D25 D45 ◇
Ershov, Yu.L. *The theory of numerations. Part I: General theory of numeration (Russian)* ◇ D25 D30 D45 G30 ◇
Hensel, G. & Putnam, H. *Normal models and the field Σ_1^** ◇ C07 C30 C57 D55 ◇
Khutoretskij, A.B. *On nonprincipal enumeration (Russian)* ◇ D25 D45 ◇
Khutoretskij, A.B. *The reducibility of computable enumerations (Russian)* ◇ D20 D25 D45 ◇
Khutoretskij, A.B. *Two existence theorems for computable numerations (Russian)* ◇ D25 D45 ◇

NOGINA, E.YU. *Correlations between certain classes of effectively topological spaces (Russian)* ◇ C57 C65 D45 ◇
SOARE, R.I. *Constructive order types on cuts* ◇ D30 D45 ◇
USPENSKIJ, V.A. *The reducibility of computable and potentially computable numerations (Russian)* ◇ D45 ◇
WAGNER, E.G. *Uniformly reflexive structures: on the nature of Goedelization and relative computability* ◇ B40 D45 D75 F40 ◇
YOUNG, P. *Toward a theory of enumerations* ◇ D10 D15 D20 D25 D45 ◇

1970

APPLEBAUM, C.H. & DEKKER, J.C.E. *Partial recursive functions and ω-functions* ◇ C57 D50 ◇
BELYAKIN, N.V. *Generalized computations and second order arithmetic (Russian)* ◇ C62 D10 D45 D65 D70 F35 ◇
CARPENTIER, M.A. *Uniformly precomplete enumeration* ◇ D45 ◇
COLLINS, D.J. *A universal semigroup (Russian)* ◇ C05 C57 D05 D40 ◇
DENISOV, S.D. & LAVROV, I.A. *Complete numerations with an infinite number of special elements (Russian)* ◇ D25 D45 ◇
FEINER, L. *Hierarchies of boolean algebras* ◇ C57 D45 G05 ◇
GANDY, R.O. & SOARE, R.I. *A problem in the theory of constructive order types* ◇ D30 D45 ◇
HAMILTON, A.G. *Bases and α-dimensions of countable vector spaces with recursive operations* ◇ C57 C60 D45 D50 ◇
KOSOVSKIJ, N.K. *Some questions in the constructive theory of normed boolean algebras (Russian)* ◇ C57 F60 G05 ◇
LACHLAN, A.H. & MADISON, E.W. *Computable fields and arithmetically definable ordered fields* ◇ C57 C60 D45 ◇
MADISON, E.W. *A note on computable real fields* ◇ C57 C60 D45 F60 ◇
MAYOH, B.H. *The relation between an object and its name: notation systems and their fixed point theorems* ◇ A05 B03 D20 D35 D45 F30 F60 ◇
SEIDENBERG, A. *Construction of the integral closure of a finite integral domain* ◇ C57 C60 F55 ◇

1971

APPLEBAUM, C.H. *ω-homomorphisms and ω-groups* ◇ D45 D50 ◇
APPLEBAUM, C.H. *Isomorphisms of ω-groups* ◇ D45 D50 ◇
CANNONITO, F.B. *A note on inverses of elementary permutations* ◇ D20 D45 F30 ◇
DEKKER, J.C.E. *Countable vector spaces with recursive operations. Part II* ◇ C57 C60 D45 D50 ◇
DEKKER, J.C.E. *Two notes on vector spaces with recursive operations* ◇ C57 C60 D45 ◇
ERSHOV, YU.L. *Computable enumerations of morphisms (Russian)* ◇ D25 D45 ◇
ERSHOV, YU.L. *La theorie des enumerations* ◇ D25 D45 ◇
ERSHOV, YU.L. *Positive equivalences (Russian)* ◇ D45 ◇
HAJKOVA, M. *The lattice of bi-numerations of arithmetic. I,II* ◇ D45 F30 ◇
KHISAMIEV, N.G. *Strongly constructive models (Russian) (Kazakh summary)* ◇ C30 C57 ◇
KHUTORETSKIJ, A.B. *On the cardinality of the upper semilattice of computable enumerations (Russian)* ◇ D25 D30 D45 ◇
KOPYTOV, V.M. *Solvability of the membership problem in finitely generated solvable matrix groups over numbered fields (Russian)* ◇ D40 D45 ◇
LACOMBE, D. *Recursion theoretic structure for relational systems* ◇ D45 D75 ◇
MADISON, E.W. *Some remarks on computable (non-archimedean) ordered fields* ◇ C57 C60 D45 ◇
MARCHENKOV, S.S. *On minimal numerations of systems of recursively enumerable sets (Russian)* ◇ D25 D45 ◇
MARCHENKOV, S.S. *On semilattices of computable numerations (Russian)* ◇ D25 D45 ◇
PERETYAT'KIN, M.G. *Strongly constructive models and numerations of the boolean algebra of recursive sets (Russian)* ◇ C57 D20 D45 G05 ◇
TANAKA, H. *Analytic well orderings and basis theorems* ◇ D45 D55 E15 ◇
VILLE, F. *Complexite des structures rigidement contenues dans une theorie du premier ordre* ◇ C50 C57 ◇

1972

CHERNOV, V.P. *Classification of spaces of operators of finite types (Russian)* ◇ D45 D65 F60 ◇
CHERNOV, V.P. *Constructive operators of finite types (Russian) (English summary)* ◇ D45 D65 F60 ◇
CULIK, K. *Algorithmization of algebras and relational structures* ◇ D20 D45 ◇
CUTLAND, N.J. *Π^1_1-models and Π^1_1-categoricity* ◇ C57 C70 D55 ◇
DENISOV, S.D. *Models of a noncontradictory formulas and the Ershov hierarchy (Russian)* ◇ C07 C57 D55 ◇
ERSHOV, YU.L. *Existence of constructivizations (Russian)* ◇ C57 D45 ◇
ERSHOV, YU.L. *Relationship between sheaf spaces and numbered sets with the C_2^* property (Russian) (English summary)* ◇ C90 D25 D45 D65 ◇
HOWARD, P.E. *A proof of a theorem of Tennenbaum* ◇ C57 C62 H15 ◇
MACINTYRE, A. *Omitting quantifier-free types in generic structures* ◇ C25 C57 C60 C75 D30 D40 ◇
MACINTYRE, A. *On algebraically closed groups* ◇ C25 C57 C60 D40 ◇
MADISON, E.W. *Real fields with characterization of the natural numbers* ◇ C57 H15 ◇
MARCHENKOV, S.S. *The computable enumerations of families of recursive functions (Russian)* ◇ D20 D45 ◇
PARIKH, R. *A note on rigid substructures* ◇ C50 C57 ◇
PIXLEY, A.F. *Local Mal'cev conditions* ◇ C05 C57 ◇

SCHNORR, C.-P. *Optimal Goedelnumberings* ◇ D20 D45 ◇

1973

ALTON, D.A. & MADISON, E.W. *Computability of boolean algebras and their extensions* ◇ C57 D45 G05 ◇
APPLEBAUM, C.H. *A result for π-groups* ◇ D45 D50 ◇
APPLEBAUM, C.H. *The recursive equivalence type of a decomposition of an ω-group: the RET of a decomposition* ◇ D45 D50 ◇
BELYAKIN, N.V. *Generalized computations over regular numerations (Russian)* ◇ D45 D65 D70 D75 ◇
CANNONITO, F.B. & GATTERDAM, R.W. *The computability of group constructions. Part I* ◇ D20 D40 D45 ◇
EMDE BOAS VAN, P. *Mostowski's universal set-algebra* ◇ C50 C57 C65 D45 E07 E20 G05 ◇
ERSHOV, YU.L. *Constructive models (Russian)* ◇ C57 D45 ◇
ERSHOV, YU.L. *Skolem functions and constructive models (Russian)* ◇ C57 C65 F50 ◇
ERSHOV, YU.L. & LAVROV, I.A. *The upper semilattice $L(\gamma)$ (Russian)* ◇ D25 D45 ◇
ERSHOV, YU.L. *Theory of numerations. Part II (Russian)* ◇ D45 D65 G30 ◇
FEINER, L. *Degrees of nonrecursive presentability* ◇ C57 D30 D45 ◇
GATTERDAM, R.W. *The computability of group constructions part II* ◇ D20 D40 D45 ◇
GONCHAROV, S.S. *Constructivizability of superatomic boolean algebras (Russian)* ◇ C57 D45 G05 ◇
GONCHAROV, S.S. & NURTAZIN, A.T. *Constructive models of complete solvable theories (Russian)* ◇ C45 C50 C57 G05 ◇
HELM, J.P. & MEYER, A.R. & YOUNG, P. *On orders of translations and enumerations* ◇ D15 D20 D25 D45 F40 ◇
MARCHENKOV, S.S. *The existence of families without positive numerations (Russian)* ◇ D20 D30 D45 ◇
PERETYAT'KIN, M.G. *A strongly constructive model without elementary submodels and extensions (Russian)* ◇ C35 C57 D45 ◇
PERETYAT'KIN, M.G. *Every recursively enumerable extension of a theory of linear order has a constructive model (Russian)* ◇ C57 C65 ◇
PERETYAT'KIN, M.G. *On complete theories with a finite number of denumerable models (Russian)* ◇ C57 ◇
SHEVYAKOV, V.S. *Formulas of the restricted predicate calculus which distinguish certain classes of models with simply computable predicates (Russian)* ◇ C57 ◇
SUTER, G.H. *Recursive elements and constructive extensions of computable local integral domains* ◇ C57 C60 D45 ◇
TANAKA, H. *Length of analytic well-orderings* ◇ D45 D55 E15 E60 ◇
V'YUGIN, V.V. *On minimal numerations of computable classes of recursively enumerable sets (Russian)* ◇ D25 D45 ◇
V'YUGIN, V.V. *On some examples of upper semilattices of computable enumerations (Russian)* ◇ D25 D45 ◇

1974

APPLEBAUM, C.H. *ω-semigroups* ◇ D45 D50 ◇
ASSER, G. *Berechenbare Graphenabbildungen II* ◇ D20 D45 ◇
BADAEV, S.A. *On incomparable enumerations (Russian)* ◇ D25 D45 ◇
BAKER, T.P. & HARTMANIS, J. *On simple Goedel numberings and translations* ◇ B75 D20 D45 ◇
BARZDINS, J. & FREJVALD, R.V. *Prognostizierung und limitaere Synthese effektiv aufzaehlbarer Klassen von Funktionen (Russian)* ◇ D15 D20 D45 ◇
BARZDINS, J. *Two theorems on the identification in the limit of τ-numbers (Russian)* ◇ D20 D45 ◇
BAUR, W. *Rekursive Algebren mit Kettenbedingungen* ◇ D45 ◇
BAUR, W. *Ueber rekursive Strukturen* ◇ C50 C57 D45 ◇
BOONE, W.W. *Between logic and group theory* ◇ C57 C60 D40 ◇
BUCHBERGER, B. *Certain decompositions of Goedel numbering and the semantics of programming languages* ◇ B75 D20 D45 ◇
BUCHBERGER, B. *On certain decompositions of Goedel numberings* ◇ D15 D20 D45 ◇
CROSSLEY, J.N. & NERODE, A. *Combinatorial functors* ◇ C57 C62 D45 D50 G30 ◇
DUSHSKIJ, V.A. *Enumerational reducibility of recursively enumerable classes of sets (Russian)* ◇ D25 D30 D45 ◇
ERSHOV, YU.L. *Theory of numerations, III: Constructive models (Russian)* ◇ B25 C57 C98 D45 F50 ◇
FREJVALD, R.V. *On the limit synthesis of numbers of general recursive functions in various computable numerations (Russian)* ◇ D20 D45 ◇
GOETZE, B.G. *Die Struktur des Halbverbandes der effektiven Numerierungen* ◇ D20 D45 ◇
GRZEGORCZYK, A. *Axiomatic theory of enumeration* ◇ D45 D75 ◇
HAMILTON, A.G. *Recursive equivalence types of vector spaces* ◇ D45 D50 ◇
HARRINGTON, L.A. *Recursively presentable prime models* ◇ C35 C50 C57 C60 ◇
KHISAMIEV, N.G. *Strongly constructive models of a decidable theory (Russian)* ◇ C15 C57 F60 ◇
KINBER, E.B. *On limiting synthesis of quasiminimal Goedel numbers (Russian)* ◇ D20 D45 ◇
LAVROV, I.A. *Certain properties of Post enumeration retracts (Russian)* ◇ D25 D45 ◇
LYNCH, N.A. *Approximations to the halting problem* ◇ D10 D20 D25 D45 ◇
MEJTUS, V.YU. & VERSHININ, K.P. *On some unsolvable problems in computable categories (Russian)* ◇ C57 D80 G30 ◇
NOGINA, E.YU. & VAJNBERG, YU.R. *Categories of effectively topological spaces (Russian)* ◇ D45 G30 ◇
NURTAZIN, A.T. *Strong and weak constructivization and computable families (Russian)* ◇ C57 D45 ◇
PODNIEKS, K.M. *Comparison of the different types of identification in the limit and prediction of functions I (Russian)* ◇ D20 D45 ◇

SAKAI, H. *On numerations of a formal system* ◇ D45 F30 ◇
SCHNORR, C.-P. *Optimal enumerations and optimal Goedel numberings* ◇ D15 D20 D25 D45 ◇
SOARE, R.I. *Isomorphisms on countable vector spaces with recursive operations* ◇ D45 D50 ◇
TAJMANOV, A.D. *On the elementary theory of topological algebras (Russian)* ◇ C57 C65 ◇
V'YUGIN, V.V. *On upper semilattices of numerations (Russian)* ◇ D25 D45 ◇

1975

ACZEL, P. *Recursive density types and Nerode extensions of arithmetic* ◇ D45 D50 F30 ◇
ASSER, G. *Rekursive Graphenabbildungen. I* ◇ D20 D45 D80 ◇
BAKER, T.P. & HARTMANIS, J. *On simple Goedel numberings and translations* ◇ B75 D20 D45 ◇
BARWISE, J. & SCHLIPF, J.S. *On recursively saturated models of arithmetic* ◇ C50 C57 C62 D80 H15 ◇
CARSTENS, H.G. *Reducing hyperarithmetic sequences* ◇ C57 D55 ◇
DENISOV, S.D. *Two-element separable enumeration of sets γ with undecidable problem $P(\gamma,\gamma)$ (Russian)* ◇ D25 D45 ◇
DOBRITSA, V.P. *Recursively numbered classes of constructive extensions and autostability of algebras (Russian)* ◇ C57 D45 ◇
ELLENTUCK, E. *Semigroups, Horn sentences and isolic structures* ◇ C05 C57 D50 E10 ◇
ERSHOV, YU.L. *The upper semilattice of numerations of a finite set (Russian)* ◇ D45 E50 ◇
FEFERMAN, S. *Impredicativity of the existence of the largest divisible subgroup of an abelian p-group* ◇ C57 C60 F65 ◇
FOWLER III, N. *Intersections of α-spaces* ◇ C57 C60 D50 ◇
FOWLER III, N. *Sum of α-spaces* ◇ C57 C60 D45 ◇
FREJVALD, R.V. *Identifiability in the limit of the numbers of general recursive functions in various computable numerations (Russian) (English summary)* ◇ D20 D25 D45 ◇
FREJVALD, R.V. *Minimal Goedel numbers and their identification in the limit* ◇ D20 D45 ◇
GOETZE, B.G. *Die Inseparabilitaetseigenschaften und die Ricesche Unentscheidbarkeitseigenschaft bei effektiven Numerierungen* ◇ D25 D45 ◇
GONCHAROV, S.S. *Autostability and computable families of constructivizations (Russian)* ◇ C50 C57 D45 ◇
GONCHAROV, S.S. *Certain properties of the constructivization of boolean algebras (Russian)* ◇ C57 D45 G05 ◇
GUHL, R. *A theorem on recursively enumerable vector spaces* ◇ C57 C60 D45 ◇
KREISEL, G. *Observations on a recent generalization of completeness theorems due to Schuette* ◇ A05 C07 C57 F05 F20 F35 F50 ◇
KREISEL, G. & MINTS, G.E. & SIMPSON, S.G. *The use of abstract language in elementary metamathematics: Some pedagogic examples* ◇ A05 C07 C57 C75 F05 F07 F20 F50 ◇
MADISON, E.W. & NELSON, GEORGE C. *Some examples of constructive and non-constructive extension of the countable atomless boolean algebra* ◇ C57 F60 G05 ◇
MARANDZHYAN, G.B. *Algorithmic languages that do not admit a mutual optimal translation (Russian) (Armenian summary)* ◇ D20 D45 ◇
METAKIDES, G. & NERODE, A. *Recursion theory and algebra* ◇ C57 C60 D25 D45 ◇
PALYUTIN, E.A. *A supplement to Ju.L.Ershov's article: "The upper semilattice of numerations of a finite set" (Russian)* ◇ D45 ◇
PINUS, A.G. *Effective linear orders (Russian)* ◇ C57 D45 E07 F15 ◇
PODNIEKS, K.M. *Comparison of the different types of identification in the limit and prediction of functions II (Russian)* ◇ D20 D45 ◇
ROBINSON, A. *Algorithms in algebra* ◇ C40 C57 C60 D45 D75 ◇
SEIDENBERG, A. *Construction of the integral closure of a finite integral domain. II* ◇ C57 C60 F55 ◇

1976

ASATRYAN, O.S. *Ueber Retrakte einer indizierten Kleeneschen Menge (Russian) (Armenian summary)* ◇ D20 D45 ◇
BARWISE, J. & SCHLIPF, J.S. *An introduction to recursively saturated and resplendent models* ◇ C15 C40 C50 C57 ◇
BOERGER, E. *On the construction of simple first-order formulae without recursive models* ◇ C57 D25 ◇
CASE, J. *Sortability and extensibility of the graphs of recursively enumerable partial and total orders* ◇ D25 D45 ◇
CROSSLEY, J.N. & NERODE, A. *Effective dimension* ◇ C57 D45 D50 ◇
DOBRITSA, V.P. & GONCHAROV, S.S. *An example of a constructive abelian group with non-constructivizable reduced subgroup (Russian)* ◇ C57 C60 D45 ◇
DOBRITSA, V.P. *On computable and strictly computable classes of constructive algebras (Russian)* ◇ C57 C60 D45 ◇
DOBRITSA, V.P. *Theorem on non-computable classes (Russian)* ◇ D45 ◇
DZGOEV, V.D. *On constructivizability of boolean algebras (Russian)* ◇ D45 ◇
FOWLER III, N. *α-decompositions of α-spaces* ◇ C57 C60 D45 D50 ◇
FRIEDMAN, H.M. *The complexity of explicit definitions* ◇ C40 C57 ◇
GONCHAROV, S.S. *Non-self-equivalent constructivization of atomic boolean algebras (Russian)* ◇ C57 D45 G05 ◇
GONCHAROV, S.S. *Restricted theories of constructive boolean algebras (Russian)* ◇ B25 C57 D35 F60 G05 ◇
HERRMANN, E. *On Lindenbaum functions of $\aleph_0$-categorical theories of finite similarity type* ◇ B25 C35 C57 ◇
LAZARD, D. *Algorithmes fondamentaux en algebre commutative* ◇ D45 ◇

MACHTEY, M. & YOUNG, P. *Simple Goedel numberings, translations, and the P-hierarchy: Preliminary report* ◇ D15 D20 D45 ◇
MORLEY, M.D. *Decidable models* ◇ C50 C57 ◇
NOGINA, E.YU. & VAJNBERG, YU.R. *Two types of continuity of computable mappings of numerated topological spaces (Russian)* ◇ D45 ◇
PINUS, A.G. *Theories of boolean algebras in a calculus with the quantifier "infinitely many exist" (Russian)* ◇ B25 C10 C35 C57 C80 G05 ◇
REMMEL, J.B. *Co-hypersimple structures* ◇ D25 D45 ◇
REMMEL, J.B. *Combinatorial functors on co-r.e. structures* ◇ D25 D30 D45 D50 ◇
SCHMERL, J.H. *Effectiveness and Vaught's gap ω two-cardinal theorem* ◇ C55 C57 ◇
SELIVANOV, V.L. *On computability of some classes of numerations (Russian)* ◇ D25 D45 ◇
SELIVANOV, V.L. *Two theorems on computable numberings (Russian)* ◇ D45 ◇
SHANIN, N.A. *On the quantifier of limiting realizability (Russian)* ◇ C57 C80 F50 ◇

1977

APT, K.R. *Recursive embeddings of partial orderings* ◇ C57 D20 G05 ◇
BADAEV, S.A. *Computable enumerations of families of general recursive functions (Russian)* ◇ D20 D45 ◇
BADAEV, S.A. *Positive numerations (Russian)* ◇ D25 D45 ◇
BIENENSTOCK, E. *Sets of degrees of computable fields* ◇ C57 C60 D30 D45 ◇
BOZOVIC, N.B. *On Markov properties of finitely presented groups* ◇ C05 C50 D45 ◇
DEKKER, J.C.E. *Planos afines con operaciones recursivas* ◇ D45 D50 ◇
DOBRITSA, V.P. *Computability of certain classes of constructive algebras (Russian)* ◇ C57 C60 D45 ◇
DROBOTUN, B.N. *Enumerations of simple models (Russian)* ◇ C57 D45 ◇
DROBOTUN, B.N. *On countable models of decidable almost categorical theories (Russian)* ◇ B25 C15 C35 C57 ◇
DZGOEV, V.D. *On the constructivization of distributed structures with relative complements (Russian)* ◇ D45 ◇
ERSHOV, YU.L. *Constructions "by finite"* ◇ D45 ◇
ERSHOV, YU.L. *Enumeration of the class* $\mathbb{C}^*_{20}$ *(Russian)* ◇ D45 D65 ◇
ERSHOV, YU.L. *Model* $\mathbb{C}$ *of partial continuous functionals* ◇ D45 D65 F10 F50 ◇
ERSHOV, YU.L. *Theory of numerations (Russian)* ◇ D45 D98 ◇
FREJVALD, R.V. & KINBER, E.B. *Identification in the limit of minimal Goedel numbers (Russian)* ◇ D20 D45 ◇
GONCHAROV, S.S. *The quantity of nonautoequivalent constructivizations (Russian)* ◇ C50 C57 D45 ◇
GUHL, R. *Two notes on recursively enumerable vector spaces* ◇ C57 C60 D45 ◇
HAY, L. & MANASTER, A.B. & ROSENSTEIN, J.G. *Concerning partial recursive similarity transformations of linearly ordered sets* ◇ C57 C65 D20 D25 D30 D45 E07 ◇
HERRMANN, E. *Ueber Lindenbaumfunktionen von* $\aleph_0$*-kategorischen Theorien endlicher Signatur* ◇ B25 C35 C57 ◇
KALANTARI, I. & RETZLAFF, A.T. *Maximal vector spaces under automorphisms of the lattice of recursively enumerable vector spaces* ◇ C57 C60 D45 ◇
KAUFMANN, M. *A rather classless model* ◇ C50 C57 C62 E65 ◇
KHISAMIEV, N.G. *On the periodical part of a strongly constructivizable abelian group (Russian)* ◇ C57 C60 D45 ◇
KINBER, E.B. *On a theory of inductive inference* ◇ D20 D45 ◇
KINBER, E.B. *On btt-degrees of sets of minimal numbers in Goedel numberings* ◇ D20 D30 D45 ◇
KINBER, E.B. *On identification in the limit of the minimal numbers for functions of effectively enumerable classes (Russian)* ◇ D20 D45 ◇
LALLEMENT, G. *Presentations de monoides et problemes algorithmiques* ◇ D40 D45 ◇
LAVROV, I.A. *Computable numberings* ◇ D25 D45 ◇
MARANDZHYAN, G.B. *On c-degrees of sets of minimal indices of algorithms (Russian) (Armenian and English summaries)* ◇ D20 D45 ◇
METAKIDES, G. *A return to constructive algebra via recursive function theory* ◇ C57 D45 ◇
METAKIDES, G. & NERODE, A. *Recursively enumerable vector spaces* ◇ C57 C60 D45 ◇
REMMEL, J.B. *Maximal and cohesive vector spaces* ◇ C57 D25 D45 ◇
SCHINZEL, B. *Decomposition of Goedelnumberings into Friedbergnumberings* ◇ D20 D45 ◇
SELIVANOV, V.L. *Enumerations of canonically calculable families of finite sets (Russian)* ◇ D45 ◇
SHANIN, N.A. *On the quantifier of limiting realizability* ◇ B55 C57 C80 F50 ◇
VUCKOVIC, V. *Recursive and recursive enumerable manifolds I,II* ◇ C57 D45 D80 ◇
WEIHRAUCH, K. *A generalized computability thesis* ◇ D45 ◇

1978

APPLEBAUM, C.H. *Some structure theorems for inverse ω-semigroups* ◇ D45 D50 ◇
BLOSHCHITSYN, V.YA. & ZAKIR'YANOV, K.KH. *Constructive abelian groups (Russian)* ◇ C57 C60 D45 ◇
BUCHBERGER, B. & ROIDER, B. *Input/output codings and transition functions in effective systems* ◇ D05 D20 D45 ◇
CALUDE, C. *On the category of recursive languages* ◇ B03 D45 F40 G30 ◇
CASE, J. & MOORE, D.J. *The complexity of total order structures* ◇ D15 D20 D45 ◇
CHEN, KEHSUN *Recursive well-founded orderings* ◇ D45 F15 ◇
DEKKER, J.C.E. *Projective bigraphs with recursive operations* ◇ D45 D50 ◇

DOBRITSA, V.P. & KHISAMIEV, N.G. & NURTAZIN, A.T. *Constructive periodic abelian groups (Russian)* ◇ C57 C60 F60 ◇
ELLENTUCK, E. *Model theoretic methods in the theory of isols* ◇ C57 D50 ◇
FOWLER III, N. *Effective inner product spaces* ◇ C57 C60 D45 D80 ◇
GONCHAROV, S.S. *Constructive models of $\aleph_1$-categorical theories (Russian)* ◇ C35 C57 ◇
GONCHAROV, S.S. *Strong constructivizability of homogeneous models (Russian)* ◇ C50 C57 D45 ◇
HECKER, H.-D. *Bemerkungen zu Enumerationssystemen und Zeitmassen* ◇ D15 D25 D45 ◇
KALANTARI, I. *Major subspaces of recursively enumerable vector spaces* ◇ C57 C60 D25 D45 ◇
KHISAMIEV, N.G. *Strongly constructive periodic abelian groups (Russian) (Kazakh summary)* ◇ C57 C60 D45 ◇
KOROL'KOV, YU.D. *Families of general recursive functions with a finite number of limit points (Russian)* ◇ D20 D45 ◇
KUSHNER, B.A. *On some systems of computable real numbers (Russian)* ◇ D45 F60 ◇
LASCAR, D. *Caractere effectif des theoremes d'approximation d'Artin (English summary)* ◇ C57 ◇
LIPSHITZ, L. & NADEL, M.E. *The additive structure of models of arithmetic* ◇ C15 C50 C57 C62 ◇
LYNCH, N.A. *Straight-line program length as a parameter for complexity measures* ◇ D15 D45 ◇
MACHTEY, M. & WINKLMANN, K. & YOUNG, P. *Simple Goedel numberings, isomorphisms, and programming properties* ◇ D05 D10 D15 D20 D45 F40 ◇
MCALOON, K. *Diagonal methods and strong cuts in models of arithmetic* ◇ C57 C62 D80 F30 ◇
METAKIDES, G. *Constructive algebra in a new frame* ◇ C57 C60 D45 ◇
MILLAR, T.S. *Foundations of recursive model theory* ◇ C50 C57 ◇
NEHRLICH, W. *Eine Bemerkung zur effektiven Fixpunktberechnung bei vollstaendigen Numerierungen* ◇ D45 ◇
NOGINA, E.YU. *Numerierte topologische Raeume (Russisch)* ◇ C57 C60 D45 ◇
NOGINA, E.YU. *On completely enumerable subsets of direct products of numbered sets (Russian)* ◇ D45 ◇
PERETYAT'KIN, M.G. *Criterion for strong constructivizability of a homogeneous model (Russian)* ◇ C35 C50 C57 D45 ◇
REMMEL, J.B. *A r-maximal vector space not contained in any maximal vector space* ◇ C57 D25 D45 ◇
REMMEL, J.B. *Realizing partial orderings by classes of co-simple sets* ◇ D25 D45 D50 ◇
REMMEL, J.B. *Recursively enumerable boolean algebras* ◇ C57 D25 D45 G05 ◇
RETZLAFF, A.T. *Simple and hyperhypersimple vector spaces* ◇ C57 D25 D45 ◇
SCHLIPF, J.S. *Toward model theory through recursive saturation* ◇ C15 C40 C50 C57 C70 ◇
SCHMERL, J.H. *A decidable $\aleph_0$-categorical theory with a non-recursive Ryll-Nardzewski function* ◇ B25 C35 C57 D30 ◇
SEIDENBERG, A. *Constructions in a polynomial ring over the ring of integers* ◇ D45 F55 ◇
SELIVANOV, V.L. *On index sets of classes of numberings (Russian)* ◇ D45 ◇
SHORE, R.A. *Controlling the dependence degree of a recursively enumerable vector space* ◇ D25 D45 D50 ◇
VALIDOV, F.I. *Minimal numerations of families of general recursive functions (Russian)* ◇ D20 D45 ◇
VERBEEK, R. & WEIHRAUCH, K. *Data representation and computational complexity* ◇ D15 D45 ◇
WIEHAGEN, R. *Characterization problems in the theory of inductive inference* ◇ D15 D20 D45 ◇

1979

BELYAKIN, N.V. *Autonomous computability (Russian)* ◇ D45 D55 D65 ◇
BERGSTRA, J.A. & TUCKER, J.V. *Equational specifications for computable data types: six hidden functions suffice and other sufficiency bounds* ◇ B75 D45 D80 ◇
BERTONI, A. & MAURI, G. & MIGLIOLI, P.A. *A characterization of abstract data as model-theoretic invariants* ◇ B75 C57 D45 D80 ◇
DEMILLO, R.A. & LIPTON, R.J. *Some connections between mathematical logic and complexity theory* ◇ C57 D15 F30 H15 ◇
DRIES VAN DEN, L. *Algorithms and bounds for polynomial rings* ◇ C57 C60 ◇
DVORNIKOV, S.G. *C-degrees of continuous everywhere defined functionals (Russian)* ◇ D45 D65 ◇
DVORNIKOV, S.G. *Precompletely enumerated sets (Russian)* ◇ D45 ◇
DZGOEV, V.D. *Cartesian degrees of constructive models (Russian)* ◇ D45 ◇
DZGOEV, V.D. *Recursive automorphisms of constructive models (Russian)* ◇ C57 D45 ◇
GIRSTMAIR, K. *Ueber konstruktive Methoden der Galoistheorie* ◇ C57 C60 F55 ◇
HARKLEROAD, L. *Recursive equivalence types on recursive manifolds* ◇ D45 D50 D80 ◇
JANTKE, K.P. *Natural properties of strategies identifying recursive functions* ◇ B75 D20 D45 ◇
KALANTARI, I. *Automorphisms of the lattice of recursively enumerable vector spaces* ◇ C57 C60 D45 ◇
KALANTARI, I. & RETZLAFF, A.T. *Recursive constructions in topological spaces* ◇ C57 C65 D45 ◇
KANDA, A. *Fully effective solutions of recursive domain equations* ◇ D45 ◇
KANDA, A. & PARK, D. *When are two effectively given domains identical?* ◇ D45 ◇
KHISAMIEV, N.G. *Criterion of the constructivizability of a direct sum of cyclic p-groups (Russian)* ◇ C57 C60 D45 ◇
KHISAMIEV, N.G. *On subgroups of finite index of abelian groups (Russian) (Kazakh summary)* ◇ C57 C60 D45 F60 ◇
KOROL'KOV, YU.D. *Families of general recursive functions without isolated points (Russian)* ◇ D20 D45 ◇
KOSOVSKIJ, N.K. *On decision procedures for invariant properties of short algorithms (Russian) (English summary)* ◇ D15 D20 D25 D45 ◇

KRUPSKIJ, V.N. *On completely enumerable sets in effectively metric spaces (Russian) (English summary)* ◇ D25 D45 ◇
KUDAJBERGENOV, K.ZH. *A theory with two strongly constructivizable models (Russian)* ◇ C57 D45 ◇
KUKIN, G.P. *Subalgebras of finitely defined Lie algebras (Russian)* ◇ C05 C57 ◇
LAVROV, I.A. *Retracts of Post's numbering and effectivization of quantifiers* ◇ D25 D45 ◇
LERMAN, M. & SCHMERL, J.H. *Theories with recursive models* ◇ C35 C57 C65 ◇
MANDERS, K.L. *The theory of all substructures of a structure: Characterisation and decision problems* ◇ B20 B25 C57 C60 ◇
MARANDZHYAN, G.B. *On the sets of minimal indices of partial recursive functions* ◇ D20 D30 D45 D55 ◇
MEAD, J. *Recursive prime models for boolean algebras* ◇ C50 C57 G05 ◇
METAKIDES, G. & NERODE, A. *Effective content of field theory* ◇ C57 C60 D45 ◇
METAKIDES, G. & REMMEL, J.B. *Recursion theory on orderings I. A model theoretic setting* ◇ C57 D45 ◇
MILLAR, T.S. *A complete, decidable theory with two decidable models* ◇ C57 ◇
REMMEL, J.B. *R-maximal boolean algebras* ◇ C57 D45 G05 ◇
RETZLAFF, A.T. *Direct summands of recursively enumerable vector spaces* ◇ C57 C60 D45 ◇
SCHADE, W. *Indexmengen rekursiver reeller Zahlen* ◇ D30 D45 ◇
SCHINZEL, B. *Classes of decompositions of a Goedelnumbering* ◇ D20 D45 ◇
SCHINZEL, B. *Uebersetzer zwischen Goedelnumerierungen* ◇ D20 D45 ◇
ZHAROV, V.G. *On the complexity of the terms of the constructive sequences of Turing machines (Russian)* ◇ D10 D45 F60 ◇

1980

ASATRYAN, O.S. *On creative equivalences (Russian) (Armenian and English summaries)* ◇ D25 D45 ◇
ASATRYAN, O.S. *On positive equivalences (Russian) (Armenian summary)* ◇ D45 ◇
BARASHKO, A.S. *On boundary sets in the theory of enumerations (Ukrainian) (English and Russian summaries)* ◇ D15 D25 D45 ◇
BERGSTRA, J.A. & TUCKER, J.V. *A characterisation of computable data types by means of a finite equational specification method* ◇ B75 D45 D80 ◇
BERGSTRA, J.A. & TUCKER, J.V. *A natural data type with a finite equational final semantics specification but no effective equational initial semantics specification* ◇ B75 D45 D80 ◇
BERGSTRA, J.A. & TUCKER, J.V. *On bounds for the specification of finite data types by means of equations and conditional equations* ◇ B75 D45 D80 ◇
BERTONI, A. & MAURI, G. & MIGLIOLI, P.A. *Towards a theory of abstract data types: a discussion on problems and tools* ◇ B75 C57 D80 ◇
DEGTEV, A.N. *A category of partial recursive functions (Russian)* ◇ D20 D45 D75 G30 ◇
DEGTEV, A.N. *On reducibility of numerations (Russian)* ◇ D25 D30 D45 ◇
DROBOTUN, B.N. & GONCHAROV, S.S. *Numerations of saturated and homogeneous models (Russian)* ◇ B25 C50 C57 D45 ◇
DZGOEV, V.D. & GONCHAROV, S.S. *Autostability of models (Russian)* ◇ C57 D45 G10 ◇
DZGOEV, V.D. *On the constructivization of certain structures (Russian)* ◇ C57 D45 ◇
ERSHOV, YU.L. *Decision problems and constructivizable models (Russian)* ◇ B25 C57 C60 C98 D45 D98 ◇
FALKINGER, J. *Reduzierbarkeit von berechenbaren Numerierungen von P_1* ◇ D20 D30 D45 ◇
FALKINGER, J. *Universalitaet von berechenbaren Numerierungen von partiell rekursiven Funktionen* ◇ D20 D30 D45 ◇
GEORGIEVA, N.V. *Theorems on the normal form of some recursive elements and mappings (Russian)* ◇ D45 D80 ◇
GONCHAROV, S.S. *A totally transcendental decidable theory without constructivizable homogeneous models (Russian)* ◇ C45 C50 C57 ◇
GONCHAROV, S.S. *Autostability of models and abelian groups (Russian)* ◇ C50 C57 C60 D45 ◇
GONCHAROV, S.S. *Computable single-valued numerations (Russian)* ◇ D20 D45 ◇
GONCHAROV, S.S. *Problem of the number of non-self-equivalent constructivizations (Russian)* ◇ C57 D45 ◇
GONCHAROV, S.S. *The problem of the number of nonautoequivalent constructivizations (Russian)* ◇ C57 D45 ◇
GONCHAROV, S.S. *Totally transcendental theory with non-constructivizable prime model (Russian)* ◇ C45 C50 C57 D45 ◇
HARNIK, V. *Game sentences, recursive saturation and definability* ◇ C40 C50 C57 C70 ◇
HAUCK, J. *Konstruktive Darstellungen in topologischen Raeumen mit rekursiver Basis* ◇ D45 D80 ◇
KUDAJBERGENOV, K.ZH. *On constructive models of undecidable theories (Russian)* ◇ C15 C35 C57 D35 ◇
LYNCH, N.A. *Straight-line program length as a parameter for complexity analysis* ◇ D15 D45 ◇
MANASTER, A.B. & ROSENSTEIN, J.G. *Two-dimensional partial orderings: Recursive model theory* ◇ C57 C65 E07 ◇
MANASTER, A.B. & ROSENSTEIN, J.G. *Two-dimensional partial orderings: Undecidability* ◇ C57 C65 D35 E07 G10 ◇
METAKIDES, G. & NERODE, A. *Recursion theory on fields and abstract dependence* ◇ C57 C60 D45 ◇
MILLAR, T.S. *Homogeneous models and decidability* ◇ C50 C57 ◇
MYERS, R.W. *Complexity of model-theoretic notions* ◇ C07 C35 C57 D55 ◇
NERODE, A. & SMITH, RICK L. *The undecidability of the lattice of recursively enumerable subspaces* ◇ D35 D45 ◇

REISER, A. & WEIHRAUCH, K. *Natural numberings and generalized computability* ◇ D45 D75 ◇
REMMEL, J.B. *On r.e. and co-r.e. vector spaces with nonextendible bases* ◇ C57 C60 D45 ◇
REMMEL, J.B. *Recursion theory on orderings II* ◇ C57 D45 ◇
REMMEL, J.B. *Recursion theory on algebraic structures with independent sets* ◇ C57 C60 D45 ◇
SCHINZEL, B. *Ueber die Kategorie der Programmbuendel (Hab.-Schr.)* ◇ D15 D20 D45 ◇
SCHINZEL, B. *Zerlegung mit Vergleichsbedingungen einer Goedelnumerierung* ◇ D20 D45 ◇
SCHLIPF, J.S. *Recursively saturated models of set theory* ◇ C50 C57 C62 E70 H20 ◇
SCHMERL, J.H. *Recursive colorings of graphs* ◇ D45 D80 ◇
SMORYNSKI, C.A. & STAVI, J. *Cofinal extension preserves recursive saturation* ◇ C50 C57 C62 ◇
SMYTH, M.B. *Computability in categories* ◇ D45 D75 G30 ◇
SPERSCHNEIDER, V. *Goedelisierung* ◇ D45 D75 ◇
STOLTENBERG-HANSEN, V. & TUCKER, J.V. *Computing roots of unity in fields* ◇ D25 D45 D80 ◇
TUCKER, J.V. *Computability and the algebra of fields: some affine constructions* ◇ B25 C57 C60 C65 D45 ◇
VISSER, A. *Numerations, λ-calculus & arithmetic* ◇ B40 D45 F30 ◇
WEIHRAUCH, K. *Berechenbarkeit auf CPO-S (Eine Vorlesung ausgearbeitet von Thomas Deil)* ◇ D45 D75 ◇

1981

ASH, C.J. & NERODE, A. *Intrinsically recursive relations* ◇ D45 D50 ◇
BARASHKO, A.S. *Exact lower bounds for indicators of the enumeration of sets (Russian)* ◇ D15 D45 ◇
BAUMSLAG, G. & CANNONITO, F.B. & MILLER III, C.F. *Computable algebra and group embeddings* ◇ D40 D45 ◇
BERGSTRA, J.A. & MEYER, J.-J.C. *Small specifications for large finite data structures* ◇ B75 D45 D80 ◇
BERTONI, A. & MAURI, G. & MIGLIOLI, P.A. *Model theoretic aspects of abstract data specification* ◇ B75 C50 C57 ◇
CLOTE, P. *A note on decidable model theory* ◇ B25 C15 C50 C57 D30 ◇
CROSSLEY, J.N. & MIRANDA, S. *A bibliography of effective algebra* ◇ A10 D45 ◇
CROSSLEY, J.N. (ED.) *Aspects of effective algebra. Proceedings of a conference at Monash University, Australia, 1-4 August, 1979* ◇ C57 C97 D45 D97 ◇
CROSSLEY, J.N. & NERODE, A. *Recursive equivalence on matroids* ◇ D45 D50 ◇
CROSSLEY, J.N. *Reminiscences of logicians. II* ◇ A10 D45 D50 ◇
DICHEV, A.V. *Computability in the sense of Moschovakis and its connection with partial recursiveness via numerations (Russian)* ◇ D45 D75 ◇
DOBRITSA, V.P. *On constructivizable abelian groups (Russian)* ◇ C57 C60 D40 D45 ◇
ELLENTUCK, E. *Galois theorems for isolated fields* ◇ C57 C60 D45 D50 ◇
GONCHAROV, S.S. *Groups with a finite number of constructivizations (Russian)* ◇ C57 C60 D45 ◇
HAUCK, J. *Berechenbarkeit in topologischen Raeumen mit rekursiver Basis* ◇ D45 ◇
HECKER, H.-D. *Zur Kompliziertheit der Beschreibung von Enumerationen rekursiv aufzaehlbarer Mengen* ◇ D15 D25 D45 ◇
HINGSTON, P. *Effective decomposition in Noetherian rings* ◇ C57 C60 D45 ◇
KALANTARI, I. *Effective content of a theorem of M.H. Stone* ◇ D45 D80 ◇
KANDA, A. *Constructive category theory* ◇ D45 D80 G30 ◇
KARR, M. *Summation in finite terms* ◇ C57 ◇
KHISAMIEV, N.G. *Criterion for constructivizability of a direct sum of cyclic p-groups (Russian) (Kazakh summary)* ◇ C57 C60 D45 ◇
KIERSTEAD, H.A. *An effective version of Dilworth's theorem* ◇ D45 D80 ◇
KIRBY, L.A.S. & MCALOON, K. & MURAWSKI, R. *Indicators, recursive saturation and expandability* ◇ C50 C57 C62 H15 ◇
KLAEREN, H.A. & SCHULZ, MARTIN *Computable algebras, word problems and canonical term algebras* ◇ D45 D80 ◇
KOTLARSKI, H. *On elementary cuts in models of arithmetic* ◇ C50 C57 C62 ◇
LACHLAN, A.H. *Full satisfaction classes and recursive saturation* ◇ C50 C57 C62 ◇
LERMAN, M. *On recursive linear orderings* ◇ C57 C65 D30 D45 D55 E07 ◇
LIN, C. *Recursively presented abelian groups: effective p-group theory. I* ◇ C57 C60 D45 ◇
LIN, C. *The effective content of Ulm's theorem* ◇ C57 C60 D45 ◇
MACINTYRE, A. *The complexity of types in field theory* ◇ C45 C50 C57 C60 ◇
MANASTER, A.B. & REMMEL, J.B. *Some recursion theoretic aspects of dense two-dimensional partial orderings* ◇ D45 ◇
MILLAR, T.S. *Counterexamples via model completions* ◇ C10 C35 C50 C57 ◇
MILLAR, T.S. *Vaught's theorem recursively revisited* ◇ C15 C57 D45 ◇
MURAWSKI, R. *A simple remark on satisfaction classes, indiscernibles and recursive saturation* ◇ C30 C50 C57 H15 ◇
NOGINA, E.YU. *The relation between separability and traceability of sets (Russian)* ◇ D45 ◇
REMMEL, J.B. *Effective structures not contained in recursively enumerable structures* ◇ C57 D45 ◇
REMMEL, J.B. *Recursive isomorphism types of recursive boolean algebras* ◇ C57 D45 G05 ◇
REMMEL, J.B. *Recursive boolean algebras with recursive atoms* ◇ C57 D45 G05 ◇
REMMEL, J.B. *Recursively categorical linear orderings* ◇ C35 C57 C65 ◇

RICHTER, L.J. *Degrees of structures*
◇ C57 D30 D45 ◇
ROCHE LA, P. *Effective Galois theory*
◇ C57 C60 D45 F60 ◇
RODRIGUEZ ARTALEJO, M. *Eine syntaktisch-algebraische Methode zur Konstruktion von Modellen*
◇ B10 C07 C30 C57 ◇
SCHAEFER, G. & WEIHRAUCH, K. *Admissible representations of effective cpo's* ◇ D45 ◇
SCHMERL, J.H. *Arborescent structures I: Recursive models*
◇ C50 C57 C65 ◇
SCHMERL, J.H. *Recursively saturated, rather classless models of Peano arithmetic*
◇ C50 C57 C62 E45 E55 ◇
SHEN', A.KH. *Some remarks on numerations that are not natural (Russian)* ◇ D20 D45 ◇
SMITH, RICK L. *Effective aspects of profinite groups*
◇ D45 ◇
SMITH, RICK L. *Effective valuation theory*
◇ C57 C60 D45 F55 ◇
SMITH, RICK L. *Two theorems on autostability in p-groups*
◇ C57 C60 ◇
SMORYNSKI, C.A. *Cofinal extensions of nonstandard models of arithmetic* ◇ C50 C57 C62 ◇
SMORYNSKI, C.A. *Elementary extensions of recursively saturated models of arithmetic* ◇ C50 C57 C62 ◇
SMORYNSKI, C.A. *Recursively saturated nonstandard models of arithmetic* ◇ C50 C57 C62 ◇
URZYCZYN, P. *The unwind property in certain algebras*
◇ C05 C57 ◇
WATNICK, R. *Constructive and recursive scattered order types* ◇ D45 ◇
WEIHRAUCH, K. *Recursion and complexity theory on CPO's* ◇ D15 D45 D75 ◇

1982

APPLEBAUM, C.H. *An introduction to ω-extensions of ω-groups* ◇ D45 D50 ◇
BALDWIN, JOHN T. *Recursion theory and abstract dependence* ◇ C57 C60 D25 D45 ◇
BERGSTRA, J.A. & KLOP, J.W. *Algebraic specifications for parametrized data types with minimal parameter and target algebras* ◇ D45 D80 ◇
BYERLY, R.E. *An invariance notion in recursion theory*
◇ D20 D45 ◇
CEGIELSKI, P. & MCALOON, K. & WILMERS, G.M. *Modeles recursivement satures de l'addition et de la multiplication des entiers naturels (English summary)*
◇ C50 C57 C62 F30 H15 ◇
COMYN, G. *Arbres infinitaires. Approximations et proprietes de calculabilite* ◇ C57 ◇
CROSSLEY, J.N. *The given* ◇ C57 D45 ◇
DUBOIS, DONALD WARD *General numeration. I: Gauged schemes. II: Division schemes* ◇ D45 ◇
DZGOEV, V.D. *Constructivizations of direct products of algebraic systems (Russian)*
◇ C30 C57 D45 G05 G10 ◇
EISENBERG, E.F. & REMMEL, J.B. *Effective isomorphisms of algebraic structures* ◇ C57 D45 ◇
FREJVALD, R.V. & KINBER, E.B. & WIEHAGEN, R. *Inductive inference and computable one-one numberings* ◇ B48 D20 D45 ◇
GOLDMAN, J. & HOMER, S. *Quadratic automata*
◇ B70 D05 D45 ◇
GONCHAROV, S.S. *Limiting equivalent constructivizations (Russian)* ◇ C57 D45 F60 ◇
HARTMANIS, J. *A note on natural complete sets and Goedel numberings* ◇ D15 D25 D45 ◇
KALANTARI, I. *Major subsets in effective topology*
◇ C57 C65 D25 D45 ◇
KALANTARI, I. & LEGGETT, A. *Simplicity in effective topology* ◇ C57 C65 D25 ◇
KNIGHT, J.F. & NADEL, M.E. *Expansions of models and Turing degrees* ◇ C50 C57 D30 ◇
KOROL'KOV, YU.D. *On the reducibility of index sets of families of general recursive functions (Russian)*
◇ D20 D45 ◇
LERMAN, M. & ROSENSTEIN, J.G. *Recursive linear orderings* ◇ C57 C65 D45 D55 E07 ◇
LERMAN, M. & REMMEL, J.B. *The universal splitting property I* ◇ D25 D45 ◇
MACINTYRE, A. *Residue fields of models of P*
◇ C57 C60 C62 H15 ◇
MAIER, W. & MENZEL, W. & SPERSCHNEIDER, V. *Embedding properties of total recursive functions*
◇ D45 ◇
MAL'TSEV, A.A. *Upper semilattices of numerations (Russian)* ◇ D45 ◇
MANN, A. *A note on recursively presented and co-recursively presented groups* ◇ C60 D40 D45 ◇
MARKER, D. *Degrees of models of true arithmetic*
◇ C57 C62 D30 H15 ◇
MCALOON, K. *On the complexity of models of arithmetic*
◇ C57 C62 ◇
METAKIDES, G. & NERODE, A. *The introduction of non-recursive methods into mathematics*
◇ C57 D45 D80 F60 ◇
MILLAR, T.S. *Type structure complexity and decidability*
◇ C50 C57 ◇
MILLER, DOUGLAS E. *Index sets and Boolean operations*
◇ D25 D45 D55 ◇
MONTAGNA, F. *Relatively precomplete numerations and arithmetic* ◇ D45 F30 ◇
MOROZOV, A.S. *Countable homogeneous boolean algebras (Russian)* ◇ C15 C50 C57 G05 ◇
MOROZOV, A.S. *Strong constructivizability of countable saturated boolean algebras (Russian)*
◇ C50 C57 G05 ◇
MULRY, P.S. *Generalized Banach-Mazur functionals in the topos of recursive sets* ◇ D20 D45 G30 ◇
NERODE, A. & REMMEL, J.B. *Recursion theory on matroids* ◇ C57 D25 D45 ◇
NOSKOV, G.A. *On conjugacy in metabelian groups (Russian)* ◇ C57 C60 ◇
PERETYAT'KIN, M.G. *Finitely axiomatizable totally transcendental theories (Russian)* ◇ C45 C57 ◇
PERETYAT'KIN, M.G. *Turing machine computations in finitely axiomatizable theories (Russian)*
◇ C57 D45 ◇
RICCARDI, G.A. *The independence of control structures in programmable numberings of the partial recursive functions* ◇ D20 D45 ◇

SCHINZEL, B. *Complexity of decomposotions of Goedel numberings* ◇ D20 D45 ◇
SCHINZEL, B. *On decomposition of Goedel numberings into Friedberg numberings* ◇ D20 D45 ◇
SELIVANOV, V.L. *On the index sets in the Kleene-Mostowski hierarchy (Russian)* ◇ D25 D45 D55 ◇
SELIVANOV, V.L. *The structure of degrees of generalized index sets (Russian)* ◇ D30 D45 ◇
SHI, NIANDONG *Creative pairs of subalgebras of recursively enumerable boolean algebras (Chinese)* ◇ C57 D25 D45 G05 ◇
SMORYNSKI, C.A. *A note on initial segment constructions in recursively saturated models of arithmetic* ◇ C50 C57 C62 ◇
SMORYNSKI, C.A. *Back-and-forth inside a recursively saturated model of arithmetic* ◇ C15 C50 C57 C62 ◇
TSUBOI, A. *On M-recursively saturated models of arithmetic* ◇ C50 C57 C62 ◇
TVERSKOJ, A.A. *Investigation of recursiveness and arithmeticity of signature functions in nonstandard models of arithmetics (Russian)* ◇ C57 C62 D45 H15 ◇
ZAKHAROV, S.D. *The algebra of enumeration operators (Russian) (English summary)* ◇ D20 D45 ◇

1983

ASH, C.J. & MILLAR, T.S. *Persistently finite, persistently arithmetic theories* ◇ C57 ◇
BARENDREGT, H.P. & LONGO, G. *Recursion theoretic operators and morphisms of numbered sets* ◇ D20 D30 D45 ◇
BERGSTRA, J.A. & KLOP, J.W. *Initial algebra specifications for parametrized data types* ◇ B75 D45 D80 G30 ◇
BERTONI, A. & MAURI, G. & MIGLIOLI, P.A. *On the power of model theory in specifying abstract data types and in capturing their recursiveness* ◇ B75 C50 C57 D20 ◇
BYERLY, R.E. *Definability of r.e. sets in a class of recursion theoretic structures* ◇ D25 D45 ◇
CROSSLEY, J.N. & REMMEL, J.B. *Undecidability and recursive equivalence I* ◇ D35 D45 D50 ◇
DOBRITSA, V.P. *Complexity of the index set of a constructive model (Russian)* ◇ C57 D45 ◇
DOBRITSA, V.P. *Some constructivizations of abelian groups (Russian)* ◇ C57 C60 D45 ◇
DOWNEY, R.G. *Abstract dependence, recursion theory, and the lattice of recursively enumerable filters* ◇ D25 D45 ◇
DOWNEY, R.G. *Nowhere simplicity in matroids* ◇ D25 D45 ◇
DOWNEY, R.G. *On a question of A.Retzlaff* ◇ C57 C60 D25 D45 ◇
GONCHAROV, S.S. *Positive numerations of families with single-valued numerations (Russian)* ◇ D45 ◇
GONCHAROV, S.S. *Universal recursively enumerable boolean algebras (Russian)* ◇ C50 C57 D45 G05 ◇
GONCHAROV, V.A. *A recursively representable Boolean algebra (Russian)* ◇ C57 D45 G05 ◇
GUICHARD, D.R. *Automorphisms of substructure lattices in recursive algebra* ◇ C57 C60 D45 G05 ◇
HARKLEROAD, L. *Manifolds allowing RET arithmetic* ◇ D45 D50 D80 ◇
HICKIN, K.K. & PHILLIPS, R.E. *Isomorphism types in wreath products and effective embeddings of periodic groups* ◇ C57 C60 D40 ◇
KALANTARI, I. & REMMEL, J.B. *Degrees of recursively enumarable topological spaces* ◇ C57 C65 D25 D45 ◇
KALANTARI, I. & LEGGETT, A. *Maximality in effective topology* ◇ C57 C65 D25 D45 ◇
KAMIMURA, T. & TANG, A. *Effectively given spaces* ◇ D45 ◇
KHISAMIEV, N.G. *Strongly constructive abelian p-groups (Russian)* ◇ C57 C60 D45 F60 ◇
KIERSTEAD, H.A. *An effective version of Hall's theorem* ◇ C57 D80 ◇
KIERSTEAD, H.A. & REMMEL, J.B. *Indiscernibles and decidable models* ◇ C30 C45 C57 C80 ◇
KINBER, E.B. *A note on limit identification of c-minimal indices (German and Russian summaries)* ◇ D20 D45 ◇
KNIGHT, J.F. *Additive structure in uncountable models for a fixed completion of P* ◇ C50 C57 C62 C75 D30 ◇
KOSSAK, R. *A certain class of models of Peano arithmetic* ◇ C50 C57 C62 ◇
KOTLARSKI, H. *On elementary cuts in models of arithmetic* ◇ C50 C57 C62 ◇
KUDAJBERGENOV, K.ZH. *The number of constructive homogeneous models of a complete decidable theory (Russian)* ◇ C50 C57 ◇
KUZ'MINA, T.M. *Reducibility by morphisms (Russian)* ◇ D20 D45 ◇
MADISON, E.W. *The existence of countable totally nonconstructive extensions of the countable atomless boolean algebras* ◇ C57 D45 G05 ◇
MILLAR, T.S. *Omitting types, type spectrums, and decidability* ◇ C07 C57 ◇
MILLAR, T.S. *Persistently finite theories with hyperarithmetic models* ◇ C15 C50 C57 D30 ◇
MOHRHERR, J. *Kleene index sets and functional m-degrees* ◇ D20 D25 D30 D45 ◇
MOROZOV, A.S. *Groups of recursive automorphisms of constructive boolean algebras (Russian)* ◇ C07 C57 D45 F60 G05 ◇
MOSES, M. *Recursive properties of isomorphism types* ◇ C57 D45 ◇
NERODE, A. & REMMEL, J.B. *Recursion theory on matroids II* ◇ D25 D45 ◇
ORLICKI, A. *On effective numberings of effective definitional schemes* ◇ D20 D45 ◇
ROY, D.K. *R.e. presented linear orders* ◇ C57 C65 D45 E07 ◇
SCHAEFER, G. & WEIHRAUCH, K. *Admissible representations of effective CPOs* ◇ B75 D45 ◇
STEINHORN, C.I. *A new omitting types theorem* ◇ C07 C15 C45 C50 C57 ◇

TSUBOI, A. *On M-recursively saturated models of Peano arithmetic (Japanese)* ◇ C50 C57 C62 ◇

1984

ABDRAZAKOV, K.T. & KHISAMIEV, N.G. *A criterion for strong constructivizability of a class of abelian p-groups (Russian)* ◇ C57 C60 D45 ◇

ASH, C.J. & DOWNEY, R.G. *Decidable subspaces and recursively enumerable subspaces* ◇ C57 C60 D35 D45 ◇

BALDWIN, JOHN T. *First-order theories of abstract dependence relations* ◇ C35 C45 C57 C60 D25 D45 ◇

BUECHLER, S. *Resplendency and recursive definability in ω-stable theories* ◇ C45 C50 C57 ◇

CARSTENS, H.G. & PAEPPINGHAUS, P. *Abstract construction of counterexamples in recursive graph theory* ◇ D45 ◇

CLOTE, P. *A recursion theoretic analysis of the clopen Ramsey theorem* ◇ D45 D55 E05 ◇

COMPTON, K.J. *On rich words* ◇ C85 D05 D45 ◇

DEGTEV, A.N. *A category of enumerated sets (Russian)* ◇ D45 G30 ◇

DENISOV, A.S. *Constructive homogeneous extensions (Russian)* ◇ C50 C57 ◇

DICHEV, A.V. *Computability in the sense of Moskovakis and its connection with partial recursivity through numbering (Russian)* ◇ D45 D75 ◇

DOWNEY, R.G. *A note on decompositions of recursively enumerable subspaces* ◇ C57 C60 D25 D45 ◇

DOWNEY, R.G. *Bases of supermaximal subspaces and Steinitz systems. I* ◇ C57 C60 D25 D45 ◇

DOWNEY, R.G. *Co-immune subspaces and complementation in V_∞* ◇ C57 C60 D45 D50 ◇

DOWNEY, R.G. *Some remarks on a theorem of Iraj Kalantari concerning convexity and recursion theory* ◇ D45 ◇

DOWNEY, R.G. & REMMEL, J.B. *The universal complementation property* ◇ C57 C60 D25 D45 ◇

DYMENT, E.Z. *Recursive metrizability of numbered topological spaces and bases of effective linear topological spaces (Russian)* ◇ C57 C65 D45 ◇

DYMENT, E.Z. *Rice numerations (Russian)* ◇ D20 D45 ◇

FREJVALD, R.V. & KINBER, E.B. & WIEHAGEN, R. *Connections between identifying functionals, standardizing operations, and computable numberings* ◇ D20 D45 ◇

GUICHARD, D.R. *A note on r-maximal subspaces of V_∞* ◇ C57 C60 D45 ◇

GUPTA, N. *Recursively presented two generated infinite p-groups* ◇ C57 C60 ◇

JOCKUSCH JR., C.G. & KALANTARI, I. *Recursively enumerable sets and van der Waerden's theorem on arithmetic progressions* ◇ C57 D25 D80 ◇

KALFA, C. *Some undecidability results in strong algebraic languages* ◇ D45 ◇

KANDA, A. *Numeration models of λ-calculus* ◇ B40 D45 ◇

KAUFMANN, M. & SCHMERL, J.H. *Saturation and simple extensions of models of Peano-arithmetic* ◇ C50 C57 C62 ◇

KHISAMIEV, N.G. *Connection between constructivizability and strong constructivizability for different classes of abelian groups (Russian)* ◇ C57 C60 D45 ◇

KIERSTEAD, H.A. & MCNULTY, G.F. & TROTTER JR., W.T. *A theory of recursive dimension of ordered sets* ◇ C65 D45 ◇

KNIGHT, J.F. & LACHLAN, A.H. & SOARE, R.I. *Two theorems on degrees of models of true arithmetic* ◇ C57 C62 D30 ◇

KOSSAK, R. *$L_{\infty\omega_1}$-elementary equivalence of ω_1-like models of PA* ◇ C50 C57 C62 C75 ◇

KOSSAK, R. *Remarks on free sets* ◇ C50 C57 C62 E05 ◇

KOTLARSKI, H. *On elementary cuts in recursively saturated models of Peano arithmetic* ◇ C50 C57 C62 ◇

KUDAJBERGENOV, K.ZH. *Autostability and extensions of constructivizations (Russian)* ◇ C50 C57 F60 ◇

KUDAJBERGENOV, K.ZH. *Constructivizability of a prime model (Russian)* ◇ C50 C57 F60 ◇

LACHLAN, A.H. *Binary homogeneous structures. I* ◇ C10 C13 C15 C45 C50 C57 ◇

LEITSCH, A. *Enumerations of subrecursive classes and self-reproducing automata. I: Effective translations and decidable indexsets. II: Index-complexity and translation-complexity* ◇ D05 D20 D45 ◇

LONGO, G. & MOGGI, E. *Cartesian closed categories of enumerations for effective type structures* ◇ D45 G30 ◇

MACINTYRE, A. & MARKER, D. *Degrees of recursively saturated models* ◇ C50 C57 C62 D30 D45 ◇

MILLAR, T.S. *Decidability and the number of countable models* ◇ C15 C57 ◇

MOROZOV, A.S. *Group $Aut_r(Q, \leqslant)$ is not constructivizable (Russian)* ◇ C07 C57 F60 ◇

MOSES, M. *Recursive linear orders with recursive successivities* ◇ C57 D45 E07 ◇

MOSES, M. *Recursive properties of isomorphism types* ◇ C57 D45 ◇

NEGRI, M. *An application of recursive saturation* ◇ C50 C57 C62 F30 ◇

NELSON, GEORGE C. *Boolean powers, recursive models, and the Horn theory of a structure* ◇ C05 C20 C30 C57 G05 ◇

NIKITIN, A.A. *Some algorithmic problems for projective planes (Russian)* ◇ C57 C65 D80 ◇

ODINTSOV, S.P. *Atomless ideals of constructive Boolean algebras (Russian)* ◇ C57 G05 ◇

ROSENSTEIN, J.G. *Recursive linear orderings (French summary)* ◇ C57 C65 D45 ◇

SCHRAM, J.M. *Recursively prime trees* ◇ C57 C65 D45 ◇

SCHWARZ, S. *Recursive automorphisms of recursive linear orderings* ◇ C57 C65 D45 ◇

SPREEN, D. *On r.e. inseparability of cpo index sets* ◇ D25 D45 ◇

TVERSKOJ, A.A. *Constructivizability of formal arithmetical structures (Russian)* ◇ C57 D45 H15 ◇

WATNICK, R. *A generalization of Tennenbaum's theorem on effectively finite recursive linear orderings* ◇ C57 C65 D25 D45 ◇

1985

ADAMOWICZ, Z. & MORALES-LUNA, G. *A recursive model for arithmetic with weak induction* ◇ C57 F30 H15 ◇

DOWNEY, R.G. & HIRD, G.R. *Automorphisms of supermaximal subspaces* ◇ C57 C60 D25 D45 ◇

DOWNEY, R.G. & KALANTARI, I. *Effective extensions of linear forms on a recursive vector space over a recursive field* ◇ C57 C60 D45 ◇

DRIES VAN DEN, L. & SMITH, RICK L. *Decidable regularly closed fields of algebraic numbers* ◇ B25 C57 C60 ◇

FRIEDMAN, H.M. & SCEDROV, A. *Arithmetic transfinite induction and recursive well-orderings* ◇ D45 F30 F50 ◇

GLASS, A.M.W. *Effective extensions of lattices-ordered groups that preserve the degree of the conjugacy and the word problems* ◇ C57 C60 D40 ◇

GLASS, A.M.W. *The word and isomorphism problems in universal algebra* ◇ C05 C57 D40 ◇

GONCHAROV, S.S. *Stongly constructive models (Russian) (English summary)* ◇ C57 ◇

KALANTARI, I. & WEITKAMP, G. *Effective topological spaces I: A definability theory. II: A hierarchy* ◇ C40 C57 C65 D45 D75 D80 ◇

KAMIMURA, T. *An effective given initial semigroup* ◇ C57 D45 ◇

KANDA, A. *Acceptable numerations of functions spaces* ◇ D45 ◇

KANDA, A. *Numeration models of* λ*-calculus* ◇ B40 B60 C57 D45 ◇

KHISAMIEV, N.G. & KHISAMIEV, Z.G. *Nonconstructivizability of the reduced part of a strongly constructive torsion-free abelian group (Russian)* ◇ C57 C60 D45 ◇

KIERSTEAD, H.A. & REMMEL, J.B. *Degrees of indiscernibles in decidable models* ◇ C30 C35 C57 D30 ◇

KOSSAK, R. *A note on satisfaction classes* ◇ C50 C57 C62 F30 ◇

KOSSAK, R. *Recursively saturated* ω_1*-like models of arithmetic* ◇ C55 C57 C62 C75 C80 E65 ◇

KOTLARSKI, H. *Bounded induction and satisfaction classes* ◇ C15 C50 C57 C62 ◇

KREITZ, C. & WEIHRAUCH, K. *Theory of representations* ◇ D45 D75 D80 ◇

LI, XIANG *Everywhere nonrecursive r.e. sets in recursively presented topological space (Chinese)* ◇ D25 D45 ◇

LI, XIANG *Turing degrees on pointwise r.e. open sets in effective Hausdorff spaces (Chinese)* ◇ D25 D30 D45 ◇

LINDSAY, P.A. *On recognizing cyclic modules effectively* ◇ C60 D45 ◇

LONGO, G. & MOGGI, E. *Strutture di tipi ed enumerazioni* ◇ D45 ◇

MADISON, E.W. *On boolean algebras and their recursive completions* ◇ C57 D45 G05 ◇

MILLAR, T.S. *Decidable Ehrenfeucht theories* ◇ B25 C15 C57 C98 D30 D35 D55 ◇

MONTAGNA, F. & SORBI, A. *Universal recursion theoretic properties of r.e. preordered structures* ◇ C57 D45 G10 ◇

MOROZOV, A.S. *Automorphisms of constructivizations of Boolean algebras (Russian)* ◇ D45 G05 ◇

MOROZOV, A.S. *Constructive Boolean algebras with almost identical automorphisms (Russian)* ◇ C07 C57 G05 ◇

NERODE, A. & REMMEL, J.B. *A survey of lattices of R.E. substructures* ◇ C57 C98 D45 D98 ◇

ORLICKI, A. *On enumerated algebras and some monads in the category of enumerated sets* ◇ D45 G30 ◇

PINUS, A.G. *Applications of boolean powers of algebraic systems (Russian)* ◇ C05 C30 C55 C57 C80 C85 D35 E50 G05 ◇

ROY, D.K. *Linear order types of nonrecursive presentability* ◇ C57 C65 D45 ◇

SCHMERL, J.H. *Recursively saturated models generated by indiscernibles* ◇ C30 C50 C57 ◇

SCHRIEBER, L. *Recursive properties of euclidean domains* ◇ C57 C60 D45 ◇

SKORNYAKOV, L.A. *Stochastic algebra (Russian)* ◇ D45 ◇

SOLOVAY, R.M. *Infinite fixed-point algebras* ◇ C57 C62 G25 ◇

TVERSKOJ, A.A. *Constructivizable and nonconstructivizable formal arithmetic structures (Russian)* ◇ C57 C62 D45 H15 ◇

WEISPFENNING, V. *Quantifier elimination for modules* ◇ C10 C35 C57 C60 ◇

D50 Recursive equivalence types of sets and structures, isols

1955

MEDVEDEV, YU.T. *On nonisomorphic recursively enumerable sets (Russian)* ◇ D25 D50 ◇

1957

DEKKER, J.C.E. *An expository account of isols* ◇ D50 ◇

1958

DEKKER, J.C.E. *Congruences in isols with a finite modulus* ◇ D50 ◇

DEKKER, J.C.E. & MYHILL, J.R. *Retraceable sets* ◇ D25 D30 D50 ◇

DEKKER, J.C.E. *The factorial function for isols* ◇ D50 ◇

MYHILL, J.R. *Recursive equivalence types and combinatorial functions* ◇ D50 ◇

1959

NERODE, A. *Some Stone spaces and recursion theory* ◇ D50 ◇

1960

DEKKER, J.C.E. & MYHILL, J.R. *Recursive equivalence types* ◇ D50 ◇

DEKKER, J.C.E. & MYHILL, J.R. *The divisibility of isols by powers of primes* ◇ D50 ◇

PAVLOVA, E.A. *The lattice of denseness of sets of natural numbers (Russian) (Moldavian summary)* ◇ D20 D50 ◇

1961

FRIEDBERG, R.M. *The uniqueness of finite division for recursive equivalence types* ◇ D50 ◇

NERODE, A. *Extensions to isols* ◇ D50 ◇

PAVLOVA, E.A. *Densities of hyperimmune sets (Russian)* ◇ D50 ◇

1962

DEKKER, J.C.E. *Infinite series of isols* ◇ D50 ◇

MYHILL, J.R. $\Omega - \Lambda$ ◇ D50 ◇

MYHILL, J.R. *Elementary properties of the group of isolic integers* ◇ D50 ◇

MYHILL, J.R. *Recursive equivalence types and combinatorial functions* ◇ D50 ◇

NERODE, A. *Arithmetically isolated sets and nonstandard models* ◇ D50 ◇

NERODE, A. *Extensions to isolic integers* ◇ D50 ◇

YATES, C.E.M. *Recursively enumerable sets and retracing functions* ◇ D25 D50 ◇

1963

DEKKER, J.C.E. *An infinite product of isols* ◇ D50 ◇

ELLENTUCK, E. *Solution of a problem of R.Friedberg* ◇ D30 D50 ◇

PAVLOVA, E.A. *On certain classes of hyperimmune sets (Russian)* ◇ D50 ◇

SANSONE, F.J. *Combinatorial functions and regressive isols* ◇ D50 ◇

1964

BARBACK, J. *Recursive functions and regressive isols* ◇ D50 ◇

DEKKER, J.C.E. *The minimum of two regressive isols* ◇ D50 ◇

DEKKER, J.C.E. *The recursive equivalence type of a class of sets* ◇ D50 ◇

ELLENTUCK, E. *Infinite products of isols* ◇ D50 ◇

1965

APPEL, K.I. & MCLAUGHLIN, T.G. *On properties of regressive sets* ◇ D50 ◇

DEKKER, J.C.E. *Closure properties of regressive functions* ◇ D50 ◇

MCLAUGHLIN, T.G. *Co-immune retraceable sets* ◇ D50 ◇

MCLAUGHLIN, T.G. *On relative coimmunity* ◇ D50 ◇

MCLAUGHLIN, T.G. *Splitting and decomposition by regressive sets* ◇ D50 ◇

NERODE, A. *Additive relations among recursive equivalence types* ◇ D50 ◇

NERODE, A. *Non-linear combinatorial functions of isols* ◇ D50 ◇

POUR-EL, M.B. *"Recursive isomorphism" and effectively extensible theories* ◇ D50 ◇

RICHTER, W.H. *Regressive sets of order n* ◇ D50 ◇

SANSONE, F.J. *A mapping of regressive isols* ◇ D50 ◇

SANSONE, F.J. *The summation of certain series of infinite regressive isols* ◇ D50 ◇

1966

BARBACK, J. *A note on regressive isols* ◇ D50 ◇

BARBACK, J. *Two notes on regressive isols* ◇ D50 ◇

DEKKER, J.C.E. *Good choice sets* ◇ D50 ◇

DEKKER, J.C.E. *Les fonctions combinatoires et les isols* ◇ D50 ◇

HAY, L. *The co-simple isols* ◇ D50 ◇

MANASTER, A.B. *Higher-order indecomposable isols* ◇ D50 ◇

MCLAUGHLIN, T.G. *Retraceable sets and recursive permutations* ◇ D50 ◇

NERODE, A. *Combinatorial series and recursive equivalence types* ◇ D50 ◇

NERODE, A. *Diophantine correct non-standard models in the isols* ◇ C62 D50 ◇

SANSONE, F.J. *On order-preserving extensions to regressive isols* ◇ D50 ◇

1967

APPEL, K.I. *No recursively enumerable set is the union of finitely many immune retraceable sets* ◇ D25 D50 ◇
APPEL, K.I. *There exist two regressive sets whose intersection is not regressive* ◇ D50 ◇
BARBACK, J. *An md-class of sets indexed by a regressive function* ◇ D50 ◇
BARBACK, J. *Double series of isols* ◇ D50 ◇
BARBACK, J. *Regressive upper bounds* ◇ D50 ◇
BREDLAU, C.E. *Regressive functions and combinatorial functions* ◇ D50 ◇
DEKKER, J.C.E. *Regressive isols* ◇ D50 ◇
ELLENTUCK, E. *Universal isols* ◇ D50 ◇
HASSETT, M.J. *A note on regressive isols* ◇ D50 ◇
HAY, L. *Elementary differences between the isols and the co-simple isols* ◇ D50 ◇
HOROWITZ, L. *Invariance of partial order of recursive equivalence types under finite division* ◇ D50 ◇
MCLAUGHLIN, T.G. *Hereditarily retraceable isols* ◇ D50 ◇
MCLAUGHLIN, T.G. *Some counterexamples in the theory of regressive sets* ◇ D50 ◇
MCLAUGHLIN, T.G. *Splitting and decomposition by regressive sets II* ◇ D25 D50 ◇
YATES, C.E.M. *Arithmetical sets and retracing functions* ◇ D30 D50 D55 ◇

1968

BARBACK, J. *On recursive sets and regressive isols* ◇ D50 ◇
CROSSLEY, J.N. *Recursive equivalence: a survey* ◇ D45 D50 ◇
DEKKER, J.C.E. *On certain vector spaces of isolic dimension* ◇ D50 D80 ◇
FERGUSON, D.C. *Infinite products of recursive equivalence types* ◇ D50 ◇
HAMILTON, A.G. *An unsolved problem in the theory of constructive order types* ◇ D45 D50 ◇
MANASTER, A.B. *Full co-ordinals of RETs* ◇ D50 E10 ◇
MARTIN, D.A. & MILLER, W. *The degrees of hyperimmune sets* ◇ D25 D50 ◇

1969

BARBACK, J. *Extensions to regressive isols* ◇ D50 ◇
BARBACK, J. *Two notes on recursive functions and regressive isols* ◇ D50 ◇
DEKKER, J.C.E. *Countable vector spaces with recursive operations. Part I* ◇ C57 C60 D45 D50 ◇
GERSTING, J.L. *A rate of growth criterion for universality of regressive isols* ◇ D50 ◇
GONSHOR, H. & RICE, H.G. *Recursive density types I* ◇ D50 ◇
GONSHOR, H. *Recursive density types II* ◇ D50 ◇
HASSETT, M.J. *Recursive equivalence types and groups* ◇ D50 ◇
HERMES, H. *Basic notions and applications of the theory of decidability* ◇ D10 D20 D25 D30 D35 D50 D98 ◇
JOCKUSCH JR., C.G. *The degrees of hyperhyperimmune sets* ◇ D30 D50 ◇
MANASTER, A.B. *Rich co-ordinals, addition isomorphisms, and RETs* ◇ D50 E10 ◇
OLIN, P. *Indefinability in the arithmetic of isolic integers* ◇ D50 H15 ◇
SANSONE, F.J. *The backward and forward summation of infinite series of isols* ◇ D50 ◇
VUCKOVIC, V. *Almost recursive sets* ◇ D25 D50 ◇

1970

APPLEBAUM, C.H. & DEKKER, J.C.E. *Partial recursive functions and ω-functions* ◇ C57 D50 ◇
BARBACK, J. & JACKSON, W.D. *On representations as an infinite series of isols* ◇ D50 ◇
CROSSLEY, J.N. *Recursive equivalence* ◇ D50 ◇
DEGTEV, A.N. *Remarks on retraceable, regressive and pointwise-decomposable sets (Russian)* ◇ D25 D50 ◇
ELLENTUCK, E. *A coding theorem for isols* ◇ D50 ◇
HAMILTON, A.G. *Bases and α-dimensions of countable vector spaces with recursive operations* ◇ C57 C60 D45 D50 ◇
HASSETT, M.J. *A mapping property of regressive isols* ◇ D50 ◇
MANASTER, A.B. & NERODE, A. *A universal embedding property of the RET's* ◇ D50 ◇

1971

APPLEBAUM, C.H. *ω-homomorphisms and ω-groups* ◇ D45 D50 ◇
APPLEBAUM, C.H. *Isomorphisms of ω-groups* ◇ D45 D50 ◇
DEKKER, J.C.E. *Countable vector spaces with recursive operations. Part II* ◇ C57 C60 D45 D50 ◇
ELLENTUCK, E. *Extensions of isolic models* ◇ D50 ◇
ELLENTUCK, E. *Uncountable suborderings of the isols* ◇ D50 ◇

1972

BARBACK, J. & JACKSON, W.D. & PARNES, M. *Analogous characterizations of finite and isolated sets* ◇ D50 ◇
BARBACK, J. *On solutions in the regressive isols* ◇ D50 ◇
BARBACK, J. *Universal regressive isols* ◇ D50 ◇
BERRY, J.W. *A note on immune sets* ◇ D30 D50 ◇
ELLENTUCK, E. *An algebraic difference between isols and cosimple isols* ◇ D50 ◇
ELLENTUCK, E. *Nonrecursive relations among isols* ◇ D50 ◇
ELLENTUCK, E. *Nonrecursive combinatorial functions* ◇ D50 ◇
ELLENTUCK, E. & MANASTER, A.B. *The decidability of a class of ∀∃ sentences in the isols* ◇ B25 D50 ◇
ELLENTUCK, E. *The positive properties of isolic integers* ◇ D50 ◇
ELLENTUCK, E. *Universal cosimple isols* ◇ D50 ◇
GERSTING, J.L. *A note on infinite series of isols* ◇ D50 ◇
HASSETT, M.J. *A characterization of the orders of regressive ω-groups* ◇ D50 ◇
HASSETT, M.J. *A closure property of regressive isols* ◇ D50 ◇
HAY, L. *A note on frame extensions* ◇ D50 ◇
JOCKUSCH JR., C.G. *Upward closure of bi-immune degrees* ◇ D30 D50 ◇

MARANDZHYAN, G.B. *The strongly effective immunity of the pivots of additively optimal recursive functions (Russian) (Armenian and English summaries)* ◇ D15 D50 ◇

MCLAUGHLIN, T.G. *Supersimple sets and the problem of extending a retracing function* ◇ D25 D50 ◇

1973

APPLEBAUM, C.H. *A result for π-groups* ◇ D45 D50 ◇

APPLEBAUM, C.H. *A stronger definition of recursively infinite set* ◇ D50 ◇

APPLEBAUM, C.H. *The recursive equivalence type of a decomposition of an ω-group: the RET of a decomposition* ◇ D45 D50 ◇

BISKUP, J. *Zufaellige Folgen und Bi-Immunitaet* ◇ D50 D80 ◇

ELLENTUCK, E. *Degrees of isolic theories* ◇ D30 D50 ◇

ELLENTUCK, E. *On the degrees of universal regressive isols* ◇ D30 D50 ◇

ELLENTUCK, E. *On the form of functions which preserve regressive isols* ◇ D50 ◇

GERSTING, J.L. *Infinite series of T-regressive isols* ◇ D50 ◇

HASSETT, M.J. *Extension pathology in regressive isols* ◇ D50 ◇

JACKSON, W.D. *A note on a theorem of C. Yates* ◇ D50 ◇

MCLAUGHLIN, T.G. *On retraceable sets with rapid growth* ◇ D25 D50 ◇

1974

APPLEBAUM, C.H. *ω-semigroups* ◇ D45 D50 ◇

APPLEBAUM, C.H. *Counterexamples in the theory of ω functions* ◇ D50 ◇

CROSSLEY, J.N. & NERODE, A. *Combinatorial functors* ◇ C57 C62 D45 D50 G30 ◇

CROSSLEY, J.N. & O'CONNOR, S. & STACEY, K. *Recursive equivalence and combinatorial functors: a bibliography* ◇ A10 D50 ◇

DEKKER, J.C.E. & ELLENTUCK, E. *Recursion relative to regressive functions* ◇ D30 D50 ◇

ELLENTUCK, E. *A Δ_2^0 theory of regressive isols* ◇ D50 ◇

FLAJOLET, P. & STEYAERT, J.-M. *Une generalisation de la notion d'ensemble immune* ◇ B75 D20 D25 D50 ◇

HAMILTON, A.G. *Recursive equivalence types of vector spaces* ◇ D45 D50 ◇

KURDYUMOV, G.L. *A certain class of immune sets (Russian)* ◇ D50 ◇

MCLAUGHLIN, T.G. *Closed basic retracing functions and hyperimmune sets* ◇ D25 D50 ◇

MCLAUGHLIN, T.G. *Degrees of unsolvability within a regressive isol* ◇ D30 D50 ◇

SOARE, R.I. *Isomorphisms on countable vector spaces with recursive operations* ◇ D45 D50 ◇

1975

ACZEL, P. *Recursive density types and Nerode extensions of arithmetic* ◇ D45 D50 F30 ◇

CROSSLEY, J.N. & NERODE, A. *Combinatorial functors* ◇ D50 ◇

CROSSLEY, J.N. & NERODE, A. *Sound functors* ◇ D50 ◇

DUJOLS, R. *Une methode de construction directe de couples d'ensembles 1-reductibles ayant des types d'equivalence recursive incomparables (English summary)* ◇ D30 D50 ◇

ELLENTUCK, E. *Hereditarily universal sets* ◇ D50 ◇

ELLENTUCK, E. *Semigroups, Horn sentences and isolic structures* ◇ C05 C57 D50 E10 ◇

FOWLER III, N. *Intersections of α-spaces* ◇ C57 C60 D50 ◇

GERSTING, J.L. *Universal pairs of regressive isols* ◇ D50 ◇

MCLAUGHLIN, T.G. *A note concerning the $\overset{*}{\vee}$ relation on Λ_R* ◇ D25 D30 D50 ◇

MCLAUGHLIN, T.G. *Trees and isols. Part I* ◇ D50 ◇

RYAN, B.F. *ω-cohesive sets* ◇ D50 ◇

TRAKHTENBROT, B.A. *On problems solvable by successive trials* ◇ D15 D50 ◇

1976

BARBACK, J. *Composite numbers and prime regressive isols* ◇ D50 ◇

BARBACK, J. *Regressive isols and comparability* ◇ D50 ◇

CATLIN, S. *ed-regressive sets of order n* ◇ D50 ◇

CROSSLEY, J.N. & NERODE, A. *Effective dimension* ◇ C57 D45 D50 ◇

DEGTEV, A.N. *Some classes of hyperimmune sets (Russian)* ◇ D50 ◇

DEKKER, J.C.E. *Equivalencia recursiva* ◇ D50 ◇

DEKKER, J.C.E. *Projective planes of infinite but isolic order* ◇ D50 D80 ◇

DUJOLS, R. *Le resultat de Karp-Myhill dans "Recursive equivalent types" de J.C.E. Dekker & J. Myhill* ◇ D50 ◇

DUJOLS, R. *Quelques remarques a propos d'un theoreme de Karp-Myhill (English summary)* ◇ D50 ◇

ELLENTUCK, E. *Decomposable isols and their degrees* ◇ D30 D50 E10 ◇

FOWLER III, N. *α-decompositions of α-spaces* ◇ C57 C60 D45 D50 ◇

GONSHOR, H. *Effective density types* ◇ D25 D50 ◇

MCLAUGHLIN, T.G. *Trees and isols II* ◇ D50 ◇

REMMEL, J.B. *Combinatorial functors on co-r.e. structures* ◇ D25 D30 D45 D50 ◇

1977

BISKUP, J. *On bi-immune isols* ◇ D50 ◇

CATLIN, S. *Pathologies in the ed-regressive sets of order 2* ◇ D50 ◇

DEKKER, J.C.E. *Planos afines con operaciones recursivas* ◇ D45 D50 ◇

DYMENT, E.Z. *Two theorems on finite unions of regressive immune sets (Russian)* ◇ D50 ◇

ELLENTUCK, E. *Boolean valued rings* ◇ C30 C60 C90 D50 ◇

ELLENTUCK, E. *Tarski semigroups* ◇ D50 ◇

GERSTING, J.L. *Infinite series of regressive isols under addition* ◇ D50 ◇

MCLAUGHLIN, T.G. *A partial comparison of two conditions on the intersection of regressive sets* ◇ D50 ◇

MCLAUGHLIN, T.G. *Degrees of unsolvability and strong forms of* $\Lambda_R + \Lambda_R \not\subseteq \Lambda_R$ ◇ D25 D30 D50 ◇
MIKHEEV, V.L. *Meager and universal regressive isols (Russian)* ◇ D50 ◇
ROJZEN, S.I. *A counterexample in the theory of automaton equivalence types (Russian)* ◇ D05 D50 ◇
ROJZEN, S.I. *An analogue of the additive system of cardinals that is connected with finite automata (Russian) (English summary)* ◇ D05 D50 ◇

1978

APPLEBAUM, C.H. *Some structure theorems for inverse* ω*-semigroups* ◇ D45 D50 ◇
BARBACK, J. & MCLAUGHLIN, T.G. *On the intersection of regressive sets* ◇ D50 ◇
DEGTEV, A.N. *Solvability of the* $\forall\exists$*-theory of a certain factor-lattice of recursively enumerable sets (Russian)* ◇ B25 D25 D50 ◇
DEKKER, J.C.E. *Projective bigraphs with recursive operations* ◇ D45 D50 ◇
DEKKER, J.C.E. *Sharply two-transitive families of permutations on an immune set* ◇ D50 ◇
ELLENTUCK, E. *Cardinals, isols, and the growth of functions* ◇ D50 E10 E25 ◇
ELLENTUCK, E. *Model theoretic methods in the theory of isols* ◇ C57 D50 ◇
GOLUNKOV, YU.V. *Precomplete classes of algorithms that preserve the membership in a set (Russian)* ◇ D20 D50 ◇
MIKHEEV, V.L. *A hierarchy of independent* ω*-processions of cosimple isols (Russian)* ◇ D50 ◇
MIKHEEV, V.L. *Infinite sums and products of isolic integers (Russian)* ◇ D50 ◇
NELSON, GEORGE C. *Isomorphism types of the hyperarithmetic sets* H_a ◇ D30 D50 D55 ◇
REMMEL, J.B. *Realizing partial orderings by classes of co-simple sets* ◇ D25 D45 D50 ◇
ROJZEN, S.I. *Unsolvability of the additive theory of automata isols over a unary alphabet (Russian)* ◇ D05 D50 ◇
ROZINAS, M.G. *Partial degrees of immune and hyperimmune sets (Russian)* ◇ D30 D50 ◇
SHORE, R.A. *Controlling the dependence degree of a recursively enumerable vector space* ◇ D25 D45 D50 ◇
SILVERSTEIN, A. *A generalization of combinatorial operators* ◇ B40 D50 ◇
SOLON, B.YA. *e-powers of hyperimmune retraceable sets (Russian)* ◇ D25 D30 D50 ◇

1979

BREDLAU, C.E. *Admissible sets and recursive equivalence types* ◇ D50 D60 ◇
ELLENTUCK, E. *Sylow subgroups of a regressive group* ◇ D50 ◇
HARKLEROAD, L. *Recursive equivalence types on recursive manifolds* ◇ D45 D50 D80 ◇
MCLAUGHLIN, T.G. *Retraceable homogeneous sets* ◇ D50 ◇
MIKHEEV, V.L. *Some remarks on isols (Russian)* ◇ D50 ◇

1980

ELLENTUCK, E. *Diagonal methods in the theory of isols* ◇ D50 ◇
GOLUNKOV, YU.V. *Precomplete classes of algorithms that preserve membership in a set II (Russian)* ◇ D25 D50 ◇
MANASTER, A.B. & REMMEL, J.B. *Co-simple higher-order indecomposable isols* ◇ D50 ◇

1981

ASH, C.J. & NERODE, A. *Intrinsically recursive relations* ◇ D45 D50 ◇
BARBACK, J. *Hyper-torre isols and an arithmetic property* ◇ D50 ◇
BARBACK, J. *On finite sums of regressive isols* ◇ D50 ◇
BARBACK, J. *On regressive isols and their differences* ◇ D50 ◇
CROSSLEY, J.N. & NERODE, A. *Recursive equivalence on matroids* ◇ D45 D50 ◇
CROSSLEY, J.N. *Reminiscences of logicians. II* ◇ A10 D45 D50 ◇
DALEY, R.P. *Retraceability, repleteness and busy beaver sets* ◇ D15 D20 D50 ◇
DEKKER, J.C.E. *Automorphisms of* ω*-cubes* ◇ D50 ◇
DEKKER, J.C.E. *Recursive equivalence types and cubes* ◇ D50 ◇
DEKKER, J.C.E. *Twilight graphs* ◇ D50 D80 ◇
ELLENTUCK, E. *Galois theorems for isolated fields* ◇ C57 C60 D45 D50 ◇
ELLENTUCK, E. *Hyper-torre isols* ◇ C62 D50 ◇
MCLAUGHLIN, T.G. *Intersection types and the terms of a regressive sum* ◇ D50 ◇
MCLAUGHLIN, T.G. *On the divergence of extension procedures in isol theory* ◇ D50 ◇
PRANK, R.K. *On the quotient lattice of the lattice of recursively enumerable sets by the immunity congruence (Russian) (Estonian and English summaries)* ◇ D25 D50 ◇
SOLON, B.YA. *PC-degrees inside an e-degree of a hyperimmune retraceable set (Russian)* ◇ D25 D30 D50 H05 ◇

1982

APPLEBAUM, C.H. *An introduction to* ω*-extensions of* ω*-groups* ◇ D45 D50 ◇
BARBACK, J. *On tame regressive isols* ◇ D50 ◇
DALEY, R.P. *Busy beaver sets: characterizations and applications* ◇ D10 D25 D50 ◇
DEKKER, J.C.E. *Automorphisms of* ω*-octahedral graphs* ◇ D50 ◇
LI, XIANG *The effective immune sets and the program index sets – a generalization and application of the recursion theorem (Chinese) (English summary)* ◇ D50 ◇
MADAN, D.B. & ROBINSON, R.W. *Monotone and 1-1 sets* ◇ D50 ◇
MCLAUGHLIN, T.G. *Regressive sets and the theory of isols* ◇ D50 D98 ◇

1983

CROSSLEY, J.N. & REMMEL, J.B. *Undecidability and recursive equivalence I* ◇ D35 D45 D50 ◇
DEKKER, J.C.E. *Isols and balanced block designs with* $\lambda = 1$ ◇ D50 D80 ◇
DEKKER, J.C.E. *Recursive equivalence types and octahedra* ◇ D50 D80 ◇
ELLENTUCK, E. *Incompatible extensions of combinatorial functions* ◇ D50 ◇
ELLENTUCK, E. *Random isols* ◇ D50 ◇
GROSS, W.F. *The inverse of a regressive object* ◇ D50 ◇
HARKLEROAD, L. *Manifolds allowing RET arithmetic* ◇ D45 D50 D80 ◇
HECK, W.S. *Large families of incomparable A-isols* ◇ D50 ◇
LI, XIANG *Effective immune sets, program index sets and effectively simple sets- generalizations and applications of the recursion theorem* ◇ D25 D50 ◇
ROLLETSCHEK, H. *Closure properties of almost-finiteness classes in recursive function theory* ◇ D50 ◇

1984

CROSSLEY, J.N. & REMMEL, J.B. *Undecidability and recursive equivalence II* ◇ D35 D50 D75 ◇
DEKKER, J.C.E. *Isolated sets and parity* ◇ D50 ◇
DOWNEY, R.G. *Co-immune subspaces and complementation in* V_∞ ◇ C57 C60 D45 D50 ◇
HARKLEROAD, L. *Fuzzy recursion, RET's and isols* ◇ D50 ◇
MAL'TSEV, A.A. *On the structure of the families of immune,hyperimmune and hyperhyperimmune sets (Russian)* ◇ D50 ◇

1985

BARBACK, J. *On hereditarily odd-even isols and a comparability of summands property* ◇ D50 ◇
FRIEDMAN, S.D. *An immune partition of the ordinals* ◇ D50 E40 E45 E55 ◇

D55 Hierarchies

1905

BAIRE, R. & BOREL, E. & HADAMARD, J. & LEBESGUE, H. *Cinq lettres sur la theorie des ensembles* ◇ A05 D55 E15 E25 F65 ◇

LEBESGUE, H. *Sur les fonctions representables analytiquement* ◇ D55 E15 ◇

1907

LEBESGUE, H. *Contribution a l'etude des correspondances de M. Zermelo* ◇ D55 E15 E25 ◇

1914

HAUSDORFF, F. *Grundzuege der Mengenlehre* ◇ D55 E10 E15 E98 ◇

LUZIN, N.N. *Sur un probleme de M.Baire* ◇ D55 E15 ◇

1916

ALEKSANDROV, P.S. *Sur la puissance des ensembles mesurables B* ◇ D55 E15 E75 ◇

1917

LUZIN, N.N. *Sur la classification de M.Baire* ◇ D55 E15 ◇

SOUSLIN, M. *Sur une definition des ensembles mesurables B sans nombres transfinis* ◇ D55 E15 E75 ◇

1918

LUZIN, N.N. & SIERPINSKI, W. *Sur quelques proprietes des ensembles mesurables (A)* ◇ D55 E15 ◇

SIERPINSKI, W. *Sur les definitions axiomatiques des ensembles mesurables (B)* ◇ D55 D70 E15 ◇

SIERPINSKI, W. *Sur un theoreme de M. Lebesgue* ◇ D55 E15 ◇

SIERPINSKI, W. *Sur une generalisation des ensembles mesurables (B)* ◇ D55 E15 ◇

SIERPINSKI, W. *Sur une propriete des fonctions representables analytiquement* ◇ D55 E15 ◇

1919

SIERPINSKI, W. *Ueber eine Verallgemeinerung der Borelschen Mengen (Polish)* ◇ D55 E15 ◇

1920

SIERPINSKI, W. *Sur les ensembles mesurables B* ◇ D55 E15 ◇

1922

KURATOWSKI, K. *Une methode d'elimination des nombres transfinis des raisonnements mathematiques* ◇ D55 E07 E15 E25 E75 ◇

ZALCWASSER, Z. *Un theoreme sur les ensembles qui sont a la fois F_σ et G_δ* ◇ D55 E15 ◇

1923

LUZIN, N.N. & SIERPINSKI, W. *Sur un ensemble non mesurable* ◇ D55 E15 ◇

1924

ALEKSANDROV, P.S. *Sur les ensembles complementaires aux ensembles (A)* ◇ D55 E15 ◇

KURATOWSKI, K. *Sur les fonctions representables analytiquement et les ensembles de premiere categorie* ◇ D55 E75 ◇

SIERPINSKI, W. *Les projections des ensembles mesurables (B) et les ensembles (A)* ◇ D55 E15 ◇

SIERPINSKI, W. *Sur la puissance des ensembles mesurables (B)* ◇ D55 E10 E15 ◇

SIERPINSKI, W. *Sur une propriete des ensembles ambigus* ◇ D55 E15 ◇

SIERPINSKI, W. *Un exemple effectif d'un ensemble mesurable (B) de classe α* ◇ D55 E15 ◇

1925

LAVRENTIEFF, M. *Sur les sous-classes de la classification de M. Baire* ◇ D55 E15 ◇

LUZIN, N.N. *Les proprietes des ensembles projectifs* ◇ D55 E15 ◇

LUZIN, N.N. *Sur le probleme de M. Emile Borel et la methode des resolvantes* ◇ D55 E15 ◇

LUZIN, N.N. *Sur les ensembles projectifs de M. Henri Lebesgue* ◇ D55 E15 ◇

LUZIN, N.N. *Sur les ensembles non mesurables B et l'emploi de la diagonale de Cantor* ◇ D55 E75 ◇

SIERPINSKI, W. *Les fonctions continues et les ensembles (A)* ◇ D55 E15 E75 ◇

SIERPINSKI, W. *Sur un ensemble ferme conduisant a un ensemble non mesurable (B)* ◇ D55 E15 ◇

SIERPINSKI, W. *Sur une classe d'ensembles* ◇ D55 E15 ◇

1926

KURATOWSKI, K. *Un theoreme concernant la puissance d'ensembles de points* ◇ D55 E15 ◇

LUZIN, N.N. *Memoire sur les ensembles analytiques et projectifs* ◇ D55 E15 ◇

SIERPINSKI, W. *Sur l'ensemble de valeurs qu'une fonction continue prend une infinite non denombrable de fois* ◇ D55 E15 ◇

SIERPINSKI, W. *Sur les ensembles hyperboreliens* ◇ D55 E15 ◇

SIERPINSKI, W. *Sur une propriete des ensembles (A)* ◇ D55 E15 ◇

1927

LUZIN, N.N. *Remarques sur les ensembles projectifs* ◇ D55 E15 ◇

LUZIN, N.N. *Sur les ensembles analytiques* ◇ D55 E15 ◇
MAZURKIEWICZ, S. *Sur une propriete des ensembles C(A)* ◇ D55 E15 ◇
SELIVANOWSKI, E. *Sur une classe d'ensembles definis par une infinite denombrable de conditions* ◇ D55 E15 ◇
SIERPINSKI, W. *Sur la puissance des ensembles d'une certaine classe* ◇ D55 E15 ◇
SIERPINSKI, W. *Sur quelques proprietes des ensembles projectifs* ◇ D55 E15 ◇
SIERPINSKI, W. *Sur un probleme de M.Hausdorff* ◇ D55 E15 ◇
SIERPINSKI, W. *Sur une classification des ensembles mesurables (B)* ◇ D55 E15 ◇
SIERPINSKI, W. *Sur une propriete characteristique des ensembles analytiques* ◇ D55 E15 ◇
SIERPINSKI, W. *Sur une propriete des complementaires analytiques* ◇ D55 E15 ◇

1928

MENGER, K. *Bemerkungen zu Grundlagenfragen. III: Ueber Potenzmengen* ◇ A05 D55 E15 E30 ◇
SELIVANOWSKI, E. *Ueber eine Klasse von effektiven Mengen (Mengen C) (Russian) (French summary)* ◇ D55 E15 ◇
SIERPINSKI, W. *Le crible de M.Lusin et l'operation (A) dans les espaces abstraits* ◇ D55 E15 ◇
SIERPINSKI, W. *Les ensembles projectifs et le crible de M. Lusin* ◇ D55 E15 ◇
SIERPINSKI, W. *Les ensembles projectifs et la propriete de Baire* ◇ D55 E15 ◇
SIERPINSKI, W. *Sur les images continues et biunivoques des complementaires analytiques* ◇ D55 E15 ◇
SIERPINSKI, W. *Sur les projections des ensembles complementaires aux ensembles (A)* ◇ D55 E15 ◇
SIERPINSKI, W. *Sur les produits des images continues des ensembles C(A)* ◇ D55 E15 ◇
SIERPINSKI, W. *Sur un ensemble analytique plan, universel pour les ensembles mesurables (B)* ◇ D55 E15 ◇

1929

LUZIN, N.N. & SIERPINSKI, W. *Sur les classes des constituantes d'un complementaire analytique* ◇ D55 E15 ◇
LUZIN, N.N. *Sur les points d'unicite d'un ensemble mesurable B* ◇ D55 E15 ◇
LUZIN, N.N. *Sur les voies de la theorie des ensembles* ◇ A05 D55 E15 E50 F55 ◇
LUZIN, N.N. *Sur un principe general de la theorie des ensembles analytiques* ◇ D55 E15 ◇
NIKODYM, O.M. *Sur les diverses classes d'ensembles* ◇ D55 E15 ◇
SIERPINSKI, W. *Contribution a la fondation de la theorie des ensembles projectifs* ◇ D55 E15 ◇
SIERPINSKI, W. *Sur l'existence de diverses classes d'ensembles* ◇ D55 E15 ◇
SIERPINSKI, W. *Sur les familles inductives et projectives d'ensembles* ◇ D55 E15 ◇

1930

BLUE, A.H. *On the structure of sets of points of classes one, two and three* ◇ D55 E15 ◇
HUREWICZ, W. *Zur Theorie der analytischen Mengen* ◇ D55 E15 ◇
KANTOROVICH, L. & LIVENSON, E. *Sur les ensembles projectifs de M. Lusin* ◇ D55 E15 ◇
LUZIN, N.N. *Analogies entre les ensembles mesurables B et les ensembles analytiques* ◇ D55 E15 ◇
LUZIN, N.N. *Lecons sur les ensembles analytiques et leurs applications* ◇ D55 E15 E98 ◇
LUZIN, N.N. *Sur le probleme de M.J.Hadamard d'uniformisation des ensembles* ◇ D55 E15 ◇
LUZIN, N.N. *Sur le probleme de M.Jacques Hadamard d'uniformisation des ensembles* ◇ D55 E15 ◇
SIERPINSKI, W. *Sur l'uniformisation des ensembles mesurables (B)* ◇ D55 E15 ◇
SIERPINSKI, W. *Sur la puissance des ensembles analytiques* ◇ D55 E15 E50 ◇

1931

KURATOWSKI, K. *Evaluation de la classe Borelienne ou projective d'un ensemble de points a l'aide des symboles logiques* ◇ D55 E15 ◇
KURATOWSKI, K. & TARSKI, A. *Les operations logiques et les ensembles projectifs* ◇ B10 D55 E15 ◇
LUZIN, N.N. *Sur une famille de complementaires analytiques* ◇ D55 E15 ◇
SIERPINSKI, W. *Les ensembles analytiques comme cribles au moyen des ensembles fermes* ◇ D55 E15 ◇
SIERPINSKI, W. *Sur certaines operations sur les ensembles fermes plans* ◇ D55 E15 ◇
SIERPINSKI, W. *Sur deux complementaires analytiques non separables B* ◇ D55 E15 ◇
SIERPINSKI, W. *Sur les cribles projectifs* ◇ D55 E15 ◇
SIERPINSKI, W. *Sur un crible universel* ◇ D55 E15 ◇
TARSKI, A. *Sur les ensembles definissables de nombres reels I* ◇ B25 B28 C10 C40 C60 C65 D55 E15 E47 F35 ◇

1932

BRAUN, S. *Sur une propriete caracteristique des ensembles* $G_{\delta\sigma\delta}$ ◇ D55 E15 ◇
KANTOROVICH, L. & LIVENSON, E. *Memoir on the analytical operations and projective sets. I* ◇ D55 E15 ◇
KURATOWSKI, K. & SZPILRAJN, E. *Sur les cribles fermes et leurs applications* ◇ D55 E15 ◇
POPRUZENKO, J. *Sur l'analyticite des ensembles (A)* ◇ D55 E15 E75 ◇
SIERPINSKI, W. *Sur les rapports entre les classifications des ensembles de MM. F.Hausdorff et Ch. de la Vallee-Poussin* ◇ D55 E15 ◇

1933

HAUSDORFF, F. *Zur Projektivitaet der* δs*-Funktionen* ◇ D55 E15 ◇
KANTOROVICH, L. & LIVENSON, E. *Memoir on the analytical operations and projective sets. II* ◇ D55 E15 ◇

KURATOWSKI, K. *Topologie I* ◇ D55 E15 E75 ◇

SZPILRAJN, E. *Sur certains invariants de l'operation (A)* ◇ D55 E15 ◇

1934

KURATOWSKI, K. & POSAMENT, T. *Sur l'isomorphie algebro-logique et les ensembles relativement Boreliens* ◇ D55 E15 ◇

LYAPUNOV, A.A. *On the separability of analytic sets (Russian) (French summary)* ◇ D55 E15 ◇

SIERPINSKI, W. *Sur la separabilite multiple des ensembles mesurables B* ◇ D55 E15 ◇

1935

LUZIN, N.N. & NOVIKOV, P.S. *Choix effectif d'un point dans un complementaire analytique arbitraire, donne par un crible* ◇ D55 E15 ◇

LUZIN, N.N. *Sur les ensembles analytiques nuls* ◇ D55 E15 E50 ◇

LYAPUNOV, A.A. *Sur la separabilite multiple des ensembles mesurables B* ◇ D55 E15 ◇

NOVIKOV, P.S. *Sur la separabilite des ensembles projectifs de seconde classe* ◇ D55 E15 ◇

SIERPINSKI, W. *Sur les transformations des ensembles par les fonctions de Baire* ◇ D55 E15 ◇

SIERPINSKI, W. *Sur un ensemble projectif de classe 2 dans l'espace des ensembles fermes plans* ◇ D55 E15 ◇

SIERPINSKI, W. *Sur une hypothese de M.Lusin* ◇ D55 E15 E50 ◇

1936

KONDO, M. *Sur un ensemble universel pour les ensembles Boreliens definis sur la famille de tous les ensembles lineaires CA* ◇ D55 E15 ◇

KURATOWSKI, K. *Sur les ensembles projectifs (Russian summary)* ◇ D55 E15 ◇

KURATOWSKI, K. *Sur les theoremes de separation dans la theorie des ensembles* ◇ D55 E15 ◇

LYAPUNOV, A.A. *Contribution a l'etude de la separabilite multiple* ◇ D55 E15 ◇

1937

BRAUN, S. *Sur l'uniformisation des ensembles fermes* ◇ D55 E15 ◇

KONDO, M. *L'uniformisation des complementaires analytiques* ◇ D55 E15 ◇

KUNUGUI, K. *Sur un theoreme d'existence dans la theorie des ensembles projectifs* ◇ D55 E15 ◇

KURATOWSKI, K. *Les suites transfinies d'ensembles et les ensembles projectifs* ◇ D55 E15 ◇

KURATOWSKI, K. *Les types d'ordre definissables et les ensembles boreliens* ◇ C75 D55 E07 E15 ◇

KURATOWSKI, K. & NEUMANN VON, J. *On some analytic sets defined by transfinite induction* ◇ D55 E15 ◇

KURATOWSKI, K. *Sur la geometrisation des types d'ordre denombrable* ◇ D55 E07 E15 ◇

KURATOWSKI, K. *Sur les suites analytiques d'ensembles* ◇ D55 E15 ◇

LYAPUNOV, A.A. *On subclasses of B-sets (Russian)* ◇ D55 E15 ◇

NOVIKOV, P.S. *Les projections des complementaires analytiques uniformes (Russian summary)* ◇ D55 E15 ◇

NOVIKOV, P.S. *On the mutual relation of the second class of projective sets and the projection of uniform analytic complements (Russian)* ◇ D55 E15 ◇

SIERPINSKI, W. *Sur un probleme de la theorie generale des ensembles concernant les familles boreliennes d'ensembles* ◇ D55 E15 ◇

1938

KONDO, M. *Sur l'uniformisation des complementaires analytiques et les ensembles projectifs de la seconde classe* ◇ D55 E15 ◇

KONDO, M. *Sur la representation parametrique reguliere des ensembles analytiques* ◇ D55 E15 ◇

KONDO, M. *Sur les operations analytiques dans la theorie des ensembles et quelques problemes qui s'y rattachent I* ◇ D55 E15 ◇

KONDO, M. *Theory of analytic sets* ◇ D55 E15 ◇

LYAPUNOV, A.A. *Sur l'uniformisation des complementaires analytiques* ◇ D55 E15 ◇

SIERPINSKI, W. *Sur un probleme concernant les ensembles projectifs* ◇ D55 E15 ◇

1939

DIENES, Z.P. *Canonic elements in the higher classes of Borel sets* ◇ D55 E15 ◇

KUNUGUI, K. *Contribution a la theorie des ensembles Boreliens et analytiques II* ◇ D55 E15 ◇

LYAPUNOV, A.A. *Separabilite multiple pour le cas de l'operation (A) (Russian) (French summary)* ◇ D55 E15 ◇

LYAPUNOV, A.A. *Sur l'uniformisation de quelques ensembles CA et A'_2 (Russian) (French summary)* ◇ D55 E15 E20 E50 ◇

NOVIKOV, P.S. *On projections of some B-sets (Russian)* ◇ D55 E15 ◇

1940

KOZLOVA, Z.I. *On some classes of A- and B-sets (Russian) (French summary)* ◇ D55 E15 ◇

KUNUGUI, K. *Contribution a la theorie des ensembles Boreliens et analytiques III* ◇ D55 E15 ◇

KUNUGUI, K. *Sur un probleme de M.E.Szpilrajn* ◇ D55 E15 ◇

MAXIMOFF, I. *Sur la separabilite d'ensembles* ◇ D55 E15 ◇

1941

KURATOWSKI, K. & SIERPINSKI, W. *Sur l'existence des ensembles projectifs non mesurables* ◇ D55 E15 ◇

YANKOV, V.A. *Sur l'uniformisation des ensembles A (Russian) (French summary)* ◇ D55 E15 ◇

1942

KONDO, M. *Sur la structure des ensembles* ◇ D55 E15 ◇

1943

KLEENE, S.C. *Recursive predicates and quantifiers* ◇ D55 ◇

1944

KLEENE, S.C. *On the forms of predicates in the theory of constructive ordinals* ◇ D55 D70 F15 ◇

KONDO, M. *La structure des fonctions projectives I* ◇ D55 E15 ◇
RIBEIRO ALBUQUERQUE, J. *Ensembles de Borel* ◇ D55 E15 ◇

1945

KELDYSH, L.V. *Sur les transformations ouvertes des ensembles* ◇ D55 E15 ◇
KLINE, S.A. *The representation of Baire's classes by transfinite sums of continuous functions* ◇ D55 E15 ◇
RIBEIRO ALBUQUERQUE, J. *Ensembles de Borel II* ◇ D55 E15 ◇

1947

ALEXIEWICZ, A. *On Hausdorff classes* ◇ D55 E15 ◇
LYAPUNOV, A.A. *On R-sets (Russian)* ◇ D55 D70 E15 ◇
LYAPUNOV, A.A. *Sur les ensembles projectifs, qui admettent des decompositions regulieres (Russian) (French summary)* ◇ D55 E15 ◇
LYAPUNOV, A.A. *Theory of R-sets (Russian)* ◇ D55 D70 E15 ◇
MOSTOWSKI, ANDRZEJ *On definable sets of positive integers* ◇ D55 F30 ◇

1948

KURATOWSKI, K. *Ensembles projectifs et ensembles singuliers* ◇ D55 E15 ◇
LYAPUNOV, A.A. *A new definition of certain classes of sets (Russian)* ◇ D55 E15 ◇
LYAPUNOV, A.A. & NOVIKOV, P.S. *Descriptive set theory (Russian)* ◇ D55 E15 ◇
MOSTOWSKI, ANDRZEJ *On a set of integers not definable by means of one-quantifier predicates* ◇ D55 ◇
POST, E.L. *Degrees of recursive unsolvability, preliminary report* ◇ D03 D30 D55 ◇
SHCHEGOL'KOV, E.A. *On the uniformization of certain B-sets (Russian)* ◇ D55 E15 ◇
SIERPINSKI, W. *L'operation du crible et les fonctions analytiques d'une suite infinie d'ensembles* ◇ D55 D70 E15 ◇
ZAHORSKI, Z. *Sur la classe de Baire des derivees approximatives d'une fonction quelconque* ◇ D55 E15 ◇

1949

LYAPUNOV, A.A. *On continuous transformations of A-sets (Russian)* ◇ D55 E15 ◇
MOSTOWSKI, ANDRZEJ *A classification of logical systems* ◇ D55 F30 ◇
MYHILL, J.R. *Note on an idea of Fitch* ◇ D55 F30 ◇
NOVIKOV, P.S. *The consistency of certain statements of the theory of sets (Russian)* ◇ D55 E15 E35 E45 ◇
SIERPINSKI, W. *Sur un probleme de M.Zarankiewicz* ◇ D55 E15 ◇
SIERPINSKI, W. *Sur un probleme de M.Lusin concernant les complementaires analytiques* ◇ D55 E15 E20 ◇

1950

ARSENIN, V.YA. & LYAPUNOV, A.A. *The theory of A-sets (Russian)* ◇ D55 D70 E15 E98 ◇
DAVIS, MARTIN D. *On the theory of recursive unsolvability* ◇ D35 D55 ◇
KREISEL, G. *Note on arithmetic models for consistent formulae of the predicate calculus I* ◇ B10 C57 D45 D55 F30 ◇
LYAPUNOV, A.A. *B-functions (Russian)* ◇ D55 E15 ◇
LYAPUNOV, A.A. *Einleitung zu "Arbeiten zur deskriptiven Mengenlehre" (Russisch)* ◇ D55 E15 ◇
LYAPUNOV, A.A. *On the equivalence of families of sets (Russian)* ◇ D55 E15 ◇
SHCHEGOL'KOV, E.A. *Elements of the theory of B-sets (Russian)* ◇ D55 E15 ◇
SIERPINSKI, W. *Les ensembles projectifs et analytiques* ◇ D55 E15 E98 ◇
WATANABE, H. *Sur une separation des ensembles analytiques plans par une courbe mesurable* ◇ D55 E15 ◇

1951

NOVIKOV, P.S. *On the consistency of some propositions on the descriptive theory of sets (Russian)* ◇ D55 E15 E35 E45 ◇
RIBEIRO ALBUQUERQUE, J. *Theory of projective sets I (Portuguese) (French summary)* ◇ D55 E15 E98 ◇
SIERPINSKI, W. *Sur quelques consequences du theoreme de M. Kondo concernant l'uniformisation des complementaires analytiques* ◇ D55 E15 ◇

1952

DAVIS, MARTIN D. *Relatively recursive functions and the extended Kleene hierarchy* ◇ D55 ◇
RIBEIRO ALBUQUERQUE, J. *Theorie des ensembles projectifs* ◇ D55 E15 ◇
RIBEIRO ALBUQUERQUE, J. *Theory of projective sets II (Portuguese) (French summary)* ◇ D55 E15 E98 ◇
RIBEIRO ALBUQUERQUE, J. *Un theoreme sur les ensembles cribles* ◇ D55 E15 ◇

1953

DAVIS, MARTIN D. *Arithmetical problems and recursively enumerable predicates* ◇ D25 D35 D55 D80 ◇
HASENJAEGER, G. *Eine Bemerkung zu Henkin's Beweis fuer die Vollstaendigkeit des Praedikatenkalkuels der ersten Stufe* ◇ B10 C07 C57 D55 ◇
KREISEL, G. *A variant to Hilbert's theory of the foundations of arithmetic* ◇ A05 B28 D30 D55 ◇
KREISEL, G. *Note on arithmetic models for consistent formulae of the predicate calculus II* ◇ B10 C57 D45 D55 F30 ◇
LYAPUNOV, A.A. *On criteria of degeneracy of R-sets (Russian)* ◇ D55 E15 ◇
LYAPUNOV, A.A. *On the classification of R-sets (Russian)* ◇ D55 D70 E15 ◇
LYAPUNOV, A.A. *R-sets (Russian)* ◇ D55 D70 E15 ◇
LYAPUNOV, A.A. *Separability and nonseparability of R-sets (Russian)* ◇ D55 E15 ◇

1954

ADDISON, J.W. *On some points of the theory of recursive functions* ◇ D20 D55 ◇
MARKWALD, W. *Zur Theorie der konstruktiven Wohlordnungen* ◇ C80 D55 F15 ◇

1955

ARSENIN, V.YA. & LYAPUNOV, A.A. & SHCHEGOL'KOV, E.A. *Arbeiten zur deskriptiven Mengenlehre* ◇ D55 D70 D97 E15 E97 ◇

HASENJAEGER, G. *On definability and derivability* ◇ B10 B15 C07 D55 ◇

KLEENE, S.C. *Arithmetical predicates and function quantifiers* ◇ D55 D70 ◇

KLEENE, S.C. *Hierarchies of number-theoretic predicates* ◇ D55 D70 ◇

KLEENE, S.C. *On the forms of the predicates in the theory of constructive ordinals II* ◇ D55 D70 F15 ◇

LACOMBE, D. *Classes recursivement fermes et fonctions majorantes* ◇ D20 D55 ◇

MOSTOWSKI, ANDRZEJ *Contributions to the theory of definable sets and functions* ◇ D55 ◇

MOSTOWSKI, ANDRZEJ *Examples of sets definable by means of two and three quantifiers* ◇ D55 ◇

ROUTLEDGE, N.A. *Concerning definable sets* ◇ D30 D55 ◇

SODNOMOV, B.S. *Consistency of the projective evaluation of some non-effective sets (Russian)* ◇ D55 E15 E35 E45 ◇

SPECTOR, C. *Recursive well-orderings* ◇ D30 D45 D55 F15 ◇

WOLFE, P. *The strict determinateness of certain infinite games* ◇ D55 E05 E15 E60 ◇

1956

KONDO, M. *Sur la nommabilite d'ensembles* ◇ D55 E15 F60 F65 ◇

KONDO, M. *Sur les analyses relatives* ◇ D55 E15 F65 ◇

KONDO, M. *Sur les nombres reels et nommables* ◇ B28 D55 E15 F65 ◇

MARKWALD, W. *Ein Satz ueber die elementararithmetischen Definierbarkeitsklassen* ◇ D25 D55 ◇

MOSTOWSKI, ANDRZEJ *Development and applications of the "projective" classification of sets of integers* ◇ C57 C80 D55 E15 E45 ◇

MOTCHANE, L. *Sur un nouveau critere de conservation de classe de Baire* ◇ D55 E15 ◇

SODNOMOV, B.S. *Consistency of the projectivity of some special sets (Russian)* ◇ D55 E15 E35 E45 ◇

1957

ADDISON, J.W. & KLEENE, S.C. *A note on function quantification* ◇ D55 ◇

ADDISON, J.W. *Hierarchies and the axiom of constructibility* ◇ D55 E15 E35 E45 ◇

GRIFFITHS, H.B. *Borel sets and countable series of operations* ◇ D55 E15 E75 ◇

LACOMBE, D. *Les ensembles recursivement ouverts ou fermes, et leurs applications a l'analyse recursive* ◇ D55 F60 ◇

LYAPUNOV, A.A. *On operations of sets admitting transfinite indexes (Russian)* ◇ D55 D70 E15 ◇

ROGERS JR., H. *Computing degrees of unsolvability* ◇ D30 D55 ◇

SPECTOR, C. *Measure theory and higher order incomparability* ◇ D30 D55 ◇

SPECTOR, C. *Recursive ordinals and predicative set theory* ◇ D20 D55 E70 F15 F65 ◇

WANG, HAO *Remarks on constructive ordinals and set theory* ◇ D55 E45 F15 ◇

1958

ADDISON, J.W. *Separation principles in the hierarchies of classical and effective descriptive set theory* ◇ D55 E15 E45 ◇

GRZEGORCZYK, A. & MOSTOWSKI, ANDRZEJ & RYLL-NARDZEWSKI, C. *The classical and the ω-complete arithmetic* ◇ C62 D55 D70 F30 ◇

KONDO, M. *Sur l'uniformisation des ensembles nommables* ◇ D55 E15 ◇

ROTHBERGER, F. *Example effectif d'un ensemble transfiniment non-projectif* ◇ D55 E15 ◇

SHOENFIELD, J.R. *Degrees of formal systems* ◇ D25 D30 D35 D55 ◇

SHOENFIELD, J.R. *The class of recursive functions* ◇ D20 D55 ◇

SIKORSKI, R. *Some examples of Borel sets* ◇ D55 E15 ◇

SPECTOR, C. *Measure-theoretic construction of imcomparable hyperdegrees* ◇ D30 D55 ◇

1959

ADDISON, J.W. *Some consequences of the axiom of constructibility* ◇ D55 E15 E45 ◇

CHOQUET, G. *Ensembles $\mathscr{K}$-analytiques et $\mathscr{K}$-Sousliniens. Cas general et cas metrique* ◇ D55 E15 ◇

KLEENE, S.C. *Quantification of number-theoretic functions* ◇ D55 ◇

KLEENE, S.C. *Recursive functionals and quantifiers of finite types I* ◇ D30 D55 D65 ◇

KONDO, M. *Sur la nommabilite d'ensembles de type superieur* ◇ D55 D65 E15 ◇

KONDO, M. *Sur la theorie projective des ensembles* ◇ D55 E15 ◇

KREISEL, G. *Analysis of Cantor-Bendixson theorem by means of the analytic hierarchy* ◇ D55 E15 ◇

KUZNETSOV, A.V. *Certain questions concerning the classification of predicates and functions (Russian)* ◇ D55 ◇

LORENZEN, P. & MYHILL, J.R. *Constructive definition of certain analytic sets of numbers* ◇ D55 D70 ◇

MOSTOWSKI, ANDRZEJ *On various degrees of constructivism* ◇ D55 F50 F60 F98 ◇

MYHILL, J.R. *Finitely representable functions* ◇ D55 ◇

ROGERS JR., H. *Computing degrees of unsolvability* ◇ D30 D55 ◇

SAMPEI, Y. *On the evaluation of the projective class of sets defined by transfinite induction* ◇ D55 D70 E15 ◇

SHCHEGOL'KOV, E.A. *On the uniformization and splitting of certain sets (Russian)* ◇ D55 E15 ◇

SHOENFIELD, J.R. *On degrees of unsolvability* ◇ D30 D55 ◇

1960

CHAUVIN, A. *Sur les modeles du calcul K_0 de Bochvar, avec ou sans egalite, et l'interpretation des paradoxes de la logique dans la theorie des ensembles elementaires arithmetiques* ◇ B50 D55 ◇

FEFERMAN, S. *Arithmetization of metamathematics in a general setting* ◇ D55 F07 F25 F30 F40 ◇
GANDY, R.O. *On a problem of Kleene's* ◇ C62 D30 D55 ◇
GANDY, R.O. & KREISEL, G. & TAIT, W.W. *Set existence* ◇ C62 D55 F35 ◇
KREISEL, G. *La predicativite* ◇ A05 D55 E45 F35 F65 ◇
SAMPEI, Y. *Note on the effective choice of a point in the complement of an analytic set* ◇ D55 E15 ◇
SPECTOR, C. *Hyperarithmetical quantifiers* ◇ D55 D70 ◇
TUGUE, T. *On predicates expressible in the 1-function quantifier forms in Kleene hierarchy with free variables of type 2* ◇ D55 D65 ◇
TUGUE, T. *Predicates recursive in a type-2 object and Kleene hierarchies* ◇ D55 D65 ◇

1961

CHAUVIN, A. *Deux modeles verifiant certains axiomes de la theorie des ensembles de Goedel, et construits dans la theorie des ensembles arithmetiques de Kleene. Construction des modeles* ◇ C62 D55 E30 E70 ◇
CHAUVIN, A. *Deux modeles verifiant certains axiomes de la theorie des ensembles de Goedel, et construits dans la theorie des ensembles arithmetiques de Kleene. Validite des axiomes* ◇ C62 D55 E30 E70 ◇
GANDY, R.O. & KREISEL, G. & TAIT, W.W. *Set existence II* ◇ C62 D55 F35 ◇
GRZEGORCZYK, A. & MOSTOWSKI, ANDRZEJ & RYLL-NARDZEWSKI, C. *Definability of sets in models of axiomatic theories* ◇ C40 D55 ◇
KERSTAN, J. *Zur topologischen Invarianz der Hausdorffschen $Q^{(\alpha)}$-Mengen* ◇ D55 E15 ◇
KONDO, M. *Sur les hyper-analyses relatives* ◇ D55 ◇
KONDO, M. *Sur les hyper-continus projectifs* ◇ D55 E15 ◇
KREIDER, D.L. & ROGERS JR., H. *Constructive versions of ordinal number classes* ◇ D55 D70 F15 ◇
KREISEL, G. *Set theoretic problems suggested by the notion of potential totality* ◇ A05 C62 D20 D55 D65 ◇
PUTNAM, H. *Uniqueness ordinals in higher constructive number classes* ◇ D55 F15 ◇
SAMPEI, Y. *On the uniformization of the complement of an analytic set* ◇ D55 E15 ◇
SAMPEI, Y. *On the uniformization of a set of class $A_{\rho\sigma}$* ◇ D55 E15 ◇
SHOENFIELD, J.R. *The problem of predicativity* ◇ D55 E15 E45 ◇
SPECTOR, C. *Inductively defined sets of natural numbers* ◇ D55 D70 ◇

1962

ADDISON, J.W. *Some problems in hierarchy theory* ◇ C40 D55 E15 ◇
ADDISON, J.W. *The theory of hierarchies* ◇ C40 C52 D55 E15 ◇
CHAUVIN, A. *Sur les classes arithmetiques constructibles* ◇ D55 ◇
CHAUVIN, A. *Sur les ensembles arithmetiques constructibles* ◇ D55 E45 F15 ◇
FEFERMAN, S. & SPECTOR, C. *Incompleteness along paths in progressions of theories* ◇ D55 F15 F30 ◇
FEFERMAN, S. *Transfinite recursive progressions of axiomatic theories* ◇ D55 F15 F30 ◇
GRZEGORCZYK, A. *A theory without recursive models* ◇ C57 C62 D55 ◇
KINO, A. & TAKEUTI, G. *On hierarchies of predicates of ordinal numbers* ◇ D55 D60 F15 ◇
KLEENE, S.C. *λ-definable functionals of finite types* ◇ B40 D30 D55 D65 ◇
KLEENE, S.C. *Turing-machine computable functionals of finite types I* ◇ D10 D30 D55 D65 ◇
KLEENE, S.C. *Turing-machine computable functionals of finite types II* ◇ D10 D30 D55 D65 ◇
KOZLOVA, Z.I. *On projective operations and the separability of projective sets (Russian)* ◇ D55 E15 ◇
KREISEL, G. *The axiom of choice and the class of hyperarithmetic functions* ◇ D55 F35 ◇
LIU, SHICHAO *Four types of general recursive well-orderings* ◇ D30 D45 D55 ◇
LIU, SHICHAO *Recursive linear orderings and hyperarithmetical functions* ◇ D30 D45 D55 ◇
MOSTOWSKI, ANDRZEJ *Representability of sets in formal systems* ◇ D55 E15 F30 F35 ◇
SHOENFIELD, J.R. *Some applications of degrees* ◇ D25 D30 D35 D55 ◇
SHOENFIELD, J.R. *The form of the negation of a predicate* ◇ D55 D65 ◇

1963

KLEENE, S.C. *Recursive functionals and quantifiers of finite types II* ◇ D30 D55 D65 ◇
KROM, MELVEN R. *Separation principles in the hierarchy theory of pure first-order logic* ◇ B10 C40 C52 D55 ◇
LYAPUNOV, A.A. *Operations on sets (Russian)* ◇ D55 D70 E15 ◇
MARKOV, A.A. *On the inversion complexity of a system of boolean functions (Russian)* ◇ B05 D55 ◇
PUTNAM, H. *A note on constructible sets of integers* ◇ D55 E45 ◇
SIERPINSKI, W. *Projective and analytic sets* ◇ D55 E15 ◇
SUZUKI, Y. *On the uniformization principle* ◇ D55 E15 ◇
TRAKHTENBROT, B.A. *On the frequency computability of functions (Russian)* ◇ D20 D55 ◇

1964

BRESSLER, D.W. & SION, M. *The current theory of analytic sets* ◇ D55 E15 ◇
BUKOVSKY, L. & VOPENKA, P. *The existence of a PCA-set of cardinality $\aleph_1$* ◇ D55 E15 E35 ◇
CLARKE, D.A. *Hierarchies of predicate of finite types* ◇ D55 D65 D70 ◇
DAVIS, MORTON *Infinite games of perfect information* ◇ D55 E15 E60 ◇
ENDERTON, H.B. *Hierarchies in recursive function theory* ◇ D30 D55 ◇
ENDERTON, H.B. & LUCKHAM, D.C. *Hierarchies over recursive well-orderings* ◇ D30 D45 D55 ◇

KOZLOVA, Z.I. *The axiom of constructibility and multiple separability and inseparability in classes of the analytic hierarchy (Russian)* ◇ D55 E15 E45 ◇
PUTNAM, H. *On hierarchies and systems of notations* ◇ D55 D70 F15 ◇
SAMPEI, Y. *On the complete basis for the* Δ^1_2 *sets* ◇ D55 ◇
SCOTT, D.S. *Invariant Borel sets* ◇ C75 D55 E15 ◇
SUZUKI, Y. *A complete classification of the* Δ^1_2*-functions* ◇ D30 D55 ◇

1965

ADDISON, J.W. *The method of alternating chains* ◇ C40 D55 D75 E15 ◇
FEFERMAN, S. *Some applications of the notions of forcing and generic sets* ◇ C62 D55 E25 E35 E40 E45 F35 ◇
GOLD, E.M. *Limiting recursion* ◇ D20 D55 ◇
HASENJAEGER, G. *Zur arithmetischen Klassifikation reeller Zahlen* ◇ D55 ◇
HENSEL, G. & PUTNAM, H. *On the notational independence of various hierarchies of degrees of unsolvability* ◇ D30 D55 F15 ◇
KREISEL, G. & SACKS, G.E. *Metarecursive sets* ◇ D30 D55 D60 ◇
KREISEL, G. *Model-theoretic invariants: applications to recursive and hyperarithmetic operations* ◇ C40 C62 D55 D60 ◇
LACHLAN, A.H. *Some notions of reducibility and productiveness* ◇ D25 D30 D55 ◇
PUTNAM, H. *Trial and error predicates and the solution to a problem of Mostowski* ◇ C57 D20 D55 ◇
SCARPELLINI, B. *A characterization of* Δ_2*-sets* ◇ D55 E15 E45 ◇

1966

AMSTISLAVSKIJ, V.I. *Recursive sieves (Russian)* ◇ D55 E15 F65 ◇
AMSTISLAVSKIJ, V.I. *Set-theoretical operations and recursive hierarchies (Russian)* ◇ D55 D70 E15 ◇
DOEPP, K. *Ueber die Bestimmbarkeit des Verhaltens von Algorithmen* ◇ D20 D45 D55 ◇
ENGELKING, R. & HOLSZTYNSKI, W. & SIKORSKI, R. *Some examples of Borel sets* ◇ D55 E15 ◇
HINMAN, P.G. *Ad astra per aspera: hierarchy schemata in recursive function theory* ◇ D55 D65 D70 E15 ◇
KOLMOGOROV, A.N. *P.S.Aleksandrov and the theory of* δs*-operations (Russian)* ◇ A10 D55 E15 ◇
LOVELAND, D.W. *The Kleene hierarchy classification of recursively random sequences* ◇ D20 D55 D80 ◇
MOSCHOVAKIS, Y.N. *Many-one degrees of the predicates* $H_a(x)$ ◇ D30 D55 ◇
ROEDDING, D. *Anzahlquantoren in der Kleene-Hierarchie* ◇ C80 D55 ◇
SACKS, G.E. *Metarecursively enumerable sets and admissible ordinals* ◇ D30 D55 D60 ◇
SACKS, G.E. *Post's problem, admissible ordinals, and regularity* ◇ D30 D55 D60 ◇
SAMPEI, Y. *On the principle of effective choice and its applications* ◇ D55 E15 E45 ◇
SIKORSKI, R. *On an analytic set (Russian summary)* ◇ D55 E15 ◇
TANAKA, H. & TUGUE, T. *A note on the effective descriptive set theory* ◇ D55 E15 ◇
TANAKA, H. *On limits of sequences of hyperarithmetical functionals and predicates* ◇ D55 ◇
TANAKA, H. *Some properties of* Σ^1_1*- and* Π^1_1*-sets in* N^N ◇ D55 ◇
YATES, C.E.M. *On the degrees of index sets* ◇ D25 D30 D55 ◇

1967

BLACKWELL, D. *Infinite games and analytic sets* ◇ D55 E15 E60 ◇
ERMOLAEVA, N.M. *Arithmetic sums of recursively-projective sets (Russian)* ◇ D55 ◇
GANDY, R.O. & SACKS, G.E. *A minimal hyperdegree* ◇ D30 D55 ◇
GANDY, R.O. *General recursive functionals of finite type and hierarchies of functions* ◇ D55 D65 ◇
JENSEN, R.B. *Modelle der Mengenlehre* ◇ C62 D55 E25 E35 E40 E45 E50 E98 ◇
KLEENE, S.C. *Mathematical logic* ◇ B98 D20 D25 D55 D98 F30 F98 ◇
MOSCHOVAKIS, Y.N. *Hyperanalytic predicates* ◇ D55 D65 ◇
OWINGS JR., J.C. *Recursion, metarecursion, and inclusion* ◇ D25 D55 D60 ◇
ROBINSON, JULIA *An introduction to hyperarithmetical functions* ◇ D55 ◇
ROGERS JR., H. *Theory of recursive functions and effective computability* ◇ D25 D30 D55 D98 F15 ◇
SACKS, G.E. *Measure-theoretic uniformity in recursion theory and set theory (summary of results)* ◇ D30 D55 D60 E25 E35 E40 E50 ◇
SACKS, G.E. *Metarecursion theory* ◇ D30 D55 D60 ◇
SOLOVAY, R.M. *A nonconstructible* Δ^1_3 *set of integers* ◇ D55 E45 E55 ◇
TANAKA, H. *A basis result for* Π^1_1 *sets of positive measure* ◇ D55 E15 ◇
TANAKA, H. *Some results in the effective descriptive set theory* ◇ D55 E15 ◇
YATES, C.E.M. *Arithmetical sets and retracing functions* ◇ D30 D50 D55 ◇

1968

ADDISON, J.W. & MOSCHOVAKIS, Y.N. *Some consequences of the axiom of definable determinateness* ◇ D55 E15 E60 ◇
AMSTISLAVSKIJ, V.I. *Expansion of recursive hierarchies and R-operations (Russian)* ◇ D55 D70 E15 ◇
BLOOM, S.L. *A note on the arithmetical hierarchy* ◇ D55 ◇
BOOLOS, G. & PUTNAM, H. *Degrees of unsolvability of constructible sets of integers* ◇ D30 D55 E45 ◇
CLEAVE, J.P. *Hyperarithmetic ultrafilters* ◇ C57 D55 E05 ◇
ERSHOV, YU.L. *On a hierarchy of sets I,II (Russian)* ◇ D30 D55 ◇

HERMAN, G.T. *Simulation of one abstract computing machine by another* ◇ D10 D25 D30 D55 ◇
KIPNIS, M.M. *The constructive classification of arithmetic predicates and the semantic bases of arithmetic (Russian)* ◇ D55 F30 F50 ◇
KOZLOVA, Z.I. *Projective sets in topological spaces of weight τ (Russian)* ◇ D55 E15 E75 ◇
MARTIN, D.A. *The axiom of determinateness and reduction principles in the analytical hierarchy* ◇ D55 E15 E60 ◇
ROGERS, C.A. & WILLMOTT, R. *On the uniformization of sets in topological spaces* ◇ D55 E15 E75 ◇
SAMPEI, Y. *A proof of Mansfield's theorem by forcing method* ◇ D55 E15 E40 E45 ◇
SHOENFIELD, J.R. *A hierarchy based on a type two object* ◇ D55 D65 ◇
SUZUKI, Y. *$\aleph_0$-standard models for set theory (Russian summary)* ◇ C62 D55 ◇
TAKAHASHI, MOTO-O *Recursive functions of ordinal numbers and Levy's hierarchy* ◇ D55 D60 E10 E45 E47 ◇
TUCKER, C. *Limit of a sequence of functions with only countably many points of discontinuity* ◇ D55 E75 ◇

1969

BLOOM, S.L. *A semi-completeness theorem* ◇ D55 E15 ◇
BOYD, R. & HENSEL, G. & PUTNAM, H. *A recursion-theoretic characterization of the ramified analytical hierarchy* ◇ D30 D55 ◇
BRUCKNER, A.M. *Some remarks on extreme derivates* ◇ D55 E15 ◇
FILIPPOV, V.P. *Certain separability and nonseparability theorems in the second class of projective sets (Russian)* ◇ D55 E15 ◇
FILIPPOV, V.P. *The multiple separability in classes of arithmetical hierarchy (Russian)* ◇ D55 ◇
GRILLIOT, T.J. *Hierarchies based on objects of finite types* ◇ D55 D65 ◇
HENSEL, G. & PUTNAM, H. *Normal models and the field Σ_1^** ◇ C07 C30 C57 D55 ◇
HEYTING, A. *What is computable ? (Dutch)* ◇ D20 D35 D55 ◇
HINMAN, P.G. *Hierarchies of effective descriptive set theory* ◇ D55 D65 D70 E15 ◇
HINMAN, P.G. *Some applications of forcing to hierarchy problems in arithmetic* ◇ D25 D30 D55 E40 ◇
JOCKUSCH JR., C.G. & MCLAUGHLIN, T.G. *Countable retracing functions and Π_2^0 predicates* ◇ D30 D55 ◇
KOZLOVA, Z.I. *Certain projective operations (Russian)* ◇ D55 E15 ◇
KRUSE, A.H. *Souslinoid and analytic sets in a general setting* ◇ D55 E15 E20 ◇
MARTIN, D.A. & SOLOVAY, R.M. *A basis theorem for Σ_3^1 sets of reals* ◇ D55 E15 E45 E55 ◇
OWINGS JR., J.C. *Π_1^1-sets, ω-sets, and metacompleteness* ◇ D55 D60 ◇
PLATEK, R.A. *Eliminating the continuum hypothesis* ◇ D55 E25 E45 E47 E50 ◇
RAO, B.V. *On discrete Borel spaces and projective sets* ◇ D55 E15 E75 ◇
ROBINSON, JULIA *Finitely generated classes of sets of natural numbers* ◇ D20 D55 E20 ◇
ROMANOV, YU.I. *Separation of projective functions (Russian)* ◇ D55 E15 ◇
SACKS, G.E. *Measure-theoretic uniformity* ◇ D30 D55 E15 E25 E35 E40 E50 ◇
SACKS, G.E. *Measure-theoretic uniformity in recursion theory and set theory* ◇ D30 D55 D60 E15 E25 E35 E40 E50 ◇
SHAPIRO, N.Z. *Real numbers and functions in the Kleene hierarchy and limits of recursive, rational functions* ◇ D55 F60 ◇
SOLOVAY, R.M. *On the cardinality of Σ_2^1 sets of reals* ◇ D55 E15 E40 E45 E50 E55 ◇
WILLMOTT, R. *On the uniformization of Souslin $\mathscr{F}$ sets* ◇ D55 E15 ◇
YATES, C.E.M. *On the degrees of index sets II* ◇ D25 D30 D55 ◇

1970

AMSTISLAVSKIJ, V.I. *Effective R-sets and transfinite extensions of recursive hierarchies* ◇ D55 D70 E15 ◇
AMSTISLAVSKIJ, V.I. *On the decomposition of a field of sets obtained by an R-operation over recursive sets (Russian)* ◇ D55 D70 E15 ◇
BARWISE, J. & FISHER, E.R. *The Shoenfield absoluteness lemma* ◇ C62 D55 E45 ◇
BLOOM, S.L. *The hyperprojective hierarchy* ◇ D55 D70 E15 ◇
CHANG, C.C. & MOSCHOVAKIS, Y.N. *The Souslin-Kleene theorem for V_κ with cofinality $(\kappa)=\omega$* ◇ D55 D70 E47 ◇
CUTLAND, N.J. *The theory of hyperarithmetic and Π_1^1 models* ◇ C70 D55 ◇
ENDERTON, H.B. & PUTNAM, H. *A note on the hyperarithmetical hierarchy* ◇ D55 ◇
ENDERTON, H.B. *The unique existential quantifier* ◇ C80 D55 ◇
ERSHOV, YU.L. *Index sets (Russian)* ◇ D25 D30 D55 ◇
ERSHOV, YU.L. *On a hierarchy of sets III (Russian)* ◇ D30 D55 F15 ◇
FRIEDMAN, H.M. *Higher set theory and mathematical practice* ◇ D55 E15 E35 E45 E60 F35 ◇
FRIEDMAN, H.M. *Iterated inductive definitions and Σ_2^1-AC* ◇ C62 D55 E25 F35 ◇
JENSEN, R.B. *Definable sets of minimal degree* ◇ D30 D55 E35 E45 E47 ◇
JENSEN, R.B. & SOLOVAY, R.M. *Some applications of almost disjoint sets* ◇ C62 D55 E05 E35 E45 E55 ◇
JOCKUSCH JR., C.G. & SOARE, R.I. *Minimal covers and arithmetical sets* ◇ D30 D55 ◇

LEVY, A. *Definability in axiomatic set theory II* ◇ D55 E35 E45 E47 E50 ◇
LYUBETSKIJ, V.A. *The existence of a nonmeasurable set of type A_2 implies the existence of an uncountable set of type CA which does not contain a perfect subset (Russian)* ◇ D55 E15 E35 ◇
MAC GIBBON, B. *Exemple d'espace $\mathcal{K}$-analytique qui n'est $\mathcal{K}$-Souslinien dans aucun espace* ◇ D55 E15 ◇
MAITRA, A. *Coanalytic sets that are not Blackwell spaces* ◇ D55 E15 ◇
MAITRA, A. & RYLL-NARDZEWSKI, C. *On the existence of two analytic non-Borel sets which are not isomorphic* ◇ D55 E15 E45 ◇
MANSFIELD, R. *Perfect subsets of definable sets of real numbers* ◇ D55 E05 E15 E45 E55 ◇
MARTIN, D.A. *Measurable cardinals and analytic games* ◇ D55 E15 E55 E60 ◇
MARTIN-LOEF, P. *On the notion of randomness* ◇ B30 D55 D80 E15 ◇
MOSCHOVAKIS, Y.N. *Determinacy and prewellorderings of the continuum* ◇ D55 E15 E25 E60 ◇
MOSCHOVAKIS, Y.N. *The Suslin-Kleene theorem for countable structures* ◇ C15 C40 D55 D70 D75 ◇
OWINGS JR., J.C. *The metarecursively enumerable sets, but not the Π^1_1-sets, can be enumerated without repetition* ◇ D55 D60 ◇
POSS, R.L. *A note on a lemma of J.W.Addison* ◇ D55 E45 ◇
RAO, B.V. *Remarks on analytic sets* ◇ D55 E15 ◇
RAO, B.V. *Remarks on generalized analytic sets and the axiom of determinateness* ◇ D55 E15 E60 ◇
SAMPEI, Y. *On the relativization of Δ^1_2 sets of reals* ◇ D55 E15 E35 E45 ◇
SILVER, J.H. *Every analytic set is Ramsey* ◇ D55 E05 E15 E55 ◇
TANAKA, H. *Notes on measure and category in recursion theory* ◇ D55 E15 ◇
TANAKA, H. *On a Π^0_1 set of positive measure* ◇ D55 E15 E45 ◇
TANAKA, H. *On analytic well-orderings* ◇ D55 E15 E45 ◇

1971

BARWISE, J. & GANDY, R.O. & MOSCHOVAKIS, Y.N. *The next admissible set* ◇ C40 C62 D55 D60 D70 E45 E47 ◇
BRAUDE, E.J. *Descriptive Baire and descriptive $\mathcal{X}$-analytic sets* ◇ D55 E15 ◇
CAMPBELL, P.J. *Suslin logic* ◇ C75 D55 E15 G05 G25 ◇
DELLACHERIE, C. *Une demonstration du theoreme de separation des ensembles analytiques* ◇ D55 E15 ◇
FELDMAN, E.D. *An extension of the hyperprojective hierarchy* ◇ D55 E15 ◇
FENSTAD, J.E. *The axiom of determinateness* ◇ D55 E55 E60 E98 ◇
FRIEDMAN, H.M. *Determinateness in the low projective hierarchy* ◇ D55 E15 E40 E45 E60 ◇
GASS, F.S. *Generalized ordinal notation* ◇ D55 F15 ◇
HINMAN, P.G. & MOSCHOVAKIS, Y.N. *Computability over the continuum* ◇ D55 D65 D70 D75 ◇
KALLIBEKOV, S. *Index sets of degrees of unsolvability (Russian)* ◇ D25 D55 ◇
KALLIBEKOV, S. *The index sets of m-degrees (Russian)* ◇ D25 D55 ◇
KRECZMAR, A. *The set of all tautologies of algorithmic logic is hyperarithmetical* ◇ B75 D35 D55 ◇
LEEDS, S. & PUTNAM, H. *An intrinsic characterisation of the hierarchy of constructible sets of integers* ◇ D55 E45 ◇
LEVY, S. *Computational equivalence* ◇ D10 D30 D55 ◇
LYUBETSKIJ, V.A. *Independence of certain propositions of descriptive set theory from Zermelo-Fraenkel set theory (Russian) (English summary)* ◇ D55 E15 E35 E45 E55 ◇
MAITRA, A. *On game-theoretic methods in the theory of Souslin sets* ◇ D55 E15 E60 ◇
MANASTER, A.B. *Some contrasts between degrees and the arithmetical hierarchy* ◇ D30 D55 ◇
MANSFIELD, R. *A Souslin operation for Π^1_2* ◇ D55 E15 E55 ◇
MOSCHOVAKIS, Y.N. *Predicative classes* ◇ C62 D55 D60 D65 D70 E70 ◇
MOSCHOVAKIS, Y.N. *Uniformization in a playful universe* ◇ D55 E15 E60 ◇
OWINGS JR., J.C. *A splitting theorem for simple Π^1_1 sets* ◇ D30 D55 D60 ◇
PLATEK, R.A. *A countable hierarchy for the superjump* ◇ D55 D65 ◇
ROEDDING, D. *Arithmetische und hyperarithmetische Praedikate I* ◇ D55 ◇
SACKS, G.E. *F-recursiveness* ◇ D30 D55 D60 D75 E40 E45 ◇
SACKS, G.E. *On the reducibility of Π^1_1 sets* ◇ D30 D55 D60 E40 ◇
SELMAN, A.L. *Arithmetical reducibilities I* ◇ D30 D55 ◇
SHOENFIELD, J.R. *Measurable cardinals* ◇ D55 E15 E45 E55 E98 ◇
SILVER, J.H. *Measurable cardinals and Δ^1_3 well-orderings* ◇ C20 C30 D55 E05 E15 E35 E45 E55 ◇
SILVER, J.H. *Some applications of model theory in set theory* ◇ C30 C55 D55 E05 E45 E55 ◇
TANAKA, H. *Analytic well orderings and basis theorems* ◇ D45 D55 E15 ◇
WILLARD, S. *Some examples in the theory of Borel sets* ◇ D55 E15 ◇
WILLMOTT, R. *Some relations between k-analytic sets and generalized Borel sets* ◇ D55 E15 ◇
YATES, C.E.M. *A note on arithmetical sets of indiscernibles* ◇ D55 E05 ◇

1972

APT, K.R. *ω-models in analytical hierarchy (Russian summary)* ◇ C62 D30 D55 ◇

Berry, J.W. *Almost recursively enumerable sets* ◇ D25 D55 ◇
Blass, A.R. *Complexity of winning strategies* ◇ D55 E60 ◇
Chikvashvili, R.I. *The index sets that are connected with effective functionals (Russian) (Georgian and English summaries)* ◇ D20 D55 ◇
Cooper, S.B. *Jump equivalence of the Δ_2^0 hyperimmune sets* ◇ D25 D30 D55 ◇
Cutland, N.J. *Π_1^1-models and Π_1^1-categoricity* ◇ C57 C70 D55 ◇
Denisov, S.D. *Models of a noncontradictory formulas and the Ershov hierarchy (Russian)* ◇ C07 C57 D55 ◇
Feldman, E.D. *Some properties of the extended hyperprojective hierarchy* ◇ D55 E15 ◇
Freiwald, R.C. *Images of Borel sets and k-analytic sets* ◇ D55 E15 ◇
Garland, S.J. *Generalized interpolation theorems* ◇ C40 C75 C85 D55 E15 ◇
Gass, F.S. *A note on Π_1^1 ordinals* ◇ D55 F15 ◇
Grilliot, T.J. *Omitting types; applications to recursion theory* ◇ C07 C62 D30 D55 D60 D75 ◇
Hughes, C.E. *Degrees of unsolvability associated with Markov algorithms* ◇ D03 D25 D55 ◇
Jockusch Jr., C.G. *Ramsey's theorem and recursion theory* ◇ D30 D55 D80 ◇
Kechris, A.S. & Moschovakis, Y.N. *Two theorems about projective sets* ◇ D55 E15 E45 E60 ◇
Krajewski, S. & Rutkowski, A. *Preliminary results and problems concerning Π_{n+1}^0 functions (Russian summary)* ◇ D55 ◇
Kreisel, G. *Which number theoretic problems can be solved in recursive progressions on Π_1^1-paths through $\mathcal{O}$?* ◇ A05 D20 D55 F15 F30 F99 ◇
Lucian, M. *Systems of notations and the constructible hierarchy* ◇ D55 E45 F15 ◇
Mathias, A.R.D. *Solution of problems of Choquet and Puritz* ◇ C20 D55 E05 E15 ◇
Moschovakis, Y.N. *The game quantifier* ◇ C75 C80 D55 D70 D75 E60 ◇
Mueller, D.W. *Randomness and extrapolation* ◇ D55 D80 ◇
Paris, J.B. *$ZF \vdash \Sigma_4^0$ determinateness* ◇ D55 E15 E60 ◇
Roedding, D. *Arithmetische und hyperarithmetische Praedikate II* ◇ D55 ◇
Selman, A.L. *Applications of forcing to the degree-theory of the arithmetical hierarchy* ◇ D30 D55 E40 ◇
Selman, A.L. *Arithmetical reducibilities II* ◇ D30 D55 ◇
Shilleto, J.R. *Minimum models of analysis* ◇ C50 C62 D55 E45 ◇
Tanaka, H. *A note on hyperdegrees* ◇ D30 D55 ◇
Tanaka, H. *A property of arithmetic sets* ◇ D55 E15 ◇
Wheeler, W.H. *Algebraically closed division rings, forcing, and the analytic hierarchy* ◇ C25 C60 D55 ◇

1973

Amstislavskij, V.I. *A comparison of the indices arising in the transfinite iteration of functions (Russian)* ◇ D55 D70 E15 ◇
Bade, W.G. *Complementation problems for the Baire classes* ◇ D55 E15 ◇
Choban, M.M. *The descriptive theory of sets* ◇ D55 E15 ◇
Devlin, K.J. *Aspects of constructibility* ◇ D55 E35 E45 E65 E98 ◇
Drabbe, J. *Hierarchie arithmetique et chaines* ◇ D55 ◇
Ershov, Yu.L. *Hierarchy of the sets of class Δ_2^0 (Russian) (English summary)* ◇ D30 D55 ◇
Firmani, B. *Su una caratterizzazione degli F_α per ogni ordinale α di $I^\wedge$ specie e di cardinalita $\leqslant \aleph_0$ (French summary)* ◇ D55 E15 ◇
Friedman, H.M. *Σ_1^1 degree determinateness fails in all boolean extensions* ◇ D30 D55 E15 E40 E60 ◇
Friedman, H.M. *Borel sets and hyperdegrees* ◇ D30 D55 E15 ◇
Galvin, F. & Prikry, K. *Borel sets and Ramsey's theorem* ◇ D55 E05 E15 ◇
Guzicki, W. *A remark on the independence of a basis hypothesis* ◇ D55 E15 E35 E45 ◇
Hay, L. *Index sets in $\underline{0}'$ (Russian)* ◇ D25 D30 D55 ◇
Hinman, P.G. *The finite levels of the hierarchy of effective R-sets* ◇ D55 D65 D70 E15 ◇
Jockusch Jr., C.G. & Soare, R.I. *Encodability of Kleene's O* ◇ D30 D55 D65 ◇
Kanovich, M.I. *On the complexity of approximation of arithmetic sets (Russian)* ◇ D55 ◇
Kechris, A.S. *Descriptive set theory* ◇ D55 E15 E45 E55 E60 ◇
Kechris, A.S. *Measure and category in effective descriptive set theory* ◇ D55 E15 E60 ◇
Keisler, H.J. & Walkoe Jr., W.J. *The diversity of quantifier prefixes* ◇ B10 C07 C13 D55 ◇
Kiselev, A.A. *Projective hierarchies on general structures (Russian)* ◇ D55 D75 E15 E45 ◇
Kiselev, A.A. *Projective hierarchies on general structures II (Russian)* ◇ D55 D75 E15 E45 E55 ◇
Lyapunov, A.A. *The method of transfinite indices in the theory of operations over sets (Russian)* ◇ D55 D70 E15 ◇
Lyapunov, A.A. *The works of P.S.Novikov in the area of descriptive set theory (Russian)* ◇ A10 D55 E15 ◇
Makkai, M. *Vaught sentences and Lindstroem's regular relations* ◇ C40 C52 C75 D55 D70 ◇
Mansfield, R. *On the possibility of a Σ_2^1-well-ordering of the Baire space* ◇ D55 E15 E40 E45 ◇
Mauldin, R.D. *The Baire order of the functions continuous almost everywhere* ◇ D55 E15 ◇
McLaughlin, T.G. *A non-enumerability theorem for infinite classes of finite structures* ◇ C13 D25 D55 ◇
Moschovakis, Y.N. *Analytical definability in a playful universe* ◇ D55 E15 E60 ◇
Nelson, George C. *Nonconstructivity of models of the reals (Russian summary)* ◇ C62 C65 D55 H05 H15 ◇

NEPEJVODA, N.N. *On a generalization of the Kleene-Mostowski hierarchy (Russian)* ◇ D55 F15 F35 ◇
OTTMANN, T. *Ketten und arithmetische Praedikate von endlichen Automaten* ◇ D05 D55 D75 ◇
POBEDIN, L.N. *Certain questions on generalized computability (Russian)* ◇ D10 D55 D75 ◇
PUTNAM, H. *Recursive functions and hierarchies* ◇ D20 D30 D35 D55 ◇
ROGERS, C.A. *Lusin's second separation theorem* ◇ D55 E15 E20 ◇
SHCHEGOL'KOV, E.A. *Uniformization of sets of certain classes (Russian)* ◇ D55 E15 ◇
STERN, J. *Reels aleatoires et ensembles de mesure nulle en theorie descriptive des ensembles* ◇ D55 E15 E35 E40 E45 E55 E60 ◇
TANAKA, H. *Π^1_1 sets of sets, hyperdegrees and related problems* ◇ D30 D55 ◇
TANAKA, H. *Length of analytic well-orderings* ◇ D45 D55 E15 E60 ◇
VAUGHT, R.L. *A Borel invariantization* ◇ C75 D55 E15 ◇
VAUGHT, R.L. *Descriptive set theory in $L_{\omega_1,\omega}$* ◇ C15 C70 C75 D55 D70 E15 ◇

1974

ACZEL, P. & HINMAN, P.G. *Recursion in the superjump* ◇ D55 D65 E55 ◇
ADDISON, J.W. *Current problems in descriptive set theory* ◇ D55 E15 ◇
AMSTISLAVSKIJ, V.I. *Recursiveness and R^c-operations (Russian)* ◇ D55 D65 D70 E15 ◇
APT, K.R. & MAREK, W. *Second order arithmetic and related topics* ◇ C62 D55 E45 F25 F35 ◇
BLASS, A.R. & CENZER, D. *Cores of Π^1_1-sets of reals* ◇ D55 E15 ◇
BOERGER, E. *Σ_3-completude de l'ensemble des types de reduction* ◇ B20 D35 D55 ◇
BOFFA, M. *Hierarchie analytique et ensembles constructibles* ◇ D35 D55 E15 E45 ◇
BOOLOS, G. *Arithmetical functions and minimalisation* ◇ D55 ◇
CASE, J. *Maximal arithmetical reducibilities* ◇ D30 D55 ◇
CENZER, D. *Analytic inductive definitions* ◇ D55 D70 ◇
CENZER, D. *Inductively defined sets of reals* ◇ D55 D70 ◇
DAVIES, R.O. *On a separation theorem of Rogers* ◇ D55 E15 ◇
DEVLIN, K.J. *An introduction to the fine structure of the constructible hierarchy* ◇ C55 D55 E45 E47 E55 E98 ◇
ELLENTUCK, E. *A new proof that analytic sets are Ramsey* ◇ D55 E05 E15 ◇
FENSTAD, J.E. & NORMANN, D. *On absolutely measurable sets* ◇ D55 E15 E40 E75 ◇
FREJVALD, R.V. & PODNIEKS, K.M. *Ueber limitaere Berechnungen auf nichtdeterministischen Turingmaschinen (Russian)* ◇ D10 D55 ◇
FRIEDMAN, H.M. *Minimality in the Δ^1_2-degrees* ◇ D30 D55 E45 ◇
FRIEDMAN, H.M. *PCA well-orderings of the line* ◇ D55 E15 E45 ◇
GANOV, V.A. *Generalized computability and the descriptive theory of sets (Russian)* ◇ B75 D10 D55 E15 ◇
GARLAND, S.J. *Second-order cardinal characterizability* ◇ B15 C40 C55 C85 D55 E10 E47 E55 ◇
GNANI, G. *Insiemi dialettici generalizzati* ◇ B60 D25 D55 ◇
GRASSIN, J. *Index sets in Ershov's hierarchy* ◇ D55 ◇
GUASPARI, D. *A note on the Kondo-Addison theorem* ◇ D55 E15 E45 ◇
GUASPARI, D. *Characterizing the constructible reals* ◇ D30 D55 E15 E45 ◇
HAY, L. *Index sets universal for differences of arithmetic sets* ◇ D25 D30 D55 ◇
JENSEN, R.B. & JOHNSBRAATEN, H. *A new construction of a non-constructible Δ^1_3 subset of ω* ◇ D55 E35 E45 ◇
KECHRIS, A.S. *On projective ordinals* ◇ D55 E10 E15 E60 ◇
KELDYSH, L.V. *Ideas of N.N.Luzin in descriptive set theory (Russian)* ◇ A10 D55 E15 ◇
KONDO, M. *Les problemes fondamentaux parus dans "Cinq lettres sur la theorie des ensembles"* ◇ D55 E15 ◇
LACHLAN, A.H. *A note on Π^0_{n+1} functions and relations* ◇ D55 ◇
LEEDS, S. & PUTNAM, H. *Solution to a problem of Gandy's* ◇ C62 D55 ◇
LOUVEAU, A. *Une demonstration topologique de theoremes de Silver et Mathias* ◇ D55 E05 E15 ◇
LUKAS, J.D. & PUTNAM, H. *Systems of notations and the ramified analytical hierarchy* ◇ D30 D55 E45 F15 F35 ◇
MAITRA, A. *On the failure of the first principle of separation for coanalytic sets* ◇ D55 E15 ◇
MANSFIELD, R. *The non-existence of Σ^1_2-well-orderings of the Cantor set* ◇ D55 E15 E40 E45 ◇
MELIKYAN, S.M. *Constructive transfinite hierarchies of pseudonumbers (Russian)* ◇ D55 F60 ◇
MOLDESTAD, J. & NORMANN, D. *2-envelopes and the analytic hierarchy* ◇ D55 D65 ◇
MOSCHOVAKIS, Y.N. *Elementary induction on abstract structures* ◇ D55 D70 D75 ◇
MOSCHOVAKIS, Y.N. *On nonmonotone inductive definability* ◇ D55 D70 D75 ◇
NELSON, GEORGE C. *Many-one reducibility within the Turing degrees of the hyperarithmetic sets $H_a(x)$* ◇ D30 D55 ◇
OTTMANN, T. *Arithmetische Praedikate ueber einem Bereich endlicher Automaten* ◇ D05 D55 D75 ◇
ROBERTSON, E.L. *Structure of complexity in the weak monadic second-order theories of the natural numbers* ◇ D15 D55 F35 ◇

SACKS, G.E. *The 1-section of a type n object* ◇ D55 D65 ◇
SHORE, R.A. *Σ_n sets which are Δ_n-incomparable (uniformly)* ◇ D30 D55 D60 E40 E47 ◇
TANAKA, H. *Some analytical rules of inference in the second-order arithmetic* ◇ C62 D55 F35 ◇
THARP, L.H. *Continuity and elementary logic* ◇ C07 C80 C95 D30 D55 ◇
VAUGHT, R.L. *Invariant sets in topology and logic* ◇ C75 D55 D70 E15 ◇
WAINER, S.S. *A hierarchy for the 1-section of any type two object* ◇ D55 D65 ◇

1975

AMSTISLAVSKIJ, V.I. *On recursive characterizations of sets obtainable by means of R^c-operations (Russian)* ◇ D55 D65 D70 E15 ◇
BARWISE, J. *Admissible sets and the interaction of model theory, recursion theory and set theory* ◇ C70 C98 D55 D60 D70 E30 E35 E45 E47 ◇
BLASS, A.R. *A forcing proof of the Kechris-Moschovakis constructibility theorem* ◇ D55 E40 E45 ◇
BOOS, W. *Lectures on large cardinal axioms* ◇ D55 E05 E45 E55 E98 ◇
BURGESS, J.P. & MILLER, DOUGLAS E. *Remarks on invariant descriptive set theory* ◇ C75 D55 E15 ◇
CARSTENS, H.G. *Reducing hyperarithmetic sequences* ◇ C57 D55 ◇
CHIKVASHVILI, R.I. *On a certain problem of Kuznecov and Trahtenbrot (Russian) (Georgian and English summaries)* ◇ D55 E15 ◇
CRISCUOLO, G. & MINICOZZI, E. & TRAUTTEUR, G. *Limiting recursion and the arithmetic hierarchy (French summary)* ◇ D20 D55 ◇
DELLACHERIE, C. *Ensembles analytiques: theoremes de separation et applications (corr ibid 544)* ◇ D55 E15 E75 ◇
DEMUTH, O. *Constructive pseudonumbers (Russian)* ◇ D55 F60 ◇
FEFERMAN, S. *A language and axioms for explicit mathematics* ◇ D55 F35 F50 F65 ◇
FELDMAN, E.D. *L-Σ^1_n transfinite induction with an application to the EHP hierarchy* ◇ D55 D70 E15 E47 ◇
GUASPARI, D. *Analytical well-orderings in $\mathbb{R}$* ◇ D55 E15 ◇
HARRINGTON, L.A. *Π^1_1 sets and Π^1_2 singletons* ◇ D55 E15 E45 ◇
HARRINGTON, L.A. & KECHRIS, A.S. *A basis result for Σ^0_3 sets of reals with an application to minimal covers* ◇ D30 D55 E15 E60 ◇
HARRINGTON, L.A. & KECHRIS, A.S. *On characterizing Spector classes* ◇ C62 D55 D60 D65 D75 E45 E55 E60 ◇
HAY, L. *Rice theorems for d.r.e. sets* ◇ D25 D30 D55 ◇
HAY, L. & MANASTER, A.B. & ROSENSTEIN, J.G. *Small recursive ordinals, many-one degrees, and the arithmetical difference hierarchy* ◇ D25 D30 D55 ◇
JEROSLOW, R. *Experimental logics and Δ^0_2-theories* ◇ B60 D55 ◇
KALLIBEKOV, S. *Hierarchy of sets for truth-table enumerability (Russian)* ◇ D30 D55 ◇
KANOVEJ, V.G. *On the independence of some propositions of descriptive set theory and second-order arithmetic (Russian)* ◇ D55 E15 E35 F35 ◇
KECHRIS, A.S. & MARTIN, D.A. *A note on universal sets for classes of countable G_δ's* ◇ D55 E15 ◇
KECHRIS, A.S. *Countable ordinals and the analytical hierarchy I* ◇ C62 D55 E10 E15 E60 ◇
KECHRIS, A.S. *The theory of countable analytical sets* ◇ D55 E15 E45 E60 ◇
MANSFIELD, R. *Omitting types: application to descriptive set theory* ◇ C75 D55 E15 E45 ◇
MARTIN, D.A. *Borel determinacy* ◇ D55 E15 E60 ◇
MATHIAS, A.R.D. *A remark on rare filters* ◇ D55 E05 E15 ◇
MAULDIN, R.D. *The Baire order of the functions continuous almost everywhere II* ◇ D55 E15 ◇
MORLEY, M.D. & SOARE, R.I. *Boolean algebras, splitting theorems and Δ^0_2 sets* ◇ D25 D55 G05 ◇
MOSCHOVAKIS, Y.N. *New methods and results in descriptive set theory* ◇ D55 E15 E55 E60 ◇
ODIFREDDI, P. *Note sugli insiemi implicitamente definibili* ◇ D55 ◇
OSTASZEWSKI, A.J. *On the descriptive set theory of the lexicographic square* ◇ D55 E15 ◇
POBEDIN, L.N. *The halt problem and theory of hierarchies (Russian)* ◇ D10 D55 ◇
RICHTER, W.H. *The least Σ^1_2- and Π^1_2-reflecting ordinals* ◇ D55 D60 D70 E45 E47 E55 ◇
SCHLIPF, J.S. *Some hyperelementary aspects of model theory* ◇ C50 C62 C70 D55 ◇
SIMPSON, S.G. *Minimal covers and hyperdegrees* ◇ D30 D55 ◇
STERN, J. *Some measure theoretic results in effective descriptive set theory* ◇ D30 D55 E15 E60 ◇
THOMASON, S.K. *The logical consequence relation of propositional tense logic* ◇ B45 D20 D55 ◇
VINNER, S. *On two complete sets in the analytical and the arithmetical hierarchies* ◇ C55 C80 D55 ◇
WAINER, S.S. *Some hierachies based on higher type quantification* ◇ C75 D55 D65 E45 ◇
WILLMOTT, R. *A form of Lusin's second separation theorem for k-analytic sets* ◇ D55 E15 ◇
YATES, C.E.M. *A general framework for simple Δ^0_2 and Σ^0_1 priority arguments* ◇ D25 D55 ◇

1976

APT, K.R. *Semantics of the infinitistic rules of proof* ◇ C62 D55 E45 F07 F35 ◇
BELYAKIN, N.V. *Iterated Kleene computability, and the superjump (Russian)* ◇ D55 D65 ◇
BOERGER, E. & HEIDLER, K. *Die m-Grade logischer Entscheidungsprobleme* ◇ D25 D35 D55 ◇
BUBIS, M.I. *On complete indexed sets (Russian)* ◇ D25 D55 ◇

BUSCH, D.R. *λ-scales, κ-Souslin sets and a new definition of analytic sets* ◇ D55 E15 E60 ◇
CARSTENS, H.G. *Δ_2^0-Mengen* ◇ D25 D30 D55 ◇
CENZER, D. *Monotone inductive definitions over the continuum* ◇ D55 D70 E15 ◇
EBBINGHAUS, H.-D. *The axiom of determinateness* ◇ D55 E15 E25 E60 ◇
FRIEDMAN, H.M. *Recursiveness in Π_1^1 paths through $\mathcal{O}$* ◇ D30 D55 F15 ◇
FRIEDMAN, H.M. *Uniformly defined descending sequences of degrees* ◇ D30 D55 F30 F35 ◇
GUASPARI, D. & HARRINGTON, L.A. *Characterizing $\mathscr{C}_3$ (the largest countable Π_3^1 set)* ◇ D55 E15 E60 ◇
HARRINGTON, L.A. & JECH, T.J. *On Σ_1 well-orderings of the universe* ◇ D55 E35 E45 E47 ◇
HAY, L. *Boolean combinations of r.e. open sets* ◇ D25 D55 E15 ◇
JOCKUSCH JR., C.G. & SIMPSON, S.G. *A degree theoretic definition of the ramified analytical hierarchy* ◇ D30 D55 E40 E45 H05 ◇
KANIEWSKI, J. *A generalization of Kondo's uniformization theorem (Russian summary)* ◇ D55 E15 ◇
KLETTE, R. *Indexmengen und Erkennung rekursiver Funktionen* ◇ D20 D25 D30 D55 ◇
LOUVEAU, A. *Determination des jeux G_ω^* (English summary)* ◇ D55 E15 E35 E60 ◇
LOUVEAU, A. *Une methode topologique pour l'etude de la propriete de Ramsey* ◇ D55 E05 E15 E50 E75 ◇
LYUBETSKIJ, V.A. *Random sequences of numbers and A_2-sets (Russian)* ◇ D55 E15 E35 ◇
MAITRA, A. & RAO, B.V. *Selection theorems for partitions of Polish spaces* ◇ D55 E15 ◇
MAREK, W. & SREBRNY, M. & ZARACH, A. (EDS.) *Set theory and hierarchy theory. A memorial tribute to Andrzej Mostowski* ◇ B97 D55 D97 E97 ◇
MARTIN, D.A. *Proof of a conjecture of Friedman* ◇ D30 D55 ◇
MYERS, D.L. *Invariant uniformization* ◇ C75 D55 E15 E35 ◇
NORMANN, D. *On a problem of S. Wainer* ◇ D55 D65 ◇
SACKS, G.E. *Countable admissible ordinals and hyperdegrees* ◇ D30 D55 D60 ◇
STAHL, S.H. *A hierarchy on the class of primitive recursive ordinal functions* ◇ D55 D60 ◇
THIELER-MEVISSEN, G. *Zur Beschreibbarkeit der hyperarithmetischen reellen Zahlen mit analysiskonformen Mitteln* ◇ D35 D55 ◇
TRAKHTENBROT, B.A. *Recursive program schemes and computable functionals* ◇ B75 D15 D55 D65 D75 ◇
WAGNER, K. *Arithmetische Operatoren* ◇ D10 D55 ◇
YASUDA, Y. *A note on the relativization of $\underset{\sim}{\Delta}_2^1$ subsets of Baire spaces* ◇ D55 E15 ◇

1977

ACZEL, P. *An introduction to inductive definitions* ◇ D55 D70 ◇
ARSENIN, V.YA. & KOZLOVA, Z.I. *A survey of A. A. Ljapunov's works on descriptive set theory (Russian)* ◇ A10 D55 E15 ◇
ARSLANOV, M.M. & NADYROV, R.F. & SOLOV'EV, V.D. *A criterion for the completeness of recursively enumerable sets, and some generalizations of a fixed point theorem (Russian)* ◇ D20 D25 D55 ◇
BURGESS, J.P. *A selector principle for Σ_1^1 equivalence relations* ◇ D55 E05 E15 E45 ◇
BURGESS, J.P. *Two selection theorems* ◇ D55 E15 ◇
BUSCH, D.R. *A problem concerning projective prewellorderings* ◇ D55 E05 E15 E60 ◇
CARSTENS, H.G. *The complexity of some combinatorial constructions* ◇ D55 D80 E05 ◇
DAVIES, R.O. & JAYNE, J.E. & OSTASZEWSKI, A.J. & ROGERS, C.A. *Theorems of Novikov type* ◇ D55 E15 ◇
DELLACHERIE, C. *Les derivations en theorie descriptive des ensembles et le theoreme de la borne* ◇ D55 E15 E75 ◇
DEMUTH, O. *A constructive analogue of the functions of the n-th Baire class* ◇ D55 E15 F60 ◇
FEFERMAN, S. *Theories of finite type related to mathematical practice* ◇ B15 D55 D65 F10 F15 F35 F98 ◇
GILLO, D. & GOETZE, B.G. & KLETTE, R. *Der iterierte Limes rekursiver Funktionen und die arithmetische Hierarchie* ◇ D20 D55 ◇
GORDEEV, L.N. *A majorizing semantics for hyperarithmetic sentences (Russian) (English summary)* ◇ D55 F15 F50 ◇
GRANT, P.W. *Strict Π_1^1-predicates on countable and cofinality ω transitive sets* ◇ C15 C40 C70 D55 D60 D70 E47 E60 ◇
GRIGORIEFF, S. *Determination des jeux boreliens et problemes logiques associes (d'apres D. Martin)* ◇ D55 E15 E60 ◇
GUZICKI, W. *On the projective class of the continuum hypothesis* ◇ D55 E47 E50 ◇
HAJEK, P. *Arithmetical complexity of some problems in computer science* ◇ D15 D55 ◇
HAJEK, P. *Experimental logics and Π_3^0 theories* ◇ B60 C62 D55 F30 ◇
HARRINGTON, L.A. & KECHRIS, A.S. *Π_2^1 singletons and $0^{\#}$* ◇ D30 D55 E45 E55 ◇
HARRINGTON, L.A. *Long projective wellorderings* ◇ C62 D55 E15 E35 E50 ◇
HINMAN, P.G. *A survey of finite-type recursion* ◇ D55 D60 D65 D98 ◇
JOHNSON, N. *Rice theorems for Σ_n^{-1} sets* ◇ D25 D30 D55 ◇
JONGH DE, D.H.J. & PARIKH, R. *Well-partial orderings and hierarchies* ◇ D55 E07 E10 ◇

KECHRIS, A.S. *Classifying projective-like hierarchies* ◇ D55 D75 E15 E45 E60 ◇
KECHRIS, A.S. *On a notion of smallness for subsets of the Baire space* ◇ D55 E15 E60 ◇
LAVORI, P. *Recursion in the extended superjump* ◇ D55 D65 ◇
LOUVEAU, A. *Boreliens a coupes $K_{\sigma\delta}$ (English summary)* ◇ D55 E15 ◇
LOUVEAU, A. *La hierarchie borelienne des ensembles Δ^1_1 (English summary)* ◇ D55 E15 ◇
LUNINA, M.A. *Luzin's arithmetic example of an analytic set that is not a Borel set (Russian)* ◇ D55 E15 ◇
MACINTYRE, J.M. *Transfinite extensions of Friedberg's completeness criterion* ◇ D30 D55 D60 ◇
MAKKAI, M. *Admissible sets and infinitary logic* ◇ C70 C98 D55 D60 D70 D98 ◇
MAKKAI, M. *An "admissible" generalization of a theorem on countable Σ^1_1 sets of reals with applications* ◇ C15 C40 C50 C70 D55 E15 ◇
MARTIN, D.A. *Descriptive set theory: projective sets* ◇ D55 E15 E55 E60 ◇
NEPEJVODA, L.K. & NEPEJVODA, N.N. *Languages without set-variables for the description of sets (Russian)* ◇ C75 D55 ◇
NORMANN, D. *Countable functionals and the analytic hierarchy* ◇ D55 D65 ◇
ODIFREDDI, P. *A note on Suzuki's chain of hyperdegrees* ◇ D30 D55 E15 ◇
PLISKO, V.E. *The nonarithmeticity of the class of realizable predicate formulas (Russian)* ◇ D35 D55 F50 ◇
RESSAYRE, J.-P. *Models with compactness properties relative to an admissible language* ◇ C50 C55 C70 C80 D55 D60 E15 ◇
SIMPSON, S.G. *Basis theorems and countable admissible ordinals* ◇ D55 D60 E15 ◇
SREBRNY, M. *Relatively constructible transitive models* ◇ C62 D55 D60 E10 E40 E45 E55 ◇
STAIGER, L. & WAGNER, K. *Recursive ω-languages* ◇ D05 D10 D55 ◇
STERN, J. *Partitions de la droite reelle en F_σ ou en G_δ (English summary)* ◇ D55 E05 E15 E35 ◇
STERN, J. *Singletons Π^1_3 et reels Δ^1_3 (English summary)* ◇ D55 E15 E35 E55 ◇
STOCKMEYER, L.J. *The polynomial-time hierarchy* ◇ D10 D15 D40 D55 ◇
TELGARSKY, R. *Topological games and analytic sets* ◇ D55 E15 E60 E75 ◇
WAGNER, K. *Arithmetische und Bairesche Operatoren* ◇ D55 E15 ◇

1978

APT, K.R. *Inductive definitions, models of comprehension and invariant definability* ◇ C40 C62 D55 D70 F35 ◇
BECKER, H. *Partially playful universes* ◇ D55 E15 E45 E60 ◇
BLACKWELL, D. *Borel-programmable functions* ◇ D55 E15 ◇
BOOK, R.V. *Simple representations of certain classes of languages* ◇ D15 D25 D55 ◇
BURGESS, J.P. *Equivalences generated by families of Borel sets* ◇ D55 E15 ◇
CENZER, D. *Parametric inductive definitions and recursive operators over the continuum* ◇ D55 D70 ◇
DAVID, R. *A Π^1_2 singleton with no sharp in a generic extension of $L^\#$* ◇ D55 E35 E45 E55 ◇
DELFINO, V. *The Victoria Delfino problems* ◇ D55 D60 D65 D70 E15 E60 ◇
DEMUTH, O. & KRYL, R. & KUCERA, A. *An application of the theory of functionals partial recursive relative to number sets in constructive mathematics (Russian)* ◇ D30 D55 F60 ◇
FITTING, M. *Elementary formal systems for hyperarithmetical relations* ◇ C75 D55 D75 ◇
GOETZE, B.G. & KLETTE, R. *Ein Normalformentheorem fuer Σ_2-Funktionen* ◇ D20 D55 ◇
GOODMAN, NICOLAS D. *The nonconstructive content of sentences of arithmetic* ◇ D20 D30 D55 F30 F50 ◇
GRASSIN, J. *Definitions inductives monotones sur le continu, dont les composantes sont denombrables (English summary)* ◇ D55 D60 D70 E15 ◇
HAJEK, P. *Some results on degrees of constructibility* ◇ D30 D55 E45 E55 ◇
HARRINGTON, L.A. & KIROUSIS, L.M. & SCHLIPF, J.S. *A generalized Kleene-Moschovakis theorem* ◇ D55 D70 D75 ◇
HARRINGTON, L.A. *Analytic determinacy and $O^\#$* ◇ D55 D65 E45 E55 E60 ◇
HARROW, K. *The bounded arithmetic hierarchy* ◇ D15 D55 ◇
HINMAN, P.G. *Recursion-theoretic hierarchies* ◇ D55 D60 D65 D70 D98 E15 ◇
HRBACEK, K. *On the complexity of analytic sets* ◇ D30 D55 D65 E15 E45 ◇
JACOBS, B.E. *The α-union theorem and generalized primitive recursion* ◇ D15 D55 D60 ◇
KANOVEJ, V.G. *A proof of a theorem of Luzin (Russian)* ◇ D55 E15 ◇
KECHRIS, A.S. *AD and projective ordinals* ◇ D55 E15 E55 E60 ◇
KECHRIS, A.S. & MOSCHOVAKIS, Y.N. (EDS.) *Cabal seminar 76–77* ◇ D55 D97 E15 E60 E97 ◇
KECHRIS, A.S. *Countable ordinals and the analytical hierarchy II* ◇ D55 E15 E60 ◇
KECHRIS, A.S. *Forcing in analysis* ◇ D55 E15 E40 E60 ◇
KECHRIS, A.S. *Minimal upper bounds for sequences of Δ^1_{2n}-degrees* ◇ D30 D55 E60 ◇
KECHRIS, A.S. & MOSCHOVAKIS, Y.N. *Notes on the theory of scales* ◇ D55 E15 E45 E60 E98 ◇
KECHRIS, A.S. & MARTIN, D.A. *On the theory of Π^1_3 sets of reals* ◇ D55 E15 E45 E60 ◇
KECHRIS, A.S. *On transfinite sequences of projective sets with an application to bold-face-Σ^1_2 equivalence relations* ◇ D55 E15 E60 ◇

KECHRIS, A.S. *On Spector classes* ◇ D55 D65 D70 D75 E15 E60 ◇
KECHRIS, A.S. *The perfect set theorem and definable wellorderings of the continuum* ◇ D55 E15 E45 E60 ◇
KISELEV, A.A. *Axiom of comparable choice and uniformizability of projective classes (Russian)* ◇ D55 E15 E25 E45 ◇
KUCERA, A. & KUSHNER, B.A. *A type of recursive isomorphism of certain concepts of constructive analysis (Russian)* ◇ D55 F60 ◇
LOUVEAU, A. *Notions elementaires de theorie descriptive effective* ◇ D55 E15 ◇
LOUVEAU, A. *Recursivity and compactness* ◇ D55 E15 ◇
LOUVEAU, A. *Relations d'equivalence co-analytiques* ◇ D55 E15 ◇
MAASS, W. *Fine structure of the constructible universe in α-and β-recursion theory* ◇ D55 D60 E45 E65 ◇
MANSFIELD, R. *A footnote to a theorem of Solovay on recursive encodability* ◇ D30 D55 ◇
MILLER, DOUGLAS E. *The invariant Π^0_α separation principle* ◇ C15 C70 C75 D55 E15 ◇
MOSCHOVAKIS, Y.N. *Inductive scales on inductive sets* ◇ D55 D70 D75 E15 E60 ◇
NELSON, GEORGE C. *Isomorphism types of the hyperarithmetic sets H_a* ◇ D30 D50 D55 ◇
NORMANN, D. *A continuous functional with noncollapsing hierarchy* ◇ D55 D65 ◇
RAO, B.V. & RAO, K.P.S.BHASKARA *On the isomorphism problem for analytic sets* ◇ D55 E15 ◇
SANCHIS, L.E. *Hyperenumeration reducibility* ◇ D30 D55 ◇
SELIVANOV, V.L. *On index sets of computable classes of finite sets (Russian)* ◇ D25 D30 D55 ◇
SIMPSON, S.G. *Sets which do not have subsets of every higher degree* ◇ D30 D55 ◇
SOLOVAY, R.M. *A Δ^1_3 coding of the subsets of $\omega\omega$* ◇ D55 E05 E15 E45 E60 ◇
SOLOVAY, R.M. *Hyperarithmetically encodable sets* ◇ D55 D60 E05 E40 E45 E55 ◇
SREBRNY, M. *Singular cardinals and analytic games* ◇ D55 E15 E45 E50 E55 E60 ◇
STAIGER, L. & WAGNER, K. *Rekursive Folgenmengen. I* ◇ D10 D55 D75 ◇
STEEL, J.R. *Forcing with tagged trees* ◇ C62 D30 D55 E40 F35 ◇
STERN, J. *Evaluation du rang de Borel de certains ensembles (English summary)* ◇ D55 E15 E50 E60 ◇
STERN, J. *Perfect set theorems for analytic and coanalytic equivalence relations* ◇ D55 E15 E60 ◇
TANAKA, H. *Recursion theory in analytical hierarchy* ◇ D25 D55 D60 E60 ◇
TROELSTRA, A.S. *Some remarks on the complexity of Henkin-Kripke models* ◇ C90 D55 F50 ◇
WAINER, S.S. *The 1-section of a nonnormal type-2 object* ◇ D55 D65 ◇
WESEP VAN, R.A. *Separation principles and the axiom of determinateness* ◇ D55 D75 E15 E60 ◇

1979

ARSLANOV, M.M. *Weak recursively enumerable degrees and computability in the limit (Russian)* ◇ D25 D55 ◇
BELYAEV, V.YA. & TAJTSLIN, M.A. *On elementary properties of existentially closed systems (Russian)* ◇ C25 C40 C60 C75 C85 D55 ◇
BELYAKIN, N.V. *Autonomous computability (Russian)* ◇ D45 D55 D65 ◇
BUDINAS, B.L. *Three linearly ordered degrees of constructibility of Δ^1_3 numbers (Russian)* ◇ D30 D55 E35 E45 ◇
BURGESS, J.P. *A reflection phenomenon in descriptive set theory* ◇ C75 D55 E15 ◇
BURGESS, J.P. *Effective enumeration of classes in a Σ^1_1 equivalence relation* ◇ D55 E15 E45 E55 ◇
DOMBROVSKIJ-KABANCHENKO, M.N. *A transfinite extension of the stepped semantic system (Russian)* ◇ D55 F50 ◇
ERIMBETOV, M.M. *On hyperarithmetical non-standard predicates (Russian)* ◇ D55 ◇
FREIWALD, R.C. & MCDOWELL, R. & MCHUGH JR., E.F. *Borel sets of exact class* ◇ D55 E15 ◇
GERLA, G. *Una generalizzazione della gerarchia di Ershov* ◇ D30 D55 ◇
HAJEK, P. *Arithmetical hierarchy and complexity of computation* ◇ D10 D15 D55 ◇
HARRINGTON, L.A. & SAMI, R.L. *Equivalence relations, projective and beyond* ◇ D55 E07 E15 E60 ◇
HINMAN, P.G. *Borel determinacy* ◇ D55 D98 E15 E60 ◇
KANOVEJ, V.G. *On descriptive forms of the countable axiom of choice (Russian)* ◇ D55 E15 E25 E35 F35 ◇
KANOVEJ, V.G. *The definability of forcing in analysis (Russian) (English summary)* ◇ D55 E40 F35 ◇
KANOVEJ, V.G. *The set of all analytically definable sets of natural numbers can be defined analytically (Russian)* ◇ C62 D55 E15 E35 E45 ◇
KECHRIS, A.S. *An overview of descriptive set theory* ◇ D55 D98 E15 E98 ◇
LEWIS, F.D. *On unsolvability in subrecursive classes of predicates* ◇ D15 D20 D30 D55 ◇
LOUVEAU, A. *Familles separantes pour les ensembles analytiques (English summary)* ◇ D55 E15 ◇
LOUVEAU, A. *Une nouvelle technique d'etude des relations d'equivalence coanalytiques* ◇ D55 E15 ◇
MARANDZHYAN, G.B. *On the sets of minimal indices of partial recursive functions* ◇ D20 D30 D45 D55 ◇
MILLER, A.W. *On the length of Borel hierarchies* ◇ D55 E15 E35 ◇
SREBRNY, M. *Constructible sets and analytic games* ◇ D55 E15 E45 E50 E55 E60 ◇
STERN, J. *Cardinalite des relations d'equivalence analytiqes et coanalytiqes* ◇ D55 E15 ◇

1980

BECKER, H. *Thin collections of sets of projective ordinals and analogs of L* ◇ D55 E15 E45 E50 E55 E60 ◇

BUDINAS, B.L. *Analytic definability of constructible real numbers (Russian)* ◇ D55 E35 E45 E50 ◇

BUDINAS, B.L. *Partial ordering of Δ^1_n-degrees of subsets of natural numbers (Russian) (English summary)* ◇ D30 D55 E35 ◇

BUSZKOWSKI, W. *Logical complexity of some classes of tree languages generated by multiple-tree-automata* ◇ D05 D55 ◇

GUASPARI, D. *Definability in models of set theory* ◇ C62 D55 E45 E47 ◇

HRBACEK, K. & SIMPSON, S.G. *On Kleene degrees of analytic sets* ◇ D30 D55 D65 E15 E35 E45 ◇

KECHRIS, A.S. & MARTIN, D.A. *Infinite games and effective descriptive set theory* ◇ D55 E15 E55 E60 E98 ◇

KECHRIS, A.S. *Recent advances in the theory of higher level projective sets* ◇ D55 E15 E60 ◇

LOUVEAU, A. *A separation theorem for Σ^1_1 sets* ◇ D55 E15 ◇

MAGIDOR, M. *Precipitous ideals and Σ^1_4 sets* ◇ D55 E05 E15 E50 E55 ◇

MOSCHOVAKIS, Y.N. *Descriptive set theory* ◇ D55 D98 E15 E98 ◇

MYERS, R.W. *Complexity of model-theoretic notions* ◇ C07 C35 C57 D55 ◇

NORMANN, D. *Recursion on the countable functionals* ◇ D30 D55 D65 D98 ◇

NORMANN, D. *The recursion theory of the continuous functionals* ◇ D55 D65 ◇

NORMANN, D. & WAINER, S.S. *The 1-section of a countable functional* ◇ D55 D65 ◇

SHINODA, J. *On the upper semi-lattice of J^s_a-degrees* ◇ D30 D55 D65 ◇

SILVER, J.H. *Counting the number of equivalence classes of Borel and coanalytic equivalence relations* ◇ D55 E15 E40 ◇

SIMPSON, S.G. *The hierarchy based on the jump operator* ◇ D30 D55 ◇

STEEL, J.R. *A note on analytic sets* ◇ D55 E15 ◇

STEEL, J.R. *Analytic sets and Borel isomorphisms* ◇ D55 E15 E60 ◇

STERN, J. *Effective partitions of the real line into Borel sets of bounded rank* ◇ D55 E15 E60 ◇

YANG, DONGPING *α-splitting theorem restricted to Δ^0_2 sets (Chinese)* ◇ D55 D60 ◇

1981

BACIU, A. & PASCU, A. *The interface between hierarchies, multiple objectives clustering and fuzzy sets* ◇ D55 E72 ◇

BUDINAS, B.L. *Construction of definable degrees of constructibility (Russian)* ◇ D30 D55 E35 E45 E47 ◇

BUDINAS, B.L. *The selector principle and analytic definability of real numbers in extensions of the constructible universe (Russian)* ◇ D30 D55 E15 E35 E45 E47 ◇

BUKHARAEV, N.R. *T-degrees of differences of recursively enumerable sets (Russian)* ◇ D25 D30 D55 ◇

CLOTE, P. *A note on the leftmost infinite branch of a recursive tree* ◇ D30 D55 ◇

EPSTEIN, R.L. & HAAS, R. & KRAMER, R.L. *Hierarchies of sets and degrees below O'* ◇ D30 D55 ◇

GIRARD, J.-Y. *Π^1_2-logic I. Dilators* ◇ D55 E10 F15 G30 ◇

GRASSIN, J. *Δ^1_1-good inductive definitions over the continuum* ◇ D55 D70 E15 ◇

HARRINGTON, L.A. & SHORE, R.A. *Definable degrees and automorphisms of $\mathscr{D}$* ◇ D30 D55 ◇

KECHRIS, A.S. *A note on Wadge degrees* ◇ D30 D55 E15 E60 ◇

KECHRIS, A.S. & MARTIN, D.A. & MOSCHOVAKIS, Y.N. (EDS.) *Cabal seminar 77 - 79* ◇ D55 D97 E15 E60 E97 ◇

KECHRIS, A.S. *Forcing with Δ perfect trees and minimal Δ-degrees* ◇ D30 D55 D75 E15 E40 E60 ◇

KECHRIS, A.S. *Homogeneous trees and projective scales* ◇ D55 E05 E15 E60 ◇

KECHRIS, A.S. *Souslin cardinals, κ-Souslin sets and the scale property in the hyperprojective hierarchy* ◇ D55 E15 E60 ◇

LERMAN, M. *On recursive linear orderings* ◇ C57 C65 D30 D45 D55 E07 ◇

MAASS, W. *A countable basis for Σ^1_2 sets and recursion theory on $\aleph_1$* ◇ D55 D60 ◇

MARTIN, D.A. *Π^1_2 monotone inductive definitions* ◇ D55 D70 D75 ◇

MOSCHOVAKIS, Y.N. *Ordinal games and playful models* ◇ D55 E15 E45 E60 ◇

MUNDICI, D. *Ergodic undefinability in set theory and recursion theory* ◇ D55 E47 E75 ◇

NORMANN, D. *Countable functionals and the projective hierarchy* ◇ D55 D65 ◇

POIZAT, B. *Degres de definissabilite arithmetique des generiques (English summary)* ◇ C25 C40 D30 D55 ◇

STEEL, J.R. *Closure properties of pointclasses* ◇ D55 D65 D75 E15 E60 ◇

STEEL, J.R. *Determinateness and the separation property* ◇ D55 D75 E15 E60 ◇

TANG, A. *Wadge reducibility and Hausdorff difference hierarchy in Pω* ◇ D30 D55 E15 ◇

YASUDA, Y. *On the existence of Cohen extensions and Σ^1_3 predicates* ◇ D55 E15 E40 ◇

YASUDA, Y. *On the existence of Cohen extensions and Σ^1_3 predicates I* ◇ D55 E15 E40 ◇

1982

ARSLANOV, M.M. *A hierarchy of degrees of unsolvability (Russian)* ◇ D30 D55 ◇

BLAIR, H.A. *The recursion-theoretic complexity of the semantics of predicate logic as a programming language* ◇ B75 D35 D55 ◇
BUDINAS, B.L. *On the selector principle and the analytic definability of constructible sets (Russian)* ◇ D30 D55 E15 E35 E45 ◇
BURGESS, J.P. *What are R-sets?* ◇ D55 D70 E15 ◇
CENZER, D. & MAULDIN, R.D. *On the Borel class of the derived set operator* ◇ D55 E15 ◇
DAVID, R. *Δ^1_3 reals* ◇ D55 E35 E45 E55 ◇
DAVID, R. *A very absolute Π^1_2 real singleton* ◇ D55 E15 E40 E45 E55 ◇
DAVID, R. *Some applications of Jensen's coding theorem* ◇ C62 D55 E40 E45 ◇
HAY, L. & MILLER, DOUGLAS E. *The Addison game played backwards: index sets in topology* ◇ D55 D80 E75 ◇
HOROWITZ, B.M. *An isomorphism type of arithmetically productive sets* ◇ D25 D55 ◇
HOROWITZ, B.M. *Arithmetical analogues of productive and universal sets* ◇ D25 D55 ◇
ISHMUKHAMETOV, SH.T. *Families of recursively enumerable sets (Russian)* ◇ D25 D55 ◇
JAYNE, J.E. & ROGERS, C.A. *The invariance of the absolute Borel classes* ◇ D55 E15 E75 ◇
KANOVEJ, V.G. *On N.N. Luzin's problems on the embeddability and decomposition of projective sets (Russian)* ◇ D55 E15 E35 E55 ◇
KECHRIS, A.S. *Effective Ramsey theorems in the projective hierarchy* ◇ D55 E05 E15 E60 ◇
KLETTE, R. *A few results on the complexity of classes of identifiable recursive function sets* ◇ D20 D55 ◇
LERMAN, M. & ROSENSTEIN, J.G. *Recursive linear orderings* ◇ C57 C65 D45 D55 E07 ◇
LOUVEAU, A. *Borel sets and the analytical hierarchy* ◇ D55 E15 E60 ◇
LOUVEAU, A. *La classification de Wadge des ensembles boreliens* ◇ D30 D55 E15 E60 ◇
MAITRA, A. *An effective selection theorem* ◇ D55 E15 ◇
MARTIN, D.A. & MOSCHOVAKIS, Y.N. & STEEL, J.R. *The extent of definable scales* ◇ D55 D70 E15 E60 ◇
MILLER, DOUGLAS E. *Index sets and Boolean operations* ◇ D25 D45 D55 ◇
NORMANN, D. *Nonobtainable continuous functionals* ◇ D55 ◇
ROSENBERG, R. *Recursively enumerable images of arithmetic sets* ◇ D25 D55 F30 ◇
SELIVANOV, V.L. *On the index sets in the Kleene-Mostowski hierarchy (Russian)* ◇ D25 D45 D55 ◇
SHORE, R.A. *Finitely generated codings and the degrees r.e. in a degree d* ◇ D30 D55 E40 ◇
SIMPSON, S.G. *Four test problems in generalized recursion theory* ◇ D25 D55 D60 E15 ◇
SORBI, A. *Σ^n_0-equivalence relations* ◇ D55 F30 ◇
WEITKAMP, G. *Analytic sets having incomparable Kleene degrees* ◇ D30 D55 D65 E15 ◇
WEITKAMP, G. *Iterating the superjump along definable prewellorderings* ◇ D55 D60 D65 ◇

1983

ADAMOWICZ, Z. *Perfect set theorems for Π^1_2 in the universe without choice* ◇ D55 E15 E25 E40 E45 E55 ◇
BELYAKIN, N.V. *A means of modeling a classical second-order arithmetic (Russian)* ◇ B15 C62 D55 F35 ◇
BUDINAS, B.L. *The selector principle for analytic equivalence relations does not imply the existence of an A_2 well ordering of the continuum (Russian)* ◇ D55 E15 E35 ◇
BUECHI, J.R. *State-strategies for games in $F_{\sigma\delta} \cap G_{\delta\sigma}$* ◇ D05 D55 E15 E60 ◇
BURGESS, J.P. *Classical hierarchies from a modern standpoint I: C-sets. II R-sets* ◇ D55 D70 E15 ◇
BURGESS, J.P. & LOCKHART, R.A. *Classical hierarchies from a modern standpoint III: BP-sets* ◇ D55 E15 ◇
BURGIN, M.S. *Inductive Turing machines (Russian)* ◇ D10 D25 D55 ◇
CENZER, D. & MAULDIN, R.D. *On the Borel class of the derived set operator. II* ◇ D55 E15 ◇
DEMUTH, O. *Arithmetic complexity of differentiation in constructive mathematics (Russian)* ◇ D55 F60 ◇
FREMLIN, D.H. & HANSELL, R.W. & JUNNILA, H.J.K. *Borel functions of bounded class* ◇ D55 E15 E35 E50 E55 E65 ◇
GUASPARI, D. *Trees, norms and scales* ◇ D55 E15 E60 E98 ◇
HAREL, D. *Recurring dominoes: making the highly undecidable highly understandable* ◇ B25 B75 D15 D55 ◇
HERRMANN, E. *Orbits of hyperhypersimple sets and the lattice of Σ^0_3 sets* ◇ D25 D55 ◇
ISHMUKHAMETOV, SH.T. *On index sets of classes of differences (Russian)* ◇ D55 ◇
KANOVEJ, V.G. *A generalization of P.S. Novikov's theorem on cross sections of Borel sets (Russian)* ◇ D55 E15 ◇
KANOVEJ, V.G. *Structure of the constituents of Π^1_1-sets (Russian)* ◇ D55 E15 E35 ◇
KECHRIS, A.S. & MARTIN, D.A. & MOSCHOVAKIS, Y.N. (EDS.) *Cabal seminar 79-81. Proceedings, Caltech-UCLA logic seminar 1979-81* ◇ D55 D97 E15 E60 E97 ◇
KECHRIS, A.S. & MARTIN, D.A. & SOLOVAY, R.M. *Introduction to Q-theory* ◇ D55 E15 E45 E60 ◇
LOUVEAU, A. *Some results in the Wadge hierarchy of Borel sets* ◇ D55 E15 ◇
MARTIN, D.A. *The largest countable this, that, and the other* ◇ D55 E15 E60 ◇
MASSERON, M. *Rungs and trees* ◇ D55 E05 F15 ◇
MILLER, A.W. *On the Borel classification of the isomorphism class of a countable model* ◇ C15 D55 E15 ◇
MOHRHERR, J. *A conjecture of Ershov for a relative hierarchy fails near $\mathfrak{D}$* ◇ D55 ◇

NORMANN, D. *General type-structures of continuous and countable functionals* ◇ D55 D65 D75 F35 ◇
ODIFREDDI, P. *Forcing and reducibilities* ◇ D30 D55 E40 F30 ◇
ODIFREDDI, P. *Forcing and reducibilities. II: Forcing in fragments of analysis. III: Forcing in fragments of set theory* ◇ D30 D55 E40 ◇
ODIFREDDI, P. *On the first order theory of the arithmetical degrees* ◇ C65 D30 D55 ◇
RAGAZ, M. *Die Unentscheidbarkeit der einstelligen unendlichwertigen Praedikatenlogik* ◇ B50 D35 D55 ◇
RAISONNIER, J. & STERN, J. *Mesurabilite et propriete de Baire (English summary)* ◇ D55 E15 ◇
SELIVANOV, V.L. *Effective analogues of A-, B- and C-sets and their application to index sets (Russian)* ◇ D25 D55 E15 ◇
SELIVANOV, V.L. *Hierarchies of hyperarithmetical sets and functions (Russian)* ◇ D55 ◇
SIMPSON, S.G. & WEITKAMP, G. *High and low Kleene degrees of coanalytic sets* ◇ D30 D55 D65 ◇
STAIGER, L. *Finite state ω-languages* ◇ D05 D55 ◇
STEEL, J.R. *Scales on Σ^1_1 sets* ◇ D55 E15 ◇
VOGEL, HELMUT *On a relationship between countable functionals and projective trees* ◇ D55 D65 ◇
WEISS, T. *Projective sets (Polish)* ◇ D55 E15 E98 ◇

1984

ADAMOWICZ, Z. *A generalization of the Shoenfield theorem on Σ^1_2 sets* ◇ D55 E40 E45 E75 ◇
BLAIR, H.A. *The intractability of validity in logic programming and dynamic logic* ◇ B75 C75 D35 D55 ◇
CENZER, D. *Monotone reducibility and the family of infinite sets* ◇ D15 D55 ◇
CLOTE, P. *A recursion theoretic analysis of the clopen Ramsey theorem* ◇ D45 D55 E05 ◇
HERRMANN, E. *Definable structures in the lattice of recursively enumerable sets* ◇ D25 D55 ◇
HOEVEN VAN DER, G.F. & MOERDIJK, I. *On choice sequences determined by spreads* ◇ D30 D55 F35 F50 G30 ◇
JOCKUSCH JR., C.G. & SHORE, R.A. *Pseudo jump operators II: Transfinite iterations, hierarchies and minimal covers* ◇ D25 D30 D55 ◇
RAISONNIER, J. *A mathematical proof of S. Shelah's theorem on the measure problem and related results* ◇ D55 E05 E15 E75 ◇
REGAN, K.W. *Arithmetical degrees of index sets for complexity cases* ◇ D15 D55 ◇
ROUHONEN, K. *On machine characterization of nonrecursive hierarchies* ◇ D10 D55 ◇
SAMI, R.L. *On Σ^1_1 equivalence relations with Borel classes of bounded rank* ◇ D55 E10 E15 E45 ◇
SCHILLING, K. *On absolutely Δ^1_2 operations* ◇ D55 E15 E75 ◇
SELIVANOV, V.L. *Index sets in the hyperarithmetical hierarchy (Russian)* ◇ D55 ◇
VETULANI, Z. *Ramified analysis and the minimal β-models of higher order arithmetics* ◇ C62 D55 E45 F35 F65 ◇
WELCH, P. *On Σ^1_3* ◇ D55 E15 E55 ◇

1985

ARTEMOV, S.N. *Nonarithmeticity of truth predicate logics of provability (Russian)* ◇ B45 D35 D55 F30 ◇
BECKER, H. *A property equivalent to the existence of scales* ◇ D55 E15 E60 ◇
CLOTE, P. *Applications of the low-basis theorem in arithmetic* ◇ C62 D20 D30 D55 F30 ◇
CLOTE, P. *On recursive trees with a unique infinite branch* ◇ D30 D55 ◇
GRIFFOR, E.R. *An application of Π^1_2-logic to descriptive set theory* ◇ D55 E15 ◇
HANSELL, R.W. & JAYNE, J.E. & ROGERS, C.A. *Separation of K-analytic sets* ◇ D55 E15 ◇
HAREL, D. *Recurring dominoes: Making the highly undecidable highly understandable* ◇ B25 B75 D05 D15 D55 ◇
KECHRIS, A.S. *Determinacy and the structure of L(R)* ◇ D55 E05 E15 E45 E60 ◇
KECHRIS, A.S. & SOLOVAY, R.M. *On the relative consistency strength of determinacy hypotheses* ◇ D55 E15 E35 E45 E60 ◇
KHOLSHCHEVNIKOVA, N.N. *Uncountable R- and N-sets (Russian)* ◇ D55 D70 E15 ◇
KOLAITIS, P.G. *Canonical forms and hierarchies in generalized recursion theories* ◇ D55 D65 D70 ◇
LERMAN, M. *Upper bounds for the arithmetical degrees* ◇ D30 D55 ◇
LOUVEAU, A. *Recursivity and capacity theory* ◇ D55 E15 ◇
MANSFIELD, R. & WEITKAMP, G. *Recursive aspects of descriptive set theory* ◇ D55 E15 E98 ◇
MARTIN, D.A. *A purely inductive proof of Borel determinacy* ◇ D55 E15 E60 ◇
MILLAR, T.S. *Decidable Ehrenfeucht theories* ◇ B25 C15 C57 C98 D30 D35 D55 ◇
RAISONNIER, J. & STERN, J. *The strength of measurability hypotheses* ◇ D55 E05 E15 ◇
SELIVANOV, V.L. *The Ershov hierarchy (Russian)* ◇ D55 ◇
USPENSKIJ, V.A. *The contribution of N.N.Luzin to the descriptive theory of sets and functions: concepts, problems, predictions (Russian)* ◇ A10 D55 E15 ◇
WEITKAMP, G. *On the existence and recursion theoretic properties of Σ^1_n-generic sets of reals* ◇ D55 D65 E15 E40 ◇

D60 Recursion theory on ordinals, admissible sets, etc.

1960

TAKEUTI, G. *On the recursive functions of ordinal numbers* ◇ D60 ◇

1961

MACHOVER, M. *The theory of transfinite recursion* ◇ C75 D60 E45 ◇

1962

KINO, A. & TAKEUTI, G. *A note on predicates of ordinal numbers* ◇ D60 ◇

KINO, A. & TAKEUTI, G. *On hierarchies of predicates of ordinal numbers* ◇ D55 D60 F15 ◇

1963

TUGUE, T. *On the partial recursive functions of ordinal numbers* ◇ D60 ◇

WANG, HAO *Computation* ◇ D20 D25 D30 D35 D60 ◇

1964

KRIPKE, S.A. *Transfinite recursion on admissible ordinals I,II* ◇ D60 ◇

TUGUE, T. *On the partial recursive functions of ordinal numbers* ◇ D60 ◇

1965

KREISEL, G. & SACKS, G.E. *Metarecursive sets* ◇ D30 D55 D60 ◇

KREISEL, G. *Model-theoretic invariants: applications to recursive and hyperarithmetic operations* ◇ C40 C62 D55 D60 ◇

TAKEUTI, G. *Recursive functions and arithmetical functions of ordinal numbers* ◇ D60 E10 E45 E55 ◇

TAKEUTI, G. *Transcendence of cardinals* ◇ D60 E35 E45 E55 E65 ◇

1966

PLATEK, R.A. *Foundations of recursion theory* ◇ D60 D75 ◇

SACKS, G.E. *Metarecursively enumerable sets and admissible ordinals* ◇ D30 D55 D60 ◇

SACKS, G.E. *Post's problem, admissible ordinals, and regularity* ◇ D30 D55 D60 ◇

1967

BARWISE, J. *Infinitary logic and admissible sets* ◇ C70 D60 ◇

OWINGS JR., J.C. *Recursion, metarecursion, and inclusion* ◇ D25 D55 D60 ◇

SACKS, G.E. *Measure-theoretic uniformity in recursion theory and set theory (summary of results)* ◇ D30 D55 D60 E25 E35 E40 E50 ◇

SACKS, G.E. *Metarecursion theory* ◇ D30 D55 D60 ◇

1968

DRISCOLL JR., G.C. *Metarecursively enumerable sets and their metadegrees* ◇ D30 D60 ◇

FRIEDMAN, H.M. & JENSEN, R.B. *Note on admissible ordinals* ◇ C70 D60 ◇

KREISEL, G. *Choice of infinitary languages by means of definability criteria: generalized recursion theory* ◇ C40 C70 C75 D60 D75 ◇

TAKAHASHI, MOTO-O *Recursive functions of ordinal numbers and Levy's hierarchy* ◇ D55 D60 E10 E45 E47 ◇

1969

BARWISE, J. *Applications of strict Π_1^1 predicates to infinitary logic* ◇ C40 C70 D60 ◇

HIRANO, M. *Some definitions for recursive functions of ordinal numbers* ◇ D60 ◇

OWINGS JR., J.C. *Π_1^1-sets, ω-sets, and metacompleteness* ◇ D55 D60 ◇

SACKS, G.E. *Measure-theoretic uniformity in recursion theory and set theory* ◇ D30 D55 D60 E15 E25 E35 E40 E50 ◇

1970

MACHTEY, M. *Admissible ordinals and intrinsic consistency* ◇ D60 ◇

MACHTEY, M. *Admissible ordinals and lattices of α-r.e. sets* ◇ D60 ◇

OHASHI, K. *On a problem of G.E.Sacks* ◇ D60 ◇

OWINGS JR., J.C. *The metarecursively enumerable sets, but not the Π_1^1-sets, can be enumerated without repetition* ◇ D55 D60 ◇

1971

BARWISE, J. & GANDY, R.O. & MOSCHOVAKIS, Y.N. *The next admissible set* ◇ C40 C62 D55 D60 D70 E45 E47 ◇

FUKUYAMA, M. *Some concepts of recursiveness and admissible ordinals* ◇ D60 ◇

GRILLIOT, T.J. *On effectively discontinuous type-2 objects* ◇ D60 D65 ◇

JENSEN, R.B. & KARP, C.R. *Primitive recursive set functions* ◇ D60 D65 E45 E47 ◇

KREISEL, G. *Some reasons for generalizing recursion theory* ◇ A05 D30 D60 D65 D70 D75 ◇

MOSCHOVAKIS, Y.N. *Predicative classes* ◇ C62 D55 D60 D65 D70 E70 ◇

OWINGS JR., J.C. *A splitting theorem for simple Π_1^1 sets* ◇ D30 D55 D60 ◇

RICHTER, W.H. *Recursively Mahlo ordinals and inductive definitions* ◇ D60 D65 D70 E55 F15 ◇
SACKS, G.E. *F-recursiveness* ◇ D30 D55 D60 D75 E40 E45 ◇
SACKS, G.E. *On the reducibility of* Π^1_1 *sets* ◇ D30 D55 D60 E40 ◇
SACKS, G.E. *Recursion in objects of finite type* ◇ D60 D65 ◇

1972

ACZEL, P. & RICHTER, W.H. *Inductive definitions and analogues of large cardinals* ◇ D60 D70 E55 ◇
FUKUYAMA, M. *Note on the recursion theory over admissible sets* ◇ D60 ◇
GRILLIOT, T.J. *Omitting types; applications to recursion theory* ◇ C07 C62 D30 D55 D60 D75 ◇
LERMAN, M. *On suborderings of the* α*-recursively enumerable* α*-degrees* ◇ D30 D60 ◇
LERMAN, M. & SACKS, G.E. *Some minimal pairs of* α*-recursively enumerable degrees* ◇ D30 D60 ◇
METAKIDES, G. α*-degrees of* α*-theories* ◇ D30 D60 ◇
NADEL, M.E. *Some Loewenheim-Skolem results for admissible sets* ◇ C62 C70 D60 E55 ◇
SACKS, G.E. & SIMPSON, S.G. *The* α*-finite injury method* ◇ D30 D60 ◇
SHORE, R.A. *Minimal* α*-degrees* ◇ D30 D60 ◇

1973

DAWSON JR., J.W. *Ordinal definability in the rank hierarchy* ◇ D60 E35 E45 E55 ◇
FUKUYAMA, M. *An introduction to recursion theory on admissible sets and admissible ordinals* ◇ D60 ◇
LERMAN, M. *Admissible ordinals and priority arguments* ◇ D60 ◇
LERMAN, M. & SIMPSON, S.G. *Maximal sets in* α*-recursion theory* ◇ D60 ◇
MACINTYRE, J.M. *Minimal* α*-recursion theoretic degrees* ◇ D30 D60 ◇
MACINTYRE, J.M. *Noninitial segments of the* α*-degrees* ◇ D30 D60 ◇
STAVI, J. *A converse of the Barwise completeness theorem* ◇ C70 D60 D70 ◇

1974

ACZEL, P. *An axiomatic approach to recursion on admissible ordinals and the Kreisel-Sacks construction of meta-recursion theory* ◇ D60 D75 ◇
ACZEL, P. & RICHTER, W.H. *Inductive definitions and reflecting properties of admissible ordinals* ◇ D60 D70 E47 E55 ◇
BARWISE, J. *Admissible sets over models of set theory* ◇ C62 D60 E30 ◇
CENZER, D. *Ordinal recursion and inductive definitions* ◇ D60 D65 D70 E45 ◇
CENZER, D. *The boundedness principle in ordinal recursion* ◇ D60 D70 ◇
CHONG, C.T. *Almost local non-*α*-recursiveness* ◇ D30 D60 ◇
FUKUYAMA, M. & NAGAOKA, K. *Strong representability of* α*-recursive functions* ◇ C70 D60 ◇
GANDY, R.O. *Inductive definitions* ◇ D60 D65 D70 E55 ◇
GREEN, J. Σ_1*-compactness for next admissible sets* ◇ C62 C70 D60 ◇
GRILLIOT, T.J. *Model theory for dissecting recursion theory* ◇ C55 C70 C75 C80 D60 D75 ◇
HARRINGTON, L.A. *The superjump and the first recursively Mahlo ordinal* ◇ D60 D65 E55 ◇
LEGGETT, A. *Maximal* α*-r.e. sets and their complements* ◇ D60 ◇
LERMAN, M. *Least upper bounds for minimal pairs of* α*-r.e.* α*-degrees* ◇ D30 D60 ◇
LERMAN, M. *Maximal* α*-r.e. sets* ◇ D60 ◇
MACHTEY, M. *Minimal degrees in generalized recursion theory* ◇ D30 D60 ◇
NADEL, M.E. *More Loewenheim-Skolem results for admissible sets* ◇ C62 C70 D60 ◇
RUTKOWSKI, A. *On the algebraic approach to the notion of* α*-recursive functions* ◇ D60 ◇
SHORE, R.A. Σ_n *sets which are* Δ_n*-incomparable (uniformly)* ◇ D30 D55 D60 E40 E47 ◇
SHORE, R.A. *Cohesive sets: countable and uncountable* ◇ D60 ◇
SIMPSON, S.G. *Degree theory on admissible ordinals* ◇ D30 D60 ◇
SIMPSON, S.G. *Post's problem for admissible sets* ◇ D30 D60 ◇

1975

BARWISE, J. *Admissible sets and the interaction of model theory, recursion theory and set theory* ◇ C70 C98 D55 D60 D70 E30 E35 E45 E47 ◇
BARWISE, J. *Admissible sets and structures. An approach to definability theory* ◇ B98 C40 C70 C98 D60 D98 E30 E98 ◇
DEVLIN, K.J. & PARIS, J.B. *Certain sequences of ordinals* ◇ D60 E10 E45 E55 E60 ◇
HARRINGTON, L.A. & KECHRIS, A.S. *On characterizing Spector classes* ◇ C62 D55 D60 D65 D75 E45 E55 E60 ◇
RICHTER, W.H. *The least* Σ^1_2*- and* Π^1_2*-reflecting ordinals* ◇ D55 D60 D70 E45 E47 E55 ◇
SHORE, R.A. *Splitting an* α*-recursively enumerable set* ◇ D30 D60 ◇
SHORE, R.A. *The irregular and non-hyperregular* α*-r.e. degrees* ◇ D30 D60 ◇

1976

ABRAMSON, F.G. & SACKS, G.E. *Uncountable Gandy ordinals* ◇ D60 D70 ◇
CHONG, C.T. *An* α*-finite injury method of the unbounded type* ◇ D30 D60 ◇
CHONG, C.T. & LERMAN, M. *Hyperhypersimple* α*-r.e. sets* ◇ D60 ◇
CHONG, C.T. *Minimal upper bounds for ascending sequences of* α*-recursively enumerable degrees* ◇ D30 D60 ◇
FUKUYAMA, M. *On the representability of* α*-recursive functions* ◇ D60 ◇

LEGGETT, A. & SHORE, R.A. *Types of simple α-recursively enumerable sets* ◇ D60 ◇
LERMAN, M. *Congruence relations, filters, ideals, and definability in lattices of α-recursively enumerable sets* ◇ D60 ◇
LERMAN, M. *Ideals of generalized finite sets in lattices of α-recursively enumerable sets* ◇ D60 ◇
LERMAN, M. *Types of simple α-recursively enumerable sets* ◇ D60 ◇
LOWENTHAL, F. *Measure and categoricity in α-recursion* ◇ D60 E75 ◇
SACKS, G.E. *Countable admissible ordinals and hyperdegrees* ◇ D30 D55 D60 ◇
SCHUETTE, K. *Primitiv-rekursive Ordinalzahlfunktionen* ◇ D60 F15 ◇
SHORE, R.A. *On the jump of an α-recursively enumerable set* ◇ D30 D60 ◇
SHORE, R.A. *The recursively enumerable α-degrees are dense* ◇ D30 D60 ◇
STAHL, S.H. *A hierarchy on the class of primitive recursive ordinal functions* ◇ D55 D60 ◇

1977

BARWISE, J. *On Moschovakis closure ordinals* ◇ D60 D65 D70 ◇
CARPENTER, A.J. *For a countable admissible ordinal α, the α-recursive functions are exactly the α-definite functions* ◇ D60 ◇
CHONG, C.T. *A recursion-theoretic characterization of constructible reals* ◇ D60 E40 E45 ◇
CHONG, C.T. *The minimal α-degree problem* ◇ D30 D60 ◇
DEVLIN, K.J. *Constructibility* ◇ D60 E35 E45 E65 E98 ◇
FENSTAD, J.E. *Between recursion theory and set theory* ◇ D60 D65 D75 ◇
FRIEDMAN, S.D. & SACKS, G.E. *Inadmissible recursion theory* ◇ D60 E65 ◇
GRANT, P.W. *Strict Π^1_1-predicates on countable and cofinality ω transitive sets* ◇ C15 C40 C70 D55 D60 D70 E47 E60 ◇
HINMAN, P.G. *A survey of finite-type recursion* ◇ D55 D60 D65 D98 ◇
JACOBS, B.E. *On generalized computational complexity* ◇ D15 D60 ◇
MAASS, W. *On minimal pairs and minimal degrees in higher recursion theory* ◇ D30 D60 D65 ◇
MACINTYRE, J.M. *Transfinite extensions of Friedberg's completeness criterion* ◇ D30 D55 D60 ◇
MAKKAI, M. *Admissible sets and infinitary logic* ◇ C70 C98 D55 D60 D70 D98 ◇
NYBERG, A.M. *Inductive operators on resolvable structures* ◇ C70 D60 D70 D75 ◇
RESSAYRE, J.-P. *Models with compactness properties relative to an admissible language* ◇ C50 C55 C70 C80 D55 D60 E15 ◇
SACKS, G.E. *RE sets higher up* ◇ D25 D60 D65 ◇
SHORE, R.A. *α-recursion theory* ◇ D30 D60 E45 ◇
SIMPSON, S.G. *Basis theorems and countable admissible ordinals* ◇ D55 D60 E15 ◇
SREBRNY, M. *Relatively constructible transitive models* ◇ C62 D55 D60 E10 E40 E45 E55 ◇
STAHL, S.H. *Primitive recursive ordinal functions with added constants* ◇ D60 ◇

1978

BARWISE, J. *Monotone quantifiers and admissible sets* ◇ C70 C80 D60 D70 ◇
CARPENTER, A.J. *For any admissible ordinal α, Kripke's α-recursive functions are exactly Machover's α-recursive functions* ◇ D60 ◇
CUTLAND, N.J. *On the form of countable admissible ordinals* ◇ C70 D60 ◇
DELFINO, V. *The Victoria Delfino problems* ◇ D55 D60 D65 D70 E15 E60 ◇
FRIEDMAN, S.D. *An introduction to β-recursion theory* ◇ D60 ◇
FRIEDMAN, S.D. *Negative solutions to Post's problem I* ◇ D30 D60 ◇
GRASSIN, J. *Definitions inductives monotones sur le continu, dont les composantes sont denombrables (English summary)* ◇ D55 D60 D70 E15 ◇
HINMAN, P.G. *Recursion-theoretic hierarchies* ◇ D55 D60 D65 D70 D98 E15 ◇
JACOBS, B.E. *The α-union theorem and generalized primitive recursion* ◇ D15 D55 D60 ◇
LEGGETT, A. *α-degrees of maximal α-r.e. sets* ◇ D30 D60 ◇
LERMAN, M. *Lattices of α-recursively enumerable sets* ◇ B25 D60 ◇
LERMAN, M. *On elementary theories of some lattices of α-recursively enumerable sets* ◇ B25 D60 E45 ◇
MAASS, W. *Contributions to α- and β-recursion theory* ◇ D25 D30 D60 E45 ◇
MAASS, W. *Fine structure of the constructible universe in α- and β-recursion theory* ◇ D55 D60 E45 E65 ◇
MAASS, W. *High α-recursively enumerable degrees* ◇ D30 D60 ◇
MAASS, W. *Inadmissibility, tame r.e. sets and the admissible collapse* ◇ D60 ◇
MAASS, W. *The uniform regular set theorem in α-recursion theory* ◇ D30 D60 E45 ◇
NORMANN, D. & STOLTENBERG-HANSEN, V. *A non-adequate admissible set with a good degree-structure* ◇ D30 D60 ◇
ODIFREDDI, P. *Ricorsivita su ordinali ammissibili* ◇ D60 ◇
PAOLA DI, R.A. *The operator gap theorem in α-recursion theory* ◇ D15 D60 ◇
SHORE, R.A. *On the $\forall\exists$-sentences of α-recursion theory* ◇ D30 D60 ◇
SHORE, R.A. *Some more minimal pairs of α-recursively enumerable degrees* ◇ D30 D60 ◇
SIMPSON, S.G. *Short course on admissible recursion theory* ◇ D30 D60 ◇
SOLOVAY, R.M. *Hyperarithmetically encodable sets* ◇ D55 D60 E05 E40 E45 E55 ◇
STARK, W.R. *A forcing approach to the strict-Π^1_1 reflection and strict-$\Pi^1_1 = \Sigma^0_1$* ◇ C70 D60 E40 ◇

STOLTENBERG-HANSEN, V. *Weakly inadmissible recursion theory* ◇ D60 D75 ◇
TANAKA, H. *Recursion theory in analytical hierarchy* ◇ D25 D55 D60 E60 ◇

1979

ABRAMSON, F.G. *Σ_1-separation* ◇ C62 C70 D60 E45 E47 ◇
ABRAMSON, F.G. *Sacks forcing does not always produce a minimal upper bound* ◇ D30 D60 E40 ◇
BREDLAU, C.E. *Admissible sets and recursive equivalence types* ◇ D50 D60 ◇
CHONG, C.T. *Σ_n-cofinalities of J_α* ◇ D60 E45 E47 ◇
CHONG, C.T. *Cones of degrees* ◇ D30 D60 E45 ◇
CHONG, C.T. *Generic sets and minimal α-degrees* ◇ D30 D60 E40 ◇
CHONG, C.T. *Major subsets of α-recursively enumerable sets* ◇ D60 ◇
FRIEDMAN, S.D. *β-recursion theory* ◇ D60 D65 ◇
FRIEDMAN, S.D. *HC of an admissible set* ◇ C70 D60 E30 ◇
GOSTANIAN, R. & HRBACEK, K. *A new proof that $\pi^1_1 < \sigma^1_1$* ◇ D60 D70 E45 ◇
GOSTANIAN, R. *The next admissible ordinal* ◇ D60 E45 ◇
JACOBS, B.E. *α-naming and α-speedup theorems* ◇ D15 D60 ◇
JACOBS, B.E. *α-speedable and non α-speedable sets* ◇ D15 D60 ◇
MAASS, W. *On α- and β-recursively enumerable degrees* ◇ D30 D60 ◇
STOLTENBERG-HANSEN, V. *Finite injury arguments in infinite computation theories* ◇ D25 D30 D60 D75 ◇
TUGUE, T. *Generalized recursion theory (Japanese)* ◇ D60 D65 D75 ◇
YANG, DONGPING *α-operator gap theorem (Chinese)* ◇ D15 D60 ◇
YANG, DONGPING *The existence of nonmitotic α-recursively enumerable sets (Chinese) (English summary)* ◇ D60 ◇

1980

CHONG, C.T. *Degree theory: from ω to singular cardinals* ◇ D30 D60 E45 ◇
CHONG, C.T. *Rich sets* ◇ D30 D60 ◇
FRIEDMAN, S.D. *Post's problem without admissibility* ◇ D30 D60 E05 E65 ◇
HOMER, S. *Two splitting theorems for β-recursion theory* ◇ D60 ◇
SHORE, R.A. *Some constructions in α-recursion theory* ◇ D30 D60 ◇
STOLTENBERG-HANSEN, V. *On computational complexity in weakly admissible structures* ◇ D15 D60 D75 ◇
YANG, DONGPING *α-splitting theorem restricted to Δ^0_2 sets (Chinese)* ◇ D55 D60 ◇

1981

FITTING, M. *Fundamentals of generalized recursion theory* ◇ D60 D65 D70 D75 ◇
FRIEDMAN, S.D. *Natural α-re degrees* ◇ D30 D60 ◇
FRIEDMAN, S.D. *Negative solutions to Post's problem II* ◇ D30 D60 ◇
FRIEDMAN, S.D. *Uncountable admissibles II: Compactness* ◇ C70 D60 E45 E47 ◇
GUERRERIO, G. *Sulle generalizzazione della ricorsivita* ◇ D60 D65 ◇
HOMER, S. & JACOBS, B.E. *Degrees of non α-speedable sets* ◇ D15 D60 ◇
MAASS, W. *A countable basis for Σ^1_2 sets and recursion theory on $\aleph_1$* ◇ D55 D60 ◇
MAASS, W. *Recursively invariant β-recursion theory* ◇ D30 D60 ◇
PAOLA DI, R.A. *A lift of a theorem of Friedberg: A Banach-Mazur functional that coincides with no α-recursive functional on the class of α-recursive functions* ◇ D60 ◇

1982

CHONG, C.T. *Double jumps of minimal degrees over cardinals* ◇ D30 D60 E45 ◇
CHONG, C.T. *Global and local admissibility* ◇ D60 ◇
FRIEDMAN, S.D. *The Turing degrees and the metadegrees have isomorphic cones* ◇ D30 D60 ◇
FRIEDMAN, S.D. *Uncountable admissibles I: Forcing* ◇ C70 D60 E40 ◇
KRANAKIS, E. *Invisible ordinals and inductive definitions* ◇ D60 D70 E45 E47 E55 ◇
KRANAKIS, E. *Reflection and partition properties of admissible ordinals* ◇ D60 E05 E45 E55 ◇
RESSAYRE, J.-P. *Bounding generalized recursive functions of ordinals by effective functors: a complement to the Girard theorem* ◇ D60 F15 ◇
SIMPSON, S.G. *Four test problems in generalized recursion theory* ◇ D25 D55 D60 E15 ◇
WEITKAMP, G. *Iterating the superjump along definable prewellorderings* ◇ D55 D60 D65 ◇
WIELE VAN DE, J. *Recursive dilators and generalized recursions* ◇ D60 D65 F15 F35 ◇

1983

CHONG, C.T. & FRIEDMAN, S.D. *Degree theory on $\aleph_\omega$* ◇ D30 D60 ◇
CHONG, C.T. *Global and local admissibility. II. Major subsets and automorphisms* ◇ D30 D60 ◇
CHONG, C.T. *Hyperhypersimple supersets in admissible recursion theory* ◇ D60 ◇
ERSHOV, YU.L. *The Σ-enumeration principle (Russian)* ◇ C62 C70 D60 E30 E65 ◇
FRIEDMAN, S.D. *Some recent developments in higher recursion theory* ◇ D60 D70 ◇
HOMER, S. *Intermediate β-r.e. degrees and the half-jump* ◇ D30 D60 ◇
HOMER, S. & SACKS, G.E. *Inverting the half-jump* ◇ D30 D60 ◇
KIROUSIS, L.M. *A selection theorem* ◇ D60 D70 ◇
KRANAKIS, E. *Definable Ramsey and definable Erdoes ordinals* ◇ C30 D60 E05 E45 E55 ◇
PAOLA DI, R.A. *The basic theory of partial α-recursive operators* ◇ D60 ◇

1984

AMBOS-SPIES, K. *An extension of the non-diamond theorem in classical and α-recursion theory* ◇ D25 D60 ◇

BAETEN, J. *Filters and ultrafilters over definable subsets of admissible ordinals* ◇ C62 D60 E05 E10 E47 ◇

CHONG, C.T. *Minimal α-hyperdegrees* ◇ D60 ◇

CHONG, C.T. *Techniques of admissible recursion theory* ◇ D60 D98 ◇

GIRARD, J.-Y. & VAUZEILLES, J. *Les premiers recursivement inaccessible et Mahlo et la theorie des dilatateurs* ◇ D60 E47 E55 F05 F15 ◇

HUMBERSTONE, I.L. *Monadic representability of certain binary relations* ◇ C52 D60 E07 ◇

KAUFMANN, M. & KRANAKIS, E. *Definable ultrapowers and ultrafilters over admissible ordinals* ◇ C20 C62 D60 E05 E45 ◇

KRANAKIS, E. & PHILLIPS, I. *Partitions and homogeneous sets for admissible ordinals* ◇ D60 E05 E45 E47 E55 ◇

KRANAKIS, E. *Stepping up lemmas in definable partitions* ◇ D60 E05 E45 E47 E55 ◇

SHINODA, J. *Countable J_a^S-admissible ordinals* ◇ D60 D65 E10 E30 E45 ◇

YANG, DONGPING *On the embedding of α-recursive presentable lattices into the α-recursive degrees below* $0'$ ◇ D30 D60 ◇

1985

CHONG, C.T. *Recursion theory on strongly Σ_2-inadmissible ordinals* ◇ D60 ◇

DAVID, R. & FRIEDMAN, S.D. *Uncountable ZF-ordinals* ◇ C62 D60 E40 E45 ◇

DIETZFELBINGER, M. & MAASS, W. *Strong reducibilities in α- and β- recursion theory* ◇ D30 D60 ◇

FRIEDMAN, S.D. *Fine structure theory and its applications* ◇ D60 E45 ◇

GIRARD, J.-Y. & RESSAYRE, J.-P. *Elements de logique Π_n^1* ◇ D60 F15 F35 ◇

KOLAITIS, P.G. *Game quantification* ◇ C40 C70 C75 C98 D60 D65 D70 E60 ◇

KRANAKIS, E. *Definable partitions and reflection properties for regular cardinals* ◇ D60 E05 E55 ◇

KRANAKIS, E. *Definable partitions and the projectum* ◇ D60 E05 E45 E55 ◇

SHINODA, J. *Absolute type 2 objects* ◇ D60 D65 E40 E45 E55 ◇

SHINODA, J. *Countable J_a^S-admissible ordinals* ◇ D60 D65 ◇

D65 Higher-type and set recursion

1957

DAVIS, MARTIN D. *Computable functionals of arbitrary finite type* ◇ D65 ◇

KLEENE, S.C. *Recursive functionals of higher finite types* ◇ D65 ◇

1959

DAVIS, MARTIN D. *Computable functionals of arbitrary finite type* ◇ D65 ◇

KLEENE, S.C. *Countable functionals* ◇ D65 ◇

KLEENE, S.C. *Recursive functionals and quantifiers of finite types I* ◇ D30 D55 D65 ◇

KONDO, M. *Sur la nommabilite d'ensembles de type superieur* ◇ D55 D65 E15 ◇

KREISEL, G. *Interpretation of analysis by means of constructive functionals of finite types* ◇ D65 F10 F30 F35 F50 ◇

1960

TUGUE, T. *On predicates expressible in the 1-function quantifier forms in Kleene hierarchy with free variables of type 2* ◇ D55 D65 ◇

TUGUE, T. *Predicates recursive in a type-2 object and Kleene hierarchies* ◇ D55 D65 ◇

1961

KONDO, M. *Sur les domaines fondamentaux des fonctionelles de type transfini* ◇ D65 ◇

KREISEL, G. *Set theoretic problems suggested by the notion of potential totality* ◇ A05 C62 D20 D55 D65 ◇

1962

KLEENE, S.C. *λ-definable functionals of finite types* ◇ B40 D30 D55 D65 ◇

KLEENE, S.C. *Herbrand-Goedel-style recursive functionals of finite types* ◇ D65 ◇

KLEENE, S.C. *Turing-machine computable functionals of finite types I* ◇ D10 D30 D55 D65 ◇

KLEENE, S.C. *Turing-machine computable functionals of finite types II* ◇ D10 D30 D55 D65 ◇

SHOENFIELD, J.R. *The form of the negation of a predicate* ◇ D55 D65 ◇

1963

HARRISON, J. *Appendix C: Equivalence of the effective operations and the hereditarily continuous functionals* ◇ D65 F50 ◇

KLEENE, S.C. *Recursive functionals and quantifiers of finite types II* ◇ D30 D55 D65 ◇

1964

CLARKE, D.A. *Hierarchies of predicate of finite types* ◇ D55 D65 D70 ◇

CURRY, H.B. *Combinatory recursive objects of all finite types* ◇ B40 D65 ◇

GRZEGORCZYK, A. *Recursive objects in all finite types* ◇ D65 F10 ◇

1966

HANATANI, Y. *Calculabilite des fonctionelles recursives primitives de type fini sur les nombres naturels* ◇ D65 F10 F35 F50 ◇

HINMAN, P.G. *Ad astra per aspera: hierarchy schemata in recursive function theory* ◇ D55 D65 D70 E15 ◇

1967

GANDY, R.O. *Computable functionals of finite type I* ◇ D65 ◇

GANDY, R.O. *General recursive functionals of finite type and hierarchies of functions* ◇ D55 D65 ◇

MAHN, F.-K. *Zu den primitiv-rekursiven Funktionen ueber einem Bereich endlicher Mengen* ◇ D20 D65 ◇

MOSCHOVAKIS, Y.N. *Hyperanalytic predicates* ◇ D55 D65 ◇

RICHTER, W.H. *Constructive transfinite number classes* ◇ D65 D70 F15 ◇

SUZUKI, Y. *Applications of the theory of β-models`* ◇ C62 D65 ◇

1968

RICHTER, W.H. *Constructively accessible ordinal numbers* ◇ D65 D70 F15 ◇

SHOENFIELD, J.R. *A hierarchy based on a type two object* ◇ D55 D65 ◇

1969

GRILLIOT, T.J. *Hierarchies based on objects of finite types* ◇ D55 D65 ◇

GRILLIOT, T.J. *Selection functions for recursive functionals* ◇ D65 ◇

HINATA, S. & TUGUE, T. *A note on continuous functionals* ◇ D65 ◇

HINMAN, P.G. *Hierarchies of effective descriptive set theory* ◇ D55 D65 D70 E15 ◇

KLEENE, S.C. *Formalized recursive functionals and formalized realizability* ◇ D65 F35 F50 ◇

MOSCHOVAKIS, Y.N. *Abstract first order computability I,II* ◇ D65 D70 D75 ◇

1970

ACZEL, P. *Representability in some systems of second order arithmetic* ◇ D65 D70 F35 ◇

BELYAKIN, N.V. *Generalized computations and second order arithmetic (Russian)* ◇ C62 D10 D45 D65 D70 F35 ◇

1971

GRILLIOT, T.J. *On effectively discontinuous type-2 objects* ◇ D60 D65 ◇

HELM, J.P. *On effectively computable operators* ◇ D65 ◇
HINMAN, P.G. & MOSCHOVAKIS, Y.N. *Computability over the continuum* ◇ D55 D65 D70 D75 ◇
JENSEN, R.B. & KARP, C.R. *Primitive recursive set functions* ◇ D60 D65 E45 E47 ◇
KREISEL, G. *Some reasons for generalizing recursion theory* ◇ A05 D30 D60 D65 D70 D75 ◇
MOSCHOVAKIS, Y.N. *Predicative classes* ◇ C62 D55 D60 D65 D70 E70 ◇
PLATEK, R.A. *A countable hierarchy for the superjump* ◇ D55 D65 ◇
RICHTER, W.H. *Recursively Mahlo ordinals and inductive definitions* ◇ D60 D65 D70 E55 F15 ◇
SACKS, G.E. *Recursion in objects of finite type* ◇ D60 D65 ◇

1972

ACZEL, P. *The ordinals of the superjump and related functionals* ◇ D65 E55 ◇
CHERNOV, V.P. *Classification of spaces of operators of finite types (Russian)* ◇ D45 D65 F60 ◇
CHERNOV, V.P. *Constructive operators of finite types (Russian) (English summary)* ◇ D45 D65 F60 ◇
ERSHOV, YU.L. *Computable functionals of finite types (Russian)* ◇ D65 ◇
ERSHOV, YU.L. *Continuous lattices and A-spaces (Russian)* ◇ D65 G10 ◇
ERSHOV, YU.L. *Everywhere-defined continuous functionals (Russian)* ◇ D65 F10 ◇
ERSHOV, YU.L. *Relationship between sheaf spaces and numbered sets with the C_2^* property (Russian) (English summary)* ◇ C90 D25 D45 D65 ◇
HAUCK, J. *Funktionale Rekursion* ◇ D20 D65 ◇
POLYAKOV, E.A. *On relative recursiveness and computability (Russian)* ◇ D30 D65 ◇

1973

AMSTISLAVSKIJ, V.I. *Computability with functionals (Russian)* ◇ D65 ◇
BELYAKIN, N.V. *Generalized computations over regular numerations (Russian)* ◇ D45 D65 D70 D75 ◇
ERSHOV, YU.L. *Theory of numerations. Part II (Russian)* ◇ D45 D65 G30 ◇
FENSTAD, J.E. *On axiomatizing recursion theory* ◇ D65 D75 ◇
HINMAN, P.G. *Degrees of continuous functionals* ◇ D30 D65 ◇
HINMAN, P.G. *The finite levels of the hierarchy of effective R-sets* ◇ D55 D65 D70 E15 ◇
JOCKUSCH JR., C.G. & SOARE, R.I. *Encodability of Kleene's O* ◇ D30 D55 D65 ◇
KECHRIS, A.S. *The structure of envelopes: a survey of recursion theory in higher types* ◇ D65 D75 ◇

1974

ACZEL, P. & HINMAN, P.G. *Recursion in the superjump* ◇ D55 D65 E55 ◇
AMSTISLAVSKIJ, V.I. *Recursiveness and R^c-operations (Russian)* ◇ D55 D65 D70 E15 ◇
BELYAKIN, N.V. *Generalized computations, and third order arithmetic (Russian)* ◇ D65 F35 ◇
CENZER, D. *Ordinal recursion and inductive definitions* ◇ D60 D65 D70 E45 ◇
ERSHOV, YU.L. *Maximal and everywhere defined functionals (Russian)* ◇ D65 ◇
GANDY, R.O. *Inductive definitions* ◇ D60 D65 D70 E55 ◇
GANDY, R.O. *Set-theoretic functions for elementary syntax* ◇ B03 D20 D65 E47 ◇
HARRINGTON, L.A. *The superjump and the first recursively Mahlo ordinal* ◇ D60 D65 E55 ◇
MOLDESTAD, J. & NORMANN, D. *2-envelopes and the analytic hierarchy* ◇ D55 D65 ◇
MOSCHOVAKIS, Y.N. *Structural characterizations of classes of relations* ◇ D65 D70 D75 ◇
NORMANN, D. *Imbedding of higher type theories* ◇ D65 D70 D75 ◇
NORMANN, D. *On abstract 1-sections* ◇ D65 D75 ◇
SACKS, G.E. *The 1-section of a type n object* ◇ D55 D65 ◇
WAINER, S.S. *A hierarchy for the 1-section of any type two object* ◇ D55 D65 ◇

1975

AMSTISLAVSKIJ, V.I. *On recursive characterizations of sets obtainable by means of R^c-operations (Russian)* ◇ D55 D65 D70 E15 ◇
HANATANI, Y. *Calculability of the primitive recursive functionals of finite type over the natural numbers* ◇ D65 F10 F35 F50 ◇
HARRINGTON, L.A. & KECHRIS, A.S. *On characterizing Spector classes* ◇ C62 D55 D60 D65 D75 E45 E55 E60 ◇
NORMANN, D. *Forcing arguments and some degree-theoretic problems in higher type recursion theory* ◇ D30 D65 E40 ◇
SCHWICHTENBERG, H. & WAINER, S.S. *Infinite terms and recursion in higher types* ◇ C75 D65 F10 ◇
TAIT, W.W. *A realizability interpretation of the theory of species* ◇ D65 F05 F35 F50 ◇
WAINER, S.S. *Some hierachies based on higher type quantification* ◇ C75 D55 D65 E45 ◇

1976

BELYAKIN, N.V. *Iterated Kleene computability, and the superjump (Russian)* ◇ D55 D65 ◇
BUSCH, D.R. *On the number of Solovay r-degrees* ◇ D30 D65 E15 ◇
CENZER, D. *Inductive definitions, positive and monotone* ◇ D65 D70 ◇
ERSHOV, YU.L. *Hereditarily effective operations (Russian)* ◇ D25 D65 ◇
HARRINGTON, L.A. & MACQUEEN, D.B. *Selection in abstract recursion theory* ◇ D65 D75 ◇
HODGES, W. *On the effectivity of some field constructions* ◇ C60 C75 D65 E35 E47 E75 ◇
LOWENTHAL, F. *Equivalence of some definitions of recursion in a higher type object* ◇ D65 ◇
MOLDESTAD, J. & NORMANN, D. *Models for recursion theory* ◇ D65 D75 ◇
NORMANN, D. *On a problem of S.Wainer* ◇ D55 D65 ◇

SAZONOV, V.YU. *Degrees of parallelism in computations* ◇ B40 B75 D30 D65 ◇
TRAKHTENBROT, B.A. *Recursive program schemes and computable functionals* ◇ B75 D15 D55 D65 D75 ◇

1977

BARWISE, J. *On Moschovakis closure ordinals* ◇ D60 D65 D70 ◇
ERSHOV, YU.L. *Enumeration of the class* $\mathbf{C}_{20}^{*}$ *(Russian)* ◇ D45 D65 ◇
ERSHOV, YU.L. *Model* $\mathbf{C}$ *of partial continuous functionals* ◇ D45 D65 F10 F50 ◇
FEFERMAN, S. *Generating schemes for partial recursively continuous functionals (summary)* ◇ D65 D70 ◇
FEFERMAN, S. *Inductive schemata and recursively continuous functionals* ◇ D65 D70 D75 ◇
FEFERMAN, S. *Recursion in total functionals of finite type* ◇ D65 ◇
FEFERMAN, S. *Theories of finite type related to mathematical practice* ◇ B15 D55 D65 F10 F15 F35 F98 ◇
FENSTAD, J.E. *Between recursion theory and set theory* ◇ D60 D65 D75 ◇
GANDY, R.O. & HYLAND, J.M.E. *Computable and recursively countable functions of higher type* ◇ D65 ◇
GIRARD, J.-Y. *Functionals and ordinoids* ◇ D65 F10 F15 ◇
HINMAN, P.G. *A survey of finite-type recursion* ◇ D55 D60 D65 D98 ◇
KECHRIS, A.S. & MOSCHOVAKIS, Y.N. *Recursion in higher types* ◇ D65 D70 D75 ◇
LAVORI, P. *Recursion in the extended superjump* ◇ D55 D65 ◇
LIH, KUOWEI *Continuous degrees* ◇ D30 D65 ◇
MAASS, W. *On minimal pairs and minimal degrees in higher recursion theory* ◇ D30 D60 D65 ◇
MOLDESTAD, J. *Computations in higher types* ◇ D65 D75 ◇
MOSCHOVAKIS, Y.N. *On the basic notions in the theory of induction* ◇ D65 D70 D75 ◇
NORMANN, D. *Countable functionals and the analytic hierarchy* ◇ D55 D65 ◇
SACKS, G.E. *RE sets higher up* ◇ D25 D60 D65 ◇
SACKS, G.E. *The k-section of a type n object* ◇ D65 ◇
VOGEL, HELMUT *Partial enumerable and finite functionals (Russian)* ◇ D65 ◇

1978

AMSTISLAVSKIJ, V.I. *An improved substitution theorem for computable functionals and some of its consequences (Russian)* ◇ D65 ◇
BERGSTRA, J.A. *The continuous functionals and* ${}^{2}E$ ◇ D65 ◇
DELFINO, V. *The Victoria Delfino problems* ◇ D55 D60 D65 D70 E15 E60 ◇
HARRINGTON, L.A. *Analytic determinacy and* $O^{\#}$ ◇ D55 D65 E45 E55 E60 ◇
HINMAN, P.G. *Recursion-theoretic hierarchies* ◇ D55 D60 D65 D70 D98 E15 ◇
HRBACEK, K. *On the complexity of analytic sets* ◇ D30 D55 D65 E15 E45 ◇
HYLAND, J.M.E. *The intrinsic recursion theory on the countable or continuous functionals* ◇ D65 D70 ◇
KECHRIS, A.S. *On Spector classes* ◇ D55 D65 D70 D75 E15 E60 ◇
KECHRIS, A.S. *Spector second order classes and reflection* ◇ D65 D70 D75 ◇
KLEENE, S.C. *Recursive functionals and quantifiers of finite types revisited I* ◇ D65 ◇
KOLAITIS, P.G. *On recursion in E and semi-Spector classes* ◇ D65 D70 D75 ◇
LIH, KUOWEI *Type two partial degrees* ◇ D30 D65 ◇
MOLDESTAD, J. *On the role of the successor function in recursion theory* ◇ D65 D75 ◇
NORMANN, D. *A continuous functional with noncollapsing hierarchy* ◇ D55 D65 ◇
NORMANN, D. *Set recursion* ◇ D65 E47 ◇
SHIMODA, H. *Recursion for type 2 objects (Japanese)* ◇ D65 ◇
WAINER, S.S. *The 1-section of a nonnormal type-2 object* ◇ D55 D65 ◇

1979

BELYAKIN, N.V. *Autonomous computability (Russian)* ◇ D45 D55 D65 ◇
DVORNIKOV, S.G. *C-degrees of continuous everywhere defined functionals (Russian)* ◇ D45 D65 ◇
FRIEDMAN, S.D. *β-recursion theory* ◇ D60 D65 ◇
HYLAND, J.M.E. *Filter spaces and continuous functionals* ◇ D65 G30 ◇
KOLAITIS, P.G. *Recursion in a quantifier vs. elementary induction* ◇ D65 D70 D75 ◇
LONGO, G. *Ricorsivita nei tipi superiori: un'introduzione alle caratterizzazioni di Ershov ed Hyland* ◇ B40 D65 G30 ◇
NORMANN, D. *A classification of higher type functionals* ◇ D65 ◇
NORMANN, D. *A jump operator in set recursion* ◇ D30 D65 E47 ◇
NORMANN, D. *A note on reflection* ◇ D65 ◇
NORMANN, D. *Degrees of functionals* ◇ D30 D65 E45 E50 ◇
NORMANN, D. *Recursion in* ${}^{3}E$ *and a splitting theorem* ◇ D30 D65 ◇
TUGUE, T. *Generalized recursion theory (Japanese)* ◇ D60 D65 D75 ◇

1980

HRBACEK, K. & SIMPSON, S.G. *On Kleene degrees of analytic sets* ◇ D30 D55 D65 E15 E35 E45 ◇
KIERSTEAD, D.P. *A semantics for Kleene's j-expressions* ◇ D65 ◇
KLEENE, S.C. *Recursive functionals and quantifiers of finite types revisited II* ◇ D65 ◇
KOLAITIS, P.G. *Recursion and nonmonotone induction in a quantifier* ◇ D65 D70 D75 ◇
NORMANN, D. *Recursion on the countable functionals* ◇ D30 D55 D65 D98 ◇
NORMANN, D. *The recursion theory of the continuous functionals* ◇ D55 D65 ◇

NORMANN, D. & WAINER, S.S. *The 1-section of a countable functional* ◇ D55 D65 ◇
SACKS, G.E. *Post's problem, absoluteness and recursion in finite types* ◇ D30 D65 ◇
SACKS, G.E. *Three aspects of recursive enumerability in higher types* ◇ D30 D65 ◇
SHINODA, J. *On the upper semi-lattice of* J^s_a*-degrees* ◇ D30 D55 D65 ◇

1981

FITTING, M. *Fundamentals of generalized recursion theory* ◇ D60 D65 D70 D75 ◇
GUERRERIO, G. *Sulle generalizzazione della ricorsivita* ◇ D60 D65 ◇
HOERNIG, K.M. *A unified approach to definability problems in the theory of higher type functionals* ◇ D65 ◇
KLEENE, S.C. *Algorithms in various contexts* ◇ D65 D75 ◇
KOLAITIS, P.G. *Model-theoretic characterizations in generalized recursion theory* ◇ D65 D70 D75 ◇
MOSCHOVAKIS, Y.N. *On the Grilliot-Harrington-MacQueen theorem* ◇ D65 ◇
NORMANN, D. *Countable functionals and the projective hierarchy* ◇ D55 D65 ◇
NORMANN, D. *The continuous functionals; computations,recursions and degrees* ◇ D30 D65 ◇
SHINODA, J. *Sections and envelopes of type 2 objects* ◇ D65 ◇
STEEL, J.R. *Closure properties of pointclasses* ◇ D55 D65 D75 E15 E60 ◇

1982

ERSHOV, YU.L. *ω-complete A-spaces (Russian)* ◇ B75 D65 ◇
KLEENE, S.C. *Recursive functionals and quantifiers of finite types revisited III* ◇ D65 ◇
NORMANN, D. *External and internal algorithms on the continuous functionals* ◇ D65 ◇
THIBAULT, M.-F. *Prerecursive categories* ◇ D20 D65 G30 ◇
WEITKAMP, G. *Analytic sets having incomparable Kleene degrees* ◇ D30 D55 D65 E15 ◇
WEITKAMP, G. *Iterating the superjump along definable prewellorderings* ◇ D55 D60 D65 ◇
WIELE VAN DE, J. *Recursive dilators and generalized recursions* ◇ D60 D65 F15 F35 ◇

1983

BOSISIO, A. *Una nota su proprieta topologiche e d'ordine e la ricorsivita generalizzata (English summary)* ◇ D65 G30 ◇
GRIFFOR, E.R. *Some consequences of AD for Kleene recursion in* ^3E ◇ D65 E45 E55 E60 ◇
HARTLEY, J.P. *The countably based functionals* ◇ D65 ◇
HRBACEK, K. *Degrees of analytic sets* ◇ D30 D65 E15 E45 ◇
KIERSTEAD, D.P. *Syntax and semantics in higher-type recursion theory* ◇ D65 ◇
LONGO, G. *Recursiveness and continuity: an introduction (Italian)* ◇ D20 D65 ◇
NORMANN, D. *Characterizing the continuous functionals* ◇ D65 H05 ◇
NORMANN, D. *General type-structures of continuous and countable functionals* ◇ D55 D65 D75 F35 ◇
NORMANN, D. *R.e. degrees of continuous functionals* ◇ D30 D65 ◇
SIMPSON, S.G. & WEITKAMP, G. *High and low Kleene degrees of coanalytic sets* ◇ D30 D55 D65 ◇
SLAMAN, T.A. *The extended plus-one hypothesis - a relative consistency result* ◇ D65 E35 ◇
SOLOV'EV, V.D. *Program schemes and effective functionals of finite type (Russian)* ◇ D65 ◇
VOGEL, HELMUT *On a relationship between countable functionals and projective trees* ◇ D55 D65 ◇

1984

GAMOVA, A.N. *Calculation of functionals of higher type (Russian)* ◇ D65 ◇
GRIFFOR, E.R. & NORMANN, D. *Effective cofinalities and admissibility in E-recursion* ◇ D65 E30 E35 E47 E50 ◇
GRIFFOR, E.R. & NORMANN, D. *The definability of* $E(\alpha)$ ◇ D65 E45 ◇
HINMAN, P.G. *Finitely approximable sets* ◇ D65 F65 ◇
LONGO, G. & MARTINI, S. *Computability in higher types and the universal domain* $P\omega$ ◇ B75 D65 ◇
LONGO, G. & MOGGI, E. *The hereditary partial effective functionals and recursion theory in higher types* ◇ D65 ◇
MIJOULE, R. *L'universalite des semi-fonctions recursives universelles* ◇ D65 G30 ◇
SHINODA, J. *Countable* J^S_a*-admissible ordinals* ◇ D60 D65 E10 E30 E45 ◇

1985

BLASS, A.R. *Kleene degrees of ultrafilters* ◇ C20 D65 E05 ◇
GIRARD, J.-Y. & NORMANN, D. *Set recursion and* Π^1_2*-logic* ◇ D65 F15 ◇
GRIFFOR, E.R. *Definability and forcing in E-recursion* ◇ D65 E40 E47 ◇
HARTLEY, J.P. *Effective discontinuity and a characterisation of the superjump* ◇ D65 ◇
KLEENE, S.C. *Unimonotone functions of finite types (Recursive functionals and quantifiers of finite types revisited IV)* ◇ D65 ◇
KOLAITIS, P.G. *Canonical forms and hierarchies in generalized recursion theories* ◇ D55 D65 D70 ◇
KOLAITIS, P.G. *Game quantification* ◇ C40 C70 C75 C98 D60 D65 D70 E60 ◇
LONGO, G. & MARTINI, S. *Computability in higher types and the universal domain* $P\omega$ *(Italian)* ◇ D65 ◇
MIHNEA, G. *On the E-recursive functions* ◇ D65 ◇
NORMANN, D. *Aspects of the continuous functionals* ◇ D65 ◇
SACKS, G.E. *Post's problem in E-recursion* ◇ D65 ◇
SHINODA, J. *Absolute type 2 objects* ◇ D60 D65 E40 E45 E55 ◇
SHINODA, J. *Countable* J^S_a*-admissible ordinals* ◇ D60 D65 ◇

SLAMAN, T.A. *Reflection and the priority method in E-recursion theory* ◇ D30 D65 E45 ◇
SLAMAN, T.A. *Reflection and forcing in E-recursion theory* ◇ D65 E40 E45 ◇
SLAMAN, T.A. *The E-recursively enumerable degrees are dense* ◇ D30 D65 E45 ◇
THOMPSON, S. *Axiomatic recursion theory and the continuous functionals* ◇ D65 D75 ◇
THOMPSON, S. *Priority arguments in the continuous r.e. degrees* ◇ D65 ◇
WEITKAMP, G. *On the existence and recursion theoretic properties of Σ_n^1-generic sets of reals* ◇ D55 D65 E15 E40 ◇
WELCH, P. *Comparing incomparable Kleene degrees* ◇ D65 E35 E45 ◇

D70 Inductive definability

1918

SIERPINSKI, W. *Sur les definitions axiomatiques des ensembles mesurables (B)* ◇ D55 D70 E15 ◇

1935

PAUC, C. *Resolution d'equations abstraites par un procede d'iteration* ◇ D70 ◇

1937

ROSSER, J.B. *Goedel theorems for non-constructive logics* ◇ D70 F30 ◇

1938

JARNIK, V. *Sur un probleme de M.Cech* ◇ D70 E10 E20 ◇

1944

KLEENE, S.C. *On the forms of predicates in the theory of constructive ordinals* ◇ D55 D70 F15 ◇

1947

LYAPUNOV, A.A. *On R-sets (Russian)* ◇ D55 D70 E15 ◇

LYAPUNOV, A.A. *Theory of R-sets (Russian)* ◇ D55 D70 E15 ◇

1948

SIERPINSKI, W. *L'operation du crible et les fonctions analytiques d'une suite infinie d'ensembles* ◇ D55 D70 E15 ◇

1950

ARSENIN, V.YA. & LYAPUNOV, A.A. *The theory of A-sets (Russian)* ◇ D55 D70 E15 E98 ◇

1953

LYAPUNOV, A.A. *On the classification of R-sets (Russian)* ◇ D55 D70 E15 ◇

LYAPUNOV, A.A. *R-sets (Russian)* ◇ D55 D70 E15 ◇

WANG, HAO *Certain predicates defined by induction schemata* ◇ D70 F30 F35 ◇

1955

ARSENIN, V.YA. & LYAPUNOV, A.A. & SHCHEGOL'KOV, E.A. *Arbeiten zur deskriptiven Mengenlehre* ◇ D55 D70 D97 E15 E97 ◇

KLEENE, S.C. *Arithmetical predicates and function quantifiers* ◇ D55 D70 ◇

KLEENE, S.C. *Hierarchies of number-theoretic predicates* ◇ D55 D70 ◇

KLEENE, S.C. *On the forms of the predicates in the theory of constructive ordinals II* ◇ D55 D70 F15 ◇

1957

LYAPUNOV, A.A. *On operations of sets admitting transfinite indexes (Russian)* ◇ D55 D70 E15 ◇

1958

BANASCHEWSKI, B. *On transfinite iteration* ◇ D70 E20 ◇

GRZEGORCZYK, A. & MOSTOWSKI, ANDRZEJ & RYLL-NARDZEWSKI, C. *The classical and the ω-complete arithmetic* ◇ C62 D55 D70 F30 ◇

1959

LORENZEN, P. & MYHILL, J.R. *Constructive definition of certain analytic sets of numbers* ◇ D55 D70 ◇

SAMPEI, Y. *On the evaluation of the projective class of sets defined by transfinite induction* ◇ D55 D70 E15 ◇

1960

SPECTOR, C. *Hyperarithmetical quantifiers* ◇ D55 D70 ◇

1961

KREIDER, D.L. & ROGERS JR., H. *Constructive versions of ordinal number classes* ◇ D55 D70 F15 ◇

SPECTOR, C. *Inductively defined sets of natural numbers* ◇ D55 D70 ◇

1963

LYAPUNOV, A.A. *Operations on sets (Russian)* ◇ D55 D70 E15 ◇

1964

CLARKE, D.A. *Hierarchies of predicate of finite types* ◇ D55 D65 D70 ◇

PUTNAM, H. *On hierarchies and systems of notations* ◇ D55 D70 F15 ◇

1965

RICHTER, W.H. *Extensions of the constructive ordinals* ◇ D70 F15 ◇

1966

AMSTISLAVSKIJ, V.I. *Set-theoretical operations and recursive hierarchies (Russian)* ◇ D55 D70 E15 ◇

HINMAN, P.G. *Ad astra per aspera: hierarchy schemata in recursive function theory* ◇ D55 D65 D70 E15 ◇

1967

RICHTER, W.H. *Constructive transfinite number classes* ◇ D65 D70 F15 ◇

1968

AMSTISLAVSKIJ, V.I. *Expansion of recursive hierarchies and R-operations (Russian)* ◇ D55 D70 E15 ◇

KUNEN, K. *Implicit definability and infinitary languages* ◇ C40 C70 C75 D70 E47 E55 ◇

RICHTER, W.H. *Constructively accessible ordinal numbers* ◇ D65 D70 F15 ◇

1969

BELYAKIN, N.V. *A variant of Richter's constructive ordinals (Russian)* ◇ D70 F15 ◇

HINMAN, P.G. *Hierarchies of effective descriptive set theory* ◇ D55 D65 D70 E15 ◇

MOSCHOVAKIS, Y.N. *Abstract computability and invariant definability* ◇ C40 D70 D75 ◇

MOSCHOVAKIS, Y.N. *Abstract first order computability I,II* ◇ D65 D70 D75 ◇

1970

ACZEL, P. *Representability in some systems of second order arithmetic* ◇ D65 D70 F35 ◇

AMSTISLAVSKIJ, V.I. *Effective R-sets and transfinite extensions of recursive hierarchies* ◇ D55 D70 E15 ◇

AMSTISLAVSKIJ, V.I. *On the decomposition of a field of sets obtained by an R-operation over recursive sets (Russian)* ◇ D55 D70 E15 ◇

BELYAKIN, N.V. *Generalized computations and second order arithmetic (Russian)* ◇ C62 D10 D45 D65 D70 F35 ◇

BLOOM, S.L. *The hyperprojective hierarchy* ◇ D55 D70 E15 ◇

CHANG, C.C. & MOSCHOVAKIS, Y.N. *The Souslin-Kleene theorem for V_κ with cofinality $(\kappa)=\omega$* ◇ D55 D70 E47 ◇

MOSCHOVAKIS, Y.N. *The Suslin-Kleene theorem for countable structures* ◇ C15 C40 D55 D70 D75 ◇

1971

BARWISE, J. & GANDY, R.O. & MOSCHOVAKIS, Y.N. *The next admissible set* ◇ C40 C62 D55 D60 D70 E45 E47 ◇

GRILLIOT, T.J. *Inductive definitions and computabilitiy* ◇ D70 D75 ◇

HINMAN, P.G. & MOSCHOVAKIS, Y.N. *Computability over the continuum* ◇ D55 D65 D70 D75 ◇

KREISEL, G. *Some reasons for generalizing recursion theory* ◇ A05 D30 D60 D65 D70 D75 ◇

MOSCHOVAKIS, Y.N. *Predicative classes* ◇ C62 D55 D60 D65 D70 E70 ◇

RICHTER, W.H. *Recursively Mahlo ordinals and inductive definitions* ◇ D60 D65 D70 E55 F15 ◇

1972

ACZEL, P. & RICHTER, W.H. *Inductive definitions and analogues of large cardinals* ◇ D60 D70 E55 ◇

MOSCHOVAKIS, Y.N. *The game quantifier* ◇ C75 C80 D55 D70 D75 E60 ◇

1973

AMSTISLAVSKIJ, V.I. *A comparison of the indices arising in the transfinite iteration of functions (Russian)* ◇ D55 D70 E15 ◇

BELYAKIN, N.V. *Generalized computations over regular numerations (Russian)* ◇ D45 D65 D70 D75 ◇

HINMAN, P.G. *The finite levels of the hierarchy of effective R-sets* ◇ D55 D65 D70 E15 ◇

LYAPUNOV, A.A. *The method of transfinite indices in the theory of operations over sets (Russian)* ◇ D55 D70 E15 ◇

MAKKAI, M. *Vaught sentences and Lindstroem's regular relations* ◇ C40 C52 C75 D55 D70 ◇

STAVI, J. *A converse of the Barwise completeness theorem* ◇ C70 D60 D70 ◇

VAUGHT, R.L. *Descriptive set theory in $L_{\omega_1,\omega}$* ◇ C15 C70 C75 D55 D70 E15 ◇

1974

AANDERAA, S.O. *Inductive definitions and their closure ordinals* ◇ D70 ◇

ACZEL, P. & RICHTER, W.H. *Inductive definitions and reflecting properties of admissible ordinals* ◇ D60 D70 E47 E55 ◇

AMSTISLAVSKIJ, V.I. *Recursiveness and R^c-operations (Russian)* ◇ D55 D65 D70 E15 ◇

CENZER, D. *Analytic inductive definitions* ◇ D55 D70 ◇

CENZER, D. *Inductively defined sets of reals* ◇ D55 D70 ◇

CENZER, D. *Ordinal recursion and inductive definitions* ◇ D60 D65 D70 E45 ◇

CENZER, D. *The boundedness principle in ordinal recursion* ◇ D60 D70 ◇

GANDY, R.O. *Inductive definitions* ◇ D60 D65 D70 E55 ◇

MOSCHOVAKIS, Y.N. *Elementary induction on abstract structures* ◇ D55 D70 D75 ◇

MOSCHOVAKIS, Y.N. *On nonmonotone inductive definability* ◇ D55 D70 D75 ◇

MOSCHOVAKIS, Y.N. *Structural characterizations of classes of relations* ◇ D65 D70 D75 ◇

NORMANN, D. *Imbedding of higher type theories* ◇ D65 D70 D75 ◇

SKORDEV, D.G. *A generalization of the theory of recursive functions (Russian)* ◇ D70 D75 ◇

VAUGHT, R.L. *Invariant sets in topology and logic* ◇ C75 D55 D70 E15 ◇

1975

ACZEL, P. *Quantifiers, games and inductive definitions* ◇ C75 C80 D70 ◇

AMSTISLAVSKIJ, V.I. *On recursive characterizations of sets obtainable by means of R^c-operations (Russian)* ◇ D55 D65 D70 E15 ◇

BARWISE, J. *Admissible sets and the interaction of model theory, recursion theory and set theory* ◇ C70 C98 D55 D60 D70 E30 E35 E45 E47 ◇

FELDMAN, E.D. *L-Σ^1_n transfinite induction with an application to the EHP hierarchy* ◇ D55 D70 E15 E47 ◇

RICHTER, W.H. *The least Σ^1_2- and Π^1_2-reflecting ordinals* ◇ D55 D60 D70 E45 E47 E55 ◇

1976

ABRAMSON, F.G. & SACKS, G.E. *Uncountable Gandy ordinals* ◇ D60 D70 ◇

BERRY, GERARD *Bottom-up computation of recursive programs (French summary)* ◇ D20 D70 ◇

CENZER, D. *Inductive definitions, positive and monotone* ◇ D65 D70 ◇
CENZER, D. *Monotone inductive definitions over the continuum* ◇ D55 D70 E15 ◇
HARNIK, V. & MAKKAI, M. *Applications of Vaught sentences and the covering theorem* ◇ C15 C40 C45 C50 C52 C75 D70 E15 ◇
HARRINGTON, L.A. & KECHRIS, A.S. *On monotone vs. nonmonotone induction* ◇ D70 ◇
NYBERG, A.M. *Uniform inductive definability and infinitary languages* ◇ C70 D70 ◇

1977

ACZEL, P. *An introduction to inductive definitions* ◇ D55 D70 ◇
BARWISE, J. *On Moschovakis closure ordinals* ◇ D60 D65 D70 ◇
FEFERMAN, S. *Generating schemes for partial recursively continuous functionals (summary)* ◇ D65 D70 ◇
FEFERMAN, S. *Inductive schemata and recursively continuous functionals* ◇ D65 D70 D75 ◇
FUKUYAMA, M. *On positive Σ inductive definitions* ◇ D70 D75 ◇
GRANT, P.W. *Strict Π^1_1-predicates on countable and cofinality ω transitive sets* ◇ C15 C40 C70 D55 D60 D70 E47 E60 ◇
KARAPETYAN, B.K. & POGOSYAN, EH.M. *Inductors and their connection with the method of empirical prediction (Russian)* ◇ B35 D70 ◇
KECHRIS, A.S. & MOSCHOVAKIS, Y.N. *Recursion in higher types* ◇ D65 D70 D75 ◇
MAKKAI, M. *Admissible sets and infinitary logic* ◇ C70 C98 D55 D60 D70 D98 ◇
MOSCHOVAKIS, Y.N. *On the basic notions in the theory of induction* ◇ D65 D70 D75 ◇
NYBERG, A.M. *Inductive operators on resolvable structures* ◇ C70 D60 D70 D75 ◇
SCHLIPF, J.S. *Ordinal spectra of first-order theories* ◇ C35 C50 C55 C62 C70 D15 D30 D70 ◇

1978

APT, K.R. *Inductive definitions, models of comprehension and invariant definability* ◇ C40 C62 D55 D70 F35 ◇
BARWISE, J. & MOSCHOVAKIS, Y.N. *Global inductive definability* ◇ C40 C70 D70 ◇
BARWISE, J. *Monotone quantifiers and admissible sets* ◇ C70 C80 D60 D70 ◇
CENZER, D. *Parametric inductive definitions and recursive operators over the continuum* ◇ D55 D70 ◇
DELFINO, V. *The Victoria Delfino problems* ◇ D55 D60 D65 D70 E15 E60 ◇
FENSTAD, J.E. *On the foundation of general recursion theory: computations versus inductive definability* ◇ D70 D75 ◇
GRASSIN, J. *Definitions inductives monotones sur le continu, dont les composantes sont denombrables (English summary)* ◇ D55 D60 D70 E15 ◇
HARRINGTON, L.A. & KIROUSIS, L.M. & SCHLIPF, J.S. *A generalized Kleene-Moschovakis theorem* ◇ D55 D70 D75 ◇
HINMAN, P.G. *Recursion-theoretic hierarchies* ◇ D55 D60 D65 D70 D98 E15 ◇
HYLAND, J.M.E. *The intrinsic recursion theory on the countable or continuous functionals* ◇ D65 D70 ◇
KECHRIS, A.S. *On Spector classes* ◇ D55 D65 D70 D75 E15 E60 ◇
KECHRIS, A.S. *Spector second order classes and reflection* ◇ D65 D70 D75 ◇
KOLAITIS, P.G. *On recursion in E and semi-Spector classes* ◇ D65 D70 D75 ◇
MOSCHOVAKIS, Y.N. *Inductive scales on inductive sets* ◇ D55 D70 D75 E15 E60 ◇

1979

BOYER, R.S. & MOORE, J.S. *A computational logic* ◇ B98 D70 ◇
GOSTANIAN, R. & HRBACEK, K. *A new proof that $\pi^1_1 < \sigma^1_1$* ◇ D60 D70 E45 ◇
KOLAITIS, P.G. *Recursion in a quantifier vs. elementary induction* ◇ D65 D70 D75 ◇

1980

BRACHO, F. *Continuously generated fixed points in $P\omega$* ◇ B40 D70 ◇
CENZER, D. & MAULDIN, R.D. *Inductive definability: measure and category* ◇ D70 E15 ◇
CENZER, D. *Non-generable formal languages* ◇ D03 D25 D30 D70 ◇
CUTLAND, N.J. *On non-monotone Σ^1_2 inductive definitions* ◇ D70 E45 ◇
KOLAITIS, P.G. *Recursion and nonmonotone induction in a quantifier* ◇ D65 D70 D75 ◇
MOLDESTAD, J. & STOLTENBERG-HANSEN, V. & TUCKER, J.V. *Finite algorithmic procedures and inductive definability* ◇ D10 D70 D75 ◇

1981

BALLO, E. *Sulle procedure induttive* ◇ D70 ◇
FITTING, M. *Fundamentals of generalized recursion theory* ◇ D60 D65 D70 D75 ◇
GRASSIN, J. *Δ^1_1-good inductive definitions over the continuum* ◇ D55 D70 E15 ◇
KOLAITIS, P.G. *Model-theoretic characterizations in generalized recursion theory* ◇ D65 D70 D75 ◇
MARTIN, D.A. *Π^1_2 monotone inductive definitions* ◇ D55 D70 D75 ◇
MCDERMOTT, M. *Inductive definitions* ◇ A05 D70 ◇

1982

BURGESS, J.P. *What are R-sets?* ◇ D55 D70 E15 ◇
FITTING, M. *A generalization of elementary formal systems* ◇ D20 D70 D75 ◇
KRANAKIS, E. *Invisible ordinals and inductive definitions* ◇ D60 D70 E45 E47 E55 ◇
MARTIN, D.A. & MOSCHOVAKIS, Y.N. & STEEL, J.R. *The extent of definable scales* ◇ D55 D70 E15 E60 ◇

1983

BURGESS, J.P. *Classical hierarchies from a modern standpoint I: C-sets. II R-sets* ◇ D55 D70 E15 ◇

FRIEDMAN, S.D. *Some recent developments in higher recursion theory* ◇ D60 D70 ◇
KIROUSIS, L.M. *A selection theorem* ◇ D60 D70 ◇
MOSCHOVAKIS, Y.N. *Scales on coinductive sets* ◇ D70 E15 E60 ◇

1984

BARUA, R. *R-sets and category* ◇ D70 E15 ◇
BLASS, A.R. & GUREVICH, Y. & KOZEN, D. *A 0-1 law for logic with a fixed-point operator* ◇ B10 B75 C80 D70 ◇

1985

KHOLSHCHEVNIKOVA, N.N. *Uncountable R- and N-sets (Russian)* ◇ D55 D70 E15 ◇
KOLAITIS, P.G. *Canonical forms and hierarchies in generalized recursion theories* ◇ D55 D65 D70 ◇
KOLAITIS, P.G. *Game quantification* ◇ C40 C70 C75 C98 D60 D65 D70 E60 ◇

D75 Abstract and axiomatic recursion theory

1960

MONTAGUE, R. *Towards a general theory of computability* ◇ D10 D75 ◇

1961

FRAISSE, R. *Une notion de recursivite relative* ◇ D20 D30 D75 ◇

PETER, R. *Ueber die Verallgemeinerung der Rekursionsbegriffe fuer abstrakte Mengen als Definitionsbereiche* ◇ D20 D75 ◇

PETER, R. *Ueber die Verallgemeinerung der Theorie der rekursiven Funktionen fuer abstrakte Mengen geeigneter Struktur als Definitionsbereiche (Russian summary)* ◇ D20 D75 ◇

1962

PETER, R. *Ueber die Verallgemeinerung der Theorie der rekursiven Funktionen fuer abstrakte Mengen geeigneter Struktur als Definitionsbereiche (Fortsetzung)* ◇ D20 D75 ◇

1963

DAVIS, MARTIN D. & PUTNAM, H. *Diophantine sets over polynomial rings* ◇ D25 D75 ◇

POLYAKOV, E.A. *Certain properties of algebras of recursive functions (Russian)* ◇ D20 D75 ◇

1964

LACHLAN, A.H. *Effective operations in a general setting* ◇ D20 D25 D75 ◇

LACOMBE, D. *Deux generalisations de la notion de recursivite* ◇ D45 D75 ◇

POLYAKOV, E.A. *Algebras of recursive functions (Russian)* ◇ D20 D75 ◇

POLYAKOV, E.A. *On some properties of algebras of recursive functions (Russian)* ◇ D20 D75 ◇

ZAROVNYJ, V.P. *The group of automatic one-to-one mappings (Russian)* ◇ D05 D10 D75 ◇

ZYKOV, A.A. *Recursively calculable functions of graphs* ◇ D20 D75 D80 ◇

1965

ADDISON, J.W. *The method of alternating chains* ◇ C40 D55 D75 E15 ◇

SCHWENKEL, F. *Rekursive Wortfunktionen ueber unendlichen Alphabeten* ◇ D20 D75 ◇

1966

LAVROV, I.A. & POLYAKOV, E.A. *Bases of algebras of recursive functions (Russian)* ◇ D20 D75 ◇

PLATEK, R.A. *Foundations of recursion theory* ◇ D60 D75 ◇

POLYAKOV, E.A. *Some properties of algebras of recursive functions (Russian)* ◇ D20 D75 ◇

1967

ENGELER, E. *Algorithmic properties of structures* ◇ B75 C75 D10 D75 ◇

LAVROV, I.A. *The use of k-order arithmetic progressions in constructing the bases of the algebra of primitive recursive functions (Russian)* ◇ D20 D75 ◇

POLYAKOV, E.A. *Several properties of the algebra of general recursive functions (Russian)* ◇ D20 D75 ◇

SHCHEGLOV, A.I. *The algebra of partially recursive functions (Russian)* ◇ D20 D75 ◇

SOKOLOV, V.A. *Isomorphisms of maximal subalgebras of R.Robinson's algebra (Russian) (English summary)* ◇ D20 D75 ◇

1968

KREISEL, G. *Choice of infinitary languages by means of definability criteria: generalized recursion theory* ◇ C40 C70 C75 D60 D75 ◇

LAMBERT JR., W.M. *A notion of effectiveness in arbitrary structures* ◇ D75 ◇

MONTAGUE, R. *Recursion theory as a branch of model theory* ◇ C40 D75 ◇

NASIBULLOV, KH.KH. *Bases in algebras of partially recursive functions (Russian)* ◇ D20 D75 ◇

POLYAKOV, E.A. *Some properties of algebras of recursive functions (Russian)* ◇ D20 D75 ◇

SHCHEGLOV, A.I. *Power of the set of maximal sub-algebras of an algebra of partial-recursive functions (Russian)* ◇ D20 D75 ◇

STRONG, H.R. *Algebraically generalized recursive function theory* ◇ D75 ◇

WAGNER, E.G. *Uniformly reflexible structures: an axiomatic approach to computability* ◇ D75 ◇

1969

BERNAYS, P. *Remark concerning the formalization of recursive definition in second order logic* ◇ B15 D75 ◇

HIRSCHELMANN, A. *Primitiv-rekursive Funktionen in Peano-Algebren* ◇ D20 D75 G25 ◇

KAPHENGST, H. *Malzew-Raeume, ein allgemeiner Begriff der rekursiven Abbildung* ◇ D20 D75 ◇

MOSCHOVAKIS, Y.N. *Abstract computability and invariant definability* ◇ C40 D70 D75 ◇

MOSCHOVAKIS, Y.N. *Abstract first order computability I,II* ◇ D65 D70 D75 ◇

PETER, R. *Ueber die Pair-schen freien Binoiden als Spezialfaelle der angeordneten freien holomorphen Mengen (Russian summary)* ◇ D20 D75 ◇

POLYAKOV, E.A. *Formal definability in algebras of recursive functions (Russian)* ◇ D20 D75 ◇

SHCHEGLOV, A.I. *A certain algebra of one-place primitive recursive functions (Russian)* ◇ D20 D75 ◇

Starke, P.H. *Abstrakte Automaten* ◇ D05 D75 D98 ◇
Tsichritzis, D.C. *Fuzzy computability* ◇ B52 D75 ◇
Wagner, E.G. *Uniformly reflexive structures: on the nature of Goedelization and relative computability* ◇ B40 D45 D75 F40 ◇

1970

Gordon, C.E. *Comparisons between some generalizations of recursion theory* ◇ D75 ◇
Herman, G.T. & Isard, S.D. *Computability over arbitrary fields* ◇ D75 ◇
Kaiser, Klaus *Zur Theorie algorithmisch abgeschlossener Modellklassen* ◇ C20 C52 D75 ◇
Koz'minykh, V.V. *On the subalgebras of R. Robinson's algebras (Russian)* ◇ D20 D75 ◇
Moschovakis, Y.N. *The Suslin-Kleene theorem for countable structures* ◇ C15 C40 D55 D70 D75 ◇
Strong, H.R. *Construction of models for algebraically generalized recursive function theory* ◇ D75 ◇

1971

Ausiello, G. *Abstract computational complexity and cycling computations* ◇ D15 D75 ◇
Friedman, H.M. *Algorithmic procedures, generalized Turing algorithms, and elementary recursion theory* ◇ A05 D10 D20 D75 F99 ◇
Friedman, H.M. *Axiomatic recursive function theory* ◇ D75 ◇
Grilliot, T.J. *Inductive definitions and computabilitiy* ◇ D70 D75 ◇
Hecker, H.-D. *Allgemeine Trennbarkeitsbegriffe bei meta-mathematischen Untersuchungen* ◇ D25 D35 D75 ◇
Hinman, P.G. & Moschovakis, Y.N. *Computability over the continuum* ◇ D55 D65 D70 D75 ◇
Kreisel, G. *Some reasons for generalizing recursion theory* ◇ A05 D30 D60 D65 D70 D75 ◇
Lacombe, D. *Recursion theoretic structure for relational systems* ◇ D45 D75 ◇
Moschovakis, Y.N. *Axioms for computation theories - first draft* ◇ D75 ◇
Sacks, G.E. *F-recursiveness* ◇ D30 D55 D60 D75 E40 E45 ◇
Wagner, E.G. *An algebraic theory of recursive definitions and recursive languages* ◇ B75 D75 G30 ◇

1972

Fahmy, M.H. *Equational calculus for primitive recursive rational-valued functions (Russian)* ◇ D20 D75 ◇
Gonshor, H. *The category of recursive functions* ◇ D20 D75 G30 ◇
Grilliot, T.J. *Omitting types; applications to recursion theory* ◇ C07 C62 D30 D55 D60 D75 ◇
Mitschke, G. *λ-definierbare Funktionen auf Peanoalgebren* ◇ B40 D75 ◇
Moschovakis, Y.N. *The game quantifier* ◇ C75 C80 D55 D70 D75 E60 ◇
Sokolov, V.A. *Certain properties of the algebra of all partially recursive functions (Russian)* ◇ D20 D75 ◇
Sokolov, V.A. *The maximal subalgebras of the algebra of all partially recursive functions (Russian) (English summary)* ◇ D20 D75 ◇
Tucker, J.V. *Algorithmic unsolvability in biological contexts* ◇ D75 D80 ◇

1973

Belyakin, N.V. *Generalized computations over regular numerations (Russian)* ◇ D45 D65 D70 D75 ◇
Fenstad, J.E. *On axiomatizing recursion theory* ◇ D65 D75 ◇
Germano, G. *A new theory of calculability* ◇ D75 ◇
Grilliot, T.J. *Implicit definability and hyperprojectivity* ◇ C40 C75 D75 ◇
Kechris, A.S. *The structure of envelopes: a survey of recursion theory in higher types* ◇ D65 D75 ◇
Kiselev, A.A. *Projective hierarchies on general structures (Russian)* ◇ D55 D75 E15 E45 ◇
Kiselev, A.A. *Projective hierarchies on general structures II (Russian)* ◇ D55 D75 E15 E45 E55 ◇
Mann, I. *Probabilistic recursive functions* ◇ D20 D75 ◇
Mikheev, V.L. *Classes of algebras of primitive recursive functions (Russian)* ◇ D20 D75 ◇
Ottmann, T. *Ketten und arithmetische Praedikate von endlichen Automaten* ◇ D05 D55 D75 ◇
Pobedin, L.N. *Certain questions on generalized computability (Russian)* ◇ D10 D55 D75 ◇
Pour-El, M.B. *Abstract computability versus analog-generability. A survey* ◇ D75 D80 F65 ◇
Pour-El, M.B. *Analog computers, digital computers, mathematical logic, differential equations - interrelations* ◇ D20 D75 D80 F65 ◇
Robinson, Julia *Axioms for number theoretic functions (Russian)* ◇ C62 D20 D75 F30 ◇
Strassen, V. *Berechnungen in partiellen Algebren endlichen Typs* ◇ D75 ◇
Strong, H.R. & Walker, S.A. *Characterizations of flowchartable recursions* ◇ D20 D75 ◇
Vuckovic, V. *Local recursive theory* ◇ D20 D25 D75 D80 ◇
Wand, M. *A concrete approach to abstract recursive definitions* ◇ D05 D75 G10 ◇

1974

Aczel, P. *An axiomatic approach to recursion on admissible ordinals and the Kreisel-Sacks construction of meta-recursion theory* ◇ D60 D75 ◇
Deutsch, M. *Reduzierungen unentscheidbarer formaler Sprachen ueber induktiv definierten Bereichen* ◇ D25 D75 ◇
Fenstad, J.E. *On axiomatizing recursion theory* ◇ D75 ◇
Gordon, C.E. *Prime and search computability, characterized as definability in certain sublanguages of constructible $L_{\omega_1,\omega}$* ◇ C75 D75 ◇
Grilliot, T.J. *Dissecting abstract recursion* ◇ D75 ◇
Grilliot, T.J. *Model theory for dissecting recursion theory* ◇ C55 C70 C75 C80 D60 D75 ◇
Grzegorczyk, A. *Axiomatic theory of enumeration* ◇ D45 D75 ◇
Moschovakis, Y.N. *Elementary induction on abstract structures* ◇ D55 D70 D75 ◇

MOSCHOVAKIS, Y.N. *On nonmonotone inductive definability* ◇ D55 D70 D75 ◇
MOSCHOVAKIS, Y.N. *Structural characterizations of classes of relations* ◇ D65 D70 D75 ◇
NORMANN, D. *Imbedding of higher type theories* ◇ D65 D70 D75 ◇
NORMANN, D. *On abstract 1-sections* ◇ D65 D75 ◇
OTTMANN, T. *Arithmetische Praedikate ueber einem Bereich endlicher Automaten* ◇ D05 D55 D75 ◇
POUR-EL, M.B. *Abstract computability and its relation to the general purpose analog computer (some connections between logic, differential equations and analog computers)* ◇ D20 D75 D80 F65 ◇
ROEVER DE, W.P. *Recursion and parameter mechanisms: An axiomatic approach* ◇ B40 B75 D75 ◇
SHMAIN, I.KH. *Extended calculus of recursive functions I (Russian)* ◇ D20 D75 F50 ◇
SKORDEV, D.G. *A generalization of the theory of recursive functions (Russian)* ◇ D70 D75 ◇
VENTURINI ZILLI, M. *On different kinds of indefinite* ◇ B40 D75 ◇

1975

BARENDREGT, H.P. *Normed uniformly reflexive structures* ◇ B40 D75 ◇
DUBINSKY, A. *Computation on arbitrary algebras* ◇ D75 ◇
ENGELER, E. *On the solvability of algorithmic problems* ◇ B75 C60 C75 D75 ◇
FENSTAD, J.E. *Computation theories: an axiomatic approach to recursion on general structures* ◇ D75 ◇
HARRINGTON, L.A. & KECHRIS, A.S. *On characterizing Spector classes* ◇ C62 D55 D60 D65 D75 E45 E55 E60 ◇
PAYNE, T.H. *Concrete computability* ◇ D75 ◇
RAYMOND, F.H. *Note sur l'algebre des fonctions (English summary)* ◇ D20 D75 ◇
REISIG, W. *Eine Verallgemeinerung des Berechenbarkeitsbegriffs durch Gleichungssysteme* ◇ D10 D75 ◇
ROBINSON, A. *Algorithms in algebra* ◇ C40 C57 C60 D45 D75 ◇
SCOTT, D.S. *Lambda calculus and recursion theory* ◇ B40 B75 D25 D75 ◇
SHEPHERDSON, J.C. *Computation over abstract structures: serial and parallel procedures and Friedman's effective definitional schemes* ◇ D75 ◇
SKORDEV, D.G. *Some topological examples of iterative combinatory spaces (Russian)* ◇ B40 D75 ◇
VENTURINI ZILLI, M. *A model with nondeterministic computation* ◇ D75 ◇

1976

BEREZIN, S.A. *An algebra of one-place primitive recursive functions with an iteration operation of general type (Russian) (English summary)* ◇ D20 D75 ◇
HARRINGTON, L.A. & MACQUEEN, D.B. *Selection in abstract recursion theory* ◇ D65 D75 ◇
LONGO, G. *On the problem of deciding equality in partial combinatory algebras and in a formal system* ◇ B40 D75 ◇
MOLDESTAD, J. & NORMANN, D. *Models for recursion theory* ◇ D65 D75 ◇
SKORDEV, D.G. *Certain combinatory spaces that are connected with the complexity of data processing (Russian)* ◇ B40 B75 D15 D75 ◇
SKORDEV, D.G. *Recursion theory on iterative combinatory spaces* ◇ B40 D75 ◇
SKORDEV, D.G. *The concept of search computability from the point of view of the theory of combinatory spaces (Russian)* ◇ B40 D75 ◇
TRAKHTENBROT, B.A. *Recursive program schemes and computable functionals* ◇ B75 D15 D55 D65 D75 ◇

1977

DEUTSCH, M. *Zum Begriff der Wortmischung als Basis fuer die Arithmetik* ◇ B28 D05 D75 ◇
FEFERMAN, S. *Inductive schemata and recursively continuous functionals* ◇ D65 D70 D75 ◇
FENSTAD, J.E. *Between recursion theory and set theory* ◇ D60 D65 D75 ◇
FUKUYAMA, M. *On positive* Σ *inductive definitions* ◇ D70 D75 ◇
HINDLEY, J.R. & MITSCHKE, G. *Some remarks about the connections between combinatory logic and axiomatic recursion theory* ◇ B40 D75 ◇
HOEHNKE, H.-J. *On partial recursive definitions and programs* ◇ D10 D75 ◇
KECHRIS, A.S. *Classifying projective-like hierarchies* ◇ D55 D75 E15 E45 E60 ◇
KECHRIS, A.S. & MOSCHOVAKIS, Y.N. *Recursion in higher types* ◇ D65 D70 D75 ◇
KRATKO, M.I. *On the axiomatic definition of the concept of computational complexity* ◇ D15 D75 ◇
KRECZMAR, A. *Programmability in fields* ◇ B75 D75 ◇
MOLDESTAD, J. *Computations in higher types* ◇ D65 D75 ◇
MOSCHOVAKIS, Y.N. *On the basic notions in the theory of induction* ◇ D65 D70 D75 ◇
NYBERG, A.M. *Inductive operators on resolvable structures* ◇ C70 D60 D70 D75 ◇
SKORDEV, D.G. *Simplification of some definitions in the theory of combinatory spaces* ◇ B40 D75 ◇
STOLTENBERG-HANSEN, V. *A regular set theorem for infinite computation theories* ◇ D25 D75 ◇

1978

BEREZIN, S.A. *Maximal subalgebras of recursive function algebras (Russian)* ◇ D20 D75 ◇
FEFERMAN, S. *Recursion theory and set theory: a marriage of convenience* ◇ D75 E30 E70 F65 ◇
FENSTAD, J.E. *On the foundation of general recursion theory: computations versus inductive definability* ◇ D70 D75 ◇
FERRARI, P.L. & LONGO, G. *Axiomatic theory of enumeration: a note on the axiom of extensionality* ◇ D75 ◇
FITTING, M. *Elementary formal systems for hyperarithmetical relations* ◇ C75 D55 D75 ◇
HARRINGTON, L.A. & KIROUSIS, L.M. & SCHLIPF, J.S. *A generalized Kleene-Moschovakis theorem* ◇ D55 D70 D75 ◇

HUPBACH, U.L. *Rekursive Funktionen in mehrsortigen Peano-Algebren* ◇ D20 D75 ◇
IVANOV, L.L. *Natural combinatory spaces* ◇ B40 D75 ◇
KECHRIS, A.S. *On Spector classes* ◇ D55 D65 D70 D75 E15 E60 ◇
KECHRIS, A.S. *Spector second order classes and reflection* ◇ D65 D70 D75 ◇
KOLAITIS, P.G. *On recursion in E and semi-Spector classes* ◇ D65 D70 D75 ◇
KOLGANOV, N.A. *Identities in an algebra of recursive functions (Russian)* ◇ D20 D75 ◇
MARCHENKOV, S.S. *A method of constructing maximal subalgebras in algebras of general recursive functions (Russian)* ◇ D20 D75 ◇
MOLDESTAD, J. *On the role of the successor function in recursion theory* ◇ D65 D75 ◇
MOSCHOVAKIS, Y.N. *Inductive scales on inductive sets* ◇ D55 D70 D75 E15 E60 ◇
SKORDEV, D.G. *A normal form theorem for recursive operators in iterative combinatory spaces* ◇ B40 D75 ◇
SOKOLOV, V.A. *"Zero" identities in the Robinson algebra (Russian)* ◇ D20 D75 ◇
STAIGER, L. & WAGNER, K. *Rekursive Folgenmengen. I* ◇ D10 D55 D75 ◇
STOLTENBERG-HANSEN, V. *Weakly inadmissible recursion theory* ◇ D60 D75 ◇
WESEP VAN, R.A. *Separation principles and the axiom of determinateness* ◇ D55 D75 E15 E60 ◇

1979

BERGSTRA, J.A. *Effective transformations on probabilistic data* ◇ D75 ◇
BERGSTRA, J.A. *Recursion theory on processes* ◇ D10 D30 D75 ◇
GALLIER, J.H. *Recursion schemes and generalized interpretations* ◇ B75 D20 D75 ◇
JUKNA, S. *On decidability of equivalence problem in some algebras of recursive functions (Russian)* ◇ B25 D20 D75 ◇
KOLAITIS, P.G. *Recursion in a quantifier vs. elementary induction* ◇ D65 D70 D75 ◇
LAMBERT, W.M. *On recursion with large objects (Spanish) (English summary)* ◇ D75 E20 E70 ◇
NIVAT, M. *Infinite words, infinite trees, infinite computations* ◇ B75 D05 D75 ◇
PETROV, V.P. & SKORDEV, D.G. *Combinatory structures* ◇ B40 D75 ◇
SCHWARTZ, DIETRICH *Beitrag zur algebraischen Rekursionstheorie* ◇ D75 ◇
SKORDEV, D.G. *The first recursion theorem for iterative combinatory spaces* ◇ B40 D75 ◇
STOLTENBERG-HANSEN, V. *Finite injury arguments in infinite computation theories* ◇ D25 D30 D60 D75 ◇
THOMAS, WILLIAM J. *A simple generalization of Turing computability* ◇ D10 D75 ◇
TUGUE, T. *Generalized recursion theory (Japanese)* ◇ D60 D65 D75 ◇
VALIANT, L.G. *Completeness classes in algebra* ◇ D15 D75 ◇

1980

DEGTEV, A.N. *A category of partial recursive functions (Russian)* ◇ D20 D45 D75 G30 ◇
FENSTAD, J.E. *General recursion theory. An axiomatic approach* ◇ D75 D98 ◇
GOLUNKOV, YU.V. *Precomplete classes of algorithms that preserve membership in a set I (Russian)* ◇ D20 D75 ◇
IVANOV, L.L. *Iterative operatory spaces* ◇ D75 ◇
IVANOV, L.L. *Some examples of iterative operatory spaces* ◇ D75 ◇
KOLAITIS, P.G. *Recursion and nonmonotone induction in a quantifier* ◇ D65 D70 D75 ◇
MOLDESTAD, J. & STOLTENBERG-HANSEN, V. & TUCKER, J.V. *Finite algorithmic procedures and inductive definability* ◇ D10 D70 D75 ◇
MOLDESTAD, J. & STOLTENBERG-HANSEN, V. & TUCKER, J.V. *Finite algorithmic procedures and computation theories* ◇ D10 D75 ◇
PAYNE, T.H. *General computability* ◇ D75 ◇
REISER, A. & WEIHRAUCH, K. *Natural numberings and generalized computability* ◇ D45 D75 ◇
SHEN', A.KH. *Axiomatic approach to the theory of algorithms and relativized computability (Russian)* ◇ D20 D75 ◇
SKORDEV, D.G. *Combinatory spaces and recursiveness in them (Russian) (English summary)* ◇ B40 D75 ◇
SKORDEV, D.G. *Semi-combinatory spaces (Russian)* ◇ B40 D75 ◇
SMYTH, M.B. *Computability in categories* ◇ D45 D75 G30 ◇
SPERSCHNEIDER, V. *Goedelisierung* ◇ D45 D75 ◇
STOLTENBERG-HANSEN, V. *On computational complexity in weakly admissible structures* ◇ D15 D60 D75 ◇
TUCKER, J.V. *Computing in algebraic systems* ◇ D10 D35 D75 D80 ◇
WEIHRAUCH, K. *Berechenbarkeit auf CPO-S (Eine Vorlesung ausgearbeitet von Thomas Deil)* ◇ D45 D75 ◇

1981

BERGSTRA, J.A. & TIURYN, J. *Algorithmic degrees of algebraic structures* ◇ D30 D75 ◇
BERGSTRA, J.A. & BROY, M. & TUCKER, J.V. & WIRSING, M. *On the power of algebraic specifications* ◇ D75 ◇
DICHEV, A.V. *Computability in the sense of Moschovakis and its connection with partial recursiveness via numerations (Russian)* ◇ D45 D75 ◇
ERSHOV, A.P. *Abstract computability on algebraic structures* ◇ B75 D75 ◇
FENSTAD, J.E. *What does axiomatic recursion theory axiomatize?* ◇ D75 ◇
FITTING, M. *Fundamentals of generalized recursion theory* ◇ D60 D65 D70 D75 ◇
GALLIER, J.H. *Recursion-closed algebraic theories* ◇ B75 D75 ◇
HIROSE, K. & NAKAYASU, F. *A representation for Spector second order classes in computation theories on two types* ◇ D75 ◇

HUWIG, H. *A definition of the P=NP-problem in categories* ◇ D15 D75 G30 ◇
IVANOV, L.L. *P-recursiveness in iterative combinatorial spaces (Russian)* ◇ B40 D75 ◇
KECHRIS, A.S. *Forcing with Δ perfect trees and minimal Δ-degrees* ◇ D30 D55 D75 E15 E40 E60 ◇
KLEENE, S.C. *Algorithms in various contexts* ◇ D65 D75 ◇
KOLAITIS, P.G. *Model-theoretic characterizations in generalized recursion theory* ◇ D65 D70 D75 ◇
LUGOWSKI, H. *Ueber die Identitaeten der Turing-Algebra I (English and Russian summaries)* ◇ C05 D10 D75 ◇
MANCA, V. *Computational formalism: Abstract combinatory view-point and related first order logical framework* ◇ B40 D75 ◇
MARTIN, D.A. *Π^1_2 monotone inductive definitions* ◇ D55 D70 D75 ◇
SHEN, BAIYING *Axiom systems for primitive recursive word arithmetic WA (Chinese)* ◇ D20 D75 ◇
STEEL, J.R. *Closure properties of pointclasses* ◇ D55 D65 D75 E15 E60 ◇
STEEL, J.R. *Determinateness and the separation property* ◇ D55 D75 E15 E60 ◇
URZYCZYN, P. *Algorithmic triviality of abstract structures* ◇ D75 ◇
WEIHRAUCH, K. *Recursion and complexity theory on CPO's* ◇ D15 D45 D75 ◇

1982

BERGSTRA, J.A. & TIURYN, J. & TUCKER, J.V. *Floyd's principle, correctness theories and program equivalence* ◇ B75 D75 ◇
ERSHOV, A.P. *Computability in arbitrary domains and bases (Russian)* ◇ B75 D75 ◇
FITTING, M. *A generalization of elementary formal systems* ◇ D20 D70 D75 ◇
HYLAND, J.M.E. *The effective topos* ◇ D75 D80 F35 F50 G30 ◇
SCOTT, D.S. *Lectures on a mathematical theory of computation* ◇ B40 B98 D75 ◇
SKORDEV, D.G. *An algebraic treatment of flow diagrams and its application to generalized recursion theory* ◇ B40 B75 D75 ◇
SKORDEV, D.G. *Applications of abstract recursion theory to studying the capability of functional programming systems (Russian)* ◇ B75 D75 ◇
SOLOV'EV, V.D. *The structure of closed classes of computable functions and predicates (Russian)* ◇ D20 D75 ◇
TORBASOVA, V.P. *Machines over models (Russian)* ◇ D05 D75 ◇

1983

CLARES RODRIGUES, B. *An introduction to W-calculability: basic operations (Spanish) (English summary)* ◇ D20 D75 ◇
DETTKI, H.J. & SCHUSTER, H. *Rekursionstheorie auf F* ◇ D75 ◇
GOLUNKOV, YU.V. *Some families of maximal subsystems of algebras in a system of algorithmic algebras of partial recursive functions and predicates. I (Russian)* ◇ D75 ◇
GUREVICH, Y. *Algebras of feasible functions* ◇ D15 D20 D75 ◇
NORMANN, D. *General type-structures of continuous and countable functionals* ◇ D55 D65 D75 F35 ◇
RICHMAN, F. *Church's thesis without tears* ◇ D20 D75 F55 F60 ◇
SAVEL'EV, A.A. *A family of subsystems of algebras for preserving an ideal (Russian)* ◇ D75 ◇
SOSKOV, I.N. *Algorithmically complete algebraic systems (Russian)* ◇ D75 ◇
SOSKOV, I.N. *Computability in algebraic systems (Russian)* ◇ D75 ◇
SZALKAI, I. *The algebraic structure of primitive recursive functions (Russian and Hungarian summaries)* ◇ D20 D75 ◇

1984

BERGSTRA, J.A. & KLOP, J.W. *The algebra of recursively defined processes and the algebra of regular processes* ◇ D05 D75 ◇
BUECHI, J.R. & MAHR, B. & SIEFKES, D. *Recursive definition and complexity of functions over arbitrary data structures* ◇ B75 D75 ◇
BUJ, D.B. & RED'KO, V.N. *Primitive program algebras I (Russian)* ◇ D20 D75 D80 ◇
COPPEY, L. *Categories de Peano et categories algorithmiques, recursivite* ◇ D75 G30 ◇
CROSSLEY, J.N. & REMMEL, J.B. *Undecidability and recursive equivalence II* ◇ D35 D50 D75 ◇
DICHEV, A.V. *Computability in the sense of Moskovakis and its connection with partial recursivity through numbering (Russian)* ◇ D45 D75 ◇
GERMANO, G. & MAZZANTI, S. *Partial closures and semantics of while: toward an iteration-based theory of data types* ◇ B75 D20 D75 D80 ◇
GOLUNKOV, YU.V. & SAVEL'EV, A.A. *A lattice of systems of algorithmic algebras of partial recursive functions and predicates (Russian)* ◇ D75 ◇
GOLUNKOV, YU.V. *On some families of maximal subsystems of algebras in a system of algorithmic algebras of partial recursive functions and predicates. II (Russian)* ◇ D20 D75 ◇
IVANOV, L.L. *Kleene-recursiveness and iterative operative spaces* ◇ D20 D75 ◇
MOSCHOVAKIS, Y.N. *Abstract recursion as a foundation for the theory of algorithms* ◇ D20 D75 ◇
SPREEN, D. & YOUNG, P. *Effective operators in a topological setting* ◇ D20 D75 D80 ◇

1985

BEZEM, M. *Isomorphism between HEO and HRO^E, ECF and ICF^E* ◇ D20 D75 ◇
JERVELL, H.R. *Recursion on homogeneous trees* ◇ D75 F15 ◇
KALANTARI, I. & WEITKAMP, G. *Effective topological spaces I: A definability theory. II: A hierarchy* ◇ C40 C57 C65 D45 D75 D80 ◇
KFOURY, A.J. & URZYCZYN, P. *Necessary and sufficient conditions for the universality of programming formalisms* ◇ B75 D75 ◇

KOSTENKO, K.I. *Classes of algorithms and computations (Russian)* ◇ D20 D75 ◇
KREITZ, C. & WEIHRAUCH, K. *Theory of representations* ◇ D45 D75 D80 ◇
RICHTER, M.M. & SZABO, M.E. *Nonstandard computation theory* ◇ D20 D75 H10 ◇
SCHAEFER, G. *A note on conjectures of Calude about the topological size of sets of partial recursive functions* ◇ D20 D75 ◇
SZALKAI, I. *On the algebraic structure of primitive recursive functions* ◇ D20 D75 ◇
THOMPSON, S. *Axiomatic recursion theory and the continuous functionals* ◇ D65 D75 ◇

D80 Applications

1940

CHURCH, A. *On the concept of a random sequence* ◇ D20 D80 E75 ◇

1947

MARKOV, A.A. *On some unsolvable problems concerning matrices (Russian)* ◇ D80 ◇

1951

MARKOV, A.A. *On an unsolvable problem concerning matrices (Russian)* ◇ D80 ◇

1953

DAVIS, MARTIN D. *Arithmetical problems and recursively enumerable predicates* ◇ D25 D35 D55 D80 ◇

FREUDENTHAL, H. *Les possibilitees des machines a calculer* ◇ D80 ◇

1954

JASKOWSKI, S. *Example of a class of systems of ordinary differential equations having no decision method for existence problems* ◇ D80 ◇

1955

FROEHLICH, A. & SHEPHERDSON, J.C. *On the factorisation of polynomials in a finite number of steps* ◇ D45 D80 ◇

MEDVEDEV, YU.T. *Degrees of difficulty of the mass problem (Russian)* ◇ D30 D80 ◇

1956

MEDVEDEV, YU.T. *On the concept of the mass problem (Russian)* ◇ D30 D80 ◇

1957

KLEENE, S.C. *Realizability* ◇ D80 F50 ◇

KREISEL, G. & LACOMBE, D. *Ensembles recursivement mesurables et ensembles recursivement ouverts ou fermes* ◇ D80 F60 ◇

MO, SHAOKUI *A sketch on model computers (Chinese)* ◇ D80 ◇

NERODE, A. *General topology and partial recursive functionals* ◇ D20 D80 ◇

RABIN, M.O. *Computable algebraic systems* ◇ C57 C60 D40 D45 D80 ◇

1958

HERMES, H. *Zum Einfachheitsprinzip in der Wahrscheinlichkeitsrechnung* ◇ A05 B48 D20 D80 ◇

MARKOV, A.A. *The insolubility of the problem of homeomorphy (Russian)* ◇ D80 ◇

MARKOV, A.A. *Unsolvability of some problems in topology (Russian)* ◇ D80 ◇

1959

LACOMBE, D. *Quelques procedes de definition en topologie recursive* ◇ D80 ◇

MEDVEDEV, YU.T. *On the concept of mass problem and its applications in the theory of recursive functions and in mathematical logic (Russian)* ◇ D30 D80 ◇

1960

KLEENE, S.C. *Realizability and Shanin's algorithm for the constructive deciphering of mathematical sentences* ◇ D80 F50 ◇

LACOMBE, D. *La theorie des fonctions recursives et ses applications (expose d'une information generale)* ◇ D10 D20 D80 D98 ◇

MARKOV, A.A. *Insolubility of the problem of homeomorphy (Russian)* ◇ D80 ◇

1962

MEDVEDEV, YU.T. *Finite problems (Russian)* ◇ D80 F50 ◇

1963

EVANS, T. *A decision problem for translations of trees* ◇ D80 ◇

GINSBURG, S. & ROSE, G.F. *Some recursively unsolvable problems in ALGOL-like languages* ◇ B70 D80 ◇

MEDVEDEV, YU.T. *Interpretation of logical formulae by means of finite problems and its relation to the realizability theory (Russian)* ◇ D80 F50 ◇

1964

ARBIB, M.A. *Brains, machines, and mathematics* ◇ D05 D10 D80 F30 ◇

BURGER, E. *Bemerkungen zu einigen Fassungen des Goedelschen Unvollstaendigkeitssatzes* ◇ D80 F30 ◇

ZYKOV, A.A. *Recursively calculable functions of graphs* ◇ D20 D75 D80 ◇

1965

GINSBURG, S. & ROSE, G.F. *Some recursively unsolvable problems in ALGOL-like languages* ◇ B75 D80 ◇

KOLMOGOROV, A.N. *Three approaches to the definition of the concept "quantity of information" (Russian)* ◇ D15 D20 D80 ◇

MULLIN, A.A. *A contribution toward computable number theory* ◇ D80 ◇

TANG, CHISUNG *A recursive theory of computer instructions (Chinese)* ◇ B75 D05 D80 ◇

1966

DEUEL, D.R. & GILL, A. *Some decision problems associated with weighted directed graphs* ◇ D05 D80 ◇

JAULIN, B. *Sur un aspect du calcul* ◇ B75 D15 D80 ◇

LOVELAND, D.W. *The Kleene hierarchy classification of recursively random sequences* ◇ D20 D55 D80 ◇
MARTIN-LOEF, P. *The definition of random sequences* ◇ D80 ◇
MEDVEDEV, YU.T. *Interpretation of logical formulae by means of finite problems (Russian)* ◇ D80 F50 ◇
MELZAK, Z.A. *An informal arithmetical approach to computability and computation III* ◇ D10 D80 ◇

1967

MCLAUGHLIN, T.G. *Some remarks on extensibility, confluence of paths, branching properties, and index sets, for certain recursively enumerable graphs* ◇ D25 D80 ◇
SCHWARTZ, J.T. (ED.) *Mathematical aspects of computer science* ◇ B35 B65 B75 B97 D10 D80 D97 ◇

1968

AGAFONOV, V.N. *Complexity of computation of pseudorandom sequences (Russian)* ◇ D15 D80 ◇
BOONE, W.W. & HAKEN, W. & POENARU, V. *On recursively unsolvable problems in topology and their classification* ◇ D35 D80 ◇
BUSKIRK VAN, J. & GILLETTE, P. *The word problem and consequences for the braid groups and mapping class groups of the 2-sphere* ◇ D40 D80 ◇
DEKKER, J.C.E. *On certain vector spaces of isolic dimension* ◇ D50 D80 ◇
ENGELER, E. *Remarks on the theory of geometrical constructions* ◇ C75 D80 ◇
KALMAR, L. *On the problem of full utilization of the technical possibilities of computers in devising appropriate approximation methods for the solution of numerical problems* ◇ D20 D80 ◇
RICHARDSON, D.B. *Some undecidable problems involving elementary functions of a real variable* ◇ D40 D80 ◇

1969

ADLER, A. *Some recursively unsolvable problems in analysis* ◇ D35 D80 ◇
CHAITIN, G.J. *On the length of programs for computing finite binary sequences: statistical considerations* ◇ D15 D80 ◇
CHAITIN, G.J. *On the simplicity and speed of programs for computing infinite sets of natural numbers* ◇ D15 D80 ◇
CLEAVE, J.P. *The primitive recursive analysis of ordinary differential equations and the complexity of their solutions* ◇ D20 D80 F60 ◇
DESTOUCHES, J.-L. *Has the theory of recursive functions application in mathematical physics?* ◇ A10 D80 ◇
KANOVICH, M.I. & KUSHNER, B.A. *Estimating the complexity of certain algorithmic problems of analysis (Russian)* ◇ D15 D20 D80 F60 ◇
KOLMOGOROV, A.N. *Logical basis of information theory and probability theory (Russian)* ◇ B30 D15 D20 D80 ◇
MARTIN-LOEF, P. *The literature on von Mises' Kollectivs revisited* ◇ A10 D80 ◇
RICHARDSON, D.B. *Solution of the identity problem for integral exponential functions* ◇ D80 ◇
SCHNORR, C.-P. *Eine Bemerkung zum Begriff der zufaelligen Folge (English summary)* ◇ D20 D80 F65 ◇

1970

JACOBS, K. *Turing-Maschinen und zufaellige 0-1-Folgen* ◇ D10 D80 ◇
LEVIN, L.A. & ZVONKIN, A.K. *The complexity of finite objects and the basing of the concepts of information and randomness on the theory of algorithms (Russian)* ◇ D20 D80 ◇
MARTIN-LOEF, P. *On the notion of randomness* ◇ B30 D55 D80 E15 ◇
MILLER, W. *Recursive function theory and numerical analysis* ◇ D20 D80 ◇
PATERSON, M.S. *Unsolvability in* 3×3 *matrices* ◇ D80 ◇
POULSEN, B.T. *The Medvedev lattice of degrees of difficulty* ◇ D30 D80 ◇
SCHNORR, C.-P. *Minimale Programmkomplexitaet und Zufaelligkeit* ◇ D15 D80 ◇
SCHNORR, C.-P. *Ueber die Zufaelligkeit und den Zufallsgrad von Folgen* ◇ D20 D80 ◇

1971

ABERTH, O. *The concepts of effective method applied to computational problems of linear algebra* ◇ D80 ◇
ASHCROFT, E.A. & MANNA, Z. & PNUELI, A. *Decidable properties of monadic functional schemas* ◇ D80 ◇
BERNHARDT, L. & LINDNER, R. & THIELE, H. *Ueber sequentiell berechenbare reelle Abbildungen (English and Russian summaries)* ◇ D80 ◇
RICHARDSON, D.B. *The simple exponential constant problem* ◇ D80 F30 ◇
ROBINSON, R.M. *Undecidability and nonperiodicity for tilings of the plane* ◇ D80 ◇
SCHNORR, C.-P. *A unified approach to the definition of random sequences* ◇ D80 ◇
SCHNORR, C.-P. *Komplexitaet von Algorithmen mit Anwendung auf die Analysis* ◇ D15 D80 ◇
SCHNORR, C.-P. *Zufaelligkeit und Wahrscheinlichkeit. Eine algorithmische Begruendung der Wahrscheinlichkeitstheorie* ◇ B30 B75 D80 ◇
SPECKER, E. *Ramsey's theorem does not hold in recursive set theory* ◇ D25 D80 E05 F60 ◇

1972

APPEL, K.I. & SCHUPP, P.E. *The conjugacy problem for the group of any tame alternating knot is solvable* ◇ D40 D80 ◇
BUEVICH, V.A. *On the algorithmic undecidability of a-completeness for the boundedly determinate functions (Russian)* ◇ D80 ◇
JOCKUSCH JR., C.G. *Ramsey's theorem and recursion theory* ◇ D30 D55 D80 ◇
KOGAN, D.I. *D-sets,* Δ*-sets and undecidable problems of discrete control (Russian) (English summary)* ◇ D80 ◇
MANASTER, A.B. & ROSENSTEIN, J.G. *Effective matchmaking (recursion theoretic aspects of a theorem of Philip Hall)* ◇ D30 D80 E05 ◇
MUELLER, D.W. *Randomness and extrapolation* ◇ D55 D80 ◇

TUCKER, J.V. *Algorithmic unsolvability in biological contexts* ◇ D75 D80 ◇

WEESE, M. *Zur Modellvollstaendigkeit und Entscheidbarkeit gewisser topologischer Raeume (Russian, English and French summaries)* ◇ B25 C35 C65 D80 ◇

1973

AANDERAA, S.O. *A proof of Higman's embedding theorem using Britton extensions of groups* ◇ C60 D40 D80 ◇

BISKUP, J. *Zufaellige Folgen und Bi-Immunitaet* ◇ D50 D80 ◇

BLUM, L. & BLUM, M. *Inductive inference: a recursion theoretic approach* ◇ B48 D15 D80 ◇

BULITKO, V.K. *Graphs with prescribed environments of the vertices (Russian)* ◇ D80 ◇

BULITKO, V.K. *On one algorithmically unsolvable problem for graphs (Russian)* ◇ D80 ◇

GLEBSKIJ, YU.V. & KOGAN, D.I. & LIOGON'KIJ, M.I. *On a certain discrete walk problem (Russian)* ◇ D80 ◇

HAKEN, W. *Connections between topological and group theoretical decision problems* ◇ B25 C60 C65 D35 D40 D80 ◇

KOGAN, D.I. *Algorithmic problems for a certain class of multi-stage games (Russian) (English summary)* ◇ D80 ◇

LAING, R. *The capabilities of some species of artificial organism* ◇ D05 D80 ◇

LEVIN, L.A. *On storage capacity for algorithms (Russian)* ◇ D20 D80 ◇

LEVIN, L.A. *On the notion of a random sequence (Russian)* ◇ D15 D80 ◇

MANASTER, A.B. & ROSENSTEIN, J.G. *Effective matchmaking and k-chromatic graphs* ◇ D80 ◇

MANIN, YU.I. *Hilbert's tenth problem (Russian)* ◇ D25 D35 D80 ◇

MYLOPOULOS, J. & TOURLAKIS, G. *Some results in computational topology* ◇ D40 D80 ◇

OREVKOV, V.P. *On the complexity of expansion of algebraic irrationalities in continued fractions (Russian)* ◇ D20 D80 ◇

PERL, J. *Anwendung von Graphenalgorithmen auf allgemeinere Problemklassen (English summary)* ◇ D80 ◇

POUR-EL, M.B. *Abstract computability versus analog-generability. A survey* ◇ D75 D80 F65 ◇

POUR-EL, M.B. *Analog computers, digital computers, mathematical logic, differential equations - interrelations* ◇ D20 D75 D80 F65 ◇

ROBINSON, JULIA *Solving diophantine equations* ◇ B25 D35 D80 H15 ◇

SEMENOV, A.L. *Algorithmic problems for power series and for context-free grammars (Russian)* ◇ D05 D80 ◇

VUCKOVIC, V. *Local recursive theory* ◇ D20 D25 D75 D80 ◇

1974

AANDERAA, S.O. & LEWIS, H.R. *Linear sampling and the ∀∃∀ case of the decision problem* ◇ B20 D03 D05 D35 D80 ◇

AITCHISON, P.W. & HERMAN, G.T. *A decision procedure using the geometry of convex sets* ◇ D10 D80 ◇

APPEL, K.I. *On the conjugacy problem for knot groups* ◇ D40 D80 ◇

BOKUT', L.A. *Undecidability of certain algorithmic problems for Lie algebras (Russian)* ◇ D40 D80 ◇

BRUDNO, A.A. *Topologische Entropie und Kompliziertheit nach A. N. Kolmogorov (Russisch)* ◇ D15 D80 ◇

CHAITIN, G.J. *Information-theoretic limitations of formal systems* ◇ D15 D80 F20 ◇

CHAITIN, G.J. *Information-theoretic computational complexity* ◇ D15 D80 ◇

HANF, W.P. *Nonrecursive tilings of the plane I* ◇ D05 D80 ◇

LEEUWEN VAN, J. *A partial solution to the reachability-problem for vector-addition systems* ◇ D80 ◇

LENDER, V.B. *On steps of solubility of lattices and degrees of idempotency of prevarieties of lattices (Russian)* ◇ C05 D80 ◇

LIPSHITZ, L. *The undecidability of the word problems for projective geometries and modular lattices* ◇ D40 D80 G10 ◇

MEJTUS, V.YU. & VERSHININ, K.P. *On some unsolvable problems in computable categories (Russian)* ◇ C57 D80 G30 ◇

MYERS, D.L. *Nonrecursive tiling of the plane II* ◇ D05 D80 ◇

PEDANOV, I.E. *The solvability of the problem of the membership of a point in an object in a geometric data-processing language (Russian)* ◇ B75 D80 ◇

PETERS, F.E. *Einfuehrung in mathematische Methoden der Informatik* ◇ B70 B98 D80 D98 ◇

PETROSYAN, G.N. *A certain basis of operators and predicates with an undecidable emptiness problem (Russian) (English summary)* ◇ B75 D05 D80 ◇

POUR-EL, M.B. *Abstract computability and its relation to the general purpose analog computer (some connections between logic, differential equations and analog computers)* ◇ D20 D75 D80 F65 ◇

RABIN, M.O. *Theoretical impediments to artificial intelligence* ◇ B75 D15 D80 ◇

WANG, HAO *Notes on a class of tiling problems* ◇ D05 D80 ◇

WANG, P.S. *The undecidability of the existence of zeros of real elementary functions* ◇ D80 ◇

ZHAROV, V.G. *On an analog of a theorem of Specker (Russian)* ◇ D20 D80 F60 ◇

1975

ASSER, G. *Rekursive Graphenabbildungen. I* ◇ D20 D45 D80 ◇

AUSIELLO, G. & PROTASI, M. *On the comparison of notions of approximation* ◇ D20 D80 ◇

BAER, ROBERT M. *Computation by assembly* ◇ D80 ◇

BAER, ROBERT M. *Self-assembly and differentiation as models of computability* ◇ D10 D80 ◇

BARWISE, J. & SCHLIPF, J.S. *On recursively saturated models of arithmetic* ◇ C50 C57 C62 D80 H15 ◇

BECK, H. *Zur Entscheidbarkeit der funktionalen Aequivalenz (English summary)* ◇ D20 D80 ◇

BURKS, A.W. *Logic, biology and automata - some historical reflections* ◇ A05 A10 D05 D10 D80 ◇
CALDWELL, J. & POUR-EL, M.B. *On a simple definition of computable function of a real variable - with applications to functions of a complex variable* ◇ D20 D80 ◇
DALEY, R.P. *Minimal program complexity of pseudo-recursive and pseudo-random sequences* ◇ D15 D80 ◇
HAVRANEK, T. *The approximation problem in computational statistics* ◇ D80 ◇
HOWARD, J.V. *Computable explanations* ◇ D80 ◇
MASLOV, S.YU. *Mutation calculi (Russian) (English summary)* ◇ D03 D25 D80 ◇
MEERSMAN, R. *A survey of techniques in applied computational complexity* ◇ D15 D80 ◇
SCHLOESSER, L. *Ueber boolesche Ianovschemata* ◇ B75 D80 ◇
SZYMANSKI, T.G. & ULLMAN, J.D. *Evaluating relational expressions with dense and sparse arguments* ◇ D15 D80 ◇

1976

ANSHEL, M. *Decision problems for HNN groups and vector addition systems* ◇ D40 D80 ◇
BEAN, D.R. *Effective coloration* ◇ D80 ◇
BEAN, D.R. *Recursive Euler and Hamilton paths* ◇ D80 ◇
BENIOFF, P.A. *Models of Zermelo Fraenkel set theory as carriers for the mathematics of physics I,II* ◇ C62 D80 E75 ◇
CARDOZA, E. & LIPTON, R.J. & MEYER, A.R. *Exponential space complete problems for Petri nets and commutative semigroups* ◇ D15 D40 D80 ◇
CHAITIN, G.J. *Algorithmic entropy of sets* ◇ D15 D25 D80 ◇
CHAITIN, G.J. *Information-theoretic characterization of recursive infinite strings* ◇ D15 D20 D80 ◇
CHERNYAKHOVSKIJ, N.P. *The expressibility of realizability, in the language of formal arithmetic (Russian)* ◇ D80 F30 F50 ◇
DAVIS, MARTIN D. & MATIYASEVICH, YU.V. & ROBINSON, JULIA *Hilbert's tenth problem: Diophantine equations: Positive aspects of a negative solution* ◇ D25 D35 D80 ◇
DEKKER, J.C.E. *Projective planes of infinite but isolic order* ◇ D50 D80 ◇
HACK, M. *The equality problem for vector addition systems is undecidable* ◇ D35 D80 ◇
HAVRANEK, T. *Statistics and computability* ◇ B75 D20 D80 ◇
HERZOG, T. & OWINGS JR., J.C. *The inequivalence of two well-known notions of randomness for binary sequence* ◇ D20 D80 ◇
IBARRA, O.H. & KIM, C.E. *A useful device for showing the solvability of some decision problems* ◇ D03 D10 D80 ◇
KUSHNER, B.A. *On Grzegorczyk's theorem on the computability of an isolated extremum (Russian)* ◇ D20 D80 F60 ◇
LEVIN, L.A. *On the principle of conservation of information in intuitionistic mathematics (Russian)* ◇ D15 D80 F50 F55 ◇
MOSTOWSKI, T. *Analytic applications of decidability theorems* ◇ B25 C60 C65 D80 ◇
PAKHOMOV, S.V. *Hierarchies of operators in constructive metric spaces (Russian) (English summary)* ◇ D80 F60 ◇
PENZIN, YU.G. *Algorithmic problems in the theory of numbers (Russian)* ◇ D80 ◇
PIGOZZI, D. *Base-undecidable properties of universal varieties* ◇ C05 D80 ◇
ROSENBLOOM, P.C. *Structural models for use in psychological research* ◇ D03 D80 ◇
SALOMAA, A. *Undecidable problems concerning growth in informationless Lindenmayer systems (German and English summaries)* ◇ D05 D80 ◇
TAJMANOV, A.D. *An algorithmic problem of number theory (Russian)* ◇ D35 D80 ◇

1977

ARAKI, T. & KASAMI, T. *Some decision problems related to the reachability problem for Petri nets* ◇ D80 G05 ◇
BROOK, T. *Order and recursion in topoi* ◇ D80 E07 G30 ◇
BUTNARIU, D. *L-fuzzy automata. Description of a neural model* ◇ D05 D80 E72 ◇
BYRNES, C. & GAUGER, M.A. *Decidability criteria for the similarity problem, with applications to the moduli of linear dynamical systems* ◇ D40 D80 ◇
CARSTENS, H.G. *The complexity of some combinatorial constructions* ◇ D55 D80 E05 ◇
CHAITIN, G.J. *Program size, oracles, and the jump operation* ◇ D15 D25 D30 D80 ◇
DASSOW, J. *Ueber Abschlusseigenschaften von biologisch motivierten Sprachen* ◇ D05 D80 ◇
DIES, J.E. *Sur l'independance algorithmique des suites infinies de Bernoulli (English summary)* ◇ D20 D80 ◇
FUCHS, P.H. & SCHNORR, C.-P. *General random sequences and learnable sequences* ◇ D20 D25 D80 ◇
JACOB, G. *Decidabilite de la finitude des demi-groups de matrices* ◇ D80 ◇
JEFFERSON, D. & SUZUKI, N. *Verification decidability of Presburger array programs* ◇ B75 D80 ◇
KRATKO, M.I. & REVIN, O. *Turing machines operating on the plane, and stable iterative systems (Russian) (English summaries)* ◇ D10 D80 ◇
LEVIN, L.A. *On a concrete method of assigning complexity measures (Russian)* ◇ D15 D80 ◇
LINDNER, R. & STAIGER, L. *Erkennungs-, mass- und informations-theoretische Eigenschaften regulaerer Folgenmengen* ◇ D80 E15 ◇
MATIYASEVICH, YU.V. *Some purely mathematical results inspired by mathematical logic* ◇ B30 D25 D35 D80 ◇
NELSON, GREG & OPPEN, D.C. *Fast decision algorithms based on union and find* ◇ D15 D80 ◇
POGORZELSKI, H.A. *Goldbach conjecture* ◇ D80 ◇

Scott, D.S. *Logic and programming languages* ◇ A10 B75 D80 ◇
Smorynski, C.A. *A note on the number of zeros of polynomials and exponential polynomials* ◇ D25 D35 D80 ◇
Vuckovic, V. *Recursive and recursive enumerable manifolds I,II* ◇ C57 D45 D80 ◇

1978

Anshel, M. *Vector groups and the equality problem for vector addition systems* ◇ D40 D80 ◇
Arnold, A. & Nivat, M. *Metric interpretations of recursive program schemes* ◇ B75 D80 ◇
Arnold, A. & Latteux, M. *Recursivite et cones rationnels fermes par intersection* ◇ D05 D80 ◇
Baxter, L.D. *The undecidability of the third order dyadic unification problem* ◇ B15 B40 D80 ◇
Bollman, D.A. & Laplaza, M.L. *Some decision problems for polynomial mappings* ◇ D80 ◇
Bulitko, V.K. *Solvable property of block-complete graphs* ◇ D80 ◇
Chaitin, G.J. & Schwartz, J.T. *A note on Monte Carlo primality tests and algorithmic information theory* ◇ D15 D80 ◇
Fowler III, N. *Effective inner product spaces* ◇ C57 C60 D45 D80 ◇
Grabowski, J. & Starke, P.H. *Petri-Netze* ◇ D05 D80 ◇
Grabowski, M. *The algebraic properties of nets from a program-theoretical point of view* ◇ D05 D80 ◇
Grigor'ev, D.Yu. *Multiplicative complexity of a pair of bilinear forms and of the polynomial multiplication* ◇ D15 D80 ◇
Gurari, E.M. & Ibarra, O.H. *An NP-complete number-theoretic problem* ◇ B25 D15 D80 ◇
Hajek, P. & Havranek, T. *Mechanizing hypothesis formation. Mathematical foundations for a general theory* ◇ A05 B35 C13 C80 C90 D80 ◇
Hughes, C.E. *The equivalence of vector addition systems to a subclass of Post canonical systems* ◇ D03 D80 ◇
Martirosyan, A.A. & Pogosyan, Eh.M. *Eine Untersuchung der Stabilitaet relativer Charakteristiken adaptiver Induktoren (Russian)* ◇ D80 ◇
McAloon, K. *Diagonal methods and strong cuts in models of arithmetic* ◇ C57 C62 D80 F30 ◇
Misercque, D. *Probleme des mariages et recursivite* ◇ D80 ◇
Morscher, E. *Inwiefern betreffen Fragen der Vollstaendigkeit und Entscheidbarkeit den Juristen* ◇ A05 D80 ◇
Pour-El, M.B. *Computer science and recursion theory* ◇ D80 D98 ◇
Pour-El, M.B. & Richards, I. *Differentiability properties of computable functions - a summary* ◇ D20 D80 F60 ◇
Robinson, R.M. *Undecidable tiling problems in the hyperbolic plane* ◇ D80 ◇
Salomaa, A. & Soittola, M. *Automata-theoretic aspects of formal power series* ◇ D05 D80 ◇
Sciore, E. & Tang, A. *Computability theory in admissible domains* ◇ B75 D80 ◇
Sims, C.C. *The role of algorithms in the teaching of algebra* ◇ D80 ◇
Singer, M.F. *The model theory of ordered differential fields* ◇ B25 C25 C35 C60 C65 D80 ◇
Starke, P.H. *Free Petri net languages* ◇ D05 D80 ◇
Uesu, T. *A system of graph grammars which generates all recursively enumerable sets of labelled graphs* ◇ D25 D80 ◇

1979

Ajtai, M. & Csirmaz, L. & Nagy, Z. *On a generalization of the game Go-moku I* ◇ D80 E60 ◇
Arbib, M.A. & Manes, E.G. *Intertwined recursion, tree transformations, and linear systems* ◇ D05 D20 D80 G30 ◇
Bergstra, J.A. & Tucker, J.V. *Equational specifications for computable data types: six hidden functions suffice and other sufficiency bounds* ◇ B75 D45 D80 ◇
Bergstra, J.A. & Tiuryn, J. *Implicit definability of algebraic structures by means of program properties* ◇ B75 D80 ◇
Bertoni, A. & Mauri, G. & Miglioli, P.A. *A characterization of abstract data as model-theoretic invariants* ◇ B75 C57 D45 D80 ◇
Boerger, E. & Kleine Buening, H. *The reachability problem for Petri nets and decision problems for Skolem arithmetic* ◇ B25 D15 D35 D80 ◇
Britton, J.L. *Integer solutions of systems of quadratic equations* ◇ B25 D35 D80 ◇
Calude, C. & Marcus, S. & Paun, G. *The universal grammar as a hypothetical brain* ◇ B65 D05 D80 ◇
Cherniavsky, J.C. *On finding test data sets for loop free programs* ◇ B75 D80 ◇
Degano, P. & Sirovich, F. *Inductive generalization and proofs of function properties* ◇ B35 B75 D20 D80 ◇
Fitting, M. *An axiomatic approach to computers* ◇ B75 D80 ◇
Grabowski, J. *The unsolvability of some Petri net language problems* ◇ D05 D80 ◇
Grunewald, F. & Segal, D. *The solubility of certain decision problems in arithmetic and algebra* ◇ D40 D80 ◇
Gurari, E.M. & Ibarra, O.H. *An NP-complete number-theoretic problem* ◇ B25 D10 D15 D25 D80 ◇
Hagihara, K. & Ito, Minoru & Kasami, T. & Taniguchi, K. *Decision problems for multivalued dependencies in relational databases* ◇ B75 D80 ◇
Harkleroad, L. *Recursive equivalence types on recursive manifolds* ◇ D45 D50 D80 ◇
Harrow, K. *Theoretical and applied computer science: antagonism or symbiosis?* ◇ B75 D10 D80 ◇
Hopcroft, J.E. & Pansiot, J.-J. *On the reachability problem for 5-dimensional vector addition systems* ◇ D80 ◇
Kleine Buening, H. *Generalized vector addition systems with finite exception sets* ◇ B75 D05 D10 D80 ◇
Kobuchi, Y. & Seki, S. *Decision promblems of locally catenative property for DIL systems* ◇ D10 D80 ◇

LANGMAACK, H. *On a theory of decision problems in programming languages* ⋄ B75 D05 D80 ⋄

MASLOV, S.YU. *Calculi with monotone deductions and their economic interpretation (Russian) (English summary)* ⋄ D05 D80 ⋄

MERZENICH, W. *A binary operation on trees and an initial algebra characterization for finite tree types* ⋄ D80 ⋄

POUR-EL, M.B. & RICHARDS, I. *A computable ordinary differential equation which possesses no computable solution* ⋄ D80 F60 ⋄

RASSIAS, G.M. *Stallings homomorphisms and the simply connectedness problem* ⋄ D40 D80 ⋄

RIEDEMANN, E.H. *The control of parallel computations by labeled Petri nets: a study in terms of multiple-firing automata and parallel program schemata (Dissertation)* ⋄ D05 D80 ⋄

SMORYNSKI, C.A. *Some rapidly growing functions* ⋄ D20 D80 F30 ⋄

STARKE, P.H. *On the languages of bounded Petri nets* ⋄ D05 D80 ⋄

STARKE, P.H. *Semilinearity and Petri nets* ⋄ D05 D80 ⋄

STILLWELL, J.C. *Unsolvability of the knot problem for surface complexes* ⋄ D80 ⋄

WEYUKER, E.J. *The applicability of program schema results to programs* ⋄ B75 D80 ⋄

1980

AVENHAUS, J. & MADLENER, K. *String matching and algorithmic problems in free groups* ⋄ D15 D40 D80 ⋄

BAKKER DE, J.W. *Mathematical theory of program correctness* ⋄ B75 D80 ⋄

BAUR, W. & RABIN, M.O. *Linear disjointness and algebraic complexity* ⋄ D15 D80 ⋄

BAUR, W. *On the elementary theory of quadruples of vector spaces* ⋄ B25 C60 D80 ⋄

BERGSTRA, J.A. & TUCKER, J.V. *A characterisation of computable data types by means of a finite equational specification method* ⋄ B75 D45 D80 ⋄

BERGSTRA, J.A. & TUCKER, J.V. *A natural data type with a finite equational final semantics specification but no effective equational initial semantics specification* ⋄ B75 D45 D80 ⋄

BERGSTRA, J.A. & TUCKER, J.V. *On bounds for the specification of finite data types by means of equations and conditional equations* ⋄ B75 D45 D80 ⋄

BERTONI, A. & MAURI, G. & MIGLIOLI, P.A. *Towards a theory of abstract data types: a discussion on problems and tools* ⋄ B75 C57 D80 ⋄

BOERGER, E. & KLEINE BUENING, H. *The reachability problem for Petri nets and decision problems for Skolem arithmetic* ⋄ B25 D15 D35 D80 ⋄

CULIK, K. *Parallel permit grammars and some graph representations of Petri nets* ⋄ D05 D80 ⋄

DAVID, R. & MITRANI, E. & TELLEZ-GIRON, R. *Emploi de cellules universelles pour la synthese de systemes asynchrones decrits par reseaux de Petri* ⋄ D05 D80 ⋄

EVEN, S. & YACOBI, Y. *Cryptocomplexity and NP-completeness* ⋄ D15 D80 ⋄

FOELDES, S. & SINGHI, N.M. *A nonconstructible projective plane* ⋄ D80 ⋄

FOELDES, S. & STEINBERG, R. *A topological space for which graph embeddability is undecidable* ⋄ D80 ⋄

FOTHI, A. & VARGA, Z. *Modeling programs via recursive functions I (Hungarian) (English summary)* ⋄ B75 D80 ⋄

FRANK, A. *The countability of the rational polynomials: A direct method* ⋄ D80 ⋄

GEHANI, N. *Generic procedures: An implementation and an undecidability result* ⋄ B75 D80 ⋄

GEORGIEVA, N.V. *Theorems on the normal form of some recursive elements and mappings (Russian)* ⋄ D45 D80 ⋄

HAUCK, J. *Konstruktive Darstellungen in topologischen Raeumen mit rekursiver Basis* ⋄ D45 D80 ⋄

INDERMARK, K. *On rational definitions in complete algebras without rank* ⋄ D80 ⋄

KLEINE BUENING, H. *Decision problems in generalized vector addition systems* ⋄ D80 ⋄

LUKAVCOVA, M. *On computable real functions* ⋄ D80 ⋄

MAYR, E.W. *Ein Algorithmus fuer das allgemeine Erreichbarkeitsproblem bei Petrinetzen und damit zusammenhaengende Probleme (Dissertation)* ⋄ D05 D80 ⋄

MUNDICI, D. *Natural limitations of algorithmic procedures in logic (Italian summary)* ⋄ C40 D05 D80 ⋄

SCHMERL, J.H. *Recursive colorings of graphs* ⋄ D45 D80 ⋄

SMITH, C.H. *Applications of classical recursion theory to computer science* ⋄ B75 D15 D20 D80 ⋄

STOLTENBERG-HANSEN, V. & TUCKER, J.V. *Computing roots of unity in fields* ⋄ D25 D45 D80 ⋄

TUCKER, J.V. *Computing in algebraic systems* ⋄ D10 D35 D75 D80 ⋄

WINKLER, P.M. *Computational characterization of abelian groups* ⋄ D05 D80 ⋄

1981

APT, K.R. *Recursive assertions and parallel programs* ⋄ B75 D80 ⋄

BAGYINSZKINE OROSZ, A. *Petri nets and their languages* ⋄ D05 D80 ⋄

BAUMSLAG, G. & DYER, E. & MILLER III, C.F. *On the integral homology of finitely presented groups* ⋄ D40 D80 ⋄

BERGSTRA, J.A. & TUCKER, J.V. *Algebraically specified programming systems and Hoare's logic* ⋄ B75 D80 ⋄

BERGSTRA, J.A. & MEYER, J.-J.C. *Small specifications for large finite data structures* ⋄ B75 D45 D80 ⋄

BERTONI, A. & MAURI, G. *On efficient computation of the coefficients of some polynomials with applications to some enumeration problems* ⋄ D15 D80 ⋄

BRUIN DE, A. *On the existence of Cook semantics* ⋄ B75 D20 D80 ⋄

CHRISTIAN, C.C. *Das rekursive Inaccessibilitaetstheorem und der Goedelsche Unvollstaendigkeitssatz in ihrer Bedeutung fuer die Informatik* ⋄ A05 D35 D80 F30 ⋄

DEKKER, J.C.E. *Twilight graphs* ⋄ D50 D80 ⋄
EVEN, S. & YACOBI, Y. *An observation concerning the complexity of problems with few solutions and its application to cryptography* ⋄ D15 D80 ⋄
FRIDMAN, G.SH. *Effective unsolvability of a discrete extremal problem (Russian)* ⋄ D80 ⋄
JOSEPH, D. & YOUNG, P. *Independence results in computer science?* ⋄ B75 D80 ⋄
JOSKO, B. *An effective retract calculus* ⋄ D80 ⋄
KALANTARI, I. *Effective content of a theorem of M.H. Stone* ⋄ D45 D80 ⋄
KANDA, A. *Constructive category theory* ⋄ D45 D80 G30 ⋄
KIERSTEAD, H.A. *An effective version of Dilworth's theorem* ⋄ D45 D80 ⋄
KIERSTEAD, H.A. & TROTTER JR., W.T. *An extremal problem in recursive combinatorics* ⋄ D80 E07 ⋄
KIERSTEAD, H.A. *Recursive colorings of highly recursive graphs* ⋄ D80 ⋄
KLAEREN, H.A. & SCHULZ, MARTIN *Computable algebras, word problems and canonical term algebras* ⋄ D45 D80 ⋄
KLOSINSKI, L.F. & SMOLARSKI, D.C. *Recognition algorithms for Fibonacci numbers* ⋄ D10 D80 ⋄
KROM, MELVEN R. *An unsolvable problem with products of matrices* ⋄ D80 ⋄
KRUPSKIJ, V.N. *On the complexity of a description of computable approximations for points of a metric space (Russian) (English summary)* ⋄ D15 D80 ⋄
MATETI, P. *A decision procedure for the correctness of a class of programs* ⋄ B75 D80 ⋄
MAYR, E.W. & MEYER, A.R. *The complexity of the finite containment problem for Petri nets* ⋄ D15 D20 D80 ⋄
MIGLIOLI, P.A. & ORNAGHI, M. *A logically justified model of computation. I,II* ⋄ B75 D80 F07 F50 ⋄
MO, SHAOKUI *From black box to computers - an analysis of the nature of computers (Chinese)* ⋄ D80 ⋄
MULLER, D.E. & SCHUPP, P.E. *Context-free languages, groups, the theory of ends, second-order logic, tiling problems, cellular automata, and vector addition systems* ⋄ B15 B25 D05 D80 ⋄
PELED, U.N. & SIMEONE, B. *A polynomial-time algorithm for recognizing threshold functions* ⋄ D15 D80 ⋄
POUR-EL, M.B. & RICHARDS, I. *The wave equation with computable initial data such that its unique solution is not computable* ⋄ D80 ⋄
RICCARDI, G.A. *The independence of control structures in abstract programming systems* ⋄ B70 D80 ⋄
SZALAS, A. *Algorithmic logic with recursive functions* ⋄ B75 D80 ⋄
TAYLOR, W. *Some universal sets of terms* ⋄ C05 D80 ⋄
TURAN, G. *On cellular graph-automata and second-order definable graph-properties* ⋄ C85 D05 D80 ⋄
VARDI, M.Y. *The decision problem for database dependencies* ⋄ B75 D80 ⋄

1982

ALBERT, J. & CULIK II, K. *Tree correspondence problems* ⋄ D80 ⋄
BENIOFF, P.A. *Quantum mechanical Hamiltonian models of Turing machines* ⋄ D10 D80 ⋄
BENIOFF, P.A. *Quantum mechanical models of Turing machines that dissipate no energy* ⋄ D10 D80 ⋄
BERGSTRA, J.A. & KLOP, J.W. *Algebraic specifications for parametrized data types with minimal parameter and target algebras* ⋄ D45 D80 ⋄
BERGSTRA, J.A. & TUCKER, J.V. *The completeness of the algebraic specification methods for computable data types* ⋄ B75 D80 ⋄
BERGSTRA, J.A. & TUCKER, J.V. *The refinement of specifications and the stability of Hoare's logic* ⋄ B75 D80 ⋄
BUDACH, L. & MEINEL, C. *Environments and Automata I., II (German and Russian summaries)* ⋄ D05 D10 D80 ⋄
CALUDE, C. & CHITESCU, I. *On Per Martin-Loef random sequences* ⋄ B75 D80 ⋄
CALUDE, C. & CHITESCU, I. *Random strings according to A.N. Kolmogorov and P. Martin-Loef classical approach* ⋄ D80 ⋄
COMYN, G. & DAUCHET, M. *Approximations of infinitary objects* ⋄ D80 ⋄
FRIEDMAN, H.M. & KO, KER-I *Computational complexity of real functions* ⋄ D15 D80 ⋄
GUREVICH, Y. & LEWIS, H.R. *The inference problem for template dependencies* ⋄ B75 D80 ⋄
HAUCK, J. *Stetigkeitseigenschaften berechenbarer Funktionale* ⋄ D80 ⋄
HAY, L. & MILLER, DOUGLAS E. *The Addison game played backwards: index sets in topology* ⋄ D55 D80 E75 ⋄
HYLAND, J.M.E. *The effective topos* ⋄ D75 D80 F35 F50 G30 ⋄
JONES, JAMES P. *Some undecidable determined games* ⋄ D80 E60 ⋄
KO, KER-I *Some negative results on the computational complexity of total variation and differentiation* ⋄ D15 D80 ⋄
KO, KER-I *The maximum value problem and NP real numbers* ⋄ D15 D80 ⋄
LAUSMAA, T. *Informational properties of a partition (Russian) (English and Estonian summaries)* ⋄ D80 E75 ⋄
MADISON, E.W. *A hierarchy of regular open sets of the Cantor space* ⋄ D80 ⋄
MANAS, M. *Algorithmically unsolvable problems in economic decision making (Czech) (English summary)* ⋄ D80 ⋄
MCNULTY, G.F. *Infinite ordered sets, a recursive perspective* ⋄ D80 ⋄
MELIKYAN, S.M. *Specker numbers and degrees of undecidability (Russian)* ⋄ D30 D80 ⋄
METAKIDES, G. & NERODE, A. *The introduction of non-recursive methods into mathematics* ⋄ C57 D45 D80 F60 ⋄
MUELLER, HEINRICH *The complexity of the vertex coloring problem for some classes of graphs restricted by cycle properties* ⋄ D15 D80 ⋄
POUR-EL, M.B. & RICHARDS, I. *Noncomputability in models of physical phenomena* ⋄ D80 ⋄

SCHMERL, J.H. *The effective version of Brooks' theorem* ◇ D80 ◇

ZASLAVSKIJ, I.D. *Logical nets and monocyclic circuits (Russian) (Armenian summary)* ◇ B70 D80 ◇

1983

ANSHEL, M. & MCALOON, K. *Reducibilities among decision problems for HNN groups, vector addition systems and subsystems of Peano arithmetic* ◇ D20 D40 D80 F30 ◇

BAEZ, J.C. *Recursivity in quantum mechanics* ◇ B51 D80 ◇

BERGSTRA, J.A. & KLOP, J.W. *Initial algebra specifications for parametrized data types* ◇ B75 D45 D80 G30 ◇

BRAMHOFF, H. & JANTZEN, M. *Notions of computability by Petri nets* ◇ B75 D20 D80 ◇

BURKHARD, HANS-DIETER *Control of Petri nets by finite automata* ◇ D05 D80 ◇

CALUDE, C. & CHITESCU, I. *On representability of P. Martin-Loef tests* ◇ D80 ◇

CALUDE, C. & CHITESCU, I. *Representability of recursive P. Martin-Loef tests* ◇ D80 ◇

CARSTENS, H.G. & PAEPPINGHAUS, P. *Recursive coloration of countable graphs* ◇ D80 ◇

DEKKER, J.C.E. *Isols and balanced block designs with $\lambda = 1$* ◇ D50 D80 ◇

DEKKER, J.C.E. *Recursive equivalence types and octahedra* ◇ D50 D80 ◇

FREUND, R. *Real functions and numbers defined by Turing machines* ◇ D10 D20 D80 ◇

HARKLEROAD, L. *Manifolds allowing RET arithmetic* ◇ D45 D50 D80 ◇

IBARRA, O.H. & LEININGER, B.S. *On the simplification and equivalence problem for straight-line programs* ◇ D80 ◇

JENSEN, C.U. *L'indecidabilite d'une classe de corps des fonctions meromorphes* ◇ D35 D80 ◇

KIERSTEAD, H.A. *An effective version of Hall's theorem* ◇ C57 D80 ◇

KO, KER-I *On the computational complexity of ordinary differential equations* ◇ D15 D80 ◇

LENARD, A. & STILLWELL, J.C. *An algorithmically unsolvable problem in analysis* ◇ D80 ◇

POUR-EL, M.B. & RICHARDS, I. *Computability and noncomputability in classical analysis* ◇ D80 ◇

POUR-EL, M.B. & RICHARDS, I. *Noncomputability in analysis and physics: a complete determination of the class of noncomputable linear operators* ◇ D80 ◇

SIEKMANN, J. & SZABO, P. *Universal unification* ◇ D80 ◇

SKANDALIS, K. *Programmability in the set of real numbers and second-order recursion* ◇ B75 D20 D80 ◇

STARKE, P.H. *Monogeneous fifo-nets and Petri nets are equivalent* ◇ D05 D80 ◇

STARKE, P.H. *On the concurrency of distributed systems* ◇ D05 D80 ◇

WINKLER, P.M. *Existence of graphs with a given set of r-neighborhoods* ◇ D80 ◇

1984

BELLENOT, S.F. *The Banach space T and the fast growing hierarchy from logic* ◇ D80 E75 ◇

BOCHERNIKOV, V.YA. & KOSOVSKIJ, N.K. *A calculus of equations of program in a model of programming language (Russian) (English summary)* ◇ D80 ◇

BUECHI, J.R. & MAHR, B. & SIEFKES, D. *Manual on REC – A language for use and cost analysis of recursion over arbitrary data structures* ◇ B75 D15 D80 ◇

BUJ, D.B. & RED'KO, V.N. *Primitive program algebras I (Russian)* ◇ D20 D75 D80 ◇

BURKHARD, HANS-DIETER & STARKE, P.H. *A note on the impact of conflict resolution to liveness and deadlock in Petri nets* ◇ D05 D80 ◇

BURKHARD, HANS-DIETER *An investigation of controls for concurrent systems by abstract control languages* ◇ D05 D80 ◇

BURR, S.A. *Some undecidable problems involving the edge-coloring and vertex-coloring of graphs* ◇ D80 ◇

CHLEBUS, B.S. & CHROBAK, M. *Probabilistic Turing machines and recursively enumerable Dedekind cuts* ◇ D10 D25 D80 ◇

COMPTON, K.J. *An undecidable problem in finite combinatorics* ◇ C13 D80 ◇

DENEF, J. & LIPSHITZ, L. *Power series solutions of algebraic differential equations* ◇ B25 C60 C65 D35 D80 ◇

FRIEDMAN, H.M. *The computational complexity of maximization and integration* ◇ D15 D80 ◇

FUERER, M. *The computational complexity of the unconstrained limited domino problem (with implications for logical decision problems)* ◇ B20 D15 D80 ◇

GERMANO, G. & MAZZANTI, S. *Partial closures and semantics of while: toward an iteration-based theory of data types* ◇ B75 D20 D75 D80 ◇

GIUSTI, M. *Some effectivity problems in polynomial ideal theory* ◇ D80 ◇

JANTKE, K.P. *Polynomial time inference of general pattern languages* ◇ D05 D15 D80 ◇

JOCKUSCH JR., C.G. & KALANTARI, I. *Recursively enumerable sets and van der Waerden's theorem on arithmetic progressions* ◇ C57 D25 D80 ◇

KRAMOSIL, I. *Recursive classification of pseudo-random sequences* ◇ D15 D80 ◇

LEFMANN, H. & VOIGT, B. *A remark on infinite arithmetic progressions* ◇ D80 E05 ◇

MCALOON, K. *Petri nets and large finite sets* ◇ D05 D20 D80 ◇

NIKITIN, A.A. *Some algorithmic problems for projective planes (Russian)* ◇ C57 C65 D80 ◇

PENTTONEN, M. *The reachability of vector addition systems and equivalent problems* ◇ D03 D05 D80 ◇

POUR-EL, M.B. & RICHARDS, I. *L^p-computability in recursive analysis* ◇ D80 F60 ◇

SPREEN, D. & YOUNG, P. *Effective operators in a topological setting* ◇ D20 D75 D80 ◇

TVERBERG, H. *On Schmerl's effective version of Brooks' theorem* ◇ D80 ◇

VERESHCHAGIN, N.K. *Zeros of linear recursive sequences (Russian)* ◇ D80 ◇

ZUBENKO, V.V. *Undecidability of strict equivalence for recursive compositions (Russian)* ◇ D80 ◇

1985

BISHOP, E.A. & BRIDGES, D.S. *Constructive analysis* ◇ D80 F55 F98 ◇

CALUDE, C. & CHITESCU, I. *A combinatorial characterization of sequential P. Martin-Loef tests* ◇ D80 ◇

CRESSWELL, M.J. *The decidable normal modal logics are not recursively enumerable* ◇ B25 B45 D25 D35 D80 ◇

CUPONA, G. & JANEVA, B. & MARKOVSKI, S. *The problem of solvability of polylinear representations of universal algebras in semigroups* ◇ C05 D80 ◇

DEUTSCH, D. *Quantum theory, the Church-Turing principle and the universal quantum computer* ◇ D20 D80 ◇

HAUCK, J. *Ein Kriterium fuer die konstrukive Loesbarkeit der Differentialgleichung* $y'=f(x,y)$ ◇ D80 ◇

JACOBSSON, C. & STOLTENBERG-HANSEN, V. *Poincare-Betti series are primitive recursive* ◇ D20 D80 ◇

KALANTARI, I. & WEITKAMP, G. *Effective topological spaces I: A definability theory. II: A hierarchy* ◇ C40 C57 C65 D45 D75 D80 ◇

KREITZ, C. & WEIHRAUCH, K. *Theory of representations* ◇ D45 D75 D80 ◇

LEHMANN, G. *Modell- und rekursionstheoretische Grundlagen psychologischer Theorienbildung* ◇ B10 C98 D80 D99 ◇

METAKIDES, G. & NERODE, A. & SHORE, R.A. *Recursive limits on the Hahn-Banach theorem* ◇ D80 ◇

MULLER, D.E. & SCHUPP, P.E. *The theory of ends, pushdown automata, and second-order logic* ◇ B15 B25 D10 D80 ◇

PAPPAS, P. *A Diophantine problem for Laurent polynomial rings* ◇ C60 D35 D80 ◇

PERES, A. *Reversible logic and quantum computers* ◇ D80 ◇

SCHMERL, J.H. *Recursion theoretic aspects of graphs and orders* ◇ D80 ◇

SIMPSON, S.G. *Recursion theoretic aspects of the dual Ramsey theorem* ◇ D80 E05 ◇

STAIGER, L. *Representable P. Martin-Loef tests* ◇ D80 ◇

VIDAKOVIC, B.D. *On some properties of the Martin-Loef's measures of randomness of finite binary words* ◇ D80 ◇

WOLFRAM, S. *Undecidability and intractability in theoretical physics* ◇ D80 ◇

D97 ∪ D98 ∪ D99 Proceedings, textbooks, surveys, and miscellaneous

1932

POLITZER, R. *Rekursive Funktionen* ◇ D20 D98 ◇

1951

PETER, R. *Rekursive Funktionen* ◇ D20 D98 ◇

1952

KLEENE, S.C. *Introduction to metamathematics* ◇ A05 B98 D98 F30 F50 F98 ◇

1954

MARKOV, A.A. *Theory of algorithms (Russian)* ◇ D03 D20 D40 D98 ◇

1955

ARSENIN, V.YA. & LYAPUNOV, A.A. & SHCHEGOL'KOV, E.A. *Arbeiten zur deskriptiven Mengenlehre* ◇ D55 D70 D97 E15 E97 ◇

HERMES, H. *Vorlesung ueber Entscheidungsprobleme in Mathematik und Logik* ◇ B25 B40 B65 D10 D20 D35 D98 ◇

1956

MCCARTHY, J. & SHANNON, C.E. (EDS.) *Automata studies* ◇ D05 D97 ◇

1957

GRZEGORCZYK, A. *Decision problems* ◇ B25 C10 D35 D98 F30 ◇

1958

DAVIS, MARTIN D. *Computability and unsolvability* ◇ D98 ◇

1959

TRAKHTENBROT, B.A. *Wieso koennen Automaten rechnen?* ◇ D10 D98 ◇

1960

KLEENE, S.C. *Mathematical logic: constructive and non-constructive operations* ◇ B98 D98 ◇

LACOMBE, D. *La theorie des fonctions recursives et ses applications (expose d'une information generale)* ◇ D10 D20 D80 D98 ◇

USPENSKIJ, V.A. *Lectures on computable functions (Russian)* ◇ D20 D98 ◇

1961

GRZEGORCZYK, A. *Fonctions recursives* ◇ D20 D98 ◇

HERMES, H. *Aufzaehlbarkeit, Entscheidbarkeit, Berechenbarkeit: Einfuehrung in die Theorie der rekursiven Funktionen* ◇ D98 F60 ◇

MCNAUGHTON, R. *The theory of automata, a survey* ◇ D05 D98 ◇

SMULLYAN, R.M. *Theory of formal systems* ◇ B98 D03 D05 D20 D25 D98 ◇

1962

DEKKER, J.C.E. (ED.) *Recursive function theory* ◇ D97 ◇

GILL, A. *Introduction to the theory of finite-state machines* ◇ D05 D10 D98 ◇

KOBRINSKIJ, N.E. & TRAKHTENBROT, B.A. *Introduction to the theory of finite automata (Russian)* ◇ D05 D98 ◇

LORENZEN, P. *Metamathematik* ◇ A05 B98 C60 D20 D35 D98 F98 ◇

NEMES, T. *Cybernetical machines (Hungarian)* ◇ D98 ◇

1963

AJZERMAN, M.A. & GUSEV, L.A. & ROZONOEHR, L.I. & SMIRNOVA, I.M. & TAL', A.A. *Logic, automata and algorithms (Russian)* ◇ B98 D05 D10 D98 ◇

BRAFFORT, P. & HIRSCHBERG, D. (EDS.) *Computer programming and formal systems* ◇ B97 D05 D97 ◇

CHURCH, A. *Logic, arithmetic, and automata* ◇ A10 B25 D05 D98 ◇

GLUSHKOV, V.M. *Theorie der abstrakten Automaten* ◇ D05 D98 ◇

SACKS, G.E. *Degrees of unsolvability* ◇ D25 D30 D98 E75 ◇

TRAKHTENBROT, B.A. *Algorithms and automatic computing machines* ◇ D10 D20 D98 ◇

1964

FENSTAD, J.E. *Algorithms in mathematics: an introduction to recursion theory and its applications I,II (Norwegian) (English summary)* ◇ D20 D98 ◇

GENENZ, J. *Reduktionstheorie nach der Methode von Kahr-Moore-Wang* ◇ B20 D35 D98 ◇

LACOMBE, D. *Theoremes de non-decidabilite* ◇ D35 D45 D98 ◇

MOORE, E.F. (ED.) *Sequential machines: selected papers* ◇ D05 D97 ◇

1965

CHOWLA, S. *The Riemann hypothesis and Hilbert's tenth problem. Mathematics and its applications Vol. 4* ◇ D35 D98 ◇

CROSSLEY, J.N. & DUMMETT, M. (EDS.) *Formal systems and recursive functions* ◇ B97 D20 D97 ◇

DAVIS, MARTIN D. (ED.) *The undecidable. Basic papers on undecidable propositions, unsolvable problems and computable functions* ◇ D20 D25 D35 D97 F30 ◇

ERSHOV, YU.L. & LAVROV, I.A. & TAJMANOV, A.D. & TAJTSLIN, M.A. *Elementary theories (Russian)* ◇ B25 C98 D35 D98 ◇

HARRISON, M.A. *Introduction to switching and automata theory* ◇ B70 D05 D98 ◇

KALMAR, L. (ED.) *Colloque sur les fondements des mathematiques, les machines mathematiques, et leurs applications* ◇ B97 D05 D10 D97 ◇
MAL'TSEV, A.I. *Algorithms and recursive functions (Russian)* ◇ D20 D98 ◇
MO, SHAOKUI *Theory of recursive functions (Chinese)* ◇ D98 ◇
MOSTOWSKI, ANDRZEJ *Thirty years of foundational studies. Lectures on the development of mathematical logic and the studies of the foundations of mathematics in 1930–1964* ◇ A10 B98 C98 D98 E98 F98 ◇

1966

CAIANIELLO, E.R. (ED.) *Automata theory* ◇ D05 D97 ◇
GINSBURG, S. *The mathematical theory of context free languages* ◇ B65 D05 D98 ◇
KORFHAGE, R. *Logic and algorithms with applications to the computer and information sciences* ◇ B98 D98 ◇
NEUMANN VON, J. *Theory of self-reproducing automata* ◇ D10 D98 ◇
OBERSCHELP, A. *Berechenbarkeit* ◇ D20 D98 ◇
OBERSCHELP, A. *Rechenmaschinen* ◇ D20 D98 ◇
STEINER, H.-G. *Verschiedene Aspekte der axiomatischen Methode im Unterricht* ◇ D35 D98 ◇

1967

CROSSLEY, J.N. (ED.) *Sets, models and recursion theory* ◇ B97 C97 D97 ◇
DAVIS, MARTIN D. *Recursive function theory* ◇ A05 D98 ◇
KLEENE, S.C. *Computability* ◇ D98 ◇
KLEENE, S.C. *Mathematical logic* ◇ B98 D20 D25 D55 D98 F30 F98 ◇
RABIN, M.O. *Mathematical theory of automata* ◇ D05 D98 ◇
ROGERS JR., H. *Some problems of definability in recursive function theory* ◇ D20 D25 D98 ◇
ROGERS JR., H. *Theory of recursive functions and effective computability* ◇ D25 D30 D55 D98 F15 ◇
SCHWARTZ, J.T. (ED.) *Mathematical aspects of computer science* ◇ B35 B65 B75 B97 D10 D80 D97 ◇
SHOENFIELD, J.R. *Mathematical logic* ◇ B98 C98 D98 E98 F98 ◇
TRAKHTENBROT, B.A. *Complexity of algorithms and computations (Russian)* ◇ D10 D15 D20 D98 ◇

1968

ARBIB, M.A. *The algebraic theory of machines, languages, and semigroups* ◇ C05 D05 D98 G30 ◇
BRAUER, W. & INDERMARK, K. *Algorithmen, rekursive Funktionen und formale Sprachen* ◇ D05 D10 D20 D98 ◇
HOTZ, G. *Automatentheorie und formale Sprachen I. Turingmaschinen und rekursive Funktionen* ◇ D05 D10 D20 D98 ◇
HU, S.-T. *Mathematical theory of switching circuits and automata* ◇ B70 D05 D98 ◇
NELSON, RAYMOND J. *Introduction to automata* ◇ D05 D10 D98 ◇
SIPALA, P. *Appunti di teoria degli algoritmi* ◇ D98 ◇

1969

ARBIB, M.A. *Theories of abstract automata* ◇ D05 D10 D98 ◇
BULLOFF, J.J. & HAHN, S.W. & HOLYOKE, T.C. *Bibliography of Kurt Goedel* ◇ A10 D99 E99 F99 ◇
HARRISON, M.A. *Lectures on linear sequential machines* ◇ D05 D98 ◇
HERMES, H. *Basic notions and applications of the theory of decidability* ◇ D10 D20 D25 D30 D35 D50 D98 ◇
HOPCROFT, J.E. & ULLMAN, J.D. *Formal languages and their relation to automata* ◇ D05 D98 ◇
HOTZ, G. & WALTER, H. *Automatentheorie und formale Sprachen II. Endliche Automaten* ◇ D05 D98 ◇
SALOMAA, A. *Theory of automata* ◇ D05 D10 D98 ◇
STARKE, P.H. *Abstrakte Automaten* ◇ D05 D75 D98 ◇

1970

BARZDINS, J. & TRAKHTENBROT, B.A. *Finite automata. Behavior and synthesis (Russian)* ◇ D05 D98 ◇
EILENBERG, S. & ELGOT, C.C. *Recursiveness* ◇ D20 D25 D98 ◇
GROSS, M. & LENTIN, A. *Introduction to formal grammars* ◇ B65 D05 D98 ◇
KOZMIDIADI, V.A. & MUCHNIK, A.A. (EDS.) *Problems of mathematical logic (Russian)* ◇ D97 ◇
LAVROV, I.A. *Logic and algorithms (Russian)* ◇ B98 D98 ◇
SHEPHERDSON, J.C. *Theory of algorithms* ◇ D98 ◇
ZAKHAROV, D.A. *Recursive functions (Russian)* ◇ D20 D98 ◇

1971

BOUCHER, C. *Lecons sur la theorie des automates mathematiques* ◇ D05 D10 D15 D98 ◇
GECSEG, F. & PEAK, I. *Algebraic theory of automata* ◇ D05 D98 ◇
GROSS, M. & LENTIN, A. *Mathematische Linguistik* ◇ B65 D05 D98 ◇
HARTMANIS, J. & HOPCROFT, J.E. *An overview of the theory of computational complexity* ◇ D15 D98 ◇
KOHAVI, Z. & PAZ, A. (EDS.) *Theory of machines and computations* ◇ D05 D97 ◇
KURKI-SUONIO, R. *A programmer's introduction to computability and formal languages* ◇ B75 D05 D20 D98 ◇
PAZ, A. *Introduction to probabilistic automata* ◇ D05 D98 ◇
PRIESE, L. *Normalformen von Markov'schen und Post'schen Algorithmen. Eine Einfuehrung in die Theorie der normierten Algorithmen* ◇ D03 D40 D98 ◇
SHOENFIELD, J.R. *Degrees of unsolvability* ◇ D25 D30 D98 ◇
YASUHARA, A. *Recursive function theory and logic* ◇ B98 D10 D20 D98 ◇

1972

ERSHOV, A.P. *Theory of program schemata* ◇ B75 D20 D98 ◇

KAIN, R.Y. *Automata theory: machines and languages* ◇ D05 D10 D98 ◇

KLOETZER, G. & RAUTENBERG, W. *Im Grenzbereich Algebra, Logik, Maschinen. 10 Jahre Forschungsarbeit der Nowosibirsker Schule A.I. Malcev's (eine Studie)* ◇ A10 B98 C05 D05 D98 ◇

LEVIN, V.I. (ED.) *Fragen der Synthese endlicher Automaten (Russian)* ◇ D05 D98 ◇

LOECKX, J. *Computability and decidability: An introduction for students of computer science* ◇ D98 ◇

LORENTS, A.A. *Elemente der konstruktiven Theorie stochastischer Automaten (Russian)* ◇ D05 D10 D98 F60 F98 ◇

MAREK, W. & ONYSZKIEWICZ, J. *Elements of logic and foundations of mathematics in problems (Polish)* ◇ B98 C98 D98 E98 ◇

PORTE, J. *La logique mathematique et le calcul mecanique* ◇ B98 D98 ◇

ROEDDING, D. *Einfuehrung in die Theorie der berechenbaren Funktionen I* ◇ D98 ◇

ROEDDING, D. *Einfuehrung in die Theorie der berechenbaren Funktionen II* ◇ D98 ◇

WALL, R.E. *Introduction to mathematical linguistics* ◇ B65 D03 D05 D98 ◇

1973

ADYAN, S.I. (ED.) *Mathematical logic, theory of algorithms and theory of sets (Russian)* ◇ D97 E97 ◇

ALFEROVA, Z. *Algorithmentheorie (Russian)* ◇ D10 D20 D98 ◇

AZRA, J.-P. & JAULIN, B. *Recursivite* ◇ D98 ◇

BEREZYUK, N.T. & FURMANOV, K.K. *Foundations of the theory of algorithms (Russian)* ◇ D20 D98 ◇

BOEHLING, K.H. & INDERMARK, K. (EDS.) *Erste Fachtagung der Gesellschaft fuer Informatik ueber Automatentheorie und formale Sprachen (Bonn, 9. - 12. Juli 1973)* ◇ D05 D97 ◇

BOONE, W.W. & CANNONITO, F.B. & LYNDON, R.C. (EDS.) *Word problems* ◇ D40 D97 ◇

ENGELER, E. *Introduction to the theory of computation* ◇ D98 ◇

ERSHOV, A.P. *The present state of the theory of program schemes (Russian)* ◇ B75 D20 D98 ◇

FISCHER, P.C. & IRLAND, M.I. *A bibliography on computational complexity* ◇ A10 D15 D99 ◇

JACOBS, K. (ED.) *Selecta mathematica II* ◇ D97 F30 ◇

JONES, N.D. *Computability theory: an introduction* ◇ D98 ◇

MARKOV, A.A. & PETRI, N.V. (EDS.) *Studies in the theory of algorithms and mathematical logic. Vol. I (Russian)* ◇ D97 ◇

NIVAT, M. (ED.) *Automata, languages and programming* ◇ B75 D05 D97 ◇

RUSTIN, R. (ED.) *Computational complexity. Courant computer science symposium* ◇ D15 D97 ◇

SALOMAA, A. *Formal languages* ◇ D05 D15 D98 ◇

1974

ARBIB, M.A. & MANES, E.G. *Machines in a category: an expository introduction* ◇ C05 D05 D98 G30 ◇

BARZDINS, J. (ED.) *Theory of algorithms and programs I (Russian)* ◇ B97 D97 ◇

BOOLOS, G. & JEFFREY, R.C. *Computability and logic* ◇ B98 D98 ◇

BRAINERD, W.S. & LANDWEBER, L.H. *Theory of computation* ◇ D98 ◇

DAVIS, MARTIN D. *Computability* ◇ B98 D98 ◇

EILENBERG, S. *Automata, languages, and machines. Vol. A* ◇ C05 D05 D98 ◇

FENSTAD, J.E. & HINMAN, P.G. (EDS.) *Generalized recursion theory* ◇ D97 ◇

GODLEVSKIJ, A.B. & LETICHEVS'KIJ, O.A. (EDS.) *Automata and algorithm theory, and mathematical logic (Russian)* ◇ B35 B97 D05 D20 D97 ◇

HAMLET, R.G. *Introduction to computation theory* ◇ D10 D15 D25 D98 ◇

HERSCHEL, R. *Einfuehrung in die Theorie der Automaten, Sprachen und Algorithmen* ◇ D05 D10 D98 ◇

JONES, JAMES P. *Recursive undecidability - an exposition* ◇ D10 D20 D35 D40 D98 ◇

KARP, R.M. (ED.) *Complexity of computation* ◇ D15 D97 ◇

KOZMIDIADI, V.A. & MASLOV, A.N. & PETRI, N.V. (EDS.) *Complexity of computations and algorithms (Russian)* ◇ D15 D20 D98 ◇

KUSHNER, B.A. & NAGORNYJ, N.M. (EDS.) *Theory of algorithms, and mathematical logic (Russian)* ◇ D97 ◇

LOECKX, J. (ED.) *Automata, languages and programming. II* ◇ D97 ◇

MANNA, Z. *Mathematical theory of computation* ◇ B75 D10 D98 ◇

PETERS, F.E. *Einfuehrung in mathematische Methoden der Informatik* ◇ B70 B98 D80 D98 ◇

SCHNORR, C.-P. *Rekursive Funktionen und ihre Komplexitaet* ◇ D10 D15 D20 D98 ◇

TAKAHASHI, S. *Methodes logiques en geometrie diophantienne* ◇ B28 C98 D98 G30 H98 ◇

1975

AGAFONOV, V.N. *Complexity of algorithms and computations. Part II (Russian)* ◇ B75 D15 D98 ◇

BARWISE, J. *Admissible sets and structures. An approach to definability theory* ◇ B98 C40 C70 C98 D60 D98 E30 E98 ◇

BARZDINS, J. (ED.) *Theory of algorithms and programs II (Russian)* ◇ B97 D97 ◇

BLOKH, A.SH. *Graph-Schemata und ihre Anwendung (Russian)* ◇ D98 ◇

BRAKHAGE, H. (ED.) *Automata theory and formal languages* ◇ D05 D97 ◇

CROSSLEY, J.N. *Reminiscenses of logicians* ◇ A05 A10 D98 ◇

FRAISSE, R. *Cours de logique mathematique. Tome 3: Recursivite et constructibilite* ◇ D25 D98 E35 E45 E98 ◇

FRIEDMAN, H.M. *One hundred and two problems in mathematical logic* ◇ B98 C98 D98 E98 F98 ◇

GINSBURG, S. *Algebraic and automata-theoretic properties of formal languages* ◇ D05 D98 ◇
LAVROV, I.A. & MAKSIMOVA, L.L. *Problems in set theory, mathematical logic and the theory of algorithms (Russian)* ◇ B98 D98 E98 ◇
MAGNARADZE, D.G. & PKHAKADZE, SH.S. (EDS.) *Studies in mathematical logic and the theory of algorithms (Russian)* ◇ B97 D97 ◇
MANASTER, A.B. *Completeness, compactness, and undecidability: an introduction to mathematical logic* ◇ B98 C07 D35 D98 ◇

1976

BOGDAN, R.J. *A selected bibliography of local induction* ◇ A10 D99 ◇
CAICEDO, X. *Algorithmen, Fibonaccizahlen und das zehnte Hilbertsche Problem (Spanish)* ◇ D35 D98 ◇
CHAPENKO, V.P. & GOBZEMIS, A.YU. & KURMIT, A.A. & LORENTS, A.A. & PETRENKO, A.F. & VASYUKEVICH, V.O. & YAKUBAJTIS, EH.A. & ZAZNOVA, N.E. *Automata theory (Russian)* ◇ D05 D98 ◇
DILLER, J. *Rekursionstheorie. Vorlesung* ◇ D98 ◇
HOPKIN, D.R. & MOSS, B.P. *Automata* ◇ D05 D10 D98 ◇
KUSHNER, B.A. & MARKOV, A.A. (EDS.) *Studies in the theory of algorithms and mathematical logic. Vol. 2 (Russian)* ◇ D97 ◇
LOECKX, J. *Algorithmentheorie* ◇ D03 D10 D98 ◇
MAREK, W. & SREBRNY, M. & ZARACH, A. (EDS.) *Set theory and hierarchy theory. A memorial tribute to Andrzej Mostowski* ◇ B97 D55 D97 E97 ◇
MICHAELSON, S. & MILNER, R. (EDS.) *Automata, languages and programming. III* ◇ D05 D97 ◇
NIVAT, M. & VIENNOT, G. (EDS.) *Journees algorithmiques. Tenues a l'Ecole Normale Superieure, Paris* ◇ D97 ◇
PETER, R. *Rekursive Funktionen in der Computer-Theorie* ◇ B75 D20 D98 ◇
POLYAKOV, E.A. & ROZINAS, M.G. *Theory of algorithms (Russian)* ◇ D20 D30 D98 ◇
SAVAGE, J.E. *The complexity of computing* ◇ B70 D10 D15 D98 ◇

1977

BARWISE, J. & MOSCHOVAKIS, Y.N. *Guide to part C: Recursion theory* ◇ D98 ◇
BARWISE, J. (ED.) *Handbook of mathematical logic* ◇ B98 C98 D98 E98 F98 H98 ◇
BARZDINS, J. (ED.) *Theory of algorithms and programs III (Russian)* ◇ B75 D97 ◇
COHORS-FRESENBORG, E. *Mathematik mit Kalkuelen und Maschinen* ◇ D03 D10 D98 ◇
ENDERTON, H.B. *Elements of recursion theory* ◇ D20 D25 D98 ◇
ERSHOV, YU.L. *Theory of numerations (Russian)* ◇ D45 D98 ◇
GANDY, R.O. & HYLAND, J.M.E. (EDS.) *Logic Colloquium 76. Proceedings of a conference held in Oxford in July 1976* ◇ B97 C97 D97 ◇
HEIDLER, K. & HERMES, H. & MAHN, F.-K. *Rekursive Funktionen* ◇ D20 D98 ◇
HENNIE III, F.C. *Introduction to computability* ◇ D15 D98 ◇
HINMAN, P.G. *A survey of finite-type recursion* ◇ D55 D60 D65 D98 ◇
HOTZ, G. *Automatentheorie und formale Sprachen* ◇ B75 D05 D97 ◇
KARPINSKI, M. (ED.) *Fundamentals of computation theory* ◇ D97 ◇
LACHLAN, A.H. & SREBRNY, M. & ZARACH, A. (EDS.) *Set theory and hierarchy theory V, Bierutowice, Poland 1976* ◇ D97 E97 ◇
MAGNARADZE, L.G. & PKHAKADZE, SH.S. (EDS.) *Studies in mathematical logic and the theory of algorithms. No. II (Russian)* ◇ B97 D97 ◇
MAKKAI, M. *Admissible sets and infinitary logic* ◇ C70 C98 D55 D60 D70 D98 ◇
MANIN, YU.I. *A course in mathematical logic* ◇ B98 C07 D98 E35 E50 F30 G12 ◇
MAREK, W. *Bibliography of Andrzej Mostowski's works* ◇ A10 D99 E99 ◇
RASIOWA, H. *A tribute to A.Mostowski* ◇ A10 D99 E99 ◇
RASIOWA, H. *In memory of Andrzej Mostowski* ◇ A10 D99 E99 ◇
SALOMAA, A. & STEINBY, M. (EDS.) *Automata,languages and programming. IV* ◇ D97 ◇
TRAKHTENBROT, B.A. *Algorithmen und Rechenautomaten* ◇ D10 D15 D40 D98 ◇
TZSCHACH, H. & WALDSCHMIDT, H. & WALTER, H.K.-G. (EDS.) *Theoretical computer science* ◇ D97 ◇
WECHSUNG, G. *Properties of complexity classes -- a short survey* ◇ D15 D98 ◇

1978

ARSLANOV, M.M. (ED.) *Algorithms and automata. Collection of scientific works (Russian)* ◇ D05 D20 D97 ◇
AUSIELLO, G. & BOEHM, C. (EDS.) *Automata, languages and programming. V* ◇ D97 ◇
DENNING, P.J. & DENNIS, J.B. & QUALITZ, J.E. *Machines, languages, and computation* ◇ B98 D05 D10 D98 ◇
FENSTAD, J.E. & GANDY, R.O. & SACKS, G.E. (EDS.) *Generalized recursion theory II* ◇ C62 C70 C80 D97 ◇
HARRISON, M.A. *Introduction to formal language theory* ◇ D05 D98 ◇
HERRMANN, E. *Der Verband der rekursiv aufzaehlbaren Mengen (Entscheidungs-Problem) (Russian and English summaries)* ◇ B25 D25 D98 ◇
HINMAN, P.G. *Recursion-theoretic hierarchies* ◇ D55 D60 D65 D70 D98 E15 ◇
KECHRIS, A.S. & MOSCHOVAKIS, Y.N. (EDS.) *Cabal seminar 76-77* ◇ D55 D97 E15 E60 E97 ◇
MACHTEY, M. & YOUNG, P. *An introduction to the general theory of algorithms* ◇ D15 D20 D98 ◇
PAUL, W.J. *Complexity theory* ◇ D15 D98 ◇
PENTTONEN, M. & ROZENBERG, G. & SALOMAA, A. *Bibliography of L systems* ◇ A10 D05 D99 ◇

POLYAKOV, E.A. (ED.) *Recursive functions (Russian)* ◇ D20 D30 D97 ◇

POUR-EL, M.B. *Computer science and recursion theory* ◇ D80 D98 ◇

SANKAPPANAVAR, H.P. *Decision problems: History and methods* ◇ A10 B25 D35 D98 ◇

TANAKA, H. *On recent recursion theory (Japanese)* ◇ D98 ◇

1979

BIANCO, E. *Informatique fondamentale. De la machine de Turing aux ordinateurs modernes* ◇ D03 D10 D98 ◇

EPSTEIN, R.L. *Degrees of unsolvability: structure and theory* ◇ D25 D30 D35 D98 ◇

ERSHOV, A.P. & KNUTH, D.E. (EDS.) *Algorithms in modern mathematics and their applications I,II* ◇ B75 B97 D97 ◇

EUDACH, L. (ED.) *Fundamentals of computation theory* ◇ D97 ◇

GAREY, M.R. & JOHNSON, D.S. *Computers and intractability.A guide to the theory of NP-completeness* ◇ D15 D98 ◇

GOLUNKOV, YU.V. (ED.) *Probabilistic methods and cybernetics No. XV (Russian)* ◇ D97 ◇

GREIBACH, S.A. *Formal languages: origins and directions* ◇ B70 D05 D98 ◇

HINMAN, P.G. *Borel determinacy* ◇ D55 D98 E15 E60 ◇

HOFSTADTER, D.R. *Goedel, Escher, Bach: an eternal golden braid* ◇ A05 B98 D99 ◇

HOPCROFT, J.E. & ULLMAN, J.D. *Introduction to automata theory, languages, and computation* ◇ D05 D10 D98 ◇

KECHRIS, A.S. *An overview of descriptive set theory* ◇ D55 D98 E15 E98 ◇

KHOMICH, V.I. & MARKOV, A.A. (EDS.) *Studies in the theory of algorithms and mathematical logic (Russian)* ◇ B97 D97 ◇

KLEENE, S.C. *Origins of recursive function theory* ◇ A10 D20 D98 ◇

KOWALTOWSKI, T. & LUCCHESI, C.L. & SIMON, IMRE & SIMON, ISTVAN & SIMON, J. *Theoretical aspects of computation (Portuguese)* ◇ D98 ◇

LEWIS, H.R. *Unsolvable classes of quantificational formulas* ◇ B20 B25 D35 D98 ◇

MALITZ, J. *Introduction to mathematical logic. Set theory, computable functions, model theory* ◇ B98 C98 D98 ◇

PETROV, B.N. & UL'YANOV, S.V. & ULANOV, G.M. *Complexity of finite objects and informational control theory (Russian)* ◇ D15 D98 ◇

ZHANG, JINWEN *Reasoning and computing* ◇ B75 D99 ◇

1980

ADYAN, S.I. & BOONE, W.W. & HIGMAN, G. (EDS.) *Word problems II. The Oxford book* ◇ D40 D97 ◇

BORGA, M. & FREGUGLIA, P. & PALLADINO, D. (EDS.) *Rassegna di matematica. Logica matematica. Matematica applicata. Didattica della matematica* ◇ B97 D97 ◇

BUCUR, I. *Special chapters of algebra (Romanian)* ◇ B80 C98 D98 ◇

CHENG, K.N. & CHONG, C.T. (EDS.) *Proceedings of the mathematical seminar* ◇ D97 ◇

CUTLAND, N.J. *Computability. An introduction to recursive function theory* ◇ B25 D98 ◇

DELLACHERIE, C. & HOFFMANN-JOERGENSEN, J. & JAYNE, J.E. & KECHRIS, A.S. & MARTIN, D.A. & ROGERS, C.A. & STONE, A.H. & TOPSOEE, F. *Analytic sets* ◇ D98 E15 E98 ◇

DRAKE, F.R. & WAINER, S.S. (EDS.) *Recursion theory: its generalizations and applications* ◇ D97 ◇

EBBINGHAUS, H.-D. *Mathematische Logik und Informatik* ◇ C13 D10 D15 D98 E70 H15 ◇

ERSHOV, YU.L. *Decision problems and constructivizable models (Russian)* ◇ B25 C57 C60 C98 D45 D98 ◇

FENSTAD, J.E. *General recursion theory. An axiomatic approach* ◇ D75 D98 ◇

GANOV, V.A. *Functionals of ordering types (Russian)* ◇ D99 ◇

GLEISER, M. *The emotional logician* ◇ A10 D99 ◇

HUET, G. & OPPEN, D.C. *Equations and rewrite rules: a survey* ◇ D03 D05 D98 ◇

KLETTE, R. & WIEHAGEN, R. *Research in the theory of inductive inference by GDR mathematicians - A survey* ◇ D98 ◇

LERMAN, M. *The degrees of unsolvability: Some recent results* ◇ D30 D98 ◇

MANIN, YU.I. *The computable and the non-computable (Russian)* ◇ D20 D25 D98 ◇

METZ, J. *Abstrakte Rechnermodelle* ◇ D05 D98 ◇

MOSCHOVAKIS, Y.N. *Descriptive set theory* ◇ D55 D98 E15 E98 ◇

NORMANN, D. *Recursion on the countable functionals* ◇ D30 D55 D65 D98 ◇

RYTTER, W. *Automata theory and complexity* ◇ D05 D15 D98 ◇

STARKE, P.H. *Petri-Netze* ◇ D05 D98 ◇

TASIK, M. *On algorithms; on recursive functions (Macedonian)* ◇ D98 ◇

1981

CROSSLEY, J.N. (ED.) *Aspects of effective algebra. Proceedings of a conference at Monash University, Australia, 1-4 August, 1979* ◇ C57 C97 D45 D97 ◇

EPSTEIN, R.L. *Initial segments of degrees below 0'* ◇ D25 D30 D35 D98 ◇

ERSHOV, A.P. & KNUTH, D.E. (EDS.) *Algorithms in modern mathematics and computer science. Proceedings, Urgench, Uzbek SSR, September 16-22, 1979* ◇ D15 D20 D97 ◇

GECSEG, F. (ED.) *Fundamentals of computation theory* ◇ D97 ◇

HERMES, H. *Recursion theory* ◇ D98 ◇

HERRMANN, E. *Die Verbandseigenschaften der rekursiv aufzaehlbaren Mengen* ◇ D25 D98 ◇

KECHRIS, A.S. & MARTIN, D.A. & MOSCHOVAKIS, Y.N. (EDS.) *Cabal seminar 77 - 79* ◇ D55 D97 E15 E60 E97 ◇

KLEENE, S.C. *The theory of recursive functions, approaching its centennial* ◇ A10 D20 D98 ◇
KOSOVSKIJ, N.K. *Elements of mathematical logic and its applications to the theory of subrecursive algorithms (Russian)* ◇ B98 D20 D98 F30 F60 F98 ◇
KRAJEWSKI, S. *Kurt Goedel and his work (Polish)* ◇ A10 C98 D98 E98 F99 ◇
LEWIS, H.R. & PAPADIMITRIOU, C.H. *Elements of the theory of computation* ◇ D98 ◇
NOLTEMEIER, H. *Informatik I: Einfuehrung in Algorithmen und Berechenbarkeit* ◇ D20 D98 ◇
SALOMAA, A. *Jewels of formal language theory* ◇ D05 D98 ◇
SAVAGE, J.A. *Space-time tradeoffs - a survey* ◇ D15 D98 ◇
SEMENOV, A.L. & USPENSKIJ, V.A. *What are the gains of the theory of algorithms: basic developments connected with the concept of algorithm and with its application in mathematics* ◇ D20 D98 ◇

1982

ARBIB, M.A. & KFOURY, A.J. & MOLL, R. *A programming approach to computability* ◇ B75 D20 D98 ◇
BUCHBERGER, B. & COLLINS, G.E. & LOOS, R. (EDS.) *Computer-Algebra - Symbolic and algebraic computation* ◇ D97 ◇
COHEN, L.J. & LOS, J. & PFEIFFER, H. & PODEWSKI, K.-P. (EDS.) *Logic, methodology and philosophy of science VI* ◇ B97 D98 ◇
COURCELLE, B. *Fundamental properties of infinite trees* ◇ D05 D98 E75 ◇
DALEN VAN, D. & LASCAR, D. & SMILEY, T.J. (EDS.) *Logic colloquium '80. Eur. Summer Meet., Prague 1980* ◇ B97 D97 ◇
DAVIS, MARTIN D. *Computability and unsolvability* ◇ D98 ◇
EMDE BOAS VAN, P. *Machine models, computational complexity and number theory* ◇ D15 D98 ◇
JOHNSON, D.S. *The NP-completeness column: an ongoing guide* ◇ D15 D98 ◇
MCLAUGHLIN, T.G. *Regressive sets and the theory of isols* ◇ D50 D98 ◇
MO, SHAOKUI *Theory of algorithms (Chinese)* ◇ D20 D98 ◇
NIELSEN, M. & SCHMIDT, ERIK MEINECHE (EDS.) *Automata, languages and programming* ◇ B75 D97 ◇
SOBOLEV, S.L. (ED.) *Mathematical logic and the theory of algorithms (Russian)* ◇ B97 C97 D97 ◇
STERN, J. (ED.) *Proceedings of the Herbrand Symposium. Logic colloquium '81, held in Marseille, France, July 1981* ◇ B97 D97 ◇

1983

ALBERT, J. & OTTMANN, T. *Automaten, Sprachen und Maschinen fuer Anwender* ◇ D05 D10 D98 ◇
DALEN VAN, D. *Algorithm and decision problems: A crash course in recursion theory* ◇ D98 ◇
DAVIS, MARTIN D. & WEYUKER, E.J. *Computability, complexity, and languages. Fundamentals of theoretical computer science* ◇ D05 D10 D20 D98 ◇
GLADKIJ, A.V. (ED.) *Mathematical logic, mathematical linguistics and theory of algorithms (Russian)* ◇ B97 D97 ◇
GLADKIJ, A.V. *Theory of algorithms (Russian)* ◇ D20 D98 ◇
KARPINSKI, M. (ED.) *Foundations of computation theory* ◇ D97 ◇
KECHRIS, A.S. & MARTIN, D.A. & MOSCHOVAKIS, Y.N. (EDS.) *Cabal seminar 79-81. Proceedings, Caltech-UCLA logic seminar 1979-81* ◇ D55 D97 E15 E60 E97 ◇
LERMAN, M. *Degrees of unsolvability. Local and global theory* ◇ D30 D98 ◇
LERMAN, M. *The structures of recursion theory* ◇ D25 D30 D98 ◇
SHAPIRO, S. *Remarks on the development of computability* ◇ A10 D98 F99 ◇

1984

ADYAN, S.I. & MAKANIN, G.S. *Studies in algorithmic questions of algebra (Russian)* ◇ D40 D98 ◇
BOERGER, E. & OBERSCHELP, W. & RICHTER, M.M. & SCHINZEL, B. & THOMAS, WOLFGANG (EDS.) *Computation and proof theory* ◇ B97 D97 F97 ◇
BOERGER, E. & HASENJAEGER, G. & ROEDDING, D. (EDS.) *Logic and machines: decision problems and complexity* ◇ B35 D97 ◇
CHONG, C.T. *Techniques of admissible recursion theory* ◇ D60 D98 ◇
GECSEG, F. & STEINBY, M. *Tree automata* ◇ A05 B25 D05 D98 ◇
LOLLI, G. & LONGO, G. & MARCJA, A. (EDS.) *Logic colloquium '82. Proceedings of the colloquium held in Florence, 23-28 August, 1982* ◇ B97 C97 D97 F97 ◇
MUELLER, GERT H. & RICHTER, M.M. (EDS.) *Models and sets* ◇ B97 C97 D97 H97 ◇
SLEZAK, P. *Minds, machines and self-reference (French and German summaries)* ◇ A05 D99 F30 ◇

1985

BOERGER, E. *Berechenbarkeit, Komplexitaet, Logik. Eine Einfuehrung in Algorithmen, Sprachen und Kalkuele unter besonderer Beruecksichtigung ihrer Komplexitaet* ◇ D98 ◇
DELON, F. & LASCAR, D. & LOUVEAU, A. & SABBAGH, G. (EDS.) *Seminaire general de logique 1982-83* ◇ B97 C98 D97 ◇
EBBINGHAUS, H.-D. & MUELLER, GERT H. & SACKS, G.E. (EDS.) *Recursion theory week* ◇ D97 ◇
GUREVICH, Y. *Monadic second-order theories* ◇ B25 C65 C85 C98 D35 D98 ◇
HARRINGTON, L.A. & MORLEY, M.D. & SCEDROV, A. & SIMPSON, S.G. *Harvey Friedman's research on the foundations of mathematics* ◇ B97 C97 D97 E97 F97 ◇
LEHMANN, G. *Modell- und rekursionstheoretische Grundlagen psychologischer Theorienbildung* ◇ B10 C98 D80 D99 ◇
NERODE, A. & REMMEL, J.B. *A survey of lattices of R.E. substructures* ◇ C57 C98 D45 D98 ◇

NERODE, A. & SHORE, R.A. (EDS.) *Recursion theory*
◇ D97 ◇

SACKS, G.E. *Some open questions in recursion theory*
◇ C13 D98 E60 ◇

Author Index

AANDERAA, S.O. & FISCHER, P.C. [1967] *The solvability of the halting problem for 2-state Post machines* (**J** 0037) ACM J 14*677-682
◇ D03 D05 ◇ REV MR 38 # 3087 Zbl 163.9 JSL 36.532 • ID 00006

AANDERAA, S.O. & COOK, S.A. [1969] *On the minimum computation time of functions* (**J** 0064) Trans Amer Math Soc 142*291-314
• TRANSL [1971] (**J** 3079) Kiber Sb Perevodov, NS 8*168-200
◇ D20 ◇ REV MR 40 # 2459 Zbl 188.334 • ID 28179

AANDERAA, S.O. & BELSNES, D. [1971] *Decision problems for tag systems* (**J** 0036) J Symb Logic 36*229-239
◇ D03 D25 D30 ◇ REV MR 46 # 8824 Zbl 251 # 02039 • ID 00008

AANDERAA, S.O. [1971] *On the decision problem for formulas in which all disjunctions are binary* (**P** 0604) Scand Logic Symp (2);1970 Oslo 1-18
◇ B20 D35 ◇ REV MR 50 # 1864 Zbl 232 # 02034 JSL 40.503 • ID 00087

AANDERAA, S.O. [1973] *A proof of Higman's embedding theorem using Britton extensions of groups* (**P** 0678) Word Probl: Decis & Burnside Probl in Group Th;1969 Irvine 1-18
◇ C60 D40 D80 ◇ REV MR 54 # 412 Zbl 265 # 20033 JSL 41.785 • ID 23990

AANDERAA, S.O. & LEWIS, H.R. [1973] *Prefix classes of Krom formulas* (**J** 0036) J Symb Logic 38*628-642
◇ B20 B25 D35 ◇ REV MR 49 # 2326 Zbl 326 # 02035 • ID 00010

AANDERAA, S.O. [1974] *Inductive definitions and their closure ordinals* (**P** 0602) Generalized Recursion Th (1);1972 Oslo 207-220
◇ D70 ◇ REV MR 52 # 13388 Zbl 305 # 02051 • ID 00093

AANDERAA, S.O. & LEWIS, H.R. [1974] *Linear sampling and the ∀∃∀ case of the decision problem* (**J** 0036) J Symb Logic 39*519-548
◇ B20 D03 D05 D35 D80 ◇ REV MR 51 # 114 Zbl 301 # 02042 • ID 03803

AANDERAA, S.O. [1974] *On k-tape versus (k - 1)-tape real time computation* (**P** 0761) Compl of Computation;1973 New York 75-96
◇ D10 D15 ◇ REV MR 56 # 10146 Zbl 316 # 68029 • ID 60006

AANDERAA, S.O. & BOERGER, E. [1979] *The Horn complexity of Boolean functions and Cook's problem* (**P** 2615) Scand Logic Symp (5);1979 Aalborg 231-256
◇ B05 D15 ◇ REV MR 83b:03048b Zbl 429 # 03022 • ID 53853

AANDERAA, S.O. & COHEN, D.E. [1980] *Modular machines, the word problem for finitely presented groups and Collins' theorem* (**P** 2634) Word Problems II;1976 Oxford 1-16
◇ D10 D30 D40 ◇ REV MR 81m:20048 Zbl 445 # 20016 • ID 80493

AANDERAA, S.O. & COHEN, D.E. [1980] *Modular machines and the Higman-Clapham-Valiev embedding theorem* (**P** 2634) Word Problems II;1976 Oxford 17-28
◇ D10 D30 D40 ◇ REV MR 81m:20049 Zbl 441 # 20023 • ID 80492

AANDERAA, S.O. & BOERGER, E. [1981] *The equivalence of Horn and network complexity for boolean functions* (**J** 1431) Acta Inf 15*303-307
◇ B05 B70 D15 ◇ REV MR 83b:03048a Zbl 477 # 94034 • ID 55613

AANDERAA, S.O. & BOERGER, E. & LEWIS, H.R. [1982] *Conservative reduction classes of Krom formulas* (**J** 0036) J Symb Logic 47*110-130
◇ B20 C13 D35 ◇ REV MR 83e:03021 Zbl 487 # 03005 • ID 35210

AANDERAA, S.O. & BOERGER, E. & GUREVICH, Y. [1982] *Prefix classes of Krom formulae with identity (German summary)* (**J** 0009) Arch Math Logik Grundlagenforsch 22*43-49
◇ B20 B25 D35 ◇ REV MR 83m:03019 Zbl 494 # 03007 • ID 33756

AANDERAA, S.O. see Vol. I, III, VI for further entries

ABDRAZAKOV, K.T. & KHISAMIEV, N.G. [1984] *A criterion for strong constructivizability of a class of abelian p-groups (Russian)* (**J** 0092) Sib Mat Zh 25/4*3-8
• TRANSL [1984] (**J** 0475) Sib Math J 25*511-515
◇ C57 C60 D45 ◇ REV MR 86g:03073 • ID 41873

ABERTH, O. [1971] *The concepts of effective method applied to computational problems of linear algebra* (**J** 0119) J Comp Syst Sci 5*17-25
◇ D80 ◇ REV Zbl 216.487 JSL 40.84 • ID 04241

ABERTH, O. see Vol. VI for further entries

ABIAN, A. [1977] *Unsolvability of the halting problem for Turing machines* (**J** 3276) Tehran J Math 1/2*1-7
◇ D10 D25 ◇ REV Zbl 417 # 03014 • ID 53251

ABIAN, A. also published under the name ABIAN, S.

ABIAN, A. see Vol. I, II, III, V for further entries

ABLAEV, F.M. [1980] *The relation of automatic measures of the complexity of languages to Loveland's complexity of word resolution (Russian)* (**J** 3937) Veroyat Met i Kibern (Kazan) 17*3-14
◇ D05 D15 ◇ REV MR 82j:03043 Zbl 474 # 68087 • ID 70284

Author Index

ABLAEV, F.M. [1983] *On the question of automaton complexity of languages and complexity of pattern recognition in the sense of Loveland (Russian)* (**J** 3937) Veroyat Met i Kibern (Kazan) 19*3-9
◇ D05 ◇ REV MR 84i:00006 Zbl 546 # 68065 • ID 43557

ABLAEV, F.M. & GABBASOV, N.Z. [1984] *Remarks on generalized reduction theorems for infinite automata (Russian)* (**J** 3937) Veroyat Met i Kibern (Kazan) 20*40-48
◇ D05 ◇ ID 46383

ABLAEV, F.M. [1985] *Possibilities of probalistic machines with respect to representation of languages in real time (Russian)* (**J** 0031) Izv Vyssh Ucheb Zaved, Mat (Kazan) 1985/7*32-40,84-85
◇ D05 D10 ◇ ID 49268

ABRAHAM, U. [1984] *A minimal model for ¬ CH: iteration of Jensen's reals* (**J** 0064) Trans Amer Math Soc 281*657-674
◇ C62 D30 E35 E40 E45 E50 E65 ◇ REV MR 85f:03044 Zbl 541 # 03029 • ID 40744

ABRAHAM, U. see Vol. III, V for further entries

ABRAMSON, F.G. & SACKS, G.E. [1976] *Uncountable Gandy ordinals* (**J** 3172) J London Math Soc, Ser 2 14*387-392
◇ D60 D70 ◇ REV MR 55 # 5419 Zbl 358 # 02052 • ID 31646

ABRAMSON, F.G. [1979] *Σ_1-separation* (**J** 0036) J Symb Logic 44*374-382
◇ C62 C70 D60 E45 E47 ◇ REV MR 82j:03057 Zbl 428 # 03029 • ID 53788

ABRAMSON, F.G. [1979] *Sacks forcing does not always produce a minimal upper bound* (**J** 0345) Adv Math 31*110-130
◇ D30 D60 E40 ◇ REV MR 81g:03054 Zbl 434 # 03030 • ID 31649

ABRAMSON, F.G. & BREJTBART, YU.YA. & LEWIS, F.D. [1980] *Complex properties of grammars* (**J** 0037) ACM J 27*484-498
◇ D05 D10 D15 ◇ REV MR 81k:68051 Zbl 422 # 68037 Zbl 475 # 68046 • ID 69036

ABRAMSON, F.G. see Vol. III, V for further entries

ACKERMANN, W. [1928] see HILBERT, D.

ACKERMANN, W. [1928] *Ueber die Erfuellbarkeit gewisser Zaehlausdruecke* (**J** 0043) Math Ann 100*638-649
◇ B25 D35 ◇ REV FdM 54.57 • ID 00109

ACKERMANN, W. [1928] *Zum Hilbertschen Aufbau der reellen Zahlen* (**J** 0043) Math Ann 99*118-133
• TRANSL [1967] (**C** 0675) From Frege to Goedel 493-507
◇ B28 D20 F15 ◇ REV FdM 54.56 • ID 00106

ACKERMANN, W. [1936] *Beitraege zum Entscheidungsproblem der mathematischen Logik* (**J** 0043) Math Ann 112*419-432
◇ B25 D35 ◇ REV Zbl 13.241 JSL 1.43 FdM 62.41 • ID 00112

ACKERMANN, W. see Vol. I, II, III, V, VI for further entries

ACKROYD, M.H. & CHIRATHAMJAREE, C. [1980] *A method for the inference of non-recursive context-free grammars* (**J** 1741) Int J Man-Mach Stud 12*379-387
◇ D05 ◇ REV MR 81g:68107 Zbl 443 # 68059 • ID 69317

ACZEL, P. & CROSSLEY, J.N. [1966] *Constructive order types III* (**J** 0009) Arch Math Logik Grundlagenforsch 9*112-116
◇ D45 F15 ◇ REV MR 35 # 5319 Zbl 199.29 • REM Part II 1966 by Crossley,J.N. • ID 00137

ACZEL, P. [1970] *Representability in some systems of second order arithmetic* (**J** 0029) Israel J Math 8*309-328
◇ D65 D70 F35 ◇ REV MR 42 # 4396 Zbl 216.6 • ID 00141

ACZEL, P. & RICHTER, W.H. [1972] *Inductive definitions and analogues of large cardinals* (**P** 2080) Conf Math Log;1970 London 1-9
◇ D60 D70 E55 ◇ REV MR 51 # 96 Zbl 272 # 02065 • ID 21302

ACZEL, P. [1972] *The ordinals of the superjump and related functionals* (**P** 2080) Conf Math Log;1970 London 336-337
◇ D65 E55 ◇ REV Zbl 227 # 02039 • ID 27366

ACZEL, P. [1974] *An axiomatic approach to recursion on admissible ordinals and the Kreisel-Sacks construction of meta-recursion theory* (**S** 1585) Rec Fct Th Newsletter
◇ D60 D75 ◇ ID 38729

ACZEL, P. & RICHTER, W.H. [1974] *Inductive definitions and reflecting properties of admissible ordinals* (**P** 0602) Generalized Recursion Th (1);1972 Oslo 301-381
◇ D60 D70 E47 E55 ◇ REV MR 52 # 13344 Zbl 318 # 02042 • ID 04129

ACZEL, P. & HINMAN, P.G. [1974] *Recursion in the superjump* (**P** 0602) Generalized Recursion Th (1);1972 Oslo 3-41
◇ D55 D65 E55 ◇ REV MR 55 # 10261 Zbl 292 # 02038 • ID 30615

ACZEL, P. [1975] *Quantifiers, games and inductive definitions* (**P** 0757) Scand Logic Symp (3);1973 Uppsala 1-14
◇ C75 C80 D70 ◇ REV MR 54 # 12477 Zbl 324 # 02009 JSL 43.373 • ID 30616

ACZEL, P. [1975] *Recursive density types and Nerode extensions of arithmetic* (**J** 0038) J Austral Math Soc 20*146-158
◇ D45 D50 F30 ◇ REV MR 51 # 10060 Zbl 308 # 02045 • ID 21303

ACZEL, P. [1977] *An introduction to inductive definitions* (**C** 1523) Handb of Math Logic 739-782
◇ D55 D70 ◇ REV MR 58 # 5109 JSL 49.975 • ID 27325

ACZEL, P. see Vol. II, III, V, VI for further entries

ADACHI, A. & KASAI, T. [1980] *A characterization of time complexity by simple loop programs* (**J** 0119) J Comp Syst Sci 20*1-17
◇ D15 D20 ◇ REV MR 82a:68076 Zbl 431 # 03027 • ID 53933

ADACHI, A. & KASAI, T. [1981] *A problem which requires $O(n^k)$ time* (**P** 3212) Math Models in Comput Systs;1981 Budapest 147-157
◇ D15 ◇ REV MR 84a:68031 Zbl 494 # 68057 • ID 38807

ADAM, ANDRAS [1971] *A description of the finite right-congruences of finitely generated free semigroups* (**J** 0049) Period Math Hung 1*135-144
◇ D40 ◇ REV MR 44 # 1751 Zbl 221 # 20075 • ID 00151

ADAM, ANDRAS [1981] *On the complexity of codes and pre-codes assigned to finite Moore automata* (**J** 0380) Acta Cybern (Szeged) 5*117-133
◇ D05 ◇ REV MR 83c:68055 Zbl 461 # 68049 • ID 69006

ADAM, ANDRAS see Vol. I, V for further entries

Adamek, J. [1974] *Free algebras and automata realizations in the language of categories* (**J** 0140) Comm Math Univ Carolinae (Prague) 15*589-602
◇ D05 G30 ◇ rev MR 50#4696 Zbl 293#18006
• id 03815

Adamek, J. & Trnkova, V. [1982] *Analyses of languages accepted by varietor machines in category* (**P** 3831) Universal Algeb & Appl;1978 Warsaw 257-272
◇ D05 ◇ rev MR 85g:18009 Zbl 513#18004 • id 37807

Adamek, J. see Vol. III, V for further entries

Adamowicz, Z. [1975] *An observation on the product of Silver's forcing* (**P** 1442) ⊧ ISILC Logic Conf;1974 Kiel 1-9
◇ D30 E40 E45 ◇ rev MR 53#12938 Zbl 326#02046
• id 27206

Adamowicz, Z. [1976] *On finite lattices of degrees of constructibility of reals* (**J** 0036) J Symb Logic 41*313-322
◇ D30 E35 E45 ◇ rev MR 58#5204 Zbl 344#02047
• id 14761

Adamowicz, Z. [1977] *Constructible semi-lattices of degrees of constructibility* (**P** 1695) Set Th & Hierarch Th (3);1976 Bierutowice 1-43
◇ D30 E35 E45 ◇ rev MR 58#21601 Zbl 369#02042
• id 51342

Adamowicz, Z. [1977] *On finite lattices of degrees of constructibility* (**J** 0036) J Symb Logic 42*349-371
◇ D30 E35 E45 ◇ rev MR 58#5203 Zbl 379#02024
• id 51857

Adamowicz, Z. [1983] *Perfect set theorems for Π^1_2 in the universe without choice* (**J** 0027) Fund Math 118*11-31
◇ D55 E15 E25 E40 E45 E55 ◇ rev Zbl 537#03031
• id 43740

Adamowicz, Z. [1984] *A generalization of the Shoenfield theorem on Σ^1_2 sets* (**J** 0027) Fund Math 123*81-90
◇ D55 E40 E45 E75 ◇ rev MR 86b:03065a • id 45774

Adamowicz, Z. [1984] *Continuous relations and generalized G_δ sets* (**J** 0027) Fund Math 123*91-107
◇ D30 E15 E40 ◇ rev MR 86b:03065b • id 45775

Adamowicz, Z. & Morales-Luna, G. [1985] *A recursive model for arithmetic with weak induction* (**J** 0036) J Symb Logic 50*49-54
◇ C57 F30 H15 ◇ rev MR 86d:03058 Zbl 576#03045
• id 39780

Adamowicz, Z. also published under the name Smolska-Adamowicz, Z.

Adamowicz, Z. see Vol. III, V, VI for further entries

Adamson, A. & Giles, R. [1979] *A game-based formal system for L_∞* (**J** 0063) Studia Logica 38*49-73
◇ B25 B50 D15 ◇ rev MR 80m:03049 Zbl 417#03008
• id 53245

Adamson, A. see Vol. III, V for further entries

Addison, J.W. [1954] *On some points of the theory of recursive functions* (0000) Diss., Habil. etc
◇ D20 D55 ◇ rem Diss. University of Wisconsin
• id 16706

Addison, J.W. & Kleene, S.C. [1957] *A note on function quantification* (**J** 0053) Proc Amer Math Soc 8*1002-1006
◇ D55 ◇ rev MR 19.934 Zbl 84.249 JSL 23.47 • id 00172

Addison, J.W. [1957] *Hierarchies and the axiom of constructibility* (**P** 1675) Summer Inst Symb Log;1957 Ithaca 355-362
◇ D55 E15 E35 E45 ◇ rev JSL 31.137 • id 29378

Addison, J.W. [1958] *Separation principles in the hierarchies of classical and effective descriptive set theory* (**J** 0027) Fund Math 46*123-135
◇ D55 E15 E45 ◇ rev MR 24#A1209 Zbl 91.52 JSL 29.60 • id 00171

Addison, J.W. [1959] *Some consequences of the axiom of constructibility* (**J** 0027) Fund Math 46*337-357
◇ D55 E15 E45 ◇ rev MR 23#A1523 Zbl 91.53 JSL 28.293 • id 00173

Addison, J.W. [1962] *Some problems in hierarchy theory* (**P** 0613) Rec Fct Th;1961 New York 123-130
◇ C40 D55 E15 ◇ rev MR 25#4997 Zbl 143.255 JSL 29.61 • id 00176

Addison, J.W. [1962] *The theory of hierarchies* (**P** 0612) Int Congr Log, Meth & Phil of Sci (1,Proc);1960 Stanford 26-37
◇ C40 C52 D55 E15 ◇ rev MR 28#1127 Zbl 133.251 JSL 29.61 • id 00175

Addison, J.W. [1965] *The method of alternating chains* (**P** 0614) Th Models;1963 Berkeley 1-16
◇ C40 D55 D75 E15 ◇ rev MR 34#1197 Zbl 199.12
• id 16215

Addison, J.W. & Moschovakis, Y.N. [1968] *Some consequences of the axiom of definable determinateness* (**J** 0054) Proc Nat Acad Sci USA 59*708-712
◇ D55 E15 E60 ◇ rev MR 36#4979 Zbl 186.253 JSL 38.334 • id 00181

Addison, J.W. [1974] *Current problems in descriptive set theory* (**P** 0693) Axiomatic Set Th;1967 Los Angeles 2*1-10
◇ D55 E15 ◇ rev MR 51#10095 Zbl 344#04001
• id 17525

Addison, J.W. see Vol. I, II, III for further entries

Adleman, L.M. & Manders, K.L. [1976] *Diophantine complexity* (**P** 1757) IEEE Symp Found of Comput Sci (17);1976 Houston 81-88
◇ D10 D15 ◇ rev MR 56#7314 • id 80512

Adleman, L.M. & Manders, K.L. [1976] *NP-complete decision problems for quadratic polynomials* (**P** 2597) ACM Symp Th of Comput (8);1976 Hershey 23-29
◇ D15 ◇ rev MR 55#11710 Zbl 381#68044 • id 69688

Adleman, L.M. & Booth, K.S. & Preparata, F.P. & Ruzzo, W.L. [1978] *Improved time and space bounds for boolean matrix multiplication* (**J** 1431) Acta Inf 11*61-70
◇ B05 D15 ◇ rev MR 80f:68038 Zbl 389#68016
• id 69046

Adleman, L.M. & Manders, K.L. [1978] *NP-complete decision problems for binary quadratics* (**J** 0119) J Comp Syst Sci 16*168-184
• transl [1980] (**J** 3079) Kiber Sb Perevodov, NS 17*124-153
◇ D15 D35 ◇ rev MR 81c:68032 Zbl 369#68030
• id 69687

Author Index

ADLEMAN, L.M. & MANDERS, K.L. [1979] *Reductions that lie* (**P** 3535) IEEE Symp Founds of Comput Sci (20);1979 San Juan 397-410
◇ D15 ◇ REV MR 82f:68045 • ID 80511

ADLEMAN, L.M. & LOUI, M.C. [1981] *Space-bounded simulation of multitape Turing machines* (**J** 0041) Math Syst Theory 14*215-222
◇ D10 D15 ◇ REV MR 82g:03071 Zbl 473 # 68045 • ID 70321

ADLER, A. [1969] *Extensions of non-standard models of number theory* (**J** 0068) Z Math Logik Grundlagen Math 15*289-290
◇ C62 D25 D35 H15 ◇ REV MR 42 # 54 Zbl 188.322 JSL 40.244 • ID 00184

ADLER, A. [1969] *Some recursively unsolvable problems in analysis* (**J** 0053) Proc Amer Math Soc 22*523-526
◇ D35 D80 ◇ REV MR 40 # 1277 Zbl 206.284 • ID 00183

ADLER, A. see Vol. I, III, V for further entries

ADLER, S. [1976] *Optimale K-bilineare Realisierung endlicher Automaten* (**J** 0947) Wiss Z Tech Univ Dresden 25*769-773
◇ D05 ◇ REV MR 55 # 5322 Zbl 412 # 68042 • ID 69050

ADLER, S. [1976] *Zur Existenz optimaler K-bilinearer Realisierungen von endlichen transitiven Automaten* (**J** 0947) Wiss Z Tech Univ Dresden 25*773-775 • ERR/ADD ibid 26*660
◇ D05 ◇ REV MR 55 # 5323 MR 57 # 9351 Zbl 412 # 68043 • ID 69051

ADLER, S. [1977] *On structured extensions of finite automata* (**J** 2716) Found Control Eng, Poznan 2*67-73
◇ D05 ◇ REV MR 56 # 11639 Zbl 412 # 68044 • ID 69048

ADRIANOPOLI, F. & LUCA DE, A. [1974] *Closure operations on measures of computational complexity* (**J** 0089) Calcolo 11*205-217
◇ D15 ◇ REV MR 55 # 11698 Zbl 311 # 68034 • ID 60068

ADRIANOPOLI, F. [1980] *Conjugated measures of computational complexity* (**J** 2095) Fund Inform, Ann Soc Math Pol, Ser 4 3*303-309
◇ D15 D20 ◇ REV MR 82g:03072 Zbl 462 # 03009 • ID 54516

ADYAN, S.I. [1955] *Algorithmic unsolvability of the problems of recognition of certain properties of groups (Russian)* (**J** 0023) Dokl Akad Nauk SSSR 103*533-535
◇ D40 ◇ REV MR 18.455 Zbl 65.9 JSL 23.54 • ID 00165

ADYAN, S.I. [1957] *Finitely generated groups and algorithms (Russian)* (**J** 0067) Usp Mat Nauk 12/3*248-249
◇ D40 ◇ REV JSL 23.54 • ID 00168

ADYAN, S.I. [1957] *Finitely presented groups and algorithms (Russian)* (**J** 0023) Dokl Akad Nauk SSSR 117*9-12
◇ B25 C60 D40 ◇ REV MR 20 # 2371 Zbl 85.251 • ID 00193

ADYAN, S.I. [1957] *Unsolvability of certain algorithmic problems in the theory of groups (Russian)* (**J** 0065) Tr Moskva Mat Obshch 6*231-298
◇ D40 ◇ REV MR 20 # 2370 Zbl 80.241 JSL 30.391 • ID 00167

ADYAN, S.I. & NOVIKOV, P.S. [1958] *Das Wortproblem fuer Halbgruppen mit einseitiger Kuerzungsregel (Russian) (German summary)* (**J** 0068) Z Math Logik Grundlagen Math 4*66-88
◇ D40 ◇ REV MR 20 # 7052 Zbl 80.241 JSL 29.57 • ID 24890

ADYAN, S.I. [1958] *On algorithmic problems in effectively complete classes of groups (Russian)* (**J** 0023) Dokl Akad Nauk SSSR 123*13-16
◇ C60 D40 ◇ REV MR 21 # 1998 Zbl 90.11 • ID 00195

ADYAN, S.I. [1959] *Undecidability of some algorithmic problems in group theory (Russian)* (**P** 0607) All-Union Math Conf (3);1956 Moskva 1*179-180
◇ D40 ◇ ID 28233

ADYAN, S.I. [1966] *Defining relations and algorithmic problems for groups and semigroups (Russian)* (**S** 0066) Tr Mat Inst Steklov 85*123pp
• TRANSL [1966] (**S** 0055) Proc Steklov Inst Math 85*iii + 152pp
◇ D40 ◇ REV MR 34 # 4340 Zbl 204.17 JSL 38.338 • ID 00197

ADYAN, S.I. & NOVIKOV, P.S. [1968] *Commutative subgroups and the conjugacy problem in free periodic groups of odd order (Russian)* (**J** 0216) Izv Akad Nauk SSSR, Ser Mat 32*1176-1190
• TRANSL [1968] (**J** 0448) Math of USSR, Izv 2*1131-1144
◇ D40 ◇ REV MR 38 # 2197 • ID 10029

ADYAN, S.I. & NOVIKOV, P.S. [1968] *Defining relations and the word problem for free periodic groups of odd order (Russian)* (**J** 0216) Izv Akad Nauk SSSR, Ser Mat 32*971-979
• TRANSL [1968] (**J** 0448) Math of USSR, Izv 2*935-942
◇ D40 ◇ REV MR 40 # 4344 Zbl 194.33 • ID 10027

ADYAN, S.I. [1973] *Burnside groups of odd exponent and irreducible systems of group identities* (**P** 0678) Word Probl: Decis & Burnside Probl in Group Th;1969 Irvine 19-37
◇ D40 ◇ REV MR 55 # 8195 Zbl 264 # 20027 JSL 41.785 • ID 27523

ADYAN, S.I. (ED.) [1973] *Mathematical logic, theory of algorithms and theory of sets (Russian)* (**X** 2027) Nauka: Moskva 276pp
◇ D97 E97 ◇ REV MR 48 # 51 • REM Dedicated to Petr Sergeevich Novikov on the occasion of his seventieth birthday • ID 70233

ADYAN, S.I. [1973] *The works of P.S.Novikov and his students on algorithmic questions of algebra (Russian)* (**S** 0066) Tr Mat Inst Steklov 133*23-32,274
• TRANSL [1973] (**S** 0055) Proc Steklov Inst Math 133*21-30
◇ D35 D40 ◇ REV MR 54 # 10424 Zbl 309 # 02051 • ID 25878

ADYAN, S.I. [1976] *Transformations of words in a semigroup presented by a system of defining relations (Russian)* (**J** 0003) Algebra i Logika 15*611-621
• TRANSL [1976] (**J** 0069) Algeb and Log 15*379-386
◇ D40 ◇ REV MR 58 # 22364 Zbl 368 # 20037 • ID 51287

ADYAN, S.I. [1980] *Classifications of periodic words and their application in group theory* (**P** 2907) Burnside Groups;1977 Bielefeld 1-40
◇ D40 ◇ REV MR 82b:20048 • ID 80509

ADYAN, S.I. & BOONE, W.W. & HIGMAN, G. (EDS.) [1980] *Word problems II. The Oxford book* (**P** 2634) Word Problems II;1976 Oxford x+578pp
◇ D40 D97 ◇ REV MR 81f:03056 Zbl 423 # 00002
• ID 53512

ADYAN, S.I. & MAKANIN, G.S. [1984] *Studies in algorithmic questions of algebra (Russian)* (**S** 0066) Tr Mat Inst Steklov 168*197-217
◇ D40 D98 ◇ REV MR 85k:20109 Zbl 549 # 20020
• ID 45771

ADYAN, S.I. see Vol. V for further entries

AE, T. [1977] *Direct or cascade product of pushdown automata* (**J** 0119) J Comp Syst Sci 14*257-263
◇ D10 ◇ REV MR 56 # 7324 Zbl 359 # 68054 • ID 50630

AGAFONOV, V.N. [1968] *Complexity of computation of pseudorandom sequences (Russian)* (**J** 0003) Algebra i Logika 7/2*4-19
• TRANSL [1968] (**J** 0069) Algeb and Log 7*72-81
◇ D15 D80 ◇ REV MR 41 # 66 Zbl 172.299 JSL 40.248
• ID 00201

AGAFONOV, V.N. & BARZDINS, J. [1974] *On sets associated with stochastic machines (Russian)* (**J** 0068) Z Math Logik Grundlagen Math 20*481-498
◇ D10 D25 ◇ REV MR 58 # 16204 Zbl 343 # 02028
• ID 60071

AGAFONOV, V.N. [1975] *Complexity of algorithms and computations. Part II (Russian)* (**X** 0913) Novosibirsk Gos Univ: Novosibirsk 146pp
◇ B75 D15 D98 ◇ REV MR 58 # 32047 Zbl 383 # 68042
• REM Part I by Trakhtenbrot: never published • ID 80516

AGASANDYAN, G.A. [1969] *Some problems of variable structure automaton theory (Russian)* (**J** 2320) Probl Peredachi Inf, Akad Nauk SSSR 5/1*71-78
• TRANSL [1969] (**J** 3419) Probl Inf Transmiss 5/1*59-64
◇ D05 ◇ REV MR 46 # 5067 Zbl 274 # 94054 • ID 60072

AGGARWAL, S.K. & HEINEN, J.A. [1979] *A general class of noncontext free grammars generating context free languages* (**J** 0194) Inform & Control 43*187-194
◇ D05 ◇ REV MR 80m:68062 Zbl 428 # 68084 • ID 69053

AGUILO, J. & FABREGAT, F. & VALDERRAMA, E. & VAQUERO, A. [1978] *State assignment for incomplete sequential machines (Spanish) (English summary)* (**J** 0234) Rev Acad Cienc Exact Fis Nat Madrid 72*623-625
◇ D05 ◇ REV Zbl 402 # 68047 • ID 69055

AGUZZI, G. [1981] *The theory of invertible algorithms (French summary)* (**J** 3441) RAIRO Inform Theor 15*253-279
◇ D03 ◇ REV MR 83d:68026 Zbl 476 # 68027 • ID 69056

AHO, A.V. & ULLMAN, J.D. [1968] *The theory of languages* (**J** 0041) Math Syst Theory 2*97-125
◇ D05 D10 D15 ◇ REV MR 38 # 6907 Zbl 165.320 JSL 36.152 • ID 00205

AHO, A.V. [1969] *Nested stack automata* (**J** 0037) ACM J 16*383-406
◇ D05 ◇ REV MR 42 # 2880 Zbl 184.286 • ID 03820

AHO, A.V. & HOPCROFT, J.E. & ULLMAN, J.D. [1969] *On the computational power of pushdown automata* (**P** 1129) Princeton Conf Inform Sci & Syst (3);1969 Princeton 150-153
◇ D10 D15 ◇ REV Zbl 286 # 68029 • ID 60075

AINHIRN, W. & MAURER, H.A. [1979] *On ε productions for terminals in EOL forms* (**J** 2702) Discr Appl Math 1*155-166
◇ D05 ◇ REV MR 81a:68077 Zbl 453 # 68041 • ID 69060

AINHIRN, W. [1980] *How to get rid of pseudoterminals* (**P** 2904) Automata, Lang & Progr (7);1980 Noordwijkerhout 1-11
◇ D05 ◇ REV MR 81j:68081 Zbl 443 # 68057 • ID 69059

AINHIRN, W. [1980] *How to get rid of pseudoterminals* (**J** 0194) Inform & Control 47*175-194
◇ D05 ◇ REV MR 83e:68094 Zbl 457 # 68080 • ID 69058

AITCHISON, P.W. & HERMAN, G.T. [1974] *A decision procedure using the geometry of convex sets* (**J** 0303) Mathematika (Univ Coll London) 21*199-206
◇ D10 D80 ◇ REV MR 54 # 9990 Zbl 295 # 02018
• ID 25757

AJTAI, M. & CSIRMAZ, L. & NAGY, Z. [1979] *On a generalization of the game Go-moku I* (**J** 0411) Studia Sci Math Hung 14*209-226
◇ D80 E60 ◇ REV MR 84j:90117a Zbl 497 # 90099 • REM Part II 1979 by Csirmaz,L. & Nagy,Zsigmond • ID 39229

AJTAI, M. [1983] *Σ^1_1-formulae on finite structures* (**J** 0073) Ann Pure Appl Logic 24*1-48
◇ C13 C62 D15 H15 ◇ REV MR 85b:03048 Zbl 519 # 03021 • ID 37538

AJTAI, M. see Vol. III, V for further entries

AJTKHOZHAEVA, E.ZH. & BARASHENKOV, V.V. [1974] *The definition of some operator scheme relations (Russian)* (**J** 0199) Zh Vychisl Mat i Mat Fiz 14*1350-1352
• TRANSL [1974] (**J** 1049) USSR Comput Math & Math Phys 14/5*251-254
◇ D05 ◇ REV MR 51 # 4709 Zbl 298 # 02027 • ID 29608

AJZENSHTEJN, M.KH. [1978] *Ordering of complexity measures according to the effectiveness (Russian)* (**C** 3177) Ehvrist Algor Optim, Vyp 2 3-20
◇ D15 D30 ◇ REV MR 83f:68034 Zbl 478 # 03022
• ID 55637

AJZENSHTEJN, M.KH. [1981] *Subrecursive reducibility and accelerability (Russian)* (**J** 0031) Izv Vyssh Ucheb Zaved, Mat (Kazan) 1981/6*11-18
• TRANSL [1981] (**J** 3449) Sov Math 25/6*11-20
◇ D15 D20 ◇ REV MR 83e:03065 Zbl 515 # 03023
• ID 35235

AJZENSHTEJN, M.KH. [1985] *Strengthened theorems on nomenclature and pseudoacceleration for the capacity measure of complexity (Russian)* (**J** 0031) Izv Vyssh Ucheb Zaved, Mat (Kazan) 1985/5*7-15,81
◇ D15 ◇ ID 48210

AJZERMAN, M.A. & GUSEV, L.A. & ROZONOEHR, L.I. & SMIRNOVA, I.M. & TAL', A.A. [1961] *The algorithmic insolubility of the problem of recognizing the representability of recursive events in finite automata (Russian)* (**J** 0011) Avtom Telemekh 22*748-755
• TRANSL [1961] (**J** 0010) Autom & Remote Control 22*646-652
◇ D05 ◇ REV MR 24 # B881 Zbl 112.357 JSL 37.410
• ID 00215

Author Index

AJZERMAN, M.A. & GUSEV, L.A. & ROZONOEHR, L.I. & SMIRNOVA, I.M. & TAL', A.A. [1963] *Logic, automata and algorithms (Russian)* (**X** 3709) Izdat Fiz-Mat Lit: Moskva 556pp
• TRANSL [1971] (**X** 0801) Academic Pr: New York xii+433pp [1967] (**X** 0814) Oldenbourg: Muenchen x+431pp (German) [1971] (**X** 1226) Academia: Prague 408pp (Czech)
◇ B98 D05 D10 D98 ◇ REV MR 29#5690 MR 35#1411 MR 43#3044 MR 48#1818 Zbl 131.8 Zbl 216.7 JSL 31.109 JSL 37.625 • ID 00220

AJZERMAN, M.A. see Vol. II for further entries

AKUSHSKY, I.Y. & BAZILEVSKIJ, YU.YA. & SHREJDER, YU.A. [1960] *Logical, recursive and operator methods for the analysis and synthesis of automata* (**P** 0696) Inform Processing (1);1959 Paris 138-144
◇ D05 D10 ◇ REV MR 27#5663 Zbl 118.254 • ID 27689

ALAD'EV, V.Z. [1971] *Certain estimates for Neumann-Moore structures (Russian) (English and Estonian summaries)* (**J** 0080) Izv Akad Nauk Ehston SSR, Fiz, Mat 20*335-342
◇ D05 ◇ REV MR 57#5706 • ID 70349

ALAD'EV, V.Z. [1972] *Computability in homogeneous structures (Estonian and Russian summaries)* (**J** 0080) Izv Akad Nauk Ehston SSR, Fiz, Mat 21*79-83
◇ D03 D05 D10 ◇ REV MR 45#9829 Zbl 231#02046 • ID 00245

ALAD'EV, V.Z. [1974] *τ_n-grammars and languages generated by them (Russian) (English summary)* (**J** 0080) Izv Akad Nauk Ehston SSR, Fiz, Mat 23/1*67-87
◇ D05 ◇ REV Zbl 277#68040 • ID 60088

ALAD'EV, V.Z. & OSIPOV, O. [1975] *About the complexity of parallel algorithms defined by homogeneous structures (Russian) (Estonian and English summaries)* (**J** 0080) Izv Akad Nauk Ehston SSR, Fiz, Mat 24*166-172
◇ D03 D05 D10 D15 ◇ REV MR 55#11699 Zbl 308#68042 • ID 60087

ALBERT, J. & MAURER, H.A. & ROZENBERG, G. [1978] *Simple EOL forms under uniform interpretation generating of languages* (**P** 1872) Automata, Lang & Progr (5);1978 Udine 1-14
◇ D05 ◇ REV MR 80d:68081 Zbl 382#68064 • ID 69070

ALBERT, J. & WEGNER, L. [1980] *Languages with homomorphic replacements* (**P** 2904) Automata, Lang & Progr (7);1980 Noordwijkerhout 19-29
◇ D05 ◇ REV MR 82b:68059 Zbl 443#68063 • ID 69064

ALBERT, J. & MAURER, H.A. & ROZENBERG, G. [1980] *Simple EOL forms under uniform interpretation generating CF languages* (**J** 2095) Fund Inform, Ann Soc Math Pol, Ser 4 3*141-156
◇ D05 ◇ REV MR 82e:68071 Zbl 459#68044 • ID 69069

ALBERT, J. & CULIK II, K. [1980] *Test sets for homomorphism equivalence on context-free languages* (**P** 2904) Automata, Lang & Progr (7);1980 Noordwijkerhout 12-18
◇ D05 ◇ REV MR 82e:68070 Zbl 443#68069 • ID 69068

ALBERT, J. & CULIK II, K. [1980] *Test sets for homomorphism equivalence on context-free languages* (**J** 0194) Inform & Control 45*273-284
◇ D05 ◇ REV MR 82e:68070 Zbl 453#68048 • ID 69067

ALBERT, J. & WEGNER, L. [1981] *Languages with homomorphic replacements* (**J** 1426) Theor Comput Sci 16*291-305
◇ D05 ◇ REV MR 82m:68115 Zbl 468#68085 • ID 69063

ALBERT, J. & MAURER, H.A. & OTTMANN, T. [1981] *On sub regular OL forms* (**J** 2095) Fund Inform, Ann Soc Math Pol, Ser 4 4*135-149
◇ D05 ◇ REV MR 83i:68103 Zbl 467#68071 • ID 69065

ALBERT, J. & CULIK II, K. & KARHUMAEKI, J. [1981] *Test sets for context free languages and algebraic systems of equations over a free monoid* (**S** 3126) Ber, Fak Inf, Univ Karlsruhe 104*27pp
◇ D05 ◇ REV Zbl 477#68082 • ID 69066

ALBERT, J. [1982] *A note on the undecidability of contextfreeness (French summary)* (**J** 3441) RAIRO Inform Theor 16*3-11
◇ D05 ◇ REV MR 84a:68071 Zbl 493#68078 • ID 38810

ALBERT, J. & CULIK II, K. [1982] *Tree correspondence problems* (**J** 0119) J Comp Syst Sci 24*167-179
◇ D80 ◇ REV MR 84d:68049 Zbl 478#68089 • ID 55658

ALBERT, J. & OTTMANN, T. [1983] *Automaten, Sprachen und Maschinen fuer Anwender* (**X** 0876) Bibl Inst: Mannheim 333pp
◇ D05 D10 D98 ◇ REV Zbl 524#68034 • ID 38236

ALBERT, J. [1984] *On the Ehrenfeucht conjecture on test sets and its dual version* (**P** 3658) Math Founds of Comput Sci (11);1984 Prague 176-184
◇ D05 ◇ REV MR 86g:68082 Zbl 554#68047 • ID 46858

ALBERTS, M. [1985] *Complexity of computations on nondeterministic Turing machines (Russian)* (**J** 0337) Mat Ezheg, Akad Nauk Latv SSR 29*112-118,252
◇ D10 D15 ◇ ID 49422

ALBERTS, M. [1985] *Space complexity of alternating Turing machines* (**P** 4647) FCT'85 Fund of Comput Th;1985 Cottbus 1-7
◇ D10 D15 ◇ ID 49068

ALBERTS, M. [1985] *Tape complexity of nondeterministic Turing machines (Russian)* (**J** 0031) Izv Vyssh Ucheb Zaved, Mat (Kazan) 1985/7*81-83,86
◇ D10 D15 ◇ ID 49269

ALBRECHT, A. & BAUERNOEPPEL, F. & JUNG, H. & SAMMLER, O. [1984] *Einfuehrende Kapitel der Kompliziertheitstheorie. Seminarberichte* (**S** 3231) Prepr NF, Sekt Math, Humboldt-Univ Berlin i+115pp
◇ D15 ◇ ID 48597

ALBRECHT, A. see Vol. I for further entries

ALEKSANDROV, P.S. [1916] *Sur la puissance des ensembles mesurables B* (**J** 0109) C R Acad Sci, Paris 162*323-325
◇ D55 E15 E75 ◇ REV FdM 46.301 • ID 38044

ALEKSANDROV, P.S. [1924] *Sur les ensembles complementaires aux ensembles (A)* (**J** 0027) Fund Math 5*160-165
◇ D55 E15 ◇ REV FdM 50.142 • ID 00266

ALEKSANDROV, P.S. see Vol. V for further entries

ALESHIN, S.V. [1972] *Finite automata and the Burnside problem for periodic groups (Russian)* (**J** 0087) Mat Zametki (Akad Nauk SSSR) 11*319-328
• TRANSL [1972] (**J** 1044) Math Notes, Acad Sci USSR 11*199-203
◇ D05 ◇ REV MR 46#265 Zbl 253#20049 • ID 00265

ALESHIN, S.V. [1977] *Ueber ein Vollstaendigkeitskriterium fuer Superpositionen von Automatenabbildungen* (**S** 2829) Rostocker Math Kolloq 3*119-132
◇ D05 ◇ REV MR 57 # 5500 Zbl 397 # 68047 • ID 69073

ALEXIEWICZ, A. [1947] *On Hausdorff classes* (**J** 0027) Fund Math 34*61-65
◇ D55 E15 ◇ REV MR 8.506 Zbl 38.33 • ID 00267

ALFEROVA, Z. [1973] *Algorithmentheorie (Russian)* (**X** 3222) Statistika: Moskva 164pp
◇ D10 D20 D98 ◇ REV Zbl 292 # 02030 • ID 60101

ALLISON, D.C.S. & HOARE, C.A.R. [1972] *Incomputability* (**J** 2331) ACM Comp Surveys 4*169-178
◇ B75 D20 ◇ REV Zbl 265 # 68028 • ID 29849

ALMAGAMBETOV, ZH.A. [1965] *On classes of axioms closed with respect to the given reduced products and powers (Russian)* (**J** 0003) Algebra i Logika 4/3*71-77
◇ B25 C20 C52 D35 ◇ REV MR 33 # 7255 Zbl 223 # 02052 • ID 00278

ALMAGAMBETOV, ZH.A. see Vol. III for further entries

ALOISIO, P. [1971] *Una nota sulle funzioni ricorsive* (**J** 0088) Ann Univ Ferrara, NS, Sez 7 16*167-168
◇ D20 ◇ REV MR 46 # 1575 Zbl 222 # 02044 • ID 00281

ALOISIO, P. [1972] *Una nota sulle funzioni parziali ricorsive* (**J** 0089) Calcolo 8*385-388
◇ D20 ◇ REV MR 48 # 5833 Zbl 243 # 02033 • ID 00283

ALOISIO, P. see Vol. II for further entries

AL'PIN, YU.A. & KUZNETSOV, S.E. [1978] *Estimates for the number of completely indecomposable and cohesive matrices and corollaries for automata theory (Russian)* (**C** 2899) Algor & Avtomaty 3-6
◇ D05 ◇ REV MR 83e:68122 Zbl 411 # 68051 • ID 69062

ALT, H. & MEHLHORN, K. [1976] *Lower bounds for the space complexity of context-free recognition* (**P** 1870) Automata, Lang & Progr (3);1976 Edinburgh 338-354
◇ D05 D15 ◇ REV Zbl 368 # 68069 • ID 51296

ALT, H. [1979] *Lower bounds on space complexity for contextfree recognition* (**J** 1431) Acta Inf 12*33-61
◇ D05 D15 ◇ REV MR 80e:68177 Zbl 395 # 68069 • ID 69074

ALTAEV, A.V. [1974] *Operators defined by γ-enumeration and related reducibilities (Russian)* (**J** 0023) Dokl Akad Nauk SSSR 216*961-963
• TRANSL [1974] (**J** 0062) Sov Math, Dokl 15*895-899
◇ D20 D30 ◇ REV MR 50 # 4268 Zbl 304 # 02015 • ID 03823

ALTON, D.A. [1971] *Recursively enumerable sets which are uniform for finite extensions* (**J** 0036) J Symb Logic 36*271-286
◇ D25 ◇ REV MR 45 # 1754 Zbl 227 # 02023 • ID 00288

ALTON, D.A. & MADISON, E.W. [1973] *Computability of boolean algebras and their extensions* (**J** 0007) Ann Math Logic 6*95-128
◇ C57 D45 G05 ◇ REV MR 49 # 4766 Zbl 272 # 02061 • ID 00289

ALTON, D.A. [1974] *Iterated quotients of the lattice of recursively enumerable sets* (**J** 3240) Proc London Math Soc, Ser 3 28*1-12
◇ D25 ◇ REV MR 50 # 12683 Zbl 278 # 02035 • ID 03824

ALTON, D.A. & LOWTHER, J.L. [1974] *Non-existence of program optimizers in an abstract setting* (**P** 3013) Progr Symp;1974 Paris 253-265
◇ B75 D15 ◇ REV MR 55 # 6941 Zbl 311 # 68035 • ID 60113

ALTON, D.A. [1975] *A characterization of R-maximal sets* (**J** 0009) Arch Math Logik Grundlagenforsch 17*35-36
◇ D25 D35 ◇ REV MR 52 # 5390 Zbl 315 # 02039 • ID 00291

ALTON, D.A. [1975] *Embedding relations in the lattice of recursively enumerable sets* (**J** 0009) Arch Math Logik Grundlagenforsch 17*37-41
◇ B25 D25 ◇ REV MR 52 # 5391 Zbl 315 # 02040 • ID 17945

ALTON, D.A. [1976] *Diversity of speed-ups and embeddability in computational complexity* (**J** 0036) J Symb Logic 41*199-214
◇ D15 ◇ REV MR 54 # 75 Zbl 375 # 68024 • ID 14800

ALTON, D.A. [1977] *"Natural" complexity measures and a subrecursive speed-up theorem* (**P** 1694) Inform Processing (7);1977 Toronto 835-838
◇ D15 ◇ REV MR 57 # 4611 Zbl 363 # 68065 • ID 80534

ALTON, D.A. [1980] *"Natural" programming languages and complexity measures for subrecursive programming languages: an abstract approach* (**P** 3021) Logic Colloq;1979 Leeds 248-285
◇ B75 D15 D20 ◇ REV MR 82i:03051 Zbl 454 # 68032 • ID 54254

AMAR, V. & PUTZOLU, G. [1966] *Generalizations of regular events* (**P** 0746) Automata Th;1964 Ravello 1-5
◇ D05 ◇ REV MR 31 # 1128 Zbl 192.80 • ID 27532

AMBARTSUMYAN, A.A. & POTEKHIN, A.I. [1977] *Standard realization of an asynchronous automaton (Russian)* (**J** 0011) Avtom Telemekh 1977/10*122-132
• TRANSL [1978] (**J** 0010) Autom & Remote Control 38*1529-1537
◇ D05 ◇ REV Zbl 416 # 94022 • ID 69077

AMBARTSUMYAN, A.A. & ZAPOL'SKIKH, E.N. [1981] *An approach to time decomposition of automata II (Russian)* (**J** 0011) Avtom Telemekh 1981/3*112-121
• TRANSL [1981] (**J** 0010) Autom & Remote Control 42*364-370
◇ D05 ◇ REV Zbl 464 # 68056 • REM Part I 1981 • ID 69076

AMBARTSUMYAN, A.A. & ZAPOL'SKIKH, E.N. [1981] *Time decomposition of automata I (Russian)* (**J** 0011) Avtom Telemekh 1981/2*135-144
• TRANSL [1981] (**J** 0010) Autom & Remote Control 42*237-244
◇ D05 ◇ REV MR 83c:68064 Zbl 464 # 68055 • REM Part II 1981 • ID 69075

AMBOS-SPIES, K. & FLEISCHHACK, H. & HAGEN, H. [1984] *p-generic sets* (**P** 4012) Automata, Lang & Progr (11);1984 Antwerpen 58-68
◇ D15 E40 ◇ REV MR 86e:03041 Zbl 556 # 03032 • ID 45088

AMBOS-SPIES, K. & JOCKUSCH JR., C.G. & SHORE, R.A. & SOARE, R.I. [1984] *An algebraic decomposition of the recursively enumerable degrees and the coincidence of several degree classes with the promptly simple degrees* (**J** 0064) Trans Amer Math Soc 281*109-128
◇ D25 ◇ REV MR 85i:03137 Zbl 539 # 03020 • ID 33576

AMBOS-SPIES, K. [1984] *An extension of the non-diamond theorem in classical and α-recursion theory* (**J** 0036) J Symb Logic 49*586-607
◇ D25 D60 ◇ REV MR 86d:03039 Zbl 533 # 03026
• ID 33573

AMBOS-SPIES, K. [1984] *Contiguous r.e. degrees* (**P** 2153) Logic Colloq;1983 Aachen 2*1-38
◇ D25 ◇ REV MR 86f:03065 Zbl 562 # 03022 • ID 40163

AMBOS-SPIES, K. [1984] *On pairs of recursively enumerable degrees* (**J** 0064) Trans Amer Math Soc 283*507-531
◇ D25 ◇ REV MR 85d:03083 Zbl 533 # 03025
Zbl 541 # 03023 • ID 33574

AMBOS-SPIES, K. [1984] *On the structure of polynomial time degrees* (**P** 3565) Symp of Th Aspects of Comput Sci (1);1984 Paris 198-208
◇ D15 ◇ REV Zbl 546 # 03021 • ID 40139

AMBOS-SPIES, K. [1984] *P-mitotic sets* (**P** 2342) Symp Rek Kombin;1983 Muenster 1-23
◇ D15 ◇ REV Zbl 546 # 03020 • ID 41157

AMBOS-SPIES, K. [1985] *Anti-mitotic recursively enumerable sets* (**J** 0068) Z Math Logik Grundlagen Math 31*461-477
◇ D25 ◇ ID 47552

AMBOS-SPIES, K. [1985] *Generators of the recursively enumerable degrees* (**P** 3342) Rec Th Week;1984 Oberwolfach 1-28
◇ D25 ◇ ID 45295

AMBOS-SPIES, K. [1985] *On the relative complexity of subproblems of intractable problems* (**P** 4241) Symp of Th Aspects of Comput Sci (2);1985 Saarbruecken
◇ D15 ◇ ID 40158

AMBOS-SPIES, K. [1985] *Sublattices of the polynomial degrees* (**J** 0194) Inform & Control 65*63-84
◇ D15 ◇ REV Zbl 556 # 03033 • ID 49212

AMIHUD, A. & CHOUEKA, Y. [1981] *Loop-programs and polynomially computable functions* (**J** 0382) Int J Comput Math 9*195-205
◇ D15 D20 ◇ REV MR 83a:68062 Zbl 456 # 68049
• ID 54323

AMOROSO, S. & COOPER, GERALD [1970] *The Garden-of-Eden theorem for finite configurations* (**J** 0053) Proc Amer Math Soc 26*158-164
◇ D05 ◇ REV MR 43 # 1760 Zbl 219 # 02025 • ID 28087

AMOROSO, S. & BLOOM, S.L. [1971] *A model of the digital computer* (**P** 1383) Symp Comput & Automata;1971 Brooklyn 539-559
◇ D10 D20 ◇ REV Zbl 262 # 68020 • ID 60117

AMOROSO, S. & COOPER, GERALD & PATT, Y.N. [1975] *Some clarifications of the concept of a Garden-of-Eden configuration* (**J** 0119) J Comp Syst Sci 10*77-82
◇ D05 ◇ REV MR 51 # 5203 Zbl 348 # 94056 • ID 60116

AMSTISLAVSKIJ, V.I. [1966] *Recursive sieves (Russian)* (**J** 0092) Sib Mat Zh 7*233-241
• TRANSL [1966] (**J** 0475) Sib Math J 7*187-193
◇ D55 E15 F65 ◇ REV MR 33 # 5477 Zbl 216.291
JSL 33.295 • ID 00303

AMSTISLAVSKIJ, V.I. [1966] *Set-theoretical operations and recursive hierarchies (Russian)* (**J** 0023) Dokl Akad Nauk SSSR 169*995-998
• TRANSL [1966] (**J** 0062) Sov Math, Dokl 7*1029-1032
◇ D55 D70 E15 ◇ REV MR 34 # 5667 Zbl 161.7
JSL 37.409 • ID 00302

AMSTISLAVSKIJ, V.I. [1968] *Expansion of recursive hierarchies and R-operations (Russian)* (**J** 0023) Dokl Akad Nauk SSSR 180*1023-1026
• TRANSL [1968] (**J** 0062) Sov Math, Dokl 9*703-706
◇ D55 D70 E15 ◇ REV MR 38 # 45 Zbl 175.271
JSL 37.409 • ID 00305

AMSTISLAVSKIJ, V.I. [1970] *Effective R-sets and transfinite extensions of recursive hierarchies* (**J** 0027) Fund Math 68*61-68
◇ D55 D70 E15 ◇ REV MR 42 # 4402 Zbl 198.26
JSL 37.409 • ID 00308

AMSTISLAVSKIJ, V.I. [1970] *On the decomposition of a field of sets obtained by an R-operation over recursive sets (Russian)* (**J** 0023) Dokl Akad Nauk SSSR 191*743-746
• TRANSL [1970] (**J** 0062) Sov Math, Dokl 11*419-422
◇ D55 D70 E15 ◇ REV MR 42 # 4401 Zbl 313 # 02024
JSL 37.409 • ID 00307

AMSTISLAVSKIJ, V.I. [1973] *A comparison of the indices arising in the transfinite iteration of functions (Russian)* (**J** 0092) Sib Mat Zh 14*699-725,909
• TRANSL [1973] (**J** 0475) Sib Math J 14*483-502
◇ D55 D70 E15 ◇ REV MR 49 # 8838 Zbl 272 # 02066
• ID 03825

AMSTISLAVSKIJ, V.I. [1973] *Computability with functionals (Russian)* (**J** 0003) Algebra i Logika 12*497-511,617
• TRANSL [1973] (**J** 0069) Algeb and Log 12*277-286
◇ D65 ◇ REV MR 50 # 6814 Zbl 291 # 02028 • ID 03826

AMSTISLAVSKIJ, V.I. [1974] *Recursiveness and R^c-operations (Russian)* (**J** 0216) Izv Akad Nauk SSSR, Ser Mat 38*1221-1237
• TRANSL [1974] (**J** 0448) Math of USSR, Izv 8*1209-1224
◇ D55 D65 D70 E15 ◇ REV MR 51 # 155
Zbl 349 # 02033 • ID 17450

AMSTISLAVSKIJ, V.I. [1975] *On recursive characterizations of sets obtainable by means of R^c-operations (Russian)* (**J** 0023) Dokl Akad Nauk SSSR 224*257-260
• TRANSL [1975] (**J** 0062) Sov Math, Dokl 16*1160-1163
◇ D55 D65 D70 E15 ◇ REV MR 52 # 13392
Zbl 349 # 02034 • ID 60124

AMSTISLAVSKIJ, V.I. [1978] *An improved substitution theorem for computable functionals and some of its consequences (Russian)* (**J** 0065) Tr Moskva Mat Obshch 37*255-269
• TRANSL [1980] (**J** 3279) Trans Moscow Math Soc 1*273-289
◇ D65 ◇ REV MR 81g:03058 Zbl 429 # 03029 • ID 53860

ANASIMOV, P.A. [1979] *A choice of a sequence of the input alphabet symbols maximizing the efficiency of the work of an automaton (Russian)* (**C** 3405) Sovrem Vopr Priklad Mat & Progr 3-9
◇ D05 ◇ REV Zbl 426 # 68037 • ID 69081

ANDERSON, MICHAEL [1967] *Note on an inequality of Tibor Rado* (**J** 0047) Notre Dame J Formal Log 8*159-160
◇ D10 ◇ REV MR 38 # 4318 Zbl 174.20 • ID 00351

ANDERSON, MICHAEL [1968] *Approximation to a decision procedure for the halting problem* (J 0047) Notre Dame J Formal Log 9*305-312
◇ D10 ◇ REV MR 40#4114 Zbl 182.19 JSL 35.480
• ID 00352

ANDERSON, MICHAEL [1969] *Note on the mortality problem for shift state trees* (J 0047) Notre Dame J Formal Log 10*275-276
◇ D05 ◇ REV MR 40#4018 Zbl 188.330 JSL 36.151
• ID 00353

ANDO, S. & ITO, T. [1974] *A complete axiom system of super-regular expressions* (P 1691) Inform Processing (6);1974 Stockholm 661-665
◇ D05 ◇ REV MR 55#12357 Zbl 301#02033 • ID 62700

ANDO, S. & ITO, T. [1975] *On applying Scott's logic to termination problems* (P 3299) Progr Kiso Riron, Algor Okeru Shomei Ron;1973/74 Kyoto 1-7
◇ B40 D05 ◇ REV MR 58#8427 Zbl 384#03007
• ID 52049

ANDREWS, P.B. [1974] *Provability in elementary type theory* (J 0068) Z Math Logik Grundlagen Math 20*411-418
◇ B15 B25 D35 F35 ◇ REV MR 52#7867
Zbl 306#02017 • ID 03832

ANDREWS, P.B. see Vol. I, III, V, VI for further entries

ANDZHAN, A.V. [1974] *Reduzierbarkeit von Funktionenklassen nach der Prognostizierung (Russisch)* (S 2587) Teor Algor & Progr (Riga) 1*214-216
◇ D20 D30 ◇ REV MR 58#21544 Zbl 341#02027
• ID 29748

ANDZHAN, A.V. [1977] *On a reducability of sets (Russian)* (S 2587) Teor Algor & Progr (Riga) 3*143-152
◇ D30 ◇ REV MR 57#5713 Zbl 376#02035 • ID 51677

ANDZHAN, A.V. [1983] *Possibilities of automata for searching one-dimensional domains (Russian)* (J 0337) Mat Ezheg, Akad Nauk Latv SSR 27*191-201
◇ D05 ◇ REV MR 85a:68046 Zbl 562#68036 • ID 38865

ANGLUIN, D. [1978] *On the complexity of minimum inference of regular sets* (J 0194) Inform & Control 39*337-350
◇ D05 D15 ◇ REV MR 80h:68025 Zbl 393#68066
• ID 69082

ANGLUIN, D. [1980] *Inductive inference of formal languages from positive data* (J 0194) Inform & Control 45*117-135
◇ D05 ◇ REV MR 82a:68135 Zbl 459#68051 • ID 80550

ANGLUIN, D. [1980] *On counting problems and the polynomial-time hierachy* (J 1426) Theor Comput Sci 12*161-173
◇ D15 ◇ REV MR 82e:68036 Zbl 499#68020 • ID 38305

ANGLUIN, D. & HOOVER, D.N. [1984] *Regular prefix relations* (J 0041) Math Syst Theory 17*167-191
◇ D05 ◇ REV MR 86c:68015 • ID 48807

ANGLUIN, D. see Vol. II for further entries

ANIKEEV, A.S. [1972] *Classification of derivable propositional formulas (Russian)* (J 0087) Mat Zametki (Akad Nauk SSSR) 11*165-174
• TRANSL [1972] (J 1044) Math Notes, Acad Sci USSR 11*106-110
◇ B15 D05 F20 F50 ◇ REV MR 45#6568
Zbl 239#02003 • ID 00375

ANISIMOV, A.V. [1971] *The group languages (Russian) (English summary)* (J 0040) Kibernetika, Akad Nauk Ukr SSR 1971/4*18-24
◇ D05 D40 ◇ REV MR 46#1134 Zbl 241#68034
• ID 60147

ANISIMOV, A.V. [1972] *Certain algorithmic questions for groups and context-free languages (Russian) (English summary)* (J 0040) Kibernetika, Akad Nauk Ukr SSR 1972/2*4-11
◇ D05 D40 ◇ REV MR 47#1329 Zbl 241#68035
• ID 60148

ANISIMOV, A.V. & SEIFERT, F.D. [1975] *Zur algebraischen Charakteristik der durch kontext-freie Sprachen definierten Gruppen (English and Russian summaries)* (J 0129) Elektr Informationsverarbeitung & Kybern 11*695-702
◇ D05 D40 ◇ REV MR 54#10425 Zbl 322#68047
• ID 25879

ANISIMOV, A.V. & LISOVIK, L.P. [1978] *Equivalence problems for finite automata mappings into a free and a commutative semigroup (Russian) (English summary)* (J 0040) Kibernetika, Akad Nauk Ukr SSR 1978/3*1-7
• TRANSL [1978] (J 0021) Cybernetics 14*321-327
◇ D05 ◇ REV MR 80m:68052 Zbl 388#03015 • ID 52269

ANISIMOV, A.V. [1979] *Church-Rosser marking transducers and decidable properties in tree processing (Russian)* (J 0023) Dokl Akad Nauk SSSR 249*1033-1035
• TRANSL [1979] (J 0062) Sov Math, Dokl 20*1356-1359
◇ D05 ◇ REV MR 80k:68016 Zbl 442#68043 • ID 69085

ANISIMOV, A.V. [1979] *Discrete backtracking converters (Russian)* (J 0040) Kibernetika, Akad Nauk Ukr SSR 1979/6*35-40
• TRANSL [1979] (J 0021) Cybernetics 15*810-816
◇ D05 ◇ REV MR 82h:68079 Zbl 475#68001 • ID 80551

ANISIMOV, A.V. [1981] *Church-Rosser transformers and decidable properties of tree processing* (P 3729) Algor in Modern Math & Comput Sci;1979 Urgench 449-457
• TRANSL [1982] (P 3803) Algor Sovrem Mat & Prilozh;1979 Urgench II*263-269
◇ D05 ◇ REV MR 84c:03077 Zbl 477.68035 • ID 34014

ANISIMOV, A.V. [1981] *Finite automaton semigroup mappings (Russian)* (J 0040) Kibernetika, Akad Nauk Ukr SSR 1981/5*1-7
• TRANSL [1981] (J 0021) Cybernetics 17*571-578
◇ D05 ◇ REV MR 84i:68092 Zbl 486#68048 • ID 38395

ANKUDINOV, G.I. [1978] *Questions of the theory of plex-languages (Russian)* (J 0040) Kibernetika, Akad Nauk Ukr SSR 1978/3*44-49
• TRANSL [1978] (J 0021) Cybernetics 14*363-369
◇ D05 ◇ REV MR 81d:68090 Zbl 384#68067 • ID 69083

ANONYMOUS [1973] *Problems* (P 0678) Word Probl: Decis & Burnside Probl in Group Th;1969 Irvine 641-646
◇ D35 D40 ◇ REV MR 53#5756 • ID 22960

ANSHEL, M. & STEBE, P. [1974] *The solvability of the conjugacy problem for certain HNN groups* (J 0015) Bull Amer Math Soc 80*266-270
◇ D40 ◇ REV MR 54#7633 Zbl 294#20036 • ID 30626

ANSHEL, M. [1976] *Conjugate powers in HNN groups* (J 0053) Proc Amer Math Soc 54*19-23
◇ D40 ◇ REV MR 52#14059 Zbl 345#20039 • ID 30622

Author Index

ANSHEL, M. & STEBE, P. [1976] *Conjugate powers in free products with amalgamation* (J 1447) Houston J Math 2*139-147
◇ B25 D40 ◇ REV MR 53 # 13416 Zbl 344 # 20032
• ID 30627

ANSHEL, M. [1976] *Decision problems for HNN groups and vector addition systems* (J 0214) Math of Comp 30*154-156
◇ D40 D80 ◇ REV MR 53 # 626 Zbl 377 # 20026
• ID 16682

ANSHEL, M. [1976] *The conjugacy problem for HNN groups and the word problem for commutative semigroups* (J 0053) Proc Amer Math Soc 61*223-224
◇ D40 ◇ REV MR 54 # 10446 Zbl 359 # 20022 • ID 25881

ANSHEL, M. [1978] *Decision problems for HNN groups and commutative semigroups* (J 1447) Houston J Math 4*137-142
◇ B25 D40 ◇ REV MR 58 # 16889 Zbl 391 # 20024
• ID 30625

ANSHEL, M. [1978] *Vector groups and the equality problem for vector addition systems* (J 0214) Math of Comp 32*614-616
◇ D40 D80 ◇ REV MR 81e:20042 Zbl 437 # 03006
• ID 30624

ANSHEL, M. & MCALOON, K. [1983] *Reducibilities among decision problems for HNN groups, vector addition systems and subsystems of Peano arithmetic* (J 0053) Proc Amer Math Soc 89*425-429
◇ D20 D40 D80 F30 ◇ REV MR 85d:03096
Zbl 523 # 20025 • ID 36994

ANSHEL, M. & DOMANSKI, B. [1985] *The complexity of Dehn's algorithm for word problems in groups* (J 2746) J Algor 6*543-549
◇ D15 D40 ◇ ID 49484

ANTELMANN, H. & BUDACH, L. & ROLLIK, H.-A. [1979] *On universal traps* (J 0129) Elektr Informationsverarbeitung & Kybern 15*123-131
◇ D05 ◇ REV MR 80i:68042 Zbl 432 # 68041 • ID 40851

APOLLONI, B. & GREGORIO DI, S. [1982] *A probabilistic analysis of a new satisfiability algorithm (French summary)* (J 3441) RAIRO Inform Theor 16*201-223
◇ B35 D15 ◇ REV MR 84a:68032 Zbl 489 # 68038
• ID 38805

APPEL, K.I. & MCLAUGHLIN, T.G. [1965] *On properties of regressive sets* (J 0064) Trans Amer Math Soc 115*83-93
◇ D50 ◇ REV MR 37 # 6176 Zbl 192.52 JSL 33.621
• ID 00407

APPEL, K.I. [1967] *No recursively enumerable set is the union of finitely many immune retraceable sets* (J 0053) Proc Amer Math Soc 18*279-281
◇ D25 D50 ◇ REV MR 34 # 7363 Zbl 192.52 JSL 33.621
• ID 00409

APPEL, K.I. [1967] *There exist two regressive sets whose intersection is not regressive* (J 0036) J Symb Logic 32*322-324
◇ D50 ◇ REV MR 35 # 6546 Zbl 192.52 JSL 33.621
• ID 00410

APPEL, K.I. & SCHUPP, P.E. [1972] *The conjugacy problem for the group of any tame alternating knot is solvable* (J 0053) Proc Amer Math Soc 33*329-336
◇ D40 D80 ◇ REV MR 45 # 3530 Zbl 243 # 20036
• ID 00411

APPEL, K.I. [1974] *On the conjugacy problem for knot groups* (J 0044) Math Z 138*273-294
◇ D40 D80 ◇ REV MR 50 # 10090 Zbl 276 # 20033
• ID 28139

APPEL, K.I. see Vol. III for further entries

APPLEBAUM, C.H. & DEKKER, J.C.E. [1970] *Partial recursive functions and ω-functions* (J 0036) J Symb Logic 35*559-568
◇ C57 D50 ◇ REV MR 43 # 3116 Zbl 217.13 • ID 00414

APPLEBAUM, C.H. [1971] *ω-homomorphisms and ω-groups* (J 0036) J Symb Logic 36*55-65
◇ D45 D50 ◇ REV MR 44 # 2603 Zbl 218 # 02041
• ID 00416

APPLEBAUM, C.H. [1971] *Isomorphisms of ω-groups* (J 0047) Notre Dame J Formal Log 12*238-248
◇ D45 D50 ◇ REV MR 46 # 3278 Zbl 188.26 • ID 00417

APPLEBAUM, C.H. [1973] *A result for π-groups* (J 0068) Z Math Logik Grundlagen Math 19*33-35
◇ D45 D50 ◇ REV MR 47 # 3156 Zbl 262 # 02046
• ID 00418

APPLEBAUM, C.H. [1973] *A stronger definition of recursively infinite set* (J 0047) Notre Dame J Formal Log 14*411-412
◇ D50 ◇ REV MR 47 # 8277 Zbl 245 # 02039 • ID 00419

APPLEBAUM, C.H. [1973] *The recursive equivalence type of a decomposition of an ω-group: the RET of a decomposition* (J 0038) J Austral Math Soc 15*172-176
◇ D45 D50 ◇ REV MR 49 # 10533 Zbl 266 # 02024
• ID 03833

APPLEBAUM, C.H. [1974] *ω-semigroups* (J 0308) Rocky Mountain J Math 4*597-620
◇ D45 D50 ◇ REV MR 50 # 9565 Zbl 295 # 02028
• ID 04135

APPLEBAUM, C.H. [1974] *Counterexamples in the theory of ω functions* (J 0017) Canad J Math 26*800-805
◇ D50 ◇ REV MR 50 # 9564 Zbl 248 # 02043 • ID 03834

APPLEBAUM, C.H. [1978] *Some structure theorems for inverse ω-semigroups* (J 0027) Fund Math 99*79-91
◇ D45 D50 ◇ REV MR 58 # 5148 Zbl 381 # 03032
• ID 30621

APPLEBAUM, C.H. [1982] *An introduction to ω-extensions of ω-groups* (J 0036) J Symb Logic 47*27-36
◇ D45 D50 ◇ REV MR 84h:03104 Zbl 496 # 03025
• ID 34294

APPLEBEE, R.C. & PAHI, B. [1970] *An unsolvable problem concerning implicational calculi* (J 0047) Notre Dame J Formal Log 11*200-202
◇ B22 D35 ◇ REV MR 44 # 74 Zbl 169.305 JSL 37.417
• ID 43922

APPLEBEE, R.C. see Vol. II for further entries

APRILE, G. & RAIMONDI, T. [1978] *Homogeneous neuron networks* (J 0141) Arch Autom & Telemech 23*359-366
◇ B70 D10 ◇ REV Zbl 378 # 94041 • ID 51831

APRILE, G. see Vol. I, II, V for further entries

APT, K.R. [1972] *ω-models in analytical hierarchy (Russian summary)* (J 0014) Bull Acad Pol Sci, Ser Math Astron Phys 20*901-904
◇ C62 D30 D55 ◇ REV MR 47 # 4777 Zbl 252 # 02045
• ID 00429

APT, K.R. & MAREK, W. [1974] *Second order arithmetic and related topics* (**J** 0007) Ann Math Logic 6*177-229
◇ C62 D55 E45 F25 F35 ◇ REV MR 51 # 12512 Zbl 299 # 02066 • ID 00432

APT, K.R. [1976] *Semantics of the infinitistic rules of proof* (**J** 0036) J Symb Logic 41*121-138
◇ C62 D55 E45 F07 F35 ◇ REV MR 55 # 10230 Zbl 328 # 02016 • ID 14789

APT, K.R. [1977] *Recursive embeddings of partial orderings* (**J** 0017) Canad J Math 29*349-359
◇ C57 D20 G05 ◇ REV MR 55 # 2673 Zbl 331 # 06002 • ID 60167

APT, K.R. [1978] *Inductive definitions, models of comprehension and invariant definability* (**J** 0029) Israel J Math 29*221-238
◇ C40 C62 D55 D70 F35 ◇ REV MR 81e:03048 Zbl 379 # 02014 • ID 29131

APT, K.R. [1981] *Recursive assertions and parallel programs* (**J** 1431) Acta Inf 15*219-232
◇ B75 D80 ◇ REV MR 82h:68009 Zbl 436 # 68009 • ID 80560

APT, K.R. see Vol. I, II, III, VI for further entries

ARAKI, T. & KASAMI, T. [1977] *Some decision problems related to the reachability problem for Petri nets* (**J** 1426) Theor Comput Sci 3*85-104
◇ D80 G05 ◇ REV MR 55 # 11666 Zbl 352 # 68083 • ID 50067

ARAKI, T. & TOKURA, N. [1981] *Flow languages equal recursively enumerable languages* (**J** 1431) Acta Inf 15*209-217
◇ D05 D25 ◇ REV MR 82h:68099 Zbl 456 # 68093 • ID 69089

ARBIB, M.A. [1961] *Turing machines, finite automata and neural nets* (**J** 0037) ACM J 8*467-475
◇ D05 D10 ◇ REV MR 25 # 4952 JSL 35.482 • ID 00452

ARBIB, M.A. [1963] *Monogenic normal systems are universal* (**J** 0038) J Austral Math Soc 3*301-306
◇ D03 D10 ◇ REV MR 27 # 3520 Zbl 137.10 • ID 00453

ARBIB, M.A. [1964] *Brains, machines, and mathematics* (**X** 0822) McGraw-Hill: New York xiv + 152pp
• TRANSL [1968] (**X** 2027) Nauka: Moskva
◇ D05 D10 D80 F30 ◇ REV MR 31 # 5766 Zbl 174.24 JSL 35.482 • ID 00454

ARBIB, M.A. [1965] *A common framework for automata theory and control theory* (**J** 4704) J Soc Ind & Appl Math, Ser Control 3*206-222
◇ D05 D10 ◇ REV MR 32 # 7269 Zbl 145.124 • ID 31323

ARBIB, M.A. & BLUM, M. [1965] *Machine dependence of degree of difficulty* (**J** 0053) Proc Amer Math Soc 16*442-447
◇ D15 ◇ REV MR 31 # 5767 Zbl 135.250 JSL 34.509 • ID 00455

ARBIB, M.A. [1966] *Simple self-reproducing universal automata* (**J** 0194) Inform & Control 9*177-189
◇ D10 ◇ REV Zbl 143.21 • ID 31324

ARBIB, M.A. [1966] *Speed-up theorems and incompleteness theorems* (**P** 0746) Automata Th;1964 Ravello 6-24
◇ D15 F30 ◇ REV Zbl 202.316 • ID 27533

ARBIB, M.A. [1967] *Automaton automorphisms* (**J** 0194) Inform & Control 11*147-154
◇ D10 ◇ REV MR 41 # 3181 Zbl 156.254 • ID 31318

ARBIB, M.A. [1967] *Some comments on self-reproducing automata* (**P** 1390) Syst & Comput Sci;1965 London ON 42-59
◇ D10 ◇ REV MR 40 # 8513 • ID 31325

ARBIB, M.A. & GIVE'ON, Y. [1968] *Algebra automata I: Parallel programming as a prolegomena for the categorical approach. II:The categorical framework for dynamic analyisis* (**J** 0194) Inform & Control 12*331-345,346-370
◇ B75 C05 D05 G30 ◇ REV MR 43 # 1761 MR 43 # 1762 Zbl 164.322 • ID 31329

ARBIB, M.A. [1968] *The algebraic theory of machines, languages, and semigroups* (**X** 0801) Academic Pr: New York 359pp
• TRANSL [1975] (**X** 3222) Statistika: Moskva 334pp
◇ C05 D05 D98 G30 ◇ REV MR 38 # 1198 Zbl 181.15
• REM Russian translation published by Arbib,M.A. & Buslenko,N.P • ID 23501

ARBIB, M.A. [1968] *The automata theory of semigroup embeddings* (**J** 0038) J Austral Math Soc 8*568-570
◇ D05 ◇ REV MR 38 # 255 • ID 31330

ARBIB, M.A. [1969] *Theories of abstract automata* (**X** 0819) Prentice Hall: Englewood Cliffs xiii + 412pp
◇ D05 D10 D98 ◇ REV MR 47 # 10159 Zbl 193.328 JSL 37.412 • ID 00456

ARBIB, M.A. [1973] *Coproducts and decomposable machines* (**J** 0119) J Comp Syst Sci 7*278-287
◇ D05 G30 ◇ REV MR 49 # 10459 Zbl 279 # 94043 • ID 31320

ARBIB, M.A. & CULIK, K. [1973] *Sequential and jumping machines and their relation to computers* (**J** 1431) Acta Inf 2*162-171
◇ B70 D05 D10 ◇ REV MR 48 # 8130 Zbl 256 # 68018 • ID 31321

ARBIB, M.A. & MANES, E.G. [1974] *Machines in a category: an expository introduction* (**J** 0163) SIAM Review 16*163-192
◇ C05 D05 D98 G30 ◇ REV MR 50 # 16156 Zbl 288 # 18005 • ID 31322

ARBIB, M.A. & MANES, E.G. [1975] *A category-theoretic approach to systems in a fuzzy world* (**J** 0154) Synthese 30*381-406
• REPR [1983] (**C** 3834) Lang, Logic and Method 199-224
◇ C05 D05 E72 G30 ◇ REV Zbl 301 # 18002 Zbl 534 # 18002 • ID 31326

ARBIB, M.A. & MANES, E.G. [1975] *Adjoint machines, state behavior machines, and duality* (**J** 0326) J Pure Appl Algebra 6*313-344
◇ D05 G30 ◇ REV MR 54 # 2753 Zbl 323 # 18002 • ID 24666

ARBIB, M.A. & MANES, E.G. [1975] *Fuzzy machines in a category* (**J** 0016) Bull Austral Math Soc 13*169-210
◇ B52 D05 G30 ◇ REV MR 53 # 10889 Zbl 318 # 18008 • ID 23098

ARBIB, M.A. & MANES, E.G. [1979] *Intertwined recursion, tree transformations, and linear systems* (**J** 0194) Inform & Control 40*144-180
◇ D05 D20 D80 G30 ◇ REV MR 80e:18001 Zbl 413 # 68053 • ID 53042

ARBIB, M.A. & KFOURY, A.J. & MOLL, R. [1982] *A programming approach to computability* (**X** 0811) Springer: Heidelberg & New York VIII*251pp
◇ B75 D20 D98 ◇ REV Zbl 497 # 68025 • ID 36641

Author Index

ARBIB, M.A. & MANES, E.G. & STEENSTRUP, M. [1983] *Port automata and the algebra of concurrent processes* (**J** 0119) J Comp Syst Sci 27*29-50
◇ D05 ◇ REV MR 86c:68022 Zbl 535 # 68025 • ID 48808

ARDEN, D.N. [1961] *Delayed-logic and finite-state machines* (**P** 0624) Switch Circ Th & Log Design (1,2);1960 Chicago;1961 Detroit 133-151
◇ D05 ◇ REV JSL 36.151 • ID 00457

ARIKAWA, S. [1969] *On the length functions of languages recognizable by linear bounded automata* (**J** 0106) Mem Fac Sci, Kyushu Univ, Ser A 23*12-27
◇ D05 ◇ REV MR 40 # 5356 Zbl 213.21 • ID 27989

ARIKAWA, S. [1970] *Elementary formal systems and formal languages - simple formal systems* (**J** 0106) Mem Fac Sci, Kyushu Univ, Ser A 24*47-75
◇ D05 ◇ REV MR 44 # 79 Zbl 206.288 • ID 00459

ARNOLD, A. & DAUCHET, M. [1978] *Caracterisation algebrique des ensembles recursivement enumerables d'arbres* (**P** 3835) AFCET-SMF Math Appliquees Colloq (1);1978 Palaiseau 2*1-13
◇ D05 ◇ REV Zbl 482 # 68076 • ID 37707

ARNOLD, A. & NIVAT, M. [1978] *Metric interpretations of recursive program schemes* (**S** 3125) Ber, Abt Inf, Univ Dortmund 74*11-22
◇ B75 D80 ◇ REV Zbl 469 # 68018 • ID 55185

ARNOLD, A. & LATTEUX, M. [1978] *Recursivite et cones rationnels fermes par intersection* (**J** 0089) Calcolo 15*382-394
◇ D05 D80 ◇ REV MR 81f:68080 Zbl 421 # 68075 • ID 53457

ARNOLD, A. & NIVAT, M. [1980] *Formal computations of nondeterministic recursive program schemes* (**J** 0041) Math Syst Theory 13*219-236
◇ B75 D20 ◇ REV MR 82m:68019 Zbl 441 # 68044 • ID 80572

ARNOLD, A. [1985] *Deterministic and nonambigous rational* ω*-languages* (**P** 4622) Autom on Infinite Words;1984 Le Mont-Dore 18-27
◇ D05 ◇ ID 49437

ARNOLD, A. see Vol. I for further entries

AROLD, D. & BERNHARDT, L. & STAIGER, L. [1972] *Ueber die Entscheidung und Aufzaehlung regulaerer Wortmengen durch Markowsche Algorithmen* (**J** 0129) Elektr Informationsverarbeitung & Kybern 8*99-114
◇ D03 D05 ◇ REV MR 47 # 7953 Zbl 253 # 02027 • ID 28932

ARON, E. & BARAK, A. [1975] *On the efficiency of ternary algorithms for multiplication and division* (**P** 1805) Int Symp Multi-Val Log (5,Proc);1975 Bloomington 331-343
◇ B50 B75 D20 ◇ REV MR 58 # 8466 • ID 35833

ARSENIN, V.YA. & LYAPUNOV, A.A. [1950] *The theory of A-sets (Russian)* (**J** 0067) Usp Mat Nauk 5/5(39)*45-108
• TRANSL [1955] (**X** 0806) Dt Verlag Wiss: Berlin iii + 108pp
◇ D55 D70 E15 E98 ◇ REV MR 12.597 Zbl 38.194
• REM Transl. in: "Arbeiten zur deskriptiven Mengenlehre" which contains also papers of Shchegol'kov,E.A and the second author • ID 00496

ARSENIN, V.YA. & LYAPUNOV, A.A. & SHCHEGOL'KOV, E.A. [1955] *Arbeiten zur deskriptiven Mengenlehre* (**X** 0806) Dt Verlag Wiss: Berlin iii + 108pp
◇ D55 D70 D97 E15 E97 ◇ REV MR 17.467 Zbl 68.270 • ID 24901

ARSENIN, V.YA. & KOZLOVA, Z.I. [1977] *A survey of A. A. Ljapunov's works on descriptive set theory (Russian)* (**J** 0052) Probl Kibern 32*15-44,246
◇ A10 D55 E15 ◇ REV MR 57 # 9432 • ID 80574

ARSENIN, V.YA. see Vol. V for further entries

ARSLANOV, M.M. [1968] *Two theorems on recursively enumerable sets (Russian)* (**J** 0003) Algebra i Logika 7/3*4-8
• TRANSL [1968] (**J** 0069) Algeb and Log 7*132-134
◇ D25 ◇ REV MR 41 # 3268 Zbl 213.17 • ID 00498

ARSLANOV, M.M. [1969] *Effectively hypersimple sets (Russian)* (**J** 0003) Algebra i Logika 8*143-153
• TRANSL [1969] (**J** 0069) Algeb and Log 8*79-85
◇ D25 ◇ REV MR 43 # 1829 Zbl 261 # 02027 • ID 00499

ARSLANOV, M.M. [1969] *The embedding of recursively enumerable sets (Russian)* (**C** 1391) Coll Candidat Works - Math, Mech, Phys (Kazan) 5-9
◇ D25 ◇ REV MR 41 # 3269 • ID 16717

ARSLANOV, M.M. [1970] *Complete hypersimple sets (Russian)* (**J** 0031) Izv Vyssh Ucheb Zaved, Mat (Kazan) 1970/4(95)*30-35
◇ D25 ◇ REV MR 43 # 4664 Zbl 261 # 02028 • ID 00500

ARSLANOV, M.M. [1976] *A theorem on density of recursively enumerable sets with respect to Turing reducibility (Russian)* (**J** 3937) Veroyat Met i Kibern (Kazan) 12-13*24-29
◇ D25 ◇ REV MR 58 # 27397 Zbl 398 # 03030 • ID 52758

ARSLANOV, M.M. & NADYROV, R.F. & SOLOV'EV, V.D. [1977] *A criterion for the completeness of recursively enumerable sets, and some generalizations of a fixed point theorem (Russian)* (**J** 0031) Izv Vyssh Ucheb Zaved, Mat (Kazan) 1977/4*3-7
◇ D20 D25 D55 ◇ REV MR 56 # 5247 Zbl 358 # 02046 • ID 50500

ARSLANOV, M.M. (ED.) [1978] *Algorithms and automata. Collection of scientific works (Russian)* (**X** 3605) Kazan Gos Univ: Kazan' 110pp
◇ D05 D20 D97 ◇ REV MR 81b:68001 Zbl 409 # 00013 • ID 48646

ARSLANOV, M.M. & SOLOV'EV, V.D. [1978] *Effectivizations of the definitions of classes of simple sets (Russian)* (**C** 2899) Algor & Avtomaty 100-108
◇ D25 ◇ REV MR 83f:03036 Zbl 412 # 03024 • ID 52955

ARSLANOV, M.M. [1979] *Weak recursively enumerable degrees and computability in the limit (Russian)* (**J** 3937) Veroyat Met i Kibern (Kazan) 15*3-9
◇ D25 D55 ◇ REV MR 81i:03061 Zbl 422 # 03015 • ID 53476

ARSLANOV, M.M. [1981] *Some generalizations of a fixed-point theorem (Russian)* (**J** 0031) Izv Vyssh Ucheb Zaved, Mat (Kazan) 1981/5*9-16
• TRANSL [1981] (**J** 3449) Sov Math 25/5*1-10
◇ D25 D30 ◇ REV MR 82j:03050 Zbl 523 # 03029 • ID 70499

ARSLANOV, M.M. [1982] *A hierarchy of degrees of unsolvability (Russian)* (**J** 3937) Veroyat Met i Kibern (Kazan) 18*10-17
◇ D30 D55 ◇ REV MR 85b:03068 Zbl 539 # 03021
• ID 40687

ARSLANOV, M.M. [1985] *A class of hypersimple incomplete sets (Russian)* (**J** 0087) Mat Zametki (Akad Nauk SSSR) 38*872-874,958
◇ D25 ◇ ID 49056

ARSLANOV, M.M. [1985] *Effectively hyperimmune sets and majorants (Russian)* (**J** 0087) Mat Zametki (Akad Nauk SSSR) 38*302-309,350
◇ D25 ◇ ID 49261

ARSLANOV, M.M. [1985] *Families of recursively enumerable sets and their degrees of undecidability (Russian)* (**J** 0031) Izv Vyssh Ucheb Zaved, Mat (Kazan) 1985/4*13-19
◇ D25 ◇ ID 46518

ARSLANOV, M.M. [1985] *Lattice properties of the degrees below 0' (Russian)* (**J** 0023) Dokl Akad Nauk SSSR 283*270-273
• TRANSL [1985] (**J** 0062) Sov Math, Dokl 32*58-62
◇ D30 ◇ ID 48590

ARTEMOV, S.N. [1985] *Nonarithmeticity of truth predicate logics of provability (Russian)* (**J** 0023) Dokl Akad Nauk SSSR 284*270-271
• TRANSL [1985] (**J** 0062) Sov Math, Dokl 32*403-405
◇ B45 D35 D55 F30 ◇ ID 48972

ARTEMOV, S.N. see Vol. II, III, VI for further entries

ARTIGUE, M. [1973] *Une generalisation des ensembles cohesifs* (**J** 2313) C R Acad Sci, Paris, Ser A-B 276*A1087-A1090
◇ D25 ◇ REV MR 47 # 3153 Zbl 255 # 02045 • ID 00502

ARTIGUE, M. [1974] *Etude d'un group de permutations des entiers naturels Lie au groupe des automorphismes du treillis des ensembles recursivement enumerables* (**P** 0776) Permutations;1972 Paris 189-209
◇ D25 ◇ REV MR 52 # 57 Zbl 286 # 02043 • ID 17222

ARTIGUE, M. see Vol. III, V, VI for further entries

ASATRYAN, O.S. [1976] *Ueber Retrakte einer indizierten Kleeneschen Menge (Russian) (Armenian summary)* (**J** 0346) Dokl Akad Nauk Armyan SSR 63*3-7
◇ D20 D45 ◇ REV MR 56 # 104 Zbl 376 # 02027
• ID 51669

ASATRYAN, O.S. [1980] *On creative equivalences (Russian) (Armenian and English summaries)* (**J** 0312) Izv Akad Nauk Armyan SSR, Ser Mat 15*145-153,163
• TRANSL [1980] (**J** 3265) Sov J Contemp Math Anal, Armen Acad Sci 15*64-73
◇ D25 D45 ◇ REV MR 82g:03080 Zbl 444 # 03025
• ID 70513

ASATRYAN, O.S. [1980] *On positive equivalences (Russian) (Armenian summary)* (**J** 0346) Dokl Akad Nauk Armyan SSR 71*136-140
◇ D45 ◇ REV MR 82k:03067 Zbl 486 # 03028 • ID 70512

ASATRYAN, O.S. [1981] *Boolean subalgebras of m-degrees of quasimaximal sets (Russian) (Armenian summary)* (**J** 0312) Izv Akad Nauk Armyan SSR, Ser Mat 72*275-279
◇ D30 ◇ REV MR 83h:03061 Zbl 498 # 03034 • ID 36069

ASATRYAN, O.S. [1981] *Two theorems on bT reducibility (Russian) (Armenian summary)* (**J** 0312) Izv Akad Nauk Armyan SSR, Ser Mat 73*67-72
◇ D25 ◇ REV MR 83d:03053 Zbl 498 # 03035 • ID 35192

ASH, C.J. & ROSENTHAL, J.W. [1980] *Some theories associated with algebraically closed fields* (**J** 0036) J Symb Logic 45*359-362
◇ C60 D35 ◇ REV MR 81e:03030 Zbl 487 # 03017
• ID 28370

ASH, C.J. & NERODE, A. [1981] *Intrinsically recursive relations* (**P** 2902) Aspects Effective Algeb;1979 Clayton 26-41
◇ D45 D50 ◇ REV MR 83a:03039 Zbl 467 # 03041
• ID 55039

ASH, C.J. & MILLAR, T.S. [1983] *Persistently finite, persistently arithmetic theories* (**J** 0053) Proc Amer Math Soc 89*487-492
◇ C57 ◇ REV Zbl 527 # 03014 • ID 37509

ASH, C.J. & DOWNEY, R.G. [1984] *Decidable subspaces and recursively enumerable subspaces* (**J** 0036) J Symb Logic 49*1137-1145
◇ C57 C60 D35 D45 ◇ ID 39867

ASH, C.J. see Vol. I, II, III, V for further entries

ASHCROFT, E.A. & MANNA, Z. & PNUELI, A. [1971] *Decidable properties of monadic functional schemas* (**P** 1058) Th Machines & Comput;1971 Haifa 3-17
◇ D80 ◇ REV MR 56 # 7283 Zbl 289 # 68036 • ID 80576

ASHCROFT, E.A. see Vol. VI for further entries

ASPVALL, B. & PLASS, M. & TARJAN, R.E. [1979] *A linear-time algorithm for testing the truth of certain quantified boolean formulas* (**J** 0232) Inform Process Lett 8*121-123
◇ B35 D15 ◇ REV MR 80b:68050 Zbl 398 # 68042
• ID 52791

ASPVALL, B. [1980] *Recognizing disguised NR(1) instances of the satisfiability problem* (**J** 2746) J Algor 1*97-103
◇ B35 D15 ◇ REV MR 82a:68051 Zbl 451 # 68037
• ID 69096

ASSER, G. [1959] *Normierte Postsche Algorithmen* (**J** 0068) Z Math Logik Grundlagen Math 5*323-333
◇ D03 ◇ REV MR 26 # 2362 Zbl 89.247 JSL 27.87
• ID 00546

ASSER, G. [1959] *Turing-Maschinen und Markowsche Algorithmen* (**J** 0068) Z Math Logik Grundlagen Math 5*346-365 • ERR/ADD ibid 7*309-310
◇ D03 D10 ◇ REV MR 23 # A3085 Zbl 89.247 JSL 27.467 • ID 00547

ASSER, G. [1959] *Zwei Rekursionstheoreme fuer normale Algorithmen (Russian, English and French summaries)* (**J** 0115) Wiss Z Humboldt-Univ Berlin, Math-Nat Reihe 9*165-168
◇ D03 ◇ REV MR 25 # 1101 Zbl 97.4 • ID 00549

ASSER, G. [1960] *Rekursive Wortfunktionen* (**J** 0068) Z Math Logik Grundlagen Math 6*258-278
◇ D20 ◇ REV MR 24 # A39 Zbl 98.244 JSL 29.199
• ID 00552

ASSER, G. & VUCKOVIC, V. [1961] *Funktionen-Algorithmen* (**J** 0068) Z Math Logik Grundlagen Math 7*1-8
◇ D03 D20 ◇ REV MR 24 # A682 Zbl 114.247 • ID 00554

ASSER, G. [1961] *Funktionen-Algorithmen und Graphschemata* (**J** 0068) Z Math Logik Grundlagen Math 7*20-27
◇ D20 ◇ REV MR 24 # A683 Zbl 117.253 JSL 28.292
• ID 40635

Author Index

Asser, G. [1963] *Ueber die algebraische Theorie der Automaten* (**C** 1424) Math & Phys-Techn Probl der Kybern 446-453
◇ D05 ◇ rev Zbl 119.128 • id 40636

Asser, G. [1966] *Ueber eine Darstellung der rekursiven Wortfunktionen in endlichen Automaten* (**J** 0068) Z Math Logik Grundlagen Math 12*1-12
◇ D05 D20 ◇ rev MR 33 # 3935 Zbl 143.14 • id 00558

Asser, G. [1969] *Malzew-Raeume* (**P** 4357) Summer Sess Th Ordered Sets & Gen Alg;1969 Cikhaj 49-56
◇ D05 ◇ rev Zbl 222 # 00003 • id 40645

Asser, G. [1974] *Berechenbare Graphenabbildungen II* (**J** 0309) Wiss Z Univ Greifswald, Math-Nat Reihe 23*29-33
◇ D20 D45 ◇ rev MR 51 # 83 Zbl 363 # 02045 • rem Part I 1975 • id 17959

Asser, G. [1975] *Rekursive Graphenabbildungen. I* (**P** 1846) Kompl, Lern- & Erkenn-Prozess;1973 Jena 1*7-17
◇ D20 D45 D80 ◇ rem Part II 1974 • id 40643

Asser, G. [1977] *Bemerkungen zum Labyrinth-Problem* (**J** 0129) Elektr Informationsverarbeitung & Kybern 13*203-216
◇ D05 ◇ rev MR 58 # 174 Zbl 362 # 94049 • id 50825

Asser, G. see Vol. I, II, III, V for further entries

Asveld, P.R.J. & Engelfriet, J. [1979] *Extended linear macro grammars, iteration grammars, and register programs* (**J** 1431) Acta Inf 11*259-285
◇ D05 D10 ◇ rev MR 80k:68055 Zbl 382 # 68061 • id 69103

Asveld, P.R.J. [1980] *On controlled iterated gsm mappings and related operations* (**J** 0060) Rev Roumaine Math Pures Appl 25*139-145
◇ D05 ◇ rev Zbl 434 # 68060 • id 69100

Asveld, P.R.J. [1980] *Space-bounded complexity classes and iterated deterministic substitution* (**J** 0194) Inform & Control 44*282-299
◇ D15 ◇ rev MR 81i:68054 Zbl 459 # 68017 • id 69099

Asveld, P.R.J. & Engelfriet, J. [1981] *A note on non-generators of full AFL's* (**X** 3205) Math Centr Amsterdam Afd Inf IW184/81*5pp
◇ D05 ◇ rev MR 83a:68086 Zbl 466 # 68065 • id 69102

Asveld, P.R.J. [1981] *Time and space complexity of inside-out macro languages* (**J** 0382) Int J Comput Math 10*3-14
◇ D15 ◇ rev MR 83a:68086 Zbl 432 # 68037 Zbl 468 # 68089 • id 69098

Auguston, M. [1975] *Universal Buchberger automata and real-time computations (Russian) (English summary)* (**S** 2587) Teor Algor & Progr (Riga) 2*183-190
◇ D10 D15 D20 ◇ rev MR 58 # 3690 Zbl 331 # 68031 • id 60211

Ausiello, G. [1971] *Abstract computational complexity and cycling computations* (**J** 0119) J Comp Syst Sci 5*118-128
◇ D15 D75 ◇ rev MR 43 # 7333 Zbl 225 # 02025 JSL 40.248 • id 00568

Ausiello, G. [1974] *Relations between semantics and complexity of recursive programs* (**P** 1869) Automata, Lang & Progr (2);1974 Saarbruecken 129-140
◇ B40 B75 D15 D20 ◇ rev MR 55 # 13868 Zbl 288 # 68006 • id 80587

Ausiello, G. [1974] *Un linguaggio di programmazione per le funzioni subricorsive* (**J** 3436) Quad, Ist Appl Calcolo, Ser 3 8*5-43
◇ B75 D20 ◇ rev Zbl 427 # 03030 • id 53717

Ausiello, G. & Protasi, M. [1975] *On the comparison of notions of approximation* (**P** 0454) Math Founds of Comput Sci (4);1975 Marianske Lazne 172-178
◇ D20 D80 ◇ rev MR 52 # 10402 Zbl 323 # 02050 • id 21700

Ausiello, G. [1976] *Difficult logical theories and their computer approximations* (**J** 1620) Asterisque 38-39*3-21
◇ B25 B35 D15 ◇ rev MR 57 # 16032 Zbl 365 # 02022 • id 51023

Ausiello, G. [1977] *On the structure and properties of NP-complete problems and their associated optimization problems* (**P** 1635) Math Founds of Comput Sci (6);1977 Tatranska Lomnica 1-16
◇ D15 ◇ rev MR 58 # 8482 Zbl 411 # 68040 • id 69107

Ausiello, G. & Boehm, C. (eds.) [1978] *Automata, languages and programming. V* (**S** 3302) Lect Notes Comput Sci VIII + 508pp
◇ D97 ◇ rev MR 80c:68003 Zbl 372 # 00024 • id 51407

Ausiello, G. & Marchetti-Spaccamela, A. & Protasi, M. [1980] *Toward a unified approach for the classification of NP-complete optimization problems* (**J** 1426) Theor Comput Sci 12*83-96
◇ D15 ◇ rev MR 81k:68027 Zbl 442 # 68029 • id 80586

Ausiello, G. & D'Atri, A. & Protasi, M. [1981] *A characterization of reductions among combinatorial problems* (**P** 4214) Anal & Design of Alg in Combin Optmiz;1979 Udine 37-63
◇ D15 ◇ rev MR 83d:68002 Zbl 484.68028 • id 46174

Ausiello, G. & D'Atri, A. & Protasi, M. [1981] *Lattice-theoretic ordering properties for NP-complete optimization problems* (**J** 2095) Fund Inform, Ann Soc Math Pol, Ser 4 4*83-94
◇ D15 ◇ rev MR 83a:68043 Zbl 467 # 68047 • id 69108

Ausiello, G. see Vol. I, VI for further entries

Auslander, L. & Winograd, S. [1980] *The multiplicative complexity of certain semilinear systems defined by polynomials* (**J** 2650) Adv Appl Math 1*257-299
◇ D15 ◇ rev MR 82g:68039 Zbl 528 # 65006 • id 80588

Austin, A.K. [1979] *A note on decision procedures for identities* (**J** 0004) Algeb Universalis 9*146-151
◇ D40 ◇ rev MR 80b:08002 Zbl 435 # 08003 • id 55814

Austin, A.K. [1983] *An elementary approach to NP-completeness* (**J** 0005) Amer Math Mon 90*398-399
◇ D15 ◇ rev MR 84m:68031 Zbl 509 # 68036 • id 36797

Austin, A.K. see Vol. III for further entries

Autebert, J.-M. [1979] *Operations de cylindre et applications sequentielles gauches inverses* (**J** 1431) Acta Inf 11*241-258
◇ D05 ◇ rev Zbl 388 # 68076 • id 69114

Autebert, J.-M. & Beauquier, J. & Boasson, L. & Nivat, M. [1979] *Quelques problemes ouverts en theorie des langages algebriques* (**J** 3441) RAIRO Inform Theor 13*363-378
◇ D05 ◇ rev MR 81a:68078 Zbl 434 # 68056 • id 69117

AUTEBERT, J.-M. [1979] *Relationships between AFDL's and cylinders* (**P** 2059) Math Founds of Comput Sci (8);1979 Olomouc 219-227
◇ D05 ◇ REV Zbl 408 # 68070 • ID 69112

AUTEBERT, J.-M. & BEAUQUIER, J. & BOASSON, L. & LATTEUX, M. [1979] *Sur la decidabilite de certaines questions liees a la linearite des langages (English summary)* (**J** 2313) C R Acad Sci, Paris, Ser A-B 288*A925-A927
◇ D05 ◇ REV MR 80e:68179 Zbl 412 # 68067 • ID 52984

AUTEBERT, J.-M. [1979] *Une note sur le cylindre des langages deterministes* (**J** 1426) Theor Comput Sci 8*395-399
◇ D05 ◇ REV MR 80g:68091 Zbl 398 # 68030 • ID 69111

AUTEBERT, J.-M. & BOASSON, L. [1980] *Generators of cones and cylinders* (**P** 4266) Form Lang Th;1979 Santa Barbara 49-87
◇ D30 ◇ REV MR 84j:68001 Zbl 545 # 68065 • ID 47197

AUTEBERT, J.-M. & BEAUQUIER, J. & BOASSON, L. [1980] *Langages sur des alphabets infinis* (**J** 2702) Discr Appl Math 2*1-20
◇ D05 ◇ REV MR 81b:68087 Zbl 443 # 68058 • ID 69113

AUTEBERT, J.-M. & BEAUQUIER, J. & BOASSON, L. [1980] *Limites de langages algebriques (English summary)* (**J** 2313) C R Acad Sci, Paris, Ser A-B 291*A555-A558
◇ D05 ◇ REV MR 81m:68057 Zbl 444 # 68065 • ID 69116

AUTEBERT, J.-M. & BEAUQUIER, J. & BOASSON, L. & LATTEUX, M. [1980] *Very small families of algebraic nonrational languages* (**P** 4266) Form Lang Th;1979 Santa Barbara 89-108
◇ D05 ◇ REV MR 84j:68001 Zbl 545 # 68065 • ID 47198

AUTEBERT, J.-M. & BEAUQUIER, J. & BOASSON, L. & LATTEUX, M. [1981] *Langages algebriques domines par des langages unaires* (**J** 0194) Inform & Control 48*49-53
◇ D05 ◇ REV MR 82h:68100 Zbl 459 # 68040 • ID 69115

AUTEBERT, J.-M. [1981] *Un resultat de discontinuite dans les familles de langages* (**P** 3475) Theor Comput Sci (5);1981 Karlsruhe 104*64-69
◇ D05 ◇ REV Zbl 457 # 68078 • ID 69110

AUTEBERT, J.-M. & BEAUQUIER, J. & BOASSON, L. [1982] *Formes de langages et de grammaires* (**J** 1431) Acta Inf 17*193-213
◇ D05 ◇ REV MR 83j:68083 Zbl 467 # 68065 • ID 69119

AUTEBERT, J.-M. & BEAUQUIER, J. & BOASSON, L. & LATTEUX, M. [1982] *Indecidabilite de la condition IRS (English summary)* (**J** 3441) RAIRO Inform Theor 16*129-138
◇ D05 ◇ REV MR 83k:68068 Zbl 493 # 68075 • ID 40334

AVALISHVILI, I. & BERISHVILI, G. [1977] *Extension of automata (Russian)* (**J** 3460) Tr Inst Kibern, Akad Nauk Gruz SSR 1*183-189
◇ D05 ◇ REV Zbl 411 # 68052 • ID 69121

AVENHAUS, J. & MADLENER, K. [1975] *$E_n - E_{n-1}$-entscheidbare Gruppen* (**P** 1449) Automata Th & Formal Lang;1975 Kaiserslautern 42-51
◇ D40 ◇ REV MR 54 # 12510 Zbl 314 # 02055 • ID 60217

AVENHAUS, J. & MADLENER, K. [1977] *Komplexitaetsuntersuchungen fuer Einrelatorgruppen* (**J** 1046) Z Angew Math Mech 57*T313-T314
◇ D40 ◇ REV MR 55 # 12495 Zbl 359 # 02044 • ID 50591

AVENHAUS, J. & MADLENER, K. [1977] *Komplexitaet bei Gruppen: Der Einbettungssatz von Higman* (**J** 1046) Z Angew Math Mech 57*T314-T315
◇ D15 D40 ◇ REV MR 57 # 85 Zbl 363 # 20030 • ID 50912

AVENHAUS, J. & MADLENER, K. [1977] *Subrekursive Komplexitaet bei Gruppen I: Gruppen mit vorgeschriebener Komplexitaet* (**J** 1431) Acta Inf 9*87-104
◇ D15 D40 ◇ REV MR 58 # 16239 Zbl 371 # 02019 • REM Part II 1978 • ID 51380

AVENHAUS, J. & MADLENER, K. [1978] *Algorithmische Probleme bei Einrelatorgruppen und ihre Komplexitaet* (**J** 0009) Arch Math Logik Grundlagenforsch 19*3-12
◇ D20 D40 ◇ REV MR 80b:03055 Zbl 396 # 03040 • ID 29160

AVENHAUS, J. & MADLENER, K. [1978] *Subrekursive Komplexitaet bei Gruppen II: Der Einbettungssatz von Higman fuer entscheidbare Gruppen* (**J** 1431) Acta Inf 9*183-193
◇ D15 D40 ◇ REV MR 80d:03044 Zbl 371 # 02020 • REM Part I 1977 • ID 51381

AVENHAUS, J. & MADLENER, K. [1979] *Complexity and independence of algorithmic problems in groups* (**P** 2935) FCT'79 Fund of Comput Th;1979 Berlin/Wendisch-Rietz 38-44
◇ D15 D40 ◇ REV Zbl 414 # 03026 • ID 53072

AVENHAUS, J. & MADLENER, K. [1980] *String matching and algorithmic problems in free groups* (**J** 0307) Rev Colomb Mat 14*1-15
◇ D15 D40 D80 ◇ REV MR 81i:20041 Zbl 431 # 03031 • ID 53937

AVENHAUS, J. & MADLENER, K. [1981] *An algorithm for the word problem in HNN extensions and the dependence of its complexity on the group representation* (**J** 3441) RAIRO Inform Theor 15*355-371
◇ D20 D40 ◇ REV MR 84g:20061 Zbl 494 # 20020 • ID 36653

AVENHAUS, J. & MADLENER, K. [1981] *P-complete problems in free groups* (**P** 3475) Theor Comput Sci (5);1981 Karlsruhe 104*42-51
◇ D15 D40 ◇ REV Zbl 457 # 68037 • ID 54403

AVENHAUS, J. & BOOK, R.V. & SQUIER, C.C. [1984] *On expressing commutativity by finite Church-Rosser presentations: a note on commutative monoids (French summary)* (**J** 3441) RAIRO Inform Theor 18*47-52
◇ D05 ◇ REV MR 85j:03057 Zbl 542 # 20038 • ID 46778

AVENHAUS, J. & MADLENER, K. [1984] *On the complexity of intersection and conjugacy problems in free groups* (**J** 1426) Theor Comput Sci 32*279-295
◇ D40 ◇ REV Zbl 555 # 20016 • ID 44086

AX, J. [1965] *The undecidability of power series fields* (**J** 0053) Proc Amer Math Soc 16*846
◇ D35 ◇ REV MR 31 # 2148 JSL 36.684 • ID 00576

AX, J. see Vol. I, III for further entries

AXT, P. [1957] *A subrecursive hierarchy and primitive recursive degrees* (**P** 1675) Summer Inst Symb Log;1957 Ithaca 344-347
◇ D20 ◇ ID 35804

Author Index

AXT, P. [1959] *On a subrecursive hierarchy and primitive recursive degrees* (J 0064) Trans Amer Math Soc 92*85-105
◇ D20 ◇ REV MR 23 # A3673 Zbl 87.11 JSL 25.167
• ID 00582

AXT, P. [1961] *Note on 3-recursive functions* (J 0068) Z Math Logik Grundlagen Math 7*97-98
◇ D20 ◇ REV MR 25 # 3821 Zbl 106.7 JSL 29.199
• ID 00583

AXT, P. [1963] *Enumeration and the Grzegorczyk hierarchy* (J 0068) Z Math Logik Grundlagen Math 9*53-65
◇ D20 ◇ REV MR 26 # 2352 Zbl 112.246 JSL 30.90
• ID 00584

AXT, P. [1963] *Relativization of a primitive recursive hierarchy* (J 0043) Math Ann 152*159-163
◇ D20 ◇ REV MR 27 # 4743 Zbl 118.251 JSL 35.480
• ID 00585

AXT, P. [1964] *On a problem of Skolem* (J 0121) Kon Norske Vidensk Selsk Forh 37*48-52
◇ D20 ◇ REV MR 31 # 2143 Zbl 178.319 • ID 00586

AXT, P. [1965] *Iteration of primitive recursion* (J 0068) Z Math Logik Grundlagen Math 11*253-255
• TRANSL [1970] (C 1540) Probl Mat Log: Slozh Algor & Kl Vychisl Funk 114-117
◇ D20 ◇ REV MR 33 # 3917 Zbl 144.2 JSL 35.479
• ID 00587

AXT, P. [1966] *Iteration of relative primitive recursion* (J 0043) Math Ann 167*53-55
• TRANSL [1970] (C 1540) Probl Mat Log: Slozh Algor & Kl Vychisl Funk 118-122
◇ D20 ◇ REV MR 35 # 50 Zbl 192.50 JSL 35.480
• ID 00588

AXT, P. & SINGLETARY, W.E. [1969] *Decision problems associated with complete deterministic normal systems* (J 0068) Z Math Logik Grundlagen Math 15*299-304
◇ D03 ◇ REV MR 43 # 1848 Zbl 209.22 • ID 00589

AXT, P. & SINGLETARY, W.E. [1969] *On deterministic normal systems* (J 0068) Z Math Logik Grundlagen Math 15*49-62
◇ D03 D25 ◇ REV MR 42 # 79 Zbl 251 # 02038
• ID 00590

AZRA, J.-P. [1971] *Relations diophantiennes et la solution negative du 10e probleme de Hilbert* (S 1567) Semin Bourbaki Exp.383*11-28
◇ D25 D35 ◇ REV MR 57 # 9665 Zbl 268 # 02030 • REM Springer: Heidelberg & New York; Lecture Notes Math 244
• ID 60221

AZRA, J.-P. & JAULIN, B. [1973] *Recursivite* (X 0834) Gauthier-Villars: Paris xix + 218pp
◇ D98 ◇ REV MR 51 # 5275 Zbl 276 # 02019 • ID 17456

BABCSANYI, I. [1980] *Semigroup-automata* (C 3483) Pap Automata Th 1980/2*79-107
◇ D05 ◇ REV MR 83g:20077 Zbl 475 # 68030 • ID 69125

BABCSANYI, I. [1981] *Rees automata (Hungarian) (English summary)* (J 0396) Mat Lapok 29*139-148
◇ D05 ◇ REV MR 83j:68061 Zbl 496 # 68039 • ID 39908

BABICHEV, A.V. & PRONINA, V.A. & TRAKHTENGERTS, EH.A. [1977] *The conversion of engendering grammars into a form observing precedence functions (Russian)* (J 2605) Programmirovanie 1977/3*43-53
• TRANSL [1977] (J 2604) Progr Comput Software 3*197-205
◇ D05 ◇ REV Zbl 414 # 68051 • ID 69127

BABICHEV, A.V. [1978] *Automata that accept languages generated by precedence grammars (Russian)* (J 2605) Programmirovanie 1978/6*20-25
• TRANSL [1978] (J 2604) Progr Comput Software 4*388-392
◇ D05 ◇ REV MR 80d:68101 Zbl 415 # 68046 • ID 69126

BABIKOV, G.V. [1965] *Normale Algorithmen und Turingmaschinen (Russian)* (J 0340) Mat Zap (Univ Sverdlovsk) 5/1*10-14
◇ D03 D10 ◇ REV Zbl 302 # 02010 • ID 60227

BABINOV, YU.P. [1979] *Class of generalized context-sensitive precedence languages (Russian)* (J 2605) Programmirovanie 1979/2*56-67
• TRANSL [1979] (J 2604) Progr Comput Software 5*117-126
◇ D05 ◇ REV MR 81b:68088 Zbl 426 # 68066 • ID 69128

BABORSKI, A. [1979] *The complexity of the original and the language of description* (J 3248) Przeglad Stat 26*259-265
◇ D15 ◇ REV MR 82j:68085 Zbl 449 # 68048 • ID 69129

BACHEM, A. [1982] *Concepts of algorithmic computation* (P 3761) Modern Appl Math; 1979 Bonn 3-49
◇ D10 D15 ◇ REV MR 83i:03064 Zbl 503 # 68025
• ID 34852

BACHMANN, P. [1980] *Theoretische Untersuchungen zur Behandlung von Kontext- Sensitivitaeten* (J 0947) Wiss Z Tech Univ Dresden 29*339
◇ D05 ◇ REV MR 81h:68062 Zbl 442 # 68074 • ID 69130

BACIU, A. & PASCU, A. [1978] *Leontief type languages* (J 0304) An Univ Timisoara, Sti Mat 16*111-118
◇ D05 ◇ REV MR 81e:68090 Zbl 463 # 68069 • ID 69131

BACIU, A. & PASCU, A. [1979] *Leontief type grammars and languages* (J 2716) Found Control Eng, Poznan 4*97-106
◇ D05 ◇ REV MR 81a:68079 Zbl 427 # 68062 • ID 69132

BACIU, A. & PASCU, A. [1980] *A Bar Hillel type property of Leontief languages* (J 0304) An Univ Timisoara, Sti Mat 18*5-9
◇ D05 ◇ REV MR 83f:68081 Zbl 471 # 68056 • ID 69133

BACIU, A. & PASCU, A. [1981] *The interface between hierarchies, multiple objectives clustering and fuzzy sets* (P 4050) Appl Syst & Cybern;1980 Acapulco 2814-2818
◇ D55 E72 ◇ REV MR 84b:00009 • ID 46297

BADAEV, S.A. [1974] *On incomparable enumerations (Russian)* (J 0092) Sib Mat Zh 15*730-738,957
• TRANSL [1974] (J 0475) Sib Math J 15*519-524
◇ D25 D45 ◇ REV MR 50 # 9554 Zbl 317 # 02044
• ID 03853

BADAEV, S.A. [1977] *Computable enumerations of families of general recursive functions (Russian)* (J 0003) Algebra i Logika 16*129-148
• TRANSL [1977] (J 0069) Algeb and Log 16*83-98
◇ D20 D45 ◇ REV MR 58 # 27390 Zbl 388 # 03018
• ID 52824

BADAEV, S.A. [1977] *Positive numerations (Russian)* (**J** 0092) Sib Mat Zh 18*483-496,717
• TRANSL [1977] (**J** 0475) Sib Math J 18*343-352
◇ D25 D45 ◇ REV MR 58 # 27398 Zbl 358 # 02047
• ID 50501

BADE, W.G. [1973] *Complementation problems for the Baire classes* (**J** 0048) Pac J Math 45*1-12
◇ D55 E15 ◇ REV MR 47 # 9261 Zbl 261 # 26005
• ID 00624

BADER, C. & MOURA, A. [1982] *A generalization of Ogden's lemma* (**J** 0037) ACM J 29*404-407
◇ D05 ◇ REV MR 83b:68077 Zbl 501 # 68039 • ID 38978

BAER, ROBERT M. [1967] *Certain directed Post systems and automata* (**J** 0068) Z Math Logik Grundlagen Math 13*151-174
◇ D03 D05 ◇ REV MR 35 # 4103 Zbl 163.9 JSL 35.158
• ID 00634

BAER, ROBERT M. [1969] *A reduction theorem for normal algorithms* (**J** 0068) Z Math Logik Grundlagen Math 15*219-222
◇ D03 ◇ REV MR 40 # 4111 Zbl 183.293 • ID 00641

BAER, ROBERT M. [1969] *Computability by normal algorithms* (**J** 0053) Proc Amer Math Soc 20*551-552
◇ D03 D20 ◇ REV MR 41 # 63 Zbl 176.278 • ID 00639

BAER, ROBERT M. [1969] *Definability by Turing machines* (**J** 0068) Z Math Logik Grundlagen Math 15*325-332
◇ D10 D35 ◇ REV MR 42 # 7513 Zbl 198.327 • ID 00640

BAER, ROBERT M. & SPANIER, E.H. [1969] *Referenced automata and metaregular families* (**J** 0119) J Comp Syst Sci 3*423-446
◇ D05 ◇ REV MR 41 # 3182 Zbl 183.296 • ID 00637

BAER, ROBERT M. [1975] *Computation by assembly* (**J** 0119) J Comp Syst Sci 11*285-294
◇ D80 ◇ REV MR 52 # 10080 Zbl 325 # 68022 • ID 28142

BAER, ROBERT M. [1975] *Self-assembly and differentiation as models of computability* (**J** 3073) Bull Math Biol 37*59-69
◇ D10 D80 ◇ REV MR 52 # 9686 Zbl 377 # 94063
• ID 51782

BAER, ROBERT M. & LEEUWEN VAN, J. [1976] *The halting problem for linear Turing assemblers* (**J** 0119) J Comp Syst Sci 13*119-135
◇ B75 D10 ◇ REV MR 54 # 9991 Zbl 342 # 02024
• ID 25759

BAETEN, J. [1984] *Filters and ultrafilters over definable subsets of admissible ordinals* (**P** 2153) Logic Colloq;1983 Aachen 1*1-8
◇ C62 D60 E05 E10 E47 ◇ REV MR 86g:03075
Zbl 562 # 03024 • ID 45395

BAETEN, J. see Vol. VI for further entries

BAETEN, J.C.M. & BERGSTRA, J.A. & KLOP, J.W. [1982] *Priority rewrite systems* (**X** 1121) Math Centr: Amsterdam CS-R 8407
◇ B75 D05 D20 ◇ ID 39331

BAEZ, J.C. [1983] *Recursivity in quantum mechanics* (**J** 0064) Trans Amer Math Soc 280*339-350
◇ B51 D80 ◇ REV MR 84i:81010 Zbl 547 # 03041
• ID 40205

BAGCHI, A. & MEYER, A.R. [1972] *Program size and economy of descriptions: Preliminary report* (**P** 1901) ACM Symp Th of Comput (4);1972 Denver 183-186
◇ D15 D25 ◇ REV Zbl 354 # 68024 • ID 50190

BAGYINSZKINE OROSZ, A. [1981] *Petri nets and their languages* (**X** 3151) Karl Marx Univ Dpt Math: Budapest 1981/1*7-9
◇ D05 D80 ◇ REV Zbl 477 # 68055 • ID 69137

BAIRE, R. & BOREL, E. & HADAMARD, J. & LEBESGUE, H. [1905] *Cinq lettres sur la theorie des ensembles* (**J** 0353) Bull Soc Math Fr 33*261-273
◇ A05 D55 E15 E25 F65 ◇ REV FdM 36.99 • ID 05473

BAIRE, R. see Vol. V for further entries

BAJAJ, C. [1985] *Geometric optimization and the polynomial hierarchy* (**P** 4672) Found of Softw Tech & Th Comput Sci (5);1985 New Delhi 176-195
◇ D15 ◇ ID 49616

BAJRASHEVA, V.R. [1982] *Degrees of automated transformations (Russian)* (**J** 3937) Veroyat Met i Kibern (Kazan) 18*17-25
◇ D05 D30 ◇ REV MR 85i:03126 Zbl 567 # 03013
• ID 44208

BAKER, B.S. & BOOK, R.V. [1974] *Reversal-bounded multi-pushdown machines* (**J** 0119) J Comp Syst Sci 8*315-332
◇ D10 D25 ◇ REV MR 51 # 12034 Zbl 309 # 68043
• ID 26099

BAKER, B.S. [1978] *Generalized syntax directed translation tree transducers, and linear space* (**J** 1428) SIAM J Comp 7*376-391
◇ D05 D15 ◇ REV MR 58 # 13952 Zbl 379 # 68052
• ID 51867

BAKER, B.S. [1978] *Tree transducers and tree languages* (**J** 0194) Inform & Control 37*241-265
◇ D05 ◇ REV MR 58 # 13951 Zbl 386 # 68071 • ID 69139

BAKER, B.S. [1979] *Composition of top-down and bottom-up tree transductions* (**J** 0194) Inform & Control 41*186-213
◇ D05 ◇ REV MR 80e:68155 Zbl 408 # 68053 • ID 69140

BAKER, J.L. [1972] *Grammars with structured vocabulary: A model for the ALGOL-68 definition* (**J** 0194) Inform & Control 20*351-395
◇ B75 D05 D25 ◇ REV MR 53 # 12089 Zbl 247 # 68027
• ID 60257

BAKER, J.L. [1975] *Factorization of Scott-style automata* (**P** 0770) Categ Th Appl to Comput & Control (1);1974 San Francisco 99-105
◇ D05 ◇ REV MR 56 # 4244 Zbl 331 # 68032 • ID 80604

BAKER, T.P. & HARTMANIS, J. [1974] *On simple Goedel numberings and translations* (**P** 1869) Automata, Lang & Progr (2);1974 Saarbruecken 301-316
◇ B75 D20 D45 ◇ REV MR 55 # 1808 Zbl 283 # 68055
• ID 62367

BAKER, T.P. & HARTMANIS, J. [1975] *On simple Goedel numberings and translations* (**J** 1428) SIAM J Comp 4*1-11
◇ B75 D20 D45 ◇ REV MR 57 # 14576 Zbl 272 # 68046
• ID 62364

BAKER, T.P. [1978] *"Natural" properties of flowchart step-counting measures* (**J** 0119) J Comp Syst Sci 16*1-22
◇ D15 ◇ REV MR 58 # 8483 Zbl 368 # 68051 • ID 51291

Author Index

BAKER, T.P. & SELMAN, A.L. [1979] *A second step toward the polynomial hierarchy* (**J** 1426) Theor Comput Sci 8*177-187
◇ D15 ◇ REV MR 80g:68052 Zbl 397 # 03023 • ID 52689

BAKER, T.P. [1979] *On "provable" analogs of 𝒫 and 𝒩𝒫* (**J** 0041) Math Syst Theory 12*213-218
◇ D15 F30 ◇ REV MR 80g:68051 Zbl 405 # 68046
• ID 80606

BAKHTURIN, YU.A. [1971] *A certain identity algorithm (Russian)* (**J** 0092) Sib Mat Zh 12*459-462
• TRANSL [1971] (**J** 0475) Sib Math J 12*328-330
◇ D40 ◇ REV MR 44 # 1717 Zbl 232 # 20061 • ID 00669

BAKKER DE, J.W. [1971] *Recursive procedures* (**X** 1121) Math Centr: Amsterdam 24*viii+108pp
◇ B40 B75 D20 ◇ REV MR 49 # 7121 Zbl 274 # 02015
JSL 40.83 • ID 28688

BAKKER DE, J.W. [1972] *Recursion, induction and symbol manipulation* (**P** 2971) MC-25 Informatica Symp;1972 Amsterdam 1*26pp
◇ D05 D20 ◇ REV MR 58 # 10380 • ID 72183

BAKKER DE, J.W. [1980] *Mathematical theory of program correctness* (**X** 0819) Prentice Hall: Englewood Cliffs xvii+505pp
◇ B75 D80 ◇ REV MR 82g:68016 Zbl 452 # 68011
• ID 81045

BAKKER DE, J.W. & ZUCKER, J.I. [1981] *Derivatives of programs* (**P** 3642) Colloq Math Log in Computer Sci;1978 Salgotarjan 321-343
◇ D20 ◇ REV MR 83g:68007 Zbl 481 # 68014 JSL 49.990
• ID 47392

BAKKER DE, J.W. see Vol. VI for further entries

BALAKRISHNAN, A.V. [1961] *Prediction theory for Markoff processes* (**J** 0048) Pac J Math 11*1171-1182
◇ D03 ◇ REV MR 25 # 5551 Zbl 106.123 • ID 00697

BALBES, R. [1973] *On free pseudo-complemented and relatively pseudo-complemented semi-lattices* (**J** 0027) Fund Math 78*119-131
◇ D40 G10 ◇ REV MR 47 # 8373 Zbl 277 # 06001
• ID 00719

BALBES, R. see Vol. III, V for further entries

BALCAR, B. & HAJEK, P. [1978] *On sequences of degrees of constructibility (solution of Friedman's Problem 75)* (**J** 0068) Z Math Logik Grundlagen Math 24*291-296
◇ D30 E45 ◇ REV MR 58 # 16300 Zbl 414 # 03032
• ID 53078

BALCAR, B. see Vol. III, V for further entries

BALCAZAR, J.L. & DIAZ, J. [1982] *A note on a theorem by Ladner* (**J** 0232) Inform Process Lett 15*84-86
◇ D15 ◇ REV MR 84d:68035 Zbl 515 # 03022 • ID 37845

BALCAZAR, J.L. [1984] *Separating, strongly separating, and collapsing relativized complexity classes* (**P** 3658) Math Founds of Comput Sci (11);1984 Prague 1-16
◇ D15 ◇ REV Zbl 554 # 68031 • ID 44957

BALCAZAR, J.L. & BOOK, R.V. & SCHOENING, U. [1984] *Sparse oracles, lowness, and highness* (**P** 3658) Math Founds of Comput Sci (11);1984 Prague 185-193
◇ D15 ◇ REV Zbl 554 # 68033 • ID 44958

BALCAZAR, J.L. & SCHOENING, U. [1985] *Bi-immune sets for complexity classes* (**J** 0041) Math Syst Theory 18*1-10
◇ D15 ◇ ID 46499

BALCAZAR, J.L. & DIAZ, J. & GABARRO, J. [1985] *On some "nonuniform" complexity measures* (**P** 4647) FCT'85 Fund of Comput Th;1985 Cottbus 18-27
◇ D15 ◇ ID 49069

BALCAZAR, J.L. [1985] *Simplicity, relativizations and nondeterminism* (**J** 1428) SIAM J Comp 14*148-157
◇ D15 ◇ REV Zbl 567 # 68027 • ID 49217

BALCER, M. & MIKA, A. & PIZON, T. [1979] *A theorem on the equivalence of the definitions of admissible words of a certain alphabet and arithmetic term expressions* (**S** 2890) Zesz Nauk, Mat Fiz, Politech Slask (Gliwice) 34*13-19
◇ D05 ◇ REV MR 81i:68121b • ID 80613

BALCER, M. & MIKA, A. & PIZON, T. [1979] *An algorithm for deciding the truth of an equation between two arbitrary arithmetic expressions* (**S** 2890) Zesz Nauk, Mat Fiz, Politech Slask (Gliwice) 34*21-27
◇ D05 ◇ REV MR 81i:68121c • ID 80612

BALCER, M. & MIKA, A. & PIZON, T. [1979] *The construction of admissible words of a certain alphabet as equivalent concepts by means of arithmetic term expressions* (**S** 2890) Zesz Nauk, Mat Fiz, Politech Slask (Gliwice) 34*5-12
◇ D05 ◇ REV MR 81i:68121a • ID 80614

BALDWIN, JOHN T. [1982] *Recursion theory and abstract dependence* (**P** 3634) Patras Logic Symp;1980 Patras 67-76
◇ C57 C60 D25 D45 ◇ REV MR 85i:03147
Zbl 519 # 03035 • ID 33739

BALDWIN, JOHN T. [1984] *First-order theories of abstract dependence relations* (**J** 0073) Ann Pure Appl Logic 26*215-243
◇ C35 C45 C57 C60 D25 D45 ◇ REV MR 85g:03056
• ID 33746

BALDWIN, JOHN T. see Vol. III, V for further entries

BALLANTYNE, A.M. & LANKFORD, D.S. [1981] *New decision algorithms for fintely presented commutative semigroups* (**J** 2687) Comp Math Appl 7*159-165
◇ D40 ◇ REV MR 83a:03038 Zbl 449.20059 • ID 35066

BALLANTYNE, A.M. & BUTLER, G.A. & LANKFORD, D.S. [1984] *A progress report on new decision algorithms for finitely presented abelian groups* (**P** 2633) Autom Deduct (7);1984 Napa 128-141
◇ D40 ◇ REV MR 86e:20040 Zbl 547 # 03009 • ID 41773

BALLANTYNE, A.M. see Vol. I for further entries

BALLO, E. [1981] *Sulle procedure induttive* (**P** 3092) Congr Naz Logica;1979 Montecatini Terme 107-118
◇ D70 ◇ ID 48364

BANACHOWSKI, L. & OKTABA, H. [1983] *A presentation of the scientific activities of the theory of computation group in the period from 1978-82* (**S** 3382) Sem-ber, Humboldt-Univ Berlin, Sekt Math 52*84-91
◇ D15 D20 ◇ ID 47882

BANACHOWSKI, L. [1983] *On the Tarjan model of pointer machines* (**S** 3382) Sem-ber, Humboldt-Univ Berlin, Sekt Math 52*1-6
◇ D10 ◇ REV MR 85f:68003 • ID 47877

BANACHOWSKI, L. see Vol. I, II for further entries

BANASCHEWSKI, B. [1958] *On transfinite iteration* (J 0027) Fund Math 46*225-229
◇ D70 E20 ◇ REV MR 21#5580 Zbl 91.54 JSL 27.236
• ID 00757

BANASCHEWSKI, B. see Vol. I, III, V for further entries

BANJEVIC, D. [1981] *On some basic properties of the Kolmogorov complexity* (J 0400) Publ Inst Math, NS (Belgrade) 30(44)*17-23
◇ D15 ◇ REV MR 84d:03050 Zbl 502#68008 • ID 34086

BANJEVIC, D. [1984] *Note on the number of sequences of given complexity* (J 0400) Publ Inst Math, NS (Belgrade) 36*107-109
◇ D15 ◇ REV Zbl 558#68036 • ID 46607

BAR-HILLEL, Y. [1962] *Some recent results in theoretical linguistics* (P 0612) Int Congr Log, Meth & Phil of Sci (1,Proc);1960 Stanford 551-557
◇ B65 D05 ◇ REV Zbl 135.8 • ID 29448

BAR-HILLEL, Y. & SHAMIR, E. [1964] *Finite state languages: formal representations and adequacy problems* (C 0635) Lang & Information 87-98
◇ B65 D05 ◇ REV Zbl 158.253 JSL 30.383 • ID 45031

BAR-HILLEL, Y. & GAIFMAN, C. & SHAMIR, E. [1964] *On categorial and phrase structure grammars* (C 0635) Lang & Information 99-115
◇ B65 D05 ◇ REV Zbl 158.253 JSL 30.383 • ID 45034

BAR-HILLEL, Y. & GAIFMAN, C. & SHAMIR, E. [1964] *On formal properties of simple phrase structure grammars* (C 0635) Lang & Information 116-150
◇ B65 D05 ◇ REV Zbl 158.253 JSL 30.383 • ID 45035

BAR-HILLEL, Y. see Vol. I, II, III, V, VI for further entries

BARAK, A. [1975] see ARON, E.

BARALT-TORRIJOS, J. & CHIARAVIGLIO, L. & GROSKY, W. [1975] *The programmatic semantics of binary predicator calculi* (J 0047) Notre Dame J Formal Log 16*591-596
◇ B40 D05 ◇ REV MR 52#7861 Zbl 283#02038
• ID 17965

BARANSKIJ, V.A. & TRAKHTMAN, A.N. [1970] *Subsemigroup graphs (Russian) (English summary)* (J 0143) Mat Chasopis (Slov Akad Ved) 20*135-140
◇ C65 D05 ◇ REV MR 47#3233 Zbl 196.41 • ID 00771

BARANSKIJ, V.A. see Vol. III for further entries

BARASHENKOV, V.V. [1974] see AJTKHOZHAEVA, E.ZH.

BARASHKO, A.S. [1974] *Tape complexity of languages that can be recognized by two-way finite-term pushdown automata (Russian) (English summary)* (J 0040) Kibernetika, Akad Nauk Ukr SSR 1974/1*40-45
◇ D10 D15 ◇ REV MR 50#6209 Zbl 283#68051
• ID 60289

BARASHKO, A.S. [1974] *The number of push-down tape reversals of a two-way push-down store automaton (Russian)* (J 0040) Kibernetika, Akad Nauk Ukr SSR 1974/4*9-14
• TRANSL [1974] (J 0021) Cybernetics 10*564-570
◇ D10 D15 ◇ REV MR 54#14436 Zbl 298#68037
• ID 60288

BARASHKO, A.S. & ROJZEN, S.I. [1975] *On redundance of the points of partial recursive functions naming the computational complexity classes (Ukrainian, English summery)* (J 0270) Dokl Akad Nauk Ukr SSR, Ser A A1975*867-869,955
◇ D15 D20 ◇ REV MR 58#180 Zbl 323#02052
• ID 30422

BARASHKO, A.S. & ROJZEN, S.I. [1976] *Kernels of partially recursive functions that name classes of computational complexity (Russian)* (J 0040) Kibernetika, Akad Nauk Ukr SSR 1976/5*10-15
• TRANSL [1976] (J 0021) Cybernetics 12*663-670
◇ D15 D20 ◇ REV MR 56#8339 Zbl 354#68070
• ID 50191

BARASHKO, A.S. [1977] *Bikernels and hierarchies of computational complexity classes (Ukrainian) (English summary)* (J 0270) Dokl Akad Nauk Ukr SSR, Ser A 1977/1*3-5
◇ D15 ◇ REV MR 58#5115 Zbl 352#68064 • ID 50065

BARASHKO, A.S. [1980] *On boundary sets in the theory of enumerations (Ukrainian) (English and Russian summaries)* (J 0270) Dokl Akad Nauk Ukr SSR, Ser A 1980/3*72-74,92
◇ D15 D25 D45 ◇ REV MR 81j:03068 Zbl 446#03032
• ID 56561

BARASHKO, A.S. [1981] *Exact lower bounds for indicators of the enumeration of sets (Russian)* (J 0087) Mat Zametki (Akad Nauk SSSR) 30*397-406,462
• TRANSL [1981] (J 1044) Math Notes, Acad Sci USSR 30*690-695
◇ D15 D45 ◇ REV MR 84j:03082 Zbl 497#03030
• ID 34674

BARBACK, J. [1964] *Recursive functions and regressive isols* (J 0132) Math Scand 15*29-42
◇ D50 ◇ REV MR 31#1189 Zbl 148.247 JSL 32.269
• ID 00773

BARBACK, J. [1966] *A note on regressive isols* (J 0047) Notre Dame J Formal Log 7*203-205
◇ D50 ◇ REV MR 35#2734 Zbl 207.308 JSL 35.156
• ID 00774

BARBACK, J. [1966] *Two notes on regressive isols* (J 0048) Pac J Math 16*407-420
◇ D50 ◇ REV MR 32#5511 Zbl 199.25 JSL 32.527
• ID 00775

BARBACK, J. [1967] *An md-class of sets indexed by a regressive function* (J 0038) J Austral Math Soc 7*301-310
◇ D50 ◇ REV MR 37#62 Zbl 157.335 JSL 35.157
• ID 00777

BARBACK, J. [1967] *Double series of isols* (J 0017) Canad J Math 19*1-15
◇ D50 ◇ REV MR 35#1473 Zbl 207.308 JSL 35.156
• ID 00776

BARBACK, J. [1967] *Regressive upper bounds* (J 0144) Rend Sem Mat Univ Padova 39*248-272
◇ D50 ◇ REV MR 37#63 Zbl 159.10 JSL 35.156
• ID 00780

BARBACK, J. [1968] *On recursive sets and regressive isols* (J 0133) Michigan Math J 15*27-32
◇ D50 ◇ REV MR 37#64 Zbl 155.16 • ID 00781

BARBACK, J. [1969] *Extensions to regressive isols* (J 0132) Math Scand 25*159-177
◇ D50 ◇ REV MR 42#1656 Zbl 212.330 • ID 00782

Author Index

BARBACK, J. [1969] *Two notes on recursive functions and regressive isols* (**J** 0064) Trans Amer Math Soc 144*77-94
◇ D50 ◇ REV MR 41 # 6691 Zbl 198.26 • ID 00783

BARBACK, J. & JACKSON, W.D. [1970] *On representations as an infinite series of isols* (**J** 0020) Compos Math 22*347-365
◇ D50 ◇ REV MR 43 # 7323 Zbl 214.17 • ID 00784

BARBACK, J. & JACKSON, W.D. & PARNES, M. [1972] *Analogous characterizations of finite and isolated sets* (**J** 0047) Notre Dame J Formal Log 13*551-555
◇ D50 ◇ REV MR 48 # 3722 Zbl 242 # 02048 • ID 00787

BARBACK, J. [1972] *On solutions in the regressive isols* (**J** 0048) Pac J Math 43*283-296
◇ D50 ◇ REV MR 47 # 4778 Zbl 257 # 02036 • ID 00790

BARBACK, J. [1972] *Universal regressive isols* (**J** 0053) Proc Amer Math Soc 36*549-551
◇ D50 ◇ REV MR 47 # 1593 Zbl 266 # 02022 • ID 00786

BARBACK, J. [1976] *Composite numbers and prime regressive isols* (**J** 0048) Pac J Math 62*49-53
◇ D50 ◇ REV MR 54 # 84 Zbl 332 # 02049 • ID 18472

BARBACK, J. [1976] *Regressive isols and comparability* (**J** 0068) Z Math Logik Grundlagen Math 22*403-412
◇ D50 ◇ REV MR 56 # 5251 Zbl 358 # 02062 • ID 23678

BARBACK, J. & MCLAUGHLIN, T.G. [1978] *On the intersection of regressive sets* (**J** 0048) Pac J Math 79*19-35
◇ D50 ◇ REV MR 80f:03047 Zbl 404 # 03034 • ID 54821

BARBACK, J. [1981] *Hyper-torre isols and an arithmetic property* (**P** 2902) Aspects Effective Algeb;1979 Clayton 53-68
◇ D50 ◇ REV MR 82k:03071 Zbl 467 # 03044 • ID 55042

BARBACK, J. [1981] *On finite sums of regressive isols* (**J** 0048) Pac J Math 97*19-28
◇ D50 ◇ REV MR 83b:03054 Zbl 473 # 03038 • ID 55367

BARBACK, J. [1981] *On regressive isols and their differences* (**P** 2902) Aspects Effective Algeb;1979 Clayton 42-52
◇ D50 ◇ REV MR 82k:03070 Zbl 467 # 03043 • ID 55041

BARBACK, J. [1982] *On tame regressive isols* (**J** 1447) Houston J Math 8*153-159
◇ D50 ◇ REV MR 84b:03061 Zbl 527 # 03026 • ID 35646

BARBACK, J. [1985] *On hereditarily odd-even isols and a comparability of summands property* (**J** 0048) Pac J Math 118*27-35
◇ D50 ◇ ID 44799

BARENDREGT, H.P. [1975] *Normed uniformly reflexive structures* (**P** 1603) λ-Calc & Comput Sci Th;1975 Roma 272-286
◇ B40 D75 ◇ REV MR 57 # 12197 Zbl 333 # 02021 • ID 27590

BARENDREGT, H.P. [1976] *The incompleteness theorems* (**X** 1051) Univ Utrecht Math Inst: Utrecht iv + 59pp
◇ D35 F30 ◇ REV MR 53 # 2611 Zbl 323 # 02057 • ID 21504

BARENDREGT, H.P. & BERGSTRA, J.A. & KLOP, J.W. & VOLKEN, H. [1978] *Degrees of sensible lambda theories* (**J** 0036) J Symb Logic 43*45-55
◇ B40 D30 ◇ REV MR 57 # 15981 Zbl 408 # 03012 • ID 31152

BARENDREGT, H.P. & LONGO, G. [1983] *Recursion theoretic operators and morphisms of numbered sets* (**J** 0027) Fund Math 119*49-62
◇ D20 D30 D45 ◇ REV MR 86b:03051 Zbl 548 # 03023 • ID 39452

BARENDREGT, H.P. see Vol. I, II, V, VI for further entries

BARNES, B.H. [1971] *A two-way automaton with fewer states than any equivalent one-way automaton* (**J** 0187) IEEE Trans Comp C-20*474-475
◇ D05 ◇ REV Zbl 218 # 02032 • ID 26282

BARON, G. & KUICH, W. [1981] *The characterization of nonexpansive grammars by rational power series* (**J** 0194) Inform & Control 48*109-118
◇ D05 ◇ REV MR 82i:68047 Zbl 481 # 68075 • ID 80630

BARRERO, A. & GONZALEZ, R.C. & THOMASON, M.G. [1981] *Equivalence and reduction of expansive tree grammars* (**J** 3191) IEEE Trans Pattern Anal & Mach Intell PAMI-3*204-206
◇ D05 ◇ REV Zbl 456 # 68095 • ID 69148

BARTH, G. [1978] *Grammars with dynamic control sets* (**P** 1872) Automata, Lang & Progr (5);1978 Udine 36-51
◇ B75 D05 ◇ REV MR 80f:68079 Zbl 382 # 68060 • ID 69150

BARTH, G. [1979] *Fast recognition of context-sensitive structures* (**J** 0373) Comp Arch Inform & Numerik 22*243-256
◇ D05 D15 ◇ REV MR 82g:68065 Zbl 407 # 68079 • ID 80635

BARTHA, M. [1981] *An algebraic definition of attributed transformations* (**P** 3165) FCT'81 Fund of Comput Th;1981 Szeged 51-60
◇ D05 ◇ REV Zbl 462 # 68049 • ID 69151

BARTHOLOMES, F. & HOTZ, G. [1970] *Homomorphismen und Reduktionen linearer Sprachen* (**X** 0811) Springer: Heidelberg & New York xii + 143pp
◇ D05 ◇ REV Zbl 201.14 • ID 23502

BARTOL, W. [1971] *On the existence of machine homomorphisms* (**J** 0014) Bull Acad Pol Sci, Ser Math Astron Phys 19*865-869
◇ D05 D10 ◇ REV Zbl 225 # 68027 • ID 00820

BARTOL, W. [1974] *Algebraic complexity of machines* (**J** 0014) Bull Acad Pol Sci, Ser Math Astron Phys 22*851-856
◇ D10 D15 ◇ REV MR 50 # 9047 Zbl 298 # 68041 • ID 60309

BARTOL, W. & RAS, Z. & SKOWRON, A. [1977] *Theory of computing systems* (**C** 4723) Math Founds of Comput Sci 2*101-165
◇ B75 D05 ◇ REV MR 57 # 8109 Zbl 357 # 68062 • ID 50447

BARUA, R. [1984] *R-sets and category* (**J** 0064) Trans Amer Math Soc 286*125-158
◇ D70 E15 ◇ REV Zbl 562 # 54053 • ID 45772

BARWISE, J. [1967] *Infinitary logic and admissible sets* (0000) Diss., Habil. etc 124pp
◇ C70 D60 ◇ REM Doctorial diss., Stanford University • ID 16732

BARWISE, J. [1969] *Applications of strict Π^1_1 predicates to infinitary logic* (**J** 0036) J Symb Logic 34*409-423
◇ C40 C70 D60 ◇ REV MR 41 # 5218 Zbl 216.3 JSL 39.335 • ID 00825

BARWISE, J. & FISHER, E.R. [1970] *The Shoenfield absoluteness lemma* (**J** 0029) Israel J Math 8*329-339 ◇ C62 D55 E45 ◇ REV MR 43 # 4660 Zbl 206.11 • ID 00828

BARWISE, J. & GANDY, R.O. & MOSCHOVAKIS, Y.N. [1971] *The next admissible set* (**J** 0036) J Symb Logic 36*108-120 ◇ C40 C62 D55 D60 D70 E45 E47 ◇ REV MR 46 # 36 Zbl 236 # 02033 • ID 00836

BARWISE, J. [1974] *Admissible sets over models of set theory* (**P** 0602) Generalized Recursion Th (1);1972 Oslo 97-122 ◇ C62 D60 E30 ◇ REV MR 53 # 2670 Zbl 355 # 02031 • ID 00911

BARWISE, J. [1975] *Admissible sets and the interaction of model theory, recursion theory and set theory* (**P** 1521) Int Congr Math (II,12);1974 Vancouver 1*229-234 ◇ C70 C98 D55 D60 D70 E30 E35 E45 E47 ◇ REV MR 55 # 2570 Zbl 342 # 02029 • ID 31017

BARWISE, J. [1975] *Admissible sets and structures. An approach to definability theory* (**X** 0811) Springer: Heidelberg & New York xiii + 394pp ◇ B98 C40 C70 C98 D60 D98 E30 E98 ◇ REV MR 54 # 12519 Zbl 316 # 02047 JSL 43.139 • ID 60316

BARWISE, J. & SCHLIPF, J.S. [1975] *On recursively saturated models of arithmetic* (**C** 0782) Model Th & Algeb (A. Robinson) 42-55 ◇ C50 C57 C62 D80 H15 ◇ REV MR 53 # 12934 Zbl 343 # 02031 • ID 23183

BARWISE, J. & SCHLIPF, J.S. [1976] *An introduction to recursively saturated and resplendent models* (**J** 0036) J Symb Logic 41*531-536 ◇ C15 C40 C50 C57 ◇ REV MR 53 # 7761 Zbl 343 # 02032 JSL 47.440 • ID 14694

BARWISE, J. & MOSCHOVAKIS, Y.N. [1977] *Guide to part C: Recursion theory* (**C** 1523) Handb of Math Logic 525-526 ◇ D98 ◇ REV MR 84g:03004b • ID 47578

BARWISE, J. (ED.) [1977] *Handbook of mathematical logic* (**X** 0809) North Holland: Amsterdam xi + 1165pp • TRANSL [1982] (**X** 2027) Nauka: Moskva ◇ B98 C98 D98 E98 F98 H98 ◇ REV MR 56 # 15351 MR 84g:03004 MR 84j:03006 Zbl 443 # 03001 JSL 49.968 JSL 49.971 JSL 49.975 JSL 49.980 • REM 3rd ed 1982. Transl. in 4 parts. Russian suppl. by Mints,G.E. & Orevkov,V.P • ID 70117

BARWISE, J. [1977] *On Moschovakis closure ordinals* (**J** 0036) J Symb Logic 42*292-296 ◇ D60 D65 D70 ◇ REV MR 58 # 5126 Zbl 367 # 02021 • ID 26461

BARWISE, J. & MOSCHOVAKIS, Y.N. [1978] *Global inductive definability* (**J** 0036) J Symb Logic 43*521-534 ◇ C40 C70 D70 ◇ REV MR 81g:03059 Zbl 395 # 03021 • ID 29280

BARWISE, J. [1978] *Monotone quantifiers and admissible sets* (**P** 1628) Generalized Recursion Th (2);1977 Oslo 1-38 ◇ C70 C80 D60 D70 ◇ REV MR 81d:03037 Zbl 453 # 03047 • ID 70698

BARWISE, J. see Vol. I, II, III, V, VI for further entries

BARZDINS, J. [1962] *On a class of Turing machines (Minsky machines) (Russian)* (**J** 0003) Algebra i Logika 1/6*42-51 ◇ D10 ◇ REV MR 27 # 2415 Zbl 163.253 JSL 32.523 • ID 00842

BARZDINS, J. [1965] *Capacity of a medium and the behavior of automata (Russian)* (**J** 0023) Dokl Akad Nauk SSSR 160*302-305 • TRANSL [1965] (**J** 0470) Sov Phys, Dokl 10*8-11 ◇ D05 ◇ REV MR 31 # 1151 Zbl 192.66 • ID 16262

BARZDINS, J. [1965] *Complexity of recognition of symmetry on Turing machines (Russian)* (**J** 0052) Probl Kibern 15*245-248 ◇ D10 D15 ◇ REV MR 36 # 1326 Zbl 255 # 02039 JSL 35.159 • ID 00843

BARZDINS, J. [1968] *Complexity of programs to determine whether natural numbers not greater than n belong to a recursively enumerable set (Russian)* (**J** 0023) Dokl Akad Nauk SSSR 182*1249-1252 • TRANSL [1968] (**J** 0062) Sov Math, Dokl 9*1251-1254 ◇ B75 D15 D25 ◇ REV MR 38 # 4307 Zbl 193.316 • ID 00844

BARZDINS, J. [1969] *On computability by probabilistic machines (Russian)* (**J** 0023) Dokl Akad Nauk SSSR 189*699-702 • TRANSL [1969] (**J** 0062) Sov Math, Dokl 10*1464-1467 ◇ D10 ◇ REV MR 41 # 9465 Zbl 217.10 • ID 28042

BARZDINS, J. [1970] *Deciphering of sequential networks in the absence of an upper limit on the number of states (Russian)* (**J** 0023) Dokl Akad Nauk SSSR 190*1048-1051 • TRANSL [1970] (**J** 0470) Sov Phys, Dokl 15*94-97 ◇ D05 ◇ REV MR 43 # 8461 Zbl 213.23 • ID 27995

BARZDINS, J. & TRAKHTENBROT, B.A. [1970] *Finite automata. Behavior and synthesis (Russian)* (**X** 2027) Nauka: Moskva 400pp • TRANSL [1973] (**X** 0809) North Holland: Amsterdam xi + 321pp ◇ D05 D98 ◇ REV MR 50 # 4174 Zbl 271 # 94032 JSL 42.111 • ID 65772

BARZDINS, J. [1970] *On the relative frequency of solution of algorithmically unsolvable mass problems (Russian)* (**J** 0023) Dokl Akad Nauk SSSR 191*967-970 • TRANSL [1970] (**J** 0062) Sov Math, Dokl 11*459-462 ◇ D20 D25 ◇ REV MR 42 # 1654 Zbl 214.18 • ID 00845

BARZDINS, J. [1971] *Complexity and accuracy of the solution of initial segments of the problem of occurrence in a recursively enumerable set (Russian)* (**J** 0023) Dokl Akad Nauk SSSR 199*262-264 • TRANSL [1971] (**J** 0062) Sov Math, Dokl 12*1054-1056 ◇ D25 ◇ REV MR 44 # 6484 Zbl 242 # 02045 • ID 26360

BARZDINS, J. & FREJVALD, R.V. [1972] *On the prediction of general recursive functions (Russian)* (**J** 0023) Dokl Akad Nauk SSSR 206*521-524 • TRANSL [1972] (**J** 0062) Sov Math, Dokl 13*1224-1228 ◇ D20 ◇ REV MR 50 # 73 Zbl 267 # 02029 • ID 03866

BARZDINS, J. [1972] *Prognostication of automata and functions* (**P** 1455) Inform Processing (5);1971 Ljubljana 81-84 ◇ D05 D20 ◇ REV MR 53 # 10563 Zbl 255 # 94031 • ID 23072

BARZDINS, J. [1973] *The frequency solution of the occurrence problem for a recursively enumerable set (Russian)* (**S** 0066) Tr Mat Inst Steklov 133*52-58,274 • TRANSL [1973] (**S** 0055) Proc Steklov Inst Math 133*49-56 ◇ D25 ◇ REV MR 50 # 6817 Zbl 295 # 02023 • ID 03868

Author Index

BARZDINS, J. [1974] *A property of limiting computable functionals (Russian)* (S 2587) Teor Algor & Progr (Riga) 1*20-24 ◇ D15 D20 ◇ REV MR 58 # 21540 Zbl 333 # 02030 • ID 60328

BARZDINS, J. [1974] *Bemerkung zur Programmsynthese anhand von rechnerischen Vorgeschichten (Russian)* (S 2587) Teor Algor & Progr (Riga) 1*145-151 ◇ D10 ◇ REV MR 58 # 25060 Zbl 328 # 94044 • ID 60330

BARZDINS, J. [1974] *Limitare Synthese von τ-Zahlen (Russian)* (S 2587) Teor Algor & Progr (Riga) 1*112-116 ◇ D15 D20 ◇ REV MR 58 # 21522 Zbl 333 # 02032 • ID 60326

BARZDINS, J. [1974] see AGAFONOV, V.N.

BARZDINS, J. & KINBER, E.B. & PODNIEKS, K.M. [1974] *On speed-up of synthesis and prognostication of functions (Russian)* (S 2587) Teor Algor & Progr (Riga) 1*117-128 ◇ D15 D20 ◇ REV MR 58 # 21523 Zbl 333 # 02033 • ID 90034

BARZDINS, J. [1974] *On synthesizing programs given by examples* (P 1511) Int Symp Th Progr;1972 Novosibirsk 56-63 ◇ D03 ◇ REV MR 56 # 1770 • ID 80638

BARZDINS, J. [1974] *Prediction and identification in the limit of finite automata (Russian)* (S 2587) Teor Algor & Progr (Riga) 1*129-144 ◇ D05 ◇ REV MR 58 # 33587 Zbl 328 # 94043 • ID 80639

BARZDINS, J. & FREJVALD, R.V. [1974] *Prognostizierung und limitaere Synthese effektiv aufzaehlbarer Klassen von Funktionen (Russian)* (S 2587) Teor Algor & Progr (Riga) 1*101-111 ◇ D15 D20 D45 ◇ REV MR 58 # 21521 Zbl 341 # 02026 • ID 29747

BARZDINS, J. (ED.) [1974] *Theory of algorithms and programs I (Russian)* (X 0895) Latv Valsts (Gos) Univ : Riga ◇ B97 D97 ◇ REM Vol.II 1975 • ID 46742

BARZDINS, J. [1974] *Two theorems on the identification in the limit of τ-numbers (Russian)* (S 2587) Teor Algor & Progr (Riga) 1*82-88 ◇ D20 D45 ◇ REV MR 58 # 21519 Zbl 333 # 02031 • ID 60327

BARZDINS, J. [1975] *Inductive derivation of automata, function and programs* (P 1521) Int Congr Math (II,12);1974 Vancouver 2*455-460 ◇ D05 D20 ◇ REV MR 54 # 9989 Zbl 356 # 62033 • ID 25756

BARZDINS, J. & FREJVALD, R.V. [1975] *The relation between predictability and identifiability in the limit (Russian) (English summary)* (S 2587) Teor Algor & Progr (Riga) 2*26-34 ◇ D20 ◇ REV MR 57 # 16009 Zbl 322 # 02040 • ID 61808

BARZDINS, J. (ED.) [1975] *Theory of algorithms and programs II (Russian)* (X 0895) Latv Valsts (Gos) Univ : Riga ◇ B97 D97 ◇ REM Vol.II 1974. Vol.III 1977 • ID 46743

BARZDINS, J. (ED.) [1977] *Theory of algorithms and programs III (Russian)* (X 0895) Latv Valsts (Gos) Univ : Riga 155pp ◇ B75 D97 ◇ REV MR 56 # 11766 • REM Vol.II 1975 • ID 70119

BARZDINS, J. see Vol. II for further entries

BASARAB, I.A. [1979] *Equivalent transformations of metalinear LA(1) grammars and unary recursive schemes (Russian)* (J 0040) Kibernetika, Akad Nauk Ukr SSR 1979/4*36-40 • TRANSL [1979] (J 0021) Cybernetics 15*476-481 ◇ D05 ◇ REV MR 82a:68137 Zbl 444 # 68071 • ID 69249

BASCA, O. [1971] *La synthese des automates finis par la methode de A.Church* (C 0640) Log, Autom, Inform 209-213 ◇ D05 ◇ REV JSL 37.625 • ID 00915

BASS, L. & YOUNG, P. [1970] *Hierarchies based on computational complexity and irregularities of class determined measured sets* (P 0641) ACM Symp Th of Comput (2);1970 Northhampton 37-41 ◇ D15 D20 ◇ ID 00916

BASS, L. & YOUNG, P. [1973] *Ordinal hierarchies and naming complexity classes* (J 0037) ACM J 20*668-686 ◇ D15 D20 ◇ REV Zbl 339 # 68038 • ID 14525

BASU, S.K. [1969] *On classes of computable functions* (P 0671) ACM Symp Th of Comput (1);1969 Marina del Rey 55-61 ◇ D20 ◇ ID 02019

BASU, S.K. [1970] *On the structure of subrecursive degrees* (J 0119) J Comp Syst Sci 4*452-464 ◇ D20 ◇ REV MR 42 # 2937 Zbl 216.290 JSL 40.87 • ID 00864

BAUDISCH, A. [1974] *Theorien abelscher Gruppen mit einem einstelligen Praedikat* (J 0027) Fund Math 83*121-127 ◇ C60 D35 ◇ REV MR 54 # 10001 Zbl 289 # 02035 • ID 00867

BAUDISCH, A. [1975] *Elementare Theorien von Halbgruppen mit Kuerzungsregeln mit einem einstelligen Praedikat* (J 0014) Bull Acad Pol Sci, Ser Math Astron Phys 23*107-109 ◇ C05 C60 D35 ◇ REV MR 51 # 10067 Zbl 307 # 02029 • ID 00868

BAUDISCH, A. see Vol. I, II, III, V for further entries

BAUER, G. & OTTO, F. [1984] *Finite complete rewriting systems and the complexity of word problem* (J 1431) Acta Inf 21*521-540 ◇ D05 D15 ◇ REV Zbl 535 # 68019 • ID 40970

BAUERNOEPPEL, F. [1984] see ALBRECHT, A.

BAUMSLAG, G. & BOONE, W.W. & NEUMANN, B.H. [1959] *Some unsolvable problems about elements and subgroups of groups* (J 0132) Math Scand 7*191-201 ◇ D35 D40 ◇ REV MR 29 # 1247 Zbl 104.7 JSL 34.506 • ID 00880

BAUMSLAG, G. [1980] *Problem areas in infinite group theory for finite group theorists* (P 3047) Santa Cruz Conf Finite Groups;1979 Santa Cruz 217-223 ◇ D40 ◇ REV MR 82c:20063 • ID 80651

BAUMSLAG, G. & CANNONITO, F.B. & MILLER III, C.F. [1981] *Computable algebra and group embeddings* (J 0032) J Algeb 69*186-212 ◇ D40 D45 ◇ REV MR 82e:20042 Zbl 497 # 20023 • ID 80653

BAUMSLAG, G. & DYER, E. & MILLER III, C.F. [1981] *On the integral homology of finitely presented groups* (J 0589) Bull Amer Math Soc (NS) 4*321-324 ◇ D40 D80 ◇ REV MR 82c:20089 Zbl 471 # 20036 • ID 80650

BAUMSLAG, G. & CANNONITO, F.B. & MILLER III, C.F. [1981] *Some recognizable properties of solvable groups* (J 0044) Math Z 178*289-295
◇ D40 ◇ REV MR 82k:20061 Zbl 455 # 20027 • ID 80652

BAUMSLAG, G. [1984] *Algorithmically insoluble problems about finitely presented solvable, Lie, and associative algebras* (P 4359) Groups-Korea;1983 Kyoungiu 1-14
◇ D40 ◇ REV Zbl 549 # 20022 • REM Part II Baumslag,G. & Gildenhuys,D. & Strebel,R. 1985 • ID 44822

BAUMSLAG, G. & GILDENHUYS, D. & STREBEL, R. [1985] *Algorithmically insoluble problems about finitely presented solvable groups, Lie and associative algebras II* (J 0032) J Algeb 97*278-285
◇ D40 ◇ REM Part I 1984 • ID 49677

BAUMSLAG, G. see Vol. III for further entries

BAUR, W. [1973] *Eine rekursiv praesentierte Gruppe mit unentscheidbarem Wortproblem* (J 0044) Math Z 131*219-222
◇ D40 ◇ REV MR 54 # 12910 Zbl 248 # 02046 • ID 30642

BAUR, W. [1974] *Rekursive Algebren mit Kettenbedingungen* (J 0068) Z Math Logik Grundlagen Math 20*37-46
◇ D45 ◇ REV MR 50 # 4269 Zbl 317 # 02050 • ID 00884

BAUR, W. [1974] *Ueber rekursive Strukturen* (J 0305) Invent Math 23*89-95
◇ C50 C57 D45 ◇ REV MR 49 # 2335 Zbl 285 # 02050 • ID 04142

BAUR, W. [1975] *Decidability and undecidability of theories of abelian groups with predicates for subgroups* (J 0020) Compos Math 31*23-30
◇ B25 C60 D35 ◇ REV MR 52 # 5399 Zbl 335 # 02032 • ID 03874

BAUR, W. [1976] *Undecidability of the theory of abelian groups with a subgroup* (J 0053) Proc Amer Math Soc 55*125-128
◇ D35 ◇ REV MR 54 # 4953 Zbl 328 # 02032 • ID 24121

BAUR, W. [1976] *Zeitlich beschraenkte Turingmaschinen und polynomiale Reduktion* (P 3196) Kompl von Entscheid Probl;1973/74 Zuerich 11-19
◇ D10 D15 ◇ REV MR 57 # 18232 Zbl 345 # 68026 • ID 60371

BAUR, W. & RABIN, M.O. [1980] *Linear disjointness and algebraic complexity* (J 3370) Enseign Math, Ser 2 26*333-344
• REPR [1982] (P 3482) Logic & Algor (Specker);1980 Zuerich 35-46
◇ D15 D80 ◇ REV MR 84a:68035 Zbl 481 # 68046 Zbl 494 # 68051 • ID 38808

BAUR, W. [1980] *On the elementary theory of quadruples of vector spaces* (J 0007) Ann Math Logic 19*243-262
◇ B25 C60 D80 ◇ REV MR 82g:03056 Zbl 453 # 03010 • ID 54140

BAUR, W. see Vol. III, V for further entries

BAVEL, Z. & MULLER, D.E. [1970] *Connectivity and reversibility in automata* (J 0037) ACM J 17*231-240
◇ D05 ◇ REV MR 43 # 1843 Zbl 212.337 • ID 00886

BAVEL, Z. & GRZYMALA-BUSSE, J.W. & SOO HONG KWANG [1984] *On the connectivity of the product of automata* (J 2095) Fund Inform, Ann Soc Math Pol, Ser 4 7*225-266
◇ D05 ◇ REV MR 86e:68074 Zbl 562 # 68044 • ID 45396

BAXTER, L.D. [1978] *The undecidability of the third order dyadic unification problem* (J 0194) Inform & Control 38*170-178
◇ B15 B40 D80 ◇ REV MR 80m:03077 Zbl 387 # 03006 • ID 52222

BAXTER, RODNEY J. [1972] *On unlimited register machines* (J 0068) Z Math Logik Grundlagen Math 18*97-102
◇ D10 ◇ REV MR 50 # 6812 Zbl 254 # 02029 • ID 00889

BAXTER, RODNEY J. see Vol. II for further entries

BAZHANOV, V.A. [1984] *Logic of quantum mechanics and the problem of its decidability (Russian)* (C 4403) Logika, Pozn, Otrazh 58-65
◇ B51 D35 ◇ ID 46738

BAZILEVSKIJ, YU.YA. [1960] see AKUSHSKY, I.Y.

BAZILEVSKIJ, YU.YA. see Vol. I for further entries

BEAN, D.R. [1976] *Effective coloration* (J 0036) J Symb Logic 41*469-480
◇ D80 ◇ REV MR 54 # 4952 Zbl 331 # 02025 • ID 14779

BEAN, D.R. [1976] *Recursive Euler and Hamilton paths* (J 0053) Proc Amer Math Soc 55*385-394
◇ D80 ◇ REV MR 54 # 4951 Zbl 327 # 05117 • ID 24120

BEATTY, J.C. [1980] *Two iteration theorems for the LL(k) languages* (J 1426) Theor Comput Sci 12*193-228
◇ D05 ◇ REV MR 81j:68085 Zbl 452 # 68084 • ID 69156

BEAUQUIER, D. [1985] *Ensembles reconnaissables de mots biinfinis. Limite et determinisme* (P 4622) Autom on Infinite Words;1984 Le Mont-Dore 28-46
◇ D05 ◇ ID 49439

BEAUQUIER, D. [1985] *Mueller automata and bi-infinite words* (P 4647) FCT'85 Fund of Comput Th;1985 Cottbus 36-43
◇ D05 ◇ ID 49070

BEAUQUIER, D. also published under the name GIRAULT-BEAUQUIER, D.

BEAUQUIER, J. [1978] *Ambiguite forte* (P 1872) Automata, Lang & Progr (5);1978 Udine 52-62
◇ D05 ◇ REV MR 80e:68180 Zbl 382 # 68065 • ID 69162

BEAUQUIER, J. [1978] *Un generateur inheremment ambigu du cone des langages algebriques* (J 3441) RAIRO Inform Theor 12*99-108 • ERR/ADD ibid 13*195
◇ D05 ◇ REV MR 80k:68056 Zbl 377 # 68043 • ID 69161

BEAUQUIER, J. [1979] *Deux familles de langages incomparables* (J 0194) Inform & Control 43*101-122
◇ D05 ◇ REV MR 81e:68091 Zbl 427 # 68064 • ID 69159

BEAUQUIER, J. [1979] *Generateurs algebriques et systemes de paires iterantes* (J 1426) Theor Comput Sci 8*293-323
◇ D05 ◇ REV MR 80i:68057 Zbl 408 # 68071 • ID 69160

BEAUQUIER, J. [1979] *Independence of linear and one-counter generators* (P 2935) FCT'79 Fund of Comput Th;1979 Berlin/Wendisch-Rietz 45-51
◇ D05 ◇ REV Zbl 416 # 68072 • ID 69164

BEAUQUIER, J. [1979] see AUTEBERT, J.-M.

BEAUQUIER, J. [1979] *Strong non-deterministic context-free languages* (P 3488) Theor Comput Sci (4);1979 Aachen 47-57
◇ D05 ◇ REV MR 82c:68047 Zbl 412 # 68066 • ID 69163

BEAUQUIER, J. [1980] see AUTEBERT, J.-M.

BEAUQUIER, J. [1981] *A remark about a substitution property* (**J** 0041) Math Syst Theory 14*189-191
◇ D05 ◇ REV MR 82f:68073 Zbl 459 # 68041 • ID 69158

BEAUQUIER, J. [1981] see AUTEBERT, J.-M.

BEAUQUIER, J. [1981] *Substitution of semi-AFL's* (**J** 1426) Theor Comput Sci 14*187-193
◇ D05 ◇ REV MR 82f:68074 Zbl 472 # 68039 • ID 69157

BEAUQUIER, J. [1982] see AUTEBERT, J.-M.

BECK, H. [1975] *Zur Entscheidbarkeit der funktionalen Aequivalenz (English summary)* (**P** 1449) Automata Th & Formal Lang;1975 Kaiserslautern 127-133
◇ D20 D80 ◇ REV MR 55 # 5424 Zbl 312 # 68049 • ID 60391

BECK, H. [1976] *Die mit Nestedstackautomaten berechenbaren Funktionen sind elementar* (**J** 0009) Arch Math Logik Grundlagenforsch 17*115-128
◇ D10 D20 ◇ REV MR 58 # 175 Zbl 365 # 02027 • ID 03879

BECK, H. [1980] *Entscheidbarkeit und Komplexitaet der Funktionsgleichheit von Programmen (Dissertation)* (**X** 3159) TH Karlsruhe Fak Informatik: Karlsruhe 220pp
◇ B75 D15 ◇ REV Zbl 476 # 68020 • ID 55588

BECKER, H. [1978] *Partially playful universes* (**C** 2908) Cabal Seminar Los Angeles 1976-77 55-90
◇ D55 E15 E45 E60 ◇ REV MR 80g:03050 Zbl 397 # 03033 • ID 52699

BECKER, H. [1980] *Thin collections of sets of projective ordinals and analogs of L* (**J** 0007) Ann Math Logic 19*205-241
◇ D55 E15 E45 E50 E55 E60 ◇ REV MR 82g:03087 Zbl 453 # 03050 • ID 54180

BECKER, H. [1985] *A property equivalent to the existence of scales* (**J** 0064) Trans Amer Math Soc 287*591-612
◇ D55 E15 E60 ◇ ID 44736

BECKER, H. see Vol. V for further entries

BECKER, J.A. & HENSON, C.W. & RUBEL, L.A. [1980] *First-order conformal invariants* (**J** 0120) Ann of Math, Ser 2 112*123-178
◇ C35 C65 D35 E35 E75 ◇ REV MR 83a:30011 Zbl 459 # 03019 • ID 54459

BECKER, J.A. & DENEF, J. & LIPSHITZ, L. [1980] *Further remarks on the elementary theory of formal power series rings* (**P** 2625) Model Th of Algeb & Arithm;1979 Karpacz 1-9
◇ C60 D35 ◇ REV MR 83a:13013 Zbl 452 # 12013 JSL 50.853 • ID 47356

BECKER, J.A. see Vol. I, III for further entries

BECKHOFF, G. [1980] *A novel approach to the state reduction problem* (**J** 2701) Digit Processes 6*305-313
◇ D05 ◇ REV MR 82m:68101 Zbl 466 # 68041 • ID 69168

BECVAR, J. [1963] *Finite and combinatorial automata. Turing automata with a programming tape* (**P** 0572) Inform Processing (2);1962 Muenchen 391-394
◇ D05 D10 ◇ REV Zbl 135.7 • ID 27612

BECVAR, J. [1965] *A universal Turing machine with a programming tape* (**P** 0797) Fonds des Math, Machines Math & Appl;1962 Tihany 11-20
◇ D10 ◇ REV Zbl 168.256 JSL 36.535 • ID 01553

BECVAR, J. [1965] *Real-time and complexity problems in automata theory* (**J** 0156) Kybernetika (Prague) 1*475-498
◇ D05 D10 D15 ◇ REV Zbl 192.87 JSL 36.346 • ID 00931

BECVAR, J. [1967] *Probleme der Komplexitaet in der Theorie der Algorithmen und Automaten* (**P** 1671) Colloq Automatenth (3);1965 Hannover 142-157
◇ A05 B75 D10 ◇ REV Zbl 174.38 • ID 31636

BECVAR, J. [1970] *Programmkomplexitaet von Funktionen und Mengen* (**P** 0577) Automatenth & Formale Sprachen;1969 Oberwolfach 317-326
◇ D15 ◇ REV Zbl 222 # 02040 • ID 26313

BECVAR, J. [1970] *Programmkomplexitaet von berechenbaren Funktionen* (**J** 1046) Z Angew Math Mech 50*T82
◇ D15 ◇ ID 31638

BECVAR, J. see Vol. I, II, VI for further entries

BEDNAREK, A.R. & WALLACE, A.D. [1967] *Finite approximants of compact totally disconnected machines* (**J** 0041) Math Syst Theory 1*209-216
◇ D05 ◇ REV MR 36 # 5919 Zbl 174.29 • ID 00936

BEDNAREK, A.R. & NORRIS, E.M. [1970] *Congruences and ideals in machines I* (**J** 0060) Rev Roumaine Math Pures Appl 15*193-199
◇ C05 D05 ◇ REV MR 41 # 7000 Zbl 213.21 • ID 27990

BEDNAREK, A.R. see Vol. V for further entries

BEESON, M.J. [1975] *The nonderivability in intuitionistic formal systems of theorems on the continuity of effective operations* (**J** 0036) J Symb Logic 40*321-346
◇ D20 F50 ◇ REV MR 51 # 10050 Zbl 316 # 02038 • ID 17547

BEESON, M.J. [1976] *Derived rules of inference related to the continuity of effective operations* (**J** 0036) J Symb Logic 41*328-336
◇ D20 F50 ◇ REV MR 54 # 7208 Zbl 333 # 02027 • ID 14764

BEESON, M.J. & SCEDROV, A. [1984] *Church's thesis, continuity, and set theory* (**J** 0036) J Symb Logic 49*630-643
◇ D20 E70 F50 ◇ REV MR 86f:03097 • ID 42422

BEESON, M.J. see Vol. V, VI for further entries

BEHMANN, H. [1922] *Beitraege zur Algebra der Logik, insbesondere zum Entscheidungsproblem* (**J** 0043) Math Ann 86*163-229
◇ B20 B25 D35 ◇ REV FdM 48.1119 • ID 00942

BEHMANN, H. [1923] *Algebra der Logik und Entscheidungsproblem* (**J** 0157) Jbuchber Dtsch Math-Ver 32*66-67,2.Abteilung
◇ B20 B25 D35 ◇ ID 00943

BEHMANN, H. [1927] *Entscheidungsproblem und Logik der Beziehungen* (**J** 0157) Jbuchber Dtsch Math-Ver 36*17-18,2.Abteilung
◇ B25 D35 ◇ REV FdM 53.41 • ID 00946

BEHMANN, H. see Vol. I, II, III, V, VI for further entries

BEICK, H.-R. & JANTKE, K.P. [1981] *Combining postulates of naturalness in inductive inference (German and Russian summaries)* (**J** 0129) Elektr Informationsverarbeitung & Kybern 17*465-484
◇ D20 ◇ REV MR 84e:03049 Zbl 459 # 03022 Zbl 526 # 03021 • ID 34385

Beick, H.-R. [1982] *Zur Konvergenzgeschwindigkeit von Strategien der induktiven Inferenz (English and Russian summaries)* (**J** 0129) Elektr Informationsverarbeitung & Kybern 18∗163–172
◇ D20 ◇ REV MR 84h:03107 • ID 34296

Bekbaev, U.D. & Khadzhiev, D. [1984] *Extension of partially defined operations II (Russian)* (**J** 0024) Dokl Akad Nauk Uzb SSR 1984/5∗6–9
◇ D20 ◇ REV Zbl 545 # 08007 • REM Part I 1984 • ID 44492

Belaga, Eh.G. [1984] *Locally synchronous complexity in the light of the trans-box method* (**P** 3565) Symp of Th Aspects of Comput Sci (1);1984 Paris 129–139
◇ D15 ◇ REV MR 85i:68002 Zbl 556 # 68016 • ID 47076

Belegradek, O.V. [1974] *Algebraically closed groups (Russian)* (**J** 0003) Algebra i Logika 13∗239–255,363
• TRANSL [1974] (**J** 0069) Algeb and Log 13∗135–143
◇ C25 C60 D40 ◇ REV MR 52 # 2859 Zbl 304 # 20019 Zbl 319 # 20039 • ID 17646

Belegradek, O.V. [1976] *On m-degrees of word problems (Russian)* (**P** 2064) All-Union Conf Math Log (4);1976 Kishinev 10
◇ D25 D40 ◇ REV Zbl 408 # 03039 • ID 43225

Belegradek, O.V. [1978] *Elementary properties of algebraically closed groups (Russian) (English summary)* (**J** 0027) Fund Math 98∗83–101
◇ C25 C60 D40 ◇ REV MR 57 # 9530 Zbl 389 # 20030 • ID 29209

Belegradek, O.V. [1978] *On m-degrees of the word problem (Russian)* (**J** 0092) Sib Mat Zh 19∗1232–1236
• TRANSL [1978] (**J** 0475) Sib Math J 19∗867–870
◇ D25 D40 ◇ REV MR 80c:03044 Zbl 408 # 03039 • ID 56277

Belegradek, O.V. [1979] *Algebraic equivalents of solvability of group-theoretic algorithmic problems (Russian)* (**J** 0092) Sib Mat Zh 20∗953–963,1164
• TRANSL [1979] (**J** 0475) Sib Math J 20∗673–680
◇ D40 ◇ REV MR 81b:20005 Zbl 426 # 20023 • ID 53673

Belegradek, O.V. [1980] *Decidable fragments of universal theories and existentially closed models (Russian)* (**J** 0092) Sib Mat Zh 21/6∗196–201,223
• TRANSL [1980] (**J** 0475) Sib Math J 21∗898–902
◇ B25 C25 D40 ◇ REV MR 82d:20033 Zbl 498 # 20024 • ID 80668

Belegradek, O.V. see Vol. III for further entries

Beletskij, M.I. [1969] *The relationship between categorical and domination grammars I,II (Russian. English summary)* (**J** 0040) Kibernetika, Akad Nauk Ukr SSR 1969/4∗129–135,1969/5∗10–14
• TRANSL [1969] (**J** 0021) Cybernetics 5∗506–512,540–545
◇ D05 ◇ REV MR 46 # 4789 • ID 00966

Belkin, V.P. & Gorbunov, V.A. [1975] *Filters in lattices of quasivarieties of algebraic systems (Russian)* (**J** 0003) Algebra i Logika 14∗373–392
• TRANSL [1975] (**J** 0069) Algeb and Log 14∗229–239
◇ C05 C60 D40 G10 ◇ REV MR 53 # 5428 Zbl 328 # 08005 • ID 22948

Bellenot, S.F. [1984] *The Banach space T and the fast growing hierarchy from logic* (**J** 0029) Israel J Math 47∗305–313
◇ D80 E75 ◇ REV MR 86c:46010 Zbl 562 # 46009 • ID 44234

Bellenot, S.F. see Vol. I for further entries

Belsnes, D. [1971] see Aanderaa, S.O.

Belsnes, D. [1971] *The immortality problem for non-erasing Turing machines* (**P** 0604) Scand Logic Symp (2);1970 Oslo 19–26
◇ D10 ◇ REV MR 49 # 7123 Zbl 238 # 02034 • ID 03886

Bel'tyukov, A.P. [1976] *An iterative description of the class $\mathfrak{E}^1$ of Grzegorczyk's hierarchy (Russian)* (**S** 0228) Zap Nauch Sem Leningrad Otd Mat Inst Steklov 60∗3–14,221
• TRANSL [1980] (**J** 1531) J Sov Math 14∗1429–1436
◇ D10 D20 ◇ REV MR 58 # 27385 Zbl 343 # 02027 • ID 60401

Bel'tyukov, A.P. [1977] *Maximum sequence of classes transformable by primitive recursion in a given class (Russian)* (**S** 0228) Zap Nauch Sem Leningrad Otd Mat Inst Steklov 68∗3–18,142
• TRANSL [1981] (**J** 1531) J Sov Math 15∗1–10
◇ D20 ◇ REV MR 58 # 21546 Zbl 358 # 02042 • ID 50496

Bel'tyukov, A.P. [1979] *A machine description and a hierarchy of initial Grzegorczyk classes (Russian) (English summary)* (**S** 0228) Zap Nauch Sem Leningrad Otd Mat Inst Steklov 88∗30–46,237
• TRANSL [1982] (**J** 1531) J Sov Math 20∗2280–2289
◇ D10 D20 ◇ REV MR 81c:03030 Zbl 429 # 03017 • ID 53848

Bel'tyukov, A.P. [1979] *Small classes based on bounded recursion (Russian)* (**S** 0716) Vychisl Tekh Vopr Kibern (Univ Leningrad) 16∗75–85,223
◇ D20 ◇ REV MR 80m:03076 Zbl 531 # 03025 • ID 70901

Bel'tyukov, A.P. [1980] *Hierarchies of complexity of computation of partial functions with values 0 and 1 (Russian)* (**J** 0087) Mat Zametki (Akad Nauk SSSR) 28∗423–431,479
• TRANSL [1980] (**J** 1044) Math Notes, Acad Sci USSR 28∗680–684
◇ D15 ◇ REV MR 82d:03066 Zbl 439 # 03015 • ID 56007

Bel'tyukov, A.P. [1982] *Lower bounds of complexity for machine models of computations (Russian) (English summary)* (**S** 0228) Zap Nauch Sem Leningrad Otd Mat Inst Steklov 118∗4–24,214
◇ D15 ◇ REV MR 83h:68054 Zbl 494 # 68058 • ID 39211

Bel'tyukov, A.P. see Vol. I, III, VI for further entries

Belyaev, V.Ya. [1978] *Subrings of finitely presented associative rings (Russian)* (**J** 0003) Algebra i Logika 17∗627–638
• TRANSL [1978] (**J** 0069) Algeb and Log 17∗407–414
◇ D40 ◇ REV MR 81i:16026 Zbl 429 # 16017 • ID 53886

Belyaev, V.Ya. & Tajtslin, M.A. [1979] *On elementary properties of existentially closed systems (Russian)* (**J** 0067) Usp Mat Nauk 34/2(206)∗39–94
• TRANSL [1979] (**J** 1399) Russ Math Surv 34/2∗43–107
◇ C25 C40 C60 C75 C85 D55 ◇ REV MR 82a:03028 Zbl 413 # 03025 • ID 32621

Belyaev, V.Ya. see Vol. III, V for further entries

Belyakin, N.V. [1961] *The universality of a computing machine with potentially infinite exterior memory (Russian)* (**J** 0052) Probl Kibern 5∗77–86
• TRANSL [1961] (**J** 1195) Probl Cybernet 5∗99–114 [1961] (**J** 0449) Probl Kybern 5∗65–77
◇ D10 ◇ REV Zbl 131.155 JSL 27.366 • ID 00974

Author Index

BELYAKIN, N.V. [1963] *Computation of effective operators on Turing machines with restricted erasure (Russian)* (J 0003) Algebra i Logika 2/1*19-23
◇ D10 ◇ REV MR 27 # 5680 Zbl 199.35 JSL 37.198
• ID 00978

BELYAKIN, N.V. [1963] *Distribution of intermediate information when computing on non-erasing Turing machines (Russian)* (J 0023) Dokl Akad Nauk SSSR 152*75-77
• TRANSL [1963] (J 0470) Sov Phys, Dokl 8*871-872 (English)
◇ D10 ◇ REV MR 32 # 7403 Zbl 171.273 • ID 00975

BELYAKIN, N.V. [1963] *On a class of Turing machines (Russian)* (J 0023) Dokl Akad Nauk SSSR 148*47-49
• TRANSL [1963] (J 0470) Sov Phys, Dokl 8*3-4 (English)
◇ D10 ◇ REV MR 27 # 2412 JSL 37.198 JSL 37.211
• ID 00976

BELYAKIN, N.V. [1963] *Simulation of Turing machines on nets (Russian)* (J 0071) Met Diskr Analiz (Novosibirsk) 1*32-41
◇ D10 ◇ REV MR 30 # 3009 JSL 37.199 • ID 00977

BELYAKIN, N.V. [1966] *Turing machines operating on a plane (Russian)* (J 0023) Dokl Akad Nauk SSSR 168*502-503
• TRANSL [1966] (J 0062) Sov Math, Dokl 7*661-662
◇ D10 ◇ REV MR 33 # 5478 Zbl 163.254 JSL 33.469
• ID 00979

BELYAKIN, N.V. [1969] *A variant of Richter's constructive ordinals (Russian)* (J 0003) Algebra i Logika 8*154-171
• TRANSL [1969] (J 0069) Algeb and Log 8*86-96
◇ D70 F15 ◇ REV MR 41 # 8223 Zbl 212.25 JSL 40.626
• ID 00982

BELYAKIN, N.V. [1970] *Generalized computations and second order arithmetic (Russian)* (J 0003) Algebra i Logika 9*375-405
• TRANSL [1970] (J 0069) Algeb and Log 9*225-243
◇ C62 D10 D45 D65 D70 F35 ◇ REV MR 46 # 32 Zbl 278 # 02036 • ID 29065

BELYAKIN, N.V. [1973] *Generalized computations over regular numerations (Russian)* (J 0003) Algebra i Logika 12*623-643,735
• TRANSL [1973] (J 0069) Algeb and Log 12*355-367
◇ D45 D65 D70 D75 ◇ REV MR 52 # 10394 Zbl 293 # 02031 • ID 21691

BELYAKIN, N.V. [1973] *On a class of recursive hierarchies (Russian)* (J 0003) Algebra i Logika 12*3-21,120
• TRANSL [1973] (J 0069) Algeb and Log 12*1-11
◇ D20 ◇ REV MR 48 # 8217 Zbl 291 # 02029 • ID 00984

BELYAKIN, N.V. [1974] *Generalized computations, and third order arithmetic (Russian)* (J 0003) Algebra i Logika 13*132-144,234
• TRANSL [1974] (J 0069) Algeb and Log 13*71-78
◇ D65 F35 ◇ REV MR 51 # 2898 Zbl 296 # 02023
• ID 17342

BELYAKIN, N.V. [1976] *Iterated Kleene computability, and the superjump (Russian)* (J 0142) Mat Sb, Akad Nauk SSSR, NS 101(143)*21-43,159
• TRANSL [1978] (J 0349) Math of USSR, Sbor 30*17-37
◇ D55 D65 ◇ REV MR 54 # 9996 Zbl 342 # 02031
• ID 25837

BELYAKIN, N.V. [1979] *Autonomous computability (Russian)* (J 0003) Algebra i Logika 18*398-407,507
• TRANSL [1979] (J 0069) Algeb and Log 18*240-247
◇ D45 D55 D65 ◇ REV MR 81j:03072 Zbl 449 # 03040
• ID 56707

BELYAKIN, N.V. [1983] *A means of modeling a classical second-order arithmetic (Russian)* (J 0003) Algebra i Logika 22*3-25
• TRANSL [1983] (J 0069) Algeb and Log 22*1-18
◇ B15 C62 D55 F35 ◇ REV MR 85h:03046 Zbl 538 # 03040 • ID 41479

BELYAKIN, N.V. see Vol. I, III, V, VI for further entries

BEN-ARI, M. [1980] *A simplified proof that regular resolution is exponential* (J 0232) Inform Process Lett 10*96-98
◇ B35 D15 F20 ◇ REV MR 81g:68062 Zbl 438 # 03054
• ID 55966

BEN-ARI, M. see Vol. I, II for further entries

BENDA, V. & BENDOVA, K. [1976] *On specific features of recognizable families of languages* (P 1401) Math Founds of Comput Sci (5);1976 Gdansk 45*81-99
◇ D05 ◇ REV Zbl 337 # 68057 • ID 33235

BENDA, V. & BENDOVA, K. [1976] *Recognizable filters and ideals* (J 0140) Comm Math Univ Carolinae (Prague) 17*251-259
◇ D05 ◇ REV MR 53 # 14984 Zbl 334 # 02020 • ID 23226

BENDA, V. & BENDOVA, K. [1977] *Characterization of recognizable families by means of regular languages* (P 1635) Math Founds of Comput Sci (6);1977 Tatranska Lomnica 53*247-252
◇ D05 ◇ REV Zbl 363 # 68090 • ID 33236

BENDA, V. & BENDOVA, K. [1977] *On families recognizable by finite branching automata* (J 0156) Kybernetika (Prague) 13*293-319
◇ D05 ◇ REV MR 57 # 4636 Zbl 374 # 94040 • ID 33237

BENDIX, P.B. & KNUTH, D.E. [1970] *Simple word problems in universal algebras* (P 0690) Comput Prob in Abstr Algeb;1967 Oxford 263-267
• REPR [1983] (C 4659) Autom of Reasoning 2*342-376
◇ C05 D40 ◇ REV MR 41 # 134 Zbl 188.49 • ID 07264

BENDOVA, K. [1976] see BENDA, V.

BENDOVA, K. [1977] see BENDA, V.

BENDOVA, K. see Vol. II, VI for further entries

BENEJAM, J.-P. & PAGET, M. [1976] *La complexite des algorithmes de Markov (English summary)* (J 2313) C R Acad Sci, Paris, Ser A-B 282*A381-A383
◇ D03 D10 D15 ◇ REV MR 53 # 111 Zbl 332 # 02041
• ID 16576

BENEJAM, J.-P. [1977] *Algebraic characterizations of the satisfiability of first-order logical formulas and the halting of programs* (J 0068) Z Math Logik Grundlagen Math 23*111-120
◇ C52 D10 G05 ◇ REV MR 58 # 10396 Zbl 383 # 03008
• ID 26472

BENEJAM, J.-P. see Vol. I, III, V for further entries

BENESOVA, M. [1969] *Real-time computable functions and almost primitive recursive functions* (J 0128) Acta Math Univ Comenianae (Bratislava) 23*121-134
◇ D15 D20 ◇ REV MR 45 # 1688 Zbl 251 # 02041
• ID 28911

BENESOVA, M. & KOREC, I. [1975] *Non-linear speed-up theorem for two register Minsky machines* (**P** 0454) Math Founds of Comput Sci (4);1975 Marianske Lazne 179-185
◇ D10 D15 ◇ REV MR 53 # 4611 Zbl 339 # 02033
• ID 60452

BENEY, J. & FRECON, L. [1980] *Langage et systeme d'ecriture de transducteurs* (**J** 2832) RAIRO Inform 14*379-394
◇ D05 ◇ REV Zbl 451 # 68048 • ID 69173

BENIOFF, P.A. [1976] *Models of Zermelo Fraenkel set theory as carriers for the mathematics of physics I,II* (**J** 0209) J Math Phys 17*618-628,629-640
◇ C62 D80 E75 ◇ REV MR 57 # 12210 Zbl 331 # 02047 Zbl 331 # 02048 • ID 30647

BENIOFF, P.A. [1980] *The computer as a physical system: a microscopic quantum mechanical Hamiltonian model of computers as represented by Turing machines* (**J** 2764) J Stat Phys 22*563-591
◇ D10 ◇ REV MR 81m:68040 • ID 80682

BENIOFF, P.A. [1982] *Quantum mechanical Hamiltonian models of Turing machines* (**J** 2764) J Stat Phys 29*515-546
◇ D10 D80 ◇ REV MR 84k:81010 Zbl 514 # 68055
• ID 39267

BENIOFF, P.A. [1982] *Quantum mechanical models of Turing machines that dissipate no energy* (**J** 2730) Phys Rev Lett 48*1581-1585
◇ D10 D80 ◇ REV MR 84i:81011 • ID 40211

BENIOFF, P.A. see Vol. I, II for further entries

BENNETT, C.H. & GILL III, J.T. [1981] *Relative to a random oracle A, $P^A \neq NP^A \neq co-NP^A$ with probability 1* (**J** 1428) SIAM J Comp 10*96-113
◇ D15 ◇ REV MR 83a:68044 Zbl 454 # 68030 • ID 69174

BENNISON, V.L. [1977] *On the problem of determining measure-independence* (**P** 3238) Conf Theoret Comput Sci;1977 Waterloo ON 100-110
◇ D15 D25 ◇ REV MR 80h:03058 Zbl 415 # 03028
• ID 53130

BENNISON, V.L. & SOARE, R.I. [1977] *Recursion theoretic characterizations of complexity theoretic properties* (**P** 3572) IEEE Symp Found of Comput Sci (18);1977 Providence 100-106
◇ D15 D20 ◇ REV MR 58 # 27391 • ID 70944

BENNISON, V.L. & SOARE, R.I. [1978] *Some lowness properties and computational complexity sequences* (**J** 1426) Theor Comput Sci 6*233-254
◇ D15 D25 D30 ◇ REV MR 58 # 183 Zbl 401 # 68021
• ID 30633

BENNISON, V.L. [1979] *Information content characterizations of complexity theoretic properties* (**P** 3488) Theor Comput Sci (4);1979 Aachen 58-66
◇ D15 D25 ◇ REV MR 82a:03034 Zbl 401 # 68022
• ID 54639

BENNISON, V.L. [1980] *Recursively enumerable complexity sequences and measure independence* (**J** 0036) J Symb Logic 45*417-438
◇ D15 D25 D30 ◇ REV MR 81j:03063 Zbl 454 # 03019
• ID 54231

BENSON, C. & MENDELSOHN, N.S. [1966] *A calculus for a certain class of word problems in groups* (**J** 1669) J Comb Th 1*202-208
◇ D40 ◇ REV MR 34 # 1380 Zbl 154.18 • ID 01568

BERARDUCCI, A. [1985] *A generalization of Goedel's recursive functions (Italian)* (**P** 4646) Atti Incontri Log Mat (2);1983/84 Siena 467-476
◇ D20 ◇ ID 49597

BERECZKI, I. [1952] *Existenz einer nichtelementaren rekursiven Funktion (Ungarisch) (Russische und deutsche Zusammenfassung)* (**P** 0662) Hungar Math Congr (1);1950 Budapest 409-417
◇ D20 ◇ REV MR 16.324 Zbl 49.8 JSL 19.298 • ID 01546

BERECZKI, I. [1952] *Loesung eines Markovschen Problems betreffs einer Ausdehnung des Begriffes der elementaren Funktion* (**J** 0001) Acta Math Acad Sci Hung 3*197-218
◇ D20 ◇ REV MR 14.937 Zbl 48.246 JSL 19.122
• ID 01048

BEREZIN, S.A. [1972] *The generation of two-place primitive recursive functions (Russian)* (**S** 0166) Mat Issl, Mold SSR 7/2*224-233,291
◇ D20 ◇ REV MR 47 # 6459 Zbl 281 # 02039 • ID 01545

BEREZIN, S.A. [1976] *An algebra of one-place primitive recursive functions with an iteration operation of general type (Russian) (English summary)* (**J** 0040) Kibernetika, Akad Nauk Ukr SSR 1976/3*12-19
• TRANSL [1976] (**J** 0021) Cybernetics 12*346-353
◇ D20 D75 ◇ REV MR 57 # 5709 Zbl 336 # 02032
• ID 60466

BEREZIN, S.A. [1978] *Maximal subalgebras of recursive function algebras (Russian)* (**J** 0040) Kibernetika, Akad Nauk Ukr SSR 1978/6*123-125
• TRANSL [1978] (**J** 0021) Cybernetics 14*935-938
◇ D20 D75 ◇ REV MR 80d:03039 Zbl 437 # 03017
• ID 55885

BEREZYUK, N.T. & FURMANOV, K.K. [1973] *Foundations of the theory of algorithms (Russian)* (**X** 2726) Aviatsion Inst: Khar'kov 163pp
◇ D20 D98 ◇ REV MR 58 # 32028 • ID 80686

BERG, E.P. & LISCHKE, G. [1977] *Zwei Saetze fuer schwache Erhaltungsmasse* (**J** 0068) Z Math Logik Grundlagen Math 23*409-410
◇ D15 D20 ◇ REV MR 80f:68041 Zbl 439 # 03016
• ID 56008

BERG, J. & CHIHARA, C.S. [1975] *Church's thesis misconstrued* (**J** 0095) Philos Stud 28*357-362
◇ A05 D20 F99 ◇ REV MR 58 # 27199 • ID 70955

BERG, J. see Vol. I, II for further entries

BERGER, R. [1966] *The undecidability of the domino problem* (**S** 0167) Mem Amer Math Soc 66*72pp
◇ D05 D10 ◇ REV MR 36 # 49 Zbl 199.308 • ID 01054

BERGMAN, GEORGE M. [1978] *Terms and cyclic permutations* (**J** 0004) Algeb Universalis 8*129-136
◇ B03 D40 ◇ REV MR 56 # 15451 Zbl 327 # 02012
• ID 50464

BERGMAN, GEORGE M. see Vol. V for further entries

BERGSTRA, J.A. [1978] *Degrees of partial functions* (**J** 0047) Notre Dame J Formal Log 19*152-154
◇ D30 ◇ REV MR 58 # 5135 Zbl 349 # 02040 • ID 27099

Author Index

BERGSTRA, J.A. [1978] see BARENDREGT, H.P.

BERGSTRA, J.A. [1978] *The continuous functionals and* ^2E (**P** 1628) Generalized Recursion Th (2);1977 Oslo 39-53 ◇ D65 ◇ REV MR 80f:03050 Zbl 453 # 03047 • ID 70958

BERGSTRA, J.A. [1979] *Effective transformations on probabilistic data* (**J** 0068) Z Math Logik Grundlagen Math 25*219-226 ◇ D75 ◇ REV MR 81d:03049 Zbl 424 # 03021 • ID 70959

BERGSTRA, J.A. & TUCKER, J.V. [1979] *Equational specifications for computable data types: six hidden functions suffice and other sufficiency bounds* (**X** 3205) Math Centr Amsterdam Afd Inf IW128/79*17pp ◇ B75 D45 D80 ◇ REV Zbl 418 # 68019 • ID 53336

BERGSTRA, J.A. & TIURYN, J. [1979] *Implicit definability of algebraic structures by means of program properties* (**P** 2935) FCT'79 Fund of Comput Th;1979 Berlin/Wendisch-Rietz 58-63 ◇ B75 D80 ◇ REV MR 81d:03048 Zbl 456 # 68025 • ID 70963

BERGSTRA, J.A. [1979] *Recursion theory on processes* (**J** 0382) Int J Comput Math 7*119-128 ◇ D10 D30 D75 ◇ REV MR 80f:68057 Zbl 404 # 03033 • ID 54820

BERGSTRA, J.A. & TUCKER, J.V. [1980] *A characterisation of computable data types by means of a finite equational specification method* (**P** 2904) Automata, Lang & Progr (7);1980 Noordwijkerhout 76-90 ◇ B75 D45 D80 ◇ REV MR 83c:68019 Zbl 417 # 68012 Zbl 449 # 68003 • ID 56751

BERGSTRA, J.A. & TUCKER, J.V. [1980] *A natural data type with a finite equational final semantics specification but no effective equational initial semantics specification* (**X** 3205) Math Centr Amsterdam Afd Inf 133*11pp ◇ B75 D45 D80 ◇ REV Zbl 421 # 68021 • ID 53454

BERGSTRA, J.A. & TUCKER, J.V. [1980] *On bounds for the specification of finite data types by means of equations and conditional equations* (**X** 3205) Math Centr Amsterdam Afd Inf 131*24pp ◇ B75 D45 D80 ◇ REV Zbl 421 # 68020 • ID 53453

BERGSTRA, J.A. & TUCKER, J.V. [1981] *Algebraically specified programming systems and Hoare's logic* (**P** 2903) Automata, Lang & Progr (8);1981 Akko 348-362 ◇ B75 D80 ◇ REV MR 83m:68012 Zbl 465 # 68003 • ID 54942

BERGSTRA, J.A. & TIURYN, J. [1981] *Algorithmic degrees of algebraic structures* (**J** 2095) Fund Inform, Ann Soc Math Pol, Ser 4 4*851-861 ◇ D30 D75 ◇ REV MR 83j:03073 Zbl 487 # 68012 • ID 35369

BERGSTRA, J.A. & BROY, M. & TUCKER, J.V. & WIRSING, M. [1981] *On the power of algebraic specifications* (**P** 3429) Math Founds of Comput Sci (10);1981 Strbske Pleso 193-204 ◇ D75 ◇ REV MR 83f:68033 Zbl 462 # 68001 • ID 40231

BERGSTRA, J.A. & MEYER, J.-J.C. [1981] *Small specifications for large finite data structures* (**J** 0382) Int J Comput Math 9*305-320 ◇ B75 D45 D80 ◇ REV MR 83h:68030 Zbl 468 # 68022 • ID 55119

BERGSTRA, J.A. & KLOP, J.W. [1982] *Algebraic specifications for parametrized data types with minimal parameter and target algebras* (**P** 3836) Automata, Lang & Progr (9);1982 Aarhus 183/81*22pp ◇ D45 D80 ◇ REV MR 83m:68036 Zbl 466 # 68019 • ID 54994

BERGSTRA, J.A. & TIURYN, J. & TUCKER, J.V. [1982] *Floyd's principle, correctness theories and program equivalence* (**J** 1426) Theor Comput Sci IW145/80*51pp ◇ B75 D75 ◇ REV MR 83j:68014 Zbl 437 # 68010 Zbl 474 # 68017 • ID 55909

BERGSTRA, J.A. [1982] see BAETEN, J.C.M.

BERGSTRA, J.A. & TUCKER, J.V. [1982] *The completeness of the algebraic specification methods for computable data types* (**J** 0194) Inform & Control 54*186-200 ◇ B75 D80 ◇ REV MR 84i:68028 Zbl 447 # 68023 • ID 56594

BERGSTRA, J.A. & TUCKER, J.V. [1982] *The refinement of specifications and the stability of Hoare's logic* (**P** 3738) Log of Progr;1981 Yorktown Heights 24-36 ◇ B75 D80 ◇ REV MR 83i:68025 Zbl 504 # 68019 • ID 39304

BERGSTRA, J.A. & KLOP, J.W. [1983] *Initial algebra specifications for parametrized data types* (**X** 1121) Math Centr: Amsterdam i+20pp • REPR [1983] (**J** 0129) Elektr Informationsverarbeitung & Kybern 19*17-31 ◇ B75 D45 D80 G30 ◇ REV MR 84j:68011 MR 85f:68068 Zbl 474 # 68019 Zbl 516 # 68019 • REM Reprinted with German and Russian summaries • ID 39980

BERGSTRA, J.A. & KLOP, J.W. [1984] *The algebra of recursively defined processes and the algebra of regular processes* (**P** 4012) Automata, Lang & Progr (11);1984 Antwerpen 82-94 ◇ D05 D75 ◇ REV MR 86e:68068 Zbl 561 # 68019 • ID 47054

BERGSTRA, J.A. see Vol. I, II, III, VI for further entries

BERISHVILI, G. [1977] see AVALISHVILI, I.

BERMAN, L. & HARTMANIS, J. [1976] *On isomorphisms and density of NP and other complete sets* (**P** 2597) ACM Symp Th of Comput (8);1976 Hershey 30-40 ◇ D15 ◇ REV MR 55 # 6946 Zbl 365 # 68045 • ID 51075

BERMAN, L. [1976] *On the structure of complete sets: almost everywhere complexity and infinitely often speedup* (**P** 1757) IEEE Symp Found of Comput Sci (17);1976 Houston 76-80 ◇ D15 ◇ REV MR 58 # 32050 • ID 80695

BERMAN, L. [1977] *Precise bounds for Presburger arithmetic and the reals with addition: preliminary report* (**P** 3572) IEEE Symp Found of Comput Sci (18);1977 Providence 95-99 ◇ B25 D15 F30 ◇ REV MR 58 # 5116 • ID 70972

BERMAN, L. [1980] *The complexity of logical theories* (**J** 1426) Theor Comput Sci 11*71-77 ◇ B25 D10 D15 F30 ◇ REV MR 82c:03061b Zbl 475 # 03017 • ID 55471

BERMAN, P. [1978] *Relationship between density and deterministic complexity of NP- complete languages* (**P** 1872) Automata, Lang & Progr (5);1978 Udine 63-71 ◇ D15 ◇ REV MR 80g:68054 Zbl 382 # 68068 • ID 80696

BERMAN, P. [1979] *Complexity of the theory of atomless boolean algebras* (**P** 2935) FCT'79 Fund of Comput Th;1979 Berlin/Wendisch-Rietz 64-70
◇ B25 D15 ◇ REV MR 81k:03033 Zbl 421 # 03028
• ID 53427

BERMAN, P. [1980] *A note on sweeping automata* (**P** 2904) Automata, Lang & Progr (7);1980 Noordwijkerhout 91-97
◇ D05 D15 ◇ REV MR 82k:68047 Zbl 479 # 68081
• ID 69187

BERMAN, P. see Vol. II for further entries

BERNARDI, C. & GNANI, G. [1973] *Un'osservazione sulle classi (English summary)* (**J** 0012) Boll Unione Mat Ital, IV Ser 8*583-585
◇ D30 ◇ REV MR 49 # 10531 Zbl 278 # 02034 • ID 03895

BERNARDI, C. [1974] *Aspetti ricorsivi degli insiemi dialettici (English summary)* (**J** 0012) Boll Unione Mat Ital, IV Ser 9*51-61
◇ D30 D35 G10 ◇ REV MR 50 # 6801 Zbl 312 # 02037
• ID 03897

BERNARDI, C. [1980] *Alcune osservazioni sugli insiemi produttivi e creativi (English summary)* (**J** 3495) Boll Unione Mat Ital, V Ser, B 17*1350-1364
◇ D25 ◇ REV MR 85k:03024 Zbl 453 # 03044 • ID 54174

BERNARDI, C. & CROCIANI, C. [1980] *Insiemi semplici e massimali in ricorsivita (English summary)* (**C** 2963) Rass di Mat 9-23
◇ D25 ◇ REV MR 82g:03075 Zbl 498 # 03027 • ID 70979

BERNARDI, C. [1981] *On the relation provable equivalence and on partitions in effectively inseparable sets* (**J** 0063) Studia Logica 40*29-37
◇ D25 D35 F30 ◇ REV MR 82m:03057 Zbl 468 # 03020
• ID 55085

BERNARDI, C. see Vol. II, III, VI for further entries

BERNAYS, P. & SCHOENFINKEL, M. [1928] *Zum Entscheidungsproblem der mathematischen Logik* (**J** 0043) Math Ann 99*342-372
◇ B20 B25 D35 ◇ REV FdM 54.56 • ID 01076

BERNAYS, P. [1958] *Remarques sur le probleme de la decision en logique elementaire* (**P** 0576) Raisonn en Math & Sci Exper;1955 Paris 39-43
◇ B10 D35 ◇ REV MR 21 # 3330 Zbl 85.251 JSL 25.285
• ID 01541

BERNAYS, P. [1969] *Remark concerning the formalization of recursive definition in second order logic* (**P** 1841) Fct Recurs & Appl;1967 Tihany 19-23
◇ B15 D75 ◇ ID 32552

BERNAYS, P. see Vol. I, II, III, V, VI for further entries

BERNHARDT, L. & LINDNER, R. & THIELE, H. [1971] *Ueber sequentiell berechenbare reelle Abbildungen (English and Russian summaries)* (**J** 0129) Elektr Informationsverarbeitung & Kybern 7*317-329
◇ D80 ◇ REV MR 48 # 1894 Zbl 268 # 02022 • ID 01102

BERNHARDT, L. [1972] see AROLD, D.

BERNHARDT, L. [1975] *Algorithmisches Konstruieren von Algorithmen* (**J** 0129) Elektr Informationsverarbeitung & Kybern 11*594
◇ B75 D20 ◇ REV Zbl 316 # 02039 • ID 60488

BERRY, GERARD [1976] *Bottom-up computation of recursive programs (French summary)* (**J** 4698) Rev Franc Autom, Inf & Rech Operat, Ser Rouge Inf Th 10/R-1*47-82
◇ D20 D70 ◇ REV MR 53 # 14960 • ID 23224

BERRY, GERARD see Vol. VI for further entries

BERRY, J.W. [1972] *A note on immune sets* (**J** 0047) Notre Dame J Formal Log 13*98-100
◇ D30 D50 ◇ REV MR 45 # 6621 Zbl 228 # 02026
• ID 01139

BERRY, J.W. [1972] *Almost recursively enumerable sets* (**J** 0064) Trans Amer Math Soc 164*241-253
◇ D25 D55 ◇ REV MR 51 # 94 Zbl 309 # 02038
• ID 03900

BERRY, J.W. [1974] *N-ary almost recursive functions* (**J** 0068) Z Math Logik Grundlagen Math 20*551-559
◇ D20 ◇ REV MR 58 # 16223 Zbl 356 # 02035 • ID 03901

BERRY, J.W. [1977] *Enumeration degrees and strong reducibilities* (**S** 1585) Rec Fct Th Newsletter 16
◇ D30 ◇ ID 31625

BERSHTEJN, L.S. & KARELIN, V.P. & MELIKHOV, A.N. & PASHKEVICH, A.P. [1969] *Graphs of regular expressions, and abstract synthesis of automata (Russian)* (**J** 0474) Avtom Vychis Tekh, Akad Nauk Latv SSR 1969/4*1-6
• TRANSL [1969] (**J** 2666) Autom Control Comput Sci 3/4*1-5
◇ D05 ◇ REV MR 46 # 1512 Zbl 254 # 94056 • ID 63857

BERSHTEJN, L.S. see Vol. V for further entries

BERSTEL, J. [1978] *Ensembles reconnaissables de nombres* (**P** 3394) Lang Algeb (1);1973 Bonascre 23-84
◇ D05 ◇ REV MR 80e:68212 Zbl 403 # 68074 • ID 69190

BERSTEL, J. [1978] *Memento sur les transductions rationnelles* (**P** 3394) Lang Algeb (1);1973 Bonascre 5-22
◇ D05 ◇ REV MR 80d:68083 Zbl 393 # 68076 • ID 69189

BERSTEL, J. [1979] *Sur la construction de mots sans carre* (**S** 2348) Semin Th Nombres Bordeaux 18*15pp
◇ D05 ◇ REV MR 82a:68156 Zbl 428 # 68090 • ID 53827

BERSTEL, J. [1979] *Sur les mots sans carre definis par un morphisme* (**P** 1873) Automata, Lang & Progr (6);1979 Graz 71*16-25
◇ D05 ◇ REV MR 81j:68104 Zbl 425 # 20046 • ID 53590

BERSTEL, J. [1979] *Transductions and context-free languages* (**X** 1079) Teubner: Leipzig 278pp
◇ D05 ◇ REV MR 80j:68056 Zbl 424 # 68040 • ID 69191

BERSTEL, J. [1984] *Some recent results on squarefree words* (**P** 3565) Symp of Th Aspects of Comput Sci (1);1984 Paris 14-25
◇ D05 ◇ REV MR 86e:68056 • ID 45397

BERTONI, A. & MAURI, G. & TORELLI, M. [1977] *Some recursively unsolvable problems relating to isolated cutpoints in probabilistic automata* (**P** 1632) Automata, Lang & Progr (4);1977 Turku SF 87-94
◇ D05 ◇ REV MR 56 # 18188 Zbl 366 # 94064 • ID 51155

BERTONI, A. & MAURI, G. & MIGLIOLI, P.A. [1979] *A characterization of abstract data as model-theoretic invariants* (**P** 1873) Automata, Lang & Progr (6);1979 Graz 26-37
◇ B75 C57 D45 D80 ◇ REV MR 82a:68027
Zbl 411 # 68033 • ID 52925

Author Index

BERTONI, A. & MAURI, G. & TORELLI, M. [1980] *Sulla complessita di alcuni problemi di conteggio* (**J** 0089) Calcolo 17*163-174
◇ D15 ◇ REV MR 83b:68031 Zbl 456 # 68048 • ID 69192

BERTONI, A. & MAURI, G. & MIGLIOLI, P.A. [1980] *Towards a theory of abstract data types: a discussion on problems and tools* (**P** 2946) Int Symp Progr (4);1980 Paris 44-58
◇ B75 C57 D80 ◇ REV MR 82c:68013 Zbl 435 # 68022 • ID 80710

BERTONI, A. & MAURI, G. & MIGLIOLI, P.A. [1981] *Model theoretic aspects of abstract data specification* (**P** 3642) Colloq Math Log in Computer Sci;1978 Salgotarjan 181-193
◇ B75 C50 C57 ◇ REV MR 83g:68007 Zbl 503 # 68013 • ID 36666

BERTONI, A. & MAURI, G. [1981] *On efficient computation of the coefficients of some polynomials with applications to some enumeration problems* (**J** 0232) Inform Process Lett 12*142-145
◇ D15 D80 ◇ REV MR 82i:68026 Zbl 462 # 68027 • ID 80711

BERTONI, A. & MAURI, G. & MIGLIOLI, P.A. [1983] *On the power of model theory in specifying abstract data types and in capturing their recursiveness* (**J** 2095) Fund Inform, Ann Soc Math Pol, Ser 4 6*127-170
◇ B75 C50 C57 D20 ◇ REV MR 84f:68014 Zbl 529 # 68008 • ID 38487

BERTONI, A. & MAURI, G. & SABADINI, N. [1984] *Non deterministic machines and their generalizations* (**P** 4559) Parallel Process;1983 Muenchen 86-97
◇ D10 D15 ◇ REV Zbl 566 # 68043 • ID 48749

BERTONI, A. see Vol. VI for further entries

BETH, E.W. [1950] *Decision problems of logic and mathematics* (**C** 0643) Philosophie 1946-48 3-18
◇ B20 B25 C10 D35 ◇ REV JSL 22.359 • ID 01156

BETH, E.W. see Vol. I, II, III, V, VI for further entries

BEYNON, W.M. [1980] *On the structure of free finite state machines* (**J** 1426) Theor Comput Sci 11*167-180
◇ D05 ◇ REV MR 83b:68067 Zbl 433 # 68048 • ID 69193

BEZEM, M. [1985] *Isomorphism between HEO and HRO^E, ECF and ICF^E* (**J** 0036) J Symb Logic 50*359
◇ D20 D75 ◇ ID 42602

BEZEM, M. see Vol. VI for further entries

BEZVERKHNIJ, V.N. & ROLLOV, EH.V. [1972] *Solution of the conjugacy problem for subsemigroups of a free semigroup (Russian)* (**C** 1435) Vopr Teor Grupp & Polugrupp 114A,115-121
◇ D40 ◇ REV MR 52 # 14093 • ID 21881

BEZVERKHNIJ, V.N. & ROLLOV, EH.V. [1974] *On subgroups of a free product of groups (Russian)* (**S** 3478) Sovrem Algebra (Leningrad) 1*16-31
◇ C05 D40 ◇ REV MR 53 # 621 Zbl 301 # 20024 • ID 43288

BEZVERKHNIJ, V.N. & ROLLOV, EH.V. [1974] *On subsemigroups of free semigroups (Russian)* (**S** 3478) Sovrem Algebra (Leningrad) 1*32-42
◇ C05 D40 ◇ REV MR 53 # 656 Zbl 301 # 20046 • ID 43291

BEZVERKHNIJ, V.N. [1977] *Loesung des Problems der Konjugiertheit von Untergruppen fuer eine Klasse von Gruppen I,II (Russian)* (**S** 3478) Sovrem Algebra (Leningrad) 6*16-23,24-32
◇ D40 ◇ REV Zbl 392 # 20021 Zbl 398 # 20046 • ID 52412

BEZVERKHNIJ, V.N. [1981] *Solvability of the inclusion problem in a class of HNN-groups (Russian)* (**C** 2264) Algor Probl Teor Grupp & Polygrupp 20-62
◇ B25 C60 D40 ◇ REV MR 84h:20020 • ID 39526

BEZVERKHNIJ, V.N. & GRINBLAT, V.A. [1981] *The root problem in Artin Groups (Russian)* (**C** 2264) Algor Probl Teor Grupp & Polygrupp 72-81
◇ B25 C60 D40 ◇ REV MR 83j:20043 • ID 39904

BEZVERKHNYAYA, I.S. [1981] *Conjugacy of finite sets of subgroups in a free product of groups (Russian)* (**C** 2264) Algor Probl Teor Grupp & Polygrupp 102-116
◇ D40 ◇ REV MR 84c:20032 • ID 39600

BIANCO, E. [1979] *Informatique fondamentale. De la machine de Turing aux ordinateurs modernes* (**X** 0804) Birkhaeuser: Basel iv+152pp
◇ D03 D10 D98 ◇ REV MR 83h:68001 Zbl 427 # 68052 • ID 53756

BIBEL, W. [1979] *Tautology testing with a generalized matrix reduction method* (**J** 1426) Theor Comput Sci 8*31-44
◇ B35 D15 F05 ◇ REV MR 80i:03023 Zbl 421 # 03011 • ID 53410

BIBEL, W. see Vol. I, II, VI for further entries

BICHEVSKIJ, YA.YA. [1973] *An estimate of the complexity of bounded mass problems (Russian) (Latvian and English summaries)* (**J** 0337) Mat Ezheg, Akad Nauk Latv SSR 13*128-139
◇ D15 ◇ REV MR 49 # 11859 Zbl 272 # 68045 • ID 60503

BIELECKI, J. [1970] *The effect of the geometry of the tape of the Turing machine on the number of instructions in the program controlling it (Polish) (Russian and English summaries)* (**J** 0141) Arch Autom & Telemech 15*499-511
◇ D10 D15 ◇ REV MR 47 # 1324 Zbl 203.300 • ID 01199

BIELECKI, J. [1971] *Selfreproduction of modular Turing machines (Polish) (Russian and English summaries)* (**J** 0141) Arch Autom & Telemech 16*161-170
◇ D10 ◇ REV Zbl 222 # 02039 • ID 26312

BIEN, Z. [1972] *Some properties of p-events* (**J** 0014) Bull Acad Pol Sci, Ser Math Astron Phys 20*779-784
◇ D05 ◇ REV MR 47 # 4486 Zbl 247 # 94038 • ID 60509

BIEN, Z. [1975] *The star measure of regular events (Russian summary)* (**J** 0014) Bull Acad Pol Sci, Ser Math Astron Phys 23*593-598
◇ D05 ◇ REV MR 53 # 10515 • ID 23049

BIEN, Z. also published under the name SOZANSKA-BIEN, Z.

BIENENSTOCK, E. [1977] *Sets of degrees of computable fields* (**J** 0029) Israel J Math 27*348-356
◇ C57 C60 D30 D45 ◇ REV MR 58 # 16233 Zbl 359 # 02030 • ID 50578

BILLINGTON, N. [1984] *Growth of groups and graded algebras* (**J** 0394) Commun Algeb 12*2579-2588 • ERR/ADD ibid 13/3*753-755
◇ C60 D40 ◇ REV MR 86e:20039ab Zbl 547 # 20026 • ID 45773

BIRD, R. [1973] *A note on definition by cases* (**J** 0068) Z Math Logik Grundlagen Math 19*207-208
◇ D20 ◇ REV MR 47 # 8275 Zbl 327 # 02032 • ID 01215

BIRD, R. [1975] *Non recursive functionals* (**J** 0068) Z Math Logik Grundlagen Math 21*41-46
◇ D20 ◇ REV MR 51 # 90 Zbl 327 # 02033 • ID 03906

BIRKHOFF, GARRETT & LIPSON, J.D. [1970] *Heterogeneous algebras* (**J** 1669) J Comb Th 8*115-133
◇ C05 D05 ◇ REV MR 40 # 4119 Zbl 211.20 • ID 01237

BIRKHOFF, GARRETT [1973] *Current trends in algebra* (**J** 0005) Amer Math Mon 80*760-782
• TRANSL [1976] (**J** 1527) Pokroky Mat Fyz Astron (Prague) 21*199-211
◇ A10 D05 D15 ◇ REV MR 48 # 32 Zbl 286 # 01013 • ID 01239

BIRKHOFF, GARRETT & LIPSON, J.D. [1974] *Universal algebra and automata* (**P** 0610) Tarski Symp;1971 Berkeley 41-52
◇ C05 D05 ◇ REV MR 50 # 11854 Zbl 312 # 94025 • ID 01240

BIRKHOFF, GARRETT see Vol. I, II, III, V for further entries

BIRYUKOV, A.P. [1966] *Solution of certain algorithmic problems for finitely determined commutative semigroups (Russian)* (**J** 0092) Sib Mat Zh 7*523-530
◇ D40 ◇ REV MR 34 # 1423 • ID 01216

BIRYUKOV, A.P. [1966] *Solvability of the problem of isomorphism for finitely defined commutative semigroups with two generators (Russian)* (**C** 1549) Tartu Mezhvuz Nauch Simp Obshchej Algeb 5-6
◇ D40 ◇ REV MR 34 # 1422 Zbl 248 # 02047 • ID 01530

BISHOP, E.A. & BRIDGES, D.S. [1985] *Constructive analysis* (**X** 0811) Springer: Heidelberg & New York xii + 477pp
◇ D80 F55 F98 ◇ ID 48575

BISHOP, E.A. see Vol. VI for further entries

BISKUP, J. [1973] *Zufaellige Folgen und Bi-Immunitaet* (**P** 1630) GI Fachtag Automatenth & Form Sprach (1);1973 Bonn 202-207
◇ D50 D80 ◇ REV MR 55 # 2528 Zbl 278 # 68041 • ID 60522

BISKUP, J. [1977] *On bi-immune isols* (**J** 0068) Z Math Logik Grundlagen Math 23*469-484
◇ D50 ◇ REV MR 58 # 10376 Zbl 383 # 03032 • ID 52014

BISKUP, J. [1977] *Ueber Projektionsmengen von Komplexitatsmassen* (**J** 0129) Elektr Informationsverarbeitung & Kybern 13*359-368
◇ D15 ◇ REV MR 56 # 17198 Zbl 368 # 68050 • ID 80732

BISKUP, J. [1978] *Path measures of Turing machine computations* (**P** 1872) Automata, Lang & Progr (5);1978 Udine 90-104
◇ D10 D15 ◇ REV Zbl 388 # 03016 • ID 52270

BISKUP, J. [1978] *The time measure of one-tape Turing machines does not have the parallel computation property* (**J** 1428) SIAM J Comp 7*115-117
◇ D10 D15 ◇ REV MR 58 # 3668 Zbl 374 # 68041 • ID 28160

BISKUP, J. [1981] *A formal approach to null values in database relations* (**P** 4199) Adv in Data Base Th;1979 Toulouse 299-341
◇ D15 ◇ REV MR 82a:68040 • ID 39963

BISKUP, J. see Vol. V for further entries

BISWAS, S. & VAISHNAVI, V.K. [1979] *Tree adjunct languages as integer set recognizers* (**J** 0382) Int J Comput Math 7*87-94
◇ D05 ◇ REV MR 80f:68080 Zbl 412 # 68068 • ID 69200

BLACKWELL, D. [1967] *Infinite games and analytic sets* (**J** 0054) Proc Nat Acad Sci USA 58*1836-1837
◇ D55 E15 E60 ◇ REV MR 36 # 4518 Zbl 224 # 90077 • ID 01250

BLACKWELL, D. [1978] *Borel-programmable functions* (**J** 2661) Ann Probab 6*321-324
◇ D55 E15 ◇ REV MR 57 # 566 Zbl 398 # 28002 • ID 80735

BLACKWELL, D. see Vol. V for further entries

BLAIR, H.A. [1982] *The recursion-theoretic complexity of the semantics of predicate logic as a programming language* (**J** 0194) Inform & Control 54*25-47
◇ B75 D35 D55 ◇ REV MR 85g:03060 Zbl 527 # 03022 • ID 37572

BLAIR, H.A. [1984] *The intractability of validity in logic programming and dynamic logic* (**P** 2989) Log of Progr; 1983 Pittsburgh 57-67
◇ B75 C75 D35 D55 ◇ REV Zbl 547 # 68036 • ID 43300

BLAIR, H.A. see Vol. I for further entries

BLANCHARD, F. & HANSEL, G. [1985] *Languages and subshifts* (**P** 4622) Autom on Infinite Words;1984 Le Mont-Dore 138-146
◇ D05 ◇ ID 49443

BLASS, A.R. [1972] *Complexity of winning strategies* (**J** 0193) Discr Math 3*295-300
◇ D55 E60 ◇ REV MR 48 # 1899 Zbl 243 # 90052 • ID 01516

BLASS, A.R. & CENZER, D. [1974] *Cores of Π^1_1-sets of reals* (**J** 0036) J Symb Logic 39*649-654
◇ D55 E15 ◇ REV MR 51 # 12526 Zbl 295 # 02038 • ID 03908

BLASS, A.R. [1975] *A forcing proof of the Kechris-Moschovakis constructibility theorem* (**J** 0053) Proc Amer Math Soc 47*195-197
◇ D55 E40 E45 ◇ REV MR 50 # 4307 Zbl 301 # 02074 • ID 03910

BLASS, A.R. & GUREVICH, Y. [1982] *On the unique satisfiability problem* (**J** 0194) Inform & Control 55*80-88
◇ D15 ◇ REV MR 85c:68020 Zbl 543 # 03027 • ID 33761

BLASS, A.R. & GUREVICH, Y. & KOZEN, D. [1984] *A 0-1 law for logic with a fixed-point operator* (1111) Preprints, Manuscr., Techn. Reports etc. CRL-TR-38-84
◇ B10 B75 C80 D70 ◇ REM Univ. of Michigan, Computing Research Lab. • ID 47716

BLASS, A.R. & GUREVICH, Y. [1984] *Equivalence relations, invariants, and normal forms* (**J** 1428) SIAM J Comp 13*682-689
◇ D15 ◇ REV MR 86g:03064a Zbl 545 # 68035 • REM Part I. Part II 1984 • ID 33771

BLASS, A.R. & GUREVICH, Y. [1984] *Equivalence relations, invariants, and normal forms II* (**P** 2342) Symp Rek Kombin;1983 Muenster 24-42
◇ D15 E15 ◇ REV MR 86g:03064b Zbl 545 # 68036 • REM Part I 1984 • ID 40312

Author Index

Blass, A.R. & Gurevich, Y. [1984] *Henkin quantifiers and complete problems* (1111) Preprints, Manuscr., Techn. Reports etc.
◇ C80 D15 ◇ rem Univ. of Michigan Technical Report • id 47720

Blass, A.R. [1985] *Kleene degrees of ultrafilters* (**P** 3342) Rec Th Week;1984 Oberwolfach 29-48
◇ C20 D65 E05 ◇ rev Zbl 573 # 03020 • id 45296

Blass, A.R. see Vol. I, III, V, VI for further entries

Blattner, M. [1973] *The unsolvability of the equality problem for sentential forms of context-free grammars* (**J** 0119) J Comp Syst Sci 7*463-468
◇ D05 ◇ rev MR 48 # 5426 Zbl 273 # 68054 • id 60543

Blattner, M. & Head, T. [1977] *Single-valued a-transducers* (J 0119) J Comp Syst Sci 15*310-327
◇ D05 ◇ rev MR 57 # 2783 Zbl 367 # 94071 • id 51214

Blattner, M. [1979] *Inherent ambiguities in families of grammars* (**P** 1873) Automata, Lang & Progr (6);1979 Graz 71*38-48
◇ D05 ◇ rev MR 82b:68060 Zbl 421 # 68069 • id 69206

Blattner, M. & Head, T. [1979] *The decidability of equivalence for deterministic finite transducers* (**J** 0119) J Comp Syst Sci 19*45-49
◇ D05 ◇ rev MR 81h:68075 Zbl 416 # 68073 • id 69207

Blattner, M. & Latteux, M. [1981] *Parikh-bounded languages* (**P** 2903) Automata, Lang & Progr (8);1981 Akko 316-323
◇ D05 ◇ rev MR 82k:68041 Zbl 462 # 68057 • id 69205

Blattner, M. & Ginsburg, S. [1982] *Position-restricted grammar forms and grammars* (**J** 1426) Theor Comput Sci 17*1-27
◇ D05 ◇ rev MR 82m:68117 Zbl 477 # 68086 • id 69208

Blikle, A.J. [1966] *Investigations in the theory of addressless computers* (**J** 0014) Bull Acad Pol Sci, Ser Math Astron Phys 14*203-208
◇ B75 D10 ◇ rev Zbl 163.397 • id 01204

Blikle, A.J. [1970] *Functions definable by means of programs* (**J** 0014) Bull Acad Pol Sci, Ser Math Astron Phys 18*391-393
◇ B75 D20 ◇ rev MR 42 # 2708 Zbl 199.310 • id 01308

Blikle, A.J. [1971] *Algorithmically definable functions: a contribution towards the semantics of programming languages* (**X** 1034) PWN: Warsaw 56pp
◇ B75 D05 D20 ◇ rev MR 52 # 2267 Zbl 221 # 68045 • id 23505

Blikle, A.J. see Vol. I, V for further entries

Blokh, A.Sh. [1974] *Ueber die Kompliziertheit von Graph-Schemata (Russisch)* (**J** 0413) Izv Akad Nauk Belor SSR, Ser Fiz-Mat 1974/4*102-103
◇ B75 D15 ◇ rev MR 50 # 6203 Zbl 325 # 94032 • id 60552

Blokh, A.Sh. [1975] *Graph-Schemata und ihre Anwendung (Russian)* (**X** 1574) Vyssheyshaya Shkola: Minsk 304pp
◇ D98 ◇ rev Zbl 334 # 68008 • id 60553

Bloom, S.L. [1968] *A note on the arithmetical hierarchy* (**J** 0047) Notre Dame J Formal Log 9*89-91
◇ D55 ◇ rev MR 38 # 4313 Zbl 185.23 • id 01319

Bloom, S.L. [1969] *A semi-completeness theorem* (**J** 0047) Notre Dame J Formal Log 10*303-308
◇ D55 E15 ◇ rev MR 44 # 1566 Zbl 184.21 • id 01320

Bloom, S.L. [1970] *The hyperprojective hierarchy* (**J** 0068) Z Math Logik Grundlagen Math 16*149-164
◇ D55 D70 E15 ◇ rev MR 42 # 1659 Zbl 162.315 • id 01321

Bloom, S.L. [1971] see Amoroso, S.

Bloom, S.L. [1971] *Some remarks on uniform halting problems* (**J** 0068) Z Math Logik Grundlagen Math 17*281-284
◇ D10 D30 ◇ rev MR 47 # 3150 Zbl 227 # 02020 • id 01331

Bloom, S.L. & Patterson, D.B. [1981] *Easy solutions are hard to find* (**P** 2923) CAAP'81 Arbres en Algeb & Progr (6);1981 Genova 135-146
◇ D15 ◇ rev MR 83b:68034 Zbl 461 # 68046 • id 69211

Bloom, S.L. [1985] *Frontiers of one-letter languages* (**J** 0380) Acta Cybern (Szeged) 7*1-18
◇ D05 ◇ rev MR 86c:68043 • id 45399

Bloom, S.L. see Vol. I, II, III, VI for further entries

Bloshchitsyn, V.Ya. & Zakir'yanov, K.Kh. [1978] *Constructive abelian groups (Russian)* (**C** 3848) Algeb & Teor Chisel ('78) 18-25
◇ C57 C60 D45 ◇ rev Zbl 485 # 03020 • id 36848

Blue, A.H. [1930] *On the structure of sets of points of classes one, two and three* (**J** 0043) Math Ann 102*624-632
◇ D55 E15 ◇ rev FdM 55.55 • id 39319

Blum, E.K. [1965] *Enumeration of recursive sets by Turing machine* (**J** 0068) Z Math Logik Grundlagen Math 11*197-201
◇ D10 D20 ◇ rev MR 33 # 3918 Zbl 144.2 • id 01338

Blum, E.K. & Lynch, N.A. [1979] *Relative complexity of operations on numeric and bit-string algebras* (**J** 0041) Math Syst Theory 13*187-207
◇ D15 D20 ◇ rev MR 83b:68052 Zbl 469 # 68046 • id 55189

Blum, L. & Blum, M. [1973] *Inductive inference: a recursion theoretic approach* (**P** 3062) IEEE Symp Switch & Automata Th (14);1973 Iowa City 200-208
◇ B48 D15 D80 ◇ rev MR 55 # 5411 • id 71122

Blum, L. & Blum, M. [1975] *Toward a mathematical theory of inductive inference* (**J** 0194) Inform & Control 28*125-155
◇ B48 D10 D15 ◇ rev MR 52 # 16109 Zbl 375 # 02028 • id 51605

Blum, L. see Vol. III for further entries

Blum, M. [1965] see Arbib, M.A.

Blum, M. [1966] *Recursive function theory and speed of computation* (**J** 0018) Canad Math Bull 9*745-750
◇ D15 D20 ◇ rev Zbl 158.249 JSL 37.199 • id 01340

Blum, M. [1967] *A machine-independent theory of the complexity of recursive functions* (**J** 0037) ACM J 14*322-336
◇ D15 ◇ rev MR 38 # 4213 Zbl 155.15 JSL 34.657 • id 01341

Blum, M. [1967] *On the size of machines* (**J** 0194) Inform & Control 11*257-265
◇ D15 ◇ rev MR 38 # 1955 Zbl 165.20 JSL 37.199 • id 01517

BLUM, M. & FISCHER, P.C. & HARTMANIS, J. [1968] *Tape reversal complexity hierarchies* (**P** 1900) IEEE Symp Switch & Automata Th (9);1968 Schenectady 373-382
◇ D10 D15 ◇ ID 33949

BLUM, M. [1971] *On effective procedures for speeding up algorithms* (**J** 0037) ACM J 18*290-305
• TRANSL [1974] (**C** 2319) Slozh Vychisl & Algor 127-149
◇ D15 ◇ REV MR 44 # 8063 Zbl 221 # 02016 • ID 01342

BLUM, M. [1973] see BLUM, L.

BLUM, M. & MARQUES, I. [1973] *On complexity properties of recursively enumerable sets* (**J** 0036) J Symb Logic 38*579-593
◇ D15 D25 ◇ REV MR 48 # 10782 Zbl 335 # 02024 • ID 01343

BLUM, M. & GILL III, J.T. [1973] *Some fruitful areas for research into complexity theory* (**P** 1584) Comput Compl - Courant Comput Sci (7);1971 New York 23-36
◇ D15 ◇ ID 27053

BLUM, M. & GILL III, J.T. [1974] *On almost everywhere complex recursive functions* (**J** 0037) ACM J 21*425-435
◇ D15 D20 ◇ REV MR 56 # 15381 Zbl 315 # 68038 • ID 62018

BLUM, M. [1975] see BLUM, L.

BLUM, M. & SAKODA, W.J. [1977] *On the capability of finite automata in 2 and 3 dimensional space* (**P** 3572) IEEE Symp Found of Comput Sci (18);1977 Providence 147-161
◇ D05 ◇ REV MR 58 # 32071 • ID 80745

BLUM, M. & CHANDRA, A.K. & WEGMAN, M.N. [1980] *Equivalence of free boolean graphs can be decided probabilistically* (**J** 0232) Inform Process Lett 10*80-82
◇ D15 ◇ REV MR 82d:68022 Zbl 444 # 68059 • ID 69212

BLUM, N. [1982] *On the power of chain rules in context free grammars* (**J** 1431) Acta Inf 17*425-433
◇ D05 ◇ REV MR 84a:68075 Zbl 493 # 68083 • ID 38811

BLUM, N. [1983] *A note on the "parallel computation thesis"* (**J** 0232) Inform Process Lett 17*203-205
◇ D15 ◇ REV MR 85d:68023 Zbl 523 # 68033 • ID 39019

BOASSON, L. & COURCELLE, B. & NIVAT, M. [1977] *A new complexity measure for languages* (**P** 3238) Conf Theoret Comput Sci;1977 Waterloo ON 130-138
◇ D05 D15 ◇ REV MR 58 # 13914 Zbl 431 # 68077 • ID 69222

BOASSON, L. & HORVATH, S. [1978] *On languages satisfying Ogden's lemma* (**J** 3441) RAIRO Inform Theor 12*201-202
◇ D05 ◇ REV MR 80j:68057 Zbl 387 # 68054 • ID 69217

BOASSON, L. [1978] *Un critere de rationalite des langages algebriques* (**P** 3394) Lang Algeb (1);1973 Bonascre 85-104
◇ D05 ◇ REV MR 80e:68181 Zbl 399 # 68072 • ID 69219

BOASSON, L. [1979] *Context-free sets of infinite words* (**P** 3488) Theor Comput Sci (4);1979 Aachen 1-9
◇ D05 ◇ REV MR 81k:68052 Zbl 402 # 68054 • ID 69216

BOASSON, L. [1979] see AUTEBERT, J.-M.

BOASSON, L. [1979] *Un langage algebrique particulier* (**J** 3441) RAIRO Inform Theor 13*203-215
◇ D05 ◇ REV MR 81g:68104 Zbl 424 # 68042 • ID 69215

BOASSON, L. & NIVAT, M. [1980] *Adherences of languages* (**J** 0119) J Comp Syst Sci 20*285-309
◇ D05 ◇ REV MR 81k:68054 Zbl 471 # 68052 • ID 69220

BOASSON, L. [1980] *Derivations et reductions dans les grammaires algebriques* (**P** 2904) Automata, Lang & Progr (7);1980 Noordwijkerhout 109-118
◇ D05 ◇ REV MR 81k:68053 Zbl 455 # 68041 • ID 69218

BOASSON, L. [1980] see AUTEBERT, J.-M.

BOASSON, L. & NIVAT, M. [1981] *Centers of languages* (**P** 3475) Theor Comput Sci (5);1981 Karlsruhe 104*245-251
◇ D05 ◇ REV Zbl 457 # 68082 • ID 69213

BOASSON, L. [1981] see AUTEBERT, J.-M.

BOASSON, L. [1981] *Some applications of CFL's over infinite alphabets* (**P** 3475) Theor Comput Sci (5);1981 Karlsruhe 104*146-151
◇ D05 ◇ REV Zbl 457 # 68083 • ID 69214

BOASSON, L. & COURCELLE, B. & NIVAT, M. [1981] *The rational index: a complexity measure for languages* (**J** 1428) SIAM J Comp 10*284-296
◇ D05 D15 ◇ REV MR 83b:68078 Zbl 469 # 68083 • ID 69221

BOASSON, L. [1982] see AUTEBERT, J.-M.

BOCHERNIKOV, V.YA. & KOSOVSKIJ, N.K. [1984] *A calculus of equations of program in a model of programming language (Russian) (English summary)* (**J** 0085) Vest Ser Mat Mekh Astron, Univ Leningrad 1984/4*5-9
◇ D80 ◇ REV MR 86g:68131 Zbl 571 # 68006 • ID 49271

BOCHMANN, D. & SIMON, F.U. [1979] *Automatengraphen und ihre analytische Behandlung* (**J** 2878) Wiss Z Tech Hochsch Karl-Marx-Stadt 21*675-684
◇ D05 ◇ REV Zbl 449 # 94033 • ID 69224

BOCHMANN, D. [1980] *Zur Dynamik und Steuerung hierarchischer Automatennetze* (**J** 2878) Wiss Z Tech Hochsch Karl-Marx-Stadt 22*249-256
◇ D05 ◇ REV Zbl 449 # 94034 • ID 69223

BOCHMANN, G.V. [1979] *Semantic equivalence of covering attribute grammars* (**J** 0435) Int J Comput & Inf Sci 8*523-539
◇ D05 ◇ REV MR 81c:68070 Zbl 426 # 68077 • ID 69225

BODNARCHUK, V.G. [1963] *Systems of equations in the algebra of events (Russian)* (**J** 0199) Zh Vychisl Mat i Mat Fiz 3*1077-1088
• TRANSL [1963] (**J** 1049) USSR Comput Math & Math Phys 3*1470-1487
◇ D05 ◇ REV MR 29 # 15 Zbl 148.250 • ID 01551

BOEHLING, K.H. [1963] *Netzwerke -Schaltwerke - Automaten. Ein Ueberblick ueber die synchrone Theorie* (**P** 1612) Schaltkreis & -werk Th (2);1961 Saarbruecken 82-106
◇ B70 D05 ◇ REV Zbl 126.129 • ID 27838

BOEHLING, K.H. [1967] *Ueber eine Darstellungstheorie sequentieller Automaten* (**P** 1671) Colloq Automatenth (3);1965 Hannover 1-25
◇ D05 ◇ REV Zbl 174.30 • ID 29428

BOEHLING, K.H. & INDERMARK, K. [1970] *Endliche Automaten I* (**X** 0876) Bibl Inst: Mannheim 100pp
◇ D05 ◇ REV Zbl 193.327 • REM Part II 1970 by Boehling,K.H. & Schuett,D. • ID 23508

BOEHLING, K.H. & SCHUETT, D. [1970] *Endliche Automaten II* (**X** 0876) Bibl Inst: Mannheim 112pp
◇ D05 ◇ REV Zbl 211.21 • REM Part I 1970 by Boehling,K.H. & Indermark,K. • ID 23510

Author Index

BOEHLING, K.H. & INDERMARK, K. (EDS.) [1973] *Erste Fachtagung der Gesellschaft fuer Informatik ueber Automatentheorie und formale Sprachen (Bonn, 9. - 12. Juli 1973)* (**X** 0811) Springer: Heidelberg & New York vii + 322pp
◇ D05 D97 ◇ REV MR 48 # 3285 • ID 80484

BOEHLING, K.H. & BRAUNMUEHL VON, B. [1974] *Komplexitaet bei Turingsmaschinen* (**X** 0876) Bibl Inst: Mannheim vii + 316pp
◇ D10 D15 D20 ◇ REV MR 53 # 7117 Zbl 343 # 02024 • ID 22965

BOEHM, C. [1962] *Macchine a indirizzi, dotate di un numero minimo di instruzioni* (**J** 0149) Atti Accad Naz Lincei Fis Mat Nat, Ser 8 32*923-930
◇ D10 ◇ REV MR 27 # 2414 Zbl 116.8 • ID 01365

BOEHM, C. & JACOPINI, G. [1962] *Nuovo tecniche di programmazione semplificanti la sintesi di macchine universali di Turing* (**J** 0149) Atti Accad Naz Lincei Fis Mat Nat, Ser 8 32*913-922
◇ B75 D10 ◇ REV MR 27 # 2413 Zbl 116.8 • ID 01363

BOEHM, C. [1964] *On a family of Turing machines and the related programming language* (**J** 0201) ICC Bull 3*185-194
◇ B75 D10 ◇ REV MR 30 # 3010 JSL 31.140 • ID 01558

BOEHM, C. & JACOPINI, G. [1966] *Flow diagrams, Turing machines and languages with only two formation rules* (**J** 0212) ACM Commun 9*366-371
◇ B75 D05 D10 ◇ ID 41121

BOEHM, C. [1978] see AUSIELLO, G.

BOEHM, C. see Vol. I, VI for further entries

BOERGER, E. [1974] *Σ_3-completude de l'ensemble des types de reduction* (**J** 0079) Logique & Anal, NS 17*89-94
◇ B20 D35 D55 ◇ REV MR 55 # 2539 Zbl 294 # 02018 • ID 28158

BOERGER, E. [1974] *Beitrag zur Reduktion des Entscheidungsproblems auf Klassen von Hornformeln mit kurzen Alternationen* (**J** 0009) Arch Math Logik Grundlagenforsch 16*67-84
◇ B20 B25 D35 ◇ REV MR 49 # 10535 Zbl 277 # 02009 • ID 01368

BOERGER, E. [1974] *Die rekursive Unloesbarkeit des zehnten Hilbertschen Problems* (**C** 4079) Mal'tsev: Algor & Rek Funkt 307-320
◇ D35 ◇ REM Anhang zur dt. Uebersetzung • ID 37198

BOERGER, E. [1975] *Recursively unsolvable algorithmic problems and related questions reexamined* (**P** 1442) ⊧ ISILC Logic Conf;1974 Kiel 10-24
◇ D03 D10 D20 ◇ REV MR 58 # 10355 Zbl 333 # 02040 • ID 27207

BOERGER, E. & HEIDLER, K. [1976] *Die m-Grade logischer Entscheidungsprobleme* (**J** 0009) Arch Math Logik Grundlagenforsch 17*105-111
◇ D25 D35 D55 ◇ REV MR 57 # 2895 Zbl 362 # 02025 • ID 03928

BOERGER, E. [1976] *On the construction of simple first-order formulae without recursive models* (**P** 1619) Coloq Log Simb;1975 Madrid 7-24
◇ C57 D25 ◇ REV MR 56 # 8348 Zbl 357 # 02010 • ID 50358

BOERGER, E. [1976] *Ueber einige Interpretationen von Registermaschinen mit Anwendungen auf Entscheidungsprobleme in der Logik, der Algorithmentheorie und der Theorie formaler Sprachen* (**P** 3198) Incont Compl Calc, Cod & Ling Form;1975 Napoli 27-46
◇ B25 D10 ◇ REV Zbl 411 # 68045 • ID 52927

BOERGER, E. [1978] *Bemerkung zu Gurevich's Arbeit ueber das Entscheidungsproblem fuer Standardklassen* (**J** 0009) Arch Math Logik Grundlagenforsch 19*111-114
◇ B20 D10 D35 ◇ REV MR 80a:03054b Zbl 402 # 03019 • ID 29169

BOERGER, E. [1978] *Ein einfacher Beweis fuer die Unentscheidbarkeit der klassischen Praedikatenlogik* (**J** 0160) Math-Phys Sem-ber, NS 25*290-299
◇ B10 D03 D10 D35 ◇ REV MR 80c:03001 Zbl 399 # 03010 • ID 52807

BOERGER, E. [1979] *A new general approach to the theory of the many-one equivalence of decision problems for algorithmic systems* (**J** 0068) Z Math Logik Grundlagen Math 25*135-162
◇ B25 D03 D05 D10 D25 D30 D40 ◇ REV MR 80f:03045 Zbl 429 # 03016 • ID 53847

BOERGER, E. & KLEINE BUENING, H. [1979] *The reachability problem for Petri nets and decision problems for Skolem arithmetic* (**P** 2615) Scand Logic Symp (5);1979 Aalborg 59-96
◇ B25 D15 D35 D80 ◇ REV MR 82b:03079 Zbl 453 # 03013 • ID 54143

BOERGER, E. [1979] see AANDERAA, S.O.

BOERGER, E. & KLEINE BUENING, H. [1980] *The r.e. complexity of decision problems for commutative semi-Thue systems with recursive rule set* (**J** 0068) Z Math Logik Grundlagen Math 26*459-469
◇ D03 D25 ◇ REV MR 82b:03074 Zbl 499 # 03025 • ID 71243

BOERGER, E. & KLEINE BUENING, H. [1980] *The reachability problem for Petri nets and decision problems for Skolem arithmetic* (**J** 1426) Theor Comput Sci 11*123-143
◇ B25 D15 D35 D80 ◇ REV MR 81h:68034 Zbl 453 # 03012 • ID 80766

BOERGER, E. [1981] *Logical description of computation processes* (**P** 3165) FCT'81 Fund of Comput Th;1981 Szeged 410-424
◇ B25 D05 D15 D35 ◇ REV MR 83e:03066 Zbl 467 # 03037 • ID 55035

BOERGER, E. [1981] see AANDERAA, S.O.

BOERGER, E. [1982] see AANDERAA, S.O.

BOERGER, E. [1982] *Note on bounded diophantine representation of subrecursive sets* (**S** 3125) Ber, Abt Inf, Univ Dortmund 11pp
◇ D20 ◇ REV Zbl 531 # 03026 • ID 37681

BOERGER, E. [1983] *From decision problems to problems of complexity* (**P** 3091) Conv Int Storica Logica;1982 San Gimignano 211-215
◇ B25 D15 ◇ ID 40032

BOERGER, E. [1983] *Undecidability versus degree complexity of decision problems for formal grammars* (**P** 4083) GTI-Worksh (1);1983 Paderborn 44-55
◇ D05 D10 D25 ◇ REV Zbl 533 # 03021 • ID 40051

BOERGER, E. & OBERSCHELP, W. & RICHTER, M.M. & SCHINZEL, B. & THOMAS, WOLFGANG (EDS.) [1984] *Computation and proof theory* (**S** 3301) Lect Notes Math 1104*viii+475pp
◇ B97 D97 F97 ◇ REV MR 85k:03002b Zbl 547#00036
• REM Proc. Log. Coll., Aachen 1983, Vol.II. Vol.I 1984 by Mueller,G.H. • ID 40062

BOERGER, E. [1984] *Decision problems in predicate logic* (**P** 3710) Logic Colloq;1982 Firenze 263-301
◇ B20 B25 D35 ◇ REV MR 85m:03008 Zbl 556#03012
• ID 40047

BOERGER, E. & HASENJAEGER, G. & ROEDDING, D. (EDS.) [1984] *Logic and machines: decision problems and complexity* (**P** 2342) Symp Rek Kombin;1983 Muenster 171*vi+456pp
◇ B35 D97 ◇ REV MR 85k:68004 Zbl 538#00005
• ID 40072

BOERGER, E. [1984] *Spektralproblem and completeness of logical decision problems* (**P** 2342) Symp Rek Kombin;1983 Muenster 333-356
◇ B25 C13 D15 ◇ REV Zbl 564#03008 • ID 40054

BOERGER, E. [1985] *Berechenbarkeit, Komplexitaet, Logik. Eine Einfuehrung in Algorithmen, Sprachen und Kalkuele unter besonderer Beruecksichtigung ihrer Komplexitaet* (**X** 0900) Vieweg: Wiesbaden xvii+469pp
◇ D98 ◇ ID 49279

BOERGER, E. [1985] *Current trends in theoretical computer science* (**P** 4084) Course on Comput Th;1984 Udine
◇ B75 D15 ◇ ID 40079

BOERGER, E. see Vol. III for further entries

BOFFA, M. [1974] *Hierarchie analytique et ensembles constructibles* (1111) Preprints, Manuscr., Techn. Reports etc. 10pp
◇ D35 D55 E15 E45 ◇ REM Semin. Theor. Descript. Ens., University Paris VII • ID 31022

BOFFA, M. [1975] *Introduction aux automates finis et aux langages reguliers* (**J** 0082) Bull Soc Math Belg 27*3-27
◇ D05 ◇ REV MR 57#11182 Zbl 362#94046 • ID 31021

BOFFA, M. see Vol. I, III, V, VI for further entries

BOGDAN, R.J. [1976] *A selected bibliography of local induction* (**C** 1702) Local Induction 329-336
◇ A10 D99 ◇ REV MR 58#27192 Zbl 319#02002
• ID 29662

BOGDAN, R.J. see Vol. II for further entries

BOGOMOLOV, A.M. & TVERDOKHLEBOV, V.A. [1978] *Synchronization of a deterministic finite automaton (Russian)* (**J** 0011) Avtom Telemekh 1978/10*200-202
• TRANSL [1979] (**J** 0010) Autom & Remote Control 39*1571-1573
◇ D05 ◇ REV MR 80e:68140 Zbl 417#68051 • ID 69226

BOKUT', L.A. [1966] *On a property of the groups of Boone (Russian)* (**J** 0003) Algebra i Logika 5/4*5-23
◇ D40 ◇ REV MR 36#5197 Zbl 189.11 JSL 33.470
• ID 01414

BOKUT', L.A. [1967] *On the Novikov groups (Russian) (English summary)* (**J** 0003) Algebra i Logika 6/1*25-38
◇ D40 ◇ REV MR 36#250 Zbl 165.318 JSL 33.623
• ID 01413

BOKUT', L.A. [1968] *Degrees of unsolvability of the conjugacy problem for finitely-presented groups (Russian)* (**J** 0003) Algebra i Logika 7/5*4-70,7/6*4-52
• TRANSL [1968] (**J** 0069) Algeb and Log 7*284-329,357-387
◇ D30 D40 ◇ REV MR 41#3574 Zbl 218#02036 JSL 27.609 • ID 01415

BOKUT', L.A. [1970] *Unsolvability of certain algorithmic problems in the class of associative rings (Russian)* (**J** 0003) Algebra i Logika 9*137-144
• TRANSL [1970] (**J** 0069) Algeb and Log 9*83-87
◇ D35 D40 ◇ REV MR 44#6492 Zbl 216.10 • ID 26247

BOKUT', L.A. [1972] *Unsolvability of the equality problem, and subalgebras of finitely presented Lie algebras (Russian)* (**J** 0216) Izv Akad Nauk SSSR, Ser Mat 36*1173-1219
• TRANSL [1972] (**J** 0448) Math of USSR, Izv 6*1153-1199
◇ D40 ◇ REV MR 48#8588 Zbl 252#02046 • ID 02021

BOKUT', L.A. [1972] *Unsolvability of the word problem for Lie algebras (Russian)* (**J** 0023) Dokl Akad Nauk SSSR 206*1277-1279
• TRANSL [1972] (**J** 0062) Sov Math, Dokl 13*1388-1391
◇ D40 ◇ REV MR 47#5071 Zbl 267#02032 • ID 60627

BOKUT', L.A. [1974] *Undecidability of certain algorithmic problems for Lie algebras (Russian)* (**J** 0003) Algebra i Logika 13*145-152,234
• TRANSL [1974] (**J** 0069) Algeb and Log 13*79-83
◇ D40 D80 ◇ REV MR 51#106 Zbl 305#02055 Zbl 356#02040 • ID 17982

BOKUT', L.A. [1980] *Decision problems for ring theory* (**P** 2634) Word Problems II;1976 Oxford 95*55-69
◇ D40 ◇ REV MR 81k:16001 Zbl 439#17001 • ID 56047

BOKUT', L.A. [1980] *Mal'cev's problem and groups with a normal form* (**P** 2634) Word Problems II;1976 Oxford 29-53
◇ D40 ◇ REV MR 82c:20064 Zbl 444#20029 • ID 80748

BOKUT', L.A. [1985] *A remark on the Borisov-Boone group (Russian)* (**J** 0092) Sib Mat Zh 26/5*43-46,204
• TRANSL [1985] (**J** 0475) Sib Math J 26*661-664
◇ D40 ◇ ID 49272

BOKUT', L.A. see Vol. III for further entries

BOLLMAN, D.A. [1967] *Formal nonassociative number theory* (**J** 0047) Notre Dame J Formal Log 8*9-16
◇ B28 D35 ◇ REV MR 39#5642 Zbl 183.13 • ID 01419

BOLLMAN, D.A. & TAPIA, M. [1972] *On the recursive unsolvability of the provability of the deduction theorem in partial propositional calculi* (**J** 0047) Notre Dame J Formal Log 13*124-128
◇ B20 D35 ◇ REV MR 46#3288 Zbl 227#02005
• ID 01420

BOLLMAN, D.A. & LAPLAZA, M.L. [1978] *Some decision problems for polynomial mappings* (**J** 1426) Theor Comput Sci 6*317-325
◇ D80 ◇ REV MR 58#16585 Zbl 383#03012 • ID 31628

BOLLMAN, D.A. see Vol. I, V, VI for further entries

BONDI, I.L. [1967] *The decidability or undecidability of certain formal logical calculi (Russian)* (**J** 0023) Dokl Akad Nauk SSSR 172*1001-1002
• TRANSL [1967] (**J** 0062) Sov Math, Dokl 8*205-206
◇ B25 D35 ◇ REV MR 35#1464 Zbl 162.21 • ID 01424

Bondi, I.L. [1971] *The solvability problem for certain formal-logical calculi (Russian)* (**S** 0208) Uch Zap, Ped Inst, Moskva 277*205-213
◇ B25 D35 ◇ REV MR 46 # 5100 • ID 02009

Bondi, I.L. [1973] *The decision problem for metric spaces (Russian)* (**J** 0031) Izv Vyssh Ucheb Zaved, Mat (Kazan) 1973/1(128)*24-27
◇ D35 ◇ REV MR 47 # 6464 Zbl 256 # 02023 • ID 02001

Bondi, I.L. [1973] *The decision problem for normed linear spaces and for Hilbert spaces (Russian)* (**J** 0031) Izv Vyssh Ucheb Zaved, Mat (Kazan) 1973/5(132)*3-10
◇ B25 C65 D35 ◇ REV MR 48 # 8218 Zbl 275 # 02044 • ID 01425

Book, R.V. & Even, S. & Greibach, S.A. & Ott, G. [1969] *Ambiguity in graphs and expressions* (**P** 1129) Princeton Conf Inform Sci & Syst (3);1969 Princeton 345-349
◇ D05 ◇ REV Zbl 294 # 94028 • ID 60637

Book, R.V. & Even, S. & Greibach, S.A. & Ott, G. [1971] *Ambiguity in graphs and expressions* (**J** 0187) IEEE Trans Comp C-20*149-153
◇ D05 ◇ REV MR 45 # 9833 • ID 02002

Book, R.V. & Ginsburg, S. [1972] *Multi-stack-counter languages* (**J** 0041) Math Syst Theory 6*37-48
◇ D05 ◇ REV MR 47 # 7956 Zbl 229 # 68030 • ID 32101

Book, R.V. [1974] *Comparing complexity classes* (**J** 0119) J Comp Syst Sci 9*213-229
◇ D10 D15 ◇ REV MR 51 # 2349 Zbl 331 # 02020 • ID 60638

Book, R.V. [1974] *On the structure of complexity classes* (**P** 1869) Automata, Lang & Progr (2);1974 Saarbruecken 437-445
◇ D10 D15 ◇ REV MR 55 # 1804 Zbl 352 # 68065 • ID 50066

Book, R.V. [1974] see Baker, B.S.

Book, R.V. [1974] *Tally languages and complexity classes* (**J** 0194) Inform & Control 26*186-193
◇ D10 D15 ◇ REV MR 49 # 10182 Zbl 287 # 68029 • ID 60639

Book, R.V. & Dobkin, D.P. & Selman, A.L. & Wrathall, C. [1977] *Inclusion complete tally languages and the Hartmanis-Berman conjecture* (**J** 0041) Math Syst Theory 11*1-8
◇ D05 ◇ REV MR 57 # 4612 Zbl 365 # 68044 • ID 31233

Book, R.V. & Greibach, S.A. & Wrathall, C. [1978] *Comparisons and reset machines* (**P** 1872) Automata, Lang & Progr (5);1978 Udine 113-124
◇ D10 D15 ◇ REV Zbl 388 # 03017 • ID 52271

Book, R.V. & Wrathall, C. [1978] *On languages accepted by time-bounded oracle machines* (**P** 4048) Allerton Conf Commun, Control & Comput (16);1978 Monticello 489-494
◇ D15 ◇ REV MR 84b:94002 Zbl 389 # 68029 • ID 46269

Book, R.V. [1978] *Simple representations of certain classes of languages* (**J** 0037) ACM J 25*23-31
◇ D15 D25 D55 ◇ REV MR 57 # 1974 Zbl 364 # 68073 • ID 50989

Book, R.V. [1979] *A remark on tally languages and complexity classes* (**J** 0194) Inform & Control 43*198-201
◇ D05 D15 ◇ REV MR 81h:68021 Zbl 441 # 68088 • ID 69228

Book, R.V. [1979] *Complexity classes of formal languages (preliminary report)* (**P** 2059) Math Founds of Comput Sci (8);1979 Olomouc 43-56
◇ D05 D15 ◇ REV MR 81i:68055 Zbl 413 # 68045 • ID 80759

Book, R.V. [1979] *On languages accepted by space-bounded oracle machines* (**J** 1431) Acta Inf 12*177-185
◇ D15 ◇ REV MR 80e:68200 Zbl 389 # 68029 Zbl 412 # 68077 • ID 69232

Book, R.V. [1979] *Polynomial space and transitive closure* (**J** 1428) SIAM J Comp 8*434-439
◇ D15 ◇ REV MR 80e:68098 Zbl 422 # 68014 • ID 53506

Book, R.V. & Brandenburg, F.-J. [1979] *Representing complexity classes by equality sets (preliminary report)* (**P** 1873) Automata, Lang & Progr (6);1979 Graz 71*49-57
◇ D15 ◇ REV Zbl 408 # 68036 • ID 69231

Book, R.V. & Greibach, S.A. & Wrathall, C. [1979] *Reset machines* (**J** 0119) J Comp Syst Sci 19*256-276
◇ D10 D15 ◇ REV MR 81e:68071 Zbl 427 # 03029 • ID 53716

Book, R.V. & Brandenburg, F.-J. [1980] *Equality sets and complexity classes* (**J** 1428) SIAM J Comp 9*729-743
◇ D15 ◇ REV MR 82h:68074 Zbl 446 # 68040 • ID 80761

Book, R.V. [1981] *Bounded query machines: on NP and PSPACE* (**J** 1426) Theor Comput Sci 15*27-39
◇ D15 ◇ REV MR 84g:68024a Zbl 473 # 68039 • ID 69227

Book, R.V. & Wrathall, C. [1981] *Bounded query machines: on NP( ) and NPQUERY( )* (**J** 1426) Theor Comput Sci 15*41-50
◇ D15 ◇ REV MR 84g:68024b Zbl 473 # 68040 • ID 69230

Book, R.V. [1981] *NTS grammars and Church-Rosser systems* (**J** 0232) Inform Process Lett 13*73-76
◇ D03 D05 ◇ REV MR 83d:68071 Zbl 476 # 68053 • ID 55592

Book, R.V. & O'Dunlaing, C. [1981] *Testing for the Church-Rosser property* (**J** 1426) Theor Comput Sci 16*223-229
◇ D03 ◇ REV MR 83h:03053 Zbl 479 # 68035 • ID 55698

Book, R.V. [1981] *The undecidability of a word problem: on a conjecture of Strong, Maggiolo-Schettini and Rosen* (**J** 0232) Inform Process Lett 12*121-122
◇ D03 D40 ◇ REV MR 82h:03035 Zbl 457 # 03036 • ID 54361

Book, R.V. & O'Dunlaing, C. [1981] *Thue congruences and the Church-Rosser property* (**J** 0136) Semigroup Forum 22*367-379
◇ D03 ◇ REV MR 83b:03046 Zbl 482 # 03017 • ID 35099

Book, R.V. & Wrathall, C. [1982] *A note on complete sets and transitive closure* (**J** 0041) Math Syst Theory 15*311-313
◇ D15 ◇ REV MR 84i:68061 Zbl 495 # 68039 • ID 40160

Book, R.V. [1982] *Confluent and other types of Thue systems* (**J** 0037) ACM J 29*171-182
◇ D03 ◇ REV MR 84c:03076 Zbl 478 # 68032 • ID 55656

Book, R.V. & Jantzen, M. & Wrathall, C. [1982] *Monadic Thue systems* (**J** 1426) Theor Comput Sci 19*231-251
◇ D03 D05 ◇ REV MR 84d:03048 Zbl 488 # 03020 • ID 34084

Book, R.V. & Wilson, Christopher B. & Xu, Meirui [1982] *Relativizing time, space, and time-space* (**J** 1428) SIAM J Comp 11*571-581
◇ D15 ◇ REV MR 84c:68028 Zbl 487 # 68038 • ID 39606

Book, R.V. [1982] *When is a monoid a group? The Church-Rosser case is tractable* (**J** 1426) Theor Comput Sci 18*325-331
◇ D03 D15 D40 ◇ REV MR 84b:20067 Zbl 489 # 68021 • ID 36598

Book, R.V. [1983] *A note on special Thue systems with a single defining relation* (**J** 0041) Math Syst Theory 16*57-60
◇ D03 ◇ REV MR 85a:03052 Zbl 505 # 03019 • ID 34802

Book, R.V. & Selman, A.L. & Xu, Meirui [1983] *Positive relativizations of complexity classes* (**J** 1428) SIAM J Comp 12*565-579
◇ D15 ◇ REV MR 85a:68055 Zbl 551 # 68043 • ID 43969

Book, R.V. & Doner, J.E. & Xu, Meirui [1983] *Refining nondeterminism in relativizations of complexity classes* (**J** 0037) ACM J 30*677-685
◇ D15 ◇ REV MR 84j:03083 • ID 34675

Book, R.V. [1983] *Thue systems and the Church-Rosser property: replacement systems, specification of formal languages, and presentations of monoids* (**P** 4384) Combin on Words:1982 Waterloo 1-38
◇ D03 ◇ REV Zbl 563 # 68062 • ID 47471

Book, R.V. & Squier, C.C. [1984] *Almost all one-rule Thue systems have decidable word problems* (**J** 0193) Discr Math 49*237-240
◇ D03 ◇ REV MR 85e:20045 Zbl 563 # 03020 • ID 39931

Book, R.V. [1984] *Homogeneous Thue systems and the Church-Rosser property* (**J** 0193) Discr Math 48*137-145
◇ D03 ◇ REV MR 85f:68042 Zbl 546 # 03019 • ID 39967

Book, R.V. & Schoening, U. [1984] *Immunity, relativizations, and nondeterminism* (**J** 1428) SIAM J Comp 13*329-337
◇ D15 ◇ REV MR 85j:68044 Zbl 558 # 68039 • ID 46256

Book, R.V. [1984] see Avenhaus, J.

Book, R.V. & Long, T.J. & Selman, A.L. [1984] *Quantitative relativizations of complexity classes* (**J** 1428) SIAM J Comp 13*461-487
◇ D15 ◇ ID 45843

Book, R.V. [1984] *Relativizations of complexity classes* (**P** 3621) Frege Konferenz (2);1984 Schwerin 296-302
◇ D15 ◇ REV MR 86e:68042 Zbl 556 # 03031 • ID 46152

Book, R.V. [1984] see Balcazar, J.L.

Book, R.V. & Otto, F. [1985] *Cancellation rules and extended word problems* (**J** 0232) Inform Process Lett 20*5-11
◇ D03 D15 ◇ REV Zbl 561 # 68030 • ID 45104

Book, R.V. & Long, T.J. & Selman, A.L. [1985] *Qualitative relativizations of complexity classes* (**J** 0119) J Comp Syst Sci 30*395-413
◇ D15 ◇ REV Zbl 569 # 03016 • ID 49234

Book, R.V. [1985] *Thue systems as rewriting systems* (**P** 4244) Rewriting Techn & Appl (1);1985 Dijon 63-94
◇ D03 ◇ ID 49759

Boolos, G. & Putnam, H. [1968] *Degrees of unsolvability of constructible sets of integers* (**J** 0036) J Symb Logic 33*497-513
◇ D30 D55 E45 ◇ REV MR 39 # 1331 Zbl 188.327 JSL 38.527 • ID 01432

Boolos, G. [1974] *Arithmetical functions and minimalisation* (**J** 0068) Z Math Logik Grundlagen Math 20*353-354
◇ D55 ◇ REV MR 51 # 10054 Zbl 298 # 02031 • ID 03932

Boolos, G. & Jeffrey, R.C. [1974] *Computability and logic* (**X** 0805) Cambridge Univ Pr: Cambridge, GB x+262pp
◇ B98 D98 ◇ REV MR 49 # 7120 MR 82d:03001 Zbl 298 # 02003 JSL 42.585 • ID 03933

Boolos, G. see Vol. I, II, III, V, VI for further entries

Boone, W.W. [1954] *Certain simple unsolvable problems of group theory I,II* (**J** 0028) Indag Math 16*231-237,492-497
◇ D40 ◇ REV MR 16.564 Zbl 55.6 Zbl 57.17 JSL 22.372 JSL 22.373 • REM Parts III,IV 1955 • ID 33578

Boone, W.W. [1955] *Certain simple unsolvable problems of group theory III,IV* (**J** 0028) Indag Math 17*252-256,571-577
◇ D40 ◇ REV MR 16.564 Zbl 67.257 JSL 22.372 • REM Parts I,II 1954. Parts V,VI 1957 • ID 33579

Boone, W.W. [1957] *Certain simple unsolvable problems of group theory V,VI* (**J** 0028) Indag Math 19*22-27,227-232
◇ D40 ◇ REV MR 20 # 5231 Zbl 67.257 JSL 22.373 • REM Parts III,IV 1955 • ID 33580

Boone, W.W. [1958] *An analysis of Turing's "The word problem in semigroups with cancellation"* (**J** 0120) Ann of Math, Ser 2 67*195-202
◇ D40 ◇ REV MR 19.1158 Zbl 84.10 JSL 24.239 • ID 01439

Boone, W.W. [1958] *The word problem* (**J** 0054) Proc Nat Acad Sci USA 44*1061-1065
◇ D40 ◇ REV MR 21 # 80 Zbl 86.247 JSL 27.241 • ID 01438

Boone, W.W. [1959] see Baumslag, G.

Boone, W.W. [1959] *The word problem* (**J** 0120) Ann of Math, Ser 2 70*207-265
◇ D40 ◇ REV MR 31 # 3485 Zbl 102.9 Zbl 86.247 JSL 27.238 • ID 01440

Boone, W.W. [1962] *Partial results regarding word problems and recursively enumerable degrees of unsolvability* (**J** 0015) Bull Amer Math Soc 68*616-623
◇ D25 D40 ◇ REV MR 26 # 3761 Zbl 118.252 JSL 28.292 • ID 01444

Boone, W.W. [1965] *Finitely presented group whose word problem has the same degree as that of an arbitrarily given Thue system (application of methods of Britton)* (**J** 0054) Proc Nat Acad Sci USA 53*265-269
◇ D03 D30 D40 ◇ REV MR 33 # 2546 Zbl 173.14 JSL 33.296 • ID 01445

Boone, W.W. & Rogers Jr., H. [1966] *On a problem of J.H.C. Whitehead and a problem of Alonzo Church* (**J** 0132) Math Scand 19*185-192
◇ D25 D30 D40 ◇ REV MR 35 # 1465 Zbl 166.265 JSL 34.506 • ID 01447

Author Index

BOONE, W.W. [1966] *Word problems and recursively enumerable degrees of unsolvability. A first paper on Thue systems* (**J** 0120) Ann of Math, Ser 2 83*520-571
◇ D03 D25 D40 ◇ REV MR 34 # 1381 JSL 33.296
• ID 01449

BOONE, W.W. [1966] *Word problems and recursively enumerable degrees of unsolvability. A sequel on finitely presented groups* (**J** 0120) Ann of Math, Ser 2 84*49-84
◇ D25 D40 ◇ REV MR 34 # 1382 Zbl 173.13 JSL 33.296
• ID 33577

BOONE, W.W. [1968] *Decision problems about algebraic and logical systems as a whole and recursively enumerable degrees of unsolvability* (**P** 0608) Logic Colloq;1966 Hannover 13-33
◇ D25 D35 D40 ◇ REV MR 38 # 5898 Zbl 237 # 02010
• REM With a supplement by Jockusch Jr.,C.G. ibid. 33-36
• ID 01453

BOONE, W.W. & HAKEN, W. & POENARU, V. [1968] *On recursively unsolvable problems in topology and their classification* (**P** 0608) Logic Colloq;1966 Hannover 37-74
◇ D35 D80 ◇ REV MR 41 # 7695 Zbl 246 # 57015
• ID 01450

BOONE, W.W. & COLLINS, D.J. & MATIYASEVICH, YU.V. [1971] *Embeddings into semigroups with only a few defining relations* (**P** 0604) Scand Logic Symp (2);1970 Oslo 27-40
◇ D03 D40 ◇ REV MR 47 # 8266 Zbl 228 # 20029
• ID 01456

BOONE, W.W. [1971] *Word problems and recursively enumerable degrees of unsolvability. An emendation* (**J** 0120) Ann of Math, Ser 2 94*389-391
◇ D25 D40 ◇ REV MR 48 # 1903 Zbl 234 # 02031
JSL 39.184 • ID 01454

BOONE, W.W. & CANNONITO, F.B. & LYNDON, R.C. (EDS.) [1973] *Word problems* (**X** 0809) North Holland: Amsterdam xii + 646pp
◇ D40 D97 ◇ REV MR 49 # 10789 Zbl 254 # 00004
• ID 80480

BOONE, W.W. & HIGMAN, G. [1974] *An algebraic characterization of groups with soluble word problem* (**J** 0038) J Austral Math Soc 18*41-53
◇ D40 ◇ REV MR 50 # 10093 Zbl 303 # 20028 • ID 03935

BOONE, W.W. [1974] *Between logic and group theory* (**P** 0709) Int Conf Th of Groups (2);1973 Canberra 90-102
◇ C57 C60 D40 ◇ REV MR 50 # 7357 Zbl 298 # 20027
• ID 04158

BOONE, W.W. & COLLINS, D.J. [1974] *Embeddings into groups with only a few defining relations* (**J** 0038) J Austral Math Soc 18*1-7
◇ D40 ◇ REV MR 52 # 578 Zbl 303 # 20027 • ID 17984

BOONE, W.W. & HIGMAN, G. [1975] *An algebraic characterization of groups with soluble order problem* (**P** 0775) Logic Colloq;1973 Bristol 53-54
◇ D40 ◇ REV MR 52 # 3344 Zbl 318 # 20019 • ID 17689

BOONE, W.W. [1980] see ADYAN, S.I.

BOOS, W. [1975] *Lectures on large cardinal axioms* (**P** 1442) ⊧ ISILC Logic Conf;1974 Kiel 25-88
◇ D55 E05 E45 E55 E98 ◇ REV MR 54 # 2467
Zbl 319 # 02064 • ID 31627

BOOS, W. see Vol. III, V for further entries

BOOTH, K.S. [1978] see ADLEMAN, L.M.

BOOTH, K.S. [1978] *Isomorphism testing for graphs, semigroups, and finite automata are polynomially equivalent problems* (**J** 1428) SIAM J Comp 7*273-279
◇ D05 D15 ◇ REV MR 58 # 3669 Zbl 381 # 68042
• ID 69235

BOOTH, T.L. [1967] *Sequential machines and automata theory* (**X** 0827) Wiley & Sons: New York 592pp
◇ D05 ◇ REV Zbl 165.23 • ID 23512

BOOTH, T.L. [1974] *Design of minimal expected processing time finite-state transducers* (**P** 1691) Inform Processing (6);1974 Stockholm 652-656
◇ D05 ◇ REV MR 54 # 9896 Zbl 299 # 68055 • ID 60651

BOREL, E. [1905] see BAIRE, R.

BOREL, E. see Vol. II, V, VI for further entries

BORGA, M. & FREGUGLIA, P. & PALLADINO, D. (EDS.) [1980] *Rassegna di matematica. Logica matematica. Matematica applicata. Didattica della matematica* (**X** 2682) Casa Ed Tilgher: Genova 100pp
◇ B97 D97 ◇ REV MR 82b:03005 Zbl 459 # 00007
• ID 54440

BORGA, M. see Vol. I, II, VI for further entries

BORGIDA, A.T. [1983] *Some formal results about stratificational grammars and their relevance to linguistics* (**J** 0041) Math Syst Theory 16*29-56
◇ B65 D05 ◇ REV MR 84c:68062 Zbl 502 # 68021
• ID 39607

BORISOV, V.V. [1969] *Simple examples of groups with unsolvable word problem (Russian)* (**J** 0087) Mat Zametki (Akad Nauk SSSR) 6*521-522
• TRANSL [1969] (**J** 1044) Math Notes, Acad Sci USSR 6*768-775
◇ D40 ◇ REV MR 41 # 5471 Zbl 211.341 • ID 01474

BORM, A.E. & ROSIER, L.E. [1985] *A note on Parikh maps, abstract languages, and decision problems* (**J** 0191) Inform Sci 35*157-166
◇ D05 ◇ ID 47711

BORODIN, A. [1969] *Complexity classes of recursive functions and the existence of complexity gaps* (**P** 0671) ACM Symp Th of Comput (1);1969 Marina del Rey 67-78
◇ D15 ◇ ID 02028

BORODIN, A. & CONSTABLE, R.L. & HOPCROFT, J.E. [1969] *Dense and non-dense families of complexity classes* (**P** 0672) IEEE Symp Switch & Automata Th (10);1969 Waterloo 7-19
◇ D15 ◇ ID 02025

BORODIN, A. & CONSTABLE, R.L. [1971] *Subrecursive programming languages II. On program size* (**J** 0119) J Comp Syst Sci 5*315-334
◇ D20 ◇ REM Part I 1972 by Constable,R.L. Part III 1971 by Constable,R.L. • ID 14666

BORODIN, A. [1972] *Computational complexity and the existence of complexity gaps* (**J** 0037) ACM J 19*158-174 • ERR/ADD ibid 19*576
◇ D15 ◇ REV MR 47 # 9888 Zbl 261 # 68024 • ID 01486

BORODIN, A. [1973] *Computational complexity: theory and practice* (**C** 1106) Curr in Th of Computing 35-89
◇ D15 ◇ REV MR 54 # 14438 • ID 21192

BORODIN, A. [1977] *On relating time and space to size and depth* (J 1428) SIAM J Comp 6*733-744
◇ D10 D15 ◇ REV MR 57 # 1966 Zbl 366 # 68039
• ID 51147

BORODIN, A. [1982] *Structured vs. general models in computational complexity* (J 3370) Enseign Math, Ser 2 28*171-189
• REPR [1982] (P 3482) Logic & Algor (Specker);1980 Zuerich 47-65
◇ B75 D15 ◇ REV MR 84d:68036b Zbl 497 # 68024 Zbl 529 # 68023 • ID 38488

BORZOV, YU.V. [1974] *Holographische Mengen (Russisch)* (S 2587) Teor Algor & Progr (Riga) 1*217-220
◇ D20 D30 ◇ REV MR 58 # 21531 Zbl 325 # 02025
• ID 30440

BORZOV, YU.V. [1980] *Four theorems on holographic sets (Russian)* (J 0337) Mat Ezheg, Akad Nauk Latv SSR 24*185-191,261
◇ D20 D30 ◇ REV MR 82j:03047 Zbl 462 # 68007
• ID 54533

BOSISIO, A. [1983] *Una nota su proprieta topologiche e d'ordine e la ricorsivita generalizzata (English summary)* (J 2100) Boll Unione Mat Ital, VI Ser, B 2*835-855
◇ D65 G30 ◇ REV MR 85b:03080 Zbl 555 # 03015
• ID 40706

BOTUSHAROV, O.I. [1983] *A recursion theory characterization of inductive inference with additional information classes* (J 2774) Koezlem MTA Szam & Autom: Kutat Intez 147*101-109
◇ B48 D20 ◇ ID 45089

BOTUSHAROV, O.I. [1985] *On inductive inference with additional information (Bulgarian summary)* (P 4391) Mat & Mat Obrazov (14);1985 Sl"nchev Bryag 330-335
◇ B48 D20 ◇ ID 48979

BOUCHER, C. [1971] *Lecons sur la theorie des automates mathematiques* (X 0811) Springer: Heidelberg & New York viii + 193pp
◇ D05 D10 D15 D98 ◇ REV MR 43 # 4573 Zbl 221 # 94067 JSL 37.759 • ID 01498

BOUDOL, G. [1979] *A new recursion induction principle* (P 3488) Theor Comput Sci (4);1979 Aachen 79-90
◇ D20 ◇ REV MR 81j:68014 Zbl 405 # 68008 • ID 80773

BOWIE, G.L. [1973] *An argument against Church's thesis* (J 0301) J Phil 70*66-73
◇ A05 D20 F99 ◇ ID 27103

BOYD, R. & HENSEL, G. & PUTNAM, H. [1969] *A recursion-theoretic characterization of the ramified analytical hierarchy* (J 0064) Trans Amer Math Soc 141*37-62
◇ D30 D55 ◇ REV MR 39 # 4003 Zbl 207.12 • ID 01510

BOYER, R.S. & MOORE, J.S. [1979] *A computational logic* (X 0801) Academic Pr: New York xiv + 397pp
◇ B98 D70 ◇ REV MR 81d:68127 Zbl 448 # 68020
• ID 56665

BOYER, R.S. & MOORE, J.S. [1984] *A mechanical proof of the Turing completeness of pure LISP* (P 3084) Autom Theor Prov After 25 Yea;1983 Denver 133-157
◇ B35 D20 ◇ REV MR 85d:68005 • ID 45265

BOYER, R.S. & MOORE, J.S. [1984] *A mechanical proof of the unsolvability of the halting problem* (J 0037) ACM J 31*441-458
◇ D10 ◇ ID 49485

BOYER, R.S. see Vol. I for further entries

BOZOVIC, I.B. & BOZOVIC, N.B. [1974] *On some unrecognizable relations among groups* (J 0389) Math Balkanica 4*71-78
◇ D30 D40 ◇ REV MR 51 # 13045 Zbl 302 # 20020
• ID 17225

BOZOVIC, N.B. [1974] see BOZOVIC, I.B.

BOZOVIC, N.B. [1977] *On Markov properties of finitely presented groups* (J 0400) Publ Inst Math, NS (Belgrade) 21(35)*29-32
◇ C05 C50 D45 ◇ REV MR 57 # 422 Zbl 374 # 02010
• ID 51534

BOZOVIC, N.B. [1979] *A note on a generalization of some undecidability results in group theory* (J 0400) Publ Inst Math, NS (Belgrade) 26(40)*61-63
◇ D35 D40 ◇ REV MR 81g:20066 Zbl 435 # 20016
• ID 55815

BOZOVIC, N.B. [1979] *On some classes of unrecognizable properties of groups* (J 0400) Publ Inst Math, NS (Belgrade) 26(40)*65-68
◇ D35 D40 ◇ REV MR 81g:20067 Zbl 435 # 20017
• ID 55816

BOZOVIC, N.B. see Vol. III for further entries

BRACHO, F. [1980] *Continuously generated fixed points in* $P\omega$ (J 2095) Fund Inform, Ann Soc Math Pol, Ser 4 3*477-489
◇ B40 D70 ◇ REV MR 82c:06015 Zbl 458 # 68004
• ID 69241

BRADFORD, R. [1971] *Cardinal addition and the axiom of choice* (J 0007) Ann Math Logic 3*111-196
◇ D35 E10 E25 ◇ REV MR 46 # 38 Zbl 287 # 02041
• ID 01574

BRADY, A.H. [1966] *The conjectured highest scoring machines for Rado's* $\Sigma(k)$ *for the value* $k=4$ (J 4305) IEEE Trans Electr Comp EC-15*802-803
◇ D10 ◇ REV Zbl 156.19 JSL 40.617 • ID 27684

BRADY, A.H. [1983] *The determination of the value of Rado's noncomputable function* $\Sigma(k)$ *for four-state Turing machines* (J 0214) Math of Comp 40*647-665
◇ D10 ◇ REV MR 84d:03049 Zbl 518 # 03013 • ID 34085

BRADY, J.M. [1976] *A programming approach to some concepts and results in the theory of computation* (J 1193) Comput J (London) 19*234-237
◇ B75 D10 D20 ◇ REV MR 55 # 13853 Zbl 329 # 68046
• ID 60686

BRADY, J.M. [1977] *Hints on proofs by recursion induction* (J 1193) Comput J (London) 20*353-355
◇ B35 D20 ◇ REV Zbl 364 # 68015 • ID 50984

BRAFFORT, P. & HIRSCHBERG, D. (EDS.) [1963] *Computer programming and formal systems* (X 0809) North Holland: Amsterdam vii + 161pp
◇ B97 D05 D97 ◇ REV MR 26 # 3225 Zbl 108.134
• ID 23565

BRAFFORT, P. see Vol. I for further entries

BRAINERD, W.S. & KNODE, R.B. [1972] *Some criteria for determining recognizability of a set* (J 0194) Inform & Control 21*171-184
◇ D05 ◇ REV MR 48 # 7664 Zbl 246 # 68009 • ID 60692

Author Index

BRAINERD, W.S. & LANDWEBER, L.H. [1974] *Theory of computation* (**X** 0827) Wiley & Sons: New York xxi+336pp
◇ D98 ◇ REV MR 53#4590 Zbl 274#68001 • ID 23513

BRAKHAGE, H. (ED.) [1975] *Automata theory and formal languages* (**X** 0811) Springer: Heidelberg & New York VIII+292pp
◇ D05 D97 ◇ REV MR 52#9658 Zbl 302#00017 • ID 48631

BRAMHOFF, H. & JANTZEN, M. [1983] *Notions of computability by Petri nets* (**P** 3850) Appl & Th of Petri Nets (3);1982 Varenna 149-165
◇ B75 D20 D80 ◇ REV MR 85e:68073 Zbl 521#68042 • ID 37475

BRANDENBURG, F.-J. [1978] *Die Zusammenhangskomplexitaet von nichtkontextfreien Grammatiken* (**S** 3180) Inform Ber, Inst Inf, Univ Bonn 20*195pp
◇ D05 D15 ◇ REV Zbl 397#68072 • ID 69242

BRANDENBURG, F.-J. [1979] see BOOK, R.V.

BRANDENBURG, F.-J. [1980] see BOOK, R.V.

BRANDENBURG, F.-J. [1980] *Multiple equality sets and Post machines* (**J** 0119) J Comp Syst Sci 21*292-316
◇ D05 D15 ◇ REV MR 83g:68116 Zbl 452#68086 • ID 54125

BRANDENBURG, F.-J. [1981] *Three write heads are as good as k* (**J** 0041) Math Syst Theory 14*1-12
◇ D05 ◇ REV MR 82a:68099 Zbl 437#68029 • ID 80785

BRANDENBURG, F.-J. [1982] *Extended Chomsky-Schuetzenberger theorems* (**P** 3836) Automata, Lang & Progr (9);1982 Aarhus 83-93
◇ B65 D05 ◇ REV MR 84a:68076 Zbl 485#68067 • ID 38812

BRANDENBURG, F.-J. [1984] *A truly morphic characterization of recursive enumerable sets* (**P** 3658) Math Founds of Comput Sci (11);1984 Prague 205-213
◇ D25 ◇ REV Zbl 551#68061 • ID 43978

BRANDSTAEDT, A. & SAALFELD, D. [1977] *Eine Hierarchie beschraenkter Rueckkehrberechnungen auf on-line Turingmaschinen* (**J** 0129) Elektr Informationsverarbeitung & Kybern 13*571-583
◇ D10 D15 D20 ◇ REV MR 57#14596 Zbl 384#68048 • ID 52104

BRANDSTAEDT, A. [1978] *On a family of complexity measures on Turing machines, defined by predicates* (**J** 0129) Elektr Informationsverarbeitung & Kybern 14*331-339
◇ D10 D15 ◇ REV MR 80g:03037 Zbl 407#68054 • ID 56233

BRANDSTAEDT, A. [1978] *Ueber den Einfluss eines zusaetzlichen Einweg-Eingabebandes bei Entscheidungen und Aufzaehlungen mit beschraenktem Regime sowie beschraenkter Rueckkehr bzw. beschraenkter dualer Rueckkehr auf einbaendrigen Turing-Maschinen* (**S** 2829) Rostocker Math Kolloq 10*11-21
◇ D10 D15 ◇ REV MR 82a:68067 Zbl 417#68030 • ID 53282

BRANDSTAEDT, A. & WECHSUNG, G. [1979] *A relation between space, return and dual return complexities* (**J** 1426) Theor Comput Sci 9*127-140
◇ D15 ◇ REV MR 80e:68138 Zbl 409#68026 • ID 69999

BRANDSTAEDT, A. [1981] *Closure properties of certain families of formal languages with respect to a generalization of cyclic closure* (**J** 3441) RAIRO Inform Theor 15*233-252
◇ D05 ◇ REV MR 82h:68101 Zbl 467#68066 • ID 80786

BRANDSTAEDT, A. [1981] *Pushdown automata with restricted use of storage symbols* (**P** 3429) Math Founds of Comput Sci (10);1981 Strbske Pleso 234-241
◇ D10 D15 ◇ REV MR 83g:68073 Zbl 464#68052 • ID 69247

BRANDSTAEDT, A. & VOGEL, J. [1982] *Kompliziertheitsbeschraenkte Checking-Stack-Baender und Raum-Zeit-Probleme (English and Russian summaries)* (**J** 0192) Wiss Z Univ Jena, Math-Nat Reihe 31*569-577
◇ D10 D15 ◇ REV MR 84f:68032 • ID 39658

BRANDSTAEDT, A. [1982] *Reversal-bounded realtime acceptors with only two tapes* (**S** 3374) Forschergeb, Univ Jena N/82/2*19pp
◇ D05 ◇ REV Zbl 479#68053 • ID 69245

BRANDSTAEDT, A. & WAGNER, K. [1983] *Reversal-bounded and visit-bounded realtime computations* (**P** 3864) FCT'83 Found of Comput Th;1983 Borgholm 26-39
◇ D10 D15 ◇ REV MR 85i:68025 Zbl 532#68052 • ID 40925

BRANDSTAEDT, A. [1983] *Space classes, intersections of languages and bounded erasing homomorphisms* (**J** 3441) RAIRO Inform Theor 17*121-130
◇ D05 D15 ◇ REV MR 85a:68086 Zbl 479#68078 • ID 69246

BRAQUELAIRE, J.P. & COURCELLE, B. [1985] *The solution of two star-height problems for regular trees* (**P** 4622) Autom on Infinite Words;1984 Le Mont-Dore 108-117
◇ D05 ◇ ID 49446

BRAUDE, E.J. [1971] *Descriptive Baire and descriptive $\mathscr{X}$-analytic sets* (**J** 3240) Proc London Math Soc, Ser 3 23*409-427
◇ D55 E15 ◇ REV MR 47#41 Zbl 222#54048 • ID 01580

BRAUER, W. & INDERMARK, K. [1968] *Algorithmen, rekursive Funktionen und formale Sprachen* (**X** 0876) Bibl Inst: Mannheim v+155pp
◇ D05 D10 D20 D98 ◇ REV MR 49#4769 • ID 04159

BRAUER, W. [1970] *Automates topologiques et ensembles reconnaissables* (**C** 4415) Semin Schuetzenberger, Lentin Nivat 1969/70 18*24pp
◇ D05 ◇ REV MR 44#2613 Zbl 216.566 • ID 48002

BRAUER, W. [1970] *Ein elementarer Beweis des Kaskadezerlegungssatzes fuer endliche Automaten* (**P** 0577) Automatenth & Formale Sprachen;1969 Oberwolfach 11-21
◇ D05 ◇ REV MR 49#10460 Zbl 213.22 • ID 27991

BRAUN, S. [1932] *Sur une propriete caracteristique des ensembles $G_{\delta\sigma\delta}$* (**J** 0027) Fund Math 18*138-147
◇ D55 E15 ◇ REV Zbl 4.203 FdM 58.86 • ID 01585

BRAUN, S. [1937] *Sur l'uniformisation des ensembles fermes* (**J** 0027) Fund Math 28*214-218
◇ D55 E15 ◇ REV Zbl 16.55 FdM 63.178 • ID 01586

BRAUN, S. see Vol. V for further entries

BRAUNMUEHL VON, B. [1974] see BOEHLING, K.H.

Braunmuehl von, B. [1977] *Zwei-Zaehler-Automaten mit gekoppelten Bewegungen* (**X** 0817) Ges Math Datenverarbeit: Bonn 101pp
◇ D10 ◇ rev Zbl 366 # 02022 • id 51105

Braunmuehl von, B. & Verbeek, R. [1979] *Finite-change automata* (**P** 3488) Theor Comput Sci (4);1979 Aachen 91-100
◇ D05 D10 ◇ rev MR 81j:03062 Zbl 404 # 68081 • id 79754

Braunmuehl von, B. & Hotzel, E. [1979] *Supercounter machines* (**P** 1873) Automata, Lang & Progr (6);1979 Graz 71*58-72
◇ D10 D15 ◇ rev Zbl 413 # 68047 • id 69248

Braunmuehl von, B. & Verbeek, R. [1983] *Input-driven languages are recognized in* log *n space* (**P** 3864) FCT'83 Found of Comput Th;1983 Borgholm 40-51
◇ D15 ◇ rev MR 85h:68029 Zbl 549 # 68082 • id 47046

Breazu, V. & Stanasila, O. [1981] *A generalization of a theorem of Kleene and applications to structured programming (Romanian) (English summary)* (**J** 3130) Bul Inst Politeh Bucuresti, Ser Electroteh 43/4*7-9
◇ B75 D05 ◇ rev MR 83j:68095 Zbl 483 # 68066 • id 39912

Breazu, V. see Vol. V for further entries

Bredlau, C.E. [1967] *Regressive functions and combinatorial functions* (**J** 0047) Notre Dame J Formal Log 8*301-310
◇ D50 ◇ rev MR 39 # 3990 Zbl 224 # 02036 JSL 38.333 • id 01588

Bredlau, C.E. [1979] *Admissible sets and recursive equivalence types* (**J** 0047) Notre Dame J Formal Log 20*355-365
◇ D50 D60 ◇ rev MR 80f:03048 Zbl 314 # 02053 • id 52647

Breidbart, S. [1978] *On splitting recursive sets* (**J** 0119) J Comp Syst Sci 17*56-64
◇ D05 D15 D20 D25 ◇ rev MR 58 # 5121 Zbl 392 # 03028 • id 52395

Brejtbart, Yu.Ya. & Kozmidiadi, V.A. [1969] *Two subclasses of Turing machines reducible to finite automata (Russian)* (**J** 0023) Dokl Akad Nauk SSSR 187*9-10
• transl [1969] (**J** 0062) Sov Math, Dokl 10*763-764
◇ D05 D10 ◇ rev MR 42 # 7435 Zbl 193.331 • id 01590

Brejtbart, Yu.Ya. [1971] *On the automaton and "zone" complexity of the predicate "to be a k-th power of an integer" (Russian)* (**J** 0023) Dokl Akad Nauk SSSR 196*16-19
• transl [1971] (**J** 0062) Sov Math, Dokl 12*10-14
◇ D05 D15 ◇ rev MR 43 # 1764 Zbl 228 # 02025 • id 29125

Brejtbart, Yu.Ya. & Kozmidiadi, V.A. [1972] *Certain subclasses of Turing machines that are reducible to finite automata (Russian) (English summary)* (**J** 0040) Kibernetika, Akad Nauk Ukr SSR 1972/1*31-41
◇ D05 D10 ◇ rev MR 46 # 1498 Zbl 253 # 02030 • id 28934

Brejtbart, Yu.Ya. [1976] *Some bounds on the complexity of predicate recognition by finite automata* (**J** 0119) J Comp Syst Sci 12*336-349
◇ D05 D15 ◇ rev MR 56 # 17199 Zbl 349 # 02028 • id 60697

Brejtbart, Yu.Ya. & Lewis, F.D. [1979] *Combined complexity classes for finite functions* (**J** 3441) RAIRO Inform Theor 13*87-98
◇ D15 ◇ rev MR 81a:68044 Zbl 414 # 68023 • id 53094

Brejtbart, Yu.Ya. [1980] see Abramson, F.G.

Brejtbart, Yu.Ya. see Vol. I for further entries

Bremer, H. [1977] *Ein vollstaendiges Problem auf der Baummaschine* (**P** 3411) Theor Comput Sci (3);1977 Darmstadt 391-406
◇ D05 D10 D15 ◇ rev MR 58 # 3670 Zbl 392 # 68053 • id 52415

Bremer, H. see Vol. VI for further entries

Brent, R.P. & Goldschlager, L.M. [1982] *Some area-time tradeoffs for VLSI* (**J** 1428) SIAM J Comp 11*737-747
◇ B70 D05 D15 ◇ rev MR 83k:68024 Zbl 492 # 68041 • id 37739

Bressler, D.W. & Sion, M. [1964] *The current theory of analytic sets* (**J** 0017) Canad J Math 16*207-230
◇ D55 E15 ◇ rev MR 29 # 1153 Zbl 126.385 • id 01592

Bressler, D.W. see Vol. V for further entries

Bridges, D.S. [1985] see Bishop, E.A.

Bridges, D.S. see Vol. II, V, VI for further entries

Britton, J.L. [1956] *Solution of the word problem for certain types of groups I, II* (**J** 0217) Proc Glasgow Math Assoc 3*45-54,68-90
◇ D40 ◇ rev MR 20 # 3205 MR 23 # A2325 Zbl 79.252 JSL 32.126 • id 02047

Britton, J.L. [1958] *The word problem for groups* (**J** 3240) Proc London Math Soc, Ser 3 8*493-506
◇ D40 ◇ rev MR 23 # A2326 Zbl 92.8 JSL 29.205 • id 01607

Britton, J.L. [1963] *The word problem* (**J** 0120) Ann of Math, Ser 2 77*16-32
◇ D40 ◇ rev MR 29 # 5891 Zbl 112.258 JSL 29.205 • id 01608

Britton, J.L. [1973] *The existence of infinite Burnside groups* (**P** 0678) Word Probl: Decis & Burnside Probl in Group Th;1969 Irvine 67-348
◇ C60 D40 ◇ rev MR 53 # 10945 Zbl 264 # 20028 JSL 41.786 • id 44434

Britton, J.L. [1979] *Integer solutions of systems of quadratic equations* (**J** 0332) Math Proc Cambridge Phil Soc 86*385-389
◇ B25 D35 D80 ◇ rev MR 80i:10080 Zbl 418 # 10019 • id 53330

Brook, T. [1977] *Order and recursion in topoi* (**X** 2005) Austral Nat Univ: Canberra ii+226pp
◇ D80 E07 G30 ◇ rev MR 58 # 16828 Zbl 362 # 18002 • id 80793

Brown, Cynthia A. & Goldberg, Allen & Purdom Jr., P.W. [1982] *Average time analyses of simplified Davis-Putnam procedures* (**J** 0232) Inform Process Lett 15*72-75
• err/add ibid 16/4*213
◇ B35 D15 ◇ rev MR 84d:68038a MR 84d:86038b Zbl 529 # 68065 • id 39699

BROY, M. & FINANCE, J.P. & QUERE, A. & REMY, J.-L. & WIRSING, M. [1979] *Methodical solution of the problem of ascending subsequences of maximum length within a given sequence* (**J** 0232) Inform Process Lett 8*224-229
◇ B75 D20 ◇ REV MR 80e:68013 Zbl 406 # 68012 • ID 56161

BROY, M. [1981] see BERGSTRA, J.A.

BROY, M. [1985] *On the Herbrand-Kleene universe for nondeterministic computations* (**J** 1426) Theor Comput Sci 36*1-19
◇ D15 ◇ ID 47239

BROY, M. see Vol. I, II for further entries

BRUCKNER, A.M. [1969] *Some remarks on extreme derivates* (**J** 0018) Canad Math Bull 12*385-388
◇ D55 E15 ◇ REV MR 40 # 1546 • ID 01652

BRUCKNER, A.M. see Vol. V for further entries

BRUCKNER, L.K. [1979] *On the Garden-of-Eden problem for one-dimensional cellular automata* (**J** 0380) Acta Cybern (Szeged) 4*259-262
◇ D10 ◇ REV MR 80j:68039 Zbl 422 # 68021 • ID 69255

BRUDNO, A.A. [1974] *Topologische Entropie und Kompliziertheit nach A. N. Kolmogorov (Russisch)* (**J** 0067) Usp Mat Nauk 29/6(180)*157-158
◇ D15 D80 ◇ REV MR 57 # 4147 Zbl 311 # 94017 • ID 60729

BRUEGGEMANN, A. & PRIESE, L. & ROEDDING, D. & SCHAETZ, R. [1984] *Modular decomposition of automata* (**P** 2342) Symp Rek Kombin;1983 Muenster 198-236
◇ D05 ◇ REV Zbl 548 # 68058 • ID 41436

BRUIN DE, A. [1981] *On the existence of Cook semantics* (**X** 3205) Math Centr Amsterdam Afd Inf IW163/81*35pp
◇ B75 D20 D80 ◇ REV Zbl 455 # 68018 • ID 54303

BRUNNER, J. & WECHLER, W. [1976] *On the behaviour of R-fuzzy automata* (**P** 1401) Math Founds of Comput Sci (5);1976 Gdansk 45*210-215
◇ B75 D05 ◇ REV Zbl 338 # 94032 • ID 40847

BRUNNER, J. & WECHLER, W. [1977] *Zur Theorie der R-fuzzy-Automaten I* (**J** 0947) Wiss Z Tech Univ Dresden 26*647-652
◇ D05 E72 ◇ REV MR 58 # 20895 Zbl 431 # 68057 • REM Part II 1978 • ID 69260

BRUNNER, J. [1978] *Zur Theorie der R-fuzzy-Automaten II* (**J** 0947) Wiss Z Tech Univ Dresden 27*693-695
◇ D05 E72 ◇ REV MR 58 # 20896 Zbl 444 # 68043 • REM Part I 1977 by Brunner,J. & Wechler,W. • ID 69261

BRUNS, G. [1976] *Free ortholattices* (**J** 0017) Canad J Math 28*977-985
◇ D40 G10 ◇ REV MR 54 # 7335 Zbl 353 # 06001 • ID 25821

BRUNS, G. see Vol. V for further entries

BRUSS, A.R. & MEYER, A.R. [1978] *On time-space classes and their relation to the theory of real addition* (**P** 1740) ACM Symp Th of Comput (10);1978 San Diego 233-239
◇ B25 D15 ◇ REV MR 81i:68056 • ID 80801

BRUSS, A.R. & MEYER, A.R. [1980] *On time-space classes and their relation to the theory of real addition* (**J** 1426) Theor Comput Sci 11*59-69
◇ B25 D15 ◇ REV MR 82c:03061a Zbl 467 # 03038 • ID 55036

BRYLL, G. & MIKLOS, S. [1978] *The theory of concrete and abstract words (Polish) (English summary)* (**S** 1454) Zesz Nauk Wyz Szk Ped Mat, Opole 20*63-76
◇ D40 ◇ REV MR 80f:03069 Zbl 394 # 03042 • ID 52519

BRYLL, G. see Vol. I, II, III for further entries

BRZOZOWSKI, J.A. [1963] *Canonical regular expressions and minimal state graphs for definite events* (**P** 0674) Symp Math Th of Automata;1962 New York 529-561
◇ D05 ◇ REV MR 30 # 5903 Zbl 116.336 • ID 27609

BRZOZOWSKI, J.A. & POAGE, J.F. [1963] *On the construction of sequential machines from regular expressions* (**J** 4305) IEEE Trans Electr Comp EC-12*402-403
◇ D05 ◇ REV Zbl 125.78 • ID 02053

BRZOZOWSKI, J.A. [1964] *Derivatives of regular expressions* (**J** 0037) ACM J 11*481-494
◇ D05 ◇ REV MR 30 # 4638 JSL 36.152 • ID 01679

BRZOZOWSKI, J.A. & CULIK, K. & GABRIELIAN, A. [1971] *Classification of noncounting events* (**J** 0119) J Comp Syst Sci 5*41-53
◇ D05 D25 ◇ REV MR 44 # 3787 Zbl 241 # 94050 • ID 60739

BRZOZOWSKI, J.A. & SIMON, IMRE [1973] *Characterizations of locally testable events* (**J** 0193) Discr Math 4*243-271
◇ D05 ◇ REV MR 47 # 7948 Zbl 255 # 94032 • ID 60738

BRZOZOWSKI, J.A. [1980] *Developments in the theory of regular languages* (**P** 3385) Inform Processing (8);1980 Tokyo & Melbourne 29-40
◇ D05 ◇ REV MR 81k:68055 Zbl 443 # 68062 • ID 69266

BRZOZOWSKI, J.A. & FICH, F.E. [1980] *Languages of R-trivial monoids* (**J** 0119) J Comp Syst Sci 20*32-49
◇ D05 ◇ REV MR 81b:68089 Zbl 446 # 68066 • ID 69265

BRZOZOWSKI, J.A. & LEISS, E. [1980] *On equations for regular languages, finite automata, and sequential networks* (**J** 1426) Theor Comput Sci 10*19-35
◇ D05 ◇ REV MR 81c:68065 Zbl 415 # 68023 • ID 69264

BRZOZOWSKI, J.A. [1980] *Open problems about regular languages* (**P** 4266) Form Lang Th;1979 Santa Barbara 23-48
◇ D05 ◇ REV MR 84j:68001 Zbl 545 # 68062 • ID 47196

BUBIS, M.I. [1976] *On complete indexed sets (Russian)* (**J** 3937) Veroyat Met i Kibern (Kazan) 12-13*30-39
◇ D25 D55 ◇ REV MR 58 # 27399 Zbl 415 # 03033 • ID 53135

BUCHBERGER, B. [1974] *Certain decompositions of Goedel numbering and the semantics of programming languages* (**P** 1511) Int Symp Th Progr;1972 Novosibirsk 152-171
◇ B75 D20 D45 ◇ REV MR 54 # 9127 Zbl 278 # 68022 • ID 25827

BUCHBERGER, B. [1974] *On certain decompositions of Goedel numberings* (**J** 0009) Arch Math Logik Grundlagenforsch 16*85-96
◇ D15 D20 D45 ◇ REV MR 52 # 5395 Zbl 284 # 02022 • ID 01680

Buchberger, B. & Roider, B. [1978] *Input/output codings and transition functions in effective systems* (**J** 1743) Int J Gen Syst 4*201-209
◇ D05 D20 D45 ◇ rev MR 80d:68063 Zbl 383 # 68046 • id 52034

Buchberger, B. & Collins, G.E. & Loos, R. (eds.) [1982] *Computer-Algebra - Symbolic and algebraic computation* (**X** 0902) Springer: Wien vii+283pp
◇ D97 ◇ rev Zbl 491 # 00019 • id 46509

Buchberger, B. see Vol. I for further entries

Bucher, W. [1981] *A note on a problem in the theory of grammatical complexity* (**J** 1426) Theor Comput Sci 14*337-344
◇ D05 ◇ rev MR 83e:68099 Zbl 469 # 68082 • id 69270

Bucher, W. & Culik II, K. & Maurer, H.A. & Wotschke, D. [1981] *Concise description of finite languages* (**J** 1426) Theor Comput Sci 14*227-246
◇ D05 ◇ rev MR 82f:68075 Zbl 469 # 68081 • id 80807

Bucher, W. & Hagauer, J. [1983] *It is decidable whether a regular language is pure context-free* (**J** 1426) Theor Comput Sci 26*233-241
◇ B65 D05 ◇ rev Zbl 529 # 68042 • id 38491

Bucher, W. & Haussler, D. & Main, M.G. [1985] *Applications of an infinite squarefree co-CFL* (**P** 4628) Automata, Lang & Progr (12);1985 Nafplion 404-412
◇ B75 D05 ◇ id 49584

Buck, R.C. [1963] *Mathematical induction and recursive definitions* (**J** 0005) Amer Math Mon 70*128-135
◇ B28 D20 ◇ rev Zbl 113.3 • id 48047

Bucur, I. [1980] *Special chapters of algebra (Romanian)* (**X** 0871) Acad Rep Soc Romania: Bucharest 335pp
• transl [1984] (**X** 0835) Reidel: Dordrecht viii+406pp
◇ B80 C98 D98 ◇ rev MR 84c:14001 MR 86f:14001 • id 46245

Bucurescu, I. [1979] *On the state minimization of incompletely specified sequential machines* (**J** 3130) Bul Inst Politeh Bucuresti, Ser Electroteh 41/4*3-10
◇ D05 ◇ rev MR 81d:68063 Zbl 433 # 68051 • id 69273

Bucurescu, I. & Pascu, A. [1981] *Fuzzy pushdown automata* (**J** 0382) Int J Comput Math 10*109-119
◇ D05 E72 ◇ rev MR 83g:68078 Zbl 474 # 68090 • id 69271

Bucurescu, I. & Pascu, A. [1981] *On the languages generated by context-sensitive fuzzy grammars* (**J** 0382) Int J Comput Math 9*278-285
◇ D05 E72 ◇ rev MR 82m:68118 Zbl 471 # 68057 • id 69272

Buda, A. [1979] *Generalized sequential machine maps* (**J** 0232) Inform Process Lett 8*38-40
◇ D05 ◇ rev MR 80f:68092 Zbl 394 # 68042 • id 69274

Buda, A. [1979] *The languages of sequential program machines are coregular (Russian) (English summary)* (**P** 3400) Mat & Mat Obrazov (8);1979 Sl"nchev Bryag 93-100
◇ D05 ◇ rev Zbl 455 # 68035 • id 69275

Buda, A. [1984] *A simple characterization of coregular tree-languages (Russian)* (**P** 4392) Mat Logika (Markova);1980 Sofia 7-19
◇ D05 ◇ id 46573

Budach, L. & Hoehnke, H.-J. [1972] *Homologie und Automaten* (**J** 1528) Sitzb Plenum & Klassen Akad Wiss DDR 11*40-41
◇ D05 ◇ id 40853

Budach, L. & Hoehnke, H.-J. [1975] *Automaten und Funktoren* (**X** 0911) Akademie Verlag: Berlin 383pp
◇ D05 ◇ rev Zbl 363 # 18006 • id 40852

Budach, L. [1975] *On the solution of the labyrinth problem for finite automata* (**J** 0129) Elektr Informationsverarbeitung & Kybern 11*661-672
◇ D05 ◇ rev MR 55 # 12346 Zbl 355 # 94060 • id 50271

Budach, L. [1977] *Environments, labyrinths and automata* (**P** 2588) FCT'77 Fund of Comput Th;1977 Poznan 56*54-64
◇ D05 ◇ rev MR 58 # 4719 Zbl 383 # 68048 • id 69278

Budach, L. [1978] *Automata and labyrinths* (**J** 0114) Math Nachr 86*195-282
◇ D05 ◇ rev MR 80k:68044 Zbl 405 # 68049 • id 69277

Budach, L. [1978] *Automaten in Labyrinthen (Russisch)* (**J** 0052) Probl Kibern 34*83-94
◇ D05 ◇ rev MR 80g:68077 Zbl 415 # 68016 • id 69276

Budach, L. [1979] see Antelmann, H.

Budach, L. [1981] *Two pebbles don't suffice* (**P** 3429) Math Founds of Comput Sci (10);1981 Strbske Pleso 578-589
◇ D05 D15 ◇ rev MR 83h:68100 Zbl 479 # 68057 • id 69279

Budach, L. & Meinel, C. [1982] *Environments and Automata I., II (German and Russian summaries)* (**J** 0129) Elektr Informationsverarbeitung & Kybern 18*3-40, 115-139
◇ D05 D10 D80 ◇ rev MR 85g:68015a MR 85g:68015b Zbl 534 # 68036 Zbl 534 # 68037 • id 40849

Budach, L. & Waack, S. [1982] *On the halting problem for automata in cones* (**J** 0129) Elektr Informationsverarbeitung & Kybern 18*489-499
◇ D05 ◇ rev MR 85a:68132 Zbl 522 # 68048 • id 36783

Budach, L. & Graw, B. [1985] *Nonuniform complexity classes, decision graphs and homological properties of posets* (**P** 4670) Comput Th (5);1984 Zaborow 7-13
◇ D15 ◇ id 49619

Budinas, B.L. [1979] *Three linearly ordered degrees of constructibility of Δ_3^1 numbers (Russian)* (**J** 0288) Vest Ser Mat Mekh, Univ Moskva 1979/5*3-6,86
• transl [1979] (**J** 0510) Moscow Univ Math Bull 34/5*1-5
◇ D30 D55 E35 E45 ◇ rev MR 82b:03094 Zbl 444 # 03028 • id 54375

Budinas, B.L. [1980] *Analytic definability of constructible real numbers (Russian)* (**J** 0087) Mat Zametki (Akad Nauk SSSR) 28*177-186,318
• transl [1980] (**J** 1044) Math Notes, Acad Sci USSR 28*551-556
◇ D55 E35 E45 E50 ◇ rev MR 83e:03075 Zbl 475 # 03029 • id 55483

Author Index

BUDINAS, B.L. [1980] *Partial ordering of Δ_n^1-degrees of subsets of natural numbers (Russian) (English summary)* (**J** 0288) Vest Ser Mat Mekh, Univ Moskva 1980/3*15-18
• TRANSL [1980] (**J** 0510) Moscow Univ Math Bull 35/3*15-18
◇ D30 D55 E35 ◇ REV MR 82b:03078 Zbl 458 # 03009
• ID 54415

BUDINAS, B.L. [1981] *Construction of definable degrees of constructibility (Russian)* (**J** 0092) Sib Mat Zh 22/1*35-46,228
• TRANSL [1981] (**J** 0475) Sib Math J 22*25-34
◇ D30 D55 E35 E45 E47 ◇ REV MR 82k:03085 Zbl 498 # 03042 • ID 71383

BUDINAS, B.L. [1981] *The selector principle and analytic definability of real numbers in extensions of the constructible universe (Russian)* (**C** 3747) Mat Log & Mat Lingvistika 13-23
◇ D30 D55 E15 E35 E45 E47 ◇ REV MR 84m:03069
• ID 35767

BUDINAS, B.L. [1982] *On the selector principle and the analytic definability of constructible sets (Russian)* (**J** 0067) Usp Mat Nauk 37/2*193-194
• TRANSL [1982] (**J** 1399) Russ Math Surv 37/2*207-208
◇ D30 D55 E15 E35 E45 ◇ REV MR 84h:03113 Zbl 515 # 03029 • ID 34305

BUDINAS, B.L. [1983] *The selector principle for analytic equivalence relations does not imply the existence of an A_2 well ordering of the continuum (Russian)* (**J** 0142) Mat Sb, Akad Nauk SSSR, NS 120(162)*164-179,286
• TRANSL [1984] (**J** 0349) Math of USSR, Sbor 48*159-172
◇ D55 E15 E35 ◇ REV MR 84f:03042 Zbl 563 # 03031
• ID 47459

BUD'KO, A.E. [1985] *Two classes of Turing machines (Russian) (English summary)* (**J** 0414) Dokl Akad Nauk Belor SSR 29/9*792-793,860
◇ D10 ◇ ID 49273

BUECHI, J.R. [1960] *Weak second-order arithmetic and finite automata* (**J** 0068) Z Math Logik Grundlagen Math 6*66-92
◇ B15 B25 C85 D05 F35 ◇ REV MR 23 # A2317 Zbl 103.247 JSL 28.100 • ID 01690

BUECHI, J.R. [1962] *Mathematische Theorie des Verhaltens endlicher Automaten* (**J** 1046) Z Angew Math Mech 42*T9-T16
◇ D05 ◇ REV Zbl 113.14 • ID 33230

BUECHI, J.R. [1962] *On a decision method in restricted second order arithmetic* (**P** 0612) Int Congr Log, Meth & Phil of Sci (1,Proc);1960 Stanford 1-11
• TRANSL [1964] (**J** 1048) Kiber Sb Perevodov 1964/8*78-90 (Russian)
◇ B25 C85 D05 F35 ◇ REV MR 32 # 1116 Zbl 147.251 JSL 28.100 • ID 01693

BUECHI, J.R. [1962] *Turing machines and the "Entscheidungsproblem"* (**J** 0043) Math Ann 148*201-213
◇ B20 D10 D35 ◇ REV MR 29 # 5719 Zbl 118.16
• ID 01691

BUECHI, J.R. [1964] *Regular canonical systems* (**J** 0009) Arch Math Logik Grundlagenforsch 6*91-111
◇ D05 D35 ◇ REV MR 29 # 6660 Zbl 129.261 JSL 31.265
• ID 01692

BUECHI, J.R. [1965] *Decision methods in the theory of ordinals* (**J** 0015) Bull Amer Math Soc 71*767-770
◇ B15 B25 C85 D05 E10 ◇ REV MR 32 # 7413
• ID 01697

BUECHI, J.R. [1965] *Transfinite automata recursions and weak second order theory of ordinals* (**P** 0623) Int Congr Log, Meth & Phil of Sci (2,Proc);1964 Jerusalem 3-23
◇ B15 B25 C85 D05 E10 ◇ REV MR 35 # 1480
• ID 01694

BUECHI, J.R. [1966] *Algebraic theory of feedback in discrete systems, part I* (**P** 0746) Automata Th;1964 Ravello 70-101
◇ C05 D05 ◇ REV MR 34 # 7293 Zbl 279 # 94042
• ID 27534

BUECHI, J.R. & LANDWEBER, L.H. [1969] *Definability in the monadic second-order theory of successor* (**J** 0036) J Symb Logic 34*166-170
◇ B15 B25 C40 C85 D05 F35 ◇ REV MR 42 # 4387 Zbl 209.22 • ID 01700

BUECHI, J.R. & LANDWEBER, L.H. [1969] *Solving sequential conditions by finite-state strategies* (**J** 0064) Trans Amer Math Soc 138*295-311
◇ D05 ◇ REV MR 43 # 5926 Zbl 182.23 JSL 37.200
• ID 01698

BUECHI, J.R. [1970] *Algorithmisches Konstruieren von Automaten und die Herstellung von Gewinnstrategien nach Cantor-Bendixson* (**P** 0577) Automatenth & Formale Sprachen;1969 Oberwolfach 385-398
◇ D05 E60 ◇ REV MR 48 # 13515 Zbl 215.606 • ID 29302

BUECHI, J.R. & HOSKEN, W.H. [1970] *Canonical systems which produce periodic sets* (**J** 0041) Math Syst Theory 4*81-90
◇ D03 ◇ REV MR 50 # 6811 Zbl 188.331 • ID 01702

BUECHI, J.R. & ZAIONTZ, C. [1983] *Deterministic automata and the monadic theory of ordinals $<\omega_2$* (**J** 0068) Z Math Logik Grundlagen Math 29*313-336
◇ B15 B25 C85 D05 E10 ◇ REV MR 85j:03008 Zbl 541 # 03004 • ID 40515

BUECHI, J.R. [1983] *State-strategies for games in $F_{\sigma\delta} \cap G_{\delta\sigma}$* (**J** 0036) J Symb Logic 48*1171-1198
◇ D05 D55 E15 E60 ◇ REV MR 85j:03012 Zbl 565 # 03009 • ID 42601

BUECHI, J.R. & SIEFKES, D. [1983] *The complete extensions of the monadic second order theory of countable ordinals* (**J** 0068) Z Math Logik Grundlagen Math 29*289-312
◇ B15 C35 C85 D05 E10 ◇ REV MR 85g:03058 Zbl 541 # 03003 • ID 33901

BUECHI, J.R. & MAHR, B. & SIEFKES, D. [1984] *Manual on REC - A language for use and cost analysis of recursion over arbitrary data structures* (1111) Preprints, Manuscr., Techn. Reports etc. 79pp
◇ B75 D15 D80 ◇ REM Techn. Univ. Berlin FB Informatik, Bericht Nr. 84-06 • ID 39989

BUECHI, J.R. & MAHR, B. & SIEFKES, D. [1984] *Recursive definition and complexity of functions over arbitrary data structures* (**P** 3621) Frege Konferenz (2);1984 Schwerin 303-308
◇ B75 D75 ◇ REV MR 86g:68039 Zbl 555 # 68019
• ID 39992

BUECHI, J.R. see Vol. I, III, V for further entries

BUECHLER, S. [1984] *Resplendency and recursive definability in ω-stable theories* (J 0029) Israel J Math 49*26-33
◇ C45 C50 C57 ◇ REV Zbl 568 # 03015 • ID 45597

BUECHLER, S. see Vol. III for further entries

BUEVICH, V.A. [1972] *On the algorithmic undecidability of a-completeness for the boundedly determinate functions (Russian)* (J 0087) Mat Zametki (Akad Nauk SSSR) 11*687-697
• TRANSL [1972] (J 1044) Math Notes, Acad Sci USSR 11*417-421
◇ D80 ◇ REV MR 47 # 1583 Zbl 251 # 02047 • ID 01713

BUEVICH, V.A. [1980] *On τ-completeness in the class of automaton mappings (Russian)* (J 0023) Dokl Akad Nauk SSSR 252*1037-1041
• TRANSL [1980] (J 0062) Sov Math, Dokl 21*856-859
◇ D05 ◇ REV MR 82i:68038 Zbl 481 # 68054 • ID 80808

BUJ, D.B. & RED'KO, V.N. [1984] *Primitive program algebras I (Russian)* (J 0040) Kibernetika, Akad Nauk Ukr SSR 1984/5*1-7,133
• TRANSL [1984] (J 0021) Cybernetics 20*607-616
◇ D20 D75 D80 ◇ REV Zbl 573 # 03018 • REM Part II 1985 • ID 48576

BUJLINA, E.M. [1984] *Degrees of unsolvability of limit-enumerable sets (Russian)* (J 3937) Veroyat Met i Kibern (Kazan) 20*13-23
◇ D30 ◇ ID 46994

BUKHARAEV, N.R. [1979] *On limit enumeration properties of complements of recursively enumerable sets (Russian)* (J 3937) Veroyat Met i Kibern (Kazan) 15*10-17
◇ D25 ◇ REV MR 81i:03063 Zbl 422 # 03016 • ID 53477

BUKHARAEV, N.R. [1981] *T-degrees of differences of recursively enumerable sets (Russian)* (J 0031) Izv Vyssh Ucheb Zaved, Mat (Kazan) 1981/5*40-49
• TRANSL [1981] (J 3449) Sov Math 15/5*40-52
◇ D25 D30 D55 ◇ REV MR 82k:03063 Zbl 479 # 03022 • ID 45961

BUKHARAEV, N.R. [1981] *r-g-simple sets, Φ M sets and their degrees of unsolvability (Russian)* (J 0031) Izv Vyssh Ucheb Zaved, Mat (Kazan) 1981/11*69-71
• TRANSL [1981] (J 3449) Sov Math 25/11*79-83
◇ D25 ◇ REV MR 83k:03046 Zbl 483 # 03027 • ID 35399

BUKHARAEV, R.G. [1985] *Foundations of the theory of probabilistic automata (Russian)* (X 2027) Nauka: Moskva 288pp
◇ D05 ◇ ID 48215

BUKHARAEVA, Z.K. & GOLUNKOV, YU.V. [1978] *Complexity and running time of normal algorithms computing boolean functions (Russian)* (J 3937) Veroyat Met i Kibern (Kazan) 14*3-20
◇ B05 B35 D03 D15 ◇ REV MR 81j:68050 Zbl 411 # 03032 • ID 52886

BUKOVSKY, L. & VOPENKA, P. [1964] *The existence of a PCA-set of cardinality* $\aleph_1$ (J 0140) Comm Math Univ Carolinae (Prague) 5*125-128
◇ D55 E15 E35 ◇ REV MR 29 # 5743 Zbl 143.259 • ID 01714

BUKOVSKY, L. see Vol. III, V, VI for further entries

BULITKO, V.K. [1973] *Graphs with prescribed environments of the vertices (Russian)* (S 0066) Tr Mat Inst Steklov 133*78-94,274
◇ D80 ◇ REV MR 55 # 7846 Zbl 295 # 05124 • ID 80810

BULITKO, V.K. [1973] *On one algorithmically unsolvable problem for graphs (Russian)* (S 3662) Upravl Sistemy (Akad Nauk SSSR, Novosibirsk) 11*61-73
◇ D80 ◇ ID 33151

BULITKO, V.K. [1978] *On computations with bounded write storage (Russian)* (J 2320) Probl Peredachi Inf, Akad Nauk SSSR 14*105-108
• TRANSL [1978] (J 3419) Probl Inf Transmiss 14/4*313-315
◇ D05 ◇ REV MR 80f:68058 Zbl 448 # 68013 • ID 33250

BULITKO, V.K. [1978] *Solvable property of block-complete graphs* (P 3663) Konf Teor Grafov;1978 Zemplinska Shirana 20-30
◇ D80 ◇ ID 33152

BULITKO, V.K. [1980] *Reducibility via Zhegalkin's linear tables (Russian)* (J 0092) Sib Mat Zh 21/3*23-31,235
• TRANSL [1980] (J 0475) Sib Math J 21*332-339
◇ D25 D30 ◇ REV MR 81h:03087 Zbl 442 # 03029 • ID 56386

BULITKO, V.K. [1985] *Boolean classes of Turing reducibilities (Russian)* (J 0216) Izv Akad Nauk SSSR, Ser Mat 49*3-31
• TRANSL [1986] (J 0448) Math of USSR, Izv 26*1-29
◇ D30 ◇ ID 44704

BULLOCK, A.M. & SCHNEIDER, H.H. [1973] *On generating the finitely satisfiable formulas* (J 0047) Notre Dame J Formal Log 14*373-376
◇ B10 C07 C13 D25 ◇ REV MR 47 # 8280 Zbl 236 # 02014 • ID 01748

BULLOCK, A.M. see Vol. I, III for further entries

BULLOFF, J.J. & HAHN, S.W. & HOLYOKE, T.C. [1969] *Bibliography of Kurt Goedel* (C 0705) Found of Math (Goedel) xi-xii
◇ A10 D99 E99 F99 ◇ ID 14484

BULNES, J. [1972] *On the speed of addition and multiplication on one-tape, off-line Turing machines* (J 0194) Inform & Control 20*415-431
◇ D10 D15 ◇ REV MR 48 # 1819 Zbl 238 # 02033 • ID 27905

BUNDY, A. [1973] *A note on omitting the replacement schema* (J 0047) Notre Dame J Formal Log 14*118-120
◇ D20 F30 ◇ REV MR 47 # 8268 Zbl 225 # 02031 • ID 01765

BUNDY, A. see Vol. I for further entries

BUNIMOVA, E.O. & KAUFMAN, V.SH. & LEVIN, V.A. [1978] *Description of language projections (Russian)* (J 2869) Vest Ser Vychisl Mat Kibern, Univ Moskva 1978/4*68-73
• TRANSL [1978] (J 3221) Moscow Univ Comp Math Cybern 1978/4*75-80
◇ D05 ◇ REV Zbl 442 # 68005 • ID 69281

BUNTING, P.W. & LEEUWEN VAN, J. & TAMARI, D. [1978] *Deciding associativity for partial multiplication tables of order 3* (J 0214) Math of Comp 32*593-605
◇ D40 ◇ REV MR 58 # 16914 Zbl 391 # 68024 • ID 52360

BURGER, E. [1964] *Bemerkungen zu einigen Fassungen des Goedelschen Unvollstaendigkeitssatzes* (J 0068) Z Math Logik Grundlagen Math 10*57-63
◇ D80 F30 ◇ REV MR 28 # 2047 Zbl 146.9 • ID 01771

BURGER, E. see Vol. V for further entries

BURGESS, J.P. & MILLER, DOUGLAS E. [1975] *Remarks on invariant descriptive set theory* (J 0027) Fund Math 90*53-75
◇ C75 D55 E15 ◇ REV MR 53 # 7784 Zbl 342 # 02049 • ID 03956

BURGESS, J.P. [1977] *A selector principle for* Σ_1^1 *equivalence relations* (J 0133) Michigan Math J 24*65-76
◇ D55 E05 E15 E45 ◇ REV MR 56 # 11792 Zbl 347 # 02046 • ID 28149

BURGESS, J.P. [1977] *Two selection theorems* (J 0465) Bull Greek Math Soc (NS) 18*121-136
◇ D55 E15 ◇ REV MR 80h:03069 • ID 71462

BURGESS, J.P. [1978] *Equivalences generated by families of Borel sets* (J 0053) Proc Amer Math Soc 69*323-326
◇ D55 E15 ◇ REV MR 57 # 16084 Zbl 403 # 04002 • ID 28148

BURGESS, J.P. [1979] *A reflection phenomenon in descriptive set theory* (J 0027) Fund Math 104*127-139
◇ C75 D55 E15 ◇ REV MR 81i:04002 Zbl 449 # 03048 • ID 56715

BURGESS, J.P. [1979] *Effective enumeration of classes in a* Σ_1^1 *equivalence relation* (J 0452) Indiana Univ Math J 28*353-364
◇ D55 E15 E45 E55 ◇ REV MR 80f:03053 Zbl 379 # 54013 • ID 71463

BURGESS, J.P. [1982] *What are R-sets?* (P 3634) Patras Logic Symp;1980 Patras 307-324
◇ D55 D70 E15 ◇ REV MR 85b:03086 Zbl 529 # 03025 • ID 37657

BURGESS, J.P. [1983] *Classical hierarchies from a modern standpoint I: C-sets. II R-sets* (J 0027) Fund Math 115*81-95,97-105
◇ D55 D70 E15 ◇ REV Zbl 515 # 28002 Zbl 515 # 28003 • REM Part III 1983 by Burgess,J.P. & Lockhart,R.A. • ID 37856

BURGESS, J.P. & LOCKHART, R.A. [1983] *Classical hierarchies from a modern standpoint III: BP-sets* (J 0027) Fund Math 115*107-118
◇ D55 E15 ◇ REV Zbl 515 # 28004 • REM Parts I,II 1983 • ID 37858

BURGESS, J.P. see Vol. II, III, V, VI for further entries

BURGIN, M.S. [1982] *Complexity of parallel algorithms and computations (Russian)* (J 0023) Dokl Akad Nauk SSSR 265*268-274
• TRANSL [1982] (J 0062) Sov Math, Dokl 26*46-51
◇ D15 ◇ REV MR 84a:68037 Zbl 515 # 68041 • ID 37862

BURGIN, M.S. [1982] *Generalized Kolmogorov complexity and duality in computational theory (Russian)* (J 0023) Dokl Akad Nauk SSSR 264*19-23
• TRANSL [1982] (J 0062) Sov Math, Dokl 25*559-564
◇ D15 ◇ REV MR 83k:68025 Zbl 511 # 03016 • ID 37389

BURGIN, M.S. [1983] *Inductive Turing machines (Russian)* (J 0023) Dokl Akad Nauk SSSR 270*1289-1293
• TRANSL [1983] (J 0062) Sov Math, Dokl 27*730-734
◇ D10 D25 D55 ◇ REV MR 85h:68016 • ID 43469

BURGIN, M.S. [1983] *Multiple computations and the Kolmogorov complexity for such processes (Russian)* (J 0023) Dokl Akad Nauk SSSR 269*793-797
• TRANSL [1983] (J 0062) Sov Math, Dokl 27*410-414
◇ D15 ◇ REV MR 84k:03109 Zbl 565 # 68035 • ID 36127

BURGIN, M.S. [1984] *Complexity measures on systems of parallel algorithms (Russian)* (J 2605) Programmirovanie 1984/1*17-28,95
• TRANSL [1984] (J 2604) Progr Comput Software 10*12-21
◇ D15 ◇ REV MR 86f:68018 Zbl 546 # 68029 • ID 48926

BURGIN, M.S. [1984] *Inductive Turing machines with multihead, and Kolmogorov's algorithm (Russian)* (J 0023) Dokl Akad Nauk SSSR 275*280-284
• TRANSL [1984] (J 0062) Sov Math, Dokl 29*189-193
◇ D10 ◇ REV MR 85f:68024 • ID 39957

BURGIN, M.S. see Vol. III, V for further entries

BURGINA, E.S. [1976] *Classes of binary grammars, binary relations and their projections (Russian) (English summary)* (J 0040) Kibernetika, Akad Nauk Ukr SSR 1976/5*24-32
• TRANSL [1976] (J 0021) Cybernetics 12*677-686
◇ D05 ◇ REV MR 57 # 18239 Zbl 362 # 68115 • ID 33006

BURGINA, E.S. [1976] *Properties of universal languages on some classes (Russian) (English summary)* (J 0040) Kibernetika, Akad Nauk Ukr SSR 1976/3*82-88
• TRANSL [1976] (J 0021) Cybernetics 12/3*417-424
◇ D05 ◇ REV MR 57 # 18239 Zbl 338 # 68056 • ID 33005

BURGINA, E.S. see Vol. I, V for further entries

BURKHARD, HANS-DIETER [1969] *Ueber Experimente an nicht-deterministischen Automaten I* (J 0129) Elektr Informationsverarbeitung & Kybern 5*347-376
◇ D05 ◇ REV MR 43 # 8470 Zbl 217.590 • ID 40804

BURKHARD, HANS-DIETER [1970] *Ueber Experimente an nicht-deterministischen Automaten II* (J 0129) Elektr Informationsverarbeitung & Kybern 6*3-14
◇ D05 ◇ REV MR 43 # 8470 Zbl 217.590 • ID 49910

BURKHARD, HANS-DIETER [1974] *Diagnose und Einstellung nicht-deterministischer Automaten bei regulaeren Unterscheidungsformen* (J 0129) Elektr Informationsverarbeitung & Kybern 10*455-469
◇ D05 ◇ REV MR 54 # 4859 Zbl 302 # 94030 • ID 40805

BURKHARD, HANS-DIETER [1975] *Zustandsidentifizierung an asynchronen Automaten* (J 0129) Elektr Informationsverarbeitung & Kybern 11*653-658
◇ D05 ◇ REV Zbl 342 # 94027 • ID 40813

BURKHARD, HANS-DIETER [1976] *Identifizierungsexperimente an asynchronen Automaten* (J 0129) Elektr Informationsverarbeitung & Kybern 12*45-59
◇ D05 ◇ REV MR 54 # 9890 Zbl 326 # 94033 • ID 40807

BURKHARD, HANS-DIETER [1976] *Zum Laengenproblem homogener Experimente an determinierten und nicht-deterministischen Automaten* (J 0129) Elektr Informationsverarbeitung & Kybern 12*301-306
◇ D05 ◇ REV MR 55 # 1820 Zbl 349 # 94061 • ID 40823

BURKHARD, HANS-DIETER [1981] *Ordered firing in Petri nets* (J 0129) Elektr Informationsverarbeitung & Kybern 17*71-86
◇ D05 ◇ REV MR 83a:68074 Zbl 535 # 68029 • ID 40839

BURKHARD, HANS-DIETER [1981] *Two pumping lemmata for Petri nets* (**J** 0129) Elektr Informationsverarbeitung & Kybern 17*349-362
◇ D05 ◇ REV MR 83a:68074 Zbl 494 # 68065 • ID 40840

BURKHARD, HANS-DIETER [1982] *Generalized identification experiments for finite deterministic automata* (**P** 3787) Discr Math;1977 Warsaw 45-51
◇ D05 ◇ REV MR 84g:68046 • ID 40837

BURKHARD, HANS-DIETER [1983] *Control of Petri nets by finite automata* (**J** 2095) Fund Inform, Ann Soc Math Pol, Ser 4 6*185-215
◇ D05 D80 ◇ REV MR 84e:68062 Zbl 531 # 68014 • ID 40841

BURKHARD, HANS-DIETER [1983] *On priorities of parallelism: Petri nets under the maximum firing strategy* (**P** 3830) Logics of Progr & Appl;1980 Poznan 86-97
◇ D05 ◇ ID 40838

BURKHARD, HANS-DIETER & STARKE, P.H. [1984] *A note on the impact of conflict resolution to liveness and deadlock in Petri nets* (**J** 2095) Fund Inform, Ann Soc Math Pol, Ser 4
◇ D05 D80 ◇ ID 40842

BURKHARD, HANS-DIETER [1984] *An investigation of controls for concurrent systems by abstract control languages* (**P** 3658) Math Founds of Comput Sci (11);1984 Prague 223-231
◇ D05 D80 ◇ REV MR 86g:68047 Zbl 566 # 68063 • ID 40843

BURKHARD, W.A. & VARAIYA, P.P. [1971] *Complexity problems in real time languages* (**J** 0191) Inform Sci 3*87-100 • TRANSL [1974] (**C** 2319) Slozh Vychisl & Algor 235-251
◇ D05 D15 ◇ REV MR 43 # 1773 Zbl 222 # 02041 • ID 26314

BURKS, A.W. & WARREN, D.W. & WRIGHT, J.B. [1954] *An analysis of a logical machine using parenthesis-free notation* (**J** 0235) Math Tables Other Aids Comp 8*53-57
◇ B03 D05 ◇ REV MR 15.833 Zbl 58.4 JSL 20.70 • ID 33359

BURKS, A.W. & MCNAUGHTON, R. & POLLMAR, C. & WARREN, D.W. & WRIGHT, J.B. [1954] *Complete decoding nets: general theory and minimality* (**J** 0514) SIAM Journ 2*201-242
◇ B70 D05 ◇ REV MR 16.1078 Zbl 59.112 JSL 21.210 • ID 31309

BURKS, A.W. & WANG, HAO [1957] *The logic of automata I,II* (**J** 0037) ACM J 4*193-218,279-297
◇ D05 ◇ REV MR 20 # 2859 JSL 30.249 • ID 01779

BURKS, A.W. & WRIGHT, J.B. [1962] *Sequence generators, graphs and formal languages* (**J** 0194) Inform & Control 5*204-212
◇ B20 D05 ◇ REV MR 26 # 4902 Zbl 109.353 JSL 29.210 • ID 01780

BURKS, A.W. & WRIGHT, J.B. [1962] *Sequence generators and digital computers* (**P** 0613) Rec Fct Th;1961 New York 139-199
◇ B20 D05 ◇ REV Zbl 145.243 JSL 29.210 • ID 02718

BURKS, A.W. [1970] *Programming and the theory of automata* (**C** 1085) Essays Cellular Automata 65-83
◇ B75 D05 ◇ REV MR 45 # 8457 Zbl 233 # 02015 • ID 27780

BURKS, A.W. [1970] *Von Neumann's self-reproducing automata* (**C** 1085) Essays Cellular Automata 3-64
◇ D10 ◇ REV MR 45 # 8457 Zbl 235 # 94041 • ID 21222

BURKS, A.W. [1975] *Logic, biology and automata - some historical reflections* (**J** 1741) Int J Man-Mach Stud 7*297-312
◇ A05 A10 D05 D10 D80 ◇ REV Zbl 318 # 94002 • ID 60788

BURKS, A.W. see Vol. I, II for further entries

BURR, S.A. [1984] *Some undecidable problems involving the edge-coloring and vertex-coloring of graphs* (**J** 0193) Discr Math 50*171-177
◇ D80 ◇ REV MR 85k:05044 Zbl 553 # 05053 • ID 43406

BURRIS, S. & SANKAPPANAVAR, H.P. [1975] *Lattice-theoretic decision problems in universal algebra* (**J** 0004) Algeb Universalis 5*163-177
◇ B25 C05 D35 ◇ REV MR 52 # 13359 Zbl 322 # 02045 • ID 21812

BURRIS, S. & LAWRENCE, J. [1982] *Two undecidability results using modified boolean powers* (**J** 0017) Canad J Math 34*500-505
◇ C05 C60 D35 G05 ◇ REV MR 83k:03051 Zbl 499 # 03029 • ID 35402

BURRIS, S. [1983] *Boolean constructions* (**P** 3841) Universal Algeb & Lattice Th (4);1982 Puebla 67-90
◇ A10 B25 C05 C30 C98 D35 G05 ◇ REV MR 85d:08010 Zbl 517 # 08001 • ID 36689

BURRIS, S. [1984] *Model companions for finitely generated universal Horn classes* (**J** 0036) J Symb Logic 49*68-74
◇ B25 C05 C10 C35 D20 ◇ REV MR 85g:03052 • ID 42428

BURRIS, S. [1985] *A simple proof of the hereditary undecidability of the theory of lattice ordered abelian groups* (**J** 0004) Algeb Universalis 20*400-401
◇ C05 C60 D35 G10 ◇ REV Zbl 575 # 06017 • ID 48302

BURRIS, S. see Vol. III, V for further entries

BURTMAN, M.I. & GASANOVA, N.P. & GUSEJNBEKOVA, A.M. [1985] *Partially recursive functions on the rationals (Russian) (Aserbaid. and English summaries)* (**J** 0135) Izv Akad Nauk Azerb SSR, Ser Fiz-Tekh Mat 6/1*42-48
◇ D20 ◇ ID 46504

BUSCH, D.R. [1976] *λ-scales, κ-Souslin sets and a new definition of analytic sets* (**J** 0036) J Symb Logic 41*373-378
◇ D55 E15 E60 ◇ REV MR 54 # 7258 Zbl 353 # 02045 • ID 14769

BUSCH, D.R. [1976] *On the number of Solovay r-degrees* (**J** 0068) Z Math Logik Grundlagen Math 22*283-286
◇ D30 D65 E15 ◇ REV MR 56 # 15389 Zbl 349 # 02038 • ID 18447

BUSCH, D.R. [1977] *A problem concerning projective prewellorderings* (**J** 0068) Z Math Logik Grundlagen Math 23*237-240
◇ D55 E05 E15 E60 ◇ REV MR 56 # 15423 Zbl 367 # 02034 • ID 26482

BUSCH, D.R. see Vol. V for further entries

BUSKIRK VAN, J. & GILLETTE, P. [1968] *The word problem and consequences for the braid groups and mapping class groups of the 2-sphere* (**J** 0064) Trans Amer Math Soc 131*277-296
◇ D40 D80 ◇ REV MR 38 # 221 Zbl 169.553 • ID 04935

Author Index

BUSZKOWSKI, W. [1978] *Undecidability of some logical extensions of Ajdukiewicz-Lambek calculus* (**J** 0063) Studia Logica 37*59-64
◇ B65 D35 ◇ REV MR 80f:03029 Zbl 394 # 03018
• ID 52495

BUSZKOWSKI, W. [1979] *Undecidability of the theory of lattice-orderable groups* (**J** 2718) Fct Approximatio, Comment Math, Poznan 7*23-28
◇ C60 D35 ◇ REV MR 80m:03078 Zbl 408 # 03010
• ID 56249

BUSZKOWSKI, W. [1980] *Logical complexity of some classes of tree languages generated by multiple-tree-automata* (**J** 0068) Z Math Logik Grundlagen Math 26*41-49
◇ D05 D55 ◇ REV MR 81c:03028 Zbl 432 # 03024
• ID 53981

BUSZKOWSKI, W. [1982] *Some decision problems in the theory of syntactic categories* (**J** 0068) Z Math Logik Grundlagen Math 28*539-548
◇ B25 D05 ◇ REV MR 84m:03014 Zbl 499 # 03010
• ID 35719

BUSZKOWSKI, W. [1985] *Algebraic models of categorial grammars* (**P** 4180) Int Congr Log, Meth & Phil of Sci (7,Pap);1983 Salzburg 403-426
◇ B65 D05 ◇ ID 48116

BUSZKOWSKI, W. [1985] *The equivalence of unidirectional Lambek categorial grammars and context-free grammars* (**J** 0068) Z Math Logik Grundlagen Math 31*369-384
◇ D05 ◇ REV Zbl 567 # 68044 • ID 47565

BUSZKOWSKI, W. see Vol. V, VI for further entries

BUTLER, G.A. [1984] see BALLANTYNE, A.M.

BUTLER, G.A. see Vol. I for further entries

BUTNARIU, D. [1977] *L-fuzzy automata. Description of a neural model* (**P** 3406) Congr Cybern & Syst (3);1975 Bucharest II*119-124
◇ D05 D80 E72 ◇ REV Zbl 428 # 68065 • ID 69287

BUTNARIU, D. see Vol. II, V for further entries

BUTZ, A.R. [1981] *Functions realized by consistent sequential machines* (**J** 0194) Inform & Control 48*147-191
◇ D05 ◇ REV MR 81i:68033 Zbl 461 # 68052 • ID 69289

BUTZBACH, P. [1973] *Une famille de congruences de Thue pour lesquelles le probleme de l'equivalence est decidable. Application a l'equivalence des grammaires separees* (**P** 0763) Automata, Lang & Progr (1);1972 Rocquencourt 3-12
◇ D03 D05 ◇ REV MR 52 # 16137 Zbl 274 # 02012
• ID 29017

BUTZBACH, P. [1978] *Sur l'equivalence des grammaires simples* (**P** 3394) Lang Algeb (1);1973 Bonascre 223-245
◇ D05 ◇ REV MR 80f:68081 Zbl 394 # 68054 • ID 69290

BUZETEANU, S. & DIMA, N. [1984] *Arithmetization of the computation of Sudan's and Knuth's functions (Romanian) (English summary)* (**J** 0197) Stud Cercet Mat Acad Romana 36*409-418
◇ D20 ◇ REV Zbl 552 # 03024 • ID 43376

BYERLY, R.E. [1982] *An invariance notion in recursion theory* (**J** 0036) J Symb Logic 47*48-66
◇ D20 D45 ◇ REV MR 83d:03052 Zbl 487 # 03021
• ID 35191

BYERLY, R.E. [1982] *Recursion theory and the lambda-calculus* (**J** 0036) J Symb Logic 47*67-83
◇ B40 D20 ◇ REV MR 83i:03066 Zbl 488 # 03023
• ID 35525

BYERLY, R.E. [1983] *Definability of r.e. sets in a class of recursion theoretic structures* (**J** 0036) J Symb Logic 48*662-669
◇ D25 D45 ◇ REV MR 84j:03085 Zbl 528 # 03031
• ID 34677

BYERLY, R.E. [1984] *Definability of recursively enumerable sets in abstract computational complexity theory* (**J** 0068) Z Math Logik Grundlagen Math 30*499-503
◇ D15 D25 ◇ REV Zbl 534 # 03016 • ID 43525

BYERLY, R.E. [1984] *Some properties of invariant sets* (**J** 0036) J Symb Logic 49*9-21
◇ D25 D30 ◇ REV MR 86a:03039 • ID 42429

BYRNES, C. & GAUGER, M.A. [1977] *Decidability criteria for the similarity problem, with applications to the moduli of linear dynamical systems* (**J** 0345) Adv Math 25*59-90
◇ D40 D80 ◇ REV MR 56 # 12002 Zbl 364 # 15006
• ID 50977

CAFFIN, R.W. & GOHEEN, H.E. & STAHL, W.R. [1963] *Simulation of a Turing machine on a digital computer* (**P** 1636) AFIPS Fall Jt Computer Conf (24);1963 Las Vegas 35-43
◇ D10 ◇ ID 29445

CAI, MAOHUA & ZHANG, JINWEN [1982] *Recursively enumerable fuzzy sets I* (**J** 3919) BUSEFAL 9*11-12
◇ D25 E72 ◇ REV Zbl 524 # 03028 • REM Part II 1982
• ID 37602

CAI, MAOHUA [1982] *Recursively enumerable fuzzy sets II* (**J** 3919) BUSEFAL 9*13-16
◇ D25 E72 ◇ REV Zbl 528 # 03025 • REM Part I 1982 by Cai,Mao Hua & Zhang,Jin Wen • ID 37640

CAI, MAOHUA [1983] *A projection theorem for recursively enumerable fuzzy sets (Chinese)* (**J** 2771) Kexue Tongbao 28*1025-1026
◇ D25 E72 ◇ ID 44088

CAIANIELLO, E.R. (ED.) [1966] *Automata theory* (**X** 0801) Academic Pr: New York xiv+342pp
◇ D05 D97 ◇ REV MR 33 # 3855 Zbl 178.290 • REM 2nd ed 1968 • ID 23516

CAICEDO, X. [1976] *Algorithmen, Fibonaccizahlen und das zehnte Hilbertsche Problem (Spanish)* (**J** 0377) Bol Mat (Bogota) 10*1-27
◇ D35 D98 ◇ REV MR 55 # 5425 Zbl 362 # 02033
• ID 50755

CAICEDO, X. see Vol. I, III, V for further entries

CALDWELL, J. & POUR-EL, M.B. [1975] *On a simple definition of computable function of a real variable - with applications to functions of a complex variable* (**J** 0068) Z Math Logik Grundlagen Math 21*1-19
◇ D20 D80 ◇ REV MR 51 # 2885 Zbl 323 # 02049
• ID 10708

CALUDE, C. [1975] *Sur une classe de distances dans un demi-groupe libre* (**J** 0070) Bull Soc Sci Math Roumanie, NS 17(65)*123-133
◇ D05 ◇ REV MR 51 # 7372 Zbl 302 # 68096 • ID 60815

CALUDE, C. [1978] *Categorical methods in computability I,II (I: Recursive, non-primitive recursive functions) (Romanian) (English summary)* (**J** 0197) Stud Cercet Mat Acad Romana 30*253-277,361-383
◇ D15 D20 ◇ REV MR 80m:03110 Zbl 381 # 03028 Zbl 405 # 03025 • ID 51896

CALUDE, C. & FANTANEANU, B. [1978] *On recursive, non-primitive recursive functions* (**J** 0070) Bull Soc Sci Math Roumanie, NS 22(70)*355-358
◇ D20 ◇ REV MR 80b:03052 Zbl 399 # 03025 • ID 52821

CALUDE, C. [1978] *On the category of recursive languages* (**J** 0517) Mathematica (Cluj) 19(24)*29-32
◇ B03 D45 F40 G30 ◇ REV MR 80b:03056 Zbl 384 # 03029 • ID 52071

CALUDE, C. & MARCUS, S. & TEVY, I. [1979] *The first example of a recursive function which is not primitive recursive* (**J** 1648) Hist Math 6*380-384
◇ A10 D20 ◇ REV MR 80i:03053 Zbl 426 # 03042 • ID 53638

CALUDE, C. & MARCUS, S. & PAUN, G. [1979] *The universal grammar as a hypothetical brain* (**P** 2950) Appl Math in Syst Th;1978 Brasov 2*93-114
◇ B65 D05 D80 ◇ REV Zbl 457 # 68093 • ID 54405

CALUDE, C. & MARCUS, S. & TEVY, I. [1980] *Recursive properties of Sudan function* (**J** 0060) Rev Roumaine Math Pures Appl 25*503-507
◇ D20 ◇ REV MR 81f:03053 Zbl 444 # 03021 • ID 71516

CALUDE, C. & VIERU, V. [1981] *A note on Knuth's iterated powers* (**J** 0230) An Univ Iasi, NS, Sect Ia 27*253-255
◇ D20 ◇ REV MR 84m:03062 Zbl 473 # 03035 • ID 55364

CALUDE, C. & VIERU, V. [1981] *An iterative normal form for partial recursive functions* (**J** 2716) Found Control Eng, Poznan 6*133-144
◇ D20 ◇ REV MR 85b:03064 Zbl 503 # 68034 • ID 36669

CALUDE, C. [1981] *Darboux property and primitive recursive functions* (**J** 0060) Rev Roumaine Math Pures Appl 26*1187-1192
◇ D20 ◇ REV MR 84h:03100 Zbl 481 # 03028 • ID 34290

CALUDE, C. & PAUN, G. [1981] *Global syntax and semantics for recursively enumerable languages* (**J** 2095) Fund Inform, Ann Soc Math Pol, Ser 4 4*245-254
◇ B03 D25 ◇ REV MR 83h:68133 Zbl 473 # 68068 • ID 55399

CALUDE, C. [1982] *Note on a hierarchy of primitive recursive functions* (**J** 0060) Rev Roumaine Math Pures Appl 27*935-936
◇ D20 ◇ REV MR 85b:03065 Zbl 495 # 03027 • ID 36885

CALUDE, C. & CHITESCU, I. [1982] *On Per Martin-Loef random sequences* (**J** 0070) Bull Soc Sci Math Roumanie, NS 26(74)*217-221
◇ B75 D80 ◇ REV MR 84g:03073 Zbl 495 # 03026 • ID 34179

CALUDE, C. & CHITESCU, I. [1982] *Random strings according to A.N. Kolmogorov and P. Martin-Loef classical approach* (**J** 2716) Found Control Eng, Poznan 7*73-85
◇ D80 ◇ REV MR 84h:60008 Zbl 521 # 03024 • ID 37070

CALUDE, C. & CHITESCU, I. [1982] *Strong noncomputability of random strings* (**J** 0382) Int J Comput Math 11*43-45
◇ D20 ◇ REV MR 83h:68066 Zbl 486 # 03026 • ID 38081

CALUDE, C. [1982] *Topological size of sets of partial recursive functions* (**J** 0068) Z Math Logik Grundlagen Math 28*455-462
◇ D15 D20 E75 ◇ REV MR 85i:03134 Zbl 495 # 03022 • ID 36883

CALUDE, C. & PAUN, G. [1983] *Independent instances for some undecidable problems (French summary)* (**J** 3441) RAIRO Inform Theor 17*49-54
◇ D20 F30 ◇ REV MR 85d:03087 Zbl 517 # 03022 • ID 37287

CALUDE, C. & CHITESCU, I. [1983] *On representability of P. Martin-Loef tests* (**J** 0156) Kybernetika (Prague) 19*42-47
◇ D80 ◇ REV Zbl 529 # 03020 • ID 37654

CALUDE, C. [1983] *On the class of independent problems related to Rice theorem* (**J** 1456) SIGACT News 2*53-57
◇ D20 D25 ◇ REV Zbl 535 # 03020 • ID 48356

CALUDE, C. & CHITESCU, I. [1983] *Representability of recursive P. Martin-Loef tests* (**J** 0156) Kybernetika (Prague) 19*526-536
◇ D80 ◇ REV MR 85h:03040 Zbl 529 # 03021 • ID 37655

CALUDE, C. & TATARAM, M. [1983] *Universal sequences of primitive recursive functions* (**J** 0060) Rev Roumaine Math Pures Appl 28*381-389
◇ D20 ◇ REV MR 85c:03016 Zbl 535 # 03017 • ID 38321

CALUDE, C. & CHITESCU, I. [1985] *A combinatorial characterization of sequential P. Martin-Loef tests* (**J** 0382) Int J Comput Math 17*53-64
◇ D80 ◇ REV Zbl 562 # 03020 • ID 47429

CALUDE, C. see Vol. II, VI for further entries

CALUGAREANU, G. [1971] *Invariants de contraction dans les groupes (Romanian and Russian summaries)* (**J** 3451) Stud Univ Cluj, Ser Math-Mech 16/1*9-27
◇ D40 ◇ REV MR 44 # 4080 Zbl 272 # 20031 • ID 02222

CAMERINI, P.M. & GALBIATI, G. & MAFFIOLI, F. [1979] *On the complexity of Steiner-like problems* (**P** 4049) Allerton Conf Commun, Control & Comput (17);1979 Monticello 969-977
◇ D05 ◇ REV MR 84b:94003 • ID 46267

CAMPBELL, P.J. [1971] *Suslin logic* (0000) Diss., Habil. etc
◇ C75 D55 E15 G05 G25 ◇ REM Ph.D.thesis, Cornell University • ID 16762

CAMPBELL, P.J. see Vol. I, V for further entries

CAMPEAU, J.O. [1959] *Simple Turing type computers* (**C** 0679) Handb of Automat, Comput & Control 2*31-01 - 31-16
◇ D10 ◇ ID 02723

CANNON, R.L. [1976] *An algebraic technique for context-sensitive parsing* (**J** 0435) Int J Comput & Inf Sci 5*257-276
◇ D05 ◇ REV MR 55 # 11722 Zbl 401 # 68063 • ID 69293

CANNONITO, F.B. [1962] *The Goedel incompleteness theorem and intelligent machines* (**P** 1197) AFIPS Spring Jt Computer Conf (21);1962 San Francisco 21*71-77
◇ B35 D10 D25 F30 ◇ REV JSL 36.693 • ID 21294

CANNONITO, F.B. [1966] *Hierarchies of computable groups and the word problem* (**J** 0036) J Symb Logic 31*376-392
◇ D40 ◇ REV MR 37 # 70 JSL 33.121 • ID 01822

Author Index

CANNONITO, F.B. & FINKELSTEIN, M. [1969] *On primitive recursive permutations and their inverses* (**J** 0036) J Symb Logic 34*634-638
◇ D20 D45 ◇ REV MR 41 # 6686 Zbl 194.312 JSL 38.655
• ID 01823

CANNONITO, F.B. [1971] *A note on inverses of elementary permutations* (**J** 0191) Inform Sci 3*355-359
◇ D20 D45 F30 ◇ REV MR 46 # 33 Zbl 229 # 02034
• ID 71525

CANNONITO, F.B. [1973] *The algebraic invariance of the word problem in groups* (**P** 0678) Word Probl: Decis & Burnside Probl in Group Th;1969 Irvine 349-364
◇ D20 D40 ◇ REV MR 53 # 7749 Zbl 274 # 02016
JSL 41.786 • ID 22996

CANNONITO, F.B. & GATTERDAM, R.W. [1973] *The computability of group constructions. Part I* (**P** 0678) Word Probl: Decis & Burnside Probl in Group Th;1969 Irvine 365-400
◇ D20 D40 D45 ◇ REV MR 56 # 5255 Zbl 274 # 02017
JSL 41.786 • REM Part II 1973 by Gatterdam,R.W. • ID 51344

CANNONITO, F.B. & GATTERDAM, R.W. [1973] *The word problem in polycyclic groups is elementary* (**J** 0020) Compos Math 27*39-45
◇ D40 ◇ REV MR 50 # 488 Zbl 279 # 20028 • ID 01827

CANNONITO, F.B. [1973] see BOONE, W.W.

CANNONITO, F.B. & GATTERDAM, R.W. [1975] *The word problem and power problem in 1-relator groups are primitive recursive* (**J** 0048) Pac J Math 61*351-359
◇ D20 D40 ◇ REV MR 53 # 5280 Zbl 335 # 02029
• ID 22912

CANNONITO, F.B. [1980] *Two decidable Markov properties over a class of solvable groups (Russian)* (**J** 0003) Algebra i Logika 19*646-658,745
• TRANSL [1980] (**J** 0069) Algeb and Log 19*419-425
◇ D40 ◇ REV MR 82j:20072 Zbl 475 # 20026 • ID 80845

CANNONITO, F.B. [1981] see BAUMSLAG, G.

CANNONITO, F.B. [1984] *On some algorithmic problems for finitely presented groups and Lie algebras* (**P** 4359) Groups-Korea;1983 Kyoungiu 21-28
◇ D40 ◇ REV Zbl 549 # 20021 • ID 44827

CANNONITO, F.B. & ROBINSON, D.J. [1984] *The word problem for finitely generated soluble groups of finite rank* (**J** 0161) Bull London Math Soc 16*43-46
◇ D40 ◇ REV MR 85i:20039 Zbl 528 # 20030 • ID 44327

CANTINI, A. [1985] *Majorizing provably recursive functions in fragments of PA* (**J** 0009) Arch Math Logik Grundlagenforsch 25*21-31
◇ D20 F15 F30 ◇ ID 49071

CANTINI, A. see Vol. I, II, III, V, VI for further entries

CANTOR, DAVID G. [1962] *On the ambiguity problem of Backus systems* (**J** 0037) ACM J 9*477-479
◇ D05 ◇ REV MR 27 # 42 Zbl 114.330 JSL 32.114
• ID 01830

CANTOR, DAVID G. & ROQUETTE, P. [1984] *On diophantine equations over the ring of all algebraic integers* (**J** 0401) J Number Th 18*1-26
◇ B25 C60 D35 ◇ REV MR 85j:11036 Zbl 538 # 12014
• ID 46787

CAPORASO, S. [1979] *Consistency proof without transfinite induction for a formal system for Turing machines* (**J** 0009) Arch Math Logik Grundlagenforsch 19*157-164
◇ D10 F05 ◇ REV MR 80j:03082 Zbl 406 # 03070
• ID 56152

CARACCIOLO, A. [1966] *Generalized Markov algorithms and automata* (**P** 0746) Automata Th;1964 Ravello 115-130
◇ D03 D05 D10 ◇ ID 27599

CARACCIOLO, A. see Vol. II, V for further entries

CARDOZA, E. & LIPTON, R.J. & MEYER, A.R. [1976] *Exponential space complete problems for Petri nets and commutative semigroups* (**P** 2597) ACM Symp Th of Comput (8);1976 Hershey 50-54
◇ D15 D40 D80 ◇ REV Zbl 374 # 20067 • ID 51569

CARPENTER, A.J. [1977] *For a countable admissible ordinal α, the α-recursive functions are exactly the α-definite functions* (**J** 0302) Rep Math Logic, Krakow & Katowice 9*3-13
◇ D60 ◇ REV MR 80g:03043 Zbl 398 # 03034 • ID 52762

CARPENTER, A.J. [1978] *For any admissible ordinal α, Kripke's α-recursive functions are exactly Machover's α-recursive functions* (**J** 3064) Abacus, Math Ass Nigeria 12*109-120
◇ D60 ◇ REV MR 82g:03085 • ID 71543

CARPENTER, A.J. see Vol. V for further entries

CARPENTIER, M.A. [1968] *Creative sequences and double sequences* (**J** 0047) Notre Dame J Formal Log 9*35-61
◇ D25 ◇ REV MR 42 # 2938 Zbl 187.276 JSL 35.590
• ID 01860

CARPENTIER, M.A. [1969] *Complete enumerations and double sequences* (**J** 0068) Z Math Logik Grundlagen Math 15*1-6
◇ D25 D45 ◇ REV MR 40 # 4101 Zbl 174.20 • ID 01861

CARPENTIER, M.A. [1970] *Uniformly precomplete enumeration* (**J** 0068) Z Math Logik Grundlagen Math 16*463-468
◇ D45 ◇ REV MR 44 # 65 Zbl 185.20 • ID 01862

CARSTENS, H.G. [1975] *Reducing hyperarithmetic sequences* (**J** 0027) Fund Math 89*5-11
◇ C57 D55 ◇ REV MR 52 # 2855 Zbl 334 # 02023
• ID 03968

CARSTENS, H.G. [1976] *Δ^0_2-Mengen* (**J** 0009) Arch Math Logik Grundlagenforsch 18*55-65
◇ D25 D30 D55 ◇ REV MR 57 # 12198 Zbl 356 # 02039
• ID 24319

CARSTENS, H.G. [1977] *The complexity of some combinatorial constructions* (**J** 0068) Z Math Logik Grundlagen Math 23*121-130
◇ D55 D80 E05 ◇ REV MR 58 # 5149 Zbl 363 # 02051
• ID 26473

CARSTENS, H.G. [1977] *The theorem of Matijasevic is provable in Peano's arithmetic by finitely many axioms* (**J** 0079) Logique & Anal, NS 20*116-121
◇ C52 D25 D35 F30 H15 ◇ REV MR 58 # 27400
Zbl 382 # 03037 • ID 51961

CARSTENS, H.G. & GOLZE, U. [1982] *Recursive paths in cross-connected trees and an application to cell spaces* (**J** 0041) Math Syst Theory 15*29-37
◇ D05 ◇ REV MR 83c:68072 Zbl 473 # 68065 • ID 39886

CARSTENS, H.G. & PAEPPINGHAUS, P. [1983] *Recursive coloration of countable graphs* (**J** 0073) Ann Pure Appl Logic 25*19-45
◇ D80 ◇ REV MR 85h:03045 Zbl 527 # 03025 • ID 37575

CARSTENS, H.G. & PAEPPINGHAUS, P. [1984] *Abstract construction of counterexamples in recursive graph theory* (**P** 2153) Logic Colloq;1983 Aachen 2*39-62
◇ D45 ◇ REV MR 86e:03043 • ID 39918

CARTWRIGHT, ROBERT & MCCARTHY, J. [1979] *Recursive programs as functions in a first order theory* (**P** 3479) Math Stud of Inform Process;1978 Kyoto 576-629
◇ B10 B75 D20 ◇ REV Zbl 407 # 68042 • ID 56231

CARTWRIGHT, ROBERT see Vol. VI for further entries

CASALEGNO, P. [1981] *Sui gradi delle funzioni parziali* (**P** 3092) Congr Naz Logica;1979 Montecatini Terme 137-144
◇ D30 ◇ ID 48377

CASALEGNO, P. [1985] *On the T-degrees of partial functions* (**J** 0036) J Symb Logic 50*580-588
◇ D30 ◇ ID 47364

CASALEGNO, P. see Vol. V for further entries

CASE, J. [1970] *Enumeration reducibility and partial degrees* (**J** 0007) Ann Math Logic 2*419-439
◇ D25 D30 ◇ REV MR 44 # 66 Zbl 223 # 02046 JSL 39.605 • ID 01887

CASE, J. [1971] *A note on degrees of self-describing Turing machines* (**J** 0037) ACM J 18*329-338
◇ D10 D25 ◇ REV MR 46 # 8823 Zbl 234 # 02023 • ID 27798

CASE, J. [1974] *Maximal arithmetical reducibilities* (**J** 0068) Z Math Logik Grundlagen Math 20*261-270
◇ D30 D55 ◇ REV MR 51 # 100 Zbl 304 # 02016 • ID 03969

CASE, J. [1974] *Periodicity in generations of automata* (**J** 0041) Math Syst Theory 8*15-32
◇ D05 ◇ REV MR 52 # 10382 Zbl 295 # 02019 • ID 21682

CASE, J. [1976] *Sortability and extensibility of the graphs of recursively enumerable partial and total orders* (**J** 0068) Z Math Logik Grundlagen Math 22*1-18
◇ D25 D45 ◇ REV MR 53 # 2654 Zbl 338 # 02021 • ID 18493

CASE, J. & SMITH, C.H. [1978] *Anomaly hierarchies of mechanized inductive inference* (**P** 1740) ACM Symp Th of Comput (10);1978 San Diego 314-319
◇ B35 D20 ◇ REV MR 80d:68047 • ID 80855

CASE, J. & MOORE, D.J. [1978] *The complexity of total order structures* (**J** 0119) J Comp Syst Sci 17*253-269
◇ D15 D20 D45 ◇ REV MR 80g:03038 Zbl 396 # 03034 • ID 52644

CASE, J. & LYNES, C. [1982] *Machine inductive inference and language identification* (**P** 3836) Automata, Lang & Progr (9);1982 Aarhus 107-115
◇ D05 D20 ◇ REV MR 83m:68131 • ID 40424

CASE, J. & SMITH, C.H. [1983] *Comparison of identification criteria for machine inductive inference* (**J** 1426) Theor Comput Sci 25*193-220
◇ B35 B48 D20 ◇ REV MR 84j:68029 Zbl 524 # 03025 • ID 37600

CASE, J. [1983] *Pseudo-extensions of computable functions* (**J** 0194) Inform & Control 56*100-111
◇ D20 ◇ REV MR 85h:03042 Zbl 538 # 68009 • ID 41498

CASPAR, K. & FENZL, W. & WEIMANN, B. [1973] *Untersuchungen ueber haltende Programme fuer Turing-Maschinen mit 2 Zeichen und bis zu 5 Befehlen* (**P** 1653) GI Jahrestag (2);1972 Karlsruhe 78*72-81
◇ D10 ◇ REV Zbl 259 # 02028 • ID 30350

CATLIN, S. [1976] *ed-regressive sets of order n* (**J** 0036) J Symb Logic 41*146-152
◇ D50 ◇ REV MR 54 # 85 Zbl 331 # 02023 • ID 14791

CATLIN, S. [1977] *Pathologies in the ed-regressive sets of order 2* (**J** 0047) Notre Dame J Formal Log 18*535-544
◇ D50 ◇ REV MR 58 # 10377 Zbl 305 # 02053 • ID 23689

CAVINESS, B.F. [1970] *On canonical forms and simplification* (**J** 0037) ACM J 17*385-396
◇ B35 D35 ◇ REV MR 43 # 7104 Zbl 193.313 • ID 03971

CAVINESS, B.F. & POLLACK, P.L. & RUBALD, C.M. [1971] *An existence lemma for canonical forms in symbolic mathematics* (**J** 0232) Inform Process Lett 1*45-46
◇ D25 F05 ◇ REV MR 45 # 1427 Zbl 221 # 68032 • ID 02214

CAZACU, C. [1969] *Machines Turing et T-schemas* (**J** 0230) An Univ Iasi, NS, Sect Ia 15*1-8
◇ D10 ◇ REV MR 40 # 7115 Zbl 188.329 • ID 02218

CAZACU, C. & CAZACU, M. [1978] *An attempt to a general theory of algorithms* (**J** 0230) An Univ Iasi, NS, Sect Ia 24*243-250
◇ B75 D20 ◇ REV MR 80e:68071 Zbl 407 # 68043 • ID 80858

CAZACU, C. [1981] *A depth-space analysis of algorithms* (**J** 0230) An Univ Iasi, NS, Sect Ia 27*239-246
◇ D15 D20 ◇ REV MR 84g:68018 Zbl 474 # 68048 • ID 69299

CAZACU, C. see Vol. II for further entries

CAZACU, M. [1978] see CAZACU, C.

CAZACU, M. [1979] *A general mathematical notion of the algorithm* (**J** 3070) Bul Inst Politeh Iasi, Sect 1 25(29)/3-4*23-26
◇ D03 D10 D20 ◇ REV MR 81g:68046 Zbl 441 # 68027 • ID 80859

CAZACU, M. [1980] *The simulation relations between the classes of algorithms* (**J** 3070) Bul Inst Politeh Iasi, Sect 1 26(30)/1-2*17-21
◇ D03 D10 D20 ◇ REV MR 82e:03039 Zbl 489 # 68025 • ID 71584

CAZANESCU, V.E. [1968] *Les notions fondamentales de la theorie des categories pour la categorie des automates sequentiels abstraits et la categorie des quasimachines* (**J** 0070) Bull Soc Sci Math Roumanie, NS 12*17-22
◇ D05 G30 ◇ REV MR 39 # 5369 Zbl 175.279 • ID 01896

CAZANESCU, V.E. [1978] *Parties algebriques d'un monoide* (**J** 0060) Rev Roumaine Math Pures Appl 23*23-28
◇ D05 ◇ REV MR 57 # 16447 Zbl 395 # 68068 • ID 69300

CAZANESCU, V.E. [1980] *On the uniqueness of the solution of a system of context-free equations* (**J** 0447) An Univ Bucuresti, Mat 29*15-20
◇ D05 ◇ REV MR 82j:68072 Zbl 479 # 68083 • ID 69301

CAZANESCU, V.E. see Vol. I for further entries

Author Index

CEGIELSKI, P. & MCALOON, K. & WILMERS, G.M. [1982] *Modeles recursivement satures de l'addition et de la multiplication des entiers naturels (English summary)* (**P** 3623) Logic Colloq;1980 Prague 57-68
◇ C50 C57 C62 F30 H15 ◇ REV MR 84i:03072 Zbl 527 # 03043 • ID 34553

CEGIELSKI, P. see Vol. I, III, VI for further entries

CELLUCCI, C. [1964] *Categorie ricorsive* (**J** 4408) Boll Unione Mat Ital, III Ser 19*300-305
◇ C57 D20 G30 ◇ REV MR 30 # 1928 Zbl 149.8 • ID 02198

CELLUCCI, C. [1985] *Proof theory and complexity* (**J** 0154) Synthese 62*173-189
◇ D15 F05 F20 F98 ◇ ID 45633

CELLUCCI, C. see Vol. III, V, VI for further entries

CELONI, J.R. & PAUL, W.J. & TARJAN, R.E. [1976] *Space bounds for a game on graphs* (**P** 2597) ACM Symp Th of Comput (8);1976 Hershey 149-160
◇ D10 D15 ◇ REV MR 56 # 4239 Zbl 365 # 05027 • ID 51064

CENZER, D. [1974] *Analytic inductive definitions* (**J** 0036) J Symb Logic 39*310-312
◇ D55 D70 ◇ REV MR 50 # 9557 Zbl 296 # 02021 • ID 03974

CENZER, D. [1974] see BLASS, A.R.

CENZER, D. [1974] *Inductively defined sets of reals* (**J** 0015) Bull Amer Math Soc 80*485-487
◇ D55 D70 ◇ REV MR 53 # 10568 Zbl 328 # 02028 • ID 14522

CENZER, D. [1974] *Ordinal recursion and inductive definitions* (**P** 0602) Generalized Recursion Th (1);1972 Oslo 221-264
◇ D60 D65 D70 E45 ◇ REV MR 55 # 2535 Zbl 284 # 02021 • ID 04278

CENZER, D. [1974] *The boundedness principle in ordinal recursion* (**J** 0027) Fund Math 81*203-212
◇ D60 D70 ◇ REV MR 52 # 13343 Zbl 328 # 02027 • ID 01914

CENZER, D. [1976] *Inductive definitions, positive and monotone* (**P** 1476) Set Th & Hierarch Th (2) (Mostowski);1975 Bierutowice 51-63
◇ D65 D70 ◇ REV MR 56 # 15390 Zbl 343 # 02029 • ID 31364

CENZER, D. [1976] *Monotone inductive definitions over the continuum* (**J** 0036) J Symb Logic 41*188-198
◇ D55 D70 E15 ◇ REV MR 55 # 90 Zbl 344 # 02032 • ID 14799

CENZER, D. [1977] *Non-generable RE sets* (**P** 2588) FCT'77 Fund of Comput Th;1977 Poznan 379-385
◇ D03 D25 ◇ REV MR 58 # 16224 Zbl 368 # 02043 • ID 51260

CENZER, D. [1978] *Parametric inductive definitions and recursive operators over the continuum* (**J** 0027) Fund Math 100*9-15
◇ D55 D70 ◇ REV MR 58 # 187 Zbl 391 # 03023 • ID 29198

CENZER, D. & MAULDIN, R.D. [1980] *Inductive definability: measure and category* (**J** 0345) Adv Math 38*55-90
◇ D70 E15 ◇ REV MR 82b:03086 Zbl 466 # 03018 • ID 54969

CENZER, D. [1980] *Non-generable formal languages* (**J** 2095) Fund Inform, Ann Soc Math Pol, Ser 4 3*95-103
◇ D03 D25 D30 D70 ◇ REV MR 82d:03068 Zbl 454 # 68087 • ID 54257

CENZER, D. & MAULDIN, R.D. [1982] *On the Borel class of the derived set operator* (**J** 0353) Bull Soc Math Fr 110*357-380
◇ D55 E15 ◇ REV MR 85b:54058 Zbl 514 # 54027 • REM Part I. Part II 1983 • ID 37461

CENZER, D. & MAULDIN, R.D. [1983] *On the Borel class of the derived set operator. II* (**J** 0353) Bull Soc Math Fr 111*367-372
◇ D55 E15 ◇ REV MR 86a:54046 Zbl 552 # 54027 • REM Part I 1982 • ID 43460

CENZER, D. [1984] *Monotone reducibility and the family of infinite sets* (**J** 0036) J Symb Logic 49*774-782
◇ D15 D55 ◇ REV MR 86a:03052 Zbl 573 # 54030 • ID 42431

CENZER, D. see Vol. V for further entries

CERNY, A. [1984] *On generalized words of Thue-Morse* (**P** 3658) Math Founds of Comput Sci (11);1984 Prague 232-239
◇ D03 ◇ REV Zbl 556 # 68042 • ID 46853

CERNY, J. [1970] *Two particular types of finite state languages* (**P** 1580) Tagung Formale Sprachen;1970 Oberwolfach 7-9
◇ D05 ◇ REV Zbl 247 # 94039 • ID 60892

CHAITIN, G.J. [1969] *On the length of programs for computing finite binary sequences: statistical considerations* (**J** 0037) ACM J 16*145-159
◇ D15 D80 ◇ REV MR 38 # 5414 Zbl 187.283 • ID 01923

CHAITIN, G.J. [1969] *On the simplicity and speed of programs for computing infinite sets of natural numbers* (**J** 0037) ACM J 16*407-422
◇ D15 D80 ◇ REV MR 39 # 2638 Zbl 187.283 • ID 01922

CHAITIN, G.J. [1974] *Information-theoretic limitations of formal systems* (**J** 0037) ACM J 21*403-424
◇ D15 D80 F20 ◇ REV MR 56 # 13775 Zbl 287 # 68027 • ID 60893

CHAITIN, G.J. [1974] *Information-theoretic computational complexity* (**J** 2745) IEEE Trans Inf Theory IT-20*10-15
◇ D15 D80 ◇ REV MR 55 # 2529 Zbl 282 # 68022 • ID 71625

CHAITIN, G.J. [1976] *Algorithmic entropy of sets* (**J** 2687) Comp Math Appl 2*233-245
◇ D15 D25 D80 ◇ REV Zbl 367 # 68036 • ID 51209

CHAITIN, G.J. [1976] *Information-theoretic characterization of recursive infinite strings* (**J** 1426) Theor Comput Sci 2*45-48
◇ D15 D20 D80 ◇ REV MR 54 # 1709 Zbl 328 # 02029 • ID 23996

CHAITIN, G.J. [1977] *Program size, oracles, and the jump operation* (**J** 0351) Osaka J Math 14*139-149
◇ D15 D25 D30 D80 ◇ REV MR 57 # 5714 Zbl 359 # 94031 • ID 50634

CHAITIN, G.J. & SCHWARTZ, J.T. [1978] *A note on Monte Carlo primality tests and algorithmic information theory* (**J** 0155) Commun Pure Appl Math 31*521-527
◇ D15 D80 ◇ REV MR 58 # 4629 Zbl 401 # 94008 • ID 54646

CHAITIN, G.J. see Vol. VI for further entries

CHAN, TATHUNG & IBARRA, O.H. [1983] *On the space and time complexity of functions computable by simple programs* (**J** 1428) SIAM J Comp 12*708-716
◇ D15 ◇ REV MR 84m:68033 Zbl 524 # 68030 • ID 39514

CHANDLER, B. & MAGNUS, W. [1982] *The history of combinatorial group theory* (**X** 0811) Springer: Heidelberg & New York viii + 234pp
◇ A10 D40 ◇ REV MR 85c:01001 Zbl 498 # 20001
• ID 38996

CHANDRA, A.K. & STOCKMEYER, L.J. [1976] *Alternation* (**P** 1757) IEEE Symp Found of Comput Sci (17);1976 Houston 98-108
◇ D10 D15 ◇ REV MR 58 # 25107 • ID 80879

CHANDRA, A.K. & STOCKMEYER, L.J. [1979] *Provably difficult combinatorial games* (**J** 1428) SIAM J Comp 8*151-174
◇ D15 ◇ REV MR 80d:68060 Zbl 421 # 68044 • ID 53456

CHANDRA, A.K. [1980] see BLUM, M.

CHANDRA, A.K. & KOZEN, D. & STOCKMEYER, L.J. [1981] *Alternation* (**J** 0037) ACM J 28*114-133
◇ D10 D15 ◇ REV MR 83g:68059 Zbl 473 # 68043
• ID 55398

CHANDRA, A.K. & HALPERN, J.Y. & MEYER, A.R. & PARIKH, R. [1985] *Equations between regular terms and application to process logic* (**J** 1428) SIAM J Comp 14*935-942
◇ B45 D05 ◇ ID 49198

CHANDRA, A.K. see Vol. I for further entries

CHANG, C.C. & MOSCHOVAKIS, Y.N. [1970] *The Souslin-Kleene theorem for V_κ with cofinality$(\kappa) = \omega$* (**J** 0048) Pac J Math 35*565-569
◇ D55 D70 E47 ◇ REV MR 43 # 56 Zbl 229 # 02051 JSL 40.245 • ID 01977

CHANG, C.C. see Vol. I, II, III, V for further entries

CHANG, NINGSAN & FU, K.S. [1979] *Parallel parsing of tree languages for syntactic pattern recognition* (**J** 3227) Pattern Recognition 11*213-222
◇ D05 ◇ REV Zbl 411 # 68074 • ID 69310

CHANG, SHIKUO [1971] *Fuzzy programs - theory and applications* (**P** 1383) Symp Comput & Automata;1971 Brooklyn 21*147-164
◇ B75 D05 ◇ REV Zbl 265 # 68018 • ID 29847

CHANG, SHIKUO see Vol. V for further entries

CHAPENKO, V.P. & GOBZEMIS, A.YU. & KURMIT, A.A. & LORENTS, A.A. & PETRENKO, A.F. & VASYUKEVICH, V.O. & YAKUBAJTIS, EH.A. & ZAZNOVA, N.E. [1976] *Automata theory (Russian)* (**J** 3188) Itogi Nauki Tekh, Ser Teor Veroyat Mat Stat Teor Kibern 13*109-188,299
• TRANSL [1977] (**J** 1531) J Sov Math 7*193-243
◇ D05 D98 ◇ REV MR 55 # 10161 Zbl 438 # 68014
• ID 81687

CHAPUT, D. & SABIDUSSI, G. [1980] *Une generalization du theoreme de Ginsburg-Rose* (**J** 0017) Canad J Math 32*567-575
◇ D05 ◇ REV MR 82a:20078 Zbl 445 # 68042 • ID 69311

CHARLESWORTH, A. [1981] *A proof of Goedel's theorem in terms of computer programs* (**J** 0497) Math Mag 54*109-121
◇ D10 F30 ◇ REV MR 83e:03001 Zbl 512 # 03006
• ID 35267

CHARLESWORTH, A. see Vol. V for further entries

CHAUVIN, A. [1960] *Sur les modeles du calcul K_0 de Bochvar, avec ou sans egalite, et l'interpretation des paradoxes de la logique dans la theorie des ensembles elementaires arithmetiques* (**J** 0150) Acad Roy Belg Bull Cl Sci (5) 46*124-131
◇ B50 D55 ◇ REV MR 22 # 7935 • ID 01998

CHAUVIN, A. [1961] *Deux modeles verifiant certains axiomes de la theorie des ensembles de Goedel, et construits dans la theorie des ensembles arithmetiques de Kleene. Construction des modeles* (**J** 0109) C R Acad Sci, Paris 253*1394-1396
◇ C62 D55 E30 E70 ◇ REV MR 24 # A1218 Zbl 112.11
• ID 01999

CHAUVIN, A. [1961] *Deux modeles verifiant certains axiomes de la theorie des ensembles de Goedel, et construits dans la theorie des ensembles arithmetiques de Kleene. Validite des axiomes* (**J** 0109) C R Acad Sci, Paris 253*1519-1521
◇ C62 D55 E30 E70 ◇ REV MR 24 # A1835 Zbl 112.11
• ID 25473

CHAUVIN, A. [1962] *Sur les classes arithmetiques constructibles* (**J** 0109) C R Acad Sci, Paris 254*3796-3798
◇ D55 ◇ REV MR 25 # 2949 Zbl 111.9 • ID 02063

CHAUVIN, A. [1962] *Sur les ensembles arithmetiques constructibles* (**J** 0109) C R Acad Sci, Paris 254*3615-3617
◇ D55 E45 F15 ◇ REV MR 25 # 2948 Zbl 113.6
• ID 02061

CHAUVIN, A. see Vol. I, V, VI for further entries

CHAZELLE, B. & MONIER, L. [1983] *Unbounded hardware is equivalent to deterministic Turing machines* (**J** 1426) Theor Comput Sci 24*123-130
◇ D10 ◇ REV MR 85a:68047 Zbl 532 # 68053 • ID 38867

CHAZELLE, B. & OTTMANN, T. & SOISALON-SOININEN, E. & WOOD, D. [1984] *The complexity and decidability of SEPARATION* (**P** 4012) Automata, Lang & Progr (11);1984 Antwerpen 119-127
◇ D15 ◇ REV Zbl 554 # 68027 • ID 47135

CHEBOTAR', K.S. [1981] *Some modifications of Knuth's algorithm for verifying cyclicity of attribute grammars (Russian)* (**J** 2605) Programmirovanie 1981/1*74-82
• TRANSL [1981] (**J** 2604) Progr Comput Software 7*58-61
◇ D05 ◇ REV MR 82j:68073 Zbl 466 # 68062 • ID 69312

CHEBOTAREVA, L.K. [1967] *On the belonging problem for free and almost free lattices (Russian)* (**J** 0092) Sib Mat Zh 8*399-405
◇ D40 G10 ◇ REV MR 35 # 2797 • ID 01898

CHEN, HUOWANG [1980] *The decision problem of families of general recursive functions (Chinese)* (**J** 3793) Jisuanjii Xuebao 3*112-118
◇ D20 ◇ ID 48506

CHEN, HUOWANG [1981] *Effectively continuous functionals (Chinese)* (**J** 0418) Shuxue Xuebao 24*801-816
◇ D20 ◇ REV MR 83m:03057 Zbl 507 # 03015 • ID 35457

CHEN, HUOWANG see Vol. V for further entries

CHEN, I-NGO & SHENG, C.L. [1970] *The decision problems of definite stochastic automata* (**J** 0946) SIAM J Control 8*124-134
◇ D05 ◇ REV MR 41 # 9668 Zbl 195.27 • ID 16307

Author Index

CHEN, JIYUAN [1984] *The satisfiability problem for simple boolean expressions belongs to P (Chinese) (English summary)* (**J** 2521) Beijing Shifan Daxue Xuebao, Ziran Kexue 1984/1*14-21
◇ B05 B25 D15 ◇ REV MR 85j:03059 • ID 44191

CHEN, KEHSUN [1978] *Recursive well-founded orderings* (**J** 0007) Ann Math Logic 13*117-147
◇ D45 F15 ◇ REV MR 80e:03047 Zbl 384 # 03027
• ID 27962

CHEN, KEJIAN [1982] *Tradeoffs in the inductive inference of nearly minimal size programs* (**J** 0194) Inform & Control 52*68-86
◇ D20 ◇ REV MR 84g:03057 Zbl 537 # 03024 • ID 34166

CHEN, ZHONGYUEH [1978] *Formalization of equivalence of recursively defined functions* (**J** 0191) Inform Sci 15*219-227
◇ D20 ◇ REV MR 81f:68020 Zbl 443 # 68023 • ID 56441

CHEN, ZHONGYUEH [1984] *On the fixpoints of nondeterministic recursive definitions* (**J** 0119) J Comp Syst Sci 29*58-79
◇ D15 ◇ ID 44089

CHENG, K.N. & CHONG, C.T. (EDS.) [1980] *Proceedings of the mathematical seminar* (**X** 1941) Nanyang Univ Publ: Singapore iii + 54pp
◇ D97 ◇ REV MR 82e:00010 Zbl 447 # 00021 • ID 48649

CHERLIN, G.L. [1980] *Rings of continuous functions: decision problems* (**P** 2625) Model Th of Algeb & Arithm;1979 Karpacz 44-91
◇ B25 C60 C65 D35 ◇ REV MR 82e:03019
Zbl 454 # 03004 • ID 54217

CHERLIN, G.L. & DRIES VAN DEN, L. & MACINTYRE, A. [1981] *Decidability and undecidability theorems for PAC-fields* (**J** 0589) Bull Amer Math Soc (NS) 4*101-104
◇ B25 C60 D35 ◇ REV MR 82g:03057 Zbl 466 # 12017
• ID 54982

CHERLIN, G.L. & SCHMITT, P.H. [1981] *Undecidable L^t theories of topological abelian groups* (**J** 0036) J Symb Logic 46*761-772
◇ C60 C90 D35 ◇ REV MR 83e:03047 Zbl 482 # 03016
• ID 35227

CHERLIN, G.L. [1984] *Definability in power series rings of nonzero characteristic* (**P** 2153) Logic Colloq;1983 Aachen 1*102-112
◇ C40 C60 D35 ◇ REV MR 86h:03059 Zbl 574 # 03017
• ID 41760

CHERLIN, G.L. [1984] *Undecidability of rational function fields in nonzero characteristic* (**P** 3710) Logic Colloq;1982 Firenze 85-95
◇ C60 C85 D35 ◇ REV MR 86f:03068 Zbl 551 # 03027
• ID 43903

CHERLIN, G.L. see Vol. I, III, V for further entries

CHERNIAK, C. [1984] *Computational complexity and the universal acceptance of logic* (**J** 0301) J Phil 81*739-758
◇ D15 ◇ REV MR 86c:03002 • ID 44377

CHERNIAVSKY, J.C. [1973] *The complexity of some non-classical logics* (**P** 3062) IEEE Symp Switch & Automata Th (14);1973 Iowa City 209-213
◇ B45 D15 F20 F50 ◇ REV MR 56 # 2801 • ID 71692

CHERNIAVSKY, J.C. [1976] *Simple programs realize exactly Presburger formulas* (**J** 1428) SIAM J Comp 5*666-677
◇ B75 D15 D20 F30 ◇ REV MR 58 # 13841
Zbl 353 # 68018 • ID 31025

CHERNIAVSKY, J.C. [1979] *On finding test data sets for loop free programs* (**J** 0232) Inform Process Lett 8*106-107
◇ B75 D80 ◇ REV MR 58 # 25063 Zbl 397 # 68005
• ID 52720

CHERNOV, V.P. [1972] *Classification of spaces of operators of finite types (Russian)* (**S** 0228) Zap Nauch Sem Leningrad Otd Mat Inst Steklov 32*148-152,158 • ERR/ADD ibid 40*161
• TRANSL [1976] (**J** 1531) J Sov Math 6*471-474
◇ D45 D65 F60 ◇ REV MR 52 # 5384a MR 52 # 5384b
Zbl 347 # 02024 • ID 17990

CHERNOV, V.P. [1972] *Constructive operators of finite types (Russian) (English summary)* (**S** 0228) Zap Nauch Sem Leningrad Otd Mat Inst Steklov 32*140-147,158 • ERR/ADD ibid 40*161
• TRANSL [1976] (**J** 1531) J Sov Math 6*465-470
◇ D45 D65 F60 ◇ REV MR 52 # 5383a MR 52 # 5383b
Zbl 347 # 02023 • ID 17991

CHERNOV, V.P. [1972] *Topological variants of the continuity theorem for mappings and related theorems theorems (Russian) (English summary)* (**S** 0228) Zap Nauch Sem Leningrad Otd Mat Inst Steklov 32*129-139,158 • ERR/ADD ibid 40*161
• TRANSL [1976] (**J** 1531) J Sov Math 6*456-464
◇ D20 F60 ◇ REV MR 52 # 5381a MR 52 # 5381b
Zbl 347 # 02022 • ID 17992

CHERNOV, V.P. see Vol. V, VI for further entries

CHERNYAKHOVSKIJ, N.P. [1976] *The expressibility of realizability, in the language of formal arithmetic (Russian)* (**S** 0554) Issl Teor Algor & Mat Logik (Moskva) 2*51-56,157
◇ D80 F30 F50 ◇ REV MR 58 # 16193 • ID 71693

CHERNYAVSKIJ, V.S. [1959] *On a certain simplification of normal algorithms (Russian)* (**P** 0607) All-Union Math Conf (3);1956 Moskva 4*91
◇ D03 ◇ ID 31285

CHERNYAVSKIJ, V.S. [1959] *On a class of normal Markov algorithms (Russian)* (**C** 1155) Log Issl (Moskva) 263-299
• TRANSL [1965] (**J** 0225) Amer Math Soc, Transl, Ser 2 48*1-35
◇ D03 ◇ REV MR 23 # A3072 Zbl 111.9 JSL 36.693
• ID 02200

CHERNYAVSKIJ, V.S. [1960] *On the reversibility of algorithms (Russian)* (**J** 0065) Tr Moskva Mat Obshch 9*425-453
• TRANSL [1964] (**J** 0225) Amer Math Soc, Transl, Ser 2 39*207-239
◇ D03 D10 ◇ REV MR 23 # A2315 Zbl 116.338
JSL 31.655 • ID 01915

CHERNYAVSKIJ, V.S. & KOZMIDIADI, V.A. [1962] *Ueber einige Begriffe der Theorie der mathematischen Maschinen (Russisch)* (**C** 4738) Vopr Teor Mat Mashin, Sbor 2 128-143
◇ D05 D10 ◇ REV MR 31 # 6738 Zbl 117.121 • ID 48077

CHESNOKOV, S.V. [1984] *Syllogisms in deterministic analysis (Russian)* (**J** 0977) Izv Akad Nauk SSSR, Tekh Kibern 1984/5*55-83
• TRANSL [1984] (**J** 0522) Engin Cybern 22/5*96-120
◇ D15 ◇ ID 46843

Chew, P. & Machtey, M. [1981] *A note on structure and looking back applied to the relative complexity of computable functions* (**J** 0119) J Comp Syst Sci 22*53-59
◇ D15 ◇ rev MR 83b:03051 Zbl 474#68063 • id 55452

Chiaraviglio, L. [1975] see Baralt-Torrijos, J.

Chien, T.Y. [1980] *Decompositions of a free monoid into disjunctive languages* (**J** 3448) Soochow J Math 5*121-127
◇ D05 ◇ rev MR 81h:20072 Zbl 426#68068 • id 69316

Chihara, C.S. [1975] see Berg, J.

Chihara, C.S. see Vol. I, V, VI for further entries

Chikvashvili, R.I. [1972] *On the theory of effective functionals (Russian) (Georgian and English summaries)* (**J** 0233) Soobshch Akad Nauk Gruz SSR 66*33-36
◇ D20 ◇ rev MR 47#1591 Zbl 234#02025 • id 02673

Chikvashvili, R.I. [1972] *The index sets that are connected with effective functionals (Russian) (Georgian and English summaries)* (**J** 0233) Soobshch Akad Nauk Gruz SSR 68*21-24
◇ D20 D55 ◇ rev MR 49#7126 Zbl 268#02029 • id 03997

Chikvashvili, R.I. [1975] *On a certain problem of Kuznecov and Trahtenbrot (Russian) (Georgian and English summaries)* (**J** 0233) Soobshch Akad Nauk Gruz SSR 79*309-312
◇ D55 E15 ◇ rev MR 52#10387 Zbl 337#04003 • id 21685

Chimev, K.N. [1984] *Separable pairs of some classes of functions (Bulgarian) (English and Russian summary)* (**J** 3171) God Vissh Ucheb Zaved, Prilozhna Mat, Sofiya 20/2*47-55
◇ D20 E20 ◇ id 49515

Chimev, K.N. [1984] *Separation of a union of sets of arguments of functions (Bulgarian) (English and Russian summary)* (**J** 3171) God Vissh Ucheb Zaved, Prilozhna Mat, Sofiya 20/2*57-65
◇ D20 E20 ◇ id 49516

Chimev, K.N. see Vol. I, II, V for further entries

Chirathamjaree, C. [1980] see Ackroyd, M.H.

Chirkov, M.K. & Eglitis, L.V. [1977] *Synthesis of partial automata by partial regular events (Russian)* (**S** 0716) Vychisl Tekh Vopr Kibern (Univ Leningrad) 14*175-183
◇ D05 ◇ rev MR 57#19168 Zbl 429#68054 • id 69480

Chirkov, M.K. [1978] *On the abstract analysis of partial automata (Russian)* (**J** 3402) Met Vychislenij, Univ Leningrad 11*178-184
◇ D05 ◇ rev MR 80a:68067 Zbl 458#68018 • id 69319

Chirkov, M.K. [1979] *On some types of incompletely specified automata* (**J** 0380) Acta Cybern (Szeged) 4*151-165
◇ D05 ◇ rev MR 80j:68043 Zbl 401#68025 • id 69320

Chirkov, M.K. see Vol. I, II for further entries

Chistov, A.L. & Grigor'ev, D.Yu. [1984] *Complexity of quantifier elimination in theory of algebraically closed fields* (**P** 3658) Math Founds of Comput Sci (11);1984 Prague 17-31
◇ C10 C60 D15 ◇ rev Zbl 562#03015 • id 44792

Chitescu, I. [1982] see Calude, C.

Chitescu, I. [1983] see Calude, C.

Chitescu, I. [1985] see Calude, C.

Chlebus, B.S. [1979] *On decidability of propositional algorithmic logic (Polish) (English summary)* (**S** 3270) Spraw Inst Inf, Uniw Warsaw 83*28pp
◇ B25 B75 D10 D15 ◇ rev Zbl 454#03006 • id 54219

Chlebus, B.S. [1980] *Decidability and definability results concerning well-orderings and some extensions of first order logic* (**J** 0068) Z Math Logik Grundlagen Math 26*529-536
◇ B15 B25 C40 C65 C80 D35 E07 ◇ rev MR 82b:03072 Zbl 445#03018 • id 56482

Chlebus, B.S. [1981] *On the computational complexity of satisfiability in propositional logic programs* (**S** 3270) Spraw Inst Inf, Uniw Warsaw 99*1-40
◇ B75 D15 ◇ rev MR 83k:68026 Zbl 508#68019 • id 38156

Chlebus, B.S. [1982] *On the computational complexity of satisfiability in propositional logics of programs* (**J** 1426) Theor Comput Sci 21*179-212
◇ B75 D15 ◇ rev MR 83k:68026 Zbl 496#68020 • id 37748

Chlebus, B.S. & Chrobak, M. [1984] *Probabilistic Turing machines and recursively enumerable Dedekind cuts* (**J** 0232) Inform Process Lett 19*167-171
◇ D10 D25 D80 ◇ rev Zbl 559#03025 • id 44829

Chlebus, B.S. [1985] *From domino tilings to a new model of computation* (**P** 4670) Comput Th (5);1984 Zaborow 24-33
◇ D20 ◇ id 49766

Chlebus, B.S. see Vol. II, III for further entries

Choban, M.M. [1973] *The descriptive theory of sets* (**J** 0137) C R Acad Bulgar Sci 26*449-452
◇ D55 E15 ◇ rev MR 58#16307 Zbl 335#54007 • id 71748

Choban, M.M. see Vol. V for further entries

Choffrut, C. [1981] *A closure property of deterministic context-free languages* (**J** 0232) Inform Process Lett 12*13-16
◇ D05 ◇ rev MR 82a:68139 Zbl 454#68096 • id 69322

Choffrut, C. [1981] *Prefix-preservation for rational partial functions is decidable* (**P** 3475) Theor Comput Sci (5);1981 Karlsruhe 159-166
◇ D05 ◇ rev Zbl 491#68053 • id 37766

Chomsky, N. & Miller, George A. [1958] *Finite state languages* (**J** 0194) Inform & Control 1*91-112
◇ B65 D05 ◇ rev MR 21#7133 Zbl 81.145 JSL 31.245 • id 02096

Chomsky, N. [1966] *Topics in the theory of generative grammar* (**X** 0873) Mouton: Paris 95pp
◇ B65 D05 ◇ id 25311

Chomsky, N. [1972] *Studies on semantics in generative grammars* (**X** 0873) Mouton: Paris 207pp
◇ B65 B98 D05 ◇ id 25314

Chomsky, N. see Vol. I, II for further entries

Chong, C.T. [1974] *Almost local non-α-recursiveness* (**J** 0036) J Symb Logic 39*552-562
◇ D30 D60 ◇ rev MR 51#5276 Zbl 298#02038 • id 04011

Author Index

Chong, C.T. [1976] *An α-finite injury method of the unbounded type* (**J** 0036) J Symb Logic 41*1-17
◇ D30 D60 ◇ REV MR 57 # 16019 Zbl 358 # 02053 • ID 14757

Chong, C.T. & Lerman, M. [1976] *Hyperhypersimple α-r.e. sets* (**J** 0007) Ann Math Logic 9*1-48
◇ D60 ◇ REV MR 52 # 7880 Zbl 317 # 02045 • ID 17999

Chong, C.T. [1976] *Minimal upper bounds for ascending sequences of α-recursively enumerable degrees* (**J** 0036) J Symb Logic 41*250-260
◇ D30 D60 ◇ REV MR 58 # 16228 Zbl 358 # 02054 • ID 14807

Chong, C.T. [1977] *A recursion-theoretic characterization of constructible reals* (**J** 0161) Bull London Math Soc 9*241-244
◇ D60 E40 E45 ◇ REV MR 58 # 21617 Zbl 407 # 03041 • ID 31058

Chong, C.T. [1977] *The minimal α-degree problem* (**J** 1735) Bull South East Asian Soc 1*44-45
◇ D30 D60 ◇ REV Zbl 418 # 03033 • ID 53318

Chong, C.T. [1979] *Σ_n-cofinalities of J_α* (**J** 0027) Fund Math 102*101-107
◇ D60 E45 E47 ◇ REV MR 80b:03078 Zbl 403 # 03033 • ID 54747

Chong, C.T. [1979] *Cones of degrees* (**J** 3926) J Singapore Nat Acad Sci 8*81-84
◇ D30 D60 E45 ◇ REV Zbl 495 # 03028 • ID 36886

Chong, C.T. [1979] *Generic sets and minimal α-degrees* (**J** 0064) Trans Amer Math Soc 254*157-169
◇ D30 D60 E40 ◇ REV MR 81c:03037 Zbl 416 # 03045 • ID 53211

Chong, C.T. [1979] *Major subsets of α-recursively enumerable sets* (**J** 0029) Israel J Math 34*106-114
◇ D60 ◇ REV MR 81k:03043 Zbl 433 # 03023 • ID 71750

Chong, C.T. [1980] *Degree theory: from ω to singular cardinals* (**P** 3236) Math Seminar;1980 Singapore 45-54
◇ D30 D60 E45 ◇ REV MR 84d:03059 Zbl 471 # 03040 • ID 55235

Chong, C.T. [1980] see Cheng, K.N.

Chong, C.T. [1980] *Rich sets* (**J** 0053) Proc Amer Math Soc 80*458-460
◇ D30 D60 ◇ REV MR 81j:03077 Zbl 481 # 03029 • ID 71751

Chong, C.T. [1981] *Hilbert's tenth problem* (**J** 2053) Math Medley 9*39-42
◇ D35 ◇ REV MR 83i:10075 Zbl 499 # 10018 • ID 38295

Chong, C.T. [1982] *Double jumps of minimal degrees over cardinals* (**J** 0036) J Symb Logic 47*329-334
◇ D30 D60 E45 ◇ REV MR 83f:03039 Zbl 526 # 03023 • ID 35280

Chong, C.T. [1982] *Global and local admissibility* (**P** 3634) Patras Logic Symp;1980 Patras 325-338
◇ D60 ◇ REV MR 85i:03153a Zbl 525 # 03034 • REM Part II 1983 • ID 38267

Chong, C.T. & Friedman, S.D. [1983] *Degree theory on $\aleph_\omega$* (**J** 0073) Ann Pure Appl Logic 24*87-97
◇ D30 D60 ◇ REV MR 84k:03114 Zbl 527 # 03027 • ID 36132

Chong, C.T. [1983] *Global and local admissibility. II. Major subsets and automorphisms* (**J** 0073) Ann Pure Appl Logic 24*99-111
◇ D30 D60 ◇ REV MR 85i:03153b Zbl 568 # 03021 • REM Part I 1982 • ID 44225

Chong, C.T. [1983] *Hyperhypersimple supersets in admissible recursion theory* (**J** 0036) J Symb Logic 48*185-192
◇ D60 ◇ REV MR 85d:03092 Zbl 575 # 03032 • ID 41104

Chong, C.T. [1984] *Minimal α-hyperdegrees* (**J** 0009) Arch Math Logik Grundlagenforsch 24*63-72
◇ D60 ◇ REV MR 85j:03076 • ID 42400

Chong, C.T. & Jockusch Jr., C.G. [1984] *Minimal degrees and 1-generic sets below 0'* (**P** 2153) Logic Colloq;1983 Aachen 2*63-77
◇ D30 ◇ REV MR 86c:03040 Zbl 574 # 03023 • ID 39942

Chong, C.T. [1984] *Techniques of admissible recursion theory* (**X** 0811) Springer: Heidelberg & New York viii+214pp
◇ D60 D98 ◇ REV Zbl 566 # 03028 • ID 44582

Chong, C.T. [1985] *Recursion theory on strongly Σ_2-inadmissible ordinals* (**P** 3342) Rec Th Week;1984 Oberwolfach 49-64
◇ D60 ◇ ID 45297

Chong, C.T. see Vol. I, II for further entries

Choquet, G. [1959] *Ensembles $\mathcal{K}$-analytiques et $\mathcal{K}$-Sousliniens. Cas general et cas metrique* (**J** 0240) Ann Inst Fourier 9*75-81
◇ D55 E15 ◇ REV MR 22 # 3692a Zbl 94.34 • ID 02221

Choquet, G. see Vol. III, V for further entries

Chottin, L. & Kierszenbaum, F. [1978] *Sur les langages d'insertions* (**P** 2599) CAAP'78 Arbres en Algeb & Progr (3);1978 Lille 250-253
◇ D05 ◇ REV Zbl 387 # 68052 • ID 69324

Chottin, L. [1979] *Strict deterministic languages and controlled rewriting systems* (**P** 1873) Automata, Lang & Progr (6);1979 Graz 71*104-117
◇ D05 ◇ REV MR 83f:68094 Zbl 438 # 68031 • ID 69323

Chou, S.M. & Fu, K.S. [1979] *Inference for transition network grammars* (**J** 3144) Comput Lang, Int J 4*83-92
◇ D05 ◇ REV Zbl 401 # 68037 • ID 69325

Choueka, Y. [1974] *Theories of automata on ω-tapes: a simplified approach* (**J** 0119) J Comp Syst Sci 8*117-141
◇ D05 ◇ REV MR 49 # 7124 Zbl 292 # 02033 • ID 03993

Choueka, Y. [1978] *Finite automata, definable sets, and regular expressions over ω^n-tapes* (**J** 0119) J Comp Syst Sci 17*81-97
◇ D05 ◇ REV MR 58 # 8514 Zbl 386 # 03018 • ID 52194

Choueka, Y. [1981] see Amihud, A.

Choueka, Y. see Vol. V for further entries

Chow, T.S. [1976] *Lower bounds on the complexity of 0-1-valued recursive functions* (**J** 0194) Inform & Control 31*17-42
◇ D15 D20 ◇ REV MR 55 # 6943 Zbl 328 # 68042 • ID 60951

Chowla, S. [1965] *The Riemann hypothesis and Hilbert's tenth problem* (**J** 0121) Kon Norske Vidensk Selsk Forh 38*62-64
◇ D35 ◇ REV MR 32 # 4101 Zbl 133.300 • ID 02224

CHOWLA, S. [1965] *The Riemann hypothesis and Hilbert's tenth problem. Mathematics and its applications Vol. 4* (**X** 0836) Gordon & Breach: New York xv + 119pp
◇ D35 D98 ◇ REV MR 31 # 2201 Zbl 133.300 • ID 49845

CHRISTEN, C. [1976] *Spektralproblem und Komplexitaetstheorie* (**P** 3196) Kompl von Entscheid Probl;1973/74 Zuerich 102-126
◇ B10 C13 D10 D15 D25 D35 ◇ REV MR 57 # 18232 Zbl 391 # 03021 • ID 52348

CHRISTIAN, C.C. [1981] *Das rekursive Inaccessibilitaetstheorem und der Goedelsche Unvollstaendigkeitssatz in ihrer Bedeutung fuer die Informatik* (**J** 3480) Informatik & Philos 11*152-168
◇ A05 D35 D80 F30 ◇ REV Zbl 475 # 03002 • ID 55456

CHRISTIAN, C.C. see Vol. I, II, V, VI for further entries

CHRISTODOULAKIS, D. [1980] *Eine einfache Basis fuer die Berechenbarkeit. Die Konstruktion der universellen Turingmaschine mit einem Zustand, zwei Symbolen und drei Baendern* (**X** 0817) Ges Math Datenverarbeit: Bonn 127*102pp
◇ D10 ◇ REV MR 83h:03056 Zbl 475 # 68021 • ID 69326

CHRISTOL, G. [1979] *Ensembles presque periodiques k-reconnaissables* (**J** 1426) Theor Comput Sci 9*141-145
◇ D05 ◇ REV MR 80e:68141 Zbl 402 # 68044 • ID 69327

CHROBAK, M. [1984] see CHLEBUS, B.S.

CHROBAK, M. [1985] *Hierarchies of one-way multihead automata languages* (**P** 4628) Automata, Lang & Progr (12);1985 Nafplion 101-110
◇ D05 ◇ ID 49558

CHU, TANIEN & SHI, HUIJAN [1980] *Some properties of quasi-disjunctive languages on a free monoid* (**J** 3448) Soochow J Math 6*59-63
◇ D05 ◇ REV MR 84e:68082 Zbl 477 # 68081 • ID 69330

CHUDNOVSKY, G.V. [1970] *Diophantine predicates (Russian)* (**J** 0067) Usp Mat Nauk 25/4*185-186
◇ D25 D35 D40 ◇ REV MR 43 # 1920 Zbl 197.281 • ID 02577

CHUDNOVSKY, G.V. see Vol. III, V for further entries

CHUNG, MOONJUNG & EVANGELIST, W.M. & SUDBOROUGH, I.H. [1985] *Complete problems for space bounded subclasses of NP* (**J** 1431) Acta Inf 22/4*379-395
◇ D15 ◇ ID 49480

CHURCH, A. [1932] *A set of postulates for the foundation of logic* (**J** 0120) Ann of Math, Ser 2 33*346-366
◇ B10 B40 D20 ◇ REV Zbl 4.145 JSL 23.23 FdM 58.997 • REM Part I. Part II 1933 • ID 02123

CHURCH, A. [1933] *A set of postulates for the foundation of logic (second paper)* (**J** 0120) Ann of Math, Ser 2 34*839-864
◇ B10 B40 D20 ◇ REV Zbl 8.289 JSL 24.94 FdM 59.52 • REM Part I 1932 • ID 02124

CHURCH, A. [1935] *A proof of freedom from contradiction* (**J** 0054) Proc Nat Acad Sci USA 21*275-281
◇ B40 D20 ◇ REV Zbl 12.241 FdM 61.55 • ID 02126

CHURCH, A. [1935] *An unsolvable problem of elementary number theory* (**J** 0015) Bull Amer Math Soc 41*332-333
◇ B40 D20 D35 ◇ REV FdM 61.62 • ID 27542

CHURCH, A. [1936] *A note on the "Entscheidungsproblem"* (**J** 0036) J Symb Logic 1*40-41 • ERR/ADD ibid 1*101-102
◇ B20 D20 D35 ◇ REV Zbl 14.385 JSL 1.74 FdM 62.1058 • ID 02132

CHURCH, A. [1936] *An unsolvable problem of elementary number theory* (**J** 0100) Amer J Math 58*345-363
• REPR [1965] (**C** 0718) The Undecidable 89-107
◇ D20 D35 ◇ REV Zbl 14.98 JSL 1.73 FdM 62.46 • ID 02127

CHURCH, A. [1936] *Mathematical logic (mimeographed notes)* (**X** 0857) Princeton Univ Pr: Princeton iii + 113pp
◇ B40 B98 D20 ◇ REV JSL 2.39 FdM 62.1048 • ID 23466

CHURCH, A. & ROSSER, J.B. [1936] *Some properties of conversion* (**J** 0064) Trans Amer Math Soc 39*472-482
◇ B40 D20 ◇ REV Zbl 14.385 JSL 1.74 FdM 62.37 • ID 02130

CHURCH, A. [1940] *On the concept of a random sequence* (**J** 0015) Bull Amer Math Soc 46*130-135
◇ D20 D80 E75 ◇ REV MR 1.149 Zbl 22.369 JSL 5.71 FdM 66.601 • ID 02141

CHURCH, A. [1941] *The calculi of λ conversion* (**X** 0857) Princeton Univ Pr: Princeton 77pp
◇ B40 D20 ◇ REV MR 3.129 Zbl 26.242 JSL 17.76 JSL 6.171 • ID 02182

CHURCH, A. & QUINE, W.V.O. [1952] *Some theorems on definability and decidability* (**J** 0036) J Symb Logic 17*179-187
◇ B15 D35 ◇ REV MR 14.233 Zbl 47.9 JSL 18.269 • ID 02144

CHURCH, A. [1957] *Binary recursive arithmetic* (**J** 3941) J Math Pures Appl, Ser 9 36*39-55
◇ D20 F30 ◇ REV MR 19.239 Zbl 77.15 JSL 23.35 • ID 02402

CHURCH, A. [1963] *Logic, arithmetic, and automata* (**P** 0677) Int Congr Math (II, 9,Proc);1962 Djursholm 23-35
◇ A10 B25 D05 D98 ◇ REV MR 31 # 65 Zbl 116.336 JSL 29.210 • ID 02403

CHURCH, A. see Vol. I, II, III, V, VI for further entries

CHURKIN, V.A. & KARGAPOLOV, M.I. & REMESLENNIKOV, V.N. & ROMAN'KOV, V.A. & ROMANOVSKIJ, N.S. [1969] *Algorithmic problems for σ-power groups (Russian)* (**J** 0003) Algebra i Logika 8*643-659
• TRANSL [1969] (**J** 0069) Algeb and Log 8*364-373
◇ D40 ◇ REV MR 44 # 293 Zbl 238 # 20050 • ID 27914

CHVALINA, J. [1976] *Autonomous automata and closures with the same endomorphism monoids* (**J** 0322) Arch Math (Brno) 12*213-224
◇ D05 ◇ REV MR 55 # 8227 Zbl 437 # 68030 • ID 69331

CHYTIL, M.P. [1973] *On changes of input/output coding I* (**J** 0140) Comm Math Univ Carolinae (Prague) 14*623-645
◇ D15 D20 ◇ REV MR 51 # 10064 Zbl 268 # 02031 • REM Part II 1974 • ID 17540

CHYTIL, M.P. [1974] *On changes of input/output coding II* (**J** 0140) Comm Math Univ Carolinae (Prague) 15*1-17
◇ D15 D20 ◇ REV MR 51 # 10064 Zbl 288 # 02028 • REM Part I 1973 • ID 17541

Author Index

CHYTIL, M.P. [1975] *On complexity of nondeterministic Turing machines computations* (**P** 0454) Math Founds of Comput Sci (4);1975 Marianske Lazne 199-205
◇ D10 D15 ◇ REV MR 53 # 12074 Zbl 322 # 68024
• ID 60963

CHYTIL, M.P. [1976] *Crossing-bounded computations and their relation to the LBA-problem* (**J** 0156) Kybernetika (Prague) 12*76-85
◇ D10 D15 ◇ REV MR 53 # 12079 Zbl 324 # 68050
• ID 23107

CHYTIL, M.P. [1976] *Return complexity of language generation* (**P** 2898) Algor Kompl, Lern-& Erkenn-Prozess;1976 Jena 17-28
◇ D10 D15 ◇ REV MR 56 # 10132 Zbl 446 # 68068
• ID 69332

CHYTIL, M.P. [1977] *Comparison of the active visiting and the crossing complexities* (**P** 1635) Math Founds of Comput Sci (6);1977 Tatranska Lomnica 272-281
◇ D10 D15 ◇ REV MR 58 # 3671 Zbl 374 # 68042
• ID 51572

CIAPALA, E. [1972] *Produkte von n-Maschinen (Polish) (Russian and English summaries)* (**J** 1929) Prace Centr Oblicz Pol Akad Nauk 84*36pp
◇ D05 ◇ REV Zbl 261 # 02019 • ID 30462

CICHON, E.A. & WAINER, S.S. [1983] *The slow-growing and the Grzegorczyk hierarchies* (**J** 0036) J Symb Logic 48*399-408
◇ D20 F15 ◇ REV MR 85e:03097 Zbl 567 # 03020
• ID 40697

CICHON, E.A. see Vol. VI for further entries

CIOBOTARU, S. [1978] *The parallel matrix grammars and the languages of Thue* (**J** 0070) Bull Soc Sci Math Roumanie, NS 22(70)*269-278
◇ D03 D05 ◇ REV MR 80k:68057 Zbl 385 # 68057
• ID 69334

CLAPHAM, C.R.J. [1964] *Finitely presented groups with word problems of arbitrary degrees of insolubility* (**J** 3240) Proc London Math Soc, Ser 3 14*633-676
◇ D25 D40 ◇ REV MR 31 # 3486 Zbl 232 # 20063 JSL 33.296 • ID 02238

CLAPHAM, C.R.J. [1967] *An embedding theorem for finitely generated groups* (**J** 3240) Proc London Math Soc, Ser 3 17*419-430
◇ D35 D40 ◇ REV MR 36 # 5199 Zbl 152.2 JSL 35.340
• ID 02239

CLAPHAM, C.R.J. [1971] *The conjugacy problem for a free product with amalgamation* (**J** 0008) Arch Math (Basel) 22*358-362
◇ D40 ◇ REV MR 46 # 5464 Zbl 239 # 20041 • ID 02240

CLARES RODRIGUES, B. [1983] *An introduction to W-calculability: basic operations (Spanish) (English summary)* (**J** 2840) Stochastica, Univ Politec Barcelona 7*111-135
◇ D20 D75 ◇ REV MR 86f:03073 Zbl 557 # 03026
• ID 44149

CLARKE, D.A. [1964] *Hierarchies of predicate of finite types* (**S** 0167) Mem Amer Math Soc 51*95pp
◇ D55 D65 D70 ◇ REV MR 31 # 35 Zbl 178.321 JSL 36.146 • ID 02246

CLARKE, J.J. [1972] *Turing machines and the mind-body problem* (**J** 0013) Brit J Phil Sci 23*1-12
◇ A05 D10 ◇ REV Zbl 315 # 02012 • ID 29800

CLAUS, V. [1970] *Zerlegungen von Semi-Thue-Systemen* (**P** 0577) Automatenth & Formale Sprachen;1969 Oberwolfach 143-153
◇ D03 ◇ REV MR 40 # 2469 Zbl 213.21 • ID 27988

CLAUS, V. & HOTZ, G. [1972] *Automatentheorie und formale Sprache III. Formale Sprachen* (**X** 0876) Bibl Inst: Mannheim 241pp
◇ D05 ◇ REV MR 53 # 1995 Zbl 235 # 68026 • REM Part II 1969 by Hotz,G. & Walter,H. • ID 23540

CLAUS, V. [1981] *The (n,k)-bounded emptiness-problem for probabilistic acceptors and related problems* (**J** 1431) Acta Inf 16*139-160
◇ D05 D10 ◇ REV MR 82j:68095 Zbl 449 # 03010
• ID 56677

CLEAVE, J.P. [1961] *Creative functions* (**J** 0068) Z Math Logik Grundlagen Math 7*205-212
◇ D25 ◇ REV MR 25 # 3829 Zbl 102.8 JSL 29.102
• ID 02261

CLEAVE, J.P. [1963] *A hierarchy of primitive recursive functions* (**J** 0068) Z Math Logik Grundlagen Math 9*331-345
• TRANSL [1970] (**C** 1540) Probl Mat Log: Slozh Algor & Kl Vychisl Funk 94-113
◇ D20 ◇ REV MR 28 # 2970 Zbl 124.3 • ID 02265

CLEAVE, J.P. [1963] *Infinite sequence of pairwise recursively inseparable, recursively enumerable sets* (**X** 3333) Unknown Publisher: See Remarks
◇ D25 ◇ REM Southampton • ID 16777

CLEAVE, J.P. & ROSE, H.E. [1967] *$\mathscr{E}_n$-arithmetic* (**P** 0691) Sets, Models & Recursion Th;1965 Leicester 297-308
• TRANSL [1974] (**C** 2319) Slozh Vychisl & Algor 18-32 (Russian)
◇ D20 F30 ◇ REV MR 39 # 2605 Zbl 162.315 • ID 02682

CLEAVE, J.P. [1968] *Hyperarithmetic ultrafilters* (**P** 0692) Summer School in Logic;1967 Leeds 223-240
◇ C57 D55 E05 ◇ REV MR 39 # 3999 Zbl 217.14
• ID 02681

CLEAVE, J.P. [1969] *The primitive recursive analysis of ordinary differential equations and the complexity of their solutions* (**J** 0119) J Comp Syst Sci 3*447-455
◇ D20 D80 F60 ◇ REV MR 40 # 4512 Zbl 263 # 34001 JSL 39.345 • ID 02266

CLEAVE, J.P. [1970] *Some properties of recursively inseparable sets* (**J** 0068) Z Math Logik Grundlagen Math 16*187-200
◇ D25 ◇ REV MR 42 # 4394 Zbl 206.281 • ID 02269

CLEAVE, J.P. [1972] *Combinatorial systems I. Cylindrical problems* (**J** 0119) J Comp Syst Sci 6*254-266
◇ D30 ◇ REV MR 47 # 8269 Zbl 256 # 02018 • REM Part II 1975 • ID 02438

CLEAVE, J.P. [1973] *Combinatorial systems III. Degrees of combinatorial problems of computing machines* (**P** 1448) Math Founds of Comput Sci (2);1973 Strbske Pleso 203-207
◇ D10 D25 D30 ◇ REV MR 56 # 11771 • REM Part II 1975
• ID 30665

CLEAVE, J.P. [1975] *Combinatorial systems II. Non-cylindrical problems* (**P** 0775) Logic Colloq;1973 Bristol 253-258
◇ D30 ◇ REV MR 52 # 5392 Zbl 325 # 02031 • REM Part I 1972. Part III 1973 • ID 18106

CLEAVE, J.P. see Vol. I, II, III for further entries

CLEAVELAND, J.C. & UZGALIS, R.C. [1977] *Grammars for programming languages* (**X** 0838) Amer Elsevier: New York xiii + 154pp
◇ B75 D05 ◇ REV MR 56 # 10086 Zbl 381 # 68009 • ID 69338

CLOTE, P. [1981] *A note on the leftmost infinite branch of a recursive tree* (**P** 2614) Open Days in Model Th & Set Th;1981 Jadwisin 93-102
◇ D30 D55 ◇ ID 33717

CLOTE, P. [1981] *A note on decidable model theory* (**P** 3404) Model Th & Arithm;1979/80 Paris 134-142
◇ B25 C15 C50 C57 D30 ◇ REV MR 83d:03046 Zbl 474 # 03020 • ID 55424

CLOTE, P. [1984] *A recursion theoretic analysis of the clopen Ramsey theorem* (**J** 0036) J Symb Logic 49*376-400
◇ D45 D55 E05 ◇ REV MR 86b:03048 Zbl 574 # 03030 • ID 42432

CLOTE, P. [1985] *Applications of the low-basis theorem in arithmetic* (**P** 3342) Rec Th Week;1984 Oberwolfach 65-88
◇ C62 D20 D30 D55 F30 ◇ REV Zbl 574 # 03053 • ID 45298

CLOTE, P. [1985] *On recursive trees with a unique infinite branch* (**J** 0053) Proc Amer Math Soc 93*335-342
◇ D30 D55 ◇ REV MR 86f:03067 Zbl 526 # 03024 • ID 43901

CLOTE, P. see Vol. I, III, V, VI for further entries

COARDOS, V. [1979] *Equations associated with a context-free grammar* (**J** 0156) Kybernetika (Prague) 15*253-260
◇ D05 ◇ REV MR 81c:68056 Zbl 418 # 68065 • ID 69339

COBHAM, A. [1956] *Reduction to a symmetric predicate* (**J** 0036) J Symb Logic 21*56-59
◇ B10 D35 ◇ REV MR 17.1173 Zbl 71.10 JSL 22.297 • ID 02278

COBHAM, A. [1963] *Some remarks concerning theories with recursively enumerable complements* (**J** 0036) J Symb Logic 28*72-74
◇ D25 D35 ◇ REV MR 30 # 3840 Zbl 124.250 JSL 30.255 • ID 02283

COBHAM, A. [1965] *The intrinsic computational difficulty of functions* (**P** 0623) Int Congr Log, Meth & Phil of Sci (2,Proc);1964 Jerusalem 24-30
◇ D15 ◇ REV MR 34 # 7376 Zbl 192.87 JSL 34.657 • ID 02284

COBHAM, A. [1972] *Uniform tag sequences* (**J** 0041) Math Syst Theory 6*164-192
◇ D03 ◇ REV MR 56 # 15230 Zbl 253 # 02029 • ID 28933

COBHAM, A. see Vol. I, III, VI for further entries

COCHET, Y. & NIVAT, M. [1971] *Une generalisation des ensembles de Dyck* (**J** 0029) Israel J Math 9*389-395
◇ D40 ◇ REV MR 43 # 1774 Zbl 215.560 • ID 02291

COCHET, Y. [1979] *Church-Rosser congruences on free semigroups* (**P** 2895) Algeb Th Semigroups;1976 Szeged 20*51-60
◇ D03 ◇ REV MR 80i:03051 Zbl 408 # 20054 • ID 56293

COCKE, J. & MINSKY, M.L. [1964] *Universality of tag systems with $P = 2$* (**J** 0037) ACM J 11*15-20
◇ D03 ◇ REV MR 30 # 5893 Zbl 149.124 JSL 36.344 • ID 02293

CODD, E.F. [1968] *Cellular automata* (**X** 0801) Academic Pr: New York ix + 122pp
◇ D05 ◇ REV Zbl 213.183 • ID 28520

COHEN, A.M. [1972] *Two algebraic proofs for regular languages* (**J** 0060) Rev Roumaine Math Pures Appl 17*183-186
◇ D05 ◇ REV MR 46 # 4790 Zbl 243 # 94051 • ID 02296

COHEN, D.E. [1980] *Degree problems for modular machines* (**J** 0036) J Symb Logic 45*510-528
◇ D10 D25 D40 ◇ REV MR 82c:03066 Zbl 474 # 03019 • ID 55423

COHEN, D.E. [1980] see AANDERAA, S.O.

COHEN, D.E. [1984] *Modular machines, undecidability and incompleteness* (**P** 2342) Symp Rek Kombin;1983 Muenster 237-247
◇ D20 D25 D35 F30 ◇ REV Zbl 549 # 03028 • ID 43120

COHEN, JACQUES & HICKEY, T. & KATCOFF, J. [1982] *Upper bounds for speedup in parallel parsing* (**J** 0037) ACM J 29*409-428
◇ D05 D15 ◇ REV MR 83b:68091 Zbl 478 # 68088 • ID 69341

COHEN, JACQUES see Vol. I for further entries

COHEN, JOEL M. [1967] *The equivalence of two concepts of categorical grammar* (**J** 0194) Inform & Control 10*475-484
◇ D05 ◇ REV MR 36 # 1243 Zbl 153.9 JSL 33.627 • ID 02297

COHEN, L.J. & LOS, J. & PFEIFFER, H. & PODEWSKI, K.-P. (EDS.) [1982] *Logic, methodology and philosophy of science VI* (**S** 3303) Stud Logic Found Math 104*856pp
◇ B97 D98 ◇ REV MR 83k:03004 Zbl 489 # 00005 • ID 36588

COHEN, L.J. see Vol. I, II for further entries

COHEN, P.E. & JOCKUSCH JR., C.G. [1975] *A lattice property of Post's simple set* (**J** 0316) Illinois J Math 19*450-453
◇ D25 ◇ REV MR 52 # 10388 Zbl 309 # 02039 • ID 21686

COHEN, P.E. see Vol. III, V for further entries

COHEN, R.S. [1972] *Rank-non-increasing transformations on transition graphs* (**J** 0194) Inform & Control 20*93-113
◇ D05 ◇ REV MR 52 # 2776 Zbl 237 # 94020 • ID 61049

COHEN, R.S. & GOLD, A.Y. [1978] *ω-computations on Turing machines* (**J** 1426) Theor Comput Sci 6*1-23
◇ D10 ◇ REV MR 57 # 5707 Zbl 368 # 68057 • ID 51293

COHEN, R.S. & GOLD, A.Y. [1978] *ω-computations on deterministic pushdown machines* (**J** 0119) J Comp Syst Sci 16*275-300
◇ D05 ◇ REV MR 82i:03050 Zbl 382 # 03025 • ID 51949

COHEN, R.S. & GOLD, A.Y. [1980] *On the complexity of ω-type Turing acceptors* (**J** 1426) Theor Comput Sci 10*249-272
◇ D10 ◇ REV MR 81d:68061 Zbl 432 # 68040 • ID 80939

COHEN, R.S. see Vol. I, II for further entries

COHN, P.M. [1973] *The word problem for free fields* (**J** 0036) J Symb Logic 38*309-314 • ERR/ADD ibid 40*69-74
◇ D40 ◇ REV MR 54 # 12511a MR 54 # 12511b Zbl 273 # 02027 • ID 02317

COHN, P.M. [1975] *Equations dans les corps gauches* (**J** 0082) Bull Soc Math Belg 27*29-39
◇ D40 ◇ REV MR 56 # 5643 Zbl 355 # 16010 • ID 80943

COHN, P.M. see Vol. III, V for further entries

COHORS-FRESENBORG, E. [1973] *Elementare und subelementare Funktionenklassen ueber binaeren Baeumen* (**P** 1630) GI Fachtag Automatenth & Form Sprach (1);1973 Bonn 220-229
◇ D15 D20 ◇ REV MR 55 # 11671 Zbl 283 # 68028 • ID 61052

COHORS-FRESENBORG, E. [1977] *Mathematik mit Kalkuelen und Maschinen* (**X** 0900) Vieweg: Wiesbaden viii + 184pp
◇ D03 D10 D98 ◇ REV MR 58 # 16215 Zbl 355 # 02028 JSL 45.380 • ID 50226

COLBOURN, C.J. [1980] *Isomorphism complete problems on matrices* (**P** 3540) West-Coast Conf Combin, Graph Th & Comput;1979 Arcata 101-107
◇ D15 ◇ REV MR 82g:68040 Zbl 442 # 68032 • ID 80945

COLBOURN, C.J. & COLBOURN, M.J. & STINSON, D.R. [1984] *The computational complexity of recognizing critical sets* (**P** 2740) SE Asian Graph Th Colloq (1);1983 Singapore 248-253
◇ D15 ◇ REV MR 85k:68035 Zbl 548 # 05014 • ID 44023

COLBOURN, M.J. [1984] see COLBOURN, C.J.

COLLINS, D.J. [1969] *On recognising Hopf groups* (**J** 0008) Arch Math (Basel) 20*235-240
◇ D40 ◇ REV MR 40 # 212 Zbl 184.41 • ID 02325

COLLINS, D.J. [1969] *Recursively enumerable degrees and the conjugacy problem* (**J** 0118) Acta Math 122*115-160
◇ D25 D40 ◇ REV MR 39 # 4001 Zbl 175.275 JSL 35.477 JSL 36.540 • ID 02323

COLLINS, D.J. [1969] *Word and conjugacy problems in groups with only a few defining relations* (**J** 0068) Z Math Logik Grundlagen Math 15*305-324
◇ D40 ◇ REV MR 41 # 8502 Zbl 188.29 • ID 02324

COLLINS, D.J. [1970] *A universal semigroup (Russian)* (**J** 0003) Algebra i Logika 9*731-740
• TRANSL [1970] (**J** 0069) Algeb and Log 9*442-446
◇ C05 C57 D05 D40 ◇ REV MR 44 # 1564 Zbl 225 # 02030 Zbl 234 # 02030 • ID 02327

COLLINS, D.J. [1970] *On recognizing properties of groups which have solvable word problem* (**J** 0008) Arch Math (Basel) 21*31-39
◇ C60 D40 ◇ REV MR 42 # 351 Zbl 195.20 JSL 39.340 • ID 02448

COLLINS, D.J. [1971] see BOONE, W.W.

COLLINS, D.J. [1971] *Truth-tables degrees and the Boone group* (**J** 0120) Ann of Math, Ser 2 94*392-396
◇ D30 D40 ◇ REV MR 48 # 1904 Zbl 234 # 02032 JSL 39.184 • ID 02328

COLLINS, D.J. [1972] *Representation of Turing reducibility by word and conjugacy problems in finitely presented groups* (**J** 0118) Acta Math 128*73-90
◇ D30 D40 ◇ REV MR 52 # 13356 Zbl 229 # 20036 • ID 21809

COLLINS, D.J. [1973] *The word, power and order problems in finitely presented groups* (**P** 0678) Word Probl: Decis & Burnside Probl in Group Th;1969 Irvine 401-420
◇ D25 D40 ◇ REV MR 53 # 2664 Zbl 267 # 02033 JSL 41.786 • ID 21611

COLLINS, D.J. [1974] see BOONE, W.W.

COLLINS, G.E. & HALPERN, J.D. [1970] *On the interpretability of arithmetic in set theory* (**J** 0047) Notre Dame J Formal Log 11*477-483
◇ B28 D35 E30 F25 F30 ◇ REV MR 45 # 4970 Zbl 212.17 • ID 02332

COLLINS, G.E. [1982] see BUCHBERGER, B.

COLLINS, G.E. see Vol. III, V for further entries

COLLINS, W.J. [1978] *Provably recursive real numbers* (**J** 0047) Notre Dame J Formal Log 19*513-522
◇ D20 F60 ◇ REV MR 82e:03040 Zbl 305 # 02054 • ID 52157

COLLINS, W.J. & YOUNG, P. [1983] *Discontinuities of provably correct operators on the provably recursive real numbers* (**J** 0036) J Symb Logic 48*913-920
◇ D20 F60 ◇ REV MR 86d:03056 Zbl 546 # 03034 • ID 43537

COMER, S.D. [1981] *The decision problem for certain nilpotent closed varieties* (**J** 0068) Z Math Logik Grundlagen Math 27*557-560
◇ B25 C05 D35 ◇ REV MR 83g:08008 Zbl 474 # 08003 • ID 55444

COMER, S.D. [1982] *Inductive domains and algebraic semantics of CF languages* (**J** 2293) Comp Linguist & Comp Lang 15*43-49
◇ D05 ◇ REV MR 83m:68150 Zbl 486 # 68089 • ID 40430

COMER, S.D. see Vol. I, III, V for further entries

COMERFORD JR., L.P. & TRUFFAULT, B. [1976] *The conjugacy problem for free products of sixth-groups with cyclic amalgamation* (**J** 0044) Math Z 149*169-181
◇ D40 ◇ REV MR 53 # 13418 Zbl 316 # 20029 • ID 23217

COMERFORD JR., L.P. & EDMUNDS, C.C. [1981] *Quadratic equations over free groups and free products* (**J** 0032) J Algeb 68*276-297
◇ D40 ◇ REV MR 82k:20060 Zbl 526.20024 • ID 80950

COMERFORD JR., L.P. [1981] *Quadratic equations over small cancellation groups* (**J** 0032) J Algeb 69*175-185
◇ D40 ◇ REV MR 82f:20060 Zbl 461 # 20016 • ID 80949

COMMENTZ-WALTER, B. [1979] *Size-depth tradeoff in monotone boolean formulae* (**J** 1431) Acta Inf 12*227-243
◇ B05 D15 ◇ REV MR 80m:68038 Zbl 395 # 94036 • ID 53164

COMMENTZ-WALTER, B. see Vol. I for further entries

COMPTON, K.J. [1984] *An undecidable problem in finite combinatorics* (**J** 0036) J Symb Logic 49*842-850
◇ C13 D80 ◇ REV MR 85i:03094 • ID 42434

COMPTON, K.J. [1984] *On rich words* (**P** 4064) Prog Graph Th;1982 Waterloo 39-61
◇ C85 D05 D45 ◇ REV Zbl 563 # 03022 • ID 47457

COMPTON, K.J. see Vol. III for further entries

COMYN, G. & DAUCHET, M. [1982] *Approximations of infinitary objects* (**P** 3836) Automata, Lang & Progr (9);1982 Aarhus 116-127
◇ D80 ◇ REV MR 83m:68052 Zbl 505 # 03013 • ID 38134

COMYN, G. [1982] *Arbres infinitaires. Approximations et proprietes de calculabilite* (**P** 4004) CAAP'82 Arbres en Algeb & Progr (7);1982 Lille 65-81
◇ C57 ◇ REV Zbl 539 # 68057 • ID 41258

Comyn, G. see Vol. VI for further entries

Constable, R.L. [1968] *Extending and refining hierarchies of computable functions* (1111) Preprints, Manuscr., Techn. Reports etc. 25*
◇ D20 ◇ REM University of Wisconsin. Computer Science • ID 21217

Constable, R.L. [1969] see Borodin, A.

Constable, R.L. [1969] *Upward and downward diagonalizations over axiomatic complexity classes* (1111) Preprints, Manuscr., Techn. Reports etc.
◇ D20 ◇ REM Cornell University. Department of Computer Science. Ithaca • ID 21218

Constable, R.L. [1971] see Borodin, A.

Constable, R.L. [1971] *Subrecursive-programming languages III: the multiple-recursive functions* (**P** 1383) Symp Comput & Automata;1971 Brooklyn 21*393-410
◇ D15 D20 ◇ REV Zbl 265 # 68019 • REM Part II 1971 by Borodin,A. & Constable,R.L. • ID 28633

Constable, R.L. [1972] *Constructive mathematics and automatic program writers* (**P** 1455) Inform Processing (5);1971 Ljubljana 229-233
◇ B75 D15 F55 ◇ REV MR 54 # 1696 Zbl 255 # 68014 • ID 61071

Constable, R.L. & Muchnick, S.S. [1972] *Subrecursive program schemata. I. Undecidable equivalence problems. II. Decidable equivalence problems* (**J** 0119) J Comp Syst Sci 6*480-537
◇ B75 D20 ◇ REV MR 54 # 14426 Zbl 257 # 68026 • ID 14667

Constable, R.L. [1972] *Subrecursive programming languages I: Efficiency and program structure* (**J** 0037) ACM J 19*526-568
◇ B75 D15 ◇ REV MR 47 # 9892 Zbl 259 # 68036 • REM Part II 1971 by Constable,R.L. & Borodin,A. • ID 46714

Constable, R.L. [1972] *The operator gap* (**J** 0037) ACM J 19*175-183
◇ D15 ◇ REV MR 47 # 3159 Zbl 229 # 68016 • ID 22030

Constable, R.L. [1973] *Two types of hierarchy theorem on axiomatic complexity classes* (**P** 1584) Comput Compl - Courant Comput Sci (7);1971 New York 37-63
◇ D15 ◇ ID 31030

Constable, R.L. & Egli, H. [1975] *Computability concepts for programming language semantics* (**P** 1618) ACM Symp Th of Comput (7);1975 Albuquerque 98-106
◇ B75 D20 ◇ REV MR 55 # 10367 MR 58 # 10367 Zbl 359 # 68017 • ID 50627

Constable, R.L. & Hunt III, H.B. & Sahni, S.K. [1980] *On the computational complexity of program scheme equivalence* (**J** 1428) SIAM J Comp 9*396-416
◇ D15 D20 ◇ REV MR 81j:68051 Zbl 447 # 68038 • ID 81617

Constable, R.L. [1980] *The role of finite automata in the development of modern computing theory* (**P** 2058) Kleene Symp;1978 Madison 61-83
◇ D05 ◇ REV MR 82k:03059 Zbl 468 # 68060 • ID 55126

Constable, R.L. see Vol. I, VI for further entries

Conway, J.H. [1971] *Regular algebra and finite machines* (**X** 1249) Chapman & Hall: London 147pp
◇ D05 ◇ REV Zbl 231 # 94041 • ID 23518

Conway, J.H. see Vol. I, III, V for further entries

Cook, C.R. [1978] *Context-free grammars with graph control* (**P** 3426) SE Conf Combin, Graph Th & Comput (9);1978 Boca Raton 203-219
◇ D05 ◇ REV MR 80e:68183 Zbl 416 # 68070 • ID 69349

Cook, S.A. [1966] *The solvability of the derivability problem for one-normal systems* (**J** 0037) ACM J 13*223-225
◇ D10 ◇ REV MR 34 # 4137 Zbl 207.312 JSL 36.344 • ID 02678

Cook, S.A. [1969] see Aanderaa, S.O.

Cook, S.A. [1971] *Characterizations of pushdown machines in terms of time-bounded computers* (**J** 0037) ACM J 18*4-18 • TRANSL [1974] (**C** 2319) Slozh Vychisl & Algor 266-288 (Russian)
◇ D10 D15 ◇ REV MR 45 # 1690 Zbl 222 # 02035 • ID 29986

Cook, S.A. [1971] *The complexity of theorem proving procedures* (**P** 0680) ACM Symp Th of Comput (3);1971 Shaker Heights 151-158
◇ D15 ◇ REV Zbl 253 # 68020 • ID 02450

Cook, S.A. & Reckhow, R.A. [1972] *Time-bounded random access machines* (**P** 1901) ACM Symp Th of Comput (4);1972 Denver 73-80
◇ D10 D15 ◇ REV MR 48 # 5416 • ID 17133

Cook, S.A. [1973] *A hierarchy for nondeterministic time complexity* (**J** 0119) J Comp Syst Sci 7*343-353
◇ D15 ◇ REV MR 49 # 2308 Zbl 278 # 68042 • ID 04019

Cook, S.A. & Reckhow, R.A. [1973] *Time bounded random access machines* (**J** 0119) J Comp Syst Sci 7*354-375
◇ D10 D15 ◇ REV MR 48 # 5416 Zbl 284 # 68038 • ID 61077

Cook, S.A. & Reckhow, R.A. [1974] *On the length of proofs in the propositional calculus: preliminary version* (**P** 1464) ACM Symp Th of Comput (6);1974 Seattle 135-148
◇ B05 D15 F20 ◇ REV MR 54 # 7215 Zbl 375 # 02004 • ID 25005

Cook, S.A. & Sethi, R. [1974] *Storage requirements for deterministic polynominal time recognizable languages* (**P** 1464) ACM Symp Th of Comput (6);1974 Seattle 33-39
◇ D15 ◇ REV MR 54 # 9166 Zbl 412 # 68078 • ID 69353

Cook, S.A. [1979] *Deterministic CEL's are accepted simultaneously in polynomial time and log squared* (**P** 3542) ACM Symp Th of Comput (11);1979 Atlanta 338-345
◇ D05 D15 ◇ REV MR 82g:68041 • ID 49846

Cook, S.A. & Reckhow, R.A. [1979] *The relative efficiency of propositional proof systems* (**J** 0036) J Symb Logic 44*36-50
◇ B05 D15 F20 ◇ REV MR 80e:03007 Zbl 408 # 03044 • ID 56282

Cook, S.A. & Rackoff, C.W. [1980] *Space lower bounds for maze threadability on restricted machines* (**J** 1428) SIAM J Comp 9*636-652
◇ D05 D15 ◇ REV MR 81k:68028 Zbl 445 # 68038 • ID 69350

COOK, S.A. [1981] *Towards a complexity theory of synchronous parallel computation* (J 3370) Enseign Math, Ser 2 27*99-124
• REPR [1982] (P 3482) Logic & Algor (Specker);1980 Zuerich 75-100
◇ B75 D15 ◇ REV MR 83b:68041a Zbl 473 # 68041 • ID 69351

COOK, S.A. [1983] *The classification of problems which have fast parallel algorithm* (P 3864) FCT'83 Found of Comput Th;1983 Borgholm 78-93
◇ D15 ◇ REV MR 85k:68036 Zbl 529 # 68014 • ID 47047

COOK, S.A. see Vol. I, V, VI for further entries

COOPER, D.C. [1969] *Program scheme equivalences and second-order logic* (J 0508) Machine Intelligence 4*3-15
◇ B15 D05 ◇ REV MR 40 # 6806 Zbl 221 # 68015 • ID 28130

COOPER, D.C. see Vol. I, III, VI for further entries

COOPER, GERALD [1970] see AMOROSO, S.

COOPER, GERALD [1975] see AMOROSO, S.

COOPER, S.B. [1972] *Degrees of unsolvability complementary between recursively enumerable degrees I* (J 0007) Ann Math Logic 4*31-73
◇ D25 D30 ◇ REV MR 45 # 3199 Zbl 248 # 02045 JSL 40.86 • ID 02361

COOPER, S.B. [1972] *Jump equivalence of the Δ_2^0 hyperimmune sets* (J 0036) J Symb Logic 37*598-600
◇ D25 D30 D55 ◇ REV MR 50 # 12690 Zbl 278 # 02037 • ID 21306

COOPER, S.B. [1972] *Minimal upper bounds for sequences of recursively enumerable degrees* (J 3172) J London Math Soc, Ser 2 5*445-450
◇ D25 D30 ◇ REV MR 50 # 9560 Zbl 256 # 02020 • ID 04020

COOPER, S.B. [1973] *Minimal degrees and the jump operator* (J 0036) J Symb Logic 38*249-271
◇ D30 ◇ REV MR 50 # 75 Zbl 309 # 02048 JSL 40.86 • ID 02363

COOPER, S.B. [1974] *An annotated bibliography for the structure of the degrees below 0' with special reference to that of the recursively enumerable degrees* (S 1585) Rec Fct Th Newsletter 5*1-15
◇ A10 D25 D30 ◇ ID 33581

COOPER, S.B. [1974] *Minimal pairs and high recursively enumerable degrees* (J 0036) J Symb Logic 39*655-660
◇ D25 ◇ REV MR 51 # 7844 Zbl 309 # 02047 • ID 04024

COOPER, S.B. [1982] *Partial degrees and the density problem* (J 0036) J Symb Logic 47*854-859
◇ D30 ◇ REV MR 84b:03058 Zbl 511 # 03019 • REM Part I. Part II 1984 • ID 33582

COOPER, S.B. [1984] *Partial degrees and the density problem. Part 2: The enumeration degrees of the Σ_2 sets are dense* (J 0036) J Symb Logic 49*503-513
◇ D30 ◇ REV MR 85j:03068 • REM Part I 1982 • ID 33583

COOPER, S.B. & MCEVOY, K. [1985] *On minimal pairs of enumeration degrees* (J 0036) J Symb Logic 50*983-1001
◇ D25 D30 ◇ ID 49556

COPPEY, L. [1984] *Categories de Peano et categories algorithmiques, recursivite* (J 3797) Diagrammes LC 1-LC 47
◇ D75 G30 ◇ REV Zbl 565 # 18004 • ID 48662

CORAY, G. [1982] *Complexite d'algorithmes modulaires* (P 3482) Logic & Algor (Specker);1980 Zuerich 101-125
◇ D15 ◇ REV MR 83m:68073 Zbl 474 # 68061 • ID 69355

CORNELIUS, B.J. & KIRBY, G. [1975] *Depth of recursion and the Ackermann function* (J 0130) BIT 15*144-150
◇ D20 ◇ REV Zbl 304 # 68017 • ID 61098

COSTE, M. [1980] see COSTE-ROY, M.-F.

COSTE, M. see Vol. III, V for further entries

COSTE-ROY, M.-F. & COSTE, M. & MAHE, L. [1980] *Contribution to the study of the natural number object in elementary topoi* (J 0326) J Pure Appl Algebra 17*35-68
◇ D20 F35 F50 G30 ◇ REV MR 81h:18005 Zbl 427 # 03056 • ID 53743

COSTE-ROY, M.-F. also published under the name COSTE, M.-F.

COSTICH, O.L. [1972] *A Medvedev characterization of sets recognized by generalized finite automata* (J 0041) Math Syst Theory 6*263-267
◇ D05 ◇ REV MR 50 # 16158 Zbl 264 # 94038 • ID 61116

COURCELLE, B. & VUILLEMIN, J. [1974] *Semantics and axiomatics of a simple recursive language (French summary)* (P 1464) ACM Symp Th of Comput (6);1974 Seattle 13-26
◇ D05 ◇ REV MR 54 # 4165 Zbl 358 # 68015 • ID 24072

COURCELLE, B. & VUILLEMIN, J. [1976] *Completeness results for the equivalence of recursive schemas* (J 0119) J Comp Syst Sci 12*179-197
◇ B75 D20 ◇ REV MR 53 # 14963 Zbl 342 # 68008 • ID 31528

COURCELLE, B. [1977] see BOASSON, L.

COURCELLE, B. [1977] *On jump-deterministic pushdown automata* (J 0041) Math Syst Theory 11*87-109
◇ D10 ◇ REV MR 57 # 4641 Zbl 365 # 02021 • ID 51022

COURCELLE, B. [1978] *A representation of trees by languages I* (J 1426) Theor Comput Sci 6*255-279
◇ D05 D10 ◇ REV MR 58 # 13955 Zbl 377 # 68040 • REM Part II 1978 • ID 46708

COURCELLE, B. [1978] *A representation of trees by languages II* (J 1426) Theor Comput Sci 7*25-55
◇ D05 D10 ◇ REV MR 58 # 13955b Zbl 428 # 68088 • REM Part I 1978 • ID 69360

COURCELLE, B. [1981] *An axiomatic approach to the Korenjak - Hopcroft algorithms* (P 2903) Automata, Lang & Progr (8);1981 Akko 393-407
◇ D05 ◇ REV MR 82k:68042 Zbl 462 # 68008 • ID 80979

COURCELLE, B. [1981] *Attribute grammars: Theory and applications* (P 2930) Formal of Progr Concepts;1981 Peniscola 75-95
◇ D05 ◇ REV MR 82g:68066 Zbl 457 # 68090 • ID 69359

COURCELLE, B. [1981] see BOASSON, L.

COURCELLE, B. [1982] *Fundamental properties of infinite trees* (P 3906) Th Found of Progr Methodol;1981 Marktoberdorf 417-471
◇ D05 D98 E75 ◇ REV MR 84e:05047 Zbl 513 # 68060 • ID 39570

COURCELLE, B. [1985] see BRAQUELAIRE, J.P.

COURT, L.M. & HIZ, H. & LAMBEK, J. [1961] *Comments* (**P** 0701) Struct of Lang & Math Aspects;1960 New York 264-265
◇ B65 D05 ◇ REV JSL 33.627 • ID 04206

COUSINEAU, G. & RIFFLET, J.M. [1975] *Langages d'interpretation des schemas recursifs* (**J** 4698) Rev Franc Autom, Inf & Rech Operat, Ser Rouge Inf Th 9/R-1*21-42
◇ B75 D05 ◇ REV MR 53 # 14988 Zbl 331 # 68007 • ID 23228

COUSOT, P. & COUSOT, R. [1979] *Constructive versions of Tarski's fixed point theorems* (**J** 0048) Pac J Math 82*43-57
◇ D20 G10 ◇ REV MR 82d:06004 Zbl 413 # 06004 • ID 80981

COUSOT, R. [1979] see COUSOT, P.

COY, W. [1976] *The logical meaning of programs of a subrecursive language* (**J** 0232) Inform Process Lett 4*121-126
◇ B75 D20 ◇ REV MR 53 # 4595 Zbl 317 # 68012 • ID 61129

COY, W. [1977] *Automata in labyrinths* (**P** 2588) FCT'77 Fund of Comput Th;1977 Poznan 56*65-71
◇ D05 ◇ REV MR 57 # 19164 Zbl 362 # 94048 • ID 50824

COY, W. see Vol. I, II for further entries

CRAIG, W. [1952] *Incompletability, with respect to validity in every finite nonempty domain of first order functional calculus* (**P** 0593) Int Congr Math (II, 6);1950 Cambridge MA 1*721
◇ C13 D35 ◇ ID 28058

CRAIG, W. [1953] *On axiomatizability within a system* (**J** 0036) J Symb Logic 18*30-32
◇ C07 D20 ◇ REV MR 14.1051 Zbl 53.201 JSL 19.62 • ID 02507

CRAIG, W. [1960] *Bases for first-order theories and subtheories* (**J** 0036) J Symb Logic 25*97-142
◇ B10 D25 F05 ◇ REV MR 24 # A1812 Zbl 108.3 JSL 37.616 • ID 02513

CRAIG, W. see Vol. I, II, III, V, VI for further entries

CREMERS, A.B. & GINSBURG, S. [1975] *Context-free grammar forms* (**J** 0119) J Comp Syst Sci 11*86-117
◇ D05 ◇ REV MR 51 # 12038 Zbl 328 # 68071 • ID 32113

CREMERS, A.B. & GINSBURG, S. & SPANIER, E.H. [1977] *The structure of context-free grammatical families* (**J** 0119) J Comp Syst Sci 15*262-279
◇ D05 ◇ REV MR 58 # 3715 Zbl 366 # 68052 • ID 32120

CRESPI-REGHIZZI, S. & GUIDA, G. & MANDRIOLI, D. [1978] *Noncounting context-free languages* (**J** 0037) ACM J 25*571-580
◇ D05 ◇ REV MR 80a:68079 Zbl 388 # 68073 • ID 69364

CRESPI-REGHIZZI, S. & GUIDA, G. & MANDRIOLI, D. [1981] *Operator precedence grammars and the noncounting property* (**J** 1428) SIAM J Comp 10*174-191
◇ D05 ◇ REV MR 83e:68100 Zbl 454 # 68090 • ID 69365

CRESSWELL, M.J. [1964] *Propositional arithmetic* (**J** 0079) Logique & Anal, NS 7*185-189
◇ B05 D03 ◇ REV MR 31 # 2129 Zbl 121.10 • ID 02531

CRESSWELL, M.J. [1977] *Categorial languages* (**J** 0063) Studia Logica 36*257-269
◇ B65 D05 ◇ REV MR 58 # 16133 Zbl 393 # 03017 • ID 52437

CRESSWELL, M.J. [1985] *The decidable normal modal logics are not recursively enumerable* (**J** 0122) J Philos Logic 14*231-233
◇ B25 B45 D25 D35 D80 ◇ ID 42657

CRESSWELL, M.J. see Vol. I, II, V for further entries

CRESTIN, J. [1978] *Langages quasirationnels* (**P** 3394) Lang Algeb (1);1973 Bonascre 123-166
◇ D05 ◇ REV MR 80b:68083 Zbl 406 # 68056 • ID 69366

CRISCUOLO, G. & MINICOZZI, E. & TRAUTTEUR, G. [1975] *Limiting recursion and the arithmetic hierarchy (French summary)* (**J** 4698) Rev Franc Autom, Inf & Rech Operat, Ser Rouge Inf Th 9/R-3*5-12
◇ D20 D55 ◇ REV MR 53 # 106 Zbl 322 # 02039 • ID 16570

CRISCUOLO, G. & GERLA, G. [1976] *tt-riducibilita e limiti ricorsivi* (**J** 0319) Matematiche (Sem Mat Catania) 31*94-103
◇ D30 ◇ REV MR 58 # 16234 Zbl 381 # 03031 • ID 51899

CRISCUOLO, G. see Vol. VI for further entries

CROCIANI, C. [1980] see BERNARDI, C.

CROCIANI, C. [1981] *Su alcune proprieta ricorsive legate al concetto di densita (English summary)* (**J** 3746) Note Math (Lecce) 1*137-153
◇ D20 ◇ REV MR 83k:03045 Zbl 484 # 03026 • ID 34875

CROSSLEY, J.N. [1965] *Constructive order types I* (**P** 0688) Logic Colloq;1963 Oxford 189-264
◇ D45 ◇ REV MR 31 # 4718 Zbl 145.10 • REM Part II 1966 • ID 04211

CROSSLEY, J.N. & DUMMETT, M. (EDS.) [1965] *Formal systems and recursive functions* (**X** 0809) North Holland: Amsterdam 320pp
◇ B97 D20 D97 ◇ REV Zbl 126.2 • ID 31683

CROSSLEY, J.N. [1966] see ACZEL, P.

CROSSLEY, J.N. [1966] *Constructive order types II* (**J** 0036) J Symb Logic 31*525-538
◇ D20 D45 F15 ◇ REV MR 35 # 5318 Zbl 156.27 • REM Part I 1965. Part III 1966 by Aczel,P.H.G. & Crossley,J.N. • ID 02555

CROSSLEY, J.N. & SCHUETTE, K. [1966] *Non-uniqueness at ω^2 in Kleene's O* (**J** 0009) Arch Math Logik Grundlagenforsch 9*95-101
◇ D45 F15 ◇ REV MR 35 # 4101 Zbl 192.55 JSL 35.336 • ID 02556

CROSSLEY, J.N. (ED.) [1967] *Sets, models and recursion theory* (**X** 0809) North Holland: Amsterdam 331pp
◇ B97 C97 D97 ◇ REV MR 36 # 24 Zbl 158.2 • ID 31684

CROSSLEY, J.N. [1968] *Recursive equivalence: a survey* (**P** 0692) Summer School in Logic;1967 Leeds 241-251
◇ D45 D50 ◇ REV MR 39 # 3991 Zbl 175.271 JSL 37.406 • ID 04212

CROSSLEY, J.N. [1969] *Constructive order types* (**X** 0809) North Holland: Amsterdam 225pp
◇ D45 ◇ REV MR 41 # 5214 Zbl 184.22 • ID 04213

CROSSLEY, J.N. [1970] *Recursive equivalence* (**J** 0161) Bull London Math Soc 2*129-151
◇ D50 ◇ REV MR 46#5125 Zbl 212.24 JSL 37.406
• ID 02558

CROSSLEY, J.N. & NERODE, A. [1974] *Combinatorial functors* (**X** 0811) Springer: Heidelberg & New York viii+146pp
◇ C57 C62 D45 D50 G30 ◇ REV MR 52#10397 Zbl 283#02036 JSL 42.586 • ID 21692

CROSSLEY, J.N. & O'CONNOR, S. & STACEY, K. [1974] *Recursive equivalence and combinatorial functors: a bibliography* (**S** 1585) Rec Fct Th Newsletter 5/Suppl.3*14pp
◇ A10 D50 ◇ ID 31686

CROSSLEY, J.N. & NERODE, A. [1975] *Combinatorial functors* (**C** 0758) Logic Colloq Boston 1972-73 1-21
◇ D50 ◇ REV MR 51#12497 Zbl 312#02036 • ID 17284

CROSSLEY, J.N. [1975] *Reminiscenses of logicians* (**P** 0765) Algeb & Log;1974 Clayton 1-62
◇ A05 A10 D98 ◇ REV MR 52#17 Zbl 312#02001
• REM Part I. Part II 1981 • ID 17606

CROSSLEY, J.N. & NERODE, A. [1975] *Sound functors* (**P** 1440) ⊧ ISILC Proof Th Symp (Schuette);1974 Kiel 500*26-43
◇ D50 ◇ REV MR 53#5278 Zbl 324#02030 • ID 22909

CROSSLEY, J.N. & NERODE, A. [1976] *Effective dimension* (**J** 0032) J Algeb 41*398-412
◇ C57 D45 D50 ◇ REV MR 54#7234 Zbl 344#02036
• ID 25025

CROSSLEY, J.N. & MIRANDA, S. [1981] *A bibliography of effective algebra* (**P** 2902) Aspects Effective Algeb;1979 Clayton 251-290
◇ A10 D45 ◇ REV MR 83g:03043 Zbl 466#03014
• ID 54965

CROSSLEY, J.N. (ED.) [1981] *Aspects of effective algebra. Proceedings of a conference at Monash University, Australia, 1-4 August, 1979* (**X** 2863) Upside Down A Book: Yarra Glen x+290pp
◇ C57 C97 D45 D97 ◇ REV MR 82h:03003 Zbl 456#00004 • ID 54309

CROSSLEY, J.N. & NERODE, A. [1981] *Recursive equivalence on matroids* (**P** 2902) Aspects Effective Algeb;1979 Clayton 69-86
◇ D45 D50 ◇ REV MR 83b:03055 Zbl 472#03035
• ID 55300

CROSSLEY, J.N. [1981] *Reminiscences of logicians. II* (**P** 2902) Aspects Effective Algeb;1979 Clayton 1-25
◇ A10 D45 D50 ◇ REV MR 82m:03060 Zbl 467#03002
• REM Part I 1975 • ID 55000

CROSSLEY, J.N. [1982] *The given* (**J** 0063) Studia Logica 41*131-139
◇ C57 D45 ◇ REV MR 85i:03149 Zbl 536#03024
• ID 37106

CROSSLEY, J.N. & REMMEL, J.B. [1983] *Undecidability and recursive equivalence I* (**P** 3669) SE Asian Conf on Log;1981 Singapore 37-54
◇ D35 D45 D50 ◇ REV Zbl 553#03030 • REM Part II 1984 • ID 43056

CROSSLEY, J.N. & REMMEL, J.B. [1984] *Undecidability and recursive equivalence II* (**P** 2153) Logic Colloq;1983 Aachen 2*79-100
◇ D35 D50 D75 ◇ REV Zbl 563#03029 • REM Part I 1983
• ID 43002

CROSSLEY, J.N. see Vol. I, II, III, V, VI for further entries

CSILLAG, P. [1947] *Eine Bemerkung zur Aufloesung der eingeschachtelten Rekursion* (**J** 0002) Acta Sci Math (Szeged) 11*169-173
◇ D20 ◇ REV MR 9.129 Zbl 30.018 JSL 13.54 • ID 02566

CSIRMAZ, L. [1979] see AJTAI, M.

CSIRMAZ, L. [1980] *Iterated grammars* (**J** 0380) Acta Cybern (Szeged) 5*43-47
◇ D05 ◇ REV MR 83f:68082 Zbl 477#68078 • ID 69367

CSIRMAZ, L. [1982] *Determinateness of program equivalence over Peano axioms* (**J** 1426) Theor Comput Sci 21*231-235
◇ B75 D20 ◇ REV MR 83k:68009 Zbl 488#68019
• ID 40293

CSIRMAZ, L. see Vol. I, III, V, VI for further entries

CSUHAJ VARJU, E. [1977] *Generative grammar forms and operators* (**J** 1458) Alkalmaz Mat Lapok 3*97-104
◇ D05 ◇ REV MR 58#19378 Zbl 408#68069 • ID 69369

CUDIA, D.F. & SINGLETARY, W.E. [1965] *Post's correspondence problem and degrees of unsolvability; degrees of unsolvability in automata and grammars* (**J** 0036) J Symb Logic 30*267-268
◇ D03 D05 D25 ◇ REV Zbl 169.312 • ID 02568

CUDIA, D.F. & SINGLETARY, W.E. [1968] *Degrees of unsolvability in formal grammars* (**J** 0037) ACM J 15*680-692
◇ D03 D25 ◇ REV MR 38#5534 Zbl 169.312 JSL 39.185
• ID 02570

CUDIA, D.F. & SINGLETARY, W.E. [1968] *The Post correspondence problem* (**J** 0036) J Symb Logic 33*418-430
◇ D03 D25 ◇ REV MR 39#5367 Zbl 169.311 JSL 39.185
• ID 02572

CUDIA, D.F. [1970] *General Problems of formal grammars* (**J** 0037) ACM J 17*31-43
◇ D05 ◇ REV MR 43#4673 Zbl 169.313 • ID 02574

CULIK, K. [1961] *Some notes on finite state languages and events represented by finite automata using labelled graphs (Czech and Russian summaries)* (**J** 0086) Cas Pestovani Mat, Ceskoslov Akad Ved 86*43-55
◇ D05 ◇ REV MR 23#B3095 JSL 35.486 • ID 02594

CULIK, K. [1963] *Some axiomatic systems for formal grammars and languages* (**P** 0572) Inform Processing (2);1962 Muenchen 313-317
◇ D05 ◇ REV Zbl 133.255 • ID 27611

CULIK, K. [1964] *Applications of graph theory to mathematical logic and linguistics* (**P** 0703) Th of Graphs & Appl;1963 Smolenice 13-20
◇ B65 D03 D05 ◇ REV MR 31#1208 Zbl 158.254
• ID 04218

CULIK, K. & HAVEL, I.M. [1967] *On multiple finite automata* (**P** 1671) Colloq Automatenth (3);1965 Hannover 158-169
◇ D05 ◇ REV Zbl 174.33 • ID 29431

CULIK, K. [1970] *On conditional context-free grammars for programming and natural languages* (**P** 0577) Automatenth & Formale Sprachen;1969 Oberwolfach 209-220
◇ B75 D05 ◇ REV Zbl 213.21 • ID 14659

CULIK, K. [1971] see BRZOZOWSKI, J.A.

CULIK, K. [1972] *Algorithmization of algebras and relational structures* (**J** 0140) Comm Math Univ Carolinae (Prague) 13*457-477
◇ D20 D45 ◇ REV Zbl 257 # 02023 • ID 29009

CULIK, K. [1973] *Main degrees of complexity of computer programs and computable functions* (**P** 3481) GI Jahrestag (3);1973 Hamburg 151-153
◇ B75 D15 D20 ◇ REV MR 48 # 1502 Zbl 269 # 68028 • ID 61172

CULIK, K. [1973] see ARBIB, M.A.

CULIK, K. [1975] *A note on comparison of Turing machines with computers* (**J** 0086) Cas Pestovani Mat, Ceskoslov Akad Ved 100*118-128
◇ B70 D10 ◇ REV MR 54 # 11853 Zbl 309 # 02034 • ID 25885

CULIK, K. & KARHUMAEKI, J. [1979] *Interactive L systems with almost interactionless behaviour* (**P** 2059) Math Founds of Comput Sci (8);1979 Olomouc 246-257
◇ D05 ◇ REV Zbl 456 # 68083 • ID 69375

CULIK, K. [1980] *Parallel permit grammars and some graph representations of Petri nets* (**J** 0049) Period Math Hung 11*105-116
◇ D05 D80 ◇ REV MR 81i:68098 Zbl 423 # 68034 • ID 69376

CULIK, K. see Vol. I, V for further entries

CULIK II, K. [1970] *N-ary grammars and the description of mapping of languages* (**J** 0156) Kybernetika (Prague) 6*99-117
◇ D05 ◇ REV MR 42 # 7444 Zbl 194.316 JSL 38.525 • ID 02597

CULIK II, K. & MAURER, H.A. & OTTMANN, T. & RUOHONEN, K. & SALOMAA, A. [1978] *Isomorphism, form equivalence and sequence equivalence of PDOL-forms* (**J** 1426) Theor Comput Sci 6*143-173
◇ D05 ◇ REV MR 57 # 11186 Zbl 368 # 68072 • ID 28340

CULIK II, K. & SALOMAA, A. [1978] *On the decidability of homomorphism equivalence for languages* (**J** 0119) J Comp Syst Sci 17*163-175
◇ D05 ◇ REV MR 80g:68093 Zbl 389 # 68042 • ID 69372

CULIK II, K. & MAURER, H.A. & OTTMANN, T. [1978] *On two-symboled complete EOL-forms* (**J** 1426) Theor Comput Sci 6*69-92
◇ D05 ◇ REV MR 57 # 4643 Zbl 369 # 68046 • ID 28339

CULIK II, K. & WOOD, D. [1978] *Speed-varying OL systems* (**J** 0191) Inform Sci 14*161-170
◇ D05 ◇ REV MR 80e:68201 Zbl 416 # 68069 • ID 69377

CULIK II, K. [1978] *The ultimate equivalence problem for DOL systems* (**J** 1431) Acta Inf 10*79-84
◇ D05 ◇ REV MR 58 # 13958 Zbl 385 # 68060 • ID 52175

CULIK II, K. & WOOD, D. [1979] *A mathematical investigation of propagating graph OL systems* (**J** 0194) Inform & Control 43*50-82
◇ D05 ◇ REV MR 81b:68099 Zbl 415 # 68039 • ID 69379

CULIK II, K. [1979] *A purely homomorphic characterization of recursively enumerable sets* (**J** 0037) ACM J 26*345-350
◇ D03 D25 ◇ REV MR 83g:03038 Zbl 395 # 68076 • ID 36005

CULIK II, K. & WOOD, D. [1979] *Doubly deterministic tabled OL systems* (**J** 0435) Int J Comput & Inf Sci 8*335-347
◇ D05 ◇ REV MR 80k:68058 Zbl 426 # 68073 • ID 69374

CULIK II, K. & RICHIER, J. [1979] *Homomorphism equivalence on ETOL languages* (**J** 0382) Int J Comput Math 7*43-51
◇ D05 ◇ REV MR 80b:68084 Zbl 401 # 68049 • ID 54641

CULIK II, K. & KARHUMAEKI, J. [1979] *Interactive L systems with almost interactionless behaviour* (**J** 0194) Inform & Control 43*83-100
◇ D05 ◇ REV MR 81i:68099 Zbl 424 # 68044 • ID 69380

CULIK II, K. & MAURER, H.A. [1979] *On simple representations of language families* (**J** 3441) RAIRO Inform Theor 13*241-250
◇ D05 ◇ REV MR 81d:68093 Zbl 432 # 68052 • ID 80996

CULIK II, K. [1979] *On the homomorphic characterizations of families of languages* (**P** 1873) Automata, Lang & Progr (6);1979 Graz 71*161-170
◇ D05 ◇ REV MR 83m:68132 Zbl 412 # 68064 • ID 69373

CULIK II, K. & DIAMOND, N.D. [1980] *A homomorphic characterization of time and space complexity classes of languages* (**J** 0382) Int J Comput Math 8*207-222
◇ D05 D10 D15 ◇ REV MR 82e:68089 Zbl 444 # 68035 • ID 80994

CULIK II, K. [1980] *Homomorphisms: decidability, equality and test sets* (**P** 4266) Form Lang Th;1979 Santa Barbara 167-194
◇ D05 ◇ REV MR 84j:68001 Zbl 545 # 68065 • ID 47202

CULIK II, K. & KARHUMAEKI, J. [1980] *On the equality sets for homomorphisms on free monoids with two generators* (**J** 3441) RAIRO Inform Theor 14*349-369
◇ D05 D40 ◇ REV MR 82f:68076 • ID 80995

CULIK II, K. & SALOMAA, A. [1980] *Test sets and checking words for homomorphism equivalence* (**J** 0119) J Comp Syst Sci 20*379-395
◇ D05 ◇ REV MR 81h:68064 Zbl 451 # 68046 • ID 69378

CULIK II, K. [1980] see ALBERT, J.

CULIK II, K. [1981] see BUCHER, W.

CULIK II, K. [1981] see ALBERT, J.

CULIK II, K. [1982] see ALBERT, J.

CULIK II, K. & WELZL, E. [1983] *Two way finite state generators* (**P** 3864) FCT'83 Found of Comput Th;1983 Borgholm 106-114
◇ D05 ◇ REV MR 85k:68026 • ID 47049

CULIK II, K. & YU, SHENG [1984] *Iterative tree automata* (**J** 1426) Theor Comput Sci 32*227-247
◇ D05 ◇ ID 44092

CULL, P. [1978] *A matrix algebra for neural nets* (**P** 3352) Appl Gen Syst Res - Devel & Trends;1977 Binghamton 5*563-573
◇ D05 ◇ REV MR 58 # 15730 Zbl 404 # 94015 • ID 69382

CUMMINGS, L.J. [1981] *Overlapping substrings and Thue's problem* (**P** 2279) Carib Conf on Combin & Comput;1981 Bridgetown 99-109
◇ D03 ◇ REV MR 84b:68096 Zbl 518 # 05009 • ID 38955

CUPONA, G. & JANEVA, B. & MARKOVSKI, S. [1985] *The problem of solvability of polylinear representations of universal algebras in semigroups* (**P** 4673) n-ary Struct (2);1983 Varna 65-70
◇ C05 D80 ◇ ID 49621

Cupona, G. see Vol. III, V for further entries

Curry, H.B. [1941] *A formalization of recursive arithmetic* (**J** 0100) Amer J Math 63*263-282
• transl [1970] (**C** 3626) Rekursiv Mat Analiz 3*437-461 (Russian)
◇ B28 D20 F30 ◇ rev MR 2.340 Zbl 25.5 JSL 7.42
• id 02612

Curry, H.B. [1964] *Combinatory recursive objects of all finite types* (**J** 0015) Bull Amer Math Soc 70*814-817
◇ B40 D65 ◇ rev MR 31 # 5795 Zbl 168.257 JSL 39.343
• id 02635

Curry, H.B. [1965] *Two examples of algorithms* (**J** 0009) Arch Math Logik Grundlagenforsch 7*29-44
◇ D03 ◇ rev MR 33 # 3919 Zbl 129.261 • id 02636

Curry, H.B. [1975] *Representation of Markov algorithms by combinators* (**C** 1856) Log Enterprise 109-119
◇ B40 D03 ◇ rev Zbl 362 # 02016 • id 50738

Curry, H.B. see Vol. I, II, VI for further entries

Cutland, N.J. [1970] *The theory of hyperarithmetic and* Π^1_1 *models* (0000) Diss., Habil. etc
◇ C70 D55 ◇ rem Ph.D. Thesis, University of Bristol
• id 16784

Cutland, N.J. [1972] Π^1_1*-models and* Π^1_1*-categoricity* (**P** 2080) Conf Math Log;1970 London 42-62
◇ C57 C70 D55 ◇ rev MR 51 # 133 Zbl 242 # 02057
• id 02662

Cutland, N.J. [1978] *On the form of countable admissible ordinals* (**S** 0019) Colloq Math (Warsaw) 38*173-174
◇ C70 D60 ◇ rev MR 58 # 21535 Zbl 381 # 03025
• id 51893

Cutland, N.J. [1980] *Computability. An introduction to recursive function theory* (**X** 0805) Cambridge Univ Pr: Cambridge, GB x+251pp
• transl [1983] (**X** 0885) Mir: Moskva 256pp
◇ B25 D98 ◇ rev MR 81i:03001 MR 84f:03001
Zbl 448 # 03029 • id 56629

Cutland, N.J. [1980] *On non-monotone* Σ^1_2 *inductive definitions* (**J** 3172) J London Math Soc, Ser 2 22*1-10
◇ D70 E45 ◇ rev MR 81m:03057 Zbl 426 # 03054
• id 55735

Cutland, N.J. see Vol. I, II, III, V for further entries

Czaja, L. [1980] *Deadlock and fairness in parallel schemas: A set-theoretic characterization and decision problems* (**J** 0232) Inform Process Lett 10*234-239
◇ B75 D05 E75 ◇ rev MR 82e:68027 Zbl 445 # 68008
• id 69386

Dahlhaus, E. [1984] *Reduction to NP-complete problems by interpretations* (**P** 2342) Symp Rek Kombin;1983 Muenster 357-365
◇ C13 D15 ◇ rev MR 86f:03060 Zbl 558 # 03019
• id 41789

Dahn, B.I. & Wolter, H. [1984] *Ordered fields with several exponential functions* (**J** 0068) Z Math Logik Grundlagen Math 30*341-348
◇ C25 C35 C65 D35 ◇ rev Zbl 504 # 12024
Zbl 527 # 03016 Zbl 548 # 03013 • id 42268

Dahn, B.I. [1984] *The limit behaviour of exponential terms* (**J** 0027) Fund Math 124*169-186
◇ C60 C65 D15 ◇ rev MR 86f:03058 • id 45383

Dahn, B.I. [1985] *Fine structure of the integral exponential functions below* 2^{2^z} (**P** 4310) Easter Conf on Model Th (3);1985 Gross Koeris 45-63
◇ D15 ◇ id 49057

Dahn, B.I. see Vol. II, III, V, VI for further entries

Dakovski, L.G. & Vojkov, G.K. [1976] *Possibilities of realizing stable natural extensions (Russian)* (**J** 0137) C R Acad Bulgar Sci 29*311-313
◇ D05 ◇ rev MR 53 # 12800 Zbl 363 # 94060 • id 83079

Dalen van, D. [1968] *Fans generated by nondeterministic automata* (**J** 0068) Z Math Logik Grundlagen Math 14*273-278
◇ D05 F55 ◇ rev MR 38 # 1961 Zbl 165.304 • id 02761

Dalen van, D. [1971] *A note on some systems of Lindenmayer* (**J** 0041) Math Syst Theory 5*128-140
◇ D03 D05 D10 D25 ◇ rev MR 49 # 6681
Zbl 218 # 02031 • id 26278

Dalen van, D. & Lascar, D. & Smiley, T.J. (eds.) [1982] *Logic colloquium '80. Eur. Summer Meet., Prague 1980* (**S** 3303) Stud Logic Found Math 108*x+342pp
◇ B97 D97 ◇ rev MR 83i:03003 Zbl 489 # 00006
• id 36589

Dalen van, D. [1983] *Algorithm and decision problems: A crash course in recursion theory* (**C** 4085) Handb Philos Log 1*409-478
◇ D98 ◇ rev Zbl 538 # 03001 • id 39627

Dalen van, D. see Vol. I, II, III, V, VI for further entries

Daley, R.P. [1971] *Recursive and pseudo-recursive sequence and the time bounded minimal-program complexity hierarchies* (**P** 2056) Princeton Conf Inform Sci & Syst (5);1971 Princeton
◇ D15 ◇ id 31695

Daley, R.P. [1973] *An example of information and computation resource trade-off* (**J** 0037) ACM J 20*687-695
◇ D15 ◇ rev MR 49 # 2307 Zbl 272 # 68042 • id 04063

Daley, R.P. [1973] *Complexity and randomness* (**P** 1584) Comput Compl - Courant Comput Sci (7);1971 New York 113-122
◇ D15 ◇ id 31696

Daley, R.P. [1973] *Minimal-program complexity of sequences with restricted resources* (**J** 0194) Inform & Control 23*301-312
◇ D15 ◇ rev MR 49 # 4301 Zbl 272 # 68041 • id 04064

Daley, R.P. [1973] *On the learning of non-recursive sequences* (**P** 2057) Princeton Conf Inform Sci & Syst (7);1973 Princeton
◇ D15 ◇ id 31697

Daley, R.P. [1973] *The process complexity in the understanding of sequences* (**P** 1448) Math Founds of Comput Sci (2);1973 Strbske Pleso 215-218
◇ D15 ◇ rev MR 53 # 7738 • id 22985

Daley, R.P. [1974] *Non-complex sequences: characterizations and examples* (**P** 1479) IEEE Symp Switch & Automata Th (15);1974 New Orleans 165-169
◇ D15 ◇ rev MR 58 # 19342 • id 32057

DALEY, R.P. [1974] *The extent and density of sequences within the minimal-program complexity hierarchies* (**J** 0119) J Comp Syst Sci 9*151-163
◇ D15 ◇ REV MR 54 # 1710 Zbl 293 # 68044 • ID 31704

DALEY, R.P. [1975] *A note on a result of Kamae* (**J** 0351) Osaka J Math 12*283-284
◇ D15 ◇ REV MR 51 # 12022 Zbl 337 # 94014 • ID 31705

DALEY, R.P. [1975] *Minimal program complexity of pseudo-recursive and pseudo-random sequences* (**J** 0041) Math Syst Theory 9*83-94
◇ D15 D80 ◇ REV MR 55 # 5407 Zbl 307 # 68033 • ID 31706

DALEY, R.P. [1976] *Noncomplex sequences: characterizations and examples* (**J** 0036) J Symb Logic 41*626-638
◇ D20 ◇ REV MR 58 # 5117 Zbl 365 # 68054 • ID 14585

DALEY, R.P. [1977] *On the inference of optimal descriptions* (**J** 1426) Theor Comput Sci 4*301-319
◇ B35 D15 ◇ REV MR 56 # 8351 Zbl 375 # 02041 • ID 31707

DALEY, R.P. [1978] *On the inference of optimal descriptions* (**P** 1962) Conf Inform Sci & Syst;1977 Baltimore
◇ D15 ◇ ID 31701

DALEY, R.P. [1978] *On the simplicity of busy beaver sets* (**J** 0068) Z Math Logik Grundlagen Math 24*207-224
◇ D10 D15 D25 D30 ◇ REV MR 80h:03061 Zbl 421 # 03031 • ID 31708

DALEY, R.P. [1978] *Quantitative and qualitative information in computations* (**P** 1962) Conf Inform Sci & Syst;1977 Baltimore
◇ D15 D30 ◇ ID 32058

DALEY, R.P. [1979] *On the simplification of constructions in degrees of unsolvability via computational complexity* (**P** 2059) Math Founds of Comput Sci (8);1979 Olomouc 258-265
◇ D10 D15 D25 ◇ REV MR 81c:68004 Zbl 463 # 03025 • ID 31703

DALEY, R.P. [1980] *Quantitative and qualitative information in computations* (**J** 0194) Inform & Control 45*236-244
◇ D15 D30 ◇ REV MR 81k:68029 Zbl 457 # 68042 • ID 54404

DALEY, R.P. [1980] *The busy beaver method* (**P** 2058) Kleene Symp;1978 Madison 333-345
◇ D10 D15 D25 ◇ REV MR 82a:03040 Zbl 501 # 03024 • ID 31702

DALEY, R.P. [1981] *Busy beaver sets and the degree of unsolvability* (**J** 0036) J Symb Logic 46*460-474
◇ D10 D15 D25 ◇ REV MR 83a:03035 Zbl 486 # 03022 • ID 33270

DALEY, R.P. [1981] *On the error correcting power of pluralism in inductive inference (Preliminary report)* (**P** 3165) FCT'81 Fund of Comput Th;1981 Szeged 90-99
◇ D10 D20 ◇ REV MR 83g:03037 Zbl 475 # 68020 • ID 55508

DALEY, R.P. [1981] *Retraceability, repleteness and busy beaver sets* (**P** 3429) Math Founds of Comput Sci (10);1981 Strbske Pleso 252-261
◇ D15 D20 D50 ◇ REV MR 83m:03049 Zbl 486 # 03021 • ID 35449

DALEY, R.P. [1982] *Busy beaver sets: characterizations and applications* (**J** 0194) Inform & Control 52*52-67
◇ D10 D25 D50 ◇ REV MR 85h:03039 Zbl 503 # 03017 • ID 44238

DALEY, R.P. & MANDERS, K.L. [1982] *The complexity of the validity problem for dynamic logic* (**J** 0194) Inform & Control 54*48-69
◇ B75 D15 ◇ REV MR 85d:03089 Zbl 521 # 03023 • ID 33516

DALEY, R.P. & SMITH, C.H. [1984] *On the complexity of inductive inference* (**P** 3658) Math Founds of Comput Sci (11);1984 Prague 255-264
◇ B75 D15 D20 ◇ ID 44831

DALIK, J. [1977] *Verbandstheoretische Eigenschaften von Sprachen II* (**J** 0322) Arch Math (Brno) 13*13-23
◇ D05 ◇ REV MR 57 # 8173b Zbl 374 # 68044 • REM Part I 1976 • ID 69394

DAMM, W. & FEHR, E. & INDERMARK, K. [1978] *Higher type recursion and self-application as control structures* (**P** 2929) Form Descr of Progr Concepts (1);1977 St.Andrews 461-489
◇ B40 B75 D20 ◇ REV MR 80e:68014 Zbl 373 # 68021 • ID 51522

DAMM, W. [1979] *An algebraic extension of the Chomsky-hierarchy* (**P** 2059) Math Founds of Comput Sci (8);1979 Olomouc 266-276
◇ B75 D05 ◇ REV Zbl 412 # 68069 • ID 52985

DAMM, W. & GUESSARIAN, I. [1981] *Combining T and level-N* (**P** 3429) Math Founds of Comput Sci (10);1981 Strbske Pleso 262-270
◇ D05 ◇ REV Zbl 477 # 68091 • ID 69395

DAMM, W. see Vol. VI for further entries

DANECKI, R. & KARPINSKI, M. [1979] *Decidability results on plane automata searching mazes* (**P** 2935) FCT'79 Fund of Comput Th;1979 Berlin/Wendisch-Rietz 84-91
◇ D05 ◇ REV MR 81d:68068 Zbl 423 # 68013 • ID 81025

DANECKI, R. [1981] *Multiple regularity and binary ETOL-systems* (**J** 2095) Fund Inform, Ann Soc Math Pol, Ser 4 4*19-34
◇ D05 ◇ REV MR 83g:68102 Zbl 467 # 68072 • ID 69398

DANECKI, R. see Vol. II for further entries

DANG HUU DAO [1978] *On the semigroups of directable automata (Preprint)* (**J** 3529) Rep Dept Numer & Comp Math, Univ Budapest 1*28pp
◇ D05 ◇ REV Zbl 409 # 68035 • ID 69409

DANG HUU DAO [1979] *On the characteristic semigroups of Mealy-automata* (**X** 3151) Karl Marx Univ Dpt Math: Budapest 1979/1*1-36
◇ D05 ◇ REV MR 82h:68077 Zbl 403 # 68056 • ID 69401

DANG HUU DAO [1980] *Further results on characteristic semigroups of Mealy-automata* (**C** 3483) Pap Automata Th 1980/2*47-78
◇ D05 ◇ REV MR 83i:68087 Zbl 461 # 68055 • ID 69402

DANG HUU DAO [1980] *Quasi-Moore-semigroups and Moore-semigroups for some classes of Mealy-automata* (**C** 3483) Pap Automata Th 1980/5*13-36
◇ D05 ◇ REV MR 83i:68086 Zbl 461 # 68056 • ID 69399

DANG HUU DAO [1981] *Mealy-automata and their characteristic semigroups* (**X** 3151) Karl Marx Univ Dpt Math: Budapest 1981/1*17-19
◇ D05 ◇ REV Zbl 461 # 68057 • ID 69400

DANG HUY RUAN & TSEJTIN, G.S. [1974] *An upper estimate of the finite-state complexity for a class of generating schemes containing complements and intersections (Russian)* (**S** 0228) Zap Nauch Sem Leningrad Otd Mat Inst Steklov 40*14-23,155
• TRANSL [1977] (**J** 1531) J Sov Math 8*254-262
◇ D05 ◇ REV MR 55 # 12486 Zbl 358 # 94065 • ID 50547

DANILOV, V.V. & PODKOPAEV, B.P. [1980] *Use of the spectral and autocorrelation properties of logic functions for synthesizing combination schemes on Maitra's cascades (Russian)* (**J** 0977) Izv Akad Nauk SSSR, Tekh Kibern 1980/1*112-120
• TRANSL [1980] (**J** 0522) Engin Cybern 18/1*95-104
◇ D05 ◇ REV Zbl 467 # 93019 • ID 55063

DANKO, W. [1974] *Not programmable function derived by a procedure (Russian summary)* (**J** 0014) Bull Acad Pol Sci, Ser Math Astron Phys 22*587-594
◇ D20 ◇ REV MR 50 # 6807 Zbl 293 # 68043 • ID 04066

DANKO, W. [1979] *Definability in algorithmic logic* (**J** 2095) Fund Inform, Ann Soc Math Pol, Ser 4 2*277-287
◇ B75 D05 D20 ◇ REV MR 82f:68031 Zbl 453 # 03026 • ID 54156

DANKO, W. [1980] *A criterion of undecidability of algorithmic theories* (**P** 3210) Math Founds of Comput Sci (9);1980 Rydzyna 205-218
◇ D35 ◇ REV MR 83m:68053 Zbl 453 # 03027 • ID 54157

DANKO, W. [1981] *A criterion of undecidability of algorithmic theories* (**J** 2095) Fund Inform, Ann Soc Math Pol, Ser 4 4*605-628
◇ B75 D35 ◇ REV MR 84j:03090 Zbl 503 # 03006 • ID 34681

DANKO, W. see Vol. II, III, VI for further entries

DANTSIN, E.YA. [1972] *Increase of the complexity of functions by an application of multiple recursion (Russian) (English summary)* (**S** 0228) Zap Nauch Sem Leningrad Otd Mat Inst Steklov 32*12-17,153
• TRANSL [1976] (**J** 1531) J Sov Math 6*353-357
◇ D20 ◇ REV MR 49 # 8841 Zbl 345 # 02027 • ID 04065

DANTSIN, E.YA. [1979] *Parameters defining the time of tautology recognition by the splitting method (Russian)* (**S** 2582) Semiotika & Inf, Akad Nauk SSSR 12*8-17
◇ B35 D15 ◇ REV MR 81i:68130 • ID 81024

DANTSIN, E.YA. [1981] *Two systems for proving tautologies based on the splitting method (Russian) (English summary)* (**S** 0228) Zap Nauch Sem Leningrad Otd Mat Inst Steklov 105*24-44,198-199
• TRANSL [1983] (**J** 1531) J Sov Math 22*1293-1305
◇ B35 D15 F20 ◇ REV MR 83i:68140 Zbl 476 # 03020 Zbl 509 # 03004 • ID 55532

DANTSIN, E.YA. see Vol. VI for further entries

DARE, V.R. & SIROMONEY, R. & SUBRAMANIAN, K.G. [1981] *Infinite arrays and infinite computations* (**P** 4258) Found of Softw Tech & Th Comput Sci (1);1981 Bangalore 281-285
◇ D20 ◇ REV MR 85a:68109 Zbl 509.68079 • ID 46481

DARE, V.R. & SIROMONEY, R. & SUBRAMANIAN, K.G. [1983] *Infinite arrays and infinite computations* (**J** 1426) Theor Comput Sci 24*195-205
◇ B75 D05 ◇ REV MR 85a:68109 Zbl 509 # 68079 • ID 38871

DARE, V.R. & SIROMONEY, R. [1985] *A generalization of the Parikh vector for finite and infinite words* (**P** 4672) Found of Softw Tech & Th Comput Sci (5);1985 New Delhi 290-302
◇ D05 ◇ ID 49648

DARE, V.R. & SIROMONEY, R. [1985] *On infinite words obtained by selective substitution grammars* (**J** 1426) Theor Comput Sci 39*281-295
◇ D05 ◇ ID 49088

DARONDEAU, P. & KOTT, L. [1984] *Towards a formal proof system for ω-rational expressions* (**J** 0232) Inform Process Lett 19*173-177
◇ D05 ◇ REV MR 86i:68087 Zbl 575 # 68082 • ID 48876

DARONDEAU, P. & KOTT, L. [1985] *A formal proof system for infinitary rational expressions* (**P** 4595) Autom Infinite Words;1984 Le Mont Dore 68-80
◇ D05 ◇ REV Zbl 575 # 68083 • ID 48877

DASSOW, J. [1974] *Kleene-Mengen und Kleene-Vollstaendigkeit (English and Russian summaries)* (**J** 0129) Elektr Informationsverarbeitung & Kybern 10*287-295
◇ D05 ◇ REV MR 54 # 4939 Zbl 293 # 94023 • ID 24112

DASSOW, J. [1974] *Kleene-Vollstaendigkeit bei stark-definiten Ereignissen* (**J** 0129) Elektr Informationsverarbeitung & Kybern 10*399-405
◇ D05 ◇ REV MR 57 # 9505 Zbl 301 # 94066 • ID 61235

DASSOW, J. [1974] *Ueber zwei abgeschlossene Mengen von Automatenoperatoren* (**J** 3465) Wiss Z Univ Rostock, Math-Nat Reihe 23*763-767
◇ D05 ◇ REV Zbl 423 # 68014 • ID 69417

DASSOW, J. [1975] *Ueber metrische Kleene-Vollstaendigkeit (English and Russian summaries)* (**J** 0129) Elektr Informationsverarbeitung & Kybern 11*557-560
◇ D05 ◇ REV MR 55 # 2399 Zbl 333 # 94026 • ID 81028

DASSOW, J. [1976] *Kleene-Mengen und trennende Mengen* (**J** 0114) Math Nachr 74*89-97
◇ B50 D05 ◇ REV MR 55 # 2400 Zbl 344 # 94027 • ID 61236

DASSOW, J. & HARNAU, W. [1976] *Spezielle Gleichungen in freien Halbgruppen und eine Anwendung auf Normalformen regulaerer Ereignisse* (**J** 3124) Beitr Algebra Geom 5*33-40
◇ D05 ◇ REV MR 55 # 12347 Zbl 355 # 20075 • ID 50258

DASSOW, J. [1977] *Einige Bemerkungen zu einem modifizierten Vollstaendigkeitsbegriff fuer Automatenabbildungen* (**S** 2829) Rostocker Math Kolloq 3*69-84
◇ D05 ◇ REV MR 58 # 9892 Zbl 397 # 68061 • ID 52722

DASSOW, J. [1977] *Some remarks on the algebra of automaton mappings* (**P** 2588) FCT'77 Fund of Comput Th;1977 Poznan 56*78-83
◇ D05 ◇ REV MR 58 # 176 Zbl 397 # 68057 • ID 72136

DASSOW, J. [1977] *Ueber Abschlusseigenschaften von biologisch motivierten Sprachen* (**S** 2829) Rostocker Math Kolloq 4*69-84
◇ D05 D80 ◇ REV MR 58 # 32092 Zbl 365 # 94074 • ID 51081

DASSOW, J. [1979] *Concentration dependent OL systems* (**J** 0382) Int J Comput Math 7*187-206
◇ D05 ◇ REV MR 81b:68090 Zbl 453 # 68042 • ID 69413

DASSOW, J. [1979] *ETOL systems and compound grammars* (**S** 2829) Rostocker Math Kolloq 11*41-46
◇ D05 ◇ REV MR 81d:68094 Zbl 453 # 68043 • ID 69414

DASSOW, J. [1979] *On the circular closure of languages* (**J** 0129) Elektr Informationsverarbeitung & Kybern 15*87-94
◇ D05 ◇ REV MR 81k:68056 Zbl 415 # 68040 • ID 69415

DASSOW, J. [1980] *On some extensions of Indian parallel context free grammars* (**J** 0380) Acta Cybern (Szeged) 4*303-310
◇ D05 ◇ REV MR 82a:68140 Zbl 441 # 68084 • ID 69411

DASSOW, J. [1980] *On Parikh-languages of L systems without interaction* (**S** 2829) Rostocker Math Kolloq 15*103-110
◇ D05 ◇ REV MR 83h:68112 Zbl 453 # 68044 • ID 69416

DASSOW, J. [1981] *Completeness problems in the structural theory of automata* (**X** 0911) Akademie Verlag: Berlin 148pp
◇ D05 ◇ REV MR 83j:68063 Zbl 481 # 68052 • ID 38460

DASSOW, J. [1981] *Ein modifizierter Vollstaendigkeitsbegriff in einer Algebra von Automatenabbildungen* (**S** 2829) Rostocker Math Kolloq 17*123-124
◇ D05 ◇ REV Zbl 463 # 68055 • ID 54589

DASSOW, J. [1981] *Equality languages and language families* (**P** 3165) FCT'81 Fund of Comput Th;1981 Szeged 100-109
◇ D05 ◇ REV MR 83g:68103 Zbl 497 # 68046 • ID 39150

DASSOW, J. [1981] *On the congruence lattice of algebras of automaton mappings* (**P** 2552) Conf Finite Algeb & Multi-Val Log;1979 Szeged 161-182
◇ D05 ◇ REV MR 83d:08001 Zbl 489 # 68047 • ID 36603

DASSOW, J. [1982] *Weitere Bemerkungen zu einem modifizierten Vollstaendigkeitsbegriff in einer Algebra von Automatenabbildungen* (**S** 2829) Rostocker Math Kolloq 21*99-110
◇ D05 ◇ REV MR 85i:68031 • ID 44365

DASSOW, J. & WOLTER, U. [1985] *Remarks on equality languages* (**J** 3289) Wiss Z Tech Hochsch Madgeburg 29*122-126
◇ A05 B20 D05 ◇ ID 49482

DATEL, V. [1976] *Anmerkungen zum Verhalten nichtdeterministischer Automaten (Tschechisch) (Russische Zusammenfassung)* (**C** 3392) Knizhnice - TU Brno, Vol A10, A12 A10*91-98
◇ D10 ◇ REV Zbl 393 # 68063 • ID 69418

D'ATRI, A. [1981] see AUSIELLO, G.

DAUCHET, M. [1977] *Grammaires transformationnelles et bimorphismes de magmoides* (**P** 3199) CAAP'77 Arbres en Algeb & Progr (2);1977 Lille 249-273
◇ D05 ◇ REV Zbl 382 # 68062 • ID 69419

DAUCHET, M. [1978] see ARNOLD, A.

DAUCHET, M. & MONGY, J. [1979] *Transformations de noyaux reconnaissables. Capacite generative des bimorphismes des forets* (**P** 2935) FCT'79 Fund of Comput Th;1979 Berlin/Wendisch-Rietz 92-98
◇ D05 ◇ REV Zbl 424 # 68045 • ID 69421

DAUCHET, M. & LATTEUX, M. [1981] *A propos d'un langage presque rationnel* (**J** 3364) C R Acad Sci, Paris, Ser 1 292*107-110
◇ D05 ◇ REV MR 82a:68072 Zbl 456 # 68091 • ID 69420

DAUCHET, M. [1982] see COMYN, G.

DAUCHET, M. & TISON, S. [1985] *Decidability of confluence for ground term rewriting systems* (**P** 4647) FCT'85 Fund of Comput Th;1985 Cottbus 80-89
◇ D05 ◇ ID 49072

DAUCHET, M. [1985] *Decidability of yield's equality for infinite regular trees* (**P** 4622) Autom on Infinite Words;1984 Le Mont-Dore 118-136
◇ D05 ◇ ID 49447

DAUSCHA, W. & NUERNBERG, G. & STARKE, P.H. & WINKLER, K.-D. [1973] *Theorie der determinierten zeitvariablen Automaten* (**J** 0129) Elektr Informationsverarbeitung & Kybern 9*455-511
◇ D05 ◇ REV MR 50 # 16149 Zbl 296 # 94031 • ID 42711

DAVID, R. [1978] *A Π^1_2 singleton with no sharp in a generic extension of $L^\#$* (**J** 0029) Israel J Math 31*343-352
◇ D55 E35 E45 E55 ◇ REV MR 80a:03064 Zbl 404 # 03038 • ID 29140

DAVID, R. & MITRANI, E. & TELLEZ-GIRON, R. [1980] *Emploi de cellules universelles pour la synthese de systemes asynchrones decrits par reseaux de Petri* (**J** 2701) Digit Processes 6*185-198
◇ D05 D80 ◇ REV Zbl 443 # 94029 • ID 69426

DAVID, R. [1982] *Δ^1_3 reals* (**J** 0007) Ann Math Logic 23*121-125
◇ D55 E35 E45 E55 ◇ REV MR 85a:03060 Zbl 519 # 03039 • ID 34808

DAVID, R. [1982] *A very absolute Π^1_2 real singleton* (**J** 0007) Ann Math Logic 23*101-120
◇ D55 E15 E40 E45 E55 ◇ REV MR 84m:03057 Zbl 519 # 03038 • ID 37544

DAVID, R. [1982] *Some applications of Jensen's coding theorem* (**J** 0007) Ann Math Logic 22*177-196
◇ C62 D55 E40 E45 ◇ REV MR 83j:03084 Zbl 489 # 03021 • ID 35373

DAVID, R. & FRIEDMAN, S.D. [1985] *Uncountable ZF-ordinals* (**P** 4046) Rec Th;1982 Ithaca 217-222
◇ C62 D60 E40 E45 ◇ REV Zbl 566 # 03033 • ID 46391

DAVID, R. see Vol. V for further entries

DAVIES, R.O. [1974] *On a separation theorem of Rogers* (**J** 0171) Proc Cambridge Phil Soc Math Phys 75*357-359
◇ D55 E15 ◇ REV MR 49 # 3868 Zbl 291 # 54047 • ID 28188

DAVIES, R.O. & JAYNE, J.E. & OSTASZEWSKI, A.J. & ROGERS, C.A. [1977] *Theorems of Novikov type* (**J** 0303) Mathematika (Univ Coll London) 24*97-114
◇ D55 E15 ◇ REV MR 57 # 1450 Zbl 359 # 54028 • ID 28189

DAVIES, R.O. see Vol. II, V for further entries

DAVIS, A.S. [1968] *Half-ring morphologies* (**P** 0692) Summer School in Logic;1967 Leeds 253-268
◇ D05 G25 ◇ REV MR 39 # 2639 Zbl 198.329 • ID 02813

DAVIS, CHARLES C. [1974] *Some semantically closed languages* (**J** 0122) J Philos Logic 3*229-240
• REPR [1977] (**P** 2116) Probl in Log & Ontology;1973 Salzburg 63-74
◇ A05 D05 ◇ REV MR 54 # 7192 MR 58 # 27426 Zbl 281 # 02004 • ID 29563

DAVIS, CHARLES C. see Vol. I, II, V for further entries

DAVIS, MARTIN D. [1950] *On the theory of recursive unsolvability* (0000) Diss., Habil. etc
◇ D35 D55 ◇ REM Ph.D. thesis, Princeton University, Princeton, NJ • ID 16786

DAVIS, MARTIN D. [1952] *Relatively recursive functions and the extended Kleene hierarchy* (**P** 0593) Int Congr Math (II, 6);1950 Cambridge MA 1*723
◇ D55 ◇ ID 28060

DAVIS, MARTIN D. [1953] *Arithmetical problems and recursively enumerable predicates* (**J** 0036) J Symb Logic 18*33-41
◇ D25 D35 D55 D80 ◇ REV MR 14.1052 Zbl 51.245 JSL 18.341 • ID 02818

DAVIS, MARTIN D. [1956] *A note on universal Turing machines* (**C** 0717) Automata Studies 167-175
• TRANSL [1974] (**C** 1902) Stud Th Automaten 195-203
◇ D10 ◇ REV MR 18.103 JSL 35.590 • ID 04177

DAVIS, MARTIN D. [1957] *Computable functionals of arbitrary finite type* (**P** 1675) Summer Inst Symb Log;1957 Ithaca 242-246
◇ D65 ◇ ID 29353

DAVIS, MARTIN D. & PUTNAM, H. [1957] *Reductions of Hilbert's 10th problem* (**P** 1675) Summer Inst Symb Log;1957 Ithaca 348-349
◇ D25 D35 ◇ REV MR 21 # 2583 • ID 29375

DAVIS, MARTIN D. [1957] *The definition of universal Turing machine* (**J** 0053) Proc Amer Math Soc 8*1125-1126
◇ D10 ◇ REV MR 20 # 2282 Zbl 84.10 JSL 35.590 • ID 02819

DAVIS, MARTIN D. [1958] *Computability and unsolvability* (**X** 0822) McGraw-Hill: New York xxv+210pp
◇ D98 ◇ REV MR 23 # A1525 Zbl 80.9 JSL 23.432 • ID 02822

DAVIS, MARTIN D. & PUTNAM, H. [1958] *Reductions of Hilbert's tenth problem* (**J** 0036) J Symb Logic 23*183-187
◇ D25 D35 ◇ REV MR 21 # 2583 Zbl 85.248 JSL 37.601 • ID 02820

DAVIS, MARTIN D. [1959] *Computable functionals of arbitrary finite type* (**P** 0634) Constructivity in Math;1957 Amsterdam 281-284
◇ D65 ◇ REV MR 21 # 6329 Zbl 85.248 • ID 04178

DAVIS, MARTIN D. & PUTNAM, H. & ROBINSON, JULIA [1961] *The decision problem for exponential diophantine equations* (**J** 0120) Ann of Math, Ser 2 74*425-436
◇ D25 D35 ◇ REV MR 24 # A3061 Zbl 111.10 JSL 35.151 • ID 02826

DAVIS, MARTIN D. [1962] *Applications of recursive function theory to number theory* (**P** 0613) Rec Fct Th;1961 New York 135-138
◇ D35 ◇ REV MR 27 # 1365 Zbl 192.53 JSL 37.602 • ID 02833

DAVIS, MARTIN D. & PUTNAM, H. [1963] *Diophantine sets over polynomial rings* (**J** 0316) Illinois J Math 7*251-256
◇ D25 D75 ◇ REV MR 26 # 4903 Zbl 113.6 JSL 37.602 • ID 04182

DAVIS, MARTIN D. [1963] *Extensions and corollaries of recent work on Hilbert's tenth problem* (**J** 0316) Illinois J Math 7*246-250
◇ D25 D35 ◇ REV MR 26 # 6046 Zbl 112.246 JSL 37.602 • ID 04184

DAVIS, MARTIN D. [1963] *Unsolvable problems: a review* (**P** 0674) Symp Math Th of Automata;1962 New York 15-22
◇ D03 D10 D35 D40 ◇ REV MR 30 # 1039 Zbl 129.257 JSL 33.297 • ID 02834

DAVIS, MARTIN D. [1965] *Introduction to Goedel's paper: On formally undecidable propositions of the Principia Mathematica and related systems I* (**C** 0718) The Undecidable 4
◇ A10 D20 F30 ◇ REV JSL 31.484 • REM Goedel's paper was published in J0124 38(1937)*173-198 • ID 04186

DAVIS, MARTIN D. (ED.) [1965] *The undecidable. Basic papers on undecidable propositions, unsolvable problems and computable functions* (**X** 0887) Raven Pr: New York 440pp
◇ D20 D25 D35 D97 F30 ◇ REV MR 32 # 7412 • ID 25888

DAVIS, MARTIN D. [1966] *Diophantine equations and recursively enumerable sets* (**P** 0746) Automata Th;1964 Ravello 146-152
◇ D25 ◇ REV MR 34 # 2524 Zbl 199.39 • ID 16234

DAVIS, MARTIN D. [1966] *Recursive functions – an introduction* (**P** 0746) Automata Th;1964 Ravello 153-163
◇ D20 D35 ◇ REV Zbl 207.305 • ID 27535

DAVIS, MARTIN D. [1967] *Recursive function theory* (**C** 0601) Encycl of Philos 7*89-95
◇ A05 D98 ◇ REV JSL 35.296 • ID 02835

DAVIS, MARTIN D. [1968] *One equation to rule them all* (**J** 0317) Trans New York Acad Sci Ser 2 30*766-773
◇ D25 D35 ◇ REV Zbl 316 # 02051 JSL 37.602 • ID 04187

DAVIS, MARTIN D. [1971] *An explicit diophantine definition of the exponential function* (**J** 0155) Commun Pure Appl Math 24*137-145
◇ D35 ◇ REV MR 42 # 7632 Zbl 222 # 10017 • ID 26329

DAVIS, MARTIN D. [1972] *On the number of the solutions of diophantine equations* (**J** 0053) Proc Amer Math Soc 35*552-554
◇ D25 D35 ◇ REV MR 46 # 3482 Zbl 275 # 02042 • ID 02836

DAVIS, MARTIN D. [1973] *Hilbert's tenth problem is unsolvable* (**J** 0005) Amer Math Mon 80*233-269
◇ D25 D35 ◇ REV MR 47 # 6465 Zbl 277 # 02008 • ID 02837

DAVIS, MARTIN D. [1974] *Computability* (**X** 1214) New York Univ Cour Inst Math: New York 248pp
◇ B98 D98 ◇ REV MR 50 # 77 Zbl 281 # 02038 • ID 21268

DAVIS, MARTIN D. & MATIYASEVICH, YU.V. & ROBINSON, JULIA [1976] *Hilbert's tenth problem: Diophantine equations: Positive aspects of a negative solution* (**P** 2957) Math Dev from Hilbert Probl;1974 DeKalb 323-378
◇ D25 D35 D80 ◇ REV MR 55 # 5522 Zbl 346 # 02026 JSL 44.116 • ID 61256

DAVIS, MARTIN D. & HERSH, R. [1977] *Hilbert's tenth problem* (J 0477) Spis Bulgar Akad Nauk 20(53)*245-252
◇ D35 ◇ REV MR 58 # 21275 • ID 81059

DAVIS, MARTIN D. [1977] *Unsolvable problems* (C 1523) Handb of Math Logic 567-594
◇ D10 D35 D40 ◇ REV MR 58 # 5109 JSL 49.975 • ID 27320

DAVIS, MARTIN D. [1982] *Computability and unsolvability* (X 0813) Dover: New York xxv+248pp
◇ D98 ◇ REV Zbl 553 # 03024 • ID 43321

DAVIS, MARTIN D. & WEYUKER, E.J. [1983] *Computability, complexity, and languages. Fundamentals of theoretical computer science* (X 0801) Academic Pr: New York xix+425pp
◇ D05 D10 D20 D98 ◇ REV MR 86b:03001 Zbl 569 # 68042 Zbl 574 # 03027 • ID 33584

DAVIS, MARTIN D. see Vol. I, II for further entries

DAVIS, MORTON [1964] *Infinite games of perfect information* (S 3513) Ann Math Stud 52*85-101
◇ D55 E15 E60 ◇ REV MR 30 # 965 Zbl 133.131 • ID 26244

DAWES, A.M. & FLORENCE, J.B. [1975] *Independent Goedel sentences and independent sets* (J 0036) J Symb Logic 40*159-166
◇ D25 F30 ◇ REV MR 52 # 10383 Zbl 324 # 02027 • ID 04287

DAWES, A.M. [1982] *Splitting theorems for speed-up related to order of enumerations* (J 0036) J Symb Logic 47*1-7
◇ D15 D25 ◇ REV MR 83c:03036 Zbl 503 # 03016 • ID 35138

DAWES, A.M. see Vol. III, V for further entries

DAWSON JR., J.W. [1973] *Ordinal definability in the rank hierarchy* (J 0007) Ann Math Logic 6*1-39 • ERR/ADD ibid 7*325
◇ D60 E35 E45 E55 ◇ REV MR 49 # 8861 Zbl 285 # 02053 • ID 02850

DAWSON JR., J.W. see Vol. III, V, VI for further entries

DAY, A. [1970] *A simple solution of the word problem for lattices* (J 0018) Canad Math Bull 13*253-254
◇ D40 G10 ◇ REV MR 42 # 2991 Zbl 206.297 • ID 02852

DAY, A. & HERRMANN, C. & WILLE, R. [1972] *On modular lattices with four generators* (J 0004) Algeb Universalis 2*317-323
◇ D40 G10 ◇ REV MR 47 # 4879 Zbl 275 # 06010 • ID 02855

DAY, A. see Vol. III for further entries

DEDEKIND, R. [1888] *Was sind und was sollen die Zahlen?* (X 0900) Vieweg: Wiesbaden xv+58pp
◇ B28 D20 E75 ◇ REV FdM 20.49 • REM 10th ed. 1965. For an English transl. see 1948 • ID 35601

DEDEKIND, R. see Vol. I, V for further entries

DEGANO, P. & SIROVICH, F. [1979] *Inductive generalization and proofs of function properties* (J 2293) Comp Linguist & Comp Lang 13*101-130
◇ B35 B75 D20 D80 ◇ REV MR 81f:68111 Zbl 449 # 68054 • ID 56755

DEGANO, P. see Vol. I for further entries

DEGTEV, A.N. [1970] *Remarks on retraceable, regressive and pointwise-decomposable sets (Russian)* (J 0003) Algebra i Logika 9*651-660
• TRANSL [1970] (J 0069) Algeb and Log 9*390-395
◇ D25 D50 ◇ REV MR 44 # 3860 Zbl 239 # 02023 • ID 02867

DEGTEV, A.N. [1971] *Hypersimple sets with retraceable complements (Russian)* (J 0003) Algebra i Logika 10*235-246
• TRANSL [1971] (J 0069) Algeb and Log 10*147-154
◇ D25 ◇ REV MR 44 # 5221 Zbl 259 # 02031 • ID 33027

DEGTEV, A.N. [1972] *Hereditary sets and truth-table reducibility (Russian)* (J 0003) Algebra i Logika 11*257-269,361
• TRANSL [1972] (J 0069) Algeb and Log 11*145-152
◇ D30 ◇ REV MR 47 # 1588 Zbl 283 # 02035 • ID 02872

DEGTEV, A.N. [1972] *The m-degrees of simple sets (Russian)* (J 0003) Algebra i Logika 11*130-139,237
• TRANSL [1972] (J 0069) Algeb and Log 11*74-80
◇ D25 ◇ REV MR 47 # 27 Zbl 287 # 02027 • ID 02874

DEGTEV, A.N. [1973] *tt and m degrees (Russian)* (J 0003) Algebra i Logika 12*143-161,243
• TRANSL [1973] (J 0069) Algeb and Log 12*78-89
◇ D25 D30 ◇ REV MR 54 # 12505 Zbl 338 # 02023 • ID 14977

DEGTEV, A.N. [1975] *Reducibility of partial recursive functions (Russian)* (J 0092) Sib Mat Zh 16*970-988,1130
• TRANSL [1975] (J 0475) Sib Math J 16*741-754
◇ D20 D25 ◇ REV MR 54 # 12506 Zbl 331 # 02021 • REM Part I. Part II 1977 • ID 61269

DEGTEV, A.N. [1976] *Minimal 1-degrees, and truth-table reducibility (Russian)* (J 0092) Sib Mat Zh 17*1014-1022,1196
• TRANSL [1976] (J 0475) Sib Math J 17*751-757
◇ D25 D30 ◇ REV MR 54 # 9997 Zbl 356 # 02037 • ID 25838

DEGTEV, A.N. [1976] *Partially ordered sets of 1-degrees, contained in recursively enumerable m-degrees (Russian)* (J 0003) Algebra i Logika 15*249-266,365
• TRANSL [1976] (J 0069) Algeb and Log 15*153-164
◇ D25 ◇ REV MR 58 # 27409 Zbl 369 # 02019 • ID 26047

DEGTEV, A.N. [1976] *Some classes of hyperimmune sets (Russian)* (C 2555) Algeb Sistemy (Irkutsk) 21-36
◇ D50 ◇ ID 33028

DEGTEV, A.N. [1977] *A family of maximal subalgebras of R. Robinson's algebra (Russian)* (J 0087) Mat Zametki (Akad Nauk SSSR) 22*511-516
• TRANSL [1977] (J 1044) Math Notes, Acad Sci USSR 22*775-778
◇ D20 G25 ◇ REV MR 58 # 191 Zbl 362 # 02029 • ID 50751

DEGTEV, A.N. [1977] *Reducibility of partial recursive functions II (Russian)* (J 0092) Sib Mat Zh 18*764-774
• TRANSL [1977] (J 0475) Sib Math J 18*541-548
◇ D20 ◇ REV MR 57 # 80 Zbl 375 # 02035 • REM Part I 1975 • ID 51612

DEGTEV, A.N. [1978] *m-degrees of supersets of simple sets (Russian)* (J 0087) Mat Zametki (Akad Nauk SSSR) 23*889-893
• TRANSL [1978] (J 1044) Math Notes, Acad Sci USSR 23*488-490
◇ D25 ◇ REV MR 80h:03062 Zbl 387 # 03014 • ID 52230

DEGTEV, A.N. [1978] *Nonerasing Turing machines (Russian)* (**C** 2595) Rekursiv Funktsii 11-17
◇ D10 ◇ REV MR 82c:03060 Zbl 504 # 03020 • ID 72215

DEGTEV, A.N. [1978] *Solvability of the ∀∃-theory of a certain factor-lattice of recursively enumerable sets (Russian)* (**J** 0003) Algebra i Logika 17*134-143,241
• TRANSL [1978] (**J** 0069) Algeb and Log 17*94-101
◇ B25 D25 D50 ◇ REV MR 80j:03055 Zbl 445 # 03020
• ID 72221

DEGTEV, A.N. [1978] *Three theorems on tt-degrees (Russian)* (**J** 0003) Algebra i Logika 17*270-281
• TRANSL [1978] (**J** 0069) Algeb and Log 17*187-194
◇ D25 D30 ◇ REV MR 80j:03056 Zbl 421 # 03032
• ID 53431

DEGTEV, A.N. & ZAKHAROV, D.A. [1979] *Recursively enumerable sets (Russian)* (**X** 0913) Novosibirsk Gos Univ: Novosibirsk 92pp
◇ D25 ◇ REV MR 82d:03069 Zbl 484 # 03022 • ID 72231

DEGTEV, A.N. [1979] *Reducibilities of tabular type in the theory of algorithms (Russian)* (**J** 0067) Usp Mat Nauk 34/3(207)*137-168,248
• TRANSL [1979] (**J** 1399) Russ Math Surv 34/3*155-192
◇ D25 D30 ◇ REV MR 81a:03045 Zbl 412 # 03027
• ID 52958

DEGTEV, A.N. [1979] *Several results on upper semilattices and m-degrees (Russian)* (**J** 0003) Algebra i Logika 18*664-679,754
• TRANSL [1979] (**J** 0069) Algeb and Log 18*420-430
◇ D25 D30 ◇ REV MR 82c:03065 Zbl 475 # 03021
• ID 72216

DEGTEV, A.N. [1980] *A category of partial recursive functions (Russian)* (**J** 0092) Sib Mat Zh 21/3*89-97,236
• TRANSL [1980] (**J** 0475) Sib Math J 21*382-388
◇ D20 D45 D75 G30 ◇ REV MR 82a:03037
Zbl 446 # 03030 • ID 56559

DEGTEV, A.N. [1980] *On reducibility of numerations (Russian)* (**J** 0142) Mat Sb, Akad Nauk SSSR, NS 112(154)*207-219
• TRANSL [1981] (**J** 0349) Math of USSR, Sbor 40*193-204
◇ D25 D30 D45 ◇ REV MR 82a:03041 Zbl 449 # 03037
• ID 56704

DEGTEV, A.N. [1981] *On (m,n)-computable sets (Russian)* (**C** 3865) Algeb Sistemy (Ivanovo) 88-99
◇ D20 ◇ REV MR 86b:03049 Zbl 536 # 03022 • ID 37104

DEGTEV, A.N. [1981] *Relations between complete sets (Russian)* (**J** 0031) Izv Vyssh Ucheb Zaved, Mat (Kazan) 1981/5*50-55
• TRANSL [1981] (**J** 3449) Sov Math 25/5*53-61
◇ D25 D30 ◇ REV MR 83g:03039 Zbl 503 # 03022
• ID 45962

DEGTEV, A.N. [1982] *Comparison of linear reducibility with other reducibilities of tabular type (Russian)* (**J** 0003) Algebra i Logika 21*511-529
• TRANSL [1982] (**J** 0069) Algeb and Log 21*339-353
◇ D20 ◇ REV MR 85k:03027 Zbl 572 # 03022 • ID 47201

DEGTEV, A.N. [1982] *Small degrees in ordinary recursion theory* (**P** 3622) Int Congr Log, Meth & Phil of Sci (6,Proc);1979 Hannover 237-240
◇ D30 ◇ REV MR 84d:03055 Zbl 505 # 03021 • ID 33586

DEGTEV, A.N. [1983] *Relations between table-type degrees (Russian)* (**J** 0003) Algebra i Logika 22*35-52
• TRANSL [1983] (**J** 0069) Algeb and Log 22*26-39
◇ D25 D30 ◇ REV MR 85k:03028 Zbl 534 # 03020
• ID 36561

DEGTEV, A.N. [1983] *Relations between reducibilities of table type (Russian)* (**J** 0003) Algebra i Logika 22*243-259
• TRANSL [1983] (**J** 0069) Algeb and Log 22*173-185
◇ D25 ◇ REV MR 85j:03069 Zbl 544 # 03017 • ID 41000

DEGTEV, A.N. [1984] *A category of enumerated sets (Russian)* (**J** 0087) Mat Zametki (Akad Nauk SSSR) 36*261-268
• TRANSL [1984] (**J** 1044) Math Notes, Acad Sci USSR 36*623-627
◇ D45 G30 ◇ REV MR 86e:03044 Zbl 568 # 03020
• ID 44247

DEGTEV, A.N. [1985] *Semilattices of disjunctive and linear degrees (Russian)* (**J** 0087) Mat Zametki (Akad Nauk SSSR) 38*310-316,350
◇ D30 ◇ ID 49262

DEHN, M. [1910] *Ueber die Topologie des dreidimensionalen Raumes* (**J** 0043) Math Ann 69*137-168
◇ D40 ◇ ID 49843

DEHN, M. [1912] *Transformationen der Kurven auf zweiseitigen Flaechen* (**J** 0043) Math Ann 72*413-421
◇ D40 ◇ ID 49844

DEIMEL JR., L.E. [1974] *Remark on the computational power of a Turing machine variant* (**J** 0232) Inform Process Lett 3*43-45
◇ D10 ◇ REV MR 50 # 12680 Zbl 295 # 02021 • ID 04074

DEKHTYAR', M.I. [1969] *Impossibility of eliminating complete enumeration in computing functions from their diagrams (Russian)* (**J** 0023) Dokl Akad Nauk SSSR 189*748-751
• TRANSL [1969] (**J** 0470) Sov Phys, Dokl 14*1146-1148
◇ D20 ◇ REV MR 42 # 1655 Zbl 253 # 02035 • ID 28938

DEKHTYAR', M.I. [1974] *On the complexity of relative computations (Russian)* (**J** 0023) Dokl Akad Nauk SSSR 214*999-1001 • ERR/ADD ibid 218*VII
• TRANSL [1974] (**J** 0062) Sov Math, Dokl 15*273-276
◇ D15 D20 ◇ REV MR 49 # 6677 MR 50 # 3642
Zbl 299 # 68032 • ID 04073

DEKHTYAR', M.I. [1975] *Reducibility with limited complexity (Russian)* (**C** 1050) Aktual Vopr Mat Log & Teor Mnozh 88-104
◇ D15 ◇ REV MR 57 # 16007 • ID 72236

DEKHTYAR', M.I. [1975] *Relative computations of predicates, and their complexity (Russian)* (**C** 1050) Aktual Vopr Mat Log & Teor Mnozh 72-87
◇ D15 ◇ REV MR 57 # 16006 • ID 72237

DEKHTYAR', M.I. [1976] *On the relativization of deterministic and nondeterministic complexity classes* (**P** 1401) Math Founds of Comput Sci (5);1976 Gdansk 255-259
◇ D15 ◇ REV Zbl 339 # 68039 • ID 61277

DEKHTYAR', M.I. [1976] *The complexity of the approximation of recursive sets (Russian) (German and English summaries)* (**J** 0129) Elektr Informationsverarbeitung & Kybern 12*115-122
◇ D15 D20 ◇ REV MR 54 # 4942 Zbl 331 # 68028
• ID 24115

DEKHTYAR', M.I. [1979] *Bounds on computational complexity and approximability of initial segments of recursive sets* (**P** 2059) Math Founds of Comput Sci (8);1979 Olomouc 277-283
◇ D15 D20 ◇ REV MR 81g:03046 Zbl 496 # 03022 • ID 72242

DEKHTYAR', M.I. [1979] *Complexity spectra of recursive sets of approximability of initial segments of complete problems* (**J** 0129) Elektr Informationsverarbeitung & Kybern 15*11-32
◇ D10 D15 D30 F20 F30 ◇ REV MR 80k:03038 Zbl 412 # 03021 • ID 52952

DEKHTYAR', M.I. [1981] *Structure of T-reducibilities with bounded complexity (Russian)* (**C** 3747) Mat Log & Mat Lingvistika 60-76
◇ D15 D30 ◇ REV MR 84h:03102 • ID 34292

DEKKER, J.C.E. [1953] *The constructivity of maximal dual ideals in certain Boolean algebras* (**J** 0048) Pac J Math 3*73-101
◇ D25 D45 G05 ◇ REV MR 14.838 Zbl 50.8 JSL 19.122 • ID 02876

DEKKER, J.C.E. [1953] *Two notes on recursively enumerable sets* (**J** 0053) Proc Amer Math Soc 4*495-501
◇ D25 ◇ REV MR 15.385 Zbl 52.250 JSL 20.73 • ID 02877

DEKKER, J.C.E. [1954] *A theorem on hypersimple sets* (**J** 0053) Proc Amer Math Soc 5*791-796
◇ D25 ◇ REV MR 16.209 Zbl 56.249 JSL 21.100 • ID 02878

DEKKER, J.C.E. [1955] *Productive sets* (**J** 0064) Trans Amer Math Soc 78*129-149
◇ D25 ◇ REV MR 16.663 Zbl 64.10 JSL 21.99 • ID 02879

DEKKER, J.C.E. [1957] *An expository account of isols* (**P** 1675) Summer Inst Symb Log;1957 Ithaca 189-200
◇ D50 ◇ REV JSL 25.356 • ID 02889

DEKKER, J.C.E. [1958] *Congruences in isols with a finite modulus* (**J** 0044) Math Z 70*113-124
◇ D50 ◇ REV MR 20 # 5133 Zbl 84.9 JSL 25.356 • ID 02884

DEKKER, J.C.E. & MYHILL, J.R. [1958] *Retraceable sets* (**J** 0017) Canad J Math 10*357-373
◇ D25 D30 D50 ◇ REV MR 20 # 5733 Zbl 82.15 JSL 27.84 • ID 02885

DEKKER, J.C.E. & MYHILL, J.R. [1958] *Some theorems on classes of recursively enumerable sets* (**J** 0064) Trans Amer Math Soc 89*25-59
◇ D25 ◇ REV MR 20 # 3780 Zbl 83.3 JSL 27.84 • ID 02887

DEKKER, J.C.E. [1958] *The factorial function for isols* (**J** 0044) Math Z 70*250-262
◇ D50 ◇ REV MR 20 # 5134 Zbl 84.10 JSL 25.356 • ID 02883

DEKKER, J.C.E. & MYHILL, J.R. [1960] *Recursive equivalence types* (**S** 0183) Publ Math Univ California 3*67-214
◇ D50 ◇ REV MR 22 # 7938 Zbl 249 # 02021 JSL 25.356 • ID 02892

DEKKER, J.C.E. & MYHILL, J.R. [1960] *The divisibility of isols by powers of primes* (**J** 0044) Math Z 73*127-133
◇ D50 ◇ REV MR 22 # 3689 Zbl 94.8 JSL 25.356 • ID 02890

DEKKER, J.C.E. [1962] *Infinite series of isols* (**P** 0613) Rec Fct Th;1961 New York 77-96
◇ D50 ◇ REV MR 26 # 16 Zbl 171.270 JSL 31.652 • ID 02894

DEKKER, J.C.E. (ED.) [1962] *Recursive function theory* (**X** 0803) Amer Math Soc: Providence vii + 247pp
◇ D97 ◇ REV MR 25 # 3828 MR 82d:03065 • REM 2nd ed 1970 • ID 23497

DEKKER, J.C.E. [1963] *An infinite product of isols* (**J** 0316) Illinois J Math 7*668-680
◇ D50 ◇ REV MR 29 # 1146 Zbl 163.253 JSL 31.652 • ID 04190

DEKKER, J.C.E. [1964] *The minimum of two regressive isols* (**J** 0044) Math Z 83*345-366
◇ D50 ◇ REV MR 28 # 3927 Zbl 122.10 JSL 32.527 • ID 02895

DEKKER, J.C.E. [1964] *The recursive equivalence type of a class of sets* (**J** 0015) Bull Amer Math Soc 70*628-632
◇ D50 ◇ REV MR 30 # 4677 Zbl 156.10 JSL 34.518 • ID 02896

DEKKER, J.C.E. [1965] *Closure properties of regressive functions* (**J** 3240) Proc London Math Soc, Ser 3 15*226-238
◇ D50 ◇ REV MR 30 # 4678 Zbl 163.252 JSL 36.539 • ID 02897

DEKKER, J.C.E. [1966] *Good choice sets* (**J** 0315) Ann Sc Norm Sup Pisa Fis Mat, Ser 3 20*367-393
◇ D50 ◇ REV MR 35 # 6551 Zbl 207.307 JSL 34.518 • ID 04191

DEKKER, J.C.E. [1966] *Les fonctions combinatoires et les isols* (**X** 0834) Gauthier-Villars: Paris iii + 77pp
◇ D50 ◇ REV MR 35 # 6552 Zbl 169.310 JSL 37.406 • ID 04192

DEKKER, J.C.E. [1967] *Regressive isols* (**P** 0691) Sets, Models & Recursion Th;1965 Leicester 272-296
◇ D50 ◇ REV Zbl 155.340 JSL 34.519 • ID 02898

DEKKER, J.C.E. [1968] *On certain vector spaces of isolic dimension* (**J** 0036) J Symb Logic 33*642
◇ D50 D80 ◇ ID 02899

DEKKER, J.C.E. [1969] *Countable vector spaces with recursive operations. Part I* (**J** 0036) J Symb Logic 34*363-387
◇ C57 C60 D45 D50 ◇ REV MR 40 # 5449 Zbl 185.20 • REM Part II 1971 • ID 02900

DEKKER, J.C.E. [1970] see APPLEBAUM, C.H.

DEKKER, J.C.E. [1971] *Countable vector spaces with recursive operations. Part II* (**J** 0036) J Symb Logic 36*477-493
◇ C57 C60 D45 D50 ◇ REV MR 45 # 4967 Zbl 231 # 02052 • REM Part I 1969 • ID 02901

DEKKER, J.C.E. [1971] *Two notes on vector spaces with recursive operations* (**J** 0047) Notre Dame J Formal Log 12*329-334
◇ C57 C60 D45 ◇ REV MR 47 # 3160 Zbl 205.308 • ID 02902

DEKKER, J.C.E. & ELLENTUCK, E. [1974] *Recursion relative to regressive functions* (**J** 0007) Ann Math Logic 6*231-257
◇ D30 D50 ◇ REV MR 51 # 101 Zbl 281 # 02048 • ID 02903

DEKKER, J.C.E. [1976] *Equivalencia recursiva* (1111) Preprints, Manuscr., Techn. Reports etc. iii + 59pp
◇ D50 ◇ REM Mimeographed Notes, Escuela de Matematica, Universidad de Costa Rica • ID 30671

DEKKER, J.C.E. [1976] *Projective planes of infinite but isolic order* (**J** 0036) J Symb Logic 41*391-404
◇ D50 D80 ◇ REV MR 54 # 7235 Zbl 332 # 02050
• ID 14771

DEKKER, J.C.E. [1977] *Planos afines con operaciones recursivas* (**J** 1680) Cienc Tecnol, Costa Rica 1*13-29
◇ D45 D50 ◇ REV MR 58 # 185 • ID 30670

DEKKER, J.C.E. [1978] *Projective bigraphs with recursive operations* (**J** 0047) Notre Dame J Formal Log 19*193-199
◇ D45 D50 ◇ REV MR 58 # 16229 Zbl 363 # 02048
• ID 30672

DEKKER, J.C.E. [1978] *Sharply two-transitive families of permutations on an immune set* (**J** 3194) J Austral Math Soc, Ser A 25*362-374
◇ D50 ◇ REV MR 80a:03058 Zbl 405 # 03021 • ID 30673

DEKKER, J.C.E. [1981] *Automorphisms of ω-cubes* (**J** 0047) Notre Dame J Formal Log 22*120-128
◇ D50 ◇ REV MR 82g:03084 Zbl 452 # 03034 • ID 54098

DEKKER, J.C.E. [1981] *Recursive equivalence types and cubes* (**P** 2902) Aspects Effective Algeb;1979 Clayton 87-121
◇ D50 ◇ REV MR 82j:03056 Zbl 467 # 03042 • ID 55040

DEKKER, J.C.E. [1981] *Twilight graphs* (**J** 0036) J Symb Logic 46*539-571
◇ D50 D80 ◇ REV MR 84h:03105 Zbl 466 # 03015
• ID 54966

DEKKER, J.C.E. [1982] *Automorphisms of ω-octahedral graphs* (**J** 0047) Notre Dame J Formal Log 23*427-434
◇ D50 ◇ REV MR 83m:03054 Zbl 464 # 03040 • ID 35453

DEKKER, J.C.E. [1983] *Isols and balanced block designs with $\lambda = 1$* (**J** 0020) Compos Math 49*75-93
◇ D50 D80 ◇ REV MR 85e:03107 Zbl 524 # 05012
• ID 38226

DEKKER, J.C.E. [1983] *Recursive equivalence types and octahedra* (**J** 3194) J Austral Math Soc, Ser A 34*101-113
◇ D50 D80 ◇ REV MR 84b:03062 Zbl 523 # 03031
• ID 35647

DEKKER, J.C.E. [1984] *Isolated sets and parity* (**J** 1680) Cienc Tecnol, Costa Rica 8*77-90
◇ D50 ◇ ID 47269

DEKKING, F.M. [1980] *Regularity and irregularity of sequences generated by automata* (**C** 4014) Semin Th Nombres 1979/80 10pp
◇ D05 ◇ REV MR 83e:68123 Zbl 438 # 10040 • ID 40170

DELFINO, V. [1978] *The Victoria Delfino problems* (**C** 2908) Cabal Seminar Los Angeles 1976-77 279-282
◇ D55 D60 D65 D70 E15 E60 ◇ REV Zbl 403 # 03036
• ID 54749

DELLACHERIE, C. [1971] *Une demonstration du theoreme de separation des ensembles analytiques* (**P** 4417) Semin Probab (5);1969/70 Strasbourg 82-85
◇ D55 E15 ◇ REV MR 52 # 3568 Zbl 217.209 • ID 46943

DELLACHERIE, C. [1975] *Ensembles analytiques: theoremes de separation et applications (corr ibid 544)* (**P** 3027) Semin Probab (9);1972/74 Strasbourg 336-372
◇ D55 E15 E75 ◇ REV MR 55 # 1331 MR 56 # 1278
Zbl 354 # 54023 • ID 50189

DELLACHERIE, C. [1977] *Les derivations en theorie descriptive des ensembles et le theoreme de la borne* (**P** 2598) Semin Probab (11);1975/76 Strasbourg 34-46 • ERR/ADD [1978] (**P** 3445) Semin Probab (12);1976/77 Strasbourg 523
◇ D55 E15 E75 ◇ REV MR 56 # 13185 Zbl 366 # 02045
• REM Erratum et addendum • ID 51128

DELLACHERIE, C. & HOFFMANN-JOERGENSEN, J. & JAYNE, J.E. & KECHRIS, A.S. & MARTIN, D.A. & ROGERS, C.A. & STONE, A.H. & TOPSOEE, F. [1980] *Analytic sets* (**X** 0801) Academic Pr: New York x+499pp
◇ D98 E15 E98 ◇ REV Zbl 451 # 04001 • ID 54044

DELLACHERIE, C. see Vol. V for further entries

DELON, F. [1978] *Definition de l'arithmetique dans la theorie des anneaux de series formelles (English summary)* (**J** 2313) C R Acad Sci, Paris, Ser A-B 286*A87-A89
◇ C60 D35 F25 F35 ◇ REV MR 82a:12018
Zbl 382 # 13010 • ID 27313

DELON, F. [1981] *Indecidabilite de la theorie des anneaux de series formelles a plusieurs indeterminees (English summary)* (**J** 0027) Fund Math 112*215-229
◇ C60 D35 ◇ REV MR 83e:03048 Zbl 515 # 12020
• ID 35228

DELON, F. & LASCAR, D. & LOUVEAU, A. & SABBAGH, G. (EDS.) [1985] *Seminaire general de logique 1982-83* (**X** 4643) Univ Paris VII, UER Math: Paris 186pp
◇ B97 C98 D97 ◇ REV MR 86f:03005 Zbl 567 # 00003
• ID 47516

DELON, F. see Vol. I, III for further entries

DEMBINSKI, P. [1972] *Equivalence of programs and their complexity* (**J** 0014) Bull Acad Pol Sci, Ser Math Astron Phys 20*309-312
◇ B75 D15 D20 ◇ REV MR 46 # 8821c Zbl 254 # 68003
• ID 02912

DEMBINSKI, P. [1972] *The machine realization of programs (Russian summary)* (**J** 0014) Bull Acad Pol Sci, Ser Math Astron Phys 20*297-301
◇ B75 D10 D20 ◇ REV MR 46 # 8821a Zbl 254 # 68001
• ID 02910

DEMBINSKI, P. [1972] *The structure of program and its semantics* (**J** 0014) Bull Acad Pol Sci, Ser Math Astron Phys 20*303-307
◇ B75 D20 ◇ REV MR 46 # 8821b Zbl 254 # 68002
• ID 02911

DEMEL, J. & DEMLOVA, M. & KOUBEK, V. [1982] *Simplicity of algebras requires to investigate almost all operations* (**J** 0140) Comm Math Univ Carolinae (Prague) 23*325-335
◇ C05 D15 ◇ REV MR 83j:08001 Zbl 518 # 08001
• ID 38370

DEMILLO, R.A. & LIPTON, R.J. [1979] *Some connections between mathematical logic and complexity theory* (**P** 3542) ACM Symp Th of Comput (11);1979 Atlanta 153-159
◇ C57 D15 F30 H15 ◇ REV MR 81h:03084 • ID 72280

DEMILLO, R.A. & EISENSTAT, S.C. & LIPTON, R.J. [1980] *Space-time trade-offs in structured programming: An improved combinatorial embedding theorem* (**J** 0037) ACM J 27*123-127
◇ D15 ◇ REV MR 81b:68009 Zbl 426 # 68046 • ID 69448

DEMILLO, R.A. see Vol. VI for further entries

DEMLOVA, M. [1982] see DEMEL, J.

DEMPSTER, J.R.H. [1968] *On the multiplying ability of two-way automata* (J 0119) J Comp Syst Sci 2*420-426
◇ D05 D10 ◇ REV MR 40#4022 Zbl 197.7 • ID 02913

DEMUTH, O. [1975] *Constructive pseudonumbers (Russian)* (J 0140) Comm Math Univ Carolinae (Prague) 16*315-331
◇ D55 F60 ◇ REV MR 52#2844 Zbl 319#02029
• ID 17635

DEMUTH, O. [1976] *The domains of definition of effective operators over general recursive functions and of constructive functions of a real variable (Russian)* (J 0140) Comm Math Univ Carolinae (Prague) 17*633-646
◇ D20 F60 ◇ REV MR 55#7755 Zbl 345#02025
• ID 29783

DEMUTH, O. [1977] *A constructive analogue of the functions of the n-th Baire class* (J 0140) Comm Math Univ Carolinae (Prague) 18*231-245
◇ D55 E15 F60 ◇ REV MR 57#5701 Zbl 359#02031
• ID 72287

DEMUTH, O. & KRYL, R. & KUCERA, A. [1978] *An application of the theory of functionals partial recursive relative to number sets in constructive mathematics (Russian)* (J 0165) Acta Univ Carolinae Math Phys (Prague) 19/1*15-60
◇ D30 D55 F60 ◇ REV MR 80a:03071 Zbl 394#03054
• ID 31808

DEMUTH, O. [1983] *Arithmetic complexity of differentiation in constructive mathematics (Russian)* (J 0140) Comm Math Univ Carolinae (Prague) 24*301-316
◇ D55 F60 ◇ REV MR 85i:03178 • ID 44288

DEMUTH, O. see Vol. I, VI for further entries

DENEF, J. [1975] *Hilbert's tenth problem for quadratic rings* (J 0053) Proc Amer Math Soc 48*214-220
◇ C60 D35 ◇ REV MR 50#12961 Zbl 324#02032
• ID 04079

DENEF, J. [1978] *Diophantine sets over $Z[T]$* (J 0053) Proc Amer Math Soc 69*148-150
◇ C60 D25 D35 ◇ REV MR 57#2899 Zbl 393#03035
• ID 52455

DENEF, J. & LIPSHITZ, L. [1978] *Diophantine sets over some rings of algebraic integers* (J 3172) J London Math Soc, Ser 2 18*385-391
◇ C60 D35 ◇ REV MR 80a:12030 Zbl 399#10049
• ID 52854

DENEF, J. [1978] *The Diophantine problem for polynomial rings and fields of rational functions* (J 0064) Trans Amer Math Soc 242*391-399
◇ C60 D35 ◇ REV MR 58#10809 Zbl 399#10048
• ID 52853

DENEF, J. [1979] *The Diophantine problem for polynomial rings of positive characteristic* (P 2627) Logic Colloq;1978 Mons 131-145
◇ C60 D35 ◇ REV MR 81h:03090 Zbl 457#12011
• ID 54394

DENEF, J. [1980] see BECKER, J.A.

DENEF, J. & LIPSHITZ, L. [1984] *Power series solutions of algebraic differential equations* (J 0043) Math Ann 267*213-238
◇ B25 C60 C65 D35 D80 ◇ REV MR 85j:12010 Zbl 518#12015 • ID 38480

DENEF, J. see Vol. III for further entries

DENENBERG, L. & LEWIS, H.R. [1984] *Logical syntax and computational complexity* (P 2153) Logic Colloq;1983 Aachen 2*101-116
◇ B25 D15 ◇ REV MR 86e:03013 Zbl 564#03009
• ID 43003

DENENBERG, L. & LEWIS, H.R. [1984] *The complexity of the satisfiability problem for Krom formulas* (J 1426) Theor Comput Sci 30*319-341
◇ B20 B25 D15 ◇ ID 43373

DENISOV, A.S. [1984] *Constructive homogeneous extensions (Russian)* (J 0092) Sib Mat Zh 25/6*60-69
• TRANSL [1984] (J 0475) Sib Math J 25*879-888
◇ C50 C57 ◇ REV Zbl 571#03012 • ID 42732

DENISOV, S.D. & LAVROV, I.A. [1970] *Complete numerations with an infinite number of special elements (Russian)* (J 0003) Algebra i Logika 9*503-509
• TRANSL [1970] (J 0069) Algeb and Log 9*301-304
◇ D25 D45 ◇ REV MR 45#56 Zbl 254#02031
• ID 72310

DENISOV, S.D. [1970] *The m-degrees of recursively enumerable sets (Russian)* (J 0003) Algebra i Logika 9*422-427
• TRANSL [1970] (J 0069) Algeb and Log 9*254-256
◇ D25 ◇ REV MR 44#3861 Zbl 239#02022 • ID 02937

DENISOV, S.D. [1972] *Models of a noncontradictory formulas and the Ershov hierarchy (Russian)* (J 0003) Algebra i Logika 11*648-655,736
• TRANSL [1972] (J 0069) Algeb and Log 11*359-362
◇ C07 C57 D55 ◇ REV MR 52#5406 Zbl 268#02028
• ID 61324

DENISOV, S.D. [1974] *Three theorems on elementary theories and tt-reducibility (Russian)* (J 0003) Algebra i Logika 13*5-8,120
• TRANSL [1974] (J 0069) Algeb and Log 13*1-2
◇ C65 D30 ◇ REV MR 50#9561 Zbl 289#02036
• ID 25962

DENISOV, S.D. [1975] *Two-element separable enumeration of sets γ with undecidable problem $P(\gamma,\gamma)$ (Russian)* (J 0003) Algebra i Logika 14*523-532
• TRANSL [1975] (J 0069) Algeb and Log 14*322-328
◇ D25 D45 ◇ REV MR 55#88 Zbl 367#02020
• ID 26019

DENISOV, S.D. [1978] *Structures of the upper semilattice of recursively enumerable m-degrees and related questions I (Russian)* (J 0003) Algebra i Logika 17*643-683,746
• TRANSL [1978] (J 0069) Algeb and Log 17*418-443
◇ D25 ◇ REV MR 81c:03035 Zbl 427#03033 • ID 53720

DENNING, P.J. & DENNIS, J.B. & QUALITZ, J.E. [1978] *Machines, languages, and computation* (X 0819) Prentice Hall: Englewood Cliffs xxii+601pp
◇ B98 D05 D10 D98 ◇ REV Zbl 492#68003 JSL 45.630
• ID 37736

DENNIS, J.B. [1978] see DENNING, P.J.

DENNIS-JONES, E.C. & WAINER, S.S. [1984] *Subrecursive hierarchies via direct limits* (P 2153) Logic Colloq;1983 Aachen 2*117-128
◇ D20 F15 G30 ◇ ID 45385

DERSHOWITZ, N. [1981] *Termination of linear rewriting systems* (P 2903) Automata, Lang & Progr (8);1981 Akko 448-458
◇ B75 D05 ◇ REV Zbl 465#68009 • ID 69446

DERSHOWITZ, N. see Vol. I for further entries

DESCHAMPS, J.-P. [1974] *Asynchronous automata and asynchronous languages* (**J** 0194) Inform & Control 24*122-143
◇ D05 ◇ REV MR 50 # 1776 Zbl 273 # 94050 • ID 61334

DESCHAMPS, J.-P. see Vol. I, II for further entries

DESTOUCHES, J.-L. [1969] *Has the theory of recursive functions application in mathematical physics?* (**P** 1841) Fct Recurs & Appl;1967 Tihany 121-125
◇ A10 D80 ◇ ID 32561

DESTOUCHES, J.-L. see Vol. I, II, VI for further entries

DETLOVS, V.K. [1953] *Normal algorithms and recursive functions (Russian)* (**J** 0023) Dokl Akad Nauk SSSR 90*723-725
◇ D03 D20 ◇ REV MR 16.436 Zbl 53.345 JSL 21.408 • ID 02970

DETLOVS, V.K. [1958] *The equivalence of normal algorithms and recursive functions (Russian)* (**S** 0066) Tr Mat Inst Steklov 52*75-139
• TRANSL [1963] (**J** 0225) Amer Math Soc, Transl, Ser 2 23*15-81
◇ D03 D20 ◇ REV MR 21 # 5 Zbl 87.252 JSL 27.362 • ID 02971

DETLOVS, V.K. see Vol. I, II, V for further entries

DETTKI, H.J. & SCHUSTER, H. [1983] *Rekursionstheorie auf F* (**X** 3176) Fernuniv Hagen: Hagen 81pp
◇ D75 ◇ REV MR 86a:03046 • ID 45386

DEUEL, D.R. & GILL, A. [1966] *Some decision problems associated with weighted directed graphs* (**J** 0374) SIAM J Appl Math 14*970-979
◇ D05 D80 ◇ REV MR 35 # 2790 Zbl 156.16 • ID 17028

DEUSSEN, P. [1967] *Some results on the set of congruence relations in a finite strongly connected automaton (German summary)* (**J** 0373) Comp Arch Inform & Numerik 2*353-367
◇ D05 ◇ REV MR 37 # 74 Zbl 166.269 • ID 17030

DEUSSEN, P. [1971] *Halbgruppen und Automaten* (**X** 0811) Springer: Heidelberg & New York 198pp
◇ D05 ◇ REV MR 49 # 8759 Zbl 226 # 94039 • ID 23520

DEUSSEN, P. [1979] *One abstract accepting algorithm for all kinds of parsers* (**P** 1873) Automata, Lang & Progr (6);1979 Graz 203-217
◇ B75 D05 ◇ REV MR 82b:68073 Zbl 412 # 68081 • ID 52988

DEUSSEN, P. see Vol. I for further entries

DEUTSCH, D. [1985] *Quantum theory, the Church-Turing principle and the universal quantum computer* (**J** 1150) Proc Roy Soc London, Ser A 400*97-117
◇ D20 D80 ◇ ID 48216

DEUTSCH, M. [1974] *Reduzierungen unentscheidbarer formaler Sprachen ueber induktiv definierten Bereichen* (**J** 0068) Z Math Logik Grundlagen Math 20*325-338
◇ D25 D75 ◇ REV MR 58 # 10363 Zbl 326 # 02031 • ID 04084

DEUTSCH, M. [1975] *Die Unentscheidbarkeit und Nichtkalkuelisierbarkeit der Mengenlehre* (**J** 0933) Math Nat Unterr 28*193-196
◇ D35 E30 ◇ REV Zbl 306 # 02003 • ID 61336

DEUTSCH, M. [1975] *Zur Darstellung koaufzaehlbarer Praedikate bei Verwendung eines einzigen unbeschraenkten Quantors* (**J** 0068) Z Math Logik Grundlagen Math 21*443-454
◇ D25 F30 ◇ REV MR 58 # 10360 Zbl 357 # 02041 • ID 04085

DEUTSCH, M. [1975] *Zur Theorie der spektralen Darstellung von Praedikaten durch Ausdruecke der Praedikatenlogik 1.Stufe* (**J** 0009) Arch Math Logik Grundlagenforsch 17*9-16
◇ B10 C13 D25 D35 ◇ REV MR 54 # 2436 Zbl 337 # 02025 • ID 02972

DEUTSCH, M. [1976] *Eine mengentheoretische Grundlegung der Theorie der Berechenbarkeit I* (**J** 0160) Math-Phys Sem-ber, NS 23*192-205
◇ D20 D25 E47 ◇ REV MR 58 # 27403a Zbl 353 # 02025 • REM Part II 1977 • ID 28192

DEUTSCH, M. [1976] *Zur Praefixoptimalitaet gewisser* $\exists\forall\exists\ldots\exists$*-Darstellungen aufzaehlbarer Praedikate* (**J** 0068) Z Math Logik Grundlagen Math 22*339-346
◇ D25 ◇ REV MR 58 # 16225 Zbl 359 # 02047 • ID 18442

DEUTSCH, M. [1977] *Eine mengentheoretische Grundlegung der Theorie der Berechenbarkeit II* (**J** 0160) Math-Phys Sem-ber, NS 24*56-70
◇ B15 D20 D35 E47 ◇ REV MR 58 # 27403b Zbl 365 # 02029 • REM Part I 1976 • ID 28193

DEUTSCH, M. [1977] *Zum Begriff der Wortmischung als Basis fuer die Arithmetik* (**J** 0068) Z Math Logik Grundlagen Math 23*241-264
◇ B28 D05 D75 ◇ REV MR 56 # 15375 Zbl 402 # 03042 • ID 26483

DEUTSCH, M. [1980] *Zur goedelisierungsfreien Darstellung der rekursiven und primitiv-rekursiven Funktionen und rekursiv aufzaehlbaren Praedikate ueber Quotiententermmengen* (**J** 0068) Z Math Logik Grundlagen Math 26*1-32
◇ D20 D25 ◇ REV MR 81g:03048 Zbl 444 # 03022 • ID 72321

DEUTSCH, M. [1981] *Registermaschinen ueber Quotiententermmengen* (**J** 0068) Z Math Logik Grundlagen Math 27*273-288
◇ D10 D20 ◇ REV MR 82j:03044 Zbl 486 # 03024 • ID 72319

DEUTSCH, M. [1981] *Zur Reduktionstheorie des Entscheidungsproblems* (**J** 0068) Z Math Logik Grundlagen Math 27*113-117
◇ B20 D35 ◇ REV MR 82f:03037 Zbl 465 # 03004 • ID 54907

DEUTSCH, M. [1982] *Zur Komplexitaetsmessung primitiv-rekursiver Funktionen ueber Quotiententermmengen* (**J** 0068) Z Math Logik Grundlagen Math 28*345-363
◇ D20 ◇ REV MR 83m:03048 Zbl 505 # 03020 • ID 35448

DEUTSCH, M. [1984] *Reductions for the satisfiability with a simple interpretation of the predicate variable* (**P** 2342) Symp Rek Kombin;1983 Muenster 285-311
◇ B20 D35 ◇ REV Zbl 547 # 03004 • ID 43185

DEUTSCH, M. see Vol. I, III, VI for further entries

DEVIDE, V. [1967] *A remark concerning the foundation of the theory of recursive functions (Serbo-Croatian summary)* (**J** 3519) Glas Mat, Ser 3 (Zagreb) 2(22)*3-8
◇ D20 ◇ REV MR 37 # 2600 Zbl 204.12 • ID 17032

DEVIDE, V. see Vol. I, II, V for further entries

DEVLIN, K.J. [1973] *Aspects of constructibility* (**S** 3301) Lect Notes Math 354*xii+240pp
◇ D55 E35 E45 E65 E98 ◇ REV MR 51 # 12527 Zbl 312 # 02054 • ID 17243

DEVLIN, K.J. [1974] *An introduction to the fine structure of the constructible hierarchy* (**P** 0602) Generalized Recursion Th (1);1972 Oslo 123-163
◇ C55 D55 E45 E47 E55 E98 ◇ REV MR 53 # 2659 Zbl 295 # 02037 • ID 15073

DEVLIN, K.J. & PARIS, J.B. [1975] *Certain sequences of ordinals* (**P** 0759) Infinite & Finite Sets (Erdoes);1973 Keszthely 1*333-360
◇ D60 E10 E45 E55 E60 ◇ REV MR 53 # 133 Zbl 324 # 02061 • ID 16656

DEVLIN, K.J. [1977] *Constructibility* (**C** 1523) Handb of Math Logic 453-489
◇ D60 E35 E45 E65 E98 ◇ REV MR 58 # 27475 JSL 49.971 • ID 27306

DEVLIN, K.J. see Vol. III, V for further entries

DEZANI-CIANCAGLINI, M. & RONCHI DELLA ROCCA, S. & SAITTA, L. [1979] *Complexity of λ-term reductions* (**J** 3441) RAIRO Inform Theor 13*257-287
◇ B40 D15 ◇ REV MR 81j:68016 Zbl 424 # 03009 • ID 81091

DEZANI-CIANCAGLINI, M. also published under the name DEZANI, M.

DEZANI-CIANCAGLINI, M. see Vol. VI for further entries

DIAMOND, N.D. [1980] see CULIK II, K.

DIAZ, J. & VERGES, M. [1982] *A conjecture in the theory of complexity (Catalan)* (**P** 3870) Congr Catala de Log Mat (1);1982 Barcelona 67
◇ D15 ◇ REV MR 84i:03003 Zbl 517 # 03011 • ID 37282

DIAZ, J. [1982] see BALCAZAR, J.L.

DIAZ, J. [1985] see BALCAZAR, J.L.

DICHEV, A.V. [1981] *An example of a set M of positive integers which is not m-equivalent to M^2* (**P** 3874) Mat & Mat Obrazov (10);1981 Sl"nchev Bryag 137-141
◇ D30 ◇ REV Zbl 549 # 03035 • ID 43127

DICHEV, A.V. [1981] *Computability in the sense of Moschovakis and its connection with partial recursiveness via numerations (Russian)* (**J** 2547) Serdica, Bulgar Math Publ 7*117-130
◇ D45 D75 ◇ REV MR 84e:03055 Zbl 497 # 03034 • ID 34390

DICHEV, A.V. [1983] *On the m-equivalence of Cartesian powers of sets of natural numbers (Russian)* (**J** 2547) Serdica, Bulgar Math Publ 9*43-48
◇ D25 D30 ◇ REV MR 86a:03041 Zbl 557 # 03025 • ID 46192

DICHEV, A.V. [1984] *Computability in the sense of Moskovakis and its connection with partial recursivity through numbering (Russian)* (**P** 4392) Mat Logika (Markova);1980 Sofia 34-46
◇ D45 D75 ◇ ID 46578

DIENES, Z.P. [1939] *Canonic elements in the higher classes of Borel sets* (**J** 0039) J London Math Soc 14*169-175
◇ D55 E15 ◇ REV MR 1.8 Zbl 21.302 FdM 65.191 • ID 03009

DIENES, Z.P. see Vol. V for further entries

DIES, J.E. [1977] *Sur l'independance algorithmique des suites infinies de Bernoulli (English summary)* (**J** 2313) C R Acad Sci, Paris, Ser A-B 285*A593-A596
◇ D20 D80 ◇ REV MR 56 # 5242 Zbl 369 # 94001 • ID 27315

DIETERICH, E. [1977] *Grobstrukturen fuer kontextfreie Grammatiken* (**P** 3411) Theor Comput Sci (3);1977 Darmstadt 96-105
◇ D05 ◇ REV MR 58 # 13960 Zbl 383 # 68059 • ID 69449

DIETZFELBINGER, M. & MAASS, W. [1985] *Strong reducibilities in α- and β- recursion theory* (**P** 3342) Rec Th Week;1984 Oberwolfach 89-120
◇ D30 D60 ◇ ID 45299

DIKOVSKIJ, A.YA. & MODINA, L. [1978] *On three unambiguousness types of context-free languages (Russian)* (**C** 3211) Mat Ling & Teor Algor 41-61
◇ D05 ◇ REV Zbl 402 # 68055 • ID 69450

DIKOVSKIJ, A.YA. [1981] *A theory of complexity of monadic recursion schemes* (**J** 3441) RAIRO Inform Theor 15*67-94
◇ D15 D20 ◇ REV MR 82d:68032 Zbl 469 # 68050 • ID 81099

DIKOVSKIJ, A.YA. [1981] *Classifications of context-free languages by complexity of derivation (Russian)* (**S** 2582) Semiotika & Inf, Akad Nauk SSSR 17*98-119
◇ D05 ◇ REV MR 84j:68048 Zbl 482 # 68068 • ID 39183

DIKOVSKIJ, A.YA. [1981] *Complexity of derivation of unambiguous context-free languages and grammars (Russian)* (**C** 3747) Mat Log & Mat Lingvistika 76-95
◇ D05 ◇ REV MR 84c:68063 • ID 39609

DILLER, J. [1976] *Rekursionstheorie. Vorlesung* (**X** 1532) Univ Muenster Inst Math Logik: Muenster iii+179pp
◇ D98 ◇ REV MR 55 # 5409 Zbl 359 # 02049 • ID 50596

DILLER, J. see Vol. I, II, III, VI for further entries

DIMA, N. [1981] *Sudan function is universal for the class of primitive recursive functions (Romanian) (English summary)* (**J** 0197) Stud Cercet Mat Acad Romana 33*59-67
◇ D20 ◇ REV MR 82k:03061 Zbl 467 # 03039 • ID 55037

DIMA, N. [1984] see BUZETEANU, S.

DIMA, N. [1984] *Topological properties of recurrent sets* (**J** 0070) Bull Soc Sci Math Roumanie, NS 28(76)*127-134
◇ D20 E07 ◇ REV MR 86g:03067 Zbl 541 # 03021 • ID 41381

DIMITRACOPOULOS, C. [1985] *A generalization of a theorem of H.Friedman* (**J** 0068) Z Math Logik Grundlagen Math 31*221-225
◇ C62 D15 F30 ◇ ID 47569

DIMITRACOPOULOS, C. see Vol. III, V, VI for further entries

DIMITROV, I. [1981] *Systematic transformation of recursive invariants in iterative algorithms (Russian)* (**P** 3874) Mat & Mat Obrazov (10);1981 Sl"nchev Bryag 248-253
◇ D20 ◇ REV Zbl 507 # 03019 • ID 37237

DIMITROV, V. & WECHLER, W. [1974] *R-fuzzy automata* (**P** 1691) Inform Processing (6);1974 Stockholm 657-660
◇ D05 E72 ◇ REV MR 53 # 15596 Zbl 302 # 94024 • ID 66008

DIMITROV, V. see Vol. II for further entries

DINCA, A. [1970] *Distribution classes, total non-ambiguity, and grammars with a finite number of states (Romanian) (French summary)* (J 0197) Stud Cercet Mat Acad Romana 22*563-573
◇ D05 ◇ REV MR 49 # 1832 Zbl 213.21 • ID 27987

DING, DECHENG [1984] *The minimal pair of r.e. recursively inseparable sets (Chinese) (English summary)* (J 4418) Nanjing Daxue Xuebao Shuxue Bannian Kan 1/2*268-271
◇ D25 ◇ REV Zbl 565 # 03019 • ID 46666

DING, DECHENG [1985] *R.e. degrees of recursively inseparable sets (Chinese)* (J 0420) Shuxue Jinzhan 14*371-372
◇ D25 ◇ ID 48507

DING, DECHENG see Vol. I, VI for further entries

DISHKANT, G.P. [1966] *Zur Frage der Entscheidbarkeit mechanischer Theorien (Russisch)* (J 0040) Kibernetika, Akad Nauk Ukr SSR 1966/2*12
◇ D35 ◇ REV Zbl 192.57 • ID 16269

DISHKANT, G.P. see Vol. II, V for further entries

DO LONG VAN [1981] *Problemes des mots et de conjugaison pour une classe de groupes de presentation finie (English summary)* (J 3364) C R Acad Sci, Paris, Ser 1 292*773-776
◇ D40 ◇ REV MR 82f:20061 Zbl 483 # 20019 • ID 81105

DO MINH THAI [1976] *Programmkomplexitaet rekursiv aufzaehlbarer Mengen* (P 2898) Algor Kompl, Lern-& Erkenn-Prozess;1976 Jena 29-33
◇ D15 D25 D30 ◇ REV MR 58 # 21532 Zbl 438 # 68011 • ID 55984

DO MINH THAI [1981] *On minimal-program complexity of complex sets* (J 0129) Elektr Informationsverarbeitung & Kybern 17*87-102
◇ D15 D25 ◇ REV MR 83a:68046 Zbl 534 # 03015 • ID 36560

DOBERKAT, E.-E. [1982] *Stochastic behavior of initial nondeterministic automata* (C 3433) Prog in Cybern & Syst Res, Vol 6 VI*331-337
◇ D05 ◇ REV MR 83i:68085 Zbl 469 # 68063 • ID 39363

DOBKIN, D.P. [1977] see BOOK, R.V.

DOBRITSA, V.P. [1975] *Recursively numbered classes of constructive extensions and autostability of algebras (Russian)* (J 0092) Sib Mat Zh 16*1148-1154,1369
• TRANSL [1975] (J 0475) Sib Math J 16*879-883
◇ C57 D45 ◇ REV MR 53 # 12917 Zbl 328 # 02030 • ID 23167

DOBRITSA, V.P. & GONCHAROV, S.S. [1976] *An example of a constructive abelian group with non-constructivizable reduced subgroup (Russian)* (P 2064) All-Union Conf Math Log (4);1976 Kishinev 33
◇ C57 C60 D45 ◇ ID 32652

DOBRITSA, V.P. [1976] *On computable and strictly computable classes of constructive algebras (Russian)* (P 2572) Material Respub Konf Molod Uchen;1976 Alma Ata 187
◇ C57 C60 D45 ◇ ID 32653

DOBRITSA, V.P. [1976] *Theorem on non-computable classes (Russian)* (J 2573) Mat Nauk (Ped Inst Alma Ata) 3*3-8
◇ D45 ◇ ID 32654

DOBRITSA, V.P. [1977] *Computability of certain classes of constructive algebras (Russian)* (J 0092) Sib Mat Zh 18*570-579
• TRANSL [1977] (J 0475) Sib Math J 18*406-413
◇ C57 C60 D45 ◇ REV MR 56 # 5244 Zbl 361 # 02065 • ID 32655

DOBRITSA, V.P. & KHISAMIEV, N.G. & NURTAZIN, A.T. [1978] *Constructive periodic abelian groups (Russian)* (J 0092) Sib Mat Zh 19*1260-1265
• TRANSL [1978] (J 0475) Sib Math J 19*886-890
◇ C57 C60 F60 ◇ REV MR 80b:20067 Zbl 421 # 20021 • ID 32600

DOBRITSA, V.P. [1981] *On constructivizable abelian groups (Russian)* (J 0092) Sib Mat Zh 22/3*208-213,239
◇ C57 C60 D40 D45 ◇ REV MR 82i:03058 Zbl 473 # 03037 • ID 55366

DOBRITSA, V.P. [1983] *Complexity of the index set of a constructive model (Russian)* (J 0003) Algebra i Logika 22*372-381
• TRANSL [1983] (J 0069) Algeb and Log 22*269-276
◇ C57 D45 ◇ REV MR 86d:03042 Zbl 537 # 03022 • ID 39619

DOBRITSA, V.P. [1983] *Some constructivizations of abelian groups (Russian)* (J 0092) Sib Mat Zh 24/2*18-25
• TRANSL [1983] (J 0475) Sib Math J 24*167-173
◇ C57 C60 D45 ◇ REV MR 85d:20062 Zbl 528 # 20038 • ID 36756

DOEMOESI, P. [1976] *On superpositions of automata* (J 0380) Acta Cybern (Szeged) 2*335-343
◇ D05 ◇ REV MR 56 # 5083 Zbl 352 # 94037 • ID 50080

DOEPP, K. [1966] *Ueber die Bestimmbarkeit des Verhaltens von Algorithmen* (J 0009) Arch Math Logik Grundlagenforsch 9*12-35
◇ D20 D45 D55 ◇ REV MR 34 # 1177 • ID 03066

DOEPP, K. [1967] *Die Unentscheidbarkeit der Grzegorczyk-Klassen* (J 0068) Z Math Logik Grundlagen Math 13*89-94
◇ D20 ◇ REV MR 37 # 6182 Zbl 341 # 02029 • ID 03067

DOEPP, K. [1971] *Automaten in Labyrinthen I,II (Russian summary)* (J 0129) Elektr Informationsverarbeitung & Kybern 7*79-94,167-189
◇ D05 ◇ REV MR 45 # 3135 MR 47 # 4772 Zbl 219 # 02024 • ID 03070

DOEPP, K. [1973] *Ein Dualitaetstheorem der Automatentheorie* (J 0129) Elektr Informationsverarbeitung & Kybern 9*211-215
◇ D05 ◇ REV MR 54 # 7140 Zbl 281 # 94030 • ID 61373

DOEPP, K. see Vol. I, III for further entries

DOERFLER, W. [1973] *Halbgruppen und Automaten* (J 0144) Rend Sem Mat Univ Padova 50*1-18
◇ D05 ◇ REV MR 50 # 13347 Zbl 284 # 20069 • ID 29932

DOERFLER, W. [1978] *The Cartesian composition of automata* (J 0041) Math Syst Theory 11*239-257
◇ D05 ◇ REV MR 58 # 9897 Zbl 385 # 68055 • ID 69390

DOERR, J. & HOTZ, G. (EDS.) [1970] *Automaten-Theorie und formale Sprachen. Bericht aus dem mathematischen Forschungsinstitut Oberwolfach 3* (X 0876) Bibl Inst: Mannheim 505pp
◇ D05 ◇ REV MR 48 # 7643 Zbl 209.25 • ID 23567

DOMANSKI, B. [1982] *The complexity of two decision problems for free groups* (J 1447) Houston J Math 8*29-38
◇ D15 D40 ◇ REV MR 83i:20027 Zbl 499 # 03031
• ID 38123

DOMANSKI, B. [1985] see ANSHEL, M.

DOMBROVSKIJ-KABANCHENKO, M.N. [1979] *A transfinite extension of the stepped semantic system (Russian)* (S 0554) Issl Teor Algor & Mat Logik (Moskva) 3*18-26
◇ D55 F50 ◇ REV MR 81j:03091 Zbl 443 # 03024
• ID 56430

DONER, J.E. [1970] *Tree acceptors and some of their applications* (J 0119) J Comp Syst Sci 4*406-451 • ERR/ADD ibid 5*453(1971))
◇ B25 C85 D05 ◇ REV MR 44 # 5179ab Zbl 212.29 JSL 37.619 • ID 03082

DONER, J.E. [1983] see BOOK, R.V.

DONER, J.E. see Vol. III, V for further entries

DONG, YUNMEI & LI, KAIDE [1965] *P grammar and its decision and analysis problems (Chinese)* (P 4564) Math Logic;1963 Xi-An 25-36
◇ D05 ◇ ID 49305

DOSEN, K. [1985] *A completeness theorem for the Lambek calculus of syntactic categories* (J 0068) Z Math Logik Grundlagen Math 31*235-241
◇ B65 C90 D05 ◇ REV Zbl 547 # 03027 • ID 47571

DOSEN, K. see Vol. II, III, VI for further entries

DOSHITA, S. & ISHIBASHI, T. & YAMASAKI, S. [1980] *Unit resolution for a subclass of the Ackermann class* (J 2794) Mem Fac Engin, Kyoto Univ 42*63-75
◇ B35 D15 ◇ REV MR 81j:03065 • ID 80151

DOSHITA, S. & YAMASAKI, S. [1983] *The satisfiability problem for a class consisting of Horn sentences and some non-Horn sentences in propositional logic* (J 0194) Inform & Control 59*1-12 • ERR/ADD ibid 60*174
◇ B20 B25 D15 ◇ REV Zbl 564 # 03010 • ID 44209

DOSHITA, S. see Vol. I for further entries

DOWNEY, R.G. [1983] *Abstract dependence, recursion theory, and the lattice of recursively enumerable filters* (J 0016) Bull Austral Math Soc 27*461-464
◇ D25 D45 ◇ REV Zbl 509 # 03021 • ID 38142

DOWNEY, R.G. [1983] *Nowhere simplicity in matroids* (J 3194) J Austral Math Soc, Ser A 35*28-45
◇ D25 D45 ◇ REV MR 85e:03103 Zbl 526 # 03027
• ID 38173

DOWNEY, R.G. [1983] *On a question of A.Retzlaff* (J 0068) Z Math Logik Grundlagen Math 29*379-384
◇ C57 C60 D25 D45 ◇ REV MR 85b:03075 Zbl 526 # 03028 • ID 38174

DOWNEY, R.G. [1984] *A note on decompositions of recursively enumerable subspaces* (J 0068) Z Math Logik Grundlagen Math 30*465-470
◇ C57 C60 D25 D45 ◇ REV MR 86b:03052 Zbl 535 # 03022 • ID 43527

DOWNEY, R.G. [1984] *Bases of supermaximal subspaces and Steinitz systems. I* (J 0036) J Symb Logic 49*1146-1159
◇ C57 C60 D25 D45 ◇ ID 39869

DOWNEY, R.G. [1984] *Co-immune subspaces and complementation in* V_∞ (J 0036) J Symb Logic 49*528-538
◇ C57 C60 D45 D50 ◇ REV MR 86d:03043 • ID 42438

DOWNEY, R.G. [1984] see ASH, C.J.

DOWNEY, R.G. [1984] *Some remarks on a theorem of Iraj Kalantari concerning convexity and recursion theory* (J 0068) Z Math Logik Grundlagen Math 30*295-302
◇ D45 ◇ REV Zbl 526 # 03029 • ID 42230

DOWNEY, R.G. & REMMEL, J.B. [1984] *The universal complementation property* (J 0036) J Symb Logic 49*1125-1136
◇ C57 C60 D25 D45 ◇ ID 39864

DOWNEY, R.G. & HIRD, G.R. [1985] *Automorphisms of supermaximal subspaces* (J 0036) J Symb Logic 50*1-9
◇ C57 C60 D25 D45 ◇ REV Zbl 572 # 03024 • ID 39783

DOWNEY, R.G. & KALANTARI, I. [1985] *Effective extensions of linear forms on a recursive vector space over a recursive field* (J 0068) Z Math Logik Grundlagen Math 31*193-200
◇ C57 C60 D45 ◇ ID 47566

DOWNEY, R.G. [1985] *The degrees of r.e. sets without the universal splitting property* (J 0064) Trans Amer Math Soc 291*337-351
◇ D25 ◇ ID 47692

DRABBE, J. [1973] *Hierarchie arithmetique et chaines* (J 0150) Acad Roy Belg Bull Cl Sci (5) 59*865-868
◇ D55 ◇ REV MR 49 # 2319 Zbl 337 # 02026 • ID 04098

DRABBE, J. see Vol. I, II, III, V, VI for further entries

DRAGALIN, A.G. [1968] *Lexicographical operator algorithms (Russian)* (S 0228) Zap Nauch Sem Leningrad Otd Mat Inst Steklov 8*46-52 • ERR/ADD ibid 16*185-187
◇ B75 D03 ◇ REV MR 39 # 4005 • ID 03112

DRAGALIN, A.G. see Vol. I, III, V, VI for further entries

DRAKE, F.R. & WAINER, S.S. (EDS.) [1980] *Recursion theory: its generalizations and applications* (S 3306) Lond Math Soc Lect Note Ser 45*vi+319pp
◇ D97 ◇ REV MR 81j:03007 Zbl 431 # 00003 • ID 53905

DRAKE, F.R. see Vol. I, II, V for further entries

DREBEN, B. [1957] *Systematic treatment of the decision problem* (P 1675) Summer Inst Symb Log;1957 Ithaca 363
◇ B20 B25 D35 ◇ ID 03611

DREBEN, B. & KAHR, A.S. & WANG, HAO [1962] *Classification of AEA formulas by letter atoms* (J 0015) Bull Amer Math Soc 68*528-532
◇ B20 B25 D35 ◇ REV MR 30 # 22 Zbl 112.5 JSL 29.101
• ID 03127

DREBEN, B. see Vol. I, II, III, VI for further entries

DRIES VAN DEN, L. [1979] *Algorithms and bounds for polynomial rings* (P 2627) Logic Colloq;1978 Mons 147-157
◇ C57 C60 ◇ REV MR 81f:03045 Zbl 461 # 13015
• ID 54499

DRIES VAN DEN, L. [1981] see CHERLIN, G.L.

DRIES VAN DEN, L. & SMITH, RICK L. [1985] *Decidable regularly closed fields of algebraic numbers* (J 0036) J Symb Logic 50*468-475
◇ B25 C57 C60 ◇ REV Zbl 574 # 12023 • ID 41808

DRIES VAN DEN, L. see Vol. I, III, V, VI for further entries

DRISCOLL JR., G.C. [1968] *Metarecursively enumerable sets and their metadegrees* (J 0036) J Symb Logic 33*389-411 ◇ D30 D60 ◇ REV MR 43 # 46 Zbl 213.17 JSL 34.115 • ID 03137

DROBOTUN, B.N. [1977] *Enumerations of simple models (Russian)* (J 0092) Sib Mat Zh 18*1002-1014,1205 • TRANSL [1977] (J 0475) Sib Math J 18*707-716 ◇ C57 D45 ◇ REV MR 58 # 16254 Zbl 382 # 03032 • ID 51956

DROBOTUN, B.N. [1977] *On countable models of decidable almost categorical theories (Russian)* (J 0403) Izv Akad Nauk Kazak SSR, Ser Fiz-Mat 1977/5*22-25,91 ◇ B25 C15 C35 C57 ◇ REV MR 58 # 27444 Zbl 374 # 02032 • ID 51556

DROBOTUN, B.N. & GONCHAROV, S.S. [1980] *Numerations of saturated and homogeneous models (Russian)* (J 0092) Sib Mat Zh 21/2*25-41,236 • TRANSL [1980] (J 0475) Sib Math J 21*164-176 ◇ B25 C50 C57 D45 ◇ REV MR 82k:03068 Zbl 441 # 03016 • ID 56069

DUBINSKY, A. [1975] *Computation on arbitrary algebras* (P 1603) λ-Calc & Comput Sci Th;1975 Roma 319-341 ◇ D75 ◇ REV MR 57 # 5867 Zbl 331 # 02018 • ID 61414

DUBITZKI, T. & WU, A. [1980] *Cellular d-graph automata with augmented memory* (J 0191) Inform Sci 22*69-77 ◇ D05 ◇ REV MR 81h:68040 Zbl 452 # 68059 • ID 69461

DUBOIS, DONALD WARD [1982] *General numeration. I: Gauged schemes. II: Division schemes* (J 0236) Rev Mat Hisp-Amer, Ser 4 42*38-50 ◇ D45 ◇ REV Zbl 541 # 05006 • ID 46427

DUBOIS, DONALD WARD see Vol. III for further entries

DUDA, W.L. [1957] *Post canonical language* (P 1675) Summer Inst Symb Log;1957 Ithaca 410-424 ◇ D03 D20 ◇ REV JSL 36.343 • ID 03156

DUDICH, A.M. & GLEBSKIJ, YU.V. & KOGAN, D.I. & LIOGON'KIJ, M.I. & MARKOV, A.A. [1965] *Algorithmen, die durch wiederholungsfreie Anwendung endlicher Automaten realisiert werden (Russisch)* (J 0052) Probl Kibern 13*241-243 ◇ D05 ◇ REV Zbl 261 # 02018 • ID 30461

DUJOLS, R. [1975] *Une methode de construction directe de couples d'ensembles 1-reductibles ayant des types d'equivalence recursive incomparables (English summary)* (J 2313) C R Acad Sci, Paris, Ser A-B 281*A1-A4 ◇ D30 D50 ◇ REV MR 52 # 5394 Zbl 312 # 02035 • ID 18121

DUJOLS, R. [1976] *Le resultat de Karp-Myhill dans "Recursive equivalent types" de J.C.E. Dekker & J. Myhill* (J 1934) Ann Sci Univ Clermont Math 13*77-80 ◇ D50 ◇ REV MR 57 # 2896 Zbl 359 # 02040 • REM The paper of Decker and Myhill was published in Univ. California Publ. Math. 3(1960)*67-213 • ID 72476

DUJOLS, R. [1976] *Quelques remarques a propos d'un theoreme de Karp-Myhill (English summary)* (J 2313) C R Acad Sci, Paris, Ser A-B 282*A1125-A1127 ◇ D50 ◇ REV MR 54 # 2439 Zbl 337 # 02027 • ID 24635

DUJOLS, R. see Vol. V for further entries

DULAC, M.-H. [1971] *Decidabilite et operations entre theories* (J 2313) C R Acad Sci, Paris, Ser A-B 273*A1113-A1114 ◇ B10 B25 D35 ◇ REV MR 45 # 3204 Zbl 224 # 02038 • ID 03165

DUMITRESCU, S. & STEFAN, G.M. [1977] *The modelling of digital logical structures as finite state systems* (J 3130) Bul Inst Politeh Bucuresti, Ser Electroteh 39/3*59-68 ◇ D05 ◇ REV MR 57 # 15782 Zbl 381 # 94034 • ID 69465

DUMMETT, M. [1965] see CROSSLEY, J.N.

DUMMETT, M. see Vol. II, VI for further entries

DUNHAM, B. [1957] *Symbolic logic and computing machines. A survey and summary* (P 1675) Summer Inst Symb Log;1957 Ithaca 164-166 ◇ D05 D10 ◇ ID 29339

DUNHAM, B. & WANG, HAO [1976] *Towards feasible solutions of the tautology problem* (J 0007) Ann Math Logic 10*117-154 ◇ B35 D15 ◇ REV MR 54 # 14464 Zbl 349 # 02006 • ID 23648

DUNHAM, B. see Vol. I for further entries

DUNHAM, C. [1968] *An uncompletable function* (J 0005) Amer Math Mon 75*1104-1105 ◇ D10 D20 ◇ REV Zbl 186.10 • ID 03175

DURIS, P. & GALIL, Z. [1984] *A time-space tradeoff for language recognition* (J 0041) Math Syst Theory 17*3-12 ◇ D05 D15 ◇ REV MR 85j:68031 Zbl 533 # 68047 • ID 46788

DURIS, P. & GALIL, Z. & PAUL, W.J. & REISCHUK, R. [1984] *Two nonlinear lower bounds for on-line computations* (J 0194) Inform & Control 60*1-11 ◇ D15 ◇ REV MR 86d:03034 • ID 44235

DURNEV, V.G. [1974] *Equations on free semigroups and groups (Russian)* (J 0087) Mat Zametki (Akad Nauk SSSR) 16*717-724 • TRANSL [1974] (J 1044) Math Notes, Acad Sci USSR 16*1024-1029 ◇ D35 ◇ REV MR 52 # 66 Zbl 334 # 02024 • ID 18124

DURNEV, V.G. [1974] *Positive formulas in free semigroups (Russian)* (J 0092) Sib Mat Zh 15*1131-1137,1182 • TRANSL [1974] (J 0475) Sib Math J 15*796-800 ◇ D35 D40 ◇ REV MR 50 # 9566 Zbl 291 # 02033 • ID 04103

DURNEV, V.G. [1977] *On the question of equations of free semigroups (Russian)* (S 2626) Vopr Teor Grupp Gomol Algeb 1*54-58 ◇ D40 ◇ REV MR 81c:20043 Zbl 411 # 20035 • ID 81127

DURNEV, V.G. see Vol. I, II, III, V for further entries

DUSHSKIJ, V.A. [1969] *Extension of partial recursive functions and functions with a recursive graph (Russian)* (J 0087) Mat Zametki (Akad Nauk SSSR) 5*261-267 • TRANSL [1969] (J 1044) Math Notes, Acad Sci USSR 5*158-161 ◇ D20 ◇ REV MR 40 # 2538 Zbl 184.18 • ID 03187

DUSHSKIJ, V.A. [1974] *Enumerational reducibility of recursively enumerable classes of sets (Russian)* (J 0068) Z Math Logik Grundlagen Math 20*19-30 ◇ D25 D30 D45 ◇ REV MR 55 # 2541 Zbl 314 # 02049 • ID 61447

DUSKE, J. [1976] *Zur Theorie kommutativer Automaten (English summary)* (J 0380) Acta Cybern (Szeged) 3*15-23
◇ D05 ◇ REV MR 54 # 10461 Zbl 355 # 94055 • ID 25882

DUSKE, J. & PARCHMANN, R. & SPECHT, J. [1979] *Szilard languages of IO-grammars* (J 0194) Inform & Control 40*319-331
◇ D05 ◇ REV MR 80g:68094 Zbl 412 # 68075 • ID 69467

DVORNIKOV, S.G. [1979] *C-degrees of continuous everywhere defined functionals (Russian)* (J 0003) Algebra i Logika 18*32-46
• TRANSL [1979] (J 0069) Algeb and Log 18*21-31
◇ D45 D65 ◇ REV MR 82k:03076 Zbl 448 # 03034
• ID 56634

DVORNIKOV, S.G. [1979] *Precompletely enumerated sets (Russian)* (J 0092) Sib Mat Zh 20*1303-1306
• TRANSL [1979] (J 0475) Sib Math J 20*927-929
◇ D45 ◇ REV MR 81a:03048 Zbl 448 # 03033 • ID 56633

D'YACHENKO, V.F. [1971] *On transformation of schemes of algorithms (Russian)* (P 1560) Tr Mezhdurn Semin Priklad Aspekt Teor Avtom;1971 Varna 1*171-178
◇ D05 D20 ◇ REV Zbl 262 # 02032 • ID 27484

DYER, C.R. [1980] *One-way bounded cellular automata* (J 0194) Inform & Control 44*261-281
◇ D05 ◇ REV MR 81g:68081 Zbl 442 # 68082 • ID 69470

DYER, C.R. [1981] *Relation of one-way parallel/sequential automata to 2-D finite-state automata* (J 0191) Inform Sci 23*25-30
◇ D05 ◇ REV MR 82a:68093 Zbl 457 # 68043 • ID 69471

DYER, C.R. & ROSENFELD, A. [1981] *Triangle cellular automata* (J 0194) Inform & Control 48*54-69
◇ D05 ◇ REV MR 82h:68078 Zbl 459 # 68023 • ID 69469

DYER, E. [1981] see BAUMSLAG, G.

DYER, E. see Vol. III for further entries

DYMENT, E.Z. [1976] *On some properties of the Medvedev lattice (Russian)* (J 0142) Mat Sb, Akad Nauk SSSR, NS 101(143)*360-379
• TRANSL [1978] (J 0349) Math of USSR, Sbor 30*321-340
◇ D30 ◇ REV MR 55 # 5421 Zbl 353 # 02019 • ID 50102

DYMENT, E.Z. [1977] *Supersets of recursively enumerable sets (Russian)* (J 0092) Sib Mat Zh 18*462-464
• TRANSL [1977] (J 0475) Sib Math J 18*331-332
◇ D25 ◇ REV MR 57 # 83 Zbl 376 # 02028 • ID 51670

DYMENT, E.Z. [1977] *Two theorems on finite unions of regressive immune sets (Russian)* (J 0087) Mat Zametki (Akad Nauk SSSR) 21*259-269
• TRANSL [1977] (J 1044) Math Notes, Acad Sci USSR 21*141-146
◇ D50 ◇ REV MR 55 # 12494 Zbl 353 # 02020 • ID 50103

DYMENT, E.Z. [1980] *Exact bounds of denumerable collections of degrees of difficulty (Russian)* (J 0087) Mat Zametki (Akad Nauk SSSR) 28*899-910,961
• TRANSL [1980] (J 1044) Math Notes, Acad Sci USSR 28*904-909
◇ D30 ◇ REV MR 82d:03070 Zbl 498 # 03036 • ID 72525

DYMENT, E.Z. [1981] *On some properties of the Q-complexity of finite objects (Russian)* (J 2320) Probl Peredachi Inf, Akad Nauk SSSR 17/4*119-123
◇ D15 ◇ REV Zbl 495 # 03023 • ID 36884

DYMENT, E.Z. [1982] *Existence of a degree of extendability of a partially recursive function, which is not a degree of separability (Russian)* (J 0087) Mat Zametki (Akad Nauk SSSR) 32*83-88
• TRANSL [1982] (J 1044) Math Notes, Acad Sci USSR 32*521-523
◇ D20 D30 ◇ REV MR 84j:03089 Zbl 529 # 03019
• ID 37653

DYMENT, E.Z. [1984] *Recursive metrizability of numbered topological spaces and bases of effective linear topological spaces (Russian)* (J 0031) Izv Vyssh Ucheb Zaved, Mat (Kazan) 1984/8*59-61
• TRANSL [1984] (J 3449) Sov Math 28/8*74-78
◇ C57 C65 D45 ◇ REV MR 86d:03044 • ID 44722

DYMENT, E.Z. [1984] *Rice numerations (Russian)* (J 0087) Mat Zametki (Akad Nauk SSSR) 36*635-646,797
◇ D20 D45 ◇ ID 45387

DYMENT, E.Z. see Vol. VI for further entries

DYMOND, P.W. [1979] *Indirect addressing and the time relationships of some models of sequential computation* (J 2687) Comp Math Appl 5*193-209
◇ D15 ◇ REV MR 81h:68024 Zbl 435 # 68036 • ID 81132

DYMOND, P.W. & TOMPA, M. [1985] *Speedups of deterministic machines by synchronous parallel machines* (J 0119) J Comp Syst Sci 30*149-161
◇ D15 ◇ ID 48217

DZGOEV, V.D. [1976] *On constructivizability of boolean algebras (Russian)* (P 2064) All-Union Conf Math Log (4);1976 Kishinev
◇ D45 ◇ ID 32627

DZGOEV, V.D. [1977] *On the constructivization of distributed structures with relative complements (Russian)* (P 2553) All-Union Algeb Conf (14);1977 Novosibirsk PART 2
◇ D45 ◇ ID 32625

DZGOEV, V.D. [1979] *Cartesian degrees of constructive models (Russian)* (P 2558) All-Union Conf Math Log (5) (Mal'tsev);1979 Novosibirsk 43-44
◇ D45 ◇ ID 32628

DZGOEV, V.D. [1979] *Recursive automorphisms of constructive models (Russian)* (P 2564) All-Union Algeb Conf (15);1979 Novosibirsk
◇ C57 D45 ◇ ID 32626

DZGOEV, V.D. & GONCHAROV, S.S. [1980] *Autostability of models (Russian)* (J 0003) Algebra i Logika 19*45-58,132
• TRANSL [1980] (J 0069) Algeb and Log 19*28-37
◇ C57 D45 G10 ◇ REV MR 82g:03082 Zbl 468 # 03023
• ID 55088

DZGOEV, V.D. [1980] *On the constructivization of certain structures (Russian)* (J 0092) Sib Mat Zh 21/1*231
◇ C57 D45 ◇ ID 32623

DZGOEV, V.D. [1982] *Constructivizations of direct products of algebraic systems (Russian)* (J 0003) Algebra i Logika 21*138-148
• TRANSL [1982] (J 0069) Algeb and Log 21*88-96
◇ C30 C57 D45 G05 G10 ◇ REV MR 85b:03076 Zbl 507 # 03013 • ID 37236

DZHANSEJTOV, K.K. [1969] *Cancellable and self-reproducing configurations (Russian) (Kazakh summary)* (J 0403) Izv Akad Nauk Kazak SSR, Ser Fiz-Mat 1969/5*58-63
◇ D10 ◇ REV MR 48 # 8134 Zbl 254 # 02030 • ID 61450

EBBINGHAUS, H.-D. [1970] *Aufzaehlbarkeit* (**S** 1415) Sel Math 2*64-113
• TRANSL [1972] (**C** 1534) Mash Turing & Rek Funk 86-149
◇ D03 D10 D20 D25 F30 ◇ REV MR 43 # 4665 Zbl 251 # 02037 • ID 28196

EBBINGHAUS, H.-D. [1970] *Turing-Maschinen und berechenbare Funktionen I,III* (**S** 1415) Sel Math 2*1-20,55-63
• TRANSL [1972] (**C** 1534) Mash Turing & Rek Funk 9-33,74-85
◇ D10 D20 ◇ REV MR 43 # 6084 MR 43 # 6086 Zbl 211.312 • REM Part II 1970 by Mahn,F.-K. • ID 28194

EBBINGHAUS, H.-D. & HERMES, H. & JACOBS, K. & MAHN, F.-K. [1972] *Turing-Maschinen und rekursive Funktionen (Russian)* (**X** 0885) Mir: Moskva 264pp
◇ D10 D20 ◇ REV MR 50 # 12679 Zbl 251 # 00011 • ID 28923

EBBINGHAUS, H.-D. & HERMES, H. [1972] *Universelle Turing-Maschinen* (**J** 0487) Math Unterricht 18*5-31
◇ D10 ◇ ID 28197

EBBINGHAUS, H.-D. [1976] *The axiom of determinateness* (**P** 1619) Coloq Log Simb;1975 Madrid 35-41
◇ D55 E15 E25 E60 ◇ REV MR 56 # 2825 Zbl 362 # 02074 • ID 50795

EBBINGHAUS, H.-D. [1980] *Mathematische Logik und Informatik* (**P** 3608) Wechselwirk Inform Math;1979 Wien 119-145
◇ C13 D10 D15 D98 E70 H15 ◇ ID 41134

EBBINGHAUS, H.-D. [1982] *Some logical aspects of complexity theory* (**P** 3607) Semin Relac Log Mat & Inform Teor;1981 Madrid 1-60
◇ D15 ◇ ID 41131

EBBINGHAUS, H.-D. [1982] *Undecidability of some domino connectability problems* (**J** 0068) Z Math Logik Grundlagen Math 28*331-336
◇ D05 ◇ REV MR 84c:03083 Zbl 495 # 03025 • ID 34018

EBBINGHAUS, H.-D. & MUELLER, GERT H. & SACKS, G.E. (EDS.) [1985] *Recursion theory week* (**S** 3301) Lect Notes Math 1141*viii+418pp
◇ D97 ◇ REV Zbl 566 # 00001 • ID 48741

EBBINGHAUS, H.-D. see Vol. I, II, III, V for further entries

EDMUNDS, C.C. [1975] *On the endomorphisms problem for free groups* (**J** 0394) Commun Algeb 3*1-20
◇ D40 ◇ REV MR 51 # 5763 Zbl 299 # 20018 • ID 15240

EDMUNDS, C.C. [1981] see COMERFORD JR., L.P.

EDMUNDSON, H.P. & TUNG, I.I. [1978] *A characterization of the recursiveness of an m-adic probabilistic language* (**J** 0194) Inform & Control 39*143-148
◇ D05 D20 ◇ REV MR 80b:68099 Zbl 387 # 68071 • ID 69477

EGLI, H. [1975] see CONSTABLE, R.L.

EGLI, H. see Vol. VI for further entries

EGLITIS, L.V. [1977] see CHIRKOV, M.K.

EGLITIS, L.V. see Vol. I for further entries

EHRENFEUCHT, A. [1957] *Two theories with axioms built by means of pleonasmus* (**J** 0036) J Symb Logic 22*36-38
◇ B25 C35 D35 ◇ REV MR 19.933 Zbl 78.244 JSL 23.445 • ID 03235

EHRENFEUCHT, A. & FEFERMAN, S. [1959] *Representability of recursively enumerable sets in formal theories* (**J** 0009) Arch Math Logik Grundlagenforsch 5*37-41
◇ D25 F30 ◇ REV MR 23 # A3088 Zbl 118.251 JSL 32.530 • ID 03237

EHRENFEUCHT, A. [1961] *Separable theories* (**J** 0014) Bull Acad Pol Sci, Ser Math Astron Phys 9*17-19
◇ D25 D35 ◇ REV MR 24 # A1825 JSL 34.127 • ID 03241

EHRENFEUCHT, A. & KREISEL, G. [1966] *Strong models of arithmetic* (**J** 0014) Bull Acad Pol Sci, Ser Math Astron Phys 14*107-110
◇ C57 C62 ◇ REV MR 36 # 2497 Zbl 137.7 • ID 03245

EHRENFEUCHT, A. & MYCIELSKI, J. [1971] *Abbreviating proofs by adding new axioms* (**J** 0015) Bull Amer Math Soc 77*366-367
• TRANSL [1974] (**C** 2319) Slozh Vychisl & Algor 172-173 (Russian)
◇ D35 F20 ◇ REV MR 43 # 1838 Zbl 216.10 • ID 29998

EHRENFEUCHT, A. [1975] *Practical decidability* (**J** 0119) J Comp Syst Sci 11*392-396
◇ B25 D15 ◇ REV MR 52 # 9683 Zbl 329 # 02020 • ID 18126

EHRENFEUCHT, A. & ROZENBERG, G. [1977] *On simplifications of PDOL systems* (**P** 3238) Conf Theoret Comput Sci;1977 Waterloo ON 81-87
◇ D05 ◇ REV MR 58 # 13962 Zbl 414 # 68046 • ID 69496

EHRENFEUCHT, A. & ROZENBERG, G. [1978] *Elementary homomorphisms and a solution of the DOL sequence equivalence problem* (**J** 1426) Theor Comput Sci 7*169-183
◇ D05 ◇ REV MR 80c:68053 Zbl 407 # 68085 • ID 56236

EHRENFEUCHT, A. & ROZENBERG, G. [1978] *On the structure of derivations in deterministic ETOL systems* (**J** 0119) J Comp Syst Sci 17*331-347
◇ D05 ◇ REV MR 80a:68080 Zbl 388 # 68066 • ID 69493

EHRENFEUCHT, A. & ROZENBERG, G. [1978] *Simplifications of homomorphisms* (**J** 0194) Inform & Control 38*298-309
◇ D05 ◇ REV MR 80a:20080 Zbl 387 # 68062 • ID 69484

EHRENFEUCHT, A. & ROZENBERG, G. [1979] *A result on the structure of ETOL languages* (**J** 2716) Found Control Eng, Poznan 4*165-171
◇ D05 ◇ REV MR 81c:68657 Zbl 436 # 68052 • ID 69486

EHRENFEUCHT, A. & ROZENBERG, G. [1979] *An observation on scattered grammars* (**J** 0232) Inform Process Lett 9*84-85
◇ D05 ◇ REV MR 80i:68058 Zbl 412 # 68073 • ID 69499

EHRENFEUCHT, A. & ROZENBERG, G. [1979] *Finding a homomorphism between two words is NP-complete* (**J** 0232) Inform Process Lett 9*86-88
◇ D05 D15 ◇ REV MR 80j:68030 Zbl 414 # 68022 • ID 69497

EHRENFEUCHT, A. & ROZENBERG, G. [1979] *On arithmetic substitutions of EDTOL languages* (**J** 2716) Found Control Eng, Poznan 4*57-62
◇ D05 ◇ REV MR 80d:68086 Zbl 421 # 68072 • ID 69485

EHRENFEUCHT, A. & ROZENBERG, G. [1979] *On the structure of polynomially bounded DOL systems* (**J** 2095) Fund Inform, Ann Soc Math Pol, Ser 4 2*187-197
◇ D05 ◇ REV MR 81f:68090 Zbl 452 # 68075 • ID 69490

EHRENFEUCHT, A. & ROZENBERG, G. & VERMEIR, D. [1979] *On ETOL systems with rank* (**J** 0119) J Comp Syst Sci 19*237-255
◇ D05 ◇ REV MR 81a:68081 Zbl 431 # 68073 • ID 69483

EHRENFEUCHT, A. & MAURER, H.A. & ROZENBERG, G. [1980] *Continuous grammars* (**J** 0194) Inform & Control 46*71-91
◇ D05 ◇ REV MR 82d:68045 Zbl 471 # 68050 • ID 81141

EHRENFEUCHT, A. & ROZENBERG, G. [1980] *DOS systems and languages* (**P** 2904) Automata, Lang & Progr (7);1980 Noordwijkerhout 134-141
◇ D05 ◇ REV MR 82a:68141 Zbl 467 # 68068 • ID 81144

EHRENFEUCHT, A. & ROZENBERG, G. [1980] *Every two equivalent DOL systems have a regular true envelope* (**J** 1426) Theor Comput Sci 10*45-52
◇ D05 ◇ REV MR 80m:68073 Zbl 471 # 68048 • ID 69494

EHRENFEUCHT, A. & ROZENBERG, G. & VERRAEDT, R. [1980] *Many-to-one simulation in EOL forms is decidable* (**J** 2702) Discr Appl Math 2*73-76
◇ D05 ◇ REV MR 81h:68065 Zbl 442 # 68071 • ID 81146

EHRENFEUCHT, A. & ROZENBERG, G. [1980] *On basic properties of DOS systems and languages* (**J** 0194) Inform & Control 47*137-153
◇ D05 ◇ REV MR 82d:68046 Zbl 469 # 68076 • ID 81143

EHRENFEUCHT, A. & ROZENBERG, G. [1980] *On the emptiness of the intersection of two DOS languages problem* (**J** 0232) Inform Process Lett 10*223-225
◇ D05 ◇ REV MR 81h:68066 Zbl 467 # 68069 • ID 81145

EHRENFEUCHT, A. & FAJTLOWICZ, S. & MALITZ, J. & MYCIELSKI, J. [1980] *Some problems on the universality of words in groups* (**J** 0004) Algeb Universalis 11*261-263
◇ D40 ◇ REV MR 81k:20045 Zbl 443 # 20032 • ID 81140

EHRENFEUCHT, A. & ROZENBERG, G. [1980] *The sequence equivalence problem is decidable for OS systems* (**J** 0037) ACM J 27*656-663
◇ D05 ◇ REV MR 81k:68057 Zbl 471 # 68047 • ID 69495

EHRENFEUCHT, A. & ROZENBERG, G. & RUOHONEN, K. [1981] *A morphic representation of complements of recursively enumerable sets* (**J** 0037) ACM J 28*706-714
◇ D05 D10 ◇ REV MR 84g:68034 Zbl 491 # 68078 • ID 37767

EHRENFEUCHT, A. & ROZENBERG, G. [1981] *FPOL systems generating counting languages* (**J** 3441) RAIRO Inform Theor 15*161-173
◇ D05 ◇ REV MR 83c:68090 Zbl 469 # 68077 • ID 69492

EHRENFEUCHT, A. & ROZENBERG, G. [1981] *On the (generalized) Post correspondence problem with lists of length 2* (**P** 2903) Automata, Lang & Progr (8);1981 Akko 408-416
◇ D03 ◇ REV MR 83c:03034 Zbl 482 # 68073 • ID 35137

EHRENFEUCHT, A. & ROZENBERG, G. [1981] *On the subword complexity of square-free DOL languages* (**J** 1426) Theor Comput Sci 16*25-32
◇ D05 ◇ REV MR 82i:68048 Zbl 481 # 68073 • ID 81142

EHRENFEUCHT, A. & ROZENBERG, G. & VERMEIR, D. [1981] *On ETOL systems with finite tree-rank* (**J** 1428) SIAM J Comp 10*40-58
◇ D05 ◇ REV MR 83c:68091 Zbl 471 # 68049 • ID 69487

EHRENFEUCHT, A. & PARIKH, R. & ROZENBERG, G. [1981] *Pumping lemmas for regular sets* (**J** 1428) SIAM J Comp 10*536-541
◇ D05 ◇ REV MR 83c:03035 Zbl 461 # 68081 • ID 69489

EHRENFEUCHT, A. & KARHUMAEKI, J. & ROZENBERG, G. [1982] *The (generalized) Post correspondence problem with lists consisting of two words is decidable* (**J** 1426) Theor Comput Sci 21*119-144
◇ D03 ◇ REV MR 84k:68035 Zbl 493 # 68076 • ID 38442

EHRENFEUCHT, A. & ROZENBERG, G. [1984] *On regularity of languages generated by copying systems* (**J** 2702) Discr Appl Math 8*313-317
◇ D05 ◇ REV MR 85j:68059 Zbl 549 # 68075 • ID 45844

EHRENFEUCHT, A. & KLEIJN, H.C.M. & ROZENBERG, G. [1985] *Adding global forbidding context to context-free grammars* (**J** 1426) Theor Comput Sci 37*337-360
◇ D05 ◇ ID 49623

EHRENFEUCHT, A. & HOOGEBOOM, H.J. & ROZENBERG, G. [1985] *On coordinated rewriting* (**P** 4647) FCT'85 Fund of Comput Th;1985 Cottbus 100-111
◇ D05 ◇ ID 49074

EHRENFEUCHT, A. see Vol. I, II, III, V, VI for further entries

EHRIG, H. & KUEHNEL, W. & PFENDER, M. [1975] *Diagram characterization of recursion* (**P** 0770) Categ Th Appl to Comput & Control (1);1974 San Francisco 137-143
◇ D20 ◇ REV MR 53 # 1999 Zbl 311 # 18001 • ID 16693

EHRIG, H. & ROSEN, B.K. [1976] *Commutativity of independent transformations on complex objects* (**S** 3145) Comput Sci Res Rep (Taiwan) 6251*52pp
◇ D03 ◇ REV Zbl 357 # 02034 • ID 50382

EHRIG, H. & MAHR, B. [1981] *Complexity of algebraic implementations for abstract data types* (**J** 0119) J Comp Syst Sci 23*223-253
◇ D15 ◇ REV MR 82m:68083 Zbl 474 # 68021 • ID 81150

EICHNER, L. [1978] *The semigroups of linearly realizable finite automata I,II* (**J** 1431) Acta Inf 10*341-367,369-390
◇ D05 ◇ REV MR 80k:68040 MR 81f:68066 Zbl 392 # 68047 Zbl 392 # 68048 • ID 69502

EICHNER, L. [1979] *The semigroups of totally linearly realizable automata* (**J** 1046) Z Angew Math Mech 59*T53-T54
◇ D05 ◇ REV MR 80k:68040 Zbl 434 # 68032 • ID 69501

EICKEL, J. [1967] *Loesung der Analyse-und Mehrdeutigkeitsprobleme durch Ueberfuehrung formaler Sprachen in sackgassenfreie formale Sprachen* (**P** 1671) Colloq Automatenth (3);1965 Hannover 234-262
◇ D05 ◇ REV Zbl 165.321 • ID 29435

EICKEL, J. [1970] *Vereinfachung der Struktur formaler Sprachen* (**P** 0577) Automatenth & Formale Sprachen;1969 Oberwolfach 255-272
◇ D05 ◇ REV MR 49 # 8429 Zbl 242 # 68033 • ID 14657

EILENBERG, S. & WRIGHT, J.B. [1967] *Automata in general algebras* (**J** 0194) Inform & Control 11*452-470
◇ D05 ◇ REV MR 36 # 6333 Zbl 175.279 • ID 31314

EILENBERG, S. & ELGOT, C.C. [1968] *Iteration and recursion* (**J** 0054) Proc Nat Acad Sci USA 61*378-379
◇ D20 D25 ◇ REV MR 39 # 2633 Zbl 193.310 JSL 35.487 • ID 03265

EILENBERG, S. & ELGOT, C.C. & SHEPHERDSON, J.C. [1969] *Sets recognized by n-tape automata* (**J** 0032) J Algeb 13*447-464 ◇ D05 ◇ REV MR 40 # 2460 Zbl 207.20 • ID 03267

EILENBERG, S. & ELGOT, C.C. [1970] *Recursiveness* (**X** 0801) Academic Pr: New York 89pp ◇ D20 D25 D98 ◇ REV MR 42 # 2939 Zbl 211.311 • ID 03270

EILENBERG, S. [1974] *Automata, languages, and machines. Vol. A* (**X** 0801) Academic Pr: New York xvi+451pp ◇ C05 D05 D98 ◇ REV MR 58 # 26604a Zbl 317 # 94045 • ID 61483

EILENBERG, S. see Vol. V for further entries

EISENBERG, E.F. & REMMEL, J.B. [1982] *Effective isomorphisms of algebraic structures* (**P** 3634) Patras Logic Symp;1980 Patras 95-122 ◇ C57 D45 ◇ REV MR 84m:03052 Zbl 525 # 03039 • ID 35753

EISENSTAT, S.C. [1980] see DEMILLO, R.A.

EJSMONT, M. [1985] *On the log-space reducibility among array languages* (**P** 4670) Comput Th (5);1984 Zaborow 80-90 ◇ D15 ◇ ID 49629

EJSMUND, J. [1971] *On a certain class of m-address machines (Polish and Russian summaries)* (**J** 0063) Studia Logica 28*131-137 ◇ D05 ◇ REV MR 48 # 5419 Zbl 254 # 94046 • ID 33138

ELGOT, C.C. [1961] *Decision problems of finite automata design and related arithmetics* (**J** 0064) Trans Amer Math Soc 98*21-51 • ERR/ADD ibid 103*558-559 ◇ B25 D03 D05 D35 F30 ◇ REV MR 25 # 2962 Zbl 111.11 JSL 34.509 • ID 03288

ELGOT, C.C. & RUTLEDGE, J.D. [1964] *RS-machines with almost blank tapes* (**J** 0037) ACM J 11*313-337 ◇ D05 D10 F30 ◇ REV MR 29 # 3376 Zbl 168.257 • ID 03289

ELGOT, C.C. & ROBINSON, A. [1964] *Random-access stored-program machines, an approach to programming languages* (**J** 0037) ACM J 11*365-399 ◇ B75 D05 D10 ◇ REV MR 30 # 4400 • ID 26135

ELGOT, C.C. & MEZEI, J.E. [1965] *On relations defined by generalized finite automata* (**J** 0284) IBM J Res Dev 9*47-68 ◇ D05 ◇ REV MR 35 # 7732 Zbl 135.7 • ID 17928

ELGOT, C.C. & RABIN, M.O. [1966] *Decidability and undecidability of extensions of second (first) order theory of (generalized) successor* (**J** 0036) J Symb Logic 31*169-181 ◇ B15 B25 C85 D05 D35 F30 F35 ◇ REV Zbl 144.245 • ID 03291

ELGOT, C.C. & ROBINSON, A. & RUTLEDGE, J.D. [1967] *Multiple control computer models* (**P** 1390) Syst & Comput Sci;1965 London ON 60-76 ◇ B70 D05 D10 ◇ REV MR 38 # 4073 • ID 26138

ELGOT, C.C. [1968] see EILENBERG, S.

ELGOT, C.C. [1969] see EILENBERG, S.

ELGOT, C.C. [1970] see EILENBERG, S.

ELGOT, C.C. [1970] *The common algebraic structure of exit-automata and machines* (**J** 0373) Comp Arch Inform & Numerik 6*349-370 ◇ D05 D10 ◇ REV Zbl 264 # 94033 • ID 61492

ELGOT, C.C. [1975] *Monadic computation and iterative algebraic theories* (**P** 0775) Logic Colloq;1973 Bristol 175-230 ◇ C05 D10 D20 ◇ REV MR 54 # 1698 Zbl 327 # 02040 • ID 61493

ELGOT, C.C. [1977] *Finite automaton from a flowchart scheme point of view* (**P** 1635) Math Founds of Comput Sci (6);1977 Tatranska Lomnica 44-51 ◇ D05 ◇ REV MR 57 # 19169 Zbl 373 # 68044 • ID 51523

ELGOT, C.C. [1982] *Selected papers* (**X** 0811) Springer: Heidelberg & New York xxiv+460pp ◇ D96 ◇ REV MR 85a:01077 Zbl 496 # 01011 • ID 38837

ELGOT, C.C. see Vol. III for further entries

ELLENTUCK, E. [1963] *Solution of a problem of R.Friedberg* (**J** 0044) Math Z 82*101-103 ◇ D30 D50 ◇ REV MR 27 # 3525 Zbl 117.13 JSL 37.611 • ID 03294

ELLENTUCK, E. [1964] *Infinite products of isols* (**J** 0048) Pac J Math 14*49-52 ◇ D50 ◇ REV MR 29 # 1147 Zbl 148.246 JSL 31.652 • ID 03295

ELLENTUCK, E. [1967] *Universal isols* (**J** 0044) Math Z 98*1-8 ◇ D50 ◇ REV MR 35 # 5315 Zbl 212.330 • ID 03298

ELLENTUCK, E. [1970] *A coding theorem for isols* (**J** 0036) J Symb Logic 35*378-382 ◇ D50 ◇ REV MR 44 # 67 Zbl 211.19 • ID 03302

ELLENTUCK, E. [1971] *Extensions of isolic models* (**J** 0068) Z Math Logik Grundlagen Math 17*323-333 ◇ D50 ◇ REV MR 45 # 60 Zbl 229 # 02038 • ID 03307

ELLENTUCK, E. [1971] *Incompleteness via simple sets* (**J** 0047) Notre Dame J Formal Log 12*255-256 ◇ D25 F30 ◇ REV MR 44 # 5222 Zbl 193.308 • ID 03306

ELLENTUCK, E. [1971] *Uncountable suborderings of the isols* (**J** 0026) Elem Math 26*277-282 ◇ D50 ◇ REV MR 48 # 5840 Zbl 269 # 02016 • ID 03319

ELLENTUCK, E. [1972] *An algebraic difference between isols and cosimple isols* (**J** 0036) J Symb Logic 37*557-561 ◇ D50 ◇ REV MR 48 # 1901 Zbl 273 # 02026 • ID 03308

ELLENTUCK, E. [1972] *Nonrecursive relations among isols* (**J** 0053) Proc Amer Math Soc 36*239-245 ◇ D50 ◇ REV MR 47 # 1594 Zbl 259 # 02035 • ID 03311

ELLENTUCK, E. [1972] *Nonrecursive combinatorial functions* (**J** 0036) J Symb Logic 37*90-95 ◇ D50 ◇ REV MR 52 # 65 Zbl 254 # 02033 • ID 03315

ELLENTUCK, E. & MANASTER, A.B. [1972] *The decidability of a class of ∀∃ sentences in the isols* (**J** 0048) Pac J Math 43*573-584 ◇ B25 D50 ◇ REV MR 48 # 5841 Zbl 242 # 02051 • ID 03309

ELLENTUCK, E. [1972] *The positive properties of isolic integers* (**J** 0036) J Symb Logic 37*114-132 ◇ D50 ◇ REV MR 52 # 13353 Zbl 247 # 02041 • ID 14990

ELLENTUCK, E. [1972] *Universal cosimple isols* (**J** 0048) Pac J Math 42*629-638 ◇ D50 ◇ REV MR 47 # 3157 Zbl 254 # 02032 • ID 03318

ELLENTUCK, E. [1973] *Degrees of isolic theories* (**J** 0047) Notre Dame J Formal Log 14*331-340 ◇ D30 D50 ◇ REV MR 51 # 10061 Zbl 258 # 02045 • ID 03321

ELLENTUCK, E. [1973] *On the degrees of universal regressive isols* (**J** 0132) Math Scand 32*145-164
◇ D30 D50 ◇ REV MR 48 # 10789 Zbl 275 # 02040 • ID 03317

ELLENTUCK, E. [1973] *On the form of functions which preserve regressive isols* (**J** 0020) Compos Math 26*283-302
◇ D50 ◇ REV MR 49 # 8842 Zbl 272 # 02067 • ID 30395

ELLENTUCK, E. [1974] *A Δ^0_2 theory of regressive isols* (**J** 0036) J Symb Logic 39*459-468
◇ D50 ◇ REV MR 51 # 102 Zbl 299 # 02052 • ID 04111

ELLENTUCK, E. [1974] *A new proof that analytic sets are Ramsey* (**J** 0036) J Symb Logic 39*163-165
◇ D55 E05 E15 ◇ REV MR 50 # 1887 Zbl 292 # 02054 • ID 03322

ELLENTUCK, E. [1974] see DEKKER, J.C.E.

ELLENTUCK, E. [1975] *Hereditarily universal sets* (**J** 0038) J Austral Math Soc 19*292-296
◇ D50 ◇ REV MR 52 # 2856 Zbl 308 # 02044 • ID 17645

ELLENTUCK, E. [1975] *Semigroups, Horn sentences and isolic structures* (**J** 0048) Pac J Math 61*87-101
◇ C05 C57 D50 E10 ◇ REV MR 53 # 5279 Zbl 324 # 02029 • ID 22911

ELLENTUCK, E. [1976] *Decomposable isols and their degrees* (**J** 0068) Z Math Logik Grundlagen Math 22*251-260
◇ D30 D50 E10 ◇ REV MR 57 # 2897 Zbl 344 # 02034 • ID 18450

ELLENTUCK, E. [1977] *Boolean valued rings* (**J** 0027) Fund Math 96*67-86
◇ C30 C60 C90 D50 ◇ REV MR 58 # 16237 Zbl 365 # 02044 • ID 27196

ELLENTUCK, E. [1977] *Tarski semigroups* (**J** 0053) Proc Amer Math Soc 65*19-23
◇ D50 ◇ REV MR 58 # 16267 Zbl 377 # 06009 • ID 51774

ELLENTUCK, E. [1978] *Cardinals, isols, and the growth of functions* (**J** 2631) J Math Kyoto Univ 18*121-130
◇ D50 E10 E25 ◇ REV MR 58 # 21549 Zbl 377 # 02030 • ID 51747

ELLENTUCK, E. [1978] *Model theoretic methods in the theory of isols* (**J** 0007) Ann Math Logic 14*273-285
◇ C57 D50 ◇ REV MR 81h:03093 Zbl 393 # 03032 • ID 29153

ELLENTUCK, E. [1979] *Sylow subgroups of a regressive group* (**J** 1447) Houston J Math 5*49-67
◇ D50 ◇ REV MR 81b:20006 Zbl 442 # 20001 • ID 81160

ELLENTUCK, E. [1980] *Diagonal methods in the theory of isols* (**J** 0068) Z Math Logik Grundlagen Math 26*193-204
◇ D50 ◇ REV MR 81i:03071 Zbl 454 # 03023 • ID 54235

ELLENTUCK, E. [1981] *Galois theorems for isolated fields* (**J** 0068) Z Math Logik Grundlagen Math 27*1-9
◇ C57 C60 D45 D50 ◇ REV MR 84a:03048 Zbl 523 # 03032 • ID 33198

ELLENTUCK, E. [1981] *Hyper-torre isols* (**J** 0036) J Symb Logic 46*1-5
◇ C62 D50 ◇ REV MR 82f:03038 Zbl 503.03020 • ID 72617

ELLENTUCK, E. [1983] *Incompatible extensions of combinatorial functions* (**J** 0036) J Symb Logic 48*752-755
◇ D50 ◇ REV MR 84j:03093 Zbl 546 # 03028 • ID 34684

ELLENTUCK, E. [1983] *Random isols* (**J** 0068) Z Math Logik Grundlagen Math 29*1-6
◇ D50 ◇ REV MR 85b:03079 Zbl 546 # 03029 • ID 40705

ELLENTUCK, E. see Vol. III, V for further entries

EMDE BOAS VAN, P. [1971] *A note on the Meyer-McCreight naming theorem in the theory of computational complexity* (**X** 1121) Math Centr: Amsterdam ZW 07/71*83pp
◇ D15 ◇ REV Zbl 329 # 68045 • ID 32513

EMDE BOAS VAN, P. [1972] *Gap and operator gap* (**X** 1121) Math Centr: Amsterdam ZN 42
◇ D15 ◇ ID 32514

EMDE BOAS VAN, P. [1973] *A comparison of the properties of complexity classes and honesty classes* (**P** 0763) Automata, Lang & Progr (1);1972 Rocquencourt 391-396
◇ D15 ◇ REV MR 51 # 4720 Zbl 263 # 68026 • ID 32515

EMDE BOAS VAN, P. [1973] *Mostowski's universal set-algebra* (**X** 1121) Math Centr: Amsterdam ZW14/73*24pp
◇ C50 C57 C65 D45 E07 E20 G05 ◇ REV Zbl 257 # 02022 • ID 29008

EMDE BOAS VAN, P. [1975] *Ten years of speedup* (**P** 0454) Math Founds of Comput Sci (4);1975 Marianske Lazne 13-29
◇ A10 D15 D20 F20 G05 ◇ REV MR 52 # 16112 Zbl 324 # 68025 • ID 21889

EMDE BOAS VAN, P. [1975] *The non-renameability of honesty classes (German summary)* (**J** 0373) Comp Arch Inform & Numerik 14*183-193
◇ D15 D20 ◇ REV MR 54 # 7224 Zbl 313 # 68037 • ID 25015

EMDE BOAS VAN, P. & VITANYI, P.M.B. [1978] *A note on the recursive enumerability of some classes of recursive languages* (**J** 0191) Inform Sci 14*89-91
◇ D05 D25 ◇ REV MR 80i:68037 Zbl 416 # 68065 • ID 31015

EMDE BOAS VAN, P. [1978] *Least fixed points and the recursion theorem* (**X** 3207) Math Centr Dep Pure Math: Amsterdam ZW78/76*10pp
◇ B75 D20 ◇ REV Zbl 333 # 68014 • ID 32518

EMDE BOAS VAN, P. [1978] *Some applications of the McCreight-Meyer algorithm in abstract complexity theory* (**J** 1426) Theor Comput Sci 7*79-98
◇ D15 ◇ REV MR 58 # 13922 • ID 31016

EMDE BOAS VAN, P. [1980] *Machine models and computational complexity* (**P** 3485) Getalth & Computers;1980 Amsterdam 17-40
◇ D15 ◇ REV Zbl 468 # 68054 • ID 69978

EMDE BOAS VAN, P. [1982] *Machine models, computational complexity and number theory* (**S** 1605) Math Centr Tracts I*7-42
◇ D15 D98 ◇ REV MR 85d:68028 Zbl 508 # 68028 • ID 38971

EMDE BOAS VAN, P. & SAVELSBERGH, M.W.P. [1984] *Bounded tiling, an alternative to satisfiability?* (**P** 3621) Frege Konferenz (2);1984 Schwerin 354-363
◇ D15 ◇ REV Zbl 561 # 68034 • ID 45402

EMDE BOAS VAN, P. see Vol. II for further entries

EMDEN VAN, M.H. [1977] *Relational equations, grammars, and programs* (**P** 3238) Conf Theoret Comput Sci;1977 Waterloo ON 191-201
◇ B75 D05 E07 ◇ REV MR 58#13847 Zbl 426#68005 • ID 53677

EMDEN VAN, M.H. see Vol. I for further entries

EMELICHEV, V.A. [1962] *Solution of some algorithmic problems for commutative semigroups (Russian)* (**J** 0023) Dokl Akad Nauk SSSR 144*261-263
• TRANSL [1962] (J 0062) Sov Math, Dokl 3*699-702
◇ D40 ◇ REV MR 25#139 • ID 03343

EMELICHEV, V.A. [1965] *An algorithm for discerning regularity of a finitely defined commutative semigroup (Russian)* (**J** 0414) Dokl Akad Nauk Belor SSR 9*713-716
◇ D40 ◇ REV MR 34#264 • ID 17933

EMELICHEV, V.A. [1966] *Regularity of a finitely determinated commutative semigroup (Russian)* (**J** 0413) Izv Akad Nauk Belor SSR, Ser Fiz-Mat 1966/2*23-30
◇ D40 ◇ REV MR 34#1424 • ID 17934

EMERSON, E.A. & SISTLA, A.P. [1984] *Deciding branching time logic: a triple exponential decision procedure for CTL* (**P** 2989) Log of Progr; 1983 Pittsburgh 176-192
◇ B75 D15 ◇ REV Zbl 559#68052 • ID 44542

EMERSON, E.A. & SISTLA, A.P. [1984] *Deciding full branching time logic* (**J** 0194) Inform & Control 61*175-201
◇ B25 B45 D15 ◇ REV MR 86h:03021 • ID 45349

EMERSON, E.A. [1985] *Automata, tableaux, and temporal logics* (**P** 4571) Log of Progr;1985 Brooklyn 79-88
◇ B45 D05 ◇ ID 49191

EMERSON, E.A. see Vol. II for further entries

ENDERTON, H.B. [1964] *Hierarchies in recursive function theory* (**J** 0064) Trans Amer Math Soc 111*457-471
◇ D30 D55 ◇ REV MR 28#2971 Zbl 222#02050 JSL 31.262 • ID 03344

ENDERTON, H.B. & LUCKHAM, D.C. [1964] *Hierarchies over recursive well-orderings* (**J** 0036) J Symb Logic 29*183-190
◇ D30 D45 D55 ◇ REV MR 36#4992 Zbl 219#02034 JSL 31.263 • ID 03345

ENDERTON, H.B. [1968] *On provable recursive functions* (**J** 0047) Notre Dame J Formal Log 9*86-88
◇ D20 F30 ◇ REV MR 39#5355 Zbl 184.18 JSL 38.526 • ID 03348

ENDERTON, H.B. & PUTNAM, H. [1970] *A note on the hyperarithmetical hierarchy* (**J** 0036) J Symb Logic 35*429-430
◇ D55 ◇ REV MR 45#65 Zbl 205.308 • ID 03349

ENDERTON, H.B. [1970] *The unique existential quantifier* (**J** 0009) Arch Math Logik Grundlagenforsch 13*52-54
◇ C80 D55 ◇ REV MR 44#1567 Zbl 272#02019 JSL 40.627 • ID 03352

ENDERTON, H.B. [1972] *Degrees of computational complexity* (**J** 0119) J Comp Syst Sci 6*389-396
◇ D15 ◇ REV MR 48#3295 Zbl 251#68029 • ID 03356

ENDERTON, H.B. [1977] *Elements of recursion theory* (**C** 1523) Handb of Math Logic 527-566
◇ D20 D25 D98 ◇ REV MR 58#5109 JSL 49.975 • ID 27309

ENDERTON, H.B. see Vol. I, II, III, V, VI for further entries

ENGELER, E. [1967] *Algorithmic properties of structures* (**J** 0041) Math Syst Theory 1*183-195
◇ B75 C75 D10 D75 ◇ REV MR 37#72 Zbl 202.8 JSL 37.197 • ID 03366

ENGELER, E. [1968] *Formal languages: automata and structures* (**X** 0906) Markham Publ: Chicago vii+81pp
◇ D05 ◇ REV MR 39#3928 Zbl 157.18 JSL 35.594 • ID 17935

ENGELER, E. [1968] *Remarks on the theory of geometrical constructions* (**P** 0637) Syntax & Semant Infinitary Lang;1967 Los Angeles 64-76
◇ C75 D80 ◇ REV Zbl 179.14 • ID 29303

ENGELER, E. [1971] *Algorithmic approximation* (**J** 0119) J Comp Syst Sci 5*67-82
◇ B75 D05 D20 ◇ REV MR 45#2950 Zbl 238#68016 JSL 39.348 • ID 04113

ENGELER, E. (ED.) [1971] *Symposium on semantics of algorithmic languages* (**X** 0811) Springer: Heidelberg & New York vi+372pp
◇ B75 B97 D05 ◇ REV MR 43#1477 Zbl 215.560 • ID 23569

ENGELER, E. [1973] *Introduction to the theory of computation* (**X** 0801) Academic Pr: New York viii+231pp
◇ D98 ◇ REV MR 51#4705 Zbl 305#68003 • ID 28203

ENGELER, E. [1975] *Algorithmic logic* (**P** 1430) Adv Course Founds Computer Sci;1974 Amsterdam 55-85
◇ B25 B75 D15 ◇ REV MR 52#12380 Zbl 314#68009 JSL 42.420 • ID 21738

ENGELER, E. [1975] *On the solvability of algorithmic problems* (**P** 0775) Logic Colloq;1973 Bristol 231-251
◇ B75 C60 C75 D75 ◇ REV MR 52#2842 Zbl 317#02051 • ID 17633

ENGELER, E. [1977] *Structural relations between programs and problems* (**P** 1704) Int Congr Log, Meth & Phil of Sci (5);1975 London ON 1*267-280
◇ B75 D15 ◇ REV MR 57#14595 Zbl 381#03027 • ID 51895

ENGELER, E. [1981] *An algorithmic model of strict finitism* (**P** 3642) Colloq Math Log in Computer Sci;1978 Salgotarjan 345-357
◇ D20 F65 ◇ REV MR 83g:68007 Zbl 498#03048 JSL 49.990 • ID 36907

ENGELER, E. [1981] *Generalized Galois theory and its application to complexity* (**J** 1426) Theor Comput Sci 13*271-293
◇ C60 D15 ◇ REV MR 82j:68018 Zbl 468#12014 • ID 81166

ENGELER, E. see Vol. I, II, III, V, VI for further entries

ENGELFRIET, J. [1978] *A hierarchy of tree transducers* (**P** 2599) CAAP'78 Arbres en Algeb & Progr (3);1978 Lille 103-106
◇ D05 ◇ REV MR 80c:68068 Zbl 386#68070 • ID 69512

ENGELFRIET, J. & SLUTZKI, G. [1979] *Bounded nesting in macro grammars* (**J** 0194) Inform & Control 42*157-193
◇ D05 ◇ REV MR 80e:68184 Zbl 453#68051 • ID 69507

ENGELFRIET, J. & ROZENBERG, G. [1979] *Equality languages and fixed point languages* (**J** 0194) Inform & Control 43*20-49
◇ D05 ◇ REV MR 81h:68067 Zbl 422#68034 • ID 69510

ENGELFRIET, J. [1979] see ASVELD, P.R.J.

ENGELFRIET, J. [1979] *Two-way automata and checking automata* (**P** 3375) Found of Comput Sci (3);1978 Amsterdam 1*1-69
◇ D05 ◇ REV MR 81i:68072 Zbl 423 # 68035 • ID 69517

ENGELFRIET, J. & ROZENBERG, G. [1980] *Fixed point languages, equality languages, and representation of recursively enumerable languages* (**J** 0037) ACM J 27*499-518
◇ D05 D25 ◇ REV MR 81f:68081 Zbl 475 # 68047 • ID 81168

ENGELFRIET, J. & FILE, G. [1980] *Formal properties of one-visit and multi-pass attribute grammars* (**P** 2904) Automata, Lang & Progr (7);1980 Noordwijkerhout 182-194
◇ D05 ◇ REV MR 82g:68067 Zbl 445 # 68062 • ID 69515

ENGELFRIET, J. [1980] *Some open questions and recent results on tree transducers and tree languages* (**P** 4266) Form Lang Th;1979 Santa Barbara 241-286
◇ D05 ◇ REV MR 84j:68001 Zbl 545 # 68062 • ID 47204

ENGELFRIET, J. & LEEUWEN VAN, J. & SCHMIDT, ERIK MEINECHE [1980] *Stack machines and classes of nonnested macro languages* (**J** 0037) ACM J 27*96-117
◇ D05 D10 ◇ REV MR 81c:68066 Zbl 428 # 68087 • ID 69514

ENGELFRIET, J. & ROZENBERG, G. & SLUTZKI, G. [1980] *Tree transducers, L systems, and two-way machines* (**J** 0119) J Comp Syst Sci 20*150-202
◇ D05 ◇ REV MR 83e:68124 Zbl 426 # 68075 • ID 69513

ENGELFRIET, J. [1981] see ASVELD, P.R.J.

ENGELFRIET, J. & FILE, G. [1981] *Passes and paths of attribute grammars* (**J** 0194) Inform & Control 49*125-169
◇ D05 ◇ REV MR 82k:68043 Zbl 471 # 68040 • ID 81169

ENGELFRIET, J. & FILE, G. [1981] *Passes, sweeps and visits* (**P** 2903) Automata, Lang & Progr (8);1981 Akko 193-207
◇ D05 D15 ◇ REV MR 82i:68003 Zbl 465 # 68040 • ID 69511

ENGELFRIET, J. & FILE, G. [1981] *The formal power of one-visit attribute grammars* (**J** 1431) Acta Inf 16*275-302
◇ D05 ◇ REV MR 83a:68088 Zbl 471 # 68044 • ID 69509

ENGELKING, R. & HOLSZTYNSKI, W. & SIKORSKI, R. [1966] *Some examples of Borel sets* (**S** 0019) Colloq Math (Warsaw) 15*271-274
◇ D55 E15 ◇ REV MR 34 # 1198 Zbl 147.227 • ID 03375

ENGELKING, R. see Vol. V for further entries

EPSTEIN, R.L. [1975] *Minimal degrees of unsolvability and the full approximation construction* (**S** 0167) Mem Amer Math Soc 3/1/162*viii+136pp
◇ D30 ◇ REV MR 52 # 2853 Zbl 315 # 02043 • ID 17644

EPSTEIN, R.L. & POSNER, D.B. [1978] *Diagonalization in degree constructions* (**J** 0036) J Symb Logic 43*280-283
◇ D30 ◇ REV MR 58 # 5139 Zbl 393 # 03031 • ID 52451

EPSTEIN, R.L. [1979] *Degrees of unsolvability: structure and theory* (**S** 3301) Lect Notes Math 759*xiv+240pp
◇ D25 D30 D35 D98 ◇ REV MR 81e:03042 Zbl 418 # 03032 • ID 53317

EPSTEIN, R.L. & HAAS, R. & KRAMER, R.L. [1981] *Hierarchies of sets and degrees below O'* (**P** 2628) Log Year;1979/80 Storrs 32-48
◇ D30 D55 ◇ REV MR 82k:03073 Zbl 467 # 03046 • ID 55044

EPSTEIN, R.L. [1981] *Initial segments of degrees below 0'* (**S** 0167) Mem Amer Math Soc 241*102pp
◇ D25 D30 D35 D98 ◇ REV MR 82m:03055 Zbl 472 # 03033 • ID 55298

EPSTEIN, R.L. see Vol. II, VI for further entries

ERIMBETOV, M.M. [1979] *On hyperarithmetical non-standard predicates (Russian)* (**C** 2065) Teor Nereg Kriv Raz Geom Post 23-31
◇ D55 ◇ REV MR 81f:03057 • ID 31750

ERIMBETOV, M.M. see Vol. II, III for further entries

ERMOLAEVA, N.M. [1967] *Arithmetic sums of recursively-projective sets (Russian)* (**J** 0023) Dokl Akad Nauk SSSR 172*1011-1013
• TRANSL [1967] (**J** 0062) Sov Math, Dokl 8*217-219
◇ D55 ◇ REV MR 34 # 7365 Zbl 217.12 • ID 28045

ERMOLAEVA, N.M. see Vol. I, II for further entries

ERNI, W. & LAPSIEN, R. [1981] *On the time and tape complexity of weak unification* (**J** 0232) Inform Process Lett 12*146-150
◇ B35 D15 ◇ REV MR 82f:68047 Zbl 469 # 68056 • ID 81184

ERSHOV, A.P. [1958] *On operator algorithms (Russian)* (**J** 0023) Dokl Akad Nauk SSSR 122*967-970
◇ B75 D03 D20 ◇ REV MR 20 # 5132 Zbl 98.244 • ID 03512

ERSHOV, A.P. [1960] *Operator algorithms. I. Basic concepts (Russian)* (**J** 0052) Probl Kibern 3*5-48
• TRANSL [1960] (**J** 1195) Probl Cybernet 3*696-763 [1960] (**J** 0449) Probl Kybern 3*3-54
◇ B75 D03 D20 ◇ REV MR 23 # B2137 Zbl 131.8 JSL 27.364 • ID 03513

ERSHOV, A.P. [1962] *Operator algorithms II. A description of basic programming constructions (Russian)* (**J** 0052) Probl Kibern 8*211-232
• TRANSL [1965] (**J** 0449) Probl Kybern 8*208-232
◇ B75 D20 ◇ REV MR 31 # 6398 Zbl 173.193 JSL 33.467 • REM Part I 1960 • ID 28635

ERSHOV, A.P. [1972] *Theory of program schemata* (**P** 1455) Inform Processing (5);1971 Ljubljana 28-45
◇ B75 D20 D98 ◇ REV MR 53 # 7091 Zbl 263 # 68007 • ID 61582

ERSHOV, A.P. [1973] *The present state of the theory of program schemes (Russian)* (**J** 0052) Probl Kibern 27*87-110,293-294
◇ B75 D20 D98 ◇ REV MR 50 # 3635 Zbl 315 # 68010 • ID 61583

ERSHOV, A.P. [1978] *Mixed computation in the class of recursive program schemata* (**J** 0380) Acta Cybern (Szeged) 4*19-23
◇ B75 D20 ◇ REV MR 80b:68020 Zbl 421 # 68017 • ID 69532

ERSHOV, A.P. & KNUTH, D.E. (EDS.) [1979] *Algorithms in modern mathematics and their applications I,II* (**X** 2652) Akad Nauk Sibirsk Otd Inst Mat: Novosibirsk 364pp,316pp
◇ B75 B97 D97 ◇ ID 45742

ERSHOV, A.P. [1981] *Abstract computability on algebraic structures* (**P** 3729) Algor in Modern Math & Comput Sci;1979 Urgench 397-420
• TRANSL [1982] (**P** 3803) Algor Sovrem Mat & Prilozh;1979 Urgench II*194-229
◇ B75 D75 ◇ REV MR 83i:03074 Zbl 477 # 68035
• ID 35528

ERSHOV, A.P. & KNUTH, D.E. (EDS.) [1981] *Algorithms in modern mathematics and computer science. Proceedings, Urgench, Uzbek SSR, September 16-22, 1979* (**S** 3302) Lect Notes Comput Sci 122*xi+487pp
• TRANSL [1982] (**X** 2652) Akad Nauk Sibirsk Otd Inst Mat: Novosibirsk 364pp; 316pp
◇ D15 D20 D97 ◇ REV MR 83e:68003 MR 85h:03004 Zbl 477 # 68035 • ID 55609

ERSHOV, A.P. [1982] *Computability in arbitrary domains and bases (Russian)* (**S** 2582) Semiotika & Inf, Akad Nauk SSSR 19*3-58
◇ B75 D75 ◇ REV MR 84h:03097 Zbl 519 # 68051
• ID 34287

ERSHOV, A.P. [1982] *Mixed computation: potential applications and problems for study* (**J** 1426) Theor Comput Sci 18*41-67
◇ B75 D20 ◇ REV MR 83b:68004 Zbl 495 # 68011
• ID 38519

ERSHOV, YU.L. & TAJTSLIN, M.A. [1963] *Undecidability of certain theories (Russian)* (**J** 0003) Algebra i Logika 2/5*37-41
◇ D35 ◇ REV MR 30 # 3018 Zbl 192.56 • ID 32024

ERSHOV, YU.L. [1964] *Unsolvability of theories of symmetric and simple finite groups (Russian)* (**J** 0023) Dokl Akad Nauk SSSR 158*777-779
• TRANSL [1964] (**J** 0062) Sov Math, Dokl 5*1309-1311
◇ C13 C60 D35 ◇ REV MR 30 # 3019 Zbl 199.30
• ID 03520

ERSHOV, YU.L. & LAVROV, I.A. & TAJMANOV, A.D. & TAJTSLIN, M.A. [1965] *Elementary theories (Russian)* (**J** 0067) Usp Mat Nauk 20/4*37-108
• TRANSL [1965] (**J** 1399) Russ Math Surv 20/4*35-105
◇ B25 C98 D35 D98 ◇ REV MR 32 # 4012 Zbl 199.30 JSL 39.603 • ID 03525

ERSHOV, YU.L. [1965] *Undecidability of certain fields (Russian)* (**J** 0023) Dokl Akad Nauk SSSR 161*27-29
• TRANSL [1965] (**J** 0062) Sov Math, Dokl 6*349-352
◇ D35 ◇ REV MR 31 # 61 Zbl 203.12 • ID 03524

ERSHOV, YU.L. [1966] *New examples of undecidable theories (Russian)* (**J** 0003) Algebra i Logika 5/5*37-47
◇ D35 ◇ REV MR 34 # 7375 Zbl 253 # 02046 • ID 03526

ERSHOV, YU.L. [1967] *Numerations of families of general recursive functions (Russian)* (**J** 0092) Sib Mat Zh 8*1015-1025
• TRANSL [1967] (**J** 0475) Sib Math J 8*771-778
◇ D20 D45 ◇ REV MR 36 # 3648 Zbl 191.304 • ID 03528

ERSHOV, YU.L. [1968] *Numbered fields* (**P** 0627) Int Congr Log, Meth & Phil of Sci (3,Proc);1967 Amsterdam 31-34
◇ C57 C60 D45 ◇ REV MR 41 # 55 • ID 37357

ERSHOV, YU.L. [1968] *On a hierarchy of sets I,II (Russian)* (**J** 0003) Algebra i Logika 7/1*47-73,7/4*15-47
• TRANSL [1968] (**J** 0069) Algeb and Log 7*25-43,212-232
◇ D30 D55 ◇ REV MR 42 # 5794 Zbl 216.9 • REM Part III 1970 • ID 21058

ERSHOV, YU.L. [1968] *On computable enumerations (Russian)* (**J** 0003) Algebra i Logika 7/5*71-99
• TRANSL [1968] (**J** 0069) Algeb and Log 7*330-346
◇ D25 D45 ◇ REV MR 43 # 47 Zbl 273 # 02024 • REM Part I. Part II 1969 by Ershov,Yu.L. & Lavrov,I.A. • ID 03531

ERSHOV, YU.L. [1968] *Restricted theories of totally ordered sets (Russian)* (**J** 0003) Algebra i Logika 7/3*38-47
• TRANSL [1968] (**J** 0069) Algeb and Log 7*153-159
◇ B25 C65 D35 E07 ◇ REV MR 41 # 44 Zbl 191.297
• ID 03532

ERSHOV, YU.L. [1969] *Completely enumerated sets (Russian)* (**J** 0092) Sib Mat Zh 10*1048-1064
• TRANSL [1969] (**J** 0475) Sib Math J 10*773-784
◇ D45 ◇ REV MR 41 # 6687 Zbl 323 # 02054 • ID 03533

ERSHOV, YU.L. [1969] *Hyperhypersimple m-degrees (Russian)* (**J** 0003) Algebra i Logika 8*523-552
• TRANSL [1969] (**J** 0069) Algeb and Log 8*298-315
◇ D25 ◇ REV MR 43 # 6088 Zbl 216.8 • ID 03537

ERSHOV, YU.L. & LAVROV, I.A. [1969] *On computable enumerations II (Russian)* (**J** 0003) Algebra i Logika 8*65-71
• TRANSL [1969] (**J** 0069) Algeb and Log 8*34-38
◇ D25 D45 ◇ REV MR 43 # 6087 Zbl 273 # 02025 • REM Part I 1968 • ID 03535

ERSHOV, YU.L. [1969] *The theory of numerations. Part I: General theory of numeration (Russian)* (**X** 0913) Novosibirsk Gos Univ: Novosibirsk 174pp
• TRANSL [1973] (**J** 0068) Z Math Logik Grundlagen Math 19*289-388
◇ D25 D30 D45 G30 ◇ REV MR 46 # 3280 Zbl 281 # 02041 • REM Part II 1973 • ID 22277

ERSHOV, YU.L. [1970] *Index sets (Russian)* (**J** 0092) Sib Mat Zh 11*326-342
• TRANSL [1970] (**J** 0475) Sib Math J 11*246-258
◇ D25 D30 D55 ◇ REV MR 41 # 8224 • ID 03542

ERSHOV, YU.L. [1970] *On a hierarchy of sets III (Russian)* (**J** 0003) Algebra i Logika 9*34-51
• TRANSL [1970] (**J** 0069) Algeb and Log 9*20-31
◇ D30 D55 F15 ◇ REV MR 45 # 8526 Zbl 233 # 02017
• REM Parts I,II 1968 • ID 03540

ERSHOV, YU.L. [1970] *On inseparable pairs (Russian)* (**J** 0003) Algebra i Logika 9*661-666
• TRANSL [1970] (**J** 0069) Algeb and Log 9*396-399
◇ D25 ◇ REV MR 44 # 3832 Zbl 263 # 02021 • ID 03541

ERSHOV, YU.L. [1971] *Computable enumerations of morphisms (Russian)* (**J** 0003) Algebra i Logika 10*247-308
• TRANSL [1971] (**J** 0069) Algeb and Log 10*155-191
◇ D25 D45 ◇ REV MR 46 # 5124 Zbl 315 # 02041
• ID 72745

ERSHOV, YU.L. [1971] *La theorie des enumerations* (**P** 0743) Int Congr Math (II,11,Proc);1970 Nice 1*223-227
◇ D25 D45 ◇ REV MR 56 # 15388 Zbl 388 # 03019
• ID 17048

ERSHOV, YU.L. [1971] *Positive equivalences (Russian)* (**J** 0003) Algebra i Logika 10*620-650
• TRANSL [1971] (**J** 0069) Algeb and Log 10*378-394
◇ D45 ◇ REV MR 46 # 7011 Zbl 276 # 02024 • ID 03545

ERSHOV, YU.L. [1972] *Computable functionals of finite types (Russian)* (J 0003) Algebra i Logika 11*367-437
• TRANSL [1972] (J 0069) Algeb and Log 11*203-242
◇ D65 ◇ REV MR 50 # 12688 Zbl 285 # 02040 • ID 61603

ERSHOV, YU.L. [1972] *Continuous lattices and A-spaces (Russian)* (J 0023) Dokl Akad Nauk SSSR 207*523-526
• TRANSL [1972] (J 0062) Sov Math, Dokl 13*1551-1555
◇ D65 G10 ◇ REV MR 51 # 2892 Zbl 275 # 54024
• ID 17338

ERSHOV, YU.L. [1972] *Elementary group theories (Russian)* (J 0023) Dokl Akad Nauk SSSR 203*1240-1243
• TRANSL [1972] (J 0062) Sov Math, Dokl 13*528-532
◇ B25 C05 C13 C60 D35 ◇ REV MR 45 # 6892
Zbl 264 # 02047 • ID 03551

ERSHOV, YU.L. [1972] *Everywhere-defined continuous functionals (Russian)* (J 0003) Algebra i Logika 11*656-665,736
• TRANSL [1972] (J 0069) Algeb and Log 11*363-368
◇ D65 F10 ◇ REV MR 50 # 12689 Zbl 285 # 02041
• ID 61604

ERSHOV, YU.L. [1972] *Existence of constructivizations (Russian)* (J 0023) Dokl Akad Nauk SSSR 204*1041-1044
• TRANSL [1972] (J 0062) Sov Math, Dokl 13*779-783
◇ C57 D45 ◇ REV MR 47 # 8288 Zbl 261 # 02017
• ID 03552

ERSHOV, YU.L. [1972] *Relationship between sheaf spaces and numbered sets with the C_2^* property (Russian) (English summary)* (S 0228) Zap Nauch Sem Leningrad Otd Mat Inst Steklov 32*18-20,154
• TRANSL [1976] (J 1531) J Sov Math 6*358-360
◇ C90 D25 D45 D65 ◇ REV MR 52 # 5382
Zbl 349 # 02030 • ID 18137

ERSHOV, YU.L. [1973] *Constructive models (Russian)* (C 0733) Izbr Vopr Algeb & Log (Mal'tsev) 111-130
◇ C57 D45 ◇ REV MR 50 # 4291 Zbl 293 # 02038
• ID 17197

ERSHOV, YU.L. [1973] *Hierarchy of the sets of class Δ_2^0 (Russian) (English summary)* (P 0793) Int Congr Log, Meth & Phil of Sci (4,Proc);1971 Bucharest 69-76
◇ D30 D55 ◇ REV MR 56 # 15394 • ID 72725

ERSHOV, YU.L. [1973] *Skolem functions and constructive models (Russian)* (J 0003) Algebra i Logika 12*644-654,735
• TRANSL [1973] (J 0069) Algeb and Log 12*368-373
◇ C57 C65 F50 ◇ REV MR 57 # 5727 Zbl 287 # 02024
• ID 30533

ERSHOV, YU.L. & LAVROV, I.A. [1973] *The upper semilattice $L(\gamma)$ (Russian)* (J 0003) Algebra i Logika 12*167-189,243-244
• TRANSL [1973] (J 0069) Algeb and Log 12*93-106
◇ D25 D45 ◇ REV MR 53 # 12910 Zbl 327 # 02038
• ID 23162

ERSHOV, YU.L. [1973] *Theory of numerations. Part II (Russian)* (X 0913) Novosibirsk Gos Univ: Novosibirsk 169pp
• TRANSL [1975] (J 0068) Z Math Logik Grundlagen Math 21*473-584
◇ D45 D65 G30 ◇ REV MR 54 # 9994 Zbl 344 # 02031
• REM Part I 1969. Part III 1974 • ID 25763

ERSHOV, YU.L. [1974] *Maximal and everywhere defined functionals (Russian)* (J 0003) Algebra i Logika 13*374-397,487
• TRANSL [1974] (J 0069) Algeb and Log 13*210-225
◇ D65 ◇ REV MR 53 # 2660 Zbl 325 # 02027 • ID 21539

ERSHOV, YU.L. [1974] *Theories of nonabelian varieties of groups* (P 0610) Tarski Symp;1971 Berkeley 255-264
◇ C05 D35 ◇ REV MR 51 # 5280 Zbl 319 # 02039
• ID 03554

ERSHOV, YU.L. [1974] *Theory of numerations, III: Constructive models (Russian)* (X 0913) Novosibirsk Gos Univ: Novosibirsk 139pp
• TRANSL [1977] (J 0068) Z Math Logik Grundlagen Math 23*289-371
◇ B25 C57 C98 D45 F50 ◇ REV MR 54 # 9995
Zbl 374 # 02028 • REM Part II 1973 • ID 25764

ERSHOV, YU.L. [1975] *The upper semilattice of numerations of a finite set (Russian)* (J 0003) Algebra i Logika 14*258-284,368
• TRANSL [1975] (J 0069) Algeb and Log 14*159-175
◇ D45 E50 ◇ REV MR 54 # 4947 Zbl 342 # 02027
• ID 24118

ERSHOV, YU.L. [1976] *Hereditarily effective operations (Russian)* (J 0003) Algebra i Logika 15*642-654,743-744
• TRANSL [1976] (J 0069) Algeb and Log 15*400-409
◇ D25 D65 ◇ REV MR 58 # 16201 Zbl 362 # 02030
• ID 50752

ERSHOV, YU.L. [1977] *Constructions "by finite"* (P 1704) Int Congr Log, Meth & Phil of Sci (5);1975 London ON 1*3-9
◇ D45 ◇ REV MR 58 # 16200 Zbl 378 # 02019 • ID 51801

ERSHOV, YU.L. [1977] *Enumeration of the class $\mathbb{C}_{20}^*$ (Russian)* (J 0003) Algebra i Logika 16*637-642,741
• TRANSL [1977] (J 0069) Algeb and Log 16*422-426
◇ D45 D65 ◇ REV MR 58 # 27406 Zbl 399 # 03031
• ID 52827

ERSHOV, YU.L. [1977] *Model $\mathbb{C}$ of partial continuous functionals* (P 1075) Logic Colloq;1976 Oxford 455-467
◇ D45 D65 F10 F50 ◇ REV MR 58 # 21541
Zbl 427 # 03037 • ID 16634

ERSHOV, YU.L. [1977] *Theory of numerations (Russian)* (X 2027) Nauka: Moskva 416pp
◇ D45 D98 ◇ REV MR 81j:03073 • ID 72721

ERSHOV, YU.L. [1980] *Decision problems and constructivizable models (Russian)* (X 2027) Nauka: Moskva 416pp
◇ B25 C57 C60 C98 D45 D98 ◇ REV MR 82h:03009
Zbl 495 # 03009 • ID 72718

ERSHOV, YU.L. [1980] *Regularly closed fields (Russian)* (J 0023) Dokl Akad Nauk SSSR 251*783-785
• TRANSL [1980] (J 0062) Sov Math, Dokl 21*510-512
◇ B25 C35 C60 D35 ◇ REV MR 82g:03050
Zbl 467 # 03026 • ID 55024

ERSHOV, YU.L. [1981] *Undecidability of regularly closed fields (Russian)* (J 0003) Algebra i Logika 20*389-394,484
• TRANSL [1981] (J 0069) Algeb and Log 20*257-260
◇ C60 D35 ◇ REV MR 84e:03052 Zbl 499 # 03018
• ID 34387

ERSHOV, YU.L. [1982] *ω-complete A-spaces (Russian)* (J 0040) Kibernetika, Akad Nauk Ukr SSR 1982/6*6-9
• TRANSL [1982] (J 0021) Cybernetics 18*701-705
◇ B75 D65 ◇ REV Zbl 517 # 68057 • ID 36697

ERSHOV, YU.L. [1983] *The Σ-enumeration principle (Russian)* (J 0023) Dokl Akad Nauk SSSR 270*786-788
• TRANSL [1983] (J 0062) Sov Math, Dokl 27*670-672
◇ C62 C70 D60 E30 E65 ◇ REV MR 85d:03074 Zbl 539 # 03016 • ID 40494

ERSHOV, YU.L. [1984] *Regularly r-closed fields with weakly universal Galois group (Russian)* (J 0003) Algebra i Logika 23*637-669
• TRANSL [1984] (J 0069) Algeb and Log 23*426-449
◇ B25 C60 D35 ◇ REV Zbl 574 # 12024 • ID 48388

ERSHOV, YU.L. [1984] *Strong nonseparability and k-heredity (Russian)* (J 0137) C R Acad Bulgar Sci 37*1139-1142
◇ D25 ◇ REV Zbl 551 # 03028 • ID 43907

ERSHOV, YU.L. see Vol. I, II, III, V, VI for further entries

ESIK, Z. [1978] *On decidability of injectivity of tree transformations* (P 2599) CAAP'78 Arbres en Algeb & Progr (3);1978 Lille 107-133
◇ D05 ◇ REV MR 58 # 19382 Zbl 384 # 68058 • ID 69536

ESIK, Z. [1979] *On functional tree transducers* (P 2935) FCT'79 Fund of Comput Th;1979 Berlin/Wendisch-Rietz 121-127
◇ D05 ◇ REV Zbl 415 # 68045 • ID 69537

ESIK, Z. [1980] *Decidability results concerning tree transducers I* (J 0380) Acta Cybern (Szeged) 5*1-20
◇ D05 ◇ REV MR 83h:68068 Zbl 456 # 68098 • REM Part II 1983 • ID 69535

ETTL, W. & LEITSCH, A. [1979] *Aufzaehlung subrekursiver Klassen und konstruierende Automaten* (J 0238) Sitzb Oesterr Akad Wiss, Math-Nat Kl, Abt 2 188*23-46
◇ D05 D20 ◇ REV MR 82c:03059 Zbl 484 # 03021 • ID 72763

ETTL, W. & LEITSCH, A. [1982] *Enumerations of subrecursive classes and generations of automata* (C 3868) Prog in Cybern & Syst Res, Vol 8 243-250
◇ D05 ◇ REV MR 84f:00031 Zbl 482 # 03019 • ID 36843

EUDACH, L. (ED.) [1979] *Fundamentals of computation theory* (X 0911) Akademie Verlag: Berlin 576pp
◇ D97 ◇ REV MR 80m:68007 Zbl 408 # 00013 • ID 48714

EVANGELIST, M. [1982] *Non-standard propositional logics and their application to complexity theory* (J 0047) Notre Dame J Formal Log 23*384-392
◇ B80 D15 F20 ◇ REV MR 83k:03014 Zbl 464 # 03028 • ID 54618

EVANGELIST, W.M. [1985] see CHUNG, MOONJUNG

EVANS, T. [1951] *The word problem for abstract algebras* (J 0039) J London Math Soc 26*64-71
◇ D40 ◇ REV MR 12.475 Zbl 42.33 JSL 34.507 • ID 03560

EVANS, T. [1954] *An embedding theorem for semigroups with cancellation* (J 0100) Amer J Math 76*399-413
◇ D40 ◇ REV MR 15.681 Zbl 55.250 JSL 20.74 • ID 03561

EVANS, T. [1963] *A decision problem for translations of trees* (J 0017) Canad J Math 15*584-590
◇ D80 ◇ REV MR 27 # 4220 Zbl 116.149 • ID 48071

EVANS, T. [1969] *An unsolvable problem concerning identities* (J 0047) Notre Dame J Formal Log 10*413-414
◇ D35 ◇ REV MR 40 # 43 Zbl 167.16 • ID 03570

EVANS, T. [1969] *Some connections between residual finiteness, finite embeddability and the word problem* (J 3172) J London Math Soc, Ser 2 1*399-403
◇ C05 D40 ◇ REV MR 40 # 2589 Zbl 184.35 • ID 03569

EVANS, T. & MANDELBERG, K.I. & NEFF, M.F. [1975] *Embedding algebras with solvable word problems in simple algebras - some Boone-Higman type theorems* (P 0775) Logic Colloq;1973 Bristol 259-277
◇ D40 ◇ REV MR 52 # 10399 Zbl 311 # 02054 • ID 21695

EVANS, T. [1978] *An algebra has a solvable word problem if and only if it is embeddable in a finitely generated simple algebra* (J 0004) Algeb Universalis 8*197-204
◇ D40 ◇ REV MR 57 # 12339 Zbl 381 # 03029 • ID 51897

EVANS, T. [1978] *Word problems* (J 0015) Bull Amer Math Soc 84*789-802
◇ D40 ◇ REV MR 58 # 16240 Zbl 389 # 03018 • ID 52307

EVANS, T. [1980] *Some solvable word problems* (P 2634) Word Problems II;1976 Oxford 87-100
◇ D40 ◇ REV MR 81k:08002 Zbl 432 # 08004 • ID 81194

EVANS, T. see Vol. II, III for further entries

EVE, J. & KURKI-SUONIO, R. [1977] *On computing the transitive closure of a relation* (J 1431) Acta Inf 8*303-314
◇ D15 E07 ◇ REV MR 57 # 4595 Zbl 349 # 68021 • ID 61616

EVEN, S. [1964] *Rational numbers and regular events* (J 4305) IEEE Trans Electr Comp EC-13*740-741
◇ D05 ◇ REV Zbl 178.331 • ID 03576

EVEN, S. [1965] *Comments on the minimization of stochastic machines* (J 4305) IEEE Trans Electr Comp EC-14*634-637
◇ D05 ◇ ID 03577

EVEN, S. [1969] see BOOK, R.V.

EVEN, S. [1971] see BOOK, R.V.

EVEN, S. & TARJAN, R.E. [1976] *A combinatorial problem which is complete in polynomial space* (J 0037) ACM J 23*710-719
◇ D15 D40 ◇ REV MR 56 # 10134 Zbl 355 # 68041 • ID 50268

EVEN, S. & YACOBI, Y. [1980] *Cryptocomplexity and NP-completeness* (P 2904) Automata, Lang & Progr (7);1980 Noordwijkerhout 195-207
◇ D15 D80 ◇ REV MR 82a:94085 Zbl 444 # 94013 • ID 69541

EVEN, S. & YACOBI, Y. [1981] *An observation concerning the complexity of problems with few solutions and its application to cryptography* (P 3575) Graphth Konzepte Inf (6);1980 Bad Honnef 270-278
◇ D15 D80 ◇ REV MR 82h:68065 Zbl 454 # 68029 • ID 81196

EVEN, S. & LONG, T.J. & YACOBI, Y. [1982] *A note on deterministic and nondeterministic time complexity* (J 0194) Inform & Control 55*117-124
◇ D15 ◇ REV MR 85e:68022 Zbl 543 # 03026 • ID 40917

EVLADENKO, V.N. [1980] *A recognition algorithm for one class of quasirecursive parametric grammars (Russian)* (J 2605) Programmirovanie 1980/5*26-30
• TRANSL [1980] (J 2604) Progr Comput Software 6*242-246
◇ D05 ◇ REV MR 84e:68084 Zbl 464 # 68083 • ID 69543

EZAWA, Y. [1980] *Structured control sets on grammars* (1111) Preprints, Manuscr., Techn. Reports etc. 21*141-150
◇ D05 ◇ REV Zbl 441 # 68079 • REM Kansai University. Osaka • ID 69545

EZAWA, Y. see Vol. II for further entries

FABIAN, R.J. & KENT, C.F. [1969] *Recursive functions defined by ordinal recursion* (J 0053) Proc Amer Math Soc 23*206-210
◇ D20 ◇ REV MR 40 # 2539 Zbl 215.323 • ID 03637

FABREGAT, F. [1978] see AGUILO, J.

FACHINI, E. & MAGGIOLO-SCHETTINI, A. [1979] *A hierarchy of primitive recursive sequence functions* (J 3441) RAIRO Inform Theor 13*49-67
◇ D20 ◇ REV MR 81i:03062 Zbl 402 # 03041 • ID 54688

FACHINI, E. & MAGGIOLO-SCHETTINI, A. [1982] *Comparing hierarchies of primitive recursive sequence functions* (J 0068) Z Math Logik Grundlagen Math 28*431-445
◇ D20 ◇ REV MR 84g:03058 Zbl 497 # 03031 • ID 38111

FACHINI, E. & NAPOLI, M. [1985] *Synchronized bottom-up tree automata and L-systems* (P 4627) CAAP'85 Arbres en Algeb & Progr (10);1985 Berlin 298-307
◇ D05 ◇ ID 49561

FAGIN, R. [1974] *Generalized first-order spectra and polynomial-time recognizable sets* (P 0761) Compl of Computation;1973 New York 43-73
◇ C13 C85 D15 ◇ REV MR 51 # 7840 Zbl 303 # 68035 • ID 18140

FAGIN, R. & KLAWE, M.M. & PIPPENGER, N.J. & STOCKMEYER, L.J. [1985] *Bounded depth, polynomial-size circuits for symmetric functions* (J 1426) Theor Comput Sci 36*239-250
◇ D15 ◇ ID 47698

FAGIN, R. see Vol. I, III, V for further entries

FAHMY, M.H. [1972] *Equational calculus for primitive recursive rational-valued functions (Russian)* (S 0228) Zap Nauch Sem Leningrad Otd Mat Inst Steklov 32*116-120,157
• TRANSL [1976] (J 1531) J Sov Math 6*444-448
◇ D20 D75 ◇ REV MR 51 # 2891 Zbl 345 # 02028 • ID 17337

FAHMY, M.H. [1977] *Equation calculi for Grzegorczyk's classes $\mathfrak{E}^n$ (Russian)* (S 0228) Zap Nauch Sem Leningrad Otd Mat Inst Steklov 68*140-141,147
• TRANSL [1981] (J 1531) J Sov Math 15*77-78
◇ D20 ◇ REV MR 58 # 21547 Zbl 358 # 02043 • ID 50497

FAHMY, M.H. [1982] *A formal proof of Ackermann's theorem (Arabic summary)* (J 0397) Proc Math Phys Soc Egypt 51*1-6
◇ D20 F30 ◇ REV MR 84f:03039 Zbl 572 # 03019 • ID 34459

FAJTLOWICZ, S. [1980] see EHRENFEUCHT, A.

FAJTLOWICZ, S. see Vol. III, V for further entries

FAL'K, V.N. & KUTEPOV, V.P. [1979] *Functional systems: theoretical and practical aspects (Russian)* (J 0040) Kibernetika, Akad Nauk Ukr SSR 1979/1*46-58
• TRANSL [1979] (J 0021) Cybernetics 15*51-66
◇ D20 ◇ REV MR 80j:68023 Zbl 443 # 68024 • ID 81984

FAL'K, V.N. see Vol. I for further entries

FALKENBERG, B. [1980] *Halteprobleme von Fang-Systemen (tag systems)* (J 0009) Arch Math Logik Grundlagenforsch 20*75-83
◇ D03 D25 ◇ REV MR 81i:03060 Zbl 436 # 03037 • ID 72790

FALKINGER, J. [1980] *Reduzierbarkeit von berechenbaren Numerierungen von P_1* (J 0068) Z Math Logik Grundlagen Math 26*445-458
◇ D20 D30 D45 ◇ REV MR 82b:03081a Zbl 449 # 03038 • ID 56705

FALKINGER, J. [1980] *Universalitaet von berechenbaren Numerierungen von partiell rekursiven Funktionen* (J 0068) Z Math Logik Grundlagen Math 26*523-528
◇ D20 D30 D45 ◇ REV MR 82b:03081b Zbl 449 # 03039 • ID 56706

FAL'KOVICH, M.A. & GALIN, A.B. [1980] *Diagnosis of memoryless digital automata using a modified technique of natural logical deduction (Russian)* (J 0011) Avtom Telemekh 1980/1*131-137
• TRANSL [1980] (J 0010) Autom & Remote Control 41*104-109
◇ B70 D05 F99 ◇ REV Zbl 439 # 94032 • ID 69607

FANG, ZHIXI [1982] *The closure property of certain classes of languages under bi-language forms (Chinese) (English summary)* (J 3793) Jisuanjii Xuebao 5*22-28
◇ D05 D15 D25 ◇ REV MR 84e:68096 • ID 39315

FANTANEANU, B. [1978] see CALUDE, C.

FARADZHEV, R.G. [1980] *Design equations for linear sequential machines* (J 0011) Avtom Telemekh 1980/9*81-90
• TRANSL [1980] (J 0010) Autom & Remote Control 41*1253-1260
◇ D05 ◇ REV MR 83b:68060 • ID 38975

FARAT, V.M. & KOZHEVNIKOVA, G.P. [1979] *Algorithms for translation of algebraic expressions into an improved parenthesis-free notation and an analysis of their effectiveness(Russian)* (C 3018) Vopr Anal Vychisl Slozhnosti Algor 23-34,47
◇ B03 D15 ◇ REV MR 82g:68044 • ID 81930

FARKAS, E. [1975] *On the evaluation of nested partial functions* (J 0129) Elektr Informationsverarbeitung & Kybern 11*264-267
◇ D15 D20 ◇ REV Zbl 327 # 68040 • ID 61632

FARMER, F.D. [1982] *Homotopy spheres in formal language* (J 0548) Stud Appl Math 66*171-179
◇ D05 ◇ REV MR 83j:68087 Zbl 511 # 55009 • ID 39909

FARRINGTON, PADDY [1984] *The first-order theory of the c-degrees* (J 0068) Z Math Logik Grundlagen Math 30*437-446
◇ D30 D35 E45 E55 F35 ◇ REV MR 86a:03057 Zbl 565 # 03025 • ID 42273

FARRINGTON, PATRICK [1982] *Constructible lattices of c-degrees* (J 0036) J Symb Logic 47*739-754
◇ D30 E40 E45 ◇ REV MR 85d:03104 Zbl 501 # 03033 • ID 36949

FARRINGTON, PATRICK [1982] *The first-order theory of the c-degrees with the #-operation* (J 0068) Z Math Logik Grundlagen Math 28*487-493
◇ D30 E45 E55 ◇ REV MR 85f:03048 Zbl 501 # 03034 • ID 36950

FARRINGTON, PATRICK [1983] *Hinges and automorphisms of the degrees of non-constructibility* (**J** 3172) J London Math Soc, Ser 2 28*193-202
◇ D30 E45 ◇ REV MR 85g:03075 Zbl 493 # 03024
• ID 37393

FASS, L.F. [1983] *Learning context-free languages from their structured sentences* (**J** 1456) SIGACT News 15/3*24-35
◇ D05 ◇ REV Zbl 528 # 68051 • ID 36763

FEAUTRIER, P. [1975] *Post grammars as a programming language description tool* (**J** 4698) Rev Franc Autom, Inf & Rech Operat, Ser Rouge Inf Th 9/R-1*43-72
◇ B75 D03 ◇ REV MR 57 # 18243 Zbl 328 # 68005
• ID 61636

FEENEY, W.J. [1954] *Certain unsolvable problems in the theory of cancellation semi-groups* (**X** 0883) Cath Univ Amer Pr: Washington
◇ D40 ◇ ID 17186

FEFERMAN, S. [1957] *Degrees of unsolvability associated with classes of formalized theories* (**J** 0036) J Symb Logic 22*161-175
◇ D25 D35 ◇ REV MR 20 # 5130 Zbl 78.6 JSL 27.85
• ID 03682

FEFERMAN, S. [1959] see EHRENFEUCHT, A.

FEFERMAN, S. [1960] *Arithmetization of metamathematics in a general setting* (**J** 0027) Fund Math 49*35-92
◇ D55 F07 F25 F30 F40 ◇ REV MR 26 # 4913
Zbl 95.243 JSL 31.269 • ID 03686

FEFERMAN, S. [1962] *Classifications of recursive functions by means of hierarchies* (**J** 0064) Trans Amer Math Soc 104*101-122
• TRANSL [1971] (**S** 1582) Matematika - Period Sb Perevodov Inostran Statej 15/6*137-158
◇ D20 ◇ REV MR 26 # 22 Zbl 106.6 JSL 30.388
• ID 03687

FEFERMAN, S. & SPECTOR, C. [1962] *Incompleteness along paths in progressions of theories* (**J** 0036) J Symb Logic 27*383-390
• TRANSL [1971] (**S** 1582) Matematika - Period Sb Perevodov Inostran Statej 15/6*159-166 (Russian)
◇ D55 F15 F30 ◇ REV MR 30 # 3012 Zbl 117.257
JSL 32.531 • ID 03688

FEFERMAN, S. [1962] *Transfinite recursive progressions of axiomatic theories* (**J** 0036) J Symb Logic 27*259-316
• TRANSL [1979] (**C** 3351) Teorem Goedel & Hipotese Cont 573-754 (Portuguese)
◇ D55 F15 F30 ◇ REV MR 30 # 3011 Zbl 117.254
JSL 32.530 • ID 03693

FEFERMAN, S. [1965] *Some applications of the notions of forcing and generic sets* (**J** 0027) Fund Math 56*325-345
• REPR [1965] (**P** 0614) Th Models;1963 Berkeley 89-95
◇ C62 D55 E25 E35 E40 E45 F35 ◇ REV
MR 31 # 1193 MR 34 # 2439 Zbl 129.264 JSL 37.612
• REM Reprint is a summary • ID 03696

FEFERMAN, S. [1975] *A language and axioms for explicit mathematics* (**P** 0765) Algeb & Log;1974 Clayton 87-139
◇ D55 F35 F50 F65 ◇ REV MR 53 # 12899
Zbl 357 # 02029 JSL 49.308 • ID 23151

FEFERMAN, S. [1975] *Impredicativity of the existence of the largest divisible subgroup of an abelian p-group* (**C** 0782) Model Th & Algeb (A. Robinson) 117-130
◇ C57 C60 F65 ◇ REV MR 53 # 5274 Zbl 316 # 02058
• ID 22905

FEFERMAN, S. [1977] *Generating schemes for partial recursively continuous functionals (summary)* (**P** 1729) Colloq Int Log;1975 Clermont-Ferrand 191-198
◇ D65 D70 ◇ REV Zbl 446 # 03035 • ID 56564

FEFERMAN, S. [1977] *Inductive schemata and recursively continuous functionals* (**P** 1075) Logic Colloq;1976 Oxford 373-392
◇ D65 D70 D75 ◇ REV MR 57 # 12195 Zbl 422 # 03023
• ID 16629

FEFERMAN, S. [1977] *Recursion in total functionals of finite type* (**J** 0020) Compos Math 35*3-22
◇ D65 ◇ REV MR 58 # 5129 Zbl 365 # 02030 • ID 27168

FEFERMAN, S. [1977] *Theories of finite type related to mathematical practice* (**C** 1523) Handb of Math Logic 913-971
◇ B15 D55 D65 F10 F15 F35 F98 ◇ REV
MR 58 # 10343 JSL 49.980 • ID 27330

FEFERMAN, S. [1978] *Recursion theory and set theory: a marriage of convenience* (**P** 1628) Generalized Recursion Th (2);1977 Oslo 55-98
◇ D75 E30 E70 F65 ◇ REV MR 81i:03074
Zbl 453 # 03047 • ID 72806

FEFERMAN, S. see Vol. I, II, III, V, VI for further entries

FEHR, E. [1978] see DAMM, W.

FEHR, E. see Vol. II, VI for further entries

FEINER, L. [1970] *Hierarchies of boolean algebras* (**J** 0036) J Symb Logic 35*365-374
◇ C57 D45 G05 ◇ REV MR 44 # 39 Zbl 222 # 02048
• ID 03711

FEINER, L. [1970] *The strong homogeneity conjecture* (**J** 0036) J Symb Logic 35*375-377
◇ D30 ◇ REV MR 44 # 3864 Zbl 251 # 02042 • ID 03713

FEINER, L. [1973] *Degrees of nonrecursive presentability* (**J** 0053) Proc Amer Math Soc 38*621-624
◇ C57 D30 D45 ◇ REV MR 48 # 5836 Zbl 266 # 02021
• ID 03714

FEJER, P.A. & SOARE, R.I. [1981] *The plus-cupping theorem for the recursively enumerable degrees* (**P** 2628) Log Year;1979/80 Storrs 49-62
◇ D25 ◇ REV MR 82j:03049 Zbl 498 # 03029 • ID 72825

FEJER, P.A. [1982] *Branching degrees above low degrees* (**J** 0064) Trans Amer Math Soc 273*157-180
◇ D25 ◇ REV MR 84a:03044 Zbl 498 # 03028 • ID 33589

FEJER, P.A. [1983] *The density of the nonbranching degrees* (**J** 0073) Ann Pure Appl Logic 24*113-130
◇ D25 ◇ REV MR 85g:03062 Zbl 521 # 03026 • ID 33590

FEJER, P.A. & SHORE, R.A. [1985] *Embeddings and extensions of embeddings in the r.e. tt and wtt - degrees* (**P** 3342) Rec Th Week;1984 Oberwolfach 121-140
◇ D25 ◇ ID 45300

FEJNBERG, V.Z. & TYSHKEVICH, R.I. [1971] *The permutability of transformations I (Russian)* (**J** 0413) Izv Akad Nauk Belor SSR, Ser Fiz-Mat 2*5-13
◇ B75 C05 D05 ◇ REV MR 45 # 2052 Zbl 224 # 20063 • ID 13755

FEJNBERG, V.Z. see Vol. V for further entries

FEKLICHEV, A.V. [1984] *Program schemes with storage and markers and their decision properties (Russian)* (**J** 0216) Izv Akad Nauk SSSR, Ser Mat 48*1060-1077
• TRANSL [1985] (**J** 0448) Math of USSR, Izv 25*375-390
◇ D20 ◇ ID 46855

FELDMAN, E.D. [1971] *An extension of the hyperprojective hierarchy* (**J** 0068) Z Math Logik Grundlagen Math 17*395-409
◇ D55 E15 ◇ REV MR 45 # 3206 Zbl 233 # 02018 • ID 03716

FELDMAN, E.D. [1972] *Some properties of the extended hyperprojective hierarchy* (**J** 0068) Z Math Logik Grundlagen Math 18*55-60
◇ D55 E15 ◇ REV MR 45 # 4973 Zbl 265 # 02030 • ID 03717

FELDMAN, E.D. & OWINGS JR., J.C. [1973] *A class of universal linear bounded automata* (**J** 0191) Inform Sci 6*187-190
◇ D10 ◇ REV MR 48 # 10707 Zbl 268 # 94043 • ID 61655

FELDMAN, E.D. [1975] *L-Σ_n^1 transfinite induction with an application to the EHP hierarchy* (**J** 0068) Z Math Logik Grundlagen Math 21*463-471
◇ D55 D70 E15 E47 ◇ REV MR 52 # 10395 Zbl 343 # 02033 • ID 04234

FELDMANN, H. & OBERQUELLE, H. & ORTLIEB, C. [1971] *Eine einfache universelle Turingmaschine in ALGOL 60 Simulation* (**J** 0373) Comp Arch Inform & Numerik 8*241-249
◇ B75 D10 ◇ REV Zbl 256 # 02016 • ID 61657

FELEA, V. [1980] *Relational VT-algebras over free theories* (**J** 0070) Bull Soc Sci Math Roumanie, NS 24(72)*239-245
◇ D05 ◇ REV MR 82j:68034 Zbl 448 # 68015 • ID 81210

FELGNER, U. [1980] *The model theory of FC-groups* (**P** 2958) Latin Amer Symp Math Log (4);1978 Santiago 163-190
◇ C35 C45 C60 D35 ◇ REV MR 81m:03045 Zbl 437 # 03014 • ID 55882

FELGNER, U. see Vol. III, V for further entries

FENG, JINGDONG [1965] *The inverse machine of Turing machines (Chinese)* (**P** 4564) Math Logic;1963 Xi-An 115
◇ D10 ◇ ID 49343

FENSTAD, J.E. [1964] *Algorithms in mathematics: an introduction to recursion theory and its applications I,II (Norwegian) (English summary)* (**J** 0311) Nordisk Mat Tidskr 12*17-35,99-111
◇ D20 D98 ◇ REV MR 33 # 39 MR 33 # 40 Zbl 125.278 • ID 17151

FENSTAD, J.E. [1965] *A universal function for the class of all primitive recursive functions* (**J** 0121) Kon Norske Vidensk Selsk Forh 38*88-93
◇ D20 ◇ REV MR 34 # 7366 Zbl 156.9 • ID 33218

FENSTAD, J.E. [1971] *Hilbert's 10th problem (Norwegian) (English summary)* (**J** 0311) Nordisk Mat Tidskr 19*5-14,60
◇ D25 D35 ◇ REV MR 48 # 85 Zbl 264 # 02043 • ID 17154

FENSTAD, J.E. [1971] *The axiom of determinateness* (**P** 0604) Scand Logic Symp (2);1970 Oslo 41-61
◇ D55 E55 E60 E98 ◇ REV MR 48 # 10806 Zbl 222 # 02076 JSL 39.331 • ID 03746

FENSTAD, J.E. [1973] *On axiomatizing recursion theory* (**S** 1626) Oslo Preprint Ser 7*i+24pp
◇ D65 D75 ◇ REV MR 52 # 13346 • ID 72859

FENSTAD, J.E. & HINMAN, P.G. (EDS.) [1974] *Generalized recursion theory* (**X** 0809) North Holland: Amsterdam VIII+460pp
◇ D97 ◇ REV MR 49 # 2306 Zbl 272 # 00006 • ID 23498

FENSTAD, J.E. & NORMANN, D. [1974] *On absolutely measurable sets* (**J** 0027) Fund Math 81*91-98
◇ D55 E15 E40 E75 ◇ REV MR 49 # 3065 Zbl 275 # 02057 • ID 03747

FENSTAD, J.E. [1974] *On axiomatizing recursion theory* (**P** 0602) Generalized Recursion Th (1);1972 Oslo 385-404
◇ D75 ◇ REV MR 54 # 79 Zbl 284 # 02019 • ID 24566

FENSTAD, J.E. [1975] *Computation theories: an axiomatic approach to recursion on general structures* (**P** 1442) ⊧ ISILC Logic Conf;1974 Kiel 143-168
◇ D75 ◇ REV MR 55 # 2538 Zbl 328 # 02025 • ID 27211

FENSTAD, J.E. [1977] *Between recursion theory and set theory* (**P** 1075) Logic Colloq;1976 Oxford 392-406
◇ D60 D65 D75 ◇ REV MR 57 # 12193 Zbl 417 # 03020 • ID 16630

FENSTAD, J.E. & GANDY, R.O. & SACKS, G.E. (EDS.) [1978] *Generalized recursion theory II* (**X** 0809) North Holland: Amsterdam vii+466pp
◇ C62 C70 C80 D97 ◇ REV MR 80a:03004 Zbl 453 # 03047 • ID 54177

FENSTAD, J.E. [1978] *On the foundation of general recursion theory: computations versus inductive definability* (**P** 1628) Generalized Recursion Th (2);1977 Oslo 99-110
◇ D70 D75 ◇ REV MR 80g:03047 Zbl 453 # 03047 • ID 72855

FENSTAD, J.E. [1980] *General recursion theory. An axiomatic approach* (**X** 0811) Springer: Heidelberg & New York xi+225pp
◇ D75 D98 ◇ REV MR 83h:03065 Zbl 439 # 03030 JSL 47.696 • ID 56021

FENSTAD, J.E. [1981] *What does axiomatic recursion theory axiomatize?* (**P** 3092) Congr Naz Logica;1979 Montecatini Terme 91-105
◇ D75 ◇ ID 33225

FENSTAD, J.E. see Vol. I, II, III, V, VI for further entries

FENZL, W. [1973] see CASPAR, K.

FERGUSON, D.C. [1968] *Infinite products of recursive equivalence types* (**J** 0036) J Symb Logic 33*221-230
◇ D50 ◇ REV MR 39 # 60 Zbl 165.316 JSL 35.590 • ID 03749

FERMENT, D. [1984] *Principality results about some matrix languages families* (**P** 4012) Automata, Lang & Progr (11);1984 Antwerpen 151-161
◇ D05 ◇ REV MR 86e:68057 Zbl 551 # 68063 • ID 47140

FERNANDEZ-PRIDA, J. [1976] *Creative and productive recursive theories (Spanish)* (**P** 1619) Coloq Log Simb;1975 Madrid 51-66
◇ D25 ◇ REV MR 56 # 8343 Zbl 362 # 02031 • ID 50753

FERNANDEZ-PRIDA, J. also published under the name PRIDA, J.F.

FERRANTE, J. & RACKOFF, C.W. [1975] *A decision procedure for the first order theory of real addition with order* (**J** 1428) SIAM J Comp 4*69-76
◇ B25 C10 C60 D15 F35 ◇ REV MR 52 # 10403 Zbl 277 # 02010 • ID 21702

FERRANTE, J. & GEISER, J.R. [1977] *An efficient decision procedure for the theory of rational order* (**J** 1426) Theor Comput Sci 4*227-233
◇ B25 D15 ◇ REV MR 56 # 13777 Zbl 372 # 02024 • ID 51432

FERRANTE, J. & RACKOFF, C.W. [1979] *The computational complexity of logical theories* (**S** 3301) Lect Notes Math 718*x+243pp
◇ B25 C98 D10 D15 ◇ REV MR 81d:03013 Zbl 404 # 03028 JSL 49.670 • ID 54815

FERRARI, P.L. & LONGO, G. [1978] *Axiomatic theory of enumeration: a note on the axiom of extensionality* (**J** 0063) Studia Logica 37*261-268
◇ D75 ◇ REV MR 80k:03049 Zbl 384 # 03023 • ID 30731

FERRARI, P.L. see Vol. I for further entries

FERRO, A. & MICALE, B. [1979] *Sulla teoria della computabilita* (**J** 1515) Archimede 31*149-163
◇ D20 D25 ◇ REV Zbl 423 # 03049 • ID 53561

FERRO, A. see Vol. I, III, V for further entries

FICH, F.E. [1980] see BRZOZOWSKI, J.A.

FILE, G. [1980] see ENGELFRIET, J.

FILE, G. [1981] see ENGELFRIET, J.

FILE, G. [1985] *Tree automata and logic programs* (**P** 4241) Symp of Th Aspects of Comput Sci (2);1985 Saarbruecken 119-130
◇ B75 D05 ◇ ID 45611

FILIPPOV, V.P. [1969] *Certain separability and nonseparability theorems in the second class of projective sets (Russian)* (**J** 0339) Uch Zap Ped Inst, Volgograd 23*153-166
◇ D55 E15 ◇ REV MR 46 # 65 • ID 17174

FILIPPOV, V.P. [1969] *The multiple separability in classes of arithmetical hierarchy (Russian)* (**J** 0339) Uch Zap Ped Inst, Volgograd 23*167-212
◇ D55 ◇ REV MR 46 # 37 • ID 17173

FILIPPOV, V.P. see Vol. V for further entries

FILOTTI, I. [1972] *On effectively levelable sets* (**S** 1585) Rec Fct Th Newsletter 2*12-13
◇ D15 ◇ ID 27054

FINANCE, J.P. [1979] see BROY, M.

FINE, B. [1974] *The structure of $PSL_2(R)$; R, the ring of integers in a Euclidian quadratic imaginary number field* (**P** 1219) Discont Groups & Riemann Surfaces;1973 College Park 145-170
◇ D40 ◇ REV MR 50 # 4776 Zbl 298 # 20036 • ID 20995

FINE, T.L. [1976] *A computational complexity viewpoint on the stability of relative frequency and on the stochastic independence* (**P** 2411) Found Probab Th, Stat Inf & Stat Th Sci;1973 London ON 1*29-40
◇ D15 ◇ REV MR 58 # 13240 Zbl 364 # 60009 • ID 50982

FINE, T.L. see Vol. I, II for further entries

FINKELSTEIN, M. [1969] see CANNONITO, F.B.

FINKELSTEIN, M. [1971] *Two conjugate primitive permutations not conjugate by a primitive recursive permutation* (**J** 0068) Z Math Logik Grundlagen Math 17*1-3
◇ D20 ◇ REV MR 43 # 6089 Zbl 228 # 02028 • ID 03791

FINSLER, P. [1926] *Formale Beweise und die Entscheidbarkeit* (**J** 0044) Math Z 25*676-682
• TRANSL [1967] (**C** 0675) From Frege to Goedel 440-445
◇ A05 D35 F25 ◇ REV FdM 52.49 • ID 03793

FINSLER, P. see Vol. I, V, VI for further entries

FIRMANI, B. [1973] *Su una caratterizzazione degli F_α per ogni ordinale α di $I^\wedge$ specie e di cardinalita $\leqslant \aleph_0$ (French summary)* (**J** 2311) Rend Mat, Ser 6 6*351-359
◇ D55 E15 ◇ REV MR 50 # 7441 Zbl 277 # 04005 • ID 29531

FISCHER, MICHAEL J. & RABIN, M.O. [1974] *Super-exponential complexity of Presburger arithmetic* (**P** 0761) Compl of Computation;1973 New York 27-41
◇ B25 D15 F20 F30 ◇ REV MR 51 # 2893 Zbl 319 # 68024 • ID 17344

FISCHER, MICHAEL J. & LYNCH, N.A. & MEYER, A.R. [1976] *Relativization of the theory of computational complexity* (**J** 0064) Trans Amer Math Soc 220*243-287
◇ D15 D20 ◇ REV MR 53 # 7742 Zbl 353 # 68059 • ID 22990

FISCHER, MICHAEL J. & MEYER, A.R. & SEIFERAS, J.I. [1978] *Separating nondeterministic time complexity classes* (**J** 0037) ACM J 25*146-167
◇ D10 D15 ◇ REV MR 57 # 4627 Zbl 366 # 68038 • ID 51146

FISCHER, MICHAEL J. & LADNER, R.E. [1979] *Propositional dynamic logic of regular programs* (**J** 0119) J Comp Syst Sci 18*194-211
◇ B75 D15 ◇ REV MR 80f:68013 Zbl 408 # 03014 • ID 56253

FISCHER, MICHAEL J. & PIPPENGER, N.J. [1979] *Relations among complexity measures* (**J** 0037) ACM J 26*361-381
◇ D05 D15 ◇ REV MR 80f:68052 Zbl 405 # 68041 • ID 82487

FISCHER, MICHAEL J. & MEYER, A.R. & PATERSON, M.S. [1982] *$\Omega(n \log n)$ lower bounds on length of Boolean formulas* (**J** 1428) SIAM J Comp 11*416-427
◇ B05 D15 ◇ REV MR 83j:03065 Zbl 488 # 94036 • ID 35363

FISCHER, MICHAEL J. see Vol. I for further entries

FISCHER, P.C. [1963] *A note on bounded-truth-table reducibility* (**J** 0053) Proc Amer Math Soc 14*875-877
◇ D30 ◇ REV MR 27 # 5691 Zbl 124.246 JSL 35.478 • ID 04299

FISCHER, P.C. [1965] *Generation of primes by a one-dimensional real-time iterative array* (**J** 0037) ACM J 12*388-394
◇ D05 ◇ REV MR 32 # 3699 Zbl 173.191 • ID 33942

FISCHER, P.C. & RUBY, S.S. [1965] *Multi-tape and infinite-state automata, a survey* (**J** 0212) ACM Commun 8*799-805
◇ D05 ◇ REV Zbl 176.281 • ID 33943

FISCHER, P.C. [1965] *On formalisms for Turing machines* (**J** 0037) ACM J 12*570-580
◇ D10 ◇ REV MR 33 # 53 Zbl 154.416 JSL 36.532 • ID 04300

FISCHER, P.C. [1965] *Theory of provably recursive functions* (**J** 0064) Trans Amer Math Soc 117*494-520
◇ D20 D30 F30 ◇ REV MR 31 # 47 Zbl 129.260 JSL 32.270 • ID 04301

FISCHER, P.C. & RUBY, S.S. [1965] *Translational methods and computational complexity* (**P** 1591) Switch Circ Th & Log Design (6);1965 Ann Arbor 173-178
• TRANSL [1970] (**C** 1540) Probl Mat Log: Slozh Algor & Kl Vychisl Funk 213-222
◇ B70 D15 D20 ◇ REV Zbl 257 # 68037 JSL 39.193 • ID 20894

FISCHER, P.C. [1966] *Turing machines with restricted memory access* (**J** 0194) Inform & Control 9*364-379
◇ D10 ◇ REV MR 33 # 7211 Zbl 145.242 • ID 33944

FISCHER, P.C. [1967] see AANDERAA, S.O.

FISCHER, P.C. [1967] *Turing machines with a schedule to keep* (**J** 0194) Inform & Control 11*138-146
◇ D10 ◇ REV Zbl 153.317 • ID 33945

FISCHER, P.C. & MEYER, A.R. & ROSENBERG, ARNOLD L. [1968] *Counter machines and counter languages* (**J** 0041) Math Syst Theory 2*265-283
• TRANSL [1970] (**C** 1540) Probl Mat Log: Slozh Algor & Kl Vychisl Funk 380-400
◇ D05 D10 D15 D20 ◇ REV MR 38 # 4233 Zbl 165.320 • ID 20875

FISCHER, P.C. & ROSENBERG, ARNOLD L. [1968] *Multitape one-way nonwriting automata* (**J** 0119) J Comp Syst Sci 2*88-101
◇ D05 ◇ REV MR 39 # 8021 Zbl 159.15 • ID 33948

FISCHER, P.C. & MEYER, A.R. [1968] *On computational speed-up* (**P** 1900) IEEE Symp Switch & Automata Th (9);1968 Schenectady 351-355
◇ D15 ◇ ID 33678

FISCHER, P.C. & MEYER, A.R. & ROSENBERG, ARNOLD L. [1968] *Real-time counter machines* (**P** 1899) IEEE Symp Switch & Automata Th (8);1967 Austin 148-154
◇ D05 ◇ ID 33947

FISCHER, P.C. [1968] *Reduction of tape reversals for off-line one-type Turing machines* (**J** 0119) J Comp Syst Sci 2*136-147
◇ D10 ◇ REV MR 39 # 6688 Zbl 199.310 • ID 33946

FISCHER, P.C. [1968] see BLUM, M.

FISCHER, P.C. [1969] *Quantificational variants on the halting problem for Turing machines* (**J** 0068) Z Math Logik Grundlagen Math 15*211-218
◇ D10 ◇ REV MR 40 # 44 Zbl 193.317 JSL 36.532 • ID 04302

FISCHER, P.C. & MEYER, A.R. & ROSENBERG, ARNOLD L. [1970] *Time-restricted sequence generation* (**J** 0119) J Comp Syst Sci 4*50-73
◇ D10 D15 ◇ REV MR 40 # 6808 Zbl 191.183 JSL 40.616 • ID 14468

FISCHER, P.C. & MEYER, A.R. [1972] *Computational speed-up by effective operators* (**J** 0036) J Symb Logic 37*55-68
◇ D15 ◇ REV MR 47 # 6457 Zbl 249 # 68018 • ID 09177

FISCHER, P.C. & WARKENTIN, J.C. [1972] *Predecessor machines and regressing functions* (**P** 1901) ACM Symp Th of Comput (4);1972 Denver 81-87
◇ B70 D20 ◇ REV MR 49 # 4291 Zbl 357 # 68068 • ID 33950

FISCHER, P.C. & MEYER, A.R. & ROSENBERG, ARNOLD L. [1972] *Real-time simulation of multihead tape units* (**J** 0037) ACM J 19*590-607
◇ D10 D15 ◇ REV MR 49 # 8452 Zbl 261 # 68027 • ID 04303

FISCHER, P.C. & IRLAND, M.I. [1973] *A bibliography on computational complexity* (**P** 1584) Comput Compl - Courant Comput Sci (7);1971 New York 187-268
◇ A10 D15 D99 ◇ ID 33953

FISCHER, P.C. [1973] *Trends in computational complexity theory* (**P** 1584) Comput Compl - Courant Comput Sci (7);1971 New York 1-22
◇ D15 ◇ ID 33951

FISCHER, P.C. [1974] *Further schemes for combining matrix algorithms* (**P** 1869) Automata, Lang & Progr (2);1974 Saarbruecken 428-436
◇ D15 ◇ REV MR 55 # 1807 Zbl 284 # 68040 • ID 33955

FISCHER, P.C. & WARKENTIN, J.C. [1974] *Predecessor machines* (**J** 0119) J Comp Syst Sci 8*190-219
◇ D05 ◇ REV MR 49 # 4291 Zbl 278 # 68048 • ID 33952

FISCHER, P.C. [1974] *Ueber Formalismen fuer Turingmaschinen* (**C** 1902) Stud Th Automaten 365-382
◇ D10 ◇ ID 33954

FISCHER, P.C. & ROBERTSON, E.L. & SAXTON, L.V. [1976] *On the sequential nature of functions* (**J** 0119) J Comp Syst Sci 13*51-73
◇ A05 D05 ◇ REV MR 57 # 18185 Zbl 342 # 68027 • ID 33956

FISCHER, P.C. & KINTALA, C.M.R. [1977] *Computations with a restricted number of nondeterministic steps* (**P** 1903) ACM Symp Th of Comput (9);1977 Boulder 178-185
◇ D15 ◇ REV MR 57 # 18251 • ID 35604

FISCHER, P.C. [1978] *Computational complexity* (**P** 1962) Conf Inform Sci & Syst;1977 Baltimore 345-350
◇ D15 ◇ ID 35603

FISCHER, R. & UHLIG, D. [1982] *Fehlerkorrigierende Realisierungen zu Booleschen Funktionen und Automatenfunktionen mit linearer Kompliziertheit* (**S** 2829) Rostocker Math Kolloq 19*91-100
◇ B05 D05 ◇ REV MR 84c:94026 Zbl 484 # 94041 • ID 42341

FISHER, A. [1982] *Formal number theory and computability* (**X** 0894) Oxford Univ Pr: Oxford xiii+190pp
◇ B98 D20 F30 F98 ◇ REV MR 85g:03001 Zbl 504 # 03002 • ID 36967

FISHER, E.R. [1970] see BARWISE, J.

FISHER, E.R. see Vol. III for further entries

FITCH, F.B. [1968] *A note on recursive relations* (**J** 0036) J Symb Logic 33*107
◇ B40 D20 ◇ REV MR 38 # 4304 Zbl 184.19 JSL 37.758 • ID 17117

FITCH, F.B. & ORGASS, R.J. [1969] *A theory of computing machines* (**J** 0178) Stud Gen 22*83-104
◇ B40 D10 ◇ REV MR 43 # 2879 • ID 14672

FITCH, F.B. see Vol. I, II, V, VI for further entries

FITTING, M. [1978] *Elementary formal systems for hyperarithmetical relations* (J 0068) Z Math Logik Grundlagen Math 24*25-30
◇ C75 D55 D75 ◇ REV MR 80a:03059 Zbl 374 # 02024 • ID 31075

FITTING, M. [1979] *An axiomatic approach to computers* (J 0105) Theoria (Lund) 45*97-113
◇ B75 D80 ◇ REV MR 82k:03077 Zbl 476 # 68038 • ID 55590

FITTING, M. [1981] *Fundamentals of generalized recursion theory* (S 3303) Stud Logic Found Math 105*xx+307pp
◇ D60 D65 D70 D75 ◇ REV MR 84c:03086 • ID 34021

FITTING, M. [1982] *A generalization of elementary formal systems* (P 3831) Universal Algeb & Appl;1978 Warsaw 89-96
◇ D20 D70 D75 ◇ REV MR 85g:03071 Zbl 508 # 03020 • ID 36940

FITTING, M. see Vol. I, II, III, V, VI for further entries

FLAJOLET, P. & STEYAERT, J.-M. [1973] *Complexite des problemes de decision relatifs aux algorithmes de tri* (P 0763) Automata, Lang & Progr (1);1972 Rocquencourt 537-548
◇ D05 D15 ◇ REV MR 54 # 4183 Zbl 298 # 68038 • ID 61727

FLAJOLET, P. & STEYAERT, J.-M. [1973] *Decision problems for multihead finite automata* (P 1448) Math Founds of Comput Sci (2);1973 Strbske Pleso 225-230
◇ D05 ◇ REV MR 53 # 12080 • ID 23108

FLAJOLET, P. & STEYAERT, J.-M. [1974] *On sets having only hard subsets* (P 1869) Automata, Lang & Progr (2);1974 Saarbruecken 446-457
◇ D15 D25 ◇ REV MR 54 # 12501 • ID 72956

FLAJOLET, P. & STEYAERT, J.-M. [1974] *Une generalisation de la notion d'ensemble immune* (J 0205) Rev Franc Autom, Inf & Rech Operat 8/R-1*37-48
◇ B75 D20 D25 D50 ◇ REV MR 50 # 1858 Zbl 283 # 02034 • ID 04370

FLAJOLET, P. & STEYAERT, J.-M. [1976] *Hierarchies de complexite et reductions entre problemes* (P 2948) Journ Algor;1975 Paris 53-72
◇ D15 ◇ REV MR 56 # 15380 Zbl 361 # 68067 • ID 72955

FLAJOLET, P. [1985] *Ambiguity and transcendence* (P 4628) Automata, Lang & Progr (12);1985 Nafplion 179-188
◇ D05 ◇ ID 49562

FLEISCHHACK, H. [1984] see AMBOS-SPIES, K.

FLEISCHMANN, K. & MAHR, B. & SIEFKES, D. [1976] *Complexity of decision problems* (1111) Preprints, Manuscr., Techn. Reports etc. 76-09*67pp
◇ D15 ◇ REM Forschungsbericht FB 20, TU Berlin • ID 33902

FLEISCHMANN, K. & MAHR, B. & SIEFKES, D. [1977] *Bounded concatenation theory as a uniform method for proving lower complexity bounds* (P 1075) Logic Colloq;1976 Oxford 471-490
◇ D15 D20 ◇ REV MR 58 # 16242 Zbl 439 # 03004 • ID 16635

FLORENCE, J.B. [1967] *Infinite subclasses of recursively enumerable classes* (J 0053) Proc Amer Math Soc 18*633-639
◇ D25 ◇ REV MR 36 # 1322 Zbl 224 # 02032 • ID 04381

FLORENCE, J.B. [1969] *Partially ordered sets representable by recursively enumerable classes* (J 0036) J Symb Logic 34*8-12
◇ D25 ◇ REV MR 40 # 39 Zbl 188.26 • ID 04383

FLORENCE, J.B. [1969] *Strong enumeration properties of recursively enumerable classes* (J 0068) Z Math Logik Grundlagen Math 15*181-192
◇ D25 ◇ REV MR 43 # 3117 Zbl 188.26 • ID 04382

FLORENCE, J.B. [1975] see DAWES, A.M.

FLORENCE, J.B. [1975] *On splitting an infinite recursively enumerable class* (J 0017) Canad J Math 27*1127-1140
◇ D25 ◇ REV MR 53 # 2655 Zbl 313 # 02022 • ID 29618

FLOYD, R.W. [1961] *A note on mathematical induction on phrase structure grammars* (J 0194) Inform & Control 4*353-358
◇ B65 D05 ◇ REV Zbl 112.113 JSL 36.693 • ID 04385

FLOYD, R.W. [1963] *Syntactic analysis and operator precedence* (J 0037) ACM J 10*316-333
◇ B75 D05 ◇ REV Zbl 133.255 JSL 35.349 • ID 04386

FLOYD, R.W. see Vol. I for further entries

FLUM, J. [1976] *El problema de las palabras en la teoria de grupos* (P 1619) Coloq Log Simb;1975 Madrid 81-90
◇ C25 C60 D40 ◇ REV MR 56 # 8701 Zbl 362 # 02032 • ID 31078

FLUM, J. see Vol. I, II, III, V for further entries

FOELDES, S. & SINGHI, N.M. [1980] *A nonconstructible projective plane* (J 2722) Geom Dedicata 9*497-499
◇ D80 ◇ REV MR 82d:03052 • ID 72988

FOELDES, S. & STEINBERG, R. [1980] *A topological space for which graph embeddability is undecidable* (J 0033) J Comb Th, Ser B 29*342-344
◇ D80 ◇ REV MR 82d:03073 Zbl 461 # 05025 • ID 72990

FOELDES, S. [1980] *On the complexity of the lattice embeddability problem* (J 0467) Utilitas Math, Canad J 17*79-84
◇ D15 D40 G10 ◇ REV MR 81i:06010 Zbl 456 # 68047 • ID 69556

FOELDES, S. & SABIDUSSI, G. [1981] *Recursive undecidability of the binding property for finitely presented equational classes* (J 0004) Algeb Universalis 12*1-4
◇ C05 D35 ◇ REV MR 82e:08007 Zbl 458 # 08005 • ID 81245

FOELDES, S. see Vol. III, V for further entries

FOERSTER, M. & PIRLING, C. & VOELKEL, L. [1980] *Zweiregister-Akzeptoren mit unterschiedlichen Druckbefehlen* (J 0129) Elektr Informationsverarbeitung & Kybern 16*325-334
◇ D05 ◇ REV MR 82d:68050 Zbl 447 # 68095 • ID 42372

FORTUNE, S. & WYLLIE, J. [1978] *Parallelism in random access machines* (P 1740) ACM Symp Th of Comput (10);1978 San Diego 114-118
◇ D10 ◇ REV MR 81i:68036 • ID 81248

FORTUNE, S. & HOPCROFT, J.E. & SCHMIDT, ERIK MEINECHE [1978] *The complexity of equivalence and containment for free single variable program schemes* (**P** 1872) Automata, Lang & Progr (5);1978 Udine 227-240
◇ B75 D15 ◇ REV MR 80d:68049 Zbl 382 # 68021
• ID 69561

FORTUNE, S. [1979] *A note on sparse complete sets* (**J** 1428) SIAM J Comp 8*431-433
◇ D15 ◇ REV MR 80f:68045 Zbl 415 # 68006 • ID 69560

FORTUNE, S. see Vol. I for further entries

FORYS, M. [1981] *On some application of F-grammars to pattern description (Polish and Russian summaries)* (**J** 2814) Podstawy Sterowania 11*51-61
◇ D05 ◇ REV MR 82i:68049 Zbl 475 # 68050 • ID 81249

FORYS, M. [1981] *On F-languages and their grammars* (**S** 2350) Zesz Nauk, Prace Mat, Uniw Krakow 22*157-162
◇ D05 ◇ REV MR 82g:68069 Zbl 475 # 68049 • ID 81250

FORYS, W. [1981] *F-languages and F-automata* (**S** 2350) Zesz Nauk, Prace Mat, Uniw Krakow 22*151-156
◇ D05 ◇ REV MR 82g:68068 Zbl 475 # 68048 • ID 81251

FORYS, W. see Vol. V for further entries

FOTHI, A. & VARGA, Z. [1980] *Modeling programs via recursive functions I (Hungarian) (English summary)* (**J** 1458) Alkalmaz Mat Lapok 6*331-336
◇ B75 D80 ◇ REV MR 83e:68006 Zbl 489 # 68008
• ID 40154

FOWLER III, N. [1975] *Intersections of α-spaces* (**J** 0038) J Austral Math Soc 20*398-418
◇ C57 C60 D50 ◇ REV MR 52 # 5397 Zbl 317 # 02049
• ID 15238

FOWLER III, N. [1975] *Sum of α-spaces* (**J** 0047) Notre Dame J Formal Log 16*379-388
◇ C57 C60 D45 ◇ REV MR 51 # 10065 Zbl 271 # 02030
• ID 17130

FOWLER III, N. [1976] *α-decompositions of α-spaces* (**J** 0036) J Symb Logic 41*483-488
◇ C57 C60 D45 D50 ◇ REV MR 55 # 92
Zbl 344 # 02035 • ID 14783

FOWLER III, N. [1978] *Effective inner product spaces* (**J** 0047) Notre Dame J Formal Log 19*693-701
◇ C57 C60 D45 D80 ◇ REV MR 80a:03056
Zbl 368 # 02044 • ID 52144

FOX, C.D. [1975] *Commutators in orderable groups* (**J** 0394) Commun Algeb 3*213-217
◇ D40 ◇ REV MR 51 # 7984 Zbl 303 # 06012 • ID 17584

FRAISSE, R. [1961] *Une notion de recursivite relative* (**P** 0633) Infinitist Meth;1959 Warsaw 323-328
◇ D20 D30 D75 ◇ REV MR 32 # 3999 JSL 32.395
• ID 04524

FRAISSE, R. [1975] *Cours de logique mathematique. Tome 3: Recursivite et constructibilite* (**X** 0834) Gauthier-Villars: Paris vi+137pp
◇ D25 D98 E35 E45 E98 ◇ REV MR 56 # 5190
Zbl 308 # 02001 • ID 61773

FRAISSE, R. [1982] *Recursivite* (**P** 3753) Penser Math;1981 Paris 126-128
◇ D20 ◇ REV MR 83j:00025 • ID 45898

FRAISSE, R. see Vol. I, II, III, V for further entries

FRANCEZ, N. & KLEBANSKY, B. & PNUELI, A. [1977] *Backtracking in recursive computations* (**J** 1431) Acta Inf 8*125-144
◇ B75 D20 ◇ REV MR 56 # 4218 Zbl 359 # 68023
• ID 81254

FRANCEZ, N. see Vol. II for further entries

FRANCO, J. & PAULL, M. [1983] *Probabilistic analysis of the Davis-Putnam procedure for solving the satisfiability problem* (**J** 2702) Discr Appl Math 5*77-87
◇ B35 D15 ◇ REV MR 84e:68038 Zbl 497 # 68021
• ID 39281

FRANK, A. [1980] *The countability of the rational polynomials: A direct method* (**J** 0005) Amer Math Mon 87*810-811
◇ D80 ◇ REV Zbl 446 # 03036 • ID 56565

FRANZEN, H. & HOFFMANN, B. [1979] *Automatic determination of data flow in extended affix grammars* (**P** 3381) GI Jahrestag (9);1979 Bonn 176-193
◇ B75 D05 ◇ REV Zbl 413 # 68078 • ID 69569

FRECON, L. [1980] see BENEY, J.

FREEDMAN, A.R. & LADNER, R.E. [1975] *Space bounds for processing contentless inputs* (**J** 0119) J Comp Syst Sci 11*118-128
◇ D10 D15 ◇ REV MR 53 # 2016 Zbl 307 # 68036
• ID 16696

FREEMAN, M. [1979] *Aspects of the upper bounds of finite input-memory and finite output-memory sequential machines* (**J** 0187) IEEE Trans Comp C-28*249-253
◇ D05 ◇ REV MR 80b:68065 Zbl 394 # 68041 • ID 69570

FREESE, R. & NATION, J.B. [1979] *Finitely presented lattices* (**J** 0053) Proc Amer Math Soc 77*174-178
◇ D40 G10 ◇ REV MR 81a:06008 Zbl 387 # 06007
• ID 81262

FREESE, R. [1980] *Free modular lattices* (**J** 0064) Trans Amer Math Soc 261*81-91
◇ D40 G10 ◇ REV MR 81k:06010 Zbl 437 # 06006
• ID 81261

FREESE, R. see Vol. III for further entries

FREGUGLIA, P. [1980] see BORGA, M.

FREGUGLIA, P. see Vol. I, V, VI for further entries

FREI-IMFELD, G. [1974] *Ueber eine Erweiterung der algebraischen Operationen* (**J** 0047) Notre Dame J Formal Log 15*279-288
◇ D20 G15 ◇ REV MR 55 # 2544 Zbl 236 # 02044
• ID 61797

FREIWALD, R.C. [1972] *Images of Borel sets and k-analytic sets* (**J** 0027) Fund Math 75*35-46
◇ D55 E15 ◇ REV MR 49 # 6199 Zbl 232 # 54048
• ID 04601

FREIWALD, R.C. & MCDOWELL, R. & MCHUGH JR., E.F. [1979] *Borel sets of exact class* (**S** 0019) Colloq Math (Warsaw) 41*187-191
◇ D55 E15 ◇ REV MR 81k:04002 Zbl 441 # 54019
• ID 73063

FREIWALD, R.C. see Vol. V for further entries

FREJDZON, R.I. [1967] *The representability of algorithmically decidable predicates by Rabin machines (Russian)* (S 0228) Zap Nauch Sem Leningrad Otd Mat Inst Steklov 4*209-218
◇ D03 D10 D20 ◇ REV MR 39#67 Zbl 167.15
• ID 04590

FREJDZON, R.I. [1968] *An index of the complexity of recursive predicates which do not depend on a standardization of the concept of algorithm (Russian)* (S 0228) Zap Nauch Sem Leningrad Otd Mat Inst Steklov 8*225-233 • ERR/ADD ibid 16*185-187
◇ D10 D15 ◇ REV MR 42#7504 Zbl 172.9 • ID 04591

FREJDZON, R.I. [1970] *A characterization of the complexity of recursive predicates (Russian)* (S 0066) Tr Mat Inst Steklov 113*79-101
• TRANSL [1970] (S 0055) Proc Steklov Inst Math 113*91-117
◇ D20 ◇ REV MR 44#6502 Zbl 233#02016 • ID 04592

FREJDZON, R.I. [1971] *Regular approximation of recursive predicates (Russian) (English summary)* (S 0228) Zap Nauch Sem Leningrad Otd Mat Inst Steklov 20*220-233,288
• TRANSL [1973] (J 1531) J Sov Math 1*139-147
◇ D20 ◇ REV MR 44#6499 Zbl 222#02042 • ID 04593

FREJDZON, R.I. [1972] *Families of recursive predicates of measure zero (Russian) (English summary)* (S 0228) Zap Nauch Sem Leningrad Otd Mat Inst Steklov 32*121-128,158
• TRANSL [1976] (J 1531) J Sov Math 6*449-455
◇ D20 F60 ◇ REV MR 49#8826 Zbl 354#02030
• ID 04594

FREJDZON, R.I. [1974] *Table approximations to recursive predicates (Russian) (English summary)* (S 0228) Zap Nauch Sem Leningrad Otd Mat Inst Steklov 40*131-135,159
• TRANSL [1977] (J 1531) J Sov Math 8*337-341
◇ D20 D25 ◇ REV MR 52#10384 Zbl 361#02052
• ID 21683

FREJDZON, R.I. [1975] *Finite approximate approach to the study of the complexity of recursive predicates (Russian)* (S 0228) Zap Nauch Sem Leningrad Otd Mat Inst Steklov 49*131-158
• TRANSL [1978] (J 1531) J Sov Math 10*604-627
◇ D05 D10 D15 ◇ REV MR 51#14640 Zbl 312#68029
• ID 61800

FREJVALD, P.V. [1982] *Increase in number of states when finite probabilistic automata are made deterministic (Russian)* (J 0474) Avtom Vychis Tekh, Akad Nauk Latv SSR 1982/3*39-42
• TRANSL [1982] (J 2666) Autom Control Comput Sci 16/3*33-35
◇ D05 ◇ REV MR 84j:68050 • ID 39190

FREJVALD, R.V. [1965] *Complexity of recognition of symmetry on Turing machines with input (Russian)* (J 0003) Algebra i Logika 4/1*47-58
◇ D10 D15 ◇ REV MR 32#2322 Zbl 146.10 JSL 35.159
• ID 04595

FREJVALD, R.V. [1966] *On the order of growth of exact time signalizers for Turing calculations (Russian)* (J 0003) Algebra i Logika 5/5*85-94
◇ D10 D15 ◇ REV MR 36#1327 Zbl 178.327 • ID 04596

FREJVALD, R.V. [1968] *Completeness up to coding of systems of functions of k-valued logic and the complexity of its determination (Russian)* (J 0023) Dokl Akad Nauk SSSR 180*803-805
• TRANSL [1968] (J 0062) Sov Math, Dokl 9*699-702
◇ B50 D15 ◇ REV MR 38#987 Zbl 199.302 • ID 04599

FREJVALD, R.V. [1972] see BARZDINS, J.

FREJVALD, R.V. [1974] *Gleichmaessige und ungleichmaessige Prognostizierbarkeit (Russian)* (S 2587) Teor Algor & Progr (Riga) 1*89-100
◇ D20 ◇ REV MR 58#21520 Zbl 341#02025 • ID 29746

FREJVALD, R.V. [1974] *Limitaere berechenbare Funktionen und Funktionale (Russian)* (S 2587) Teor Algor & Progr (Riga) 1*6-19
◇ D20 ◇ REV MR 58#21542 Zbl 341#02024 • ID 29745

FREJVALD, R.V. [1974] *Limitaere Berechnungen auf stochastischen Automaten (Russisch)* (S 2587) Teor Algor & Progr (Riga) 1*32-47
◇ D10 D20 ◇ REV MR 58#27386 Zbl 338#94034
• ID 61803

FREJVALD, R.V. [1974] *On the limit synthesis of numbers of general recursive functions in various computable numerations (Russian)* (J 0023) Dokl Akad Nauk SSSR 219*812-814
• TRANSL [1974] (J 0062) Sov Math, Dokl 15*1681-1683
◇ D20 D45 ◇ REV MR 51#12492 Zbl 322#02042
• ID 17281

FREJVALD, R.V. [1974] see BARZDINS, J.

FREJVALD, R.V. & PODNIEKS, K.M. [1974] *Ueber limitaere Berechnungen auf nichtdeterministischen Turingmaschinen (Russian)* (S 2587) Teor Algor & Progr (Riga) 1*25-31
◇ D10 D55 ◇ REV MR 58#21518 Zbl 341#02023
• ID 29744

FREJVALD, R.V. [1975] *Fast computation by probabilistic Turing machines (Russian) (English summary)* (S 2587) Teor Algor & Progr (Riga) 2*201-205
◇ D10 D15 ◇ REV MR 58#3677 Zbl 346#94027
• ID 61804

FREJVALD, R.V. [1975] *Functions computable in the limit by probabilistic machines* (P 1755) Math Founds of Comput Sci (3);1974 Jadwisin 77-87
◇ D10 D20 ◇ REV Zbl 309#94063 JSL 42.422 • ID 44490

FREJVALD, R.V. [1975] *Identifiability in the limit of the numbers of general recursive functions in various computable numerations (Russian) (English summary)* (S 2587) Teor Algor & Progr (Riga) 2*3-25
◇ D20 D25 D45 ◇ REV MR 57#16008 Zbl 322#02043
• ID 61809

FREJVALD, R.V. [1975] *Minimal Goedel numbers and their identification in the limit* (P 0454) Math Founds of Comput Sci (4);1975 Marianske Lazne 219-225
◇ D20 D45 ◇ REV MR 54#12499 Zbl 329#02016
• ID 61805

FREJVALD, R.V. [1975] *On complexity and optimality of the computation in the limit (Russian) (English summary)* (S 2587) Teor Algor & Progr (Riga) 2*155-173
◇ D15 D20 ◇ REV MR 58#3676 Zbl 322#02041
• ID 61807

FREJVALD, R.V. [1975] see BARZDINS, J.

Frejvald, R.V. & Kinber, E.B. [1977] *Identification in the limit of minimal Goedel numbers (Russian)* (**S** 2587) Teor Algor & Progr (Riga) 3*3-34
◇ D20 D45 ◇ REV MR 58 # 16218 Zbl 376 # 02033
• ID 51675

Frejvald, R.V. [1977] *On sets recognizable by a probabilistic machine with isolated cut-point and by no deterministic machine (Russian)* (**J** 0031) Izv Vyssh Ucheb Zaved, Mat (Kazan) 1977/1(176)*100-107
◇ D05 D10 ◇ REV MR 56 # 13793 Zbl 371 # 94068
• ID 51403

Frejvald, R.V. [1977] *Probabilistic machines can use less running time* (**P** 1694) Inform Processing (7);1977 Toronto 839-842
◇ D10 D15 ◇ REV MR 57 # 18219 Zbl 367 # 94079
• ID 51215

Frejvald, R.V. & Ikaunieks, E. [1977] *Some advantages of nondeterministic machines over probabilistic ones (Russian)* (**J** 0031) Izv Vyssh Ucheb Zaved, Mat (Kazan) 1977/2(177)*118-123
• TRANSL [1977] (**J** 3449) Sov Math 21/2*91-95
◇ D10 D15 ◇ REV MR 58 # 33612 Zbl 371 # 94069
• ID 51404

Frejvald, R.V. [1978] *Effektive Operationen und limitar berechenbare Funktionale (Russisch)* (**J** 0068) Z Math Logik Grundlagen Math 24*193-206
◇ D20 ◇ REV MR 80c:03048 Zbl 399 # 03026 • ID 52822

Frejvald, R.V. [1978] *Recognition of languages with high probability on different classes of automata (Russian)* (**J** 0023) Dokl Akad Nauk SSSR 239*60-62
• TRANSL [1978] (**J** 0062) Sov Math, Dokl 19*295-298
◇ D05 D10 ◇ REV MR 57 # 11188 Zbl 401 # 68032
• ID 69578

Frejvald, R.V. [1979] *Fast probabilistic verification of number multiplication (Russian)* (**J** 0474) Avtom Vychis Tekh, Akad Nauk Latv SSR 1979/1*40-43,93
• TRANSL [1979] (**J** 2666) Autom Control Comput Sci 13/1*37-39
◇ D10 D15 ◇ REV MR 81h:68026 Zbl 422 # 68015
• ID 81270

Frejvald, R.V. [1979] *Fast probabilistic algorithms* (**P** 2059) Math Founds of Comput Sci (8);1979 Olomouc 57-69
◇ D10 D15 ◇ REV MR 82a:68089 Zbl 408 # 68035
• ID 81272

Frejvald, R.V. [1979] *Finite identification of general recursive functions by probabilistic strategies* (**P** 2935) FCT'79 Fund of Comput Th;1979 Berlin/Wendisch-Rietz 138-145
◇ D10 D20 ◇ REV MR 81j:03074 Zbl 422 # 03012
• ID 53473

Frejvald, R.V. & Wiehagen, R. [1979] *Inductive inference with additional information* (**J** 0129) Elektr Informationsverarbeitung & Kybern 15*179-185
◇ D20 ◇ REV MR 81e:03036 Zbl 437 # 03018 • ID 55886

Frejvald, R.V. [1979] *Language recognition using finite probabilistic multitape and multihead automata (Russian)* (**J** 2320) Probl Peredachi Inf, Akad Nauk SSSR 15/3*99-106
• TRANSL [1979] (**J** 3419) Probl Inf Transmiss 15*235-241
◇ D05 ◇ REV MR 81f:68092 Zbl 433 # 68063 • ID 69579

Frejvald, R.V. [1979] *On shortening the recognition time of some word sets by using a random number generator (Russian)* (**J** 0052) Probl Kibern 36*209-224,280
◇ D05 D15 ◇ REV MR 81i:68083 Zbl 447 # 68046
• ID 81269

Frejvald, R.V. [1979] *Running time for deterministic and nondeterministic Turing machines (Russian)* (**J** 0337) Mat Ezheg, Akad Nauk Latv SSR 23*158-165,275
◇ D10 D15 ◇ REV MR 81d:03042 Zbl 462 # 68028
• ID 73050

Frejvald, R.V. [1981] *Probabilistic two-way machines* (**P** 3429) Math Founds of Comput Sci (10);1981 Strbske Pleso 33-45
◇ D05 D10 D15 ◇ REV MR 83e:03059 Zbl 486 # 68045
• ID 35232

Frejvald, R.V. [1981] *Two-way finite probabilistic automata and tape-bounded Turing machines (Russian)* (**J** 0023) Dokl Akad Nauk SSSR 256*1326-1329
• TRANSL [1981] (**J** 0062) Sov Math, Dokl 23*198-202
◇ D05 D10 ◇ REV MR 83h:68080 Zbl 467 # 68073
• ID 69582

Frejvald, R.V. & Kinber, E.B. & Wiehagen, R. [1982] *Inductive inference and computable one-one numberings* (**J** 0068) Z Math Logik Grundlagen Math 28*463-479
◇ B48 D20 D45 ◇ REV MR 85g:03065 Zbl 541 # 03025
• ID 43888

Frejvald, R.V. [1983] *Space and reversal complexity of probabilistic one-way Turing machines* (**P** 3864) FCT'83 Found of Comput Th;1983 Borgholm 159-170
◇ D10 D15 ◇ REV MR 85f:68025 Zbl 539 # 68040
• ID 39961

Frejvald, R.V. [1983] *Undecidability of the emptiness problem for probabilistic finite multitape machines (Russian)* (**C** 3798) Mat Log, Mat Ling & Teor Algor 69-74
◇ D05 ◇ REV MR 85e:03102 • ID 40707

Frejvald, R.V. & Kinber, E.B. & Wiehagen, R. [1984] *Connections between identifying functionals, standardizing operations, and computable numberings* (**J** 0068) Z Math Logik Grundlagen Math 30*145-164
◇ D20 D45 ◇ REV MR 85i:03135 Zbl 559 # 03028
• ID 42395

Frejvald, R.V. & Kinber, E.B. & Wiehagen, R. [1984] *On the power of probabilistic strategies in inductive inference* (**J** 1426) Theor Comput Sci 28*111-133
◇ B48 D20 ◇ REV MR 85j:03064 Zbl 555 # 68014
• ID 42396

Frejvald, R.V. [1985] *Space and reversal complexity of probabilistic one-way Turing machines* (**P** 3083) FCT'83 Found of Comput Th (Sel Pap);1983 Borgholm 39-50
◇ D10 ◇ REV MR 85f:68025 Zbl 557 # 03024 • ID 46191

Frejvald, R.V. see Vol. I, II for further entries

Fremlin, D.H. & Hansell, R.W. & Junnila, H.J.K. [1983] *Borel functions of bounded class* (**J** 0064) Trans Amer Math Soc 277*835-849
◇ D55 E15 E35 E50 E55 E65 ◇ REV MR 84d:54067 Zbl 518 # 54032 • ID 38382

Fremlin, D.H. see Vol. V for further entries

FRENKEL', V.I. [1963] *Algorithmic problems in partially ordered groups (Russian)* (**J** 0023) Dokl Akad Nauk SSSR 152*67-70
• TRANSL [1963] (**J** 0062) Sov Math, Dokl 4*1266-1269
◇ C57 C60 D40 ◇ REV MR 27 # 3715 MR 31 # 1194
• ID 20969

FRENKEL', V.I. [1964] *On the effective partial ordering of finitely defined groups (Russian)* (**J** 0092) Sib Mat Zh 5*651-670
◇ C57 C60 ◇ REV MR 29 # 2290 • ID 04602

FRENKEL', V.I. [1965] *The unsolvability of some algorithmic problems in groups, given by a system of generators and defining inequalities (Russian)* (**J** 0092) Sib Mat Zh 6*1144-1162
◇ D40 ◇ REV MR 34 # 227 • ID 04603

FREUDENTHAL, H. [1953] *Les possibilitees des machines a calculer* (**J** 0082) Bull Soc Math Belg 6*14-22
◇ D80 ◇ REV MR 16.555 Zbl 57.107 • ID 04607

FREUDENTHAL, H. [1953] *Machines pensantes* (**X** 1623) Univ Paris VI Inst Poincare: Paris 24
◇ A05 D05 ◇ REV MR 15.258 Zbl 53.345 • ID 28215

FREUDENTHAL, H. see Vol. I, II, VI for further entries

FREUND, R. [1983] *Real functions and numbers defined by Turing machines* (**J** 1426) Theor Comput Sci 23*287-304
◇ D10 D20 D80 ◇ REV MR 85b:03061 Zbl 512 # 68064
• ID 37557

FREY, T. [1964] *Ueber die Konstruktion nichtvollstaendiger Automaten* (**J** 0001) Acta Math Acad Sci Hung 15*375-381
◇ D05 ◇ REV MR 29 # 3377 Zbl 168.13 • ID 04615

FREY, T. [1964] *Ueber die Konstruktion endlicher Automaten* (**J** 0001) Acta Math Acad Sci Hung 15*383-398
◇ D05 ◇ REV MR 29 # 3378 Zbl 133.256 • ID 04614

FREY, T. [1969] *Ueber die Reduktion und Konstruktion nicht vollstaendiger /partieller/ Automaten* (**P** 1841) Fct Recurs & Appl;1967 Tihany 49-70
◇ D05 ◇ ID 32556

FRIDMAN, A.A. [1960] *On the relation between the word problem and the conjugacy problem in finitely defined groups (Russian)* (**J** 0065) Tr Moskva Mat Obshch 9*329-356
◇ D40 ◇ REV MR 31 # 1195 Zbl 114.254 JSL 34.507
• ID 04623

FRIDMAN, A.A. [1962] *Degrees of unsolvability of the word problem for finitely presented groups (Russian)* (**J** 0023) Dokl Akad Nauk SSSR 147*805-808
• TRANSL [1962] (**J** 0062) Sov Math, Dokl 3*1733-1737
◇ D25 D40 ◇ REV MR 29 # 5890 JSL 33.296 • ID 24965

FRIDMAN, A.A. [1967] *Unloesbarkeitsgrade des Wortproblems fuer endlich-definierte Gruppen (Russisch)* (**X** 2027) Nauka: Moskva 190pp
◇ D30 D40 ◇ REV MR 40 # 5711 Zbl 218 # 02042
• ID 26284

FRIDMAN, A.A. [1973] *A solution of the conjugacy problem in a class of groups (Russian)* (**S** 0066) Tr Mat Inst Steklov 133*233-242,276
• TRANSL [1973] (**S** 0055) Proc Steklov Inst Math 133*235-244
◇ D40 ◇ REV MR 49 # 424 Zbl 295 # 20041 • ID 50254

FRIDMAN, EH.I. & PENZIN, YU.G. [1972] *The elementary and the universal theory of the ordered group of integers with maximal subgroups (Russian)* (**C** 3549) Algebra, Vyp 1 (Irkutsk) 80-86
◇ B20 B25 D35 ◇ REV MR 56 # 11776 • ID 77209

FRIDMAN, EH.I. [1972] *Undecidability of the elementary theory of abelian torsion-free groups with a finite set of serving subgroups (Russian)* (**C** 3549) Algebra, Vyp 1 (Irkutsk) 97-100
◇ C60 D35 ◇ REV MR 57 # 2901 • ID 73069

FRIDMAN, EH.I. [1973] *Positive element-subgroup theories (Russian)* (**C** 1443) Algebra, Vyp 2 (Irkutsk) 161-163
◇ B25 D35 ◇ REV MR 53 # 7754 • ID 22999

FRIDMAN, EH.I. & SLOBODSKOJ, A.M. [1975] *Theories of Abelian groups with predicates specifying a subgroup (Russian)* (**J** 0003) Algebra i Logika 14*572-575
• TRANSL [1975] (**J** 0069) Algeb and Log 14*353-355
◇ C60 D35 ◇ REV MR 55 # 12512 Zbl 347 # 02037
• ID 52191

FRIDMAN, EH.I. & SLOBODSKOJ, A.M. [1976] *Undecidable universal theories of lattices of subgroups of abelian groups (Russian)* (**J** 0003) Algebra i Logika 15*227-234,246
• TRANSL [1976] (**J** 0069) Algeb and Log 15*142-146
◇ D35 ◇ REV MR 57 # 9515 Zbl 348 # 02042 • ID 26045

FRIDMAN, EH.I. see Vol. III, VI for further entries

FRIDMAN, G.SH. [1981] *Effective unsolvability of a discrete extremal problem (Russian)* (**C** 2355) Metody Anal Mnogormernoj Ehkon Inf 49-53
◇ D80 ◇ REV MR 84f:90102 Zbl 514 # 05035 • ID 39706

FRIDMAN, G.SH. see Vol. I for further entries

FRIED, M. & SACERDOTE, G.S. [1976] *Solving diophantine problems over all residue class fields of a number field and all finite fields* (**J** 0120) Ann of Math, Ser 2 104*203-233
◇ B25 C10 C13 C60 D35 ◇ REV MR 58 # 10722
Zbl 376 # 02042 • ID 51684

FRIED, M. see Vol. III for further entries

FRIEDBERG, R.M. [1957] *A criterion for completeness of degrees of unsolvability* (**J** 0036) J Symb Logic 22*159-160
◇ D30 ◇ REV MR 20 # 4488 Zbl 78.6 JSL 25.165
• ID 04634

FRIEDBERG, R.M. [1957] *The existence of a maximal set* (**P** 1675) Summer Inst Symb Log;1957 Ithaca 407-409
◇ D25 ◇ ID 29388

FRIEDBERG, R.M. [1957] *The fine structure of degrees of unsolvability of recursively enumerable sets* (**P** 1675) Summer Inst Symb Log;1957 Ithaca 404-406
◇ D25 ◇ REV Zbl 201.330 JSL 28.166 • ID 04641

FRIEDBERG, R.M. [1957] *Two recursively enumerable sets of incomparable degrees of unsolvability* (**J** 0054) Proc Nat Acad Sci USA 43*236-238
◇ D25 ◇ REV MR 18.867 Zbl 80.243 JSL 23.225
• ID 04635

FRIEDBERG, R.M. [1958] *Three theorems on recursive enumeration. I. Decomposition II. Maximal set. III. Enumeration without duplication* (**J** 0036) J Symb Logic 23*309-316
◇ D25 ◇ REV MR 22 # 13 Zbl 88.16 JSL 25.165
• ID 04638

FRIEDBERG, R.M. [1958] *Un contre-example relatif aux fonctionelles recursives* (**J** 0109) C R Acad Sci, Paris 247*852-854
◇ D20 ◇ REV MR 20#6355 Zbl 81.13 JSL 24.171
• ID 04637

FRIEDBERG, R.M. [1958] *4-quantifier completeness: A Banach-Mazur functional not uniformly partial recursive* (**J** 0014) Bull Acad Pol Sci, Ser Math Astron Phys 6*1-5
◇ D20 ◇ REV MR 20#3071 Zbl 84.249 JSL 24.52
• ID 04636

FRIEDBERG, R.M. & ROGERS JR., H. [1959] *Reducibility and completeness for sets of integers* (**J** 0068) Z Math Logik Grundlagen Math 5*117-125
◇ D30 ◇ REV MR 22#3682 Zbl 108.6 JSL 25.362
• ID 04639

FRIEDBERG, R.M. [1961] *The uniqueness of finite division for recursive equivalence types* (**J** 0044) Math Z 75*3-7
◇ D50 ◇ REV MR 24#A1214 Zbl 93.12 JSL 25.363
• ID 04642

FRIEDE, D. [1978] *Ueber deterministische kontextfreie Sprachen und rekursiven Abstieg* (1111) Preprints, Manuscr., Techn. Reports etc. IFI-HH-B-49*99pp
◇ D05 ◇ REV Zbl 432#68056 • REM Fachbereich Informatik. Bericht. Universitaet Hamburg • ID 69594

FRIEDE, D. [1979] *Partitioned LL(k) grammars* (**P** 1873) Automata, Lang & Progr (6);1979 Graz 245-255
◇ D05 ◇ REV MR 81i:68100 Zbl 417#68068 • ID 69593

FRIEDE, D. [1979] *Transition diagrams and strict deterministic grammars* (**P** 3488) Theor Comput Sci (4);1979 Aachen 113-123
◇ D05 ◇ REV MR 81k:68058 Zbl 401#68056 • ID 69592

FRIEDMAN, E.P. [1973] *Equivalence problems in monadic recursion schemes* (**P** 3062) IEEE Symp Switch & Automata Th (14);1973 Iowa City 26-33
◇ D20 ◇ REV MR 56#7291 • ID 81282

FRIEDMAN, E.P. [1974] *Relationships between monadic recursion schemes and deterministic context-free languages* (**P** 1479) IEEE Symp Switch & Automata Th (15);1974 New Orleans 43-51
◇ D05 D20 ◇ REV MR 55#6958 • ID 81283

FRIEDMAN, E.P. [1978] *A note on non-singular deterministic pushdown automata* (**J** 1426) Theor Comput Sci 7*333-339
◇ D10 ◇ REV MR 80b:68075 Zbl 393#68079 • ID 52477

FRIEDMAN, E.P. & GREIBACH, S.A. [1978] *On equivalence and subclass containment problems for deterministic context-free languages* (**J** 0232) Inform Process Lett 7*287-290
◇ D05 ◇ REV MR 80a:68097 Zbl 392#68069 • ID 69595

FRIEDMAN, E.P. & GREIBACH, S.A. [1979] *Monadic recursion schemes: The effect of constants* (**J** 0119) J Comp Syst Sci 18*254-266
◇ B75 D20 ◇ REV MR 80e:68202 Zbl 411#03010
• ID 52864

FRIEDMAN, E.P. & GREIBACH, S.A. [1979] *Superdeterministic DPDAs: The method of accepting does affect decision problems* (**J** 0119) J Comp Syst Sci 19*79-117
◇ D05 D10 D15 ◇ REV MR 81h:68077 Zbl 419#68100
• ID 53400

FRIEDMAN, E.P. & GREIBACH, S.A. [1980] *Superdeterministic PDAs: a subcase with a decidable inclusion problem* (**J** 0037) ACM J 27*675-700
◇ D10 ◇ REV MR 83c:68059 Zbl 462#68030 • ID 39048

FRIEDMAN, H.M. & JENSEN, R.B. [1968] *Note on admissible ordinals* (**P** 0637) Syntax & Semant Infinitary Lang;1967 Los Angeles 77-79
◇ C70 D60 ◇ REV Zbl 199.305 • ID 04643

FRIEDMAN, H.M. [1970] *Higher set theory and mathematical practice* (**J** 0007) Ann Math Logic 2*325-357
◇ D55 E15 E35 E45 E60 F35 ◇ REV MR 44#1556 Zbl 215.327 • ID 03773

FRIEDMAN, H.M. [1970] *Iterated inductive definitions and* Σ^1_2-AC (**P** 0603) Intuitionism & Proof Th;1968 Buffalo 435-442
◇ C62 D55 E25 F35 ◇ REV MR 44#1555 Zbl 216.6
• ID 14546

FRIEDMAN, H.M. [1971] *Algorithmic procedures, generalized Turing algorithms, and elementary recursion theory* (**P** 0638) Logic Colloq;1969 Manchester 361-389
◇ A05 D10 D20 D75 F99 ◇ REV MR 46#3275 Zbl 221#02018 • ID 04648

FRIEDMAN, H.M. [1971] *Axiomatic recursive function theory* (**P** 0638) Logic Colloq;1969 Manchester 113-137
◇ D75 ◇ REV MR 43#1830 Zbl 221#02019 • ID 04647

FRIEDMAN, H.M. [1971] *Determinateness in the low projective hierarchy* (**J** 0027) Fund Math 72*79-95
◇ D55 E15 E40 E45 E60 ◇ REV MR 45#8518 Zbl 244#02015 JSL 39.599 • ID 04649

FRIEDMAN, H.M. [1973] Σ^1_1 *degree determinateness fails in all boolean extensions* (1111) Preprints, Manuscr., Techn. Reports etc.
◇ D30 D55 E15 E40 E60 ◇ REM Mimeographed Notes, SUNY at Buffalo, NY • ID 21340

FRIEDMAN, H.M. [1973] *Borel sets and hyperdegrees* (**J** 0036) J Symb Logic 38*405-409
◇ D30 D55 E15 ◇ REV MR 49#30 Zbl 335#02028
• ID 04653

FRIEDMAN, H.M. [1974] *Minimality in the* Δ^1_2*-degrees* (**J** 0027) Fund Math 81*183-192
◇ D30 D55 E45 ◇ REV MR 52#61 Zbl 305#02050
• ID 18148

FRIEDMAN, H.M. [1974] *PCA well-orderings of the line* (**J** 0036) J Symb Logic 39*79-80
◇ D55 E15 E45 ◇ REV MR 50#107 Zbl 324#02059
• ID 04658

FRIEDMAN, H.M. [1975] *One hundred and two problems in mathematical logic* (**J** 0036) J Symb Logic 40*113-129
◇ B98 C98 D98 E98 F98 ◇ REV MR 51#5254 Zbl 318#02002 • ID 04296

FRIEDMAN, H.M. [1976] *On decidability of equational theories* (**J** 0326) J Pure Appl Algebra 7*1-3
◇ B25 C05 D35 ◇ REV MR 53#5285 Zbl 323#02056
• ID 22920

FRIEDMAN, H.M. [1976] *Recursiveness in* Π^1_1 *paths through* $\mathcal{O}$ (**J** 0053) Proc Amer Math Soc 54*311-315
◇ D30 D55 F15 ◇ REV MR 53#2663 Zbl 349#02037
• ID 21610

FRIEDMAN, H.M. [1976] *The complexity of explicit definitions* (**J** 0345) Adv Math 20*18-29
◇ C40 C57 ◇ REV MR 53 # 5287 Zbl 355 # 02010
• ID 22924

FRIEDMAN, H.M. [1976] *Uniformly defined descending sequences of degrees* (**J** 0036) J Symb Logic 41*363-367
◇ D30 D55 F30 F35 ◇ REV MR 55 # 2540
Zbl 366 # 02030 • ID 14767

FRIEDMAN, H.M. [1978] *Classically and intuitionistically provably recursive functions* (**P** 1864) Higher Set Th;1977 Oberwolfach 21-27
◇ B15 D20 E70 F30 F50 ◇ REV MR 80b:03093
Zbl 396 # 03045 • ID 31762

FRIEDMAN, H.M. & KO, KER-I [1982] *Computational complexity of real functions* (**J** 1426) Theor Comput Sci 20*323-352
◇ D15 D80 ◇ REV MR 83j:03103 Zbl 498 # 03047
• ID 35385

FRIEDMAN, H.M. [1984] *On the spectra of universal relational sentences* (**J** 0194) Inform & Control 62*205-209
◇ C13 D15 ◇ ID 46357

FRIEDMAN, H.M. [1984] *The computational complexity of maximization and integration* (**J** 0345) Adv Math 53*80-98
◇ D15 D80 ◇ REV MR 86c:03037 Zbl 563 # 03023
• ID 45281

FRIEDMAN, H.M. & SCEDROV, A. [1985] *Arithmetic transfinite induction and recursive well-orderings* (**J** 0345) Adv Math 56*283-294
◇ D45 F30 F50 ◇ ID 47284

FRIEDMAN, H.M. see Vol. I, III, V, VI for further entries

FRIEDMAN, JOYCE [1962] *A decision procedure for computations of finite automata* (**J** 0037) ACM J 9*315-323
◇ D05 ◇ REV MR 26 # 6021 Zbl 106.110 JSL 30.248
• ID 04665

FRIEDMAN, JOYCE & WARREN, D.S. [1978] *A parsing method for Montague grammars* (**J** 2130) Linguist Philos 2*347-372
◇ B65 D05 ◇ REV Zbl 388 # 68081 • ID 69596

FRIEDMAN, JOYCE see Vol. I, III, VI for further entries

FRIEDMAN, S.D. & SACKS, G.E. [1977] *Inadmissible recursion theory* (**J** 0015) Bull Amer Math Soc 83*255-256
◇ D60 E65 ◇ REV MR 58 # 21536 Zbl 376 # 02030
• ID 51672

FRIEDMAN, S.D. [1978] *An introduction to β-recursion theory* (**P** 1628) Generalized Recursion Th (2);1977 Oslo 111-126
◇ D60 ◇ REV MR 80b:03057 Zbl 453 # 03047 • ID 73119

FRIEDMAN, S.D. [1978] *Negative solutions to Post's problem I* (**P** 1628) Generalized Recursion Th (2);1977 Oslo 127-133
◇ D30 D60 ◇ REV MR 80c:03047 Zbl 453 # 03048 • REM Part II 1981 • ID 73118

FRIEDMAN, S.D. [1979] *β-recursion theory* (**J** 0064) Trans Amer Math Soc 255*173-200
◇ D60 D65 ◇ REV MR 81g:03055 Zbl 435 # 03035
JSL 46.664 • ID 55795

FRIEDMAN, S.D. [1979] *HC of an admissible set* (**J** 0036) J Symb Logic 44*95-102
◇ C70 D60 E30 ◇ REV MR 81i:03054 Zbl 406 # 03056
• ID 56139

FRIEDMAN, S.D. [1980] *Post's problem without admissibility* (**J** 0345) Adv Math 35*30-49
◇ D30 D60 E05 E65 ◇ REV MR 81g:03056
Zbl 453 # 03046 JSL 46.664 • ID 54176

FRIEDMAN, S.D. [1981] *Natural α-re degrees* (**P** 2628) Log Year;1979/80 Storrs 63-66
◇ D30 D60 ◇ REV MR 84a:03050 Zbl 471 # 03041
• ID 55236

FRIEDMAN, S.D. [1981] *Negative solutions to Post's problem II* (**J** 0120) Ann of Math, Ser 2 113*25-43
◇ D30 D60 ◇ REV MR 82k:03075 Zbl 461 # 03008 • REM Part I 1978 • ID 54485

FRIEDMAN, S.D. [1981] *Uncountable admissibles II: Compactness* (**J** 0029) Israel J Math 40*129-149
◇ C70 D60 E45 E47 ◇ REV MR 83k:03058b
Zbl 522 # 03035 • REM Part I 1982 • ID 33348

FRIEDMAN, S.D. [1982] *The Turing degrees and the metadegrees have isomorphic cones* (**P** 3634) Patras Logic Symp;1980 Patras 145-157
◇ D30 D60 ◇ REV MR 84g:03070 Zbl 515 # 03027
• ID 34177

FRIEDMAN, S.D. [1982] *Uncountable admissibles I: Forcing* (**J** 0064) Trans Amer Math Soc 270*61-73
◇ C70 D60 E40 ◇ REV MR 83k:03058a Zbl 493 # 03025
• REM Part II 1981 • ID 33350

FRIEDMAN, S.D. [1983] see CHONG, C.T.

FRIEDMAN, S.D. [1983] *Some recent developments in higher recursion theory* (**J** 0036) J Symb Logic 48*629-642
◇ D60 D70 ◇ REV Zbl 549 # 03032 • ID 43125

FRIEDMAN, S.D. [1985] *An immune partition of the ordinals* (**P** 3342) Rec Th Week;1984 Oberwolfach 141-147
◇ D50 E40 E45 E55 ◇ ID 45301

FRIEDMAN, S.D. [1985] *Fine structure theory and its applications* (**P** 4046) Rec Th;1982 Ithaca 259-269
◇ D60 E45 ◇ REV Zbl 573 # 03024 • ID 46395

FRIEDMAN, S.D. [1985] see DAVID, R.

FRIEDMAN, S.D. see Vol. III, V for further entries

FRIEDRICH, U. [1972] *Entscheidbarkeit der monadischen Nachfolgerarithmetik mit endlichen Automaten ohne Analysetheorem (Russian, English and French summaries)* (**J** 0115) Wiss Z Humboldt-Univ Berlin, Math-Nat Reihe 21*503-504
◇ B25 C85 D05 F35 ◇ REV MR 48 # 3724
Zbl 248 # 02051 • ID 04668

FRISH, I. [1965] *On stop-conditions in the definitions of constructive languages* (**J** 0068) Z Math Logik Grundlagen Math 11*61-73
◇ D05 ◇ REV MR 30 # 1930 • ID 04676

FRISH, I. [1968] *Grammars with partial orderings of the rules* (**J** 0194) Inform & Control 12*415-425
◇ D05 ◇ REV MR 39 # 5270 Zbl 172.300 • ID 15001

FRIZEN, D.G. & STARKOV, M.A. & TROFIMOV, O.E. [1976] *An automaton for calculation of the number of objects (Russian)* (**J** 0011) Avtom Telemekh 1976/6*134-140
• TRANSL [1976] (**J** 0010) Autom & Remote Control 37*927-933
◇ D05 D15 ◇ REV MR 55 # 7634 • ID 90038

FROEHLICH, A. & SHEPHERDSON, J.C. [1955] *Effective procedures in field theory* (**J** 0354) Phil Trans Roy Soc London, Ser A 248*407-432
◇ C57 C60 D45 ◇ REV MR 17.570 Zbl 70.35 JSL 24.169 • ID 17092

FROEHLICH, A. & SHEPHERDSON, J.C. [1955] *On the factorisation of polynomials in a finite number of steps* (**J** 0044) Math Z 62*331-334
◇ D45 D80 ◇ REV MR 17.119 Zbl 64.249 JSL 24.169 • ID 04679

FROEHLICH, R. [1976] *Kontextisolierte Regelgrammatiken* (**J** 2878) Wiss Z Tech Hochsch Karl-Marx-Stadt 18*489-501
◇ D05 ◇ REV MR 55 # 11724 Zbl 417 # 68057 • ID 69565

FROLOV, E.B. [1979] *Connectivity of the graph of a finite automaton with a large set of states (Russian)* (**J** 0011) Avtom Telemekh 1979/4*94-100
• TRANSL [1979] (**J** 0010) Autom & Remote Control 40*558-563
◇ D05 ◇ REV MR 80h:68046 Zbl 418 # 68057 • ID 69597

FROLOV, G.D. & KRINITSKIJ, N.A. & MIRONOV, G.A. [1974] *Certain criteria for the complexity of systems (Russian)* (**C** 3138) Tsifr Vychislit Tekhn & Progr 8*4-8
◇ D15 ◇ REV MR 52 # 4706 Zbl 307 # 02023 • ID 63140

FROUGNY, C. [1979] *Langages tres simples generateurs* (**J** 3441) RAIRO Inform Theor 13*69-86
◇ D05 ◇ REV MR 81d:68097 Zbl 405 # 68063 • ID 69598

FROUGNY, C. [1985] *Context-free grammars with cancellation properties* (**J** 1426) Theor Comput Sci 39*3-13
◇ D05 ◇ ID 49274

FU, K.S. [1979] see CHOU, S.M.

FU, K.S. [1979] see CHANG, NINGSAN

FU, K.S. see Vol. II, V for further entries

FUCHS, P.H. [1975] *Statistical characterization of learnable sequences* (**P** 1449) Automata Th & Formal Lang;1975 Kaiserslautern 52-56
◇ D20 ◇ REV MR 53 # 7739 Zbl 313 # 68035 • ID 22986

FUCHS, P.H. & SCHNORR, C.-P. [1977] *General random sequences and learnable sequences* (**J** 0036) J Symb Logic 42*329-340
◇ D20 D25 D80 ◇ REV MR 58 # 13937 Zbl 376 # 02026 • ID 51668

FUCHS, P.H. see Vol. III for further entries

FUERER, M. [1976] *Polynomiale Transformationen und Auswahlaxiom* (**P** 3196) Kompl von Entscheid Probl;1973/74 Zuerich 86-101
◇ D15 E25 ◇ REV MR 57 # 18232 Zbl 342 # 68029 • ID 61859

FUERER, M. [1976] *Simulation von Turingmaschinen mit logischen Netzen* (**P** 3196) Kompl von Entscheid Probl;1973/74 Zuerich 163-181
◇ B70 D10 D15 ◇ REV MR 57 # 18232 Zbl 344 # 02029 • ID 61860

FUERER, M. [1980] *The complexity of the inequivalence problem for regular expressions with intersection* (**P** 2904) Automata, Lang & Progr (7);1980 Noordwijkerhout 234-245
◇ D05 D15 ◇ REV MR 82g:68042 Zbl 444 # 68072 • ID 69026

FUERER, M. [1981] *Alternation and the Ackermann case of the decision problem* (**J** 3370) Enseign Math, Ser 2 27*137-162
• REPR [1982] (**P** 3482) Logic & Algor (Specker);1980 Zuerich 161-186
◇ B20 B25 D15 ◇ REV MR 83d:03015a Zbl 479 # 03005 Zbl 502 # 03025 • ID 35169

FUERER, M. [1982] *The complexity of Presburger arithmetic with bounded quantifier alternation depth* (**J** 1426) Theor Comput Sci 18*105-111
◇ B25 D15 ◇ REV MR 83b:03049 Zbl 484 # 03003 • ID 35102

FUERER, M. [1984] *The computational complexity of the unconstrained limited domino problem (with implications for logical decision problems)* (**P** 2342) Symp Rek Kombin;1983 Muenster 312-319
◇ B20 D15 D80 ◇ REV Zbl 549 # 68037 • ID 43160

FUKSMAN, A.L. [1976] *Diagnostic and other simplified forms of push-down automata (Russian)* (**J** 2605) Programmirovanie 1976/5*39-47
• TRANSL [1976] (**J** 2604) Progr Comput Software 2*356-363
◇ D20 ◇ REV MR 58 # 8503 Zbl 397 # 68059 • ID 69600

FUKUYAMA, M. [1968] *A note on incomparable degrees* (**J** 0350) Sci Rep Tokyo Kyoiku Daigaku Sect A 9*257-259
◇ D25 D30 ◇ REV MR 37 # 2605 Zbl 172.297 • ID 17106

FUKUYAMA, M. [1968] *A remark on minimal degrees* (**J** 0350) Sci Rep Tokyo Kyoiku Daigaku Sect A 9*255-256
◇ D30 ◇ REV MR 37 # 6183 Zbl 172.297 • ID 17105

FUKUYAMA, M. [1971] *Some concepts of recursiveness and admissible ordinals* (**J** 0090) J Math Soc Japan 23*435-451
◇ D60 ◇ REV MR 46 # 7006 Zbl 216.8 • ID 04717

FUKUYAMA, M. [1972] *Note on the recursion theory over admissible sets* (**J** 0350) Sci Rep Tokyo Kyoiku Daigaku Sect A 11*154-166
◇ D60 ◇ REV MR 50 # 12686 Zbl 263 # 02022 • ID 17107

FUKUYAMA, M. [1973] *An introduction to recursion theory on admissible sets and admissible ordinals* (**J** 0091) Sugaku 25*120-133
◇ D60 ◇ REV MR 58 # 27404 • ID 73147

FUKUYAMA, M. & NAGAOKA, K. [1974] *Strong representability of α-recursive functions* (**S** 1459) Mem School Sci & Engin, Waseda Univ 38*119-128
◇ C70 D60 ◇ REV MR 53 # 12907 Zbl 351 # 02032 • ID 23159

FUKUYAMA, M. [1976] *On the representability of α-recursive functions* (**S** 1459) Mem School Sci & Engin, Waseda Univ 39*67-78
◇ D60 ◇ REV MR 56 # 2809 Zbl 361 # 02053 • ID 50688

FUKUYAMA, M. [1977] *On positive Σ inductive definitions* (**S** 1459) Mem School Sci & Engin, Waseda Univ 41*121-134
◇ D70 D75 ◇ REV MR 80g:03048 Zbl 428 # 03041 • ID 53800

FURMANOV, K.K. [1973] see BEREZYUK, N.T.

FURST, M. & SAXE, J.B. & SIPSER, M. [1981] *Parity, circuits, and the polynomial-time hierarchy* (**P** 4235) IEEE Symp Found of Comp Sci (22);1981 Nashville 260-270
◇ D15 ◇ REV MR 84a:68004 • ID 45816

FURST, M. & LIPTON, R.J. & STOCKMEYER, L.J. [1983] *Pseudorandom number generation and space complexity* (**P** 3864) FCT'83 Found of Comput Th;1983 Borgholm 171-176
◇ D10 D15 ◇ REV MR 85m:68010 Zbl 549 # 68032 • ID 47052

FURST, M. & SAXE, J.B. & SIPSER, M. [1984] *Parity, circuits, and the polynomial-time hierarchy* (**J** 0041) Math Syst Theory 17*13-27
◇ D15 ◇ REV MR 86e:68048 Zbl 534 # 94008 • ID 38364

GABARRO, J. [1984] *Pushdown space complexity and related full-A.F.L.s* (**P** 3565) Symp of Th Aspects of Comput Sci (1);1984 Paris 250-259
◇ D10 D15 ◇ REV MR 86c:68024 Zbl 542 # 68064 • ID 47098

GABARRO, J. [1985] see BALCAZAR, J.L.

GABBASOV, N.Z. & MURTAZINA, T. [1978] *An improvement of the estimate in Rabin's reduction theorem (Russian)* (**C** 2899) Algor & Avtomaty 7-10
◇ D05 ◇ REV MR 82e:68055 Zbl 411 # 68049 • ID 69603

GABBASOV, N.Z. [1984] see ABLAEV, F.M.

GABBAY, D.M. [1972] *Sufficient conditions for the undecidability of intuitionistic theories with applications* (**J** 0036) J Symb Logic 37*375-384
◇ D35 F50 ◇ REV MR 47 # 8278 Zbl 266 # 02026 • ID 04731

GABBAY, D.M. [1973] *The undecidability of intuitionistic theories of algebraically closed fields and real closed fields* (**J** 0036) J Symb Logic 38*86-92
◇ B25 C60 C90 D35 F50 ◇ REV MR 47 # 6466 Zbl 266 # 02027 • ID 61880

GABBAY, D.M. [1976] *On Kreisel's notion of validity in Post systems* (**J** 0063) Studia Logica 35*285-295
◇ B50 D03 F50 ◇ REV MR 55 # 12477 Zbl 363 # 02027 • ID 50854

GABBAY, D.M. [1977] *Undecidability of intuitionistic theories formulated with the apartness relation* (**J** 0027) Fund Math 97*57-69
◇ D35 F50 ◇ REV MR 57 # 5690 Zbl 365 # 02012 • ID 27197

GABBAY, D.M. see Vol. I, II, III, VI for further entries

GABRIELIAN, A. [1971] see BRZOZOWSKI, J.A.

GABRIELIAN, A. & GINSBURG, S. [1973] *Structured storage AFA* (**J** 0187) IEEE Trans Comp C-22*534-537
◇ B75 D05 ◇ REV MR 50 # 9438 Zbl 252 # 68034 • ID 32105

GABRIELIAN, A. & GINSBURG, S. [1974] *Grammar schemata* (**J** 0037) ACM J 21*213-226
◇ D05 ◇ REV MR 51 # 14654 Zbl 279 # 68059 • ID 32108

GABRIELIAN, A. [1981] *Pure grammars and pure languages* (**J** 0382) Int J Comput Math 9*3-16
◇ B05 D03 D05 D10 ◇ REV MR 82f:68077 Zbl 454 # 68097 • ID 54258

GACS, P. [1974] *On the symmetry of algorithmic information (Russian)* (**J** 0023) Dokl Akad Nauk SSSR 218*1265-1267 • TRANSL [1974] (**J** 0062) Sov Math, Dokl 15*1477-1480
◇ D20 ◇ REV MR 53 # 7611 Zbl 314 # 94019 • ID 61885

GACS, P. & LOVASZ, L. [1977] *Some remarks on generalized spectra* (**J** 0068) Z Math Logik Grundlagen Math 23*547-554
◇ B15 C13 D15 ◇ REV MR 58 # 10398 Zbl 398 # 03025 • ID 52753

GACS, P. [1981] *On the relation between descriptional complexity and algorithmic probability* (**P** 4235) IEEE Symp Found of Comp Sci (22);1981 Nashville 296-303
◇ D15 ◇ REV MR 84a:68004 • ID 45811

GACS, P. [1983] *On the relation between descriptional complexity and algorithmic probability* (**J** 1426) Theor Comput Sci 22*71-93
◇ D15 ◇ REV MR 84h:60010 Zbl 562 # 68035 • ID 46432

GACS, P. see Vol. II for further entries

GAIFMAN, C. [1964] see BAR-HILLEL, Y.

GAIFMAN, H. & SHAMIR, E. [1980] *Roots of the hardest context free language and other constructs* (**J** 1456) SIGACT News 12/3*45-51
◇ D05 ◇ REV Zbl 453 # 68049 • ID 69605

GAIFMAN, H. see Vol. I, II, III, V, VI for further entries

GAINES, B.R. & KOHOUT, L.J. [1975] *Possible automata* (**P** 1805) Int Symp Multi-Val Log (5,Proc);1975 Bloomington 183-196
◇ B50 D05 ◇ ID 35823

GAINES, B.R. & KOHOUT, L.J. [1976] *The logic of automata* (**J** 1743) Int J Gen Syst 2*191-208
◇ D05 ◇ REV Zbl 327 # 94056 • ID 61898

GAINES, B.R. see Vol. I, II, V for further entries

GAL, L.N. [1958] *A note on direct products* (**J** 0036) J Symb Logic 23*1-6
◇ B25 C30 D35 ◇ REV MR 20 # 5121 Zbl 92.6 JSL 36.541 • ID 04757

GALBIATI, G. [1979] see CAMERINI, P.M.

GALIL, Z. [1974] *Two way deterministic pushdown automaton languages and some open problems in the theory of computation* (**P** 1479) IEEE Symp Switch & Automata Th (15);1974 New Orleans 170-177
◇ D05 D10 D20 ◇ REV MR 54 # 2440 • ID 24030

GALIL, Z. [1975] *Functional schemas with nested predicates* (**J** 0194) Inform & Control 27*349-368
◇ B75 D05 ◇ REV MR 52 # 4720 Zbl 302 # 68097 • ID 17692

GALIL, Z. [1975] *Two fast simulations which imply some fast string matching and palindrome-recognition algorithms* (**J** 0232) Inform Process Lett 4*85-87
◇ D05 D10 D15 ◇ REV MR 54 # 9158 Zbl 328 # 68047 • ID 61901

GALIL, Z. [1976] *Real-time algorithms for string-matching and palindrome recognition* (**P** 2597) ACM Symp Th of Comput (8);1976 Hershey 161-173
◇ D15 ◇ REV MR 56 # 10123 Zbl 365 # 68032 • ID 51074

GALIL, Z. [1977] *On resolution with clauses of bounded size* (**J** 1428) SIAM J Comp 6*444-459
◇ B35 D15 ◇ REV MR 56 # 13813 Zbl 368 # 68085 • ID 51297

GALIL, Z. [1977] *On the complexity of regular resolution and the Davis-Putnam procedure* (**J** 1426) Theor Comput Sci 4*23-46
◇ B35 D15 ◇ REV MR 56 # 4269 Zbl 385 # 68048 • ID 52174

GALIL, Z. & SEIFERAS, J.I. [1977] *Real-time recognition of substring repetition and reversal* (**J** 0041) Math Syst Theory 11*111-146
◇ D10 D15 ◇ REV MR 57 # 4669 Zbl 349 # 68035 • ID 50717

GALIL, Z. & SEIFERAS, J.I. [1978] *A linear-time on-line recognition algorithm for "palstar"* (**J** 0037) ACM J 25*102-111
◇ D10 D15 ◇ REV MR 58 # 3693 Zbl 365 # 68058 • ID 51078

GALIL, Z. [1978] *Palindrome recognition in real time by a multitape Turing machine* (**J** 0119) J Comp Syst Sci 16*140-157
◇ D10 D15 ◇ REV MR 58 # 3650 Zbl 386 # 03020 • ID 52196

GALIL, Z. [1984] see DURIS, P.

GALIN, A.B. [1980] see FAL'KOVICH, M.A.

GALLAIRE, H. [1969] *Recognition time for context-free languages by on-line Turing machines* (**J** 0194) Inform & Control 15*288-295
◇ D05 D10 D15 ◇ REV MR 40 # 2471 Zbl 186.13 • ID 04759

GALLEGO DE, M.S. [1978] *The abstract family of languages of categorical type* (**J** 0962) Ann Soc Sci Bruxelles, Ser 1 92*249-262
◇ D05 ◇ REV MR 80d:68088 Zbl 447 # 68086 • ID 31643

GALLEGO DE, M.S. [1979] *A characterization of context-free languages* (**J** 0962) Ann Soc Sci Bruxelles, Ser 1 93*155-158
◇ D05 ◇ REV MR 81b:68091 Zbl 459 # 68046 • ID 69609

GALLIER, J.H. [1979] *Recursion schemes and generalized interpretations* (**P** 1873) Automata, Lang & Progr (6);1979 Graz 71*256-270
◇ B75 D20 D75 ◇ REV Zbl 409 # 68014 • ID 69611

GALLIER, J.H. [1981] *Recursion-closed algebraic theories* (**J** 0119) J Comp Syst Sci 23*69-105
◇ B75 D75 ◇ REV MR 83h:68020 Zbl 472 # 68006 • ID 69610

GALLIER, J.H. & PELIN, A. [1984] *Solving word problems in free algebras using complexity functions* (**P** 2633) Autom Deduct (7);1984 Napa 476-495
◇ D40 ◇ REV MR 86e:03015 Zbl 546 # 08003 • ID 43548

GALLIER, J.H. see Vol. I, III for further entries

GALVIN, F. [1967] *Reduced products, Horn sentences, and decision problems* (**J** 0015) Bull Amer Math Soc 73*59-64
◇ B20 B25 C20 C30 C40 D35 ◇ REV MR 35 # 39 Zbl 155.349 JSL 33.477 • ID 04765

GALVIN, F. [1970] *Horn sentences* (**J** 0007) Ann Math Logic 1*389-422
◇ B20 C20 C30 C40 D35 ◇ REV MR 48 # 3729 Zbl 206.278 JSL 38.651 • ID 04766

GALVIN, F. & PRIKRY, K. [1973] *Borel sets and Ramsey's theorem* (**J** 0036) J Symb Logic 38*193-198
◇ D55 E05 E15 ◇ REV MR 49 # 2399 Zbl 276 # 04003 • ID 04769

GALVIN, F. see Vol. III, V for further entries

GALYUSHKOV, B.S. [1978] *Compound grammars (Russian)* (**C** 3211) Mat Ling & Teor Algor 15-34
◇ D05 ◇ REV Zbl 395 # 68075 • ID 69608

GALYUSHKOV, B.S. [1981] *Semi-context-free grammars (Russian)* (**C** 3747) Mat Log & Mat Lingvistika 38-50
◇ D05 ◇ REV MR 83m:68136 • ID 40425

GAMOVA, A.N. [1984] *Calculation of functionals of higher type (Russian)* (**X** 2235) VINITI: Moskva 3463-84
◇ D65 ◇ ID 46992

GANDY, R.O. [1960] *On a problem of Kleene's* (**J** 0015) Bull Amer Math Soc 66*501-502
◇ C62 D30 D55 ◇ REV MR 23 # A64 Zbl 97.246 JSL 29.104 • ID 04774

GANDY, R.O. & KREISEL, G. & TAIT, W.W. [1960] *Set existence* (**J** 0014) Bull Acad Pol Sci, Ser Math Astron Phys 8*577-583
◇ C62 D55 F35 ◇ REV MR 28 # 2964a JSL 27.232 • REM Part I. Part II 1961 • ID 15061

GANDY, R.O. & KREISEL, G. & TAIT, W.W. [1961] *Set existence II* (**J** 0014) Bull Acad Pol Sci, Ser Math Astron Phys 9*881-882
◇ C62 D55 F35 ◇ REV MR 28 # 2964b JSL 27.232 • REM Part I 1960 • ID 22330

GANDY, R.O. & SACKS, G.E. [1967] *A minimal hyperdegree* (**J** 0027) Fund Math 61*215-223
◇ D30 D55 ◇ REV MR 37 # 1246 Zbl 204.15 • ID 04777

GANDY, R.O. [1967] *Computable functionals of finite type I* (**P** 0691) Sets, Models & Recursion Th;1965 Leicester 202-242
◇ D65 ◇ REV MR 36 # 2495 Zbl 155.340 JSL 35.157 • ID 04776

GANDY, R.O. [1967] *General recursive functionals of finite type and hierarchies of functions* (**J** 0179) Ann Fac Sci Clermont 35*5-24
◇ D55 D65 ◇ REV MR 43 # 1841 • ID 04779

GANDY, R.O. & SOARE, R.I. [1970] *A problem in the theory of constructive order types* (**J** 0036) J Symb Logic 35*119-121
◇ D30 D45 ◇ REV MR 45 # 1755 Zbl 195.18 • ID 04780

GANDY, R.O. [1971] see BARWISE, J.

GANDY, R.O. [1974] *Inductive definitions* (**P** 0602) Generalized Recursion Th (1);1972 Oslo 265-299
◇ D60 D65 D70 E55 ◇ REV MR 53 # 7744 Zbl 318 # 02041 • ID 22992

GANDY, R.O. [1974] *Set-theoretic functions for elementary syntax* (**P** 0693) Axiomatic Set Th;1967 Los Angeles 2*103-126
◇ B03 D20 D65 E47 ◇ REV MR 51 # 12524 Zbl 323 # 02067 • ID 17242

GANDY, R.O. & HYLAND, J.M.E. [1977] *Computable and recursively countable functions of higher type* (**P** 1075) Logic Colloq;1976 Oxford 407-438
◇ D65 ◇ REV MR 58 # 10368 Zbl 422 # 03022 • ID 16631

GANDY, R.O. & HYLAND, J.M.E. (EDS.) [1977] *Logic Colloquium 76. Proceedings of a conference held in Oxford in July 1976* (**S** 3303) Stud Logic Found Math 87*x+612pp
◇ B97 C97 D97 ◇ REV MR 57 # 53 Zbl 409 # 00002 • ID 16612

GANDY, R.O. [1978] see FENSTAD, J.E.

GANDY, R.O. [1980] *Church's thesis and principles for mechanisms* (**P** 2058) Kleene Symp;1978 Madison 123-148
◇ A05 D05 D10 D20 F99 ◇ REV MR 82h:03036 Zbl 465 # 03022 • ID 54925

GANDY, R.O. [1984] *Some relations between classes of low computational complexity* (**J** 0161) Bull London Math Soc 16*127-134
◇ D20 ◇ REV MR 85j:03060 Zbl 567 # 03018 • ID 45245

GANDY, R.O. see Vol. I, III, V, VI for further entries

GANOV, V.A. [1974] *Generalized computability and the descriptive theory of sets (Russian)* (**J** 0092) Sib Mat Zh 15*1242-1261,1429
• TRANSL [1974] (**J** 0475) Sib Math J 15*873-887
◇ B75 D10 D55 E15 ◇ REV MR 51 # 158 Zbl 315 # 02042 • ID 17498

GANOV, V.A. [1980] *Functionals of ordering types (Russian)* (**J** 0092) Sib Mat Zh 21
◇ D99 ◇ ID 33266

GANOV, V.A. [1980] *Generalized computability and jump operation (Russian)* (**C** 3646) Uchebnoe Posobie
◇ D30 ◇ ID 33910

GANOV, V.A. see Vol. VI for further entries

GAO, HENGSHAN [1973] *The decision problem of modal predicate calculi. I: On Slomson's reductions (Chinese)* (**J** 3816) Zhongguo Kexue Jishu Daxue Xuebao 3/2*49-54
◇ B45 D35 ◇ REV MR 58 # 27330 Zbl 349 # 02016 • REM Part II 1976 • ID 46707

GAO, HENGSHAN [1973] *Two remarks on the decision problem (Chinese)* (**J** 2771) Kexue Tongbao 18*259-260
◇ D35 ◇ REV MR 58 # 27420 • ID 73224

GAO, HENGSHAN [1976] *The decision problem of modal predicate calculi. II: On Slomson's reductions (Chinese)* (**J** 0418) Shuxue Xuebao 19*276-280
◇ B45 C13 D35 ◇ REV MR 58 # 27330 Zbl 349 # 02016 • REM Part I 1973 • ID 61914

GAO, HENGSHAN see Vol. I, II, III, V, VI for further entries

GAREY, M.R. & JOHNSON, D.S. & TARJAN, R.E. [1976] *The planar Hamiltonian circuit problem is NP-complete* (**J** 1428) SIAM J Comp 5*704-714
◇ D15 ◇ REV MR 56 # 2867 Zbl 346 # 05110 • ID 81304

GAREY, M.R. & JOHNSON, D.S. [1979] *Computers and intractability.A guide to the theory of NP-completeness* (**X** 0994) Freeman: San Francisco x+338pp
• TRANSL [1982] (**X** 0885) Mir: Moskva 416pp
◇ D15 D98 ◇ REV MR 80g:68056 Zbl 411 # 68039 JSL 48.498 • ID 81305

GARFUNKEL, S. & SHANK, H.S. [1971] *On the undecidability of finite planar graphs* (**J** 0036) J Symb Logic 36*121-125
◇ C13 D35 ◇ REV MR 44 # 2607 Zbl 222 # 02056 • ID 04793

GARFUNKEL, S. & SHANK, H.S. [1972] *On the indecidability of finite planar cubic graphs* (**J** 0036) J Symb Logic 37*595-597
◇ C13 D35 ◇ REV MR 47 # 4781 Zbl 269 # 02020 • ID 04795

GARFUNKEL, S. & SCHMERL, J.H. [1974] *The undecidability of theories of groupoids with an extra predicate* (**J** 0053) Proc Amer Math Soc 42*286-289
◇ D35 ◇ REV MR 48 # 3725 Zbl 253 # 02048 Zbl 273 # 02032 • ID 04797

GARLAND, S.J. [1972] *Generalized interpolation theorems* (**J** 0036) J Symb Logic 37*343-351
◇ C40 C75 C85 D55 E15 ◇ REV MR 51 # 10045 Zbl 267 # 02037 • ID 04799

GARLAND, S.J. [1974] *Second-order cardinal characterizability* (**P** 0693) Axiomatic Set Th;1967 Los Angeles 2*127-146
◇ B15 C40 C55 C85 D55 E10 E47 E55 ◇ REV MR 54 # 4982 Zbl 319 # 02065 • ID 24144

GASANOV, R.G. [1983] *On a mathematical model of algorithms (Russian) (English and Azerbaijani summaries)* (**J** 0135) Izv Akad Nauk Azerb SSR, Ser Fiz-Tekh Mat 1983/4*155-158
◇ D20 ◇ REV MR 85j:03058 Zbl 553 # 03028 • ID 43330

GASANOVA, N.P. [1985] see BURTMAN, M.I.

GASARCH, W.I. & HOMER, S. [1983] *Relativizations comparing NP and exponential time* (**J** 0194) Inform & Control 58*88-100
◇ D15 ◇ REV MR 85h:03041 Zbl 562 # 68034 • ID 43308

GASS, F.S. [1971] *Generalized ordinal notation* (**J** 0047) Notre Dame J Formal Log 12*104-114
◇ D55 F15 ◇ REV MR 44 # 1561 Zbl 221 # 02028 • ID 04808

GASS, F.S. [1972] *A note on Π^1_1 ordinals* (**J** 0047) Notre Dame J Formal Log 13*103-104
◇ D55 F15 ◇ REV MR 47 # 1590 Zbl 228 # 02020 • ID 04809

GASS, F.S. see Vol. VI for further entries

GATTERDAM, R.W. [1973] *The computability of group constructions part II* (**J** 0016) Bull Austral Math Soc 8*27-60
◇ D20 D40 D45 ◇ REV MR 56 # 5256 Zbl 243 # 02036 • REM Part I 1973 by Cannonito,F.B. & Gatterdam,R.W. • ID 04810

GATTERDAM, R.W. [1973] see CANNONITO, F.B.

GATTERDAM, R.W. [1973] *The Higman theorem for primitive-recursive groups - a preliminary report* (**P** 0678) Word Probl: Decis & Burnside Probl in Group Th;1969 Irvine 421-425
◇ D20 D40 ◇ REV MR 53 # 632 Zbl 263 # 02024 JSL 41.786 • ID 16688

GATTERDAM, R.W. [1973] *The Higman theorem for $E^n(A)$ computable groups* (**P** 0779) Conf Group Th;1972 Racine 71-74
◇ D20 D40 ◇ REV MR 52 # 585 Zbl 261 # 02032 • ID 18154

GATTERDAM, R.W. [1975] see CANNONITO, F.B.

GAUGER, M.A. [1977] see BYRNES, C.

GAVRILOV, S. [1978] *Relations between different forms of relative computability of everywhere defined functions (Russian)* (**C** 2595) Rekursiv Funktsii 3-10
◇ D30 ◇ REV MR 82d:03071 Zbl 504 # 03020 • ID 73275

GECSEG, F. & PEAK, I. [1965] *Automata with isomorphic semi-groups* (**J** 0002) Acta Sci Math (Szeged) 26*43-47
◇ D05 ◇ REV MR 31 # 2151 Zbl 139.9 • ID 04820

GECSEG, F. & PEAK, I. [1971] *Algebraic theory of automata* (**X** 0928) Akad Kiado: Budapest xiii+326pp
◇ D05 D98 ◇ REV Zbl 246 # 94029 • ID 61940

GECSEG, F. [1976] *On products of abstract automata* (**J** 0002) Acta Sci Math (Szeged) 38*21-43
◇ D05 ◇ REV MR 53 # 10518 Zbl 342 # 94024 • ID 27234

GECSEG, F. (ED.) [1981] *Fundamentals of computation theory* (**X** 0811) Springer: Heidelberg & New York XI+471pp
◇ D97 ◇ REV MR 83c:68002 Zbl 459 # 00019 • ID 48715

GECSEG, F. & STEINBY, M. [1984] *Tree automata* (**X** 0928) Akad Kiado: Budapest 235pp
◇ A05 B25 D05 D98 ◇ REV MR 86c:68061 Zbl 537 # 68056 • ID 41334

GECSEG, F. see Vol. III for further entries

GEHANI, N. [1980] *Generic procedures: An implementation and an undecidability result* (**J** 3144) Comput Lang, Int J 5*155-161
◇ B75 D80 ◇ REV Zbl 444 # 68009 • ID 69627

GEISER, J.R. [1977] see FERRANTE, J.

GEISER, J.R. see Vol. I, III, VI for further entries

GENENZ, J. [1964] *Reduktionstheorie nach der Methode von Kahr-Moore-Wang* (**X** 3333) Unknown Publisher: See Remarks 88pp
◇ B20 D35 D98 ◇ REM Muenster • ID 21347

GENG, SUYUN & ZHANG, LIANG [1984] *A note on the formal definition of "Turing reduction" (Chinese) (English summary)* (**J** 2521) Beijing Shifan Daxue Xuebao, Ziran Kexue 1984/4*50-53
◇ D30 ◇ REV MR 86a:68030 • ID 44156

GENRICH, H.J. [1975] *Extended simple regular expressions* (**P** 0454) Math Founds of Comput Sci (4);1975 Marianske Lazne 231-237
◇ B75 D05 ◇ REV MR 53 # 12124 Zbl 337 # 02012 • ID 61944

GENRICH, H.J. & THIELER-MEVISSEN, G. [1976] *The calculus of facts* (**P** 1401) Math Founds of Comput Sci (5);1976 Gdansk 588-595
◇ B45 D15 ◇ REV Zbl 341 # 68039 • ID 61945

GENRICH, H.J. & LAUTENBACH, K. & THIAGARAJAN, P.S. [1980] *Elements of general net theory* (**P** 2973) Net Th & Appl;1979 Hamburg 21-163
◇ B70 B75 D05 ◇ REV MR 82a:68112 • ID 81319

GENRICH, H.J. see Vol. I for further entries

GEORGEFF, M.P. [1981] *Interdependent translation schemes* (**J** 0119) J Comp Syst Sci 22*198-219
◇ D05 ◇ REV MR 82f:68078 Zbl 458 # 68031 • ID 81324

GEORGIEVA, N.V. [1975] *A certain reflexive recursive scheme (Russian)* (**J** 0137) C R Acad Bulgar Sci 28*723-725
◇ D20 ◇ REV MR 51 # 12491 Zbl 339 # 02035 • ID 17280

GEORGIEVA, N.V. [1976] *Another simplification of the recursion scheme* (**J** 0009) Arch Math Logik Grundlagenforsch 18*1-3
◇ D20 ◇ REV MR 58 # 5118 Zbl 343 # 02026 • ID 23708

GEORGIEVA, N.V. [1976] *Classes of one-argument recursive functions* (**J** 0068) Z Math Logik Grundlagen Math 22*127-130
◇ D20 ◇ REV MR 57 # 81 Zbl 339 # 02036 • ID 18495

GEORGIEVA, N.V. [1976] *Hierarchies of one argument recursive functions I* (**J** 0137) C R Acad Bulgar Sci 29*919-922
◇ D20 ◇ REV MR 55 # 5412 Zbl 358 # 02060 • ID 31113

GEORGIEVA, N.V. [1978] *An extension of the decidable class of equations considered by Goodstein and Lee* (**J** 0068) Z Math Logik Grundlagen Math 24*399-404
◇ D20 ◇ REV MR 80b:03053 Zbl 401 # 03028 • ID 73324

GEORGIEVA, N.V. [1980] *Theorems on the normal form of some recursive elements and mappings (Russian)* (**J** 0137) C R Acad Bulgar Sci 33*1577-1580
◇ D45 D80 ◇ REV MR 82h:06020 Zbl 468 # 03028 • ID 55093

GEORGIEVA, N.V. [1984] *A theorem on representability of general recursive functions* (**P** 4392) Mat Logika (Markova);1980 Sofia 27-33
◇ D20 ◇ ID 46963

GEORGIEVA, N.V. see Vol. I, II, III, VI for further entries

GERASIMOV, V.N. [1976] *Distributive lattices of subspaces and the equality problem for algebras with a single relation (Russian)* (**J** 0003) Algebra i Logika 15*384-435
• TRANSL [1976] (**J** 0069) Algeb and Log 15*238-274
◇ D40 ◇ REV MR 56 # 5618 Zbl 372 # 08001 • ID 26051

GERGELY, T. [1981] *Algebraic representation of language hierarchies* (**J** 0380) Acta Cybern (Szeged) 5*307-323
◇ D05 G15 ◇ REV MR 83i:68106 Zbl 475 # 68045 • ID 69615

GERGELY, T. see Vol. I, II, III for further entries

GERHARD, J.A. & PETRICH, M. [1981] *The word problem for orthogroups* (**J** 0017) Canad J Math 33*893-900
◇ D40 ◇ REV MR 82m:20063 Zbl 472 # 20021 • ID 81328

GERLA, G. [1976] see CRISCUOLO, G.

GERLA, G. [1979] *Una generalizzazione della gerarchia di Ershov* (**J** 3495) Boll Unione Mat Ital, V Ser, B 16*765-778
◇ D30 D55 ◇ REV MR 80j:03062 Zbl 417 # 03018 • ID 53255

GERLA, G. see Vol. II, III, V for further entries

GERMANO, G. [1971] *Incompleteness and truth definitions* (**J** 0105) Theoria (Lund) 37*86-90
◇ C40 D35 F30 ◇ REV MR 50 # 4274 Zbl 243 # 02038 • ID 04907

GERMANO, G. & MAGGIOLO-SCHETTINI, A. [1972] *Equivalence of partial recursivity and computability by algorithms without concluding formulas* (**J** 0089) Calcolo 8*273-291
◇ D03 D20 ◇ REV MR 50 # 9549 Zbl 242 # 02043 • ID 04908

GERMANO, G. & MAGGIOLO-SCHETTINI, A. [1973] *A flow diagram composition of Markov's normal algorithms without concluding formulas* (**J** 0130) BIT 13*301-312
◇ B75 D03 D20 ◇ REV MR 49 # 11854 Zbl 283 # 68026 • ID 61989

GERMANO, G. [1973] *A new theory of calculability* (**P** 0763) Automata, Lang & Progr (1);1972 Rocquencourt 33-36
◇ D75 ◇ ID 33893

GERMANO, G. & MAGGIOLO-SCHETTINI, A. [1973] *Markov's normal algorithms without concluding formulas* (**P** 0793) Int Congr Log, Meth & Phil of Sci (4,Proc);1971 Bucharest 29,VIII-4 IX
◇ D03 ◇ ID 33891

GERMANO, G. & MAGGIOLO-SCHETTINI, A. [1973] *Quelques characterisations des fonctions recursives partielles* (**J** 2313) C R Acad Sci, Paris, Ser A-B 276*A1325-A1327
◇ D20 ◇ REV MR 51 # 91 Zbl 324 # 02025 • ID 18159

GERMANO, G. & MAGGIOLO-SCHETTINI, A. [1974] *Sequence-to-sequence functions. A systematic study* (**J** 1089) IBM Res Rep
◇ B75 D20 ◇ ID 33894

GERMANO, G. & MAGGIOLO-SCHETTINI, A. [1975] *Sequence-to-sequence recursiveness* (**J** 0232) Inform Process Lett 4*1-6
◇ B75 D20 G30 ◇ REV MR 52 # 7884 Zbl 311 # 02047 • ID 18161

GERMANO, G. & MAGGIOLO-SCHETTINI, A. [1976] *A language for Markov's algorithms composition* (**J** 0380) Acta Cybern (Szeged) 3*31-35
◇ D03 ◇ REV MR 56 # 10104 Zbl 345 # 02021 • ID 29780

GERMANO, G. & MAGGIOLO-SCHETTINI, A. [1976] *Recursivity, sequence recursivity, stack recursivity and semantics of programs* (**P** 1401) Math Founds of Comput Sci (5);1976 Gdansk 52-64
◇ B75 D15 D20 ◇ REV Zbl 339 # 68003 • ID 61990

GERMANO, G. & MAGGIOLO-SCHETTINI, A. [1977] *Sequence-to-sequence partial recursive functions* (**P** 1704) Int Congr Log, Meth & Phil of Sci (5);1975 London ON 3*3-4
◇ B75 D20 ◇ ID 33900

GERMANO, G. & MAGGIOLO-SCHETTINI, A. [1979] *Computable stack functions for semantics of stack programs* (**J** 0119) J Comp Syst Sci 19*133-144
◇ B75 D20 ◇ REV MR 81b:68060 Zbl 428 # 68057 • ID 53823

GERMANO, G. & MAGGIOLO-SCHETTINI, A. [1981] *Sequence recursiveness without cylindrification and limited register machines* (**J** 1426) Theor Comput Sci 15*213-221
◇ B75 D15 D20 ◇ REV MR 83f:68007 Zbl 464 # 68050 • ID 40169

GERMANO, G. & MAZZANTI, S. [1984] *Partial closures and semantics of while: toward an iteration-based theory of data types* (**P** 2153) Logic Colloq;1983 Aachen 2*163-174
◇ B75 D20 D75 D80 ◇ REV Zbl 553 # 68011 • ID 40182

GERMANO, G. see Vol. I, III, VI for further entries

GERSTING, J.L. [1969] *A rate of growth criterion for universality of regressive isols* (**J** 0048) Pac J Math 31*669-677
◇ D50 ◇ REV MR 41 # 59 Zbl 209.20 • ID 04914

GERSTING, J.L. [1972] *A note on infinite series of isols* (**J** 0308) Rocky Mountain J Math 2*661-666
◇ D50 ◇ REV MR 47 # 4779 Zbl 266 # 02023 • ID 04915

GERSTING, J.L. [1973] *Infinite series of T-regressive isols* (**J** 0047) Notre Dame J Formal Log 14*519-526
◇ D50 ◇ REV MR 48 # 10790 Zbl 232 # 02032 • ID 04916

GERSTING, J.L. [1975] *Universal pairs of regressive isols* (**J** 0047) Notre Dame J Formal Log 16*409-414
◇ D50 ◇ REV MR 52 # 7883 Zbl 262 # 02044 • ID 04917

GERSTING, J.L. [1977] *Infinite series of regressive isols under addition* (**J** 0047) Notre Dame J Formal Log 18*299-304
◇ D50 ◇ REV MR 56 # 11773 Zbl 314 # 02054 • ID 23621

GHIRON, H. [1962] *Rules to manipulate regular expressions of finite automata* (**J** 0072) IRE Trans Electr Comp EC-11*574-575
◇ D05 ◇ REV Zbl 137.10 • ID 04922

GILDENHUYS, D. [1985] see BAUMSLAG, G.

GILES, R. [1979] see ADAMSON, A.

GILES, R. see Vol. I, II, V for further entries

GILL, A. [1962] *Introduction to the theory of finite-state machines* (**X** 0822) McGraw-Hill: New York 207pp
◇ D05 D10 D98 ◇ REV MR 34 # 8891 Zbl 158.258 • ID 23523

GILL, A. [1966] see DEUEL, D.R.

GILL, A. see Vol. I for further entries

GILL III, J.T. [1973] see BLUM, M.

GILL III, J.T. [1974] *Computational complexity of probabilistic Turing machines* (**P** 1464) ACM Symp Th of Comput (6);1974 Seattle 91-95
◇ D10 D15 ◇ REV MR 54 # 1711 Zbl 357 # 68056 • ID 50443

GILL III, J.T. [1974] see BLUM, M.

GILL III, J.T. & MORRIS, P.H. [1974] *On subcreative sets and s-reducibility* (**J** 0036) J Symb Logic 39*669-677
◇ D25 D30 ◇ REV MR 51 # 95 Zbl 296 # 02020 • ID 04931

GILL III, J.T. & SIMON, I. [1976] *Ink, dirty-tape Turing machines, and quasicomplexity measures* (**P** 1870) Automata, Lang & Progr (3);1976 Edinburgh 285-306
◇ D10 D15 ◇ REV Zbl 356 # 02033 • ID 50305

GILL III, J.T. [1977] *Computational complexity of probabilistic Turing machines* (**J** 1428) SIAM J Comp 6*675-695
◇ D10 D15 ◇ REV MR 57 # 4616 Zbl 366 # 02024 • ID 51107

GILL III, J.T. [1981] see BENNETT, C.H.

GILLETTE, P. [1968] see BUSKIRK VAN, J.

GILLO, D. & GOETZE, B.G. & KLETTE, R. [1977] *Der iterierte Limes rekursiver Funktionen und die arithmetische Hierarchie* (**J** 0068) Z Math Logik Grundlagen Math 23*265-272
◇ D20 D55 ◇ REV MR 58 # 16235 Zbl 382 # 03036 • ID 26486

GINSBERG, S. & GOLDSTINE, J. & SPANIER, E.H. [1983] *On the equality of grammatical families* (**J** 0119) J Comp Syst Sci 26*171-196
◇ B65 D05 ◇ REV MR 85g:68037 Zbl 535 # 68037 • ID 43937

GINSBURG, S. [1958] *On the length of the smallest uniform experiment which distinguishes the terminal states of a machine* (**J** 0037) ACM J 5*266-280
◇ D05 ◇ REV MR 22 # 10882 Zbl 88.344 • ID 32068

GINSBURG, S. [1959] *A synthesis technique for minimal state sequential machines* (**J** 0072) IRE Trans Electr Comp EC-8*13-24
◇ D05 ◇ ID 32069

GINSBURG, S. [1959] *A technique for the reduction of a given machine to a minimal state machine* (J 0072) IRE Trans Electr Comp EC-8*346-355
◇ D05 ◇ ID 32070

GINSBURG, S. [1960] *Connective properties preserved in minimal state machines* (J 0037) ACM J 7*311-325
◇ D05 ◇ REV Zbl 97.5 • ID 32072

GINSBURG, S. [1960] *Some remarks on abstract machines* (J 0064) Trans Amer Math Soc 96*400-440
◇ D05 ◇ REV Zbl 97.5 JSL 37.411 • ID 04961

GINSBURG, S. [1960] *Synthesis of minimal state machines* (J 0072) IRE Trans Electr Comp EC-9*441-449
◇ D05 ◇ ID 32071

GINSBURG, S. [1961] *Compatibility of states in input-independent machines* (J 0037) ACM J 8*400-403
◇ D05 ◇ REV MR 23 # B2133 Zbl 104.7 • ID 32075

GINSBURG, S. [1961] *Distinguishability of a semigroup by a machine* (J 0053) Proc Amer Math Soc 12*661-668
◇ D05 ◇ REV MR 31 # 1312 Zbl 118.254 • ID 32076

GINSBURG, S. [1961] *Sets of tapes accepted by different types of automata* (J 0037) ACM J 8*81-86
◇ D05 ◇ REV Zbl 103.346 • ID 32074

GINSBURG, S. & ROSE, G.F. [1962] *A comparison of the work done by generalized sequential machines and Turing machines* (J 0064) Trans Amer Math Soc 103*394-402
◇ D05 D10 ◇ REV MR 25 # 1990 Zbl 114.249 JSL 37.411 • ID 04963

GINSBURG, S. [1962] *An introduction to mathematical machine theory* (X 0832) Addison-Wesley: Reading ix + 148pp
◇ D05 D10 ◇ REV MR 26 # 3222 Zbl 102.338 • ID 32121

GINSBURG, S. [1962] *Examples of abstract machines* (J 0072) IRE Trans Electr Comp EC-11*132-135
◇ D05 ◇ REV MR 26 # 3556 Zbl 163.255 • ID 32077

GINSBURG, S. & ROSE, G.F. [1963] *Operations which preserve definability in languages* (J 0037) ACM J 10*175-195
◇ D05 ◇ REV MR 28 # 747 • ID 32080

GINSBURG, S. & SPANIER, E.H. [1963] *Quotients of context-free languages* (J 0037) ACM J 10*487-492
◇ D05 ◇ REV MR 29 # 1108 Zbl 148.8 JSL 34.135 • ID 04967

GINSBURG, S. & ROSE, G.F. [1963] *Some recursively unsolvable problems in ALGOL-like languages* (J 0037) ACM J 10*29-47
◇ B70 D80 ◇ REV MR 28 # 748 Zbl 134.13 • ID 32078

GINSBURG, S. & GREIBACH, S.A. & HIBBARD, T.N. [1964] *Solvability of machine mappings of regular sets to regular sets* (J 0037) ACM J 11*302-312
◇ D05 ◇ ID 04969

GINSBURG, S. & HIBBARD, T.N. & ULLIAN, J.S. [1965] *Sequences in context free languages* (J 0316) Illinois J Math 9*321-337
◇ D05 ◇ REV MR 34 # 2467 Zbl 134.13 JSL 37.197 • ID 04973

GINSBURG, S. & ROSE, G.F. [1965] *Some recursively unsolvable problems in ALGOL-like languages* (C 1180) Readings Math Psychology 2*172-190
◇ B75 D80 ◇ REV Zbl 148.250 • ID 32079

GINSBURG, S. & ROSE, G.F. [1966] *A characterization of machine mappings* (J 0017) Canad J Math 18*381-388
◇ D05 ◇ REV MR 32 # 9165 Zbl 143.19 JSL 33.468 • ID 04987

GINSBURG, S. & ULLIAN, J.S. [1966] *Ambiguity in context free languages* (J 0037) ACM J 13*62-89
◇ D05 ◇ REV MR 36 # 2446 Zbl 139.122 JSL 33.301 • ID 04985

GINSBURG, S. & SPANIER, E.H. [1966] *Bounded regular sets* (J 0053) Proc Amer Math Soc 17*1043-1049
◇ D05 ◇ REV MR 34 # 1194 Zbl 147.253 • ID 04978

GINSBURG, S. & GREIBACH, S.A. [1966] *Deterministic context free languages* (J 0194) Inform & Control 9*620-648
◇ D05 ◇ REV MR 34 # 7301 Zbl 145.8 JSL 33.302 • ID 05327

GINSBURG, S. & SPANIER, E.H. [1966] *Finite-turn pushdown automata* (J 0946) SIAM J Control 4*429-543
◇ D10 ◇ REV MR 34 # 4138 Zbl 147.253 • ID 22308

GINSBURG, S. & GREIBACH, S.A. [1966] *Mappings which preserve context sensitive languages* (J 0194) Inform & Control 9*563-582
◇ D05 ◇ REV MR 34 # 5595 Zbl 179.317 • ID 15004

GINSBURG, S. & ULLIAN, J.S. [1966] *Preservation of unambiguity and inherent ambiguity in context-free languages* (J 0037) ACM J 13*364-368
◇ D05 ◇ REV MR 34 # 1119 Zbl 143.16 JSL 33.301 • ID 04989

GINSBURG, S. & ROSE, G.F. [1966] *Preservation of languages by transducers* (J 0194) Inform & Control 9*153-176
• ERR/ADD ibid 12*549-552
◇ D05 ◇ REV MR 38 # 4234 Zbl 186.13 • ID 19405

GINSBURG, S. & SPANIER, E.H. [1966] *Semigroups, Presburger formulas, and languages* (J 0048) Pac J Math 16*285-296
◇ D05 F30 ◇ REV MR 32 # 9172 Zbl 143.16 JSL 34.137 • ID 04980

GINSBURG, S. [1966] *The mathematical theory of context free languages* (X 0822) McGraw-Hill: New York xii + 232pp
◇ B65 D05 D98 ◇ REV MR 35 # 2692 Zbl 184.284 JSL 33.300 • ID 04982

GINSBURG, S. & HARRISON, M.A. [1967] *Bracketed context free languages* (J 0119) J Comp Syst Sci 1*1-23
◇ D05 ◇ REV MR 38 # 4236 Zbl 153.8 • ID 32085

GINSBURG, S. & GREIBACH, S.A. & HARRISON, M.A. [1967] *One-way stack automata* (J 0037) ACM J 14*389-418
◇ D05 D10 ◇ REV MR 39 # 5262 Zbl 171.148 • ID 32084

GINSBURG, S. & GREIBACH, S.A. & HARRISON, M.A. [1967] *Stack automata and compiling* (J 0037) ACM J 14*172-201
◇ B75 D05 D10 ◇ REV MR 39 # 1241 Zbl 153.11 • ID 04991

GINSBURG, S. & ROSE, G.F. [1968] *A note on preservation of languages by transducers* (J 0194) Inform & Control 12*549-552
◇ D05 ◇ REV MR 38 # 4235 Zbl 165.23 • ID 32089

GINSBURG, S. & SPANIER, E.H. [1968] *Control sets on grammars* (J 0041) Math Syst Theory 2*159-177
◇ D05 ◇ REV MR 38 # 4237 Zbl 157.336 • ID 32086

GINSBURG, S. & SPANIER, E.H. [1968] *Derivation-bounded languages* (J 0119) J Comp Syst Sci 2*228-250
◇ D05 ◇ REV MR 39 # 2546 Zbl 176.167 • ID 32090

GINSBURG, S. & HARRISON, M.A. [1968] *On the elimination of endmarkers* (**J** 0194) Inform & Control 12*103-115
◇ D05 ◇ REV Zbl 175.278 • ID 32087

GINSBURG, S. & HARRISON, M.A. [1968] *One-way nondeterministic real-time list-storage languages* (**J** 0037) ACM J 15*428-446
◇ D05 ◇ REV MR 38 # 6896 Zbl 207.16 • ID 32088

GINSBURG, S. & PARTEE, B. [1969] *A mathematical model of transformational grammars* (**J** 0194) Inform & Control 15*297-334
◇ B65 D35 ◇ REV MR 40 # 8533 Zbl 207.15 • ID 32093

GINSBURG, S. & GREIBACH, S.A. [1969] *Abstract families of languages* (**S** 0167) Mem Amer Math Soc 87*1-32
◇ D05 ◇ REV MR 45 # 6547a • ID 32091

GINSBURG, S. & GREIBACH, S.A. & HOPCROFT, J.E. [1969] *Pre-AFL* (**S** 0167) Mem Amer Math Soc 86*41-51
◇ D05 ◇ REV MR 45 # 6547c • ID 32092

GINSBURG, S. & GREIBACH, S.A. & HOPCROFT, J.E. [1969] *Studies in abstract families of languages* (**S** 0167) Mem Amer Math Soc 87*51pp
◇ D05 ◇ REV MR 40 # 8534 Zbl 194.314 • ID 32123

GINSBURG, S. & HARRISON, M.A. [1970] *On the closure of AFL under reversal* (**J** 0194) Inform & Control 17*395-409
◇ D05 ◇ REV MR 43 # 1776 Zbl 225 # 68042 • ID 32098

GINSBURG, S. & ROSE, G.F. [1970] *On the existence of generators for certain AFL* (**J** 0191) Inform Sci 2*431-446
◇ D05 ◇ REV MR 44 # 2539 Zbl 204.319 • ID 32097

GINSBURG, S. & GREIBACH, S.A. [1970] *Principal AFL* (**J** 0119) J Comp Syst Sci 4*308-338
◇ D05 ◇ REV MR 44 # 3808 Zbl 198.31 • ID 32096

GINSBURG, S. & SPANIER, E.H. [1970] *Substitution in families of languages* (**J** 0191) Inform Sci 2*83-110
◇ D05 ◇ REV Zbl 211.314 • ID 32095

GINSBURG, S. & HOPCROFT, J.E. [1970] *Two-way balloon automata and AFL* (**J** 0037) ACM J 17*3-13
◇ D05 ◇ REV MR 43 # 3047 Zbl 198.31 • ID 32094

GINSBURG, S. & SPANIER, E.H. [1971] *AFL with the semilinear property* (**J** 0119) J Comp Syst Sci 5*365-396
◇ D05 ◇ REV MR 49 # 4316 Zbl 235 # 68029 • ID 32099

GINSBURG, S. & HOPCROFT, J.E. [1971] *Images of AFL under certain families of homomorphisms* (**J** 0041) Math Syst Theory 5*216-227
◇ D05 ◇ REV MR 45 # 8475 Zbl 222 # 68035 • ID 32100

GINSBURG, S. & GOLDSTINE, J. [1971] *Intersection-closed full AFL and the recursively enumerable languages* (**P** 0680) ACM Symp Th of Comput (3);1971 Shaker Heights 121-131
◇ D05 D25 ◇ REV MR 48 # 5429 Zbl 251 # 68044 • ID 62030

GINSBURG, S. [1972] see BOOK, R.V.

GINSBURG, S. & GREIBACH, S.A. [1972] *Multitape AFA* (**J** 0037) ACM J 19*193-221
◇ D05 ◇ REV MR 45 # 9873 Zbl 241 # 68031 • ID 32102

GINSBURG, S. & GOLDSTINE, J. [1972] *On the largest full sub-AFL of an AFL* (**J** 0041) Math Syst Theory 6*241-242
◇ D05 ◇ REV Zbl 239 # 68013 • ID 32103

GINSBURG, S. & GOLDSTINE, J. [1973] *Intersection-closed full AFL and the recursively enumerable languages* (**J** 0194) Inform & Control 22*201-231
◇ D05 D25 ◇ REV MR 48 # 5429 Zbl 267 # 68035 • ID 32106

GINSBURG, S. & GREIBACH, S.A. [1973] *On AFL generators for finitely encoded AFA* (**J** 0119) J Comp Syst Sci 7*1-27
◇ D05 ◇ REV MR 48 # 5430 Zbl 249 # 68025 • ID 32104

GINSBURG, S. [1973] see GABRIELIAN, A.

GINSBURG, S. [1974] see GABRIELIAN, A.

GINSBURG, S. & SPANIER, E.H. [1974] *On incomparable AFL* (**J** 0119) J Comp Syst Sci 9*88-108
◇ D05 ◇ REV MR 49 # 8431 Zbl 289 # 68038 • ID 32109

GINSBURG, S. & ROVAN, B. [1974] *On the periodicity of word-length in DOL languages* (**J** 0194) Inform & Control 26*34-44
◇ D05 ◇ REV MR 50 # 9441 Zbl 288 # 68036 • ID 32110

GINSBURG, S. & ROSE, G.F. [1974] *The equivalence of stack-counter acceptors and quasi-realtime stack-counter acceptors* (**J** 0119) J Comp Syst Sci 8*243-269
◇ D05 D10 D15 ◇ REV MR 48 # 12881 Zbl 275 # 68017 • ID 32107

GINSBURG, S. [1975] *Algebraic and automata-theoretic properties of formal languages* (**X** 0809) North Holland: Amsterdam xii + 313pp
◇ D05 D98 ◇ REV MR 56 # 1816 Zbl 325 # 68002 JSL 41.788 • ID 44415

GINSBURG, S. [1975] see CREMERS, A.B.

GINSBURG, S. & SPANIER, E.H. [1975] *Substitution on grammar forms* (**J** 1431) Acta Inf 5*377-386
◇ D05 ◇ REV MR 53 # 12102 Zbl 373 # 68046 • ID 32114

GINSBURG, S. & ROZENBERG, G. [1975] *Tol systems and control sets* (**J** 0194) Inform & Control 27*109-125
◇ D05 ◇ REV MR 51 # 4733 Zbl 294 # 68027 • ID 32111

GINSBURG, S. & GOLDSTINE, J. & GREIBACH, S.A. [1975] *Uniformly erasable AFL* (**J** 0119) J Comp Syst Sci 10*165-182
◇ D05 ◇ REV MR 52 # 12421 Zbl 325 # 68042 • ID 32112

GINSBURG, S. & MAURER, H.A. [1976] *On strongly equivalent context-free grammar forms* (**J** 0373) Comp Arch Inform & Numerik 16*281-290
◇ D05 ◇ REV MR 53 # 12103 Zbl 376 # 68046 • ID 32115

GINSBURG, S. & LYNCH, N.A. [1976] *Size complexity in context-free grammatical families* (**J** 0037) ACM J 23*582-598
◇ D05 D15 ◇ REV MR 58 # 13966 Zbl 333 # 68057 • ID 32117

GINSBURG, S. & GOLDSTINE, J. & GREIBACH, S.A. [1976] *Some uniformly erasable families of languages* (**J** 1426) Theor Comput Sci 2*29-44
◇ D05 ◇ REV MR 53 # 14992 Zbl 343 # 68033 • ID 32116

GINSBURG, S. & LYNCH, N.A. [1977] *Derivation complexity in context-free grammar forms* (**J** 1428) SIAM J Comp 6*123-138
◇ D05 ◇ REV MR 55 # 11726 Zbl 346 # 68035 • ID 32118

GINSBURG, S. & MAURER, H.A. [1977] *On quasi-interpretations of grammar forms* (**J** 0373) Comp Arch Inform & Numerik 19*141-147
◇ D05 ◇ REV MR 56 # 13803 Zbl 363 # 68109 • ID 32119

GINSBURG, S. [1977] see CREMERS, A.B.

GINSBURG, S. & LEONG, B. & MAYER, O. & WOTSCHKE, D. [1979] *On strict interpretations of grammar forms* (**J** 0041) Math Syst Theory 12*233-252
◇ D05 ◇ REV MR 80f:68083 Zbl 412 # 68071 • ID 52986

GINSBURG, S. [1980] *Methods for specifying families of formal languages - past-present-future* (**P** 4266) Form Lang Th;1979 Santa Barbara 1-21
◇ D05 ◇ REV MR 84j:68001 Zbl 545 # 68065 • ID 47195

GINSBURG, S. & KINTALA, C.M.R. [1981] *Strict interpretations of deterministic pushdown acceptors* (**J** 0041) Math Syst Theory 14*229-240
◇ D10 ◇ REV MR 82f:68079 Zbl 471 # 68062 • ID 81342

GINSBURG, S. [1982] see BLATTNER, M.

GINSBURG, S. see Vol. V for further entries

GINZBURG, A. [1968] *Algebraic theory of automata* (**X** 0801) Academic Pr: New York 165pp
◇ D05 ◇ REV MR 39 # 4009 Zbl 195.25 • ID 04996

GINZBURG, A. see Vol. V for further entries

GIRARD, J.-Y. [1977] *Functionals and ordinoids* (**P** 1729) Colloq Int Log;1975 Clermont-Ferrand 59-71
◇ D65 F10 F15 ◇ REV MR 80h:03067 Zbl 441 # 03012 • ID 56065

GIRARD, J.-Y. [1981] *Π^1_2-logic I. Dilators* (**J** 0007) Ann Math Logic 21*75-219
◇ D55 E10 F15 G30 ◇ REV MR 83i:03093 Zbl 496 # 03037 • ID 35541

GIRARD, J.-Y. & VAUZEILLES, J. [1984] *Les premiers recursivement inaccessible et Mahlo et la theorie des dilatateurs* (**J** 0009) Arch Math Logik Grundlagenforsch 24*167-191
◇ D60 E47 E55 F05 F15 ◇ REV Zbl 556 # 03043 • ID 42404

GIRARD, J.-Y. & RESSAYRE, J.-P. [1985] *Elements de logique Π^1_n* (**P** 4046) Rec Th;1982 Ithaca 389-445
◇ D60 F15 F35 ◇ REV Zbl 573 # 03029 • ID 46368

GIRARD, J.-Y. & NORMANN, D. [1985] *Set recursion and Π^1_2-logic* (**J** 0073) Ann Pure Appl Logic 28*255-286
◇ D65 F15 ◇ REV Zbl 575 # 03034 • ID 42597

GIRARD, J.-Y. see Vol. I, II, III, V, VI for further entries

GIRAULT-BEAUQUIER, D. [1984] *Bilimites de langages reconaissables* (**J** 1426) Theor Comput Sci 33*335-342
◇ D05 ◇ REV Zbl 555 # 68044 • ID 44153

GIRAULT-BEAUQUIER, D. [1984] *Some results about finite and infinite behaviours of a pushdown automation* (**P** 4012) Automata, Lang & Progr (11);1984 Antwerpen 187-195
◇ D10 ◇ REV Zbl 554 # 68057 • ID 47146

GIRAULT-BEAUQUIER, D. also published under the name BEAUQUIER, D.

GIRE, F. & NIVAT, M. [1984] *Relations rationnelles infinitaires* (**J** 0089) Calcolo 21*91-125
◇ D05 ◇ REV Zbl 552 # 68064 • ID 48120

GIRSTMAIR, K. [1979] *Ueber konstruktive Methoden der Galoistheorie* (**J** 0504) Manuscr Math 26*423-441
◇ C57 C60 F55 ◇ REV MR 80b:12016 Zbl 408 # 12024 • ID 56289

GIULIANO, J.A. [1972] *Writing stack acceptors* (**J** 0119) J Comp Syst Sci 6*168-204
◇ D05 D10 ◇ REV MR 52 # 9689 Zbl 242 # 68023 • ID 05003

GIUSTI, M. [1984] *Some effectivity problems in polynomial ideal theory* (**P** 4013) Symb & Algeb Comput;1984 Cambridge 159-171
◇ D80 ◇ ID 47041

GIVE'ON, Y. [1968] see ARBIB, M.A.

GLADKIJ, A.V. [1961] *On simple Dyck words (Russian)* (**J** 0092) Sib Mat Zh 2*36-45
◇ D40 ◇ REV MR 23 # A3089 Zbl 118.265 • ID 05004

GLADKIJ, A.V. [1963] *Configuration characteristics of languages (Russian)* (**J** 0052) Probl Kibern 10*251-260
◇ D05 ◇ REV MR 32 # 2287 JSL 35.339 • ID 05006

GLADKIJ, A.V. [1963] *On the recognition of replaceability in recursive languages (Russian)* (**J** 0003) Algebra i Logika 2/3*5-22
◇ D05 ◇ REV MR 28 # 3933 Zbl 143.16 • ID 05005

GLADKIJ, A.V. [1965] *Algorithmic undecidability of inherent ambiguity in context-free languages (Russian)* (**J** 0003) Algebra i Logika 4/4*53-63
◇ D05 D10 ◇ REV MR 34 # 2376 JSL 35.339 • ID 05007

GLADKIJ, A.V. [1965] *Certain decision problems for context-free grammars (Russian)* (**J** 0003) Algebra i Logika 4/1*3-13 • ERR/ADD ibid 4/5*93
◇ D05 ◇ REV MR 34 # 45 Zbl 143.16 JSL 35.339 • ID 05008

GLADKIJ, A.V. [1973] *Formale Grammatiken und Sprachen (Russisch)* (**X** 2027) Nauka: Moskva 368pp
◇ D05 D10 D15 ◇ REV MR 50 # 6214 Zbl 262 # 68029 • ID 62037

GLADKIJ, A.V. (ED.) [1983] *Mathematical logic, mathematical linguistics and theory of algorithms (Russian)* (**X** 1434) Kalinin Gos Univ: Kalinin 116pp
◇ B97 D97 ◇ REV MR 84k:03005 • ID 34944

GLADKIJ, A.V. [1983] *Theory of algorithms (Russian)* (**X** 1434) Kalinin Gos Univ: Kalinin 60pp
◇ D20 D98 ◇ REV MR 85m:03028 • ID 45285

GLADKIJ, A.V. see Vol. I, II, V for further entries

GLADSTONE, M.D. [1965] *Some ways of constructing a propositional calculus of any required degree of unsolvability* (**J** 0064) Trans Amer Math Soc 118*192-210
◇ B20 D25 D35 ◇ REV MR 31 # 26 Zbl 168.12 JSL 34.505 • ID 05010

GLADSTONE, M.D. [1967] *A reduction of the recursion scheme* (**J** 0036) J Symb Logic 32*505-508
◇ D20 ◇ REV MR 37 # 59 Zbl 157.17 JSL 35.591 • ID 05012

GLADSTONE, M.D. [1968] *A single-axiom implicational calculus of given unsolvability* (**J** 0068) Z Math Logik Grundlagen Math 14*193-204
◇ B20 D25 D30 D35 ◇ REV MR 37 # 6180 Zbl 164.318 • ID 05013

GLADSTONE, M.D. [1971] *Simplification of the recursion scheme* (J 0036) J Symb Logic 36*653-665
◇ D20 ◇ REV MR 46 # 5121 Zbl 248 # 02040 • ID 05015

GLADSTONE, M.D. see Vol. I, III for further entries

GLASS, A.M.W. [1974] *The word problem for lattice ordered groups* (J 3420) Proc Edinburgh Math Soc, Ser 2 19*217-219
◇ D40 ◇ REV MR 51 # 3313 Zbl 323 # 06042 • ID 17368

GLASS, A.M.W. & GUREVICH, Y. [1983] *The word problem for lattice ordered groups* (J 0064) Trans Amer Math Soc 280*127-138
◇ D40 ◇ REV MR 85d:06015 Zbl 527 # 06009 • ID 33767

GLASS, A.M.W. [1984] *The isomorphism problem and undecidable properties for finitely presented lattice-ordered groups* (P 2167) Orders: Descr & Roles;1982 L'Arbresle 157-170
◇ C60 D40 G10 ◇ REV Zbl 549 # 06010 • ID 44713

GLASS, A.M.W. & MADDEN, J.J. [1984] *The word problem versus the isomorphism problem* (J 3172) J London Math Soc, Ser 2 30*53-61
◇ D40 ◇ REV Zbl 551 # 20018 • ID 44094

GLASS, A.M.W. [1985] *Effective extensions of lattices-ordered groups that preserve the degree of the conjugacy and the word problems* (P 4663) Ordered Algeb Struct;1982 Cincinnati 89-98
◇ C57 C60 D40 ◇ ID 49075

GLASS, A.M.W. [1985] *The word and isomorphism problems in universal algebra* (P 4178) Universal Algeb & Lattice Th;1984 Charleston 123-128
◇ C05 C57 D40 ◇ ID 47863

GLASS, A.M.W. see Vol. III, V for further entries

GLEBSKIJ, YU.V. [1965] see DUDICH, A.M.

GLEBSKIJ, YU.V. [1970] *A certain class of word sets (Russian) (English summary)* (J 1572) Izv Vyssh Ucheb Zaved, Radiofizika (Moskva) 13*1256-1258
◇ D05 ◇ REV MR 45 # 1765 • ID 05036

GLEBSKIJ, YU.V. & KOGAN, D.I. [1971] *Additively controllable systems and languages: certain algorithmic problems (Russian) (English summary)* (J 0040) Kibernetika, Akad Nauk Ukr SSR 1971/4*25-29
◇ D05 ◇ REV MR 46 # 4792 Zbl 241 # 68032 • ID 62043

GLEBSKIJ, YU.V. & KOGAN, D.I. & LIOGON'KIJ, M.I. [1973] *On a certain discrete walk problem (Russian)* (J 0499) Uch Zap Univ, Gor'kij 166*186-194
◇ D80 ◇ REV MR 56 # 5257 • ID 73406

GLEBSKIJ, YU.V. & GORDON, E.I. [1974] *Asynchronous automata with delays, and logical languages (Russian) (English summary)* (J 0011) Avtom Telemekh 1974/12*143-148
• TRANSL [1974] (J 0010) Autom & Remote Control 35*2002-2007
◇ D05 ◇ REV MR 56 # 11644 Zbl 307 # 94042 • ID 62045

GLEBSKIJ, YU.V. see Vol. I, III for further entries

GLEISER, M. [1980] *The emotional logician* (J 2697) Datamation 26*221-223
◇ A10 D99 ◇ REV MR 82a:01033 • ID 81351

GLINERT, E.P. [1971] *On restricted Turing computability* (J 0041) Math Syst Theory 5*331-343
◇ D20 ◇ REV MR 48 # 1896 Zbl 226 # 02038 • ID 05038

GLUKHOV, M.M. [1964] *Algorithimic solvability of the word problem for completely free modular lattices (Russian)* (J 0092) Sib Mat Zh 5*1027-1034
◇ D40 G10 ◇ REV MR 35 # 5367 Zbl 161.13 • ID 05044

GLUKHOV, M.M. [1970] *On free products and algorithm problems in R-varieties of universal algebras (Russian)* (J 0023) Dokl Akad Nauk SSSR 193*514-517
• TRANSL [1970] (J 0062) Sov Math, Dokl 11*957-960
◇ C05 D40 ◇ REV MR 45 # 146 Zbl 219 # 08004
• ID 05045

GLUKHOV, M.M. [1971] *Free expansions and algorithmic problems in R-varieties of universal algebras (Russian)* (J 0142) Mat Sb, Akad Nauk SSSR, NS 85(127)*307-338
• TRANSL [1971] (J 0349) Math of USSR, Sbor 14*297-328
◇ C05 D40 ◇ REV MR 45 # 147 Zbl 245 # 08002
• ID 05046

GLUKHOV, M.M. [1974] *Some algorithmic problems and free products in R-varieties of linear Ω-algebras (Russian)* (J 0092) Sib Mat Zh 15*229
• TRANSL [1974] (J 0475) Sib Math J 15*166
◇ C05 D40 ◇ REV Zbl 289 # 08003 • ID 30021

GLUSHKOV, V.M. [1960] *On an algorithm of synthesis of abstract automata (Russian) (English summary)* (J 0265) Ukr Mat Zh, Akad Nauk Ukr SSR 12*147-156
◇ D05 ◇ REV MR 24 # B2522 Zbl 101.250 JSL 33.629
• ID 21999

GLUSHKOV, V.M. [1960] *Ueber eine Methode der Analyse abstrakter Automaten (Ukrainisch)* (J 0270) Dokl Akad Nauk Ukr SSR, Ser A 1960*1151-1154
◇ D05 ◇ REV MR 23 # B1073 Zbl 94.8 • ID 47983

GLUSHKOV, V.M. [1961] *Abstract automata and the decomposition of free semigroups (Russian)* (J 0023) Dokl Akad Nauk SSSR 136*765-767
• TRANSL [1961] (J 0062) Sov Math, Dokl 2*121-124
◇ D05 ◇ REV MR 26 # 12 Zbl 103.248 • ID 05048

GLUSHKOV, V.M. [1961] *The abstract theory of automata (Russian)* (J 0067) Usp Mat Nauk 16/5*3-62 • ERR/ADD ibid 17/2*270
• TRANSL [1961] (J 1399) Russ Math Surv 16/5*1-53
◇ D05 ◇ REV MR 25 # 1976 Zbl 104.354 • ID 47998

GLUSHKOV, V.M. [1963] *Theorie der abstrakten Automaten* (X 0806) Dt Verlag Wiss: Berlin 103pp
◇ D05 D98 ◇ REV MR 29 # 4691 Zbl 128.13 • ID 24873

GLUSHKOV, V.M. & LETICHEVS'KIJ, O.A. [1973] *The theory of discrete transformators (Russian)* (C 0733) Izbr Vopr Algeb & Log (Mal'tsev) 5-39
◇ D05 ◇ REV MR 49 # 10184 Zbl 301 # 94062 • ID 62054

GLUSHKOV, V.M. & TSEJTLIN, G.E. & YUSHCHENKO, E.L. [1975] *Automata theory, and some questions of the synthesis of language processor structures (Russian) (English summary)* (J 0040) Kibernetika, Akad Nauk Ukr SSR 1975/5*1-20
• TRANSL [1975] (J 0021) Cybernetics 11*671-691
◇ D05 ◇ REV MR 58 # 33595 Zbl 364 # 68065 • ID 50987

GLUSHKOV, V.M. see Vol. I, II, VI for further entries

GNANI, G. [1973] see BERNARDI, C.

GNANI, G. [1974] *Insiemi dialettici generalizzati* (J 0319) Matematiche (Sem Mat Catania) 29*263-273
◇ B60 D25 D55 ◇ REV MR 55 # 5416 Zbl 343 # 02018
• ID 62056

GOBZEMIS, A.YU. [1976] see CHAPENKO, V.P.

GODDARD, L. [1962] *Proof-making* (**J** 0094) Mind 71*74-80
◇ A05 D35 ◇ ID 31088

GODDARD, L. [1965] *An augmented modal logic* (**J** 0047) Notre Dame J Formal Log 6*81-98
◇ B45 D35 ◇ REV MR 36 # 1303 Zbl 245 # 02024
• ID 05062

GODDARD, L. & ROUTLEY, R. [1966] *Use, mention and quotation* (**J** 0273) Australasian J Phil 44*1-49
◇ A05 D35 ◇ ID 31090

GODDARD, L. see Vol. I, II, VI for further entries

GODLEVSKIJ, A.B. [1973] *Certain special cases of the stopping problem and of the functional equivalence of automata (Russian) (English summary)* (**J** 0040) Kibernetika, Akad Nauk Ukr SSR 1973/4*90-97
◇ D05 ◇ REV MR 48 # 5421 Zbl 268 # 02025 • ID 62068

GODLEVSKIJ, A.B. & LETICHEVS'KIJ, O.A. (EDS.) [1974] *Automata and algorithm theory, and mathematical logic (Russian)* (**X** 2522) Akad Nauk Inst Kibernet: Kiev 98pp
◇ B35 B97 D05 D20 D97 ◇ REV MR 55 # 1829
• ID 80465

GODLEVSKIJ, A.B. [1974] *Ueber einen Spezialfall des Problems der Funktionalaequivalenz diskreter Transformer (Russisch)* (**J** 0040) Kibernetika, Akad Nauk Ukr SSR 1974/3*32-35
◇ D05 ◇ REV Zbl 301 # 94063 • ID 62069

GOEDEL, K. [1931] *Ueber formal unentscheidbare Saetze der "Principia Mathematica" und verwandter Systeme I* (**J** 0124) Monatsh Math-Phys 38*173-198
• TRANSL [1965] (**C** 0718) The Undecidable 5-38 [1967] (**C** 0675) From Frege to Goedel 596-616 [1962] (**X** 1323) Oliver & Boyd: Edinburgh vii + 72pp
◇ B28 B30 D20 D35 F25 F30 F35 ◇ REV MR 27 # 1373 Zbl 124.4 Zbl 2.1 FdM 57.54 • ID 15052

GOEDEL, K. [1933] *Zum Entscheidungsproblem des logischen Funktionenkalkuels* (**J** 0124) Monatsh Math-Phys 40*433-443
◇ B10 B20 B25 C13 D35 ◇ REV Zbl 8.289 FdM 59.865
• ID 20885

GOEDEL, K. [1965] *On undecidable propositions of formal mathematical systems* (**C** 0718) The Undecidable 41-71
◇ D20 D35 F30 ◇ REM Publ. as mimeographed lecture notes in 1934 • ID 21353

GOEDEL, K. see Vol. I, II, III, V, VI for further entries

GOESSEL, M. [1978] *A generalized principle of automata superposition relative to a pair of boolean functions (Russian) (English summary)* (**J** 0040) Kibernetika, Akad Nauk Ukr SSR 1978/3*33-37
• TRANSL [1978] (**J** 0021) Cybernetics 14*352-357
◇ B05 D05 ◇ REV MR 80m:68049 Zbl 393 # 94041
• ID 69601

GOETZE, B.G. [1974] *Die Struktur des Halbverbandes der effektiven Numerierungen* (**J** 0068) Z Math Logik Grundlagen Math 20*183-188
◇ D20 D45 ◇ REV MR 50 # 12681 Zbl 301 # 02035
• ID 05090

GOETZE, B.G. [1974] *Unentscheidbarkeitsgrade rekursiver Funktionen* (**J** 0068) Z Math Logik Grundlagen Math 20*189-191
◇ D20 ◇ REV MR 50 # 12691 Zbl 301 # 02034 • ID 05089

GOETZE, B.G. [1975] *Die Inseparabilitaetseigenschaften und die Ricesche Unentscheidbarkeitseigenschaft bei effektiven Numerierungen* (**J** 0129) Elektr Informationsverarbeitung & Kybern 11*583-586
◇ D25 D45 ◇ REV Zbl 328 # 02024 • ID 62076

GOETZE, B.G. & KLETTE, R. [1975] *Limes-rekursive Funktionen* (**J** 0129) Elektr Informationsverarbeitung & Kybern 11*586-589
◇ D20 ◇ REV Zbl 329 # 02018 • ID 62999

GOETZE, B.G. & KLETTE, R. [1975] *Some properties of limit recursive functions* (**P** 1755) Math Founds of Comput Sci (3);1974 Jadwisin 88-90
◇ D20 ◇ REV Zbl 331 # 02022 JSL 42.422 • ID 62072

GOETZE, B.G. [1976] *The structure of the lattice of recursive sets* (**J** 0068) Z Math Logik Grundlagen Math 22*187-191
◇ D20 ◇ REV MR 56 # 2805 Zbl 328 # 02023 • ID 18477

GOETZE, B.G. [1977] see GILLO, D.

GOETZE, B.G. & KLETTE, R. [1978] *Ein Normalformentheorem fuer Σ_2-Funktionen* (**J** 0129) Elektr Informationsverarbeitung & Kybern 14*251-256
◇ D20 D55 ◇ REV MR 80e:03043 Zbl 403 # 03030
• ID 54744

GOETZE, B.G. & NEHRLICH, W. [1978] *Loop programs and classes of primitive recursive functions* (**P** 1707) Math Founds of Comput Sci (7);1978 Zakopane 232-238
◇ D20 ◇ REV MR 80g:68035 Zbl 385 # 03032 • ID 52140

GOETZE, B.G. & NEHRLICH, W. [1980] *The structure of LOOP programs and subrecursive hierarchies* (**J** 0068) Z Math Logik Grundlagen Math 26*255-278
◇ D15 D20 ◇ REV MR 81j:03066 Zbl 439 # 03018
• ID 56010

GOETZE, B.G. & NEHRLICH, W. [1981] *The number of loops necessary and sufficient for computing functions (German and Russian summaries)* (**J** 0129) Elektr Informationsverarbeitung & Kybern 17*363-376
◇ D15 ◇ REV MR 84e:03048 Zbl 504 # 68026 • ID 34384

GOGUEN, J.A. & THATCHER, J.W. & WAGNER, E.G. & WRIGHT, J.B. [1975] *Factorisations, congruences, and the decomposition of automata and systemi* (**P** 1755) Math Founds of Comput Sci (3);1974 Jadwisin 33-45
◇ D05 ◇ REV MR 58 # 25071 Zbl 306 # 18005 • ID 33358

GOGUEN, J.A. see Vol. I, II, III, V for further entries

GOHEEN, H.E. [1963] see CAFFIN, R.W.

GOLD, A.Y. [1978] see COHEN, R.S.

GOLD, A.Y. [1980] see COHEN, R.S.

GOLD, E.M. [1965] *Limiting recursion* (**J** 0036) J Symb Logic 30*28-48
◇ D20 D55 ◇ REV MR 39 # 1326 Zbl 203.12 JSL 36.342
• ID 05097

GOLD, E.M. [1971] *Universal goal-seekers* (**J** 0194) Inform & Control 18*395-403
◇ D05 D10 D20 ◇ REV MR 46 # 7009 Zbl 218 # 68021
• ID 05098

GOLDBERG, ALLEN [1982] see BROWN, CYNTHIA A.

GOLDFARB, W.D. & LEWIS, H.R. [1973] *The decision problem for formulas with a small number of atomic subformulas* (J 0036) J Symb Logic 38*471-480
◇ B20 D35 ◇ REV MR 49 # 2328 Zbl 276 # 02029
• ID 08111

GOLDFARB, W.D. & LEWIS, H.R. [1975] *Skolem reduction classes* (J 0036) J Symb Logic 40*62-68
◇ B20 D35 ◇ REV MR 58 # 193 Zbl 306 # 02043
• ID 05117

GOLDFARB, W.D. [1981] *On the Goedel class with identity* (J 0036) J Symb Logic 46*354-364
◇ B20 B25 D35 ◇ REV MR 82g:03079 Zbl 472 # 03009
• ID 55275

GOLDFARB, W.D. [1981] *The undecidability of the second-order unification problem* (J 1426) Theor Comput Sci 13*225-230
◇ B15 D35 F20 ◇ REV MR 82c:03067 Zbl 457 # 03006
• ID 54331

GOLDFARB, W.D. [1984] *The unsolvability of the Goedel class with identity* (J 0036) J Symb Logic 49*1237-1252
◇ B20 D35 ◇ REV MR 86g:03015a • ID 42450

GOLDFARB, W.D. [1984] *The Goedel class with identity is unsolvable* (J 0589) Bull Amer Math Soc (NS) 10*113-115
◇ B20 D35 ◇ REV MR 85f:03006 Zbl 534 # 03005
• ID 36555

GOLDFARB, W.D. see Vol. I, III, VI for further entries

GOLDMAN, J. & HOMER, S. [1982] *Quadratic automata* (J 0119) J Comp Syst Sci 24*180-196
◇ B70 D05 D45 ◇ REV MR 84i:03079 Zbl 492 # 68046
• ID 34558

GOLDSCHLAGER, L.M. [1977] *The monotone and planar circuit value problems are log space complete for P* (J 1456) SIGACT News 9/2*25-29
◇ B70 D10 D15 ◇ REV Zbl 356 # 94042 • ID 50338

GOLDSCHLAGER, L.M. [1982] see BRENT, R.P.

GOLDSTINE, J. [1971] see GINSBURG, S.

GOLDSTINE, J. [1972] see GINSBURG, S.

GOLDSTINE, J. [1972] *Substitution and bounded languages* (J 0119) J Comp Syst Sci 6*9-29
◇ D05 ◇ REV MR 46 # 8477 Zbl 232 # 68030 • ID 05120

GOLDSTINE, J. [1973] see GINSBURG, S.

GOLDSTINE, J. [1975] see GINSBURG, S.

GOLDSTINE, J. [1976] see GINSBURG, S.

GOLDSTINE, J. [1980] *Formal languages and their relation to automata: what Hopcroft & Ullman didn't tell us* (P 4266) Form Lang Th;1979 Santa Barbara 109-140
◇ D05 ◇ REV MR 84j:68001 Zbl 545 # 68042 • ID 47199

GOLDSTINE, J. [1983] see GINSBERG, S.

GOLUNKOV, YU.V. [1978] see BUKHARAEVA, Z.K.

GOLUNKOV, YU.V. [1978] *Precomplete classes of algorithms that preserve the membership in a set (Russian)* (J 0031) Izv Vyssh Ucheb Zaved, Mat (Kazan) 1978/7(194)*93-96
• TRANSL [1978] (J 3449) Sov Math 22/7*77-79
◇ D20 D50 ◇ REV MR 58 # 27411 Zbl 413 # 68034
• ID 73480

GOLUNKOV, YU.V. & KNYAZEV, E.A. [1979] *Completeness of systems of commands with a finite set of memory states (Russian)* (J 3937) Veroyat Met i Kibern (Kazan) 15*18-33
◇ D05 ◇ REV MR 81h:68035 Zbl 422 # 68025 • ID 81364

GOLUNKOV, YU.V. (ED.) [1979] *Probabilistic methods and cybernetics No. XV (Russian)* (X 3605) Kazan Gos Univ: Kazan' 95pp
◇ D97 ◇ REV MR 81d:03003 • ID 70050

GOLUNKOV, YU.V. [1980] *Precomplete classes of algorithms that preserve membership in a set I (Russian)* (J 3937) Veroyat Met i Kibern (Kazan) 16*21-40
◇ D20 D75 ◇ REV MR 83h:68052 Zbl 446 # 68022 • REM Part II 1980 • ID 39209

GOLUNKOV, YU.V. [1980] *Precomplete classes of algorithms that preserve membership in a set II (Russian)* (J 3937) Veroyat Met i Kibern (Kazan) 17*35-51
◇ D25 D50 ◇ REV MR 82m:03054 Zbl 496 # 68022 • REM Part I 1980 • ID 73479

GOLUNKOV, YU.V. [1983] *Some families of maximal subsystems of algebras in a system of algorithmic algebras of partial recursive functions and predicates. I (Russian)* (J 0031) Izv Vyssh Ucheb Zaved, Mat (Kazan) 1983/12*13-20
• TRANSL [1983] (J 3449) Sov Math 27/12*13-22
◇ D75 ◇ REV MR 85g:03061 Zbl 539 # 03024 • REM Part II 1984 • ID 44010

GOLUNKOV, YU.V. & SAVEL'EV, A.A. [1984] *A lattice of systems of algorithmic algebras of partial recursive functions and predicates (Russian)* (J 0031) Izv Vyssh Ucheb Zaved, Mat (Kazan) 1984/11*57-59
◇ D75 ◇ ID 44717

GOLUNKOV, YU.V. [1984] *On some families of maximal subsystems of algebras in a system of algorithmic algebras of partial recursive functions and predicates. II (Russian)* (J 0031) Izv Vyssh Ucheb Zaved, Mat (Kazan) 1984/2*20-26
• TRANSL [1984] (J 3449) Sov Math 28/2*26-35
◇ D20 D75 ◇ REV Zbl 539 # 03025 • REM Part I 1983
• ID 41233

GOLUNKOV, YU.V. & SAVEL'EV, A.A. [1985] *Systems of algorithmic algebras preserving ideals in a lattice of recursively enumerable sets (Russian) (English summary)* (J 0040) Kibernetika, Akad Nauk Ukr SSR 1985/5*123-125,136
◇ D25 ◇ ID 49058

GOLUNKOV, YU.V. see Vol. II for further entries

GOLZE, U. [1982] see CARSTENS, H.G.

GONCHAROV, S.S. [1973] *Constructivizability of superatomic boolean algebras (Russian)* (J 0003) Algebra i Logika 12*31-40,120
• TRANSL [1973] (J 0069) Algeb and Log 12*17-22
◇ C57 D45 G05 ◇ REV MR 47 # 8265 Zbl 278 # 02039
• ID 05125

GONCHAROV, S.S. & NURTAZIN, A.T. [1973] *Constructive models of complete solvable theories (Russian)* (J 0003) Algebra i Logika 12*125-142,243
• TRANSL [1973] (J 0069) Algeb and Log 12*67-77
◇ C45 C50 C57 G05 ◇ REV MR 53 # 2667
Zbl 278 # 02038 • ID 21613

GONCHAROV, S.S. [1975] *Autostability and computable families of constructivizations (Russian)* (**J** 0003) Algebra i Logika 14*647-680,727
• TRANSL [1975] (**J** 0069) Algeb and Log 14*392-409
◇ C50 C57 D45 ◇ REV MR 55 # 10267 Zbl 367 # 02023
• ID 26031

GONCHAROV, S.S. [1975] *Certain properties of the constructivization of boolean algebras (Russian)* (**J** 0092) Sib Mat Zh 16*264-278,420
• TRANSL [1975] (**J** 0475) Sib Math J 16*203-214
◇ C57 D45 G05 ◇ REV MR 52 # 2846 Zbl 309 # 02052
• ID 17637

GONCHAROV, S.S. [1976] see DOBRITSA, V.P.

GONCHAROV, S.S. [1976] *Non-self-equivalent constructivization of atomic boolean algebras (Russian)* (**J** 0087) Mat Zametki (Akad Nauk SSSR) 19*853-858
• TRANSL [1976] (**J** 1044) Math Notes, Acad Sci USSR 19*500-503
◇ C57 D45 G05 ◇ REV MR 55 # 12490 Zbl 357 # 02043
• ID 50391

GONCHAROV, S.S. [1976] *Restricted theories of constructive boolean algebras (Russian)* (**J** 0092) Sib Mat Zh 17*797-812
• TRANSL [1976] (**J** 0475) Sib Math J 17*601-611
◇ B25 C57 D35 F60 G05 ◇ REV MR 55 # 93
Zbl 361 # 02066 • ID 52189

GONCHAROV, S.S. [1977] *The quantity of nonautoequivalent constructivizations (Russian)* (**J** 0003) Algebra i Logika 16*257-282
• TRANSL [1977] (**J** 0069) Algeb and Log 16*169-185
◇ C50 C57 D45 ◇ REV MR 81h:03067 Zbl 405 # 03016
• ID 54891

GONCHAROV, S.S. [1978] *Constructive models of $\aleph_1$-categorical theories (Russian)* (**J** 0087) Mat Zametki (Akad Nauk SSSR) 23*885-888
• TRANSL [1978] (**J** 1044) Math Notes, Acad Sci USSR 23*486-487
◇ C35 C57 ◇ REV MR 80g:03029 Zbl 385 # 03025
• ID 52133

GONCHAROV, S.S. [1978] *Strong constructivizability of homogeneous models (Russian)* (**J** 0003) Algebra i Logika 17*363-388
• TRANSL [1978] (**J** 0069) Algeb and Log 17*247-263
◇ C50 C57 D45 ◇ REV MR 82a:03026 Zbl 441 # 03015
• ID 56068

GONCHAROV, S.S. [1980] *A totally transcendental decidable theory without constructivizable homogeneous models (Russian)* (**J** 0003) Algebra i Logika 19*137-149
• TRANSL [1980] (**J** 0069) Algeb and Log 19*85-93
◇ C45 C50 C57 ◇ REV MR 83d:03047 Zbl 468 # 03024
• ID 55089

GONCHAROV, S.S. [1980] *Autostability of models and abelian groups (Russian)* (**J** 0003) Algebra i Logika 19*23-44
• TRANSL [1980] (**J** 0069) Algeb and Log 19*13-27
◇ C50 C57 C60 D45 ◇ REV MR 82g:03081
Zbl 468 # 03022 • ID 55087

GONCHAROV, S.S. [1980] see DZGOEV, V.D.

GONCHAROV, S.S. [1980] *Computable single-valued numerations (Russian)* (**J** 0003) Algebra i Logika 19*507-551
• TRANSL [1980] (**J** 0069) Algeb and Log 19*325-356
◇ D20 D45 ◇ REV MR 82m:03059 Zbl 514 # 03029
• ID 37410

GONCHAROV, S.S. [1980] see DROBOTUN, B.N.

GONCHAROV, S.S. [1980] *Problem of the number of non-self-equivalent constructivizations (Russian)* (**J** 0003) Algebra i Logika 19*621-639
• TRANSL [1980] (**J** 0069) Algeb and Log 19*401-414
◇ C57 D45 ◇ REV MR 82k:03046 Zbl 476 # 03046
• ID 55558

GONCHAROV, S.S. [1980] *The problem of the number of nonautoequivalent constructivizations (Russian)* (**J** 0023) Dokl Akad Nauk SSSR 251*271-274
• TRANSL [1980] (**J** 0062) Sov Math, Dokl 21*411-414
◇ C57 D45 ◇ REV MR 81i:03070 Zbl 476 # 03045
• ID 73488

GONCHAROV, S.S. [1980] *Totally transcendental theory with non-constructivizable prime model (Russian)* (**J** 0092) Sib Mat Zh 21/1*44-51
• TRANSL [1980] (**J** 0475) Sib Math J 21*32-37
◇ C45 C50 C57 D45 ◇ REV MR 81b:03039
Zbl 463 # 03005 • ID 54545

GONCHAROV, S.S. [1981] *Groups with a finite number of constructivizations (Russian)* (**J** 0023) Dokl Akad Nauk SSSR 256*269-272
• TRANSL [1981] (**J** 0062) Sov Math, Dokl 23*58-61
◇ C57 C60 D45 ◇ REV MR 82g:03083 Zbl 496 # 20021
• ID 37746

GONCHAROV, S.S. [1982] *Limiting equivalent constructivizations (Russian)* (**C** 3953) Mat Log & Teor Algor 4-12
◇ C57 D45 F60 ◇ REV MR 85d:03122 Zbl 543 # 03017
• ID 40426

GONCHAROV, S.S. [1983] *Positive numerations of families with single-valued numerations (Russian)* (**J** 0003) Algebra i Logika 22*481-488
• TRANSL [1983] (**J** 0069) Algeb and Log 22*345-350
◇ D45 ◇ REV Zbl 572 # 03023 • ID 44262

GONCHAROV, S.S. [1983] *Universal recursively enumerable boolean algebras (Russian)* (**J** 0092) Sib Mat Zh 24/6*36-43
• TRANSL [1983] (**J** 0475) Sib Math J 24*852-858
◇ C50 C57 D45 G05 ◇ REV MR 86b:03054
Zbl 522 # 03026 • ID 43384

GONCHAROV, S.S. [1985] *Stongly constructive models (Russian) (English summary)* (**P** 4310) Easter Conf on Model Th (3);1985 Gross Koeris 103-114
◇ C57 ◇ REV Zbl 575 # 03024 • ID 48825

GONCHAROV, S.S. see Vol. III for further entries

GONCHAROV, V.A. [1983] *A recursively representable Boolean algebra (Russian)* (**C** 4019) Kraevye Zadach Dlya Diff Uravnenij & Priloz Mekh & Tekh 43-46
◇ C57 D45 G05 ◇ REV MR 86b:03054 Zbl 565 # 03020
• ID 45205

GONCZAROWSKI, J. & WARMUTH, M.K. [1985] *Applications of scheduling theory to formal language theory* (**J** 1426) Theor Comput Sci 37*217-243
◇ D05 D15 ◇ ID 49076

GONCZAROWSKI, J. [1985] *Decidable properties of monadic recursive schemas with a depth parameter* (J 1431) Acta Inf 22/3*277-310
◇ D20 ◇ REV Zbl 564 # 68057 • ID 48983

GONSHOR, H. & RICE, H.G. [1969] *Recursive density types I* (J 0064) Trans Amer Math Soc 140*493-503
◇ D50 ◇ REV MR 39 # 1327 Zbl 183.293 • REM Part II 1969 • ID 05137

GONSHOR, H. [1969] *Recursive density types II* (J 0064) Trans Amer Math Soc 140*505-509
◇ D50 ◇ REV MR 40 # 4102 Zbl 198.293 • REM Part I 1969 by Gonshor,H. & Rice,G • ID 05139

GONSHOR, H. [1972] *The category of recursive functions* (J 0027) Fund Math 75*87-94
◇ D20 D75 G30 ◇ REV MR 46 # 3281 Zbl 235 # 08005 • ID 05141

GONSHOR, H. [1976] *Effective density types* (J 0047) Notre Dame J Formal Log 17*303-307
◇ D25 D50 ◇ REV MR 54 # 4949 Zbl 292 # 02040 • ID 18172

GONSHOR, H. see Vol. I, V for further entries

GONZALEZ, R.C. [1981] see BARRERO, A.

GOODMAN, NICOLAS D. [1978] *The nonconstructive content of sentences of arithmetic* (J 0036) J Symb Logic 43*497-501
◇ D20 D30 D55 F30 F50 ◇ REV MR 81e:03045 Zbl 398 # 03046 • ID 29277

GOODMAN, NICOLAS D. see Vol. II, V, VI for further entries

GOODSTEIN, R.L. [1954] *The relatively exponential, logarithmic and circular functions in recursive function theory* (J 0118) Acta Math 92*171-190
◇ D20 F30 F60 ◇ REV MR 16.783 Zbl 58.250 • ID 05175

GOODSTEIN, R.L. [1973] *On limiting the applications of the uniqueness rules in the equation calculus* (J 0068) Z Math Logik Grundlagen Math 19*115-116
◇ D20 ◇ REV MR 47 # 3152 Zbl 298 # 02037 • ID 05196

GOODSTEIN, R.L. [1975] *Hilbert's tenth problem and the independence of recursive difference* (J 3172) J London Math Soc, Ser 2 10*175-176
◇ D25 D35 ◇ REV MR 51 # 10055 Zbl 324 # 02031 • ID 17545

GOODSTEIN, R.L. see Vol. I, II, III, V, VI for further entries

GORAJ, A. & MIRKOWSKA, M. & PALUSZKIEWICZ, A. [1970] *On the notion of the description of the programm (Russian summary)* (J 0014) Bull Acad Pol Sci, Ser Math Astron Phys 18*499-505
◇ B75 D05 ◇ REV MR 42 # 7514 Zbl 222 # 02053 • ID 28470

GORALCIK, P. & GORALCIKOVA, A. & KOUBEK, V. [1980] *Testing of properties of finite algebras* (P 2904) Automata, Lang & Progr (7);1980 Noordwijkerhout 273-281
◇ C13 D15 D40 ◇ REV MR 82b:03088 Zbl 449 # 68033 • ID 73522

GORALCIKOVA, A. [1980] see GORALCIK, P.

GORBUNOV, V.A. [1975] see BELKIN, V.P.

GORBUNOV, V.A. see Vol. III for further entries

GORDEEV, L.N. [1977] *A majorizing semantics for hyperarithmetic sentences (Russian) (English summary)* (S 0228) Zap Nauch Sem Leningrad Otd Mat Inst Steklov 68*30-37,143
• TRANSL [1981] (J 1531) J Sov Math 15*16-21
◇ D55 F15 F50 ◇ REV MR 58 # 27410 Zbl 359 # 02032 • ID 50580

GORDEEV, L.N. see Vol. III, V, VI for further entries

GORDON, C.E. [1970] *Comparisons between some generalizations of recursion theory* (J 0020) Compos Math 22*333-346
◇ D75 ◇ REV MR 44 # 2601 Zbl 235 # 02037 • ID 05198

GORDON, C.E. [1974] *Prime and search computability, characterized as definability in certain sublanguages of constructible* $L_{\omega_1,\omega}$ (J 0064) Trans Amer Math Soc 197*391-408
◇ C75 D75 ◇ REV MR 54 # 4945 Zbl 341 # 02030 • ID 05199

GORDON, C.E. see Vol. V, VI for further entries

GORDON, D. [1979] *Complexity classes of provable recursive functions* (J 0119) J Comp Syst Sci 18*294-303
◇ D15 D20 F30 ◇ REV MR 80e:68105 Zbl 423 # 03045 • ID 53557

GORDON, D. & SHAMIR, E. [1983] *Computation of recursive functionals using minimal initial segments* (J 1426) Theor Comput Sci 23*305-315
◇ D20 ◇ REV MR 84h:03106 Zbl 531 # 03022 • ID 34295

GORDON, D. see Vol. V for further entries

GORDON, E.I. [1974] see GLEBSKIJ, YU.V.

GORDON, E.I. see Vol. I, III, V for further entries

GORDON, H.G. [1976] *Complete degrees of finite-state transformability* (J 0194) Inform & Control 32*169-187
◇ D05 D20 ◇ REV MR 54 # 4862 Zbl 342 # 94022 • ID 24081

GORLOV, V.V. [1973] *Congruences on closed Post classes (Russian)* (J 0087) Mat Zametki (Akad Nauk SSSR) 13*725-734
• TRANSL [1973] (J 1044) Math Notes, Acad Sci USSR 13*434-438
◇ B50 D03 ◇ REV MR 48 # 10781 Zbl 261 # 02037 • ID 05201

GORLOV, V.V. see Vol. II for further entries

GORN, S. [1963] *Processors for infinite codes of the Shannon-Fano type* (P 0674) Symp Math Th of Automata;1962 New York 223-240
◇ B75 D05 ◇ REV MR 30 # 5897 Zbl 116.96 JSL 35.487 • ID 05204

GORN, S. see Vol. II for further entries

GORYAGA, A.V. & KIRKINSKIJ, A.S. [1975] *Decidability of the conjugacy problem does not carry over to finite extensions of groups (Russian)* (J 0003) Algebra i Logika 14*393-406
• TRANSL [1975] (J 0069) Algeb and Log 14*240-248
◇ D40 ◇ REV MR 54 # 2813 Zbl 358 # 20046 • ID 24671

GOSTANIAN, R. & HRBACEK, K. [1979] *A new proof that* $\pi^1_1 < \sigma^1_1$ (J 0068) Z Math Logik Grundlagen Math 25*407-408
◇ D60 D70 E45 ◇ REV MR 81g:03057 Zbl 424 # 03027 • ID 73545

GOSTANIAN, R. [1979] *The next admissible ordinal* (J 0007) Ann Math Logic 17*171-203
◇ D60 E45 ◇ REV MR 81c:03039 Zbl 424 # 03024
• ID 73542

GOSTANIAN, R. see Vol. III, V for further entries

GOSTEV, YU.G. [1981] *Atomary languages and grammars. On the theory of data structure families (Russian)* (J 0040) Kibernetika, Akad Nauk Ukr SSR 1981/2*20-25
• TRANSL [1981] (J 0021) Cybernetics 17*176-181
◇ B05 D05 ◇ REV MR 82k:68044 Zbl 469 # 68030
• ID 81380

GRABOWSKI, J. [1976] *Unpredictable sets* (J 0129) Elektr Informationsverarbeitung & Kybern 12*439-447
◇ D20 ◇ REV MR 58 # 27417 Zbl 335 # 02030 • ID 62144

GRABOWSKI, J. & STARKE, P.H. [1978] *Petri-Netze* (J 3920) ZKI Inf, Akad Wiss DDR 3/1978*1-88
◇ D05 D80 ◇ ID 42725

GRABOWSKI, J. [1979] *The unsolvability of some Petri net language problems* (J 0232) Inform Process Lett 9*60-63
◇ D05 D80 ◇ REV MR 80j:68017 Zbl 414 # 68032
• ID 53097

GRABOWSKI, M. [1978] *The algebraic properties of nets from a program-theoretical point of view* (J 2095) Fund Inform, Ann Soc Math Pol, Ser 4 2*83-101
◇ D05 D80 ◇ REV MR 80b:68021 Zbl 404 # 68045
• ID 69622

GRABOWSKI, M. see Vol. I, III for further entries

GRANDJEAN, E. [1976] *Sur les classes de relations recursives definies par la methode de Trahtenbrot (English summary)* (J 2313) C R Acad Sci, Paris, Ser A-B 283*A127-A129
◇ D20 ◇ REV MR 54 # 7225 Zbl 341 # 02037 • ID 25016

GRANDJEAN, E. [1976] *Une notion de recursivite relative qui correspond a la recursivite au sens de Trahtenbrot (English summary)* (J 2313) C R Acad Sci, Paris, Ser A-B 282*A251-A253
◇ D20 D30 ◇ REV MR 57 # 84 Zbl 326 # 02029
• ID 27259

GRANDJEAN, E. [1983] *Complexity of the first-order theory of almost all finite structures* (J 0194) Inform & Control 57*180-204
◇ B25 C13 D15 ◇ REV MR 85j:03061 Zbl 542 # 03018
• ID 40384

GRANDJEAN, E. [1984] *Spectre des formules du premier order et complexite algorithmique* (C 4356) Gen Log Semin Paris 1982/83 139-152
◇ C13 C98 D15 ◇ ID 48072

GRANDJEAN, E. [1984] *The spectra of first-order sentences and computational complexity* (J 1428) SIAM J Comp 13*356-373
◇ C13 D15 ◇ REV MR 86g:68055 Zbl 535 # 03014
• ID 45011

GRANDJEAN, E. [1984] *Universal quantifiers and time complexity of random access machines* (P 2342) Symp Rek Kombin;1983 Muenster 366-379
◇ C13 D15 ◇ REV Zbl 548 # 03017 • ID 41788

GRANDJEAN, E. [1985] *Universal quantifiers and time complexity of random access machines* (J 0041) Math Syst Theory 18*171-187
◇ C13 D10 D15 ◇ REV Zbl 578 # 03022 • ID 42738

GRANDY, R.E. [1966] *A note on the recursive unsolvability of primitive recursive arithmetic* (J 0053) Proc Amer Math Soc 17*146-147
◇ D35 F30 ◇ REV MR 32 # 7404 Zbl 148.247 • ID 05303

GRANDY, R.E. see Vol. I, II, III for further entries

GRANT, P.W. [1977] *Recognition of EOL languages in less than quartic time* (J 0232) Inform Process Lett 6*174-175
◇ D05 D15 ◇ REV MR 56 # 10163 Zbl 381 # 68069
• ID 31080

GRANT, P.W. [1977] *Strict Π^1_1-predicates on countable and cofinality ω transitive sets* (J 0036) J Symb Logic 42*161-173
◇ C15 C40 C70 D55 D60 D70 E47 E60 ◇ REV MR 58 # 21537 Zbl 369 # 02020 • ID 26439

GRANT, P.W. [1980] *Some more independence results in complexity theory* (J 1426) Theor Comput Sci 12*119-126
◇ D15 ◇ REV MR 81m:03051 Zbl 447 # 68041 • ID 56595

GRANT, P.W. see Vol. III for further entries

GRASSELLI, J. [1975] *On the tenth Hilbert problem (Slovene) (English summary)* (J 2310) Obz Mat Fiz, Ljubljana 22*1-3
◇ D35 ◇ REV MR 56 # 2911 Zbl 294 # 02021 • ID 62165

GRASSIN, J. [1972] *Une generalisation du theoreme de Myhill-Sheperdson aux combinaisons boolennes d'ouverts dont une cle est recursivement enumerable* (J 2313) C R Acad Sci, Paris, Ser A-B 275*A1023-A1026
◇ D25 ◇ REV MR 46 # 7007 Zbl 255 # 02047 • ID 05306

GRASSIN, J. [1974] *Index sets in Ershov's hierarchy* (J 0036) J Symb Logic 39*97-104
◇ D55 ◇ REV MR 50 # 4266 Zbl 287 # 02030 JSL 39.97
• ID 05307

GRASSIN, J. [1978] *Definitions inductives monotones sur le continu, dont les composantes sont denombrables (English summary)* (J 2313) C R Acad Sci, Paris, Ser A-B 287*A911-A913
◇ D55 D60 D70 E15 ◇ REV MR 80c:03049 Zbl 428 # 03039 • ID 32137

GRASSIN, J. [1981] *Δ^1_1-good inductive definitions over the continuum* (J 0068) Z Math Logik Grundlagen Math 27*11-16
◇ D55 D70 E15 ◇ REV MR 82j:03058 Zbl 499 # 03037
• ID 73595

GRAW, B. [1985] see BUDACH, L.

GREEN, J. [1974] *Σ_1-compactness for next admissible sets* (J 0036) J Symb Logic 39*105-116
◇ C62 C70 D60 ◇ REV MR 51 # 5262 Zbl 286 # 02059
• ID 05312

GREEN, J. see Vol. III, V for further entries

GREEN, M.W. [1964] *A lower bound on Rado's sigma function for binary Turing machines* (P 0550) Switch Circ Th & Log Design (5);1964 Princeton 91-94
◇ D10 ◇ REV JSL 40.617 • ID 14693

GREGORIO DI, S. [1982] see APOLLONI, B.

GREGUSOVA, L. & KOREC, I. [1979] *Small universal Minsky machines* (P 2059) Math Founds of Comput Sci (8);1979 Olomouc 308-316
◇ D10 ◇ REV MR 81g:03042 Zbl 412 # 03019 • ID 52950

GREIBACH, S.A. [1963] *The undecidability of the ambiguity problem for minimal linear grammars* (J 0194) Inform & Control 6*119-125
◇ D05 ◇ REV MR 28 # 3886 Zbl 115.370 JSL 32.114
• ID 05323

GREIBACH, S.A. [1964] see GINSBURG, S.

GREIBACH, S.A. [1965] *A new normal-form theorem for context-free phrase structure grammars* (J 0037) ACM J 12*42-52
◇ D05 ◇ REV MR 30 # 2960 Zbl 135.184 JSL 34.658
• ID 05324

GREIBACH, S.A. [1966] see GINSBURG, S.

GREIBACH, S.A. [1966] *The unsolvability of the recognition of linear context-free languages* (J 0037) ACM J 13*582-587
◇ D05 ◇ REV MR 34 # 5596 Zbl 148.9 JSL 36.693
• ID 05325

GREIBACH, S.A. [1967] *A note on pushdown store automata and regular systems* (J 0053) Proc Amer Math Soc 18*263-268
◇ D10 ◇ REV MR 34 # 8894 Zbl 183.17 JSL 33.302
• ID 05328

GREIBACH, S.A. [1967] see GINSBURG, S.

GREIBACH, S.A. [1968] *A note on undecidable properties of formal languages* (J 0041) Math Syst Theory 2*1-6
◇ D05 ◇ REV MR 37 # 1202 Zbl 157.19 JSL 40.245
• ID 05329

GREIBACH, S.A. [1969] see GINSBURG, S.

GREIBACH, S.A. [1969] see BOOK, R.V.

GREIBACH, S.A. [1970] see GINSBURG, S.

GREIBACH, S.A. [1971] see BOOK, R.V.

GREIBACH, S.A. [1972] see GINSBURG, S.

GREIBACH, S.A. [1972] *Syntactic operators on full semi-AFL's* (J 0119) J Comp Syst Sci 6*30-76
◇ D05 ◇ REV MR 46 # 6655 Zbl 269 # 68046 • ID 05330

GREIBACH, S.A. [1973] see GINSBURG, S.

GREIBACH, S.A. [1974] *Some restrictions on w-grammars* (J 0435) Int J Comput & Inf Sci 3*289-327
◇ D05 ◇ REV MR 50 # 3648 Zbl 288 # 68035 • ID 26101

GREIBACH, S.A. [1975] see GINSBURG, S.

GREIBACH, S.A. [1976] see GINSBURG, S.

GREIBACH, S.A. [1978] see BOOK, R.V.

GREIBACH, S.A. [1978] see FRIEDMAN, E.P.

GREIBACH, S.A. [1978] *One way finite visit automata* (J 1426) Theor Comput Sci 6*175-221
◇ D10 ◇ REV MR 58 # 8521 Zbl 368 # 68059 • ID 51295

GREIBACH, S.A. [1978] *Visits, crosses, and reversals for nondeterministic off-line machines* (J 0194) Inform & Control 36*174-216
◇ D10 D15 ◇ REV MR 80g:68102 Zbl 375 # 02027
• ID 51604

GREIBACH, S.A. [1979] *Formal languages: origins and directions* (P 3535) IEEE Symp Founds of Comput Sci (20);1979 San Juan 66-90
◇ B70 D05 D98 ◇ REV MR 82h:68103 • ID 81398

GREIBACH, S.A. [1979] *Linearity is polynomially decidable for realtime pushdown store automata* (J 0194) Inform & Control 42*27-37
◇ D10 D15 ◇ REV MR 80e:68106 Zbl 413 # 68086
• ID 53043

GREIBACH, S.A. [1979] see FRIEDMAN, E.P.

GREIBACH, S.A. [1979] see BOOK, R.V.

GREIBACH, S.A. [1980] see FRIEDMAN, E.P.

GREIBACH, S.A. [1981] *Formal languages: origins and directions* (J 3789) Ann Hist of Comp 3*14-41
◇ A10 D05 ◇ REV MR 83e:01049 • ID 40115

GRIFFITHS, H.B. [1957] *Borel sets and countable series of operations* (J 0027) Fund Math 44*115-155
◇ D55 E15 E75 ◇ REV MR 20 # 1294 Zbl 79.8
• ID 05337

GRIFFITHS, T.V. [1968] *Some remarks on derivations in general rewriting system* (J 0194) Inform & Control 12*27-54
• TRANSL [1981] (J 3079) Kiber Sb Perevodov, NS 18*75-99
◇ D03 ◇ REV MR 38 # 4238 Zbl 162.319 • ID 15003

GRIFFOR, E.R. [1983] *Some consequences of AD for Kleene recursion in* ^3E (J 0068) Z Math Logik Grundlagen Math 29*485-492
◇ D65 E45 E55 E60 ◇ REV MR 85k:03030
Zbl 549 # 03034 • ID 43126

GRIFFOR, E.R. & NORMANN, D. [1984] *Effective cofinalities and admissibility in E-recursion* (J 0027) Fund Math 123*151-161
◇ D65 E30 E35 E47 E50 ◇ REV MR 86f:03072
• ID 44080

GRIFFOR, E.R. & NORMANN, D. [1984] *The definability of* $E(\alpha)$ (J 0036) J Symb Logic 49*437-442
◇ D65 E45 ◇ REV MR 85i:03155 MR 86a:03048
Zbl 575 # 03033 • ID 42453

GRIFFOR, E.R. [1985] *An application of* Π^1_2*-logic to descriptive set theory* (P 3342) Rec Th Week;1984 Oberwolfach 148-158
◇ D55 E15 ◇ ID 45302

GRIFFOR, E.R. [1985] *Definability and forcing in E-recursion* (J 0132) Math Scand 57*5-28
◇ D65 E40 E47 ◇ ID 49429

GRIGORAS, G. [1980] *On the completeness in abstract complexity classes of languages* (J 0230) An Univ Iasi, NS, Sect Ia 26*173-177
◇ D15 ◇ REV MR 82e:68075 Zbl 443 # 68033 • ID 81403

GRIGORAS, G. [1981] *On a generalization of gamma-reducibility* (J 0060) Rev Roumaine Math Pures Appl 26*855-858
◇ D15 ◇ REV MR 83b:68046 Zbl 482 # 68041 • ID 37705

GRIGORCHUK, R.I. [1985] *A relationship between algorithmic problems and entropy characteristics of groups (Russian)* (J 0023) Dokl Akad Nauk SSSR 284*24-29
• TRANSL [1985] (J 0062) Sov Math, Dokl 32*356-360
◇ D15 D40 ◇ ID 48984

GRIGOR'EV, D.YU. [1976] *An application of separability and independence notions for proving lower bounds of circuit complexity (Russian) (English summary)* (S 0228) Zap Nauch Sem Leningrad Otd Mat Inst Steklov 60*38-48
• TRANSL [1980] (J 1531) J Sov Math 14*1450-1457
◇ D15 ◇ REV MR 58 # 32054 Zbl 341 # 94020 • ID 32124

GRIGOR'EV, D.YU. [1976] *Kolmogorov algorithms are stronger than Turing machines (Russian)* (**S** 0228) Zap Nauch Sem Leningrad Otd Mat Inst Steklov 60*29-37,221 • TRANSL [1980] (**J** 1531) J Sov Math 14*1445-1450 ◇ D10 D15 ◇ REV MR 58 # 27374 Zbl 345 # 02023 • ID 29781

GRIGOR'EV, D.YU. [1977] *Imbedding theorems for Turing machines of different dimensions and Kolmogorov's algorithms (Russian)* (**J** 0023) Dokl Akad Nauk SSSR 234*15-18 • TRANSL [1977] (**J** 0062) Sov Math, Dokl 18*588-592 ◇ D10 ◇ REV MR 58 # 5113 Zbl 386 # 03019 • ID 52195

GRIGOR'EV, D.YU. [1978] *Multiplicative complexity of a pair of bilinear forms and of the polynomial multiplication* (**P** 1707) Math Founds of Comput Sci (7);1978 Zakopane 250-256 ◇ D15 D80 ◇ REV MR 80d:68052 Zbl 381 # 68045 • ID 32129

GRIGOR'EV, D.YU. [1979] *Time bounds of multidimensional Turing machines (Russian)* (**S** 0228) Zap Nauch Sem Leningrad Otd Mat Inst Steklov 88*47-55,237 • TRANSL [1982] (**J** 1531) J Sov Math 20*2290-2295 ◇ D10 D15 ◇ REV MR 81a:68058 Zbl 429 # 03020 • ID 31777

GRIGOR'EV, D.YU. [1981] *Complexity of "wild" matrix problems and of the isomorphism of algebras and graphs (Russian) (English summary)* (**S** 0228) Zap Nauch Sem Leningrad Otd Mat Inst Steklov 105*10-17,198 ◇ D15 D40 ◇ REV MR 82k:03060 Zbl 478 # 68042 • ID 73627

GRIGOR'EV, D.YU. [1984] see CHISTOV, A.L.

GRIGOR'EV, D.YU. see Vol. I, V for further entries

GRIGORIEFF, S. [1975] *Minimalite des reels definis par "forcing" sur certaines familles d'arbres de suites finies d'entiers (English summary)* (**J** 2313) C R Acad Sci, Paris, Ser A-B 281*A301-A304 ◇ D30 E40 E45 ◇ REV MR 52 # 93 Zbl 316 # 02062 • ID 18177

GRIGORIEFF, S. [1977] *Determination des jeux boreliens et problemes logiques associes (d'apres D. Martin)* (**S** 1567) Semin Bourbaki Exp.478*14pp ◇ D55 E15 E60 ◇ REV MR 55 # 10272 Zbl 359 # 02070 • REM Springer: Heidelberg & New York; Lecture Notes Math 567 • ID 50617

GRIGORIEFF, S. see Vol. V for further entries

GRILLIOT, T.J. [1969] *Hierarchies based on objects of finite types* (**J** 0036) J Symb Logic 34*177-182 ◇ D55 D65 ◇ REV MR 44 # 75 Zbl 185.23 JSL 36.147 • ID 05343

GRILLIOT, T.J. [1969] *Selection functions for recursive functionals* (**J** 0047) Notre Dame J Formal Log 10*225-234 ◇ D65 ◇ REV MR 42 # 65 Zbl 185.23 JSL 38.653 • ID 05344

GRILLIOT, T.J. [1971] *Inductive definitions and computabilitiy* (**J** 0064) Trans Amer Math Soc 158*309-317 ◇ D70 D75 ◇ REV MR 46 # 3276 Zbl 223 # 02043 JSL 38.654 • ID 05345

GRILLIOT, T.J. [1971] *On effectively discontinuous type-2 objects* (**J** 0036) J Symb Logic 36*245-248 ◇ D60 D65 ◇ REV MR 45 # 66 Zbl 224 # 02034 • ID 05346

GRILLIOT, T.J. [1972] *Omitting types; applications to recursion theory* (**J** 0036) J Symb Logic 37*81-89 ◇ C07 C62 D30 D55 D60 D75 ◇ REV MR 49 # 8839 Zbl 244 # 02016 JSL 40.87 • ID 05347

GRILLIOT, T.J. [1973] *Implicit definability and hyperprojectivity* (**J** 0287) Scripta Math 29*151-155 ◇ C40 C75 D75 ◇ REV MR 50 # 9558 Zbl 259 # 02032 • ID 05348

GRILLIOT, T.J. [1974] *Dissecting abstract recursion* (**P** 0602) Generalized Recursion Th (1);1972 Oslo 405-420 ◇ D75 ◇ REV MR 54 # 80 Zbl 281 # 02045 • ID 23961

GRILLIOT, T.J. [1974] *Model theory for dissecting recursion theory* (**P** 0602) Generalized Recursion Th (1);1972 Oslo 421-428 ◇ C55 C70 C75 C80 D60 D75 ◇ REV MR 53 # 12929 Zbl 281 # 02046 • ID 23178

GRILLIOT, T.J. see Vol. III for further entries

GRINBLAT, V.A. [1970] *On the applicability problem for normal algorithms (Russian)* (**J** 0789) Uch Zap Mat Ped Inst, Tula 2*260-266 ◇ D03 ◇ REV MR 53 # 12904 • ID 23156

GRINBLAT, V.A. [1981] see BEZVERKHNIJ, V.N.

GRINDLINGER, E.I. [1964] *Solution of the word problem for a class of groups by Dehn's algorithm, and of the conjugacy problem by means of a generalization of Dehn's algorithm (Russian)* (**J** 0023) Dokl Akad Nauk SSSR 154*507-509 • TRANSL [1964] (**J** 0062) Sov Math, Dokl 5*110-112 ◇ D40 ◇ REV MR 28 # 3080 • ID 05351

GRINDLINGER, E.I. [1964] *Solution of the isomorphism problem for a class of semi-groups (Russian)* (**J** 0092) Sib Mat Zh 5*788-792 ◇ D40 ◇ REV MR 29 # 3527 Zbl 178.326 • ID 49847

GRINDLINGER, E.I. [1964] *The word problem for a class of semigroups with a finite number of defining relations(Russian)* (**J** 0092) Sib Mat Zh 5*77-85 ◇ D40 ◇ REV MR 28 # 3079 Zbl 178.326 • ID 05352

GRINDLINGER, E.I. [1966] *On the unsolvability of the word problem for a class of semigroups with a solvable isomorphism problem (Russian)* (**J** 0023) Dokl Akad Nauk SSSR 171*519-520 • TRANSL [1966] (**J** 0062) Sov Math, Dokl 7*1502-1503 ◇ D40 ◇ REV MR 34 # 5903 Zbl 207.313 JSL 33.469 • ID 19305

GRINDLINGER, E.I. & GRINDLINGER, M.D. [1968] *The word problem for a class of semigroups (Russian)* (**J** 0789) Uch Zap Mat Ped Inst, Tula 1968*119-121 ◇ D40 ◇ REV MR 43 # 7529 • ID 22223

GRINDLINGER, E.I. & GRINDLINGER, M.D. [1970] *An algorithm for the solution of the word problem for certain semigroups (Russian)* (**J** 0031) Izv Vyssh Ucheb Zaved, Mat (Kazan) 1970/9(100)*45-47 ◇ D40 ◇ REV MR 44 # 1718 Zbl 223 # 20066 • ID 05355

GRINDLINGER, E.I. [1970] *The word, divisibility and occurence problem for quotient semigroups of free products (Russian)* (**J** 0789) Uch Zap Mat Ped Inst, Tula 2*303-310 ◇ D40 ◇ REV MR 53 # 13443 • ID 23219

GRINDLINGER, M.D. [1960] *Dehn's algorithm for the word problem* (**J** 0155) Commun Pure Appl Math 13*67-83 ◇ D40 ◇ REV MR 23 # A1693 • ID 05315

GRINDLINGER, M.D. [1963] *Solution of the isomorphism problem for a certain class of groups (Russian)* (J 0226) Uch Zap Ped Inst, Ivanovo 31*59-61
◇ D40 ◇ REV MR 47 # 343 • ID 05359

GRINDLINGER, M.D. [1964] *On Magnus's generalized word problem (Russian)* (J 0092) Sib Mat Zh 5*955-957
◇ D40 ◇ REV MR 29 # 3526 Zbl 192.67 • ID 16260

GRINDLINGER, M.D. [1964] *Solution of the conjugacy problem for a class of groups, coinciding with their anti-centers, by means of the generalized Dehn algorithm (Russian)* (J 0023) Dokl Akad Nauk SSSR 158*1254-1256
• TRANSL [1964] (J 0062) Sov Math, Dokl 5*1384-1386
◇ D40 ◇ REV MR 30 # 4819 • ID 05356

GRINDLINGER, M.D. [1965] *On the word problem and the conjugacy problem (Russian)* (J 0216) Izv Akad Nauk SSSR, Ser Mat 29*245-268
◇ D40 ◇ REV MR 30 # 4820 • ID 05357

GRINDLINGER, M.D. [1968] *Strengthening of a theorem of J.S.Smetanich (Russian)* (J 0789) Uch Zap Mat Ped Inst, Tula 1968*115-118
◇ D25 D40 ◇ REV MR 44 # 4081 • ID 22244

GRINDLINGER, M.D. [1968] see GRINDLINGER, E.I.

GRINDLINGER, M.D. [1970] see GRINDLINGER, E.I.

GRINDLINGER, M.D. see Vol. III for further entries

GRODZKI, Z. [1968] *On the equivalence of Markov normal algorithms* (J 0014) Bull Acad Pol Sci, Ser Math Astron Phys 16*333-336
◇ D03 ◇ REV MR 38 # 49 Zbl 159.12 • ID 05371

GRODZKI, Z. [1970] *The k-machines* (J 0014) Bull Acad Pol Sci, Ser Math Astron Phys 18*399-403
◇ D03 D05 ◇ REV Zbl 203.302 • ID 18178

GRODZKI, Z. [1970] *The degree of complexity of the sets of computations I* (J 0014) Bull Acad Pol Sci, Ser Math Astron Phys 18*609-612
◇ D20 ◇ REV MR 43 # 1844 Zbl 205.314 • REM Part II 1971 • ID 05372

GRODZKI, Z. [1971] *The degree of complexity of the sets of computations II* (J 0014) Bull Acad Pol Sci, Ser Math Astron Phys 19*539-543
◇ D20 ◇ REV MR 46 # 1573 Zbl 234 # 94051 • REM Part I 1970 • ID 05373

GRODZKI, Z. [1974] *The Kolmogorov's complexity of the computation of the k-machines* (J 0302) Rep Math Logic, Krakow & Katowice 3*19-29
◇ D03 D15 D20 ◇ REV MR 54 # 4934 Zbl 301 # 68055 • ID 21934

GRODZKI, Z. [1975] *Generalized Markov normal algorithms* (J 0302) Rep Math Logic, Krakow & Katowice 5*27-35
◇ D03 ◇ REV MR 57 # 9506 Zbl 339 # 02029 • ID 21898

GRODZKI, Z. & ZURAWIECKI, J. [1976] *The (k,m)-computation sets* (J 0302) Rep Math Logic, Krakow & Katowice 6*79-86
◇ D03 D05 ◇ REV MR 57 # 15785 Zbl 353 # 68069 • ID 21914

GROSCHE, G. [1971] *Syntaktische Untersuchungen zu mathematischen Sprachen I,II,III* (J 0129) Elektr Informationsverarbeitung & Kybern 7*3-29,107-128,153-165
◇ D05 ◇ REV MR 46 # 1586 MR 47 # 33 Zbl 215.320 Zbl 219 # 02009 • ID 05380

GROSKY, W. [1975] see BARALT-TORRIJOS, J.

GROSS, M. [1966] *Sur certains procedes de definition de langages formels* (P 0746) Automata Th;1964 Ravello 181-200
◇ D05 ◇ REV Zbl 192.71 • ID 27536

GROSS, M. & LENTIN, A. [1967] *Notions sur les grammaires formelles* (X 0834) Gauthier-Villars: Paris 198pp
◇ D05 ◇ REV MR 37 # 2556 Zbl 217.538 JSL 34.298 • ID 25383

GROSS, M. & LENTIN, A. [1970] *Introduction to formal grammars* (X 0811) Springer: Heidelberg & New York xi+231pp
◇ B65 D05 D98 ◇ REV Zbl 191.310 JSL 36.346 • ID 05384

GROSS, M. & LENTIN, A. [1971] *Mathematische Linguistik* (X 0811) Springer: Heidelberg & New York x+286pp
◇ B65 D05 D98 ◇ REV MR 50 # 15451 Zbl 212.27 • ID 05386

GROSS, W.F. [1983] *The inverse of a regressive object* (J 0036) J Symb Logic 48*804-815
◇ D50 ◇ REV MR 85d:03090 Zbl 536 # 03019 • ID 37102

GROSS, W.F. see Vol. III, V, VI for further entries

GROSZEK, M.J. & SLAMAN, T.A. [1983] *Independence results on the global structure of the Turing degrees* (J 0064) Trans Amer Math Soc 277*579-588
◇ D30 E35 ◇ REV MR 85b:03069 Zbl 528 # 03027 • ID 37641

GRUNEWALD, F. & SEGAL, D. [1979] *The solubility of certain decision problems in arithmetic and algebra* (J 0589) Bull Amer Math Soc (NS) 1*915-918
◇ D40 D80 ◇ REV MR 81b:10014 Zbl 431 # 20029 • ID 81410

GRUNEWALD, F. & SEGAL, D. [1980] *Some general algorithms I: Arithmetic groups. II: Nilpotent groups* (J 0120) Ann of Math, Ser 2 112*531-583,585-617
◇ D40 ◇ REV MR 82d:20048 Zbl 457 # 20047 Zbl 457 # 20048 • ID 81411

GRUNEWALD, F. & SEGAL, D. [1985] *Decision problems concerning S-arithmetic groups* (J 0036) J Symb Logic 50*743-772
◇ D40 ◇ ID 47377

GRUNEWALD, F. see Vol. III for further entries

GRZEGORCZYK, A. [1951] *Undecidability of some topological theories* (J 0027) Fund Math 38*137-152
◇ D35 ◇ REV MR 13.898 Zbl 45.2 JSL 18.73 • ID 05390

GRZEGORCZYK, A. [1953] *Some classes of recursive functions* (X 1034) PWN: Warsaw 46pp
• TRANSL [1970] (C 1540) Probl Mat Log: Slozh Algor & Kl Vychisl Funk 9-49
◇ D20 ◇ REV MR 15.667 Zbl 52.249 JSL 20.71 • ID 23469

GRZEGORCZYK, A. [1955] *Computable functionals* (J 0027) Fund Math 42*168-202
◇ D20 ◇ REV MR 19.238 Zbl 66.260 JSL 24.50 • ID 05393

GRZEGORCZYK, A. [1955] *On the definition of computable functionals* (J 0027) Fund Math 42*232-239
◇ D20 F60 ◇ REV MR 19.238 Zbl 67.3 JSL 24.50 • ID 05392

GRZEGORCZYK, A. [1956] *Some proofs of undecidability of arithmetic* (**J** 0027) Fund Math 43*166-177
⋄ D35 F30 ⋄ REV MR 18.552 Zbl 72.5 JSL 23.46
• ID 05395

GRZEGORCZYK, A. [1957] *Decision problems* (**X** 1034) PWN: Warsaw 142pp
⋄ B25 C10 D35 D98 F30 ⋄ REV MR 22#7936
Zbl 89.246 • ID 24864

GRZEGORCZYK, A. [1957] *On the definitions of computable real continuous functions* (**J** 0027) Fund Math 44*61-71
⋄ D20 F60 ⋄ REV MR 19.723 Zbl 79.248 • ID 05396

GRZEGORCZYK, A. & MOSTOWSKI, ANDRZEJ & RYLL-NARDZEWSKI, C. [1958] *The classical and the ω-complete arithmetic* (**J** 0036) J Symb Logic 23*188-206
⋄ C62 D55 D70 F30 ⋄ REV MR 21#4908 Zbl 84.248
JSL 27.80 • ID 05398

GRZEGORCZYK, A. & MOSTOWSKI, ANDRZEJ & RYLL-NARDZEWSKI, C. [1961] *Definability of sets in models of axiomatic theories* (**J** 0014) Bull Acad Pol Sci, Ser Math Astron Phys 9*163-167
⋄ C40 D55 ⋄ REV MR 29#1138 Zbl 99.9 JSL 34.126
• ID 05402

GRZEGORCZYK, A. [1961] *Fonctions recursives* (**X** 0834) Gauthier-Villars: Paris 100pp
⋄ D20 D98 ⋄ REV MR 26#14 Zbl 101.250 JSL 31.481
• ID 05401

GRZEGORCZYK, A. [1962] *A theory without recursive models* (**J** 0014) Bull Acad Pol Sci, Ser Math Astron Phys 10*63-69
⋄ C57 C62 D55 ⋄ REV MR 25#3820 Zbl 113.6
JSL 28.102 • ID 05406

GRZEGORCZYK, A. [1962] *An example of two weak essentially undecidable theories F and F** (**J** 0014) Bull Acad Pol Sci, Ser Math Astron Phys 10*5-9
⋄ D35 ⋄ REV MR 25#3819 Zbl 118.252 JSL 27.358
• ID 05407

GRZEGORCZYK, A. [1964] *Recursive objects in all finite types* (**J** 0027) Fund Math 54*73-93
⋄ D65 F10 ⋄ REV MR 28#3926 Zbl 196.14 JSL 39.343
• ID 05411

GRZEGORCZYK, A. [1974] *Axiomatic theory of enumeration* (**P** 0602) Generalized Recursion Th (1);1972 Oslo 429-436
⋄ D45 D75 ⋄ REV MR 52#10389 Zbl 309#02037
• ID 21688

GRZEGORCZYK, A. see Vol. I, II, III, V, VI for further entries

GRZYMALA-BUSSE, J.W. [1971] *Operation-preserving functions and autonomous factors of finite automata* (**J** 0119) J Comp Syst Sci 5*465-474
⋄ D05 ⋄ REV MR 44#5229 Zbl 224#94058 • ID 05419

GRZYMALA-BUSSE, J.W. [1984] see BAVEL, Z.

GUAN, JIWEN [1965] *Theory of linear inner automata (Chinese)* (**P** 4564) Math Logic;1963 Xi-An 65-68
⋄ D05 ⋄ ID 49313

GUASPARI, D. [1974] *A note on the Kondo-Addison theorem* (**J** 0036) J Symb Logic 39*567-570
⋄ D55 E15 E45 ⋄ REV MR 52#58 Zbl 299#02049
• ID 05420

GUASPARI, D. [1974] *Characterizing the constructible reals* (**J** 0014) Bull Acad Pol Sci, Ser Math Astron Phys 22*357-358
⋄ D30 D55 E15 E45 ⋄ REV MR 50#9587
Zbl 314#02070 • ID 05421

GUASPARI, D. [1975] *Analytical well-orderings in* $\mathbb{R}$ (**P** 0775) Logic Colloq;1973 Bristol 317-346
⋄ D55 E15 ⋄ REV MR 52#2849 Zbl 317#02081
• ID 17640

GUASPARI, D. & HARRINGTON, L.A. [1976] *Characterizing* $\mathscr{C}_3$ *(the largest countable* Π^1_3 *set)* (**J** 0053) Proc Amer Math Soc 57*127-129
⋄ D55 E15 E60 ⋄ REV MR 53#5303 Zbl 336#04001
• ID 22937

GUASPARI, D. [1980] *Definability in models of set theory* (**J** 0036) J Symb Logic 45*9-19
⋄ C62 D55 E45 E47 ⋄ REV MR 83k:03065
Zbl 453#03053 • ID 54183

GUASPARI, D. [1983] *Trees, norms and scales* (**C** 3847) Surveys in Set Th 135-161
⋄ D55 E15 E60 E98 ⋄ REV Zbl 549#03040 • ID 43128

GUASPARI, D. see Vol. II, III, VI for further entries

GUCCIONE, S. & LO SARDO, P. [1974] *Effective procedures and randomness* (**J** 2834) Scientia (Milano) 109*489-498
⋄ B75 D20 ⋄ REV MR 56#1386 • ID 81419

GUCCIONE, S. see Vol. I, II for further entries

GUERRERIO, G. [1981] *Sulle generalizzazione della ricorsivita* (**P** 3092) Congr Naz Logica;1979 Montecatini Terme 119-136
⋄ D60 D65 ⋄ ID 48718

GUESSARIAN, I. [1981] see DAMM, W.

GUESSARIAN, I. [1983] *Pushdown tree automata* (**J** 0041) Math Syst Theory 16*237-263
⋄ D10 ⋄ REV MR 85c:68053 Zbl 524#68047 • ID 39090

GUESSARIAN, I. see Vol. I, V for further entries

GUHL, R. [1975] *A theorem on recursively enumerable vector spaces* (**J** 0047) Notre Dame J Formal Log 16*357-362
⋄ C57 C60 D45 ⋄ REV MR 52#2857 Zbl 254#02034
• ID 05433

GUHL, R. [1977] *Two notes on recursively enumerable vector spaces* (**J** 0047) Notre Dame J Formal Log 18*295-298
⋄ C57 C60 D45 ⋄ REV MR 56#5252 Zbl 305#02057
• ID 24226

GUICHARD, D.R. [1983] *Automorphisms of substructure lattices in recursive algebra* (**J** 0073) Ann Pure Appl Logic 25*47-58
⋄ C57 C60 D45 G05 ⋄ REV MR 85e:03104 • ID 40309

GUICHARD, D.R. [1984] *A note on r-maximal subspaces of* V_∞ (**J** 0073) Ann Pure Appl Logic 26*1-9
⋄ C57 C60 D45 ⋄ REV MR 85j:03075 Zbl 522#03027
• ID 43386

GUICHARD, D.R. see Vol. III for further entries

GUIDA, G. [1978] see CRESPI-REGHIZZI, S.

GUIDA, G. [1981] see CRESPI-REGHIZZI, S.

GUITART, R. [1974] *Remarques sur les machines et les structures* (**J** 0306) Cah Topol & Geom Differ 15*113-144
◇ D05 G30 ◇ REV MR 52 # 5761 Zbl 319 # 18003
• ID 18180

GUITART, R. see Vol. III, V for further entries

GUMIN, H. [1963] *Digital computers, mathematical logic and principal limitations of computability* (**P** 0572) Inform Processing (2);1962 Muenchen 29-32
◇ B75 D35 ◇ REV Zbl 143.184 • ID 27610

GUMIN, H. see Vol. I for further entries

GUPTA, N. [1984] *Recursively presented two generated infinite p-groups* (**J** 0044) Math Z 188*89-90
◇ C57 C60 ◇ REV MR 86f:20032 Zbl 553 # 20015
• ID 39718

GURARI, E.M. & IBARRA, O.H. [1978] *An NP-complete number-theoretic problem* (**P** 1740) ACM Symp Th of Comput (10);1978 San Diego 205-215
◇ B25 D15 D80 ◇ REV MR 81c:68034 • ID 42545

GURARI, E.M. & IBARRA, O.H. [1979] *An NP-complete number-theoretic problem* (**J** 0037) ACM J 26*567-581
◇ B25 D10 D15 D25 D80 ◇ REV Zbl 407 # 68053
• ID 56232

GURARI, E.M. & IBARRA, O.H. [1979] *Simple counter machines and number-theoretic problems* (**J** 0119) J Comp Syst Sci 19*145-162
◇ D05 D10 ◇ REV MR 81i:68086 Zbl 426 # 68036
• ID 53681

GURARI, E.M. & IBARRA, O.H. [1979] *Some decision problems concerning sequential transducers and checking automata* (**J** 0119) J Comp Syst Sci 18*18-34
◇ D05 D10 ◇ REV MR 80h:68051 Zbl 404 # 68057
• ID 54862

GURARI, E.M. & IBARRA, O.H. [1979] *The complexity of the equivalence problem for counter machines, semilinear sets, and simple programs* (**P** 3542) ACM Symp Th of Comput (11);1979 Atlanta 142-152
◇ D05 D15 ◇ REV MR 81d:68057 • ID 81432

GURARI, E.M. & IBARRA, O.H. [1980] *Path systems: constructions, solutions and applications* (**J** 1428) SIAM J Comp 9*348-374
◇ D05 D10 ◇ REV MR 82f:68048 Zbl 447 # 68049
• ID 81430

GURARI, E.M. & IBARRA, O.H. [1981] *The complexity of the equivalence problem for two characterizations of Presburger sets* (**J** 1426) Theor Comput Sci 13*295-314
◇ B25 D05 D15 F30 ◇ REV MR 82m:68084
Zbl 454 # 03005 • ID 54218

GURARI, E.M. & IBARRA, O.H. [1981] *The complexity of decision problems for finite-turn multicounter machines* (**J** 0119) J Comp Syst Sci 22*220-229
◇ D10 D15 ◇ REV MR 82j:68031b Zbl 458 # 68011
• ID 54436

GURARI, E.M. & IBARRA, O.H. [1981] *The complexity of the equivalence problem for simple programs* (**J** 0037) ACM J 28*535-560
◇ D15 D20 ◇ REV MR 82g:68043 Zbl 462.68023
• ID 81429

GURARI, E.M. & IBARRA, O.H. [1981] *Two-way counter machines and Diophantine equations* (**P** 4235) IEEE Symp Found of Comp Sci (22);1981 Nashville 45-52
◇ D05 D35 ◇ REV MR 84a:68004 Zbl 496 # 03020
• ID 45817

GURARI, E.M. [1982] *The equivalence problem for deterministic two-way sequential transducers is decidable* (**J** 1428) SIAM J Comp 11*448-452
◇ D05 ◇ REV MR 83i:68081 Zbl 486 # 68037 • ID 39361

GURARI, E.M. & IBARRA, O.H. [1982] *Two-way counter machines and Diophantine equations* (**J** 0037) ACM J 29*863-873
◇ D05 D35 ◇ REV MR 84a:68089 Zbl 496 # 03020
• ID 47210

GUREVICH, G.A. [1972] *On the conjugacy problem for groups with one defining relation (Russian)* (**J** 0023) Dokl Akad Nauk SSSR 207*18-20
• TRANSL [1972] (**J** 0062) Sov Math, Dokl 13*1436-1439
◇ D40 ◇ REV MR 47 # 5123 • ID 05448

GUREVICH, G.A. [1973] *On the conjugacy problem for groups with a single defining relation (Russian)* (**S** 0066) Tr Mat Inst Steklov 133*109-120,275
• TRANSL [1973] (**S** 0055) Proc Steklov Inst Math 133*108-120
◇ D40 ◇ REV MR 49 # 2956 Zbl 291 # 20045 • ID 50255

GUREVICH, I.B. [1974] *The noncomputability within a class of local algorithms of certain predicates connected with the minimization of Boolean functions (Russian)* (**J** 0040) Kibernetika, Akad Nauk Ukr SSR 1974/2*24-30
• TRANSL [1974] (**J** 0021) Cybernetics 10*213-218
◇ B35 D20 ◇ REV MR 53 # 7628 Zbl 286 # 02038
• ID 62232

GUREVICH, I.B. see Vol. I for further entries

GUREVICH, Y. [1965] *Existential interpretation (Russian)* (**J** 0003) Algebra i Logika 4/4*71-85
◇ B20 B25 C60 D35 F25 ◇ REV MR 34 # 46
Zbl 294 # 02023 • REM Part II 1982 • ID 05452

GUREVICH, Y. [1966] *Certain algorithmic questions of the theory of classes of algebraic systems (Russian)* (**C** 1549) Tartu Mezhvuz Nauch Simp Obshchej Algeb 38-41
◇ B25 D35 ◇ REV MR 34 # 29 Zbl 248 # 02048 • ID 05457

GUREVICH, Y. [1966] *Effective recognition of satisfiability of formulae of the restricted predicate calculus (Russian)* (**J** 0003) Algebra i Logika 5/2*25-55
◇ B20 B25 D35 ◇ REV MR 35 # 4096 Zbl 198.324
• ID 05455

GUREVICH, Y. [1966] *On the decision problem for pure restricted predicate logic (Russian)* (**J** 0023) Dokl Akad Nauk SSSR 166*1032-1034
• TRANSL [1966] (**J** 0062) Sov Math, Dokl 7*217-219
◇ B20 C13 D35 ◇ REV MR 34 # 47 Zbl 158.252
• ID 05458

GUREVICH, Y. [1966] *The decision problem for the restricted predicate calculus (Russian)* (**J** 0023) Dokl Akad Nauk SSSR 168*510-511
• TRANSL [1966] (**J** 0062) Sov Math, Dokl 7*669-670
◇ B20 C13 D35 ◇ REV MR 34 # 60 Zbl 163.253
• ID 05456

Gurevich, Y. [1966] *The word problem for certain classes of semigroups (Russian)* (**J** 0003) Algebra i Logika 5/5*25–35 ◇ D40 ◇ REV MR 34 # 5904 Zbl 178.325 • ID 05453

Gurevich, Y. [1967] *A contribution to the elementary theory of lattice ordered abelian groups and K-lineals (Russian)* (**J** 0023) Dokl Akad Nauk SSSR 175*1213–1215 • TRANSL [1967] (**J** 0062) Sov Math, Dokl 8*987–989 ◇ B25 C60 D35 ◇ REV MR 36 # 1373 Zbl 189.326 • ID 33753

Gurevich, Y. [1967] *Hereditary undecidability of a class of lattice-ordered abelian groups (Russian)(English summary)* (**J** 0003) Algebra i Logika 6/1*45–62 ◇ D35 ◇ REV MR 36 # 92 Zbl 165.318 • ID 05459

Gurevich, Y. [1969] *A decision problem for decision problems (Russian)* (**J** 0003) Algebra i Logika 8*640–642 • TRANSL [1969] (**J** 0069) Algeb and Log 8*362–363 ◇ B25 D35 ◇ REV MR 44 # 71 Zbl 211.313 • ID 05461

Gurevich, Y. [1969] *The decision problem for the logic of predicates and operations (Russian)* (**J** 0003) Algebra i Logika 8*284–308 • TRANSL [1969] (**J** 0069) Algeb and Log 8*160–174 ◇ B25 D35 ◇ REV MR 41 # 8205 Zbl 198.325 • ID 31104

Gurevich, Y. [1970] *Minsky machines and the case* $\forall\exists\forall$ & $\exists^{\infty}$ *of the decision problem (Russian)* (**J** 0340) Mat Zap (Univ Sverdlovsk) 7/3*77–83 ◇ B20 D05 D35 ◇ REV MR 44 # 2614 Zbl 317 # 02052 • ID 22243

Gurevich, Y. & Koryakov, I.O. [1972] *Remarks on R.Berger's paper on the domino problem (Russian)* (**J** 0092) Sib Mat Zh 13*459–463 • TRANSL [1972] (**J** 0475) Sib Math J 13*319–321 ◇ D10 ◇ REV MR 45 # 4971 Zbl 241 # 02013 • ID 05462

Gurevich, Y. & Turashvili, T.V. [1973] *A strengthening of a certain result of J. Suranyi (Russian) (Georgian and English summaries)* (**J** 0233) Soobshch Akad Nauk Gruz SSR 70*289–292 ◇ B20 D35 ◇ REV MR 54 # 10003 Zbl 296 # 02024 • ID 25843

Gurevich, Y. [1976] *Semi-conservative reduction* (**J** 0009) Arch Math Logik Grundlagenforsch 18*23–25 ◇ B20 D25 D35 ◇ REV MR 57 # 2875 Zbl 351 # 02011 • ID 23714

Gurevich, Y. [1976] *The decision problem for standard classes* (**J** 0036) J Symb Logic 41*460–464 ◇ B20 B25 D35 ◇ REV MR 53 # 10572 Zbl 339 # 02045 • ID 14777

Gurevich, Y. [1979] *Modest theory of short chains I* (**J** 0036) J Symb Logic 44*481–490 ◇ B25 C65 C85 D35 E07 ◇ REV MR 81a:03038a Zbl 464 # 03013 • REM Part II 1979 by Gurevich,Y. & Shelah,S. • ID 54603

Gurevich, Y. & Shelah, S. [1979] *Modest theory of short chains II* (**J** 0036) J Symb Logic 44*491–502 ◇ B25 C65 C85 D35 E07 E50 ◇ REV MR 81a:03038b Zbl 464 # 03014 • REM Part I 1979 by Gurevich,Y. • ID 54604

Gurevich, Y. & Harrington, L.A. [1982] *Automata, trees, and games* (**P** 3714) ACM Symp Th of Comput (14);1982 60–65 ◇ B25 C85 D05 ◇ ID 33759

Gurevich, Y. [1982] *Existential interpretation II* (**J** 0009) Arch Math Logik Grundlagenforsch 22*103–120 ◇ B20 D35 F25 ◇ REV MR 84b:03059 Zbl 493 # 03020 • REM Part I 1965 • ID 33757

Gurevich, Y. & Shelah, S. [1982] *Monadic theory of order and topology in ZFC* (**J** 0007) Ann Math Logic 23*179–198 ◇ B25 C65 C85 D35 E07 ◇ REV MR 85d:03080 Zbl 516 # 03007 • ID 33758

Gurevich, Y. [1982] see Blass, A.R.

Gurevich, Y. [1982] see Aanderaa, S.O.

Gurevich, Y. & Lewis, H.R. [1982] *The inference problem for template dependencies* (**J** 0194) Inform & Control 55*69–79 ◇ B75 D80 ◇ REV MR 85e:68013 Zbl 548 # 68095 • ID 33760

Gurevich, Y. [1983] *Algebras of feasible functions* (**P** 4238) IEEE Symp Found of Comput Sci (24);1983 Tuscon 210–214 ◇ D15 D20 D75 ◇ ID 47724

Gurevich, Y. [1983] *Decision problem for separated distributive lattices* (**J** 0036) J Symb Logic 48*193–196 ◇ D35 G10 ◇ REV MR 84f:03041 Zbl 508 # 03016 • ID 33763

Gurevich, Y. & Shelah, S. [1983] *Interpreting second-order logic in the monadic theory of order* (**J** 0036) J Symb Logic 48*816–828 ◇ B15 C65 C85 D35 E07 E50 F25 ◇ REV MR 85f:03007 Zbl 559 # 03008 • ID 33765

Gurevich, Y. & Shelah, S. [1983] *Random models and the Goedel case of the decision problem* (**J** 0036) J Symb Logic 48*1120–1124 ◇ B20 B25 C13 D35 ◇ REV MR 85d:03019 Zbl 534 # 03006 • ID 33766

Gurevich, Y. & Magidor, M. & Shelah, S. [1983] *The monadic theory of* ω_2 (**J** 0036) J Symb Logic 48*387–398 ◇ B15 B25 C65 C85 D35 E10 E35 ◇ REV MR 84i:03076 Zbl 549 # 03010 • ID 33764

Gurevich, Y. [1983] see Glass, A.M.W.

Gurevich, Y. & Lewis, H.R. [1984] *A logic for constant-depth circuits* (**J** 0194) Inform & Control 61*65–74 ◇ B75 D15 ◇ REV MR 86e:03008 • ID 47721

Gurevich, Y. [1984] see Blass, A.R.

Gurevich, Y. & Stockmeyer, L.J. & Vishkin, U. [1984] *Solving NP-hard problems on graphs that are almost trees and an application to facility location problems* (**J** 0037) ACM J 31*459–473 ◇ D15 ◇ ID 33772

Gurevich, Y. & Lewis, H.R. [1984] *The word problem for cancellation semigroups with zero* (**J** 0036) J Symb Logic 49*184–191 ◇ D40 ◇ REV MR 85k:20170 • ID 42456

Gurevich, Y. [1984] *Toward logic tailored for computational complexity* (**P** 2153) Logic Colloq;1983 Aachen 2*175–216 ◇ B75 C13 D15 ◇ REV MR 86f:03061 • ID 43005

Gurevich, Y. [1985] *Monadic second-order theories* (**C** 4183) Model-Theor Log 479–509 ◇ B25 C65 C85 C98 D35 D98 ◇ ID 48332

GUREVICH, Y. & SHELAH, S. [1985] *The decision problem for branching time logic* (**J** 0036) J Symb Logic 50*668-681
◇ B25 B45 C65 D05 D25 ◇ ID 47372

GUREVICH, Y. see Vol. I, II, III, V, VI for further entries

GUSEJNBEKOVA, A.M. [1985] see BURTMAN, M.I.

GUSEV, L.A. [1961] see AJZERMAN, M.A.

GUSEV, L.A. [1963] see AJZERMAN, M.A.

GUSEV, L.A. see Vol. V for further entries

GUZICKI, W. [1973] *A remark on the independence of a basis hypothesis* (**J** 0027) Fund Math 78*189-191
◇ D55 E15 E35 E45 ◇ REV MR 48 # 1921 Zbl 273 # 02047 • ID 30498

GUZICKI, W. [1977] *On the projective class of the continuum hypothesis* (**P** 1695) Set Th & Hierarch Th (3);1976 Bierutowice 181-185
◇ D55 E47 E50 ◇ REV MR 58 # 231 Zbl 364 # 02038 • ID 50968

GUZICKI, W. see Vol. III, V, VI for further entries

GYURIS, L. [1969] *On the connection of Glushkovian microprogram-algebras and logical schemas of the algorithms* (**P** 1841) Fct Recurs & Appl;1967 Tihany 43-48
◇ B75 D20 ◇ ID 32555

HAAS, R. [1981] see EPSTEIN, R.L.

HACK, M. [1976] *The equality problem for vector addition systems is undecidable* (**J** 1426) Theor Comput Sci 2*77-95
◇ D35 D80 ◇ REV MR 55 # 12496 Zbl 357 # 68038 • ID 50440

HADAMARD, J. [1905] see BAIRE, R.

HADAMARD, J. see Vol. I, V, VI for further entries

HAEUSSLER, A.F. [1976] *Polynomial beschraenkte nichtdeterministische Turingmaschinen und die Vollstaendigkeit des aussagelogischen Erfuellungsproblems* (**P** 3196) Kompl von Entscheid Probl;1973/74 Zuerich 20-35
◇ B05 D10 D15 ◇ REV Zbl 383 # 03025 • ID 52007

HAFNER, I. [1976] *Regressive recursion* (**J** 0389) Math Balkanica 6*75-77
◇ D20 ◇ REV MR 80g:03039 Zbl 418 # 03030 • ID 53315

HAFNER, I. see Vol. I, V, VI for further entries

HAGAUER, J. [1982] *On form-equivalence of deterministic pure grammar forms* (**J** 1426) Theor Comput Sci 18*69-87
◇ D05 ◇ REV MR 83b:68079 Zbl 485 # 68064 • ID 38979

HAGAUER, J. [1983] see BUCHER, W.

HAGEN, H. [1984] see AMBOS-SPIES, K.

HAGIHARA, K. & ITO, MINORU & KASAMI, T. & TANIGUCHI, K. [1979] *Decision problems for multivalued dependencies in relational databases* (**J** 1428) SIAM J Comp 8*247-264
◇ B75 D80 ◇ REV MR 80c:68021 Zbl 408 # 68025 • ID 69628

HAHN, S.W. [1969] see BULLOFF, J.J.

HAIMO, F. & SINGER, M.F. & TRETKOFF, M. [1980] *Remarks on analytic continuation* (**J** 0161) Bull London Math Soc 12*9-12
◇ D40 ◇ REV MR 81j:30009 Zbl 447 # 30003 • ID 81455

HAIMO, F. see Vol. III for further entries

HAJEK, P. [1974] *Degrees of dependence in the theory of semisets* (**J** 0027) Fund Math 82*11-24
◇ D30 E40 E45 E70 ◇ REV MR 51 # 10090 Zbl 302 # 02022 • ID 05528

HAJEK, P. [1977] *Arithmetical complexity of some problems in computer science* (**P** 1635) Math Founds of Comput Sci (6);1977 Tatranska Lomnica 282-287
◇ D15 D55 ◇ REV MR 57 # 18222 Zbl 354 # 68073 • ID 31129

HAJEK, P. [1977] *Experimental logics and Π^0_3 theories* (**J** 0036) J Symb Logic 42*515-522
◇ B60 C62 D55 F30 ◇ REV MR 58 # 16243 Zbl 428 # 03043 • ID 26853

HAJEK, P. & HAVRANEK, T. [1978] *Mechanizing hypothesis formation. Mathematical foundations for a general theory* (**X** 0811) Springer: Heidelberg & New York xv+396pp • TRANSL [1984] (**X** 2027) Nauka: Moskva 278pp
◇ A05 B35 C13 C80 C90 D80 ◇ REV MR 82f:03017 MR 86e:03022 Zbl 371 # 02002 • ID 31131

HAJEK, P. [1978] see BALCAR, B.

HAJEK, P. [1978] *Some results on degrees of constructibility* (**P** 1864) Higher Set Th;1977 Oberwolfach 55-71
◇ D30 D55 E45 E55 ◇ REV MR 80d:03055b Zbl 421 # 03041 • ID 53440

HAJEK, P. [1979] *Arithmetical hierarchy and complexity of computation* (**J** 1426) Theor Comput Sci 8*227-237
◇ D10 D15 D55 ◇ REV MR 80b:03051 Zbl 402 # 03038 • ID 54685

HAJEK, P. see Vol. I, II, III, V, VI for further entries

HAJKOVA, M. [1971] *The lattice of bi-numerations of arithmetic. I,II* (**J** 0140) Comm Math Univ Carolinae (Prague) 12*81-104,281-306
◇ D45 F30 ◇ REV MR 44 # 1565 MR 45 # 64 Zbl 216.289 Zbl 232 # 02025 • ID 05529

HAJKOVA, M. see Vol. VI for further entries

HAKEN, A. [1985] *The intractability of resolution* (**J** 1426) Theor Comput Sci 39*297-308
◇ B35 D15 D35 ◇ ID 49032

HAKEN, W. [1968] see BOONE, W.W.

HAKEN, W. [1973] *Connections between topological and group theoretical decision problems* (**P** 0678) Word Probl: Decis & Burnside Probl in Group Th;1969 Irvine 427-441
◇ B25 C60 C65 D35 D40 D80 ◇ REV MR 53 # 1594 Zbl 265 # 02033 JSL 41.786 • ID 29825

HALL, P. [1958] *Some word-problems* (**J** 0039) J London Math Soc 33*482-496
◇ D40 ◇ REV MR 21 # 1331 Zbl 198.29 • ID 05582

HALL, P. see Vol. III for further entries

HALL JR., M. [1949] *The word problem for semigroups with two generators* (**J** 0036) J Symb Logic 14*115-118
◇ D40 ◇ REV MR 11.1 Zbl 34.13 JSL 14.259 • ID 05581

HALL PARTEE, B. [1978] *Fundamentals of mathematics for linguistics* (**X** 0835) Reidel: Dordrecht xxiv+242pp
◇ B65 D05 D10 ◇ REV Zbl 404 # 68073 • ID 54863

HALL PARTEE, B. also published under the name PARTEE, B.

HALPERN, J.D. [1970] see COLLINS, G.E.

HALPERN, J.D. see Vol. III, V for further entries

HALPERN, J.Y. [1985] see CHANDRA, A.K.

HALPERN, J.Y. see Vol. II for further entries

HAMBLIN, C.L. [1973] *Language types and logical theorems* (**J** 0194) Inform & Control 22*183-187
◇ B05 D05 ◇ REV MR 48 # 3304 Zbl 256 # 68037
• ID 62308

HAMBLIN, C.L. see Vol. I, II, III for further entries

HAMILTON, A.G. [1968] *An unsolved problem in the theory of constructive order types* (**J** 0036) J Symb Logic 33*565-567
◇ D45 D50 ◇ REV MR 39 # 3992 Zbl 165.317 • ID 05624

HAMILTON, A.G. [1970] *Bases and α-dimensions of countable vector spaces with recursive operations* (**J** 0036) J Symb Logic 35*85-96
◇ C57 C60 D45 D50 ◇ REV MR 44 # 2604 Zbl 195.18
• ID 05625

HAMILTON, A.G. [1974] *Recursive equivalence types of vector spaces* (**J** 0038) J Austral Math Soc 18*376-384
◇ D45 D50 ◇ REV MR 53 # 12915 Zbl 301 # 02038
• ID 23165

HAMILTON, A.G. see Vol. I, II for further entries

HAMILTON, W.L. & MERTENS, J.R. [1975] *Reproduction in tessellation structures* (**J** 0119) J Comp Syst Sci 10*248-252
◇ D05 ◇ REV MR 51 # 14651 Zbl 304 # 94048 • ID 62311

HAMLET, R.G. [1972] *A patent problem for abstract programming languages; machine-independent computations* (**P** 1901) ACM Symp Th of Comput (4);1972 Denver 193-197
◇ B75 D10 D20 ◇ REV Zbl 363 # 68043 • ID 50919

HAMLET, R.G. [1974] *Introduction to computation theory* (**X** 3224) Intext Educ Publ: New York xvi + 197pp
◇ D10 D15 D25 D98 ◇ REV Zbl 305 # 68004 • ID 62312

HANAK, J. [1966] *On real-time Turing machines* (**J** 0322) Arch Math (Brno) 2*79-92
◇ D10 D15 ◇ REV MR 35 # 2741 Zbl 245 # 02035
• ID 05627

HANATANI, Y. [1966] *Calculabilite des fonctionelles recursives primitives de type fini sur les nombres naturels* (**J** 0260) Ann Jap Ass Phil Sci 3*19-30
◇ D65 F10 F35 F50 ◇ REV MR 33 # 5479 Zbl 143.13
• ID 24779

HANATANI, Y. [1975] *Calculability of the primitive recursive functionals of finite type over the natural numbers* (**P** 1440) ⊧ ISILC Proof Th Symp (Schuette);1974 Kiel 152-163
◇ D65 F10 F35 F50 ◇ REV MR 53 # 5273
Zbl 317 # 02033 • ID 22904

HANATANI, Y. see Vol. VI for further entries

HANF, W.P. [1965] *Model-theoretic methods in the study of elementary logic* (**P** 0614) Th Models;1963 Berkeley 132-145
◇ B25 C07 D20 D25 D35 F25 ◇ REV MR 35 # 1457
Zbl 166.258 JSL 34.127 • ID 05635

HANF, W.P. [1974] *Nonrecursive tilings of the plane I* (**J** 0036) J Symb Logic 39*283-285
◇ D05 D80 ◇ REV MR 51 # 110 Zbl 299 # 02054 • REM Part II 1974 by Myers,D.L. • ID 05636

HANF, W.P. see Vol. I, III, V for further entries

HANSEL, G. [1985] see BLANCHARD, F.

HANSEL, G. see Vol. II for further entries

HANSELL, R.W. [1983] see FREMLIN, D.H.

HANSELL, R.W. & JAYNE, J.E. & ROGERS, C.A. [1985] *Separation of K-analytic sets* (**J** 0303) Mathematika (Univ Coll London) 32*147-190
◇ D55 E15 ◇ ID 49732

HANSELL, R.W. see Vol. V for further entries

HANSON, DAVID R. [1976] *A simple variant of the boundary-tag algorithm for the allocation of coroutine environments* (**J** 0232) Inform Process Lett 4*109-112
◇ D03 ◇ REV Zbl 321 # 68014 • ID 62322

HANSON, N.R. [1961] *The Goedel theorem* (**J** 0047) Notre Dame J Formal Log 2*94-110,228
◇ D20 D35 F30 ◇ REV Zbl 121.254 JSL 27.471
• ID 05642

HARAGUCHI, M. [1979] *The theoretical foundation of programming by examples* (**J** 0106) Mem Fac Sci, Kyushu Univ, Ser A 33*237-255
◇ B75 D20 ◇ REV MR 80i:68069 Zbl 426 # 03043
• ID 53639

HARAN, D. [1984] *The undecidability of pseudo-real-closed fields* (**J** 0504) Manuscr Math 49*91-108
◇ B25 C60 D35 ◇ REV MR 85j:03072 Zbl 578 # 12020
• ID 44141

HARAN, D. see Vol. III for further entries

HARDOUIN-DUPARC, J. [1974] *Paradis terrestre dans l'automate cellulaire de Conway* (**J** 0205) Rev Franc Autom, Inf & Rech Operat 8/R-3*63-71
◇ D05 ◇ REV MR 51 # 12442 Zbl 311 # 94043 • ID 62331

HAREL, D. [1983] *Recurring dominoes: making the highly undecidable highly understandable* (**P** 3864) FCT'83 Found of Comput Th;1983 Borgholm 177-194
◇ B25 B75 D15 D55 ◇ REV MR 85e:68003
Zbl 531 # 68002 • ID 38572

HAREL, D. & PELEG, D. [1984] *On static logics, dynamic logics, and complexity classes* (**J** 0194) Inform & Control 60*86-102
◇ B75 D15 ◇ ID 44121

HAREL, D. & PATERSON, M.S. [1984] *Undecidability of PDL with* $L=\{a^{2^i} \mid i \geqslant 0\}$ (**J** 0119) J Comp Syst Sci 29*359-365
◇ B75 D35 ◇ REV MR 86g:68056 Zbl 553 # 03029
• ID 43336

HAREL, D. [1985] *Recurring dominoes: Making the highly undecidable highly understandable* (**S** 3358) Ann Discrete Math 24*51-72
◇ B25 B75 D05 D15 D55 ◇ ID 41208

HAREL, D. see Vol. I, II, III for further entries

HARKLEROAD, L. [1979] *Recursive equivalence types on recursive manifolds* (**J** 0047) Notre Dame J Formal Log 20*1-31
◇ D45 D50 D80 ◇ REV MR 83d:03054 Zbl 332 # 02048
• ID 52398

HARKLEROAD, L. [1983] *Manifolds allowing RET arithmetic* (**J** 0047) Notre Dame J Formal Log 24*482-484
◇ D45 D50 D80 ◇ REV MR 86a:03044 Zbl 534 # 03021
• ID 36562

HARKLEROAD, L. [1984] *Fuzzy recursion, RET's and isols* (**J** 0068) Z Math Logik Grundlagen Math 30*425-436
◇ D50 ◇ REV MR 86b:03058 Zbl 534 # 03023 • ID 43529

HARNAU, W. [1976] see DASSOW, J.

HARNAU, W. see Vol. II for further entries

HARNIK, V. & MAKKAI, M. [1976] *Applications of Vaught sentences and the covering theorem* (**J** 0036) J Symb Logic 41*171-187
◇ C15 C40 C45 C50 C52 C75 D70 E15 ◇ REV MR 56 # 5265 Zbl 333 # 02013 • ID 14797

HARNIK, V. [1980] *Game sentences, recursive saturation and definability* (**J** 0036) J Symb Logic 45*35-46
◇ C40 C50 C57 C70 ◇ REV MR 81h:03076 Zbl 445 # 03014 • ID 56478

HARNIK, V. see Vol. III, V, VI for further entries

HARPER, L.H. [1975] *A note on some classes of boolean functions* (**J** 0548) Stud Appl Math 54*161-164
◇ B05 D15 ◇ REV MR 56 # 5068 Zbl 307 # 02007 • ID 62346

HARRINGTON, L.A. [1974] *Recursively presentable prime models* (**J** 0036) J Symb Logic 39*305-309
◇ C35 C50 C57 C60 ◇ REV MR 50 # 4292 Zbl 332 # 02055 JSL 49.671 • ID 05664

HARRINGTON, L.A. [1974] *The superjump and the first recursively Mahlo ordinal* (**P** 0602) Generalized Recursion Th (1);1972 Oslo 43-52
◇ D60 D65 E55 ◇ REV MR 52 # 13347 Zbl 292 # 02039 • ID 21802

HARRINGTON, L.A. [1975] *Π^1_1 sets and Π^1_2 singletons* (**J** 0053) Proc Amer Math Soc 52*356-360
◇ D55 E15 E45 ◇ REV MR 51 # 10096 Zbl 356 # 02048 • ID 17524

HARRINGTON, L.A. & KECHRIS, A.S. [1975] *A basis result for Σ^0_3 sets of reals with an application to minimal covers* (**J** 0053) Proc Amer Math Soc 53*445-448
◇ D30 D55 E15 E60 ◇ REV MR 53 # 2683 Zbl 376 # 02054 • ID 21631

HARRINGTON, L.A. & KECHRIS, A.S. [1975] *On characterizing Spector classes* (**J** 0036) J Symb Logic 40*19-24
◇ C62 D55 D60 D65 D75 E45 E55 E60 ◇ REV MR 55 # 5420 Zbl 312 # 02033 • ID 05665

HARRINGTON, L.A. [1976] see GUASPARI, D.

HARRINGTON, L.A. & JECH, T.J. [1976] *On Σ_1 well-orderings of the universe* (**J** 0036) J Symb Logic 41*167-170
◇ D55 E35 E45 E47 ◇ REV MR 53 # 7780 Zbl 376 # 02053 • ID 14795

HARRINGTON, L.A. & KECHRIS, A.S. [1976] *On monotone vs. nonmonotone induction* (**J** 0015) Bull Amer Math Soc 82*888-890
◇ D70 ◇ REV MR 55 # 10259 Zbl 363 # 02044 • ID 30724

HARRINGTON, L.A. & MACQUEEN, D.B. [1976] *Selection in abstract recursion theory* (**J** 0036) J Symb Logic 41*153-158
◇ D65 D75 ◇ REV MR 56 # 105 Zbl 409 # 03029 • ID 14792

HARRINGTON, L.A. & KECHRIS, A.S. [1977] *Π^1_2 singletons and $0^{\#}$* (**J** 0027) Fund Math 95*167-171
◇ D30 D55 E45 E55 ◇ REV MR 57 # 117 Zbl 384 # 03044 • ID 26528

HARRINGTON, L.A. [1977] *Long projective wellorderings* (**J** 0007) Ann Math Logic 12*1-24
◇ C62 D55 E15 E35 E50 ◇ REV MR 57 # 5752 Zbl 384 # 03033 • ID 24274

HARRINGTON, L.A. & KIROUSIS, L.M. & SCHLIPF, J.S. [1978] *A generalized Kleene-Moschovakis theorem* (**J** 0053) Proc Amer Math Soc 68*209-213
◇ D55 D70 D75 ◇ REV MR 57 # 16020 Zbl 347 # 02027 • ID 30951

HARRINGTON, L.A. [1978] *Analytic determinacy and $0^{\#}$* (**J** 0036) J Symb Logic 43*685-693
◇ D55 D65 E45 E55 E60 ◇ REV MR 80b:03065 Zbl 398 # 03039 JSL 49.665 • ID 52767

HARRINGTON, L.A. & SAMI, R.L. [1979] *Equivalence relations, projective and beyond* (**P** 2627) Logic Colloq;1978 Mons 247-264
◇ D55 E07 E15 E60 ◇ REV MR 82d:03080 Zbl 521 # 03031 • ID 73861

HARRINGTON, L.A. & SHORE, R.A. [1981] *Definable degrees and automorphisms of $\mathscr{D}$* (**J** 0589) Bull Amer Math Soc (NS) 4*97-100
◇ D30 D55 ◇ REV MR 82g:03077 Zbl 452 # 03033 • ID 54097

HARRINGTON, L.A. [1982] see GUREVICH, Y.

HARRINGTON, L.A. & SHELAH, S. [1982] *The undecidability of the recursively enumerable degrees* (**J** 0589) Bull Amer Math Soc (NS) 6*79-80
◇ D25 D35 ◇ REV MR 83i:03067 Zbl 518 # 03016 • ID 35526

HARRINGTON, L.A. & MORLEY, M.D. & SCEDROV, A. & SIMPSON, S.G. [1985] *Harvey Friedman's research on the foundations of mathematics* (**X** 0809) North Holland: Amsterdam xvi + 408pp
◇ B97 C97 D97 E97 F97 ◇ ID 49810

HARRINGTON, L.A. see Vol. III, V, VI for further entries

HARRISON, J. [1963] *Appendix C: Equivalence of the effective operations and the hereditarily continuous functionals* (**C** 4220) Rep Sem Found Anal 2*C1-3
◇ D65 F50 ◇ ID 49833

HARRISON, J. [1968] *Recursive pseudo-well-orderings* (**J** 0064) Trans Amer Math Soc 131*526-543
◇ D20 D45 ◇ REV MR 39 # 5366 Zbl 186.11 JSL 37.197 • ID 05674

HARRISON, M.A. [1965] *Introduction to switching and automata theory* (**X** 0822) McGraw-Hill: New York 499pp
◇ B70 D05 D98 ◇ REV MR 32 # 3963 • ID 23525

HARRISON, M.A. [1967] see GINSBURG, S.

HARRISON, M.A. [1968] see GINSBURG, S.

HARRISON, M.A. [1969] *Lectures on linear sequential machines* (**X** 0801) Academic Pr: New York x + 210pp
◇ D05 D98 ◇ REV MR 44 # 2531 Zbl 212.340 • ID 23526

HARRISON, M.A. [1970] see GINSBURG, S.

Harrison, M.A. & Schkolnick, M. [1971] *A grammatical characterization of one-way nondeterministic stack languages* (**J** 0037) ACM J 18*148-172
◇ D10 ◇ REV MR 44 # 8081 Zbl 274 # 68020 • ID 62357

Harrison, M.A. [1978] *Introduction to formal language theory* (**X** 0832) Addison-Wesley: Reading xiv + 594pp
◇ D05 D98 ◇ REV MR 80h:68060 Zbl 411 # 68058 • ID 69631

Harrison, M.A. & Yehudai, A. [1981] *Eliminating null rules in linear time* (**J** 1193) Comput J (London) 24*156-161
◇ D05 D15 ◇ REV MR 82e:68076 Zbl 456 # 68089 • ID 81495

Harrison, M.A. see Vol. I, II for further entries

Harrop, R. [1961] *On the recursivity of finite sets* (**J** 0068) Z Math Logik Grundlagen Math 7*136-140
◇ D20 ◇ REV MR 25 # 1988 Zbl 109.241 JSL 33.115 • ID 05680

Harrop, R. [1964] *A relativization procedure for propositional calculi, with an application to a generalized form of Post's theorem* (**J** 3240) Proc London Math Soc, Ser 3 14*595-617
◇ B22 D35 ◇ REV MR 30 # 16 Zbl 158.251 JSL 32.125 • ID 05682

Harrop, R. [1965] *Some generalizations and applications of a relativization procedure for propositional calculi* (**P** 0688) Logic Colloq;1963 Oxford 12-41
◇ B22 B25 D35 ◇ REV MR 35 # 2737 Zbl 158.252 JSL 32.125 • ID 05684

Harrop, R. see Vol. I, III, V, VI for further entries

Harrow, K. [1975] *Small Grzegorczyk classes and limited minimum* (**J** 0068) Z Math Logik Grundlagen Math 21*417-426
◇ D20 ◇ REV MR 53 # 7741 Zbl 316 # 02043 • ID 05688

Harrow, K. [1978] *The bounded arithmetic hierarchy* (**J** 0194) Inform & Control 36*102-117
◇ D15 D55 ◇ REV MR 57 # 16010 Zbl 374 # 02019 • ID 30717

Harrow, K. [1979] *Equivalence of some hierarchies of primitive recursive functions* (**J** 0068) Z Math Logik Grundlagen Math 25*411-418
◇ D20 ◇ REV MR 80k:03039 Zbl 446 # 03029 • ID 56558

Harrow, K. [1979] *Theoretical and applied computer science: antagonism or symbiosis?* (**J** 0005) Amer Math Mon 86*253-260
◇ B75 D10 D80 ◇ REV MR 80c:68001 Zbl 416 # 68031 • ID 69655

Hart, J. [1969] *$\mathcal{E}^0$-arithmetic* (**J** 0068) Z Math Logik Grundlagen Math 15*237
◇ C62 D20 ◇ REV MR 40 # 40 Zbl 157.336 • ID 05689

Hartley, J.P. [1983] *The countably based functionals* (**J** 0036) J Symb Logic 48*458-474
◇ D65 ◇ REV MR 84k:03116 Zbl 548 # 03024 • ID 36134

Hartley, J.P. [1985] *Effective discontinuity and a characterisation of the superjump* (**J** 0036) J Symb Logic 50*349-358
◇ D65 ◇ ID 42552

Hartmanis, J. & Stearns, R.E. [1962] *Some dangers in state reduction of sequential machines* (**J** 0194) Inform & Control 5*252-260
◇ D05 ◇ REV MR 27 # 1360 Zbl 105.323 • ID 05691

Hartmanis, J. [1963] *Regularity preserving modifications of regular expressions* (**J** 0194) Inform & Control 6*55-69
◇ D05 ◇ REV JSL 31.265 • ID 05693

Hartmanis, J. & Stearns, R.E. [1964] *Computational complexity of recursive sequences* (**P** 0550) Switch Circ Th & Log Design (5);1964 Princeton 82-90
◇ D15 ◇ REV Zbl 131.154 JSL 32.121 • ID 22001

Hartmanis, J. & Lewis, P.M. & Stearns, R.E. [1965] *Classification of computation by time and memory requirements* (**P** 0573) Inform Processing (3);1965 New York 1*31-35
◇ D15 ◇ REV JSL 37.624 • ID 28468

Hartmanis, J. & Lewis II, P.M. & Stearns, R.E. [1965] *Hierarchies of memory limited computations* (**P** 1591) Switch Circ Th & Log Design (6);1965 Ann Arbor 179-190
◇ D10 D15 ◇ REV Zbl 229 # 02033 JSL 37.624 • ID 43996

Hartmanis, J. & Lewis II, P.M. & Stearns, R.E. [1965] *Memory bounds for recognition of context-free and context-sensitive languages* (**P** 1591) Switch Circ Th & Log Design (6);1965 Ann Arbor 191-202 • ERR/ADD ibid 190
◇ B65 D05 D15 ◇ REV Zbl 272 # 68054 JSL 37.625 • ID 43989

Hartmanis, J. & Stearns, R.E. [1965] *On the computational complexity of algorithms* (**J** 0064) Trans Amer Math Soc 117*285-306
◇ D15 ◇ REV MR 30 # 1040 Zbl 156.256 JSL 32.120 • ID 05694

Hartmanis, J. [1967] *Context free languages and Turing machine computation* (**P** 0737) Math Aspects Comput Sci;1966 New York 42-51
◇ D05 D10 ◇ REV MR 38 # 4239 Zbl 189.291 JSL 37.759 • ID 05696

Hartmanis, J. [1967] *On memory requirements for context free language recognition* (**J** 0037) ACM J 14*663-665 • TRANSL [1970] (**C** 1540) Probl Mat Log: Slozh Algor & Kl Vychisl Funk 339-343
◇ D05 ◇ REV MR 39 # 3930 Zbl 155.18 • ID 20874

Hartmanis, J. [1968] *Computational complexity of one-tape Turing machine computations* (**J** 0037) ACM J 15*325-339 • TRANSL [1970] (**C** 1540) Probl Mat Log: Slozh Algor & Kl Vychisl Funk 282-300
◇ D10 D15 D20 ◇ REV MR 40 # 5352 Zbl 162.317 • ID 20878

Hartmanis, J. [1968] see Blum, M.

Hartmanis, J. [1968] *Tape-reversal bounded Turing machine computations* (**J** 0119) J Comp Syst Sci 2*117-135
◇ D10 D15 ◇ REV MR 39 # 5265 Zbl 259 # 68020 • ID 62363

Hartmanis, J. [1969] *On the complexity of undecidable problems in automata theory* (**J** 0037) ACM J 16*160-167
◇ D05 D15 ◇ REV MR 41 # 9662 Zbl 182.24 • ID 05697

Hartmanis, J. & Hopcroft, J.E. [1970] *What makes some language theory problems undecidable* (**J** 0119) J Comp Syst Sci 4*368-376
◇ D05 D30 ◇ REV MR 43 # 3053 Zbl 198.30 JSL 40.245 • ID 05698

HARTMANIS, J. & HOPCROFT, J.E. [1971] *An overview of the theory of computational complexity* (**J** 0037) ACM J 18*444-475
◇ D15 D98 ◇ REV MR 44 # 5226 Zbl 289 # 68011
• ID 05702

HARTMANIS, J. [1971] *Size arguments in the study of computation speeds* (**P** 1383) Symp Comput & Automata;1971 Brooklyn 21*7-18
◇ D15 ◇ REV Zbl 265 # 68029 • ID 29850

HARTMANIS, J. & LEWIS II, P.M. [1971] *The use of lists in the study of undecidable problems in automata theory* (**J** 0119) J Comp Syst Sci 5*54-66
• TRANSL [1974] (**J** 3079) Kiber Sb Perevodov, NS 11*117-130
◇ D05 ◇ REV MR 43 # 1839 Zbl 288 # 02019 JSL 39.347
• ID 05700

HARTMANIS, J. [1974] see BAKER, T.P.

HARTMANIS, J. & HUNT III, H.B. [1974] *The LBA problem and its importance in the theory of computing* (**P** 0761) Compl of Computation;1973 New York 1-26
◇ D05 D15 ◇ REV MR 50 # 11845 Zbl 301 # 68056
• ID 62368

HARTMANIS, J. [1975] see BAKER, T.P.

HARTMANIS, J. [1976] *On effective speed-up and long proofs of trivial theorems in formal theories (French summary)* (**J** 4698) Rev Franc Autom, Inf & Rech Operat, Ser Rouge Inf Th 10/R-1*29-38
◇ B25 B35 D15 D20 F20 ◇ REV MR 54 # 6550
Zbl 399 # 03042 • ID 24160

HARTMANIS, J. [1976] see BERMAN, L.

HARTMANIS, J. [1978] *Feasible computations and provable complexity properties* (**X** 2378) SIAM: Philadelphia vii + 62pp
◇ D15 ◇ REV MR 82c:68021 Zbl 383 # 68043 • ID 81502

HARTMANIS, J. & IMMERMAN, N. & MAHANEY, S.R. [1978] *One-way log-tape reductions* (**P** 3578) IEEE Symp Found of Comput Sci (19);1978 Ann Arbor 65-72
◇ D10 D15 ◇ REV MR 80e:68133 • ID 81501

HARTMANIS, J. [1979] *Relations between diagonalization, proof systems, and complexity gaps* (**J** 1426) Theor Comput Sci 8*239-253
◇ D15 D20 ◇ REV MR 81g:68066 Zbl 398 # 68013
• ID 52790

HARTMANIS, J. & MAHANEY, S.R. [1981] *Languages simultaneously complete for one-way and two-way log-tape automata* (**J** 1428) SIAM J Comp 10*383-390
◇ D05 D15 ◇ REV MR 83m:68147 Zbl 471 # 68059
• ID 40428

HARTMANIS, J. [1981] *Observations about the development of theoretical computer science* (**J** 3789) Ann Hist of Comp 3*42-51
◇ A10 D05 D15 ◇ REV MR 83e:01056 • ID 40116

HARTMANIS, J. [1982] *A note on natural complete sets and Goedel numberings* (**J** 1426) Theor Comput Sci 17*75-89
◇ D15 D25 D45 ◇ REV MR 84h:03098 Zbl 483 # 03026
• ID 34288

HARTMANIS, J. & YESHA, Y. [1983] *Computation times of NP sets of different densities* (**P** 3851) Automata, Lang & Progr (10);1983 Barcelona 319-330
◇ D15 ◇ REV MR 85g:68021 Zbl 533 # 03023 • ID 36547

HARTMANIS, J. [1983] *On sparse sets in NP-P* (**J** 0232) Inform Process Lett 16*55-60
◇ D15 ◇ REV MR 85d:68029 Zbl 501 # 68014 • ID 39021

HARTMANIS, J. [1983] *On Goedel speed-up and succinctness of language representation* (**J** 1426) Theor Comput Sci 26*335-342
◇ D15 F20 ◇ REV MR 85e:68039 Zbl 526 # 68034
• ID 38195

HARTMANIS, J. & YESHA, Y. [1984] *Computation times of NP sets of different densities* (**J** 1426) Theor Comput Sci 34*17-32
◇ D15 ◇ REV MR 86d:68026 • ID 48884

HARTMANIS, J. & IMMERMAN, N. [1985] *On complete problems for NP ∩ CoNP* (**P** 4628) Automata, Lang & Progr (12);1985 Nafplion 250-259
◇ D15 ◇ ID 49568

HARTMANIS, J. & IMMERMAN, N. & SEWELSON, V. [1985] *Sparse sets in NP - P: EXPTIME versus NEXPTIME* (**J** 0194) Inform & Control 65*158-181
◇ D15 ◇ ID 49565

HASENJAEGER, G. [1953] *Eine Bemerkung zu Henkin's Beweis fuer die Vollstaendigkeit des Praedikatenkalkuels der ersten Stufe* (**J** 0036) J Symb Logic 18*42-48
◇ B10 C07 C57 D55 ◇ REV MR 14.1052 Zbl 51.5
JSL 31.268 • ID 05723

HASENJAEGER, G. [1955] *On definability and derivability* (**P** 1589) Math Interpr of Formal Systs;1954 Amsterdam 15-25
◇ B10 B15 C07 D55 ◇ REV MR 17.699 Zbl 68.246
JSL 24.171 • ID 27719

HASENJAEGER, G. [1958] *Formales und produktives Schliessen* (**J** 0160) Math-Phys Sem-ber, NS 6*184-194
◇ A05 D35 ◇ REV MR 31 # 5 Zbl 231 # 02003 JSL 29.59
• ID 05724

HASENJAEGER, G. [1965] *Zur arithmetischen Klassifikation reeller Zahlen* (**J** 0096) Acta Philos Fenn 18*13-19
◇ D55 ◇ REV MR 32 # 5506 Zbl 199.32 • ID 05730

HASENJAEGER, G. [1983] *Exponential diophantine description of a small universal Turing machine (UTM) using Matiyasevich's 1975/80 "masking"* (**P** 4083) GTI-Worksh (1);1983 Paderborn 105-116
◇ D10 D15 D35 ◇ ID 47069

HASENJAEGER, G. [1984] see BOERGER, E.

HASENJAEGER, G. [1984] *Universal Turing machines (UTM) and Jones-Matiyasevich-masking* (**P** 2342) Symp Rek Kombin;1983 Muenster 248-253
◇ D10 D35 ◇ ID 43371

HASENJAEGER, G. see Vol. I, II, III, V, VI for further entries

HASHIGUCHI, K. [1979] *A decision procedure for the order of regular events* (**J** 1426) Theor Comput Sci 8*69-72
◇ D05 ◇ REV MR 80g:68065 Zbl 419 # 68088 • ID 69632

HASHIGUCHI, K. [1982] *Limitedness theorem on finite automata with distance functions* (**J** 0119) J Comp Syst Sci 24*233-244
◇ D05 ◇ REV MR 83i:68082 Zbl 513 # 68051 • ID 39362

HASKELL, R. [1975] *Efficient implementation of a class of recursively defined functions* (J 1193) Comput J (London) 18*23-29
◇ D15 ◇ REV Zbl 296 # 68047 • ID 62372

HASSETT, M.J. [1967] *A note on regressive isols* (J 0044) Math Z 97*425-426
◇ D50 ◇ REV MR 35 # 2735 Zbl 189.10 • ID 05739

HASSETT, M.J. [1969] *Recursive equivalence types and groups* (J 0036) J Symb Logic 34*13-20
◇ D50 ◇ REV MR 39 # 2634 Zbl 185.21 • ID 05740

HASSETT, M.J. [1970] *A mapping property of regressive isols* (J 0316) Illinois J Math 14*478-487
◇ D50 ◇ REV MR 42 # 1657 Zbl 201.12 • ID 05741

HASSETT, M.J. [1972] *A characterization of the orders of regressive ω-groups* (J 0038) J Austral Math Soc 14*379-382
◇ D50 ◇ REV MR 47 # 4780 Zbl 247 # 02042 • ID 29507

HASSETT, M.J. [1972] *A closure property of regressive isols* (J 0308) Rocky Mountain J Math 2*1-24
◇ D50 ◇ REV MR 45 # 1761 Zbl 238 # 02036 • ID 05742

HASSETT, M.J. [1973] *Extension pathology in regressive isols* (J 0020) Compos Math 26*31-40
◇ D50 ◇ REV MR 47 # 3158 Zbl 257 # 02037 • ID 05743

HATCHER, W.S. & RUS, T. [1976] *Context-free algebras* (J 0383) J Cybern 6*65-77
◇ D05 ◇ REV MR 57 # 14621 Zbl 362 # 68107 • ID 32204

HATCHER, W.S. & HODGSON, B.R. [1981] *Complexity bounds on proofs* (J 0036) J Symb Logic 46*255-258
◇ D25 F20 ◇ REV MR 82j:03075 Zbl 469 # 03041 • ID 55168

HATCHER, W.S. see Vol. I, II, III, V, VI for further entries

HAUCK, J. [1972] *Funktionale Rekursion* (J 0068) Z Math Logik Grundlagen Math 18*31-36
◇ D20 D65 ◇ REV MR 45 # 4968 Zbl 272 # 02056 • ID 05755

HAUCK, J. [1972] *Zur berechenbaren gleichmaessigen Finitheit rekursiver Funktionale* (J 0115) Wiss Z Humboldt-Univ Berlin, Math-Nat Reihe 21*527-529
◇ D20 F60 ◇ REV MR 49 # 2321 Zbl 251 # 02036 • ID 05754

HAUCK, J. [1975] *Turingmaschinen und berechenbare reelle Funktionen* (J 0115) Wiss Z Humboldt-Univ Berlin, Math-Nat Reihe 24*797-799
◇ D10 D20 F60 ◇ REV MR 58 # 5130 Zbl 327 # 02030 • ID 62385

HAUCK, J. [1980] *Konstruktive Darstellungen in topologischen Raeumen mit rekursiver Basis* (J 0068) Z Math Logik Grundlagen Math 26*565-576
◇ D45 D80 ◇ REV MR 83i:03072 Zbl 455 # 03025 • ID 54286

HAUCK, J. [1980] *Stetigkeitseigenschaften berechenbarer reeller Funktionen* (J 0068) Z Math Logik Grundlagen Math 26*69-76
◇ D20 F60 ◇ REV MR 82b:03107 Zbl 475 # 03035 • ID 55489

HAUCK, J. [1981] *Berechenbarkeit in topologischen Raeumen mit rekursiver Basis* (J 0068) Z Math Logik Grundlagen Math 27*473-480
◇ D45 ◇ REV MR 82j:03078 Zbl 473 # 03051 • ID 55379

HAUCK, J. [1982] *Stetigkeitseigenschaften berechenbarer Funktionale* (J 0068) Z Math Logik Grundlagen Math 28*377-383
◇ D80 ◇ REV MR 83m:03068 Zbl 509 # 03033 • ID 35467

HAUCK, J. [1985] *Ein Kriterium fuer die konstrukive Loesbarkeit der Differentialgleichung* $y' = f(x,y)$ (J 0068) Z Math Logik Grundlagen Math 31*357-362
◇ D80 ◇ REV Zbl 567 # 03030 • ID 47563

HAUCK, J. see Vol. I, III, VI for further entries

HAUSCHILD, K. [1966] *Ueber die Unentscheidbarkeit nichtassoziativer Schiefringe (English and French summaries)* (J 0115) Wiss Z Humboldt-Univ Berlin, Math-Nat Reihe 15*681-683
◇ D35 ◇ REV MR 36 # 2501 Zbl 199.31 • ID 05763

HAUSCHILD, K. & RAUTENBERG, W. [1970] *Universelle Interpretierbarkeit in Verbaenden (Russian, English and French summaries)* (J 0115) Wiss Z Humboldt-Univ Berlin, Math-Nat Reihe 19*575-577
◇ D35 F25 G10 ◇ REV MR 48 # 8220 Zbl 275 # 02043 • ID 05774

HAUSCHILD, K. & RAUTENBERG, W. [1971] *Interpretierbarkeit in der Gruppentheorie* (J 0004) Algeb Universalis 1*136-151
◇ C60 D35 F25 ◇ REV MR 46 # 39 Zbl 242 # 02052 • ID 05780

HAUSCHILD, K. & RAUTENBERG, W. [1971] *Interpretierbarkeit und Entscheidbarkeit in der Graphentheorie I* (J 0068) Z Math Logik Grundlagen Math 17*47-55
◇ B25 C65 D35 F25 ◇ REV MR 44 # 6491 Zbl 231 # 02058 • REM Part II 1972 by Hauschild,K. & Herre,H. & Rautenberg,W • ID 27450

HAUSCHILD, K. & HERRE, H. & RAUTENBERG, W. [1972] *Interpretierbarkeit und Entscheidbarkeit in der Graphentheorie II* (J 0068) Z Math Logik Grundlagen Math 18*457-480
◇ B25 C65 D35 F25 ◇ REV MR 48 # 3737 Zbl 284 # 02023 • REM Part I 1971 • ID 05787

HAUSCHILD, K. [1972] *Universalitaet in der Ringtheorie* (J 0115) Wiss Z Humboldt-Univ Berlin, Math-Nat Reihe 21*505-506
◇ C60 D35 F25 ◇ REV MR 48 # 8221 Zbl 245 # 02041 • ID 05790

HAUSCHILD, K. [1974] *Rekursive Unentscheidbarkeit der Theorie der pythagoraeischen Koerper* (J 0027) Fund Math 82*191-197
◇ D35 ◇ REV MR 50 # 9567 Zbl 299 # 02056 • ID 16923

HAUSCHILD, K. [1981] *Zum Vergleich von Haertigquantor und Rescherquantor* (J 0068) Z Math Logik Grundlagen Math 27*255-264
◇ B25 C10 C55 C80 D35 ◇ REV MR 82h:03033 Zbl 503.03011 • ID 73915

HAUSCHILD, K. [1982] *Model-theoretic properties of cause-and-effect structures* (J 0140) Comm Math Univ Carolinae (Prague) 23*541-555
◇ B45 C52 D35 ◇ REV MR 84e:03023 Zbl 522 # 03005 • ID 34362

HAUSCHILD, K. see Vol. I, III, V, VI for further entries

HAUSDORFF, F. [1914] *Grundzuege der Mengenlehre* (X 2636) Veit: Leipzig viii+476pp
• REPR [1965] (X 0848) Chelsea: New York viii+476pp
◇ D55 E10 E15 E98 ◇ REV MR 25 # 4999 Zbl 41.20 FdM 45.123 • ID 23280

HAUSDORFF, F. [1933] *Zur Projektivitaet der δs-Funktionen* (J 0027) Fund Math 20*100-104
◇ D55 E15 ◇ REV Zbl 7.241 FdM 59.885 • ID 05798

HAUSDORFF, F. see Vol. III, V for further entries

HAUSMANN, D. & KORTE, B. [1980] *The relative strength of oracles for independence systems* (P 3164) Fct Anal, Num Anal & Optimization (Unger);1979 Bonn 195-211
◇ D15 ◇ REV MR 81m:68030 Zbl 471 # 05019 • ID 55254

HAUSSLER, D. & ZEIGER, H.P. [1980] *Very special languages and representations of recursively enumerable languages via computation histories* (J 0194) Inform & Control 47*201-212
◇ D05 ◇ REV MR 83e:68103 Zbl 457 # 68086 • ID 40167

HAUSSLER, D. [1981] *Model completeness of an algebra of languages* (J 0053) Proc Amer Math Soc 83*371-374
◇ B25 C35 D05 ◇ REV MR 82k:03048 Zbl 478 # 03014 • ID 55629

HAUSSLER, D. [1983] *Insertion languages* (J 0191) Inform Sci 31*77-89
◇ D05 ◇ REV MR 85m:68015 Zbl 544 # 68049 • ID 45929

HAUSSLER, D. & WARMUTH, M.K. [1984] *On the complexity of iterated shuffle* (J 0119) J Comp Syst Sci 28*345-358
◇ B75 D15 ◇ REV MR 86f:68016 Zbl 549 # 68039 • ID 45482

HAUSSLER, D. [1985] see BUCHER, W.

HAVEL, I.M. [1967] see CULIK, K.

HAVEL, I.M. [1967] *On the modification of normal algorithms of Markov* (P 1671) Colloq Automatenth (3);1965 Hannover 183-189
◇ D03 ◇ REV Zbl 174.22 • ID 29432

HAVEL, I.M. [1968] *Regular expressions over generalized alphabet and design of logical nets* (J 0156) Kybernetika (Prague) 4*516-537
◇ D05 ◇ REV MR 40 # 1215 Zbl 169.315 JSL 38.524 • ID 05805

HAVEL, I.M. [1971] *Weak complexity measures* (J 1456) SIGACT News 8*21-30
◇ D15 ◇ ID 31135

HAVEL, I.M. [1973] *Ueber das zehnte Hilbertsche Problem (Tschechisch)* (J 1527) Pokroky Mat Fyz Astron (Prague) 18*185-192
◇ D35 ◇ REV MR 56 # 15382 Zbl 264 # 02044 • ID 27516

HAVEL, I.M. [1975] *Finite branching automata: automata theory motivated by problem solving* (P 1755) Math Founds of Comput Sci (3);1974 Jadwisin 53-61
◇ D05 ◇ REV Zbl 306 # 68039 • ID 62405

HAVEL, I.M. see Vol. I for further entries

HAVRANEK, T. [1974] *An application of logical-probabilistic expressions to the realization of stochastic automata* (J 0156) Kybernetika (Prague) 10*241-257
◇ B70 B80 D05 ◇ REV MR 49 # 12214 Zbl 283 # 94020 • ID 31143

HAVRANEK, T. [1975] *The approximation problem in computational statistics* (P 0454) Math Founds of Comput Sci (4);1975 Marianske Lazne 258-265
◇ D80 ◇ REV MR 55 # 1822 Zbl 324 # 62005 • ID 32354

HAVRANEK, T. [1976] *Statistics and computability* (J 0156) Kybernetika (Prague) 12*303-315
◇ B75 D20 D80 ◇ REV MR 56 # 4230 Zbl 375 # 62001 • ID 32355

HAVRANEK, T. [1978] *Enumeration calculi and rank methods* (J 1741) Int J Man-Mach Stud 10*59-65
◇ B50 D10 ◇ REV MR 80g:68118a Zbl 404 # 68094 • ID 31145

HAVRANEK, T. [1978] see HAJEK, P.

HAVRANEK, T. see Vol. I, II, III for further entries

HAWRUSIK, F.M. & VENKATARAMAN, K.N. & YASUHARA, A. [1983] *A view of computability on term algebras* (J 0119) J Comp Syst Sci 26*410-471
◇ B75 D20 ◇ REV MR 85i:68016 Zbl 514 # 68053 • ID 44356

HAY, L. [1965] *On creative sets and indices of partial recursive functions* (J 0064) Trans Amer Math Soc 120*359-367
◇ D20 D25 ◇ REV MR 33 # 3928 Zbl 163.252 JSL 39.186 • ID 05810

HAY, L. [1966] *Isomorphism types of index sets of partial recursive functions* (J 0053) Proc Amer Math Soc 17*106-110
◇ D20 D25 D30 ◇ REV MR 32 # 4000 Zbl 163.252 JSL 39.186 • ID 05811

HAY, L. [1966] *The co-simple isols* (J 0120) Ann of Math, Ser 2 83*231-256
◇ D50 ◇ REV MR 34 # 5671 Zbl 189.289 JSL 37.407 • ID 05812

HAY, L. [1967] *Elementary differences between the isols and the co-simple isols* (J 0064) Trans Amer Math Soc 127*427-441
◇ D50 ◇ REV MR 35 # 4100 Zbl 189.289 JSL 37.408 • ID 05813

HAY, L. [1969] *Index sets of finite classes of recursively enumerable sets* (J 0036) J Symb Logic 34*39-44
◇ D25 D30 ◇ REV MR 40 # 4103 Zbl 193.310 JSL 39.186 • ID 05814

HAY, L. [1972] *A discrete chain of degrees of index sets* (J 0036) J Symb Logic 37*139-149
◇ D25 D30 ◇ REV MR 47 # 8270 Zbl 256 # 02019 • ID 05816

HAY, L. [1972] *A note on frame extensions* (J 0036) J Symb Logic 37*543-545
◇ D50 ◇ REV MR 47 # 6462 Zbl 262 # 02045 • ID 05815

HAY, L. [1973] *Discrete ω-sequences of index sets (1)* (J 0064) Trans Amer Math Soc 183*293-311
◇ D25 D30 ◇ REV MR 50 # 1859 Zbl 285 # 02037 • ID 05817

HAY, L. [1973] *Index sets in $\underline{0}'$ (Russian)* (J 0003) Algebra i Logika 12*713-729
• TRANSL [1973] (J 0069) Algeb and Log 12*713-729
◇ D25 D30 D55 ◇ REV MR 55 # 2530 Zbl 288 # 02023 • ID 25961

HAY, L. [1973] *The class of recursively enumerable subsets of a recursively enumerable set* (J 0048) Pac J Math 46*167-183
◇ D25 D30 ◇ REV MR 52 # 13342 Zbl 226 # 02039 • ID 05819

HAY, L. [1973] *The halting problem relativized to complements* (J 0053) Proc Amer Math Soc 41*583-587
◇ D25 D30 ◇ REV MR 48 # 5837 Zbl 248 # 02044
• ID 05818

HAY, L. [1974] *A noninitial segment of index sets* (J 0036) J Symb Logic 39*209-224
◇ D25 D30 ◇ REV MR 50 # 6813 Zbl 292 # 02037
• ID 05820

HAY, L. [1974] *Index sets universal for differences of arithmetic sets* (J 0068) Z Math Logik Grundlagen Math 20*239-254
◇ D25 D30 D55 ◇ REV MR 55 # 2531 Zbl 311 # 02052
• ID 05821

HAY, L. [1975] *Rice theorems for d.r.e. sets* (J 0017) Canad J Math 27*352-365
◇ D25 D30 D55 ◇ REV MR 55 # 2532 Zbl 309 # 02040
• ID 30703

HAY, L. & MANASTER, A.B. & ROSENSTEIN, J.G. [1975] *Small recursive ordinals, many-one degrees, and the arithmetical difference hierarchy* (J 0007) Ann Math Logic 8*297-343
◇ D25 D30 D55 ◇ REV MR 53 # 7748 Zbl 309 # 02050
• ID 05823

HAY, L. [1975] *Spectra and halting problems* (J 0068) Z Math Logik Grundlagen Math 21*167-176
◇ C13 D10 D25 D30 ◇ REV MR 51 # 10056
Zbl 309 # 02049 • ID 05822

HAY, L. [1976] *Boolean combinations of r.e. open sets* (J 0036) J Symb Logic 41*235-238
◇ D25 D55 E15 ◇ REV MR 55 # 89 Zbl 339 # 02039
• ID 14803

HAY, L. & MANASTER, A.B. & ROSENSTEIN, J.G. [1977] *Concerning partial recursive similarity transformations of linearly ordered sets* (J 0048) Pac J Math 71*57-70
◇ C57 C65 D20 D25 D30 D45 E07 ◇ REV
MR 56 # 2806 Zbl 409 # 03027 • ID 30705

HAY, L. [1978] *Convex subsets of* 2^n *and bounded truth-table reducibility* (J 0193) Discr Math 21*31-46
◇ B05 D30 ◇ REV MR 81c:03031 Zbl 377 # 02036
• ID 30704

HAY, L. & JOHNSON, N. [1979] *Extensional characterization of index sets* (J 0068) Z Math Logik Grundlagen Math 25*227-234
◇ D25 ◇ REV MR 81e:03040 Zbl 433 # 03022 • ID 73959

HAY, L. & MILLER, DOUGLAS E. [1982] *A topological analog to the Rice-Shapiro index theorem* (J 0036) J Symb Logic 47*824-832
◇ D20 E15 ◇ REV MR 85a:03058 Zbl 518 # 03015
• ID 34806

HAY, L. [1982] *On the recursion-theoretic complexity of relative succinctness of representations of languages* (J 0194) Inform & Control 52*2-7
◇ D05 D30 ◇ REV MR 84m:03061 Zbl 512 # 03020
• ID 35761

HAY, L. & MILLER, DOUGLAS E. [1982] *The Addison game played backwards: index sets in topology* (P 3634) Patras Logic Symp;1980 Patras 231-237
◇ D55 D80 E75 ◇ REV MR 84d:03060 Zbl 524 # 03035
• ID 34093

HAY, L. see Vol. II for further entries

HAYASHI, T. & MIYANO, S. [1984] *Alternating finite automata on ω-words* (J 1426) Theor Comput Sci 32*321-330
◇ B75 D05 ◇ REV MR 86e:68079 Zbl 544 # 68042
• ID 45928

HAYASHI, T. & MIYANO, S. [1985] *Finite tree automata of infinte trees* (J 3957) Bull Inf & Cybern (Kyushu Univ) 21/3-4*71-82
◇ D05 ◇ ID 47693

HAYASHI, T. see Vol. II for further entries

HEAD, T. [1977] see BLATTNER, M.

HEAD, T. [1979] *Codeterministic Lindenmayer schemes and systems* (J 0119) J Comp Syst Sci 19*203-210
◇ D05 ◇ REV MR 81e:68100 Zbl 434 # 68058 • ID 69633

HEAD, T. [1979] see BLATTNER, M.

HEAD, T. [1984] *Adherence equivalence is decidable for DOL languages* (P 3565) Symp of Th Aspects of Comput Sci (1);1984 Paris 241-249
◇ B75 D05 ◇ REV MR 86a:68053 Zbl 543 # 68060
• ID 47097

HEAD, T. [1985] *The adherence of languages as topological spaces* (P 4622) Autom on Infinite Words;1984 Le Mont-Dore 147-163
◇ D05 ◇ ID 49448

HEBEISEN, F. [1978] *Charakterisierung der Aufzaehlungsreduzierbarkeit* (J 0009) Arch Math Logik Grundlagenforsch 19*89-95
◇ D30 ◇ REV MR 80e:03048 Zbl 496 # 03024 • ID 29167

HEBEISEN, F. [1979] *Masstheoretische Ergebnisse fuer WT-Grade* (J 0068) Z Math Logik Grundlagen Math 25*33-36
◇ D30 ◇ REV MR 80e:03049 Zbl 437 # 03021 • ID 55889

HEBEISEN, F. [1979] *Ueber Halbordnungen von WT-Graden in e-Graden* (J 0068) Z Math Logik Grundlagen Math 25*209-212
◇ D30 ◇ REV MR 80j:03057 Zbl 497 # 03032 • ID 73975

HECK, W.S. [1983] *Large families of incomparable A-isols* (J 0036) J Symb Logic 48*250-252
◇ D50 ◇ REV MR 85a:03057 Zbl 534 # 03022 • ID 34805

HECKER, H.-D. [1969] *Definierbarkeit und Unentscheidbarkeit* (J 0068) Z Math Logik Grundlagen Math 15*81-84
◇ D20 D35 ◇ REV MR 38 # 4309 Zbl 177.11 • ID 05849

HECKER, H.-D. [1969] *Riegeralgebren und rekursive Untrennbarkeit* (P 4112) Th Ordered Sets & Gen Algeb;1969 Cikhaj
◇ D25 D35 ◇ ID 42913

HECKER, H.-D. [1971] *Allgemeine Trennbarkeitsbegriffe bei meta-mathematischen Untersuchungen* (J 0068) Z Math Logik Grundlagen Math 17*205-217
◇ D25 D35 D75 ◇ REV MR 46 # 1576 Zbl 222 # 02054
• ID 05851

HECKER, H.-D. [1971] *Rekursive Untrennbarkeit bei elementaren Theorien* (J 0068) Z Math Logik Grundlagen Math 17*443-464
◇ D25 D35 ◇ REV MR 48 # 3726 Zbl 231 # 02057
• ID 05850

HECKER, H.-D. [1973] *Kompliziertheitsmasse bei Turingmaschinen* (S 1536) Schr Weiterbildungszentr MKR (Dresden) 1973/2*44-55
◇ D10 D15 ◇ ID 42914

HECKER, H.-D. [1975] *Eine Bemerkung zur Berechnungskompliziertheit auf Registermaschinen* (J 0129) Elektr Informationsverarbeitung & Kybern 11*437-438
◇ D10 D15 ◇ REV MR 52 # 4705 Zbl 318 # 68036
• ID 62442

HECKER, H.-D. [1976] *Zur Programmkomplexitaet rekursiv aufzaehlbarer Mengen* (J 0068) Z Math Logik Grundlagen Math 22*239-244
◇ D15 D25 ◇ REV MR 58 # 5122 Zbl 359 # 02036
• ID 18452

HECKER, H.-D. [1977] *Ein Zusammenhang zwischen Universalitaet und Komplexitaet* (J 1670) Mitt Math Ges DDR 23
◇ D15 ◇ ID 42915

HECKER, H.-D. [1978] *Bemerkungen zu Enumerationssystemen und Zeitmassen* (S 2829) Rostocker Math Kolloq 10*43-48
◇ D15 D25 D45 ◇ REV Zbl 412 # 03030 • ID 52961

HECKER, H.-D. [1980] *Enumerationen in speziellen Standardklassen rekursiv-aufzaehlbarer Mengen* (J 0068) Z Math Logik Grundlagen Math 26*165-180
◇ D25 ◇ REV MR 81j:03075 Zbl 439 # 03022 • ID 56013

HECKER, H.-D. [1981] *Zur Kompliziertheit der Beschreibung von Enumerationen rekursiv aufzaehlbarer Mengen* (S 2829) Rostocker Math Kolloq 18*123-128
◇ D15 D25 D45 ◇ REV MR 83f:03034 Zbl 478 # 03023
• ID 55638

HECKER, H.-D. [1984] *Abstrakte Tempomasse und speed-up-Theoreme fuer Enumerationen rekursiv-aufzaehlbarer Mengen* (J 0068) Z Math Logik Grundlagen Math 30*269-281
◇ D15 D25 ◇ REV MR 86f:03062 Zbl 557 # 03022
• ID 42227

HECKER, H.-D. [1984] *Das Kompressionstheorem fuer Tempomasse* (J 0068) Z Math Logik Grundlagen Math 30*283-288
◇ D15 D25 ◇ REV MR 86f:03063 Zbl 557 # 03023
• ID 42228

HEFFERNAN, J.D. [1978] *Discussion: some doubts about "Turing machine arguments"* (J 0153) Phil of Sci (East Lansing) 45*638-647
◇ A05 D10 ◇ REV MR 80b:03050 • ID 73994

HEIDLER, K. [1976] *Berechenbarkeit und Entscheidbarkeit auf der Grundlage der Programmiersprache Basic* (J 0487) Math Unterricht 22*88-91
◇ B75 D20 ◇ ID 28270

HEIDLER, K. [1976] see BOERGER, E.

HEIDLER, K. [1976] *Die Unentscheidbarkeit des Eck-Dominoproblems* (J 0160) Math-Phys Sem-ber, NS 23*237-250
◇ D05 ◇ REV MR 55 # 5427 Zbl 349 # 02041 • ID 28268

HEIDLER, K. & HERMES, H. & MAHN, F.-K. [1977] *Rekursive Funktionen* (X 0876) Bibl Inst: Mannheim vii + 236PP
◇ D20 D98 ◇ REV MR 58 # 5108 Zbl 372 # 02021
JSL 46.165 • ID 32181

HEIDLER, K. [1978] *Die Unentscheidbarkeit des Halteproblems auf der Grundlage der Programmiersprache Basic* (J 0160) Math-Phys Sem-ber, NS 25*52-78
◇ B75 D10 ◇ REV MR 57 # 11144 Zbl 407 # 68056
• ID 28271

HEILBRUNNER, S. [1980] *An algorithm for the solution of fixed-point equations for infinite words* (J 3441) RAIRO Inform Theor 14*131-141
◇ D05 ◇ REV MR 81g:68110 Zbl 433 # 68062 • ID 69634

HEILBRUNNER, S. [1983] *A metatheorem for undecidable properties of formal languages and its application to LRR and LLR grammars and languages* (J 1426) Theor Comput Sci 23*49-68
◇ D05 ◇ REV MR 84g:68060 Zbl 507 # 68044 • ID 39491

HEINDORF, L. [1984] *Regular ideals and boolean pairs* (J 0068) Z Math Logik Grundlagen Math 30*547-560
◇ C30 C85 D35 G05 ◇ REV MR 86g:03063
Zbl 558 # 03016 • ID 39661

HEINDORF, L. see Vol. III, V for further entries

HEINEN, J.A. [1979] see AGGARWAL, S.K.

HEINTZ, JOOS [1976] *Untere Schranken fuer die Komplexitaet logischer Entscheidungsprobleme* (P 3196) Kompl von Entscheid Probl;1973/74 Zuerich 127-137
◇ D10 D15 ◇ REV MR 57 # 18232 Zbl 343 # 02035
• ID 62451

HEINTZ, JOOS [1979] *Definability bounds of first order theories of algebraically closed fields* (P 2935) FCT'79 Fund of Comput Th;1979 Berlin/Wendisch-Rietz 160-166
◇ B25 C40 C60 D10 D15 ◇ REV MR 81j:03064
Zbl 439 # 03003 • ID 55995

HEINTZ, JOOS [1983] *Definability and fast quantifier elimination in algebraically closed fields* (J 1426) Theor Comput Sci 24*239-277 • ERR/ADD ibid 39*343
◇ B25 C10 C40 C60 D10 D15 ◇ REV MR 85a:68062
Zbl 546 # 03017 • ID 38863

HELLER, A. [1985] *Dominical categories and recursion theory* (P 4646) Atti Incontri Log Mat (2);1983/84 Siena 339-344
◇ D20 ◇ ID 49603

HELLER, A. see Vol. III for further entries

HELLER, H. [1981] *Relativized polynomial hierarchies extending two levels* (X 3160) TU Muenchen Fak Math: Muenchen 70pp
◇ D15 ◇ REV Zbl 516 # 03021 • ID 37256

HELLER, H. & ZACHOS, S. [1984] *A new characterization of BPP* (P 4010) Found of Softw Tech & Th Comput Sci (4);1984 Bangalore 179-187
◇ D15 ◇ REV Zbl 552 # 68050 • ID 43466

HELLER, H. [1984] *On relativized polynomial and exponential computations* (J 1428) SIAM J Comp 13*717-725
◇ D15 ◇ REV Zbl 563 # 03025 • ID 47458

HELLER, H. [1984] *Relativized polynomial hierachies extending two levels* (J 0041) Math Syst Theory 17*71-84
◇ D15 ◇ REV MR 86e:68043 Zbl 543 # 03028 • ID 40920

HELM, J.P. [1971] *On effectively computable operators* (J 0068) Z Math Logik Grundlagen Math 17*231-244
◇ D65 ◇ REV MR 44 # 3865 Zbl 245 # 02036 • ID 05876

HELM, J.P. & YOUNG, P. [1971] *On size vs. efficiency for programs admitting speed-ups* (**J** 0036) J Symb Logic 36*21-27
• TRANSL [1974] (**C** 2319) Slozh Vychisl & Algor 150-159
◇ D15 ◇ REV MR 46 # 5119 Zbl 258 # 68023 • ID 05877

HELM, J.P. & MEYER, A.R. & YOUNG, P. [1973] *On orders of translations and enumerations* (**J** 0048) Pac J Math 46*185-195
◇ D15 D20 D25 D45 F40 ◇ REV MR 48 # 1898 Zbl 276 # 02022 • ID 05879

HEMMERLING, A. [1976] *Turing-Maschinen und Zellularraeume in der Ebene (English and Russian summaries)* (**J** 0129) Elektr Informationsverarbeitung & Kybern 12*379-402
◇ D10 ◇ REV MR 54 # 4863 Zbl 339 # 02032 • ID 24082

HEMMERLING, A. [1977] *Allgemeine Untersuchungen zur Entscheidbarkeit und Trennbarkeit bezueglich Kodierungen* (**J** 0129) Elektr Informationsverarbeitung & Kybern 13*295-309
◇ D05 ◇ REV MR 58 # 10455 Zbl 363 # 02050 • ID 50877

HEMMERLING, A. [1977] *Vergleich einiger Kodierungen der natuerlichen Zahlen bezueglich der regulaeren Entscheidbarkeit von Zahlenmengen* (**J** 0129) Elektr Informationsverarbeitung & Kybern 13*217-230
◇ D05 D20 ◇ REV MR 56 # 8337 Zbl 361 # 02049 • ID 50684

HEMMERLING, A. [1977] *Zur verallgemeinerten Realisierbarkeit von Pseudomusterfunktionen durch Zellularraeume* (**J** 0129) Elektr Informationsverarbeitung & Kybern 13*523-528
◇ D10 D20 ◇ REV MR 58 # 26607 Zbl 364 # 02020 • ID 50950

HEMMERLING, A. [1978] *Realisierung der Verbandsoperationen fuer Regularitaetsklassen von Kodierungen* (**J** 0129) Elektr Informationsverarbeitung & Kybern 14*145-152
◇ D20 ◇ REV MR 58 # 16217 Zbl 379 # 02013 • ID 51846

HEMMERLING, A. [1979] *Concentration of multidimensional tape-bounded systems of Turing automata and cellular spaces* (**P** 2935) FCT'79 Fund of Comput Th;1979 Berlin/Wendisch-Rietz 167-174
◇ D05 D10 ◇ REV MR 82c:68030 Zbl 417 # 68034 • ID 42902

HEMMERLING, A. [1979] *Systeme von Turing-Automaten und Zellularraeume auf rahmbaren Pseudomustermengen* (**J** 0129) Elektr Informationsverarbeitung & Kybern 15*47-72
◇ D10 ◇ REV MR 80m:68048 Zbl 428 # 68058 • ID 53824

HEMMERLING, A. [1979] *Zur Raumkompliziertheit mehrdimensionaler Zellularraeume und Turing-Automaten (Russian and English summaries)* (**J** 0129) Elektr Informationsverarbeitung & Kybern 15*143-158
◇ D05 D10 ◇ REV MR 80m:68051 Zbl 408 # 68044 • ID 42901

HEMMERLING, A. [1982] *On the computational equivalence of synchronous and asynchronous cellular spaces (German and Russian summaries)* (**J** 0129) Elektr Informationsverarbeitung & Kybern 18*423-434
◇ D05 ◇ REV MR 84i:68086 Zbl 511 # 68076 • ID 42904

HEMMERLING, A. [1982] *Zur Raumkompliziertheit von Absuchprozessen auf endlichen Graphen* (**S** 2829) Rostocker Math Kolloq 19*77-90
◇ D05 D15 ◇ REV MR 83i:68069 Zbl 484 # 68039 • ID 42903

HEMMERLING, A. [1983] *D ≠ ND fuer mehrdimensionale Turing-Automaten mit sublogarithmischer Raumschranke* (**J** 0129) Elektr Informationsverarbeitung & Kybern 19*85-94
◇ D10 D15 ◇ REV Zbl 538 # 68040 • ID 41502

HEMMERLING, A. & KRIEGEL, K. [1984] *On searching of special classes of mazes and finite embedded graphs* (**P** 3658) Math Founds of Comput Sci (11);1984 Prague 291-300
◇ D05 ◇ REV Zbl 555 # 68022 • ID 42905

HEMMERLING, A. & MURAWSKI, G. [1984] *Zur Raumkompliziertheit mehrdimensionaler Turing-Automaten* (**J** 0068) Z Math Logik Grundlagen Math 30*233-258
◇ D10 D15 ◇ ID 42226

HENKE VON, F.W. & INDERMARK, K. & WEIHRAUCH, K. [1973] *Hierarchies of primitive recursive wordfunctions and transductions defined by automata* (**P** 0763) Automata, Lang & Progr (1);1972 Rocquencourt 549-561
◇ D05 D20 ◇ REV MR 55 # 1840 Zbl 357 # 02035 • ID 50383

HENKE VON, F.W. & WEIHRAUCH, K. [1973] *Klassifizierung von primitiv-rekursiven Transformationen und Automatentransduktionen* (**P** 1653) GI Jahrestag (2);1972 Karlsruhe 78*63-71
◇ D05 D20 ◇ REV Zbl 261 # 02024 • ID 30467

HENKE VON, F.W. [1974] *Primitiv-rekursive Transformationen* (**X** 0817) Ges Math Datenverarbeit: Bonn 51pp
◇ D20 ◇ REV MR 51 # 107 Zbl 324 # 02026 • ID 17466

HENKE VON, F.W. & INDERMARK, K. & ROSE, G.F. & WEIHRAUCH, K. [1975] *On primitive recursive wordfunctions (German summary)* (**J** 0373) Comp Arch Inform & Numerik 15*217-234
◇ D20 ◇ REV MR 54 # 2434 Zbl 317 # 02041 • ID 24630

HENKIN, L. [1957] *Sums of squares* (**P** 1675) Summer Inst Symb Log;1957 Ithaca 284-291
◇ C57 C60 ◇ REV JSL 31.128 • ID 05913

HENKIN, L. & SUPPES, P. & TARSKI, A. (EDS.) [1959] *The axiomatic method, with special reference to geometry and physics* (**X** 0809) North Holland: Amsterdam xi + 488pp
◇ B30 B97 C65 D35 ◇ REV Zbl 88.244 • ID 22387

HENKIN, L. see Vol. I, II, III, V, VI for further entries

HENNIE III, F.C. [1965] *One-tape, off-line Turing machine computations* (**J** 0194) Inform & Control 8*553-578
• TRANSL [1970] (**C** 1540) Probl Mat Log: Slozh Algor & Kl Vychisl Funk 223-248
◇ D10 ◇ REV MR 32 # 9171 Zbl 231 # 02048 JSL 33.119 • ID 05937

HENNIE III, F.C. [1966] *Online Turing machine computations* (**J** 4305) IEEE Trans Electr Comp 15*35-44
• TRANSL [1970] (**C** 1540) Probl Mat Log: Slozh Algor & Kl Vychisl Funk 249-270
◇ D10 ◇ REV Zbl 143.14 • ID 20869

HENNIE III, F.C. [1977] *Introduction to computability* (**X** 0832) Addison-Wesley: Reading ix + 374pp
◇ D15 D98 ◇ REV MR 56 # 11767 Zbl 365 # 02020 • ID 51021

HENSEL, G. & PUTNAM, H. [1965] *On the notational independence of various hierarchies of degrees of unsolvability* (**J** 0036) J Symb Logic 30*69-86
◇ D30 D55 F15 ◇ REV MR 37 # 5096 Zbl 137.9 JSL 32.124 • ID 05946

HENSEL, G. [1969] see BOYD, R.

HENSEL, G. & PUTNAM, H. [1969] *Normal models and the field* Σ_1^* (**J** 0027) Fund Math 64*231-240
◇ C07 C30 C57 D55 ◇ REV MR 39 # 5357 Zbl 193.302 • ID 05948

HENSON, C.W. [1972] *Countable homogeneous relational structures and* $\aleph_0$*-categorical theories* (**J** 0036) J Symb Logic 37*494-500
◇ C10 C15 C35 C50 D35 ◇ REV MR 48 # 94 Zbl 259 # 02040 • ID 05954

HENSON, C.W. & JOCKUSCH JR., C.G. & RUBEL, L.A. & TAKEUTI, G. [1977] *First-order topology* (**J** 0202) Diss Math (Warsaw) 143*40pp
◇ B25 C65 C75 D35 H05 ◇ REV MR 55 # 5434 Zbl 399 # 03019 • ID 30718

HENSON, C.W. [1980] see BECKER, J.A.

HENSON, C.W. see Vol. I, III, V, VI for further entries

HERINGER, H.J. [1972] *Formale Logik und Grammatik* (**X** 0877) Niemeyer: Tuebingen vi + 104pp
◇ B65 B98 D05 ◇ REV MR 52 # 13299 Zbl 337 # 68003 • ID 21764

HERMAN, G.T. [1968] *Simulation of one abstract computing machine by another* (**J** 0212) ACM Commun 11*802-813
◇ D10 D25 D30 D55 ◇ ID 14808

HERMAN, G.T. [1968] *The halting problem of one state Turing machine with n-dimensional tape* (**J** 0068) Z Math Logik Grundlagen Math 14*185-191
◇ D10 ◇ REV MR 37 # 3923 Zbl 169.311 • ID 05982

HERMAN, G.T. [1969] *A new hierarchy of elementary functions* (**J** 0053) Proc Amer Math Soc 20*557-562
◇ D20 ◇ REV MR 40 # 4110 • ID 05985

HERMAN, G.T. [1969] *A simple solution of the uniform halting problem* (**J** 0036) J Symb Logic 34*639-640
◇ D10 ◇ REV MR 41 # 6695 Zbl 188.330 • ID 05983

HERMAN, G.T. [1969] *On hierarchies of elementary functions* (**P** 1841) Fct Recurs & Appl;1967 Tihany 105-120
◇ D20 ◇ ID 32560

HERMAN, G.T. [1969] *The uniform halting problem for generalized one-state Turing-machines* (**J** 0194) Inform & Control 15*353-367
◇ D10 ◇ REV MR 40 # 4024 Zbl 185.25 • ID 05986

HERMAN, G.T. [1969] *The unsolvability of the uniform halting problem for two state Turing-machines* (**J** 0036) J Symb Logic 34*161-165
◇ D10 ◇ REV MR 40 # 4116 Zbl 181.13 • ID 05984

HERMAN, G.T. & ISARD, S.D. [1970] *Computability over arbitrary fields* (**J** 3172) J London Math Soc, Ser 2 2*71-79
◇ D75 ◇ REV MR 40 # 2544 Zbl 188.25 • ID 05987

HERMAN, G.T. [1971] *Strong computability and variants of the uniform halting problem* (**J** 0068) Z Math Logik Grundlagen Math 17*115-131
◇ D10 ◇ REV MR 44 # 1569 Zbl 218 # 02033 • ID 05989

HERMAN, G.T. [1971] *The equivalence of different hierarchies of elementary functions* (**J** 0068) Z Math Logik Grundlagen Math 17*219-224
• TRANSL [1974] (**C** 2319) Slozh Vychisl & Algor 7-17
◇ D20 ◇ REV MR 44 # 6494 Zbl 222 # 02043 • ID 05991

HERMAN, G.T. [1971] *When is a sequential machine the realization of another* (**J** 0041) Math Syst Theory 5*115-127
◇ D05 ◇ REV MR 46 # 1487 Zbl 218 # 94023 • ID 05990

HERMAN, G.T. & JACKOWSKI, J.A. [1973] *A decision procedure using discrete geometry* (**J** 0193) Discr Math 5*131-144
◇ D10 ◇ REV MR 48 # 5832 Zbl 259 # 02029 • ID 05992

HERMAN, G.T. [1974] see AITCHISON, P.W.

HERMANN, G. [1926] *Die Frage der endlich vielen Schritte in der Theorie der Polynomideale* (**J** 0043) Math Ann 95*736-788
◇ C57 C60 D45 F55 ◇ REV FdM 52.127 • ID 05997

HERMES, H. [1937] *Definite Begriffe und berechenbare Zahlen* (**J** 2074) Sem-ber, Muenster 10*110-123
◇ D05 D10 D20 F60 ◇ REV JSL 6.35 FdM 63.823 • ID 32166

HERMES, H. [1951] *Zum Begriff der Axiomatisierbarkeit* (**J** 0114) Math Nachr 4*343-347
◇ A05 B30 D25 ◇ REV MR 12.578 Zbl 42.7 JSL 22.83 • ID 06000

HERMES, H. [1955] *Vorlesung ueber Entscheidungsprobleme in Mathematik und Logik* (**X** 0910) Aschendorffsche Verlagsbuchh: Muenster ii + 140pp
◇ B25 B40 B65 D10 D20 D35 D98 ◇ REV MR 17.569 Zbl 67.249 • ID 06008

HERMES, H. [1958] *Zum Einfachheitsprinzip in der Wahrscheinlichkeitsrechnung* (**J** 0076) Dialectica 12*317-331
◇ A05 B48 D20 D80 ◇ REV MR 21 # 350 Zbl 90.348 • ID 14545

HERMES, H. [1961] *Aufzaehlbarkeit, Entscheidbarkeit, Berechenbarkeit: Einfuehrung in die Theorie der rekursiven Funktionen* (**X** 0811) Springer: Heidelberg & New York x + 246pp
• TRANSL [1965] (**X** 0811) Springer: Heidelberg & New York ix + 245pp (English) [1975] (**X** 0905) Boringhieri: Torino 318pp (Italian)
◇ D98 F60 ◇ REV MR 26 # 1252 MR 41 # 8225 MR 49 # 8825 MR 80b:03049 Zbl 383 # 03023 JSL 31.254 • ID 20981

HERMES, H. [1964] *Unentscheidbarkeit der Arithmetik* (**J** 0160) Math-Phys Sem-ber, NS 11*20-34
◇ D35 F30 ◇ REV MR 29 # 3370 Zbl 127.9 JSL 33.469 • ID 06013

HERMES, H. & ROEDDING, D. [1965] *A method for producing reduction types in the restricted lower predicate calculus* (**P** 0688) Logic Colloq;1963 Oxford 42-47
◇ D35 ◇ REV MR 35 # 57 Zbl 213.18 JSL 40.517 • ID 06011

HERMES, H. [1968] *Praedikatenlogik und Theorie der rekursiven Funktionen* (**C** 0552) Phil Contemp - Chroniques 254-265
◇ B10 D20 ◇ ID 14956

HERMES, H. [1969] *Basic notions and applications of the theory of decidability* (**P** 0630) Aspects Math Log;1968 Varenna 1-54
◇ D10 D20 D25 D30 D35 D50 D98 ◇ REV MR 41 # 3271 Zbl 212.331 • ID 06019

HERMES, H. [1969] *On the notion of constructivity* (**P** 1060) Constr Aspects Fund Thm Algeb;1967 Zuerich 115-129
◇ D20 F65 ◇ REV MR 41 # 1529 Zbl 179.316 • ID 20948

HERMES, H. [1970] *Entscheidungsproblem und Dominospiele* (**S** 1415) Sel Math 2*114–140
• TRANSL [1972] (**C** 1534) Mash Turing & Rek Funk 150–182
◇ D05 D10 D35 ◇ REV MR 44 # 6489 Zbl 233 # 02019
• ID 27783

HERMES, H. [1971] *A simplified proof for the unsolvability of the decision problem in the case ∀∃∀* (**P** 0638) Logic Colloq;1969 Manchester 307–310
◇ B20 D35 ◇ REV MR 43 # 1840 Zbl 221 # 02032
• ID 06021

HERMES, H. [1972] *Die Unloesbarkeit des zehnten Hilbertschen Problems* (**J** 3370) Enseign Math, Ser 2 18*47–56
◇ D25 D35 ◇ REV MR 49 # 2322 Zbl 238 # 02038
• ID 06022

HERMES, H. [1972] see EBBINGHAUS, H.-D.

HERMES, H. [1976] *Domino games* (**P** 1619) Coloq Log Simb;1975 Madrid 103–114
◇ D05 D35 ◇ REV MR 56 # 5207 Zbl 361 # 02059
• ID 50694

HERMES, H. [1976] *Hilbert's tenth problem* (**P** 1619) Coloq Log Simb;1975 Madrid 127–136
◇ A10 D25 D35 ◇ REV MR 56 # 15347 Zbl 414 # 03028
• ID 53074

HERMES, H. [1977] see HEIDLER, K.

HERMES, H. [1978] *Die Unloesbarkeit des zehnten Hilbertschen Problems* (**J** 0933) Math Nat Unterr 31*260–263
◇ A10 D25 D35 ◇ REV Zbl 384 # 03028 • ID 52070

HERMES, H. [1981] *Recursion theory* (**C** 2617) Modern Log Survey 173–195
◇ D98 ◇ REV MR 82f:03002 Zbl 464 # 03001 • ID 42769

HERMES, H. see Vol. I, II, III, V, VI for further entries

HERRE, H. [1972] see HAUSCHILD, K.

HERRE, H. [1973] *Unentscheidbarkeit in der Graphentheorie* (**J** 0014) Bull Acad Pol Sci, Ser Math Astron Phys 21*201–208
◇ D35 ◇ REV MR 50 # 1865 Zbl 279 # 05101 • ID 06028

HERRE, H. & WOLTER, H. [1975] *Entscheidbarkeit von Theorien in Logiken mit verallgemeinerten Quantoren* (**J** 0068) Z Math Logik Grundlagen Math 21*229–246
◇ B25 C10 C55 C80 D35 ◇ REV MR 53 # 2623 Zbl 318 # 02049 • ID 14283

HERRE, H. & PINUS, A.G. [1978] *Zum Entscheidungsproblem fuer Theorien in Logiken mit monadischen verallgemeinerten Quantoren* (**J** 0068) Z Math Logik Grundlagen Math 24*375–384
◇ B25 C10 C55 C80 D35 ◇ REV MR 58 # 16246 Zbl 397 # 03010 • ID 52676

HERRE, H. [1981] *Miscellaneous results and problems in extended model theory* (**P** 2623) Worksh Extended Model Th;1980 Berlin 20–65
◇ B25 C65 C80 C98 D35 ◇ REV MR 84h:03095 Zbl 522 # 03026 • ID 34285

HERRE, H. [1984] *Bemerkungen zum Begriff der praktischen Berechenbarkeit* (**J** 0309) Wiss Z Univ Greifswald, Math-Nat Reihe 33*65
◇ D15 D20 ◇ ID 46645

HERRE, H. see Vol. I, III, V, VI for further entries

HERRMANN, C. [1972] see DAY, A.

HERRMANN, C. [1973] *On the equational theory of submodule lattices* (**P** 0457) Lattice Th;1973 Houston 105–118
◇ C05 D40 G10 ◇ REV MR 53 # 10668 Zbl 313 # 06004
• ID 23092

HERRMANN, C. [1975] *Concerning M.M.Gluhov's paper on the word problem for free modular lattices* (**J** 0004) Algeb Universalis 5*445
◇ D40 G10 ◇ REV MR 52 # 214 Zbl 346 # 06008
• ID 24484

HERRMANN, C. & HUHN, A.P. [1975] *Zum Wortproblem fuer freie Untermodulverbaende* (**J** 0008) Arch Math (Basel) 26*449–453
◇ D40 ◇ REV MR 53 # 5389 Zbl 343 # 06012 • ID 06037

HERRMANN, C. [1983] *On the word problem for the modular lattice with four free generators* (**J** 0043) Math Ann 265*513–527
◇ C05 D40 G10 ◇ REV MR 84m:06014 Zbl 506 # 06004
• ID 39455

HERRMANN, C. see Vol. III for further entries

HERRMANN, E. [1976] *On Lindenbaum functions of $\aleph_0$-categorical theories of finite similarity type* (**J** 0014) Bull Acad Pol Sci, Ser Math Astron Phys 24*17–21
◇ B25 C35 C57 ◇ REV MR 53 # 7758 Zbl 333 # 02042
• ID 18190

HERRMANN, E. [1977] *A note on no hh-simple recursive enumerable sets (Russian summary)* (**J** 0014) Bull Acad Pol Sci, Ser Math Astron Phys 25*333–336
◇ D25 ◇ REV MR 56 # 15386 Zbl 358 # 02048 • ID 26558

HERRMANN, E. [1977] *Ueber Lindenbaumfunktionen von $\aleph_0$-kategorischen Theorien endlicher Signatur* (**J** 0115) Wiss Z Humboldt-Univ Berlin, Math-Nat Reihe 26*637–646
◇ B25 C35 C57 ◇ REV MR 80d:03028 Zbl 423 # 03005
• ID 53517

HERRMANN, E. [1978] *Der Verband der rekursiv aufzaehlbaren Mengen (Entscheidungs-Problem) (Russian and English summaries)* (**S** 3382) Sem-ber, Humboldt-Univ Berlin, Sekt Math 10*v+304pp
◇ B25 D25 D98 ◇ REV MR 81e:03041 Zbl 415 # 03029
• ID 53131

HERRMANN, E. [1979] *The lattice structures of the recursively enumerable sets* (**P** 2935) FCT'79 Fund of Comput Th;1979 Berlin/Wendisch-Rietz 175–181
◇ D25 ◇ REV MR 81i:03064 Zbl 422 # 03018 • ID 53479

HERRMANN, E. [1981] *Die Verbandseigenschaften der rekursiv aufzaehlbaren Mengen* (**S** 3382) Sem-ber, Humboldt-Univ Berlin, Sekt Math viii+275pp
◇ D25 D98 ◇ REV MR 84b:03057 Zbl 472 # 03030
• ID 55295

HERRMANN, E. [1983] *Definable Boolean pairs in the lattice of recursively enumerable sets* (**P** 1601) Easter Conf on Model Th (1);1983 Diedrichshagen 42–67
◇ D25 D35 ◇ REV MR 84i:03008 Zbl 517 # 03013
• ID 37283

HERRMANN, E. [1983] *Orbits of hyperhypersimple sets and the lattice of Σ_3^0 sets* (**J** 0036) J Symb Logic 48*693–699
◇ D25 D55 ◇ REV MR 85a:03055 Zbl 532 # 03018
• ID 33592

HERRMANN, E. [1984] *Classes of simple sets, filter properties and their mutual position* (**S** 3382) Sem-ber, Humboldt-Univ Berlin, Sekt Math 60*60-72
◇ D25 ◇ REV Zbl 551 # 03025 • ID 42360

HERRMANN, E. [1984] *Definable structures in the lattice of recursively enumerable sets* (**J** 0036) J Symb Logic 49*1190-1197
◇ D25 D55 ◇ REV MR 86c:03039 • ID 42349

HERRMANN, E. [1984] *The undecidability of the elementary theory of the lattice of recursively enumerable sets* (**P** 3621) Frege Konferenz (2);1984 Schwerin 66-72
◇ D25 D35 ◇ REV MR 85m:03006 • ID 45388

HERRMANN, E. see Vol. I, III for further entries

HERSCHEL, R. [1974] *Einfuehrung in die Theorie der Automaten, Sprachen und Algorithmen* (**X** 0814) Oldenbourg: Muenchen 226pp
◇ D05 D10 D98 ◇ REV MR 57 # 18167 Zbl 305 # 68002 • ID 62496

HERSH, R. [1977] see DAVIS, MARTIN D.

HERSH, R. see Vol. I, V for further entries

HERZOG, T. & OWINGS JR., J.C. [1976] *The inequivalence of two well-known notions of randomness for binary sequence* (**J** 0068) Z Math Logik Grundlagen Math 22*385-389
◇ D20 D80 ◇ REV MR 56 # 2802 Zbl 353 # 02026 • ID 50109

HESSE, W. [1979] *A correspondence between W-grammars and formal systems of logic and its application to formal language description* (**J** 2293) Comp Linguist & Comp Lang 13*19-30
◇ D03 D05 ◇ REV MR 81d:68098 Zbl 429 # 68041 • ID 81548

HEYTING, A. [1969] *What is computable ? (Dutch)* (**J** 3077) Nieuw Arch Wisk, Ser 3 17*1-7
◇ D20 D35 D55 ◇ REV MR 40 # 4104 Zbl 185.19 • ID 06067

HEYTING, A. see Vol. I, II, III, VI for further entries

HIBBARD, T.N. [1964] see GINSBURG, S.

HIBBARD, T.N. [1965] see GINSBURG, S.

HIBBARD, T.N. & ULLIAN, J.S. [1966] *The independence of inherent ambiguity from complementedness among context-free languages* (**J** 0037) ACM J 13*588-593
◇ D05 ◇ REV MR 34 # 2377 Zbl 154.258 JSL 33.301 • ID 06070

HICKEY, T. [1982] see COHEN, JACQUES

HICKIN, K.K. & MACINTYRE, A. [1980] *Algebraically closed groups: embeddings and centralizers* (**P** 2634) Word Problems II;1976 Oxford 141-155
◇ C25 C60 D40 ◇ REV MR 82c:20065 Zbl 444 # 20025 • ID 81551

HICKIN, K.K. [1981] *Bounded HNN presentations* (**J** 0032) J Algeb 71*422-434
◇ D25 D40 ◇ REV MR 82j:20059 Zbl 497 # 20011 • ID 81550

HICKIN, K.K. & PHILLIPS, R.E. [1983] *Isomorphism types in wreath products and effective embeddings of periodic groups* (**J** 0064) Trans Amer Math Soc 277*765-778
◇ C57 C60 D40 ◇ REV MR 85i:20034 Zbl 516 # 20015 • ID 38580

HICKIN, K.K. see Vol. II, III, V for further entries

HIGMAN, G. & NEUMANN, B.H. & NEUMANN, H. [1949] *Embedding theorems for groups* (**J** 0039) J London Math Soc 24*247-254
◇ C60 D40 ◇ REV MR 11.322 Zbl 34.301 • ID 90239

HIGMAN, G. [1961] *Subgroups of finitely presented groups* (**J** 1150) Proc Roy Soc London, Ser A 262*455-475
◇ C60 D25 D45 ◇ REV MR 24 # A152 Zbl 104.21 JSL 29.204 • ID 22083

HIGMAN, G. [1974] see BOONE, W.W.

HIGMAN, G. [1975] see BOONE, W.W.

HIGMAN, G. [1980] see ADYAN, S.I.

HIGMAN, G. see Vol. III, V for further entries

HILBERT, D. [1900] *Mathematische Probleme* (**J** 1109) Nachr Akad Wiss Goettingen, Math-Phys Kl 1900*253-297
• TRANSL [1900] (**J** 0152) Enseign Math 2*349-354 (French, fragment) [1901] (**J** 0767) Rev Gen Sci Pur Appl 12*168-174 (French, fragment) [1902] (**P** 1484) Int Congr Math (2);1900 Paris 58-114 (French) [1902] (**J** 0015) Bull Amer Math Soc 8*437-479 (English) [1976] (**P** 2957) Math Dev from Hilbert Probl;1974 DeKalb 1*1-35 (English) • REPR [1901] (**J** 3975) Arch Math & Phys 1*44-63,213-237
◇ A05 D35 E30 E50 F25 ◇ REV FdM 31.68 • ID 06078

HILBERT, D. [1926] *Ueber das Unendliche* (**J** 0043) Math Ann 95*161-190
• TRANSL [1964] (**C** 1105) Phil of Math. Sel Readings 134-151 (English) [1967] (**C** 0675) From Frege to Goedel 367-392 (English) [1967] (**C** 2141) Filos Matematica 161-183 (Spanish) • REPR [1927] (**J** 0157) Jbuchber Dtsch Math-Ver 36*201-215
◇ A05 B28 D20 E10 E30 ◇ REV FdM 53.41 • REM Reprint is a shortened version • ID 45196

HILBERT, D. & ACKERMANN, W. [1928] *Grundzuege der theoretischen Logik* (**X** 0811) Springer: Heidelberg & New York viii + 120pp
• TRANSL [1950] (**X** 0848) Chelsea: New York xii + 172pp (English) [1950] (**X** 1876) Kexue Chubanshe: Beijing
◇ B25 B98 D35 ◇ REV MR 50 # 4230 Zbl 239 # 02001 JSL 15.59 JSL 16.52 JSL 25.158 JSL 3.83 FdM 54.55 • REM 4th ed. 1959;viii + 188pp. • ID 00107

HILBERT, D. see Vol. I, II, V, VI for further entries

HILTON, A.M. [1963] *Logic, computing machines, and automation* (**X** 1354) Spartan Books : Sutton xxi + 427pp
◇ A05 B98 D05 ◇ REV MR 28 # 741 Zbl 109.101 JSL 38.341 • ID 23532

HINATA, S. & TUGUE, T. [1969] *A note on continuous functionals* (**J** 0260) Ann Jap Ass Phil Sci 3*138-145
◇ D65 ◇ REV MR 40 # 4108 Zbl 188.327 JSL 39.606 • ID 13716

HINATA, S. see Vol. VI for further entries

HINDLEY, J.R. & MITSCHKE, G. [1977] *Some remarks about the connections between combinatory logic and axiomatic recursion theory* (**J** 0009) Arch Math Logik Grundlagenforsch 18*99-103
◇ B40 D75 ◇ REV MR 57 # 9498 Zbl 365 # 02028 • ID 24332

HINDLEY, J.R. see Vol. I, II, VI for further entries

HINGSTON, P. [1981] *Effective decomposition in Noetherian rings* (**P** 2902) Aspects Effective Algeb;1979 Clayton 122-127
◇ C57 C60 D45 ◇ REV MR 83a:03040 Zbl 473 # 03036
• ID 55365

HINMAN, P.G. [1966] *Ad astra per aspera: hierarchy schemata in recursive function theory* (0000) Diss., Habil. etc
◇ D55 D65 D70 E15 ◇ REM Diss., University of California, Berkeley • ID 20914

HINMAN, P.G. [1969] *Hierarchies of effective descriptive set theory* (**J** 0064) Trans Amer Math Soc 142*111-140
◇ D55 D65 D70 E15 ◇ REV MR 42 # 74 Zbl 191.305 JSL 37.758 • ID 06103

HINMAN, P.G. [1969] *Some applications of forcing to hierarchy problems in arithmetic* (**J** 0068) Z Math Logik Grundlagen Math 15*341-352
◇ D25 D30 D55 E40 ◇ REV MR 43 # 6096 Zbl 191.306
• ID 06104

HINMAN, P.G. & MOSCHOVAKIS, Y.N. [1971] *Computability over the continuum* (**P** 0638) Logic Colloq;1969 Manchester 77-105
◇ D55 D65 D70 D75 ◇ REV MR 43 # 4675
Zbl 234 # 02028 • ID 06105

HINMAN, P.G. [1973] *Degrees of continuous functionals* (**J** 0036) J Symb Logic 38*393-395
◇ D30 D65 ◇ REV MR 49 # 2316 Zbl 281 # 02047
• ID 06107

HINMAN, P.G. [1973] *The finite levels of the hierarchy of effective R-sets* (**J** 0027) Fund Math 79*1-10
◇ D55 D65 D70 E15 ◇ REV MR 52 # 10396
Zbl 285 # 02039 • ID 06108

HINMAN, P.G. [1974] see FENSTAD, J.E.

HINMAN, P.G. [1974] see ACZEL, P.

HINMAN, P.G. [1977] *A survey of finite-type recursion* (**P** 1695) Set Th & Hierarch Th (3);1976 Bierutowice 187-209
◇ D55 D60 D65 D98 ◇ REV MR 58 # 5131
Zbl 366 # 02026 • ID 51109

HINMAN, P.G. [1978] *Recursion-theoretic hierarchies* (**X** 0811) Springer: Heidelberg & New York xii+480pp
◇ D55 D60 D65 D70 D98 E15 ◇ REV MR 82b:03084
Zbl 371 # 02017 JSL 48.497 • ID 51378

HINMAN, P.G. [1979] *Borel determinacy* (**J** 0005) Amer Math Mon 86*114-115
◇ D55 D98 E15 E60 ◇ REV Zbl 404 # 03037 • ID 54824

HINMAN, P.G. [1984] *Finitely approximable sets* (**P** 2153) Logic Colloq;1983 Aachen 2*233-258
◇ D65 F65 ◇ ID 43007

HINMAN, P.G. & ZACHOS, S. [1985] *Probabilistic machines, oracles, and quantifiers* (**P** 3342) Rec Th Week;1984 Oberwolfach 159-192
◇ C80 D15 ◇ ID 45303

HINMAN, P.G. see Vol. V for further entries

HIRAI, T. [1973] *Total undecidability on the class of general recursive functions with a given property* (**J** 0407) Comm Math Univ St Pauli (Tokyo) 22/1*43-48
◇ D20 ◇ REV MR 50 # 4265 Zbl 281 # 02040 • ID 06131

HIRANO, M. [1969] *Some definitions for recursive functions of ordinal numbers* (**J** 0350) Sci Rep Tokyo Kyoiku Daigaku Sect A 10*135-141
◇ D60 ◇ REV MR 41 # 3272 Zbl 215.323 • ID 06134

HIRD, G.R. [1985] see DOWNEY, R.G.

HIROSE, K. & SEKI, S. [1963] *A remark on the theory of automata* (**J** 0407) Comm Math Univ St Pauli (Tokyo) 11*115-119
◇ D05 ◇ REV MR 27 # 3483 Zbl 118.254 • ID 37162

HIROSE, K. [1964] *A theorem on incomparable degrees of recursive unsolvability* (**J** 0407) Comm Math Univ St Pauli (Tokyo) 13*27-33
◇ D25 D30 ◇ REV MR 31 # 5797 • ID 06135

HIROSE, K. [1965] *On complete degrees* (**J** 0081) Proc Japan Acad 41*875-877
◇ D30 ◇ REV MR 36 # 3651 • ID 06137

HIROSE, K. [1965] *On the degree of unsolvability* (**J** 0091) Sugaku 17*72-83
◇ D30 ◇ REV MR 35 # 5316 • ID 06136

HIROSE, K. [1968] *A conjecture on Hilbert's 10th problem* (**J** 0407) Comm Math Univ St Pauli (Tokyo) 17*31-34
◇ D30 D35 ◇ REV MR 38 # 5627 Zbl 205.310 JSL 37.604
• ID 06139

HIROSE, K. [1968] *An investigation on degrees of unsolvability* (**J** 0090) J Math Soc Japan 20*609-633
◇ D30 ◇ REV MR 37 # 5095 Zbl 195.304 • ID 06138

HIROSE, K. & IIDA, S. [1973] *A proof of negative answer to Hilbert's 10th problem* (**J** 0081) Proc Japan Acad 49*10-12
◇ D25 D35 ◇ REV MR 49 # 2325 Zbl 279 # 02028
• ID 06140

HIROSE, K. [1973] *On Hilbert's tenth problem (negative solution)* (**J** 0091) Sugaku 25*1-9
◇ D25 D35 ◇ REV MR 58 # 27750 • ID 81564

HIROSE, K. & NAKAYASU, F. [1981] *A representation for Spector second order classes in computation theories on two types* (**P** 3201) Logic Symposia;1979/80 Hakone 31-47
◇ D75 ◇ REV MR 84a:03051 Zbl 495 # 03030 • ID 33416

HIROSE, K. see Vol. II, III, V for further entries

HIROSE, S. & YONEDA, M. [1985] *On the Chomsky and Stanley's homomorphic characterization of context-free languages* (**J** 1426) Theor Comput Sci 36*109-112
◇ D05 ◇ ID 47259

HIRSCHBERG, D. [1963] see BRAFFORT, P.

HIRSCHELMANN, A. [1969] *Primitiv-rekursive Funktionen in Peano-Algebren* (**X** 0817) Ges Math Datenverarbeit: Bonn ii+37pp
◇ D20 D75 G25 ◇ REV MR 39 # 5358 Zbl 185.19
• ID 24936

HIRSCHELMANN, A. see Vol. III for further entries

HIRSCHFELD, J. [1974] *Models of arithmetic and the semiring of recursive functions* (**P** 1083) Victoria Symp Nonstand Anal;1972 Victoria 369*99-105
◇ C62 D20 H15 ◇ REV MR 58 # 5189 Zbl 281 # 02056
• ID 21188

HIRSCHFELD, J. [1975] *Models of arithmetic and recursive functions* (**J** 0029) Israel J Math 20*111-126
◇ C62 D20 H15 ◇ REV MR 52 # 2858 Zbl 311 # 02050
• ID 06142

HIRSCHFELD, J. see Vol. I, II, III, V, VI for further entries

HIZ, H. [1961] see COURT, L.M.

HIZ, H. see Vol. I, VI for further entries

HO, C. [1964] *On completely reversible Turing machines (Russian)* (**J** 0023) Dokl Akad Nauk SSSR 157*1307-1310
• TRANSL [1964] (**J** 0470) Sov Phys, Dokl 9*636-638
◇ D10 ◇ REV MR 30 # 17 Zbl 171.272 • ID 05829

HOARE, C.A.R. [1971] *Procedures and parameters: an axiomatic approach* (**C** 0628) Symp Semant of Algor Lang 102-116
◇ D05 ◇ REV MR 43 # 4297 Zbl 221 # 68020 • ID 28134

HOARE, C.A.R. [1972] see ALLISON, D.C.S.

HOARE, C.A.R. see Vol. I, II for further entries

HODES, H.T. [1978] *Uniform upper bounds on ideals of Turing degrees* (**J** 0036) J Symb Logic 43*601-612
◇ D30 ◇ REV MR 58 # 10369 Zbl 389 # 03017 • ID 29286

HODES, H.T. [1980] *Jumping through the transfinite: The master code hierarchy of Turing degrees* (**J** 0036) J Symb Logic 45*204-220
◇ D30 E40 E45 ◇ REV MR 81m:03052 Zbl 441 # 03014
• ID 56067

HODES, H.T. [1981] *Upper bounds on locally countable admissible initial segments of a Turing degree hierarchy* (**J** 0036) J Symb Logic 46*753-760
◇ D30 E40 E45 ◇ REV MR 84a:03049 Zbl 483 # 03028
• ID 33271

HODES, H.T. [1982] *Jumping to a uniform upper bound* (**J** 0053) Proc Amer Math Soc 85*600-602
◇ D30 ◇ REV MR 83j:03071 Zbl 498 # 03037 • ID 35367

HODES, H.T. [1983] *A minimal upper bound on a sequence of Turing degrees which represents that sequence* (**J** 0048) Pac J Math 108*115-119
◇ D30 ◇ REV MR 85i:03143 Zbl 546 # 03023 • ID 43526

HODES, H.T. [1983] *More about uniform upper bounds on ideals of Turing degrees* (**J** 0036) J Symb Logic 48*441-457
◇ D30 ◇ REV MR 85e:03100 Zbl 514 # 03027 • ID 37409

HODES, H.T. [1984] *Finite level Borel games and a problem concerning the jump hierarchy* (**J** 0036) J Symb Logic 49*1301-1318
◇ D30 E45 E60 ◇ ID 42462

HODES, H.T. [1984] *The modal theory of pure identity and some related decision problems* (**J** 0068) Z Math Logik Grundlagen Math 30*415-423
◇ B25 B45 D35 ◇ REV MR 86d:03018 Zbl 534 # 03007
• ID 41142

HODES, H.T. see Vol. II for further entries

HODGES, A. [1983] *Alan Turing: The Enigma* (**X** 1078) Simon & Schuster: New York xii + 587pp
◇ A10 D10 ◇ REV Zbl 541 # 68001 JSL 50.1065
• ID 41396

HODGES, W. [1976] *On the effectivity of some field constructions* (**J** 3240) Proc London Math Soc, Ser 3 32*133-162
◇ C60 C75 D65 E35 E47 E75 ◇ REV MR 55 # 10252
Zbl 325 # 12105 • ID 30711

HODGES, W. [1984] *Finite extensions of finite groups* (**P** 2153) Logic Colloq;1983 Aachen 1*193-206
◇ C13 C50 C60 D35 ◇ REV MR 86g:03059 • ID 39733

HODGES, W. see Vol. I, II, III, V, VI for further entries

HODGSON, B.R. [1981] see HATCHER, W.S.

HODGSON, B.R. & KENT, C.F. [1982] *An arithmetical characterization of NP* (**J** 1426) Theor Comput Sci 21*255-267
◇ D15 ◇ REV MR 84k:03110 Zbl 498 # 03023 • ID 36128

HODGSON, B.R. [1982] *On direct products of automaton decidable theories* (**J** 1426) Theor Comput Sci 19*331-335
◇ B25 D05 ◇ REV MR 83m:03046 Zbl 493 # 03002
• ID 35446

HODGSON, B.R. & KENT, C.F. [1983] *A normal form for arithmetical representation of NP-sets* (**J** 0119) J Comp Syst Sci 27*378-388
◇ D15 ◇ REV MR 85m:68011 Zbl 535 # 03016 • ID 38320

HODGSON, B.R. [1983] *Decidabilite par automate fini* (**J** 2660) Ann Sci Math Quebec 7*39-57
◇ B25 D05 D35 ◇ REV MR 84m:03016 Zbl 531 # 03007
• ID 35721

HOEHNKE, H.-J. [1972] see BUDACH, L.

HOEHNKE, H.-J. [1972] *Superposition partieller Funktionen* (**C** 0722) Stud Algeb & Anwendgn 7-26
◇ C05 D20 ◇ REV MR 50 # 62 Zbl 275 # 08007 • ID 29729

HOEHNKE, H.-J. [1975] see BUDACH, L.

HOEHNKE, H.-J. [1977] *On partial recursive definitions and programs* (**P** 2588) FCT'77 Fund of Comput Th;1977 Poznan 260-274
◇ D10 D75 ◇ REV MR 58 # 25095 Zbl 391 # 68010
• ID 81583

HOEHNKE, H.-J. [1981] *On K-recursive definitions and bicategories* (**P** 3642) Colloq Math Log in Computer Sci;1978 Salgotarjan 485-490
◇ D05 G30 ◇ REV MR 83g:68007 Zbl 491 # 68052
• ID 47136

HOEHNKE, H.-J. see Vol. III for further entries

HOERNIG, K.M. [1981] *A unified approach to definability problems in the theory of higher type functionals* (**X** 1663) Ludwig-Maximilians-Univ: Muenchen 53pp
◇ D65 ◇ REV Zbl 516 # 03024 • ID 37257

HOERNIG, K.M. see Vol. I for further entries

HOEVEN VAN DER, G.F. & MOERDIJK, I. [1984] *On choice sequences determined by spreads* (**J** 0036) J Symb Logic 49*908-916
◇ D30 D55 F35 F50 G30 ◇ ID 42464

HOEVEN VAN DER, G.F. see Vol. III, V, VI for further entries

HOFFMANN, B. [1979] see FRANZEN, H.

HOFFMANN-JOERGENSEN, J. [1980] see DELLACHERIE, C.

HOFFMANN-JOERGENSEN, J. see Vol. V for further entries

HOFSTADTER, D.R. [1979] *Goedel, Escher, Bach: an eternal golden braid* (**X** 2671) Basic Books: New York xxi + 777pp
◇ A05 B98 D99 ◇ REV MR 80j:03009 Zbl 457 # 03001
JSL 48.864 • ID 74228

HOLLAND, W.C. & MCCLEARY, S.H. [1979] *Solvability of the word problem in free lattice-ordered groups* (**J** 1447) Houston J Math 5*99-105
◇ D40 ◇ REV MR 80f:06018 Zbl 387 # 06011 • ID 54852

HOLLAND, W.C. see Vol. III, V for further entries

HOLLERER, W.O. & VOLLMAR, R. [1975] *On "forgetful" cellular automata* (J 0119) J Comp Syst Sci 11*237-251
◇ D05 ◇ REV MR 52 # 7741 Zbl 329 # 94026 • ID 62590

HOLSZTYNSKI, W. [1966] see ENGELKING, R.

HOLSZTYNSKI, W. see Vol. III for further entries

HOLYOKE, T.C. [1969] see BULLOFF, J.J.

HOMER, S. [1980] *Two splitting theorems for β-recursion theory* (J 0007) Ann Math Logic 18*137-151
◇ D60 ◇ REV MR 81m:03055 Zbl 471 # 03042 • ID 55237

HOMER, S. & JACOBS, B.E. [1981] *Degrees of non α-speedable sets* (J 0068) Z Math Logik Grundlagen Math 27*539-548
◇ D15 D60 ◇ REV MR 83e:03072 Zbl 503 # 03021 • ID 33201

HOMER, S. [1982] see GOLDMAN, J.

HOMER, S. [1983] *Intermediate β-r.e. degrees and the half-jump* (J 0036) J Symb Logic 48*790-796
◇ D30 D60 ◇ REV MR 84m:03066 Zbl 529 # 03023 • ID 35802

HOMER, S. & SACKS, G.E. [1983] *Inverting the half-jump* (J 0064) Trans Amer Math Soc 278*317-331
◇ D30 D60 ◇ REV MR 84k:03115 Zbl 523 # 03033 • ID 36133

HOMER, S. & MAASS, W. [1983] *Oracle dependent properties of the lattice of NP sets* (J 1426) Theor Comput Sci 24*279-289
◇ D15 ◇ REV MR 85f:68030 Zbl 543 # 03024 • ID 33606

HOMER, S. [1983] see GASARCH, W.I.

HOMER, S. [1985] *Minimal polynomial degrees of nonrecursive sets* (P 3342) Rec Th Week;1984 Oberwolfach 193-202
◇ D15 ◇ ID 45305

HONDA, N. & NASU, M. [1969] *Mappings induced by PGSM mappings and some recursively unsolvable problems of finite probabilistic automata* (J 0194) Inform & Control 15*250-273
◇ D05 E72 ◇ REV MR 40 # 8528 • ID 26103

HONDA, N. & IGARASHI, Y. [1974] *Deterministic multitape automata computations* (J 0119) J Comp Syst Sci 8*167-189
◇ D10 D15 ◇ REV MR 48 # 12898 Zbl 277 # 68023 • ID 62674

HONDA, N. & NASU, M. [1976] *A completeness property of one-dimensional tessellation automata* (J 0119) J Comp Syst Sci 12*36-48
◇ D05 ◇ REV MR 53 # 12796 Zbl 339 # 02031 • ID 64164

HONDA, N. & OYAMAGUCHI, M. [1978] *The decidability of equivalence for deterministic stateless pushdown automata* (J 0194) Inform & Control 38*367-376
◇ D10 ◇ REV MR 80f:68094 Zbl 393 # 68078 • ID 52476

HONDA, N. & INAGAKI, Y. & OYAMAGUCHI, M. [1980] *On the equivalence problem for two DPDA's, one of which is real-time* (P 3385) Inform Processing (8);1980 Tokyo & Melbourne 53-57
◇ D10 D15 ◇ REV MR 81j:68009 Zbl 443 # 68037 • ID 56443

HONDA, N. & INAGAKI, Y. & OYAMAGUCHI, M. [1980] *The equivalence problem for real-time strict deterministic languages* (J 0194) Inform & Control 45*90-115
◇ D05 D15 ◇ REV MR 81f:68093 Zbl 444 # 68038 • ID 69754

HONDA, N. & INAGAKI, Y. & OYAMAGUCHI, M. [1981] *The equivalence problem for two dpda's, one of which is a finite-turn or one-counter machine* (J 0119) J Comp Syst Sci 23*366-382
◇ D05 D10 ◇ REV MR 83b:68061 Zbl 473 # 68046 • ID 38977

HONDA, N. & INAGAKI, Y. & IZUMI, H. [1984] *A complete axiom system for algebra of closed-regular expression* (P 4012) Automata, Lang & Progr (11);1984 Antwerpen 260-269
◇ D05 ◇ REV Zbl 554 # 68050 • ID 47159

HONG, JIAWEI [1981] *On similarity and duality of computation (Chinese)* (J 1024) Zhongguo Kexue 24*141-152
◇ D15 ◇ REV MR 83i:68052a Zbl 468 # 68058 • ID 39354

HONG, JIAWEI [1984] *A tradeoff theorem for space and reversal* (J 1426) Theor Comput Sci 32*221-224
◇ D10 ◇ REV MR 85j:68032 Zbl 545 # 68037 • ID 46789

HONG, JIAWEI [1984] *On similarity and duality of computation I* (J 0194) Inform & Control 62*109-128
◇ D20 ◇ ID 46385

HONG, JIAWEI see Vol. VI for further entries

HOOGEBOOM, H.J. [1985] see EHRENFEUCHT, A.

HOOPER, P.K. [1966] *Monogenic Post normal systems of arbitrary degree* (J 0037) ACM J 13*359-363
◇ D03 ◇ REV MR 33 # 3846 Zbl 168.12 JSL 34.508 • ID 06212

HOOPER, P.K. [1966] *The immortality problem for Post normal systems* (J 0037) ACM J 13*594-599
◇ D03 ◇ REV MR 34 # 1192 Zbl 173.12 • ID 06211

HOOPER, P.K. [1966] *The undecidability of the Turing machine immortality problem* (J 0036) J Symb Logic 31*219-234
◇ D10 ◇ REV MR 33 # 7261 Zbl 173.12 JSL 36.150 • ID 06210

HOOVER, D.N. [1984] see ANGLUIN, D.

HOOVER, D.N. see Vol. I, II, III for further entries

HOPCROFT, J.E. & ULLMAN, J.D. [1967] *An approach to a unified theory of automata* (J 0432) Bell Syst Tech J 46*1793-1829
◇ D05 D10 ◇ REV MR 38 # 4218 Zbl 155.343 • ID 06214

HOPCROFT, J.E. & WEINER, P. [1967] *Modular decomposition of synchronous sequential machines* (J 4305) IEEE Trans Electr Comp EC-16*233-239
◇ D05 ◇ ID 20899

HOPCROFT, J.E. & ULLMAN, J.D. [1967] *Nonerasing stack automata* (J 0119) J Comp Syst Sci 1*166-186
◇ D05 D10 ◇ REV Zbl 166.5 • ID 19019

HOPCROFT, J.E. & ULLMAN, J.D. [1968] *Decidable and undecidable questions about automata* (J 0037) ACM J 15*317-324
◇ D05 D10 ◇ REV MR 38 # 5530 Zbl 155.343 • ID 06216

HOPCROFT, J.E. & ULLMAN, J.D. [1968] *Relations between time and tape complexities* (J 0037) ACM J 15*414-427
◇ D05 D10 D15 ◇ REV MR 38 # 4219 Zbl 169.311 JSL 38.343 • ID 15006

HOPCROFT, J.E. [1969] see BORODIN, A.

HOPCROFT, J.E. & ULLMAN, J.D. [1969] *Formal languages and their relation to automata* (**X** 0832) Addison-Wesley: Reading x+242pp
• TRANSL [1979] (**X** 1876) Kexue Chubanshe: Beijing
◇ D05 D98 ◇ REV MR 38 # 5533 Zbl 196.17 • ID 23533

HOPCROFT, J.E. [1969] see AHO, A.V.

HOPCROFT, J.E. [1969] see GINSBURG, S.

HOPCROFT, J.E. [1970] see GINSBURG, S.

HOPCROFT, J.E. [1970] see HARTMANIS, J.

HOPCROFT, J.E. [1971] see HARTMANIS, J.

HOPCROFT, J.E. [1971] see GINSBURG, S.

HOPCROFT, J.E. & ULLMAN, J.D. [1974] *Some results on tape-bounded Turing machines (Russian)* (**C** 2319) Slozh Vychisl & Algor 252-265
◇ D10 D15 ◇ REV Zbl 289 # 68014 • ID 62598

HOPCROFT, J.E. & PAUL, W.J. & VALIANT, L.G. [1977] *On time versus space* (**J** 0037) ACM J 24*332-337
◇ D15 ◇ REV MR 56 # 1798 Zbl 358 # 68082 • REM Part I. Part II 1981 by Paul,W. & Reischuk,R. • ID 46716

HOPCROFT, J.E. [1978] see FORTUNE, S.

HOPCROFT, J.E. & ULLMAN, J.D. [1979] *Introduction to automata theory, languages, and computation* (**X** 0832) Addison-Wesley: Reading x+418pp
◇ D05 D10 D98 ◇ REV MR 83j:68002 Zbl 426 # 68001 • ID 39857

HOPCROFT, J.E. & PANSIOT, J.-J. [1979] *On the reachability problem for 5-dimensional vector addition systems* (**J** 1426) Theor Comput Sci 8*135-159
◇ D80 ◇ REV MR 80h:68022 Zbl 466 # 68048 • ID 54998

HOPKIN, D.R. & MOSS, B.P. [1976] *Automata* (**X** 0843) Macmillan : New York & London vi+170PP
◇ D05 D10 D98 ◇ REV Zbl 372 # 94035 • ID 32394

HORADAM, K.J. [1981] *A quick test for nonisomorphism of one-relator groups* (**J** 0053) Proc Amer Math Soc 81*195-200
◇ D40 ◇ REV MR 82j:20069 Zbl 426 # 20022 • ID 81590

HORADAM, K.J. [1981] *The word problem and related results for graph product groups* (**J** 0053) Proc Amer Math Soc 82*157-164
◇ D40 ◇ REV MR 82e:20043 Zbl 439 # 20024 • ID 81591

HOREJS, J. [1964] *Note on definition of recursiveness* (**J** 0068) Z Math Logik Grundlagen Math 10*119-120
◇ D20 ◇ REV MR 28 # 3928 Zbl 159.9 JSL 33.115 • ID 06221

HOREJS, J. [1965] *On generalizations of a theorem on recursive sets* (**J** 0322) Arch Math (Brno) 1*221-227
◇ D20 D25 ◇ REV MR 34 # 1178 Zbl 203.12 • ID 06222

HOREJS, J. [1967] *Automata without internal memory (Czech) (Russian and English summaries)* (**J** 0086) Cas Pestovani Mat, Ceskoslov Akad Ved 92*193-205
◇ D05 ◇ REV Zbl 168.256 JSL 35.486 • ID 19018

HOREJS, J. [1969] *Recursive functions computable within $C\bar{f}\log\bar{f}$ (Czech summary)* (**J** 0156) Kybernetika (Prague) 5*384-399
◇ D15 ◇ REV MR 44 # 1570 Zbl 181.305 • ID 06223

HOROWITZ, B.M. [1978] *Sets completely creative via recursive permutations* (**J** 0068) Z Math Logik Grundlagen Math 24*445-452
◇ D20 D25 ◇ REV MR 80c:03043 Zbl 393 # 03030 • ID 52450

HOROWITZ, B.M. [1980] *Constructively nonpartial recursive functions* (**J** 0047) Notre Dame J Formal Log 21*273-276
◇ D20 D25 ◇ REV MR 81e:03037 Zbl 394 # 03043 • ID 53801

HOROWITZ, B.M. [1982] *An isomorphism type of arithmetically productive sets* (**J** 0068) Z Math Logik Grundlagen Math 28*211-214
◇ D25 D55 ◇ REV MR 83m:03055b Zbl 515 # 03024 • ID 35455

HOROWITZ, B.M. [1982] *Arithmetical analogues of productive and universal sets* (**J** 0068) Z Math Logik Grundlagen Math 28*203-210
◇ D25 D55 ◇ REV MR 83m:03055a Zbl 515 # 03026 • ID 35454

HOROWITZ, B.M. [1982] *Elementary formal systems as a framework for relative recursion theory* (**J** 0047) Notre Dame J Formal Log 23*39-52
◇ D20 D25 D30 ◇ REV MR 84g:03059 Zbl 442 # 03030 • ID 55094

HOROWITZ, L. [1967] *Invariance of partial order of recursive equivalence types under finite division* (**J** 0028) Indag Math 29*76-82
◇ D50 ◇ REV MR 35 # 2736 Zbl 149.247 • ID 06237

HORVATH, S. [1977] *Complexity of sequence encodings* (**P** 2588) FCT'77 Fund of Comput Th;1977 Poznan 399-404
◇ D15 D20 ◇ REV MR 58 # 10381 Zbl 371 # 02015 • ID 51376

HORVATH, S. [1978] see BOASSON, L.

HORVATH, S. [1984] *A generating system for partial recursive functions on N** (**J** 2774) Koezlem MTA Szam & Autom: Kutat Intez 158*314-318
◇ D20 ◇ ID 45090

HORVATH, S. [1985] *On defining the Blum complexity of partial recursive sequence functions* (**J** 0129) Elektr Informationsverarbeitung & Kybern 21*229-231
◇ D15 D20 ◇ ID 49632

HOSCH, F.A. & LANDWEBER, L.H. [1973] *Finite delay solutions for sequential conditions* (**P** 0763) Automata, Lang & Progr (1);1972 Rocquencourt 45-60
◇ D05 ◇ REV MR 56 # 15232 • ID 81592

HOSKEN, W.H. [1970] see BUECHI, J.R.

HOSKEN, W.H. [1972] *An asymmetric regular set* (**J** 0130) BIT 12*115-117
◇ D05 ◇ REV MR 48 # 8135 Zbl 263 # 94020 • ID 62601

HOSKEN, W.H. [1972] *Some Post canonical systems in one letter* (**J** 0130) BIT 12*509-515
◇ D03 ◇ REV MR 55 # 12485 Zbl 305 # 02047 • ID 06239

HOTZ, G. [1966] *Eindeutigkeit und Mehrdeutigkeit formaler Sprachen* (**J** 0129) Elektr Informationsverarbeitung & Kybern 2*235-246
◇ D05 ◇ REV MR 35 # 1414 Zbl 177.17 JSL 35.348 • ID 06253

HOTZ, G. [1967] *Homomorphie und Aequivalenz formaler Sprachen* (**P** 1671) Colloq Automatenth (3);1965 Hannover 204-211
◇ D05 ◇ REV Zbl 165.321 • ID 29433

HOTZ, G. [1968] *Automatentheorie und formale Sprachen I. Turingmaschinen und rekursive Funktionen* (**X** 0876) Bibl Inst: Mannheim 174pp
◇ D05 D10 D20 D98 ◇ REV MR 39 # 5359 Zbl 212.27 • REM Part II 1969 by Hotz,G. & Walter,H. • ID 06254

HOTZ, G. & WALTER, H. [1969] *Automatentheorie und formale Sprachen II. Endliche Automaten* (**X** 0876) Bibl Inst: Mannheim 226pp
◇ D05 D98 ◇ REV Zbl 212.27 • REM Part I 1968. Part III 1972 by Claus,V. & Hotz,G • ID 23536

HOTZ, G. [1970] see DOERR, J.

HOTZ, G. [1970] see BARTHOLOMES, F.

HOTZ, G. [1972] see CLAUS, V.

HOTZ, G. [1974] *Komplexitaetsmasse fuer Ausdruecke (English summary)* (**P** 1869) Automata, Lang & Progr (2);1974 Saarbruecken 398-412
◇ D15 ◇ REV MR 54 # 14440 Zbl 284 # 68042 • ID 81596

HOTZ, G. [1975] *Strukturelle Verwandtschaften von Semi-Thue-Systemen* (**P** 0770) Categ Th Appl to Comput & Control (1);1974 San Francisco 174-179
◇ D03 ◇ REV MR 52 # 4723 Zbl 305 # 68057 • ID 62606

HOTZ, G. [1977] *Automatentheorie und formale Sprachen* (**J** 1670) Mitt Math Ges DDR 1977/4*23-63
◇ B75 D05 D97 ◇ REV MR 48 # 7643 Zbl 411 # 68046 • ID 52928

HOTZ, G. [1979] *Ueber die Darstellbarkeit des syntaktischen Monoides kontextfreier Sprachen* (**J** 3441) RAIRO Inform Theor 13*337-345
◇ D05 ◇ REV MR 81c:68058 Zbl 428 # 68085 • ID 53826

HOTZEL, E. [1979] see BRAUNMUEHL VON, B.

HOWARD, J.V. [1975] *Computable explanations* (**J** 0068) Z Math Logik Grundlagen Math 21*215-224
◇ D80 ◇ REV MR 58 # 5102 Zbl 326 # 60036 • ID 06259

HOWARD, P.E. [1972] *A proof of a theorem of Tennenbaum* (**J** 0068) Z Math Logik Grundlagen Math 18*111-112
◇ C57 C62 H15 ◇ REV MR 46 # 3293 Zbl 251 # 02053 • ID 06260

HOWARD, P.E. see Vol. III, V for further entries

HOWARD, W.A. & POUR-EL, M.B. [1964] *A structural criterion for recursive enumeration without repetition* (**J** 0068) Z Math Logik Grundlagen Math 10*105-114
◇ D25 ◇ REV MR 29 # 5720 Zbl 132.247 JSL 38.155 • ID 16368

HOWARD, W.A. [1981] *Computability of ordinal recursion of type level two* (**P** 3146) Constr Math;1980 Las Cruces 87-104
◇ D20 F10 F15 F35 F50 ◇ REV MR 83e:03091 Zbl 463 # 03032 • ID 54572

HOWARD, W.A. see Vol. V, VI for further entries

HRBACEK, K. [1978] *On the complexity of analytic sets* (**J** 0068) Z Math Logik Grundlagen Math 24*419-425
◇ D30 D55 D65 E15 E45 ◇ REV MR 80b:03066 Zbl 411 # 03040 JSL 49.665 • ID 52894

HRBACEK, K. [1979] see GOSTANIAN, R.

HRBACEK, K. & SIMPSON, S.G. [1980] *On Kleene degrees of analytic sets* (**P** 2058) Kleene Symp;1978 Madison 347-352
◇ D30 D55 D65 E15 E35 E45 ◇ REV MR 82d:03081 Zbl 464 # 03042 JSL 49.665 • ID 74290

HRBACEK, K. [1983] *Degrees of analytic sets* (**J** 0068) Z Math Logik Grundlagen Math 29*75-82
◇ D30 D65 E15 E45 ◇ REV MR 85d:03086 Zbl 549 # 03039 • ID 41103

HRBACEK, K. see Vol. I, III, V for further entries

HROMKOVIC, J. [1981] *Closure properties of the family of languages recognized by one-way two-head deterministic finite state automata* (**P** 3429) Math Founds of Comput Sci (10);1981 Strbske Pleso 304-313
◇ D05 ◇ REV MR 83e:68125 Zbl 472 # 68047 • ID 40172

HROMKOVIC, J. [1985] *Fooling a two-way nondeterministic multihead automaton with reversal number restriction* (**J** 1431) Acta Inf 22*589-594
◇ D05 ◇ REV Zbl 565 # 68078 • ID 49077

HROMKOVIC, J. [1985] *On the power of alternation in automata theory* (**J** 0119) J Comp Syst Sci 31*28-39
◇ D05 D10 ◇ ID 49486

HROMKOVIC, J. see Vol. II for further entries

HSIA, PEI & YEH, R.T. [1973] *Finite automata with markers* (**P** 0763) Automata, Lang & Progr (1);1972 Rocquencourt 443-451
◇ D05 ◇ REV MR 56 # 11648 Zbl 274 # 02013 • ID 29018

HSIA, PEI & YEH, R.T. [1975] *Marker automata* (**J** 0191) Inform Sci 8*71-88
◇ D05 ◇ REV MR 52 # 7203 Zbl 341 # 94029 • ID 62617

HU, S.-T. [1968] *Mathematical theory of switching circuits and automata* (**X** 0926) Univ Calif Pr: Berkeley 261pp
◇ B70 D05 D98 ◇ REV MR 39 # 5244 Zbl 174.291 • ID 23541

HU, SHIHUA [1956] *The primitive recursiveness of a kind of recursion (Chinese) (English summary)* (**J** 0418) Shuxue Xuebao 6/1*93-104
◇ D20 ◇ REV MR 18 # 104 Zbl 75.235 • ID 48509

HU, SHIHUA & LOH, CHUNGWAN [1958] *Normal forms of general recursive functions (Chinese)* (**J** 0418) Shuxue Xuebao 8*507-520
• TRANSL [1958] (**J** 1153) Kexue Jilu (Beijing) 2*134-139 (**J** 1024) Zhongguo Kexue 9*889-896 • REPR [1962] (**J** 0419) Chinese Math Acta 1*110-117
◇ D20 ◇ REV MR 21 # 3332 JSL 37.612 • ID 22151

HU, SHIHUA [1960] *Kernel functions and normal forms of recursive functions (Chinese)* (**J** 1153) Kexue Jilu (Beijing) 4*99-101
• TRANSL (**J** 1024) Zhongguo Kexue 9*876-888 • REPR [1962] (**J** 0419) Chinese Math Acta 1*97-109
◇ D20 ◇ REV MR 23 # A798b Zbl 93.13 • ID 22053

HU, SHIHUA & LOH, CHUNGWAN [1960] *Kernel functions. Theory of recursive algorithms II (Chinese) (English summary)* (**J** 0418) Shuxue Xuebao 10*89-97
• TRANSL [1960] (**J** 1024) Zhongguo Kexue 9*876-888 [1962] (**J** 0419) Chinese Math Acta 1*97-109
◇ D10 D20 ◇ REV MR 22 # 3684b Zbl 168.255 JSL 37.612 • REM Part I 1960. Part III 1960 • ID 21127

HU, SHIHUA [1960] *Normal forms of recursive functions. Theory of recursive algorithms III (Chinese) (English summary)* (**J** 0418) Shuxue Xuebao 10*98-103
• TRANSL [1960] (**J** 1024) Zhongguo Kexue 9*889-896
◇ D20 ◇ REV MR 22#3684c Zbl 168.255 JSL 37.612
• REM Part II 1960 • ID 21129

HU, SHIHUA [1960] *Recursive algorithms. Theory of recursive algorithms I (Chinese) (English summary)* (**J** 0418) Shuxue Xuebao 10*66-88
• TRANSL [1960] (**J** 1024) Zhongguo Kexue 9*843-875
◇ D10 D20 ◇ REV MR 22#3684a JSL 37.612 • REM Part II 1960 • ID 25479

HU, SHIHUA [1960] *Theory of recursive algorithms* (**J** 1153) Kexue Jilu (Beijing) 4*91-98
• TRANSL [1960] (**J** 1024) Zhongguo Kexue 9*843-875 [1962] (**J** 0419) Chinese Math Acta 1*64-96
◇ D20 ◇ REV MR 23#A798a Zbl 93.13 • ID 22052

HU, SHIHUA & HUANG, ZULIANG [1963] *Addition and multiplication (Chinese)* (**J** 0420) Shuxue Jinzhan 6*371-378
◇ D20 F30 ◇ ID 48690

HU, SHIHUA & YANG, DONGPING [1964] *On primitive recursiveness (Chinese)* (**J** 0418) Shuxue Xuebao 14*607-618
• TRANSL [1964] (**J** 0419) Chinese Math Acta 5*653-665
◇ D20 ◇ REV MR 33#5480 Zbl 158.250 • ID 06280

HU, SHIHUA & YANG, DONGPING [1965] *About primitive recursivity (Chinese)* (**P** 4564) Math Logic;1963 Xi-An 114
◇ D20 ◇ ID 49339

HU, SHIHUA [1973] *The description problem of algorithmic languages (Chinese)* (**X** 4565) Jixuan Jishu Yanjiushuo, Zhongguo Kexueyuan: Beijing 47pp
◇ B75 D20 ◇ ID 48686

HU, SHIHUA see Vol. I, II for further entries

HU, ZHAOGUANG [1982] *The proof to Mizumoto and Tanaka's problem* (**J** 3919) BUSEFAL 9*66-77
◇ D30 E72 ◇ REV Zbl 521#03040 • ID 37077

HUANG, WENGI & NERODE, A. [1985] *Application of pure recursion theory in recursive analysis (Chinese)* (**J** 0418) Shuxue Xuebao 28/5*625-635
◇ D20 F60 ◇ ID 49393

HUANG, ZULIANG [1963] see HU, SHIHUA

HUBER-DYSON, V. [1964] *On the decision problem for theories of finite models* (**J** 0029) Israel J Math 2*55-70
◇ B25 C13 D35 ◇ REV MR 31#2149 Zbl 143.249
• ID 03214

HUBER-DYSON, V. [1969] *On the decision problem for extensions of a decidable theory* (**J** 0027) Fund Math 64*7-40
◇ B25 C13 C60 D35 ◇ REV MR 40#5431 Zbl 193.312
• ID 03215

HUBER-DYSON, V. [1974] *A family of groups with nice word problems* (**J** 0038) J Austral Math Soc 17*414-425
◇ D40 ◇ REV MR 50#13290 Zbl 304#02018 • ID 04105

HUBER-DYSON, V. [1977] *Talking about free groups in naturally enriched languages* (**J** 0394) Commun Algeb 5*1163-1191
◇ C60 D40 ◇ REV MR 56#11775 Zbl 365#02045
• ID 51046

HUBER-DYSON, V. [1981] *A reduction of the open sentence problem for finite groups* (**J** 0161) Bull London Math Soc 13*331-338
◇ C13 C60 D40 ◇ REV MR 82j:03052 Zbl 441#20002
• ID 54396

HUBER-DYSON, V. [1982] *Decision problems in group theory* (**P** 3808) Rect Trends in Math;1982 Reinhardsbrunn 174-182
◇ B25 D40 ◇ REV MR 84d:20032 Zbl 508#20018
• ID 39680

HUBER-DYSON, V. [1982] *Finiteness conditions and the word problem* (**P** 3886) Groups-St.Andrews;1981 St.Andrews 244-251
◇ D40 ◇ REV MR 84i:20032 Zbl 494#20019 • ID 36652

HUBER-DYSON, V. & JONES, JAMES P. & SHEPHERDSON, J.C. [1982] *Some diophantine forms of Goedel's theorem* (**J** 0009) Arch Math Logik Grundlagenforsch 22*51-60
◇ D35 F30 ◇ REV MR 83k:03073 Zbl 494#03043
• ID 35411

HUBER-DYSON, V. [1982] *Symmetric groups and the open sentence problem* (**P** 3634) Patras Logic Symp;1980 Patras 159-169
◇ C13 C60 D35 ◇ REV MR 84j:20032 Zbl 514#20025
• ID 37444

HUBER-DYSON, V. see Vol. III, VI for further entries

HUET, G. [1980] *Confluent reductions: abstract properties and applications to term rewriting systems* (**J** 0037) ACM J 27*797-821
◇ D03 D05 ◇ REV MR 82a:68090 Zbl 458#68007
• ID 69015

HUET, G. & OPPEN, D.C. [1980] *Equations and rewrite rules: a survey* (**P** 4266) Form Lang Th;1979 Santa Barbara 349-405
◇ D03 D05 D98 ◇ REV MR 84j:68001 Zbl 545#68065
• ID 47206

HUET, G. see Vol. I, III, VI for further entries

HUGHES, C.E. & OVERBEEK, R.A. & SINGLETARY, W.E. [1971] *The many-one equivalence of some general combinatorial decision problems* (**J** 0015) Bull Amer Math Soc 77*467-472
◇ D03 D05 D10 D25 D30 ◇ REV MR 43#7324 Zbl 216.9 • ID 06293

HUGHES, C.E. [1972] *Degrees of unsolvability associated with Markov algorithms* (**J** 0435) Int J Comput & Inf Sci 1*355-365
◇ D03 D25 D55 ◇ REV MR 49#7129 Zbl 298#02040
• ID 06296

HUGHES, C.E. & SINGLETARY, W.E. [1973] *Combinatorial systems with axiom* (**J** 0047) Notre Dame J Formal Log 14*354-360
◇ D03 D05 D10 D25 D30 ◇ REV MR 48#10791 Zbl 258#02043 • ID 06298

HUGHES, C.E. [1973] *Many-one degrees associated with problems of tag* (**J** 0036) J Symb Logic 38*1-17
◇ D03 D10 D25 D40 ◇ REV MR 48#5834 Zbl 272#02062 JSL 38.1 • ID 06297

HUGHES, C.E. [1973] *Many-one degrees associated with semi-Thue systems* (**J** 0119) J Comp Syst Sci 7*497-505
◇ D03 D25 ◇ REV MR 48#5838 Zbl 284#02020
• ID 06300

HUGHES, C.E. [1974] *Single premise Post canonical forms defined over one-letter alphabets* (J 0036) J Symb Logic 39*489-495
◇ D03 ◇ REV MR 51 # 98 Zbl 298 # 02029 • ID 06301

HUGHES, C.E. & SINGLETARY, W.E. [1975] *Combinatorial systems defined over one- and two-letter alphabets* (J 0009) Arch Math Logik Grundlagenforsch 17*25-33
◇ D03 ◇ REV MR 53 # 105 Zbl 336 # 02035 • ID 06306

HUGHES, C.E. [1975] *Sets derived by deterministic systems with axiom* (J 0068) Z Math Logik Grundlagen Math 21*71-80
◇ D03 D25 ◇ REV MR 56 # 2810 Zbl 307 # 02025
• ID 06305

HUGHES, C.E. [1975] *The general decision problem for Markov algorithms with axiom* (J 0047) Notre Dame J Formal Log 16*208-216
◇ D03 ◇ REV MR 52 # 2854 Zbl 232 # 02029 • ID 06302

HUGHES, C.E. [1976] *A reduction class containing formulas with one monadic predicate and one binary function symbol* (J 0036) J Symb Logic 41*45-49
◇ B20 D35 ◇ REV MR 54 # 4955 Zbl 332 # 02051
• ID 14752

HUGHES, C.E. [1976] *Two variable implicational calculi of prescribed many-one degrees of unsolvability* (J 0036) J Symb Logic 41*39-44
◇ B20 D25 ◇ REV MR 53 # 12911 Zbl 339 # 02044
• ID 14753

HUGHES, C.E. & SINGLETARY, W.E. [1977] *The one-one equivalence of some general combinatorial decision problems* (J 0047) Notre Dame J Formal Log 18*305-309
◇ D30 D40 ◇ REV MR 57 # 9513 Zbl 314 # 02052
• ID 24228

HUGHES, C.E. [1978] *The equivalence of vector addition systems to a subclass of Post canonical systems* (J 0232) Inform Process Lett 7*201-205
◇ D03 D80 ◇ REV MR 80i:68038 Zbl 391 # 68020
• ID 52359

HUGHES, C.E. [1980] *Derivatives and quotients of prefix-free context-free languages* (J 0194) Inform & Control 45*229-235
◇ D05 D25 ◇ REV MR 83b:68080 Zbl 453 # 68050
• ID 54207

HUGHES, C.E. & STRAIGHT, D.W. [1980] *Word problems for bidirectional single-premise Post systems* (J 0047) Notre Dame J Formal Log 21*501-508
◇ D03 D25 ◇ REV MR 81h:03089 Zbl 416 # 03042
• ID 53932

HUGHES, C.E. see Vol. II for further entries

HUGILL, D.F. [1969] *Initial segments of Turing degrees* (J 3240) Proc London Math Soc, Ser 3 19*1-16
◇ D30 ◇ REV MR 38 # 5621 Zbl 196.14 • ID 06310

HUHN, A.P. [1975] see HERRMANN, C.

HUHN, A.P. see Vol. III for further entries

HULE, H. & MAURER, H.A. & OTTMANN, T. [1976] *OL forms* (1111) Preprints, Manuscr., Techn. Reports etc.
◇ D05 ◇ REM Bericht 48 Inst. Angewandte Informatik & Formale Beschreibungsverfahren, Univ. Karlsruhe, Nov. 1976 • ID 28329

HULE, H. & MAURER, H.A. & OTTMANN, T. [1978] *Good OL forms* (J 1431) Acta Inf 9*345-353
◇ D05 ◇ REV MR 58 # 13970 Zbl 374 # 68048 • ID 28337

HULE, H. see Vol. III for further entries

HUMBERSTONE, I.L. [1984] *Monadic representability of certain binary relations* (J 0016) Bull Austral Math Soc 29*365-376
◇ C52 D60 E07 ◇ REV MR 85f:04001 Zbl 531 # 04004
• ID 37693

HUMBERSTONE, I.L. see Vol. II, III for further entries

HUMMITZSCH, P. [1971] *Die Entscheidbarkeit der endlichen Ununterscheidbarkeit endlicher Automaten* (J 0068) Z Math Logik Grundlagen Math 17*315-322
◇ D05 ◇ REV MR 46 # 40 Zbl 279 # 94045 • ID 06324

HUMMITZSCH, P. [1971] *Zur schwachen Aequivalenz von endlichen Automaten (Russian and English summaries)* (J 0129) Elektr Informationsverarbeitung & Kybern 7*467-483
◇ D05 ◇ REV MR 48 # 1824 Zbl 243 # 94044 • ID 06323

HUNT III, H.B. [1974] see HARTMANIS, J.

HUNT III, H.B. & SZYMANSKI, T.G. [1976] *Dichotomization, reachability, and the forbidden subgraph problem* (P 2597) ACM Symp Th of Comput (8);1976 Hershey 126-134
◇ B75 D05 D15 ◇ REV MR 56 # 13806 Zbl 365 # 68031
• ID 51073

HUNT III, H.B. & ROSENKRANTZ, D.J. [1978] *Polynomial algorithms for deterministic pushdown automata* (J 1428) SIAM J Comp 7*405-412
◇ D10 D15 ◇ REV MR 80f:68053 Zbl 386 # 68072
• ID 69877

HUNT III, H.B. [1979] *Observations on the complexity of regular expression problems* (J 0119) J Comp Syst Sci 19*222-236
◇ D05 D15 ◇ REV MR 81h:68028 Zbl 453 # 68015
• ID 81616

HUNT III, H.B. [1980] see CONSTABLE, R.L.

HUNT III, H.B. & STEARNS, R.E. [1981] *On the equivalence and containment problems for unambiguous regular expressions, grammars, and automata* (P 4235) IEEE Symp Found of Comp Sci (22);1981 Nashville 74-81
◇ D05 ◇ REV MR 84a:68004 • ID 45823

HUNT III, H.B. & ROSENKRANTZ, D.J. [1983] *The complexity of monadic recursion schemes: executability problems, nesting depth, and applications* (J 1426) Theor Comput Sci 27*3-38
◇ D15 D20 ◇ REV Zbl 537 # 68039 • ID 41329

HUNT III, H.B. [1984] *Terminating Turing machine computations and the complexity and/or decidability of correspondence problems, grammars, and program schemes* (J 0037) ACM J 31*299-318
◇ D10 D15 D20 ◇ ID 49571

HUNT III, H.B. & ROSENKRANTZ, D.J. [1984] *The complexity of monadic recursion schemes: Exponential time bounds* (J 0119) J Comp Syst Sci 28*395-419
◇ D15 D20 ◇ REV Zbl 543 # 68034 • ID 40971

HUNT III, H.B. & ROSENKRANTZ, D.J. [1985] *Testing for grammatical coverings* (J 1426) Theor Comput Sci 38/2-3*323-341
◇ D05 ◇ ID 49282

HUNYADVARI, L. [1979] *The L-fuzzy Kleene theorem* (**J** 3955) Ann Univ Budapest, Sect Comp 1979/2*39-48
◇ D05 ◇ REV MR 83i:68053 Zbl 488 # 68042 • ID 39379

HUNYADVARI, L. [1982] *A characteristic of L-fuzzy languages generated by L-fuzzy grammars of type 3 (Russian)* (**S** 0716) Vychisl Tekh Vopr Kibern (Univ Leningrad) 18*147-152,228
◇ B65 D05 ◇ REV MR 84c:68077 Zbl 534 # 68056
• ID 39631

HUNYADVARI, L. [1982] *Some properties of L-regular languages (Russian)* (**S** 0716) Vychisl Tekh Vopr Kibern (Univ Leningrad) 18*133-147,228
◇ D05 ◇ REV MR 84d:68065 Zbl 534 # 68055 • ID 39732

HUPBACH, U.L. [1978] *Rekursive Funktionen in mehrsortigen Peano-Algebren* (**J** 0129) Elektr Informationsverarbeitung & Kybern 14*491-506
◇ D20 D75 ◇ REV MR 80g:03042 Zbl 399 # 03036
• ID 52832

HUREWICZ, W. [1930] *Zur Theorie der analytischen Mengen* (**J** 0027) Fund Math 15*4-17
◇ D55 E15 ◇ REV FdM 56.845 • ID 06346

HUREWICZ, W. see Vol. V for further entries

HURLEY, B. [1980] *A note on the word problem for groups* (**J** 0131) Quart J Math, Oxford Ser 2 31*329-334
◇ D40 ◇ REV MR 81m:20052 Zbl 441 # 20022 • ID 81620

HURWITZ, R.D. [1976] *On the conjugacy problem in a free product with commuting subgroups* (**J** 0043) Math Ann 221*1-8
◇ D40 ◇ REV MR 54 # 414 Zbl 319 # 20041 • ID 23992

HURWITZ, R.D. [1980] *On cyclic subgroups and the conjugacy problem* (**J** 0053) Proc Amer Math Soc 79*1-8
◇ D40 ◇ REV MR 81k:20048 Zbl 441 # 20024 • ID 81622

HURWITZ, R.D. [1984] *A survey of the conjugacy problem* (**C** 3510) Contrib Group Theory (Lyndon) 278-298
◇ D40 ◇ REV MR 86f:20036 Zbl 549 # 20023 • ID 44134

HUTCHINSON, G. [1973] *Recursively unsolvable word problems of modular lattices and diagram chasing* (**J** 0032) J Algeb 26*385-399
◇ D40 G10 ◇ REV MR 48 # 10925 Zbl 272 # 06010
• ID 06349

HUTCHINSON, G. [1977] *Embedding and unsolvability theorems for modular lattices* (**J** 0004) Algeb Universalis 7*47-84
◇ D40 G10 ◇ REV MR 56 # 198 Zbl 376 # 06015
• ID 24174

HUTCHINSON, G. see Vol. III for further entries

HUWIG, H. [1976] *A machine independent description of complexity classes, definable by nondeterministic as well as deterministic Turing machines with primitive recursive tape or time bounds* (**P** 1401) Math Founds of Comput Sci (5);1976 Gdansk 345-351
◇ D10 D15 ◇ REV Zbl 337 # 68028 • ID 62663

HUWIG, H. [1981] *A definition of the P=NP-problem in categories* (**P** 3165) FCT'81 Fund of Comput Th;1981 Szeged 146-153
◇ D15 D75 G30 ◇ REV MR 83h:68056 Zbl 475 # 03016
• ID 55470

HUWIG, H. [1982] *Ein Modell des P=NP-Problems mit einer positiven Loesung (English summary)* (**J** 1431) Acta Inf 17*221-243
◇ D15 G30 ◇ REV MR 84f:03037 Zbl 496 # 03023
• ID 34457

HUYNH, D.T. [1979] *On complexity measures which are induced by probability distributions* (**P** 2935) FCT'79 Fund of Comput Th;1979 Berlin/Wendisch-Rietz 437-442
◇ D15 D20 ◇ REV MR 81b:68046 Zbl 435 # 03032
• ID 55792

HUYNH, D.T. [1983] *Commutative grammars: The complexity of uniform word problems* (**J** 0194) Inform & Control 57*21-39
◇ D05 D15 ◇ REV MR 85j:68060 Zbl 541 # 68044
• ID 41403

HUYNH, D.T. [1984] *Deciding the inequivalence of context-free grammars with 1-letter alphabet is Σ_2^p-complete* (**J** 1426) Theor Comput Sci 33*305-326
◇ D05 D15 ◇ REV Zbl 556 # 68040 • ID 44154

HUYNH, D.T. [1985] *Complexity of the word problem for commutative semigroups of fixed dimension* (**J** 1431) Acta Inf 22/4*421-432
◇ D40 ◇ REV Zbl 564 # 20033 • ID 49539

HUYNH, D.T. see Vol. III for further entries

HUZINO, S. [1959] *On the existence of Sheffer stroke class in the sequential machines* (**J** 0106) Mem Fac Sci, Kyushu Univ, Ser A 13*53-68
◇ B05 D05 ◇ REV MR 21 # 4917 Zbl 114.331 • ID 06355

HUZINO, S. [1959] *Some properties of convolution machines and σ-composite machines* (**J** 0106) Mem Fac Sci, Kyushu Univ, Ser A 13*69-93
◇ D05 ◇ REV MR 21 # 4918 Zbl 114.329 • ID 06356

HUZINO, S. [1959] *Turing transformation and strong computability of Turing computers* (**J** 0106) Mem Fac Sci, Kyushu Univ, Ser A 13*173-195
◇ D10 ◇ REV MR 23 # A2316 Zbl 97.4 • ID 06354

HUZINO, S. [1960] *On some sequential equations* (**J** 0106) Mem Fac Sci, Kyushu Univ, Ser A 14*50-62
◇ D05 ◇ REV MR 22 # 5573 Zbl 116.8 • ID 06358

HUZINO, S. [1960] *On the pseudo universal Turing computer, its structure and programming* (**J** 0106) Mem Fac Sci, Kyushu Univ, Ser A 14*207-316
◇ D10 ◇ REV MR 23 # A3086 Zbl 103.347 • ID 06357

HUZINO, S. [1962] *Finite automata and Asser's function algorithms* (**J** 0068) Z Math Logik Grundlagen Math 8*77-80
◇ D03 D05 D20 ◇ REV MR 26 # 3603 Zbl 192.79
• ID 06359

HUZINO, S. & YONEYAMA, N. [1962] *On a proof of Shepherdson's theorem* (**J** 0106) Mem Fac Sci, Kyushu Univ, Ser A 16*88-93
◇ D05 D10 ◇ REV MR 26 # 4515 Zbl 168.12 JSL 33.628
• ID 18055

HUZINO, S. [1966] *On the simulation of real-time Turing machines by a modified Shepherdson-Sturgis' machine* (**J** 0106) Mem Fac Sci, Kyushu Univ, Ser A 20*16-26
◇ D05 D10 D15 ◇ REV MR 33 # 5493 Zbl 199.41
JSL 33.628 • ID 06361

HUZINO, S. [1966] *Simulatability of finite automata by Shepherdson and Sturgis' machines* (J 0106) Mem Fac Sci, Kyushu Univ, Ser A 20*1-15
◇ D05 D10 ◇ REV MR 33 # 5492 Zbl 199.42 JSL 33.628 • ID 06360

HUZINO, S. & SHIBATA, R. [1977] *An elementary proof of R. W. Ritchie's theorem on the set of squares* (J 0106) Mem Fac Sci, Kyushu Univ, Ser A 31*9-14
◇ D05 ◇ REV MR 56 # 1818 Zbl 358 # 02041 • ID 50495

HUZINO, S. & IIMORI, S. [1979] *On LISP-realization of partial recursive functions* (J 0106) Mem Fac Sci, Kyushu Univ, Ser A 33*127-138
◇ B75 D20 ◇ REV MR 80c:68036 Zbl 401 # 03017 • ID 81633

HYLAND, J.M.E. [1977] see GANDY, R.O.

HYLAND, J.M.E. [1978] *The intrinsic recursion theory on the countable or continuous functionals* (P 1628) Generalized Recursion Th (2);1977 Oslo 135-145
◇ D65 D70 ◇ REV MR 80g:03045 Zbl 453 # 03047 • ID 74350

HYLAND, J.M.E. [1979] *Filter spaces and continuous functionals* (J 0007) Ann Math Logic 16*101-143
◇ D65 G30 ◇ REV MR 81e:03047 Zbl 415 # 03037 • ID 53139

HYLAND, J.M.E. [1982] *The effective topos* (P 3638) Brouwer Centenary Symp;1981 Noordwijkerhout 165-216
◇ D75 D80 F35 F50 G30 ◇ REV MR 84m:03101 Zbl 522 # 03055 • ID 35795

HYLAND, J.M.E. see Vol. III, V, VI for further entries

IBARRA, O.H. [1971] *Characterizations of some tape and time complexity classes of Turing machines in terms of multihead and auxiliary stack automata* (J 0119) J Comp Syst Sci 5*88-117
◇ D10 D15 ◇ REV MR 44 # 1519 Zbl 255 # 68012 JSL 39.188 • ID 06366

IBARRA, O.H. [1972] *A note concerning nondeterministic tape complexities* (J 0037) ACM J 19*608-612
◇ D10 D15 ◇ REV MR 48 # 7659 Zbl 245 # 94044 • ID 62669

IBARRA, O.H. [1974] *A hierarchy theorem for polynomial-space recognition* (J 1428) SIAM J Comp 3*184-187
◇ D10 D15 ◇ REV MR 55 # 6949 Zbl 294 # 02013 • ID 62668

IBARRA, O.H. & MELSON, R.T. [1974] *Some results concerning automata on two-dimensional tapes* (J 0382) Int J Comput Math 4*269-279
◇ D10 ◇ REV MR 51 # 2887 Zbl 358 # 94066 • ID 17380

IBARRA, O.H. & SAHNI, S.K. [1975] *Hierarchies of Turing machines with restricted tape alphabet size* (J 0119) J Comp Syst Sci 11*56-67
◇ D10 D15 ◇ REV MR 51 # 14643 Zbl 307 # 68037 • ID 62670

IBARRA, O.H. & KIM, C.E. [1976] *A useful device for showing the solvability of some decision problems* (J 0119) J Comp Syst Sci 13*153-160
◇ D03 D10 D80 ◇ REV MR 56 # 10149 Zbl 338 # 68046 • ID 62671

IBARRA, O.H. [1977] *Counter machines and number-theoretic problems* (P 4047) Allerton Conf Commun, Control & Comput (15);1977 Monticello 202-210
◇ D10 D35 ◇ REV MR 84b:94001 • ID 46258

IBARRA, O.H. [1978] see GURARI, E.M.

IBARRA, O.H. [1978] *Reversal-bounded multicounter machines and their decision problems* (J 0037) ACM J 25*116-133
◇ D10 ◇ REV MR 57 # 1970 Zbl 365 # 68059 • ID 51079

IBARRA, O.H. [1979] see GURARI, E.M.

IBARRA, O.H. [1979] *Restricted one-counter machines with undecidable universe problems* (J 0041) Math Syst Theory 13*181-186
◇ D10 ◇ REV MR 80m:68044 Zbl 428 # 03038 • ID 53797

IBARRA, O.H. [1980] see GURARI, E.M.

IBARRA, O.H. & LEININGER, B.S. [1981] *Characterizations of Presburger functions* (J 1428) SIAM J Comp 10*22-39
◇ D20 F30 ◇ REV MR 82k:68008 Zbl 471 # 68032 • ID 55261

IBARRA, O.H. & ROSIER, L.E. [1981] *On restricted one-counter machines* (J 0041) Math Syst Theory 14*241-245
◇ D05 ◇ REV MR 82f:68086 Zbl 471 # 68061 • ID 81630

IBARRA, O.H. & ROSIER, L.E. [1981] *On the decidability of equivalence for deterministic pushdown transducers* (J 0232) Inform Process Lett 13*89-93
◇ D10 ◇ REV MR 84j:68052 Zbl 473 # 68077 • ID 39197

IBARRA, O.H. [1981] see GURARI, E.M.

IBARRA, O.H. & MORAN, S. [1982] *On some decision problems for RAM programs* (J 0119) J Comp Syst Sci 24*69-81
◇ D10 ◇ REV MR 83g:68069 Zbl 475 # 68019 • ID 69638

IBARRA, O.H. [1982] see GURARI, E.M.

IBARRA, O.H. [1982] *2DST mappings on languages and related problems* (J 1426) Theor Comput Sci 19*219-227
◇ D05 ◇ REV MR 83h:68129 Zbl 485 # 68072 • ID 39223

IBARRA, O.H. & LEININGER, B.S. [1983] *On the simplification and equivalence problem for straight-line programs* (J 0037) ACM J 30*641-653
◇ D80 ◇ REV MR 85e:68026 • ID 39994

IBARRA, O.H. [1983] see CHAN, TATHUNG

IBARRA, O.H. & ROSIER, L.E. [1983] *Simple programming languages and restricted classes of Turing machines* (J 1426) Theor Comput Sci 26*197-220
◇ B75 D10 ◇ REV MR 85f:68031 Zbl 537 # 68045 • ID 39965

IBARRA, O.H. & MORAN, S. [1983] *Some time-space tradeoff results concerning single-tape and offline TMs* (J 1428) SIAM J Comp 12*388-394
◇ D10 D15 ◇ REV MR 84f:68039 Zbl 512 # 68036 • ID 39660

IBARRA, O.H. & KIM, S.M. [1984] *A characterization of systolic binary tree automata and applications* (J 1431) Acta Inf 21*193-207
◇ B75 D05 ◇ REV Zbl 535 # 68028 • ID 44100

IBARRA, O.H. & KIM, S.M. & ROSIER, L.E. [1984] *Space and time efficient simulations and characterizations of some restricted classes of PDAS* (**P** 4012) Automata, Lang & Progr (11);1984 Antwerpen 247-259
◇ D10 D15 ◇ REV MR 86e:68031 Zbl 568 # 68066
• ID 47156

IBRAGIMOV, M.YA. [1976] *Common fixed points (Russian)* (**J** 3937) Veroyat Met i Kibern (Kazan) 12-13*40-45
◇ D20 ◇ REV MR 58 # 27392 Zbl 398 # 03028 • ID 52756

IDT, J. [1984] *Automates a pile sur des alphabets infinis* (**P** 3565) Symp of Th Aspects of Comput Sci (1);1984 Paris 260-273
◇ D05 ◇ REV MR 86b:68024 Zbl 542 # 68067 • ID 47099

IGARASHI, S. [1964] *A formalization of the descriptions of languages and the related problems in a Gentzen-type formal system* (**J** 0996) RAAG Res Notes 80*44pp
◇ B75 D05 ◇ REV Zbl 199.40 • ID 16226

IGARASHI, S. see Vol. VI for further entries

IGARASHI, Y. [1974] see HONDA, N.

IGARASHI, Y. [1978] *Tape bounds for some subclasses of deterministic context-free languages* (**J** 0194) Inform & Control 37*321-333
◇ D10 D15 ◇ REV MR 58 # 3727 Zbl 376 # 68050
• ID 51710

IGARASHI, Y. see Vol. I for further entries

IHRIG, A.H. [1965] *The Post-Lineal theorems for arbitrary recursively enumerable degrees of unsolvability* (**J** 0047) Notre Dame J Formal Log 6*54-72
◇ D25 ◇ REV MR 33 # 7250 Zbl 132.248 JSL 32.529
• ID 06376

IIDA, S. [1973] see HIROSE, K.

IIDA, S. see Vol. V for further entries

IIMORI, S. [1979] see HUZINO, S.

IKAUNIEKS, E. [1977] see FREJVALD, R.V.

ILYUSHKIN, V.A. [1972] *On the complexity of a grammatical description of context-free languages (Russian)* (**J** 0023) Dokl Akad Nauk SSSR 203*1244-1245
• TRANSL [1972] (**J** 0062) Sov Math, Dokl 13*533-535
◇ D05 ◇ REV MR 47 # 7962 Zbl 299 # 68050 • ID 62677

IMMERMAN, N. [1978] see HARTMANIS, J.

IMMERMAN, N. [1979] *Length of predicate calculus formulas as a new complexity measure* (**P** 3535) IEEE Symp Founds of Comput Sci (20);1979 San Juan 337-347
◇ D15 ◇ REV MR 82b:68035 • ID 81637

IMMERMAN, N. [1981] *Number of quantifiers is better than number of tape cells* (**J** 0119) J Comp Syst Sci 22*384-406
◇ C07 D15 ◇ REV MR 84e:68039 Zbl 486 # 03019
• ID 38079

IMMERMAN, N. [1982] *Upper and lower bounds for first order expressibility* (**J** 0119) J Comp Syst Sci 25*76-98
◇ D15 ◇ REV MR 84a:68041 Zbl 503 # 68032 • ID 36668

IMMERMAN, N. [1983] *Languages which capture complexity classes* (**P** 4195) ACM Symp Th of Comput (15); 347-354
◇ D15 ◇ ID 47722

IMMERMAN, N. [1985] see HARTMANIS, J.

INAGAKI, Y. [1980] see HONDA, N.

INAGAKI, Y. [1981] see HONDA, N.

INAGAKI, Y. [1984] see HONDA, N.

INDERMARK, K. [1968] see BRAUER, W.

INDERMARK, K. [1970] see BOEHLING, K.H.

INDERMARK, K. [1973] see BOEHLING, K.H.

INDERMARK, K. [1973] see HENKE VON, F.W.

INDERMARK, K. [1975] see HENKE VON, F.W.

INDERMARK, K. & REISIG, W. [1976] *On recursively definable relations* (**P** 1913) Lattice Th;1974 Szeged 149-169
◇ D20 D25 ◇ REV MR 55 # 10260 Zbl 365 # 02032
• ID 51033

INDERMARK, K. [1978] see DAMM, W.

INDERMARK, K. [1980] *On rational definitions in complete algebras without rank* (**S** 1642) Schr Inf Angew Math, Ber (Aachen) 64*46pp
◇ D80 ◇ REV Zbl 472 # 68007 • ID 55328

INDERMARK, K. see Vol. III for further entries

INOUE, K. & NAKAMURA, A. [1977] *Some properties of two-dimensional on-line tessellation acceptors* (**J** 0191) Inform Sci 13*95-121
◇ D05 D15 ◇ REV Zbl 371 # 94067 • ID 51402

INOUE, K. & TAKANAMI, I. [1978] *A note on closure properties of the classes of sets accepted by tape-bounded two-dimensional Turing machines* (**J** 0464) Syst-Comp-Controls 9/3*11-19
◇ D10 ◇ REV MR 81i:68071 • ID 81639

INOUE, K. & TAKANAMI, I. [1978] *A note on closure properties of the classes of sets accepted by tape-bounded two-dimensional Turing machines* (**J** 0191) Inform Sci 15*143-158
◇ D10 ◇ REV MR 80f:68059a Zbl 436 # 68031 • ID 81642

INOUE, K. & TAKANAMI, I. [1978] *Closure properties of the classes of sets accepted by automata on a two-dimensional tape under two operations* (**J** 0464) Syst-Comp-Controls 9/6*10-17
◇ D10 ◇ REV MR 81i:68075 • ID 81638

INOUE, K. & TAKANAMI, I. [1978] *Cyclic closure properties of automata on a two-dimensional tape* (**J** 0191) Inform Sci 15*229-242 • ERR/ADD ibid 21*261-262
◇ D10 ◇ REV MR 80e:68144 Zbl 436 # 68032 • ID 69644

INOUE, K. & TAKANAMI, I. [1979] *Closure properties of three-way and four-way tape-bounded two-dimensional Turing machines* (**J** 0191) Inform Sci 18*247-265
◇ D10 ◇ REV MR 80f:68059b Zbl 442 # 68036 • ID 81641

INOUE, K. & TAKANAMI, I. [1979] *Three-way tape-bounded two-dimensional Turing machines* (**J** 0191) Inform Sci 17*195-220
◇ D10 ◇ REV MR 80e:68134 Zbl 442 # 68035 • ID 81643

INOUE, K. & TAKANAMI, I. [1980] *A note on decision problems for three-way two-dimensional finite automata* (**J** 0232) Inform Process Lett 10*245-248
◇ D05 ◇ REV MR 81h:68036 Zbl 454 # 68046 • ID 54256

INOUE, K. & TAKANAMI, I. [1980] *A note on deterministic three-way tape-bounded two-dimensional Turing machines* (**J** 0191) Inform Sci 20*41-55 • ERR/ADD ibid 21*261-262
◇ D10 ◇ REV MR 80i:68039 Zbl 452 # 68064 • ID 69640

INOUE, K. & ITO, A. & TAKANAMI, I. & TANIGUCHI, H. [1982] *Two-dimensional alternating Turing machines with only universal states* (**J** 0194) Inform & Control 55*193-221
◇ D10 ◇ REV MR 85i:68017 Zbl 557 # 68037 • ID 44360

INOUE, K. & TAKANAMI, I. & TANIGUCHI, H. [1983] *Two-dimensional alternating Turing machines* (**J** 1426) Theor Comput Sci 27*61-83
◇ D10 D15 ◇ REV MR 85i:03127 Zbl 539 # 68039 • ID 44210

INOUE, K. & ITO, A. & TAKANAMI, I. & TANIGUCHI, H. [1985] *A space-hierarchy result on two-dimensional alternating Turing machines with only universal states* (**J** 0191) Inform Sci 35*79-90
◇ D10 D15 ◇ REV Zbl 563 # 68045 • ID 44719

INOUE, K. & MATSUNO, H. & TAKANAMI, I. & TANIGUCHI, H. [1985] *Alernating simple multihead finite automata* (**J** 1426) Theor Comput Sci 36*291-298
◇ D05 ◇ ID 47715

INOUE, K. & MATSUNO, H. & TAKANAMI, I. & TANIGUCHI, H. [1985] *Alternating one-way multihead finite automata with only universal states* (**J** 4641) Mem Fac Sci Kochi Univ Ser A Math 6*25-31
◇ D05 ◇ REV MR 86e:68033 • ID 48913

INOUE, K. see Vol. I, VI for further entries

ION, I.D. [1963] *On the connection between the theory of algorithms and the abstract theory of automata* (**J** 0060) Rev Roumaine Math Pures Appl 8*673-682
◇ D05 D20 ◇ REV MR 33 # 5494 Zbl 135.251 • ID 06393

ION, I.D. [1968] *Sur l'axiomatisation de l'algebre des evenements reguliers* (**J** 0070) Bull Soc Sci Math Roumanie, NS 12*41-45
◇ C05 D05 ◇ REV MR 39 # 2640 Zbl 174.290 • ID 06394

IRLAND, M.I. [1973] see FISCHER, P.C.

ISARD, S.D. [1970] see HERMAN, G.T.

ISARD, S.D. [1977] *A finitely axiomatizable undecidable extension of K* (**J** 0105) Theoria (Lund) 43*195-202
◇ B45 D05 D35 ◇ REV MR 58 # 27335 Zbl 391 # 03011 • ID 52338

ISHIBASHI, T. [1980] see DOSHITA, S.

ISHMUKHAMETOV, SH.T. [1982] *Families of recursively enumerable sets (Russian)* (**J** 3937) Veroyat Met i Kibern (Kazan) 18*46-53
◇ D25 D55 ◇ REV MR 85e:03098 Zbl 539 # 03019 • ID 40701

ISHMUKHAMETOV, SH.T. [1983] *On index sets of classes of differences (Russian)* (**J** 0031) Izv Vyssh Ucheb Zaved, Mat (Kazan) 1983/3(250)*78-79
• TRANSL [1983] (**J** 3449) Sov Math 27/3*99-102
◇ D55 ◇ REV MR 85h:03043 Zbl 522 # 03030 • ID 37789

ISHMUKHAMETOV, SH.T. [1985] *Differences of recursively enumerable sets (Russian)* (**J** 0031) Izv Vyssh Ucheb Zaved, Mat (Kazan) 1985/8*3-12,84
◇ D25 ◇ ID 49434

ISKANDER, A.A. [1970] *Word problem for ringoids of numerical functions* (**P** 1417) Open House for Algeb;1970 Aarhus 122-140
◇ D40 ◇ REV MR 43 # 141 Zbl 234 # 08005 • ID 28682

ISKANDER, A.A. [1971] *Word problem for ringoids of numerical functions* (**J** 0064) Trans Amer Math Soc 158*399-408
◇ D40 ◇ REV MR 43 # 6095 Zbl 221 # 08014 • ID 06452

ISKANDER, A.A. see Vol. III for further entries

ISTRAIL, S. [1978] *Tag systems generating Thue irreducible sequences* (**J** 0232) Inform Process Lett 7*129-131
◇ D03 ◇ REV MR 58 # 5110 Zbl 385 # 03031 • ID 52139

ISTRAIL, S. [1979] *On complements of some bounded context-sensitive languages* (**J** 0194) Inform & Control 42*283-289
◇ D05 ◇ REV MR 82b:68063 Zbl 424 # 68043 • ID 81672

ISTRAIL, S. [1982] *Generalization of the Ginsburg-Rice Schuetzenberger fixed-point theorem for context-sensitive and recursive-enumerable languages* (**J** 1426) Theor Comput Sci 18*333-341
◇ D05 ◇ REV MR 83h:68115 Zbl 477 # 68089 • ID 39221

ITKIN, V.E. [1972] *Term-logical equivalence of schemes of programs (Russian)* (**J** 0040) Kibernetika, Akad Nauk Ukr SSR 1972/1*5-27
• TRANSL [1972] (**J** 0021) Cybernetics 8*5-28
◇ B75 D05 ◇ REV MR 52 # 7186 Zbl 242 # 02038 • ID 26352

ITO, A. [1982] see INOUE, K.

ITO, A. [1985] see INOUE, K.

ITO, MAKOTO [1965] *Boolean recursive functions and closure algebra* (**P** 0614) Th Models;1963 Berkeley 431-432
◇ B25 B45 D20 ◇ REV Zbl 147.256 • ID 27530

ITO, MAKOTO see Vol. I, II, III, VI for further entries

ITO, MINORU [1979] see HAGIHARA, K.

ITO, MINORU see Vol. II for further entries

ITO, T. [1974] see ANDO, S.

ITO, T. [1975] see ANDO, S.

ITO, T. see Vol. VI for further entries

ITZHAIK, Y. & YEHUDAI, A. [1983] *On containment problems for finite-turn languages* (**P** 3864) FCT'83 Found of Comput Th;1983 Borgholm 219-231
◇ D10 ◇ REV MR 85e:68003 Zbl 539 # 68070 • ID 47055

IVANOV, A.A. [1983] *Decidability of theories in a certain calculus (Russian)* (**J** 0087) Mat Zametki (Akad Nauk SSSR) 33*617-625
• TRANSL [1983] (**J** 1044) Math Notes, Acad Sci USSR 33*317-321
◇ B25 C10 C55 C60 C80 D35 ◇ REV MR 85c:03013 Zbl 524 # 03023 • ID 37599

IVANOV, A.A. see Vol. III, V, VI for further entries

IVANOV, A.G. [1979] *Theorems on the time hierarchy for random access machines (Russian) (English summary)* (**S** 0228) Zap Nauch Sem Leningrad Otd Mat Inst Steklov 88*62-72,238
• TRANSL [1982] (**J** 1531) J Sov Math 20*2299-2304
◇ D10 D15 ◇ REV MR 81a:68056 Zbl 429 # 03018 • ID 53849

IVANOV, L.L. [1978] *Natural combinatory spaces* (**J** 2547) Serdica, Bulgar Math Publ 4*296-310
◇ B40 D75 ◇ REV MR 80m:03081 Zbl 443 # 03020 • ID 56426

IVANOV, L.L. [1980] *Iterative operatory spaces* (**J** 0137) C R Acad Bulgar Sci 33*735-738
◇ D75 ◇ REV MR 82d:03076a Zbl 449 # 03042 • ID 56709

IVANOV, L.L. [1980] *Some examples of iterative operatory spaces* (**J** 0137) C R Acad Bulgar Sci 33*877-879
◇ D75 ◇ REV MR 82d:03076c Zbl 457 # 03046 • ID 54371

IVANOV, L.L. [1981] *P-recursiveness in iterative combinatorial spaces (Russian)* (**J** 2547) Serdica, Bulgar Math Publ 7*281-297
◇ B40 D75 ◇ REV MR 83h:03066 Zbl 499 # 03039
• ID 36073

IVANOV, L.L. [1984] *Kleene-recursiveness and iterative operative spaces* (**P** 4117) Conf Math Log (Markov);1980 Sofia 47-62
◇ D20 D75 ◇ ID 41125

IVANOV, L.L. see Vol. VI for further entries

IVANOV, N.N. & MIKHAJLOV, G.I. & RUDNEV, V.V. & TAL', A.A. [1984] *Finite automata: equivalence and behaviour (Russian)* (**X** 2027) Nauka: Moskva 192pp
◇ D05 ◇ REV Zbl 534 # 68035 • ID 38361

IVSHIN, V.YU. [1975] *Complete complexity classes of computable functions (Russian)* (**C** 1050) Aktual Vopr Mat Log & Teor Mnozh 33-48
◇ D15 ◇ REV MR 56 # 8340 • ID 74429

IWAMOTO, S. [1977] *A class of inverse theorems on recursive programming with monotonicity* (**J** 3195) J Oper Res Soc Japan 20*94-112
◇ D20 ◇ REV MR 56 # 4846 Zbl 367 # 90124 • ID 51212

IWATA, S. & KASAI, T. [1985] *Gradually intractable problems and non-deterministic log-space lower bounds* (**J** 0041) Math Syst Theory 18*153-170
◇ D15 ◇ ID 47712

IZUMI, H. [1984] see HONDA, N.

IZUMI, M. & TAKAHASHI, N. & TANAKA, H. [1984] *Some fine hierarchies on relativized time-bounded complexity classes* (**P** 3668) Log & Founds of Math;1983 Kyoto 53-78
◇ D15 ◇ ID 42933

JACKOWSKI, J.A. [1973] see HERMAN, G.T.

JACKSON, W.D. [1970] see BARBACK, J.

JACKSON, W.D. [1972] see BARBACK, J.

JACKSON, W.D. [1973] *A note on a theorem of C. Yates* (**J** 0047) Notre Dame J Formal Log 14*100-102
◇ D50 ◇ REV MR 48 # 1905 Zbl 247 # 02039 • ID 06485

JACOB, G. [1977] *Decidabilite de la finitude des demi-groups de matrices* (**P** 3411) Theor Comput Sci (3);1977 Darmstadt 259-269
◇ D80 ◇ REV MR 58 # 22363 Zbl 367 # 68051 • ID 51210

JACOBS, B.E. [1977] *On generalized computational complexity* (**J** 0036) J Symb Logic 42*57-58
◇ D15 D60 ◇ REV MR 58 # 21538 Zbl 384 # 03022
• ID 23744

JACOBS, B.E. [1978] *The α-union theorem and generalized primitive recursion* (**J** 0064) Trans Amer Math Soc 237*63-81
◇ D15 D55 D60 ◇ REV MR 80b:03058 Zbl 381 # 03030
• ID 51898

JACOBS, B.E. [1979] *α-naming and α-speedup theorems* (**J** 0047) Notre Dame J Formal Log 20*241-261
◇ D15 D60 ◇ REV MR 80k:03046 Zbl 363 # 02046
• ID 50873

JACOBS, B.E. [1979] *α-speedable and non α-speedable sets* (**J** 0017) Canad J Math 31*282-299
◇ D15 D60 ◇ REV MR 80k:03047 Zbl 368 # 02041
• ID 51258

JACOBS, B.E. [1981] see HOMER, S.

JACOBS, B.E. see Vol. II for further entries

JACOBS, K. [1970] *Turing-Maschinen und zufaellige 0-1-Folgen* (**S** 1415) Sel Math 2*141-167
• TRANSL [1972] (**C** 1534) Mash Turing & Rek Funk 183-215
◇ D10 D80 ◇ REV MR 44 # 6413 Zbl 219 # 02023
• ID 28085

JACOBS, K. [1972] see EBBINGHAUS, H.-D.

JACOBS, K. (ED.) [1973] *Selecta mathematica II* (**X** 0811) Springer: Heidelberg & New York xi+185pp
◇ D97 F30 ◇ ID 23474

JACOBSSON, C. & STOLTENBERG-HANSEN, V. [1985] *Poincare-Betti series are primitive recursive* (**J** 3172) J London Math Soc, Ser 2 31*1-9
◇ D20 D80 ◇ ID 49264

JACOPINI, G. [1962] see BOEHM, C.

JACOPINI, G. [1966] see BOEHM, C.

JACOPINI, G. [1971] *Una molto semplice caratterizzazione delle funzioni calcolabili operanti su simboli* (**J** 3434) Pubbl Ist Appl Calcolo, Ser 3 65*102-105
◇ D20 ◇ REV Zbl 298 # 68039 • ID 62721

JACOPINI, G. & MENTRASTI, P. [1975] *Funzioni ricorsive di sequenza* (**J** 3434) Pubbl Ist Appl Calcolo, Ser 3 104*39pp
◇ D20 ◇ REV Zbl 423 # 03048 • ID 53560

JACOPINI, G. see Vol. VI for further entries

JAERVI, T. [1970] *On control sets induced by grammars* (**J** 3994) Ann Acad Sci Fennicae Ser A I 480*7pp
◇ D05 ◇ REV MR 45 # 6537 Zbl 218 # 68016 • ID 06495

JA'JA', J. [1979] *On the complexity of bilinear forms with commutativity* (**P** 3542) ACM Symp Th of Comput (11);1979 Atlanta 197-208
◇ D15 ◇ REV MR 81d:68058 • ID 81686

JA'JA', J. see Vol. III, V for further entries

JAMBU-GIRAUDET, M. [1980] *Theorie des modeles de groupes d'automorphismes d'ensembles totalement ordonnes 2-homogenes (English summary)* (**J** 2313) C R Acad Sci, Paris, Ser A-B 290*A1037-A1039
◇ C07 C60 C62 C65 D35 E07 F25 F35 ◇ REV MR 81f:03046 Zbl 453 # 03037 • ID 54167

JAMBU-GIRAUDET, M. [1981] *Interpretations d'arithmetiques dans des groupes et des treillis* (**P** 3404) Model Th & Arithm;1979/80 Paris 143-153
◇ C07 C62 C65 D35 E07 F25 F30 G10 ◇ REV MR 83g:03064 Zbl 476 # 03023 • ID 55535

JAMBU-GIRAUDET, M. see Vol. III, V for further entries

JAMIESON, D.W. & TEUTSCH, R.J. [1974] *Hockett on effective computability* (**J** 0293) Found Lang 11*287-293
◇ B65 D20 ◇ REV MR 57 # 9467 • ID 79293

JANEVA, B. [1985] see CUPONA, G.

JANICKI, R. [1980] *Some remarks on deterministic Mazurkiewicz algorithms and languages associated with them* (**J** 2095) Fund Inform, Ann Soc Math Pol, Ser 4 3*65-75
◇ D05 ◇ REV MR 82g:68070 Zbl 445 # 68023 • ID 81690

JANICZAK, A. [1953] *Undecidability of some simple formalized theories* (J 0027) Fund Math 40*131-139
◇ B25 D35 ◇ REV MR 15.669 Zbl 52.252 JSL 22.217
• ID 06508

JANICZAK, A. [1954] *On the reducibility of decision problems* (S 0019) Colloq Math (Warsaw) 3*33-36
◇ D10 D20 D25 D30 ◇ REV MR 15.925 Zbl 56.12
JSL 21.100 • ID 06510

JANICZAK, A. [1954] *Some remarks on partially recursive functions* (S 0019) Colloq Math (Warsaw) 3*37-38
◇ D20 ◇ REV MR 15.925 Zbl 56.12 JSL 21.100 • ID 06509

JANICZAK, A. see Vol. III for further entries

JANIGA, L. [1981] *Another hierarchy defined by multihead finite automata* (P 3429) Math Founds of Comput Sci (10);1981 Strbske Pleso 314-320
◇ D05 ◇ REV MR 83e:68052 Zbl 466#68069 • ID 40162

JANSSEN, D. & MONIEN, B. [1977] *Ueber die Komplexitaet der Fehlerdiagnose bei Systemen* (J 1046) Z Angew Math Mech 57*T315-T317
◇ B70 D15 ◇ REV MR 56#4238 Zbl 381#03047
• ID 51915

JANSSENS, D. & ROZENBERG, G. [1981] *Decision problems for node label controlled graph grammars* (J 0119) J Comp Syst Sci 22*144-177
◇ D05 ◇ REV MR 83c:68076 Zbl 466#68067 • ID 39053

JANTKE, K.P. [1976] *Zur Kompliziertheit der konsistenten Erkennung von Klassen allgemein-rekursiver Funktionen* (P 2898) Algor Kompl, Lern-& Erkenn-Prozess;1976 Jena 45-52
◇ D15 D20 ◇ REV MR 58#5119 Zbl 439#03017
• ID 56009

JANTKE, K.P. [1978] *Universal methods for identification of total recursive functions* (S 2829) Rostocker Math Kolloq 10*63-69
◇ D20 ◇ REV MR 81c:03032 Zbl 437#03019 • ID 55887

JANTKE, K.P. [1979] *Automatic synthesis of programs and inductive inference of functions* (P 2935) FCT'79 Fund of Comput Th;1979 Berlin/Wendisch-Rietz 219-225
◇ D20 ◇ REV Zbl 437#03020 • ID 55888

JANTKE, K.P. [1979] *Natural properties of strategies identifying recursive functions* (J 0129) Elektr Informationsverarbeitung & Kybern 15*487-496
◇ B75 D20 D45 ◇ REV MR 81d:68014 Zbl 448#68006
• ID 56663

JANTKE, K.P. [1981] see BEICK, H.-R.

JANTKE, K.P. [1983] *The recursive power of algebraic semantics* (S 3231) Prepr NF, Sekt Math, Humboldt-Univ Berlin 57*33pp
◇ B75 D20 ◇ REV Zbl 521#68023 • ID 37474

JANTKE, K.P. [1984] *Polynomial time inference of general pattern languages* (P 3565) Symp of Th Aspects of Comput Sci (1);1984 Paris 314-325
◇ D05 D15 D80 ◇ REV MR 86b:68025 Zbl 545#68078
• ID 41210

JANTKE, K.P. see Vol. II for further entries

JANTZEN, M. [1981] *The power of synchronizing operations on strings* (J 1426) Theor Comput Sci 14*127-154
◇ D05 D25 ◇ REV MR 82e:68029 Zbl 477#68034
• ID 81694

JANTZEN, M. & KUDLEK, M. [1982] *Homomorphic images of sentential form languages defined by semi-Thue systems* (P 3767) Found of Softw Tech & Th Comput Sci (2);1982 Bangalore 126-135
◇ D03 D05 ◇ REV MR 84b:68098 Zbl 542#68060
• ID 38957

JANTZEN, M. [1982] see BOOK, R.V.

JANTZEN, M. [1983] see BRAMHOFF, H.

JANTZEN, M. & KUDLEK, M. [1984] *Homomorphic images of sentential form languages defined by semi-Thue systems* (J 1426) Theor Comput Sci 33*13-43
◇ D03 ◇ REV Zbl 542#68059 • ID 45430

JANTZEN, M. [1984] *Thue systems and the Church-Rosser property* (P 3658) Math Founds of Comput Sci (11);1984 Prague 80-95
◇ D03 ◇ REV Zbl 553#03025 • ID 43324

JANTZEN, M. [1985] *A note on a special one-rule semi-Thue system* (J 0232) Inform Process Lett 21*135-140
◇ D03 ◇ ID 49475

JANTZEN, M. [1985] *Extending regular expressions with iterated shuffle* (J 1426) Theor Comput Sci 38/2-3*223-247
◇ D05 ◇ ID 49275

JARNIK, V. [1938] *Sur un probleme de M.Cech* (J 0953) Vest Cesk Spol Nauk 1938*7pp
◇ D70 E10 E20 ◇ REV FdM 64.933 • ID 41088

JARZABEK, S. [1972] *Product k-machines and their classification (Polish) (Russian and English summaries)* (J 1929) Prace Centr Oblicz Pol Akad Nauk 86*40pp
◇ D05 ◇ REV Zbl 261#02020 • ID 30463

JASKOWSKI, S. [1948] *Sur le probleme de decision de la topologie et de la theorie des groups* (S 0019) Colloq Math (Warsaw) 1*176-178
◇ D35 D40 ◇ REV Zbl 37.296 • ID 06545

JASKOWSKI, S. [1954] *Example of a class of systems of ordinary differential equations having no decision method for existence problems* (J 0014) Bull Acad Pol Sci, Ser Math Astron Phys 2*155-157
◇ D80 ◇ REV MR 16.103 Zbl 56.13 JSL 28.103 • ID 06552

JASKOWSKI, S. [1956] *Indecidability of first order sentences in the theory of free groupoids* (J 0027) Fund Math 43*36-45
◇ D35 D40 ◇ REV MR 18.271 Zbl 75.6 JSL 23.445
• ID 06553

JASKOWSKI, S. see Vol. I, II, V, VI for further entries

JAULIN, B. [1966] *Sur un aspect du calcul* (J 0392) Math Sci Hum 14*1-8
◇ B75 D15 D80 ◇ REV MR 36#53 • ID 06561

JAULIN, B. [1973] see AZRA, J.-P.

JAYNE, J.E. [1977] see DAVIES, R.O.

JAYNE, J.E. [1980] see DELLACHERIE, C.

JAYNE, J.E. & ROGERS, C.A. [1982] *The invariance of the absolute Borel classes* (P 3888) Convex Anal & Optim (Ioffe);1980 London 118-151
◇ D55 E15 E75 ◇ REV MR 83g:54052 Zbl 492#54022
• ID 37743

JAYNE, J.E. [1985] see HANSELL, R.W.

JAYNE, J.E. see Vol. V for further entries

JAZAYERI, M. [1981] *A simpler construction for showing the intrinsically exponential complexity of the circularity problem for attribute grammars* (**J** 0037) ACM J 28*715-720
◇ D05 D15 ◇ REV MR 83k:68030 Zbl 468 # 68056
• ID 40302

JEANROND, H. [1980] *Deciding unique termination of permutative rewriting systems: choose your term algebra carefully* (**P** 3063) Autom Deduct (5);1980 Les Arcs 335-355
◇ D05 ◇ REV MR 81i:68006 Zbl 438 # 68004 • ID 55983

JEANROND, H. see Vol. I for further entries

JECH, T.J. [1976] see HARRINGTON, L.A.

JECH, T.J. see Vol. III, V for further entries

JEDRZEJOWICZ, J. [1976] *One-one degrees of Turing machines decision problems* (**P** 1401) Math Founds of Comput Sci (5);1976 Gdansk 45*285-289
◇ D10 D25 ◇ REV Zbl 341 # 02034 • ID 29754

JEDRZEJOWICZ, J. [1979] *Inverse derivability and confluence problems of deterministic combinatorial systems* (**P** 2935) FCT'79 Fund of Comput Th;1979 Berlin/Wendisch-Rietz 226-229
◇ D25 ◇ REV MR 81d:03045 Zbl 417 # 03015 • ID 53252

JEDRZEJOWICZ, J. [1979] *One-one degrees of combinatorial systems decision problems* (**J** 3293) Bull Acad Pol Sci, Ser Math 27*819-821
◇ D25 ◇ REV MR 82k:03066 Zbl 464 # 03038 • ID 74493

JEDRZEJOWICZ, J. [1981] *Undecidable problems associated with combinatorial systems and their one-one degrees of unsolvability* (**J** 0068) Z Math Logik Grundlagen Math 27*453-462
◇ D20 D25 ◇ REV MR 83c:03039 Zbl 473 # 03034
• ID 55363

JEFFERSON, D. & SUZUKI, N. [1977] *Verification decidability of Presburger array programs* (**P** 3238) Conf Theoret Comput Sci;1977 Waterloo ON 202-212
◇ B75 D80 ◇ REV MR 58 # 13880 Zbl 411 # 68017
• ID 52924

JEFFREY, R.C. [1974] see BOOLOS, G.

JEFFREY, R.C. see Vol. I, II for further entries

JENSEN, C.U. [1982] *Sur une classe de corps indecidables (English summary)* (**J** 3364) C R Acad Sci, Paris, Ser 1 295*507-509
◇ C60 D35 ◇ REV MR 83m:12046 Zbl 514 # 03020
• ID 37407

JENSEN, C.U. [1983] *L'indecidabilite d'une classe de corps des fonctions meromorphes* (**J** 2128) C R Math Acad Sci, Soc Roy Canada 5*69-74
◇ D35 D80 ◇ REV MR 84g:03062 Zbl 515 # 12019
• ID 34169

JENSEN, C.U. [1984] *Theorie des modeles pour des anneaux de fonctions entieres et des corps de fonctions meromorphes* (**S** 3521) Mem Soc Math Fr 16*23-40
◇ C60 C65 D35 F35 ◇ REV Zbl 562 # 12025 • ID 39754

JENSEN, C.U. see Vol. III, V for further entries

JENSEN, F.V. & LARSEN, K.G. [1985] *Recursively defined domains and their induction principles* (**P** 4672) Found of Softw Tech & Th Comput Sci (5);1985 New Delhi 225-245
◇ B75 D20 ◇ ID 49753

JENSEN, F.V. see Vol. I, II, III, V for further entries

JENSEN, R.B. [1967] *Modelle der Mengenlehre* (**S** 3301) Lect Notes Math 37*viii+176pp
◇ C62 D55 E25 E35 E40 E45 E50 E98 ◇ REV MR 36 # 4982 Zbl 191.299 JSL 40.92 • ID 06596

JENSEN, R.B. [1968] see FRIEDMAN, H.M.

JENSEN, R.B. [1970] *Definable sets of minimal degree* (**P** 1072) Math Log & Founds of Set Th;1968 Jerusalem 122-128
◇ D30 D55 E35 E45 E47 ◇ REV MR 46 # 5130
Zbl 245 # 02055 • ID 28527

JENSEN, R.B. & SOLOVAY, R.M. [1970] *Some applications of almost disjoint sets* (**P** 1072) Math Log & Founds of Set Th;1968 Jerusalem 84-104
◇ C62 D55 E05 E35 E45 E55 ◇ REV MR 44 # 6482
Zbl 222 # 02077 • ID 22245

JENSEN, R.B. & KARP, C.R. [1971] *Primitive recursive set functions* (**P** 0693) Axiomatic Set Th;1967 Los Angeles 1*143-176
◇ D60 D65 E45 E47 ◇ REV MR 43 # 7317
Zbl 221 # 02027 JSL 40.505 • ID 06600

JENSEN, R.B. & JOHNSBRAATEN, H. [1974] *A new construction of a non-constructible Δ^1_3 subset of ω* (**J** 0027) Fund Math 81*279-290
◇ D55 E35 E45 ◇ REV MR 54 # 7253 Zbl 289 # 02048
• ID 62767

JENSEN, R.B. see Vol. III, V, VI for further entries

JEROSLOW, R. [1975] *Experimental logics and Δ^0_2-theories* (**J** 0122) J Philos Logic 4*253-267
◇ B60 D55 ◇ REV MR 58 # 21558 Zbl 319 # 02024
• ID 28273

JEROSLOW, R. see Vol. III, VI for further entries

JERVELL, H.R. [1985] *Recursion on homogeneous trees* (**J** 0068) Z Math Logik Grundlagen Math 31*295-298
◇ D75 F15 ◇ REV Zbl 567 # 03027 • ID 47555

JERVELL, H.R. see Vol. I, II, III, V, VI for further entries

JIRKU, P. [1975] *Towards an integrated theory of formal and natural languages* (**J** 0156) Kybernetika (Prague) 11*91-100
◇ B65 D05 ◇ REV MR 54 # 4912 Zbl 304 # 68068
• ID 24088

JIRKU, P. see Vol. I for further entries

JOCKUSCH JR., C.G. [1968] *Semirecursive sets and positive reducibility* (**J** 0064) Trans Amer Math Soc 131*420-436
◇ D25 D30 ◇ REV MR 36 # 3649 Zbl 198.324 • ID 06618

JOCKUSCH JR., C.G. [1968] *Supplement to Boone's "Algebraic systems"* (**P** 0608) Logic Colloq;1966 Hannover 34-36
◇ D25 D40 ◇ REV MR 38 # 5899 • ID 06617

JOCKUSCH JR., C.G. [1968] *Uniformly introreducible sets* (**J** 0036) J Symb Logic 33*521-536
◇ D25 D30 ◇ REV MR 38 # 5619 Zbl 165.19 • ID 06619

JOCKUSCH JR., C.G. & MCLAUGHLIN, T.G. [1969] *Countable retracing functions and Π^0_2 predicates* (**J** 0048) Pac J Math 30*67-93
◇ D30 D55 ◇ REV MR 42 # 4403 Zbl 181.306 • ID 06622

JOCKUSCH JR., C.G. [1969] *Relationships between reducibilities* (**J** 0064) Trans Amer Math Soc 142*229-237
◇ D25 D30 ◇ REV MR 39 # 6747 Zbl 188.26 • ID 06620

JOCKUSCH JR., C.G. [1969] *The degrees of bi-immune sets* (**J** 0068) Z Math Logik Grundlagen Math 15*135-140
◇ D30 ◇ REV MR 39 # 5360 Zbl 184.20 • ID 06621

JOCKUSCH JR., C.G. [1969] *The degrees of hyperhyperimmune sets* (**J** 0036) J Symb Logic 34*489-493
◇ D30 D50 ◇ REV MR 40 # 5445 Zbl 181.306 • ID 06624

JOCKUSCH JR., C.G. & SOARE, R.I. [1970] *Minimal covers and arithmetical sets* (**J** 0053) Proc Amer Math Soc 25*856-859
◇ D30 D55 ◇ REV MR 42 # 67 Zbl 205.10 • ID 06625

JOCKUSCH JR., C.G. & SOARE, R.I. [1971] *A minimal pair of Π^0_1 classes* (**J** 0036) J Symb Logic 36*66-78
◇ D25 D30 ◇ REV MR 44 # 68 Zbl 219 # 02033
• ID 06627

JOCKUSCH JR., C.G. & SOARE, R.I. [1972] *Π^0_1 classes and degrees of theories* (**J** 0064) Trans Amer Math Soc 173*33-56
◇ D25 D30 ◇ REV MR 47 # 4775 Zbl 262 # 02041
• ID 06631

JOCKUSCH JR., C.G. [1972] *A reducibility arising from the Boone groups* (**J** 0132) Math Scand 31*262-266
◇ D25 D30 D40 ◇ REV MR 48 # 78 Zbl 263 # 02023
• ID 06636

JOCKUSCH JR., C.G. [1972] *Degrees in which the recursive sets are uniformly recursive* (**J** 0017) Canad J Math 24*1092-1099
◇ D25 D30 ◇ REV MR 48 # 83 Zbl 221 # 02029
• ID 06630

JOCKUSCH JR., C.G. & SOARE, R.I. [1972] *Degrees of members of Π^0_1 classes* (**J** 0048) Pac J Math 40*605-616
◇ D25 D30 F30 ◇ REV MR 46 # 8827 Zbl 232 # 02031
• ID 06633

JOCKUSCH JR., C.G. [1972] *Ramsey's theorem and recursion theory* (**J** 0036) J Symb Logic 37*268-280
◇ D30 D55 D80 ◇ REV MR 51 # 12495 Zbl 262 # 02042
• ID 06629

JOCKUSCH JR., C.G. [1972] *Upward closure of bi-immune degrees* (**J** 0068) Z Math Logik Grundlagen Math 18*285-287
◇ D30 D50 ◇ REV MR 46 # 1574 Zbl 257 # 02033
• ID 06635

JOCKUSCH JR., C.G. [1973] *An application of Σ^0_4 determinacy to the degrees of unsolvability* (**J** 0036) J Symb Logic 38*293-294
◇ D30 ◇ REV MR 48 # 10786 Zbl 279 # 02026 • ID 06640

JOCKUSCH JR., C.G. & SOARE, R.I. [1973] *Encodability of Kleene's O* (**J** 0036) J Symb Logic 38*437-440
◇ D30 D55 D65 ◇ REV MR 51 # 97 Zbl 279 # 02025
• ID 17322

JOCKUSCH JR., C.G. & SOARE, R.I. [1973] *Post's problem and his hypersimple set* (**J** 0036) J Symb Logic 38*446-452
◇ D25 ◇ REV MR 49 # 2313 Zbl 279 # 02023 • ID 06638

JOCKUSCH JR., C.G. [1973] *Upward closure and cohesive degrees* (**J** 0029) Israel J Math 15*332-335
◇ D30 ◇ REV MR 50 # 76 Zbl 279 # 02024 • ID 06637

JOCKUSCH JR., C.G. [1974] *Π^0_1 classes and boolean combinations of recursively enumerable sets* (**J** 0036) J Symb Logic 39*95-96
◇ D25 D30 ◇ REV MR 49 # 8834 Zbl 286 # 02045
• ID 06641

JOCKUSCH JR., C.G. [1975] see COHEN, P.E.

JOCKUSCH JR., C.G. [1975] *Recursiveness of initial segments of Kleene's O* (**J** 0027) Fund Math 87*161-167
◇ D20 F15 ◇ REV MR 58 # 189 Zbl 338 # 02024
• ID 06642

JOCKUSCH JR., C.G. & SIMPSON, S.G. [1976] *A degree theoretic definition of the ramified analytical hierarchy* (**J** 0007) Ann Math Logic 10*1-32
◇ D30 D55 E40 E45 H05 ◇ REV MR 58 # 10370 Zbl 333 # 02039 • ID 18212

JOCKUSCH JR., C.G. & PATERSON, M.S. [1976] *Completely autoreducible degrees* (**J** 0068) Z Math Logik Grundlagen Math 22*571-575
◇ D25 D30 ◇ REV MR 56 # 107 Zbl 384 # 03026
• ID 52068

JOCKUSCH JR., C.G. [1977] see HENSON, C.W.

JOCKUSCH JR., C.G. & SOLOVAY, R.M. [1977] *Fixed points of jump preserving automorphisms of degrees* (**J** 0029) Israel J Math 26*91-94
◇ D30 ◇ REV MR 55 # 5422 Zbl 372 # 02023 • ID 26091

JOCKUSCH JR., C.G. [1977] *Simple proofs of some theorems on high degrees of unsolvability* (**J** 0017) Canad J Math 29*1072-1080
◇ D30 ◇ REV MR 57 # 16023 Zbl 342 # 02030 • ID 50230

JOCKUSCH JR., C.G. & POSNER, D.B. [1978] *Double jumps of minimal degrees* (**J** 0036) J Symb Logic 43*715-724
◇ D30 ◇ REV MR 80d:03042 Zbl 411 # 03034 • ID 52888

JOCKUSCH JR., C.G. [1980] *Degrees of generic sets* (**P** 3021) Logic Colloq;1979 Leeds 110-139
◇ D30 E40 ◇ REV MR 83i:03070 Zbl 457 # 03042
• ID 54367

JOCKUSCH JR., C.G. [1980] *Fine degrees of word problems of cancellation semigroups* (**J** 0068) Z Math Logik Grundlagen Math 26*93-95
◇ D25 D40 ◇ REV MR 81j:03071 Zbl 443 # 03019
• ID 56425

JOCKUSCH JR., C.G. & POSNER, D.B. [1981] *Automorphism bases for degrees of unsolvability* (**J** 0029) Israel J Math 40*150-164
◇ D30 ◇ REV MR 83j:03072 Zbl 486 # 03025 • ID 35368

JOCKUSCH JR., C.G. [1981] *Three easy constructions of recursively enumerable sets* (**P** 2628) Log Year;1979/80 Storrs 83-91
◇ D25 ◇ REV MR 83a:03036 Zbl 472 # 03031 • ID 55296

JOCKUSCH JR., C.G. & SHORE, R.A. [1983] *Pseudo jump operators I: The r.e. case* (**J** 0064) Trans Amer Math Soc 275*599-609
◇ D25 D30 ◇ REV MR 84c:03081 MR 86e:03030 Zbl 514 # 03028 • REM Part II 1984 • ID 33593

JOCKUSCH JR., C.G. [1984] see AMBOS-SPIES, K.

JOCKUSCH JR., C.G. [1984] see CHONG, C.T.

JOCKUSCH JR., C.G. & SHORE, R.A. [1984] *Pseudo jump operators II: Transfinite iterations, hierarchies and minimal covers* (**J** 0036) J Symb Logic 49*1205-1236
◇ D25 D30 D55 ◇ REV MR 86g:03072 Zbl 574 # 03026
• REM Part I 1983 • ID 33594

JOCKUSCH JR., C.G. & KALANTARI, I. [1984] *Recursively enumerable sets and van der Waerden's theorem on arithmetic progressions* (J 0048) Pac J Math 115*143-153
◇ C57 D25 D80 ◇ REV MR 86d:03045 Zbl 571 # 03017 • ID 39945

JOCKUSCH JR., C.G. & MOHRHERR, J. [1985] *Embedding the diamond lattice in the recursively enumerable truth-table degrees* (J 0053) Proc Amer Math Soc 94*123-128
◇ D25 ◇ REV Zbl 556 # 03038 • ID 44708

JOCKUSCH JR., C.G. [1985] *Genericity for recursively enumerable sets* (P 3342) Rec Th Week;1984 Oberwolfach 203-232
◇ D25 E40 ◇ ID 45306

JOCKUSCH JR., C.G. & SHORE, R.A. [1985] *REA operators, R.E. degrees and minimal covers* (P 4046) Rec Th;1982 Ithaca 3-11
◇ D30 ◇ ID 39949

JOHANSEN, P. [1979] *The generating function of the number of subpatterns of a DOL sequence* (J 1426) Theor Comput Sci 8*57-68
◇ D05 ◇ REV MR 80h:68061 Zbl 419 # 68098 • ID 53399

JOHNSBRAATEN, H. [1974] see JENSEN, R.B.

JOHNSBRAATEN, H. see Vol. V for further entries

JOHNSON, D.S. [1976] see GAREY, M.R.

JOHNSON, D.S. [1979] see GAREY, M.R.

JOHNSON, D.S. [1981] *The NP-completeness column: an ongoing guide* (J 2746) J Algor 2*393-405
◇ D15 ◇ REV MR 83f:68039 Zbl 494 # 68047 • REM This is a series of papers: See also 1982ff. • ID 39358

JOHNSON, D.S. [1982] *The NP-completeness column: an ongoing guide* (J 2746) J Algor 3*88-99,182-195,288-300,381-395
◇ D15 D98 ◇ REV MR 83i:68072 Zbl 494 # 68048 Zbl 502 # 68007 • REM This is a series of papers: See also 1981 & 1983 • ID 39359

JOHNSON, D.S. [1983] *The NP-completeness column: an ongoing guide* (J 2746) J Algor 4*87-100,189-203,286-300,397-411
◇ D15 ◇ REV MR 84e:68040 MR 84g:68026 MR 85g:68028 MR 85g:68029a Zbl 509 # 68035 • REM This is a series of papers: See also 1982 & 1984 • ID 46386

JOHNSON, D.S. [1984] *The NP-completeness column: an ongoing guide* (J 2746) J Algor 5*147-160,284-299,433-447,595-609
◇ D15 ◇ REV MR 85g:68029b MR 86a:68038 Zbl 547 # 68048 • REM This is a series of papers: See also 1983 & 1985 • ID 47875

JOHNSON, D.S. [1985] *The NP-completeness column: an ongoing guide* (J 2746) J Algor 6*145-159,291-305,434-451
◇ D15 ◇ REV Zbl 562 # 68032 • REM This is a series of papers: See also 1984 • ID 47876

JOHNSON, N. [1977] *Rice theorems for Σ_n^{-1} sets* (J 0017) Canad J Math 29*794-805
◇ D25 D30 D55 ◇ REV MR 55 # 12493 Zbl 342 # 02026 • ID 50101

JOHNSON, N. [1978] *Classifications of generalized index sets of open classes* (J 0036) J Symb Logic 43*694-714
◇ D25 D30 ◇ REV MR 80h:03063 Zbl 449 # 03035 • ID 56702

JOHNSON, N. [1979] see HAY, L.

JONES, JAMES P. [1969] *Effectively retractable theories and degrees of undecidability* (J 0036) J Symb Logic 34*597-604
◇ D25 ◇ REV MR 42 # 4399 Zbl 195.305 • ID 06696

JONES, JAMES P. [1969] *Independent recursive axiomatizability in arithmetic* (J 0053) Proc Amer Math Soc 23*107-113
◇ D25 F30 ◇ REV MR 41 # 1534 Zbl 186.8 • ID 06695

JONES, JAMES P. [1974] *Recursive undecidability - an exposition* (J 0005) Amer Math Mon 81*724-738
◇ D10 D20 D35 D40 D98 ◇ REV MR 50 # 9568 Zbl 326 # 02034 • ID 06697

JONES, JAMES P. & SATO, D. & WADA, H. & WIENS, D. [1976] *Diophantine representation of the set of prime numbers* (J 0005) Amer Math Mon 83*449-464
◇ D25 ◇ REV MR 54 # 2615 Zbl 336 # 02037 • ID 24054

JONES, JAMES P. [1978] *Three universal representations of recursively enumerable sets* (J 0036) J Symb Logic 43*335-351
◇ D25 F20 ◇ REV MR 58 # 16226 Zbl 414 # 03025 • ID 29266

JONES, JAMES P. [1979] *Diophantine representation of Mersenne and Fermat primes* (J 0399) Acta Arith, Pol Akad Nauk 35*209-221
◇ D25 D35 ◇ REV MR 81a:10020 Zbl 414 # 03029 • ID 53075

JONES, JAMES P. [1980] *Undecidable Diophantine equations* (J 0589) Bull Amer Math Soc (NS) 3*859-862
◇ D25 D35 ◇ REV MR 81k:10094 Zbl 442 # 03028 • ID 56385

JONES, JAMES P. [1981] *Classification of quantifier prefixes over Diophantine equations* (J 0068) Z Math Logik Grundlagen Math 27*403-410
◇ D25 D35 ◇ REV MR 83a:03037 Zbl 472 # 03034 • ID 55299

JONES, JAMES P. & MATIYASEVICH, YU.V. [1982] *A new representation for the symmetric binomial coefficient and its applications (French summary)* (J 2660) Ann Sci Math Quebec 6*81-97 • ERR/ADD ibid 6*223
◇ D25 ◇ REV MR 84g:03060a Zbl 499 # 03028 • ID 34168

JONES, JAMES P. & MATIYASEVICH, YU.V. [1982] *Exponential diophantine representation of recursively enumerable sets (French summary)* (P 3708) Herbrand Symp Logic Colloq;1981 Marseille 159-177
◇ D20 D25 D35 ◇ REV MR 85i:03138 Zbl 499 # 03027 • ID 38122

JONES, JAMES P. [1982] see HUBER-DYSON, V.

JONES, JAMES P. [1982] *Some undecidable determined games* (J 3184) Int J Game Theory 11*63-70
◇ D80 E60 ◇ REV MR 84b:90107 Zbl 498 # 90090 • ID 38973

JONES, JAMES P. [1982] *Universal diophantine equation* (J 0036) J Symb Logic 47*549-571
◇ D25 ◇ REV MR 84e:10070 Zbl 492 # 03018 • ID 38095

JONES, JAMES P. & MATIYASEVICH, YU.V. [1984] *Register machine proof of the theorem on exponential diophantine representation of enumerable sets* (J 0036) J Symb Logic 49*818-829
◇ D10 D25 D35 ◇ REV MR 85i:03139 JSL 51.478 • ID 42466

JONES, JAMES P. see Vol. VI for further entries

JONES, N.D. [1968] *Classes of automata and transitive closure* (J 0194) Inform & Control 13*207-229
◇ D10 D15 ◇ REV MR 39 # 8023 Zbl 184.23 • ID 06698

JONES, N.D. [1969] *Context-free languages and rudimentary attributes* (J 0041) Math Syst Theory 3*102-109
◇ D05 ◇ REV MR 41 # 9671 Zbl 179.22 • ID 06699

JONES, N.D. & SELMAN, A.L. [1972] *Turing machines and the spectra of first-order formulas with equality* (P 1901) ACM Symp Th of Comput (4);1972 Denver 157-167
◇ C13 D10 D15 ◇ REV Zbl 381 # 03026 • ID 51894

JONES, N.D. [1973] *Computability theory: an introduction* (X 0801) Academic Pr: New York xiv + 154pp
◇ D98 ◇ REV MR 51 # 88 Zbl 291 # 02025 • ID 15233

JONES, N.D. & SELMAN, A.L. [1974] *Turing machines and the spectra of first-order formulas* (J 0036) J Symb Logic 39*139-150
◇ C13 D10 D15 ◇ REV MR 58 # 5164 Zbl 288 # 02021 • ID 06700

JONES, N.D. [1975] *Space-bounded reducibility among combinatorial problems* (J 0119) J Comp Syst Sci 11*68-85 • ERR/ADD ibid 15*241
◇ D15 ◇ REV MR 53 # 2020 Zbl 317 # 02039 • ID 29647

JONGH DE, D.H.J. & PARIKH, R. [1977] *Well-partial orderings and hierarchies* (J 0028) Indag Math 39*195-207
◇ D55 E07 E10 ◇ REV MR 56 # 5371 Zbl 435 # 06004 • ID 26064

JONGH DE, D.H.J. see Vol. I, II, III, VI for further entries

JONSSON, B. & MCNULTY, G.F. & QUACKENBUSH, R.W. [1975] *The ascending and descending varietal chains of a variety* (J 0017) Canad J Math 27*25-31
◇ B25 C05 D35 ◇ REV MR 50 # 12860 Zbl 305 # 08003 • ID 06740

JONSSON, B. see Vol. III, V for further entries

JOSEPH, D. & YOUNG, P. [1981] *A survey of some recent results on computational complexity in weak theories of arithmetic* (P 3429) Math Founds of Comput Sci (10);1981 Strbske Pleso 46-60
◇ D15 F30 ◇ REV MR 84j:03111 Zbl 481 # 03025 • ID 34701

JOSEPH, D. & YOUNG, P. [1981] *Independence results in computer science?* (J 0119) J Comp Syst Sci 23*205-222 • ERR/ADD ibid 24*378
◇ B75 D80 ◇ REV MR 82m:68060 MR 84c:68017 Zbl 474 # 68046 • ID 55451

JOSEPH, D. [1983] *Polynomial time computations in models of ET* (J 0119) J Comp Syst Sci 26*311-338
◇ D15 F30 ◇ REV MR 84k:68034 Zbl 542 # 03019 • ID 39245

JOSEPH, D. & YOUNG, P. [1985] *Some remarks on witness functions for polynomial and noncomplete sets in NP* (J 1426) Theor Comput Sci 39*225-237
◇ D15 ◇ ID 49060

JOSHI, A.K. & KOSARAJU, S.RAO & YAMADA, H. [1972] *String adjunct grammars I: Local and distributed adjunction. II: Equational representation, null symbols, and linguistic relevance* (J 0194) Inform & Control 21*93-116,235-260
◇ D05 ◇ REV MR 49 # 4319 MR 49 # 4320 Zbl 245 # 68027 Zbl 245 # 68028 • ID 14998

JOSHI, A.K. & LEVY, L.S. & YUEH, KANG [1980] *Local constraints in programming languages I. Syntax* (J 1426) Theor Comput Sci 12*265-290
◇ B75 D05 ◇ REV MR 82f:68080 Zbl 441 # 68086 • ID 81738

JOSHI, A.K. see Vol. II for further entries

JOSKO, B. [1981] *An effective retract calculus* (P 3475) Theor Comput Sci (5);1981 Karlsruhe 104*184-194
◇ D80 ◇ REV Zbl 461 # 03010 • ID 54487

JOUANNAUD, J.-P. [1983] *Confluent and coherent equational term rewriting systems application to proofs in abstract data types* (P 3889) CAAP'83 Arbres en Algeb & Progr (8);1983 L'Aquila 269-283
◇ B75 D03 ◇ REV MR 85e:68050 Zbl 522 # 68013 • ID 37055

JOUANNAUD, J.-P. see Vol. I, II for further entries

JUERGENSEN, H. [1984] *Komplexitaet von Erzeugen in Algebren* (J 0380) Acta Cybern (Szeged) 6*371-379
◇ D40 ◇ REV Zbl 552 # 68049 • ID 44266

JUKNA, S. [1979] *On decidability of equivalence problem in some algebras of recursive functions (Russian)* (J 2574) Litov Mat Sb (Vil'nyus) 19/3*133-136
◇ B25 D20 D75 ◇ ID 33258

JUKNA, S. [1982] *A principle for obtaining lower bounds of arithmetical complexity (Russian) (English and Lithuanian summaries)* (J 3939) Mat Logika Primen (Akad Nauk Litov SSR) 2*108-112
◇ D15 ◇ REV MR 85i:03129 Zbl 516 # 03034 • ID 45156

JUKNA, S. [1982] *Arithmetical representations of machine complexity classes (Russian) (English and Lithuanian summaries)* (J 3939) Mat Logika Primen (Akad Nauk Litov SSR) 2*92-107
◇ D15 D20 ◇ REV MR 85i:03130 Zbl 515 # 68045 • ID 45162

JUKNA, S. see Vol. III, VI for further entries

JUNG, H. & WIEHAGEN, R. [1977] *Rekursionstheoretische Charakterisierung von erkennbaren Klassen rekursiver Funktionen (English and Russian summaries)* (J 0129) Elektr Informationsverarbeitung & Kybern 13*385-397
◇ D20 ◇ REV MR 56 # 11770 Zbl 364 # 02022 • ID 50952

JUNG, H. [1981] *Relationships between probabilistic and deterministic tape complexity* (P 3429) Math Founds of Comput Sci (10);1981 Strbske Pleso 339-346
◇ D15 ◇ REV MR 83j:03066 Zbl 462 # 68031 • ID 35364

JUNG, H. [1984] see ALBRECHT, A.

JUNG, H. [1985] *On probabilistic time and space* (P 4628) Automata, Lang & Progr (12);1985 Nafplion 310-317
◇ D15 ◇ ID 49573

JUNNILA, H.J.K. [1983] see FREMLIN, D.H.

JUNNILA, H.J.K. see Vol. V for further entries

JURA, A. [1980] *Some remarks on nonexistence of an algorithm for finding all ideals of a given finite index in a finitely presented semigroup* (J 1008) Demonstr Math (Warsaw) 13*573-578
◇ D40 ◇ REV MR 82f:20083 Zbl 465 # 20056 • ID 81749

KADANE, J.B. & SIMON, H.A. [1976] *Problems of computational complexity in artificial intelligence* (**P** 3356) Algor & Complex;1976 Pittsburgh 281-299
◇ B75 D15 ◇ REV MR 56 # 10144 Zbl 385 # 68049
• ID 69920

KAHR, A.S. [1962] see DREBEN, B.

KAHR, A.S. & MOORE, E.F. & WANG, HAO [1962] *Entscheidungsproblem reduced to the ∀∃∀ case* (**J** 0054) Proc Nat Acad Sci USA 48*365-377
◇ B20 D35 ◇ REV MR 30 # 21 Zbl 102.8 JSL 27.225
• ID 19074

KAHR, A.S. [1963] *Improved reductions of the "Entscheidungsproblem" to subclasses of ∀∃∀ formulas* (**P** 0674) Symp Math Th of Automata;1962 New York 57-70
◇ D35 ◇ REV MR 29 # 4689 • ID 06793

KAIN, R.Y. [1972] *Automata theory: machines and languages* (**X** 0822) McGraw-Hill: New York xvii+301pp
◇ D05 D10 D98 ◇ REV Zbl 248 # 68004 • ID 23542

KAISER, KLAUS [1970] *Zur Theorie algorithmisch abgeschlossener Modellklassen* (**J** 0041) Math Syst Theory 4*160-167
◇ C20 C52 D75 ◇ REV MR 42 # 4384 Zbl 202.8
• ID 06800

KAISER, KLAUS see Vol. I, III for further entries

KALANDARISHVILI, N.G. [1978] *An estimate of the algorithm of analysis in the language of equivalent transformations (Russian) (Georgian and English summaries)* (**S** 2043) Issl Mat Log & Teor Algor (Tbilisi) 1978*4-7
◇ D05 ◇ REV MR 80d:68039 • ID 81759

KALANTARI, I. & RETZLAFF, A.T. [1977] *Maximal vector spaces under automorphisms of the lattice of recursively enumerable vector spaces* (**J** 0036) J Symb Logic 42*481-491
◇ C57 C60 D45 ◇ REV MR 58 # 21539 Zbl 383 # 03033
• ID 26850

KALANTARI, I. [1978] *Major subspaces of recursively enumerable vector spaces* (**J** 0036) J Symb Logic 43*293-303
◇ C57 C60 D25 D45 ◇ REV MR 81f:03058
Zbl 403 # 03034 • ID 29262

KALANTARI, I. [1979] *Automorphisms of the lattice of recursively enumerable vector spaces* (**J** 0068) Z Math Logik Grundlagen Math 25*385-401
◇ C57 C60 D45 ◇ REV MR 81b:03047 Zbl 424 # 03023
• ID 74612

KALANTARI, I. & RETZLAFF, A.T. [1979] *Recursive constructions in topological spaces* (**J** 0036) J Symb Logic 44*609-625
◇ C57 C65 D45 ◇ REV MR 81f:03059 Zbl 427 # 03035
• ID 53722

KALANTARI, I. [1981] *Effective content of a theorem of M.H. Stone* (**P** 2902) Aspects Effective Algeb;1979 Clayton 128-146
◇ D45 D80 ◇ REV MR 83a:03041 Zbl 476 # 03044
• ID 55556

KALANTARI, I. [1982] *Major subsets in effective topology* (**P** 3634) Patras Logic Symp;1980 Patras 77-94
◇ C57 C65 D25 D45 ◇ REV MR 84g:03064
Zbl 521 # 03029 • ID 34171

KALANTARI, I. & LEGGETT, A. [1982] *Simplicity in effective topology* (**J** 0036) J Symb Logic 47*169-183
◇ C57 C65 D25 ◇ REV MR 83m:03053 Zbl 516 # 03023
• ID 35452

KALANTARI, I. & REMMEL, J.B. [1983] *Degrees of recursively enumerable topological spaces* (**J** 0036) J Symb Logic 48*610-622
◇ C57 C65 D25 D45 ◇ REV MR 85i:03151
Zbl 532 # 03021 • ID 33595

KALANTARI, I. & LEGGETT, A. [1983] *Maximality in effective topology* (**J** 0036) J Symb Logic 48*100-112
◇ C57 C65 D25 D45 ◇ REV MR 84k:03112
Zbl 532 # 03020 • ID 36130

KALANTARI, I. [1984] see JOCKUSCH JR., C.G.

KALANTARI, I. & WEITKAMP, G. [1985] *Effective topological spaces I: A definability theory. II: A hierarchy* (**J** 0073) Ann Pure Appl Logic 29*1-27,207-224
◇ C40 C57 C65 D45 D75 D80 ◇ REV Zbl 569 # 03018
• ID 47472

KALANTARI, I. [1985] see DOWNEY, R.G.

KALFA, C. [1984] *Some undecidability results in strong algebraic languages* (**J** 0036) J Symb Logic 49*951-954
◇ D45 ◇ REV MR 85k:03029 • ID 42467

KALFA, C. see Vol. III for further entries

KALICKI, J. [1954] *An undecidable problem in the algebra of truth-tables* (**J** 0036) J Symb Logic 19*172-176
◇ B05 D35 ◇ REV MR 16.324 Zbl 58.246 JSL 20.283
• ID 06815

KALICKI, J. see Vol. I, II, III for further entries

KALISZ, E. [1980] *The conversion of PRIM-URM conflicts in context-free grammars into PRIM-PRIM conflicts (Romanian, English summery)* (**J** 3130) Bul Inst Politeh Bucuresti, Ser Electroteh 42/3*89-94
◇ D05 ◇ REV MR 82g:68071 Zbl 462 # 68051 • ID 81762

KALLIBEKOV, S. [1971] *Index sets of degrees of unsolvability (Russian)* (**J** 0003) Algebra i Logika 10*316-326
• TRANSL [1971] (**J** 0069) Algeb and Log 10*198-204
◇ D25 D55 ◇ REV MR 45 # 3202 Zbl 318 # 02045
• ID 74624

KALLIBEKOV, S. [1971] *The index sets of m-degrees (Russian)* (**J** 0092) Sib Mat Zh 12*1292-1300
• TRANSL [1971] (**J** 0475) Sib Math J 12*931-937
◇ D25 D55 ◇ REV MR 45 # 59 Zbl 264 # 02040
• ID 06829

KALLIBEKOV, S. [1973] *On degrees of recursively enumerable sets (Russian)* (**J** 0092) Sib Mat Zh 14*421-426,463
• TRANSL [1973] (**J** 0475) Sib Math J 14*290-293
◇ D25 ◇ REV MR 48 # 8216 Zbl 275 # 02039 • ID 06830

KALLIBEKOV, S. [1973] *Truth tabular degrees of recursively enumerable sets (Russian)* (**J** 0087) Mat Zametki (Akad Nauk SSSR) 14*697-702
• TRANSL [1973] (**J** 1044) Math Notes, Acad Sci USSR 14*958-961
◇ D25 ◇ REV MR 49 # 2314 Zbl 318 # 02046 • ID 06809

KALLIBEKOV, S. [1975] *Hierarchy of sets for truth-table enumerability (Russian)* (**C** 1050) Aktual Vopr Mat Log & Teor Mnozh 155-163
◇ D30 D55 ◇ REV MR 57 # 5719 • ID 74626

KALLIBEKOV, S. [1983] *Index sets and degrees of some classes of recursively enumerable sets (Russian)* (**X** 1700) Fan: Tashkent 84pp
◇ D25 D30 ◇ ID 46369

KALMAR, L. [1930] *Ein Beitrag zum Entscheidungsproblem* (**J** 0460) Acta Univ Szeged, Sect Mat 5*222-236
◇ B25 D35 ◇ REV Zbl 4.146 FdM 57.1321 • ID 06836

KALMAR, L. [1932] *Zum Entscheidungsproblem der mathematischen Logik* (**P** 0653) Int Congr Math (II, 4);1932 Zuerich 2*337-338
◇ B20 B25 D35 ◇ REV FdM 58.70 • ID 06837

KALMAR, L. [1937] *Zur Reduktion des Entscheidungsproblems* (**J** 4510) Norsk Mat Tidsskr 19*121-130
◇ B20 D35 ◇ REV Zbl 17.337 JSL 3.46 FdM 63.823 • REM A note by Skolem ibid 19*130-133 (Norwegian) • ID 06842

KALMAR, L. [1937] *Zurueckfuehrung des Entscheidungsproblems auf den Fall von Formeln mit einer einzigen, binaeren Funktionsvariablen* (**J** 0020) Compos Math 4*137-144
◇ B20 D35 ◇ REV Zbl 15.338 JSL 2.48 FdM 62.1058 • ID 06841

KALMAR, L. [1939] *On the reduction of the decision problem I: Ackermann prefix, a single binary predicate* (**J** 0036) J Symb Logic 4*1-9
◇ B20 D35 ◇ REV Zbl 20.195 JSL 4.127 FdM 65.27 • REM Part II 1947 by Kalmar,L. & Suranyi,J. • ID 06843

KALMAR, L. [1943] *Ein einfaches Beispiel fuer ein unentscheidbares arithmetisches Problem (Hungarian) (German summary)* (**J** 0461) Mat Fiz Lapok 50*1-23
◇ D20 F30 ◇ REV MR 8.558 JSL 9.24 • ID 19071

KALMAR, L. & SURANYI, J. [1947] *On the reduction of the decision problem II: Goedel prefix, a single binary predicate* (**J** 0036) J Symb Logic 12*65-73
◇ B20 D35 ◇ REV MR 11.303 Zbl 29.99 JSL 13.48 • REM Part I 1939. Part III 1950 • ID 06846

KALMAR, L. [1949] *On unsolvable mathematical problems* (**P** 0682) Int Congr Philos (10);1948 Amsterdam 756-758
◇ A05 D35 ◇ REV MR 10.423 Zbl 31.194 JSL 14.130 • ID 20803

KALMAR, L. [1950] *Contribution to the reduction theory of the decision problem I. Prefix* (x_1) (x_2) $(Ex_3)\ldots(Ex_{n-1})$ (x_n), *a single binary predicate* (**J** 0001) Acta Math Acad Sci Hung 1*64-73
◇ B20 D35 ◇ REV MR 12.661 Zbl 39.246 JSL 17.73 • REM Part II 1950 by Suranyi,J. • ID 06854

KALMAR, L. & SURANYI, J. [1950] *On the reduction of the decision problem III: Pepis prefix, a single binary predicate* (**J** 0036) J Symb Logic 15*161-173
◇ B20 D35 ◇ REV MR 12.661 Zbl 41.352 JSL 16.215 • REM Part II 1947 • ID 06851

KALMAR, L. [1951] *Contributions to the reduction theory of the decision problem. III Prefix* $(x_1)(Ex_2)\ldots(Ex_{n-2})(x_{n-1})(x_n)$, *a single binary predicate* (**J** 0001) Acta Math Acad Sci Hung 2*19-38
◇ B10 D35 ◇ REV MR 13.715 Zbl 44.3 JSL 18.264 • REM Part II 1950 by Suranyi,J. Part IV 1951 • ID 42204

KALMAR, L. [1951] *Contribution to the reduction theory of the decision problem IV. Reduction to the case of a finite set of individuals* (**J** 0001) Acta Math Acad Sci Hung 2*125-142
◇ D35 ◇ REV MR 14.713 Zbl 45.2 JSL 18.264 • REM Part III 1951. Part V 1951 by Suranyi,J. • ID 06856

KALMAR, L. [1952] *Another proof to the Markov-Post theorem* (**J** 0001) Acta Math Acad Sci Hung 3*1-27
◇ D03 D20 D40 ◇ REV MR 14.528 Zbl 47.11 JSL 23.447 • ID 06857

KALMAR, L. [1952] *Reduction of the decision problem to the satisfiability question of logical formulae on a finite set (Hungarian)* (**P** 0662) Hungar Math Congr (1);1950 Budapest 163-190
◇ D35 ◇ REV MR 14.1051 Zbl 48.246 JSL 20.72 • ID 19070

KALMAR, L. [1955] *The solution of a problem of K.Schroeter, concerning the definition of general recursive functions* (**J** 0462) Mat Fiz Oszt Koezlem, Acad Sci Hung 5*103-127
◇ D20 ◇ REV MR 19.237 Zbl 163.8 JSL 25.164 • ID 19069

KALMAR, L. [1955] *Ueber ein Problem, betreffend die Definition des Begriffes der allgemein-rekursiven Funktion* (**J** 0068) Z Math Logik Grundlagen Math 1*93-96
◇ D20 ◇ REV MR 17.225 Zbl 68.247 JSL 23.227 • ID 06858

KALMAR, L. [1956] *A direct proof of the unsolvability of the decision problem by means of a general recursive algorithm (Hungarian)* (**J** 0462) Mat Fiz Oszt Koezlem, Acad Sci Hung 6*1-25
◇ B20 D35 ◇ REV MR 20 # 4 Zbl 75.5 JSL 24.173 • ID 16972

KALMAR, L. [1956] *Ein direkter Beweis fuer die allgemein-rekursive Unloesbarkeit des Entscheidungsproblems des Praedikatenkalkuels der ersten Stufe mit Identitaet* (**J** 0068) Z Math Logik Grundlagen Math 2*1-14
◇ B10 D35 ◇ REV MR 18.369 Zbl 75.5 JSL 27.86 • ID 06861

KALMAR, L. [1957] *On Church's hypothesis as foundation for studies related to so-called unsolvable mathematical problems (Hungarian)* (**J** 0462) Mat Fiz Oszt Koezlem, Acad Sci Hung 7*19-38
◇ D20 D35 ◇ REV MR 20 # 3781 Zbl 124.249 • ID 06862

KALMAR, L. [1957] *Ueber arithmetische Funktionen von unendlich vielen Variablen, welche an jeder Stelle bloss von einer endlichen Anzahl von Variablen abhaengig sind* (**S** 0019) Colloq Math (Warsaw) 5*1-5
◇ D20 E20 ◇ REV MR 20 # 4489 Zbl 81.241 JSL 35.152 • ID 06863

KALMAR, L. [1959] *An argument against the plausibility of Church's thesis* (**P** 0634) Constructivity in Math;1957 Amsterdam 72-80
◇ A05 D20 F99 ◇ REV MR 21 # 5567 Zbl 88.249 JSL 33.471 • ID 06864

KALMAR, L. [1959] *On a hypothesis used in investigations of so-called unsolvable arithmetical problems (Russian)* (**P** 0607) All-Union Math Conf (3);1956 Moskva 4*227-231
◇ A05 D35 ◇ REV Zbl 124.250 • ID 31286

KALMAR, L. [1961] *A practical infinitistic computer* (**P** 0633) Infinitist Meth;1959 Warsaw 347-362
◇ B70 D10 ◇ REV MR 26 # 2042 Zbl 119.13 JSL 34.510 • ID 06865

KALMAR, L. (ED.) [1965] *Colloque sur les fondements des mathematiques, les machines mathematiques, et leurs applications* (**X** 0928) Akad Kiado: Budapest 320pp
◇ B97 D05 D10 D97 ◇ REV MR 32 # 5493 Zbl 148.1
• ID 48630

KALMAR, L. [1968] *On the problem of full utilization of the technical possibilities of computers in devising appropriate approximation methods for the solution of numerical problems* (**J** 0070) Bull Soc Sci Math Roumanie, NS 12*75-79
◇ D20 D80 ◇ REV Zbl 186.10 • ID 06871

KALMAR, L. [1969] *R.Peter's work in the theory of recursive functions* (**P** 1841) Fct Recurs & Appl;1967 Tihany 1-11
◇ A10 D20 ◇ ID 32550

KALMAR, L. see Vol. I, III, V, VI for further entries

KALORKOTI, K.A. [1982] *Decision problems in group theory* (**J** 3240) Proc London Math Soc, Ser 3 44*312-332
◇ D25 D40 ◇ REV MR 84f:20036 Zbl 485 # 20022
• ID 39695

KALUZHNIN, L.A. [1959] *On the algorithmization of mathematical problems (Russian)* (**J** 0052) Probl Kibern 2*51-67
◇ B75 D20 ◇ REV MR 24 # B2071 JSL 27.363 • ID 19068

KALUZHNIN, L.A. see Vol. I, II, V for further entries

KAMAE, T. [1973] *On Kolmogorov's complexity and information* (**J** 0351) Osaka J Math 10*305-307
◇ D15 D20 ◇ REV MR 49 # 4302 Zbl 273 # 94024
• ID 62849

KAMAE, T. see Vol. I for further entries

KAMBAYASHI, Y. & YAJIMA, S. [1972] *Finite memory machines satisfying the lower bound of memory* (**J** 0194) Inform & Control 20*150-157
◇ D05 ◇ REV MR 48 # 1826 Zbl 237 # 94014 • ID 06873

KAMEDA, T. [1968] *Generalized transition matrix of a sequential machine and its applications* (**J** 0194) Inform & Control 12*259-275
◇ D05 ◇ REV MR 41 # 5120 Zbl 157.337 • ID 15002

KAMEDA, T. & VOLLMAR, R. [1970] *Note on tape reversal complexity of languages* (**J** 0194) Inform & Control 17*203-215
◇ D10 ◇ REV MR 42 # 7446 Zbl 196.18 • ID 16318

KAMEDA, T. & VOLLMAR, R. [1970] *Zur Umkehrkomplexitaet von Sprachen* (**P** 0577) Automatenth & Formale Sprachen;1969 Oberwolfach 327-339
• TRANSL [1974] (**C** 2319) Slozh Vychisl & Algor 222-234
◇ D05 D10 D15 ◇ REV MR 49 # 8424 Zbl 289 # 68035
• ID 29299

KAMEDA, T. [1972] *Pushdown automata with counters* (**J** 0119) J Comp Syst Sci 6*138-150
◇ D10 ◇ REV MR 55 # 1824 Zbl 242 # 68022 • ID 06875

KAMIMURA, T. & TANG, A. [1983] *Effectively given spaces* (**P** 3851) Automata, Lang & Progr (10);1983 Barcelona 385-396
◇ D45 ◇ REV MR 85m:03030 Zbl 535 # 03021 • ID 38324

KAMIMURA, T. [1985] *An effective given initial semigroup* (**J** 1431) Acta Inf 203-227
◇ C57 D45 ◇ REV Zbl 571 # 20057 • ID 48019

KANDA, A. [1979] *Fully effective solutions of recursive domain equations* (**P** 2059) Math Founds of Comput Sci (8);1979 Olomouc 326-336
◇ D45 ◇ REV MR 81e:68054 Zbl 404 # 68008 • ID 81767

KANDA, A. & PARK, D. [1979] *When are two effectively given domains identical?* (**P** 3488) Theor Comput Sci (4);1979 Aachen 170-181
◇ D45 ◇ REV MR 81j:68058 Zbl 405 # 03020 • ID 54895

KANDA, A. [1981] *Constructive category theory* (**P** 3429) Math Founds of Comput Sci (10);1981 Strbske Pleso 563-577
◇ D45 D80 G30 ◇ REV MR 83e:03068 Zbl 481 # 03039
• ID 35237

KANDA, A. [1984] *Numeration models of λ-calculus* (**P** 4007) CAAP'84 Arbres en Algeb & Progr (9);1984 Bordeaux 155-168
◇ B40 D45 ◇ ID 45525

KANDA, A. [1985] *Acceptable numerations of functions spaces* (**J** 0068) Z Math Logik Grundlagen Math 31*503-508
◇ D45 ◇ ID 47800

KANDA, A. [1985] *Numeration models of λ-calculus* (**J** 0068) Z Math Logik Grundlagen Math 31*209-220
◇ B40 B60 C57 D45 ◇ ID 47568

KANDRI-RODY, A. & KAPUR, D. & NARENDRAN, P. [1985] *An ideal-theoretic approach to word problems and unification problems over finitely presented commutative algebras* (**P** 4244) Rewriting Techn & Appl (1);1985 Dijon 345-364
◇ C05 D40 ◇ ID 49636

KANELLAKIS, P.C. [1980] *On the computational complexity of cardinality constraints in relational databases* (**J** 0232) Inform Process Lett 11*98-101
◇ B75 D15 ◇ REV MR 82m:68041 Zbl 465 # 68058
• ID 81776

KANIEWSKI, J. [1976] *A generalization of Kondo's uniformization theorem (Russian summary)* (**J** 0014) Bull Acad Pol Sci, Ser Math Astron Phys 24*393-398
◇ D55 E15 ◇ REV MR 54 # 2468 Zbl 333 # 04001
• ID 24651

KANIEWSKI, J. see Vol. V for further entries

KANNAN, R. [1981] *A circuit-size lower bound* (**P** 4235) IEEE Symp Found of Comp Sci (22);1981 Nashville 304-309
◇ D15 ◇ REV MR 84a:68004 • ID 45819

KANNAN, R. [1981] *Towards separating nondeterministic time from deterministic time* (**P** 4235) IEEE Symp Found of Comp Sci (22);1981 Nashville 235-243
◇ D15 ◇ REV MR 84a:68004 • ID 45818

KANNAN, R. [1982] *Circuit-size lower bounds and nonreducibility to sparse sets* (**J** 0194) Inform & Control 55*40-56
◇ D15 ◇ REV MR 85e:68023 Zbl 537 # 94027 • ID 40004

KANOVEJ, V.G. [1974] *On degrees of constructibility and descriptive properties of the set of real numbers in an initial model and in its extensions (Russian)* (**J** 0023) Dokl Akad Nauk SSSR 216*728-729
• TRANSL [1974] (**J** 0062) Sov Math, Dokl 15*866-868
◇ C62 D30 E40 E45 ◇ REV MR 51 # 143
Zbl 327 # 02051 • ID 17445

KANOVEJ, V.G. [1975] *On the independence of some propositions of descriptive set theory and second-order arithmetic (Russian)* (J 0023) Dokl Akad Nauk SSSR 223∗552-554
• TRANSL [1975] (J 0062) Sov Math, Dokl 16∗937-940
◇ D55 E15 E35 F35 ◇ REV MR 52 # 10416 Zbl 363 # 02067 • ID 50893

KANOVEJ, V.G. [1975] *The majorization of initial segments of degrees of constructivity (Russian)* (J 0087) Mat Zametki (Akad Nauk SSSR) 17∗939-946
• TRANSL [1975] (J 1044) Math Notes, Acad Sci USSR 17∗563-567
◇ D30 E40 E45 ◇ REV MR 56 # 11784 Zbl 344 # 02046
• ID 62865

KANOVEJ, V.G. [1978] *A proof of a theorem of Luzin (Russian)* (J 0087) Mat Zametki (Akad Nauk SSSR) 23∗61-66
• TRANSL [1978] (J 1044) Math Notes, Acad Sci USSR 23∗35-37
◇ D55 E15 ◇ REV MR 58 # 16308 Zbl 414 # 04003
• ID 74664

KANOVEJ, V.G. [1979] *On descriptive forms of the countable axiom of choice (Russian)* (C 2581) Issl Neklass Log & Teor Mnozh 3-136
◇ D55 E15 E25 E35 F35 ◇ REV MR 83c:03045 Zbl 439 # 03033 • ID 56024

KANOVEJ, V.G. [1979] *The definability of forcing in analysis (Russian) (English summary)* (J 0288) Vest Ser Mat Mekh, Univ Moskva 1979/2∗3-13,101
• TRANSL [1979] (J 0510) Moscow Univ Math Bull 34/2∗1-12
◇ D55 E40 F35 ◇ REV MR 80j:03085 Zbl 419 # 03032
• ID 53375

KANOVEJ, V.G. [1979] *The set of all analytically definable sets of natural numbers can be defined analytically (Russian)* (J 0216) Izv Akad Nauk SSSR, Ser Mat 43∗1259-1293
• TRANSL [1980] (J 0448) Math of USSR, Izv 15∗469-500
◇ C62 D55 E15 E35 E45 ◇ REV MR 82a:03043 Zbl 427 # 03045 • ID 54105

KANOVEJ, V.G. [1982] *On N.N. Luzin's problems on the embeddability and decomposition of projective sets (Russian)* (J 0087) Mat Zametki (Akad Nauk SSSR) 32∗23-39,124
• TRANSL [1982] (J 1044) Math Notes, Acad Sci USSR 32∗490-499
◇ D55 E15 E35 E55 ◇ REV MR 84g:03075 Zbl 559 # 03030 • ID 34181

KANOVEJ, V.G. [1983] *A generalization of P.S. Novikov's theorem on cross sections of Borel sets (Russian)* (J 0087) Mat Zametki (Akad Nauk SSSR) 33∗289-292,319
• TRANSL [1983] (J 1044) Math Notes, Acad Sci USSR 33∗144-146
◇ D55 E15 ◇ REV MR 84h:04003 Zbl 539 # 28001
• ID 34332

KANOVEJ, V.G. [1983] *Structure of the constituents of Π^1_1-sets (Russian)* (J 0092) Sib Mat Zh 24/2∗56-76
• TRANSL [1983] (J 0475) Sib Math J 24∗198-215
◇ D55 E15 E35 ◇ REV MR 85j:04002 Zbl 518 # 03018
• ID 37518

KANOVEJ, V.G. see Vol. I, V, VI for further entries

KANOVICH, M.I. [1969] *Estimates of the reduction complexity of algorithms (Russian)* (S 0228) Zap Nauch Sem Leningrad Otd Mat Inst Steklov 16∗77-80
• TRANSL [1969] (J 0521) Semin Math, Inst Steklov 16∗38-39
◇ B75 D15 ◇ REV MR 41 # 6693 Zbl 318 # 02035
• ID 06897

KANOVICH, M.I. & KUSHNER, B.A. [1969] *Estimating the complexity of certain algorithmic problems of analysis (Russian)* (S 0228) Zap Nauch Sem Leningrad Otd Mat Inst Steklov 16∗81-90
• TRANSL [1969] (J 0521) Semin Math, Inst Steklov 16∗40-44
◇ D15 D20 D80 F60 ◇ REV MR 42 # 77 Zbl 267 # 02022 • ID 29887

KANOVICH, M.I. [1969] *On the decision complexity of algorithms (Russian)* (J 0023) Dokl Akad Nauk SSSR 186∗1008-1009
• TRANSL [1969] (J 0062) Sov Math, Dokl 10∗700-701
◇ B75 D15 ◇ REV MR 41 # 6692 Zbl 272 # 02054
• ID 06893

KANOVICH, M.I. & PETRI, N.V. [1969] *Some theorems on the complexity of normal algorithms and computations (Russian)* (J 0023) Dokl Akad Nauk SSSR 184∗1275-1276
• TRANSL [1969] (J 0062) Sov Math, Dokl 10∗233-234
◇ D03 D15 ◇ REV MR 39 # 4006 Zbl 181.313 • ID 06891

KANOVICH, M.I. [1970] *Complexity of resolution of a recursively enumerable set as a criterion of its universality (Russian)* (J 0023) Dokl Akad Nauk SSSR 194∗500-505
• TRANSL [1970] (J 0062) Sov Math, Dokl 11∗1224-1228
◇ D25 ◇ REV MR 43 # 1831 Zbl 299 # 02045 • ID 06899

KANOVICH, M.I. [1970] *On the complexity of enumeration and decision of predicates (Russian)* (J 0023) Dokl Akad Nauk SSSR 190∗23-26
• TRANSL [1970] (J 0062) Sov Math, Dokl 11∗17-20
◇ D15 ◇ REV MR 41 # 6694 Zbl 255 # 02049 • ID 06894

KANOVICH, M.I. [1970] *On the decision complexity of recursively enumerable sets (Russian)* (J 0023) Dokl Akad Nauk SSSR 192∗721-723
• TRANSL [1970] (J 0062) Sov Math, Dokl 11∗704-706
◇ D15 D25 ◇ REV MR 42 # 68 Zbl 272 # 02055
• ID 06898

KANOVICH, M.I. [1971] *On complexity of Boolean function minimzation (Russian)* (J 0023) Dokl Akad Nauk SSSR 198∗35-38
• TRANSL [1971] (J 0062) Sov Math, Dokl 12∗720-724
◇ B35 D15 ◇ REV MR 43 # 7334 Zbl 235 # 02026
• ID 06900

KANOVICH, M.I. [1971] *On domains of definition of optimal algorithms (Russian)* (J 0023) Dokl Akad Nauk SSSR 198∗283-285
• TRANSL [1971] (J 0062) Sov Math, Dokl 12∗773-776
◇ B75 D15 ◇ REV MR 44 # 5169 Zbl 235 # 02025
• ID 33374

KANOVICH, M.I. [1972] *Complexity of bounded decidability of semirecursively enumerable sets (Russian)* (J 0023) Dokl Akad Nauk SSSR 203∗1246-1248
• TRANSL [1972] (J 0062) Sov Math, Dokl 13∗536-538
◇ D15 D25 ◇ REV MR 45 # 4978 Zbl 269 # 02013
• ID 06901

KANOVICH, M.I. [1972] *On the universality of strongly undecidable sets (Russian)* (**J** 0023) Dokl Akad Nauk SSSR 204*533-535
• TRANSL [1972] (**J** 0062) Sov Math, Dokl 13*687-690
◇ D15 D20 D25 D30 ◇ REV MR 46 # 3279 • ID 06902

KANOVICH, M.I. [1973] *Irreducibility of languages of the stepped semantic system (Russian)* (**J** 0023) Dokl Akad Nauk SSSR 212*800-803
• TRANSL [1973] (**J** 0062) Sov Math, Dokl 14*1459-1463
◇ D15 D30 F50 ◇ REV MR 50 # 4275 Zbl 299 # 02050
• ID 06903

KANOVICH, M.I. [1973] *Komplexitaet der beschraenkten Entscheidbarkeit von Algorithmen (Russian)* (**S** 0554) Issl Teor Algor & Mat Logik (Moskva) 1*3-41
◇ D05 D15 D20 ◇ REV MR 49 # 2298 Zbl 284 # 02012
• ID 30516

KANOVICH, M.I. [1973] *On the complexity of approximation of arithmetic sets (Russian)* (**J** 0023) Dokl Akad Nauk SSSR 211*1038-1041
• TRANSL [1973] (**J** 0062) Sov Math, Dokl 14*1153-1158
◇ D55 ◇ REV MR 49 # 31 Zbl 293 # 02029 • ID 06904

KANOVICH, M.I. [1974] *"Complex" and "simple" numbers (Russian)* (**J** 0023) Dokl Akad Nauk SSSR 218*276-277
• TRANSL [1974] (**J** 0062) Sov Math, Dokl 15*1316-1318
◇ D20 ◇ REV MR 51 # 7838 Zbl 326 # 02028 • ID 18217

KANOVICH, M.I. [1974] *A certain extension of the step-by-step semantic system of A. A. Markov (Russian)* (**C** 1450) Teor Algor & Mat Logika (Markov) 62-70
◇ D15 F50 ◇ REV MR 57 # 16000 Zbl 299 # 02051
• ID 62868

KANOVICH, M.I. [1974] *A theorem on speeding up in formal systems (Russian)* (**C** 2319) Slozh Vychisl & Algor 186-189
◇ B35 D15 ◇ REV Zbl 307 # 68034 • ID 62867

KANOVICH, M.I. [1974] *Complexity of the limit of Specker sequences (Russian)* (**J** 0023) Dokl Akad Nauk SSSR 214*1020-1023
• TRANSL [1974] (**J** 0062) Sov Math, Dokl 15*299-303
◇ D15 D20 F60 ◇ REV MR 49 # 4765 Zbl 299 # 02040
• ID 07592

KANOVICH, M.I. [1975] *A hierachical semantic system with set variables (Russian)* (**J** 0023) Dokl Akad Nauk SSSR 221*1256-1259
• TRANSL [1975] (**J** 0062) Sov Math, Dokl 16*504-509
◇ B20 B65 C40 D05 F50 ◇ REV MR 52 # 5385
Zbl 325 # 02030 • ID 18218

KANOVICH, M.I. [1975] *Dekker's construction and effective nonrecursiveness (Russian)* (**J** 0023) Dokl Akad Nauk SSSR 222*1028-1030
• TRANSL [1975] (**J** 0062) Sov Math, Dokl 16*719-721
◇ D20 D25 ◇ REV MR 52 # 10390 Zbl 327 # 02036
• ID 21689

KANOVICH, M.I. [1975] *On sets of complex-programmed numbers* (**P** 0454) Math Founds of Comput Sci (4);1975 Marianske Lazne 271-272
◇ D15 ◇ REV MR 54 # 11846 Zbl 316 # 02040 • ID 62886

KANOVICH, M.I. & KUSHNER, B.A. [1976] *Complexity of algorithms, and Specker sequences (Russian)* (**S** 0554) Issl Teor Algor & Mat Logik (Moskva) 2*73-83,160
◇ D15 D20 ◇ REV MR 58 # 16203 • ID 74681

KANOVICH, M.I. [1976] *The relation between upper and lower bounds for the reduction complexity of enumerable sets (Russian)* (**S** 0554) Issl Teor Algor & Mat Logik (Moskva) 2*66-72
◇ D15 D25 ◇ REV MR 58 # 19351 • ID 81783

KANOVICH, M.I. [1977] *Complex properties of context-sensitive languages (Russian)* (**J** 0023) Dokl Akad Nauk SSSR 233/3*289-292
• TRANSL [1977] (**J** 0062) Sov Math, Dokl 18*383-387
◇ D03 D05 D15 ◇ REV MR 58 # 19352 Zbl 371 # 02014
• ID 51375

KANOVICH, M.I. [1977] *On computability of Kolmogorov complexity* (**P** 2588) FCT'77 Fund of Comput Th;1977 Poznan 421-422
◇ D15 ◇ REV Zbl 356 # 68063 • ID 33376

KANOVICH, M.I. [1977] *On the precision of a complexity criterion for nonrecursiveness and universality (Russian)* (**J** 0023) Dokl Akad Nauk SSSR 232*1249-1252
• TRANSL [1977] (**J** 0062) Sov Math, Dokl 18*232-236
◇ D15 D20 D25 D30 ◇ REV MR 57 # 16014
Zbl 383 # 03029 • ID 52011

KANOVICH, M.I. [1978] *An estimate of the complexity of arithmetic incompleteness (Russian)* (**J** 0023) Dokl Akad Nauk SSSR 238*1283-1286
• TRANSL [1978] (**J** 0062) Sov Math, Dokl 19*206-210
◇ D15 F20 F30 ◇ REV MR 81i:03091 Zbl 394 # 03041
• ID 52518

KANOVICH, M.I. [1978] *Complexity of recognition of invariant properties (Russian)* (**C** 3211) Mat Ling & Teor Algor 69-76
◇ D03 D15 ◇ REV Zbl 395 # 03027 • ID 52578

KANOVICH, M.I. [1979] *A complexity version of Goedel's incompleteness theorem* (**P** 2935) FCT'79 Fund of Comput Th;1979 Berlin/Wendisch-Rietz 542-543
◇ D15 F20 F30 ◇ REV MR 81e:03035 Zbl 412 # 03014
• ID 52945

KANOVICH, M.I. [1979] *The complexity of search for short programs (Russian)* (**S** 2582) Semiotika & Inf, Akad Nauk SSSR 12*147-148
◇ D15 D20 ◇ ID 33377

KANOVICH, M.I. [1981] *Complexity of a fixed point and of a shift algorithm (Russian)* (**C** 3747) Mat Log & Mat Lingvistika 96-111
◇ D15 D20 ◇ REV MR 84a:03043 • ID 34897

KANOVICH, M.I. [1982] *On the complexity of the problem of separating recursively enumerable sets (Russian)* (**J** 0023) Dokl Akad Nauk SSSR 267*1300-1304
• TRANSL [1982] (**J** 0062) Sov Math, Dokl 26*766-769
◇ D25 D30 ◇ REV MR 84f:03038 Zbl 573 # 03013
• ID 34458

KANOVICH, M.I. [1982] *Truth-table reducibilities of problems of the extension of partial recursive functions (Russian)* (**J** 0023) Dokl Akad Nauk SSSR 264*294-298
• TRANSL [1982] (**J** 0062) Sov Math, Dokl 25*631-635
◇ D20 D30 ◇ REV MR 85j:03070 Zbl 527 # 03035
• ID 37581

KANOVICH, M.I. [1983] *Complexity and convergence of algorithmic mass problems (Russian)* (**J** 0023) Dokl Akad Nauk SSSR 272*289-293
• TRANSL [1983] (**J** 0062) Sov Math, Dokl 28*369-373
◇ D15 D30 ◇ REV MR 85i:03131 Zbl 563 # 03026
• ID 44214

KANOVICH, M.I. [1983] *Effectiveness of the translation theorem and the recursion theorem (Russian)* (**C** 3798) Mat Log, Mat Ling & Teor Algor 39-51
◇ D20 ◇ REV MR 86a:03040 • ID 48786

KANOVICH, M.I. [1983] *On the implicativity of the lattice of truth-table degrees of algorithmic problems (Russian)* (**J** 0023) Dokl Akad Nauk SSSR 270*1046-1050
• TRANSL [1983] (**J** 0062) Sov Math, Dokl 27*699-703
◇ D30 ◇ REV MR 84m:03064 Zbl 561 # 03024 • ID 35762

KANOVICH, M.I. [1983] *Reducibility via general recursive operators (Russian)* (**J** 0023) Dokl Akad Nauk SSSR 273*793-796
• TRANSL [1983] (**J** 0062) Sov Math, Dokl 28*706-710
◇ D30 ◇ REV MR 85b:03070 Zbl 566 # 03027 • ID 40695

KANOVICH, M.I. [1984] *On the independence of invariant propositions (Russian)* (**J** 0023) Dokl Akad Nauk SSSR 276*27-31
• TRANSL [1984] (**J** 0062) Sov Math, Dokl 29*425-429
◇ B30 D20 ◇ ID 45178

KANOVICH, M.I. [1984] *The solution of Roger's problem on the relation between the strong and weak recursion theorem (Russian)* (**J** 0023) Dokl Akad Nauk SSSR 279*1040-1044
• TRANSL [1984] (**J** 0062) Sov Math, Dokl 30*762-764
◇ D20 ◇ ID 46667

KANOVICH, M.I. see Vol. I, III for further entries

KANTOROVICH, L. & LIVENSON, E. [1930] *Sur les ensembles projectifs de M. Lusin* (**J** 0109) C R Acad Sci, Paris 190*1113-1115
◇ D55 E15 ◇ REV FdM 56.87 • ID 39457

KANTOROVICH, L. & LIVENSON, E. [1932] *Memoir on the analytical operations and projective sets. I* (**J** 0027) Fund Math 18*214-279
◇ D55 E15 ◇ REV Zbl 4.294 FdM 58.83 • REM Part II 1933 • ID 08214

KANTOROVICH, L. & LIVENSON, E. [1933] *Memoir on the analytical operations and projective sets. II* (**J** 0027) Fund Math 20*54-97
◇ D55 E15 ◇ REV Zbl 7.241 FdM 59.884 • REM Part I 1932 • ID 08216

KANTOROVICH, L. see Vol. V for further entries

KAPHENGST, H. [1959] *Eine abstrakte programmgesteuerte Rechenmaschine* (**J** 0068) Z Math Logik Grundlagen Math 5*366-379
◇ D10 ◇ REV Zbl 135.390 • ID 06910

KAPHENGST, H. [1969] *Malzew-Raeume, ein allgemeiner Begriff der rekursiven Abbildung* (**J** 0068) Z Math Logik Grundlagen Math 15*63-76
◇ D20 D75 ◇ REV MR 39 # 3993 Zbl 188.329 • ID 06912

KAPHENGST, H. & REICHEL, HORST [1977] *Initial algebraic semantics for non context-free languages* (**P** 2588) FCT'77 Fund of Comput Th;1977 Poznan 56*120-126
◇ B75 D05 G30 ◇ REV MR 58 # 25075 Zbl 395 # 68070 • ID 52601

KAPHENGST, H. see Vol. I for further entries

KAPUR, D. & NARENDRAN, P. [1983] *The Knuth-Bendix completion procedure and Thue systems* (**P** 3893) Found of Softw Tech & Th Comput Sci (3);1983 Bangalore 363-385
◇ D03 ◇ REV MR 85e:68040 Zbl 535 # 68012 • ID 38340

KAPUR, D. & KRISHNAMOORTHY, M.S. & MCNAUGHTON, R. & NARENDRAN, P. [1985] *An $O(|T|^3)$ algorithm for testing the Church-Rosser property of Thue systems* (**J** 1426) Theor Comput Sci 35*109-114
◇ D03 D15 ◇ REV Zbl 563 # 03018 • ID 45091

KAPUR, D. [1985] see KANDRI-RODY, A.

KAPUR, D. & KRISHNAMOORTHY, M.S. & MCNAUGHTON, R. & NARENDRAN, P. [1985] *The Church-Rosser property and special Thue systems* (**J** 1426) Theor Comput Sci 39*123-133
◇ D03 ◇ ID 49062

KAPUR, D. & NARENDRAN, P. [1985] *The Knuth-Bendix completion procedure and Thue systems* (**J** 1428) SIAM J Comp 14*1052-1072
◇ D03 ◇ ID 49276

KAPUR, D. see Vol. I, VI for further entries

KARACUBA, A.A. [1960] *Solution to a problem in the theory of finite automata (Russian)* (**J** 0067) Usp Mat Nauk 15/3*157-159
◇ D05 ◇ REV MR 22 # 10880 Zbl 97.5 • ID 47981

KARAPETYAN, B.K. & POGOSYAN, EH.M. [1977] *Inductors and their connection with the method of empirical prediction (Russian)* (**S** 0507) Vychisl Sist (Akad Nauk SSSR Novosibirsk) 69*102-112
◇ B35 D70 ◇ REV Zbl 449 # 68055 • ID 56756

KARASEK, J. [1976] *Nondeterministic and deterministic k-machines* (**J** 0014) Bull Acad Pol Sci, Ser Math Astron Phys 24*295-298
◇ D03 ◇ REV MR 55 # 9606 Zbl 329 # 68052 • ID 18219

KARELIN, V.P. [1969] see BERSHTEJN, L.S.

KARGAPOLOV, M.I. [1969] see CHURKIN, V.A.

KARGAPOLOV, M.I. see Vol. I, III for further entries

KARHUMAEKI, J. [1979] see CULIK, K.

KARHUMAEKI, J. [1979] see CULIK II, K.

KARHUMAEKI, J. [1980] see CULIK II, K.

KARHUMAEKI, J. [1981] *Generalized Parikh mappings and homomorphisms* (**P** 2903) Automata, Lang & Progr (8);1981 Akko 324-332
◇ D05 ◇ REV MR 82k:68026 Zbl 462 # 68056 • ID 81786

KARHUMAEKI, J. [1981] *On strongly cube-free ω-words generated by binary morphisms* (**P** 3165) FCT'81 Fund of Comput Th;1981 Szeged 182-189
◇ D05 ◇ REV MR 83k:20066 Zbl 462 # 68055 • ID 40278

KARHUMAEKI, J. [1981] see ALBERT, J.

KARHUMAEKI, J. [1982] see EHRENFEUCHT, A.

KARHUMAEKI, J. & LINNA, M. [1983] *A note on morphic characterisation of languages* (**J** 2702) Discr Appl Math 5*243-246
◇ B65 D05 ◇ REV MR 84b:68099 Zbl 499 # 68031 • ID 38958

KARMAZIN, V.N. [1976] *Mitteilung ueber minimale universelle Turingmaschinen (Russisch)* (**J** 0040) Kibernetika, Akad Nauk Ukr SSR 1976/2*139-140
◇ D10 ◇ REV Zbl 328 # 02022 • ID 62888

KAROL', A.M. [1978] *Complexity of the set of constructive real numbers (Russian)* (**C** 3211) Mat Ling & Teor Algor 77-82
◇ D15 F60 ◇ REV Zbl 395 # 03028 • ID 52579

KARP, C.R. [1971] see JENSEN, R.B.

KARP, C.R. see Vol. III, V for further entries

KARP, R.M. [1972] *Reducibility among combinatorial problems* (**P** 0551) Compl of Computer Computation;1972 Yorktown Heights 85-103
◇ D15 ◇ REV MR 51 # 14644 JSL 40.618 • ID 14692

KARP, R.M. (ED.) [1974] *Complexity of computation* (**X** 0803) Amer Math Soc: Providence VIII + 166pp
◇ D15 D97 ◇ REV MR 50 # 3631 Zbl 293 # 00022 • ID 48712

KARP, R.M. & LIPTON, R.J. [1982] *Turing machines that take advice* (**J** 3370) Enseign Math, Ser 2 28*191-209
• REPR [1982] (**P** 3482) Logic & Algor (Specker);1980 Zuerich 255-273
◇ D15 ◇ REV MR 84k:68036 Zbl 494 # 68061 • ID 39248

KARPINSKI, M. [1973] *Free structure tree automata I: Equivalence. II: Non-deterministic and deterministic regularity. III: Normalized climbing automata* (**J** 0014) Bull Acad Pol Sci, Ser Math Astron Phys 21*441-446,447-450,567-572
◇ D05 ◇ REV MR 58 # 9908 Zbl 282 # 02011 Zbl 282 # 02012 Zbl 282 # 02013 • REM Part IV 1974 • ID 06937

KARPINSKI, M. [1974] *Free structure tree automata IV: sequential representation* (**J** 0014) Bull Acad Pol Sci, Ser Math Astron Phys 22*87-91
◇ D05 ◇ REV MR 51 # 2351 Zbl 282 # 02014 • REM Parts I,II,III 1973 • ID 29084

KARPINSKI, M. [1974] *Probabilistic climbing and sinking languages* (**J** 0014) Bull Acad Pol Sci, Ser Math Astron Phys 22*1057-1061
◇ D05 ◇ REV MR 52 # 5234 Zbl 306 # 68040 • ID 06940

KARPINSKI, M. [1975] *Almost deterministic ω-automata with existential output condition* (**J** 0053) Proc Amer Math Soc 53*449-452
◇ D05 ◇ REV MR 55 # 10253 Zbl 332 # 02043 • ID 62891

KARPINSKI, M. [1975] *Decision algorithm for Havel's branching automata* (**P** 0454) Math Founds of Comput Sci (4);1975 Marianske Lazne 273-279
◇ D05 ◇ REV MR 53 # 5196 Zbl 335 # 02019 • ID 21664

KARPINSKI, M. [1976] *Multiplicity functions on ω-automata* (**P** 1401) Math Founds of Comput Sci (5);1976 Gdansk 596-601
◇ D05 ◇ REV Zbl 344 # 02028 • ID 62894

KARPINSKI, M. (ED.) [1977] *Fundamentals of computation theory* (**X** 0811) Springer: Heidelberg & New York xi + 542pp
◇ D97 ◇ REV MR 57 # 11147 Zbl 351 # 00018 • ID 80451

KARPINSKI, M. [1979] see DANECKI, R.

KARPINSKI, M. [1980] *The ultimate equivalence of iterated homomorphisms is recursively unsolvable* (**X** 0817) Ges Math Datenverarbeit: Bonn i + 12pp
◇ D05 ◇ REV MR 82j:68027 Zbl 426 # 28072 • ID 81787

KARPINSKI, M. [1982] *Decidability of "Skolem matrix emptiness problem" entails constructability of exact regular expression* (**J** 1426) Theor Comput Sci 17*99-102
◇ D05 ◇ REV MR 84k:03039 Zbl 498 # 03008 • ID 34975

KARPINSKI, M. (ED.) [1983] *Foundations of computation theory* (**S** 3302) Lect Notes Comput Sci 158*xi + 517pp
◇ D97 ◇ REV MR 85e:68003 Zbl 513 # 00010 • ID 39975

KARR, M. [1981] *Summation in finite terms* (**J** 0037) ACM J 28*305-350
◇ C57 ◇ REV MR 82m:12018 Zbl 494 # 68044 • ID 81790

KARR, M. see Vol. VI for further entries

KASAI, T. [1977] *Computational complexity of multitape Turing machines and random access machines* (**J** 0390) Publ Res Inst Math Sci (Kyoto) 13*469-496
◇ D10 D15 ◇ REV MR 57 # 8149 Zbl 385 # 68047 • ID 52173

KASAI, T. [1980] see ADACHI, A.

KASAI, T. [1981] see ADACHI, A.

KASAI, T. [1985] see IWATA, S.

KASAMI, T. & TANIGUCHI, K. [1971] *Some decision problems for two-dimensional nonwriting automata (Japanese)* (**J** 0979) Denshi Tsushin Gakkai Ronbunshi, Sect A-D 54-C/7*578-585
• TRANSL [1971] (**J** 0464) Syst-Comp-Controls 2/4*42-48
◇ D05 ◇ REV MR 48 # 8211 • ID 13389

KASAMI, T. [1977] see ARAKI, T.

KASAMI, T. [1979] see HAGIHARA, K.

KASHAPOVA, F.R. [1984] *Constructive set theory with types, and consistency with Church's thesis (Russian)* (**J** 0288) Vest Ser Mat Mekh, Univ Moskva 1984/4*72-75
• TRANSL [1984] (**J** 0510) Moscow Univ Math Bull 39/4*470-474
◇ D20 E70 F35 F50 ◇ REV MR 85j:03105 • ID 44311

KASHAPOVA, F.R. see Vol. VI for further entries

KASHEF, R.S. & ROWAN, J.H. [1973] *A universal programmable cellular array* (**P** 3244) Hawaii Int Conf Syst Sci (6);1973 Honolulu 264-267
◇ D10 ◇ REV Zbl 354 # 94042 • ID 50195

KASHINTSEV, E.V. [1970] *Graphs and the word problem for finitely presented semigroups (Russian)* (**J** 0789) Uch Zap Mat Ped Inst, Tula 2*290-302
◇ D40 ◇ REV MR 52 # 14097 • ID 21882

KASHINTSEV, E.V. [1970] *On the word problem (Russian)* (**J** 0789) Uch Zap Mat Ped Inst, Tula 2*185-214
◇ D40 ◇ REV MR 53 # 627 • ID 16683

KASHINTSEV, E.V. [1978] *An algorithm for the solution of the conjugacy problem for certain semigroups (Russian)* (**C** 2595) Rekursiv Funktsii 18-26
◇ D40 ◇ REV MR 82h:20068 Zbl 504 # 03020 • ID 81795

KASHINTSEV, E.V. [1978] *On the word problem for special semigroups (Russian)* (**J** 0216) Izv Akad Nauk SSSR, Ser Mat 42*1401-1416,1440
• TRANSL [1978] (**J** 0448) Math of USSR, Izv 13*663-676
◇ D40 ◇ REV MR 80c:20080 Zbl 401 # 20051 • ID 81796

KASTANAS, I.G. [1984] *The jump inversion theorem for Q_{2n+1}-degrees* (**J** 0053) Proc Amer Math Soc 90*422-424
◇ D30 E15 E60 ◇ REV MR 85h:03050 Zbl 545 # 03027 • ID 43508

KASTANAS, I.G. see Vol. V for further entries

KATCOFF, J. [1982] see COHEN, JACQUES

KATERINOCHKINA, N.N. [1970] *The equivalence of certain computational equipment (Russian) (English summary)* (**J** 0040) Kibernetika, Akad Nauk Ukr SSR 1970/5*27-31
◇ D10 D15 D20 ◇ REV MR 45 # 1766 Zbl 249 # 02018
• ID 06946

KATERINOCHKINA, N.N. see Vol. I, II for further entries

KATS, B.E. [1978] *Regular Buechi systems and time-space trade-off in language recognition (Russian)* (**C** 3211) Mat Ling & Teor Algor 83-104
◇ D15 ◇ REV Zbl 395 # 03029 • ID 52580

KATSEFF, H.P. & SIPSER, M. [1981] *Several results in program size complexity* (**J** 1426) Theor Comput Sci 15*291-309
◇ D15 ◇ REV MR 82i:68028 Zbl 459 # 68014 • ID 81801

KAUFMAN, V.SH. [1978] see BUNIMOVA, E.O.

KAUFMANN, M. [1977] *A rather classless model* (**J** 0053) Proc Amer Math Soc 62*330-333
◇ C50 C57 C62 E65 ◇ REV MR 57 # 16058
Zbl 359 # 02054 • ID 31400

KAUFMANN, M. & KRANAKIS, E. [1984] *Definable ultrapowers and ultrafilters over admissible ordinals* (**J** 0068) Z Math Logik Grundlagen Math 30*97-118
◇ C20 C62 D60 E05 E45 ◇ REV MR 86e:03037
Zbl 519 # 03022 • ID 41225

KAUFMANN, M. & SCHMERL, J.H. [1984] *Saturation and simple extensions of models of Peano-arithmetic* (**J** 0073) Ann Pure Appl Logic 27*109-136
◇ C50 C57 C62 ◇ REV MR 85j:03051 Zbl 557 # 03021
• ID 39638

KAUFMANN, M. see Vol. I, III, V, VI for further entries

KAZANOVICH, YA.B. [1970] *A classification of primitive recursive functions using Turing machines (Russian)* (**J** 0052) Probl Kibern 22*95-106,297
• TRANSL [1972] (**J** 0471) Syst Th Res 22*93-104
◇ D10 D20 ◇ REV MR 45 # 6622 Zbl 241 # 02011
• ID 06978

KECHRIS, A.S. & MOSCHOVAKIS, Y.N. [1972] *Two theorems about projective sets* (**J** 0029) Israel J Math 12*391-399
◇ D55 E15 E45 E60 ◇ REV MR 48 # 1900
Zbl 257 # 02034 • ID 06986

KECHRIS, A.S. [1973] *Descriptive set theory* (1111) Preprints, Manuscr., Techn. Reports etc.
◇ D55 E15 E45 E55 E60 ◇ REM Lecture Notes, MIT Cambridge, MA • ID 21379

KECHRIS, A.S. [1973] *Measure and category in effective descriptive set theory* (**J** 0007) Ann Math Logic 5*337-384
◇ D55 E15 E60 ◇ REV MR 51 # 5308 Zbl 277 # 02019
• ID 06988

KECHRIS, A.S. [1973] *The structure of envelopes: a survey of recursion theory in higher types* (**X** 0865) MIT Pr: Cambridge, MA 28pp
◇ D65 D75 ◇ REM Logic seminar notes • ID 38725

KECHRIS, A.S. [1974] *On projective ordinals* (**J** 0036) J Symb Logic 39*269-282
◇ D55 E10 E15 E60 ◇ REV MR 53 # 2684
Zbl 292 # 02056 • ID 06989

KECHRIS, A.S. [1975] see HARRINGTON, L.A.

KECHRIS, A.S. & MARTIN, D.A. [1975] *A note on universal sets for classes of countable G_δ's* (**J** 0303) Mathematika (Univ Coll London) 22*43-45
◇ D55 E15 ◇ REV MR 54 # 8593 Zbl 307 # 54040
• ID 30723

KECHRIS, A.S. [1975] *Countable ordinals and the analytical hierarchy I* (**J** 0048) Pac J Math 60*223-227
◇ C62 D55 E10 E15 E60 ◇ REV MR 52 # 7900
Zbl 287 # 02042 • REM Part II 1978 • ID 14831

KECHRIS, A.S. [1975] *The theory of countable analytical sets* (**J** 0064) Trans Amer Math Soc 202*259-297
◇ D55 E15 E45 E60 ◇ REV MR 54 # 7259
Zbl 317 # 02082 • ID 06990

KECHRIS, A.S. [1976] see HARRINGTON, L.A.

KECHRIS, A.S. [1977] see HARRINGTON, L.A.

KECHRIS, A.S. [1977] *Classifying projective-like hierarchies* (**J** 0465) Bull Greek Math Soc (NS) 18*254-275
◇ D55 D75 E15 E45 E60 ◇ REV MR 80m:03083
Zbl 417 # 03022 • ID 53258

KECHRIS, A.S. [1977] *On a notion of smallness for subsets of the Baire space* (**J** 0064) Trans Amer Math Soc 229*191-207
◇ D55 E15 E60 ◇ REV MR 56 # 8369 Zbl 401 # 03022
• ID 26243

KECHRIS, A.S. & MOSCHOVAKIS, Y.N. [1977] *Recursion in higher types* (**C** 1523) Handb of Math Logic 681-737
◇ D65 D70 D75 ◇ REV MR 58 # 5109 JSL 49.975
• ID 27324

KECHRIS, A.S. [1978] *AD and projective ordinals* (**C** 2908) Cabal Seminar Los Angeles 1976-77 91-132
◇ D55 E15 E55 E60 ◇ REV MR 80j:03069
Zbl 434 # 03036 • ID 55740

KECHRIS, A.S. & MOSCHOVAKIS, Y.N. (EDS.) [1978] *Cabal seminar 76-77* (**X** 0811) Springer: Heidelberg & New York iii + 282pp
◇ D55 D97 E15 E60 E97 ◇ REV MR 80b:03004
Zbl 379 # 00001 JSL 50.849 • ID 51832

KECHRIS, A.S. [1978] *Countable ordinals and the analytical hierarchy II* (**J** 0007) Ann Math Logic 15*193-223
◇ D55 E15 E60 ◇ REV MR 81b:03050 Zbl 449 # 03047
• REM Part I 1975 • ID 56714

KECHRIS, A.S. [1978] *Forcing in analysis* (**P** 1864) Higher Set Th;1977 Oberwolfach 277-302
◇ D55 E15 E40 E60 ◇ REV MR 80c:03051
Zbl 391 # 03024 • ID 52351

KECHRIS, A.S. [1978] *Minimal upper bounds for sequences of Δ^1_{2n}-degrees* (**J** 0036) J Symb Logic 43*502-507
◇ D30 D55 E60 ◇ REV MR 81e:03043 Zbl 405 # 03019
• ID 29278

KECHRIS, A.S. & MOSCHOVAKIS, Y.N. [1978] *Notes on the theory of scales* (**C** 2908) Cabal Seminar Los Angeles 1976-77 1-53
◇ D55 E15 E45 E60 E98 ◇ REV MR 83b:03059
Zbl 397 # 03032 • ID 52698

KECHRIS, A.S. & MARTIN, D.A. [1978] *On the theory of Π^1_3 sets of reals* (**J** 0015) Bull Amer Math Soc 84*149-151
◇ D55 E15 E45 E60 ◇ REV MR 57 # 5753
Zbl 385 # 03037 • ID 30726

KECHRIS, A.S. [1978] *On transfinite sequences of projective sets with an application to bold-face-Σ^1_2 equivalence relations* (**P** 1897) Logic Colloq;1977 Wroclaw 155-160
◇ D55 E15 E60 ◇ REV MR 80j:03070 Zbl 448 # 03040 • ID 56640

KECHRIS, A.S. [1978] *On Spector classes* (**C** 2908) Cabal Seminar Los Angeles 1976-77 245-277
◇ D55 D65 D70 D75 E15 E60 ◇ REV MR 81b:03053 Zbl 405 # 03022 • ID 54897

KECHRIS, A.S. [1978] *Spector second order classes and reflection* (**P** 1628) Generalized Recursion Th (2);1977 Oslo 147-183
◇ D65 D70 D75 ◇ REV MR 81i:03072 Zbl 453 # 03047 • ID 74736

KECHRIS, A.S. [1978] *The perfect set theorem and definable wellorderings of the continuum* (**J** 0036) J Symb Logic 43*630-634
◇ D55 E15 E45 E60 ◇ REV MR 80b:03067 Zbl 401 # 03023 • ID 74744

KECHRIS, A.S. [1979] *An overview of descriptive set theory* (**C** 3825) Semin Init Analyse (18) Paris 1978/79 Exp.4*35pp
◇ D55 D98 E15 E98 ◇ REV MR 84h:03114 Zbl 496 # 03026 • ID 34337

KECHRIS, A.S. [1980] see DELLACHERIE, C.

KECHRIS, A.S. & MARTIN, D.A. [1980] *Infinite games and effective descriptive set theory* (**P** 4522) Anal Sets 404-470
◇ D55 E15 E55 E60 E98 ◇ REV MR 82m:03063 • ID 40463

KECHRIS, A.S. [1980] *Recent advances in the theory of higher level projective sets* (**P** 2058) Kleene Symp;1978 Madison 149-166
◇ D55 E15 E60 ◇ REV MR 82b:03090 Zbl 451 # 03015 • ID 54030

KECHRIS, A.S. [1981] *A note on Wadge degrees* (**C** 2922) Cabal Seminar Los Angeles 1977-79 165-168
◇ D30 D55 E15 E60 ◇ REV MR 82j:03061 Zbl 485 # 03033 • ID 74733

KECHRIS, A.S. & MARTIN, D.A. & MOSCHOVAKIS, Y.N. (EDS.) [1981] *Cabal seminar 77 - 79* (**S** 3301) Lect Notes Math 839*v+274pp
◇ D55 D97 E15 E60 E97 ◇ REV MR 82c:03002 Zbl 456 # 00005 JSL 50.849 • ID 54310

KECHRIS, A.S. [1981] *Forcing with Δ perfect trees and minimal Δ-degrees* (**J** 0036) J Symb Logic 46*803-816
◇ D30 D55 D75 E15 E40 E60 ◇ REV MR 83e:03080 Zbl 485 # 03025 • ID 34923

KECHRIS, A.S. [1981] *Homogeneous trees and projective scales* (**C** 2922) Cabal Seminar Los Angeles 1977-79 33-73
◇ D55 E05 E15 E60 ◇ REV MR 82j:03060 Zbl 485 # 03029 • ID 74734

KECHRIS, A.S. [1981] *Souslin cardinals, κ-Souslin sets and the scale property in the hyperprojective hierarchy* (**C** 2922) Cabal Seminar Los Angeles 1977-79 127-146
◇ D55 E15 E60 ◇ REV MR 82k:03079 Zbl 485 # 03032 • ID 74732

KECHRIS, A.S. & SOLOVAY, R.M. & STEEL, J.R. [1981] *The axiom of determinacy and the prewellordering property* (**C** 2922) Cabal Seminar Los Angeles 1977-79 101-125
◇ D30 E15 E45 E60 ◇ REV MR 83f:03042 Zbl 485 # 03031 • ID 35283

KECHRIS, A.S. [1982] *Effective Ramsey theorems in the projective hierarchy* (**P** 3708) Herbrand Symp Logic Colloq;1981 Marseille 179-187
◇ D55 E05 E15 E60 ◇ REV MR 85k:03031 Zbl 507 # 03023 • ID 37240

KECHRIS, A.S. & MARTIN, D.A. & MOSCHOVAKIS, Y.N. (EDS.) [1983] *Cabal seminar 79-81. Proceedings, Caltech-UCLA logic seminar 1979-81* (**S** 3301) Lect Notes Math 1019*284pp
◇ D55 D97 E15 E60 E97 ◇ REV Zbl 511 # 00005 • ID 36754

KECHRIS, A.S. & MARTIN, D.A. & SOLOVAY, R.M. [1983] *Introduction to Q-theory* (**C** 3875) Cabal Seminar Los Angeles 1979-81 199-282
◇ D55 E15 E45 E60 ◇ ID 40468

KECHRIS, A.S. [1985] *Determinacy and the structure of L(R)* (**P** 4046) Rec Th;1982 Ithaca 271-283
◇ D55 E05 E15 E45 E60 ◇ REV Zbl 573 # 03027 • ID 40477

KECHRIS, A.S. & SOLOVAY, R.M. [1985] *On the relative consistency strength of determinacy hypotheses* (**J** 0064) Trans Amer Math Soc 290*179-211
◇ D55 E15 E35 E45 E60 ◇ ID 45614

KECHRIS, A.S. see Vol. V for further entries

KEISLER, H.J. & WALKOE JR., W.J. [1973] *The diversity of quantifier prefixes* (**J** 0036) J Symb Logic 38*79-85
◇ B10 C07 C13 D55 ◇ REV MR 51 # 12472 Zbl 259 # 02007 • ID 07033

KEISLER, H.J. see Vol. I, II, III, V, VI for further entries

KEKLIKOGLOU, I. [1981] *Verallgemeinerte kommutative Sprachen (English summary)* (**P** 3475) Theor Comput Sci (5);1981 Karlsruhe 70-75
◇ D05 ◇ REV Zbl 456 # 68092 • ID 46466

KELDYSH, L.V. [1945] *Sur les transformations ouvertes des ensembles* (**J** 0023) Dokl Akad Nauk SSSR 49*622-624
◇ D55 E15 ◇ REV MR 8.16 Zbl 60.402 • ID 24914

KELDYSH, L.V. [1974] *Ideas of N.N.Luzin in descriptive set theory (Russian)* (**J** 0067) Usp Mat Nauk 29/5*183-196
• TRANSL [1974] (**J** 1399) Russ Math Surv 29/5*179-193
◇ A10 D55 E15 ◇ REV MR 52 # 7809 Zbl 304 # 01013 • ID 18221

KELDYSH, L.V. see Vol. V for further entries

KELMANS, A.K. [1985] *The existence of a "most complex" problem in the class of problems verifiable in nonpolynomial time (Russian)* (**J** 0023) Dokl Akad Nauk SSSR 282*1299-1303
◇ D15 ◇ ID 48591

KEMPNER, A.J. [1936] *Remarks on "unsolvable" problems* (**J** 0005) Amer Math Mon 43*467-473
◇ D20 D35 ◇ REV Zbl 15.145 JSL 2.41 FdM 62.1060 • ID 41309

KENT, C.F. [1962] *Constructive analogues of the group of permutations of the natural numbers* (**J** 0064) Trans Amer Math Soc 104*347-362
◇ C57 D45 ◇ REV MR 25 # 3826 Zbl 105.248 JSL 34.517 • ID 07059

KENT, C.F. [1969] see FABIAN, R.J.

KENT, C.F. [1969] *Reducing ordinal recursion* (J 0053) Proc Amer Math Soc 22*690-696
◇ D20 ◇ REV MR 42 # 69 Zbl 317 # 02042 • ID 07060

KENT, C.F. [1982] see HODGSON, B.R.

KENT, C.F. [1983] see HODGSON, B.R.

KENT, C.F. see Vol. I, VI for further entries

KENZHEBAEV, S. [1966] *Certain undecidable rings (Russian) (Kazakh summary)* (J 0403) Izv Akad Nauk Kazak SSR, Ser Fiz-Mat 1966/1*25-30
◇ C60 D35 ◇ REV MR 34 # 1346 Zbl 199.31 • ID 16220

KENZHEBAEV, S. [1966] *Undecidability of the elementary theory of a certain ring (Russian)* (J 0468) Uch Zap Ped Inst (Alma Ata) 23*24-25
◇ D35 ◇ REV MR 37 # 2604 • ID 07062

KERSTAN, J. [1961] *Zur topologischen Invarianz der Hausdorffschen $Q^{(\alpha)}$-Mengen* (J 0068) Z Math Logik Grundlagen Math 7*259-277
◇ D55 E15 ◇ REV MR 25 # 4491 Zbl 123.158 • ID 07066

KESEL'MAN, D.YA. [1974] *The elementary theories of graphs and abelian loops (Russian)* (J 0087) Mat Zametki (Akad Nauk SSSR) 16*957-968
• TRANSL [1974] (J 1044) Math Notes, Acad Sci USSR 16*1167-1171
◇ C65 D35 ◇ REV MR 51 # 7959 Zbl 311 # 02063
• ID 17588

KESEL'MAN, D.YA. [1982] *Decidability of theories of certain classes of elimination graphs* (J 0040) Kibernetika, Akad Nauk Ukr SSR 1982/3i*34-37,133
• TRANSL [1982] (J 0021) Cybernetics 18*305-310
◇ B25 C65 D35 ◇ REV MR 84m:05076 • ID 39437

KFOURY, A.J. [1982] see ARBIB, M.A.

KFOURY, A.J. & URZYCZYN, P. [1985] *Necessary and sufficient conditions for the universality of programming formalisms* (J 1431) Acta Inf 22*347-377
◇ B75 D75 ◇ REV Zbl 552 # 68008 • ID 48239

KFOURY, A.J. see Vol. I, II for further entries

KHACHATRYAN, V.E. & PETROSYAN, G.N. & PODLOVCHENKO, R.I. [1972] *Interpretations of schemes of algorithms, and various types of equivalence relations between schemes (Russian) (Armenian and English summaries)* (J 0312) Izv Akad Nauk Armyan SSR, Ser Mat 7*140-151
◇ B75 D20 ◇ REV MR 46 # 10240 Zbl 243 # 02023
• ID 10575

KHACHIYAN, L.G. & TARASOV, S.P. [1980] *Bounds of solutions and algorithmic complexity of systems of convex diophantine inequalities (Russian)* (J 0023) Dokl Akad Nauk SSSR 255*296-300
• TRANSL [1980] (J 0062) Sov Math, Dokl 22*700-704
◇ D15 ◇ REV MR 82f:68052 Zbl 467 # 90048 • ID 82952

KHACHIYAN, L.G. & KOZLOV, M.K. & TARASOV, S.P. [1980] *The polynomial solvability of convex quadratic programming (Russian)* (J 0199) Zh Vychisl Mat i Mat Fiz 20*1319-1323
• TRANSL [1980] (J 1049) USSR Comput Math & Math Phys 20/5*223-228
◇ D10 D15 ◇ REV MR 82c:90070 Zbl 475 # 90068
• ID 38400

KHADZHIEV, D. [1984] see BEKBAEV, U.D.

KHAKHANYAN, V.KH. [1980] *Comparative strength of variants of Church's thesis at the level of set theory (Russian)* (J 0023) Dokl Akad Nauk SSSR 252*1070-1074
• TRANSL [1980] (J 0062) Sov Math, Dokl 21*894-898
◇ D20 E70 F50 ◇ REV MR 81g:03069 Zbl 482 # 03026
• ID 73755

KHAKHANYAN, V.KH. [1980] *The consistency of intuitionistic set theory with Church's principle and the uniformization principle (Russian)* (J 0288) Vest Ser Mat Mekh, Univ Moskva 1980/5*3-7
• TRANSL [1980] (J 0510) Moscow Univ Math Bull 35/5*1-5
◇ D20 E35 E70 F25 F50 ◇ REV MR 81h:03109 MR 82b:03106 Zbl 445 # 03031 • ID 56495

KHAKHANYAN, V.KH. [1980] *The consistency of intuitionistic set theory with formal mathematical analysis (Russian)* (J 0023) Dokl Akad Nauk SSSR 253*48-52
• TRANSL [1980] (J 0062) Sov Math, Dokl 22*46-50
◇ D20 E35 E70 F25 F35 F50 ◇ REV MR 81h:03109 Zbl 531 # 03036 • ID 73754

KHAKHANYAN, V.KH. [1981] *The consistency of some intuitionistic and constructive principles with a set theory* (J 0063) Studia Logica 40*237-248
◇ D20 E35 E70 F50 ◇ REV MR 84e:03061 Zbl 491 # 03019 • ID 36853

KHAKHANYAN, V.KH. [1983] *Set theory and Church's thesis (Russian)* (C 3807) Issl Neklass Log & Formal Sist 198-208
◇ D20 E35 E70 F30 F35 F50 ◇ REV MR 85d:03117
• ID 41114

KHARLAMPOVICH, O.G. [1981] *A finitely presented solvable group with undecidable word problem (Russian)* (J 0216) Izv Akad Nauk SSSR, Ser Mat 45*852-873,928
• TRANSL [1982] (J 0448) Math of USSR, Izv 19*151-169
◇ D40 ◇ REV MR 82m:20036 Zbl 485 # 20023 • ID 81494

KHARLAMPOVICH, O.G. [1983] *The universal theory of the class of finite nilpotent groups is undecidable (Russian)* (J 0087) Mat Zametki (Akad Nauk SSSR) 33*499-516
• TRANSL [1983] (J 1044) Math Notes, Acad Sci USSR 33*254-263
◇ C60 D35 ◇ REV MR 85b:20046 Zbl 516 # 20018
• ID 39367

KHARLAMPOVICH, O.G. see Vol. III for further entries

KHASIN, L.S. [1969] *Complexity bounds for the realization of monotonic symmetrical functions by means of formulas in the basis* $\vee$, $\wedge$, $\neg$ *(Russian)* (J 0023) Dokl Akad Nauk SSSR 189*752-755
• TRANSL [1969] (J 0470) Sov Phys, Dokl 14*1149-1151
◇ B05 D20 ◇ REV MR 41 # 6621 Zbl 206.290 • ID 07080

KHASIN, L.S. see Vol. I, II for further entries

KHISAMIEV, N.G. [1967] *Unsolvability of the elementary theory of a free lattice (Russian)* (J 0003) Algebra i Logika 6/5*45-48
◇ D35 G10 ◇ REV MR 37 # 3921 Zbl 209.23 • ID 06145

KHISAMIEV, N.G. [1971] *Strongly constructive models (Russian) (Kazakh summary)* (J 0403) Izv Akad Nauk Kazak SSR, Ser Fiz-Mat 1971/3*59-63
◇ C30 C57 ◇ REV MR 45 # 4960 Zbl 305 # 02062
• ID 06146

KHISAMIEV, N.G. [1974] *Strongly constructive models of a decidable theory (Russian)* (**J** 0403) Izv Akad Nauk Kazak SSR, Ser Fiz-Mat 1974/1*83-84,94
◇ C15 C57 F60 ◇ REV MR 50#6824 Zbl 275#02051 • ID 06147

KHISAMIEV, N.G. [1977] *On the periodical part of a strongly constructivizable abelian group (Russian)* (**C** 2557) Teor & Priklad Zad Mat & Mekh 299-303
◇ C57 C60 D45 ◇ REV MR 80c:03046 • ID 32598

KHISAMIEV, N.G. [1978] see DOBRITSA, V.P.

KHISAMIEV, N.G. [1978] *Strongly constructive periodic abelian groups (Russian) (Kazakh summary)* (**J** 0403) Izv Akad Nauk Kazak SSR, Ser Fiz-Mat 1978/1*58-62,92
◇ C57 C60 D45 ◇ REV MR 80b:03040 Zbl 396#20034 • ID 32599

KHISAMIEV, N.G. [1979] *Criterion of the constructivizability of a direct sum of cyclic p-groups (Russian)* (**P** 2558) All-Union Conf Math Log (5) (Mal'tsev);1979 Novosibirsk 157
◇ C57 C60 D45 ◇ REV MR 82h:20042 • ID 32602

KHISAMIEV, N.G. [1979] *On subgroups of finite index of abelian groups (Russian) (Kazakh summary)* (**J** 0403) Izv Akad Nauk Kazak SSR, Ser Fiz-Mat 1979/3*43-47,89
◇ C57 C60 D45 F60 ◇ REV MR 81b:20037 Zbl 416#20049 • ID 32601

KHISAMIEV, N.G. [1981] *Criterion for constructivizability of a direct sum of cyclic p-groups (Russian) (Kazakh summary)* (**J** 0403) Izv Akad Nauk Kazak SSR, Ser Fiz-Mat 1981/1*51-55,86
◇ C57 C60 D45 ◇ REV MR 82h:20042 Zbl 457#20001 • ID 81571

KHISAMIEV, N.G. [1983] *Strongly constructive abelian p-groups (Russian)* (**J** 0003) Algebra i Logika 22*198-217
• TRANSL [1983] (**J** 0069) Algeb and Log 22*142-158
◇ C57 C60 D45 F60 ◇ REV MR 85e:03105 Zbl 568#20052 • ID 40239

KHISAMIEV, N.G. [1984] see ABDRAZAKOV, K.T.

KHISAMIEV, N.G. [1984] *Connection between constructivizability and strong constructivizability for different classes of abelian groups (Russian)* (**J** 0003) Algebra i Logika 23*319-335,363
• TRANSL [1984] (**J** 0069) Algeb and Log 23*220-233
◇ C57 C60 D45 ◇ REV MR 86h:03080 • ID 42697

KHISAMIEV, N.G. & KHISAMIEV, Z.G. [1985] *Nonconstructivizability of the reduced part of a strongly constructive torsion-free abelian group (Russian)* (**J** 0003) Algebra i Logika 24*108-118,123
• TRANSL [1985] (**J** 0069) Algeb and Log 24*69-76
◇ C57 C60 D45 ◇ ID 49409

KHISAMIEV, N.G. see Vol. III for further entries

KHISAMIEV, Z.G. [1984] *Multiple m-reducibility of index sets (Russian)* (**J** 0092) Sib Mat Zh 25/4*192-198
• TRANSL [1984] (**J** 0475) Sib Math J 25*665-671
◇ D25 D30 ◇ REV MR 86a:03042 • ID 45476

KHISAMIEV, Z.G. [1985] see KHISAMIEV, N.G.

KHMELEVSKIJ, YU.I. [1964] *The solution of certain systems of word equations (Russian)* (**J** 0023) Dokl Akad Nauk SSSR 156*749-751
• TRANSL [1964] (**J** 0062) Sov Math, Dokl 5*724-727
◇ D40 ◇ REV MR 29#1145 Zbl 207.13 • ID 06158

KHMELEVSKIJ, YU.I. [1966] *Word equations without coefficients (Russian)* (**J** 0023) Dokl Akad Nauk SSSR 171*1047-1049
• TRANSL [1966] (**J** 0062) Sov Math, Dokl 7*1611-1613
◇ D40 ◇ REV MR 35#1685 Zbl 207.13 • ID 06159

KHMELEVSKIJ, YU.I. [1967] *Solution of word equations in three unknowns (Russian)* (**J** 0023) Dokl Akad Nauk SSSR 177*1023-1025
• TRANSL [1967] (**J** 0062) Sov Math, Dokl 8*1554-1556
◇ D40 ◇ REV MR 36#3899 Zbl 207.13 • ID 06160

KHMELEVSKIJ, YU.I. [1969] *On Hilbert's tenth problem (Russian)* (**C** 3724) Probl Gil'berta 141-153
◇ D35 ◇ REV MR 40#7228 Zbl 213.288 • ID 21047

KHMELEVSKIJ, YU.I. [1971] *Equations in free semigroups (Russian)* (**S** 0066) Tr Mat Inst Steklov 107*286pp
• TRANSL [1971] (**S** 0055) Proc Steklov Inst Math 107*iii+270pp
◇ D40 ◇ REV MR 51#5808 Zbl 224#02037 • ID 62571

KHMELEVSKIJ, YU.I. see Vol. III for further entries

KHODZHAEV, D. [1967] *On the complexity of calculations on Turing machines with an oracle (Russian)* (**J** 0430) Vopr Kibern (Akad Nauk Uzb SSR) 12*69-76
◇ D10 D15 ◇ REV MR 38#4317 • ID 06172

KHODZHAYANTS, M.YU. [1979] *Some characteristics of enumeration operators (Russian)* (**J** 0312) Izv Akad Nauk Armyan SSR, Ser Mat 14*142-149,156
• TRANSL [1979] (**J** 3265) Sov J Contemp Math Anal, Armen Acad Sci 14/2*64-72
◇ D15 D25 ◇ REV MR 81c:03033 Zbl 418#03031 • ID 74223

KHODZHAYANTS, M.YU. [1979] *Structure of e-degrees (Russian)* (**J** 0346) Dokl Akad Nauk Armyan SSR 68*35-38
◇ D25 D30 ◇ REV MR 80k:03043 Zbl 412#03026 • ID 74224

KHODZHAYANTS, M.YU. [1980] *Structure of e-degrees (Russian) (Armenian and English summaries)* (**J** 0312) Izv Akad Nauk Armyan SSR, Ser Mat 15*165-175,246
◇ D25 D30 ◇ REV MR 82a:03039 Zbl 497#03033 • ID 74222

KHODZHAYANTS, M.YU. [1981] *e-degrees, T-degrees and axiomatic theories (Russian) (Armenian summary)* (**J** 0346) Dokl Akad Nauk Armyan SSR 73*73-77
◇ D30 ◇ REV MR 83c:03038 Zbl 519#03034 • ID 35140

KHOLSHCHEVNIKOVA, N.N. [1985] *Uncountable R- and N-sets (Russian)* (**J** 0087) Mat Zametki (Akad Nauk SSSR) 38*270-277,349
◇ D55 D70 E15 ◇ ID 49292

KHOLSHCHEVNIKOVA, N.N. see Vol. V for further entries

KHOMICH, V.I. [1970] *The complexity of the algorithms that are connected with the realization of logico-arithmetical and propositional formulae (Russian)* (**J** 0023) Dokl Akad Nauk SSSR 191*1004-1006
• TRANSL [1970] (**J** 0062) Sov Math, Dokl 11*500-502
◇ D15 ◇ REV MR 42#76 Zbl 212.332 • ID 48031

KHOMICH, V.I. [1975] *Weakly and strongly nonrealizable propositional formulae (Russian)* (**J** 0068) Z Math Logik Grundlagen Math 21*267-288
◇ B20 D15 D20 F50 ◇ REV MR 51#10048 Zbl 309#02021 • ID 17549

KHOMICH, V.I. & MARKOV, A.A. (EDS.) [1979] *Studies in the theory of algorithms and mathematical logic (Russian)* (**X** 2027) Nauka: Moskva 135pp
◇ B97 D97 ◇ REV MR 81b:03003 • ID 70052

KHOMICH, V.I. see Vol. I, II, VI for further entries

KHRAPCHENKO, V.M. [1973] *A quadratic estimate from below for complexity that is based on the continuity of the second derivative (Russian)* (**J** 0052) Probl Kibern 26*203-206,327
◇ D15 ◇ REV MR 50 # 12466 Zbl 282 # 68026 • ID 62615

KHRAPCHENKO, V.M. see Vol. I for further entries

KHUDYAKOV, V.V. & SOLON, B.YA. [1971] *On certain algebras of primitive recursive functions (Russian)* (**P** 2586) All-Union Algeb Conf (11);1971 Kishinev
◇ D20 ◇ ID 90057

KHUTORETSKIJ, A.B. [1965] *On recursive isomorphisms (Russian)* (**J** 0003) Algebra i Logika 4/3*85-88
◇ D20 D45 ◇ REV MR 33 # 3921 Zbl 192.50 JSL 34.117 • ID 19027

KHUTORETSKIJ, A.B. [1969] *On nonprincipal enumeration (Russian)* (**J** 0003) Algebra i Logika 8*726-732
• TRANSL [1969] (**J** 0069) Algeb and Log 8*412-415
◇ D25 D45 ◇ REV MR 45 # 1756 Zbl 259 # 02030 • ID 06351

KHUTORETSKIJ, A.B. [1969] *The reducibility of computable enumerations (Russian)* (**J** 0003) Algebra i Logika 8*251-264
• TRANSL [1969] (**J** 0069) Algeb and Log 8*145-151
◇ D20 D25 D45 ◇ REV MR 42 # 66 Zbl 317 # 02043 • ID 06352

KHUTORETSKIJ, A.B. [1969] *Two existence theorems for computable numerations (Russian)* (**J** 0003) Algebra i Logika 8*483-492
• TRANSL [1969] (**J** 0069) Algeb and Log 8*277-282
◇ D25 D45 ◇ REV MR 44 # 6485 Zbl 298 # 02034 • ID 06350

KHUTORETSKIJ, A.B. [1971] *On the cardinality of the upper semilattice of computable enumerations (Russian)* (**J** 0003) Algebra i Logika 10*561-569
• TRANSL [1971] (**J** 0069) Algeb and Log 10*348-352
◇ D25 D30 D45 ◇ REV MR 46 # 1577 Zbl 298 # 02035 • ID 06353

KIERCZAK, J. [1985] *Rough grammars* (**J** 2095) Fund Inform, Ann Soc Math Pol, Ser 4 8*73-81
◇ D05 ◇ ID 47258

KIERSTEAD, D.P. [1980] *A semantics for Kleene's j-expressions* (**P** 2058) Kleene Symp;1978 Madison 353-366
◇ D65 ◇ REV MR 82f:03039 Zbl 469 # 03031 • ID 55158

KIERSTEAD, D.P. [1983] *Syntax and semantics in higher-type recursion theory* (**J** 0064) Trans Amer Math Soc 276*67-105
◇ D65 ◇ REV MR 84k:03117 Zbl 506 # 03010 • ID 36135

KIERSTEAD, H.A. [1981] *An effective version of Dilworth's theorem* (**J** 0064) Trans Amer Math Soc 268*63-77
◇ D45 D80 ◇ REV MR 82k:03069 Zbl 485 # 03019 • ID 74800

KIERSTEAD, H.A. & TROTTER JR., W.T. [1981] *An extremal problem in recursive combinatorics* (**P** 3866) SE Conf Combin, Graph Th & Comput (12);1981 Baton Rouge 143-153
◇ D80 E07 ◇ REV MR 84d:06003 Zbl 489 # 05001 • ID 36594

KIERSTEAD, H.A. [1981] *Recursive colorings of highly recursive graphs* (**J** 0017) Canad J Math 33*1279-1290
◇ D80 ◇ REV MR 84b:05045 Zbl 458 # 05034 • ID 38900

KIERSTEAD, H.A. [1983] *An effective version of Hall's theorem* (**J** 0053) Proc Amer Math Soc 88*124-128
◇ C57 D80 ◇ REV MR 84g:03065 Zbl 533 # 03028 • ID 34172

KIERSTEAD, H.A. & REMMEL, J.B. [1983] *Indiscernibles and decidable models* (**J** 0036) J Symb Logic 48*21-32
◇ C30 C45 C57 C80 ◇ REV MR 84k:03103 Zbl 523 # 03030 • ID 36120

KIERSTEAD, H.A. & MCNULTY, G.F. & TROTTER JR., W.T. [1984] *A theory of recursive dimension of ordered sets* (**J** 3457) Order 1*67-82
◇ C65 D45 ◇ REV MR 86a:06003 Zbl 562 # 06001 • ID 45164

KIERSTEAD, H.A. & REMMEL, J.B. [1985] *Degrees of indiscernibles in decidable models* (**J** 0064) Trans Amer Math Soc 289*41-57
◇ C30 C35 C57 D30 ◇ ID 44599

KIERSTEAD, H.A. see Vol. III for further entries

KIERSZENBAUM, F. [1978] see CHOTTIN, L.

KILMISTER, C.W. [1967] *Language, logic and mathematics* (**X** 1165) Hodder & Stroughton: London v+124pp
◇ A05 B98 D35 ◇ REV MR 37 # 1223 Zbl 162.11 • ID 22338

KIM, C.E. [1976] see IBARRA, O.H.

KIM, S.M. [1984] see IBARRA, O.H.

KIMBALL, J. [1967] *Predicates definable over transformational derivations by intersection with regular languages* (**J** 0194) Inform & Control 11*177-195
◇ B65 D05 ◇ REV MR 39 # 3932 Zbl 156.254 JSL 34.137 • ID 07088

KIMURA, NAOKI & TAMURA, T. [1954] *On decompositions of a commutative semigroup* (**J** 0536) Kodai Math Sem Rep (Kyoto) 4*109-112
◇ D40 ◇ REV MR 16.670 Zbl 58.15 • ID 13357

KINBER, E.B. [1972] *Frequency calculations of general recursive predicates and frequency enumeration of sets (Russian)* (**J** 0023) Dokl Akad Nauk SSSR 205*23-25
• TRANSL [1972] (**J** 0062) Sov Math, Dokl 13*873-876
◇ D20 ◇ REV MR 46 # 3283 Zbl 264 # 02039 • ID 07090

KINBER, E.B. [1974] *Frequency enumeration of sets (Russian)* (**J** 0003) Algebra i Logika 13*398-419,487
• TRANSL [1974] (**J** 0069) Algeb and Log 13*226-237
◇ D25 ◇ REV MR 52 # 13354 Zbl 319 # 02031 • ID 62960

KINBER, E.B. [1974] *Frequency calculations with small number of mistakes (Russian)* (**S** 2587) Teor Algor & Progr (Riga) 1*206-213
◇ D20 ◇ REV MR 58 # 21524 Zbl 335 # 02021 • ID 90032

KINBER, E.B. [1974] *On limiting synthesis of quasiminimal Goedel numbers (Russian)* (**S** 2587) Teor Algor & Progr (Riga) 1*221-223
◇ D20 D45 ◇ REV MR 58 # 21525 Zbl 335 # 02022 • ID 90033

KINBER, E.B. [1974] see BARZDINS, J.

KINBER, E.B. [1974] *Ueber Frequenzberechnungen auf unendlichen Auswahlmengen (Russisch)* (S 2587) Teor Algor & Progr (Riga) 1*48-67
◇ D20 D25 ◇ REV MR 58 # 21512 Zbl 335 # 02020
• ID 62962

KINBER, E.B. [1975] *On comparison of limit identification and standardization of general recursive functions (Russian) (English summary)* (S 2587) Teor Algor & Progr (Riga) 2*45-56
◇ D20 ◇ REV MR 58 # 3624 Zbl 318 # 02039 • ID 62964

KINBER, E.B. [1975] *On frequency real-time computations (Russian) (English summary)* (S 2587) Teor Algor & Progr (Riga) 2*174-182
◇ D10 D15 D20 ◇ REV MR 58 # 3679 Zbl 335 # 02023
• ID 62963

KINBER, E.B. [1976] *Frequenzberechnungen auf endlichen Automaten (Russian) (English summary)* (J 0040) Kibernetika, Akad Nauk Ukr SSR 1976/2*7-15
◇ D05 D20 ◇ REV Zbl 333 # 94024 • ID 62958

KINBER, E.B. [1977] see FREJVALD, R.V.

KINBER, E.B. [1977] *On a theory of inductive inference* (P 2588) FCT'77 Fund of Comput Th;1977 Poznan 435-440
◇ D20 D45 ◇ REV MR 58 # 16220 Zbl 368 # 02045
• ID 51262

KINBER, E.B. [1977] *On btt-degrees of sets of minimal numbers in Goedel numberings* (J 0068) Z Math Logik Grundlagen Math 23*201-212
◇ D20 D30 D45 ◇ REV MR 58 # 21545 Zbl 382 # 03029
• ID 26479

KINBER, E.B. [1977] *On identification in the limit of the minimal numbers for functions of effectively enumerable classes (Russian)* (S 2587) Teor Algor & Progr (Riga) 3*35-56
◇ D20 D45 ◇ REV MR 58 # 16219 Zbl 376 # 02034
• ID 51676

KINBER, E.B. [1977] *On speed-up of limiting identification of recursive functions by changing the order of questions (Russian) (English and German summary)* (J 0129) Elektr Informationsverarbeitung & Kybern 13*369-383
◇ D15 D20 ◇ REV MR 56 # 15383 Zbl 364 # 02021
• ID 50951

KINBER, E.B. [1977] *On Turing degrees of hypersimple Post sets (Russian)* (J 0337) Mat Ezheg, Akad Nauk Latv SSR 21*164-170
◇ D25 D30 ◇ REV MR 58 # 10361 Zbl 415 # 03030
• ID 53132

KINBER, E.B. [1980] *On inclusion problem for deterministic multitape automata* (J 0232) Inform Process Lett 11*144-146
◇ D10 ◇ REV MR 82f:68055 Zbl 447 # 68047 • ID 81853

KINBER, E.B. [1982] see FREJVALD, R.V.

KINBER, E.B. [1983] *A class of multitape automata with a decidable equivalence problem (Russian)* (J 2605) Programmirovanie 1983/3*16-24
• TRANSL [1983] (J 2604) Progr Comput Software 9*121-127
◇ D05 ◇ REV MR 85j:68033 Zbl 541 # 68033 • ID 46791

KINBER, E.B. [1983] *A note on limit identification of c-minimal indices (German and Russian summaries)* (J 0129) Elektr Informationsverarbeitung & Kybern 19*459-463
◇ D20 D45 ◇ REV MR 85j:03065 Zbl 565 # 03018
• ID 46779

KINBER, E.B. [1983] *The inclusion problem for some classes of deterministic multitape automata* (J 1426) Theor Comput Sci 26*1-24
◇ D05 ◇ REV MR 85e:68020 Zbl 523 # 68047 • ID 37000

KINBER, E.B. [1984] see FREJVALD, R.V.

KINO, A. & TAKEUTI, G. [1962] *A note on predicates of ordinal numbers* (J 0090) J Math Soc Japan 14*367-378
◇ D60 ◇ REV MR 26 # 3605 Zbl 147.252 JSL 33.294
• ID 07098

KINO, A. & TAKEUTI, G. [1962] *On hierarchies of predicates of ordinal numbers* (J 0090) J Math Soc Japan 14*199-232
◇ D55 D60 F15 ◇ REV JSL 33.293 • ID 07097

KINO, A. [1968] *On provably recursive functions and ordinal recursive functions* (J 0090) J Math Soc Japan 20*456-476
◇ D20 F05 F30 F35 ◇ REV MR 38 # 43 Zbl 195.303
• ID 07103

KINO, A. see Vol. III, V, VI for further entries

KINTALA, C.M.R. [1977] see FISCHER, P.C.

KINTALA, C.M.R. [1981] see GINSBURG, S.

KIPNIS, M.M. [1968] *The constructive classification of arithmetic predicates and the semantic bases of arithmetic (Russian)* (S 0228) Zap Nauch Sem Leningrad Otd Mat Inst Steklov 8*53-65
• TRANSL [1968] (J 0521) Semin Math, Inst Steklov 8*22-27
◇ D55 F30 F50 ◇ REV MR 43 # 1836 Zbl 262 # 02035
• ID 27486

KIPNIS, M.M. [1970] *Invariant properties of systems of formulae of elementary axiomatic theories (Russian)* (J 0216) Izv Akad Nauk SSSR, Ser Mat 34*963-976
• TRANSL [1970] (J 0448) Math of USSR, Izv 4*965-978
◇ D25 D35 ◇ REV MR 42 # 7483 Zbl 234 # 02033
• ID 07109

KIPNIS, M.M. see Vol. I, VI for further entries

KIRBY, G. [1975] see CORNELIUS, B.J.

KIRBY, L.A.S. & MCALOON, K. & MURAWSKI, R. [1981] *Indicators, recursive saturation and expandability* (J 0027) Fund Math 114*127-139
◇ C50 C57 C62 H15 ◇ REV MR 84i:03117
Zbl 488 # 03038 • ID 34593

KIRBY, L.A.S. see Vol. III, V, VI for further entries

KIREEVSKIJ, N.N. [1934] *On the problem of the solvability of the decision problem (Russian)* (J 4717) Izv Akad Nauk SSSR 1934*1493-1499
◇ B25 D35 ◇ REV FdM 60.850 • ID 07112

KIREEVSKIJ, N.N. see Vol. III, VI for further entries

KIRKINSKIJ, A.S. [1975] see GORYAGA, A.V.

KIRKINSKIJ, A.S. & REMESLENNIKOV, V.N. [1975] *The isomorphism problem for solvable groups (Russian)* (J 0087) Mat Zametki (Akad Nauk SSSR) 18*437-443
• TRANSL [1975] (J 1044) Math Notes, Acad Sci USSR 18*849-852
◇ D40 ◇ REV MR 53 # 628 Zbl 351 # 20012 • ID 16684

KIROUSIS, L.M. [1978] see HARRINGTON, L.A.

KIROUSIS, L.M. [1983] *A selection theorem* (**J** 0036) J Symb Logic 48*585-594
◇ D60 D70 ◇ REV MR 85g:03066 Zbl 533 # 03029
• ID 36548

KISELEV, A.A. [1973] *Projective hierarchies on general structures (Russian)* (**J** 0226) Uch Zap Ped Inst, Ivanovo 123*100-163
◇ D55 D75 E15 E45 ◇ REV MR 57 # 5720a • REM Part I. Part II 1973 • ID 32586

KISELEV, A.A. [1973] *Projective hierarchies on general structures II (Russian)* (**J** 0226) Uch Zap Ped Inst, Ivanovo 125*29-45
• ERR/ADD ibid 130*92
◇ D55 D75 E15 E45 E55 ◇ REV MR 57 # 5720b • REM Part I 1973 • ID 32587

KISELEV, A.A. [1978] *Axiom of comparable choice and uniformizability of projective classes (Russian)* (**J** 0003) Algebra i Logika 17*144-168,241
• TRANSL [1978] (**J** 0069) Algeb and Log 17*101-119
◇ D55 E15 E25 E45 ◇ REV MR 80k:03052 Zbl 419 # 03029 Zbl 427 # 030038 • ID 32588

KISELEV, A.A. see Vol. I, III for further entries

KLAEREN, H.A. & SCHULZ, MARTIN [1981] *Computable algebras, word problems and canonical term algebras* (**P** 3475) Theor Comput Sci (5);1981 Karlsruhe 104*203-213
◇ D45 D80 ◇ REV Zbl 461 # 03009 • ID 54486

KLASSEN, V.P. [1970] *Inclusion problem for a certain class of groups (Russian)* (**J** 0003) Algebra i Logika 9*306-312
• TRANSL [1970] (**J** 0069) Algeb and Log 9*183-186
◇ D40 ◇ REV MR 43 # 7493 Zbl 234 # 20018 • ID 07135

KLAWE, M.M. [1985] see FAGIN, R.

KLEBANSKY, B. [1977] see FRANCEZ, N.

KLEENE, S.C. [1936] *A note on recursive functions* (**J** 0015) Bull Amer Math Soc 42*544-546 • ERR/ADD [1936] (**J** 0036) J Symb Logic 1*119
◇ D20 ◇ REV Zbl 15.50 JSL 1.119 FdM 62.45 • ID 07164

KLEENE, S.C. [1936] *General recursive functions of natural numbers* (**J** 0043) Math Ann 112*727-742
• REPR [1965] (**C** 0718) The Undecidable 237-252,253
◇ D20 D25 ◇ REV Zbl 14.194 JSL 2.38 JSL 31.485 FdM 62.44 • ID 07163

KLEENE, S.C. [1943] *Recursive predicates and quantifiers* (**J** 0064) Trans Amer Math Soc 53*41-73
• REPR [1965] (**C** 0718) The Undecidable 255-287
◇ D55 ◇ REV MR 4.126 JSL 31.485 JSL 8.32 • REM ERR&ADD in Reprint p.254, and in "Introduction to metamathematics", p.527 • ID 07167

KLEENE, S.C. [1944] *On the forms of predicates in the theory of constructive ordinals* (**J** 0100) Amer J Math 66*41-58
◇ D55 D70 F15 ◇ REV MR 5.197 Zbl 61.10 JSL 11.127
• REM ERR in Kleene,S.C.: Introduction to metamathematics, p.527. Part II 1955 • ID 07168

KLEENE, S.C. [1950] *A symmetric form of Goedel's theorem* (**J** 0028) Indag Math 12*244-246
◇ D25 ◇ REV MR 12.71 Zbl 38.31 JSL 16.147 • ID 07170

KLEENE, S.C. [1952] *Introduction to metamathematics* (**X** 0809) North Holland: Amsterdam x+550pp
• TRANSL [1957] (**X** 1656) Izdat Inostr Lit: Moskva 526pp (Russian) [1984] (**X** 1876) Kexue Chubanshe: Beijing xii+234pp,x+235-688pp (Chinese) [1974] (**X** 1781) Tecnos: Madrid (Spanish)
◇ A05 B98 D98 F30 F50 F98 ◇ REV MR 14.525 Zbl 47.7 JSL 19.215 JSL 25.280 JSL 33.290 JSL 35.350 JSL 38.333 • REM Co-publisher: Wolters-Noordhoff; 8th revised ed. 1980. Chinese transl. in 2 parts • ID 07173

KLEENE, S.C. & POST, E.L. [1954] *The upper semi-lattice of degrees of recursive unsolvability* (**J** 0120) Ann of Math, Ser 2 59*379-407 • ERR/ADD [1959] (**J** 0064) Trans Amer Math Soc 91*52
◇ D30 ◇ REV MR 15.772 Zbl 57.247 JSL 21.407
• ID 07174

KLEENE, S.C. [1955] *Arithmetical predicates and function quantifiers* (**J** 0064) Trans Amer Math Soc 79*312-340
• ERR/ADD [1957] (**J** 0053) Proc Amer Math Soc 8*1006
◇ D55 D70 ◇ REV MR 17.4 Zbl 66.257 JSL 21.409
• ID 07177

KLEENE, S.C. [1955] *Hierarchies of number-theoretic predicates* (**J** 0015) Bull Amer Math Soc 61*193-213 • ERR/ADD [1957] (**J** 0053) Proc Amer Math Soc 8*1006
◇ D55 D70 ◇ REV MR 17.4 Zbl 66.259 JSL 21.411
• ID 07176

KLEENE, S.C. [1955] *On the forms of the predicates in the theory of constructive ordinals II* (**J** 0100) Amer J Math 77*405-428
• ERR/ADD [1959] (**J** 0064) Trans Amer Math Soc 91*51
◇ D55 D70 F15 ◇ REV MR 17.5 Zbl 67.252 JSL 21.410
• REM Part I 1944 • ID 07178

KLEENE, S.C. [1956] *A note on computable functionals* (**J** 0028) Indag Math 18*275-280
◇ D20 ◇ REV MR 19.238 Zbl 74.250 JSL 24.51 • ID 07180

KLEENE, S.C. [1956] *Representations of events in nerve nets and finite automata* (**C** 0717) Automata Studies 3-41
• TRANSL [1956] (**C** 3340) Avtomaty (Moskva) 15-67 (Russian) [1974] (**C** 1902) Stud Th Automaten 3-55,423-424 (German)
◇ D05 ◇ REV MR 17.1040 JSL 23.59 JSL 28.175
• ID 07181

KLEENE, S.C. [1957] see ADDISON, J.W.

KLEENE, S.C. [1957] *Realizability* (**P** 1675) Summer Inst Symb Log;1957 Ithaca 1*100-104
• REPR [1959] (**P** 0634) Constructivity in Math;1957 Amsterdam 285-289
◇ D80 F50 ◇ REV MR 21 # 2590 Zbl 88.249 JSL 27.242
• REM ERR in Kleene,S.C. & Vesley,R.E.: The foundations of intuitionistic mathematics, 192 • ID 22336

KLEENE, S.C. [1957] *Recursive functionals of higher finite types* (**P** 1675) Summer Inst Symb Log;1957 Ithaca 1*148-154
• ERR/ADD ibid 3*429
◇ D65 ◇ REV JSL 27.359 • ID 07187

KLEENE, S.C. [1958] *Extension of an effectively generated class of functions by enumeration* (**S** 0019) Colloq Math (Warsaw) 6*68-78 • ERR/ADD [1963] (**J** 0064) Trans Amer Math Soc 108*141
◇ D20 ◇ REV MR 22 # 9443 Zbl 85.246 JSL 25.279
• ID 07182

KLEENE, S.C. [1959] *Countable functionals* (**P** 0634) Constructivity in Math;1957 Amsterdam 81-100 • ERR/ADD [1963] (**J** 0064) Trans Amer Math Soc 108*141 ◇ D65 ◇ REV MR 22 # 3686 Zbl 100.249 JSL 27.359 • ID 07183

KLEENE, S.C. [1959] *Quantification of number-theoretic functions* (**J** 0020) Compos Math 14*23-40 ◇ D55 ◇ REV MR 21 # 2586 Zbl 85.247 JSL 27.82 • ID 07184

KLEENE, S.C. [1959] *Recursive functionals and quantifiers of finite types I* (**J** 0064) Trans Amer Math Soc 91*1-52 • ERR/ADD [1963] (**J** 0064) Trans Amer Math Soc 108*142 ◇ D30 D55 D65 ◇ REV MR 21 # 1273 Zbl 88.13 JSL 27.82 • REM Part II 1963 • ID 07185

KLEENE, S.C. [1960] *Mathematical logic: constructive and non-constructive operations* (**P** 0660) Int Congr Math (II, 8);1958 Edinburgh 137-153 • ERR/ADD [1963] (**J** 0064) Trans Amer Math Soc 108*142 ◇ B98 D98 ◇ REV MR 22 # 5569 Zbl 126.19 JSL 27.78 • ID 07186

KLEENE, S.C. [1960] *Realizability and Shanin's algorithm for the constructive deciphering of mathematical sentences* (**J** 0079) Logique & Anal, NS 3*154-165 ◇ D80 F50 ◇ REV JSL 27.243 • REM ERR in Kleene,S.C. & Vesley,R.E.: The foundations of intuitionistic mathematics, p.192-193 • ID 07188

KLEENE, S.C. [1962] *λ-definable functionals of finite types* (**J** 0027) Fund Math 50*281-303 ◇ B40 D30 D55 D65 ◇ REV MR 32 # 4003 Zbl 100.250 JSL 29.104 • ID 07191

KLEENE, S.C. [1962] *Herbrand-Goedel-style recursive functionals of finite types* (**P** 0613) Rec Fct Th;1961 New York 49-75 ◇ D65 ◇ REV MR 25 # 4992 Zbl 171.269 • ID 33701

KLEENE, S.C. [1962] *Turing-machine computable functionals of finite types I* (**P** 0612) Int Congr Log, Meth & Phil of Sci (1,Proc);1960 Stanford 38-45 ◇ D10 D30 D55 D65 ◇ REV MR 32 # 4001 Zbl 192.53 JSL 35.588 • REM Part II 1962 • ID 07189

KLEENE, S.C. [1962] *Turing-machine computable functionals of finite types II* (**J** 3240) Proc London Math Soc, Ser 3 12*245-258 ◇ D10 D30 D55 D65 ◇ REV MR 32 # 4002 Zbl 192.53 JSL 35.588 • REM Part I 1962 • ID 07190

KLEENE, S.C. [1963] *Recursive functionals and quantifiers of finite types II* (**J** 0064) Trans Amer Math Soc 108*106-142 ◇ D30 D55 D65 ◇ REV MR 27 # 3521 Zbl 121.13 JSL 36.146 • REM ERR in Kleene,S.C. & Vesley,R.E.:The foundations of intuitionistic mathematics, p.193. Part I 1959 • ID 07192

KLEENE, S.C. [1967] *Computability* (**C** 1842) Phil of Sci Today 36-45 ◇ D98 ◇ ID 32370

KLEENE, S.C. [1967] *Mathematical logic* (**X** 0827) Wiley & Sons: New York xiii+398pp • TRANSL [1967] (**X** 3636) Tokyo Tosho: Tokyo 200pp+266pp [1971] (**X** 0850) Colin: Paris [1973] (**X** 0885) Mir: Moskva 480pp ◇ B98 D20 D25 D55 D98 F30 F98 ◇ REV MR 36 # 25 Zbl 149.243 JSL 35.438 • ID 45895

KLEENE, S.C. [1969] *Formalized recursive functionals and formalized realizability* (**S** 0167) Mem Amer Math Soc 89*106pp ◇ D65 F35 F50 ◇ REV MR 39 # 5319 Zbl 184.20 • ID 24941

KLEENE, S.C. [1969] *On the normal form theorem* (**P** 1841) Fct Recurs & Appl;1967 Tihany 71-83 • ERR/ADD [1969] (**S** 0167) Mem Amer Math Soc 89*105 ◇ D20 ◇ ID 32557

KLEENE, S.C. [1978] *Recursive functionals and quantifiers of finite types revisited I* (**P** 1628) Generalized Recursion Th (2);1977 Oslo 185-222 • ERR/ADD [1982] (**P** 3634) Patras Logic Symp;1980 Patras 38 ◇ D65 ◇ REV MR 80k:03048 • REM Part II 1980 • ID 30930

KLEENE, S.C. [1979] *Origins of recursive function theory* (**P** 3535) IEEE Symp Founds of Comput Sci (20);1979 San Juan 371-382 • TRANSL [1982] (**P** 3803) Algor Sovrem Mat & Prilozh;1979 Urgench II*270-308 • REPR [1981] (**J** 3789) Ann Hist of Comp 3*52-671 ◇ A10 D20 D98 ◇ REV MR 82d:03003 MR 84h:03005 Zbl 565 # 03002 • REM Revised reprint • ID 74851

KLEENE, S.C. [1980] *Recursive functionals and quantifiers of finite types revisited II* (**P** 2058) Kleene Symp;1978 Madison 1-29 • ERR/ADD [1982] (**P** 3634) Patras Logic Symp;1980 Patras 38 ◇ D65 ◇ REV MR 83j:03076 Zbl 453 # 03048 • REM Part I 1978. Part III 1982 • ID 54178

KLEENE, S.C. [1981] *Algorithms in various contexts* (**P** 3729) Algor in Modern Math & Comput Sci;1979 Urgench 355-360 • TRANSL [1982] (**P** 3803) Algor Sovrem Mat & Prilozh;1979 Urgench 2*139-146 ◇ D65 D75 ◇ REV MR 84d:03051 Zbl 477 # 68035 • ID 34087

KLEENE, S.C. [1981] *The theory of recursive functions, approaching its centennial* (**J** 0589) Bull Amer Math Soc (NS) 5*43-61 ◇ A10 D20 D98 ◇ REV MR 82k:03058 Zbl 486 # 03023 • ID 74850

KLEENE, S.C. [1982] *Recursive functionals and quantifiers of finite types revisited III* (**P** 3634) Patras Logic Symp;1980 Patras 1-40 • ERR/ADD [1982] (**J** 0194) Inform & Control 54*3ff ◇ D65 ◇ REV MR 84g:03071 Zbl 572 # 03027 • REM ADD & CORR in footnotes 10, 12 (paper of M. Davis). Part II 1980. Part IV 1985 • ID 34178

KLEENE, S.C. [1985] *Unimonotone functions of finite types (Recursive functionals and quantifiers of finite types revisited IV)* (**P** 4046) Rec Th;1982 Ithaca 119-138 ◇ D65 ◇ REM Part III 1982 • ID 45926

KLEENE, S.C. see Vol. I, II, VI for further entries

KLEIJN, H.C.M. [1985] see EHRENFEUCHT, A.

KLEIMAN, J.G. [1979] *Identities and some algorithmic problems in groups (Russian)* (**J** 0023) Dokl Akad Nauk SSSR 244*814-818 • TRANSL [1979] (**J** 0062) Sov Math, Dokl 20*115-119 ◇ D40 ◇ REV MR 80b:20044 Zbl 415 # 20026 • ID 81867

KLEIMAN, J.G. [1982] *On identities in groups (Russian)* (**J** 0065) Tr Moskva Mat Obshch 44*62-108
◇ C52 C60 D40 ◇ REV MR 84e:20040 • ID 39318

KLEINBERG, E.M. [1970] *Recursion theory and formal deducibility* (**J** 0036) J Symb Logic 35*556-558
◇ D25 F07 F40 ◇ REV MR 43 # 4666 Zbl 221 # 02017 • ID 07211

KLEINBERG, E.M. [1973] *A characterization of determinacy for Turing degree games* (**J** 0027) Fund Math 80*287-291
◇ D30 E60 ◇ REV MR 48 # 10787 Zbl 276 # 02047 • ID 07221

KLEINBERG, E.M. see Vol. I, III, V for further entries

KLEINE BUENING, H. & OTTMANN, T. [1977] *Kleine universelle mehrdimensionale Turingmaschinen* (**J** 0129) Elektr Informationsverarbeitung & Kybern 13*179-201
◇ D10 ◇ REV MR 56 # 15378 Zbl 363 # 02041 • ID 28334

KLEINE BUENING, H. [1979] *Generalized vector addition systems with finite exception sets* (**P** 2935) FCT'79 Fund of Comput Th;1979 Berlin/Wendisch-Rietz 237-242
◇ B75 D05 D10 D80 ◇ REV MR 81b:68061 Zbl 418 # 03029 • ID 53314

KLEINE BUENING, H. [1979] see BOERGER, E.

KLEINE BUENING, H. [1980] *Decision problems in generalized vector addition systems* (**J** 2095) Fund Inform, Ann Soc Math Pol, Ser 4 3*497-511
◇ D80 ◇ REV MR 83c:68034 Zbl 453 # 03011 • ID 54141

KLEINE BUENING, H. [1980] see BOERGER, E.

KLEINE BUENING, H. [1981] *Classes of functions over binary trees* (**P** 3165) FCT'81 Fund of Comput Th;1981 Szeged 199-204
◇ B75 D05 D15 D20 ◇ REV MR 83h:03055 Zbl 466 # 68037 • ID 54997

KLEINE BUENING, H. [1981] *Some undecidable theories with monadic predicates and without equality* (**J** 0009) Arch Math Logik Grundlagenforsch 21*137-148
◇ B20 D35 ◇ REV MR 83k:03052 Zbl 475 # 03022 • ID 55476

KLEINE BUENING, H. [1982] *Note on the $E_1^* - E_2^*$ problem* (**J** 0068) Z Math Logik Grundlagen Math 28*277-284
◇ D20 ◇ REV MR 84i:03081 Zbl 515 # 03025 • ID 34560

KLEINE BUENING, H. [1983] *A classification of an iterative hierarchy* (**J** 0009) Arch Math Logik Grundlagenforsch 23*175-186
◇ D20 ◇ REV MR 85f:03038b Zbl 544 # 03016 • ID 40727

KLEINE BUENING, H. [1983] *Durch syntaktische Rekursion definierte Klassen* (**J** 0068) Z Math Logik Grundlagen Math 29*169-175
◇ D20 ◇ REV MR 84h:03101 Zbl 572 # 03018 • ID 34291

KLEINE BUENING, H. [1984] *Complexity of loop-problems in normed networks* (**P** 2342) Symp Rek Kombin;1983 Muenster 254-269
◇ D05 D15 ◇ REV Zbl 533 # 68037 • ID 47032

KLESHCHEV, A.S. [1980] *Relational model of computations (Russian)* (**J** 2605) Programmirovanie 1980/4*20-29
• TRANSL [1980] (**J** 2604) Progr Comput Software 6*180-188
◇ D05 ◇ REV MR 82f:68016 Zbl 464 # 68039 • ID 81869

KLETTE, R. [1975] see GOETZE, B.G.

KLETTE, R. [1976] *Erkennung allgemein rekursiver Funktionen (English and Russian summaries)* (**J** 0129) Elektr Informationsverarbeitung & Kybern 12*227-243
◇ D20 ◇ REV MR 54 # 2435 Zbl 329 # 02017 • ID 24028

KLETTE, R. [1976] *Indexmengen und Erkennung rekursiver Funktionen* (**J** 0068) Z Math Logik Grundlagen Math 22*231-238
◇ D20 D25 D30 D55 ◇ REV MR 58 # 21526 Zbl 356 # 02036 • ID 18453

KLETTE, R. [1977] see GILLO, D.

KLETTE, R. [1978] see GOETZE, B.G.

KLETTE, R. & LINDNER, R. [1979] *Zweidimensional arbeitende Vektormaschinen und ihr Leistungsvermoegen bei der Loesung von Entscheidungsproblemen der Aussagenlogik* (**J** 0129) Elektr Informationsverarbeitung & Kybern 15*37-46
◇ B35 D05 D15 ◇ REV MR 80g:68081 Zbl 414 # 68024 • ID 53095

KLETTE, R. & WIEHAGEN, R. [1980] *Research in the theory of inductive inference by GDR mathematicians - A survey* (**J** 0191) Inform Sci 22*149-169
◇ D98 ◇ REV MR 83j:03068 Zbl 459 # 03021 • ID 54461

KLETTE, R. [1982] *A few results on the complexity of classes of identifiable recursive function sets* (**P** 3787) Discr Math;1977 Warsaw 135-142
◇ D20 D55 ◇ REV MR 85d:03084 Zbl 541 # 03019 • ID 41102

KLINE, S.A. [1945] *The representation of Baire's classes by transfinite sums of continuous functions* (**J** 0039) J London Math Soc 20*4-7
◇ D55 E15 ◇ REV MR 7.377 Zbl 60.143 • ID 07237

KLOETZER, G. & RAUTENBERG, W. [1972] *Im Grenzbereich Algebra, Logik, Maschinen. 10 Jahre Forschungsarbeit der Nowosibirsker Schule A.I. Malcev's (eine Studie)* (**J** 1670) Mitt Math Ges DDR 1972/1-2*43-102
◇ A10 B98 C05 D05 D98 ◇ REV Zbl 243 # 02001 • ID 30031

KLOP, J.W. [1978] see BARENDREGT, H.P.

KLOP, J.W. [1982] see BERGSTRA, J.A.

KLOP, J.W. [1982] see BAETEN, J.C.M.

KLOP, J.W. [1983] see BERGSTRA, J.A.

KLOP, J.W. [1984] see BERGSTRA, J.A.

KLOP, J.W. see Vol. VI for further entries

KLOPOTOWSKI, J. [1978] *Transition operators for (Z,Q)-machines* (**J** 1008) Demonstr Math (Warsaw) 11*499-508
◇ D05 ◇ REV MR 80f:68069 Zbl 393 # 68064 • ID 81873

KLOSINSKI, L.F. & SMOLARSKI, D.C. [1981] *Recognition algorithms for Fibonacci numbers* (**J** 1905) Fibonacci Quart 19*57-61
◇ D10 D80 ◇ REV MR 82i:10013 Zbl 455 # 68025 • ID 54306

KLOSS, B.M. [1964] *The definition of complexity of algorithms (Russian)* (**J** 0023) Dokl Akad Nauk SSSR 157*38-40
• TRANSL [1964] (**J** 0062) Sov Math, Dokl 5*880-882
◇ D15 D20 ◇ REV MR 29 # 1144 Zbl 127.10 JSL 34.509 • ID 07239

KLOSS, B.M. see Vol. II for further entries

KLUPP, H. & SCHNORR, C.-P. [1977] *A universally hard set of formulae with respect to non-deterministic Turing acceptors* (J 0232) Inform Process Lett 6*35-37
◇ B25 D10 D15 ◇ REV MR 56#4241 Zbl 353#68064 • ID 50134

KNIGHT, J.F. & NADEL, M.E. [1982] *Expansions of models and Turing degrees* (J 0036) J Symb Logic 47*587-604
◇ C50 C57 D30 ◇ REV MR 83k:03039 Zbl 527#03013 • ID 36173

KNIGHT, J.F. & NADEL, M.E. [1982] *Models of arithmetic and closed ideals* (J 0036) J Symb Logic 47*833-840
◇ C62 D30 ◇ REV MR 85d:03072 Zbl 518#03032 • ID 37529

KNIGHT, J.F. [1983] *Additive structure in uncountable models for a fixed completion of P* (J 0036) J Symb Logic 48*623-628
◇ C50 C57 C62 C75 D30 ◇ REV MR 85h:03034 Zbl 549#03026 • ID 42733

KNIGHT, J.F. [1983] *Degrees of types and independent sequences* (J 0036) J Symb Logic 48*1074-1081
◇ C35 C45 D30 ◇ REV MR 86b:03039 Zbl 541#03015 • ID 40363

KNIGHT, J.F. & LACHLAN, A.H. & SOARE, R.I. [1984] *Two theorems on degrees of models of true arithmetic* (J 0036) J Symb Logic 49*425-436
◇ C57 C62 D30 ◇ REV MR 85i:03144 Zbl 576#03044 • ID 33597

KNIGHT, J.F. see Vol. III, V for further entries

KNODE, R.B. [1972] see BRAINERD, W.S.

KNUTH, D.E. [1970] see BENDIX, P.B.

KNUTH, D.E. [1971] *Examples of formal semantics* (C 0628) Symp Semant of Algor Lang 212-235
◇ D05 ◇ REV MR 49#1834 Zbl 221#68014 • ID 28129

KNUTH, D.E. [1979] see ERSHOV, A.P.

KNUTH, D.E. [1981] see ERSHOV, A.P.

KNUTH, D.E. see Vol. V for further entries

KNYAZEV, E.A. [1979] see GOLUNKOV, YU.V.

KO, KER-I & MOORE, D.J. [1981] *Completeness, approximation and density* (J 1428) SIAM J Comp 10*787-796
◇ D15 ◇ REV MR 83c:68046 Zbl 468#68051 • ID 39046

KO, KER-I [1982] see FRIEDMAN, H.M.

KO, KER-I [1982] *Some negative results on the computational complexity of total variation and differentiation* (J 0194) Inform & Control 53*21-31
◇ D15 D80 ◇ REV MR 85e:68031 Zbl 521#03046 • ID 37081

KO, KER-I [1982] *The maximum value problem and NP real numbers* (J 0119) J Comp Syst Sci 24*15-35
◇ D15 D80 ◇ REV MR 84e:68041 Zbl 481#03038 • ID 36831

KO, KER-I [1983] *On self-reducibility and weak P-selectivity* (J 0119) J Comp Syst Sci 26*209-221
◇ D15 ◇ REV MR 85e:03094 Zbl 519#68062 • ID 36734

KO, KER-I [1983] *On the computational complexity of ordinary differential equations* (J 0194) Inform & Control 58*157-194
◇ D15 D80 ◇ REV MR 86d:03035 Zbl 541#03035 • ID 41390

KO, KER-I [1983] *On the definitions of some complexity classes of real numbers* (J 0041) Math Syst Theory 16*95-109
◇ D15 ◇ REV MR 85e:03151 Zbl 529#03016 • ID 37651

KO, KER-I [1984] *Reducibilities on real numbers* (J 1426) Theor Comput Sci 31*101-123
◇ D15 D30 ◇ REV Zbl 542#03033 • ID 43694

KO, KER-I [1985] *Nonlevelable sets and immune sets in the accepting density hierachy in NP* (J 0041) Math Syst Theory 18*189-205
◇ D15 ◇ ID 49540

KO, KER-I & SCHOENING, U. [1985] *On circuit-size complexity and the low hierarchy in NP* (J 1428) SIAM J Comp 14*41-51
◇ D15 ◇ ID 45431

KO, KER-I [1985] *On some natural complete operators* (J 1426) Theor Comput Sci 37*1-30
◇ D15 ◇ ID 47713

KOBAYASHI, K. [1971] *Structural complexity of context-free languages* (J 0194) Inform & Control 18*299-310
◇ D05 ◇ REV MR 43#4681 Zbl 217.539 • ID 07265

KOBAYASHI, K. [1974] *A note on extending equivalence theories of algorithms* (J 0232) Inform Process Lett 3*54-56
◇ B75 D20 D25 ◇ REV MR 52#2843 Zbl 314#02046 • ID 17634

KOBAYASHI, K. & TAKAHASHI, MASAKO & YAMASAKI, H. [1984] *Characterization of ω-regular languages by first-order formulas* (J 1426) Theor Comput Sci 28*315-327
◇ D05 ◇ REV Zbl 551#68069 • ID 43985

KOBAYASHI, K. & TAKAHASHI, M. & YAMASAKI, H. [1985] *Logical formulas and four subclasses of ω-regular languages* (P 4622) Autom on Infinite Words;1984 Le Mont-Dore 81-88
◇ D05 ◇ REV Zbl 575#03029 • ID 48828

KOBAYASHI, K. [1985] *On proving time constructibility of functions* (J 1426) Theor Comput Sci 35*215-225
◇ D15 ◇ REV Zbl 566#68042 • ID 48748

KOBRINSKIJ, N.E. & TRAKHTENBROT, B.A. [1959] *Ueber den Aufbau einer allgemeinen Theorie der logischen Netze (Russian)* (C 1155) Log Issl (Moskva) 352-378
◇ D05 ◇ REV Zbl 109.99 • ID 48037

KOBRINSKIJ, N.E. & TRAKHTENBROT, B.A. [1962] *Introduction to the theory of finite automata (Russian)* (X 3709) Izdat Fiz-Mat Lit: Moskva 404pp
• TRANSL [1965] (X 0809) North Holland: Amsterdam x+337pp
◇ D05 D98 ◇ REV MR 32#3914 Zbl 104.355 JSL 29.97 JSL 33.466 • ID 07268

KOBUCHI, Y. & SEKI, S. [1979] *Decision promblems of locally catenative property for DIL systems* (J 0194) Inform & Control 43*266-279
◇ D10 D80 ◇ REV MR 81h:68069 Zbl 426#68074 • ID 69658

KOBZEV, G.N. [1973] *btt-reducibility I,II (Russian)* (J 0003) Algebra i Logika 12*190-204,244,433-444,492
• TRANSL [1973] (J 0069) Algeb and Log 12*107-115,242-248
◇ D25 D30 ◇ REV MR 53#5275 Zbl 309#02042 Zbl 309#02043 • ID 22906

KOBZEV, G.N. [1973] *Pointwise decomposable sets (Russian)* (**J** 0087) Mat Zametki (Akad Nauk SSSR) 13*893-898
• ERR/ADD ibid 14*944
• TRANSL [1973] (**J** 1044) Math Notes, Acad Sci USSR 13*533-536
◇ D20 D25 D30 ◇ REV MR 48 # 8214 Zbl 292 # 02036
• ID 07269

KOBZEV, G.N. [1974] *On complete Btt-degrees* (**J** 0003) Algebra i Logika 13*22-25,120
• TRANSL [1974] (**J** 0069) Algeb and Log 13*10-12
◇ D25 D30 ◇ REV MR 51 # 7843 Zbl 309 # 02044
• ID 25964

KOBZEV, G.N. [1975] *r-separated sets (Russian) (Georgian summary)* (**S** 2043) Issl Mat Log & Teor Algor (Tbilisi) 1975*19-30
◇ D25 ◇ REV MR 57 # 12192 • ID 74910

KOBZEV, G.N. [1976] *About 1-degrees* (**S** 1585) Rec Fct Th Newsletter 11*7-8
◇ D30 ◇ ID 37195

KOBZEV, G.N. [1976] *Relationship between recursively enumerable tt- and w-degrees (Russian) (English summary)* (**J** 0233) Soobshch Akad Nauk Gruz SSR 84*585-587
◇ D25 ◇ REV Zbl 363 # 02047 • ID 50874

KOBZEV, G.N. [1977] *Maximal m-degrees (Russian) (Georgian and English summaries)* (**J** 0233) Soobshch Akad Nauk Gruz SSR 85*325-327
◇ D25 ◇ REV MR 56 # 5249 Zbl 353 # 02023 • ID 50106

KOBZEV, G.N. [1977] *Recursively enumerable bw-degrees (Russian)* (**J** 0087) Mat Zametki (Akad Nauk SSSR) 21*839-846
• TRANSL [1977] (**J** 1044) Math Notes, Acad Sci USSR 21*473-477
◇ D25 ◇ REV MR 57 # 5715 Zbl 402 # 03039 • ID 54686

KOBZEV, G.N. [1978] *On the semi-lattice of tt-degrees (Russian) (English summary)* (**J** 0233) Soobshch Akad Nauk Gruz SSR 90*281-283
◇ D25 D30 ◇ REV MR 80e:03045a Zbl 389 # 03016
• ID 52305

KOBZEV, G.N. [1978] *On tt-degrees of recursively enumerable Turing degrees (Russian)* (**J** 0142) Mat Sb, Akad Nauk SSSR, NS 106(148)*507-514
• TRANSL [1978] (**J** 0349) Math of USSR, Sbor 35*173-180
◇ D25 ◇ REV MR 80e:03045b Zbl 386 # 03022 • REM Part I. Part II 1979 • ID 52198

KOBZEV, G.N. [1979] *tt-degrees of recursively enumerable Turing degrees II (Russian)* (**J** 0003) Algebra i Logika 18*415-425
• TRANSL [1979] (**J** 0069) Algeb and Log 18*252-259
◇ D25 D30 ◇ REV MR 81k:03039 Zbl 441 # 03013 • REM Part I 1978 • ID 56066

KOERBER, P. & OTTMANN, T. [1974] *Simulation endlicher Automaten durch Ketten aus einfachen Bausteinautomaten* (**J** 0129) Elektr Informationsverarbeitung & Kybern 10*133-148
◇ D05 ◇ REV MR 51 # 15246 Zbl 358 # 94061 • ID 28326

KOGALOVSKIJ, S.R. [1966] *On the properties preserved under algebraic constructions (Russian)* (**C** 1549) Tartu Mezhvuz Nauch Simp Obshchej Algeb 44-51
◇ B25 C30 C52 C60 D35 D45 ◇ REV MR 34 # 5679 Zbl 248 # 02050 • ID 63026

KOGALOVSKIJ, S.R. see Vol. I, III, V, VI for further entries

KOGAN, D.I. [1965] see DUDICH, A.M.

KOGAN, D.I. [1971] see GLEBSKIJ, YU.V.

KOGAN, D.I. [1972] *D-sets, Δ-sets and undecidable problems of discrete control (Russian) (English summary)* (**J** 1572) Izv Vyssh Ucheb Zaved, Radiofizika (Moskva) 15*358-364
◇ D80 ◇ REV MR 46 # 3282 • ID 07301

KOGAN, D.I. [1973] *Algorithmic problems for a certain class of multi-stage games (Russian) (English summary)* (**P** 2894) All-Union Conf Game Th (2);1971 Vil'nyus 174-176
◇ D80 ◇ REV MR 56 # 15395 Zbl 277 # 90103 • ID 74936

KOGAN, D.I. [1973] see GLEBSKIJ, YU.V.

KOGAN, D.I. see Vol. I, III for further entries

KOHAVI, Z. & PAZ, A. (EDS.) [1971] *Theory of machines and computations* (**X** 0801) Academic Pr: New York 430pp
◇ D05 D97 ◇ REV Zbl 266 # 94019 • ID 66289

KOHOUT, L.J. [1975] see GAINES, B.R.

KOHOUT, L.J. [1976] see GAINES, B.R.

KOHOUT, L.J. see Vol. I, II, V for further entries

KOKORIN, A.I. & MART'YANOV, V.I. [1973] *Universal extended theories (Russian)* (**C** 1443) Algebra, Vyp 2 (Irkutsk) 107-113
◇ B25 C52 D35 ◇ REV MR 53 # 5286 • ID 22922

KOKORIN, A.I. & PINUS, A.G. [1978] *Decidability problems of extended theories (Russian)* (**J** 0067) Usp Mat Nauk 33/2(200)*49-84
• TRANSL [1978] (**J** 1399) Russ Math Surv 33/2*53-96
◇ B25 C85 C98 D35 ◇ REV MR 58 # 5153 Zbl 395 # 03008 Zbl 434 # 03014 • ID 55718

KOKORIN, A.I. see Vol. III for further entries

KOLAITIS, P.G. [1978] *On recursion in E and semi-Spector classes* (**C** 2908) Cabal Seminar Los Angeles 1976-77 209-244
◇ D65 D70 D75 ◇ REV MR 80j:03064b Zbl 405 # 03023
• ID 54898

KOLAITIS, P.G. [1979] *Recursion in a quantifier vs. elementary induction* (**J** 0036) J Symb Logic 44*235-259
◇ D65 D70 D75 ◇ REV MR 80j:03064a Zbl 426 # 03047
• ID 53643

KOLAITIS, P.G. [1980] *Recursion and nonmonotone induction in a quantifier* (**P** 2058) Kleene Symp;1978 Madison 367-389
◇ D65 D70 D75 ◇ REV MR 83j:03078 Zbl 468 # 03025
• ID 55090

KOLAITIS, P.G. [1981] *Model-theoretic characterizations in generalized recursion theory* (**P** 2628) Log Year;1979/80 Storrs 104-119
◇ D65 D70 D75 ◇ REV MR 82j:03059 Zbl 499 # 03038
• ID 74942

KOLAITIS, P.G. [1985] *Canonical forms and hierarchies in generalized recursion theories* (**P** 4046) Rec Th;1982 Ithaca 139-170
◇ D55 D65 D70 ◇ REV Zbl 573 # 03017 • ID 46370

KOLAITIS, P.G. [1985] *Game quantification* (**C** 4183) Model-Theor Log 365-421
◇ C40 C70 C75 C98 D60 D65 D70 E60 ◇ ID 48331

KOLETSOS, G. [1985] *Church-Rosser theorem for typed functional systems* (**J** 0036) J Symb Logic 50*782-790
◇ D05 ◇ ID 47379

KOLETSOS, G. see Vol. II, VI for further entries

KOLGANOV, N.A. [1978] *Identities in an algebra of recursive functions (Russian)* (C 2595) Rekursiv Funktsii 27-36
◇ D20 D75 ◇ REV MR 82d:03067 Zbl 504 # 03020
• ID 74946

KOLMOGOROV, A.N. [1953] *On the concept of algorithm (Russian)* (J 0067) Usp Mat Nauk 8/4(56)*175-176
◇ B25 D20 ◇ REV Zbl 51.245 • ID 43578

KOLMOGOROV, A.N. & USPENSKIJ, V.A. [1958] *On the definition of an algorithm (Russian)* (J 0067) Usp Mat Nauk 13/4(82)*3-28
• TRANSL [1963] (J 0225) Amer Math Soc, Transl, Ser 2 29*217-245
◇ D20 ◇ REV MR 20 # 5735 Zbl 90.11 JSL 38.655
• ID 19135

KOLMOGOROV, A.N. [1965] *Three approaches to the definition of the concept "quantity of information" (Russian)* (J 2320) Probl Peredachi Inf, Akad Nauk SSSR 1/1*3-11
◇ D15 D20 D80 ◇ REV MR 32 # 2273 Zbl 271 # 94018
• ID 63035

KOLMOGOROV, A.N. [1966] *P.S.Aleksandrov and the theory of δs-operations (Russian)* (J 0067) Usp Mat Nauk 21/4*275-278
• TRANSL [1966] (J 1399) Russ Math Surv 21/4*247-250
◇ A10 D55 E15 ◇ REV MR 33 # 2783 Zbl 163.250
• ID 24780

KOLMOGOROV, A.N. [1969] *Logical basis of information theory and probability theory (Russian)* (J 2320) Probl Peredachi Inf, Akad Nauk SSSR 5/3*3-7
• TRANSL [1969] (J 3419) Probl Inf Transmiss 5/3*1-4
◇ B30 D15 D20 D80 ◇ REV MR 47 # 3105
Zbl 265 # 94010 • ID 63036

KOLMOGOROV, A.N. & VARPAKHOVSKIJ, F.I. [1971] *The solution of Hilbert's tenth problem (Bulgarian)* (J 0477) Spis Bulgar Akad Nauk 14(47)
◇ D35 ◇ REV MR 58 # 5150 • REM Transl. from Russian
• ID 79667

KOLMOGOROV, A.N. see Vol. I, V, VI for further entries

KOLODIJ, A.N. [1974] *Two-sided nondeterministic automata (Russian) (English summary)* (J 0040) Kibernetika, Akad Nauk Ukr SSR 1974/5*40-45
• TRANSL [1974] (J 0021) Cybernetics 10*778-785
◇ D05 ◇ REV MR 53 # 2558 Zbl 312 # 94030 • ID 21501

KOLYADA, K.V. [1984] *Completeness of regular mappings (Russian)* (J 0052) Probl Kibern 41*41-47
◇ D05 ◇ REV MR 86e:08004 • ID 48908

KOMJATH, P. & SZABO, Z. [1984] *Orientation problems on sequences by recursive functions* (X 3151) Karl Marx Univ Dpt Math: Budapest 1984/2*129-138
◇ D20 ◇ REV Zbl 569 # 03017 • ID 49236

KOMJATH, P. see Vol. V for further entries

KONDO, M. [1936] *Sur un ensemble universel pour les ensembles Boreliens definis sur la famille de tous les ensembles lineaires CA* (J 0081) Proc Japan Acad 12*307-309
◇ D55 E15 ◇ REV Zbl 17.159 FdM 62.1175 • ID 37170

KONDO, M. [1937] *L'uniformisation des complementaires analytiques* (J 0081) Proc Japan Acad 13*287-291
◇ D55 E15 ◇ REV Zbl 19.297 FdM 63.933 • ID 37174

KONDO, M. [1938] *Sur l'uniformisation des complementaires analytiques et les ensembles projectifs de la seconde classe* (J 2307) Japan J Math 15*197-230
◇ D55 E15 ◇ REV Zbl 22.123 FdM 64.1008 • ID 27560

KONDO, M. [1938] *Sur la representation parametrique reguliere des ensembles analytiques* (J 0027) Fund Math 31*29-46
◇ D55 E15 ◇ REV Zbl 19.158 FdM 64.184 • ID 37175

KONDO, M. [1938] *Sur les operations analytiques dans la theorie des ensembles et quelques problemes qui s'y rattachent I* (J 0438) J Hokkaido Univ Educ Ser 1 7*1-38
◇ D55 E15 ◇ REV Zbl 20.349 FdM 64.1007 • REM Part II 1941 • ID 37176

KONDO, M. [1938] *Theory of analytic sets* (X 3552) Iwanami Shoten: Tokyo
◇ D55 E15 ◇ ID 33398

KONDO, M. [1939] *On the enumerable sets* (J 3631) Tokyo School of Physics J 567*1-6
◇ D25 ◇ REV JSL 15.63 • ID 33399

KONDO, M. [1942] *Sur la structure des ensembles* (J 0081) Proc Japan Acad 18*57-64
◇ D55 E15 ◇ REV MR 7.277 Zbl 60.129 • ID 37179

KONDO, M. [1944] *La structure des fonctions projectives I* (J 0081) Proc Japan Acad 20*439-445
◇ D55 E15 ◇ REV Zbl 60.129 • ID 37180

KONDO, M. [1956] *Sur la nommabilite d'ensembles* (J 0109) C R Acad Sci, Paris 242*1841-1843
◇ D55 E15 F60 F65 ◇ REV MR 17.933 Zbl 70.278
JSL 22.299 • ID 07330

KONDO, M. [1956] *Sur les analyses relatives* (J 0109) C R Acad Sci, Paris 242*2084-2087
◇ D55 E15 F65 ◇ REV MR 17.933 Zbl 70.278
JSL 22.299 • ID 07332

KONDO, M. [1956] *Sur les nombres reels et nommables* (J 0109) C R Acad Sci, Paris 242*1945-1948
◇ B28 D55 E15 F65 ◇ REV MR 17.933 Zbl 70.278
JSL 22.299 • ID 07331

KONDO, M. [1958] *Sur l'uniformisation des ensembles nommables* (J 0109) C R Acad Sci, Paris 246*2712-2715
◇ D55 E15 ◇ REV MR 20 # 2286 Zbl 81.47 • ID 07333

KONDO, M. [1959] *Sur la nommabilite d'ensembles de type superieur* (J 0109) C R Acad Sci, Paris 248*3099-3101
◇ D55 D65 E15 ◇ REV MR 21 # 4922 Zbl 88.264
• ID 07335

KONDO, M. [1959] *Sur la theorie projective des ensembles* (J 0109) C R Acad Sci, Paris 248*2940-2942
◇ D55 E15 ◇ REV MR 21 # 4921 Zbl 88.264 • ID 07334

KONDO, M. [1961] *Sur les domaines fondamentaux des fonctionelles de type transfini* (J 0109) C R Acad Sci, Paris 252*3711-3713
◇ D65 ◇ REV MR 23 # A2307 Zbl 105.6 • ID 07339

KONDO, M. [1961] *Sur les hyper-analyses relatives* (J 0109) C R Acad Sci, Paris 253*51-53
◇ D55 ◇ REV MR 23 # A2312 Zbl 105.7 • ID 07338

KONDO, M. [1961] *Sur les hyper-continus projectifs* (J 0109) C R Acad Sci, Paris 253*930-932
◇ D55 E15 ◇ REV MR 24 # A2539 Zbl 101.42 • ID 07340

KONDO, M. [1974] *Les problemes fondamentaux parus dans "Cinq lettres sur la theorie des ensembles"* (J 0463) Proc Fac Sci Tokai Univ 9*21-35
◇ D55 E15 ◇ REV MR 49 # 2380 Zbl 357 # 04018
• ID 07348

KONDO, M. [1976] *Peut-on prolonger decimalement tout nombre naturel?* (J 0463) Proc Fac Sci Tokai Univ 11(1)*3-6
◇ A05 D15 ◇ REV MR 54 # 12496 Zbl 359 # 02033
• ID 50581

KONDO, M. [1979] *Syntactic structure of mathematical words (Japanese)* (P 4108) Found of Math;1979 Kyoto 30-50
◇ B65 D05 ◇ ID 47657

KONDO, M. see Vol. I, II, V, VI for further entries

KONIKOWSKA, B. [1971] *Continuous machines, τ-computations and τ-computable sets* (J 0014) Bull Acad Pol Sci, Ser Math Astron Phys 19*525-530
◇ D10 D20 ◇ REV MR 46 # 1133a Zbl 226 # 02035
• ID 07349

KONIKOWSKA, B. [1971] *Continuous machines* (J 1929) Prace Centr Oblicz Pol Akad Nauk 40*18pp
◇ D10 ◇ REV MR 49 # 1824 Zbl 269 # 68030 • ID 63043

KONIKOWSKA, B. [1971] *On some properties of continuous machines* (J 0014) Bull Acad Pol Sci, Ser Math Astron Phys 19*531-538
◇ D10 D20 ◇ REV MR 46 # 1133b Zbl 226 # 02036
• ID 07350

KONIKOWSKA, B. [1971] *Properties of continuous machines* (J 1929) Prace Centr Oblicz Pol Akad Nauk 41*22pp
◇ D10 D20 ◇ REV MR 46 # 1133b Zbl 269 # 68031
• ID 63044

KONIKOWSKA, B. [1972] *Formalisierung des Begriffs der stetigen Maschine (Polish) (Russian and English summaries)* (J 1929) Prace Centr Oblicz Pol Akad Nauk 72*44pp
◇ D15 D20 ◇ REV Zbl 262 # 68019 • ID 63045

KOPIEKI, R. & SARALSKI, B. & WALIGORA, G. [1975] *The research of Jaskowski on decidability theory of first order sentences I* (J 0063) Studia Logica 34*201-214
◇ A10 B20 B25 D35 ◇ REV MR 54 # 86 Zbl 315 # 02044
• ID 23964

KOPYTOV, V.M. [1971] *Solvability of the membership problem in finitely generated solvable matrix groups over numbered fields (Russian)* (J 0003) Algebra i Logika 10*169-182
• TRANSL [1971] (J 0069) Algeb and Log 10*108-116
◇ D40 D45 ◇ REV MR 45 # 355 Zbl 248 # 20043
• ID 07365

KOPYTOV, V.M. see Vol. III for further entries

KOREC, I. [1969] *A complexity valuation of the partial recursive functions following the expectation of the length of their computations on Minsky machines* (J 0128) Acta Math Univ Comenianae (Bratislava) 23*53-112
◇ D10 D15 ◇ REV MR 47 # 9882 Zbl 215.322 • ID 28019

KOREC, I. [1972] *The mathematical expectation of the time for computations of partial recursive functions on Minsky machines (Russian) (German, English, and French summaries)* (J 0115) Wiss Z Humboldt-Univ Berlin, Math-Nat Reihe 21*541-543
◇ D10 D15 ◇ REV MR 48 # 10196 Zbl 256 # 02017
• ID 07369

KOREC, I. [1973] *Creative sets with prescribed density properties* (P 2276) Conf Algeb Th Automata;1973 Szeged 21*145-158
◇ D25 ◇ REV Zbl 357 # 02037 • ID 50385

KOREC, I. [1975] see BENESOVA, M.

KOREC, I. [1977] *Decidability (undecidability) of equivalence of Minsky machines with components consisting of at most seven (eight) instructions* (P 1635) Math Founds of Comput Sci (6);1977 Tatranska Lomnica 324-332
◇ D05 D10 ◇ REV MR 56 # 18183 Zbl 361 # 02060
• ID 50695

KOREC, I. & PROCHAZKA, J. [1977] *Real-time computability of* $[\ |\alpha|X\ ]$ *and* $[X^{\alpha}]$ (J 0129) Elektr Informationsverarbeitung & Kybern 13*13-26
◇ D15 ◇ REV MR 56 # 17203 Zbl 358 # 68083 • ID 50541

KOREC, I. [1979] see GREGUSOVA, L.

KOREC, I. [1980] *Densities of first order theories* (J 0128) Acta Math Univ Comenianae (Bratislava) 36*219-227
◇ D25 ◇ REV MR 82m:03058 Zbl 522 # 03017 • ID 74981

KOREC, I. see Vol. III, V, VI for further entries

KORFHAGE, R. [1966] *Logic and algorithms with applications to the computer and information sciences* (X 0827) Wiley & Sons: New York xii + 194pp
◇ B98 D98 ◇ REV MR 35 # 5279 Zbl 148.7 JSL 36.344
• ID 07376

KOROL'KOV, YU.D. [1978] *Families of general recursive functions with a finite number of limit points (Russian)* (J 0003) Algebra i Logika 17*169-177,241-242
• TRANSL [1978] (J 0069) Algeb and Log 17*120-127
◇ D20 D45 ◇ REV MR 80k:03040 Zbl 421 # 03034
• ID 53433

KOROL'KOV, YU.D. [1979] *Families of general recursive functions without isolated points (Russian)* (J 0087) Mat Zametki (Akad Nauk SSSR) 26*747-755,814
• TRANSL [1979] (J 1044) Math Notes, Acad Sci USSR 26*867-872
◇ D20 D45 ◇ REV MR 81a:03041 Zbl 423 # 03050
• ID 53562

KOROL'KOV, YU.D. [1982] *On the reducibility of index sets of families of general recursive functions (Russian)* (J 0092) Sib Mat Zh 23/1*190-193,223
◇ D20 D45 ◇ REV MR 83j:03069 Zbl 503 # 03019
• ID 35365

KORSHUNOV, A.D. [1969] *Comparison of the complexity of the largest and shortest disjunctive normal forms and a lower estimate of the number of irredundant disjunctive normal forms for almost all Boolean functions (Russian)* (J 0040) Kibernetika, Akad Nauk Ukr SSR 1969/4*1-11
• TRANSL [1969] (J 0021) Cybernetics 5*357-369
◇ B05 D20 ◇ REV MR 46 # 3215 Zbl 195.29 • ID 16309

KORSHUNOV, A.D. [1970] *The invariant properties of finite automata (Russian)* (J 0071) Met Diskr Analiz (Novosibirsk) 16*51-76
◇ D05 ◇ REV MR 44 # 3875 Zbl 212.338 • ID 07381

KORSHUNOV, A.D. [1974] *A survey of certain trends in automata theory (Russian)* (J 0071) Met Diskr Analiz (Novosibirsk) 25*19-55,62
◇ D05 ◇ REV MR 51 # 5209 Zbl 306 # 94034 • ID 17295

KORSHUNOV, A.D. [1978] *Enumeration of finite automata (Russian)* (J 0052) Probl Kibern 34*5-82,272
◇ D05 ◇ REV MR 80a:68050 Zbl 415 # 03027 • ID 53129

KORSHUNOV, A.D. see Vol. I for further entries

KORTAS, K. & KUBIAK, W. [1984] *The quasilinear complete number problems in NQL* (J 2716) Found Control Eng, Poznan 9/2*83-92
◇ D15 ◇ REV MR 86c:68025 • ID 48811

KORTE, B. [1980] see HAUSMANN, D.

KORTE, B. & SCHRADER, R. [1981] *A survey on oracle techniques* (P 3429) Math Founds of Comput Sci (10);1981 Strbske Pleso 61-77
◇ D15 ◇ REV MR 83e:68036 Zbl 462 # 68009 • ID 69659

KORTELAINEN, J. [1982] *On language families generated by commutative languages* (J 0446) Ann Acad Sci Fennicae, Ser A I, Diss 44*60pp
◇ D05 ◇ REV MR 83m:68138 Zbl 503 # 68054 • ID 40427

KORYAKOV, I.O. [1972] see GUREVICH, Y.

KOSARAJU, S.RAO [1972] see JOSHI, A.K.

KOSARAJU, S.RAO [1974] *Regularity preserving functions* (J 1456) SIGACT News 6/2*16-17 • ERR/ADD ibid 6*22
◇ D05 ◇ REV MR 53 # 12083 MR 56 # 10150 • ID 23110

KOSOVSKIJ, N.K. [1970] *Some questions in the constructive theory of normed boolean algebras (Russian)* (S 0066) Tr Mat Inst Steklov 113*2-38
• TRANSL [1970] (S 0055) Proc Steklov Inst Math 113*1-41
◇ C57 F60 G05 ◇ REV MR 44 # 5988 Zbl 229 # 02032 • ID 28600

KOSOVSKIJ, N.K. [1971] *Algorithmic sequences from the initial class of the Grzegorczyk hierarchy (Russian) (English summary)* (S 0228) Zap Nauch Sem Leningrad Otd Mat Inst Steklov 20*60-65,284
• TRANSL [1973] (J 1531) J Sov Math 1*36-40
◇ D20 ◇ REV MR 45 # 6623 Zbl 222 # 02028 • ID 07388

KOSOVSKIJ, N.K. [1971] *Diophantine representations of the sequence of solutions of the Pell equation (Russian) (English summary)* (S 0228) Zap Nauch Sem Leningrad Otd Mat Inst Steklov 20*49-59,283
• TRANSL [1973] (J 1531) J Sov Math 1*28-35
◇ D35 ◇ REV MR 45 # 226 Zbl 222 # 02052 • ID 07389

KOSOVSKIJ, N.K. [1972] *Properties of the solutions of equations in a free semigroup (Russian) (English summary)* (S 0228) Zap Nauch Sem Leningrad Otd Mat Inst Steklov 32*21-28,154
• TRANSL [1976] (J 1531) J Sov Math 6*361-367
◇ D20 D25 D40 ◇ REV MR 49 # 8844 Zbl 365 # 02033 • ID 07391

KOSOVSKIJ, N.K. [1972] *Recognition of invariant properties of algorithms (Russian) (English summary)* (S 0228) Zap Nauch Sem Leningrad Otd Mat Inst Steklov 32*29-34,154
• TRANSL [1976] (J 1531) J Sov Math 6*368-373
◇ D20 ◇ REV MR 49 # 8831 Zbl 351 # 02027 • ID 07390

KOSOVSKIJ, N.K. [1974] *Solutions of systems consisting of word equations and inequalities in lengths of words (Russian) (English summary)* (S 0228) Zap Nauch Sem Leningrad Otd Mat Inst Steklov 40*24-29,156
• TRANSL [1977] (J 1531) J Sov Math 8*262-265
◇ D20 D25 ◇ REV MR 51 # 10063 Zbl 365 # 02034 • ID 17542

KOSOVSKIJ, N.K. & VINOGRADOV, A.K. [1975] *Hierarchy of Diophantine representations of primitive recursive predicates (Russian)* (S 0716) Vychisl Tekh Vopr Kibern (Univ Leningrad) 12*99-107,162
◇ D15 D20 ◇ REV MR 58 # 182 Zbl 448 # 03030 • ID 56630

KOSOVSKIJ, N.K. [1975] *Possibilities for the operations of one-place summation and one-place restricted multiplication (Russian) (English summary)* (S 0228) Zap Nauch Sem Leningrad Otd Mat Inst Steklov 49*3-6,176
• TRANSL [1978] (J 1531) J Sov Math 10*493-495
◇ D20 ◇ REV MR 51 # 12490 Zbl 346 # 02017 • ID 17287

KOSOVSKIJ, N.K. [1977] *On classes of functions defined by addition (Russian)* (S 0716) Vychisl Tekh Vopr Kibern (Univ Leningrad) 14*54-60
◇ D20 ◇ REV MR 58 # 27393 Zbl 429 # 03025 • ID 53856

KOSOVSKIJ, N.K. [1978] *On Kolmogorov's subrecursive algorithmic complexity (Russian)* (J 0085) Vest Ser Mat Mekh Astron, Univ Leningrad 1978/4*54-59,150
• TRANSL [1983] (J 3945) Vest Math Univ Leningrad 11*301-307
◇ D15 D20 ◇ REV MR 80a:68041 Zbl 397 # 03024 • ID 81923

KOSOVSKIJ, N.K. [1978] *The complexity of the solvability of boolean functional equations (Russian)* (S 0716) Vychisl Tekh Vopr Kibern (Univ Leningrad) 15*104-111
◇ B05 D15 G05 ◇ REV MR 80d:68062 Zbl 534 # 94007 • ID 90060

KOSOVSKIJ, N.K. [1979] *On decision procedures for invariant properties of short algorithms (Russian) (English summary)* (S 0228) Zap Nauch Sem Leningrad Otd Mat Inst Steklov 88*73-77,238-239
• TRANSL [1982] (J 1531) J Sov Math 20*2304-2307
◇ D15 D20 D25 D45 ◇ REV MR 82b:03075 Zbl 432 # 03025 • ID 53982

KOSOVSKIJ, N.K. [1980] *An example of a sequence of symbols with large subrecursive Kolmogorov's complexity of initial segments (Russian)* (S 2582) Semiotika & Inf, Akad Nauk SSSR 14*108-114
◇ D15 ◇ REV MR 82a:68077 Zbl 455 # 03023 • ID 54284

KOSOVSKIJ, N.K. [1981] *Elements of mathematical logic and its applications to the theory of subrecursive algorithms (Russian)* (X 0938) Leningrad Univ: Leningrad 192pp
◇ B98 D20 D98 F30 F60 F98 ◇ REV MR 83m:03013 Zbl 479 # 03001 • ID 55663

KOSOVSKIJ, N.K. [1983] *Polynomial lower bounds for the complexity of establishing the solvability of logical-arithmetical equations (Russian) (English summary)* (J 3955) Ann Univ Budapest, Sect Comp 4*67-73
◇ D15 ◇ REV MR 85i:03132 Zbl 541 # 68016 • ID 44215

KOSOVSKIJ, N.K. [1984] see BOCHERNIKOV, V.YA.

KOSOVSKIJ, N.K. see Vol. VI for further entries

KOSSAK, R. [1983] *A certain class of models of Peano arithmetic* (J 0036) J Symb Logic 48*311-320
◇ C50 C57 C62 ◇ REV MR 84j:03076 Zbl 514 # 03036 • ID 34668

KOSSAK, R. [1984] *$L_{\infty\omega_1}$-elementary equivalence of ω_1-like models of PA* (J 0027) Fund Math 123*123-131
◇ C50 C57 C62 C75 ◇ REV MR 86f:03108 Zbl 545 # 03018 • ID 41820

KOSSAK, R. [1984] *Remarks on free sets* (**S** 3382) Sem-ber, Humboldt-Univ Berlin, Sekt Math 60*78-86
◇ C50 C57 C62 E05 ◇ REV MR 86h:03066 Zbl 562 # 03017 • ID 44634

KOSSAK, R. [1985] *A note on satisfaction classes* (**J** 0047) Notre Dame J Formal Log 26*1-8
◇ C50 C57 C62 F30 ◇ REV MR 86c:03055 Zbl 562 # 03040 • ID 42582

KOSSAK, R. [1985] *Recursively saturated* ω_1*-like models of arithmetic* (**J** 0047) Notre Dame J Formal Log 26*413-422
◇ C55 C57 C62 C75 C80 E65 ◇ REV Zbl 552 # 03021 Zbl 571 # 03014 • ID 47534

KOSSAK, R. see Vol. III, VI for further entries

KOSTENKO, K.I. [1985] *Classes of algorithms and computations (Russian)* (**J** 0023) Dokl Akad Nauk SSSR 280*33-37
• TRANSL [1985] (**J** 0062) Sov Math, Dokl 31*25-29
◇ D20 D75 ◇ ID 45389

KOSTYRKO, V.F. [1964] *The reduction class* $\forall\exists^n\forall$ *(Russian)* (**J** 0003) Algebra i Logika 3/5-6*45-55
◇ B20 D35 ◇ REV MR 31 # 27 Zbl 178.323 • ID 19124

KOSTYRKO, V.F. [1966] *On the AEA reduction class (Russian) (English summary)* (**J** 0040) Kibernetika, Akad Nauk Ukr SSR 1966/1*17-22
◇ B20 D35 ◇ REV MR 34 # 4136 Zbl 145.242 • ID 19123

KOSTYRKO, V.F. [1971] *The reduction class* $\forall x \forall y \exists z F(x,y,z) \wedge \forall^m \mathfrak{A}(F)$ *(Russian) (English summary)* (**J** 0040) Kibernetika, Akad Nauk Ukr SSR 1971/5*1-3
◇ B20 C13 D35 ◇ REV MR 46 # 1550 Zbl 231 # 02055 • ID 19122

KOSTYRKO, V.F. [1971] *Undecidability of the elementary* $\exists$*-theory of groupoids (Russian)* (**C** 3567) Teor Kibernetika Kiev 1967/71 41-47
◇ D35 ◇ REV MR 58 # 21560 • ID 75002

KOSTYRKO, V.F. see Vol. III for further entries

KOTLARSKI, H. [1981] *On elementary cuts in models of arithmetic* (**J** 3293) Bull Acad Pol Sci, Ser Math 29*419-423
◇ C50 C57 C62 ◇ REV MR 83b:03075 Zbl 471 # 03053 • ID 55248

KOTLARSKI, H. [1983] *On elementary cuts in models of arithmetic* (**J** 0027) Fund Math 115*27-31
◇ C50 C57 C62 ◇ REV MR 84f:03031 Zbl 515 # 03038 • ID 34451

KOTLARSKI, H. [1984] *On elementary cuts in recursively saturated models of Peano arithmetic* (**J** 0027) Fund Math 120*205-222
◇ C50 C57 C62 ◇ REV MR 86f:03056 Zbl 572 # 03016 • ID 45761

KOTLARSKI, H. [1985] *Bounded induction and satisfaction classes* (**P** 4310) Easter Conf on Model Th (3);1985 Gross Koeris 143-167
◇ C15 C50 C57 C62 ◇ ID 49911

KOTLARSKI, H. see Vol. III, V for further entries

KOTT, L. [1984] see DARONDEAU, P.

KOTT, L. [1985] see DARONDEAU, P.

KOUBEK, V. [1980] see GORALCIK, P.

KOUBEK, V. [1982] see DEMEL, J.

KOUBEK, V. see Vol. V for further entries

KOWALCZYK, W. [1982] *A sufficient condition for the consistency of P= NP with Peano arithmetic* (**J** 2095) Fund Inform, Ann Soc Math Pol, Ser 4 5*233-245
◇ D15 F25 F30 ◇ REV MR 84b:68043 Zbl 524 # 03026 • ID 37601

KOWALCZYK, W. [1984] *Some connections between presentability of complexity classes and the power of formal systems of reasoning* (**P** 3658) Math Founds of Comput Sci (11);1984 Prague 364-369
◇ B75 D15 ◇ REV Zbl 563 # 03024 • ID 44838

KOWALCZYK, W. [1985] *On the effectiveness of some operations on algorithms* (**P** 4670) Comput Th (5);1984 Zaborow 127-133
◇ D20 ◇ ID 49604

KOWALTOWSKI, T. & LUCCHESI, C.L. & SIMON, IMRE & SIMON, ISTVAN & SIMON, J. [1979] *Theoretical aspects of computation (Portuguese)* (**X** 2233) Inst Mat Pura Apl: Rio Janeiro xii+292pp
◇ D98 ◇ REV MR 84g:68001 • ID 39488

KOZEN, D. [1976] *On parallelism in Turing machines* (**P** 1757) IEEE Symp Found of Comput Sci (17);1976 Houston 89-97
◇ D10 D15 ◇ REV MR 57 # 4621 • ID 81929

KOZEN, D. [1977] *Lower bounds for natural proof systems* (**P** 3572) IEEE Symp Found of Comput Sci (18);1977 Providence 254-266
◇ D15 F20 ◇ REV MR 58 # 13931 • ID 81928

KOZEN, D. [1978] *Indexing of subrecursive classes* (**P** 1740) ACM Symp Th of Comput (10);1978 San Diego 287-295
◇ D15 D20 ◇ REV MR 80d:03040 • ID 75028

KOZEN, D. [1980] *Complexity of boolean algebras* (**J** 1426) Theor Comput Sci 10*221-247
◇ B25 D15 G05 ◇ REV MR 81e:03008 Zbl 428 # 03036 • ID 53795

KOZEN, D. [1980] *Indexings of subrecursive classes* (**J** 1426) Theor Comput Sci 11*277-301
◇ D15 D20 ◇ REV MR 82a:03035 Zbl 435 # 03033 • ID 55793

KOZEN, D. [1981] see CHANDRA, A.K.

KOZEN, D. [1981] *Positive first-order logic is NP-complete* (**J** 0284) IBM J Res Dev 25*327-332
◇ B25 D15 ◇ REV MR 83b:03050 Zbl 481 # 03026 • ID 35103

KOZEN, D. [1984] see BLASS, A.R.

KOZEN, D. see Vol. II, III, VI for further entries

KOZHEVNIKOVA, G.P. [1979] see FARAT, V.M.

KOZLOV, G.T. [1970] *Unsolvability of the elementary theory of lattices of subgroups of finite abelian p-groups (Russian)* (**J** 0003) Algebra i Logika 9*167-171
• TRANSL [1970] (**J** 0069) Algeb and Log 9*104-107
◇ C60 D35 ◇ REV MR 38 # 4298 MR 43 # 7329 Zbl 222 # 02055 • ID 07424

KOZLOV, G.T. [1972] *Undecidability of the theory of abelian groups with a chain of subgroups (Russian)* (**C** 3549) Algebra, Vyp 1 (Irkutsk) 21-23
◇ D35 ◇ REV MR 57 # 2902 • ID 75030

KOZLOV, G.T. see Vol. III for further entries

KOZLOV, M.K. [1980] see KHACHIYAN, L.G.

KOZLOVA, Z.I. [1940] *On some classes of A- and B-sets (Russian) (French summary)* (J 0216) Izv Akad Nauk SSSR, Ser Mat 4*479-500
◇ D55 E15 ◇ REV MR 3.225 Zbl 24.302 FdM 66.205 • ID 32604

KOZLOVA, Z.I. [1962] *On projective operations and the separability of projective sets (Russian)* (J 0216) Izv Akad Nauk SSSR, Ser Mat 26*223-260
◇ D55 E15 ◇ REV MR 27 # 2434 Zbl 109.152 • ID 07428

KOZLOVA, Z.I. [1964] *The axiom of constructibility and multiple separability and inseparability in classes of the analytic hierarchy (Russian)* (J 0092) Sib Mat Zh 5*1239-1258
◇ D55 E15 E45 ◇ REV MR 30 # 1936 Zbl 199.28 JSL 38.529 • ID 07429

KOZLOVA, Z.I. [1968] *Projective sets in topological spaces of weight τ (Russian)* (J 0023) Dokl Akad Nauk SSSR 183*37-40
• TRANSL [1968] (J 0062) Sov Math, Dokl 9*1326-1329
◇ D55 E15 E75 ◇ REV MR 38 # 4326 Zbl 198.552 • ID 07430

KOZLOVA, Z.I. [1969] *Certain projective operations (Russian)* (J 0339) Uch Zap Ped Inst, Volgograd 23*3-59
◇ D55 E15 ◇ REV MR 44 # 3878 • ID 07434

KOZLOVA, Z.I. [1977] see ARSENIN, V.YA.

KOZLOVA, Z.I. see Vol. V for further entries

KOZMIDIADI, V.A. [1962] *Sets enumerable and solvable by automata (Russian)* (J 0023) Dokl Akad Nauk SSSR 142*1005-1006
• TRANSL [1962] (J 0062) Sov Math, Dokl 3*216-218
◇ D05 ◇ REV MR 32 # 1124 Zbl 121.15 • ID 07438

KOZMIDIADI, V.A. [1962] see CHERNYAVSKIJ, V.S.

KOZMIDIADI, V.A. & MARCHENKOV, S.S. [1969] *On multihead automata (Russian)* (J 0052) Probl Kibern 21*127-158
• TRANSL [1969] (J 0471) Syst Th Res 21*124-156
◇ D05 D10 ◇ REV MR 44 # 3876 Zbl 243 # 02032 JSL 38.655 • ID 63107

KOZMIDIADI, V.A. [1969] see BREJTBART, YU.YA.

KOZMIDIADI, V.A. [1970] *A certain generalization of finite automata that produces a hierarchy analogous to A.Grzegorczyk's classification of primitively recursive functions (Russian)* (J 0052) Probl Kibern 23*127-170,303
◇ D05 D10 D20 ◇ REV MR 45 # 1758 Zbl 267 # 02026 • ID 07439

KOZMIDIADI, V.A. & MUCHNIK, A.A. (EDS.) [1970] *Problems of mathematical logic (Russian)* (X 0885) Mir: Moskva 432pp
◇ D97 ◇ REV MR 54 # 44 Zbl 223 # 02029 • ID 70166

KOZMIDIADI, V.A. [1972] see BREJTBART, YU.YA.

KOZMIDIADI, V.A. & MASLOV, A.N. & PETRI, N.V. (EDS.) [1974] *Complexity of computations and algorithms (Russian)* (X 0885) Mir: Moskva 389pp
◇ D15 D20 D98 ◇ REV MR 51 # 2325 • ID 80476

KOZ'MINYKH, V.V. [1968] *One-place primitively recursive functions (Russian)* (J 0003) Algebra i Logika 7/1*75-90
• TRANSL [1968] (J 0069) Algeb and Log 7*44-53
◇ D20 G15 ◇ REV MR 40 # 2540 Zbl 248 # 02041 • ID 07440

KOZ'MINYKH, V.V. [1970] *On the subalgebras of R. Robinson's algebras (Russian)* (J 0003) Algebra i Logika 9*672-690
• TRANSL [1970] (J 0069) Algeb and Log 9*403-415
◇ D20 D75 ◇ REV MR 44 # 3866 Zbl 264 # 02038 • ID 27511

KOZ'MINYKH, V.V. [1972] *Representation of partial recursive functions in the form of superpositions (Russian)* (J 0003) Algebra i Logika 11*270-294
• TRANSL [1972] (J 0069) Algeb and Log 11*153-167
◇ D20 ◇ REV MR 47 # 26 Zbl 264 # 02037 • ID 27509

KOZ'MINYKH, V.V. [1974] *Representation of partial recursive functions with certain conditions in the form of superpositions (Russian)* (J 0003) Algebra i Logika 13*420-424,487-488
• TRANSL [1974] (J 0069) Algeb and Log 13*238-240
◇ D20 ◇ REV MR 53 # 107 Zbl 319 # 02030 • ID 16573

KRAJEWSKI, S. & RUTKOWSKI, A. [1972] *Preliminary results and problems concerning Π^0_{n+1} functions (Russian summary)* (J 0014) Bull Acad Pol Sci, Ser Math Astron Phys 20*805-809
◇ D55 ◇ REV MR 48 # 1897 Zbl 248 # 02042 • ID 07443

KRAJEWSKI, S. [1981] *Kurt Goedel and his work (Polish)* (J 0519) Wiad Mat, Ann Soc Math Pol, Ser 2 23*161-187
◇ A10 C98 D98 E98 F99 ◇ REV MR 84j:01068 Zbl 535 # 01011 • ID 38331

KRAJEWSKI, S. see Vol. II, III, V, VI for further entries

KRAJNEV, V.A. [1980] *Words not containing sequential subwords are equal with respect to the frequency structure (Russian)* (J 0071) Met Diskr Analiz (Novosibirsk) 34*27-37
◇ D40 ◇ REV MR 83d:05014 Zbl 455 # 05013 • ID 54290

KRAL, J. [1970] *A modification of substitution theorem with some applications* (P 0577) Automatenth & Formale Sprachen;1969 Oberwolfach 227-241
◇ D05 ◇ REV MR 42 # 7447 Zbl 221 # 68043 • ID 14658

KRAMER, R.L. [1981] see EPSTEIN, R.L.

KRAMOSIL, I. [1975] *A probabilistic approach to automaton-environment systems* (J 0156) Kybernetika (Prague) 11*173-206
◇ D05 ◇ REV MR 56 # 2704 Zbl 308 # 94045 • ID 63115

KRAMOSIL, I. [1979] *A note on computational complexity of a statistical deducibility testing procedure* (P 2059) Math Founds of Comput Sci (8);1979 Olomouc 337-345
◇ B35 D15 ◇ REV MR 81e:68113 Zbl 417 # 68079 • ID 53285

KRAMOSIL, I. [1980] *Computational complexity of a statistical verification procedure for propositional calculus* (P 3897) Inform Th (3);1980 Liblice 123-130
◇ B35 D15 ◇ REV Zbl 529 # 68067 • ID 38493

KRAMOSIL, I. [1983] *On extremum-searching approximate probabilistic algorithms* (J 0156) Kybernetika (Prague) 19*365-373
◇ D15 D20 ◇ REV MR 85c:68027 Zbl 543 # 03029 • ID 40922

KRAMOSIL, I. [1984] *Recursive classification of pseudo-random sequences* (J 0156) Kybernetika (Prague) 20*Suppl.,34pp
◇ D15 D80 ◇ REV Zbl 544 # 60006 • ID 41037

KRAMOSIL, I. see Vol. I, II, III, VI for further entries

KRANAKIS, E. [1982] *Invisible ordinals and inductive definitions* (J 0068) Z Math Logik Grundlagen Math 28*137-148
◇ D60 D70 E45 E47 E55 ◇ REV MR 84b:03064 Zbl 494 # 03032 • ID 35649

KRANAKIS, E. [1982] *Reflection and partition properties of admissible ordinals* (J 0007) Ann Math Logic 22*213-242
◇ D60 E05 E45 E55 ◇ REV MR 83m:03056 Zbl 494 # 03030 • ID 35456

KRANAKIS, E. [1983] *Definable Ramsey and definable Erdoes ordinals* (J 0009) Arch Math Logik Grundlagenforsch 23*115-128
◇ C30 D60 E05 E45 E55 ◇ REV MR 86b:03067 Zbl 535 # 03028 • ID 38326

KRANAKIS, E. [1984] see KAUFMANN, M.

KRANAKIS, E. & PHILLIPS, I. [1984] *Partitions and homogeneous sets for admissible ordinals* (P 2153) Logic Colloq;1983 Aachen 1*235-260
◇ D60 E05 E45 E47 E55 ◇ REV Zbl 572 # 03031 • ID 45390

KRANAKIS, E. [1984] *Stepping up lemmas in definable partitions* (J 0036) J Symb Logic 49*22-31
◇ D60 E05 E45 E47 E55 ◇ REV MR 86a:03049 • ID 42470

KRANAKIS, E. [1985] *Definable partitions and reflection properties for regular cardinals* (J 0047) Notre Dame J Formal Log 26*408-412
◇ D60 E05 E55 ◇ ID 47533

KRANAKIS, E. [1985] *Definable partitions and the projectum* (J 0068) Z Math Logik Grundlagen Math 31*351-355
◇ D60 E05 E45 E55 ◇ REV Zbl 567 # 03021 • ID 47562

KRANAKIS, E. see Vol. III, V for further entries

KRATKO, M.I. [1963] *On the reducibility of the combinatorial problem of Post to certain mass problems in the theory of finite automata (Russian)* (S 0507) Vychisl Sist (Akad Nauk SSSR Novosibirsk) 9*71-73
◇ D03 D05 ◇ REV MR 31 # 50 • ID 42975

KRATKO, M.I. [1964] *Algorithmic unsolvability of the problem of recognition of completeness for finite automata (Russian)* (J 0023) Dokl Akad Nauk SSSR 155*35-37
• TRANSL [1964] (J 0062) Sov Math, Dokl 5*338-340
◇ D05 ◇ REV MR 28 # 3936 Zbl 171.276 • ID 07453

KRATKO, M.I. [1964] *On the existence of non-recursive bases of finite automata (Russian)* (J 0003) Algebra i Logika 3/2*33-44
◇ D03 D05 ◇ REV MR 30 # 1938 Zbl 166.269 • ID 07452

KRATKO, M.I. [1964] *The algorithmic unsolvability of a problem in the theory of finite automata (Russian)* (J 0071) Met Diskr Analiz (Novosibirsk) 2*37-41
◇ D05 ◇ REV MR 29 # 3379 • ID 07451

KRATKO, M.I. [1965] *A class of Post calculi (Russian)* (J 0023) Dokl Akad Nauk SSSR 165*994-995
• TRANSL [1965] (J 0062) Sov Math, Dokl 6*1544-1545
◇ D03 ◇ REV MR 32 # 7414 Zbl 192.64 JSL 32.393 • ID 19117

KRATKO, M.I. [1966] *Formal Post calculi and finite automata (Russian)* (J 0052) Probl Kibern 17*41-65
◇ D03 D05 ◇ REV MR 37 # 5098 Zbl 217.10 JSL 32.393 • ID 19115

KRATKO, M.I. [1977] *On the axiomatic definition of the concept of computational complexity* (P 1704) Int Congr Log, Meth & Phil of Sci (5);1975 London ON
◇ D15 D75 ◇ ID 43170

KRATKO, M.I. & REVIN, O. [1977] *Turing machines operating on the plane, and stable iterative systems (Russian) (English summaries)* (J 0040) Kibernetika, Akad Nauk Ukr SSR 1977/5*146-147
• TRANSL [1977] (J 0021) Cybernetics 13*787-788
◇ D10 D80 ◇ REV MR 57 # 11967 Zbl 385 # 68050 • ID 69664

KRATKO, M.I. see Vol. V for further entries

KRECZMAR, A. [1971] *The set of all tautologies of algorithmic logic is hyperarithmetical* (J 0014) Bull Acad Pol Sci, Ser Math Astron Phys 19*781-783
◇ B75 D35 D55 ◇ REV MR 46 # 10248 Zbl 221 # 02030 • ID 28106

KRECZMAR, A. [1972] *Degree of recursive unsolvability of algorithmic logic (Russian summary)* (J 0014) Bull Acad Pol Sci, Ser Math Astron Phys 20*615-617
◇ B75 D30 D35 ◇ REV MR 47 # 4776 Zbl 245 # 02038 • ID 07466

KRECZMAR, A. [1977] *Programmability in fields* (J 2095) Fund Inform, Ann Soc Math Pol, Ser 4 1*195-230
◇ B75 D75 ◇ REV MR 57 # 8124 Zbl 364 # 68025 • ID 50985

KRECZMAR, A. see Vol. II, III for further entries

KREIDER, D.L. & ROGERS JR., H. [1961] *Constructive versions of ordinal number classes* (J 0064) Trans Amer Math Soc 100*325-369
◇ D55 D70 F15 ◇ REV MR 27 # 1381 Zbl 100.12 JSL 31.134 • ID 07467

KREIDER, D.L. & RITCHIE, R.W. [1964] *Predictably computable functionals and definition by recursion* (J 0068) Z Math Logik Grundlagen Math 10*65-80
◇ D10 D20 ◇ REV MR 35 # 6556 Zbl 131.7 JSL 33.298 • ID 07469

KREIDER, D.L. & RITCHIE, R.W. [1966] *A basis theorem for a class of two-way automata* (J 0068) Z Math Logik Grundlagen Math 12*243-255
◇ D10 D25 ◇ REV MR 34 # 5681 Zbl 207.305 • ID 07473

KREIDER, D.L. & RITCHIE, R.W. [1966] *A universal two-way automaton* (J 0009) Arch Math Logik Grundlagenforsch 9*49-58
◇ D05 ◇ REV MR 35 # 2738 Zbl 173.14 • ID 07471

KREISEL, G. [1950] *Note on arithmetic models for consistent formulae of the predicate calculus I* (J 0027) Fund Math 37*265-285
◇ B10 C57 D45 D55 F30 ◇ REV MR 12.790 JSL 18.180 • REM Part II 1953 • ID 07481

KREISEL, G. [1953] *A variant to Hilbert's theory of the foundations of arithmetic* (J 0013) Brit J Phil Sci 4*107-129,357
◇ A05 B28 D30 D55 ◇ REV MR 15.670 JSL 22.304 • ID 07487

KREISEL, G. [1953] *Note on arithmetic models for consistent formulae of the predicate calculus II* (P 0645) Int Congr Philos (11);1953 Bruxelles 14*39-49
◇ B10 C57 D45 D55 F30 ◇ REV MR 15.668 Zbl 53.200 JSL 21.403 • REM Part I 1951 • ID 20815

KREISEL, G. & LACOMBE, D. & SHOENFIELD, J.R. [1957] *Effective operations and partial recursive functionals* (**P** 1675) Summer Inst Symb Log;1957 Ithaca 364-365
◇ D20 F60 ◇ REV JSL 31.261 • ID 07520

KREISEL, G. & LACOMBE, D. [1957] *Ensembles recursivement mesurables et ensembles recursivement ouverts ou fermes* (**J** 0109) C R Acad Sci, Paris 245∗1106-1109
◇ D80 F60 ◇ REV MR 22 # 3680 Zbl 79.9 JSL 31.113 • ID 07490

KREISEL, G. & LACOMBE, D. & SHOENFIELD, J.R. [1957] *Fonctionnelles recursivement definissables et fonctionnelles recursives* (**J** 0109) C R Acad Sci, Paris 245∗399-402
◇ D20 F60 ◇ REV MR 19.521 Zbl 78.7 JSL 23.48 • ID 07494

KREISEL, G. [1957] *Sums of squares* (**P** 1675) Summer Inst Symb Log;1957 Ithaca 313-320
◇ C57 C60 D20 F07 F99 ◇ REV JSL 31.128 • ID 29369

KREISEL, G. [1958] *Mathematical significance of consistency proofs* (**J** 0036) J Symb Logic 23∗155-182
◇ C57 C60 F05 F25 F50 ◇ REV MR 22 # 6710 Zbl 88.15 JSL 31.129 • ID 07514

KREISEL, G. [1959] *Analysis of Cantor-Bendixson theorem by means of the analytic hierarchy* (**J** 0014) Bull Acad Pol Sci, Ser Math Astron Phys 7∗621-626
◇ D55 E15 ◇ REV MR 22 # 9444 Zbl 93.14 JSL 35.334 • ID 07515

KREISEL, G. [1959] *Interpretation of analysis by means of constructive functionals of finite types* (**P** 0634) Constructivity in Math;1957 Amsterdam 101-128
◇ D65 F10 F30 F35 F50 ◇ REV MR 21 # 5568 Zbl 134.10 JSL 36.169 • ID 07499

KREISEL, G. & LACOMBE, D. & SHOENFIELD, J.R. [1959] *Partial recursive functionals and effective operations* (**P** 0634) Constructivity in Math;1957 Amsterdam 290-297
◇ D20 F60 ◇ REV MR 21 # 7159 JSL 31.261 • ID 07508

KREISEL, G. [1960] *La predicativite* (**J** 0353) Bull Soc Math Fr 88∗371-391
◇ A05 D55 E45 F35 F65 ◇ REV MR 23 # A800 Zbl 131.6 JSL 27.79 • ID 07519

KREISEL, G. [1960] see GANDY, R.O.

KREISEL, G. & TAIT, W.W. [1961] *Finite definability of number-theoretic functions and parametric completeness of equational calculi* (**J** 0068) Z Math Logik Grundlagen Math 7∗28-38
◇ B20 D20 ◇ REV MR 24 # A1822 Zbl 116.5 JSL 32.270 • ID 07524

KREISEL, G. [1961] see GANDY, R.O.

KREISEL, G. [1961] *Set theoretic problems suggested by the notion of potential totality* (**P** 0633) Infinitist Meth;1959 Warsaw 103-140
◇ A05 C62 D20 D55 D65 ◇ REV MR 26 # 3599 • ID 07523

KREISEL, G. [1962] *The axiom of choice and the class of hyperarithmetic functions* (**J** 0028) Indag Math 24∗307-319
◇ D55 F35 ◇ REV MR 25 # 3838 Zbl 108.8 JSL 35.333 • ID 07526

KREISEL, G. & SACKS, G.E. [1965] *Metarecursive sets* (**J** 0036) J Symb Logic 30∗318-338
◇ D30 D55 D60 ◇ REV MR 35 # 4097 Zbl 156.252 JSL 33.622 • ID 16397

KREISEL, G. [1965] *Model-theoretic invariants: applications to recursive and hyperarithmetic operations* (**P** 0614) Th Models;1963 Berkeley 190-205
◇ C40 C62 D55 D60 ◇ REV MR 33 # 7257 Zbl 225 # 02040 • ID 33546

KREISEL, G. [1966] see EHRENFEUCHT, A.

KREISEL, G. [1968] *Choice of infinitary languages by means of definability criteria: generalized recursion theory* (**P** 0637) Syntax & Semant Infinitary Lang;1967 Los Angeles 139-151
◇ C40 C70 C75 D60 D75 ◇ REV Zbl 177.10 • ID 07540

KREISEL, G. [1971] *Some reasons for generalizing recursion theory* (**P** 0638) Logic Colloq;1969 Manchester 139-198
◇ A05 D30 D60 D65 D70 D75 ◇ REV MR 48 # 46 Zbl 219 # 02027 JSL 40.230 • ID 07548

KREISEL, G. [1972] *Which number theoretic problems can be solved in recursive progressions on Π^1_1-paths through $\mathcal{O}$?* (**J** 0036) J Symb Logic 37∗311-334
◇ A05 D20 D55 F15 F30 F99 ◇ REV MR 51 # 5273 Zbl 255 # 02048 • ID 07553

KREISEL, G. [1974] *A notion of mechanistic theory* (**J** 0154) Synthese 29∗11-26
• REPR [1976] (**C** 2953) Log & Probab in Quant Mech 3-18
◇ A05 B30 D10 ◇ REV Zbl 307 # 02028 Zbl 335 # 02031 • ID 63130

KREISEL, G. [1975] *Observations on a recent generalization of completeness theorems due to Schuette* (**P** 1440) ⊧ ISILC Proof Th Symp (Schuette);1974 Kiel 164-181
◇ A05 C07 C57 F05 F20 F35 F50 ◇ REV MR 54 # 73 Zbl 324 # 02020 • ID 23955

KREISEL, G. & MINTS, G.E. & SIMPSON, S.G. [1975] *The use of abstract language in elementary metamathematics: Some pedagogic examples* (**C** 0758) Logic Colloq Boston 1972-73 38-131
◇ A05 C07 C57 C75 F05 F07 F20 F50 ◇ REV Zbl 318 # 02003 • ID 27698

KREISEL, G. [1975] *Was hat die Logik in den letzten 25 Jahren fur die Mathematik geleistet?* (**J** 2688) Conceptus (Wien) 9∗40-45
◇ A05 B98 C75 D40 ◇ REV MR 58 # 21356 • ID 75092

KREISEL, G. see Vol. I, II, III, V, VI for further entries

KREITZ, C. & WEIHRAUCH, K. [1982] *Complexity theory on real numbers and functions* (**P** 3862) Theor Comput Sci (6);1983 Dortmund 165-174
◇ D15 ◇ REV MR 84b:68003 Zbl 501 # 03025 • ID 36945

KREITZ, C. & WEIHRAUCH, K. [1985] *Theory of representations* (**J** 1426) Theor Comput Sci 38∗35-53
◇ D45 D75 D80 ◇ ID 48592

KREITZ, C. see Vol. VI for further entries

KREJNIN, YA.L. [1975] *The basis and classification of recursive functions (Russian)* (**C** 1050) Aktual Vopr Mat Log & Teor Mnozh 164-170
◇ D20 ◇ REV MR 57 # 5710 • ID 75073

KREJNIN, YA.L. see Vol. V for further entries

KREMPA, J. & PETTOROSSI, A. & SKOWRON, A. [1983] *𝔉-computable numbers* (**P** 2943) Symp Math Found Comp Sci; 1982 Diedrichshagen 24-39
◇ D15 ◇ REV MR 85f:68003 Zbl 537 # 68044 • ID 44242

KREMPA, J. see Vol. I, II, V for further entries

KRICHEVSKIJ, R.E. [1978] *Digital enumeration of binary dictionaries (Russian)* (**J** 0023) Dokl Akad Nauk SSSR 239*1044-1047
• TRANSL [1978] (**J** 0062) Sov Math, Dokl 19*469-473
◇ D15 ◇ REV MR 80a:03057 Zbl 398 # 68017 • ID 75096

KRICHEVSKIJ, R.E. see Vol. II for further entries

KRIEGEL, K. [1984] see HEMMERLING, A.

KRIEGEL, K. see Vol. I for further entries

KRINITSKIJ, N.A. [1974] see FROLOV, G.D.

KRINITSKIJ, N.A. [1984] *Algorithms around us (Russian)* (**X** 2027) Nauka: Moskva 224pp
◇ D20 ◇ REV MR 86d:00020 • REM 2nd ed. • ID 45117

KRIPKE, S.A. [1962] *The undecidability of monadic modal quantification theory* (**J** 0068) Z Math Logik Grundlagen Math 8*113-116
• TRANSL [1974] (**C** 4138) Feys: Modal Logika
◇ B45 D35 ◇ REV MR 28 # 2975 Zbl 111.11 JSL 31.277
• ID 07555

KRIPKE, S.A. [1964] *Transfinite recursion on admissible ordinals I,II* (**J** 0036) J Symb Logic 29*161-162
◇ D60 ◇ ID 38731

KRIPKE, S.A. & POUR-EL, M.B. [1967] *Deduction-preserving "recursive isomorphisms" between theories* (**J** 0015) Bull Amer Math Soc 73*145-148
◇ B30 D25 D35 F30 ◇ REV MR 35 # 6548 Zbl 174.20
• ID 31215

KRIPKE, S.A. & POUR-EL, M.B. [1967] *Deduction-preserving "recursive isomorphisms" between theories* (**J** 0027) Fund Math 61*141-163
◇ B30 D25 D35 F30 ◇ REV MR 40 # 5447 Zbl 174.20
• ID 31216

KRIPKE, S.A. see Vol. II, III, V, VI for further entries

KRISHNAMOORTHY, M.S. [1985] see KAPUR, D.

KROM, MELVEN R. [1963] *Separation principles in the hierarchy theory of pure first-order logic* (**J** 0036) J Symb Logic 28*222-236
◇ B10 C40 C52 D55 ◇ REV MR 31 # 2133 Zbl 137.9 JSL 31.503 • ID 07571

KROM, MELVEN R. [1967] *The decision problem for segregated formulas in first-order logic* (**J** 0132) Math Scand 21*233-240
◇ B20 B25 D35 ◇ REV MR 39 # 1286 Zbl 169.310
• ID 07574

KROM, MELVEN R. [1970] *The decision problem for formulas in prenex conjunctive normal form with binary disjunctions* (**J** 0036) J Symb Logic 35*210-216
◇ B20 D03 D35 ◇ REV MR 43 # 55 Zbl 207.13
• ID 07576

KROM, MELVEN R. [1981] *An unsolvable problem with products of matrices* (**J** 0041) Math Syst Theory 14*335-337
◇ D80 ◇ REV MR 83k:03053 Zbl 476 # 03025 • ID 35403

KROM, MELVEN R. see Vol. I, III, V for further entries

KRUPSKIJ, V.N. [1979] *On completely enumerable sets in effectively metric spaces (Russian) (English summary)* (**J** 0288) Vest Ser Mat Mekh, Univ Moskva 1979/5*20-25,86
• TRANSL [1979] (**J** 0510) Moscow Univ Math Bull 34/5*22-28
◇ D25 D45 ◇ REV MR 82a:03042 Zbl 445 # 03021
• ID 56485

KRUPSKIJ, V.N. [1981] *On the complexity of a description of computable approximations for points of a metric space (Russian) (English summary)* (**J** 0288) Vest Ser Mat Mekh, Univ Moskva 1981/5*3-9
• TRANSL [1981] (**J** 0510) Moscow Univ Math Bull 36/5*1-9
◇ D15 D80 ◇ REV MR 83e:03073 Zbl 466 # 03012
• ID 46473

KRUPSKIJ, V.N. [1982] *Simultaneous approximability of real numbers (Russian)* (**J** 0023) Dokl Akad Nauk SSSR 267*45-48
• TRANSL [1982] (**J** 0062) Sov Math, Dokl 26*545-547
◇ D15 ◇ REV MR 85b:11063 Zbl 525 # 03031 • ID 38264

KRUSE, A.H. [1960] *Some developments in the theory of numerations* (**J** 0064) Trans Amer Math Soc 97*523-553
◇ D45 E10 E25 E50 ◇ REV MR 23 # A803 Zbl 135.253
• ID 07579

KRUSE, A.H. [1962] *Constructive methods of numeration* (**J** 0068) Z Math Logik Grundlagen Math 8*57-70
◇ D45 E10 E25 ◇ REV MR 26 # 4917 Zbl 143.22
• ID 07582

KRUSE, A.H. [1969] *Souslinoid and analytic sets in a general setting* (**X** 0803) Amer Math Soc: Providence 127pp
◇ D55 E15 E20 ◇ REV MR 44 # 82 Zbl 204.314
• ID 22237

KRUSE, A.H. see Vol. III, V for further entries

KRYAUCHYUKAS, V.YU. [1979] *Diophantine representation of perfect numbers (Russian)* (**S** 0228) Zap Nauch Sem Leningrad Otd Mat Inst Steklov 88*78-89
• TRANSL [1982] (**J** 1531) J Sov Math 20*2307-2313
◇ D35 ◇ REV Zbl 435 # 10006 • ID 36647

KRYAZHOVSKIKH, G.V. [1980] *Approximability of finitely presented algebras (Russian)* (**J** 0092) Sib Mat Zh 21/5*58-62
• TRANSL [1980] (**J** 0475) Sib Math J 21*688-691
◇ C60 D40 ◇ REV MR 82f:08004 Zbl 457 # 17001
• ID 54395

KRYL, R. [1978] see DEMUTH, O.

KRYL, R. see Vol. VI for further entries

KRYLOV, S.M. [1982] *Models of universal discrete-analog computers on the basis of Turing machines (Russian) (English summary)* (**J** 4526) Ehlektr Modelirovanie 1982/3*6-10
◇ D10 ◇ REV MR 84a:68044 • ID 38803

KRYNICKI, M. [1977] *Henkin quantifier and decidability* (**P** 1629) Symp Math Log;1974 Oulo;1975 Helsinki 2*89-90
◇ B25 C10 C80 D35 ◇ ID 33547

KRYNICKI, M. see Vol. I, III, V, VI for further entries

KRYUKOV, YU.A. [1971] *Turing machines with three states and two symbols and with one state and n symbols (Russian) (English summary)* (**J** 0040) Kibernetika, Akad Nauk Ukr SSR 1971/1*12-13
◇ D10 ◇ REV MR 45 # 1695 Zbl 259 # 02027 • ID 30349

KUBIAK, W. [1984] see KORTAS, K.

KUBINETS, M.V. [1972] *Recognition of a self-crossing of a plane trajectory by a Kolmogorov algorithm (Russian)* (**S** 0228) Zap Nauch Sem Leningrad Otd Mat Inst Steklov 32*35-44,154
• TRANSL [1976] (**J** 1531) J Sov Math 6*374-382
◇ D10 D15 ◇ REV MR 49 # 8830 Zbl 345 # 02022
• ID 19161

KUBINSKI, T. [1962] *On extension of the theory of syntactic categories (Polish summary)* (**J** 0481) Acta Univ Wroclaw 12*19-36
◇ D05 ◇ REV MR 37 # 6166 • ID 07594

KUBINSKI, T. see Vol. I, II, V, VI for further entries

KUCERA, A. & KUSHNER, B.A. [1978] *A type of recursive isomorphism of certain concepts of constructive analysis (Russian)* (**J** 0140) Comm Math Univ Carolinae (Prague) 19*97-105
◇ D55 F60 ◇ REV MR 58 # 173 Zbl 398 # 03049
• ID 52777

KUCERA, A. [1978] see DEMUTH, O.

KUCERA, A. [1982] *On recursive measure of classes of recursive sets* (**J** 0140) Comm Math Univ Carolinae (Prague) 23*117-121
◇ D20 D30 ◇ REV MR 83m:03050 Zbl 493 # 03035
• ID 35450

KUCERA, A. [1985] *Measure, Π^0_1-classes and complete extensions of PA* (**P** 3342) Rec Th Week;1984 Oberwolfach 245-259
◇ D30 E15 E75 F30 ◇ ID 45308

KUCERA, A. see Vol. VI for further entries

KUDAJBERGENOV, K.ZH. [1979] *A theory with two strongly constructivizable models (Russian)* (**J** 0003) Algebra i Logika 18*176-185,253
• TRANSL [1979] (**J** 0069) Algeb and Log 18*111-117
◇ C57 D45 ◇ REV MR 81f:03038 Zbl 448 # 03028
• ID 75139

KUDAJBERGENOV, K.ZH. [1980] *On constructive models of undecidable theories (Russian)* (**J** 0092) Sib Mat Zh 21/5*155-158,192
◇ C15 C35 C57 D35 ◇ REV MR 82h:03040
Zbl 454 # 03011 • ID 54224

KUDAJBERGENOV, K.ZH. [1983] *The number of constructive homogeneous models of a complete decidable theory (Russian)* (**J** 0087) Mat Zametki (Akad Nauk SSSR) 34*135-143
• TRANSL [1983] (**J** 1044) Math Notes, Acad Sci USSR 34*552-557
◇ C50 C57 ◇ REV MR 85e:03075 Zbl 537 # 03021
• ID 40506

KUDAJBERGENOV, K.ZH. [1984] *Autostability and extensions of constructivizations (Russian)* (**J** 0092) Sib Mat Zh 25/5*72-78
• TRANSL [1984] (**J** 0475) Sib Math J 25*743-749
◇ C50 C57 F60 ◇ REV MR 86b:03040 • ID 41816

KUDAJBERGENOV, K.ZH. [1984] *Constructivizability of a prime model (Russian)* (**J** 0092) Sib Mat Zh 25/4*93-98
• TRANSL [1984] (**J** 0475) Sib Math J 25*584-588
◇ C50 C57 F60 ◇ REV MR 86d:03030 • ID 41871

KUDAJBERGENOV, K.ZH. see Vol. III for further entries

KUDLEK, M. [1979] *Context free normal systems* (**P** 2059) Math Founds of Comput Sci (8);1979 Olomouc 346-353
◇ D03 ◇ REV MR 81i:68104 Zbl 408 # 68066 • ID 56297

KUDLEK, M. [1982] see JANTZEN, M.

KUDLEK, M. [1984] see JANTZEN, M.

KUDLEK, M. see Vol. II for further entries

KUDRYAVTSEV, V.B. [1962] *A completeness theorem for a class of automata without feedback (Russian)* (**J** 0052) Probl Kibern 8*91-115
◇ D05 ◇ REV MR 30 # 28 • ID 07603

KUDRYAVTSEV, V.B. [1973] *The functional properties of logical nets* (**J** 0114) Math Nachr 55*187-211
◇ D05 ◇ REV MR 47 # 10157 Zbl 269 # 94022 • ID 07608

KUDRYAVTSEV, V.B. [1973] *The function system $\mathscr{P}_\Sigma$ (Russian)* (**J** 0023) Dokl Akad Nauk SSSR 210*521-522
• TRANSL [1973] (**J** 0062) Sov Math, Dokl 14*755-757
◇ D05 ◇ REV MR 48 # 3645 Zbl 309 # 02064 • ID 19157

KUDRYAVTSEV, V.B. [1974] *On the functional system $\mathscr{P}_\Sigma$ (Russian)* (**J** 0199) Zh Vychisl Mat i Mat Fiz 14*198-208,269
• TRANSL [1974] (**J** 1049) USSR Comput Math & Math Phys 14/1*194-203
◇ D05 ◇ REV MR 49 # 2369 Zbl 282 # 94037 • ID 19156

KUDRYAVTSEV, V.B. [1980] *Functional systems (Russian)* (**X** 0898) Moskov Gos Univ: Moskva 158pp
◇ B50 C05 D05 ◇ REV Zbl 491 # 03024 • ID 36855

KUDRYAVTSEV, V.B. see Vol. I, II, V for further entries

KUEHNEL, W. [1975] see EHRIG, H.

KUEHNEL, W. & MESEGUER, J. & PFENDER, M. & SOLS, I. [1975] *Primitive recursive algebraic theories with applications to program schemes* (**J** 0306) Cah Topol & Geom Differ 16*271-273
◇ B75 D20 G30 ◇ REV Zbl 353 # 02016 • ID 50099

KUEHNEL, W. & MESEGUER, J. & PFENDER, M. & SOLS, I. [1977] *Primitive recursive algebraic theories and program schemes* (**J** 0016) Bull Austral Math Soc 17*207-233
◇ B75 D20 G30 ◇ REV MR 57 # 6138 Zbl 311 # 18003
Zbl 354 # 18001 • ID 50186

KUICH, W. [1981] see BARON, G.

KUKIN, G.P. [1977] *Problem of equality for Lie algebras (Russian)* (**J** 0092) Sib Mat Zh 18*1194-1197,1208
• TRANSL [1977] (**J** 0475) Sib Math J 18*849-851
◇ D40 ◇ REV MR 56 # 15721 Zbl 383 # 17006 • ID 52031

KUKIN, G.P. [1978] *Algorithmic problems for solvable Lie algebras (Russian)* (**J** 0003) Algebra i Logika 17*402-415
• TRANSL [1978] (**J** 0069) Algeb and Log 17*270-278
◇ D40 ◇ REV MR 80h:17017 Zbl 445 # 17010 • ID 56513

KUKIN, G.P. [1979] *Subalgebras of finitely defined Lie algebras (Russian)* (**J** 0003) Algebra i Logika 18*311-327
• TRANSL [1979] (**J** 0069) Algeb and Log 18*190-201
◇ C05 C57 ◇ REV MR 81k:17010 Zbl 445 # 17011
• ID 56514

KUKIN, G.P. [1983] *The equality problem and free products of Lie algebras and of associative algebras (Russian)* (**J** 0092) Sib Mat Zh 24/2*85-96
• TRANSL [1983] (**J** 0475) Sib Math J 24*221-231
◇ D40 ◇ REV MR 84m:17012 Zbl 517 # 17009 • ID 36724

KULAGINA, O.S. [1958] *A method of determining grammatical concepts on the basis of set theory (Russian)* (**J** 0052) Probl Kibern 1*203-214
• TRANSL [1960] (**J** 1195) Probl Cybernet 1*228-242 [1962] (**J** 0449) Probl Kybern 1*239-254
◇ B65 D05 E75 ◇ REV Zbl 83.155 JSL 35.339 • ID 21973

KUNEN, K. [1968] *Implicit definability and infinitary languages* (**J** 0036) J Symb Logic 33*446-451
◇ C40 C70 C75 D70 E47 E55 ◇ REV MR 38 # 5597 Zbl 195.302 JSL 35.341 • ID 07631

KUNEN, K. see Vol. I, II, III, V for further entries

KUNUGUI, K. [1937] *Sur un theoreme d'existence dans la theorie des ensembles projectifs* (**J** 0027) Fund Math 29*167-181
◇ D55 E15 ◇ REV Zbl 17.345 FdM 63.931 • ID 07642

KUNUGUI, K. [1939] *Contribution a la theorie des ensembles Boreliens et analytiques II* (**J** 0438) J Hokkaido Univ Educ Ser 1 8*1-24
◇ D55 E15 ◇ REV MR 1.301 Zbl 21.112 • REM Part I 1939. Part III 1940 • ID 07643

KUNUGUI, K. [1940] *Contribution a la theorie des ensembles Boreliens et analytiques III* (**J** 0438) J Hokkaido Univ Educ Ser 1 8*79-108
◇ D55 E15 ◇ REV MR 1.301 Zbl 60.129 FdM 66.1222 • REM Part II 1939 • ID 07644

KUNUGUI, K. [1940] *Sur un probleme de M.E.Szpilrajn* (**J** 0081) Proc Japan Acad 16*73-78
◇ D55 E15 ◇ REV MR 1.302 Zbl 23.115 FdM 66.205 • ID 07645

KUNUGUI, K. see Vol. V for further entries

KUNZE, J. [1967] *Selektive Graphenschemata* (**J** 0068) Z Math Logik Grundlagen Math 13*101-122
◇ D05 ◇ REV MR 35 # 6557 Zbl 148.248 • ID 07647

KURATOWSKI, K. [1922] *Une methode d'elimination des nombres transfinis des raisonnements mathematiques* (**J** 0027) Fund Math 3*76-108
◇ D55 E07 E15 E25 E75 ◇ REV FdM 48.205 • ID 07652

KURATOWSKI, K. [1924] *Sur les fonctions representables analytiquement et les ensembles de premiere categorie* (**J** 0027) Fund Math 5*75-86
◇ D55 E75 ◇ REV FdM 50.181 • ID 07654

KURATOWSKI, K. [1926] *Un theoreme concernant la puissance d'ensembles de points* (**J** 0459) C R Soc Sci Lett Varsovie Cl 3 19*25-35
◇ D55 E15 ◇ REV FdM 57.1335 • ID 39828

KURATOWSKI, K. [1931] *Evaluation de la classe Borelienne ou projective d'un ensemble de points a l'aide des symboles logiques* (**J** 0027) Fund Math 17*249-272
◇ D55 E15 ◇ REV Zbl 3.105 FdM 57.92 • ID 07658

KURATOWSKI, K. & TARSKI, A. [1931] *Les operations logiques et les ensembles projectifs* (**J** 0027) Fund Math 17*240-248
• TRANSL [1956] (**C** 1159) Tarski: Logic, Semantics, Metamathematics 143-151
◇ B10 D55 E15 ◇ REV Zbl 3.105 JSL 34.99 FdM 57.92 • ID 07656

KURATOWSKI, K. & SZPILRAJN, E. [1932] *Sur les cribles fermes et leurs applications* (**J** 0027) Fund Math 18*160-170
◇ D55 E15 ◇ REV Zbl 4.204 FdM 58.85 • ID 07659

KURATOWSKI, K. [1933] *Topologie I* (**X** 1034) PWN: Warsaw xi + 450pp
• TRANSL [1933] (**X** 0801) Academic Pr: New York xx + 560pp
◇ D55 E15 E75 ◇ REV MR 19.873 MR 36 # 840 Zbl 158.408 Zbl 49.397 FdM 59.563 • REM 4th ed. 1958; 2nd engl. ed. 1966. Vol.II 1950 • ID 28504

KURATOWSKI, K. & POSAMENT, T. [1934] *Sur l'isomorphie algebro-logique et les ensembles relativement Boreliens* (**J** 0027) Fund Math 22*281-286
◇ D55 E15 ◇ REV Zbl 9.205 FdM 60.42 • ID 07662

KURATOWSKI, K. [1936] *Sur les ensembles projectifs (Russian summary)* (**J** 0142) Mat Sb, Akad Nauk SSSR, NS 1(43)*713-714
◇ D55 E15 ◇ REV Zbl 16.156 FdM 62.1175 • ID 20865

KURATOWSKI, K. [1936] *Sur les theoremes de separation dans la theorie des ensembles* (**J** 0027) Fund Math 26*183-191
◇ D55 E15 ◇ REV Zbl 14.54 FdM 62.1174 • ID 07664

KURATOWSKI, K. [1937] *Les suites transfinies d'ensembles et les ensembles projectifs* (**J** 0027) Fund Math 28*186-196
◇ D55 E15 ◇ REV Zbl 17.344 FdM 64.833 • ID 07665

KURATOWSKI, K. [1937] *Les types d'ordre definissables et les ensembles boreliens* (**J** 0027) Fund Math 29*97-100
◇ C75 D55 E07 E15 ◇ REV Zbl 17.49 JSL 3.48 FdM 64.833 • ID 07668

KURATOWSKI, K. & NEUMANN VON, J. [1937] *On some analytic sets defined by transfinite induction* (**J** 0120) Ann of Math, Ser 2 38*521-525
◇ D55 E15 ◇ REV Zbl 17.344 FdM 63.177 • ID 31806

KURATOWSKI, K. [1937] *Sur la geometrisation des types d'ordre denombrable* (**J** 0027) Fund Math 28*167-185
◇ D55 E07 E15 ◇ REV Zbl 17.343 FdM 63.28 • ID 07666

KURATOWSKI, K. [1937] *Sur les suites analytiques d'ensembles* (**J** 0027) Fund Math 29*54-59
◇ D55 E15 ◇ REV Zbl 17.343 FdM 63.933 • ID 07667

KURATOWSKI, K. & SIERPINSKI, W. [1941] *Sur l'existence des ensembles projectifs non mesurables* (**J** 0477) Spis Bulgar Akad Nauk 61*207-212
◇ D55 E15 ◇ ID 31807

KURATOWSKI, K. [1948] *Ensembles projectifs et ensembles singuliers* (**J** 0027) Fund Math 35*131-140
◇ D55 E15 ◇ REV MR 10.358 Zbl 31.290 • ID 07669

KURATOWSKI, K. see Vol. V for further entries

KURDUMOV, G.I. [1980] *An algorithm-theoretic method for the study of uniform random networks* (**C** 2970) Multicomp Random Systs 471-503
◇ D05 D10 ◇ REV MR 82j:68033 Zbl 441 # 68054 • ID 81976

KURDYUMOV, G.L. [1972] *Ueber Abbildungen mit rekursiven Graphen (Russisch)* (**J** 1460) Tr Inst Ehlekt Mashinostr, Moskva 24*3-13
◇ D20 ◇ REV Zbl 264 # 02041 • ID 27515

KURDYUMOV, G.L. [1974] *A certain class of immune sets (Russian)* (**J** 1460) Tr Inst Ehlekt Mashinostr, Moskva 30*3-18
◇ D50 ◇ REV MR 53 # 12916 • ID 23166

KUR'EROV, YU.N. [1981] *Equiprobable canonical calculi (Russian)* (S 2582) Semiotika & Inf, Akad Nauk SSSR 17*90-97
◇ B35 D03 ◇ REV MR 84m:03060 Zbl 484 # 03018 • ID 35760

KUR'EROV, YU.N. [1985] *On the embedding of an Ackermann class in a graph grammar (Russian) (English and Lithuanian summaries)* (J 3939) Mat Logika Primen (Akad Nauk Litov SSR) 1985/4*89-99,141
◇ D05 D20 ◇ ID 49034

KUR'EROV, YU.N. see Vol. I, VI for further entries

KURKI-SUONIO, R. [1971] *A programmer's introduction to computability and formal languages* (X 2329) Auerbach: Pennsauken 140pp
◇ B75 D05 D20 D98 ◇ REV Zbl 265 # 68001 • ID 29846

KURKI-SUONIO, R. [1977] see EVE, J.

KURMIT, A.A. [1972] *On the question of the enumerability and representability of sets of words by means of finite automata (Russian)* (J 0474) Avtom Vychis Tekh, Akad Nauk Latv SSR 1972/3*10-17
• TRANSL [1972] (J 2666) Autom Control Comput Sci 6/3*10-17
◇ D05 ◇ REV MR 47 # 3148 Zbl 239 # 94064 • ID 07678

KURMIT, A.A. [1976] see CHAPENKO, V.P.

KURMIT, A.A. [1979] *Generation of classes of information - lossless automata by means of substitutions of a set of states (Russian)* (J 0474) Avtom Vychis Tekh, Akad Nauk Latv SSR 1979/2*44-49
• TRANSL [1979] (J 2666) Autom Control Comput Sci 13/2*37-41
◇ D05 ◇ REV MR 81h:68038 Zbl 447 # 68054 • ID 69023

KURMIT, A.A. see Vol. I for further entries

KURODA, S.-Y. [1964] *Classes of languages and linear-bounded automata* (J 0194) Inform & Control 7*207-223
◇ D05 ◇ REV MR 29 # 6968 JSL 32.116 • ID 07692

KURTZ, S.A. [1983] *Notions of weak genericity* (J 0036) J Symb Logic 48*764-770
◇ D30 E40 ◇ REV MR 85a:03056 Zbl 549 # 03042 • ID 33598

KURTZ, S.A. [1983] *On the random oracle hypthesis* (J 0194) Inform & Control 57*40-47
◇ D15 ◇ REV MR 85j:03062 Zbl 549 # 68038 • ID 45246

KURTZ, S.A. [1985] *Sparse sets in $NP-P$: relativizations* (J 1428) SIAM J Comp 14*113-119
◇ D15 ◇ REV Zbl 574 # 03020 • ID 45392

KUSABA, T. [1973] *On Hilbert's tenth problem (affirmative cases)* (J 0091) Sugaku 25*10-19
◇ D25 D35 ◇ REV MR 58 # 27751 • ID 81981

KUSHNER, B.A. [1969] see KANOVICH, M.I.

KUSHNER, B.A. [1970] *Some mass problems connected with the integration of constructive functions (Russian)* (S 0066) Tr Mat Inst Steklov 113*39-72
• TRANSL [1970] (S 0055) Proc Steklov Inst Math 113*42-83
◇ D30 F60 ◇ REV MR 44 # 6917 Zbl 285 # 02033 • ID 07698

KUSHNER, B.A. [1973] *A certain problem of Mostowski (Russian)* (S 0554) Issl Teor Algor & Mat Logik (Moskva) 1*257-261
◇ D20 F60 ◇ REV MR 49 # 28 Zbl 285 # 02030 • ID 07702

KUSHNER, B.A. [1973] *Computationally complex real numbers (Russian)* (J 0068) Z Math Logik Grundlagen Math 19*447-452
◇ D20 F60 ◇ REV MR 49 # 4764 Zbl 296 # 02019 • ID 07699

KUSHNER, B.A. & NAGORNYJ, N.M. (EDS.) [1974] *Theory of algorithms, and mathematical logic (Russian)* (X 2265) Akad Nauk Vychis Tsentr: Moskva 216pp
◇ D97 ◇ REV MR 52 # 7873 Zbl 286 # 00009 • ID 70187

KUSHNER, B.A. [1976] see KANOVICH, M.I.

KUSHNER, B.A. [1976] *On Grzegorczyk's theorem on the computability of an isolated extremum (Russian)* (S 0554) Issl Teor Algor & Mat Logik (Moskva) 2*112-121
◇ D20 D80 F60 ◇ REV MR 58 # 16209 • ID 75199

KUSHNER, B.A. & MARKOV, A.A. (EDS.) [1976] *Studies in the theory of algorithms and mathematical logic. Vol. 2 (Russian)* (X 2265) Akad Nauk Vychis Tsentr: Moskva 160pp
◇ D97 ◇ REV MR 56 # 74 • ID 70127

KUSHNER, B.A. [1978] see KUCERA, A.

KUSHNER, B.A. [1978] *On some systems of computable real numbers (Russian)* (C 3211) Mat Ling & Teor Algor 105-112
◇ D45 F60 ◇ REV Zbl 411 # 03056 • ID 52910

KUSHNER, B.A. see Vol. VI for further entries

KUTEPOV, V.P. [1979] see FAL'K, V.N.

KUZENKO, V.F. & SHKIL'NYAK, S.S. & ZUBENKO, V.V. [1980] *Syntactical definitions and the problem of reorientation (Russian)* (S 2653) Prepr Inst Kib, Akad Nauk Ukr SSR 3*45pp
◇ D05 ◇ REV MR 82g:68081 • ID 83230

KUZ'MIN, V.A. [1965] *Realization of functions of the algebra of logic by means of automata, normal algorithms and Turing machines (Russian)* (J 0052) Probl Kibern 13*75-96
◇ B05 D03 D05 D10 ◇ REV MR 33 # 7212 Zbl 263 # 02020 • ID 29866

KUZ'MINA, T.M. [1981] *Structure of the m-degrees of the index sets of families of partial recursive functions (Russian)* (J 0003) Algebra i Logika 20*55-68,123-124
• TRANSL [1981] (J 0069) Algeb and Log 20*37-48
◇ D20 D30 ◇ REV MR 83h:03060 Zbl 498 # 03032 • ID 36068

KUZ'MINA, T.M. [1983] *Reducibility by morphisms (Russian)* (J 3937) Veroyat Met i Kibern (Kazan) 19*29-39
◇ D20 D45 ◇ REV MR 85b:03071 Zbl 566 # 03026 • ID 40699

KUZNETSOV, A.V. [1950] *On primitive recursive functions of large oscillation (Russian)* (J 0023) Dokl Akad Nauk SSSR 71*233-236
◇ D20 ◇ REV MR 11.635 Zbl 37.297 JSL 17.270 • ID 19147

KUZNETSOV, A.V. & TRAKHTENBROT, B.A. [1955] *Investigation of partial recursive operators by means of the theory of Baire's space (Russian)* (J 0023) Dokl Akad Nauk SSSR 105*897-900
◇ D20 D25 D30 ◇ REV MR 17.1039 Zbl 66.261 JSL 22.301 • ID 19741

KUZNETSOV, A.V. [1959] *Certain questions concerning the classification of predicates and functions (Russian)* (P 0607) All-Union Math Conf (3);1956 Moskva 4*86-87
◇ D55 ◇ ID 30601

KUZNETSOV, A.V. [1959] *On the equivalence and functional completeness problems (Russian)* (**P** 0607) All-Union Math Conf (3);1956 Moskva 2*145-146
◇ B25 D35 ◇ ID 28393

KUZNETSOV, A.V. [1963] *Undecidability of the general problems of completeness solvability and equivalence for propositional calculi (Russian)* (**J** 0003) Algebra i Logika 2/4*47-66
• TRANSL [1966] (**J** 0225) Amer Math Soc, Transl, Ser 2 59*56-72
◇ B22 D35 ◇ REV MR 28 # 3935 Zbl 166.263 JSL 37.756 • ID 19144

KUZNETSOV, A.V. see Vol. I, II, III, VI for further entries

KUZNETSOV, O.P. [1961] *Asynchronous logical networks and the representation of non-regular events* (**P** 1607) Hungar Math Congr (2);1960 Budapest V32-V37
◇ D05 ◇ ID 29313

KUZNETSOV, O.P. [1963] *On a class of regular events (Russian)* (**C** 4103) Struktur Teor Relej Ustrojstv 100-109
◇ D05 ◇ REV Zbl 217.10 JSL 33.629 • ID 43232

KUZNETSOV, O.P. [1965] *Representation of regular events in asynchronous automata (Russian)* (**J** 0011) Avtom Telemekh 26*1086-1093
• TRANSL [1965] (**J** 0010) Autom & Remote Control 26*1073-1079
◇ D05 ◇ REV MR 32 # 9166 Zbl 234 # 02022 • ID 27797

KUZNETSOV, O.P. [1965] *Ueber eine Klasse von regulaeren Ereignissen (Russisch)* (**P** 1578) IFAC Symp Teor Relej Ustroj & Kon Avtom;1962 Moskva 211-214
◇ D05 ◇ REV Zbl 217.10 • ID 28040

KUZNETSOV, O.P. [1972] *Ueber die Kompliziertheit der Berechnungen in eindimensionalen iterativen Strukturen (Russian)* (**P** 3117) All-Union Conf Autom Contr-Eng Cybern (4);1968 Tbilisi 24-32
◇ D05 D10 ◇ REV MR 55 # 4776 Zbl 271 # 68037 • ID 63245

KUZNETSOV, S.E. [1978] see AL'PIN, YU.A.

KUZNETSOV, S.E. see Vol. I for further entries

KWASOWIEC, W. [1969] *Some properties of machines (Russian)* (**J** 0485) Algorytmy, Pol Akad Nauk 5*21-24
• TRANSL [1970] (**J** 0068) Z Math Logik Grundlagen Math 16*399-404
◇ B70 D05 ◇ REV MR 40 # 8311 Zbl 185.25 • ID 07727

LACAVA, F. [1985] *Undecidability of the theory of L-algebras (Italian) (English summary)* (**J** 2099) Boll Unione Mat Ital, VI Ser, A 4*133-136
◇ D35 ◇ REV Zbl 566 # 03004 • ID 44805

LACAVA, F. see Vol. III for further entries

LACHLAN, A.H. [1962] *Multiple recursion* (**J** 0068) Z Math Logik Grundlagen Math 8*81-107
◇ D20 ◇ REV MR 26 # 3602 Zbl 107.11 • ID 07735

LACHLAN, A.H. [1964] *Effective operations in a general setting* (**J** 0036) J Symb Logic 29*163-178
◇ D20 D25 D75 ◇ REV MR 34 # 7367 Zbl 219 # 02028 JSL 31.654 • ID 07740

LACHLAN, A.H. [1964] *Standard classes of recursively enumerable sets* (**J** 0068) Z Math Logik Grundlagen Math 10*23-42
◇ D25 ◇ REV MR 28 # 4994 Zbl 282 # 02016 • ID 07736

LACHLAN, A.H. [1965] *Effective inseparability for sequences of sets* (**J** 0053) Proc Amer Math Soc 16*647-653
◇ D25 ◇ REV MR 32 # 42 Zbl 204.13 • ID 07741

LACHLAN, A.H. [1965] *On a problem of G.E. Sacks* (**J** 0053) Proc Amer Math Soc 16*972-979
◇ D25 D30 ◇ REV MR 32 # 43 Zbl 156.9 JSL 32.125 • ID 07739

LACHLAN, A.H. [1965] *On recursive enumeration without repetition* (**J** 0068) Z Math Logik Grundlagen Math 11*209-220 • ERR/ADD ibid 13*99-100
◇ D25 ◇ REV MR 32 # 7405 Zbl 158.251 JSL 38.155 • ID 19171

LACHLAN, A.H. [1965] *Some notions of reducibility and productiveness* (**J** 0068) Z Math Logik Grundlagen Math 11*17-44
◇ D25 D30 D55 ◇ REV MR 30 # 3014 Zbl 158.251 JSL 35.478 • ID 07738

LACHLAN, A.H. [1966] *A note on universal sets* (**J** 0036) J Symb Logic 31*573-574
◇ D25 ◇ REV MR 35 # 2732 Zbl 239 # 02021 JSL 32.395 • ID 07745

LACHLAN, A.H. [1966] *Lower bounds for pairs of r.e.degrees* (**J** 3240) Proc London Math Soc, Ser 3 16*537-569
◇ D25 ◇ REV MR 34 # 4126 Zbl 156.9 JSL 37.611 • ID 07744

LACHLAN, A.H. [1966] *On the indexing of classes of recursively enumerable sets* (**J** 0036) J Symb Logic 31*10-22
◇ D25 ◇ REV MR 34 # 2451 Zbl 143.12 • ID 07742

LACHLAN, A.H. [1966] *The impossibility of finding relative complements for recursively enumerable degrees* (**J** 0036) J Symb Logic 31*434-454
◇ D25 ◇ REV MR 34 # 5673 Zbl 156.10 JSL 36.539 • ID 07743

LACHLAN, A.H. [1967] *The priority method I* (**J** 0068) Z Math Logik Grundlagen Math 13*1-10
◇ D25 ◇ REV MR 35 # 1469 Zbl 153.316 • ID 07747

LACHLAN, A.H. [1968] *Complete recursively enumerable sets* (**J** 0053) Proc Amer Math Soc 19*99-102
◇ D25 ◇ REV MR 36 # 4985 Zbl 155.340 • ID 07749

LACHLAN, A.H. [1968] *Degrees of recursively enumerable sets which have no maximal supersets* (**J** 0036) J Symb Logic 33*431-443
◇ D25 ◇ REV MR 38 # 4314 Zbl 197.5 • ID 07750

LACHLAN, A.H. [1968] *Distributive initial segments of the degrees of unsolvability* (**J** 0068) Z Math Logik Grundlagen Math 14*457-472
◇ D30 ◇ REV MR 38 # 5620 Zbl 176.278 • ID 07748

LACHLAN, A.H. [1968] *On the lattice of recursively enumerable sets* (**J** 0064) Trans Amer Math Soc 130*1-37
◇ B25 C10 D25 ◇ REV MR 37 # 2594 Zbl 281 # 02042 JSL 35.153 • ID 07751

LACHLAN, A.H. [1968] *The elementary theory of recursively enumerable sets* (**J** 0025) Duke Math J 35*123-146
◇ B25 D25 ◇ REV MR 37 # 2593 Zbl 281 # 02043 JSL 35.153 • ID 07752

LACHLAN, A.H. [1969] *Initial segments of one-one degrees* (**J** 0048) Pac J Math 29*351-366
◇ D30 ◇ REV MR 41 # 5215 Zbl 182.16 • ID 07753

LACHLAN, A.H. & MADISON, E.W. [1970] *Computable fields and arithmetically definable ordered fields* (**J** 0053) Proc Amer Math Soc 24*803-807
◇ C57 C60 D45 ◇ REV MR 40 # 7110 Zbl 229 # 02039 • ID 07756

LACHLAN, A.H. [1970] *Initial segments of many-one degrees* (**J** 0017) Canad J Math 22*75-85
◇ D30 ◇ REV MR 41 # 1531 Zbl 229 # 02036 • ID 07755

LACHLAN, A.H. [1970] *On some games which are relevant to the theory of recursively enumerable sets* (**J** 0120) Ann of Math, Ser 2 91*291-310
◇ D25 E60 ◇ REV MR 44 # 1562 Zbl 276 # 02023 JSL 39.345 • ID 07754

LACHLAN, A.H. [1971] *Solution to a problem of Spector* (**J** 0017) Canad J Math 23*247-256
◇ D20 D30 ◇ REV MR 47 # 6460 Zbl 272 # 02064 • ID 07758

LACHLAN, A.H. [1972] *Embedding nondistributive lattices in the recursively enumerable degrees* (**P** 2080) Conf Math Log;1970 London 255*149-177
◇ D25 ◇ REV MR 51 # 12494 Zbl 256 # 02021 • ID 17283

LACHLAN, A.H. [1972] *Recursively enumerable many-one degrees* (**J** 0003) Algebra i Logika 11*326-358,362 • REPR [1972] (**J** 0069) Algeb and Log 11*186-202
◇ D25 ◇ REV MR 46 # 8826 Zbl 287 # 02028 • ID 19168

LACHLAN, A.H. [1972] *Two theorems of many-one degrees of recursively enumerable sets* (**J** 0003) Algebra i Logika 11*216-229
• REPR [1972] (**J** 0069) Algeb and Log 11*127-132
◇ D25 ◇ REV MR 46 # 8825 Zbl 309 # 02046 • ID 19170

LACHLAN, A.H. [1973] *The priority method for the construction of recursively enumerable sets* (**P** 0713) Cambridge Summer School Math Log;1971 Cambridge GB 299-310
◇ D25 ◇ REV MR 49 # 29 Zbl 279 # 02021 • ID 07760

LACHLAN, A.H. [1974] *A note on Π^0_{n+1} functions and relations* (**J** 0014) Bull Acad Pol Sci, Ser Math Astron Phys 22*863-866
◇ D55 ◇ REV MR 54 # 83 Zbl 449 # 03041 • ID 19166

LACHLAN, A.H. [1975] *Uniform enumeration operations* (**J** 0036) J Symb Logic 40*401-409
◇ D25 D30 ◇ REV MR 52 # 62 Zbl 316 # 02048 • ID 18243

LACHLAN, A.H. [1975] *Wtt-complete sets are not necessarily tt-complete* (**J** 0053) Proc Amer Math Soc 48*429-434
◇ D25 D30 ◇ REV MR 50 # 9555 Zbl 311 # 02048 • ID 07764

LACHLAN, A.H. [1976] *A recursively enumerable degree which will not split over all lesser ones* (**J** 0007) Ann Math Logic 9*307-365
◇ D25 ◇ REV MR 53 # 12912 Zbl 357 # 02040 • ID 18244

LACHLAN, A.H. & LEBEUF, R. [1976] *Countable initial segments of the degrees of unsolvability* (**J** 0036) J Symb Logic 41*289-300
◇ D30 ◇ REV MR 53 # 7746 Zbl 361 # 02054 • ID 14759

LACHLAN, A.H. & SREBRNY, M. & ZARACH, A. (EDS.) [1977] *Set theory and hierarchy theory V, Bierutowice, Poland 1976* (**S** 3301) Lect Notes Math 619*viii+358pp
◇ D97 E97 ◇ REV MR 57 # 56 Zbl 354 # 00007 • ID 50140

LACHLAN, A.H. [1979] *Bounding minimal pairs* (**J** 0036) J Symb Logic 44*626-642
◇ D25 ◇ REV MR 80k:03041 Zbl 428 # 03037 • ID 53796

LACHLAN, A.H. [1980] *Decomposition of recursively enumerable degrees* (**J** 0053) Proc Amer Math Soc 79*629-634
◇ D25 ◇ REV MR 81i:03065 Zbl 451 # 03013 • ID 54028

LACHLAN, A.H. & SOARE, R.I. [1980] *Not every finite lattice is embeddable in the recursively enumerable degrees* (**J** 0345) Adv Math 37*74-82
◇ D25 ◇ REV MR 82a:03038 Zbl 439 # 03024 • ID 56015

LACHLAN, A.H. [1981] *Full satisfaction classes and recursive saturation* (**J** 0018) Canad Math Bull 24*295-297
◇ C50 C57 C62 ◇ REV MR 84c:03066b Zbl 471 # 03055 • ID 55250

LACHLAN, A.H. [1984] *Binary homogeneous structures. I* (**S** 3382) Sem-ber, Humboldt-Univ Berlin, Sekt Math 60*100-156
◇ C10 C13 C15 C45 C50 C57 ◇ REV MR 86h:03048 Zbl 552 # 03020 • ID 43365

LACHLAN, A.H. [1984] see KNIGHT, J.F.

LACHLAN, A.H. see Vol. I, II, III, V, VI for further entries

LACOMBE, D. [1954] *Sur le semi-reseau constitue par les degres d'indecidabilite recursive* (**J** 0109) C R Acad Sci, Paris 239*1108-1109
◇ D30 ◇ REV MR 16.555 Zbl 58.7 JSL 23.226 • ID 07765

LACOMBE, D. [1955] *Classes recursivement fermes et fonctions majorantes* (**J** 0109) C R Acad Sci, Paris 240*716-718
◇ D20 D55 ◇ REV MR 16.662 Zbl 67.2 JSL 24.52 • ID 07767

LACOMBE, D. [1957] see KREISEL, G.

LACOMBE, D. [1957] *Les ensembles recursivement ouverts ou fermes, et leurs applications a l'analyse recursive* (**J** 0109) C R Acad Sci, Paris 245*1040-1043
◇ D55 F60 ◇ REV MR 21 # 5572 Zbl 78.7 JSL 24.53 • ID 07768

LACOMBE, D. [1959] see KREISEL, G.

LACOMBE, D. [1959] *Quelques procedes de definition en topologie recursive* (**P** 0634) Constructivity in Math;1957 Amsterdam 129-158
◇ D80 ◇ REV MR 22 # 3687 Zbl 89.7 JSL 31.133 • ID 07770

LACOMBE, D. [1960] *La theorie des fonctions recursives et ses applications (expose d'une information generale)* (**J** 0353) Bull Soc Math Fr 88*393-468
◇ D10 D20 D80 D98 ◇ REV MR 23 # A60 Zbl 156.252 JSL 38.526 • ID 07771

LACOMBE, D. [1964] *Deux generalisations de la notion de recursivite* (**J** 0109) C R Acad Sci, Paris 258*3141-3143
◇ D45 D75 ◇ REV MR 28 # 3929 Zbl 154.256 JSL 36.536 • ID 07772

LACOMBE, D. [1964] *Deux generalisations de la notion de recursivite relative* (**J** 0109) C R Acad Sci, Paris 258*3410-3413
◇ D30 ◇ REV MR 28 # 3930 Zbl 154.256 JSL 36.536 • ID 43455

LACOMBE, D. [1964] *Theoremes de non-decidabilite* (S 1567) Semin Bourbaki Exp.266*41pp
◇ D35 D45 D98 ◇ REV MR 31 # 62 Zbl 154.257
• ID 42978

LACOMBE, D. [1971] *Recursion theoretic structure for relational systems* (P 0638) Logic Colloq;1969 Manchester 3-17
◇ D45 D75 ◇ REV MR 43 # 1823 Zbl 225 # 02027
JSL 40.454 • ID 22225

LACOMBE, D. see Vol. I, VI for further entries

LADNER, R.E. [1973] *A completely mitotic nonrecursive r.e. degree* (J 0064) Trans Amer Math Soc 184*479-507
◇ D25 ◇ REV MR 53 # 2656 Zbl 309 # 02045 • ID 07775

LADNER, R.E. [1973] *Mitotic recursively enumerable sets* (J 0036) J Symb Logic 38*199-211
◇ D25 ◇ REV MR 49 # 7127 Zbl 286 # 02042 • ID 07774

LADNER, R.E. & LYNCH, N.A. & SELMAN, A.L. [1974] *Comparison of polynominal-time reducibilities* (P 1464) ACM Symp Th of Comput (6);1974 Seattle 110-121
◇ D15 D20 ◇ REV MR 54 # 6554 Zbl 381 # 68041
• ID 24767

LADNER, R.E. & LYNCH, N.A. & SELMAN, A.L. [1975] *A comparison of polynominal time reducibilities* (J 1426) Theor Comput Sci 1*103-123
◇ D15 ◇ REV MR 52 # 16116 Zbl 321 # 68039 • ID 31232

LADNER, R.E. [1975] see FREEDMAN, A.R.

LADNER, R.E. & SASSO JR., L.P. [1975] *The weak truth table degrees of recursively enumerable sets* (J 0007) Ann Math Logic 8*429-448
◇ D25 ◇ REV MR 52 # 63 Zbl 324 # 02028 • ID 18245

LADNER, R.E. & LYNCH, N.A. [1976] *Relativization of questions about log space computability* (J 0041) Math Syst Theory 10*19-32
◇ D10 D15 ◇ REV MR 54 # 7222 Zbl 341 # 68036
• ID 25012

LADNER, R.E. [1977] *Application of model theoretic games to discrete linear orders and finite automata* (J 0194) Inform & Control 33*281-303
• TRANSL [1980] (J 3079) Kiber Sb Perevodov, NS 17*164-191
◇ B15 B25 C07 C13 C65 C85 D05 E07 E60 ◇ REV MR 58 # 10387 Zbl 387 # 68037 • ID 52252

LADNER, R.E. [1977] *The computational complexity of provability in systems of modal propositional logic* (J 1428) SIAM J Comp 6*467-480
◇ B45 D15 ◇ REV MR 56 # 8326 Zbl 373 # 02025
• ID 51489

LADNER, R.E. [1979] see FISCHER, MICHAEL J.

LADNER, R.E. [1980] *Complexity theory with emphasis on the complexity of logical theories* (P 3021) Logic Colloq;1979 Leeds 286-319
◇ B25 C98 D05 D10 D15 ◇ REV MR 82h:03037
Zbl 444 # 03018 • ID 75296

LADNER, R.E. & LIPTON, R.J. & STOCKMEYER, L.J. [1984] *Alternation bounded auxiliary pushdown automata* (J 0194) Inform & Control 62*93-108
◇ D10 D15 ◇ ID 46387

LADNER, R.E. & NORMAN, J.K. [1985] *Solitaire automata* (J 0119) J Comp Syst Sci 30*116-129
◇ D10 ◇ REV Zbl 564 # 90108 • ID 48241

LADRIERE, J. [1961] *Expression de la recursion primitive dans le calcul-λK* (J 0079) Logique & Anal, NS 4*23-54
◇ B40 D20 ◇ REV JSL 30.91 • ID 07778

LADRIERE, J. see Vol. I, II, III, VI for further entries

LADZIANSKA, Z. [1974] *Poproduct of lattices and Sorkin's theorem* (J 0143) Mat Chasopis (Slov Akad Ved) 24*247-251
◇ C05 D40 G10 ◇ REV MR 50 # 6938 • ID 07779

LAEMMEL, A.E. [1963] *Application of lattice-ordered semigroups to codes and finite-state transducers* (P 0674) Symp Math Th of Automata;1962 New York 241-256
◇ B70 D05 ◇ REV MR 30 # 5898 Zbl 122.128 • ID 07780

LAGANA, M.R. & LEONI, G. & PINZANI, R. & SPRUGNOLI, R. [1975] *Improvements in the execution of Markov algorithms (Italian summary)* (J 0012) Boll Unione Mat Ital, IV Ser 11*473-489
◇ D03 ◇ REV MR 52 # 7885 Zbl 323 # 68030 • ID 18247

LAING, R. [1973] *A note on Maxwell's demon and universal computation* (J 0383) J Cybern 3*76-78
◇ D10 ◇ REV MR 51 # 2888 Zbl 279 # 68043 • ID 17379

LAING, R. [1973] *The capabilities of some species of artificial organism* (J 0383) J Cybern 3*16-25
◇ D05 D80 ◇ REV MR 50 # 11857 Zbl 278 # 68051
• ID 63282

LAING, R. [1976] *Automaton introspection* (J 0119) J Comp Syst Sci 13*172-183
◇ D05 ◇ REV MR 55 # 12360 Zbl 336 # 94031 • ID 63281

LAING, R. [1978] *Anomalies of self-description* (J 0154) Synthese 38*373-387
◇ A05 D10 ◇ REV Zbl 384 # 03021 • ID 52063

LAKINA, N.I. [1974] *On the question of programming of Turing machines (Russian)* (J 0413) Izv Akad Nauk Belor SSR, Ser Fiz-Mat 1974/4*104-107,142
◇ D10 ◇ REV MR 50 # 11837 Zbl 295 # 02020 • ID 07804

LALLEMENT, G. [1973] *The role of regular and inverse semigroups in the theory of finite state machines and languages* (P 0760) Symp Inverse Semigroups & General;1973 DeKalb 44-63
◇ D05 ◇ REV MR 53 # 683 Zbl 381 # 20052 • ID 16690

LALLEMENT, G. [1977] *Presentations de monoides et problemes algorithmiques* (S 2250) Semin Dubreil: Algebre 29(1975/76)*136-144
◇ D40 D45 ◇ REV MR 58 # 28227 Zbl 367 # 02025
• ID 51185

LAMBEK, J. [1961] see COURT, L.M.

LAMBEK, J. [1961] *How to program an infinite abacus* (J 0018) Canad Math Bull 4*295-302
◇ D10 D20 ◇ REV MR 24 # A2532 Zbl 112.9 JSL 31.514
• ID 07818

LAMBEK, J. see Vol. I, III, V, VI for further entries

LAMBERT, W.M. [1979] *On recursion with large objects (Spanish) (English summary)* (J 1680) Cienc Tecnol, Costa Rica 3/1*11-24
◇ D75 E20 E70 ◇ REV MR 82k:03080 Zbl 471 # 03044
• ID 55239

LAMBERT JR., W.M. [1968] *A notion of effectiveness in arbitrary structures* (J 0036) J Symb Logic 33*577-602
◇ D75 ◇ REV MR 39 # 5361 Zbl 165.20 • ID 07831

LAMBERT JR., W.M. see Vol. I, VI for further entries

LANDWEBER, L.H. [1969] *Decision problems for ω-automata* (**J** 0041) Math Syst Theory 3*376-384
◇ D05 D10 ◇ REV MR 41 # 5219 Zbl 182.24 • ID 07842

LANDWEBER, L.H. [1969] see BUECHI, J.R.

LANDWEBER, L.H. [1969] *Synthesis algorithms for sequential machines* (**P** 0594) Inform Processing (4);1968 Edinburgh 1*300-304
◇ B70 D05 ◇ REV Zbl 213.22 • ID 27992

LANDWEBER, L.H. & ROBERTSON, E.L. [1972] *Recursive properties of abstract complexity classes* (**J** 0037) ACM J 19*296-308
◇ D15 ◇ REV MR 46 # 35 Zbl 261 # 68025 • ID 07843

LANDWEBER, L.H. [1973] see HOSCH, F.A.

LANDWEBER, L.H. [1974] see BRAINERD, W.S.

LANDWEBER, L.H. & LIPTON, R.J. & ROBERTSON, E.L. [1979] *On the stucture of sets in NP and other complexity classes* (**P** 4049) Allerton Conf Commun, Control & Comput (17);1979 Monticello 661-669
◇ D15 ◇ REV MR 84b:94003 • ID 46282

LANDWEBER, L.H. & LIPTON, R.J. & ROBERTSON, E.L. [1981] *On the structure of sets in NP and other complexity classes* (**J** 1426) Theor Comput Sci 15*181-200
◇ D15 ◇ REV MR 84g:68028 Zbl 482 # 68042 • ID 37706

LANDWEBER, P.S. [1964] *Decision problems of phrase-structure grammars* (**J** 4305) IEEE Trans Electr Comp EC-13*354-362
◇ B65 D05 ◇ REV Zbl 133.256 JSL 32.115 • ID 07846

LANGE, E.E. [1979] *On the solution of a class of logical-combinatorial problems by a permutational method (Russian)* (**S** 0764) Teor Konech Avtom & Prilozh (Riga) 10*35-46
◇ D05 ◇ REV Zbl 449 # 05026 • ID 56744

LANGE, K.-J. [1984] *Nondeterministic logspace reductions* (**P** 3658) Math Founds of Comput Sci (11);1984 Prague 378-388
◇ D15 ◇ REV MR 86g:68059 Zbl 567 # 03016 • ID 46839

LANGE, K.-J. & WELZL, E. [1985] *String grammars with disconnecting* (**P** 4647) FCT'85 Fund of Comput Th;1985 Cottbus 249-256
◇ D05 ◇ ID 49079

LANGMAACK, H. & SCHMIDT, GUNTHER [1970] *Klassen unwesentlich verschiedener Ableitungen als Verbaende* (**P** 0577) Automatenth & Formale Sprachen;1969 Oberwolfach 169-182
◇ D03 D40 G10 ◇ REV MR 51 # 4716 Zbl 209.310 • ID 29296

LANGMAACK, H. [1979] *On a theory of decision problems in programming languages* (**P** 3479) Math Stud of Inform Process;1978 Kyoto 75*538-558
◇ B75 D05 D80 ◇ REV MR 81i:68041 Zbl 463 # 68019 • ID 69672

LANKFORD, D.S. [1981] see BALLANTYNE, A.M.

LANKFORD, D.S. [1984] see BALLANTYNE, A.M.

LANKFORD, D.S. see Vol. I for further entries

LAPLAZA, M.L. [1978] see BOLLMAN, D.A.

LAPLAZA, M.L. see Vol. I, V, VI for further entries

LAPSIEN, R. [1981] see ERNI, W.

LARSEN, K.G. [1985] see JENSEN, F.V.

LASCAR, D. [1978] *Caractere effectif des theoremes d'approximation d'Artin (English summary)* (**J** 2313) C R Acad Sci, Paris, Ser A-B 287*A907-A910
◇ C57 ◇ REV MR 80c:14006 Zbl 424 # 13005 • ID 82025

LASCAR, D. [1982] see DALEN VAN, D.

LASCAR, D. [1985] see DELON, F.

LASCAR, D. see Vol. I, III, V, VI for further entries

LATTEUX, M. [1978] *Mots infinis et langages commutatifs* (**J** 3441) RAIRO Inform Theor 12*185-192
◇ D05 ◇ REV MR 80b:68091 Zbl 387 # 68051 • ID 69673

LATTEUX, M. [1978] see ARNOLD, A.

LATTEUX, M. [1979] see AUTEBERT, J.-M.

LATTEUX, M. [1980] see AUTEBERT, J.-M.

LATTEUX, M. [1981] see DAUCHET, M.

LATTEUX, M. [1981] see AUTEBERT, J.-M.

LATTEUX, M. [1981] see BLATTNER, M.

LATTEUX, M. [1982] see AUTEBERT, J.-M.

LATTEUX, M. & THIERRIN, G. [1983] *Semidiscrete context-free languages* (**J** 0382) Int J Comput Math 14*3-18
◇ D05 ◇ REV MR 85e:68042 Zbl 514 # 68072 • ID 40013

LATTEUX, M. & ROZENBERG, G. [1984] *Commutative one-counter languages are regular* (**J** 0119) J Comp Syst Sci 29*54-57
◇ D05 ◇ REV MR 86g:68092 Zbl 549 # 68074 • ID 44103

LATTEUX, M. & LEGUY, B. & RATOANDROMANANA, B. [1985] *The family of one-counter languages is closed under quotient* (**J** 1431) Acta Inf 22*579-588
◇ D05 ◇ ID 49081

LAURINOLLI, T. [1978] *Bounded quantification and relations recognizable by finite automata* (**J** 1431) Acta Inf 10*67-78
◇ D05 ◇ REV MR 58 # 19370 Zbl 381 # 68067 • ID 69674

LAUSMAA, T. [1982] *Informational properties of a partition (Russian) (English and Estonian summaries)* (**J** 0080) Izv Akad Nauk Ehston SSR, Fiz, Mat 31*390-398,473
◇ D80 E75 ◇ REV MR 84i:68071 Zbl 518 # 94010 • ID 40166

LAUTEMANN, C. [1983] *BPP and the polynomial hierarchy* (**J** 0232) Inform Process Lett 17*215-217
◇ D15 ◇ REV MR 85j:03063 Zbl 515 # 68042 • ID 37863

LAUTENBACH, K. [1980] see GENRICH, H.J.

LAUTENBACH, K. see Vol. I for further entries

LAVORI, P. [1977] *Recursion in the extended superjump* (**J** 0316) Illinois J Math 21*752-758
◇ D55 D65 ◇ REV MR 58 # 5132 • ID 75381

LAVRENTIEFF, M. [1925] *Sur les sous-classes de la classification de M. Baire* (**J** 0109) C R Acad Sci, Paris 180*111-114
◇ D55 E15 ◇ REV FdM 51.166 • ID 41584

LAVRENTIEFF, M. see Vol. V for further entries

LAVROV, I.A. [1962] *Undecidability of elementary theories of certain rings (Russian)* (**J** 0003) Algebra i Logika 1/3*39-45
◇ C60 D35 ◇ REV MR 28 # 3934 • ID 07884

LAVROV, I.A. [1963] *Effective inseparability of the sets of identically true formulae and finitely refutable formulae for certain elementary theories (Russian)* (**J** 0003) Algebra i Logika 2/1*5-18
◇ C13 D25 D35 ◇ REV MR 28#1132 Zbl 199.32
• ID 07885

LAVROV, I.A. [1965] see ERSHOV, YU.L.

LAVROV, I.A. & POLYAKOV, E.A. [1966] *Bases of algebras of recursive functions (Russian)* (**J** 0092) Sib Mat Zh 7*1059-1067
• TRANSL [1966] (**J** 0475) Sib Math J 7*843-849
◇ D20 D75 ◇ REV MR 34#2452 Zbl 189.9 • ID 19188

LAVROV, I.A. [1967] *The use of k-order arithmetic progressions in constructing the bases of the algebra of primitive recursive functions (Russian)* (**J** 0023) Dokl Akad Nauk SSSR 172*279-282
• TRANSL [1967] (**J** 0062) Sov Math, Dokl 8*83-86
◇ D20 D75 ◇ REV MR 37#60 Zbl 235#02040
• ID 07886

LAVROV, I.A. [1968] *Answer to a question of P.R.Young (Russian)* (**J** 0003) Algebra i Logika 7/2*48-54
• TRANSL [1968] (**J** 0069) Algeb and Log 7*98-101
◇ D25 ◇ REV MR 39#2635 Zbl 208.19 • ID 07887

LAVROV, I.A. [1969] see ERSHOV, YU.L.

LAVROV, I.A. [1970] see DENISOV, S.D.

LAVROV, I.A. [1970] *Logic and algorithms (Russian)* (**X** 0913) Novosibirsk Gos Univ: Novosibirsk 173pp
◇ B98 D98 ◇ REV MR 48#8188 • ID 07889

LAVROV, I.A. [1973] see ERSHOV, YU.L.

LAVROV, I.A. [1974] *Certain properties of Post enumeration retracts (Russian)* (**J** 0003) Algebra i Logika 13*662-675,720
• TRANSL [1974] (**J** 0069) Algeb and Log 13*379-387
◇ D25 D45 ◇ REV MR 54#78 Zbl 316#02045
• ID 23959

LAVROV, I.A. & MAKSIMOVA, L.L. [1975] *Problems in set theory, mathematical logic and the theory of algorithms (Russian)* (**X** 2027) Nauka: Moskva 240pp
◇ B98 D98 E98 ◇ REV MR 52#13300 Zbl 307#02001
• REM 2nd ed. 1984; 224pp • ID 21765

LAVROV, I.A. [1977] *Computable numberings* (**P** 1704) Int Congr Log, Meth & Phil of Sci (5);1975 London ON 1*195-206
◇ D25 D45 ◇ REV MR 58#21533 Zbl 386#03023
• ID 52199

LAVROV, I.A. [1979] *Computability of partial functions and enumerability of sets in Peano's arithmitic (Russian)* (**J** 0092) Sib Mat Zh 20*1269-1274,1407
• TRANSL [1979] (**J** 0475) Sib Math J 20*900-904
◇ D20 D25 F30 ◇ REV MR 81d:03044 Zbl 431#03030
• ID 53936

LAVROV, I.A. [1979] *Retracts of Post's numbering and effectivization of quantifiers* (**P** 1705) Scand Logic Symp (4);1976 Jyvaeskylae 287-292
◇ D25 D45 ◇ REV MR 80k:03042 Zbl 399#03030
• ID 52826

LAVROV, I.A. see Vol. I, II, V for further entries

LAWLER, E.L. [1975] *Complexity of combinatorial computations* (**C** 3277) Topics Combin Optim 87-95
◇ D15 ◇ REV Zbl 355#68038 • ID 50266

LAWLER, E.L. & LENSTRA, J.K. [1982] *Machine scheduling with precedence constraints* (**P** 3780) Ordered Sets;1981 Banff 655-676
◇ D15 ◇ REV MR 83h:68049 Zbl 474#68057 • ID 43071

LAWLER, E.L. see Vol. I for further entries

LAWRENCE, J. [1982] see BURRIS, S.

LAWRENCE, J. see Vol. III for further entries

LAZARD, D. [1976] *Algorithmes fondamentaux en algebre commutative* (**J** 1620) Asterisque 38-39*131-138
◇ D45 ◇ REV MR 55#12713 Zbl 353#13002 • ID 50129

LAZDINYA, G.K. [1981] *Probabilistic supercounter machines (Russian)* (**J** 0031) Izv Vyssh Ucheb Zaved, Mat (Kazan) 1981/5*83-84
• TRANSL [1981] (**J** 3449) Sov Math 25/5*98-99
◇ D05 ◇ REV MR 83d:68047 Zbl 547.68049 • ID 39615

LEBESGUE, H. [1905] see BAIRE, R.

LEBESGUE, H. [1905] *Sur les fonctions representables analytiquement* (**J** 0416) J Mathematiques (6)1*139-216
◇ D55 E15 ◇ REV FdM 36.453 • ID 27561

LEBESGUE, H. [1907] *Contribution a l'etude des correspondances de M. Zermelo* (**J** 0353) Bull Soc Math Fr 35*202-212
◇ D55 E15 E25 ◇ REV FdM 38.96 • ID 07905

LEBESGUE, H. see Vol. I, V for further entries

LEBEUF, R. [1976] see LACHLAN, A.H.

LECERF, Y. [1963] *Machines de Turing reversibles. Recursive insolublite en $n \in N$ de l'equation $u = \theta^n u$, ou θ est un "isomorphisme de codes"* (**J** 0109) C R Acad Sci, Paris 257*2597-2600
◇ D10 D20 ◇ REV MR 31#66 Zbl 192.69 • ID 28597

LECERF, Y. [1963] *Recursive insolubilite de l'equation generale de diagonalisation de deux monomorphismes de monoides libres $\varphi x = \psi x$* (**J** 0109) C R Acad Sci, Paris 257*2940-2943
◇ D10 D20 ◇ REV MR 31#2144 Zbl 192.69 • ID 16257

LECONTE, M. [1985] *Kth power free-codes* (**P** 4622) Autom on Infinite Words;1984 Le Mont-Dore 172-187
◇ D05 ◇ ID 49449

LEDGARD, H.F. [1972] *Embedding Markov normal algorithms within the λ-calculus* (**J** 0382) Int J Comput Math 3*131-140
◇ B40 D03 D15 ◇ REV Zbl 261#68026 • ID 63344

LEDGARD, H.F. see Vol. VI for further entries

LEE, C.Y. [1960] *Automata and finite automata* (**J** 0432) Bell Syst Tech J 39*1267-1295
◇ D05 D10 ◇ REV MR 22#3633 JSL 36.534 • ID 07956

LEE, C.Y. [1961] *Categorizing automata by W-machine programs* (**J** 0037) ACM J 8*384-399
◇ D05 ◇ REV MR 23#A1530 Zbl 104.357 JSL 33.628
• ID 07957

LEE, C.Y. [1963] *A Turing machine which prints its own code script* (**P** 0674) Symp Math Th of Automata;1962 New York 155-164
◇ D10 ◇ REV MR 30#1041 Zbl 116.337 • ID 07958

LEE, E.T. [1983] *Algorithms for finding Chomsky and Greibach normal forms for a fuzzy context-free grammar using an algebraic approach* (**J** 1429) Kybernetes 12*125-134
◇ D05 ◇ REV MR 84i:68131 Zbl 509#68073 • ID 40175

LEE, E.T. see Vol. II for further entries

LEEDS, S. & PUTNAM, H. [1971] *An intrinsic characterisation of the hierarchy of constructible sets of integers* (**P** 0638) Logic Colloq;1969 Manchester 311-350
◇ D55 E45 ◇ REV MR 54#7232 Zbl 234#02027
• ID 07967

LEEDS, S. & PUTNAM, H. [1974] *Solution to a problem of Gandy's* (**J** 0027) Fund Math 81*99-106
◇ C62 D55 ◇ REV MR 55#91 Zbl 325#02028 • ID 07969

LEEUW DE, K. & MOORE, E.F. & SHANNON, C.E. & SHAPIRO, N.Z. [1956] *Computability by probabilistic machines* (**C** 0717) Automata Studies 183-212
• TRANSL [1974] (**C** 1902) Stud Th Automaten 211-246
◇ D05 ◇ REV MR 18.104 JSL 35.481 • ID 12014

LEEUWEN VAN, J. [1974] *A partial solution to the reachability-problem for vector-addition systems* (**P** 1464) ACM Symp Th of Comput (6);1974 Seattle 303-309
◇ D80 ◇ REV MR 54#1725 Zbl 362#68105 • ID 23997

LEEUWEN VAN, J. [1976] see BAER, ROBERT M.

LEEUWEN VAN, J. [1977] *Recursively enumerable languages and van Wijngaarden grammars* (**J** 0028) Indag Math 39*29-39
◇ D05 D25 ◇ REV MR 56#10170 Zbl 355#68051
• ID 26061

LEEUWEN VAN, J. [1978] see BUNTING, P.W.

LEEUWEN VAN, J. [1980] see ENGELFRIET, J.

LEFMANN, H. & VOIGT, B. [1984] *A remark on infinite arithmetic progressions* (**J** 0193) Discr Math 52*285-286
◇ D80 E05 ◇ REV Zbl 551#10044 • ID 44423

LEGGETT, A. [1974] *Maximal α-r.e. sets and their complements* (**J** 0007) Ann Math Logic 6*293-357
◇ D60 ◇ REV MR 50#12687 Zbl 282#02017 • ID 07975

LEGGETT, A. & SHORE, R.A. [1976] *Types of simple α-recursively enumerable sets* (**J** 0036) J Symb Logic 41*681-694
◇ D60 ◇ REV MR 58#5127 Zbl 353#02022 • ID 14590

LEGGETT, A. [1978] *α-degrees of maximal α-r.e. sets* (**J** 0036) J Symb Logic 43*456-473
◇ D30 D60 ◇ REV MR 58#5128 Zbl 397#03029
• ID 28282

LEGGETT, A. [1982] see KALANTARI, I.

LEGGETT, A. [1983] see KALANTARI, I.

LEGGETT JR., E.W. & MOORE, D.J. [1981] *Optimization problems and the polynomial hierarchy* (**J** 1426) Theor Comput Sci 15*279-289
◇ D15 ◇ REV MR 82m:68085 Zbl 459#68016 • ID 82044

LEGUY, B. [1985] see LATTEUX, M.

LEHMANN, D.J. [1971] *Lr(k) grammars and deterministic languages* (**J** 0029) Israel J Math 10*526-530
◇ D05 ◇ REV MR 45#6548 Zbl 227#68038 • ID 07976

LEHMANN, D.J. see Vol. II, III for further entries

LEHMANN, G. [1985] *Modell- und rekursionstheoretische Grundlagen psychologischer Theorienbildung* (**X** 0811) Springer: Heidelberg & New York xxii+297pp
◇ B10 C98 D80 D99 ◇ ID 42737

LEININGER, B.S. [1981] see IBARRA, O.H.

LEININGER, B.S. [1983] see IBARRA, O.H.

LEISS, E. [1979] *A note on infinite graphs with actions* (**P** 3424) SE Conf Combin, Graph Th & Comput (10);1979 Boca Raton II*667-678
◇ D05 E20 ◇ REV MR 81g:05062 Zbl 427#68059
• ID 53757

LEISS, E. [1979] *On tractable unrestricted regular expressions* (1111) Preprints, Manuscr., Techn. Reports etc. 62*23pp
◇ D05 ◇ REV Zbl 432#68055 • REM University of Kentucky, Department of Computer Science. Lexington
• ID 54005

LEISS, E. [1980] see BRZOZOWSKI, J.A.

LEISS, E. [1981] *On generalized language equations* (**J** 1426) Theor Comput Sci 14*63-77
◇ D05 ◇ REV MR 82b:68066 Zbl 454#68105 • ID 82048

LEISS, E. [1981] *Succinct representation of regular languages by boolean automata* (**J** 1426) Theor Comput Sci 13*323-330
◇ D05 ◇ REV MR 82e:68090 Zbl 458#68017 • ID 82047

LEISS, E. [1985] *Succinct representation of regular languages by Boolean automata.II* (**J** 1426) Theor Comput Sci 38*133-136
◇ D05 ◇ ID 48599

LEITSCH, A. [1978] *Unsolvability in systems of constructing automata* (**P** 3428) Prog in Cybern & Syst Res;1978 Wien III*117-124
◇ D05 ◇ REV Zbl 393#03029 • ID 52449

LEITSCH, A. [1979] see ETTL, W.

LEITSCH, A. [1980] *Complexity of index sets and translating functions* (**J** 2095) Fund Inform, Ann Soc Math Pol, Ser 4 3*181-188
◇ D15 D20 ◇ REV MR 83h:03057 Zbl 454#03020
• ID 54232

LEITSCH, A. [1982] see ETTL, W.

LEITSCH, A. [1984] *Enumerations of subrecursive classes and self-reproducing automata. I: Effective translations and decidable indexsets. II: Index-complexity and translation-complexity* (**J** 0238) Sitzb Oesterr Akad Wiss, Math-Nat Kl, Abt 2 193*19-44,135-158
◇ D05 D20 D45 ◇ REV Zbl 575#03031 • ID 44807

LEITSCH, A. see Vol. I for further entries

LEIVANT, D. [1982] *Unprovability of theorems of complexity theory in weak number theories* (**J** 1426) Theor Comput Sci 18*259-268
◇ D15 F30 ◇ REV MR 84g:03098 Zbl 482#03025
• ID 34202

LEIVANT, D. [1985] *Syntactic translations and provably recursive functions* (**J** 0036) J Symb Logic 50*682-688
◇ D20 F30 F50 ◇ ID 39403

LEIVANT, D. see Vol. I, II, VI for further entries

LEMPEL, A. & ZIV, J. [1976] *On the complexity of finite sequences* (**J** 2745) IEEE Trans Inf Theory IT-22*75-81
◇ D15 D20 ◇ REV MR 52#10234 Zbl 337#94013
• ID 63372

LENARD, A. & STILLWELL, J.C. [1983] *An algorithmically unsolvable problem in analysis* (**J** 0053) Proc Amer Math Soc 88*129-130
◇ D80 ◇ REV MR 85a:58098 Zbl 516#58043 • ID 38585

Lender, V.B. [1974] *On steps of solubility of lattices and degrees of idempotency of prevarieties of lattices (Russian)* (**J** 0142) Mat Sb, Akad Nauk SSSR, NS 95(137)*445-460 • ERR/ADD [1976] (**J** 0142) Mat Sb, Akad Nauk SSSR, NS 99(141)*477 • TRANSL [1974] (**J** 0349) Math of USSR, Sbor 24*435-449 [1976] (**J** 0349) Math of USSR, Sbor 28*430
◇ C05 D80 ◇ REV MR 50 # 6940 MR 53 # 7874 Zbl 335 # 06002 Zbl 335 # 06003 • ID 63374

Lenstra, J.K. [1982] see Lawler, E.L.

Lentin, A. [1967] see Gross, M.

Lentin, A. [1970] see Gross, M.

Lentin, A. [1971] see Gross, M.

Lentin, A. [1973] *Equations in free monoids* (**P** 0763) Automata, Lang & Progr (1);1972 Rocquencourt 67-85
◇ D40 ◇ REV MR 51 # 13087 Zbl 269 # 20057 • ID 17224

Leong, B. & Seiferas, J.I. [1977] *New real-time simulations of multihead tape units* (**P** 1903) ACM Symp Th of Comput (9);1977 Boulder 239-248
◇ D10 D15 ◇ REV MR 58 # 8490 Zbl 454 # 68033 • ID 82050

Leong, B. [1979] see Ginsburg, S.

Leoni, G. [1975] see Lagana, M.R.

Leoni, G. & Sprugnoli, R. [1975] *The compilation of Pointer Markov Algorithms* (**P** 3525) Int Comput Symp;1975 Antibes 129-135
◇ D03 D15 ◇ REV MR 58 # 3629 Zbl 324 # 68014 • ID 63376

Leoni, G. & Sprugnoli, R. [1977] *Some relations between Markov algorithms and formal languages* (**J** 0089) Calcolo 14*261-284
◇ D03 D05 ◇ REV MR 58 # 21513 Zbl 374 # 68043 • ID 75461

Lepistoe, T. [1971] *On commutative languages* (**J** 3994) Ann Acad Sci Fennicae Ser A I 486*6pp
◇ D05 ◇ REV MR 44 # 2615 Zbl 212.336 • ID 08015

Lerman, M. [1969] *Some nondistributive lattices as initial segments of the degrees of unsolvability* (**J** 0036) J Symb Logic 34*85-98
◇ D30 ◇ REV MR 40 # 1276 Zbl 191.304 • ID 08019

Lerman, M. [1970] *Recursive functions modulo co-r-maximal sets* (**J** 0064) Trans Amer Math Soc 148*429-444
◇ D25 D30 ◇ REV MR 42 # 70 Zbl 198.25 • ID 08021

Lerman, M. [1970] *Turing degrees and many-one degrees of maximal sets* (**J** 0036) J Symb Logic 35*29-40
◇ D25 D30 ◇ REV MR 43 # 4674 Zbl 198.25 • ID 08020

Lerman, M. [1971] *Initial segments of the degrees of unsolvability* (**J** 0120) Ann of Math, Ser 2 93*365-389
◇ D30 ◇ REV MR 46 # 7008 Zbl 193.310 • ID 08023

Lerman, M. [1971] *Some theorems on r-maximal sets and major subsets of recursively enumerable sets* (**J** 0036) J Symb Logic 36*193-215
◇ D25 D30 ◇ REV MR 45 # 3200 Zbl 242 # 02047 • ID 08022

Lerman, M. [1972] *On suborderings of the α-recursively enumerable α-degrees* (**J** 0007) Ann Math Logic 4*369-392
◇ D30 D60 ◇ REV MR 48 # 5835 Zbl 262 # 02038 • ID 08026

Lerman, M. & Sacks, G.E. [1972] *Some minimal pairs of α-recursively enumerable degrees* (**J** 0007) Ann Math Logic 4*415-442
◇ D30 D60 ◇ REV MR 55 # 12491 Zbl 262 # 02040 • ID 08024

Lerman, M. [1973] *Admissible ordinals and priority arguments* (**P** 0713) Cambridge Summer School Math Log;1971 Cambridge GB 311-344
◇ D60 ◇ REV MR 52 # 59 Zbl 293 # 02028 • ID 17223

Lerman, M. & Simpson, S.G. [1973] *Maximal sets in α-recursion theory* (**J** 0029) Israel J Math 14*236-247
◇ D60 ◇ REV MR 47 # 8271 Zbl 261 # 02030 • ID 08027

Lerman, M. [1974] *Least upper bounds for minimal pairs of α-r.e. α-degrees* (**J** 0036) J Symb Logic 39*49-56
◇ D30 D60 ◇ REV MR 50 # 9559 Zbl 286 # 02044 • ID 08029

Lerman, M. [1974] *Maximal α-r.e. sets* (**J** 0064) Trans Amer Math Soc 188*341-386
◇ D60 ◇ REV MR 48 # 10785 Zbl 333 # 02036 • ID 08030

Lerman, M. [1976] *Congruence relations, filters, ideals, and definability in lattices of α-recursively enumerable sets* (**J** 0036) J Symb Logic 41*405-418
◇ D60 ◇ REV MR 54 # 2437 Zbl 333 # 02037 JSL 41.405 • ID 14763

Lerman, M. [1976] see Chong, C.T.

Lerman, M. [1976] *Ideals of generalized finite sets in lattices of α-recursively enumerable sets* (**J** 0068) Z Math Logik Grundlagen Math 22*347-352
◇ D60 ◇ REV MR 57 # 16021 Zbl 352 # 02032 • ID 18441

Lerman, M. [1976] *Types of simple α-recursively enumerable sets* (**J** 0036) J Symb Logic 41*419-426
◇ D60 ◇ REV MR 55 # 12492 Zbl 333 # 02038 • ID 14772

Lerman, M. & Shore, R.A. & Soare, R.I. [1978] *r-maximal major subsets* (**J** 0029) Israel J Math 31*1-18
◇ D25 ◇ REV MR 81g:03049 Zbl 384 # 03025 • ID 52067

Lerman, M. [1978] *Lattices of α-recursively enumerable sets* (**P** 1628) Generalized Recursion Th (2);1977 Oslo 223-238
◇ B25 D60 ◇ REV MR 80e:03054 Zbl 453 # 03047 • ID 30739

Lerman, M. [1978] *On elementary theories of some lattices of α-recursively enumerable sets* (**J** 0007) Ann Math Logic 14*227-272
◇ B25 D60 E45 ◇ REV MR 80i:03056 Zbl 391 # 03022 • ID 29152

Lerman, M. & Schmerl, J.H. [1979] *Theories with recursive models* (**J** 0036) J Symb Logic 44*59-76
◇ C35 C57 C65 ◇ REV MR 81g:03036 Zbl 423 # 03038 • ID 53550

Lerman, M. & Soare, R.I. [1980] *d-simple sets, small sets, and degree classes* (**J** 0048) Pac J Math 87*135-155
◇ D25 ◇ REV MR 82g:03076 Zbl 467 # 03040 • ID 55038

Lerman, M. & Soare, R.I. [1980] *A decidable fragment of the elementary theory of the lattice of recursively enumerable sets* (**J** 0064) Trans Amer Math Soc 257*1-37
◇ B25 C10 D25 ◇ REV MR 81c:03034 Zbl 439 # 03023 • ID 56014

LERMAN, M. [1980] *The degrees of unsolvability: Some recent results* (**P** 3021) Logic Colloq;1979 Leeds 140-157
◇ D30 D98 ◇ REV MR 82i:03055 Zbl 457 # 03041
• ID 54366

LERMAN, M. [1981] *On recursive linear orderings* (**P** 2628) Log Year;1979/80 Storrs 132-142
◇ C57 C65 D30 D45 D55 E07 ◇ REV MR 82i:03059 Zbl 467 # 03045 • ID 55043

LERMAN, M. & ROSENSTEIN, J.G. [1982] *Recursive linear orderings* (**P** 3634) Patras Logic Symp;1980 Patras 123-126
◇ C57 C65 D45 D55 E07 ◇ REV MR 84j:03092 Zbl 511 # 03018 • ID 34683

LERMAN, M. & REMMEL, J.B. [1982] *The universal splitting property I* (**P** 3623) Logic Colloq;1980 Prague 181-207
◇ D25 D45 ◇ REV MR 84d:03054 Zbl 497 # 03036 • REM Part II 1984 • ID 34090

LERMAN, M. [1983] *Degrees of unsolvability. Local and global theory* (**X** 0811) Springer: Heidelberg & New York xiii+307pp
◇ D30 D98 ◇ REV MR 85h:03044 Zbl 542 # 03023 JSL 50.549 • ID 33601

LERMAN, M. [1983] *The structures of recursion theory* (**P** 3669) SE Asian Conf on Log;1981 Singapore 77-95
◇ D25 D30 D98 ◇ REV MR 85b:03060 Zbl 539 # 03022
• ID 33602

LERMAN, M. & SHORE, R.A. & SOARE, R.I. [1984] *The elementary theory of the recursively enumerable degrees is not* $\aleph_0$ *categorical* (**J** 0345) Adv Math 53*301-320
◇ C35 D25 ◇ REV MR 86d:03040 Zbl 559 # 03027
• ID 33603

LERMAN, M. & REMMEL, J.B. [1984] *The universal splitting property. II* (**J** 0036) J Symb Logic 49*137-150
◇ D25 ◇ REV MR 86f:03066 • REM Part I 1982 • ID 42471

LERMAN, M. [1985] *On the ordering of classes in high/low hierarchies* (**P** 3342) Rec Th Week;1984 Oberwolfach 260-270
◇ D25 ◇ ID 45309

LERMAN, M. [1985] *The embedding problem for the recursively enumerable degrees* (**P** 4046) Rec Th;1982 Ithaca 13-20
◇ D25 ◇ ID 46371

LERMAN, M. [1985] *Upper bounds for the arithmetical degrees* (**J** 0073) Ann Pure Appl Logic 29*225-254
◇ D30 D55 ◇ ID 47477

LERMAN, M. see Vol. I, III for further entries

LETICHEVS'KIJ, O.A. [1969] *Functional equivalence of discrete transformers I (Russian) (English summary)* (**J** 0040) Kibernetika, Akad Nauk Ukr SSR 1969/2*5-16
◇ D05 ◇ REV MR 45 # 8446 Zbl 238 # 02031 • REM Part II 1970 • ID 63391

LETICHEVS'KIJ, O.A. [1970] *Functional equivalence of discrete transformers. II (Russian)* (**J** 0040) Kibernetika, Akad Nauk Ukr SSR 1970/2*14-28
◇ D05 ◇ REV Zbl 238 # 02032 • REM Part I 1969. Part III 1972 • ID 27904

LETICHEVS'KIJ, O.A. [1972] *Functional equivalence of discrete transformers. III (Russian)* (**J** 0040) Kibernetika, Akad Nauk Ukr SSR 1972/1*1-4
◇ D05 ◇ REV Zbl 242 # 02040 • REM Part II 1971 • ID 26354

LETICHEVS'KIJ, O.A. [1972] *Optimization of the strategy of a control processor (Russian) (English summary)* (**J** 0040) Kibernetika, Akad Nauk Ukr SSR 1972/2*1-3
◇ D05 ◇ REV MR 46 # 4785 Zbl 242 # 94046 • ID 63392

LETICHEVS'KIJ, O.A. [1973] *Practical methods for recognizing the equivalence of discrete transformers and program schemes (Russian) (English summary)* (**J** 0040) Kibernetika, Akad Nauk Ukr SSR 1973/4*15-26
◇ B75 D05 ◇ REV MR 49 # 4312 Zbl 277 # 94037
• ID 63394

LETICHEVS'KIJ, O.A. [1973] *The equivalence of automata with respect to cancellative semigroups (Russian)* (**J** 0052) Probl Kibern 27*195-212,294 • ERR/ADD ibid 28(1974)*280
◇ D05 D40 ◇ REV MR 49 # 7383 Zbl 358 # 94062
• ID 08038

LETICHEVS'KIJ, O.A. [1973] see GLUSHKOV, V.M.

LETICHEVS'KIJ, O.A. [1974] see GODLEVSKIJ, A.B.

LETICHEVS'KIJ, O.A. [1976] *Acceleration of the iteration of monotone operators (Russian)* (**J** 0040) Kibernetika, Akad Nauk Ukr SSR 1976/4*1-7
• TRANSL [1976] (**J** 0021) Cybernetics 12*495-501
◇ D15 ◇ REV MR 57 # 18193 Zbl 341 # 02021 • ID 29742

LETICHEVS'KIJ, O.A. & SMIKUN, L.B. [1976] *On a class of groups with solvable problem of automata equivalence (Russian)* (**J** 0023) Dokl Akad Nauk SSSR 227*36-38
• TRANSL [1976] (**J** 0062) Sov Math, Dokl 17*341-344
◇ D05 D40 ◇ REV MR 53 # 10964 Zbl 359 # 02034
• ID 23102

LETICHEVS'KIJ, O.A. see Vol. I for further entries

LEVENSHTEJN, V.I. [1962] *The inversion of finite automata (Russian)* (**J** 0023) Dokl Akad Nauk SSSR 147*1300-1303
• TRANSL [1962] (**J** 0470) Sov Phys, Dokl 7*1081-1084
◇ D05 ◇ REV MR 27 # 3526 Zbl 178.330 • ID 08039

LEVIEN, R.E. [1963] *Set-theoretic formalizations of computational algorithms, computable functions, and general-purpose computers* (**P** 0674) Symp Math Th of Automata;1962 New York 101-123
◇ B75 D10 D20 E75 ◇ REV MR 30 # 1628 Zbl 147.150
• ID 27607

LEVIN, A.G. [1978] *Generating an input sequence which brings an asynchronous circuit to a given state (Russian)* (**J** 0474) Avtom Vychis Tekh, Akad Nauk Latv SSR 1978/3*28-31
◇ B70 D05 ◇ REV MR 80e:68157 Zbl 382 # 94043
• ID 69020

LEVIN, A.G. see Vol. I for further entries

LEVIN, L.A. & ZVONKIN, A.K. [1970] *The complexity of finite objects and the basing of the concepts of information and randomness on the theory of algorithms (Russian)* (**J** 0067) Usp Mat Nauk 25/6*85-127
• TRANSL [1970] (**J** 1399) Russ Math Surv 25/6*83-124
◇ D20 D80 ◇ REV MR 46 # 7004 Zbl 222 # 02027
• ID 14459

LEVIN, L.A. [1973] *On storage capacity for algorithms (Russian)* (**J** 0023) Dokl Akad Nauk SSSR 212*804-805
• TRANSL [1973] (**J** 0062) Sov Math, Dokl 14*1464-1466
◇ D20 D80 ◇ REV MR 49 # 2309 • ID 08047

LEVIN, L.A. [1973] *On the notion of a random sequence (Russian)* (**J** 0023) Dokl Akad Nauk SSSR 212*548-550
• TRANSL [1973] (**J** 0062) Sov Math, Dokl 14*1413-1416
◇ D15 D80 ◇ REV MR 51 # 2346 Zbl 312 # 94006
• ID 63406

LEVIN, L.A. [1973] *Universal problems of full search (Russian)* (**J** 2320) Probl Peredachi Inf, Akad Nauk SSSR 9/3*115-116
◇ D15 ◇ REV MR 49 # 4799 Zbl 313 # 02026 • ID 29623

LEVIN, L.A. [1974] *Bezeichnung rekursiver Funktionen (Russisch)* (**C** 2319) Slozh Vychisl & Algor 174-185
◇ D15 ◇ REV Zbl 298 # 02030 • ID 63403

LEVIN, L.A. [1976] *On the principle of conservation of information in intuitionistic mathematics (Russian)* (**J** 0023) Dokl Akad Nauk SSSR 227*1293-1296
• TRANSL [1976] (**J** 0062) Sov Math, Dokl 17*601-605
◇ D15 D80 F50 F55 ◇ REV MR 58 # 21509 Zbl 358 # 02033 • ID 50487

LEVIN, L.A. [1977] *On a concrete method of assigning complexity measures (Russian)* (**J** 0023) Dokl Akad Nauk SSSR 234*536-539
• TRANSL [1977] (**J** 0062) Sov Math, Dokl 18*727-731
◇ D15 D80 ◇ REV MR 56 # 15376 Zbl 382 # 03044
• ID 51968

LEVIN, L.A. see Vol. II, VI for further entries

LEVIN, V.A. [1972] *Infinite-valued logic and transient processes in finite automata (Russian)* (**J** 0474) Avtom Vychis Tekh, Akad Nauk Latv SSR 1972/6*1-9
• TRANSL [1972] (**J** 2666) Autom Control Comput Sci 6/6*1-8
◇ B50 D05 ◇ REV MR 49 # 4681 Zbl 245 # 94023
• ID 63410

LEVIN, V.A. [1974] *Equations in infinite-valued logic, and transition processes in finite automata (Russian)* (**J** 0474) Avtom Vychis Tekh, Akad Nauk Latv SSR 1974/5*12-15
• TRANSL [1974] (**J** 2666) Autom Control Comput Sci 8/5*12-15
◇ B50 D05 ◇ REV MR 50 # 16166 Zbl 295 # 94060
• ID 63412

LEVIN, V.A. [1974] *Transients in simple asynchronous automata with memory (Russian)* (**J** 0474) Avtom Vychis Tekh, Akad Nauk Latv SSR 1974/2*24-28
• TRANSL [1974] (**J** 2666) Autom Control Comput Sci 8/2*19-23
◇ D10 ◇ REV MR 50 # 4160 Zbl 315 # 94044 • ID 63409

LEVIN, V.A. [1978] see BUNIMOVA, E.O.

LEVIN, V.A. [1980] *Some new data on logical determinants and their applications to dynamics of automata (Russian)* (**J** 0474) Avtom Vychis Tekh, Akad Nauk Latv SSR 1980/5*14-23
• TRANSL [1980] (**J** 2666) Autom Control Comput Sci 14/5*12-20
◇ D05 ◇ REV MR 83d:68049 • ID 39620

LEVIN, V.A. see Vol. I, II for further entries

LEVIN, V.I. (ED.) [1972] *Fragen der Synthese endlicher Automaten (Russian)* (**X** 2230) Zinatne: Riga 200pp
◇ D05 D98 ◇ REV MR 55 # 12342 Zbl 243 # 00020
• ID 48632

LEVIN, V.I. [1977] *Logical determinants and automata with continuous time. I: Theory of logical determinants. II: Automata without memory. III: Automata with memory (Russian)* (**J** 0977) Izv Akad Nauk SSSR, Tekh Kibern 1977/3*113-122,1977/4*124-134,1977/5*134-142
• TRANSL [1977] (**J** 0522) Engin Cybern 15/3*90-99,15/4*88-97,15/5*76-84
◇ D05 D10 ◇ REV MR 80a:68051 MR 81b:94068 Zbl 364 # 94054 Zbl 364 # 94055 Zbl 374 # 94039
• ID 82062

LEVINSON, H. [1972] *On the genera of graphs of group presentations. II* (**J** 0033) J Comb Th, Ser B 12*205-225
◇ D40 ◇ REV MR 46 # 7080 Zbl 228 # 05106 • REM Part I 1970 • ID 08049

LEVITZ, H. [1978] *An ordinal bound for the set of polynomial functions with exponentiation* (**J** 0004) Algeb Universalis 8*233-243
◇ D15 D20 E07 E10 F15 F30 ◇ REV MR 81f:04003 Zbl 376 # 06001 • ID 30732

LEVITZ, H. see Vol. I, III, V, VI for further entries

LEVY, A. [1970] *Definability in axiomatic set theory II* (**P** 1072) Math Log & Founds of Set Th;1968 Jerusalem 129-145
◇ D55 E35 E45 E47 E50 ◇ REV MR 42 # 2936 Zbl 218 # 02056 • REM Part I 1965 • ID 20962

LEVY, A. see Vol. I, II, III, V, VI for further entries

LEVY, L.S. [1980] see JOSHI, A.K.

LEVY, S. [1971] *Computational equivalence* (**C** 1566) Semin IRIA Log & Automates 1971 127-136
◇ D10 D30 D55 ◇ REV Zbl 276 # 02021 • ID 29041

LEWIS, A.A. [1985] *On effectively computable realizations of choice functions* (**J** 3914) Math Soc Sci 10*43-80
◇ D20 ◇ ID 48600

LEWIS, A.A. [1985] *The minimum degree of recursively representable choice functions* (**J** 3914) Math Soc Sci 10*179-188
◇ D20 D30 ◇ ID 48981

LEWIS, A.A. see Vol. I, V for further entries

LEWIS, F.D. [1970] *The classification of unsolvable problems in automata theory* (1111) Preprints, Manuscr., Techn. Reports etc. 49*
◇ D05 D25 ◇ REM Cornell University, Department of Computer Science. Ithaca • ID 21223

LEWIS, F.D. [1971] *Classes of recursive functions and their index sets* (**J** 0068) Z Math Logik Grundlagen Math 17*291-294
◇ D20 ◇ REV MR 45 # 3203 Zbl 229 # 02035 • ID 08108

LEWIS, F.D. [1971] *The enumerability and invariance of complexity classes* (**J** 0119) J Comp Syst Sci 5*286-303
◇ D15 D25 ◇ REV MR 43 # 4677 Zbl 215.47 JSL 39.349
• ID 08109

LEWIS, F.D. [1976] *On computational reducibility* (**J** 0119) J Comp Syst Sci 12*122-131
◇ D15 ◇ REV MR 53 # 4616 Zbl 343 # 02025 • ID 21659

LEWIS, F.D. [1976] *Subrecursive reducibilities and completeness (Preliminary version)* (**P** 4736) Conf Inform Sci & Syst;1976 Baltimore 36-41
◇ D15 D20 ◇ REV MR 58 # 19355 • ID 82069

LEWIS, F.D. [1979] see BREJTBART, YU.YA.

LEWIS, F.D. [1979] *On unsolvability in subrecursive classes of predicates* (**J** 0047) Notre Dame J Formal Log 20*55-67
◇ D15 D20 D30 D55 ◇ REV MR 80h:03060 Zbl 316 # 02050 • ID 52397

LEWIS, F.D. [1980] see ABRAMSON, F.G.

LEWIS, F.D. [1981] *A note on context free languages, complexity classes, and diagonalization* (**J** 0041) Math Syst Theory 14*223-227
◇ D05 D15 ◇ REV MR 82f:03035 Zbl 473 # 68038 • ID 55397

LEWIS, F.D. [1981] *Stateless Turing machines and fixed points* (**J** 0435) Int J Comput & Inf Sci 10*215-218
◇ D10 D15 D20 ◇ REV MR 84a:03042 Zbl 468 # 68059 • ID 55125

LEWIS, H.R. [1973] see AANDERAA, S.O.

LEWIS, H.R. [1973] see GOLDFARB, W.D.

LEWIS, H.R. [1974] see AANDERAA, S.O.

LEWIS, H.R. [1975] *Description of restricted automata by first-order formulae* (**J** 0041) Math Syst Theory 9*97-104
◇ D05 ◇ REV MR 52 # 2848 Zbl 317 # 02038 • ID 17639

LEWIS, H.R. [1975] see GOLDFARB, W.D.

LEWIS, H.R. [1976] *Krom formulas with one dyadic predicate letter* (**J** 0036) J Symb Logic 41*341-362
◇ B20 D05 D35 ◇ REV MR 53 # 10573 Zbl 381 # 03012 • ID 14766

LEWIS, H.R. [1977] *A measure of complexity for combinatorial decision problems of the tiling variety* (**P** 3238) Conf Theoret Comput Sci;1977 Waterloo ON 94-99
◇ B25 D10 D15 ◇ REV MR 58 # 10388 Zbl 414 # 03008 • ID 53054

LEWIS, H.R. [1978] *Complexity of solvable cases of the decision problem for the predicate calculus* (**P** 3578) IEEE Symp Found of Comput Sci (19);1978 Ann Arbor 35-47
◇ B20 B25 D15 ◇ REV MR 80e:03041 • ID 75527

LEWIS, H.R. [1978] *Renaming a set of clauses as a Horn set* (**J** 0037) ACM J 25*134-135
◇ B20 D15 ◇ REV MR 57 # 8151 Zbl 365 # 68082 • ID 82072

LEWIS, H.R. [1979] *Satisfiability problems for propositional calculi* (**J** 0041) Math Syst Theory 13*45-53
◇ B05 D15 ◇ REV MR 80m:03025 Zbl 428 # 03035 • ID 53794

LEWIS, H.R. [1979] *Unsolvable classes of quantificational formulas* (**X** 0832) Addison-Wesley: Reading xvii + 198pp
◇ B20 B25 D35 D98 ◇ REV MR 81i:03069 Zbl 423 # 03003 JSL 47.221 • ID 53515

LEWIS, H.R. [1980] *Complexity results for classes of quantificational formulas* (**J** 0119) J Comp Syst Sci 21*317-353
• TRANSL [1983] (**J** 3079) Kiber Sb Perevodov, NS 20*64-106
◇ B20 B25 D10 D15 ◇ REV MR 82m:03052 Zbl 471 # 03034 • ID 36546

LEWIS, H.R. & PAPADIMITRIOU, C.H. [1981] *Elements of the theory of computation* (**X** 0819) Prentice Hall: Englewood Cliffs xiv + 466pp
◇ D98 ◇ REV Zbl 464 # 68001 JSL 49.989 • ID 47391

LEWIS, H.R. [1982] see AANDERAA, S.O.

LEWIS, H.R. & PAPADIMITRIOU, C.H. [1982] *Symmetric space-bounded computation* (**J** 1426) Theor Comput Sci 19*161-187
◇ D15 ◇ REV MR 84h:68034 Zbl 491 # 68045 • ID 37764

LEWIS, H.R. [1982] see GUREVICH, Y.

LEWIS, H.R. [1984] see GUREVICH, Y.

LEWIS, H.R. [1984] see DENENBERG, L.

LEWIS, H.R. see Vol. I for further entries

LEWIS, P.M. [1965] see HARTMANIS, J.

LEWIS II, P.M. [1965] see HARTMANIS, J.

LEWIS II, P.M. [1971] see HARTMANIS, J.

LEWIS II, P.M. see Vol. I, II for further entries

LI, KAIDE [1965] see DONG, YUNMEI

LI, MING [1985] *Lower bounds by Kolmogorov-complexity (extended abstract)* (**P** 4628) Automata, Lang & Progr (12);1985 Nafplion 383-393
◇ D15 ◇ ID 49578

LI, XIANG [1979] *A recursively unsolvable problem concerning the mu operator (Chinese) (English summary)* (**J** 2754) Huazhong Gongxueyuan Xuebao 7/1*24-27
◇ D25 ◇ REV MR 83f:03035 • ID 35277

LI, XIANG [1979] *The productiveness of Blum's measure of effective computational complexity (Chinese)* (**J** 3793) Jisuanjii Xuebao 2*28-34
◇ D15 D25 ◇ ID 48517

LI, XIANG [1981] *Study on the theory of computability and computational complexity of real numbers, I: Unsolvable problems in the constructive continuum* (**J** 2684) J Huazhong Inst Tech (Engl Ed) 3/1*1-14
◇ D20 ◇ REV MR 83k:03056 • REM Transl. from Chinese • ID 35404

LI, XIANG [1982] *The effective immune sets and the program index sets - a generalization and application of the recursion theorem (Chinese) (English summary)* (**J** 3793) Jisuanjii Xuebao 5*273-279
◇ D50 ◇ REV MR 84i:03082 • ID 34561

LI, XIANG [1983] *Effective immune sets, program index sets and effectively simple sets- generalizations and applications of the recursion theorem* (**P** 3669) SE Asian Conf on Log;1981 Singapore 97-106
◇ D25 D50 ◇ REV MR 85f:03039 • ID 40729

LI, XIANG [1985] *Everywhere nonrecursive r.e. sets in recursively presented topological space (Chinese)* (**J** 3187) Shuxue Niankan, Xi A 6*383-386
◇ D25 D45 ◇ ID 48684

LI, XIANG [1985] *On relativized nondeterministic polynomial-time bounded combinations* (**J** 0382) Int J Comput Math 17*151-153
◇ D15 ◇ REV Zbl 558 # 68041 • ID 46635

LI, XIANG [1985] *Turing degrees on pointwise r.e. open sets in effective Hausdorff spaces (Chinese)* (**J** 2771) Kexue Tongbao 30*1557
◇ D25 D30 D45 ◇ ID 48683

LI, XIANG see Vol. II for further entries

LIBUS, M. & OSTROWSKA, M. [1982] *Remarks on L.J.Stockmeyer's note "The complexity of decision problems in automata theory and logic" (Polish) (English summary)* (S 3291) Zesz Nauk Mat Fiz Chem (Uniw Gdansk) 5*37-52
◇ D05 D10 D15 ◇ REV Zbl 518 # 68029 • ID 38383

LICHTENSTEIN, D. [1982] *Planar formulae and their uses* (J 1428) SIAM J Comp 11*329-343
◇ D15 ◇ REV MR 83k:68032 Zbl 478 # 68043 • ID 55657

LIEBERHERR, K.J. & SPECKER, E. [1979] *Complexity of partial satisfaction* (P 3535) IEEE Symp Founds of Comput Sci (20);1979 San Juan 132-139
◇ B05 D15 ◇ REV MR 83a:68036 • ID 38843

LIEBERHERR, K.J. & SPECKER, E. [1981] *Complexity of partial satisfaction* (J 0037) ACM J 28*411-421
◇ B05 D15 ◇ REV MR 83a:03033 Zbl 456 # 68078
• ID 69677

LIEBERHERR, K.J. & VAVASIS, S.A. [1982] *Analysis of polynomial approximation algorithms for constant expressions* (P 3862) Theor Comput Sci (6);1983 Dortmund 187-197
◇ B05 D15 ◇ REV MR 84b:68003 Zbl 495 # 68028
• ID 46285

LIEPE, W. & WIEHAGEN, R. [1976] *Charakteristische Eigenschaften von erkennbaren Klassen rekursiver Funktionen* (J 0129) Elektr Informationsverarbeitung & Kybern 12*421-438
◇ D20 ◇ REV MR 55 # 2527 Zbl 344 # 02030 • ID 66084

LIFSCHITZ, V. [1967] *Constructive groups (Russian)* (S 0228) Zap Nauch Sem Leningrad Otd Mat Inst Steklov 4*86-95
• TRANSL [1967] (J 0521) Semin Math, Inst Steklov 4*32-35
◇ D45 F60 ◇ REV MR 38 # 4281 Zbl 198.27 • ID 08139

LIFSCHITZ, V. [1967] *Deductive general validity and reduction classes (Russian)* (S 0228) Zap Nauch Sem Leningrad Otd Mat Inst Steklov 4*69-77
◇ B20 D35 ◇ REV MR 38 # 5604 Zbl 165.19 • ID 08138

LIFSCHITZ, V. [1967] *Some reduction classes and undecidable theories (Russian)* (S 0228) Zap Nauch Sem Leningrad Otd Mat Inst Steklov 4*65-68
• TRANSL [1967] (J 0521) Semin Math, Inst Steklov 4*24-25
◇ D35 ◇ REV MR 38 # 5603 Zbl 237 # 02013 • ID 08137

LIFSCHITZ, V. [1967] *The decision problem for some constructive theories of equality (Russian)* (S 0228) Zap Nauch Sem Leningrad Otd Mat Inst Steklov 4*78-85
• TRANSL [1967] (J 0521) Semin Math, Inst Steklov 4*29-31
◇ B25 D35 F50 ◇ REV MR 39 # 65 Zbl 165.19 • ID 33665

LIFSCHITZ, V. [1972] *Sufficient conditions for the solvability of the word problem in microprogram semigroups (Russian)* (J 0496) Tr Inst Inzh Zheleznod Transp, Moskva 413*58-75,167
◇ D40 ◇ REV MR 48 # 446 • ID 08145

LIFSCHITZ, V. see Vol. I, VI for further entries

LIH, KUOWEI [1977] *Continuous degrees* (J 0406) Bull Inst Math, Acad Sin (Taipei) 5*171-180
◇ D30 D65 ◇ REV MR 58 # 5133 Zbl 385 # 03035
• ID 30729

LIH, KUOWEI [1978] *Type two partial degrees* (J 0036) J Symb Logic 43*623-629
◇ D30 D65 ◇ REV MR 80d:03043 Zbl 397 # 03027
• ID 52693

LILLO DE, N.J. [1978] *A note on Turing machine regularity and primitive recursion* (J 0047) Notre Dame J Formal Log 19*289-294
◇ D10 D20 ◇ REV MR 57 # 12191 Zbl 305 # 02048
• ID 51316

LILLO DE, N.J. see Vol. I, III, V for further entries

LIN, C. [1981] *Recursively presented abelian groups: effective p-group theory. I* (J 0036) J Symb Logic 46*617-624
◇ C57 C60 D45 ◇ REV MR 82j:03054 Zbl 499 # 03032
• ID 75552

LIN, C. [1981] *The effective content of Ulm's theorem* (P 2902) Aspects Effective Algeb;1979 Clayton 147-160
◇ C57 C60 D45 ◇ REV MR 83f:03038 Zbl 499 # 03033
• ID 35279

LIN, S. & RADO, T. [1965] *Computer studies of Turing machine problems* (J 0037) ACM J 12*196-212
◇ D10 ◇ REV MR 33 # 3847 Zbl 137.10 JSL 40.617
• ID 14466

LIN, YUCAI [1984] *The decision problems about the periodic solutions of the domino problems* (J 4719) Chinese Ann Math Ser B 5*721-726
◇ D03 ◇ REV Zbl 547 # 03006 • ID 43187

LINDNER, C.C. [1971] *Finite partial cyclic triple systems can be finitely embedded* (J 0004) Algeb Universalis 1*93-96
◇ C05 C13 D35 ◇ REV MR 46 # 1617 Zbl 219 # 05011
• ID 08168

LINDNER, C.C. [1972] *Finite embedding theorems for partial Latin squares, quasi-groups and loops* (J 0164) J Comb Th, Ser A 13*339-345
◇ D40 ◇ REV MR 47 # 3200 Zbl 246 # 05012 • ID 08170

LINDNER, C.C. see Vol. III for further entries

LINDNER, R. [1971] see BERNHARDT, L.

LINDNER, R. & WAGNER, K. [1974] *The axiomatisation of a certain sequential propositional calculus (Russian)* (J 0052) Probl Kibern 28*43-80,278 • ERR/ADD ibid 29*248
◇ B60 B70 D05 D10 ◇ REV MR 51 # 89
Zbl 313 # 94021 • ID 24790

LINDNER, R. & WERNER, GUENTER [1976] *Eine vergleichende Analyse von Stopstrategien fuer allgemeinrekursive Prognosen (English and Russian summaries)* (J 0129) Elektr Informationsverarbeitung & Kybern 12*275-280
◇ D20 ◇ REV MR 54 # 7226 Zbl 333 # 02034 • ID 25017

LINDNER, R. & STAIGER, L. [1977] *Erkennungs-, mass- und informations-theoretische Eigenschaften regulaerer Folgenmengen* (J 0068) Z Math Logik Grundlagen Math 23*283-287
◇ D80 E15 ◇ REV MR 58 # 3770 Zbl 384 # 68061
• ID 26489

LINDNER, R. [1979] see KLETTE, R.

LINDNER, R. see Vol. II, VI for further entries

LINDSAY, P.A. [1985] *On recognizing cyclic modules effectively* (J 0394) Commun Algeb 13*1579-1595
◇ C60 D45 ◇ ID 46372

LINNA, M. [1970] *Set of schemata of c-valid equations between regular expressions is independent of the basic alphabet* (J 0498) Ann Univ Turku, Ser A I 144*7pp
◇ D05 ◇ REV MR 43 # 7336 Zbl 256 # 68033 • ID 08178

LINNA, M. [1973] *Finite power property of regular languages* (**P** 0763) Automata, Lang & Progr (1);1972 Rocquencourt 87-98
◇ D05 ◇ REV MR 51 # 7379 Zbl 276 # 02020 • ID 29040

LINNA, M. [1975] *On ω-words and ω-computations* (**J** 0498) Ann Univ Turku, Ser A I 168*53pp
◇ D05 ◇ REV MR 56 # 1822 Zbl 319 # 68046 • ID 82078

LINNA, M. [1979] *Two decidability results for deterministic pushdown automata* (**J** 0119) J Comp Syst Sci 18*92-107
◇ D05 D10 ◇ REV MR 81d:68070 Zbl 399 # 03023 • ID 52819

LINNA, M. [1983] see KARHUMAEKI, J.

LIOGON'KIJ, M.I. [1965] see DUDICH, A.M.

LIOGON'KIJ, M.I. [1973] see GLEBSKIJ, YU.V.

LIOGON'KIJ, M.I. see Vol. I, III for further entries

LIPSCHUTZ, S. [1964] *An extension of Greendliger's results on the word problem* (**J** 0053) Proc Amer Math Soc 15*37-43
◇ D40 ◇ REV MR 28 # 4018 Zbl 199.57 • ID 08183

LIPSCHUTZ, S. [1971] *Note on independent equation problem in groups* (**J** 0008) Arch Math (Basel) 22*113-116
◇ D40 ◇ REV MR 44 # 4083 Zbl 235 # 20033 • ID 08185

LIPSCHUTZ, S. [1973] *On the word problem and T-fourth-groups* (**P** 0678) Word Probl: Decis & Burnside Probl in Group Th;1969 Irvine 443-451
◇ D40 ◇ REV MR 53 # 8267 Zbl 265 # 20034 JSL 41.786 • ID 23042

LIPSCHUTZ, S. [1980] *Groups with solvable conjugacy problems* (**J** 0316) Illinois J Math 24*192-195
◇ D40 ◇ REV MR 81k:20049 Zbl 429 # 20037 • ID 82079

LIPSHITZ, L. [1974] *The undecidability of the word problems for projective geometries and modular lattices* (**J** 0064) Trans Amer Math Soc 193*171-180
◇ D40 D80 G10 ◇ REV MR 51 # 295 Zbl 288 # 02026 • ID 17484

LIPSHITZ, L. [1977] *Undecidable existential problems for addition and divisibility in algebraic number rings II* (**J** 0053) Proc Amer Math Soc 64*122-128
◇ D35 ◇ REV MR 58 # 27375b Zbl 365 # 02019 • REM Part I 1978 • ID 31824

LIPSHITZ, L. [1978] see DENEF, J.

LIPSHITZ, L. & NADEL, M.E. [1978] *The additive structure of models of arithmetic* (**J** 0053) Proc Amer Math Soc 68*331-336
◇ C15 C50 C57 C62 ◇ REV MR 58 # 10424 Zbl 383 # 03047 • ID 31825

LIPSHITZ, L. [1978] *The diophantine problem for addition and divisibility* (**J** 0064) Trans Amer Math Soc 235*271-283
◇ B25 D35 ◇ REV MR 57 # 9666 Zbl 374 # 02025 • ID 31822

LIPSHITZ, L. [1978] *Undecidable existential problems for addition and divisibility in algebraic number rings* (**J** 0064) Trans Amer Math Soc 241*121-128
◇ D35 ◇ REV MR 58 # 27375a Zbl 385 # 03012 • REM Part I. Part II 1977 • ID 31823

LIPSHITZ, L. [1980] see BECKER, J.A.

LIPSHITZ, L. [1981] *Some remarks on the diophantine problem for addition and divisibility* (**J** 3133) Bull Soc Math Belg, Ser B 33*41-52
◇ B25 D15 ◇ REV MR 82k:03011 Zbl 497 # 03007 • ID 75573

LIPSHITZ, L. [1984] see DENEF, J.

LIPSHITZ, L. see Vol. I, III for further entries

LIPSON, J.D. [1970] see BIRKHOFF, GARRETT

LIPSON, J.D. [1974] see BIRKHOFF, GARRETT

LIPTON, R.J. [1976] see CARDOZA, E.

LIPTON, R.J. & MILLER, R.E. & SNYDER, LAWRENCE [1977] *Synchronization and computing capabilities of linear asynchronous structures* (**J** 0119) J Comp Syst Sci 14*49-72
◇ B40 B70 D05 ◇ REV MR 55 # 7632 Zbl 368 # 02038 • ID 51255

LIPTON, R.J. & ZALCSTEIN, Y. [1977] *Word problems solvable in logspace* (**J** 0037) ACM J 24*522-526
◇ D15 D40 ◇ REV MR 56 # 4234 Zbl 359 # 68049 • ID 50628

LIPTON, R.J. [1978] *Model theoretic aspects of computational complexity* (**P** 3578) IEEE Symp Found of Comput Sci (19);1978 Ann Arbor 193-200
◇ C62 D15 F30 F65 H15 ◇ REV MR 80e:03042 • ID 75578

LIPTON, R.J. [1979] *On the consistency of P = NP and fragments of arithmetic* (**P** 2935) FCT'79 Fund of Comput Th;1979 Berlin/Wendisch-Rietz 269-278
◇ D15 F30 ◇ REV MR 81k:03034 Zbl 419 # 03023 • ID 53366

LIPTON, R.J. [1979] see LANDWEBER, L.H.

LIPTON, R.J. [1979] see DEMILLO, R.A.

LIPTON, R.J. [1980] see DEMILLO, R.A.

LIPTON, R.J. [1981] see LANDWEBER, L.H.

LIPTON, R.J. [1982] see KARP, R.M.

LIPTON, R.J. [1983] see FURST, M.

LIPTON, R.J. [1984] see LADNER, R.E.

LIPTON, R.J. see Vol. VI for further entries

LISCHKE, G. [1974] *Eine Charakterisierung der rekursiven Funktionen mit endlichen Niveaumengen (𝔎-Funktionen)* (**J** 0068) Z Math Logik Grundlagen Math 20*465-472
◇ D20 ◇ REV MR 53 # 108 Zbl 307 # 02026 • ID 08192

LISCHKE, G. [1975] *Flussbildmasse - Ein Versuch zur Definition natuerlicher Kompliziertheitsmasse* (**J** 0129) Elektr Informationsverarbeitung & Kybern 11*423-436,606
◇ D15 ◇ REV MR 52 # 4708 Zbl 315 # 68037 • ID 42622

LISCHKE, G. [1975] *Kompliziertheitsmasse, die gewisse Funktionalgleichungen erfuellen* (**P** 1846) Kompl, Lern- & Erkenn-Prozess;1973 Jena 1*84-85
◇ D15 ◇ ID 42625

LISCHKE, G. [1975] *Ueber die Erfuellung gewisser Erhaltungssaetze durch Kompliziertheitsmasse* (**J** 0068) Z Math Logik Grundlagen Math 21*159-166
◇ D15 D20 ◇ REV MR 52 # 67 Zbl 339 # 02038 • ID 08193

LISCHKE, G. [1976] *Natuerliche Kompliziertheitsmasse und Erhaltungssaetze I* (J 0068) Z Math Logik Grundlagen Math 22*413-418
◇ D15 ◇ REV MR 57 # 4624a Zbl 346 # 02020 • REM Part II 1977 • ID 24285

LISCHKE, G. [1977] *Natuerliche Kompliziertheitsmasse und Erhaltungssaetze II* (J 0068) Z Math Logik Grundlagen Math 23*193-200
◇ D15 ◇ REV MR 57 # 4624b Zbl 346 # 02021 Zbl 371 # 02021 • REM Part I 1976 • ID 26478

LISCHKE, G. [1977] see BERG, E.P.

LISCHKE, G. [1978] *Complexity measures defined by Mazurkiewicz-algorithms* (P 1707) Math Founds of Comput Sci (7);1978 Zakopane 326-332
◇ D15 ◇ REV MR 80d:68054 Zbl 383 # 68044 • ID 52033

LISCHKE, G. [1979] *Some considerations about complexity for Mazurkiewicz-algorithms* (J 0129) Elektr Informationsverarbeitung & Kybern 15*549-557
◇ D15 ◇ REV MR 82b:03080 Zbl 426 # 68008 • ID 53678

LISCHKE, G. [1981] *Two types of properties for complexity measures* (J 0232) Inform Process Lett 12*123-126
◇ D15 ◇ REV MR 82f:68050 Zbl 474 # 68060 • ID 69678

LISOVIK, L.P. [1975] *Generalized finite operators, and the word problem of the definitor algebra of regular events (Russian)* (J 0040) Kibernetika, Akad Nauk Ukr SSR 1975/1*25-28
• TRANSL [1975] (J 0021) Cybernetics 11*28-33
◇ D05 ◇ REV MR 54 # 9160 Zbl 314 # 94033 • ID 63466

LISOVIK, L.P. [1977] *The word problem for an algebra of regular events over semigroups (Russian) (English summary)* (J 0040) Kibernetika, Akad Nauk Ukr SSR 1977/2*15-16
◇ D05 ◇ REV MR 58 # 5147 Zbl 355 # 94066 • ID 75584

LISOVIK, L.P. [1978] see ANISIMOV, A.V.

LISOVIK, L.P. [1978] *The comparison of finitely many-valued transformations of finite automata (Russian)* (S 2653) Prepr Inst Kib, Akad Nauk Ukr SSR 3*3-15
◇ D05 ◇ REV MR 81b:68100 • ID 82092

LISOVIK, L.P. [1979] *Equivalence problem for finitely ambiguous finite automata over semigroups (Russian)* (J 0040) Kibernetika, Akad Nauk Ukr SSR 1979/4*24-27
• TRANSL [1979] (J 0021) Cybernetics 15*463-467
◇ D05 ◇ REV MR 82b:68045 Zbl 455 # 68036 • ID 54308

LISOVIK, L.P. [1979] *The identity problem for regular events over the direct product of a free and a cyclic semigroup (Russian) (English Summary)* (J 0270) Dokl Akad Nauk Ukr SSR, Ser A 1979*410-413,492
◇ D05 ◇ REV MR 80i:20037 Zbl 402 # 68058 • ID 82093

LISOVIK, L.P. & RED'KO, V.N. [1980] *Regular events in semigroups (Russian)* (J 0052) Probl Kibern 37*155-184,239
◇ D05 D40 ◇ REV MR 82b:20101 Zbl 475 # 68034 • ID 82591

LISOVIK, L.P. [1980] *Strict sets and finite semigroup coverings (Russian)* (J 0040) Kibernetika, Akad Nauk Ukr SSR 1980/1*12-16,149
• TRANSL [1980] (J 0021) Cybernetics 16*13-17
◇ D40 ◇ REV MR 82e:20080 Zbl 491 # 20048 • ID 82090

LISOVIK, L.P. [1981] *Program schemata over cyclic interpretations (Russian)* (J 2605) Programmirovanie 1981/1*29-33,95
• TRANSL [1981] (J 2604) Progr Comput Software 7*21-24
◇ D05 ◇ REV MR 83c:68017 Zbl 469 # 68020 • ID 39045

LISOVIK, L.P. [1981] *Recursive-deterministic abstract automata (Russian)* (J 0040) Kibernetika, Akad Nauk Ukr SSR 1981/2*36-41,149
• TRANSL [1981] (J 0021) Cybernetics 17*195-201
◇ D05 ◇ REV MR 82k:68023 Zbl 466 # 68071 • ID 82089

LISOVIK, L.P. [1983] *Finite transformers on marked trees and quasi-identities in a free semigroup (Russian)* (J 0052) Probl Kibern 40*19-41
◇ D05 ◇ REV MR 85d:20080 Zbl 516 # 20034 • ID 38976

LISOVIK, L.P. [1983] *Minimal undecidable identity problem for finite-automaton mappings* (J 0040) Kibernetika, Akad Nauk Ukr SSR 1983/2*11-14
• TRANSL [1983] (J 0021) Cybernetics 19*160-165
◇ D05 ◇ REV MR 85f:68026 Zbl 532 # 68058 • ID 39964

LISOVIK, L.P. [1984] *Construction of decidable singular theories of two successor functions with an extra predicate (Russian)* (J 0003) Algebra i Logika 23*266-277
• TRANSL [1984] (J 0069) Algeb and Log 23*181-189
◇ B25 D05 ◇ ID 42703

LISOVIK, L.P. [1984] *Monadic second-order theories of two successor functions with an additional predicate (Russian)* (J 0270) Dokl Akad Nauk Ukr SSR, Ser A 8*80-82
◇ B15 B25 C10 C85 D05 ◇ REV MR 86c:03008 • ID 48409

LISOVIK, L.P. see Vol. III for further entries

LITMAN, A. [1976] *On the monadic theory of ω_1 without AC* (J 0029) Israel J Math 23*251-266
◇ B25 C85 D05 E10 E25 ◇ REV MR 54 # 12526 Zbl 348 # 02041 • ID 18261

LITMAN, A. see Vol. III, V for further entries

LITOW, B. & SUDBOROUGH, I.H. [1978] *On non-erasing oracle tapes in space bounded reducibility* (J 1456) SIGACT News 10/2*53-57
◇ D10 D15 ◇ REV Zbl 391 # 68027 • ID 52363

LITOW, B.E. [1985] *On efficient deterministic simulation of Turing machine computations below log-space* (J 0041) Math Syst Theory 18*11-18
◇ D10 D15 ◇ ID 46500

LITVINTSEVA, Z.K. [1970] *On the complexity of individual identity problems in semigroups (Russian)* (J 0003) Algebra i Logika 9*172-199
• TRANSL [1970] (J 0069) Algeb and Log 9*108-125
◇ D15 D40 ◇ REV MR 43 # 7330 Zbl 251 # 02046 • ID 08196

LITVINTSEVA, Z.K. [1970] *On the complexity of some problems for groups and semi-groups (Russian)* (J 0023) Dokl Akad Nauk SSSR 191*989-992
• TRANSL [1970] (J 0062) Sov Math, Dokl 11*485-488
◇ D15 D40 ◇ REV MR 42 # 1884 Zbl 215.109 • ID 28011

LITVINTSEVA, Z.K. [1970] *The conjugacy problem for finitely presented groups (Russian)* (S 1427) Dal'nevostoch Mat Sb, Khabarovsk 1*54-71
◇ D40 ◇ REV MR 52 # 10400 • ID 21698

LIU, SHICHAO [1960] *A theorem on general recursive functions* (**J** 0053) Proc Amer Math Soc 11*184-187
◇ D20 ◇ REV MR 24 # A42 Zbl 93.12 JSL 29.104
• ID 08201

LIU, SHICHAO [1960] *An enumeration of the primitive recursive functions without repetition* (**J** 0261) Tohoku Math J 12*400-402
◇ D20 ◇ REV MR 23 # A63 Zbl 106.6 • ID 08199

LIU, SHICHAO [1960] *An example of general recursive well-ordering which is not primitive recursive* (**J** 0261) Tohoku Math J (2)12*233-234
◇ D20 D45 ◇ REV MR 23 # A799 Zbl 95.10 • ID 08200

LIU, SHICHAO [1960] *Proof of a conjecture of Routledge* (**J** 0053) Proc Amer Math Soc 11*967-969
◇ D20 ◇ REV MR 22 # 7941 Zbl 97.3 • ID 08198

LIU, SHICHAO [1962] *A generalized concept of primitive recursion and its application to deriving general recursive functions* (**C** 2368) Chow,Hung-Ching 60th Anniv Vol 93-98
◇ D20 ◇ REV MR 31 # 51 Zbl 149.8 • ID 42976

LIU, SHICHAO [1962] *Four types of general recursive well-orderings* (**J** 0047) Notre Dame J Formal Log 3*75-78
◇ D30 D45 D55 ◇ REV MR 29 # 12 JSL 30.255
• ID 08203

LIU, SHICHAO [1962] *Recursive linear orderings and hyperarithmetical functions* (**J** 0047) Notre Dame J Formal Log 3*129-132
◇ D30 D45 D55 ◇ REV MR 29 # 13 Zbl 156.10
JSL 31.137 • ID 08202

LIU, SHICHAO [1963] *A note an many-one reducibility* (**J** 0036) J Symb Logic 28*35-42
◇ D30 ◇ REV MR 30 # 3020a Zbl 129.4 JSL 31.512
• ID 08204

LIU, SHICHAO [1963] *On many-one degrees* (**J** 0036) J Symb Logic 28*143-153
◇ D30 ◇ REV MR 30 # 3020b Zbl 192.52 JSL 31.512
• ID 08205

LIU, SHICHAO [1967] *Application of a general method for dealing with many-one degrees* (**C** 2383) Chow,Hung-Ching 65th Anniv Vol 97-113
◇ D30 ◇ REV MR 37 # 2606 Zbl 189.10 • ID 42953

LIU, SHICHAO see Vol. I, II, V, VI for further entries

LIU, ZHENHONG [1983] *NP-complete problems and some approximate algorithms I,II (Chinese)* (**J** 4632) Qufu Shiyuan Xuebao 1983/2*1-12,1983/3*5-15
◇ D15 ◇ ID 49581

LIVCHAK, A.B. [1985] *On polynomial computability (Russian)* (**J** 0031) Izv Vyssh Ucheb Zaved, Mat (Kazan) 1985/1*66-67
◇ D15 ◇ ID 46724

LIVCHAK, A.B. see Vol. III, V, VI for further entries

LIVENSON, E. [1930] see KANTOROVICH, L.

LIVENSON, E. [1932] see KANTOROVICH, L.

LIVENSON, E. [1933] see KANTOROVICH, L.

LIVENSON, E. see Vol. V for further entries

LO SARDO, P. [1974] see GUCCIONE, S.

LOBODA, M. [1976] *Generalized canonical systems for generation of regular sets* (**J** 2814) Podstawy Sterowania 6*233-236
◇ D05 ◇ REV MR 56 # 18184 Zbl 331 # 94018 • ID 63491

LOCKHART, J. [1981] *Decision problems in classes of group presentations with uniformly solvable word problem* (**J** 0008) Arch Math (Basel) 37*1-6
◇ D40 ◇ REV MR 83h:20033 Zbl 443 # 20031 • ID 54397

LOCKHART, J. [1982] *Markov-type properties* (**J** 0053) Proc Amer Math Soc 85*305-309
◇ D40 ◇ REV MR 83i:20028 Zbl 494 # 20018 • ID 39348

LOCKHART, R.A. [1983] see BURGESS, J.P.

LOEB, M.H. [1970] *A model theoretic characterization of effective operations* (**J** 0036) J Symb Logic 35*217-222 • ERR/ADD ibid 39*225
◇ C62 D20 ◇ REV MR 43 # 48 MR 50 # 74
Zbl 223 # 02044 Zbl 289 # 02029 • ID 22235

LOEB, M.H. & WAINER, S.S. [1970] *Hierarchies of number-theoretic functions I,II* (**J** 0009) Arch Math Logik Grundlagenforsch 13*39-51,97-113 • ERR/ADD ibid 14*198-199
◇ D20 ◇ REV MR 44 # 1568 MR 47 # 6461
Zbl 222 # 02049 • ID 08232

LOEB, M.H. & WAINER, S.S. [1974] *Hierarchies of number-theoretic functions (Russian)* (**C** 2319) Slozh Vychisl & Algor 33-64
◇ D20 F15 ◇ REV Zbl 288 # 02024 • ID 63480

LOEB, M.H. [1976] *Embedding first order predicate logic in fragments of intuitionistic logic* (**J** 0036) J Symb Logic 41*705-718
◇ B10 D35 F50 ◇ REV MR 56 # 79 Zbl 358 # 02012
• ID 14746

LOEB, M.H. see Vol. I, II, III, VI for further entries

LOECKX, J. [1972] *Computability and decidability: An introduction for students of computer science* (**X** 0811) Springer: Heidelberg & New York vi+76pp
◇ D98 ◇ REV MR 55 # 13849 Zbl 237 # 68009 • ID 23475

LOECKX, J. (ED.) [1974] *Automata, languages and programming. II* (**X** 0811) Springer: Heidelberg & New York VIII+611pp
◇ D97 ◇ REV MR 54 # 9164 Zbl 277 # 00027 • ID 48710

LOECKX, J. [1976] *Algorithmentheorie* (**X** 0811) Springer: Heidelberg & New York xiv+223pp
◇ D03 D10 D98 ◇ REV MR 57 # 4554 Zbl 355 # 68031
• ID 50264

LOEFGREN, L. [1963] *Self-repair as a computability concept in the theory of automata* (**P** 0674) Symp Math Th of Automata;1962 New York 205-222
◇ D10 ◇ REV MR 30 # 2990 Zbl 116.341 • ID 27608

LOEFGREN, L. [1966] *Explicability of sets and transfinite automata* (**P** 0746) Automata Th;1964 Ravello 251-268
◇ D05 E25 E50 E55 ◇ REV MR 34 # 4069
Zbl 276 # 02051 • ID 27537

LOH, CHUNGWAN [1958] see HU, SHIHUA

LOH, CHUNGWAN [1960] see HU, SHIHUA

LOLLI, G. [1983] *Complessita delle teorie* (**P** 3829) Atti Incontri Log Mat (1);1982 Siena 159-182
◇ B25 D15 ◇ REV MR 84k:03006 Zbl 525 # 03030
• ID 38263

LOLLI, G. & LONGO, G. & MARCJA, A. (EDS.) [1984] *Logic colloquium '82. Proceedings of the colloquium held in Florence, 23-28 August, 1982* (**S** 3303) Stud Logic Found Math 112*viii+358pp
◇ B97 C97 D97 F97 ◇ REV MR 85g:03006 Zbl 538#00003 • ID 41493

LOLLI, G. [1985] *Foundational problems from computation theory* (**J** 0154) Synthese 62*275-288
◇ D20 ◇ ID 45608

LOLLI, G. see Vol. I, II, III, V, VI for further entries

LONG, T.J. [1981] *On γ-reducibility versus polynomial time many-one reducibility* (**J** 1426) Theor Comput Sci 14*91-101
◇ D15 ◇ REV MR 82e:68041 Zbl 454#68031 • ID 54253

LONG, T.J. [1982] *A note on sparse-oracles for NP* (**J** 0119) J Comp Syst Sci 24*224-232
◇ D15 ◇ REV MR 83i:03065 Zbl 486#68033 • ID 35524

LONG, T.J. [1982] see EVEN, S.

LONG, T.J. [1982] *Strong nondeterministic polynomial-time reducibilities* (**J** 1426) Theor Comput Sci 21*1-25
◇ D15 ◇ REV MR 84j:68024 Zbl 521#03028 • ID 37072

LONG, T.J. [1984] see BOOK, R.V.

LONG, T.J. [1985] *On restricting the size of oracles compared with restricting access to oracles* (**J** 1428) SIAM J Comp 14*585-597
◇ D15 ◇ ID 47714

LONG, T.J. [1985] see BOOK, R.V.

LONG, T.J. see Vol. V for further entries

LONGO, G. [1974] *I problemi di decisione e la loro complessita* (**J** 3436) Quad, Ist Appl Calcolo, Ser 3 8*87-108
◇ B25 B35 D15 F20 ◇ REV Zbl 427#03010 • ID 53697

LONGO, G. [1976] *On the problem of deciding equality in partial combinatory algebras and in a formal system* (**J** 0063) Studia Logica 35*363-375
◇ B40 D75 ◇ REV MR 56#8330 Zbl 353#02009 • ID 50092

LONGO, G. [1978] see FERRARI, P.L.

LONGO, G. [1979] *Ricorsivita nei tipi superiori: un'introduzione alle caratterizzazioni di Ershov ed Hyland* (**J** 2038) Rend Sem Mat, Torino 37/3*1-29
◇ B40 D65 G30 ◇ REV MR 81i:03073 Zbl 437#03024 • ID 55892

LONGO, G. [1983] see BARENDREGT, H.P.

LONGO, G. [1983] *Recursiveness and continuity: an introduction (Italian)* (**P** 3829) Atti Incontri Log Mat (1);1982 Siena 215-223
◇ D20 D65 ◇ REV MR 84k:03006 • ID 44707

LONGO, G. & MOGGI, E. [1984] *Cartesian closed categories of enumerations for effective type structures* (**P** 3090) Semant of Data Types;1984 Sophia-Antipolis 235-255
◇ D45 G30 ◇ REV Zbl 564#03037 • ID 48226

LONGO, G. & MARTINI, S. [1984] *Computability in higher types and the universal domain Pω* (**P** 3565) Symp of Th Aspects of Comput Sci (1);1984 Paris 186-197
◇ B75 D65 ◇ REV MR 86e:03045 • ID 45405

LONGO, G. [1984] see LOLLI, G.

LONGO, G. & MOGGI, E. [1984] *The hereditary partial effective functionals and recursion theory in higher types* (**J** 0036) J Symb Logic 49*1319-1332
◇ D65 ◇ ID 42472

LONGO, G. & MARTINI, S. [1985] *Computability in higher types and the universal domain Pω (Italian)* (**P** 4646) Atti Incontri Log Mat (2);1983/84 Siena 501-506
◇ D65 ◇ ID 49712

LONGO, G. & MOGGI, E. [1985] *Strutture di tipi ed enumerazioni* (**P** 4646) Atti Incontri Log Mat (2);1983/84 Siena 507-512
◇ D45 ◇ ID 49606

LONGO, G. see Vol. III, V, VI for further entries

LOOS, R. [1982] see BUCHBERGER, B.

LORENTS, A.A. [1967] *Certain problems in the theory of finite probabilistic automata (Russian)* (**J** 0474) Avtom Vychis Tekh, Akad Nauk Latv SSR 1967/5*57-80
◇ D05 F50 ◇ REV MR 45#6543 Zbl 169.317 • ID 08270

LORENTS, A.A. [1968] *Some problems in the constructive theory of finite probabilistic automata (Russian) (German summary)* (**J** 0068) Z Math Logik Grundlagen Math 14*413-447
◇ D05 F65 ◇ REV MR 40#1211 Zbl 169.317 • ID 08275

LORENTS, A.A. [1969] *Questions of the reducibility of finite probabilistic automata (Russian)* (**J** 0474) Avtom Vychis Tekh, Akad Nauk Latv SSR 1969/1*4-13
◇ D05 ◇ REV MR 45#9869 Zbl 254#94057 • ID 08272

LORENTS, A.A. [1969] *Synthesis of stable finite probabilistic automata (Russian)* (**J** 0474) Avtom Vychis Tekh, Akad Nauk Latv SSR 1969/4*90-91
• TRANSL [1969] (**J** 2666) Autom Control Comput Sci 3/4*83-84
◇ D05 ◇ REV MR 46#1524 • ID 08271

LORENTS, A.A. [1969] *The economy of states of finite probabilistic automata (Russian)* (**J** 0474) Avtom Vychis Tekh, Akad Nauk Latv SSR 1969/2*1-9
◇ D05 ◇ REV MR 45#9870 Zbl 254#94058 • ID 08273

LORENTS, A.A. [1972] *Elemente der konstruktiven Theorie stochastischer Automaten (Russian)* (**X** 2230) Zinatne: Riga 235pp
◇ D05 D10 D98 F60 F98 ◇ REV MR 57#11977a Zbl 252#94036 • ID 63510

LORENTS, A.A. [1976] see CHAPENKO, V.P.

LORENTS, A.A. see Vol. VI for further entries

LORENTS, P.P. [1978] *The Lob-Wainer hierarchy and general recursive functions (Russian)* (**C** 2595) Rekursiv Funktsii 37-54
◇ D20 ◇ REV MR 82b:03076 Zbl 504#03019 • ID 75661

LORENTS, P.P. [1981] *Generation of hierarchies of recursive functions and the solution of problems "A" and "B" of Loeb and Wainer using the method of correcting fundamental sequences (Russian) (Estonian and English summaries)* (**S** 0393) Uch Zap Univ, Tartu 556*15-26
◇ D20 ◇ REV MR 83a:03034 Zbl 503#03015 • ID 35065

LORENTS, P.P. see Vol. VI for further entries

LORENZEN, P. & MYHILL, J.R. [1959] *Constructive definition of certain analytic sets of numbers* (**J** 0036) J Symb Logic 24*37-49
◇ D55 D70 ◇ REV MR 22#14 Zbl 91.13 JSL 33.295 • ID 08296

LORENZEN, P. [1962] *Metamathematik* (**X** 0876) Bibl Inst: Mannheim 173pp
• TRANSL [1967] (**X** 0834) Gauthier-Villars: Paris 162pp (French) [1971] (**X** 1781) Tecnos: Madrid (Spanish)
◇ A05 B98 C60 D20 D35 D98 F98 ◇ REV MR 28 # 3932 Zbl 105.246 JSL 31.106 • ID 08303

LORENZEN, P. see Vol. I, II, V, VI for further entries

LOS, J. [1982] see COHEN, L.J.

LOS, J. see Vol. I, II, III, V, VI for further entries

LOSEV, G.F. [1982] *Local algorithms for computing information with nonfixed memory (Russian)* (**J** 0023) Dokl Akad Nauk SSSR 264∗547-550
• TRANSL [1982] (**J** 0062) Sov Math, Dokl 25∗681-684
◇ D10 ◇ REV Zbl 515 # 03021 • ID 37844

LOSEV, G.F. see Vol. I for further entries

LOUI, M.C. [1981] *A space bound for one-tape multidimensional Turing machines* (**J** 1426) Theor Comput Sci 15∗311-320
◇ D10 D15 ◇ REV MR 82i:68052 Zbl 459 # 68018
• ID 82117

LOUI, M.C. [1981] see ADLEMAN, L.M.

LOUI, M.C. [1982] *Simulations among multidimensional Turing machines* (**J** 1426) Theor Comput Sci 21∗145-161
◇ D10 ◇ REV MR 83m:68094 Zbl 486 # 68038 • ID 40420

LOUVEAU, A. [1974] *Une demonstration topologique de theoremes de Silver et Mathias* (**J** 0247) Bull Sci Math, Ser 2 98∗97-102
◇ D55 E05 E15 ◇ REV MR 55 # 7788 Zbl 311 # 54043
• ID 15025

LOUVEAU, A. [1976] *Determination des jeux G^*_ω (English summary)* (**J** 2313) C R Acad Sci, Paris, Ser A-B 282∗A495-A497
◇ D55 E15 E35 E60 ◇ REV MR 53 # 5300 Zbl 321 # 90056 • ID 22934

LOUVEAU, A. [1976] *Une methode topologique pour l'etude de la propriete de Ramsey* (**J** 0029) Israel J Math 23∗97-116
◇ D55 E05 E15 E50 E75 ◇ REV MR 54 # 100 Zbl 333 # 54022 • ID 18266

LOUVEAU, A. [1977] *Boreliens a coupes $K_{\sigma\delta}$ (English summary)* (**J** 2313) C R Acad Sci, Paris, Ser A-B 285∗A309-A312
◇ D55 E15 ◇ REV MR 56 # 131 Zbl 369 # 04006
• ID 27311

LOUVEAU, A. [1977] *La hierarchie borelienne des ensembles Δ^1_1 (English summary)* (**J** 2313) C R Acad Sci, Paris, Ser A-B 285∗A601-A604
◇ D55 E15 ◇ REV MR 56 # 5299 Zbl 363 # 04008
• ID 27317

LOUVEAU, A. [1978] *Notions elementaires de theorie descriptive effective* (**C** 1692) Semin Init Analyse (17) Paris 1977/78 10pp
◇ D55 E15 ◇ REV MR 80f:03054 Zbl 397 # 03031
• ID 52697

LOUVEAU, A. [1978] *Recursivity and compactness* (**P** 1864) Higher Set Th;1977 Oberwolfach 303-337
◇ D55 E15 ◇ REV MR 80j:03071 Zbl 385 # 03038
• ID 52146

LOUVEAU, A. [1978] *Relations d'equivalence co-analytiques* (**C** 1525) Semin Init Analyse (9-10) Paris 1969/71 16/19∗8pp
◇ D55 E15 ◇ REV MR 58 # 5242 Zbl 429 # 04001
• ID 75682

LOUVEAU, A. [1979] *Familles separantes pour les ensembles analytiques (English summary)* (**J** 2313) C R Acad Sci, Paris, Ser A-B 288∗A391-A394
◇ D55 E15 ◇ REV MR 80i:04004 Zbl 432 # 04001
• ID 75680

LOUVEAU, A. [1979] *Une nouvelle technique d'etude des relations d'equivalence coanalytiques* (**C** 2050) Semin Th des Ensembles GMS Paris 1976/78 35-42
◇ D55 E15 ◇ REV MR 84k:03121 Zbl 512 # 03026
• ID 36514

LOUVEAU, A. [1980] *A separation theorem for Σ^1_1 sets* (**J** 0064) Trans Amer Math Soc 260∗363-378
◇ D55 E15 ◇ REV MR 81j:04001 Zbl 455 # 03021 Zbl 463 # 03029 • ID 54569

LOUVEAU, A. [1982] *Borel sets and the analytical hierarchy* (**P** 3708) Herbrand Symp Logic Colloq;1981 Marseille 209-215
◇ D55 E15 E60 ◇ REV MR 85j:03081 Zbl 498 # 03040
• ID 36905

LOUVEAU, A. [1982] *La classification de Wadge des ensembles boreliens* (**X** 1623) Univ Paris VI Inst Poincare: Paris 7∗11pp
◇ D30 D55 E15 E60 ◇ REV Zbl 549 # 04003 • ID 43141

LOUVEAU, A. [1983] *Some results in the Wadge hierarchy of Borel sets* (**C** 3875) Cabal Seminar Los Angeles 1979-81 28-55
◇ D55 E15 ◇ REV Zbl 535 # 03026 • ID 45238

LOUVEAU, A. [1985] *Recursivity and capacity theory* (**P** 4046) Rec Th;1982 Ithaca 285-301
◇ D55 E15 ◇ ID 46388

LOUVEAU, A. [1985] see DELON, F.

LOUVEAU, A. see Vol. V for further entries

LOVASZ, L. [1977] see GACS, P.

LOVASZ, L. [1980] *Efficient algorithms: an approach by formal logic* (**C** 3494) Stud on Math Progr. Math Meth Oper Res, Vol 1 1∗119-126
◇ B35 C13 D15 ◇ REV MR 82e:68042 Zbl 427 # 68050
• ID 53755

LOVASZ, L. see Vol. III, V for further entries

LOVELAND, D.W. [1966] *The Kleene hierarchy classification of recursively random sequences* (**J** 0064) Trans Amer Math Soc 125∗497-510
◇ D20 D55 D80 ◇ REV MR 34 # 7377 Zbl 189.11 JSL 36.537 • ID 08356

LOVELAND, D.W. & REDDY, C.R. [1978] *Presburger arithmetic with bounded quantifier alternation* (**P** 1740) ACM Symp Th of Comput (10);1978 San Diego 320-325
◇ B25 C10 D15 F20 F30 ◇ REV MR 80d:68057
• ID 31164

LOVELAND, D.W. see Vol. I for further entries

LOWENTHAL, F. [1976] *Equivalence of some definitions of recursion in a higher type object* (**J** 0036) J Symb Logic 41*427-435
◇ D65 ◇ REV MR 53 # 12909 Zbl 366 # 02028 • ID 14773

LOWENTHAL, F. [1976] *Measure and categoricity in α-recursion* (**P** 1476) Set Th & Hierarch Th (2) (Mostowski);1975 Bierutowice 185-201
◇ D60 E75 ◇ REV MR 57 # 12194 Zbl 375 # 02038 • ID 23805

LOWTHER, J.L. [1974] see ALTON, D.A.

LUBIW, A. [1981] *Some NP-complete problems similar to graph isomorphism* (**J** 1428) SIAM J Comp 10*11-21
◇ D15 ◇ REV MR 82f:03036 Zbl 454 # 68025 • ID 75698

LUCA DE, A. [1974] see ADRIANOPOLI, F.

LUCA DE, A. [1975] *Complexity and information theory* (**C** 3294) Coding & Complexity 207-269
◇ D15 ◇ REV MR 58 # 9786 Zbl 357 # 94023 • ID 50453

LUCA DE, A. see Vol. V for further entries

LUCCHESI, C.L. [1979] see KOWALTOWSKI, T.

LUCCHESI, C.L. see Vol. VI for further entries

LUCHKIN, V.D. [1966] *On the ranks of configurations of contex-free languages (Russian)* (**J** 0003) Algebra i Logika 5/3*59-70
◇ D05 ◇ REV JSL 35.339 • ID 19292

LUCIAN, M. [1972] *Systems of notations and the constructible hierarchy* (**X** 0858) Harvard Univ Pr: Cambridge
◇ D55 E45 F15 ◇ ID 28788

LUCKHAM, D.C. [1964] see ENDERTON, H.B.

LUCKHAM, D.C. see Vol. I, VI for further entries

LUCKHARDT, H. [1979] *A limit for higher recursion theory* (**J** 0068) Z Math Logik Grundlagen Math 25*475-479
◇ D20 ◇ REV MR 82c:03068 Zbl 421 # 03037 • ID 53436

LUCKHARDT, H. [1984] *Obere Komplexitaetsschranken fuer TAUT- Entscheidungen* (**P** 3621) Frege Konferenz (2);1984 Schwerin 331-337
◇ B05 D15 ◇ REV MR 86d:03036 • ID 45393

LUCKHARDT, H. see Vol. I, II, III, VI for further entries

LUE, YIZHONG [1980] *Discussion on a computing complex problem* (**J** 4567) Comput Rev 6*68-69
◇ D15 ◇ ID 48519

LUE, YIZHONG see Vol. III for further entries

LUGOWSKI, H. [1979] *Ueber Normalformen von Elementen freier Turing-Algebren* (**J** 0598) Wiss Z Paed Hochsch Potsdam 23*7-11
◇ D10 ◇ REV MR 80j:03091 Zbl 428 # 68064 • ID 53825

LUGOWSKI, H. [1981] *Ueber die Identitaeten der Turing-Algebra I (English and Russian summaries)* (**J** 0598) Wiss Z Paed Hochsch Potsdam 25*779-784
◇ C05 D10 D75 ◇ REV MR 84k:03108a Zbl 519 # 68065 • REM Part II 1983 • ID 36125

LUGOWSKI, H. [1983] *Ueber die Identitaeten der Turing-Algebra II (English and Russian summaries)* (**J** 0598) Wiss Z Paed Hochsch Potsdam 27*9-16
◇ C05 D10 ◇ REV MR 84k:03108b Zbl 542 # 68039 • REM Part I 1981 • ID 36126

LUKAS, J.D. & PUTNAM, H. [1974] *Systems of notations and the ramified analytical hierarchy* (**J** 0036) J Symb Logic 39*243-253
◇ D30 D55 E45 F15 F35 ◇ REV MR 51 # 12496 Zbl 295 # 02027 • ID 08387

LUKAVCOVA, M. [1980] *On computable real functions* (**J** 0156) Kybernetika (Prague) 16*240-247
◇ D80 ◇ REV MR 82d:03098 Zbl 451 # 68040 • ID 54056

LUNINA, M.A. [1977] *Luzin's arithmetic example of an analytic set that is not a Borel set (Russian)* (**J** 0087) Mat Zametki (Akad Nauk SSSR) 22*525-534
• TRANSL [1977] (**J** 1044) Math Notes, Acad Sci USSR 22*783-788
◇ D55 E15 ◇ REV MR 58 # 5234 Zbl 358 # 28001 • ID 75722

LUO, LIBO [1983] *The τ-theory for free groups is undecidable* (**J** 0036) J Symb Logic 48*700-703
◇ C35 C60 D35 ◇ REV MR 85j:03073 Zbl 527 # 03021 • ID 44346

LUO, LIBO see Vol. III, V for further entries

LUPANOV, O.B. [1974] *Methods to obtain complexity and calculability estimations for individual functions (Russian)* (**J** 0071) Met Diskr Analiz (Novosibirsk) 25*3-18
◇ D10 D15 D20 ◇ REV MR 51 # 4722 Zbl 299 # 94032 • ID 63532

LUPANOV, O.B. see Vol. I for further entries

LUZIN, N.N. [1914] *Sur un probleme de M.Baire* (**J** 0109) C R Acad Sci, Paris 158*A1258-A1261
◇ D55 E15 ◇ REV FdM 45.632 • ID 27672

LUZIN, N.N. [1917] *Sur la classification de M.Baire* (**J** 0109) C R Acad Sci, Paris 164*91-94
◇ D55 E15 ◇ REV FdM 46.390 • ID 27562

LUZIN, N.N. & SIERPINSKI, W. [1918] *Sur quelques proprietes des ensembles mesurables (A)* (**J** 3977) Bull Int Acad Sci Cracovie, Cl Math Nat 1918*35-48
◇ D55 E15 ◇ REV FdM 46.296 • ID 38036

LUZIN, N.N. & SIERPINSKI, W. [1923] *Sur un ensemble non mesurable* (**J** 0416) J Mathematiques (9)2*53-72
◇ D55 E15 ◇ REV FdM 49.701 • ID 27564

LUZIN, N.N. [1925] *Les proprietes des ensembles projectifs* (**J** 0109) C R Acad Sci, Paris 188*1817-1819
◇ D55 E15 ◇ REV FdM 51.169 • ID 41587

LUZIN, N.N. [1925] *Sur le probleme de M. Emile Borel et la methode des resolvantes* (**J** 0109) C R Acad Sci, Paris 181*279-281
◇ D55 E15 ◇ REV FdM 51.169 • ID 41588

LUZIN, N.N. [1925] *Sur les ensembles projectifs de M. Henri Lebesgue* (**J** 0109) C R Acad Sci, Paris 180*A1572-A1575
◇ D55 E15 ◇ REV FdM 51.169 • ID 27563

LUZIN, N.N. [1925] *Sur les ensembles non mesurables B et l'emploi de la diagonale de Cantor* (**J** 0109) C R Acad Sci, Paris 181*95-96
◇ D55 E75 ◇ REV FdM 51.169 • ID 38627

LUZIN, N.N. [1926] *Memoire sur les ensembles analytiques et projectifs* (**J** 1404) Mat Sb, Akad Nauk SSSR 33*237-290
◇ D55 E15 ◇ REV FdM 52.199 • ID 41575

Luzin, N.N. [1927] *Remarques sur les ensembles projectifs* (**J** 0109) C R Acad Sci, Paris 185*835-837
◇ D55 E15 ◇ rev FdM 53.173 • id 41559

Luzin, N.N. [1927] *Sur les ensembles analytiques* (**J** 0027) Fund Math 10*1-95
◇ D55 E15 ◇ rev FdM 53.171 • id 38628

Luzin, N.N. & Sierpinski, W. [1929] *Sur les classes des constituantes d'un complementaire analytique* (**J** 0109) C R Acad Sci, Paris 189*794-796
◇ D55 E15 ◇ rev FdM 55.54 • id 39296

Luzin, N.N. [1929] *Sur les points d'unicite d'un ensemble mesurable B* (**J** 0109) C R Acad Sci, Paris 189*422-425
◇ D55 E15 ◇ rev FdM 55.53 • id 39294

Luzin, N.N. [1929] *Sur les voies de la theorie des ensembles* (**P** 0741) Int Congr Math (II, 3);1928 Bologna 1*295-299
◇ A05 D55 E15 E50 F55 ◇ rev FdM 55.657 • id 16811

Luzin, N.N. [1929] *Sur un principe general de la theorie des ensembles analytiques* (**J** 0109) C R Acad Sci, Paris 189*390-392
◇ D55 E15 ◇ rev FdM 55.53 • id 39291

Luzin, N.N. [1930] *Analogies entre les ensembles mesurables B et les ensembles analytiques* (**J** 0027) Fund Math 16*48-76
◇ D55 E15 ◇ rev FdM 56.846 • id 08417

Luzin, N.N. [1930] *Lecons sur les ensembles analytiques et leurs applications* (**X** 0834) Gauthier-Villars: Paris xv+328pp
• transl [1953] (**X** 0704) Gos Izdat Tekhn-Teor Lit: Moskva 359pp (Russian) • last ed [1972] (**X** 0848) Chelsea: New York xvi+328pp
◇ D55 E15 E98 ◇ rev MR 16.21 Zbl 51.291 FdM 56.85 • id 23405

Luzin, N.N. [1930] *Sur le probleme de M.J.Hadamard d'uniformisation des ensembles* (**J** 0109) C R Acad Sci, Paris 190*349-351
◇ D55 E15 ◇ rev FdM 56.87 • id 39463

Luzin, N.N. [1930] *Sur le probleme de M.Jacques Hadamard d'uniformisation des ensembles* (**J** 0517) Mathematica (Cluj) 4*54-66
◇ D55 E15 ◇ rev FdM 56.847 • id 39538

Luzin, N.N. [1931] *Sur une famille de complementaires analytiques* (**J** 0027) Fund Math 17*4-7
◇ D55 E15 ◇ rev FdM 57.92 • id 08418

Luzin, N.N. & Novikov, P.S. [1935] *Choix effectif d'un point dans un complementaire analytique arbitraire, donne par un crible* (**J** 0027) Fund Math 25*559-560
◇ D55 E15 ◇ rev Zbl 12.344 FdM 61.223 • id 08419

Luzin, N.N. [1935] *Sur les ensembles analytiques nuls* (**J** 0027) Fund Math 25*109-131
◇ D55 E15 E50 ◇ rev Zbl 12.344 FdM 61.222 • id 28418

Luzin, N.N. see Vol. V for further entries

Lyapin, E.S. [1970] *Intersections of independent subsemigroups of a semigroup (Russian)* (**J** 0031) Izv Vyssh Ucheb Zaved, Mat (Kazan) 1970/4(95)*67-73
◇ D40 ◇ rev MR 43 # 7526 Zbl 206.305 • id 08218

Lyapunov, A.A. [1934] *On the separability of analytic sets (Russian) (French summary)* (**J** 0023) Dokl Akad Nauk SSSR 2*276-280
◇ D55 E15 ◇ rev Zbl 9.105 FdM 60.40 • id 43676

Lyapunov, A.A. [1935] *Sur la separabilite multiple des ensembles mesurables B* (**J** 0459) C R Soc Sci Lett Varsovie Cl 3 28*118-119
◇ D55 E15 ◇ rev Zbl 15.8 FdM 62.1175 • id 43678

Lyapunov, A.A. [1936] *Contribution a l'etude de la separabilite multiple* (**J** 0142) Mat Sb, Akad Nauk SSSR, NS 1(43)*503-510
◇ D55 E15 ◇ rev Zbl 15.297 FdM 62.234 • id 08116

Lyapunov, A.A. [1937] *On subclasses of B-sets (Russian)* (**J** 0216) Izv Akad Nauk SSSR, Ser Mat 419-426
◇ D55 E15 ◇ rev FdM 63.1564 • id 43679

Lyapunov, A.A. [1938] *Sur l'uniformisation des complementaires analytiques* (**J** 0142) Mat Sb, Akad Nauk SSSR, NS 3(45)*219-223 • err/add ibid 5(47)*445)
◇ D55 E15 ◇ rev Zbl 18.347 FdM 64.1009 • id 08117

Lyapunov, A.A. [1939] *Separabilite multiple pour le cas de l'operation (A) (Russian) (French summary)* (**J** 0216) Izv Akad Nauk SSSR, Ser Mat 1939*539-552
◇ D55 E15 ◇ rev MR 1.302 Zbl 24.303 FdM 65.1168 • id 24922

Lyapunov, A.A. [1939] *Sur l'uniformisation de quelques ensembles CA et A_2' (Russian) (French summary)* (**J** 0216) Izv Akad Nauk SSSR, Ser Mat 1939*41-52
◇ D55 E15 E20 E50 ◇ rev Zbl 24.302 FdM 65.1168 • id 43794

Lyapunov, A.A. [1947] *On R-sets (Russian)* (**J** 0023) Dokl Akad Nauk SSSR 58*1887-1890
◇ D55 D70 E15 ◇ rev MR 9.339 • id 08120

Lyapunov, A.A. [1947] *Sur les ensembles projectifs, qui admettent des decompositions regulieres (Russian) (French summary)* (**J** 0142) Mat Sb, Akad Nauk SSSR, NS 20(62)*179-196
◇ D55 E15 ◇ rev MR 9.83 • id 08119

Lyapunov, A.A. [1947] *Theory of R-sets (Russian)* (**J** 0067) Usp Mat Nauk 2/3(19)*191
◇ D55 D70 E15 ◇ id 43688

Lyapunov, A.A. [1948] *A new definition of certain classes of sets (Russian)* (**J** 0023) Dokl Akad Nauk SSSR 59*847-848
◇ D55 E15 ◇ rev MR 9.417 Zbl 31.290 • id 08121

Lyapunov, A.A. & Novikov, P.S. [1948] *Descriptive set theory (Russian)* (**C** 4134) Mat SSSR za 30 Let 243-255
◇ D55 E15 ◇ rev Zbl 41.20 • id 43710

Lyapunov, A.A. [1949] *On continuous transformations of A-sets (Russian)* (**J** 0216) Izv Akad Nauk SSSR, Ser Mat 13*61-64
◇ D55 E15 ◇ rev MR 10.518 Zbl 40.165 • id 08122

Lyapunov, A.A. [1950] *B-functions (Russian)* (**J** 0067) Usp Mat Nauk 5/5(39)*109-119
• transl [1955] (**X** 0806) Dt Verlag Wiss: Berlin iii+108pp
◇ D55 E15 ◇ rev MR 12.597 Zbl 38.195 • rem Transl. in: "Arbeiten zur deskriptiven Mengenlehre" which contains also papers of Arsenin,V.Ya. & Shchegol'kov,E.A. • id 21021

Lyapunov, A.A. [1950] *Einleitung zu "Arbeiten zur deskriptiven Mengenlehre" (Russisch)* (**J** 0067) Usp Mat Nauk 5/5(39)*11-13
◇ D55 E15 ◇ rev MR 17.467 Zbl 68.270 • id 28543

Lyapunov, A.A. [1950] *On the equivalence of families of sets (Russian)* (**J** 0067) Usp Mat Nauk 5/6*143-144
◇ D55 E15 ◇ rev MR 12.597 Zbl 39.48 • id 21022

LYAPUNOV, A.A. [1950] see ARSENIN, V.YA.

LYAPUNOV, A.A. [1953] *On criteria of degeneracy of R-sets (Russian)* (J 0216) Izv Akad Nauk SSSR, Ser Mat 17*563-578
◇ D55 E15 ◇ REV MR 15.690 Zbl 51.291 • ID 08126

LYAPUNOV, A.A. [1953] *On the classification of R-sets (Russian)* (J 0142) Mat Sb, Akad Nauk SSSR, NS 32(74)*255-262
◇ D55 D70 E15 ◇ REV MR 16.226 Zbl 50.55 Zbl 53.485 • ID 08127

LYAPUNOV, A.A. [1953] *R-sets (Russian)* (S 0066) Tr Mat Inst Steklov 40*67pp
◇ D55 D70 E15 ◇ REV MR 16.226 Zbl 53.364 • ID 08128

LYAPUNOV, A.A. [1953] *Separability and nonseparability of R-sets (Russian)* (J 0142) Mat Sb, Akad Nauk SSSR, NS 32(74)*515-532
◇ D55 E15 ◇ REV MR 14.1069 Zbl 51.291 • ID 08125

LYAPUNOV, A.A. [1955] see ARSENIN, V.YA.

LYAPUNOV, A.A. [1957] *On operations of sets admitting transfinite indexes (Russian)* (J 0065) Tr Moskva Mat Obshch 6*195-230
◇ D55 D70 E15 ◇ REV MR 19.521 Zbl 81.46 • ID 08129

LYAPUNOV, A.A. [1963] *Operations on sets (Russian)* (J 0003) Algebra i Logika 2/2*47-56
◇ D55 D70 E15 ◇ REV MR 27 # 3556 Zbl 163.248 • ID 08130

LYAPUNOV, A.A. [1973] *The method of transfinite indices in the theory of operations over sets (Russian)* (S 0066) Tr Mat Inst Steklov 133*132-148,275
• TRANSL [1973] (S 0055) Proc Steklov Inst Math 133*133-149
◇ D55 D70 E15 ◇ REV MR 51 # 2921 Zbl 294 # 04003 • ID 17359

LYAPUNOV, A.A. [1973] *The works of P.S.Novikov in the area of descriptive set theory (Russian)* (S 0066) Tr Mat Inst Steklov 133*11-22,274
• TRANSL [1973] (S 0055) Proc Steklov Inst Math 133*9-19
◇ A10 D55 E15 ◇ REV MR 51 # 13168 Zbl 294 # 04002 • ID 17231

LYAPUNOV, A.A. see Vol. V for further entries

LYNCH, J.F. [1982] *Complexity classes and theories of finite models* (J 0041) Math Syst Theory 15*127-144
◇ C13 D15 ◇ REV MR 84e:03042 Zbl 484 # 03020 • ID 34379

LYNCH, J.F. see Vol. III, V for further entries

LYNCH, N.A. [1974] *Approximations to the halting problem* (J 0119) J Comp Syst Sci 9*143-150
◇ D10 D20 D25 D45 ◇ REV MR 53 # 2657 Zbl 299 # 02042 • ID 21538

LYNCH, N.A. [1974] see LADNER, R.E.

LYNCH, N.A. [1975] *"Helping": several formalizations* (J 0036) J Symb Logic 40*555-566
◇ D15 D20 ◇ REV MR 52 # 13341 Zbl 342 # 02025 • ID 14811

LYNCH, N.A. [1975] see LADNER, R.E.

LYNCH, N.A. [1975] *On reducibility to complex or sparse sets* (J 0037) ACM J 22*341-345
◇ D15 ◇ REV MR 52 # 56 Zbl 311 # 68037 • ID 17221

LYNCH, N.A. [1976] *Complexity-class-encoding sets* (J 0119) J Comp Syst Sci 13*100-118
◇ D10 D15 D25 ◇ REV MR 53 # 14980 Zbl 355 # 68039 • ID 50267

LYNCH, N.A. [1976] see FISCHER, MICHAEL J.

LYNCH, N.A. [1976] see LADNER, R.E.

LYNCH, N.A. [1976] see GINSBURG, S.

LYNCH, N.A. [1977] see GINSBURG, S.

LYNCH, N.A. [1978] *Log space machines with multiple oracle tapes* (J 1426) Theor Comput Sci 6*25-39
◇ D10 D15 ◇ REV MR 57 # 8153 Zbl 368 # 68058 • ID 51294

LYNCH, N.A. [1978] *Straight-line program length as a parameter for complexity measures* (P 1740) ACM Symp Th of Comput (10);1978 San Diego 150-161
◇ D15 D45 ◇ REV MR 82e:68009 • ID 82127

LYNCH, N.A. [1979] see BLUM, E.K.

LYNCH, N.A. [1980] *Straight-line program length as a parameter for complexity analysis* (J 0119) J Comp Syst Sci 21*251-280
◇ D15 D45 ◇ REV MR 83e:68040 Zbl 458 # 68008 • ID 69682

LYNCH, N.A. [1982] *Accessibility of values as a determinant of relative complexity in algebras* (J 0119) J Comp Syst Sci 24*101-113
◇ D15 ◇ REV MR 83k:68033 Zbl 491 # 68044 • ID 37763

LYNDON, R.C. [1973] see BOONE, W.W.

LYNDON, R.C. see Vol. I, III, V for further entries

LYNES, C. [1982] see CASE, J.

LYNN, D.S. [1972] *New results for Rado's sigma function for binary Turing machines* (J 0187) IEEE Trans Comp C-21*894-896
◇ D10 D20 ◇ REV Zbl 248 # 02038 • ID 63540

LYUBETSKIJ, V.A. [1970] *The existence of a nonmeasurable set of type A_2 implies the existence of an uncountable set of type CA which does not contain a perfect subset (Russian)* (J 0023) Dokl Akad Nauk SSSR 195*548-550
• TRANSL [1970] (J 0062) Sov Math, Dokl 11*1513-1515
◇ D55 E15 E35 ◇ REV MR 43 # 4662 Zbl 258 # 02063 • ID 08210

LYUBETSKIJ, V.A. [1971] *Independence of certain propositions of descriptive set theory from Zermelo-Fraenkel set theory (Russian) (English summary)* (J 0288) Vest Ser Mat Mekh, Univ Moskva 26/2*78-82
• TRANSL [1971] (J 0510) Moscow Univ Math Bull 26/2*116-119
◇ D55 E15 E35 E45 E55 ◇ REV MR 44 # 83 Zbl 218 # 02060 • ID 08211

LYUBETSKIJ, V.A. [1976] *Random sequences of numbers and A_2-sets (Russian)* (C 3271) Issl Teor Mnozh & Neklass Logik 96-122
◇ D55 E15 E35 ◇ REV MR 58 # 27485 Zbl 405 # 03026 • ID 54901

LYUBETSKIJ, V.A. see Vol. I, III, V for further entries

LYUBICH, YU.I. [1964] *Estimates of the number of states that arise in the determinization of a nondeterministic autonomous automaton (Russian)* (**J** 0023) Dokl Akad Nauk SSSR 155*41-43
• TRANSL [1964] (**J** 0062) Sov Math, Dokl 5*345-348
◇ D05 ◇ REV MR 28 # 3938 Zbl 131.10 • ID 08219

MAASS, W. [1976] *Eine Funktionalinterpretation der praedikativen Analysis* (**J** 0009) Arch Math Logik Grundlagenforsch 18*27-46
◇ D20 F10 F35 F65 ◇ REV MR 58 # 21503 Zbl 381 # 03043 • ID 23711

MAASS, W. [1977] *On minimal pairs and minimal degrees in higher recursion theory* (**J** 0009) Arch Math Logik Grundlagenforsch 18*169-186
◇ D30 D60 D65 ◇ REV MR 58 # 5136 Zbl 371 # 02018 • ID 24331

MAASS, W. [1978] *Contributions to α- and β-recursion theory* (**X** 2797) Minerva Publ: Muenchen iii + 107pp
◇ D25 D30 D60 E45 ◇ REV MR 81k:03044 Zbl 368 # 02042 • REM Habilitationsschrift • ID 31878

MAASS, W. [1978] *Fine structure of the constructible universe in α- and β-recursion theory* (**P** 1864) Higher Set Th;1977 Oberwolfach 669*339-359
◇ D55 D60 E45 E65 ◇ REV MR 80g:03044 Zbl 397 # 03030 • ID 31877

MAASS, W. [1978] *High α-recursively enumerable degrees* (**P** 1628) Generalized Recursion Th (2);1977 Oslo 239-269
◇ D30 D60 ◇ REV MR 80b:03059 Zbl 453 # 03047 • ID 31879

MAASS, W. [1978] *Inadmissibility, tame r.e. sets and the admissible collapse* (**J** 0007) Ann Math Logic 13*149-170
◇ D60 ◇ REV MR 80a:03060 Zbl 385 # 03034 JSL 46.665 • ID 27963

MAASS, W. [1978] *The uniform regular set theorem in α-recursion theory* (**J** 0036) J Symb Logic 43*270-279
◇ D30 D60 E45 ◇ REV MR 81b:03051 Zbl 403 # 03032 • ID 29260

MAASS, W. [1979] *On α- and β-recursively enumerable degrees* (**J** 0007) Ann Math Logic 16*205-231
◇ D30 D60 ◇ REV MR 81m:03056 Zbl 441 # 03017 JSL 46.666 • ID 56070

MAASS, W. [1981] *A countable basis for Σ^1_2 sets and recursion theory on $\aleph_1$* (**J** 0053) Proc Amer Math Soc 82*267-270
◇ D55 D60 ◇ REV MR 83k:03060 Zbl 465 # 03026 • ID 54929

MAASS, W. [1981] *Recursively invariant β-recursion theory* (**J** 0007) Ann Math Logic 21*27-73
◇ D30 D60 ◇ REV MR 83k:03059 Zbl 482 # 03021 • ID 35405

MAASS, W. & SHORE, R.A. & STOB, M. [1981] *Splitting properties and jump classes* (**J** 0029) Israel J Math 39*210-224
◇ D25 ◇ REV MR 84i:03083 Zbl 469 # 03026 • ID 55153

MAASS, W. [1982] *Recursively enumerable generic sets* (**J** 0036) J Symb Logic 47*809-823
◇ D25 E40 ◇ REV MR 84e:03051 Zbl 498 # 03026 • ID 33604

MAASS, W. [1983] *Characterization of recursively enumerable sets with supersets effectively isomorphic to all recursively enumerable sets* (**J** 0064) Trans Amer Math Soc 279*311-336
◇ D25 ◇ REV MR 85e:03099 Zbl 546 # 03024 • ID 33605

MAASS, W. [1983] see HOMER, S.

MAASS, W. & STOB, M. [1983] *The intervals of the lattice of recursively enumerable sets determined by major subsets* (**J** 0073) Ann Pure Appl Logic 24*189-212
◇ D25 ◇ REV MR 85j:03066 Zbl 538 # 03037 • ID 33607

MAASS, W. [1984] *On the orbits of hyperhypersimple sets* (**J** 0036) J Symb Logic 49*51-62
◇ D25 ◇ REV MR 85k:03025 • ID 42475

MAASS, W. [1985] *Combinatorial lower bound arguments for deterministic and nondeterministic Turing machines* (**J** 0064) Trans Amer Math Soc 292*675-693
◇ D10 D15 ◇ ID 49265

MAASS, W. [1985] *Major subsets and automorphisms of recursively enumerable sets* (**P** 4046) Rec Th;1982 Ithaca 21-32
◇ D25 ◇ ID 46373

MAASS, W. [1985] see DIETZFELBINGER, M.

MAASS, W. [1985] *Variations on promptly simple sets* (**J** 0036) J Symb Logic 50*138-148
◇ D25 ◇ ID 42531

MAASS, W. see Vol. VI for further entries

MAC GIBBON, B. [1970] *Exemple d'espace $\mathscr{K}$-analytique qui n'est $\mathscr{K}$-Souslinien dans aucun espace* (**J** 0247) Bull Sci Math, Ser 2 94*3-4
◇ D55 E15 ◇ REV MR 42 # 1964 Zbl 193.227 • ID 19284

MACHOVER, M. [1961] *The theory of transfinite recursion* (**J** 0015) Bull Amer Math Soc 67*575-578
◇ C75 D60 E45 ◇ REV MR 26 # 17 Zbl 103.246 JSL 35.335 • ID 08457

MACHOVER, M. [1966] *Contextual determinacy in Lesniewski's grammar (Polish and Russian sumaries)* (**J** 0063) Studia Logica 19*47-57
◇ D05 ◇ REV MR 34 # 5640 Zbl 299 # 02059 • ID 63544

MACHOVER, M. see Vol. I, II, III, V for further entries

MACHTEY, M. [1970] *Admissible ordinals and intrinsic consistency* (**J** 0036) J Symb Logic 35*389-399
◇ D60 ◇ REV MR 45 # 1759 Zbl 218 # 02034 • ID 08460

MACHTEY, M. [1970] *Admissible ordinals and lattices of α-r.e. sets* (**J** 0007) Ann Math Logic 2*379-417
◇ D60 ◇ REV MR 43 # 6090 Zbl 252 # 02042 • ID 08461

MACHTEY, M. [1971] *Classification of computable functions by primitive recursive classes* (**P** 0680) ACM Symp Th of Comput (3);1971 Shaker Heights 251-257
◇ D15 D20 ◇ REV Zbl 263 # 02019 • ID 29865

MACHTEY, M. [1972] *Augmented loop languages and classes of computable functions* (**J** 0119) J Comp Syst Sci 6*603-624
◇ D05 D20 ◇ REV MR 53 # 10565 Zbl 312 # 68028 • ID 14526

MACHTEY, M. [1973] *A notion of helping and pseudo-complementation in lattices of honest subrecursive classes* (**P** 3062) IEEE Symp Switch & Automata Th (14);1973 Iowa City 181-189
◇ D15 D20 ◇ REV MR 55 # 5413 • ID 75750

MACHTEY, M. [1974] *Minimal degrees in generalized recursion theory* (**J** 0068) Z Math Logik Grundlagen Math 20*133-148
◇ D30 D60 ◇ REV MR 50 # 4267 Zbl 335 # 02025
• ID 08462

MACHTEY, M. [1974] *The honest subrecursive classes are a lattice* (**J** 0194) Inform & Control 24*247-263
◇ D15 D20 ◇ REV MR 55 # 10256 Zbl 291 # 02026
• ID 63548

MACHTEY, M. [1975] *Helping and the meet of pairs of honest subrecursive classes* (**J** 0194) Inform & Control 28*76-89
◇ D20 ◇ REV MR 51 # 12025 Zbl 301 # 68057 • ID 17506

MACHTEY, M. [1975] *On the density of honest sub-recursive classes* (**J** 0119) J Comp Syst Sci 10*183-199
◇ D20 ◇ REV MR 52 # 13349 Zbl 336 # 02031 • ID 21803

MACHTEY, M. [1976] *Minimal pairs of polynomial degrees with subexponential complexity* (**J** 1426) Theor Comput Sci 2*73-76
◇ D15 ◇ REV MR 53 # 7747 Zbl 332 # 68040 • ID 22994

MACHTEY, M. & YOUNG, P. [1976] *Simple Goedel numberings, translations, and the P-hierarchy: Preliminary report* (**P** 2597) ACM Symp Th of Comput (8);1976 Hershey 236-243
◇ D15 D20 D45 ◇ REV MR 55 # 5414 Zbl 383 # 03027
• ID 52009

MACHTEY, M. & YOUNG, P. [1978] *An introduction to the general theory of algorithms* (**X** 0809) North Holland: Amsterdam vii+264pp
◇ D15 D20 D98 ◇ REV MR 81k:68001 Zbl 376 # 68027
JSL 46.877 • ID 51709

MACHTEY, M. & WINKLMANN, K. & YOUNG, P. [1978] *Simple Goedel numberings, isomorphisms, and programming properties* (**J** 1428) SIAM J Comp 7*39-60
◇ D05 D10 D15 D20 D45 F40 ◇ REV MR 58 # 178
Zbl 412 # 03020 • ID 52951

MACHTEY, M. [1981] see CHEW, P.

MACHTEY, M. & YOUNG, P. [1981] *Remarks on recursion versus diagonalization and exponentially difficult problems* (**J** 0119) J Comp Syst Sci 22*442-453
◇ D15 D20 ◇ REV MR 83g:68063 Zbl 468 # 68043
• ID 55123

MACINTYRE, A. [1971] *On the elementary theory of Banach algebras* (**J** 0007) Ann Math Logic 3*239-269
◇ C60 C65 D35 ◇ REV MR 48 # 87 Zbl 286 # 02049
• ID 08464

MACINTYRE, A. [1972] *Omitting quantifier-free types in generic structures* (**J** 0036) J Symb Logic 37*512-520
◇ C25 C57 C60 C75 D30 D40 ◇ REV MR 49 # 37
Zbl 273 # 02038 • ID 08467

MACINTYRE, A. [1972] *On algebraically closed groups* (**J** 0120) Ann of Math, Ser 2 96*53-97
◇ C25 C57 C60 D40 ◇ REV MR 47 # 6477
Zbl 254 # 20021 • ID 08468

MACINTYRE, A. [1973] *The word problem for division rings* (**J** 0036) J Symb Logic 38*428-436
◇ D40 ◇ REV MR 49 # 2323 Zbl 286 # 02046 • ID 08470

MACINTYRE, A. & SIMMONS, H. [1975] *Algebraic properties of number theories* (**J** 0029) Israel J Math 22*7-27
◇ C25 C40 C52 C62 D25 H15 ◇ REV MR 53 # 2671
Zbl 356 # 02043 • ID 08474

MACINTYRE, A. [1975] *Dense embeddings. I. A theorem of Robinson in a general setting* (**C** 0782) Model Th & Algeb (A. Robinson) 200-219
◇ B25 C10 C35 C60 D40 ◇ REV MR 53 # 10574
Zbl 327 # 02049 • ID 23079

MACINTYRE, A. [1979] *Combinatorial problems for skew fields I. Analogue of Britton's lemma, and results of Adjan-Rabin type* (**J** 3240) Proc London Math Soc, Ser 3 39*211-236
◇ D40 ◇ REV MR 81h:03092 Zbl 433 # 16018 • ID 75762

MACINTYRE, A. [1980] see HICKIN, K.K.

MACINTYRE, A. [1981] see CHERLIN, G.L.

MACINTYRE, A. [1981] *The complexity of types in field theory* (**P** 2628) Log Year;1979/80 Storrs 143-156
◇ C45 C50 C57 C60 ◇ REV MR 83g:03044
Zbl 499 # 03015 • ID 90377

MACINTYRE, A. [1982] *Residue fields of models of P* (**P** 3622) Int Congr Log, Meth & Phil of Sci (6,Proc);1979 Hannover 193-206
◇ C57 C60 C62 H15 ◇ REV MR 86b:03042
Zbl 514 # 03021 • ID 37408

MACINTYRE, A. & MARKER, D. [1984] *Degrees of recursively saturated models* (**J** 0064) Trans Amer Math Soc 282*539-554
◇ C50 C57 C62 D30 D45 ◇ REV MR 85e:03106
Zbl 557 # 03046 • ID 40310

MACINTYRE, A. see Vol. I, II, III, V, VI for further entries

MACINTYRE, J.M. [1973] *Minimal α-recursion theoretic degrees* (**J** 0036) J Symb Logic 38*18-28
◇ D30 D60 ◇ REV MR 48 # 80 Zbl 335 # 02026
• ID 08476

MACINTYRE, J.M. [1973] *Noninitial segments of the α-degrees* (**J** 0036) J Symb Logic 38*368-388
◇ D30 D60 ◇ REV MR 49 # 2317 Zbl 335 # 02027
• ID 08477

MACINTYRE, J.M. [1977] *Transfinite extensions of Friedberg's completeness criterion* (**J** 0036) J Symb Logic 42*1-10
◇ D30 D55 D60 ◇ REV MR 58 # 10371 Zbl 382 # 03030
• ID 24345

MACQUEEN, D.B. [1976] see HARRINGTON, L.A.

MADAN, D.B. & ROBINSON, R.W. [1982] *Monotone and 1-1 sets* (**J** 3194) J Austral Math Soc, Ser A 33*62-75
◇ D50 ◇ REV MR 83k:03047 Zbl 513 # 03018 • ID 35400

MADATYAN, K.A. [1980] *On correction of the set of algorithms of pattern recognition by the schemes of functional elements (Russian)* (**J** 0023) Dokl Akad Nauk SSSR 255*286-290
• TRANSL [1980] (**J** 0062) Sov Math, Dokl 22*687-691
◇ D20 ◇ REV MR 83g:68125 Zbl 475 # 68055 • ID 39152

MADDEN, J.J. [1984] see GLASS, A.M.W.

MADDUX, R. [1980] *The equational theory of CA_3 is undecidable* (**J** 0036) J Symb Logic 45*311-316
◇ C05 D35 G15 ◇ REV MR 81e:03060 Zbl 435 # 03010
• ID 55771

MADDUX, R. see Vol. V, VI for further entries

MADISON, E.W. [1968] *Computable algebraic structures and nonstandard arithmetic* (**J** 0064) Trans Amer Math Soc 130*38-54
◇ C57 C60 C62 D45 ◇ REV MR 36 # 2498 Zbl 176.275 • ID 08500

MADISON, E.W. [1968] *Structures elementarily closed relative to a model for arithmetic* (**J** 0036) J Symb Logic 33*101-104
◇ C57 C60 C62 ◇ REV MR 38 # 5599 Zbl 191.296 • ID 08499

MADISON, E.W. [1970] *A note on computable real fields* (**J** 0036) J Symb Logic 35*239-241
◇ C57 C60 D45 F60 ◇ REV MR 42 # 7500 Zbl 291 # 02031 • ID 24826

MADISON, E.W. [1970] see LACHLAN, A.H.

MADISON, E.W. [1971] *Some remarks on computable (non-archimedean) ordered fields* (**J** 3172) J London Math Soc, Ser 2 4*304-308
◇ C57 C60 D45 ◇ REV MR 45 # 6624 Zbl 258 # 02044 • ID 63572

MADISON, E.W. [1972] *Real fields with characterization of the natural numbers* (**J** 0047) Notre Dame J Formal Log 13*211-218
◇ C57 H15 ◇ REV MR 45 # 4963 Zbl 197.278 • ID 08502

MADISON, E.W. [1973] see ALTON, D.A.

MADISON, E.W. & NELSON, GEORGE C. [1975] *Some examples of constructive and non-constructive extension of the countable atomless boolean algebra* (**J** 3172) J London Math Soc, Ser 2 11*325-336
◇ C57 F60 G05 ◇ REV MR 52 # 89 Zbl 375 # 06009 • ID 18269

MADISON, E.W. [1982] *A hierarchy of regular open sets of the Cantor space* (**J** 0001) Acta Math Acad Sci Hung 40*139-145
◇ D80 ◇ REV MR 84d:03057 Zbl 466 # 03016 • ID 54967

MADISON, E.W. [1983] *The existence of countable totally nonconstructive extensions of the countable atomless boolean algebras* (**J** 0036) J Symb Logic 48*167-170
◇ C57 D45 G05 ◇ REV MR 84g:03066 Zbl 523 # 06020 • ID 34173

MADISON, E.W. [1985] *On boolean algebras and their recursive completions* (**J** 0068) Z Math Logik Grundlagen Math 31*481-486
◇ C57 D45 G05 ◇ ID 47796

MADLENER, K. [1975] see AVENHAUS, J.

MADLENER, K. [1977] see AVENHAUS, J.

MADLENER, K. [1978] see AVENHAUS, J.

MADLENER, K. [1979] see AVENHAUS, J.

MADLENER, K. [1980] see AVENHAUS, J.

MADLENER, K. [1981] see AVENHAUS, J.

MADLENER, K. [1984] see AVENHAUS, J.

MAEHARA, S. [1957] *General recursive functions in the number-theoretic formal system* (**J** 0260) Ann Jap Ass Phil Sci 1*119-130
◇ D20 F30 ◇ REV MR 23 # A61 Zbl 109.9 JSL 27.90 • ID 08518

MAEHARA, S. see Vol. I, II, III, VI for further entries

MAFFIOLI, F. [1979] see CAMERINI, P.M.

MAGGIOLO-SCHETTINI, A. [1972] see GERMANO, G.

MAGGIOLO-SCHETTINI, A. [1973] see GERMANO, G.

MAGGIOLO-SCHETTINI, A. [1974] see GERMANO, G.

MAGGIOLO-SCHETTINI, A. [1975] see GERMANO, G.

MAGGIOLO-SCHETTINI, A. [1976] see GERMANO, G.

MAGGIOLO-SCHETTINI, A. [1977] see GERMANO, G.

MAGGIOLO-SCHETTINI, A. [1979] see FACHINI, E.

MAGGIOLO-SCHETTINI, A. [1979] see GERMANO, G.

MAGGIOLO-SCHETTINI, A. [1981] see GERMANO, G.

MAGGIOLO-SCHETTINI, A. [1982] see FACHINI, E.

MAGGIOLO-SCHETTINI, A. see Vol. I for further entries

MAGIDOR, M. & MORAN, G. [1970] *Probabilistic tree automata and context free languages* (**J** 0029) Israel J Math 8*340-348
◇ D05 ◇ REV MR 42 # 9104 Zbl 207.20 • ID 08540

MAGIDOR, M. [1980] *Precipitous ideals and Σ^1_4 sets* (**J** 0029) Israel J Math 35*109-134
◇ D55 E05 E15 E50 E55 ◇ REV MR 81k:03048 Zbl 445 # 03023 JSL 50.239 • ID 56487

MAGIDOR, M. [1983] see GUREVICH, Y.

MAGIDOR, M. see Vol. I, III, V for further entries

MAGNARADZE, D.G. & PKHAKADZE, SH.S. (EDS.) [1975] *Studies in mathematical logic and the theory of algorithms (Russian)* (**X** 1052) Tbilisi Univ: Tbilisi 152pp
◇ B97 D97 ◇ REV MR 53 # 2614 • REM Part I. Part II 1977 • ID 21507

MAGNARADZE, L.G. & PKHAKADZE, SH.S. (EDS.) [1977] *Studies in mathematical logic and the theory of algorithms. No. II (Russian)* (**X** 1052) Tbilisi Univ: Tbilisi 39pp
◇ B97 D97 ◇ REV MR 56 # 2779 • REM Part I 1975. Part III 1978 • ID 70125

MAGNARADZE, L.G. see Vol. I for further entries

MAGNUS, W. [1931] *Untersuchungen ueber einige unendliche diskontinuierliche Gruppen* (**J** 0043) Math Ann 105*52-74
◇ D40 ◇ REV Zbl 2.113 • ID 49841

MAGNUS, W. [1932] *Das Identitaetsproblem fuer Gruppen mit einer definierenden Relation* (**J** 0043) Math Ann 106*295-307
◇ D40 ◇ REV Zbl 4.97 FdM 58.125 • ID 08547

MAGNUS, W. [1982] see CHANDLER, B.

MAHANEY, S.R. [1978] see HARTMANIS, J.

MAHANEY, S.R. [1980] *Sparse complete sets for NP: solution of a conjecture of Berman and Hartmanis* (**P** 3577) IEEE Symp Found of Comput Sci (21);1980 Syracuse 54-60
◇ D15 ◇ REV MR 81k:68033 Zbl 493 # 68043 • ID 82137

MAHANEY, S.R. [1981] see HARTMANIS, J.

MAHANEY, S.R. [1981] *On the number of P-isomorphism classes of NP-complete sets* (**P** 4235) IEEE Symp Found of Comp Sci (22);1981 Nashville 271-278
◇ D15 ◇ REV MR 84a:68004 • ID 45815

MAHANEY, S.R. [1982] *Sparse complete sets for NP: solution of a conjecture of Berman and Hartmanis* (**J** 0119) J Comp Syst Sci 25*130-143
◇ D15 ◇ REV MR 84d:68042 Zbl 493 # 68043 • ID 38439

MAHANEY, S.R. & YOUNG, P. [1985] *Reductions among polynomial isomorphism types* (**J** 1426) Theor Comput Sci 39*207-224
◇ D15 ◇ ID 49063

MAHE, L. [1980] see COSTE-ROY, M.-F.

MAHE, L. see Vol. I, V for further entries

MAHN, F.-K. [1967] *Zu den primitiv-rekursiven Funktionen ueber einem Bereich endlicher Mengen* (**J** 0009) Arch Math Logik Grundlagenforsch 10*30-33
◇ D20 D65 ◇ REV MR 35 # 5314 Zbl 265 # 02027 • ID 08555

MAHN, F.-K. [1969] *Primitiv-rekursive Funktionen auf Termmengen* (**J** 0009) Arch Math Logik Grundlagenforsch 12*54-65
◇ D20 ◇ REV Zbl 265 # 02028 • ID 63588

MAHN, F.-K. [1970] *Turing-Maschinen und berechenbare Funktionen II* (**S** 1415) Sel Math 2*21-54
• TRANSL [1972] (**C** 1534) Mash Turing & Rek Funk 34-73
◇ D10 D20 ◇ REV MR 43 # 6085 Zbl 211.312 • REM Parts I,III 1970 by Ebbinghaus,H.-D • ID 27339

MAHN, F.-K. [1972] see EBBINGHAUS, H.-D.

MAHN, F.-K. [1977] see HEIDLER, K.

MAHR, B. [1976] see FLEISCHMANN, K.

MAHR, B. [1977] see FLEISCHMANN, K.

MAHR, B. [1981] see EHRIG, H.

MAHR, B. & SIEFKES, D. [1981] *Relating uniform and nonuniform models of computation* (**P** 3380) GI Jahrestag (11) & ECI Conf (3);1981 Muenchen 41-48
◇ D15 ◇ REV MR 83j:68043 Zbl 484 # 68035 • ID 36618

MAHR, B. [1984] see BUECHI, J.R.

MAIBAUM, T.S.E. [1978] *Pumping lemmas for term languages* (**J** 0119) J Comp Syst Sci 17*319-330
◇ D05 ◇ REV MR 82e:68081 Zbl 388 # 68071 • ID 82140

MAIDA, A. [1982] *Una revisitazione della logica classica in termini di grammatiche generative (English summary)* (**J** 0971) Rend Accad Sci Napoli Fis Mat, Ser 4 48*209-216
◇ D05 ◇ REV MR 84i:03080 Zbl 512 # 68057 • ID 34559

MAIER, W. [1981] *Graphen total rekursiver Funktionen* (**S** 3126) Ber, Fak Inf, Univ Karlsruhe 111pp
◇ D15 D20 ◇ REV Zbl 475 # 03019 • ID 55473

MAIER, W. & MENZEL, W. & SPERSCHNEIDER, V. [1982] *Embedding properties of total recursive functions* (**J** 0068) Z Math Logik Grundlagen Math 28*565-575
◇ D45 ◇ REV MR 84c:03079 Zbl 501 # 03026 • ID 34016

MAIN, M.G. [1985] see BUCHER, W.

MAITRA, A. [1970] *Coanalytic sets that are not Blackwell spaces* (**J** 0027) Fund Math 67*251-254
◇ D55 E15 ◇ REV MR 42 # 1662 Zbl 207.485 • ID 08558

MAITRA, A. & RYLL-NARDZEWSKI, C. [1970] *On the existence of two analytic non-Borel sets which are not isomorphic* (**J** 0014) Bull Acad Pol Sci, Ser Math Astron Phys 18*177-178
◇ D55 E15 E45 ◇ REV MR 42 # 3743 Zbl 197.489 JSL 49.665 • ID 47389

MAITRA, A. [1971] *On game-theoretic methods in the theory of Souslin sets* (**J** 0027) Fund Math 70*179-185
◇ D55 E15 E60 ◇ REV MR 44 # 6928 Zbl 218 # 54028 • ID 08559

MAITRA, A. [1974] *On the failure of the first principle of separation for coanalytic sets* (**J** 0053) Proc Amer Math Soc 46*299-301
◇ D55 E15 ◇ REV MR 50 # 8478 Zbl 313 # 04001 • ID 63591

MAITRA, A. & RAO, B.V. [1976] *Selection theorems for partitions of Polish spaces* (**J** 0027) Fund Math 93*47-56
◇ D55 E15 ◇ REV MR 58 # 12908 Zbl 343 # 54008 • ID 26502

MAITRA, A. [1982] *An effective selection theorem* (**J** 0036) J Symb Logic 47*388-394
◇ D55 E15 ◇ REV MR 83m:03060 • ID 35460

MAITRA, A. see Vol. V for further entries

MAKANIN, G.S. [1977] *The problem of the solvability of equations in a free semigroup (Russian)* (**J** 0142) Mat Sb, Akad Nauk SSSR, NS 103(145)*147-236,319
• TRANSL [1977] (**J** 0349) Math of USSR, Sbor 32*129-198
◇ B03 B25 D40 ◇ REV MR 57 # 9874 Zbl 371 # 20047 • ID 51395

MAKANIN, G.S. [1977] *The problem of solvability of equations in a free semigroup (Russian)* (**J** 0023) Dokl Akad Nauk SSSR 233/2*287-290
• TRANSL [1977] (**J** 0062) Sov Math, Dokl 18*330-334
◇ B03 B25 D40 ◇ REV MR 58 # 5997 Zbl 379 # 20046 • ID 51863

MAKANIN, G.S. [1979] *Identification of the rank of equations in a free semigroup (Russian)* (**J** 0216) Izv Akad Nauk SSSR, Ser Mat 43*547-602,734
• TRANSL [1980] (**J** 0448) Math of USSR, Izv 14*499-545
◇ D40 ◇ REV MR 80h:20084 Zbl 409 # 20044 • ID 82144

MAKANIN, G.S. [1980] *Equations in a free semigroup (Russian)* (**P** 1959) Int Congr Math (II,13);1978 Helsinki 1*263-268
• TRANSL [1981] (**J** 0225) Amer Math Soc, Transl, Ser 2 117*1-6
◇ B03 B25 D40 ◇ REV MR 82b:20081 Zbl 436 # 20039 • ID 82143

MAKANIN, G.S. [1984] see ADYAN, S.I.

MAKANIN, G.S. [1985] *On the decidability of the theory of a free group (Russian)* (**P** 4647) FCT'85 Fund of Comput Th;1985 Cottbus 279-284
◇ B25 C60 C98 D35 ◇ REV Zbl 574 # 20001 • ID 48835

MAKANIN, G.S. see Vol. I, III for further entries

MAKAREVSKIJ, A.YA. & STOTSKAYA, E.D. [1969] *Representability in deterministic multi-tape automata (Russian) (English summary)* (**J** 0040) Kibernetika, Akad Nauk Ukr SSR 1969/4*29-38
• TRANSL [1969] (**J** 0021) Cybernetics 5*390-399
◇ D05 D10 ◇ REV MR 46 # 1510 Zbl 227 # 94033 • ID 08561

MAKAROV, I.T. [1979] *A theorem on nonprovability of lower time bounds for a class of functions (Russian)* (**J** 3937) Veroyat Met i Kibern (Kazan) 15*48-50
◇ D15 ◇ REV MR 82c:03062 Zbl 422 # 03014 • ID 53475

MAKAROV, S.V. [1963] *On the realization of stochastic matrices by finite automata (Russian)* (**S** 0507) Vychisl Sist (Akad Nauk SSSR Novosibirsk) 9*65-70
◇ D05 ◇ REV MR 30 # 3022 • ID 08563

MAKAROV, S.V. [1964] *Turing machines and finite automata (Russian)* (**J** 0092) Sib Mat Zh 5*102-108
◇ D05 D10 ◇ REV MR 29 # 28 • ID 08564

MAKAROV, V.P. [1982] *On theory of abstract algorithms* (**J** 0338) Nauch-Tekh Inf, Ser 2, Akad Nauk SSSR 1982/9*35-40
• TRANSL [1982] (**J** 2667) Autom Doc Math Linguist 16/5*61-75
◇ D20 ◇ REV Zbl 539 # 68020 • ID 44269

MAKKAI, M. [1973] *Vaught sentences and Lindstroem's regular relations* (**P** 0713) Cambridge Summer School Math Log;1971 Cambridge GB 622-660
◇ C40 C52 C75 D55 D70 ◇ REV MR 50 # 4285 Zbl 268 # 02009 • ID 20987

MAKKAI, M. [1976] see HARNIK, V.

MAKKAI, M. [1977] *Admissible sets and infinitary logic* (**C** 1523) Handb of Math Logic 233-281
◇ C70 C98 D55 D60 D70 D98 ◇ REV MR 58 # 10395 JSL 49.968 • ID 24202

MAKKAI, M. [1977] *An "admissible" generalization of a theorem on countable Σ^1_1 sets of reals with applications* (**J** 0007) Ann Math Logic 11*1-30
◇ C15 C40 C50 C70 D55 E15 ◇ REV MR 58 # 10408 Zbl 376 # 02031 • ID 23657

MAKKAI, M. see Vol. III, V, VI for further entries

MAKSIMOVA, L.L. [1975] see LAVROV, I.A.

MAKSIMOVA, L.L. see Vol. I, II, III, V for further entries

MALENGE, J.P. & RIX, H. [1981] *Indecidable...pas sur! (English summary)* (**J** 2832) RAIRO Inform 15*99-102
◇ D05 ◇ REV MR 83b:68058 Zbl 456 # 68051 • ID 38945

MALINOVSKIJ, V.I. [1968] *An equivalence problem in a certain class of address algorithms (Russian) (English summary)* (**J** 0040) Kibernetika, Akad Nauk Ukr SSR 1968/5*33-39
◇ D20 ◇ REV MR 45 # 8065 • ID 08622

MALITZ, J. [1979] *Introduction to mathematical logic. Set theory, computable functions, model theory* (**X** 0811) Springer: Heidelberg & New York xii + 198pp
◇ B98 C98 D98 ◇ REV MR 81h:03002 Zbl 407 # 03001 JSL 49.672 • ID 56167

MALITZ, J. [1980] see EHRENFEUCHT, A.

MALITZ, J. see Vol. I, III, V for further entries

MAL'TSEV, A.A. [1981] *The structure of an m-jump (Russian)* (**J** 0092) Sib Mat Zh 22/4*129-135,230
• TRANSL [1981] (**J** 0475) Sib Math J 22*583-589
◇ D30 ◇ REV MR 82m:03056 • ID 75895

MAL'TSEV, A.A. [1982] *Upper semilattices of numerations (Russian)* (**J** 0092) Sib Mat Zh 23/4*122-136,220
• TRANSL [1982] (**J** 0475) Sib Math J 23*545-556
◇ D45 ◇ REV MR 84e:03053 Zbl 506 # 03009 • ID 34388

MAL'TSEV, A.A. [1984] *On the structure of the families of immune, hyperimmune and hyperhyperimmune sets (Russian)* (**J** 0142) Mat Sb, Akad Nauk SSSR, NS 124(166)*307-319
• TRANSL [1985] (**J** 0349) Math of USSR, Sbor 52*301-314
◇ D50 ◇ ID 45477

MAL'TSEV, A.A. [1985] *Structure of the semilattice of tt 1-degrees (Russian)* (**J** 0092) Sib Mat Zh 26/2*132-139,223
• TRANSL [1985] (**J** 0475) Sib Math J 26*264-270
◇ D30 ◇ ID 45609

MAL'TSEV, A.I. [1960] *On free soluble groups (Russian)* (**J** 0023) Dokl Akad Nauk SSSR 130*495-498
• TRANSL [1960] (**J** 0062) Sov Math, Dokl 1*65-68 [1971] (**C** 2621) Mal'tsev: Metamath of Algeb Syst 119-123
◇ C60 D35 ◇ REV MR 22 # 8056 Zbl 97.248 JSL 30.99
• ID 19257

MAL'TSEV, A.I. [1960] *On the undecidability of the elementary theory of certain fields (Russian)* (**J** 0092) Sib Mat Zh 1*71-77
• TRANSL [1965] (**J** 0225) Amer Math Soc, Transl, Ser 2 48*36-43 [1971] (**C** 2621) Mal'tsev: Metamath of Algeb Syst 138-146
◇ D35 ◇ REV MR 23 # A3094 Zbl 118.253 JSL 30.395
• ID 19258

MAL'TSEV, A.I. [1960] *Some correspondences between rings and groups (Russian)* (**J** 0142) Mat Sb, Akad Nauk SSSR, NS 50(92)*257-266
• TRANSL [1965] (**J** 0225) Amer Math Soc, Transl, Ser 2 45*221-231 [1971] (**C** 2621) Mal'tsev: Metamath of Algeb Syst 124-137
◇ C60 D35 ◇ REV MR 22 # 9448 Zbl 100.14 JSL 30.393
• ID 08609

MAL'TSEV, A.I. [1961] *Constructive algebra I (Russian)* (**J** 0067) Usp Mat Nauk 16/3*3-60
• TRANSL [1961] (**J** 1399) Russ Math Surv 16/3*77-129 [1971] (**C** 2621) Mal'tsev: Metamath of Algeb Syst 148-214
◇ C57 C98 F60 F98 ◇ REV MR 27 # 1362 Zbl 129.259 JSL 31.647 • ID 28706

MAL'TSEV, A.I. [1961] *Effective inseparability of the set of identically true from the set of finitely refutable formulae of certain elementary theories (Russian)* (**J** 0023) Dokl Akad Nauk SSSR 139*802-805
• TRANSL [1961] (**J** 0062) Sov Math, Dokl 2*1005-1008 [1971] (**C** 2621) Mal'tsev: Metamath of Algeb Syst 248-254
◇ D35 ◇ REV MR 25 # 17 Zbl 117.12 JSL 30.394
• ID 19249

MAL'TSEV, A.I. [1961] *Remark on the paper "on the undecidability of the elementary theories of certain fields" (Russian)* (**J** 0092) Sib Mat Zh 2*639
• TRANSL [1965] (**J** 0225) Amer Math Soc, Transl, Ser 2 48*43-44 [1971] (**C** 2621) Mal'tsev: Metamath of Algeb Syst 146
◇ D35 ◇ REV Zbl 161.8 JSL 30.395 • ID 19254

MAL'TSEV, A.I. [1961] *Undecidability of the elementary theory of finite groups (Russian)* (**J** 0023) Dokl Akad Nauk SSSR 138*771-774
• TRANSL [1961] (**J** 0062) Sov Math, Dokl 2*714-717 [1971] (**C** 2621) Mal'tsev: Metamath of Algeb Syst 215-220
◇ C13 D35 ◇ REV MR 27 # 3550 Zbl 119.252 JSL 30.394
• ID 19251

MAL'TSEV, A.I. [1962] *On recursive abelian groups (Russian)* (J 0023) Dokl Akad Nauk SSSR 146*1009-1012
• TRANSL [1962] (J 0062) Sov Math, Dokl 3*1431-1434 [1971] (C 2621) Mal'tsev: Metamath of Algeb Syst 282-286
◇ C57 C60 D45 ◇ REV MR 27 # 1363 Zbl 156.11 JSL 31.649 • ID 41272

MAL'TSEV, A.I. [1962] *Strongly related models and recursively complete algebras (Russian)* (J 0023) Dokl Akad Nauk SSSR 145*276-279
• TRANSL [1962] (J 0062) Sov Math, Dokl 3*987-991 [1971] (C 2621) Mal'tsev: Metamath of Algeb Syst 255-261
◇ C57 C60 C62 D45 ◇ REV MR 26 # 1254 Zbl 132.247 JSL 31.649 • ID 19247

MAL'TSEV, A.I. [1963] *Complete enumeration of a set (Russian)* (J 0003) Algebra i Logika 2/2*4-29
• TRANSL [1971] (C 2621) Mal'tsev: Metamath of Algeb Syst 287-312
◇ D45 ◇ REV MR 34 # 1179 Zbl 163.8 JSL 35.336 • ID 19241

MAL'TSEV, A.I. [1964] *On the theory of computable families of objects (Russian)* (J 0003) Algebra i Logika 3/4*5-31
• TRANSL [1971] (C 2621) Mal'tsev: Metamath of Algeb Syst 353-378
◇ D45 ◇ REV MR 34 # 48 Zbl 199.34 JSL 35.336 • ID 19240

MAL'TSEV, A.I. [1965] *Algorithms and recursive functions (Russian)* (X 2027) Nauka: Moskva 391pp
• TRANSL [1970] (X 0812) Wolters-Noordhoff : Groningen 372pp [1974] (X 0911) Akademie Verlag: Berlin xiv+336pp (German) [1974] (X 0900) Vieweg: Wiesbaden xiv+336pp (German)
◇ D20 D98 ◇ REV MR 34 # 2453 MR 41 # 8233 Zbl 178.324 Zbl 198.25 JSL 34.290 • ID 19237

MAL'TSEV, A.I. [1965] *Positive and negative numerations (Russian)* (J 0023) Dokl Akad Nauk SSSR 160*278-280
• TRANSL [1965] (J 0062) Sov Math, Dokl 6*75-77 [1971] (C 2621) Mal'tsev: Metamath of Algeb Syst 379-383
◇ D20 D45 ◇ REV MR 32 # 1119 JSL 35.336 • ID 19239

MAL'TSEV, A.I. [1966] *Identical relations on varieties of quasigroups (Russian)* (J 0142) Mat Sb, Akad Nauk SSSR, NS 69(111)*3-12
• TRANSL [1969] (J 0225) Amer Math Soc, Transl, Ser 2 82*225-235 [1971] (C 2621) Mal'tsev: Metamath of Algeb Syst 384-395
◇ C05 D40 ◇ REV MR 34 # 61 Zbl 202.312 JSL 40.640 • ID 08612

MAL'TSEV, A.I. see Vol. I, II, III, V for further entries

MAL'TSEV, I.A. [1972] *Congruences and automorphisms in cells of Post algebras (Russian)* (J 0003) Algebra i Logika 11*666-672,737
• TRANSL [1972] (J 0069) Algeb and Log 11*369-373
◇ C05 D03 G20 ◇ REV MR 48 # 5856 Zbl 281 # 02059 • ID 08616

MAL'TSEV, I.A. [1972] *Some properties of cellular subalgebras of a Post algebra and their basic cells (Russian)* (J 0003) Algebra i Logika 11*571-587,615
• TRANSL [1972] (J 0069) Algeb and Log 11*315-325
◇ D03 G20 ◇ REV MR 48 # 5855 Zbl 281 # 02058 • ID 08617

MAL'TSEV, I.A. see Vol. I, II for further entries

MAN, V.D. [1976] *Die Aequivalenzeigenschaft der Konfigurationen der n-dimensionalen iterativen Automaten* (P 2898) Algor Kompl, Lern-& Erkenn-Prozess;1976 Jena 1976*109-120
◇ D05 ◇ REV MR 58 # 9931 Zbl 428 # 68063 • ID 69011

MAN, V.D. [1978] *Berechnung von Wortfunktionen auf deterministischen n-dimensionalen iterativen Automaten* (J 0129) Elektr Informationsverarbeitung & Kybern 14*341-360
◇ D05 D20 ◇ REV MR 80a:68054 Zbl 394 # 68036 • ID 69010

MANAS, M. [1982] *Algorithmically unsolvable problems in economic decision making (Czech) (English summary)* (J 2711) Ekonom Mat Obzor (Prague) 18*137-141
◇ D80 ◇ REV MR 83m:90053 Zbl 488 # 90052 • ID 36750

MANASTER, A.B. [1966] *Higher-order indecomposable isols* (J 0064) Trans Amer Math Soc 125*363-383
◇ D50 ◇ REV MR 37 # 67 Zbl 209.20 JSL 33.295 • ID 08640

MANASTER, A.B. [1968] *Full co-ordinals of RETs* (J 0048) Pac J Math 26*547-553
◇ D50 E10 ◇ REV MR 39 # 57 Zbl 172.9 • ID 08641

MANASTER, A.B. [1969] *Rich co-ordinals, addition isomorphisms, and RETs* (J 0036) J Symb Logic 34*45-52
◇ D50 E10 ◇ REV MR 39 # 6749 Zbl 209.21 JSL 36.342 • ID 08642

MANASTER, A.B. & NERODE, A. [1970] *A universal embedding property of the RET's* (J 0036) J Symb Logic 35*51-59
◇ D50 ◇ REV MR 44 # 2605 Zbl 198.25 • ID 09887

MANASTER, A.B. [1971] *Some contrasts between degrees and the arithmetical hierarchy* (J 0036) J Symb Logic 36*301-304
◇ D30 D55 ◇ REV MR 44 # 3867 Zbl 225 # 02029 • ID 08643

MANASTER, A.B. & ROSENSTEIN, J.G. [1972] *Effective matchmaking (recursion theoretic aspects of a theorem of Philip Hall)* (J 3240) Proc London Math Soc, Ser 3 25*615-654
◇ D30 D80 E05 ◇ REV MR 47 # 3161 Zbl 251 # 05001 • ID 08644

MANASTER, A.B. [1972] see ELLENTUCK, E.

MANASTER, A.B. & ROSENSTEIN, J.G. [1973] *Effective matchmaking and k-chromatic graphs* (J 0053) Proc Amer Math Soc 39*371-378
◇ D80 ◇ REV MR 49 # 4838 Zbl 274 # 02018 • ID 08646

MANASTER, A.B. [1975] *Completeness, compactness, and undecidability: an introduction to mathematical logic* (X 0819) Prentice Hall: Englewood Cliffs vi+154pp
◇ B98 C07 D35 D98 ◇ REV MR 53 # 12857 Zbl 306 # 02001 JSL 42.320 • ID 23120

MANASTER, A.B. [1975] see HAY, L.

MANASTER, A.B. [1977] see HAY, L.

MANASTER, A.B. & REMMEL, J.B. [1980] *Co-simple higher-order indecomposable isols* (J 0068) Z Math Logik Grundlagen Math 26*279-288
◇ D50 ◇ REV MR 82h:03041 Zbl 436 # 03047 • ID 77738

MANASTER, A.B. & ROSENSTEIN, J.G. [1980] *Two-dimensional partial orderings: Recursive model theory* (J 0036) J Symb Logic 45*121-132
◇ C57 C65 E07 ◇ REV MR 81d:03047a Zbl 468 # 03009 • ID 55074

MANASTER, A.B. & ROSENSTEIN, J.G. [1980] *Two-dimensional partial orderings: Undecidability* (J 0036) J Symb Logic 45*133-143
◇ C57 C65 D35 E07 G10 ◇ REV MR 81d:03047b Zbl 468 # 03008 • ID 55073

MANASTER, A.B. & REMMEL, J.B. [1981] *Partial orderings of fixed finite dimension: Model companions and density* (J 0036) J Symb Logic 46*789-802
◇ C25 C65 D35 E07 ◇ REV MR 83b:06002 Zbl 491 # 03012 • ID 33275

MANASTER, A.B. & REMMEL, J.B. [1981] *Some decision problems for subtheories of two-dimensional partial orderings* (P 2628) Log Year;1979/80 Storrs 202-214
◇ B25 C10 C65 D35 ◇ REV MR 82h:03011 Zbl 486 # 03010 • ID 75934

MANASTER, A.B. & REMMEL, J.B. [1981] *Some recursion theoretic aspects of dense two-dimensional partial orderings* (P 2902) Aspects Effective Algeb;1979 Clayton 161-188
◇ D45 ◇ REV MR 84b:03060 Zbl 491 # 03013 • ID 35645

MANCA, V. [1981] *Computational formalism: Abstract combinatory view-point and related first order logical framework* (J 2095) Fund Inform, Ann Soc Math Pol, Ser 4 4*3-18
◇ B40 D75 ◇ REV MR 84i:03088 Zbl 468 # 03026 • ID 55091

MANCA, V. see Vol. III for further entries

MANDELBERG, K.I. [1975] see EVANS, T.

MANDERS, K.L. [1976] see ADLEMAN, L.M.

MANDERS, K.L. [1978] see ADLEMAN, L.M.

MANDERS, K.L. [1979] see ADLEMAN, L.M.

MANDERS, K.L. [1979] *The theory of all substructures of a structure: Characterisation and decision problems* (J 0036) J Symb Logic 44*583-598
◇ B20 B25 C57 C60 ◇ REV MR 81i:03047 Zbl 429 # 03007 • ID 53838

MANDERS, K.L. [1980] *Computational complexity of decision problems* (P 2625) Model Th of Algeb & Arithm;1979 Karpacz 211-227
◇ D15 ◇ REV MR 82m:03053 Zbl 444 # 03019 • ID 75940

MANDERS, K.L. [1982] see DALEY, R.P.

MANDERS, K.L. see Vol. I, III, V for further entries

MANDRIOLI, D. [1978] see CRESPI-REGHIZZI, S.

MANDRIOLI, D. [1981] see CRESPI-REGHIZZI, S.

MANES, E.G. [1974] see ARBIB, M.A.

MANES, E.G. [1975] see ARBIB, M.A.

MANES, E.G. [1979] see ARBIB, M.A.

MANES, E.G. [1983] see ARBIB, M.A.

MANES, E.G. see Vol. II for further entries

MANIN, YU.I. [1973] *Hilbert's tenth problem (Russian)* (J 1452) Itogi Nauki Tekh, Ser Sovrem Probl Mat 1*5-37
• TRANSL [1975] (J 1531) J Sov Math 3*161-184
◇ D25 D35 D80 ◇ REV MR 53 # 7938 Zbl 292 # 02005 • ID 23038

MANIN, YU.I. [1977] *A course in mathematical logic* (X 0811) Springer: Heidelberg & New York xiii + 286pp
◇ B98 C07 D98 E35 E50 F30 G12 ◇ REV MR 56 # 15345 Zbl 383 # 03002 • REM Translated from Russian • ID 51984

MANIN, YU.I. [1980] *The computable and the non-computable (Russian)* (X 2643) Sovet Radio: Moskva 128pp
◇ D20 D25 D98 ◇ REV MR 82i:03002 Zbl 471 # 03003 • ID 55198

MANIN, YU.I. see Vol. I, II, V, VI for further entries

MANN, A. [1982] *A note on recursively presented and co-recursively presented groups* (J 0161) Bull London Math Soc 14*112-118
◇ C60 D40 D45 ◇ REV MR 84d:20033 Zbl 483 # 20020 • ID 39712

MANN, I. [1973] *Probabilistic recursive functions* (J 0064) Trans Amer Math Soc 177*447-467
◇ D20 D75 ◇ REV MR 48 # 1281 Zbl 267 # 02030 • ID 08659

MANNA, Z. [1971] see ASHCROFT, E.A.

MANNA, Z. [1974] *Mathematical theory of computation* (X 0822) McGraw-Hill: New York x + 448pp
◇ B75 D10 D98 ◇ REV MR 53 # 4601 Zbl 353 # 68066 JSL 44.122 • ID 50135

MANNA, Z. & SHAMIR, A. [1976] *The theoretical aspects of the optimal fixedpoint* (J 1428) SIAM J Comp 5*414-426
◇ D20 ◇ REV MR 55 # 13861 Zbl 358 # 68017 • ID 50536

MANNA, Z. & SHAMIR, A. [1978] *The convergence of functions to fixed points of recursive definitions* (J 1426) Theor Comput Sci 6*109-141
◇ D20 ◇ REV MR 57 # 14583 Zbl 401 # 03016 • ID 82172

MANNA, Z. see Vol. I, II for further entries

MANOHAR, R.P. & TREMBLAY, J.P. [1975] *Discrete mathematical structures with applications to computer science* (X 0822) McGraw-Hill: New York xvi + 606pp
◇ D05 D20 G05 G10 ◇ REV MR 52 # 13052 Zbl 313 # 68001 • ID 21747

MANSFIELD, A. [1983] *On the computational complexity of a merge recognition problem* (J 2702) Discr Appl Math 5*119-122
◇ D15 ◇ REV MR 83k:68035 Zbl 496 # 68028 • ID 40329

MANSFIELD, R. [1970] *Perfect subsets of definable sets of real numbers* (J 0048) Pac J Math 35*451-457
◇ D55 E05 E15 E45 E55 ◇ REV MR 43 # 6100 Zbl 251 # 02060 JSL 40.462 • ID 28913

MANSFIELD, R. [1971] *A Souslin operation for Π^1_2* (J 0029) Israel J Math 9*367-379
◇ D55 E15 E55 ◇ REV MR 45 # 6629 Zbl 295 # 02039 • ID 08668

MANSFIELD, R. [1973] *On the possibility of a Σ^1_2-well-ordering of the Baire space* (J 0036) J Symb Logic 38*396-398
◇ D55 E15 E40 E45 ◇ REV MR 49 # 4775 Zbl 276 # 02046 • ID 08673

MANSFIELD, R. [1974] *The non-existence of Σ^1_2-well-orderings of the Cantor set* (J 0027) Fund Math 86*279-282
◇ D55 E15 E40 E45 ◇ REV MR 52 # 7903 Zbl 317 # 02080 • ID 08675

MANSFIELD, R. [1975] *Omitting types: application to descriptive set theory* (**J** 0053) Proc Amer Math Soc 47*198-200
◇ C75 D55 E15 E45 ◇ REV MR 50#6851 Zbl 302#02027 • ID 08674

MANSFIELD, R. [1978] *A footnote to a theorem of Solovay on recursive encodability* (**P** 1897) Logic Colloq;1977 Wroclaw 195-198
◇ D30 D55 ◇ REV MR 80h:03064 Zbl 449#03044 • ID 56711

MANSFIELD, R. & WEITKAMP, G. [1985] *Recursive aspects of descriptive set theory* (**X** 0815) Clarendon Pr: Oxford vii+144pp
◇ D55 E15 E98 ◇ ID 45094

MANSFIELD, R. see Vol. I, III, V for further entries

MARANDZHYAN, G.B. [1969] *Certain properties of asymptotically optimal recursive functions (Russian) (Armenian and English summaries)* (**J** 0312) Izv Akad Nauk Armyan SSR, Ser Mat 4*3-22
◇ D15 D20 ◇ REV MR 42#5790 Zbl 181.12 • ID 08681

MARANDZHYAN, G.B. [1969] *Hierarchies of recursive functions and asymptotic optimality (Russian) (Armenian summary)* (**J** 0346) Dokl Akad Nauk Armyan SSR 48*193-197
◇ D15 D20 ◇ REV MR 41#3274 Zbl 253#02037 • ID 08680

MARANDZHYAN, G.B. [1969] *On complexity scales of natural numbers (Russian)* (**P** 2534) Konf Molod Special Vychisl (1);1969 Erevan 3*79-88
◇ D15 D20 ◇ ID 43265

MARANDZHYAN, G.B. [1971] *Lattices of blocks of recursive functions (Russian) (Armenian summary)* (**J** 0346) Dokl Akad Nauk Armyan SSR 52*7-9
◇ D15 D20 ◇ REV MR 44#1563 Zbl 224#02028 • ID 08682

MARANDZHYAN, G.B. [1972] *The strongly effective immunity of the pivots of additively optimal recursive functions (Russian) (Armenian and English summaries)* (**J** 0312) Izv Akad Nauk Armyan SSR, Ser Mat 7*391-398,478
◇ D15 D50 ◇ REV MR 48#8213 Zbl 271#02027 • ID 08683

MARANDZHYAN, G.B. [1973] *On algorithms of minimal complexity (Russian)* (**J** 0023) Dokl Akad Nauk SSSR 213*787-788
• TRANSL [1973] (**J** 0062) Sov Math, Dokl 14*1797-1799
◇ D15 ◇ REV MR 49#2315 Zbl 295#68046 • ID 08685

MARANDZHYAN, G.B. [1973] *The complexities of the representation of natural numbers by means of recursive functions (Russian)* (**S** 0554) Issl Teor Algor & Mat Logik (Moskva) 1*42-96
◇ D15 D20 ◇ REV MR 49#7131 Zbl 283#02033 • ID 08684

MARANDZHYAN, G.B. [1975] *Algorithmic languages that do not admit a mutual optimal translation (Russian) (Armenian summary)* (**J** 0346) Dokl Akad Nauk Armyan SSR 61*193-197
◇ D20 D45 ◇ REV MR 54#4936 Zbl 325#68005 • ID 24109

MARANDZHYAN, G.B. [1976] *On weakly positive degrees of the sets of minimal indices of algorithms (Russian)* (**P** 2064) All-Union Conf Math Log (4);1976 Kishinev 85
◇ D20 D30 ◇ ID 43269

MARANDZHYAN, G.B. [1977] *On c-degrees of sets of minimal indices of algorithms (Russian) (Armenian and English summaries)* (**J** 0312) Izv Akad Nauk Armyan SSR, Ser Mat 12*130-137
◇ D20 D45 ◇ REV MR 58#5137 Zbl 429#03026 • ID 53857

MARANDZHYAN, G.B. [1979] *On the sets of minimal indices of partial recursive functions* (**P** 2059) Math Founds of Comput Sci (8);1979 Olomouc 372-374
◇ D20 D30 D45 D55 ◇ REV MR 81f:03052 Zbl 437#03023 • ID 55891

MARCHENKOV, S.S. [1969] *Elimination of recursion schemes in Grzegorczyk's class $\mathscr{E}^2$ (Russian)* (**J** 0087) Mat Zametki (Akad Nauk SSSR) 5*561-568
• TRANSL [1969] (**J** 1044) Math Notes, Acad Sci USSR 5*336-340
◇ D20 ◇ REV MR 40#5446 Zbl 179.18 • ID 08687

MARCHENKOV, S.S. [1969] see KOZMIDIADI, V.A.

MARCHENKOV, S.S. [1970] *Multiple recursions that are limited in the class of primitively recursive functions (Russian) (English summary)* (**J** 0040) Kibernetika, Akad Nauk Ukr SSR 1970/6*53-59
◇ D20 ◇ REV MR 45#4974 Zbl 257#02030 • ID 08688

MARCHENKOV, S.S. [1971] *On minimal numerations of systems of recursively enumerable sets (Russian)* (**J** 0023) Dokl Akad Nauk SSSR 198*530-532
• TRANSL [1971] (**J** 0062) Sov Math, Dokl 12*843-846
◇ D25 D45 ◇ REV MR 44#6483 Zbl 255#02044 • ID 08690

MARCHENKOV, S.S. [1971] *On semilattices of computable numerations (Russian)* (**J** 0023) Dokl Akad Nauk SSSR 198*766-768
• TRANSL [1971] (**J** 0062) Sov Math, Dokl 12*886-888
◇ D25 D45 ◇ REV MR 44#3868 Zbl 306#02032 • ID 08689

MARCHENKOV, S.S. [1972] *Bounded recursions (Russian)* (**J** 0389) Math Balkanica 2*124-142
◇ D20 ◇ REV MR 48#5839 Zbl 257#02031 • ID 08693

MARCHENKOV, S.S. [1972] *The computable enumerations of families of recursive functions (Russian)* (**J** 0003) Algebra i Logika 11*588-607,616
• TRANSL [1972] (**J** 0069) Algeb and Log 11*326-336
◇ D20 D45 ◇ REV MR 50#12682 Zbl 282#02015 Zbl 306#02033 • ID 08691

MARCHENKOV, S.S. [1973] *The existence of families without positive numerations (Russian)* (**J** 0087) Mat Zametki (Akad Nauk SSSR) 13*597-604
• TRANSL [1973] (**J** 1044) Math Notes, Acad Sci USSR 13*360-363
◇ D20 D30 D45 ◇ REV MR 48#10783 Zbl 306#02033 • ID 08694

MARCHENKOV, S.S. [1975] *The existence of recursively enumerable minimal truth-table degrees (Russian)* (**J** 0003) Algebra i Logika 14*422-429
• TRANSL [1975] (**J** 0069) Algeb and Log 14*257-261
◇ D25 ◇ REV MR 53#12913 Zbl 378#02020 • ID 23164

MARCHENKOV, S.S. [1976] *On the congruence of the upper semilattices of recursively enumerable m-powers and tabular powers (Russian)* (**J** 0087) Mat Zametki (Akad Nauk SSSR) 20*19-26
• TRANSL [1976] (**J** 1044) Math Notes, Acad Sci USSR 20*567-570
◇ D25 D30 ◇ REV MR 54#9998 Zbl 372#02022
• ID 25839

MARCHENKOV, S.S. [1976] *One class of partial sets (Russian)* (**J** 0087) Mat Zametki (Akad Nauk SSSR) 20*473-478
• TRANSL [1976] (**J** 1044) Math Notes, Acad Sci USSR 20*823-825
◇ D25 ◇ REV MR 58#184 Zbl 396#03035 • ID 52645

MARCHENKOV, S.S. [1976] *Tabular powers of maximal sets (Russian)* (**J** 0087) Mat Zametki (Akad Nauk SSSR) 20*373-381
• TRANSL [1976] (**J** 1044) Math Notes, Acad Sci USSR 20*766-770
◇ D25 ◇ REV MR 55#2533 Zbl 378#02021 • ID 51803

MARCHENKOV, S.S. [1977] *On recursively enumerable minimal btt-degrees (Russian)* (**J** 0142) Mat Sb, Akad Nauk SSSR, NS 103(145)*550-562
• TRANSL [1977] (**J** 0349) Math of USSR, Sbor 32*477-487
◇ D25 ◇ REV MR 56#15392 Zbl 364#02026 • ID 50956

MARCHENKOV, S.S. [1978] *A method of constructing maximal subalgebras in algebras of general recursive functions (Russian)* (**J** 0003) Algebra i Logika 17*581-595,622
• TRANSL [1978] (**J** 0069) Algeb and Log 17*383-392
◇ D20 D75 ◇ REV MR 81a:03042 Zbl 431#03029
• ID 75986

MARCHENKOV, S.S. & MATROSOV, V.L. [1979] *Complexity of algorithms and computations (Russian)* (**J** 3188) Itogi Nauki Tekh, Ser Teor Veroyat Mat Stat Teor Kibern 16*103-149
• TRANSL [1981] (**J** 1531) J Sov Math 15*140-165
◇ D10 D15 D25 ◇ REV Zbl 454#03018 • ID 54230

MARCHENKOV, S.S. [1979] *On the quasi-Peano property of recursive functions (Russian)* (**J** 0052) Probl Kibern 35*199-204,208
◇ D20 ◇ REV MR 80j:03054 Zbl 454#03021 • ID 54233

MARCHENKOV, S.S. [1980] *A basis with respect to superposition in the class of functions that are elementary in the sense of Kalmar (Russian)* (**J** 0087) Mat Zametki (Akad Nauk SSSR) 27*321-332,492
• TRANSL [1980] (**J** 1044) Math Notes, Acad Sci USSR 27*161-166
◇ D20 ◇ REV MR 81e:03039 Zbl 483#03025 • ID 75984

MARCHENKOV, S.S. [1980] *Existence of superposition bases in countable primitive-recursively closed classes of one-place functions (Russian)* (**J** 0087) Mat Zametki (Akad Nauk SSSR) 27*877-883,989
• TRANSL [1980] (**J** 1044) Math Notes, Acad Sci USSR 27*422-425
◇ D20 ◇ REV MR 81j:03067 Zbl 439#03019 • ID 54278

MARCHENKOV, S.S. [1982] *On the complexity of exponent calculation (Russian)* (**J** 0087) Mat Zametki (Akad Nauk SSSR) 31*457-463,475
• TRANSL [1982] (**J** 1044) Math Notes, Acad Sci USSR 31*234-237
◇ D15 ◇ REV MR 83j:68052 Zbl 508#03018 • ID 37680

MARCHENKOV, S.S. [1982] *Undecidability of the positive ∀∃-theory of a free semigroup (Russian)* (**J** 0092) Sib Mat Zh 23/1*196-198,223
◇ B20 D10 D35 ◇ REV MR 83e:03067 • ID 35236

MARCHENKOV, S.S. see Vol. I, II for further entries

MARCHETTI-SPACCAMELA, A. [1980] see AUSIELLO, G.

MARCJA, A. [1984] see LOLLI, G.

MARCJA, A. see Vol. I, II, III for further entries

MARCUS, S. [1963] *Typologie des langues et modeles logiques* (**J** 0001) Acta Math Acad Sci Hung 14*269-281
◇ D05 ◇ REV MR 28#17 • ID 08712

MARCUS, S. [1964] *Grammar and finite automata (Romanian)* (**X** 0871) Acad Rep Soc Romania: Bucharest 255pp
◇ B65 D05 ◇ REV MR 30#4615 Zbl 129.118 • ID 16804

MARCUS, S. [1964] *Langues completement adequates et langues regulieres* (**J** 0068) Z Math Logik Grundlagen Math 10*7-13
◇ D05 ◇ REV MR 28#2036 Zbl 124.18 • ID 08710

MARCUS, S. [1964] *Sur un modele de H.B. Curry pour le langage mathematique* (**J** 0109) C R Acad Sci, Paris 258*1954-1956
◇ D05 ◇ REV MR 28#3937 Zbl 132.399 • ID 08713

MARCUS, S. [1965] *Sur la notion de projectivite* (**J** 0068) Z Math Logik Grundlagen Math 11*181-192
◇ D05 ◇ REV MR 30#4616 Zbl 131.364 • ID 08716

MARCUS, S. [1969] *Contextual grammars* (**J** 0060) Rev Roumaine Math Pures Appl 14*1525-1534
◇ D05 ◇ REV MR 41#6636 Zbl 193.324 • ID 14660

MARCUS, S. [1979] see CALUDE, C.

MARCUS, S. [1980] see CALUDE, C.

MARCUS, S. see Vol. V for further entries

MAREK, W. & ONYSZKIEWICZ, J. [1972] *Elements of logic and foundations of mathematics in problems (Polish)* (**X** 1034) PWN: Warsaw 278pp
• TRANSL [1982] (**X** 0835) Reidel: Dordrecht viii+276pp
◇ B98 C98 D98 E98 ◇ REV MR 84h:03001 Zbl 288#02001 Zbl 574#03001 • ID 34216

MAREK, W. [1974] see APT, K.R.

MAREK, W. & SREBRNY, M. & ZARACH, A. (EDS.) [1976] *Set theory and hierarchy theory. A memorial tribute to Andrzej Mostowski* (**X** 0811) Springer: Heidelberg & New York xiii+345pp
◇ B97 D55 D97 E97 ◇ REV MR 54#2412 Zbl 324#00007 • ID 24424

MAREK, W. [1977] *Bibliography of Andrzej Mostowski's works* (**J** 0063) Studia Logica 36*3-8
◇ A10 D99 E99 ◇ REV MR 56#65b Zbl 357#01046
• ID 50347

MAREK, W. see Vol. I, III, V, VI for further entries

MARGENSTERN, M. [1981] *Le theoreme de Matiyassevitch et resultats connexes* (**P** 3404) Model Th & Arithm;1979/80 Paris 198-241
◇ D25 D35 ◇ REV MR 83g:03041 Zbl 481#10016
• ID 36008

MARGENSTERN, M. see Vol. VI for further entries

MARGOLIS, S.W. & PIN, J.-E. [1984] *Languages and inverse semigroups* (**P** 4012) Automata, Lang & Progr (11);1984 Antwerpen 337-346
◇ D05 ◇ REV Zbl 566 # 68061 • ID 47167

MARINESCU, D. [1978] *A universal Turing machine with three tapes for normal algorithms* (**J** 3071) Bul Univ Brasov, Ser C 20*87-100
◇ D03 D10 ◇ REV MR 81a:03040 Zbl 442 # 68040 • ID 76044

MARINKOVIC, I. [1977] *The unsolvability of the emptiness problem of the two deterministic finite transducer mappings* (**J** 0042) Mat Vesn, Drust Mat Fiz Astron Serb 1(14)(29)*49-50
◇ D05 ◇ REV MR 57 # 11969 Zbl 374 # 68054 • ID 51573

MARKER, D. [1982] *Degrees of models of true arithmetic* (**P** 3708) Herbrand Symp Logic Colloq;1981 Marseille 233-242
◇ C57 C62 D30 H15 ◇ REV MR 85i:03109 Zbl 522 # 03058 • ID 37802

MARKER, D. [1984] see MACINTYRE, A.

MARKER, D. see Vol. III for further entries

MARKOV, A.A. [1947] *On some unsolvable problems concerning matrices (Russian)* (**J** 0023) Dokl Akad Nauk SSSR 57*539-542
◇ D80 ◇ REV MR 9.221 JSL 13.53 • ID 08767

MARKOV, A.A. [1947] *On the impossibility of certain algorithms in the theory of associative systems I (Russian)* (**J** 0023) Dokl Akad Nauk SSSR 55*583-586
◇ D03 D40 ◇ REV MR 8.558 Zbl 29.101 JSL 13.52 • REM Part II 1947 • ID 19385

MARKOV, A.A. [1947] *On the representation of recursive functions (Russian)* (**J** 0023) Dokl Akad Nauk SSSR 58*1891-1892
◇ D20 ◇ REV MR 9.403 Zbl 34.7 JSL 14.67 • ID 19388

MARKOV, A.A. [1947] *The impossibility of certain algorithms in the theory of associative systems II (Russian)* (**J** 0023) Dokl Akad Nauk SSSR 58*353-356
◇ D03 D40 ◇ REV MR 9.321 Zbl 30.194 JSL 13.53 • REM Part I 1947 • ID 19386

MARKOV, A.A. [1949] *On the representation of recursive functions* (**J** 0216) Izv Akad Nauk SSSR, Ser Mat 13*417-424
• TRANSL [1951] (**X** 0803) Amer Math Soc: Providence 13pp
◇ D20 ◇ REV MR 11.151 Zbl 37.297 JSL 15.66 JSL 17.72 • ID 19383

MARKOV, A.A. [1951] *On an unsolvable problem concerning matrices (Russian)* (**J** 0023) Dokl Akad Nauk SSSR 78*1089-1092
◇ D80 ◇ REV MR 13.97 Zbl 42.246 JSL 13.53 JSL 17.152 • ID 19387

MARKOV, A.A. [1951] *Selected works (Russian)* (**X** 0899) Akad Nauk SSSR : Moskva 720pp
◇ D96 ◇ REV MR 14.344 Zbl 54.3 • ID 45548

MARKOV, A.A. [1951] *The impossibility of certain algorithms in the theory of associative systems (Russian)* (**J** 0023) Dokl Akad Nauk SSSR 77*19-20
◇ D03 D35 D40 ◇ REV MR 12.661 Zbl 43.11 JSL 16.215 • ID 19379

MARKOV, A.A. [1951] *The impossibility of algorithms for the recognition of certain properties of associative systems (Russian)* (**J** 0023) Dokl Akad Nauk SSSR 77*953-956
◇ D03 D40 ◇ REV MR 13.4 Zbl 43.11 JSL 17.151 • ID 19380

MARKOV, A.A. [1951] *The theory of algorithms (Russian)* (**S** 0066) Tr Mat Inst Steklov 38*176-189
• TRANSL [1960] (**J** 0225) Amer Math Soc, Transl, Ser 2 15(2)*1-14
◇ D03 D20 ◇ REV MR 13.811 Zbl 49.151 Zbl 49.293 JSL 18.340 • ID 19381

MARKOV, A.A. [1952] *On unsolvable algorithmic problems (Russian)* (**J** 0142) Mat Sb, Akad Nauk SSSR, NS 31*34-42
◇ D03 D20 D25 D35 D40 ◇ REV MR 14.233 Zbl 48.3 JSL 18.341 • ID 19378

MARKOV, A.A. [1952] *Theory of algorithms (Russian) (Hungarian summary)* (**P** 0662) Hungar Math Congr (1);1950 Budapest 191-203
◇ D03 D20 ◇ REV MR 16.436 Zbl 49.151 Zbl 49.293 JSL 20.73 • ID 45549

MARKOV, A.A. [1954] *Theory of algorithms (Russian)* (**S** 0066) Tr Mat Inst Steklov 42*375pp
• TRANSL [1961] (**X** 2737) Israel Progr Sci Transl: Jerusalem iv+444pp [1960] (**X** 1876) Kexue Chubanshe: Beijing 452pp • LAST ED [1984] (**X** 2027) Nauka: Moskva 432pp
◇ D03 D20 D40 D98 ◇ REV MR 17.1038 Zbl 58.5 JSL 22.77 • ID 24898

MARKOV, A.A. [1958] *The insolubility of the problem of homeomorphy (Russian)* (**J** 0023) Dokl Akad Nauk SSSR 121*218-220
◇ D80 ◇ REV MR 20 # 4260 Zbl 92.7 • ID 49842

MARKOV, A.A. [1958] *Unsolvability of some problems in topology (Russian)* (**J** 0023) Dokl Akad Nauk SSSR 123*978-980
◇ D80 ◇ REV MR 21 # 2224 Zbl 90.12 JSL 37.197 • ID 08769

MARKOV, A.A. [1958] *Zum Problem der Darstellbarkeit von Matrizen (Russisch)* (**J** 0068) Z Math Logik Grundlagen Math 4*157-168
◇ D40 ◇ REV MR 23 # A2324 Zbl 92.7 JSL 31.653 • ID 19371

MARKOV, A.A. [1960] *Insolubility of the problem of homeomorphy (Russian)* (**P** 0660) Int Congr Math (II, 8);1958 Edinburgh 300-306
◇ D80 ◇ REV MR 22 # 5962 Zbl 119.252 JSL 27.99 • ID 19370

MARKOV, A.A. [1962] *Computable invariants (Russian)* (**J** 0023) Dokl Akad Nauk SSSR 146*1017-1020
• TRANSL [1962] (**J** 0062) Sov Math, Dokl 3*1440-1443
◇ D05 D20 ◇ REV MR 31 # 1196 Zbl 139.7 • ID 21159

MARKOV, A.A. [1962] *Sur les invariants calculables* (**P** 1606) Colloq Math (Pascal);1962 Clermont-Ferrand 117-120
◇ D35 D40 D45 ◇ REV MR 44 # 6490 • ID 16839

MARKOV, A.A. [1963] *Indistinguishability by invariants in the theory of associative calculi (Russian)* (**J** 0216) Izv Akad Nauk SSSR, Ser Mat 27*907-936
◇ D05 D20 ◇ REV MR 34 # 49 Zbl 139.6 • ID 08771

MARKOV, A.A. [1963] *On certain algorithms related to systems of words (Russian)* (**J** 0216) Izv Akad Nauk SSSR, Ser Mat 27*101-160
◇ D05 D40 ◇ REV MR 27 # 2426 Zbl 117.14 • ID 08772

MARKOV, A.A. [1963] *On the inversion complexity of a system of boolean functions (Russian)* (J 0023) Dokl Akad Nauk SSSR 150*477-479
• TRANSL [1963] (J 0062) Sov Math, Dokl 4*694-696
◇ B05 D55 ◇ REV MR 27 # 1369 Zbl 171.279 • ID 08773

MARKOV, A.A. [1964] *Normal algorithms which compute Boolean functions (Russian)* (J 0023) Dokl Akad Nauk SSSR 157*262-264
• TRANSL [1964] (J 0062) Sov Math, Dokl 5*922-924
◇ B05 D03 ◇ REV MR 30 # 3841 Zbl 127.10 • ID 08774

MARKOV, A.A. [1965] see DUDICH, A.M.

MARKOV, A.A. [1967] *An example of an independent system of words which cannot be included in a finite complete system (Russian)* (J 0087) Mat Zametki (Akad Nauk SSSR) 1*87-90
• TRANSL [1967] (J 1044) Math Notes, Acad Sci USSR 1*56-58
◇ D05 ◇ REV MR 35 # 1481 Zbl 154.7 • ID 08775

MARKOV, A.A. [1967] *Normal algorithms connected with computation of boolean functions (Russian)* (J 0216) Izv Akad Nauk SSSR, Ser Mat 31*161-208
• TRANSL [1967] (J 0448) Math of USSR, Izv 1*151-194
◇ B05 D03 ◇ REV MR 35 # 1474 Zbl 164.318 • ID 08776

MARKOV, A.A. & PETRI, N.V. (EDS.) [1973] *Studies in the theory of algorithms and mathematical logic. Vol. I (Russian)* (X 2265) Akad Nauk Vychis Tsentr: Moskva 287pp
◇ D97 ◇ REV MR 48 # 5809 • ID 70221

MARKOV, A.A. [1974] *On the language $\mathfrak{A}_1$ (Russian)* (J 0023) Dokl Akad Nauk SSSR 214*279-282
• TRANSL [1974] (J 0062) Sov Math, Dokl 15*125-128
◇ D03 D35 F50 ◇ REV MR 49 # 2288 Zbl 308 # 02033
• ID 63719

MARKOV, A.A. [1976] see KUSHNER, B.A.

MARKOV, A.A. [1979] see KHOMICH, V.I.

MARKOV, A.A. see Vol. I, II, III, V, VI for further entries

MARKOVSKI, S. [1985] see CUPONA, G.

MARKOVSKI, S. see Vol. III for further entries

MARKSHAJTIS, G.N. [1980] *Solvability of the word problem for certain groups (Russian) (Lithuanian and English summaries)* (J 2574) Litov Mat Sb (Vil'nyus) 20/2*87-90,208-209
• TRANSL [1980] (J 3283) Lith Math J 20*129-131
◇ D40 ◇ REV MR 82c:20067 Zbl 445 # 20017 • ID 82182

MARKWALD, W. [1954] *Zur Theorie der konstruktiven Wohlordnungen* (J 0043) Math Ann 127*135-149
◇ C80 D55 F15 ◇ REV MR 15.771 Zbl 56.47 JSL 20.283
• ID 08779

MARKWALD, W. [1955] *Zur Eigenschaft primitiv-rekursiver Funktionen, unendlich viele Werte anzunehmen* (J 0027) Fund Math 42*166-167
◇ D20 ◇ REV MR 17.225 Zbl 66.261 JSL 23.48 • ID 08780

MARKWALD, W. [1956] *Ein Satz ueber die elementararithmetischen Definierbarkeitsklassen* (J 1114) Arch Phil 5*398-406
• REPR [1956] (J 0009) Arch Math Logik Grundlagenforsch 2*78-86
◇ D25 D55 ◇ REV MR 18.1 Zbl 71.12 JSL 22.374
• ID 08781

MARKWALD, W. see Vol. I, II for further entries

MARQUES, I. [1973] see BLUM, M.

MARQUES, I. [1975] *On degrees of unsolvability and complexity properties* (J 0036) J Symb Logic 40*529-540
◇ D15 D20 D30 ◇ REV MR 57 # 16024 Zbl 346 # 02022
• ID 14813

MARQUES, I. [1975] *On speedability of recursively enumerable sets* (J 0068) Z Math Logik Grundlagen Math 21*199-214
◇ D15 D25 ◇ REV MR 51 # 10057 Zbl 309 # 02041
• ID 24808

MARTIC, B. [1981] *Iterative systems and diagram algorithms* (J 0068) Z Math Logik Grundlagen Math 27*561-564
◇ D20 ◇ REV MR 83e:03060 Zbl 472 # 03029 • ID 55294

MARTIC, B. see Vol. II for further entries

MARTIN, D.A. [1963] *A theorem on hyperhypersimple sets* (J 0036) J Symb Logic 28*273-278
◇ D25 ◇ REV MR 31 # 2145 Zbl 207.305 JSL 31.139
• ID 08792

MARTIN, D.A. [1966] *Classes of recursively enumerable sets and degrees of unsolvability* (J 0068) Z Math Logik Grundlagen Math 12*295-310
◇ D25 D30 ◇ REV MR 37 # 68 Zbl 181.305 JSL 32.528
• ID 08793

MARTIN, D.A. [1966] *Completeness, the recursion theorem, and effectively simple sets* (J 0053) Proc Amer Math Soc 17*838-842
◇ D25 ◇ REV MR 36 # 45 Zbl 216.290 • ID 08794

MARTIN, D.A. [1966] *On a question of G.E.Sacks* (J 0036) J Symb Logic 31*66-69
◇ D25 D30 ◇ REV MR 34 # 4127 Zbl 154.6 JSL 32.528
• ID 08795

MARTIN, D.A. [1968] *The axiom of determinateness and reduction principles in the analytical hierarchy* (J 0015) Bull Amer Math Soc 74*687-689
◇ D55 E15 E60 ◇ REV MR 37 # 2607 JSL 38.334
• ID 08796

MARTIN, D.A. & MILLER, W. [1968] *The degrees of hyperimmune sets* (J 0068) Z Math Logik Grundlagen Math 14*159-166
◇ D25 D50 ◇ REV MR 37 # 3922 Zbl 216.291 • ID 08797

MARTIN, D.A. & SOLOVAY, R.M. [1969] *A basis theorem for Σ^1_3 sets of reals* (J 0120) Ann of Math, Ser 2 89*138-159
◇ D55 E15 E45 E55 ◇ REV MR 41 # 53 Zbl 176.276
• ID 08799

MARTIN, D.A. & POUR-EL, M.B. [1970] *Axiomatizable theories with few axiomatizable extensions* (J 0036) J Symb Logic 35*205-209
◇ D25 ◇ REV MR 43 # 6094 Zbl 209.12 • ID 08804

MARTIN, D.A. [1970] *Measurable cardinals and analytic games* (J 0027) Fund Math 66*287-291
◇ D55 E15 E55 E60 ◇ REV MR 41 # 3283 Zbl 216.14
• ID 08801

MARTIN, D.A. [1975] see KECHRIS, A.S.

MARTIN, D.A. [1975] *Borel determinacy* (J 0120) Ann of Math, Ser 2 102*363-371
◇ D55 E15 E60 ◇ REV MR 53 # 7785 Zbl 336 # 02049 JSL 49.1425 • ID 23023

MARTIN, D.A. [1976] *Proof of a conjecture of Friedman* (**J** 0053) Proc Amer Math Soc 55*129
◇ D30 D55 ◇ REV MR 53 # 10571 Zbl 325 # 02029
• ID 23078

MARTIN, D.A. [1977] *Descriptive set theory: projective sets* (**C** 1523) Handb of Math Logic 783-815
◇ D55 E15 E55 E60 ◇ REV MR 58 # 5109 JSL 49.975
• ID 27326

MARTIN, D.A. [1978] see KECHRIS, A.S.

MARTIN, D.A. [1980] see DELLACHERIE, C.

MARTIN, D.A. [1980] see KECHRIS, A.S.

MARTIN, D.A. [1981] Π^1_2 *monotone inductive definitions* (**C** 2922) Cabal Seminar Los Angeles 1977-79 215-233
◇ D55 D70 D75 ◇ REV MR 83g:03047 Zbl 495 # 03029
• ID 36009

MARTIN, D.A. [1981] see KECHRIS, A.S.

MARTIN, D.A. & MOSCHOVAKIS, Y.N. & STEEL, J.R. [1982] *The extent of definable scales* (**J** 0589) Bull Amer Math Soc (NS) 6*435-440
◇ D55 D70 E15 E60 ◇ REV MR 83j:03081 Zbl 509 # 03025 • ID 33519

MARTIN, D.A. [1983] see KECHRIS, A.S.

MARTIN, D.A. [1983] *The largest countable this, that, and the other* (**C** 3875) Cabal Seminar Los Angeles 1979-81 97-106
◇ D55 E15 E60 ◇ REV Zbl 571 # 03019 • ID 45240

MARTIN, D.A. [1985] *A purely inductive proof of Borel determinacy* (**P** 4046) Rec Th;1982 Ithaca 303-308
◇ D55 E15 E60 ◇ ID 46399

MARTIN, D.A. see Vol. III, V for further entries

MARTIN, R.M. [1949] *A note on nominalism and recursive functions* (**J** 0036) J Symb Logic 14*27-31
◇ A05 D20 ◇ REV MR 10.668 Zbl 41.352 JSL 15.153
• ID 08816

MARTIN, R.M. see Vol. I, II, V for further entries

MARTIN-LOEF, P. [1966] *The definition of random sequences* (**J** 0194) Inform & Control 9*602-619
◇ D80 ◇ REV MR 36 # 6228 JSL 40.450 • ID 14557

MARTIN-LOEF, P. [1969] *The literature on von Mises' Kollectivs revisited* (**J** 0105) Theoria (Lund) 35*12-37
◇ A10 D80 ◇ REV MR 39 # 2186 Zbl 198.231 JSL 40.450
• ID 14556

MARTIN-LOEF, P. [1970] *On the notion of randomness* (**P** 0603) Intuitionism & Proof Th;1968 Buffalo 73-78
• TRANSL [1974] (**C** 2319) Slozh Vychisl & Algor 364-369
◇ B30 D55 D80 E15 ◇ REV MR 43 # 1237 Zbl 203.299 JSL 40.450 • ID 20933

MARTIN-LOEF, P. see Vol. I, III, VI for further entries

MARTINI, S. [1984] see LONGO, G.

MARTINI, S. [1985] see LONGO, G.

MARTIROSYAN, A.A. & POGOSYAN, EH.M. [1978] *Eine Untersuchung der Stabilitaet relativer Charakteristiken adaptiver Induktoren (Russian)* (**J** 0346) Dokl Akad Nauk Armyan SSR 67*212-215
◇ D80 ◇ REV Zbl 422 # 68040 • ID 53507

MARTIROSYAN, A.A. see Vol. II for further entries

MART'YANOV, V.I. [1973] see KOKORIN, A.I.

MART'YANOV, V.I. [1977] *Extended universal theories of the integers (Russian)* (**J** 0003) Algebra i Logika 16*588-602,624
• TRANSL [1977] (**J** 0069) Algeb and Log 16*395-405
◇ B25 D35 F30 ◇ REV MR 58 # 27421 Zbl 394 # 03038
• ID 29230

MART'YANOV, V.I. [1978] *Undecidability of the theory of abelian groups with an automorphism (Russian)* (**J** 0087) Mat Zametki (Akad Nauk SSSR) 23*515-520
• TRANSL [1978] (**J** 1044) Math Notes, Acad Sci USSR 23*281-284
◇ D35 ◇ REV MR 58 # 21561 Zbl 383 # 06006 • ID 90298

MART'YANOV, V.I. [1982] *Undecidability of the theory of Boolean algebras with automorphism (Russian)* (**J** 0092) Sib Mat Zh 23/3*147-154
• TRANSL [1982] (**J** 0475) Sib Math J 23*408-415
◇ D35 G05 ◇ REV MR 83m:03052 Zbl 503 # 03018
• ID 36922

MART'YANOV, V.I. see Vol. I, III for further entries

MARUSCIAC, I. & NADIU, G.S. [1969] *A universal Turing machine for normal algorithms (Romanian) (French summary)* (**J** 0197) Stud Cercet Mat Acad Romana 21*85-95
◇ D03 D10 ◇ REV MR 42 # 4404 Zbl 176.278 • ID 08837

MASHURYAN, A.S. [1974] *Recursive definitions on induction models (Russian) (Armenian summary)* (**J** 0346) Dokl Akad Nauk Armyan SSR 59*199-204
◇ C13 D20 ◇ REV MR 51 # 2899 Zbl 309 # 02060
• ID 17343

MASHURYAN, A.S. [1979] *On a class of primitive recursive functions (Russian)* (**J** 0346) Dokl Akad Nauk Armyan SSR 69*209-212
◇ C13 D20 ◇ REV Zbl 431 # 03028 • ID 53934

MASHURYAN, A.S. [1980] *On the class of primitive recursive functions definable on finite models (Russian) (Armenian summary)* (**J** 0346) Dokl Akad Nauk Armyan SSR 71*209-211
◇ C13 D20 ◇ REV MR 82i:03052 Zbl 494 # 03026
• ID 76106

MASHURYAN, A.S. see Vol. II for further entries

MASLOV, A.N. [1972] *Probabilistic Turing machines and recursive functions (Russian)* (**J** 0023) Dokl Akad Nauk SSSR 203*1018-1020
• TRANSL [1972] (**J** 0470) Sov Phys, Dokl 17*318-320
◇ D10 D20 ◇ REV Zbl 268 # 02027 • ID 63761

MASLOV, A.N. [1974] see KOZMIDIADI, V.A.

MASLOV, A.N. [1974] *The hierarchy of indexed languages of an arbitrary level (Russian)* (**J** 0023) Dokl Akad Nauk SSSR 217*1013-1016
• TRANSL [1974] (**J** 0062) Sov Math, Dokl 15*1170-1174
◇ D05 ◇ REV MR 51 # 2363 Zbl 316 # 68042 • ID 63762

MASLOV, S.YU. [1962] *Transformation of arbitrary canonical calculi into canonical calculi of special types (Russian)* (**J** 0023) Dokl Akad Nauk SSSR 147*779-782
• TRANSL [1962] (**J** 0062) Sov Math, Dokl 3*1708-1711
◇ D03 D05 ◇ REV MR 26 # 1245 Zbl 143.256 JSL 32.527
• ID 08839

MASLOV, S.YU. [1963] *Some methods for the definition of sets in generating bases (Russian)* (J 0023) Dokl Akad Nauk SSSR 153*266-269
• TRANSL [1963] (J 0062) Sov Math, Dokl 4*1649-1652
◇ D05 ◇ REV MR 28 # 4990 Zbl 192.63 JSL 32.392
• ID 19341

MASLOV, S.YU. [1963] *Strong representability of sets by calculi (Russian)* (J 0023) Dokl Akad Nauk SSSR 152*272-274
• TRANSL [1963] (J 0062) Sov Math, Dokl 4*1292-1295
◇ D03 D20 ◇ REV MR 27 # 4753 Zbl 217.9 JSL 32.391
• ID 08840

MASLOV, S.YU. [1964] *On the "tag problem" of E.L.Post (Russian)* (S 0066) Tr Mat Inst Steklov 72*57-68
• TRANSL [1971] (J 0225) Amer Math Soc, Transl, Ser 2 97*1-14
◇ D03 ◇ REV MR 34 # 4129 Zbl 178.326 JSL 32.526
• ID 19343

MASLOV, S.YU. [1964] *Some properties of E.L.Post's apparatus of canonical systems (Russian)* (S 0066) Tr Mat Inst Steklov 72*5-56
• TRANSL [1971] (J 0225) Amer Math Soc, Transl, Ser 2 98*91-161
◇ D03 D05 D10 ◇ REV MR 34 # 4128 Zbl 192.64 JSL 32.524 • ID 19339

MASLOV, S.YU. & MINTS, G.E. & OREVKOV, V.P. [1965] *Unsolvability in the constructive predicate calculus of certain classes of formulas containing only monadic predicate variables (Russian)* (J 0023) Dokl Akad Nauk SSSR 163*295-297
• TRANSL [1965] (J 0062) Sov Math, Dokl 6*918-920
◇ B20 D35 F50 ◇ REV MR 33 # 52 Zbl 173.10 JSL 35.143 • ID 19335

MASLOV, S.YU. [1967] *Representation of enumerable sets by means of local calculi (Russian)* (S 0066) Tr Mat Inst Steklov 93*43-49
◇ D05 D25 ◇ REV MR 36 # 6291 Zbl 162.23 JSL 39.188
• ID 08844

MASLOV, S.YU. [1967] *The concept of strict representability in the general theory of calculi (Russian)* (S 3713) Probl Konstr Napravleniya Mat 4*3-42
• TRANSL [1970] (S 1500) Probl Constr Math 4*1-50
◇ D05 D20 ◇ REV MR 36 # 6290 Zbl 193.313 JSL 39.187
• ID 08843

MASLOV, S.YU. [1972] *Deduction search in calculi of general type (Russian) (English summary)* (S 0228) Zap Nauch Sem Leningrad Otd Mat Inst Steklov 32*59-65,156
• TRANSL [1976] (J 1531) J Sov Math 6*395-400
◇ B35 D03 F07 ◇ REV MR 49 # 8845 Zbl 344 # 02026
• ID 08850

MASLOV, S.YU. & RUSAKOV, E.D. [1972] *Probabilistic canonical calculi (Russian) (English summary)* (S 0228) Zap Nauch Sem Leningrad Otd Mat Inst Steklov 32*66-76,155 • ERR/ADD ibid 40*161
• TRANSL [1976] (J 1531) J Sov Math 6*401-409
◇ B48 D03 ◇ REV MR 49 # 8829 Zbl 344 # 02027
• ID 19328

MASLOV, S.YU. [1975] *Mutation calculi (Russian) (English summary)* (S 0228) Zap Nauch Sem Leningrad Otd Mat Inst Steklov 49*7-30,176
◇ D03 D25 D80 ◇ REV MR 52 # 7876 Zbl 317 # 92020
• ID 18281

MASLOV, S.YU. [1976] *Absorption relation on regular sets (Russian) (English summary)* (S 0228) Zap Nauch Sem Leningrad Otd Mat Inst Steklov 60*65-74
• TRANSL [1980] (J 1531) J Sov Math 14*1468-1475
◇ D05 ◇ REV MR 58 # 32011 Zbl 355 # 94061 • ID 50272

MASLOV, S.YU. [1978] *Macroevolution as deduction process* (J 0154) Synthese 39*417-434
◇ D03 ◇ REV MR 80b:92018 Zbl 397 # 92008 • ID 82192

MASLOV, S.YU. [1979] *Calculi with monotone deductions and their economic interpretation (Russian) (English summary)* (S 0228) Zap Nauch Sem Leningrad Otd Mat Inst Steklov 88*90-105,239-240
• TRANSL [1982] (J 1531) J Sov Math 20*2314-2321
◇ D05 D80 ◇ REV MR 81g:03030 Zbl 429 # 03042
• ID 53873

MASLOV, S.YU. see Vol. I, III, VI for further entries

MASLOVA, T.M. [1979] *Bounded m-reducibilities (Russian)* (J 3937) Veroyat Met i Kibern (Kazan) 15*51-60
◇ D25 D30 ◇ REV MR 81i:03066 Zbl 422 # 03019
• ID 53480

MASSERON, M. [1983] *Rungs and trees* (J 0036) J Symb Logic 48*847-863
◇ D55 E05 F15 ◇ REV MR 85e:03145 Zbl 567 # 03026
• ID 40760

MATETI, P. [1981] *A decision procedure for the correctness of a class of programs* (J 0037) ACM J 28*215-232
◇ B75 D80 ◇ REV MR 82g:68017 Zbl 464 # 68018
• ID 69697

MATHIAS, A.R.D. [1972] *Solution of problems of Choquet and Puritz* (P 2080) Conf Math Log;1970 London 204-210
◇ C20 D55 E05 E15 ◇ REV MR 51 # 166 Zbl 232 # 02036 • ID 17477

MATHIAS, A.R.D. [1975] *A remark on rare filters* (P 0759) Infinite & Finite Sets (Erdoes);1973 Keszthely 3*1095-1097
◇ D55 E05 E15 ◇ REV MR 51 # 10098 Zbl 342 # 02050
• ID 17523

MATHIAS, A.R.D. see Vol. I, III, V for further entries

MATIYASEVICH, YU.V. [1967] *Simple examples of undecidable associative calculi (Russian)* (J 0023) Dokl Akad Nauk SSSR 173*1264-1266
• TRANSL [1967] (J 0062) Sov Math, Dokl 8*555-557
◇ D03 ◇ REV MR 36 # 50 Zbl 189.11 JSL 33.469
• ID 19326

MATIYASEVICH, YU.V. [1967] *Simple examples of unsolvable canonical calculi (Russian)* (S 0066) Tr Mat Inst Steklov 93*50-88
• TRANSL [1967] (S 0055) Proc Steklov Inst Math 93*61-110
◇ D03 ◇ REV MR 36 # 50 Zbl 193.318 • ID 33031

MATIYASEVICH, YU.V. [1968] *Arithmetic representations of powers (Russian)* (S 0228) Zap Nauch Sem Leningrad Otd Mat Inst Steklov 8*159-165
• TRANSL [1968] (J 0521) Semin Math, Inst Steklov 8*75-78
◇ D25 D35 ◇ REV MR 39 # 62 Zbl 235 # 10012 JSL 37.605 • ID 08879

MATIYASEVICH, YU.V. [1968] *The connection between Hilbert's tenth problem and systems of equations between words and lengths (Russian)* (**S** 0228) Zap Nauch Sem Leningrad Otd Mat Inst Steklov 8*132-144
• TRANSL [1968] (**J** 0521) Semin Math, Inst Steklov 8*61-67
◇ D25 D35 D40 ◇ REV MR 40 # 41 Zbl 212.333 JSL 37.604 • ID 08878

MATIYASEVICH, YU.V. [1968] *Two reductions of Hilbert's tenth problem (Russian)* (**S** 0228) Zap Nauch Sem Leningrad Otd Mat Inst Steklov 8*145-158
• TRANSL [1968] (**J** 0521) Semin Math, Inst Steklov 8*68-74
◇ D25 D35 ◇ REV MR 40 # 42 Zbl 212.333 JSL 37.604
• ID 33034

MATIYASEVICH, YU.V. [1970] *Enumerable sets are diophantine (Russian)* (**J** 0023) Dokl Akad Nauk SSSR 191*279-282
• TRANSL [1970] (**J** 0062) Sov Math, Dokl 11*354-357
◇ D25 D35 ◇ REV MR 41 # 3390 Zbl 212.334 JSL 37.605
• ID 19324

MATIYASEVICH, YU.V. [1970] *Solution of the tenth problem of Hilbert* (**J** 0396) Mat Lapok 21*83-87
◇ B28 D25 D35 ◇ REV MR 46 # 3447 Zbl 223 # 02041
• ID 82198

MATIYASEVICH, YU.V. [1971] *Diophantine representation of the set of prime numbers (Russian)* (**J** 0023) Dokl Akad Nauk SSSR 196*770-773
• TRANSL [1971] (**J** 0062) Sov Math, Dokl 12*249-254
◇ D25 D35 ◇ REV MR 43 # 1921 Zbl 222 # 10018 JSL 37.607 • ID 26330

MATIYASEVICH, YU.V. [1971] *Diophantine representation of enumerable predicates (Russian)* (**J** 0216) Izv Akad Nauk SSSR, Ser Mat 35*3-30
• TRANSL [1971] (**J** 0448) Math of USSR, Izv 5*1-28
◇ D25 D35 ◇ REV MR 43 # 54 Zbl 219 # 02035 JSL 39.605 • ID 28089

MATIYASEVICH, YU.V. [1971] *Diophantine representation of recursively enumerable predicates* (**P** 0743) Int Congr Math (II,11,Proc);1970 Nice 1*235-238
◇ D25 D35 ◇ REV MR 54 # 2444 Zbl 235 # 02039 JSL 37.606 • ID 27826

MATIYASEVICH, YU.V. [1971] *Diophantine representation of recursively enumerable predicates* (**P** 0604) Scand Logic Symp (2);1970 Oslo 171-177
◇ D25 D35 ◇ REV MR 50 # 6820 Zbl 223 # 02042
• ID 76156

MATIYASEVICH, YU.V. [1971] see BOONE, W.W.

MATIYASEVICH, YU.V. [1971] *Real-time recognition of the inclusion relation (Russian) (English summary)* (**S** 0228) Zap Nauch Sem Leningrad Otd Mat Inst Steklov 20*104-114,285
• TRANSL [1973] (**J** 1531) J Sov Math 1*64-70
◇ D10 D15 ◇ REV MR 45 # 3208 Zbl 222 # 02051
• ID 08881

MATIYASEVICH, YU.V. [1972] *Arithmetic representation of enumerable sets with a small number of quantifiers (Russian) (English summary)* (**S** 0228) Zap Nauch Sem Leningrad Otd Mat Inst Steklov 32*77-84,156
• TRANSL [1976] (**J** 1531) J Sov Math 6*410-416
◇ D25 ◇ REV MR 49 # 8835 Zbl 354 # 02032 • ID 08885

MATIYASEVICH, YU.V. [1972] *Diophantine sets (Russian)* (**J** 0067) Usp Mat Nauk 27/5*185-222
• TRANSL [1972] (**J** 1399) Russ Math Surv 27/5*124-164
◇ D25 D35 ◇ REV MR 56 # 109 Zbl 269 # 02019
• ID 30367

MATIYASEVICH, YU.V. [1972] *Diophantine representation of enumerable predicates (Russian)* (**J** 0087) Mat Zametki (Akad Nauk SSSR) 12*115-120
• TRANSL [1972] (**J** 1044) Math Notes, Acad Sci USSR 12*501-506
◇ D25 D35 ◇ REV MR 48 # 3727 Zbl 219 # 02035
• ID 76150

MATIYASEVICH, YU.V. [1973] *On recursive unsolvability of Hilbert's tenth problem* (**P** 0793) Int Congr Log, Meth & Phil of Sci (4,Proc);1971 Bucharest 89-110
◇ D25 D35 ◇ REV MR 57 # 5711 • ID 76154

MATIYASEVICH, YU.V. [1974] *Existence of noneffectivizable estimates in the theory of exponentially diophantine equations (Russian) (English summary)* (**S** 0228) Zap Nauch Sem Leningrad Otd Mat Inst Steklov 40*77-93,157
• TRANSL [1977] (**J** 1531) J Sov Math 8*299-311
◇ D25 D35 ◇ REV MR 51 # 10225 Zbl 361 # 02057
• ID 17510

MATIYASEVICH, YU.V. & ROBINSON, JULIA [1974] *Two universal three-quantifier representations of enumerable sets (Russian)* (**C** 1450) Teor Algor & Mat Logika (Markov) 112-123,216
◇ D25 D35 ◇ REV MR 53 # 10566 Zbl 327 # 02035
• ID 23074

MATIYASEVICH, YU.V. & ROBINSON, JULIA [1975] *Reduction of an arbitrary diophantine equation to one in 13 unknowns* (**J** 0399) Acta Arith, Pol Akad Nauk 27*521-553
◇ D25 D35 ◇ REV MR 52 # 8033 Zbl 279 # 10019
• ID 18283

MATIYASEVICH, YU.V. [1976] *A new proof of the theorem on exponential diophantine representation of enumerable sets (Russian) (English summary)* (**S** 0228) Zap Nauch Sem Leningrad Otd Mat Inst Steklov 60*75-92,223
• TRANSL [1980] (**J** 1531) J Sov Math 14*1475-1486
◇ D25 ◇ REV MR 58 # 27402 Zbl 346 # 02025 • ID 63791

MATIYASEVICH, YU.V. [1976] see DAVIS, MARTIN D.

MATIYASEVICH, YU.V. [1977] *Primes are non-negative values of a polynomial in 10 variables (Russian) (English summery)* (**S** 0228) Zap Nauch Sem Leningrad Otd Mat Inst Steklov 68*62-82,144-145
• TRANSL [1981] (**J** 1531) J Sov Math 15*33-44
◇ D25 ◇ REV MR 58 # 21534 Zbl 357 # 10034 • ID 50431

MATIYASEVICH, YU.V. [1977] *Some purely mathematical results inspired by mathematical logic* (**P** 1704) Int Congr Log, Meth & Phil of Sci (5);1975 London ON 1*121-127
◇ B30 D25 D35 D80 ◇ REV MR 58 # 5508 Zbl 377 # 02001 • ID 51718

MATIYASEVICH, YU.V. [1979] *Algorithmic unsolvability of exponential Diophantine equations with three indeterminates (Russian)* (**S** 0554) Issl Teor Algor & Mat Logik (Moskva) 3*69-78,133
• TRANSL [1984] (**S** 3489) Sel Math Sov 3*223-232 (English)
◇ D25 D35 ◇ REV MR 81f:03055 Zbl 425 # 10019
• ID 53587

MATIYASEVICH, YU.V. [1981] *What should we do having proved a decision problem to be unsovable?* (P 3729) Algor in Modern Math & Comput Sci;1979 Urgench 441-443
• TRANSL [1982] (P 3803) Algor Sovrem Mat & Prilozh;1979 Urgench II*254-256
◇ D35 ◇ REV MR 84j:03091 Zbl 477 # 68035 • ID 34682

MATIYASEVICH, YU.V. [1982] see JONES, JAMES P.

MATIYASEVICH, YU.V. [1984] see JONES, JAMES P.

MATIYASEVICH, YU.V. [1984] *Studies in certain algorithmic problems of algebra and number theory (Russian)* (S 0066) Tr Mat Inst Steklov 168*218-235
◇ D03 D05 D35 ◇ REV MR 85k:01040 • ID 45770

MATIYASEVICH, YU.V. see Vol. I, VI for further entries

MATOS, A.B. & PORTO, A.G. [1980] *Ackermann and the superpowers* (J 1456) SIGACT News 12/3*90-95
◇ D20 ◇ REV Zbl 455 # 03015 • ID 54276

MATROSOV, V.L. [1975] *An analytic description of the complexity classes of computable functions (Russian)* (C 1050) Aktual Vopr Mat Log & Teor Mnozh 3-12
◇ D15 D20 ◇ REV MR 58 # 27381 • ID 76157

MATROSOV, V.L. [1975] *The closure of certain complexity classes with respect to recursive operations (Russian)* (C 1050) Aktual Vopr Mat Log & Teor Mnozh 13-32
◇ D15 D20 ◇ REV MR 56 # 8341 • ID 76159

MATROSOV, V.L. [1976] *Complexity classes and classes of computable signaling functions (Russian)* (J 0023) Dokl Akad Nauk SSSR 226*513-515
• TRANSL [1976] (J 0062) Sov Math, Dokl 17*137-139
◇ D15 D20 ◇ REV MR 58 # 21527 Zbl 354 # 68072
• ID 76158

MATROSOV, V.L. [1976] *Signalling computable functions for certain improvement of the complexity measure (Russian)* (J 2744) Izv Sev-Kavk Nauch Tsentra, Estestv (Rostov nD) 1*57-61,118
◇ D15 ◇ REV MR 55 # 5415 • ID 76160

MATROSOV, V.L. [1979] see MARCHENKOV, S.S.

MATROSOV, V.L. [1981] *Complexity of computable functions for a generalized storage measure (Russian)* (J 0087) Mat Zametki (Akad Nauk SSSR) 29*895-905
• TRANSL [1981] (J 1044) Math Notes, Acad Sci USSR 29*457-462
◇ D15 D20 ◇ REV MR 84h:68028 Zbl 481 # 68050
• ID 38458

MATSUMOTO, K. [1965] *Word problem for free lattice (Japanese)* (S 2609) Mem Nara Tech College 1*53-59
◇ D40 G10 ◇ ID 90095

MATSUMOTO, K. see Vol. I, II, III, VI for further entries

MATSUNO, H. [1985] see INOUE, K.

MATTHEWS, J. [1966] *The conjugacy problem in wreath products and free metaabelian groups* (J 0064) Trans Amer Math Soc 121*329-339
◇ D40 ◇ REV MR 33 # 1351 Zbl 136.276 • ID 08897

MATVEEVA, G.S. [1965] *On a theorem of Rabin concerning the complexity of computable functions (Russian)* (J 0092) Sib Mat Zh 6*546-555
◇ D15 ◇ REV MR 33 # 2540 Zbl 253 # 02036 JSL 34.133
• ID 19322

MAULDIN, R.D. [1973] *The Baire order of the functions continuous almost everywhere* (J 0053) Proc Amer Math Soc 41*535-540
◇ D55 E15 ◇ REV MR 48 # 2319 Zbl 274 # 26002 • REM Part I. Part II 1975 • ID 08906

MAULDIN, R.D. [1975] *The Baire order of the functions continuous almost everywhere II* (J 0053) Proc Amer Math Soc 51*371-377
◇ D55 E15 ◇ REV MR 51 # 8345 Zbl 306 # 26004 • REM Part I 1973 • ID 17569

MAULDIN, R.D. [1980] see CENZER, D.

MAULDIN, R.D. [1982] see CENZER, D.

MAULDIN, R.D. [1983] see CENZER, D.

MAULDIN, R.D. see Vol. V for further entries

MAURER, H.A. [1976] see GINSBURG, S.

MAURER, H.A. [1976] see HULE, H.

MAURER, H.A. & SALOMAA, A. & WOOD, D. [1977] *EOL forms* (J 1431) Acta Inf 8*75-96
• REPR [1978] (J 0119) J Comp Syst Sci 16*345-361
◇ B75 D03 D05 ◇ REV MR 56 # 4255 MR 58 # 3735 Zbl 348 # 68046 Zbl 376 # 68053 • ID 31590

MAURER, H.A. [1977] see GINSBURG, S.

MAURER, H.A. & OTTMANN, T. & SALOMAA, A. [1977] *On the form equivalence of L-forms* (J 1426) Theor Comput Sci 4*199-225
◇ D05 ◇ REV MR 57 # 4655 Zbl 358 # 68111 • ID 28336

MAURER, H.A. [1978] see HULE, H.

MAURER, H.A. & ROZENBERG, G. [1978] *Increasing the similarity of EOL form interpretations* (J 0194) Inform & Control 38*330-342
◇ D05 ◇ REV MR 80d:68090 Zbl 386 # 68068 • ID 69705

MAURER, H.A. [1978] see CULIK II, K.

MAURER, H.A. & SALOMAA, A. & WOOD, D. [1978] *On good EOL forms* (J 1428) SIAM J Comp 7*158-166
◇ D03 D05 ◇ REV MR 57 # 18258 Zbl 375 # 68034
• ID 31593

MAURER, H.A. & SALOMAA, A. & WOOD, D. [1978] *Relative goodness of EOL forms* (J 3441) RAIRO Inform Theor 12*291-304
◇ B75 D05 ◇ REV MR 80b:68092 Zbl 388 # 68067
• ID 69702

MAURER, H.A. [1978] see ALBERT, J.

MAURER, H.A. & SALOMAA, A. & WOOD, D. [1978] *Uniform interpretations of L-forms* (J 0194) Inform & Control 36*157-173
◇ D03 D05 ◇ REV MR 57 # 8187 Zbl 367 # 68053
• ID 31594

MAURER, H.A. & SALOMAA, A. & WOOD, D. [1979] *Context-dependent L forms* (J 0194) Inform & Control 42*97-118
◇ B75 D05 ◇ REV MR 80e:68188 Zbl 416 # 68068
• ID 69700

MAURER, H.A. [1979] see AINHIRN, W.

MAURER, H.A. & PENTTONEN, M. & SALOMAA, A. & WOOD, D. [1979] *On non context-free grammar forms* (**J** 0041) Math Syst Theory 12*297-324
◇ B75 D05 ◇ REV MR 80h:68062 Zbl 415 # 68036
• ID 69704

MAURER, H.A. [1979] see CULIK II, K.

MAURER, H.A. & ROZENBERG, G. & SALOMAA, A. & WOOD, D. [1979] *Pure interpretations of EOL forms* (**J** 3441) RAIRO Inform Theor 13*347-362
◇ B75 D05 ◇ REV MR 81h:68073 Zbl 434 # 68059
• ID 69706

MAURER, H.A. [1980] see EHRENFEUCHT, A.

MAURER, H.A. & SALOMAA, A. & WOOD, D. [1980] *On generators and generative capacity of EOL forms* (**J** 1431) Acta Inf 13*87-107
◇ D05 ◇ REV MR 81f:68085 Zbl 446 # 68061 • ID 69703

MAURER, H.A. & SALOMAA, A. & WOOD, D. [1980] *Pure grammars* (**J** 0194) Inform & Control 44*47-72
◇ D05 ◇ REV MR 81d:68103 Zbl 446 # 68063 • ID 69699

MAURER, H.A. [1980] see ALBERT, J.

MAURER, H.A. & SALOMAA, A. & WOOD, D. [1981] *Completeness of context-free grammar forms* (**J** 0119) J Comp Syst Sci 23*1-10
◇ B75 D05 ◇ REV MR 83e:68113 Zbl 477 # 68085
• ID 69707

MAURER, H.A. [1981] see BUCHER, W.

MAURER, H.A. [1981] see ALBERT, J.

MAURER, H.A. & ROZENBERG, G. [1981] *Sub context-free L forms* (**J** 0382) Int J Comput Math 9*25-41
◇ B75 D05 ◇ REV MR 81m:68060 Zbl 455 # 68040
• ID 69701

MAURER, H.A. & SALOMAA, A. & WOOD, D. [1982] *Dense hierarchies of grammatical families* (**J** 0037) ACM J 29*118-126
◇ D05 ◇ REV MR 83i:68110 Zbl 491 # 68077 • ID 39364

MAURI, G. [1977] see BERTONI, A.

MAURI, G. [1979] see BERTONI, A.

MAURI, G. [1980] see BERTONI, A.

MAURI, G. [1981] see BERTONI, A.

MAURI, G. [1983] see BERTONI, A.

MAURI, G. [1984] see BERTONI, A.

MAURI, G. see Vol. VI for further entries

MAXIMOFF, I. [1940] *Sur la separabilite d'ensembles* (**J** 0174) Bull Sect Sci Acad Roumaine 22*384-389
◇ D55 E15 ◇ REV MR 1.301 Zbl 23.115 FdM 66.1221
• ID 08912

MAXIMOFF, I. see Vol. V for further entries

MAYER, O. [1973] *Die Darstellung indizierter Sprachen durch Ausdruecke* (**P** 1630) GI Fachtag Automatenth & Form Sprach (1);1973 Bonn 166-174
◇ D05 ◇ REV MR 55 # 9610 Zbl 347 # 68047 • ID 63798

MAYER, O. [1979] see GINSBURG, S.

MAYOH, B.H. [1965] *Unsolvable problems in the theory of computable numbers* (**P** 0688) Logic Colloq;1963 Oxford 272-279
◇ D20 D35 F60 ◇ REV MR 35 # 56 Zbl 192.59
• ID 08924

MAYOH, B.H. [1967] *Groups and semigroups with solvable word problems* (**J** 0053) Proc Amer Math Soc 18*1038-1039
◇ C57 C60 D40 ◇ REV MR 37 # 4151 Zbl 153.349
JSL 36.541 • ID 08925

MAYOH, B.H. [1968] *Semi-effective numberings and definitions of the computable numbers* (**J** 0009) Arch Math Logik Grundlagenforsch 11*113-125
◇ D25 D30 F60 ◇ REV MR 40 # 4106 Zbl 179.19
• ID 08926

MAYOH, B.H. [1970] *The relation between an object and its name: notation systems and their fixed point theorems* (**P** 0785) Scand Logic Symp (1);1968 Aabo 77-95
◇ A05 B03 D20 D35 D45 F30 F60 ◇ REV
MR 52 # 2850 Zbl 318 # 02048 • ID 17641

MAYOH, B.H. see Vol. I, II, VI for further entries

MAYR, E.W. [1980] *Ein Algorithmus fuer das allgemeine Erreichbarkeitsproblem bei Petrinetzen und damit zusammenhaengende Probleme (Dissertation)* (**X** 3160) TU Muenchen Fak Math: Muenchen 59pp
◇ D05 D80 ◇ REV Zbl 476 # 68046 • ID 55591

MAYR, E.W. & MEYER, A.R. [1981] *The complexity of the finite containment problem for Petri nets* (**J** 0037) ACM J 28*561-576
◇ D15 D20 D80 ◇ REV MR 82f:68061 Zbl 462 # 68020
• ID 54534

MAYR, E.W. & MEYER, A.R. [1982] *The complexity of the word problems for commutative semigroups and polynomial ideals* (**J** 0345) Adv Math 46*305-329
◇ D15 D40 ◇ REV MR 84g:20099 Zbl 506 # 03007
• ID 36961

MAYS, W. [1958] *Cybernetic models and thought processes* (**P** 1177) Int Congr Cybern (1);1956 Namur 103-110
◇ D05 ◇ REV JSL 25.163 • ID 22139

MAYS, W. see Vol. I for further entries

MAZURKIEWICZ, A.W. [1972] *Iteratively computable relations* (**J** 0014) Bull Acad Pol Sci, Ser Math Astron Phys 20*793-798
◇ D15 D20 ◇ REV MR 47 # 9875 Zbl 254 # 68004
• ID 63804

MAZURKIEWICZ, A.W. [1972] *Recursive algorithms and formal languages* (**J** 0014) Bull Acad Pol Sci, Ser Math Astron Phys 20*799-803
◇ D05 D20 ◇ REV Zbl 259 # 68033 • ID 63805

MAZURKIEWICZ, A.W. see Vol. I for further entries

MAZURKIEWICZ, S. [1927] *Sur une propriete des ensembles C(A)* (**J** 0027) Fund Math 10*172-174
◇ D55 E15 ◇ REV FdM 53.173 • ID 08935

MAZURKIEWICZ, S. see Vol. V for further entries

MAZZANTI, S. [1984] see GERMANO, G.

MAZZANTI, S. see Vol. I for further entries

McAloon, K. [1978] *Diagonal methods and strong cuts in models of arithmetic* (**P** 1897) Logic Colloq;1977 Wroclaw 171–181
◇ C57 C62 D80 F30 ◇ rev MR 80k:03068 Zbl 458 # 03021 • id 54427

McAloon, K. [1980] *Progressions transfinies de theories axiomatiques, formes combinatoires du theoreme d'incompletude et fonctions recursives a croissance rapide* (**J** 1620) Asterisque 73*41–58
◇ C62 D20 F15 F30 ◇ rev MR 82g:03098b Zbl 462 # 03017 JSL 48.483 • id 54524

McAloon, K. [1981] see Kirby, L.A.S.

McAloon, K. [1982] see Cegielski, P.

McAloon, K. [1982] *On the complexity of models of arithmetic* (**J** 0036) J Symb Logic 47*403–415
◇ C57 C62 ◇ rev MR 84h:03084 Zbl 519 # 03056 • id 34279

McAloon, K. [1983] see Anshel, M.

McAloon, K. [1984] *Petri nets and large finite sets* (**J** 1426) Theor Comput Sci 32*173–183
◇ D05 D20 D80 ◇ rev MR 86g:68135 Zbl 569 # 68046 • id 49247

McAloon, K. see Vol. I, III, V, VI for further entries

McAndrew, M.H. [1963] *A partition problem* (**J** 0214) Math of Comp 17*291–295
◇ D05 ◇ rev MR 29 # 2194 Zbl 114.263 • id 08940

McBeth, R. [1980] *A generalization of Ackermann's function* (**J** 0068) Z Math Logik Grundlagen Math 26*509–516
◇ D20 ◇ rev MR 82c:03063 Zbl 455 # 03016 • id 54277

McBeth, R. [1980] *Exponential polynomials of linear height* (**J** 0068) Z Math Logik Grundlagen Math 26*399–404
◇ D20 E07 ◇ rev MR 82g:03074 Zbl 466 # 04001 • id 54978

McBeth, R. [1981] *A note on exponential polynomials and prime factors* (**J** 0068) Z Math Logik Grundlagen Math 27*213–214
◇ D20 F15 ◇ rev MR 82i:03053 Zbl 471 # 04002 • id 55251

McBeth, R. see Vol. V, VI for further entries

McCarthy, J. & Shannon, C.E. (eds.) [1956] *Automata studies* (**X** 0857) Princeton Univ Pr: Princeton ix + 285pp
◇ D05 D97 ◇ rev MR 18.103 Zbl 74.112 • rem 2nd ed 1957, 3rd ed 1960, 4th ed 1965. • id 23558

McCarthy, J. [1956] *The inversion of functions defined by Turing machines* (**C** 0717) Automata Studies 177–181 • transl [1974] (**C** 1902) Stud Th Automaten 205–210
◇ D10 D20 ◇ rev MR 18.103 JSL 35.481 • id 08953

McCarthy, J. [1960] *Recursive functions of symbolic expressions and their computation by machine. Part I* (**J** 0212) ACM Commun 3*184–195
◇ B75 D20 ◇ rev Zbl 101.104 JSL 33.117 • id 08954

McCarthy, J. [1979] see Cartwright, Robert

McCarthy, J. see Vol. I, II for further entries

McCawley, J.D. [1970] *English is a VSO language* (**J** 0242) Language (Baltimore) 46*286–299
◇ B65 D05 ◇ id 08960

McCawley, J.D. see Vol. I, II for further entries

McCleary, S.H. [1967] *Primitive recursive computations* (**J** 0047) Notre Dame J Formal Log 8*311–317
◇ D20 ◇ rev MR 39 # 3994 Zbl 189.288 • id 08961

McCleary, S.H. [1979] see Holland, W.C.

McCleary, S.H. [1980] *A solution of the word problem in free normal-valued lattice-ordered groups* (**P** 2983) Ordered Groups;1978 Boise 107–129
◇ D40 ◇ rev MR 82h:06026 Zbl 446 # 68063 • id 82212

McCleary, S.H. [1982] *The word problem in free normal valued lattice-ordered groups: A solution and practical shortcuts* (**J** 0004) Algeb Universalis 14*317-348
◇ D40 ◇ rev MR 83f:06027 Zbl 453 # 06012 • id 69709

McCleary, S.H. see Vol. V for further entries

McCool, J. [1968] *Elements of finite order in free product sixth-groups* (**J** 0126) Glasgow Math J 9*128–145
◇ D40 ◇ rev MR 38 # 2196 Zbl 197.19 • id 08969

McCool, J. [1969] *The order problem and the power problem for free product sixth-groups* (**J** 0126) Glasgow Math J 10*1–9
◇ D40 ◇ rev MR 39 # 2852 Zbl 195.306 • id 08970

McCool, J. [1970] *Unsolvable problems in groups with solvable word problem* (**J** 0017) Canad J Math 22*836–838
◇ D40 ◇ rev MR 42 # 353 Zbl 216.87 • id 08971

McCool, J. [1971] *The power problem for groups with one defining relator* (**J** 0053) Proc Amer Math Soc 28*427–430
◇ D40 ◇ rev MR 43 # 326 Zbl 223 # 20036 • id 08972

McCreight, E.M. & Meyer, A.R. [1969] *Classes of computable functions defined by bounds on computation: preliminary report* (**P** 0671) ACM Symp Th of Comput (1);1969 Marina del Rey 79–88
◇ D15 D20 ◇ id 08975

McCreight, E.M. & Meyer, A.R. [1969] *Properties of bounds on computation* (**P** 1129) Princeton Conf Inform Sci & Syst (3);1969 Princeton 154–156
◇ D15 D20 ◇ id 22112

McCreight, E.M. & Meyer, A.R. [1971] *Computationally complex and pseudo-random zero-one valued functions* (**P** 1058) Th Machines & Comput;1971 Haifa 19–42
◇ D15 ◇ rev MR 53 # 4617 • id 21661

McCulloch, W.S. & Pitts, W. [1943] *A logical calculus of the ideas immanent in nervous activity* (**J** 0515) Bull Math Biophys 5*115–133
◇ B60 D05 ◇ rev MR 6.12 JSL 9.49 • id 08977

McDermott, M. [1981] *Inductive definitions* (**J** 0079) Logique & Anal, NS 24*315–330
◇ A05 D70 ◇ rev MR 83k:03008 Zbl 519 # 03005 • id 36153

McDowell, R. [1979] see Freiwald, R.C.

McEvoy, K. [1985] *Jumps of quasi-minimal enumeration degrees* (**J** 0036) J Symb Logic 50*839–848
◇ D30 ◇ id 47384

McEvoy, K. [1985] see Cooper, S.B.

McHugh Jr., E.F. [1979] see Freiwald, R.C.

McKenzie, R. [1971] *Negative solution of the decision problem for sentences true in every subalgebra of* $\langle N, + \rangle$ (**J** 0036) J Symb Logic 36*607–609
◇ D35 F30 ◇ rev MR 45 # 6625 Zbl 248 # 02052 • id 08997

McKenzie, R. & Thompson, R.J. [1973] *An elementary construction of unsolvable word problems in group theory* (**P** 0678) Word Probl: Decis & Burnside Probl in Group Th;1969 Irvine 457-478
◇ D40 ◇ REV MR 53 # 629 Zbl 286 # 02047 JSL 41.786
• ID 16686

McKenzie, R. [1975] *On spectra, and the negative solution of the decision problem for identities having a finite nontrivial model* (**J** 0036) J Symb Logic 40*186-196
◇ B20 C05 C13 D35 ◇ REV MR 51 # 12499
Zbl 316 # 02052 • ID 09009

McKenzie, R. see Vol. I, III, V for further entries

McLaughlin, T.G. [1962] *A note on contraproduction domains* (**J** 0132) Math Scand 11*175-178
◇ D20 D25 ◇ REV MR 27 # 4741 • ID 09038

McLaughlin, T.G. [1962] *On an extension of a theorem of Friedberg* (**J** 0047) Notre Dame J Formal Log 3*270-273
◇ D25 ◇ REV MR 26 # 4900 • ID 09039

McLaughlin, T.G. [1963] *A remark on semiproductive sets* (**J** 0045) Monatsh Math 67*97-100
◇ D25 D30 ◇ REV MR 27 # 2432 • ID 09041

McLaughlin, T.G. [1963] *A theorem on productive functions* (**J** 0053) Proc Amer Math Soc 14*444
◇ D25 ◇ REV MR 26 # 4901 Zbl 203.11 JSL 37.196
• ID 09040

McLaughlin, T.G. [1964] *A note on pseudo doubly creative pairs* (**J** 0047) Notre Dame J Formal Log 5*24-26
◇ D25 ◇ REV MR 30 # 3842 • ID 09044

McLaughlin, T.G. [1964] *On contraproductive sets which are not productive* (**J** 0068) Z Math Logik Grundlagen Math 10*49-52
◇ D25 ◇ REV MR 28 # 4995 • ID 09042

McLaughlin, T.G. [1964] *Some observations on quasicohesive sets* (**J** 0133) Michigan Math J 11*83-87
◇ D25 ◇ REV MR 28 # 2972 Zbl 122.11 JSL 32.270
• ID 09043

McLaughlin, T.G. [1965] *Co-immune retraceable sets* (**J** 0015) Bull Amer Math Soc 71*523-525
◇ D50 ◇ REV MR 30 # 3843 Zbl 221 # 02026 JSL 32.123
• ID 09045

McLaughlin, T.G. [1965] *On a class of complete simple sets* (**J** 0018) Canad Math Bull 8*33-37
◇ D25 ◇ REV MR 32 # 44 • ID 09047

McLaughlin, T.G. [1965] see Appel, K.I.

McLaughlin, T.G. [1965] *On relative coimmunity* (**J** 0048) Pac J Math 15*1319-1327
◇ D50 ◇ REV MR 32 # 7406 Zbl 224 # 02033 • ID 09048

McLaughlin, T.G. [1965] *Splitting and decomposition by regressive sets* (**J** 0133) Michigan Math J 12*499-505
◇ D50 ◇ REV MR 32 # 4004 Zbl 221 # 02024 • REM Part I. Part II 1967 • ID 09046

McLaughlin, T.G. [1965] *Strong reducibility on hypersimple sets* (**J** 0047) Notre Dame J Formal Log 6*229-234
◇ D25 ◇ REV MR 33 # 1234 Zbl 143.12 • ID 09049

McLaughlin, T.G. [1966] *Retraceable sets and recursive permutations* (**J** 0053) Proc Amer Math Soc 17*427-429
◇ D50 ◇ REV MR 32 # 5507 Zbl 192.52 JSL 33.114
• ID 09052

McLaughlin, T.G. [1966] *Two remarks on indecomposable number sets* (**J** 0068) Z Math Logik Grundlagen Math 12*187-190
◇ D20 D25 ◇ REV MR 33 # 5481 Zbl 221 # 02023
• ID 09053

McLaughlin, T.G. [1967] *Hereditarily retraceable isols* (**J** 0015) Bull Amer Math Soc 73*113-115
◇ D50 ◇ REV MR 34 # 4130 Zbl 165.316 JSL 33.114
• ID 09054

McLaughlin, T.G. [1967] *Some counterexamples in the theory of regressive sets* (**J** 0068) Z Math Logik Grundlagen Math 13*81-87
◇ D50 ◇ REV MR 34 # 7368 Zbl 227 # 02022 JSL 34.296 JSL 35.296 • ID 09055

McLaughlin, T.G. [1967] *Some remarks on extensibility, confluence of paths, branching properties, and index sets, for certain recursively enumerable graphs* (**J** 0316) Illinois J Math 11*257-279
◇ D25 D80 ◇ REV MR 35 # 1470 Zbl 227 # 02014
JSL 34.518 • ID 09057

McLaughlin, T.G. [1967] *Splitting and decomposition by regressive sets II* (**J** 0017) Canad J Math 19*291-311
◇ D25 D50 ◇ REV MR 35 # 51 Zbl 221 # 02025 • REM Part I 1965 • ID 09056

McLaughlin, T.G. [1968] *A theorem on intermediate reducibilities* (**J** 0053) Proc Amer Math Soc 19*87-90
◇ D30 ◇ REV MR 36 # 2502 Zbl 295 # 02024 • ID 09058

McLaughlin, T.G. [1969] see Jockusch Jr., C.G.

McLaughlin, T.G. [1971] *The family of all recursively enumerable classes of finite sets* (**J** 0064) Trans Amer Math Soc 155*127-136
◇ D25 ◇ REV MR 43 # 1832 Zbl 221 # 02022 • ID 09059

McLaughlin, T.G. [1972] *Complete index sets of recursively enumerable families* (**J** 0020) Compos Math 24*83-91
◇ D25 ◇ REV MR 47 # 28 Zbl 306 # 02040 • ID 09061

McLaughlin, T.G. [1972] *Supersimple sets and the problem of extending a retracing function* (**J** 0048) Pac J Math 41*485-494
◇ D25 D50 ◇ REV MR 48 # 3719 Zbl 255 # 02043
• ID 09060

McLaughlin, T.G. [1973] *A non-enumerability theorem for infinite classes of finite structures* (**P** 0678) Word Probl: Decis & Burnside Probl in Group Th;1969 Irvine 479-481
◇ C13 D25 D55 ◇ REV MR 55 # 5418 Zbl 267 # 02031
JSL 41.786 • ID 29894

McLaughlin, T.G. [1973] *On retraceable sets with rapid growth* (**J** 0053) Proc Amer Math Soc 40*573-576
◇ D25 D50 ◇ REV MR 49 # 4767 Zbl 274 # 02014
• ID 09062

McLaughlin, T.G. [1974] *Closed basic retracing functions and hyperimmune sets* (**J** 0068) Z Math Logik Grundlagen Math 20*49-52 • ERR/ADD ibid 22/3*287
◇ D25 D50 ◇ REV MR 49 # 10532 MR 53 # 10567
Zbl 298 # 02032 • ID 19314

McLaughlin, T.G. [1974] *Degrees of unsolvability within a regressive isol* (**J** 0027) Fund Math 86*29-40
◇ D30 D50 ◇ REV MR 51 # 10062 Zbl 271 # 02029
• ID 09063

McLaughlin, T.G. [1975] *A note concerning the $\overset{*}{\vee}$ relation on Λ_R* (J 0068) Z Math Logik Grundlagen Math 21*177-179
◇ D25 D30 D50 ◇ REV MR 53#5276 Zbl 311#02049 • ID 09064

McLaughlin, T.G. [1975] *Trees and isols. Part I* (J 0308) Rocky Mountain J Math 5*401-418
◇ D50 ◇ REV MR 54#2438 Zbl 338#02025 • REM Part II 1976 • ID 14973

McLaughlin, T.G. [1976] *Trees and isols II* (J 0068) Z Math Logik Grundlagen Math 22*45-78
◇ D50 ◇ REV MR 55#2542 Zbl 338#02026 • REM Part I 1975 • ID 18489

McLaughlin, T.G. [1977] *A partial comparison of two conditions on the intersection of regressive sets* (J 0009) Arch Math Logik Grundlagenforsch 18*159-167
◇ D50 ◇ REV MR 58#10378 Zbl 385#03046 • ID 24330

McLaughlin, T.G. [1977] *Degrees of unsolvability and strong forms of $\Lambda_R + \Lambda_R \nsubseteq \Lambda_R$* (J 0047) Notre Dame J Formal Log 18*545-566
◇ D25 D30 D50 ◇ REV MR 58#21550 Zbl 383#03031 • ID 24296

McLaughlin, T.G. [1978] see Barback, J.

McLaughlin, T.G. [1978] *On the relations between some rate-of-growth conditions* (J 0053) Proc Amer Math Soc 69*151-155
◇ D15 ◇ REV MR 58#16227 Zbl 382#03038 • ID 51962

McLaughlin, T.G. [1979] *Retraceable homogeneous sets* (J 0027) Fund Math 103*223-229
◇ D50 ◇ REV MR 80i:03054 Zbl 422#03020 • ID 53481

McLaughlin, T.G. [1981] *Intersection types and the terms of a regressive sum* (P 2902) Aspects Effective Algeb;1979 Clayton 189-195
◇ D50 ◇ REV MR 82k:03072 Zbl 472#03036 • ID 55301

McLaughlin, T.G. [1981] *On the divergence of extension procedures in isol theory* (J 0053) Proc Amer Math Soc 83*769-773
◇ D50 ◇ REV MR 83a:03043 Zbl 481#03027 • ID 35067

McLaughlin, T.G. [1982] *Regressive sets and the theory of isols* (S 3310) Lect Notes Pure Appl Math 66*viii+371pp
◇ D50 D98 ◇ REV MR 83i:03073 Zbl 484#03025 • ID 35527

McLaughlin, T.G. see Vol. I, V for further entries

McNaughton, R. [1954] see Burks, A.W.

McNaughton, R. & Yamada, H. [1960] *Regular expressions and state graphs for automata* (J 0072) IRE Trans Electr Comp EC-9*39-47
◇ D05 ◇ REV Zbl 156.255 JSL 32.390 • ID 19312

McNaughton, R. [1961] *The theory of automata, a survey* (C 4192) Adv in Comput 2*379-421
◇ D05 D98 ◇ REV MR 24#B2520 Zbl 133.98 JSL 37.760 • ID 44062

McNaughton, R. & Yamada, H. [1964] *Regular expressions and state graphs for automata* (C 0681) Sequential Machines Sel Pap 157-174
◇ D05 ◇ REV Zbl 156.255 JSL 32.390 • ID 19310

McNaughton, R. [1966] *Testing and generating infinite sequences by a finite automaton* (J 0194) Inform & Control 9*521-530
◇ D05 ◇ REV MR 35#4105 Zbl 212.339 JSL 37.618 • ID 09074

McNaughton, R. [1971] *A decision procedure for generalized sequential mapability-onto of regular sets* (P 0680) ACM Symp Th of Comput (3);1971 Shaker Heights 206-218
◇ D05 ◇ REV Zbl 257#68080 • ID 63848

McNaughton, R. [1974] *Algebraic decision procedures for local testability* (J 0041) Math Syst Theory 8*60-76
◇ D05 ◇ REV MR 52#13361 Zbl 287#02022 • ID 21815

McNaughton, R. & Seiferas, J.I. [1976] *Regularity preserving relations* (J 1426) Theor Comput Sci 2*147-154
◇ D05 ◇ REV MR 53#12111 Zbl 353#94044 • ID 23112

McNaughton, R. & Narendran, P. [1984] *The undecidability of the preperfectness of Thue systems* (J 1426) Theor Comput Sci 31*165-174
◇ D03 ◇ REV MR 85k:68028 Zbl 545#03022 • ID 43503

McNaughton, R. [1985] see Kapur, D.

McNaughton, R. see Vol. I, II, III, V, VI for further entries

McNulty, G.F. [1975] see Jonsson, B.

McNulty, G.F. [1976] *The decision problem for equational bases of algebras* (J 0007) Ann Math Logic 10*193-259
◇ C05 D35 ◇ REV MR 55#5428 Zbl 376#08005 JSL 47.903 • ID 24259

McNulty, G.F. [1976] *Undecidable properties of finite sets of equations* (J 0036) J Symb Logic 41*589-604
◇ C05 D35 ◇ REV MR 58#5154 Zbl 375#02040 JSL 47.903 • ID 14582

McNulty, G.F. [1982] *Infinite ordered sets, a recursive perspective* (P 3780) Ordered Sets;1981 Banff 299-330
◇ D80 ◇ REV MR 83h:06009 Zbl 513#03021 • ID 37225

McNulty, G.F. [1984] see Kierstead, H.A.

McNulty, G.F. see Vol. I, III, V for further entries

Mead, J. [1979] *Recursive prime models for boolean algebras* (S 0019) Colloq Math (Warsaw) 41*25-33
◇ C50 C57 G05 ◇ REV MR 80j:03050 Zbl 445#03016 • ID 56480

Mead, J. see Vol. III for further entries

Medina, A. [1966] *Logical analysis of organization in finite automata (Spanish summary)* (J 0516) Rev Mexicana Fis 15*1-54
◇ D05 ◇ REV MR 33#3936 • ID 09075

Medvedev, Yu.T. [1955] *Degrees of difficulty of the mass problem (Russian)* (J 0023) Dokl Akad Nauk SSSR 104*501-504
◇ D30 D80 ◇ REV MR 17.448 Zbl 65.3 JSL 21.320 • ID 09077

Medvedev, Yu.T. [1955] *On nonisomorphic recursively enumerable sets (Russian)* (J 0023) Dokl Akad Nauk SSSR 102*211-214
◇ D25 D50 ◇ REV MR 18.272 Zbl 64.288 JSL 21.101 • ID 19308

MEDVEDEV, YU.T. [1956] *On the class of events able to be represented in a finite automation (Russian)* (**J** 1884) Avtomatika, Akad Nauk Ukr SSR Appendix 2
◇ D05 ◇ ID 45551

MEDVEDEV, YU.T. [1956] *On the concept of the mass problem (Russian)* (**J** 0067) Usp Mat Nauk 11/5(71)*231-232
◇ D30 D80 ◇ ID 45550

MEDVEDEV, YU.T. [1959] *On the concept of mass problem and its applications in the theory of recursive functions and in mathematical logic (Russian)* (**P** 0607) All-Union Math Conf (3);1956 Moskva 1*183
◇ D30 D80 ◇ ID 28237

MEDVEDEV, YU.T. [1962] *Finite problems (Russian)* (**J** 0023) Dokl Akad Nauk SSSR 142*1015-1018
• TRANSL [1962] (**J** 0062) Sov Math, Dokl 3*227-230
◇ D80 F50 ◇ REV MR 24 # A3067 Zbl 286 # 02028 JSL 38.330 • ID 09078

MEDVEDEV, YU.T. [1963] *Interpretation of logical formulae by means of finite problems and its relation to the realizability theory (Russian)* (**J** 0023) Dokl Akad Nauk SSSR 148*771-774
• TRANSL [1963] (**J** 0062) Sov Math, Dokl 4*180-183
◇ D80 F50 ◇ REV MR 26 # 4904 Zbl 218 # 02048 JSL 38.330 • ID 09079

MEDVEDEV, YU.T. [1966] *Interpretation of logical formulae by means of finite problems (Russian)* (**J** 0023) Dokl Akad Nauk SSSR 169*20-23
• TRANSL [1966] (**J** 0062) Sov Math, Dokl 7*857-860
◇ D80 F50 ◇ REV MR 38 # 37 Zbl 192.38 JSL 38.330
• ID 09080

MEDVEDEV, YU.T. [1969] *A method for proving the unsolvability of algorithmic problems (Russian)* (**J** 0023) Dokl Akad Nauk SSSR 185*1232-1235
• TRANSL [1969] (**J** 0062) Sov Math, Dokl 10*495-498
◇ D35 F50 ◇ REV MR 40 # 4112 Zbl 193.316 • ID 09081

MEDVEDEV, YU.T. [1972] *Locally finitary algorithmic problems (Errata ibid 204*1286) (Russian)* (**J** 0023) Dokl Akad Nauk SSSR 203*285-288
• TRANSL [1972] (**J** 0062) Sov Math, Dokl 13*382-386
◇ D20 F30 F50 ◇ REV MR 47 # 24 Zbl 262 # 02034
• ID 63851

MEDVEDEV, YU.T. see Vol. II, VI for further entries

MEERSMAN, R. [1975] *A survey of techniques in applied computational complexity* (**J** 2749) J Comp Appl Math 1*39-46
◇ D15 D80 ◇ REV MR 51 # 9572 Zbl 324 # 65023 Zbl 349 # 65030 • ID 82224

MEERSMAN, R. & ROZENBERG, G. & VERMEIR, D. [1979] *Persistent ETOL systems* (**J** 0191) Inform Sci 18*189-212
◇ D05 ◇ REV MR 80e:68204 Zbl 442 # 68072 • ID 69716

MEHLHORN, K. [1973] *On the size of sets of computable functions* (**P** 3062) IEEE Symp Switch & Automata Th (14);1973 Iowa City 190-196
◇ D15 D20 ◇ REV MR 55 # 2526 • ID 76263

MEHLHORN, K. [1974] *Polynomial and abstract subrecursive classes* (**P** 1464) ACM Symp Th of Comput (6);1974 Seattle 96-109
◇ D15 D20 ◇ REV MR 56 # 5245 Zbl 363 # 68063
• ID 50920

MEHLHORN, K. [1974] *The "almost all" theory of subrecursive degrees is decidable* (**P** 1869) Automata, Lang & Progr (2);1974 Saarbruecken 317-325
◇ B25 C80 D20 ◇ REV MR 55 # 7752 Zbl 284 # 68041
• ID 63853

MEHLHORN, K. [1976] see ALT, H.

MEHLHORN, K. [1976] *Polynomial and abstract subrecursive classes* (**J** 0119) J Comp Syst Sci 12*147-178
◇ D15 D20 ◇ REV MR 54 # 76 Zbl 329 # 68049
• ID 23957

MEINEL, C. [1982] *Embedding of the poset of Turing degrees in the poset* $[\mathfrak{U}, \leqslant]$ *(German and Russian summaries)* (**J** 0129) Elektr Informationsverarbeitung & Kybern 18*339-344
◇ D10 D30 ◇ REV MR 85c:03017 Zbl 527 # 03020
• ID 37570

MEINEL, C. [1982] see BUDACH, L.

MEINHARDT, D. & WAGNER, K. [1982] *Eine Bemerkung zu einer Arbeit von Monien ueber Kopfzahlhierarchien fuer Zweiwegautomaten (English and Russian summaries)* (**J** 0129) Elektr Informationsverarbeitung & Kybern 18*69-74
◇ D05 ◇ REV MR 84e:68050 Zbl 493 # 68047 • ID 38441

MEINHARDT, D. [1984] *Tape reversal bounded Turing machines with an auxiliary pushdown or an auxiliary counter* (**P** 3621) Frege Konferenz (2);1984 Schwerin 338-344
◇ D10 ◇ REV Zbl 555 # 68020 • ID 45404

MEISSNER, H.G. [1976] *Limes-Erkennung von regulaeren und linearen Sprachen mit Optimalitaetseigenschaften* (**J** 0129) Elektr Informationsverarbeitung & Kybern 12*525-549
◇ D05 ◇ REV MR 56 # 7349 Zbl 358 # 68068 • ID 82227

MEISSNER, H.G. [1976] *Ueber die Fortsetzbarkeit von sequentiellen Baum-Operatoren mit endlichem Gewicht* (**J** 0129) Elektr Informationsverarbeitung & Kybern 12*403-414
◇ B75 D05 ◇ REV MR 56 # 2698 Zbl 336 # 94029
• ID 82226

MEISSNER, H.G. [1977] *Zu einigen Begriffen und Resultaten aus der Theorie der Baumautomaten* (**S** 2829) Rostocker Math Kolloq 3*85-102
◇ D05 ◇ REV MR 57 # 14606 Zbl 403 # 68059 • ID 69717

MEJTUS, V.YU. & VERSHININ, K.P. [1974] *On some unsolvable problems in computable categories (Russian)* (**J** 0023) Dokl Akad Nauk SSSR 216*42-43
• TRANSL [1974] (**J** 0062) Sov Math, Dokl 15*752-754
◇ C57 D80 G30 ◇ REV MR 49 # 10536 Zbl 304 # 02019
• ID 09089

MELIKHOV, A.N. [1969] see BERSHTEJN, L.S.

MELIKHOV, A.N. see Vol. II, V for further entries

MELIKYAN, S.M. [1974] *Constructive transfinite hierarchies of pseudonumbers (Russian)* (**J** 0346) Dokl Akad Nauk Armyan SSR 59*257-260
◇ D55 F60 ◇ REV MR 53 # 7737 Zbl 307 # 02024
• ID 22984

MELIKYAN, S.M. [1982] *Specker numbers and degrees of undecidability (Russian)* (**J** 0023) Dokl Akad Nauk SSSR 263*1308-1310
• TRANSL [1982] (**J** 0062) Sov Math, Dokl 25*532-534
◇ D30 D80 ◇ REV MR 83g:03040 Zbl 512 # 03021
• ID 36007

MELIKYAN, S.M. see Vol. VI for further entries

MELLISH, M. [1978] *Some prediction algorithms for nonrecursive sequences* (J 0194) Inform & Control 37*234-239
◇ D20 ◇ REV MR 58 # 181 Zbl 383 # 03028 • ID 52010

MEL'NICHUK, I.L. [1984] *Unsolvability of problems of equality and divisibility in certain varieties of semigroups (Russian)* (J 0003) Algebra i Logika 23*430-438
• TRANSL [1984] (J 0069) Algeb and Log 23*297-302
◇ C05 D40 ◇ REV MR 86d:20065 • ID 44845

MELSON, R.T. [1974] see IBARRA, O.H.

MELVILLE, R. [1981] *An improved simulation result for ink-bounded Turing machines* (J 0119) J Comp Syst Sci 22*98-105
◇ D10 ◇ REV MR 83a:68063 Zbl 456 # 68050 • ID 69718

MELZAK, Z.A. [1961] *An informal arithmetical approach to computability and computation* (J 0018) Canad Math Bull 4*279-293
◇ D10 ◇ REV MR 27 # 1364 Zbl 112.357 JSL 31.514
• REM Part II 1964 • ID 09095

MELZAK, Z.A. [1964] *An informal arithmetical approach to computability and computation II* (J 0018) Canad Math Bull 7*183-200
◇ D10 D15 ◇ REV MR 31 # 4727 Zbl 112.357
Zbl 203.161 • REM Part I 1961. Part III 1966 • ID 09096

MELZAK, Z.A. [1966] *An informal arithmetical approach to computability and computation III* (J 0018) Canad Math Bull 9*593-609
◇ D10 D80 ◇ REV MR 35 # 2739 Zbl 216.291 • REM Part II 1964 • ID 09097

MELZI, G. [1976] *I supporti fisici dell'inferenza formale. Vol. I* (X 1364) Vita e Pensiero: Milano xv + 243pp
◇ B10 B75 D05 ◇ REV MR 55 # 75 Zbl 376 # 94023
• ID 76279

MENDELSOHN, N.S. [1966] see BENSON, C.

MENDELSON, E. [1963] *On some recent criticism of Church's thesis* (J 0047) Notre Dame J Formal Log 4*201-205
◇ A05 D20 F99 ◇ REV MR 29 # 3369 Zbl 126.20
JSL 33.471 • ID 09116

MENDELSON, E. see Vol. I, II, III, V for further entries

MENGER, K. [1928] *Bemerkungen zu Grundlagenfragen. III: Ueber Potenzmengen* (J 0157) Jbuchber Dtsch Math-Ver 37*303-308
◇ A05 D55 E15 E30 ◇ REV FdM 54.95 • REM Parts II,IV 1928 • ID 31949

MENGER, K. see Vol. I, II, V, VI for further entries

MENTRASTI, P. [1975] see JACOPINI, G.

MENTRASTI, P. [1976] *Sulle classi di funzioni elementari inferiori in una variabile (English summary)* (J 2311) Rend Mat, Ser 6 9*37-56
◇ D20 ◇ REV MR 54 # 7227 Zbl 339 # 02034 • ID 25019

MENTRASTI, P. & PROTASI, M. [1982] *Extended primitive recursive functions (French summary)* (J 3441) RAIRO Inform Theor 16*73-84
◇ D20 ◇ REV MR 84d:03052 Zbl 498 # 03024 • ID 34088

MENZEL, W. [1982] see MAIER, W.

MENZEL, W. & SPERSCHNEIDER, V. [1982] *Universal automata with uniform bounds on simulation time* (J 0194) Inform & Control 52*19-35
◇ D20 ◇ REV MR 84h:68040 Zbl 501 # 03023 • ID 37777

MENZEL, W. & SPERSCHNEIDER, V. [1984] *Recursively enumerable extensions of R_1 by finite functions* (P 2342) Symp Rek Kombin;1983 Muenster 62-76
◇ D20 D25 ◇ REV Zbl 545 # 03025 • ID 43506

MERCANTI, F. & PIAZZESE, F. [1981] *Neuromachines as Turing machines* (S 2990) Rend Semin Mat Brescia 5*103-114
◇ D10 ◇ REV Zbl 475 # 68023 • ID 69719

MERCANTI, F. [1981] *On a neuromachine equivalent to the universal Turing machine* (S 2990) Rend Semin Mat Brescia 6*9-14
◇ D10 ◇ REV Zbl 475 # 68022 • ID 69720

MERTENS, J.R. [1975] see HAMILTON, W.L.

MERZENICH, W. [1979] *A binary operation on trees and an initial algebra characterization for finite tree types* (J 1431) Acta Inf 11*149-168
◇ D80 ◇ REV MR 80d:68077 Zbl 379 # 02016 • ID 52181

MESEGUER, J. [1975] see KUEHNEL, W.

MESEGUER, J. [1977] see KUEHNEL, W.

MESEGUER, J. see Vol. I, III, V for further entries

MESKIN, S. [1974] *A finitely generated residually finite group with an unsolvable word problem* (J 0053) Proc Amer Math Soc 43*8-10
◇ D40 ◇ REV MR 49 # 425 Zbl 261 # 20020 • ID 09165

METAKIDES, G. [1972] *α-degrees of α-theories* (J 0036) J Symb Logic 37*677-682
◇ D30 D60 ◇ REV MR 48 # 81 Zbl 327 # 02037
• ID 09166

METAKIDES, G. & NERODE, A. [1975] *Recursion theory and algebra* (P 0765) Algeb & Log;1974 Clayton 209-219
◇ C57 C60 D25 D45 ◇ REV MR 51 # 7798
Zbl 306 # 02038 • ID 17296

METAKIDES, G. [1977] *A return to constructive algebra via recursive function theory* (J 4135) Dialexeis 2*112-126
◇ C57 D45 ◇ ID 33933

METAKIDES, G. & NERODE, A. [1977] *Recursively enumerable vector spaces* (J 0007) Ann Math Logic 11*147-171
◇ C57 C60 D45 ◇ REV MR 56 # 5253 Zbl 389 # 03019
JSL 48.880 • ID 23661

METAKIDES, G. [1978] *Constructive algebra in a new frame* (J 4373) GMS Math Inst Patras 4*13-25
◇ C57 C60 D45 ◇ ID 33934

METAKIDES, G. & NERODE, A. [1979] *Effective content of field theory* (J 0007) Ann Math Logic 17*289-320
◇ C57 C60 D45 ◇ REV MR 82b:03082 Zbl 469 # 03028
JSL 48.880 • ID 55155

METAKIDES, G. & REMMEL, J.B. [1979] *Recursion theory on orderings I. A model theoretic setting* (J 0036) J Symb Logic 44*383-402
◇ C57 D45 ◇ REV MR 80m:03080 Zbl 471 # 03035 • REM Part II 1980 by Remmel,J.B. • ID 55230

METAKIDES, G. & NERODE, A. [1980] *Recursion theory on fields and abstract dependence* (**J** 0032) J Algeb 65*36-59
◇ C57 C60 D45 ◇ REV MR 81k:03041 Zbl 469 # 03029 JSL 48.880 • ID 55156

METAKIDES, G. & NERODE, A. [1982] *The introduction of non-recursive methods into mathematics* (**P** 3638) Brouwer Centenary Symp;1981 Noordwijkerhout 319-335
◇ C57 D45 D80 F60 ◇ REV MR 85d:03121 Zbl 511 # 03020 • ID 37391

METAKIDES, G. & NERODE, A. & SHORE, R.A. [1985] *Recursive limits on the Hahn-Banach theorem* (**P** 4271) E.Bishop-Reflection on Him & Research;1983 San Diego 85-91
◇ D80 ◇ ID 45612

METAKIDES, G. see Vol. I, III, V for further entries

METIVIER, Y. [1985] *Calcul de longueurs de chaines de reecriture dans le monoide libre* (**J** 1426) Theor Comput Sci 35*71-87
◇ D05 ◇ REV Zbl 562 # 03019 • ID 47428

METRA, I. [1971] *Comparison of the number of states of probabilistic and deterministic automata that represent given events (Russian)* (**J** 0474) Avtom Vychis Tekh, Akad Nauk Latv SSR 1971/5*94-96
◇ D05 ◇ REV MR 45 # 8465 Zbl 252 # 94032 • ID 63886

METROPOLIS, N. & ROTA, G.-C. [1974] *Significance arithmetic on the algebra of binary strings* (**C** 1213) Stud in Num Anal (Lanczos) 241-251
◇ D15 F65 G05 ◇ REV MR 50 # 6815 Zbl 311 # 65031 • ID 21270

METROPOLIS, N. see Vol. VI for further entries

METZ, J. [1980] *Abstrakte Rechnermodelle* (**X** 0911) Akademie Verlag: Berlin 144pp
◇ D05 D98 ◇ REV MR 82j:68001 Zbl 455 # 68004 • ID 54302

MEY VANDER, J.E. & PIERCE, J.C. & SINGLETARY, W.E. [1973] *Tutor - a Turing machine simulator* (**J** 0191) Inform Sci 5*265-278
◇ D10 ◇ REV Zbl 248 # 68017 • ID 64503

MEYER, A.R. & RITCHIE, R.W. [1967] *Computational complexity and program structure* (1111) Preprints, Manuscr., Techn. Reports etc. RC1817
◇ B75 D15 ◇ REM IBM, Armonk • ID 21176

MEYER, A.R. & RITCHIE, D.M. [1967] *The complexity of loop programs* (**P** 1384) ACM Nat Conf (22);1967 465-469
◇ B75 D15 ◇ ID 28655

MEYER, A.R. & RITCHIE, D.M. [1968] *Classification of functions by computational complexity* (**P** 1130) Hawaii Int Conf Syst Sci (1);1968 Honolulu 17-19
◇ D15 D20 ◇ ID 22109

MEYER, A.R. [1968] see FISCHER, P.C.

MEYER, A.R. [1969] see MCCREIGHT, E.M.

MEYER, A.R. [1970] see FISCHER, P.C.

MEYER, A.R. [1971] see MCCREIGHT, E.M.

MEYER, A.R. & RITCHIE, D.M. [1972] *A classification of the recursive functions* (**J** 0068) Z Math Logik Grundlagen Math 18*71-82
◇ D20 ◇ REV MR 45 # 4975 Zbl 247 # 02037 • ID 21445

MEYER, A.R. [1972] see FISCHER, P.C.

MEYER, A.R. & MOLL, R. [1972] *Honest bounds for complexity classes of recursive functions* (**P** 1126) IEEE Symp Switch & Automata Th (3);1962 College Park 61-66
◇ D15 D20 ◇ REV MR 50 # 9552 • ID 22116

MEYER, A.R. [1972] see BAGCHI, A.

MEYER, A.R. [1973] see HELM, J.P.

MEYER, A.R. & STOCKMEYER, L.J. [1973] *Word problems requiring exponential time: preliminary report* (**P** 1482) ACM Symp Th of Comput (5);1973 Austin 1-9
◇ D15 D40 ◇ REV MR 54 # 6557 Zbl 359 # 68050 • ID 24771

MEYER, A.R. & MOLL, R. [1974] *Honest bounds for complexity classes of recursive functions* (**J** 0036) J Symb Logic 39*127-138
◇ D15 D20 ◇ REV MR 50 # 9552 Zbl 322 # 02038 • ID 09386

MEYER, A.R. [1975] *The inherent computational complexity of theories of ordered sets* (**P** 1521) Int Congr Math (II,12);1974 Vancouver 2*477-482
◇ B25 D15 ◇ REV MR 56 # 103 Zbl 361 # 02061 • ID 27549

MEYER, A.R. [1975] *Weak monadic second order theory of successor is not elementary recursive* (**C** 0758) Logic Colloq Boston 1972-73 132-154
• TRANSL [1975] (**J** 3079) Kiber Sb Perevodov, NS 12*62-77
◇ B25 D10 D15 D20 ◇ REV MR 52 # 13358 Zbl 326 # 02036 • ID 21811

MEYER, A.R. [1976] see CARDOZA, E.

MEYER, A.R. [1976] see FISCHER, MICHAEL J.

MEYER, A.R. [1978] see BRUSS, A.R.

MEYER, A.R. [1978] see FISCHER, MICHAEL J.

MEYER, A.R. & WINKLMANN, K. [1979] *The fundamental theorem of complexity theory (Preliminary version)* (**P** 3375) Found of Comput Sci (3);1978 Amsterdam 1*98-112
◇ B75 D10 D15 ◇ REV MR 81b:68053 Zbl 409 # 68029 • ID 69721

MEYER, A.R. [1980] see BRUSS, A.R.

MEYER, A.R. [1981] see MAYR, E.W.

MEYER, A.R. & MIRKOWSKA, G. & STREETT, R.S. [1981] *The deducibility problem in propositional dynamic logic* (**P** 3497) Log of Progr;1979 Zuerich 12-22
• REPR [1981] (**P** 2903) Automata, Lang & Progr (8);1981 Akko 238-248
◇ B75 D25 D35 ◇ REV MR 83i:03046 MR 85h:03029 Zbl 466 # 68024 • REM The reprint is a revised version • ID 54996

MEYER, A.R. [1982] see FISCHER, MICHAEL J.

MEYER, A.R. [1982] see MAYR, E.W.

MEYER, A.R. [1985] see CHANDRA, A.K.

MEYER, A.R. see Vol. I, II, VI for further entries

MEYER, J.-J.C. [1981] see BERGSTRA, J.A.

MEYER, J.-J.C. see Vol. II for further entries

MEYER, R.K. [1968] *An undecidability result in the theory of relevant implication* (**J** 0068) Z Math Logik Grundlagen Math 14*255-262
◇ B46 D35 ◇ REV MR 38 # 4287 Zbl 172.297 • ID 09181

MEYER, R.K. & ROUTLEY, R. [1973] *An undecidable relevant logic* (**J** 0068) Z Math Logik Grundlagen Math 19*389-397
◇ B46 D35 D40 ◇ REV MR 49 # 32 Zbl 301 # 02024
• ID 09201

MEYER, R.K. see Vol. I, II, III, V, VI for further entries

MEYER AUF DER HEIDE, F. & ROLLIK, A. [1981] *Random access machines and straight-line programs* (**P** 3165) FCT'81 Fund of Comput Th;1981 Szeged 259-264
◇ D10 ◇ REV MR 83e:68050 • ID 40161

MEYER AUF DER HEIDE, F. [1985] *Lower bounds for solving linear Diophantine equations on random access machines* (**J** 0037) ACM J 32*929-937
◇ D10 D15 ◇ ID 49278

MEZEI, J.E. [1963] *Structure of monoids with applications to automata* (**P** 0674) Symp Math Th of Automata;1962 New York 267-299
◇ D05 ◇ REV MR 30 # 5900 Zbl 124.250 • ID 09206

MEZEI, J.E. & WRIGHT, J.B. [1965] *Generalized ALGOL-like languages* (**J** 1089) IBM Res Rep RC-1528
◇ B75 D05 ◇ ID 31317

MEZEI, J.E. [1965] see ELGOT, C.C.

MEZEI, J.E. & WRIGHT, J.B. [1967] *Algebraic automata and context-free sets* (**J** 0194) Inform & Control 11*3-29
◇ D05 ◇ REV MR 38 # 3091 Zbl 155.343 • ID 31311

MEZNIK, I. [1972] *G-machines and generable sets* (**J** 0194) Inform & Control 20*499-509
◇ D05 ◇ REV MR 48 # 3301 Zbl 239 # 94063 • ID 63915

MICALE, B. [1979] see FERRO, A.

MICALE, B. see Vol. III for further entries

MICALI, S. [1981] *Two-way deterministic finite automata are exponentially more succinct than sweeping automata* (**J** 0232) Inform Process Lett 12*103-105
◇ D05 D15 ◇ REV MR 82d:68034 Zbl 471 # 68039
• ID 82242

MICALI, S. see Vol. VI for further entries

MICHAELSON, S. & MILNER, R. (EDS.) [1976] *Automata, languages and programming. III* (**X** 1261) Edinburgh Univ Pr: Edinburgh VI + 559pp
◇ D05 D97 ◇ REV Zbl 339 # 00025 • ID 48706

MICHEL, M. [1984] *Algebre de machines et logique temporelle* (**P** 3565) Symp of Th Aspects of Comput Sci (1);1984 Paris 287-298
◇ B45 D05 ◇ REV MR 86a:68026 Zbl 542 # 68021
• ID 43712

MICHEL, M. see Vol. II for further entries

MICHEL, P. [1981] *Borne superieure de la complexite de la theorie de N muni de la relation de divisibilite* (**P** 3404) Model Th & Arithm;1979/80 Paris 242-250
◇ B25 D15 ◇ REV MR 83e:03063 Zbl 488 # 03021
• ID 35234

MICHELIS DE, G. [1974] *Recursive functions not dependent on the computational rules* (**P** 3388) Int Comput Symp;1973 Davos 25-31
◇ D20 ◇ REV Zbl 269 # 68027 • ID 61264

MICHELIS DE, G. [1974] *Sulla computazione di funzioni ricorsive (English summary)* (**J** 0089) Calcolo 11*17-31
◇ D20 ◇ REV MR 52 # 5388 Zbl 282 # 68024 • ID 18295

MIGLIOLI, P.A. [1979] see BERTONI, A.

MIGLIOLI, P.A. [1980] see BERTONI, A.

MIGLIOLI, P.A. & ORNAGHI, M. [1981] *A logically justified model of computation. I,II* (**J** 2095) Fund Inform, Ann Soc Math Pol, Ser 4 4*151-172,277-341
◇ B75 D80 F07 F50 ◇ REV MR 83e:03090
Zbl 473 # 68017 Zbl 483 # 68032 • ID 55396

MIGLIOLI, P.A. [1981] see BERTONI, A.

MIGLIOLI, P.A. [1983] see BERTONI, A.

MIGLIOLI, P.A. see Vol. I, II, III, VI for further entries

MIGNOTTE, M. [1978] *Some effective results about linear recursive sequences* (**P** 1872) Automata, Lang & Progr (5);1978 Udine 322-329
◇ D20 ◇ REV MR 80d:10019 Zbl 387 # 10007 • ID 82246

MIHAILESCU, E.G. [1967] *Decision problem in the classical logic* (**J** 0047) Notre Dame J Formal Log 8*239-253
◇ B25 D35 ◇ REV MR 38 # 3124 Zbl 174.21 • ID 09234

MIHAILESCU, E.G. see Vol. I, II for further entries

MIHNEA, G. [1985] *On the E-recursive functions* (**J** 0070) Bull Soc Sci Math Roumanie, NS 29*261-274
◇ D65 ◇ ID 49608

MIJOULE, R. [1984] *L'universalite des semi-fonctions recursives universelles* (**J** 3797) Diagrammes 12*M1-M12
◇ D65 G30 ◇ REV Zbl 564 # 18002 • ID 48236

MIJOULE, R. see Vol. I, III, V, VI for further entries

MIKA, A. [1979] see BALCER, M.

MIKENBERG, A.M. [1964] *Complete systems of nondecreasing general recursive functions (Russian)* (**J** 0023) Dokl Akad Nauk SSSR 154*517-519
• TRANSL [1964] (**J** 0062) Sov Math, Dokl 5*121-123
◇ D20 ◇ REV MR 28 # 2046 Zbl 158.250 • ID 09244

MIKHAJLOV, G.I. [1984] see IVANOV, N.N.

MIKHAJLOVA, K.A. [1958] *The occurence problem for direct products of groups (Russian)* (**J** 0023) Dokl Akad Nauk SSSR 119*1103-1105
◇ D40 ◇ REV MR 20 # 6454 Zbl 84.253 JSL 36.540
• ID 21032

MIKHAJLOVA, K.A. [1959] *The occurence problem for free products of groups (Russian)* (**J** 0023) Dokl Akad Nauk SSSR 127*746-748
◇ D40 ◇ REV MR 21 # 5670 • ID 09237

MIKHAJLOVA, K.A. [1966] *The occurence problem for direct products of groups (Russian)* (**J** 0142) Mat Sb, Akad Nauk SSSR, NS 70*241-251
◇ D40 ◇ REV MR 33 # 2707 JSL 36.540 • ID 21033

MIKHEEV, V.L. [1973] *Classes of algebras of primitive recursive functions (Russian)* (**J** 0087) Mat Zametki (Akad Nauk SSSR) 14*143-156
• TRANSL [1973] (**J** 1044) Math Notes, Acad Sci USSR 14*638-645
◇ D20 D75 ◇ REV MR 49 # 2311 Zbl 285 # 02035
• ID 09239

MIKHEEV, V.L. [1977] *Meager and universal regressive isols (Russian)* (**J** 0092) Sib Mat Zh 18*358-367,479
• TRANSL [1977] (**J** 0475) Sib Math J 18*257-264
◇ D50 ◇ REV MR 56 # 8344 Zbl 366 # 02033 • ID 51116

MIKHEEV, V.L. [1978] *A hierarchy of independent ω-processions of cosimple isols (Russian)* (**J** 0003) Algebra i Logika 17*56-78,122
• TRANSL [1978] (**J** 0069) Algeb and Log 17*40-58
◇ D50 ◇ REV MR 80e:03053 Zbl 397 # 03028 • ID 52694

MIKHEEV, V.L. [1978] *Infinite sums and products of isolic integers (Russian)* (**J** 0087) Mat Zametki (Akad Nauk SSSR) 23*471-485
• TRANSL [1978] (**J** 1044) Math Notes, Acad Sci USSR 23*255-262
◇ D50 ◇ REV MR 57 # 16026 Zbl 413 # 03026 • ID 76358

MIKHEEV, V.L. [1979] *Some remarks on isols (Russian)* (**J** 0031) Izv Vyssh Ucheb Zaved, Mat (Kazan) 1979/3(202)*76-79
• TRANSL [1979] (**J** 3449) Sov Math 23/3*54-56
◇ D50 ◇ REV MR 80j:03061 Zbl 426 # 03045 • ID 53641

MIKLOS, S. [1978] see BRYLL, G.

MIKOLAJCZAK, B. [1980] *On the simulation problem of a set of automata* (**J** 0947) Wiss Z Tech Univ Dresden 29*383-385
◇ D05 ◇ REV MR 81g:68083 • ID 82249

MIKOLAJCZAK, B. [1981] *On the time computational complexity of several problems of simulation of a set of finite automata* (**P** 2334) Probl Rechnerarchitektur;1980 Altenberg 54-61
◇ D05 D15 ◇ REV MR 83g:68067b • ID 39148

MIKULECKY, P. [1975] *On configurations in cellular automata* (**P** 1755) Math Founds of Comput Sci (3);1974 Jadwisin 62-68
◇ D05 ◇ REV Zbl 309 # 94061 • ID 63936

MILGRAM, D.L. & ROSENFELD, A. [1972] *Array automata and array grammars* (**P** 1455) Inform Processing (5);1971 Ljubljana 69-74
◇ D05 D10 ◇ REV MR 53 # 7135 Zbl 244 # 68032 • ID 63941

MILGRAM, D.L. & ROSENFELD, A. [1973] *Parallel/sequential array automata* (**J** 0232) Inform Process Lett 2*43-46
◇ D05 ◇ REV MR 48 # 8141 Zbl 285 # 68026 • ID 11526

MILGRAM, D.L. [1976] *A region crossing problem for array-bounded automata* (**J** 0194) Inform & Control 31*147-152
◇ D05 D10 ◇ REV MR 54 # 9161 Zbl 328 # 94047 • ID 25834

MILLAR, T.S. [1978] *Foundations of recursive model theory* (**J** 0007) Ann Math Logic 13*45-72
◇ C50 C57 ◇ REV MR 80a:03051 Zbl 432 # 03018 JSL 49.671 • ID 27960

MILLAR, T.S. [1979] *A complete, decidable theory with two decidable models* (**J** 0036) J Symb Logic 44*307-312
◇ C57 ◇ REV MR 81b:03040 Zbl 421 # 03026 JSL 49.671 • ID 53425

MILLAR, T.S. [1980] *Homogeneous models and decidability* (**J** 0048) Pac J Math 91*407-418
◇ C50 C57 ◇ REV MR 83i:03056 Zbl 467 # 03007 • ID 55005

MILLAR, T.S. [1981] *Counterexamples via model completions* (**P** 2628) Log Year;1979/80 Storrs 215-229
◇ C10 C35 C50 C57 ◇ REV MR 84b:03052 Zbl 493 # 03009 • ID 35640

MILLAR, T.S. [1981] *Vaught's theorem recursively revisited* (**J** 0036) J Symb Logic 46*397-411
◇ C15 C57 D45 ◇ REV MR 82d:03053 Zbl 493.03008 • ID 76379

MILLAR, T.S. [1982] *Type structure complexity and decidability* (**J** 0064) Trans Amer Math Soc 271*73-81
◇ C50 C57 ◇ REV MR 83h:03046 Zbl 493 # 03010 • ID 36061

MILLAR, T.S. [1983] *Omitting types, type spectrums, and decidability* (**J** 0036) J Symb Logic 48*171-181
◇ C07 C57 ◇ REV MR 85a:03049 Zbl 516 # 03017 • ID 34800

MILLAR, T.S. [1983] *Persistently finite theories with hyperarithmetic models* (**J** 0064) Trans Amer Math Soc 278*91-99
◇ C15 C50 C57 D30 ◇ REV MR 84m:03053 Zbl 525 # 03027 • ID 35754

MILLAR, T.S. [1983] see ASH, C.J.

MILLAR, T.S. [1984] *Decidability and the number of countable models* (**J** 0073) Ann Pure Appl Logic 27*137-153
◇ C15 C57 ◇ REV MR 86f:03052 Zbl 563 # 03014 • ID 39678

MILLAR, T.S. [1985] *Decidable Ehrenfeucht theories* (**P** 4046) Rec Th;1982 Ithaca 311-321
◇ B25 C15 C57 C98 D30 D35 D55 ◇ REV Zbl 573 # 03003 • ID 46374

MILLAR, T.S. see Vol. III for further entries

MILLER, A.W. [1979] *On the length of Borel hierarchies* (**J** 0007) Ann Math Logic 16*233-267
◇ D55 E15 E35 ◇ REV MR 80m:04003 Zbl 415 # 03038 • ID 53140

MILLER, A.W. [1983] *On the Borel classification of the isomorphism class of a countable model* (**J** 0047) Notre Dame J Formal Log 24*22-34
◇ C15 D55 E15 ◇ REV MR 84c:03055 Zbl 487 # 03013 • ID 35705

MILLER, A.W. see Vol. III, V for further entries

MILLER, DAVID P. [1981] *High recursively enumerable degrees and the anticupping property* (**P** 2628) Log Year;1979/80 Storrs 230-245
◇ D25 ◇ REV MR 82j:03048 Zbl 498 # 03033 • ID 76396

MILLER, DOUGLAS E. [1975] see BURGESS, J.P.

MILLER, DOUGLAS E. [1978] *The invariant Π^0_α separation principle* (**J** 0064) Trans Amer Math Soc 242*185-204
◇ C15 C70 C75 D55 E15 ◇ REV MR 81b:03054 Zbl 409 # 03030 • ID 33521

MILLER, DOUGLAS E. [1982] see HAY, L.

MILLER, DOUGLAS E. [1982] *Index sets and Boolean operations* (**J** 0053) Proc Amer Math Soc 84*568-572
◇ D25 D45 D55 ◇ REV MR 83j:03070 Zbl 498 # 03025 • ID 35366

MILLER, DOUGLAS E. & REMMEL, J.B. [1984] *Effectively nowhere simple sets* (**J** 0036) J Symb Logic 49*129-136
◇ D25 ◇ REV MR 85i:03140 • ID 42483

MILLER, DOUGLAS E. see Vol. III, V for further entries

MILLER, GARY L. [1979] *Graph isomorphism, general remarks* (**J** 0119) J Comp Syst Sci 18*128-142
◇ D15 ◇ REV MR 80h:68056 Zbl 403 # 03029 • ID 54743

MILLER, GEORGE A. [1958] see CHOMSKY, N.

MILLER, R.E. [1977] see LIPTON, R.J.

MILLER, W. [1968] see MARTIN, D.A.

MILLER, W. [1970] *Recursive function theory and numerical analysis* (**J** 0119) J Comp Syst Sci 4*465-472
◇ D20 D80 ◇ REV MR 42 # 4398 Zbl 302 # 65054 JSL 39.346 • ID 09259

MILLER III, C.F. [1968] *On Britton's theorem A* (**J** 0053) Proc Amer Math Soc 19*1151-1154
◇ D40 ◇ REV MR 38 # 244 Zbl 185.51 • ID 09249

MILLER III, C.F. [1971] *On group theoretic decision problems and their classification* (**S** 3513) Ann Math Stud 68*viii + 106pp
◇ D25 D40 ◇ REV MR 46 # 9147 Zbl 277 # 20054 • ID 09250

MILLER III, C.F. [1973] *Decision problems in algebraic classes of groups (a survey)* (**P** 0678) Word Probl: Decis & Burnside Probl in Group Th;1969 Irvine 507-523
◇ D40 ◇ REV MR 53 # 3121 Zbl 288 # 20052 JSL 41.786 • ID 21654

MILLER III, C.F. [1973] *Some connections between Hilbert's 10th problem and the theory of groups* (**P** 0678) Word Probl: Decis & Burnside Probl in Group Th;1969 Irvine 483-506
◇ D35 D40 ◇ REV MR 58 # 10385 Zbl 286 # 02048 JSL 41.786 • ID 63944

MILLER III, C.F. [1981] see BAUMSLAG, G.

MILLER III, C.F. [1981] *The word problem in quotients of a group* (**P** 2902) Aspects Effective Algeb;1979 Clayton 246-250
◇ D40 ◇ REV MR 82i:20048 Zbl 481 # 20022 • ID 38454

MILNER, R. [1969] *Program schemes and recursive function theory* (**J** 0508) Machine Intelligence 5*39-58
◇ B75 D10 D20 ◇ REV MR 47 # 8267 Zbl 221 # 68017 • ID 09269

MILNER, R. [1976] see MICHAELSON, S.

MILNER, R. [1976] *Models of LCF* (**C** 3517) Found of Comput Sci, Vol 2, Part 2 49-63
◇ B40 D20 ◇ REV MR 58 # 19304 Zbl 364 # 02018 • ID 50948

MILNER, R. see Vol. I, II, VI for further entries

MINICOZZI, E. [1975] see CRISCUOLO, G.

MINICOZZI, E. see Vol. I, II for further entries

MINSKY, M.L. [1956] *Some universal elements for finite automata* (**C** 0717) Automata Studies 117-128
◇ D05 ◇ REV MR 17.1170 JSL 35.480 • ID 09302

MINSKY, M.L. [1961] *Recursive unsolvability of Post's problem of "tag" and other topics in the theory of Turing machines* (**J** 0120) Ann of Math, Ser 2 74*437-455
◇ D03 D10 ◇ REV MR 25 # 3825 Zbl 105.8 JSL 31.654 • ID 09303

MINSKY, M.L. [1962] *Size and structure of universal Turing machines using tag systems* (**P** 0613) Rec Fct Th;1961 New York 229-238
◇ D03 D10 ◇ REV MR 26 # 21 Zbl 192.66 JSL 31.655 • ID 09304

MINSKY, M.L. [1964] see COCKE, J.

MINSKY, M.L. [1967] *Computation: finite and infinite machines* (**X** 0819) Prentice Hall: Englewood Cliffs xvii + 317pp
• TRANSL [1971] (**X** 0808) Kohlhammer: Stuttgart 383pp (German) [1971] (**X** 0885) Mir: Moskva 364pp (Russian)
◇ D05 D10 D20 ◇ REV MR 50 # 9050 Zbl 195.24 • ID 23547

MINTS, G.E. [1965] see MASLOV, S.YU.

MINTS, G.E. [1971] *Quantifier-free and one-quantifier systems (Russian) (English summary)* (**S** 0228) Zap Nauch Sem Leningrad Otd Mat Inst Steklov 20*115-133,285
• TRANSL [1973] (**J** 1531) J Sov Math 1*71-84
◇ B20 D20 F30 ◇ REV MR 44 # 6486 Zbl 222 # 02022 • ID 09286

MINTS, G.E. [1974] *Note on equation systems (Russian)* (**C** 1450) Teor Algor & Mat Logika (Markov) 124-126
◇ D20 ◇ REV MR 58 # 21552 Zbl 295 # 02022 • ID 63948

MINTS, G.E. [1975] see KREISEL, G.

MINTS, G.E. & TYUGU, E.KH. [1982] *The completeness of structural synthesis rules (Russian)* (**J** 0023) Dokl Akad Nauk SSSR 263*291-295
• TRANSL [1982] (**J** 0062) Sov Math, Dokl 25*343-346
◇ B20 B75 D20 F50 ◇ REV MR 84m:03013 Zbl 517 # 68058 • ID 35718

MINTS, G.E. see Vol. I, II, III, V, VI for further entries

MIR, H. [1976] *Two families of context-sensitive languages generated by controlled operator grammars* (**J** 3395) Libyan J Sci 6B*79-82
◇ D05 ◇ REV Zbl 403 # 68070 • ID 69012

MIRANDA, S. [1981] see CROSSLEY, J.N.

MIRKIN, B.G. [1966] *An algorithm for constructing a base in a language of regular expressions (Russian)* (**J** 0977) Izv Akad Nauk SSSR, Tekh Kibern 1966/5*113-119
• TRANSL [1966] (**J** 0522) Engin Cybern 5*110-116
◇ D05 ◇ REV JSL 36.694 • ID 09309

MIRKIN, B.G. [1966] *On a language of pseudo-regular expressions (Russian) (English summary)* (**J** 0040) Kibernetika, Akad Nauk Ukr SSR 1966/6*8-11
◇ D05 ◇ REV MR 35 # 5260 Zbl 192.80 JSL 34.658 • ID 19416

MIRKIN, B.G. [1966] *On the theory of multitape automata (Russian) (English summary)* (**J** 0040) Kibernetika, Akad Nauk Ukr SSR 1966/5*12-18
◇ D05 ◇ REV MR 36 # 1328 Zbl 149.250 • ID 09308

MIRKOWSKA, G. [1981] see MEYER, A.R.

MIRKOWSKA, G. see Vol. II, VI for further entries

MIRKOWSKA, M. [1970] see GORAJ, A.

MIRONOV, G.A. [1974] see FROLOV, G.D.

MISERCQUE, D. [1978] *Probleme des mariages et recursivite* (**J** 3824) Bull Soc Math Belg, Ser A 30*111-121
◇ D80 ◇ REV MR 84h:03108 Zbl 495 # 03031 • ID 34338

MISERCQUE, D. see Vol. I, III, VI for further entries

MITCHELL, R. [1966] *A generalization of productive sets* (**J** 0036) J Symb Logic 31*455-459
◇ D25 ◇ REV MR 36 # 4986 Zbl 192.51 • ID 09321

MITRANI, E. [1980] see DAVID, R.

MITSCHKE, G. [1972] *λ-definierbare Funktionen auf Peanoalgebren* (**J** 0009) Arch Math Logik Grundlagenforsch 15*31-35
◇ B40 D75 ◇ REV MR 48 # 3720 Zbl 252 # 02041 • ID 09330

MITSCHKE, G. [1977] see HINDLEY, J.R.

MITSCHKE, G. see Vol. III, VI for further entries

MIYANO, S. [1983] *Remarks on multihead pushdown automata and multihead stack automata* (**J** 0119) J Comp Syst Sci 27*116-124
◇ D10 ◇ REV MR 85k:68027 Zbl 516 # 68044 • ID 38590

MIYANO, S. [1984] see HAYASHI, T.

MIYANO, S. [1985] see HAYASHI, T.

MIZUMOTO, M. & TANAKA, K. & TOYODA, J. [1972] *General formulation of formal grammars* (**J** 0191) Inform Sci 4*87
◇ B65 D05 ◇ REV MR 48 # 10211 • ID 42020

MIZUMOTO, M. see Vol. II, V for further entries

MO, SHAOKUI [1955] *On the definition of primitive recursive functions (Chinese) (English summary)* (**J** 0418) Shuxue Xuebao 5*109-115
◇ D20 ◇ REV MR 17.225 JSL 25.182 • ID 09350

MO, SHAOKUI [1956] *On the explicit form of general recursive functions (Chinese) (English summary)* (**J** 0418) Shuxue Xuebao 6*548-564
◇ D20 ◇ REV MR 20 # 3070 JSL 25.183 • ID 09351

MO, SHAOKUI [1957] *A sketch on model computers (Chinese)* (**J** 2771) Kexue Tongbao 9*1-11
◇ D80 ◇ ID 48524

MO, SHAOKUI [1958] *On the construction of number-theoretic functions (Chinese)* (**J** 2804) Nanjing Daxue Xuebao, Ziran Kexue 1958/1*27-31
◇ D20 ◇ ID 47123

MO, SHAOKUI [1959] *The application of recursive functions on programming (Chinese)* (**J** 2804) Nanjing Daxue Xuebao, Ziran Kexue 1959/4*1-10
◇ D20 ◇ ID 47124

MO, SHAOKUI [1963] *On general recursive functions (Chinese)* (**J** 2804) Nanjing Daxue Xuebao, Ziran Kexue 1963/3*71-86
◇ D20 ◇ ID 47129

MO, SHAOKUI [1964] *Studies on number-theoretic operators (Chinese)* (**J** 2804) Nanjing Daxue Xuebao, Ziran Kexue 1964/8*1-23
• LAST ED [1966] (**J** 4197) Acta Sci Nat, Pars Math, Mech & Astronom 2/1*24-41
◇ D20 ◇ ID 47137

MO, SHAOKUI & SHEN, BAIYING [1965] *Systems of primitive recursive arithmetic (Chinese)* (**P** 4564) Math Logic;1963 Xi-An 99-109
◇ D20 F30 ◇ ID 48522

MO, SHAOKUI [1965] *Theory of recursive functions (Chinese)* (**X** 0740) Kexue Jishu Chubanshe: Shanghai
◇ D98 ◇ ID 47175

MO, SHAOKUI & YE, DAXING [1979] *On recursion schemas defining operators (Chinese) (English summary)* (**J** 2754) Huazhong Gongxueyuan Xuebao 7/4*7,1-9
◇ D20 ◇ REV MR 82i:03054 • ID 76486

MO, SHAOKUI [1981] *From black box to computers - an analysis of the nature of computers (Chinese)* (**J** 4430) Ziran Zazhi 4*403-408
◇ D80 ◇ ID 48526

MO, SHAOKUI [1982] *Theory of algorithms (Chinese)* (**X** 1876) Kexue Chubanshe: Beijing
◇ D20 D98 ◇ ID 47179

MO, SHAOKUI [1983] *A note on a regular language (Chinese) (English summary)* (**J** 2804) Nanjing Daxue Xuebao, Ziran Kexue 1983/4*614-615
◇ D05 ◇ REV MR 85k:68047 Zbl 542 # 68066 • ID 48523

MO, SHAOKUI see Vol. I, II, III, V, VI for further entries

MODINA, L. [1978] see DIKOVSKIJ, A.YA.

MOERDIJK, I. [1984] see HOEVEN VAN DER, G.F.

MOERDIJK, I. see Vol. III, V, VI for further entries

MOGGI, E. [1984] see LONGO, G.

MOGGI, E. [1985] see LONGO, G.

MOGGI, E. see Vol. V, VI for further entries

MOHRHERR, J. [1983] *A conjecture of Ershov for a relative hierarchy fails near $\mathfrak{O}$* (**J** 0003) Algebra i Logika 22*232-235
• TRANSL [1983] (**J** 0069) Algeb and Log 22*169-171
◇ D55 ◇ REV Zbl 544 # 03019 • ID 41006

MOHRHERR, J. [1983] *Kleene index sets and functional m-degrees* (**J** 0036) J Symb Logic 48*829-840
◇ D20 D25 D30 D45 ◇ REV MR 84j:03086 Zbl 538 # 03038 • ID 33610

MOHRHERR, J. [1984] *Density of a final segment of the truth-table degrees* (**J** 0048) Pac J Math 115*409-419
◇ D30 ◇ REV MR 86a:03043 Zbl 534 # 03018 • ID 43685

MOHRHERR, J. [1985] see JOCKUSCH JR., C.G.

MOISEEVA, G.N. [1972] *Ueber die Synthese von Automaten auf der Basis von Algorithmengraphen mit parallelen Zweigen (Russisch)* (**S** 0716) Vychisl Tekh Vopr Kibern (Univ Leningrad) 9*115-125
◇ D05 ◇ REV Zbl 253 # 94022 • ID 63973

MOISIL, G.C. [1955] *Contribution a l'etude algebrique des mecanismes automatiques (Romanian) (Russian and French summaries)* (**J** 0518) Bul Sti Mat-Fiz, Acad Romina 7*183-230
◇ B70 D05 ◇ REV MR 18.784 Zbl 67.100 JSL 37.414 • ID 19392

MOISIL, G.C. [1967] *Theorie structurelle des automates finis* (**X** 0834) Gauthier-Villars: Paris 337pp
◇ D05 ◇ REV Zbl 168.260 • ID 23550

MOISIL, G.C. [1973] *Sur la possibilite de modeler le fini par l'infini* (**J** 2293) Comp Linguist & Comp Lang 9*163-173
◇ D05 D10 ◇ REV MR 52 # 13024 • ID 82274

MOISIL, G.C. see Vol. I, II, V, VI for further entries

MOITRA, A. [1981] *Relation between algebraic specification and Turing machines* (**P** 4258) Found of Softw Tech & Th Comput Sci (1);1981 Bangalore 241-249
◇ B75 D10 ◇ REV Zbl 511 # 68029 • ID 46468

MOKATSYAN, A.A. [1977] *Some properties of relativizated Kolmogorov complexities (Russian)* (**J** 0346) Dokl Akad Nauk Armyan SSR 64*77-80
◇ D15 D30 ◇ REV MR 58 # 27394 Zbl 383 # 03026 • ID 52008

MOKATSYAN, A.A. [1982] *ω-mitotic but not btt-mitotic enumerable sets (Russian) (Armenian summary)* (**S** 0422) Tr Vychisl Tsentra Akad Nauk Armyan SSR & Univ Erevan 10*143-151
◇ D25 ◇ REV MR 85i:03145 • ID 44218

MOLDESTAD, J. & NORMANN, D. [1974] *2-envelopes and the analytic hierarchy* (**S** 1626) Oslo Preprint Ser 21
◇ D55 D65 ◇ ID 28313

MOLDESTAD, J. & NORMANN, D. [1976] *Models for recursion theory* (**J** 0036) J Symb Logic 41*719-729
◇ D65 D75 ◇ REV MR 56 # 106 Zbl 352 # 02034 • ID 14573

MOLDESTAD, J. [1977] *Computations in higher types* (**S** 3301) Lect Notes Math 574*iv+203pp
◇ D65 D75 ◇ REV Zbl 352 # 02033 • ID 28298

MOLDESTAD, J. [1978] *On the role of the successor function in recursion theory* (**P** 1628) Generalized Recursion Th (2);1977 Oslo 283-301
◇ D65 D75 ◇ REV MR 80g:03046 Zbl 453 # 03047 • ID 76498

MOLDESTAD, J. & STOLTENBERG-HANSEN, V. & TUCKER, J.V. [1980] *Finite algorithmic procedures and inductive definability* (**J** 0132) Math Scand 46*62-76
◇ D10 D70 D75 ◇ REV MR 82d:03077a Zbl 448 # 03037 • ID 56637

MOLDESTAD, J. & STOLTENBERG-HANSEN, V. & TUCKER, J.V. [1980] *Finite algorithmic procedures and computation theories* (**J** 0132) Math Scand 46*77-94
◇ D10 D75 ◇ REV MR 82d:03077b Zbl 419 # 68078 • ID 30577

MOLDESTAD, J. see Vol. III for further entries

MOLL, K.R. [1980] *Left context precedence grammars* (**J** 1431) Acta Inf 14*317-335
◇ D05 ◇ REV MR 82j:68079 Zbl 431 # 68081 • ID 82276

MOLL, R. [1972] see MEYER, A.R.

MOLL, R. [1974] see MEYER, A.R.

MOLL, R. [1976] *An operator embedding theorem for complexity classes of recursive functions* (**J** 1426) Theor Comput Sci 1*193-198
◇ D15 ◇ REV MR 54 # 77 Zbl 326 # 68031 • ID 23958

MOLL, R. [1982] see ARBIB, M.A.

MONGY, J. [1979] see DAUCHET, M.

MONIEN, B. [1973] *Relationships between pushdown automata and tape-bounded Turing machines* (**P** 0763) Automata, Lang & Progr (1);1972 Rocquencourt 575-583
◇ D10 D15 ◇ REV MR 51 # 14647 Zbl 268 # 68032 • ID 63989

MONIEN, B. [1974] *Beschreibung von Zeitkomplexitaetsklassen bei Turingmaschinen durch andere Automatenmodelle* (**J** 0129) Elektr Informationsverarbeitung & Kybern 10*37-51
◇ D10 D15 ◇ REV MR 50 # 3646 Zbl 286 # 68026 • ID 63987

MONIEN, B. [1975] *About the deterministic simulation of nondeterministic* (log *n*)*-tape bounded Turing machines* (**P** 1449) Automata Th & Formal Lang;1975 Kaiserslautern 118-126
◇ D10 D15 ◇ REV MR 55 # 1813 Zbl 341 # 68032 • ID 63988

MONIEN, B. [1976] *Transformational methods and their application to complexity problems* (**J** 1431) Acta Inf 6*95-108 • ERR/ADD ibid 8*383-384
◇ D10 D15 ◇ REV MR 53 # 7124 MR 57 # 8154 Zbl 329 # 02015 • ID 63986

MONIEN, B. [1977] *A recursive and a grammatical characterization of the exponential-time languages* (**J** 1426) Theor Comput Sci 3*61-74
◇ D05 D10 D15 ◇ REV MR 55 # 2545 Zbl 355 # 68056 • ID 76503

MONIEN, B. [1977] *The LBA-problem and the transformability of the class* $\mathfrak{E}^2$ (**P** 3411) Theor Comput Sci (3);1977 Darmstadt 339-350
◇ D05 D15 ◇ REV MR 57 # 14599 Zbl 365 # 68046 • ID 51076

MONIEN, B. [1977] see JANSSEN, D.

MONIEN, B. & SUDBOROUGH, I.H. [1980] *The interface between language theory and complexity theory* (**P** 4266) Form Lang Th;1979 Santa Barbara 287-324
◇ D05 D15 ◇ REV MR 84j:68001 Zbl 545 # 68065 • ID 47205

MONIEN, B. [1980] *Two-way multihead automata over a one-letter alphabet* (**J** 3441) RAIRO Inform Theor 14*67-82
◇ D05 ◇ REV MR 81i:68077 Zbl 442 # 68039 • ID 82278

MONIEN, B. & SUDBOROUGH, I.H. [1981] *Bounding the bandwidth of NP-complete problems* (**P** 3575) Graphth Konzepte Inf (6);1980 Bad Honnef 279-292
◇ D15 ◇ REV MR 82h:68067 Zbl 454 # 68072 • ID 82279

MONIEN, B. [1981] *On the LBA problem* (**P** 3165) FCT'81 Fund of Comput Th;1981 Szeged 265-280
◇ D05 ◇ REV MR 83g:68064 Zbl 474 # 68064 • ID 39147

MONIEN, B. & SUDBOROUGH, I.H. [1981] *Time and space bounded complexity classes and bandwidth constrained problems* (**P** 3429) Math Founds of Comput Sci (10);1981 Strbske Pleso 78-93
◇ D15 ◇ ID 46199

MONIEN, B. & SUDBOROUGH, I.H. [1982] *On eliminating nondeterminism from Turing machines which use less than logarithm worktape space* (**J** 1426) Theor Comput Sci 21*237-253
◇ D10 D15 ◇ REV MR 84e:68029 Zbl 493 # 68046 • ID 39300

MONIEN, B. [1984] *Deterministic two-way one-head pushdown automata are very powerful* (**J** 0232) Inform Process Lett 18*239-242
◇ D10 ◇ REV MR 85j:68034 Zbl 549 # 68041 • ID 46793

MONIER, L. [1983] see CHAZELLE, B.

MONK, L.G. [1975] *Elementary-recursive decision procedures* (0000) Diss., Habil. etc 89pp
◇ B25 C60 D20 ◇ REM Univ. Berkeley • ID 90301

MONTAGNA, F. [1980] *The undecidability of the first-order theory of diagonalizable algebras* (J 0063) Studia Logica 39*355-359
◇ D35 G25 ◇ REV MR 82i:03077b Zbl 463 # 03016 • ID 54556

MONTAGNA, F. [1982] *Relatively precomplete numerations and arithmetic* (J 0122) J Philos Logic 11*419-430
◇ D45 F30 ◇ REV MR 84f:03052 Zbl 498 # 03046 • ID 34471

MONTAGNA, F. & SORBI, A. [1985] *Universal recursion theoretic properties of r.e. preordered structures* (J 0036) J Symb Logic 50*397-406
◇ C57 D45 G10 ◇ ID 41807

MONTAGNA, F. see Vol. II, III, V, VI for further entries

MONTAGUE, R. & TARSKI, A. [1957] *Independent recursive axiomatizability* (P 1675) Summer Inst Symb Log;1957 Ithaca 270
◇ B30 D25 ◇ ID 29361

MONTAGUE, R. [1960] *Towards a general theory of computability* (J 0154) Synthese 12*429-438
◇ D10 D75 ◇ REV MR 24 # A1213 JSL 29.63 • ID 09419

MONTAGUE, R. [1968] *Recursion theory as a branch of model theory* (P 0627) Int Congr Log, Meth & Phil of Sci (3,Proc);1967 Amsterdam 63-86
◇ C40 D75 ◇ REV MR 42 # 7505 Zbl 247 # 02040 JSL 38.158 • ID 09432

MONTAGUE, R. [1970] *Universal grammar* (J 0105) Theoria (Lund) 36*373-398
• REPR [1974] (C 4062) Montague: Formal Philos 222-246
◇ A05 B65 D05 ◇ REV MR 46 # 6988 Zbl 243 # 02002 JSL 47.210 • ID 09433

MONTAGUE, R. see Vol. I, II, III, V, VI for further entries

MOON, B.A. [1973] *A Markov algorithm interpreter* (J 2734) Int J Math Educ Sci Technol 4*205-220
◇ D03 D15 ◇ REV Zbl 262 # 68016 • ID 64018

MOORE, D.J. [1978] see CASE, J.

MOORE, D.J. [1981] see KO, KER-I

MOORE, D.J. [1981] see LEGGETT JR., E.W.

MOORE, E.F. [1952] *A simplified universal Turing Machine* (P 0700) ACM Proc Conf;1952 Pittsburgh 50-55
◇ D10 ◇ REV MR 16.633 JSL 19.57 • ID 42246

MOORE, E.F. [1956] see LEEUW DE, K.

MOORE, E.F. [1956] *Gedanken-experiments on sequential machines* (C 0717) Automata Studies 129-153
• TRANSL [1974] (C 1902) Stud Th Automaten 151-179
◇ D05 ◇ REV MR 17.1140 JSL 23.60 • ID 09462

MOORE, E.F. [1962] see KAHR, A.S.

MOORE, E.F. (ED.) [1964] *Sequential machines: selected papers* (X 0832) Addison-Wesley: Reading iv+266pp
◇ D05 D97 ◇ REV Zbl 147.241 • ID 23552

MOORE, J.S. [1979] see BOYER, R.S.

MOORE, J.S. [1979] *A mechanical proof of the termination of Takeuchi's function* (J 0232) Inform Process Lett 9*176-181
◇ B35 D20 ◇ REV MR 81f:68113 Zbl 411 # 68079 • ID 52930

MOORE, J.S. [1984] see BOYER, R.S.

MOORE, J.S. see Vol. I for further entries

MORAGA, C. [1978] *Comments on a method of Karpovsky* (J 0194) Inform & Control 39*243-246
◇ B70 D15 ◇ REV MR 80d:94042 Zbl 388 # 94023 • ID 52289

MORAGA, C. see Vol. II for further entries

MORALES-LUNA, G. [1985] see ADAMOWICZ, Z.

MORAN, G. [1970] see MAGIDOR, M.

MORAN, G. see Vol. V for further entries

MORAN, S. & PAZ, A. [1981] *Non deterministic polynomial optimization problems and their approximations* (J 1426) Theor Comput Sci 15*251-277
◇ D15 ◇ REV MR 82i:68029 Zbl 459 # 68015 • ID 69765

MORAN, S. & PAZ, A. [1981] *Nondeterministic polynomial optimization problems and their approximation* (P 4214) Anal & Design of Alg in Combin Optmiz;1979 Udine 1-35
◇ D15 ◇ REV MR 82i:68029 Zbl 459 # 68015 • ID 46175

MORAN, S. [1981] *Some results on relativized deterministic and nondeterministic time hierarchies* (J 0119) J Comp Syst Sci 22*1-8
◇ D15 ◇ REV MR 83b:68059 Zbl 482 # 68040 • ID 38974

MORAN, S. [1982] see IBARRA, O.H.

MORAN, S. [1983] see IBARRA, O.H.

MORAVEK, J. [1973] *Computational optimality of a dynamic programming method* (P 1448) Math Founds of Comput Sci (2);1973 Strbske Pleso 267-270
◇ B40 D15 ◇ REV MR 53 # 2023 • ID 82295

MORGAN, C.G. [1976] *Some undecidability results for construction problems in tessellation automata* (J 2695) Cybernetica 19*133-139
◇ D05 ◇ REV MR 54 # 4867 Zbl 329 # 94027 • ID 31882

MORGAN, C.G. see Vol. I, II, III, VI for further entries

MORGENSTERN, J. [1973] *Algorithmes lineaires tangents et complexite* (J 2313) C R Acad Sci, Paris, Ser A-B 277*A367-A369
◇ D15 ◇ REV MR 48 # 10198 Zbl 266 # 68021 • ID 64033

MORITA, K. & SUGATA, K. & UMEO, H. [1977] *Computational complexity of L(m, n) tape-bounded two-dimensional Turing machines (Japanese)* (J 0979) Denshi Tsushin Gakkai Ronbunshi, Sect A-D 60-D/11*982-989
• TRANSL [1977] (J 0464) Syst-Comp-Controls 8/6*17-24
◇ D10 D15 ◇ REV MR 80f:68060 • ID 82301

MORITA, K. & SUGATA, K. & UMEO, H. [1982] *Deterministic one-way simulation of two-way real-time cellular automata and its related problems* (J 0232) Inform Process Lett 14*158-161
◇ D05 D15 ◇ REV MR 83h:68077 Zbl 488 # 68041 • ID 39219

MORIYA, T. [1982] *Pebble machines and tree walking machines* (J 0191) Inform Sci 27*99-119
◇ D05 ◇ REV MR 84b:68064 Zbl 496 # 68042 • ID 38952

MORLEY, M.D. & SOARE, R.I. [1975] *Boolean algebras, splitting theorems and Δ_2^0 sets* (**J** 0027) Fund Math 90*45-52
◇ D25 D55 G05 ◇ REV MR 53 # 2658 Zbl 326 # 02030
• ID 09499

MORLEY, M.D. [1976] *Decidable models* (**J** 0029) Israel J Math 25*233-240
◇ C50 C57 ◇ REV MR 56 # 15405 Zbl 361 # 02067
• ID 26084

MORLEY, M.D. [1985] see HARRINGTON, L.A.

MORLEY, M.D. see Vol. III, V for further entries

MOROZOV, A.S. [1982] *Countable homogeneous boolean algebras (Russian)* (**J** 0003) Algebra i Logika 21*269-282
• TRANSL [1982] (**J** 0069) Algeb and Log 21*181-190
◇ C15 C50 C57 G05 ◇ REV MR 85a:06023 Zbl 539 # 03011 • ID 39816

MOROZOV, A.S. [1982] *Decidability of theories of Boolean algebras with a distinguished ideal (Russian)* (**J** 0092) Sib Mat Zh 23/1*199-201,223
◇ D35 G05 ◇ REV MR 84i:03086 Zbl 499 # 03030
• ID 34563

MOROZOV, A.S. [1982] *Strong constructivizability of countable saturated boolean algebras (Russian)* (**J** 0003) Algebra i Logika 21*193-203
• TRANSL [1982] (**J** 0069) Algeb and Log 21*130-137
◇ C50 C57 G05 ◇ REV MR 85b:03052 Zbl 526 # 03016
• ID 38169

MOROZOV, A.S. [1983] *Groups of recursive automorphisms of constructive boolean algebras (Russian)* (**J** 0003) Algebra i Logika 22*138-158
• TRANSL [1983] (**J** 0069) Algeb and Log 22*95-112
◇ C07 C57 D45 F60 G05 ◇ REV Zbl 549 # 03031
• ID 43123

MOROZOV, A.S. [1984] *Group $Aut_r(Q, \leqslant)$ is not constructivizable (Russian)* (**J** 0087) Mat Zametki (Akad Nauk SSSR) 36*473-478
• TRANSL [1984] (**J** 1044) Math Notes, Acad Sci USSR 36*733-736
◇ C07 C57 F60 ◇ REV MR 86g:20001 Zbl 574 # 03028
• ID 41757

MOROZOV, A.S. [1985] *Automorphisms of constructivizations of Boolean algebras (Russian)* (**J** 0092) Sib Mat Zh 26/4*98-110,204
• TRANSL [1985] (**J** 0475) Sib Math J 26*555-565
◇ D45 G05 ◇ ID 48610

MOROZOV, A.S. [1985] *Constructive Boolean algebras with almost identical automorphisms (Russian)* (**J** 0087) Mat Zametki (Akad Nauk SSSR) 37*478-482,599
• TRANSL [1985] (**J** 1044) Math Notes, Acad Sci USSR 37*266-268
◇ C07 C57 G05 ◇ ID 46375

MORRIS, P.H. [1974] see GILL III, J.T.

MORRIS, P.H. [1976] *A reducibility condition for recursiveness* (**J** 0053) Proc Amer Math Soc 60*270-272
◇ D20 D30 ◇ REV MR 54 # 4943 Zbl 358 # 02049
• ID 24722

MORRIS JR., J.H. [1971] *Another recursion induction principle* (**J** 0212) ACM Commun 14*351-354
◇ B75 D20 ◇ REV MR 45 # 57 Zbl 226 # 68026 • ID 09502

MORSCHER, E. [1978] *Inwiefern betreffen Fragen der Vollstaendigkeit und Entscheidbarkeit den Juristen* (**P** 1850) Strukt & Entsch des Rechts;1976 Salzburg 99-118
◇ A05 D80 ◇ ID 32393

MORSCHER, E. see Vol. I, II for further entries

MOSCHOVAKIS, Y.N. [1964] *Recursive metric spaces* (**J** 0027) Fund Math 55*215-238
◇ C57 C65 D45 F60 ◇ REV MR 32 # 45 Zbl 221 # 02015 JSL 31.651 • ID 09514

MOSCHOVAKIS, Y.N. [1966] *Many-one degrees of the predicates $H_a(x)$* (**J** 0048) Pac J Math 18*329-342
◇ D30 D55 ◇ REV MR 37 # 1247 Zbl 147.252 JSL 32.529
• ID 09515

MOSCHOVAKIS, Y.N. [1966] *Notation systems and recursive ordered fields* (**J** 0020) Compos Math 17*40-71
◇ C57 C60 D45 F60 ◇ REV MR 31 # 5798 Zbl 143.13 JSL 31.650 • ID 21265

MOSCHOVAKIS, Y.N. [1967] *Hyperanalytic predicates* (**J** 0064) Trans Amer Math Soc 129*249-282
◇ D55 D65 ◇ REV MR 38 # 4308 Zbl 159.11 JSL 36.147
• ID 09516

MOSCHOVAKIS, Y.N. [1968] see ADDISON, J.W.

MOSCHOVAKIS, Y.N. [1969] *Abstract computability and invariant definability* (**J** 0036) J Symb Logic 34*605-633
◇ C40 D70 D75 ◇ REV MR 42 # 5791 Zbl 218 # 02039
• ID 09517

MOSCHOVAKIS, Y.N. [1969] *Abstract first order computability I,II* (**J** 0064) Trans Amer Math Soc 138*427-464,465-504
◇ D65 D70 D75 ◇ REV MR 39 # 5362 Zbl 218 # 02038 JSL 37.758 • ID 19462

MOSCHOVAKIS, Y.N. [1970] *Determinacy and prewellorderings of the continuum* (**P** 1072) Math Log & Founds of Set Th;1968 Jerusalem 24-62
◇ D55 E15 E25 E60 ◇ REV MR 43 # 6082 Zbl 209.15
• ID 22217

MOSCHOVAKIS, Y.N. [1970] see CHANG, C.C.

MOSCHOVAKIS, Y.N. [1970] *The Suslin-Kleene theorem for countable structures* (**J** 0025) Duke Math J 37*341-352
◇ C15 C40 D55 D70 D75 ◇ REV MR 42 # 7509 Zbl 207.12 • ID 09518

MOSCHOVAKIS, Y.N. [1971] *Axioms for computation theories - first draft* (**P** 0638) Logic Colloq;1969 Manchester 199-255
◇ D75 ◇ REV MR 43 # 7325 Zbl 243 # 02034 • ID 22216

MOSCHOVAKIS, Y.N. [1971] see HINMAN, P.G.

MOSCHOVAKIS, Y.N. [1971] *Predicative classes* (**P** 0693) Axiomatic Set Th;1967 Los Angeles 1*247-264
◇ C62 D55 D60 D65 D70 E70 ◇ REV MR 43 # 7314 Zbl 234 # 02029 JSL 40.506 • ID 09520

MOSCHOVAKIS, Y.N. [1971] see BARWISE, J.

MOSCHOVAKIS, Y.N. [1971] *Uniformization in a playful universe* (**J** 0015) Bull Amer Math Soc 77*731-736
◇ D55 E15 E60 ◇ REV MR 44 # 2609 Zbl 232 # 04002
• ID 09519

MOSCHOVAKIS, Y.N. [1972] *The game quantifier* (**J** 0053) Proc Amer Math Soc 31*245-250
◇ C75 C80 D55 D70 D75 E60 ◇ REV MR 44 # 3871 Zbl 243 # 02035 JSL 38.653 • ID 09521

MOSCHOVAKIS, Y.N. [1972] see KECHRIS, A.S.

MOSCHOVAKIS, Y.N. [1973] *Analytical definability in a playful universe* (**P** 0793) Int Congr Log, Meth & Phil of Sci (4,Proc);1971 Bucharest 77-85
◇ D55 E15 E60 ◇ REV MR 58 # 27486 • ID 26209

MOSCHOVAKIS, Y.N. [1974] *Elementary induction on abstract structures* (**X** 0809) North Holland: Amsterdam x + 218pp
◇ D55 D70 D75 ◇ REV MR 53 # 2661 Zbl 307 # 02003 JSL 44.124 • ID 21540

MOSCHOVAKIS, Y.N. [1974] *On nonmonotone inductive definability* (**J** 0027) Fund Math 82*39-83
◇ D55 D70 D75 ◇ REV MR 50 # 6853 Zbl 306 # 02039 • ID 09522

MOSCHOVAKIS, Y.N. [1974] *Structural characterizations of classes of relations* (**P** 0602) Generalized Recursion Th (1);1972 Oslo 53-79
◇ D65 D70 D75 ◇ REV MR 54 # 82 Zbl 295 # 02026 • ID 24569

MOSCHOVAKIS, Y.N. [1975] *New methods and results in descriptive set theory* (**P** 1521) Int Congr Math (II,12);1974 Vancouver 1*251-257
◇ D55 E15 E55 E60 ◇ REV MR 56 # 2827 Zbl 344 # 02052 • ID 64058

MOSCHOVAKIS, Y.N. [1977] see BARWISE, J.

MOSCHOVAKIS, Y.N. [1977] *On the basic notions in the theory of induction* (**P** 1704) Int Congr Log, Meth & Phil of Sci (5);1975 London ON 1*207-236
◇ D65 D70 D75 ◇ REV MR 58 # 10364 Zbl 375 # 02039 • ID 51616

MOSCHOVAKIS, Y.N. [1977] see KECHRIS, A.S.

MOSCHOVAKIS, Y.N. [1978] see KECHRIS, A.S.

MOSCHOVAKIS, Y.N. [1978] see BARWISE, J.

MOSCHOVAKIS, Y.N. [1978] *Inductive scales on inductive sets* (**C** 2908) Cabal Seminar Los Angeles 1976-77 185-192
◇ D55 D70 D75 E15 E60 ◇ REV MR 81m:03059 Zbl 398 # 03036 • ID 52764

MOSCHOVAKIS, Y.N. [1980] *Descriptive set theory* (**X** 0809) North Holland: Amsterdam xii + 637pp
◇ D55 D98 E15 E98 ◇ REV MR 82e:03002 Zbl 433 # 03025 JSL 46.874 • ID 76581

MOSCHOVAKIS, Y.N. [1981] see KECHRIS, A.S.

MOSCHOVAKIS, Y.N. [1981] *On the Grilliot-Harrington-MacQueen theorem* (**P** 2628) Log Year;1979/80 Storrs 246-267
◇ D65 ◇ REV MR 84b:03063 Zbl 572 # 03026 • ID 35648

MOSCHOVAKIS, Y.N. [1981] *Ordinal games and playful models* (**C** 2922) Cabal Seminar Los Angeles 1977-79 169-201
◇ D55 E15 E45 E60 ◇ REV MR 84h:03115 Zbl 473 # 03040 • ID 55369

MOSCHOVAKIS, Y.N. [1982] see MARTIN, D.A.

MOSCHOVAKIS, Y.N. [1983] see KECHRIS, A.S.

MOSCHOVAKIS, Y.N. [1983] *Scales on coinductive sets* (**C** 3875) Cabal Seminar Los Angeles 1979-81 77-85
◇ D70 E15 E60 ◇ REV Zbl 549 # 03047 • ID 43130

MOSCHOVAKIS, Y.N. [1984] *Abstract recursion as a foundation for the theory of algorithms* (**P** 2153) Logic Colloq;1983 Aachen 2*289-362
◇ D20 D75 ◇ ID 43010

MOSCHOVAKIS, Y.N. see Vol. V for further entries

MOSES, M. [1983] *Recursive properties of isomorphism types* (**J** 3194) J Austral Math Soc, Ser A 34*269-286
◇ C57 D45 ◇ REV MR 84g:03067 Zbl 532 # 03022 Zbl 546 # 03025 • ID 34174

MOSES, M. [1984] *Recursive linear orders with recursive successivities* (**J** 0073) Ann Pure Appl Logic 27*253-264
◇ C57 D45 E07 ◇ REV Zbl 572 # 03025 • ID 39682

MOSES, M. [1984] *Recursive properties of isomorphism types* (**J** 0016) Bull Austral Math Soc 29*419-421
◇ C57 D45 ◇ REV Zbl 532 # 03022 • ID 46602

MOSHCHENSKIJ, V.A. [1969] *On the question of the complexity of Turing computations (Russian)* (**J** 0414) Dokl Akad Nauk Belor SSR 13*876-878 • ERR/ADD ibid 14*88
◇ D10 D15 D20 ◇ REV MR 40 # 7114 Zbl 251 # 68028 • ID 09508

MOSHCHENSKIJ, V.A. [1971] *A certain estimate of the state sequences of Turing machines (Russian) (English summary)* (**J** 0040) Kibernetika, Akad Nauk Ukr SSR 1971/4*39-43
◇ D10 ◇ REV MR 46 # 8781 Zbl 248 # 02037 • ID 09509

MOSHCHENSKIJ, V.A. [1971] *The estimation of certain functions that characterize the performance of Turing machines (Russian) (English summary)* (**J** 0040) Kibernetika, Akad Nauk Ukr SSR 1971/1*34-40
◇ D10 D15 ◇ REV MR 46 # 4781 Zbl 257 # 02029 • ID 28999

MOSHCHENSKIJ, V.A. [1972] *Zur Analyse von Turing-berechnungen (Russian)* (**J** 0413) Izv Akad Nauk Belor SSR, Ser Fiz-Mat 1972/2*47-55
◇ D10 D15 ◇ REV MR 46 # 6649 Zbl 289 # 02019 • ID 29984

MOSHCHENSKIJ, V.A. [1973] *Lectures on mathematical logic (Russian)* (**X** 1212) Belor Gos Univ: Minsk 159pp
◇ B98 D10 D20 ◇ REV MR 50 # 6775 Zbl 275 # 02001 • ID 21271

MOSHCHENSKIJ, V.A. [1977] *A method of analysis for Turing computations. I (Russian) (English summary)* (**J** 0040) Kibernetika, Akad Nauk Ukr SSR 1977/3*86-94
• TRANSL [1977] (**J** 0021) Cybernetics 13*402-412
◇ D10 D15 ◇ REV MR 57 # 14600 Zbl 356 # 02034 • ID 50306

MOSHCHENSKIJ, V.A. [1980] *Turing a-computations and essential complexity of a-inversions of binary words (Russian)* (**J** 0071) Met Diskr Analiz (Novosibirsk) 35*83-92,105
◇ D10 ◇ REV MR 82j:68023 Zbl 487 # 68039 • ID 82302

MOSHCHENSKIJ, V.A. see Vol. I, II, V for further entries

MOSS, B.P. [1976] see HOPKIN, D.R.

MOSS, B.P. see Vol. II, V for further entries

MOSTOWSKI, A.WLODZIMIERZ [1966] *Computational algorithms for deciding some problems for nilpotent groups* (**J** 0027) Fund Math 59*137-152
◇ D40 ◇ REV MR 37 # 293 Zbl 143.37 JSL 35.476 • ID 09580

MOSTOWSKI, A.WLODZIMIERZ [1966] *On the decidability of some problems in special classes of groups* (**J** 0027) Fund Math 59*123-135
◇ D40 ◇ REV MR 37#292 Zbl 143.37 JSL 35.476 • ID 09581

MOSTOWSKI, A.WLODZIMIERZ [1973] *Uniform algorithm for deciding group-theoretic problems* (**P** 0678) Word Probl: Decis & Burnside Probl in Group Th;1969 Irvine 525-551
◇ D30 D40 ◇ REV MR 53#7750 Zbl 269#02017 JSL 41.786 • ID 22997

MOSTOWSKI, A.WLODZIMIERZ [1978] *Recursively enumerable degrees of programming problems* (**J** 2095) Fund Inform, Ann Soc Math Pol, Ser 4 1*365-377
◇ B75 D03 D25 ◇ REV MR 58#5138 Zbl 386#68053 • ID 69730

MOSTOWSKI, A.WLODZIMIERZ [1979] *A note concerning the complexity of a decision problem for positive formulas in SkS* (**P** 2952) CAAP'79 Arbres en Algeb & Progr (4);1979 Lille 173-180
◇ B15 B25 D15 ◇ REV MR 80m:03028 • ID 76601

MOSTOWSKI, A.WLODZIMIERZ [1980] *Finite automata on infinite trees and subtheories of SkS* (**P** 3057) CAAP'80 Arbres en Algeb & Progr (5);1980 Lille 228-240
◇ B25 C85 D05 ◇ REV MR 83e:03061 Zbl 489#03012 • ID 34829

MOSTOWSKI, A.WLODZIMIERZ [1980] *Nearly deterministic automata acceptation of infinite trees and a complexity of weak theory of SkS* (**P** 3057) CAAP'80 Arbres en Algeb & Progr (5);1980 Lille 54-62
◇ D05 ◇ REV Zbl 489#03011 • ID 37205

MOSTOWSKI, A.WLODZIMIERZ [1981] *The complexity of automata and subtheories of monadic second order arithmetics* (**P** 3165) FCT'81 Fund of Comput Th;1981 Szeged 453-466
◇ B15 B25 D05 D15 F35 ◇ REV MR 83g:03036 Zbl 469#03025 • ID 55152

MOSTOWSKI, A.WLODZIMIERZ [1985] *Regular expressions for infinite trees and a standard form of automata* (**P** 4670) Comput Th (5);1984 Zaborow 157-168
◇ D05 ◇ ID 49639

MOSTOWSKI, A.WLODZIMIERZ see Vol. I, II for further entries

MOSTOWSKI, ANDRZEJ [1947] *On definable sets of positive integers* (**J** 0027) Fund Math 34*81-112
◇ D55 F30 ◇ REV MR 9.129 Zbl 31.194 JSL 13.112 • ID 09537

MOSTOWSKI, ANDRZEJ [1948] *On a set of integers not definable by means of one-quantifier predicates* (**J** 0283) Ann Soc Pol Math 21*114-119
◇ D55 ◇ REV MR 10.175 Zbl 31.339 JSL 15.135 • ID 09540

MOSTOWSKI, ANDRZEJ [1949] *A classification of logical systems* (**J** 4716) Stud Philos, Leopolis (Poznan) 4*237-274
◇ D55 F30 ◇ REV JSL 17.274 • ID 09547

MOSTOWSKI, ANDRZEJ [1953] *A lemma concerning recursive functions and its applications* (**J** 0014) Bull Acad Pol Sci, Ser Math Astron Phys 1*277-280
◇ D20 ◇ REV MR 15.667 Zbl 52.12 JSL 19.299 • ID 09554

MOSTOWSKI, ANDRZEJ [1953] *On a system of axioms which has no recursively enumerable arithmetic model* (**J** 0027) Fund Math 40*56-61
◇ C57 D45 E30 E70 ◇ REV MR 15.667 Zbl 53.3 JSL 23.45 • ID 09553

MOSTOWSKI, ANDRZEJ & ROBINSON, R.M. & TARSKI, A. [1953] *Undecidability and essential undecidability in arithmetic* (**C** 4472) Undecidable Th 39-74
◇ D35 F30 ◇ REV Zbl 53.4 JSL 24.167 • ID 42176

MOSTOWSKI, ANDRZEJ & ROBINSON, R.M. & TARSKI, A. [1953] *Undecidable theories* (**X** 0809) North Holland: Amsterdam xi+98pp
◇ D35 F25 F30 ◇ REV MR 15.384 Zbl 53.4 • REM 2nd ed. 1968; 3rd ed. 1971 • ID 28815

MOSTOWSKI, ANDRZEJ [1955] *A formula with no recursively enumerable model* (**J** 0027) Fund Math 42*125-140
◇ C57 D45 ◇ REV MR 17.225 Zbl 67.251 JSL 23.45 • ID 09558

MOSTOWSKI, ANDRZEJ [1955] *Contributions to the theory of definable sets and functions* (**J** 0027) Fund Math 42*271-275
◇ D55 ◇ REV MR 17.816 Zbl 67.252 JSL 25.365 • ID 09556

MOSTOWSKI, ANDRZEJ [1955] *Examples of sets definable by means of two and three quantifiers* (**J** 0027) Fund Math 42*259-270
◇ D55 ◇ REV MR 17.815 Zbl 66.262 JSL 25.364 • ID 09557

MOSTOWSKI, ANDRZEJ [1956] *Concerning a problem of H.Scholz* (**J** 0068) Z Math Logik Grundlagen Math 2*210-214
◇ C13 D20 ◇ REV MR 19.240 Zbl 74.249 JSL 24.241 • ID 09561

MOSTOWSKI, ANDRZEJ [1956] *Development and applications of the "projective" classification of sets of integers* (**P** 0575) Int Congr Math (II, 7);1954 Amsterdam 3*280-288
◇ C57 C80 D55 E15 E45 ◇ REV MR 19.238 Zbl 75.235 JSL 23.44 • ID 09560

MOSTOWSKI, ANDRZEJ [1957] *On computable sequences* (**J** 0027) Fund Math 44*37-51
◇ D20 F60 ◇ REV MR 19.934 Zbl 79.247 JSL 25.367 • ID 09564

MOSTOWSKI, ANDRZEJ [1957] *On recursive models of formalised arithmetic* (**J** 0014) Bull Acad Pol Sci, Ser Math Astron Phys 5*705-710
◇ C57 C62 D45 ◇ REV MR 20#7 Zbl 81.12 JSL 23.45 • ID 09562

MOSTOWSKI, ANDRZEJ [1958] see GRZEGORCZYK, A.

MOSTOWSKI, ANDRZEJ [1959] *On various degrees of constructivism* (**P** 0634) Constructivity in Math;1957 Amsterdam 178-194
◇ D55 F50 F60 F98 ◇ REV MR 22#3688 Zbl 88.250 JSL 35.575 • ID 09568

MOSTOWSKI, ANDRZEJ [1961] see GRZEGORCZYK, A.

MOSTOWSKI, ANDRZEJ [1962] *L'espace des modeles d'une theorie formalisee et quelques-unes de ses applications* (**P** 1606) Colloq Math (Pascal);1962 Clermont-Ferrand 7*107-116
◇ B50 C07 C40 C57 C80 C85 C90 D45 ◇ REV MR 47#4785 JSL 40.501 • ID 09574

MOSTOWSKI, ANDRZEJ [1962] *Representability of sets in formal systems* (**P** 0613) Rec Fct Th;1961 New York V*29-48
◇ D55 E15 F30 F35 ◇ REV MR 26#27 Zbl 158.8
• ID 33705

MOSTOWSKI, ANDRZEJ [1965] *Thirty years of foundational studies. Lectures on the development of mathematical logic and the studies of the foundations of mathematics in 1930-1964* (**J** 0096) Acta Philos Fenn 17*1-180
• REPR [1966] (**X** 1096) Blackwell: Oxford 180pp
◇ A10 B98 C98 D98 E98 F98 ◇ REV MR 33#18 MR 33#5445 Zbl 146.245 JSL 33.111 • ID 09578

MOSTOWSKI, ANDRZEJ [1972] *A transfinite sequence of ω-models* (**J** 0036) J Symb Logic 37*96-102
◇ C62 D30 ◇ REV MR 48#3721 Zbl 246#02038
• ID 09588

MOSTOWSKI, ANDRZEJ [1979] *Foundational studies. Selected works Vol. I,II* (**X** 0809) North Holland: Amsterdam xlvi+635pp,viii+605pp
◇ B96 C96 D96 E96 ◇ REV MR 81i:01018 Zbl 425#01021 • ID 82308

MOSTOWSKI, ANDRZEJ see Vol. I, II, III, V, VI for further entries

MOSTOWSKI, T. [1976] *Analytic applications of decidability theorems* (**J** 0519) Wiad Mat, Ann Soc Math Pol, Ser 2 20*1-6
◇ B25 C60 C65 D80 ◇ REV MR 56#5518 • ID 82309

MOSTOWSKI, T. see Vol. III for further entries

MOSZNER, Z. [1967] *On the theory of relations (Polish)* (**X** 2881) Wydawn Szkol Ped: Warsaw 207pp
◇ D03 E07 E98 ◇ REV MR 38#2028 • REM 2nd ed. 1974
• ID 76611

MOSZNER, Z. [1973] *Structure de l'automate plein, reduit et inversible* (**J** 0529) Aequationes Math 9*46-59
◇ D05 ◇ REV MR 47#3585 Zbl 263#94016 • ID 09592

MOSZNER, Z. see Vol. V for further entries

MOTCHANE, L. [1956] *Sur un nouveau critere de conservation de classe de Baire* (**J** 0109) C R Acad Sci, Paris 242*605-608
◇ D55 E15 ◇ REV MR 17.1063 Zbl 70.56 • ID 09593

MOTCHANE, L. see Vol. V for further entries

MOURA, A. [1982] see BADER, C.

MROWKA, S. [1957] *Recursive families of sets* (**J** 0027) Fund Math 44*186-191
◇ D20 ◇ REV MR 20#4490 Zbl 80.243 • ID 09608

MROWKA, S. see Vol. V for further entries

MUCHNICK, S.S. [1972] see CONSTABLE, R.L.

MUCHNICK, S.S. [1976] *Computational complexity of multiple recursive schemata* (**J** 1428) SIAM J Comp 5*427-451
◇ D15 D20 ◇ REV MR 56#8352 Zbl 342#68026
• ID 64100

MUCHNICK, S.S. [1976] *The vectorized Grzegorczyk hierarchy* (**J** 0068) Z Math Logik Grundlagen Math 22*441-480
◇ D20 ◇ REV MR 58#5120 Zbl 362#02028 • ID 23685

MUCHNIK, A.A. [1956] *On separability of recursively enumerable sets (Russian)* (**J** 0023) Dokl Akad Nauk SSSR 109*29-32
◇ D25 ◇ REV MR 18.866 Zbl 71.246 JSL 23.224
• ID 19460

MUCHNIK, A.A. [1956] *On the unsolvability of the problem of reducibility in the theory of algorithms (Russian)* (**J** 0023) Dokl Akad Nauk SSSR 108*194-197
◇ D25 D30 D35 ◇ REV MR 18.457 Zbl 70.247 JSL 22.218 • ID 09612

MUCHNIK, A.A. [1957] *Negative solution of the reducibility problem of Post (Russian)* (**J** 0067) Usp Mat Nauk 12/2(74)*215-216
◇ D25 ◇ ID 45566

MUCHNIK, A.A. [1958] *Isomorphism of systems of recursively enumerable sets with effective properties (Russian)* (**J** 0065) Tr Moskva Mat Obshch 7*407-412
• TRANSL [1963] (**J** 0225) Amer Math Soc, Transl, Ser 2 23*7-13
◇ D25 D30 ◇ REV MR 21#4098 JSL 32.393 • ID 19457

MUCHNIK, A.A. [1958] *Solution of Post's reduction problem and of certain other problems in the theory of algorithms I (Russian)* (**J** 0065) Tr Moskva Mat Obshch 7*391-405
• TRANSL [1963] (**J** 0225) Amer Math Soc, Transl, Ser 2 29*197-215
◇ D25 D30 ◇ REV MR 22#5570 Zbl 85.250 JSL 30.90
• ID 19458

MUCHNIK, A.A. [1959] *Solution of the problem of reducibility of Post (Russian)* (**P** 0607) All-Union Math Conf (3);1956 Moskva 1*184
◇ D25 ◇ ID 45565

MUCHNIK, A.A. [1963] *On strong and weak reducibility of algorithmic problems (Russian)* (**J** 0092) Sib Mat Zh 4*1328-1341
◇ D30 ◇ REV MR 34#50 Zbl 156.16 • ID 45433

MUCHNIK, A.A. [1965] *On the reducibility of problems of the solvability of enumerable sets to problems of separability (Russian)* (**J** 0216) Izv Akad Nauk SSSR, Ser Mat 29*717-724
◇ D25 D30 ◇ REV MR 31#3328 Zbl 133.250 • ID 09613

MUCHNIK, A.A. [1970] see KOZMIDIADI, V.A.

MUCHNIK, A.A. [1970] *Ueber zwei Zugaenge zur Klassifikation rekursiver Funktionen (Russisch)* (**C** 1540) Probl Mat Log: Slozh Algor & Kl Vychisl Funk 123-138
◇ D20 ◇ REV Zbl 242#02044 • ID 26358

MUCHNIK, A.A. [1983] *Supplement of the translator to the paper "On alternation. I, II" (Russian)* (**J** 3079) Kiber Sb Perevodov, NS 20*141-158
◇ B45 D05 D10 D15 F50 ◇ REV Zbl 545#68041 • REM For the papers see Chandra,A.K. et al. 1976 and 1981
• ID 43519

MUCHNIK, A.A. [1985] *Games on infinite trees and automata with dead ends. A new proof of decidability for the monadic theory with two successor functions (Russian)* (**S** 2582) Semiotika & Inf, Akad Nauk SSSR 24*16-40,142
◇ B15 B25 C85 D05 ◇ REV Zbl 576#03010 • ID 47256

MUCHNIK, A.A. see Vol. I, II for further entries

MUELLER, D.W. [1972] *Randomness and extrapolation* (**P** 1520) Berkeley Symp Math Stat & Probab (6);1970 Berkeley 2*1-31
◇ D55 D80 ◇ REV MR 54#9988 • ID 25755

MUELLER, D.W. see Vol. I, V for further entries

MUELLER, GERT H. [1965] *Charakterisierung einer Klasse von rekursiven Funktionen* (**P** 0797) Fonds des Math, Machines Math & Appl;1962 Tihany 45-51
◇ D20 ◇ REV JSL 38.156 • ID 09620

MUELLER, GERT H. [1970] *Rekursionsformen in der Żahlentheorie* (**P** 0577) Automatenth & Formale Sprachen;1969 Oberwolfach 399-440
◇ D20 ◇ REV MR 48 # 1902 Zbl 279 # 02020 JSL 38.156 • ID 09621

MUELLER, GERT H. & RICHTER, M.M. (EDS.) [1984] *Models and sets* (**S** 3301) Lect Notes Math 1103*viii+484pp
◇ B97 C97 D97 H97 ◇ REV MR 85k:03002a Zbl 547 # 00008 • REM Logic Colloquium;1983 Aachen, Vol.I. Vol.II 1984 by Boerger,E. • ID 41750

MUELLER, GERT H. [1985] see EBBINGHAUS, H.-D.

MUELLER, GERT H. see Vol. I, II, III, V, VI for further entries

MUELLER, HEINRICH [1982] *The complexity of the vertex coloring problem for some classes of graphs restricted by cycle properties* (**S** 3126) Ber, Fak Inf, Univ Karlsruhe 4/82*13pp
◇ D15 D80 ◇ REV Zbl 477 # 05036 • ID 55598

MUELLER, HORST [1970] *Ueber die mit Stackautomaten berechenbaren Funktionen* (**J** 0009) Arch Math Logik Grundlagenforsch 13*60-73
◇ D10 D20 ◇ REV MR 45 # 70 Zbl 255 # 02040 • ID 09623

MUELLER, HORST [1971] *Endliche Automaten und Labyrinthe (English and Russian summaries)* (**J** 0129) Elektr Informationsverarbeitung & Kybern 7*261-264
◇ D05 ◇ REV MR 47 # 4773 Zbl 224 # 94057 • ID 09624

MUELLER, HORST [1971] *Stackautomaten in Labyrinthen* (**J** 0009) Arch Math Logik Grundlagenforsch 14*127-134
◇ D10 ◇ REV MR 50 # 4264 Zbl 234 # 02021 • ID 09625

MUELLER, HORST [1972] *Endliche Automaten* (**J** 0487) Math Unterricht 18*43-57
◇ D05 ◇ ID 28300

MUELLER, HORST [1972] *Jede mit Stackautomaten berechenbare Funktion ist elementar* (**J** 0009) Arch Math Logik Grundlagenforsch 15*19-30
◇ D10 D15 D20 ◇ REV MR 47 # 8273 Zbl 255 # 02041 • ID 09626

MUELLER, HORST [1977] *A one-symbol printing automaton escaping from every labyrinth* (**J** 0373) Comp Arch Inform & Numerik 19*95-110
◇ D05 ◇ REV MR 58 # 20907 Zbl 361 # 94063 • ID 28301

MUELLER, HORST [1985] *Weak Petri net computers for Ackermann functions* (**J** 0129) Elektr Informationsverarbeitung & Kybern 21*236-246
◇ D05 D20 ◇ ID 49641

MUELLER, HORST see Vol. III, V for further entries

MUENTEFERING, P. & STARKE, P.H. [1972] *Ueber reduzible Ereignisse* (**J** 0129) Elektr Informationsverarbeitung & Kybern 8*187-196
◇ D05 D10 ◇ REV MR 57 # 19183 Zbl 242 # 02041 • ID 26355

MUIR, A. & WARNER, M.W. [1984] *Lattice valued relations and automata* (**J** 2702) Discr Appl Math 7*65-78
◇ D05 E07 G10 ◇ REV MR 85e:68081 Zbl 547 # 68052 • ID 40043

MUIR, A. see Vol. V for further entries

MULLER, D.E. [1956] *Complexity in electronic switching circuits* (**J** 0072) IRE Trans Electr Comp EC-5*15-19
◇ B70 D15 ◇ REV JSL 25.300 • ID 16979

MULLER, D.E. [1970] see BAVEL, Z.

MULLER, D.E. & SCHUPP, P.E. [1981] *Context-free languages, groups, the theory of ends, second-order logic, tiling problems, cellular automata, and vector addition systems* (**J** 0589) Bull Amer Math Soc (NS) 4*331-334
◇ B15 B25 D05 D80 ◇ REV MR 82m:03051 Zbl 484 # 03019 • ID 76648

MULLER, D.E. & SCHUPP, P.E. [1983] *Groups, the theory of ends, and context-free languages* (**J** 0119) J Comp Syst Sci 26*295-310
◇ D05 D40 ◇ REV MR 84k:20016 Zbl 537 # 20011 • ID 39206

MULLER, D.E. & SCHUPP, P.E. [1985] *Alternating automata on infinite objects, determinacy and Rabin's Theorem* (**P** 4595) Autom Infinite Words;1984 Le Mont Dore 100-107
◇ B25 C85 D05 E60 ◇ ID 48421

MULLER, D.E. & SCHUPP, P.E. [1985] *The theory of ends, pushdown automata, and second-order logic* (**J** 1426) Theor Comput Sci 37*51-75
◇ B15 B25 D10 D80 ◇ ID 47646

MULLER, D.E. see Vol. I for further entries

MULLIN, A.A. [1963] *On a theorem equivalent to Post's fundamental theorem of recursive function theory* (**J** 0068) Z Math Logik Grundlagen Math 9*203-205
◇ D25 ◇ REV MR 27 # 2404 JSL 36.343 • ID 09640

MULLIN, A.A. [1965] *A contribution toward computable number theory* (**J** 0068) Z Math Logik Grundlagen Math 11*117-119
◇ D80 ◇ REV MR 33 # 1276 Zbl 254 # 10008 • ID 33193

MULLIN, A.A. see Vol. I, V, VI for further entries

MULRY, P.S. [1982] *Generalized Banach-Mazur functionals in the topos of recursive sets* (**J** 0326) J Pure Appl Algebra 26*71-83
◇ D20 D45 G30 ◇ REV MR 84e:03079 Zbl 491 # 03017 • ID 34411

MUNDICI, D. [1980] *Natural limitations of algorithmic procedures in logic (Italian summary)* (**J** 0149) Atti Accad Naz Lincei Fis Mat Nat, Ser 8 69*101-105
◇ C40 D05 D80 ◇ REV MR 84e:03007 Zbl 518 # 03005 • ID 33528

MUNDICI, D. [1981] *Complexity of Craig's interpolation* (**P** 2614) Open Days in Model Th & Set Th;1981 Jadwisin 185-204
◇ C40 D15 ◇ REV Zbl 507 # 03025 • ID 33726

MUNDICI, D. [1981] *Craig's interpolation theorem in computation theory* (**J** 0149) Atti Accad Naz Lincei Fis Mat Nat, Ser 8 70*6-11
◇ C40 D15 ◇ REV Zbl 523 # 03027 • ID 33529

MUNDICI, D. [1981] *Ergodic undefinability in set theory and recursion theory* (**J** 0053) Proc Amer Math Soc 82*107-111
◇ D55 E47 E75 ◇ REV MR 82h:03057 Zbl 472 # 03039 • ID 55303

MUNDICI, D. [1981] *Irreversibility, uncertainty, relativity and computer limitations (Italian and Russian summaries)* (J 2775) Nuovo Cimento B, Ser 2 61*297-305
◇ A05 D15 ◇ REV MR 84b:03012 • ID 35613

MUNDICI, D. [1982] *Complexity of Craig's interpolation* (J 2095) Fund Inform, Ann Soc Math Pol, Ser 4 5*261-278
◇ C40 D15 ◇ REV MR 84h:03081 Zbl 507 # 03025 • ID 34339

MUNDICI, D. [1983] *A lower bound for the complexity of Craig's interpolants in sentential logic* (J 0009) Arch Math Logik Grundlagenforsch 23*27-36
◇ B20 C40 D15 ◇ REV MR 85c:03003 Zbl 511 # 03004 • ID 37383

MUNDICI, D. [1983] *Natural limitations of decision procedures for arithmetic with bounded quantifiers* (J 0009) Arch Math Logik Grundlagenforsch 23*37-54
◇ B25 D10 D15 F30 ◇ REV MR 84k:03040 Zbl 523 # 03028 • ID 34976

MUNDICI, D. [1984] *Δ-tautologies, uniform and nonuniform upper bounds in computation theory (Italian summary)* (J 0149) Atti Accad Naz Lincei Fis Mat Nat, Ser 8 75*99-101
◇ B75 C40 D15 ◇ REV MR 86d:03037 Zbl 568 # 03018 Zbl 568 # 03019 • ID 44693

MUNDICI, D. [1984] *NP and Craig's interpolation theorem* (P 3710) Logic Colloq;1982 Firenze 345-358
◇ C40 D15 ◇ REV MR 86e:03042 • ID 41841

MUNDICI, D. [1984] *Tautologies with a unique Craig interpolant, uniform vs. nonuniform complexity* (J 0073) Ann Pure Appl Logic 27*265-273
◇ C40 D15 ◇ REV MR 86g:03056 • ID 44069

MUNDICI, D. see Vol. I, II, III, V for further entries

MURAWSKI, G. [1984] see HEMMERLING, A.

MURAWSKI, R. [1981] *A simple remark on satisfaction classes, indiscernibles and recursive saturation* (J 2718) Fct Approximatio, Comment Math, Poznan 11*149-151
◇ C30 C50 C57 H15 ◇ REV MR 84f:03059 Zbl 481 # 03022 • ID 34478

MURAWSKI, R. [1981] see KIRBY, L.A.S.

MURAWSKI, R. [1984] *Mathematical incompleteness of arithmetic (Polish)* (J 0519) Wiad Mat, Ann Soc Math Pol, Ser 2 26*47-58
◇ D35 F30 ◇ ID 39679

MURAWSKI, R. see Vol. III, V for further entries

MURSKIJ, V.L. [1971] *Nondiscernible properties of finite systems of identity relations (Russian)* (J 0023) Dokl Akad Nauk SSSR 196*520-522
• TRANSL [1971] (J 0062) Sov Math, Dokl 12*183-186
◇ C05 D35 D40 ◇ REV MR 44 # 4084 Zbl 238 # 02039 JSL 47.903 • ID 19453

MURSKIJ, V.L. see Vol. I, II, III for further entries

MURTAZINA, T. [1978] see GABBASOV, N.Z.

MUZALEWSKI, M. [1975] *On the decidability of the identities problem in some classes of algebras* (J 0014) Bull Acad Pol Sci, Ser Math Astron Phys 23*7-9
◇ D40 ◇ REV MR 54 # 10107 Zbl 311 # 02055 • ID 09661

MUZALEWSKI, M. see Vol. III for further entries

MUZYUKINA, G.I. [1973] *mu -degrees of unsolvability of functions (Russian)* (J 0226) Uch Zap Ped Inst, Ivanovo 125*84-93
◇ D30 ◇ REV MR 57 # 5716 • ID 76704

MYASNIKOV, A.G. & REMESLENNIKOV, V.N. [1982] *Classification of nilpotent powered groups according to elementary properties (Russian)* (C 3953) Mat Log & Teor Algor 56-87
◇ C60 D40 ◇ REV MR 85h:20042 Zbl 516 # 20021 • ID 38582

MYASNIKOV, A.G. see Vol. III for further entries

MYCIELSKI, J. [1971] see EHRENFEUCHT, A.

MYCIELSKI, J. [1980] see EHRENFEUCHT, A.

MYCIELSKI, J. [1983] *The meaning of the conjecture $P \neq NP$ for mathematical logic* (J 0005) Amer Math Mon 90*129-130
◇ B30 D15 ◇ REV MR 85f:03001 Zbl 513 # 03028 • ID 37228

MYCIELSKI, J. see Vol. I, III, V, VI for further entries

MYERS, D.L. [1974] *Nonrecursive tiling of the plane II* (J 0036) J Symb Logic 39*286-294
◇ D05 D80 ◇ REV MR 51 # 111 Zbl 299 # 02055 • REM Part I 1974 by Hanf,W.P. • ID 09697

MYERS, D.L. [1976] *Invariant uniformization* (J 0027) Fund Math 91*65-72
◇ C75 D55 E15 E35 ◇ REV MR 53 # 10582 Zbl 358 # 02068 • ID 30751

MYERS, D.L. see Vol. I, III for further entries

MYERS, R.W. [1980] *Complexity of model-theoretic notions* (J 0047) Notre Dame J Formal Log 21*656-658
◇ C07 C35 C57 D55 ◇ REV MR 81j:03070 Zbl 416 # 03043 • ID 55957

MYHILL, J.R. [1949] *Note on an idea of Fitch* (J 0036) J Symb Logic 14*175-176
◇ D55 F30 ◇ REV MR 11.151 Zbl 37.297 JSL 14.260 • ID 09698

MYHILL, J.R. [1950] *A system which can define its own truth* (J 0027) Fund Math 37*190-192
◇ B30 D20 F25 F30 ◇ REV MR 13.97 Zbl 41.150 JSL 21.319 • ID 09699

MYHILL, J.R. [1953] *Three contributions to recursive function theory* (P 0645) Int Congr Philos (11);1953 Bruxelles 14*50-59
◇ D25 ◇ REV MR 15.667 Zbl 52.249 JSL 20.176 • ID 16870

MYHILL, J.R. [1955] *Creative sets* (J 0068) Z Math Logik Grundlagen Math 1*97-108
◇ D25 ◇ REV MR 17.118 Zbl 65.1 JSL 22.73 • ID 09710

MYHILL, J.R. & SHEPHERDSON, J.C. [1955] *Effective operations on partial recursive functions* (J 0068) Z Math Logik Grundlagen Math 1*310-317
◇ D20 ◇ REV MR 17.1039 Zbl 68.247 JSL 22.303 • ID 09711

MYHILL, J.R. [1956] *A problem on recursively enumerable sets (problem 8)* (J 0036) J Symb Logic 21*215
◇ D25 ◇ ID 33302

MYHILL, J.R. [1956] *A problem on recursively enumerable supersets (problem 9)* (J 0036) J Symb Logic 21*215
◇ D25 ◇ REM See also 1959 by Lorenzen,P. & Myhill,J.R. • ID 33303

MYHILL, J.R. [1956] *Solution of a problem of Tarski* (J 0036) J Symb Logic 21*49-51
◇ D35 ◇ REV MR 17.816 Zbl 71.10 JSL 23.445 • ID 09713

MYHILL, J.R. [1958] *Recursive equivalence types and combinatorial functions* (J 0015) Bull Amer Math Soc 64*373-376
◇ D50 ◇ REV MR 21 # 7 Zbl 85.248 JSL 25.356 JSL 31.510 • ID 09716

MYHILL, J.R. [1958] see DEKKER, J.C.E.

MYHILL, J.R. [1958] *The foundations of mathematics, II. The theory of recursive functions* (C 0742) Phil Mid-Century 130-139
◇ D20 ◇ ID 09715

MYHILL, J.R. [1959] see LORENZEN, P.

MYHILL, J.R. [1959] *Finitely representable functions* (P 0634) Constructivity in Math;1957 Amsterdam 195-207
◇ D55 ◇ REV MR 21 # 4104 Zbl 91.13 JSL 38.157 • ID 09714

MYHILL, J.R. [1959] *Recursive digraphs, splinters, and cylinders* (J 0043) Math Ann 138*211-218
◇ D25 ◇ REV MR 22 # 2546 Zbl 87.251 JSL 25.361 • ID 09718

MYHILL, J.R. [1960] see DEKKER, J.C.E.

MYHILL, J.R. [1961] *Category methods in recursion theory* (J 0048) Pac J Math 11*1479-1486
◇ D30 ◇ REV MR 25 # 14 Zbl 112.246 JSL 38.654 • ID 09720

MYHILL, J.R. [1961] *Note on degrees of partial functions* (J 0053) Proc Amer Math Soc 12*519-521
◇ D30 ◇ REV MR 23 # A3091 Zbl 101.12 JSL 37.408 • ID 09721

MYHILL, J.R. [1962] $\Omega - \Lambda$ (P 0613) Rec Fct Th;1961 New York 97-104
◇ D50 ◇ REV MR 26 # 15 Zbl 149.247 JSL 33.619 • ID 19446

MYHILL, J.R. [1962] *Elementary properties of the group of isolic integers* (J 0044) Math Z 78*126-130
◇ D50 ◇ REV MR 25 # 1102 Zbl 100.248 JSL 38.333 • ID 09724

MYHILL, J.R. [1962] *Recursive equivalence types and combinatorial functions* (P 0612) Int Congr Log, Meth & Phil of Sci (1,Proc);1960 Stanford 46-55
◇ D50 ◇ REV MR 27 # 2405 Zbl 149.246 JSL 31.510 • ID 09723

MYHILL, J.R. see Vol. I, II, III, V, VI for further entries

MYLOPOULOS, J. & TOURLAKIS, G. [1973] *Some results in computational topology* (J 0037) ACM J 20*439-455
◇ D40 D80 ◇ REV MR 48 # 10230 Zbl 298 # 68065 • ID 13653

NABEBIN, A.A. [1976] *Multitape automata in a unary alphabet (Russian)* (S 2850) Tr Ehnerg Inst Moskva 292*7-11
◇ D05 D10 ◇ REV MR 58 # 27387 • ID 76721

NABEBIN, A.A. [1977] *Expressibility in restricted second-order arithmetic (Russian)* (J 0092) Sib Mat Zh 18*830-837,957 • TRANSL [1977] (J 0475) Sib Math J 18*588-593
◇ B15 C40 D05 F35 ◇ REV MR 58 # 16250 Zbl 385 # 03047 • ID 52155

NABIALEK, I. [1973] *Some properties of τ-computable functions* (J 1929) Prace Centr Oblicz Pol Akad Nauk 131*10pp
◇ D20 ◇ REV Zbl 282 # 68023 • ID 64133

NABIALEK, I. [1974] *Rational τ-computable functions* (J 0014) Bull Acad Pol Sci, Ser Math Astron Phys 22*381-383
◇ D20 ◇ REV MR 49 # 10185 Zbl 306 # 68023 • ID 64131

NABIALEK, I. & ZAKOWSKI, W. [1975] *On Z-computability of functions and sets of functions* (J 1929) Prace Centr Oblicz Pol Akad Nauk 173*8pp
◇ D15 D20 ◇ REV Zbl 304 # 68050 • ID 64132

NABIALEK, I. [1977] *Bases of aggregable sets* (J 1008) Demonstr Math (Warsaw) 10*77-84
◇ D15 D20 ◇ REV MR 57 # 2900 Zbl 375 # 02019 • ID 51596

NADEL, M.E. [1972] *Some Loewenheim-Skolem results for admissible sets* (J 0029) Israel J Math 12*427-432
◇ C62 C70 D60 E55 ◇ REV MR 47 # 3143 Zbl 262 # 02053 • ID 09743

NADEL, M.E. [1974] *More Loewenheim-Skolem results for admissible sets* (J 0029) Israel J Math 18*53-64
◇ C62 C70 D60 ◇ REV MR 51 # 2901 Zbl 309 # 02058 • ID 16930

NADEL, M.E. [1978] see LIPSHITZ, L.

NADEL, M.E. [1982] see KNIGHT, J.F.

NADEL, M.E. see Vol. III, V, VI for further entries

NADIU, G.S. [1966] *On some theorems of recursion of normal algorithms* (J 0197) Stud Cercet Mat Acad Romana 18*1071-1078
◇ D03 ◇ REV MR 36 # 3653 Zbl 166.265 • ID 09747

NADIU, G.S. [1969] see MARUSCIAC, I.

NADIU, G.S. see Vol. I, III, VI for further entries

NADYROV, R.F. [1977] see ARSLANOV, M.M.

NAGAOKA, K. [1974] see FUKUYAMA, M.

NAGAOKA, K. see Vol. II for further entries

NAGASAKA, K. [1973] *On minimal-program complexity measure* (P 3244) Hawaii Int Conf Syst Sci (6);1973 Honolulu 477-479
◇ D15 ◇ REV Zbl 356 # 94022 • ID 50337

NAGASHIMA, T. [1968] *On elementary functions of natural numbers* (J 0531) Hitotsubashi J Arts Sci (Tokyo) 9*50-58
◇ D20 ◇ REV MR 38 # 1001 • ID 09757

NAGASHIMA, T. [1975] *On a certain class of recursive functions* (J 0531) Hitotsubashi J Arts Sci (Tokyo) 16*72-81
◇ D20 ◇ REV MR 52 # 13355 • ID 21808

NAGASHIMA, T. see Vol. II, III, V, VI for further entries

NAGEL, E. & NEWMAN, J.R. [1958] *Goedel's proof* (X 0924) New York Univ Pr: New York ix + 118pp
◇ B28 D35 F30 F98 ◇ REV MR 20 # 1625 Zbl 86.246 JSL 24.222 • ID 09765

NAGEL, E. & NEWMAN, J.R. [1967] *Goedel's proof* (C 0721) Contemp Readings in Log Th 51-71
◇ D35 F30 ◇ REV JSL 21.374 • ID 09763

NAGEL, E. see Vol. I, II, VI for further entries

NAGORNYJ, N.M. [1953] *On strengthening the reduction theorem of the theory of algorithm (Russian)* (J 0023) Dokl Akad Nauk SSSR 90*341-342
◇ D03 ◇ REV MR 16.436 Zbl 53.345 JSL 19.218
• ID 09773

NAGORNYJ, N.M. [1958] *Beispiel einer Gruppe mit nicht rekursivem Zentrum (Russisch)* (J 0068) Z Math Logik Grundlagen Math 4*304-308
◇ D45 ◇ REV MR 21 # 5669 Zbl 92.10 JSL 31.653
• ID 19444

NAGORNYJ, N.M. [1958] *On a minimal alphabet of algorithms over a given alphabet (Russian)* (S 0066) Tr Mat Inst Steklov 52*66-74
• TRANSL [1963] (J 0225) Amer Math Soc, Transl, Ser 2 29*153-161
◇ B75 D05 ◇ REV MR 20 # 6356 Zbl 87.253 JSL 29.108
• ID 21255

NAGORNYJ, N.M. [1958] *Some generalisations of the concept of normal algorithm (Russian)* (S 0066) Tr Mat Inst Steklov 52*7-65
• TRANSL [1963] (J 0225) Amer Math Soc, Transl, Ser 2 29*91-151
◇ D03 D20 ◇ REV MR 21 # 6 Zbl 87.252 JSL 27.360
• ID 16958

NAGORNYJ, N.M. [1960] *On the investigation of the isomorphisms of associative calculi (Russian) (German summary)* (J 0068) Z Math Logik Grundlagen Math 6*319-324
◇ D05 D20 ◇ REV MR 32 # 52 Zbl 156.13 JSL 29.57
• ID 09774

NAGORNYJ, N.M. [1961] *On the realization of functions in alphabets by algorithms of certain classes (Russian)* (J 0023) Dokl Akad Nauk SSSR 140*52-55
• TRANSL [1961] (J 0062) Sov Math, Dokl 2*1156-1159
◇ D05 ◇ REV MR 25 # 16 Zbl 199.37 • ID 09775

NAGORNYJ, N.M. [1973] *Separability with respect to invariants (Russian)* (S 0554) Issl Teor Algor & Mat Logik (Moskva) 1*205-210
◇ D03 D25 ◇ REV MR 49 # 2324 Zbl 289 # 02025
• ID 09779

NAGORNYJ, N.M. [1974] see KUSHNER, B.A.

NAGORNYJ, N.M. [1974] *Ueber einige Verfahren zur Beschreibung der Arbeit von Systemen wechselwirkender Rechenmaschinen (Russian)* (C 1450) Teor Algor & Mat Logika (Markov) 127-142
◇ D05 ◇ REV Zbl 302 # 68072 • ID 64143

NAGORNYJ, N.M. [1976] *Normal algorithms and first order languages (Russian)* (S 0554) Issl Teor Algor & Mat Logik (Moskva) 2*46-50,157
◇ B10 D03 ◇ REV MR 58 # 16205 • ID 76745

NAGORNYJ, N.M. see Vol. VI for further entries

NAGY, Z. [1979] see AJTAI, M.

NAGY, Z. see Vol. V for further entries

NAKAMURA, A. [1970] *On a propositional calculus whose decision problem is recursively unsolvable* (J 0111) Nagoya Math J 38*145-152
◇ B22 D35 ◇ REV MR 41 # 3277 Zbl 202.13 • ID 09788

NAKAMURA, A. [1970] *On the undecidability of monadic modal predicate logic* (J 0068) Z Math Logik Grundlagen Math 16*257-260
◇ B45 D35 ◇ REV MR 43 # 6059 Zbl 207.5 • ID 09789

NAKAMURA, A. [1975] *On causal ω^2-systems* (J 0119) J Comp Syst Sci 10*253-265
◇ D05 ◇ REV MR 52 # 13028 Zbl 302 # 94021 • ID 21743

NAKAMURA, A. [1977] see INOUE, K.

NAKAMURA, A. & ONO, H. [1977] *Two-dimensional finite automata and their application to the decision problem of monadic first-order arithmetic $A[P, F(x), G(x)]$* (P 2032) Stud on Polyautomata;1976 Zushi & Kyoto 51-71
◇ D05 D35 F30 F35 ◇ ID 32225

NAKAMURA, A. & ONO, H. [1979] *Undecidability of the first-order arithmetic $A[P(x),2x,x+1]$* (J 0119) J Comp Syst Sci 18*243-253
◇ D05 D35 F30 F35 ◇ REV MR 80k:03045 Zbl 414 # 03006 • ID 53052

NAKAMURA, A. & ONO, H. [1980] *On the size of refutation Kripke models for some linear modal and tense logics* (J 0063) Studia Logica 39*325-333
◇ B45 D15 ◇ REV MR 82i:03030 Zbl 466 # 03008
• ID 54959

NAKAMURA, A. & ONO, H. [1981] *Undecidability of extensions of the monadic first-order theory of successor and two-dimensional finite automata* (P 3201) Logic Symposia;1979/80 Hakone 155-174
◇ B15 B25 D05 D35 ◇ REV MR 83k:03054 Zbl 474 # 03023 • ID 55427

NAKAMURA, A. & ONO, H. [1983] *Pictures of functions and their acceptability by automata* (J 1426) Theor Comput Sci 23*37-48
◇ D10 D20 ◇ REV MR 84c:68081 Zbl 504 # 68027
• ID 37166

NAKAMURA, A. see Vol. I, II, III, VI for further entries

NAKAYASU, F. [1981] see HIROSE, K.

NAPOLI, M. [1985] see FACHINI, E.

NARASIMHAN, R. & SRINIVASAN, C.V. [1959] *On the synthesis of finite sequential machines* (J 0545) Proc Indian Acad Sci, Sect A 50*68-82
◇ D05 ◇ REV MR 23 # B1069 Zbl 87.12 • ID 47978

NARENDRAN, P. [1983] see KAPUR, D.

NARENDRAN, P. [1984] see MCNAUGHTON, R.

NARENDRAN, P. [1985] see KAPUR, D.

NARENDRAN, P. [1985] see KANDRI-RODY, A.

NARUSHIMA, H. & NOJIMA, S. & OOHARA, S. [1974] *A tag type automaton with a double reading head* (J 0463) Proc Fac Sci Tokai Univ 9*15-20
◇ D03 D05 ◇ REV MR 49 # 7122 Zbl 364 # 02019
• ID 09987

NARUSHIMA, H. see Vol. I, V for further entries

NASH, B. [1973] *Reachability problems in vector addition systems* (J 0005) Amer Math Mon 80*292-295
◇ D05 ◇ REV MR 47 # 7964 Zbl 273 # 68034 • ID 64161

NASIBULLOV, KH.KH. [1967] *Some bases of the Robinson algebra (Russian)* (J 0040) Kibernetika, Akad Nauk Ukr SSR 1967/2*24-26
• TRANSL [1967] (J 0021) Cybernetics 3/2*18-20
◇ D20 ◇ REV MR 42 # 4395 Zbl 235 # 02041 • ID 09816

NASIBULLOV, KH.KH. [1968] *Bases in algebras of partially recursive functions (Russian)* (J 0003) Algebra i Logika 7/4*87-105
• TRANSL [1968] (J 0069) Algeb and Log 7*255-267
◇ D20 D75 ◇ REV MR 41 # 56 Zbl 215.323 • ID 09817

NASIBULLOV, KH.KH. & POLYAKOV, E.A. [1969] *Certain recursive schemes (Russian)* (J 0226) Uch Zap Ped Inst, Ivanovo 44*6-10
◇ D20 ◇ REV MR 47 # 1587 • ID 16366

NASIBULLOV, KH.KH. [1969] *Recursive functions of large scope (Russian)* (J 0092) Sib Mat Zh 10*105-115
• TRANSL [1969] (J 0475) Sib Math J 10*74-81
◇ D20 ◇ REV MR 38 # 5622 Zbl 231 # 02049 • ID 27445

NASIBULLOV, KH.KH. [1981] *Some problems connected with lower classes of the Grzegorczyk hierarchy (Russian)* (C 3865) Algeb Sistemy (Ivanovo) 169-173
◇ D20 ◇ ID 45158

NASLIN, P. [1958] *Circuits a relais et automatismes a sequences* (X 0856) Dunod: Paris xii+229pp
◇ B70 D05 ◇ REV Zbl 93.313 JSL 24.187 JSL 37.627
• ID 09819

NASU, M. [1969] see HONDA, N.

NASU, M. [1976] see HONDA, N.

NASYROV, I.R. [1982] *A reduction theorem for automata over trees and context-free grammars (Russian)* (J 3937) Veroyat Met i Kibern (Kazan) 18*60-70
◇ D05 ◇ REV MR 85e:68051 Zbl 537 # 68055 • ID 40021

NASYROV, I.R. [1982] *Representability of languages in deterministic and nondeterministic automata with a countable number of states (Russian)* (J 0031) Izv Vyssh Ucheb Zaved, Mat (Kazan) 1982/7*79-80
• TRANSL [1982] (J 3449) Sov Math 26/7*104-106
◇ D05 D10 ◇ REV MR 84d:68092 Zbl 491 # 68049
• ID 39716

NASYROV, I.R. [1984] *Representation of languages in Turing machines and automata with a countable number of states (Russian)* (J 3937) Veroyat Met i Kibern (Kazan) 20*87-94
◇ D10 ◇ ID 46389

NATION, J.B. [1979] see FREESE, R.

NAUMOVIC, J. [1983] *A classification of the one-argument primitive recursive functions* (J 0009) Arch Math Logik Grundlagenforsch 23*161-174
◇ D20 ◇ REV MR 85f:03038a Zbl 544 # 03015 • ID 40725

NAZARYAN, G.A. [1975] *On the realization of boolean functions in algorithmic languages (Russian)* (S 0422) Tr Vychisl Tsentra Akad Nauk Armyan SSR & Univ Erevan 8*36-56
◇ B05 B75 D03 ◇ REV MR 56 # 5243 Zbl 426 # 03063
• ID 53659

NAZARYAN, G.A. [1976] *Complexity classes of sets of boolean functions (Russian) (Armenian summary)* (J 0346) Dokl Akad Nauk Armyan SSR 63*257-263
◇ B05 D15 ◇ REV MR 58 # 10359 Zbl 356 # 94050
• ID 76769

NAZARYAN, G.A. [1978] *Ueber eine Synthese von Algorithmen approximativ berechenbarer boolescher Funktionen (Russian)* (J 0346) Dokl Akad Nauk Armyan SSR 66*15-21
◇ B35 D15 D20 ◇ REV MR 80c:65040 Zbl 383 # 94036
• ID 52040

NAZARYAN, G.A. [1979] *Ueber disjunkte Zerlegungen von Mengen boolescher Funktionen (Russian)* (J 0346) Dokl Akad Nauk Armyan SSR 69*88-94
◇ B05 D15 ◇ REV Zbl 421 # 03027 • ID 53426

NAZARYAN, G.A. [1980] *The relations of some complexity characteristics of sets of boolean functions (Russian) (Armenian summary)* (J 0346) Dokl Akad Nauk Armyan SSR 71*217-220
◇ B05 D15 ◇ REV MR 82g:03073 Zbl 473 # 68035
• ID 76768

NAZARYAN, G.A. [1982] *Realization of Boolean functions in algorithmic languages under constraints on the running time of the algorithms (Russian) (Armenian summary)* (S 0422) Tr Vychisl Tsentra Akad Nauk Armyan SSR & Univ Erevan 10*41-52
◇ B05 D15 ◇ REV MR 85a:03076 • ID 34748

NAZARYAN, G.A. see Vol. I for further entries

NEANDER, J. [1963] *Aequivalenzkalkuel und orientierte Graphen* (P 1612) Schaltkreis & -werk Th (2);1961 Saarbruecken 129-152
◇ B70 D05 ◇ ID 27839

NEFF, M.F. [1975] see EVANS, T.

NEGRESCU, I. & PAVALOIU, I. [1965] *A property of matrix schemes of algorithms (Romanian)* (J 0197) Stud Cercet Mat Acad Romana 17*1405-1409
◇ D20 ◇ REV MR 37 # 73 • ID 09841

NEGRI, M. [1984] *An application of recursive saturation* (J 2099) Boll Unione Mat Ital, VI Ser, A 3*449-451
◇ C50 C57 C62 F30 ◇ REV MR 86f:03109
Zbl 562 # 03039 • ID 44715

NEGRI, M. see Vol. VI for further entries

NEHRLICH, W. [1978] *Eine Bemerkung zur effektiven Fixpunktberechnung bei vollstaendigen Numerierungen* (S 2829) Rostocker Math Kolloq 10*83-86
◇ D45 ◇ REV MR 81g:03052 Zbl 412 # 03029 • ID 52960

NEHRLICH, W. [1978] see GOETZE, B.G.

NEHRLICH, W. [1980] see GOETZE, B.G.

NEHRLICH, W. [1981] see GOETZE, B.G.

NEKVINDA, M. [1972] *On a certain event recognizable in real time (Czech summary)* (J 0156) Kybernetika (Prague) 8*149-153
◇ D05 D15 ◇ REV MR 48 # 5423 Zbl 235 # 02033
• ID 27821

NEKVINDA, M. [1973] *On the complexity of events recognizable in real time* (J 0156) Kybernetika (Prague) 9*1-10
◇ D05 D15 ◇ REV MR 48 # 7662 Zbl 264 # 94039
• ID 64180

NELSON, EVELYN [1980] *Categorical and topological aspects of formal languages* (**J** 0041) Math Syst Theory 13*255-273
◇ D05 G30 ◇ REV MR 82a:68149 Zbl 471 # 68053
• ID 69732

NELSON, EVELYN [1983] *Iterative algebras* (**J** 1426) Theor Comput Sci 25*67-94
◇ D05 ◇ REV MR 84e:08005 Zbl 533 # 03014 • ID 36538

NELSON, EVELYN see Vol. I, III, V for further entries

NELSON, GEORGE C. [1973] *Nonconstructivity of models of the reals (Russian summary)* (**J** 0014) Bull Acad Pol Sci, Ser Math Astron Phys 21*1067-1071
◇ C62 C65 D55 H05 H15 ◇ REV MR 51 # 10066 Zbl 293 # 02042 • ID 17539

NELSON, GEORGE C. [1974] *Many-one reducibility within the Turing degrees of the hyperarithmetic sets* $H_a(x)$ (**J** 0064) Trans Amer Math Soc 191*1-44
◇ D30 D55 ◇ REV MR 50 # 1861 Zbl 313 # 02023
• ID 19487

NELSON, GEORGE C. [1975] see MADISON, E.W.

NELSON, GEORGE C. [1978] *Isomorphism types of the hyperarithmetic sets* H_a (**J** 0047) Notre Dame J Formal Log 19*1-19
◇ D30 D50 D55 ◇ REV MR 58 # 16236 Zbl 364 # 02024
• ID 27070

NELSON, GEORGE C. [1984] *Boolean powers, recursive models, and the Horn theory of a structure* (**J** 0048) Pac J Math 114*207-220
◇ C05 C20 C30 C57 G05 ◇ REV MR 85k:03021 Zbl 505 # 03014 • ID 39839

NELSON, GEORGE C. see Vol. III, VI for further entries

NELSON, GREG & OPPEN, D.C. [1977] *Fast decision algorithms based on union and find* (**P** 3572) IEEE Symp Found of Comput Sci (18);1977 Providence 114-119
◇ D15 D80 ◇ REV MR 58 # 13895 • ID 82344

NELSON, GREG see Vol. I, III for further entries

NELSON, RAYMOND J. [1968] *Introduction to automata* (**X** 0827) Wiley & Sons: New York xii+400pp
◇ D05 D10 D98 ◇ REV MR 37 # 5061 Zbl 165.22 JSL 36.151 • ID 09856

NELSON, RAYMOND J. see Vol. I for further entries

NEMES, T. [1962] *Cybernetical machines (Hungarian)* (**X** 0928) Akad Kiado: Budapest 259pp
◇ D98 ◇ REV Zbl 104.356 • ID 48000

NEPEJVODA, L.K. & NEPEJVODA, N.N. [1977] *Languages without set-variables for the description of sets (Russian)* (**S** 2579) Teor Mnozhestv & Topol (Izhevsk) 1*70-75
◇ C75 D55 ◇ REV Zbl 479 # 03027 • ID 55689

NEPEJVODA, N.N. [1973] *On a generalization of the Kleene-Mostowski hierarchy (Russian)* (**J** 0023) Dokl Akad Nauk SSSR 212*295-298
• TRANSL [1973] (**J** 0062) Sov Math, Dokl 14*1365-1368
◇ D55 F15 F35 ◇ REV MR 49 # 4760 Zbl 336 # 02036
• ID 09867

NEPEJVODA, N.N. [1977] see NEPEJVODA, L.K.

NEPEJVODA, N.N. see Vol. I, II, III, V, VI for further entries

NEPOMNYASHCHIJ, V.A. [1965] *On algorithms realized by re-application of finite automata (Russian)* (**J** 0071) Met Diskr Analiz (Novosibirsk) 5*77-82
◇ D05 ◇ REV MR 33 # 3937 Zbl 192.79 • ID 09869

NEPOMNYASHCHIJ, V.A. [1966] *On a basis for recursively enumerable sets (Russian)* (**J** 0023) Dokl Akad Nauk SSSR 170*1262-1264
• TRANSL [1966] (**J** 0062) Sov Math, Dokl 7*1369-1372
◇ D25 ◇ REV MR 35 # 5312 Zbl 163.251 • ID 09870

NEPOMNYASHCHIJ, V.A. [1966] *On certain automata capable of computing a basis for recursively enumerable sets (Russian)* (**J** 0003) Algebra i Logika 5/5*69-83
◇ D05 D25 ◇ REV MR 35 # 58 Zbl 192.67 • ID 09871

NEPOMNYASHCHIJ, V.A. [1970] *Rudimentary predicates and Turing calculations (Russian)* (**J** 0023) Dokl Akad Nauk SSSR 195*282-284
• TRANSL [1970] (**J** 0062) Sov Math, Dokl 11*1462-1465
◇ D10 D20 ◇ REV MR 43 # 7326 Zbl 223 # 02031
• ID 22215

NEPOMNYASHCHIJ, V.A. [1970] *The rudimentary interpretation of two-tape Turing computations (Russian) (English summary)* (**J** 0040) Kibernetika, Akad Nauk Ukr SSR 1970/2*29-35
◇ D10 D20 ◇ REV MR 52 # 7879 Zbl 245 # 02034
• ID 18311

NEPOMNYASHCHIJ, V.A. [1971] *The completeness of operations in operator algorithms (Russian)* (**J** 0023) Dokl Akad Nauk SSSR 199*780-782
• TRANSL [1971] (**J** 0062) Sov Math, Dokl 12*1185-1188
◇ D20 ◇ REV MR 45 # 6627 Zbl 235 # 02024 • ID 64198

NEPOMNYASHCHIJ, V.A. [1972] *Conditions for the algorithmic completeness of systems of operations* (**P** 1455) Inform Processing (5);1971 Ljubljana 52-55
◇ D05 ◇ REV MR 53 # 12902 • ID 23154

NEPOMNYASHCHIJ, V.A. [1973] *Rudimentary simulation of nondeterministic Turing calculations (Russian) (English summary)* (**J** 0040) Kibernetika, Akad Nauk Ukr SSR 1973/2*23-34
◇ D10 ◇ REV MR 49 # 7125 Zbl 267 # 02027 • ID 09872

NEPOMNYASHCHIJ, V.A. [1974] *Criteria for the algorithmic completeness of the systems of operators* (**P** 1511) Int Symp Th Progr;1972 Novosibirsk 172-186
◇ D20 ◇ REV MR 57 # 16001 • ID 76812

NEPOMNYASHCHIJ, V.A. [1978] *Examples of predicates inexpressible by s-rudimentary formulas (Russian) (English summary)* (**J** 0040) Kibernetika, Akad Nauk Ukr SSR 1978/2*44-46
• TRANSL [1978] (**J** 0021) Cybernetics 14*198-200
◇ D05 D20 ◇ REV MR 81h:03083 Zbl 391 # 03017
• ID 76815

NEPOMNYASHCHIJ, V.A. & VOELKEL, L. [1978] *Zur Vollstaendigkeit von Befehlssystemen* (**J** 0129) Elektr Informationsverarbeitung & Kybern 14*43-48
◇ D20 ◇ REV MR 80a:03052 Zbl 381 # 68047 • ID 51923

NERODE, A. [1957] *General topology and partial recursive functionals* (**P** 1675) Summer Inst Symb Log;1957 Ithaca 247-251
◇ D20 D80 ◇ ID 29354

Nerode, A. [1958] *Linear automaton transformations* (**J** 0053) Proc Amer Math Soc 9*541-544
◇ D05 ◇ REV MR 24 # B1726 Zbl 89.334 JSL 28.173
• ID 09874

Nerode, A. [1959] *Some Stone spaces and recursion theory* (**J** 0025) Duke Math J 26*397-406
◇ D50 ◇ REV MR 21 # 7165 Zbl 114.247 • ID 09875

Nerode, A. [1961] *Extensions to isols* (**J** 0120) Ann of Math, Ser 2 73*362-403
◇ D50 ◇ REV MR 24 # A1215 Zbl 101.12 JSL 25.359
• ID 09878

Nerode, A. [1962] *Arithmetically isolated sets and nonstandard models* (**P** 0613) Rec Fct Th;1961 New York 105-116
◇ D50 ◇ REV MR 26 # 1251 Zbl 178.319 JSL 32.269
• ID 09880

Nerode, A. [1962] *Extensions to isolic integers* (**J** 0120) Ann of Math, Ser 2 75*419-448
◇ D50 ◇ REV MR 25 # 3830 Zbl 106.8 JSL 32.268
• ID 09879

Nerode, A. [1965] *Additive relations among recursive equivalence types* (**J** 0043) Math Ann 159*329-343
◇ D50 ◇ REV MR 32 # 4005 Zbl 168.11 JSL 35.589
• ID 09882

Nerode, A. [1965] *Non-linear combinatorial functions of isols* (**J** 0044) Math Z 86*410-424
◇ D50 ◇ REV MR 34 # 5672 JSL 33.619 • ID 09883

Nerode, A. [1966] *Combinatorial series and recursive equivalence types* (**J** 0027) Fund Math 58*113-141
◇ D50 ◇ REV MR 34 # 1180 Zbl 168.11 JSL 33.620
• ID 09885

Nerode, A. [1966] *Diophantine correct non-standard models in the isols* (**J** 0120) Ann of Math, Ser 2 84*421-432
◇ C62 D50 ◇ REV MR 34 # 2465 Zbl 158.251 JSL 33.619
• ID 09884

Nerode, A. [1970] see Manaster, A.B.

Nerode, A. [1974] see Crossley, J.N.

Nerode, A. [1975] see Crossley, J.N.

Nerode, A. [1975] see Metakides, G.

Nerode, A. [1976] see Crossley, J.N.

Nerode, A. [1977] see Metakides, G.

Nerode, A. [1979] see Metakides, G.

Nerode, A. [1980] see Metakides, G.

Nerode, A. & Shore, R.A. [1980] *Reducibility orderings: theories, definability and automorphisms* (**J** 0007) Ann Math Logic 18*61-89
◇ C62 D30 D35 ◇ REV MR 81k:03040 Zbl 494 # 03028
• ID 76822

Nerode, A. & Shore, R.A. [1980] *Second order logic and first order theories of reducibility orderings* (**P** 2058) Kleene Symp;1978 Madison 181-200
◇ B10 B15 D30 D35 F35 F40 ◇ REV MR 82g:03078 Zbl 465 # 03024 • ID 54927

Nerode, A. & Smith, Rick L. [1980] *The undecidability of the lattice of recursively enumerable subspaces* (**P** 3006) Brazil Conf Math Log (3);1979 Recife 245-252
◇ D35 D45 ◇ REV MR 82i:03060 Zbl 451 # 03014
• ID 54029

Nerode, A. [1981] see Ash, C.J.

Nerode, A. [1981] see Crossley, J.N.

Nerode, A. & Remmel, J.B. [1982] *Recursion theory on matroids* (**P** 3634) Patras Logic Symp;1980 Patras 41-65
◇ C57 D25 D45 ◇ REV MR 86b:03055a Zbl 526 # 03026
• REM Part I. Part II 1983 • ID 38172

Nerode, A. [1982] see Metakides, G.

Nerode, A. & Remmel, J.B. [1983] *Recursion theory on matroids II* (**P** 3669) SE Asian Conf on Log;1981 Singapore 133-184
◇ D25 D45 ◇ REV MR 86b:03055b Zbl 526 # 03026
• REM Part I 1982 • ID 43057

Nerode, A. & Remmel, J.B. [1985] *A survey of lattices of R.E. substructures* (**P** 4046) Rec Th;1982 Ithaca 323-375
◇ C57 C98 D45 D98 ◇ REV Zbl 573 # 03015 • ID 46402

Nerode, A. [1985] see Huang, Wengi

Nerode, A. & Remmel, J.B. [1985] *Generic objects in recursion theory* (**P** 3342) Rec Th Week;1984 Oberwolfach 271-314
◇ D25 E40 ◇ ID 45310

Nerode, A. & Shore, R.A. (eds.) [1985] *Recursion theory* (**S** 3304) Proc Symp Pure Math vii + 528pp
◇ D97 ◇ REV MR 86f:03004 Zbl 556 # 00008 • REM Proceedings of the AMS-ASL summer institute 1982
• ID 46160

Nerode, A. [1985] see Metakides, G.

Nerode, A. see Vol. III for further entries

Neuber, S. & Starke, P.H. [1967] *Ueber Homomorphie und Reduktion bei nicht-deterministischen Automaten* (**J** 0129) Elektr Informationsverarbeitung & Kybern 3*351-362
◇ D05 ◇ REV MR 39 # 1244 • ID 42683

Neumann, B.H. [1949] see Higman, G.

Neumann, B.H. [1959] see Baumslag, G.

Neumann, B.H. [1968] *Lectures on topics in the theory of infinite groups* (**X** 0925) Tata Inst Fund Res: Bombay iii + 267 + ivpp
◇ C05 C60 C98 D40 ◇ REV MR 42 # 1881 Zbl 237 # 20001 • ID 24829

Neumann, B.H. [1973] *The isomorphism problem for algebraically closed groups* (**P** 0678) Word Probl: Decis & Burnside Probl in Group Th;1969 Irvine 553-562
◇ C25 C60 D40 ◇ REV MR 54 # 2767 Zbl 262 # 20046 JSL 41.786 • ID 24064

Neumann, B.H. see Vol. III, V for further entries

Neumann, H. [1949] see Higman, G.

Neumann, H. see Vol. I, II, III for further entries

Neumann von, J. [1937] see Kuratowski, K.

Neumann von, J. [1966] *Theory of self-reproducing automata* (**X** 1285) Univ Ill Pr: Urbana xix + 388pp
• TRANSL [1971] (**X** 0885) Mir: Moskva 382pp (Russian)
◇ D10 D98 ◇ REV Zbl 247 # 94029 • ID 64203

Neumann von, J. see Vol. I, II, V, VI for further entries

Newman, J.R. [1958] see Nagel, E.

Newman, J.R. [1967] see Nagel, E.

Newman, J.R. see Vol. I, VI for further entries

NGO THE KHANH [1980] *Simple deterministic machines and their languages* (J 2293) Comp Linguist & Comp Lang 14*209-242
◇ D05 ◇ REV MR 83d:68050 Zbl 485 # 68074 • ID 39625

NGO THE KHANH [1983] *Prefix-free languages and simple deterministic machines (Hungarian)* (J 2845) Tanulmanyok 141*183pp
◇ D05 ◇ REV MR 85e:68045 • ID 40018

NGUEN HYU NGI [1966] *Estimates for sequences of states of Turing machines (Russian)* (J 0052) Probl Kibern 17*67-89
◇ D10 ◇ REV MR 37 # 5097 Zbl 199.36 • ID 09938

NGUYEN XUAN MY [1980] *Some classes of semi-Thue systems (German and Russian summaries)* (J 0129) Elektr Informationsverarbeitung & Kybern 16*3-10
◇ D03 ◇ REV MR 81j:68094 Zbl 443 # 68064 • ID 82352

NICOLESCU, R. [1973] *Some remarks on the theory of normal algorithms* (J 0447) An Univ Bucuresti, Mat 22*83-87
◇ D03 ◇ REV MR 50 # 1855 Zbl 292 # 02031 • ID 09942

NIELSEN, M. & SCHMIDT, ERIK MEINECHE (EDS.) [1982] *Automata, languages and programming* (S 3302) Lect Notes Comput Sci 140*vii + 614pp
◇ B75 D97 ◇ REV MR 83k:68002 Zbl 477 # 00027 • REM Proceedings of the 1982 conference, Aarhus • ID 40060

NIGIYAN, S.A. [1980] *Reducibility to free schemes (Russian)* (J 2605) Programmirovanie 1980/4*14-19,94
• TRANSL [1980] (J 2604) Progr Comput Software 6*175-180
◇ D05 ◇ REV MR 82f:68011 Zbl 469 # 68019 • ID 82354

NIGIYAN, S.A. [1983] *Solvability of algorithmic problems* (C 4581) Priklad Mat, Vyp 2 2*82-90,143
◇ D20 ◇ ID 48601

NIGMATULLIN, R.G. [1977] *The complexity of languages of type* $\cup M$ *(Russian)* (J 0199) Zh Vychisl Mat i Mat Fiz 17*1278-1284,1335
• TRANSL [1977] (J 1049) USSR Comput Math & Math Phys 17/5*174-181
◇ D15 ◇ REV MR 57 # 14601 Zbl 404 # 68051 • ID 54859

NIGMATULLIN, R.G. [1981] *The problem of lower bounds of complexity and the theory of NP-completeness (Russian)* (J 0031) Izv Vyssh Ucheb Zaved, Mat (Kazan) 1981/5(228)*17-25
• TRANSL [1981] (J 3449) Sov Math 25/5*11-20
◇ D15 ◇ REV MR 83e:68041 Zbl 478 # 68037 • ID 69738

NIGMATULLIN, R.G. see Vol. I for further entries

NIJHOLT, A. [1980] *Context-free grammars: covers, normal forms, and parsing* (X 0811) Springer: Heidelberg & New York vii + 253pp
◇ D05 ◇ REV MR 82f:68081 Zbl 477 # 68077 • ID 82355

NIJHOLT, A. [1982] *The equivalence problem for LL- and LR-regular grammars* (J 0119) J Comp Syst Sci 24*149-161
◇ D05 ◇ REV MR 83i:68114 Zbl 465 # 68038 • ID 39365

NIKITCHENKO, N.S. [1984] *Computable compositions and universal imperative program logics (Russian)* (J 2605) Programmirovanie 6*3-14,95
◇ B75 D20 ◇ REV MR 86d:68047 Zbl 551 # 68028 • ID 44847

NIKITIN, A.A. [1984] *Some algorithmic problems for projective planes (Russian)* (J 0003) Algebra i Logika 23*512-529
• TRANSL [1984] (J 0069) Algeb and Log 23*347-358
◇ C57 C65 D80 ◇ ID 48426

NIKODYM, O.M. [1929] *Sur les diverses classes d'ensembles* (J 0027) Fund Math 14*145-204
◇ D55 E15 ◇ REV FdM 55.657 • ID 09964

NIKODYM, O.M. see Vol. V for further entries

NISHIMURA, T. [1959] *On Goedel's theorem (Japanese)* (J 0091) Sugaku 11*1-12
• TRANSL [1961] (J 0090) J Math Soc Japan 13*1-12
◇ D35 F30 ◇ REV MR 27 # 2416 Zbl 100.11 JSL 29.106 JSL 34.649 • ID 09973

NISHIMURA, T. see Vol. I, II, III, V, VI for further entries

NISHIZAWA, T.W. [1970] *A proof of the equivalence of computability to recursiveness with no arithmetization* (J 0106) Mem Fac Sci, Kyushu Univ, Ser A 24*100-103
◇ D20 ◇ REV MR 42 # 7507 Zbl 267 # 02028 • ID 09977

NIVAT, M. [1971] *Congruence de Thue et t-languages* (J 0411) Studia Sci Math Hung 6*243-249
◇ D03 D05 D40 ◇ REV MR 45 # 1764 Zbl 242 # 68040 • ID 09978

NIVAT, M. [1971] see COCHET, Y.

NIVAT, M. (ED.) [1973] *Automata, languages and programming* (X 0809) North Holland: Amsterdam 638pp
◇ B75 D05 D97 ◇ REV MR 50 # 12420 Zbl 254 # 00022 • ID 23570

NIVAT, M. [1973] *Congruences parfaites et quasi-parfaites* (S 2250) Semin Dubreil: Algebre 7*9pp
◇ D03 ◇ REV MR 52 # 13574 Zbl 338 # 02018 • ID 64219

NIVAT, M. & VIENNOT, G. (EDS.) [1976] *Journees algorithmiques. Tenues a l'Ecole Normale Superieure, Paris* (J 1620) Asterisque 38-39*ii + 293pp
◇ D97 ◇ REV MR 54 # 12421 • ID 48647

NIVAT, M. [1977] see BOASSON, L.

NIVAT, M. [1978] see ARNOLD, A.

NIVAT, M. [1979] *Infinite words, infinite trees, infinite computations* (P 3375) Found of Comput Sci (3);1978 Amsterdam 2*1-52
◇ B75 D05 D75 ◇ REV MR 82e:68015 Zbl 423 # 68012 • ID 69743

NIVAT, M. [1979] see AUTEBERT, J.-M.

NIVAT, M. [1980] see BOASSON, L.

NIVAT, M. [1980] see ARNOLD, A.

NIVAT, M. [1981] see BOASSON, L.

NIVAT, M. & PERRIN, D. (EDS.) [1984] *Automata on infinite words* (S 3302) Lect Notes Comput Sci 194*v + 216pp
◇ D05 ◇ REV MR 86h:68006 Zbl 563 # 00019 • ID 49470

NIVAT, M. [1984] see GIRE, F.

NIVAT, M. see Vol. V for further entries

NIWINSKI, D. [1983] *Fixed-point semantics for algebraic (tree) grammars* (S 3382) Sem-ber, Humboldt-Univ Berlin, Sekt Math 52*69-83
◇ D05 ◇ REV MR 85f:68003 Zbl 539 # 68012 • ID 47881

NIWINSKI, D. [1984] *Fixed-point characterization of context free ∞-languages* (**J** 0194) Inform & Control 61*247-276
◇ D05 ◇ ID 45403

NIWINSKI, D. [1985] *Equational μ-calculus* (**P** 4670) Comput Th (5);1984 Zaborow 169-176
◇ D20 ◇ ID 49725

NOGINA, E.YU. [1966] *On effective topological spaces* (**J** 0023) Dokl Akad Nauk SSSR 169*28-31
◇ C57 C65 ◇ REV Zbl 154.7 • ID 49889

NOGINA, E.YU. [1969] *Correlations between certain classes of effectively topological spaces (Russian)* (**J** 0087) Mat Zametki (Akad Nauk SSSR) 5*483-495
• TRANSL [1969] (**J** 1044) Math Notes, Acad Sci USSR 5*288-294
◇ C57 C65 D45 ◇ REV MR 40 # 4082 Zbl 188.328
• ID 09983

NOGINA, E.YU. & VAJNBERG, YU.R. [1974] *Categories of effectively topological spaces (Russian)* (**C** 2577) Issl Formaliz Yazyk & Neklass Log 253-273
◇ D45 G30 ◇ REV MR 56 # 8336 • ID 79581

NOGINA, E.YU. & VAJNBERG, YU.R. [1976] *Two types of continuity of computable mappings of numerated topological spaces (Russian)* (**S** 0554) Issl Teor Algor & Mat Logik (Moskva) 2*84-99,159
◇ D45 ◇ REV MR 58 # 16214 • ID 79580

NOGINA, E.YU. [1978] *Numerierte topologische Raeume (Russisch)* (**J** 0068) Z Math Logik Grundlagen Math 24*141-176
◇ C57 C60 D45 ◇ REV MR 80e:03051 Zbl 415 # 03034
• ID 53136

NOGINA, E.YU. [1978] *On completely enumerable subsets of direct products of numbered sets (Russian)* (**C** 3211) Mat Ling & Teor Algor 130-132
◇ D45 ◇ REV Zbl 419 # 03027 • ID 53370

NOGINA, E.YU. [1981] *The relation between separability and traceability of sets (Russian)* (**C** 3747) Mat Log & Mat Lingvistika 135-144
◇ D45 ◇ REV MR 84e:03054 • ID 34389

NOJIMA, S. [1974] see NARUSHIMA, H.

NOJIMA, S. see Vol. I for further entries

NOLIN, L. [1966] *Organigrammes et machines de Turing* (**P** 0746) Automata Th;1964 Ravello 295-303
◇ A05 D10 D20 ◇ REV Zbl 253 # 02033 • ID 28936

NOLIN, L. [1969] *Formalisation des notions de machine et de programme* (**X** 0834) Gauthier-Villars: Paris xiv+218pp
◇ B75 D10 D20 ◇ REV Zbl 309 # 68004 • ID 64226

NOLIN, L. [1974] *Algorithmes universels* (**J** 0205) Rev Franc Autom, Inf & Rech Operat 8/R-1*5-18
◇ B75 D15 ◇ REV MR 51 # 4711 Zbl 291 # 68016
• ID 64225

NOLIN, L. [1975] *Algorithmes seriels, algorithmes paralleles* (**P** 2974) New Concepts & Tech in Parall Inform Process;1973 Capri 167-186
◇ B40 B75 D20 ◇ REV MR 58 # 25098 • ID 82368

NOLIN, L. [1975] *Theorie des algorithmes et semantique des langages de programmation* (**P** 1475) Conv Inform Teor & Conv Strutt Corpi Algeb;1973 Roma 283-303
◇ B40 B75 D20 ◇ REV MR 54 # 1700 Zbl 332 # 68011
• ID 23994

NOLIN, L. see Vol. I, VI for further entries

NOLTEMEIER, H. [1981] *Informatik I: Einfuehrung in Algorithmen und Berechenbarkeit* (**X** 3223) Hanser: Muenchen 148pp
◇ D20 D98 ◇ REV MR 86f:68001a Zbl 477 # 68002 • REM Part II 1984 by Laue,R. & Noltemeier,H. • ID 69744

NORGELA, S.A. [1976] *On approximating reduction classes of CPC by decidable classes (Russian) (English summary)* (**S** 0228) Zap Nauch Sem Leningrad Otd Mat Inst Steklov 60*103-108,224
• TRANSL [1980] (**J** 1531) J Sov Math 14*1493-1496
◇ B20 B25 D35 ◇ REV MR 58 # 27422 Zbl 342 # 02033
• ID 32671

NORGELA, S.A. [1977] *On recursive nonseparability of the strategies of deduction-search in the classical predicate calculus (Russian)* (**J** 2574) Litov Mat Sb (Vil'nyus) 17/3*146-147
◇ B35 D35 ◇ ID 32680

NORGELA, S.A. [1978] *Herbrand strategies of deduction-search in predicate calculus I (Russian) (English and Lithuanian summaries)* (**J** 2574) Litov Mat Sb (Vil'nyus) 18/4*95-100,201
• TRANSL [1978] (**J** 3283) Lith Math J 18*513-517
◇ B25 B35 D35 F07 ◇ REV MR 80e:03050
Zbl 394 # 03017 • REM Part II 1979 • ID 32674

NORGELA, S.A. see Vol. I, III, VI for further entries

NORMAN, J.K. [1985] see LADNER, R.E.

NORMANN, D. [1974] *Imbedding of higher type theories* (**S** 1626) Oslo Preprint Ser 16*
◇ D65 D70 D75 ◇ ID 27558

NORMANN, D. [1974] see FENSTAD, J.E.

NORMANN, D. [1974] *On abstract 1-sections* (**J** 0154) Synthese 27*259-263
◇ D65 D75 ◇ REV MR 58 # 5134 Zbl 327 # 02039
• ID 64228

NORMANN, D. [1974] see MOLDESTAD, J.

NORMANN, D. [1975] *Forcing arguments and some degree-theoretic problems in higher type recursion theory* (**S** 1626) Oslo Preprint Ser 17
◇ D30 D65 E40 ◇ ID 28312

NORMANN, D. [1976] see MOLDESTAD, J.

NORMANN, D. [1976] *On a problem of S.Wainer* (**S** 1626) Oslo Preprint Ser 10
◇ D55 D65 ◇ ID 28317

NORMANN, D. [1977] *Countable functionals and the analytic hierarchy* (**S** 1626) Oslo Preprint Ser 17
◇ D55 D65 ◇ ID 28321

NORMANN, D. [1978] *A continuous functional with noncollapsing hierarchy* (**J** 0036) J Symb Logic 43*487-491
◇ D55 D65 ◇ REV MR 80b:03063 Zbl 393 # 03007
• ID 29275

NORMANN, D. & STOLTENBERG-HANSEN, V. [1978] *A non-adequate admissible set with a good degree-structure* (**P** 1628) Generalized Recursion Th (2);1977 Oslo 321-329
◇ D30 D60 ◇ REV MR 80b:03060 Zbl 453 # 03047
• ID 30969

NORMANN, D. [1978] *Set recursion* (**P** 1628) Generalized Recursion Th (2);1977 Oslo 303-320
◇ D65 E47 ◇ REV MR 80j:03063 Zbl 453 # 03047
• ID 28318

NORMANN, D. [1979] *A classification of higher type functionals* (**P** 2615) Scand Logic Symp (5);1979 Aalborg 301-308
◇ D65 ◇ REV MR 82e:03044 Zbl 427 # 03036 • ID 53723

NORMANN, D. [1979] *A jump operator in set recursion* (**J** 0068) Z Math Logik Grundlagen Math 25*251-264
◇ D30 D65 E47 ◇ REV MR 83c:03042 Zbl 414 # 03027 JSL 47.902 • ID 28319

NORMANN, D. [1979] *A note on reflection* (**J** 0132) Math Scand 45*5-12
◇ D65 ◇ REV MR 81k:03046 Zbl 428 # 03040 • ID 53799

NORMANN, D. [1979] *Degrees of functionals* (**J** 0007) Ann Math Logic 16*269-304
◇ D30 D65 E45 E50 ◇ REV MR 84g:03072 Zbl 421 # 03036 JSL 48.212 • ID 53435

NORMANN, D. [1979] *Recursion in ^3E and a splitting theorem* (**P** 1705) Scand Logic Symp (4);1976 Jyvaeskylae 275-285
◇ D30 D65 ◇ REV MR 81b:03052 Zbl 398 # 03035
• ID 52763

NORMANN, D. [1980] *Recursion on the countable functionals* (**X** 0811) Springer: Heidelberg & New York viii + 191pp
◇ D30 D55 D65 D98 ◇ REV MR 82d:03075 Zbl 446 # 03034 JSL 49.668 • ID 56563

NORMANN, D. [1980] *The recursion theory of the continuous functionals* (**P** 3021) Logic Colloq;1979 Leeds 171-183
◇ D55 D65 ◇ REV MR 82h:03045 Zbl 439 # 03028
• ID 56019

NORMANN, D. & WAINER, S.S. [1980] *The 1-section of a countable functional* (**J** 0036) J Symb Logic 45*549-562
◇ D55 D65 ◇ REV MR 82f:03040 Zbl 439 # 03029
• ID 56020

NORMANN, D. [1981] *Countable functionals and the projective hierarchy* (**J** 0036) J Symb Logic 46*209-215
◇ D55 D65 ◇ REV MR 83j:03077 Zbl 459 # 03020
• ID 54460

NORMANN, D. [1981] *The continuous functionals; computations,recursions and degrees* (**J** 0007) Ann Math Logic 21*1-26
◇ D30 D65 ◇ REV MR 83b:03056 Zbl 472 # 03037 JSL 49.668 • ID 35105

NORMANN, D. [1982] *External and internal algorithms on the continuous functionals* (**P** 3634) Patras Logic Symp;1980 Patras 137-144
◇ D65 ◇ REV MR 85g:03067 Zbl 574 # 03035 • ID 43891

NORMANN, D. [1982] *Nonobtainable continuous functionals* (**P** 3622) Int Congr Log, Meth & Phil of Sci (6,Proc);1979 Hannover 241-249
◇ D55 ◇ REV MR 85d:03093 Zbl 499 # 03036 • ID 41105

NORMANN, D. [1983] *Characterizing the continuous functionals* (**J** 0036) J Symb Logic 48*965-969
◇ D65 H05 ◇ REV MR 85d:03094 Zbl 536 # 03026
• ID 37108

NORMANN, D. [1983] *General type-structures of continuous and countable functionals* (**J** 0068) Z Math Logik Grundlagen Math 29*177-192
◇ D55 D65 D75 F35 ◇ REV MR 85d:03095 Zbl 529 # 03022 • ID 37656

NORMANN, D. [1983] *R.e. degrees of continuous functionals* (**J** 0009) Arch Math Logik Grundlagenforsch 23*79-98
◇ D30 D65 ◇ REV MR 85g:03068 Zbl 527 # 03028
• ID 37576

NORMANN, D. [1984] see GRIFFOR, E.R.

NORMANN, D. [1984] *The infinite - a mathematical necessity* (**J** 3075) Normat 32*63-70
◇ A05 D20 F30 ◇ REV MR 85j:03001 Zbl 536 # 03013
• ID 37098

NORMANN, D. [1985] *Aspects of the continuous functionals* (**P** 4046) Rec Th;1982 Ithaca 171-176
◇ D65 ◇ ID 46376

NORMANN, D. [1985] see GIRARD, J.-Y.

NORMANN, D. see Vol. V for further entries

NORRIS, E.M. [1970] see BEDNAREK, A.R.

NOSKOV, G.A. [1982] *On conjugacy in metabelian groups (Russian)* (**J** 0087) Mat Zametki (Akad Nauk SSSR) 31*495-507,653
• TRANSL [1982] (**J** 1044) Math Notes, Acad Sci USSR 31*252-258
◇ C57 C60 ◇ REV MR 83i:20029 Zbl 492 # 20018
• ID 39350

NOSKOV, G.A. [1983] *Elementary theory of a finitely generated commutative ring (Russian)* (**J** 0087) Mat Zametki (Akad Nauk SSSR) 33*23-29,157
• TRANSL [1983] (**J** 1044) Math Notes, Acad Sci USSR 33*12-15
◇ C60 D35 ◇ REV MR 84g:03063 Zbl 524 # 13014
• ID 34170

NOSKOV, G.A. [1983] *The elementary theory of a finitely generated almost solvable group (Russian)* (**J** 0216) Izv Akad Nauk SSSR, Ser Mat 47*498-517
• TRANSL [1983] (**J** 0448) Math of USSR, Izv 22*465-482
◇ B25 C60 D35 ◇ REV MR 85d:20029 Zbl 521 # 20019
• ID 38972

NOVIKOV, P.S. [1935] see LUZIN, N.N.

NOVIKOV, P.S. [1935] *Sur la separabilite des ensembles projectifs de seconde classe* (**J** 0027) Fund Math 25*459-466
◇ D55 E15 ◇ REV Zbl 12.344 FdM 61.222 • ID 10014

NOVIKOV, P.S. [1937] *Les projections des complementaires analytiques uniformes (Russian summary)* (**J** 0142) Mat Sb, Akad Nauk SSSR, NS 2(44)*3-16
◇ D55 E15 ◇ REV Zbl 17.158 FdM 63.178 • ID 20900

NOVIKOV, P.S. [1937] *On the mutual relation of the second class of projective sets and the projection of uniform analytic complements (Russian)* (**J** 0216) Izv Akad Nauk SSSR, Ser Mat 1*231-252
◇ D55 E15 ◇ REV FdM 63.1564 • ID 45567

NOVIKOV, P.S. [1939] *On projections of some B-sets (Russian)* (**J** 0023) Dokl Akad Nauk SSSR 23*863-864
◇ D55 E15 ◇ REV MR 1.302 Zbl 61.95 FdM 65.1169
• ID 10015

NOVIKOV, P.S. [1948] see LYAPUNOV, A.A.

NOVIKOV, P.S. [1949] *On the axiom of complete induction (Russian)* (J 0023) Dokl Akad Nauk SSSR 64*457-459
◇ B25 B28 D35 F30 ◇ REV MR 11.304 Zbl 38.151 JSL 14.256 • ID 19476

NOVIKOV, P.S. [1949] *The consistency of certain statements of the theory of sets (Russian)* (J 0067) Usp Mat Nauk 4/2(30)*171
◇ D55 E15 E35 E45 ◇ REV MR 11.304 • ID 10019

NOVIKOV, P.S. [1951] *On the consistency of some propositions on the descriptive theory of sets (Russian)* (S 0066) Tr Mat Inst Steklov 38*279-316
• TRANSL [1963] (J 0225) Amer Math Soc, Transl, Ser 2 29*51-89
◇ D55 E15 E35 E45 ◇ REV MR 14.234 Zbl 53.364 JSL 19.123 • ID 19475

NOVIKOV, P.S. [1952] *Algorithmic unsolvability of the word problem in group theory (Russian)* (J 0067) Usp Mat Nauk 7/5(51)*197
◇ D40 ◇ REV Zbl 68.13 • ID 45568

NOVIKOV, P.S. [1952] *On algorithmic unsolvability of the word problem (Russian)* (J 0023) Dokl Akad Nauk SSSR 85*709-712
◇ D40 ◇ REV MR 14.618 Zbl 47.249 JSL 19.58 • ID 19474

NOVIKOV, P.S. [1954] *Unsolvability of the conjugacy problem in group theory (Russian)* (J 0216) Izv Akad Nauk SSSR, Ser Mat 18*485-524
◇ D40 ◇ REV MR 17.706 Zbl 57.15 JSL 23.52 • ID 19473

NOVIKOV, P.S. [1955] *On the algorithmic insolvability of the word problem in group theory (Russian)* (S 0066) Tr Mat Inst Steklov 44*143pp
• TRANSL [1958] (J 0225) Amer Math Soc, Transl, Ser 2 9*1-122
◇ D40 ◇ REV MR 17.706 Zbl 68.13 JSL 23.50 • ID 19472

NOVIKOV, P.S. [1956] *On the unsolvability of the word problem for groups and some other problems of algebra (Russian) (English summary)* (J 0022) Cheskoslov Mat Zh 6(81)*450-454
◇ D40 ◇ REV MR 22 # 12133 Zbl 75.235 JSL 29.56 • ID 19471

NOVIKOV, P.S. [1958] see ADYAN, S.I.

NOVIKOV, P.S. [1958] *Ueber einige algorithmische Probleme der Gruppentheorie* (J 0157) Jbuchber Dtsch Math-Ver 61*88-92
◇ D35 D40 ◇ REV MR 21 # 6387 Zbl 85.251 • ID 10021

NOVIKOV, P.S. [1959] *On the unsolvability of some problems in algebra I (Russian)* (P 0607) All-Union Math Conf (3);1956 Moskva 2*65-66
◇ D35 D40 ◇ ID 28389

NOVIKOV, P.S. [1968] see ADYAN, S.I.

NOVIKOV, P.S. see Vol. I, II, III, V, VI for further entries

NOVOTNY, M. [1966] *Ueber endlich charakterisierbare Sprachen* (J 0086) Cas Pestovani Mat, Ceskoslov Akad Ved 91*92-94
◇ D05 ◇ REV Zbl 142.248 JSL 35.487 • ID 10037

NOVOTNY, M. [1974] *On some operators reducing generalized grammars* (J 0194) Inform & Control 26*225-235
◇ D05 ◇ REV MR 50 # 15457 Zbl 293 # 68061 • ID 31959

NOVOTNY, M. [1974] *Sets constructed by acceptors* (J 0194) Inform & Control 26*116-133
◇ D05 D25 ◇ REV MR 54 # 14458 Zbl 292 # 94033 • ID 82374

NOVOTNY, M. see Vol. II, III, V for further entries

NOWACZYK, A. [1978] *Categorial languages and variable-binding operators* (J 0063) Studia Logica 37*27-39
◇ A05 D05 ◇ REV MR 81h:03055 Zbl 414 # 03003 • ID 53049

NOWAK, J. [1969] *A theory of finite many-output automata defined by matrices (Polish and Russian summaries)* (J 0063) Studia Logica 24*55-81
◇ D05 ◇ REV MR 42 # 1601 Zbl 298 # 94056 • ID 33130

NOWAK, J. [1970] *A simplification of the basic theorem on automata (Polish and Russian summaries)* (J 0063) Studia Logica 26*35-44
◇ D05 ◇ REV MR 57 # 19184 Zbl 298 # 94057 • ID 33134

NOWAK, J. [1973] *On a certain congruence of automata with respect to connecting and coupling with retardation (Polish and Russian summaries)* (J 0063) Studia Logica 31*159-165
◇ D05 ◇ REV MR 49 # 772 Zbl 292 # 94030 • ID 33140

NOWAK, J. [1973] *Remarks on the notion of equivalence of automata (Polish and Russian summaries)* (J 0063) Studia Logica 31*153-158
◇ D05 ◇ REV MR 58 # 15741 Zbl 292 # 94029 • ID 33139

NOWAKOWSKI, R. [1965] *A structural theory of matrix-defined finite automata (Polish) (Russian and English summaries)* (J 0063) Studia Logica 16*75-116
◇ D05 ◇ REV MR 32 # 1075 JSL 32.391 • ID 19470

NOWAKOWSKI, R. [1970] *Two definitions of the equivalence of automata (Polish and Russian summaries)* (J 0063) Studia Logica 26*7-17
◇ D05 ◇ REV MR 54 # 4868 • ID 33133

NOWAKOWSKI, R. [1973] *On the homomorphisms of graphs of automata* (S 4733) Prace Inst Mat, Politech Wroclaw, Ser Stud Mater 5*15-24
◇ D05 ◇ REV Zbl 259 # 94047 • ID 64237

NOWAKOWSKI, R. [1973] *On the products of graphs of automata* (S 4733) Prace Inst Mat, Politech Wroclaw, Ser Stud Mater 5*3-14
◇ D05 ◇ REV MR 57 # 11970 Zbl 259 # 94046 • ID 64236

NOWAKOWSKI, R. [1973] *Some properties of connection of graphs of automata* (S 4733) Prace Inst Mat, Politech Wroclaw, Ser Stud Mater 5*25-28
◇ D05 ◇ REV Zbl 259 # 94048 • ID 64238

NOZAKI, A. [1969] *On the notion of universality of Turing machine (Czech summary)* (J 0156) Kybernetika (Prague) 5*29-43
◇ D10 ◇ REV MR 41 # 3187 Zbl 167.15 • ID 10038

NOZAKI, A. see Vol. I, II for further entries

NUERNBERG, G. [1973] see DAUSCHA, W.

NURMEEV, N.N. [1976] *Computation of boolean functions by Turing machines (Russian)* (J 3937) Veroyat Met i Kibern (Kazan) 12-13*60-76
◇ B05 D10 D15 ◇ REV MR 58 # 27389 Zbl 397 # 68043 • ID 52721

NURTAZIN, A.T. [1973] see GONCHAROV, S.S.

NURTAZIN, A.T. [1974] *Strong and weak constructivization and computable families (Russian)* (**J** 0003) Algebra i Logika 13*311-323,364
• TRANSL [1974] (**J** 0069) Algeb and Log 13*177-184
◇ C57 D45 ◇ REV MR 52#2851 Zbl 302#02014
• ID 17642

NURTAZIN, A.T. [1978] see DOBRITSA, V.P.

NURTAZIN, A.T. see Vol. III, V for further entries

NYBERG, A.M. [1976] *Uniform inductive definability and infinitary languages* (**J** 0036) J Symb Logic 41*109-120
◇ C70 D70 ◇ REV MR 54#7201 Zbl 374#02023
• ID 14788

NYBERG, A.M. [1977] *Inductive operators on resolvable structures* (**P** 1629) Symp Math Log;1974 Oulo;1975 Helsinki 91-100
◇ C70 D60 D70 D75 ◇ ID 28324

NYBERG, A.M. see Vol. I, III, V for further entries

OBERQUELLE, H. [1971] see FELDMANN, H.

OBERSCHELP, A. [1960] *Ein Satz ueber die Unloesbarkeitsgrade der Mengen von natuerlichen Zahlen* (**J** 0535) Abh Braunschweig Wiss Ges 12*1-3
◇ D25 ◇ REV MR 23#A3680 Zbl 109.9 JSL 32.124
• ID 10044

OBERSCHELP, A. [1961] *Ueber die Unentscheidbarkeit gewisser Axiommengen* (**J** 0009) Arch Math Logik Grundlagenforsch 5*112
◇ D35 ◇ REV MR 25#1098 Zbl 124.4 • ID 10045

OBERSCHELP, A. [1966] *Berechenbarkeit* (**C** 4260) Fischer Lexikon Math 2 30-57
◇ D20 D98 ◇ ID 47105

OBERSCHELP, A. [1966] *Rechenmaschinen* (**C** 4260) Fischer Lexikon Math 2 260-285
◇ D20 D98 ◇ ID 47109

OBERSCHELP, A. see Vol. I, II, III, V for further entries

OBERSCHELP, W. [1958] *Varianten von Turingmaschinen* (**J** 0009) Arch Math Logik Grundlagenforsch 4*53-62
◇ D10 ◇ REV MR 20#4494 Zbl 84.10 JSL 36.534
• ID 10053

OBERSCHELP, W. [1984] see BOERGER, E.

OBERSCHELP, W. see Vol. I, III for further entries

OCHAKOVSKAYA, O.N. [1975] *Complexity of the derivation of equivalent Janov operator schemes (Russian) (English summary)* (**J** 0040) Kibernetika, Akad Nauk Ukr SSR 1975/6*43-50
• TRANSL [1975] (**J** 0021) Cybernetics 11*891-898
◇ B75 D15 ◇ REV MR 58#32063 Zbl 331#68008
• ID 64248

OCHRANOVA-DOLEZELOVA, R. [1968] *Real-time decidability, computability, countability and generability* (**J** 0068) Z Math Logik Grundlagen Math 14*283-288
◇ D10 D15 ◇ REV MR 37#3875 • ID 10063

O'CONNOR, S. [1974] see CROSSLEY, J.N.

ODIFREDDI, P. [1975] *Note sugli insiemi implicitamente definibili* (**J** 2038) Rend Sem Mat, Torino 34*327-332
◇ D55 ◇ REV MR 56#5250 Zbl 356#02038 • ID 50310

ODIFREDDI, P. [1977] *A note on Suzuki's chain of hyperdegrees* (**J** 0047) Notre Dame J Formal Log 18*589-590
◇ D30 D55 E15 ◇ REV MR 58#21548 Zbl 351#02033
• ID 23692

ODIFREDDI, P. [1978] *Ricorsivita su ordinali ammissibili* (**J** 3285) Boll Unione Mat Ital, V Ser, A 15*495-516
◇ D60 ◇ REV MR 80a:03061 Zbl 393#03034 • ID 52454

ODIFREDDI, P. [1981] *Insiemi subcreativi (English summary)* (**J** 2038) Rend Sem Mat, Torino 39/3*69-73
◇ D15 D25 ◇ REV MR 84j:03087 Zbl 514#03024
• ID 34678

ODIFREDDI, P. [1981] *Strong reducibilities* (**J** 0589) Bull Amer Math Soc (NS) 4*37-86
◇ D25 D30 ◇ REV MR 82k:03064 Zbl 484#03024
• ID 76927

ODIFREDDI, P. [1981] *Trees and degrees* (**C** 2922) Cabal Seminar Los Angeles 1977-79 235-271
◇ D30 ◇ REV MR 82h:03038 Zbl 495#03024 • ID 76928

ODIFREDDI, P. [1983] *Forcing and reducibilities* (**J** 0036) J Symb Logic 48*288-310
◇ D30 D55 E40 F30 ◇ REV MR 85m:03034a Zbl 526#03025 • REM Part I. Parts II,III 1983 • ID 38171

ODIFREDDI, P. [1983] *Forcing and reducibilities. II: Forcing in fragments of analysis. III: Forcing in fragments of set theory* (**J** 0036) J Symb Logic 48*724-743,1013-1034
◇ D30 D55 E40 ◇ REV MR 85m:03034 Zbl 574#03024 Zbl 574#03025 • REM Part I 1983 • ID 33611

ODIFREDDI, P. [1983] *On the first order theory of the arithmetical degrees* (**J** 0053) Proc Amer Math Soc 87*505-507
◇ C65 D30 D55 ◇ REV MR 84d:03056 Zbl 554#03023
• ID 34091

ODIFREDDI, P. [1985] *Global properties (automorphisms and definability) of m-degrees (Italian) (English summary)* (**J** 2099) Boll Unione Mat Ital, VI Ser, A 4*71-76
◇ D30 ◇ ID 44816

ODIFREDDI, P. [1985] *Rapidly growing recursive functions and associated ordinals (Italian)* (**P** 4646) Atti Incontri Log Mat (2);1983/84 Siena 393-429
◇ D20 ◇ ID 49610

ODIFREDDI, P. [1985] *The structure of m-degrees* (**P** 3342) Rec Th Week;1984 Oberwolfach 315-332
◇ D30 ◇ ID 45311

ODINTSOV, S.P. [1984] *Atomless ideals of constructive Boolean algebras (Russian)* (**J** 0003) Algebra i Logika 23*278-295,362
• TRANSL [1984] (**J** 0069) Algeb and Log 23*190-203
◇ C57 G05 ◇ REV MR 86f:03069 • ID 44901

O'DONNELL, M.J. [1977] *Subtree replacement systems: a unifying theory for recursive equations, LISP, lucid and combinatory logic* (**P** 1903) ACM Symp Th of Comput (9);1977 Boulder 295-305
◇ B40 B75 D05 D20 ◇ REV MR 57#9507 • ID 76932

O'DONNELL, M.J. [1979] *A practical programming theorem which is independent of Peano arithmetic* (**P** 2539) Frege Konferenz (1);1979 Jena 284-298
◇ B40 B75 D20 F05 F30 ◇ REV MR 82i:03066
• ID 76931

O'Donnell, M.J. [1979] *A programming language theorem which is independent of Peano arithmetic* (**P** 3542) ACM Symp Th of Comput (11);1979 Atlanta 176-188
◇ B75 D15 F30 ◇ REV MR 81d:68015 • ID 82386

O'Donnell, M.J. see Vol. I, VI for further entries

O'Dunlaing, C. [1981] see Book, R.V.

O'Dunlaing, C. [1983] *Infinite regular Thue systems* (**J** 1426) Theor Comput Sci 25∗171-192
◇ D03 ◇ REV MR 84f:03035 Zbl 512 # 03018 • ID 34455

O'Dunlaing, C. [1983] *Undecidable questions related to Church-Rosser Thue systems* (**J** 1426) Theor Comput Sci 23∗339-345
◇ D03 ◇ REV MR 84m:68024 Zbl 512 # 03019 • ID 36513

Oganesyan, G.U. [1978] *A class of semigroups with a decidable word problem (Russian)* (**J** 0087) Mat Zametki (Akad Nauk SSSR) 24∗259-265,303
• TRANSL [1978] (**J** 1044) Math Notes, Acad Sci USSR 24∗640-643
◇ D40 ◇ REV MR 80i:03055 Zbl 389 # 20048 • ID 52321

Oganesyan, G.U. [1984] *The isomorphism problem for semigroups with one defining relation (Russian)* (**J** 0087) Mat Zametki (Akad Nauk SSSR) 35∗685-690
• TRANSL [1984] (**J** 1044) Math Notes, Acad Sci USSR 35∗360-363
◇ D40 ◇ REV MR 85m:20085 Zbl 549 # 20035 • ID 45845

Ohashi, K. [1964] *A stronger form of a theorem of Friedberg* (**J** 0047) Notre Dame J Formal Log 5∗10-12
◇ D25 ◇ REV MR 30 # 1042 Zbl 143.253 • ID 10072

Ohashi, K. [1964] *Enumeration of some classes of recursively enumerable sets* (**J** 0068) Z Math Logik Grundlagen Math 10∗1-6
◇ D25 ◇ REV MR 28 # 4996 Zbl 154.257 • ID 10071

Ohashi, K. [1970] *On a problem of G.E.Sacks* (**J** 0036) J Symb Logic 35∗46-50
◇ D60 ◇ REV MR 45 # 1760 Zbl 205.10 • ID 10073

Ohashi, K. see Vol. VI for further entries

Okee, J. [1975] *A semantical proof of the undecidability of the monadic intuitionistic predicate calculus of first order* (**J** 0047) Notre Dame J Formal Log 16∗552-554
◇ D35 F50 ◇ REV MR 52 # 5370 Zbl 258 # 02032
• ID 18315

Okee, J. see Vol. VI for further entries

Oktaba, H. [1983] see Banachowski, L.

Olin, P. [1969] *Indefinability in the arithmetic of isolic integers* (**J** 0048) Pac J Math 29∗175-186
◇ D50 H15 ◇ REV MR 43 # 49 Zbl 184.20 • ID 10092

Olin, P. see Vol. III, V for further entries

Omanadze, R.Sh. [1976] *On one kind of reducibility (Russian) (English summary)* (**J** 0233) Soobshch Akad Nauk Gruz SSR 83∗281-284
◇ D30 ◇ REV MR 55 # 12488 Zbl 334 # 02022 • ID 32657

Omanadze, R.Sh. [1976] *The completeness of recursively enumerable sets (Russian) (Georgian and English summaries)* (**J** 0233) Soobshch Akad Nauk Gruz SSR 81∗529-532
◇ D25 ◇ REV MR 54 # 4944 Zbl 339 # 02040 • ID 24117

Omanadze, R.Sh. [1978] *On some generalizations of the notion of set productivity (Russian)* (**J** 0031) Izv Vyssh Ucheb Zaved, Mat (Kazan) 1978/9(196)∗84-88
• TRANSL [1978] (**J** 3449) Sov Math 22/9∗65-68
◇ D25 ◇ REV MR 80e:03046 Zbl 412 # 03023 • ID 32661

Omanadze, R.Sh. [1978] *On the reducibility on the class of recursive enumerable sets (Russian) (English summary)* (**J** 0233) Soobshch Akad Nauk Gruz SSR 91∗549-552
◇ D25 D30 ◇ REV MR 81e:03044 Zbl 399 # 03029
• ID 32658

Omanadze, R.Sh. [1979] *On some properties of Q-reducibility (Russian)* (**P** 2558) All-Union Conf Math Log (5) (Mal'tsev);1979 Novosibirsk
◇ D25 D30 ◇ ID 32660

Omanadze, R.Sh. [1979] *On Q-reducibility (Russian)* (**J** 0233) Soobshch Akad Nauk Gruz SSR 95∗29-32
◇ D25 D30 ◇ REV MR 81b:03045 Zbl 415 # 03032
• ID 32659

Omanadze, R.Sh. [1980] *On bounded Q-reducibility (Russian) (Georgian and English summaries)* (**J** 0233) Soobshch Akad Nauk Gruz SSR 100∗57-60
◇ D25 D30 ◇ REV MR 82h:03039 Zbl 468 # 03019
• ID 55084

Omanadze, R.Sh. [1984] *Upper semilattice of recursively enumerable Q-degrees (Russian)* (**J** 0003) Algebra i Logika 23∗175-184
• TRANSL [1984] (**J** 0069) Algeb and Log 23∗124-130
◇ D25 ◇ ID 44817

Omarov, A.I. [1984] *Elementary theory of D-degrees (Russian)* (**J** 0003) Algebra i Logika 23∗530-537
• TRANSL [1984] (**J** 0069) Algeb and Log 23∗358-363
◇ C20 C30 C50 D30 G10 ◇ ID 46564

Omarov, A.I. see Vol. III for further entries

Ono, H. [1974] *A formal system of partial recursive functions* (**J** 0390) Publ Res Inst Math Sci (Kyoto) 10∗271-291
◇ D20 ◇ REV MR 52 # 10398 Zbl 309 # 02036 • ID 21694

Ono, H. [1977] see Nakamura, A.

Ono, H. [1979] see Nakamura, A.

Ono, H. [1980] see Nakamura, A.

Ono, H. [1981] see Nakamura, A.

Ono, H. [1983] see Nakamura, A.

Ono, H. see Vol. I, II, III, V, VI for further entries

Onyszkiewicz, J. [1972] see Marek, W.

Onyszkiewicz, J. see Vol. III, V for further entries

Oohara, S. [1974] see Narushima, H.

Oppen, D.C. [1973] *Elementary bounds for Presburger arithmetic* (**P** 1482) ACM Symp Th of Comput (5);1973 Austin 34-37
◇ B25 C10 D15 F30 ◇ REV MR 54 # 4187
Zbl 306 # 02044 • ID 24077

Oppen, D.C. [1977] see Nelson, Greg

Oppen, D.C. [1978] *A $2^{2^{2^{pn}}}$ upper bound on the complexity of Presburger arithmetic* (**J** 0119) J Comp Syst Sci 16∗323-332
◇ B25 C10 D15 F30 ◇ REV MR 57 # 18224
Zbl 381 # 03021 • ID 51889

OPPEN, D.C. [1980] *Complexity, convexity and combinations of theories* (**J** 1426) Theor Comput Sci 12*291-302
◇ B25 B35 D15 ◇ REV MR 82a:03013 Zbl 437 # 03007
• ID 55875

OPPEN, D.C. [1980] see HUET, G.

OPPEN, D.C. [1980] *Reasoning about recursively defined data structures* (**J** 0037) ACM J 27*403-411
◇ B25 D15 ◇ REV MR 82a:68177 Zbl 477 # 68025
• ID 82395

OREJAS, F. [1976] *The priority method in degree theory (Spanish)* (**P** 1619) Coloq Log Simb;1975 Madrid 137-159
◇ D25 ◇ REV MR 56 # 108 Zbl 367 # 02022 • ID 51182

OREVKOV, V.P. [1965] see MASLOV, S.YU.

OREVKOV, V.P. [1967] *The undecidability of a class of formulas containing just one single place predicate variable in modal calculus (Russian)* (**S** 0228) Zap Nauch Sem Leningrad Otd Mat Inst Steklov 4*168-173
• TRANSL [1967] (**J** 0521) Semin Math, Inst Steklov 4*67-70
◇ B45 D35 ◇ REV MR 41 # 1514 Zbl 153.316 • ID 10149

OREVKOV, V.P. [1968] *Two undecidable classes of formulas in classical predicate calculus (Russian)* (**S** 0228) Zap Nauch Sem Leningrad Otd Mat Inst Steklov 8*202-210 • ERR/ADD ibid 20*292-294
• TRANSL [1968] (**J** 0521) Semin Math, Inst Steklov 8*98-102
◇ B20 D35 ◇ REV MR 42 # 5776 Zbl 182.18 • ID 10153

OREVKOV, V.P. [1971] *On biconjunctive reduction classes (Russian) (English summary)* (**S** 0228) Zap Nauch Sem Leningrad Otd Mat Inst Steklov 20*170-174,287
• TRANSL [1973] (**J** 1531) J Sov Math 1*106-109
◇ B20 D35 ◇ REV MR 45 # 4944 Zbl 222 # 02034
• ID 10157

OREVKOV, V.P. [1972] *Undecidable classes of formulas for the constructive predicate calculus I (Russian)* (**S** 0066) Tr Mat Inst Steklov 121*100-108,165
• TRANSL [1972] (**S** 0055) Proc Steklov Inst Math 121*111-119
◇ D35 F50 ◇ REV MR 49 # 7132 Zbl 286 # 02031
• ID 10161

OREVKOV, V.P. [1973] *On the complexity of expansion of algebraic irrationalities in continued fractions (Russian)* (**S** 0066) Tr Mat Inst Steklov 129*24-29,267
• TRANSL [1973] (**S** 0055) Proc Steklov Inst Math 129*20-24
◇ D20 D80 ◇ REV MR 55 # 12647 Zbl 359 # 10049
• ID 50619

OREVKOV, V.P. see Vol. I, II, III, VI for further entries

ORGASS, R.J. [1969] see FITCH, F.B.

ORGASS, R.J. see Vol. VI for further entries

ORLICKI, A. [1983] *On effective numberings of effective definitional schemes* (**J** 2095) Fund Inform, Ann Soc Math Pol, Ser 4 6*235-245
◇ D20 D45 ◇ REV MR 86b:03056 Zbl 527 # 03024
• ID 37574

ORLICKI, A. [1985] *On enumerated algebras and some monads in the category of enumerated sets* (**J** 2095) Fund Inform, Ann Soc Math Pol, Ser 4 8/3-4*285-307
◇ D45 G30 ◇ ID 49477

ORLOV, V.A. [1973] *A simple proof of the algorithmic undecidability of certain problems on the completeness of automaton bases (Russian) (English summary)* (**J** 0040) Kibernetika, Akad Nauk Ukr SSR 1973/4*109-113
◇ D05 ◇ REV MR 48 # 5831 Zbl 278 # 94033 • ID 10169

ORLOV, V.A. [1973] *Ueber die Kompliziertheit der Realisierung beschraenkt-determinierter Operatoren durch Schemata in Automatenbasen (Russisch)* (**J** 0052) Probl Kibern 26*141-182
◇ D03 D05 ◇ REV Zbl 268 # 94044 • ID 64297

ORLOVSKIJ, E.S. [1958] *Some questions in the theory of algorithms (Russian)* (**S** 0066) Tr Mat Inst Steklov 52*140-171
• TRANSL [1970] (**J** 0225) Amer Math Soc, Transl, Ser 2 94*1-36
◇ D20 ◇ REV MR 20 # 6357 Zbl 87.253 JSL 29.108
• ID 21254

ORLOVSKIJ, E.S. [1959] *Algorithmic operators in the narrow sense (Russian)* (**P** 0607) All-Union Math Conf (3);1956 Moskva 4*87
◇ D20 ◇ ID 31281

ORMAN, G. [1978] *An automata realization of the functions associated with a derivation* (**J** 3071) Bul Univ Brasov, Ser C 20*107-113
◇ D05 D10 ◇ REV MR 81e:68103 Zbl 442 # 68077
• ID 82403

ORMAN, G. [1979] *Control Turing machines for some derivational functions* (**J** 3071) Bul Univ Brasov, Ser C 21*103-112
◇ D10 ◇ REV MR 83e:68048 Zbl 465 # 68039 • ID 69751

ORMAN, G. [1980] *An automatic realization of $\xi(h)$-function (Romanian summary)* (**J** 3071) Bul Univ Brasov, Ser C 22*81-88
◇ D05 D10 ◇ REV MR 83i:68115 • ID 39368

ORNAGHI, M. [1981] see MIGLIOLI, P.A.

ORNAGHI, M. see Vol. I, II, III, VI for further entries

ORPONEN, P. [1983] *Complexity classes of alternating machines with oracles* (**P** 3851) Automata, Lang & Progr (10);1983 Barcelona 573-584
◇ D15 ◇ REV Zbl 521 # 68043 • ID 37476

ORPONEN, P. & SCHOENING, U. [1984] *The structure of polynomial complexity cores* (**P** 3658) Math Founds of Comput Sci (11);1984 Prague 452-458
◇ D15 ◇ REV Zbl 556 # 68014 • ID 46164

ORPONEN, P. & RUSSO, D.A. & SCHOENING, U. [1985] *Polynomial levelability and maximal complexity cores* (**P** 4628) Automata, Lang & Progr (12);1985 Nafplion 435-444
◇ D15 ◇ ID 49586

ORTLIEB, C. [1971] see FELDMANN, H.

OSIPOV, O. [1975] see ALAD'EV, V.Z.

OSIPOVA, V.A. [1968] *On the word problem for finitely presented semigroups (Russian)* (**J** 0023) Dokl Akad Nauk SSSR 178*1017-1020
• TRANSL [1968] (**J** 0062) Sov Math, Dokl 9*237-240
◇ D40 ◇ REV MR 37 # 1453 • ID 10177

OSIPOVA, V.A. [1972] *On equations with one unknown in semigroups with a bounded measure of overlap of the defining words (Russian)* (J 0023) Dokl Akad Nauk SSSR 203*1252-1254
• TRANSL [1972] (J 0062) Sov Math, Dokl 13*542-545
◇ D40 ◇ REV MR 45 # 6953 Zbl 275 # 20103 • ID 10178

OSIPOVA, V.A. [1973] *An algorithm for recognizing the solvability of equations with one unknown in semigroups with a measure of overlap of the defining words that is less than 1/3 (Russian)* (J 0142) Mat Sb, Akad Nauk SSSR, NS 92(134)*3-33,165
• TRANSL [1973] (J 0349) Math of USSR, Sbor 21*1-32
◇ C05 D40 ◇ REV MR 48 # 4161 Zbl 284 # 20060
• ID 29930

OSIPOVA, V.A. [1973] *On the conjugacy problem in semigroups (Russian)* (S 0066) Tr Mat Inst Steklov 133*169-182,275
◇ D40 ◇ REV MR 48 # 4162 Zbl 294 # 20058 • ID 10179

OSIPOVA, V.A. see Vol. V for further entries

OSTASZEWSKI, A.J. [1975] *On the descriptive set theory of the lexicographic square* (J 0027) Fund Math 87*261-281
◇ D55 E15 ◇ REV MR 52 # 1650 Zbl 303 # 54015
• ID 10191

OSTASZEWSKI, A.J. [1977] see DAVIES, R.O.

OSTASZEWSKI, A.J. see Vol. III, V for further entries

OSTROUKHOV, D.A. [1969] *Estimation of the complexity of normal algorithms (Russian)* (J 0023) Dokl Akad Nauk SSSR 184*1292-1293
• TRANSL [1969] (J 0062) Sov Math, Dokl 10*251-253
◇ D03 ◇ REV MR 40 # 7042 Zbl 236 # 02029 • ID 27859

OSTROUKHOV, D.A. [1970] *On the coding of natural numbers with the aid of schemes of normal algorithms (Russian)* (J 0068) Z Math Logik Grundlagen Math 16*347-352
◇ D03 D20 ◇ REV MR 45 # 3209 Zbl 222 # 02026
• ID 19538

OSTROUKHOV, D.A. [1973] *Linearization of constructive sequences of normal algorithms (Russian)* (S 0554) Issl Teor Algor & Mat Logik (Moskva) 1*97-133
◇ D03 F60 ◇ REV MR 49 # 2299 Zbl 284 # 02011
• ID 10192

OSTROUKHOV, D.A. [1973] *The complexity of terms of sequences of normal algorithms (Russian)* (C 2978) Vychisl Mat (Kiev) 55-59
◇ D03 D20 ◇ REV MR 57 # 16002 • ID 77035

OSTROUKHOV, D.A. see Vol. I for further entries

OSTROWSKA, M. [1982] see LIBUS, M.

OTT, G. [1969] see BOOK, R.V.

OTT, G. [1971] see BOOK, R.V.

OTTMANN, T. [1973] *Ketten und arithmetische Praedikate von endlichen Automaten* (P 1630) GI Fachtag Automatenth & Form Sprach (1);1973 Bonn 2*74-80
◇ D05 D55 D75 ◇ REV MR 55 # 10254 Zbl 273 # 02023
• ID 28325

OTTMANN, T. [1973] *Ueber Moeglichkeiten zur Simulation endlicher Automaten durch eine Art sequentieller Netzwerke aus einfachen Bausteinen* (J 0068) Z Math Logik Grundlagen Math 19*223-238
◇ D05 ◇ REV MR 47 # 8219 Zbl 305 # 94049 • ID 10198

OTTMANN, T. [1974] *Arithmetische Praedikate ueber einem Bereich endlicher Automaten* (J 0009) Arch Math Logik Grundlagenforsch 16*159-176
◇ D05 D55 D75 ◇ REV MR 55 # 7751 Zbl 352 # 02030
• ID 28327

OTTMANN, T. [1974] see KOERBER, P.

OTTMANN, T. [1975] *Eine universelle Turingmaschine mit zweidimensionalem Band, 7 Buchstaben und 2 Zustaenden (English and Russian summaries)* (J 0129) Elektr Informationsverarbeitung & Kybern 11*27-38
◇ D10 ◇ REV MR 52 # 16130 Zbl 312 # 02030 • ID 21890

OTTMANN, T. [1975] *Mit regulaeren Grundbegriffen definierbare Praedikate* (J 0373) Comp Arch Inform & Numerik 14*213-223
◇ D05 ◇ REV MR 56 # 2804 Zbl 312 # 02029 • ID 28331

OTTMANN, T. [1975] *Some classes of nets of finite automata* (1111) Preprints, Manuscr., Techn. Reports etc.
◇ D05 ◇ REM Bericht 29 Inst. Angewandte Informatik & Formale Beschreibungsverfahren, Univ. Karlsruhe, Jan. 1975
• ID 28328

OTTMANN, T. [1976] see HULE, H.

OTTMANN, T. [1976] *Rekursive Prozeduren und partiell rekursive Funktionen* (J 0160) Math-Phys Sem-ber, NS 23*206-227
◇ D20 ◇ REV MR 55 # 2525 Zbl 345 # 02029 • ID 28332

OTTMANN, T. [1977] see KLEINE BUENING, H.

OTTMANN, T. [1977] *Lokale Simulierbarkeit zweidimensionaler Turingmaschinen* (J 0129) Elektr Informationsverarbeitung & Kybern 13*465-471
◇ D10 ◇ REV MR 56 # 15379 Zbl 365 # 02025 • ID 28335

OTTMANN, T. [1977] see MAURER, H.A.

OTTMANN, T. [1978] *Eine einfache universelle Menge endlicher Automaten* (J 0068) Z Math Logik Grundlagen Math 24*55-61
◇ D05 ◇ REV MR 57 # 9349 Zbl 373 # 94029 • ID 28333

OTTMANN, T. [1978] see HULE, H.

OTTMANN, T. [1978] see CULIK II, K.

OTTMANN, T. [1981] see ALBERT, J.

OTTMANN, T. & SALOMAA, A. & WOOD, D. [1981] *Sub-regular grammar forms* (J 0232) Inform Process Lett 12*184-187
◇ D05 ◇ REV Zbl 477 # 68084 • ID 47324

OTTMANN, T. [1983] see ALBERT, J.

OTTMANN, T. [1984] see CHAZELLE, B.

OTTO, F. [1984] *Conjugacy in monoids with a special Church-Rosser presentation is decidable* (J 0136) Semigroup Forum 29*223-240
◇ C05 D03 D05 D40 ◇ REV Zbl 551 # 20044 • ID 43942

OTTO, F. [1984] see BAUER, G.

OTTO, F. [1984] *Some undecidability results for nonmonadic Church-Rosser Thue systems* (J 1426) Theor Comput Sci 33*261-278
◇ D03 ◇ REV MR 86f:68014 Zbl 563 # 03019 • ID 44155

OTTO, F. & WRATHALL, C. [1985] *A note on Thue systems with a single defining relation* (J 0041) Math Syst Theory 18*135-143
◇ D03 ◇ ID 47707

OTTO, F. [1985] see BOOK, R.V.

OTTO, F. [1985] *Deciding algebraic properties of monoids presented by finite Church-Rosser Thue systems* (**P** 4244) Rewriting Techn & Appl (1);1985 Dijon 95-106
◇ D03 ◇ ID 49770

OTTO, F. [1985] *Elements of finite order for finite monadic Church-Rosser Thue systems* (**J** 0064) Trans Amer Math Soc 291*629-637
◇ D03 ◇ ID 48212

OVERBEEK, R.A. [1971] see HUGHES, C.E.

OVERBEEK, R.A. [1973] *The representation of many-one degrees by decision problems of Turing machines* (**J** 3240) Proc London Math Soc, Ser 3 26*167-183
◇ D10 D25 ◇ REV MR 47 # 3151 Zbl 253 # 02041 • ID 64322

OVERBEEK, R.A. [1973] *The representation of many-one degrees by the word problem for Thue systems* (**J** 3240) Proc London Math Soc, Ser 3 26*184-192
◇ D03 D30 D40 ◇ REV MR 47 # 3147 Zbl 253 # 02042 • ID 10202

OVERBEEK, R.A. see Vol. I for further entries

OWINGS JR., J.C. [1967] *Recursion, metarecursion, and inclusion* (**J** 0036) J Symb Logic 32*173-179
◇ D25 D55 D60 ◇ REV MR 36 # 46 Zbl 204.12 • ID 10203

OWINGS JR., J.C. [1969] *Π^1_1-sets, ω-sets, and metacompleteness* (**J** 0036) J Symb Logic 34*194-204
◇ D55 D60 ◇ REV MR 43 # 50 Zbl 188.27 • ID 19535

OWINGS JR., J.C. [1970] *Commutativity and common fixed points in recursion theory* (**J** 0053) Proc Amer Math Soc 24*385-387
◇ D20 ◇ REV MR 41 # 5216 Zbl 205.10 • ID 10205

OWINGS JR., J.C. [1970] *The metarecursively enumerable sets, but not the Π^1_1-sets, can be enumerated without repetition* (**J** 0036) J Symb Logic 35*223-229
◇ D55 D60 ◇ REV MR 44 # 76 Zbl 225 # 02026 • ID 19534

OWINGS JR., J.C. [1971] *A splitting theorem for simple Π^1_1 sets* (**J** 0036) J Symb Logic 36*433-438
◇ D30 D55 D60 ◇ REV MR 46 # 3284 Zbl 287 # 02025 • ID 19533

OWINGS JR., J.C. [1973] see FELDMAN, E.D.

OWINGS JR., J.C. [1973] *Diagonalization and the recursion theorem* (**J** 0047) Notre Dame J Formal Log 14*95-99
◇ B40 D20 F30 ◇ REV MR 48 # 84 Zbl 247 # 02038 • ID 10206

OWINGS JR., J.C. [1975] *Splitting a context-sensitive set* (**J** 0119) J Comp Syst Sci 10*83-87
◇ D05 D25 ◇ REV MR 50 # 12684 Zbl 312 # 68047 • ID 10207

OWINGS JR., J.C. [1976] see HERZOG, T.

OWINGS JR., J.C. see Vol. V for further entries

OYAMAGUCHI, M. [1978] see HONDA, N.

OYAMAGUCHI, M. [1980] see HONDA, N.

OYAMAGUCHI, M. [1981] see HONDA, N.

OYAMAGUCHI, M. [1984] *Some remarks on subclass containment problems for several classes of dpda's* (**J** 0232) Inform Process Lett 19*9-12
◇ D10 ◇ REV MR 86d:68020 Zbl 548 # 68078 • ID 48882

PAEPPINGHAUS, P. [1983] see CARSTENS, H.G.

PAEPPINGHAUS, P. [1984] see CARSTENS, H.G.

PAEPPINGHAUS, P. see Vol. I, II, VI for further entries

PAGER, D. [1970] *On the efficiency of algorithms* (**J** 0037) ACM J 17*708-714
◇ D15 ◇ REV MR 43 # 1846 Zbl 206.286 • ID 10243

PAGER, D. [1970] *The categorization of tag systems in terms of decidability* (**J** 3172) J London Math Soc, Ser 2 2*473-480
◇ D03 ◇ REV MR 43 # 4679 Zbl 199.311 • ID 10242

PAGER, D. see Vol. I for further entries

PAGET, M. [1976] see BENEJAM, J.-P.

PAGET, M. [1978] *Proprietes de complexite pour une famille d'algorithmes de Markov* (**J** 3441) RAIRO Inform Theor 12*15-32
◇ B75 D03 D10 D15 ◇ REV MR 58 # 13932 Zbl 368 # 68056 • ID 51292

PAHI, B. [1970] see APPLEBEE, R.C.

PAHI, B. see Vol. I, II for further entries

PAIR, C. & QUERE, A. [1968] *Definition et etude des bilangages reguliers* (**J** 0194) Inform & Control 13*565-593
◇ D05 ◇ REV MR 40 # 4035 Zbl 181.16 • ID 10259

PAIR, C. & QUERE, A. [1970] *Sur les fonctions recursives primitives de ramifications* (**J** 0001) Acta Math Acad Sci Hung 21*437
◇ D20 ◇ REV MR 42 # 2940 Zbl 207.11 • ID 10257

PAIR, C. see Vol. III for further entries

PAKHOMOV, S.V. [1972] *Properties of graphs of functions in the Grzegorczyk hierarchy (Russian) (English summary)* (**S** 0228) Zap Nauch Sem Leningrad Otd Mat Inst Steklov 32*105-107,157
• TRANSL [1976] (**J** 1531) J Sov Math 6*434-436
◇ D20 ◇ REV MR 48 # 10788 Zbl 347 # 02025 • ID 10253

PAKHOMOV, S.V. [1974] *A simple syntactic definition of all classes of the Grzegorczyk hierarchy (Russian)* (**S** 0228) Zap Nauch Sem Leningrad Otd Mat Inst Steklov 40*127-130,159
• TRANSL [1977] (**J** 1531) J Sov Math 8*334-337
◇ D20 ◇ REV MR 51 # 93 Zbl 354 # 02031 • ID 15194

PAKHOMOV, S.V. [1976] *Hierarchies of operators in constructive metric spaces (Russian) (English summary)* (**S** 0228) Zap Nauch Sem Leningrad Otd Mat Inst Steklov 60*183-193,226
• TRANSL [1980] (**J** 1531) J Sov Math 14*1547-1554
◇ D80 F60 ◇ REV MR 58 # 21515 Zbl 338 # 02016 • ID 64344

PAKHOMOV, S.V. [1977] *How to prove that some classes of simple primitive recursive functions are distinct (Russian)* (**S** 0228) Zap Nauch Sem Leningrad Otd Mat Inst Steklov 68*115-122,145-146
• TRANSL [1981] (**J** 1531) J Sov Math 15*63-67
◇ D20 ◇ REV MR 58 # 27395 Zbl 358 # 02045 • ID 50499

PAKHOMOV, S.V. [1979] *Machine-independent description of some machine complexity classes (Russian) (English summary)* (S 0228) Zap Nauch Sem Leningrad Otd Mat Inst Steklov 88*176-185,244
• TRANSL [1982] (J 1531) J Sov Math 20*2358-2363
◇ D10 D15 ◇ REV MR 81a:68057 Zbl 429 # 03019 • ID 53850

PAKHOMOV, S.V. see Vol. VI for further entries

PALASINSKI, M. [1982] *On the word problem for BCK-algebras* (J 0352) Math Jap 27*335-344
◇ D40 G25 ◇ REV MR 83j:06019 Zbl 494 # 03049 • ID 36878

PALASINSKI, M. see Vol. III for further entries

PALLADINO, D. [1980] see BORGA, M.

PALLADINO, D. see Vol. III, V for further entries

PALTANEA, R. [1980] *Example of a grammar that generates the formulas of propositional calculus* (J 3071) Bul Univ Brasov, Ser C 22*93-100
◇ B05 D05 ◇ REV MR 83e:03062 • ID 35233

PALUSZKIEWICZ, A. [1970] see GORAJ, A.

PALYUTIN, E.A. [1971] *Boolean algebras with a categorical theory in a weak second order logic (Russian)* (J 0003) Algebra i Logika 10*523-534
• TRANSL [1971] (J 0069) Algeb and Log 10*325-331
◇ B15 C35 C85 D35 G05 ◇ REV MR 46 # 3302 Zbl 248 # 02055 • ID 64365

PALYUTIN, E.A. [1975] *A supplement to Ju.L.Ershov's article: "The upper semilattice of numerations of a finite set" (Russian)* (J 0003) Algebra i Logika 14*284-287,368
• TRANSL [1975] (J 0069) Algeb and Log 14*176-178
◇ D45 ◇ REV MR 54 # 4948 Zbl 342 # 02028 • ID 24725

PALYUTIN, E.A. see Vol. I, II, III, V for further entries

PAN, LUQUAN [1985] *On reduced Thue systems* (J 0041) Math Syst Theory 18*145-151
◇ D03 ◇ ID 47708

PANSIOT, J.-J. [1979] see HOPCROFT, J.E.

PANSIOT, J.-J. [1981] *A note on Post's correspondence problem* (J 0232) Inform Process Lett 12*233
◇ D03 ◇ REV MR 83h:03054 Zbl 485 # 03018 • ID 36065

PANSIOT, J.-J. [1983] *Hierarchie et fermeture de certaines classes de tag-systemes (English summary)* (J 1431) Acta Inf 20*179-196
◇ D03 ◇ REV MR 85j:68062 Zbl 507 # 68046 • ID 46797

PANSIOT, J.-J. [1984] *Bornes inferieures sur la complexite des facteurs des mots infinis engendres par morphismes iteres* (P 3565) Symp of Th Aspects of Comput Sci (1);1984 Paris 230-240
◇ D05 ◇ REV MR 86b:68030 Zbl 543 # 68061 • ID 47096

PANSIOT, J.-J. [1984] *Complexite des facteurs des mots infinis engendres par morphismes iteres* (P 4012) Automata, Lang & Progr (11);1984 Antwerpen 380-389
◇ D05 ◇ REV Zbl 554 # 68053 • ID 47173

PANSIOT, J.-J. [1985] *On various classes of infinite words obtained by iterated mappings* (P 4622) Autom on Infinite Words;1984 Le Mont-Dore 188-197
◇ D05 ◇ ID 49451

PAOLA DI, R.A. [1966] *On sets represented by the same formula in distinct consistent axiomatizable Rosser theories* (J 0048) Pac J Math 18*455-456
◇ D25 F30 ◇ REV MR 37 # 6181 Zbl 203.11 • ID 03047

PAOLA DI, R.A. [1966] *Pseudo-complements and ordinal logics based on consistency statements* (J 0036) J Symb Logic 31*359-364
◇ D25 F15 F30 ◇ REV MR 38 # 2022 Zbl 212.329 JSL 37.406 • ID 03046

PAOLA DI, R.A. [1966] *Some properties of pseudo-complements of recursively enumerable sets* (J 0064) Trans Amer Math Soc 121*296-308
◇ D25 D30 F30 ◇ REV MR 33 # 3920 Zbl 203.11 JSL 37.406 • ID 03045

PAOLA DI, R.A. [1967] *A survey of Soviet work in the theory of computer programming* (1111) Preprints, Manuscr., Techn. Reports etc. 146pp
◇ B25 B75 D35 G30 ◇ REM The Rand Corporation, RM-5424-PR, October 1967 • ID 30681

PAOLA DI, R.A. [1967] *Some theorems on extensions of arithmetic* (J 0036) J Symb Logic 32*180-189
◇ D25 D35 F15 F30 ◇ REV MR 36 # 2493 • ID 03048

PAOLA DI, R.A. [1968] *A note on diminishing the undecidable region of a recursively enumerable set* (J 0025) Duke Math J 35*399-405
◇ D25 ◇ REV MR 37 # 2601 Zbl 197.5 • ID 03049

PAOLA DI, R.A. [1969] *Random sets in subrecursive hierarchies* (J 0037) ACM J 16*621-630
◇ D20 ◇ REV MR 40 # 2545 Zbl 188.27 JSL 40.249 • ID 03050

PAOLA DI, R.A. [1969] *The recursive unsolvability of the decision problem for the class of definite formulas* (J 0037) ACM J 16*324-327
◇ D35 ◇ REV Zbl 182.332 • ID 30682

PAOLA DI, R.A. [1971] *The relational data file and the decision problem for classes of proper formulas* (P 1681) Symp Inform Storage & Retrieval;1971 College Park 95-105
◇ B20 B75 D35 ◇ ID 30683

PAOLA DI, R.A. [1978] *The operator gap theorem in α-recursion theory* (J 0009) Arch Math Logik Grundlagenforsch 19*115-129
◇ D15 D60 ◇ REV MR 81c:03038 Zbl 412 # 03031 • ID 52962

PAOLA DI, R.A. [1981] *A lift of a theorem of Friedberg: A Banach-Mazur functional that coincides with no α-recursive functional on the class of α-recursive functions* (J 0036) J Symb Logic 46*216-232
◇ D60 ◇ REV MR 82h:03043 Zbl 471 # 03039 • ID 55234

PAOLA DI, R.A. [1983] *The basic theory of partial α-recursive operators* (J 3526) Ann Mat Pura Appl, Ser 4 134*169-199
◇ D60 ◇ REV MR 85i:03154 • ID 44227

PAOLA DI, R.A. [1985] *Creativity and effective inseparability in dominical categories* (P 4646) Atti Incontri Log Mat (2);1983/84 Siena 477-478
◇ D25 G30 ◇ ID 49600

PAOLA DI, R.A. see Vol. I, III, VI for further entries

PAPADIMITRIOU, C.H. [1976] *On the algebraic complexity of sets of functions* (J 0465) Bull Greek Math Soc (NS) 17*50-58
◇ D15 ◇ REV MR 58 # 19359 Zbl 379 # 68038 • ID 51865

Papadimitriou, C.H. [1981] see Lewis, H.R.

Papadimitriou, C.H. [1982] see Lewis, H.R.

Papadimitriou, C.H. & Zachos, S. [1982] *Two remarks on the power of counting* (**P** 3862) Theor Comput Sci (6);1983 Dortmund 269-275
◇ D15 ◇ rev MR 84b:68003 Zbl 506 # 68039 • id 46286

Pappas, P. [1985] *A Diophantine problem for Laurent polynomial rings* (**J** 0053) Proc Amer Math Soc 93*713-718
◇ C60 D35 D80 ◇ rev MR 86d:03041 Zbl 532 # 10032 • id 44466

Pappas, P. see Vol. III for further entries

Parchmann, R. [1979] see Duske, J.

Parikh, R. [1961] *Language generating devices* (**S** 1082) Quart Prog Rep MIT Res Lab Electron 1961*199-212
◇ D05 ◇ id 21225

Parikh, R. [1966] *On context-free languages* (**J** 0037) ACM J 13*570-581
◇ D05 ◇ rev MR 34 # 8901 Zbl 154.258 • id 10278

Parikh, R. [1966] *Some generalisations of the notion of well ordering* (**J** 0068) Z Math Logik Grundlagen Math 12*333-340
◇ D20 D45 E07 F35 ◇ rev MR 34 # 4131 Zbl 202.310 • id 10279

Parikh, R. [1971] *Existence and feasibility in arithmetic* (**J** 0036) J Symb Logic 36*494-508
◇ A05 D15 F30 F65 H15 ◇ rev MR 46 # 3287 Zbl 243 # 02037 • id 10282

Parikh, R. [1972] *A note on rigid substructures* (**J** 0053) Proc Amer Math Soc 33*520-522
◇ C50 C57 ◇ rev MR 45 # 3186 Zbl 273 # 02034 • id 10285

Parikh, R. [1977] see Jongh de, D.H.J.

Parikh, R. [1978] *Effectiveness* (1111) Preprints, Manuscr., Techn. Reports etc.
◇ D15 ◇ rem Technical Memo, Lab for Computer Science, MIT • id 32423

Parikh, R. [1981] see Ehrenfeucht, A.

Parikh, R. [1985] see Chandra, A.K.

Parikh, R. see Vol. I, II, III, VI for further entries

Paris, J.B. [1972] *ZF* $\vdash$ Σ_4^0 *determinateness* (**J** 0036) J Symb Logic 37*661-667
◇ D55 E15 E60 ◇ rev MR 51 # 159 Zbl 264 # 02053 • id 19500

Paris, J.B. [1975] see Devlin, K.J.

Paris, J.B. [1977] *Measure and minimal degrees* (**J** 0007) Ann Math Logic 11*203-216
◇ D30 E75 ◇ rev MR 56 # 15393 Zbl 359 # 02037 • id 23665

Paris, J.B. & Wilkie, A.J. [1981] *Models of arithmetic and the rudimentary sets* (**J** 3133) Bull Soc Math Belg, Ser B 33*157-169
◇ C62 D20 ◇ rev MR 82k:03074 Zbl 499 # 03021 • id 79972

Paris, J.B. see Vol. I, II, III, V, VI for further entries

Park, D. [1979] see Kanda, A.

Park, D. [1981] *Concurrency and automata on infinite sequences* (**P** 3475) Theor Comput Sci (5);1981 Karlsruhe 104*167-183
◇ D05 ◇ rev Zbl 457 # 68049 • id 69757

Park, D. see Vol. III, V for further entries

Parnes, M. [1972] see Barback, J.

Parnes, M. see Vol. I for further entries

Parsons, C. [1968] *Hierarchies of primitive recursive functions* (**J** 0068) Z Math Logik Grundlagen Math 14*357-376
◇ D20 ◇ rev MR 39 # 66 Zbl 172.297 JSL 36.538 • id 10307

Parsons, C. see Vol. I, II, V, VI for further entries

Partee, B. [1969] see Ginsburg, S.

Partee, B. also published under the name Hall Partee, B.

Partis, M.T. [1963] *Commutative partially ordered recursive arithmetics* (**J** 0132) Math Scand 13*199-216
◇ D20 F60 ◇ rev MR 29 # 4675 JSL 34.117 • id 10312

Partis, M.T. [1967] *Limited universal and existential quantifiers in commutative partially ordered recursive arithmetics* (**J** 0047) Notre Dame J Formal Log 8*17-23
◇ D20 F60 ◇ rev MR 39 # 63 Zbl 211.312 • id 10313

Pascu, A. [1974] *On the recursiveness of context-sensitive languages* (**J** 0517) Mathematica (Cluj) 16(39)*293-297
◇ D05 ◇ rev MR 58 # 25145 Zbl 377 # 68052 • id 51779

Pascu, A. [1978] see Baciu, A.

Pascu, A. [1979] see Baciu, A.

Pascu, A. [1980] see Baciu, A.

Pascu, A. [1981] see Bucurescu, I.

Pascu, A. [1981] see Baciu, A.

Pashkevich, A.P. [1969] see Bershtejn, L.S.

Passy, S. [1980] *Structured programs for Turing machines* (**J** 0232) Inform Process Lett 10*63-67
◇ D10 ◇ rev MR 81h:68080 Zbl 455 # 68033 • id 69758

Passy, S. see Vol. II for further entries

Paterson, M.S. [1970] *Unsolvability in* 3×3 *matrices* (**J** 0548) Stud Appl Math 49*105-107
◇ D80 ◇ rev MR 41 # 62 Zbl 186.11 • id 10320

Paterson, M.S. [1972] *Tape bounds for time-bounded Turing machines* (**J** 0119) J Comp Syst Sci 6*116-124
• transl [1974] (**C** 2319) Slozh Vychisl & Algor 213-221 (Russian)
◇ D10 D15 ◇ rev MR 46 # 4797 Zbl 289 # 02020 • id 29985

Paterson, M.S. & Valiant, L.G. [1973] *Deterministic one-counter automata* (**P** 1630) GI Fachtag Automatenth & Form Sprach (1);1973 Bonn 104-115
◇ D05 D15 ◇ rev MR 55 # 6960 Zbl 341 # 94030 • id 65868

Paterson, M.S. [1975] *Complexity of monotone networks for boolean matrix product* (**J** 1426) Theor Comput Sci 1*13-20
• transl [1978] (**J** 3079) Kiber Sb Perevodov, NS 15*28-37 (Russian)
◇ B70 D15 ◇ rev Zbl 307 # 68031 • id 69759

PATERSON, M.S. & VALIANT, L.G. [1975] *Deterministic one-counter automata* (**J** 0119) J Comp Syst Sci 10*340-350 ◇ D05 ◇ REV MR 52#55 Zbl 307#68038 • ID 17220

PATERSON, M.S. [1976] see JOCKUSCH JR., C.G.

PATERSON, M.S. [1982] see FISCHER, MICHAEL J.

PATERSON, M.S. [1984] see HAREL, D.

PATERSON, M.S. see Vol. I, II for further entries

PATRIKEEV, V.L. [1971] *The computability of the translation functions of Janov's operator schemes (Russian)* (**C** 3567) Teor Kibernetika Kiev 1967/71 3-16 ◇ B75 D20 ◇ REV MR 58#16206 • ID 77162

PATRIZIA, M. [1974] *Funzioni elementari* (**J** 3436) Quad, Ist Appl Calcolo, Ser 3 8*109-136 ◇ D20 ◇ REV Zbl 434#03027 • ID 55731

PATT, Y.N. [1975] see AMOROSO, S.

PATT, Y.N. see Vol. II for further entries

PATTERSON, D.B. [1981] see BLOOM, S.L.

PATTON, T.E. [1963] *On n-adic representation of numbers* (**J** 0036) J Symb Logic 28*161-163 ◇ D03 D25 ◇ REV MR 32#7407 Zbl 199.27 • ID 10324

PATTON, T.E. [1965] *Church's theorem on the decision problem* (**J** 0047) Notre Dame J Formal Log 6*147-153 ◇ D35 ◇ REV MR 34#5644 Zbl 147.251 • ID 10326

PATTON, T.E. see Vol. I for further entries

PAUC, C. [1935] *Resolution d'equations abstraites par un procede d'iteration* (**J** 0109) C R Acad Sci, Paris 200*2047-2050 ◇ D70 ◇ REV FdM 61.980 • ID 40835

PAUL, W.J. [1976] see CELONI, J.R.

PAUL, W.J. [1977] see HOPCROFT, J.E.

PAUL, W.J. [1978] *Complexity theory* (**X** 0823) Teubner: Stuttgart 247pp ◇ D15 D98 ◇ REV MR 80c:68033 Zbl 382#68044 • ID 82446

PAUL, W.J. [1979] *Kolmogorov complexity and lower bounds* (**P** 2935) FCT'79 Fund of Comput Th;1979 Berlin/Wendisch-Rietz 325-334 ◇ D15 ◇ REV MR 81d:68079 Zbl 415#68012 • ID 69764

PAUL, W.J. [1979] *On time hierarchies* (**J** 0119) J Comp Syst Sci 19*197-202 ◇ D15 ◇ REV MR 81b:03043 Zbl 428#68055 • ID 77166

PAUL, W.J. & PRAUSS, E.J. & REISCHUK, R. [1980] *On alternation* (**J** 1431) Acta Inf 14*243-255 • TRANSL [1983] (**J** 3079) Kiber Sb Perevodov, NS 20*107-122 ◇ D10 D15 ◇ REV MR 82e:68048a Zbl 437#68025 • REM Part I. Part II 1980 • ID 55911

PAUL, W.J. & REISCHUK, R. [1980] *On alternation II. A graph-theoretic approach to determinism versus nondeterminism* (**J** 1431) Acta Inf 14*391-403 • TRANSL [1983] (**J** 3079) Kiber Sb Perevodov, NS 20*123-140 (Russian) ◇ D10 D15 ◇ REV MR 82e:68048b Zbl 437#68025 • REM Part I 1980 by Paul,W.J. & Prauss,E.J. & Reischuk,R. • ID 82448

PAUL, W.J. [1981] *On heads versus tapes* (**P** 4235) IEEE Symp Found of Comp Sci (22);1981 Nashville 68-73 ◇ D10 D15 ◇ REV MR 84a:68004 • ID 45821

PAUL, W.J. & REISCHUK, R. [1981] *On time versus space II* (**J** 0119) J Comp Syst Sci 22*312-327 ◇ D15 ◇ REV MR 83e:68043 Zbl 462#68029 • REM Part I 1977 by Hopcroft,J. & Paul,W. & Valiant,L. • ID 69761

PAUL, W.J. [1982] *On-line simulation of k+1 tapes by k tapes requires nonlinear time* (**J** 0194) Inform & Control 53*1-8 ◇ D10 D15 ◇ REV MR 85c:68015 Zbl 536#68049 • ID 39063

PAUL, W.J. [1984] *On heads versus tapes* (**J** 1426) Theor Comput Sci 28*1-12 ◇ D10 D15 ◇ REV MR 85m:68009 Zbl 536#68050 • ID 45012

PAUL, W.J. [1984] see DURIS, P.

PAUL, W.J. see Vol. I for further entries

PAULL, M. [1983] see FRANCO, J.

PAUN, G. [1978] *An infinite hierarchy of contextual languages with choice* (**J** 0070) Bull Soc Sci Math Roumanie, NS 22(70)*425-429 ◇ D05 ◇ REV MR 80c:68058 Zbl 392#68067 • ID 82449

PAUN, G. [1979] *On the family of finite index matrix languages* (**J** 0119) J Comp Syst Sci 18*267-280 ◇ D05 ◇ REV MR 80k:68060 Zbl 411#68062 • ID 52929

PAUN, G. [1979] see CALUDE, C.

PAUN, G. [1981] see CALUDE, C.

PAUN, G. [1981] *The languages of propositional calculus and the Chomsky hierarchy (Romanian) (English summary)* (**J** 0197) Stud Cercet Mat Acad Romana 33*299-309 ◇ B65 D05 ◇ REV MR 83e:68117 Zbl 516#68056 • ID 46246

PAUN, G. [1981] *Thue languages and matrix grammars* (**J** 2716) Found Control Eng, Poznan 6*273-278 ◇ D03 D05 ◇ REV MR 84a:03041 Zbl 495#68062 • ID 35578

PAUN, G. [1983] see CALUDE, C.

PAUN, G. & TATARAM, M. [1985] *Classes of mappings having context sensitive graphs* (**J** 0060) Rev Roumaine Math Pures Appl 30*273-288 ◇ D05 ◇ ID 48602

PAUN, G. see Vol. II for further entries

PAVALOIU, I. [1965] see NEGRESCU, I.

PAVLENKO, V.A. [1981] *The combinatorial problem of Post with two pairs of words (Russian) (English summary)* (**J** 0270) Dokl Akad Nauk Ukr SSR, Ser A 1981/7*9-11 ◇ D03 ◇ REV MR 83a:03032 Zbl 466#03013 • ID 54964

PAVLENKO, V.A. [1982] *The Post combinatorial problem for two pairs of words (Russian)* (**S** 2651) Prepr Inst Prikl Mat, Akad Nauk SSSR 1982*65pp ◇ D03 ◇ REV MR 83j:03064 • ID 35362

PAVLIDIS, T. [1972] *Linear and context-free graph grammars* (**J** 0037) ACM J 19*11-22 ◇ D05 ◇ REV MR 46#3357 Zbl 229#68027 • ID 10329

PAVLOTSKAYA, L.M. [1973] *Solvability of the halting problem for certain classes of Turing machines (Russian)* (**J** 0087) Mat Zametki (Akad Nauk SSSR) 13*899-909
• TRANSL [1973] (**J** 1044) Math Notes, Acad Sci USSR 13*537-541
◇ D10 ◇ REV MR 48 # 76 Zbl 284 # 02017 • ID 10330

PAVLOTSKAYA, L.M. [1975] *The minimal number of different vertex codes in the graph of the universal Turing machine (Russian)* (**J** 0071) Met Diskr Analiz (Novosibirsk) 27*52-60,74
◇ D10 ◇ REV MR 53 # 2653 Zbl 322 # 02036 • ID 21537

PAVLOTSKAYA, L.M. [1978] *Sufficient conditions for the solvability of the halting problem for Turing machines (Russian)* (**J** 0052) Probl Kibern 33*91-118,232
◇ D10 ◇ REV MR 58 # 5114 Zbl 426 # 68025 • ID 53680

PAVLOV, R.D. [1971] *Nonsolvability of certain algorithmic problems of group theory in minimal alphabets (Russian)* (**J** 0137) C R Acad Bulgar Sci 24*855-858
◇ D40 ◇ REV MR 45 # 4972 Zbl 251 # 02045 • ID 10332

PAVLOV, R.D. [1971] *On the problem of recognition of group properties (Russian)* (**J** 0087) Mat Zametki (Akad Nauk SSSR) 10*169-180
• TRANSL [1971] (**J** 1044) Math Notes, Acad Sci USSR 10*524-530
◇ D40 ◇ REV MR 45 # 3536 Zbl 222 # 20010 • ID 10331

PAVLOV, R.D. [1971] *The impossibility of certain algorithms for the recognition of group properties in minimal alphabets* (**J** 0255) God Fak Mat & Mekh, Univ Sofiya 66*191-217
◇ D40 ◇ REV MR 58 # 16241 Zbl 326 # 20003 • ID 77181

PAVLOV, R.D. [1976] *Group theoretic algorithmic problems in minimal alphabets* (**P** 3208) Mat & Mat Obrazov (3);1974 Burgas 115-117
◇ D40 ◇ REV Zbl 355 # 20001 • ID 50253

PAVLOV, R.D. [1979] *On the problem of recognizing group properties in bounded alphabets* (**J** 2547) Serdica, Bulgar Math Publ 5*252-271
◇ D40 ◇ REV MR 82a:20043 Zbl 463 # 20031 • ID 54584

PAVLOV, R.D. [1979] *On the problem of recognizing homomorphisms of finitely presented groups* (**J** 2547) Serdica, Bulgar Math Publ 5*370-373
◇ D40 ◇ REV MR 82d:20034 Zbl 446 # 20019 • ID 82450

PAVLOVA, E.A. [1960] *The lattice of denseness of sets of natural numbers (Russian) (Moldavian summary)* (**J** 0967) Izv Akad Nauk Mold SSR, Ser Fiz-Tekh Mat 1960/10*31-38
◇ D20 D50 ◇ REV MR 32 # 46 • ID 42943

PAVLOVA, E.A. [1961] *Densities of hyperimmune sets (Russian)* (**J** 0023) Dokl Akad Nauk SSSR 139*814-817
• TRANSL [1961] (**J** 0062) Sov Math, Dokl 2*1017-1019
◇ D50 ◇ REV MR 32 # 47 Zbl 119.13 • ID 10333

PAVLOVA, E.A. [1963] *On certain classes of hyperimmune sets (Russian)* (**J** 0967) Izv Akad Nauk Mold SSR, Ser Fiz-Tekh Mat 1963/11*26-33
◇ D50 ◇ REV MR 34 # 5674 Zbl 235 # 02036 • ID 27824

PAVLOVA, E.A. [1965] *Certain arithmetic and algebraic properties of a system of densities (Russian)* (**C** 4212) Issl Algeb & Mat Anal 55-64
◇ D20 ◇ REV MR 35 # 1471 • ID 45440

PAWLAK, Z. [1968] *On the notion of a computer* (**P** 0627) Int Congr Log, Meth & Phil of Sci (3,Proc);1967 Amsterdam 255-267
◇ D10 ◇ REV MR 39 # 6752 Zbl 184.26 • ID 10336

PAWLAK, Z. [1969] *Programmierte Maschinen* (**J** 0485) Algorytmy, Pol Akad Nauk 5*5-19
◇ D10 ◇ REV MR 40 # 8310 Zbl 283 # 68001 • ID 64415

PAWLAK, Z. see Vol. I, II, V for further entries

PAYNE, T.H. [1971] *Effectively minimizing effective fixed-points* (**J** 0053) Proc Amer Math Soc 30*561-562
◇ D20 ◇ REV MR 44 # 2602 Zbl 232 # 02033 • ID 10338

PAYNE, T.H. [1972] *Sequences having an effectve fixed-point property* (**J** 0064) Trans Amer Math Soc 165*227-237
◇ D20 ◇ REV MR 52 # 10391 Zbl 319 # 02032 • ID 21690

PAYNE, T.H. [1973] *Effective extendability and fixed-points* (**J** 0047) Notre Dame J Formal Log 14*123-124
◇ D20 ◇ REV MR 58 # 21529 Zbl 247 # 02044 • ID 29508

PAYNE, T.H. [1975] *Computability of finite linear configurations* (**J** 0047) Notre Dame J Formal Log 16*354-356
◇ D20 ◇ REV MR 52 # 10380 Zbl 292 # 02034 • ID 10340

PAYNE, T.H. [1975] *Concrete computability* (**J** 0047) Notre Dame J Formal Log 16*238-244
◇ D75 ◇ REV MR 54 # 4946 Zbl 301 # 02036 • ID 10339

PAYNE, T.H. [1980] *General computability* (**J** 0047) Notre Dame J Formal Log 21*277-292
◇ D75 ◇ REV MR 81k:03047 Zbl 454 # 03024 • ID 54236

PAYNE, T.H. see Vol. III, V for further entries

PAZ, A. [1971] *Introduction to probabilistic automata* (**X** 0801) Academic Pr: New York 248pp
◇ D05 D98 ◇ REV Zbl 234 # 94005 • ID 23553

PAZ, A. [1971] see KOHAVI, Z.

PAZ, A. & SALOMAA, A. [1973] *Integral sequential word functions and growth equivalence of Lindenmayer systems* (**J** 0194) Inform & Control 23*313-343
◇ D05 ◇ REV MR 48 # 3309 Zbl 273 # 68056 • ID 31489

PAZ, A. [1981] see MORAN, S.

PEAK, I. [1964] *Automaten und Halbgruppen I* (**J** 0002) Acta Sci Math (Szeged) 25*193-201
◇ D05 ◇ REV MR 30 # 1011 Zbl 139.8 • REM Part II 1965
• ID 46700

PEAK, I. [1965] *Automata and semi-groups II (Russian)* (**J** 0002) Acta Sci Math (Szeged) 26*49-54
◇ D05 ◇ REV MR 31 # 3336 Zbl 139.9 • REM Part I 1964
• ID 10343

PEAK, I. [1965] see GECSEG, F.

PEAK, I. [1965] *Certain extensions of semi-simple automata (Russian)* (**J** 0057) Publ Math (Univ Debrecen) 12*25-29
◇ D05 ◇ REV MR 32 # 5521 • ID 10342

PEAK, I. [1971] see GECSEG, F.

PEAK, I. see Vol. V for further entries

PECKEL, J. [1978] *A deterministic subclass of context-free languages* (**J** 0086) Cas Pestovani Mat, Ceskoslov Akad Ved 103*43-52
◇ D05 D15 ◇ REV MR 57 # 8193 Zbl 369 # 68033
• ID 51358

PECUCHET, J.-P. [1984] *Automates Boustrophedon, semi-groupe de Birget et monoide inversif libre* (**P** 4012) Automata, Lang & Progr (11);1984 Antwerpen 390
◇ D05 ◇ ID 47178

PECUCHET, J.-P. [1985] *Automates boustrophedon sur des mots infinis* (**P** 4622) Autom on Infinite Words;1984 Le Mont-Dore 47-54
◇ D05 ◇ ID 49452

PECUCHET, J.-P. [1985] *Automates boustrophedon et mots infinis (French) (English summary)* (**J** 1426) Theor Comput Sci 35*115-122
◇ D05 ◇ ID 46525

PEDANOV, I.E. [1974] *The solvability of the problem of the membership of a point in an object in a geometric data-processing language (Russian)* (**J** 0199) Zh Vychisl Mat i Mat Fiz 14*450-460
• TRANSL [1974] (**J** 1049) USSR Comput Math & Math Phys 14/2*174-183
◇ B75 D80 ◇ REV MR 52 # 2291 Zbl 282 # 02020
• ID 29087

PELED, U.N. & SIMEONE, B. [1981] *A polynomial-time algorithm for recognizing threshold functions* (**J** 3401) Meth Oper Res 40*397-400
◇ D15 D80 ◇ REV Zbl 466 # 68031 • ID 69766

PELEG, D. [1984] see HAREL, D.

PELEG, D. see Vol. II, V for further entries

PELIN, A. [1984] see GALLIER, J.H.

PELZ, E. [1984] *Les debuts de la methode de priorite et les theorems de Friedberg-Muchnick* (**C** 4356) Gen Log Semin Paris 1982/83 45-64
◇ D25 ◇ ID 48088

PENNER, V. [1973] *Ueber eine Hierarchie von push-down-entscheidbaren Mengen* (**P** 1630) GI Fachtag Automatenth & Form Sprach (1);1973 Bonn 254-262
◇ D10 D20 ◇ REV MR 56 # 7329 Zbl 273 # 68038
• ID 64438

PENTTONEN, M. & ROZENBERG, G. & SALOMAA, A. [1978] *Bibliography of L systems* (**J** 1426) Theor Comput Sci 5*339-354
◇ A10 D05 D99 ◇ REV MR 58 # 19391 Zbl 374 # 68047
• ID 31591

PENTTONEN, M. [1979] see MAURER, H.A.

PENTTONEN, M. [1984] *The reachability of vector addition systems and equivalent problems* (**J** 0498) Ann Univ Turku, Ser A I 186*80-86
◇ D03 D05 D80 ◇ REV MR 85j:68035 Zbl 551 # 68048
• ID 43972

PENZIN, YU.G. [1972] see FRIDMAN, EH.I.

PENZIN, YU.G. [1973] *The undecidability of fields of rational functions over fields of characteristic 2 (Russian)* (**J** 0003) Algebra i Logika 12*205-210,244
• TRANSL [1973] (**J** 0069) Algeb and Log 12*116-119
◇ D35 ◇ REV MR 52 # 10705 Zbl 282 # 02019 • ID 21735

PENZIN, YU.G. [1976] *Algorithmic problems in the theory of numbers (Russian)* (**C** 2555) Algeb Sistemy (Irkutsk) 122-148
◇ D80 ◇ ID 32593

PENZIN, YU.G. [1976] *Undecidability of a theory of the integers with addition and predicate mutually disjoint (Russian)* (**C** 2555) Algeb Sistemy (Irkutsk) 149-153
◇ D35 F30 ◇ ID 32592

PENZIN, YU.G. [1978] *Unsolvable theories of a ring of continuous functions (Russian)* (**C** 2556) Algor Vopr Algeb Sist 142-147
◇ D35 ◇ ID 32594

PENZIN, YU.G. see Vol. III, VI for further entries

PEPIS, J. [1936] *Beitraege zur Reduktionstheorie des logischen Entscheidungsproblemes* (**J** 0460) Acta Univ Szeged, Sect Mat 8*7-41
◇ B20 D35 ◇ REV Zbl 14.98 JSL 2.84 FdM 62.1059
• ID 10360

PEPIS, J. [1937] *Ueber das Entscheidungsproblem des engeren logischen Funktionskalkuels (Polish) (German summary)* (**S** 0281) Arch Towarz Nauk Lwow, Sect 3 7/8*1-172
◇ B20 B25 D35 ◇ REV Zbl 19.97 JSL 4.93 FdM 63.823
• ID 19493

PEPIS, J. [1938] *Ein Verfahren der mathematischen Logik* (**J** 0036) J Symb Logic 3*61-76
◇ B20 D35 ◇ REV JSL 3.161 FdM 64.29 • ID 10361

PEPIS, J. [1938] *Untersuchungen ueber das Entscheidungsproblem der mathematischen Logik* (**J** 0027) Fund Math 30*257-348
◇ B20 D35 ◇ REV Zbl 18.385 JSL 3.160 FdM 64.29
• ID 10362

PERES, A. [1985] *Reversible logic and quantum computers* (**J** 4619) Phys Rev A, Ser 3 32/6*3266-3276
◇ D80 ◇ ID 49326

PERETYAT'KIN, M.G. [1971] *Strongly constructive models and numerations of the boolean algebra of recursive sets (Russian)* (**J** 0003) Algebra i Logika 10*535-557
• TRANSL [1971] (**J** 0069) Algeb and Log 10*332-345
◇ C57 D20 D45 G05 ◇ REV MR 46 # 5126
Zbl 311 # 02051 • ID 10368

PERETYAT'KIN, M.G. [1973] *A strongly constructive model without elementary submodels and extensions (Russian)* (**J** 0003) Algebra i Logika 12*312-322,364
• TRANSL [1973] (**J** 0069) Algeb and Log 12*178-183
◇ C35 C57 D45 ◇ REV MR 52 # 7892 Zbl 298 # 02046
• ID 18330

PERETYAT'KIN, M.G. [1973] *Every recursively enumerable extension of a theory of linear order has a constructive model (Russian)* (**J** 0003) Algebra i Logika 12*211-219,244
• TRANSL [1973] (**J** 0069) Algeb and Log 12*120-124
◇ C57 C65 ◇ REV MR 54 # 7243 Zbl 298 # 02045
• ID 25033

PERETYAT'KIN, M.G. [1973] *On complete theories with a finite number of denumerable models (Russian)* (**J** 0003) Algebra i Logika 12*550-576,618
• TRANSL [1973] (**J** 0069) Algeb and Log 12*310-326
◇ C57 ◇ REV MR 50 # 6827 Zbl 298 # 02047 • ID 10370

PERETYAT'KIN, M.G. [1978] *Criterion for strong constructivizability of a homogeneous model (Russian)* (**J** 0003) Algebra i Logika 17*436-454,491
• TRANSL [1978] (**J** 0069) Algeb and Log 17*290-301
◇ C35 C50 C57 D45 ◇ REV MR 81j:03051
Zbl 431 # 03021 • ID 53927

PERETYAT'KIN, M.G. [1982] *Finitely axiomatizable totally transcendental theories (Russian)* (**C** 3953) Mat Log & Teor Algor 88-135
◇ C45 C57 ◇ REV MR 85i:03103 Zbl 524#03017 • ID 37596

PERETYAT'KIN, M.G. [1982] *Turing machine computations in finitely axiomatizable theories (Russian)* (**J** 0003) Algebra i Logika 21*410-441
• TRANSL [1982] (**J** 0069) Algeb and Log 21*272-295
◇ C57 D45 ◇ REV MR 85c:03015 Zbl 567#03014
• ID 39943

PERETYAT'KIN, M.G. see Vol. III, VI for further entries

PERI, C. [1984] *Algorithms on sets of words that are isomorphic to Markov algorithms (Italian) (English summary)* (**S** 2990) Rend Semin Mat Brescia 8*111-126
◇ D03 D20 ◇ ID 47273

PERKINS, P. [1967] *Unsolvable problems for equational theories* (**J** 0047) Notre Dame J Formal Log 8*175-185
◇ C05 D35 ◇ REV MR 38#4310 Zbl 197.282 • ID 10372

PERKINS, P. [1972] *An unsolvable provability problem for one variable groupoid equations* (**J** 0047) Notre Dame J Formal Log 13*359-362
◇ C05 D35 ◇ REV MR 50#12696 Zbl 238#02040
• ID 10374

PERL, J. [1973] *Anwendung von Graphenalgorithmen auf allgemeinere Problemklassen (English summary)* (**J** 0373) Comp Arch Inform & Numerik 11*235-247
◇ D80 ◇ REV MR 52#7956 Zbl 262#05128 • ID 18329

PERL, J. [1978] *Entropie von Problemen* (**P** 3148) Graphth Konzepte Inf (3);1977 Linz 179-190
◇ D15 ◇ REV MR 80a:68045 Zbl 395#68047 • ID 69768

PERLES, M.A. & RABIN, M.O. & SHAMIR, E. [1963] *The theory of definite automata* (**J** 4305) IEEE Trans Electr Comp EC-12*233-243
◇ D05 ◇ REV MR 27#3484 Zbl 158.10 • ID 10380

PERLES, M.A. see Vol. V for further entries

PERRAUD, J. [1977] *Sur les conditions de petite simplification et l'algorithme de Dehn (English summary)* (**J** 2313) C R Acad Sci, Paris, Ser A-B 284*A659-A662
◇ D40 ◇ REV MR 55#513 Zbl 359#02045 • ID 50592

PERRAUD, J. [1980] *Sur la condition de petite simplification C'(1/6) dans un produit libre amalgame (English summary)* (**J** 2313) C R Acad Sci, Paris, Ser A-B 291*A247-A250
◇ D40 ◇ REV MR 82c:20068 Zbl 461#20010 • ID 82466

PERRAUD, J. [1980] *Sur le probleme des mots des quotients de groupes et produits libres* (**J** 0353) Bull Soc Math Fr 108*285-331
◇ D40 ◇ REV MR 82h:20043 Zbl 449#20047 • ID 82465

PERRIN, D. [1983] *Varietes de semigroupes et mots infinis* (**P** 3851) Automata, Lang & Progr (10);1983 Barcelona 610-616
◇ C05 D05 ◇ REV MR 85f:68046 Zbl 519#68070
• ID 39971

PERRIN, D. [1984] see NIVAT, M.

PERRIN, D. [1984] *Recent results on automata and infinite words* (**P** 3658) Math Founds of Comput Sci (11);1984 Prague 134-148
◇ D05 ◇ ID 44850

PERRIN, D. [1985] *An introduction to finite automata on infinite words* (**P** 4622) Autom on Infinite Words;1984 Le Mont-Dore 2-17
◇ D05 ◇ REV MR 86h:68006 • ID 49453

PERRIN, D. [1985] *Words over a partially commutative alphabet* (**P** 4626) Combin Algor on Words;1984 Maratea 329-340
◇ D05 ◇ ID 49454

PERROT, J.F. [1975] *Une theorie algebrique des automates finis monogenes* (**P** 1475) Conv Inform Teor & Conv Strutt Corpi Algeb;1973 Roma XV*201-244
◇ D05 ◇ REV MR 54#10462 Zbl 327#20027 • ID 25883

PETER, R. [1934] *Ueber den Zusammenhang der verschiedenen Begriffe der rekursiven Funktionen* (**J** 0043) Math Ann 110*612-632
◇ D20 ◇ REV Zbl 10.241 FdM 60.851 • ID 10393

PETER, R. [1935] *Konstruktion nichtrekursiver Funktionen* (**J** 0043) Math Ann 111*42-60
◇ D20 ◇ REV Zbl 11.3 FdM 61.52 • ID 10394

PETER, R. [1935] *Zur Theorie der rekursiven Funktionen (Hungarian) (German summary)* (**J** 0461) Mat Fiz Lapok 42*25-49
◇ D20 ◇ REV Zbl 12.2 FdM 61.974 • ID 40812

PETER, R. [1936] *Ueber die mehrfache Rekursion* (**J** 0043) Math Ann 113*489-527
◇ D20 ◇ REV Zbl 15.339 JSL 2.57 FdM 62.1055
• ID 10395

PETER, R. [1937] *Ueber rekursive Funktionen der zweiten Stufe* (**P** 1608) Int Congr Math (II, 5);1936 Oslo 2*267
◇ D20 ◇ REV FdM 63.31 • ID 27935

PETER, R. [1940] *Contribution to recursive number theory* (**J** 0460) Acta Univ Szeged, Sect Mat 9*233-238
◇ D20 F30 ◇ REV MR 1.132 Zbl 22.194 JSL 5.70 FdM 66.32 • ID 10396

PETER, R. [1950] *Zusammenhang der mehrfachen und transfiniten Rekursionen* (**J** 0036) J Symb Logic 15*248-272
◇ D20 ◇ REV MR 12.469 Zbl 40.294 JSL 16.216
• ID 10398

PETER, R. [1951] *Probleme der Hilbertschen Theorie der hoeheren Stufen von rekursiven Funktionen* (**J** 0001) Acta Math Acad Sci Hung 2*247-274
◇ D20 ◇ REV MR 14.713 Zbl 45.4 JSL 18.263 • ID 10401

PETER, R. [1951] *Rekursive Funktionen* (**X** 0928) Akad Kiado: Budapest 206pp
• TRANSL [1951] (**X** 1876) Kexue Chubanshe: Beijing [1954] (**X** 1656) Izdat Inostr Lit: Moskva [1967] (**X** 0801) Academic Pr: New York 300pp
◇ D20 D98 ◇ REV MR 13.421 MR 36#2496 Zbl 43.248 Zbl 77.13 JSL 16.280 JSL 23.362 • REM 2nd enl. ed. 1957;278pp. X0911 • ID 46603

PETER, R. [1952] *Transfinite Rekursionen (Grundlagenforschung und rekursive Funktionen)* (**P** 0662) Hungar Math Congr (1);1950 Budapest 419-428
◇ D20 ◇ REV MR 15.493 Zbl 48.247 JSL 20.73 • ID 10400

PETER, R. [1953] *Rekursive Definitionen, wobei fruehere Funktionswerte von variabler Anzahl verwendet werden* (**J** 0057) Publ Math (Univ Debrecen) 3*33-70
◇ D20 ◇ REV MR 15.771 Zbl 58.6 JSL 20.176 • ID 10403

PETER, R. [1955] *Ein neuer Beweis fuer die Tatsache, dass die Klasse der primitiv-rekursiven Funktionen umfassender als die Klasse der elementaren Funktionen ist* (**J** 0068) Z Math Logik Grundlagen Math 1*29-36
◇ D20 ◇ REV MR 16.987 Zbl 68.247 JSL 20.282 • ID 10404

PETER, R. [1956] *Die beschraenkt-rekursiven Funktionen und die Ackermannsche Majorisierungsmethode* (**J** 0057) Publ Math (Univ Debrecen) 4*362-375
◇ D20 ◇ REV MR 18.104 Zbl 70.8 JSL 22.376 • ID 10405

PETER, R. [1957] *The boundedly recursive functions of Grzegorczyk and the majorisation method of Ackermann (Hungarian)* (**J** 0396) Mat Lapok 8*93-99
◇ D20 ◇ REV MR 20 # 5732 Zbl 102.8 • ID 42820

PETER, R. [1958] *Graphschemata und rekursive Funktionen* (**J** 0076) Dialectica 12*373-393
◇ D20 ◇ REV MR 21 # 1274 JSL 27.83 • ID 10407

PETER, R. [1959] *Rekursivitaet und Konstruktivitaet* (**P** 0634) Constructivity in Math;1957 Amsterdam 226-233
◇ A05 D20 F60 ◇ REV MR 21 # 4916 Zbl 85.250 JSL 33.471 • ID 10410

PETER, R. [1959] *Ueber die Partiell-Rekursivitaet der durch Graphschemata definierten zahlentheoretischen Funktionen* (**J** 0006) Ann Univ Budapest, Sect Math 2*41-48
◇ B75 D20 ◇ REV MR 22 # 7940 JSL 27.83 • ID 10409

PETER, R. [1960] *Zu einem Rekursionsschema von Hu Shih-Hua* (**J** 0006) Ann Univ Budapest, Sect Math 3-4*227-232
◇ D20 ◇ REV MR 24 # A2533 Zbl 103.247 • ID 10411

PETER, R. [1961] *Primitiv-rekursive Wortbeziehungen in der Programmierungssprache "Algol 60" (Russian summary)* (**J** 1487) Publ Math Inst Acad Sci Hung 6*137-144
◇ B75 D20 ◇ REV MR 27 # 2137 Zbl 109.352 • ID 29316

PETER, R. [1961] *Ueber die Verallgemeinerung der Rekursionsbegriffe fuer abstrakte Mengen als Definitionsbereiche* (**P** 0633) Infinitist Meth;1959 Warsaw 329-335
◇ D20 D75 ◇ REV MR 26 # 19 Zbl 121.255 • ID 10408

PETER, R. [1961] *Ueber die Verallgemeinerung der Theorie der rekursiven Funktionen fuer abstrakte Mengen geeigneter Struktur als Definitionsbereiche (Russian summary)* (**J** 0001) Acta Math Acad Sci Hung 12*271-314
◇ D20 D75 ◇ REV MR 25 # 15 Zbl 107.9 JSL 40.620 • REM Part II 1962 • ID 10412

PETER, R. [1962] *Ueber die Rekursivitaet einiger Uebersetzungs-Transformationen I,II (II: Verwendung einer Linearisierungsweise des Kantorowitsch-schen Ausdrucks-graphen) (Russian summaries)* (**J** 1487) Publ Math Inst Acad Sci Hung 7*69-78,373-384
◇ D20 ◇ REV MR 32 # 48 MR 32 # 49 Zbl 112.8 • ID 10414

PETER, R. [1962] *Ueber die Verallgemeinerung der Theorie der rekursiven Funktionen fuer abstrakte Mengen geeigneter Struktur als Definitionsbereiche (Fortsetzung)* (**J** 0001) Acta Math Acad Sci Hung 13*1-24
◇ D20 D75 ◇ REV MR 26 # 3601 Zbl 107.9 • REM Part I 1961 • ID 10416

PETER, R. [1963] *Programmierung und partiell-rekursive Funktionen* (**J** 0001) Acta Math Acad Sci Hung 14*373-401
◇ B75 D20 ◇ REV MR 28 # 23 Zbl 119.252 • ID 10419

PETER, R. [1963] *Ueber die Rekursivitaet der Begriffe der mathematischen Grammatiken (Russian summary)* (**J** 0462) Mat Fiz Oszt Koezlem, Acad Sci Hung 8*213-228
◇ D05 D20 ◇ REV MR 29 # 4678 Zbl 119.361 • ID 10417

PETER, R. [1963] *Ueber die Primitiv-Rekursivitaet einiger den Aufbau von Formeln charakterisierenden Wortfunktionen* (**J** 0001) Acta Math Acad Sci Hung 14*149-172
◇ D20 ◇ REV MR 29 # 3372 Zbl 117.14 • ID 10418

PETER, R. [1965] *Programmierung und partiell-rekursive Funktionen* (**P** 0797) Fonds des Math, Machines Math & Appl;1962 Tihany 131-132
◇ B75 D20 ◇ ID 27605

PETER, R. [1965] *Ueber die sequenzielle Berechenbarkeit von rekursiven Wortfunktionen durch Kellerspeicher* (**J** 0001) Acta Math Acad Sci Hung 16*231-253
◇ D10 D20 ◇ REV MR 35 # 52 Zbl 192.66 • ID 10420

PETER, R. [1965] *Zum Beitrag von F. Schwenkel "Rekursive Wortfunktionen ueber unendlichen Alphabeten"* (**J** 0068) Z Math Logik Grundlagen Math 11*377-378
◇ D20 ◇ REV MR 34 # 2454 Zbl 125.278 JSL 40.622 • ID 10422

PETER, R. [1969] *Automatische Programmierung zur Berechnung der partiell-rekursiven Funktionen* (**J** 0411) Studia Sci Math Hung 4*447-463
◇ B75 D20 ◇ REV MR 40 # 2541 Zbl 188.25 • ID 10423

PETER, R. [1969] *Ueber die Pair-schen freien Binoiden als Spezialfaelle der angeordneten freien holomorphen Mengen (Russian summary)* (**J** 0014) Bull Acad Pol Sci, Ser Math Astron Phys 17*181-184
• REPR [1970] (**J** 0001) Acta Math Acad Sci Hung 21*297-313
◇ D20 D75 ◇ REV MR 39 # 6750 MR 42 # 5798 Zbl 184.18 • ID 10424

PETER, R. [1969] *Ueber zweistufig definierte Sprachen* (**P** 1841) Fct Recurs & Appl;1967 Tihany 12-18
◇ B15 D20 ◇ ID 32551

PETER, R. [1972] *Zur Frage der Rekursivitaet der im "Algol 68" verwendeten zweistufigen Grammatik* (**J** 0006) Ann Univ Budapest, Sect Math 15*89-101
◇ B75 D20 ◇ REV MR 48 # 5436 • ID 10427

PETER, R. [1973] *Veranschaulichung einer sequenziellen Berechnung der rekursiven Wortfunktionen durch "eisenbahnrangierende Graphen"* (**J** 0049) Period Math Hung 3*183-187
◇ D20 ◇ REV MR 57 # 12190 Zbl 264 # 02036 • ID 27508

PETER, R. [1973] *Zur Rekursivitaet der mathematischen Grammatiken* (**J** 2293) Comp Linguist & Comp Lang 9*193-216
◇ D20 ◇ REV Zbl 342 # 68049 • ID 64462

PETER, R. [1975] *Die Rekursivitaet der Programmierungssprache "Lisp 1.5" in Spezialfaellen der angeordneten freien homomorphen Mengen (Russian summary)* (**J** 0380) Acta Cybern (Szeged) 2*183-201
◇ B75 D20 ◇ REV MR 53 # 12062 Zbl 324 # 68004 • ID 23106

PETER, R. [1976] *Rekursive Funktionen in der Computer-Theorie* (**X** 0928) Akad Kiado: Budapest 190pp
• TRANSL [1981] (**X** 0827) Wiley & Sons: New York 179pp [1981] (**X** 4433) Horwood: Chichester 179pp
◇ B75 D20 D98 ◇ REV MR 55 # 6926 MR 82m:68013 Zbl 323 # 68001 Zbl 464 # 68002 JSL 43.154 • ID 64463

PETER, R. also published under the name POLITZER, R.

PETER, R. see Vol. I, III, V, VI for further entries

PETERS, F.E. [1974] *Einfuehrung in mathematische Methoden der Informatik* (**X** 0876) Bibl Inst: Mannheim 342pp
◇ B70 B98 D80 D98 ◇ REV MR 53 # 2499 Zbl 319 # 94001 • ID 64464

PETERS JR., P.S. & RITCHIE, R.W. [1973] *Nonfiltering and local-filtering transformational grammars* (**C** 2282) Approach Nat Language 180-194
◇ D05 ◇ REV Zbl 298 # 68055 • ID 64465

PETRENKO, A.F. [1976] see CHAPENKO, V.P.

PETRESCO, J. [1962] *Algorithmes de decision et de construction dans les groupes libres* (**J** 0044) Math Z 79*32-43
◇ B25 D40 ◇ REV MR 28 # 5106 Zbl 104.243 • ID 10432

PETRESCO, J. [1969] *Pregroupes de mots et probleme des mots* (**S** 1057) Semin Dubreil: Alg Th Nombr 2/17*11pp
◇ D40 ◇ REV MR 40 # 7338 Zbl 208.32 • ID 28645

PETRI, C.A. [1967] *Grundsaetzliches zur Beschreibung diskreter Prozesse* (**P** 1671) Colloq Automatenth (3);1965 Hannover 121-140
◇ B70 D05 ◇ REV Zbl 157.337 • ID 29429

PETRI, N.V. [1969] *Algorithms connected with predicates and Boolean functions (Russian)* (**J** 0023) Dokl Akad Nauk SSSR 185*37-99
• TRANSL [1969] (**J** 0062) Sov Math, Dokl 10*294-297
◇ B35 D15 ◇ REV MR 40 # 4113 Zbl 193.315 • ID 10436

PETRI, N.V. [1969] see KANOVICH, M.I.

PETRI, N.V. [1969] *The complexity of algorithms and their operating time (Russian)* (**J** 0023) Dokl Akad Nauk SSSR 186*30-31
• TRANSL [1969] (**J** 0062) Sov Math, Dokl 10*547-549
◇ B35 D15 D20 ◇ REV MR 40 # 7115 Zbl 272 # 02053 • ID 10438

PETRI, N.V. [1969] *Two theorems on the complexity of algorithms and computations (Russian)* (**S** 0228) Zap Nauch Sem Leningrad Otd Mat Inst Steklov 16*165-174
• TRANSL [1969] (**J** 0521) Semin Math, Inst Steklov 16*84-89
◇ D15 D25 ◇ REV MR 41 # 8234 Zbl 318 # 02036 • ID 10437

PETRI, N.V. [1973] see MARKOV, A.A.

PETRI, N.V. [1974] see KOZMIDIADI, V.A.

PETRI, N.V. [1979] *Unsolvability of the problem of recognition of cancelling iterative nets (Russian)* (**S** 0554) Issl Teor Algor & Mat Logik (Moskva) 3*90-98,133
◇ D10 ◇ REV MR 81g:03043 Zbl 431 # 68054 • ID 77250

PETRI, N.V. see Vol. I, V, VI for further entries

PETRICH, M. [1981] see GERHARD, J.A.

PETROSYAN, G.N. [1972] see KHACHATRYAN, V.E.

PETROSYAN, G.N. [1974] *A certain basis of operators and predicates with an undecidable emptiness problem (Russian) (English summary)* (**J** 0040) Kibernetika, Akad Nauk Ukr SSR 1974/5*23-28
◇ B75 D05 D80 ◇ REV MR 53 # 4626 Zbl 301 # 02030 • ID 21662

PETROV, B.N. & UL'YANOV, S.V. & ULANOV, G.M. [1979] *Complexity of finite objects and informational control theory (Russian)* (**C** 4703) Tekh Kibernetika, Vyp 11 77-147,192
◇ D15 D98 ◇ REV MR 82h:93064 • ID 71239

PETROV, V.P. & SKORDEV, D.G. [1979] *Combinatory structures* (**J** 2547) Serdica, Bulgar Math Publ 5*128-148
◇ B40 D75 ◇ REV MR 81j:03079 Zbl 438 # 03047 • ID 55959

PETTOROSSI, A. [1979] *On the definition of hierarchies of infinite sequential computations* (**P** 2935) FCT'79 Fund of Comput Th;1979 Berlin/Wendisch-Rietz 335-341
◇ B40 D20 ◇ REV MR 81j:68021 Zbl 419 # 68077 • ID 53394

PETTOROSSI, A. [1983] see KREMPA, J.

PETTOROSSI, A. see Vol. VI for further entries

PFEIFFER, H. [1982] see COHEN, L.J.

PFEIFFER, H. see Vol. II, V, VI for further entries

PFENDER, M. [1975] see EHRIG, H.

PFENDER, M. [1975] see KUEHNEL, W.

PFENDER, M. [1977] see KUEHNEL, W.

PFENDER, M. see Vol. V, VI for further entries

PHAN DINH DIEU [1971] *On a class of stochastic languages* (**J** 0068) Z Math Logik Grundlagen Math 17*421-425
◇ D05 ◇ REV MR 45 # 1706 Zbl 242 # 94047 • ID 61364

PHAN DINH DIEU [1971] *On the languages representable by finite probabilistic automata* (**J** 0068) Z Math Logik Grundlagen Math 17*427-442
◇ D05 D10 ◇ REV MR 45 # 1707 Zbl 242 # 94048 • ID 61365

PHAN DINH DIEU [1976] *A note on iterative arrays of finite automata over the one-letter alphabet* (**J** 0129) Elektr Informationsverarbeitung & Kybern 12*551-555
◇ D05 ◇ REV MR 56 # 13808 Zbl 352 # 94040 • ID 50081

PHAN DINH DIEU see Vol. VI for further entries

PHILLIPS, I. [1984] see KRANAKIS, E.

PHILLIPS, R.E. [1983] see HICKIN, K.K.

PHILLIPS, R.E. see Vol. III for further entries

PIAZZESE, F. [1981] see MERCANTI, F.

PIECZKOWSKI, A. [1968] *Undecidability of the homogeneous formulas of degree 3 of the predicate calculus (Polish and Russian summaries)* (**J** 0063) Studia Logica 22*7-16
◇ B20 D35 ◇ REV MR 38 # 4311 Zbl 315 # 02045 • ID 10474

PIECZKOWSKI, A. see Vol. I, II, VI for further entries

PIERCE, J.C. [1973] see MEY VANDER, J.E.

PIGOZZI, D. [1974] *The join of equational theories* (**S** 0019) Colloq Math (Warsaw) 30*15-25
◇ B25 C05 D35 ◇ REV MR 50 # 4270 Zbl 319 # 02037 • ID 10495

PIGOZZI, D. [1976] *Base-undecidable properties of universal varieties* (J 0004) Algeb Universalis 6*193-223 ◇ C05 D80 ◇ REV MR 55#7757 Zbl 356#08005 JSL 47.904 • ID 23901

PIGOZZI, D. [1976] *The universality of the variety of quasigroups* (J 3194) J Austral Math Soc, Ser A 21*194-219 ◇ B25 C05 C52 D35 ◇ REV MR 52#13582 Zbl 323#20073 • ID 21868

PIGOZZI, D. see Vol. III for further entries

PILIPOSYAN, A.G. [1973] *Certain transformations of logical schemes (Russian) (Armenian summary)* (J 0346) Dokl Akad Nauk Armyan SSR 56*193-197 ◇ D05 ◇ REV MR 48#12886 Zbl 287#02019 • ID 30531

PIN, J.-E. [1984] see MARGOLIS, S.W.

PIN, J.-E. [1985] *Star-free ω-languages and first order logic* (P 4622) Autom on Infinite Words;1984 Le Mont-Dore 56-67 ◇ D05 ◇ ID 49457

PINUS, A.G. [1971] *A remark on the paper by Ju.M.Vazhenin: "Elementary properties of transformation semigroups of ordered sets" (Russian)* (J 0003) Algebra i Logika 10*327-328 • TRANSL [1971] (J 0069) Algeb and Log 10*205 ◇ B25 C65 D35 E07 ◇ REV MR 45#1804 Zbl 262#20079 • REM For the paper see Vazhenin 1970 • ID 19598

PINUS, A.G. [1972] *On the theory of convex subsets (Russian)* (J 0092) Sib Mat Zh 13*218-224 • TRANSL [1972] (J 0475) Sib Math J 13*157-161 ◇ B25 C65 C85 D35 ◇ REV MR 45#3205 Zbl 255#02052 • ID 10513

PINUS, A.G. [1973] *A weak second order theory of fixed sets (Russian)* (C 1443) Algebra, Vyp 2 (Irkutsk) 154-160 ◇ B15 D35 ◇ REV MR 54#4956 • ID 24124

PINUS, A.G. [1975] *Effective linear orders (Russian)* (J 0092) Sib Mat Zh 16*1246-1254,1371 • TRANSL [1975] (J 0475) Sib Math J 16*956-962 ◇ C57 D45 E07 F15 ◇ REV MR 53#137 Zbl 333#02035 • ID 16662

PINUS, A.G. [1976] *Theories of boolean algebras in a calculus with the quantifier "infinitely many exist" (Russian)* (J 0092) Sib Mat Zh 17*1417-1421,1440 • TRANSL [1976] (J 0475) Sib Math J 17*1035-1038 ◇ B25 C10 C35 C57 C80 G05 ◇ REV MR 56#84 Zbl 353#02006 • ID 50089

PINUS, A.G. [1978] see KOKORIN, A.I.

PINUS, A.G. [1978] see HERRE, H.

PINUS, A.G. [1983] *Calculus with the quantifier of elementary equivalence (Russian)* (J 0092) Sib Mat Zh 24/3(139)*136-141 • TRANSL [1983] (J 0475) Sib Math J 24*428-432 ◇ C40 C55 C80 D35 ◇ REV MR 85b:03056 Zbl 536#03020 • ID 37103

PINUS, A.G. [1983] *The operation of Cartesian product* (J 0031) Izv Vyssh Ucheb Zaved, Mat (Kazan) 1983/8*51-53 • TRANSL [1983] (J 0062) Sov Math, Dokl 27*62-65 ◇ B25 C05 D35 ◇ REV MR 85b:08009 Zbl 539#08004 • ID 39357

PINUS, A.G. [1985] *Applications of boolean powers of algebraic systems (Russian)* (J 0092) Sib Mat Zh 26/3*117-125,225 • TRANSL [1985] (J 0475) Sib Math J 26*400-407 ◇ C05 C30 C55 C57 C80 C85 D35 E50 G05 ◇ ID 47255

PINUS, A.G. see Vol. I, III, V, VI for further entries

PINZANI, R. [1975] see LAGANA, M.R.

PIPPENGER, N.J. [1979] see FISCHER, MICHAEL J.

PIPPENGER, N.J. [1980] *On another boolean matrix* (J 1426) Theor Comput Sci 11*49-56 ◇ B70 D15 ◇ REV MR 81f:94044 Zbl 429#94038 • ID 69774

PIPPENGER, N.J. [1981] *Computational complexity of algebraic functions* (J 0119) J Comp Syst Sci 22*454-470 ◇ D15 ◇ REV MR 82m:12001 Zbl 469#68045 • ID 82486

PIPPENGER, N.J. [1985] see FAGIN, R.

PIRICKA-KELEMENOUA, A. [1972] *Directable automata and directly subdefinite events (Czech summary)* (J 0156) Kybernetika (Prague) 8*395-403 ◇ D05 ◇ REV MR 52#16915 Zbl 249#94015 • ID 21895

PIRLING, C. [1980] see FOERSTER, M.

PISANSKI, T. [1978] *Computability and solvability (Slovanian) (English summary)* (J 2310) Obz Mat Fiz, Ljubljana 25*41-54 ◇ D10 ◇ REV MR 80m:03004 Zbl 366#02005 • ID 51088

PITTS, W. [1943] see MCCULLOCH, W.S.

PIXLEY, A.F. [1972] *Local Mal'cev conditions* (J 0018) Canad Math Bull 15*559-568 ◇ C05 C57 ◇ REV MR 46#8942 Zbl 254#08009 • ID 31207

PIXLEY, A.F. see Vol. III for further entries

PIZON, T. [1979] see BALCER, M.

PIZZI, C. [1981] *"Since", "even if", "as if"* (C 3515) Ital Studies in Phil of Sci 73-87 ◇ B65 D35 ◇ REV MR 82d:03043 • ID 77349

PIZZI, C. see Vol. II for further entries

PKHAKADZE, SH.S. [1975] *A certain class of abbreviating symbols I (Russian) (Georgian summary)* (S 2043) Issl Mat Log & Teor Algor (Tbilisi) 1975*31-102 ◇ B75 D15 ◇ REV MR 56#8342 • ID 77268

PKHAKADZE, SH.S. [1975] see MAGNARADZE, D.G.

PKHAKADZE, SH.S. [1977] see MAGNARADZE, L.G.

PKHAKADZE, SH.S. [1983] *An example of an intuitively computable everywhere defined function, and Church's thesis (Russian) (English and Georgian summaries)* (J 3112) Annot Dokl, Inst Prikl Mat, Tbilisi 17*64-76 ◇ D20 ◇ REV MR 86b:03046 • ID 45335

PKHAKADZE, SH.S. [1984] *An example of an intuitively computable everywhere defined function and Church's thesis (Russian)* (X 1052) Tbilisi Univ: Tbilisi 70pp ◇ D20 ◇ ID 45610

PKHAKADZE, SH.S. see Vol. I, V for further entries

PLA I CARRERA, J. [1982] *On the R-representability of primitive recursive functions (Catalan)* (P 3870) Congr Catala de Log Mat (1);1982 Barcelona 101-105
◇ D20 F30 ◇ REV MR 84i:03003 Zbl 521 # 03025
• ID 37071

PLA I CARRERA, J. see Vol. I, II, III, V, VI for further entries

PLAISTED, D.A. [1977] *Sparse complex polynomials and polynomial reducibility* (J 0119) J Comp Syst Sci 14*210-221
◇ D15 ◇ REV MR 56 # 7319 Zbl 359 # 65043 • ID 50625

PLAISTED, D.A. [1980] *The application of multivariate polynomials to inference rules and partial tests for unsatisfiability* (J 1428) SIAM J Comp 9*698-705
◇ B35 D15 F20 ◇ REV MR 82g:68045 Zbl 448 # 68022
• ID 56667

PLAISTED, D.A. [1984] *Complete problems in the first-order predicate calculus* (J 0119) J Comp Syst Sci 29*8-35
◇ B10 D15 ◇ REV MR 86b:03045 • ID 44042

PLAISTED, D.A. [1985] *The undecidability of self-embedding for term rewriting systems* (J 0232) Inform Process Lett 20*61-64
◇ D05 ◇ REV MR 86f:68022 • ID 48927

PLAISTED, D.A. see Vol. I, II for further entries

PLASS, M. [1979] see ASPVALL, B.

PLATEK, M. [1974] *Questions of graphs and automata in a generative description of language* (J 2817) Prague Bull Math Linguist 21*27-64
◇ D05 ◇ REV MR 54 # 14459 Zbl 367 # 68056 • ID 51211

PLATEK, M. [1984] *Recognizing of languages by composition of deterministic pushdown transducers* (J 2817) Prague Bull Math Linguist 41*1-13
◇ D10 ◇ REV MR 86e:68058 Zbl 567 # 68049 • ID 48914

PLATEK, R.A. [1966] *Foundations of recursion theory* (0000) Diss., Habil. etc 215pp
◇ D60 D75 ◇ REM Ph.D. thesis, Stanford University, Stanford, CA • ID 19596

PLATEK, R.A. [1969] *Eliminating the continuum hypothesis* (J 0036) J Symb Logic 34*219-225
◇ D55 E25 E45 E47 E50 ◇ REV MR 41 # 1528 Zbl 206.11 JSL 36.166 • ID 10526

PLATEK, R.A. [1970] *A note on the cardinality of the Medvedev lattice* (J 0053) Proc Amer Math Soc 25*917
◇ D30 ◇ REV MR 41 # 6690 Zbl 205.11 • ID 10528

PLATEK, R.A. [1970] *A note on the failure of the relativized enumeration theorem in recursive function theory* (J 0053) Proc Amer Math Soc 25*915-916
◇ D20 ◇ REV MR 41 # 8226 Zbl 205.11 • ID 42981

PLATEK, R.A. [1971] *A countable hierarchy for the superjump* (P 0638) Logic Colloq;1969 Manchester 257-271
◇ D55 D65 ◇ REV MR 44 # 6495 Zbl 234 # 02026
• ID 10529

PLATEK, R.A. see Vol. III, V for further entries

PLISKO, V.E. [1977] *The nonarithmeticity of the class of realizable predicate formulas (Russian)* (J 0216) Izv Akad Nauk SSSR, Ser Mat 41*483-502
• TRANSL [1977] (J 0448) Math of USSR, Izv 11*453-471
◇ D35 D55 F50 ◇ REV MR 57 # 16031 Zbl 373 # 02032
• ID 51967

PLISKO, V.E. see Vol. I, VI for further entries

PLOTKIN, G.D. [1977] *LCF considered as a programming language* (J 1426) Theor Comput Sci 5*223-255
◇ B75 D05 ◇ REV MR 81e:68016 Zbl 369 # 68006
• ID 82497

PLOTKIN, G.D. see Vol. I, VI for further entries

PLOTKIN, J.M. & ROSENTHAL, J.W. [1982] *The expected complexity of analytic tableaux analysis in propositional calculus* (J 0047) Notre Dame J Formal Log 23*409-426
◇ B05 B35 D15 F20 ◇ REV MR 83k:03044
Zbl 464 # 03011 • ID 35398

PLOTKIN, J.M. see Vol. II, III, V for further entries

PLOTNIKOVA, N.A. [1983] *Calculation with oracles* (J 0092) Sib Mat Zh 24/1*146-151,193
• TRANSL [1983] (J 0475) Sib Math J 24*119-124
◇ D10 D30 ◇ REV MR 85e:03093 • ID 40689

PLYASUNOV, A.V. [1979] *Splinters and Turing degrees (Russian)* (J 0087) Mat Zametki (Akad Nauk SSSR) 25*307-310,319
• TRANSL [1979] (J 1044) Math Notes, Acad Sci USSR 25*158-159
◇ D25 ◇ REV MR 80g:03040 Zbl 421 # 03033 • ID 53432

PNUELI, A. [1971] see ASHCROFT, E.A.

PNUELI, A. [1977] see FRANCEZ, N.

PNUELI, A. & SLUTZKI, G. [1981] *Automatic programming of finite state linear programs* (J 1428) SIAM J Comp 10*519-535
◇ D05 ◇ REV MR 82m:68103 Zbl 462 # 68002 • ID 82501

PNUELI, A. see Vol. I, II for further entries

POAGE, J.F. [1963] see BRZOZOWSKI, J.A.

POBEDIN, L.N. [1973] *Certain questions on generalized computability (Russian)* (J 0003) Algebra i Logika 12*220-231,244
• TRANSL [1973] (J 0069) Algeb and Log 12*125-131
◇ D10 D55 D75 ◇ REV MR 52 # 13352 Zbl 288 # 02022
• ID 64570

POBEDIN, L.N. [1975] *The halt problem and theory of hierarchies (Russian)* (J 0003) Algebra i Logika 14*186-203,240
• TRANSL [1975] (J 0069) Algeb and Log 14*112-123
◇ D10 D55 ◇ REV MR 54 # 12508 Zbl 324 # 02024
• ID 25999

PODEWSKI, K.-P. [1982] see COHEN, L.J.

PODEWSKI, K.-P. see Vol. III, V for further entries

PODKOLZIN, A.S. [1978] *On universal homogeneous structures (Russian)* (J 0052) Probl Kibern 34*109-131
◇ D05 ◇ REV MR 80g:68073 Zbl 417 # 68038 • ID 53283

PODKOPAEV, B.P. [1980] see DANILOV, V.V.

PODLOVCHENKO, R.I. [1970] *Schemes of algorithms that are defined on situations (Russian)* (J 0052) Probl Kibern 23*213-246
• TRANSL [1973] (J 0471) Syst Th Res 23*225-258
◇ D05 D20 ◇ REV MR 46 # 4776 Zbl 267 # 02020
• ID 29884

PODLOVCHENKO, R.I. [1972] see KHACHATRYAN, V.E.

PODLOVCHENKO, R.I. [1973] *R-schemes and equivalence relations between them (Russian)* (J 0052) Probl Kibern 27*213-237,294-295
◇ D05 ◇ REV MR 49 # 11856 Zbl 311 # 68046 • ID 10577

PODLOVCHENKO, R.I. [1973] *A complete system of similar transformations of R-schemes (Russian)* (J 0052) Probl Kibern 27*239-250,295
◇ D05 ◇ REV MR 49 # 11857 Zbl 302 # 68032 • ID 10578

PODLOVCHENKO, R.I. [1974] *Non-determined algorithm schemata or R-schemata* (P 1511) Int Symp Th Progr;1972 Novosibirsk 5*86-110
◇ D05 ◇ REV MR 55 # 6927 Zbl 283 # 68010 • ID 64582

PODNIEKS, K.M. [1974] *Comparison of the different types of identification in the limit and prediction of functions I (Russian)* (S 2587) Teor Algor & Progr (Riga) 1*68-81
◇ D20 D45 ◇ REV MR 58 # 16221 Zbl 332 # 02045 • REM Part II 1975 • ID 64587

PODNIEKS, K.M. [1974] see BARZDINS, J.

PODNIEKS, K.M. [1974] see FREJVALD, R.V.

PODNIEKS, K.M. [1975] *Comparison of the different types of identification in the limit and prediction of functions II (Russian)* (S 2587) Teor Algor & Progr (Riga) 2*35-44,206-207
◇ D20 D45 ◇ REV MR 57 # 16011 Zbl 332 # 02046 • REM Part I 1974 • ID 77390

PODNIEKS, K.M. [1975] *Probabilistic synthesis of enumerated classes of functions (Russian)* (J 0023) Dokl Akad Nauk SSSR 223*1071-1074
• TRANSL [1975] (J 0062) Sov Math, Dokl 16*1042-1045
◇ D20 ◇ REV MR 58 # 21530 Zbl 341 # 02028 • ID 29750

PODNIEKS, K.M. [1975] *Probabilistic prediction of function values (Russian) (English summary)* (S 2587) Teor Algor & Progr (Riga) 2*57-76
◇ D20 ◇ REV MR 58 # 3683 Zbl 318 # 02040 • ID 64586

PODNIEKS, K.M. [1975] *The double-incompleteness theorem (Russian) (English summary)* (S 2587) Teor Algor & Progr (Riga) 2*191-200,209
◇ D35 F30 ◇ REV MR 57 # 16036 Zbl 347 # 02021 • ID 64585

PODNIEKS, K.M. [1977] *Forecasting strategies of limited complexity (Russian) (English Summary)* (S 2587) Teor Algor & Progr (Riga) 3*89-102,154
◇ D15 D20 ◇ REV MR 57 # 14602 • ID 82506

PODNIEKS, K.M. [1981] *Prediction of the following value of a function (Russian)* (J 0031) Izv Vyssh Ucheb Zaved, Mat (Kazan) 1981/5*71-77
• TRANSL [1981] (J 3449) Sov Math 25/5*83-91
◇ D10 D20 ◇ REV MR 84d:68051 Zbl 484 # 68074 • ID 36622

PODNIEKS, K.M. see Vol. VI for further entries

POENARU, V. [1963] *Expose sommaire de la theorie des algorithmes (d'apres A.A. Markov)* (J 0082) Bull Soc Math Belg 15*271-302
◇ D03 D20 ◇ REV MR 30 # 4679 Zbl 116.8 • ID 10579

POENARU, V. [1968] see BOONE, W.W.

POGORZELSKI, H.A. [1962] *A note on an arithmetization of a word system in a denumerable alphabet* (J 0068) Z Math Logik Grundlagen Math 8*247-249
◇ D05 D20 ◇ REV MR 26 # 3577 Zbl 117.253 JSL 29.200 • ID 10583

POGORZELSKI, H.A. [1962] *Exponential chains of natural numbers and Vuckovic's recursion functions I (Polish) (Russian and English summaries)* (J 0051) Commentat Math, Ann Soc Math Pol, Ser 1 7*19-34
◇ D20 ◇ REV MR 32 # 5508 Zbl 199.26 • ID 10581

POGORZELSKI, H.A. [1962] *Word arithmetic: theory of primitive words* (J 0068) Z Math Logik Grundlagen Math 8*251-255
◇ D05 D20 ◇ REV MR 26 # 3578 Zbl 127.9 JSL 29.200 • ID 10582

POGORZELSKI, H.A. [1964] *Commutative recursive word arithmetic in the alphabet of prime numbers* (J 0047) Notre Dame J Formal Log 5*13-23
◇ D05 D20 ◇ REV MR 32 # 1120 Zbl 133.255 JSL 31.271 • ID 10586

POGORZELSKI, H.A. [1964] *Primitive words in an infinite abstract alphabet* (J 0068) Z Math Logik Grundlagen Math 10*193-198
◇ D05 D20 ◇ REV MR 28 # 3924 Zbl 129.5 JSL 31.271 • ID 10584

POGORZELSKI, H.A. [1964] *Skolem arithmetics on certain concrete word systems* (J 0132) Math Scand 14*93-105
◇ D05 ◇ REV MR 30 # 1050 Zbl 126.20 JSL 31.271 • ID 10585

POGORZELSKI, H.A. [1965] *Nonconcatenative abstract Skolem arithmetics I,II,III* (J 0068) Z Math Logik Grundlagen Math 11*89-92,249-252,373-376
◇ D05 ◇ REV MR 31 # 2187 MR 31 # 3318 MR 32 # 1121 Zbl 203.12 JSL 35.150 • ID 24793

POGORZELSKI, H.A. [1969] *Goldbach sentences in abstract arithmetics* $\mathscr{A}^k(A)$ *I* (J 0127) J Reine Angew Math 237*65-96
◇ D20 F30 ◇ REV MR 40 # 4107 Zbl 179.348 • REM Part II 1970 • ID 19592

POGORZELSKI, H.A. [1977] *Goldbach conjecture* (J 0127) J Reine Angew Math 292*1-12
◇ D80 ◇ REV MR 58 # 27414 Zbl 351 # 02030 • ID 77391

POGORZELSKI, H.A. see Vol. I, VI for further entries

POGOSYAN, EH.M. [1974] *1-inductors and some of their properties (Russian) (Armenian summary)* (J 0346) Dokl Akad Nauk Armyan SSR 58*10-14
◇ D15 ◇ REV MR 49 # 6699 Zbl 305 # 94042 • ID 82508

POGOSYAN, EH.M. [1975] *Inductive inference with feedback (Russian) (Armenian summary)* (J 0346) Dokl Akad Nauk Armyan SSR 60*193-197
◇ B48 D15 ◇ REV MR 56 # 8345 Zbl 327 # 94008 • ID 77403

POGOSYAN, EH.M. & SARKISYAN, O.A. [1975] *On a formalization of inductive generalization (Russian) (Armenian summary)* (S 0422) Tr Vychisl Tsentra Akad Nauk Armyan SSR & Univ Erevan 8*157-175
◇ D20 ◇ REV MR 57 # 86 Zbl 397 # 68095 • ID 52723

POGOSYAN, EH.M. [1977] see KARAPETYAN, B.K.

POGOSYAN, EH.M. [1978] see MARTIROSYAN, A.A.

POGOSYAN, EH.M. [1978] *Training as a variety of induction inference (Russian)* (J 0977) Izv Akad Nauk SSSR, Tekh Kibern 1978/3*62-69
• TRANSL [1978] (J 0522) Engin Cybern 16/3*46-52
◇ D20 ◇ REV MR 82j:68091 Zbl 387 # 68075 • ID 82509

POGOSYAN, EH.M. see Vol. II for further entries

POHL, H.-J. [1968] *Ueber die Reduzierung der Anzahl von Eingabesignalen von Automaten* (J 0068) Z Math Logik Grundlagen Math 14*93-96
◇ D05 ◇ REV MR 36#4922 Zbl 169.315 • ID 10600

POHL, H.-J. [1969] *Darstellbarkeit von Ereignissen in Z-endlichen Automaten* (J 0068) Z Math Logik Grundlagen Math 15*93-95
◇ D05 ◇ REV MR 39#4008 Zbl 175.280 • ID 10601

POIGNE, A. [1978] *Context-free rewriting* (S 3125) Ber, Abt Inf, Univ Dortmund 74*85-91
◇ D03 ◇ REV Zbl 473#68069 • ID 55400

POIGNE, A. [1981] *Context-free languages of infinite words as least fixpoints* (P 3165) FCT'81 Fund of Comput Th;1981 Szeged 301-310
◇ D05 ◇ REV MR 84e:68091 Zbl 482#68069 • ID 39308

POIGNE, A. see Vol. V, VI for further entries

POIZAT, B. [1981] *Degres de definissabilite arithmetique des generiques (English summary)* (J 3364) C R Acad Sci, Paris, Ser 1 293*289-291
◇ C25 C40 D30 D55 ◇ REV MR 83f:03031 Zbl 472#03022 • ID 55288

POIZAT, B. see Vol. I, III, V for further entries

POLITZER, R. [1932] *Rekursive Funktionen* (P 0653) Int Congr Math (II, 4);1932 Zuerich 2*336-337
◇ D20 D98 ◇ ID 10392

POLITZER, R. also published under the name PETER, R.

POLJAK, S. & TURZIK, D. [1982] *A polynomial algorithm for constructing a large bipartite subgraph, with an application to a satisfiability problem* (J 0017) Canad J Math 34*519-524
◇ B05 D15 ◇ REV MR 83j:05048 Zbl 471#68041 • ID 39883

POLLACK, P.L. [1971] see CAVINESS, B.F.

POLLMAR, C. [1954] see BURKS, A.W.

POLYAKOV, E.A. [1963] *Certain properties of algebras of recursive functions (Russian)* (J 0226) Uch Zap Ped Inst, Ivanovo 34*46-51
◇ D20 D75 ◇ REV MR 33#5482 • ID 10614

POLYAKOV, E.A. [1964] *Algebras of recursive functions (Russian)* (J 0003) Algebra i Logika 3/1*41-56
◇ D20 D75 ◇ REV MR 29#17 Zbl 236#02034 JSL 37.408 • ID 10615

POLYAKOV, E.A. [1964] *On some properties of algebras of recursive functions (Russian)* (J 0003) Algebra i Logika 3/3*39-57
◇ D20 D75 ◇ REV MR 31#52 Zbl 236#02035 JSL 37.408 • ID 10616

POLYAKOV, E.A. [1966] see LAVROV, I.A.

POLYAKOV, E.A. [1966] *Some properties of algebras of recursive functions (Russian)* (J 0092) Sib Mat Zh 7*720-723
• TRANSL [1966] (J 0475) Sib Math J 7*574-576
◇ D20 D75 ◇ REV MR 34#1190 Zbl 207.305 • ID 10617

POLYAKOV, E.A. [1967] *Several properties of the algebra of general recursive functions (Russian)* (J 0040) Kibernetika, Akad Nauk Ukr SSR 1967/2*20-23
• TRANSL [1967] (J 0021) Cybernetics 3/2*15-17
◇ D20 D75 ◇ REV MR 42#5792 Zbl 235#02042 • ID 64611

POLYAKOV, E.A. [1968] *Recursive subsets of sets of recursive functions (Russian)* (J 0023) Dokl Akad Nauk SSSR 183*1262-1264
• TRANSL [1968] (J 0062) Sov Math, Dokl 9*1548-1550
◇ D20 ◇ REV MR 39#3996 Zbl 179.18 • ID 24974

POLYAKOV, E.A. [1968] *Some problems in the theory of recursive functions (Russian)* (J 0003) Algebra i Logika 7/2*77-84
• TRANSL [1968] (J 0069) Algeb and Log 7*117-121
◇ D20 ◇ REV MR 39#3995 Zbl 249#02020 • ID 10618

POLYAKOV, E.A. [1968] *Some properties of algebras of recursive functions (Russian)* (J 0023) Dokl Akad Nauk SSSR 178*296-298
• TRANSL [1968] (J 0062) Sov Math, Dokl 9*88-90
◇ D20 D75 ◇ REV MR 36#4987 Zbl 249#02019 • ID 28888

POLYAKOV, E.A. [1969] see NASIBULLOV, KH.KH.

POLYAKOV, E.A. [1969] *Formal definability in algebras of recursive functions (Russian)* (J 0226) Uch Zap Ped Inst, Ivanovo 61*183-187
◇ D20 D75 ◇ REV MR 48#3716 • ID 10620

POLYAKOV, E.A. [1969] *Precomplete classes of primitive recursive functions (Russian)* (J 0226) Uch Zap Ped Inst, Ivanovo 44*3-5
◇ D20 ◇ REV MR 47#1586 • ID 10619

POLYAKOV, E.A. [1972] *On relative recursiveness and computability (Russian)* (P 2585) All-Union Conf Math Log (2);1972 Moskva 41
◇ D30 D65 ◇ ID 90050

POLYAKOV, E.A. & ROZINAS, M.G. [1974] *Some algebraic problems in the theory of reducibilities of sets (Russian)* (P 1590) All-Union Conf Math Log (3);1974 Novosibirsk 176-177
◇ D30 ◇ ID 90045

POLYAKOV, E.A. & ROZINAS, M.G. [1976] *Functional degrees (Russian)* (P 2064) All-Union Conf Math Log (4);1976 Kishinev 119
◇ D30 ◇ ID 90046

POLYAKOV, E.A. & ROZINAS, M.G. [1976] *Theory of algorithms (Russian)* (X 2594) Ivanovo Gos Univ: Ivanovo 87pp
◇ D20 D30 D98 ◇ REV MR 57#5717 Zbl 394#03039 • ID 52516

POLYAKOV, E.A. & ROZINAS, M.G. [1977] *Enumeration reducibilities (Russian)* (J 0092) Sib Mat Zh 18*838-845,957
• TRANSL [1977] (J 0475) Sib Math J 18*594-599
◇ D30 ◇ REV MR 56#8346 Zbl 377#02029 • ID 51746

POLYAKOV, E.A. & ROZINAS, M.G. [1978] *Correlations between different forms of relative computability of functions (Russian)* (J 0142) Mat Sb, Akad Nauk SSSR, NS 107(149)*134-145,160
• TRANSL [1979] (J 0349) Math of USSR, Sbor 35*425-436
◇ D20 D30 ◇ REV MR 80e:68135 Zbl 412#03022 • ID 82513

POLYAKOV, E.A. (ED.) [1978] *Recursive functions (Russian)* (**X** 2594) Ivanovo Gos Univ: Ivanovo 100pp
◇ D20 D30 D97 ◇ REV MR 81m:03007 Zbl 504 # 03020
• ID 70034

POLYAKOV, E.A. [1978] *Some remarks on the relative recursiveness (Russian)* (**C** 2595) Rekursiv Funktsii 61-70
◇ D30 ◇ REV MR 82i:03056 Zbl 504 # 03020 • ID 90051

POLYAKOV, E.A. [1980] *The theory of recursive functionals and effective operations (Russian)* (**J** 0031) Izv Vyssh Ucheb Zaved, Mat (Kazan) 1980/5*37-39
• TRANSL [1980] (**J** 3449) Sov Math 24/5*40-43
◇ D20 ◇ REV MR 81j:03078 Zbl 436 # 03038 • ID 77425

POLYAKOV, E.A. [1981] *Partial recursive functions with a recursive graph (Russian)* (**C** 3865) Algeb Sistemy (Ivanovo) 174-175
◇ D20 ◇ REV MR 85k:03026 Zbl 541 # 03020 • ID 41380

POLYAKOV, E.A. see Vol. V for further entries

POPOV, S.V. [1977] *Undecidable interval arithmetic (Russian) (English Summary)* (**S** 2651) Prepr Inst Prikl Mat, Akad Nauk SSSR 95*55pp
◇ B55 D35 ◇ REV MR 58 # 10389 • ID 77439

POPOV, S.V. [1981] *Nondecidable intermediate calculus (Russian)* (**J** 0003) Algebra i Logika 20*654-706,728
• TRANSL [1981] (**J** 0069) Algeb and Log 20*424-461
◇ B55 D35 ◇ REV MR 84h:03103 Zbl 528 # 03028
JSL 50.1081 • ID 34293

POPOV, S.V. see Vol. I, III, VI for further entries

POPRUZENKO, J. [1932] *Sur l'analyticite des ensembles (A)* (**J** 0027) Fund Math 18*77-84
◇ D55 E15 E75 ◇ REV Zbl 4.203 FdM 58.86 • ID 10656

POPRUZENKO, J. see Vol. V for further entries

PORTE, J. [1958] *Systemes de Post, algorithmes de Markov* (**J** 2695) Cybernetica 1*114-149
◇ D03 ◇ REV MR 22 # 4636 Zbl 173.9 JSL 24.239
• ID 33679

PORTE, J. [1958] *Une simplification de la theorie de Turing* (**P** 1177) Int Congr Cybern (1);1956 Namur 251-280
◇ D10 D20 ◇ REV MR 23 # A3084 JSL 25.162 • ID 42543

PORTE, J. [1960] *Quelques pseudo-paradoxes de la "calculabilite effective"* (**P** 1184) Int Congr Cybern (2);1958 Namur 332-334
◇ A05 D20 ◇ REV JSL 33.471 • ID 21985

PORTE, J. [1972] *La logique mathematique et le calcul mecanique* (**S** 0889) Notas Logica Mat 8*105pp
◇ B98 D98 ◇ REV MR 51 # 5253 Zbl 279 # 02001
JSL 24.70 • ID 17471

PORTE, J. see Vol. I, II, VI for further entries

PORTO, A.G. [1980] see MATOS, A.B.

POSAMENT, T. [1934] see KURATOWSKI, K.

POSNER, D.B. [1978] see EPSTEIN, R.L.

POSNER, D.B. [1978] see JOCKUSCH JR., C.G.

POSNER, D.B. [1980] *A survey of non-r.e. degrees ‹ = 0'* (**P** 3021) Logic Colloq;1979 Leeds 52-109
◇ D30 ◇ REV MR 83i:03071 Zbl 475 # 03020 • ID 55474

POSNER, D.B. [1981] see JOCKUSCH JR., C.G.

POSNER, D.B. & ROBINSON, R.W. [1981] *Degrees joining to 0'* (**J** 0036) J Symb Logic 46*714-722
◇ D25 D30 ◇ REV MR 83c:03040 Zbl 517 # 03014
• ID 33279

POSNER, D.B. [1981] *The upper semilattice of degrees $\leqslant 0'$ is complemented* (**J** 0036) J Symb Logic 46*705-713
◇ D25 D30 ◇ REV MR 84i:03084 Zbl 517 # 03015
• ID 33278

POSS, R.L. [1970] *A note on a lemma of J.W.Addison* (**J** 0047) Notre Dame J Formal Log 11*337-339
◇ D55 E45 ◇ REV MR 44 # 6496 Zbl 185.15 JSL 38.334
• ID 10678

POSS, R.L. see Vol. V for further entries

POST, E.L. [1936] *Finite combinatory processes - formulation I* (**J** 0036) J Symb Logic 1*103-105
• REPR [1965] (**C** 0718) The Undecidable 289-291
◇ D03 D20 D40 ◇ REV Zbl 15.193 JSL 2.43
FdM 62.1060 • ID 10682

POST, E.L. [1943] *Formal reductions of the general combinatorial decision problem* (**J** 0100) Amer J Math 65*197-215
◇ D03 D40 ◇ REV MR 4.209 JSL 8.50 • ID 10684

POST, E.L. [1944] *Recursively enumerable sets of positive integers and their decision problems* (**J** 0015) Bull Amer Math Soc 50*284-316
• REPR [1965] (**C** 0718) The Undecidable 305-337
◇ D03 D25 D40 ◇ REV MR 6.29 JSL 10.18 JSL 31.485
• ID 16994

POST, E.L. [1946] *A variant of a recursively unsolvable problem* (**J** 0015) Bull Amer Math Soc 52*264-268
◇ D03 D05 D40 ◇ REV MR 7.405 JSL 12.55 • ID 10685

POST, E.L. [1946] *Note on a conjecture of Skolem* (**J** 0036) J Symb Logic 11*73-74
◇ D20 ◇ REV MR 8.307 JSL 12.28 • ID 10686

POST, E.L. [1947] *Recursive unsolvability of a problem of Thue* (**J** 0036) J Symb Logic 12*1-11
• REPR [1965] (**C** 0718) The Undecidable 293-303
◇ D03 D10 D40 ◇ REV MR 8.558 JSL 12.90 • ID 10687

POST, E.L. [1948] *Degrees of recursive unsolvability, preliminary report* (**J** 0015) Bull Amer Math Soc 54*641-642
◇ D03 D30 D55 ◇ ID 10688

POST, E.L. [1954] see KLEENE, S.C.

POST, E.L. [1965] *Absolutely unsolvable problems and relatively undecidable propositions* (**C** 0718) The Undecidable 338-433
◇ D35 ◇ ID 15053

POST, E.L. see Vol. I, II for further entries

POTEKHIN, A.I. [1977] see AMBARTSUMYAN, A.A.

POULSEN, B.T. [1970] *The Medvedev lattice of degrees of difficulty* (**S** 3462) Var Publ Ser, Aarhus Univ 12*34pp
◇ D30 D80 ◇ REV MR 43 # 4672 Zbl 237 # 02012
• ID 27876

POUR-EL, M.B. [1960] *A comparison of five "computable" operators* (**J** 0068) Z Math Logik Grundlagen Math 6*325-340
◇ D20 ◇ REV MR 23 # A2319 Zbl 95.246 • ID 10699

POUR-EL, M.B. [1964] see HOWARD, W.A.

POUR-EL, M.B. [1964] *Goedel numberings versus Friedberg numberings* (**J** 0053) Proc Amer Math Soc 15*252-256
◇ D25 D45 ◇ REV MR 30 # 4680 Zbl 168.254 • ID 10700

POUR-EL, M.B. [1965] *"Recursive isomorphism" and effectively extensible theories* (**J** 0015) Bull Amer Math Soc 71*551-555
◇ D50 ◇ REV MR 31 # 59 • ID 10703

POUR-EL, M.B. & PUTNAM, H. [1965] *Recursively enumerable classes and their application to recursive sequences of formal theories* (**J** 0009) Arch Math Logik Grundlagenforsch 8*104-121
◇ D25 ◇ REV MR 34 # 7370 Zbl 242 # 02046 JSL 38.155 • ID 10701

POUR-EL, M.B. [1967] see KRIPKE, S.A.

POUR-EL, M.B. [1968] *Effectively extensible theories* (**J** 0036) J Symb Logic 33*56-68
◇ B30 D25 ◇ REV MR 38 # 3148 Zbl 179.19 • ID 10705

POUR-EL, M.B. [1968] *Independent axiomatization and its relation to the hypersimple set* (**J** 0068) Z Math Logik Grundlagen Math 14*449-456
◇ C07 D25 ◇ REV MR 40 # 4105 Zbl 182.9 JSL 38.654 • ID 10704

POUR-EL, M.B. [1969] *A recursion-theoretic view of axiomatizable theories* (**P** 1841) Fct Recurs & Appl;1967 Tihany 26-42
◇ C07 D25 D35 ◇ ID 32554

POUR-EL, M.B. [1969] *Independent axiomatization and its relation to the hypersimple set* (**P** 1841) Fct Recurs & Appl;1967 Tihany 24-25
◇ C07 D25 D35 ◇ REV Zbl 182.9 • ID 32553

POUR-EL, M.B. [1970] *A recursion-theoretic view of axiomatizable theories* (**J** 0076) Dialectica 24*267-276
◇ C07 D25 D35 ◇ REV Zbl 274 # 02001 • ID 29012

POUR-EL, M.B. [1970] see MARTIN, D.A.

POUR-EL, M.B. [1973] *Abstract computability versus analog-generability. A survey* (**P** 0713) Cambridge Summer School Math Log;1971 Cambridge GB 345-360
◇ D75 D80 F65 ◇ REV MR 49 # 17 Zbl 272 # 02068 • ID 10709

POUR-EL, M.B. [1973] *Analog computers, digital computers, mathematical logic, differential equations - interrelations* (**P** 1753) Int Congr AICA Hybrid Comput (7); 122-124
◇ D20 D75 D80 F65 ◇ ID 31218

POUR-EL, M.B. [1974] *Abstract computability and its relation to the general purpose analog computer (some connections between logic, differential equations and analog computers)* (**J** 0064) Trans Amer Math Soc 199*1-28
◇ D20 D75 D80 F65 ◇ REV MR 50 # 78 Zbl 296 # 02022 • ID 10706

POUR-EL, M.B. [1975] see CALDWELL, J.

POUR-EL, M.B. [1978] *Computer science and recursion theory* (**P** 2905) ACM Annual Conf;1978 Washington 15-20
◇ D80 D98 ◇ REV MR 81d:68003 • ID 82521

POUR-EL, M.B. & RICHARDS, I. [1978] *Differentiability properties of computable functions - a summary* (**J** 0380) Acta Cybern (Szeged) 4*123-125
◇ D20 D80 F60 ◇ REV MR 80d:03059 Zbl 397 # 03038 • ID 52704

POUR-EL, M.B. & RICHARDS, I. [1979] *A computable ordinary differential equation which possesses no computable solution* (**J** 0007) Ann Math Logic 17*61-90
◇ D80 F60 ◇ REV MR 81k:03064 Zbl 424 # 68028 JSL 47.900 • ID 77493

POUR-EL, M.B. & RICHARDS, I. [1981] *The wave equation with computable initial data such that its unique solution is not computable* (**J** 0345) Adv Math 39*215-239
◇ D80 ◇ REV MR 83e:03101 Zbl 465 # 35054 JSL 47.900 • ID 54937

POUR-EL, M.B. & RICHARDS, I. [1982] *Noncomputability in models of physical phenomena* (**J** 2736) Int J Theor Phys 21*553-555
◇ D80 ◇ REV Zbl 493 # 35057 • ID 38435

POUR-EL, M.B. & RICHARDS, I. [1983] *Computability and noncomputability in classical analysis* (**J** 0064) Trans Amer Math Soc 275*539-560
◇ D80 ◇ REV MR 84e:03056 Zbl 513 # 03031 • ID 34391

POUR-EL, M.B. & RICHARDS, I. [1983] *Noncomputability in analysis and physics: a complete determination of the class of noncomputable linear operators* (**J** 0345) Adv Math 48*44-74
◇ D80 ◇ REV MR 84j:03114 Zbl 519 # 03045 • ID 34704

POUR-EL, M.B. & RICHARDS, I. [1984] *L^p-computability in recursive analysis* (**J** 0053) Proc Amer Math Soc 92*93-97
◇ D80 F60 ◇ REV MR 86f:03102 Zbl 558 # 03030 • ID 45841

PRANK, R.K. [1978] *On congruence relations in the lattice of recursively enumerable sets (Russian) (English summary)* (**S** 3468) Tr Mat & Mekh (Tartu) 22(464)*28-36
◇ D25 ◇ REV MR 80g:03041 Zbl 409 # 03024 • ID 56327

PRANK, R.K. [1979] *Expressibility in the elementary theory of recursive sets with realizability logic (Russian) (English summary)* (**S** 3468) Tr Mat & Mekh (Tartu) 25(500)*119-129
◇ B60 D20 F50 ◇ REV MR 81a:03043 Zbl 421 # 03038 • ID 53437

PRANK, R.K. [1980] *Semantics of realizability for a language with variables for recursively enumerable sets (Russian)* (**J** 2852) Tr Vychisl Tsentra, Univ Tartu 43*112-131
◇ B60 D25 F35 F50 ◇ REV MR 81h:03111 • ID 77510

PRANK, R.K. [1981] *Expressibility in the elementary theory of recursively enumerable sets with realizability logic (Russian)* (**J** 0003) Algebra i Logika 20*427-439,484-485
• TRANSL [1981] (**J** 0069) Algeb and Log 20*282-291
◇ B60 D25 F50 ◇ REV MR 83i:03068 • ID 35552

PRANK, R.K. [1981] *On the quotient lattice of the lattice of recursively enumerable sets by the immunity congruence (Russian) (Estonian and English summaries)* (**S** 0393) Uch Zap Univ, Tartu 556*11-14
◇ D25 D50 ◇ REV MR 82k:03062 Zbl 522 # 03028 • ID 77509

PRATHER, R.E. [1970] *On categories of infinite automata* (**J** 0041) Math Syst Theory 4*295-305
◇ D10 ◇ REV MR 53 # 8186 Zbl 208.21 • ID 23039

PRATHER, R.E. [1971] *An algebraic proof of the Paull-Unger theorem* (**J** 0187) IEEE Trans Comp C-20*578-580
◇ D05 ◇ REV MR 43 # 3122 Zbl 222 # 94058 • ID 10724

PRATHER, R.E. [1975] *A convenient cryptomorphic version of recursive function theory* (**J** 0194) Inform & Control 27*178-195
◇ D20 ◇ REV MR 52 # 5398 Zbl 306 # 02031 • ID 18345

PRATHER, R.E. [1977] *Structured Turing machines* (**J** 0194) Inform & Control 35*159-171
◇ D10 ◇ REV MR 57 # 1971 Zbl 365 # 02024 • ID 51025

PRATHER, R.E. see Vol. I for further entries

PRATT, V.R. & STOCKMEYER, L.J. [1976] *A characterization of the power of vector machines* (**J** 0119) J Comp Syst Sci 12*198-221
◇ D10 D15 ◇ REV MR 55 # 1827 Zbl 342 # 68033 • ID 64658

PRATT, V.R. [1979] *Axioms or algorithms* (**P** 2059) Math Founds of Comput Sci (8);1979 Olomouc 160-169
◇ B25 B75 D15 ◇ REV MR 81d:68023 Zbl 404 # 68044 • ID 54858

PRATT, V.R. see Vol. I, II for further entries

PRAUSS, E.J. [1980] see PAUL, W.J.

PREPARATA, F.P. [1978] see ADLEMAN, L.M.

PREPARATA, F.P. see Vol. I, II for further entries

PRESIC, S.B. [1979] *On quasi-algebras and the word problem* (**J** 0400) Publ Inst Math, NS (Belgrade) 26(40)*255-268
◇ C05 D40 ◇ REV MR 81g:08010 Zbl 431 # 08004 • ID 69789

PRESIC, S.B. see Vol. I, II, III for further entries

PRESTEL, A. & ZIEGLER, M. [1975] *Erblich euklidische Koerper* (**J** 0127) J Reine Angew Math 274/275*196-205
◇ C60 D35 ◇ REV MR 52 # 363 Zbl 307 # 12103 • ID 18349

PRESTEL, A. [1979] *Entscheidbarkeit mathematischer Theorien* (**J** 0157) Jbuchber Dtsch Math-Ver 81*177-188
◇ B25 C60 C98 D35 ◇ REV MR 80j:03001 Zbl 426 # 03013 • ID 53609

PRESTEL, A. see Vol. I, III, V for further entries

PRIDA, J.F. [1982] *A non-standard study of the theory of relative recursivity (Spanish)* (**J** 0264) Collect Math (Barcelona) 33*201-214
◇ D20 D30 H05 ◇ REV MR 85b:03066 Zbl 536 # 03023 • ID 37105

PRIDA, J. F. also published under the name FERNANDEZ-PRIDA, J.

PRIDA, J.F. see Vol. VI for further entries

PRIDE, S.J. [1977] *The isomorphism problem for two-generator one-relator groups with torsion is solvable* (**J** 0064) Trans Amer Math Soc 227*109-139
◇ C60 D40 ◇ REV MR 55 # 3092 Zbl 356 # 20037 • ID 26231

PRIESE, L. [1971] *Normalformen von Markov'schen und Post'schen Algorithmen. Eine Einfuehrung in die Theorie der normierten Algorithmen* (**X** 1532) Univ Muenster Inst Math Logik: Muenster 124pp
◇ D03 D40 D98 ◇ REV MR 57 # 5702 Zbl 246 # 02029 • ID 64671

PRIESE, L. & ROEDDING, D. [1974] *A combinatorial approach to selfcorrection* (**J** 0383) J Cybern 4*7-25
◇ D05 D10 ◇ REV MR 56 # 11657 Zbl 336 # 94027 • ID 14988

PRIESE, L. [1976] *On a simple combinatorial structure sufficient for sub-lying nontrivial self-reproduction* (**J** 0383) J Cybern 6*101-137
◇ D03 D10 ◇ REV MR 57 # 9361 Zbl 359 # 02048 • ID 50595

PRIESE, L. [1976] *Reversible Automaten und einfache universelle 2-dimensionale Thue-Systeme* (**J** 0068) Z Math Logik Grundlagen Math 22*353-384
◇ D03 D05 D10 ◇ REV MR 58 # 5111 Zbl 391 # 03020 • ID 18440

PRIESE, L. [1978] *A note on asynchronous cellular automata* (**J** 0119) J Comp Syst Sci 17*237-252
◇ D03 ◇ REV MR 80b:68069 Zbl 396 # 68039 • ID 52664

PRIESE, L. [1979] *Towards a precise characterization of the complexity of universal and nonuniversal Turing machines* (**J** 1428) SIAM J Comp 8*508-523
◇ D10 ◇ REV MR 81k:03035 Zbl 426 # 68024 • ID 53679

PRIESE, L. [1979] *Ueber ein 2-dimensionales Thue-System mit zwei Regeln und unentscheidbarem Wortproblem* (**J** 0068) Z Math Logik Grundlagen Math 25*179-192
◇ D03 ◇ REV MR 81d:03041 Zbl 419 # 03022 • ID 53365

PRIESE, L. [1979] *Ueber eine minimale universelle Turing-Maschine* (**P** 3488) Theor Comput Sci (4);1979 Aachen 244-259
◇ D10 ◇ REV MR 81j:68063 Zbl 404 # 68056 • ID 54861

PRIESE, L. [1984] see BRUEGGEMANN, A.

PRIKRY, K. [1973] see GALVIN, F.

PRIKRY, K. see Vol. III, V for further entries

PRINOTH, R. [1978] *Starke Faerbbarkeit in Petri-Netzen* (**J** 3531) Ber Ges Math Datenverarb, Bonn 117*85pp
◇ D05 D15 ◇ REV MR 81k:68044 Zbl 412 # 68048 • ID 69793

PROCHAZKA, J. [1977] see KOREC, I.

PROCHNOW, D. [1974] *Einfach-rekursive Programmschemata (English and Russian summaries)* (**J** 0129) Elektr Informationsverarbeitung & Kybern 10*519-541
◇ B75 D20 ◇ REV MR 54 # 1702 Zbl 329 # 02013 • ID 64683

PRONINA, V.A. [1977] see BABICHEV, A.V.

PROSKURIN, A.V. [1979] *Positive rudimentarity of the graphs of Ackermann and Grzegorczyk functions (Russian) (English summary)* (**S** 0228) Zap Nauch Sem Leningrad Otd Mat Inst Steklov 88*186-191,244-245
• TRANSL [1982] (**J** 1531) J Sov Math 20*2363-2366
◇ D20 ◇ REV MR 81b:03044 Zbl 429 # 03023 • ID 53854

PROTASI, M. [1975] see AUSIELLO, G.

PROTASI, M. [1980] see AUSIELLO, G.

PROTASI, M. [1981] see AUSIELLO, G.

PROTASI, M. [1982] see MENTRASTI, P.

PUDLAK, P. [1975] *Polynomially complete problems in the logic of automated discovery* (**P** 0454) Math Founds of Comput Sci (4);1975 Marianske Lazne 32*358-361
◇ B35 D15 ◇ REV MR 52 # 12404 Zbl 318 # 68055 • ID 21740

PUDLAK, P. [1975] *The observational predicate calculus and complexity of computations* (**J** 0140) Comm Math Univ Carolinae (Prague) 16*395-398
◇ B10 D15 ◇ REV MR 52 # 7202 Zbl 311 # 02021
• ID 29600

PUDLAK, P. & SPRINGSTEEL, F. [1979] *Complexity in mechanized hypothesis formation* (**J** 1426) Theor Comput Sci 8*203-225
◇ B35 D15 ◇ REV MR 81b:68054 Zbl 404 # 68097
• ID 54869

PUDLAK, P. [1984] *Bounds for Hodes-Specker theorem* (**P** 2342) Symp Rek Kombin;1983 Muenster 421-445
◇ B05 D15 ◇ REV Zbl 551 # 03022 • ID 43900

PUDLAK, P. see Vol. I, III, V, VI for further entries

PURDOM JR., P.W. [1982] see BROWN, CYNTHIA A.

PUTNAM, H. [1957] *Decidability and essential undecidability* (**J** 0036) J Symb Logic 22*39-54
◇ B25 D35 ◇ REV MR 19.626 Zbl 78.245 Zbl 79.454 JSL 23.446 • ID 10827

PUTNAM, H. [1957] see DAVIS, MARTIN D.

PUTNAM, H. [1958] see DAVIS, MARTIN D.

PUTNAM, H. [1960] *An unsolvable problem in number theory* (**J** 0036) J Symb Logic 25*220-232
◇ D35 ◇ REV MR 28 # 2048 Zbl 108.7 JSL 37.601
• ID 10831

PUTNAM, H. & SMULLYAN, R.M. [1960] *Exact separation of recursively enumerable sets within theories* (**J** 0053) Proc Amer Math Soc 11*574-577
◇ D25 D35 ◇ REV MR 22 # 10907 JSL 25.362 • ID 12540

PUTNAM, H. [1961] see DAVIS, MARTIN D.

PUTNAM, H. [1961] *Uniqueness ordinals in higher constructive number classes* (**C** 0622) Essays Found of Math (Fraenkel) 190-206
◇ D55 F15 ◇ REV MR 29 # 4686 Zbl 143.254 JSL 31.135
• ID 10835

PUTNAM, H. [1963] *A note on constructible sets of integers* (**J** 0047) Notre Dame J Formal Log 4*270-273
◇ D55 E45 ◇ REV MR 30 # 3844 Zbl 192.43 JSL 36.339
• ID 10837

PUTNAM, H. [1963] see DAVIS, MARTIN D.

PUTNAM, H. [1964] *Minds and machines* (**C** 0618) Minds & Machines 72-97
◇ A05 D10 ◇ REV JSL 36.177 • ID 10830

PUTNAM, H. [1964] *On families of sets represented in theories* (**J** 0009) Arch Math Logik Grundlagenforsch 6*66-70
◇ C40 D25 F30 ◇ REV MR 31 # 3329 Zbl 126.21
• ID 10838

PUTNAM, H. [1964] *On hierarchies and systems of notations* (**J** 0053) Proc Amer Math Soc 15*44-50
◇ D55 D70 F15 ◇ REV MR 28 # 1126 Zbl 237 # 02011 JSL 31.136 • ID 10839

PUTNAM, H. [1965] see HENSEL, G.

PUTNAM, H. [1965] see POUR-EL, M.B.

PUTNAM, H. [1965] *Trial and error predicates and the solution to a problem of Mostowski* (**J** 0036) J Symb Logic 30*49-57
◇ C57 D20 D55 ◇ REV MR 33 # 3923 JSL 36.342
• ID 10840

PUTNAM, H. [1968] see BOOLOS, G.

PUTNAM, H. [1969] see BOYD, R.

PUTNAM, H. [1969] see HENSEL, G.

PUTNAM, H. [1970] see ENDERTON, H.B.

PUTNAM, H. [1971] see LEEDS, S.

PUTNAM, H. [1973] *Recursive functions and hierarchies* (**J** 0005) Amer Math Mon 80*68-86
◇ D20 D30 D35 D55 ◇ REV MR 50 # 71 Zbl 268 # 02024 • ID 10842

PUTNAM, H. [1974] see LEEDS, S.

PUTNAM, H. [1974] see LUKAS, J.D.

PUTNAM, H. see Vol. I, II, III, V, VI for further entries

PUTZOLU, G. [1966] see AMAR, V.

QUACKENBUSH, R.W. [1975] see JONSSON, B.

QUACKENBUSH, R.W. see Vol. III, V for further entries

QUALITZ, J.E. [1978] see DENNING, P.J.

QUERE, A. [1968] see PAIR, C.

QUERE, A. [1970] see PAIR, C.

QUERE, A. [1979] see BROY, M.

QUINE, W.V.O. [1949] *On decidability and completeness* (**J** 0154) Synthese 7*441-446
◇ B25 D35 F30 ◇ REV MR 12.70 JSL 16.76 • ID 10881

QUINE, W.V.O. [1952] see CHURCH, A.

QUINE, W.V.O. [1966] *Church's theorem on the decision problem* (**C** 0587) Quine: Sel Logic Papers 212-219
◇ D35 ◇ REV MR 34 # 7333 • ID 32229

QUINE, W.V.O. [1969] *The limits of decision* (**P** 1571) Int Congr Philos (14);1968 Wien 3*57-62
◇ D35 ◇ ID 32230

QUINE, W.V.O. see Vol. I, II, III, V, VI for further entries

RABIN, M.O. [1957] *Computable algebraic systems* (**P** 1675) Summer Inst Symb Log;1957 Ithaca 134-138
◇ C57 C60 D40 D45 D80 ◇ REV JSL 32.412 • ID 10933

RABIN, M.O. [1957] *Effective computability of winning strategies* (**S** 3513) Ann Math Stud 39*147-157
◇ D20 D25 ◇ REV MR 20 # 263 Zbl 78.329 JSL 23.224
• ID 42167

RABIN, M.O. & SCOTT, D.S. [1957] *Remarks on finite automata* (**P** 1675) Summer Inst Symb Log;1957 Ithaca 106-112
◇ D05 ◇ REV Zbl 158.9 • ID 29327

RABIN, M.O. [1957] *Two-way finite automata* (**P** 1675) Summer Inst Symb Log;1957 Ithaca 366-369
◇ D05 ◇ ID 29381

RABIN, M.O. [1958] *On recursively enumerable and arithmetic models of set theory* (**J** 0036) J Symb Logic 23*408-416
◇ C57 C62 D45 ◇ REV MR 22 # 10908 Zbl 95.246 JSL 28.167 • ID 10930

RABIN, M.O. [1958] *Recursive unsolvability of group theoretic problems* (**J** 0120) Ann of Math, Ser 2 67*172-194
◇ D40 ◇ REV MR 22 # 1611 Zbl 79.248 JSL 23.55
• ID 10926

RABIN, M.O. & SCOTT, D.S. [1959] *Finite automata and their decision problems* (**J** 0284) IBM J Res Dev 3*114-125
◇ D05 ◇ REV MR 21 # 2559 JSL 25.163 • ID 10927

RABIN, M.O. [1960] *Computable algebra, general theory and theory of computable fields* (**J** 0064) Trans Amer Math Soc 95*341-360
◇ C57 C60 D45 ◇ REV MR 22 # 4639 Zbl 156.12
• ID 33612

RABIN, M.O. [1960] *Degree of difficulty of computing a function and a partial ordering of recursive sets* (1111) Preprints, Manuscr., Techn. Reports etc.
◇ D15 D20 ◇ REM Hebrew University, Department of Mathematics. Jerusalem • ID 27543

RABIN, M.O. [1962] *Diophantine equations and non-standard models of arithmetic* (**P** 0612) Int Congr Log, Meth & Phil of Sci (1,Proc);1960 Stanford 151-158
◇ C62 D25 H15 ◇ REV MR 27 # 3540 • ID 20907

RABIN, M.O. [1963] *Probabilistic automata* (**J** 0194) Inform & Control 6*230-245
◇ D05 ◇ ID 10936

RABIN, M.O. [1963] *Real time computation* (**J** 0029) Israel J Math 1*203-211
• TRANSL [1970] (**C** 1540) Probl Mat Log: Slozh Algor & Kl Vychisl Funk 156-167
◇ D05 D10 D15 ◇ REV MR 29 # 1148 Zbl 156.256 JSL 31.657 • ID 10939

RABIN, M.O. [1963] see PERLES, M.A.

RABIN, M.O. & WANG, HAO [1963] *Words in the history of a Turing machine with a fixed input* (**J** 0037) ACM J 10*526-527
◇ D10 ◇ REV MR 28 # 4998 Zbl 192.67 JSL 34.508
• ID 10937

RABIN, M.O. [1965] *A simple method for undecidability proofs and some applications* (**P** 0623) Int Congr Log, Meth & Phil of Sci (2,Proc);1964 Jerusalem 58-68
◇ D35 F25 ◇ REV MR 36 # 4976 Zbl 192.55 JSL 36.150
• ID 16270

RABIN, M.O. [1966] see ELGOT, C.C.

RABIN, M.O. [1966] *Lectures on classical and probabilistic automata* (**P** 0746) Automata Th;1964 Ravello 309-313
◇ D05 ◇ REV Zbl 192.75 • ID 10941

RABIN, M.O. [1967] *Mathematical theory of automata* (**P** 0737) Math Aspects Comput Sci;1966 New York 153-175
◇ D05 D98 ◇ REV MR 39 # 1243 Zbl 189.13 JSL 40.520
• ID 16807

RABIN, M.O. [1968] *Decidability of second-order theories and automata on infinite trees* (**J** 0015) Bull Amer Math Soc 74*1025-1029
◇ B15 B25 C85 D05 ◇ REV MR 38 # 44 Zbl 313 # 02029
• ID 29625

RABIN, M.O. [1969] *Decidability of second order theories and automata on infinite trees* (**J** 0064) Trans Amer Math Soc 141*1-35
• TRANSL [1971] (**J** 3079) Kiber Sb Perevodov, NS 8*72-116
◇ B15 B25 C85 D05 ◇ REV MR 40 # 30
Zbl 221 # 02031 JSL 37.618 • ID 10942

RABIN, M.O. [1970] *Weakly definable relations and special automata* (**P** 1072) Math Log & Founds of Set Th;1968 Jerusalem 1-23
◇ B15 B25 C40 C85 D05 ◇ REV MR 43 # 3121
Zbl 214.22 JSL 40.622 • ID 22233

RABIN, M.O. [1971] *Decidability and definability in second-order theories* (**P** 0743) Int Congr Math (II,11,Proc);1970 Nice 1*239-244
◇ B15 B25 C40 C85 D05 ◇ REV MR 54 # 12512
Zbl 226 # 02041 JSL 40.623 • ID 14690

RABIN, M.O. [1972] *Automata on infinite objects and Church's problem* (**X** 0803) Amer Math Soc: Providence iii + 22pp
◇ B15 B25 C85 D05 ◇ REV MR 48 # 75
Zbl 315 # 02037 JSL 40.623 • ID 10943

RABIN, M.O. [1974] see FISCHER, MICHAEL J.

RABIN, M.O. [1974] *Theoretical impediments to artificial intelligence* (**P** 1691) Inform Processing (6);1974 Stockholm 615-619
◇ B75 D15 D80 ◇ REV MR 54 # 9191 Zbl 296 # 68054
• ID 64724

RABIN, M.O. [1977] *Decidable theories* (**C** 1523) Handb of Math Logic 595-629
◇ B25 C98 D20 ◇ REV MR 58 # 5109 JSL 49.975
• ID 27321

RABIN, M.O. [1980] see BAUR, W.

RABIN, M.O. see Vol. I, III, V, VI for further entries

RACKOFF, C.W. [1974] *On the complexity of the theories of weak direct products: a preliminary report* (**P** 1464) ACM Symp Th of Comput (6);1974 Seattle 149-160
◇ B25 C13 C30 C60 D15 ◇ REV MR 54 # 4957
Zbl 358 # 68080 • ID 24125

RACKOFF, C.W. [1975] see FERRANTE, J.

RACKOFF, C.W. [1976] *On the complexity of the theories of weak direct powers* (**J** 0036) J Symb Logic 41*561-573
◇ B25 C13 C30 C60 D15 ◇ REV MR 58 # 5155
Zbl 383 # 03011 • ID 14579

RACKOFF, C.W. [1979] see FERRANTE, J.

RACKOFF, C.W. [1980] see COOK, S.A.

RACKOFF, C.W. & SEIFERAS, J.I. [1981] *Limitations on separating nondeterministic complexity classes* (**J** 1428) SIAM J Comp 10*742-745
◇ D15 ◇ REV MR 84f:68035 Zbl 468 # 68052 • ID 69862

RACKOFF, C.W. [1982] *Relativized questions involving probabilistic algorithms* (**J** 0037) ACM J 29*261-268
◇ D15 ◇ REV MR 83j:03067 Zbl 477 # 68037 • ID 69863

RADENSKY, A.A. [1983] *Transformations in context-free languages* (**J** 0137) C R Acad Bulgar Sci 36*185-187
◇ D05 ◇ REV MR 84i:68135 Zbl 563 # 68066 • ID 47482

RADO, T. [1962] *On non-computable functions* (**J** 0432) Bell Syst Tech J 41*877-884
◇ D10 D20 ◇ REV MR 24 # A3063 JSL 32.524 • ID 10964

RADO, T. [1963] *On a simple source for non-computable functions* (**P** 0674) Symp Math Th of Automata;1962 New York 75-81
◇ D10 D20 ◇ REV MR 30 # 1043 Zbl 132.248 JSL 33.524
• ID 10965

RADO, T. [1965] see LIN, S.

RADO, T. see Vol. I, V for further entries

RADZISZOWSKI, S. [1978] *Programmability and P= NP conjecture* (**J** 2095) Fund Inform, Ann Soc Math Pol, Ser 4 2*71-82
◇ D15 ◇ REV MR 80a:68046 Zbl 404 # 68043 • ID 69864

RADZISZOWSKI, S. [1979] *Logic and complexity of synchronous parallel computations* (**J** 1927) Prace Inst Podstaw Inf, Pol Akad Nauk 353*31pp
◇ D15 ◇ REV Zbl 473 # 68018 • ID 69865

RADZISZOWSKI, S. [1981] *Logic and complexity of synchronous parallel computations* (**P** 3642) Colloq Math Log in Computer Sci;1978 Salgotarjan 675-698
◇ B75 D15 ◇ REV MR 83g:68007 Zbl 512 # 68035 • ID 47145

RAGAZ, M. [1983] *Die Unentscheidbarkeit der einstelligen unendlichwertigen Praedikatenlogik* (**J** 0009) Arch Math Logik Grundlagenforsch 23*129-139
◇ B50 D35 D55 ◇ REV MR 85e:03059a Zbl 533 # 03007 • ID 36532

RAGAZ, M. see Vol. II, V for further entries

RAIMONDI, T. [1978] see APRILE, G.

RAISONNIER, J. & STERN, J. [1983] *Mesurabilite et propriete de Baire (English summary)* (**J** 3364) C R Acad Sci, Paris, Ser 1 296*323-326
◇ D55 E15 ◇ REV MR 84g:03077 Zbl 549 # 03038 • ID 34182

RAISONNIER, J. [1984] *A mathematical proof of S. Shelah's theorem on the measure problem and related results* (**J** 0029) Israel J Math 48*48-56
◇ D55 E05 E15 E75 ◇ REV MR 86g:03082b • ID 44806

RAISONNIER, J. & STERN, J. [1985] *The strength of measurability hypotheses* (**J** 0029) Israel J Math 50*337-349
◇ D55 E05 E15 ◇ ID 48248

RAISONNIER, J. see Vol. V for further entries

RAJLICH, V. [1971] *Absolutely parallel grammars and two-way deterministic finite-state transducers* (**P** 0680) ACM Symp Th of Comput (3);1971 Shaker Heights 132-137
◇ D05 ◇ REV Zbl 245 # 68029 • ID 64740

RAJLICH, V. [1972] *Absolutely parallel grammars and two-way finite-state transducers* (**J** 0119) J Comp Syst Sci 6*324-342
◇ D05 ◇ REV MR 45 # 9878 Zbl 246 # 68013 • ID 64739

RAKHMATULIN, N.A. [1975] *Automata-theoretic characteristics for the spectra of formulas of finite levels (Russian)* (**C** 1050) Aktual Vopr Mat Log & Teor Mnozh 105-127
◇ C13 D05 D10 ◇ REV MR 58 # 5165 • ID 77649

RANDELL, B. [1972] *On Alan Turing and the origins of digital computers* (**J** 0508) Machine Intelligence 7*3-20
◇ A10 D10 ◇ REV Zbl 249 # 01010 • ID 28874

RANEY, G.N. [1958] *Sequential functions* (**J** 0037) ACM J 5*177-180
◇ D05 ◇ REV MR 22 # 12016 Zbl 88.18 • ID 47980

RANGEL, J.L. [1974] *The equivalence problem for regular expressions over one letter is elementary* (**P** 1479) IEEE Symp Switch & Automata Th (15);1974 New Orleans 24-27
◇ D05 D20 ◇ REV MR 54 # 4188 • ID 24078

RAO, B.V. [1969] *On discrete Borel spaces and projective sets* (**J** 0015) Bull Amer Math Soc 75*614-617
◇ D55 E15 E75 ◇ REV MR 39 # 4014 Zbl 175.7 • ID 16953

RAO, B.V. [1970] *Remarks on analytic sets* (**J** 0027) Fund Math 66*237-239
◇ D55 E15 ◇ REV MR 43 # 451 Zbl 191.303 • ID 10990

RAO, B.V. [1970] *Remarks on generalized analytic sets and the axiom of determinateness* (**J** 0027) Fund Math 69*125-129
◇ D55 E15 E60 ◇ REV MR 44 # 62 Zbl 211.18 • ID 10991

RAO, B.V. [1976] see MAITRA, A.

RAO, B.V. & RAO, K.P.S.BHASKARA [1978] *On the isomorphism problem for analytic sets* (**J** 0014) Bull Acad Pol Sci, Ser Math Astron Phys 26*767-769
◇ D55 E15 ◇ REV MR 81g:54043 Zbl 399 # 04002 • ID 29179

RAO, B.V. see Vol. V for further entries

RAO, K.P.S.BHASKARA [1978] see RAO, B.V.

RAO, K.P.S.BHASKARA see Vol. V for further entries

RAOULT, J.-C. [1981] *Finiteness results on rewriting systems (French summary)* (**J** 3441) RAIRO Inform Theor 15*373-391
◇ D05 ◇ REV MR 83m:68056 Zbl 491 # 03015 • ID 40412

RAS, Z. [1971] *Deductive systems of computing machines* (**J** 0014) Bull Acad Pol Sci, Ser Math Astron Phys 19*517-524
◇ B05 D05 ◇ REV MR 49 # 4333 Zbl 227 # 02018 • ID 27358

RAS, Z. [1971] *On a relationship between the propositional calculus and a grammar (Russian summary)* (**J** 0014) Bull Acad Pol Sci, Ser Math Astron Phys 19*635-637
◇ B05 D05 ◇ REV MR 47 # 1584 Zbl 227 # 02019 • ID 10995

RAS, Z. [1971] *Semi-Thue systems as deductive systems of certain computing machines (Russian summary)* (**J** 0014) Bull Acad Pol Sci, Ser Math Astron Phys 19*631-633
◇ D03 ◇ REV MR 46 # 8820 Zbl 251 # 02040 • ID 10994

RAS, Z. [1972] *On a relationship between certain grammars and enumerable first order predicate calculi (Russian summary)* (**J** 0014) Bull Acad Pol Sci, Ser Math Astron Phys 20*95-99
◇ B20 D05 ◇ REV MR 47 # 1585 Zbl 237 # 02009 • ID 10996

RAS, Z. [1977] see BARTOL, W.

RASIOWA, H. [1973] *Formalized ω^+ valued algorithmic systems (Russian summary)* (**J** 0014) Bull Acad Pol Sci, Ser Math Astron Phys 21*559-565
◇ B50 B75 D15 ◇ REV MR 48 # 8198 Zbl 277 # 68025 • ID 19572

RASIOWA, H. [1974] *On ω^+-valued algorithmic logic and related problems* (**J** 1929) Prace Centr Oblicz Pol Akad Nauk 150*23pp
◇ B75 D05 ◇ REV Zbl 313 # 02027 • ID 29624

RASIOWA, H. [1977] *A tribute to A.Mostowski* (**P** 1075) Logic Colloq;1976 Oxford 139-144
◇ A10 D99 E99 ◇ REV MR 58 # 4949 Zbl 441 # 01015 • ID 16617

RASIOWA, H. [1977] *In memory of Andrzej Mostowski* (J 0063) Studia Logica 36*1-3
◇ A10 D99 E99 ◇ REV MR 56#65a Zbl 356#01020
• ID 31454

RASIOWA, H. [1977] *Many-valued algorithmic logic as a tool to investigate programs* (P 1894) Int Symp Multi-Val Log (5,Inv Pap);1975 Bloomington 77-102
◇ B75 D03 G20 ◇ REV MR 58#8455 Zbl 386#03009
• ID 52185

RASIOWA, H. see Vol. I, II, III, V, VI for further entries

RASSIAS, G.M. [1979] *Stallings homomorphisms and the simply connectedness problem* (J 2713) Eleutheria, Math J Sem Zervos (Athens) 1979/2*459-462
◇ D40 D80 ◇ REV MR 82i:57001 • ID 82581

RATOANDROMANANA, B. [1985] see LATTEUX, M.

RAUTENBERG, W. [1961] *Unentscheidbarkeit der euklidischen Inzidenzgeometrie* (J 0068) Z Math Logik Grundlagen Math 7*12-15
◇ D35 ◇ REV MR 25#1099 Zbl 166.263 JSL 29.58
• ID 11040

RAUTENBERG, W. [1962] *Ueber metatheoretische Eigenschaften einiger geometrischer Theorien* (J 0068) Z Math Logik Grundlagen Math 8*5-41
◇ B30 C65 D35 ◇ REV MR 26#4883 Zbl 112.247
• ID 11041

RAUTENBERG, W. [1967] *Elementare Schemata nichtelementarer Axiome* (J 0068) Z Math Logik Grundlagen Math 13*329-366
◇ B25 C52 C60 C65 C85 D35 ◇ REV MR 36#1316 Zbl 207.296 • ID 11046

RAUTENBERG, W. [1968] *Unterscheidbarkeit endlicher geordneter Mengen mit gegebener Anzahl von Quantoren* (J 0068) Z Math Logik Grundlagen Math 14*267-272
◇ B20 C07 C13 C65 D35 ◇ REV MR 37#2595 Zbl 169.7 • ID 11048

RAUTENBERG, W. [1970] see HAUSCHILD, K.

RAUTENBERG, W. [1971] see HAUSCHILD, K.

RAUTENBERG, W. [1972] see KLOETZER, G.

RAUTENBERG, W. [1972] see HAUSCHILD, K.

RAUTENBERG, W. & ZIEGLER, M. [1975] *Recursive inseparability in graph theory* (J 0369) Notices Amer Math Soc 22*523
◇ C13 D35 ◇ ID 49852

RAUTENBERG, W. see Vol. I, II, III, VI for further entries

RAYMOND, F.H. [1975] *Note sur l'algebre des fonctions (English summary)* (J 4698) Rev Franc Autom, Inf & Rech Operat, Ser Rouge Inf Th 9/R-3*25-49
◇ D20 D75 ◇ REV MR 54#9141 Zbl 362#68058
• ID 25830

RAYNA, G. [1974] *Degrees of finite-state transformability* (J 0194) Inform & Control 24*144-154
• TRANSL [1977] (J 3079) Kiber Sb Perevodov, NS 14*95-106
◇ D05 D30 ◇ REV MR 57#5515 Zbl 277#02007
• ID 29527

RECKHOW, R.A. [1972] see COOK, S.A.

RECKHOW, R.A. [1973] see COOK, S.A.

RECKHOW, R.A. [1974] see COOK, S.A.

RECKHOW, R.A. [1979] see COOK, S.A.

REDDY, C.R. [1978] see LOVELAND, D.W.

REDDY, C.R. see Vol. I for further entries

RED'KO, V.N. [1963] *On commutative closing of events (Ukrainian) (Russian and English summaries)* (J 0270) Dokl Akad Nauk Ukr SSR, Ser A 1963*1156-1159
◇ D05 ◇ REV MR 29#4692 Zbl 163.256 • ID 26652

RED'KO, V.N. [1964] *On the determinate aggregate of relationships of the algebra of regular events (Ukrainian)* (J 0265) Ukr Mat Zh, Akad Nauk Ukr SSR 16*120-126
◇ D05 ◇ REV MR 31#3284 • ID 26653

RED'KO, V.N. [1976] *Definitor-theoretic aspects of languages (Russian)* (J 0052) Probl Kibern 31*43-52,237
◇ D03 D05 D25 ◇ REV MR 55#13889 Zbl 414#68048
• ID 82590

RED'KO, V.N. [1980] see LISOVIK, L.P.

RED'KO, V.N. [1984] see BUJ, D.B.

REEDY, A. & SAVITCH, W.J. [1975] *The Turing degree of the inherent ambiguity problem for context-free languages* (J 1426) Theor Comput Sci 1*77-91
◇ D05 D30 ◇ REV MR 52#10381 Zbl 313#68064
• ID 21680

REGAN, K.W. [1984] *Arithmetical degrees of index sets for complexity cases* (P 2342) Symp Rek Kombin;1983 Muenster 118-130
◇ D15 D55 ◇ REV Zbl 548#03019 • ID 43198

REICHBACH, J. [1965] *On characterization and undecidability of the first-order functional calculus* (J 0537) Yokohama Math J 13*11-30
◇ D35 ◇ REV MR 34#4108 • ID 11069

REICHBACH, J. [1980] *Decidability of mathematical sciences and their undecidability* (J 1550) Creation Math 12*
◇ B25 D35 ◇ ID 32005

REICHBACH, J. see Vol. I, II, III, VI for further entries

REICHEL, HORST [1977] see KAPHENGST, H.

REICHEL, HORST see Vol. III for further entries

REIF, J.H. [1984] *The complexity of two-player games of incomplete information* (J 0119) J Comp Syst Sci 29*274-301
◇ D15 E60 ◇ REV MR 86c:68026 Zbl 551#90100
• ID 44064

REIF, J.H. see Vol. II for further entries

REISCHER, C. & SIMOVICI, D.A. [1970] *On the reduced form of the linear boolean automata* (J 0230) An Univ Iasi, NS, Sect Ia 16*479-485
◇ B70 D05 ◇ REV Zbl 213.23 • ID 27996

REISCHER, C. see Vol. I, II for further entries

REISCHUK, R. [1980] see PAUL, W.J.

REISCHUK, R. [1981] see PAUL, W.J.

REISCHUK, R. [1984] see DURIS, P.

REISER, A. & WEIHRAUCH, K. [1980] *Natural numberings and generalized computability* (J 0129) Elektr Informationsverarbeitung & Kybern 16*11-20
◇ D45 D75 ◇ REV MR 82b:03083 Zbl 446#03033
• ID 56562

REISIG, W. [1975] *Eine Verallgemeinerung des Berechenbarkeitsbegriffs durch Gleichungssysteme* (S 3180) Inform Ber, Inst Inf, Univ Bonn 3*111pp
◇ D10 D75 ◇ REV MR 54 # 12502 Zbl 374 # 02020
• ID 51544

REISIG, W. [1976] see INDERMARK, K.

REMESLENNIKOV, V.N. [1969] see CHURKIN, V.A.

REMESLENNIKOV, V.N. & SOKOLOV, V.G. [1970] *Some properties of a Magnus embedding (Russian)* (J 0003) Algebra i Logika 9*566-578
• TRANSL [1970] (J 0069) Algeb and Log 9*342-349
◇ C60 D40 ◇ REV MR 45 # 2001 Zbl 247 # 20026
• ID 11095

REMESLENNIKOV, V.N. [1973] *Example of a finitely presented group in the variety $\mathfrak{A}^5$ with the unsolvable word problem (Russian)* (J 0003) Algebra i Logika 12*577-602,618
• TRANSL [1973] (J 0069) Algeb and Log 12*327-346
◇ D40 ◇ REV MR 51 # 8266 Zbl 288 # 02027 • ID 17570

REMESLENNIKOV, V.N. [1975] see KIRKINSKIJ, A.S.

REMESLENNIKOV, V.N. [1979] *An algorithmic problem for nilpotent groups and rings (Russian)* (J 0092) Sib Mat Zh 20*1077-1081,1167
• TRANSL [1979] (J 0475) Sib Math J 20*761-764
◇ D40 ◇ REV MR 81g:03051 Zbl 426 # 20024 • ID 77726

REMESLENNIKOV, V.N. & ROMANOVSKIJ, N.S. [1980] *Algorithmic problems for solvable groups* (P 2634) Word Problems II;1976 Oxford 337-346
◇ D40 ◇ REV MR 81h:20044 Zbl 432 # 20030 • ID 82598

REMESLENNIKOV, V.N. [1982] see MYASNIKOV, A.G.

REMESLENNIKOV, V.N. & ROMAN'KOV, V.A. [1983] *Model-theoretic and algorithmic questions of group theory (Russian)* (J 1501) Itogi Nauki Tekh, Ser Algeb, Topol, Geom 21*3-79
• TRANSL [1985] (J 1531) J Sov Math 31*2887-2939
◇ B25 C60 C98 D15 D30 D40 ◇ REV MR 85f:20002 Zbl 563 # 20032 Zbl 573 # 20031 • ID 39954

REMESLENNIKOV, V.N. see Vol. III for further entries

REMMEL, J.B. [1976] *Co-hypersimple structures* (J 0036) J Symb Logic 41*611-625
◇ D25 D45 ◇ REV MR 58 # 5123 Zbl 345 # 02032
• ID 14584

REMMEL, J.B. [1976] *Combinatorial functors on co-r.e. structures* (J 0007) Ann Math Logic 10*261-287
◇ D25 D30 D45 D50 ◇ REV MR 55 # 2543
Zbl 345 # 02033 • ID 24260

REMMEL, J.B. [1977] *Maximal and cohesive vector spaces* (J 0036) J Symb Logic 42*400-418
◇ C57 D25 D45 ◇ REV MR 57 # 5712 Zbl 429 # 03027
• ID 53858

REMMEL, J.B. [1978] *A r-maximal vector space not contained in any maximal vector space* (J 0036) J Symb Logic 43*430-441
◇ C57 D25 D45 ◇ REV MR 58 # 16230 Zbl 409 # 03028
• ID 29270

REMMEL, J.B. [1978] *Realizing partial orderings by classes of co-simple sets* (J 0048) Pac J Math 76*169-184
◇ D25 D45 D50 ◇ REV MR 58 # 10379 Zbl 393 # 03033
• ID 52453

REMMEL, J.B. [1978] *Recursively enumerable boolean algebras* (J 0007) Ann Math Logic 15*75-107
◇ C57 D25 D45 G05 ◇ REV MR 80e:03052
Zbl 413 # 03027 • ID 29157

REMMEL, J.B. [1979] *R-maximal boolean algebras* (J 0036) J Symb Logic 44*533-548
◇ C57 D45 G05 ◇ REV MR 81b:03049 Zbl 439 # 03026
• ID 56017

REMMEL, J.B. [1979] see METAKIDES, G.

REMMEL, J.B. [1980] see MANASTER, A.B.

REMMEL, J.B. [1980] *On r.e. and co-r.e. vector spaces with nonextendible bases* (J 0036) J Symb Logic 45*20-34
◇ C57 C60 D45 ◇ REV MR 81b:03048 Zbl 471 # 03038
• ID 55233

REMMEL, J.B. [1980] *Recursion theory on orderings II* (J 0036) J Symb Logic 45*317-333
◇ C57 D45 ◇ REV MR 81e:03046 Zbl 471 # 03036
JSL 51.229 • REM Part I 1979 by Metakides,G. & Remmel,J.B.
• ID 55231

REMMEL, J.B. [1980] *Recursion theory on algebraic structures with independent sets* (J 0007) Ann Math Logic 18*153-191
◇ C57 C60 D45 ◇ REV MR 81j:03076 Zbl 471 # 03037
• ID 55232

REMMEL, J.B. [1981] *Effective structures not contained in recursively enumerable structures* (P 2902) Aspects Effective Algeb;1979 Clayton 206-225
◇ C57 D45 ◇ REV MR 83a:03029 Zbl 497 # 03035
• ID 35064

REMMEL, J.B. [1981] see MANASTER, A.B.

REMMEL, J.B. [1981] *Recursive isomorphism types of recursive boolean algebras* (J 0036) J Symb Logic 46*572-594
◇ C57 D45 G05 ◇ REV MR 83a:03042 Zbl 543 # 03031
• ID 33280

REMMEL, J.B. [1981] *Recursive boolean algebras with recursive atoms* (J 0036) J Symb Logic 46*595-616
◇ C57 D45 G05 ◇ REV MR 82j:03055 Zbl 543 # 03032
• ID 77739

REMMEL, J.B. [1981] *Recursively categorical linear orderings* (J 0053) Proc Amer Math Soc 83*387-391
◇ C35 C57 C65 ◇ REV MR 82j:03037 Zbl 493 # 03022
• ID 77729

REMMEL, J.B. [1982] see EISENBERG, E.F.

REMMEL, J.B. [1982] see NERODE, A.

REMMEL, J.B. [1982] see LERMAN, M.

REMMEL, J.B. [1983] see KALANTARI, I.

REMMEL, J.B. [1983] see KIERSTEAD, H.A.

REMMEL, J.B. [1983] see NERODE, A.

REMMEL, J.B. [1983] see CROSSLEY, J.N.

REMMEL, J.B. [1984] see MILLER, DOUGLAS E.

REMMEL, J.B. [1984] see DOWNEY, R.G.

REMMEL, J.B. [1984] see LERMAN, M.

REMMEL, J.B. [1984] see CROSSLEY, J.N.

REMMEL, J.B. [1985] see NERODE, A.

REMMEL, J.B. [1985] see KIERSTEAD, H.A.

REMY, J.-L. [1979] see BROY, M.

RENNIE, M.K. [1968] *A function which bounds truth-tabular calculations in S5* (**J** 0079) Logique & Anal, NS 11*425-439
◇ B25 B45 D15 ◇ REV MR 43 # 6062 Zbl 181.6
• ID 11098

RENNIE, M.K. see Vol. I, II for further entries

RESSAYRE, J.-P. [1977] *Models with compactness properties relative to an admissible language* (**J** 0007) Ann Math Logic 11*31-55
◇ C50 C55 C70 C80 D55 D60 E15 ◇ REV MR 57 # 5735 Zbl 376 # 02032 JSL 47.439 • ID 23658

RESSAYRE, J.-P. [1982] *Bounding generalized recursive functions of ordinals by effective functors: a complement to the Girard theorem* (**P** 3708) Herbrand Symp Logic Colloq;1981 Marseille 251-279
◇ D60 F15 ◇ REV MR 86f:03071 Zbl 513 # 03020
• ID 37224

RESSAYRE, J.-P. [1985] see GIRARD, J.-Y.

RESSAYRE, J.-P. see Vol. III, V, VI for further entries

RESTIVO, A. & REUTENAUER, C. [1984] *Cancellation, pumping and permutation in formal languages* (**P** 4012) Automata, Lang & Progr (11);1984 Antwerpen 414-422
◇ D05 ◇ ID 47184

RESTIVO, A. & SALEMI, S. [1985] *Overlap free words on two symbols* (**P** 4622) Autom on Infinite Words;1984 Le Mont-Dore 198-206
◇ D05 ◇ ID 49461

RESTIVO, A. & SALEMI, S. [1985] *Some decision results on nonrepetitive words* (**P** 4626) Combin Algor on Words;1984 Maratea 289-295
◇ D05 ◇ ID 49462

RETZLAFF, A.T. [1977] see KALANTARI, I.

RETZLAFF, A.T. [1978] *Simple and hyperhypersimple vector spaces* (**J** 0036) J Symb Logic 43*260-269
◇ C57 D25 D45 ◇ REV MR 81g:03053 Zbl 399 # 03033
• ID 29259

RETZLAFF, A.T. [1979] *Direct summands of recursively enumerable vector spaces* (**J** 0068) Z Math Logik Grundlagen Math 25*363-372
◇ C57 C60 D45 ◇ REV MR 81a:03049 Zbl 444 # 03024
• ID 77762

RETZLAFF, A.T. [1979] see KALANTARI, I.

REUSCH, B. [1969] *Lineare Automaten* (**X** 0876) Bibl Inst: Mannheim 149pp
◇ D05 ◇ REV MR 41 # 5124 Zbl 209.30 • ID 23554

REUSCH, B. see Vol. I for further entries

REUTENAUER, C. [1984] see RESTIVO, A.

REVESZ, G. [1969] *On certain formal grammars and syntatic analysis without blind path* (**P** 1841) Fct Recurs & Appl;1967 Tihany 84-98
◇ D05 ◇ ID 32558

REVESZ, G. [1975] *Phrase-structure grammars and dual pushdown automata (Hungarian) (English summary)* (**J** 1458) Alkalmaz Mat Lapok 1*397-404
◇ D05 D10 ◇ REV MR 57 # 11179 Zbl 361 # 68100
• ID 50716

REVESZ, G. [1976] *A note on the relation of Turing machines to phrase structure grammars* (**J** 2293) Comp Linguist & Comp Lang 11*11-16
◇ D05 D10 ◇ REV Zbl 404 # 68080 • ID 69870

REVESZ, G. [1976] *Multicontrol Turing machines* (**J** 0380) Acta Cybern (Szeged) 3*173-177
◇ D10 ◇ REV MR 56 # 2803 Zbl 363 # 02042 • ID 50869

REVESZ, G. [1976] *Multicontrol Turing machines* (**P** 2898) Algor Kompl, Lern-& Erkenn-Prozess;1976 Jena 71-77
◇ D10 ◇ REV MR 56 # 10153 Zbl 447 # 68045 • ID 69871

REVESZ, G. see Vol. VI for further entries

REVIN, O. [1977] see KRATKO, M.I.

REYNOLDS, J.C. [1969] *A generalized resolution principle based upon context-free grammars* (**P** 0594) Inform Processing (4);1968 Edinburgh 2*1405-1411
◇ B35 D05 ◇ REV Zbl 211.315 • ID 46610

REYNOLDS, J.C. [1975] *On the interpretation of Scott's domains* (**P** 1475) Conv Inform Teor & Conv Strutt Corpi Algeb;1973 Roma 123-135
◇ B75 D05 G10 ◇ REV MR 54 # 1704 Zbl 318 # 68024
• ID 23995

REYNOLDS, J.C. see Vol. I, VI for further entries

REYNVAAN, C. & SCHNORR, C.-P. [1977] *Ueber Netzwerkgroessen hoeherer Ordnung und die mittlere Anzahl der in Netzwerken benutzten Operationen* (**P** 3411) Theor Comput Sci (3);1977 Darmstadt 368-390
◇ B70 D15 ◇ REV MR 57 # 19141 Zbl 386 # 94027
• ID 69872

RHODES, J. [1984] *Algebraic and topological theory of languages and computation I: Theorems for arbitrary languages generalizing the theorems of Eilenberg, Kleene, Schuetzenberger, and Straubing* (**P** 3565) Symp of Th Aspects of Comput Sci (1);1984 Paris 299-304
◇ D05 ◇ REV MR 85i:68002 Zbl 569 # 68062 • ID 47100

RIBEIRO ALBUQUERQUE, J. [1944] *Ensembles de Borel* (**J** 0050) Port Math 4*161-198
◇ D55 E15 ◇ REV MR 7.196 • REM Part I. Part II 1945
• ID 00252

RIBEIRO ALBUQUERQUE, J. [1945] *Ensembles de Borel II* (**J** 0050) Port Math 4*217-224
◇ D55 E15 ◇ REV MR 7.196 • REM Part I 1944 • ID 00253

RIBEIRO ALBUQUERQUE, J. [1951] *Theory of projective sets I (Portuguese) (French summary)* (**J** 0082) Bull Soc Math Belg 1*345-400
◇ D55 E15 E98 ◇ REV MR 17.467 Zbl 49.40 • REM Part II 1952 • ID 00254

RIBEIRO ALBUQUERQUE, J. [1952] *Theorie des ensembles projectifs* (**J** 0050) Port Math 11*11-33
◇ D55 E15 ◇ REV MR 14.146 Zbl 46.280 • ID 00256

RIBEIRO ALBUQUERQUE, J. [1952] *Theory of projective sets II (Portuguese) (French summary)* (**J** 0082) Bull Soc Math Belg 2*5-44
◇ D55 E15 E98 ◇ REV MR 17.467 Zbl 49.40 • REM Part I 1951 • ID 00401

RIBEIRO ALBUQUERQUE, J. [1952] *Un theoreme sur les ensembles cribles* (**J** 0050) Port Math 11*95-103
◇ D55 E15 ◇ REV MR 14.147 Zbl 49.317 • ID 00255

RIBEIRO ALBUQUERQUE, J. see Vol. V for further entries

RICCARDI, G.A. [1981] *The independence of control structures in abstract programming systems* (**J** 0119) J Comp Syst Sci 22*107-143
◇ B70 D80 ◇ REV MR 82m:03061 Zbl 467 # 68009 • ID 77770

RICCARDI, G.A. [1982] *The independence of control structures in programmable numberings of the partial recursive functions* (**J** 0068) Z Math Logik Grundlagen Math 28*285-296
◇ D20 D45 ◇ REV MR 83m:03058 Zbl 527 # 03023 • ID 35458

RICE, H.G. [1953] *Classes of recursively enumerable sets and their decision problems* (**J** 0064) Trans Amer Math Soc 74*358-366
◇ D25 ◇ REV MR 14.713 Zbl 53.3 JSL 19.121 • ID 11150

RICE, H.G. [1956] *On completely recursively enumerable classes and their key arrays* (**J** 0036) J Symb Logic 21*304-308
◇ D25 ◇ REV MR 18.369 Zbl 72.5 JSL 23.48 • ID 11152

RICE, H.G. [1956] *Recursive and recursively enumerable orders* (**J** 0064) Trans Amer Math Soc 83*277-300
◇ D20 D25 D45 ◇ REV MR 18.712 Zbl 79.246 JSL 22.375 • ID 11153

RICE, H.G. [1957] *On the relative density of sets of integers* (**J** 0053) Proc Amer Math Soc 8*320-321
◇ D20 D30 ◇ REV MR 19.3 Zbl 79.247 • ID 11154

RICE, H.G. [1965] *Recursion and iteration* (**J** 0212) ACM Commun 8*114-115
◇ D20 ◇ REV Zbl 129.103 • ID 26658

RICE, H.G. [1969] see GONSHOR, H.

RICE, H.G. see Vol. VI for further entries

RICHALET, J. [1965] *Calcul operationnel dans un anneau fini* (**J** 0109) C R Acad Sci, Paris 260*3541-3543
◇ D05 ◇ REV MR 31 # 1197 • ID 11156

RICHARD, D. [1982] *La theorie sans egalite du successeur et de la coprimarite des entiers naturels est indecidable. Le predicat de primarite est definissable dans le langage de cette theorie (English summary)* (**J** 3364) C R Acad Sci, Paris, Ser 1 294*143-146
◇ D35 F30 ◇ REV MR 84c:03084 Zbl 486 # 03027 • ID 34019

RICHARD, D. [1984] *The arithmetics as theories of two orders (English and French summaries)* (**P** 2167) Orders: Descr & Roles;1982 L'Arbresle 287-311
◇ B28 C62 D35 F30 ◇ REV MR 85k:06001 MR 86h:03102 Zbl 555 # 03026 • ID 41779

RICHARD, D. see Vol. I, III, V, VI for further entries

RICHARDS, I. [1978] see POUR-EL, M.B.

RICHARDS, I. [1979] see POUR-EL, M.B.

RICHARDS, I. [1981] see POUR-EL, M.B.

RICHARDS, I. [1982] see POUR-EL, M.B.

RICHARDS, I. [1983] see POUR-EL, M.B.

RICHARDS, I. [1984] see POUR-EL, M.B.

RICHARDSON, D.B. [1968] *Some undecidable problems involving elementary functions of a real variable* (**J** 0036) J Symb Logic 33*514-520
◇ D40 D80 ◇ REV MR 39 # 1330 Zbl 175.274 • ID 11162

RICHARDSON, D.B. [1969] *Solution of the identity problem for integral exponential functions* (**J** 0068) Z Math Logik Grundlagen Math 15*333-340
◇ D80 ◇ REV MR 41 # 6678 Zbl 184.23 • ID 11163

RICHARDSON, D.B. [1971] *The simple exponential constant problem* (**J** 0068) Z Math Logik Grundlagen Math 17*133-136
◇ D80 F30 ◇ REV MR 44 # 3872 Zbl 231 # 02053 • ID 11164

RICHARDSON, D.B. [1976] *Continuous self-reproduction* (**J** 0119) J Comp Syst Sci 12*6-12
◇ D10 ◇ REV MR 53 # 2559 Zbl 328 # 94049 • ID 64819

RICHARDSON, D.B. see Vol. III, VI for further entries

RICHIER, J. [1979] see CULIK II, K.

RICHMAN, F. [1983] *Church's thesis without tears* (**J** 0036) J Symb Logic 48*797-803
◇ D20 D75 F55 F60 ◇ REV MR 84j:03084 Zbl 527 # 03036 • ID 34676

RICHMAN, F. see Vol. VI for further entries

RICHTER, L.J. [1979] *On automorphisms of the degrees that preserve jumps* (**J** 0029) Israel J Math 32*27-31
◇ D30 ◇ REV MR 80h:03065 Zbl 395 # 03030 • ID 52581

RICHTER, L.J. [1981] *Degrees of structures* (**J** 0036) J Symb Logic 46*723-731
◇ C57 D30 D45 ◇ REV MR 83d:03048 Zbl 512 # 03024 • ID 33281

RICHTER, M.M. [1984] see BOERGER, E.

RICHTER, M.M. [1984] see MUELLER, GERT H.

RICHTER, M.M. & SZABO, M.E. [1985] *Nonstandard computation theory* (**C** 4213) Algeb, Combin & Log in Comput Sci
◇ D20 D75 H10 ◇ ID 39589

RICHTER, M.M. see Vol. I, II, III, V, VI for further entries

RICHTER, W.H. [1965] *Extensions of the constructive ordinals* (**J** 0036) J Symb Logic 30*193-211
◇ D70 F15 ◇ REV MR 36 # 2500 Zbl 134.12 JSL 36.341 • ID 11170

RICHTER, W.H. [1965] *Regressive sets of order n* (**J** 0044) Math Z 86*372-374
◇ D50 ◇ REV MR 35 # 6554 JSL 38.525 • ID 11169

RICHTER, W.H. [1967] *Constructive transfinite number classes* (**J** 0015) Bull Amer Math Soc 73*261-265
◇ D65 D70 F15 ◇ REV MR 34 # 7372 Zbl 155.341 JSL 36.341 • ID 11171

RICHTER, W.H. [1968] *Constructively accessible ordinal numbers* (**J** 0036) J Symb Logic 33*43-55
◇ D65 D70 F15 ◇ REV MR 38 # 4305 Zbl 197.5 JSL 36.341 • ID 11172

RICHTER, W.H. [1971] *Recursively Mahlo ordinals and inductive definitions* (**P** 0638) Logic Colloq;1969 Manchester 273-288
◇ D60 D65 D70 E55 F15 ◇ REV MR 43 # 7331 Zbl 252 # 02024 • ID 11173

RICHTER, W.H. [1972] see ACZEL, P.

RICHTER, W.H. [1974] see ACZEL, P.

RICHTER, W.H. [1975] *The least Σ^1_2- and Π^1_2-reflecting ordinals* (**P** 1442) ⊧ ISILC Logic Conf;1974 Kiel 568-578 ◇ D55 D60 D70 E45 E47 E55 ◇ REV MR 54 # 81 Zbl 354 # 02035 • ID 23962

RIEDEMANN, E.H. [1979] *The control of parallel computations by labeled Petri nets: a study in terms of multiple-firing automata and parallel program schemata (Dissertation)* (**S** 3125) Ber, Abt Inf, Univ Dortmund 4*290 ◇ D05 D80 ◇ REV Zbl 449 # 68023 • ID 69013

RIEGER, L. [1961] *On a critique of Church's thesis concerning general recursive functions in arithmetic (Czech)* (**J** 0086) Cas Pestovani Mat, Ceskoslov Akad Ved 86*480-481 ◇ A05 D20 F99 ◇ REV JSL 35.489 • ID 22008

RIEGER, L. [1963] *Kleene's normal form for computable functions (Czech) (Russian and English summaries)* (**J** 0086) Cas Pestovani Mat, Ceskoslov Akad Ved 88*349-363 ◇ D20 ◇ REV MR 31 # 4719 Zbl 163.251 • ID 11191

RIEGER, L. see Vol. I, II, III, V, VI for further entries

RIEKSTINS, J. [1973] *The solvability of the halting problem for two-cycle Turing machines (Russian) (Latvian and English summaries)* (**J** 0337) Mat Ezheg, Akad Nauk Latv SSR 13*140-168 ◇ D10 ◇ REV MR 49 # 2310 Zbl 284 # 02018 • ID 11195

RIFFLET, J.M. [1975] see COUSINEAU, G.

RIGUET, J. [1953] *Sur les rapports entre les concepts de machine de multipole et de structure algebrique* (**J** 0109) C R Acad Sci, Paris 237*425-427 ◇ D03 D05 ◇ REV MR 15.559 Zbl 53.17 JSL 23.62 • ID 11198

RIGUET, J. [1956] *Algorithmes de Markov et theorie des machines* (**J** 0109) C R Acad Sci, Paris 242*435-437 ◇ D03 D05 ◇ REV Zbl 70.9 JSL 23.62 • ID 11199

RIGUET, J. see Vol. V for further entries

RINE, D.C. [1973] *A correspondence between control logic of associative memories and Markov algorithms* (**J** 2820) Proc West Virginia Acad Sci 45*335-345 ◇ B70 D03 ◇ REV MR 50 # 6202 Zbl 287 # 68031 • ID 64836

RINE, D.C. [1975] *Representation and design of production systems* (**J** 0302) Rep Math Logic, Krakow & Katowice 4*43-66 ◇ B70 D03 G20 ◇ REV MR 55 # 12487 Zbl 343 # 02023 • ID 21936

RINE, D.C. see Vol. I, II for further entries

RIPS, E. [1982] *Another characterization of finitely generated groups with a solvable word problem* (**J** 0161) Bull London Math Soc 14*43-44 ◇ D40 ◇ REV MR 82m:20038 Zbl 481 # 20021 • ID 82613

RIPS, E. see Vol. III for further entries

RITCHIE, D.M. [1967] see MEYER, A.R.

RITCHIE, D.M. [1968] see MEYER, A.R.

RITCHIE, D.M. [1972] see MEYER, A.R.

RITCHIE, R.W. [1963] *Classes of predictably computable functions* (**J** 0064) Trans Amer Math Soc 106*139-173 • TRANSL [1970] (**C** 1540) Probl Mat Log: Slozh Algor & Kl Vychisl Funk 50-93 ◇ D15 D20 ◇ REV MR 28 # 2045 Zbl 107.10 JSL 28.252 • ID 11205

RITCHIE, R.W. [1964] see KREIDER, D.L.

RITCHIE, R.W. [1965] *Classes of recursive functions based on Ackermann's function* (**J** 0048) Pac J Math 15*1027-1044 ◇ D20 ◇ REV MR 33 # 1235 Zbl 133.249 JSL 31.654 • ID 11207

RITCHIE, R.W. [1966] see KREIDER, D.L.

RITCHIE, R.W. [1967] see MEYER, A.R.

RITCHIE, R.W. & YOUNG, P. [1968] *Strong representability of partial functions in arithmetic theories* (**J** 0191) Inform Sci 1*189-204 ◇ D20 F30 ◇ REV MR 41 # 3276 • ID 11208

RITCHIE, R.W. & SPRINGSTEEL, F. [1972] *Language recognition by marking automata* (**J** 0194) Inform & Control 20*313-330 ◇ D05 D10 ◇ REV MR 48 # 3649 Zbl 242 # 68032 • ID 64842

RITCHIE, R.W. [1973] see PETERS JR., P.S.

RITCHIE, R.W. see Vol. I for further entries

RITTER, W.E. [1966] *Notation systems and an effective fixed point property* (**J** 0053) Proc Amer Math Soc 17*390-395 ◇ D20 D30 F15 ◇ REV MR 34 # 1181 Zbl 216.290 JSL 40.626 • ID 11210

RITTER, W.E. [1967] *Representability of partial recursive functions in formal theories* (**J** 0053) Proc Amer Math Soc 18*647-651 ◇ D20 F30 ◇ REV MR 37 # 6177 Zbl 207.305 • ID 11211

RIX, H. [1981] see MALENGE, J.P.

ROBERTSON, E.L. [1971] *Complexity classes of partial recursive functions* (**P** 0680) ACM Symp Th of Comput (3);1971 Shaker Heights 258-266 ◇ D15 D20 ◇ REV MR 49 # 8832 Zbl 263 # 68025 • ID 28674

ROBERTSON, E.L. [1972] see LANDWEBER, L.H.

ROBERTSON, E.L. [1974] *Complexity classes of partial recursive functions* (**J** 0119) J Comp Syst Sci 9*69-87 ◇ D15 D20 ◇ REV MR 49 # 8832 Zbl 315 # 68039 • ID 64850

ROBERTSON, E.L. [1974] *Structure of complexity in the weak monadic second-order theories of the natural numbers* (**P** 1464) ACM Symp Th of Comput (6);1974 Seattle 161-171 ◇ D15 D55 F35 ◇ REV MR 54 # 11850 Zbl 361 # 68071 • ID 25884

ROBERTSON, E.L. [1976] see FISCHER, P.C.

ROBERTSON, E.L. [1979] see LANDWEBER, L.H.

ROBERTSON, E.L. [1981] see LANDWEBER, L.H.

ROBINSON, A. [1961] *Model theory and non-standard arithmetic* (**P** 0633) Infinitist Meth;1959 Warsaw 265-302 • REPR [1979] (**C** 4594) Sel Pap Robinson 1*167-204 ◇ C35 C50 C57 C60 C62 H15 ◇ REV MR 26 # 32 Zbl 126.11 JSL 35.149 • ID 11249

ROBINSON, A. [1964] see ELGOT, C.C.

ROBINSON, A. [1967] see ELGOT, C.C.

ROBINSON, A. [1975] *Algorithms in algebra* (**C** 0782) Model Th & Algeb (A. Robinson) 14-40
• REPR [1979] (**C** 4594) Sel Pap Robinson 1*504-520
◇ C40 C57 C60 D45 D75 ◇ REV MR 53 # 5298 Zbl 318 # 02033 • ID 22933

ROBINSON, A. also published under the name ROBINSOHN, A.

ROBINSON, A. see Vol. I, II, III, V for further entries

ROBINSON, D.J. [1984] *Decision problems for infinite soluble groups* (**P** 4359) Groups-Korea;1983 Kyoungiu 111-117
◇ D35 D40 ◇ REV MR 86f:20037 Zbl 555 # 20024 • ID 44852

ROBINSON, D.J. [1984] see CANNONITO, F.B.

ROBINSON, JULIA [1949] *Definability and decision problems in arithmetic* (**J** 0036) J Symb Logic 14*98-114
◇ D35 F30 ◇ REV MR 11.151 Zbl 34.8 JSL 15.68 • ID 11292

ROBINSON, JULIA [1950] *General recursive functions* (**J** 0053) Proc Amer Math Soc 1*703-718
◇ D20 ◇ REV MR 12.469 Zbl 41.151 JSL 16.280 • ID 11293

ROBINSON, JULIA [1952] *Existential definability in arithmetic* (**P** 0593) Int Congr Math (II, 6);1950 Cambridge MA 1*728-729
◇ D25 ◇ ID 28063

ROBINSON, JULIA [1952] *Existential definability in arithmetic* (**J** 0064) Trans Amer Math Soc 72*437-449
◇ D25 ◇ REV MR 14.4 Zbl 47.248 JSL 20.182 • ID 49836

ROBINSON, JULIA [1955] *A note on primitive recursive functions* (**J** 0053) Proc Amer Math Soc 6*667-670
◇ D20 ◇ REV MR 17.447 Zbl 67.2 JSL 22.376 • ID 11294

ROBINSON, JULIA [1959] *The undecidability of algebraic rings and fields* (**J** 0053) Proc Amer Math Soc 10*950-957
◇ C60 D35 ◇ REV MR 22 # 3691 Zbl 100.15 JSL 29.57 • ID 11295

ROBINSON, JULIA [1961] see DAVIS, MARTIN D.

ROBINSON, JULIA [1962] *On the decision problem for algebraic rings* (**C** 0595) Stud Math Anal & Rel Topics (Polya) 297-304
◇ C60 D35 ◇ REV MR 26 # 3609 Zbl 117.12 JSL 35.475 • ID 11296

ROBINSON, JULIA [1962] *The undecidability of exponential diophantine equations* (**P** 0612) Int Congr Log, Meth & Phil of Sci (1,Proc);1960 Stanford 12-13
◇ D35 ◇ REV MR 29 # 5724 Zbl 178.324 JSL 35.152 • ID 11297

ROBINSON, JULIA [1965] *The decision problem for fields* (**P** 0614) Th Models;1963 Berkeley 299-311
◇ B25 C60 D35 ◇ REV MR 34 # 62 Zbl 274 # 02020 • ID 11298

ROBINSON, JULIA [1967] *An introduction to hyperarithmetical functions* (**J** 0036) J Symb Logic 32*325-342
◇ D55 ◇ REV MR 37 # 71 Zbl 153.8 • ID 11299

ROBINSON, JULIA [1968] *Finite generation of recursively enumerable sets* (**J** 0053) Proc Amer Math Soc 19*1480-1486
◇ D25 ◇ REV MR 39 # 1328 Zbl 182.16 • ID 11301

ROBINSON, JULIA [1968] *Recursive functions of one variable* (**J** 0053) Proc Amer Math Soc 19*815-820
◇ D20 D40 ◇ REV MR 37 # 6178 Zbl 165.316 JSL 35.476 • ID 11300

ROBINSON, JULIA [1969] *Diophantine decision problems* (**C** 0596) Stud in Number Th 76-116
◇ D25 D35 ◇ REV MR 39 # 5364 Zbl 269 # 02018 JSL 37.603 • ID 11304

ROBINSON, JULIA [1969] *Finitely generated classes of sets of natural numbers* (**J** 0053) Proc Amer Math Soc 21*608-614
◇ D20 D55 E20 ◇ REV MR 40 # 7111 Zbl 184.19 • ID 11302

ROBINSON, JULIA [1969] *Unsolvable diophantine problems* (**J** 0053) Proc Amer Math Soc 22*534-538
◇ D25 D35 ◇ REV MR 39 # 5363 Zbl 182.19 JSL 37.603 • ID 11303

ROBINSON, JULIA [1971] *Hilbert's tenth problem* (**P** 1469) AMS Numb Th Summer Inst;1969 Stony Brook 191-194
◇ D35 ◇ REV MR 47 # 4782 Zbl 238 # 02037 • ID 27906

ROBINSON, JULIA [1973] *Axioms for number theoretic functions (Russian)* (**C** 0733) Izbr Vopr Algeb & Log (Mal'tsev) 253-263
◇ C62 D20 D75 F30 ◇ REV MR 48 # 8224 Zbl 279 # 02035 • ID 29547

ROBINSON, JULIA [1973] *Solving diophantine equations* (**P** 0793) Int Congr Log, Meth & Phil of Sci (4,Proc);1971 Bucharest 63-67
◇ B25 D35 D80 H15 ◇ REV MR 58 # 215 • ID 77822

ROBINSON, JULIA [1974] see MATIYASEVICH, YU.V.

ROBINSON, JULIA [1975] see MATIYASEVICH, YU.V.

ROBINSON, JULIA [1976] see DAVIS, MARTIN D.

ROBINSON, R.M. [1947] *Primitive recursive functions* (**J** 0015) Bull Amer Math Soc 53*925-942
◇ D20 ◇ REV MR 9.221 Zbl 34.291 JSL 13.113 • REM Part I. Part II 1955 • ID 28764

ROBINSON, R.M. [1948] *Recursion and double recursion* (**J** 0015) Bull Amer Math Soc 54*987-992
◇ D20 ◇ REV MR 10.229 Zbl 34.292 JSL 14.191 • ID 11308

ROBINSON, R.M. [1951] *Undecidable rings* (**J** 0064) Trans Amer Math Soc 70*137-159
◇ C60 D35 ◇ REV MR 12.791 Zbl 42.245 JSL 17.268 • ID 11311

ROBINSON, R.M. [1952] *An essentially undecidable axiom system* (**P** 0593) Int Congr Math (II, 6);1950 Cambridge MA 1*729-730
◇ D35 F30 ◇ ID 11309

ROBINSON, R.M. [1953] see MOSTOWSKI, ANDRZEJ

ROBINSON, R.M. [1955] *Primitive recursive functions II* (**J** 0053) Proc Amer Math Soc 6*663-666
◇ D20 ◇ REV MR 17.447 Zbl 67.2 JSL 22.375 • REM Part I 1947 • ID 11313

ROBINSON, R.M. [1956] *Arithmetical representation of recursively enumerable sets* (**J** 0036) J Symb Logic 21*162-187
◇ D25 F30 ◇ REV MR 18.272 Zbl 73.252 JSL 24.170
• ID 11314

ROBINSON, R.M. [1963] *Undecidability of the elementary theory of the field of rational functions of one variable with rational coefficients (Russian)* (**J** 0003) Algebra i Logika 2/4*5-11
◇ D35 ◇ REV MR 29 # 26 Zbl 192.57 JSL 31.254
• ID 11317

ROBINSON, R.M. [1964] *The undecidability of pure transcendental extension of real fields* (**J** 0068) Z Math Logik Grundlagen Math 10*275-282
◇ D35 ◇ REV MR 30 # 3021 Zbl 221 # 02034 JSL 31.254
• ID 11319

ROBINSON, R.M. [1971] *Undecidability and nonperiodicity for tilings of the plane* (**J** 0305) Invent Math 12*177-209
◇ D80 ◇ REV MR 45 # 6626 Zbl 197.468 • ID 11320

ROBINSON, R.M. [1972] *Some representations of diophantine sets* (**J** 0036) J Symb Logic 37*572-578
◇ D25 D35 ◇ REV MR 50 # 9556 Zbl 264 # 02042
• ID 11321

ROBINSON, R.M. [1978] *Undecidable tiling problems in the hyperbolic plane* (**J** 0305) Invent Math 44*259-264
◇ D80 ◇ REV MR 81h:03091 Zbl 354 # 50006 • ID 31983

ROBINSON, R.M. see Vol. I, III, V, VI for further entries

ROBINSON, R.W. [1967] *Simplicity of recursively enumerable sets* (**J** 0036) J Symb Logic 32*162-172
◇ D25 ◇ REV MR 36 # 1323 Zbl 204.12 JSL 35.153
• ID 11323

ROBINSON, R.W. [1967] *Two theorems on hyperhypersimple sets* (**J** 0064) Trans Amer Math Soc 128*531-538
◇ D25 ◇ REV MR 35 # 6549 Zbl 153.315 JSL 35.153
• ID 11322

ROBINSON, R.W. [1968] *A dichotomy of the recursively enumerable sets* (**J** 0068) Z Math Logik Grundlagen Math 14*339-356
◇ D25 ◇ REV MR 38 # 5623 Zbl 217.12 • ID 11324

ROBINSON, R.W. [1971] *Interpolation and embedding in the recursively enumerable degrees* (**J** 0120) Ann of Math, Ser 2 93*285-314
◇ D25 ◇ REV MR 43 # 51 Zbl 259 # 02033 • ID 11325

ROBINSON, R.W. [1971] *Jump restricted interpolation in the recursively enumerable degrees* (**J** 0120) Ann of Math, Ser 2 93*586-596
◇ D25 ◇ REV MR 47 # 1592 Zbl 259 # 02034 • ID 11326

ROBINSON, R.W. [1981] see POSNER, D.B.

ROBINSON, R.W. [1982] see MADAN, D.B.

ROCHE LA, P. [1981] *Effective Galois theory* (**J** 0036) J Symb Logic 46*385-392
◇ C57 C60 D45 F60 ◇ REV MR 82j:03053
Zbl 464 # 03039 • ID 75255

RODRIGUEZ ARTALEJO, M. [1981] *Eine syntaktisch-algebraische Methode zur Konstruktion von Modellen* (**J** 0068) Z Math Logik Grundlagen Math 27*59-71
◇ B10 C07 C30 C57 ◇ REV MR 82e:03027
Zbl 481 # 03016 • ID 77840

RODRIGUEZ ARTALEJO, M. see Vol. I, III for further entries

ROEDDING, D. [1964] *Ueber die Eliminierbarkeit von Definitionsschemata in der Theorie der recursiven Funktionen* (**J** 0068) Z Math Logik Grundlagen Math 10*315-330
◇ D20 ◇ REV MR 30 # 18 Zbl 199.25 • ID 11335

ROEDDING, D. [1965] see HERMES, H.

ROEDDING, D. [1965] *Darstellungen der (im Kalmar-Csillagschen Sinne) elementaren Funktionen* (**J** 0009) Arch Math Logik Grundlagenforsch 7*139-158
◇ D20 ◇ REV Zbl 217.11 • REM Part II 1966 • ID 28043

ROEDDING, D. [1965] *Einige aequivalente Praezisierungen des intuitiven Berechenbarkeitsbegriffs* (**J** 0487) Math Unterricht 11/2*21-38
◇ D20 ◇ REV MR 32 # 2325 • ID 11336

ROEDDING, D. [1966] *Anzahlquantoren in der Kleene-Hierarchie* (**J** 0009) Arch Math Logik Grundlagenforsch 9*61-65
◇ C80 D55 ◇ REV MR 38 # 2023 Zbl 161.7 JSL 33.472
• ID 24963

ROEDDING, D. [1966] *Der Entscheidbarkeitsbegriff in der mathematischen Logik* (**J** 0178) Stud Gen 19*516-522
◇ B25 D10 D20 D25 D35 ◇ REV Zbl 192.20 • ID 16282

ROEDDING, D. [1966] *Ueber Darstellungen der elementaren Funktionen II* (**J** 0009) Arch Math Logik Grundlagenforsch 9*36-48
◇ D20 ◇ REV MR 33 # 7252 Zbl 217.12 • REM Part I 1965
• ID 11338

ROEDDING, D. [1967] *Primitiv-rekursive Funktionen ueber einen Bereich endlicher Mengen* (**J** 0009) Arch Math Logik Grundlagenforsch 10*13-29
◇ D20 ◇ REV MR 35 # 5313 Zbl 189.9 • ID 11340

ROEDDING, D. [1968] *Klassen rekursiver Funktionen* (**P** 0692) Summer School in Logic;1967 Leeds 159-222
◇ D20 ◇ REV MR 38 # 5624 Zbl 212.328 JSL 37.196
• ID 11341

ROEDDING, D. [1970] *Reduktionstypen der Praedikatenlogik. Nach einer einstuendigen Vorlesung ausgearbeitet von Egon Boerger* (**X** 1532) Univ Muenster Inst Math Logik: Muenster 59pp
◇ D35 ◇ REV MR 57 # 2903 Zbl 267 # 02034 • ID 29896

ROEDDING, D. [1971] *Arithmetische und hyperarithmetische Praedikate I* (**X** 1532) Univ Muenster Inst Math Logik: Muenster 75pp
◇ D55 ◇ REV Zbl 252 # 02043 • REM Part II 1972 • ID 27756

ROEDDING, D. [1972] *Arithmetische und hyperarithmetische Praedikate II* (**X** 1532) Univ Muenster Inst Math Logik: Muenster 93pp
◇ D55 ◇ REV Zbl 252 # 02044 • REM Part I 1971 • ID 27757

ROEDDING, D. & SCHWICHTENBERG, H. [1972] *Bemerkungen zum Spektralproblem* (**J** 0068) Z Math Logik Grundlagen Math 18*1-12
◇ B15 C13 D10 D15 ◇ REV MR 46 # 5128
Zbl 242 # 02049 • ID 11343

ROEDDING, D. [1972] *Einfuehrung in die Theorie der berechenbaren Funktionen I* (**X** 1532) Univ Muenster Inst Math Logik: Muenster iii + 113pp
◇ D98 ◇ REV MR 56 # 15384a • REM Part II 1972
• ID 77836

ROEDDING, D. [1972] *Einfuehrung in die Theorie der berechenbaren Funktionen II* (**X** 1532) Univ Muenster Inst Math Logik: Muenster ii + 117pp
◇ D98 ◇ REV MR 56 # 15384b • REM Part I 1972 • ID 77835

ROEDDING, D. [1972] *Registermaschinen* (**J** 0487) Math Unterricht 18*32-41
◇ D10 ◇ ID 14507

ROEDDING, D. [1974] see PRIESE, L.

ROEDDING, D. & ROEDDING, W. [1979] *Network of finite automata* (**P** 3432) Prog in Cybern & Syst Res;1972 Wien 139-158
◇ D05 ◇ REV Zbl 416 # 68050 • ID 41438

ROEDDING, D. [1983] *Modular decomposition of automata* (**P** 3864) FCT'83 Found of Comput Th;1983 Borgholm 394-412
◇ D05 ◇ REV MR 85f:68073 Zbl 523 # 68044 • ID 39983

ROEDDING, D. [1984] see BOERGER, E.

ROEDDING, D. [1984] see BRUEGGEMANN, A.

ROEDDING, D. [1984] *Some logical problems connected with a modular decomposition theory of automata* (**P** 2153) Logic Colloq;1983 Aachen 2*365-388
◇ D05 ◇ REV MR 85k:03002b • ID 41437

ROEDDING, D. see Vol. III for further entries

ROEDDING, W. [1979] see ROEDDING, D.

ROEDDING, W. see Vol. I, V, VI for further entries

ROEVER DE, W.P. [1974] *Recursion and parameter mechanisms: An axiomatic approach* (**P** 1869) Automata, Lang & Progr (2);1974 Saarbruecken 14*34-65
◇ B40 B75 D75 ◇ REV MR 55 # 1788 Zbl 302 # 68019 • ID 64875

ROGERS, C.A. & WILLMOTT, R. [1968] *On the uniformization of sets in topological spaces* (**J** 0118) Acta Math 120*1-52
◇ D55 E15 E75 ◇ REV MR 38 # 6014 Zbl 167.208 • ID 11352

ROGERS, C.A. [1973] *Lusin's second separation theorem* (**J** 3172) J London Math Soc, Ser 2 6*491-503
◇ D55 E15 E20 ◇ REV MR 51 # 14008 Zbl 256 # 04003 • ID 64880

ROGERS, C.A. [1977] see DAVIES, R.O.

ROGERS, C.A. [1980] see DELLACHERIE, C.

ROGERS, C.A. [1982] see JAYNE, J.E.

ROGERS, C.A. [1985] see HANSELL, R.W.

ROGERS, C.A. see Vol. V for further entries

ROGERS JR., H. [1956] *Certain logical reduction and decision problems* (**J** 0120) Ann of Math, Ser 2 64*264-284
◇ B20 D35 ◇ REV MR 18.271 Zbl 74.14 JSL 22.217 • ID 11355

ROGERS JR., H. [1957] *Computing degrees of unsolvability* (**P** 1675) Summer Inst Symb Log;1957 Ithaca 277-283
◇ D30 D55 ◇ REV JSL 25.363 • ID 11361

ROGERS JR., H. [1958] *Goedel numberings of partial recursive functions* (**J** 0036) J Symb Logic 23*331-341
◇ D20 D45 F40 ◇ REV MR 21 # 2585 Zbl 88.16 JSL 29.146 • ID 11357

ROGERS JR., H. [1959] *Computing degrees of unsolvability* (**J** 0043) Math Ann 138*125-140
◇ D30 D55 ◇ REV MR 22 # 5571 Zbl 86.11 JSL 25.363 • ID 11358

ROGERS JR., H. [1959] *Recursive functions over well-ordered partial orderings* (**J** 0053) Proc Amer Math Soc 10*847-853
◇ D20 D45 ◇ REV MR 22 # 2547 Zbl 95.10 JSL 27.83 • ID 11360

ROGERS JR., H. [1959] see FRIEDBERG, R.M.

ROGERS JR., H. [1959] *The present theory of Turing machine computability* (**J** 0514) SIAM Journ 7*114-130
• REPR [1969] (**C** 0569) Phil of Math Oxford Readings 130-146
◇ D10 ◇ REV MR 20 # 6359 Zbl 103.247 JSL 31.513 • ID 11359

ROGERS JR., H. [1961] see KREIDER, D.L.

ROGERS JR., H. [1965] *On universal functions* (**J** 0053) Proc Amer Math Soc 16*39-44
◇ D20 ◇ REV MR 30 # 1932 Zbl 287 # 02023 JSL 31.513 • ID 11363

ROGERS JR., H. [1966] see BOONE, W.W.

ROGERS JR., H. [1967] *Some problems of definability in recursive function theory* (**P** 0691) Sets, Models & Recursion Th;1965 Leicester 183-201
◇ D20 D25 D98 ◇ REV MR 36 # 6286 Zbl 165.316 • ID 33613

ROGERS JR., H. [1967] *Theory of recursive functions and effective computability* (**X** 0822) McGraw-Hill: New York xx + 482pp
• TRANSL [1972] (**X** 0885) Mir: Moskva 624pp
◇ D25 D30 D55 D98 F15 ◇ REV MR 37 # 61 MR 50 # 4262 Zbl 183.14 JSL 24.70 JSL 36.141 • ID 24823

ROGERS JR., H. see Vol. I, III for further entries

ROGOZHIN, YU.V. [1975] *The immortality problem for Post machines (Russian)* (**J** 0040) Kibernetika, Akad Nauk Ukr SSR 1975/2*1-6
• TRANSL [1975] (**J** 0021) Cybernetics 11*177-182
◇ D10 ◇ REV MR 54 # 12497 Zbl 308 # 02042 • ID 64885

ROGOZHIN, YU.V. [1975] *Unsolvability of the immortality problem for Turing machines with three states (Russian) (English summary insert)* (**J** 0040) Kibernetika, Akad Nauk Ukr SSR 1975/1*41-43
• TRANSL [1975] (**J** 0021) Cybernetics 11*46-59
◇ D10 ◇ REV MR 53 # 10564 Zbl 306 # 02030 • ID 23073

ROGOZHIN, YU.V. [1980] *Universal one-place partial recursive functions (Russian)* (**S** 0166) Mat Issl, Mold SSR 59*36-43,155
◇ D10 D20 ◇ REV MR 82j:03046 Zbl 457 # 03037 • ID 54362

ROGOZHIN, YU.V. [1982] *Seven universal Turing machines (Russian)* (**S** 0166) Mat Issl, Mold SSR 69*76-90
◇ D10 ◇ REV MR 84i:68078 Zbl 515 # 03020 • ID 37843

ROGUSKI, S. [1974] *Degrees of nonconstructibility in Cohen's model* (**J** 0014) Bull Acad Pol Sci, Ser Math Astron Phys 22*1193-1194
◇ D30 E40 E45 ◇ REV MR 54 # 4976 Zbl 306 # 02063 • ID 11367

ROGUSKI, S. see Vol. III, V for further entries

ROHLEDER, H. [1961] *Zum Ausmultiplizieren der Klammern beim Verfahren von Nelson* (J 1046) Z Angew Math Mech 41*77-78
◇ B35 D05 ◇ REV Zbl 103.248 • ID 20926

ROHLEDER, H. see Vol. I, II for further entries

ROIDER, B. [1978] see BUCHBERGER, B.

ROJZEN, S.I. [1975] see BARASHKO, A.S.

ROJZEN, S.I. [1976] see BARASHKO, A.S.

ROJZEN, S.I. [1977] *A counterexample in the theory of automaton equivalence types (Russian)* (C 2925) Diskr Sist, Formal Yazyki & Slozhn Algor ('77) 27-42
◇ D05 D50 ◇ REV MR 58 # 21517 • ID 77869

ROJZEN, S.I. [1977] *An analogue of the additive system of cardinals that is connected with finite automata (Russian) (English summary)* (J 0040) Kibernetika, Akad Nauk Ukr SSR 1977/2*22-29
• TRANSL [1977] (J 0021) Cybernetics 13*166-174
◇ D05 D50 ◇ REV MR 57 # 9362 Zbl 363 # 02040
• ID 50867

ROJZEN, S.I. [1978] *Unsolvability of the additive theory of automata isols over a unary alphabet (Russian)* (C 3722) Diskr Sist, Formal Yazyki & Slozhn Algor ('78) 66-74
◇ D05 D50 ◇ REV MR 81a:03047 • ID 77868

ROLLETSCHEK, H. [1983] *Closure properties of almost-finiteness classes in recursive function theory* (J 0036) J Symb Logic 48*756-763
◇ D50 ◇ REV MR 84j:03088 Zbl 528 # 03026 • ID 34679

ROLLIK, A. [1981] see MEYER AUF DER HEIDE, F.

ROLLIK, H.-A. [1979] see ANTELMANN, H.

ROLLOV, EH.V. [1972] see BEZVERKHNIJ, V.N.

ROLLOV, EH.V. [1974] see BEZVERKHNIJ, V.N.

ROLLOV, EH.V. [1980] *Subsemigroups of a class of semigroups (Russian)* (S 3478) Sovrem Algebra (Leningrad) 1980*113-118
◇ D40 ◇ REV MR 82e:20068 Zbl 453 # 20048 • ID 82638

ROMAN'KOV, V.A. [1969] see CHURKIN, V.A.

ROMAN'KOV, V.A. [1977] *Unsolvability of the endomorphic reducibility problem in free nilpotent groups and in free rings (Russian)* (J 0003) Algebra i Logika 16*457-471
• TRANSL [1977] (J 0069) Algeb and Log 16*310-320
◇ D40 ◇ REV MR 81b:20027 Zbl 411 # 20021 • ID 29227

ROMAN'KOV, V.A. [1979] *Equations in free metabelian groups (Russian)* (J 0092) Sib Mat Zh 20*671-673,694
• TRANSL [1979] (J 0475) Sib Math J 20*469-471
◇ D40 ◇ REV MR 80k:20040 Zbl 419 # 20030
Zbl 427 # 20025 • ID 82640

ROMAN'KOV, V.A. [1979] *Universal theory of nilpotent groups (Russian)* (J 0087) Mat Zametki (Akad Nauk SSSR) 25*487-495,635
• TRANSL [1979] (J 1044) Math Notes, Acad Sci USSR 25*253-258
◇ B25 C60 D35 ◇ REV MR 80j:03058 Zbl 419 # 20031
• ID 53388

ROMAN'KOV, V.A. [1983] see REMESLENNIKOV, V.N.

ROMAN'KOV, V.A. see Vol. III for further entries

ROMANOV, YU.I. [1969] *Separation of projective functions (Russian)* (J 0339) Uch Zap Ped Inst, Volgograd 23*147-152
◇ D55 E15 ◇ REV MR 44 # 1579 • ID 11385

ROMANOV, YU.I. see Vol. V for further entries

ROMANOVSKIJ, N.S. [1969] see CHURKIN, V.A.

ROMANOVSKIJ, N.S. [1974] *Some algorithmic problems for solvable groups (Russian)* (J 0003) Algebra i Logika 13*26-34,121
• TRANSL [1974] (J 0069) Algeb and Log 13*13-16
◇ D40 ◇ REV MR 50 # 10092 Zbl 292 # 20026 • ID 11386

ROMANOVSKIJ, N.S. [1980] see REMESLENNIKOV, V.N.

ROMANOVSKIJ, N.S. [1980] *The elementary theory of an almost polycyclic group (Russian)* (J 0142) Mat Sb, Akad Nauk SSSR, NS 111(153)*135-143,160
• TRANSL [1981] (J 0349) Math of USSR, Sbor 39*125-132
◇ C60 D35 ◇ REV MR 81d:03035 Zbl 424 # 20031
• ID 77873

ROMANOVSKIJ, N.S. [1980] *The embedding problem for abelian-by-nilpotent groups (Russian)* (J 0092) Sib Mat Zh 21/2*170-174,239
• TRANSL [1980] (J 0475) Sib Math J 21*273-276
◇ D40 ◇ REV MR 82c:20069 • ID 82642

ROMANOVSKIJ, N.S. [1982] *The word problem for centrally metabelian groups (Russian)* (J 0092) Sib Mat Zh 23/4*201-205,222
◇ D40 ◇ REV MR 83m:20050 Zbl 496 # 20022 • ID 40411

ROMANOVS'KIJ, V.YU. [1976] *On the solvability of the equivalence problem for a subclass of unary recursion schemes (Russian) (English summary)* (J 0270) Dokl Akad Nauk Ukr SSR, Ser A 1976*885-886
◇ D20 ◇ REV MR 58 # 8456 Zbl 338 # 02020 • ID 64891

ROMANOVS'KIJ, V.YU. [1978] *Undecidable properties of unary recursive schemes over a free group (Ukrainian) (English and Russian summaries)* (J 0270) Dokl Akad Nauk Ukr SSR, Ser A 1978*73-76,96
◇ D05 ◇ REV MR 58 # 5156 Zbl 382 # 03031 • ID 51955

ROMANOVS'KIJ, V.YU. [1980] *Solvability of the equivalence of linear unary recursive schemes with individual constants (Ukrainian) (English and Russian summaries)* (J 0270) Dokl Akad Nauk Ukr SSR, Ser A 1*71-74,96
◇ D20 ◇ REV MR 81d:68060 Zbl 418 # 68050 • ID 53337

ROMOV, B.A. [1971] *The uniprimitive foundations of the maximal subalgebras of Post algebras (Russian) (English summary)* (J 0040) Kibernetika, Akad Nauk Ukr SSR 1971/6*21-30
◇ B50 D03 G20 ◇ REV MR 47 # 8666 Zbl 265 # 02011
• ID 11387

ROMOV, B.A. [1975] *Algorithmically decidable problems that are connected with expressibility in Post algebras of finite degree (Russian)* (C 2962) Mat Voprosy Teor Intell Mashin 71-83
◇ D40 G20 ◇ REV MR 58 # 21598 Zbl 424 # 03033
• ID 77878

ROMOV, B.A. see Vol. II, III, V for further entries

RONCHI DELLA ROCCA, S. [1979] see DEZANI-CIANCAGLINI, M.

RONCHI DELLA ROCCA, S. see Vol. I, VI for further entries

ROOTSELAAR VAN, B. [1962] *Algebraische Kennzeichnung freier Wortarithmetiken* (**J** 0020) Compos Math 15*156-168
◇ D05 ◇ REV MR 29#5722 Zbl 117.14 JSL 30.509
• ID 11395

ROOTSELAAR VAN, B. [1962] *Die Struktur der rekursiven Wortarithmetik des Herrn v. Vucovic* (**J** 0028) Indag Math 24*192-200
◇ D05 ◇ REV MR 26#3597 Zbl 199.26 JSL 29.201
• ID 11394

ROOTSELAAR VAN, B. see Vol. I, II, VI for further entries

ROQUETTE, P. [1984] see CANTOR, DAVID G.

ROQUETTE, P. see Vol. I, III for further entries

ROSE, A. [1958] *Applications of logical computers to the construction of electrical control tables for signalling frames* (**J** 0068) Z Math Logik Grundlagen Math 4*222-243
◇ B70 D05 ◇ REV MR 21#1926 Zbl 93.142 • ID 11444

ROSE, A. [1959] *A note on the representation of general recursive functions and the μ-quantifier* (**J** 0171) Proc Cambridge Phil Soc Math Phys 55*145-148
◇ D20 ◇ REV MR 26#4899 Zbl 88.15 • ID 11449

ROSE, A. [1962] *Remarque sur la machine universelle de Turing* (**J** 0109) C R Acad Sci, Paris 255*2044-2045
◇ D10 ◇ REV MR 26#20 Zbl 253#02032 • ID 11464

ROSE, A. [1962] *Sur les applications de la logique polyvalente a la construction des machines Turing* (**J** 0109) C R Acad Sci, Paris 255*1836-1838
◇ B50 D10 ◇ REV MR 25#4994 Zbl 173.11 • ID 11465

ROSE, A. see Vol. I, II, V, VI for further entries

ROSE, G.F. [1962] see GINSBURG, S.

ROSE, G.F. [1962] *Output completeness in sequential machines* (**J** 0053) Proc Amer Math Soc 13*611-614
◇ D05 ◇ REV MR 25#1989 Zbl 109.241 JSL 31.140
• ID 11503

ROSE, G.F. & ULLIAN, J.S. [1963] *Approximation of functions on the integers* (**J** 0048) Pac J Math 13*693-701
◇ D20 D25 ◇ REV MR 27#4742 Zbl 129.260 • ID 11504

ROSE, G.F. [1963] see GINSBURG, S.

ROSE, G.F. [1965] see GINSBURG, S.

ROSE, G.F. [1966] see GINSBURG, S.

ROSE, G.F. [1968] see GINSBURG, S.

ROSE, G.F. [1970] see GINSBURG, S.

ROSE, G.F. & WEIHRAUCH, K. [1973] *A characterization of the classes L_1 and R_1 of primitive recursive word functions* (**P** 1630) GI Fachtag Automatenth & Form Sprach (1);1973 Bonn 263-266
◇ D05 D20 ◇ REV MR 56#15385 Zbl 283#68031
• ID 64909

ROSE, G.F. [1974] see GINSBURG, S.

ROSE, G.F. [1975] see HENKE VON, F.W.

ROSE, G.F. see Vol. VI for further entries

ROSE, H.E. [1961] *On the consistency and undecidability of recursive arithmetic* (**J** 0068) Z Math Logik Grundlagen Math 7*124-135
◇ D35 F30 ◇ REV MR 25#3833 Zbl 109.10 • ID 11507

ROSE, H.E. [1967] see CLEAVE, J.P.

ROSE, H.E. [1972] *$\mathcal{E}^{\alpha}$-arithmetic and transfinite induction* (**J** 0036) J Symb Logic 37*19-30
◇ D20 F30 ◇ REV MR 54#12500 Zbl 262#02043
• ID 11511

ROSE, H.E. [1984] *Subrecursion. Functions and hierarchies* (**X** 0815) Clarendon Pr: Oxford xiii+191pp
◇ D20 ◇ REV MR 86g:03004 Zbl 539#03018 • ID 45450

ROSE, H.E. see Vol. I, II, VI for further entries

ROSEN, B.K. [1976] see EHRIG, H.

ROSEN, B.K. see Vol. V, VI for further entries

ROSENBERG, ARNOLD L. [1967] *A machine realization of the linear context-free languages* (**J** 0194) Inform & Control 10*175-188
◇ D05 ◇ ID 26680

ROSENBERG, ARNOLD L. [1967] *Multitape finite automata with rewind instructions* (**J** 0119) J Comp Syst Sci 1*299-315
◇ D05 ◇ REV MR 38#1958 Zbl 245#94037 • ID 26679

ROSENBERG, ARNOLD L. [1968] see FISCHER, P.C.

ROSENBERG, ARNOLD L. [1970] see FISCHER, P.C.

ROSENBERG, ARNOLD L. [1972] see FISCHER, P.C.

ROSENBERG, R. [1982] *Recursively enumerable images of arithmetic sets* (**J** 0068) Z Math Logik Grundlagen Math 28*189-201
◇ D25 D55 F30 ◇ REV MR 84e:03074 Zbl 535#03019
• ID 34406

ROSENBLOOM, P.C. [1976] *Structural models for use in psychological research* (**P** 1792) Struct Learning;1968 Philadelphia II*171-178
◇ D03 D80 ◇ REV MR 56#17927 • ID 82652

ROSENBLOOM, P.C. see Vol. I, II, VI for further entries

ROSENFELD, A. [1972] see MILGRAM, D.L.

ROSENFELD, A. [1973] see MILGRAM, D.L.

ROSENFELD, A. [1981] see DYER, C.R.

ROSENFELD, A. see Vol. V for further entries

ROSENKRANTZ, D.J. [1967] *Matrix equations and normal forms for context-free grammars* (**J** 0037) ACM J 14*501-507
◇ D05 ◇ REV MR 38#3100 Zbl 148.251 • ID 26688

ROSENKRANTZ, D.J. [1978] see HUNT III, H.B.

ROSENKRANTZ, D.J. [1983] see HUNT III, H.B.

ROSENKRANTZ, D.J. [1984] see HUNT III, H.B.

ROSENKRANTZ, D.J. [1985] see HUNT III, H.B.

ROSENSTEIN, J.G. [1968] *Initial segments of degrees* (**J** 0048) Pac J Math 24*163-172
◇ D30 ◇ REV MR 37#1241 Zbl 153.316 • ID 11529

ROSENSTEIN, J.G. [1972] see MANASTER, A.B.

ROSENSTEIN, J.G. [1973] see MANASTER, A.B.

ROSENSTEIN, J.G. [1975] see HAY, L.

ROSENSTEIN, J.G. [1977] see HAY, L.

ROSENSTEIN, J.G. [1980] see MANASTER, A.B.

ROSENSTEIN, J.G. [1982] see LERMAN, M.

ROSENSTEIN, J.G. [1984] *Recursive linear orderings (French summary)* (**P** 2167) Orders: Descr & Roles;1982 L'Arbresle 465–475
◇ C57 C65 D45 ◇ REV MR 86d:06002 Zbl 554 # 06001 • ID 41775

ROSENSTEIN, J.G. see Vol. III, V for further entries

ROSENTHAL, J.W. [1980] see ASH, C.J.

ROSENTHAL, J.W. [1982] see PLOTKIN, J.M.

ROSENTHAL, J.W. see Vol. III, V for further entries

ROSIER, L.E. [1981] see IBARRA, O.H.

ROSIER, L.E. [1983] see IBARRA, O.H.

ROSIER, L.E. [1984] see IBARRA, O.H.

ROSIER, L.E. [1985] see BORM, A.E.

ROSSER, J.B. [1936] *Extensions of some theorems of Goedel and Church* (**J** 0036) J Symb Logic 1*87–91
◇ D25 D35 F30 ◇ REV JSL 2.52 FdM 62.1058 • ID 11542

ROSSER, J.B. [1936] see CHURCH, A.

ROSSER, J.B. [1937] *Goedel theorems for non-constructive logics* (**J** 0036) J Symb Logic 2*129–137 • ERR/ADD ibid 2*IV
◇ D70 F30 ◇ REV Zbl 17.242 JSL 3.50 FdM 63.824 • ID 11544

ROSSER, J.B. [1939] *An informal exposition of proofs of Goedel's theorems and Church's theorem* (**J** 0036) J Symb Logic 4*53–60
◇ A05 B20 D35 F30 ◇ REV Zbl 22.292 JSL 4.165 FdM 65.27 • ID 11547

ROSSER, J.B. [1955] *Deux esquisses de logique* (**X** 0834) Gauthier-Villars: Paris 69pp
◇ B10 B40 D20 E30 ◇ REV MR 16.661 Zbl 64.7 • ID 23483

ROSSER, J.B. [1955] *Logique combinatoire et λ-conversion* (**C** 4157) Rosser:Deux Esquisses Log 3–31
◇ B40 D20 ◇ REV JSL 22.293 • ID 42147

ROSSER, J.B. [1957] *The relation between Turing machines and actual computing machines* (**P** 1675) Summer Inst Symb Log;1957 Ithaca 105
◇ D10 ◇ ID 29326

ROSSER, J.B. see Vol. I, II, III, V, VI for further entries

ROTA, G.-C. [1974] see METROPOLIS, N.

ROTA, G.-C. see Vol. I, III, VI for further entries

ROTHBERGER, F. [1958] *Example effectif d'un ensemble transfiniment non-projectif* (**J** 0017) Canad J Math 10*554–560
◇ D55 E15 ◇ REV MR 20 # 3789 Zbl 85.39 • ID 11580

ROTHBERGER, F. see Vol. V for further entries

ROTMAN, J.R. [1965] *The theory of groups. An introduction* (**X** 0802) Allyn & Bacon: London xiii + 305pp
◇ D40 ◇ REV MR 34 # 4338 Zbl 123.20 JSL 32.127 • ID 11591

ROUHONEN, K. [1984] *On machine characterization of nonrecursive hierarchies* (**J** 0498) Ann Univ Turku, Ser A I 186*87–101
◇ D10 D55 ◇ REV MR 85h:03047 Zbl 562 # 03023 • ID 43310

ROUTLEDGE, N.A. [1953] *Ordinal recursion* (**J** 0171) Proc Cambridge Phil Soc Math Phys 49*175–182
◇ D20 F15 ◇ REV MR 14.714 Zbl 52.13 JSL 24.69 • ID 11604

ROUTLEDGE, N.A. [1955] *Concerning definable sets* (**J** 0027) Fund Math 41*6–11
◇ D30 D55 ◇ REV MR 16.555 Zbl 57.8 JSL 24.69 • ID 11605

ROUTLEDGE, N.A. see Vol. I for further entries

ROUTLEY, R. [1966] see GODDARD, L.

ROUTLEY, R. [1973] see MEYER, R.K.

ROUTLEY, R. see Vol. I, II, III, V, VI for further entries

ROVAN, B. [1969] *Bounded push down automata (Czech summary)* (**J** 0156) Kybernetika (Prague) 5*261–265
◇ D05 ◇ REV MR 40 # 1220 Zbl 184.285 • ID 30843

ROVAN, B. [1974] see GINSBURG, S.

ROVAN, B. [1981] *A framework for studying grammars* (**P** 3429) Math Founds of Comput Sci (10);1981 Strbske Pleso 473–482
◇ D05 ◇ REV MR 83e:68118 Zbl 491 # 68076 • ID 40168

ROWAN, J.H. [1973] see KASHEF, R.S.

ROWICKI, A. [1965] *Remarks on order of simple processes* (**J** 0014) Bull Acad Pol Sci, Ser Math Astron Phys 13*55–60
◇ D10 D15 ◇ REV MR 30 # 5516 Zbl 163.398 • ID 30845

ROWICKI, A. [1967] *On the notion of occupancy of the tree* (**J** 0014) Bull Acad Pol Sci, Ser Math Astron Phys 15*283–288
◇ D05 D15 ◇ REV MR 35 # 7731 Zbl 171.149 • ID 11628

ROWICKI, A. [1969] *On minimal and maximal orders of simple processes* (**J** 0068) Z Math Logik Grundlagen Math 15*359–384
◇ D05 ◇ REV MR 44 # 3873 Zbl 194.484 • ID 11630

ROWICKI, A. [1969] *Ueber ein Kompliziertheitsmass einfacher Prozesse* (**J** 0129) Elektr Informationsverarbeitung & Kybern 5*147–152
◇ D05 ◇ REV MR 40 # 7051 Zbl 185.25 • ID 11629

ROWICKI, A. see Vol. I for further entries

ROY, D.K. [1983] *R.e. presented linear orders* (**J** 0036) J Symb Logic 48*369–376
◇ C57 C65 D45 E07 ◇ REV MR 85b:03077 Zbl 525 # 03032 • ID 38265

ROY, D.K. [1985] *Linear order types of nonrecursive presentability* (**J** 0068) Z Math Logik Grundlagen Math 31*495–501
◇ C57 C65 D45 ◇ ID 47798

ROZENBERG, G. [1967] *Decision problems for quasi-uniform events (Russian summary)* (**J** 0014) Bull Acad Pol Sci, Ser Math Astron Phys 15*745–752
◇ D05 ◇ REV MR 37 # 1249 Zbl 174.34 • ID 11513

ROZENBERG, G. [1968] *Some remarks on Rabin and Scott's notion of multitape automaton (Russian summary)* (**J** 0014) Bull Acad Pol Sci, Ser Math Astron Phys 16*215–218
◇ D05 ◇ REV MR 37 # 5099 Zbl 205.313 • ID 11638

ROZENBERG, G. [1969] *p-automata and p-events* (**J** 0014) Bull Acad Pol Sci, Ser Math Astron Phys 17*565–570
◇ D05 ◇ REV MR 41 # 1448 Zbl 211.314 • ID 30848

ROZENBERG, G. [1969] *Finite memory address machines are universal* (**J** 0014) Bull Acad Pol Sci, Ser Math Astron Phys 17*401-403
◇ D05 ◇ REV MR 40 # 7043 Zbl 186.12 • ID 11639

ROZENBERG, G. [1969] *On the introduction of orderings into the grammars of Chomsky's hierarchy* (**J** 0014) Bull Acad Pol Sci, Ser Math Astron Phys 17*559-563
◇ B65 D05 ◇ REV MR 41 # 6638 Zbl 193.326 • ID 30849

ROZENBERG, G. [1971] *The unsolvability of the isomorphism problem for address machines* (**J** 0060) Rev Roumaine Math Pures Appl 16*1553-1558
◇ D05 ◇ REV MR 46 # 1571 Zbl 254 # 02035 • ID 11640

ROZENBERG, G. [1972] *The equivalence problem for deterministic TOL-systems is undecidable* (**J** 0232) Inform Process Lett 1*201-204
◇ D03 ◇ REV Zbl 267 # 68033 • ID 64949

ROZENBERG, G. [1975] see GINSBURG, S.

ROZENBERG, G. & SALOMAA, A. [1976] *Context-free grammars with graph-controlled tables* (**J** 0119) J Comp Syst Sci 13*90-99
◇ D05 ◇ REV MR 55 # 6980 Zbl 338 # 68061 • ID 31583

ROZENBERG, G. & RUOHONEN, K. & SALOMAA, A. [1976] *Developmental systems with fragmentation* (**J** 0382) Int J Comput Math 5*177-191
◇ D05 ◇ REV MR 55 # 11731 Zbl 343 # 68040 • ID 31584

ROZENBERG, G. & SALOMAA, A. [1976] *The mathematical theory of L-systems* (**J** 1937) Adv Inform Syst Sci 6*161-206
◇ D05 ◇ REV MR 57 # 11196 Zbl 365 # 68072 • ID 31495

ROZENBERG, G. & SALOMAA, A. [1977] *New squeezing mechanisms for L systems* (**J** 0191) Inform Sci 12*187-201
◇ A05 B75 D05 ◇ REV MR 80e:68205 Zbl 363 # 68095 • ID 31588

ROZENBERG, G. [1977] see EHRENFEUCHT, A.

ROZENBERG, G. [1978] see PENTTONEN, M.

ROZENBERG, G. [1978] see EHRENFEUCHT, A.

ROZENBERG, G. [1978] see MAURER, H.A.

ROZENBERG, G. [1978] see ALBERT, J.

ROZENBERG, G. [1979] see EHRENFEUCHT, A.

ROZENBERG, G. [1979] see ENGELFRIET, J.

ROZENBERG, G. [1979] see MEERSMAN, R.

ROZENBERG, G. [1979] see MAURER, H.A.

ROZENBERG, G. [1980] *A survey of results and open problems in the mathematical theory of L systems* (**P** 4266) Form Lang Th;1979 Santa Barbara 195-240
◇ D05 ◇ REV MR 84j:68001 Zbl 545 # 68065 • ID 47203

ROZENBERG, G. [1980] see EHRENFEUCHT, A.

ROZENBERG, G. [1980] see ENGELFRIET, J.

ROZENBERG, G. [1980] see ALBERT, J.

ROZENBERG, G. [1981] see EHRENFEUCHT, A.

ROZENBERG, G. [1981] see JANSSENS, D.

ROZENBERG, G. [1981] *On subwords of formal languages* (**P** 3165) FCT'81 Fund of Comput Th;1981 Szeged 328-333
◇ D05 ◇ REV MR 83i:68118 Zbl 481 # 68074 • ID 39370

ROZENBERG, G. [1981] see MAURER, H.A.

ROZENBERG, G. [1982] see EHRENFEUCHT, A.

ROZENBERG, G. [1984] see LATTEUX, M.

ROZENBERG, G. [1984] see EHRENFEUCHT, A.

ROZENBERG, G. [1985] see EHRENFEUCHT, A.

ROZENBLAT, B.V. [1979] *Positive theories of free inverse semigroups (Russian)* (**J** 0092) Sib Mat Zh 20*1282-1293,1408
• TRANSL [1979] (**J** 0475) Sib Math J 20*910-918
◇ B20 C05 D35 D40 ◇ REV MR 80m:03079 Zbl 431 # 20045 • ID 53949

ROZENBLAT, B.V. & VAZHENIN, YU.M. [1981] *Decidability of the positive theory of a free countably generated semigroup (Russian)* (**J** 0142) Mat Sb, Akad Nauk SSSR, NS 116(158)*120-127
• TRANSL [1983] (**J** 0349) Math of USSR, Sbor 44*109-116
◇ B25 C05 D35 ◇ REV MR 83b:03014 Zbl 472 # 20020 • ID 35085

ROZENBLAT, B.V. see Vol. I, III for further entries

ROZINAS, M.G. [1972] *Algebra of multiple primitive recursive functions (Russian)* (**J** 0226) Uch Zap Ped Inst, Ivanovo 117*95-111
◇ D20 ◇ ID 90039

ROZINAS, M.G. [1974] *Partial degrees and r-degrees (Russian)* (**J** 0092) Sib Mat Zh 15*1323-1331,1431
• TRANSL [1974] (**J** 0475) Sib Math J 15*935-941
◇ D30 ◇ REV MR 58 # 10372 Zbl 358 # 02059 • ID 50513

ROZINAS, M.G. [1974] see POLYAKOV, E.A.

ROZINAS, M.G. [1976] see POLYAKOV, E.A.

ROZINAS, M.G. [1976] *Jump operation for certain kinds of reducibilities (Russian)* (**X** 2235) VINITI: Moskva 3185-76*31pp
◇ D30 ◇ REV Zbl 348 # 02040 • REM An abstract in J0092 18(1977)/1*234 • ID 90041

ROZINAS, M.G. [1977] see POLYAKOV, E.A.

ROZINAS, M.G. [1977] *Minimal non-recursively enumerable pm-degrees (Russian)* (**X** 2235) VINITI: Moskva 2998-77*7pp
◇ D30 ◇ ID 90042

ROZINAS, M.G. [1978] see POLYAKOV, E.A.

ROZINAS, M.G. [1978] *On the semilattice of e-degrees (Russian)* (**C** 2595) Rekursiv Funktsii 71-84
◇ D30 ◇ REV MR 82i:03057 Zbl 504 # 03020 • ID 90043

ROZINAS, M.G. [1978] *Partial degrees of immune and hyperimmune sets (Russian)* (**J** 0092) Sib Mat Zh 19*866-870,955
• TRANSL [1978] (**J** 0475) Sib Math J 19*613-616
◇ D30 D50 ◇ REV MR 58 # 10373 Zbl 394 # 03040 • ID 52517

ROZINAS, M.G. [1979] *On the representation of e-degrees in the form of greatest lower bound of two incomparable e-degrees (Russian)* (**P** 2558) All-Union Conf Math Log (5) (Mal'tsev);1979 Novosibirsk 129
◇ D30 ◇ ID 90044

ROZINAS, M.G. & SOLON, B.YA. [1979] *Weakly semirecursive sets (Russian)* (J 0031) Izv Vyssh Ucheb Zaved, Mat (Kazan) 1979/12*48-50
• TRANSL [1979] (J 3449) Sov Math 23/12*50-52
◇ D20 D25 D30 ◇ REV MR 81d:03046 Zbl 436 # 03043 • ID 55639

ROZINAS, M.G. [1981] *Expansion of partial degrees into r-degrees (Russian)* (C 3865) Algeb Sistemy (Ivanovo) 195-197
◇ D30 ◇ REV MR 85f:03040 Zbl 538 # 03039 • ID 40733

ROZONOEHR, L.I. [1961] see AJZERMAN, M.A.

ROZONOEHR, L.I. [1963] see AJZERMAN, M.A.

ROZONOEHR, L.I. see Vol. II for further entries

RUBALD, C.M. [1971] see CAVINESS, B.F.

RUBEL, L.A. [1977] see HENSON, C.W.

RUBEL, L.A. [1980] see BECKER, J.A.

RUBEL, L.A. see Vol. I, III, V, VI for further entries

RUBIN, M. [1976] *The theory of boolean algebras with a distinguished subalgebra is undecidable* (J 1934) Ann Sci Univ Clermont Math 13*129-134
◇ D35 G05 G25 ◇ REV MR 57 # 5721 Zbl 354 # 02036 • ID 50176

RUBIN, M. [1983] *A Boolean algebra with few subalgebras, interval algebras and retractiveness* (J 0064) Trans Amer Math Soc 278*65-89
◇ C55 C80 D35 E50 E65 G05 ◇ REV MR 85a:06024 Zbl 524 # 06020 • ID 33531

RUBIN, M. see Vol. I, III, V for further entries

RUBY, S.S. [1965] see FISCHER, P.C.

RUDNEV, V.V. [1984] see IVANOV, N.N.

RUMELY, R. [1980] *Undecidability and definability for the theory of global fields* (J 0064) Trans Amer Math Soc 262*195-217
◇ C40 C60 D35 ◇ REV MR 81m:03053 Zbl 472 # 03010 • ID 55276

RUMSAS, A. [1983] *On the complexity of computations over the ring of integers (Russian) (English summary)* (J 3939) Mat Logika Primen (Akad Nauk Litov SSR) 3*101-103
◇ D15 ◇ REV Zbl 558 # 03020 • ID 45166

RUOHONEN, K. [1972] *Hilbert's tenth problem (Finnish) (English summary)* (J 1108) Arkhimedes (Helsinki) 1972*71-100
◇ D25 D35 ◇ REV MR 58 # 190 Zbl 258 # 02047 • ID 64969

RUOHONEN, K. [1976] see ROZENBERG, G.

RUOHONEN, K. [1978] see CULIK II, K.

RUOHONEN, K. [1979] *On some decidability problems for HDOL systems with nonsingular Parikh matrices* (J 1426) Theor Comput Sci 9*377-384
◇ D05 ◇ REV MR 81i:68114 Zbl 414 # 68044 • ID 53098

RUOHONEN, K. [1979] *On the decidability of the OL-DOL equivalence problem* (J 0194) Inform & Control 40*301-318
◇ D05 ◇ REV MR 80e:68197 Zbl 412 # 68062 • ID 52983

RUOHONEN, K. [1980] *Hilbert's tenth problem (Swedish) (English summary)* (J 3075) Normat 28*145-154,180
◇ D25 D35 ◇ REV MR 81m:10031 Zbl 445 # 10001 • ID 56510

RUOHONEN, K. [1981] see EHRENFEUCHT, A.

RUOHONEN, K. [1981] *The decidability of the DOL-DTOL equivalence problem* (J 0119) J Comp Syst Sci 22*42-52
◇ B75 D05 ◇ REV MR 83c:68098 Zbl 491 # 68048 • ID 37765

RUOHONEN, K. [1983] *On some variants of Post's correspondence problem* (J 1431) Acta Inf 19*357-367
◇ D03 D05 ◇ REV MR 85d:03082 Zbl 537 # 68080 • ID 41101

RUOHONEN, K. [1984] *A note on off-line machines with "Brownian" input heads* (J 2702) Discr Appl Math 9*69-75
◇ D10 ◇ REV MR 86a:68025 Zbl 553 # 03026 • ID 43327

RUOHONEN, K. [1985] *Reversible machines and Post's correspondence problem for biprefix morphisms (German and Russian summaries)* (J 0129) Elektr Informationsverarbeitung & Kybern 21*579-595
◇ D03 D05 ◇ ID 49645

RUS, T. [1964] *Ueber ein formales System I,II (Rumaenisch)* (J 0197) Stud Cercet Mat Acad Romana 15*459-470,595-615
◇ B10 D20 ◇ REV MR 32 # 3991 Zbl 199.3 • ID 16208

RUS, T. [1967] *Algebraic treatment of formalized languages (Romanian)* (J 0197) Stud Cercet Mat Acad Romana 19*259-272
◇ D05 G15 ◇ REV MR 39 # 2549 Zbl 149.249 • ID 30866

RUS, T. [1967] *The algebra of formalized languages (Romanian)* (J 0197) Stud Cercet Mat Acad Romana 19*1309-1324
◇ D05 G15 ◇ REV MR 40 # 5362 Zbl 227 # 68034 • ID 30867

RUS, T. [1976] see HATCHER, W.S.

RUS, T. [1980] *HAS-Hierarchy; a natural tool for language specification* (J 2095) Fund Inform, Ann Soc Math Pol, Ser 4 3*269-293
◇ D05 ◇ REV MR 82k:68046 Zbl 452 # 68087 • ID 82675

RUS, T. see Vol. I for further entries

RUSAKOV, E.D. [1972] see MASLOV, S.YU.

RUSIECKI, A. [1975] *Computability of recursive-defined functions by executing the program, equivalent to their definitions (Polish) (English and Russian summaries)* (J 1929) Prace Centr Oblicz Pol Akad Nauk 184*76pp
◇ B75 D05 ◇ REV Zbl 315 # 02038 • ID 29811

RUSSO, D.A. [1985] see ORPONEN, P.

RUSTIN, R. (ED.) [1973] *Computational complexity. Courant computer science symposium* (X 4582) Algorithms: New York 268pp
◇ D15 D97 ◇ ID 48713

RUTKOWSKI, A. [1972] see KRAJEWSKI, S.

RUTKOWSKI, A. [1974] *On the algebraic approach to the notion of α-recursive functions* (J 0014) Bull Acad Pol Sci, Ser Math Astron Phys 22*993-995
◇ D60 ◇ REV MR 51 # 10059 Zbl 298 # 02039 • ID 11697

RUTKOWSKI, A. see Vol. I, II, III, V for further entries

RUTLEDGE, J.D. [1964] see ELGOT, C.C.

RUTLEDGE, J.D. [1967] see ELGOT, C.C.

RUTLEDGE, J.D. see Vol. II for further entries

RUZICKA, M. [1982] *Input-output systems, their types and applications for the automata theory* (J 0156) Kybernetika (Prague) 18*131-144
◇ B75 D05 ◇ REV MR 83m:68097 Zbl 495 # 93004
• ID 38524

RUZZO, W.L. [1978] see ADLEMAN, L.M.

RUZZO, W.L. [1981] *On uniform circuit complexity* (J 0119) J Comp Syst Sci 22*365-383
◇ D10 D15 ◇ REV MR 82j:68024 Zbl 462 # 68013
• ID 69879

RYAN, B.F. [1975] *ω-cohesive sets* (J 0064) Trans Amer Math Soc 202*161-171
◇ D50 ◇ REV MR 51 # 103 Zbl 301 # 02039 • ID 17605

RYAN, W.J. [1978] *Goedel's second incompleteness theorem for general recursive arithmetic* (J 0068) Z Math Logik Grundlagen Math 24*457-459
◇ D20 F30 ◇ REV MR 80b:03092 Zbl 427 # 03031
• ID 53718

RYAN, W.J. see Vol. I, VI for further entries

RYLL-NARDZEWSKI, C. [1958] see GRZEGORCZYK, A.

RYLL-NARDZEWSKI, C. [1961] see GRZEGORCZYK, A.

RYLL-NARDZEWSKI, C. [1970] see MAITRA, A.

RYLL-NARDZEWSKI, C. see Vol. I, III, V for further entries

RYSTSOV, I.K. [1983] *Polynomial complete problems in automata theory* (J 0232) Inform Process Lett 16*147-151
◇ D05 D15 ◇ REV MR 84g:68031 Zbl 508 # 68030
• ID 38157

RYTTER, W. [1980] *Automata theory and complexity* (S 3270) Spraw Inst Inf, Uniw Warsaw 91*40pp
◇ D05 D15 D98 ◇ REV Zbl 471 # 68030 • ID 69880

RYTTER, W. [1982] *A note on two-way nondeterministic pushdown automata* (J 0232) Inform Process Lett 15*5-9
◇ D10 ◇ REV MR 83m:68098 Zbl 496 # 68035 • ID 40423

RYTTER, W. [1983] *Remarks on trace languages* (S 3382) Sem-ber, Humboldt-Univ Berlin, Sekt Math 52*119-123
◇ D05 ◇ REV MR 85f:68003 Zbl 534 # 68054 • ID 38362

RYTTER, W. [1985] *On the recognition of context-free languages* (P 4670) Comput Th (5);1984 Zaborow 318-325
◇ D05 ◇ ID 49646

RYTTER, W. [1985] *The complexity of two-way pushdown automata and recursive programs* (P 4626) Combin Algor on Words;1984 Maratea 341-356
◇ D10 D15 ◇ ID 49464

SAALFELD, D. [1977] see BRANDSTAEDT, A.

SABADINI, N. [1984] see BERTONI, A.

SABBAGH, G. [1971] *Logique mathematique V. Decidabilite et fonctions recursives* (C 1495) Encycl Universalis 10*71-73
◇ B25 D20 D35 ◇ REV JSL 38.341 • REM Part IV 1971
• ID 28610

SABBAGH, G. [1976] *Caracterisation algebrique des groupes de type fini ayant un probleme de mots resoluble (Theoreme de Boone-Higman, travaux de B. H. Neumann et Macintyre)* (S 1567) Semin Bourbaki Exp.457*61-80
◇ D40 ◇ REV MR 56 # 5258 Zbl 366 # 20023 • REM Springer: Heidelberg & New York; Lecture Notes Math 514
• ID 51142

SABBAGH, G. [1985] see DELON, F.

SABBAGH, G. see Vol. I, II, III, V for further entries

SABEL'FEL'D, V.K. [1979] *Polynomial estimate of the complexity of the recognition of logic-term equivalence (Russian)* (J 0023) Dokl Akad Nauk SSSR 249*793-796
• TRANSL [1979] (J 0470) Sov Phys, Dokl 24*954-956
◇ B75 D15 D20 ◇ REV MR 81g:68072 Zbl 443 # 68031
• ID 82682

SABEL'FEL'D, V.K. [1980] *The logic-termal equivalence is polynomial-time decidable* (J 0232) Inform Process Lett 10*57-62
◇ B75 D15 ◇ REV MR 81a:68039 Zbl 443 # 68032
• ID 69882

SABEL'FEL'D, V.K. [1981] *Tree equivalence of linear recursive schemata is polynomial-time decidable* (J 0232) Inform Process Lett 13*147-153
◇ B75 D15 ◇ REV MR 83e:68044 Zbl 479 # 68052
• ID 55700

SABIDUSSI, G. [1980] see CHAPUT, D.

SABIDUSSI, G. [1981] see FOELDES, S.

SACERDOTE, G.S. [1972] *On a problem of Boone* (J 0132) Math Scand 31*111-117
◇ C60 D30 D35 D40 ◇ REV MR 47 # 6871
Zbl 255 # 20022 • ID 30879

SACERDOTE, G.S. [1972] *Some undecidable problems in group theory* (J 0053) Proc Amer Math Soc 36*231-238
◇ D40 ◇ REV MR 47 # 8660 Zbl 259 # 02036 • ID 11735

SACERDOTE, G.S. [1976] *A characterization of the subgroups of finitely presented groups* (J 0015) Bull Amer Math Soc 82*609-611
◇ C60 D40 ◇ REV MR 54 # 2814 Zbl 377 # 20028
• ID 24067

SACERDOTE, G.S. [1976] see FRIED, M.

SACERDOTE, G.S. [1977] *Subgroups of finitely presented groups* (J 3240) Proc London Math Soc, Ser 3 35*193-212
◇ D40 ◇ REV MR 56 # 5728 Zbl 373 # 20041 • ID 51518

SACERDOTE, G.S. [1977] *The Boone-Higman theorem and the conjugacy problem* (J 0032) J Algeb 49*212-221
◇ D40 ◇ REV MR 56 # 8350 Zbl 401 # 20030 • ID 78045

SACERDOTE, G.S. see Vol. III for further entries

SACKS, G.E. [1961] *A minimal degree less than O'* (J 0015) Bull Amer Math Soc 67*416-419
◇ D30 ◇ REV MR 23 # A3676 Zbl 101.12 JSL 34.295
• ID 11740

SACKS, G.E. [1961] *On suborderings of degrees of recursive unsolvability* (J 0068) Z Math Logik Grundlagen Math 7*46-56
◇ D30 ◇ REV MR 24 # A1820 Zbl 118.252 JSL 29.203
• ID 11741

SACKS, G.E. [1963] *Degrees of unsolvability* (J 0120) Ann of Math, Ser 2 55*xii + 174pp
◇ D25 D30 D98 E75 ◇ REV MR 32 # 4013 Zbl 143.253
JSL 29.202 • REM 2nd edition 1966;171pp • ID 21256

SACKS, G.E. [1963] *On the degrees less than 0'* (J 0120) Ann of Math, Ser 2 77*211-231 • ERR/ADD ibid 78*204
◇ D25 D30 ◇ REV MR 26 # 3604 Zbl 118.251 JSL 29.60
• ID 19607

SACKS, G.E. [1963] *Recursive enumerability and the jump operator* (**J** 0064) Trans Amer Math Soc 108*223-239
◇ D25 D30 ◇ REV MR 27#5681 Zbl 118.251 JSL 29.204
• ID 11742

SACKS, G.E. [1964] *A maximal set which is not complete* (**J** 0133) Michigan Math J 11*193-205
◇ D25 ◇ REV MR 29#3368 Zbl 135.249 JSL 32.528
• ID 11744

SACKS, G.E. [1964] *A simple set which is not effectively simple* (**J** 0053) Proc Amer Math Soc 15*51-55
◇ D25 ◇ REV MR 28#24 Zbl 154.257 • ID 11743

SACKS, G.E. [1964] *The recursively enumerable degrees are dense* (**J** 0120) Ann of Math, Ser 2 80*300-312
◇ D25 ◇ REV MR 29#3367 Zbl 135.7 JSL 34.294
• ID 11745

SACKS, G.E. [1965] see KREISEL, G.

SACKS, G.E. [1966] *Metarecursively enumerable sets and admissible ordinals* (**J** 0015) Bull Amer Math Soc 72*59-64
◇ D30 D55 D60 ◇ REV MR 35#6542 Zbl 149.247
JSL 34.115 • ID 11746

SACKS, G.E. [1966] *Post's problem, admissible ordinals, and regularity* (**J** 0064) Trans Amer Math Soc 124*1-23
◇ D30 D55 D60 ◇ REV MR 34#1183 Zbl 149.247
JSL 34.115 • ID 11747

SACKS, G.E. [1967] see GANDY, R.O.

SACKS, G.E. [1967] *Measure-theoretic uniformity in recursion theory and set theory (summary of results)* (**J** 0015) Bull Amer Math Soc 73*169-174
◇ D30 D55 D60 E25 E35 E40 E50 ◇ REV
MR 35#4098 Zbl 164.315 • REM Summary. See alsa 1969
• ID 16821

SACKS, G.E. [1967] *Metarecursion theory* (**P** 0691) Sets, Models & Recursion Th;1965 Leicester 243-263
◇ D30 D55 D60 ◇ REV MR 40#7109 Zbl 189.10
JSL 34.115 • ID 11748

SACKS, G.E. [1967] *On a theorem of Lachlan and Martin* (**J** 0053) Proc Amer Math Soc 18*140-141
◇ D25 D30 ◇ REV MR 34#7373 Zbl 154.6 JSL 32.529
• ID 11749

SACKS, G.E. [1969] *Measure-theoretic uniformity* (**C** 0705) Found of Math (Goedel) 51-57
◇ D30 D55 E15 E25 E35 E40 E50 ◇ REV
MR 39#4004 Zbl 172.295 • ID 14481

SACKS, G.E. [1969] *Measure-theoretic uniformity in recursion theory and set theory* (**J** 0064) Trans Amer Math Soc 142*381-420
◇ D30 D55 D60 E15 E25 E35 E40 E50 ◇ REV
MR 40#7108 Zbl 209.16 • REM For a summary see 1967
• ID 16822

SACKS, G.E. [1971] *F-recursiveness* (**P** 0638) Logic Colloq;1969 Manchester 289-303
◇ D30 D55 D60 D75 E40 E45 ◇ REV MR 43#6097
Zbl 298#02036 • ID 11750

SACKS, G.E. [1971] *On the reducibility of Π_1^1 sets* (**J** 0345) Adv Math 7*57-82
◇ D30 D55 D60 E40 ◇ REV MR 45#4976
Zbl 223#02045 • ID 11752

SACKS, G.E. [1971] *Recursion in objects of finite type* (**P** 0743) Int Congr Math (II,11,Proc);1970 Nice 1*251-254
◇ D60 D65 ◇ REV MR 56#15391 Zbl 226#02040
JSL 39.343 • ID 11753

SACKS, G.E. [1972] see LERMAN, M.

SACKS, G.E. & SIMPSON, S.G. [1972] *The α-finite injury method* (**J** 0007) Ann Math Logic 4*343-367
◇ D30 D60 ◇ REV MR 51#5277 Zbl 262#02037
• ID 11754

SACKS, G.E. [1974] *The 1-section of a type n object* (**P** 0602) Generalized Recursion Th (1);1972 Oslo 81-96
◇ D55 D65 ◇ REV MR 53#2662 Zbl 287#02026
• ID 14512

SACKS, G.E. [1976] *Countable admissible ordinals and hyperdegrees* (**J** 0345) Adv Math 20*213-262
◇ D30 D55 D60 ◇ REV MR 55#2536 Zbl 439#03027
• ID 56018

SACKS, G.E. [1976] see ABRAMSON, F.G.

SACKS, G.E. [1977] see FRIEDMAN, S.D.

SACKS, G.E. [1977] *RE sets higher up* (**P** 1704) Int Congr Log, Meth & Phil of Sci (5);1975 London ON 1*173-194
◇ D25 D60 D65 ◇ REV MR 57#12196 Zbl 376#02029
• ID 51671

SACKS, G.E. [1977] *The k-section of a type n object* (**J** 0100) Amer J Math 99*901-917
◇ D65 ◇ REV MR 58#21543 Zbl 373#02033 • ID 51497

SACKS, G.E. [1978] see FENSTAD, J.E.

SACKS, G.E. [1980] *Post's problem, absoluteness and recursion in finite types* (**P** 2058) Kleene Symp;1978 Madison 201-222
◇ D30 D65 ◇ REV MR 82g:03086 Zbl 472#03032
• ID 55297

SACKS, G.E. [1980] *Three aspects of recursive enumerability in higher types* (**P** 3021) Logic Colloq;1979 Leeds 184-214
◇ D30 D65 ◇ REV MR 83d:03056 Zbl 469#03030
• ID 55157

SACKS, G.E. [1983] see HOMER, S.

SACKS, G.E. [1985] *Post's problem in E-recursion* (**P** 4046) Rec Th;1982 Ithaca 177-193
◇ D65 ◇ ID 46377

SACKS, G.E. [1985] see EBBINGHAUS, H.-D.

SACKS, G.E. [1985] *Some open questions in recursion theory* (**P** 3342) Rec Th Week;1984 Oberwolfach 333-342
◇ C13 D98 E60 ◇ ID 42739

SACKS, G.E. see Vol. III, V for further entries

SAGER, H. [1979] *The nature of reduction in Turing machines and its relation to reduction in automata* (**S** 2787) Math Colloq, Univ Cape Town 12*43-54
◇ D05 D10 ◇ REV MR 81g:03044 Zbl 429#68057
• ID 69883

SAHNI, S.K. [1975] see IBARRA, O.H.

SAHNI, S.K. [1980] see CONSTABLE, R.L.

SAITTA, L. [1979] see DEZANI-CIANCAGLINI, M.

SAJO, A. [1984] *On subword complexity functions* (**J** 2702) Discr Appl Math 8*209-212
◇ D05 ◇ REV MR 86a:68059 Zbl 562#68060 • ID 48792

SAKAI, H. [1974] *On numerations of a formal system* (**J** 0260) Ann Jap Ass Phil Sci 4*227-230
◇ D45 F30 ◇ REV MR 56#100 Zbl 291#02023 • ID 65025

SAKAI, H. see Vol. I, VI for further entries

SAKAROVITCH, JACQUES [1977] *Sur les groupes infinis, considerес comme monoides syntaxiques de langages formels* (**S** 2250) Semin Dubreil: Algebre 29(1975/76)*168-179
◇ D05 D40 ◇ REV MR 58#32114 Zbl 395#68067 • ID 69885

SAKAROVITCH, JACQUES [1981] *Sur une propriete d'iteration des langages algebriques deterministes* (**J** 0041) Math Syst Theory 14*247-288
◇ D05 ◇ REV MR 82f:68032 Zbl 472#68042 • ID 82692

SAKODA, W.J. [1977] see BLUM, M.

SALEMI, S. [1985] see RESTIVO, A.

SALOMAA, A. [1964] *Axiom systems for regular expressions of finite automata* (**J** 0498) Ann Univ Turku, Ser A I 75*29pp
◇ D05 ◇ REV MR 29#5726 Zbl 127.10 • ID 11789

SALOMAA, A. [1964] *On the reducibility of events represented in automata* (**J** 3994) Ann Acad Sci Fennicae Ser A I 353*16pp
◇ D05 ◇ REV MR 29#5725 Zbl 168.260 • ID 11788

SALOMAA, A. [1965] *On probabilistic automata with one input letter* (**J** 0498) Ann Univ Turku, Ser A I 85*16pp
◇ D05 ◇ REV MR 32#9170 Zbl 133.258 • ID 31472

SALOMAA, A. [1966] *Axiomatization of an algebra of events realizable by logical network (Russian)* (**J** 0052) Probl Kibern 17*237-246
◇ D05 ◇ REV MR 35#6559 Zbl 216.289 • ID 11792

SALOMAA, A. [1966] *Two complete axiom systems for the algebra of regular events* (**J** 0037) ACM J 13*158-169
◇ D05 G05 ◇ REV MR 32#7411 Zbl 149.249 • ID 11793

SALOMAA, A. [1967] *On m-adic probabilistic automata* (**J** 0194) Inform & Control 10*215-219
◇ D05 ◇ REV MR 35#6494 Zbl 155.347 • ID 11794

SALOMAA, A. [1968] *On events represented by probabilistic automata of different types* (**J** 0017) Canad J Math 20*242-251
◇ D05 ◇ REV MR 36#7444 Zbl 157.22 • ID 31473

SALOMAA, A. [1968] *On finite automata with a time-variant structure* (**J** 0194) Inform & Control 13*85-98
◇ D05 ◇ REV MR 40#1205 Zbl 193.332 • ID 31476

SALOMAA, A. [1968] *On finite time-variant automata with monitors of different types* (**J** 0498) Ann Univ Turku, Ser A I 118/3*12pp
◇ D05 ◇ REV MR 40#1206 Zbl 203.303 • ID 31477

SALOMAA, A. [1968] *On languages accepted by probabilistic and time-variant automata* (**P** 1935) Princeton Conf Inform Sci & Syst (2);1968 Princeton II*184-188
◇ D05 ◇ ID 31474

SALOMAA, A. [1968] *On regular expressions and regular canonical systems* (**J** 0041) Math Syst Theory 2*341-355
◇ D05 ◇ REV MR 38#5630 Zbl 177.19 • ID 31478

SALOMAA, A. & TIXIER, V. [1968] *Two complete axiom systems for the extended language of regular expressions* (**J** 0187) IEEE Trans Comp C-17*700-701
◇ D05 ◇ REV MR 38#4243 Zbl 174.290 • ID 31475

SALOMAA, A. [1969] *On grammars with restricted use of productions* (**J** 3994) Ann Acad Sci Fennicae Ser A I 454*32pp
◇ D05 ◇ REV MR 41#3250 Zbl 193.325 • ID 11796

SALOMAA, A. [1969] *On the index of a context-free grammar and language* (**J** 0194) Inform & Control 14*474-477
◇ D05 ◇ REV MR 39#5276 Zbl 181.310 • ID 31479

SALOMAA, A. [1969] *Probabilistic and weighted grammars* (**J** 0194) Inform & Control 15*529-544
◇ D05 ◇ REV MR 42#7452 Zbl 188.32 • ID 31481

SALOMAA, A. [1969] *Theory of automata* (**X** 0869) Pergamon Pr: Oxford 263pp
◇ D05 D10 D98 ◇ REV MR 41#6631 Zbl 193.329 • ID 11795

SALOMAA, A. [1970] *On some families of formal languages obtained by regulated derivations* (**J** 3994) Ann Acad Sci Fennicae Ser A I 479*18pp
◇ D05 ◇ REV MR 43#1779 Zbl 217.226 • ID 31480

SALOMAA, A. [1970] *Periodically time-variant context-free grammars* (**J** 0194) Inform & Control 17*294-311
◇ D05 ◇ REV MR 42#7453 Zbl 222#68032 • ID 31482

SALOMAA, A. [1971] *Grammars with control languages* (**X** 1051) Univ Utrecht Math Inst: Utrecht 7*8pp
◇ D05 ◇ ID 31484

SALOMAA, A. [1971] *The generative capacity of transformational grammars of Ginsburg and Partee* (**J** 0194) Inform & Control 18*227-232
◇ D05 ◇ REV MR 43#3054 Zbl 225#68043 • ID 31483

SALOMAA, A. [1972] *Matrix grammars with a leftmost restriction* (**J** 0194) Inform & Control 20*143-149
◇ D05 ◇ REV MR 49#1837 Zbl 241#68033 • ID 31485

SALOMAA, A. [1972] *On a homomorphic characterization of recursively enumerable languages* (**J** 3994) Ann Acad Sci Fennicae Ser A I 525*10pp
◇ D25 ◇ REV MR 50#6223 Zbl 253#68010 • ID 31486

SALOMAA, A. [1973] *Developmental languages: a new type of formal languages* (**J** 2083) Ann Univ Turku Ser B 126*183-189
◇ D05 ◇ ID 31581

SALOMAA, A. [1973] *Formal languages* (**X** 0801) Academic Pr: New York xiii+322pp
◇ D05 D15 D98 ◇ REV MR 55#11661 Zbl 262#68025 JSL 42.583 • ID 23556

SALOMAA, A. [1973] see PAZ, A.

SALOMAA, A. [1973] *L-systems: a device in biologically motivated automata theory* (**P** 1448) Math Founds of Comput Sci (2);1973 Strbske Pleso 147-151
◇ D05 ◇ REV MR 53#7151 • ID 31492

SALOMAA, A. [1973] *On exponential growth in Lindenmayer systems* (**J** 0028) Indag Math 35*23-30
◇ D05 D15 ◇ REV MR 47#6134 Zbl 267#68032 • ID 31487

SALOMAA, A. [1973] *On sentential forms of context-free grammars* (**J** 1431) Acta Inf 2*40-49
◇ D05 ◇ REV MR 48#3311 Zbl 264#68029 • ID 31488

SALOMAA, A. [1973] *On some recent problems concerning developmental languages* (**P** 1630) GI Fachtag Automatenth & Form Sprach (1);1973 Bonn 2*23-34
◇ B75 D05 ◇ REV MR 55 # 6981 Zbl 278 # 68070
• ID 31491

SALOMAA, A. [1974] *L systems* (**P** 1939) L Systems;1974 Aarhus 15*338pp
◇ D05 ◇ REV MR 53 # 1996 • ID 31497

SALOMAA, A. [1974] *Recent results on L-systems* (**P** 1938) Conf Biol Motiv Automata Th;1974 McClean 38-45
◇ D05 ◇ ID 31496

SALOMAA, A. [1974] *Solution of a decision problem concerning unary Lindenmayer systems* (**J** 0193) Discr Math 9*71-77
◇ D05 ◇ REV MR 49 # 8435 • ID 31494

SALOMAA, A. [1975] *Formal power series and growth functions of Lindenmayer systems* (**P** 0454) Math Founds of Comput Sci (4);1975 Marianske Lazne 32*101-113
◇ D05 ◇ REV MR 53 # 2051 Zbl 325 # 68044 • ID 31582

SALOMAA, A. [1975] *On some decidability problems concerning developmental languages* (**P** 0757) Scand Logic Symp (3);1973 Uppsala 144-153
◇ B75 D05 ◇ REV MR 52 # 7218 Zbl 315 # 68065
JSL 43.373 • ID 31490

SALOMAA, A. [1976] see ROZENBERG, G.

SALOMAA, A. [1976] *L systems: a parallel way of looking at formal languages. New ideas and recent developments* (**S** 1605) Math Centr Tracts 82*65-107
◇ D05 ◇ REV Zbl 358 # 68112 • ID 31585

SALOMAA, A. [1976] *Recent results on L systems* (**P** 1401) Math Founds of Comput Sci (5);1976 Gdansk 115-123
◇ D05 ◇ REV Zbl 338 # 68052 • ID 31587

SALOMAA, A. [1976] *Sequential and parallel rewriting* (**P** 1940) Form Lang & Progr;1975 Madrid 111-129
◇ D05 ◇ REV MR 56 # 7354 Zbl 356 # 68090 • ID 31586

SALOMAA, A. [1976] *Undecidable problems concerning growth in informationless Lindenmayer systems (German and English summaries)* (**J** 0129) Elektr Informationsverarbeitung & Kybern 12*331-335
◇ D05 D80 ◇ REV MR 54 # 4203 Zbl 332 # 68051
• ID 24079

SALOMAA, A. & STEINBY, M. (EDS.) [1977] *Automata,languages and programming. IV* (**X** 0811) Springer: Heidelberg & New York X+569pp
◇ D97 ◇ REV MR 56 # 4210 Zbl 349 # 00016 • ID 48711

SALOMAA, A. [1977] see MAURER, H.A.

SALOMAA, A. [1977] *Formal power series and language theory* (**X** 1941) Nanyang Univ Publ: Singapore 23pp
◇ D05 ◇ ID 31589

SALOMAA, A. [1977] see ROZENBERG, G.

SALOMAA, A. & SOITTOLA, M. [1978] *Automata-theoretic aspects of formal power series* (**X** 0811) Springer: Heidelberg & New York x+171pp
◇ D05 D80 ◇ REV MR 58 # 3698 Zbl 377 # 68039
• ID 31592

SALOMAA, A. [1978] see PENTTONEN, M.

SALOMAA, A. [1978] see CULIK II, K.

SALOMAA, A. [1978] see MAURER, H.A.

SALOMAA, A. [1979] see MAURER, H.A.

SALOMAA, A. [1980] *Morphisms on free monoids and language theory* (**P** 4266) Form Lang Th;1979 Santa Barbara 141-166
◇ D05 ◇ REV MR 84j:68001 Zbl 545 # 68065 • ID 47200

SALOMAA, A. [1980] see MAURER, H.A.

SALOMAA, A. [1980] see CULIK II, K.

SALOMAA, A. [1981] see MAURER, H.A.

SALOMAA, A. [1981] *Jewels of formal language theory* (**X** 3581) Comput Sci Press: Rockville ix+144pp
◇ D05 D98 ◇ REV MR 83f:68091 Zbl 487 # 68064
• ID 40234

SALOMAA, A. [1981] see OTTMANN, T.

SALOMAA, A. [1982] see MAURER, H.A.

SALOMAA, A. [1985] *Computation and automata* (**X** 0805) Cambridge Univ Pr: Cambridge, GB xiii+282pp
◇ D05 ◇ REV Zbl 565 # 68046 • ID 48219

SALOMAA, A. see Vol. I, II, V for further entries

SALOMON, K.B. [1975] *The decidability of a mapping problem for generalized sequential machines with final states* (**J** 0119) J Comp Syst Sci 10*200-218
◇ D05 ◇ REV MR 52 # 2292 Zbl 302 # 94022 • ID 65035

SALOVAARA, S. [1967] *On set theoretical foundations of system theory. A study of the state concept* (**J** 1191) Acta Polytech Scand, Math Comp Sci 15*78pp
◇ D05 E75 ◇ REV Zbl 308 # 93001 JSL 35.597 • ID 21977

SALWICKI, A. [1975] *Procedures, formal computations and models* (**P** 1755) Math Founds of Comput Sci (3);1974 Jadwisin 464-484
◇ B75 D20 ◇ REV Zbl 323 # 68009 JSL 42.423 • ID 44496

SALWICKI, A. see Vol. II, V for further entries

SAMI, R.L. [1979] see HARRINGTON, L.A.

SAMI, R.L. [1984] *On Σ^1_1 equivalence relations with Borel classes of bounded rank* (**J** 0036) J Symb Logic 49*1273-1283
◇ D55 E10 E15 E45 ◇ REV MR 86g:03077 • ID 42499

SAMI, R.L. see Vol. III, V for further entries

SAMMLER, O. [1984] see ALBRECHT, A.

SAMOJLENKO, L.G. [1969] *On a class of grammars of direct components (Russian)* (**J** 0040) Kibernetika, Akad Nauk Ukr SSR 1969/2*94-96
◇ D05 ◇ REV JSL 35.349 • ID 11804

SAMOJLENKO, L.G. [1969] *Some subclasses of immediate-constitutent grammars which are weakly equivalent to context-free grammars (Russian)* (**J** 0040) Kibernetika, Akad Nauk Ukr SSR 1969/4*48-55
• TRANSL [1969] (**J** 0021) Cybernetics 5*411-420
◇ D05 ◇ REV MR 46 # 4799 • ID 11805

SAMOJLENKO, L.G. see Vol. VI for further entries

SAMPEI, Y. [1959] *On the evaluation of the projective class of sets defined by transfinite induction* (**J** 0407) Comm Math Univ St Pauli (Tokyo) 7*21-26,132
◇ D55 D70 E15 ◇ REV MR 24 # A47 Zbl 102.285
• ID 11808

SAMPEI, Y. [1960] *Note on the effective choice of a point in the complement of an analytic set* (J 0407) Comm Math Univ St Pauli (Tokyo) 9*91–95
◇ D55 E15 ◇ REV MR 25 # 2963 Zbl 112.11 JSL 35.146
• ID 11809

SAMPEI, Y. [1961] *On the uniformization of the complement of an analytic set* (J 0407) Comm Math Univ St Pauli (Tokyo) 10*57–62
◇ D55 E15 ◇ REV MR 26 # 4920 Zbl 117.258 JSL 35.146
• ID 11811

SAMPEI, Y. [1961] *On the uniformization of a set of class* $A_{\rho\sigma}$ (J 0407) Comm Math Univ St Pauli (Tokyo) 10*67–73
◇ D55 E15 ◇ REV MR 26 # 4921 Zbl 125.281 JSL 35.146
• ID 11810

SAMPEI, Y. [1964] *On the complete basis for the* Δ^1_2 *sets* (J 0407) Comm Math Univ St Pauli (Tokyo) 13*81–88
◇ D55 ◇ REV MR 32 # 7416 Zbl 171.270 JSL 40.243
• ID 11812

SAMPEI, Y. [1966] *On the principle of effective choice and its applications* (J 0407) Comm Math Univ St Pauli (Tokyo) 15*29–42 • ERR/ADD ibid 17*104
◇ D55 E15 E45 ◇ REV MR 35 # 1486 MR 39 # 5347 Zbl 178.320 JSL 40.243 • ID 19651

SAMPEI, Y. [1968] *A proof of Mansfield's theorem by forcing method* (J 0407) Comm Math Univ St Pauli (Tokyo) 17*99–103
◇ D55 E15 E40 E45 ◇ REV MR 40 # 1283 Zbl 211.16 JSL 20.462 JSL 40.462 • ID 11813

SAMPEI, Y. [1970] *On the relativization of* Δ^1_2 *sets of reals* (J 0407) Comm Math Univ St Pauli (Tokyo) 18*149–151
◇ D55 E15 E35 E45 ◇ REV MR 43 # 1828 Zbl 246 # 02046 • ID 11814

SAMPEI, Y. see Vol. I, V for further entries

SANCHIS, L.E. [1967] *Functionals defined by recursion* (J 0047) Notre Dame J Formal Log 8*161–174
◇ D20 F10 ◇ REV MR 39 # 1294 Zbl 183.14 • ID 11825

SANCHIS, L.E. [1977] *Data types as lattices: Retractions, closures and projections* (J 3441) RAIRO Inform Theor 11*329–344
◇ B40 B75 D25 G10 ◇ REV MR 58 # 13876 Zbl 394 # 03019 • ID 52496

SANCHIS, L.E. [1978] *Hyperenumeration reducibility* (J 0047) Notre Dame J Formal Log 19*405–415
◇ D30 D55 ◇ REV MR 58 # 10374 Zbl 336 # 02034
• ID 51498

SANCHIS, L.E. [1979] *Reducibilities in two models for combinatory logic* (J 0036) J Symb Logic 44*221–234
◇ B40 D30 ◇ REV MR 80k:03044 Zbl 417 # 03016
• ID 53253

SANCHIS, L.E. see Vol. I, III, VI for further entries

SANDRING, S. & STARKE, P.H. [1982] *A note on liveness in generalized Petri nets* (J 2095) Fund Inform, Ann Soc Math Pol, Ser 4 5*217–232
◇ D05 ◇ REV MR 83m:68105 Zbl 504 # 68034 • ID 42745

SANKAPPANAVAR, H.P. [1975] see BURRIS, S.

SANKAPPANAVAR, H.P. [1977] *On the decision problem of the congruence lattices of pseudocomplemented semilattices* (P 1076) Latin Amer Symp Math Log (3);1976 Campinas 255–266
◇ D35 G10 ◇ REV MR 57 # 5723 Zbl 361 # 02062
• ID 16609

SANKAPPANAVAR, H.P. [1978] *Decision problems: History and methods* (P 1800) Brazil Conf Math Log (1);1977 Campinas 241–291
◇ A10 B25 D35 D98 ◇ REV MR 81a:03011 Zbl 385 # 03011 • ID 52119

SANKAPPANAVAR, H.P. see Vol. III for further entries

SANSONE, F.J. [1963] *Combinatorial functions and regressive isols* (J 0048) Pac J Math 13*703–707
◇ D50 ◇ REV MR 32 # 5509 Zbl 121.255 JSL 33.113
• ID 11837

SANSONE, F.J. [1965] *A mapping of regressive isols* (J 0316) Illinois J Math 9*726–735
◇ D50 ◇ REV MR 32 # 5510 Zbl 148.246 JSL 33.114
• ID 11838

SANSONE, F.J. [1965] *The summation of certain series of infinite regressive isols* (J 0053) Proc Amer Math Soc 16*1135–1140
◇ D50 ◇ REV MR 37 # 1243 Zbl 192.51 JSL 33.114
• ID 11839

SANSONE, F.J. [1966] *On order-preserving extensions to regressive isols* (J 0133) Michigan Math J 13*353–355
◇ D50 ◇ REV MR 35 # 6555 Zbl 192.51 JSL 33.114
• ID 11840

SANSONE, F.J. [1969] *The backward and forward summation of infinite series of isols* (J 0132) Math Scand 24*217–220
◇ D50 ◇ REV MR 43 # 1833 Zbl 211.19 • ID 11841

SANTHA, M. [1983] *Constructions d'oracles pour la hierarchie polynomiale relativisee (English summary)* (J 3364) C R Acad Sci, Paris, Ser 1 297*377–380
◇ D15 ◇ REV MR 85b:03067 Zbl 547 # 03029 • ID 40683

SANTHA, M. [1984] *La hierarchie polynomiale avec oracle* (C 4356) Gen Log Semin Paris 1982/83 153–177
◇ D15 ◇ ID 48094

SANTOS, E.S. [1969] *Probabilistic Turing machines and computability* (J 0053) Proc Amer Math Soc 22*704–710
◇ D10 ◇ REV MR 40 # 2468 Zbl 186.12 • ID 11842

SANTOS, E.S. [1971] *Computability by probabilistic Turing machines* (J 0064) Trans Amer Math Soc 159*165–184
◇ D10 D20 ◇ REV MR 43 # 7270 Zbl 246 # 02030
• ID 65060

SANTOS, E.S. [1976] *Fuzzy and probabilistic programs* (J 0191) Inform Sci 10*331–345
◇ B52 B75 D10 ◇ REV MR 58 # 19335 Zbl 334 # 68013
• ID 65059

SAOUDI, A. [1984] *Infinitary tree languages recognized by* ω*-automata* (J 0232) Inform Process Lett 18*15–19
◇ D05 ◇ REV MR 85e:03092 Zbl 539 # 68072 • ID 40686

SARALSKI, B. [1975] see KOPIEKI, R.

SARKISYAN, A.D. [1979] *Realization of arithmetic functions on iterative networks (Russian) (Armenian summary)* (J 0346) Dokl Akad Nauk Armyan SSR 68*267–272
◇ D05 D15 ◇ REV MR 81c:03029 Zbl 408 # 94027
• ID 78135

Sarkisyan, A.D. [1982] *Classes of functions defined by their calculation time on iterative networks (Russian) (Armenian summary)* (**S** 0422) Tr Vychisl Tsentra Akad Nauk Armyan SSR & Univ Erevan 10*9-28
◇ D10 D15 ◇ REV MR 85i:68018 • ID 44364

Sarkisyan, G.Z. [1978] *Effective computability of arithmetic predicates and functions on the basis of schemes of functional elements (Russian) (English and Armenian summaries)* (**J** 0312) Izv Akad Nauk Armyan SSR, Ser Mat 13*128-139,172
◇ D15 ◇ REV MR 80a:68047 Zbl 382 # 68050 • ID 82712

Sarkisyan, G.Z. [1978] *Ueber eine Klasse arithmetischer Funktionen, die mit Hilfe von Schemata aus Funktionalelementen berechenbar sind (Russian) (Armenian summary)* (**J** 0346) Dokl Akad Nauk Armyan SSR 66*257-262
◇ D15 D20 ◇ REV MR 81d:03043 Zbl 389 # 10040 • ID 52319

Sarkisyan, O.A. [1975] see Pogosyan, Eh.M.

Sarkisyan, O.A. [1976] *On the connection between algorithmic problems in groups and semigroups (Russian)* (**J** 0023) Dokl Akad Nauk SSSR 227*1305-1307
• TRANSL [1976] (**J** 0062) Sov Math, Dokl 17*615-617
◇ D40 ◇ REV MR 53 # 7751 Zbl 359 # 02046 • ID 50593

Sarkisyan, O.A. [1979] *Beziehungen zwischen Identitaets- und Teilbarkeitsproblemen in Gruppen und Halbgruppen (Russisch)* (**J** 0216) Izv Akad Nauk SSSR, Ser Mat 43*909-921
• TRANSL [1980] (**J** 0448) Math of USSR, Izv 15*161-171
◇ D40 ◇ REV MR 80k:20056 Zbl 412 # 20051 • ID 52978

Sarkisyan, O.A. [1981] *Word and divisibility problems in semigroups and groups without cycles (Russian)* (**J** 0216) Izv Akad Nauk SSSR, Ser Mat 45*1424-1440
• TRANSL [1982] (**J** 0448) Math of USSR, Izv 19*643-656
◇ C05 D40 ◇ REV MR 83e:20066 Zbl 517 # 20033 • ID 40133

Sarkisyan, R.A. [1980] *Algorithmic questions for linear algebraic groups I,II (Russian)* (**J** 0142) Mat Sb, Akad Nauk SSSR, NS 113(155)*179-216,350,400-436,495
• TRANSL [1982] (**J** 0349) Math of USSR, Sbor 41*149-179,329-359
◇ D40 ◇ REV MR 83a:20049 Zbl 446 # 20025 Zbl 459 # 20037 • ID 38857

Sasso Jr., L.P. [1970] *A cornucopia of minimal degrees* (**J** 0036) J Symb Logic 35*383-388
◇ D30 ◇ REV MR 44 # 69 Zbl 219 # 02029 • ID 11855

Sasso Jr., L.P. [1973] *A minimal partial degree* $\leqslant 0'$ (**J** 0053) Proc Amer Math Soc 38*388-392
◇ D30 ◇ REV MR 47 # 29 Zbl 232 # 02030 • ID 11856

Sasso Jr., L.P. [1974] *A minimal degree not realizing least possible jump* (**J** 0036) J Symb Logic 39*571-574
◇ D30 ◇ REV MR 50 # 12692 Zbl 318 # 02044 • ID 11857

Sasso Jr., L.P. [1974] *Deficiency sets and bounded information reducibilities* (**J** 0064) Trans Amer Math Soc 200*267-290
◇ D25 D30 ◇ REV MR 57 # 9510 Zbl 449 # 03034 • ID 56701

Sasso Jr., L.P. [1975] *A survey of partial degrees* (**J** 0036) J Symb Logic 40*130-140
◇ D30 ◇ REV MR 52 # 64 Zbl 318 # 02043 • ID 11858

Sasso Jr., L.P. [1975] see Ladner, R.E.

Sato, D. [1976] see Jones, James P.

Savage, J.A. [1981] *Space-time tradeoffs - a survey* (**P** 3212) Math Models in Comput Systs;1981 Budapest 171-181
◇ D15 D98 ◇ REV MR 83m:68086 Zbl 467 # 68048 • ID 69894

Savage, J.E. [1976] *The complexity of computing* (**X** 0827) Wiley & Sons: New York xiii+391pp
◇ B70 D10 D15 D98 ◇ REV MR 58 # 13936 Zbl 391 # 68025 • ID 52361

Savage, J.E. & Swamy, S. [1983] *Space-time tradeoffs for linear recursion* (**J** 0041) Math Syst Theory 16*9-27
◇ D15 ◇ REV MR 84b:68051 Zbl 502 # 68006 • ID 38951

Savel'ev, A.A. [1983] *A family of subsystems of algebras for preserving an ideal (Russian)* (**J** 3937) Veroyat Met i Kibern (Kazan) 19*99-106
◇ D75 ◇ REV MR 84k:03111 • ID 36129

Savel'ev, A.A. [1984] see Golunkov, Yu.V.

Savel'ev, A.A. [1985] see Golunkov, Yu.V.

Savelsbergh, M.W.P. [1984] see Emde Boas van, P.

Savitch, W.J. [1970] *Relationship between nondeterministic and deterministic tape complexities* (**J** 0119) J Comp Syst Sci 4*177-192
◇ D10 D15 ◇ REV MR 42 # 1605 Zbl 188.335 JSL 39.346 • ID 11868

Savitch, W.J. [1972] *Maze recognizing automata* (**P** 1901) ACM Symp Th of Comput (4);1972 Denver 151-156
◇ D10 D15 ◇ REV Zbl 358 # 68087 • ID 50542

Savitch, W.J. [1973] *Maze recognizing automata and nondeterministic tape complexity* (**J** 0119) J Comp Syst Sci 7*389-403
◇ D10 D15 ◇ REV MR 49 # 4314 Zbl 273 # 02022 • ID 30491

Savitch, W.J. [1973] *Nondeterministic finite automata revisited* (**P** 3244) Hawaii Int Conf Syst Sci (6);1973 Honolulu 249-251
◇ D05 D15 ◇ REV Zbl 361 # 68080 • ID 50715

Savitch, W.J. [1975] see Reedy, A.

Savitch, W.J. [1976] *Three hardest problems* (**C** 3517) Found of Comput Sci, Vol 2, Part 2 111-149
◇ D15 ◇ REV MR 58 # 3685 Zbl 357 # 68053 • ID 50442

Savitch, W.J. & Vitanyi, P.M.B. [1977] *Linear time simulation of multihead Turing machines with head-to-head jumps* (**P** 1632) Automata, Lang & Progr (4);1977 Turku SF 453-464
◇ D10 D15 ◇ REV MR 56 # 10155 Zbl 353 # 68074 • ID 82717

Savitch, W.J. [1977] *Recursive Turing machines* (**J** 0382) Int J Comput Math 6*3-31
◇ D10 D15 ◇ REV MR 56 # 11768 Zbl 355 # 02029 • ID 50227

Savitch, W.J. [1978] *Parallel and nondeterministic time complexity classes(Preliminary report)* (**P** 1872) Automata, Lang & Progr (5);1978 Udine 411-424
◇ D15 ◇ REV Zbl 382 # 68046 • ID 69897

Savitch, W.J. [1978] *The influence of the machine model on computational complexity* (**P** 3387) Interface Comput Sci & Oper Res;1976 Amsterdam 1-32
◇ D15 ◇ REV MR 81f:68001 Zbl 407 # 68055 • ID 69896

SAVITCH, W.J. & STIMSON, M.J. [1979] *Hierarchies of recursive computations* (J 0382) Int J Comput Math 7*271-286
◇ D10 D15 ◇ REV MR 81c:68038 Zbl 437 # 68024
• ID 55910

SAVITCH, W.J. & VERMEIR, D. [1981] *On the amount of nondeterminism in pushdown automata* (J 2095) Fund Inform, Ann Soc Math Pol, Ser 4 4*401-418
◇ D10 ◇ REV MR 83i:68127 Zbl 528 # 68034 • ID 39371

SAVITCH, W.J. [1983] *A note on relativized log space* (J 0041) Math Syst Theory 16*229-235
◇ D15 ◇ REV MR 85c:68021 Zbl 524 # 68031 • ID 39069

SAVITCH, W.J. & VITANYI, P.M.B. [1984] *On the power of real-time two-way multihead finite automata with jumps* (J 0232) Inform Process Lett 19*31-35
◇ D05 ◇ REV MR 85j:68036 Zbl 539 # 68042 • ID 46795

SAXE, J.B. [1981] see FURST, M.

SAXE, J.B. [1984] see FURST, M.

SAXTON, L.V. [1975] *An extension of the class of on-line machines* (P 3243) Manitoba Conf Num Math (4);1974 Winnipeg 329-341
◇ D10 D15 ◇ REV MR 51 # 9578 Zbl 361 # 02050
• ID 50685

SAXTON, L.V. [1976] see FISCHER, P.C.

SAZONOV, V.YU. [1976] *Degrees of parallelism in computations* (P 1401) Math Founds of Comput Sci (5);1976 Gdansk 45*517-523
◇ B40 B75 D30 D65 ◇ REV Zbl 341 # 02035 • ID 29755

SAZONOV, V.YU. [1976] *Expressibility of functions in D.Scott's LFC language (Russian)* (J 0003) Algebra i Logika 15*308-330,366
• TRANSL [1976] (J 0069) Algeb and Log 15*192-206
◇ B40 D20 ◇ REV MR 55 # 12480 Zbl 415 # 03011
• ID 26049

SAZONOV, V.YU. [1976] *Functionals computable in series and in parallel (Russian)* (J 0092) Sib Mat Zh 17*648-672,717
• TRANSL [1976] (J 0475) Sib Math J 17*498-516
◇ B40 D20 ◇ REV MR 54 # 7211 Zbl 342 # 02016
• ID 25002

SAZONOV, V.YU. [1980] *A logical approach to the problem "P=NP?"* (P 3210) Math Founds of Comput Sci (9);1980 Rydzyna 562-575
◇ D15 ◇ REV MR 83j:68055 Zbl 453 # 03042 • ID 54172

SAZONOV, V.YU. [1980] *Polynomial computability and recursivity in finite domains* (J 0129) Elektr Informationsverarbeitung & Kybern 16*319-323
◇ D15 ◇ REV MR 83c:68053 Zbl 455 # 03018 • ID 54279

SAZONOV, V.YU. see Vol. VI for further entries

SCARPELLINI, B. [1962] *Die Nichtaxiomatisierbarkeit des unendlichwertigen Praedikatenkalkuels von Lukasiewicz* (J 0036) J Symb Logic 27*159-170
◇ B50 D35 ◇ REV MR 27 # 3503 Zbl 112.245 JSL 29.145
• ID 11877

SCARPELLINI, B. [1963] *Zwei unentscheidbare Probleme der Analysis* (J 0068) Z Math Logik Grundlagen Math 9*265-289
◇ D25 D35 ◇ REV MR 27 # 4744 • ID 11876

SCARPELLINI, B. [1965] *A characterization of Δ_2-sets* (J 0064) Trans Amer Math Soc 117*441-450
◇ D55 E15 E45 ◇ REV MR 30 # 3007 Zbl 192.43
JSL 37.193 • ID 11879

SCARPELLINI, B. [1984] *Complexity of subcases of Presburger arithmetic* (J 0064) Trans Amer Math Soc 284*203-218
◇ B25 D15 F20 F30 ◇ REV MR 86c:03010
Zbl 548 # 03018 • ID 43196

SCARPELLINI, B. [1985] *Lower bound results on lengths of second-order formulas* (J 0073) Ann Pure Appl Logic 29*29-58
◇ B15 C13 D10 F20 F35 ◇ ID 47473

SCARPELLINI, B. see Vol. I, III, V, VI for further entries

SCEDROV, A. [1984] see BEESON, M.J.

SCEDROV, A. [1985] see FRIEDMAN, H.M.

SCEDROV, A. [1985] see HARRINGTON, L.A.

SCEDROV, A. see Vol. I, II, III, V, VI for further entries

SCHADE, W. [1979] *Indexmengen rekursiver reeller Zahlen* (J 0068) Z Math Logik Grundlagen Math 25*103-110
◇ D30 D45 ◇ REV MR 80d:03041 Zbl 471 # 03050
• ID 55245

SCHAEFER, G. & WEIHRAUCH, K. [1981] *Admissible representations of effective cpo's* (P 3429) Math Founds of Comput Sci (10);1981 Strbske Pleso 544-553
◇ D45 ◇ REV MR 83f:03041 Zbl 485 # 68041 • ID 35282

SCHAEFER, G. & WEIHRAUCH, K. [1983] *Admissible representations of effective CPOs* (J 1426) Theor Comput Sci 26*131-147
◇ B75 D45 ◇ REV MR 85f:68054 • ID 39976

SCHAEFER, G. [1985] *A note on conjectures of Calude about the topological size of sets of partial recursive functions* (J 0068) Z Math Logik Grundlagen Math 31*279-280
◇ D20 D75 ◇ ID 47576

SCHAEFER, T.J. [1978] *The complexity of satisfiability problems* (P 1740) ACM Symp Th of Comput (10);1978 San Diego 216-226
◇ B05 D15 ◇ REV MR 80d:68058 • ID 82718

SCHAETZ, R. [1984] see BRUEGGEMANN, A.

SCHEIN, B.M. [1969] *On some problems in the theory of partial automata (Czech summary)* (J 0156) Kybernetika (Prague) 5*44-49
◇ D05 ◇ REV MR 40 # 4121 • ID 48082

SCHEIN, B.M. see Vol. III, V for further entries

SCHIEK, H. [1973] *Equations over groups* (P 0678) Word Probl: Decis & Burnside Probl in Group Th;1969 Irvine 563-567
◇ D40 ◇ REV MR 55 # 3096 Zbl 264 # 20026 JSL 41.786
• ID 44405

SCHILLING, K. [1984] *On absolutely Δ^1_2 operations* (J 0027) Fund Math 121*239-250
◇ D55 E15 E75 ◇ REV MR 86d:54059 Zbl 562 # 03025
• ID 44440

SCHILLING, K. see Vol. V for further entries

SCHINZEL, B. [1977] *Decomposition of Goedelnumberings into Friedbergnumberings* (**J** 0068) Z Math Logik Grundlagen Math 23*393-399
◇ D20 D45 ◇ REV MR 58 # 10386 Zbl 439 # 03020
• ID 56011

SCHINZEL, B. [1977] *Struktur von Programmbuendeln* (**P** 3411) Theor Comput Sci (3);1977 Darmstadt 226-233
◇ D20 G30 ◇ REV MR 58 # 3750 Zbl 363 # 68042
• ID 82721

SCHINZEL, B. [1979] *Classes of decompositions of a Goedelnumbering* (**P** 2935) FCT'79 Fund of Comput Th;1979 Berlin/Wendisch-Rietz 397-403
◇ D20 D45 ◇ REV MR 81k:03037a Zbl 424 # 03022
• ID 78175

SCHINZEL, B. [1979] *Uebersetzer zwischen Goedelnumerierungen* (**P** 2539) Frege Konferenz (1);1979 Jena 401-413
◇ D20 D45 ◇ REV MR 83e:03070 • ID 35239

SCHINZEL, B. [1980] *Ueber die Kategorie der Programmbuendel (Hab.-Schr.)* (**X** 3155) TH Darmstadt Fachb Informatik: Darmstadt 139pp
◇ D15 D20 D45 ◇ REV Zbl 476 # 03047 • ID 55559

SCHINZEL, B. [1980] *Zerlegung mit Vergleichsbedingungen einer Goedelnumerierung* (**J** 0068) Z Math Logik Grundlagen Math 26*215-226
◇ D20 D45 ◇ REV MR 81k:03037b Zbl 439 # 03021
• ID 56012

SCHINZEL, B. [1982] *Complexity of decomposotions of Goedel numberings* (**J** 2095) Fund Inform, Ann Soc Math Pol, Ser 4 5*15-33
◇ D20 D45 ◇ REV MR 83j:03074 Zbl 505 # 03025
• ID 35370

SCHINZEL, B. [1982] *On decomposition of Goedel numberings into Friedberg numberings* (**J** 0036) J Symb Logic 47*267-274
◇ D20 D45 ◇ REV MR 83e:03069 Zbl 514 # 03025 Zbl 533 # 03027 • ID 35238

SCHINZEL, B. [1984] see BOERGER, E.

SCHKOLNICK, M. [1971] see HARRISON, M.A.

SCHLIPF, J.S. [1975] see BARWISE, J.

SCHLIPF, J.S. [1975] *Some hyperelementary aspects of model theory* (0000) Diss., Habil. etc 165pp
◇ C50 C62 C70 D55 ◇ REM Ph.D. thesis, University of Wisconsin, Madison, WI, USA • ID 19103

SCHLIPF, J.S. [1976] see BARWISE, J.

SCHLIPF, J.S. [1977] *Ordinal spectra of first-order theories* (**J** 0036) J Symb Logic 42*492-505
◇ C35 C50 C55 C62 C70 D15 D30 D70 ◇ REV MR 58 # 27446 Zbl 411 # 03035 • ID 26851

SCHLIPF, J.S. [1978] see HARRINGTON, L.A.

SCHLIPF, J.S. [1978] *Toward model theory through recursive saturation* (**J** 0036) J Symb Logic 43*183-206
◇ C15 C40 C50 C57 C70 ◇ REV MR 58 # 10399 Zbl 409 # 03019 • ID 29252

SCHLIPF, J.S. [1980] *Recursively saturated models of set theory* (**J** 0053) Proc Amer Math Soc 80*135-142
◇ C50 C57 C62 E70 H20 ◇ REV MR 81h:03101 Zbl 455 # 03022 • ID 54283

SCHLIPF, J.S. see Vol. III for further entries

SCHLOESSER, L. [1975] *Ueber boolesche Ianovschemata* (**S** 3180) Inform Ber, Inst Inf, Univ Bonn 9*67pp
◇ B75 D80 ◇ REV MR 54 # 1705 Zbl 357 # 02030
• ID 50378

SCHMERL, J.H. [1974] see GARFUNKEL, S.

SCHMERL, J.H. [1976] *Effectiveness and Vaught's gap ω two-cardinal theorem* (**J** 0053) Proc Amer Math Soc 58*237-240
◇ C55 C57 ◇ REV MR 55 # 5433 Zbl 357 # 02049
• ID 50397

SCHMERL, J.H. [1978] *A decidable $\aleph_0$-categorical theory with a non-recursive Ryll-Nardzewski function* (**J** 0027) Fund Math 98*121-125
◇ B25 C35 C57 D30 ◇ REV MR 80g:03031 Zbl 372 # 02025 • ID 29211

SCHMERL, J.H. [1979] see LERMAN, M.

SCHMERL, J.H. [1980] *Decidability and $\aleph_0$-categoricity of theories of partially ordered sets* (**J** 0036) J Symb Logic 45*585-611
◇ B25 C15 C35 C65 D35 ◇ REV MR 84g:03037 Zbl 441 # 03007 • ID 56060

SCHMERL, J.H. [1980] *Recursive colorings of graphs* (**J** 0017) Canad J Math 32*821-830
◇ D45 D80 ◇ REV MR 81m:03054 Zbl 438 # 05030
• ID 78188

SCHMERL, J.H. [1981] *Arborescent structures I: Recursive models* (**P** 2902) Aspects Effective Algeb;1979 Clayton 226-231
◇ C50 C57 C65 ◇ REV MR 83g:03033a Zbl 475 # 03013
• REM Part II 1981 • ID 55467

SCHMERL, J.H. [1981] *Recursively saturated, rather classless models of Peano arithmetic* (**P** 2628) Log Year;1979/80 Storrs 268-282
◇ C50 C57 C62 E45 E55 ◇ REV MR 83b:03039 Zbl 469 # 03051 • ID 55178

SCHMERL, J.H. [1982] *The effective version of Brooks' theorem* (**J** 0017) Canad J Math 34*1036-1046
◇ D80 ◇ REV MR 84j:05058 Zbl 477 # 05035 • ID 39140

SCHMERL, J.H. [1984] see KAUFMANN, M.

SCHMERL, J.H. [1985] *Recursion theoretic aspects of graphs and orders* (**P** 4583) Graphs & Order;1984 Banff 467-484
◇ D80 ◇ REV Zbl 565 # 05028 • ID 48661

SCHMERL, J.H. [1985] *Recursively saturated models generated by indiscernibles* (**J** 0047) Notre Dame J Formal Log 26*99-105
◇ C30 C50 C57 ◇ REV Zbl 556 # 03030 • ID 42589

SCHMERL, J.H. see Vol. I, III, V for further entries

SCHMERL, U.R. [1982] *Ueber die schwach und die stark wachsende Hierarchie zahlentheoretischer Funktionen* (**J** 1944) Sitzb, Akad Wiss, Bayern, Math-Nat Kl 1981*1-8
◇ D20 ◇ REV MR 84e:03050 Zbl 501 # 03028 • ID 34386

SCHMERL, U.R. see Vol. III, VI for further entries

SCHMIDT, DIANA [1976] *Built-up systems of fundamental sequences and hierarchies of number theoretic functions* (**J** 0009) Arch Math Logik Grundlagenforsch 18*47-53
• ERR/ADD ibid 18*145-146
◇ D20 E10 F15 ◇ REV MR 57 # 16025 Zbl 358 # 02061
• ID 24318

SCHMIDT, DIANA [1981] *An algebraic characterisation of P* (**S** 3126) Ber, Fak Inf, Univ Karlsruhe 2/81*11pp
◇ D15 ◇ REV Zbl 452 # 68081 • ID 69899

SCHMIDT, DIANA [1983] *An alternative definition of NP* (**J** 2010) Bull Europ Assoc Th Comput Sci 21*57-67
◇ D15 ◇ ID 39433

SCHMIDT, DIANA [1984] *Limitations on separating nondeterministic and deterministic complexity classes* (**S** 3126) Ber, Fak Inf, Univ Karlsruhe 1/84*10pp
◇ D15 ◇ ID 39436

SCHMIDT, DIANA [1984] *The complement of one complexity class in another* (**P** 2342) Symp Rek Kombin;1983 Muenster 77-87
◇ D15 ◇ REV Zbl 551 # 03023 • ID 39424

SCHMIDT, DIANA [1985] *The recursion-theoretic structure of complexity classes* (**J** 1426) Theor Comput Sci 38*143-156
◇ D15 ◇ ID 39438

SCHMIDT, DIANA see Vol. V, VI for further entries

SCHMIDT, ERIK MEINECHE [1978] see FORTUNE, S.

SCHMIDT, ERIK MEINECHE [1980] see ENGELFRIET, J.

SCHMIDT, ERIK MEINECHE [1982] see NIELSEN, M.

SCHMIDT, GUNTHER [1970] see LANGMAACK, H.

SCHMIDT, GUNTHER see Vol. I, II, V for further entries

SCHMIDTKE, K. [1972] *Classification of abstract computers with respect to the rho-inclusion (Polish) (Russian and English summaries)* (**J** 1929) Prace Centr Oblicz Pol Akad Nauk 88*58pp
◇ D05 ◇ REV Zbl 267 # 02025 • ID 29892

SCHMIDTKE, K. [1973] *Classification of computing machines with finite computations with respect to* ρ *-inclusion* (**J** 0014) Bull Acad Pol Sci, Ser Math Astron Phys 21*71-77
◇ D05 ◇ REV MR 51 # 9563 • ID 28413

SCHMIDTKE, K. [1974] *Classification of computing machines with infinite computations with respect to* ρ *-inclusion (Russian summary)* (**J** 0014) Bull Acad Pol Sci, Ser Math Astron Phys 22*71-77
◇ D10 ◇ REV MR 55 # 10168 Zbl 289 # 68019 • ID 28414

SCHMITT, A.A. [1969] *Zur Theorie der nichtdeterministischen und unvollstaendigen Automaten* (**J** 0373) Comp Arch Inform & Numerik 4*56-74
◇ D05 ◇ REV MR 39 # 8027 Zbl 213.22 • ID 27993

SCHMITT, A.A. [1970] *Die Zustands-Komplexitaetsklassen von Turingmaschinen* (**P** 0577) Automatenth & Formale Sprachen;1969 Oberwolfach 341-349
• TRANSL [1970] (**J** 0194) Inform & Control 17*217-225
◇ D10 D15 ◇ REV MR 42 # 7439 Zbl 215.322 Zbl 222 # 02038 • ID 28018

SCHMITT, A.A. [1971] *Automaten. Algorithmen. Gehirne* (**X** 0916) Suhrkamp: Frankfurt 159pp
◇ D05 D10 ◇ REV Zbl 218 # 00008 • ID 26268

SCHMITT, P.H. [1981] see CHERLIN, G.L.

SCHMITT, P.H. [1984] *Undecidable theories of valued Abelian groups* (**S** 3521) Mem Soc Math Fr 16*67-76
◇ C60 D35 ◇ REV Zbl 555 # 20034 • ID 46136

SCHMITT, P.H. see Vol. II, III for further entries

SCHNEIDER, H.H. [1973] see BULLOCK, A.M.

SCHNEIDER, H.H. see Vol. I, III, V for further entries

SCHNORR, C.-P. [1969] *Eine Bemerkung zum Begriff der zufaelligen Folge (English summary)* (**J** 0982) Z Wahrscheinltheor & Verw Geb 14*27-35
◇ D20 D80 F65 ◇ REV MR 41 # 9315 Zbl 188.28 • ID 12929

SCHNORR, C.-P. & WALTER, H. [1969] *Pullbackkonstruktionen bei Semi-Thuesystemen (English and Russian summaries)* (**J** 0129) Elektr Informationsverarbeitung & Kybern 5*27-36
◇ D03 ◇ REV MR 39 # 6753 Zbl 177.18 • ID 12927

SCHNORR, C.-P. [1970] *Minimale Programmkomplexitaet und Zufaelligkeit* (**P** 1580) Tagung Formale Sprachen;1970 Oberwolfach 39-42
◇ D15 D80 ◇ REV Zbl 219 # 02021 • ID 28084

SCHNORR, C.-P. [1970] *Ueber die Zufaelligkeit und den Zufallsgrad von Folgen* (**P** 0577) Automatenth & Formale Sprachen;1969 Oberwolfach 351-367
◇ D20 D80 ◇ REV MR 49 # 1827 Zbl 214.18 • ID 12930

SCHNORR, C.-P. [1971] *A unified approach to the definition of random sequences* (**J** 0041) Math Syst Theory 5*246-258
• TRANSL [1974] (**C** 2319) Slozh Vychisl & Algor 370-387
◇ D80 ◇ REV MR 50 # 6808 Zbl 227 # 62005 • ID 12931

SCHNORR, C.-P. [1971] *Komplexitaet von Algorithmen mit Anwendung auf die Analysis* (**J** 0009) Arch Math Logik Grundlagenforsch 14*54-68
◇ D15 D80 ◇ REV MR 43 # 7328 Zbl 219 # 02020 • ID 12932

SCHNORR, C.-P. [1971] *Zufaelligkeit und Wahrscheinlichkeit. Eine algorithmische Begruendung der Wahrscheinlichkeitstheorie* (**X** 0811) Springer: Heidelberg & New York iv+212pp
◇ B30 B75 D80 ◇ REV MR 54 # 2328 Zbl 232 # 60001 • ID 23998

SCHNORR, C.-P. [1972] *Optimal Goedelnumberings* (**P** 1455) Inform Processing (5);1971 Ljubljana 71*56-58
◇ D20 D45 ◇ REV Zbl 255 # 02042 • ID 28970

SCHNORR, C.-P. [1973] *Process complexity and effective random tests* (**J** 0119) J Comp Syst Sci 7*376-388
◇ D15 ◇ REV MR 48 # 3713 Zbl 273 # 68036 • ID 12933

SCHNORR, C.-P. [1974] *Optimal enumerations and optimal Goedel numberings* (**J** 0041) Math Syst Theory 8*182-191
◇ D15 D20 D25 D45 ◇ REV MR 52 # 10385 Zbl 316 # 02044 • ID 21684

SCHNORR, C.-P. [1974] *Rekursive Funktionen und ihre Komplexitaet* (**X** 0823) Teubner: Stuttgart 191pp
◇ D10 D15 D20 D98 ◇ REV MR 57 # 2892 Zbl 299 # 02043 • ID 28795

SCHNORR, C.-P. & STUMPF, G. [1975] *A characterization of complexity sequences* (**J** 0068) Z Math Logik Grundlagen Math 21*47-56
◇ D10 D20 ◇ REV MR 51 # 12027 Zbl 309 # 02033 • ID 12934

SCHNORR, C.-P. [1976] *The combinational complexity of equivalence* (**J** 1426) Theor Comput Sci 1*289-295
• TRANSL [1979] (**J** 3079) Kiber Sb Perevodov, NS 16*74-81
◇ B05 D15 ◇ REV Zbl 333 # 68032 • ID 69903

SCHNORR, C.-P. [1976] *The network complexity and the Turing machine complexity of finite functions* (**J** 1431) Acta Inf 7*95-107
◇ D10 D15 ◇ REV MR 54 # 9883 Zbl 338 # 02019 • ID 65119

SCHNORR, C.-P. [1977] see KLUPP, H.

SCHNORR, C.-P. [1977] see FUCHS, P.H.

SCHNORR, C.-P. [1977] *The network complexity and the breadth of Boolean functions* (**P** 1075) Logic Colloq;1976 Oxford 491-504
◇ B05 B70 D15 ◇ REV MR 58 # 8494 Zbl 418 # 68048 • ID 16638

SCHNORR, C.-P. [1977] see REYNVAAN, C.

SCHNORR, C.-P. [1978] *Satisfiability is quasilinear complete in NQL* (**J** 0037) ACM J 25*136-145
◇ D10 D15 ◇ REV MR 58 # 8495 Zbl 364 # 68056 • ID 50986

SCHNORR, C.-P. [1980] *A 3n-lower bound on the network complexity of boolean functions* (**J** 1426) Theor Comput Sci 10*83-92
◇ B70 D15 ◇ REV MR 81f:94046 Zbl 438 # 68012 • ID 69904

SCHNORR, C.-P. [1981] *On self-transformable combinatorial problems* (**P** 4021) Math Progr;1979 Oberwolfach 14*225-243
◇ D15 ◇ REV MR 83a:68056 Zbl 449.90071 • ID 38866

SCHNORR, C.-P. see Vol. I for further entries

SCHOENFELD, W. [1979] *An undecidability result for relation algebras* (**J** 0036) J Symb Logic 44*111-115
◇ D35 G15 ◇ REV MR 80f:03074 Zbl 407 # 03019 • ID 56185

SCHOENFELD, W. see Vol. I, III, V, VI for further entries

SCHOENFINKEL, M. [1928] see BERNAYS, P.

SCHOENFINKEL, M. see Vol. I, VI for further entries

SCHOENHAGE, A. [1970] *Universelle Turing-Speicherung* (**P** 0577) Automatenth & Formale Sprachen;1969 Oberwolfach 369-383
◇ D10 ◇ REV MR 51 # 14628 Zbl 215.322 • ID 28017

SCHOENHAGE, A. [1980] *Storage modification machines* (**J** 1428) SIAM J Comp 9*490-508
◇ D10 ◇ REV MR 82b:68040 Zbl 454 # 68034 • ID 82743

SCHOENHAGE, A. [1982] *Random access machines and Presburger arithmetic* (**P** 3482) Logic & Algor (Specker);1980 Zuerich 353-363
◇ B25 D15 ◇ REV MR 83c:03037 Zbl 496 # 03021 • ID 35139

SCHOENHAGE, A. see Vol. V for further entries

SCHOENING, U. [1981] *A note on complete sets for the polynomial-time hierarchy* (**J** 1456) SIGACT News 13/1*30-34
◇ D15 ◇ REV Zbl 483 # 68046 • ID 36812

SCHOENING, U. [1981] *Untersuchungen zur Struktur von NP und verwandten Komplexitaetsklassen mit Hilfe verschiedener polynomieller Reduktionen* (**X** 3503) Univ Stuttgart Inst Informatik: Stuttgart 78pp
◇ D15 ◇ REV Zbl 516 # 03036 • ID 37265

SCHOENING, U. [1982] *A uniform approach to obtain diagonal sets in complexity classes* (**J** 1426) Theor Comput Sci 18*95-103
◇ D15 ◇ REV MR 83b:68055 Zbl 485 # 68039 • ID 38970

SCHOENING, U. [1982] *On NP-decomposable sets* (**J** 1456) SIGACT News 14/1*18-20
◇ D15 ◇ REV Zbl 565 # 68045 • ID 48669

SCHOENING, U. [1983] *A low and a high hierarchy within NP* (**J** 0119) J Comp Syst Sci 27*14-28
◇ D15 ◇ REV Zbl 515 # 68046 • ID 37866

SCHOENING, U. [1983] *On the structure of* Δ_2^P (**J** 0232) Inform Process Lett 16*209-211
◇ D15 ◇ REV MR 85a:03054 Zbl 533 # 03024 • ID 34804

SCHOENING, U. [1984] see BOOK, R.V.

SCHOENING, U. [1984] *Minimal pairs for P* (**J** 1426) Theor Comput Sci 31*41-48
◇ D15 ◇ REV MR 86d:68028 Zbl 543 # 03025 • ID 40913

SCHOENING, U. [1984] *Robust algorithms: a different approach to oracles* (**P** 4012) Automata, Lang & Progr (11);1984 Antwerpen 448-453
◇ D15 ◇ REV Zbl 554 # 68034 • ID 47188

SCHOENING, U. [1984] see BALCAZAR, J.L.

SCHOENING, U. [1984] see ORPONEN, P.

SCHOENING, U. [1985] see BALCAZAR, J.L.

SCHOENING, U. [1985] see KO, KER-I

SCHOENING, U. [1985] see ORPONEN, P.

SCHRADER, R. [1981] see KORTE, B.

SCHRAM, J.M. [1984] *Recursively prime trees* (**J** 2761) J Recreational Math 16*281-288
◇ C57 C65 D45 ◇ REV MR 85m:11008 • ID 44732

SCHREIBER, P. [1966] *Ueber die Entbehrlichkeit von Hilfsbuchstaben bei der Berechnung mehrstelliger Wortfunktionen durch Markowsche Algorithmen* (**J** 0068) Z Math Logik Grundlagen Math 12*241-242
◇ D03 D20 ◇ REV MR 34 # 2455 Zbl 202.311 • ID 12977

SCHREIBER, P. [1967] *Normale Algorithmen ohne abbrechende Regeln* (**J** 0068) Z Math Logik Grundlagen Math 13*189-191
◇ D03 ◇ REV MR 35 # 6558 Zbl 204.17 • ID 12978

SCHREIBER, P. [1971] *Theseus im Labyrinth als Turingmaschine* (**J** 0068) Z Math Logik Grundlagen Math 17*57-60
◇ D10 ◇ REV MR 46 # 5120 Zbl 216.7 • ID 12981

SCHREIBER, P. see Vol. I, II, III, VI for further entries

SCHRIEBER, L. [1985] *Recursive properties of euclidean domains* (**J** 0073) Ann Pure Appl Logic 29*59-77
◇ C57 C60 D45 ◇ REV Zbl 574 # 03029 • ID 47474

SCHROEDER, M.E. [1970] *Hierarchien primitiv-rekursiver Funktionen im Transfiniten* (**J** 0009) Arch Math Logik Grundlagenforsch 13*114-133
◇ D20 E10 E47 ◇ REV MR 43 # 7332 Zbl 244 # 02014 • ID 12984

SCHROEDER, M.E. see Vol. V for further entries

SCHROEPPEL, R. & SHAMIR, A. [1981] *A $T=O(2^{n/2})$, $S=O(2^{n/4})$ algorithm for certain NP-complete problems* (J 1428) SIAM J Comp 10*456-464
◇ D15 ◇ REV MR 83a:90116 Zbl 462 # 68015 • ID 69905

SCHUBERT, L.K. [1974] *Iterated limiting recursion and the program minimization problem* (J 0037) ACM J 21*436-445
◇ B75 D15 D20 ◇ REV MR 57 # 2893 Zbl 352 # 68060 • ID 78250

SCHUBERT, L.K. [1974] *Representative samples of programmable functions* (J 0194) Inform & Control 25*30-44
◇ B75 D15 D20 ◇ REV MR 58 # 32065 Zbl 304 # 68027 • ID 65136

SCHUETT, D. [1970] see BOEHLING, K.H.

SCHUETTE, K. [1933] *Untersuchungen zum Entscheidungsproblem der mathematischen Logik* (J 0043) Math Ann 109*572-603
◇ B20 B25 D35 ◇ REV Zbl 9.2 FdM 60.21 • ID 12997

SCHUETTE, K. [1951] *Eine Bemerkung ueber quasirekursive Funktionen* (J 1114) Arch Phil 4*223-224
• REPR [1950] (J 0009) Arch Math Logik Grundlagenforsch 1*63-64
◇ D20 ◇ REV MR 14.527 Zbl 45.5 JSL 18.75 • ID 19695

SCHUETTE, K. [1954] *Kennzeichnung von Ordnungszahlen durch rekursiv erklaerte Funktionen* (J 0043) Math Ann 127*15-32
◇ D20 F15 ◇ REV MR 15.689 Zbl 55.49 JSL 19.217 • ID 13005

SCHUETTE, K. [1966] see CROSSLEY, J.N.

SCHUETTE, K. [1976] *Primitiv-rekursive Ordinalzahlfunktionen* (J 1944) Sitzb, Akad Wiss, Bayern, Math-Nat Kl 1975*143-153
◇ D60 F15 ◇ REV MR 54 # 12503 Zbl 358 # 02055 • ID 31510

SCHUETTE, K. see Vol. I, II, III, V, VI for further entries

SCHUETZENBERGER, M.-P. [1961] *A remark on finite transducers* (J 0194) Inform & Control 4*185-196
◇ D05 ◇ REV MR 26 # 1235 Zbl 119.139 JSL 34.297 • ID 13023

SCHUETZENBERGER, M.-P. [1961] *On the definition of a family of automata* (J 0194) Inform & Control 4*245-270
◇ D05 ◇ REV MR 24 # B1725 Zbl 104.7 JSL 34.296 • ID 13024

SCHUETZENBERGER, M.-P. [1962] *Finite counting automata* (J 0194) Inform & Control 5*91-107
◇ D05 ◇ REV MR 27 # 4720 Zbl 118.125 JSL 35.296 • ID 13025

SCHUETZENBERGER, M.-P. [1963] *Certain elementary families of automata* (P 0674) Symp Math Th of Automata;1962 New York 139-153
◇ D05 ◇ REV MR 29 # 5696 JSL 34.296 • ID 13026

SCHUETZENBERGER, M.-P. [1963] *On context-free languages and push-down automata* (J 0194) Inform & Control 6*246-264
◇ D05 D10 ◇ REV Zbl 123.125 JSL 34.297 • ID 13027

SCHUETZENBERGER, M.-P. [1966] *On a family of sets related to McNaughton's L-language* (P 0746) Automata Th;1964 Ravello 320-324
◇ D05 ◇ REV MR 36 # 2448 Zbl 192.79 • ID 27539

SCHUETZENBERGER, M.-P. [1973] *A propos du relation rationelles fonctionelles* (P 0763) Automata, Lang & Progr (1);1972 Rocquencourt 103-114
◇ D05 ◇ REV MR 52 # 7205 Zbl 283 # 94018 • ID 65149

SCHUETZENBERGER, M.-P. see Vol. I, VI for further entries

SCHULER, P.F. [1975] *A note on degrees of context-sensitivity* (J 1431) Acta Inf 5*387-394
◇ D05 ◇ REV MR 54 # 1731 Zbl 312 # 02032 • ID 29101

SCHULER, P.F. [1975] *WCS-analysis of the context-sensitive* (J 1431) Acta Inf 4*359-371
◇ D05 D25 ◇ REV MR 52 # 12434 Zbl 306 # 02037 • ID 21742

SCHULZ, MARTIN [1981] see KLAEREN, H.A.

SCHUPP, P.E. [1968] *On Dehn's algorithm and the conjugacy problem* (J 0043) Math Ann 178*119-130
◇ D40 ◇ REV MR 38 # 5901 Zbl 164.19 • ID 13046

SCHUPP, P.E. [1970] *A note on recursively enumerable predicates in groups* (J 0027) Fund Math 66*61-63
◇ B25 D25 D40 ◇ REV MR 40 # 4345 Zbl 193.317 • ID 13048

SCHUPP, P.E. [1970] *On the conjugacy problem for certain quotient groups of free products* (J 0043) Math Ann 186*123-129
◇ D40 ◇ REV MR 41 # 5475 Zbl 182.35 • ID 13047

SCHUPP, P.E. [1972] see APPEL, K.I.

SCHUPP, P.E. [1973] *A survey of small cancellation theory* (P 0678) Word Probl: Decis & Burnside Probl in Group Th;1969 Irvine 569-589
◇ C60 D40 ◇ REV MR 54 # 415 Zbl 292 # 20034 JSL 41.786 • ID 44402

SCHUPP, P.E. [1981] see MULLER, D.E.

SCHUPP, P.E. [1983] see MULLER, D.E.

SCHUPP, P.E. [1985] see MULLER, D.E.

SCHURMANN, A. [1971] *Functions computable by a computer (Polish and Russian summaries)* (J 0063) Studia Logica 27*57-72
◇ D05 D10 D20 ◇ REV MR 47 # 6458 Zbl 253 # 68006 • ID 13049

SCHUSTER, H. [1983] see DETTKI, H.J.

SCHUSTER, P. [1976] *Probleme, die zum Erfuellungsproblem der Aussagenlogik polynomial aequivalent sind* (P 3196) Kompl von Entscheid Probl;1973/74 Zuerich 36-48
◇ B05 D15 ◇ REV MR 57 # 18232 Zbl 386 # 68048 • ID 69906

SCHWABHAEUSER, W. & SZMIELEW, W. & TARSKI, A. [1983] *Metamathematische Methoden in der Geometrie* (X 0811) Springer: Heidelberg & New York viii + 482pp
◇ B30 C65 C98 D35 ◇ REV MR 85e:03004 Zbl 564 # 51001 • ID 40225

SCHWABHAEUSER, W. see Vol. I, III, V, VI for further entries

SCHWARTZ, DIETRICH [1978] *Kanonische Abbildungen und Eilenberg-Maschinen* (J 0068) Z Math Logik Grundlagen Math 24*177-186
◇ D05 ◇ REV MR 80c:03042 Zbl 391 # 68031 • ID 78294

SCHWARTZ, DIETRICH [1979] *Beitrag zur algebraischen Rekursionstheorie* (J 0114) Math Nachr 90*249-256
◇ D75 ◇ REV MR 80m:03082 Zbl 467 # 03036 • ID 55034

SCHWARTZ, DIETRICH see Vol. I, II, III, V for further entries

SCHWARTZ, J.T. (ED.) [1967] *Mathematical aspects of computer science* (**X** 0803) Amer Math Soc: Providence vi+226pp
◇ B35 B65 B75 B97 D10 D80 D97 ◇ REV MR 38 # 2975 Zbl 165.2 • ID 23572

SCHWARTZ, J.T. [1978] see CHAITIN, G.J.

SCHWARTZ, J.T. see Vol. I, III, V for further entries

SCHWARTZ, T. [1969] *A simple treatment of Church's theorem on the decision problem* (**J** 0079) Logique & Anal, NS 12*153-156
◇ D35 ◇ REV MR 44 # 73 Zbl 193.311 • ID 13069

SCHWARTZ, T. see Vol. I, II for further entries

SCHWARZ, S. [1984] *Recursive automorphisms of recursive linear orderings* (**J** 0073) Ann Pure Appl Logic 26*69-73
◇ C57 C65 D45 ◇ REV MR 85c:03018 Zbl 571 # 03018 • ID 39829

SCHWARZ, S. [1984] *The quotient semilattice of the recursively enumerable degrees modulo the cappable degrees* (**J** 0064) Trans Amer Math Soc 283*315-328
◇ D25 ◇ REV MR 85i:03141 Zbl 512 # 03023 • ID 33614

SCHWARZ, S. see Vol. V for further entries

SCHWEIZER, B. & SKLAR, A. [1969] *A grammar of functions I* (**J** 0529) Aequationes Math 2*62-85
◇ D05 ◇ REV MR 38 # 2227 Zbl 164.335 • REM Part II 1969 • ID 46701

SCHWEIZER, B. & SKLAR, A. [1969] *A grammar of functions II* (**J** 0529) Aequationes Math 3*15-43
◇ D05 ◇ REV MR 42 # 4663 Zbl 179.38 • REM Part I 1969 • ID 13072

SCHWEIZER, B. see Vol. I, V for further entries

SCHWENKEL, F. [1965] *Rekursive Wortfunktionen ueber unendlichen Alphabeten* (**J** 0068) Z Math Logik Grundlagen Math 11*133-147 • ERR/ADD ibid 11*379-380
◇ D20 D75 ◇ REV MR 30 # 3845 Zbl 125.278 JSL 40.621 • ID 13074

SCHWICHTENBERG, H. [1969] *Rekursionsformeln und die Grzegorczyk-Hierarchie* (**J** 0009) Arch Math Logik Grundlagenforsch 12*85-97
◇ D20 ◇ REV MR 40 # 7113 Zbl 213.18 JSL 35.480 • ID 13075

SCHWICHTENBERG, H. [1971] *Eine Klassifikation der ε_0-rekursiven Funktionen* (**J** 0068) Z Math Logik Grundlagen Math 17*61-74
◇ D20 ◇ REV MR 44 # 77 Zbl 232 # 02028 • ID 13076

SCHWICHTENBERG, H. [1972] see ROEDDING, D.

SCHWICHTENBERG, H. [1972] *Beweistheoretische Charakterisierung einer Erweiterung der Grzegorczyk-Hierarchie* (**J** 0009) Arch Math Logik Grundlagenforsch 15*129-145
◇ D20 F15 F30 ◇ REV MR 48 # 3717 Zbl 257 # 02021 • ID 13077

SCHWICHTENBERG, H. & WAINER, S.S. [1975] *Infinite terms and recursion in higher types* (**P** 1440) ⊧ ISILC Proof Th Symp (Schuette);1974 Kiel 341-364
◇ C75 D65 F10 ◇ REV MR 54 # 7231 Zbl 341 # 02033 • ID 25022

SCHWICHTENBERG, H. see Vol. I, III, V, VI for further entries

SCIORE, E. & TANG, A. [1978] *Computability theory in admissible domains* (**P** 1740) ACM Symp Th of Comput (10);1978 San Diego 95-104
◇ B75 D80 ◇ REV MR 80f:68056 • ID 82757

SCIORE, E. see Vol. V, VI for further entries

SCOGNAMIGLIO, G. [1963] *Un metodo di calcolo dei prodotti delle matrici booleane elementari* (**J** 2328) Ann Pont Ist Sup Sci Lett Napoli 13*413-429
◇ B05 D03 ◇ REV Zbl 265 # 06013 • ID 29833

SCOTT, D.S. [1957] see RABIN, M.O.

SCOTT, D.S. [1959] see RABIN, M.O.

SCOTT, D.S. [1960] *On a theorem of Rabin* (**J** 0028) Indag Math 22*481-484
◇ C35 C57 C62 D35 ◇ REV MR 25 # 1985 • ID 11909

SCOTT, D.S. [1961] *On constructing models for arithmetic* (**P** 0633) Infinitist Meth;1959 Warsaw 235-255
◇ C20 C57 C62 H15 ◇ REV MR 27 # 2423 Zbl 126.12 JSL 38.336 • ID 24930

SCOTT, D.S. [1964] *Invariant Borel sets* (**J** 0027) Fund Math 56*117-128
◇ C75 D55 E15 ◇ REV MR 30 # 3027 Zbl 152.213 • ID 11916

SCOTT, D.S. [1967] *Some definitional suggestions for automata theory* (**J** 0119) J Comp Syst Sci 1*187-212
◇ D05 ◇ REV Zbl 164.321 JSL 40.615 • ID 11918

SCOTT, D.S. [1975] *Lambda calculus and recursion theory* (**P** 0757) Scand Logic Symp (3);1973 Uppsala 154-193
◇ B40 B75 D25 D75 ◇ REV MR 52 # 5372 Zbl 322 # 02023 JSL 43.373 • ID 18375

SCOTT, D.S. [1977] *Logic and programming languages* (**J** 0212) ACM Commun 20*634-641
◇ A10 B75 D80 ◇ REV MR 56 # 10114 Zbl 355 # 68019 • ID 50263

SCOTT, D.S. [1982] *Lectures on a mathematical theory of computation* (**P** 3906) Th Found of Progr Methodol;1981 Marktoberdorf 145-292
◇ B40 B98 D75 ◇ REV MR 85g:68043 Zbl 516 # 68064 • ID 38593

SCOTT, D.S. see Vol. I, II, III, V, VI for further entries

SCOTT, E.A. [1984] *A finitely presented simple group with unsolvable conjugacy problem* (**J** 0032) J Algeb 90*333-353
◇ D40 ◇ REV MR 86f:20029c Zbl 544 # 20029 • ID 44271

SEBELIK, J. [1982] *Horn clause programs and recursive functions defined by systems of equations* (**J** 0156) Kybernetika (Prague) 18*106-120
◇ D20 ◇ REV MR 84c:03078 Zbl 489 # 03013 • ID 34015

SEDOL, YA.YA. [1964] *The free product of associative calculi with common subalphabet, and some related questions (Russian)* (**J** 0023) Dokl Akad Nauk SSSR 158*1034-1037 • TRANSL [1964] (**J** 0062) Sov Math, Dokl 5*1362-1365
◇ D05 D40 ◇ REV MR 29 # 4687 Zbl 239 # 02020 • ID 11930

SEEBOLD, P. [1985] *Generalized Thue-Morse sequences* (**P** 4647) FCT'85 Fund of Comput Th;1985 Cottbus 402-411
◇ D03 ◇ ID 49084

SEEBOLD, P. [1985] *Overlap-free sequences* (**P** 4622) Autom on Infinite Words;1984 Le Mont-Dore 207-215
◇ D05 ◇ ID 49466

SEESE, D.G. [1972] *Entscheidbarkeits- und Definierbarkeitsfragen der Theorie "netzartiger" Graphen I (Russian) (English and French summaries)* (**J** 0115) Wiss Z Humboldt-Univ Berlin, Math-Nat Reihe 21*513-517
◇ C80 C85 D35 ◇ REV MR 49 # 7133 Zbl 254 # 02036 • ID 11931

SEESE, D.G. [1975] *Ein Unentscheidbarkeitskriterium* (**J** 0115) Wiss Z Humboldt-Univ Berlin, Math-Nat Reihe 24*772-780
◇ D35 ◇ REV MR 58 # 5158 Zbl 331 # 02026 • ID 65177

SEESE, D.G. [1977] *Second order logic, generalized quantifiers and decidability* (**J** 0014) Bull Acad Pol Sci, Ser Math Astron Phys 25*725-732
◇ B15 B25 C55 C65 C80 C85 D35 ◇ REV MR 57 # 87 Zbl 383 # 03010 • ID 27132

SEESE, D.G. [1978] *A remark to the undecidability of well-orderings with the Haertig quantifier* (**J** 0014) Bull Acad Pol Sci, Ser Math Astron Phys 26*951
◇ C55 C65 C80 D35 ◇ REV MR 80b:03014 Zbl 408 # 03031 • ID 56270

SEESE, D.G. [1978] *Decidability of ω-trees with bounded sets (German and Russian summaries)* (**X** 2888) ZI Math Mech Akad Wiss DDR: Berlin 52pp
◇ B25 C65 C85 D05 ◇ REV MR 58 # 194 Zbl 386 # 03004 • ID 52180

SEESE, D.G. [1978] *Ueber unentscheidbare Erweiterungen von SC* (**J** 0068) Z Math Logik Grundlagen Math 24*63-71
◇ D35 ◇ REV MR 57 # 5724 Zbl 375 # 02043 • ID 51620

SEESE, D.G. & TUSCHIK, H.-P. & WEESE, M. [1982] *Undecidable theories in stationary logic* (**J** 0053) Proc Amer Math Soc 84*563-567
◇ C55 C65 C80 D35 E05 E75 ◇ REV MR 84c:03071 Zbl 515 # 03002 • ID 33532

SEESE, D.G. see Vol. I, II, III, V for further entries

SEGAL, D. [1979] see GRUNEWALD, F.

SEGAL, D. [1980] see GRUNEWALD, F.

SEGAL, D. [1985] see GRUNEWALD, F.

SEGAL, D. see Vol. III for further entries

SEIBT, H. [1983] *On isomorphic partial recursive definitions (German and Russian summaries)* (**J** 0598) Wiss Z Paed Hochsch Potsdam 27*25-27
◇ D20 ◇ REV MR 84i:68043 • ID 40152

SEIDENBERG, A. [1970] *Construction of the integral closure of a finite integral domain* (**J** 0059) Rend Sem Mat Fis Milano 40*100-120
◇ C57 C60 F55 ◇ REV MR 45 # 3396 Zbl 218 # 14023 • REM Part II 1975 • ID 41695

SEIDENBERG, A. [1975] *Construction of the integral closure of a finite integral domain. II* (**J** 0053) Proc Amer Math Soc 52*368-372
◇ C57 C60 F55 ◇ REV MR 54 # 12741 Zbl 333 # 13004 • REM Part I 1970 • ID 41698

SEIDENBERG, A. [1978] *Constructions in a polynomial ring over the ring of integers* (**J** 0100) Amer J Math 100*685-706
◇ D45 F55 ◇ REV MR 81g:13016 Zbl 416 # 13013 • ID 53225

SEIDENBERG, A. see Vol. III, VI for further entries

SEIDL, H. [1985] *A quadratic regularity test for nondeleting macro S grammars* (**P** 4647) FCT'85 Fund of Comput Th;1985 Cottbus 422-430
◇ D05 ◇ ID 49086

SEIFERAS, J.I. [1974] *Observations on nondeterministic multidimensional iterative arrays* (**P** 1464) ACM Symp Th of Comput (6);1974 Seattle 276-289
◇ D10 D15 ◇ REV MR 55 # 1854 Zbl 358 # 68088 • ID 50543

SEIFERAS, J.I. [1976] see MCNAUGHTON, R.

SEIFERAS, J.I. [1977] *Iterative arrays with direct central control* (**J** 1431) Acta Inf 8*177-192
◇ D10 D15 ◇ REV MR 58 # 3699 Zbl 337 # 94035 • ID 65192

SEIFERAS, J.I. [1977] *Linear-time computation by nondeterministic multidimensional iterative arrays* (**J** 1428) SIAM J Comp 6*487-504
◇ D10 D15 ◇ REV MR 56 # 7330 Zbl 368 # 68049 • ID 51290

SEIFERAS, J.I. [1977] see LEONG, B.

SEIFERAS, J.I. [1977] see GALIL, Z.

SEIFERAS, J.I. [1978] see GALIL, Z.

SEIFERAS, J.I. [1978] see FISCHER, MICHAEL J.

SEIFERAS, J.I. [1981] see RACKOFF, C.W.

SEIFERAS, J.I. see Vol. V for further entries

SEIFERT, F.D. [1975] see ANISIMOV, A.V.

SEIFERT, F.D. [1976] *Eine Klassifizierung endlich erzeugbarer Gruppen durch formale Sprachen* (**J** 0068) Z Math Logik Grundlagen Math 22*419-424
◇ D05 ◇ REV MR 58 # 28175 Zbl 358 # 68122 • ID 24286

SEKI, S. [1963] see HIROSE, K.

SEKI, S. [1979] see KOBUCHI, Y.

SEKI, S. see Vol. I, V, VI for further entries

SELIVANOV, V.L. [1976] *Enumerations of families of general recursive function (Russian)* (**J** 0003) Algebra i Logika 15*205-226,246
• TRANSL [1976] (**J** 0069) Algeb and Log 15*128-141
◇ D20 D25 ◇ REV MR 57 # 16012 Zbl 348 # 02038 • ID 26041

SELIVANOV, V.L. [1976] *On computability of some classes of numerations (Russian)* (**J** 3937) Veroyat Met i Kibern (Kazan) 12-13*157-170
◇ D25 D45 ◇ REV MR 58 # 27396 Zbl 398 # 03031 • ID 52759

SELIVANOV, V.L. [1976] *Two theorems on computable numberings (Russian)* (**J** 0003) Algebra i Logika 15*470-484,488
• TRANSL [1976] (**J** 0069) Algeb and Log 15*297-306
◇ D45 ◇ REV MR 56 # 15387 Zbl 358 # 02050 • ID 26054

SELIVANOV, V.L. [1977] *Enumerations of canonically calculable families of finite sets (Russian)* (**J** 0092) Sib Mat Zh 18*1373-1380,1437
• TRANSL [1977] (**J** 0475) Sib Math J 18*973-979
◇ D45 ◇ REV MR 58 # 21555 Zbl 384 # 03024 • ID 52066

SELIVANOV, V.L. [1978] *On index sets of classes of numberings (Russian)* (**J** 3937) Veroyat Met i Kibern (Kazan) 14*90-103
◇ D45 ◇ REV MR 81k:03042 Zbl 411 # 03037 • ID 52891

SELIVANOV, V.L. [1978] *On index sets of computable classes of finite sets (Russian)* (C 2899) Algor & Avtomaty 95-99
◇ D25 D30 D55 ◇ REV MR 82b:03077 Zbl 412 # 03025
• ID 52956

SELIVANOV, V.L. [1978] *Some remarks about classes of recursively enumerable sets (Russian)* (J 0092) Sib Mat Zh 19*153-160,238
• TRANSL [1978] (J 0475) Sib Math J 19*109-114
◇ D25 ◇ REV MR 81c:03036 Zbl 387 # 03013 • ID 52229

SELIVANOV, V.L. [1979] *Structures of the degrees of unsolvability of index sets (Russian)* (J 0003) Algebra i Logika 18*463-480,508-509
• TRANSL [1979] (J 0069) Algeb and Log 18*286-299
◇ D30 ◇ REV MR 81i:03067 Zbl 439 # 03025 • ID 56016

SELIVANOV, V.L. [1982] *A class of reducibilities in the theory of recursive functions (Russian)* (J 3937) Veroyat Met i Kibern (Kazan) 18*83-100
◇ D30 ◇ REV MR 85e:03101 Zbl 543 # 03030 • ID 40704

SELIVANOV, V.L. [1982] *On the index sets in the Kleene-Mostowski hierarchy (Russian)* (C 3953) Mat Log & Teor Algor 135-158
◇ D25 D45 D55 ◇ REV MR 85h:03048 Zbl 522 # 03029
• ID 37788

SELIVANOV, V.L. [1982] *The structure of degrees of generalized index sets (Russian)* (J 0003) Algebra i Logika 21*472-491
• TRANSL [1982] (J 0069) Algeb and Log 21*316-330
◇ D30 D45 ◇ REV MR 85m:03031 Zbl 572 # 03021
• ID 44853

SELIVANOV, V.L. [1983] *Effective analogues of A-, B- and C-sets and their application to index sets (Russian)* (J 3937) Veroyat Met i Kibern (Kazan) 19*112-128
◇ D25 D55 E15 ◇ ID 44858

SELIVANOV, V.L. [1983] *Hierarchies of hyperarithmetical sets and functions (Russian)* (J 0003) Algebra i Logika 22*666-692
• TRANSL [1983] (J 0069) Algeb and Log 22*473-491
◇ D55 ◇ REV Zbl 536 # 03025 • ID 37107

SELIVANOV, V.L. [1984] *Index sets in the hyperarithmetical hierarchy (Russian)* (J 0092) Sib Mat Zh 25/3*164-181
• TRANSL [1984] (J 0475) Sib Math J 25*474-488
◇ D55 ◇ REV MR 86a:03045 • ID 45283

SELIVANOV, V.L. [1984] *On the hierarchy of limit computations (Russian)* (J 0092) Sib Mat Zh 25/5*146-156
• TRANSL [1984] (J 0475) Sib Math J 25*798-806
◇ D20 D25 ◇ REV MR 86f:03070 • ID 44084

SELIVANOV, V.L. [1985] *The Ershov hierarchy (Russian)* (J 0092) Sib Mat Zh 26/1*134-149
• TRANSL [1985] (J 0475) Sib Math J 26/1*105-133
◇ D55 ◇ ID 44469

SELIVANOWSKI, E. [1927] *Sur une classe d'ensembles definis par une infinite denombrable de conditions* (J 0109) C R Acad Sci, Paris 184*1311-1313
◇ D55 E15 ◇ REV FdM 53.173 • ID 41560

SELIVANOWSKI, E. [1928] *Ueber eine Klasse von effektiven Mengen (Mengen C) (Russian) (French summary)* (J 1404) Mat Sb, Akad Nauk SSSR 35*379-412
◇ D55 E15 ◇ REV FdM 54.94 • ID 39107

SELIVANOWSKI, E. see Vol. V for further entries

SELMAN, A.L. [1971] *Arithmetical reducibilities I* (J 0068) Z Math Logik Grundlagen Math 17*335-350
◇ D30 D55 ◇ REV MR 46 # 3285 Zbl 229 # 02037 • REM Part II 1972 • ID 11962

SELMAN, A.L. [1972] *Applications of forcing to the degree-theory of the arithmetical hierarchy* (J 3240) Proc London Math Soc, Ser 3 25*586-602
◇ D30 D55 E40 ◇ REV MR 47 # 3155 Zbl 251 # 02043
• ID 11963

SELMAN, A.L. [1972] *Arithmetical reducibilities II* (J 0068) Z Math Logik Grundlagen Math 18*83-92
◇ D30 D55 ◇ REV MR 46 # 3286 Zbl 238 # 02035 • REM Part I 1971 • ID 11965

SELMAN, A.L. [1972] see JONES, N.D.

SELMAN, A.L. [1973] *Sets of formulas valid in finite structures* (J 0064) Trans Amer Math Soc 177*491-504
◇ C13 D30 ◇ REV MR 47 # 8272 Zbl 276 # 02025
• ID 11966

SELMAN, A.L. [1974] see LADNER, R.E.

SELMAN, A.L. [1974] *Relativized halting problems* (J 0068) Z Math Logik Grundlagen Math 20*193-198
◇ D10 D25 D30 ◇ REV MR 52 # 13350 Zbl 298 # 02043
• ID 11967

SELMAN, A.L. [1974] see JONES, N.D.

SELMAN, A.L. [1975] see LADNER, R.E.

SELMAN, A.L. [1977] see BOOK, R.V.

SELMAN, A.L. [1978] *Polynomial time enumeration reducibility* (J 1428) SIAM J Comp 7*440-457
◇ D15 ◇ REV MR 80d:68059 Zbl 386 # 68055 • ID 69911

SELMAN, A.L. [1979] see BAKER, T.P.

SELMAN, A.L. [1979] *P-selective sets, tally languages, and the behaviour of polynomial time reducibilities on NP (preliminary report)* (P 1873) Automata, Lang & Progr (6);1979 Graz 546-555
◇ D05 D15 ◇ REV MR 81j:68055 Zbl 422 # 03013
• ID 53474

SELMAN, A.L. [1979] *P-selective sets, tally languages, and the behaviour of polynomial time reducibilities on NP* (J 0041) Math Syst Theory 13*55-65
◇ D05 D15 ◇ REV MR 81a:68086 Zbl 405 # 03018
• ID 53133

SELMAN, A.L. [1981] *Some observations on NP real numbers and P-selective sets* (J 0119) J Comp Syst Sci 23*326-332
◇ D15 ◇ REV MR 83f:68046 Zbl 486 # 03020 • ID 38080

SELMAN, A.L. [1982] *Analogues of semirecursive sets and effective reducibilities to the study of NP complexity* (J 0194) Inform & Control 52*36-51
◇ D15 ◇ REV MR 84g:03056 Zbl 504 # 03022 • ID 34165

SELMAN, A.L. [1982] *Reductions on NP and p-selective sets* (J 1426) Theor Comput Sci 19*287-304
◇ D15 ◇ REV MR 84b:68048 Zbl 489 # 03016 • ID 37207

SELMAN, A.L. [1983] see BOOK, R.V.

SELMAN, A.L. [1984] see BOOK, R.V.

SELMAN, A.L. [1985] see BOOK, R.V.

SELMAN, A.L. see Vol. II, III for further entries

SEMENOV, A.L. [1973] *Algorithmic problems for power series and for context-free grammars (Russian)* (**J** 0023) Dokl Akad Nauk SSSR 212*50-52
• TRANSL [1973] (**J** 0062) Sov Math, Dokl 14*1319-1322
◇ D05 D80 ◇ REV MR 48 # 10216 Zbl 322 # 68052
• ID 11973

SEMENOV, A.L. [1977] *Presburgerness of predicates regular in two number systems (Russian)* (**J** 0092) Sib Mat Zh 18*403-418,479
• TRANSL [1977] (**J** 0475) Sib Math J 18*289-300
◇ B10 D20 F30 ◇ REV MR 56 # 8349 Zbl 369 # 02023
• ID 51323

SEMENOV, A.L. [1980] *An interpretation of free algebras in free groups (Russian)* (**J** 0023) Dokl Akad Nauk SSSR 252*1329-1332
• TRANSL [1980] (**J** 0062) Sov Math, Dokl 21*952-955
◇ D20 D35 ◇ REV MR 82b:08011 Zbl 482 # 20022
• ID 82771

SEMENOV, A.L. & USPENSKIJ, V.A. [1981] *What are the gains of the theory of algorithms: basic developments connected with the concept of algorithm and with its application in mathematics* (**P** 3729) Algor in Modern Math & Comput Sci;1979 Urgench 100-234
• TRANSL [1982] (**P** 3803) Algor Sovrem Mat & Prilozh;1979 Urgench 1*99-342
◇ D20 D98 ◇ REV MR 83h:68048 Zbl 477 # 68035
• ID 39208

SEMENOV, A.L. see Vol. III, VI for further entries

SENDOV, B. & SKORDEV, D.G. [1961] *On equations in words (Russian) (German summary)* (**J** 0068) Z Math Logik Grundlagen Math 7*289-297
◇ D05 ◇ REV MR 31 # 57 Zbl 119.13 • ID 12435

SENDOV, B. see Vol. VI for further entries

SEPER, K. [1966] *A note on normalizability of Ter-Zaharjan's quasi-normal algorithms and of Markov's branching, iteration and union operations* (**J** 3519) Glas Mat, Ser 3 (Zagreb) 1(21)*133-137
◇ D03 ◇ REV Zbl 166.5 • ID 31504

SEPER, K. [1979] *Algorithmic constructions inspired by Caporaso* (**J** 0400) Publ Inst Math, NS (Belgrade) 25(39)*210-218
◇ D05 D20 ◇ REV MR 81a:68040 Zbl 415 # 03050
• ID 53152

SEPER, K. see Vol. II, VI for further entries

SETH, S.C. & STECKELBERG, J.M. [1977] *On a relation between algebraic programs and Turing machines* (**J** 0232) Inform Process Lett 6*180-183
◇ B75 D10 ◇ REV MR 58 # 25085 Zbl 368 # 68032
• ID 51289

SETHI, R. [1974] see COOK, S.A.

SETHI, R. see Vol. V, VI for further entries

SEWELSON, V. [1985] see HARTMANIS, J.

SHABUNIN, L.V. [1973] *The undecidability of certain formal systems of combinatory logic (Russian)* (**J** 0288) Vest Ser Mat Mekh, Univ Moskva 28/5*36-40
• TRANSL [1973] (**J** 0510) Moscow Univ Math Bull 28/5-6*29-33
◇ B40 D35 ◇ REV MR 49 # 8846 Zbl 281 # 02034
• ID 11733

SHABUNIN, L.V. [1974] *Some algorithmic problems of calculi of combinatory logic (Russian) (English summary)* (**J** 0288) Vest Ser Mat Mekh, Univ Moskva 29/6*36-41
• TRANSL [1974] (**J** 0510) Moscow Univ Math Bull 29/5-6*72-77
◇ B40 D35 ◇ REV MR 52 # 2834 Zbl 298 # 02018
• ID 17628

SHABUNIN, L.V. [1975] *Combinatory calculi I,II (Russian)* (**J** 0288) Vest Ser Mat Mekh, Univ Moskva 30/1*12-17,30/2*10-14
• TRANSL [1975] (**J** 0510) Moscow Univ Math Bull 30/1-2*9-13,79-83
◇ B40 D35 ◇ REV MR 52 # 7863 MR 52 # 7864 Zbl 314 # 02040 Zbl 314 # 02041 • ID 18369

SHABUNIN, L.V. see Vol. VI for further entries

SHAKUOV, S.N. [1976] *Linear acceleration of the operating time of single-tape Turing machines (Russian)* (**J** 0023) Dokl Akad Nauk SSSR 230*792-794
• TRANSL [1976] (**J** 0062) Sov Math, Dokl 17*1407-1409
◇ D10 D15 ◇ REV MR 54 # 7223 Zbl 361 # 02051
• ID 25014

SHAKUOV, S.N. [1977] *Fast Turing computations and their linear speedup (Russian)* (**J** 0023) Dokl Akad Nauk SSSR 236*556-557
• TRANSL [1977] (**J** 0062) Sov Math, Dokl 18*1250-1252
◇ D10 D15 ◇ REV MR 57 # 5708 Zbl 391 # 68026
• ID 52362

SHAMIR, A. [1976] *The fixedpoints of recursive definitions* (**X** 3249) Weizmann Inst Sci: Rehovot
◇ D20 ◇ REV Zbl 423 # 03051 • REM Thesis for the degree of doctor of philosophy • ID 53563

SHAMIR, A. [1976] see MANNA, Z.

SHAMIR, A. [1978] see MANNA, Z.

SHAMIR, A. [1981] see SCHROEPPEL, R.

SHAMIR, E. [1963] *On sequential languages and two classes of regular events. Introduction* (**J** 0146) Z Phonetik Sprachwiss Kommunikation 16*389-390
◇ D05 ◇ REV JSL 37.200 • ID 12007

SHAMIR, E. [1963] see PERLES, M.A.

SHAMIR, E. [1964] see BAR-HILLEL, Y.

SHAMIR, E. [1965] *On sequential languages and two classes of regular events* (**J** 0146) Z Phonetik Sprachwiss Kommunikation 18*61-69
◇ D05 ◇ REV MR 32 # 1122 JSL 37.200 • ID 12009

SHAMIR, E. & SNIR, M. [1980] *On the depth complexity of formulas* (**J** 0041) Math Syst Theory 13*301-322
◇ D15 ◇ REV MR 81i:68067 Zbl 445 # 68031 • ID 56522

SHAMIR, E. [1980] see GAIFMAN, H.

SHAMIR, E. [1983] see GORDON, D.

SHANIN, N.A. [1955] *On some logical problems of arithmetic (Russian)* (**S** 0066) Tr Mat Inst Steklov 43*112pp
◇ D20 F30 F50 ◇ REV MR 19.4 Zbl 68.12 JSL 22.79
• ID 19648

SHANIN, N.A. [1976] *On the quantifier of limiting realizability (Russian)* (**S** 0228) Zap Nauch Sem Leningrad Otd Mat Inst Steklov 60*209-220,227
• TRANSL [1980] (**J** 1531) J Sov Math 14*1565-1572
◇ C57 C80 F50 ◇ REV MR 58 # 27418 Zbl 344 # 02025
• ID 65058

SHANIN, N.A. [1977] *On the quantifier of limiting realizability* (**P** 3269) Set Th Found Math (Kurepa);1977 Beograd 127
◇ B55 C57 C80 F50 ◇ REV MR 58 # 21516
Zbl 361 # 02046 • ID 53812

SHANIN, N.A. [1979] *On canonical recursive functions and operations (Russian) (English summary)* (**S** 0228) Zap Nauch Sem Leningrad Otd Mat Inst Steklov 88*218-235,246-247
• TRANSL [1982] (**J** 1531) J Sov Math 20*2381-2390
◇ D20 ◇ REV MR 81a:03044 Zbl 429 # 03024 • ID 53855

SHANIN, N.A. see Vol. I, V, VI for further entries

SHANK, H.S. [1971] see GARFUNKEL, S.

SHANK, H.S. [1971] *Records of Turing machines* (**J** 0041) Math Syst Theory 5*50-55
◇ D10 ◇ REV MR 46 # 7005 Zbl 214.19 • ID 12011

SHANK, H.S. [1972] see GARFUNKEL, S.

SHANK, H.S. see Vol. III for further entries

SHANNON, C.E. [1953] *Computers and automata* (**J** 4711) IRE Proc 41*1234-1241
◇ B70 D05 ◇ REV MR 15.902 JSL 19.140 • ID 16860

SHANNON, C.E. [1956] *A universal Turing machine with two internal states* (**C** 0717) Automata Studies 157-165
• TRANSL [1974] (**C** 1902) Stud Th Automaten 183-193
◇ D10 ◇ REV MR 18.103 JSL 36.532 • ID 12012

SHANNON, C.E. [1956] see MCCARTHY, J.

SHANNON, C.E. [1956] see LEEUW DE, K.

SHAPIRO, E.Y. [1984] *Alternation and the computational complexity of logic programs* (**J** 2551) J Log Progr 1/1*19-33
◇ D10 D15 ◇ REV MR 86g:68061 • ID 45467

SHAPIRO, H.S. [1959] *Numbers and functions computable by means of rational recurrence formulae* (**J** 0155) Commun Pure Appl Math 12*513-522
◇ D20 ◇ REV MR 22 # 10905 Zbl 98.244 • ID 12017

SHAPIRO, N.Z. [1956] see LEEUW DE, K.

SHAPIRO, N.Z. [1956] *Degrees of computability* (**J** 0064) Trans Amer Math Soc 82*281-299
◇ D25 D30 ◇ REV MR 19.2 Zbl 70.246 JSL 23.48
• ID 12018

SHAPIRO, N.Z. [1963] *Functions which remain partial recursive under all similarity transformations* (**J** 0036) J Symb Logic 28*17-19
◇ D20 ◇ REV MR 31 # 2146 Zbl 149.246 JSL 32.527
• ID 12019

SHAPIRO, N.Z. [1969] *Real numbers and functions in the Kleene hierarchy and limits of recursive, rational functions* (**J** 0036) J Symb Logic 34*207-214
◇ D55 F60 ◇ REV MR 45 # 67 Zbl 185.23 • ID 12020

SHAPIRO, N.Z. see Vol. VI for further entries

SHAPIRO, S. [1977] *On Church's thesis* (**P** 1704) Int Congr Log, Meth & Phil of Sci (5);1975 London ON 4*23-24
◇ A05 D20 F99 ◇ ID 32450

SHAPIRO, S. [1981] *Understanding Church's thesis* (**J** 0122) J Philos Logic 10*353-365
◇ A05 D20 F99 ◇ REV MR 82j:03006 • ID 78442

SHAPIRO, S. [1982] *Acceptable notation* (**J** 0047) Notre Dame J Formal Log 23*14-20
◇ D20 ◇ REV MR 83m:03047 Zbl 452 # 68055 • ID 35447

SHAPIRO, S. [1983] *Remarks on the development of computability* (**J** 2028) Hist & Phil Log 4*203-220
◇ A10 D98 F99 ◇ REV MR 85c:01037 Zbl 529 # 03015
• ID 37650

SHAPIRO, S. see Vol. I, II, III, VI for further entries

SHARONOV, V.I. & ZAMOV, N.K. [1970] *Amplifications of formulae of predicate calculus (Russian)* (**J** 0468) Uch Zap Ped Inst (Alma Ata) 130/3*54-59
◇ B10 B20 D35 ◇ REV MR 44 # 2573 Zbl 223 # 02010
• ID 14357

SHARONOV, V.I. see Vol. I for further entries

SHATROVA, N.A. [1983] *Degree of complexity of an algorithm for a class of groups (Russian)* (**C** 4017) Algeb Dejstviya & Uporyadochennosti 142-151
◇ D30 D40 ◇ REV MR 85h:20036 Zbl 551 # 20019
• ID 43468

SHATROVA, N.P. [1981] *Upper bound of the degree of complexity of an algorithm for solving the conjugacy problem for a class of groups (Russian)* (**S** 3478) Sovrem Algebra (Leningrad) 1981*105-127
◇ D15 D40 ◇ REV MR 84b:20039 Zbl 486 # 20024
• ID 38392

SHAY, M. & YOUNG, P. [1978] *Characterizing the orders changed by program translators* (**J** 0048) Pac J Math 76*485-490
◇ B75 D20 D25 ◇ REV MR 80a:03053 Zbl 392 # 03029
• ID 32052

SHAYAKHMETOV, T.K. [1968] *Undecidability of certain theories with a supplemental predicate (Russian) (Kazakh summary)* (**J** 0429) Vest Akad Nauk Kazak SSR 24/3*48-50
◇ D35 ◇ REV MR 37 # 1245 Zbl 223 # 02409 • ID 24948

SHCHEGLOV, A.I. [1967] *The algebra of partially recursive functions (Russian)* (**J** 0003) Algebra i Logika 6/6*33-48
◇ D20 D75 ◇ REV MR 37 # 3919 Zbl 236 # 02037
• ID 11888

SHCHEGLOV, A.I. [1968] *Power of the set of maximal sub-algebras of an algebra of partial-recursive functions (Russian)* (**J** 0003) Algebra i Logika 7/3*119-121
• TRANSL [1968] (**J** 0069) Algeb and Log 7*201-202
◇ D20 D75 ◇ REV MR 41 # 57 Zbl 236 # 02038
• ID 11889

SHCHEGLOV, A.I. [1969] *A certain algebra of one-place primitive recursive functions (Russian)* (**J** 0226) Uch Zap Ped Inst, Ivanovo 44*11-16
◇ D20 D75 ◇ REV MR 47 # 1595 • ID 11890

SHCHEGLOV, A.I. see Vol. V for further entries

SHCHEGOL'KOV, E.A. [1948] *On the uniformization of certain B-sets (Russian)* (**J** 0023) Dokl Akad Nauk SSSR 59*1065-1068
◇ D55 E15 ◇ REV MR 9.417 Zbl 35.323 • ID 11891

SHCHEGOL'KOV, E.A. [1950] *Elements of the theory of B-sets (Russian)* (J 0067) Usp Mat Nauk 5/5(39)*14-44
• TRANSL [1955] (X 0806) Dt Verlag Wiss: Berlin iii+108pp
◇ D55 E15 ◇ REV MR 12.597 Zbl 38.194 • REM Transl. in: "Arbeiten zur deskriptiven Mengenlehre" which contains also papers of Arsenin,V.Ya & Lyapunov,A.A. • ID 11892

SHCHEGOL'KOV, E.A. [1955] see ARSENIN, V.YA.

SHCHEGOL'KOV, E.A. [1959] *On the uniformization and splitting of certain sets (Russian)* (J 0023) Dokl Akad Nauk SSSR 124*783-785
◇ D55 E15 ◇ REV MR 21 # 2593 • ID 11894

SHCHEGOL'KOV, E.A. [1973] *Uniformization of sets of certain classes (Russian)* (S 0066) Tr Mat Inst Steklov 133*251-262,276
• TRANSL [1973] (S 0055) Proc Steklov Inst Math 133*255-265
◇ D55 E15 ◇ REV Zbl 296 # 04004 • ID 65096

SHCHEGOL'KOV, E.A. see Vol. I, V for further entries

SHCHEPIN, G.G. [1965] *On the imbedding problem for the nilpotent product of finitely presented groups (Russian)* (J 0023) Dokl Akad Nauk SSSR 160*294-297
• TRANSL [1965] (J 0062) Sov Math, Dokl 6*94-97
◇ D40 ◇ REV MR 30 # 4825 Zbl 132.14 • ID 11896

SHEKHTMAN, V.B. [1978] *An undecidable superintuitionistic propositional calculus (Russian)* (J 0023) Dokl Akad Nauk SSSR 240*549-552
• TRANSL [1978] (J 0062) Sov Math, Dokl 19*656-660
◇ B55 D35 ◇ REV MR 58 # 10330 Zbl 417 # 03010 JSL 50.1081 • ID 78453

SHEKHTMAN, V.B. [1982] *Undecidable propositional calculi (Russian)* (S 2874) Vopr Kibern, Akad Nauk SSSR 75*74-116
◇ B22 B55 D35 ◇ REV Zbl 499 # 03003 • ID 38116

SHEKHTMAN, V.B. see Vol. II, III for further entries

SHELAH, S. [1975] *The monadic theory of order* (J 0120) Ann of Math, Ser 2 102*379-419
◇ B25 C65 C85 D35 E50 ◇ REV MR 58 # 10390 Zbl 345 # 02034 • ID 15024

SHELAH, S. [1979] see GUREVICH, Y.

SHELAH, S. [1982] see GUREVICH, Y.

SHELAH, S. [1982] see HARRINGTON, L.A.

SHELAH, S. [1983] see GUREVICH, Y.

SHELAH, S. [1985] see GUREVICH, Y.

SHELAH, S. see Vol. I, II, III, V, VI for further entries

SHEN', A.KH. [1979] *The priority method and separation problems (Russian)* (J 0023) Dokl Akad Nauk SSSR 248*1309-1313
• TRANSL [1979] (J 0062) Sov Math, Dokl 20*1159-1163
◇ D25 D30 ◇ REV MR 81b:03046 Zbl 448 # 03031
• ID 56631

SHEN', A.KH. [1980] *Axiomatic approach to the theory of algorithms and relativized computability (Russian)* (J 0288) Vest Ser Mat Mekh, Univ Moskva 1980/2*27-29,102
• TRANSL [1980] (J 0510) Moscow Univ Math Bull 35/2*29-32
◇ D20 D75 ◇ REV MR 82b:03087 Zbl 428 # 68056
• ID 78514

SHEN', A.KH. [1981] *Some remarks on numerations that are not natural (Russian)* (C 3747) Mat Log & Mat Lingvistika 162-165
◇ D20 D45 ◇ REV MR 83k:03057 • ID 34876

SHEN', A.KH. see Vol. VI for further entries

SHEN, BAIYING [1965] see MO, SHAOKUI

SHEN, BAIYING [1981] *Axiom systems for primitive recursive word arithmetic WA (Chinese)* (J 0418) Shuxue Xuebao 24*717-724
◇ D20 D75 ◇ REV MR 83m:03066 Zbl 524 # 03027
• ID 35465

SHEN, BAIYING [1984] *Explicit representations for inverse functions of number-theoretic functions (Chinese) (English summary)* (J 2804) Nanjing Daxue Xuebao, Ziran Kexue 1984*203-210
◇ D20 ◇ REV Zbl 571 # 03016 • ID 44150

SHEN, BAIYING [1984] *Inverse functions of number-theoretic functions III (Chinese)* (J 3187) Shuxue Niankan, Xi A 5*483-494
◇ D20 ◇ REV MR 86g:03069 Zbl 545 # 03024 • REM English summary in J4719 4*535. Parts I,II 1982 by Mo,Shaokui & Shen,Baiying • ID 43505

SHEN, BAIYING [1984] *Primitive recursive arithmetic in the second class $A^0(D)$ (Chinese)* (J 0418) Shuxue Xuebao 27*345-363
◇ D20 F30 ◇ REV Zbl 564 # 03036 • ID 46417

SHEN, BAIYING [1985] *Primitive recursive arithmetic in the first class A^o. I (Chinese)* (J 0418) Shuxue Xuebao 28*294-307
◇ D20 F30 ◇ ID 46288

SHEN, BAIYING see Vol. I, II, V, VI for further entries

SHENG, C.L. [1970] see CHEN, I-NGO

SHENG, C.L. see Vol. I, II for further entries

SHEPHERDSON, J.C. [1951] *Inverses and zero divisors in matrix rings* (J 3240) Proc London Math Soc, Ser 3 1*71-85
• ERR/ADD ibid 1*II
◇ B25 C60 D40 ◇ REV MR 13.7 Zbl 43.17 • ID 12066

SHEPHERDSON, J.C. [1955] see MYHILL, J.R.

SHEPHERDSON, J.C. [1955] see FROEHLICH, A.

SHEPHERDSON, J.C. [1959] *The reduction of two-way automata to one-way automata* (J 0284) IBM J Res Dev 3*198-200
◇ D05 D10 ◇ REV MR 21 # 2560 JSL 25.163 • ID 12070

SHEPHERDSON, J.C. [1961] *Representability of recursively enumerable sets in formal theories* (J 0009) Arch Math Logik Grundlagenforsch 5*119-127
◇ D25 F30 ◇ REV MR 23 # A3674 Zbl 113.243 JSL 34.117 • ID 12071

SHEPHERDSON, J.C. & STURGIS, H.E. [1963] *Computability of recursive functions* (J 0037) ACM J 10*217-255
◇ D10 D20 ◇ REV MR 27 # 1359 Zbl 118.254 JSL 32.122
• ID 12073

SHEPHERDSON, J.C. [1965] *Machine configuration and word problems of given degree of unsolvability* (J 0068) Z Math Logik Grundlagen Math 11*149-175
◇ D05 D30 D35 D40 ◇ REV MR 30 # 4681 Zbl 161.8 JSL 33.120 • ID 12076

SHEPHERDSON, J.C. [1969] see EILENBERG, S.

SHEPHERDSON, J.C. [1970] *Theory of algorithms* (J 3434) Pubbl Ist Appl Calcolo, Ser 3 51*59pp
◇ D98 ◇ REV Zbl 309 # 02032 • ID 65253

SHEPHERDSON, J.C. [1975] *Computation over abstract structures: serial and parallel procedures and Friedman's effective definitional schemes* (P 0775) Logic Colloq;1973 Bristol 445-513
◇ D75 ◇ REV MR 55 # 7753 Zbl 325 # 02026 • ID 30441

SHEPHERDSON, J.C. [1976] *On the definition of computable function of a real variable* (J 0068) Z Math Logik Grundlagen Math 22*391-402
◇ D20 F60 ◇ REV MR 56 # 102 Zbl 359 # 02029 • ID 23677

SHEPHERDSON, J.C. [1982] see HUBER-DYSON, V.

SHEPHERDSON, J.C. see Vol. I, II, III, V, VI for further entries

SHEVYAKOV, V.S. [1973] *Formulas of the restricted predicate calculus which distinguish certain classes of models with simply computable predicates (Russian)* (J 0023) Dokl Akad Nauk SSSR 210*285-287
• TRANSL [1973] (J 0062) Sov Math, Dokl 14*743-745
◇ C57 ◇ REV MR 49 # 2346 Zbl 329 # 02004 • ID 11994

SHI, HUIJAN [1980] see CHU, TANIEN

SHI, NIANDONG [1982] *Creative pairs of subalgebras of recursively enumerable boolean algebras (Chinese)* (J 0418) Shuxue Xuebao 25*737-745
◇ C57 D25 D45 G05 ◇ REV MR 85b:03078 Zbl 524 # 03030 • ID 37603

SHI, NIANDONG see Vol. II for further entries

SHIBATA, R. [1977] see HUZINO, S.

SHIELDS, M.W. [1985] *Deterministic asynchronous automata* (P 4622) Autom on Infinite Words;1984 Le Mont-Dore 89-98
◇ D05 ◇ ID 49468

SHILLETO, J.R. [1972] *Minimum models of analysis* (J 0036) J Symb Logic 37*48-54
◇ C50 C62 D55 E45 ◇ REV MR 51 # 7872 Zbl 247 # 02053 JSL 39.601 • ID 29513

SHIMODA, H. [1978] *Recursion for type 2 objects (Japanese)* (P 4109) B-Val Anal & Nonstand Anal;1978 Kyoto 101-116
◇ D65 ◇ ID 47670

SHINODA, J. [1980] *On the upper semi-lattice of J^s_a-degrees* (J 0111) Nagoya Math J 80*75-106
◇ D30 D55 D65 ◇ REV MR 83a:03044 Zbl 445 # 03022 • ID 56486

SHINODA, J. [1981] *Sections and envelopes of type 2 objects* (P 3201) Logic Symposia;1979/80 Hakone 175-188
◇ D65 ◇ REV MR 83g:03046 Zbl 478 # 03025 • ID 55640

SHINODA, J. [1984] *Countable J^S_a-admissible ordinals* (P 3668) Log & Founds of Math;1983 Kyoto 79-91
◇ D60 D65 E10 E30 E45 ◇ ID 42934

SHINODA, J. [1985] *Absolute type 2 objects* (P 3342) Rec Th Week;1984 Oberwolfach 343-356
◇ D60 D65 E40 E45 E55 ◇ ID 45312

SHINODA, J. [1985] *Countable J^S_a-admissible ordinals* (J 0111) Nagoya Math J 99*1-10
◇ D60 D65 ◇ ID 48593

SHINODA, J. see Vol. V for further entries

SHIRSHOV, A.I. [1962] *Some algorithmic problems for ε-algebras (Russian)* (J 0092) Sib Mat Zh 3*132-137
◇ D40 ◇ REV MR 32 # 1222 Zbl 143.256 • ID 12375

SHIRSHOV, A.I. [1962] *Some algorithm problems for Lie algebras (Russian)* (J 0092) Sib Mat Zh 3*292-296
◇ D40 ◇ REV MR 32 # 1231 Zbl 104.260 • ID 12374

SHIRSHOV, A.I. see Vol. I for further entries

SHIRVANYAN, V.L. [1980] *The word problem for groups with a recursive set of defining relations of the form $A^n=1$ (Russian) (Armenian summary)* (J 0346) Dokl Akad Nauk Armyan SSR 71*193-197
◇ D40 ◇ REV MR 82j:20073 Zbl 471 # 20022 • ID 82796

SHIRVANYAN, V.L. see Vol. V for further entries

SHKIL'NYAK, S.S. [1980] see KUZENKO, V.F.

SHKIRA, V.V. [1973] *Universal functions for certain classes of recursive functions and sets (Russian)* (S 0066) Tr Mat Inst Steklov 133*243-250,276
• TRANSL [1973] (S 0055) Proc Steklov Inst Math 133*245-253
◇ D20 D25 ◇ REV MR 48 # 3718 Zbl 299 # 02044 • ID 22291

SHMAIN, I.KH. [1974] *Extended calculus of recursive functions I (Russian)* (C 2577) Issl Formaliz Yazyk & Neklass Log 50-81
◇ D20 D75 F50 ◇ REV MR 57 # 16028 • ID 78543

SHMAIN, I.KH. see Vol. VI for further entries

SHOENFIELD, J.R. [1957] see KREISEL, G.

SHOENFIELD, J.R. [1957] *Quasicreative sets* (J 0053) Proc Amer Math Soc 8*964-967
◇ D25 ◇ REV MR 19.723 Zbl 80.244 JSL 25.166 • ID 12088

SHOENFIELD, J.R. [1957] *The non-enumerability of degrees of unsolvability* (P 1675) Summer Inst Symb Log;1957 Ithaca 213
◇ D30 ◇ ID 29349

SHOENFIELD, J.R. [1958] *Degrees of formal systems* (J 0036) J Symb Logic 23*389-392
◇ D25 D30 D35 D55 ◇ REV MR 22 # 3678 Zbl 93.13 JSL 27.85 • ID 12089

SHOENFIELD, J.R. [1958] *The class of recursive functions* (J 0053) Proc Amer Math Soc 9*690-692
◇ D20 D55 ◇ REV MR 20 # 2281 Zbl 87.252 JSL 24.238 • ID 12090

SHOENFIELD, J.R. [1959] *On degrees of unsolvability* (J 0120) Ann of Math, Ser 2 69*644-653
◇ D30 D55 ◇ REV MR 21 # 4097 Zbl 119.251 JSL 29.203 • ID 12092

SHOENFIELD, J.R. [1959] see KREISEL, G.

SHOENFIELD, J.R. [1960] *An uncountable set of incomparable degrees* (J 0053) Proc Amer Math Soc 11*61-62
◇ D30 ◇ REV MR 22 # 7937 Zbl 109.241 JSL 29.203 • ID 12094

SHOENFIELD, J.R. [1960] *Degrees of models* (J 0036) J Symb Logic 25*233-237
◇ C57 D30 D35 ◇ REV MR 25 # 2957 Zbl 105.248 JSL 22.623 JSL 33.623 • ID 12098

SHOENFIELD, J.R. [1961] *The problem of predicativity* (**C** 0622) Essays Found of Math (Fraenkel) 132-139
◇ D55 E15 E45 ◇ REV MR 29 # 2177 JSL 34.515
• ID 21160

SHOENFIELD, J.R. [1961] *Undecidable and creative theories* (**J** 0027) Fund Math 49*171-179
◇ D25 D35 ◇ REV MR 24 # A44 Zbl 96.243 JSL 32.123
• ID 12096

SHOENFIELD, J.R. [1962] *Some applications of degrees* (**P** 0612) Int Congr Log, Meth & Phil of Sci (1,Proc);1960 Stanford 56-59
◇ D25 D30 D35 D55 ◇ REV MR 32 # 2331 JSL 37.610
• ID 12095

SHOENFIELD, J.R. [1962] *The form of the negation of a predicate* (**P** 0613) Rec Fct Th;1961 New York 131-134
◇ D55 D65 ◇ REV MR 26 # 18 Zbl 143.13 JSL 33.116
• ID 12097

SHOENFIELD, J.R. [1965] *Applications of model theory to degrees of unsolvability* (**P** 0614) Th Models;1963 Berkeley 359-363
◇ C50 D25 D30 ◇ REV MR 34 # 53 Zbl 192.52
JSL 37.610 • ID 12099

SHOENFIELD, J.R. [1966] *A theorem on minimal degrees* (**J** 0036) J Symb Logic 31*539-544
◇ D30 ◇ REV MR 34 # 5676 Zbl 202.309 JSL 32.529
• ID 12100

SHOENFIELD, J.R. [1967] *Mathematical logic* (**X** 0832) Addison-Wesley: Reading viii + 344pp
• TRANSL [1975] (**X** 2027) Nauka: Moskva 527pp
◇ B98 C98 D98 E98 F98 ◇ REV MR 37 # 1224
MR 53 # 87 Zbl 155.11 JSL 40.234 • ID 22384

SHOENFIELD, J.R. [1968] *A hierarchy based on a type two object* (**J** 0064) Trans Amer Math Soc 134*103-108
◇ D55 D65 ◇ REV MR 41 # 8227 Zbl 191.305 JSL 36.340
• ID 12101

SHOENFIELD, J.R. [1971] *Degrees of unsolvability* (**X** 0838) Amer Elsevier: New York vii + 111pp
• TRANSL [1977] (**X** 2027) Nauka: Moskva 192pp
◇ D25 D30 D98 ◇ REV MR 49 # 4768 MR 57 # 5718
Zbl 245 # 02037 JSL 40.452 • ID 12103

SHOENFIELD, J.R. [1971] *Measurable cardinals* (**P** 0638) Logic Colloq;1969 Manchester 19-49
◇ D55 E15 E45 E55 E98 ◇ REV MR 44 # 64
Zbl 268 # 02047 JSL 40.93 • ID 12105

SHOENFIELD, J.R. [1975] *The decision problem for recursively enumerable degrees* (**J** 0015) Bull Amer Math Soc 81*973-977
◇ B25 D25 ◇ REV MR 52 # 7882 Zbl 339 # 02043
• ID 18383

SHOENFIELD, J.R. [1976] *Degrees of classes of RE sets* (**J** 0036) J Symb Logic 41*695-696
◇ D25 D30 ◇ REV MR 58 # 5140 Zbl 366 # 02029
• ID 14592

SHOENFIELD, J.R. see Vol. I, III, V, VI for further entries

SHORE, R.A. [1972] *Minimal α-degrees* (**J** 0007) Ann Math Logic 4*393-414
◇ D30 D60 ◇ REV MR 51 # 5278 Zbl 262 # 02039
• ID 12112

SHORE, R.A. [1974] *Σ_n sets which are Δ_n-incomparable (uniformly)* (**J** 0036) J Symb Logic 39*295-304
◇ D30 D55 D60 E40 E47 ◇ REV MR 54 # 7229
Zbl 308 # 02043 • ID 19679

SHORE, R.A. [1974] *Cohesive sets: countable and uncountable* (**J** 0053) Proc Amer Math Soc 44*442-445
◇ D60 ◇ REV MR 49 # 7128 Zbl 291 # 02027 • ID 12113

SHORE, R.A. [1975] *Splitting an α-recursively enumerable set* (**J** 0064) Trans Amer Math Soc 204*65-77
◇ D30 D60 ◇ REV MR 52 # 60 Zbl 306 # 02034
• ID 12115

SHORE, R.A. [1975] *The irregular and non-hyperregular α-r.e. degrees* (**J** 0029) Israel J Math 22*28-41
◇ D30 D60 ◇ REV MR 55 # 2537 Zbl 374 # 02021
• ID 12116

SHORE, R.A. [1976] *On the jump of an α-recursively enumerable set* (**J** 0064) Trans Amer Math Soc 217*351-363
◇ D30 D60 ◇ REV MR 54 # 12504 Zbl 343 # 02030
• ID 18384

SHORE, R.A. [1976] *The recursively enumerable α-degrees are dense* (**J** 0007) Ann Math Logic 9*123-155
◇ D30 D60 ◇ REV MR 52 # 2852 Zbl 374 # 02022
• ID 17643

SHORE, R.A. [1976] see LEGGETT, A.

SHORE, R.A. [1977] *α-recursion theory* (**C** 1523) Handb of Math Logic 653-680
◇ D30 D60 E45 ◇ REV MR 58 # 5109 JSL 49.975
• ID 27323

SHORE, R.A. [1977] *Determining automorphisms of the recursively enumerable sets* (**J** 0053) Proc Amer Math Soc 65*318-325
◇ D25 ◇ REV MR 56 # 5248 Zbl 364 # 02023 • ID 32239

SHORE, R.A. [1978] see LERMAN, M.

SHORE, R.A. [1978] *Controlling the dependence degree of a recursively enumerable vector space* (**J** 0036) J Symb Logic 43*13-22
◇ D25 D45 D50 ◇ REV MR 58 # 10365 Zbl 399 # 03032
• ID 29238

SHORE, R.A. [1978] *Nowhere simple sets and the lattice of recursively enumerable sets* (**J** 0036) J Symb Logic 43*322-330
◇ D25 ◇ REV MR 58 # 10362 Zbl 398 # 03029 • ID 29264

SHORE, R.A. [1978] *On the $\forall\exists$-sentences of α-recursion theory* (**P** 1628) Generalized Recursion Th (2);1977 Oslo 331-353
◇ D30 D60 ◇ REV MR 80e:03055 Zbl 453 # 03047
• ID 78559

SHORE, R.A. [1978] *Some more minimal pairs of α-recursively enumerable degrees* (**J** 0068) Z Math Logik Grundlagen Math 24*409-418
◇ D30 D60 ◇ REV MR 80b:03061 Zbl 416 # 03044
• ID 53210

SHORE, R.A. [1979] *The homogeneity conjecture* (**J** 0054) Proc Nat Acad Sci USA 76*4218-4219
◇ D30 ◇ REV MR 81a:03046 Zbl 412 # 03028 • ID 52959

SHORE, R.A. [1980] *$\mathscr{L}^*(K)$ and other lattices of recursively enumerable sets* (**J** 0053) Proc Amer Math Soc 80*143-146
◇ D25 ◇ REV MR 81h:03088 Zbl 444 # 03023 • ID 78557

SHORE, R.A. [1980] see NERODE, A.

SHORE, R.A. [1980] *Some constructions in α-recursion theory* (**P** 3021) Logic Colloq;1979 Leeds 158-170
◇ D30 D60 ◇ REV MR 82e:03043 Zbl 461 # 03007
• ID 54484

SHORE, R.A. [1981] see HARRINGTON, L.A.

SHORE, R.A. [1981] see MAASS, W.

SHORE, R.A. [1981] *The degrees of unsolvability: global results* (**P** 2628) Log Year;1979/80 Storrs 283-301
◇ C40 D30 D35 F35 ◇ REV MR 82k:03065
Zbl 474 # 03022 • ID 55426

SHORE, R.A. [1981] *The theory of the degrees below O'* (**J** 3172) J London Math Soc, Ser 2 24*1-14
◇ D25 D30 D35 F30 ◇ REV MR 83m:03051
Zbl 469 # 03027 JSL 50.550 • ID 55154

SHORE, R.A. [1982] *Finitely generated codings and the degrees r.e. in a degree d* (**J** 0053) Proc Amer Math Soc 84*256-263
◇ D30 D55 E40 ◇ REV MR 84g:03061 Zbl 498 # 03031
• ID 33615

SHORE, R.A. [1982] *On homogeneity and definability in the first order theory of the Turing degrees* (**J** 0036) J Symb Logic 47*8-16
◇ D30 ◇ REV MR 84a:03046 Zbl 521 # 03027 • ID 33616

SHORE, R.A. [1982] *The Turing and truth-table degrees are not elementarily equivalent* (**P** 3623) Logic Colloq;1980 Prague 231-237
◇ D30 ◇ REV MR 84c:03082 Zbl 542 # 03021 • ID 33617

SHORE, R.A. [1983] see JOCKUSCH JR., C.G.

SHORE, R.A. [1984] see AMBOS-SPIES, K.

SHORE, R.A. [1984] see JOCKUSCH JR., C.G.

SHORE, R.A. [1984] *The arithmetic and Turing degrees are not elementarily equivalent* (**J** 0009) Arch Math Logik Grundlagenforsch 24*137-140
◇ D30 ◇ REV MR 86c:03041 • ID 42402

SHORE, R.A. [1984] *The degrees of unsolvability: the ordering of functions by relative computability* (**P** 4313) Int Congr Math (II,14);1983 Warsaw 1*337-345
◇ D30 ◇ ID 48595

SHORE, R.A. [1984] see LERMAN, M.

SHORE, R.A. [1985] see FEJER, P.A.

SHORE, R.A. [1985] see NERODE, A.

SHORE, R.A. [1985] see METAKIDES, G.

SHORE, R.A. [1985] see JOCKUSCH JR., C.G.

SHORE, R.A. [1985] *The structure of the degrees of unsolvability* (**P** 4046) Rec Th;1982 Ithaca 33-51
◇ D30 ◇ REV Zbl 573 # 03016 • ID 46348

SHORE, R.A. see Vol. V for further entries

SHREJDER, YU.A. [1960] see AKUSHSKY, I.Y.

SHREJDER, YU.A. [1971] *Equality, resemblance, and order (Russian)* (**X** 2027) Nauka: Moskva 254pp
• TRANSL [1975] (**X** 0885) Mir: Moskva 279pp (English)
◇ D10 E07 ◇ REV MR 52 # 10430 MR 52 # 10431
• ID 78575

SHREJDER, YU.A. see Vol. II, III, V for further entries

SHU, YONGCHANG & WANG, YIZHI [1981] *Fuzzy languages and fuzzy grammars (Chinese)* (**J** 3732) Mohu Shuxue 1/2*113-123
◇ B52 D05 ◇ REV MR 84g:68071 • ID 47233

SHUKURYAN, S.K. [1976] *Solvability of the equivalence problem in a class of multitape multihead automata and flow charts over memory (Russian) (English summary insert)* (**J** 0040) Kibernetika, Akad Nauk Ukr SSR 1976/4*12-16
• TRANSL [1976] (**J** 0021) Cybernetics 12*507-512
◇ D05 ◇ REV MR 58 # 3701 Zbl 343 # 02022 • ID 65580

SHUKURYAN, S.K. [1976] *Some decidable cases of the special problem of functional equivalence for x-y-automata (Russian)* (**J** 0346) Dokl Akad Nauk Armyan SSR 63*27-32
◇ D05 ◇ REV MR 58 # 20913 Zbl 357 # 68063 • ID 50448

SHURYGIN, V.A. [1966] *Nontrivial constructive mappings of certain sets (Russian)* (**J** 0023) Dokl Akad Nauk SSSR 168*40-42 • ERR/ADD ibid 171*510
• TRANSL [1966] (**J** 0062) Sov Math, Dokl 7*604-607
◇ D20 D45 F60 ◇ REV MR 34 # 57 Zbl 219 # 02019
• REM Transl-err ibid 7*below table of contents • ID 12786

SHURYGIN, V.A. [1967] *Constructive sets with equality and their mappings (Russian)* (**J** 0023) Dokl Akad Nauk SSSR 173*54-57
• TRANSL [1967] (**J** 0062) Sov Math, Dokl 8*348-351
◇ D03 F60 ◇ REV MR 35 # 1478 Zbl 234 # 02020
• ID 12761

SHURYGIN, V.A. [1968] *Complete constructive sets with equality, and some of their properties (Russian)* (**S** 0228) Zap Nauch Sem Leningrad Otd Mat Inst Steklov 8*272-280
• TRANSL [1968] (**J** 0521) Semin Math, Inst Steklov 8*133-135
◇ D03 F60 ◇ REV MR 40 # 5448 Zbl 233 # 02012
• ID 12762

SHURYGIN, V.A. [1970] *Constructive sets with equality and their mappings (Russian)* (**S** 0066) Tr Mat Inst Steklov 113*173-259
• TRANSL [1970] (**S** 0055) Proc Steklov Inst Math 113*195-287
◇ D03 F60 ◇ REV MR 45 # 58 Zbl 247 # 02032 • ID 12763

SHURYGIN, V.A. [1974] *Einige Eigenschaften der Kompliziertheit konstruktiver reeller Zahlen (Russian)* (**C** 1450) Teor Algor & Mat Logika (Markov) 177-194
◇ D15 F60 ◇ REV MR 58 # 16211 Zbl 311 # 02045
• ID 29609

SHURYGIN, V.A. [1977] *On estimates of the complexity of algorithmic problems in constructive analysis (Russian)* (**J** 0023) Dokl Akad Nauk SSSR 233*1064-1067
• TRANSL [1977] (**J** 0062) Sov Math, Dokl 18*577-581
◇ D15 D20 F60 ◇ REV MR 57 # 16003 Zbl 378 # 02018
• ID 51800

SHURYGIN, V.A. [1979] *Bounds on the complexity of normal algorithms (Russian)* (**J** 0023) Dokl Akad Nauk SSSR 244*1097-1101
• TRANSL [1979] (**J** 0062) Sov Math, Dokl 20*195-199
◇ D03 D15 ◇ REV MR 80h:03059 Zbl 418 # 68042
• ID 69917

SHURYGIN, V.A. see Vol. VI for further entries

SIEFKES, D. [1970] *Decidable theories I. Buechi's monadic second order successor arithmetic* (S 3301) Lect Notes Math 120*xii+130pp
◇ B15 B25 C85 D05 F35 F98 ◇ REV MR 44 # 6488 Zbl 399 # 03011 • ID 12145

SIEFKES, D. [1970] *Decidable extensions of monadic second order successor arithmetic* (P 0577) Automatenth & Formale Sprachen;1969 Oberwolfach 441-472
◇ B15 B25 C85 D05 F35 ◇ REV MR 56 # 8354 Zbl 213.19 • ID 27984

SIEFKES, D. [1971] *Undecidable extensions of monadic second order successor arithmetic* (J 0068) Z Math Logik Grundlagen Math 17*385-394
◇ B15 B25 C85 D35 F35 ◇ REV MR 45 # 1763 Zbl 193.312 • ID 12146

SIEFKES, D. [1975] *The recursive sets in certain monadic second order fragments of arithmetic* (J 0009) Arch Math Logik Grundlagenforsch 17*71-80
◇ D20 F35 ◇ REV MR 52 # 13362 Zbl 325 # 02033 • ID 12147

SIEFKES, D. [1976] see FLEISCHMANN, K.

SIEFKES, D. [1977] see FLEISCHMANN, K.

SIEFKES, D. [1977] *Degrees of circuit complexity* (P 2588) FCT'77 Fund of Comput Th;1977 Poznan 56*522-531
◇ D15 ◇ REV MR 58 # 8496 Zbl 368 # 94049 • ID 33903

SIEFKES, D. [1978] *An axiom system for the weak monadic second order theory of two successors* (J 0029) Israel J Math 30*264-284
◇ B15 B25 C85 D05 ◇ REV MR 80a:03015 Zbl 397 # 03009 • ID 29136

SIEFKES, D. [1981] see MAHR, B.

SIEFKES, D. [1983] see BUECHI, J.R.

SIEFKES, D. [1984] see BUECHI, J.R.

SIEFKES, D. see Vol. I, III, V for further entries

SIEKMANN, J. & SZABO, P. [1983] *Universal unification* (P 3858) Adequate Modeling of Syst;1982 Bad Honnef 102-141
◇ D80 ◇ REV MR 85f:03011 • ID 40644

SIEKMANN, J. see Vol. I, III for further entries

SIERPINSKI, W. [1918] *Sur les definitions axiomatiques des ensembles mesurables (B)* (J 3977) Bull Int Acad Sci Cracovie, Cl Math Nat 1918*29-34
◇ D55 D70 E15 ◇ REV FdM 46.295 • ID 38030

SIERPINSKI, W. [1918] see LUZIN, N.N.

SIERPINSKI, W. [1918] *Sur un theoreme de M. Lebesgue* (J 3977) Bull Int Acad Sci Cracovie, Cl Math Nat 1918*168-172
◇ D55 E15 ◇ REV FdM 46.298 • ID 38039

SIERPINSKI, W. [1918] *Sur une generalisation des ensembles mesurables (B)* (J 3977) Bull Int Acad Sci Cracovie, Cl Math Nat 1918*161-167
◇ D55 E15 ◇ REV FdM 46.297 • ID 38038

SIERPINSKI, W. [1918] *Sur une propriete des fonctions representables analytiquement* (J 3977) Bull Int Acad Sci Cracovie, Cl Math Nat 1918*179-184
◇ D55 E15 ◇ REV FdM 46.298 • ID 38041

SIERPINSKI, W. [1919] *Ueber eine Verallgemeinerung der Borelschen Mengen (Polish)* (J 0611) Pol Tow Mat, Prace Mat-Fiz 30*89-94
◇ D55 E15 ◇ REV FdM 47.181 • ID 41759

SIERPINSKI, W. [1920] *Sur les ensembles mesurables B* (J 0109) C R Acad Sci, Paris 171*24-26
◇ D55 E15 ◇ REV FdM 47.179 • ID 41749

SIERPINSKI, W. [1923] see LUZIN, N.N.

SIERPINSKI, W. [1924] *Les projections des ensembles mesurables (B) et les ensembles (A)* (J 0027) Fund Math 5*155-159
◇ D55 E15 ◇ REV FdM 50.142 • ID 12160

SIERPINSKI, W. [1924] *Sur la puissance des ensembles mesurables (B)* (J 0027) Fund Math 5*166-171
◇ D55 E10 E15 ◇ REV FdM 50.141 • ID 12158

SIERPINSKI, W. [1924] *Sur une propriete des ensembles ambigus* (J 0027) Fund Math 6*1-5
◇ D55 E15 ◇ REV FdM 50.141 • ID 12159

SIERPINSKI, W. [1924] *Un exemple effectif d'un ensemble mesurable (B) de classe α* (J 0027) Fund Math 6*39-44
◇ D55 E15 ◇ REV FdM 50.142 • ID 12156

SIERPINSKI, W. [1925] *Les fonctions continues et les ensembles (A)* (J 0027) Fund Math 7*155-158
◇ D55 E15 E75 ◇ REV FdM 51.168 • ID 12161

SIERPINSKI, W. [1925] *Sur un ensemble ferme conduisant a un ensemble non mesurable (B)* (J 0027) Fund Math 7*198-202
◇ D55 E15 ◇ REV FdM 51.169 • ID 41589

SIERPINSKI, W. [1925] *Sur une classe d'ensembles* (J 0027) Fund Math 7*237-243
◇ D55 E15 ◇ REV FdM 51.166 • ID 41583

SIERPINSKI, W. [1926] *Sur l'ensemble de valeurs qu'une fonction continue prend une infinite non denombrable de fois* (J 0027) Fund Math 8*370-373
◇ D55 E15 ◇ REV FdM 52.201 • ID 41576

SIERPINSKI, W. [1926] *Sur les ensembles hyperboreliens* (J 0459) C R Soc Sci Lett Varsovie Cl 3 19*16-25
◇ D55 E15 ◇ REV FdM 57.1334 • ID 39827

SIERPINSKI, W. [1926] *Sur une propriete des ensembles (A)* (J 0027) Fund Math 8*362-369
◇ D55 E15 ◇ REV FdM 52.201 • ID 12162

SIERPINSKI, W. [1927] *Sur la puissance des ensembles d'une certaine classe* (J 0027) Fund Math 9*45-49
◇ D55 E15 ◇ REV FdM 53.172 • ID 12168

SIERPINSKI, W. [1927] *Sur quelques proprietes des ensembles projectifs* (J 0109) C R Acad Sci, Paris 185*833-835
◇ D55 E15 ◇ REV FdM 53.172 • ID 41558

SIERPINSKI, W. [1927] *Sur un probleme de M.Hausdorff* (J 0027) Fund Math 10*427-430
◇ D55 E15 ◇ REV FdM 53.173 • ID 12165

SIERPINSKI, W. [1927] *Sur une classification des ensembles mesurables (B)* (J 0027) Fund Math 10*320-327
◇ D55 E15 ◇ REV FdM 53.172 • ID 12164

SIERPINSKI, W. [1927] *Sur une propriete characteristique des ensembles analytiques* (J 0027) Fund Math 10*169-171
◇ D55 E15 ◇ REV FdM 53.172 • ID 12166

SIERPINSKI, W. [1927] *Sur une propriete des complementaires analytiques* (J 0176) Bull Acad Pol Sci, Ser Sci Math 1927*449-458
◇ D55 E15 ◇ REV FdM 53.172 • ID 41557

SIERPINSKI, W. [1928] *Le crible de M.Lusin et l'operation (A) dans les espaces abstraits* (J 0027) Fund Math 11*16-18
◇ D55 E15 ◇ REV FdM 54.93 • ID 12172

SIERPINSKI, W. [1928] *Les ensembles projectifs et le crible de M. Lusin* (J 0027) Fund Math 12*1-3
◇ D55 E15 ◇ REV FdM 54.99 • ID 39194

SIERPINSKI, W. [1928] *Les ensembles projectifs et la propriete de Baire* (J 0459) C R Soc Sci Lett Varsovie Cl 3 20*477-480
◇ D55 E15 ◇ REV FdM 57.1334 • ID 39825

SIERPINSKI, W. [1928] *Sur les images continues et biunivoques des complementaires analytiques* (J 0027) Fund Math 12*211-213
◇ D55 E15 ◇ REV FdM 54.94 • ID 12169

SIERPINSKI, W. [1928] *Sur les projections des ensembles complementaires aux ensembles (A)* (J 0027) Fund Math 11*117-122
◇ D55 E15 ◇ REV FdM 54.93 • ID 12174

SIERPINSKI, W. [1928] *Sur les produits des images continues des ensembles C(A)* (J 0027) Fund Math 11*123-126
◇ D55 E15 ◇ REV FdM 54.93 • ID 12173

SIERPINSKI, W. [1928] *Sur un ensemble analytique plan, universel pour les ensembles mesurables (B)* (J 0027) Fund Math 12*75-77
◇ D55 E15 ◇ REV FdM 54.97 • ID 12171

SIERPINSKI, W. [1929] *Contribution a la fondation de la theorie des ensembles projectifs* (J 0459) C R Soc Sci Lett Varsovie Cl 3 21*219-233
◇ D55 E15 ◇ REV FdM 57.1333 • ID 39815

SIERPINSKI, W. [1929] *Sur l'existence de diverses classes d'ensembles* (J 0027) Fund Math 14*82-91
◇ D55 E15 ◇ REV FdM 55.54 • ID 12178

SIERPINSKI, W. [1929] see LUZIN, N.N.

SIERPINSKI, W. [1929] *Sur les familles inductives et projectives d'ensembles* (J 0027) Fund Math 13*228-239
◇ D55 E15 ◇ REV FdM 55.54 • ID 12176

SIERPINSKI, W. [1930] *Sur l'uniformisation des ensembles mesurables (B)* (J 0027) Fund Math 16*136-139
◇ D55 E15 ◇ REV FdM 56.87 • ID 12184

SIERPINSKI, W. [1930] *Sur la puissance des ensembles analytiques* (J 0027) Fund Math 15*128-130
◇ D55 E15 E50 ◇ REV FdM 56.85 • ID 12183

SIERPINSKI, W. [1931] *Les ensembles analytiques comme cribles au moyen des ensembles fermes* (J 0027) Fund Math 17*77-91
◇ D55 E15 ◇ REV Zbl 3.107 FdM 57.92 • ID 12185

SIERPINSKI, W. [1931] *Sur certaines operations sur les ensembles fermes plans* (J 0459) C R Soc Sci Lett Varsovie Cl 3 24*57-77
◇ D55 E15 ◇ REV Zbl 3.154 FdM 57.1335 • ID 39830

SIERPINSKI, W. [1931] *Sur deux complementaires analytiques non separables B* (J 0027) Fund Math 17*296-297
◇ D55 E15 ◇ REV Zbl 3.109 FdM 57.93 • ID 12187

SIERPINSKI, W. [1931] *Sur les cribles projectifs* (J 0027) Fund Math 17*30-31
◇ D55 E15 ◇ REV Zbl 3.107 FdM 57.91 • ID 12186

SIERPINSKI, W. [1931] *Sur un crible universel* (J 0027) Fund Math 17*1-3
◇ D55 E15 ◇ REV Zbl 3.107 FdM 57.91 • ID 12188

SIERPINSKI, W. [1932] *Sur les rapports entre les classifications des ensembles de MM. F.Hausdorff et Ch. de la Vallee-Poussin* (J 0027) Fund Math 19*257-264
◇ D55 E15 ◇ REV Zbl 5.390 FdM 58.82 • ID 12189

SIERPINSKI, W. [1934] *Sur la separabilite multiple des ensembles mesurables B* (J 0027) Fund Math 23*292-303
◇ D55 E15 ◇ REV Zbl 10.55 FdM 60.41 • ID 12195

SIERPINSKI, W. [1935] *Sur les transformations des ensembles par les fonctions de Baire* (J 0027) Fund Math 25*98-101
◇ D55 E15 ◇ REV Zbl 13.6 FdM 61.224 • ID 12198

SIERPINSKI, W. [1935] *Sur un ensemble projectif de classe 2 dans l'espace des ensembles fermes plans* (J 0027) Fund Math 25*261-263
◇ D55 E15 ◇ REV Zbl 12.57 FdM 61.222 • ID 12197

SIERPINSKI, W. [1935] *Sur une hypothese de M.Lusin* (J 0027) Fund Math 25*132-135
◇ D55 E15 E50 ◇ REV Zbl 13.7 FdM 61.222 • ID 12196

SIERPINSKI, W. [1937] *Sur un probleme de la theorie generale des ensembles concernant les familles boreliennes d'ensembles* (J 0027) Fund Math 29*206-208
◇ D55 E15 ◇ REV Zbl 17.60 FdM 63.932 • ID 12200

SIERPINSKI, W. [1938] *Sur un probleme concernant les ensembles projectifs* (J 0027) Fund Math 30*61-64
◇ D55 E15 ◇ REV Zbl 18.247 FdM 64.1006 • ID 12201

SIERPINSKI, W. [1941] see KURATOWSKI, K.

SIERPINSKI, W. [1948] *L'operation du crible et les fonctions analytiques d'une suite infinie d'ensembles* (J 0459) C R Soc Sci Lett Varsovie Cl 3 41*47-62
◇ D55 D70 E15 ◇ REV MR 13.217 Zbl 39.47 • ID 12225

SIERPINSKI, W. [1949] *Sur un probleme de M.Zarankiewicz* (J 0459) C R Soc Sci Lett Varsovie Cl 3 42*1-3
◇ D55 E15 ◇ REV MR 14.146 Zbl 41.376 • ID 12233

SIERPINSKI, W. [1949] *Sur un probleme de M.Lusin concernant les complementaires analytiques* (J 0027) Fund Math 36*44-47
◇ D55 E15 E20 ◇ REV MR 11.675 Zbl 38.197 • ID 12239

SIERPINSKI, W. [1950] *Les ensembles projectifs et analytiques* (X 0834) Gauthier-Villars: Paris 80pp
◇ D55 E15 E98 ◇ REV MR 14.627 Zbl 39.47 • ID 21089

SIERPINSKI, W. [1951] *Sur quelques consequences du theoreme de M. Kondo concernant l'uniformisation des complementaires analytiques* (J 0459) C R Soc Sci Lett Varsovie Cl 3 44*56-62
◇ D55 E15 ◇ REV MR 14.960 Zbl 45.168 • ID 43590

SIERPINSKI, W. [1963] *Projective and analytic sets* (J 0287) Scripta Math 26*187-195
◇ D55 E15 ◇ REV MR 26#6057 Zbl 114.33 • ID 12273

SIERPINSKI, W. see Vol. II, V, VI for further entries

Sikdar, K. [1981] *On the complexity classes and optimal algorithms for parallel evaluation of polynomials* (**P** 4258) Found of Softw Tech & Th Comput Sci (1);1981 Bangalore 09-38
◇ D15 ◇ REV Zbl 511 # 68025 • ID 46480

Sikorski, R. [1958] *Some examples of Borel sets* (**S** 0019) Colloq Math (Warsaw) 5*170-171
◇ D55 E15 ◇ REV MR 21 # 3336 Zbl 85.39 • ID 12301

Sikorski, R. [1966] *On an analytic set (Russian summary)* (**J** 0014) Bull Acad Pol Sci, Ser Math Astron Phys 14*15-16
◇ D55 E15 ◇ REV MR 33 # 2784 • ID 12315

Sikorski, R. [1966] see Engelking, R.

Sikorski, R. see Vol. I, II, III, V, VI for further entries

Silberger, D.M. [1980] *Universal terms of complexity three* (**J** 0004) Algeb Universalis 11*393-395
◇ D05 ◇ REV MR 82b:08006 Zbl 533 # 20036 • ID 82809

Silberger, D.M. see Vol. V for further entries

Silver, J.H. [1970] *Every analytic set is Ramsey* (**J** 0036) J Symb Logic 35*60-64
◇ D55 E05 E15 E55 ◇ REV MR 48 # 10807 Zbl 216.13 • ID 12317

Silver, J.H. [1971] *Measurable cardinals and Δ^1_3 well-orderings* (**J** 0120) Ann of Math, Ser 2 94*414-446
◇ C20 C30 D55 E05 E15 E35 E45 E55 ◇ REV MR 45 # 8517 Zbl 259 # 02054 JSL 39.330 • ID 19669

Silver, J.H. [1971] *Some applications of model theory in set theory* (**J** 0007) Ann Math Logic 3*45-110
◇ C30 C55 D55 E05 E45 E55 ◇ REV MR 53 # 12950 Zbl 215.324 JSL 39.597 • REM This paper is most of the author's 1966 Berkeley dissertation • ID 12319

Silver, J.H. [1980] *Counting the number of equivalence classes of Borel and coanalytic equivalence relations* (**J** 0007) Ann Math Logic 18*1-28
◇ D55 E15 E40 ◇ REV MR 81d:03051 Zbl 517 # 03018 • ID 78616

Silver, J.H. see Vol. III, V for further entries

Silverstein, A. [1978] *A generalization of combinatorial operators* (**J** 0047) Notre Dame J Formal Log 19*639-645
◇ B40 D50 ◇ REV MR 84d:03058 Zbl 363 # 02049 • ID 52141

Simeone, B. [1981] see Peled, U.N.

Simmons, H. [1973] *The word problem for absolute presentations* (**J** 3172) J London Math Soc, Ser 2 6*275-280
◇ D40 ◇ REV MR 49 # 7360 Zbl 253 # 02044 • ID 12329

Simmons, H. [1975] see Macintyre, A.

Simmons, H. [1980] *The word and torsion problems for commutative Thue systems* (**P** 2634) Word Problems II;1976 Oxford 395-400
◇ D03 D15 D40 ◇ REV MR 81k:20081 Zbl 432 # 03023 • ID 53980

Simmons, H. see Vol. II, III, VI for further entries

Simon, F.U. [1979] see Bochmann, D.

Simon, H.-U. [1979] *Word problems for groups and contextfree recognition* (**P** 2935) FCT'79 Fund of Comput Th;1979 Berlin/Wendisch-Rietz 417-422
◇ D15 D40 ◇ REV MR 81d:20024 Zbl 413 # 68044 • ID 53040

Simon, H.A. [1976] see Kadane, J.B.

Simon, H.A. see Vol. I for further entries

Simon, I. [1976] see Gill III, J.T.

Simon, Imre [1973] see Brzozowski, J.A.

Simon, Imre [1979] see Kowaltowski, T.

Simon, Istvan [1979] see Kowaltowski, T.

Simon, J. [1977] *On the difference between one and many (Preliminary version)* (**P** 1632) Automata, Lang & Progr (4);1977 Turku SF 480-491
◇ D10 D15 ◇ REV MR 56 # 17207 Zbl 364 # 68066 • ID 50988

Simon, J. [1977] *Polynomially bounded quantification over higher types and a new hierarchy of the elementary sets* (**P** 1076) Latin Amer Symp Math Log (3);1976 Campinas 267-281
◇ B15 D10 D15 D20 ◇ REV MR 58 # 179 Zbl 393 # 03028 • ID 16610

Simon, J. [1979] see Kowaltowski, T.

Simon, J. [1981] *Division in idealized unit cost RAMs* (**J** 0119) J Comp Syst Sci 22*421-441
◇ D10 D15 ◇ REV MR 82i:68030 Zbl 476 # 68037 • ID 82815

Simon, J. [1981] *On tape-bounded probabilistic Turing machine acceptors* (**J** 1426) Theor Comput Sci 16*75-91
◇ D10 D15 ◇ REV MR 82i:68032 Zbl 473 # 68044 • ID 69921

Simovici, D.A. [1970] see Reischer, C.

Simovici, D.A. [1979] *Polylocal languages* (**P** 2539) Frege Konferenz (1);1979 Jena 449-457
◇ D05 ◇ REV MR 82g:68077 • ID 82817

Simovici, D.A. [1980] *Computing of graphs of relations using generative grammars* (**J** 3441) RAIRO Inform Theor 14*279-299
◇ D05 ◇ REV MR 82d:68048 Zbl 486 # 68073 • ID 82818

Simovici, D.A. [1980] *Context-sensitive languages and Ackermann's function* (**J** 2716) Found Control Eng, Poznan 5*91-103
◇ D05 D20 ◇ REV MR 82c:68052 Zbl 463 # 68071 • ID 69922

Simovici, D.A. [1982] *Several remarks on the complexity of graph computations* (**J** 0230) An Univ Iasi, NS, Sect Ia 28*89-96
◇ D10 D15 ◇ REV MR 85b:03062 Zbl 549 # 68040 • ID 40679

Simovici, D.A. see Vol. I, II for further entries

Simpson, S.G. [1972] see Sacks, G.E.

Simpson, S.G. [1973] see Lerman, M.

Simpson, S.G. [1974] *Degree theory on admissible ordinals* (**P** 0602) Generalized Recursion Th (1);1972 Oslo 165-193
◇ D30 D60 ◇ REV MR 58 # 27405 Zbl 319 # 02033 • ID 12340

SIMPSON, S.G. [1974] *Post's problem for admissible sets* (**P** 0602) Generalized Recursion Th (1);1972 Oslo 437-441
◇ D30 D60 ◇ REV MR 52 # 7881 Zbl 345 # 02030
• ID 18390

SIMPSON, S.G. [1975] *Minimal covers and hyperdegrees* (**J** 0064) Trans Amer Math Soc 209*45-64
◇ D30 D55 ◇ REV MR 52 # 13351 Zbl 316 # 02049
• ID 12341

SIMPSON, S.G. [1975] see KREISEL, G.

SIMPSON, S.G. [1976] see JOCKUSCH JR., C.G.

SIMPSON, S.G. [1977] *Basis theorems and countable admissible ordinals* (**P** 1729) Colloq Int Log;1975 Clermont-Ferrand 161-165
◇ D55 D60 E15 ◇ REV MR 80d:03045 Zbl 448 # 03035
• ID 56635

SIMPSON, S.G. [1977] *Degrees of unsolvability: A survey of results* (**C** 1523) Handb of Math Logic 631-652
◇ D25 D30 D35 F35 ◇ REV MR 58 # 5109 JSL 49.975
• ID 27322

SIMPSON, S.G. [1977] *First order theory of the degrees of recursive unsolvability* (**J** 0120) Ann of Math, Ser 2 105*121-139
◇ D30 D35 F35 ◇ REV MR 55 # 5423 Zbl 349 # 02035
• ID 30798

SIMPSON, S.G. [1978] *Sets which do not have subsets of every higher degree* (**J** 0036) J Symb Logic 43*135-138
◇ D30 D55 ◇ REV MR 81f:03054 Zbl 402 # 03040
• ID 29250

SIMPSON, S.G. [1978] *Short course on admissible recursion theory* (**P** 1628) Generalized Recursion Th (2);1977 Oslo 355-390
◇ D30 D60 ◇ REV MR 80f:03049 Zbl 453 # 03047
• ID 32237

SIMPSON, S.G. [1980] see HRBACEK, K.

SIMPSON, S.G. [1980] *The hierarchy based on the jump operator* (**P** 2058) Kleene Symp;1978 Madison 267-276
◇ D30 D55 ◇ REV MR 82a:03044 Zbl 498 # 03039
• ID 78648

SIMPSON, S.G. [1982] *Four test problems in generalized recursion theory* (**P** 3622) Int Congr Log, Meth & Phil of Sci (6,Proc);1979 Hannover 263-270
◇ D25 D55 D60 E15 ◇ REV MR 84m:03067
Zbl 525 # 03033 • ID 38266

SIMPSON, S.G. & WEITKAMP, G. [1983] *High and low Kleene degrees of coanalytic sets* (**J** 0036) J Symb Logic 48*356-368
◇ D30 D55 D65 ◇ REV MR 85g:03064 Zbl 529 # 03018
• ID 37652

SIMPSON, S.G. [1985] see HARRINGTON, L.A.

SIMPSON, S.G. [1985] *Recursion theoretic aspects of the dual Ramsey theorem* (**P** 3342) Rec Th Week;1984 Oberwolfach 357-371
◇ D80 E05 ◇ REV Zbl 574 # 03031 • ID 45316

SIMPSON, S.G. see Vol. I, III, V, VI for further entries

SIMS, C.C. [1978] *The role of algorithms in the teaching of algebra* (**P** 3278) Topics in Algeb;1978 Canberra 95-107
◇ D80 ◇ REV Zbl 396 # 03041 • ID 52651

SINGER, M.F. [1978] *The model theory of ordered differential fields* (**J** 0036) J Symb Logic 43*82-91
◇ B25 C25 C35 C60 C65 D80 ◇ REV MR 80a:03044
Zbl 396 # 03031 • ID 29244

SINGER, M.F. [1980] see HAIMO, F.

SINGER, M.F. see Vol. I, III for further entries

SINGHI, N.M. [1980] see FOELDES, S.

SINGLETARY, W.E. [1964] *A complex of problems proposed by Post* (**J** 0015) Bull Amer Math Soc 70*105-109 • ERR/ADD ibid 70*826
◇ B20 D03 D25 D30 D35 ◇ REV MR 28 # 2040
MR 29 # 5714 Zbl 171.274 JSL 31.273 • ID 12350

SINGLETARY, W.E. [1965] see CUDIA, D.F.

SINGLETARY, W.E. [1967] *Recursive unsolvability of a complex of problems proposed by Post* (**J** 0434) J Fac Sci Univ Tokyo, Sect 1 14*25-58
◇ D03 D25 D35 ◇ REV MR 36 # 6288 Zbl 167.16
• ID 12352

SINGLETARY, W.E. [1967] *The equivalence of some general combinatorial decision problems* (**J** 0015) Bull Amer Math Soc 73*446-451
◇ D03 D25 ◇ REV MR 35 # 1477 Zbl 169.312 • ID 12353

SINGLETARY, W.E. [1968] see CUDIA, D.F.

SINGLETARY, W.E. [1969] see AXT, P.

SINGLETARY, W.E. [1971] see HUGHES, C.E.

SINGLETARY, W.E. [1973] see HUGHES, C.E.

SINGLETARY, W.E. [1973] see MEY VANDER, J.E.

SINGLETARY, W.E. [1974] *Many-one degrees associated with partial propositional calculi* (**J** 0047) Notre Dame J Formal Log 15*335-343
◇ B20 D25 ◇ REV MR 50 # 1862 Zbl 232 # 02009
• ID 12356

SINGLETARY, W.E. [1975] see HUGHES, C.E.

SINGLETARY, W.E. [1977] see HUGHES, C.E.

SINGLETARY, W.E. see Vol. I, II for further entries

SINGLETON, R.C. [1962] *A test for linear separability as applied to self-organizing machines* (**P** 0578) Self Organizing Systs;1962 Chicago 503-524
◇ D05 ◇ ID 12357

SINTZOFF, M. [1967] *Existence of a van Wijngaarden syntax for every recursively enumerable set* (**J** 0962) Ann Soc Sci Bruxelles, Ser 1 81*115-118
◇ D03 D05 D25 ◇ REV MR 38 # 4306 Zbl 165.318
• ID 12360

SION, M. [1964] see BRESSLER, D.W.

SION, M. see Vol. V for further entries

SIPALA, P. [1968] *Appunti di teoria degli algoritmi* (**X** 4342) Univ Degli Stud Trieste: Trieste 89pp
◇ D98 ◇ REV MR 40 # 7117 • ID 48079

SIPSER, M. [1978] *Halting space-bounded computations* (**P** 3578) IEEE Symp Found of Comput Sci (19);1978 Ann Arbor 73-74
◇ D10 D15 ◇ REV MR 80e:68137 Zbl 423 # 68011
• ID 82824

SIPSER, M. [1980] *Halting space-bounded computations* (J 1426) Theor Comput Sci 10*335-338
◇ D10 D15 ◇ REV MR 81c:68037 Zbl 423#68011
• ID 69924

SIPSER, M. [1981] see FURST, M.

SIPSER, M. [1981] see KATSEFF, H.P.

SIPSER, M. [1982] *On relativization and the existence of complete sets* (P 3836) Automata, Lang & Progr (9);1982 Aarhus 523-531
◇ D15 ◇ REV MR 83m:68088 Zbl 515#68040 • ID 37861

SIPSER, M. [1984] *A topological view of some problems in complexity theory* (P 3658) Math Founds of Comput Sci (11);1984 Prague 567-572
◇ D15 ◇ ID 46889

SIPSER, M. [1984] see FURST, M.

SIROMONEY, R. [1981] see DARE, V.R.

SIROMONEY, R. [1983] see DARE, V.R.

SIROMONEY, R. [1985] see DARE, V.R.

SIROVICH, F. [1979] see DEGANO, P.

SISTLA, A.P. [1984] see EMERSON, E.A.

SISTLA, A.P. & VARDI, M.Y. & WOLPER, P. [1985] *The complementation problem for Buechi automata with applications to temporal logic (extended abstract)* (P 4628) Automata, Lang & Progr (12);1985 Nafplion 465-474
◇ B45 D05 ◇ ID 49529

SISTLA, A.P. see Vol. II for further entries

SKANDALIS, K. [1977] *Finite lattices of degrees of definability* (J 0014) Bull Acad Pol Sci, Ser Math Astron Phys 25*217-219
◇ D30 E35 E45 ◇ REV MR 56#125 Zbl 379#02025
• ID 26555

SKANDALIS, K. [1983] *Programmability in the set of real numbers and second-order recursion* (J 2095) Fund Inform, Ann Soc Math Pol, Ser 4 6*257-274
◇ B75 D20 D80 ◇ REV MR 84k:03113 Zbl 561#03022
• ID 36131

SKANDALIS, K. [1984] *Programmable real numbers and functions* (J 2095) Fund Inform, Ann Soc Math Pol, Ser 4 7*27-56
◇ D20 ◇ REV MR 85m:68028 Zbl 561#03023 • ID 45167

SKARBEK, W. & ZEMBRZUSKI, K. [1972] *Computable real functions and their relation to the analog computer* (J 1929) Prace Centr Oblicz Pol Akad Nauk 83*40pp
◇ D20 F60 ◇ REV Zbl 263#02018 • ID 29864

SKARBEK, W. & ZEMBRZUSKI, K. [1972] *Some properties of continuous machines* (J 0014) Bull Acad Pol Sci, Ser Math Astron Phys 20*583-589
◇ D10 D20 ◇ REV MR 47#2862 Zbl 257#94032
• ID 65340

SKLAR, A. [1969] see SCHWEIZER, B.

SKLAR, A. see Vol. I, V for further entries

SKOBELEV, V.G. [1984] *Enumeration of automata possessing homing sequences (Russian) (English summary)* (J 0040) Kibernetika, Akad Nauk Ukr SSR 1984/5*120-122,136
◇ D05 ◇ ID 48596

SKOBELEV, V.G. see Vol. V for further entries

SKOLEM, T.A. [1919] *Untersuchungen ueber die Axiome des Klassenkalkuels und ueber Produktions- und Summationsprobleme, welche gewisse Klassen von Aussagen betreffen* (J 0974) Norsk Vid-Akad Oslo Mat-Natur Kl Skr 3*37pp
• REPR [1970] (C 1098) Skolem: Select Works in Logic 67-101
◇ B20 B25 C10 D35 G05 ◇ ID 12383

SKOLEM, T.A. [1923] *Begruendung der elementaren Arithmetik durch die rekurrierende Denkweise ohne Anwendung scheinbarer Veraenderlichen mit unendlichem Ausdehnungsbereich* (J 1145) Vidensk Selsk Kristiana Skrifter Ser 1 6*1-38
• TRANSL [1967] (C 0675) From Frege to Goedel 303-333 (English) [1970] (C 1098) Skolem: Select Works in Logic 153-188
◇ A05 B28 D20 F30 ◇ REV Zbl 228#02001 • ID 38689

SKOLEM, T.A. [1936] *Einige Reduktionen des Entscheidungsproblems* (J 0752) Avh Norske Vid-Akad Oslo I 1936/6*17pp
• REPR [1970] (C 1098) Skolem: Select Works in Logic 395-409
◇ B20 D35 ◇ REV Zbl 15.338 FdM 62.1059 • ID 40914

SKOLEM, T.A. [1937] *Eine Bemerkung zum Entscheidungsproblem* (P 1608) Int Congr Math (II, 5);1936 Oslo 2*268-270
◇ B20 B25 D35 ◇ REV JSL 3.57 FdM 63.31 • ID 27938

SKOLEM, T.A. [1940] *Einfacher Beweis der Unmoeglichkeit eines allgemeinen Loesungsverfahren fuer arithmetische Probleme* (J 0121) Kon Norske Vidensk Selsk Forh 13*1-4
• REPR [1970] (C 1098) Skolem: Select Works in Logic 451-454
◇ D35 F30 ◇ REV MR 2.210 Zbl 61.10 JSL 9.21 FdM 66.32 • ID 12398

SKOLEM, T.A. [1944] *A note on recursive arithmetic* (J 0121) Kon Norske Vidensk Selsk Forh 17*107-109
◇ D20 F30 ◇ REV MR 8.4 JSL 11.26 • ID 12402

SKOLEM, T.A. [1944] *Remarks on recursive functions and relations* (J 0121) Kon Norske Vidensk Selsk Forh 17*89-92
• REPR [1970] (C 1098) Skolem: Select Works in Logic 483-486
◇ D20 ◇ REV MR 8.4 JSL 11.26 • ID 28452

SKOLEM, T.A. [1944] *Some remarks on recursive arithmetic* (J 0121) Kon Norske Vidensk Selsk Forh 17*103-106
• REPR [1970] (C 1098) Skolem: Select Works in Logic 487-493
◇ D20 F30 ◇ REV MR 8.4 JSL 11.26 • ID 12401

SKOLEM, T.A. [1944] *Some remarks on the comparison between recursive functions* (J 0121) Kon Norske Vidensk Selsk Forh 17*126-129
• REPR [1970] (C 1098) Skolem: Select Works in Logic 495-498
◇ D20 ◇ REV MR 8.4 JSL 11.26 • ID 12403

SKOLEM, T.A. [1953] *Some considerations concerning recursive functions* (J 0132) Math Scand 1*213-221
• REPR [1970] (C 1098) Skolem: Select Works in Logic 553-561
◇ D20 ◇ REV MR 15.667 Zbl 52.248 JSL 21.98 • ID 12410

SKOLEM, T.A. [1954] *Remarks on "elementary" arithmetic functions* (**J** 0121) Kon Norske Vidensk Selsk Forh 27/6*6pp
• REPR [1970] (**C** 1098) Skolem: Select Works in Logic 575-580
◇ D20 ◇ REV MR 16.324 Zbl 58.5 • ID 12412

SKOLEM, T.A. [1954] *Results in investigations in the foundations (Norwegian)* (**P** 0788) Skand Mat Kongr (12);1953 Lund 273-289
◇ A05 B98 D03 F55 ◇ REV MR 16.553 Zbl 56.245 • ID 21133

SKOLEM, T.A. [1959] *Some remarks on the constructions of functions by substitution* (**J** 0121) Kon Norske Vidensk Selsk Forh 32*49-56
◇ D20 ◇ REV MR 23 # A2318 Zbl 96.243 • ID 12424

SKOLEM, T.A. [1962] *Proof of some theorems on recursively enumerable sets* (**J** 0047) Notre Dame J Formal Log 3*65-74 • ERR/ADD ibid 4*44-47
• REPR [1970] (**C** 1098) Skolem: Select Works in Logic 689-698,699-702
◇ D25 ◇ REV MR 27 # 3513a Zbl 113.6 • ID 12429

SKOLEM, T.A. [1962] *Recursive enumeration of some classes of primitive recursive functions and a majorisation theorem* (**J** 0121) Kon Norske Vidensk Selsk Forh 35*142
• REPR [1970] (**C** 1098) Skolem: Select Works in Logic 681-687
◇ D20 D25 ◇ REV MR 27 # 3512 Zbl 106.239 JSL 38.526 • ID 12428

SKOLEM, T.A. see Vol. I, II, III, V, VI for further entries

SKORDEV, D.G. [1961] see SENDOV, B.

SKORDEV, D.G. [1962] *One more example of a recursively complete arithmetic operation (Bulgarian) (German summary)* (**J** 0255) God Fak Mat & Mekh, Univ Sofiya 57*1-7
◇ D20 ◇ REV MR 33 # 41 Zbl 147.251 • ID 12437

SKORDEV, D.G. [1963] *Computable and μ-recursive operators (Bulgarian) (Russian and German summaries)* (**J** 1156) Izv Bulgar Akad Nauk Mat Inst 7*5-43
◇ D20 ◇ REV MR 27 # 3516 • ID 22056

SKORDEV, D.G. [1964] *On the concept of a recursively complete arithmetic operation (Bulgarian) (Russian summary)* (**J** 0255) God Fak Mat & Mekh, Univ Sofiya 59*117-137
◇ D20 ◇ REV MR 34 # 54 Zbl 156.9 • ID 12439

SKORDEV, D.G. [1965] *A class of primitive recursive functions (Russian) (English summary)* (**J** 0255) God Fak Mat & Mekh, Univ Sofiya 60*105-111
◇ D20 ◇ REV MR 38 # 5625 Zbl 162.21 • ID 12440

SKORDEV, D.G. [1970] *Some simple examples of universal functions (Russian)* (**J** 0023) Dokl Akad Nauk SSSR 190*45-46
• TRANSL [1970] (**J** 0062) Sov Math, Dokl 11*41-43
◇ D20 ◇ REV MR 41 # 1532 Zbl 197.281 • ID 12442

SKORDEV, D.G. [1973] *Some examples of universal functions that are recursively definable by means of small systems of inequalities (Russian)* (**S** 0554) Issl Teor Algor & Mat Logik (Moskva) 1*134-177
◇ D20 ◇ REV MR 49 # 2312 Zbl 291 # 02032 • ID 12443

SKORDEV, D.G. [1974] *A generalization of the theory of recursive functions (Russian)* (**J** 0023) Dokl Akad Nauk SSSR 219*1079-1082
• TRANSL [1974] (**J** 0062) Sov Math, Dokl 15*1756-1760
◇ D70 D75 ◇ REV MR 54 # 7230 Zbl 317 # 02046 • ID 29652

SKORDEV, D.G. [1974] *Recursively complete operations on words (Russian)* (**J** 0137) C R Acad Bulgar Sci 27*449-452
◇ D20 ◇ REV MR 49 # 8833 Zbl 333 # 02029 • ID 12444

SKORDEV, D.G. [1975] *Some topological examples of iterative combinatory spaces (Russian)* (**J** 0137) C R Acad Bulgar Sci 28*1575-1578
◇ B40 D75 ◇ REV MR 55 # 7754 Zbl 341 # 02032 • ID 29752

SKORDEV, D.G. [1976] *Certain combinatory spaces that are connected with the complexity of data processing (Russian)* (**J** 0137) C R Acad Bulgar Sci 29*7-10
◇ B40 B75 D15 D75 ◇ REV MR 53 # 12905 Zbl 358 # 02056 • ID 23157

SKORDEV, D.G. [1976] *On Turing computable operators* (**J** 0255) God Fak Mat & Mekh, Univ Sofiya 67*103-112
◇ D20 ◇ REV MR 55 # 12497 Zbl 362 # 02026 • ID 50748

SKORDEV, D.G. [1976] *Recursion theory on iterative combinatory spaces* (**J** 0014) Bull Acad Pol Sci, Ser Math Astron Phys 24*23-31
◇ B40 D75 ◇ REV MR 54 # 4950 Zbl 328 # 02026 • ID 18392

SKORDEV, D.G. [1976] *Some methods for building up the theory of recursive functions* (**P** 3208) Mat & Mat Obrazov (3);1974 Burgas 41-53
◇ D20 D25 ◇ REV Zbl 369 # 02022 • ID 51322

SKORDEV, D.G. [1976] *The concept of search computability from the point of view of the theory of combinatory spaces (Russian)* (**J** 2547) Serdica, Bulgar Math Publ 2*343-349
◇ B40 D75 ◇ REV MR 57 # 2894 Zbl 437 # 03025 • ID 55893

SKORDEV, D.G. [1977] *Simplification of some definitions in the theory of combinatory spaces* (**J** 0137) C R Acad Bulgar Sci 30*947-950
◇ B40 D75 ◇ REV MR 58 # 5084 Zbl 364 # 02025 • ID 50955

SKORDEV, D.G. [1978] *A normal form theorem for recursive operators in iterative combinatory spaces* (**J** 0068) Z Math Logik Grundlagen Math 24*115-124
◇ B40 D75 ◇ REV MR 80b:03062 Zbl 405 # 03024 • ID 54899

SKORDEV, D.G. [1979] *Algebraic generalization of a result of Bohm and Jacopini* (**J** 0137) C R Acad Bulgar Sci 32*151-154
◇ D05 ◇ REV MR 81d:03050 Zbl 454 # 68035 • ID 78678

SKORDEV, D.G. [1979] see PETROV, V.P.

SKORDEV, D.G. [1979] *The first recursion theorem for iterative combinatory spaces* (**J** 0068) Z Math Logik Grundlagen Math 25*69-77
◇ B40 D75 ◇ REV MR 80g:03049 Zbl 412 # 03032 • ID 52963

SKORDEV, D.G. [1980] *Combinatory spaces and recursiveness in them (Russian) (English summary)* (**P** 4117) Conf Math Log (Markov);1980 Sofia 456pp
◇ B40 D75 ◇ REV Zbl 537 # 03029 • ID 41127

SKORDEV, D.G. [1980] *Semi-combinatory spaces (Russian)* (**J** 0137) C R Acad Bulgar Sci 33*739-742
◇ B40 D75 ◇ REV MR 82d:03076b Zbl 452 # 03037
• ID 54101

SKORDEV, D.G. [1982] *An algebraic treatment of flow diagrams and its application to generalized recursion theory* (**P** 3831) Universal Algeb & Appl;1978 Warsaw 277-287
◇ B40 B75 D75 ◇ REV MR 85f:68007 Zbl 518 # 03017
• ID 37517

SKORDEV, D.G. [1982] *Applications of abstract recursion theory to studying the capability of functional programming systems (Russian)* (**C** 4413) Math Th & Pract Software Syst 7-16
◇ B75 D75 ◇ ID 41128

SKORDEV, D.G. see Vol. II, III, V, VI for further entries

SKORNYAKOV, L.A. [1985] *Stochastic algebra (Russian)* (**J** 0031) Izv Vyssh Ucheb Zaved, Mat (Kazan) 1985/7*3-11,84
◇ D45 ◇ ID 49352

SKORNYAKOV, L.A. see Vol. III, V for further entries

SKOWRON, A. [1972] *Durch festprogrammierte Maschinen akzeptierte Mengen (Polish) (Russian and English summaries)* (**J** 1929) Prace Centr Oblicz Pol Akad Nauk 67*16pp
◇ D10 ◇ REV Zbl 261 # 02021 • ID 30464

SKOWRON, A. [1972] *Equivalence of generalized machines* (**J** 1929) Prace Centr Oblicz Pol Akad Nauk 94*14pp
◇ D10 ◇ REV Zbl 261 # 02022 • ID 30465

SKOWRON, A. [1973] *Functions computable by machines (Polish) (Russian and English summaries)* (**J** 1929) Prace Centr Oblicz Pol Akad Nauk 105*44pp
◇ D10 ◇ REV Zbl 261 # 02023 • ID 30466

SKOWRON, A. [1973] *Machines with input and output* (**P** 1448) Math Founds of Comput Sci (2);1973 Strbske Pleso 317-319
◇ D05 D10 ◇ REV MR 54 # 4940 • ID 24113

SKOWRON, A. [1973] *Stored program machines* (**J** 0014) Bull Acad Pol Sci, Ser Math Astron Phys 21*459-465
◇ D10 D20 ◇ REV MR 58 # 10357 Zbl 287 # 68038
• ID 78689

SKOWRON, A. [1977] see BARTOL, W.

SKOWRON, A. [1983] see KREMPA, J.

SKRZYPKOWSKI, T. [1973] *On the inclusion of relational machines (Polish) (Russian and English summaries)* (**J** 1929) Prace Centr Oblicz Pol Akad Nauk 126*54pp
◇ D05 ◇ REV Zbl 277 # 68028 • ID 65352

SKYUM, S. [1975] *Confusion in the Garden of Eden* (**J** 0053) Proc Amer Math Soc 50*332-336
◇ D05 ◇ REV MR 52 # 7206 Zbl 271 # 02024 • ID 29610

SKYUM, S. & VALIANT, L.G. [1981] *A complexity theory based on Boolean algebra* (**P** 4235) IEEE Symp Found of Comp Sci (22);1981 Nashville 244-253
◇ D15 G05 ◇ REV MR 84a:68004 • ID 45822

SLAMAN, T.A. [1983] see GROSZEK, M.J.

SLAMAN, T.A. [1983] *The extended plus-one hypothesis - a relative consistency result* (**J** 0111) Nagoya Math J 92*107-120
◇ D65 E35 ◇ REV MR 85b:03081 Zbl 549 # 03033
• ID 40710

SLAMAN, T.A. [1985] *Reflection and the priority method in E-recursion theory* (**P** 3342) Rec Th Week;1984 Oberwolfach 372-404
◇ D30 D65 E45 ◇ REV Zbl 574 # 03032 • ID 45317

SLAMAN, T.A. [1985] *Reflection and forcing in E-recursion theory* (**J** 0073) Ann Pure Appl Logic 29*79-106
◇ D65 E40 E45 ◇ REV Zbl 574 # 03034 • ID 47710

SLAMAN, T.A. [1985] *The E-recursively enumerable degrees are dense* (**P** 4046) Rec Th;1982 Ithaca 195-213
◇ D30 D65 E45 ◇ REV Zbl 574 # 03033 • ID 46378

SLEZAK, P. [1984] *Minds, machines and self-reference (French and German summaries)* (**J** 0076) Dialectica 38*17-34
◇ A05 D99 F30 ◇ REV MR 85b:03011 • ID 40543

SLEZAK, P. see Vol. VI for further entries

SLISENKO, A.O. [1971] *A property of enumerable sets containing "complexly deducible" formulas (Russian) (English summary)* (**S** 0228) Zap Nauch Sem Leningrad Otd Mat Inst Steklov 20*200-207,288
• TRANSL [1973] (**J** 1531) J Sov Math 1*126-131
◇ D25 F20 ◇ REV MR 44 # 6501 Zbl 222 # 02045
• ID 12465

SLISENKO, A.O. [1973] *Identification of the symmetry predicate by means of multihead Turing machines with input (Russian)* (**S** 0066) Tr Mat Inst Steklov 129*30-202,267
• TRANSL [1973] (**S** 0055) Proc Steklov Inst Math 129*25-208
◇ D10 D15 ◇ REV MR 55 # 5410 Zbl 306 # 02029
• ID 65367

SLISENKO, A.O. [1977] *A simplified proof of the real-time recognizability of parlindromes on Turing machines (Russian) (English summary)* (**S** 0228) Zap Nauch Sem Leningrad Otd Mat Inst Steklov 68*123-139,146-147
• TRANSL [1981] (**J** 1531) J Sov Math 15*68-77
◇ D10 D15 ◇ REV MR 58 # 19361 Zbl 359 # 02035
• ID 82827

SLISENKO, A.O. [1978] *Models of computation based on address organisation of memory (Russian)* (**P** 2591) All-Union Symp Artif Intel & Autom of Invest Math;1978 Kiev 94-96
◇ D10 ◇ ID 90028

SLISENKO, A.O. [1979] *Complexity problems in the theory of computations (Russian) (English summary)* (**X** 2593) Nauch Sov Probl Kompl Kibern Akad Nauk SSSR: Moskva 32pp
◇ D15 ◇ ID 90030

SLISENKO, A.O. [1982] *Context-free grammars as a tool for describing polynomial-time subclasses of hard problems* (**J** 0232) Inform Process Lett 14*52-56
◇ D05 ◇ REV MR 83h:68064 Zbl 486 # 68034 • ID 39213

SLISENKO, A.O. [1984] *Linguistic considerations in devising effective algorithms* (**P** 4313) Int Congr Math (II,14);1983 Warsaw 1*347-357
◇ B35 D20 ◇ ID 48589

SLISENKO, A.O. see Vol. I, VI for further entries

SLOBODSKOJ, A.M. [1975] see FRIDMAN, EH.I.

SLOBODSKOJ, A.M. [1976] see FRIDMAN, EH.I.

SLOBODSKOJ, A.M. [1981] *Unsolvability of the universal theory of finite groups (Russian)* (J 0003) Algebra i Logika 20*207-230,251
• TRANSL [1981] (J 0069) Algeb and Log 20*139-156
◇ C13 C60 D35 D40 ◇ REV MR 83h:03062 Zbl 519 # 03006 • ID 36070

SLOBODSKOJ, A.M. see Vol. III, VI for further entries

SLOMSON, A. [1969] *An undecidable two sorted predicate calculus* (J 0036) J Symb Logic 34*21-23
◇ B10 D35 ◇ REV MR 39 # 2636 Zbl 187.277 • ID 12473

SLOMSON, A. see Vol. I, III, V for further entries

SLUTZKI, G. [1979] see ENGELFRIET, J.

SLUTZKI, G. [1980] see ENGELFRIET, J.

SLUTZKI, G. [1981] see PNUELI, A.

SLYAKHOVA, N.I. & TYUPA, V.G. [1961] *On algorithms of Turing type (Ukrainian) (Russian summary)* (J 4406) Zb Prats Obchis Mat Tekhn 1*31-44
◇ D10 ◇ REV MR 33 # 3849 Zbl 111.9 • ID 48042

SMART, J.J.C. [1961] *Goedel's theorem, Church's theorem and mechanism* (J 0154) Synthese 13*105-110
◇ A05 B10 D35 F30 ◇ REV Zbl 104.242 • ID 31260

SMART, J.J.C. see Vol. I for further entries

SMIKUN, L.B. [1974] *The problem of the equivalence of automata with respect to the direct product of groups (Russian) (English summary)* (J 0040) Kibernetika, Akad Nauk Ukr SSR 1974/6*145-146
• TRANSL [1974] (J 0021) Cybernetics 10*1076-1077
◇ D05 ◇ REV MR 53 # 12918 • ID 23168

SMIKUN, L.B. [1976] see LETICHEVS'KIJ, O.A.

SMIKUN, L.B. [1978] *Degrees of unsolvability of algorithmic problems connected with automata operations on groups (Russian)* (J 0040) Kibernetika, Akad Nauk Ukr SSR 1978/5*9-12
• TRANSL [1978] (J 0021) Cybernetics 14*656-659
◇ D05 D30 ◇ REV MR 80d:03037 Zbl 438 # 03046
• ID 78734

SMILEY, T.J. [1982] see DALEN VAN, D.

SMILEY, T.J. see Vol. I, II, VI for further entries

SMIRNOVA, I.M. [1961] see AJZERMAN, M.A.

SMIRNOVA, I.M. [1963] see AJZERMAN, M.A.

SMIRNOVA, I.M. see Vol. V for further entries

SMITH, C.H. [1978] see CASE, J.

SMITH, C.H. [1980] *Applications of classical recursion theory to computer science* (P 3021) Logic Colloq;1979 Leeds 236-247
◇ B75 D15 D20 D80 ◇ REV MR 82b:03089 Zbl 464 # 68038 • ID 78748

SMITH, C.H. [1983] see CASE, J.

SMITH, C.H. [1984] see DALEY, R.P.

SMITH, C.H. see Vol. II for further entries

SMITH, D.D. [1972] *Non-recursiveness of the set of finite sets of equations whose theories are one-based* (J 0047) Notre Dame J Formal Log 13*135-138
◇ C05 D35 ◇ REV MR 45 # 8525 Zbl 227 # 02024
• ID 12526

SMITH, L.W. & YAU, S.S. [1972] *Generation of regular expressions for automata by the integral of regular expressions* (J 1193) Comput J (London) 15*222-228
◇ D05 ◇ REV MR 52 # 16916 Zbl 251 # 94045 • ID 65396

SMITH, RICK L. [1980] see NERODE, A.

SMITH, RICK L. [1981] *Effective aspects of profinite groups* (J 0036) J Symb Logic 46*851-863
◇ D45 ◇ REV MR 83g:03045 Zbl 519 # 03036 • ID 33282

SMITH, RICK L. [1981] *Effective valuation theory* (P 2902) Aspects Effective Algeb;1979 Clayton 232-245
◇ C57 C60 D45 F55 ◇ REV MR 83b:03053 Zbl 479 # 03023 • ID 55685

SMITH, RICK L. [1981] *Two theorems on autostability in p-groups* (P 2628) Log Year;1979/80 Storrs 302-311
◇ C57 C60 ◇ REV MR 83h:03064 Zbl 488 # 03024
• ID 36072

SMITH, RICK L. [1985] see DRIES VAN DEN, L.

SMITH, RICK L. see Vol. V, VI for further entries

SMOLARSKI, D.C. [1981] see KLOSINSKI, L.F.

SMOLSKA-ADAMOWICZ, Z. [1977] *On finite lattices of the degrees of constructibility of reals* (P 1639) Set Th & Hierarch Th (1);1974 Karpacz 31-49
◇ D30 E35 E45 ◇ REV MR 58 # 10445 Zbl 386 # 03025
• ID 52201

SMOLSKA-ADAMOWICZ, Z. also published under the name ADAMOWICZ, Z.

SMORYNSKI, C.A. [1973] *Elementary intuitionistic theories* (J 0036) J Symb Logic 38*102-134
◇ B25 C90 D35 F50 ◇ REV MR 48 # 5842 Zbl 261 # 02033 • ID 12535

SMORYNSKI, C.A. [1977] *A note on the number of zeros of polynomials and exponential polynomials* (J 0036) J Symb Logic 42*99-106
◇ D25 D35 D80 ◇ REV MR 58 # 5124 Zbl 367 # 02019
• ID 24358

SMORYNSKI, C.A. [1978] *Avoiding self-referential statements* (J 0053) Proc Amer Math Soc 70*181-184
◇ D25 F30 ◇ REV MR 57 # 16015 Zbl 392 # 03037
• ID 28384

SMORYNSKI, C.A. [1979] *Some rapidly growing functions* (J 2789) Math Intell 2*149-154
◇ D20 D80 F30 ◇ REV MR 82c:03064 Zbl 453 # 03049
• ID 54179

SMORYNSKI, C.A. & STAVI, J. [1980] *Cofinal extension preserves recursive saturation* (P 2625) Model Th of Algeb & Arithm;1979 Karpacz 338-345
◇ C50 C57 C62 ◇ REV MR 82g:03062 Zbl 467 # 03059
• ID 55057

SMORYNSKI, C.A. [1981] *Cofinal extensions of nonstandard models of arithmetic* (J 0047) Notre Dame J Formal Log 22*133-144
◇ C50 C57 C62 ◇ REV MR 82g:03063 Zbl 481 # 03045
• ID 78761

SMORYNSKI, C.A. [1981] *Elementary extensions of recursively saturated models of arithmetic* (J 0047) Notre Dame J Formal Log 22*193-203
◇ C50 C57 C62 ◇ REV MR 82g:03064 Zbl 503.03032
• ID 78760

SMORYNSKI, C.A. [1981] *Recursively saturated nonstandard models of arithmetic* (J 0036) J Symb Logic 46*259-286
• ERR/ADD ibid 47*493-494
◇ C50 C57 C62 ◇ REV MR 82i:03045 MR 84b:03054 Zbl 501 # 03044 Zbl 549 # 03059 • ID 78758

SMORYNSKI, C.A. [1982] *A note on initial segment constructions in recursively saturated models of arithmetic* (J 0047) Notre Dame J Formal Log 23*393-408
◇ C50 C57 C62 ◇ REV MR 83j:03058 Zbl 519 # 03055 • ID 35359

SMORYNSKI, C.A. [1982] *Back-and-forth inside a recursively saturated model of arithmetic* (P 3623) Logic Colloq;1980 Prague 273-278
◇ C15 C50 C57 C62 ◇ REV MR 83m:03041 Zbl 503 # 03033 • ID 35443

SMORYNSKI, C.A. [1982] *The finite inseparability of the first-order theory of diagonalisable algebras* (J 0063) Studia Logica 41*347-349
◇ B45 C13 D35 F30 ◇ REV Zbl 542 # 03024 • ID 43687

SMORYNSKI, C.A. [1982] *The varieties of arboreal experience* (J 2789) Math Intell 4*182-189
◇ D20 F30 ◇ REV MR 84i:03110 Zbl 521 # 03041 • ID 34588

SMORYNSKI, C.A. see Vol. II, III, V, VI for further entries

SMULLYAN, R.M. [1958] *Undecidability and recursive inseparability* (J 0068) Z Math Logik Grundlagen Math 4*143-147
◇ D25 D35 ◇ REV MR 20 # 5734 Zbl 99.9 JSL 25.165 • ID 12539

SMULLYAN, R.M. [1960] see PUTNAM, H.

SMULLYAN, R.M. [1960] *Theories with effectively inseparable nuclei* (J 0068) Z Math Logik Grundlagen Math 6*219-224
◇ D25 D35 ◇ REV MR 23 # A3675 Zbl 95.9 • ID 12541

SMULLYAN, R.M. [1961] *Elementary formal systems* (J 0090) J Math Soc Japan 13*38-44
◇ D03 D25 ◇ REV MR 24 # A685 Zbl 100.10 JSL 34.117 • ID 12544

SMULLYAN, R.M. [1961] *Extended canonical systems* (J 0053) Proc Amer Math Soc 12*440-442
◇ B45 D03 D25 ◇ REV MR 23 # A3092 Zbl 101.11 JSL 32.524 • ID 12542

SMULLYAN, R.M. [1961] *Monadic elementary formal systems* (J 0068) Z Math Logik Grundlagen Math 7*81-83
◇ D03 D25 ◇ REV MR 25 # 3824 Zbl 218 # 02030 • ID 12543

SMULLYAN, R.M. [1961] *Theory of formal systems* (S 3513) Ann Math Stud xi+142pp
• TRANSL [1981] (X 2027) Nauka: Moskva 208pp
◇ B98 D03 D05 D20 D25 D98 ◇ REV MR 22 # 12042 MR 27 # 2409 MR 83h:03002 Zbl 529 # 03014 Zbl 97.245 JSL 30.88 • ID 21025

SMULLYAN, R.M. [1962] *On Post's canonical systems* (J 0036) J Symb Logic 27*55-57
◇ D03 D25 ◇ REV MR 33 # 3924 Zbl 134.8 JSL 33.623 • ID 12548

SMULLYAN, R.M. [1963] *Creativity and effective inseparability* (J 0064) Trans Amer Math Soc 109*135-145
◇ D25 ◇ REV MR 27 # 3518 Zbl 117.254 JSL 30.391 • ID 12549

SMULLYAN, R.M. [1963] *Pseudo-uniform reducibility* (J 0090) J Math Soc Japan 15*129-133
◇ D20 D25 D30 D35 ◇ REV MR 27 # 2411 Zbl 118.249 • ID 12546

SMULLYAN, R.M. [1964] *Effectively simple sets* (J 0053) Proc Amer Math Soc 15*893-894
◇ D25 ◇ REV MR 31 # 4720 Zbl 192.51 • ID 12550

SMULLYAN, R.M. see Vol. I, II, III, V, VI for further entries

SMYTH, M.B. [1980] *Computability in categories* (P 2904) Automata, Lang & Progr (7);1980 Noordwijkerhout 609-620
◇ D45 D75 G30 ◇ REV MR 82c:68055 Zbl 471 # 68031 • ID 55260

SMYTH, M.B. see Vol. I, VI for further entries

SNIR, M. [1980] see SHAMIR, E.

SNIR, M. see Vol. I for further entries

SNYDER, LAWRENCE [1977] see LIPTON, R.J.

SOARE, R.I. [1969] *A note on degrees of subsets* (J 0036) J Symb Logic 34*256
◇ D30 ◇ REV Zbl 188.27 • ID 12567

SOARE, R.I. [1969] *Cohesive sets and recursively enumerable Dedekind cuts* (J 0048) Pac J Math 31*215-231
◇ D25 D30 ◇ REV MR 40 # 2542 Zbl 172.9 JSL 36.148 • ID 12566

SOARE, R.I. [1969] *Constructive order types on cuts* (J 0036) J Symb Logic 34*285-289
◇ D30 D45 ◇ REV Zbl 182.17 • ID 12568

SOARE, R.I. [1969] *Recursion theory and Dedekind cuts* (J 0064) Trans Amer Math Soc 140*271-294
◇ D25 D30 F60 ◇ REV MR 39 # 3997 Zbl 181.305 JSL 36.148 • ID 12565

SOARE, R.I. [1969] *Sets with no subset of higher degree* (J 0036) J Symb Logic 34*53-56
◇ D30 ◇ REV MR 41 # 8228 Zbl 182.16 • ID 12564

SOARE, R.I. [1970] see GANDY, R.O.

SOARE, R.I. [1970] see JOCKUSCH JR., C.G.

SOARE, R.I. [1971] see JOCKUSCH JR., C.G.

SOARE, R.I. [1972] see JOCKUSCH JR., C.G.

SOARE, R.I. [1972] *The Friedberg-Muchnik theorem re-examined* (J 0017) Canad J Math 24*1070-1078
◇ D25 ◇ REV MR 49 # 8836 Zbl 225 # 02028 • ID 12569

SOARE, R.I. [1973] see JOCKUSCH JR., C.G.

SOARE, R.I. [1974] *Automorphisms of the lattice of recursively enumerable sets* (J 0015) Bull Amer Math Soc 80*53-58
◇ D25 ◇ REV MR 51 # 10058 Zbl 281 # 02044 • ID 17544

SOARE, R.I. [1974] *Automorphisms of the lattice of recursively enumerable sets I: Maximal sets* (J 0120) Ann of Math, Ser 2 100*80-120
◇ D25 ◇ REV MR 50 # 12685 Zbl 279 # 02022 • REM Part II 1982 • ID 12570

SOARE, R.I. [1974] *Isomorphisms on countable vector spaces with recursive operations* (J 0038) J Austral Math Soc 18*230-235
◇ D45 D50 ◇ REV MR 58 # 186 Zbl 301 # 02037 • ID 65409

SOARE, R.I. [1975] see MORLEY, M.D.

SOARE, R.I. [1976] *The infinite injury priority method* (J 0036) J Symb Logic 41*513-530
◇ D25 ◇ REV MR 55 # 2534 Zbl 329 # 02019 • ID 14787

SOARE, R.I. [1977] *Computational complexity, speedable and levelable sets* (J 0036) J Symb Logic 42*545-563
◇ D15 ◇ REV MR 58 # 5125 Zbl 401 # 68020 • ID 26857

SOARE, R.I. [1977] see BENNISON, V.L.

SOARE, R.I. [1978] see LERMAN, M.

SOARE, R.I. [1978] *Recursively enumerable sets and degrees* (J 0015) Bull Amer Math Soc 84*1149-1181
◇ D25 ◇ REV MR 81g:03050 Zbl 401 # 03018 • ID 78783

SOARE, R.I. [1978] see BENNISON, V.L.

SOARE, R.I. [1980] see LERMAN, M.

SOARE, R.I. [1980] *Constructions in the recursively enumerable degrees* (P 2129) Rec Th & Comput Complex (CIME);1979 Bressanone
◇ D25 ◇ ID 33619

SOARE, R.I. [1980] *Fundamental methods for constructing recursively enumerable degrees* (P 3021) Logic Colloq;1979 Leeds 1-51
◇ D25 ◇ REV MR 82j:03051 Zbl 451 # 03012 • ID 54027

SOARE, R.I. [1980] see LACHLAN, A.H.

SOARE, R.I. [1980] *Recursive enumerability* (P 1959) Int Congr Math (II,13);1978 Helsinki 1*275-280
◇ D25 ◇ REV MR 81k:03038 Zbl 434 # 03028 • ID 55732

SOARE, R.I. [1981] see FEJER, P.A.

SOARE, R.I. [1982] *Automorphisms of the lattice of recursively enumerable sets Part II:Low sets* (J 0007) Ann Math Logic 22*69-107
◇ D25 ◇ REV MR 83k:03048 Zbl 526 # 03022 • REM Part I 1974 • ID 33620

SOARE, R.I. [1982] *Computational complexity of recursively enumerable sets* (J 0194) Inform & Control 52*8-18
◇ D15 D25 ◇ REV MR 85b:03063 Zbl 512 # 03022 • ID 33621

SOARE, R.I. & STOB, M. [1982] *Relative recursive enumerability* (P 3708) Herbrand Symp Logic Colloq;1981 Marseille 299-324
◇ D25 ◇ REV MR 86b:03050 Zbl 513 # 03019 • ID 33624

SOARE, R.I. [1984] see AMBOS-SPIES, K.

SOARE, R.I. [1984] see LERMAN, M.

SOARE, R.I. [1984] see KNIGHT, J.F.

SOARE, R.I. [1985] *Tree arguments in recursion theory and the 0'''-priority method* (P 4046) Rec Th;1982 Ithaca 53-106
◇ D25 ◇ REV Zbl 573 # 03014 • ID 46379

SOARE, R.I. see Vol. I, III for further entries

SOBOLEV, S.K. [1977] *On finite-dimensional superintuitionistic logics (Russian)* (J 0216) Izv Akad Nauk SSSR, Ser Mat 41*963-986,1199
• TRANSL [1977] (J 0448) Math of USSR, Izv 11*909-935
◇ B55 D05 F50 G10 G25 ◇ REV MR 58 # 10333 Zbl 368 # 02062 • ID 51279

SOBOLEV, S.K. [1977] *The intuitionistic propositional calculus with quantifiers (Russian)* (J 0087) Mat Zametki (Akad Nauk SSSR) 22*69-76
• TRANSL [1977] (J 1044) Math Notes, Acad Sci USSR 22*528-532
◇ B55 C80 C90 D35 F50 ◇ REV MR 56 # 15371 Zbl 365 # 02013 • ID 51014

SOBOLEV, S.K. see Vol. II for further entries

SOBOLEV, S.L. (ED.) [1982] *Mathematical logic and the theory of algorithms (Russian)* (X 2642) Nauka: Novosibirsk 176pp
◇ B97 C97 D97 ◇ REV MR 84i:03007 Zbl 539 # 00002 • ID 34492

SODNOMOV, B.S. [1955] *Consistency of the projective evaluation of some non-effective sets (Russian)* (J 0067) Usp Mat Nauk 10/1(63)*155-158
◇ D55 E15 E35 E45 ◇ REV MR 17.570 Zbl 64.10 JSL 21.406 • ID 12627

SODNOMOV, B.S. [1956] *Consistency of the projectivity of some special sets (Russian)* (J 4162) Uch Zap Ped Inst, Buryat 10*3-10
◇ D55 E15 E35 E45 ◇ ID 45667

SODNOMOV, B.S. see Vol. V for further entries

SOIL, A. [1977] *Remarks on uniform tag sequences* (J 0230) An Univ Iasi, NS, Sect Ia 23*415-418
◇ D03 ◇ REV MR 58 # 16216 Zbl 372 # 02020 • ID 51428

SOISALON-SOININEN, E. [1984] see CHAZELLE, B.

SOITTOLA, M. [1978] see SALOMAA, A.

SOKOLOV, V.A. [1967] *Isomorphisms of maximal subalgebras of R.Robinson's algebra (Russian) (English summary)* (J 0003) Algebra i Logika 6/3*91-99
◇ D20 D75 ◇ REV MR 36 # 4989 Zbl 236 # 02036 • ID 27861

SOKOLOV, V.A. [1972] *Certain properties of the algebra of all partially recursive functions (Russian)* (S 0166) Mat Issl, Mold SSR 7/1(23)*133-149
◇ D20 D75 ◇ REV MR 47 # 37 Zbl 261 # 02025 • ID 30468

SOKOLOV, V.A. [1972] *The maximal subalgebras of the algebra of all partially recursive functions (Russian) (English summary)* (J 0040) Kibernetika, Akad Nauk Ukr SSR 1972/1*70-73
◇ D20 D75 ◇ REV MR 46 # 5122 Zbl 261 # 02026 • ID 12632

SOKOLOV, V.A. [1975] *A problem in the class of computable functions with the superposition operation (Russian)* (J 2868) Vest Univ Yaroslavl' 9*111-114
◇ D20 ◇ REV MR 57 # 79 • ID 78829

SOKOLOV, V.A. [1978] *"Zero" identities in the Robinson algebra (Russian)* (C 3177) Ehvrist Algor Optim, Vyp 2 137-146
◇ D20 D75 ◇ REV Zbl 468 # 03018 • ID 55083

SOKOLOV, V.A. see Vol. II for further entries

SOKOLOV, V.G. [1970] see REMESLENNIKOV, V.N.

SOKOLOV, V.G. [1971] *An algorithm for the solution of the word problem for a certain class of solvable groups (Russian)* (J 0092) Sib Mat Zh 12*1405-1410
• TRANSL [1971] (J 0475) Sib Math J 12*1016-1020
◇ D40 ◇ REV MR 46 # 1913 • ID 12633

SOLDATOVA, V.V. [1969] *Solution of the word problem for a certain class of groups (Russian)* (**J** 0226) Uch Zap Ped Inst, Ivanovo 44*17-25
◇ D40 ◇ REV MR 46 # 9168 • ID 12635

SOLDATOVA, V.V. see Vol. III for further entries

SOLNTSEV, S.V. [1976] *Two remarks on the stopping problem for cellular growth models (Russian)* (**J** 2320) Probl Peredachi Inf, Akad Nauk SSSR 12/1*63-69
• TRANSL [1976] (**J** 3419) Probl Inf Transmiss 12/1*43-49
◇ D05 ◇ REV MR 54 # 14892 Zbl 366 # 94061 • ID 51154

SOLOMON, M.K. [1978] *Some results on measure independent Goedel speed-ups* (**J** 0036) J Symb Logic 43*667-672
◇ D15 F20 ◇ REV MR 80b:03091 Zbl 398 # 03027
• ID 52755

SOLOMONOFF, R.J. [1978] *Complexity-based induction systems: Comparisons and convergence theorems* (**J** 2745) IEEE Trans Inf Theory IT-24*422-432
◇ B30 D10 ◇ REV MR 58 # 20912 Zbl 382 # 60003
• ID 51980

SOLON, B.YA. [1971] see KHUDYAKOV, V.V.

SOLON, B.YA. [1976] *On non-minimal pm-degrees and pc-degrees (Russian)* (**P** 2064) All-Union Conf Math Log (4);1976 Kishinev 139
◇ D30 ◇ ID 90054

SOLON, B.YA. [1977] *Reducibility according to computability and e-interreducible sets (Russian)* (**X** 2235) VINITI: Moskva 2999-77
◇ D30 ◇ ID 90053

SOLON, B.YA. [1978] *e-powers of hyperimmune retraceable sets (Russian)* (**J** 0092) Sib Mat Zh 19*172-179,239
• TRANSL [1978] (**J** 0475) Sib Math J 19*172-179
◇ D25 D30 D50 ◇ REV MR 57 # 16018 Zbl 396 # 03036
• ID 52646

SOLON, B.YA. [1978] *Quasiminimal pF-degrees (Russian)* (**C** 2595) Rekursiv Funktsii 85-98
◇ D30 ◇ REV MR 82d:03072 Zbl 504 # 03020 • ID 90055

SOLON, B.YA. [1979] *e-degrees of productive sets (Russian)* (**P** 2558) All-Union Conf Math Log (5) (Mal'tsev);1979 Novosibirsk 142
◇ D25 ◇ ID 90056

SOLON, B.YA. [1979] see ROZINAS, M.G.

SOLON, B.YA. [1981] *PC-degrees inside an e-degree of a hyperimmune retraceable set (Russian)* (**C** 3865) Algeb Sistemy (Ivanovo) 203-217
◇ D25 D30 D50 H05 ◇ REV MR 85g:03063
Zbl 549 # 03029 • ID 43887

SOLOVAY, R.M. [1967] *A nonconstructible Δ^1_3 set of integers* (**J** 0064) Trans Amer Math Soc 127*50-75
◇ D55 E45 E55 ◇ REV MR 35 # 2748 Zbl 161.7
JSL 36.340 • ID 12640

SOLOVAY, R.M. [1969] see MARTIN, D.A.

SOLOVAY, R.M. [1969] *On the cardinality of Σ^1_2 sets of reals* (**C** 0705) Found of Math (Goedel) 58-73
◇ D55 E15 E40 E45 E50 E55 ◇ REV MR 43 # 3115
Zbl 188.325 JSL 39.330 • ID 22230

SOLOVAY, R.M. [1970] see JENSEN, R.B.

SOLOVAY, R.M. [1976] *On sets Cook-reducible to sparse sets* (**J** 1428) SIAM J Comp 5*646-652
◇ D15 ◇ REV MR 56 # 7321 Zbl 367 # 02018 • ID 51178

SOLOVAY, R.M. [1977] see JOCKUSCH JR., C.G.

SOLOVAY, R.M. [1977] *On random r.e. sets* (**P** 1076) Latin Amer Symp Math Log (3);1976 Campinas 283-307
◇ D15 D25 ◇ REV MR 57 # 16016 Zbl 366 # 02025
• ID 16611

SOLOVAY, R.M. [1978] *A Δ^1_3 coding of the subsets of $\omega\omega$* (**C** 2908) Cabal Seminar Los Angeles 1976-77 133-150
◇ D55 E05 E15 E45 E60 ◇ REV MR 80g:03054
Zbl 392 # 03032 • ID 52399

SOLOVAY, R.M. [1978] *Hyperarithmetically encodable sets* (**J** 0064) Trans Amer Math Soc 239*99-122
◇ D55 D60 E05 E40 E45 E55 ◇ REV MR 58 # 10375
Zbl 411 # 03039 • ID 52893

SOLOVAY, R.M. [1981] see KECHRIS, A.S.

SOLOVAY, R.M. [1983] see KECHRIS, A.S.

SOLOVAY, R.M. [1985] *Infinite fixed-point algebras* (**P** 4046) Rec Th;1982 Ithaca 473-486
◇ C57 C62 G25 ◇ REV Zbl 573 # 03030 • ID 46419

SOLOVAY, R.M. [1985] see KECHRIS, A.S.

SOLOVAY, R.M. see Vol. II, III, V, VI for further entries

SOLOV'EV, S.YU. [1980] *On a constructive procedure of grammatical inference (Russian)* (**J** 2869) Vest Ser Vychisl Mat Kibern, Univ Moskva 1980/3*44-48,70
◇ D05 ◇ REV MR 82g:68078 Zbl 442 # 68081 • ID 82849

SOLOV'EV, V.D. [1974] *Q-reducibility and hypersimple sets (Russian)* (**J** 3937) Veroyat Met i Kibern (Kazan) 10-11*121-128
◇ D25 D30 ◇ REV MR 57 # 16017 Zbl 314 # 02050
• ID 65453

SOLOV'EV, V.D. [1976] *Some generalizations of the notions of reducibility and creativity (Russian)* (**J** 0031) Izv Vyssh Ucheb Zaved, Mat (Kazan) 1976/3(166)*65-72
• TRANSL [1976] (**J** 3449) Sov Math 20/3*56-62
◇ D25 D30 ◇ REV MR 54 # 9999 Zbl 349 # 02029
• ID 25840

SOLOV'EV, V.D. [1976] *Superhypersimple sets (Russian)* (**J** 0031) Izv Vyssh Ucheb Zaved, Mat (Kazan) 1976/2(165)*108-110
• TRANSL [1976] (**J** 3449) Sov Math 20/2*101-103
◇ D25 ◇ REV MR 56 # 2807 Zbl 348 # 02037 • ID 65454

SOLOV'EV, V.D. [1977] see ARSLANOV, M.M.

SOLOV'EV, V.D. [1978] see ARSLANOV, M.M.

SOLOV'EV, V.D. [1982] *The structure of closed classes of computable functions and predicates (Russian)* (**J** 0031) Izv Vyssh Ucheb Zaved, Mat (Kazan) 1982/12*51-56
• TRANSL [1982] (**J** 3449) Sov Math 26/12*53-60
◇ D20 D75 ◇ REV MR 84f:03040 Zbl 511 # 03017
• ID 37390

SOLOV'EV, V.D. [1983] *Program schemes and effective functionals of finite type (Russian)* (**J** 3937) Veroyat Met i Kibern (Kazan) 19*129-133
◇ D65 ◇ REV MR 84i:00006 Zbl 543 # 68035 • ID 40975

SOLS, I. [1975] see KUEHNEL, W.

SOLS, I. [1976] *Logic, foundations, geometry and automata in the realm of topoi* (P 1619) Coloq Log Simb;1975 Madrid 159-176
◇ D05 E70 G30 ◇ REV MR 56 # 15419 Zbl 381 # 18006 • ID 78853

SOLS, I. [1977] see KUEHNEL, W.

SOLS, I. see Vol. I, V for further entries

SONTAG, E.D. [1975] *On some questions of rationality and decidability* (J 0119) J Comp Syst Sci 11*375-381
◇ D05 ◇ REV MR 54 # 4941 Zbl 357 # 68064 • ID 24114

SONTAG, E.D. [1985] *Real addition and the polynomial hierarchy* (J 0232) Inform Process Lett 20*115-120
◇ D15 ◇ REV Zbl 575 # 03030 • ID 48866

SOO HONG KWANG [1984] see BAVEL, Z.

SOPRUNOV, S.F. [1974] *On the power of a real-closed field (Russian) (English summary)* (J 0288) Vest Ser Mat Mekh, Univ Moskva 29/4*70-73
• TRANSL [1974] (J 0510) Moscow Univ Math Bull 29/3-4*98-101
◇ C20 C60 D35 ◇ REV MR 51 # 5294 Zbl 298 # 02060 • ID 17402

SOPRUNOV, S.F. see Vol. III, V for further entries

SORBI, A. [1982] *Σ_0^n-equivalence relations* (J 0063) Studia Logica 41*351-358
◇ D55 F30 ◇ REV MR 85j:03071 Zbl 539 # 03023 • REM The title of this article contains a misprint: Σ_0^n should be Σ_n^0 • ID 41231

SORBI, A. [1982] *Numerazioni positive, r.e. classi e formule (English summary)* (J 3768) Boll Unione Mat Ital, VI Ser, D 1*79-88
• REPR [1982] (J 2100) Boll Unione Mat Ital, VI Ser, B 1/1*403-411
◇ D20 F30 ◇ REV MR 83e:03071 MR 84c:03085 Zbl 502 # 03023 Zbl 508 # 03022 • ID 46829

SORBI, A. [1985] see MONTAGNA, F.

SORBI, A. see Vol. VI for further entries

SORENSEN, R.A. [1984] *Unique alternative guessing* (J 0079) Logique & Anal, NS 27*77-85
◇ B45 D20 ◇ REV MR 85h:03024 • ID 43297

SORKIN, YU.I. [1961] *Algorithmic solvability of isomorphism problems (Russian)* (J 0023) Dokl Akad Nauk SSSR 137*804-806
• TRANSL [1961] (J 0470) Sov Phys, Dokl 6*294-295
◇ D40 ◇ REV MR 25 # 4998 Zbl 112.9 • ID 21098

SOSKOV, I.N. [1983] *Algorithmically complete algebraic systems (Russian)* (J 0137) C R Acad Bulgar Sci 36*729-731
◇ D75 ◇ REV Zbl 531 # 03039 • ID 37686

SOSKOV, I.N. [1983] *Computability in algebraic systems (Russian)* (J 0137) C R Acad Bulgar Sci 36*301-304
◇ D75 ◇ REV MR 85b:03082 • ID 40712

SOSKOV, I.N. [1984] *Simply calculable functions on a base set (Russian)* (P 4392) Mat Logika (Markova);1980 Sofia 112-138
◇ D20 ◇ ID 46579

SOUSLIN, M. [1917] *Sur une definition des ensembles mesurables B sans nombres transfinis* (J 0109) C R Acad Sci, Paris 164*88-91
◇ D55 E15 E75 ◇ REV FdM 46.296 • ID 38033

SOUSLIN, M. see Vol. V for further entries

SOZANSKA-BIEN, Z. [1970] *Some properties of m-address machines (Polish and Russian summaries)* (J 0063) Studia Logica 26*19-34
◇ D10 ◇ REV MR 44 # 1521 Zbl 253 # 94029 • ID 65465

SOZANSKA-BIEN, Z. also published under the name BIEN, Z.

SPANIER, E.H. [1963] see GINSBURG, S.

SPANIER, E.H. [1966] see GINSBURG, S.

SPANIER, E.H. [1968] see GINSBURG, S.

SPANIER, E.H. [1969] see BAER, ROBERT M.

SPANIER, E.H. [1970] see GINSBURG, S.

SPANIER, E.H. [1971] see GINSBURG, S.

SPANIER, E.H. [1974] see GINSBURG, S.

SPANIER, E.H. [1975] see GINSBURG, S.

SPANIER, E.H. [1977] see CREMERS, A.B.

SPANIER, E.H. [1983] see GINSBERG, S.

SPECHT, J. [1979] see DUSKE, J.

SPECKER, E. [1971] *Ramsey's theorem does not hold in recursive set theory* (P 0638) Logic Colloq;1969 Manchester 439-442
◇ D25 D80 E05 F60 ◇ REV MR 43 # 4667 Zbl 285 # 02038 • ID 29960

SPECKER, E. & WICK, G. [1976] *Laengen von Formeln* (P 3196) Kompl von Entscheid Probl;1973/74 Zuerich 182-217
◇ D15 ◇ REV MR 57 # 18232 Zbl 388 # 94019 • ID 52287

SPECKER, E. [1979] see LIEBERHERR, K.J.

SPECKER, E. [1981] see LIEBERHERR, K.J.

SPECKER, E. see Vol. I, II, III, V, VI for further entries

SPECTOR, C. [1955] *Recursive well-orderings* (J 0036) J Symb Logic 20*151-163
◇ D30 D45 D55 F15 ◇ REV MR 17.570 Zbl 67.3 JSL 21.412 • ID 12672

SPECTOR, C. [1956] *On degrees of recursive unsolvability* (J 0120) Ann of Math, Ser 2 64*581-592
◇ D30 ◇ REV MR 18.552 Zbl 74.13 JSL 22.374 • ID 12673

SPECTOR, C. [1957] *Measure theory and higher order incomparability* (P 1675) Summer Inst Symb Log;1957 Ithaca 265-266
◇ D30 D55 ◇ ID 29359

SPECTOR, C. [1957] *Recursive ordinals and predicative set theory* (P 1675) Summer Inst Symb Log;1957 Ithaca 377-382
◇ D20 D55 E70 F15 F65 ◇ REV JSL 31.138 • ID 12676

SPECTOR, C. [1958] *Measure-theoretic construction of imcomparable hyperdegrees* (J 0036) J Symb Logic 23*280-288
◇ D30 D55 ◇ REV MR 22 # 3681 Zbl 85.249 JSL 27.242 • ID 12674

SPECTOR, C. [1960] *Hyperarithmetical quantifiers* (J 0027) Fund Math 48*313-320
◇ D55 D70 ◇ REV MR 22 # 10904 Zbl 98.243 JSL 31.137 • ID 12675

SPECTOR, C. [1961] *Inductively defined sets of natural numbers* (**P** 0633) Infinitist Meth;1959 Warsaw 97-102
◇ D55 D70 ◇ REV MR 25 # 4991 Zbl 116.7 JSL 34.295 JSL 35.295 • ID 12677

SPECTOR, C. [1962] see FEFERMAN, S.

SPECTOR, C. see Vol. VI for further entries

SPEHNER, J.-C. [1974] *Deux algorithmes relatifs aux sous-monoides d'un monoide libre* (**J** 2313) C R Acad Sci, Paris, Ser A-B 278*A1335-A1338
◇ D05 ◇ REV MR 50 # 514 Zbl 284 # 20066 • ID 29931

SPENCER, J.H. [1983] *Large numbers and unprovable theorems* (**J** 0005) Amer Math Mon 90*669-675
◇ D20 F30 ◇ REV MR 85a:03073 Zbl 539 # 03042 • ID 34745

SPENCER, J.H. see Vol. V, VI for further entries

SPERSCHNEIDER, V. [1980] *Goedelisierung* (**S** 3126) Ber, Fak Inf, Univ Karlsruhe 29/80*58pp
◇ D45 D75 ◇ REV Zbl 452 # 03035 • ID 54099

SPERSCHNEIDER, V. [1982] see MAIER, W.

SPERSCHNEIDER, V. [1982] see MENZEL, W.

SPERSCHNEIDER, V. [1984] see MENZEL, W.

SPERSCHNEIDER, V. [1984] *The length-problem* (**P** 2342) Symp Rek Kombin;1983 Muenster 88-102
◇ D20 D25 ◇ REV Zbl 544 # 03018 • ID 41003

SPIVAK, M.A. [1965] *Algorithm for abstract synthesis of automata for an expanded language of regular expressions (Russian)* (**J** 0977) Izv Akad Nauk SSSR, Tekh Kibern 1965/1*51-57
• TRANSL [1965] (**J** 0522) Engin Cybern 1*43-49
◇ D05 ◇ REV MR 32 # 5467 Zbl 133.257 JSL 37.620 • ID 12693

SPIVAK, M.A. see Vol. V for further entries

SPREEN, D. & YOUNG, P. [1984] *Effective operators in a topological setting* (**P** 2153) Logic Colloq;1983 Aachen 2*437-452
◇ D20 D75 D80 ◇ REV Zbl 563 # 03030 • ID 43014

SPREEN, D. [1984] *On r.e. inseparability of cpo index sets* (**P** 2342) Symp Rek Kombin;1983 Muenster 103-117
◇ D25 D45 ◇ REV Zbl 545 # 03026 • ID 43507

SPRINGSTEEL, F. [1972] see RITCHIE, R.W.

SPRINGSTEEL, F. [1979] see PUDLAK, P.

SPRINGSTEEL, F. [1981] *Complexity of hypothesis formation problems* (**J** 1741) Int J Man-Mach Stud 15*319-332
◇ B35 D15 ◇ REV MR 82m:68091 • ID 82856

SPRUGNOLI, R. [1975] see LAGANA, M.R.

SPRUGNOLI, R. [1975] see LEONI, G.

SPRUGNOLI, R. [1977] see LEONI, G.

SQUIER, C.C. [1984] see BOOK, R.V.

SQUIER, C.C. [1984] see AVENHAUS, J.

SREBRNY, M. [1976] see MAREK, W.

SREBRNY, M. [1977] *Relatively constructible transitive models* (**J** 0027) Fund Math 96*161-172
◇ C62 D55 D60 E10 E40 E45 E55 ◇ REV MR 58 # 16273 Zbl 383 # 03036 • ID 26534

SREBRNY, M. [1977] see LACHLAN, A.H.

SREBRNY, M. [1978] *Singular cardinals and analytic games* (**P** 1864) Higher Set Th;1977 Oberwolfach 391-422
◇ D55 E15 E45 E50 E55 E60 ◇ REV MR 80j:03072 Zbl 413 # 03030 • ID 53024

SREBRNY, M. [1979] *Constructible sets and analytic games* (**J** 0519) Wiad Mat, Ann Soc Math Pol, Ser 2 22*30-34
◇ D55 E15 E45 E50 E55 E60 ◇ REV MR 82a:03050 • ID 78903

SREBRNY, M. see Vol. III, V, VI for further entries

SRINIVASAN, C.V. [1959] see NARASIMHAN, R.

STACEY, K. [1974] see CROSSLEY, J.N.

STAHL, G. [1981] *Character and acceptability of Church's thesis* (**J** 0302) Rep Math Logic, Krakow & Katowice 11*63-67
◇ D20 ◇ REV MR 84d:03013 Zbl 506 # 03011 • ID 34053

STAHL, G. see Vol. I, II, V, VI for further entries

STAHL, S.H. [1976] *A hierarchy on the class of primitive recursive ordinal functions* (**J** 0017) Canad J Math 28*1205-1209
◇ D55 D60 ◇ REV MR 54 # 7233 Zbl 327 # 02034 • ID 25024

STAHL, S.H. [1977] *Primitive recursive ordinal functions with added constants* (**J** 0036) J Symb Logic 42*77-82
◇ D60 ◇ REV MR 58 # 16231 Zbl 371 # 02016 • ID 23749

STAHL, S.H. see Vol. III, V for further entries

STAHL, W.R. [1963] see CAFFIN, R.W.

STAIGER, L. [1972] see AROLD, D.

STAIGER, L. & WAGNER, K. [1974] *Automatentheoretische und automatenfreie Charakterisierungen topologischer Klassen regulaerer Folgenmengen* (**J** 0129) Elektr Informationsverarbeitung & Kybern 10*379-392
◇ D05 ◇ REV MR 57 # 11971 Zbl 301 # 94069 • ID 65483

STAIGER, L. & WAGNER, K. [1975] *Finite automata acceptation of infinite sequences* (**P** 1755) Math Founds of Comput Sci (3);1974 Jadwisin 69-72
◇ D05 E15 ◇ REV Zbl 312 # 94029 • ID 65959

STAIGER, L. [1976] *Regulaere Nullmengen* (**J** 0129) Elektr Informationsverarbeitung & Kybern 12*307-311
◇ D05 E75 ◇ REV MR 54 # 4818 Zbl 345 # 94030 • ID 42788

STAIGER, L. & WAGNER, K. [1976] *Zur Theorie der abstrakten Familien von ω-Sprachen (ω-AFL)* (**P** 2898) Algor Kompl, Lern-& Erkenn-Prozess;1976 Jena 79-91
◇ D05 ◇ REV MR 56 # 13812 Zbl 447 # 68088 • ID 40916

STAIGER, L. [1977] *Empty-storage-acceptance of ω-languages* (**P** 2588) FCT'77 Fund of Comput Th;1977 Poznan 516-521
◇ D05 D10 ◇ REV MR 58 # 25149 Zbl 367 # 02016 • ID 51176

STAIGER, L. [1977] see LINDNER, R.

STAIGER, L. & WAGNER, K. [1977] *Recursive ω-languages* (**P** 2588) FCT'77 Fund of Comput Th;1977 Poznan 532-537
◇ D05 D10 D55 ◇ REV MR 58 # 3752 Zbl 367 # 02017 • ID 51177

STAIGER, L. & WAGNER, K. [1978] *Rekursive Folgenmengen. I* (**J** 0068) Z Math Logik Grundlagen Math 24*523-538
◇ D10 D55 D75 ◇ REV MR 80f:03044 Zbl 421 # 03035 • ID 53434

STAIGER, L. [1981] *Complexity and entropy* (**P** 3429) Math Founds of Comput Sci (10);1981 Strbske Pleso 508-514
◇ D15 ◇ REV MR 83e:03064 Zbl 471 # 68029 • ID 69931

STAIGER, L. [1983] *Finite state ω-languages* (**J** 0119) J Comp Syst Sci 27*434-488
◇ D05 D55 ◇ REV MR 84m:68076 Zbl 479 # 68075 Zbl 479 # 68076 • ID 55702

STAIGER, L. [1985] *Representable P. Martin-Loef tests* (**J** 0156) Kybernetika (Prague) 21*235-243
◇ D80 ◇ ID 48982

STAIGER, L. see Vol. V for further entries

STANASILA, O. [1981] see BREAZU, V.

STANASILA, O. see Vol. V for further entries

STANCIULESCU, F.S. [1965] *Einige Bemerkungen zur Sequentiallogik der endlichen Automaten* (**J** 0068) Z Math Logik Grundlagen Math 11*57-60
◇ D05 ◇ REV MR 30 # 3846 Zbl 127.11 • ID 20927

STANKIEWICZ, E. & ZAKOWSKI, W. [1978] *The (Z, Q)-systems* (**J** 1008) Demonstr Math (Warsaw) 11*557-565
◇ D05 ◇ REV MR 80f:68071 Zbl 391 # 68034 • ID 82863

STANULOV, N. [1971] *On the information structure of a class of finite automata* (**P** 1560) Tr Mezhdurn Semin Priklad Aspekt Teor Avtom;1971 Varna 1971*82-89
◇ D05 ◇ REV Zbl 277 # 94032 • ID 65487

STAPLES, J. [1983] *Two-level expression representation for faster evaluation* (**P** 3908) Graph-Gram & Appl to Comput Sci (2);1982 Neunkirchen 392-404
◇ B35 B40 D03 ◇ REV Zbl 522 # 68036 • ID 37059

STAPLES, J. see Vol. VI for further entries

STARK, W.R. [1978] *A forcing approach to the strict-Π^1_1 reflection and strict-$\Pi^1_1 = \Sigma^0_1$* (**J** 0068) Z Math Logik Grundlagen Math 24*467-479
◇ C70 D60 E40 ◇ REV MR 81c:03025 Zbl 412 # 03033 • ID 52964

STARK, W.R. see Vol. I, II, III, V for further entries

STARKE, P.H. [1960] *Bemerkungen zu der von Asser entwickelten Version der Turing-Maschine* (**J** 0068) Z Math Logik Grundlagen Math 6*106-108
◇ D10 ◇ REV MR 23 # A3083 Zbl 97.4 • ID 13119

STARKE, P.H. [1963] *Ueber die Darstellbarkeit von Ereignissen in nicht-initialen Automaten* (**J** 0068) Z Math Logik Grundlagen Math 9*315-319
◇ D05 ◇ REV MR 28 # 2976 Zbl 129.263 • ID 13120

STARKE, P.H. [1964] *Einige Probleme der Automatentheorie I, II* (**S** 4163) Math in Schule 2*190-200,277-288
◇ D05 ◇ ID 42628

STARKE, P.H. [1965] *Die Imitation endlicher Medwedjew-Automaten durch Nervennetze* (**J** 0068) Z Math Logik Grundlagen Math 11*241-248
◇ D05 ◇ REV MR 39 # 3923 Zbl 163.11 • ID 13121

STARKE, P.H. [1965] *Einfuehrung in die Theorie der Nervennetze* (**J** 4068) Dt Z fuer Phil 13*64-85
◇ D05 ◇ ID 42629

STARKE, P.H. [1965] *Theorie stochastischer Automaten I,II (English and Russian summaries)* (**J** 0129) Elektr Informationsverarbeitung & Kybern 1*5-32,71-98
◇ D05 ◇ REV MR 36 # 6236 Zbl 149.10 • ID 28799

STARKE, P.H. [1966] *Eine Bemerkung ueber homogene Experimente* (**J** 0129) Elektr Informationsverarbeitung & Kybern 2*257-259
◇ D05 ◇ REV Zbl 166.270 • ID 42633

STARKE, P.H. [1966] *Einige Bemerkungen ueber nicht-deterministische Automaten* (**J** 0129) Elektr Informationsverarbeitung & Kybern 2*61-82
◇ D05 ◇ REV MR 36 # 6238 Zbl 163.11 • ID 42630

STARKE, P.H. [1966] *Stochastische Ereignisse und Wortmengen* (**J** 0068) Z Math Logik Grundlagen Math 12*61-68
◇ D05 ◇ REV MR 33 # 5409 Zbl 163.11 • ID 13122

STARKE, P.H. [1966] *Stochastische Ereignisse und stochastische Operatoren* (**J** 0129) Elektr Informationsverarbeitung & Kybern 2*177-190
◇ D05 ◇ REV MR 36 # 4925 Zbl 149.13 • ID 42631

STARKE, P.H. [1966] *Theory of stochastic automata* (**J** 0156) Kybernetika (Prague) 2*475-482
◇ D05 ◇ REV MR 35 # 4060 Zbl 168.13 • ID 42632

STARKE, P.H. [1967] *Huellenoperationen fuer nicht-deterministische Automaten* (**J** 0129) Elektr Informationsverarbeitung & Kybern 3*281-294
◇ D05 ◇ REV MR 37 # 6135 Zbl 168.13 • ID 42682

STARKE, P.H. [1967] *Ueber Experimente an Automaten* (**J** 0068) Z Math Logik Grundlagen Math 13*67-80
◇ D05 ◇ REV MR 35 # 60 Zbl 162.28 • ID 13123

STARKE, P.H. [1967] see NEUBER, S.

STARKE, P.H. & THIELE, H. [1967] *Zufaellige Zustaende in stochastischen Automaten* (**J** 0129) Elektr Informationsverarbeitung & Kybern 3*25-37
◇ D05 ◇ REV MR 36 # 7445 Zbl 163.10 • ID 13124

STARKE, P.H. [1968] *Aequivalenz und Reduktion bei stochastischen Automaten* (**J** 0947) Wiss Z Tech Univ Dresden 17*1118-1122
◇ D05 ◇ REV MR 40 # 1212 • ID 42688

STARKE, P.H. [1968] *Die Reduktion von stochastischen Automaten* (**J** 0129) Elektr Informationsverarbeitung & Kybern 4*93-100
◇ D05 ◇ REV MR 38 # 6906 Zbl 157.23 • ID 13126

STARKE, P.H. [1969] *Abstrakte Automaten* (**X** 0806) Dt Verlag Wiss: Berlin 392pp
• TRANSL [1971] (**X** 0809) North Holland: Amsterdam 419pp (English)
◇ D05 D75 D98 ◇ REV MR 43 # 1769 Zbl 182.21 JSL 37.413 • ID 22029

STARKE, P.H. [1969] *Schwache Homomorphismen fuer stochastische Automaten* (**J** 0068) Z Math Logik Grundlagen Math 15*421-429
◇ D05 ◇ REV MR 42 # 1613 • ID 13127

STARKE, P.H. [1969] *Ueber die Minimalisierung von stochastischen Rabin-Automaten* (**J** 0129) Elektr Informationsverarbeitung & Kybern 5*153-170
◇ D05 ◇ REV MR 44 # 3806 Zbl 184.287 • ID 42689

STARKE, P.H. & THIELE, H. [1970] *On asynchronous stochastic automata* (**J** 0194) Inform & Control 17*265-293
◇ D05 ◇ REV MR 43 # 3051 Zbl 248 # 94068 • ID 42692

STARKE, P.H. & THIELE, H. [1970] *Ueber asynchrone stochastische Automaten* (**P** 0577) Automatenth & Formale Sprachen;1969 Oberwolfach 131-142
◇ D05 ◇ REV MR 48 # 8150 Zbl 248 # 94067 • ID 42691

STARKE, P.H. [1970] *Ueber regulaere nicht-deterministische Operatoren* (**J** 0129) Elektr Informationsverarbeitung & Kybern 6*229-237
◇ D05 ◇ REV MR 43 # 7271 Zbl 209.29 • ID 42690

STARKE, P.H. [1970] *Ueber Minima von nicht-deterministischen Automaten* (**J** 0115) Wiss Z Humboldt-Univ Berlin, Math-Nat Reihe 19*663-664
◇ D05 ◇ REV MR 48 # 1834 Zbl 286 # 94049 • ID 42694

STARKE, P.H. [1971] *Einige Bemerkungen ueber asynchrone stochastische Automaten* (**J** 3994) Ann Acad Sci Fennicae Ser A I 491*21pp
◇ D05 ◇ REV MR 46 # 5086 Zbl 248 # 94069 • ID 42698

STARKE, P.H. [1971] *Ueber die Transformation zweiseitig unendlicher Folgen durch determinierte Automaten* (**J** 0129) Elektr Informationsverarbeitung & Kybern 7*425-436
◇ D05 ◇ REV MR 48 # 1829 Zbl 241 # 94047 • ID 42696

STARKE, P.H. [1972] *Allgemeine Probleme und Methoden in der Automatentheorie* (**J** 0129) Elektr Informationsverarbeitung & Kybern 8*489-517
◇ D05 ◇ REV MR 47 # 6403 Zbl 257 # 94026 • ID 65496

STARKE, P.H. [1972] *Das Analyse-Synthese-Problem in der Automatentheorie* (**J** 1528) Sitzb Plenum & Klassen Akad Wiss DDR 11*531-533
◇ D05 ◇ ID 42705

STARKE, P.H. [1972] *Ueber die Minimisierung von Automaten mit halbgeordnetem Ausgabealphabet (Russian, English and French summaries)* (**J** 0115) Wiss Z Humboldt-Univ Berlin, Math-Nat Reihe 21*531-533
◇ D05 ◇ REV MR 48 # 10719 Zbl 256 # 94053 • ID 13128

STARKE, P.H. [1972] *Ueber die Experimentmengen determinierter Automaten* (**J** 0129) Elektr Informationsverarbeitung & Kybern 8*67-76
◇ D05 ◇ REV MR 47 # 6402 Zbl 247 # 94036 • ID 42702

STARKE, P.H. [1972] see MUENTEFERING, P.

STARKE, P.H. [1972] *Ueber Finalexperimente an determinierten Automaten* (**J** 0129) Elektr Informationsverarbeitung & Kybern 8*617-622
◇ D05 ◇ REV MR 57 # 19188 Zbl 337 # 94032 • ID 42707

STARKE, P.H. [1973] *Grammatiken und Sprachen* (**S** 1536) Schr Weiterbildungszentr MKR (Dresden) 2*1-20
◇ D05 ◇ ID 42709

STARKE, P.H. [1973] *On the sequential relations of time-variant automata* (**P** 1448) Math Founds of Comput Sci (2);1973 Strbske Pleso 163-168
◇ D05 ◇ REV MR 53 # 2031 • ID 42712

STARKE, P.H. [1973] see DAUSCHA, W.

STARKE, P.H. [1973] *Ueber diagnostische Strategien an determinierten Automaten* (**J** 0068) Z Math Logik Grundlagen Math 19*271-276
◇ D05 ◇ REV MR 52 # 5387 Zbl 317 # 94047 • ID 13129

STARKE, P.H. & THIELE, H. [1973] *Ueber die Experimentmengen nicht-deterministischer Automaten (Russian)* (**J** 0474) Avtom Vychis Tekh, Akad Nauk Latv SSR 1973/6*7-13
◇ D05 ◇ REV Zbl 285 # 94031 • ID 42708

STARKE, P.H. [1973] *Ueber die Experimentmengen schwach-initialer Automaten* (**J** 3920) ZKI Inf, Akad Wiss DDR 2/1973*41-45
◇ D05 ◇ ID 42710

STARKE, P.H. [1974] *Das allgemeine Diagnose-Problem* (**S** 1536) Schr Weiterbildungszentr MKR (Dresden) 1974/6*14-36
◇ D05 ◇ ID 42713

STARKE, P.H. [1974] *On diagnosing experiments with nondeterministic automata with final states* (**J** 0129) Elektr Informationsverarbeitung & Kybern 10*471-480
◇ D05 ◇ REV MR 53 # 15594 Zbl 299 # 94046 • ID 42715

STARKE, P.H. [1974] *Theorie der sequentiellen Automaten* (**C** 2318) Entwicklung Math in DDR 701-705
◇ D05 ◇ ID 42714

STARKE, P.H. [1975] *Application of two-tape automata to analysis and synthesis of nondeterministic generalized sequential machines* (**J** 4553) Knizh Odborn Ved Spis Vys Uch Tech, Brno B-56*189-194
◇ B70 D05 ◇ REV MR 52 # 16906 Zbl 323 # 94035 • ID 42717

STARKE, P.H. [1975] *On the representation of relations by deterministic and nondeterministic multitape automata* (**P** 0454) Math Founds of Comput Sci (4);1975 Marianske Lazne 114-124
◇ D05 ◇ REV MR 53 # 9722 Zbl 317 # 94051 • ID 42718

STARKE, P.H. [1975] *Ueber die Darstellbarkeit von Relationen in Mehrbandautomaten* (**J** 3920) ZKI Inf, Akad Wiss DDR 1/1975*6-8
◇ D05 ◇ ID 42716

STARKE, P.H. [1975] *Ueber die Darstellbarkeit von Relationen in Mehrbandautomaten* (**J** 0129) Elektr Informationsverarbeitung & Kybern 11*580
◇ D05 ◇ REV MR 56 # 7331 Zbl 316 # 94054 • ID 65498

STARKE, P.H. [1976] *Closedness properties and decision problems for finite multi-tape automata* (**J** 0156) Kybernetika (Prague) 12*61-75
◇ D05 ◇ REV MR 55 # 11721 Zbl 344 # 94028 • ID 65497

STARKE, P.H. [1976] *Decision problems for multi-tape automata* (**P** 1401) Math Founds of Comput Sci (5);1976 Gdansk 124-136
◇ D05 ◇ REV Zbl 337 # 02024 • ID 65499

STARKE, P.H. [1976] *Entscheidungsprobleme fuer autonome Mehrbandautomaten* (**J** 0068) Z Math Logik Grundlagen Math 22*131-140
◇ D05 ◇ REV MR 58 # 177 Zbl 328 # 02020 • ID 18484

STARKE, P.H. [1976] *Mehrbandakzeptoren* (**S** 1536) Schr Weiterbildungszentr MKR (Dresden) 17*14-51
◇ D05 ◇ ID 42719

STARKE, P.H. [1976] *On the diagonals of n-regular relations* (**J** 0129) Elektr Informationsverarbeitung & Kybern 12*281-288 • ERR/ADD ibid 13*147-148(1977))
◇ D05 ◇ REV MR 54 # 14444 Zbl 333 # 68056 • ID 42721

STARKE, P.H. [1976] *Ueber die Darstellbarkeit von Relationen in Mehrbandautomaten* (**J** 0129) Elektr Informationsverarbeitung & Kybern 12/1-2*61-81
◇ D05 ◇ REV MR 56 # 7331 Zbl 328 # 02019 • ID 14827

STARKE, P.H. [1977] *Analyse und Synthese von asynchronen ND-Automaten* (**J** 1426) Theor Comput Sci 3*261-266
◇ D05 ◇ REV MR 55 # 12366 Zbl 365 # 94072 • ID 42720

STARKE, P.H. [1977] *Closedness properties of multihead languages* (**J** 2716) Found Control Eng, Poznan 2*51-63
◇ D05 ◇ REV MR 57 # 11199 Zbl 379 # 68053 • ID 42723

STARKE, P.H. [1977] *Multitape automata and languages* (**J** 2845) Tanulmanyok 63*13-26
◇ D05 ◇ REV MR 57 # 18234 • ID 42722

STARKE, P.H. [1978] *Free Petri net languages* (**S** 3382) Sem-ber, Humboldt-Univ Berlin, Sekt Math 7*41pp
◇ D05 ◇ REV Zbl 408 # 68065 • ID 42724

STARKE, P.H. [1978] *Free Petri net languages* (**P** 1707) Math Founds of Comput Sci (7);1978 Zakopane 506-515
• ERR/ADD [1983] (**J** 4385) NEWS 13*11-12
◇ D05 D80 ◇ REV MR 81i:68117 Zbl 391 # 68038
• ID 42747

STARKE, P.H. [1978] see GRABOWSKI, J.

STARKE, P.H. [1979] *On the languages of bounded Petri nets* (**P** 2059) Math Founds of Comput Sci (8);1979 Olomouc 425-433 • ERR/ADD [1983] (**J** 4385) NEWS 13*11-12
• REPR [1979] (**J** 0129) Elektr Informationsverarbeitung & Kybern 15*355-365
◇ D05 D80 ◇ REV MR 81d:68109 MR 82c:68036
Zbl 412 # 68049 Zbl 426 # 68071 • ID 42749

STARKE, P.H. [1979] *Semilinearity and Petri nets* (**P** 2935) FCT'79 Fund of Comput Th;1979 Berlin/Wendisch-Rietz 423-429
◇ D05 D80 ◇ REV MR 81h:68050 Zbl 419 # 68091
• ID 42750

STARKE, P.H. [1980] *Beitraege zur Theorie der Mehrbandautomaten* (**J** 1670) Mitt Math Ges DDR 1980/2-3*111-112
◇ D05 ◇ ID 42751

STARKE, P.H. [1980] *Petri-Netze* (**X** 0806) Dt Verlag Wiss: Berlin 192pp
◇ D05 D98 ◇ REV MR 85e:68078 Zbl 449 # 68020
• ID 42726

STARKE, P.H. [1980] *Remarks on Reusch's nondeterminism problem* (**J** 2010) Bull Europ Assoc Th Comput Sci 10*40-48
◇ D05 ◇ ID 42752

STARKE, P.H. [1981] *A note on conflicts in Petri nets* (**J** 2010) Bull Europ Assoc Th Comput Sci 14*26-33
◇ D05 ◇ ID 42753

STARKE, P.H. [1981] *Processes in Petri nets* (**P** 3165) FCT'81 Fund of Comput Th;1981 Szeged 350-359
◇ D05 ◇ REV MR 83h:68089 Zbl 481 # 68058 • ID 42728

STARKE, P.H. [1981] *Processes in Petri nets* (**J** 0129) Elektr Informationsverarbeitung & Kybern 17*389-416
◇ D05 ◇ REV MR 83j:68068 Zbl 504 # 68035 • ID 42727

STARKE, P.H. [1982] see SANDRING, S.

STARKE, P.H. [1982] *A CE-system and its reachability graph* (**J** 4385) NEWS 12*13
◇ D05 ◇ ID 42756

STARKE, P.H. [1982] *Graph grammars and Petri net processes* (**S** 1536) Schr Weiterbildungszentr MKR (Dresden) 57*75-86
◇ D05 ◇ REV Zbl 533 # 68064 • ID 42754

STARKE, P.H. [1982] *Praedikat-Transitions-Netze* (**C** 2502) Petri-Netze & Anwendungen 3-11
◇ D05 ◇ ID 42746

STARKE, P.H. [1983] *Graph grammars for Petri net processes* (**J** 0129) Elektr Informationsverarbeitung & Kybern 19*199-233
◇ D05 ◇ REV MR 85h:68057 Zbl 533 # 68064 • ID 42730

STARKE, P.H. [1983] *Monogeneous fifo-nets and Petri nets are equivalent* (**J** 2010) Bull Europ Assoc Th Comput Sci 21*68-77
◇ D05 D80 ◇ ID 42757

STARKE, P.H. [1983] *On the concurrency of distributed systems* (**S** 3382) Sem-ber, Humboldt-Univ Berlin, Sekt Math 52*133-142
◇ D05 D80 ◇ REV MR 85f:68003 Zbl 546 # 68040
• ID 42731

STARKE, P.H. [1984] see BURKHARD, HANS-DIETER

STARKE, P.H. [1984] *An uninvited address to "serializers" and "nontransitivists"* (**J** 1456) SIGACT News 16*11-12
◇ D05 ◇ ID 42761

STARKE, P.H. [1984] *Multiprocessor systems and their concurrency* (**P** 3658) Math Founds of Comput Sci (11);1984 Prague 516-525
◇ B75 D05 ◇ ID 42759

STARKE, P.H. [1984] *Multiprocessor systems and their concurrency (German and Russian summaries)* (**J** 0129) Elektr Informationsverarbeitung & Kybern 20*207-227
◇ B75 D05 ◇ REV MR 86e:68041 Zbl 552 # 68058
• ID 42758

STARKE, P.H. [1984] *Ueber die Nebenlaeufigkeit von Multiprozessorsystemen* (**P** 4367) INFO'84;1984 Dresden 2*30-33
◇ D05 ◇ ID 42760

STARKOV, M.A. [1976] see FRIZEN, D.G.

STATMAN, R. [1977] *Herbrand's theorem and Gentzen's notion of a direct proof* (**C** 1523) Handb of Math Logic 897-912
◇ D15 F05 F07 F20 ◇ REV MR 58 # 10343 JSL 49.980
• ID 27329

STATMAN, R. [1977] *The typed λ-calculus is not elementary recursive* (**P** 3572) IEEE Symp Found of Comput Sci (18);1977 Providence 90-94
◇ B40 D15 D20 ◇ REV MR 58 # 27356 • ID 78954

STATMAN, R. [1979] *Intuitionistic propositional logic is polynomial-space complete* (**J** 1426) Theor Comput Sci 9*67-72
◇ B40 D15 F20 F50 ◇ REV MR 80m:68042
Zbl 411 # 03049 • ID 52903

STATMAN, R. [1979] *The typed λ-calculus is not elementary recursive* (**J** 1426) Theor Comput Sci 9*73-81
◇ B40 D15 D20 ◇ REV MR 80m:68043 Zbl 411 # 03050
• ID 52904

STATMAN, R. [1981] *Number theoretic functions computable by polymorphic programs* (**P** 4235) IEEE Symp Found of Comp Sci (22);1981 Nashville 279-282
◇ D20 ◇ REV MR 84a:68004 • ID 45814

STATMAN, R. [1982] *Completeness, invariance and λ-definability* (J 0036) J Symb Logic 47*17-26
◇ B40 D15 D20 ◇ REV MR 84e:03020 Zbl 487#03006
• ID 34359

STATMAN, R. see Vol. I, VI for further entries

STAVI, J. [1973] *A converse of the Barwise completeness theorem* (J 0036) J Symb Logic 38*594-612
◇ C70 D60 D70 ◇ REV MR 51#12473 Zbl 308#02015
• ID 13133

STAVI, J. [1980] see SMORYNSKI, C.A.

STAVI, J. see Vol. I, II, III, V, VI for further entries

STEARNS, R.E. [1962] see HARTMANIS, J.

STEARNS, R.E. [1964] see HARTMANIS, J.

STEARNS, R.E. [1965] see HARTMANIS, J.

STEARNS, R.E. [1981] see HUNT III, H.B.

STEBE, P. [1974] see ANSHEL, M.

STEBE, P. [1976] see ANSHEL, M.

STECKELBERG, J.M. [1977] see SETH, S.C.

STEEL, J.R. [1975] *Descending sequences of degrees* (J 0036) J Symb Logic 40*59-61
◇ C62 D30 ◇ REV MR 56#8347 Zbl 349#02036
• ID 13142

STEEL, J.R. [1978] *Forcing with tagged trees* (J 0007) Ann Math Logic 15*55-74
◇ C62 D30 D55 E40 F35 ◇ REV MR 81c:03044
Zbl 404#03020 • ID 29156

STEEL, J.R. [1980] *A note on analytic sets* (J 0053) Proc Amer Math Soc 80*655-657
◇ D55 E15 ◇ REV MR 82b:03092 Zbl 462#03012
• ID 54519

STEEL, J.R. [1980] *Analytic sets and Borel isomorphisms* (J 0027) Fund Math 108*83-88
◇ D55 E15 E60 ◇ REV MR 82b:03091 Zbl 463#03028
JSL 49.665 • ID 54568

STEEL, J.R. [1981] *Closure properties of pointclasses* (C 2922) Cabal Seminar Los Angeles 1977-79 147-163
◇ D55 D65 D75 E15 E60 ◇ REV MR 84b:03066
Zbl 496#03033 • ID 35650

STEEL, J.R. [1981] *Determinateness and the separation property* (J 0036) J Symb Logic 46*41-44
◇ D55 D75 E15 E60 ◇ REV MR 83d:03058
Zbl 487#03031 • ID 33283

STEEL, J.R. [1981] see KECHRIS, A.S.

STEEL, J.R. [1982] *A classification of jump operators* (J 0036) J Symb Logic 47*347-358
◇ D30 E45 E60 ◇ REV MR 84i:03085 Zbl 524#03029
• ID 34562

STEEL, J.R. [1982] see MARTIN, D.A.

STEEL, J.R. [1983] *Scales on Σ^1_1 sets* (C 3875) Cabal Seminar Los Angeles 1979-81 72-76
◇ D55 E15 ◇ REV Zbl 567#03022 • ID 45242

STEEL, J.R. see Vol. III, V for further entries

STEEL JR., T.B. (ED.) [1966] *Formal language description languages for computer programming* (X 0809) North Holland: Amsterdam x+330pp
◇ B75 D05 ◇ REV Zbl 152.157 • ID 23573

STEEL JR., T.B. see Vol. II for further entries

STEENSTRUP, M. [1983] see ARBIB, M.A.

STEFAN, G.M. [1977] see DUMITRESCU, S.

STEFAN, T. [1984] *A logical approach to regular languages* (J 0060) Rev Roumaine Math Pures Appl 29*433-438
◇ D05 ◇ REV MR 85m:68019 Zbl 574#68061 • ID 44201

STEFANI, S. [1974] *Gradi di insolubilita e limiti* (J 0088) Ann Univ Ferrara, NS, Sez 7 19*51-64
◇ D30 ◇ REV MR 53#110 Zbl 302#02012 • ID 16575

STEFANI, S. [1980] *Recursive functions with measurable oracles: An approach to a probabilistic recursion theory* (J 3495) Boll Unione Mat Ital, V Ser, B 17*634-649
◇ D20 D30 ◇ REV MR 82h:03046 Zbl 454#03022
• ID 54234

STEFANI, S. [1982] *Recursive functions with measurable oracles: the semilattice of the effectively splitting distributions (Italian summary)* (J 2100) Boll Unione Mat Ital, VI Ser, B 1*327-345
◇ D20 D30 ◇ REV MR 84m:03065 Zbl 502#03024
• ID 35764

STEGMUELLER, W. [1959] *Unvollstaendigkeit und Unentscheidbarkeit. Die metamathematischen Resultate von Goedel, Church, Kleene, Rosser und ihre erkenntnistheoretische Bedeutung* (X 0902) Springer: Wien iii+114pp
◇ A05 D35 F30 F98 ◇ REV MR 22#2538 Zbl 86.244
• REM 2nd ed. 1970; iii+114pp • ID 13145

STEGMUELLER, W. see Vol. I, II, III, V, VI for further entries

STEINBERG, R. [1980] see FOELDES, S.

STEINBY, M. [1977] see SALOMAA, A.

STEINBY, M. [1984] *Some decidable properties of Σ-rational and Σ-algebraic tree transformations* (J 0498) Ann Univ Turku, Ser A I 186*102-109
◇ D05 ◇ REV Zbl 536#68073 • ID 36778

STEINBY, M. [1984] see GECSEG, F.

STEINER, H.-G. [1966] *Verschiedene Aspekte der axiomatischen Methode im Unterricht* (P 1672) Repercussions Rech Math sur Enseig;1965 Echternach 31-70
◇ D35 D98 ◇ ID 29420

STEINER, H.-G. see Vol. I, VI for further entries

STEINHORN, C.I. [1983] *A new omitting types theorem* (J 0053) Proc Amer Math Soc 89*480-486
◇ C07 C15 C45 C50 C57 ◇ REV MR 84j:03071
Zbl 529#03009 • ID 34663

STEINHORN, C.I. see Vol. III, V for further entries

STENDER, P.V. & TARTAKOVSKIJ, V.A. [1968] *On the word problem in semigroups (Russian)* (J 0142) Mat Sb, Akad Nauk SSSR, NS 75(117)*15-38
• TRANSL [1968] (J 0349) Math of USSR, Sbor 4*13-32
◇ D40 ◇ REV MR 37#2879 Zbl 174.306 • ID 28472

STENGER, H.-J. [1984] *Algebraic characterisations of NTIME(F) and NTIME(F,A)* (J 3441) RAIRO Inform Theor 18*365-385
◇ B20 B35 D15 D20 ◇ REV Zbl 547#68080 • ID 43303

STENLUND, S. [1970] *Combinators as effectively calculable functions* (**C** 0735) Logic & Value (Dahlquist) 67–74
◇ B40 D20 ◇ REV MR 58#21490 • ID 78994

STENLUND, S. see Vol. II, VI for further entries

STEPHAN, B.J. [1975] *Compactness and recursive enumerability in intensional logic* (**J** 0068) Z Math Logik Grundlagen Math 21*343–346
◇ B45 D25 ◇ REV MR 52#46 Zbl 312#02025 • ID 13174

STERN, J. [1973] *Reels aleatoires et ensembles de mesure nulle en theorie descriptive des ensembles* (**J** 2313) C R Acad Sci, Paris, Ser A-B 276*A1249-A1252
◇ D55 E15 E35 E40 E45 E55 E60 ◇ REV MR 47#3175 Zbl 261#02040 • ID 13176

STERN, J. [1975] *Some measure theoretic results in effective descriptive set theory* (**J** 0029) Israel J Math 20*97–110
◇ D30 D55 E15 E60 ◇ REV MR 52#7904 Zbl 322#02061 • ID 13180

STERN, J. [1977] *Partitions de la droite reelle en F_σ ou en G_δ (English summary)* (**J** 2313) C R Acad Sci, Paris, Ser A-B 284*A921-A922
◇ D55 E05 E15 E35 ◇ REV MR 55#5445 Zbl 362#04007 • ID 27300

STERN, J. [1977] *Singletons Π^1_3 et reels Δ^1_3 (English summary)* (**J** 2313) C R Acad Sci, Paris, Ser A-B 284*A831-A833
◇ D55 E15 E35 E55 ◇ REV MR 55#5444 Zbl 368#02065 • ID 27298

STERN, J. [1978] *Evaluation du rang de Borel de certains ensembles (English summary)* (**J** 2313) C R Acad Sci, Paris, Ser A-B 286*A855-A857
◇ D55 E15 E50 E60 ◇ REV MR 81e:03049 Zbl 377#04007 • ID 31515

STERN, J. [1978] *Perfect set theorems for analytic and coanalytic equivalence relations* (**P** 1897) Logic Colloq;1977 Wroclaw 277–284
◇ D55 E15 E60 ◇ REV MR 82a:03046 Zbl 449#03046 • ID 56713

STERN, J. [1979] *Cardinalite des relations d'equivalence analytiqes et coanalytiqes* (**C** 2050) Semin Th des Ensembles GMS Paris 1976/78 21–34
◇ D55 E15 ◇ REV MR 84k:03120a Zbl 521#03032 • ID 36138

STERN, J. [1980] *Effective partitions of the real line into Borel sets of bounded rank* (**J** 0007) Ann Math Logic 18*29–60
◇ D55 E15 E60 ◇ REV MR 81d:03052 Zbl 522#03032 • ID 79002

STERN, J. (ED.) [1982] *Proceedings of the Herbrand Symposium. Logic colloquium '81, held in Marseille, France, July 1981* (**S** 3303) Stud Logic Found Math 107*xi+384pp
◇ B97 D97 ◇ REV MR 85f:03003 Zbl 489#00007 • ID 36590

STERN, J. [1982] *Quelques aspects du probleme P=NP* (**J** 2107) Publ Dep Math, Lyon, NS 1/B*15–26
◇ D15 ◇ REV MR 84i:68009 Zbl 534#68029 • ID 45053

STERN, J. [1983] see RAISONNIER, J.

STERN, J. [1985] *Complexty of some problems from the theory of automata* (**J** 0194) Inform & Control 66*163–176
◇ D05 D15 ◇ ID 49090

STERN, J. [1985] see RAISONNIER, J.

STERN, J. see Vol. I, III, V for further entries

STEYAERT, J.-M. [1973] see FLAJOLET, P.

STEYAERT, J.-M. [1974] see FLAJOLET, P.

STEYAERT, J.-M. [1976] see FLAJOLET, P.

STILLWELL, J.C. [1972] *Decidability of the "almost all" theory of degrees* (**J** 0036) J Symb Logic 37*501–506
◇ B25 C80 D30 ◇ REV MR 50#1863 Zbl 287#02029 • ID 13190

STILLWELL, J.C. [1977] *Concise survey of mathematical logic* (**J** 3194) J Austral Math Soc, Ser A 24*139–161
◇ B98 C98 D10 D35 ◇ REV MR 57#5655 Zbl 393#03002 • ID 52422

STILLWELL, J.C. [1979] *Unsolvability of the knot problem for surface complexes* (**J** 0016) Bull Austral Math Soc 20*131–137
◇ D80 ◇ REV MR 80j:03060 Zbl 397#57005 • ID 79030

STILLWELL, J.C. [1982] *The word problem and the isomorphism problem for groups* (**J** 0589) Bull Amer Math Soc (NS) 6*33–56
◇ D40 ◇ REV MR 82m:20039 Zbl 483#20018 • ID 82890

STILLWELL, J.C. [1983] see LENARD, A.

STILLWELL, J.C. [1983] *Efficient computation in groups and simplicial complexes* (**J** 0064) Trans Amer Math Soc 276*715–727
◇ D15 D40 ◇ REV MR 84h:03099 Zbl 519#20031 • ID 34289

STILLWELL, J.C. see Vol. I, II for further entries

STIMSON, M.J. [1979] see SAVITCH, W.J.

STINSON, D.R. [1984] see COLBOURN, C.J.

STOB, M. [1981] see MAASS, W.

STOB, M. [1982] *Index sets and degrees of unsolvability* (**J** 0036) J Symb Logic 47*241–248
◇ D30 ◇ REV MR 83i:03069 Zbl 532#03019 • ID 33623

STOB, M. [1982] *Invariance of properties under automorphisms of the lattice of recursively enumerable sets* (**J** 0048) Pac J Math 100*445–471
◇ D25 ◇ REV MR 83k:03049 Zbl 534#03017 • ID 33622

STOB, M. [1982] see SOARE, R.I.

STOB, M. [1983] see MAASS, W.

STOB, M. [1983] *WTT-degrees and T-degrees of recursively enumerable sets* (**J** 0036) J Symb Logic 48*921–930
◇ D25 ◇ REV MR 85b:03072 Zbl 563#03028 • ID 33625

STOB, M. [1985] *Major subsets and the lattice of recursively enumerable sets* (**P** 4046) Rec Th;1982 Ithaca 107–116
◇ D25 ◇ ID 46380

STOCKMEYER, L.J. [1973] see MEYER, A.R.

STOCKMEYER, L.J. [1976] see PRATT, V.R.

STOCKMEYER, L.J. [1976] see CHANDRA, A.K.

STOCKMEYER, L.J. [1977] *The polynomial-time hierarchy* (**J** 1426) Theor Comput Sci 3*1–22
◇ D10 D15 D40 D55 ◇ REV MR 55#11716 Zbl 353#02024 • ID 50107

STOCKMEYER, L.J. [1979] see CHANDRA, A.K.

STOCKMEYER, L.J. [1980] *Difficult computational problems* (**S** 0780) Jbuch Ueberblick Math 1980*61-74
◇ D15 ◇ REV MR 84f:68038 Zbl 468 # 68053 • ID 55124

STOCKMEYER, L.J. [1981] see CHANDRA, A.K.

STOCKMEYER, L.J. [1983] see FURST, M.

STOCKMEYER, L.J. [1984] see LADNER, R.E.

STOCKMEYER, L.J. [1984] see GUREVICH, Y.

STOCKMEYER, L.J. [1985] see FAGIN, R.

STOCKMEYER, L.J. see Vol. I for further entries

STOEVA, S.P. & TOPENCHAROV, V.V. [1985] *Fuzzy-topological automata* (**J** 2720) Fuzzy Sets Syst 16*65-74
◇ D05 ◇ ID 46415

STOKLOSA, J. [1981] *Properties of* (α, k)*-computations* (**J** 2716) Found Control Eng, Poznan 6*163-171
◇ D05 ◇ REV MR 84a:68045 Zbl 491 # 68047 • ID 38804

STOLTENBERG-HANSEN, V. [1977] *A regular set theorem for infinite computation theories* (**S** 1626) Oslo Preprint Ser 15
◇ D25 D75 ◇ ID 30987

STOLTENBERG-HANSEN, V. [1978] see NORMANN, D.

STOLTENBERG-HANSEN, V. [1978] *Weakly inadmissible recursion theory* (**P** 1628) Generalized Recursion Th (2);1977 Oslo 391-405
◇ D60 D75 ◇ REV MR 80e:03056 Zbl 453 # 03047
• ID 30970

STOLTENBERG-HANSEN, V. [1979] *Finite injury arguments in infinite computation theories* (**J** 0007) Ann Math Logic 16*57-80
◇ D25 D30 D60 D75 ◇ REV MR 81h:03094
Zbl 417 # 03019 • ID 53256

STOLTENBERG-HANSEN, V. & TUCKER, J.V. [1980] *Computing roots of unity in fields* (**J** 0161) Bull London Math Soc 12*463-471
◇ D25 D45 D80 ◇ REV MR 82d:03074 Zbl 448 # 03032
• ID 56632

STOLTENBERG-HANSEN, V. [1980] see MOLDESTAD, J.

STOLTENBERG-HANSEN, V. [1980] *On computational complexity in weakly admissible structures* (**J** 0036) J Symb Logic 45*353-358
◇ D15 D60 D75 ◇ REV MR 82a:03036 Zbl 435 # 03031
• ID 55791

STOLTENBERG-HANSEN, V. [1985] see JACOBSSON, C.

STONE, A.H. [1980] see DELLACHERIE, C.

STONE, A.H. see Vol. V for further entries

STORK, H.-G. [1979] *Remarks on the satisfiability problem of propositional logic* (**J** 3359) Appl Comp Sci, Ber Prakt Inf 13*31-43
◇ B05 D15 ◇ REV MR 81b:68058 Zbl 398 # 03024
• ID 52752

STOSS, H.-J. [1970] *k-Band-Simulation von k-Kopf-Turing-Maschinen* (**J** 0373) Comp Arch Inform & Numerik 6*309-317
◇ D10 ◇ REV MR 44 # 3534 Zbl 222 # 02036 • ID 26309

STOSS, H.-J. [1971] *Zwei-Band Simulation von Turingmaschinen* (**J** 0373) Comp Arch Inform & Numerik 7*222-235
• TRANSL [1974] (**C** 2319) Slozh Vychisl & Algor 199-212
◇ D10 ◇ REV MR 45 # 1432 Zbl 222 # 02037 • ID 26310

STOSS, H.-J. [1974] *k-Band-Simulation von k-Kopf-Turing-Maschinen (Russian)* (**C** 2319) Slozh Vychisl & Algor 190-198
◇ D10 ◇ REV Zbl 289 # 02018 • ID 29983

STOTSKAYA, E.D. [1969] see MAKAREVSKIJ, A.YA.

STOTSKAYA, E.D. [1971] *Multitape deterministic automata without end markers (Russian)* (**J** 0011) Avtom Telemekh 1971/9*105-111
• TRANSL [1972] (**J** 0010) Autom & Remote Control 33*1435-1439
◇ D10 ◇ REV MR 51 # 2889 Zbl 234 # 94052 • ID 17334

STRAIGHT, D.W. [1979] *Domino f-sets* (**J** 0068) Z Math Logik Grundlagen Math 25*235-249
◇ D10 D20 ◇ REV MR 80h:03057 Zbl 453 # 68018
• ID 79050

STRAIGHT, D.W. [1980] see HUGHES, C.E.

STRASSEN, V. [1973] *Berechnungen in partiellen Algebren endlichen Typs* (**J** 0373) Comp Arch Inform & Numerik 11*181-196
◇ D75 ◇ REV MR 53 # 2025 Zbl 265 # 68030 • ID 29851

STRASSEN, V. [1984] *Algebraische Berechnungskomplexitaet* (**C** 3444) Perspectives in Math 509-550
◇ D15 ◇ ID 44726

STRASSEN, V. see Vol. II for further entries

STREBEL, R. [1985] see BAUMSLAG, G.

STREETT, R.S. [1981] see MEYER, A.R.

STREETT, R.S. [1982] *Global process logic is undecidable* (**P** 3767) Found of Softw Tech & Th Comput Sci (2);1982 Bangalore 96-105
◇ B75 D35 ◇ REV MR 84c:03042 Zbl 537 # 03008
• ID 34935

STREETT, R.S. see Vol. II for further entries

STRIGIN, YU.D. [1973] *Hierarchies of general recursive operators and general recursive types of degrees of noncomputability (Russian)* (**J** 0023) Dokl Akad Nauk SSSR 210*537-540
• TRANSL [1973] (**J** 0062) Sov Math, Dokl 14*776-780
◇ D20 D30 ◇ REV MR 57 # 9512 Zbl 289 # 02028
• ID 29995

STRIGIN, YU.D. [1973] *The hierarchy of general recursive functionals (Russian)* (**J** 0023) Dokl Akad Nauk SSSR 210*282-284
• TRANSL [1973] (**J** 0062) Sov Math, Dokl 14*739-742
◇ D20 D30 ◇ REV MR 57 # 9511 Zbl 289 # 02027
• ID 29993

STRNAD, P. [1968] *On-line Turing machine recognition* (**J** 0194) Inform & Control 12*442-452 • ERR/ADD ibid 13*508
• TRANSL [1970] (**C** 1540) Probl Mat Log: Slozh Algor & Kl Vychisl Funk 271-281
◇ D10 D15 ◇ REV MR 40 # 4117 Zbl 172.10 • ID 19682

STRNAD, P. [1969] *On optimum time bounds for recognition of some sets of words by on-line Turing machines* (**J** 0156) Kybernetika (Prague) 5*266-279
◇ D10 D15 ◇ REV MR 48 # 1516 Zbl 184.23 • ID 13228

STRNAD, P. [1972] *A hierarchy of optimal recognitions (Russian) (German, English and French summaries)* (**J** 0115) Wiss Z Humboldt-Univ Berlin, Math-Nat Reihe 21*545-546
◇ D15 D20 ◇ REV MR 54 # 9992 Zbl 252 # 68020
• ID 25761

STRONG, H.R. [1968] *Algebraically generalized recursive function theory* (J 0284) IBM J Res Dev 12*465-475
◇ D75 ◇ REV MR 41 # 6689 Zbl 185.19 • ID 13233

STRONG, H.R. [1970] *Construction of models for algebraically generalized recursive function theory* (J 0036) J Symb Logic 35*401-409
◇ D75 ◇ REV MR 45 # 3201 Zbl 218 # 02035 • ID 13234

STRONG, H.R. [1970] *Depth-bounded computation* (J 0119) J Comp Syst Sci 4*1-14
• TRANSL [1974] (C 2319) Slozh Vychisl & Algor 349-363
◇ D15 D20 ◇ REV MR 40 # 5363 Zbl 313 # 02025
• ID 29621

STRONG, H.R. & WALKER, S.A. [1973] *Characterizations of flowchartable recursions* (J 0119) J Comp Syst Sci 7*404-447
◇ D20 D75 ◇ REV MR 48 # 10186 Zbl 266 # 68011
• ID 13997

STUMPF, G. [1975] see SCHNORR, C.-P.

STURGIS, H.E. [1963] see SHEPHERDSON, J.C.

SU, ZHE [1980] *The significance of the "street vendor's load" problem (Chinese)* (J 2772) Shuxue de Shijian yu Renshi 2*33-36
◇ D15 ◇ REV MR 82j:68025 • ID 82912

SUBRAMANIAN, K.G. [1981] see DARE, V.R.

SUBRAMANIAN, K.G. [1983] see DARE, V.R.

SUDAN, G. [1927] *Sur le nombre transfini ω^{ω}* (J 0494) Bull Math Soc Sci Roumanie 30*11-30
◇ D20 E10 ◇ REV FdM 53.171 • ID 12706

SUDAN, G. see Vol. V for further entries

SUDBOROUGH, I.H. [1974] *Bounded-reversal multihead finite automata languages* (J 0194) Inform & Control 25*317-328
◇ D05 ◇ REV MR 53 # 4648 Zbl 282 # 68033 • ID 65576

SUDBOROUGH, I.H. [1975] *On tape-bounded complexity classes and multihead finite automata* (J 0119) J Comp Syst Sci 10*62-76
◇ D05 D10 D15 ◇ REV MR 51 # 87 Zbl 299 # 68031
• ID 15186

SUDBOROUGH, I.H. [1976] *On deterministic context-free languages, multihead automata, and the power of an auxiliary pushdown store* (P 2597) ACM Symp Th of Comput (8);1976 Hershey 141-148
◇ D05 D10 D15 ◇ REV MR 55 # 9615 Zbl 365 # 68077
• ID 51080

SUDBOROUGH, I.H. [1976] *One-way multihead writing finite automata* (J 0194) Inform & Control 30*1-20
◇ D05 ◇ REV MR 53 # 4649 Zbl 337 # 02023 • ID 65577

SUDBOROUGH, I.H. [1977] *Some remarks on multihead automata* (J 3441) RAIRO Inform Theor 11*181-195
◇ D10 D15 ◇ REV MR 57 # 14608 Zbl 369 # 68035
• ID 51359

SUDBOROUGH, I.H. [1978] see LITOW, B.

SUDBOROUGH, I.H. [1978] *On the tape complexity of deterministic context-free languages* (J 0037) ACM J 25*405-414
◇ D05 D10 D15 ◇ REV MR 80h:68065 Zbl 379 # 68054
• ID 51868

SUDBOROUGH, I.H. [1980] see MONIEN, B.

SUDBOROUGH, I.H. [1981] see MONIEN, B.

SUDBOROUGH, I.H. [1982] see MONIEN, B.

SUDBOROUGH, I.H. [1983] *Bandwidth constraints on problems complete for polynomial time* (J 1426) Theor Comput Sci 26*25-52
◇ D15 ◇ REV MR 86c:68027 Zbl 535 # 68021 • ID 48812

SUDBOROUGH, I.H. [1985] see CHUNG, MOONJUNG

SUGATA, K. [1977] see MORITA, K.

SUGATA, K. [1982] see MORITA, K.

SUNDBLAD, Y. [1971] *The Ackermann functions. A theoretical, computational and formula manipulative study* (J 0130) BIT 11*107-119
◇ D20 ◇ REV MR 44 # 2384 Zbl 221 # 68033 • ID 28135

SUPPES, P. [1959] see HENKIN, L.

SUPPES, P. see Vol. I, II, III, V for further entries

SURANYI, J. [1943] *Zur Reduktion des Entscheidungsproblems des logischen Funktionskalkuels (Hungarian) (German summary)* (J 0461) Mat Fiz Lapok 50*51-74
◇ B20 D35 ◇ REV MR 9.129 JSL 9.22 • ID 19712

SURANYI, J. [1947] see KALMAR, L.

SURANYI, J. [1949] *Reduction of the decision problem to formulas containing a bounded number of quantifiers only* (P 0682) Int Congr Philos (10);1948 Amsterdam 759-762
◇ B20 D35 ◇ REV MR 11.303 Zbl 34.153 JSL 14.131
• ID 20805

SURANYI, J. [1950] *Contributions to the reduction theory of the decision problem II: Three universal, one existential quantifier* (J 0001) Acta Math Acad Sci Hung 1*261-271
◇ B20 D35 ◇ REV MR 13.715 Zbl 41.351 JSL 18.264
• REM Part I 1950 by Kalmar,L. Part III 1951 by Kalmar,L.
• ID 12740

SURANYI, J. [1950] see KALMAR, L.

SURANYI, J. [1951] *Contributions to the reduction theory of the decision problem V: Ackermann prefix with three universal quantifiers* (J 0001) Acta Math Acad Sci Hung 2*325-335
◇ B20 D35 ◇ REV MR 14.344 Zbl 45.2 JSL 18.265 • REM Part IV 1951 by Kalmar,L. • ID 12741

SURANYI, J. [1955] *On the reduction theory of the decision problem of symbolic logic (Hungarian) (Russian and English summaries)* (J 0396) Mat Lapok 6*180-197
◇ B20 D35 ◇ REV MR 17.447 JSL 22.296 • ID 12742

SURANYI, J. [1959] *Reduktionstheorie des Entscheidungsproblems im Praedikatenkalkuel der ersten Stufe* (X 0928) Akad Kiado: Budapest 216pp
• LAST ED (X 0806) Dt Verlag Wiss: Berlin 212pp
◇ B20 D35 ◇ REV MR 21 # 7156 Zbl 91.11 JSL 25.274
• ID 12743

SURANYI, J. [1971] *Reduction of the decision problem of the first order predicate calculus to reflexive and symmetrical binary predicates* (J 0049) Period Math Hung 1*97-106
◇ B20 D35 ◇ REV MR 44 # 2569 Zbl 242 # 02050
• ID 12744

SURANYI, J. see Vol. V for further entries

SUSSMANN, H.J. [1971] *Hilbert's tenth problem* (J 0307) Rev Colomb Mat 9*iv+69pp
◇ D25 D35 ◇ REV MR 43 # 4669 • ID 28475

SUTER, G.H. [1973] *Recursive elements and constructive extensions of computable local integral domains* (**J** 0036) J Symb Logic 38*272-290
◇ C57 C60 D45 ◇ REV MR 52#54 Zbl 276#02028 • ID 12784

SUZUKI, N. [1977] see JEFFERSON, D.

SUZUKI, Y. [1959] *Enumeration of recursive sets* (**J** 0036) J Symb Logic 24*311
◇ D20 ◇ REV MR 26#1253 Zbl 96.245 JSL 33.115 • ID 12788

SUZUKI, Y. [1963] *On the uniformization principle* (**P** 1127) Symp Founds of Math;1962 Katada 137-144
◇ D55 E15 ◇ REV JSL 36.687 • ID 33698

SUZUKI, Y. [1964] *A complete classification of the Δ^1_2-functions* (**J** 0015) Bull Amer Math Soc 70*246-253
◇ D30 D55 ◇ REV MR 28#2042 Zbl 166.263 JSL 36.688 • ID 12789

SUZUKI, Y. [1967] *Applications of the theory of β-models* (**J** 0407) Comm Math Univ St Pauli (Tokyo) 16*57-68
◇ C62 D65 ◇ REV MR 37#1240 Zbl 216.296 • ID 12790

SUZUKI, Y. [1968] *$\aleph_0$-standard models for set theory (Russian summary)* (**J** 0014) Bull Acad Pol Sci, Ser Math Astron Phys 16*265-267
◇ C62 D55 ◇ REV MR 39#51 Zbl 206.11 • ID 12791

SUZUKI, Y. see Vol. III, V for further entries

SVENONIUS, L. [1979] *Two kinds of extensions of primitive recursive arithmetic* (**P** 1705) Scand Logic Symp (4);1976 Jyvaeskylae 49-94
◇ D20 F30 F35 ◇ REV MR 81e:03058 Zbl 399#03047 • ID 52843

SVENONIUS, L. see Vol. I, III for further entries

SWAMY, S. [1983] see SAVAGE, J.E.

SWANSON, J.W. [1966] *A reduction theorem for normal algorithms* (**J** 0036) J Symb Logic 31*86-97
◇ D03 ◇ REV MR 33#7253 Zbl 135.250 JSL 32.123 • ID 12801

SWANSON, J.W. [1967] *A variant of Turing machines requiring print instructions only* (**J** 0079) Logique & Anal, NS 10*200-206
◇ D10 ◇ REV MR 37#7223 Zbl 207.312 JSL 34.134 • ID 12802

SYMES, D. [1972] *The computation of finite functions* (**P** 1901) ACM Symp Th of Comput (4);1972 Denver 177-182
◇ D15 D20 ◇ REV Zbl 357#68059 • ID 50444

SYRKIN, G.I. [1967] *A test for the validity of the translation theorem in the theory of normal algorithms (Russian)* (**J** 0023) Dokl Akad Nauk SSSR 173*270-272
• TRANSL [1967] (**J** 0062) Sov Math, Dokl 8*377-380
◇ D03 ◇ REV MR 35#1479 Zbl 174.22 • ID 12816

SZABO, M.E. [1985] see RICHTER, M.M.

SZABO, M.E. see Vol. I, II, VI for further entries

SZABO, P. [1978] *The undecidability of the D_A-unification problem* (**X** 3159) TH Karlsruhe Fak Informatik: Karlsruhe 19pp
◇ B35 D35 ◇ REV Zbl 401#68066 • ID 54642

SZABO, P. [1983] see SIEKMANN, J.

SZABO, P. see Vol. I, III for further entries

SZABO, Z. [1982] *On the ability of some inductive inferential strategies* (**X** 3151) Karl Marx Univ Dpt Math: Budapest 1982/2*213-221
◇ D20 ◇ REV Zbl 538#03036 • ID 41478

SZABO, Z. [1984] see KOMJATH, P.

SZALAS, A. [1981] *Algorithmic logic with recursive functions* (**J** 2095) Fund Inform, Ann Soc Math Pol, Ser 4 4*975-995
◇ B75 D80 ◇ REV MR 84h:68060 Zbl 494#68029 • ID 36656

SZALKAI, I. [1983] *The algebraic structure of primitive recursive functions (Russian and Hungarian summaries)* (**J** 2774) Koezlem MTA Szam & Autom: Kutat Intez 29*75-91
◇ D20 D75 ◇ REV MR 85i:03136 • ID 44216

SZALKAI, I. [1985] *On the algebraic structure of primitive recursive functions* (**J** 0068) Z Math Logik Grundlagen Math 31*551-556
◇ D20 D75 ◇ REV Zbl 572#03020 • ID 47806

SZCZERBA, L.W. & TARSKI, A. [1965] *Metamathematical properties of some affine geometries* (**P** 0623) Int Congr Log, Meth & Phil of Sci (2,Proc);1964 Jerusalem 166-178
◇ B25 B30 C52 C65 D35 ◇ REV MR 35#843 Zbl 149.385 JSL 36.333 • ID 12831

SZCZERBA, L.W. [1971] *Undecidability of elementary Pasch-Free geometry (Russian summary)* (**J** 0014) Bull Acad Pol Sci, Ser Math Astron Phys 19*469-474
◇ D35 ◇ REV MR 46#1580 Zbl 224#50001 • ID 12835

SZCZERBA, L.W. & TARSKI, A. [1979] *Metamathematical discussion of some affine geometries* (**J** 0027) Fund Math 104*155-192
◇ B25 B30 C35 C52 C65 D35 ◇ REV MR 81a:03012 Zbl 497#03008 • ID 79177

SZCZERBA, L.W. see Vol. I, III, VI for further entries

SZEKELY, D.L. [1966] *On general purpose unifying automata* (**J** 0047) Notre Dame J Formal Log 7*305-322
◇ D10 ◇ REV Zbl 192.21 • ID 12840

SZMIELEW, W. & TARSKI, A. [1952] *Mutual interpretability of some essentially undecidable theories* (**P** 0593) Int Congr Math (II, 6);1950 Cambridge MA 734
◇ D35 F25 ◇ ID 12847

SZMIELEW, W. [1983] see SCHWABHAEUSER, W.

SZMIELEW, W. see Vol. I, III, V for further entries

SZPILRAJN, E. [1932] see KURATOWSKI, K.

SZPILRAJN, E. [1933] *Sur certains invariants de l'operation (A)* (**J** 0027) Fund Math 21*229-235
◇ D55 E15 ◇ REV Zbl 8.110 FdM 59.93 • ID 12855

SZPILRAJN, E. also published under the name MARCZEWSKI, E.and SZPILRAJN-MARCZEWSKI, E.

SZPILRAJN, E. see Vol. V for further entries

SZWAST, W. [1985] *On some properties of Horn's spectra (Polish) (English summary)* (**S** 1454) Zesz Nauk Wyz Szk Ped Mat, Opole 23*5-9
◇ C13 D15 ◇ REV Zbl 574#03021 • ID 48831

SZYMANSKI, T.G. & ULLMAN, J.D. [1975] *Evaluating relational expressions with dense and sparse arguments* (**P** 1513) IEEE Symp Founds of Comput Sci (16);1975 Berkeley 90-97
◇ D15 D80 ◇ REV MR 54#9155 • ID 25832

SZYMANSKI, T.G. [1976] see HUNT III, H.B.

TAERNLUND, S.-A. [1977] *Horn clause computability* (J 0130) BIT 17*215-226
◇ B75 D20 ◇ REV MR 58 # 10383 Zbl 359 # 02042
• ID 50589

TAERNLUND, S.-A. see Vol. I for further entries

TAIT, W.W. [1960] see GANDY, R.O.

TAIT, W.W. [1961] see KREISEL, G.

TAIT, W.W. [1961] *Nested recursion* (J 0043) Math Ann 143*236-250
◇ D20 ◇ REV MR 23 # A1529 Zbl 111.10 JSL 28.103
• ID 13279

TAIT, W.W. [1961] see GANDY, R.O.

TAIT, W.W. [1975] *A realizability interpretation of the theory of species* (C 0758) Logic Colloq Boston 1972-73 240-251
◇ D65 F05 F35 F50 ◇ REV MR 52 # 5380
Zbl 328 # 02014 • ID 18413

TAIT, W.W. see Vol. I, III, V, VI for further entries

TAJMANOV, A.D. [1958] *On classes of models, closed under direct product (Russian)* (J 0067) Usp Mat Nauk 13/3*231-232
◇ C30 C52 D35 ◇ ID 13265

TAJMANOV, A.D. [1965] see ERSHOV, YU.L.

TAJMANOV, A.D. [1966] *On formulae of Horn-type (Russian)* (J 0216) Izv Akad Nauk SSSR, Ser Mat 30*523-524
◇ B20 D35 ◇ REV MR 34 # 1160 Zbl 156.251 • ID 19765

TAJMANOV, A.D. [1974] *On the elementary theory of topological algebras (Russian)* (J 0027) Fund Math 81*331-342
◇ C57 C65 ◇ REV MR 50 # 4290 Zbl 292 # 13011
• ID 13277

TAJMANOV, A.D. [1976] *An algorithmic problem of number theory (Russian)* (J 0429) Vest Akad Nauk Kazak SSR 9*61-62
◇ D35 D80 ◇ REV MR 58 # 16249 • ID 79191

TAJMANOV, A.D. see Vol. I, III, V for further entries

TAJTSLIN, M.A. [1962] *Effective inseparability of the set of identically true and the set of finitely refutable formulas of the elementary theory of lattices (Russian)* (J 0003) Algebra i Logika 1/3*24-38
◇ C13 D25 D35 G10 ◇ REV MR 28 # 1131 Zbl 199.31
• ID 19772

TAJTSLIN, M.A. [1962] *Undecidability of the elementary theory of commutative semigroups with cancellation (Russian)* (J 0092) Sib Mat Zh 3*308-309
◇ D35 ◇ REV MR 26 # 6 Zbl 156.12 • ID 19771

TAJTSLIN, M.A. [1963] *Undecidability of elementary theories of certain classes of finite commutative associative rings (Russian)* (J 0003) Algebra i Logika 2/3*29-51
◇ C13 D35 ◇ REV MR 28 # 1128 Zbl 192.56 • ID 13254

TAJTSLIN, M.A. [1963] see ERSHOV, YU.L.

TAJTSLIN, M.A. [1964] *On elementary theories of free nilpotent algebras (Russian)* (J 0003) Algebra i Logika 3/5-6*57-63
◇ B25 C60 D35 ◇ REV MR 32 # 5519 Zbl 231 # 02059
• ID 13255

TAJTSLIN, M.A. [1965] see ERSHOV, YU.L.

TAJTSLIN, M.A. [1965] *On the elementary theory of classical Lie algebras (Russian)* (J 0023) Dokl Akad Nauk SSSR 164*1243-1245
• TRANSL [1965] (J 0062) Sov Math, Dokl 6*1373-1376
◇ C60 D35 ◇ REV MR 33 # 1338 Zbl 231 # 02060
• ID 65641

TAJTSLIN, M.A. [1965] *On the theory of finite rings with division (Russian)* (J 0003) Algebra i Logika 4/4*103-114
◇ C13 D35 ◇ REV MR 35 # 1641 Zbl 216.292 • ID 13257

TAJTSLIN, M.A. [1966] *On elementary theories of commutative semigroups with cancellation (Russian)* (J 0003) Algebra i Logika 5/1*51-69
◇ B25 C60 D40 ◇ REV MR 33 # 5486 Zbl 253 # 02047
• ID 13258

TAJTSLIN, M.A. [1967] *Some further examples of undecidable theories (Russian) (English summary)* (J 0003) Algebra i Logika 6/3*105-111
◇ D35 ◇ REV MR 37 # 69 Zbl 209.23 • ID 13259

TAJTSLIN, M.A. [1968] *Elementary lattice theories for ideals in polynomial rings (Russian)* (J 0003) Algebra i Logika 7/2*94-97
• TRANSL [1968] (J 0069) Algeb and Log 7*127-129
◇ C60 D35 G10 ◇ REV MR 38 # 4312 Zbl 216.292
• ID 13260

TAJTSLIN, M.A. [1968] *On simple ideals in polynomial rings (Russian)* (J 0003) Algebra i Logika 7/6*64-66
• TRANSL [1968] (J 0069) Algeb and Log 7*394-395
◇ D35 ◇ REV MR 40 # 5607 Zbl 218 # 02043 • ID 27378

TAJTSLIN, M.A. [1968] *On the isomorphism problem for commutative semigroups (Russian)* (J 0092) Sib Mat Zh 9*375-401
• TRANSL [1968] (J 0475) Sib Math J 9*286-304
◇ C05 C07 C60 D40 ◇ REV MR 37 # 330
Zbl 242 # 20069 • ID 32029

TAJTSLIN, M.A. [1969] *Equivalence of automata with respect to a commutative semigroup (Russian)* (J 0003) Algebra i Logika 8*553-600
• TRANSL [1969] (J 0069) Algeb and Log 8*316-342
◇ D05 ◇ REV MR 44 # 3874 Zbl 223 # 02030 • ID 13261

TAJTSLIN, M.A. [1970] *On elementary theories of lattices of subgroups (Russian)* (J 0003) Algebra i Logika 9*473-483
• TRANSL [1970] (J 0069) Algeb and Log 9*285-290
◇ B25 C60 D35 G10 ◇ REV MR 44 # 1739
Zbl 221 # 02033 • ID 28107

TAJTSLIN, M.A. [1974] *On the isomorphism problem for commutative semigroups (Russian)* (J 0142) Mat Sb, Akad Nauk SSSR, NS 93(135)*103-128,152
• TRANSL [1974] (J 0349) Math of USSR, Sbor 22*104-128
◇ D40 ◇ REV MR 48 # 8663 Zbl 294 # 20054 • ID 13262

TAJTSLIN, M.A. [1979] see BELYAEV, V.YA.

TAJTSLIN, M.A. [1980] *The isomorphism problem for commutative semigroups solved positively (Russian)* (C 2620) Teor Model & Primen 75-81
◇ C05 D40 ◇ REV MR 82h:20079 Zbl 525 # 20049
• ID 41671

TAJTSLIN, M.A. see Vol. I, II, III for further entries

TAKAHASHI, H. [1977] *Information transmission in one-dimensional cellular space and the maximum invariant set* (J 0194) Inform & Control 33*35-55
◇ D10 ◇ REV MR 55 # 12369 Zbl 364 # 94087 • ID 51000

TAKAHASHI, H. [1977] *Undecidable equations about the maximum invariant set* (**J** 0194) Inform & Control 33*1-34
◇ D10 ◇ REV MR 55 # 12368 Zbl 364 # 94086 • ID 50999

TAKAHASHI, M. [1985] see KOBAYASHI, K.

TAKAHASHI, MASAKO [1973] *Primitive transformations of regular sets and recognizable sets* (**P** 0763) Automata, Lang & Progr (1);1972 Rocquencourt 475-480
◇ D05 ◇ REV MR 51 # 5216 Zbl 281 # 94037 • ID 65649

TAKAHASHI, MASAKO [1984] see KOBAYASHI, K.

TAKAHASHI, MOTO-O [1968] *Ackermann's model and recursive predicates* (**J** 0081) Proc Japan Acad 44*41-42
◇ D20 ◇ REV MR 37 # 2603 Zbl 191.304 • ID 13292

TAKAHASHI, MOTO-O [1968] *Recursive functions of ordinal numbers and Levy's hierarchy* (**J** 0407) Comm Math Univ St Pauli (Tokyo) 17*21-29
◇ D55 D60 E10 E45 E47 ◇ REV MR 39 # 6751 Zbl 181.11 • ID 13293

TAKAHASHI, MOTO-O see Vol. I, II, III, V, VI for further entries

TAKAHASHI, N. [1984] see IZUMI, M.

TAKAHASHI, S. [1973] *A non-standard treatment of infinitely near points* (**C** 2976) Numb Th, Algeb Geom & Comm Algeb (Akizuki) 231-241
◇ D35 H15 H20 ◇ REV MR 55 # 5632 Zbl 291 # 02040 • ID 65648

TAKAHASHI, S. [1974] *Methodes logiques en geometrie diophantienne* (**X** 0893) Pr Univ Montreal: Montreal 48*178pp
◇ B28 C98 D98 G30 H98 ◇ REV MR 51 # 424 Zbl 325 # 02038 • ID 17488

TAKAHASHI, S. see Vol. I, III for further entries

TAKANAMI, I. [1978] see INOUE, K.

TAKANAMI, I. [1979] see INOUE, K.

TAKANAMI, I. [1980] see INOUE, K.

TAKANAMI, I. [1982] see INOUE, K.

TAKANAMI, I. [1983] see INOUE, K.

TAKANAMI, I. [1985] see INOUE, K.

TAKAOKA, T. [1973] *Two measures over language space (Japanese)* (**J** 0979) Denshi Tsushin Gakkai Ronbunshi, Sect A-D 56-D/5*305-311
• TRANSL [1973] (**J** 0464) Syst-Comp-Controls 4/3*30-36
◇ D05 ◇ REV MR 49 # 8438 • ID 13302

TAKAOKA, T. see Vol. II for further entries

TAKEUCHI, K. [1956] *The word problem of free algebras* (**J** 0091) Sugaku 8*219-229
◇ D40 ◇ REV MR 20 # 897 JSL 34.302 • ID 43282

TAKEUCHI, K. [1969] *The word problem for free distributive lattices* (**J** 0090) J Math Soc Japan 21*330-333
◇ D40 G10 ◇ REV MR 39 # 4066 Zbl 208.295 • ID 13305

TAKEUTI, G. [1960] *On the recursive functions of ordinal numbers* (**J** 0090) J Math Soc Japan 12*119-128
◇ D60 ◇ REV MR 23 # A1524 Zbl 95.11 JSL 27.88 • ID 13315

TAKEUTI, G. [1962] see KINO, A.

TAKEUTI, G. [1965] *Recursive functions and arithmetical functions of ordinal numbers* (**P** 0623) Int Congr Log, Meth & Phil of Sci (2,Proc);1964 Jerusalem 179-196
◇ D60 E10 E45 E55 ◇ REV MR 35 # 53 Zbl 199.28 • ID 13329

TAKEUTI, G. [1965] *Transcendence of cardinals* (**J** 0036) J Symb Logic 30*1-7
◇ D60 E35 E45 E55 E65 ◇ REV MR 33 # 2529 Zbl 192.42 • ID 13330

TAKEUTI, G. [1977] see HENSON, C.W.

TAKEUTI, G. see Vol. I, II, III, V, VI for further entries

TAL', A.A. [1961] see AJZERMAN, M.A.

TAL', A.A. [1963] see AJZERMAN, M.A.

TAL', A.A. [1984] see IVANOV, N.N.

TALJA, J. [1983] *On the complexity-relativized strong reducibilities* (**J** 0063) Studia Logica 42*259-267
◇ D15 D30 ◇ REV MR 85i:03146 Zbl 549 # 03030 • ID 42350

TALJA, J. see Vol. II, III for further entries

TAMARI, D. [1954] *Une contribution aux theories de communication: machines de Turing et problemes de mot* (**J** 0154) Synthese 9*205-227
◇ D10 D40 ◇ REV MR 17.755 JSL 31.139 • ID 13346

TAMARI, D. [1970] *The equivalence of associativity and word problems* (**P** 0629) Conf Universal Algeb;1969 Kingston 171-189
◇ C05 D40 ◇ REV MR 41 # 7020 Zbl 225 # 08005 • ID 13347

TAMARI, D. [1973] *The associativity problem for monoids and the word problem for semigroups and groups* (**P** 0678) Word Probl: Decis & Burnside Probl in Group Th;1969 Irvine 591-607
◇ D40 ◇ REV MR 52 # 14098 Zbl 264 # 20044 • ID 21883

TAMARI, D. [1978] see BUNTING, P.W.

TAMARI, D. see Vol. I, V for further entries

TAMURA, S. [1980] *On the word problem for lo-semigroups* (**P** 3011) Symp Semigroups (4);1980 Yamaguchi 97-102
◇ D40 ◇ REV MR 82g:06026 Zbl 461 # 06012 • ID 82949

TAMURA, S. see Vol. I, II, III for further entries

TAMURA, T. [1954] see KIMURA, NAOKI

TAMURA, T. see Vol. V for further entries

TANAKA, E. [1977] *The Turing machine constructed by trainable threshold elements* (**J** 2338) IEEE Trans Syst Man & Cybern SMC-7*881-886
◇ B70 D10 ◇ REV MR 58 # 26616 Zbl 365 # 94075 • ID 51082

TANAKA, H. & TUGUE, T. [1966] *A note on the effective descriptive set theory* (**J** 0407) Comm Math Univ St Pauli (Tokyo) 15*19-28
◇ D55 E15 ◇ REV MR 35 # 63 Zbl 178.320 JSL 39.344 • ID 13250

TANAKA, H. [1966] *On limits of sequences of hyperarithmetical functionals and predicates* (**J** 0407) Comm Math Univ St Pauli (Tokyo) 14*105-121
◇ D55 ◇ REV MR 36 # 1325 Zbl 148.247 JSL 39.344 • ID 13360

TANAKA, H. [1966] *Some properties of Σ_1^1- and Π_1^1-sets in N^N* (J 0081) Proc Japan Acad 42*304-307
◇ D55 ◇ REV MR 34#5683 Zbl 178.321 • ID 31269

TANAKA, H. [1967] *A basis result for Π_1^1 sets of positive measure* (J 0407) Comm Math Univ St Pauli (Tokyo) 16*115-127
◇ D55 E15 ◇ REV MR 38#4315 Zbl 189.289 • ID 31270

TANAKA, H. [1967] *Some results in the effective descriptive set theory* (J 0390) Publ Res Inst Math Sci (Kyoto) 3*11-52
◇ D55 E15 ◇ REV MR 37#3926 Zbl 178.320 • ID 13361

TANAKA, H. [1970] *Notes on measure and category in recursion theory* (J 0260) Ann Jap Ass Phil Sci 3(5)*231-241
◇ D55 E15 ◇ REV MR 44#78 Zbl 314#02048 • ID 13363

TANAKA, H. [1970] *On a Π_1^0 set of positive measure* (J 0111) Nagoya Math J 38*139-144
◇ D55 E15 E45 ◇ REV MR 43#4676 Zbl 218#02040
• ID 13362

TANAKA, H. [1970] *On analytic well-orderings* (J 0036) J Symb Logic 35*198-203
◇ D55 E15 E45 ◇ REV MR 42#5796 Zbl 222#02047
JSL 38.155 • ID 13364

TANAKA, H. [1971] *Analytic well orderings and basis theorems* (J 0091) Sugaku 23*177-192
◇ D45 D55 E15 ◇ REV MR 58#27412 • ID 79236

TANAKA, H. [1972] *A note on hyperdegrees* (J 0407) Comm Math Univ St Pauli (Tokyo) 20*23-25
◇ D30 D55 ◇ REV MR 47#8274 Zbl 256#02022
• ID 13365

TANAKA, H. [1972] *A property of arithmetic sets* (J 0053) Proc Amer Math Soc 31*521-524
◇ D55 E15 ◇ REV MR 44#3870 Zbl 251#02044
• ID 13366

TANAKA, H. [1973] *Π_1^1 sets of sets, hyperdegrees and related problems* (J 0090) J Math Soc Japan 25*609-621
◇ D30 D55 ◇ REV MR 49#2320 Zbl 261#02029
• ID 13368

TANAKA, H. [1973] *Length of analytic well-orderings* (J 0407) Comm Math Univ St Pauli (Tokyo) 21*7-10
◇ D45 D55 E15 E60 ◇ REV MR 50#12693
Zbl 273#02029 • ID 13367

TANAKA, H. [1974] *Some analytical rules of inference in the second-order arithmetic* (J 0407) Comm Math Univ St Pauli (Tokyo) 23/1*71-81
◇ C62 D55 F35 ◇ REV MR 53#10559 Zbl 357#02012
• ID 23068

TANAKA, H. [1978] *On recent recursion theory (Japanese)* (P 4109) B-Val Anal & Nonstand Anal;1978 Kyoto 65-86
◇ D98 ◇ ID 47668

TANAKA, H. [1978] *Recursion theory in analytical hierarchy* (J 0407) Comm Math Univ St Pauli (Tokyo) 27*113-132
◇ D25 D55 D60 E60 ◇ REV MR 81j:03069
Zbl 418#03034 • ID 53319

TANAKA, H. [1984] see IZUMI, M.

TANAKA, H. see Vol. I, II, III, V, VI for further entries

TANAKA, K. [1972] see MIZUMOTO, M.

TANAKA, K. see Vol. I, II, V for further entries

TANG, A. [1978] see SCIORE, E.

TANG, A. [1981] *Wadge reducibility and Hausdorff difference hierarchy in $P\omega$* (P 3368) Continuous Lattices;1979 Bremen 360-371
◇ D30 D55 E15 ◇ REV MR 83d:06001 Zbl 464#04002
• ID 46091

TANG, A. [1983] see KAMIMURA, T.

TANG, A. see Vol. V, VI for further entries

TANG, CHISUNG [1965] *A recursive theory of computer instructions (Chinese)* (J 1024) Zhongguo Kexue 14*1229-1232
◇ B75 D05 D80 ◇ REV MR 35#1475 Zbl 148.253
• ID 24855

TANG, TONGGAO [1981] *A note on the relative recursion theorem with a functional index (Chinese)* (J 3735) Fudan Xuebao, Ziran Kexue 20*356-358
◇ D20 ◇ REV MR 83h:03059 Zbl 567#03019 • ID 34842

TANG, TONGGAO see Vol. I, II, VI for further entries

TANG, ZHISONG [1965] *Recursiveness of systems of computer instructions (Chinese)* (J 0418) Shuxue Xuebao 15/6*842-860
• TRANSL [1965] (J 0419) Chinese Math Acta 7*577-597
◇ D20 ◇ REV Zbl 156.253 • ID 48531

TANG, ZHISONG see Vol. I, II for further entries

TANIGUCHI, H. [1982] see INOUE, K.

TANIGUCHI, H. [1983] see INOUE, K.

TANIGUCHI, H. [1985] see INOUE, K.

TANIGUCHI, K. [1971] see KASAMI, T.

TANIGUCHI, K. [1979] see HAGIHARA, K.

TAO, RENJI [1965] *Reduction of automata, universal automata and some problems regarding the behaviour of automata (Chinese)* (P 4564) Math Logic;1963 Xi-An 47-56
◇ D05 ◇ ID 48532

TAO, RENJI [1978] *Automata and its reduction (Chinese)* (J 3793) Jisuanjii Xuebao 1*11-32
◇ D05 ◇ ID 48533

TAO, RENJI [1978] *Some problems regarding the behaviour of automata (Chinese)* (J 3793) Jisuanjii Xuebao 1*81-92
◇ D05 ◇ ID 48534

TAO, RENJI [1979] *The reversibility of finite automata (Chinese)* (X 1876) Kexue Chubanshe: Beijing 300pp
◇ D05 ◇ ID 48537

TAO, RENJI [1979] *Universal automata (Chinese)* (J 3793) Jisuanjii Xuebao 2*14-27
◇ D05 ◇ ID 48535

TAO, RENJI [1981] *On the computational power of automata with time or space bounded by Ackermann's or superexponential functions* (J 1426) Theor Comput Sci 16*115-148
◇ D10 D15 D20 ◇ REV MR 82m:68093 Zbl 475#03018
• ID 55472

TAO, RENJI [1982] *A lower bound of kn^2 on time-complexity for one-tape automata on computations* (J 3766) Zhongguo Kexue, Xi A 25*866-876
◇ D05 D15 ◇ REV MR 84c:68034 Zbl 494#68059
• ID 48536

TAPIA, M. [1972] see BOLLMAN, D.A.

TARASOV, S.P. [1980] see KHACHIYAN, L.G.

TARJAN, R.E. [1976] see EVEN, S.

TARJAN, R.E. [1976] see CELONI, J.R.

TARJAN, R.E. [1976] see GAREY, M.R.

TARJAN, R.E. [1978] *Complexity of monotone networks for computing conjunctions* (**S** 3358) Ann Discrete Math 2*121-133
◇ B05 B70 D15 ◇ REV MR 81f:68058 Zbl 383 # 94038 • ID 69946

TARJAN, R.E. [1979] see ASPVALL, B.

TARSKI, A. [1931] see KURATOWSKI, K.

TARSKI, A. [1931] *Sur les ensembles definissables de nombres reels I* (**J** 0027) Fund Math 17*210-239
• TRANSL [1956] (**C** 1159) Tarski: Logic, Semantics, Metamathematics 110-142 (English)
◇ B25 B28 C10 C40 C60 C65 D55 E15 E47 F35 ◇ REV Zbl 75.4 JSL 34.99 FdM 57.60 • ID 16924

TARSKI, A. [1952] see SZMIELEW, W.

TARSKI, A. [1953] *A general method in proofs of undecidability* (**C** 4472) Undecidable Th 3-35
◇ D35 F25 ◇ REV JSL 24.167 • ID 42175

TARSKI, A. [1953] see MOSTOWSKI, ANDRZEJ

TARSKI, A. [1953] *Undecidability of the elementary theory of groups* (**C** 4472) Undecidable Th 77-87
◇ D35 ◇ REV Zbl 53.4 JSL 24.167 • ID 42177

TARSKI, A. [1957] see MONTAGUE, R.

TARSKI, A. [1959] see HENKIN, L.

TARSKI, A. [1965] see SZCZERBA, L.W.

TARSKI, A. [1979] see SZCZERBA, L.W.

TARSKI, A. [1983] see SCHWABHAEUSER, W.

TARSKI, A. also published under the name TAJTELBAUM, A.

TARSKI, A. see Vol. I, II, III, V, VI for further entries

TARTAKOVSKIJ, V.A. [1947] *On the process of extinction (Russian)* (**J** 0023) Dokl Akad Nauk SSSR 58*1605-1608
◇ D40 ◇ REV MR 9.321 Zbl 37.11 • ID 13473

TARTAKOVSKIJ, V.A. [1947] *On the problem of equivalence for certain types of groups (Russian)* (**J** 0023) Dokl Akad Nauk SSSR 58*1909-1910
◇ D40 ◇ REV MR 10.500 Zbl 37.151 • ID 13474

TARTAKOVSKIJ, V.A. [1949] *Application of the sieve method to the solution of the word problem for certain types of groups (Russian)* (**J** 0142) Mat Sb, Akad Nauk SSSR, NS 25(67)*251-274
◇ D40 ◇ REV MR 11.493 Zbl 34.163 • ID 49848

TARTAKOVSKIJ, V.A. [1949] *Solution of the word problem for groups with a k-reduced basis for k>6 (Russian)* (**J** 0216) Izv Akad Nauk SSSR, Ser Mat 13*483-494
◇ D40 ◇ REV MR 11.493 Zbl 35.295 • ID 49849

TARTAKOVSKIJ, V.A. [1949] *The sieve method in group theory (Russian)* (**J** 0142) Mat Sb, Akad Nauk SSSR, NS 25(67)*3-50
◇ D40 ◇ REV MR 11.493 Zbl 34.15 • ID 49850

TARTAKOVSKIJ, V.A. [1952] *On primitive composition (Russian)* (**J** 0142) Mat Sb, Akad Nauk SSSR, NS 30(72)*39-52
◇ D40 ◇ REV MR 11.493 Zbl 47.257 • ID 49851

TARTAKOVSKIJ, V.A. [1968] see STENDER, P.V.

TASIK, M. [1980] *On algorithms; on recursive functions (Macedonian)* (**X** 3760) Univ Kiril et Metodij, Mat Fak: Skopje v+117pp
◇ D98 ◇ REV MR 83i:03063 Zbl 491 # 03016 • ID 34851

TATARAM, M. [1983] see CALUDE, C.

TATARAM, M. [1985] see PAUN, G.

TAUTS, A. [1983] *Parallelizing of recursive computations (Russian) (German and Estonian summaries)* (**J** 0080) Izv Akad Nauk Ehston SSR, Fiz, Mat 32*121-127
◇ D20 ◇ REV Zbl 516 # 03020 • ID 37255

TAUTS, A. see Vol. I, II, III, V, VI for further entries

TAYLOR, W. [1977] *Equational logic* (**P** 2112) Contrib to Universal Algeb;1975 Szeged 465-501
◇ C05 C98 D40 ◇ REV MR 57 # 12341 Zbl 421 # 08003 • ID 69948

TAYLOR, W. [1981] *Some universal sets of terms* (**J** 0064) Trans Amer Math Soc 267*595-607
◇ C05 D80 ◇ REV MR 82h:08005 Zbl 483 # 04003 • ID 82957

TAYLOR, W. see Vol. III, V for further entries

TCHUENTE, M. [1983] *Computation of Boolean functions on networks of binary automata* (**J** 0119) J Comp Syst Sci 26*269-277
◇ D05 ◇ REV MR 84i:68088 Zbl 508 # 94025 • ID 40171

TEIXEIRA, M.T. [1961] *Recursive functions and the foundations of mathematics (Portuguese)* (**J** 0084) Gaz Mat (Lisboa) 22/84-85*12-16
◇ D20 ◇ REV MR 25 # 4995 Zbl 147.250 • ID 13516

TEIXEIRA, M.T. see Vol. VI for further entries

TELGARSKY, R. [1977] *Topological games and analytic sets* (**J** 1447) Houston J Math 3*549-553
◇ D55 E15 E60 E75 ◇ REV MR 58 # 24215 Zbl 388 # 90093 • REM Part I. Part II 1980 by Ostaszewski,A.J. & Telgarsky,R. • ID 32018

TELGARSKY, R. see Vol. V for further entries

TELLEZ-GIRON, R. [1980] see DAVID, R.

TENNENBAUM, S. [1963] *Degree of unsolvability and the rate of growth of functions* (**P** 0674) Symp Math Th of Automata;1962 New York 71-73
◇ D25 ◇ REV MR 29 # 4679 Zbl 199.25 JSL 32.524 • ID 13518

TENNENBAUM, S. see Vol. V for further entries

TER-AKOPOV, A.K. [1975] *Fast response optimization of discrete transformers with respect to automata composition (Russian) (English summary)* (**J** 0040) Kibernetika, Akad Nauk Ukr SSR 1975/5*27-31
• TRANSL [1975] (**J** 0021) Cybernetics 11*697-701
◇ D05 ◇ REV MR 57 # 19189 Zbl 326 # 94030 • ID 65681

TER-ZAKHARYAN, N.P. [1973] *On the language of multi-place recursive functions (Russian)* (**J** 0023) Dokl Akad Nauk SSSR 210*541-542
• TRANSL [1973] (**J** 0062) Sov Math, Dokl 14*781-783
◇ D20 ◇ REV MR 47 # 8276 Zbl 291 # 02030 • ID 13522

TER-ZAKHARYAN, N.P. [1973] *Some entropy properties of algorithmic languages (Russian)* (S 0554) Issl Teor Algor & Mat Logik (Moskva) 1*178-204
◇ D20 ◇ REV MR 48 # 10780 Zbl 284 # 02013 • ID 30517

TER-ZAKHARYAN, N.P. [1975] *Quantitative characteristics of the language of many-argument recursive functions (Russian)* (S 0422) Tr Vychisl Tsentra Akad Nauk Armyan SSR & Univ Erevan 8*9-35
◇ D20 ◇ REV MR 58 # 27415 Zbl 397 # 03025 • ID 52691

TER-ZAKHARYAN, N.P. [1976] *Some nonoptimal languages of recursive functions (Russian) (English and Armenian summaries)* (J 0312) Izv Akad Nauk Armyan SSR, Ser Mat 11*256-262,289
◇ D20 ◇ REV MR 55 # 5408 Zbl 359 # 02043 • ID 50590

TER-ZAKHARYAN, N.P. [1982] *The possibilities of coding messages in various languages of formal arithmetic (Russian) (Armenian summary)* (S 0422) Tr Vychisl Tsentra Akad Nauk Armyan SSR & Univ Erevan 10*93-102
◇ D20 ◇ REV MR 85a:03074 • ID 34746

TERLOUW, J. [1982] *On definition trees of ordinal recursive functionals: reduction of the recursion orders by means of type level raising* (J 0036) J Symb Logic 47*395-402
◇ D20 ◇ REV MR 83d:03055 Zbl 388 # 03030 • ID 35193

TERLOUW, J. see Vol. VI for further entries

TERZILER, M. [1983] *Une recherche sur les fonctions primitives recursives* (J 0650) Ege Ueniv Fen Fak Derg, Ser A 6*47-54
◇ D20 ◇ REV Zbl 522 # 03025 • ID 43380

TERZILER, M. see Vol. I for further entries

TETRUASHVILI, M.R. [1970] *On the word problem for a certain class of finitely determined semigroups (Russian) (Georgian and English summaries)* (J 0233) Soobshch Akad Nauk Gruz SSR 59*541-544
◇ D40 ◇ REV MR 44 # 6803 Zbl 223 # 20062 • ID 13523

TETRUASHVILI, M.R. [1973] *A realization of the Magnus algorithm on a Turing machine, and an upper bound of the complexity of the computations (Russian) (Georgian and English summaries)* (J 0233) Soobshch Akad Nauk Gruz SSR 70*541-544
◇ D10 D15 D40 ◇ REV MR 50 # 6816 Zbl 275 # 20068 • ID 13524

TETRUASHVILI, M.R. [1975] *Estimation of the complexity of a certain reduction (Russian)* (S 2043) Issl Mat Log & Teor Algor (Tbilisi) 1975*125-134
◇ D15 D40 ◇ REV MR 56 # 2811 • ID 79291

TETRUASHVILI, M.R. [1975] *The complexity of Turing computations in terms of Magnus's algorithm (Russian) (Georgian summary)* (S 2043) Issl Mat Log & Teor Algor (Tbilisi) 1975*111-123
◇ D15 D40 ◇ REV MR 57 # 464 • ID 82964

TETRUASHVILI, M.R. [1978] *The problem of conjugacy for one class of groups and the computational complexity (Russian)* (J 0954) Tr Inst Prikl Mat, Tbilisi 5*250-258
◇ D15 D40 ◇ REV MR 80e:68125 Zbl 441 # 20025 • ID 33961

TETRUASHVILI, M.R. [1979] *The conjugacy problem for groups with one defining relation and the complexity of Turing calculations* (P 2539) Frege Konferenz (1);1979 Jena 477-482
◇ D15 D40 ◇ REV MR 82b:20050 • ID 82965

TETRUASHVILI, M.R. [1982] *On the conjugacy problem for finitely presented subgroups of finite index of a finitely presented group with unsolvable conjugacy problem (Russian) (English and Georgian summaries)* (J 0954) Tr Inst Prikl Mat, Tbilisi 11*100-113
◇ D15 D40 ◇ REV MR 85k:20110 • ID 45338

TETRUASHVILI, M.R. [1984] *Computational complexity of recognizing word equality in semigroups of a certain class (Russian) (English and Georgian summaries)* (J 0233) Soobshch Akad Nauk Gruz SSR 114*33-36
◇ D15 D40 ◇ REV MR 86e:68044 • ID 44856

TETRUASHVILI, M.R. [1984] *The computational complexity of the theory of abelian groups with a given number of generators* (P 3621) Frege Konferenz (2);1984 Schwerin 371-375
◇ B25 D15 ◇ ID 45394

TEUTSCH, R.J. [1974] see JAMIESON, D.W.

TEVY, I. [1979] see CALUDE, C.

TEVY, I. [1980] see CALUDE, C.

THARP, L.H. [1974] *Continuity and elementary logic* (J 0036) J Symb Logic 39*700-716
◇ C07 C80 C95 D30 D55 ◇ REV MR 51 # 49 Zbl 299 # 02011 • ID 13528

THARP, L.H. see Vol. I, III, V, VI for further entries

THATCHER, J.W. [1963] *The construction of a self-describing Turing machine* (P 0674) Symp Math Th of Automata;1962 New York 165-171
◇ D10 ◇ REV MR 30 # 1044 Zbl 116.337 • ID 13529

THATCHER, J.W. [1966] *Decision problems for multiple successor arithmetics* (J 0036) J Symb Logic 31*182-190
◇ D35 ◇ REV MR 34 # 4139 Zbl 144.1 • ID 13530

THATCHER, J.W. [1967] *Characterizing derivation trees of context free grammars through a generalization of finite automata theory* (J 0119) J Comp Syst Sci 1*317-322
◇ D05 ◇ REV MR 38 # 2024 Zbl 155.18 • ID 13531

THATCHER, J.W. & WRIGHT, J.B. [1968] *Generalized finite automata theory with an application to a decision problem of second order logic* (J 0041) Math Syst Theory 2*57-81
◇ B15 B25 D05 ◇ REV MR 37 # 75 Zbl 157.22 JSL 37.619 • ID 28596

THATCHER, J.W. [1970] *Self-describing Turing machines and self-reproducing cellular automata* (C 1085) Essays Cellular Automata 103-131
◇ D05 D10 ◇ REV Zbl 235 # 02034 • ID 27822

THATCHER, J.W. [1970] *Universality in the von Neumann cellular model* (C 1085) Essays Cellular Automata 132-186
◇ D05 D10 ◇ REV Zbl 235 # 02035 • ID 27823

THATCHER, J.W. [1975] see GOGUEN, J.A.

THATCHER, J.W. see Vol. V for further entries

THIAGARAJAN, P.S. [1980] see GENRICH, H.J.

THIBAULT, M.-F. [1982] *Prerecursive categories* (J 0326) J Pure Appl Algebra 24*79-93
◇ D20 D65 G30 ◇ REV MR 83f:03040 Zbl 485 # 18005 • ID 35281

THIELE, H. [1967] see STARKE, P.H.

THIELE, H. [1970] see STARKE, P.H.

THIELE, H. [1971] see BERNHARDT, L.

THIELE, H. [1973] see STARKE, P.H.

THIELE, H. [1983] *A classification of propositional process logics on the basis of theory of automata* (**S** 3382) Sem-ber, Humboldt-Univ Berlin, Sekt Math 52*160-173
◇ B75 D05 ◇ REV MR 85f:68003 Zbl 547 # 68035
• ID 43298

THIELE, H. see Vol. I, II for further entries

THIELER-MEVISSEN, G. [1976] see GENRICH, H.J.

THIELER-MEVISSEN, G. [1976] *Zur Beschreibbarkeit der hyperarithmetischen reellen Zahlen mit analysiskonformen Mitteln* (**S** 0478) Bonn Math Schr 75*41pp
◇ D35 D55 ◇ REV MR 56 # 11772 Zbl 353 # 02021
• ID 50104

THIERRIN, G. [1983] see LATTEUX, M.

THOMAS, WILLIAM J. [1973] *Doubts about some standard arguments for Church's Thesis* (**P** 0580) Int Congr Log, Meth & Phil of Sci (4,Sel Pap);1971 Bucharest 55-62
◇ A05 D10 D20 F99 ◇ REV MR 57 # 2869
Zbl 286 # 02009 • ID 65709

THOMAS, WILLIAM J. [1979] *A simple generalization of Turing computability* (**J** 0047) Notre Dame J Formal Log 20*95-102
◇ D10 D75 ◇ REV MR 80h:03068 Zbl 397 # 03022
• ID 52688

THOMAS, WILLIAM J. see Vol. I for further entries

THOMAS, WOLFGANG [1975] *A note on undecidable extensions of monadic second order successor arithmetic* (**J** 0009) Arch Math Logik Grundlagenforsch 17*43-44
◇ D35 F35 ◇ REV MR 56 # 110 Zbl 325 # 02032
• ID 13578

THOMAS, WOLFGANG [1978] *The theory of successor with an extra predicate* (**J** 0043) Math Ann 237*121-132
◇ B25 D35 ◇ REV MR 80b:03054 Zbl 369 # 02025
• ID 31271

THOMAS, WOLFGANG [1979] *Star-free regular sets of ω-sequences* (**J** 0194) Inform & Control 42*148-156
◇ C07 D05 ◇ REV MR 80h:68052 Zbl 411 # 03031
• ID 52885

THOMAS, WOLFGANG [1981] *A combinatorial approach to the theory of ω-automata* (**J** 0194) Inform & Control 48*261-283
◇ B25 D05 ◇ REV MR 84d:68054 Zbl 478 # 03020
• ID 55635

THOMAS, WOLFGANG [1982] *Classifying regular events in symbolic logic* (**J** 0119) J Comp Syst Sci 25*360-376
◇ D05 ◇ REV MR 85a:03053 Zbl 503 # 68055 • ID 34803

THOMAS, WOLFGANG [1984] *An application of the Ehrenfeucht-Fraisse game in formal language theory* (**S** 3521) Mem Soc Math Fr 16*11-21
◇ C07 D05 ◇ REV Zbl 558 # 68064 • ID 39756

THOMAS, WOLFGANG [1984] see BOERGER, E.

THOMAS, WOLFGANG see Vol. I, II, III, V for further entries

THOMASON, M.G. [1981] see BARRERO, A.

THOMASON, M.G. see Vol. II for further entries

THOMASON, S.K. [1967] *The forcing method and the upper semilattice of hyperdegrees* (**J** 0064) Trans Amer Math Soc 129*38-57
◇ D30 E40 ◇ REV MR 36 # 2503 Zbl 168.251 • ID 13585

THOMASON, S.K. [1969] *A note on non-distributive sublattices of degrees and hyperdegrees* (**J** 0017) Canad J Math 21*147-148
◇ D30 E40 ◇ REV MR 39 # 58 Zbl 169.310 • ID 13586

THOMASON, S.K. [1970] *A theorem on initial segments of degrees* (**J** 0036) J Symb Logic 35*41-45
◇ D30 ◇ REV MR 44 # 6493 Zbl 196.14 • ID 13589

THOMASON, S.K. [1970] *On initial segments of hyperdegrees* (**J** 0036) J Symb Logic 35*189-197
◇ D30 D35 E40 ◇ REV MR 44 # 5225 Zbl 206.281
• ID 13588

THOMASON, S.K. [1970] *Sublattices and initial segments of the degrees of unsolvability* (**J** 0017) Canad J Math 22*569-581
◇ D30 ◇ REV MR 42 # 73 Zbl 206.282 • ID 13587

THOMASON, S.K. [1971] *Sublattices of the recursively enumerable degrees* (**J** 0068) Z Math Logik Grundlagen Math 17*273-280
◇ D25 ◇ REV MR 45 # 8523 Zbl 219 # 02030 • ID 13591

THOMASON, S.K. [1975] *The logical consequence relation of propositional tense logic* (**J** 0068) Z Math Logik Grundlagen Math 21*29-40
◇ B45 D20 D55 ◇ REV MR 51 # 67 Zbl 324 # 02014
• ID 13597

THOMASON, S.K. see Vol. II, III for further entries

THOMPSON, R.J. [1973] see MCKENZIE, R.

THOMPSON, R.J. [1980] *Embeddings into finitely generated simple groups which preserve the word problem* (**P** 2634) Word Problems II;1976 Oxford 401-441
◇ D40 ◇ REV MR 81k:20050 Zbl 431 # 20030 • ID 82971

THOMPSON, S. [1985] *Axiomatic recursion theory and the continuous functionals* (**J** 0036) J Symb Logic 50*442-450
◇ D65 D75 ◇ ID 42557

THOMPSON, S. [1985] *Priority arguments in the continuous r.e. degrees* (**J** 0036) J Symb Logic 50*661-667
◇ D65 ◇ ID 47371

THOMPSON, S. see Vol. V for further entries

THUE, A. [1913] *Ueber die gegenseitige Lage gleicher Teile gewisser Zeichenreihen* (**J** 1145) Vidensk Selsk Kristiana Skrifter Ser 1 9*67pp
◇ D03 D40 ◇ REV FdM 44.462 • ID 22087

THUE, A. [1914] *Probleme ueber Veraenderungen von Zeichenreihen nach gegebenen Regeln* (**J** 1145) Vidensk Selsk Kristiana Skrifter Ser 1 10*34pp
◇ D03 D40 ◇ REV FdM 45.333 • ID 22088

THURAISINGHAM, M.B. [1982] *Representation of one-one degrees by decision problems for system functions* (**J** 0119) J Comp Syst Sci 24*373-377
◇ D25 ◇ REV MR 84a:03045 Zbl 507 # 03016 • ID 35579

THURAISINGHAM, M.B. [1983] *Cylindrical decision problems for system functions* (**J** 0047) Notre Dame J Formal Log 24*188-198
◇ D20 D25 ◇ REV MR 85b:03073 Zbl 508 # 03017
• ID 36938

THURAISINGHAM, M.B. [1983] *Some elementary closure properties of n-cylinders* (**J** 0047) Notre Dame J Formal Log 24∗242–254
◇ D20 D25 ◇ REV MR 85b:03074 Zbl 508 # 03020 • ID 36939

THURAISINGHAM, M.B. [1983] *The concept of n-cylinder and its relationship to simple sets* (**J** 0047) Notre Dame J Formal Log 24∗328–336
◇ D25 ◇ REV MR 85d:03085 Zbl 515 # 03020 • ID 38141

THURAISINGHAM, M.B. [1984] *System functions and their decision problems* (**J** 0068) Z Math Logik Grundlagen Math 30∗119–128
◇ D25 ◇ REV MR 86f:03064 • ID 42216

TICHY, P. [1969] *Intension in terms of Turing machines (Polish and Russian summaries)* (**J** 0063) Studia Logica 24∗7–25
◇ A05 D10 ◇ REV MR 45 # 4950 Zbl 246 # 02010 • ID 13605

TICHY, P. see Vol. I, II for further entries

TIMOSHENKO, E.I. [1973] *Algorithmic problems for metabelian group (Russian)* (**J** 0003) Algebra i Logika 12∗232–240
• TRANSL [1973] (**J** 0069) Algeb and Log 12∗132–137
◇ D40 ◇ REV MR 54 # 10426 Zbl 281 # 20030 • ID 25931

TIMOSHENKO, E.I. see Vol. III for further entries

TINHOFER, G. [1978] *On the simultaneous isomorphism of special relations (isomorphism of automata)* (**P** 3148) Graphth Konzepte Inf (3);1977 Linz 205–213
◇ C13 D05 ◇ REV MR 58 # 20915 Zbl 389 # 68033 • ID 52324

TISON, S. [1985] see DAUCHET, M.

TITARENKO, L.N. [1973] *The complexity of recognizing functional properties of normal algorithms (Russian)* (**C** 2978) Vychisl Mat (Kiev) 60–63
◇ D03 D15 ◇ REV MR 57 # 16004 • ID 79366

TITGEMEYER, D. [1965] *Einfuehrung in die Theorie der Kalkuele und Algorithmen* (**J** 0487) Math Unterricht 11/2∗5–20
◇ D03 D05 D10 ◇ REV MR 32 # 51 • ID 13618

TITGEMEYER, D. [1965] *Untersuchungen ueber die Struktur des Kleene-Post'schen Halbverbandes der Grade der rekursiven Unloesbarkeit* (**J** 0009) Arch Math Logik Grundlagenforsch 8∗45–62
◇ D30 ◇ REV MR 32 # 5520 Zbl 168.12 JSL 35.155 • ID 13619

TITOV, N.N. [1985] *On the finitely generatedness of some classes of recursively enumerable sets (Russian)* (**J** 0040) Kibernetika, Akad Nauk Ukr SSR 1985/2∗100–102
◇ D25 ◇ ID 46522

TITS, J. [1969] *Le probleme des mots dans les groupes de Coxeter* (**P** 1062) Conv Teor Gruppi & Cont Polari;1967/68 Roma 1∗175–185
◇ D40 ◇ REV MR 40 # 7339 Zbl 206.30 • ID 20945

TITS, J. see Vol. III, V for further entries

TIURYN, J. [1979] see BERGSTRA, J.A.

TIURYN, J. [1981] see BERGSTRA, J.A.

TIURYN, J. [1982] see BERGSTRA, J.A.

TIURYN, J. see Vol. II, III, V for further entries

TIXIER, V. [1968] see SALOMAA, A.

TKACHEV, G.A. [1977] *Complexity of realization of one sequence of functions of k-valued logic (Russian)* (**J** 2869) Vest Ser Vychisl Mat Kibern, Univ Moskva 1977/1∗45–57
• TRANSL [1977] (**J** 3221) Moscow Univ Comp Math Cybern 1977/1∗36–45
◇ B50 B70 D15 ◇ REV Zbl 447 # 94031 • ID 69960

TKACHEV, G.A. see Vol. II for further entries

TLYUSTEN, V.SH. [1978] *Solvability of a uniform halting problem in the class of Post machines with two states (Russian) (English summary)* (**J** 0040) Kibernetika, Akad Nauk Ukr SSR 1978/1∗24–30
• TRANSL [1978] (**J** 0021) Cybernetics 14∗24–31
◇ D10 ◇ REV MR 58 # 5112 Zbl 382 # 03026 • ID 51950

TOCA, A. [1979] *Ω-automata* (**J** 0517) Mathematica (Cluj) 21(44)∗209–213
◇ D05 ◇ REV MR 82c:68037 Zbl 424 # 68032 • ID 82977

TOGER, A.V. [1971] *On the complexity of some functional classes (Russian)* (**J** 0023) Dokl Akad Nauk SSSR 199∗789–791
• TRANSL [1971] (**J** 0062) Sov Math, Dokl 12∗1197–1200
◇ D15 ◇ REV MR 44 # 5228 Zbl 321 # 68037 • ID 19746

TOGER, A.V. [1976] *Asymptotic behavior of the complexity of the approximate computation of functions that satisfy a Lipschitz condition (Russian)* (**P** 2961) Mat Progr & Smezhnye Vopr;1974 Drogobych 141–149
◇ D15 ◇ REV MR 58 # 16213 • ID 79378

TOKURA, N. [1981] see ARAKI, T.

TOMESCU, I. [1967] *On some combinatorical problems in the theory of classification (Romanian)* (**J** 0197) Stud Cercet Mat Acad Romana 19∗1385–1393
◇ D05 ◇ REV MR 40 # 4132 Zbl 203.14 • ID 13638

TOMESCU, I. see Vol. I for further entries

TOMPA, M. [1981] *An extension of Savitch's theorem to small space bounds* (**J** 0232) Inform Process Lett 12∗106–108
◇ D15 ◇ REV MR 82d:68035 Zbl 457 # 68040 • ID 82988

TOMPA, M. [1985] see DYMOND, P.W.

TONOYAN, R.N. [1965] *Logical schemes for algorithms and their equivalent transforms (Russian)* (**J** 0052) Probl Kibern 14∗161–188
◇ B75 D05 ◇ REV MR 34 # 4074 Zbl 267 # 02019 • ID 29883

TOPENCHAROV, V.V. [1985] see STOEVA, S.P.

TOPENCHAROV, V.V. see Vol. V for further entries

TOPSOEE, F. [1980] see DELLACHERIE, C.

TORBASOVA, V.P. [1982] *Machines over models (Russian)* (**S** 3909) Mat Met Opt & Upravleniya Slozh Sist (Kalinin) 1982∗3–13
◇ D05 D75 ◇ REV Zbl 549 # 68054 • ID 43161

TORBASOVA, V.P. see Vol. I for further entries

TORELLI, M. [1977] see BERTONI, A.

TORELLI, M. [1980] see BERTONI, A.

TOURLAKIS, G. [1973] see MYLOPOULOS, J.

TOURLAKIS, G. [1984] *An inductive number-theoretic characterization of NP* (**J** 0232) Inform Process Lett 19∗245–247
◇ D15 ◇ REV Zbl 563 # 68044 • ID 44481

Tourneau le, J.J. [1968] *Decision problems related to the concept of operation* (**X** 0926) Univ Calif Pr: Berkeley
◇ C05 D35 ◇ ID 27136

Tourneau le, J.J. see Vol. III for further entries

Tovey, C.A. [1984] *A simplified NP-complete satisfiability problem* (**J** 2702) Discr Appl Math 8*85-89
◇ D15 ◇ REV MR 85f:68033 Zbl 534#68028 • ID 38360

Townsend, R. [1974] *A decidability result in algebraic language theory* (**J** 0041) Math Syst Theory 8*225-227
◇ D05 ◇ REV MR 53#4651 Zbl 361#68110 • ID 21663

Toyoda, J. [1972] see Mizumoto, M.

Trakhtenbrot, B.A. [1950] *The impossibility of an algorithm for the decision problem in finite domains (Russian)* (**J** 0023) Dokl Akad Nauk SSSR 70*569-572
• TRANSL [1963] (**J** 0225) Amer Math Soc, Transl, Ser 2 23*1-5
◇ C13 D35 ◇ REV MR 11.488 Zbl 38.150 JSL 15.229 • ID 19743

Trakhtenbrot, B.A. [1953] *On recursively separability (Russian)* (**J** 0023) Dokl Akad Nauk SSSR 88*953-956
◇ B10 C13 D35 ◇ REV MR 16.436 Zbl 50.8 JSL 19.60 • ID 19742

Trakhtenbrot, B.A. [1955] see Kuznetsov, A.V.

Trakhtenbrot, B.A. [1955] *Tabular representation of recursive operators (Russian)* (**J** 0023) Dokl Akad Nauk SSSR 101*417-420
◇ D20 ◇ REV MR 18.457 Zbl 64.11 JSL 21.207 • ID 20966

Trakhtenbrot, B.A. [1956] *Signalizing functions and matrix operators (Russian)* (**J** 4167) Uch Zap Ped Inst, Penza 4*75-87
◇ B70 D20 ◇ REV MR 20#4487 • ID 45723

Trakhtenbrot, B.A. [1956] *The definition of a finite set and the deductive incompleteness of the theory of sets (Russian)* (**J** 0216) Izv Akad Nauk SSSR, Ser Mat 20*569-582
• TRANSL [1964] (**J** 0225) Amer Math Soc, Transl, Ser 2 39*177-192
◇ B28 D35 E30 ◇ REV MR 18.269 Zbl 71.246 JSL 27.236 • ID 19739

Trakhtenbrot, B.A. [1957] *On operators realizable in logical nets (Russian)* (**J** 0023) Dokl Akad Nauk SSSR 112*1005-1007
◇ D05 ◇ REV MR 19.888 Zbl 78.306 JSL 27.252 • ID 19738

Trakhtenbrot, B.A. [1958] *Synthesis of logical nets whose operators are described by means of the calculus of one-place predicates (Russian)* (**J** 0023) Dokl Akad Nauk SSSR 118*646-649
◇ B70 D05 ◇ REV MR 20#5142 Zbl 84.11 JSL 28.254 • ID 19737

Trakhtenbrot, B.A. [1959] *Descriptive classifications in recursive arithmetics (Russian)* (**P** 0607) All-Union Math Conf (3);1956 Moskva 1*185
◇ D20 F30 ◇ ID 45721

Trakhtenbrot, B.A. [1959] *On effective operators and properties related to their continuousness (Russian)* (**P** 0607) All-Union Math Conf (3);1956 Moskva 2*147-148
◇ D20 ◇ ID 28397

Trakhtenbrot, B.A. [1959] see Kobrinskij, N.E.

Trakhtenbrot, B.A. [1959] *Wieso koennen Automaten rechnen?* (**X** 0806) Dt Verlag Wiss: Berlin 101pp
◇ D10 D98 ◇ REV MR 23#B527 Zbl 87.126 JSL 27.224 • ID 13665

Trakhtenbrot, B.A. [1960] *Algorithms and machine solutions of problems (Russian)* (**X** 3709) Izdat Fiz-Mat Lit: Moskva 119pp
• TRANSL [1977] (**X** 0885) Mir: Moskva 109pp (Spanish)
◇ B25 B35 D20 ◇ REV MR 22#10906 Zbl 80.114 JSL 47.702 • ID 42812

Trakhtenbrot, B.A. [1961] *Finite automata and the logic of one-place predicates (Russian)* (**J** 0023) Dokl Akad Nauk SSSR 140*326-329
• TRANSL [1961] (**J** 0470) Sov Phys, Dokl 6*753-755
◇ B20 D05 F35 ◇ REV MR 26#1246 Zbl 115.7 JSL 29.100 • ID 19734

Trakhtenbrot, B.A. [1962] *Finite automata and the logic of one-place predicates (Russian)* (**J** 0092) Sib Mat Zh 3*103-131
◇ B20 D05 F35 ◇ REV MR 26#4908 Zbl 115.7 JSL 29.98 • ID 19733

Trakhtenbrot, B.A. [1962] see Kobrinskij, N.E.

Trakhtenbrot, B.A. [1963] *Algorithms and automatic computing machines* (**X** 1004) Heath: Lexington vi+101pp
◇ D10 D20 D98 ◇ REV JSL 28.104 JSL 29.147 • ID 19732

Trakhtenbrot, B.A. [1963] *On the frequency computability of functions (Russian)* (**J** 0003) Algebra i Logika 2/1*25-32
◇ D20 D55 ◇ REV MR 29#14 Zbl 192.50 JSL 39.606 • ID 13666

Trakhtenbrot, B.A. [1964] *Turing computations with logarithmic delay (Russian)* (**J** 0003) Algebra i Logika 3/4*33-48
◇ D10 D15 ◇ REV MR 32#1125 JSL 33.118 • ID 19731

Trakhtenbrot, B.A. [1965] *Optimal computations and the frequency occurrence of Ablonskij (Russian)* (**J** 0003) Algebra i Logika 4/5*79-93
◇ D15 ◇ REV MR 33#2493 JSL 34.134 • ID 19729

Trakhtenbrot, B.A. [1966] *Normed signalizers for Turing computations (Russian)* (**J** 0003) Algebra i Logika 5/6*61-70
◇ D10 D15 ◇ REV MR 37#7224 Zbl 253#02031 • ID 28935

Trakhtenbrot, B.A. [1967] *Complexity of algorithms and computations (Russian)* (**X** 0913) Novosibirsk Gos Univ: Novosibirsk 258pp
◇ D10 D15 D20 D98 ◇ REV JSL 35.337 • ID 19728

Trakhtenbrot, B.A. [1969] *On the complexity of reduction algorithms in Novikov-Boone constructions (Russian)* (**J** 0003) Algebra i Logika 8*93-128
• TRANSL [1969] (**J** 0069) Algeb and Log 8*50-71
◇ D15 D40 ◇ REV MR 44#2608 Zbl 199.312 JSL 37.607 • ID 19727

Trakhtenbrot, B.A. [1970] see Barzdins, J.

Trakhtenbrot, B.A. [1970] *On autoreducibility (Russian)* (**J** 0023) Dokl Akad Nauk SSSR 192*1224-1227
• TRANSL [1970] (**J** 0062) Sov Math, Dokl 11*814-817
◇ D30 ◇ REV MR 43#52 Zbl 216.289 JSL 38.527 • ID 19726

TRAKHTENBROT, B.A. [1973] *A formalization of certain concepts in terms of complexity of computations* (**P** 0793) Int Congr Log, Meth & Phil of Sci (4,Proc);1971 Bucharest 205-213
◇ D15 ◇ REV MR 57 # 82 • ID 79437

TRAKHTENBROT, B.A. [1973] *Autoreducible and nonautoreducible predicates and sets (Russian)* (**S** 0554) Issl Teor Algor & Mat Logik (Moskva) 1*211-234
◇ D25 D30 ◇ REV MR 49 # 8837 Zbl 258 # 02036
• ID 19724

TRAKHTENBROT, B.A. [1973] *Frequency computations (Russian)* (**S** 0066) Tr Mat Inst Steklov 133*221-232,276
• TRANSL [1973] (**S** 0055) Proc Steklov Inst Math 133*223-234
◇ D10 D15 D20 ◇ REV MR 54 # 12498 Zbl 292 # 02032
• ID 50028

TRAKHTENBROT, B.A. [1974] *Bemerkungen ueber die Kompliziertheit von Berechnungen auf stochastischen Automaten (Russisch)* (**C** 1450) Teor Algor & Mat Logika (Markov) 159-176
• TRANSL [1979] (**C** 2897) Algeb Model, Kateg & Gruppoide 165-178
◇ D10 D15 ◇ REV MR 58 # 25129 Zbl 304 # 02014
• ID 65771

TRAKHTENBROT, B.A. [1975] *On problems solvable by successive trials* (**P** 0454) Math Founds of Comput Sci (4);1975 Marianske Lazne 32*125-137
◇ D15 D50 ◇ REV MR 52 # 16126 Zbl 357 # 68060
• ID 50445

TRAKHTENBROT, B.A. [1976] *Recursive program schemes and computable functionals* (**P** 1401) Math Founds of Comput Sci (5);1976 Gdansk 137-152
◇ B75 D15 D55 D65 D75 ◇ REV Zbl 352 # 02028
• ID 50027

TRAKHTENBROT, B.A. [1977] *Algorithmen und Rechenautomaten* (**X** 0806) Dt Verlag Wiss: Berlin 208pp
◇ D10 D15 D40 D98 ◇ REV MR 58 # 10356
Zbl 355 # 68032 • REM Uebersetzung aus dem Russischen: G.Asser & H.-D.Hecker & L.Voelkel • ID 50265

TRAKHTENBROT, B.A. [1977] *Frequency algorithms and computations* (**P** 1635) Math Founds of Comput Sci (6);1977 Tatranska Lomnica 148-161
◇ D15 D20 ◇ REV MR 58 # 19336 Zbl 392 # 68030
• ID 52414

TRAKHTENBROT, B.A. see Vol. I, II, VI for further entries

TRAKHTENBROT, M.B. [1976] *Relationship between classes of monotonic functions* (**J** 1426) Theor Comput Sci 2*225-247
◇ B75 D15 D20 ◇ REV MR 54 # 4172 Zbl 352 # 68052
• ID 24076

TRAKHTENGERTS, EH.A. [1977] see BABICHEV, A.V.

TRAKHTMAN, A.N. [1970] see BARANSKIJ, V.A.

TRAUB, J. [1978] *Recent results and open problems in analytic computational complexity* (**C** 3507) Math Models & Numer Meth 3*269-272
◇ D15 ◇ REV MR 80j:68034 Zbl 416 # 68036 • ID 69968

TRAUTTEUR, G. [1975] see CRISCUOLO, G.

TREMBLAY, J.P. [1975] see MANOHAR, R.P.

TRETKOFF, M. [1980] see HAIMO, F.

T"RKALANOV, K.D. [1970] *A certain class of partially ordered semigroups with a solvable inequality problem (Russian)* (**J** 0137) C R Acad Bulgar Sci 23*129-132
◇ D05 D40 ◇ REV MR 43 # 130 Zbl 247 # 02034
• ID 13643

T"RKALANOV, K.D. [1971] *A solution of the inequality in a certain class of partially ordered semigroups by the method of Osipova (Russian and French summmaries)* (**S** 1002) Nauch Trud Vissh Ped Inst, Plovdiv 9/2*37-43
◇ B25 D40 ◇ REV MR 45 # 5047 • ID 13644

T"RKALANOV, K.D. & ZHELEVA, S. [1971] *On the problem of inequality for partially ordered semigroups (Bulgarian) (Russian and German summaries)* (**S** 1002) Nauch Trud Vissh Ped Inst, Plovdiv 9/3*25-32
◇ B25 D40 ◇ REV MR 45 # 5048 • ID 13645

T"RKALANOV, K.D. [1974] *On the recognition of the applicability of a certain algorithm for divisibility (Bulgarian) (Russian and French summaries)* (**S** 1441) Nauch Trud, Univ Plovdiv 12*163-168
◇ D40 ◇ REV MR 53 # 5281 • ID 22914

T"RKALANOV, K.D. [1976] *On the o-isomorphism and broken-line problems, and the applicability of the algorithm for divisibility (Russian) (English summary)* (**J** 0270) Dokl Akad Nauk Ukr SSR, Ser A 1976*502-505,574
◇ D40 ◇ REV MR 54 # 7666 Zbl 338 # 02027 • ID 65821

TRNKOVA, V. [1980] *General theory of relational automata* (**J** 2095) Fund Inform, Ann Soc Math Pol, Ser 4 3*189-233
◇ D05 ◇ REV MR 82e:68058 Zbl 453 # 68022 • ID 83008

TRNKOVA, V. [1982] see ADAMEK, J.

TRNKOVA, V. see Vol. II, III, V for further entries

TROELSTRA, A.S. [1978] *Some remarks on the complexity of Henkin-Kripke models* (**J** 0028) Indag Math 40*296-302
◇ C90 D55 F50 ◇ REV MR 58 # 21575 Zbl 389 # 03024
• ID 27952

TROELSTRA, A.S. see Vol. I, II, III, V, VI for further entries

TROFIMOV, O.E. [1976] see FRIZEN, D.G.

TROFIMOV, V.I. [1981] *Growth functions of some classes of languages* (**J** 0040) Kibernetika, Akad Nauk Ukr SSR 1981/6*i,9-12,149
• TRANSL [1981] (**J** 0021) Cybernetics 17*727-731
◇ D05 ◇ REV MR 84d:68088 Zbl 512 # 68054 • ID 39713

TROTTER JR., W.T. [1981] see KIERSTEAD, H.A.

TROTTER JR., W.T. [1984] see KIERSTEAD, H.A.

TROTTER JR., W.T. see Vol. V for further entries

TRUFFAULT, B. [1968] *Sur le probleme des mots pour les groupes de Greendlinger* (**J** 2313) C R Acad Sci, Paris, Ser A-B 267*A1-A3
◇ D40 ◇ REV MR 37 # 6359 Zbl 169.32 • ID 13690

TRUFFAULT, B. [1976] see COMERFORD JR., L.P.

TRUSS, J.K. [1978] *A note on increasing sequences of constructibility degrees* (**P** 1864) Higher Set Th;1977 Oberwolfach 473-476
◇ D30 E45 ◇ REV MR 80d:03055a Zbl 389 # 03021
• ID 52310

TRUSS, J.K. see Vol. III, V for further entries

TSEJTIN, G.S. [1956] *Associative calculus with unsolvable equivalence problem (Russian)* (**J** 0023) Dokl Akad Nauk SSSR 107*370-371
◇ D03 D40 ◇ REV MR 18.103 Zbl 70.9 JSL 22.219
• ID 01902

TSEJTIN, G.S. [1956] *On the problem of recognition of properties of associative calculi (Russian)* (**J** 0023) Dokl Akad Nauk SSSR 107*209-212
◇ D03 D40 ◇ REV MR 18.456 Zbl 70.9 JSL 22.219
• ID 01900

TSEJTIN, G.S. [1958] *Associative calculus with unsolvable equivalence problem (Russian)* (**S** 0066) Tr Mat Inst Steklov 52*172-189
• TRANSL [1970] (**J** 0225) Amer Math Soc, Transl, Ser 2 94*73-92
◇ D03 D40 ◇ REV MR 20 # 6358 Zbl 87.253 JSL 30.254
• ID 03595

TSEJTIN, G.S. [1959] *A simple example of an associative calculus with an unsolvable equivalence problem (Russian)* (**P** 0607) All-Union Math Conf (3);1956 Moskva 1*187-188
◇ D03 D40 ◇ ID 28242

TSEJTIN, G.S. [1959] *On the problem of recognizing properties of associative calculi (Russian)* (**P** 0607) All-Union Math Conf (3);1956 Moskva 1*189
◇ D03 D40 ◇ ID 28244

TSEJTIN, G.S. [1959] *Uniform recursiveness of algorithmic operators on general recursive functions and a canonical representation for constructive functions of a real argument (Russian)* (**P** 0607) All-Union Math Conf (3);1956 Moskva 1*188-189
◇ D20 F60 ◇ ID 28243

TSEJTIN, G.S. [1964] *A method of presenting the theory of algorithms and enumerable sets (Russian)* (**S** 0066) Tr Mat Inst Steklov 72*69-98
• TRANSL [1972] (**J** 0225) Amer Math Soc, Transl, Ser 2 99*1-39
◇ D20 D25 ◇ REV MR 34 # 5668 Zbl 173.11 JSL 35.478
• ID 22013

TSEJTIN, G.S. [1970] *On upper bounds of recursively enumerable sets of constructive real numbers (Russian)* (**S** 0066) Tr Mat Inst Steklov 113*102-172
• TRANSL [1970] (**S** 0055) Proc Steklov Inst Math 113*119-194
◇ D25 F60 ◇ REV MR 48 # 77 Zbl 229 # 02031 • ID 01909

TSEJTIN, G.S. [1971] *Lower estimate of the number of steps for an inverting normal algorithm and other similar algorithms (Russian) (English summary)* (**S** 0228) Zap Nauch Sem Leningrad Otd Mat Inst Steklov 20*243-262,289
• TRANSL [1973] (**J** 1531) J Sov Math 1*154-168
◇ D03 D15 ◇ REV MR 44 # 6498 Zbl 222 # 02025
• ID 26304

TSEJTIN, G.S. [1971] *Reduced form of normal algorithms and a linear acceleration theorem (Russian) (English summary)* (**S** 0228) Zap Nauch Sem Leningrad Otd Mat Inst Steklov 20*234-242,289
• TRANSL [1973] (**J** 1531) J Sov Math 1*148-153
◇ D03 D15 ◇ REV MR 44 # 6497 Zbl 222 # 02024
• ID 26303

TSEJTIN, G.S. [1974] see DANG HUY RUAN

TSEJTIN, G.S. see Vol. I, VI for further entries

TSEJTLIN, G.E. & YUSHCHENKO, E.L. [1974] *Ueber die Darstellung von Sprachen in Bobsleigh-Automaten (Russisch)* (**J** 0040) Kibernetika, Akad Nauk Ukr SSR 1974/6*40-51
◇ D05 ◇ REV Zbl 311 # 68038 • ID 60875

TSEJTLIN, G.E. [1975] see GLUSHKOV, V.M.

TSEJTLIN, G.E. [1977] *The system of algorithmic algebras and some control schemes in homogeneous structures (Russian)* (**S** 0507) Vychisl Sist (Akad Nauk SSSR Novosibirsk) 70*29-40
◇ D05 G25 ◇ REV MR 80b:68012 Zbl 409 # 68016
• ID 56352

TSEJTLIN, G.E. & YUSHCHENKO, E.L. [1978] *Theory of parametric models of languages and networks of parallel automata. Program transformations* (**P** 3247) Transform de Progr;1978 Paris 81-94
◇ D05 ◇ REV MR 80b:68097 Zbl 419 # 68099 • ID 33148

TSEJTLIN, G.E. see Vol. I for further entries

TSETLIN, M.L. [1961] *Certain problems in the behavior of finite automata* (**J** 0023) Dokl Akad Nauk SSSR 139*830-833
• TRANSL [1961] (**J** 0470) Sov Phys, Dokl 6*670-673
◇ D05 ◇ REV MR 24 # B480 Zbl 109.242 • ID 48041

TSICHRITZIS, D.C. [1969] *Fuzzy computability* (**P** 1129) Princeton Conf Inform Sci & Syst (3);1969 Princeton 157-161
◇ B52 D75 ◇ REV Zbl 286 # 68028 • ID 65798

TSICHRITZIS, D.C. [1971] *A note on comparison of subrecursive hierarchies* (**J** 0232) Inform Process Lett 1*42-44
◇ D20 ◇ REV MR 46 # 5123 • ID 13703

TSICHRITZIS, D.C. [1973] *A model for iterative computation* (**J** 0191) Inform Sci 5*187-197
◇ D20 ◇ REV MR 47 # 9886 Zbl 277 # 68019 • ID 13704

TSICHRITZIS, D.C. see Vol. II for further entries

TSINMAN, L.L. [1971] *Certain examples and theorems from the theory of recursive functions (Russian)* (**S** 0208) Uch Zap, Ped Inst, Moskva 277*218-222
◇ D20 D25 ◇ REV MR 46 # 34 • ID 71810

TSINMAN, L.L. see Vol. I, III, VI for further entries

TSUBOI, A. [1982] *On M-recursively saturated models of arithmetic* (**J** 2606) Tsukuba J Math 6*305-318
◇ C50 C57 C62 ◇ REV MR 85i:03110 Zbl 543 # 03018
• ID 37188

TSUBOI, A. [1983] *On M-recursively saturated models of Peano arithmetic (Japanese)* (**P** 4113) Found of Math;1982 Kyoto 158-177
◇ C50 C57 C62 ◇ REV Zbl 543 # 03018 • ID 47682

TSUBOI, A. see Vol. III, VI for further entries

TUCKER, C. [1968] *Limit of a sequence of functions with only countably many points of discontinuity* (**J** 0053) Proc Amer Math Soc 19*118-122
◇ D55 E75 ◇ REV MR 36 # 2112 Zbl 157.203 • ID 13705

TUCKER, J.V. [1972] *Algorithmic unsolvability in biological contexts* (**P** 2080) Conf Math Log;1970 London 348-350
◇ D75 D80 ◇ REV Zbl 236 # 02040 • ID 27863

TUCKER, J.V. [1979] see BERGSTRA, J.A.

TUCKER, J.V. [1980] see BERGSTRA, J.A.

TUCKER, J.V. [1980] *Computability and the algebra of fields: some affine constructions* (J 0036) J Symb Logic 45*103-120 ◇ B25 C57 C60 C65 D45 ◇ REV MR 82a:03027 Zbl 481 # 03020 • ID 79494

TUCKER, J.V. [1980] *Computing in algebraic systems* (P 3021) Logic Colloq;1979 Leeds 215-235 ◇ D10 D35 D75 D80 ◇ REV MR 82h:03047 Zbl 415 # 03035 • ID 53137

TUCKER, J.V. [1980] see STOLTENBERG-HANSEN, V.

TUCKER, J.V. [1980] see MOLDESTAD, J.

TUCKER, J.V. [1981] see BERGSTRA, J.A.

TUCKER, J.V. [1982] see BERGSTRA, J.A.

TUCKER, J.V. see Vol. I, VI for further entries

TUGUE, T. [1960] *On predicates expressible in the 1-function quantifier forms in Kleene hierarchy with free variables of type 2* (J 0081) Proc Japan Acad 36*10-14 ◇ D55 D65 ◇ REV MR 22 # 3683 Zbl 97.248 JSL 33.115 • ID 13707

TUGUE, T. [1960] *Predicates recursive in a type-2 object and Kleene hierarchies* (J 0407) Comm Math Univ St Pauli (Tokyo) 8*97-117 ◇ D55 D65 ◇ REV MR 22 # 1515 Zbl 97.247 JSL 33.115 • ID 13708

TUGUE, T. [1963] *On the partial recursive functions of ordinal numbers* (P 1127) Symp Founds of Math;1962 Katada 1-49 ◇ D60 ◇ REV MR 29 # 4677 Zbl 147.252 • ID 22115

TUGUE, T. [1964] *On the partial recursive functions of ordinal numbers* (J 0090) J Math Soc Japan 16*1-31 ◇ D60 ◇ REV MR 29 # 4677 Zbl 147.252 • ID 13709

TUGUE, T. [1966] see TANAKA, H.

TUGUE, T. [1969] see HINATA, S.

TUGUE, T. [1979] *Generalized recursion theory (Japanese)* (P 4108) Found of Math;1979 Kyoto 51-74 ◇ D60 D65 D75 ◇ ID 47658

TUGUE, T. see Vol. I, V for further entries

TULIPANI, S. [1982] *A use of the method of interpretations for decidability or undecidability of measure spaces* (J 0004) Algeb Universalis 15*228-232 ◇ B25 C65 D35 ◇ REV MR 84i:03087 Zbl 518 # 03003 • ID 34564

TULIPANI, S. [1985] *An algorithm to determine for any prime p, a polynomial-sized Horn sentence which expresses "the cardinality is not p"* (J 0036) J Symb Logic 50*1062-1064 ◇ B35 D15 ◇ ID 49361

TULIPANI, S. see Vol. I, III, V for further entries

TUNG, I.I. [1978] see EDMUNDSON, H.P.

TURAKAINEN, P. [1978] *On characterization of recursively enumerable languages in terms of linear languages and VW-grammars* (J 0028) Indag Math 40*145-153 ◇ D05 D25 ◇ REV MR 57 # 11200 Zbl 369 # 68049 • ID 29188

TURAKAINEN, P. [1981] *On nonstochastic languages and homomorphic images of stochastic languages* (J 0191) Inform Sci 24*229-253 ◇ D05 ◇ REV MR 82i:68053 Zbl 483 # 68069 • ID 36813

TURAKAINEN, P. [1982] *A homomorphic characterisation of principal semiAFLs without using intersection with regular sets* (J 0191) Inform Sci 27*141-149 ◇ D05 ◇ REV MR 84b:68104 Zbl 506 # 68063 • ID 38961

TURAKAINEN, P. [1982] *Rational stochastic automata in formal language theory* (P 3787) Discr Math;1977 Warsaw 31-44 ◇ D05 ◇ REV MR 84j:68053 Zbl 541 # 68054 • ID 39193

TURAN, G. [1981] *On cellular graph-automata and second-order definable graph-properties* (P 3165) FCT'81 Fund of Comput Th;1981 Szeged 384-393 ◇ C85 D05 D80 ◇ REV MR 83h:68076 Zbl 479 # 68059 • ID 55701

TURAN, G. see Vol. III for further entries

TURAN, P. [1971] *On the work of Alan Baker* (P 0743) Int Congr Math (II,11,Proc);1970 Nice 1*3-5 ◇ B25 D35 ◇ REV MR 54 # 2394 JSL 37.606 • ID 13720

TURASHVILI, T.V. [1973] see GUREVICH, Y.

TURASHVILI, T.V. [1974] *A reduction of the decidability problem of first order predicate logic to a class with asymetric and irreflexive two-place predicates (Russian) (Georgian and English summaries)* (J 0233) Soobshch Akad Nauk Gruz SSR 75*297-300 ◇ D35 ◇ REV MR 50 # 6821 Zbl 329 # 02021 • ID 13721

TURASHVILI, T.V. [1975] *The decidability problem of first order predicate logic (Russian)* (J 1477) Tr Vychisl Tsentr, Akad Nauk Gruz SSR 15*146-157 ◇ B20 D35 ◇ REV MR 54 # 2415 • ID 24005

TURASHVILI, T.V. [1977] *On the undecidable minimal classes of first order predicate logic (Russian) (English summary)* (J 0233) Soobshch Akad Nauk Gruz SSR 86*589-591 ◇ B20 D35 ◇ REV MR 56 # 15361 Zbl 362 # 02009 • ID 50731

TURCAT, C. & VERDILLON, A. [1976] *Recursion and testing of combinational circuits* (J 0187) IEEE Trans Comp C-25*652-654 ◇ B70 D20 ◇ REV MR 54 # 9884 Zbl 333 # 94014 • ID 65810

TURCHIN, V.F. [1972] *Equivalent transformations of recursive functions that are described in the language REFAL (Russian)* (P 3049) Teor Yazykov & Metody Postroe Sist Progr 31-42 ◇ B75 D20 ◇ REV MR 50 # 9045 • ID 83019

TURING, A.M. [1936] *On computable numbers, with an application to the "Entscheidungsproblem"* (J 1910) Proc London Math Soc, Ser 2 42*230-265 • ERR/ADD ibid 43*544-546 ◇ D05 D10 D20 D35 F60 ◇ REV Zbl 16.97 JSL 2.42 FdM 62.1059 • ID 13723

TURING, A.M. [1937] *Computability and λ-definability* (J 0036) J Symb Logic 2*153-163 ◇ A05 B40 D05 D10 D20 F99 ◇ REV Zbl 18.193 JSL 3.89 FdM 63.824 • ID 13725

TURING, A.M. [1939] *Systems of logic based on ordinals* (J 1910) Proc London Math Soc, Ser 2 45*161-228 ◇ B40 D25 D30 F15 F30 ◇ REV Zbl 21.97 JSL 4.128 FdM 65.1102 • ID 13726

TURING, A.M. [1950] *The word problem in semi-groups with cancellation* (J 0120) Ann of Math, Ser 2 52*491-505 ◇ D40 ◇ REV MR 12.239 Zbl 37.301 JSL 17.74 • ID 13731

TURING, A.M. [1954] *Solvable and unsolvable problems* (J 4681) Sci News 31*7-23
◇ B25 D10 D35 ◇ REV JSL 20.74 • ID 42433

TURING, A.M. [1964] *Computing machinery and intelligence* (C 0618) Minds & Machines 4-30
◇ A05 B75 D10 ◇ ID 13732

TURING, A.M. see Vol. I, VI for further entries

TURQUETTE, A.R. [1950] *Goedel and the synthetic a priori* (J 0301) J Phil 47*125-129
◇ A05 D35 ◇ REV JSL 15.221 • ID 30571

TURQUETTE, A.R. [1968] *Dualizable quasi-strokes for m-state automata* (J 0191) Inform Sci 1*131-142
◇ B50 D05 ◇ REV MR 42 # 46 • ID 13749

TURQUETTE, A.R. see Vol. I, II for further entries

TURZIK, D. [1982] see POLJAK, S.

TUSCHIK, H.-P. [1982] see SEESE, D.G.

TUSCHIK, H.-P. see Vol. I, II, III, V for further entries

TUZOV, V.A. [1970] *Graph-schemes with associative calculus (Russian)* (J 0199) Zh Vychisl Mat i Mat Fiz 10*132-145
• TRANSL [1970] (J 1049) USSR Comput Math & Math Phys 10/1*172-190
◇ B70 D05 ◇ REV MR 43 # 1780 Zbl 212.25 • ID 27857

TUZOV, V.A. [1970] *The equivalence of logical schemes with permutation operators (Russian) (English summary)* (J 0040) Kibernetika, Akad Nauk Ukr SSR 1970/6*33-38
◇ B75 D05 ◇ REV MR 45 # 1684 Zbl 231 # 02042
• ID 27440

TUZOV, V.A. [1971] *Optimal schemes of algorithms (Russian)* (J 0199) Zh Vychisl Mat i Mat Fiz 11*1282-1289
• TRANSL [1971] (J 1049) USSR Comput Math & Math Phys 11/5*232-241
◇ D15 D20 ◇ REV MR 45 # 2952 Zbl 242 # 02039
• ID 26353

TVERBERG, H. [1984] *On Schmerl's effective version of Brooks' theorem* (J 0033) J Comb Th, Ser B 37*27-30
◇ D80 ◇ REV MR 86b:03057 • ID 44143

TVERBERG, H. see Vol. V for further entries

TVERDOKHLEBOV, V.A. [1978] see BOGOMOLOV, A.M.

TVERSKOJ, A.A. [1982] *Investigation of recursiveness and arithmeticity of signature functions in nonstandard models of arithmetics (Russian)* (J 0023) Dokl Akad Nauk SSSR 262*1325-1328
• TRANSL [1982] (J 0062) Sov Math, Dokl 25*249-253
◇ C57 C62 D45 H15 ◇ REV MR 83k:03079 Zbl 496 # 03047 • ID 35413

TVERSKOJ, A.A. [1984] *Constructivizability of formal arithmetical structures (Russian)* (C 4091) Algeb & Diskret Mat (Riga) 134-136
◇ C57 D45 H15 ◇ REV MR 85j:03053 • ID 44226

TVERSKOJ, A.A. [1985] *Constructivizable and nonconstructivizable formal arithmetic structures (Russian)* (J 0067) Usp Mat Nauk 40/6*159-160
◇ C57 C62 D45 H15 ◇ ID 49425

TVERSKOJ, A.A. see Vol. III, V, VI for further entries

TYSHKEVICH, R.I. [1971] see FEJNBERG, V.Z.

TYSHKEVICH, R.I. see Vol. III for further entries

TYUGU, E.KH. [1982] see MINTS, G.E.

TYUPA, V.G. [1961] see SLYAKHOVA, N.I.

TZSCHACH, H. & WALDSCHMIDT, H. & WALTER, H.K.-G. (EDS.) [1977] *Theoretical computer science* (X 0811) Springer: Heidelberg & New York VII+418pp
◇ D97 ◇ REV MR 57 # 11148 Zbl 348 # 00019 • ID 48716

UESU, T. [1978] *A system of graph grammars which generates all recursively enumerable sets of labelled graphs* (J 2606) Tsukuba J Math 2*11-26
◇ D25 D80 ◇ REV MR 80h:68067 Zbl 458 # 68027
• ID 54437

UESU, T. [1979] *A complete system of grammars for plane graphs* (J 2606) Tsukuba J Math 3*129-160
◇ D05 ◇ REV MR 81d:68110 Zbl 425 # 05022 • ID 37189

UESU, T. see Vol. I, III, V, VI for further entries

UHLIG, D. [1982] see FISCHER, R.

UHLIG, D. see Vol. I for further entries

UKKONEN, E. [1982] *Structure preserving elimination of null productions from context-free grammars* (J 1426) Theor Comput Sci 17*43-54
◇ D05 ◇ REV MR 82m:68126 Zbl 479 # 68072 • ID 69003

UKKONEN, E. [1982] *The equivalence problem for some non-real-time deterministic pushdown automata* (J 0037) ACM J 29*1166-1181
◇ D10 ◇ REV MR 83j:68058 Zbl 489 # 68075 • ID 39907

UKKONEN, E. [1983] *Two results on polynomial time truth-table reductions to sparse sets* (J 1428) SIAM J Comp 12*580-587
◇ D15 ◇ REV MR 85g:68023 Zbl 532 # 68051 • ID 36822

ULANOV, G.M. [1979] see PETROV, B.N.

ULLIAN, J.S. [1960] *Splinters of recursive functions* (J 0036) J Symb Logic 25*33-38
◇ D25 D30 ◇ REV MR 24 # A1216 Zbl 112.8 JSL 31.138
• ID 13771

ULLIAN, J.S. [1961] *A theorem on maximal sets* (J 0047) Notre Dame J Formal Log 2*222-223
◇ D25 ◇ REV MR 25 # 3823 Zbl 154.257 JSL 27.244
• ID 13772

ULLIAN, J.S. [1963] see ROSE, G.F.

ULLIAN, J.S. [1965] see GINSBURG, S.

ULLIAN, J.S. [1966] see GINSBURG, S.

ULLIAN, J.S. [1966] *Failure of a conjecture about context free languages* (J 0194) Inform & Control 9*61-65
◇ D05 ◇ REV MR 32 # 9174 Zbl 155.342 Zbl 242 # 68047 JSL 32.266 • ID 13773

ULLIAN, J.S. [1966] see HIBBARD, T.N.

ULLIAN, J.S. [1967] *Partial algorithm problems for context free languages* (J 0194) Inform & Control 11*80-101
◇ D05 ◇ REV MR 37 # 1205 Zbl 155.342 JSL 37.196
• ID 13774

ULLIAN, J.S. [1971] *Three theorems concerning principal AFLs* (J 0119) J Comp Syst Sci 5*304-314
◇ D05 ◇ REV MR 43 # 5761 Zbl 217.227 • ID 30568

ULLIAN, J.S. see Vol. II for further entries

ULLMAN, J.D. [1967] see HOPCROFT, J.E.

ULLMAN, J.D. [1968] see HOPCROFT, J.E.

ULLMAN, J.D. [1968] see AHO, A.V.

ULLMAN, J.D. [1969] see HOPCROFT, J.E.

ULLMAN, J.D. [1969] *Halting stack automata* (J 0037) ACM J 16*550-563
◇ D10 ◇ REV MR 40 # 1207 • ID 13776

ULLMAN, J.D. [1969] see AHO, A.V.

ULLMAN, J.D. [1974] see HOPCROFT, J.E.

ULLMAN, J.D. [1975] see SZYMANSKI, T.G.

ULLMAN, J.D. [1975] *NP-complete scheduling problems* (J 0119) J Comp Syst Sci 10*384-393
◇ D15 ◇ REV MR 52 # 12406 Zbl 313 # 68054 • ID 65831

ULLMAN, J.D. [1979] see HOPCROFT, J.E.

ULLMAN, J.D. see Vol. V for further entries

UL'YANOV, S.V. [1979] see PETROV, B.N.

UMEO, H. [1977] see MORITA, K.

UMEO, H. [1982] see MORITA, K.

UMEZAWA, T. [1967] *On uniform transition of states in a finite automaton* (J 1005) Rep Fac Sci, Shizuoka Univ 2*16-22
◇ D05 ◇ ID 31274

UMEZAWA, T. see Vol. I, II, III, V, VI for further entries

UMIRBAEV, U.U. [1984] *Equality problem for center-by-metabelian Lie-algebras (Russian)* (J 0003) Algebra i Logika 23*305-318
• TRANSL [1984] (J 0069) Algeb and Log 23*209-219
◇ B25 C60 D40 ◇ REV MR 86g:17009 • ID 42699

URBAN, J. [1966] *Die Minimalisierung der zur sequentiellen Berechnung der partiell-rekursiven Wortfunktionen notwendigen Kellerspeicher* (J 0001) Acta Math Acad Sci Hung 17*335-358
◇ D10 D20 ◇ REV MR 35 # 1476 Zbl 192.66 • ID 13788

URBANO, R.H. [1963] *Boolean matrices and the stability of neural nets* (P 0572) Inform Processing (2);1962 Muenchen 755-757
◇ B05 D05 ◇ REV Zbl 121.344 JSL 35.348 • ID 13791

URPONEN, T. [1971] *On axiom systems for regular expressions and on equations involving languages* (J 0498) Ann Univ Turku, Ser A I 145*52
◇ D05 ◇ REV MR 44 # 6419 Zbl 252 # 02039 • ID 27754

URPONEN, T. [1972] *On regular expressions possessing the empty word property* (J 0498) Ann Univ Turku, Ser A I 154*12pp
◇ D05 ◇ REV MR 47 # 4774 Zbl 247 # 02035 • ID 13792

URPONEN, T. [1972] *On regular expressions over one letter and on commutative languages* (J 3994) Ann Acad Sci Fennicae Ser A I 517*24pp
◇ D05 ◇ REV MR 49 # 6696 Zbl 252 # 02040 • ID 27755

URQUHART, A.I.F. [1981] *Decidability and the finite model property* (J 0122) J Philos Logic 10*367-370
◇ B22 B25 D35 ◇ REV MR 83i:03038 Zbl 465 # 03005 • ID 54908

URQUHART, A.I.F. [1981] *The decision problem for equational theories* (J 1447) Houston J Math 7*587-589
◇ B25 C05 D35 ◇ REV MR 83h:03063 Zbl 496 # 03004 • ID 36071

URQUHART, A.I.F. [1984] *The undecidability of entailment and relevant implication* (J 0036) J Symb Logic 49*1059-1073
◇ B46 D35 G10 ◇ ID 40056

URQUHART, A.I.F. see Vol. I, II, VI for further entries

URSIC, S. [1984] *A linear characterization of NP-complete problems* (P 2633) Autom Deduct (7);1984 Napa 80-100
◇ D15 ◇ REV MR 86g:68062 Zbl 548 # 90049 • ID 44461

URSINI, A. [1985] *Decision problems for classes of diagonalizable algebras* (J 0063) Studia Logica 44*87-90
◇ B25 B45 C05 D35 ◇ ID 47525

URSINI, A. see Vol. I, II, III, VI for further entries

URZYCZYN, P. [1981] *Algorithmic triviality of abstract structures* (J 2095) Fund Inform, Ann Soc Math Pol, Ser 4 4*819-849
◇ D75 ◇ REV MR 83j:03075 Zbl 501 # 03027 • ID 35371

URZYCZYN, P. [1981] *The unwind property in certain algebras* (J 0194) Inform & Control 50*91-109
◇ C05 C57 ◇ REV MR 84c:08005 Zbl 491 # 03018 • ID 36852

URZYCZYN, P. [1985] see KFOURY, A.J.

URZYCZYN, P. see Vol. II, III for further entries

USPENSKIJ, V.A. [1953] *The Goedel theorem and the theory of algorithms (Russian)* (J 0067) Usp Mat Nauk 8/4(56)*176-178
◇ D20 D35 F30 ◇ REV Zbl 51.245 • ID 45765

USPENSKIJ, V.A. [1953] *Theorem of Goedel and theory of algorithms (Russian)* (J 0023) Dokl Akad Nauk SSSR 91*737-740
◇ D20 D25 F30 ◇ REV MR 17.4 Zbl 52.251 JSL 19.218 • ID 16353

USPENSKIJ, V.A. [1955] *On calculable operations (Russian)* (J 0023) Dokl Akad Nauk SSSR 103*773-776
◇ D20 ◇ REV MR 18.369 Zbl 65.2 JSL 22.76 • ID 16351

USPENSKIJ, V.A. [1955] *Systems of enumerable sets and their enumeration (Russian)* (J 0023) Dokl Akad Nauk SSSR 105*1155-1158
◇ D25 D45 ◇ REV Zbl 67.3 JSL 22.220 • ID 16352

USPENSKIJ, V.A. [1956] *Calculable operations and the notion of a program (Russian)* (J 0067) Usp Mat Nauk 11/4*172-176
◇ D20 D45 ◇ REV Zbl 70.10 JSL 23.49 • ID 16349

USPENSKIJ, V.A. [1957] *Some notes on recursively enumerable sets (Russian) (English summary)* (J 0068) Z Math Logik Grundlagen Math 3*157-170
• TRANSL [1963] (J 0225) Amer Math Soc, Transl, Ser 2 23*89-101
◇ D25 ◇ REV MR 19.1032 Zbl 121.15 Zbl 80.242 JSL 23.49 • ID 16348

USPENSKIJ, V.A. [1958] see KOLMOGOROV, A.N.

USPENSKIJ, V.A. [1959] *Computable operations, computable operators and constructively continuous functions (Russian)* (P 0607) All-Union Math Conf (3);1956 Moskva 1*185
◇ D20 F60 ◇ ID 28239

USPENSKIJ, V.A. [1959] *On algorithmic reducibility (Russian)* (P 0607) All-Union Math Conf (3);1956 Moskva 2*66-69
◇ D30 ◇ REV JSL 23.225 • ID 16350

USPENSKIJ, V.A. [1959] *The concept of program and computable operators (Russian)* (**P** 0607) All-Union Math Conf (3);1956 Moskva 1*186
◇ D20 D45 ◇ ID 28240

USPENSKIJ, V.A. [1960] *Lectures on computable functions (Russian)* (**X** 3709) Izdat Fiz-Mat Lit: Moskva 492pp
• TRANSL [1966] (**X** 0740) Kexue Jishu Chubanshe: Shanghai 412pp [1966] (**X** 0859) Hermann: Paris 416pp
◇ D20 D98 ◇ REV MR 22 # 12043 MR 33 # 2542
• ID 33627

USPENSKIJ, V.A. [1969] *The reducibility of computable and potentially computable numerations (Russian)* (**J** 0087) Mat Zametki (Akad Nauk SSSR) 6*3-9
• TRANSL [1969] (**J** 1044) Math Notes, Acad Sci USSR 6*461-464
◇ D45 ◇ REV MR 41 # 58 Zbl 184.19 • ID 13804

USPENSKIJ, V.A. [1981] see SEMENOV, A.L.

USPENSKIJ, V.A. [1983] *Post's machine (Russian)* (**X** 0885) Mir: Moskva 88pp
• TRANSL [1985] (**X** 0885) Mir: Moskva 83pp (Portuguese)
◇ D20 ◇ ID 45220

USPENSKIJ, V.A. [1985] *The contribution of N.N.Luzin to the descriptive theory of sets and functions: concepts, problems, predictions (Russian)* (**J** 0067) Usp Mat Nauk 40/3*85-116,240
• TRANSL [1985] (**J** 1399) Russ Math Surv 40/3*97-134
◇ A10 D55 E15 ◇ ID 47706

USPENSKIJ, V.A. see Vol. I, V, VI for further entries

USTYAN, A.E. [1970] *Examples of semigroups with an unsolvable word problem (Russian)* (**J** 0789) Uch Zap Mat Ped Inst, Tula 1970*267-275
◇ D40 ◇ REV MR 52 # 14099 • ID 21884

USTYAN, A.E. [1970] *On the isomorphism problem for finitely presented semigroups (Russian)* (**J** 0789) Uch Zap Mat Ped Inst, Tula 2*276-289
◇ D40 ◇ REV MR 52 # 14100 • ID 21885

USTYAN, A.E. [1972] *On the word problem for finitely-generated semigroups (Russian)* (**J** 0092) Sib Mat Zh 13*198-210
• TRANSL [1972] (**J** 0475) Sib Math J 13*141-150
◇ D40 ◇ REV MR 45 # 6957 Zbl 244 # 20073 • ID 13805

USTYAN, A.E. [1972] *The connection between the occurence problem and the word problem in semigroups (Russian)* (**C** 1435) Vopr Teor Grupp & Polugrupp 106-108
◇ D40 ◇ REV MR 52 # 14101 • ID 21886

UZGALIS, R.C. [1977] see CLEAVELAND, J.C.

VAEAENAENEN, J. [1979] *A new incompleteness in arithmetic (Finnish) (English summary)* (**J** 1108) Arkhimedes (Helsinki) 31*30-37
◇ A10 C62 D25 E05 F30 ◇ REV MR 80c:03058 Zbl 398 # 03021 • ID 52749

VAEAENAENEN, J. see Vol. I, III, V for further entries

VAISER, A. [1975] *Complexity of computation and stability of separation of languages by finite probabilistic automata* (**C** 3264) Sist Upravleniya 172-181
◇ D05 D15 ◇ REV Zbl 346 # 94025 • ID 65859

VAISER, A. [1975] *Remarks on time and space functions of probabilistic machines* (**C** 3264) Sist Upravleniya 182-196
◇ D10 D15 ◇ REV Zbl 346 # 94026 • ID 65857

VAISER, A. [1976] *Stochastic languages and the complexity of computations on probabilistic Turing machines (Russian) (English summary)* (**J** 0040) Kibernetika, Akad Nauk Ukr SSR 1976/1*21-25
• TRANSL [1976] (**J** 0021) Cybernetics 12*22-27
◇ D10 D15 ◇ REV MR 58 # 19398 Zbl 337 # 94037
• ID 65858

VAISHNAVI, V.K. [1979] see BISWAS, S.

VAJNBERG, YU.R. [1974] see NOGINA, E.YU.

VAJNBERG, YU.R. [1976] see NOGINA, E.YU.

VAJNBERG, YU.R. see Vol. VI for further entries

VALDERRAMA, E. [1978] see AGUILO, J.

VALIANT, L.G. [1973] see PATERSON, M.S.

VALIANT, L.G. [1974] *The decidability of equivalence for deterministic finite-turn pushdown automata JA JC* (**P** 1464) ACM Symp Th of Comput (6);1974 Seattle 27-32
◇ D10 ◇ REV MR 53 # 15002 Zbl 358 # 68090 • ID 23230

VALIANT, L.G. [1974] *The equivalence problem for deterministic finite-turn pushdown automata* (**J** 0194) Inform & Control 25*123-133
◇ D10 ◇ REV MR 52 # 12412 Zbl 285 # 68025 • ID 21741

VALIANT, L.G. [1975] see PATERSON, M.S.

VALIANT, L.G. [1977] see HOPCROFT, J.E.

VALIANT, L.G. [1979] *Completeness classes in algebra* (**P** 3542) ACM Symp Th of Comput (11);1979 Atlanta 249-261
◇ D15 D75 ◇ REV MR 83e:68046 • ID 40155

VALIANT, L.G. [1979] *Negative results on counting* (**P** 3488) Theor Comput Sci (4);1979 Aachen 38-46
◇ D15 ◇ REV MR 81j:68056 • ID 83027

VALIANT, L.G. [1979] *The complexity of enumeration and reliability problems* (**J** 1428) SIAM J Comp 8*410-421
◇ D15 ◇ REV MR 80f:68055 Zbl 419 # 68082 • ID 53397

VALIANT, L.G. [1981] see SKYUM, S.

VALIANT, L.G. [1982] *Reducibility by algebraic projections* (**J** 3370) Enseign Math, Ser 2 28*253-268
• REPR [1982] (**P** 3482) Logic & Algor (Specker);1980 Zuerich 365-380
◇ D15 ◇ REV MR 84d:68046b Zbl 474 # 68062 • ID 38440

VALIANT, L.G. [1984] *An algebraic approach to computational complexity* (**P** 4313) Int Congr Math (II,14);1983 Warsaw 2*1637-1643
◇ D15 ◇ ID 48604

VALIDOV, F.I. [1978] *Minimal numerations of families of general recursive functions (Russian)* (**J** 0031) Izv Vyssh Ucheb Zaved, Mat (Kazan) 1978/9(196)*13-24
• TRANSL [1978] (**J** 3449) Sov Math 22/9*10-18
◇ D20 D45 ◇ REV MR 80f:03046 Zbl 398 # 03032
• ID 52760

VALIDOV, F.I. [1984] *Recursively enumerable sets and discrete families of general recursive functions (Russian)* (**J** 0031) Izv Vyssh Ucheb Zaved, Mat (Kazan) 1984/4*6-11
• TRANSL [1984] (**J** 3449) Sov Math 28/4*6-11
◇ D25 ◇ REV MR 85j:03067 Zbl 556 # 03037 • ID 45842

VALIEV, M.K. [1968] *A theorem of G.Higman (Russian)* (**J** 0003) Algebra i Logika 7/3*9-22
• TRANSL [1968] (**J** 0069) Algeb and Log 7*135-143
◇ C60 D35 D40 ◇ REV MR 41 # 1849 Zbl 209.331
• ID 13814

VALIEV, M.K. [1969] *The complexity of the word problem for finitely presented groups (Russian)* (**J** 0003) Algebra i Logika 8*5-43
• TRANSL [1969] (**J** 0069) Algeb and Log 8*2-21
◇ D20 D40 ◇ REV MR 46 # 5460 Zbl 193.317 JSL 37.607
• ID 13815

VALIEV, M.K. [1970] *Certain estimates of the computing time on Turing machines with input (Russian) (English summary)* (**J** 0040) Kibernetika, Akad Nauk Ukr SSR 1970/6*26-32
◇ D10 D15 ◇ REV MR 46 # 10243 Zbl 257 # 02028
• ID 28998

VALIEV, M.K. [1975] *On polynomial reducibility of the word problem under embedding of recursively presented groups in finitely presented groups* (**P** 0454) Math Founds of Comput Sci (4);1975 Marianske Lazne 32*432-438
◇ D15 D40 ◇ REV MR 54 # 413 Zbl 323 # 02055
• ID 23991

VALIEV, M.K. [1977] *Real time computations with restrictions on tape alphabet* (**P** 1635) Math Founds of Comput Sci (6);1977 Tatranska Lomnica 532-536
◇ D10 D15 ◇ REV MR 57 # 8157 Zbl 365 # 02023
• ID 51024

VALIEV, M.K. [1980] *Decision complexity of variants of propositional dynamic logic* (**P** 3210) Math Founds of Comput Sci (9);1980 Rydzyna 656-664
◇ B25 B75 D15 ◇ REV MR 81k:68003 Zbl 451 # 03003
• ID 54018

VALIEV, M.K. see Vol. II for further entries

VALK, R. [1983] *Infinite behaviour of Petri nets* (**J** 1426) Theor Comput Sci 25*311-341
◇ D05 ◇ REV MR 85e:68079 Zbl 559 # 68057 • ID 40042

VANDERVEKEN, D.R. [1976] *The Lesniewski-Curry theory of syntactical categories and the categorially open functors* (**J** 0063) Studia Logica 35*191-201
◇ B15 D05 G30 ◇ REV MR 56 # 5233 Zbl 356 # 02013
• ID 50285

VANDERVEKEN, D.R. see Vol. II, VI for further entries

VAQUERO, A. [1978] see AGUILO, J.

VARAIYA, P.P. [1971] see BURKHARD, W.A.

VARDI, M.Y. [1981] *The decision problem for database dependencies* (**J** 0232) Inform Process Lett 12*251-254
◇ B75 D80 ◇ REV MR 84a:03047 Zbl 482 # 68094
• ID 35580

VARDI, M.Y. [1985] see SISTLA, A.P.

VARGA, Z. [1980] see FOTHI, A.

VARPAKHOVSKIJ, F.I. [1971] see KOLMOGOROV, A.N.

VARTAPETOV, E.A. [1968] *Coding of abstract automata by normal codes I (Russian) (English summary)* (**J** 0040) Kibernetika, Akad Nauk Ukr SSR 1968/6*7-15
• TRANSL [1968] (**J** 0021) Cybernetics 4/6*7-15
◇ D05 ◇ REV MR 45 # 8393 Zbl 217.589 JSL 35.348
• REM Part II 1969 • ID 13831

VARTAPETOV, E.A. [1969] *Coding of abstract automata by normal codes II (Russian) (English summary)* (**J** 0040) Kibernetika, Akad Nauk Ukr SSR 1969/4*21-28
• TRANSL [1969] (**J** 0021) Cybernetics 5*381-389
◇ D05 ◇ REV MR 45 # 8394 Zbl 217.589 • REM Part I 1968
• ID 46713

VASIL'EV, EH.S. [1973] *The elementary theories of complete torsion-free abelian groups with p-adic topology (Russian)* (**J** 0087) Mat Zametki (Akad Nauk SSSR) 14*201-208
• TRANSL [1973] (**J** 1044) Math Notes, Acad Sci USSR 14*673-677
◇ B25 C20 C60 D35 ◇ REV MR 49 # 2329
Zbl 306 # 02053 • ID 13840

VASIL'EV, EH.S. see Vol. III for further entries

VASILEVSKIJ, M.P. [1973] *The detection of the disrepair of automata (Russian) (English summary)* (**J** 0040) Kibernetika, Akad Nauk Ukr SSR 1973/4*98-108
◇ D05 ◇ REV MR 48 # 10722 Zbl 268 # 94047 • ID 65895

VASYUKEVICH, V.O. [1976] see CHAPENKO, V.P.

VASYUKEVICH, V.O. [1977] *Ueber die Formalisierung der Loesung von Problemen der statischen Analyse diskreter Automaten mit moeglichen Funktionsstoerungen (Russian)* (**J** 0474) Avtom Vychis Tekh, Akad Nauk Latv SSR 1977/1*12-16
◇ D05 ◇ REV Zbl 348 # 94038 • ID 65896

VAUGHT, R.L. [1957] *Sentences true in all constructive models* (**P** 1675) Summer Inst Symb Log;1957 Ithaca 341-343
◇ C13 C57 D35 ◇ REV MR 24 # A40 Zbl 108.8
JSL 31.132 • REM See also 1960 • ID 24868

VAUGHT, R.L. [1960] *Sentences true in all constructive models* (**J** 0036) J Symb Logic 25*39-53
◇ C13 C57 D35 ◇ REV MR 24 # A40 Zbl 108.8
JSL 31.132 • REM See also 1957 • ID 24869

VAUGHT, R.L. [1962] *On a theorem of Cobham concerning undecidable theories* (**P** 0612) Int Congr Log, Meth & Phil of Sci (1,Proc);1960 Stanford 14-25
◇ D25 D35 ◇ REV MR 28 # 32 Zbl 178.323 JSL 34.126
• ID 13853

VAUGHT, R.L. [1973] *A Borel invariantization* (**J** 0015) Bull Amer Math Soc 79*1292-1295
◇ C75 D55 E15 ◇ REV MR 48 # 10818 Zbl 296 # 54037
• ID 13909

VAUGHT, R.L. [1973] *Descriptive set theory in* $L_{\omega_1,\omega}$ (**P** 0713) Cambridge Summer School Math Log;1971 Cambridge GB 574-598
◇ C15 C70 C75 D55 D70 E15 ◇ REV MR 53 # 12868
Zbl 308 # 02054 JSL 47.217 • ID 13864

VAUGHT, R.L. [1974] *Invariant sets in topology and logic* (**J** 0027) Fund Math 82*269-294
◇ C75 D55 D70 E15 ◇ REV MR 51 # 167
Zbl 309 # 02068 • ID 17478

VAUGHT, R.L. see Vol. I, II, III, V for further entries

VAUZEILLES, J. [1984] see GIRARD, J.-Y.

VAUZEILLES, J. see Vol. V, VI for further entries

VAVASIS, S.A. [1982] see LIEBERHERR, K.J.

VAZHENIN, YU.M. [1974] *On the elementary theory of free inverse semigroups* (**J** 0136) Semigroup Forum 9*189-195
◇ B25 C60 D35 ◇ REV MR 55 # 2547 Zbl 299 # 20050
• ID 79687

VAZHENIN, YU.M. [1974] *The elementary theories of symmetric groups and semigroups (Russian)* (**J** 0031) Izv Vyssh Ucheb Zaved, Mat (Kazan) 1974/1(140)*15-20
◇ D35 ◇ REV MR 50 # 4271 Zbl 297 # 20049 • ID 13914

VAZHENIN, YU.M. [1981] see ROZENBLAT, B.V.

VAZHENIN, YU.M. [1981] *Sur la liaison entre problemes combinatoires et algorithmiques* (**J** 1426) Theor Comput Sci 16*33-41
◇ B25 D05 D15 D35 ◇ REV MR 83g:03042
Zbl 469 # 03004 • ID 55132

VAZHENIN, YU.M. see Vol. I, III, V for further entries

VAZIRANI, U.V. & VAZIRANI, V.V. [1983] *A natural encoding scheme proved probabilistic polynomial complete* (**J** 1426) Theor Comput Sci 24*291-300
◇ D15 ◇ REV MR 85e:03095 Zbl 525 # 68025 • ID 40692

VAZIRANI, V.V. [1983] see VAZIRANI, U.V.

VELIKANOV, K.M. [1974] *Specialization of the form of deduction in the extended calculus of recursive functions (Russian)* (**C** 2577) Issl Formaliz Yazyk & Neklass Log 82-87
◇ D20 ◇ REV MR 57 # 16029 • ID 79694

VELINOV, YU.P. [1982] *Polycategories and recursiveness* (**P** 2185) Symp on n-ary Structures;1982 Skopje 47-56
◇ D20 ◇ REV MR 85i:20005 Zbl 561 # 18002 • ID 45221

VELINOV, YU.P. see Vol. V for further entries

VELOSO, P.A.S. [1977] *Some bounds on quasi-initialised finite automata* (**P** 3182) Int Comput Symp;1977 Liege 389-393
◇ D05 ◇ REV MR 57 # 15793 Zbl 391 # 68030 • ID 52364

VELOSO, P.A.S. see Vol. II for further entries

VENKATARAMAN, K.N. [1983] see HAWRUSIK, F.M.

VENTURINI ZILLI, M. [1974] *On different kinds of indefinite* (**J** 0089) Calcolo 11*67-77
◇ B40 D75 ◇ REV MR 54 # 70 Zbl 294 # 02016 • ID 23952

VENTURINI ZILLI, M. [1975] *A model with nondeterministic computation* (**P** 1603) λ-Calc & Comput Sci Th;1975 Roma 287-296
◇ D75 ◇ REV MR 57 # 14594 Zbl 342 # 02032 • ID 65907

VENTURINI ZILLI, M. see Vol. I, VI for further entries

VERBEEK, R. [1973] *Erweiterungen subrekursiver Programmiersprachen* (**P** 1630) GI Fachtag Automatenth & Form Sprach (1);1973 Bonn 311-318
◇ B75 D15 D20 ◇ REV MR 55 # 4768 Zbl 279 # 68019
• ID 65908

VERBEEK, R. & WEIHRAUCH, K. [1977] *Data presentation and computational complexity* (**P** 3238) Conf Theoret Comput Sci;1977 Waterloo ON 111-119
◇ D05 D15 ◇ REV MR 81b:68059 Zbl 426 # 03040
• ID 53636

VERBEEK, R. & WEIHRAUCH, K. [1978] *Data representation and computational complexity* (**J** 1426) Theor Comput Sci 7*99-116
◇ D15 D45 ◇ REV MR 80j:68035 Zbl 421 # 03029
• ID 53428

VERBEEK, R. [1978] *Primitiv-rekursive Grzegorczyk-Hierarchien* (**S** 3180) Inform Ber, Inst Inf, Univ Bonn 22*137pp
◇ D20 ◇ REV MR 81h:03086 Zbl 421 # 03030 • ID 53429

VERBEEK, R. [1979] see BRAUNMUEHL VON, B.

VERBEEK, R. [1979] *On the naturalness of the Grzegorczyk hierarchy* (**P** 2935) FCT'79 Fund of Comput Th;1979 Berlin/Wendisch-Rietz 483-490
◇ D15 D20 ◇ REV MR 82e:03041 Zbl 426 # 03041
• ID 53637

VERBEEK, R. [1981] *Time-space trade-offs for general recursion* (**P** 4235) IEEE Symp Found of Comp Sci (22);1981 Nashville 228-234
◇ D15 ◇ REV MR 84a:68004 • ID 45855

VERBEEK, R. [1983] see BRAUNMUEHL VON, B.

VERDILLON, A. [1976] see TURCAT, C.

VERESHCHAGIN, N.K. [1984] *Zeros of linear recursive sequences (Russian)* (**J** 0023) Dokl Akad Nauk SSSR 278*1036-1039
• TRANSL [1984] (**J** 0062) Sov Math, Dokl 30*502-505
◇ D80 ◇ REV MR 86c:68064 • ID 44151

VERGES, M. [1982] see DIAZ, J.

VERKHOZINA, M.I. [1973] *The undecidability of the separation problem for positive fragments of logical calculi (Russian)* (**C** 1443) Algebra, Vyp 2 (Irkutsk) 69-83
◇ D35 F50 ◇ REV MR 54 # 4959 • ID 24127

VERMEIR, D. [1979] see EHRENFEUCHT, A.

VERMEIR, D. [1979] see MEERSMAN, R.

VERMEIR, D. [1981] see SAVITCH, W.J.

VERMEIR, D. [1981] see EHRENFEUCHT, A.

VERRAEDT, R. [1980] see EHRENFEUCHT, A.

VERSHININ, K.P. [1974] see MEJTUS, V.YU.

VERSHININ, K.P. see Vol. I, III, V for further entries

VERSHININ, V.A. [1975] *On the question of superposability of constructive sets with equality (Russian)* (**J** 0023) Dokl Akad Nauk SSSR 223*781-784
• TRANSL [1975] (**J** 0062) Sov Math, Dokl 16*976-980
◇ D25 F60 ◇ REV MR 53 # 7743 Zbl 338 # 02022
• ID 22991

VERSHININ, V.A. see Vol. VI for further entries

VETULANI, Z. [1984] *Ramified analysis and the minimal β-models of higher order arithmetics* (**J** 0027) Fund Math 121*1-15
◇ C62 D55 E45 F35 F65 ◇ REV Zbl 576 # 03021
• ID 44142

VETULANI, Z. see Vol. III, V for further entries

VIDAKOVIC, B.D. [1985] *On some properties of the Martin-Loef's measures of randomness of finite binary words* (**P** 4661) Algeb & Log;1984 Zagreb 171-176
◇ D80 ◇ ID 49066

VIDAL-NAQUET, G. [1971] *Programmes formels et logique du second ordre* (**J** 2313) C R Acad Sci, Paris, Ser A-B 273*A1268-A1270
◇ D20 ◇ REV MR 52 # 12389 Zbl 241 # 02012 • ID 21739

Vidal-Naquet, G. [1973] *Quelques applications des automates a arbres infinis* (**P** 0763) Automata, Lang & Progr (1);1972 Rocquencourt 115-122
◇ B25 D05 D35 ◇ rev MR 57 # 18236 Zbl 264 # 02045 • id 27517

Vidal Abascal, E. [1967] *Undecidability in mathematics* (**J** 3100) Acta Cient Compostelana 4*55-60
◇ D35 ◇ rev Zbl 337 # 02029 • id 65916

Viennot, G. [1976] see Nivat, M.

Vier, L.C. [1972] *Church's thesis in northern Dutch constructivism* (**S** 3303) Stud Logic Found Math x+321pp
◇ D20 F50 ◇ id 49913

Vieru, V. [1981] see Calude, C.

Ville, F. [1971] *Complexite des structures rigidement contenues dans une theorie du premier ordre* (**J** 2313) C R Acad Sci, Paris, Ser A-B 272*A561-A563
◇ C50 C57 ◇ rev MR 44 # 1552 Zbl 218 # 02047 • id 13946

Ville, F. see Vol. III, V for further entries

Vincenzi, A. [1983] *Experiments as abstract machines (Italian)* (**P** 3829) Atti Incontri Log Mat (1);1982 Siena 235-238
◇ D20 ◇ rev MR 84k:03006 • id 44821

Vincenzi, A. see Vol. II, III for further entries

Vinner, S. [1975] *On two complete sets in the analytical and the arithmetical hierarchies* (**J** 0009) Arch Math Logik Grundlagenforsch 17*81-84
◇ C55 C80 D55 ◇ rev MR 52 # 5346 Zbl 317 # 02048 • id 13949

Vinner, S. see Vol. I, III, V for further entries

Vinogradov, A.K. [1975] see Kosovskij, N.K.

Vinogradov, A.P. [1977] *On the existence of a majorizing local algorithm in effectively describable problems of computing information (Russian)* (**J** 0023) Dokl Akad Nauk SSSR 235*745-748
• transl [1977] (**J** 0062) Sov Math, Dokl 18*1014-1018
◇ D15 D20 ◇ rev MR 58 # 3663 Zbl 386 # 03021 • id 52197

Vinogradov, A.P. see Vol. I for further entries

Viragh, J. [1980] *Deterministic ascending tree automata. I* (**J** 0380) Acta Cybern (Szeged) 5*33-41
◇ D05 ◇ rev MR 83c:68063 Zbl 454 # 68044 • id 39050

Vishkin, U. [1984] see Gurevich, Y.

Vishkin, U. see Vol. I for further entries

Visser, A. [1980] *Numerations, λ-calculus & arithmetic* (**C** 3050) Essays Combin Log, Lambda Calc & Formalism (Curry) 259-284
◇ B40 D45 F30 ◇ rev MR 83c:03041 Zbl 469 # 03006 • id 35141

Visser, A. see Vol. II, III, VI for further entries

Vitanyi, P.M.B. [1977] see Savitch, W.J.

Vitanyi, P.M.B. [1978] see Emde Boas van, P.

Vitanyi, P.M.B. [1980] *On the power of real-time Turing machines under varying specifications* (**P** 2904) Automata, Lang & Progr (7);1980 Noordwijkerhout 658-671
◇ D10 D15 ◇ rev MR 81j:68065 Zbl 422 # 68017 • id 83077

Vitanyi, P.M.B. [1980] *Real-time Turing machines under varying specifications* (**X** 3205) Math Centr Amsterdam Afd Inf IW140*30pp
◇ D10 D15 ◇ rev Zbl 432 # 68038 • rem Part II 1980 • id 69992

Vitanyi, P.M.B. [1980] *Real-time Turing machines under varying specifications II* (**X** 3205) Math Centr Amsterdam Afd Inf 147*9pp
◇ D10 D15 ◇ rev Zbl 445 # 68035 • rem Part I 1980 • id 69991

Vitanyi, P.M.B. [1984] see Savitch, W.J.

Vitanyi, P.M.B. [1984] *On the simulation of many storage heads by one* (**J** 1426) Theor Comput Sci 34*157-168
◇ D10 ◇ rev MR 86d:68021 • id 48883

Vitanyi, P.M.B. [1984] *The simple roots of real-time computation hierarchies* (**P** 4012) Automata, Lang & Progr (11);1984 Antwerpen 486-489
◇ D15 ◇ id 47190

Vitanyi, P.M.B. [1985] *An $n^{1.618}$ lower bound on the time to simulate one queue or two pushdown stores by one tape* (**J** 0232) Inform Process Lett 21*147-152
◇ D10 D15 ◇ id 49543

Voelkel, L. [1973] *Untersuchungen ueber die Kompliziertheit von Berechnungs- und Tabellenprogrammen* (**J** 0129) Elektr Informationsverarbeitung & Kybern 9*137-159
◇ D15 D20 ◇ rev MR 54 # 9156 Zbl 277 # 68021 • id 65934

Voelkel, L. [1976] *Staerke-Relationen fuer URM-Befehlssysteme* (**J** 0129) Elektr Informationsverarbeitung & Kybern 12*245-257
◇ D10 D15 D20 ◇ rev MR 55 # 1817 Zbl 332 # 02044 • id 65933

Voelkel, L. [1977] *Spracherkennung durch Registermaschinen* (**S** 1536) Schr Weiterbildungszentr MKR (Dresden) 27*
◇ D10 ◇ id 42368

Voelkel, L. [1978] see Nepomnyashchij, V.A.

Voelkel, L. [1979] *Language recognition by linear bounded and copy programs* (**P** 2935) FCT'79 Fund of Comput Th;1979 Berlin/Wendisch-Rietz 491-495
◇ D05 ◇ rev MR 81b:68101 Zbl 413 # 68084 • id 42370

Voelkel, L. [1980] see Foerster, M.

Vogel, Helmut [1977] *Ausgezeichnete Folgen fuer praedikative Ordinalzahlen und praedikativ-rekursive Funktionen* (**J** 0068) Z Math Logik Grundlagen Math 23*435-438
◇ D20 F15 ◇ rev MR 58 # 16232 Zbl 453 # 03043 • id 54173

Vogel, Helmut [1977] *Partial enumerable and finite functionals (Russian)* (**J** 0003) Algebra i Logika 16*109-119,125
• transl [1977] (**J** 0069) Algeb and Log 16*75-81
◇ D65 ◇ rev MR 58 # 27407 Zbl 398 # 03037 • id 52765

Vogel, Helmut [1983] *On a relationship between countable functionals and projective trees* (**J** 0027) Fund Math 119*169-183
◇ D55 D65 ◇ rev MR 85g:03069 Zbl 569 # 03019 • id 43893

Vogel, Helmut see Vol. VI for further entries

VOGEL, J. & WAGNER, K. [1981] *On a class of automata accepting exactly the languages which are elementary in the sense of KALMAR* (**P** 3212) Math Models in Comput Systs;1981 Budapest
◇ D05 D20 ◇ ID 40923

VOGEL, J. & WAGNER, K. [1981] *Two-way automata with more than one storage medium* (**S** 3374) Forschergeb, Univ Jena N/81/52*19pp
◇ D10 D15 ◇ REV Zbl 481 # 68051 • ID 38459

VOGEL, J. [1982] see BRANDSTAEDT, A.

VOGEL, J. & WAGNER, K. [1985] *Two-way automata with more than one storage medium* (**J** 1426) Theor Comput Sci 39*267-280
◇ D05 ◇ ID 49093

VOGLER, H. [1985] *Iterated linear control and iterated one-turn pushdowns* (**P** 4647) FCT'85 Fund of Comput Th;1985 Cottbus 474-484
◇ D10 ◇ ID 49094

VOIGT, B. [1984] see LEFMANN, H.

VOIGT, B. see Vol. V for further entries

VOJKOV, G.K. [1976] see DAKOVSKI, L.G.

VOLGER, H. [1983] *A new hierarchy of elementary recursive decision problems* (**J** 3401) Meth Oper Res 45*509-519
◇ B25 D15 D20 ◇ REV MR 85i:68019 Zbl 531 # 03006 • ID 37672

VOLGER, H. [1983] *Turing machines with linear alternation, theories of bounded concatenation and the decision problem of first order theories* (**J** 1426) Theor Comput Sci 23*333-337
◇ B25 D10 D15 ◇ REV MR 84m:68042 Zbl 538 # 03035 • ID 39717

VOLGER, H. [1984] *Rudimentary relations and Turing machines with linear alternation* (**P** 2342) Symp Rek Kombin;1983 Muenster 131-136
◇ D10 D15 D20 ◇ REV Zbl 558 # 68042 • ID 39721

VOLGER, H. [1984] *The role of rudimentary relations in complexity theory* (**S** 3521) Mem Soc Math Fr 16*41-51
◇ D10 D15 ◇ REV Zbl 558 # 68043 • ID 46636

VOLGER, H. see Vol. III, V for further entries

VOLKEN, H. [1978] see BARENDREGT, H.P.

VOLKEN, H. see Vol. VI for further entries

VOLLMAR, R. [1970] see KAMEDA, T.

VOLLMAR, R. [1973] *Ueber Turingmaschinen mit variablem Speicher* (**J** 0129) Elektr Informationsverarbeitung & Kybern 9*3-13
◇ D10 ◇ REV MR 56 # 8338 Zbl 271 # 02025 • ID 29915

VOLLMAR, R. [1975] see HOLLERER, W.O.

VOLLMAR, R. [1975] *On Turing machines with variable structure* (**J** 0191) Inform Sci 8*259-270
◇ D10 ◇ REV MR 51 # 2890 Zbl 312 # 02031 • ID 17336

VOL'VACHEV, R.T. [1975] *Undecidability of the isomorphism and the conjugacy problem of commutative matrix groups and algebras (Russian)* (**J** 0413) Izv Akad Nauk Belor SSR, Ser Fiz-Mat 1975/5*98-101,141
◇ D25 D40 ◇ REV MR 52 # 13357 Zbl 367 # 02024 • ID 21810

VOL'VACHEV, R.T. see Vol. III for further entries

VOPENKA, P. [1964] see BUKOVSKY, L.

VOPENKA, P. see Vol. III, V, VI for further entries

VUCKOVIC, V. [1959] *Partially ordered recursive arithmetics* (**J** 0132) Math Scand 7*305-320
◇ D20 F30 ◇ REV MR 22 # 7939 Zbl 93.12 JSL 28.251 • ID 13892

VUCKOVIC, V. [1960] *Rekursive Wortarithmetik* (**J** 4706) Publ Inst Math (Belgrade) 14*9-60
◇ D20 ◇ REV MR 23 # A3087 Zbl 96.244 JSL 29.200 • ID 13893

VUCKOVIC, V. [1961] *Basic theorems on Turing algorithms* (**J** 0400) Publ Inst Math, NS (Belgrade) 1(15)*31-65
◇ D10 ◇ REV MR 34 # 4133 Zbl 118.16 • ID 13895

VUCKOVIC, V. [1961] see ASSER, G.

VUCKOVIC, V. [1961] *Turing algorithms* (**J** 0068) Z Math Logik Grundlagen Math 7*106-116
◇ D10 ◇ REV MR 26 # 1249 Zbl 109.10 • ID 13894

VUCKOVIC, V. [1962] *Einfuehrung von $\Sigma_f(x)$ und $\Pi_f(x)$ in der rekursiven Gitterpunktarithmetik* (**J** 1156) Izv Bulgar Akad Nauk Mat Inst 6*15-25
◇ D20 F30 ◇ REV MR 26 # 1250 JSL 28.251 • ID 22057

VUCKOVIC, V. [1962] *On some possibilities in the foundations of recursive arithmetics of words (English) (Serbo-Croatian summary)* (**J** 0371) Glas Mat-Fiz Astron, Ser 2 (Zagreb) 17*145-157
◇ D05 ◇ REV MR 29 # 3371 Zbl 117.14 JSL 36.549 • ID 43487

VUCKOVIC, V. [1962] *Recursive arithmetic of words and finite automata (Serbo-Croatian) (English summary)* (**J** 1179) Filoz Jugos Chas (Belgrade) 6*41-66
◇ D05 ◇ REV JSL 36.549 • ID 43493

VUCKOVIC, V. [1964] *On a class of regular sets* (**J** 0047) Notre Dame J Formal Log 5*113-124
◇ D05 ◇ REV MR 36 # 1328 Zbl 133.257 JSL 36.549 • ID 13896

VUCKOVIC, V. [1967] *Creative and weakly creative sequences of r.e. sets* (**J** 0053) Proc Amer Math Soc 18*478-483
◇ D25 ◇ REV MR 35 # 4099 Zbl 204.14 JSL 34.296 • ID 13901

VUCKOVIC, V. [1967] *Mathematics of incompleteness and undecidability* (**J** 0068) Z Math Logik Grundlagen Math 13*123-150
◇ D35 F30 ◇ REV MR 35 # 5303 Zbl 183.10 JSL 37.195 • ID 13899

VUCKOVIC, V. [1967] *Recursive models for three-valued propositional calculi with classical implication* (**J** 0047) Notre Dame J Formal Log 8*148-153
◇ B50 C57 C90 F30 ◇ REV Zbl 262 # 02021 • ID 27478

VUCKOVIC, V. [1969] *Almost recursive sets* (**J** 0053) Proc Amer Math Soc 23*114-119
◇ D25 D50 ◇ REV MR 41 # 5217 Zbl 185.20 JSL 38.525 • ID 13902

VUCKOVIC, V. [1970] *Effective enumerability of some families of partially recursive functions connected with computable functionals* (**J** 0068) Z Math Logik Grundlagen Math 16*113-121
◇ D20 D25 ◇ REV MR 43 # 1834 Zbl 199.308 • ID 13904

VUCKOVIC, V. [1970] *Recursive word-functions over infinite alphabets* (J 0068) Z Math Logik Grundlagen Math 16*123-138
◇ D20 ◇ REV MR 43 # 4670 Zbl 198.324 • ID 13903

VUCKOVIC, V. [1973] *Local recursive theory* (J 0047) Notre Dame J Formal Log 14*237-246
◇ D20 D25 D75 D80 ◇ REV MR 48 # 79 Zbl 245 # 02040 • ID 13906

VUCKOVIC, V. [1974] *Almost recursivity and partial degrees* (J 0068) Z Math Logik Grundlagen Math 20*419-426
◇ D25 D30 ◇ REV MR 52 # 10392 Zbl 355 # 02030 • ID 13907

VUCKOVIC, V. [1977] *Recursive and recursive enumerable manifolds I,II* (J 0047) Notre Dame J Formal Log 18*265-291,383-405
◇ C57 D45 D80 ◇ REV MR 56 # 2808 Zbl 306 # 02035 Zbl 306 # 02036 • ID 23633

VUCKOVIC, V. [1982] *Relativized cylindrification* (J 0068) Z Math Logik Grundlagen Math 28*167-172
◇ D20 D30 ◇ REV MR 84c:03080 Zbl 528 # 03024 • ID 34017

VUCKOVIC, V. see Vol. I, II, V, VI for further entries

VUILLEMIN, J. [1974] see COURCELLE, B.

VUILLEMIN, J. [1976] see COURCELLE, B.

VUILLEMIN, J. see Vol. II for further entries

V'YUGIN, V.V. [1972] *On discrete families of recursively enumerable sets (Russian)* (J 0003) Algebra i Logika 11*243-256,361
• TRANSL [1972] (J 0069) Algeb and Log 11*137-144
◇ D25 ◇ REV MR 47 # 1589 Zbl 272 # 02063 • ID 13953

V'YUGIN, V.V. [1973] *On minimal numerations of computable classes of recursively enumerable sets (Russian)* (J 0023) Dokl Akad Nauk SSSR 212*273-275
• TRANSL [1973] (J 0062) Sov Math, Dokl 14*1338-1340
◇ D25 D45 ◇ REV MR 48 # 10784 Zbl 298 # 02033 • ID 13955

V'YUGIN, V.V. [1973] *On some examples of upper semilattices of computable enumerations (Russian)* (J 0003) Algebra i Logika 12*512-529,617
• TRANSL [1973] (J 0069) Algeb and Log 12*287-296
◇ D25 D45 ◇ REV MR 51 # 12493 Zbl 305 # 02059 • ID 17282

V'YUGIN, V.V. [1974] *On upper semilattices of numerations (Russian)* (J 0023) Dokl Akad Nauk SSSR 217*749-751
• TRANSL [1974] (J 0062) Sov Math, Dokl 15*1110-1113
◇ D25 D45 ◇ REV MR 50 # 9562 Zbl 305 # 02060 • ID 13956

V'YUGIN, V.V. [1974] *Segments of recursively enumerable M-degrees (Russian)* (J 0003) Algebra i Logika 13*635-654,719
• TRANSL [1974] (J 0069) Algeb and Log 13*361-373
◇ D25 ◇ REV MR 53 # 5277 Zbl 361 # 02055 • ID 22908

V'YUGIN, V.V. [1976] *On Turing invariant sets (Russian)* (J 0023) Dokl Akad Nauk SSSR 229*790-793
• TRANSL [1976] (J 0062) Sov Math, Dokl 17*1090-1094
◇ D30 ◇ REV MR 54 # 10000 Zbl 359 # 02041 • ID 25841

V'YUGIN, V.V. [1981] *Algorithmic entropy (complexity) of finite objects and its application to the definition of randomness and quantity of information (Russian)* (S 2582) Semiotika & Inf, Akad Nauk SSSR 16*14-43
◇ D15 ◇ REV MR 83e:94019 • ID 40174

WAACK, S. [1981] *Tape complexity of word problems* (J 3437) Rep, Akad Wiss DDR, Inst Math 04/81*99pp
◇ D15 D40 ◇ REV MR 82m:20040 Zbl 468 # 03021 • ID 55086

WAACK, S. [1981] *Tape complexity of word problems* (P 3165) FCT'81 Fund of Comput Th;1981 Szeged 467-471
◇ D15 D40 ◇ REV MR 83k:03055 Zbl 498 # 03038 • ID 33707

WAACK, S. [1982] see BUDACH, L.

WAACK, S. see Vol. I for further entries

WADA, H. [1976] see JONES, JAMES P.

WAERDEN VAN DER, B.L. [1930] *Eine Bemerkung ueber die Unzerlegbarkeit von Polynomen* (J 0043) Math Ann 102*738-739
◇ B25 C60 D45 F55 ◇ REV FdM 56.825 • ID 38703

WAERDEN VAN DER, B.L. see Vol. V for further entries

WAGNER, E.G. [1968] *Bounded action machines: toward an abstract theory of computer structure* (J 0119) J Comp Syst Sci 2*13-75
◇ D05 ◇ REV MR 43 # 5764 Zbl 162.483 • REM Part II 1972 • ID 13964

WAGNER, E.G. [1968] *Uniformly reflexible structures: an axiomatic approach to computability* (J 0191) Inform Sci 1*343-362
◇ D75 ◇ REV MR 41 # 6688 • ID 21040

WAGNER, E.G. [1969] *Uniformly reflexive structures: on the nature of Goedelization and relative computability* (J 0064) Trans Amer Math Soc 144*1-41
◇ B40 D45 D75 F40 ◇ REV MR 40 # 2543 Zbl 265 # 02029 • ID 13965

WAGNER, E.G. [1971] *An algebraic theory of recursive definitions and recursive languages* (P 0680) ACM Symp Th of Comput (3);1971 Shaker Heights 12-23
◇ B75 D75 G30 ◇ REV Zbl 252 # 02048 • ID 27760

WAGNER, E.G. [1972] *Bounded action machines II. The basic structure of tapeless computers (German summary)* (J 0373) Comp Arch Inform & Numerik 9*211-232
◇ D05 ◇ REV MR 47 # 1328 Zbl 257 # 68042 • REM Part I 1968 • ID 13966

WAGNER, E.G. [1975] see GOGUEN, J.A.

WAGNER, E.G. see Vol. II, V for further entries

WAGNER, K. [1972] *Zur Axiomatisierung eines sequentiellen Aussagenkalkuels (Russian, English and French summaries)* (J 0115) Wiss Z Humboldt-Univ Berlin, Math-Nat Reihe 21*471-472
◇ B22 D05 ◇ REV MR 48 # 8212 Zbl 251 # 94032 • ID 13969

WAGNER, K. [1973] *Die Modellierung der Arbeit von Turingmaschinen mit n-dimensionalem Band durch Turingmaschinen mit eindimensionalem Band (English and Russian summaries)* (J 0129) Elektr Informationsverarbeitung & Kybern 9*121-135
◇ D10 ◇ REV MR 50 # 72 Zbl 268 # 02026 • ID 13970

WAGNER, K. [1973] *Universelle Turingmaschinen mit n-dimensionalem Band (English and Russian summaries)* (**J** 0129) Elektr Informationsverarbeitung & Kybern 9*423-431
◇ D10 ◇ REV MR 50 # 9553 Zbl 318 # 02038 • ID 13971

WAGNER, K. [1974] see STAIGER, L.

WAGNER, K. [1974] see LINDNER, R.

WAGNER, K. [1974] *Zellulare Berechenbarkeit von Funktionen ueber n-dimensionalen Zeichensystemen* (**J** 0129) Elektr Informationsverarbeitung & Kybern 10*259-269
◇ D10 ◇ REV MR 51 # 9574 Zbl 296 # 94033 • ID 65964

WAGNER, K. [1975] *A hierarchy of regular sequence sets* (**P** 0454) Math Founds of Comput Sci (4);1975 Marianske Lazne 445-449
◇ D05 ◇ REV MR 57 # 5520 Zbl 327 # 94062 • ID 40901

WAGNER, K. [1975] *Akzeptierbarkeitsgrade regulaerer Folgenmengen* (**P** 4263) Diskr Math;1974 Berlin 626-630
◇ D05 ◇ REV Zbl 327 # 94063 • ID 40911

WAGNER, K. [1975] see STAIGER, L.

WAGNER, K. [1975] *Turing-Berechenbarkeit in linearer Zeit* (**J** 0380) Acta Cybern (Szeged) 2*235-248
◇ D10 D15 ◇ REV MR 55 # 10255 Zbl 322 # 02035 • ID 65960

WAGNER, K. [1976] *Arithmetische Operatoren* (**J** 0068) Z Math Logik Grundlagen Math 22*553-570
◇ D10 D55 ◇ REV MR 58 # 27408 Zbl 352 # 02031 • ID 50030

WAGNER, K. [1976] *Eine Axiomatisierung der Theorie der regulaeren Folgenmengen (English and Russian summaries)* (**J** 0129) Elektr Informationsverarbeitung & Kybern 12*337-354
◇ D05 ◇ REV MR 54 # 9163 Zbl 335 # 94028 • ID 25835

WAGNER, K. [1976] see STAIGER, L.

WAGNER, K. [1977] *Arithmetische und Bairesche Operatoren* (**J** 0068) Z Math Logik Grundlagen Math 23*181-191
◇ D55 E15 ◇ REV MR 58 # 5141 Zbl 359 # 02038 • ID 26477

WAGNER, K. & WECHSUNG, G. [1977] *Complexity hierarchies of oracles* (**P** 1635) Math Founds of Comput Sci (6);1977 Tatranska Lomnica 543-548
◇ D15 ◇ REV MR 58 # 10358 Zbl 361 # 68073 • ID 79795

WAGNER, K. [1977] *Eine topologische Charakterisierung einiger Klassen regulaerer Folgenmengen* (**J** 0129) Elektr Informationsverarbeitung & Kybern 13*473-487
◇ D05 ◇ REV MR 58 # 27388 Zbl 379 # 94070 • ID 79790

WAGNER, K. [1977] see STAIGER, L.

WAGNER, K. [1978] see STAIGER, L.

WAGNER, K. [1979] *Bounded recursion and complexity classes* (**P** 2059) Math Founds of Comput Sci (8);1979 Olomouc 492-498
◇ D15 D20 ◇ REV MR 81h:03085 Zbl 409 # 03021 • ID 56324

WAGNER, K. [1979] *On ω-regular sets* (**J** 0194) Inform & Control 43*123-177
◇ D05 ◇ REV MR 81f:68070 Zbl 434 # 68061 • ID 55757

WAGNER, K. & WECHSUNG, G. [1980] *Kompliziertheitshierarchien* (**J** 0192) Wiss Z Univ Jena, Math-Nat Reihe 29*229-250
◇ D15 ◇ REV MR 81h:68033 Zbl 436 # 68030 • ID 83096

WAGNER, K. [1981] see VOGEL, J.

WAGNER, K. [1982] see MEINHARDT, D.

WAGNER, K. [1983] see BRANDSTAEDT, A.

WAGNER, K. [1984] *Compact descriptions and the counting polynomial time hierarchy* (**P** 3621) Frege Konferenz (2);1984 Schwerin 383-392
◇ D15 ◇ REV MR 85m:03006 Zbl 554 # 68032 • ID 40931

WAGNER, K. [1984] *The complexity of problems concerning graphs with regularities* (**P** 3658) Math Founds of Comput Sci (11);1984 Prague 544-552
◇ D15 ◇ REV Zbl 548 # 68039 • ID 40937

WAGNER, K. [1984] *The complexity of combinatorial problems with compactly described instances* (**S** 3374) Forschergeb, Univ Jena 84/23/29pp
◇ D15 ◇ REV Zbl 548 # 68041 • ID 43248

WAGNER, K. [1985] see VOGEL, J.

WAGNER, K. see Vol. II, V for further entries

WAINER, S.S. [1970] *A classification of the ordinal recursive functions* (**J** 0009) Arch Math Logik Grundlagenforsch 13*136-153
• TRANSL [1974] (**C** 2319) Slozh Vychisl & Algor 65-84 (Russian)
◇ D20 D25 ◇ REV MR 45 # 3207 Zbl 228 # 02027 • ID 13973

WAINER, S.S. [1970] see LOEB, M.H.

WAINER, S.S. [1972] *Ordinal recursion, and a refinement of the extended Grzegorczyk hierarchy* (**J** 0036) J Symb Logic 37*281-292
◇ D20 ◇ REV MR 48 # 82 Zbl 261 # 02031 • ID 13974

WAINER, S.S. [1974] *A hierarchy for the 1-section of any type two object* (**J** 0036) J Symb Logic 39*88-94
◇ D55 D65 ◇ REV MR 50 # 12694 Zbl 299 # 02048 • ID 13975

WAINER, S.S. [1974] see LOEB, M.H.

WAINER, S.S. [1975] see SCHWICHTENBERG, H.

WAINER, S.S. [1975] *Some hierachies based on higher type quantification* (**P** 0775) Logic Colloq;1973 Bristol 305-316
◇ C75 D55 D65 E45 ◇ REV MR 58 # 5142 Zbl 311 # 02053 • ID 29611

WAINER, S.S. [1978] *The 1-section of a nonnormal type-2 object* (**P** 1628) Generalized Recursion Th (2);1977 Oslo 407-417
◇ D55 D65 ◇ REV MR 80j:03065 Zbl 453 # 03047 • ID 79802

WAINER, S.S. [1980] see DRAKE, F.R.

WAINER, S.S. [1980] see NORMANN, D.

WAINER, S.S. [1983] see CICHON, E.A.

WAINER, S.S. [1984] see DENNIS-JONES, E.C.

WAINER, S.S. [1985] *Subrecursive ordinals* (**P** 3342) Rec Th Week;1984 Oberwolfach 405-418
◇ D20 F15 ◇ ID 45318

WAINER, S.S. [1985] *The "slow-growing" Π_2^1 approach to hierarchies* (**P** 4046) Rec Th;1982 Ithaca 487-502
◇ D20 F15 ◇ ID 46381

WAJS, R. [1973] *On certain properties of relational machines* (**J** 1929) Prace Centr Oblicz Pol Akad Nauk 129*42pp
◇ D05 ◇ REV Zbl 274 # 94053 • ID 65971

WAKULICZ-DEJA, ALICJA [1973] *Eigenschaften von Maschinen, die durch partielle Uebergangsfunktionen charakterisiert werden* (**J** 1929) Prace Centr Oblicz Pol Akad Nauk 97*18pp
◇ D05 ◇ REV Zbl 261 # 94046 • ID 65972

WALAT, A. [1971] *Some properties of k-machines (Russian summary)* (**J** 0014) Bull Acad Pol Sci, Ser Math Astron Phys 19*639-643
◇ D03 D05 ◇ REV MR 46 # 1572 Zbl 236 # 94035 • ID 13990

WALDSCHMIDT, H. [1977] see TZSCHACH, H.

WALIGORA, G. [1975] see KOPIEKI, R.

WALIGORSKI, S. [1969] *Algebraic theory of automata (Polish)* (**J** 0485) Algorytmy, Pol Akad Nauk 6*1-131
◇ D05 ◇ REV MR 41 # 3190 Zbl 195.25 • ID 16306

WALIGORSKI, S. see Vol. I, II for further entries

WALKER, S.A. [1973] see STRONG, H.R.

WALKOE JR., W.J. [1973] see KEISLER, H.J.

WALKOE JR., W.J. see Vol. I, III for further entries

WALL, R.E. [1972] *Introduction to mathematical linguistics* (**X** 0819) Prentice Hall: Englewood Cliffs xiv+337pp
◇ B65 D03 D05 D98 ◇ REV MR 50 # 3659 Zbl 352 # 68085 JSL 39.615 • ID 13999

WALLACE, A.D. [1967] see BEDNAREK, A.R.

WALLACE, A.D. see Vol. V for further entries

WALSH, T.R.S. [1982] *The busy beaver on a one-way infinite tape* (**J** 1456) SIGACT News 14/1*38-43
◇ D10 ◇ REV Zbl 565 # 68048 • ID 48670

WALSH, T.R.S. see Vol. I for further entries

WALTER, H. [1969] see HOTZ, G.

WALTER, H. [1969] see SCHNORR, C.-P.

WALTER, H.K.-G. [1977] see TZSCHACH, H.

WAND, M. [1973] *A concrete approach to abstract recursive definitions* (**P** 0763) Automata, Lang & Progr (1);1972 Rocquencourt 331-341
◇ D05 D75 G10 ◇ REV MR 51 # 3013 Zbl 278 # 68066 • ID 17362

WAND, M. [1975] *An algebraic formulation of the Chomsky hierarchy* (**P** 0770) Categ Th Appl to Comput & Control (1);1974 San Francisco 209-213
◇ D05 ◇ REV MR 56 # 1834 Zbl 305 # 68056 • ID 32526

WAND, M. see Vol. II, VI for further entries

WANG, HAO [1953] *Certain predicates defined by induction schemata* (**J** 0036) J Symb Logic 18*49-59
◇ D70 F30 F35 ◇ REV MR 14.936 Zbl 51.5 JSL 30.99 • ID 14021

WANG, HAO [1957] *A variant to Turing's theory of computing machines* (**J** 0037) ACM J 4*63-92
◇ D10 ◇ REV MR 20 # 4492 JSL 28.288 • ID 14028

WANG, HAO [1957] *Remarks on constructive ordinals and set theory* (**P** 1675) Summer Inst Symb Log;1957 Ithaca 383-390
◇ D55 E45 F15 ◇ ID 29384

WANG, HAO [1957] *Symbolic representations of calculating machines* (**P** 1675) Summer Inst Symb Log;1957 Ithaca 181-188
◇ B70 D05 D10 F30 ◇ REV Zbl 145.407 JSL 27.103 • ID 14033

WANG, HAO [1957] see BURKS, A.W.

WANG, HAO [1957] *Universal Turing machines: an exercise in coding* (**J** 0068) Z Math Logik Grundlagen Math 3*69-80
◇ D10 ◇ REV MR 20 # 4493 Zbl 90.10 JSL 28.288 • ID 14027

WANG, HAO [1962] see DREBEN, B.

WANG, HAO [1962] see KAHR, A.S.

WANG, HAO [1963] *Computation* (**C** 1009) Wang: Survey Math Logic 82-125
◇ D20 D25 D30 D35 D60 ◇ REV JSL 29.105 • ID 14047

WANG, HAO [1963] *Dominoes and the AEA case of the decision problem* (**P** 0674) Symp Math Th of Automata;1962 New York 23-55
◇ B20 B25 D05 D35 ◇ REV MR 29 # 4688 Zbl 137.10 • ID 14040

WANG, HAO [1963] *Tag systems and lag systems* (**J** 0043) Math Ann 152*65-74
◇ D03 ◇ REV MR 29 # 27 Zbl 131.246 JSL 36.344 • ID 14041

WANG, HAO [1963] see RABIN, M.O.

WANG, HAO [1965] *Logic and computers* (**J** 0005) Amer Math Mon 72*135-140
◇ A05 B35 D03 D10 ◇ REV MR 30 # 4639 Zbl 123.335 JSL 31.264 • ID 14058

WANG, HAO [1965] *Remarks on machines, sets, and the decision problem* (**P** 0688) Logic Colloq;1963 Oxford 304-320
◇ B20 B25 D03 D05 D10 D35 E30 ◇ REV MR 39 # 6729 Zbl 133.254 • ID 14057

WANG, HAO [1974] *Notes on a class of tiling problems* (**J** 0027) Fund Math 82*295-305
◇ D05 D80 ◇ REV MR 51 # 109 Zbl 301 # 02043 • ID 14066

WANG, HAO [1976] see DUNHAM, B.

WANG, HAO see Vol. I, II, III, V, VI for further entries

WANG, JIE [1983] *A note on a theorem about generating sets of recursively enumerable sets (Chinese)* (**J** 3942) Zhongshan Daxue Xuebao, Ziran Kexue 3*98
◇ D25 ◇ ID 44231

WANG, JIE [1985] *A necessary and sufficient condition for the existence for a given B of an A such that $P^A = NP^B$ (Chinese)* (**J** 2771) Kexue Tongbao 30*792-793
◇ D15 ◇ ID 48560

WANG, JUENTIN [1973] *On the representation of generative grammars as first-order theories* (**P** 0580) Int Congr Log, Meth & Phil of Sci (4,Sel Pap);1971 Bucharest 302-316
◇ B10 D05 ◇ REV MR 57 # 8200 Zbl 296 # 68079
• ID 65982

WANG, P.S. [1974] *The undecidability of the existence of zeros of real elementary functions* (**J** 0037) ACM J 21*586-589
◇ D80 ◇ REV MR 51 # 117 Zbl 289 # 68017 • ID 15174

WANG, YIZHI [1981] see SHU, YONGCHANG

WARKENTIN, J.C. [1972] see FISCHER, P.C.

WARKENTIN, J.C. [1974] see FISCHER, P.C.

WARMUTH, M.K. [1984] see HAUSSLER, D.

WARMUTH, M.K. [1985] see GONCZAROWSKI, J.

WARNER, M.W. [1984] see MUIR, A.

WARNER, M.W. see Vol. I, V for further entries

WARREN, D.S. [1978] see FRIEDMAN, JOYCE

WARREN, D.S. see Vol. VI for further entries

WARREN, D.W. [1954] see BURKS, A.W.

WASILEWSKA, A. [1980] *On the Gentzen-type formalizations* (**J** 0068) Z Math Logik Grundlagen Math 26*439-444
◇ B25 D05 F07 ◇ REV MR 82g:03102 Zbl 471 # 03021
• ID 55216

WASILEWSKA, A. [1985] *Monadic second-order definability as a common characterization of finite automata, certain classes of programs and logics* (**J** 2095) Fund Inform, Ann Soc Math Pol, Ser 4 8/3-4*309-320
◇ B15 B75 C40 D05 ◇ ID 49467

WASILEWSKA, A. see Vol. I, II, III, VI for further entries

WATANABE, H. [1950] *Sur une separation des ensembles analytiques plans par une courbe mesurable* (**J** 0081) Proc Japan Acad 26/7*17-20
◇ D55 E15 ◇ REV MR 14.960 Zbl 40.312 • ID 14087

WATANABE, H. see Vol. V for further entries

WATANABE, O. [1983] *The time-precision tradeoff problem on on-line probabilistic Turing machines* (**J** 1426) Theor Comput Sci 24*105-117
◇ D10 D15 ◇ REV MR 85c:68016 Zbl 509 # 68045
• ID 39066

WATANABE, O. [1985] *On one-one polynomial time equivalence relations* (**J** 1426) Theor Comput Sci 38/2-3*157-165
◇ D15 ◇ ID 49283

WATANABE, S. [1963] *Periodicity of Post's normal process of tag* (**P** 0674) Symp Math Th of Automata;1962 New York 83-99
◇ D03 ◇ REV MR 30 # 1045 Zbl 129.257 JSL 33.298
• ID 14090

WATANABE, S. see Vol. II, III, V for further entries

WATNICK, R. [1981] *Constructive and recursive scattered order types* (**P** 2628) Log Year;1979/80 Storrs 312-326
◇ D45 ◇ REV MR 82h:03042 Zbl 468 # 03027 • ID 55092

WATNICK, R. [1984] *A generalization of Tennenbaum's theorem on effectively finite recursive linear orderings* (**J** 0036) J Symb Logic 49*563-569
◇ C57 C65 D25 D45 ◇ REV MR 85i:03152 • ID 42511

WEBB, J.C. [1980] *Mechanism, mentalism, and metamathematics. An essay on finitism* (**X** 0835) Reidel: Dordrecht xiii+277pp
◇ A05 D20 F99 ◇ REV MR 83j:03015 Zbl 454 # 03001 JSL 51.472 • ID 54214

WEBB, J.C. [1983] *Goedel's theorems and Church's thesis: A prologue to mechanism* (**C** 3834) Lang, Logic and Method 309-353
◇ A05 D20 F99 ◇ REV Zbl 537 # 03002 • ID 43711

WEBER VON, S. [1978] *Zur Definition eines speziellen algorithmischen Fehlers mittels Turingmaschinen* (**S** 2829) Rostocker Math Kolloq 10*115-121
◇ D10 ◇ REV Zbl 433 # 68034 • ID 69997

WEBER VON, S. [1980] *Zur Definition eines algorithmischen Fehlers 2. Art mittels Turingmaschinen* (**J** 0947) Wiss Z Tech Univ Dresden 29*388-389
◇ D10 ◇ REV Zbl 442 # 68041 • ID 69998

WECHLER, W. [1974] see DIMITROV, V.

WECHLER, W. [1975] *R-fuzzy automata with a time-variant structure* (**P** 1755) Math Founds of Comput Sci (3);1974 Jadwisin 28*73-76
◇ D05 E72 ◇ REV Zbl 313 # 94024 • ID 40845

WECHLER, W. [1975] *R-fuzzy grammars* (**P** 0454) Math Founds of Comput Sci (4);1975 Marianske Lazne 32*450-456
◇ D05 ◇ REV MR 54 # 1733 Zbl 321 # 68058 • ID 40846

WECHLER, W. [1976] see BRUNNER, J.

WECHLER, W. [1977] see BRUNNER, J.

WECHLER, W. [1978] *The concept of fuzziness in automata and language theory* (**X** 0911) Akademie Verlag: Berlin vii+141pp
◇ B52 D05 E72 ◇ REV MR 80g:68105 Zbl 401 # 94048
• ID 54647

WECHLER, W. [1979] *Fuzzy sets and languages* (**C** 3514) Adv Fuzzy Sets & Appl 263-278
◇ D05 E72 ◇ REV MR 81a:68087 • ID 40844

WECHLER, W. see Vol. V for further entries

WECHSUNG, G. [1972] *Quasisequentielle Wortfunktionen* (**J** 0115) Wiss Z Humboldt-Univ Berlin, Math-Nat Reihe 21*553-554
◇ D05 ◇ REV MR 48 # 10723 Zbl 252 # 94031 • ID 66010

WECHSUNG, G. [1972] *Ueber die Gruppe der eineindeutigen laengentreuen sequentiellen Funktionen* (**J** 0129) Elektr Informationsverarbeitung & Kybern 8*335-352
◇ D05 ◇ REV MR 47 # 8741 Zbl 261 # 94049 • ID 42613

WECHSUNG, G. [1973] *Funktionen, die von pushdown-Automaten berechnet werden* (**J** 0380) Acta Cybern (Szeged) 2*115-134
◇ D10 ◇ REV MR 53 # 4629 Zbl 301 # 94068 • ID 42615

WECHSUNG, G. [1973] *Isomorphe Darstellungen der Kleeneschen Algebra der regulaeren Mengen* (**J** 1670) Mitt Math Ges DDR 1973/2-3*161-171
◇ D05 ◇ REV Zbl 301 # 94067 • ID 66012

WECHSUNG, G. [1973] *Quasisequentielle Funktionen* (**J** 0380) Acta Cybern (Szeged) 2*24-33
◇ D05 E20 ◇ REV MR 47 # 8742 Zbl 256 # 94055
• ID 42614

WECHSUNG, G. [1975] *Eine algebraische Charakterisierung der linearen Sprachen* (**J** 0129) Elektr Informationsverarbeitung & Kybern 11*19-25
◇ D05 ◇ REV MR 52 # 4734 Zbl 304 # 68077 • ID 66011

WECHSUNG, G. [1975] *Minimale and optimale "Blumsche Masse" (English and Russian summaries)* (**J** 0129) Elektr Informationsverarbeitung & Kybern 11*673-679
◇ D15 D20 ◇ REV MR 53 # 12906 Zbl 323 # 02051 • ID 23158

WECHSUNG, G. [1975] *The axiomatization problem of a theory of linear languages* (**P** 1755) Math Founds of Comput Sci (3);1974 Jadwisin 298-302
◇ D05 ◇ REV Zbl 312 # 68043 JSL 42.422 • ID 66013

WECHSUNG, G. [1976] *Kompliziertheitstheoretische Charakterisierung der kontextfreien und linearen Sprachen* (**J** 0129) Elektr Informationsverarbeitung & Kybern 12*289-300
◇ D05 D15 ◇ REV MR 56 # 7358 Zbl 346 # 68024 • ID 42616

WECHSUNG, G. [1977] see WAGNER, K.

WECHSUNG, G. [1977] *Properties of complexity classes -- a short survey* (**P** 1635) Math Founds of Comput Sci (6);1977 Tatranska Lomnica 177-191
◇ D15 D98 ◇ REV MR 58 # 8499 Zbl 361 # 68077 • ID 83120

WECHSUNG, G. [1979] *A crossing measure for 2-tape Turing machines* (**P** 2059) Math Founds of Comput Sci (8);1979 Olomouc 508-516
◇ D10 D15 ◇ REV MR 81e:03034 Zbl 404 # 68055 • ID 54860

WECHSUNG, G. [1979] see BRANDSTAEDT, A.

WECHSUNG, G. [1979] *The oscillation complexity and a hierarchy of context-free languages* (**P** 2935) FCT'79 Fund of Comput Th;1979 Berlin/Wendisch-Rietz 508-515
◇ D05 ◇ REV Zbl 412 # 68040 • ID 42618

WECHSUNG, G. [1980] *A note on the return complexity (German and Russian summaries)* (**J** 0129) Elektr Informationsverarbeitung & Kybern 16*139-146
◇ D15 ◇ REV MR 83a:68059 Zbl 453 # 68017 • ID 38868

WECHSUNG, G. [1980] see WAGNER, K.

WECHSUNG, G. [1985] *On sparse complete sets* (**J** 0068) Z Math Logik Grundlagen Math 31*281-287
◇ D15 ◇ REV Zbl 553 # 03027 • ID 47577

WECHSUNG, G. [1985] *On sparse complete sets* (**J** 0129) Elektr Informationsverarbeitung & Kybern 21*253-254
◇ D15 ◇ ID 49614

WECHSUNG, G. [1985] *On the Boolean closure of NP* (**P** 4647) FCT'85 Fund of Comput Th;1985 Cottbus 485-493
◇ D15 ◇ ID 49067

WECHSUNG, G. see Vol. I, III for further entries

WEEG, G.P. [1965] *The automorphism group of the direct product of strongly related automata* (**J** 0037) ACM J 12*187-195
◇ D05 G20 ◇ REV MR 31 # 4728 Zbl 125.279 • ID 14104

WEESE, M. [1972] *Zur Modellvollstaendigkeit und Entscheidbarkeit gewisser topologischer Raeume (Russian, English and French summaries)* (**J** 0115) Wiss Z Humboldt-Univ Berlin, Math-Nat Reihe 21*477-485
◇ B25 C35 C65 D80 ◇ REV MR 48 # 5852 Zbl 255 # 02053 • ID 14105

WEESE, M. [1976] *The universality of boolean algebras with the Haertig quantifier* (**P** 1476) Set Th & Hierarch Th (2) (Mostowski);1975 Bierutowice 291-296
◇ C55 C80 D35 F25 G05 ◇ REV MR 54 # 12478 Zbl 331 # 02027 • ID 23814

WEESE, M. [1977] *The undecidability of well-ordering with the Haertig quantifier (Russian summary)* (**J** 0014) Bull Acad Pol Sci, Ser Math Astron Phys 25*89-91
◇ C55 C65 C80 D35 E07 ◇ REV MR 55 # 7737 Zbl 315 # 02046 • ID 26551

WEESE, M. [1981] *Decidability with respect to Haertig quantifier and Rescher quantifier* (**J** 0068) Z Math Logik Grundlagen Math 27*569-576
◇ B25 C55 C80 D35 ◇ REV MR 84i:03075 Zbl 503 # 03012 • ID 33324

WEESE, M. [1982] see SEESE, D.G.

WEESE, M. [1984] *The theory of Boolean algebras extended by a group of automorphisms* (**S** 3382) Sem-ber, Humboldt-Univ Berlin, Sekt Math 60*218-222
◇ C07 D35 G05 ◇ REV MR 86h:03110 Zbl 561 # 03005 • ID 42386

WEESE, M. [1984] *Undecidable extensions of the theory of Boolean algebras* (**S** 3231) Prepr NF, Sekt Math, Humboldt-Univ Berlin 89*30pp
◇ D35 G05 ◇ ID 42384

WEESE, M. see Vol. I, II, III, V for further entries

WEGENER, I. [1981] *An improved complexity hierarchy on the depth of boolean functions* (**J** 1431) Acta Inf 15*147-152
◇ B05 B70 D15 ◇ REV MR 82i:94036 Zbl 431 # 94058 • ID 69392

WEGENER, I. [1981] *Boolean functions whose monotone complexity is of size $n^2 \log n$* (**P** 3475) Theor Comput Sci (5);1981 Karlsruhe 104*22-31
◇ B05 D15 ◇ REV Zbl 456 # 94023 • ID 66300

WEGENER, I. see Vol. I, II for further entries

WEGMAN, M.N. [1980] see BLUM, M.

WEGNER, L. [1979] *Bracketed two-level grammars -- a decidable and practical approach to language definitions* (**P** 1873) Automata, Lang & Progr (6);1979 Graz 668-682
◇ D05 ◇ REV MR 82e:68088 Zbl 411 # 68061 • ID 83126

WEGNER, L. [1980] see ALBERT, J.

WEGNER, L. [1981] see ALBERT, J.

WEICKER, R. [1971] *Tabulator-Turingmaschinen und Komplexitaet (English summary)* (**J** 0373) Comp Arch Inform & Numerik 7*264-274 • ERR/ADD ibid 9*165-167
◇ D10 D15 ◇ REV MR 46 # 3277 Zbl 242 # 02042 • ID 14122

WEICKER, R. [1974] *Turing machines with associative memory access* (**P** 1869) Automata, Lang & Progr (2);1974 Saarbruecken 14*458-472
◇ D10 D15 ◇ REV MR 55 # 1828 Zbl 286 # 68025 • ID 66034

WEIHRAUCH, K. [1973] see ROSE, G.F.

WEIHRAUCH, K. [1973] see HENKE VON, F.W.

WEIHRAUCH, K. [1974] *Teilklassen primitiv-rekursiver Wortfunktionen* (**X** 0817) Ges Math Datenverarbeit: Bonn 49pp
◇ D20 ◇ REV MR 51 # 108 Zbl 293 # 02026 • ID 17465

WEIHRAUCH, K. [1975] see HENKE VON, F.W.

WEIHRAUCH, K. [1975] *Program schemata with polynomial bounded counters* (**J** 0232) Inform Process Lett 3*91-96
◇ B75 D15 ◇ REV Zbl 302 # 68031 • ID 66036

WEIHRAUCH, K. [1977] *A generalized computability thesis* (**P** 2588) FCT'77 Fund of Comput Th;1977 Poznan 538-542
◇ D45 ◇ REV MR 57 # 16022 Zbl 368 # 02039 • ID 51256

WEIHRAUCH, K. [1977] see VERBEEK, R.

WEIHRAUCH, K. [1978] see VERBEEK, R.

WEIHRAUCH, K. [1980] *Berechenbarkeit auf CPO-S (Eine Vorlesung ausgearbeitet von Thomas Deil)* (**S** 1642) Schr Inf Angew Math, Ber (Aachen) 63*101pp
◇ D45 D75 ◇ REV Zbl 475 # 03023 • ID 55477

WEIHRAUCH, K. [1980] see REISER, A.

WEIHRAUCH, K. [1981] see SCHAEFER, G.

WEIHRAUCH, K. [1981] *Recursion and complexity theory on CPO's* (**P** 3475) Theor Comput Sci (5);1981 Karlsruhe 195-202
◇ D15 D45 D75 ◇ REV Zbl 464 # 03041 • ID 46482

WEIHRAUCH, K. [1982] see KREITZ, C.

WEIHRAUCH, K. [1983] see SCHAEFER, G.

WEIHRAUCH, K. [1985] see KREITZ, C.

WEIHRAUCH, K. [1985] *Type 2 recursion theory* (**J** 1426) Theor Comput Sci 38*17-33
◇ D20 ◇ ID 48598

WEIHRAUCH, K. see Vol. VI for further entries

WEIMANN, B. [1973] see CASPAR, K.

WEINBAUM, C.M. [1966] *Visualizing the word problem, with an application to sixth groups* (**J** 0048) Pac J Math 16*557-578
◇ D40 ◇ REV MR 35 # 241 Zbl 146.33 • ID 14123

WEINBAUM, C.M. [1971] *The word and conjugacy problems for the knot group of any tame, prime, alternating knot* (**J** 0053) Proc Amer Math Soc 30*22-26
◇ D40 ◇ REV MR 43 # 4895 Zbl 228 # 55004 • ID 14124

WEINBERG, G.M. [1967] *Computing machines* (**C** 0601) Encycl of Philos 2*168-173
◇ B35 D10 ◇ REV JSL 35.298 • ID 14125

WEINER, P. [1967] see HOPCROFT, J.E.

WEINER, P. see Vol. I for further entries

WEISPFENNING, V. [1976] *Negative-existentially complete structures and definability in free extensions* (**J** 0036) J Symb Logic 41*95-108
◇ C25 C40 C60 C75 D40 G05 ◇ REV MR 54 # 2451 Zbl 335 # 02035 • ID 14747

WEISPFENNING, V. [1985] *Quantifier elimination for modules* (**J** 0009) Arch Math Logik Grundlagenforsch 25*1-11
◇ C10 C35 C57 C60 ◇ ID 48460

WEISPFENNING, V. [1985] *The complexity of elementary problems in archimedian ordered groups* (**P** 4601) EUROCAL;1985 Linz 2*87-88
◇ B25 C10 D15 D40 ◇ ID 48461

WEISPFENNING, V. see Vol. II, III for further entries

WEISS, M. [1967] *Axiomatische Untersuchungen zur elementaren Theorie der freien Halbgruppen mit Substitution als undefiniertem Grundbegriff* (**J** 0068) Z Math Logik Grundlagen Math 13*265-280
◇ B03 D05 ◇ REV MR 35 # 6540 Zbl 229 # 20056 • ID 14130

WEISS, M. see Vol. I, V for further entries

WEISS, T. [1983] *Projective sets (Polish)* (**J** 0519) Wiad Mat, Ann Soc Math Pol, Ser 2 25*51-64
◇ D55 E15 E98 ◇ REV MR 85g:03073 • ID 43895

WEITKAMP, G. [1982] *Analytic sets having incomparable Kleene degrees* (**J** 0036) J Symb Logic 47*860-868
◇ D30 D55 D65 E15 ◇ REV MR 85d:03091 Zbl 527 # 03029 • ID 37577

WEITKAMP, G. [1982] *Iterating the superjump along definable prewellorderings* (**J** 0068) Z Math Logik Grundlagen Math 28*385-394
◇ D55 D60 D65 ◇ REV MR 84g:03069 Zbl 523 # 03034 • ID 34176

WEITKAMP, G. [1983] see SIMPSON, S.G.

WEITKAMP, G. [1985] see KALANTARI, I.

WEITKAMP, G. [1985] *On the existence and recursion theoretic properties of Σ^1_n-generic sets of reals* (**J** 0068) Z Math Logik Grundlagen Math 31*97-108
◇ D55 D65 E15 E40 ◇ ID 42291

WEITKAMP, G. [1985] see MANSFIELD, R.

WELCH, L.V. [1984] *A hierarchy of families of recursively enumerable degrees* (**J** 0036) J Symb Logic 49*1160-1170
◇ D25 ◇ ID 42512

WELCH, P. [1984] *On Σ^1_3* (**P** 2153) Logic Colloq;1983 Aachen 1*473-484
◇ D55 E15 E55 ◇ REV Zbl 574 # 03037 • ID 45409

WELCH, P. [1985] *Comparing incomparable Kleene degrees* (**J** 0036) J Symb Logic 50*55-58
◇ D65 E35 E45 ◇ ID 42522

WELCH, P. see Vol. V, VI for further entries

WELSH, D.J.A. [1982] *Problems in computational complexity* (**P** 4033) Appl of Combin;1981 Milton Keynes 6*75-85
◇ D15 ◇ REV MR 83k:68041 • ID 40333

WELSH, D.J.A. [1983] *Randomised algorithms* (**J** 2702) Discr Appl Math 5*133-145
◇ D15 ◇ REV MR 84e:68046 Zbl 506 # 68040 • ID 39303

WELZL, E. [1983] see CULIK II, K.

WELZL, E. [1984] *Encoding graphs by derivations and implications for the theory of graph grammars* (**P** 4012) Automata, Lang & Progr (11);1984 Antwerpen 503-513
◇ D05 ◇ REV Zbl 557 # 68050 • ID 47192

WELZL, E. [1985] see LANGE, K.-J.

WERNER, GEORGES [1970] *Quelques remarques sur la complexite des algorithmes (English summary)* (**J** 3954) Rev Franc Inf & Rech Operat 4/R-2*33-50
◇ D15 D20 ◇ REV MR 45 # 6628 Zbl 209.33 • ID 14140

WERNER, GEORGES [1971] *Propriete d'invariance des classes de fonctions de complexite bornee* (**J** 2313) C R Acad Sci, Paris, Ser A-B 273*A133-A136
◇ D15 D20 ◇ REV MR 49 # 11861 Zbl 221 # 02020 • ID 24209

WERNER, GEORGES [1974] *Sous-classes recursivement enumerables d'une classe de complexite* (**P** 0787) C R Journ Soc Math France;1974 Montpellier 369-375
◇ D15 D20 ◇ REV MR 52 # 5389 • ID 18426

WERNER, GUENTER [1975] *Prognose von Folgen* (**J** 0129) Elektr Informationsverarbeitung & Kybern 11*649-653
◇ D20 ◇ REV Zbl 325 # 02024 • ID 30439

WERNER, GUENTER [1976] see LINDNER, R.

WESEP VAN, R.A. [1978] *Separation principles and the axiom of determinateness* (**J** 0036) J Symb Logic 43*77-81
◇ D55 D75 E15 E60 ◇ REV MR 81e:03050 Zbl 396 # 03042 • ID 29243

WESEP VAN, R.A. [1978] *Wadge degrees and descriptive set theory* (**C** 2908) Cabal Seminar Los Angeles 1976-77 151-170
◇ D30 E15 E60 ◇ REV MR 80i:03058 Zbl 393 # 03037 • ID 52457

WESEP VAN, R.A. see Vol. V for further entries

WETTE, E. [1974] *The refutation of number theory. I* (**J** 0286) Int Logic Rev 10*8pp
◇ D10 F30 ◇ REM Without use of any natural language. Printed in two colours • ID 31553

WETTE, E. [1976] *On the formalization of productive logic* (**J** 0286) Int Logic Rev 13*23-33
◇ B60 D20 ◇ REV Zbl 362 # 02018 • ID 31554

WETTE, E. see Vol. I, III, V, VI for further entries

WEYUKER, E.J. [1979] *The applicability of program schema results to programs* (**J** 0435) Int J Comput & Inf Sci 8*387-403
◇ B75 D80 ◇ REV MR 81d:68026 Zbl 427 # 68017 • ID 83144

WEYUKER, E.J. [1983] see DAVIS, MARTIN D.

WHEELER, W.H. [1972] *Algebraically closed division rings, forcing, and the analytic hierarchy* (0000) Diss., Habil. etc
◇ C25 C60 D55 ◇ REM Doct. diss., Yale University • ID 19104

WHEELER, W.H. see Vol. I, III for further entries

WHITMAN, P.M. [1961] *Status of word problems for lattices* (**P** 0620) Lattice Th;1959 Monterey 2*17-21
◇ D40 G10 ◇ REV MR 23 # A1560 • ID 16231

WHITMAN, P.M. see Vol. V for further entries

WICHMANN, B. [1976] *Ackermann's function: A study in the efficiency of calling procedures* (**J** 0130) BIT 16*103-110
◇ D15 D20 ◇ REV MR 54 # 11851 Zbl 323 # 68010 • ID 66079

WICK, G. [1976] see SPECKER, E.

WIEDMER, E. [1980] *Computing with infinite objects* (**J** 1426) Theor Comput Sci 10*133-155
◇ D10 D20 ◇ REV MR 81c:68039 Zbl 473 # 68042 • ID 83148

WIEDMER, E. see Vol. I, VI for further entries

WIEHAGEN, R. [1975] *Dechiffrierung von ND-Automaten* (**J** 0129) Elektr Informationsverarbeitung & Kybern 11*39-60
◇ D05 ◇ REV MR 52 # 13031 Zbl 311 # 94044 • ID 42392

WIEHAGEN, R. [1975] *Inductive inference of recursive functions* (**P** 0454) Math Founds of Comput Sci (4);1975 Marianske Lazne 32*462-464
◇ D20 ◇ REV MR 53 # 109 Zbl 346 # 02018 • ID 16574

WIEHAGEN, R. [1976] see LIEPE, W.

WIEHAGEN, R. [1976] *Limes-Erkennung rekursiver Funktionen durch spezielle Strategien* (**J** 0129) Elektr Informationsverarbeitung & Kybern 12*93-99
◇ D20 ◇ REV MR 54 # 7228 Zbl 346 # 02019 • ID 25020

WIEHAGEN, R. [1976] *Primitiv-rekursive Erkennung* (**P** 2898) Algor Kompl, Lern-& Erkenn-Prozess;1976 Jena 121-131
◇ D20 ◇ REV MR 56 # 5246 Zbl 423 # 03046 • ID 53558

WIEHAGEN, R. [1977] *Identification of formal languages* (**P** 1635) Math Founds of Comput Sci (6);1977 Tatranska Lomnica 571-579
◇ B75 D05 ◇ REV MR 58 # 8547 Zbl 375 # 68032 • ID 42393

WIEHAGEN, R. [1977] see JUNG, H.

WIEHAGEN, R. [1978] *Characterization problems in the theory of inductive inference* (**P** 1872) Automata, Lang & Progr (5);1978 Udine 494-508
◇ D15 D20 D45 ◇ REV MR 80e:03044 Zbl 389 # 03015 • ID 52304

WIEHAGEN, R. [1979] see FREJVALD, R.V.

WIEHAGEN, R. [1980] see KLETTE, R.

WIEHAGEN, R. [1982] see FREJVALD, R.V.

WIEHAGEN, R. [1984] see FREJVALD, R.V.

WIELE VAN DE, J. [1982] *Recursive dilators and generalized recursions* (**P** 3708) Herbrand Symp Logic Colloq;1981 Marseille 325-332
◇ D60 D65 F15 F35 ◇ REV MR 85i:03184 Zbl 499 # 03035 • ID 38124

WIENS, D. [1976] see JONES, JAMES P.

WIETLISBACH, M.N. [1981] *Zur Komplexitaet von Entscheidungsalgorithmen, die auf dem Herbrand'schen Satz und regulaerer Resolution beruhen* (**X** 2710) Eidgen Techn Hochsch: Zuerich 112pp
◇ B35 D15 ◇ REV Zbl 516 # 03037 • ID 37266

WIJNGAARDEN VAN, A. [1974] *The generative power of two-level grammars* (**P** 1869) Automata, Lang & Progr (2);1974 Saarbruecken 9-16
◇ D05 ◇ REV MR 55 # 6987 Zbl 291 # 68029 • ID 66092

WILKIE, A.J. [1980] *Applications of complexity theory to Σ_0-definability problems in arithmetic* (**P** 2625) Model Th of Algeb & Arithm;1979 Karpacz 363-369
◇ C62 D15 F30 ◇ REV MR 82b:03085 Zbl 483 # 03024 • ID 79968

WILKIE, A.J. [1981] see PARIS, J.B.

WILKIE, A.J. see Vol. I, III, V, VI for further entries

WILKS, Y. [1975] *Putnam and Clarke and mind and body* (J 0013) Brit J Phil Sci 26*213-225
◇ A05 D10 ◇ REV Zbl 382 # 03007 • ID 51931

WILLARD, S. [1971] *Some examples in the theory of Borel sets* (J 0027) Fund Math 71*187-191
◇ D55 E15 ◇ REV MR 45 # 4347 Zbl 222 # 54046 • ID 14209

WILLARD, S. see Vol. V for further entries

WILLE, R. [1972] see DAY, A.

WILLE, R. see Vol. V for further entries

WILLIAMS, J.H. [1981] *Formal representations for recursively defined functional programs* (P 2930) Formal of Progr Concepts;1981 Peniscola 460-470
◇ B75 D20 ◇ REV MR 82j:68009 Zbl 543 # 68005 • ID 83151

WILLIS, D.G. [1970] *Computational complexity and probability constructions* (J 0037) ACM J 17*241-259
◇ D10 D15 ◇ REV MR 43 # 4579 Zbl 233 # 68013 • ID 14225

WILLMOTT, R. [1968] see ROGERS, C.A.

WILLMOTT, R. [1969] *On the uniformization of Souslin $\mathscr{F}$ sets* (J 0053) Proc Amer Math Soc 22*148-155
◇ D55 E15 ◇ REV MR 39 # 2935 Zbl 176.519 • ID 14226

WILLMOTT, R. [1971] *Some relations between k-analytic sets and generalized Borel sets* (J 0027) Fund Math 71*263-271
◇ D55 E15 ◇ REV MR 45 # 4980 Zbl 229 # 04003 • ID 14227

WILLMOTT, R. [1975] *A form of Lusin's second separation theorem for k-analytic sets* (J 3172) J London Math Soc, Ser 2 12*213-218
◇ D55 E15 ◇ REV MR 54 # 12527 Zbl 317 # 54049 • ID 79991

WILLMOTT, R. see Vol. II, V for further entries

WILMERS, G.M. [1982] see CEGIELSKI, P.

WILMERS, G.M. see Vol. III, V, VI for further entries

WILSON, CHRISTOPHER B. [1982] see BOOK, R.V.

WINETT, J.M. [1962] *An α-state finite automaton for multiplication by α* (J 0072) IRE Trans Electr Comp EC-11*412-414
◇ D05 ◇ ID 14234

WINKLER, K.-D. [1973] see DAUSCHA, W.

WINKLER, P.M. [1980] *Classification of algebraic structures by work space* (J 0004) Algeb Universalis 11*320-333
◇ C05 D15 ◇ REV MR 83c:08007 Zbl 455 # 08003 • ID 39039

WINKLER, P.M. [1980] *Computational characterization of abelian groups* (P 2058) Kleene Symp;1978 Madison 423-425
◇ D05 D80 ◇ REV MR 82b:20051 • ID 83154

WINKLER, P.M. [1983] *Existence of graphs with a given set of r-neighborhoods* (J 0033) J Comb Th, Ser B 34*165-176
◇ D80 ◇ REV MR 85e:05135 Zbl 517 # 05056 • ID 39854

WINKLER, P.M. [1983] *Polynomial hyperforms* (J 0004) Algeb Universalis 17*101-109
◇ C05 D10 ◇ REV Zbl 558 # 08003 • ID 46630

WINKLER, P.M. see Vol. III for further entries

WINKLMANN, K. [1978] see MACHTEY, M.

WINKLMANN, K. [1979] see MEYER, A.R.

WINKLMANN, K. see Vol. II for further entries

WINKOWSKI, J. [1972] *Composed abstract machines (Russian summary)* (J 0014) Bull Acad Pol Sci, Ser Math Astron Phys 20*407-411
◇ D05 ◇ REV MR 46 # 8822 Zbl 293 # 68045 • ID 14235

WINKOWSKI, J. see Vol. I for further entries

WINOGRAD, S. [1980] see AUSLANDER, L.

WINSKEL, G. [1984] *A new definition of morphism on Petri nets* (P 3565) Symp of Th Aspects of Comput Sci (1);1984 Paris 140-150
◇ D05 ◇ REV MR 85i:68002 Zbl 559 # 68058 • ID 47090

WINSKEL, G. see Vol. II for further entries

WIRSING, M. [1977] *Das Entscheidungsproblem der Klasse von Formeln, die hoechstens zwei Primformeln enthalten* (J 0504) Manuscr Math 22*13-25
◇ B20 B25 D35 ◇ REV MR 57 # 12199 Zbl 365 # 02035 • ID 51036

WIRSING, M. [1978] *Kleine unentscheidbare Klassen der Praedikatenlogik mit Identitaet und Funktionszeichen* (J 0009) Arch Math Logik Grundlagenforsch 19*97-109
◇ B20 D35 ◇ REV MR 80a:03054a Zbl 398 # 03005 • ID 29168

WIRSING, M. [1979] see BROY, M.

WIRSING, M. [1979] *Small universal Post systems* (J 0068) Z Math Logik Grundlagen Math 25*559-564
◇ D03 ◇ REV MR 80m:03075 Zbl 467 # 03035 • ID 55033

WIRSING, M. [1981] see BERGSTRA, J.A.

WIRSING, M. see Vol. II for further entries

WISNIEWSKI, K. [1979] *A notion of the acceptance of infinite sequences by finite automata* (J 3293) Bull Acad Pol Sci, Ser Math 27*331-332
◇ D05 E15 ◇ REV MR 80k:68033 Zbl 415 # 03026 • ID 53128

WISNIEWSKI, K. see Vol. V for further entries

WOEHL, K. [1979] *Zur Komplexitaet der Presburger Arithmetik und des Aequivalenz-Problems einfacher Programme* (P 3488) Theor Comput Sci (4);1979 Aachen 310-318
◇ B25 B75 D15 F20 F30 ◇ REV MR 81k:03015 Zbl 419 # 03024 • ID 53367

WOJCIECHOWSKI, J. [1984] *Classes of transfinite sequences accepted by nondeterministic finite automata* (J 2095) Fund Inform, Ann Soc Math Pol, Ser 4 7*191-223
◇ D05 ◇ REV Zbl 553 # 68048 • ID 43439

WOJCIECHOWSKI, J. [1985] *Finite automata on transfinite sequences and regular expressions* (J 2095) Fund Inform, Ann Soc Math Pol, Ser 4 8/3-4*379-396
◇ D05 ◇ ID 49479

WOLFE, P. [1955] *The strict determinateness of certain infinite games* (**J** 0048) Pac J Math 5*841-847
◇ D55 E05 E15 E60 ◇ REV MR 17.506 Zbl 66.380
• ID 14269

WOLFRAM, S. [1985] *Undecidability and intractability in theoretical physics* (**J** 2730) Phys Rev Lett 54*735-738
◇ D80 ◇ REV MR 86d:03046 • ID 44471

WOLPER, P. [1985] see SISTLA, A.P.

WOLPER, P. see Vol. II for further entries

WOLTER, H. [1975] see HERRE, H.

WOLTER, H. [1984] see DAHN, B.I.

WOLTER, H. [1984] *Some remarks on exponential functions in ordered fields* (**P** 1545) Easter Conf on Model Th (2);1984 Wittenberg 229-243
◇ C25 C65 D35 ◇ REV MR 86h:03065 Zbl 568 # 03009
• ID 44686

WOLTER, H. see Vol. I, III, V, VI for further entries

WOLTER, U. [1985] see DASSOW, J.

WOOD, D. [1976] *Iterated a-NGSM maps and Γ systems* (**J** 0194) Inform & Control 32*1-26
◇ D05 ◇ REV MR 54 # 4206 Zbl 346 # 68036 • ID 66145

WOOD, D. [1977] see MAURER, H.A.

WOOD, D. [1978] see MAURER, H.A.

WOOD, D. [1978] see CULIK II, K.

WOOD, D. [1979] see CULIK II, K.

WOOD, D. [1979] see MAURER, H.A.

WOOD, D. [1980] see MAURER, H.A.

WOOD, D. [1981] see MAURER, H.A.

WOOD, D. [1981] see OTTMANN, T.

WOOD, D. [1982] see MAURER, H.A.

WOOD, D. [1984] see CHAZELLE, B.

WOTSCHKE, D. [1979] see GINSBURG, S.

WOTSCHKE, D. [1981] see BUCHER, W.

WRATHALL, C. [1977] *Complete sets and the polynomial-time hierarchy* (**J** 1426) Theor Comput Sci 3*23-33
◇ D15 ◇ REV MR 55 # 11717 Zbl 366 # 02031 • ID 51114

WRATHALL, C. [1977] see BOOK, R.V.

WRATHALL, C. [1978] see BOOK, R.V.

WRATHALL, C. [1978] *Rudimentary predicates and relative computation* (**J** 1428) SIAM J Comp 7*194-209
◇ D10 D15 ◇ REV MR 58 # 3754 Zbl 375 # 68030
• ID 51640

WRATHALL, C. [1979] see BOOK, R.V.

WRATHALL, C. [1981] see BOOK, R.V.

WRATHALL, C. [1982] see BOOK, R.V.

WRATHALL, C. [1985] see OTTO, F.

WRIGHT, J.B. [1954] see BURKS, A.W.

WRIGHT, J.B. [1962] see BURKS, A.W.

WRIGHT, J.B. [1965] see MEZEI, J.E.

WRIGHT, J.B. [1967] see MEZEI, J.E.

WRIGHT, J.B. [1967] see EILENBERG, S.

WRIGHT, J.B. [1968] see THATCHER, J.W.

WRIGHT, J.B. [1972] *Characterization of recursively enumerable sets* (**J** 0036) J Symb Logic 37*507-511
◇ D25 ◇ REV MR 47 # 3154 Zbl 262 # 02033 • ID 14331

WRIGHT, J.B. [1975] see GOGUEN, J.A.

WRIGHT, J.B. see Vol. I, III, V for further entries

WU, A. [1980] see DUBITZKI, T.

WU, YUNZENG [1979] *Two problems in contemporary mathematical logic - CH and P= ?NP (Chinese)* (**J** 4452) Zhexue Yanjiu
◇ D15 E50 ◇ ID 49441

WU, YUNZENG see Vol. I, II for further entries

WUETHRICH, H.R. [1976] *Ein Entscheidungsverfahren fuer die Theorie der reell abgeschlossenen Koerper* (**P** 3196) Kompl von Entscheid Probl;1973/74 Zuerich 138-162
• TRANSL [1981] (**J** 3079) Kiber Sb Perevodov, NS 18*100-124
◇ B25 C60 D15 ◇ REV MR 57 # 18232 Zbl 363 # 02052
• ID 43103

WYLLIE, J. [1978] see FORTUNE, S.

XU, MEIRUI [1982] see BOOK, R.V.

XU, MEIRUI [1983] see BOOK, R.V.

YABANZHI, G.G. [1981] *The word problem for some groups of the variety N_2A (Russian)* (**J** 0967) Izv Akad Nauk Mold SSR, Ser Fiz-Tekh Mat 1981/1*39-44,94
◇ D40 ◇ REV MR 82m:20037 Zbl 467 # 20031 • ID 81679

YABLON, P. [1975] *A generalized propositional calculus* (**J** 0047) Notre Dame J Formal Log 16*295-297
◇ B60 D35 ◇ REV MR 51 # 5259 Zbl 236 # 02021
• ID 17472

YABLONSKIJ, S.V. [1980] *On some results in the theory of function systems (Russian)* (**P** 1959) Int Congr Math (II,13);1978 Helsinki 2*963-971
• TRANSL [1981] (**J** 0225) Amer Math Soc, Transl, Ser 2 117*39-46
◇ B50 D20 G25 ◇ REV MR 81j:03038 Zbl 473 # 03019
• ID 55350

YABLONSKIJ, S.V. see Vol. I, II, V for further entries

YACOBI, Y. [1980] see EVEN, S.

YACOBI, Y. [1981] see EVEN, S.

YACOBI, Y. [1982] see EVEN, S.

YAGZHEV, A.V. [1980] *Algorithmic problem of recognizing automorphisms among endomorphisms of free associative algebras of finite rank (Russian)* (**J** 0092) Sib Mat Zh 21/1*193-199,238
• TRANSL [1980] (**J** 0475) Sib Math J 21*142-146
◇ D40 ◇ REV MR 81d:08006 Zbl 433 # 16025 • ID 54049

YAJIMA, S. [1972] see KAMBAYASHI, Y.

YAKU, T. [1973] *The constructibility of a configuration in a cellular automaton* (**J** 0119) J Comp Syst Sci 7*481-496
◇ D05 ◇ REV MR 48 # 10724 Zbl 271 # 94037 • ID 66167

YAKU, T. [1976] *Surjectivity of nondeterministic parallel maps induced by nondeterministic cellular automata* (**J** 0119) J Comp Syst Sci 12*1-5
◇ D05 ◇ REV MR 54#4871 Zbl 338#94031 • ID 66168

YAKU, T. [1985] *Wiring a Turing machine in the cellular automaton and the surjectivity problem for a parallel map* (**J** 0463) Proc Fac Sci Tokai Univ 20*51-72
◇ D05 D10 ◇ ID 48214

YAKUBAJTIS, EH.A. [1976] see CHAPENKO, V.P.

YAKUBAJTIS, EH.A. see Vol. I for further entries

YAMADA, H. [1960] see MCNAUGHTON, R.

YAMADA, H. [1962] *Real-time computation and recursive functions not real-time computable* (**J** 0072) IRE Trans Electr Comp EC-11*753-760 • ERR/ADD ibid EC-12*400
• TRANSL [1970] (**C** 1540) Probl Mat Log: Slozh Algor & Kl Vychisl Funk 139-155
◇ D15 ◇ REV MR 27#2141 Zbl 124.250 JSL 31.656
• ID 18008

YAMADA, H. [1964] see MCNAUGHTON, R.

YAMADA, H. [1972] see JOSHI, A.K.

YAMASAKI, H. [1979] *On multitape automata* (**P** 2059) Math Founds of Comput Sci (8);1979 Olomouc 533-541
◇ D05 ◇ REV MR 81e:68067 Zbl 414#68029 • ID 53096

YAMASAKI, H. [1984] see KOBAYASHI, K.

YAMASAKI, H. [1985] see KOBAYASHI, K.

YAMASAKI, S. [1980] see DOSHITA, S.

YAMASAKI, S. [1983] see DOSHITA, S.

YAMASAKI, S. see Vol. I for further entries

YANG, DONGPING [1964] see HU, SHIHUA

YANG, DONGPING [1965] see HU, SHIHUA

YANG, DONGPING [1965] *On creative pair and productive pair (Chinese)* (**J** 0420) Shuxue Jinzhan 8/4*414-416
◇ D25 ◇ ID 48547

YANG, DONGPING [1966] *Primitive recursive simple sets and their hierarchy (Chinese)* (**J** 0420) Shuxue Jinzhan 9*97-101
◇ D20 D25 ◇ REV MR 36#3652 • ID 18010

YANG, DONGPING [1979] *α-operator gap theorem (Chinese)* (**J** 3793) Jisuanjii Xuebao 2*163-173
◇ D15 D60 ◇ ID 48548

YANG, DONGPING [1979] *The existence of nonmitotic α-recursively enumerable sets (Chinese) (English summary)* (**J** 0418) Shuxue Xuebao 22*195-203
◇ D60 ◇ REV MR 80h:03066 Zbl 398#03033 • ID 52761

YANG, DONGPING [1980] *α-splitting theorem restricted to Δ^0_2 sets (Chinese)* (**J** 0418) Shuxue Xuebao 23*730-739
◇ D55 D60 ◇ REV MR 82h:03044 Zbl 530#03023
• ID 80155

YANG, DONGPING [1984] *On the embedding of α-recursive presentable lattices into the α-recursive degrees below $0'$* (**J** 0036) J Symb Logic 49*488-502
◇ D30 D60 ◇ REV MR 85m:03032 • ID 42516

YANKOV, V.A. [1941] *Sur l'uniformisation des ensembles A (Russian) (French summary)* (**J** 0023) Dokl Akad Nauk SSSR 30*597-598
◇ D55 E15 ◇ REV MR 3.225 Zbl 24.385 • ID 24916

YANKOV, V.A. see Vol. II, III, VI for further entries

YANOV, YU.I. [1958] *On logical schemata of algorithms (Russian)* (**J** 0052) Probl Kibern 1*75-127
◇ B75 D20 ◇ REV MR 24#B1735 Zbl 85.340 JSL 27.362
• ID 00381

YANOV, YU.I. [1962] *On identical transformations of regular expressions (Russian)* (**J** 0023) Dokl Akad Nauk SSSR 147*327-330
• TRANSL [1962] (**J** 0062) Sov Math, Dokl 3*1630-1634
◇ D05 ◇ REV MR 26#29 Zbl 129.262 • ID 06519

YANOV, YU.I. [1964] *Invariant operations over events (Russian)* (**J** 0052) Probl Kibern 12*253-258
◇ D05 ◇ REV MR 32#2289 Zbl 264#02034 • ID 27506

YANOV, YU.I. [1966] *Certain subalgebras of events having no finite complete systems of identities (Russian)* (**J** 0052) Probl Kibern 17*255-258
◇ C05 D05 ◇ REV MR 35#5263 Zbl 199.44 • ID 16225

YANOV, YU.I. [1968] *The logical transformations of schemes of algorithms (Russian)* (**J** 0052) Probl Kibern 20*201-216
◇ D20 ◇ REV MR 44#3771 Zbl 195.306 • ID 06521

YANOV, YU.I. [1973] *Ueber das Problem aequivalenter Umformungen (Russian)* (**J** 1670) Mitt Math Ges DDR 1973/2-3*47-58
◇ D05 ◇ REV Zbl 267#02021 • ID 29885

YANOV, YU.I. [1975] *On some semantic characteristics of Turing machines (Russian)* (**J** 0023) Dokl Akad Nauk SSSR 224*301-304
• TRANSL [1975] (**J** 0062) Sov Math, Dokl 16*1213-1217
◇ D10 ◇ REV MR 52#7878 Zbl 334#02021 • ID 18207

YANOV, YU.I. [1977] *A convolution method for the solution of properties of formal systems (Russian)* (**S** 2651) Prepr Inst Prikl Mat, Akad Nauk SSSR 77/11*41pp
◇ D10 D20 ◇ REV MR 58#16202 • ID 74465

YANOV, YU.I. [1977] *Computations in a class of programs (Russian)* (**J** 0052) Probl Kibern 32*237-245,247
◇ B75 D20 F30 ◇ REV MR 58#10382 Zbl 415#03009
• ID 74466

YANOV, YU.I. [1977] *Equivalent transformations of computational trees (Russian) (English Summary)* (**S** 2651) Prepr Inst Prikl Mat, Akad Nauk SSSR 76*48pp
◇ B75 D20 ◇ REV MR 58#5146 • ID 74467

YANOV, YU.I. [1978] *Several theorems on convolutions (Russian) (English Summary)* (**S** 2651) Prepr Inst Prikl Mat, Akad Nauk SSSR 95*73pp
◇ D10 ◇ REV MR 80g:03036 Zbl 416#68045 • ID 74464

YANOV, YU.I. [1980] *The complete limited system of equivalent transformation rules for programs computing total functions (Russian)* (**J** 0052) Probl Kibern 37*215-239
◇ B75 D20 ◇ REV Zbl 468#03017 • ID 55082

YANOV, YU.I. see Vol. I, II, III for further entries

YAP, C.K. [1983] *Some consequences of non-uniform conditions on uniform classes* (**J** 1426) Theor Comput Sci 26*287-300
◇ D15 ◇ REV MR 85i:03133 Zbl 541#68017 • ID 41402

YASUDA, Y. [1976] *A note on the relativization of $\underset{\sim}{\Delta}^1_2$ subsets of Baire spaces* (**J** 3630) Bull Dept of Lib Arts (Numazu) 3*19-22
◇ D55 E15 ◇ ID 33404

YASUDA, Y. [1981] *On the existence of Cohen extensions and Σ_3^1 predicates* (J 3630) Bull Dept of Lib Arts (Numazu) 8*13-18
◇ D55 E15 E40 ◇ ID 33409

YASUDA, Y. [1981] *On the existence of Cohen extensions and Σ_3^1 predicates I* (P 4153) B-Val Anal & Nonstand Anal;1981 Kyoto 15-26
◇ D55 E15 E40 ◇ ID 37193

YASUDA, Y. see Vol. III, V, VI for further entries

YASUHARA, A. [1967] *A remark on Post normal systems* (J 0037) ACM J 14*167-171
◇ D03 ◇ REV MR 35 # 4104 Zbl 173.12 JSL 33.116
• ID 18026

YASUHARA, A. [1970] *The solvability of the word problem for certain semigroups* (J 0053) Proc Amer Math Soc 26*645-650
◇ D40 ◇ REV MR 42 # 3156 Zbl 213.20 • ID 18027

YASUHARA, A. [1971] *Recursive function theory and logic* (X 0801) Academic Pr: New York xv+338pp
◇ B98 D10 D20 D98 ◇ REV MR 47 # 1582
Zbl 254 # 02002 JSL 40.619 • ID 18028

YASUHARA, A. [1974] *Some non-recursive classes of Thue systems with solvable word problem* (J 0068) Z Math Logik Grundlagen Math 20*121-132
◇ D03 D40 ◇ REV MR 50 # 9551 Zbl 383 # 03024
• ID 18029

YASUHARA, A. [1983] see HAWRUSIK, F.M.

YATES, C.E.M. [1962] *Recursively enumerable sets and retracing functions* (J 0068) Z Math Logik Grundlagen Math 8*331-345
◇ D25 D50 ◇ REV MR 26 # 3598 Zbl 111.9 JSL 32.394
• ID 18037

YATES, C.E.M. [1965] *Three theorems on the degrees of recursively enumerable sets* (J 0025) Duke Math J 32*461-468
◇ D25 ◇ REV MR 31 # 4721 Zbl 134.8 JSL 32.394
• ID 18038

YATES, C.E.M. [1966] *A minimal pair of recursively enumerable degrees* (J 0036) J Symb Logic 31*159-168
◇ D25 ◇ REV MR 34 # 5677 Zbl 143.254 JSL 37.611
• ID 18041

YATES, C.E.M. [1966] *On the degrees of index sets* (J 0064) Trans Amer Math Soc 121*309-328
◇ D25 D30 D55 ◇ REV MR 32 # 2326 Zbl 143.254
JSL 39.344 • REM Part II 1969 • ID 18039

YATES, C.E.M. [1967] *Arithmetical sets and retracing functions* (J 0068) Z Math Logik Grundlagen Math 13*193-204
◇ D30 D50 D55 ◇ REV MR 37 # 6184 Zbl 162.315
JSL 40.453 • ID 18043

YATES, C.E.M. [1967] *Recursively enumerable degrees and the degrees less than* $0^{(1)}$ (P 0691) Sets, Models & Recursion Th;1965 Leicester 264-271
◇ D25 D30 ◇ REV MR 36 # 2504 Zbl 204.15 JSL 35.589
• ID 18042

YATES, C.E.M. [1969] *On the degrees of index sets II* (J 0064) Trans Amer Math Soc 135*249-266
◇ D25 D30 D55 ◇ REV MR 39 # 2637 Zbl 185.22
JSL 39.344 • REM Part I 1966 • ID 18044

YATES, C.E.M. [1970] *Initial segments of the degrees of unsolvability I: a survey* (P 1072) Math Log & Founds of Set Th;1968 Jerusalem 63-83
◇ D30 ◇ REV MR 42 # 4400 Zbl 207.12 • REM Part II 1970
• ID 20961

YATES, C.E.M. [1970] *Initial segments of the degrees of unsolvability - part II; minimal degrees* (J 0036) J Symb Logic 35*243-266
◇ D30 ◇ REV MR 43 # 53 Zbl 219 # 02031 • REM Part I 1970 • ID 18045

YATES, C.E.M. [1971] *A note on arithmetical sets of indiscernibles* (P 0638) Logic Colloq;1969 Manchester 443-451
◇ D55 E05 ◇ REV MR 43 # 4668 Zbl 219 # 02032
• ID 14508

YATES, C.E.M. [1972] *Degrees of unsolvability* (1111) Preprints, Manuscr., Techn. Reports etc.
◇ D30 ◇ REM Lecture Notes, University of Co... • ID 18046

YATES, C.E.M. [1972] *Initial segments and implications for the structure of degrees* (P 2080) Conf Math Log;1970 London 305-335
◇ D30 ◇ REV MR 50 # 9563 Zbl 236 # 02032 • ID 18047

YATES, C.E.M. [1974] *Prioric games and minimal degrees below* $\underline{0}^{(1)}$ (J 0027) Fund Math 82*217-237
◇ D25 D30 E60 ◇ REV MR 53 # 10570 Zbl 298 # 02042
• ID 18048

YATES, C.E.M. [1975] *A general framework for simple Δ_2^0 and Σ_1^0 priority arguments* (P 1521) Int Congr Math (II,12);1974 Vancouver 1*269-273
◇ D25 D55 ◇ REV MR 54 # 12507 Zbl 369 # 02018
• ID 51318

YATES, C.E.M. [1976] *Banach-Mazur games, comeager sets and degrees of unsolvability* (J 0332) Math Proc Cambridge Phil Soc 79*195-220
◇ D30 E60 E75 ◇ REV MR 53 # 10569 Zbl 344 # 02033
• ID 23077

YAU, S.S. [1972] see SMITH, L.W.

YE, DAXING [1979] see MO, SHAOKUI

YEH, R.T. [1971] *Some structural properties of generalized automata and algebras* (J 0041) Math Syst Theory 5*306-318
◇ C05 D05 ◇ REV MR 46 # 5218 Zbl 226 # 94040
• ID 18049

YEH, R.T. [1973] see HSIA, PEI

YEH, R.T. [1975] see HSIA, PEI

YEH, R.T. see Vol. II, V for further entries

YEHUDAI, A. [1980] *The decidability of equivalence for a family of linear grammars* (J 0194) Inform & Control 47*122-136
◇ D05 ◇ REV MR 82d:68049 Zbl 469 # 68074 • ID 66310

YEHUDAI, A. [1981] see HARRISON, M.A.

YEHUDAI, A. [1983] see ITZHAIK, Y.

YESHA, Y. [1983] see HARTMANIS, J.

YESHA, Y. [1983] *On certain polynomial-time truth-table reducibilities of complete sets to sparse sets* (J 1428) SIAM J Comp 12*411-425
◇ D15 ◇ REV MR 85g:68024 Zbl 545 # 03023 • ID 43504

YESHA, Y. [1984] see HARTMANIS, J.

YNTEMA, M.K. [1964] *A detailed argument for the Post-Linial theorems* (J 0047) Notre Dame J Formal Log 5*37-50
◇ B20 D35 ◇ REV MR 30 # 1031 Zbl 168.256 JSL 31.117 • ID 18050

YNTEMA, M.K. [1967] *Inclusion relations among families of context-free languages* (J 0194) Inform & Control 10*572-597
◇ D05 ◇ REV Zbl 207.314 • ID 18051

YOELI, M. [1965] *Lattice-ordered semigroups, graphs, and automata* (J 0514) SIAM Journ 13*411-422
◇ D05 ◇ REV MR 31 # 2344 Zbl 128.251 • ID 22318

YOELI, M. see Vol. I for further entries

YOKOMORI, T. [1980] *Stochastic characterizations of EOL languages* (J 0194) Inform & Control 45*26-33
◇ D05 ◇ REV MR 81f:68094 Zbl 443 # 68056 • ID 66311

YONEDA, M. [1985] see HIROSE, S.

YONEYAMA, N. [1962] see HUZINO, S.

YOSHIMOTO, Y. [1981] *The proof of sufficiency of McNaughton's condition on generalized sequentialmachine mappability-onto of regular sets* (J 0381) J Tsuda College (Tokyo) 13*57-77
◇ D05 ◇ REV MR 84f:03036 • ID 34456

YOUNG, P. [1964] *A note on pseudo-creative sets and cylinders* (J 0048) Pac J Math 14*749-753
◇ D25 D30 ◇ REV MR 29 # 18 Zbl 208.18 JSL 35.335 • ID 18059

YOUNG, P. [1964] *On reducibility by recursive functions* (J 0053) Proc Amer Math Soc 15*889-892
◇ D30 ◇ REV MR 31 # 3330 Zbl 134.8 • ID 18060

YOUNG, P. [1965] *On semi-cylinders, splinters, and bounded-truth-table reducibility* (J 0064) Trans Amer Math Soc 115*329-339
◇ D25 D30 ◇ REV MR 35 # 54 Zbl 163.252 JSL 35.335 • ID 18061

YOUNG, P. [1966] *A theorem on recursively enumerable classes and splinters* (J 0053) Proc Amer Math Soc 17*1050-1056
◇ D25 ◇ REV MR 34 # 7371 Zbl 207.306 • ID 15058

YOUNG, P. [1966] *Linear orderings under one-one reducibility* (J 0036) J Symb Logic 31*70-85
◇ D30 ◇ REV MR 38 # 48 Zbl 163.252 • ID 18062

YOUNG, P. [1967] *On pseudo-creative sets, splinters, and bounded truth-table reducibility* (J 0068) Z Math Logik Grundlagen Math 13*25-31
◇ D25 D30 ◇ REV MR 34 # 2457 Zbl 207.306 JSL 35.335 • ID 18063

YOUNG, P. [1968] *An effective operator, continuous but not partial recursive* (J 0053) Proc Amer Math Soc 19*103-108
◇ D20 ◇ REV MR 37 # 5093 Zbl 155.16 JSL 35.477 • ID 18064

YOUNG, P. [1968] see RITCHIE, R.W.

YOUNG, P. [1969] *Toward a theory of enumerations* (J 0037) ACM J 16*328-348
• TRANSL [1971] (J 3079) Kiber Sb Perevodov, NS 8*201-231 (Russian)
◇ D10 D15 D20 D25 D45 ◇ REV MR 39 # 2542 Zbl 231 # 02050 • ID 18065

YOUNG, P. [1970] see BASS, L.

YOUNG, P. [1971] *A note on dense and nondense families of complexity classes* (J 0041) Math Syst Theory 5*66-70
◇ D15 ◇ REV Zbl 216.291 • ID 26262

YOUNG, P. [1971] *A note on "axioms" for computational complexity and computation of finite functions* (J 0194) Inform & Control 19*377-386
◇ D15 ◇ REV MR 47 # 9889 Zbl 256 # 68014 • ID 66182

YOUNG, P. [1971] see HELM, J.P.

YOUNG, P. [1971] *Speed-up changing the order in which sets are enumerated* (J 0041) Math Syst Theory 5*148-156
• ERR/ADD ibid 7*352
◇ D15 D20 D25 ◇ REV MR 46 # 4783 MR 55 # 2546 Zbl 282 # 68017 • ID 18067

YOUNG, P. [1973] *Easy constructions in complexity theory: Gap and speed-up theorems* (J 0053) Proc Amer Math Soc 37*555-563
◇ D15 D20 ◇ REV MR 47 # 1323 Zbl 328 # 68044 • ID 18068

YOUNG, P. [1973] see HELM, J.P.

YOUNG, P. [1973] see BASS, L.

YOUNG, P. [1976] see MACHTEY, M.

YOUNG, P. [1977] *Optimization among provably equivalent programs* (J 0037) ACM J 24*693-700
◇ D15 ◇ REV MR 57 # 4585 Zbl 401 # 68016 • ID 66313

YOUNG, P. [1978] see MACHTEY, M.

YOUNG, P. [1978] see SHAY, M.

YOUNG, P. [1981] see JOSEPH, D.

YOUNG, P. [1981] see MACHTEY, M.

YOUNG, P. [1983] see COLLINS, W.J.

YOUNG, P. [1984] see SPREEN, D.

YOUNG, P. [1985] *Goedel theorems, exponential difficulty and undecidability of arithmetic theories: an exposition* (P 4046) Rec Th;1982 Ithaca 503-522
◇ D15 D35 F20 F30 ◇ REV Zbl 577 # 03023 • ID 46382

YOUNG, P. [1985] see MAHANEY, S.R.

YOUNG, P. [1985] see JOSEPH, D.

YOUNGER, D.H. [1967] *Recognition and parsing of context-free languages in time n^3 (Russian)* (J 0194) Inform & Control 10*189-208
• TRANSL [1970] (C 1540) Probl Mat Log: Slozh Algor & Kl Vychisl Funk 344-362
◇ D05 D15 ◇ REV Zbl 149.248 JSL 39.193 • ID 16894

YU, SHENG [1984] see CULIK II, K.

YUEH, KANG [1980] see JOSHI, A.K.

YUKAMI, T. [1984] *Some results on speed-up* (J 0260) Ann Jap Ass Phil Sci 6*195-205
◇ D15 F20 F30 ◇ REV MR 86c:03050 Zbl 545 # 03032 • ID 43511

YUKAMI, T. see Vol. I, VI for further entries

YUSHCHENKO, E.L. [1974] see TSEJTLIN, G.E.

YUSHCHENKO, E.L. [1975] see GLUSHKOV, V.M.

YUSHCHENKO, E.L. [1978] see TSEJTLIN, G.E.

YUSHCHENKO, E.L. see Vol. I for further entries

ZACHOS, S. [1982] *Robustness of probabilistic computational complexity classes under definitional perturbations* (J 0194) Inform & Control 54*143-154
◇ D15 ◇ REV MR 85e:68024 Zbl 529 # 68024 • ID 38489

ZACHOS, S. [1982] see PAPADIMITRIOU, C.H.

ZACHOS, S. [1984] see HELLER, H.

ZACHOS, S. [1985] see HINMAN, P.G.

ZAHORSKI, Z. [1948] *Sur la classe de Baire des derivees approximatives d'une fonction quelconque* (J 0283) Ann Soc Pol Math 21*306-323
◇ D55 E15 ◇ REV MR 11.89 Zbl 36.316 • ID 18091

ZAIONTZ, C. [1983] see BUECHI, J.R.

ZAIONTZ, C. see Vol. I, III for further entries

ZAK, S. [1979] *A Turing machine oracle hierarchy* (P 2059) Math Founds of Comput Sci (8);1979 Olomouc 542-551
◇ D10 D15 ◇ REV Zbl 413 # 68046 • ID 53041

ZAK, S. [1979] *A Turing machine space hierarchy* (J 0156) Kybernetika (Prague) 15*100-121
◇ D10 D15 ◇ REV MR 80i:68040 Zbl 421 # 68050 • ID 83222

ZAK, S. [1980] *A Turing machine oracle hierarchy I,II* (J 0140) Comm Math Univ Carolinae (Prague) 21*11-26,27-39
◇ D10 D15 ◇ REV MR 81k:03036 Zbl 429 # 68049 Zbl 429 # 68050 • ID 53894

ZAK, S. [1983] *A Turing machine time hierarchy* (J 1426) Theor Comput Sci 26*327-333
◇ D10 D15 ◇ REV MR 85j:68037 Zbl 525 # 68026 • ID 46796

ZAKHAROV, D.A. [1970] *Recursive functions (Russian)* (X 0913) Novosibirsk Gos Univ: Novosibirsk 206pp
◇ D20 D98 ◇ REV MR 57 # 16013 • ID 80199

ZAKHAROV, D.A. [1979] see DEGTEV, A.N.

ZAKHAROV, D.A. see Vol. III for further entries

ZAKHAROV, S.D. [1982] *The algebra of enumeration operators (Russian) (English summary)* (J 0288) Vest Ser Mat Mekh, Univ Moskva 1982/5*7-11,89
• TRANSL [1982] (J 0510) Moscow Univ Math Bull 37/5*7-11
◇ D20 D45 ◇ REV MR 84g:03068 Zbl 541 # 03024 • ID 34175

ZAKHAROV, S.D. [1984] *e- and s-degrees (Russian)* (J 0003) Algebra i Logika 23*395-406,478
• TRANSL [1984] (J 0069) Algeb and Log 23*273-281
◇ D30 ◇ ID 44818

ZAKIR'YANOV, K.KH. [1978] see BLOSHCHITSYN, V.YA.

ZAKOWSKI, W. [1973] *Multidimensional continuous machines* (J 1929) Prace Centr Oblicz Pol Akad Nauk 120*12pp
◇ D20 ◇ REV Zbl 272 # 68039 • ID 66200

ZAKOWSKI, W. [1973] *On some properties of N-dimensional simple continuous machines* (J 1929) Prace Centr Oblicz Pol Akad Nauk 128*18pp
◇ D20 ◇ REV Zbl 275 # 68010 • ID 66201

ZAKOWSKI, W. [1974] *On τ-computability of functions and sets of functions of n real variables* (J 0014) Bull Acad Pol Sci, Ser Math Astron Phys 22*959-962
◇ D20 ◇ REV Zbl 303 # 68030 • ID 18098

ZAKOWSKI, W. [1974] *On some properties of N-dimensional simple continuous machines* (J 0014) Bull Acad Pol Sci, Ser Math Astron Phys 22*81-85
◇ D20 ◇ REV MR 55 # 4790 Zbl 283 # 68029 • ID 18099

ZAKOWSKI, W. [1975] *On τ-computability and almost τ-computability of functions of n real variables* (J 1008) Demonstr Math (Warsaw) 7*543-553
◇ D20 ◇ REV MR 51 # 5274 Zbl 295 # 68050 • ID 66199

ZAKOWSKI, W. [1975] see NABIALEK, I.

ZAKOWSKI, W. [1975] *The (Z,Q)-machines* (J 0014) Bull Acad Pol Sci, Ser Math Astron Phys 23*205-208
◇ D20 ◇ REV MR 53 # 4631 Zbl 318 # 68037 • ID 18100

ZAKOWSKI, W. [1976] *Controlled (Z,Q)-machines and generalized (Z,Q)-computable sets (Russian summary)* (J 0014) Bull Acad Pol Sci, Ser Math Astron Phys 24*129-133
◇ D20 ◇ REV MR 53 # 2652 Zbl 329 # 68048 • ID 21536

ZAKOWSKI, W. [1977] *A generalization of the notions of a machine and computability* (J 0194) Inform & Control 33*166-176
◇ D10 D20 ◇ REV MR 54 # 14446 Zbl 346 # 68025 • ID 66198

ZAKOWSKI, W. [1978] *On some mathematical models of computing machines (Polish)* (P 3349) Nat Math Conf (9);1978 Isfahan 377-386
◇ D10 ◇ REV MR 84m:68044 • ID 39524

ZAKOWSKI, W. [1978] see STANKIEWICZ, E.

ZAKOWSKI, W. see Vol. V for further entries

ZAKREVSKIJ, A.D. [1965] *Algorithms of minimalization of weakly defined Boolean functions (Russian) (English summary)* (J 0040) Kibernetika, Akad Nauk Ukr SSR 1965/2*53-60
◇ B35 D20 ◇ REV MR 34 # 5588 Zbl 192.86 • ID 16244

ZAKREVSKIJ, A.D. see Vol. I for further entries

ZALCSTEIN, Y. [1977] see LIPTON, R.J.

ZALCWASSER, Z. [1922] *Un theoreme sur les ensembles qui sont a la fois F_σ et G_δ* (J 0027) Fund Math 3*44-45
◇ D55 E15 ◇ REV FdM 48.206 • ID 14343

ZAMA, N. [1966] *A generalization of algorithms and one of its applications I* (J 0407) Comm Math Univ St Pauli (Tokyo) 15*109-116
◇ D10 ◇ REV MR 37 # 76 Zbl 153.317 • ID 14344

ZAMA, N. [1968] *On some algebraic formulation of algorithms* (J 0407) Comm Math Univ St Pauli (Tokyo) 17*53-62
◇ D20 ◇ REV MR 38 # 5628 Zbl 199.310 • ID 14345

ZAMOV, N.K. [1970] see SHARONOV, V.I.

ZAMOV, N.K. see Vol. I, VI for further entries

ZAMYATIN, A.P. [1978] *A non-abelian variety of groups has an undecidable elementary theory(Russian)* (J 0003) Algebra i Logika 17*20-27
• TRANSL [1978] (J 0069) Algeb and Log 17*13-17
◇ C05 C60 D35 ◇ REV MR 80c:03045 Zbl 408 # 03011 • ID 90332

ZAMYATIN, A.P. see Vol. III for further entries

ZAPOL'SKIKH, E.N. [1981] see AMBARTSUMYAN, A.A.

ZARACH, A. [1976] see MAREK, W.

ZARACH, A. [1977] see LACHLAN, A.H.

ZARACH, A. see Vol. III, V, VI for further entries

ZAROVNYJ, V.P. [1964] *The group of automatic one-to-one mappings (Russian)* (J 0023) Dokl Akad Nauk SSSR 156*1266-1269
• TRANSL [1964] (J 0062) Sov Math, Dokl 5*801-804
◇ D05 D10 D75 ◇ REV MR 30 # 3024 Zbl 131.9
• ID 14367

ZASHEV, J.A. [1984] *Basic recursion theory in partially ordered models of some fragments of the combinatory logic* (J 0137) C R Acad Bulgar Sci 37*561-564
◇ B40 D20 ◇ REV MR 85j:03077 Zbl 544 # 03020
• ID 41008

ZASHEV, J.A. see Vol. VI for further entries

ZASLAVSKIJ, I.D. [1964] *Graph schemes with memory (Russian)* (S 0066) Tr Mat Inst Steklov 72*99-192
• TRANSL [1971] (J 0225) Amer Math Soc, Transl, Ser 2 98*163-288
◇ B70 B75 D05 D10 ◇ REV MR 34 # 5589 Zbl 199.38
• ID 16224

ZASLAVSKIJ, I.D. [1969] *On Shannon pseudofunctions (Russian)* (S 0228) Zap Nauch Sem Leningrad Otd Mat Inst Steklov 16*65-76
• TRANSL [1969] (J 0521) Semin Math, Inst Steklov 16*32-37
◇ D05 D15 F60 ◇ REV MR 42 # 78 Zbl 199.309
• ID 27441

ZASLAVSKIJ, I.D. [1969] *The axiomatic determination of constructive objects and operations (Russian) (Armenian and English summaries)* (J 0312) Izv Akad Nauk Armyan SSR, Ser Mat 4*153-181
◇ D20 F50 ◇ REV MR 42 # 2942 Zbl 176.278 • ID 14375

ZASLAVSKIJ, I.D. [1974] *Recursive extrapolators (Russian)* (C 1450) Teor Algor & Mat Logika (Markov) 55-61
◇ D20 ◇ REV MR 57 # 9509 Zbl 309 # 02035 • ID 66217

ZASLAVSKIJ, I.D. [1979] *The realization of three-valued logical functions by means of recursive and Turing operators (Russian)* (S 0554) Issl Teor Algor & Mat Logik (Moskva) 3*52-61
◇ B50 D10 D20 ◇ REV MR 81i:03031 Zbl 478 # 03007
• ID 55622

ZASLAVSKIJ, I.D. [1982] *Logical nets and monocyclic circuits (Russian) (Armenian summary)* (S 0422) Tr Vychisl Tsentra Akad Nauk Armyan SSR & Univ Erevan 10*29-40
◇ B70 D80 ◇ REV MR 85a:94039 • ID 38876

ZASLAVSKIJ, I.D. see Vol. II, VI for further entries

ZASTAVKA, Z. [1968] *On algorithms with memory elements (Russian) (Czech summary)* (J 0156) Kybernetika (Prague) 4*201-225
◇ D10 ◇ REV MR 38 # 4316 Zbl 176.280 JSL 36.346
• ID 19786

ZAZNOVA, N.E. [1976] see CHAPENKO, V.P.

ZDEBSKAYA, G.V. [1979] *On some properties of computation in the limit (Russian)* (J 3937) Veroyat Met i Kibern (Kazan) 15*34-39
◇ D20 D30 ◇ REV MR 81i:03068 Zbl 422 # 03017
• ID 53478

ZDEBSKAYA, G.V. [1980] *On weakly hypersimple sets (Russian)* (J 3937) Veroyat Met i Kibern (Kazan) 16*45-50
◇ D25 ◇ REV MR 83k:03050 Zbl 446 # 03031 • ID 56560

ZDEBSKAYA, G.V. [1982] *Recursively enumerable sets and limit computability (Russian)* (J 0031) Izv Vyssh Ucheb Zaved, Mat (Kazan) 1982/10*19-27
• TRANSL [1982] (J 3449) Sov Math 26/10*21-30
◇ D20 D25 ◇ REV MR 85i:03142 Zbl 563 # 03027
• ID 44217

ZEIGER, H.P. [1980] see HAUSSLER, D.

ZEIGLER, B.P. [1973] *Every discrete input machine is linearly simulatable* (J 0119) J Comp Syst Sci 7*161-167
◇ D05 ◇ REV MR 47 # 10158 Zbl 258 # 94036 • ID 66225

ZEILBERGER, D. [1981] *Enumeration of words by their number of mistakes* (J 0193) Discr Math 34*89-91
◇ D40 ◇ REV MR 82d:05016 Zbl 461 # 05005 • ID 54495

ZELEZNIKAR, A. [1962] *Some arithmetic normal algorithms (Serbo-Croatian summary)* (J 0371) Glas Mat-Fiz Astron, Ser 2 (Zagreb) 17*159-170
◇ D03 ◇ REV MR 29 # 2186 Zbl 118.254 • ID 14381

ZELEZNIKAR, A. [1963] *Some algorithm theory and its applicability (Serbo-Croatian summary)* (J 0371) Glas Mat-Fiz Astron, Ser 2 (Zagreb) 18*141-151
◇ D03 D05 ◇ REV MR 29 # 2187 Zbl 122.11 • ID 14383

ZELEZNIKAR, A. [1967] *Overlapping algorithms* (J 0041) Math Syst Theory 1*325-345
◇ B75 D03 ◇ REV MR 39 # 68 Zbl 143.314 • ID 14384

ZELEZNIKAR, A. see Vol. I for further entries

ZELINKA, B. [1966] *Sur le PS-isomorphisme des langues* (J 0068) Z Math Logik Grundlagen Math 12*263-265
◇ D05 ◇ REV MR 37 # 43 Zbl 143.15 • ID 14385

ZELINKA, B. see Vol. I, V for further entries

ZEMANEK, H. [1962] *Automaten und Denkprozesse* (C 1018) Digit Inf.-wandler 1-66
◇ A05 D10 ◇ REV JSL 30.382 • ID 14402

ZEMBRZUSKI, K. [1972] see SKARBEK, W.

ZEMBRZUSKI, K. [1974] *On D-machines (Russian summary)* (J 0014) Bull Acad Pol Sci, Ser Math Astron Phys 22*697-702
◇ D05 ◇ REV MR 50 # 1856 Zbl 302 # 68067 • ID 14404

ZEMBRZUSKI, K. [1974] *On D-machine representation of continuous functions (Russian summary)* (J 0014) Bull Acad Pol Sci, Ser Math Astron Phys 22*703-706
◇ D05 ◇ REV MR 50 # 1857 Zbl 302 # 68068 • ID 14405

ZEMKE, F. [1977] *P.R.-regulated systems of notation and the subrecursive hierarchy equivalence property* (J 0064) Trans Amer Math Soc 234*89-118
◇ D20 F15 ◇ REV MR 58 # 16222 Zbl 366 # 02032
• ID 51115

ZERMELO, E. [1931] *Ueber Stufen der Quantifikation und die Logik des Unendlichen* (J 0157) Jbuchber Dtsch Math-Ver 41*85-88,2.Abt.
◇ A05 C75 D35 E07 E30 ◇ REV FdM 58.60 • ID 14412

ZERMELO, E. see Vol. I, III, V for further entries

ZEUGMANN, T. [1983] *A-posteriori characterizations in inductive inference of recursive functions (French, German and Russian summaries)* (J 0129) Elektr Informationsverarbeitung & Kybern 19*559-594
◇ D15 ◇ REV Zbl 542 # 03020 • ID 43684

ZEUGMANN, T. [1984] *On the nonboundability of total effective operators* (**J** 0068) Z Math Logik Grundlagen Math 30*169-172
◇ D20 ◇ REV MR 85g:03070 Zbl 548 # 03026 • ID 42219

ZEUGMANN, T. [1984] *Recursive operators versus recursive functions with respect to the generation of classes of functions having a fastest program* (**P** 4175) Algeb & Log Found Progr 75-85
◇ D15 D20 ◇ REV MR 86c:03038 Zbl 565 # 03016 • ID 44474

ZEUGMANN, T. see Vol. II for further entries

ZHANG, GUOQIANG [1983] *NP-completeness and restricted partitions (Chinese) (English summary)* (**J** 2521) Beijing Shifan Daxue Xuebao, Ziran Kexue 1983/6*1-6
◇ D15 ◇ REV MR 85k:68039 • ID 44276

ZHANG, JINWEN [1979] *Reasoning and computing* (**P** 4588) Logic in China;1978 Beijing 243-252
◇ B75 D99 ◇ ID 48549

ZHANG, JINWEN [1982] see CAI, MAOHUA

ZHANG, JINWEN see Vol. II, III, V, VI for further entries

ZHANG, LIANG [1984] see GENG, SUYUN

ZHAROV, V.G. [1972] *The complexity of the terms of constructive sequences of normal algorithms (Russian)* (**J** 0023) Dokl Akad Nauk SSSR 203*746-748
• TRANSL [1972] (**J** 0062) Sov Math, Dokl 13*445-448
◇ D03 D15 ◇ REV MR 47 # 25 Zbl 329 # 02014 • ID 66216

ZHAROV, V.G. [1974] *On an analog of a theorem of Specker (Russian)* (**J** 0023) Dokl Akad Nauk SSSR 215*526-528
• TRANSL [1974] (**J** 0062) Sov Math, Dokl 15*538-541
◇ D20 D80 F60 ◇ REV MR 50 # 6810 Zbl 301 # 02031 • ID 66215

ZHAROV, V.G. [1974] *The complexity of the universal algorithm (Russian)* (**C** 1450) Teor Algor & Mat Logika (Markov) 34-54,213
◇ D03 D15 ◇ REV MR 53 # 7740 Zbl 308 # 02040 • ID 22987

ZHAROV, V.G. [1976] *Codes of words that belong to an enumerable set (Russian)* (**S** 0554) Issl Teor Algor & Mat Logik (Moskva) 2*57-65,158
◇ D15 D25 ◇ REV MR 58 # 16207 • ID 80254

ZHAROV, V.G. [1976] *On lower and upper estimates for the complexity of the terms of constructive sequences of normal algorithms (Russian)* (**J** 0023) Dokl Akad Nauk SSSR 228*1029-1030
• TRANSL [1976] (**J** 0062) Sov Math, Dokl 17*861-863
◇ D03 D15 ◇ REV MR 57 # 4631 Zbl 362 # 68078 • ID 50819

ZHAROV, V.G. [1979] *On the complexity of the terms of the constructive sequences of Turing machines (Russian)* (**S** 0554) Issl Teor Algor & Mat Logik (Moskva) 3*39-52
◇ D10 D45 F60 ◇ REV MR 81g:03045 Zbl 454 # 03016 • ID 54228

ZHEGALKIN, I.I. [1939] *Sur l'Entscheidungsproblem (Russian) (French summary)* (**J** 0142) Mat Sb, Akad Nauk SSSR, NS 6(48)*185-198
◇ B20 B25 D35 ◇ REV MR 1.322 Zbl 22.193 JSL 5.69 FdM 65.27 • ID 17903

ZHEGALKIN, I.I. see Vol. I, III, V, VI for further entries

ZHELEVA, S. [1971] see T"RKALANOV, K.D.

ZHOU, CHAOCHEN [1979] *Program schemes and predicate calculus (Chinese)* (**J** 3793) Jisuanjii Xuebao 2*174-189
◇ B75 D20 ◇ ID 48555

ZHU, SHANGYONG [1981] *Fuzzy tree grammars and fuzzy forest grammars (Chinese)* (**J** 3732) Mohu Shuxue 1/2*45-64
◇ D05 ◇ REV MR 84g:68072 • ID 47234

ZHU, SHANGYONG [1983] *Error-correcting fuzzy tree grammar. I (Chinese) (English summary)* (**J** 3732) Mohu Shuxue 3/1*15-26
◇ D05 ◇ REV MR 84m:68074 Zbl 532 # 68078 • REM Part II 1983 • ID 48993

ZHU, SHANGYONG [1983] *Error-correcting fuzzy tree grammars. II (Chinese) (English summary)* (**J** 3732) Mohu Shuxue 3/4*19-22
◇ D05 ◇ REV MR 85g:68040 • REM Part I 1983 • ID 43954

ZHUK, I.K. [1973] *The word problem for a certain class of groups (Russian)* (**J** 0414) Dokl Akad Nauk Belor SSR 17*1081-1084,1163
◇ D40 ◇ REV MR 50 # 489 Zbl 297 # 20045 • ID 14446

ZHUKOV, S.A. [1977] *Some asymptotic estimates of the complexity of partial recursive functions (Russian)* (**C** 3017) Vopr Sist Progr 39-45
◇ D15 D20 ◇ REV MR 80f:03043 • ID 80263

ZHURAVLEV, YU.I. [1964] *Estimate of complexity of local algorithms for some extremum problems on finite sets (Russian)* (**J** 0023) Dokl Akad Nauk SSSR 158*1018-1021
• TRANSL [1964] (**J** 0062) Sov Math, Dokl 5*1343-1347
◇ D15 ◇ REV MR 29 # 3374 Zbl 156.19 • ID 14456

ZHURAVLEV, YU.I. see Vol. I, V for further entries

ZIEGLER, M. [1975] see PRESTEL, A.

ZIEGLER, M. [1975] *Gruppen mit vorgeschriebenem Wortproblem* (**J** 0043) Math Ann 219*43-51
◇ D40 ◇ REV MR 53 # 12914 Zbl 302 # 02013 • ID 14981

ZIEGLER, M. [1975] see RAUTENBERG, W.

ZIEGLER, M. [1976] *A language for topological structures which satisfies a Lindstroem-theorem* (**J** 0015) Bull Amer Math Soc 82*568-570
◇ B25 C40 C50 C65 C80 C90 C95 D35 ◇ REV MR 54 # 4971 Zbl 338 # 02007 • ID 24137

ZIEGLER, M. [1976] *Ein rekursiv aufzaehlbarer btt-Grad, der nicht zum Wortproblem einer Gruppe gehoert* (**J** 0068) Z Math Logik Grundlagen Math 22*165-168
◇ D25 D40 ◇ REV MR 56 # 2812 Zbl 366 # 02034 • ID 18480

ZIEGLER, M. [1980] *Algebraisch abgeschlossene Gruppen (English summary)* (**P** 2634) Word Problems II;1976 Oxford 449-576
◇ C25 C60 D30 D40 ◇ REV MR 82b:20004 Zbl 451 # 20001 • ID 54050

ZIEGLER, M. [1982] *Einige unentscheidbare Koerpertheorien* (**J** 3370) Enseign Math, Ser 2 28*269-280
• REPR [1982] (**P** 3482) Logic & Algor (Specker);1980 Zuerich 381-392
◇ C60 D35 ◇ REV MR 83m:03040 Zbl 499 # 03002 Zbl 519 # 12018 JSL 50.552 • ID 35441

ZIEGLER, M. see Vol. III for further entries

ZIELONKA, W. [1979] *On the equivalence of Lambek's syntactic calculus and categorial calculi* (**P** 2935) FCT'79 Fund of Comput Th;1979 Berlin/Wendisch-Rietz 537-541
◇ D05 ◇ REV MR 84d:03023 Zbl 427 # 03018 • ID 53705

ZIELONKA, W. [1981] *Axiomatizability of Ajdukiewicz-Lambek calculus by means of cancellation schemes* (**J** 0068) Z Math Logik Grundlagen Math 27*215-224
◇ B65 D05 ◇ REV MR 82h:68107 Zbl 467 # 03020
• ID 55018

ZIMAND, M. [1983] *Complexity of probabilistic algorithms* (**J** 2716) Found Control Eng, Poznan 8*33-49
◇ D15 ◇ REV MR 85e:03096 Zbl 546 # 68030 • ID 40694

ZIV, J. [1976] see LEMPEL, A.

ZUBENKO, V.V. [1980] see KUZENKO, V.F.

ZUBENKO, V.V. [1984] *Undecidability of strict equivalence for recursive compositions (Russian)* (**C** 4734) Model & Sist Obrabotki Inform, Vyp 3 38-42,117
◇ D80 ◇ ID 47274

ZUBENKO, V.V. see Vol. III for further entries

ZUCKER, J.I. [1981] see BAKKER DE, J.W.

ZUCKER, J.I. see Vol. I, II, VI for further entries

ZURAWIECKI, J. [1976] see GRODZKI, Z.

ZURAWIECKI, J. see Vol. V for further entries

ZVONKIN, A.K. [1970] see LEVIN, L.A.

ZVONKIN, A.K. see Vol. I for further entries

ZYKIN, G.P. [1963] *Remark on a theorem of Hao Wang (Russian)* (**J** 0003) Algebra i Logika 2/1*33-35
◇ D10 ◇ REV MR 27 # 5682 Zbl 171.273 JSL 36.534
• ID 14463

ZYKOV, A.A. [1959] *Remarks in connection with the reduction theorem for logical calculi (Russian)* (**P** 0607) All-Union Math Conf (3);1956 Moskva 4*85-86
◇ B20 D35 ◇ ID 30599

ZYKOV, A.A. [1964] *Recursively calculable functions of graphs* (**P** 0703) Th of Graphs & Appl;1963 Smolenice 99-106
◇ D20 D75 D80 ◇ REV MR 31 # 92 Zbl 173.265
• ID 14464

ZYKOV, A.A. see Vol. I, III for further entries

Source Index

Journals

J 0001 Acta Math Acad Sci Hung • H
Acta Mathematica Academiae Scientiarum Hungaricae [1950–1982] ISSN 0001-5954
• CONT AS (J 4729) Acta Math Hung

J 0002 Acta Sci Math (Szeged) • H
Acta Scientiarum Mathematicarum [1947ff] ISSN 0001-6969
• CONT OF (J 0460) Acta Univ Szeged, Sect Mat

J 0003 Algebra i Logika • SU
Algebra i Logika (Algebra and Logic) [1962ff] ISSN 0373-9252
• TRANSL IN (J 0069) Algeb and Log

J 0004 Algeb Universalis • CDN
Algebra Universalis [1970ff] ISSN 0002-5240

J 0005 Amer Math Mon • USA
American Mathematical Monthly [1894ff] ISSN 0002-9890

J 0006 Ann Univ Budapest, Sect Math • H
Annales Universitatis Scientiarum Budapestinensis. Sectio Mathematica [1958ff] ISSN 0524-9007

J 0007 Ann Math Logic • NL
Annals of Mathematical Logic [1970–1982] ISSN 0003-4843
• CONT AS (J 0073) Ann Pure Appl Logic

J 0008 Arch Math (Basel) • CH
*Archiv der Mathematik * Archives of Mathematics * Archives Mathematiques* [1948ff] ISSN 0003-889X

J 0009 Arch Math Logik Grundlagenforsch • D
Archiv fuer Mathematische Logik und Grundlagenforschung [1950ff] ISSN 0003-9268

J 0010 Autom & Remote Control • USA
Automation and Remote Control [1958ff] ISSN 0005-1179
• TRANSL OF (J 0011) Avtom Telemekh

J 0011 Avtom Telemekh • SU
Avtomatika i Telemekhanika (Automation and Telemechanics) [1934ff] ISSN 0005-2310
• TRANSL IN (J 0010) Autom & Remote Control

J 0012 Boll Unione Mat Ital, IV Ser • I
Bolletino della Unione Matematica Italiana. Serie IV [1968–1975] ISSN 0041-7084
• CONT OF (J 4408) Boll Unione Mat Ital, III Ser • CONT AS (J 3285) Boll Unione Mat Ital, V Ser, A & (J 3495) Boll Unione Mat Ital, V Ser, B

J 0013 Brit J Phil Sci • GB
British Journal for the Philosophy of Science [1950ff] ISSN 0007-0882

J 0014 Bull Acad Pol Sci, Ser Math Astron Phys • PL
Bulletin de l'Academie Polonaise des Sciences. Serie des Sciences Mathematiques, Astronomiques et Physiques [1953–1978] ISSN 0001-4117
• CONT AS (J 3293) Bull Acad Pol Sci, Ser Math

J 0015 Bull Amer Math Soc • USA
Bulletin of the American Mathematical Society [1894–1978] ISSN 0002-9904
• CONT AS (J 0589) Bull Amer Math Soc (NS)

J 0016 Bull Austral Math Soc • AUS
Bulletin of the Australian Mathematical Society [1969ff] ISSN 0004-9727

J 0017 Canad J Math • CDN
*Canadian Journal of Mathematics * Journal Canadien de Mathematiques* [1949ff] ISSN 0008-414X

J 0018 Canad Math Bull • CDN
*Canadian Mathematical Bulletin * Bulletin Canadien de Mathematiques* [1958ff] ISSN 0008-4395

J 0020 Compos Math • NL
Compositio Mathematica [1933ff] ISSN 0010-437X

J 0021 Cybernetics • USA
Cybernetics [1965ff] ISSN 0011-4235
• TRANSL OF (J 0040) Kibernetika, Akad Nauk Ukr SSR

J 0022 Cheskoslov Mat Zh • CS
*Cheskoslovatskij Matematicheskij Zhurnal * Czechoslovak Mathematical Journal* [1951ff] ISSN 0011-4642
• REM From Vol. 19 (1969) on the title is only: Czechoslovak Mathematical Journal

J 0023 Dokl Akad Nauk SSSR • SU
Doklady Akademii Nauk SSSR (Reports of the Academy of Sciences of the USSR) [1933ff] ISSN 0002-3264
• TRANSL IN (J 0062) Sov Math, Dokl & (J 0470) Sov Phys, Dokl

J 0024 Dokl Akad Nauk Uzb SSR • SU
*Doklady Akademii Nauk Uzb SSR (DAN Uzb SSR) * UzSSR Fanlar Akademijasining Dokladlari (Reports of the Academy of Sciences of the Uzb SSR)* [1944ff] ISSN 0134-4307

J 0025 Duke Math J • USA
Duke Mathematical Journal [1935ff] ISSN 0012-7094

J 0026 Elem Math • CH
*Elemente der Mathematik * Revue de Mathematiques Elementaires * Rivista di Matematica Elementare* [1946ff] ISSN 0013-6018

J 0027 Fund Math • PL
Fundamenta Mathematicae [1920ff] ISSN 0016-2736

J 0028 Indag Math • NL
*Indagationes Mathematicae * Nederlandse Akademie van Wetenschappen. Proceedings* [1939ff] ISSN 0019-3577, ISSN 0023-3358
• REM Until 1950 part of Koninklijke Nederlandsche Akademie van Wetenschappen, Proceedings of the Section of Sciences; vol n+41 with separate pagination. Since 1951 same as Koninklijke Nederlandse Akademie van Wetenschappen, Proceedings of the Section of Sciences, Series A; vol n+41. Before 1951 page numbers in Proceedings and Indagationes different. Since 1951 the same page numbers as Proceedings Series A.

J 0029 Israel J Math • IL
Israel Journal of Mathematics [1963ff] ISSN 0021-2172
• CONT OF (**J** 0493) Bull Res Counc Israel Sect F

J 0031 Izv Vyssh Ucheb Zaved, Mat (Kazan) • SU
Izvestiya Vysshikh Uchebnykh Zavedenij. Matematika (Proceedings of the University. Mathematics) [1957ff] ISSN 0021-3446
• TRANSL IN (**J** 3449) Sov Math

J 0032 J Algeb • USA
Journal of Algebra [1964ff] ISSN 0021-8693

J 0033 J Comb Th, Ser B • USA
Journal of Combinatorial Theory. Series B [1971ff] ISSN 0095-8956
• CONT OF (**J** 1669) J Comb Th

J 0036 J Symb Logic • USA
The Journal of Symbolic Logic [1936ff] ISSN 0022-4812

J 0037 ACM J • USA
Journal of the ACM (=Association for Computing Machinery) [1954ff] ISSN 0004-5411

J 0038 J Austral Math Soc • AUS
Journal of the Australian Mathematical Society [1959-1975] ISSN 0004-9735
• CONT AS (**J** 3194) J Austral Math Soc, Ser A

J 0039 J London Math Soc • GB
The Journal of the London Mathematical Society [1926-1968]
• CONT AS (**J** 3172) J London Math Soc, Ser 2

J 0040 Kibernetika, Akad Nauk Ukr SSR • SU
Kibernetika. Akademiya Nauk Ukrainskoj SSR (Cybernetics. Academy of Sciences of the Ukrainian SSR) [1965ff] ISSN 0023-1274
• TRANSL IN (**J** 0021) Cybernetics

J 0041 Math Syst Theory • D
Mathematical Systems Theory. An International Journal [1967ff] ISSN 0025-5661

J 0042 Mat Vesn, Drust Mat Fiz Astron Serb • YU
Matematichki Vesnik. Drushtvo Matematichara, Fizichara i Astromoma SR Serbije, SFR Jugoslavija (Mathematical Publications. Society Serbe of Mathematicians, Physicists and Astronomers) [1964ff] ISSN 0025-5165
• CONT OF (**J** 4277) Vesn Drusht Mat Fiz Serbije

J 0043 Math Ann • D
Mathematische Annalen [1868ff] ISSN 0025-5831

J 0044 Math Z • D
Mathematische Zeitschrift [1918ff] ISSN 0025-5874

J 0045 Monatsh Math • A
Monatshefte fuer Mathematik [1943ff] ISSN 0026-9255
• CONT OF (**J** 0124) Monatsh Math-Phys

J 0046 Nieuw Arch Wisk • NL
Nieuw Archief voor Wiskunde [1875-1893]
• CONT AS (**J** 1793) Nieuw Arch Wisk, Ser 2

J 0047 Notre Dame J Formal Log • USA
Notre Dame Journal of Formal Logic [1960ff] ISSN 0029-4527

J 0048 Pac J Math • USA
Pacific Journal of Mathematics [1951ff] ISSN 0030-8730

J 0049 Period Math Hung • H
Periodica Mathematica Hungarica [1971ff] ISSN 0031-5303

J 0050 Port Math • P
Portugaliae Mathematica [1937ff] ISSN 0032-5155

J 0051 Commentat Math, Ann Soc Math Pol, Ser 1 • PL
*Annales Societatis Mathematicae Polonae. Series I. Commentationes Mathematicae * Roczniki Polskiego Towarzystwa Matematycznego. Seria I. Prace Matematyczne.* [1955ff] ISSN 0079-368X
• CONT OF (**J** 0611) Pol Tow Mat, Prace Mat-Fiz

J 0052 Probl Kibern • SU
Problemy Kibernetiki. Glavnaya Redaktsiya Fiziko-Matematicheskoj Literatury [1958ff] ISSN 0555-277X
• TRANSL IN (**J** 0471) Syst Th Res & (**J** 0449) Probl Kybern & (**J** 1195) Probl Cybernet

J 0053 Proc Amer Math Soc • USA
Proceedings of the American Mathematical Society [1950ff] ISSN 0002-9939

J 0054 Proc Nat Acad Sci USA • USA
Proceedings of the National Academy of Sciences of the United States of America [1915ff] ISSN 0027-8424

J 0056 Publ Dep Math, Lyon • F
Publications du Departement de Mathematiques. Faculte des Sciences de Lyon. [1964-1981] ISSN 0076-1656
• CONT AS (**J** 2107) Publ Dep Math, Lyon, NS

J 0057 Publ Math (Univ Debrecen) • H
Publicationes Mathematicae [1949ff] ISSN 0033-3883

J 0059 Rend Sem Mat Fis Milano • I
Rendiconti del Seminario Matematico e Fisico di Milano. Sotto gli Auspici dell'Universita e del Politecnico [1927ff]

J 0060 Rev Roumaine Math Pures Appl • RO
Revue Roumaine de Mathematiques Pures et Appliquees. Academia Republicii Socialiste Romania [1956ff] ISSN 0035-3965

J 0062 Sov Math, Dokl • USA
Soviet Mathematics. Doklady. [1960ff] ISSN 0038-5573
• TRANSL OF (**J** 0023) Dokl Akad Nauk SSSR

J 0063 Studia Logica • PL
Studia Logica [1953ff] ISSN 0039-3215

J 0064 Trans Amer Math Soc • USA
Transactions of the American Mathematical Society [1900ff] ISSN 0002-9947

J 0065 Tr Moskva Mat Obshch • SU
Trudy Moskovskogo Matematicheskogo Obshchestva (Publications of the Moscow Mathematical Society) [1952ff] ISSN 0134-8663
• TRANSL IN (**J** 3279) Trans Moscow Math Soc

J 0067 Usp Mat Nauk • SU
Uspekhi Matematicheskikh Nauk (Advances in Mathematical Sciences) [1936ff] ISSN 0042-1316
• TRANSL IN (**J** 1399) Russ Math Surv

J 0068 Z Math Logik Grundlagen Math • DDR
Zeitschrift fuer Mathematische Logik und Grundlagen der Mathematik [1955ff] ISSN 0044-3050

J 0069 Algeb and Log • USA
Algebra and Logic [1968ff] ISSN 0002-5232
• TRANSL OF (**J** 0003) Algebra i Logika

J 0070 Bull Soc Sci Math Roumanie, NS • RO
Bulletin Mathematique de la Societe des Sciences Mathematiques de la Republique Socialiste de Roumanie. Nouvelle Serie. [1957ff] ISSN 0007-4691
• CONT OF (**J** 0494) Bull Math Soc Sci Roumanie

J 0071 Met Diskr Analiz (Novosibirsk) • SU
Metody Diskretnogo Analiza. Sbornik Trudov (Methods of Discrete Analysis. Collected Papers) [1963ff] ISSN 0419-4160, ISSN 0136-1228

J 0072 IRE Trans Electr Comp • USA
Transactions on Electronic Computers. IRE (= Institute of Radio Engineers) [1952-1962]
• CONT AS (**J** 4305) IEEE Trans Electr Comp

J 0073 Ann Pure Appl Logic • NL
Annals of Pure and Applied Logic [1983ff] ISSN 0168-0072
• CONT OF (**J** 0007) Ann Math Logic

J 0076 Dialectica • CH
Dialectica. International Review of Philosophy of Knowledge [1947ff] ISSN 0012-2017

J 0077 Proc London Math Soc • GB
Proceedings of the London Mathematical Society [1865-1903]
• CONT AS (**J** 1910) Proc London Math Soc, Ser 2

J 0079 Logique & Anal, NS • B
Logique et Analyse. Nouvelle Serie. Publication Trimestrielle du Centre National Belge de Recherche de Logique [1958ff] ISSN 0024-5836

J 0080 Izv Akad Nauk Ehston SSR, Fiz, Mat • SU
*Izvestiya Akademiya Nauk Ehstonskoj SSR. Fizika. Matematika * Eesti NSV Teaduste Akadeemia Toimetised. Fueuesika-Matemaatika (Proceedings of the Academy of Sciences of the Estonian SSR. Physics. Mathematics)* [1956ff] ISSN 0002-3140

J 0081 Proc Japan Acad • J
Proceedings of the Japan Academy [1925-1977] ISSN 0021-4280
• CONT AS (**J** 3239) Proc Japan Acad, Ser A

J 0082 Bull Soc Math Belg • B
Bulletin de la Societe Mathematique de Belgique [1948-1976] ISSN 0037-9476
• CONT AS (**J** 3133) Bull Soc Math Belg, Ser B & (**J** 3824) Bull Soc Math Belg, Ser A

J 0084 Gaz Mat (Lisboa) • P
Gazeta de Matematica. Jornal dos Concorrentes ao Exame de Aptidao e dos Estudantes de Matematica das Escolas Superiores [1940ff]

J 0085 Vest Ser Mat Mekh Astron, Univ Leningrad • SU
Vestnik Leningradskogo Universiteta. Seriya Matematika, Mekhanika, Astronomiya (Publications of the Leningrad University. Series: Mathematics, Mechanics, Astronomy) [1946ff] ISSN 0024-0850
• TRANSL IN (**J** 3945) Vest Math Univ Leningrad

J 0086 Cas Pestovani Mat, Ceskoslov Akad Ved • CS
Casopis pro Pestovani Matematiky. Ceskoslovenska Akademie Ved (Journal for the Cultivation of Mathematics. Czechoslovak Academy of Sciences) [1872ff]

J 0087 Mat Zametki (Akad Nauk SSSR) • SU
Matematicheskie Zametki (Mathematical Notes) [1967ff] ISSN 0025-567X
• TRANSL IN (**J** 1044) Math Notes, Acad Sci USSR

J 0088 Ann Univ Ferrara, NS, Sez 7 • I
Annali dell'Universita di Ferrara. Nuova Serie. Sezione 7. Scienze Matematiche [1966ff]

J 0089 Calcolo • I
Calcolo [1964ff] ISSN 0008-0624

J 0090 J Math Soc Japan • J
Journal of the Mathematical Society of Japan [1885ff] ISSN 0025-5645

J 0091 Sugaku • J
Sugaku (Mathematics) [1947ff] ISSN 0039-470X

J 0092 Sib Mat Zh • SU
Sibirskij Matematicheskij Zhurnal. Akademiya Nauk SSSR. Sibirskoe Otdelenie (Siberian Mathematical Journal. Academy of Sciences of the USSR. Siberian Section) [1960ff] ISSN 0037-4474
• TRANSL IN (**J** 0475) Sib Math J

J 0094 Mind • GB
Mind. A Quarterly Review of Philosophy [1876ff] ISSN 0026-4423

J 0095 Philos Stud • NL
Philosophical Studies. An International Journal for Philosophy in the Analytic Tradition [1950ff] ISSN 0031-8116

J 0096 Acta Philos Fenn • SF
Acta Philosophica Fennica [1948ff] ISSN 0355-1792

J 0100 Amer J Math • USA
American Journal of Mathematics [1878ff] ISSN 0002-9327

J 0105 Theoria (Lund) • S
Theoria. A Swedish Journal of Philosophy [1934ff] ISSN 0040-5825

J 0106 Mem Fac Sci, Kyushu Univ, Ser A • J
*Memoirs of the Faculty of Science. Kyushu University. Series A. Mathematics * Kyushu Daigaku Rigakubu Kiyo. A. Sugaku* [1940ff] ISSN 0373-6385

J 0109 C R Acad Sci, Paris • F
Academie des Sciences de Paris. Comptes Rendus Hebdomadaires des Seances [1835-1965]
• CONT AS (**J** 2313) C R Acad Sci, Paris, Ser A-B

J 0111 Nagoya Math J • J
Nagoya Sugaku Zashi (Nagoya Mathematical Journal) [1950ff] ISSN 0027-7630

J 0114 Math Nachr • DDR
Mathematische Nachrichten [1948ff] ISSN 0025-584X

J 0115 Wiss Z Humboldt-Univ Berlin, Math-Nat Reihe • DDR
Wissenschaftliche Zeitschrift der Humboldt-Universitaet Berlin. Mathematisch-Naturwissenschaftliche Reihe [1951ff] ISSN 0043-6852

J 0116 Electr & Comm Japan • USA
*Electronics and Communications in Japan * Scripta Electronica Japonica* [1963ff] ISSN 0036-9683
• TRANSL OF (**J** 0979) Denshi Tsushin Gakkai Ronbunshi, Sect A-D

J 0118 Acta Math • S
Acta Mathematica [1882ff] ISSN 0001-5962

J 0119 J Comp Syst Sci • USA
Journal of Computer and System Sciences [1967ff] ISSN 0022-0000

J 0120 Ann of Math, Ser 2 • USA
Annals of Mathematics. 2nd Series [1899ff] ISSN 0003-486X

J 0121 Kon Norske Vidensk Selsk Forh • N
Kongelige Norske Videnskabers Selskabs. Forhandlinger. (Proceedings of the Royal Scandinavian Society of Sciences) [1926ff] ISSN 0368-6302

J 0122 J Philos Logic • NL
Journal of Philosophical Logic [1972ff] ISSN 0022-3611

J 0124 Monatsh Math-Phys • A
Monatshefte fuer Mathematik und Physik [1900-1942]
• CONT AS (**J** 0045) Monatsh Math

J 0126 Glasgow Math J • GB
Glasgow Mathematical Journal [1967ff] ISSN 0017-0895
• CONT OF (**J** 0217) Proc Glasgow Math Assoc

J 0127 J Reine Angew Math • D
Journal fuer die Reine und Angewandte Mathematik [1826ff] ISSN 0075-4102

J 0128 Acta Math Univ Comenianae (Bratislava) • CS
Universitas Comeniana. Acta Facultatis Rerum Naturalium. Mathematica [1956ff]

J 0129 Elektr Informationsverarbeitung & Kybern • DDR
Elektronische Informationsverarbeitung und Kybernetik [1965ff] ISSN 0013-5712

J 0130 BIT • DK
BIT. Nordisk Tidskrift for Informationsbehandling (BIT. Scandinavian Journal for Informatics) [1961ff] ISSN 0006-3835

J 0131 Quart J Math, Oxford Ser 2 • GB
The Quarterly Journal of Mathematics. Oxford Second Series [1950ff] ISSN 0033-5606
• CONT OF (**J** 1138) Quart J Math, Oxford Ser

J 0132 Math Scand • DK
Mathematica Scandinavica [1953ff] ISSN 0025-5521

J 0133 Michigan Math J • USA
The Michigan Mathematical Journal [1952ff] ISSN 0026-2285

J 0135 Izv Akad Nauk Azerb SSR, Ser Fiz-Tekh Mat • SU
Izvestiya Akademii Nauk Azerbajdzhanskoj SSR. Seriya Fiziko-Tekhnicheskikh i Matematicheskikh Nauk (Proceedings of the Academy of Sciences of the Azerbaijan SSR. Series: Physical-Technical and Mathematical Sciences) [1958ff] ISSN 0002-3108

J 0136 Semigroup Forum • D
Semigroup Forum [1970ff] ISSN 0037-1912

J 0137 C R Acad Bulgar Sci • BG
Doklady Bolgarskoi Akademii Nauk (Comptes Rendus de l'Academie Bulgare des Sciences) [1948ff] ISSN 0001-3978

J 0140 Comm Math Univ Carolinae (Prague) • CS
Commentationes Mathematicae Universitatis Carolinae [1960ff] ISSN 0010-2628

J 0141 Arch Autom & Telemech • PL
Archiwum Automatyki i Telemechaniki (Archives of Automation and Telemechanics) [1956ff] ISSN 0004-072X

J 0142 Mat Sb, Akad Nauk SSSR, NS • SU
Matematicheskij Sbornik. Novaya Seriya. Akademiya Nauk SSSR i Moskovskoe Matematicheskoe Obshchestvo (Mathematical Collected Articles. New Series. Academy of Sciences of the USSR and Moskovian Mathematical Society) [1936ff] ISSN 0025-5157
• CONT OF (**J** 1404) Mat Sb, Akad Nauk SSSR • TRANSL IN (**J** 0349) Math of USSR, Sbor

J 0143 Mat Chasopis (Slov Akad Ved) • CS
Matematicky Chasopis (Journal of Mathematics) [1967-1975] ISSN 0025-5173
• CONT OF (**J** 4713) Mat Fyz Chasopis (Slov Akad Ved)
• CONT AS (**J** 1522) Math Slovaca

J 0144 Rend Sem Mat Univ Padova • I
Rendiconti del Seminario Matematico dell'Universita di Padova [1930ff] ISSN 0041-8994

J 0146 Z Phonetik Sprachwiss Kommunikation • DDR
Zeitschrift fuer Phonetik, Sprachwissenschaft und Kommunikationsforschung [1948ff] ISSN 0044-331X

J 0149 Atti Accad Naz Lincei Fis Mat Nat, Ser 8 • I
Atti della Accademia Nazionale dei Lincei. Rendiconti. Classe di Scienze Fisiche, Matematiche e Naturali. Serie VIII [1946ff] ISSN 0001-4435

J 0150 Acad Roy Belg Bull Cl Sci (5) • B
*Academie Royale des Sciences, des Lettres et des Beaux Arts de Belgique. Bulletin de la Classe des Sciences. Cinquieme Serie * Koninklijke Academie voor Wetenschappen. Mededeelingen van de Afdeeling Wetenschappen. 5. Serie* [1915ff] ISSN 0001-4141

J 0152 Enseign Math • CH
L'Enseignement Mathematique: Revue Internationale [1899-1954]
• CONT AS (**J** 3370) Enseign Math, Ser 2

J 0153 Phil of Sci (East Lansing) • USA
Philosophy of Science [1934ff] ISSN 0031-8248

J 0154 Synthese • NL
Synthese. An International Journal for Epistemology, Methodology and Philosophy of Science [1936ff] ISSN 0039-7857

J 0155 Commun Pure Appl Math • USA
Communications on Pure and Applied Mathematics [1939ff] ISSN 0010-3640

J 0156 Kybernetika (Prague) • CS
Kybernetika (Cybernetics) [1965ff] ISSN 0023-5954
• REL PUBL (**J** 3524) Kybernetika Suppl (Prague)

J 0157 Jbuchber Dtsch Math-Ver • D
Jahresbericht der Deutschen Mathematiker-Vereinigung [1890ff] ISSN 0012-0456

J 0160 Math-Phys Sem-ber, NS • D
Mathematisch-Physikalische Semesterberichte: Zur Pflege des Zusammenhangs von Schule und Universitaet. Neue Folge [1950–1979] ISSN 0340–4897
• CONT AS (J 2790) Math Sem-ber

J 0161 Bull London Math Soc • GB
The Bulletin of the London Mathematical Society [1926ff] ISSN 0024–6093

J 0163 SIAM Review • USA
Review. SIAM (= Society for Industrial and Applied Mathematics) [1959ff] ISSN 0036–1445

J 0164 J Comb Th, Ser A • USA
Journal of Combinatorial Theory. Series A [1971ff] ISSN 0097–3165
• CONT OF (J 1669) J Comb Th

J 0165 Acta Univ Carolinae Math Phys (Prague) • CS
Acta Universitatis Carolinae. Mathematica et Physica [1959ff] ISSN 0001–7140

J 0171 Proc Cambridge Phil Soc Math Phys • GB
Proceedings of the Cambridge Philosophical Society. Mathematical and Physical Sciences [1843–1974] ISSN 0008–1981
• CONT AS (J 0332) Math Proc Cambridge Phil Soc

J 0174 Bull Sect Sci Acad Roumaine • RO
Academie Roumaine. Bulletin de la Section Scientifique [1918–1948]
• CONT AS (J 0518) Bul Sti Mat-Fiz, Acad Romina

J 0176 Bull Acad Pol Sci, Ser Sci Math • PL
Bulletin International de l'Academie Polonaise des Sciences et des Lettres. Classe des Sciences Mathematiques et Naturelles. Serie A Sciences Mathematiques [1926–1939]

J 0178 Stud Gen • D
Studium Generale: Zeitschrift fuer die Einheit der Wissenschaften. Zusammenhang ihrer Begriffsbildungen und Forschungsmethoden. [1947–1971] ISSN 0039–4149

J 0179 Ann Fac Sci Clermont • F
Universite de Clermont. Faculte des Sciences. Annales [1952–1972]
• CONT AS (J 1934) Ann Sci Univ Clermont Math

J 0185 Unterrichtsbl Math Nat • D
Unterrichtsblaetter fuer Mathematik und Naturwissenschaften [1895–1947]
• CONT AS (J 0933) Math Nat Unterr

J 0187 IEEE Trans Comp • USA
Transactions on Computers. IEEE (= Institute of Electrical and Electronics Engineers) [1968ff] ISSN 0018–9340
• CONT OF (J 4305) IEEE Trans Electr Comp

J 0191 Inform Sci • USA
Information Sciences. An International Journal [1957ff] ISSN 0020–0255

J 0192 Wiss Z Univ Jena, Math-Nat Reihe • DDR
Wissenschaftliche Zeitschrift der Friedrich-Schiller-Universitaet Jena. Mathematisch-Naturwissenschaftliche Reihe. [1951ff] ISSN 0043–6836

J 0193 Discr Math • NL
Discrete Mathematics [1971ff] ISSN 0012–365X

J 0194 Inform & Control • USA
Information and Control [1958ff] ISSN 0019–9958

J 0197 Stud Cercet Mat Acad Romana • RO
Studii si Cercetari Matematice. Academia Republicii Socialiste Romania. (Mathematische Studien und Untersuchungen. Akademie der Sozialistischen Republik Rumaenien) [1950ff] ISSN 0567–6401
• CONT OF (J 0524) Disq Math Phys

J 0198 Bul Inst Politeh Iasi NS • RO
*Buletinul Institutului Politehnic din Iasi. Serie Noua * Bulletin de l'Ecole Polytechnique de Jassy. Nouveau Serie* [1946–1976]
• CONT AS (J 3070) Bul Inst Politeh Iasi, Sect 1

J 0199 Zh Vychisl Mat i Mat Fiz • SU
Zhurnal Vychislitel'noj Matematiki i Matematicheskoj Fiziki (Journal of Computational Mathematical and Mathematical Physics) [1961ff] ISSN 0044–4669
• TRANSL IN (J 1049) USSR Comput Math & Math Phys

J 0201 ICC Bull • I
ICC (International Computation Centre). Bulletin [1962ff] ISSN 0536–1222

J 0202 Diss Math (Warsaw) • PL
*Dissertationes Mathematicae. Polska Akademia Nauk, Instytut Matematyczny * Rozprawy Matematyczne* [1952ff] ISSN 0012–3862

J 0205 Rev Franc Autom, Inf & Rech Operat • F
Revue Francaise d'Automatique, Informatique et Recherche Operationelle (RAIRO). Series: Bleue, Jaune, Rouge, Verte [1972–1976] ISSN 0399–0559
• CONT OF (J 3954) Rev Franc Inf & Rech Operat • CONT AS (J 4698) Rev Franc Autom, Inf & Rech Operat, Ser Rouge Inf Th & (J 2831) RAIRO Autom & (J 2832) RAIRO Inform
• REM In 1975 the Serie Rouge split into: Serie Rouge Analyse Numerique & J4698

J 0209 J Math Phys • USA
Journal of Mathematical Physics [1960ff] ISSN 0022–2488

J 0211 Rev Gen Sci Pur Appl & Bull Soc Philomatique • F
Revue Generale des Sciences Pures et Appliquees et Bulletin de la Societe Philomatique. [1953–1959]
• CONT OF (J 0767) Rev Gen Sci Pur Appl

J 0212 ACM Commun • USA
Communications of ACM (= Association for Computing Machinery) [1958ff] ISSN 0001–0782

J 0214 Math of Comp • USA
Mathematics of Computation [1960ff] ISSN 0025–5718
• CONT OF (J 0235) Math Tables Other Aids Comp

J 0216 Izv Akad Nauk SSSR, Ser Mat • SU
Izvestiya Akademii Nauk SSSR. Seriya Matematicheskaya (Proceedings of the Academy of Sciences of the USSR. Mathematical Series) [1937ff] ISSN 0373–2436
• CONT OF (J 4717) Izv Akad Nauk SSSR • TRANSL IN (J 0448) Math of USSR, Izv

J 0217 Proc Glasgow Math Assoc • GB
Glasgow Mathematical Association. Proceedings [1952–1966]
• CONT AS (J 0126) Glasgow Math J

J 0224 God Vissh Tekh Ucheb Zaved Mat, Sofiya • BG
Godishnik na Visshite Tekhnicheski Uchebni Zavedeniya Matematika (Annuaire des Ecoles Techniques Superieures:Mathematiques) [1964–1973] ISSN 0436–1083
• CONT AS (J 3171) God Vissh Ucheb Zaved, Prilozhna Mat, Sofiya

J 0225 Amer Math Soc, Transl, Ser 2 • USA
American Mathematical Society. Translations. Series 2 [1955ff] ISSN 0065-9290

J 0226 Uch Zap Ped Inst, Ivanovo • SU
Ivanovskij Gosudarstvennyj Pedagogicheskij Institut imeni D.A.Furmanova. Uchenye Zapiski (Furmanov-Institute of Education in Ivanovo. Scientific Notes) [1941-1973] ISSN 0444-9681
• CONT AS (**J** 3536) Uch Zap Univ, Ivanovo

J 0229 Ann Mat Pura Appl, Ser 3 • I
Annali di Matematica Pura ed Applicata [1898-1923]
• CONT AS (**J** 3526) Ann Mat Pura Appl, Ser 4

J 0230 An Univ Iasi, NS, Sect Ia • RO
Analele Stiintifice ale Universitatii Al.I. Cuza din Iasi. (Serie Noua) Sectiunea 1a: Matematica (Wissenschaftliche Annalen der Al.I. Cuza Universitaet Iasi. (Neue Serie) Sektion 1a: Mathematik) [1955ff] ISSN 0041-9109

J 0232 Inform Process Lett • NL
Information Processing Letters. Devoted to the Rapid Publication of Short Contributions to Information Processing [1971ff] ISSN 0020-0190

J 0233 Soobshch Akad Nauk Gruz SSR • SU
*Soobshcheniya Akademii Nauk Gruzinskoj SSR * Sakaharth SSR Mecnierebatha Akademia Moambe (Communications of the Academy of Sciences of the Georgian SSR)* [1940ff] ISSN 0002-3167

J 0234 Rev Acad Cienc Exact Fis Nat Madrid • E
Revista de la Real Academia de Ciencias Exactas, Fisicas y Naturales de Madrid [1904ff] ISSN 0034-0596

J 0235 Math Tables Other Aids Comp • USA
Mathematical Tables and other Aids to Computation [1945-1959]
• CONT AS (**J** 0214) Math of Comp

J 0236 Rev Mat Hisp-Amer, Ser 4 • E
Revista Matematica Hispano-Americana. 4a Serie. Real Sociedad Matematica Espanola. [1941ff] ISSN 0373-0999
• CONT OF (**J** 3993) Rev Mat Hisp-Amer, Ser 2

J 0237 Stud Univ Cluj, Ser Math Phys Chem • RO
Studia Universitatis Babes-Bolyai. Series Mathematica-Physica-Chemia [1956-1963]
• CONT AS (**J** 0355) Studia Univ Babes-Bolyai, Math-Phys (Cluj) Kozlemenyei. Termeszettuclomanyi Sorozat

J 0238 Sitzb Oesterr Akad Wiss, Math-Nat Kl, Abt 2 • A
Oesterreichische Akademie der Wissenschaften. Mathematisch-Naturwissenschaftliche Klasse. Sitzungsberichte. Abteilung II. Mathematische, Physikalische und Technische Wissenschaft [1846ff] ISSN 0029-8816

J 0240 Ann Inst Fourier • F
Annales de l'Institut Fourier [1949ff] ISSN 0373-0956

J 0242 Language (Baltimore) • USA
Language [1924ff] ISSN 0097-8507

J 0247 Bull Sci Math, Ser 2 • F
Bulletin des Sciences Mathematiques, Serie 2 [1870ff] ISSN 0007-4497

J 0249 J Math Pures Appl • F
Journal de Mathematiques Pures et Appliquees [1836-1921]
• CONT AS (**J** 3941) J Math Pures Appl, Ser 9

J 0255 God Fak Mat & Mekh, Univ Sofiya • BG
Godishnik na Sofijskiya Universitet. Fakultet po Matematika i Mekhanika (Annuaire de l'Universite de Sofia. Faculte de Mathematiques.)

J 0259 Nyt Tidsskr Mat • DK
Nyt Tidsskrift for Matematik (New Journal for Mathematics) [1890-1918]
• CONT AS (**J** 4510) Norsk Mat Tidsskr

J 0260 Ann Jap Ass Phil Sci • J
Annals of the Japan Association for Philosophy of Science [1956ff]

J 0261 Tohoku Math J • J
Tohoku Mathematical Journal (Tohoku Sugaku Zashi) [1911ff] ISSN 0040-8735

J 0264 Collect Math (Barcelona) • E
Collectanea Mathematica [1948ff] ISSN 0010-0757

J 0265 Ukr Mat Zh, Akad Nauk Ukr SSR • SU
Ukrainskij Matematicheskij Zhurnal. Akademiya Nauk Ukrainskoj SSR. Institut Matematiki (Ukrainian Mathematical Journal. Academy of Sciences of the Ukrainian SSR. Institute of Mathematics) [1949ff] ISSN 0041-6053
• TRANSL IN (**J** 3281) Ukr Math J

J 0270 Dokl Akad Nauk Ukr SSR, Ser A • SU
*Doklady Akademii Nauk Ukrainskoj SSR. Seriya A. Fiziko-Matematicheskie i Tekhnicheskie Nauki * Dopovidi Akademii Nauk Uk'rainskoj RSR. Seriya A. Fiziko-Matematichni Ta Tekhnichi Nauki (Reports of the Academy of Sciences of the Ukrainian SSR. Series A. Physical-Mathematical and Engineering Sciences)* [1939ff] ISSN 0002-3531, ISBN 0201-8446

J 0273 Australasian J Phil • AUS
Australasian Journal of Philosophy [1947ff] ISSN 0004-8402
• CONT OF (**J** 4731) Australasian J Psych & Phil

J 0283 Ann Soc Pol Math • PL
*Societe Polonaise de Mathematique. Annales. * Rocznik i Polskiego Towarzystwa Matematycznego* [1922-1952]
• CONT AS (**J** 1405) Ann Pol Math

J 0284 IBM J Res Dev • USA
IBM (= International Business Machines) Journal of Research and Development [1957ff] ISSN 0018-8646

J 0286 Int Logic Rev • I
*International Logic Review. * Rassegna Internazionale di Logica* [1970ff] ISSN 0048-6779

J 0287 Scripta Math • USA
Scripta Mathematica. [1932ff] ISSN 0036-9713

J 0288 Vest Ser Mat Mekh, Univ Moskva • SU
Vestnik Moskovskogo Universiteta. Seriya I. Matematika, Mekhanika (Publications of the Moscow University. Series I. Mathematics. Mechanics) [1946ff] ISSN 0201-7385, ISSN 0579-9368
• TRANSL IN (**J** 0510) Moscow Univ Math Bull & Moscow University. Mechanics Bulletin

J 0293 Found Lang • NL
Foundations of Language. International Journal of Language and Philosophy [1965-1976] ISSN 0015-900X

J 0301 J Phil • USA
The Journal of Philosophy [1904ff] ISSN 0022-362X

J 0302 Rep Math Logic, Krakow & Katowice • PL
Reports on Mathematical Logic. The Jagiellonian University of Cracow. The Silesian University of Katowice [1973ff] ISSN 0083-4432
• CONT OF (S 0458) Zesz Nauk, Prace Log, Uniw Krakow

J 0303 Mathematika (Univ Coll London) • GB
Mathematika. A Journal of Pure and Applied Mathematics [1954ff] ISSN 0025-5793

J 0304 An Univ Timisoara, Sti Mat • RO
Analele Universitatii din Timisoara. Stiinte Matematice. Facultatea de Stiinte Matematice ale Naturii. (Annalen der Universitaet Timisoara. Mathematische Wissenschaften. Fakultaet fuer mathematische Wissenschaften der Natur.) [1963ff] ISSN 0563-5608
• CONT OF (J 1152) Inst Ped Timisoara Lucr Sti Mat Fiz

J 0305 Invent Math • D
Inventiones Mathematicae [1966ff] ISSN 0020-9910

J 0306 Cah Topol & Geom Differ • F
Cahiers de Topologie et Geometrie Differentielle [1959ff] ISSN 0008-0004

J 0307 Rev Colomb Mat • CO
Revista Colombiana de Matematicas [1967ff] ISSN 0034-7426
• CONT OF (J 0348) Rev Mat Elementales

J 0308 Rocky Mountain J Math • USA
The Rocky Mountain Journal of Mathematics [1971ff] ISSN 0035-7596

J 0309 Wiss Z Univ Greifswald, Math-Nat Reihe • DDR
Wissenschaftliche Zeitschrift der Ernst Moritz Arndt-Universitaet Greifswald. Mathematisch-Naturwissenschaftliche Reihe [1951ff] ISSN 0075-7512

J 0311 Nordisk Mat Tidskr • N
Nordisk Matematisk Tidskrift (Scandinavian Mathematical Journal) [1953-1978] ISSN 0029-1412
• CONT OF (J 4510) Norsk Mat Tidsskr • CONT AS (J 3075) Normat

J 0312 Izv Akad Nauk Armyan SSR, Ser Mat • SU
Izvestiya Akademii Nauk Armyanskoj SSR. Seriya Matematika (Proceedings of the Academy of Sciences of the Armenian SSR. Series: Mathematics) [1965ff] ISSN 0002-3043
• TRANSL IN (J 3265) Sov J Contemp Math Anal, Armen Acad Sci

J 0315 Ann Sc Norm Sup Pisa Fis Mat, Ser 3 • I
Annali della Scuola Normale Superiore di Pisa. Classe di Science. Fisiche e Matematiche. Seria III [1947-1973] ISSN 0036-9918
• CONT OF (J 1568) Ann Sc Norm Sup Pisa, Fis Mat, Ser 2
• CONT AS (J 4702) Ann Sc Norm Sup Pisa Fis Mat, Ser 4

J 0316 Illinois J Math • USA
Illinois Journal of Mathematics [1957ff] ISSN 0019-2082

J 0317 Trans New York Acad Sci Ser 2 • USA
Transactions of the New York Academy of Sciences. Series 2 [1938ff] ISSN 0028-7113

J 0319 Matematiche (Sem Mat Catania) • I
Le Matematiche [1946ff]

J 0322 Arch Math (Brno) • CS
Archivum Mathematicum [1965ff] ISSN 0044-8753

J 0326 J Pure Appl Algebra • NL
Journal of Pure and Applied Algebra [1971ff] ISSN 0022-4049

J 0332 Math Proc Cambridge Phil Soc • GB
Mathematical Proceedings of the Cambridge Philosophical Society [1975ff] ISSN 0305-0041
• CONT OF (J 0171) Proc Cambridge Phil Soc Math Phys

J 0337 Mat Ezheg, Akad Nauk Latv SSR • SU
Latvijskij Matematicheskij Ezhegodnik. Latvijskij Ordena Trudovogo Krasnogo Znameni Gosudarstvennyj Universitet imeni P.Stuchki. Akademiya Nauk Latvijskoj SSR (Latvian Mathematical Yearbook) [1965ff] ISSN 0458-8223
• TRANSL IN Latvian Mathematical Yearbook

J 0338 Nauch-Tekh Inf, Ser 2, Akad Nauk SSSR • SU
Nauchno-Tekhnicheskaya Informatsiya. Seriya 2. Gosudarstvennyj Komitet SSSR po Nauke i Tekhnike. Akademiya Nauk SSSR. Vsesoyuznyj Institut Nauchnoj i Tekhnicheskoj Informatsii. Informatsionnye Protsesy i Sistemy (Scientific Technical Information. Series 2) [1967ff]
• TRANSL IN (J 2667) Autom Doc Math Linguist

J 0339 Uch Zap Ped Inst, Volgograd • SU
Uchenye Zopiski Volgogradskogo Gosudarstvennogo Pedagogicheskogo Instituta im. A.S. Serafimovicha (Scientific Notes of the Serafimovich-Institute of Education of Volgograd) [1948ff]

J 0340 Mat Zap (Univ Sverdlovsk) • SU
Matematicheskie Zapiski (Mathematical Notes) ISSN 0076-5368

J 0345 Adv Math • USA
Advances in Mathematics [1964ff] ISSN 0001-8708
• REL PUBL (S 3105) Adv Math, Suppl Stud

J 0346 Dokl Akad Nauk Armyan SSR • SU
Doklady Akademii Nauk Armyanskoj SSR (Reports of the Academy of Sciences of the Armenian SSR) [1944ff] ISSN 0321-1339

J 0348 Rev Mat Elementales • CO
Revista de Matematicas Elementales [1952-1966]
• CONT AS (J 0307) Rev Colomb Mat

J 0349 Math of USSR, Sbor • USA
Mathematics of the USSR, Sbornik [1967ff] ISSN 0025-5734
• TRANSL OF (J 0142) Mat Sb, Akad Nauk SSSR, NS

J 0350 Sci Rep Tokyo Kyoiku Daigaku Sect A • J
Tokyo Kyoiku Daigaku (Tokyo University of Education. Science Reports. Section A.)

J 0351 Osaka J Math • J
Osaka Journal of Mathematics [1964ff] ISSN 0030-6126
• CONT OF (J 1770) Osaka Math J

J 0352 Math Jap • J
Mathematica Japonica [1948ff] ISSN 0025-5513

J 0353 Bull Soc Math Fr • F
Bulletin de la Societe Mathematique de France [1873ff] ISSN 0037-9484

J 0354 Phil Trans Roy Soc London, Ser A • GB
Philosophical Transactions of the Royal Society of London. Series A. Mathematical and Physical Sciences. ISSN 0080-4614

Source Index

J 0355 Studia Univ Babes-Bolyai, Math-Phys (Cluj) • RO
Studia Universitatis Babes-Bolyai. Mathematica - Physica [1964-1970]
• CONT OF (J 0237) Stud Univ Cluj, Ser Math Phys Chem
• CONT AS (J 3451) Stud Univ Cluj, Ser Math-Mech

J 0369 Notices Amer Math Soc • USA
Notices of the American Mathematical Society [1953ff] ISSN 0002-9920

J 0371 Glas Mat-Fiz Astron, Ser 2 (Zagreb) • YU
Glasnik Matematichko-Fizichki i Astronomichki. Ser II (Publications of Mathematics, Physics, Astronomy. Ser II) [1946-1965]
• CONT AS (J 3519) Glas Mat, Ser 3 (Zagreb)

J 0373 Comp Arch Inform & Numerik • A
*Computing: Archiv fuer Informatik und Numerik * Computing: Archives for Informatics and Numerical Computation* [1966ff] ISSN 0010-485X

J 0374 SIAM J Appl Math • USA
Journal on Applied Mathematics. SIAM (= Society for Industrial and Applied Mathematics) [1966ff] ISSN 0036-1399
• CONT OF (J 0514) SIAM Journ

J 0377 Bol Mat (Bogota) • CO
Boletin de Matematicas [1967ff]

J 0380 Acta Cybern (Szeged) • H
Acta Cybernetica. Forum Centrale Publicationum Cyberneticarum Hungaricum [1969ff] ISSN 0324-721X

J 0381 J Tsuda College (Tokyo) • J
Journal of Tsuda College [1969ff]

J 0382 Int J Comput Math • GB
International Journal of Computer Mathematics. Section A: Programming Languages; Theory and Methods. Section B: Computational Methods [1964ff] ISSN 0020-7160

J 0383 J Cybern • USA
Journal of Cybernetics. Transactions of the American Society for Cybernetics [1971ff] ISSN 0022-0280
• REM Includes Translations from Appropriate Russian and Japanese Journals

J 0384 Rend Mat, Ser 5 • I
Rendiconti di Matematica. Serie 5 [1940-1967]
• CONT AS (J 2311) Rend Mat, Ser 6

J 0386 Proc Edinburgh Math Soc • GB
Proceedings of the Edinburgh Mathematical Society [1883-1926]
• CONT AS (J 3420) Proc Edinburgh Math Soc, Ser 2

J 0389 Math Balkanica • YU
Mathematica Balkanica: Unitas Mathematicorum Balkanica [1971ff]

J 0390 Publ Res Inst Math Sci (Kyoto) • J
Publications of the Research Institute for Mathematical Sciences [1965ff] ISSN 0034-5318
• REM Vols 1-4 Issued as: Kyoto Univ. Research Institute for Mathematical Sciences. Publications. Series A.

J 0392 Math Sci Hum • F
Mathematiques et Sciences Humaines [1962ff] ISSN 0025-5815

J 0394 Commun Algeb • USA
Communications in Algebra [1974ff] ISSN 0092-7872

J 0396 Mat Lapok • H
Matematikai Lapok (Mathematical Papers) [1949ff] ISSN 0025-519X
• CONT OF (J 0461) Mat Fiz Lapok

J 0397 Proc Math Phys Soc Egypt • ET
Proceedings of the Mathematical and Physical Society of Egypt [1937ff] ISSN 0076-5317

J 0399 Acta Arith, Pol Akad Nauk • PL
Acta Arithmetica. Academia Scientiarum Polona. Institutum Mathematicum [1937ff]

J 0400 Publ Inst Math, NS (Belgrade) • YU
Institut Mathematique. Publications de l'Institut Mathematique. Nouvelle Serie [1961ff] ISSN 0522-828X
• CONT OF (J 4706) Publ Inst Math (Belgrade)

J 0401 J Number Th • USA
Journal of Number Theory [1968ff] ISSN 0022-314X

J 0403 Izv Akad Nauk Kazak SSR, Ser Fiz-Mat • SU
Izvestiya Akademii Nauk Kazakhskoj SSR. Seriya Fiziko-Matematicheskaya (Proceedings of the Academy of Sciences of the Kazakh SSR. Series: Physics & Mathematics) [1963ff] ISSN 0002-3191

J 0406 Bull Inst Math, Acad Sin (Taipei) • RC
Chung Yang Yen Chui y Uan Shu Hs Ueh Yen Chiu So (Bulletin of the Institute of Mathematics. Academia Sinica.) [1973ff]

J 0407 Comm Math Univ St Pauli (Tokyo) • J
Commentarii Mathematici Universitatis Sancti Pauli [1952ff] ISSN 0010-258X

J 0411 Studia Sci Math Hung • H
Studia Scientiarum Mathematicarum Hungaria. Auxilio Consilii Instituti Mathematici. Academiae Scientiarum Hungaricae [1966ff] ISSN 0081-6906

J 0413 Izv Akad Nauk Belor SSR, Ser Fiz-Mat • SU
*Vestsi Akademii Navuk BeSSR. Seriya Fizika-Matematychnykh Navuk * Izvestiya Akademii Nauk BSSR. Seriya Fiziko-Matematicheskikh Nauk (Proceedings of the Academy of Sciences of the Byelorussian SSR. Series: Physics, Mathematics)* [1964ff] ISSN 0002-3574

J 0414 Dokl Akad Nauk Belor SSR • SU
Doklady Akademii Nauk BSSR (Reports of the Academy of Sciences of the BSSR) [1957ff] ISSN 0002-354X

J 0416 J Mathematiques • F
Journal de Mathematiques [1836ff] ISSN 0021-7824

J 0418 Shuxue Xuebao • TJ
Shuxue Xuebao (Acta Mathematica Sinica) [1951ff]
• TRANSL IN (J 0419) Chinese Math Acta
• REM In 1951 published as: Journal of the Chinese Mathematical Society (N.S.)

J 0419 Chinese Math Acta • USA
Chinese Mathematics. Acta [1962-1967]
• TRANSL OF (J 0418) Shuxue Xuebao

J 0420 Shuxue Jinzhan • TJ
Shuxue Jinzhan (Advances in Mathematics) [1955ff] ISSN 0559-9326

J 0429 Vest Akad Nauk Kazak SSR • SU
Vestnik Akademii Nauk Kazakhskoj SSR (Publications of the Academy of Sciences of the Kazakh SSR) [1944ff] ISSN 0002-3213

J 0430 Vopr Kibern (Akad Nauk Uzb SSR) • SU
Voprosy Kibernetiki i Vychislitel'noj Matematiki (Problems of Cybernetics and Numerical Mathematics) [1966ff] ISSN 0507-3502

J 0432 Bell Syst Tech J • USA
The Bell System Technical Journal. [1922ff] ISSN 0005-8580

J 0434 J Fac Sci Univ Tokyo, Sect 1 • J
Journal of the Faculty of Science. University of Tokyo. Section 1 Mathematics, Astronomy, Physics, Chemistry [1925-1970] ISSN 0040-8980
• CONT AS (**J** 2332) J Fac Sci, Univ Tokyo, Sect 1 A

J 0435 Int J Comput & Inf Sci • USA
International Journal of Computer and Information Sciences [1972ff] ISSN 0091-7036

J 0438 J Hokkaido Univ Educ Ser 1 • J
Journal of Hokkaido University of Education. Faculty of Science. Series 1. Mathematics [1930ff] ISSN 0018-3482

J 0446 Ann Acad Sci Fennicae, Ser A I, Diss • SF
Annales Academiae Scientiarum Fennicae. Series AI. Mathematica. Dissertationes [1975ff] ISSN 0066-1953
• CONT OF (**J** 3994) Ann Acad Sci Fennicae Ser A I

J 0447 An Univ Bucuresti, Mat • RO
Analele Universitatii Bucuresti. Matematica (Annalen der Universitaet Bukarest. Mathematik) [1951ff]

J 0448 Math of USSR, Izv • USA
Mathematics of the USSR, Izvestiya [1967ff] ISSN 0025-5726
• TRANSL OF (**J** 0216) Izv Akad Nauk SSSR, Ser Mat

J 0449 Probl Kybern • DDR
Probleme der Kybernetik [1958-1965]
• CONT AS (**J** 0471) Syst Th Res • TRANSL OF (**J** 0052) Probl Kibern

J 0452 Indiana Univ Math J • USA
Indiana University Mathematics Journal [1971ff] ISSN 0022-2518
• CONT OF (**J** 4732) J Math Mech (Indiana Univ)

J 0459 C R Soc Sci Lett Varsovie Cl 3 • PL
*Societe des Sciences et des Lettres de Varsovie. Comptes Rendus des Seances. Classe III: Sciences Mathematiques et Physiques * Towarzystwo Naukowe Warszawskie. Sprawozdania z Posiedze. Wydzialu III: Nauk Matematyczno-Fizycznych (Warschauer Sitzungsberichte)* [1908-1950]

J 0460 Acta Univ Szeged, Sect Mat • H
Acta Litterarum ac Scientiarum Regiae Universitatis Hungaricae Francisco-Josephinae, Sectio Scientiarum Mathematicarum [1922-1946]
• CONT AS (**J** 0002) Acta Sci Math (Szeged)

J 0461 Mat Fiz Lapok • H
Matematikai es Fizikai Lapok (Mathematical and Physical Papers) [1892-1948]
• CONT AS (**J** 0396) Mat Lapok

J 0462 Mat Fiz Oszt Koezlem, Acad Sci Hung • H
Magyar Tudomanyos Akademia. Matematikai es Fizikai Tudomanyok Osztalyanak Koezlemenyek. (Hungarian Academy of Sciences. Bulletin of the Mathematical and Physical Sciences) [1952-1974]
• CONT AS (**J** 1458) Alkalmaz Mat Lapok

J 0463 Proc Fac Sci Tokai Univ • J
Proceedings of the Faculty of Science of Tokai University [1966ff]

J 0464 Syst-Comp-Controls • USA
Systems - Computers - Controls [1970ff] ISSN 0096-8765
• TRANSL OF (**J** 0979) Denshi Tsushin Gakkai Ronbunshi, Sect A-D

J 0465 Bull Greek Math Soc (NS) • GR
Bulletin of the Greek Mathematical Society.New Series (Hellenike Mathematike Hetaireia. Deltion. Nea Seira.) [1960ff] ISSN 0072-7466
• CONT OF (**J** 1699) Bull Soc Math Grece

J 0467 Utilitas Math, Canad J • CDN
Utilitas Mathematica. A Canadian Journal of Applied Mathematics, Computer Science, and Statistics [1927ff] ISSN 0315-3681

J 0468 Uch Zap Ped Inst (Alma Ata) • SU
Kazakhskij Gosudarstvennyj Pedagogicheskij Institut im. Abaya. Uchenye Zapiski (Abaya-Institute of Education of Kazakhstan. Scientific Notes)

J 0470 Sov Phys, Dokl • USA
Soviet Physics. Doklady. [1956ff] ISSN 0038-5689
• TRANSL OF (**J** 0023) Dokl Akad Nauk SSSR

J 0471 Syst Th Res • USA
Systems Theory Research [1966ff] ISSN 0082-1255
• CONT OF (**J** 1195) Probl Cybernet & (**J** 0449) Probl Kybern
• TRANSL OF (**J** 0052) Probl Kibern

J 0474 Avtom Vychis Tekh, Akad Nauk Latv SSR • SU
Avtomatika i Vychislitel'naya Tekhnika. Akademiya Nauk Latvijskoj SSR (Automation and Computer Science. Academy of Sciences of the Latvian SSR) [1967ff] ISSN 0572-4538
• TRANSL IN (**J** 2666) Autom Control Comput Sci

J 0475 Sib Math J • USA
Siberian Mathematical Journal [1966ff] ISSN 0037-4466
• TRANSL OF (**J** 0092) Sib Mat Zh

J 0477 Spis Bulgar Akad Nauk • BG
B"lgarski Akademija na Naukite. Spisanie (Bulgarian Academy of Sciences. Journal) ISSN 0015-3265

J 0481 Acta Univ Wroclaw • PL
Acta Universitatis Wratislaviensis

J 0483 Uch Zap, Univ Riga • SU
*Uchenye Zapiski Latvijskogo Gosudarstvennogo Universiteta imeni Petra Stutski. * Zinatniskie Raksti. Latvijas Valsts Universitate. (Scientific Notes of the Latvian State University.)*
• REM This journal consists of two series: Teoriya Algoritmov i Programm & Teoreticheskie Voprosy Avtomaticheskikh Sistem Upravleniya

J 0485 Algorytmy, Pol Akad Nauk • PL
Algorytmy (Algorithms) [1962ff] ISSN 0065-6240

J 0487 Math Unterricht • D
Der Mathematikunterricht: Beitraege zu seiner Wissenschaftlichen und Methodischen Gestaltung [1955ff] ISSN 0025-5807

J 0493 Bull Res Counc Israel Sect F • IL
Research Council of Israel. Bulletin. Section F. Mathematics and Physics [1952-1962]
• CONT AS (**J** 0029) Israel J Math

J 0494 Bull Math Soc Sci Roumanie • RO
Societatea Romana de Stiinte, Sectia Mathematica. Bulletin Mathematiques de la Societe Roumaine des Sciences [1908-1956]
• CONT AS (**J** 0070) Bull Soc Sci Math Roumanie, NS

J 0496 Tr Inst Inzh Zheleznod Transp, Moskva • SU
Moskovskij Institut Inzhenerov Zheleznodorozhnogo Transporta. Trudy (Institute for Railway Engineers, Moscow. Publications)

J 0497 Math Mag • USA
Mathematics Magazine [1947ff] ISSN 0025-570X
• CONT OF (**J** 1737) Nat Math Magazine (Louisiana)

J 0498 Ann Univ Turku, Ser A I • SF
Annales Universitatis Turkuensis. Series A.I: Astronomica, Chemica, Physica, Mathematica (Turun Yliopiston Julkaisuja. Sarja A.1.: Astronomica, Chemica, Physica, Mathematica) [1957ff] ISSN 0082-7002

J 0499 Uch Zap Univ, Gor'kij • SU
Uchenye Zapiski Gor'kovskogo Gosudarstvennogo Universiteta imeni N.I.Lobachevskogo (Scientific Notes of the Lobachevskij University of Gor'kij)

J 0504 Manuscr Math • D
Manuscripta Mathematica [1969ff] ISSN 0025-2611

J 0508 Machine Intelligence • GB
Machine Intelligence [1967ff] ISSN 0541-6418

J 0510 Moscow Univ Math Bull • USA
Moscow University Mathematics Bulletin [1969ff] ISSN 0027-1322
• TRANSL OF (**J** 0288) Vest Ser Mat Mekh, Univ Moskva

J 0514 SIAM Journ • USA
Journal. SIAM (= Society for Industrial and Applied Mathematics) [1953-1965]
• CONT AS (**J** 0374) SIAM J Appl Math

J 0515 Bull Math Biophys • GB
Bulletin of Mathematical Biophysics [1939-1971] ISSN 0007-4985
• CONT AS (**J** 3073) Bull Math Biol

J 0516 Rev Mexicana Fis • MEX
Revista Mexicana de Fisica [1952ff] ISSN 0035-001X

J 0517 Mathematica (Cluj) • RO
Mathematica. Revue d'Analyse Numerique et de Theorie de l'Approximation [1929ff] ISSN 0025-5505

J 0518 Bul Sti Mat-Fiz, Acad Romina • RO
*Academia Republicii Populare Romine. Buletinul Stiintific. Sectia de Stiinte Matematice si Fizice * Academie de la Republique Populaire Roumainie. Bulletin Scientifique. Section des Sciences Mathematiques et Physiques * Akademija Rumynskoi Respubliki. Nachnyi Vestnik. Otdelenie Matematicheskih i Fizicheskih Nauk* [1949-1965] ISSN 0515-1333
• CONT OF (**J** 0174) Bull Sect Sci Acad Roumaine

J 0519 Wiad Mat, Ann Soc Math Pol, Ser 2 • PL
*Annales Societatis Mathematicae Polonae. Seria 2. Wiadomosci Matematyczne * Roczniki Polskiego Towarzystwa Matematycznego. Seria 2. Wiadomosci Matematyczne* [1955ff] ISSN 0079-3698
• CONT OF (**J** 4710) Pol Tow Mat, Wiad Mat

J 0521 Semin Math, Inst Steklov • USA
Seminars in Mathematics. V.A.Steklov Mathematical Institute Leningrad [1967ff]
• TRANSL OF (**S** 0228) Zap Nauch Sem Leningrad Otd Mat Inst Steklov

J 0522 Engin Cybern • USA
Engineering Cybernetics. Soviet Journal of Computer and System Sciences. Essential Serials in Electronics and Systems Science [1963ff] ISSN 0013-788X
• TRANSL OF (**J** 0977) Izv Akad Nauk SSSR, Tekh Kibern

J 0524 Disq Math Phys • RO
Disquisitiones Mathematicae et Physicae [1940-1949]
• CONT AS (**J** 0197) Stud Cercet Mat Acad Romana

J 0529 Aequationes Math • CH
Aequationes Mathematicae [1968ff] ISSN 0001-9054

J 0531 Hitotsubashi J Arts Sci (Tokyo) • J
Hitotsubashi Journal of Arts & Sciences [1960ff] ISSN 0073-2788

J 0535 Abh Braunschweig Wiss Ges • D
Abhandlungen der Braunschweigischen Wissenschaftlichen Gesellschaft [1949ff] ISSN 0068-0737

J 0536 Kodai Math Sem Rep (Kyoto) • J
Kodai Mathematical Seminar Reports [1949-1975] ISSN 0023-2599

J 0537 Yokohama Math J • J
The Yokohama Mathematical Journal [1953ff] ISSN 0044-0523

J 0545 Proc Indian Acad Sci, Sect A • IND
Indian Academy of Sciences. Proceedings. Section A: Physical Sciences [1931ff] ISSN 0019-428X

J 0548 Stud Appl Math • USA
Studies in Applied Mathematics [1969ff] ISSN 0022-2526

J 0589 Bull Amer Math Soc (NS) • USA
Bulletin of the American Mathematical Society. New Series [1979ff] ISSN 0273-0979
• CONT OF (**J** 0015) Bull Amer Math Soc

J 0598 Wiss Z Paed Hochsch Potsdam • DDR
Wissenschaftliche Zeitschrift der Paedagogischen Hochschule Karl Liebknecht zu Potsdam [1954ff]

J 0611 Pol Tow Mat, Prace Mat-Fiz • PL
Polskie Towarzystwo Matematyczene. Prace Matematyczno-Fizyczne (Mathematische und Physikalische Abhandlungen) [1887-1954]
• CONT AS (**J** 0051) Commentat Math, Ann Soc Math Pol, Ser 1 & (**J** 2095) Fund Inform, Ann Soc Math Pol, Ser 4

J 0650 Ege Ueniv Fen Fak Derg, Ser A • TR
Ege Ueniversitesi Fen fakueltesi Dergisi. Seri A: Matematik, Astronomi, Fizik. Kimya, Jeologi ve Jeofizik (Ege University Faculty of Science Journal)

J 0752 Avh Norske Vid-Akad Oslo I • N
Avhandlinger. Norske Videnskaps-Akademi i Oslo. I: Matematisk-Naturvidenskapelig Klasse (Proceedings of the Scandinavian Academy of Sciences. Mathematical and Natural Sciences Class)

J 0767 Rev Gen Sci Pur Appl • F
Revue Generale des Sciences Pures et Appliquees [1890-1952]
• CONT AS (**J** 0211) Rev Gen Sci Pur Appl & Bull Soc Philomatique

J 0789 Uch Zap Mat Ped Inst, Tula • SU
Tul'skij Gosudarstvennyj Pedagogicheskij Institut im. L.N.Tolstogo. Uchenye Zapiski Matematicheskikh Kafedr (Tolstoi State Institute of Education in Tula. Scientific Notes of the Mathematical Faculties)

J 0933 Math Nat Unterr • D
Der Mathematische und Naturwissenschaftliche Unterricht [1948ff] ISSN 0025-5866
• CONT OF (**J** 0185) Unterrichtsbl Math Nat

J 0946 SIAM J Control • USA
Journal on Control. SIAM (= Society for Industrial and Applied Mathematics) [1966-1975] ISSN 0036-1402
• CONT OF (**J** 4704) J Soc Ind & Appl Math, Ser Control
• CONT AS (**J** 4705) SIAM J Control & Optim

J 0947 Wiss Z Tech Univ Dresden • DDR
Wissenschaftliche Zeitschrift der Technischen Universitaet Dresden [1951ff] ISSN 0043-6925

J 0953 Vest Cesk Spol Nauk • CS
Vestnik Ceska Spolechnost Nauk V Praze. (Publications of the Czech Society of Sciences. Prague) [1918-1952]

J 0954 Tr Inst Prikl Mat, Tbilisi • SU
*Tbilisskij Gosudarstvennyj Universitet. Institut Prikladnoj Matematiki. Trudy * Thbilisis Sahelmcipho Universiteti Gamoqenebithi Mathematikis Instituti Shromebi (State University of Tbilisi. Institute of Applied Mathematics. Publications)* [1969ff] ISSN 0082-2191

J 0962 Ann Soc Sci Bruxelles, Ser 1 • B
Annales de la Societe Scientifique de Bruxelles. Serie I. Sciences Mathematiques, Astronomiques et Physiques [1875ff] ISSN 0037-959X

J 0967 Izv Akad Nauk Mold SSR, Ser Fiz-Tekh Mat • SU
*Izvestiya Akademii Nauk Moldavskoj SSR. Seriya Fiziko-Tekhnicheskikh i Matematicheskikh Nauk * Buletinul Akademiei de Shtiince a RSS Moldovensht (Proceedings of the Academy of Sciences of the Moldavian SSR. Series: Physical-Technical and Mathematical Sciences)* [1951ff] ISSN 0321-169X

J 0971 Rend Accad Sci Napoli Fis Mat, Ser 4 • I
Societa Nazionale di Scienze, Lettere ed Arti in Napoli. Rendiconto dell'Accademia delle Scienze Fisiche e Matematiche. Serie IV.

J 0974 Norsk Vid-Akad Oslo Mat-Natur Kl Skr • N
Norske Videnskaps - Akademi i Oslo. Matematisk-Naturvidenskapelig Klasse. Skrifter. (Monographs of the Scandinavian Academy of Sciences. Mathematical and Natural Sciences Class) [1929ff] ISSN 0029-2338
• CONT OF (**J** 1145) Vidensk Selsk Kristiana Skrifter Ser 1

J 0977 Izv Akad Nauk SSSR, Tekh Kibern • SU
Izvestiya Akademii Nauk SSSR. Tekhnicheskaya Kibernetika. Otdelenie Mekhaniki i Protsessov Upravlenija (Proceedings of the Academy of Sciences of the USSR. Engineering Cybernetics. Department of Mechanics and Control Processes) [1963ff] ISSN 0002-3388
• TRANSL IN (**J** 0522) Engin Cybern

J 0979 Denshi Tsushin Gakkai Ronbunshi, Sect A-D • J
Denshi Tsushin Gakkai Ronbunshi. Sect. A-D (Reports of the University of Electro-Communications) [1949ff] ISSN 0493-4253
• TRANSL IN (**J** 0116) Electr & Comm Japan & (**J** 0464) Syst-Comp-Controls

J 0982 Z Wahrscheinltheor & Verw Geb • D
Zeitschrift fuer Wahrscheinlichkeitstheorie und Verwandte Gebiete [1962ff] ISSN 0044-3719

J 0996 RAAG Res Notes • J
Research Notes and Memoranda of Applied Geometry for Prevenient Natural Philosophy. Post RAAG-Reports.

J 1005 Rep Fac Sci, Shizuoka Univ • J
Reports of the Faculty of Science. Shizuoka University. [1965ff] ISSN 0583-0923

J 1008 Demonstr Math (Warsaw) • PL
Demonstratio Mathematica [1969ff] ISSN 0420-1213

J 1021 Itogi Nauki Ser Mat • SU
Itogi Nauki. Seriya Matematiki (Progress in Science. Mathematical Series) [1962-1971] ISSN 0579-1731
• CONT AS (**J** 1488) Itogi Nauki Tekh, Ser Probl Geom & (**J** 1501) Itogi Nauki Tekh, Ser Algeb, Topol, Geom & (**J** 1452) Itogi Nauki Tekh, Ser Sovrem Probl Mat & (**J** 3188) Itogi Nauki Tekh, Ser Teor Veroyat Mat Stat Teor Kibern & (**J** 4387) Itogi Nauki Tekh, Ser Tekh Kibern • TRANSL IN (**J** 1531) J Sov Math
• REM J1531 contains only selected translations

J 1024 Zhongguo Kexue • TJ
Zhongguo Kexue (Scientia Sinica) [1950-1981]
• CONT AS (**J** 3766) Zhongguo Kexue, Xi A

J 1044 Math Notes, Acad Sci USSR • USA
Mathematical Notes of the Academy of Sciences of the USSR [1967ff] ISSN 0001-4346
• TRANSL OF (**J** 0087) Mat Zametki (Akad Nauk SSSR)

J 1046 Z Angew Math Mech • DDR
Zeitschrift fuer Angewandte Mathematik und Mechanik: Ingeneurwissenschaftliche Forschungsarbeiten. Applied Mathematics and Mechanics [1921ff] ISSN 0044-2267

J 1048 Kiber Sb Perevodov • SU
Kiberneticheskij Sbornik: Sbornik Perevodov. (Collected Articles on Cybernetics: Collected Translations) [1960-1964]
• CONT AS (**J** 3079) Kiber Sb Perevodov, NS

J 1049 USSR Comput Math & Math Phys • GB
USSR Computational Mathematics and Mathematical Physics [1962ff] ISSN 0041-5553
• TRANSL OF (**J** 0199) Zh Vychisl Mat i Mat Fiz

J 1077 IEEE Proc • USA
Proceedings. IEEE (= Institute of Electrical and Electronics Engineers) [1963ff] ISSN 0018-9219
• CONT OF (**J** 4711) IRE Proc

J 1089 IBM Res Rep • USA
IBM (= International Business Machines) Research Report

J 1108 Arkhimedes (Helsinki) • SF
Arkhimedes [1949ff] ISSN 0004-1920

J 1109 Nachr Akad Wiss Goettingen, Math-Phys Kl • D
Nachrichten der Akademie der Wissenschaften in Goettingen. II. Mathematisch-Physikalische Klasse [1893ff] ISSN 0065-5295

J 1114 Arch Phil • D
Archiv fuer Philosophie [1911-1930, 1947-1964]

J 1138 Quart J Math, Oxford Ser • GB
The Quarterly Journal of Mathematics. Oxford Series [1930-1949]
• CONT AS (**J** 0131) Quart J Math, Oxford Ser 2

J 1145 Vidensk Selsk Kristiana Skrifter Ser 1 • N
Videnskaps Selskapet i Kristiana. Skrifter Utgit. 1 Matematisk-Naturvidenskapelig Klasse [?-1928]
• CONT AS (J 0974) Norsk Vid-Akad Oslo Mat-Natur Kl Skr

J 1150 Proc Roy Soc London, Ser A • GB
Proceedings of the Royal Society of London. Series A. Mathematical and Physical Sciences. [1905ff] ISSN 0080-4630

J 1152 Inst Ped Timisoara Lucr Sti Mat Fiz • RO
Lucrarile Stiintifice ale Institutului Pedagogic Timisoara. Matematica-Fizica (Wissenschaftliche Arbeiten des Paedagogischen Instituts Timisoara. Mathematik-Physik) [1958-1962]
• CONT AS (J 0304) An Univ Timisoara, Sti Mat

J 1153 Kexue Jilu (Beijing) • TJ
Kexue Jilu (Science Record) [1957ff] ISSN 0559-1244

J 1156 Izv Bulgar Akad Nauk Mat Inst • BG
B"lgarska Akademiya na Naukite. Otdelenie Za Fiziko-Matematicheski i Tekhnicheski Nauki. Izvestiya na Matematicheski Institut (Bulgarian Academy of Sciences. Department of Physics, Mathematics & Engineering. Reports of the Mathematical Institute) [1957-1974]
• CONT AS (J 2547) Serdica, Bulgar Math Publ

J 1179 Filoz Jugos Chas (Belgrade) • YU
Filozofija: Jugoslovenski Chasopis Za Filozofiju (Philosophy: Yugoslavian Journal of Philosophy) [1957ff] ISSN 0015-1866

J 1191 Acta Polytech Scand, Math Comp Sci • SF
Acta Polytechnica Scandinavica, Mathematics and Computer Science Series ISSN 0355-2713

J 1193 Comput J (London) • GB
The Computer Journal [1958ff] ISSN 0010-4620

J 1195 Probl Cybernet • GB
Problems of Cybernetics [1958-1965]
• CONT AS (J 0471) Syst Th Res • TRANSL OF (J 0052) Probl Kibern

J 1399 Russ Math Surv • GB
Russian Mathematical Surveys [1946ff] ISSN 0036-0279
• TRANSL OF (J 0067) Usp Mat Nauk

J 1404 Mat Sb, Akad Nauk SSSR • SU
Matematicheskij Sbornik. Akademiya Nauk SSSR i Moskovskoe Matematicheskoe Obshchestvo (Mathematical Collected Articles. New Series. Academy of Sciences of the USSR and the Moscovian Mathematical Society) [1866-1935]
• CONT AS (J 0142) Mat Sb, Akad Nauk SSSR, NS

J 1405 Ann Pol Math • PL
Annales Polonici Mathematici. Polska Akademia Nauk, Instytut Matematyczny [1953ff] ISSN 0066-2216
• CONT OF (J 0283) Ann Soc Pol Math

J 1426 Theor Comput Sci • NL
Theoretical Computer Science [1975ff] ISSN 0304-3975

J 1428 SIAM J Comp • USA
Journal on Computing. SIAM (= Society for Industrial and Applied Mathematics) [1972ff] ISSN 0097-5397

J 1429 Kybernetes • GB
Kybernetes: An International Journal of Cybernetics and General Systems [1972ff] ISSN 0368-492X

J 1431 Acta Inf • D
Acta Informatica [1971ff] ISSN 0001-5903

J 1447 Houston J Math • USA
Houston Journal of Mathematics [1975ff] ISSN 0362-1588

J 1452 Itogi Nauki Tekh, Ser Sovrem Probl Mat • SU
Itogi Nauki i Tekhniki: Seriya Sovremennye Problemy Matematiki (Progress in Science and Technology: Series on Current Problems in Mathematics) [1972ff]
• CONT OF (J 1021) Itogi Nauki Ser Mat • TRANSL IN (J 1531) J Sov Math

J 1456 SIGACT News • USA
SIGACT (= ACM Special Interest Group on Automata and Computability Theory). News [1969ff]

J 1458 Alkalmaz Mat Lapok • H
Alkalmazott Matematikai Lapok (Papers in Applied Mathematics) [1951ff] ISSN 0133-3399
• CONT OF (J 0462) Mat Fiz Oszt Koezlem, Acad Sci Hung

J 1460 Tr Inst Ehlekt Mashinostr, Moskva • SU
Trudy MIEM (= Moskovskij Institut Ehlektronnogo Mashinostroeniya) (Institute of Electronic Engineering, Moscow. Publications of the MIEM) ISSN 0579-8671

J 1477 Tr Vychisl Tsentr, Akad Nauk Gruz SSR • SU
Trudy Vychislitel'nogo Tsentra. Akademiya Nauk Gruzinskoj SSR Sakharthvelos SSR Mecnierebatha Akademis Gamothvlithi Centris Shromebi. (Publications of the Computational Centre. Academy of Sciences of the Georgian SSR) [1960ff] ISSN 0568-4900

J 1487 Publ Math Inst Acad Sci Hung • H
Publications of the Mathematical Institute of the Hungarian Academy of Sciences [1956ff]

J 1488 Itogi Nauki Tekh, Ser Probl Geom • SU
Itogi Nauki i Tekhnike. Seriya Problemy Geometrii (Progress in Science and Technology. Series Problems in Geometry) [1972ff] ISSN 0202-7461
• CONT OF (J 1021) Itogi Nauki Ser Mat • TRANSL IN (J 1531) J Sov Math

J 1501 Itogi Nauki Tekh, Ser Algeb, Topol, Geom • SU
Itogi Nauki i Tekhniki. Seriya Algebra, Topologiya, Geometriya. (Progress in Science and Technology. Series Algebra, Topology, Geometry) [1972ff] ISSN 0202-7445
• CONT OF (J 1021) Itogi Nauki Ser Mat • TRANSL IN (J 1531) J Sov Math & (C 4688) Prog in Math, Vol 12
• REM C4688 is Volume 1968

J 1515 Archimede • I
Archimede. Rivista per gli Insegnanti e i Cultori di Matematiche Pure e Applicate [1949ff] ISSN 0003-8369

J 1522 Math Slovaca • CS
Mathematica Slovaca [1976ff] ISSN 0025-5173
• CONT OF (J 0143) Mat Chasopis (Slov Akad Ved)

J 1527 Pokroky Mat Fyz Astron (Prague) • CS
Pokroky Matematiky, Fyziky a Astronomie (Progress in Mathematics, Physics and Astronomy) [1956ff] ISSN 0032-2423

J 1528 Sitzb Plenum & Klassen Akad Wiss DDR • DDR
Sitzungsberichte des Plenums und der Klassen der Akademie der Wissenschaften der DDR

J 1531 J Sov Math • USA
Journal of Soviet Mathematics [1973ff] ISSN 0090-4104
• TRANSL OF (J 1021) Itogi Nauki Ser Mat & (S 0228) Zap Nauch Sem Leningrad Otd Mat Inst Steklov & Problemy Matimaticheskogo Analiza & (J 1452) Itogi Nauki Tekh, Ser Sovrem Probl Mat & (J 1501) Itogi Nauki Tekh, Ser Algeb, Topol, Geom & (J 3188) Itogi Nauki Tekh, Ser Teor Veroyat Mat Stat Teor Kibern & (J 1488) Itogi Nauki Tekh, Ser Probl Geom
• REM This contains selected translations from each of the Russian Journals listed

J 1550 Creation Math • IL
Creation in Mathematics [1970ff]

J 1561 IEEE Trans Syst & Sci Cybern • USA
Transactions on Systems Science and Cybernetics. IEEE (= Institute of Electrical and Electronics Engineers) [1945-1970] ISSN 0018-9472
• CONT AS (J 2338) IEEE Trans Syst Man & Cybern

J 1568 Ann Sc Norm Sup Pisa, Fis Mat, Ser 2 • I
Annali della R. Scuola Normale Superiore di Pisa. Serie 2 Scienze Fisiche e Matematiche [1932-1946]
• CONT OF (J 1908) Ann Sc Norm Sup Pisa, Fis Mat • CONT AS (J 0315) Ann Sc Norm Sup Pisa Fis Mat, Ser 3

J 1572 Izv Vyssh Ucheb Zaved, Radiofizika (Moskva) • SU
Izvestiya Vysshikh Uchebnykh Zavednii. Radiofizika (Proceedings of the Universities. Radiophysics) [1958ff] ISSN 0021-3462
• TRANSL IN (J 4330) Radiophys & Quant Electr

J 1620 Asterisque • F
Asterisque [1973ff] ISSN 0303-1179

J 1648 Hist Math • USA
Historia Mathematica. International Journal of History of Mathematics [1974ff] ISSN 0315-0860

J 1669 J Comb Th • USA
Journal of Combinatorial Theory [1966-1970] ISSN 0021-9800
• CONT AS (J 0164) J Comb Th, Ser A & (J 0033) J Comb Th, Ser B

J 1670 Mitt Math Ges DDR • DDR
Mitteilungen der Mathematischen Gesellschaft der DDR

J 1680 Cienc Tecnol, Costa Rica • CR
Ciencia y Tecnologia, Revista de la Universidad de Costa Rica [1976ff] ISSN 0378-052X

J 1699 Bull Soc Math Grece • GR
(Bulletin de la Societe Mathematique Grece) [1921-1959]
• CONT AS (J 0465) Bull Greek Math Soc (NS)

J 1735 Bull South East Asian Soc • SGP
Bulletin of the South East Asian Society [1977ff]

J 1737 Nat Math Magazine (Louisiana) • USA
National Mathematics Magazine [1926-1946]
• CONT AS (J 0497) Math Mag

J 1741 Int J Man-Mach Stud • USA
International Journal of Man-Machine Studies [1969ff] ISSN 0020-7373

J 1743 Int J Gen Syst • USA
International Journal of General Systems: Methodology, Applications, Education [1974ff] ISSN 0308-1079

J 1770 Osaka Math J • J
Osaka Mathematical Journal [1949-1963]
• CONT AS (J 0351) Osaka J Math

J 1793 Nieuw Arch Wisk, Ser 2 • NL
Nieuw Archief voor Wiskunde. Reeks 2 [1894-1952]
• CONT OF (J 0046) Nieuw Arch Wisk • CONT AS (J 3077) Nieuw Arch Wisk, Ser 3

J 1884 Avtomatika, Akad Nauk Ukr SSR • SU
Avtomatika. Akademiya Nauk Ukrainskoj SSR. Nauchno-tekhnicheskij Kompleks "Institut Kibernetiki imeni V.M.Glushkova". Nauchno-tekhnicheskij Zhurnal (Automation. Academy of Sciences of the Ukrainian SSR. Scientific-Technical Complex: Glushkov Institute of Cybernetics) [1956ff] ISSN 0572-2691
• TRANSL IN (J 2548) Sov Autom Control

J 1905 Fibonacci Quart • USA
The Fibonacci Quarterly. [1963ff] ISSN 0015-0517

J 1908 Ann Sc Norm Sup Pisa, Fis Mat • I
Annali della Reale Scuola Normale Superiore Universitaria di Pisa. Scienze Fisiche e Matematiche [1913-1931]
• CONT AS (J 1568) Ann Sc Norm Sup Pisa, Fis Mat, Ser 2

J 1910 Proc London Math Soc, Ser 2 • GB
Proceedings of the London Mathematical Society. Serie 2 [1904-1951]
• CONT OF (J 0077) Proc London Math Soc • CONT AS (J 3240) Proc London Math Soc, Ser 3

J 1927 Prace Inst Podstaw Inf, Pol Akad Nauk • PL
Prace Instytut Podstaw Informatyki Polskiej Akademii Nauk (Reports. Institute of Computer Science Polish Academy of Science)

J 1929 Prace Centr Oblicz Pol Akad Nauk • PL
Polska Akademija Nauk. Centrum Obliczeniowe. Prace. (Polish Academy of Sciences. Computation Centre. Reports) ISSN 0079-3175

J 1934 Ann Sci Univ Clermont Math • F
Annales Scientifiques de l'Universite de Clermont-Ferrand II, Section Mathematiques (Clermont Ferrand) [1973ff] ISSN 0069-472X
• CONT OF (J 0179) Ann Fac Sci Clermont

J 1937 Adv Inform Syst Sci • USA
Advances in Information Systems Science [1969ff] ISSN 0065-2784

J 1944 Sitzb, Akad Wiss, Bayern, Math-Nat Kl • D
Bayerische Akademie der Wissenschaften. Sitzungsberichte [1931ff] ISSN 0340-7586

J 2010 Bull Europ Assoc Th Comput Sci • D
Bulletin of the European Association for Theoretic Computer Science [1978ff]

J 2028 Hist & Phil Log • GB
History and Philosophy of Logic [1980ff] ISSN 0144-5340

J 2038 Rend Sem Mat, Torino • I
Rendiconti del Seminario Matematico (gia "Conferenze di Fisica e di Matematica"). Universita e Politecnico di Torino

J 2053 Math Medley • SGP
Mathematical Medley [1975ff]

J 2074 Sem-ber, Muenster • D
Semesterbericht Muenster [1931-1939]

J 2083 Ann Univ Turku Ser B • SF
Annales Universitatis Turkuensis. Ser B (Yliopisto Turku. Jukaisuja. Sarja B. Humaniora) [1923ff] ISSN 0082-6987

J 2095 Fund Inform, Ann Soc Math Pol, Ser 4 • PL
*Fundamenta Informaticae. Annales Societatis Mathematicae Polonae. Series 4 * Roczniki Polskiego Towarzystwa Matematycznego. Seria 4* [1977ff] ISSN 0324-8429
• CONT OF (**J** 0611) Pol Tow Mat, Prace Mat-Fiz

J 2099 Boll Unione Mat Ital, VI Ser, A • I
Bolletino della Unione Matematica Italiana. Serie VI. A [1982ff] ISSN 0041-7084
• CONT OF (**J** 3285) Boll Unione Mat Ital, V Ser, A

J 2100 Boll Unione Mat Ital, VI Ser, B • I
Bolletino della Unione Matematica Italiana. Serie VI. B [1982ff] ISSN 0041-7084
• CONT OF (**J** 3495) Boll Unione Mat Ital, V Ser, B

J 2107 Publ Dep Math, Lyon, NS • F
Publications du Departement de Mathematiques. Nouvelle Serie. Faculte des Sciences de Lyon. [1982ff] ISSN 0076-1656
• CONT OF (**J** 0056) Publ Dep Math, Lyon

J 2128 C R Math Acad Sci, Soc Roy Canada • CDN
*Comptes Rendus Mathematiques de l'Academie des Sciences. La Societe Royale du Canada * Mathematical Reports of the Academy of Sciences* [1979ff] ISSN 0706-1994

J 2130 Linguist Philos • NL
Linguistics and Philosophy [1977ff] ISSN 0165-0157

J 2293 Comp Linguist & Comp Lang • H
Computational Linguistics and Computer Languages

J 2307 Japan J Math • J
Japanese Journal of Mathematics [1924-1974]
• CONT AS (**J** 2347) Japan J Math, NS

J 2310 Obz Mat Fiz, Ljubljana • YU
Obzornik za Matematiko in Fiziko (Mathematical and Physical Reviews) [1951ff] ISSN 0473-7466

J 2311 Rend Mat, Ser 6 • I
Rendiconti di Matematica. Serie 6 [1968-1980] ISSN 0034-4427
• CONT OF (**J** 0384) Rend Mat, Ser 5

J 2313 C R Acad Sci, Paris, Ser A-B • F
Academie des Sciences de Paris. Comptes Rendus Hebdomadaires des Seances. Serie A: Sciences Mathematiques, Serie B: Sciences Physiques [1966-1980] ISSN 0001-4036
• CONT OF (**J** 0109) C R Acad Sci, Paris • CONT AS (**J** 3364) C R Acad Sci, Paris, Ser 1 & (**J** 2314) C R Acad Sci, Paris, Ser 2

J 2314 C R Acad Sci, Paris, Ser 2 • F
Comptes Rendus des Seances de l'Academie des Sciences. Serie II. Mecanique-Physique, Chimie, Sciences de la Terre, Sciences de l'Univers [1981ff]
• CONT OF (**J** 2313) C R Acad Sci, Paris, Ser A-B

J 2320 Probl Peredachi Inf, Akad Nauk SSSR • SU
Problemy Peredachi Informatsii. Zhurnal Akademii Nauk SSSR (Probleme der Informationsuebertragung) [1965ff] ISSN 0555-2923
• TRANSL IN (**J** 3419) Probl Inf Transmiss

J 2328 Ann Pont Ist Sup Sci Lett Napoli • I
Pontifico Istituto Superiore di Scienze et Lettere "S.Chiara" di Napoli Annali [1952ff]

J 2331 ACM Comp Surveys • USA
Computing Surveys: The Survey and Tutorial Journal of the ACM (= Association for Computing Machinery) [1969ff] ISSN 0360-0300

J 2332 J Fac Sci, Univ Tokyo, Sect 1 A • J
Journal of the Faculty of Science. University of Tokyo. Section 1 A. Mathematics [1971ff] ISSN 0040-8980
• CONT OF (**J** 0434) J Fac Sci Univ Tokyo, Sect 1

J 2338 IEEE Trans Syst Man & Cybern • USA
Transactions on Systems, Man and Cybernetics. IEEE (= Institute of Electrical and Electronics Engineers) [1971ff] ISSN 0018-9472
• CONT OF (**J** 1561) IEEE Trans Syst & Sci Cybern

J 2345 Ueberblicke Math • D
Ueberblicke Math [1967-1974]
• CONT AS (**S** 0780) Jbuch Ueberblick Math

J 2347 Japan J Math, NS • J
Japanese Journal of Mathematics. New Series [1975ff] ISSN 0075-3432
• CONT OF (**J** 2307) Japan J Math

J 2521 Beijing Shifan Daxue Xuebao, Ziran Kexue • TJ
Beijing Shifan Daxue Xuebao. Ziran Kexue Ban (Journal of Natural Sciences of Beijing Normal University. Natural Science Edition)

J 2547 Serdica, Bulgar Math Publ • BG
Serdica. Bulgaricae Mathematicae Publicationes [1975ff] ISSN 0204-4110
• CONT OF (**J** 1156) Izv Bulgar Akad Nauk Mat Inst

J 2548 Sov Autom Control • USA
Soviet Automatic Control. Essential Serials in Electronics and Cybernetics [1968ff] ISSN 0038-5328
• TRANSL OF (**J** 1884) Avtomatika, Akad Nauk Ukr SSR

J 2551 J Log Progr • USA
Journal of Logic Programming [1984ff] ISSN 0743-1066

J 2573 Mat Nauk (Ped Inst Alma Ata) • SU
Matematicheskie Nauki (Mathematical Sciences)

J 2574 Litov Mat Sb (Vil'nyus) • SU
*Litovskij Matematicheskij Sbornik * Lietuvos Matematikos Rinkinys (Lithuanian Mathematical Collected Articles)* [1961ff] ISSN 0132-2818
• TRANSL IN (**J** 3283) Lith Math J

J 2604 Progr Comput Software • USA
Programming and Computer Software [1975ff] ISSN 0361-7688
• TRANSL OF (**J** 2605) Programmirovanie

J 2605 Programmirovanie • SU
Programmirovanie (Programming) [1975ff] ISSN 0132-3474
• TRANSL IN (**J** 2604) Progr Comput Software

J 2606 Tsukuba J Math • J
Tsukuba Journal of Mathematics [1977ff] ISSN 0387-4982

J 2631 J Math Kyoto Univ • J
Journal of Mathematics of Kyoto University [1961ff] ISSN 0023-608X

J 2650 Adv Appl Math • USA
Advances in Applied Mathematics. [1980ff] ISSN 0196-8858

J 2660 Ann Sci Math Quebec • CDN
Les Annales des Sciences Mathematiques du Quebec. [1977ff] ISSN 0707-9109

J 2661 Ann Probab • USA
The Annals of Probability. An Official Journal of the Institute of Mathematical Statistics. [1973ff] ISSN 0091-1798

J 2666 Autom Control Comput Sci • USA
Automatic Control and Computer Sciences [1969ff] ISSN 0146-4116
• TRANSL OF (**J** 0474) Avtom Vychis Tekh, Akad Nauk Latv SSR

J 2667 Autom Doc Math Linguist • USA
Automatic Documentation and Mathematical Linguistics [1967ff] ISSN 0005-1055
• TRANSL OF (**J** 0338) Nauch-Tekh Inf, Ser 2, Akad Nauk SSSR

J 2684 J Huazhong Inst Tech (Engl Ed) • TJ
Journal of Huazhong Institute of Technology. English Edition [1979-1981]
• CONT AS (**J** 3218) J Huazhong Univ Sci Tech (Engl Ed)
• TRANSL OF (**J** 2754) Huazhong Gongxueyuan Xuebao

J 2687 Comp Math Appl • USA
Computers and Mathematics with Applications. [1975ff] ISSN 0097-4943

J 2688 Conceptus (Wien) • A
Conceptus. Zeitschrift fuer Philosophie [1967ff] ISSN 0010-5155

J 2695 Cybernetica • B
Cybernetica. Revue Trimestrielle de l'Association Internationale de Cybernetique [1958ff] ISSN 0011-4227

J 2697 Datamation • USA
Datamation [1957ff] ISSN 0011-6963

J 2701 Digit Processes • CH
Digital Processes. An International Journal on the Theory and Design of Digital Systems [1975ff] ISSN 0301-4185

J 2702 Discr Appl Math • NL
Discrete Applied Mathematics [1979ff] ISSN 0166-218X

J 2711 Ekonom Mat Obzor (Prague) • CS
Ekonomicko-Matematicky Obzor (Economical & Mathematical Review) [1965ff] ISSN 0013-3027

J 2713 Eleutheria, Math J Sem Zervos (Athens) • GR
Eleutheria. Mathematical Journal of the Seminar of P. Zervos. (Liberty) [1978ff]

J 2716 Found Control Eng, Poznan • PL
Foundations of Control Engineering. Institute of Control Engineering. Technical University of Poznan [1975ff] ISSN 0324-8747

J 2718 Fct Approximatio, Comment Math, Poznan • PL
Functiones et Approximatio. Commentarii Mathematici [1974ff]

J 2720 Fuzzy Sets Syst • NL
Fuzzy Sets and Systems [1978ff] ISSN 0165-0114

J 2722 Geom Dedicata • NL
Geometriae Dedicata [1973ff] ISSN 0046-5755

J 2730 Phys Rev Lett • USA
Physical Review Letters [1958ff] ISSN 0031-9007

J 2734 Int J Math Educ Sci Technol • GB
International Journal of Mathematical Education in Science and Technology [1970ff] ISSN 0020-739X

J 2736 Int J Theor Phys • USA
International Journal of Theoretical Physics [1968ff] ISSN 0020-7748

J 2744 Izv Sev-Kavk Nauch Tsentra, Estestv (Rostov nD) • SU
Izvestiya Severo-Kavkazskogo Nauchnogo Tsentra Vysshej Shkoly. Estestvennye Nauki (Proceedings of the Scientific University Centre of the North-Caucasus. Natural Sciences)

J 2745 IEEE Trans Inf Theory • USA
Transactions on Information Theory. IEEE (= Institute of Electrical and Electronics Engineers) [1955ff] ISSN 0018-9448

J 2746 J Algor • USA
Journal of Algorithms [1980ff] ISSN 0196-6774

J 2749 J Comp Appl Math • NL
Journal of Computational and Applied Mathematics [1975ff] ISSN 0377-0427

J 2754 Huazhong Gongxueyuan Xuebao • TJ
*Huazhong Gongxueyuan Xuebao * Zhongguo Kexue Shuxue Zhuanji (Journal Huazhong (Central China) University of Science and Technology)*
• TRANSL IN (**J** 2684) J Huazhong Inst Tech (Engl Ed) & (**J** 3218) J Huazhong Univ Sci Tech (Engl Ed)

J 2761 J Recreational Math • USA
Journal of Recreational Mathematics [1968ff] ISSN 0022-412X

J 2764 J Stat Phys • USA
Journal of Statistical Physics [1969ff] ISSN 0022-4715

J 2771 Kexue Tongbao • TJ
Kexue Tongbao (Science Bulletin) [1950ff] ISSN 0023-074X
• TRANSL IN (**J** 3769) Sci Bull, Foreign Lang Ed

J 2772 Shuxue de Shijian yu Renshi • TJ
Shuxue de Shijian yu Renshi (Mathematics in Practice and Theory)

J 2774 Koezlem MTA Szam & Autom: Kutat Intez • H
Koezlemenyek. Magyar Tudomanyos Akademia Szamitastechnikai es Automatizalasi Kutato Intezet Budapest (Bulletin of the Hungarian Academy of Sciences Budapest, Research Institut of Computer Science and Automatization) [1968ff]

J 2775 Nuovo Cimento B, Ser 2 • I
Il Nuovo Cimento. B. Serie II

J 2789 Math Intell • D
The Mathematical Intelligencer [1978ff] ISSN 0343-6993

J 2790 Math Sem-ber • D
Mathematische Semesterberichte [1980ff] ISSN 0720-728X
• CONT OF (**J** 0160) Math-Phys Sem-ber, NS

J 2794 Mem Fac Engin, Kyoto Univ • J
Memoirs of the Faculty of Engineering. Kyoto University [1914ff] ISSN 0023-6063

J 2804 Nanjing Daxue Xuebao, Ziran Kexue • TJ
Nanjing Daxue Xuebao. Ziran Kexue Ban (Nanjing University Journal. Natural Sciences Edition) [1957ff]

J 2814 Podstawy Sterowania • PL
Podstawy Sterowania. Zaklad Systemow Automatyki Kompleksowej (Foundations of Cybernetics. Institute of Complex Automation Systems)

J 2817 Prague Bull Math Linguist • CS
The Prague Bulletin of Mathematical Linguistics [1964ff] ISSN 0032-6585

J 2820 Proc West Virginia Acad Sci • USA
Proceedings. West Virginia Academy of Science [1952ff] ISSN 0096-4263

J 2831 RAIRO Autom • F
RAIRO Automatique. RAIRO (= Revue Francaise d'Automatique, d'Informatique et de Recherche Operationnelle). Series Automatique [1977ff] ISSN 0399-0524
• CONT OF (**J** 0205) Rev Franc Autom, Inf & Rech Operat Ser Jaune

J 2832 RAIRO Inform • F
RAIRO Informatique. Revue Francaise d'Automatique, d'Informatique et de Recherche Operationnelle. Series Informatique [1977ff] ISSN 0399-0532
• CONT OF (**J** 0205) Rev Franc Autom, Inf & Rech Operat Ser Bleue

J 2834 Scientia (Milano) • I
Scientia. Rivista Internazionale di Sintesi Scientifica. [1907ff] ISSN 0036-8687

J 2840 Stochastica, Univ Politec Barcelona • E
Stochastica. Revista de Matematica Pura y Aplicada del Departamento de Matematicas y Estadistica de la Escuela Tecnica Superior de Arquitectura [1975ff]

J 2845 Tanulmanyok • H
Tanulmanyok. Szamitastechnikai es Automatizalasi Kutato Intezete (Studies. Research Institut for Computer Science and Automatization)

J 2852 Tr Vychisl Tsentra, Univ Tartu • SU
Trudy Vychislitel'nogo Tsentra. Tartuskij Gosudarstvennyj Universitet (Publications of the Computational Centre of the State University of Tartu) [1893ff]

J 2868 Vest Univ Yaroslavl' • SU
Vestnik Yaroslavskogo Universiteta (Publications of the Yaroslavl' University)

J 2869 Vest Ser Vychisl Mat Kibern, Univ Moskva • SU
Vestnik Moskovskogo Universiteta. Nauchnyj Zhurnal. Seriya XV. Vychislitel'naya Matematika i Kibernetika (Publications of the Moscow University. Scientific Journal. Series XV. Computational Mathematics and Cybernetics) [1946ff] ISSN 0201-7385, ISSN 0137-0782
• TRANSL IN (**J** 3221) Moscow Univ Comp Math Cybern

J 2878 Wiss Z Tech Hochsch Karl-Marx-Stadt • DDR
Wissenschaftliche Zeitschrift der Technischen Hochschule Karl-Marx-Stadt

J 2887 Zbor Radova, NS • YU
Zbornik Radova. Nova Serija. (Collected Papers. New Series)

J 3064 Abacus, Math Ass Nigeria • WAN
Abacus. The Journal of the Mathematical Association of Nigeria [1960ff] ISSN 0001-3099

J 3070 Bul Inst Politeh Iasi, Sect 1 • RO
Buletinul Institutului Politehnic din Iasi. Sectia 1. Matematica, Mecanica Teoretica, Fizica (Bulletin des Polytechnischen Instituts Jassy. Sektion 1. Mathematik, Theoretische Mechanik, Physik) [1977ff] ISSN 0304-5188
• CONT OF (**J** 0198) Bul Inst Politeh Iasi NS

J 3071 Bul Univ Brasov, Ser C • RO
Buletinul Universitatii din Brasov. Seria C. Matematica, Fizica, Chimie, Stiinte Naturale (Bulletin der Universitaet Brasov. Serie C. Mathematik, Physik, Chemie, Naturwissenschaften)

J 3073 Bull Math Biol • GB
Bulletin of Mathematical Biology [1973ff] ISSN 0092-8240
• CONT OF (**J** 0515) Bull Math Biophys

J 3075 Normat • N
Normat. Nordisk Matematisk Tidskrift (Normat. Scandinavian Mathematical Journal) [1979ff]
• CONT OF (**J** 0311) Nordisk Mat Tidskr

J 3077 Nieuw Arch Wisk, Ser 3 • NL
Nieuw Archief voor Wiskunde. Derde Serie [1953-1982] ISSN 0028-9825
• CONT OF (**J** 1793) Nieuw Arch Wisk, Ser 2 • CONT AS (**J** 3929) Nieuw Arch Wisk, Ser 4

J 3079 Kiber Sb Perevodov, NS • SU
Kiberneticheskij Sbornik. Novaya Seriya. Sbornik Perevodov (Collected Articles on Cybernetics: New Series. Collected Translations) [1965ff] ISSN 0453-8382
• CONT OF (**J** 1048) Kiber Sb Perevodov

J 3100 Acta Cient Compostelana • E
Acta Cientifica Compostelana. Revista de la Universidad de Santiago de Compostela [1964ff] ISSN 0567-7378

J 3112 Annot Dokl, Inst Prikl Mat, Tbilisi • SU
Annotatsii Dokladov. Seminar Instituta Prikladnoj Matematiki Tbilisskogo Universiteta (Annotations of Papers. Seminar of the Institute of Applied Mathematics. University of Tbilisi) [1969ff] ISSN 0082-2191

J 3124 Beitr Algebra Geom • DDR
Beitraege zur Algebra und Geometrie [1971ff] ISSN 0440-1298

J 3130 Bul Inst Politeh Bucuresti, Ser Electroteh • RO
Buletinul Institutului Politehnic "Gheorghe Gheorghiu-Dej" Bucuresti. Seria Electrotehnica (Bulletin des Polytechnischen Instituts "Gheorghe Gheorghiu-Dej" Bukarest. Serie Elektrotechnik)

J 3133 Bull Soc Math Belg, Ser B • B
Bulletin de la Societe Mathematique de Belgique. Serie B [1977ff] ISSN 0037-9476
• CONT OF (**J** 0082) Bull Soc Math Belg

J 3137 Shuxue Niankan • TJ
Shuxue Niankan (Chinese Annals of Mathematics) [1980-1982]
• CONT AS (**J** 3187) Shuxue Niankan, Xi A & (**J** 4719) Chinese Ann Math Ser B

J 3144 Comput Lang, Int J • GB
Computer Languages. An International Journal [1975ff] ISSN 0096-0551

J 3171 God Vissh Ucheb Zaved, Prilozhna Mat, Sofiya • BG
Godishnik na Visshite Uchebni Zavedeniya. Prilozhna Matematika (Annuaire de l'Universite, Mathematiques Appliquees) [1974ff]
• CONT OF (**J** 0224) God Vissh Tekh Ucheb Zaved Mat, Sofiya

J 3172 J London Math Soc, Ser 2 • GB
Journal of the London Mathematical Society. 2nd Series [1969ff] ISSN 0024-6107
• CONT OF (**J** 0039) J London Math Soc

J 3184 Int J Game Theory • D
International Journal of Game Theory [1971ff] ISSN 0020-7276

J 3187 Shuxue Niankan, Xi A • TJ
Shuxue Niankan. Xi A (Chinese Annals of Mathematics. Series A) [1983ff]
• CONT OF (**J** 3137) Shuxue Niankan

J 3188 Itogi Nauki Tekh, Ser Teor Veroyat Mat Stat Teor Kibern • SU
Itogi Nauki i Tekhniki. Seriya Teoriya Veroyatnostej, Matematicheskaya Statistika, Teoreticheskaya Kibernetika (Progress in Science and Technology. Series Probability Theory, Mathematical Statistics, Theoretical Cybernetics) [1972ff] ISSN 0202-7488
• CONT OF (**J** 1021) Itogi Nauki Ser Mat • TRANSL IN (**J** 1531) J Sov Math

J 3191 IEEE Trans Pattern Anal & Mach Intell • USA
Transactions on Pattern Analysis and Machine Intelligence. IEEE (= Institute of Electrical and Electronics Engineers) [1979ff] ISSN 0162-8828

J 3194 J Austral Math Soc, Ser A • AUS
Journal of the Australian Mathematical Society. Series A [1976ff] ISSN 0263-6115
• CONT OF (**J** 0038) J Austral Math Soc

J 3195 J Oper Res Soc Japan • J
Journal of the Operations Research Society of Japan [1965ff]

J 3218 J Huazhong Univ Sci Tech (Engl Ed) • TJ
Journal of Huazhong University of Science and Technology [1982ff]
• CONT OF (**J** 2684) J Huazhong Inst Tech (Engl Ed) • TRANSL OF (**J** 2754) Huazhong Gongxueyuan Xuebao

J 3220 Mitt Ges Math Datenverarb Bonn • D
Mitteilungen der Gesellschaft fuer Mathematik und Datenverarbeitung

J 3221 Moscow Univ Comp Math Cybern • USA
Moscow University Computational Mathematics and Cybernetics ISSN 0278-6419
• TRANSL OF (**J** 2869) Vest Ser Vychisl Mat Kibern, Univ Moskva

J 3227 Pattern Recognition • GB
Pattern Recognition. The Journal of the Pattern Recognition Society [1968ff] ISSN 0031-3203

J 3239 Proc Japan Acad, Ser A • J
Proceedings of the Japan Academy. Series A. Mathematical Sciences [1978ff] ISSN 0386-2194

J 3240 Proc London Math Soc, Ser 3 • GB
Proceedings of the London Mathematical Society. 3rd Series [1951ff] ISSN 0024-6115
• CONT OF (**J** 1910) Proc London Math Soc, Ser 2

J 3248 Przeglad Stat • PL
Przeglad Statystyczny (Revue Statistique) [1954ff] ISSN 0033-2372

J 3265 Sov J Contemp Math Anal, Armen Acad Sci • USA
Soviet Journal of Contemporary Mathematical Analysis. Armenian Academy of Sciences [1979ff] ISSN 0735-2719
• TRANSL OF (**J** 0312) Izv Akad Nauk Armyan SSR, Ser Mat

J 3276 Tehran J Math • IR
Tehran Journal of Mathematics

J 3279 Trans Moscow Math Soc • USA
Transactions of the Moscow Mathematical Society ISSN 0077-1554
• TRANSL OF (**J** 0065) Tr Moskva Mat Obshch

J 3281 Ukr Math J • USA
Ukrainian Mathematical Journal [1967ff] ISSN 0041-5995
• TRANSL OF (**J** 0265) Ukr Mat Zh, Akad Nauk Ukr SSR

J 3283 Lith Math J • USA
Lithuanian Mathematical Journal. [1975ff] ISSN 0363-1672
• TRANSL OF (**J** 2574) Litov Mat Sb (Vil'nyus)
• REM Only selected translations of J2574

J 3285 Boll Unione Mat Ital, V Ser, A • I
Bollettino della Unione Matematica Italiana. Serie V. A [1976-1981] ISSN 0041-7084
• CONT OF (**J** 0012) Boll Unione Mat Ital, IV Ser • CONT AS (**J** 2099) Boll Unione Mat Ital, VI Ser, A

J 3289 Wiss Z Tech Hochsch Madgeburg • DDR
Wissenschaftliche Zeitschrift der Technischen Hochschule Otto von Guericke. Magdeburg [1957ff] ISSN 0541-8933

J 3293 Bull Acad Pol Sci, Ser Math • PL
Bulletin de l'Academie Polonaise des Sciences. Serie des Sciences Mathematiques [1979-1982] ISSN 0001-4117
• CONT OF (**J** 0014) Bull Acad Pol Sci, Ser Math Astron Phys
• CONT AS (**J** 3417) Bull Pol Acad Sci, Math

J 3359 Appl Comp Sci, Ber Prakt Inf • D
Applied Computer Science. Berichte zur praktischen Informatik

J 3364 C R Acad Sci, Paris, Ser 1 • F
Comptes Rendus des Seances de l'Academie des Sciences. Serie I: Science Mathematique [1981ff] ISSN 0151-0509
• CONT OF (**J** 2313) C R Acad Sci, Paris, Ser A-B

J 3370 Enseign Math, Ser 2 • CH
L'Enseignement Mathematique. Revue Internationale. Serie 2 [1955ff] ISSN 0013-8584
• CONT OF (**J** 0152) Enseign Math

J 3395 Libyan J Sci • LAR
The Libyan Journal of Science [1971ff]

J 3396 Mat Model Teor Ehlektr Tsepej • SU
Matematicheskoe Modelirovanie i Teoriya Ehlektricheskikh Tsepej (Mathematical Modelling and Theory of Electric Circuits) [1962-1978]
• CONT AS (**J** 4526) Ehlektr Modelirovanie

J 3401 Meth Oper Res • D
Methods of Operations Research [1963ff] ISSN 0078-5318

J 3402 Met Vychislenij, Univ Leningrad • SU
Metody Vychislenij. Leningradskij Ordena Lenina Gosurdarstvennyj Universitet imeni A.A. Zhdanova (Method of Computation. Zhdanov-University, Leningrad) [1963ff] ISSN 0543-6273

J 3417 Bull Pol Acad Sci, Math • PL
Bulletin of the Polish Academy of Sciences. Mathematics [1983ff] ISSN 0001-4117
• CONT OF (**J** 3293) Bull Acad Pol Sci, Ser Math

J 3419 Probl Inf Transmiss • USA
Problems of Information Transmission [1965ff] ISSN 0032-9460
• TRANSL OF (**J** 2320) Probl Peredachi Inf, Akad Nauk SSSR

J 3420 Proc Edinburgh Math Soc, Ser 2 • GB
Proceedings of the Edinburgh Mathematical Society. Series 2. [1927ff] ISSN 0013-0915
• CONT OF (**J** 0386) Proc Edinburgh Math Soc

J 3434 Pubbl Ist Appl Calcolo, Ser 3 • I
Pubblicazioni. Serie III. Istituto per le Applicazioni del Calcolo "Mauro Picone" (IAC) Consiglio Nazionale delle Ricerche

J 3436 Quad, Ist Appl Calcolo, Ser 3 • I
Quaderni Serie III. Istituto Applicazione Calcolo

J 3437 Rep, Akad Wiss DDR, Inst Math • DDR
Report. Akademie der Wissenschaften der DDR. Institut fuer Mathematik.

J 3441 RAIRO Inform Theor • F
Revue Francaise d'Automatique, d'Informatique et de Recherche Operationnelle (RAIRO), Informatique Theorique [1977ff] ISSN 0399-0540
• CONT OF (**J** 4698) Rev Franc Autom, Inf & Rech Operat, Ser Rouge Inf Th

J 3448 Soochow J Math • RC
Soochow Journal of Mathematics [1979ff]

J 3449 Sov Math • USA
Soviet Mathematics [1974ff] ISSN 0197-7156
• TRANSL OF (**J** 0031) Izv Vyssh Ucheb Zaved, Mat (Kazan)

J 3450 Studia Univ Babes-Bolyai, Math (Cluj) • RO
Studia Universitatis Babes-Bolyai. Mathematica [1976ff]
• CONT OF (**J** 3451) Stud Univ Cluj, Ser Math-Mech

J 3451 Stud Univ Cluj, Ser Math-Mech • RO
Studia Universitatis Babes-Bolyai, Series Mathematica-Mechanica [1971-1975] ISSN 0370-8659
• CONT OF (**J** 0355) Studia Univ Babes-Bolyai, Math-Phys (Cluj) • CONT AS (**J** 3450) Studia Univ Babes-Bolyai, Math (Cluj)

J 3457 Order • NL
Order [1984ff]

J 3460 Tr Inst Kibern, Akad Nauk Gruz SSR • SU
Trudy Instituta Kibernetiki. Akademiya Nauk Gruzinskoj SSR (Publications of the Institute of Cybernetics. Academy of Sciences of the Georgian SSR) [1963ff] ISSN 0515-9970

J 3465 Wiss Z Univ Rostock, Math-Nat Reihe • DDR
Wissenschaftliche Zeitschrift der Wilhelm-Pieck-Universitaet Rostock. Mathematisch-Naturwissenschaftliche Reihe

J 3480 Informatik & Philos • A
Informatik und Philosophie

J 3495 Boll Unione Mat Ital, V Ser, B • I
Bolletino della Unione Matematica Italiana. Serie V. B [1976-1981] ISSN 0041-7084
• CONT OF (**J** 0012) Boll Unione Mat Ital, IV Ser • CONT AS (**J** 2100) Boll Unione Mat Ital, VI Ser, B

J 3519 Glas Mat, Ser 3 (Zagreb) • YU
Glasnik Matematichki. Serija 3 (Publications of Mathematics. Series 3) [1966ff] ISSN 0017-095X
• CONT OF (**J** 0371) Glas Mat-Fiz Astron, Ser 2 (Zagreb)

J 3524 Kybernetika Suppl (Prague) • CS
Kybernetika. Supplement. (Prague) (Cybernetics. Supplement. (Prague))
• REL PUBL (**J** 0156) Kybernetika (Prague)

J 3526 Ann Mat Pura Appl, Ser 4 • I
Annali di Matematica Pura ed Applicata. Serie Quarta. Sotto gli Auspici del Consiglio Nazionale delle Ricerche ISSN 0003-4622
• CONT OF (**J** 0229) Ann Mat Pura Appl, Ser 3

J 3529 Rep Dept Numer & Comp Math, Univ Budapest • H
Reports of the Department of Numerical and Computer Mathematics of L. Eoetvoes University, Budapest. (Automata Theoretic Letters, No. 1.)

J 3531 Ber Ges Math Datenverarb, Bonn • D
Berichte der Gesellschaft fuer Mathematik und Datenverarbeitung. (Reports of the Society for Mathematics and Data Processing) ISSN 0533-9480

J 3536 Uch Zap Univ, Ivanovo • SU
Ivanovskij Gosudarstvennyj Universitet. Uchenye Zapiski (State University of Ivanovo. Scientific Notes) [1974ff]
• CONT OF (**J** 0226) Uch Zap Ped Inst, Ivanovo

J 3630 Bull Dept of Lib Arts (Numazu) • J
Bulletin of the Department of Liberal Arts [1974ff]

J 3631 Tokyo School of Physics J • J
Tokyo Buturi Gakko Zassi (Tokyo School of Physics Journal) [1891-1944]

J 3732 Mohu Shuxue • TJ
Mohu Shuxue (Fuzzy Mathematics) [1981ff]

J 3735 Fudan Xuebao, Ziran Kexue • TJ
Fudan Xuebao. Ziran Kexue Ban (Fudan Journal. Natural Science)

J 3746 Note Math (Lecce) • I
Note di Matematica. Pubblicazione Semestrale [1981ff]

J 3766 Zhongguo Kexue, Xi A • TJ
*Zhongguo Kexue. Xi A. Mathematical, Physical, Astronomical & Technical Sciences * Scientia Sinica. Series A* [1982ff]
• CONT OF (**J** 1024) Zhongguo Kexue

J 3768 Boll Unione Mat Ital, VI Ser, D • I
Bolletino della Unione Matematica Italiana. Serie VI. D. Algebra e Geometria [1982ff] ISSN 0041-7084

J 3769 Sci Bull, Foreign Lang Ed • TJ
Kexue Tongbao. Foreign Language Edition (Science Bulletin) [1980ff]
• TRANSL OF (**J** 2771) Kexue Tongbao

J 3789 Ann Hist of Comp • USA
Annals of the History of Computing [1979ff] ISSN 0164-1239

J 3793 Jisuanjii Xuebao • TJ
Jisuanji Xuebao (Chinese Journal of Computers) [1978ff]

J 3797 Diagrammes • F
Diagrammes

J 3816 Zhongguo Kexue Jishu Daxue Xuebao • TJ
Zhongguo Kexue Jishu Daxue Xuebao (Journal of China University of Science and Technology)

J 3824 Bull Soc Math Belg, Ser A • B
Bulletin de la Societe Mathematique de Belgique. Serie A [1977ff]
• CONT OF (**J** 0082) Bull Soc Math Belg

J 3914 Math Soc Sci • NL
Mathematical Social Sciences [1980ff] ISSN 0165-4896

J 3919 BUSEFAL • F
BUSEFAL (= Bulletin pour les Sous-Ensembles Flous et leurs Applications)

J 3920 ZKI Inf, Akad Wiss DDR • DDR
Informationen ZKI (= Zentralinstitut fuer Kybernetik und Informationsprozesse) der Akademie der Wissenschaften der DDR

J 3926 J Singapore Nat Acad Sci • SGP
Journal of the Singapore National Academy of Science

J 3929 Nieuw Arch Wisk, Ser 4 • NL
Nieuw Archief voor Wiskunde. Vierde Serie. Uitgegeven door het Wiskundid Genootschap te Amsterdam [1983ff] ISSN 0028-9825
• CONT OF (J 3077) Nieuw Arch Wisk, Ser 3

J 3937 Veroyat Met i Kibern (Kazan) • SU
Veroyatnostnye Metody i Kibernetika (Probabilistic Methods and Cybernetics)

J 3939 Mat Logika Primen (Akad Nauk Litov SSR) • SU
Matematicheskaya Logika i Ee Primeneniya (Mathematical Logic and its Applications) [1981ff]

J 3941 J Math Pures Appl, Ser 9 • F
Journal de Mathematiques Pures et Appliquees. Neuvieme Serie [1922ff] ISSN 0021-7824
• CONT OF (J 0249) J Math Pures Appl

J 3942 Zhongshan Daxue Xuebao, Ziran Kexue • TJ
Zhongshan Daxue Xuebao. Ziran Kexue Ban (Acta Scientiarum Naturalium Universitatis Sunyatseni)

J 3945 Vest Math Univ Leningrad • USA
Vestnik Leningrad University. Mathematics [1968ff] ISSN 0146-924X
• TRANSL OF (J 0085) Vest Ser Mat Mekh Astron, Univ Leningrad

J 3954 Rev Franc Inf & Rech Operat • F
Association Francaise pour la Cybernetique Economique et Technique. Revue Francaise d'Informatique et de Recherche Operationnelle [1967-1971]
• CONT AS (J 0205) Rev Franc Autom, Inf & Rech Operat

J 3955 Ann Univ Budapest, Sect Comp • H
Annales Universitatis Scientiarum Budapestinensis. Sectio Computatorica

J 3957 Bull Inf & Cybern (Kyushu Univ) • J
Bulletin of Informatics and Cybernetics. Research Association of Statistical Sciences [1982ff] ISSN 0286-522X

J 3975 Arch Math & Phys • D
Archiv der Mathematik und Physik mit Besonderer Ruecksicht auf die Beduerfnisse der Lehrer an Hoeheren Unterrichtsanstalten. 3.Reihe [1841ff]

J 3977 Bull Int Acad Sci Cracovie, Cl Math Nat • PL
Bulletin International de l'Academie des Sciences de Cracovie. Classe de Mathematiques-Sciences Naturelles

J 3993 Rev Mat Hisp-Amer, Ser 2 • E
Revista Matematica Hispano-Americana. Serie 2 ISSN 0373-0999
• CONT AS (J 0236) Rev Mat Hisp-Amer, Ser 4

J 3994 Ann Acad Sci Fennicae Ser A I • SF
Annales Academiae Scientiarum Fennicae. Serie A I [1941-1974] ISSN 0066-1953
• CONT AS (J 0446) Ann Acad Sci Fennicae, Ser A I, Diss

J 3996 Boll Unione Mat Ital • I
Bollettino della Unione Matematica Italiana [1922-1945]
• CONT AS (J 4408) Boll Unione Mat Ital, III Ser

J 4068 Dt Z fuer Phil • D
Deutsche Zeitschrift fuer Philosophie [1953ff] ISSN 0012-1045

J 4135 Dialexeis • GR
Dialexeis (Reports of the Greek Mathematical Society)

J 4162 Uch Zap Ped Inst, Buryat • SU
Uchenye Zapiski Buryat-Mongol. Ped. Instituta (Scientific Notes of the Buryat-Mongolian Institute of Education)

J 4167 Uch Zap Ped Inst, Penza • SU
Penza. Gosudarstvennyj Pedagogicheskij Institut. Uchenye Zapiski (Penza. State Institute of Education. Scientific Notes) [1953ff] ISSN 0553-6286

J 4197 Acta Sci Nat, Pars Math, Mech & Astronom • TJ
Acta Scientiarum Naturalium Scholarum Superiorum Sinensium, Pars Mathematica, Mechanica et Astronomica

J 4277 Vesn Drusht Mat Fiz Serbije • YU
*Vesnik Drushtva Matematichara i Fizichara Narodne Republike Serbije * Vestnik Obshchestva Matematikov i Fizikov N.R. Serbie * Bulletin de la Societe des Mathematiciens et Physiciens de la R.P. Serbie (Publications of the Society of Mathematicians and Physicists of the P.R. Serbia)* [1949-1963]
• CONT AS (J 0042) Mat Vesn, Drust Mat Fiz Astron Serb

J 4305 IEEE Trans Electr Comp • USA
Transactions of Electronic Computers. IEEE (= Institute of Electrical and Electronics Engineers) [1963-1967]
• CONT OF (J 0072) IRE Trans Electr Comp • CONT AS (J 0187) IEEE Trans Comp

J 4330 Radiophys & Quant Electr • SU
Radiophysics and Quantum Electronics
• TRANSL OF (J 1572) Izv Vyssh Ucheb Zaved, Radiofizika (Moskva)

J 4373 GMS Math Inst Patras • GR
Greek Mathematical Society, Mathematical Institut of Patras

J 4385 NEWS • D
NEWS - Special Interest Group "Petri Nets and Related System Models". Newsletter

J 4387 Itogi Nauki Tekh, Ser Tekh Kibern • SU
Itogi Nauki i Tekhniki. Seriya Tekhnicheskaya Kibernetika. Gosudarstvennyj Komitet SSSR po Nauke i Tekhnike. Akademiya Nauk SSSR. Vsesoyuznyj Institut Nauchnoj i Tekhnicheskoj Informatsii. (Progress in Science and Technology. Series: Engineering Cybernetics) [1972ff]
• CONT OF (J 1021) Itogi Nauki Ser Mat

J 4406 Zb Prats Obchis Mat Tekhn • SU
Zbirnik Prats'z Obchislyuv. Matematika i Tekhnika (Collected Periodic Papers. Mathematics and Technology)

J 4408 Boll Unione Mat Ital, III Ser • I
Bolletino della Unione Matematica Italiana, Ser III [1946-1967]
• CONT OF (J 3996) Boll Unione Mat Ital • CONT AS (J 0012) Boll Unione Mat Ital, IV Ser

J 4418 Nanjing Daxue Xuebao Shuxue Bannian Kan • TJ
Nanjing Daxue Xuebao Shuxue Bannian Kan (Nanjing University Journal Mathematical Biquarterly)

J 4430 Ziran Zazhi • TJ
Ziran Zazhi (Nature Journal) [1978ff]

Source Index

J 4452 Zhexue Yanjiu • TJ
Zhexue Yanjiu (Philosophical Research. Studies on Philosophy)

J 4510 Norsk Mat Tidsskr • N
Norsk Matematisk Tidsskrift (Scandinavian Mathematical Journal) [1919-1952]
• CONT OF (**J** 0259) Nyt Tidsskr Mat • CONT AS (**J** 0311) Nordisk Mat Tidskr

J 4526 Ehlektr Modelirovanie • SU
Ehlektronnoe Modelirovanie (Electronic Modelling) [1979ff]
• CONT OF (**J** 3396) Mat Model Teor Ehlektr Tsepej

J 4553 Knizh Odborn Ved Spis Vys Uch Tech, Brno • CS
Knizhnice Odborn. Ved. Spisu Vysoke Ucheni Tech v Brne (Bulletin Technical University Brno)

J 4567 Comput Rev
Computer Review

J 4619 Phys Rev A, Ser 3 • USA
Physical Review. A. General Physics. 3rd Series

J 4632 Qufu Shiyuan Xuebao • TJ
Qufu Shiyuan Xuebao

J 4641 Mem Fac Sci Kochi Univ Ser A Math • J
Kochi University. Faculty of Science. Memoirs. Series A. Mathematics

J 4681 Sci News • GB
Science News

J 4698 Rev Franc Autom, Inf & Rech Operat, Ser Rouge Inf Th • F
Revue Francaise d'Automatique, d'Informatique et de Recherche Operationnelle (RAIRO). Serie Rouge Informatique Theorique [1975-1976] ISSN 0399-0540
• CONT OF (**J** 0205) Rev Franc Autom, Inf & Rech Operat Ser Rouge • CONT AS (**J** 3441) RAIRO Inform Theor

J 4702 Ann Sc Norm Sup Pisa Fis Mat, Ser 4 • I
Annali della Scuola Normale Superiore di Pisa. Classe di Science. Fisiche e Matematiche. Seria IV [1974ff]
• CONT OF (**J** 0315) Ann Sc Norm Sup Pisa Fis Mat, Ser 3

J 4704 J Soc Ind & Appl Math, Ser Control • USA
Journal of the Society for Industrial and Applied Mathematics. Series A: Control [1963-1965]
• CONT AS (**J** 0946) SIAM J Control

J 4705 SIAM J Control & Optim • USA
Journal on Control and Optimization. SIAM (= Society for Industrial and Applied Mathematics) [1976ff]
• CONT OF (**J** 0946) SIAM J Control

J 4706 Publ Inst Math (Belgrade) • YU
Academie Serbe des Sciences. Publications de l'Institut Mathematique [1948-1960]
• CONT AS (**J** 0400) Publ Inst Math, NS (Belgrade)

J 4710 Pol Tow Mat, Wiad Mat • PL
Polskie Towarzystwo Matematyczne. Wiadomosci Matematiczne [1899-1954]
• CONT AS (**J** 0519) Wiad Mat, Ann Soc Math Pol, Ser 2

J 4711 IRE Proc • USA
Proceedings. IRE (= Institute of Radio Engineers) [1911-1962]
• CONT AS (**J** 1077) IEEE Proc

J 4713 Mat Fyz Chasopis (Slov Akad Ved) • CS
Matematicky-Fyzikalny Chasopis (Journal of Mathematical Physics) [1951-1966]
• CONT AS (**J** 0143) Mat Chasopis (Slov Akad Ved)

J 4716 Stud Philos, Leopolis (Poznan) • PL
Studia Philosophica. Commentarii Societatis Polonorum. Leopolis [1935ff]

J 4717 Izv Akad Nauk SSSR • SU
Izvestiya Akademii Nauk SSSR (Bulletin de l'Academie des Sciences Mathematiques et Naturelles. Leningrad) [?-1936]
• CONT AS (**J** 0216) Izv Akad Nauk SSSR, Ser Mat

J 4719 Chinese Ann Math Ser B • TJ
Shuxue Niankan. Xi B (Chinese Annals of Mathematics. Series B) [1983ff]
• CONT OF (**J** 3137) Shuxue Niankan

J 4729 Acta Math Hung • H
Acta Mathematica Hungarica [1983ff] ISSN 0236-5294
• CONT OF (**J** 0001) Acta Math Acad Sci Hung

J 4731 Australasian J Psych & Phil • AUS
The Australasian Journal of Psychology and Philosophy [1923-1946]
• CONT AS (**J** 0273) Australasian J Phil

J 4732 J Math Mech (Indiana Univ) • USA
Journal of Mathematics and Mechanics [1962-1970]
• CONT OF (**J** 0452) Indiana Univ Math J

Series

S 0019 Colloq Math (Warsaw) • PL
Colloquium Mathematicum [1947ff] • PUBL Academie Polonaise des Sciences, Institut Mathematique: Warsaw • ISSN 0010-1354

S 0055 Proc Steklov Inst Math • USA
Proceedings of the Steklov Institute of Mathematics [1967ff] • PUBL (**X** 0803) Amer Math Soc: Providence • TRANSL OF (**S** 0066) Tr Mat Inst Steklov • ISSN 0081-5438

S 0066 Tr Mat Inst Steklov • SU
Trudy Ordena Lenina Matematicheskogo Instituta imeni V.A.Steklova. Akademiya Nauk SSSR (Proceedings of the Mathematical Steklov-Institute of the Academy of Sciences SSSR) [1938ff] • PUBL (**X** 0899) Akad Nauk SSSR : Moskva • CONT OF (**S** 1644) Tr Fiz-Mat Inst Steklov • TRANSL IN (**S** 0055) Proc Steklov Inst Math

S 0166 Mat Issl, Mold SSR • SU
Matematicheskie Issledovaniya. Akademiya Nauk Moldavskoj SSR. Ordena Trudovogo Krasnogo Znameni Institut Matematiki s Vychislitel'nym Tsentrom (Mathematical Studies) [1966ff] • PUBL (**X** 2741) Shtiintsa: Kishinev • ISSN 0542-9994

S 0167 Mem Amer Math Soc • USA
Memoirs of the American Mathematical Society [1950-1974] • PUBL (**X** 0803) Amer Math Soc: Providence • CONT AS (**S** 2450) Mem Amer Math Soc, NS • ISSN 0065-9266

S 0183 Publ Math Univ California • USA
University of California Publications in Mathematics PUBL (**X** 0926) Univ Calif Pr: Berkeley

S 0208 Uch Zap, Ped Inst, Moskva • SU
Uchenye Zapiski Moskovskogo Gosudarstvennogo Pedagogicheskogo Instituta imeni V.I.Lenina (Scientific Notes of the Moscow State Institute of Education) [1950ff] • PUBL (**X** 2802) Moskov Ped Inst: Moskva

S 0228 Zap Nauch Sem Leningrad Otd Mat Inst Steklov • SU
Zapiski Nauchnykh Seminarov Leningradskogo Otdeleniya Ordena Lenina Matematicheskogo Instituta imeni V.A.Steklova Akademii Nauk SSSR (LOMI) (Reports of the Scientific Seminars of the Leningrad Steklov Institute of Mathematics) PUBL (**X** 2641) Nauka: Leningrad • TRANSL IN (**J** 1531) J Sov Math & (**J** 0521) Semin Math, Inst Steklov • REM Transl. in J0521 up to vol. 19

S 0281 Arch Towarz Nauk Lwow, Sect 3 • PL
Archivum Towarzystwa Naukowego we Lwowie. Dzial 3. Matematyczno-Przyrodniczy (Archive of the Scientific Society of Lwow. Section 3. Mathematics and Natural Sciences) [1920-1939] • PUBL Tow Nauk: Lwow

S 0393 Uch Zap Univ, Tartu • SU
*Uchenye Zapiski Tartuskogo Gosudarstvennogo Universiteta. * Tartu Riikliku Uelikooli Toimetised. * Acta et Commentationes Universitatis Tartuensis (Scientific Notes of the Tartu State University.)* [1961ff] • PUBL (**X** 2463) Tartusk Gos Univ: Tartu

S 0422 Tr Vychisl Tsentra Akad Nauk Armyan SSR & Univ Erevan • SU
Trudy Vychislitel'nogo Tsentra Akademij Nauk Armyanskoj SSR i Erevanskogo Gosudarstvennogo Universiteta. Matematicheskie Voprosy Kibernetiki i Vychislitel'noj Tekhniki (Publications of the Computing Centre of the Academy of Sciences of the Armyan SSR and of the Erevan State University. Mathematical Problems of Cybernetics and Computational Techniques) [1963ff] • PUBL (**X** 2225) Akad Nauk Armyan SSR : Erevan

S 0458 Zesz Nauk, Prace Log, Uniw Krakow • PL
Zeszyty Naukowe Uniwersytetu Jagiellonskiego Prace z Logiki (Scientific Papers. Jagielleonian University. Reports on Logic) [1965-1972] • PUBL (**X** 1034) PWN: Warsaw • CONT AS (**J** 0302) Rep Math Logic, Krakow & Katowice

S 0478 Bonn Math Schr • D
Bonner Mathematische Schriften [1957ff] • PUBL (**X** 0908) Univ Math Inst: Bonn • ISSN 0524-045X

S 0507 Vychisl Sist (Akad Nauk SSSR Novosibirsk) • SU
Vychislitel'nye Sistemy. Sbornik Trudov. Akademiya Nauk SSSR. Sibirskoe Otdelenie (Computer Systems. Collected Articles) [1962ff] • PUBL (**X** 2652) Akad Nauk Sibirsk Otd Inst Mat: Novosibirsk • ISSN 0568-661X

S 0554 Issl Teor Algor & Mat Logik (Moskva) • SU
Issledovaniya po Teorii Algorifmov i Matematicheskoj Logike (Studies in the Theory of Algorithms and Mathematical Logic) [1973,1976,1979] • ED: MARKOV, A.A. & PETRI, N.V. (VOL 1). MARKOV, A.A. & KUSHNER, B.A. (VOL 2). MARKOV, A.A. & KHOMICH, V.I. (VOL 3) • PUBL (**X** 2265) Akad Nauk Vychis Tsentr: Moskva 287pp,160pp,133pp • ISSN 0302-9085

S 0716 Vychisl Tekh Vopr Kibern (Univ Leningrad) • SU
Vychislitel'naya Tekhnika i Voprosy Kibernetiki (Computer Technology and Questions of Cybernetics) [1962ff] • ED: CHIRKOV, M.K. & MASLOV, S.P. & TSAR'KOVA, Z.I. (V 8). BRUSENTSOVA, N.P. & SHAUMAN, A.M. (V 7, V 15 - V 19) • PUBL (**X** 0938) Leningrad Univ: Leningrad • ISSN 0507-536X

S 0764 Teor Konech Avtom & Prilozh (Riga) • SU
Teoriya Konechnykh Avtomatov i ee Prilozheniya (Theory of Finite Automata and its Applications) [1972ff] • PUBL (**X** 2230) Zinatne: Riga • LC-No 74-304069

Source Index

S 0780 Jbuch Ueberblick Math • D
Jahrbuch Ueberblicke Mathematik [1975ff] • PUBL (**X** 0876) Bibl Inst: Mannheim
• CONT OF (**J** 2345) Ueberblicke Math

S 0889 Notas Logica Mat • RA
Notas de Logica Matematica. [1963ff] • PUBL Inst Mat, Univ Nacional del Sur: Bahia Blanca
• ISSN 0078-2017

S 1002 Nauch Trud Vissh Ped Inst, Plovdiv • BG
Vissh Pedagogicheski Institut Plovdiv. Nauchni Trudove (Institute of Education of Plovdiv. Scientific Papers) [1963-1972] • PUBL Ped Inst: Plovdiv
• CONT AS (**S** 1441) Nauch Trud, Univ Plovdiv

S 1057 Semin Dubreil: Alg Th Nombr • F
Seminaire P.Dubreil: Algebre et Theorie des Nombres [1946-1970] • ED: DUBREIL, P. • PUBL (**X** 1623) Univ Paris VI Inst Poincare: Paris
• CONT AS (**S** 2250) Semin Dubreil: Algebre

S 1082 Quart Prog Rep MIT Res Lab Electron • USA
MIT Research Laboratory of Electronics Quarterly Progress Report

S 1415 Sel Math • D
Selecta Mathematica [1969ff] • PUBL (**X** 0811) Springer: Heidelberg & New York
• ISSN 0586-4313

S 1427 Dal'nevostoch Mat Sb, Khabarovsk • SU
Dal'nevostochnyj Matematicheskij Sbornik. Ministerstvo Prosveshcheniya RSFSR Khabarovskij Gosudarstvennyj Institut (Far-Eastern Mathematical Collected Articles. Ministry of Education of the RSFSR (Russian Soviet Federative Socialist Republic). State University of Khabarovsk) [1970ff]
• PUBL Gos Univ: Khabarovsk

S 1441 Nauch Trud, Univ Plovdiv • BG
Nauchni Trudove na Plovdivski Universitet "Paissi Khilendarski" (Scientific Papers of the University "Paissi Khilendarski", Plovdiv) [1973ff] • PUBL Univ Plovdiv: Plovdiv
• CONT OF (**S** 1002) Nauch Trud Vissh Ped Inst, Plovdiv

S 1454 Zesz Nauk Wyz Szk Ped Mat, Opole • PL
Zeszyty Naukowe Wyzszej Szkoly Pedagogicznej W Opolu Matematyka (Scientific Papers. University of Education in Opole. Mathematics) [1961ff] • PUBL Wyzsza Szkola Pedagogiczna: Opole
• ISSN 0078-5431

S 1459 Mem School Sci & Engin, Waseda Univ • J
Memoirs of the School of Sciences and Engineering. Waseda University [1922ff] • PUBL Waseda Univ.: Tokyo
• ISSN 0369-1950

S 1500 Probl Constr Math • USA
Problems in the Constructive Trend in Mathematics. [1970ff]
• ED: OREVKOV, V.P. & SHANIN, N.A. • SER (**S** 0055) Proc Steklov Inst Math • PUBL (**X** 0803) Amer Math Soc: Providence
• TRANSL OF (**S** 3713) Probl Konstr Napravleniya Mat
• REM Volumes 1-3 of S3713 not translated into English.

S 1536 Schr Weiterbildungszentr MKR (Dresden) • DDR
Schriftenreihe des Weiterbildungszentrum fuer Mathematik, Kybernetik und Rechentechnik PUBL Tech Univ Dresden: Dresden

S 1562 Prace Inst Mat Fis, Politech Wroclaw, Ser Stud Mater • PL
Prace Naukowe. Instytut Matematyki i Fizyki Teoretycznej. Politechnika Wroclawska. Seria Studia i Materialy. (Scientific Publications. Institute of Mathematics and Theoretical Physics. Technical University of Wroclaw) [?-1969] • PUBL Politechnika Wroclawska: Wroclaw
• CONT AS (**S** 4733) Prace Inst Mat, Politech Wroclaw, Ser Stud Mater

S 1567 Semin Bourbaki • F
Seminaire Bourbaki [1948ff] • SER (**S** 3301) Lect Notes Math, (**J** 1620) Asterisque • PUBL (**X** 0811) Springer: Heidelberg & New York, (**X** 2244) Soc Math France: Paris, (**X** 1623) Univ Paris VI Inst Poincare: Paris Secretariat Mathematique: Paris

S 1582 Matematika - Period Sb Perevodov Inostran Statej • SU
Matematika. Periodicheskij Sbornik Perevodov Inostrannykh Statej (Mathematics. Series of Collections of Translations of Foreign Articles) [1957ff] • PUBL (**X** 0885) Mir: Moskva
• REM Does not appear any more

S 1585 Rec Fct Th Newsletter • GB
Recursive Function Theory: Newsletter
• REM Obtainable from Barry Cooper, The British Logic Colloquium, Leeds Univ., LS2 9JT Leeds, GB, for Europe, Israel, and Japan, and from Iraj Kalantari, Dept. of Math., Western Illinois Univ., 61455 Macomb, IL, USA

S 1605 Math Centr Tracts • NL
Mathematical Centre Tracts [1963-1983] • PUBL (**X** 1121) Math Centr: Amsterdam

S 1626 Oslo Preprint Ser • N
Oslo Preprint Series [1970ff] • PUBL (**X** 2786) Univ Oslo Mat Inst: Oslo

S 1642 Schr Inf Angew Math, Ber (Aachen) • D
Schriften zur Informatik und Angewandten Mathematik. Bericht PUBL (**X** 3215) TH Aachen Math Nat Fak: Aachen

S 1644 Tr Fiz-Mat Inst Steklov • SU
Trudy Fiziko-Matematicheskogo Instituta imeni V.A.Steklova Akademiya Nauk SSSR (Travaux de l'Institut Physico-Mathematique V.A.Stekloff de l'Academie des Sciences SSSR) [1930-1937] • PUBL (**X** 0899) Akad Nauk SSSR : Moskva
• CONT AS (**S** 0066) Tr Mat Inst Steklov

S 1802 Colloq Int CNRS • F
Colloques Internationaux du Centre National de la Recherche Scientifique (CNRS) PUBL (**X** 0999) CNRS Inst B Pascal: Paris

S 1926 Banach Cent Publ • PL
Banach Center Publications. Polish Academy of Science. Institut of Mathematics [1976ff] • PUBL (**X** 1034) PWN: Warsaw
• ISSN 0137-6934

S 1956 Tagungsbericht, Oberwolfach • D
Tagungsbericht. Mathematisches Forschungsinstitut Oberwolfach PUBL (**X** 0876) Bibl Inst: Mannheim

S 2043 Issl Mat Log & Teor Algor (Tbilisi) • SU
Issledovaniya po Matematicheskoj Logike i Teorii Algoritmov (Studies in Mathematical Logic and the Theory of Algorithms) [1975,1977,1978] • ED: PKHAKADZE, SH.S. & MAGNARADZE, L.G. • PUBL (**X** 1052) Tbilisi Univ: Tbilisi 152pp,39pp,54pp
• LC-No 75-404082

S 2073 Actualites Sci Indust • F
Actualites Scientifiques et Industrielles PUBL (**X** 0859) Hermann: Paris

S 2250 Semin Dubreil: Algebre • F
Seminaire: Algebre. [1971ff] • ED: DUBREIL, P. & ARIBAUD, F. & MALLIAVIN, M.-P. • PUBL (**X** 1623) Univ Paris VI Inst Poincare: Paris
• CONT OF (**S** 1057) Semin Dubreil: Alg Th Nombr

S 2308 Symp Kyoto Univ Res Inst Math Sci (RIMS) • J
Surikaisekikenkyusho Kokyuroku (Kyoto University. Research Institute for Mathematical Sciences (RIMS). Proceedings of Symposia) PUBL (**X** 2441) Kyoto Univ Res Inst Math Sci: Kyoto

S 2348 Semin Th Nombres Bordeaux • F
Seminaire de Theorie des Nombres. 1972–1973 PUBL Univ Bourdeaux: Valence

S 2350 Zesz Nauk, Prace Mat, Uniw Krakow • PL
Zeszyty Naukowe Uniwersytetu Jagiellonskiego. Prace Matematyczne (Scientific Papers. Jagiellonian University. Reports on Mathematics) [1954–1984] • PUBL (**X** 2451) Uniw Jagiell Inst Mat: Krakow • ALT PUBL (**X** 1034) PWN: Warsaw • ISSN 0083-4386

S 2450 Mem Amer Math Soc, NS • USA
Memoirs of the American Mathematical Society. New Series [1975ff] • PUBL (**X** 0803) Amer Math Soc: Providence
• CONT OF (**S** 0167) Mem Amer Math Soc
• ISSN 0065-9266

S 2579 Teor Mnozhestv & Topol (Izhevsk) • SU
Teoriya Mnozhestv i Topologiya (Set Theory and Topology) [1977,1979,1982] • ED: GRYZLOV, A.A. • PUBL (**X** 4562) Udmurtskij Gos Univ: Izhevsk 114pp,116pp,116pp
• REM Title of Vol.2: Sovremennaya Topologiya i Teoriya Mnozhestv, Vyp.2

S 2582 Semiotika & Inf, Akad Nauk SSSR • SU
Semiotika i Informatika. Gosudarstvennyj Komitet SSSR po Nauke i Tekhnike. Akademiya Nauk SSSR. Vsesoyuznyj Institut Nauchnoj i Tekhnicheskoj Informatsii (Semiotics and Information Science) ED: MIKHAJLOV, A.I. • PUBL (**X** 2235) VINITI: Moskva

S 2587 Teor Algor & Progr (Riga) • SU
Teoriya Algoritmov i Programm (Theory of Algorithms and Programs) [1974,1975,1977] • ED: BARZDIN', YA.M. (VOL.2 & 3) & IKAUNIEKS, E. & FREIVALDS, R. (VOL.2) • SER (**J** 0483) Uch Zap, Univ Riga 210 (Vol.1), 233 (Vol.2) • PUBL (**X** 0895) Latv Valsts (Gos) Univ : Riga

S 2609 Mem Nara Tech College • J
Memoirs of Nara Technical College [1965ff]

S 2626 Vopr Teor Grupp Gomol Algeb • SU
Voprosy Teoriya Grupp i Gomologicheskoj Algebry (Questions in the Theory of Groups and Homological Algebra) [1977,1979] • PUBL (**X** 2766) Yaroslav Gos Univ: Yaroslavl'

S 2651 Prepr Inst Prikl Mat, Akad Nauk SSSR • SU
Preprint. Akademiya Nauk SSSR. Institut Prikladnoj Matematiki. (Preprint. Academy of Sciences of the USSR. Institute of Applied Mathematics) PUBL Akad Nauk SSSR, Inst Prikl Mat: Moskva

S 2653 Prepr Inst Kib, Akad Nauk Ukr SSR • SU
Preprint. Akademiya Nauk Ukrainskoj SSR. Institut Kibernetiki. (Preprint. Academy of Sciences of the Ukrainian SSR. Institute of Cybernetics) [1971ff] • PUBL (**X** 2522) Akad Nauk Inst Kibernet: Kiev

S 2787 Math Colloq, Univ Cape Town • ZA
Mathematics Colloquium University of Cape Town. PUBL Univ Cape Town, Dep Math: Rondebosch

S 2829 Rostocker Math Kolloq • DDR
Rostocker Mathematisches Kolloquium. PUBL Wilhelm-Pieck-Univ Rostock, Sekt Math: Rostock

S 2850 Tr Ehnerg Inst Moskva • SU
Trudy Moskovskogo Ordena Lenina Ehnergeticheskogo Instituta Tematicheskij Sbornik (Proceedings of the Moscow Institute of Energetics) PUBL Moskovsk Ehnergetichesk Instituta: Moskva

S 2874 Vopr Kibern, Akad Nauk SSSR • SU
Akademiya Nauk SSSR Nauchnyj Sovet po Kompleksnoj Probleme. "Kibernetika". Voprosy Kibernetiki (Problems of Cybernetics. Academy of Sciences. Scientific Council for Complexity Problems. Cybernetics) PUBL Akad Nauk SSSR Nauch Sovet Komplek Probl Kibern: Moskva

S 2890 Zesz Nauk, Mat Fiz, Politech Slask (Gliwice) • PL
Zeszyty Naukowe Politechniki Slaskiej. Matematyka, Fizyka (Scientific Papers. Silesian Technical University. Mathematics. Physics) [1961ff] • PUBL Politech Slask: Gliwice
• ISSN 0072-470X

S 2990 Rend Semin Mat Brescia • I
Rendiconti del Seminario Matematico di Brescia PUBL (**X** 1364) Vita e Pensiero: Milano

S 3105 Adv Math, Suppl Stud • USA
Advances in Mathematics. Supplementary Studies. [1965ff]
• PUBL (**X** 0801) Academic Pr: New York
• REL PUBL (**J** 0345) Adv Math

S 3125 Ber, Abt Inf, Univ Dortmund • D
Bericht. Abteilung Informatik. Universitaet Dortmund, Dortmund PUBL Univ Dortmund, Abt Inf: Dortmund

S 3126 Ber, Fak Inf, Univ Karlsruhe • D
Bericht. Institut fuer Informatik II PUBL (**X** 3159) TH Karlsruhe Fak Informatik: Karlsruhe

S 3145 Comput Sci Res Rep (Taiwan) • RC
Computer Science Research Report

S 3180 Inform Ber, Inst Inf, Univ Bonn • D
Informatik Berichte PUBL (**X** 1512) Univ Bonn Inst Informatik: Bonn

S 3231 Prepr NF, Sekt Math, Humboldt-Univ Berlin • DDR
Preprint (Neue Folge). Humboldt-Universitaet zu Berlin. Sektion Mathematik PUBL (**X** 2219) Humboldt-Univ Berlin: Berlin

S 3270 Spraw Inst Inf, Uniw Warsaw • PL
Sprawozdania Instytutu Informatyki Uniwersytetu Warszawskiego (Reports. Institute of Informatics. University of Warsaw.) PUBL Uniwersytet Warszawska: Warsaw

S 3291 Zesz Nauk Mat Fiz Chem (Uniw Gdansk) • PL
Zeszyty Naukowe Wydzialu Matematyki, Fizyki i Chemii. Matematyka (Scientific Papers of the Faculty of Mathematics, Physics & Chemistry) PUBL Uniwersytet Gdanski: Gdansk

S 3301 Lect Notes Math • D
Lecture Notes in Mathematics [1964ff] • PUBL (**X** 0811) Springer: Heidelberg & New York
• ISSN 0075-8434

S 3302 Lect Notes Comput Sci • D
Lecture Notes in Computer Science [1973ff] • PUBL (**X** 0811) Springer: Heidelberg & New York
• ISSN 0302-9743

S 3303 Stud Logic Found Math • NL
Studies in Logic and the Foundations of Mathematics [1954ff] • PUBL (**X** 0809) North Holland: Amsterdam
• ISSN 0049-237X

S 3304 Proc Symp Pure Math • USA
Proceedings of Symposia in Pure Mathematics PUBL (**X** 0803) Amer Math Soc: Providence

S 3305 Symposia Matematica • I
Symposia Matematica [1969ff] • PUBL (**X** 3604) INDAM: Roma
• ISSN 0082-0725

S 3306 Lond Math Soc Lect Note Ser • GB
London Mathematical Society Lecture Note Series [1971ff] • PUBL (**X** 0805) Cambridge Univ Pr: Cambridge, GB
• ISSN 0076-0552

S 3307 Synth Libr • NL
Synthese Library. Studies in Epistemology, Logic, Methodology, and Philosophy of Science [1959ff] • PUBL (**X** 0835) Reidel: Dordrecht
• ISSN 0082-1128

S 3308 Univ Western Ontario Ser in Philos of Sci • NL
University of Western Ontario Series in Philosophy of Science [1972ff] • PUBL (**X** 0835) Reidel: Dordrecht

S 3309 AFIPS Conference Proc • USA
AFIPS Conference Proceedings PUBL (**X** 1354) Spartan Books : Sutton

S 3310 Lect Notes Pure Appl Math • USA
Lecture Notes in Pure and Applied Mathematics [1971ff] • PUBL (**X** 1684) Dekker: New York
• ISSN 0075-8469

S 3311 Boston St Philos Sci • NL
Boston Studies in the Philosophy of Science [1963ff] • PUBL (**X** 0835) Reidel: Dordrecht
• ISSN 0068-0346

S 3312 Coll Math Soc Janos Bolyai • H
Colloquia Mathematica Societatis Janos Bolyai PUBL (**X** 0809) North Holland: Amsterdam

S 3313 Contemp Math • USA
Contemporary Mathematics [1980ff] • PUBL (**X** 0803) Amer Math Soc: Providence
• ISSN 0271-4132

S 3314 Lect Notes Econ & Math Syst • D
Lecture Notes in Economics and Mathematical Systems [1968ff] • PUBL (**X** 0811) Springer: Heidelberg & New York
• ISSN 0075-8442

S 3358 Ann Discrete Math • NL
Annals of Discrete Mathematics [1977ff] • PUBL (**X** 0809) North Holland: Amsterdam

S 3374 Forschergeb, Univ Jena • DDR
Forschungsergebnisse. Friedrich-Schiller-Universitaet PUBL (**X** 2211) Schiller Univ: Jena

S 3382 Sem-ber, Humboldt-Univ Berlin, Sekt Math • DDR
Seminarberichte. Humboldt-Universitaet zu Berlin, Sektion Mathematik PUBL (**X** 2219) Humboldt-Univ Berlin: Berlin

S 3412 Prace Inst Mat, Politech Wroclaw, Ser Konf • PL
Prace Naukowe Instytutu Matematyki Politechniki Wroclaw. Serija: Konferencje (Scientific Papers of the Institute of Mathematics of Wroclaw Technical University. Series: Conferences) PUBL Politechnika Wroclawska: Wroclaw

S 3462 Var Publ Ser, Aarhus Univ • DK
Various Publications Series [1962ff] • PUBL (**X** 1599) Aarhus Univ Mat Inst: Aarhus
• ISSN 0065-0188

S 3468 Tr Mat & Mekh (Tartu) • SU
*Matemaatika - ja Mekhaanika-Alaseid Toeid. * Trudy po Matematike i Mekhanike. (Works about Mathematics and Mechanics.)* SER (**S** 0393) Uch Zap Univ, Tartu • PUBL (**X** 2463) Tartusk Gos Univ: Tartu

S 3478 Sovrem Algebra (Leningrad) • SU
Sovremennaya Algebra (Modern Algebra) [1974ff] • ED: LYAPIN, E.S. • PUBL Leningrad Ped Inst im. Gartsena: Leningrad

S 3489 Sel Math Sov • CH
Selecta Mathematica Sovietica [1981ff] • PUBL (**X** 0804) Birkhaeuser: Basel
• ISSN 0272-9903

S 3513 Ann Math Stud • USA
Annals of Mathematics Studies [1940ff] • PUBL (**X** 0857) Princeton Univ Pr: Princeton

S 3521 Mem Soc Math Fr • F
Memoire de la Mathematique de France. Supplement au Bulletin [1927ff] • PUBL (**X** 0834) Gauthier-Villars: Paris
• REM Since 1981: Nouvelle Serie

S 3662 Upravl Sistemy (Akad Nauk SSSR, Novosibirsk) • SU
Upravljaemye Sistemy. Sbornik Trudov (Controlled Systems. Collected Articles) [1968ff] • PUBL (**X** 2652) Akad Nauk Sibirsk Otd Inst Mat: Novosibirsk
• ISSN 0566-7275

S 3713 Probl Konstr Napravleniya Mat • SU
Problemy Konstruktivnogo Napravleniya v Matematike. Sbornik Rabot (Problems of the Constructive Direction in Mathematics. Collected Papers) ED: OREVKOV, V.P. & SHANIN, N.A. • SER (**S** 0066) Tr Mat Inst Steklov • PUBL (**X** 2641) Nauka: Leningrad
• TRANSL IN (**S** 1500) Probl Constr Math
• REM Volumes 1-3 not translated into English

S 3726 Congressus Numerantium • CDN
Congressus Numerantium [1970ff] • PUBL (**X** 2420) Utilitas Mathematica Publ: Winnipeg

S 3909 Mat Met Opt & Upravleniya Slozh Sist (Kalinin) • SU
Matematicheskie Metody Optimizatsii i Upravleniya v Slozhnykh Sistemakh (Mathematical Methods of Optimization and Control in Complex Systems) [1981,1982] • ED: ABRAMOV, YU.A. • PUBL (**X** 1434) Kalinin Gos Univ: Kalinin 180pp,196pp

S 4163 Math in Schule • DDR
Mathematik in der Schule [1963ff] • PUBL (**X** 1036) Volk & Wissen: Berlin
• ISSN 0465-3750

S 4733 Prace Inst Mat, Politech Wroclaw, Ser Stud Mater • PL
Politechnika Wroclawska Instytut Matematyki. Prace Naukowe. Studia i Materialy (Scientific Publications. Institute of Mathematics. Technical University of Wroclaw) [1970ff]
• PUBL Politechnika Wroclawska: Wroclaw
• CONT OF (**S** 1562) Prace Inst Mat Fis, Politech Wroclaw, Ser Stud Mater

Proceedings

P 0454 Math Founds of Comput Sci (4);1975 Marianske Lazne • CS
[1975] *Mathematical Foundations of Computer Science. Proceedings of the 4th Symposium* ED: BECVAR, J. • SER (S 3302) Lect Notes Comput Sci 32 • PUBL (X 0811) Springer: Heidelberg & New York x+476pp
• DAT&PL 1975 Sep; Marianske Lazne, CS • ISBN 3-540-07389-2, LC-No 75-22406

P 0457 Lattice Th;1973 Houston • USA
[1973] *Proceedings of the University of Houston Lattice Theory Conference* ED: FAJTLOWICZ, S. & KAISER, K. • PUBL (X 2379) Univ Houston Dept Math: Houston viii+632pp
• DAT&PL 1973 Mar; Houston, TX, USA • LC-No 75-622429

P 0550 Switch Circ Th & Log Design (5);1964 Princeton • USA
[1964] *Proceedings of 5th Annual Symposium on Switching Theory and Logical Design* PUBL (X 2179) IEEE: New York
• DAT&PL 1964 Nov; Princeton, NJ, USA

P 0551 Compl of Computer Computation;1972 Yorktown Heights • USA
[1972] *Proceedings of a Symposium on the Complexity of Computer Computations* ED: MILLER, R.E. & THATCHER, J.W. & BOHLINGER, J.D. • SER IBM Research Symposia Series
• PUBL (X 1332) Plenum Publ: New York x+225pp • ALT PUBL (X 2471) IBM: Armonk
• DAT&PL 1972 Mar; Yorktown Heights, NY, USA • ISBN 0-306-30707-3, LC-No 72-85736

P 0572 Inform Processing (2);1962 Muenchen • D
[1963] *Information Processing '62. Proccedings of IFIP Congress* ED: POPPLEWELL, C.M. • PUBL (X 0809) North Holland: Amsterdam xiv+780pp
• DAT&PL 1962 Aug; Muenchen, D • LC-No 76-462349

P 0573 Inform Processing (3);1965 New York • USA
[1965-1966] *Information Processing '65. Proceedings of IFIP Congress* ED: KALENICH, W.A. • PUBL (X 0843) Macmillan : New York & London Vol 1: xv+1-304, Vol 2: viii+305-648
• ALT PUBL (X 1354) Spartan Books : Sutton
• DAT&PL 1965 May; New York, NY, USA • LC-No 65-24118

P 0575 Int Congr Math (II, 7);1954 Amsterdam • NL
[1954-1957] *Proceedings of the International Congress of Mathematicians 1954* ED: GERRETSEN, J.C.H. & GROOT DE, J.
• PUBL (X 0809) North Holland: Amsterdam 3 Vols: 582pp,440pp,560pp • ALT PUBL (X 1317) Noordhoff: Groningen 1954-1957 & (X 3602) Kraus: Vaduz 1967
• DAT&PL 1954 Sep; Amsterdam, NL • LC-No 52-1808

P 0576 Raisonn en Math & Sci Exper;1955 Paris • F
[1958] *La Raisonnement en Mathematiques et en Sciences Experimentales.* SER (S 1802) Colloq Int CNRS 70 • PUBL (X 0999) CNRS Inst B Pascal: Paris 140pp
• DAT&PL 1955 Sep; Paris, F • LC-No 63-24106

P 0577 Automatenth & Formale Sprachen;1969 Oberwolfach • D
[1970] *Automatentheorie und Formale Sprachen.* ED: DOERR, J. & HOTZ, G. • SER (S 1956) Tagungsbericht, Oberwolfach 3 • PUBL (X 0876) Bibl Inst: Mannheim 505pp
• DAT&PL 1969 Oct; Oberwolfach, D • LC-No 76-857074

P 0578 Self Organizing Systs;1962 Chicago • USA
[1962] *Self Organizing Systems* ED: YOVITIS, M.C. & JACOBI, G.T. & GOLDSTEIN, G.D. • PUBL (X 1354) Spartan Books : Sutton ix+563pp
• DAT&PL 1962 -?-; Chicago, IL, USA • LC-No 62-20444

P 0580 Int Congr Log, Meth & Phil of Sci (4,Sel Pap);1971 Bucharest • RO
[1973] *Logic, Language and Probability.* ED: BOGDAN, R.J. & NIINILUOTO, I. • SER (S 3307) Synth Libr • PUBL (X 0835) Reidel: Dordrecht x+323pp
• DAT&PL 1971 Aug; Bucharest, RO • ISBN 90-277-0312-4, LC-No 72-95892 • REL PUBL (P 0793) Int Congr Log, Meth & Phil of Sci (4,Proc);1971 Bucharest
• REM A Selection of Papers Contributed to Sections IV, VI and XI of P0793

P 0593 Int Congr Math (II, 6);1950 Cambridge MA • USA
[1952] *Proceedings of the International Congress of Mathematicians* ED: GRAVES, L.M & HILLE, E. & SMITH, P.A. & ZARISKI, O. • PUBL (X 0803) Amer Math Soc: Providence 2 Vols: 769pp,461pp • ALT PUBL (X 3602) Kraus: Vaduz 1967 2 Vols
• DAT&PL 1950 Aug; Cambridge, MA, USA • LC-No 52-1808

P 0594 Inform Processing (4);1968 Edinburgh • GB
[1969] *Information Processing '68. Proceedings of IFIP Congress* ED: MORRELL, A.J.H. • PUBL (X 0809) North Holland: Amsterdam 2 Vols: xxvi+xi+1650pp • ALT PUBL (X 2696) Humanities Pr: Atlantic Highlands
• DAT&PL 1968 Aug; Edinburgh, GB • ISBN 0-7204-2032-6, LC-No 76-462349
• REM Vol 1: Mathematics, Software. Vol 2: Hardware, Applications

P 0602 Generalized Recursion Th (1);1972 Oslo • N
[1974] *Generalized Recursion Theory* ED: FENSTAD, J.E.& HINMAN, P.G. • SER (S 3303) Stud Logic Found Math 79 • PUBL (X 0809) North Holland: Amsterdam viii+456pp
• ALT PUBL (X 0838) Amer Elsevier: New York
• DAT&PL 1972 Jun; Oslo, N • ISBN 0-444-10545-X, LC-No 73-81531

P 0603 Intuitionism & Proof Th;1968 Buffalo • USA
[1970] *Intuitionism and Proof Theory* ED: KINO, A. & MYHILL, J. & VESLEY, R.E. • PUBL (X 0809) North Holland: Amsterdam viii+516pp
• DAT&PL 1968 Aug,Buffalo, NY, USA • ISBN 0-7204-2257-4, LC-No 77-97196

P 0604 Scand Logic Symp (2);1970 Oslo • N
[1971] *Proceedings of the 2nd Scandinavian Logic Symposium* ED: FENSTAD, J.E. • SER (S 3303) Stud Logic Found Math 63 • PUBL (X 0809) North Holland: Amsterdam ii+405pp • DAT&PL 1970 Jun; Oslo, N • ISBN 0-7204-2259-0, LC-No 71-153401

P 0607 All-Union Math Conf (3);1956 Moskva • SU
[1959] *Trudy 3'go Vsesoyuznogo Matematicheskogo S'ezda (Proceedings of the 3rd All Union Mathematical Conference)* ED: NIKOL'SKIJ, S.M. & ABRAMOV, A.A. & BOLTYANSKIJ, V.G. • PUBL (X 0899) Akad Nauk SSSR : Moskva 4 Vols • DAT&PL 1956 Jun; Moskva, SU

P 0608 Logic Colloq;1966 Hannover • D
[1968] *Contributions to Mathematical Logic. Proceedings of the Logic Colloquium* ED: SCHMIDT, H.A. & SCHUETTE, K. & THIELE, H.-J. • SER (S 3303) Stud Logic Found Math • PUBL (X 0809) North Holland: Amsterdam ix+298pp • DAT&PL 1966 Aug; Hannover, D • LC-No 68-24434

P 0610 Tarski Symp;1971 Berkeley • USA
[1974] *Proceedings of the Tarski Symposium. An International Symposium held to Honor Alfred Tarski on the Occasion of His 70th Birthday* ED: HENKIN, L. & ADDISON, J. & CHANG, C.C. & CRAIG, W. & SCOTT, D.S. & VAUGHT, R. • SER (S 3304) Proc Symp Pure Math 25 • PUBL (X 0803) Amer Math Soc: Providence ix+498pp • DAT&PL 1971 Jun; Berkeley, CA, USA • ISBN 0-8218-1425-7, LC-No 74-8666 • REM Corrected Reprint 1979; xx+498pp

P 0612 Int Congr Log, Meth & Phil of Sci (1,Proc);1960 Stanford • USA
[1962] *Proceedings of the 1st International Congress for Logic, Methodology and Philosophy of Science* ED: NAGEL, E. & SUPPES, P. & TARSKI, A. • PUBL (X 1355) Stanford Univ Pr: Stanford ix+661pp • DAT&PL 1960 Aug; Stanford, CA, USA • LC-No 62-9620 • TRANSL IN [1965] (P 2251) Mat Log & Primen;1960 Stanford

P 0613 Rec Fct Th;1961 New York • USA
[1962] *Recursive Function Theory* ED: DEKKER, J.C.E. • SER (S 3304) Proc Symp Pure Math 5 • PUBL (X 0803) Amer Math Soc: Providence vii+247pp • DAT&PL 1961 Apr; New York, NY, USA • ISBN 0-8218-1405-2, LC-No 50-1183 • REM 2nd ed. 1970

P 0614 Th Models;1963 Berkeley • USA
[1965] *The Theory of Models.* ED: ADDISON, J.W. & HENKIN, L. & TARSKI, A. • SER (S 3303) Stud Logic Found Math • PUBL (X 0809) North Holland: Amsterdam xv+494pp • DAT&PL 1963 Jun; Berkeley, CA, USA • LC-No 66-7051

P 0620 Lattice Th;1959 Monterey • USA
[1961] *Lattice Theory* ED: DILWORTH, R.P. • SER (S 3304) Proc Symp Pure Math 2 • PUBL (X 0803) Amer Math Soc: Providence viii+208pp • DAT&PL 1959 Apr; Monterey, CA, USA • LC-No 50-1183

P 0623 Int Congr Log, Meth & Phil of Sci (2,Proc);1964 Jerusalem • IL
[1965] *Proceedings of the 2nd International Congress for Logic, Methodology and Philosophy of Science* ED: BAR-HILLEL, Y. • SER (S 3303) Stud Logic Found Math • PUBL (X 0809) North Holland: Amsterdam viii+440pp • DAT&PL 1964 Aug; Jerusalem, IL • ISBN 0-7204-2235-3, LC-No 66-7008 • REM 2nd ed. 1972

P 0624 Switch Circ Th & Log Design (1,2);1960 Chicago;1961 Detroit • USA
[1961] *Switching Circuit Theory and Logical Design. Proceedings of the 2nd Annual Symposium and Papers from the 1st Annual Symposium* ED: LEDLEY, R.S. • PUBL American Institute of Electrical Engineers: New York xi+341pp • DAT&PL 1960 Oct; Chicago, IL, USA, 1961 Oct; Detroit, MI, USA

P 0627 Int Congr Log, Meth & Phil of Sci (3,Proc);1967 Amsterdam • NL
[1968] *Proceedings of the 3rd International Congress for Logic, Methodology and Philosophy of Science* ED: ROOTSELAAR VAN, B. & STAAL, J.F. • SER (S 3303) Stud Logic Found Math • PUBL (X 0809) North Holland: Amsterdam xii+553pp • DAT&PL 1967 Aug; Amsterdam, NL • ISBN 0-444-85423-1, LC-No 68-29768

P 0629 Conf Universal Algeb;1969 Kingston • CDN
[1970] *Proceedings of the Conference on Universal Algebra* ED: WENZEL, G.H. • SER Queen's Papers in Pure and Applied Mathematics 25 • PUBL (X 0997) Queen's Univ: Kingston v+275pp • DAT&PL 1969 Oct; Kingston, ON, CDN • LC-No 76-586989

P 0630 Aspects Math Log;1968 Varenna • I
[1969] *Aspects of Mathematical Logic. Centro Internazionale Matematico Estivo (CIME)* ED: CASARI, E. • PUBL (X 0860) Cremonese: Firenze ii+285pp • DAT&PL 1968 Sep; Varenna, I • LC-No 70-477028

P 0633 Infinitist Meth;1959 Warsaw • PL
[1961] *Infinitistic Methods. Proceedings of the Symposium on Foundations of Mathematics* PUBL (X 1034) PWN: Warsaw 362pp • ALT PUBL (X 0869) Pergamon Pr: Oxford 1961 • DAT&PL 1959 Sep; Warsaw, PL • LC-No 61-11351

P 0634 Constructivity in Math;1957 Amsterdam • NL
[1959] *Constructivity in Mathematics* ED: HEYTING, A. • SER (S 3303) Stud Logic Found Math • PUBL (X 0809) North Holland: Amsterdam viii+298pp • DAT&PL 1957 Aug; Amsterdam, NL • LC-No 63-522

P 0637 Syntax & Semant Infinitary Lang;1967 Los Angeles • USA
[1968] *The Syntax and Semantics of Infinitary Languages* ED: BARWISE, K.J. • SER (S 3301) Lect Notes Math 72 • PUBL (X 0811) Springer: Heidelberg & New York iv+268pp • DAT&PL 1967 Dec; Los Angeles, CA, USA • ISBN 3-540-04242-3, LC-No 68-57175

P 0638 Logic Colloq;1969 Manchester • GB
[1971] *Logic Colloquium '69* ED: GANDY, R.O. & YATES, C.M.E. • SER (S 3303) Stud Logic Found Math 61 • PUBL (X 0809) North Holland: Amsterdam xiv+457pp • DAT&PL 1969 Aug; Manchester, GB • ISBN 0-7204-2261-2, LC-No 71-146188

P 0641 ACM Symp Th of Comput (2);1970 Northhampton • USA
[1970] *2nd Annual ACM Symposium on the Theory of Computation (Association for Computing Machinery)* PUBL (**X** 2205) ACM: New York
• DAT&PL 1970 May; Northhampton, MA, USA • LC-No 82-642181

P 0645 Int Congr Philos (11);1953 Bruxelles • B
[1953] *Actes du 11eme Congres International de Philosophie. * Proceedings of the 11th International Congress of Philosophy* PUBL (**X** 1313) Nauwelaerts: Louvain • ALT PUBL (**X** 0809) North Holland: Amsterdam 1953;14 Vols
• DAT&PL 1953 Aug; Bruxelles, B

P 0653 Int Congr Math (II, 4);1932 Zuerich • CH
[1932] *Verhandlungen des Internationalen Mathematiker-Kongresses Zuerich* ED: SAX, W. • PUBL (**X** 1268) Orell Fuessli: Zuerich 2 Vols: 335pp,365pp
• DAT&PL 1932 Sep; Zuerich, CH • LC-No 52-1808

P 0660 Int Congr Math (II, 8);1958 Edinburgh • GB
[1960] *Proceedings of the International Congress of Mathematicians* ED: TODD, J.A. • PUBL (**X** 0805) Cambridge Univ Pr: Cambridge, GB lxiv+573pp
• DAT&PL 1958 Aug; Edinburgh, GB • LC-No 52-1808

P 0662 Hungar Math Congr (1);1950 Budapest • H
[1952] *Comptes Rendus du 1ier Congres de Mathematiciens Hongrois* PUBL (**X** 0928) Akad Kiado: Budapest 789pp
• DAT&PL 1950 Aug; Budapest, H

P 0671 ACM Symp Th of Comput (1);1969 Marina del Rey • USA
[1969] *Conference Records of the Association for Computing Machinery Symposium on the Theory of Computation.* PUBL (**X** 2205) ACM: New York 272pp
• DAT&PL 1969 May; Marina del Rey, CA, USA • LC-No 82-642181

P 0672 IEEE Symp Switch & Automata Th (10);1969 Waterloo • CDN
[1969] *IEEE Conference Records of the 1969 Annual Symposium on Switching and Automata Theory: Papers Presented at the 10th Annual Symposium.* PUBL (**X** 2179) IEEE: New York 276pp
• DAT&PL 1969 Oct; Waterloo, ON, CDN • LC-No 72-181681

P 0674 Symp Math Th of Automata;1962 New York • USA
[1963] *Proceedings of the Symposium on Mathematical Theory of Automata* ED: FOX, J. • SER Microwave Research Institute Symposia Series 12 • PUBL (**X** 2039) Poly Inst New York: Brooklyn xix+640pp • ALT PUBL (**X** 0827) Wiley & Sons: New York
• DAT&PL 1962 Apr; New York, NY, USA • LC-No 63-11286

P 0677 Int Congr Math (II, 9,Proc);1962 Djursholm • S
[1963] *Proceedings of the International Congress of Mathematicians* ED: STENSTROEM, V. • PUBL (**X** 1163) Almqvist & Wiksell: Stockholm 1+595pp
• DAT&PL 1962 Aug; Djursholm, S • 52-1808

P 0678 Word Probl: Decis & Burnside Probl in Group Th;1969 Irvine • USA
[1973] *Word Problems. Decision Problems and the Burnside Problem in Group Theory* ED: BOONE, W.W. & CANNONITO, F.B. & LYNDON, R.C. • SER (**S** 3303) Stud Logic Found Math 71 • PUBL (**X** 0809) North Holland: Amsterdam xii+646pp
• DAT&PL 1969 Sep; Irvine, CA, USA • ISBN 0-7204-2271-X, LC-No 70-146190

P 0680 ACM Symp Th of Comput (3);1971 Shaker Heights • USA
[1971] *Conference Records of the 3rd Association for Computing Machinery Symposium on the Theory of Computing* PUBL (**X** 2205) ACM: New York 266pp
• DAT&PL 1971 May; Shaker Heights, OH, USA • LC-No 82-642181

P 0682 Int Congr Philos (10);1948 Amsterdam • NL
[1949] *Library of the 10th International Congress of Philosophy* ED: BETH, E.W. & POS, H.J. & HOLLAK, H.J.A. • PUBL (**X** 0809) North Holland: Amsterdam Vol 1, L.J.Veen: Amsterdam Vol 2
• DAT&PL 1948 Aug; Amsterdam, NL • LC-No 50-35721

P 0688 Logic Colloq;1963 Oxford • GB
[1965] *Formal Systems and Recursive Functions. Proceedings of the 8th Logic Colloquium* ED: CROSSLEY, J.N. & DUMMETT, M.A.E. • SER (**S** 3303) Stud Logic Found Math • PUBL (**X** 0809) North Holland: Amsterdam 320pp
• DAT&PL 1963 Jul; Oxford, GB • LC-No 66-2289

P 0690 Comput Prob in Abstr Algeb;1967 Oxford • GB
[1970] *Computational Problems in Abstract Algebra.* ED: LEECH, J. • PUBL (**X** 0869) Pergamon Pr: Oxford x+402pp
• DAT&PL 1967 Aug; Oxford, GB • ISBN 0-08-012975-7, LC-No 75-84072

P 0691 Sets, Models & Recursion Th;1965 Leicester • GB
[1967] *Sets, Models and Recursion Theory. Proceedings of the Summer School in Mathematical Logic and 10th Logic Colloquium* ED: CROSSLEY, J.N. • SER (**S** 3303) Stud Logic Found Math • PUBL (**X** 0809) North Holland: Amsterdam v+331pp • ALT PUBL (**X** 0838) Amer Elsevier: New York 1967
• DAT&PL 1965 Aug; Leicester, GB • ISBN 0-7204-2242-6, ISBN 0-444-10696-0, LC-No 67-21973
• REM 2nd ed. 1974

P 0692 Summer School in Logic;1967 Leeds • GB
[1968] *Proceedings of the Summer School in Logic* ED: LOEB, M.H. • SER (**S** 3301) Lect Notes Math 70 • PUBL (**X** 0811) Springer: Heidelberg & New York iv+331pp
• DAT&PL 1967 Aug; Leeds, GB • ISBN 3-540-04240-7, LC-No 68-56951

P 0693 Axiomatic Set Th;1967 Los Angeles • USA
[1971-1974] *Axiomatic Set Theory* ED: SCOTT, D. (V 1) & JECH, T.J. (V 2) • SER (**S** 3304) Proc Symp Pure Math 13 • PUBL (**X** 0803) Amer Math Soc: Providence 2 Vols: vi+474pp,viii+222pp
• DAT&PL 1967 Jul; Los Angeles, CA, USA • ISBN 0-8218-0245-3 (V1), ISBN 0-8218-0246-1 (V2), LC-No 78-125172

P 0696 Inform Processing (1);1959 Paris • F
[1960] *Information Processing '59 (Unesco)* PUBL Unesco: Paris 520pp • ALT PUBL (**X** 0814) Oldenbourg: Muenchen 1960 & Butterworth: London 1960
• DAT&PL 1959 Jun; Paris, F • LC-No 76-462349

P 0700 ACM Proc Conf;1952 Pittsburgh • USA
[1952] *Proceedings of the Association for Computing Machinery* PUBL (**X** 2205) ACM: New York 305pp
• DAT&PL 1952 May; Pittsburgh, PA, USA • LC-No 53-3390

P 0701 Struct of Lang & Math Aspects;1960 New York • USA
[1961] *Structure of Language and Its Mathematical Aspects* ED: JAKOBSON, R. • SER Proc Symp Appl Math 12 • PUBL (**X** 0803) Amer Math Soc: Providence vi+279pp
• DAT&PL 1960 Apr; New York, NY, USA • LC-No 50-1183

P 0703 Th of Graphs & Appl;1963 Smolenice • CS
[1964] *Theory of Graphs and Its Applications* ED: FIEDLER, M.
• PUBL (**X** 1226) Academia: Prague 234pp
• DAT&PL 1963 Jun; Smolenice, CS • LC-No 64-23026

P 0709 Int Conf Th of Groups (2);1973 Canberra • AUS
[1974] *Proceedings of the 2nd International Conference on the Theory of Groups* ED: NEWMAN, M.F. • SER (**S** 3301) Lect Notes Math 372 • PUBL (**X** 0811) Springer: Heidelberg & New York 740+viipp
• DAT&PL 1973 Aug; Canberra, ACT, AUS • ISBN 3-540-06845-7, LC-No 74-13872

P 0713 Cambridge Summer School Math Log;1971 Cambridge GB • GB
[1973] *Cambridge Summer School in Mathematical Logic* ED: ROGERS, H. & MATHIAS, A.R.D. • SER (**S** 3301) Lect Notes Math 337 • PUBL (**X** 0811) Springer: Heidelberg & New York ix+660pp
• DAT&PL 1971 Aug; Cambridge, GB • ISBN 3-540-05569-X, LC-No 73-12410

P 0737 Math Aspects Comput Sci;1966 New York • USA
[1967] *Proceedings of Symposia in Applied Mathematics. Vol 19: Mathematical Aspects of Computer Science* ED: SCHWARTZ, J.T. • PUBL (**X** 0803) Amer Math Soc: Providence v+224pp
• DAT&PL 1966, Apr; New York, NY, USA • LC-No 67-16554

P 0741 Int Congr Math (II, 3);1928 Bologna • I
[1929-1932] *Atti del Congresso Internazionale dei Matematici* PUBL (**X** 1375) Zanichelli: Bologna 6 Vols: 338pp,365pp,472pp,429pp,494pp,554pp
• DAT&PL 1928 Sep; Bologna, I • LC-No 52-1808

P 0743 Int Congr Math (II,11,Proc);1970 Nice • F
[1971] *Actes du Congres International de Mathematiciens 1970* ED: BERGER, M. & DIEUDONNE, J. & LERAY, J. & LIONS, J.-L. & MALLIAVIN, M.P. & SERRE, J.-P. • PUBL (**X** 0834) Gauthier-Villars: Paris 3 Vols: xxxiii+532pp,959pp,iii+371pp
• DAT&PL 1970 Sep; Nice, F • REL PUBL (**P** 1158) Int Congr Math (II,11,Comm Ind);1970 Nice
• REM Vol 1: Documents.Medailles Fields.Conferences Generales.Logique.Algebre. Vol 2: Geometrie et Topologie. Analyse. Vol 3: Mathematiques Appliquees.Historie et Enseignement.

P 0746 Automata Th;1964 Ravello • I
[1966] *Automata Theory. International School of Physics* ED: CAIANIELLO, E.R. • PUBL (**X** 0801) Academic Pr: New York xiv+342pp
• DAT&PL 1964 Jun; Ravello, I • LC-No 65-22775
• REM 2nd ed. 1968

P 0757 Scand Logic Symp (3);1973 Uppsala • S
[1975] *Proceedings of the 3rd Scandinavian Logic Symposium* ED: KANGER, S. • SER (**S** 3303) Stud Logic Found Math 82
• PUBL (**X** 0809) North Holland: Amsterdam vii+214pp
• ALT PUBL (**X** 0838) Amer Elsevier: New York
• DAT&PL 1973 Apr; Uppsala, S • ISBN 0-444-10679-0, LC-No 74-80113

P 0759 Infinite & Finite Sets (Erdoes);1973 Keszthely • H
[1975] *Infinite and Finite Sets. Dedicated to Paul Erdoes on His 60th Birthday* ED: HAJNAL, A. & RADO, R. & SOS, V.T.
• SER (**S** 3312) Coll Math Soc Janos Bolyai 10 • PUBL (**X** 3725) Bolyai Janos Mat Tars: Budapest 3 Vols: 1555pp
• ALT PUBL (**X** 0809) North Holland: Amsterdam
• DAT&PL 1973 Jun; Keszthely, H • ISBN 0-7204-2814-9, LC-No 75-321560

P 0760 Symp Inverse Semigroups & General;1973 DeKalb • USA
[1973] *Proceedings of a Symposium on Inverse Semigroups and Their Generalizations* ED: MCALLISTER, D. & MCFADDEN, B.
• PUBL Northern Illinois Univ: DeKalb 185pp
• DAT&PL 1973 Feb; DeKalb, IL, USA

P 0761 Compl of Computation;1973 New York • USA
[1974] *Complexity of Computation. Proceedings of a Symposium in Applied Mathematics of the American Mathematical Society (AMS) and the Society for Industrial and Applied Mathematics (SIAM)* ED: KARP, R.M. • SER SIAM-AMS Proceedings: 7 • PUBL (**X** 0803) Amer Math Soc: Providence viii+166pp
• DAT&PL 1973 Apr; New York, NY, USA • ISBN 0-8218-1327-7, LC-No 74-22062

P 0763 Automata, Lang & Progr (1);1972 Rocquencourt • F
[1973] *Automata, Languages and Programming. Proceedings of a Symposium Organised by IRIA (Institut de Recherche d'Informatique et d'Automatique)* ED: NIVAT, M. • PUBL (**X** 0809) North Holland: Amsterdam 638pp • ALT PUBL (**X** 0838) Amer Elsevier: New York
• DAT&PL 1972 Jul; Versailles-Rocquencourt, F • ISBN 0-444-10426-7, LC-No 72-93498
• REM Also Abbreviated as ICALP 72

P 0765 Algeb & Log;1974 Clayton • AUS
[1975] *Algebra and Logic. Papers from the 1974 Summer Research Institute of the Australian Mathematical Society* ED: CROSSLEY, J.N. • SER (**S** 3301) Lect Notes Math 450 • PUBL (**X** 0811) Springer: Heidelberg & New York viii+307pp
• DAT&PL 1974 Jan; Clayton, Vic, AUS • ISBN 3-540-07152-0, LC-No 75-9903

P 0770 Categ Th Appl to Comput & Control (1);1974 San Francisco • USA
[1975] *Proceedings of the 1st International Symposium on Category Theory Applied to Computation and Control* ED: MANES, E.G. • SER (**S** 3302) Lect Notes Comput Sci 25
• PUBL (**X** 0811) Springer: Heidelberg & New York x+245pp • ALT PUBL University of Massachussetts, Maths Department and Department of Computer Science: Amhurst
• DAT&PL 1974 Feb; San Francisco, CA, USA • ISBN 3-540-07142-3, LC-No 74-34481

P 0775 Logic Colloq;1973 Bristol • GB
[1975] *Logic Colloquium '73* ED: ROSE, H.E. & SHEPHERDSON, J.C. • SER (**S** 3303) Stud Logic Found Math 80 • PUBL (**X** 0809) North Holland: Amsterdam viii+513pp • ALT PUBL (**X** 0838) Amer Elsevier: New York
• DAT&PL 1973 Jul; Bristol, GB • ISBN 0-444-10642-1, LC-No 74-79302

Source Index

P 0776 Permutations;1972 Paris • F
[1974] *Permutations. Actes du Colloque sur les Permutations* SER Mathematiques et Sciences de L'Homme 20 • PUBL (**X** 0834) Gauthier-Villars: Paris xiv+289pp • ALT PUBL (**X** 0873) Mouton: Paris
• DAT&PL 1972 Jul; Paris, F • ISBN 2-04-009681-7, ISBN 2-7193-0897-8, LC-No 74-189927

P 0779 Conf Group Th;1972 Racine • USA
[1973] *Conference on Group Theory* ED: GATTERDAM, R.W. & WESTON, K.W. • SER (**S** 3301) Lect Notes Math 319 • PUBL (**X** 0811) Springer: Heidelberg & New York iv+188pp
• DAT&PL 1972 Jun; Racine, WI, USA • ISBN 3-540-06205-X, LC-No 73-76679

P 0785 Scand Logic Symp (1);1968 Aabo • S
[1970] *Proceedings of the 1st Scandinavian Logic Symposium* SER Filosofiska Studier 8 • PUBL (**X** 0882) Univ Filos Foeren: Uppsala 171pp
• DAT&PL 1968 Sep; Aabo, S • LC-No 72-186670

P 0787 C R Journ Soc Math France;1974 Montpellier • F
[1974] *Comptes Rendus des Journees Mathematiques de la Societe Mathematique de France* SER Cahiers Mathematiques de Montpellier 3 • PUBL (**X** 2358) Univ Languedoc Ma: Montpellier viii+379pp
• DAT&PL 1974 Apr; Montpellier, F

P 0788 Skand Mat Kongr (12);1953 Lund • S
[1954] *12te Skandinaviska Matematikerkongressen (Comptes-Rendus du 12eme Congres des Mathematiciens Scandinaves)* PUBL Hakan Oh Issons Boktryckeri: Lund xvi+337pp
• DAT&PL 1953 Aug; Lund, S • LC-No 55-58514

P 0793 Int Congr Log, Meth & Phil of Sci (4,Proc);1971 Bucharest • RO
[1973] *Proceedings of the 4th International Congress for Logic, Methodology and Philosophy of Science* ED: SUPPES, P. & HENKIN, L. & MOISIL, G.C. & JOJA, A. • SER (**S** 3303) Stud Logic Found Math 74 • PUBL (**X** 0809) North Holland: Amsterdam x+981pp • ALT PUBL (**X** 0838) Amer Elsevier: New York & (**X** 1034) PWN: Warsaw
• DAT&PL 1971 Aug; Bucharest, RO • ISBN 0-444-10491-7, LC-No 72-88505 • REL PUBL (**P** 0580) Int Congr Log, Meth & Phil of Sci (4,Sel Pap);1971 Bucharest

P 0797 Fonds des Math, Machines Math & Appl;1962 Tihany • H
[1965] *Colloque sur les Fondements des Mathematiques, les Machines Mathematiques, et leurs Applications* ED: KALMAR, L. • SER Collection de Logique Mathematique, Serie A 19 • PUBL (**X** 0928) Akad Kiado: Budapest 320pp • ALT PUBL (**X** 0834) Gauthier-Villars: Paris & (**X** 1313) Nauwelaerts: Louvain
• DAT&PL 1962 Sep; Tihany, H

P 1058 Th Machines & Comput;1971 Haifa • IL
[1971] *Theory of Machines and Computations* ED: KOHAVI, Z. & PAZ, A. • PUBL (**X** 0801) Academic Pr: New York xii+416pp
• DAT&PL 1971 Aug; Haifa, IL • LC-No 72-176298

P 1060 Constr Aspects Fund Thm Algeb;1967 Zuerich • CH
[1969] *Constructive Aspects of the Fundamental Theorem of Algebra. Proceedings of a Symposium Conducted at the IBM Research Laboratory.* ED: DEJON, B. & HENRICI, P. • PUBL (**X** 0827) Wiley & Sons: New York vii+337pp
• DAT&PL 1967 Jun; Zuerich, CH • ISBN 0-471-20300-9, LC-No 69-19380

P 1062 Conv Teor Gruppi & Cont Polari;1967/68 Roma • I
[1969] *Convegni Teoria dei Gruppi e dei Continui Polari. Teoria dei Polari Continui.* SER (**S** 3305) Symposia Matematica 1 • PUBL (**X** 3604) INDAM: Roma iii+445pp • ALT PUBL (**X** 0801) Academic Pr: New York
• DAT&PL 1967 Dec; Roma, I, 1968 Apr; Roma, I • LC-No 77-497037

P 1072 Math Log & Founds of Set Th;1968 Jerusalem • IL
[1970] *Mathematical Logic and the Foundations of Set Theory* ED: BAR-HILLEL, Y. • SER (**S** 3303) Stud Logic Found Math • PUBL (**X** 0809) North Holland: Amsterdam 145pp
• DAT&PL 1968 Nov; Jerusalem, IL • ISBN 0-7204-2255-8, LC-No 73-97195

P 1075 Logic Colloq;1976 Oxford • GB
[1977] *Logic Colloquium 76* ED: GANDY, R.O. & HYLAND, J.M.E. • SER (**S** 3303) Stud Logic Found Math 87 • PUBL (**X** 0809) North Holland: Amsterdam x+612pp • ALT PUBL (**X** 0838) Amer Elsevier: New York
• DAT&PL 1976 Jul; Oxford, GB • ISBN 0-7204-0691-9, LC-No 77-8943

P 1076 Latin Amer Symp Math Log (3);1976 Campinas • BR
[1977] *Non-Classical Logic, Model Theory and Computability. 3rd Latin American Symposium on Mathematical Logic* ED: ARRUDA, A.I. & COSTA DA, N.C.A & CHUAQUI, R. • SER (**S** 3303) Stud Logic Found Math 89 • PUBL (**X** 0809) North Holland: Amsterdam xviii+307pp • ALT PUBL (**X** 0838) Amer Elsevier: New York
• DAT&PL 1976 Jul; Campinas, BR • ISBN 0-7204-0752-4, LC-No 77-7366

P 1083 Victoria Symp Nonstand Anal;1972 Victoria • AUS
[1974] *Victoria Symposium on Non-Standard Analysis* ED: HURD, A. & LOEB, P. • SER (**S** 3301) Lect Notes Math 369 • PUBL (**X** 0811) Springer: Heidelberg & New York xviii+339pp
• DAT&PL 1972 May; Victoria, VIC, AUS • ISBN 3-540-06656-X, LC-No 73-22552

P 1126 IEEE Symp Switch & Automata Th (3);1962 College Park • USA
[1972] *Proceedings of the 3rd Annual Switching and Automata Theory Symposium* PUBL (**X** 2179) IEEE: New York
• DAT&PL 1962 Oct; College Park, MA, USA • LC-No 72-181681

P 1127 Symp Founds of Math;1962 Katada • J
[1963] *Proceedings of the Symposium on the Foundations of Mathematics* ED: TAKEUTI, G. • PUBL Tokyo University of Education: Tokyo vi+144pp
• DAT&PL 1962 Oct; Katada, J

P 1129 Princeton Conf Inform Sci & Syst (3);1969 Princeton • USA
[1969] *Proceedings of the 3rd Annual Conference on Information Science and Systems* ED: THOMAS, J.B. & VALKENBURG VAN, M.E. & WEINER, P. • PUBL (**X** 2188) Princeton Univ Dept Elect Eng & Comp Sci: Princeton xiii+550pp
• DAT&PL 1969 Mar; Princeton, NJ, USA

P 1130 Hawaii Int Conf Syst Sci (1);1968 Honolulu • USA
[1968] *Proceedings of the 1st Hawaii International Conference on System Sciences* ED: KINARIAWALA, B.K. & KUO, F.F. • PUBL University of Hawaii Press: Honolulu xv+816pp • ALT PUBL (**X** 1777) Western Periodicals: Hollywood 1968
• DAT&PL 1968 -?-; Honolulu, HI, USA • LC-No 68-19464

P 1158 Int Congr Math (II,11,Comm Ind);1970 Nice • F
[1970] *Congres International des Mathematiciens 1970. Les 265 Communications Individuels* PUBL (**X** 0834) Gauthier-Villars: Paris vii + 290pp
• DAT&PL 1970 Sep; Nice, F • LC-No 72-374601 • REL PUBL (**P** 0743) Int Congr Math (II,11,Proc);1970 Nice

P 1177 Int Congr Cybern (1);1956 Namur • B
[1958] *1ier Congres International de Cybernetique: Actes* PUBL (**X** 0834) Gauthier-Villars: Paris 924pp • ALT PUBL Association Internationale de Cybernetique: Namur 1958
• DAT&PL 1956 Jun; Namur, B • LC-No 59-4013

P 1184 Int Congr Cybern (2);1958 Namur • B
[1960] *2eme Congres International de Cybernetique: Actes* PUBL Association Internationale de Cybernetique: Namur xxxi + 1002pp
• DAT&PL 1958 Sep; Namur, B • LC-No 59-4013

P 1197 AFIPS Spring Jt Computer Conf (21);1962 San Francisco • USA
[1962] *AFIPS Conference Proceedings: 1962: Spring Joint Computer Conference* SER (**S** 3309) AFIPS Conference Proc 21 • PUBL (**X** 1354) Spartan Books : Sutton xv + 392pp • ALT PUBL (**X** 0874) Nat Pr Books : Palo Alto
• DAT&PL 1962 May; San Francisco, CA, USA • LC-No 55-44701

P 1219 Discont Groups & Riemann Surfaces;1973 College Park • USA
[1974] *Discontinuous Groups and Riemanian Surfaces* ED: GREENBERG, L. • SER (**S** 3513) Ann Math Stud 79 • PUBL (**X** 0857) Princeton Univ Pr: Princeton ix + 443pp • ALT PUBL (**X** 1015) Univ Toronto Pr: Toronto
• DAT&PL 1973 May; College Park, MD, USA • ISBN 0-691-08138-7, LC-No 73-16783

P 1383 Symp Comput & Automata;1971 Brooklyn • USA
[1971] *Proceedings of the Symposium on Computers and Automata* ED: FOX, J. • SER Microwave Research Institute Symposia Series 21 • PUBL (**X** 2039) Poly Inst New York: Brooklyn xxiv + 653pp • ALT PUBL (**X** 0820) Intersci Publ: New York
• DAT&PL 1971 Apr; Brooklyn, NY, USA • LC-No 70-185593

P 1384 ACM Nat Conf (22);1967 • USA
[1967] *Proceedings of the 22nd National Conference ACM* PUBL (**X** 2205) ACM: New York
• DAT&PL 1967 -?-; -?- • LC-No 64-25615

P 1390 Syst & Comput Sci;1965 London ON • CDN
[1967] *Systems and Computer Science* ED: HART, J.F. & TAKASU, S. • PUBL (**X** 1015) Univ Toronto Pr: Toronto x + 249pp
• DAT&PL 1965 Sep; London, ON, CDN • Lc-No 68-114245

P 1401 Math Founds of Comput Sci (5);1976 Gdansk • PL
[1976] *Mathematical Foundations of Computer Science. Proceedings of the 5th Symposium* ED: MAZURKIEWICZ, A. • SER (**S** 3302) Lect Notes Comput Sci 45 • PUBL (**X** 0811) Springer: Heidelberg & New York xi + 606pp
• DAT&PL 1976 Sep; Gdansk, PL • ISBN 3-540-07854-1, LC-No 76-25494

P 1417 Open House for Algeb;1970 Aarhus • DK
[1970] *Papers from the 'Open House for Algebraists'* SER (**S** 3462) Var Publ Ser, Aarhus Univ 17 • PUBL (**X** 1599) Aarhus Univ Mat Inst: Aarhus iv + 161pp
• DAT&PL 1970 Jun; Aarhus, DK

P 1430 Adv Course Founds Computer Sci;1974 Amsterdam • NL
[1975] *Foundations of Computer Science. Advanced Course* ED: BAKKER DE, J.W. • SER (**S** 1605) Math Centr Tracts 63 • PUBL (**X** 1121) Math Centr: Amsterdam 215pp
• DAT&PL 1974 May; Amsterdam, NL • ISBN 90-6196-111-4, LC-No 76-363070

P 1440 ⊧ ISILC Proof Th Symp (Schuette);1974 Kiel • D
[1975] ⊧ *ISILC Proof Theory Symposium. Dedicated to Kurt Schuette on the Occasion of His 65th Birthday. Proceedings of the International Summer Institute and Logic Colloquium* ED: DILLER, J. & MUELLER, GERT H. • SER (**S** 3301) Lect Notes Math 500 • PUBL (**X** 0811) Springer: Heidelberg & New York viii + 383pp
• DAT&PL 1974 Jul; Kiel, D • ISBN 3-540-07533-X, LC-No 75-40482 • REL PUBL (**P** 1442) ⊧ ISILC Logic Conf;1974 Kiel
• REM This Volume Contains Only the Proof Theory Part of the Conference.

P 1442 ⊧ ISILC Logic Conf;1974 Kiel • D
[1975] ⊧ *ISILC Logic Conference. Proceedings of the International Summer Institute and Logic Colloquium* ED: MUELLER, GERT H. & OBERSCHELP, A. & POTTHOFF, K. • SER (**S** 3301) Lect Notes Math 499 • PUBL (**X** 0811) Springer: Heidelberg & New York iv + 651pp
• DAT&PL 1974 Jul; Kiel, D • ISBN 3-540-07534-8, LC-No 75-40431 • REL PUBL (**P** 1440) ⊧ ISILC Proof Th Symp (Schuette);1974 Kiel

P 1448 Math Founds of Comput Sci (2);1973 Strbske Pleso • CS
[1973] *Mathematical Foundations of Computer Science. Proceedings of the 2nd Symposium* ED: HAVEL, I.M. • PUBL (**X** 1773) Vydat Slov Akad: Bratislava 338pp
• DAT&PL 1973 Sep; Strbske Pleso, CS

P 1449 Automata Th & Formal Lang;1975 Kaiserslautern • D
[1975] *Automata Theory and Formal Languages. 2nd GI-Conference (Gesellschaft fuer Informatik)* ED: BRAKHAGE, H. • SER (**S** 3302) Lect Notes Comput Sci 33 • PUBL (**X** 0811) Springer: Heidelberg & New York viii + 292pp
• DAT&PL 1975 May; Kaiserslautern, D • ISBN 3-540-07407-4, LC-No 75-28494

P 1455 Inform Processing (5);1971 Ljubljana • YU
[1972] *Information Processing '71. Proceedings of IFIP Congress* ED: FREIMAN, C.V. & GRIFFITH, J.E. & ROSENFELD, J.L. • PUBL (**X** 0809) North Holland: Amsterdam 2 Vols: xviii + 1621pp • ALT PUBL (**X** 2403) AFIPS Pr: Montvale
• DAT&PL 1971 Aug; Ljubljana, YU • ISBN 0-7204-2063-6, LC-No 76-184997
• REM Vol 1: Foundations and Systems. Vol 2: Applications

P 1464 ACM Symp Th of Comput (6);1974 Seattle • USA
[1974] *Proceedings of 6th Annual ACM Symposium on Theory of Computing (Association for Computing Machinery)* PUBL (**X** 2205) ACM: New York iv + 347pp
• DAT&PL 1974 Apr; Seattle, WA, USA • LC-No 82-642181

P 1469 AMS Numb Th Summer Inst;1969 Stony Brook • USA
[1971] *Proceedings of the 1969 Summer Institute on Number Theory, Analytic Number Theory, Diophantine Problems and Analytic Number Theory* ED: LEWIS, D.J. • SER (**S** 3304) Proc Symp Pure Math 20 • PUBL (**X** 0803) Amer Math Soc: Providence xiii + 451pp
• DAT&PL 1969 Jul; Stony Brook, NY, USA • ISBN 0-8218-1420-6, LC-No 76-125938

P 1475 Conv Inform Teor & Conv Strutt Corpi Algeb;1973 Roma • I
[1975] *Convegno di Informatica Teoretica. Convegno di Strutture in Corpi Algebrici* SER (**S** 3305) Symposia Matematica 15 • PUBL (**X** 3604) INDAM: Roma 604pp
• ALT PUBL (**X** 0801) Academic Pr: New York
• DAT&PL 1973 Feb; Rome, I, 1973 Apr; Rome, I

P 1476 Set Th & Hierarch Th (2) (Mostowski);1975 Bierutowice • PL
[1976] *Set Theory and Hierarchy Theory. A Memorial Tribute to Andrzej Mostowski. Proceedings of the 2nd Conference on Set Theory and Hierarchy Theory* ED: MAREK, W. & SREBRNY, M. & ZARACH, A. • SER (**S** 3301) Lect Notes Math 537
• PUBL (**X** 0811) Springer: Heidelberg & New York xiii + 345pp
• DAT&PL 1975 Sep; Bierutowice, PL • ISBN 3-540-07856-8, LC-No 76-26536

P 1479 IEEE Symp Switch & Automata Th (15);1974 New Orleans • USA
[1974] *Proceedings of IEEE Annual Symposium on Switching and Automata Theory. Conference Record of the 15th Annual Symposium* PUBL (**X** 2179) IEEE: New York v + 211pp
• DAT&PL 1974 Oct; New Orleans, LA, USA • LC-No 75-304938

P 1482 ACM Symp Th of Comput (5);1973 Austin • USA
[1973] *5th Annual ACM Symposium on Theory of Computing (Association for Computing Machinery)* PUBL (**X** 2205) ACM: New York iv + 277pp
• DAT&PL 1973 May; Austin, TX, USA • LC-No 82-642181

P 1484 Int Congr Math (2);1900 Paris • F
[1902] *Comptes Rendus du 2eme Congres International des Mathematiciens. Proces Verbaux et Communications* ED: DUPORCQ, E. • PUBL (**X** 0834) Gauthier-Villars: Paris 455pp
• DAT&PL 1900 Aug; Paris, F

P 1511 Int Symp Th Progr;1972 Novosibirsk • SU
[1974] *International Symposium on Theoretical Programming* ED: ERSHOV, A.P. & NEPOMNYASHCHIJ, V.A. • SER (**S** 3302) Lect Notes Comput Sci 5 • PUBL (**X** 0811) Springer: Heidelberg & New York vi + 407pp
• DAT&PL 1972 Aug; Novosibirsk, SU • ISBN 3-540-06720-5, LC-No 74-176124

P 1513 IEEE Symp Founds of Comput Sci (16);1975 Berkeley • USA
[1975] *16th Annual IEEE Symposium on Foundations of Computer Science* PUBL (**X** 2179) IEEE: New York 193pp
• DAT&PL 1975 Oct; Berkeley, CA, USA • LC-No 80-646634

P 1520 Berkeley Symp Math Stat & Probab (6);1970 Berkeley • USA
[1972] *Proceedings of the 6th Berkeley Symposium on Mathematical Statistics and Probability* ED: LECAM, L.M. & NEYMAN, J. & SCOTT, E.L. • PUBL (**X** 0926) Univ Calif Pr: Berkeley 5 Vols:
xiii + 760pp,xlix + 605pp,xx + 711pp,xvi + 353pp,xvi + 369pp
• ALT PUBL (**X** 0805) Cambridge Univ Pr: Cambridge, GB
• DAT&PL 1970 Jun; Berkeley, CA, USA • ISBN 0-520-01964-4 (V1), ISBN 0-520-02184-3 (V2), ISBN 0-520-02185-1 (V3), ISBN 0-520-02187-8 (V4), ISBN 0-520-02188-6 (V5), ISBN 0-520-02189-4 (V6), LC-No 49-8189
• REM Vol 1: Theory of Statistics. Vol 2: Probability Theory. Vol 3: Probability Theory. Vol 4: Biology and Health. Vol 5: Darwinian, Neo-Darwinian, and Non Darwinian Evolution.

P 1521 Int Congr Math (II,12);1974 Vancouver • CDN
[1975] *Proceedings of the International Congress of Mathematicians* ED: JAMES, R.D. • PUBL Canadian Mathematical Congress: Montreal 2 Vols:
xlix + 552pp,viii + 600pp
• DAT&PL 1974 Aug; Vancouver, BC, CDN • ISBN 0-919558-04-6, LC-No 74-34533

P 1545 Easter Conf on Model Th (2);1984 Wittenberg • DDR
[1984] *Proceedings of the 2nd Easter Conference on Model Theory* SER (**S** 3382) Sem-ber, Humboldt-Univ Berlin, Sekt Math 60 • PUBL (**X** 2219) Humboldt-Univ Berlin: Berlin ii + 243pp
• DAT&PL 1984 Apr; Wittenberg, DDR

P 1560 Tr Mezhdurn Semin Priklad Aspekt Teor Avtom;1971 Varna • BG
[1971] *Trudy Mezhdurnarodnogo Seminara po Prikladnym Aspektam Teorij Avtomatov (Proceedings of the International Seminar on Applied Aspects of the Automata Theory.)* PUBL (**X** 2237) Publ Bulg Acad Sci: Sofia 252pp
• DAT&PL 1971 May; Varna, BG

P 1571 Int Congr Philos (14);1968 Wien • A
[1968-1971] *Akten des 14. Internationalen Kongresses fuer Philosophie (Proceedings of the 14th International Congress of Philosophy)* PUBL (**X** 1279) Herder: Freiburg 6 Vols
• DAT&PL 1968 Sep; Wien, A • LC-No 78-352552

P 1578 IFAC Symp Teor Relej Ustroj & Kon Avtom;1962 Moskva • SU
[1965] *Trudy Mezhdunarodnogo Simpoziuma po Teorii Relejnykh Ustrojstv i Konechnykh Avtomatov (IFAK) (Arbeiten des Internationalen Symposiums ueber die Theorie der Relaisvorrichtungen und Endlichen Automaten (IFAC))* PUBL (**X** 2027) Nauka: Moskva 403pp
• DAT&PL 1962 -?-; Moskva, SU • LC-No 66-41255

P 1580 Tagung Formale Sprachen;1970 Oberwolfach • D
[1970] *Tagung ueber Formale Sprachen* ED: HOTZ, G. & SCHNORR, C.P. • SER (**J** 3220) Mitt Ges Math Datenverarb Bonn 8 • PUBL (**X** 0817) Ges Math Datenverarbeit: Bonn iii + 58pp
• DAT&PL 1970 Aug; Oberwolfach, D • LC-No 72-334973

P 1584 Comput Compl - Courant Comput Sci (7);1971 New York • USA
[1973] *Computational Complexity. Courant Computer Science Symposium* ED: RUSTIN, R. • PUBL Algorithmics Press: New York 268pp
• DAT&PL 1971 Oct; New York, NY, USA

P 1589 Math Interpr of Formal Systs;1954 Amsterdam • NL
[1955] *Mathematical Interpretations of Formal Systems* SER (**S** 3303) Stud Logic Found Math 10 • PUBL (**X** 0809) North Holland: Amsterdam viii + 113pp
• DAT&PL 1954 Sep; Amsterdam, NL • LC-No 56-3127
• REM 2nd ed. 1971

P 1590 All-Union Conf Math Log (3);1974 Novosibirsk • SU
[1974] *3 Vsesoyuznaya Konferentsiya po Matematicheskoj Logike. Tezitsy Doklady i Soobshcheniya (3rd All-Union Conference on Mathematical Logic)*
• DAT&PL 1974 -?-; Novosibirsk, SU • LC-No 75-558515

P 1591 Switch Circ Th & Log Design (6);1965 Ann Arbor • USA
[1965] *Conference Record of the 6th Annual Symposium on Switching Circuit Theory and Logical Design* PUBL (**X** 2179) IEEE: New York
• DAT&PL 1965 Oct; Ann Arbor, MI, USA

P 1601 Easter Conf on Model Th (1);1983 Diedrichshagen • DDR
[1983] *Proceedings of the 1st Easter Conference on Model Theory* ED: DAHN, B.I. • SER (**S** 3382) Sem-ber, Humboldt-Univ Berlin, Sekt Math 49 • PUBL (**X** 2219) Humboldt-Univ Berlin: Berlin 154pp
• DAT&PL 1983 Apr; Diedrichshagen, DDR

P 1603 λ-Calc & Comput Sci Th;1975 Roma • I
[1975] *λ Calculus and Computer Science Theory* ED: BOEHM, C. • SER (**S** 3302) Lect Notes Comput Sci 37 • PUBL (**X** 0811) Springer: Heidelberg & New York xii+370pp
• DAT&PL 1975 Mar; Roma, I • ISBN 3-540-07416-3, LC-No 75-33375

P 1606 Colloq Math (Pascal);1962 Clermont-Ferrand • F
[1962] *Actes du Colloque de Mathematiques Reuni a Clermont a l'Occasion du Tricentenaire de la Mort de Blaise Pascal* SER (**J** 0179) Ann Fac Sci Clermont 7,8 • PUBL Univ Clermont, Fac. Sci.: Clermont 2 Vols: 123pp,189pp
• DAT&PL 1962 Jun; Clermont-Ferrand, F
• REM Vol 1: Introduction et Logique Mathematique. Vol 2: Calcul des Probabilites, Analyse Numerique et Calcul Automatique, Geometrie et Physique Mathematique

P 1607 Hungar Math Congr (2);1960 Budapest • H
[1961] *Magyar Matematikai Kongresszus (2nd Hungarian Mathematical Congress. Abstracts)* PUBL (**X** 0928) Akad Kiado: Budapest 7 Vols
• DAT&PL 1960 Aug; Budapest, H

P 1608 Int Congr Math (II, 5);1936 Oslo • N
[1937] *Comptes Rendus du Congres International des Mathematiciens* PUBL A.M. Broeggers Boktrykkeri a/S: Oslo 2 Vols: 316pp,vv+289pp • ALT PUBL (**X** 3602) Kraus: Vaduz 1967 2 Vols
• DAT&PL 1936 Jul; Oslo, N • LC-No 52-1808

P 1612 Schaltkreis & -werk Th (2);1961 Saarbruecken • D
[1963] *2. Colloquium ueber Schaltkreis- und Schaltwerk-Theorie* ED: DOERR, J. & PESCHL, E. & UNGER, H.
• SER International Series of Numerical Mathematics 4
• PUBL (**X** 0804) Birkhaeuser: Basel 152pp
• DAT&PL 1961 Oct; Saarbruecken, D • LC-No 63-48116

P 1618 ACM Symp Th of Comput (7);1975 Albuquerque • USA
[1975] *Proceedings of the 7th Annual ACM Symposium on the Theory of Computing (Association for Computing Machinery)* PUBL (**X** 2205) ACM: New York 265pp
• DAT&PL 1975 May; Albuquerque, NM, USA • LC-No 82-642181

P 1619 Coloq Log Simb;1975 Madrid • E
[1976] *Coloquio Sobre Logica Simbolica* PUBL Centro Calculo Univ. Complutense: Madrid 176pp
• DAT&PL 1975 Feb; Madrid, E • LC-No 77-555677

P 1628 Generalized Recursion Th (2);1977 Oslo • N
[1978] *Generalized Recursion Theory II* ED: FENSTAD, J.E. & GANDY, R.O. & SACKS, G.E. • SER (**S** 3303) Stud Logic Found Math 94 • PUBL (**X** 0809) North Holland: Amsterdam vii+466pp • ALT PUBL (**X** 0838) Amer Elsevier: New York
• DAT&PL 1977 Jun; Oslo, N • ISBN 0-444-85163-1, LC-No 78-5366

P 1629 Symp Math Log;1974 Oulo;1975 Helsinki • SF
[1977] *Proceedings of the Symposia on Mathematical Logic* ED: MIETTINEN, S. & VAEAENAENEN, J. • PUBL University of Helsinki, Department of Philosophy: Helsinki iv+103pp
• DAT&PL 1974 -?-; Oulo, SF, 1975 -?-; Helsinki, SF

P 1630 GI Fachtag Automatenth & Form Sprach (1);1973 Bonn • D
[1973] *1. GI-Fachtagung ueber Automatentheorie und Formale Sprachen. (Gesellschaft fuer Informatik)* ED: BOEHLING, K.H. & INDERMARK, K. • SER (**S** 3302) Lect Notes Comput Sci 2
• PUBL (**X** 0811) Springer: Heidelberg & New York vii+322pp
• DAT&PL 1973 Jul; Bonn, D • ISBN 3-540-06527-X, LC-No 74-649668

P 1632 Automata, Lang & Progr (4);1977 Turku SF • SF
[1977] *Automata, Languages and Programming. 4th Colloquium* ED: SALOMAA, A. & STEINBY, M. • SER (**S** 3302) Lect Notes Comput Sci 52 • PUBL (**X** 0811) Springer: Heidelberg & New York x+569pp
• DAT&PL 1977 Jul; Turku, SF • ISBN 3-540-08342-1, LC-No 78-337949
• REM Also Abbreviated as ICALP 77

P 1635 Math Founds of Comput Sci (6);1977 Tatranska Lomnica • CS
[1977] *Mathematical Foundations of Computer Science. Proceedings of the 6th Symposium* ED: GRUSKA, J. • SER (**S** 3302) Lect Notes Comput Sci 53 • PUBL (**X** 0811) Springer: Heidelberg & New York xi+595pp
• DAT&PL 1977 Sep; Tatranska Lomnica, CS • ISBN 3-540-08353-7, LC-No 77-10135

P 1636 AFIPS Fall Jt Computer Conf (24);1963 Las Vegas • USA
[1963] *AFIPS Conference Proceedings: 1963: Fall Joint Computer Conference* SER (**S** 3309) AFIPS Conference Proc 24 • PUBL (**X** 1354) Spartan Books : Sutton vii+647pp • ALT PUBL (**X** 2508) Cleaver-Hume Pr: London 1963
• DAT&PL 1963 Nov; Las Vegas, NV, USA • LC-No 55-44701

P 1639 Set Th & Hierarch Th (1);1974 Karpacz • PL
[1977] *Set Theory and Hierarchy Theory. Proceedings of the 1st Colloquium in Set Theory and Hierarchy Theory* SER (**S** 3412) Prace Inst Mat, Politech Wroclaw, Ser Konf 14/1 • PUBL Polytechnical Edition: Wroclaw 123pp
• DAT&PL 1974 Sep; Karpacz, PL

P 1653 GI Jahrestag (2);1972 Karlsruhe • D
[1973] *GI - 2. Jahrestagung (Gesellschaft fuer Informatik)* ED: DEUSSEN, P. • SER (**S** 3314) Lect Notes Econ & Math Syst 78
• PUBL (**X** 0811) Springer: Heidelberg & New York xi+576pp
• DAT&PL 1972 Oct; Karlsruhe, D • ISBN 3-540-06127-4, LC-No 72-96727

P 1671 Colloq Automatenth (3);1965 Hannover • D
[1967] *3. Colloquium ueber Automatentheorie. Vortragsauszuege* ED: HAENDLER, W. & PESCHL, E. & UNGER, H. • SER International Series of Numerical Mathematics 6
• PUBL (**X** 0804) Birkhaeuser: Basel iv+316pp
• DAT&PL 1965 Oct; Hannover, D • LC-No 85-212889

P 1672 Repercussions Rech Math sur Enseig;1965 Echternach • L
[1966] *Les Repercussions de la Recherche Mathematique sur l'Enseignement. Textes Originaux des Conferences Faites au Seminaire Organise par la CIEM* PUBL Institut Grand-Ducal, Section des Sciences Naturelles, Physiques et Mathematiques: Luxembourg;288pp
• DAT&PL 1965 -?-; Echternach, L

P 1675 Summer Inst Symb Log;1957 Ithaca • USA
[1957] *Summaries of Talks Presented at the Summer Institute for Symbolic Logic* PUBL Institute for Defense Analyses, Communications Research Division: Princeton; xvi+427pp
• DAT&PL 1957 Jul; Ithaca, NY, USA • LC-No 65-4418
• REM 2nd ed. 1960

P 1681 Symp Inform Storage & Retrieval;1971 College Park • USA
[1971] *Proceedings of the Symposium on Information Storage and Retrieval* ED: MINKER, J. & ROSENFELD, S. • PUBL (**X** 2205) ACM: New York viii+285pp
• DAT&PL 1971 Apr; College Park, MD, USA • LC-No 72-178466

P 1691 Inform Processing (6);1974 Stockholm • S
[1974] *Information Processing '74. Proceedings of IFIP Congress* ED: ROSENFELD, J.L. • PUBL (**X** 0809) North Holland: Amsterdam xxi+1107pp • ALT PUBL (**X** 0838) Amer Elsevier: New York
• DAT&PL 1974 Aug; Stockholm, S • ISBN 0-444-10689-8, LC-No 74-76063

P 1694 Inform Processing (7);1977 Toronto • CDN
[1977] *Information Processing '77. Proceedings of IFIP Congress* ED: GILCHRIST, B. • SER IFIP Congress Series 7 • PUBL (**X** 0809) North Holland: Amsterdam xix+1004pp
• DAT&PL 1977 Aug; Toronto, ON, CDN • ISBN 0-7204-0755-9, LC-No 77-80624

P 1695 Set Th & Hierarch Th (3);1976 Bierutowice • PL
[1977] *Set Theory and Hierarchy Theory V. Proceedings of the 3rd Conference on Set Theory and Hierarchy Theory* ED: LACHLAN, A. & SREBRNY, M. & ZARACH, A. • SER (**S** 3301) Lect Notes Math 619 • PUBL (**X** 0811) Springer: Heidelberg & New York viii+358pp
• DAT&PL 1976 Sep; Bierutowice, PL • ISBN 3-540-08521-1, LC-No 78-309663

P 1704 Int Congr Log, Meth & Phil of Sci (5);1975 London ON • CDN
[1977] *Proceedings of 5th International Congress of Logic, Methodology and Philosophy of Science* ED: BUTTS, R.E. & HINTIKKA, J. • SER (**S** 3308) Univ Western Ontario Ser in Philos of Sci 9-12 • PUBL (**X** 0835) Reidel: Dordrecht 4 Vols: x+406pp, x+427pp, x+321pp, x+336pp
• DAT&PL 1975 Aug; London, ON, CDN • ISBN 90-277-0708-1 (V1), ISBN 90-277-0710-3 (V2), ISBN 90-277-0829-0 (V3), ISBN 90-277-0831-2 (V4), ISBN 90-277-0706-5 (Set of the 4 Vols), LC-No 77-22429 (V1), LC-No 77-22431 (V2), LC-No 77-22432 (V3), LC-No 77-22433 (V4)
• REM Vol 1: Logic, Foundations of Mathematics, and Computability Theory. Vol 2: Foundational problems in the Special Sciences. Vol 3: Basic Problems in Methodology and Linguistics. Vol 4: Historical and Philosophical Dimensions of Logic, Methodology and Philosophy of Science.

P 1705 Scand Logic Symp (4);1976 Jyvaeskylae • SF
[1979] *Essays on Mathematical and Philosophical Logic. Proceedings of the 4th Scandinavian Logic Symposium and of the 1st Soviet-Finnish Logic Conference* ED: HINTIKKA, J. & NIINILUOTO, I. & SAARINEN, E. • SER (**S** 3307) Synth Libr 122 • PUBL (**X** 0835) Reidel: Dordrecht viii+462pp
• DAT&PL 1976 Jun; Jyvaeskylae, SF • ISBN 90-277-0879-7, LC-No 78-14736

P 1707 Math Founds of Comput Sci (7);1978 Zakopane • PL
[1978] *Mathematical Foundations of Computer Science. Proceedings of the 7th Symposium* ED: WINKOWOSKI, J • SER (**S** 3302) Lect Notes Comput Sci 64 • PUBL (**X** 0811) Springer: Heidelberg & New York ix+551pp
• DAT&PL 1978 Sep; Zakopane, PL • ISBN 3-540-08921-7, LC-No 78-14457

P 1729 Colloq Int Log;1975 Clermont-Ferrand • F
[1977] *Colloque International de Logique* SER (**S** 1802) Colloq Int CNRS 249 • PUBL (**X** 0999) CNRS Inst B Pascal: Paris 224pp
• DAT&PL 1975 Jul; Clermont-Ferrand, F • ISBN 2-222-02019-0, LC-No 78-367483

P 1740 ACM Symp Th of Comput (10);1978 San Diego • USA
[1978] *Conference Record of the 10th Annual ACM Symposium on Theory of Computing (Association for Computing Machinery)* PUBL (**X** 2205) ACM: New York 346pp
• DAT&PL 1978 May; San Diego, CA, USA • LC-No 79-101797

P 1753 Int Congr AICA Hybrid Comput (7); • USA
[1973] *Proceedings of the 7th International Congress of AICA on Hybrid Computation, Part 1*

P 1755 Math Founds of Comput Sci (3);1974 Jadwisin • PL
[1975] *Mathematical Foundations of Computer Science. Proceedings of the 3rd Symposium* ED: BLIKLE, A. • SER (**S** 3302) Lect Notes Comput Sci 28 • PUBL (**X** 0811) Springer: Heidelberg & New York 484pp
• DAT&PL 1974 Jun; Jadwisin, PL • ISBN 3-540-07162-8, LC-No 75-9642

P 1757 IEEE Symp Found of Comput Sci (17);1976 Houston • USA
[1976] *17th Annual IEEE Symposium on Foundations of Computing* PUBL (**X** 2179) IEEE: New York v+276pp
• DAT&PL 1976 Oct; Houston, TX, USA • LC-No 80-646634

P 1792 Struct Learning;1968 Philadelphia • USA
[1973-1976] *Structural Learning* ED: SCANDURA, J.M. • SER Structural Learning Series • PUBL (**X** 0836) Gordon & Breach: New York 2 Vols
• DAT&PL 1968 Apr; Philadelphia, PA, USA • ISBN 0-677-15110-1 (V2), LC-No 73-76710

P 1800 Brazil Conf Math Log (1);1977 Campinas • BR
[1978] *Proceedings of 1st Brazilian Conference on Mathematical Logic* ED: ARRUDA, A.I. & CHAQUI, R. & COSTA DA, N.C.A. • SER (**S** 3310) Lect Notes Pure Appl Math 39 • PUBL (**X** 1684) Dekker: New York xii+303pp
• DAT&PL 1977 Jul; Campinas, BR • LC-No 78-14488

P 1805 Int Symp Multi-Val Log (5,Proc);1975 Bloomington • USA
[1975] *Proceedings of the 1975 International Symposium on Multiple-Valued Logic* PUBL (**X** 2179) IEEE: New York iv+475pp
• DAT&PL 1975 May; Bloomington, IN, USA • LC-No 76-370321 • REL PUBL (**P** 1894) Int Symp Multi-Val Log (5,Inv Pap);1975 Bloomington

P 1841 Fct Recurs & Appl;1967 Tihany • H
[1969] *Les Fonctions Recursives et leurs Applications* PUBL (**X** 0999) CNRS Inst B Pascal: Paris • ALT PUBL (**X** 3725) Bolyai Janos Mat Tars: Budapest
• DAT&PL 1967 Sep; Tihany, H

P 1846 Kompl, Lern- & Erkenn-Prozess;1973 Jena • DDR
[1975] *Kompliziertheit, Lern- und Erkennungsprozesse* SER Wissenschaftliche Beitraege Schiller-Univ Jena 75 • PUBL (**X** 2211) Schiller Univ: Jena 2 Vols
• DAT&PL 1973 Oct; Jena, DDR

P 1850 Strukt & Entsch des Rechts;1976 Salzburg • A
[1978] *Strukturierungen und Entscheidungen im Rechtsdenken: Notation, Terminologie und Datenverarbeitung in der Rechtslogik* ED: TAMMELO, I. & SCHREINER, H. • SER Forschungen aus Staat und Recht 43 • PUBL (**X** 0902) Springer: Wien viii+313pp
• DAT&PL 1976 Nov; Salzburg, A • ISBN 3-211-81470-1, LC-No 78-1539

P 1864 Higher Set Th;1977 Oberwolfach • D
[1978] *Higher Set Theory* ED: MUELLER, GERT H. & SCOTT, D.S. • SER (**S** 3301) Lect Notes Math 669 • PUBL (**X** 0811) Springer: Heidelberg & New York xii+476pp
• DAT&PL 1977 Apr; Oberwolfach, D • ISBN 3-540-08926-8, LC-No 79-312135

P 1869 Automata, Lang & Progr (2);1974 Saarbruecken • D
[1974] *Automata, Languages and Programming: 2nd Colloquium* ED: LOECKX, J. • SER (**S** 3302) Lect Notes Comput Sci 14 • PUBL (**X** 0811) Springer: Heidelberg & New York viii+611pp
• DAT&PL 1974 Jul; Saarbruecken, D • ISBN 3-540-06841-4, LC-No 74-180345
• REM Also Abbreviated as ICALP 74

P 1870 Automata, Lang & Progr (3);1976 Edinburgh • GB
[1976] *Automata, Languages and Programming. 3rd Colloquium* ED: MICHAELSON, S. & MILNER, R. • PUBL (**X** 1261) Edinburgh Univ Pr: Edinburgh vi+559pp
• DAT&PL 1976 Jul; Edinburgh, GB • ISBN 0-85224-308-1, LC-No 77-359145
• REM Also Abbreviated as ICALP 76

P 1872 Automata, Lang & Progr (5);1978 Udine • I
[1978] *Automata, Languages and Programming. 5th Colloquium* ED: AUSIELLO, G. & BOEHM, C. • SER (**S** 3302) Lect Notes Comput Sci 62 • PUBL (**X** 0811) Springer: Heidelberg & New York viii+508pp
• DAT&PL 1978 Jul; Udine, I • ISBN 3-540-08860-1, LC-No 79-303999
• REM Also Abbreviated as ICALP 78

P 1873 Automata, Lang & Progr (6);1979 Graz • A
[1979] *Automata, Languages and Programming. 6th Colloquium* ED: MAURER, H.A. • SER (**S** 3302) Lect Notes Comput Sci 71 • PUBL (**X** 0811) Springer: Heidelberg & New York ix+682pp
• DAT&PL 1979 Jul; Graz, A • ISBN 3-540-09510-1, LC-No 79-15859
• REM Also Abbreviated as ICALP 79

P 1894 Int Symp Multi-Val Log (5,Inv Pap);1975 Bloomington • USA
[1977] *Modern Uses of Multiple-Valued Logic. Invited Papers from the 5th International Symposium on Multiple-Valued Logic* ED: DUNN, J.M. & EPSTEIN, G. • SER Epistime 2
• PUBL (**X** 0835) Reidel: Dordrecht x+338pp
• DAT&PL 1975 May; Bloomington, IN, USA • ISBN 90-277-0747-2, LC-No 77-23098 • REL PUBL (**P** 1805) Int Symp Multi-Val Log (5,Proc);1975 Bloomington
• REM With a Bibliography of Many-Valued Logic

P 1897 Logic Colloq;1977 Wroclaw • PL
[1978] *Logic Colloquium 77* ED: MACINTYRE, A. & PACHOLSKI, L. & PARIS, J. • SER (**S** 3303) Stud Logic Found Math 96 • PUBL (**X** 0809) North Holland: Amsterdam x+311pp
• DAT&PL 1977 Aug; Wroclaw, PL • ISBN 0-444-85178-X, LC-No 78-13396

P 1899 IEEE Symp Switch & Automata Th (8);1967 Austin • USA
[1968] *IEEE Conference Record of 1967 8th Annual Symposium on Switching and Automata Theory* PUBL (**X** 2179) IEEE: New York 335pp
• DAT&PL 1967 Oct; Austin, TX, USA • LC-No 72-181681

P 1900 IEEE Symp Switch & Automata Th (9);1968 Schenectady • USA
[1968] *IEEE Conference Record of 1968 9th Annual Symposium on Switching and Automata Theory* PUBL (**X** 2179) IEEE: New York 448pp
• DAT&PL 1968 Oct; Schenectady, NY, USA • LC-No 72-181681

P 1901 ACM Symp Th of Comput (4);1972 Denver • USA
[1972] *Proceedings of the 4th Annual ACM Symposium on the Theory of Computing. Spring Joint Computer Conference. (Association for Computing Machinery)* SER (**S** 3309) AFIPS Conference Proc 40 • PUBL (**X** 2205) ACM: New York 263pp
• DAT&PL 1972 May; Denver, CO, USA • LC-No 82-642181

P 1903 ACM Symp Th of Comput (9);1977 Boulder • USA
[1977] *Conference Records of the 9th Annual ACM Symposium on the Theory of Computing (Association for Computing Machinery)* PUBL (**X** 2205) ACM: New York v+314pp
• DAT&PL 1977 May; Boulder, CO, USA • LC-No 82-642181

P 1913 Lattice Th;1974 Szeged • H
[1976] *Lattice Theory* ED: HUHN, A.P. & SCHMIDT, E.T. • SER (**S** 3312) Coll Math Soc Janos Bolyai 14 • PUBL (**X** 0809) North Holland: Amsterdam 462pp • ALT PUBL (**X** 3725) Bolyai Janos Mat Tars: Budapest
• DAT&PL 1974 Aug; Szeged, H • ISBN 0-7204-0498-3, ISBN 963-8021-18-7, LC-No 77-484237

P 1935 Princeton Conf Inform Sci & Syst (2);1968 Princeton • USA
[1968] *Proceedings of the 2nd Princeton Conference on Information Science and Systems* PUBL (**X** 2188) Princeton Univ Dept Elect Eng & Comp Sci: Princeton xii+507pp
• DAT&PL 1968 Mar; Princeton, NJ, USA
• REM Published in Co-Operation with the IEEE Group on Circuit Theory

P 1938 Conf Biol Motiv Automata Th;1974 McClean • USA
[1974] *Proceedings of the 1974 Conference on Biologically Motivated Automata Theory* PUBL (**X** 2179) IEEE: New York v+221pp
• DAT&PL 1974 Jun; McClean, VA, USA • LC-No 75-302460

P 1939 L Systems;1974 Aarhus • DK
[1974] *L Systems* ED: ROZENBERG, G.& SALOMAA, A. • SER (S 3302) Lect Notes Comput Sci 15 • PUBL (X 0811) Springer: Heidelberg & New York vi+338pp
• DAT&PL 1974 Jan; Aarhus, DK • ISBN 3-540-06867-8, LC-No 74-16417

P 1940 Form Lang & Progr;1975 Madrid • E
[1976] *Formal Languages and Programming* ED: AGUILAR, R. • PUBL (X 0809) North Holland: Amsterdam ix+129pp
• ALT PUBL (X 0838) Amer Elsevier: New York
• DAT&PL 1975 Apr; Madrid, E • ISBN 0-444-11084-4, LC-No 76-373380

P 1959 Int Congr Math (II,13);1978 Helsinki • SF
[1980] *Proceedings of the International Congress of Mathematicians* ED: LEHTO, O. • PUBL Academia Scientiarum Fennica: Helsinki 2 Vols: 1022pp
• DAT&PL 1978 Aug; Helsinki, SF • ISBN 951-41-0352-1

P 1962 Conf Inform Sci & Syst;1977 Baltimore • USA
[1978] *Proceedings of a Conference on Information Sciences and Systems* PUBL (X 1291) Johns Hopkins Univ Pr: Baltimore
• DAT&PL 1977 Mar; Baltimore, MD, USA

P 2032 Stud on Polyautomata;1976 Zushi & Kyoto • J
[1977] *Studies on Polyautomata* ED: NISHIO, H.
• DAT&PL 1976 Aug; Zushi, J, 1977 Feb; Kyoto, J

P 2056 Princeton Conf Inform Sci & Syst (5);1971 Princeton • USA
[1971] *Proceedings of the 5th Annual Princeton Conference on Information Sciences and Systems* ED: VALKENBURG VAN, M.E. ET AL. • PUBL (X 2188) Princeton Univ Dept Elect Eng & Comp Sci: Princeton
• DAT&PL 1971 Mar; Princeton, NJ, USA

P 2057 Princeton Conf Inform Sci & Syst (7);1973 Princeton • USA
[1973] *Proceedings of the 7th Annual Princeton Conference on Information Sciences and Systems* ED: PAVLIDIS, T. • PUBL (X 2188) Princeton Univ Dept Elect Eng & Comp Sci: Princeton xv+584pp
• DAT&PL 1973 Mar; Princeton, NJ, USA

P 2058 Kleene Symp;1978 Madison • USA
[1980] *The Kleene Symposium* ED: BARWISE, K.J. & KEISLER, H.J. & KUNEN, K. • SER (S 3303) Stud Logic Found Math 101 • PUBL (X 0809) North Holland: Amsterdam xx+425pp
• DAT&PL 1978 Jun; Madison, WI, USA • ISBN 0-444-85345-6, LC-No 79-20792

P 2059 Math Founds of Comput Sci (8);1979 Olomouc • CS
[1979] *Mathematical Foundations of Computer Science. Proceedings of the 8th Symposium* ED: BECVAR, J. • SER (S 3302) Lect Notes Comput Sci 74 • PUBL (X 0811) Springer: Heidelberg & New York ix+580pp
• DAT&PL 1979 Sep; Olomouc, CS • ISBN 3-540-09526-8, LC-No 79-17801

P 2064 All-Union Conf Math Log (4);1976 Kishinev • SU
[1976] *4 Vsesoyuznaya Konferentsiya po Matematicheskoj Logike. Tezitsy Doklady i Soobshcheniya (4th All-Union Conference on Mathematical Logic)* ED: KUZNETSOV, A.V.
• PUBL (X 2741) Shtiintsa: Kishinev 170pp
• DAT&PL 1976 -?-; Kishinev, SU • LC-No 78-410667

P 2080 Conf Math Log;1970 London • GB
[1972] *Conference on Mathematical Logic - London '70* ED: HODGES, W. • SER (S 3301) Lect Notes Math 255 • PUBL (X 0811) Springer: Heidelberg & New York viii+351pp
• DAT&PL 1970 Aug; London, GB • ISBN 3-540-05744-7, LC-No 70-189457

P 2112 Contrib to Universal Algeb;1975 Szeged • H
[1977] *Contributions to Universal Algebra* ED: CSAKANY, B. & SCHMIDT, J. • SER (S 3312) Coll Math Soc Janos Bolyai 17
• PUBL (X 0809) North Holland: Amsterdam 607pp • ALT PUBL (X 3725) Bolyai Janos Mat Tars: Budapest
• DAT&PL 1975 Aug; Szeged, H • ISBN 963-8021-01-2, ISBN 0-7204-0725-7, LC-No 79-350398

P 2116 Probl in Log & Ontology;1973 Salzburg • A
[1977] *Problems in Logic and Ontology. Internationales Forschungszentrum Salzburg. Forschungsgespraeche.* ED: MORSCHER, E. & CZERMAK, J. & WEINGARTNER, P. • PUBL (X 2596) Akad Druck-& Verlagsanstalt: Graz 310pp
• DAT&PL 1973 Sep; Salzburg, A • ISBN 3-201-01021-9, LC-No 80-487240
• REM Reprint of Vol. 3/3*171-343 of J0122

P 2129 Rec Th & Comput Complex (CIME);1979 Bressanone • I
[1980] *Proceedings of the CIME Summer School on Recursion Theory and Computational Complexity* ED: LOLLI, G. • PUBL Liguori Ed: Napoli 236pp
• DAT&PL 1979 Jun; Bressanone, I • LC-No 84-111950, ISBN 88-207-0913-9

P 2153 Logic Colloq;1983 Aachen • D
[1984] *Proceedings of the Logic Colloquium, Part 1, 2* ED: MUELLER, GERT H. & RICHTER, M.M. (V1). BOERGER, E. & OBERSCHELP, W. & RICHTER, M.M. & SCHINZEL, B. & THOMAS, W. (V2) • SER (S 3301) Lect Notes Math 1103,1104
• PUBL (X 0811) Springer: Heidelberg & New York 2 Vols: viii+484pp,viii+475pp
• DAT&PL 1983 Jul; Aachen, D • ISBN 3-540-13900-1 (V1), ISBN 3-540-13901-X (V2), LC-No 84-26704
• REM Vol 1: Models and Sets. Vol 2: Computation and Proof Theory.

P 2167 Orders: Descr & Roles;1982 L'Arbresle • F
[1984] *Orders: Description and Roles in Set Theory, Lattices, Ordered Groups, Topology, Theory of Models and Relations, Combinatorics, Effectiveness, Social Sciences. Proceedings of a Conference on Ordered Sets and Their Applications* ED: POUZET, M. & RICHARD, D. • SER (S 3358) Ann Discrete Math 23, North Holland Mathematics Studies 99 • PUBL (X 0809) North Holland: Amsterdam xxviii+548pp
• DAT&PL 1982 Jul; L'Arbresle, F • ISBN 0-444-87601-4, LC-No 84-13749

P 2185 Symp on n-ary Structures;1982 Skopje • YU
[1982] *Proceedings of the Symposium on n-ary Structures* ED: POPOV, B. & CUPONA, G. & TRPENOVSKI, B. • PUBL Makedon Akad Nauk: Skopje x+289pp
• DAT&PL 1982 Jan; Skopje, YU

P 2251 Mat Log & Primen;1960 Stanford • USA
[1965] *Matematicheskaya Logika i Ee Primeneniya: Sbornik Statei (Mathematical Logic and Its Applications. Logic, Methodology and Philosophy of Science)* ED: MAL'TSEV, A.I. & NAGEL, E. & SUPPES, P. & TARSKI, A. • PUBL (**X** 0885) Mir: Moskva 341pp
• DAT&PL 1960 Aug; Stanford, CA, USA
• TRANSL OF [1962] (**P** 0612) Int Congr Log, Meth & Phil of Sci (1,Proc);1960 Stanford
• REM Only Parts of P0612 are translated.

P 2276 Conf Algeb Th Automata;1973 Szeged • H
[1973] *Proceedings of a Conference on the Algebraic Theory of Automata* PUBL (**X** 3725) Bolyai Janos Mat Tars: Budapest 80pp
• DAT&PL 1973 Aug; Szeged, H

P 2279 Carib Conf on Combin & Comput;1981 Bridgetown • USA
[1981] *Proceedings of the 3rd Caribbean Conference on Combinatorics and Computing* ED: CADOGAN, C.C. • PUBL Univ West Indies: Bridgetown vi+189pp
• DAT&PL 1981 Jan; Bridgetown, USA

P 2334 Probl Rechnerarchitektur;1980 Altenberg • DDR
[1981] *Probleme der Rechnerarchitektur* SER Weiterbildungszentrum fuer Math Kybern & Rechentechn, Inf.verarb. 51 • PUBL TU Dresden: Dresden 82pp
• DAT&PL 1980 Dec; Altenberg, DDR

P 2342 Symp Rek Kombin;1983 Muenster • D
[1984] *Logic and Machines: Decision Problems and Complexity. Proceedings of the Symposium Rekursive Kombinatorik* ED: BOERGER, E. & HASENJAEGER, G. & ROEDDING, D. • SER (**S** 3302) Lect Notes Comput Sci 171
• PUBL (**X** 0811) Springer: Heidelberg & New York vi+456pp
• DAT&PL 1983 May; Muenster, D • ISBN 3-540-13331-3, LC-No 84-55

P 2411 Found Probab Th, Stat Inf & Stat Th Sci;1973 London ON • CDN
[1976] *Foundations of Probability Theory, Statistical Inference and Statistical Theories of Science* ED: HARPER, W.L. & HOOKER, C.A. • SER (**S** 3308) Univ Western Ontario Ser in Philos of Sci 6 • PUBL (**X** 0835) Reidel: Dordrecht 3 Vols: x+308pp,x+455pp,xii+241pp
• DAT&PL 1973 May; London,ON, CDN • ISBN 90-277-0616-6, ISBN 90-277-0617-4 (V1), ISBN 90-277-0618-2, ISBN 90-277-0619-0 (V2), ISBN 90-277-0620-4, ISBN 90-277-0621-2 (V3), LC-No 75-34354 (V1), LC-No 75-38667 (V2), LC-No 75-33879 (V3), ISBN 90-277-0614-X (V1-V3)
• REM Vol.1: Foundations and Philosophy of Epistemic Applications of Probability Theory. Vol.2: Foundations and Philosophy of Statistical Inference. Vol.3: Foundations and Philosophy of Statistical Theories in the Physical Sciences.

P 2534 Konf Molod Special Vychisl (1);1969 Erevan • SU
[1969] *Trudy 1 Konferentsiya Molodych Spetsialistov Vychisleniya (Proceedings of the 1st Conference of Young Computer Specialists)* PUBL (**X** 2225) Akad Nauk Armyan SSR : Erevan
• DAT&PL 1969 Apr; Erevan, SU

P 2539 Frege Konferenz (1);1979 Jena • DDR
[1979] *"Begriffsschrift". Jenaer Frege-Konferenz* ED: BOLCK, F. • PUBL (**X** 2211) Schiller Univ: Jena iii+548pp
• DAT&PL 1979 May; Jena, DDR

P 2552 Conf Finite Algeb & Multi-Val Log;1979 Szeged • H
[1981] *Proceedings of the Conference on Finite Algebra and Multiple-Valued Logic* ED: CSAKANY, B. & ROSENBERG, J.G.
• SER (**S** 3312) Coll Math Soc Janos Bolyai 28 • PUBL (**X** 0809) North Holland: Amsterdam 880pp • ALT PUBL (**X** 3725) Bolyai Janos Mat Tars: Budapest
• DAT&PL 1979 Aug; Szeged, H • ISBN 0-444-85439-8, LC-No 81-214217

P 2553 All-Union Algeb Conf (14);1977 Novosibirsk • SU
[1977] *14 Vsesoyuznyj Algebraicheskij Kollokviyum (14th All-Union Algebraic Conference)*
• DAT&PL 1977 -?-; Novosibirsk, SU

P 2558 All-Union Conf Math Log (5) (Mal'tsev);1979 Novosibirsk • SU
[1979] *5 Vsesoyuznaya Konferentsiya po Matematicheskoj Logike. Tezitsy Doklady i Soobshcheniya (5th All-Union Conference on Mathematical Logic. Dedicated to the 70th Anniversary of the Academician A.I. Mal'tsev)*
• DAT&PL 1979 -?-; Novosibirsk, SU

P 2564 All-Union Algeb Conf (15);1979 Novosibirsk • SU
[1979] *15 Vsesoyuznyj Algebraicheskij Kollokvium (15th All-Union Algebraic Conference)*
• DAT&PL 1979 -?-; Novosibirsk, SU

P 2572 Material Respub Konf Molod Uchen;1976 Alma Ata • SU
[1976] *Materialy Respublikanskoj Konferentsij Molodykh Uchenykh. (Materials of the Republican Conference of Young Scientists)* PUBL (**X** 2443) Nauka: Alma-Ata
• DAT&PL 1976 -?-; Alma-Ata, SU

P 2585 All-Union Conf Math Log (2);1972 Moskva • SU
[1972] *2 Vsesoyuznaya Konferentsiya po Matematicheskoj Logike. Tezitsy Doklady i Soobshcheniya (On Relative Recursiveness and Computability. 2nd All-Union Conference on Mathematical Logic)*
• DAT&PL 1972 -?-; Moskva, SU

P 2586 All-Union Algeb Conf (11);1971 Kishinev • SU
[1971] *11 Vsesoyuznyj Algebraicheskij Kollokvium. Rezyume Soobshchenij i Dokladov (11th All-Union Algebraic Conference. Abstracts of Reports and Papers)* PUBL Redaktsionno-Izdat Otdel Akad Nauk Moldav SSR: Kishinev 354pp
• DAT&PL 1971 -?-; Kishinev, SU • LC-No 72-308891

P 2588 FCT'77 Fund of Comput Th;1977 Poznan • PL
[1977] *Fundamentals of Computation Theory. Proceedings of the International FCT '77 - Conference* ED: KARPINSKI, M.
• SER (**S** 3302) Lect Notes Comput Sci 56 • PUBL (**X** 0811) Springer: Heidelberg & New York 542pp
• DAT&PL 1977 Sep; Poznan-Kornik, PL • ISBN 3-540-08442-8, LC-No 77-14022

P 2591 All-Union Symp Artif Intel & Autom of Invest Math;1978 Kiev • SU
[1978] *Vsesoyuznyj Simpozium "Iskusstvennyj Intellekt i Avtomatizatsiya Issledovanij v Matematike. Tezitsy Dokladov i Soobshchenij (Proceedings of the All Union Symposium on Artificial Intelligence and Automation of Investigations in Mathematics. * Abstracts of Reports and Communications)* PUBL (**X** 2522) Akad Nauk Inst Kibernet: Kiev 115pp
• DAT&PL 1978 Nov; Kiev, SU • LC-No 80-450294

P 2597 ACM Symp Th of Comput (8);1976 Hershey • USA
[1976] *Proceedings of the 8th Annual ACM Symposium on Theory of Computing (Association for Computing Machinery)* PUBL (**X** 2205) ACM: New York iv+246pp
• DAT&PL 1976 May; Hershey, PA, USA • LC-No 82-642181

Source Index

P 2598 Semin Probab (11);1975/76 Strasbourg • F
[1977] *Seminaire de Probabilites XI. Universite de Strasbourg.* ED: DELLACHERIE, C. & MEYER, P.A. & WEIL, M. • SER (S 3301) Lect Notes Math 581 • PUBL (X 0811) Springer: Heidelberg & New York v+573pp
• DAT&PL 1975 -?-; Strasbourg, F • ISBN 3-540-08145-3, LC-No 67-29618

P 2599 CAAP'78 Arbres en Algeb & Progr (3);1978 Lille • F
[1978] *Les Arbres en Algebre et en Programmation. 3eme Colloquium* PUBL Universite de Lille I: Lille ii+253pp
• DAT&PL 1978 Feb; Lille, F

P 2614 Open Days in Model Th & Set Th;1981 Jadwisin • PL
[1981] *Proceedings of the International Conference "Open Days in Model Theory and Set Theory"* ED: GUZICKI, W. & MAREK, W. & PELC, A. & RAUSZER, C. • PUBL Selbstverlag 333pp
• DAT&PL 1981 Sep; Jadwisin, PL
• REM Available from John Derrick, School of Mathematics, University of Leeds, GB, or from Cecylia Rauszer, Institute of Mathematics, University of Warsaw, PL

P 2615 Scand Logic Symp (5);1979 Aalborg • DK
[1979] *Proceedings of the 5th Scandinavian Logic Symposium* ED: JENSEN, F.V. & MAYOH, B.H. & MOLLER, K.K. • PUBL (X 2646) Aalborg Univ Pr: Aalborg vii+361pp
• DAT&PL 1979 Jan; Aalborg, DK • ISBN 87-7307-037-8, LC-No 80-464603

P 2623 Worksh Extended Model Th;1980 Berlin • DDR
[1981] *Workshop on Extended Model Theory* ED: HERRE, H. • SER (J 3437) Rep, Akad Wiss DDR, Inst Math 1981/3 • PUBL (X 2655) Akad Wiss DDR Inst Math: Berlin ii+160pp
• DAT&PL 1980 Nov; Berlin, DDR

P 2625 Model Th of Algeb & Arithm;1979 Karpacz • PL
[1980] *Model Theory of Algebra and Arithmetic. Proceedings of the Conference on Applications of Logic to Algebra and Arithmetic* ED: PACHOLSKI, L. & WIERZEJEWSKI, J. & WILKIE, A.J. • SER (S 3301) Lect Notes Math 834 • PUBL (X 0811) Springer: Heidelberg & New York vi+410pp
• DAT&PL 1979 Sep; Karpacz, PL • ISBN 3-540-10269-8, LC-No 82-131180

P 2627 Logic Colloq;1978 Mons • B
[1979] *Logic Colloquium '78* ED: BOFFA, M. & DALEN VAN, D. & MCALOON, K. • SER (S 3303) Stud Logic Found Math 97 • PUBL (X 0809) North Holland: Amsterdam x+434pp
• DAT&PL 1978 Aug; Mons, B • LC-No 79-21152

P 2628 Log Year;1979/80 Storrs • USA
[1981] *Logic Year 1979-80. The University of Connecticut* ED: LERMAN, M. & SCHMERL, J.H. & SOARE, R.I. • SER (S 3301) Lect Notes Math 859 • PUBL (X 0811) Springer: Heidelberg & New York viii+326pp
• DAT&PL 1979 Nov; Storrs, CT, USA • ISBN 3-540-10708-8, LC-No 81-5628

P 2633 Autom Deduct (7);1984 Napa • USA
[1984] *7th International Conference on Automated Deduction* ED: SHOSTAK, R.E. • SER (S 3302) Lect Notes Comput Sci 170 • PUBL (X 0811) Springer: Heidelberg & New York vi+508pp
• DAT&PL 1984 May; Napa, CA, USA • ISBN 3-540-96022-8, LC-No 84-5441

P 2634 Word Problems II;1976 Oxford • GB
[1980] *Word Problems Vol 2. The Oxford Book* ED: ADYAN, S.I. & BOONE, W.W. & HIGMAN, G. • SER (S 3303) Stud Logic Found Math 95 • PUBL (X 0809) North Holland: Amsterdam x+578pp
• DAT&PL 1976 Jun; Oxford, GB • ISBN 0-444-85343-X, LC-No 79-15276

P 2740 SE Asian Graph Th Colloq (1);1983 Singapore • SGP
[1984] *Graph Theory. Proceedings of the 1st Southeast Asian Graph Theory Colloquium* ED: KOH, K.M. & YAP, H.P. • SER (S 3301) Lect Notes Math 1073 • PUBL (X 0811) Springer: Heidelberg & New York xiii+335pp
• DAT&PL 1983 May; Singapore, SGP • ISBN 3-540-13368-2, LC-No 84-13997

P 2894 All-Union Conf Game Th (2);1971 Vil'nyus • SU
[1973] *Uspekhi Teorij Igr. Trudy 2 Vsesoyuznoj Konferentsij po Teorij Igr (Advances in Game Theory. Proceedings of the 2nd All-Union Conference)* ED: VILKAS, E. • PUBL Izdatel'stvo Mintis: Vil'nyus 332pp
• DAT&PL 1971 Jun; Vil'nyus, SU • LC-No 74-343644

P 2895 Algeb Th Semigroups;1976 Szeged • H
[1979] *Algebraic Theory of Semigroups. Proceedings of the 6th Algebraic Conference* ED: POLLAK, G. • SER (S 3312) Coll Math Soc Janos Bolyai 20 • PUBL (X 0809) North Holland: Amsterdam 753pp • ALT PUBL (X 3725) Bolyai Janos Mat Tars: Budapest
• DAT&PL 1976 Aug; Szeged, H • ISBN 0-444-85282-4, LC-No 79-315517

P 2898 Algor Kompl, Lern-& Erkenn-Prozess;1976 Jena • DDR
[1976] *Algorithmische Kompliziertheit, Lern- und Erkennungsprozesse. 2tes Internationales Symposium* ED: BOLCK, F. • PUBL (X 2211) Schiller Univ: Jena 132pp
• DAT&PL 1976 Oct; Jena, DDR

P 2902 Aspects Effective Algeb;1979 Clayton • AUS
[1981] *Aspects of Effective Algebra. Proceedings of a Conference at Monash University, Australia* ED: CROSSLEY, J.N. • PUBL (X 2863) Upside Down A Book: Yarra Glen x+290pp
• DAT&PL 1979 Aug; Clayton, Vic, AUS • ISBN 0-949865-01-X

P 2903 Automata, Lang & Progr (8);1981 Akko • IL
[1981] *Automata, Languages and Programming. 8th Colloquium* ED: EVEN, S. & KARIV, O. • SER (S 3302) Lect Notes Comput Sci 115 • PUBL (X 0811) Springer: Heidelberg & New York viii+552pp
• DAT&PL 1981 Jul; Akko, IL • ISBN 3-540-10843-2, LC-No 81-9053
• REM Also Abbreviated as ICALP 81

P 2904 Automata, Lang & Progr (7);1980 Noordwijkerhout • NL
[1980] *Automata, Languages and Programming. 7th Colloquium* ED: BAKKER DE, J.W. & LEEUWEN VAN, J. • SER (S 3302) Lect Notes Comput Sci 85 • PUBL (X 0811) Springer: Heidelberg & New York viii+671pp
• DAT&PL 1980 Jul; Noordwijkerhout, NL • ISBN 3-540-10003-2, LC-No 81-156919
• REM Also Abbreviated as ICALP 80

P 2905 ACM Annual Conf;1978 Washington • USA
[1978] *ACM 78. Proceedings of the 1978 Annual Conference of the Association for Computing Machinery* PUBL (**X** 2205) ACM: New York 2 Vols:xxii + 501pp;xxiii + 502-990pp
• DAT&PL 1978 Dec; Washington, DC, USA • ISBN 0-89791-000-1 (Set), LC-No 77-649466

P 2907 Burnside Groups;1977 Bielefeld • D
[1980] *Burnside Groups.* SER (**S** 3301) Lect Notes Math 806 • PUBL (**X** 0811) Springer: Heidelberg & New York ii + 274pp
• DAT&PL 1977 Jun; Bielefeld, D • ISBN 3-540-10006-7, LC-No 80-17687

P 2923 CAAP'81 Arbres en Algeb & Progr (6);1981 Genova • I
[1981] *CAAP '81. Les Arbres en Algebre et en Programmation. 6eme Colloque* ED: ASTESIANO, E. & BOEHM, C. • SER (**S** 3302) Lect Notes Comput Sci 112 • PUBL (**X** 0811) Springer: Heidelberg & New York vi + 364pp
• DAT&PL 1981 Mar; Genova, I • ISBN 3-540-10828-9, LC-No 81-8959

P 2929 Form Descr of Progr Concepts (1);1977 St.Andrews • CDN
[1978] *Formal Description of Programming Concepts. Proceedings of the IFIP Working Conference* ED: NEUHOLD, E.J. • PUBL (**X** 0809) North Holland: Amsterdam xviii + 648pp
• DAT&PL 1977 Aug; St. Andrews, NB, CDN • ISBN 0-444-85107-0, LC-No 81-455275

P 2930 Formal of Progr Concepts;1981 Peniscola • E
[1981] *Formalization of Programming Concepts.* ED: DIAZ, J. & RAMOS, I. • SER (**S** 3302) Lect Notes Comput Sci 107 • PUBL (**X** 0811) Springer: Heidelberg & New York vii + 478pp
• DAT&PL 1981 Apr; Peniscola, E • ISBN 3-540-10699-5, LC-No 81-5715

P 2935 FCT'79 Fund of Comput Th;1979 Berlin/Wendisch-Rietz • DDR
[1979] *Fundamentals of Computation Theory - FCT '79. Proceedings of the Conference on Algebraic, Arithmetic, and Categorical Methods in Computation Theory* ED: BUDACH, L. • SER Mathematical Research - Mathematische Forschung 2 • PUBL (**X** 0911) Akademie Verlag: Berlin 576pp
• DAT&PL 1979 Sep; Berlin/Wendisch-Rietz, DDR • LC-No 82-460828

P 2943 Symp Math Found Comp Sci; 1982 Diedrichshagen • DDR
[1983] *Symposium on Mathematical Foundations of Computer Science* SER (**S** 3382) Sem-ber, Humboldt-Univ Berlin, Sekt Math 52 • PUBL (**X** 2219) Humboldt-Univ Berlin: Berlin iv + 173pp
• DAT&PL 1982 Dec; Diedrichshagen, DDR • LC-No 84-214921

P 2946 Int Symp Progr (4);1980 Paris • F
[1980] *Proceedings of the 4th International Symposium on Programming* ED: ROBINET, B. • SER (**S** 3302) Lect Notes Comput Sci 83 • PUBL (**X** 0811) Springer: Heidelberg & New York viii + 341pp
• DAT&PL 1980 Apr; Paris, F • ISBN 3-540-09981-6, LC-No 80-13593

P 2948 Journ Algor;1975 Paris • F
[1976] *Journees Algorithmiques. Tenues a l'Ecole Normale Superieure, Paris* ED: NIVAT, M. & VIENNOT, G. • SER (**J** 1620) Asterisque 38-39 • PUBL (**X** 2244) Soc Math France: Paris ii + 293pp
• DAT&PL 1975 Dec; Paris, F • LC-No 77-556712

P 2950 Appl Math in Syst Th;1978 Brasov • RO
[1979] *Proceedings of the International Symposium on Applications of Mathematics in System Theory.* ED: SIMIONESCU, C.L. & BENKOE, I. • PUBL University of Brasov, Balkan Union of Mathematicians, Association of Scientists of Romania, Mathematical Society of Romania: Brasov; 2 Vols: ii + 339pp, 376pp
• DAT&PL 1978 Dec; Brasov, RO • LC-No 81-102730

P 2952 CAAP'79 Arbres en Algeb & Progr (4);1979 Lille • F
[1979] *Les Arbres en Algebre et en Programmation. 4eme Colloquium* PUBL Universite de Lille I: Lille iv + 327pp
• DAT&PL 1979 Feb; Lille, F

P 2957 Math Dev from Hilbert Probl;1974 DeKalb • USA
[1976] *Mathematical Developments Arising from Hilbert Problems* ED: BROWDER, F.E. • SER (**S** 3304) Proc Symp Pure Math 28 • PUBL (**X** 0803) Amer Math Soc: Providence xii + 628pp
• DAT&PL 1974 May; DeKalb, IL, USA • LC-No 76-20437

P 2958 Latin Amer Symp Math Log (4);1978 Santiago • RCH
[1980] *Mathematical Logic in Latin America. Proceedings of the 4th Latin American Symposium on Mathematical Logic* ED: ARRUDA, A.I. & CHUAQUI, R. & COSTA DA, N.C.A. • SER (**S** 3303) Stud Logic Found Math 99 • PUBL (**X** 0809) North Holland: Amsterdam xii + 392pp
• DAT&PL 1978 Dec; Santiago, RCH • ISBN 0-444-85402-9, LC-No 79-20797

P 2961 Mat Progr & Smezhnye Vopr;1974 Drogobych • SU
[1976] *Matematicheskoe Programmirovanie i Smezhnye Voprosy (Mathematical Programming and Related Questions. Proceedings of the 7th Winter School. Theory of Functions and Functional Analysis)* ED: MITYAGIN, B.S. • PUBL Central Ekonom-Mat Inst Akad Nauk SSSR:Moskva 161pp
• DAT&PL 1974 -?-; Drogobych, SU

P 2971 MC-25 Informatica Symp;1972 Amsterdam • NL
[1971] *MC-25 Informatica Symposium* SER (**S** 1605) Math Centr Tracts 37 • PUBL (**X** 1121) Math Centr: Amsterdam vii + 190pp
• DAT&PL 1972 Jan; Amsterdam, NL • LC-No 73-152705

P 2973 Net Th & Appl;1979 Hamburg • D
[1980] *Net Theory and Applications. Proceedings of the Advanced Course on General Net Theory of Processes and Systems* ED: BRAUER, W. • SER (**S** 3302) Lect Notes Comput Sci 84 • PUBL (**X** 0811) Springer: Heidelberg & New York xiii + 537pp
• DAT&PL 1979 Oct; Hamburg, D • ISBN 3-540-10001-6, LC-No 80-17287

P 2974 New Concepts & Tech in Parall Inform Process;1973 Capri • I
[1975] *New Concepts and Technologies in Parallel Information Processing. Proceedings of the NATO Advanced Study Institute* ED: CAIANIELLO, E.R. • SER NATO Advanced Study Institute, Series E 9 • PUBL (**X** 1317) Noordhoff: Groningen ix + 401pp
• DAT&PL 1973 Jun; Capri, I • ISBN 90-286-0553-3, LC-No 75-319021

Source Index

P 2983 Ordered Groups;1978 Boise • USA
[1980] *Ordered Groups.* ED: SMITH, J.E. & KENNY, G.O. & BALL, R.N. • SER (**S** 3310) Lect Notes Pure Appl Math 62 • PUBL (**X** 1684) Dekker: New York xi + 174pp
• DAT&PL 1978 Oct; Boise, ID, USA • ISBN 0-8247-6943-0, LC-No 80-24251

P 2989 Log of Progr; 1983 Pittsburgh • USA
[1984] *Logics of Programs* ED: CLARKE, E. & KOZEN, D.
• SER (**S** 3302) Lect Notes Comput Sci 164 • PUBL (**X** 0811) Springer: Heidelberg & New York vi + 527pp
• DAT&PL 1983 Jan; Pittsburgh, PA, USA • ISBN 3-540-12896-4, LC-No 84-3123

P 3006 Brazil Conf Math Log (3);1979 Recife • BR
[1980] *Proceedings of the 3rd Brazilian Conference on Mathematical Logic* ED: ARRUDA, A.I. & COSTA DA, N.C.A. & SETTE, A.M. • PUBL (**X** 2836) Soc Brasil Log: Sao Paulo vi + 336pp
• DAT&PL 1979 Dec; Recife, BR

P 3011 Symp Semigroups (4);1980 Yamaguchi • J
[1980] *Proceedings of the 4th Symposium on Semigroups. Semigroup Theory and Its Related Fields* ED: MURATA, K.
• PUBL Yamaguchi University: Yamaguchi ii + 131pp
• DAT&PL 1980 Oct; Yamaguchi, J

P 3013 Progr Symp;1974 Paris • F
[1974] *Programming Symposium. Proceedings. Colloque sur la Programmation* ED: ROBINET, B. • SER (**S** 3302) Lect Notes Comput Sci 19 • PUBL (**X** 0811) Springer: Heidelberg & New York v + 425pp
• DAT&PL 1974 Apr; Paris, F • ISBN 3-540-06859-7, LC-No 75-19256

P 3021 Logic Colloq;1979 Leeds • GB
[1980] *Recursion Theory: Its Generalisations and Applications. Proceedings of the Logic Colloquium '79* ED: DRAKE, F.R. & WAINER, S.S. • SER (**S** 3306) Lond Math Soc Lect Note Ser 45 • PUBL (**X** 0805) Cambridge Univ Pr: Cambridge, GB vi + 319pp
• DAT&PL 1979 Aug; Leeds, GB • ISBN 0-521-23543-X, LC-No 82-180570

P 3027 Semin Probab (9);1972/74 Strasbourg • F
[1975] *Seminaire de Probabilites. IX. 1ere Partie: Questions de Theorie des Flots. 1972/73. 2nde Partie: Exposes 1973/1974* ED: MEYER, P.A. • SER (**S** 3301) Lect Notes Math 465
• PUBL (**X** 0811) Springer: Heidelberg & New York iv + 589pp
• DAT&PL 1972 -?-; Strasbourg, F • ISBN 3-540-07178-4, LC-No 67-29618

P 3047 Santa Cruz Conf Finite Groups;1979 Santa Cruz • USA
[1980] *The Santa Cruz Conference on Finite Groups* ED: COOPERSTEIN, B. & MASON, G. • SER (**S** 3304) Proc Symp Pure Math 37 • PUBL (**X** 0803) Amer Math Soc: Providence xviii + 634pp
• DAT&PL 1979 Jun; Santa Cruz, CA, USA • ISBN 0-8218-1440-0, LC-No 80-26879

P 3049 Teor Yazykov & Metody Postroe Sist Progr • SU
[1972] *Teoriya Yazykov i Metody Postroeniya Sistem Programmirovaniya (Theory of Languages and Methods of Construction of Programming Systems. Proceedings of a Symposium)* PUBL (**X** 2522) Akad Nauk Inst Kibernet: Kiev 416pp

P 3057 CAAP'80 Arbres en Algeb & Progr (5);1980 Lille • F
[1980] *Les Arbres en Algebre et en Programmation. 5eme Colloquium* PUBL Universite de Lille I: Lille vi + 268pp
• DAT&PL 1980 Feb; Lille, F

P 3062 IEEE Symp Switch & Automata Th (14);1973 Iowa City • USA
[1973] *14th Annual IEEE Symposium on Switching and Automata Theory* PUBL (**X** 2179) IEEE: New York v + 213pp
• DAT&PL 1973 Oct; Iowa City, IA, USA • LC-No 80-646635

P 3063 Autom Deduct (5);1980 Les Arcs • F
[1980] *5th Conference on Automated Deduction* ED: BIBEL, W. & KOWALSKI, R. • SER (**S** 3302) Lect Notes Comput Sci 87
• PUBL (**X** 0811) Springer: Heidelberg & New York vii + 385pp
• DAT&PL 1980 Jul; Les Arcs, F • ISBN 3-540-10009-1, LC-No 80-18708

P 3083 FCT'83 Found of Comput Th (Sel Pap);1983 Borgholm • S
[1985] *Topics in the Theory of Computation. Selected Papers of the International Conference on Foundations of Computation Theory, FCT '83* ED: KARPINSKI, M. & LEEUWEN VAN, J.
• SER (**S** 3358) Ann Discrete Math 24, North-Holland Math Stud 102 • PUBL (**X** 0809) North Holland: Amsterdam ix + 187pp
• DAT&PL 1983 Aug; Borgholm, S • LC-No 84-21089

P 3084 Autom Theor Prov After 25 Yea;1983 Denver • USA
[1984] *Automated Theorem Proving After 25 Years. Proceedings of the Special Session on Automatic Theorem Proving, 89th Annual Meeting of the American Mathematical Society (AMS)* ED: BLEDSOE, W.W. & LOVELAND, D.W. • SER (**S** 3313) Contemp Math 29 • PUBL (**X** 0803) Amer Math Soc: Providence ix + 360pp
• DAT&PL 1983 Jan; Denver, CO, USA • ISBN 0-8218-5027-X, LC-No 84-9226

P 3090 Semant of Data Types;1984 Sophia-Antipolis • F
[1984] *Semantics of Data Types* ED: KAHN, G. & MACQUEEN, D.B. & PLOTKIN, G. • SER (**S** 3302) Lect Notes Comput Sci 173 • PUBL (**X** 0811) Springer: Heidelberg & New York vi + 391pp
• DAT&PL 1984 Jun; Sophia-Antipolis, F • ISBN 3-540-13346-1, LC-No 84-10575

P 3091 Conv Int Storica Logica;1982 San Gimignano • I
[1983] *Atti del Convegno Internazionale di Storica della Logica* ED: ABRUSCI, V.M. & CASARI, E. & MUGNAI, M.
• PUBL Cooperativa Libraria, Universitaria Ed: Bologna x + 401pp
• DAT&PL 1982 Dec; San Gimignano, I • LC-No 84-181758

P 3092 Congr Naz Logica;1979 Montecatini Terme • I
[1981] *Atti del Congresso Nazionale di Logica* ED: BERNINI, S.
• PUBL (**X** 1732) Bibliopolis: Napoli 735pp
• DAT&PL 1979 Oct; Montecatini Terme, I • LC-No 81-198713

P 3117 All-Union Conf Autom Contr-Eng Cybern (4);1968 Tbilisi • SU
[1972] *Avtomaty, Gibridnye Upravlyayushchie Mashiny. Trudy 4 Vsesoyuznogo Soveshchaniya po Avtomaticheskomu Upravleniyu (Tekhnicheskoj Kibernetike). Tom 3 (Automata, Hybrid and Control Machines. Proceedings of the 4th All-Union Conference on Automatic Control-Engineering Cybernetics. Vol.3)* ED: TRAPEZNIKOV, V.A. & GAVRILOV, M.A. & AJZERMAN, M.A. • PUBL (**X** 2027) Nauka: Moskva 268pp
• DAT&PL 1968 Sep; Tbilisi, SU • LC-No 72-338109

P 3146 Constr Math;1980 Las Cruces • USA
[1981] *Constructive Mathematics. Proceedings of the New Mexico State University Conference* ED: RICHMAN, F. • SER (**S** 3301) Lect Notes Math 873 • PUBL (**X** 0811) Springer: Heidelberg & New York vii+347pp
• DAT&PL 1980 Aug; Las Cruces, NM, USA • ISBN 3-540-10850-5, LC-No 81-9345

P 3148 Graphth Konzepte Inf (3);1977 Linz • A
[1978] *Datenstrukturen, Graphen, Algorithmen. Ergebnisse des Workshops WG 77. 3. Fachtagung ueber Graphentheoretische Konzepte der Informatik* ED: MUEHLBACHER, J. • SER (**J** 3359) Appl Comp Sci, Ber Prakt Inf 8 • PUBL (**X** 3223) Hanser: Muenchen 364pp
• DAT&PL 1977 Jun; Linz, A • ISBN 3-446-12526-4, LC-No 78-370096

P 3164 Fct Anal, Num Anal & Optimization (Unger);1979 Bonn • D
[1980] *Functional Analysis, Numerical Analysis and Optimization. Special Topics of Applied Mathematics. Proceedings of the Seminar held at the Gesellschaft fuer Mathematik und Datenverarbeitung (GMD)* ED: FREHSE, J. & PALLASCHKE, D. & TROTTENBERG, U. • PUBL (**X** 0809) North Holland: Amsterdam viii+242pp
• DAT&PL 1979 Oct; Bonn, D • ISBN 0-444-86035-5, LC-No 80-17794

P 3165 FCT'81 Fund of Comput Th;1981 Szeged • H
[1981] *Fundamentals of Computation Theory. Proceedings of the 1981 International FCT-Conference* ED: GECSEG, F. • SER (**S** 3302) Lect Notes Comput Sci 117 • PUBL (**X** 0811) Springer: Heidelberg & New York xi+471pp
• DAT&PL 1981 Aug; Szeged, H • ISBN 3-540-10854-8, LC-No 81-13533

P 3182 Int Comput Symp;1977 Liege • B
[1977] *International Computing Symposium 1977* ED: MORLET, E. & RIBBENS, D. • PUBL (**X** 0809) North Holland: Amsterdam xi+610pp
• DAT&PL 1977 Apr; Liege, B • ISBN 0-7204-0741-9, LC-No 77-10016

P 3196 Kompl von Entscheid Probl;1973/74 Zuerich • CH
[1976] *Komplexitaet von Entscheidungsproblemen. Ein Seminar* ED: SPECKER, E. & STRASSEN, V. • SER (**S** 3302) Lect Notes Comput Sci 43 • PUBL (**X** 0811) Springer: Heidelberg & New York ii+217pp
• DAT&PL 1973 -?-; Zuerich, CH • ISBN 3-540-07805-3, LC-No 76-25088

P 3198 Incont Compl Calc, Cod & Ling Form;1975 Napoli • I
[1976] *Atti dell' Incontro su Complessita die Calcolo, Codici e Linguaggi Formali.*
• DAT&PL 1975 -?-; Napoli, I

P 3199 CAAP'77 Arbres en Algeb & Progr (2);1977 Lille • F
[1977] *Les Arbres en Algebre et en Programmation. 2eme Colloquium* PUBL Universite de Lille I: Lille 273pp
• DAT&PL 1977 Feb; Lille, F

P 3201 Logic Symposia;1979/80 Hakone • J
[1981] *Logic Symposia Hakone 1979,1980* ED: MUELLER, GERT H. & TAKEUTI, G. & TUGUE, T. • SER (**S** 3301) Lect Notes Math 891 • PUBL (**X** 0811) Springer: Heidelberg & New York xi+394pp
• DAT&PL 1979 Mar; Hakone, J, 1980 Feb; Hakone, J • ISBN 3-540-11161-1, LC-No 81-18424

P 3208 Mat & Mat Obrazov (3);1974 Burgas • BG
[1976] *Matematika i Matematichesko Obrazovanie. Doklady na 3a Proletna Konferentsiya na B"lgarskoto Matematichesko Druzhestvo (Mathematics and Mathematical Education. Proceedings of the 3rd Spring Conference of the Bulgarian Mathematical Society)* ED: SENDOV, B. & VACHOV, D. • PUBL (**X** 2237) Publ Bulg Acad Sci: Sofia 322pp
• DAT&PL 1974 Apr; Burgas, BG • LC-No 76-512734

P 3210 Math Founds of Comput Sci (9);1980 Rydzyna • PL
[1980] *Mathematical Foundations of Computer Science. Proceedings of the 9th Symposium* ED: DEMBINSKI, P. • SER (**S** 3302) Lect Notes Comput Sci 88 • PUBL (**X** 0811) Springer: Heidelberg & New York viii+723pp
• DAT&PL 1980 Sep; Rydzyna, PL • ISBN 3-540-10027-X, LC-No 80-20087

P 3212 Math Models in Comput Systs;1981 Budapest • H
[1981] *Mathematical Models in Computer Systems. Proceedings of the 3rd Hungarian Computer Science Conference* ED: ARATO, M. & VARGA, L. • PUBL (**X** 0928) Akad Kiado: Budapest 371pp
• DAT&PL 1981 Jan; Budapest, H • LC-No 82-108964

P 3236 Math Seminar;1980 Singapore • SGP
[1980] *Proceedings of the Mathematical Seminar* ED: CHENG, K.N. & CHONG, C.T. • PUBL (**X** 1941) Nanyang Univ Publ: Singapore iii+54pp
• DAT&PL 1980 Jun; Singapore, SGP

P 3238 Conf Theoret Comput Sci;1977 Waterloo ON • CDN
[1977] *Proceedings of a Conference on Theoretical Computer Science* PUBL University of Waterloo, Computer Science Department: Waterloo iv+283pp
• DAT&PL 1977 Aug; Waterloo, ON, CDN

P 3243 Manitoba Conf Num Math (4);1974 Winnipeg • CDN
[1975] *Proceedings of the 4th Manitoba Conference on Numerical Mathematics* ED: HARTNELL, B.L. & WILLIAMS, H.C. • SER (**S** 3726) Congressus Numerantium 12 • PUBL (**X** 2420) Utilitas Mathematica Publ: Winnipeg 410pp
• DAT&PL 1974 Oct; Winnipeg, MB, CDN • LC-No 79-302019

P 3244 Hawaii Int Conf Syst Sci (6);1973 Honolulu • USA
[1973] *Proceedings of the 6th Hawaii International Conference on System Sciences* ED: LEW, A. • PUBL (**X** 1777) Western Periodicals: Hollywood xx+533pp
• DAT&PL 1973 Jan; Honolulu, HI, USA • LC-No 72-180444

P 3247 Transform de Progr;1978 Paris • F
[1978] *Transformations de Programmes. Actes du 3eme Colloque International sur la Programmation* ED: ROBINET, B. • PUBL (**X** 1422) Bordas: Paris xxii+426pp
• DAT&PL 1978 Mar; Paris, F • LC-No 79-371673

P 3269 Set Th Found Math (Kurepa);1977 Beograd • YU
[1977] *Set Theory. Foundations of Mathematics* SER (**J** 2887) Zbor Radova, NS 2(10) • PUBL (**X** 3727) Beograd Mat Inst: Belgrade 152pp
• DAT&PL 1977 Aug; Beograd, YU • LC-No 79-373865

P 3278 Topics in Algeb;1978 Canberra • AUS
[1978] *Topics in Algebra. Proceedings of the 18th Summer Research Institute of the Australian Mathematical Society* ED: NEWMAN, M.F. • SER (S 3301) Lect Notes Math 697 • PUBL (**X** 0811) Springer: Heidelberg & New York ix+229pp
• DAT&PL 1978 Jan; Canberra, ACT, AUS • ISBN 3-540-09103-3, LC-No 78-20866

P 3299 Progr Kiso Riron, Algor Okeru Shomei Ron;1973/74 Kyoto • J
[1975] *"Program no Kiso Riron" Kenkyu Shugo Oyobi Tanki Kyodo Kenkyu "Algorithm ni Okeru Shomei Ron". Hokoku Shu (The 3rd Symposium on Basic Theory of Programs and the Workshop for Proof Theory about Algorithms. Proceedings)* ED: IGARASHI, S. • SER (S 2308) Symp Kyoto Univ Res Inst Math Sci (RIMS) 236 • PUBL (**X** 2441) Kyoto Univ Res Inst Math Sci: Kyoto 215pp
• DAT&PL 1973 May; Kyoto, J, 1974 Nov; Kyoto, J

P 3342 Rec Th Week;1984 Oberwolfach • D
[1985] *Recursion Theory Week* ED: EBBINGHAUS, H.-D. & MUELLER, GERT H. & SACKS, G.E. • SER (S 3301) Lect Notes Math 1141 • PUBL (**X** 0811) Springer: Heidelberg & New York viii+418pp
• DAT&PL 1984 Apr; Oberwolfach, D • ISBN 3-540-15673-9, LC-No 85-20808

P 3349 Nat Math Conf (9);1978 Isfahan • IR
[1978] *Proceedings of the 9th National Mathematics Conference* PUBL Univ Isfahan: Isfahan vi+386pp
• DAT&PL 1978 Mar; Isfahan, IR

P 3352 Appl Gen Syst Res - Devel & Trends;1977 Binghamton • USA
[1978] *Applied General Systems Research: Recent Developments and Trends* ED: KLIR, G.J. • SER NATO Conference Series. II: Systems Science 5 • PUBL (**X** 1332) Plenum Publ: New York xvii+1001pp
• DAT&PL 1977 Aug; Binghamton, NY, USA • ISBN 0-306-32845-3, LC-No 77-26044

P 3356 Algor & Complex;1976 Pittsburgh • USA
[1976] *Algorithms and Complexity. New Directions and Recent Results. Poceedings of a Symposium on New Directions and Recent Results in Algorithms and Complexity* ED: TRAUB, J.F. • PUBL (**X** 0801) Academic Pr: New York ix+523pp
• DAT&PL 1976 Apr; Pittsburgh, PA, USA • LC-No 76-49046

P 3368 Continuous Lattices;1979 Bremen • D
[1981] *Continuous Lattices. Proceedings of the Conference on Topological and Categorical Aspects of Continuous Lattices (Workshop IV)* ED: BANASCHEWSKI, B. & HOFFMANN, R.-E. • SER (S 3301) Lect Notes Math 871 • PUBL (**X** 0811) Springer: Heidelberg & New York x+413pp
• DAT&PL 1979 Nov; Bremen, D • ISBN 3-540-10848-3, LC-No 81-9291

P 3375 Found of Comput Sci (3);1978 Amsterdam • NL
[1979] *Foundations of Computer Science. III. Part 1, 2* ED: BAKKER DE, J.W. & LEEUWEN VAN, J. • SER (S 1605) Math Centr Tracts 108,109 • PUBL (**X** 1121) Math Centr: Amsterdam 2 Vol: iii+112pp,i+164pp
• DAT&PL 1978 Aug; Amsterdam, NL • ISBN 90-6196-176-9 (V 1), ISBN 90-6196-177-7 (V 2)
• REM Vol 1: Automata, Data Structures, Complexity. Vol 2: Languages, Logic, Semantics.

P 3380 GI Jahrestag (11) & ECI Conf (3);1981 Muenchen • D
[1981] *GI - 11. Jahrestagung (Gesellschaft fuer Informatik). In Verbindung mit 3rd Conference of the European Co-Operation in Informatics (ECI)* ED: BRAUER, W. • SER Informatik-Fachberichte 50 • PUBL (**X** 0811) Springer: Heidelberg & New York xiv+617pp
• DAT&PL 1981 Oct; Muenchen, D • ISBN 3-540-10884-X, LC-No 82-122474

P 3381 GI Jahrestag (9);1979 Bonn • D
[1979] *GI - 9. Jahrestagung (Gesellschaft fuer Informatik)* ED: BOEHLING, K.H. & SPIES, P.P. • SER Informatik-Fachberichte 19 • PUBL (**X** 0811) Springer: Heidelberg & New York xii+690pp
• DAT&PL 1979 Oct; Bonn, D • ISBN 3-540-09664-7

P 3385 Inform Processing (8);1980 Tokyo & Melbourne • J
[1980] *Information Processing '80. Proceedings of IFIP Congress* ED: LAVINGTON, S. • SER IFIP Congress Series 8
• PUBL (**X** 0809) North Holland: Amsterdam xiii+1070pp
• DAT&PL 1980 Oct; Tokyo, J, 1980 Oct; Melbourne, AUS
• ISBN 0-444-86034-7, LC-No 76-462349

P 3387 Interface Comput Sci & Oper Res;1976 Amsterdam • NL
[1978] *Interfaces Between Computer Science and Operations Research* ED: LENSTRA, J.K. & RINNOY KAN, A.H.G. & EMDE BOAS VAN, P. • SER (S 1605) Math Centr Tracts 99
• PUBL (**X** 1121) Math Centr: Amsterdam ix+231pp
• DAT&PL 1976 Sep; Amsterdam, NL • ISBN 90-6196-170-X, LC-No 79-302334

P 3388 Int Comput Symp;1973 Davos • CH
[1974] *International Computing Symposium 1973* ED: GUENTHER, A. & LEVRAT, B. & LIPPS, H. • PUBL (**X** 0809) North Holland: Amsterdam 635pp • ALT PUBL (**X** 0838) Amer Elsevier: New York
• DAT&PL 1973 Sep; Davos, CH • LC-No 73-91449

P 3394 Lang Algeb (1);1973 Bonascre • F
[1978] *Langages Algebriques. Actes des Premieres Journees d'Informatique Theorique* ED: CRESTIN, J.-P. & NIVAT, M.
• PUBL Ecole Nationale Superieure de Techniques Avancees 281pp
• DAT&PL 1973 Apr; Bonascre, F • ISBN 2-7225-0424-3

P 3400 Mat & Mat Obrazov (8);1979 Sl"nchev Bryag • BG
[1979] *Matematika i Matematichesko Obrazovanie (Mathematics and Mathematical Education. Proceedings of the 8th Spring Conference of the Union of Bulgarian Mathematicians)* ED: PENKOV, B. • PUBL (**X** 2237) Publ Bulg Acad Sci: Sofia 698pp
• DAT&PL 1979 Apr; Sl"nchev Bryag (Sunny Beach), BG
• LC-No 80-643123
• REM Bulgarian. Russian and English Summaries

P 3404 Model Th & Arithm;1979/80 Paris • F
[1981] *Model Theory and Arithmetic. Comptes Rendus d'une Action Thematique Programmee du CNRS sur la Theorie des Modeles et l'Arithmetique* ED: BERLINE, C. & MCALOON, K. & RESSAYRE, J.-P. • SER (S 3301) Lect Notes Math 890 • PUBL (X 0811) Springer: Heidelberg & New York vi+306pp
• DAT&PL 1979 -?-; Paris, F • ISBN 3-540-11159-X, LC-No 82-137454

P 3406 Congr Cybern & Syst (3);1975 Bucharest • RO
[1977] *Modern Trends in Cybernetics and Systems. Vol 1-3. Proceeedings of the 3rd International Congress of Cybernetics and Systems* ED: ROSE, J. & BILCIU, C. • PUBL (X 0811) Springer: Heidelberg & New York 3 Vols: 3326pp
• DAT&PL 1975 Aug; Bucharest, RO • ISBN 3-540-08199-2
• REM Vol 1: Proceedings of Official Meetings, Symposia and Section 1

P 3411 Theor Comput Sci (3);1977 Darmstadt • D
[1977] *Theoretical Computer Science. 3rd GI Conference (Gesellschaft fuer Informatik)* ED: TZSCHACH, H. & WALDSCHMIDT, H. & WALTER, H.K.-G. • SER (S 3302) Lect Notes Comput Sci 48 • PUBL (X 0811) Springer: Heidelberg & New York vii+418pp
• DAT&PL 1977 Mar; Darmstadt, D • ISBN 3-540-08138-0, LC-No 77-3607

P 3424 SE Conf Combin, Graph Th & Comput (10);1979 Boca Raton • USA
[1979] *Proceedings of the 10th Southeastern Conference on Combinatorics, Graph Theory and Computing* ED: HOFFMAN, F. & MCCARTHY, D. & MULLIN, R.C. & STANTON, R.G. • SER (S 3726) Congressus Numerantium 23,24 • PUBL (X 2420) Utilitas Mathematica Publ: Winnipeg 2 Vols: xi+xx+939pp
• DAT&PL 1979 Apr; Boca Raton, FL, USA • ISBN 0-919628-23-0 (V1), ISBN 0-919628-24-9 (V2)

P 3426 SE Conf Combin, Graph Th & Comput (9);1978 Boca Raton • USA
[1978] *Proceedings of the 9th Southeastern Conference on Combinatorics, Graph Theory and Computing* ED: HOFFMAN, F. & MCCARTHY, D. & MULLIN, R.C. & STANTON, R.G. • SER (S 3726) Congressus Numerantium 21 • PUBL (X 2420) Utilitas Mathematica Publ: Winnipeg x+694pp
• DAT&PL 1978 Jan; Boca Raton, FL, USA • ISBN 0-919628-21-4

P 3428 Prog in Cybern & Syst Res;1978 Wien • A
[1978] *Progress in Cybernetics and Systems Research. Vol. 3: General Systems Methodology, Fuzzy Mathematics and Fuzzy Systems, Biocybernetics and Theoretical Neurobiology.* ED: TRAPPL, R. & KLIR, G.J. & RICCIARDI, L. • PUBL (X 0827) Wiley & Sons: New York xiv+674pp • ALT PUBL (X 2437) Hemisphere Publ: Washington
• DAT&PL 1978 -?-; Wien, A • LC-No 75-6641
• REM At least 4 Vols

P 3429 Math Founds of Comput Sci (10);1981 Strbske Pleso • CS
[1981] *Mathematical Foundations of Computer Science 1981. Proceedings of the 10th Symposium* ED: GRUSKA, J. & CHYTIL, M. • SER (S 3302) Lect Notes Comput Sci 118 • PUBL (X 0811) Springer: Heidelberg & New York xi+589pp
• DAT&PL 1981 Aug; Strbske Pleso, CS • ISBN 3-540-10856-4, LC-No 81-9302

P 3432 Prog in Cybern & Syst Res;1972 Wien • A
[1979] *Progress in Cybernetics and Systems Research. Vol. 5: Organization and Management, Organic Problem-Solving in Management, System Approach in Urban and Regional Planning, Computer Performance, Control and Evaluation, Computer Linguistics.* ED: TRAPPL, R. & HANIKA, F. DE P. & PICHLER, F.R. • PUBL (X 0827) Wiley & Sons: New York xv+683pp • ALT PUBL (X 2437) Hemisphere Publ: Washington
• DAT&PL 1972 -?-; Wien, A • ISBN 0-470-26553-1, LC-No 75-6641

P 3445 Semin Probab (12);1976/77 Strasbourg • F
[1978] *Semininaire de Probabilites XII. 1976/77. 1re Partie: Martingales et Integrales Stochastiques. 2de Partie: Exposes Supplementaires* ED: DELLACHERIE, C. & MEYER, P.A. & WEIL, M. • SER (S 3301) Lect Notes Math 649 • PUBL (X 0811) Springer: Heidelberg & New York viii+805pp
• DAT&PL 1976 -?-; Strasbourg, F • ISBN 3-540-08761-3, LC-No 67-29618

P 3475 Theor Comput Sci (5);1981 Karlsruhe • D
[1981] *Theoretical Computer Science. 5th GI-Conference (Gesellschaft fuer Informatik)* ED: DEUSSEN, P. • SER (S 3302) Lect Notes Comput Sci 104 • PUBL (X 0811) Springer: Heidelberg & New York vii+261pp
• DAT&PL 1981 Mar; Karlsruhe, D • ISBN 3-540-10576-X, LC-No 81-4579

P 3479 Math Stud of Inform Process;1978 Kyoto • J
[1979] *Mathematical Studies of Information Processing* ED: BLUM, E.K. & PAUL, M. & TAKASU, S. • SER (S 3302) Lect Notes Comput Sci 75 • PUBL (X 0811) Springer: Heidelberg & New York viii+629pp
• DAT&PL 1978 Aug; Kyoto, J • ISBN 3-540-09541-1, LC-No 80-494262

P 3481 GI Jahrestag (3);1973 Hamburg • D
[1973] *GI - 3. Jahrestagung (Gesellschaft fuer Informatik)* ED: BRAUER, W. • SER (S 3302) Lect Notes Comput Sci 1 • PUBL (X 0811) Springer: Heidelberg & New York xi+508pp
• DAT&PL 1973 Oct; Hamburg, D • ISBN 3-540-06473-7

P 3482 Logic & Algor (Specker);1980 Zuerich • CH
[1982] *Logic and Algorithmic. An International Symposium Held in Honour of Ernst Specker* ED: ENGELER, E. & LAEUCHLI, H. & STRASSEN, V. • SER Monogr l'Enseign Math 30 • PUBL (X 3718) Enseign Math, Univ Geneve: Geneve 392pp
• DAT&PL 1980 Feb; Zuerich, CH • LC-No 82-184564

P 3485 Getalth & Computers;1980 Amsterdam • NL
[1980] *Studieweek "Getaltheorie en Computers" (Number Theory and Computers.)* PUBL (X 1121) Math Centr: Amsterdam v+347pp
• DAT&PL 1980 Sep; Amsterdam, NL

P 3488 Theor Comput Sci (4);1979 Aachen • D
[1979] *Theoretical Computer Science. 4th GI Conference (Gesellschaft fuer Informatik)* ED: WEIHRAUCH, K. • SER (S 3302) Lect Notes Comput Sci 67 • PUBL (X 0811) Springer: Heidelberg & New York vii+324pp
• DAT&PL 1979 Mar; Aachen D • ISBN 3-540-09118-1, LC-No 79-9707

P 3497 Log of Progr;1979 Zuerich • CH
[1981] *Logic of Programs* ED: ENGELER, E. • SER (**S** 3302) Lect Notes Comput Sci 125 • PUBL (**X** 0811) Springer: Heidelberg & New York v+245pp
• DAT&PL 1979 May; Zuerich, CH • ISBN 3-540-11160-3, LC-No 82-137449

P 3525 Int Comput Symp;1975 Antibes • F
[1975] *International Computing Symposium 1975* ED: GELENBE, E. & POTIER, D. • PUBL (**X** 0809) North Holland: Amsterdam vii+266pp • ALT PUBL (**X** 0838) Amer Elsevier: New York
• DAT&PL 1975 Jun; Antibes, F • ISBN 0-7204-2839-4, LC-No 75-38988

P 3535 IEEE Symp Founds of Comput Sci (20);1979 San Juan • PRI
[1979] *20th Annual IEEE Symposium on Foundations of Computer Science* PUBL (**X** 2179) IEEE: New York vii+431pp
• DAT&PL 1979 Oct; San Juan, PRI • LC-No 80-646634
• REM IEEE Publication No. 79CH1471-26

P 3540 West-Coast Conf Combin, Graph Th & Comput;1979 Arcata • USA
[1980] *Proceedings of the West Coast Conference on Combinatorics, Graph Theory and Computing* ED: CHINN, P.Z. & McCARTHY, D. • SER (**S** 3726) Congressus Numerantium 26 • PUBL (**X** 2420) Utilitas Mathematica Publ: Winnipeg vi+322pp
• DAT&PL 1979 Sep; Arcata, CA, USA • ISBN 0-919628-26-5

P 3542 ACM Symp Th of Comput (11);1979 Atlanta • USA
[1979] *Conference Record of the 11th Annual ACM Symposium on Theory of Computing (Association for Computing Machinery)* PUBL (**X** 2205) ACM: New York vii+368pp
• DAT&PL 1979 Apr; Atlanta, GA, USA • ISBN 0-89791-003-6, LC-No 82-642181

P 3565 Symp of Th Aspects of Comput Sci (1);1984 Paris • F
[1984] *STACS 84. Symposium of Theoretical Aspects of Computer Science by AFCET (Association Francaise pour la Cybernetique Economique et Technique) and GI (Gesellschaft fuer Informatik)* ED: FONTET, M. & MEHLHORN, K. • SER (**S** 3302) Lect Notes Comput Sci 166 • PUBL (**X** 0811) Springer: Heidelberg & New York vi+338pp
• DAT&PL 1984 Apr; Paris, F • ISBN 3-540-12920-0, LC-No 84-5299

P 3572 IEEE Symp Found of Comput Sci (18);1977 Providence • USA
[1977] *18th Annual IEEE Symposium on Foundations of Computer Science* PUBL (**X** 2179) IEEE: New York v+269pp
• DAT&PL 1977 Oct; Providence, RI, USA • LC-No 80-646634

P 3575 Graphth Konzepte Inf (6);1980 Bad Honnef • D
[1981] *Graph-Theoretic Concepts in Computer Science. Proceedings of the 6th International Workshop WG 80 on Graphentheoretische Konzepte der Informatik* ED: NOLTEMEIER, H. • SER (**S** 3302) Lect Notes Comput Sci 100 • PUBL (**X** 0811) Springer: Heidelberg & New York x+403pp
• DAT&PL 1980 Jun; Bad Honnef, D • ISBN 3-540-10291-4, LC-No 81-265

P 3577 IEEE Symp Found of Comput Sci (21);1980 Syracuse • USA
[1980] *21th Annual IEEE Symposium on Foundations of Computer Science* PUBL (**X** 2179) IEEE: New York vi+421pp
• DAT&PL 1980 Oct; Syracuse, NY, USA • LC-No 80-646634

P 3578 IEEE Symp Found of Comput Sci (19);1978 Ann Arbor • USA
[1978] *19th Annual IEEE Symposium on Foundations of Computer Science* PUBL (**X** 2179) IEEE: New York v+290pp
• DAT&PL 1978 Oct; Ann Arbor, MI, USA • LC-No 80-646634

P 3607 Semin Relac Log Mat & Inform Teor;1981 Madrid • E
[1982] *Seminario sobre Relaciones entre la Logica Matematica y la Informatica Teorica* SER Ser Univ Complutense Madrid, Fac Mat • PUBL Univ Madrid Fac Mat 152pp
• DAT&PL 1981 Mar; Madrid, E

P 3608 Wechselwirk Inform Math;1979 Wien • A
[1980] *Wechselwirkungen zwischen Informatik und Mathematik. Kongressband fuer die Tagung Mathematik und Informatik* ED: DOERFLER, W. & SCHAUER, H. • SER Schriftenr Oesterreich Computer Ges 9 • PUBL (**X** 0814) Oldenbourg: Muenchen 235pp
• DAT&PL May 1979; Wien, A • ISBN 3-486-24451-5, LC-No 81-111866

P 3621 Frege Konferenz (2);1984 Schwerin • DDR
[1984] *Frege Konferenz* ED: WECHSUNG, G. • SER Mathematical Research - Mathematische Forschungen 20
• PUBL (**X** 0911) Akademie Verlag: Berlin 408pp
• DAT&PL 1984 Sep; Schwerin, DDR • LC-No 84-248695

P 3622 Int Congr Log, Meth & Phil of Sci (6,Proc);1979 Hannover • D
[1982] *Proceedings of the 6th International Congress for Logic, Methodology and Philosophy of Science* ED: COHEN, L.J. & LOS, J. & PFEIFFER, H. & PODEWSKI, K.-P. • SER (**S** 3303) Stud Logic Found Math 104 • PUBL (**X** 0809) North Holland: Amsterdam xiv+842pp
• DAT&PL 1979 Aug; Hannover, D • LC-No 80-12713

P 3623 Logic Colloq;1980 Prague • CS
[1982] *Logic Colloquium '80* ED: DALEN VAN, D. & LASCAR, D. & SMILEY, T.J. • SER (**S** 3303) Stud Logic Found Math 108 • PUBL (**X** 0809) North Holland: Amsterdam x+342pp
• ISBN 0-444-86465-2, LC-No 82-12611

P 3634 Patras Logic Symp;1980 Patras • GR
[1982] *Patras Logic Symposium* ED: METAKIDES, G. • SER (**S** 3303) Stud Logic Found Math 109 • PUBL (**X** 0809) North Holland: Amsterdam ix+391pp
• DAT&PL 1980 Aug; Patras, GR • ISBN 0-444-86476-8, LC-No 82-14107

P 3638 Brouwer Centenary Symp;1981 Noordwijkerhout • NL
[1982] *The L.E.J. Brouwer Centenary Symposium* ED: TROELSTRA, A.S. & DALEN VAN, D. • SER (**S** 3303) Stud Logic Found Math 110 • PUBL (**X** 0809) North Holland: Amsterdam x+524pp
• DAT&PL 1981 Jun; Noordwijkerhout, NL • ISBN 0-444-86494-6, LC-No 82-239358

P 3642 Colloq Math Log in Computer Sci;1978 Salgotarjan • H
[1981] *Proceedings of the Colloquium on Mathematical Logic in Computer Science* ED: DOEMALKI, B. & GERGELY, T. • SER (S 3312) Coll Math Soc Janos Bolyai 26 • PUBL (**X** 0809) North Holland: Amsterdam 758pp • ALT PUBL (**X** 3725) Bolyai Janos Mat Tars: Budapest
• DAT&PL 1978 Sep; Salgotarjan, H • ISBN 0-444-85440-1, LC-No 82-101557

P 3658 Math Founds of Comput Sci (11);1984 Prague • CS
[1984] *Mathematical Foundations of Computer Science 1984. Proceedings of the 11th Symposium* ED: CHYTIL, M.P. & KOUBEK, V. • SER (S 3302) Lect Notes Comput Sci 176
• PUBL (**X** 0811) Springer: Heidelberg & New York 581pp
• DAT&PL 1984 Sep; Prague, CS • ISBN 3-540-13372-0, LC-No 83-13980

P 3663 Konf Teor Grafov;1978 Zemplinska Shirana • SU
[1978] *Konferentsiya z Teorie Grafov (Conference on Graph Theory)*
• DAT&PL 1978 -?-; Zemplinska Shirana, SU

P 3668 Log & Founds of Math;1983 Kyoto • J
[1984] *Logic and the Foundations of Mathematics* ED: TUGUE, T. • SER (S 2308) Symp Kyoto Univ Res Inst Math Sci (RIMS) 516 • PUBL (**X** 2441) Kyoto Univ Res Inst Math Sci: Kyoto ii+195pp
• DAT&PL 1983 Oct; Kyoto, J

P 3669 SE Asian Conf on Log;1981 Singapore • SGP
[1983] *Southeast Asian Conference on Logic* ED: CHONG, CHI TAT & WICKS, M.J. • SER (S 3303) Stud Logic Found Math 111 • PUBL (**X** 0809) North Holland: Amsterdam xiv+210pp
• DAT&PL 1981 Nov; Singapore, SGP • ISBN 0-444-86706-6, LC-No 83-11458

P 3708 Herbrand Symp Logic Colloq;1981 Marseille • F
[1982] *Proceedings of the Herbrand Symposium Logic. Colloquium '81* ED: STERN, J. • SER (S 3303) Stud Logic Found Math 107 • PUBL (**X** 0809) North Holland: Amsterdam xi+384pp
• DAT&PL 1981 Jul; Marseille, F • ISBN 0-444-86417-2, LC-No 82-6433

P 3710 Logic Colloq;1982 Firenze • I
[1984] *Proceedings of the Logic Colloquium '82* ED: LOLLI, G. & LONGO, G. & MARCJA, A. • SER (S 3303) Stud Logic Found Math 112 • PUBL (**X** 0809) North Holland: Amsterdam viii+358pp
• DAT&PL 1982 Aug; Firenze, I • ISBN 0-444-86876-3, LC-No 84-1630

P 3714 ACM Symp Th of Comput (14);1982 • USA
[1982] *Symposium on Theory of Computing* PUBL (**X** 2205) ACM: New York
• LC-No 83-641095

P 3729 Algor in Modern Math & Comput Sci;1979 Urgench • SU
[1981] *Algorithms in Modern Mathematics and Computer Science* ED: ERSHOV, A.P. & KNUTH, D.E. • SER (S 3302) Lect Notes Comput Sci 122 • PUBL (**X** 0811) Springer: Heidelberg & New York xi+487pp
• DAT&PL 1979 Sep; Urgench, SU • ISBN 3-540-11157-3, LC-NO 81-18418
• TRANSL IN [1982] (**P** 3803) Algor Sovrem Mat & Prilozh;1979 Urgench

P 3738 Log of Progr;1981 Yorktown Heights • USA
[1982] *Logics of Programs. Papers Presented at the Workshop* ED: KOZEN, D. • SER (S 3302) Lect Notes Comput Sci 131
• PUBL (**X** 0811) Springer: Heidelberg & New York vi+429pp
• DAT&PL 1981 May; Yorktown Heights, NY, USA • ISBN 3-540-11212-X, LC-No 82-3219

P 3753 Penser Math;1981 Paris • F
[1982] *Penser les Mathematiques. Lectures from a Seminar on Philosophy and Mathematics* ED: GUENARD, F. & LELIEVRE, G. • SER Collection Points: Serie Sciences 29 • PUBL (**X** 1349) Seuil: Paris 277pp
• DAT&PL 1981; Paris, F • ISBN 2-02-006061-2, LC-No 82-144038

P 3761 Modern Appl Math; 1979 Bonn • D
[1982] *Modern Applied Mathematics. Optimization and Operations Research. Papers based on Lectures Presented at the Summer School on Optimization and Operations Research* ED: KORTE, B. • PUBL (**X** 0809) North Holland: Amsterdam ix+693pp
• DAT&PL 1979 Sep; Bonn, D • ISBN 0-444-86134-3, LC-No 81-9487

P 3767 Found of Softw Tech & Th Comput Sci (2);1982 Bangalore • IND
[1982] *Foundations of Software Technology and Theoretical Computer Science. Proceedings of the 2nd Conference* ED: JOSEPH, M. & SHYAMASUNDAR, R.K. • PUBL (**X** 0925) Tata Inst Fund Res: Bombay v+428pp
• DAT&PL 1982 Dec; Bangalore, IND

P 3780 Ordered Sets;1981 Banff • CDN
[1982] *Ordered Sets. Proceedings of a NATO Advanced Study Institute* ED: RIVAL, I. • SER NATO Advanced Study Inst. Series C 83 • PUBL (**X** 0835) Reidel: Dordrecht xviii+966pp
• DAT&PL 1981 Aug; Banff, CDN • ISBN 90-277-1396-0, LC-No 82-544

P 3787 Discr Math;1977 Warsaw • PL
[1982] *Discrete Mathematics* ED: KULIKOWSKI, J.L. & YABLONSKIJ, S.V. & ZHURAVLEV, JU.I. & MICHALEWICZ, M.
• SER (S 1926) Banach Cent Publ 7 • PUBL (**X** 1034) PWN: Warsaw 224pp
• DAT&PL 1977 -?-; Warsaw, PL • ISBN 83-01-01494-6, LC-No 82-207902

P 3803 Algor Sovrem Mat & Prilozh;1979 Urgench • SU
[1982] *Algoritmy v Sovremennoj Matematike i ee Prilozheniyakh. Chast' I, II (Algorithms in Modern Mathematics and Computer Science Part I, II)* ED: ERSHOV, A.P. & KNUTH, D.E. • PUBL (**X** 2652) Akad Nauk Sibirsk Otd Inst Mat: Novosibirsk 2 Vols: 364pp,316pp
• DAT&PL 1979 Sep; Urgench, SU • LC-No 84-128439
• TRANSL OF [1981] (**P** 3729) Algor in Modern Math & Comput Sci;1979 Urgench
• REM With Additions to the Original Collection

P 3808 Rect Trends in Math;1982 Reinhardsbrunn • D
[1982] *Recent Trends in Mathematics* ED: KURKE, H. & MECKE, J. & TRIEBEL, H. & THIELE, R. • SER Teubner-Texte zur Mathematik 50 • PUBL (**X** 0823) Teubner: Stuttgart 336pp
• DAT&PL 1982 Oct; Reinhardsbrunn, D • LC-No 83-168766

P 3829 Atti Incontri Log Mat (1);1982 Siena • I
[1983] *Atti Degli Incontri di Logica Matematica* ED: BERNARDI, C. • PUBL (**X** 3812) Univ Siena, Dip Mat: Siena 398pp
• DAT&PL 1982 Jan; Siena, I, 1982 Apr; Siena, I, 1982 Jun; Siena, I

P 3830 Logics of Progr & Appl;1980 Poznan • PL
[1983] *Logics of Programs and Their Applications* ED: SALWICKI, A. • SER (**S** 3302) Lect Notes Comput Sci 148 • PUBL (**X** 0811) Springer: Heidelberg & New York vi+324pp
• DAT&PL 1980 Aug; Poznan, PL • ISBN 3-540-11981-7, LC-No 83-158934

P 3831 Universal Algeb & Appl;1978 Warsaw • PL
[1982] *Universal Algebra and Applications* ED: TRACZYK, T. • SER (**S** 1926) Banach Cent Publ 9 • PUBL (**X** 1034) PWN: Warsaw 454pp
• DAT&PL 1978 Feb-Sep; Warsaw, PL • ISBN 83-01-02145-4, LC-No 83-167533

P 3835 AFCET-SMF Math Appliquees Colloq (1);1978 Palaiseau • F
[1978] *1er Colloque AFCET-SMF de Mathematiques Appliquees. Tome I, II, III. Association Francaise pour la Cybernetique Economique et Technique (AFCET) et Societe Mathematique de France (SMF)* PUBL Ecole Polytechnique: Palaiseau 3 Vols: 375pp, 323pp, 85pp
• DAT&PL 1978 -?-; Palaiseau, F

P 3836 Automata, Lang & Progr (9);1982 Aarhus • DK
[1982] *Automata, Languages and Programming. 9th Colloquium* ED: NIELSEN, M. & SCHMIDT, E.M. • SER (**S** 3302) Lect Notes Comput Sci 140 • PUBL (**X** 0811) Springer: Heidelberg & New York vii+614pp
• DAT&PL 1982 Jul; Aarhus, DK • ISBN 3-540-11576-5, LC-No 83-10430
• REM Also Abbreviated as ICALP '82

P 3841 Universal Algeb & Lattice Th (4);1982 Puebla • MEX
[1983] *Universal Algebra and Lattice Theory* ED: FREESE, R.S. & GRACIA, O.C. • SER (**S** 3301) Lect Notes Math 1004 • PUBL (**X** 0811) Springer: Heidelberg & New York vi+308pp
• DAT&PL 1982 Jan; Puebla, MEX • ISBN 3-540-12329-6, LC-No 83-14637

P 3850 Appl & Th of Petri Nets (3);1982 Varenna • I
[1983] *Applications and Theory of Petri Nets. Selected Papers from the 3rd European Workshop* ED: PAGNONI, A. & ROZENBERG, G. • SER Informatik-Fachberichte 66 • PUBL (**X** 0811) Springer: Heidelberg & New York vi+315pp
• DAT&PL 1982 Sep; Varenna, I • ISBN 3-540-12309-1, LC-No 83-6632

P 3851 Automata, Lang & Progr (10);1983 Barcelona • E
[1983] *Automata, Languages and Programming. 10th Colloquium* ED: DIAZ, J. • SER (**S** 3302) Lect Notes Comput Sci 154 • PUBL (**X** 0811) Springer: Heidelberg & New York viii+734pp
• DAT&PL 1983 Jul; Barcelona, E • ISBN 3-540-12317-2, LC-No 83-10435
• REM Also Abbreviated as ICALP 83

P 3858 Adequate Modeling of Syst;1982 Bad Honnef • D
[1983] *Adequate Modeling of Systems. Proceedings of the International Working Conference on Model Realism* ED: WEDDE, H. • PUBL (**X** 0811) Springer: Heidelberg & New York xi+336pp
• DAT&PL 1982 Apr; Bad Honnef, D • ISBN 3-540-12567-1, LC-No 83-10335

P 3862 Theor Comput Sci (6);1983 Dortmund • D
[1982] *Theoretical Computer Science. 6th GI Conference. (Gesellschaft fuer Informatik)* ED: CREMERS, A.B. & KRIEGEL, H.P. • SER (**S** 3302) Lect Notes Comput Sci 145 • PUBL (**X** 0811) Springer: Heidelberg & New York x+365pp
• DAT&PL 1983 Jan; Dortmund, D • ISBN 3-540-11973-6, LC-No 82-19673

P 3864 FCT'83 Found of Comput Th;1983 Borgholm • S
[1983] *Foundation of Computation Theory. 1983 FCT-Conference* ED: KARPINSKI, M. • SER (**S** 3302) Lect Notes Comput Sci 158 • PUBL (**X** 0811) Springer: Heidelberg & New York xi+517pp
• DAT&PL 1983 Aug; Borgholm, S • ISBN 3-540-12689-9, LC-No 83-16736

P 3866 SE Conf Combin, Graph Th & Comput (12);1981 Baton Rouge • USA
[1981] *Proceedings of the 12th Southeastern Conference on Combinatorics, Graph Theory and Computing. Vol. 1,2* ED: HOFFMAN, F. & REID, K.B. & MULLIN, R.C. & STANTON, R.G. • SER (**S** 3726) Congressus Numerantium 32,33 • PUBL (**X** 2420) Utilitas Mathematica Publ: Winnipeg 2 Vols: iv+394pp, 415pp
• DAT&PL 1981 Mar; Baton Rouge, LA, USA

P 3870 Congr Catala de Log Mat (1);1982 Barcelona • E
[1982] *1er Congres Catala de Logica Matematica. Actes* PUBL Univ Politecnica & Univ Barcelona: Barcelona 130pp
• DAT&PL 1982 -?-; Barcelona, E

P 3874 Mat & Mat Obrazov (10);1981 Sl"nchev Bryag • BG
[1981] *Matematika i Matematichesko Obrazovanie (Mathematics and Mathematical Education. 10th Spring Conference of the Union of Bulgarian Mathematicans)* PUBL (**X** 2237) Publ Bulg Acad Sci: Sofia 436pp
• DAT&PL 1981 Apr; Sl"nchev Bryag (Sunny Beach), BG
• LC-No 80-643123

P 3886 Groups-St.Andrews;1981 St.Andrews • GB
[1982] *Groups - St.Andrews 1981* ED: CAMPBELL, C.M. & ROBERTSON, E.F. • SER (**S** 3306) Lond Math Soc Lect Note Ser 71 • PUBL (**X** 0805) Cambridge Univ Pr: Cambridge, GB viii+360pp
• DAT&PL 1981 Jul; St.Andrews, GB • ISBN 0-521-28974-2, LC-No 82-4427

P 3888 Convex Anal & Optim (Ioffe);1980 London • GB
[1982] *Convex Analysis and Optimization. An Expanded Version of Papers Presented at the Colloquium Held in Honour of Alexander D.Ioffe* ED: AUBIN, J.P. & VINTER, R.B. • SER Research Notes in Math 57 • PUBL (**X** 1330) Pitman Publ: Belmont & London vii+210pp
• DAT&PL 1980 Feb; London, GB • ISBN 0-273-08547-6, LC-No 81-21060

P 3889 CAAP'83 Arbres en Algeb & Progr (8);1983 L'Aquila • I
[1983] *CAAP'83. Les Arbres en Algebre et en Programmation. 8eme Colloque* ED: AUSIELLO, G. & PROTASI, M. • SER (S 3302) Lect Notes Comput Sci 159 • PUBL (X 0811) Springer: Heidelberg & New York vi+416pp
• DAT&PL 1983 Mar; L'Aquila, I • ISBN 3-540-12727-5, LC-No 83-16909

P 3893 Found of Softw Tech & Th Comput Sci (3);1983 Bangalore • IND
[1983] *Foundations of Software Technology and Theoretical Computer Science. Proceedings of the 3rd Conference* PUBL (X 0925) Tata Inst Fund Res: Bombay 613pp
• DAT&PL 1983 Dec; Bangalore, IND

P 3897 Inform Th (3);1980 Liblice • CS
[1980] *Proceedings of the 3rd Czechoslovak-Soviet-Hungarian Seminar on Information Theory* ED: DRIML, M. & VISEK, J.A.
• PUBL (X 1226) Academia: Prague 224pp
• DAT&PL 1980 Jun; Liblice, CS

P 3906 Th Found of Progr Methodol;1981 Marktoberdorf • D
[1982] *Theoretical Foundations of Programming Methodology* ED: BROY, M. & SCHMIDT, G. • SER NATO Adv Study Inst Ser C 91 • PUBL (X 0835) Reidel: Dordrecht xiii+658pp
• DAT&PL 1981 -?-; Marktoberdorf, D • LC-No 82-12347

P 3908 Graph-Gram & Appl to Comput Sci (2);1982 Neunkirchen • D
[1983] *Graph-Grammars and Their Application to Computer Science.* ED: EHRIG, H. & NAGL, M. & ROZENBERG, G. • SER (S 3302) Lect Notes Comput Sci 153 • PUBL (X 0811) Springer: Heidelberg & New York vii+452pp
• DAT&PL 1982 Oct; Neunkirchen, D • ISBN 3-540-12310-5, LC-No 83-6677

P 4004 CAAP'82 Arbres en Algeb & Progr (7);1982 Lille • F
[1982] *Les Arbres en Algebre et en Programmation. 7eme Colloquium* PUBL Universite de Lille I: Lille iv+321pp
• DAT&PL 1982 Mar; Lille, F

P 4007 CAAP'84 Arbres en Algeb & Progr (9);1984 Bordeaux • F
[1984] *Les Arbres en Algebre et en Programmation. 9eme Colloquium* PUBL (X 0805) Cambridge Univ Pr: Cambridge, GB vi+326pp
• DAT&PL 1984 Mar; Bordeaux, F

P 4010 Found of Softw Tech & Th Comput Sci (4);1984 Bangalore • IND
[1984] *Foundations of Software Technology and Theoretical Computer Science. Proceedings of the 4th Conference* ED: JOSEPH, M. & SHYAMASUNDAR, R. • SER (S 3302) Lect Notes Comput Sci 181 • PUBL (X 0811) Springer: Heidelberg & New York viii+468pp
• DAT&PL 1984 Dec; Bangalore, IND • ISBN 3-540-13883-8, LC-No 84-23528

P 4012 Automata, Lang & Progr (11);1984 Antwerpen • B
[1984] *Automata, Languages and Programming. 11th Colloquium* ED: PAREDAENS, J. • SER (S 3302) Lect Notes Comput Sci 172 • PUBL (X 0811) Springer: Heidelberg & New York viii+527pp
• DAT&PL 1984 Jul; Antwerpen, B • ISBN 3-540-13345-3, LC-No 84-10577
• REM Also Abbreviated as ICALP '84

P 4013 Symb & Algeb Comput;1984 Cambridge • GB
[1984] *EUROSAM 84. International Symposium on Symbolic and Algebraic Computation* ED: FITCH, J. • SER (S 3302) Lect Notes Comput Sci 174 • PUBL (X 0811) Springer: Heidelberg & New York xi+396pp
• DAT&PL 1984 Jul; Cambridge, GB • ISBN 3-540-13350-X, LC-No 84-10620

P 4021 Math Progr;1979 Oberwolfach • D
[1981] *Mathematical Programming at Oberwolfach* ED: KOENIG, H. & KORTE, B. & RITTER, K. • SER Math Program Stud 14 • PUBL (X 0809) North Holland: Amsterdam viii+257pp
• DAT&PL 1979 May; Oberwolfach, D • LC-No 80-27351

P 4033 Appl of Combin;1981 Milton Keynes • GB
[1982] *Applications of Combinatorics. Proceedings of a Conference on Combinatorics and Its Applications* ED: WILSON, R.J. • SER Shiva Math Series 6 • PUBL Shiva Publishing: Nantwich ii+114pp • ALT PUBL (X 0804) Birkhaeuser: Basel
• DAT&PL 1981 Nov; Milton Keynes, GB • ISBN 0-906812-14-3, LC-No 83-200957

P 4046 Rec Th;1982 Ithaca • USA
[1985] *Recursion Theory. Proceedings of the AMS-ASL Summer Institute* ED: NERODE, A. & SHORE, R.A. • SER (S 3304) Proc Symp Pure Math 42 • PUBL (X 0803) Amer Math Soc: Providence vii+528pp
• DAT&PL 1982 Jun; Ithaca, NY, USA • ISBN 0-8218-1447-8, LC-No 84-18525

P 4047 Allerton Conf Commun, Control & Comput (15);1977 Monticello • USA
[1977] *15th Annual Allerton Conference on Communication, Control and Computing* PUBL (X 1285) Univ Ill Pr: Urbana xii+761pp
• DAT&PL 1977 Sep; Monticello, IL, USA

P 4048 Allerton Conf Commun, Control & Comput (16);1978 Monticello • USA
[1978] *16th Annual Allerton Conference on Communication, Control and Computing* PUBL (X 1285) Univ Ill Pr: Urbana xv+994pp
• DAT&PL 1978 Oct; Monticello, IL, USA

P 4049 Allerton Conf Commun, Control & Comput (17);1979 Monticello • USA
[1979] *17th Annual Allerton Conference on Communication, Control and Computing* PUBL (X 1285) Univ Ill Pr: Urbana xiv+1036pp
• DAT&PL 1979 Oct; Monticello, IL, USA

P 4050 Appl Syst & Cybern;1980 Acapulco • MEX
[1981] *Applied Systems and Cybernetics, Vol. I, II, III, IV, V, VI* ED: LASKER, G.E. • PUBL (X 0869) Pergamon Pr: Oxford 6 Vols: xxxii+3294pp
• DAT&PL 1980 Dec; Acapulco, MEX • ISBN 0-08-027196-0, LC-No 81-13765

P 4064 Prog Graph Th;1982 Waterloo • CDN
[1984] *Progress in Graph Theory*
• DAT&PL 1982 Jul; Waterloo, ON, CDN

P 4083 GTI-Worksh (1);1983 Paderborn • D
[1983] *Report on the 1st GTI-Workshop* ED: PRIESE, L.
• DAT&PL 1983 -?-; Paderborn, D

P 4084 Course on Comput Th;1984 Udine • I
[1985] *Proceedings des Unesco Kurses "Course on Computation Theory"*
• DAT&PL 1984 Sep; Udine, I

P 4108 Found of Math;1979 Kyoto • J
[1979] *Sugaku Kisoron (Foundations of Mathematics)* ED: TUGUE, T. • SER (S 2308) Symp Kyoto Univ Res Inst Math Sci (RIMS) 362 • PUBL (X 2441) Kyoto Univ Res Inst Math Sci: Kyoto ii + 185pp
• DAT&PL 1979 May; Kyoto, J

P 4109 B-Val Anal & Nonstand Anal;1978 Kyoto • J
[1978] *Boole Daisuchi no Kaisekigaku to Chojun Kaiseki (B-Valued Analysis and Nonstandard Analysis)* ED: NAMBA, K. • SER (S 2308) Symp Kyoto Univ Res Inst Math Sci (RIMS) 336 • PUBL (X 2441) Kyoto Univ Res Inst Math Sci: Kyoto ii + 149pp
• DAT&PL 1978 May; Kyoto, J
• REM All Papers in Japanese

P 4112 Th Ordered Sets & Gen Algeb;1969 Cikhaj • CS
[1969] *Proceedings of the Summer Session on the Theory of Ordered Sets and General Algebra* PUBL (X 2229) Univ Purkyne: Brno 126pp
• DAT&PL 1969 -?-; Cikhaj, CS

P 4113 Found of Math;1982 Kyoto • J
[1983] *Foundations of Mathematics* ED: SHINODA, J. • SER (S 2308) Symp Kyoto Univ Res Inst Math Sci (RIMS) 480
• PUBL (X 2441) Kyoto Univ Res Inst Math Sci: Kyoto iii + 236pp
• DAT&PL 1982 Oct; Kyoto, J

P 4117 Conf Math Log (Markov);1980 Sofia • BG
[1984] *Matematicheskaya Logika: Trudy Konferentsij po Matematicheskoj Logike, Posvyashchennoj Pamiati A.A.Markova 1903-1979 (Proceedings of a Conference on Mathematical Logic)* ED: SKORDEV, D. • PUBL Bulgar Akad Nauk 172pp
• DAT&PL 1980 Sep; Sofia, BG • LC-No 84-215513

P 4153 B-Val Anal & Nonstand Anal;1981 Kyoto • J
[1981] *Boolean-Algebra-Valued Analysis and Nonstandard Analysis* ED: NANBA, K. • SER (S 2308) Symp Kyoto Univ Res Inst Math Sci (RIMS) 441 • PUBL (X 2441) Kyoto Univ Res Inst Math Sci: Kyoto ii + 158pp
• DAT&PL 1981 May; Kyoto, J

P 4175 Algeb & Log Found Progr • DDR
[1984] *Algebraic and Logical Foundations of Programming* SER Weiterbildungszentrum fuer Math Kybern & Rechentechn, Inf.verarb. 67 • PUBL TU Dresden: Dresden

P 4178 Universal Algeb & Lattice Th;1984 Charleston • USA
[1985] *Universal Algebra and Lattice Theory* ED: COMER, S.D.
• SER (S 3301) Lect Notes Math 1149 • PUBL (X 0811) Springer: Heidelberg & New York vi + 282pp
• DAT&PL 1984 Jul; Charleston, SC, USA • ISBN 3-540-15691-7, LC-No 85-17329

P 4180 Int Congr Log, Meth & Phil of Sci (7,Pap);1983 Salzburg • A
[1985] *Foundations of Logic and Linguistics. Problems and Their Solutions. Papers from the 7th International Congress of Logic, Methodology and Philosophy of Science* ED: DORN, G. & WEINGARTNER, P. • PUBL (X 1332) Plenum Publ: New York xi + 715pp
• DAT&PL 1983 Jul; Salzburg, A • ISBN 0-306-41916-5, LC-No 84-26518

P 4195 ACM Symp Th of Comput (15); • USA
[1983] *Proceedings of the 15th ACM Symposium on Theory of Computing (Association for Computing Machinery)*
• LC-No 83-641095

P 4199 Adv in Data Base Th;1979 Toulouse • F
[1981] *Advances in Data Base Theory, Vol 1* ED: GALLAIRE, H. & MINKER, J. & NICOLAS, J.M. • PUBL (X 1332) Plenum Publ: New York xiii + 432pp
• DAT&PL 1979 Dec; Toulouse, F • ISBN 0-306-40629-2,LC-No 81-116229

P 4214 Anal & Design of Alg in Combin Optmiz;1979 Udine • I
[1981] *Analysis and Design of Algorithms in Combinatorial Optimization* ED: AUSIELLO, G. & LUCERTINI, M. • PUBL (X 0902) Springer: Wien vi + 209pp
• DAT&PL 1979 Sep; Udine, I • ISBN 3-211-81626-7

P 4235 IEEE Symp Found of Comp Sci (22);1981 Nashville • USA
[1981] *22nd Annual IEEE Symposium on Foundations of Computer Science* PUBL (X 2179) IEEE: New York x + 429pp
• DAT&PL 1981 Oct; Nashville, TN, USA • LC-No 80-646634

P 4238 IEEE Symp Found of Comput Sci (24);1983 Tuscon • USA
[1983] *24th Annual IEEE Symposium on Foundations of Computer Science* PUBL (X 2179) IEEE: New York
• DAT&PL 1983 Nov; Tuscon, AZ, USA • LC-No 80-646634

P 4241 Symp of Th Aspects of Comput Sci (2);1985 Saarbruecken • D
[1985] *STACS 85. Proceedings of the 2nd Annual Symposium on Theoretical Aspects of Computer Science* ED: MEHLHORN, K. • SER (S 3302) Lect Notes Comput Sci 182 • PUBL (X 0811) Springer: Heidelberg & New York vii + 374pp
• DAT&PL 1985 Jan; Saarbruecken, D • ISBN 3-540-13912-5, LC-No 84-26882

P 4244 Rewriting Techn & Appl (1);1985 Dijon • F
[1985] *Rewriting Techniques and Applications. Papers from the 1st Conference* ED: JOUANNAUD, J.-P. • SER (S 3302) Lect Notes Comput Sci 202 • PUBL (X 0811) Springer: Heidelberg & New York vi + 441pp
• DAT&PL 1985 May; Dijon, F • ISBN 3-540-15976-2, LC-No 85-22164

P 4258 Found of Softw Tech & Th Comput Sci (1);1981 Bangalore • IND
[1981] *Foundations of Software Technology and Theoretical Computer Science. Proceedings of the 1st Conference* ED: SHYAMAZUNDER, R.K. & JOSEPH, M. • PUBL (X 0925) Tata Inst Fund Res: Bombay viii + 311pp
• DAT&PL 1981 Dec; Bangalore, IND

P 4263 Diskr Math;1974 Berlin • D
[1975] *Diskrete Mathematik*
• DAT&PL 1974 -?-; Berlin

P 4266 Form Lang Th;1979 Santa Barbara • USA
[1980] *Formal Language Theory. Perspectives and Open Problems* ED: BOOK, R.V. • PUBL (X 0801) Academic Pr: New York xiii + 454pp
• DAT&PL 1979 Dec; Santa Barbara, CA, USA • ISBN 0-12-115350-9, LC-No 80-22435

P 4271 E.Bishop-Reflection on Him & Research;1983 San Diego • USA
[1985] *Errett Bishop: Reflections on Him and His Research* ED: ROSENBLATT, M. • SER (**S** 3313) Contemp Math 39 • PUBL (**X** 0803) Amer Math Soc: Providence xvii+91pp • DAT&PL 1983 Sep; San Diego, CA, USA • ISBN 0-8218-5040-7, LC-No 85-752

P 4310 Easter Conf on Model Th (3);1985 Gross Koeris • DDR
[1985] *Proceedings of the 3rd Easter Conference on Model Theory* SER Seminarbericht 70 • PUBL (**X** 2219) Humboldt-Univ Berlin: Berlin iv+226pp • DAT&PL 1985 Apr; Gross Koeris, DDR

P 4313 Int Congr Math (II,14);1983 Warsaw • PL
[1984] *Proceedings of the International Congress of Mathematicians, Vol 1,2* ED: CIESIELSKI, Z. & OLECH, C. • PUBL (**X** 1034) PWN: Warsaw lxii+1730pp • ALT PUBL (**X** 0809) North Holland: Amsterdam • DAT&PL 1983 Aug; Warsaw, PL • ISBN 83-01-05523-5, LC-No 84-13788

P 4357 Summer Sess Th Ordered Sets & Gen Alg;1969 Cikhaj • YU
[1969] *Summer Session on the Theory of Ordered Sets and General Algebra* PUBL (**X** 2229) Univ Purkyne: Brno 126pp • DAT&PL 1969 Sep; Cikhaj, YU

P 4359 Groups-Korea;1983 Kyoungiu • ROK
[1984] *Groups - Korea 1983* ED: KIM, A.C. & NEUMANN, B.H. • SER (**S** 3301) Lect Notes Math 1098 • PUBL (**X** 0811) Springer: Heidelberg & New York viii+183pp • DAT&PL 1983 Aug; Kyoungiu, ROK • LC-No 84-26864, ISBN 3-540-13890-0

P 4367 INFO'84;1984 Dresden • DDR
INFO '84. SER Tagungsberichte 2 • DAT&PL 1984 Feb; Dresden, DDR

P 4384 Combin on Words:1982 Waterloo • CDN
[1983] *Combinatorics on Words. Progress and Perspectives.* ED: CUMMINGS, L. • PUBL (**X** 0801) Academic Pr: New York x+405pp • DAT&PL 1982 Aug; Waterloo, CDN • LC-No 83-21327, ISBN 0-12-198820-1

P 4391 Mat & Mat Obrazov (14);1985 Sl"nchev Bryag • BG
[1985] *Matematika i Matematichesko Obrazovanie. Doklady na 14a Proletna Konferentsiya na S"juza na Matematicite v B"lgariya (Mathematics and Mathematical Education. Proceedings of the 14th Spring Conference of the Union of Bulgarian Mathematicians)* PUBL (**X** 2237) Publ Bulg Acad Sci: Sofia • DAT&PL 1985 Apr; Sl"nchev Bryag (Sunny Beach), BG • LC-No 80-643123

P 4392 Mat Logika (Markov);1980 Sofia • BG
identical with P 4117

P 4417 Semin Probab (5);1969/70 Strasbourg • F
[1971] *Seminaire Probabilites V, 1969/70* SER (**S** 3301) Lect Notes Math 191 • PUBL (**X** 0811) Springer: Heidelberg & New York iv+372pp • DAT&PL 1969 -?-; Strasbourg, F • ISBN 3-540-05397-2

P 4522 Anal Sets • GB
[1980] *Analytic Sets. Lectures Delivered at a Conference* PUBL (**X** 0801) Academic Pr: New York x+499pp • DAT&PL 1978 Jul; london, GB • ISBN 0-12-593150-6

P 4559 Parallel Process;1983 Muenchen • D
[1984] *WOPPLOT 83. Parallel Processing: Logic, Organisation and Technology* SER Lect Notes Phys 196 • PUBL (**X** 0811) Springer: Heidelberg & New York v+189pp • DAT&PL 1983 Jun; Muenchen, D • ISBN 3-540-12917-0

P 4564 Math Logic;1963 Xi-An • TJ
[1965] *1963 Nian Cueanguo Shuli Luoji Zhuanye Xueshu Huiyi Lunwen Xuanji (Mathematical Logic. Proceedings of the National Symposium)* PUBL Defence Industry Press: Beijing • DAT&PL 1963 Oct; Xi-An, TJ

P 4571 Log of Progr;1985 Brooklyn • USA
Logic of Programs ED: PARIKH, R. • SER (**S** 3302) Lect Notes Comput Sci 193 • PUBL (**X** 0811) Springer: Heidelberg & New York vi+424pp • DAT&PL 1985 Jun; Brooklyn, NY, USA • ISBN 3-540-15648-8

P 4583 Graphs & Order;1984 Banff • CDN
Graphs and Order. The Role of Graphs in the Theory of Order Sets and Its Applications ED: RIVAL, I. • SER NATO ASI Ser, Ser C 147 • PUBL (**X** 0835) Reidel: Dordrecht xix+796pp • DAT&PL 1984 May; Banff, CDN

P 4588 Logic in China;1978 Beijing • TJ
(Issue of the Symposium on Logic in China) PUBL Jilin People Press • DAT&PL 1978 May; Beijing, TJ

P 4595 Autom Infinite Words;1984 Le Mont Dore • F
Automata on Infinite Words SER (**S** 3302) Lect Notes Comput Sci 192 • PUBL (**X** 0811) Springer: Heidelberg & New York iv+216pp • DAT&PL 1984 May; Le Mont Dore, F • ISBN 3-540-15641-0

P 4601 EUROCAL;1985 Linz • A
[1985] *Proceedings of European Conference on Computer Algebra (EUROCAL)* ED: BUCHBERGER, B. (V1) & CAVINESS, B.F. (V2) • SER (**S** 3302) Lect Notes Comput Sci 203,204 • PUBL (**X** 0811) Springer: Heidelberg & New York vi+233pp, xvi+650pp • DAT&PL 1985 Apr; Linz, A • ISBN 3-540-15983-5 (V1), ISBN 3-540-15984-3 (V2) • REM Vol 1: Invited Lectures. Vol 2: Research Contributions

P 4622 Autom on Infinite Words;1984 Le Mont-Dore • F
[1985] *Automata on Infinite Words* ED: NIVAT, M. & PERRIN, D. • SER (**S** 3302) Lect Notes Comput Sci 192 • PUBL (**X** 0811) Springer: Heidelberg & New York v+216pp • DAT&PL 1984 May; Le Mont-Dore, F • ISBN 3-540-15641-0

P 4626 Combin Algor on Words;1984 Maratea
[1985] *Combinatorial Algorithms on Words* ED: APOSTOLICO, A. & GALIL, Z. • SER NATO Adv Sci Inst,Ser F 12 • PUBL (**X** 0811) Springer: Heidelberg & New York viii+361pp • DAT&PL 1984 Jun; Maratea • ISBN 3-540-15227-X

P 4627 CAAP'85 Arbres en Algeb & Progr (10);1985 Berlin • D
[1985] *Mathematical Foundations of Software Development, Vol 1. Colloquium on Trees in Algebra and Programming (CAAP '85)* ED: EHRIG, H. & FLOYD, C. & NIVAT, M. & THATCHER, J. • SER (**S** 3302) Lect Notes Comput Sci 185 • PUBL (**X** 0811) Springer: Heidelberg & New York xiii+418pp • DAT&PL 1985 Mar; Berlin, D • ISBN 3-540-15198-2

P 4628 Automata, Lang & Progr (12);1985 Nafplion • GR
[1985] *Automata, Languages and Programming.* ED: BRAUER, W. • SER (**S** 3302) Lect Notes Comput Sci 194 • PUBL (**X** 0811) Springer: Heidelberg & New York viii+520pp
• DAT&PL 1985 Jul; Nafplion, GR • ISBN 3-540-15650-X

P 4646 Atti Incontri Log Mat (2);1983/84 Siena • I
[1985] *Atti degli Incontri di Logica Matematica* ED: BERNARDI, C. & PAGLI, P. • PUBL (**X** 3812) Univ Siena, Dip Mat: Siena 648pp
• DAT&PL 1983 Jan; Siena, I, 1983 Apr; Siena, I, 1984 Jan; Siena, I, 1984 Apr; Siena, I

P 4647 FCT'85 Fund of Comput Th;1985 Cottbus • DDR
[1985] *Fundamentals of Computation Theory. FCT '85* ED: BUDACH, L. • SER (**S** 3302) Lect Notes Comput Sci 199
• PUBL (**X** 0811) Springer: Heidelberg & New York xii+542pp
• DAT&PL 1985 Sep; Cottbus, DDR • ISBN 3-540-15689-5

P 4661 Algeb & Log;1984 Zagreb • YU
[1985] *Proceedings of the Conference 'Algebra and Logic'* ED: STOJAKOVIC, Z. • PUBL (**X** 4030) Univ Novom Sadu, Inst Mat: Novi Sad vi+193pp
• DAT&PL 1984 Jun; Zagreb, YU

P 4663 Ordered Algeb Struct;1982 Cincinnati • USA
[1985] *Ordered Algebraic Structures. Proceedings of a Special Session of the 88th Annuanl Meeting of the American Mathematical Society (AMS)* ED: POWELL, W.B. & TSINAKIS, T. • SER (**S** 3310) Lect Notes Pure Appl Math 99 • PUBL (**X** 1684) Dekker: New York xii+196pp
• DAT&PL 1982 -?-; Cincinnati, OH, USA • ISBN 0-8247-7342-X

P 4670 Comput Th (5);1984 Zaborow • SU
[1985] *Computation Theory. Proceedings of the 5th Symposium* ED: SKOWRON, A. • SER (**S** 3302) Lect Notes Comput Sci 208 • PUBL (**X** 0811) Springer: Heidelberg & New York vii+397pp
• DAT&PL 1984 Dec; Zaborow, SU • ISBN 3-540-16066-3

P 4672 Found of Softw Tech & Th Comput Sci (5);1985 New Delhi • IND
[1985] *Foundations of Software Technology and Theoretical Computer Science. Proceedings of the 5th Conference* ED: MAHESHWARI, S.N. • SER (**S** 3302) Lect Notes Comput Sci 206 • PUBL (**X** 0811) Springer: Heidelberg & New York x+522pp
• DAT&PL 1985 Dec; New Delhi, IND • ISBN 3-540-16042-6

P 4673 n-ary Struct (2);1983 Varna • BG
[1985] *Proceedings of the 2nd International Symposium, n-ary Structures* ED: TOPENCHAROV, V.V. & ARNAOUDOV, YA.N.
• PUBL Center of Applied Math: Sofia x+266pp
• DAT&PL 1983 Sep; Varna, BG

P 4736 Conf Inform Sci & Syst;1976 Baltimore • USA
[1976] *Proceedings of the 1976 Conference on Information Sciences and Systems. 10th Annual Meeting* PUBL (**X** 1291) Johns Hopkins Univ Pr: Baltimore xiii+601pp
• DAT&PL 1976 Mar; Baltimore, MD, USA

Collection volumes

C 0552 Phil Contemp - Chroniques • I
[1968] *La Philosophie Contemporaine: Chroniques (Contemporary Philosophy: A Survey)* ED: KLIBANSKY, R. • PUBL (**X** 1319) La Nuova Italia: Firenze 4 Vols • LC-No 68-55649

C 0569 Phil of Math Oxford Readings • GB
[1969] *The Philosophy of Mathematics.* ED: HINTIKKA, K.J.J. • SER Oxford Readings in Philosophy • PUBL (**X** 0894) Oxford Univ Pr: Oxford v+186pp • ISBN 0-19-875011-0, LC-No 71-441791

C 0587 Quine: Sel Logic Papers • USA
[1966] *Selected Logic Papers of Willard van Orman Quine* PUBL (**X** 0981) Random House: New York x+250pp • LC-No 66-11147

C 0595 Stud Math Anal & Rel Topics (Polya) • USA
[1962] *Studies in Mathematical Analysis and Related Topics. Essays in Honor of George Polya* ED: SZEGOE, G. & LOEWNER, C. ET AL. • SER Stanford Studies in Mathematics and Statistics 4 • PUBL (**X** 1355) Stanford Univ Pr: Stanford xxi+447pp • LC-No 62-15265

C 0596 Stud in Number Th • USA
[1969] *Studies in Number Theory* ED: LEVEQUE, W.J. • SER MAA Studies in Mathematics 6 • PUBL (**X** 1298) Math Ass Amer: Washington vii+212pp • ALT PUBL (**X** 0819) Prentice Hall: Englewood Cliffs • ISBN 0-13-541359-1, LC-No 75-76868

C 0601 Encycl of Philos • USA
[1967] *Encyclopedia of Philosophy* ED: EDWARDS, P. • PUBL (**X** 0843) Macmillan : New York & London 8 Vols • LC-No 67-10059

C 0618 Minds & Machines • USA
[1964] *Minds and Machines.* ED: ANDERSON, A.R. • SER Contemporary Prospects in Philosophy Series • PUBL (**X** 0819) Prentice Hall: Englewood Cliffs viii+114pp • ISBN 0-13-583393-0, LC-No 64-11553

C 0622 Essays Found of Math (Fraenkel) • IL
[1961] *Essays on the Foundations of Mathematics: Dedicated to A.A.Fraenkel on His 70th Anniversary* ED: BAR-HILLEL, Y. & POZNANSKI, E.I.J. & RABIN, M.O. & ROBINSON, A. • PUBL (**X** 1299) Magnes Pr: Jerusalem x+351pp • ALT PUBL (**X** 0809) North Holland: Amsterdam & (**X** 0833) Acad Pr: Jerusalem & (**X** 0894) Oxford Univ Pr: Oxford • LC-No 63-753

C 0628 Symp Semant of Algor Lang • D
[1971] *Symposium on Semantics of Algorithmic Languages* ED: ENGELER, E. • SER (**S** 3301) Lect Notes Math 188 • PUBL (**X** 0811) Springer: Heidelberg & New York vi+372pp • ISBN 3-540-05377-8, LC-No 78-151406

C 0635 Lang & Information • IL
[1964] *Language and Information. Selected Essays on Their Theory and Application* ED: BAR-HILLEL, Y. • SER Adives International Series, Addison-Wesley Series in Logic • PUBL (**X** 0833) Acad Pr: Jerusalem x+388pp • ALT PUBL (**X** 0832) Addison-Wesley: Reading • LC-No 64-55085

C 0640 Log, Autom, Inform • RO
[1971] *Logique, Automatique, Informatique* ED: MOISIL, G.C. • PUBL (**X** 0871) Acad Rep Soc Romania: Bucharest 456pp • LC-No 77-880254

C 0643 Philosophie 1946-48 • F
[1950] *Philosophie. Chronique des Annees d'apres la Guerre 1946-1948* ED: BAYER, R. • SER (**S** 2073) Actualites Sci Indust 1104 (V 12), 1110 (V 14) • PUBL (**X** 0859) Hermann: Paris at least 14 Vols • REM V 12: Histoire de la Philosohie Metaphysique, Philosophie des Valeurs. V 13: Philosophie des Sciences. V 14: Psychologie, Phenomenologie et Existentialisme.

C 0675 From Frege to Goedel • USA
[1967] *From Frege to Goedel: A Source Book in Mathematical Logic 1879-1931* ED: HEIJENOORT VAN, J. • PUBL (**X** 0858) Harvard Univ Pr: Cambridge x+660pp • ALT PUBL (**X** 0894) Oxford Univ Pr: Oxford • LC-No 67-10905 • REM 2nd ed. 1971; xi+660pp. Some Articles have been Reprinted 1970; iv+117pp

C 0679 Handb of Automat, Comput & Control • GB
[1958,1959] *Handbook of Automation, Computation and Control* ED: GRABLE, E.M. & RAMO, S. & WOOLDRIDGE, D.E. • PUBL (**X** 1249) Chapman & Hall: London 2 Vols • ALT PUBL (**X** 0827) Wiley & Sons: New York • LC-No 58-10800 • REM Vol 1: Control Fundamentals. Vol 2: Computers & Data Processing

C 0681 Sequential Machines Sel Pap • USA
[1964] *Sequential Machines: Selected Papers* ED: MOORE, E.F. • SER Addison-Wesley Series in Computer Science and Information Processing • PUBL (**X** 0832) Addison-Wesley: Reading v+266pp • LC-No 63-14678

C 0705 Found of Math (Goedel) • D
[1969] *Foundations of Mathematics. Symposium Papers Commemorating the 60th Birthday of Kurt Goedel.* ED: BULLOFF, J.J. & HOLYOKE, T.C. & HAHN, S.W. • PUBL (**X** 0811) Springer: Heidelberg & New York xii+195pp • ISBN 3-540-04490-6, LC-No 68-28757

C 0717 Automata Studies • USA
[1956] *Automata Studies* ED: SHANNON, C.E. & MCCARTHY, J.E. • SER (**S** 3513) Ann Math Stud 34 • PUBL (**X** 0857) Princeton Univ Pr: Princeton viii+285pp
• LC-No 56-7637
• TRANSL IN [1974] (**C** 1902) Stud Th Automaten
• REM 4th ed. 1965

C 0718 The Undecidable • USA
[1965] *The Undecidable. Basic Papers on Undecidable Propositions, Unsolvable Problems and Computable Functions* ED: DAVIS, M. • PUBL (**X** 0887) Raven Pr: New York 440pp
• LC-No 65-3996

C 0721 Contemp Readings in Log Th • USA
[1967] *Contemporary Readings in Logical Theory* ED: COPI, I.M. & GOULD, J.A. • PUBL (**X** 0843) Macmillan : New York & London 342pp
• LC-No 67-15535

C 0722 Stud Algeb & Anwendgn • DDR
[1972] *Studien zur Algebra und ihre Anwendungen. Mit Anwendungen in der Mathematik, Physik und Rechentechnik* ED: HOEHNKE, H.J. • SER Schriftenreihe des Zentralinstituts fuer Mathematik und Mechanik bei der Akademie der Wissenschaften der DDR; 16 • PUBL (**X** 0911) Akademie Verlag: Berlin vi+154pp
• LC-No 73-336816

C 0733 Izbr Vopr Algeb & Log (Mal'tsev) • SU
[1973] *Izbrannye Voprosy Algebry i Logiki: Sbornik Posvyashch Pamyati A.I.Mal'tsev (Selected Questions of Algebra and Logic: Volume Dedicated to the Memory of A.I.Mal'tsev)* ED: ERSHOV, YU.L. & KARGAPOLOV, M.I. & MERZLYAKOV, YU.I. & SMIRNOV, D.M. & SHIRSHOV, A.L. • PUBL (**X** 2642) Nauka: Novosibirsk 339pp
• LC-No 73-360316

C 0735 Logic & Value (Dahlquist) • S
[1970] *Logic and Value. Essays Dedicated to Thorild Dahlquist on His 50th Birthday* ED: PAULI, T. • SER Filosofiska Studier 9 • PUBL (**X** 0882) Univ Filos Foeren: Uppsala vi+247pp
• LC-No 72-186683

C 0742 Phil Mid-Century • I
[1958-1959] *Philosophy in the Mid-Century. A Survey * La Philosophie au Milieu du Vingtieme Siecle. Chroniques* ED: KLIBANSKY, R. • PUBL (**X** 1319) La Nuova Italia: Firenze 4 Vols

C 0758 Logic Colloq Boston 1972-73 • D
[1975] *Logic Colloquium. Symposium on Logic Held at Boston, MA, USA, 1972-1973* ED: PARIKH, R.J. • SER (**S** 3301) Lect Notes Math 453 • PUBL (**X** 0811) Springer: Heidelberg & New York iv+251pp
• ISBN 3-540-07155-5, LC-No 75-11528

C 0782 Model Th & Algeb (A. Robinson) • D
[1975] *Model Theory and Algebra. A Memorial Tribute to Abraham Robinson.* ED: SARACINO, D.H. & WEISPFENNING, V.B. • SER (**S** 3301) Lect Notes Math 498 • PUBL (**X** 0811) Springer: Heidelberg & New York 436pp
• ISBN 3-540-07538-0, LC-No 75-40483

C 1009 Wang: Survey Math Logic • TJ
[1963] *Wang,H.: Survey of Mathematical Logic* PUBL (**X** 1876) Kexue Chubanshe: Beijing • ALT PUBL (**X** 0809) North Holland: Amsterdam 1963

C 1018 Digit Inf.-wandler • D
[1962] *Digitale Informationswandler* ED: HOFFMANN, W. ET AL. • PUBL (**X** 0900) Vieweg: Wiesbaden vii+740pp
• ALT PUBL (**X** 0820) Intersci Publ: New York
• LC-No 62-16102

C 1050 Aktual Vopr Mat Log & Teor Mnozh • SU
[1975] *Aktual'nye Voprosy Matematicheskoj Logiki i Teorii Mnozhestv (Current Questions in Mathematical Logic and Set Theory)* ED: SHCHELGOL'KOV, E.A. & BOKSHTEJN, M.F. & PETRI, N.V. • PUBL (**X** 2802) Moskov Ped Inst: Moskva
• LC-No 76-530761

C 1085 Essays Cellular Automata • USA
[1970] *Essays on Cellular Automata* ED: BURKS, A.W. • PUBL (**X** 1285) Univ Ill Pr: Urbana xxvi+375pp
• ISBN 0-252-00023-4, LC-No 71-83547

C 1098 Skolem: Select Works in Logic • N
[1970] *Skolem,T.A.: Selected Works in Logic* ED: FENSTAD, J.E. • SER Scandinavian University Books • PUBL (**X** 1554) Universitesforlaget: Oslo 732pp
• LC-No 74-485971

C 1105 Phil of Math. Sel Readings • USA
[1964] *Philosophy of Mathematics. Selected Readings.* ED: BENACERRAF, P. & PUTNAM, H. • SER Prentice Hall Philosophy Series • PUBL (**X** 0819) Prentice Hall: Englewood Cliffs vii+563pp • ALT PUBL (**X** 1096) Blackwell: Oxford
• LC-No 64-13252

C 1106 Curr in Th of Computing • USA
[1973] *Currents in the Theory of Computing* ED: AHO, A.V. • SER Prentice-Hall Series in Automatic Computation • PUBL (**X** 0819) Prentice Hall: Englewood Cliffs x+245pp
• ISBN 0-13-195651-5, LC-No 72-12477

C 1155 Log Issl (Moskva) • SU
[1959] *Logicheskie Issledovaniya (Logical Investigations)* PUBL (**X** 0899) Akad Nauk SSSR : Moskva
• LC-No 60-40373

C 1159 Tarski: Logic, Semantics, Metamathematics • GB
[1956] *Logic, Semantics, Metamathematics. Papers from 1923 to 1938 by Alfred Tarski* PUBL (**X** 0815) Clarendon Pr: Oxford 471pp
• LC-No 56-4171
• TRANSL IN [1964] (**C** 1186) Tarski: Logique, Semantique, Metamath
• REM Translations of Several Articles

C 1180 Readings Math Psychology • USA
[1963-1965] *Readings in Mathematical Psychology* ED: LUCE, R.D. & BUSH, R.R. & GALANTER, E. • PUBL (**X** 0827) Wiley & Sons: New York 2 Vols Psychology.
• LC-No 63-14066

C 1186 Tarski: Logique, Semantique, Metamath • F
[1964] *Tarski,A.: Logique, Semantique, Metamathematique. 1923-1944. Tome 1* PUBL (**X** 0850) Colin: Paris
• TRANSL OF [1956] (**C** 1159) Tarski: Logic, Semantics, Metamathematics
• REM Revised and extended French ed. 1972

C 1213 Stud in Num Anal (Lanczos) • USA
[1974] *Studies in Numerical Analysis. Papers in Honour of Cornelius Lanczos* ED: SCAIFE, B.K.P. • PUBL (**X** 0801) Academic Pr: New York xxii+333pp
• ISBN 0-12-621150-7, LC-No 72-12280

C 1391 Coll Candidat Works - Math, Mech, Phys (Kazan) • SU
[1969] *(Collection of Candidates Works in Exact Sciences: Mathematics, Mechanics, Physics)* PUBL (**X** 3605) Kazan Gos Univ: Kazan'
• LC-No 73-316643

C 1424 Math & Phys-Techn Probl der Kybern • DDR
[1963] *Mathematische und Physikalisch-Technische Probleme der Kybernetik* PUBL (**X** 0911) Akademie Verlag: Berlin
• LC-No 64-56602

C 1435 Vopr Teor Grupp & Polugrupp • SU
[1972] *Voprosy Teorii Grupp i Polugrupp (Questions in the Theory of Groups and Semigroups)* ED: GRINDLINGER, M.D. & GRINDLINGER, E.I. • PUBL Tul'skij Gosudarstvennyj Pedagogicheskij Institut imeni L.N.Tolstoj: Tula;200pp

C 1443 Algebra, Vyp 2 (Irkutsk) • SU
[1973] *Algebra. Vyp 2* ED: FRIDMAN, E.I. • PUBL (**X** 1006) Irkutsk Gos Univ: Irkutsk 164pp
• LC-No 74-645580 • REL PUBL (**C** 3549) Algebra, Vyp 1 (Irkutsk)

C 1450 Teor Algor & Mat Logika (Markov) • SU
[1974] *Teoriya Algorifmov i Matematicheskaya Logika: Sbornik Statej (Theory of Algorithms and Mathematical Logic. Collection of Articles Dedicated to Andrej Andrejevich Markov)* ED: KUSHNER, B.A. & NAGORNYJ, N.M. • PUBL (**X** 2265) Akad Nauk Vychis Tsentr: Moskva 216pp
• LC-No 75-554732

C 1495 Encycl Universalis • F
[1968-1976] *Encyclopaedia Universalis* PUBL (**X** 2524) Encyclopaedia Universalis: Paris 20 Vols
• ISBN 2-85229-281-5, LC-No 75-516014

C 1523 Handb of Math Logic • NL
[1977 & 2nd ed. 1978] *Handbook of Mathematical Logic* ED: BARWISE, J. • SER (**S** 3303) Stud Logic Found Math 90
• PUBL (**X** 0809) North Holland: Amsterdam xi+1165pp
• ALT PUBL (**X** 0838) Amer Elsevier: New York
• ISBN 0-7204-2285-X, LC-No 76-26032
• TRANSL IN [1982] (**C** 1920) Spravochnaya Kniga po Mat Logike, Chast 1-4

C 1525 Semin Init Analyse (9-10) Paris 1969/71 • F
[1970ff] *Seminaire Choquet 9e-10e Annees: 1969-1971. Inititation a l'Analyse* ED: CHOQUET, G. • PUBL (**X** 1623) Univ Paris VI Inst Poincare: Paris

C 1534 Mash Turing & Rek Funk • SU
[1972] *Mashiny Tyuringa i Rekursivnye Funktsii (Turing Machines and Recursive Functions)* PUBL (**X** 0885) Mir: Moskva 264pp
• TRANSL IN Turing Machines and Recursive Functions

C 1540 Probl Mat Log: Slozh Algor & Kl Vychisl Funk • SU
[1970] *Problemy Matematicheskoj Logiki. Slozhnost' Algoritmov i Klassy Vychislimykh Funktsii. Sbornik Peredov (Probleme der Mathematischen Logik. Kompliziertheit von Algorithmen und Klassen Berechenbarer Funktionen. Sammlung von Uebersetzungen.)* ED: KOZMIDIADI, V.A. & MUCHNIK, A.A. • SER Biblioteka "Kiberneticheskogo Sbornika" • PUBL (**X** 0885) Mir: Moskva 431pp

C 1549 Tartu Mezhvuz Nauch Simp Obshchej Algeb • SU
[1966] *Tartuskij Gosudarstvennyj Universitet Mezhvuzovskij Nauchnyj Simpozium po Obshchej Algebre (Wissenschaftliches Interhochschul-Symposium ueber Allgemeine Algebra)* ED: GABOVICH, E. ET AL. • PUBL (**X** 2463) Tartusk Gos Univ: Tartu 232pp

C 1566 Semin IRIA Log & Automates 1971 • F
[1971] *Seminaires IRIA Logiques et Automates* PUBL (**X** 2732) INRIA: Le Chesnay Cedex 205pp
• LC-No 75-509899

C 1692 Semin Init Analyse (17) Paris 1977/78 • F
[1978] *Seminiare d'Initiation a l'Analyse: G.Choquet-M.Rogalski- J.Saint-Raymond. 18e Annee. 1978/1979* SER Publ. Math. Univ. Pierre et Marie Curie • PUBL (**X** 1623) Univ Paris VI Inst Poincare: Paris

C 1702 Local Induction • NL
[1976] *Local Induction* ED: BOGDAN, R.J. • SER (**S** 3307) Synth Libr 93 • PUBL (**X** 0835) Reidel: Dordrecht xiv+340pp
• ISBN 90-277-0649-2, LC-No 75-34922

C 1842 Phil of Sci Today • USA
[1967] *Philosophy of Science Today.* ED: MORGENBESSER, S.
• PUBL (**X** 2671) Basic Books: New York xvi+208pp
• LC-No 67-17391

C 1856 Log Enterprise • USA
[1975] *The Logical Enterprise* ED: MARCUS, R.B. & ANDERSON, A.R. & MARTIN, R.M. • PUBL (**X** 0875) Yale Univ Pr: New Haven x+261pp
• ISBN 0-300-01790-1, LC-No 74-20084

C 1902 Stud Th Automaten • D
[1974] *Studien zur Theorie der Automaten* ED: SHANNON, C.E. & MCCARTHY, J. • SER R & B Studien 1 • PUBL Rogner & Bernhard: Muenchen xxxii+452pp
• TRANSL OF [1956] (**C** 0717) Automata Studies

C 1920 Spravochnaya Kniga po Mat Logike, Chast 1-4 • SU
[1982] *Spravochnaya Kniga po Matematicheskoj Logike. Chast. 1-4 (Handbook of Mathematical Logic. Part 1-4)* ED: ERSHOV, YU.L. & PALYUTIN, E.A. & TAJMANOV, A.D. • PUBL (**X** 2027) Nauka: Moskva 4 Vols: 392pp, 375pp, 360pp, 391pp
• TRANSL OF [1977] (**C** 1523) Handb of Math Logic
• REM The Translation Contains a New Supplement by Palyutin,E.A. Part 1: Model Theory. Part 2: Set Theory. Part 3: Recursion Theory. Part 4: Proof Theory and Constructive Mathematics.

C 2050 Semin Th des Ensembles GMS Paris 1976/78 • F
[1979] *Theorie des Ensembles: Seminaire GMS* SER Publications Mathematiques de l'Universite Paris VII 5
• PUBL Universite de Paris VII, UER de Mathematiques: Paris 228pp
• DAT&PL 1976-77; Paris, F, 1977-78; Paris, F

C 2065 Teor Nereg Kriv Raz Geom Post • SU
[1979] *Teoriya Neregulyarnykh Krivykh v Razlichnykh Geometricheskykh Postranstvakh (Theorie der Nichtregulaeren Kurven in Verschiedenen Geometrischen Raeumen)* ED: TAJTSLIN, M.A. • PUBL (**X** 2769) Kazakh Gos Univ: Alma-Ata 136pp

C 2141 Filos Matematica • I
[1967] *La Filosofia della Matematica* ED: CELLUCCI, C.
• PUBL Laterza: Bari 320pp
• LC-No 68-120024

C 2264 Algor Probl Teor Grupp & Polygrupp • SU
[1981] *Algoritmicheskie Problemy Teorii Grupp i Polygrupp (Algorithmic Problems of the Theory of Groups and Semigroups)* ED: GRINDLINGER, M.D. • PUBL Tulisk Gos Ped Inst: Tula 134pp

C 2282 Approach Nat Language • NL
[1973] *Approaches to Natural Language* ED: HINTIKKA, K.J.J. & MORAVCSIK, J.M.E. & SUPPES, P. • SER (S 3307) Synth Libr • PUBL (**X** 0835) Reidel: Dordrecht viii + 526pp
• LC-No 72-179892

C 2318 Entwicklung Math in DDR • DDR
[1974] *Entwicklung der Mathematik in der DDR: Zum 25. Jahrestag der Gruendung der Deutschen Demokratischen Republik* ED: SACHS, H. & AHRENS, H. ET AL. • PUBL (**X** 0806) Dt Verlag Wiss: Berlin xx + 756pp
• LC-No 75-569431

C 2319 Slozh Vychisl & Algor • SU
[1974] *Slozhnost' Vychislennij i Algorifmov. Sbornik Peredov (Die Kompliziertheit von Berechnungen und Algorithmen. Sammlung von Uebersetzungen)* ED: KOZMIDIADI, V.A. & MASLOV, A.N. & PETRI, N.V. • PUBL (**X** 0885) Mir: Moskva 392pp
• LC-No 75-568382

C 2355 Metody Anal Mnogormernoj Ehkon Inf • SU
[1981] *Metody Analiza Mnogomernoj Ehkonomicheskoj Informatsii (Methods of Analysis of Multidimensional Economic Information)* ED: MIRKIN, B.G. • PUBL (**X** 2642) Nauka: Novosibirsk 207pp
• LC-No 82-152452

C 2368 Chow,Hung-Ching 60th Anniv Vol • RC
[1962] *Hung-Ching Chow 60th Anniversary Volume* PUBL (**X** 2445) Acad Sin Inst Math: Taipei iii + 134pp

C 2383 Chow,Hung-Ching 65th Anniv Vol • RC
[1967] *Hung-Ching Chow 65th Anniversary Volume* PUBL (**X** 2445) Acad Sin Inst Math: Taipei vi + 204pp
• LC-No 74-899243

C 2502 Petri-Netze & Anwendungen • DDR
[1982] *Petri-Netze und Ihre Anwendungen* PUBL Techn Hochschule, Sekt Automatisierungstechnik: Karl-Marx-Stadt

C 2555 Algeb Sistemy (Irkutsk) • SU
[1976] *Algebraicheskie Sistemy (Algebraic Systems)* ED: KOKORIN, A.I. • PUBL (**X** 1006) Irkutsk Gos Univ: Irkutsk 170pp
• LC-No 79-410179

C 2556 Algor Vopr Algeb Sist • SU
[1978] *Algoritmicheskie Voprosy Algebraicheskikh Sistem (Algorithmic Questions of Algebraic Systems)* ED: KOKORIN, A.I. • PUBL (**X** 1006) Irkutsk Gos Univ: Irkutsk 217pp

C 2557 Teor & Priklad Zad Mat & Mekh • SU
[1977] *Teoreticheskie i Prikladnye Zadachi Matematiki i Mekhaniki. Sbornik Trudov (Theoretical and Applied Problems in Mathematics and Mechanics. Work Collection)* ED: AMANOV, T.I. • PUBL Nauka Kazakh SSR:Alma-Ata 322pp

C 2577 Issl Formaliz Yazyk & Neklass Log • SU
[1974] *Issledovaniya po Formalizovannym Yazykam i Neklassicheskim Logikam (Investigations on Formalized Languages and Non-Classical Logics)* ED: BOCHVAR, D.A. • PUBL (**X** 2027) Nauka: Moskva 275pp
• LC-No 75-554105

C 2581 Issl Neklass Log & Teor Mnozh • SU
[1979] *Issledovaniya po Neklassicheskim Logikam i Teorii Mnozhestv (Investigations on Non-Classical Logics and Set Theory)* ED: MIKHAJLOV, A.I. ET AL. • PUBL (**X** 2027) Nauka: Moskva 374pp
• LC-No 80-475529

C 2595 Rekursiv Funktsii • SU
[1978] *Rekursivnye Funktsii (Recursive Functions)* ED: POLYAKOV, E.A. • PUBL (**X** 2594) Ivanovo Gos Univ: Ivanovo 100pp

C 2617 Modern Log Survey • NL
[1981] *Modern Logic - A Survey. Historical, Philosophical and Mathematical Aspects of Modern Logic and Its Applications.* ED: AGAZZI, E. • SER (S 3307) Synth Libr 149 • PUBL (**X** 0835) Reidel: Dordrecht viii + 475pp
• ISBN 90-277-1137-2, LC-No 80-22027

C 2620 Teor Model & Primen • SU
[1980] *Teoriya Modelej i Ee Primeneniya (Theory of Models and Its Applications)* ED: BAJZHANOV, B.S. & TAJTSLIN, M.A. • PUBL (**X** 2769) Kazakh Gos Univ: Alma-Ata 83pp

C 2621 Mal'tsev: Metamath of Algeb Syst • NL
[1971] *The Metamathematics of Algebraic Systems. Mal'tsev,A.I: Collected Papers, 1936-1967* ED: WELLS III, B.F. • SER (S 3303) Stud Logic Found Math 66 • PUBL (**X** 0809) North Holland: Amsterdam xviii + 494pp
• LC-No 73-157020

C 2897 Algeb Model, Kateg & Gruppoide • DDR
[1979] *Algebraische Modelle, Kategorien und Gruppoide* ED: HOEHNKE, H.-J. • SER Studien zur Algebra und ihre Anwendungen 7 • PUBL (**X** 0911) Akademie Verlag: Berlin vi + 178pp
• LC-No 80-475994

C 2899 Algor & Avtomaty • SU
[1978] *Algoritmy i Avtomaty. Sbornik Nauchnykh Trudov (Algorithms and Automata. Collection of Scientific Works)* ED: ARSLANOV, M.M. • PUBL (**X** 3605) Kazan Gos Univ: Kazan' 110pp

C 2908 Cabal Seminar Los Angeles 1976-77 • USA
[1978] *Cabal Seminar 76-77. Proceedings of the Caltech-UCLA Logic Seminar 1976-77* ED: KECHRIS, A.S. & MOSCHOVAKIS, Y.N. • SER (S 3301) Lect Notes Math 689 • PUBL (**X** 0811) Springer: Heidelberg & New York iii + 282pp
• ISBN 3-540-09086-X, LC-No 78-24063

C 2922 Cabal Seminar Los Angeles 1977-79 • USA
[1981] *Cabal Seminar 77-79. Proceedings of the Caltech-UCLA Logic Seminar 1977-79* ED: KECHRIS, A.S. & MARTIN, D.A. & MOSCHOVAKIS, Y.N. • SER (S 3301) Lect Notes Math 839 • PUBL (**X** 0811) Springer: Heidelberg & New York iv + 274pp
• ISBN 3-540-10288-4, LC-No 81-1059

C 2925 Diskr Sist, Formal Yazyki & Slozhn Algor ('77) • SU
[1977] *Diskretnye Sistemy, Formalnye Yazyki i Slozhnost' Algoritmov (Discrete Systems, Formal Languages and Complexity of Algorithms)* ED: BOGOMOLOV, A.M. & LIPSKAYA, V.A. • PUBL (**X** 2522) Akad Nauk Inst Kibernet: Kiev 84pp
• LC-No 78-421552

C 2953 Log & Probab in Quant Mech • NL
[1976] *Logic and Probability in Quantum Mechanics* ED: SUPPES, P. • SER (S 3307) Synth Libr 78 • PUBL (**X** 0835) Reidel: Dordrecht xv + 541pp

C 2962 Mat Voprosy Teor Intell Mashin • SU
[1975] *Matematicheskie Voprosy Teorii Intellektual'nykh Mashin. Sbornik Trudov (Mathematical Questions of the Theory of Machine Intelligence. Collection of Papers)* ED: KAPITONOVA, YU.V. & ASEL'DEROV, Z.M. • PUBL (**X** 2522) Akad Nauk Inst Kibernet: Kiev 84pp
• LC-No 77-507084

C 2963 Rass di Mat • I
[1980] *Rassegna di Matematica. Logica Matematica. Matematica Applicata. Didattica della Matematica (Mathematics Review. Mathematical Logic. Applied Mathematics. Teaching of Mathematics)* ED: BORGA, M. & FREGUGLIA, P. & PALLADINO, D. • PUBL (**X** 2682) Casa Ed Tilgher: Genova 100pp

C 2970 Multicomp Random Systs • USA
[1980] *Multicomponent Random Systems* ED: DOBRUSHIN, R.L. & SINAJ, YA.G. & GRIFFEATH, D. • SER Advances in Probability and Related Topics 6 • PUBL (**X** 1684) Dekker: New York xi+606pp
• ISBN 0-8247-6831-0, LC-No 80-17688
• TRANSL OF Mnogokomponentnye Sluchajnye Sistemy

C 2976 Numb Th, Algeb Geom & Comm Algeb (Akizuki) • J
[1973] *Number Theory, Algebraic Geometry and Commutative Algebra. In Honor of Yasuo Akizuki.* ED: KUSUNOKI, Y. & MIZOHATA, S. & NAGATA, M. & TODA, H. & YAMAGUTI, M. & YOSHIZAWA, H. • PUBL (**X** 2465) Kinokuniya Company: Tokyo vi+528pp
• LC-No 74-159266

C 2978 Vychisl Mat (Kiev) • SU
[1973] *Vychislitel'naya Matematika (Numerical Analysis)* ED: KIRO, S.N. & DUDYAK, M.V. • PUBL (**X** 2522) Akad Nauk Inst Kibernet: Kiev 81pp

C 3017 Vopr Sist Progr • SU
[1977] *Voprosy Sistemnogo Programmirovaniya (Questions of Systems Programming)* ED: YUSHCHENKO, E.L. & KULYABKO, P.P. • PUBL (**X** 2522) Akad Nauk Inst Kibernet: Kiev 100pp
• LC-No 78-422428

C 3018 Vopr Anal Vychisl Slozhnosti Algor • SU
[1979] *Voprosy Analiza Vychislitel'noj Slozhnosti Algoritmov (Questions of the Computational Complexity of Algorithms)* ED: ISHCHUK, V.A. • SER (**S** 2653) Prepr Inst Kib, Akad Nauk Ukr SSR 79/14 • PUBL (**X** 2522) Akad Nauk Inst Kibernet: Kiev 48pp

C 3050 Essays Combin Log, Lambda Calc & Formalism (Curry) • USA
[1980] *To H.B. Curry: Essays on Combinatory Logic, Lambda Calculus and Formalism* ED: SELDIN, J.P. & HINDLEY, J.R. • PUBL (**X** 0801) Academic Pr: New York xxv+606pp
• ISBN 0-12-349050-2, LC-No 80-40139

C 3138 Tsifr Vychislit Tekhn & Progr • SU
[1974] *Tsifrovaya Vychislitel'naya Tekhnika i Programmirovanie. Sbornik Statej. Vypusk 8 (Digitale Rechentechnik und Programmierung. Artikelsammlung. Nr. 8)* ED: KITOV, A.I. ET AL. • PUBL (**X** 2643) Sovet Radio: Moskva 192pp

C 3177 Ehvrist Algor Optim, Vyp 2 • SU
[1978] *Ehvristicheskie Algoritmy Optimizatsii. Mezhvuzovskij Tematicheskij Sbornik. Vypusk 2. (Heuristic Optimization Algorithms. Interuniversity Thematic Work Collection. No. 2.)* ED: MAMATOV, YU.A. • PUBL (**X** 2766) Yaroslav Gos Univ: Yaroslavl' 167pp

C 3211 Mat Ling & Teor Algor • SU
[1978] *Matematicheskaya Lingvistika i Teoriya Algoritmov. Mezhvuzovskij Tematicheskij Sbornik (Mathematical Linguistics and Theory of Algorithms. Interuniversity Thematic Collection)* ED: GLADKIJ, A.V. • PUBL (**X** 1434) Kalinin Gos Univ: Kalinin 155pp
• LC-No 80-458139

C 3264 Sist Upravleniya • SU
[1975] *Sistemy Upravleniya (Control Systems)* PUBL (**X** 3606) Tomsk Univ: Tomsk

C 3271 Issl Teor Mnozh & Neklass Logik • SU
[1976] *Issledovaniya po Teorij Mnozhestv i Neklassicheskim Logikam. Sbornik Trudov (Studies in Set Theory and Nonclassical Logics. Collection of Papers)* ED: BOCHVAR, D.A. & GRISHIN, V.N. • PUBL (**X** 2027) Nauka: Moskva 328pp
• LC-No 77-501571

C 3277 Topics Combin Optim • A
[1975] *Topics in Combinatorial Optimization* ED: RINALDI, S. • SER CISM. International Centre for Mechanical Sciences. Courses and Lectures; 175 • PUBL (**X** 0902) Springer: Wien 186pp
• LC-No 78-313826

C 3294 Coding & Complexity • A
[1975] *Coding and Complexity* ED: LONGO, G. • SER CISM. International Centre for Mechanical Sciences. Courses and Lectures; 216 • PUBL (**X** 0902) Springer: Wien vii+334pp
• ISBN 3-211-81341-1, LC-No 77-375125

C 3340 Avtomaty (Moskva) • SU
[1956] *Avtomaty (Automata)* PUBL (**X** 1656) Izdat Inostr Lit: Moskva

C 3351 Teorem Goedel & Hipotese Cont • P
[1979] *O Teorema de Goedel e a Hipotese do Continuo* PUBL (**X** 2719) Fundacao Calouste Gulbenkian: Lisbon

C 3392 Knizhnice - TU Brno, Vol A10, A12 • CS
[1976] *Knizhnice. Publications of Technical and Scientific Papers of the Technical University in Brno. Vol. A 10 and Vol A 12* PUBL (**X** 2229) Univ Purkyne: Brno 2 Vols: 299pp,211pp

C 3405 Sovrem Vopr Priklad Mat & Progr • SU
[1979] *Sovremennye Voprosy Prikladnoj Matematiki i Programmirovaniya. Matematicheskie Nauki. Mezhvuzovskij Sbornik (Modern Questions of Applied Mathematics and Programming. Mathematical Sciences. Interuniversity Work Collection)* PUBL (**X** 2741) Shtiintsa: Kishinev 176pp

C 3433 Prog in Cybern & Syst Res, Vol 6 • A
[1982] *Progress in Cybernetics and Systems Research. Vol. 6 Cybernetics in Biology and Medicine, Systems Analysis, Systems Engineering Methodology, Mathematical Systems Theory.* ED: PICHLER, F.R. & TRAPPL, R. • PUBL (**X** 0822) McGraw-Hill: New York xv+398pp • ALT PUBL (**X** 2437) Hemisphere Publ: Washington
• ISBN 0-89116-194-5, LC-No 75-6641

C 3444 Perspectives in Math • CH
[1984] *Perspectives in Mathematics. Anniversary of Oberwolfach 1984* ED: JAEGER, W. & MOSER, J. & REMMERT, R. • PUBL (**X** 0804) Birkhaeuser: Basel 587pp
• ISBN 3-7643-1624-1, LC-No 84-16922

C 3483 Pap Automata Th • H
[1980] *Papers on Automata Theory. Vol 1, 2* ED: PEAK, I. • PUBL (**X** 3151) Karl Marx Univ Dpt Math: Budapest 2 Vols: 108pp,123pp
• LC-No 81-180200

C 3494 Stud on Math Progr. Math Meth Oper Res, Vol 1 • H
[1980] *Studies on Mathematical Programming. Mathematical Methods of Operation Research. Vol 1* ED: PREKOPA, A. • PUBL (**X** 0928) Akad Kiado: Budapest 200pp • ISBN 963-05-1854-6, LC-No 80-496383

C 3507 Math Models & Numer Meth • PL
[1978] *(Mathematical Models and Numerical Methods)* ED: TIKHONOV, A.N. & KUHNERT, F. & KUZNECOV, N.N. & MOSZYNSKI, K. & WAKULICZ, A. • SER (**S** 1926) Banach Cent Publ 3 • PUBL (**X** 1034) PWN: Warsaw 391pp • LC-No 78-318876

C 3510 Contrib Group Theory (Lyndon) • USA
[1984] *Contributions to Group Theory. Papers Dedicated to Lyndon, R.C. on His 65th Birthday* ED: APPEL, K.I. & RATELIFFE, J.G. & SCHUPP, P.E. • SER (**S** 3313) Contemp Math 33 • PUBL (**X** 0803) Amer Math Soc: Providence xi+519pp • ISBN 0-8218-5035-0, LC-No 84-18454

C 3514 Adv Fuzzy Sets & Appl • NL
[1979] *Advances in Fuzzy Set Theory and Applications* ED: GUPTA, M.M. & RAGADE, R.K. & YAGER, R.R. • PUBL (**X** 0809) North Holland: Amsterdam xv+753pp • ISBN 0-444-85372-3, LC-No 79-17151

C 3515 Ital Studies in Phil of Sci • NL
[1981] *Italian Studies in the Philosophy of Science* ED: CHIARA SCABIA DALLA, M.L. • SER (**S** 3311) Boston St Philos Sci 47 • PUBL (**X** 0835) Reidel: Dordrecht xi+525pp • ISBN 90-277-0735-9, LC-No 80-16665

C 3517 Found of Comput Sci, Vol 2, Part 2 • NL
[1976] *Foundations of Computer Science Vol 2, Part 2.* ED: APT, K.R. & BAKKER DE, J.W. • SER (**S** 1605) Math Centr Tracts 82 • PUBL (**X** 1121) Math Centr: Amsterdam 149pp • ISBN 90-6196-141-6, LC-No 77-368695

C 3549 Algebra, Vyp 1 (Irkutsk) • SU
[1972] *Algebra. Vyp. 1* ED: KOKORIN, A.I. & PENZIN, YU.G. • PUBL (**X** 1006) Irkutsk Gos Univ: Irkutsk 135pp • LC-No 74-645580 • REL PUBL (**C** 1443) Algebra, Vyp 2 (Irkutsk)

C 3567 Teor Kibernetika Kiev 1967/71 • SU
[1971] *Teoreticheskaya Kibernetika (Theoretical Cybernetics.)* ED: GLUSHKOV, V.M. • PUBL (**X** 2522) Akad Nauk Inst Kibernet: Kiev 196pp • LC-No 73-325460 • REM Proceedings of Seminars: Automata Theory (1970/1971), Computer Theory (1967/1969), Complex Systems and Simulation (1971)

C 3626 Rekursiv Mat Analiz • SU
[1970] *Rekursivnyj Matematicheskij Analiz (Recursive Mathematical Analysis)* ED: MINTS, G.E. • SER Prilozhenije 3 • PUBL (**X** 2027) Nauka: Moskva

C 3646 Uchebnoe Posobie • SU
[1980] *Uchebnoe Posobie (Textbook)* PUBL Altaysk Gos Univ: Barnaul

C 3722 Diskr Sist, Formal Yazyki & Slozhn Algor ('78) • SU
[1978] *Diskretnye Sistemy, Formal'nye Yazyki i Slozhnost' Algoritmov (Discrete Systems, Formal Languages and Complexity of Algorithms)* ED: SPERANSKIJ, D.V. & LIPSKAYA, V.A. • PUBL (**X** 2522) Akad Nauk Inst Kibernet: Kiev 95pp

C 3724 Probl Gil'berta • SU
[1969] *Problemy Gil'berta (Hilbert's Problems)* PUBL (**X** 2027) Nauka: Moskva 240pp • LC-No 75-444471

C 3747 Mat Log & Mat Lingvistika • SU
[1981] *Matematicheskaya Logika i Matematicheskaya Lingvistika (Mathematical Logic and Mathematical Linguistics)* ED: GLADKIJ, M. • PUBL (**X** 1434) Kalinin Gos Univ: Kalinin 172pp

C 3798 Mat Log, Mat Ling & Teor Algor • SU
[1983] *Matematicheskaya Logika, Matematicheskaya Lingvistike i Teoriya Algoritmov (Mathematical Logic, Mathematical Linguistics and Theory of Algorithms)* ED: GLADKIJ, A.V. • PUBL (**X** 1434) Kalinin Gos Univ: Kalinin 116pp • LC-No 84-157035

C 3807 Issl Neklass Log & Formal Sist • SU
[1983] *Issledovaniya po Neklassicheskim Logikam i Formal'nym Sistemam (Studies in Nonclassical Logics and Formal Systems)* ED: MIKHAJLOV, A.I. • PUBL (**X** 2027) Nauka: Moskva 360pp • LC-No 83-181942

C 3825 Semin Init Analyse (18) Paris 1978/79 • F
[1979] *Seminiare d'Initiation a l'Analyse: G.Choquet-M.Rogalski- J.Saint-Raymond. 18e Annee. 1978/1979* SER Publ. Math. Univ. Piere et Marie Curie 29 • PUBL (**X** 1623) Univ Paris VI Inst Poincare: Paris 188pp

C 3834 Lang, Logic and Method • NL
[1983] *Language, Logic, and Method* ED: COHEN, R.S. & WARTOFSKY, M.W. • SER (**S** 3311) Boston St Philos Sci 31 • PUBL (**X** 0835) Reidel: Dordrecht viii+464pp • LC-No 82-7558

C 3847 Surveys in Set Th • GB
[1983] *Surveys in Set Theory* ED: MATHIAS, A.R.D. • SER (**S** 3306) Lond Math Soc Lect Note Ser 87 • PUBL (**X** 0805) Cambridge Univ Pr: Cambridge, GB vii+247pp • LC-No 83-10106

C 3848 Algeb & Teor Chisel ('78) • SU
[1978] *Algebra i Teoriya Chisel. Tematicheskij Sbornik Nauchnykh Trudov. Professorsko-prepodatel'skogo Sostava i Aspirantov Vysshikh Uchebnykh Zavedenij Ministerstva Proveshcheniya Kazakhskoj SSR (Algebra and Number Theory. Thematic Work Collection)* ED: DZOZ, N.A. • PUBL Kazakh. Ped. Inst. im. Abaya 136p

C 3865 Algeb Sistemy (Ivanovo) • SU
[1981] *Algebraicheskie Sistemy. Mezhvuzovskij Sbornik Nauchnykh Trudov (Algebraic Systems. Interuniversitary Collection of Scientific Works)* ED: MOLDAVANSKIJ, D.I. • PUBL (**X** 2594) Ivanovo Gos Univ: Ivanovo 234pp

C 3868 Prog in Cybern & Syst Res, Vol 8 • USA
[1982] *Progress in Cybernetics and Systems Research. Vol. VIII: General Systems Methodology; Mathematical Systems Theory; Fuzzy Sets* ED: TRAPPL, R. & KLIR, G.J. & PICHLER, F.R. • PUBL (**X** 2437) Hemisphere Publ: Washington xiii+529pp • ALT PUBL (**X** 0822) McGraw-Hill: New York xiii+529pp • ISBN 0-89116-237-2, LC-No 75-6641

C 3875 Cabal Seminar Los Angeles 1979-81 • USA
[1983] *Cabal Seminar 79-81. Proceeings of the Caltech-UCLA Logic Seminar 1979-81* ED: KECHRIS, A.S. & MARTIN, D.A. & MOSCHOVAKIS, Y.N. • SER (S 3301) Lect Notes Math 1019 • PUBL (X 0811) Springer: Heidelberg & New York 284pp • ISBN 3-540-12688-0, LC-No 83-16866

C 3953 Mat Log & Teor Algor • SU
[1982] *Matematicheskaya Logika i Teoriya Algoritmov (Mathematical Logic and the Theory of Algorithms)* ED: SOBOLEV, S.L. • SER Trudy Inst Mat 2 • PUBL (X 2642) Nauka: Novosibirsk 176pp

C 4014 Semin Th Nombres 1979/80 • F
[1980] *Seminaire de Theorie des Nombres 1979-1980* PUBL Univ Bordeaux I, UER Math Inf:Valence 258pp
• LC-No 85-648844

C 4017 Algeb Dejstviya & Uporyadochennosti • SU
[1983] *Algebraicheskie Dejstviya i Uporyadochennosti (Algebraic Actions and Orderings)* ED: LYAPIN, E.S. • PUBL Leningrad Gos Ped Inst: Leningrad 164pp

C 4019 Kraevye Zadach Dlya Diff Uravnenij & Priloz Mekh & Tekh • SU
[1983] *Kraevye Zadachi Dlya Differentsial'nykh Uravnenij i Ikh Prilozheniya v Mekhanike i Tekhnike (Boundary Value Problems for Differential Equations and Their Applications in Mechanics and Technology)* ED: ZHAUTYKOV, O.A. • PUBL (X 2443) Nauka: Alma-Ata 167pp

C 4062 Montague: Formal Philos • USA
[1974] *Formal Philosophy. Selected Papers of Richard Montague* ED: THOMASON, R.H. • PUBL (X 0875) Yale Univ Pr: New Haven 369pp
• ISBN 0-300-01527-5, LC-No 73-77159

C 4079 Mal'tsev: Algor & Rek Funkt • SU
[1974] *Mal'tsev: Algorithmen und Rekursive Funktionen* PUBL (X 0900) Vieweg: Wiesbaden xiv+336pp • ALT PUBL (X 0911) Akademie Verlag: Berlin 1974 xiv+336pp
• LC-No 66-93980
• TRANSL IN Algorithms and Recursive Functions • TRANSL OF Algoritmy i Rekursivnye Funktsii

C 4085 Handb Philos Log • NL
[1983 & 1984] *Handbook of Philosophical Logic* ED: GABBAY, D. & GUENTHNER, F. • SER (S 3307) Synth Libr 164,165
• PUBL (X 0835) Reidel: Dordrecht 2 Vols: xi+493pp,xi+776pp
• ISBN 90-277-1542-4 (V1), ISBN 90-277-1604-8 (V2), LC-No 83-4277

C 4091 Algeb & Diskret Mat (Riga) • SU
[1984] *Algebra i Diskretnaya Matematika (Algebra and Discrete Mathematics)* ED: IKAUNIEKS, EH.A. • PUBL (X 0895) Latv Valsts (Gos) Univ : Riga 164pp

C 4103 Struktur Teor Relej Ustrojstv • SU
[1963] *Strukturnaya Teoria Relejnykh Ustrojstv* PUBL (X 0899) Akad Nauk SSSR : Moskva
• LC-No 64-44665

C 4134 Mat SSSR za 30 Let • SU
[1948] *Matematika v SSSR za 30 Let 1917-1947 (Mathematics in the USSR for 30 Years 1917-1947)* PUBL OGIZ: Moskva & Leningrad

C 4138 Feys: Modal Logika • SU
[1974] *Feys,R.: Modal'naya Logika (Modal Logic)* ED: MINTS, G.E. • PUBL (X 2027) Nauka: Moskva 520pp
• TRANSL OF Modal Logic
• REM Contains translations of Kripke,S.A. 1959 ID 07554, 1962 ID 07555, 1963 ID 07558, 1965 ID 07559 and of Schuette,K. 1968 ID 24822

C 4157 Rosser:Deux Esquisses Log • F
[1955] *Rosser,J.B.: Deux Esquisses de Logique.* SER Collection de Logique Mathematique, Ser A 7 • PUBL (X 0834) Gauthier-Villars: Paris 67pp • ALT PUBL (X 1313) Nauwelaerts: Louvain
• LC-No 55-3012

C 4183 Model-Theor Log • D
[1985] *Model-Theoretic Logics* ED: BARWISE, J. & FEFERMAN, S. • PUBL (X 0811) Springer: Heidelberg & New York xvii+893pp
• ISBN 3-540-90936-2, LC-No 83-20277

C 4192 Adv in Comput • USA
[1960-1962] *Advances in Computers. Vol 1, 2, 3* ED: ALT, F. (V1-3) & BOOTH, A.D. & MEAGHER, R.E. (V1-2) & RUBINOFF, M. (V3) • PUBL (X 0801) Academic Pr: New York 3 Vols: x+316pp, xiii+434pp, xiii+361pp
• LC-No 59-15761

C 4212 Issl Algeb & Mat Anal • SU
[1965] *Issledovaniya po Algebre i Matematicheskomu Analizu (Studies in Algebra and Mathematical Analysis)* PUBL Karta Moldovenyaske: Kishinev 159pp

C 4213 Algeb, Combin & Log in Comput Sci • NL
[1985] *Algebra, Combinatorics, and Logic in Computer Science* PUBL (X 0809) North Holland: Amsterdam

C 4220 Rep Sem Found Anal • USA
[1963] *Reports of the Seminar on Foundation of Analysis* ED: KREISEL, G. • PUBL Stanford Univ: Stanford
• REM Mimeographed

C 4260 Fischer Lexikon Math 2 • D
[1966] *Fischer Lexikon "Mathematik 2"* ED: BEHNKE, H. & TIETZ, H. • PUBL (X 1265) Fischer: Stuttgart

C 4356 Gen Log Semin Paris 1982/83 • F
[1984] *General Logic Seminar. Paris 1982-83* ED: DELON, F. & LASCAR, D. & LOUVEAU, A. & SABBAGH, G. • SER Publ Math Univ Paris VII • PUBL Univ Paris VII, UER Math: Paris iii+186pp

C 4403 Logika, Pozn, Otrazh • SU
[1984] *Logika, Poznanie, Otrazhenie - Sverdlovsk (Logik, Erkenntnis, Reflektion - Sverdlovsk)*

C 4413 Math Th & Pract Software Syst • SU
[1982] *(Mathematical Theory and Practice of Software Systems)* ED: ERSHOV, A.P. • PUBL -?-: Novosibirsk

C 4415 Semin Schuetzenberger, Lentin Nivat 1969/70 • F
[1970] *Seminaire M.P. Schuetzenberger, A. Lentin et M. Nivat 1969/70: Problemes Mathematiques de la Theorie des Automates, Exposes 1 a 25* PUBL Faculte des Sciences de Paris, Secret Mathematiques: Paris

C 4472 Undecidable Th • NL
[1953] *Undecidable Theories* SER (S 3303) Stud Logic Found Math • PUBL (X 0809) North Holland: Amsterdam

Source Index

C 4581 Priklad Mat, Vyp 2 • SU
[1983] *Prikladnaya Matematika. Mezhvuzovskij Sbornik Nauchnykh Trudov Vyp 2 (Applied Mathematics, Vol. 2)* ED: TONOYAN, R.N. • PUBL (**X** 3559) Erevan Univ: Erevan 148pp

C 4594 Sel Pap Robinson • USA
[1979] *Selected Papers of Abraham Robinson* ED: KEISLER, H.J. & KOERNER, S. & LUXEMBURG, W.A.J. & YOUNG, A.D. • PUBL (**X** 0875) Yale Univ Pr: New Haven xxxvii + 694pp,xlv + 582pp • ALT PUBL (**X** 0809) North Holland: Amsterdam
• REM Vol 1: Model Theory and Algebra. Vol 2: Nonstandard Analysis and Philosophy

C 4659 Autom of Reasoning • D
[1983] *Automation of Reasoning. Vol 1, 2* ED: SIEKMANN, J. & WRIGHTSON, G. • PUBL (**X** 0811) Springer: Heidelberg & New York xii + 525pp,xii + 637pp
• ISBN 3-540-12043-2 (V1), ISBN 3-540-12044-0 (V2)
• REM Vol 1: Classical Papers on Computational Logic 1957-1966. Vol 2: Classical Papers on Computational Logic 1967-1970

C 4688 Prog in Math, Vol 12 • USA
[1972] *Progress in Mathematics. Vol 12: Algebra and Geometry* ED: GAMKRELIDZE, R.V. • PUBL (**X** 1332) Plenum Publ: New York ix + 254pp
• TRANSL OF [1972] (**J** 1501) Itogi Nauki Tekh, Ser Algeb, Topol, Geom 1968

C 4703 Tekh Kibernetika, Vyp 11 • SU
[1979] *Tekhnicheskaya Kibernetika. Tom 11 (Engineering Cybernetics Vol 11)* ED: PETROV, B.N. • PUBL (**X** 2235) VINITI: Moskva 192pp

C 4723 Math Founds of Comput Sci • PL
[1977] *Mathematical Foundations of Computer Science* ED: MAZURKIEWICZ, A. & PAWLAK, Z. • SER (**S** 1926) Banach Cent Publ 2 • PUBL (**X** 1034) PWN: Warsaw 259pp

C 4734 Model & Sist Obrabotki Inform, Vyp 3 • SU
[1984] *Modeli i Sistemy Obrabotki Informatsii. Vyp. 3 (Models and Systems of Information Processing. Vol. 3)* ED: BUBLIK, B.N. • PUBL (**X** 2645) Vishcha Shkola: Kiev 120pp

C 4738 Vopr Teor Mat Mashin, Sbor 2 • SU
[1962] *Voprosy Teorii Matematicheskikh Mashin. Sbornik 2 (Questions in the Theory of Mathematical Machines. Collection 2)* ED: BAZHILEVSKIJ, YU.YA. • PUBL (**X** 0704) Gos Izdat Tekhn-Teor Lit: Moskva 239pp

Publishers

X 0704 *Gosudarstvennoe Izdatel'stvo Tekhnicheskij-Teoreticheskoj Literatury (GITTL) (State Publisher of Technical-Theoretical Literature)* (Moskva, SU)

X 0740 *Shanghai Kexue Jishu Chubanshe (Scientific and Technical Press)* (Shanghai, TJ)

X 0801 *Academic Press* (New York, NY, USA & London, GB) ISBN 0-12

X 0802 *Allyn & Bacon* (London, GB & Boston, MA, USA & Spit Junction, NSW, AUS) ISBN 0-205, ISBN 0-695

X 0803 *American Mathematical Society* (Providence, RI, USA) ISBN 0-8218

X 0804 *Birkhaeuser Verlag* (Basel, CH & Stuttgart, D & Cambridge, MA, USA) ISBN 3-7643

X 0805 *The Cambridge University Press.* (Cambridge, GB & New York, NY, USA & Melbourne, Vic, AUS) ISBN 0-521

X 0806 *VEB Deutscher Verlag der Wissenschaften* (Berlin, DDR) ISBN 3-326

X 0807 *Duke University Press* (Durham, NC, USA) ISBN 0-8223

X 0808 *W. Kohlhammer* (Stuttgart, D & Koeln, D & Berlin, D & Mainz, D) ISBN 3-17

X 0809 *North-Holland Publishing Company.* (Amsterdam, NL & Oxford, GB) ISBN 0-7204 • REL PUBL (**X** 0838) Amer Elsevier: New York

X 0811 *Springer-Verlag* (Heidelberg, D & Berlin, D & New York, NY, USA & Tokyo, J) ISBN 3-540, ISBN 0-387 • REL PUBL (**X** 1231) Barth: Leipzig & (**X** 0902) Springer: Wien

X 0812 *Wolters-Noordhoff* (Groningen, NL) ISBN 90-01 • REL PUBL (**X** 1317) Noordhoff: Groningen

X 0813 *Dover Publications* (New York, NY, USA) ISBN 0-486

X 0814 *R.Oldenbourg Verlag* (Muenchen, D & Wien, A) ISBN 3-486

X 0815 *The Clarendon Press* (Oxford, GB) ISBN 0-19 • REL PUBL (**X** 0894) Oxford Univ Pr: Oxford • REM This Imprint is Used for Academic Books Published by X0894.

X 0817 *Gesellschaft fuer Mathematik und Datenverarbeitung* (Bonn, D) ISBN 3-88457

X 0819 *Prentice Hall* (Englewood Cliffs, NJ, USA & Brookvale, NSW, AUS & Scarborough, ON, CDN) ISBN 0-13 • REL PUBL (**X** 2040) Winthrop: Cambridge

X 0820 *Interscience Publishers* (New York, NY, USA & Chichester, GB) ISBN 0-470 • REL PUBL (**X** 0827) Wiley & Sons: New York

X 0822 *McGraw-Hill Book Company* (New York, NY, USA & Roseville, NSW, AUS & Isando, ZA & Maidenhead, GB & Singapore, SGP & Scarborough, CDN & Sao Paulo, BR) ISBN 0-07 • REM Member Firms: 1) CRM/McGraw-Hill, Del Mar, CA, USA. 2) CTB/McGraw-Hill, Monterey, CA, USA. 3) Edutronics/McGraw-Hill, Los Angeles, CA, USA. 4) Instruto/McGraw-Hill, Paoli, PA, USA. 5) McGraw-Hill Continuing Education Center, Washington, DC, USA. 6) McGraw-Hill International Book Company, Singapore, SGP. 7) Sheperd's/McGraw-Hill, Colorado Springs, CO, USA. 8) McGraw-Hill do Brasil, Sao Paulo, BR.

X 0823 *B.G.Teubner* (Stuttgart, D) ISBN 3-519 • REM See also X1079

X 0824 *The Free Press* (New York, NY, USA) ISBN 0-02 • REL PUBL (**X** 0843) Macmillan : New York & London

X 0827 *J.Wiley & Sons* (New York, NY, USA & Chichester, GB & Rexdale, ON, CDN & Auckland, NZ) ISBN 0-471 • REL PUBL (**X** 0942) Norton: New York & (**X** 0820) Intersci Publ: New York & (**X** 0880) Ronald Press: New York & (**X** 2737) Israel Progr Sci Transl: Jerusalem

X 0832 *Addison-Wesley Publishing Co.* (Reading, MA, USA & London, GB & Don Mills, ON, CDN & North Ryde, NSW, AUS) ISBN 0-201 • REL PUBL (**X** 0867) Benjamin: Reading

X 0833 *Jerusalem Academic Press* (Jerusalem, IL)

X 0834 *Gauthier-Villars Editeur* (Paris, F) ISBN 2-04

X 0835 *D.Reidel Publishing Company* (Dordrecht, NL & Hingham, MA, USA) ISBN 90-277

X 0836 *Gordon & Breach, Science Publishers* (New York, NY, USA & London, GB & Paris, F) ISBN 0-677

X 0838 *American Elsevier Publishing Co.* (New York, NY, USA & Amsterdam, NL & London, GB) ISBN 0-444, ISBN 0-525, ISBN 0-87690 • REL PUBL (**X** 0809) North Holland: Amsterdam

X 0842 *University of Wisconsin Press* (Madison, WI, USA & London, GB) ISBN 0-299 • REL PUBL (**X** 3828) Amer Univ Publ Group: London

X 0843 *Macmillan Publishing Company* (New York, NY, USA & Melbourne, Vic, AUS & London, GB & Toronto, Ont, CDN) ISBN 0-02 • REL PUBL (**X** 2375) Macmillan Journal: London & (**X** 0824) Free Press: New York

X 0845 *University of Notre Dame Press* (Notre Dame, IN, USA & London, GB) ISBN 0-268

X 0848 *Chelsea Publishing Company* (New York, NY, USA) ISBN 0-8284

X 0850 *Armand Colin, Editeur* (Paris, F) ISBN 2-200

X 0856 *Dunod, Editeur* (Paris, F) ISBN 2-04

X 0857 *Princeton University Press* (Princeton, NJ, USA & Guildford, GB) ISBN 0-691

X 0858 *Harvard University Press* (Cambridge, MA, USA & London, GB) ISBN 0-674

X 0859 *Hermann, Editeurs des Sciences et des Arts* (Paris, F) ISBN 2-7056

X 0860 *Cremonese Edizioni* (Firenze, I) ISBN 88-7083

X 0865 *The MIT Press* (Cambridge, MA, USA & London, GB) ISBN 0-262

X 0867 *W.A. Benjamin* (Reading, MA, USA) ISBN 0-8053 • REL PUBL (**X** 0832) Addison-Wesley: Reading

X 0868 *Penguin Books* (Harmondsworth, GB & New York, NY, USA & Ringwood, Vic, AUS & & Auckland, NZ & Markham, CDN) ISBN 0-14

X 0869 *Pergamon Press* (Oxford, GB & Elmsford, NY, USA & Rushcutters Bay, NSW, AUS & & Willowdale, ON, CDN & Paris, F) ISBN 0-08 • REL PUBL (**X** 0900) Vieweg: Wiesbaden

X 0870 *Physica-Verlag Rudolf Liebing* (Wuerzburg, D & Wien, A) ISBN 3-7908

X 0871 *Academiei Republicii Socialiste Romania Editura (RSR)* (Bucharest, RO)

X 0873 *Mouton et Cie.* (Paris, F) ISBN 2-7193

X 0874 *National Press Books* (Palo Alto, CA, USA) ISBN 0-87484

X 0875 *Yale University Press* (New Haven, CT, USA & London, GB) ISBN 0-300

X 0876 *Bibliographisches Institut* (Mannheim, D & Wien, A & Zuerich, CH) ISBN 3-411

X 0877 *Max Niemeyer Verlag* (Tuebingen, D & Halle, DDR) ISBN 3-484

X 0880 *Ronald Press & Co.* (New York, NY, USA) ISBN 0-8260 • REL PUBL (**X** 0827) Wiley & Sons: New York

X 0882 *Uppsala Universitet, Filosofiska Foereningen och Filosofiska Institutionen* (Uppsala, S)

X 0883 *The Catholic University of America Press* (Washington, DC, USA & Baltimore, MD, USA) ISBN 0-8132

X 0885 *Izdatel'stvo Mir* (Moskva, SU)

X 0887 *Raven Press* (New York, NY, USA) ISBN 0-89004, ISBN 0-911216

X 0893 *Les Presses de l'Universite de Montreal* (Montreal, PQ, CDN) ISBN 2-7606

X 0894 *Oxford University Press* (Oxford, GB & London, GB & Melbourne, Vic, AUS & Don Mills, ON, CDN & Nairobi, EAK & Auckland, NZ & Petaling Jaya, MAL & New York, NY, USA & Karachi, PAK & Harare, ZW) ISBN 0-19 • REL PUBL (**X** 0815) Clarendon Pr: Oxford

X 0895 *Latvijas Valsts Universitate* (Riga, SU)

X 0898 *Izdatel'stvo Moskovskogo Gosudarstvennogo Universiteta* (Moskva, SU)

X 0899 *Izdatel'stvo Akademii Nauk SSSR* (Moskva, SU)

X 0900 *Vieweg, Friedrich & Sohn Verlagsgesellschaft* (Wiesbaden, D) ISBN 3-528 • REL PUBL (**X** 0869) Pergamon Pr: Oxford

X 0902 *Springer-Verlag* (Wien, A) ISBN 3-211 • REL PUBL (**X** 0811) Springer: Heidelberg & New York

X 0903 *Vandenhoeck & Ruprecht* (Goettingen, D) ISBN 3-525 • REM Member Firms: 1) E. Klotz Verlag, Goettingen, D. 2) Verlag der Medizinischen Psychologie Goettingen, Dr D.& Dr A. Ruprecht, Goettingen, D

X 0905 *Boringhieri Editore* (Torino, I) ISBN 88-339

X 0906 *Markham Publishing Company. A Rand McNally College Publ. Co.* (Chicago, IL, USA) ISBN 0-8140

X 0908 *Universitaet Bonn, Mathematisches Institut* (Bonn, D)

X 0909 *Cedam* (Padova, I)

X 0910 *Aschendorffsche Verlagsbuchhandlung* (Muenster, D) ISBN 3-402

X 0911 *Akademie Verlag* (Berlin, DDR)

X 0913 *Novosibirskij Gosudarstvennyj Universitet* (Novosibirsk, SU)

X 0916 *Suhrkamp Verlag* (Frankfurt, D) ISBN 3-518

X 0918 *Ernst Klett Verlag* (Stuttgart, D) ISBN 3-12 • REL PUBL (**X** 1255) Deuticke: Wien

X 0924 *New York University Press* (New York, NY, USA) ISBN 0-8147

X 0925 *Tata Institute of Fundamental Research* (Bombay, IND)

X 0926 *University of California Press* (Berkeley, CA, USA & London, GB) ISBN 0-520 • REL PUBL (**X** 1291) Johns Hopkins Univ Pr: Baltimore

X 0928 *Akademiai Kiado, Publishing House of the Hungarian Academy of Sciences.* (Budapest, H) ISBN 963-05

X 0932 *Otto Salle Verlag* (Frankfurt, D) ISBN 3-7935

X 0938 *Izdatel'stvo Leningradskogo Universiteta* (Leningrad, SU)

X 0942 *W.W.Norton & Co.* (New York, NY, USA & London, GB) ISBN 0-393 • REL PUBL (**X** 0827) Wiley & Sons: New York

X 0981 *Random House* (New York, NY, USA & Mississauga, ON, CDN) ISBN 0-394

X 0994 *W.H.Freeman and Co.* (San Francisco, CA, USA & Oxford, GB) ISBN 0-7167

X 0997 *Queen's University* (Kingston, ON, CDN) ISBN 0-88911, ISBN 0-9690334

X 0999 *Centre National de la Recherche Scientifique (CNRS), Institut Blaise Pascal* (Paris, F)

X 1004 *D.C.Heath & Co.* (Lexington, MA, USA & Toronto, ON, CDN) ISBN 0-669

X 1006 *Irkutskij Gosudarstvennyj Universitet* (Irkutsk, SU)

X 1015 *University of Toronto Press* (Toronto, ON, CDN & Buffalo, NY, USA) ISBN 0-8020

X 1034 *Panstwowe Wydawnictwo Naukowe (PWN)* (Warsaw, PL) ISBN 83-01

X 1036 *Volk und Wissen, Volkseigener Verlag Berlin* (Berlin, DDR) ISBN 3-353

X 1051 *Rijksuniversiteit Utrecht, Mathematisch Instituut.* (Utrecht, NL)

X 1052 *Izdatel'stvo Tbilisskogo Universiteta* (Tbilisi, SU)

X 1078 *Simon and Schuster* (New York, NY, USA) ISBN 0-671

X 1079 *B.G.Teubner Verlagsgesellschaft* (Leipzig, DDR) ISBN 3-519 • REM See also X0823

X 1080 *Scottish Academic Press* (Edinburgh, GB) ISBN 0-7073

X 1096 *Basil Blackwell* (Oxford, GB) ISBN 0-631, ISBN 0-632, ISBN 0-86286, ISBN 0-86793 • REM Also: Blackwell Scientific Publications: Oxford, GB & Carlton, Vic, AUS

X 1121 *Mathematisch Centrum* (Amsterdam, NL) ISBN 90-6196

X 1163 *Almqvist & Wiksell Foerlag* (Stockholm, S & Bromma, S & Goeteborg, S & Malmoe, S) ISBN 91-20

X 1165 *Hodder & Stroughton Educational* (London, GB & Sevenoaks, GB & Glenfield, NZ & Don Mills, ON, CDN & Lane Cove, NSW, AUS & Ashwood, Vic, AUS & Brisbane, Qld, AUS & Adelaide, SA, AUS & West Perth, WA, AUS) ISBN 0-340 • REM Formerly: The English University Press: London, GB

X 1172 *Les Editions de l'Office Central de Librairie S.A.R.L. (O.C.D.L.)* (Paris, F) ISBN 2-7043

X 1174 *Walter de Gruyter* (Berlin, D) ISBN 3-11 • REM Member Firms: 1) de Gruyter: Hawthorne, NY, USA. 2) Mouton Publishers: Berlin, D

X 1212 *Izdatel'stvo Belorusskogo Gosudarstvennogo Universiteta* (Minsk, SU)

X 1214 *New York University, Courant Institute of Mathematical Science.* (New York, NY, USA)

X 1226 *Academia* (Prague, CS) • REM Publishing House of the Czechoslovak Academy of Science

X 1231 *Johann Ambrosius Barth* (Leipzig, DDR) ISBN 3-335 • REL PUBL (**X** 0811) Springer: Heidelberg & New York

X 1234 *G.Bell & Sons* (London, GB) ISBN 0-7135

X 1249 *Chapman & Hall* (London, GB) ISBN 0-412

X 1255 *Franz Deuticke, Verlagsgesellschaft* (Wien, A) ISBN 3-7005 • REL PUBL (**X** 0918) Klett: Stuttgart

X 1261 *Edinburgh University Press* (Edinburgh, GB) ISBN 0-85224

X 1265 *Gustav Fischer Verlag* (Stuttgart, D) ISBN 3-437

X 1268 *Orell Fuessli Verlag* (Zuerich, CH) ISBN 3-280

X 1275 *Anton Hain K.G. Meisenheim Verlag* (Koenigstein, D & Meisenheim, D) ISBN 3-445 • REL PUBL (**X** 1588) Athenaeum/Hain/Hanstein: Koenigstein

X 1277 *Hayden Book Company* (Rochelle Park, NY, USA) ISBN 0-8104 • REL PUBL (**X** 1354) Spartan Books : Sutton

X 1279 *Herder & Co.* (Freiburg, D & Roma, I) ISBN 3-451

X 1285 *University of Illinois Press* (Urbana, IL, USA & London, GB) ISBN 0-252 • REL PUBL (**X** 3828) Amer Univ Publ Group: London

X 1286 *Indiana University Press* (Bloomington, IN, USA & London, GB) ISBN 0-253 • REL PUBL (**X** 3828) Amer Univ Publ Group: London

X 1291 *Johns Hopkins University Press* (Baltimore, MD, USA) ISBN 0-8018

X 1298 *Mathematical Association of America* (Washington, DC, USA) ISBN 0-88385

X 1299 *Magnes Press* (Jerusalem, IL) ISBN 965-223

X 1313 *Nauwelaerts, Beatrice* (Louvain, B) ISBN 2-900014

X 1317 *P.Noordhoff International Publishing* (Groningen, NL & Leiden, NL) ISBN 90-01 • REL PUBL (**X** 0812) Wolters-Noordhoff : Groningen & (**X** 1352) Sijthoff: Leiden • REM Now: Sijthoff & Noordhoff International Publishers: Leiden, NL

X 1319 *La Nuova Italia Editrice* (Firenze, I) ISBN 88-221

X 1323 *Oliver & Boyd* (Edinburgh, GB) ISBN 0-05

X 1330 *Pitman Publishing* (Belmont, CA, USA & London, GB & Toronto, ON, CDN & & Carlton, Vic, AUS & Johannesburg, ZA & Wellington, NZ & & Auckland, NZ) ISBN 0-273, ISBN 0-8224, ISBN 0-915092, ISBN 0-272

X 1332 *Plenum Publishing Corporation* (New York, NY, USA & London, GB) ISBN 0-306 • REM Also Called Plenum Press

X 1349 *Editions du Seuil* (Paris, F) ISBN 2-02

X 1352 *A.W.Sijthoff International Publishing Co.* (Leiden, NL) ISBN 90-218, ISBN 90-286 • REL PUBL (**X** 1317) Noordhoff: Groningen

X 1354 *Spartan Books* (Sutton, GB & Rochelle Park, NJ, USA) ISBN 0-905532 • REL PUBL (**X** 1277) Hayden: Rochelle Park

X 1355 *Stanford University Press* (Stanford, CA, USA) ISBN 0-8047

X 1358 *University of Texas Press* (Austin, TX, USA & London, GB) ISBN 0-292 • REL PUBL (**X** 3828) Amer Univ Publ Group: London

X 1364 *Vita e Pensiero, Publicazioni della Universita Cattolica* (Milano, I) ISBN 88-343

X 1367 *University of Washington Press* (Seattle, WA, USA & London, GB) ISBN 0-295 • REL PUBL (**X** 3828) Amer Univ Publ Group: London

X 1375 *Nicola Zanichelli Editore* (Bologna, I) ISBN 88-08

X 1422 *Editions Bordas* (Paris, F) ISBN 2-04

X 1434 *Kalininskij Gosudarstvennyj Universitet* (Kalinin, SU)

X 1493 *C.W.K.Gleerup Bokfoerlag a.B.* (Lund, S) ISBN 91-40

X 1512 *Rheinische Friedrich-Wilhelms-Universitaet Bonn, Institut fuer Informatik.* (Bonn, D)

X 1532 *Westfaelische Wilhelms-Universitaet Muenster, Institut fuer Mathematische Logik und Grundlagenforschung* (Muenster, D)

X 1553 *Verlag Dokumentation Saur* (Muenchen, D & London, GB) ISBN 3-7940, ISBN 3-598 • REL PUBL (**X** 2797) Minerva Publ: Muenchen • REM Member Firm: Zell Publishers: Oxford, GB

X 1554 *Universitetsforlaget* (Oslo, N & Bergen, N & Tromsoe, N & New York, NY, USA) ISBN 82-00 • REM Also: UB-Forlaget: Oslo, N

X 1574 *Vyssheyshaya Shkola* (Minsk, SU)

X 1588 *Verlagsgruppe Athenaeum/ Hain/ Hanstein* (Koenigstein/Ts, D) • REL PUBL (**X** 1275) Hain: Koenigstein & (**X** 2665) Athenaeum: Frankfurt

X 1599 *Aarhus Universitet, Matematisk Institut* (Aarhus, DK) ISBN 87-87436

X 1623 *Universite de Paris VI, Institut Henri Poincare* (Paris, F) • REM The University Has Now Been Divided Into Smaller Units. The Mathematics Department Is Now: Universite de Paris VII-Pierre et Marie Curie, Secretariat Mathematique

X 1656 *Izdatel'stvo Inostr. Lit.* (Moskva, SU)

X 1663 *Ludwig-Maximilians-Universitaet, Pressereferat* (Muenchen, D) ISBN 3-922480

X 1684 *Marcel Dekker* (New York, NY, USA & Basel, CH) ISBN 0-8247 • REL PUBL (**X** 2442) Dekker Journal: New York

X 1700 *Fan* (Tashkent, SU)

X 1732 *Bibliopolis* (Napoli, I) ISBN 88-7088

X 1758 *Ars Polona* (Warsaw, PL)

X 1761 *Hekdosis Hellinikis Mathematikes Hetaireias-Greek Mathematical Society* (Athens, GR)

X 1763 *Royal Society* (London, GB) ISBN 0-85403

X 1773 *VEDA - Vydavatelstvo Slovenskej Akademie Vied* (Bratislava, CS)

X 1777 *Western Periodicals Company* (North Hollywood, CA, USA)

X 1781 *Tecnos* (Madrid, E) ISBN 84-309

X 1821 *Real Academia de Ciencias Exactas, Fisicas y Naturales* (Madrid, E)

X 1876 *Kexue Chubanshe (Science Press)* (Beijing, TJ)

X 1941 *Nanyang University Publications* (Singapore, MAL)

X 2005 *Australian National University* (Canberra, ACT, AUS) ISBN 0-909851

X 2027 *Izdatel'stvo Nauka* (Moskva, SU & Alma-Ata, SU & Leningrad, SU & Novosibirsk, SU)

X 2039 *Polytechnic Institute of New York* (Brooklyn, NY, USA)

X 2040 *Winthrop* (Cambridge, MA, USA) ISBN 0-87626 • REL PUBL (**X** 0819) Prentice Hall: Englewood Cliffs

X 2091 *Consejo Superior de Investigaciones Cientificas.* (Madrid, E) ISBN 84-00

X 2121 *Accademia Nazionale dei Lincei* (Roma, I) ISBN 88-218

X 2152 *Tokai University Press* (Tokyo, J) ISBN 4-486

X 2179 *IEEE (Institute of Electrical and Electronics Engineers* (New York, NY, USA & Long Beach, CA, USA & Piscataway, CA, USA) ISBN 0-87942 • REM Section: IEEE Computer Society: Long Beach, CA, USA. Section: IEEE United States Activities Commitee: Piscataway, CA, USA

X 2188 *Princeton University, Department of Electrical Engineering and Computer Science* (Princeton, NJ, USA)

X 2189 *Universitatea "Babes-Bolyai", Biblioteca Centrala Universitara* (Cluj-Napoca, RO)

X 2190 *Linguistic Society of America* (Arlington, VA, USA)

X 2197 *Tohoku University, Mathematical Institute* (Sendai, J)

X 2198 *Universidad de Barcelona, Facultad de Ciencias, Seminario Matematico* (Barcelona, E)

X 2199 *Izdatel'stvo Naukova Dumka* (Kiev, SU)

X 2201 *Australasian Association of Philosophy.* (Bundoora, Vic, AUS) ISBN 0-9592545

X 2203 *Philosophy of Science Association* (East Lansing, MI, USA) ISBN 0-917586

X 2204 *American Institute of Physics* (New York, NY, USA) ISBN 0-88318

X 2205 *(ACM) Association for Computing Machinery* (New York, NY, USA) ISBN 0-89791

X 2207 *Universitatea "Al. I. Cuza" din Iasi* (Jassy (Iasi), RO)

X 2208 *Taylor & Francis* (London, GB) ISBN 0-85066

X 2209 *Izdatel'stvo Akademii Nauk Gruzinskoj SSR* (Tbilisi, SU)

X 2211 *Friedrich Schiller Universitaet* (Jena, DDR)

X 2214 *Dialectica* (Bienne/Biel, CH)

X 2219 *Humboldt-Universitaet zu Berlin* (Berlin, DDR)

X 2224 *Matematisk Institut* (Bergen, N)

X 2225 *Izdatel'stvo Akademii Nauk Armyanskoj SSR* (Erevan, SU)

X 2227 *Scuola Normale Superiore di Pisa* (Pisa, I) ISBN 88-7642

X 2229 *Universita J.E.Purkyne* (Brno, CS) • REM Formerly Named Masarykova Universita

X 2230 *Isdevnieciba Zinatne* (Riga, SU)

X 2233 *Instituto de Matematica Pura et Aplicada (IMPA)* (Rio de Janeiro, BR) ISBN 85-244

X 2235 *Vsesoyuznyj Institut Nauchnoj i Tekhnicheskoj Informatsii (VINITI), Gosudarstvennyj Komitet SSSR po Nauke: Tekhnike (GKNT SSSR), Akademiya Nauk (AN) SSSR* (Moskva, SU)

X 2237 *Izdatelstvo na Bulgarskata Akademia na Naukite (Publishing House of the Bulgarian Academy of Sciences.)* (Sofia, BG)

X 2238 *Sociedad Colombiana de Matematicas* (Bogota, CO) ISBN 958-95081

X 2241 *New York Academy of Sciences* (New York, NY, USA) ISBN 0-89072, ISBN 0-89766

X 2242 *Osaka University, Department of Mathematics* (Osaka, J)

X 2243 *University of Osaka Prefecture, Department of Mathematics* (Osaka, J)

X 2244 *Societe Mathematique de France* (Paris, F) ISBN 2-85629

X 2248 *Drushtvo Matematichara i Fizichara Sr Hrvatske.* (Zagreb, YU)

X 2256 *Casa Editrice Felice le Monnier* (Firenze, I)

X 2257 *John Benjamins B.V.* (Amsterdam, NL & Philadelphia, PA, USA) ISBN 90-272

X 2258 *Periodika* (Tallinn, SU)

X 2261 *Ministerstvo Vysshego i Srednego Spetsial'nogo Obrazovaniya* (Moskva, SU)

Miscellaneous Indexes

External classifications

This index complements the Subject Index at the beginning of this volume; it lists the items which, in addition to classifications in the present volume, have classifications *external to this volume*. These items are ordered by external classification code and within each code by author (the first alphabetically in the case of multi-author items), year and identification number (thus an item with, for example, two external classifications occurs twice in this listing). This index provides another way to search the bibliography. With it, the user can easily identify those items in this volume classified also in some area external to this volume.

B03

Bergman, George M. [1978] 50464
Burks, A.W. [1954] 33359
Calude, C. [1978] 52071
Calude, C. [1981] 55399
Farat, V.M. [1979] 81930
Gandy, R.O. [1974] 17242
Makanin, G.S. [1977] 51395
Makanin, G.S. [1977] 51863
Makanin, G.S. [1980] 82143
Mayoh, B.H. [1970] 17641
Weiss, M. [1967] 14130

B05

Aanderaa, S.O. [1979] 53853
Aanderaa, S.O. [1981] 55613
Adleman, L.M. [1978] 69046
Bukharaeva, Z.K. [1978] 52886
Chen, Jiyuan [1984] 44191
Commentz-Walter, B. [1979] 53164
Cook, S.A. [1974] 25005
Cook, S.A. [1979] 56282
Cresswell, M.J. [1964] 02531
Fischer, Michael J. [1982] 35363
Fischer, R. [1982] 42341
Gabrielian, A. [1981] 54258
Goessel, M. [1978] 69601
Gostev, Yu.G. [1981] 81380
Haeussler, A.F. [1976] 52007
Hamblin, C.L. [1973] 62308
Harper, L.H. [1975] 62346
Hay, L. [1978] 30704
Huzino, S. [1959] 06355
Kalicki, J. [1954] 06815
Khasin, L.S. [1969] 07080
Korshunov, A.D. [1969] 16309
Kosovskij, N.K. [1978] 90060
Kuz'min, V.A. [1965] 29866
Lewis, H.R. [1979] 53794
Lieberherr, K.J. [1979] 38843
Lieberherr, K.J. [1981] 69677
Lieberherr, K.J. [1982] 46285
Luckhardt, H. [1984] 45393
Markov, A.A. [1963] 08773
Markov, A.A. [1964] 08774
Markov, A.A. [1967] 08776
Nazaryan, G.A. [1975] 53659
Nazaryan, G.A. [1976] 76769
Nazaryan, G.A. [1979] 53426
Nazaryan, G.A. [1980] 76768
Nazaryan, G.A. [1982] 34748
Nurmeev, N.N. [1976] 52721
Paltanea, R. [1980] 35233
Plotkin, J.M. [1982] 35398
Poljak, S. [1982] 39883
Pudlak, P. [1984] 43900
Ras, Z. [1971] 10995
Ras, Z. [1971] 27358
Schaefer, T.J. [1978] 82718
Schnorr, C.-P. [1976] 69903
Schnorr, C.-P. [1977] 16638
Schuster, P. [1976] 69906
Scognamiglio, G. [1963] 29833
Stork, H.-G. [1979] 52752
Tarjan, R.E. [1978] 69946
Urbano, R.H. [1963] 13791
Wegener, I. [1981] 66300
Wegener, I. [1981] 69392

B10

Bernays, P. [1958] 01541
Blass, A.R. [1984] 47716
Boerger, E. [1978] 52807
Bullock, A.M. [1973] 01748
Cartwright, Robert [1979] 56231
Christen, C. [1976] 52348
Church, A. [1932] 02123
Church, A. [1933] 02124
Cobham, A. [1956] 02278
Craig, W. [1960] 02513
Deutsch, M. [1975] 02972
Dulac, M.-H. [1971] 03165
Goedel, K. [1933] 20885
Hasenjaeger, G. [1953] 05723
Hasenjaeger, G. [1955] 27719
Hermes, H. [1968] 14956
Kalmar, L. [1951] 42204
Kalmar, L. [1956] 06861
Keisler, H.J. [1973] 07033
Kreisel, G. [1950] 07481
Kreisel, G. [1953] 20815
Krom, Melven R. [1963] 07571
Kuratowski, K. [1931] 07656
Lehmann, G. [1985] 42737
Loeb, M.H. [1976] 14746
Melzi, G. [1976] 76279
Nagornyj, N.M. [1976] 76745
Nerode, A. [1980] 54927
Plaisted, D.A. [1984] 44042
Pudlak, P. [1975] 29600
Rodriguez Artalejo, M. [1981] 77840
Rosser, J.B. [1955] 23483
Rus, T. [1964] 16208
Semenov, A.L. [1977] 51323
Sharonov, V.I. [1970] 14357
Slomson, A. [1969] 12473
Smart, J.J.C. [1961] 31260
Trakhtenbrot, B.A. [1953] 19742
Wang, Juentin [1973] 65982

B15

Andrews, P.B. [1974] 03832
Anikeev, A.S. [1972] 00375
Baxter, L.D. [1978] 52222
Belyakin, N.V. [1983] 41479
Bernays, P. [1969] 32552
Buechi, J.R. [1960] 01690
Buechi, J.R. [1965] 01694
Buechi, J.R. [1965] 01697
Buechi, J.R. [1969] 01700
Buechi, J.R. [1983] 33901
Buechi, J.R. [1983] 40515
Chlebus, B.S. [1980] 56482
Church, A. [1952] 02144
Cooper, D.C. [1969] 28130
Deutsch, M. [1977] 28193
Elgot, C.C. [1966] 03291
Feferman, S. [1977] 27330
Friedman, H.M. [1978] 31762
Gacs, P. [1977] 52753
Garland, S.J. [1974] 24144
Goldfarb, W.D. [1981] 54331
Gurevich, Y. [1983] 33764
Gurevich, Y. [1983] 33765
Hasenjaeger, G. [1955] 27719
Ladner, R.E. [1977] 52252
Lisovik, L.P. [1984] 48409
Mostowski, A.Wlodzimierz [1979] 76601
Mostowski, A.Wlodzimierz [1981] 55152
Muchnik, A.A. [1985] 47256
Muller, D.E. [1981] 76648
Muller, D.E. [1985] 47646
Nabebin, A.A. [1977] 52155
Nakamura, A. [1981] 55427
Nerode, A. [1980] 54927
Palyutin, E.A. [1971] 64365
Peter, R. [1969] 32551

Pinus, A.G. [1973] 24124
Rabin, M.O. [1968] 29625
Rabin, M.O. [1969] 10942
Rabin, M.O. [1970] 22233
Rabin, M.O. [1971] 14690
Rabin, M.O. [1972] 10943
Roedding, D. [1972] 11343
Scarpellini, B. [1985] 47473
Seese, D.G. [1977] 27132
Siefkes, D. [1970] 12145
Siefkes, D. [1970] 27984
Siefkes, D. [1971] 12146
Siefkes, D. [1978] 29136
Simon, J. [1977] 16610
Thatcher, J.W. [1968] 28596
Vanderveken, D.R. [1976] 50285
Wasilewska, A. [1985] 49467

B20

Aanderaa, S.O. [1971] 00087
Aanderaa, S.O. [1973] 00010
Aanderaa, S.O. [1974] 03803
Aanderaa, S.O. [1982] 33756
Aanderaa, S.O. [1982] 35210
Behmann, H. [1922] 00942
Behmann, H. [1923] 00943
Bernays, P. [1928] 01076
Beth, E.W. [1950] 01156
Boerger, E. [1974] 01368
Boerger, E. [1974] 28158
Boerger, E. [1978] 29169
Boerger, E. [1984] 40047
Bollman, D.A. [1972] 01420
Buechi, J.R. [1962] 01691
Burks, A.W. [1962] 01780
Burks, A.W. [1962] 02718
Church, A. [1936] 02132
Dassow, J. [1985] 49482
Denenberg, L. [1984] 43373
Deutsch, M. [1981] 54907
Deutsch, M. [1984] 43185
Doshita, S. [1983] 44209
Dreben, B. [1957] 03611
Dreben, B. [1962] 03127
Fridman, Eh.I. [1972] 77209
Fuerer, M. [1981] 35169
Fuerer, M. [1984] 43160
Galvin, F. [1967] 04765
Galvin, F. [1970] 04766
Genenz, J. [1964] 21347
Gladstone, M.D. [1965] 05010
Gladstone, M.D. [1968] 05013
Goedel, K. [1933] 20885
Goldfarb, W.D. [1973] 08111
Goldfarb, W.D. [1975] 05117
Goldfarb, W.D. [1981] 55275
Goldfarb, W.D. [1984] 36555
Goldfarb, W.D. [1984] 42450
Gurevich, Y. [1965] 05452
Gurevich, Y. [1966] 05455
Gurevich, Y. [1966] 05456
Gurevich, Y. [1966] 05458
Gurevich, Y. [1970] 22243
Gurevich, Y. [1973] 25843
Gurevich, Y. [1976] 14777
Gurevich, Y. [1976] 23714
Gurevich, Y. [1982] 33757
Gurevich, Y. [1983] 33766
Hermes, H. [1971] 06021
Hughes, C.E. [1976] 14752
Hughes, C.E. [1976] 14753
Kahr, A.S. [1962] 19074
Kalmar, L. [1932] 06837
Kalmar, L. [1937] 06841
Kalmar, L. [1937] 06842
Kalmar, L. [1939] 06843
Kalmar, L. [1947] 06846
Kalmar, L. [1950] 06851
Kalmar, L. [1950] 06854
Kalmar, L. [1956] 16972
Kanovich, M.I. [1975] 18218
Khomich, V.I. [1975] 17549
Kleine Buening, H. [1981] 55476
Kopieki, R. [1975] 23964
Kostyrko, V.F. [1964] 19124
Kostyrko, V.F. [1966] 19123
Kostyrko, V.F. [1971] 19122
Kreisel, G. [1961] 07524
Krom, Melven R. [1967] 07574
Krom, Melven R. [1970] 07576
Lewis, H.R. [1976] 14766
Lewis, H.R. [1978] 75527
Lewis, H.R. [1978] 82072
Lewis, H.R. [1979] 53515
Lewis, H.R. [1980] 36546
Lifschitz, V. [1967] 08138
Manders, K.L. [1979] 53838
Marchenkov, S.S. [1982] 35236
Maslov, S.Yu. [1965] 19335
McKenzie, R. [1975] 09009
Mints, G.E. [1971] 09286
Mints, G.E. [1982] 35718
Mundici, D. [1983] 37383
Norgela, S.A. [1976] 32671
Orevkov, V.P. [1968] 10153
Orevkov, V.P. [1971] 10157
Paola di, R.A. [1971] 30683
Pepis, J. [1936] 10360
Pepis, J. [1937] 19493
Pepis, J. [1938] 10361
Pepis, J. [1938] 10362
Pieczkowski, A. [1968] 10474
Ras, Z. [1972] 10996
Rautenberg, W. [1968] 11048
Rogers Jr., H. [1956] 11355
Rosser, J.B. [1939] 11547
Rozenblat, B.V. [1979] 53949
Schuette, K. [1933] 12997
Sharonov, V.I. [1970] 14357
Singletary, W.E. [1964] 12350
Singletary, W.E. [1974] 12356
Skolem, T.A. [1919] 12383
Skolem, T.A. [1936] 40914
Skolem, T.A. [1937] 27938
Stenger, H.-J. [1984] 43303
Suranyi, J. [1943] 19712
Suranyi, J. [1949] 20805
Suranyi, J. [1950] 12740
Suranyi, J. [1951] 12741
Suranyi, J. [1955] 12742
Suranyi, J. [1959] 12743
Suranyi, J. [1971] 12744
Tajmanov, A.D. [1966] 19765
Trakhtenbrot, B.A. [1961] 19734
Trakhtenbrot, B.A. [1962] 19733
Turashvili, T.V. [1975] 24005
Turashvili, T.V. [1977] 50731
Wang, Hao [1963] 14040
Wang, Hao [1965] 14057
Wirsing, M. [1977] 51036
Wirsing, M. [1978] 29168
Yntema, M.K. [1964] 18050
Zhegalkin, I.I. [1939] 17903
Zykov, A.A. [1959] 30599

B22

Applebee, R.C. [1970] 43922
Harrop, R. [1964] 05682
Harrop, R. [1965] 05684
Kuznetsov, A.V. [1963] 19144
Nakamura, A. [1970] 09788
Shekhtman, V.B. [1982] 38116
Urquhart, A.I.F. [1981] 54908
Wagner, K. [1972] 13969

B25

Aanderaa, S.O. [1973] 00010
Aanderaa, S.O. [1982] 33756
Ackermann, W. [1928] 00109
Ackermann, W. [1936] 00112
Adamson, A. [1979] 53245
Adyan, S.I. [1957] 00193
Almagambetov, Zh.A. [1965] 00278
Alton, D.A. [1975] 17945
Andrews, P.B. [1974] 03832
Anshel, M. [1976] 30627
Anshel, M. [1978] 30625
Ausiello, G. [1976] 51023
Baur, W. [1975] 03874
Baur, W. [1980] 54140
Behmann, H. [1922] 00942
Behmann, H. [1923] 00943
Behmann, H. [1927] 00946
Belegradek, O.V. [1980] 80668
Berman, L. [1977] 70972
Berman, L. [1980] 55471
Berman, P. [1979] 53427
Bernays, P. [1928] 01076
Beth, E.W. [1950] 01156
Bezverkhnij, V.N. [1981] 39526
Bezverkhnij, V.N. [1981] 39904
Boerger, E. [1974] 01368
Boerger, E. [1976] 52927
Boerger, E. [1979] 53847
Boerger, E. [1979] 54143
Boerger, E. [1980] 80766
Boerger, E. [1981] 55035
Boerger, E. [1983] 40032
Boerger, E. [1984] 40047
Boerger, E. [1984] 40054
Bondi, I.L. [1967] 01424
Bondi, I.L. [1971] 02009
Bondi, I.L. [1973] 01425
Britton, J.L. [1979] 53330
Bruss, A.R. [1978] 80801
Bruss, A.R. [1980] 55036
Buechi, J.R. [1960] 01690
Buechi, J.R. [1962] 01693
Buechi, J.R. [1965] 01694
Buechi, J.R. [1965] 01697

Buechi, J.R. [1969] 01700
Buechi, J.R. [1983] 40515
Burris, S. [1975] 21812
Burris, S. [1983] 36689
Burris, S. [1984] 42428
Buszkowski, W. [1982] 35719
Cantor, David G. [1984] 46787
Chen, Jiyuan [1984] 44191
Cherlin, G.L. [1980] 54217
Cherlin, G.L. [1981] 54982
Chlebus, B.S. [1979] 54219
Chlebus, B.S. [1980] 56482
Church, A. [1963] 02403
Clote, P. [1981] 55424
Comer, S.D. [1981] 55444
Cresswell, M.J. [1985] 42657
Cutland, N.J. [1980] 56629
Degtev, A.N. [1978] 72221
Denef, J. [1984] 38480
Denenberg, L. [1984] 43003
Denenberg, L. [1984] 43373
Doner, J.E. [1970] 03082
Doshita, S. [1983] 44209
Dreben, B. [1957] 03611
Dreben, B. [1962] 03127
Dries van den, L. [1985] 41808
Drobotun, B.N. [1977] 51556
Drobotun, B.N. [1980] 56069
Dulac, M.-H. [1971] 03165
Ehrenfeucht, A. [1957] 03235
Ehrenfeucht, A. [1975] 18126
Elgot, C.C. [1961] 03288
Elgot, C.C. [1966] 03291
Ellentuck, E. [1972] 03309
Emerson, E.A. [1984] 45349
Engeler, E. [1975] 21738
Ershov, Yu.L. [1965] 03525
Ershov, Yu.L. [1968] 03532
Ershov, Yu.L. [1972] 03551
Ershov, Yu.L. [1974] 25764
Ershov, Yu.L. [1980] 55024
Ershov, Yu.L. [1980] 72718
Ershov, Yu.L. [1984] 48388
Ferrante, J. [1975] 21702
Ferrante, J. [1977] 51432
Ferrante, J. [1979] 54815
Fischer, Michael J. [1974] 17344
Fridman, Eh.I. [1972] 77209
Fridman, Eh.I. [1973] 22999
Fried, M. [1976] 51684
Friedman, H.M. [1976] 22920
Friedrich, U. [1972] 04668
Fuerer, M. [1981] 35169
Fuerer, M. [1982] 35102
Gabbay, D.M. [1973] 61880
Gal, L.N. [1958] 04757
Galvin, F. [1967] 04765
Gecseg, F. [1984] 41334
Goedel, K. [1933] 20885
Goldfarb, W.D. [1981] 55275
Goncharov, S.S. [1976] 52189
Grandjean, E. [1983] 40384
Grzegorczyk, A. [1957] 24864
Gurari, E.M. [1978] 42545
Gurari, E.M. [1979] 56232
Gurari, E.M. [1981] 54218
Gurevich, Y. [1965] 05452
Gurevich, Y. [1966] 05455
Gurevich, Y. [1966] 05457
Gurevich, Y. [1967] 33753
Gurevich, Y. [1969] 05461
Gurevich, Y. [1969] 31104
Gurevich, Y. [1976] 14777
Gurevich, Y. [1979] 54603
Gurevich, Y. [1979] 54604
Gurevich, Y. [1982] 33758
Gurevich, Y. [1982] 33759
Gurevich, Y. [1983] 33764
Gurevich, Y. [1983] 33766
Gurevich, Y. [1985] 47372
Gurevich, Y. [1985] 48332
Haken, W. [1973] 29825
Hanf, W.P. [1965] 05635
Haran, D. [1984] 44141
Harel, D. [1983] 38572
Harel, D. [1985] 41208
Harrop, R. [1965] 05684
Hartmanis, J. [1976] 24160
Hauschild, K. [1971] 27450
Hauschild, K. [1972] 05787
Hauschild, K. [1981] 73915
Haussler, D. [1981] 55629
Heintz, Joos [1979] 55995
Heintz, Joos [1983] 38863
Henson, C.W. [1977] 30718
Hermes, H. [1955] 06008
Herre, H. [1975] 14283
Herre, H. [1978] 52676
Herre, H. [1981] 34285
Herrmann, E. [1976] 18190
Herrmann, E. [1977] 53517
Herrmann, E. [1978] 53131
Ackermann, W. [1928] 00107
Hodes, H.T. [1984] 41142
Hodgson, B.R. [1982] 35446
Hodgson, B.R. [1983] 35721
Huber-Dyson, V. [1964] 03214
Huber-Dyson, V. [1969] 03215
Huber-Dyson, V. [1982] 39680
Ito, Makoto [1965] 27530
Ivanov, A.A. [1983] 37599
Janiczak, A. [1953] 06508
Jonsson, B. [1975] 06740
Jukna, S. [1979] 33258
Kalmar, L. [1930] 06836
Kalmar, L. [1932] 06837
Kesel'man, D.Ya. [1982] 39437
Kireevskij, N.N. [1934] 07112
Klupp, H. [1977] 50134
Kogalovskij, S.R. [1966] 63026
Kokorin, A.I. [1973] 22922
Kokorin, A.I. [1978] 55718
Kolmogorov, A.N. [1953] 43578
Kopieki, R. [1975] 23964
Kozen, D. [1980] 53795
Kozen, D. [1981] 35103
Krom, Melven R. [1967] 07574
Krynicki, M. [1977] 33547
Kuznetsov, A.V. [1959] 28393
Lachlan, A.H. [1968] 07751
Lachlan, A.H. [1968] 07752
Ladner, R.E. [1977] 52252
Ladner, R.E. [1980] 75296
Lerman, M. [1978] 29152
Lerman, M. [1978] 30739
Lerman, M. [1980] 56014
Lewis, H.R. [1977] 53054
Lewis, H.R. [1978] 75527
Lewis, H.R. [1979] 53515
Lewis, H.R. [1980] 36546
Lifschitz, V. [1967] 33665
Lipshitz, L. [1978] 31822
Lipshitz, L. [1981] 75573
Lisovik, L.P. [1984] 42703
Lisovik, L.P. [1984] 48409
Litman, A. [1976] 18261
Lolli, G. [1983] 38263
Longo, G. [1974] 53697
Loveland, D.W. [1978] 31164
Macintyre, A. [1975] 23079
Makanin, G.S. [1977] 51395
Makanin, G.S. [1977] 51863
Makanin, G.S. [1980] 82143
Makanin, G.S. [1985] 48835
Manaster, A.B. [1981] 75934
Manders, K.L. [1979] 53838
Mart'yanov, V.I. [1977] 29230
Mehlhorn, K. [1974] 63853
Meyer, A.R. [1975] 21811
Meyer, A.R. [1975] 27549
Michel, P. [1981] 35234
Mihailescu, E.G. [1967] 09234
Millar, T.S. [1985] 46374
Monk, L.G. [1975] 90301
Mostowski, A.Wlodzimierz [1979] 76601
Mostowski, A.Wlodzimierz [1980] 34829
Mostowski, A.Wlodzimierz [1981] 55152
Mostowski, T. [1976] 82309
Muchnik, A.A. [1985] 47256
Muller, D.E. [1981] 76648
Muller, D.E. [1985] 47646
Muller, D.E. [1985] 48421
Mundici, D. [1983] 34976
Nakamura, A. [1981] 55427
Norgela, S.A. [1976] 32671
Norgela, S.A. [1978] 32674
Noskov, G.A. [1983] 38972
Novikov, P.S. [1949] 19476
Oppen, D.C. [1973] 24077
Oppen, D.C. [1978] 51889
Oppen, D.C. [1980] 55875
Oppen, D.C. [1980] 82395
Paola di, R.A. [1967] 30681
Pepis, J. [1937] 19493
Petresco, J. [1962] 10432
Pigozzi, D. [1974] 10495
Pigozzi, D. [1976] 21868
Pinus, A.G. [1971] 19598
Pinus, A.G. [1972] 10513
Pinus, A.G. [1976] 50089
Pinus, A.G. [1983] 39357
Pratt, V.R. [1979] 54858
Prestel, A. [1979] 53609
Putnam, H. [1957] 10827
Quine, W.V.O. [1949] 10881
Rabin, M.O. [1968] 29625
Rabin, M.O. [1969] 10942
Rabin, M.O. [1970] 22233
Rabin, M.O. [1971] 14690
Rabin, M.O. [1972] 10943
Rabin, M.O. [1977] 27321

Burris, S. [1984] 42428
Dahn, B.I. [1984] 42268
Drobotun, B.N. [1977] 51556
Ehrenfeucht, A. [1957] 03235
Ershov, Yu.L. [1980] 55024
Felgner, U. [1980] 55882
Goncharov, S.S. [1978] 52133
Harrington, L.A. [1974] 05664
Haussler, D. [1981] 55629
Henson, C.W. [1972] 05954
Herrmann, E. [1976] 18190
Herrmann, E. [1977] 53517
Kierstead, H.A. [1985] 44599
Knight, J.F. [1983] 40363
Kudajbergenov, K.Zh. [1980] 54224
Lerman, M. [1979] 53550
Lerman, M. [1984] 33603
Luo, Libo [1983] 44346
Macintyre, A. [1975] 23079
Millar, T.S. [1981] 35640
Myers, R.W. [1980] 55957
Palyutin, E.A. [1971] 64365
Peretyat'kin, M.G. [1973] 18330
Peretyat'kin, M.G. [1978] 53927
Pinus, A.G. [1976] 50089
Remmel, J.B. [1981] 77729
Robinson, A. [1961] 11249
Schlipf, J.S. [1977] 26851
Schmerl, J.H. [1978] 29211
Schmerl, J.H. [1980] 56060
Scott, D.S. [1960] 11909
Singer, M.F. [1978] 29244
Szczerba, L.W. [1979] 79177
Weese, M. [1972] 14105
Weispfenning, V. [1985] 48460

C40

Addison, J.W. [1962] 00175
Addison, J.W. [1962] 00176
Addison, J.W. [1965] 16215
Apt, K.R. [1978] 29131
Barwise, J. [1969] 00825
Barwise, J. [1971] 00836
Barwise, J. [1975] 60316
Barwise, J. [1976] 14694
Barwise, J. [1978] 29280
Belyaev, V.Ya. [1979] 32621
Buechi, J.R. [1969] 01700
Cherlin, G.L. [1984] 41760
Chlebus, B.S. [1980] 56482
Friedman, H.M. [1976] 22924
Galvin, F. [1967] 04765
Galvin, F. [1970] 04766
Garland, S.J. [1972] 04799
Garland, S.J. [1974] 24144
Germano, G. [1971] 04907
Grant, P.W. [1977] 26439
Grilliot, T.J. [1973] 05348
Grzegorczyk, A. [1961] 05402
Harnik, V. [1976] 14797
Harnik, V. [1980] 56478
Heintz, Joos [1979] 55995
Heintz, Joos [1983] 38863
Kalantari, I. [1985] 47472
Kanovich, M.I. [1975] 18218
Kolaitis, P.G. [1985] 48331
Kreisel, G. [1965] 33546
Kreisel, G. [1968] 07540
Krom, Melven R. [1963] 07571
Kunen, K. [1968] 07631
Macintyre, A. [1975] 08474
Makkai, M. [1973] 20987
Makkai, M. [1977] 23657
Montague, R. [1968] 09432
Moschovakis, Y.N. [1969] 09517
Moschovakis, Y.N. [1970] 09518
Mostowski, Andrzej [1962] 09574
Mundici, D. [1980] 33528
Mundici, D. [1981] 33529
Mundici, D. [1981] 33726
Mundici, D. [1982] 34339
Mundici, D. [1983] 37383
Mundici, D. [1984] 41841
Mundici, D. [1984] 44069
Mundici, D. [1984] 44693
Nabebin, A.A. [1977] 52155
Pinus, A.G. [1983] 37103
Poizat, B. [1981] 55288
Putnam, H. [1964] 10838
Rabin, M.O. [1970] 22233
Rabin, M.O. [1971] 14690
Robinson, A. [1975] 22933
Rumely, R. [1980] 55276
Schlipf, J.S. [1978] 29252
Shore, R.A. [1981] 55426
Tarski, A. [1931] 16924
Wasilewska, A. [1985] 49467
Weispfenning, V. [1976] 14747
Ziegler, M. [1976] 24137

C45

Baldwin, John T. [1984] 33746
Buechler, S. [1984] 45597
Felgner, U. [1980] 55882
Goncharov, S.S. [1973] 21613
Goncharov, S.S. [1980] 54545
Goncharov, S.S. [1980] 55089
Harnik, V. [1976] 14797
Kierstead, H.A. [1983] 36120
Knight, J.F. [1983] 40363
Lachlan, A.H. [1984] 43365
Macintyre, A. [1981] 90377
Peretyat'kin, M.G. [1982] 37596
Steinhorn, C.I. [1983] 34663

C50

Barwise, J. [1975] 23183
Barwise, J. [1976] 14694
Baur, W. [1974] 04142
Bertoni, A. [1981] 36666
Bertoni, A. [1983] 38487
Bozovic, N.B. [1977] 51534
Buechler, S. [1984] 45597
Cegielski, P. [1982] 34553
Clote, P. [1981] 55424
Denisov, A.S. [1984] 42732
Drobotun, B.N. [1980] 56069
Emde Boas van, P. [1973] 29008
Goncharov, S.S. [1973] 21613
Goncharov, S.S. [1975] 26031
Goncharov, S.S. [1977] 54891
Goncharov, S.S. [1978] 56068
Goncharov, S.S. [1980] 54545
Goncharov, S.S. [1980] 55087
Goncharov, S.S. [1980] 55089
Goncharov, S.S. [1983] 43384
Harnik, V. [1976] 14797
Harnik, V. [1980] 56478
Harrington, L.A. [1974] 05664
Henson, C.W. [1972] 05954
Hodges, W. [1984] 39733
Kaufmann, M. [1977] 31400
Kaufmann, M. [1984] 39638
Kirby, L.A.S. [1981] 34593
Knight, J.F. [1982] 36173
Knight, J.F. [1983] 42733
Kossak, R. [1983] 34668
Kossak, R. [1984] 41820
Kossak, R. [1984] 44634
Kossak, R. [1985] 42582
Kotlarski, H. [1981] 55248
Kotlarski, H. [1983] 34451
Kotlarski, H. [1984] 45761
Kotlarski, H. [1985] 49911
Kudajbergenov, K.Zh. [1983] 40506
Kudajbergenov, K.Zh. [1984] 41816
Kudajbergenov, K.Zh. [1984] 41871
Lachlan, A.H. [1981] 55250
Lachlan, A.H. [1984] 43365
Lipshitz, L. [1978] 31825
Macintyre, A. [1981] 90377
Macintyre, A. [1984] 40310
Makkai, M. [1977] 23657
Mead, J. [1979] 56480
Millar, T.S. [1978] 27960
Millar, T.S. [1980] 55005
Millar, T.S. [1981] 35640
Millar, T.S. [1982] 36061
Millar, T.S. [1983] 35754
Morley, M.D. [1976] 26084
Morozov, A.S. [1982] 38169
Morozov, A.S. [1982] 39816
Murawski, R. [1981] 34478
Negri, M. [1984] 44715
Omarov, A.I. [1984] 46564
Parikh, R. [1972] 10285
Peretyat'kin, M.G. [1978] 53927
Ressayre, J.-P. [1977] 23658
Robinson, A. [1961] 11249
Schlipf, J.S. [1975] 19103
Schlipf, J.S. [1977] 26851
Schlipf, J.S. [1978] 29252
Schlipf, J.S. [1980] 54283
Schmerl, J.H. [1981] 55178
Schmerl, J.H. [1981] 55467
Schmerl, J.H. [1985] 42589
Shilleto, J.R. [1972] 29513
Shoenfield, J.R. [1965] 12099
Smorynski, C.A. [1980] 55057
Smorynski, C.A. [1981] 78758
Smorynski, C.A. [1981] 78760
Smorynski, C.A. [1981] 78761
Smorynski, C.A. [1982] 35359
Smorynski, C.A. [1982] 35443
Steinhorn, C.I. [1983] 34663
Tsuboi, A. [1982] 37188
Tsuboi, A. [1983] 47682
Ville, F. [1971] 13946
Ziegler, M. [1976] 24137

C52

Addison, J.W. [1962] 00175
Almagambetov, Zh.A. [1965] 00278
Benejam, J.-P. [1977] 26472
Carstens, H.G. [1977] 51961
Harnik, V. [1976] 14797
Hauschild, K. [1982] 34362
Humberstone, I.L. [1984] 37693
Kaiser, Klaus [1970] 06800
Kleiman, J.G. [1982] 39318
Kogalovskij, S.R. [1966] 63026
Kokorin, A.I. [1973] 22922
Krom, Melven R. [1963] 07571
Macintyre, A. [1975] 08474
Makkai, M. [1973] 20987
Pigozzi, D. [1976] 21868
Rautenberg, W. [1967] 11046
Szczerba, L.W. [1965] 12831
Szczerba, L.W. [1979] 79177
Tajmanov, A.D. [1958] 13265

C55

Devlin, K.J. [1974] 15073
Garland, S.J. [1974] 24144
Grilliot, T.J. [1974] 23178
Hauschild, K. [1981] 73915
Herre, H. [1975] 14283
Herre, H. [1978] 52676
Ivanov, A.A. [1983] 37599
Kossak, R. [1985] 47534
Pinus, A.G. [1983] 37103
Pinus, A.G. [1985] 47255
Ressayre, J.-P. [1977] 23658
Rubin, M. [1983] 33531
Schlipf, J.S. [1977] 26851
Schmerl, J.H. [1976] 50397
Seese, D.G. [1977] 27132
Seese, D.G. [1978] 56270
Seese, D.G. [1982] 33532
Silver, J.H. [1971] 12319
Vinner, S. [1975] 13949
Weese, M. [1976] 23814
Weese, M. [1977] 26551
Weese, M. [1981] 33324

C60

Aanderaa, S.O. [1973] 23990
Abdrazakov, K.T. [1984] 41873
Adyan, S.I. [1957] 00193
Adyan, S.I. [1958] 00195
Ash, C.J. [1980] 28370
Ash, C.J. [1984] 39867
Baldwin, John T. [1982] 33739
Baldwin, John T. [1984] 33746
Baudisch, A. [1974] 00867
Baudisch, A. [1975] 00868
Baur, W. [1975] 03874
Baur, W. [1980] 54140
Becker, J.A. [1980] 47356
Belegradek, O.V. [1974] 17646
Belegradek, O.V. [1978] 29209
Belkin, V.P. [1975] 22948
Belyaev, V.Ya. [1979] 32621
Bezverkhnij, V.N. [1981] 39526
Bezverkhnij, V.N. [1981] 39904
Bienenstock, E. [1977] 50578
Billington, N. [1984] 45773
Bloshchitsyn, V.Ya. [1978] 36848
Boone, W.W. [1974] 04158
Britton, J.L. [1973] 44434
Burris, S. [1982] 35402
Burris, S. [1985] 48302
Buszkowski, W. [1979] 56249
Cantor, David G. [1984] 46787
Cherlin, G.L. [1980] 54217
Cherlin, G.L. [1981] 35227
Cherlin, G.L. [1981] 54982
Cherlin, G.L. [1984] 41760
Cherlin, G.L. [1984] 43903
Chistov, A.L. [1984] 44792
Collins, D.J. [1970] 02448
Dahn, B.I. [1984] 45383
Dekker, J.C.E. [1969] 02900
Dekker, J.C.E. [1971] 02901
Dekker, J.C.E. [1971] 02902
Delon, F. [1978] 27313
Delon, F. [1981] 35228
Denef, J. [1975] 04079
Denef, J. [1978] 52455
Denef, J. [1978] 52853
Denef, J. [1978] 52854
Denef, J. [1979] 54394
Denef, J. [1984] 38480
Dobritsa, V.P. [1976] 32652
Dobritsa, V.P. [1976] 32653
Dobritsa, V.P. [1977] 32655
Dobritsa, V.P. [1978] 32600
Dobritsa, V.P. [1981] 55366
Dobritsa, V.P. [1983] 36756
Downey, R.G. [1983] 38174
Downey, R.G. [1984] 39864
Downey, R.G. [1984] 39869
Downey, R.G. [1984] 42438
Downey, R.G. [1984] 43527
Downey, R.G. [1985] 39783
Downey, R.G. [1985] 47566
Dries van den, L. [1979] 54499
Dries van den, L. [1985] 41808
Ellentuck, E. [1977] 27196
Ellentuck, E. [1981] 33198
Engeler, E. [1975] 17633
Engeler, E. [1981] 81166
Ershov, Yu.L. [1964] 03520
Ershov, Yu.L. [1968] 37357
Ershov, Yu.L. [1972] 03551
Ershov, Yu.L. [1980] 55024
Ershov, Yu.L. [1980] 72718
Ershov, Yu.L. [1981] 34387
Ershov, Yu.L. [1984] 48388
Feferman, S. [1975] 22905
Felgner, U. [1980] 55882
Ferrante, J. [1975] 21702
Flum, J. [1976] 31078
Fowler III, N. [1975] 15238
Fowler III, N. [1975] 17130
Fowler III, N. [1976] 14783
Fowler III, N. [1978] 52144
Frenkel', V.I. [1963] 20969
Frenkel', V.I. [1964] 04602
Fridman, Eh.I. [1972] 73069
Fridman, Eh.I. [1975] 52191
Fried, M. [1976] 51684
Froehlich, A. [1955] 17092
Gabbay, D.M. [1973] 61880
Girstmair, K. [1979] 56289
Glass, A.M.W. [1984] 44713
Glass, A.M.W. [1985] 49075
Goncharov, S.S. [1980] 55087
Goncharov, S.S. [1981] 37746
Guhl, R. [1975] 05433
Guhl, R. [1977] 24226
Guichard, D.R. [1983] 40309
Guichard, D.R. [1984] 43386
Gupta, N. [1984] 39718
Gurevich, Y. [1965] 05452
Gurevich, Y. [1967] 33753
Haken, W. [1973] 29825
Hamilton, A.G. [1970] 05625
Haran, D. [1984] 44141
Harrington, L.A. [1974] 05664
Hauschild, K. [1971] 05780
Hauschild, K. [1972] 05790
Heintz, Joos [1979] 55995
Heintz, Joos [1983] 38863
Henkin, L. [1957] 05913
Hermann, G. [1926] 05997
Hickin, K.K. [1980] 81551
Hickin, K.K. [1983] 38580
Higman, G. [1949] 90239
Higman, G. [1961] 22083
Hingston, P. [1981] 55365
Hodges, W. [1976] 30711
Hodges, W. [1984] 39733
Huber-Dyson, V. [1969] 03215
Huber-Dyson, V. [1977] 51046
Huber-Dyson, V. [1981] 54396
Huber-Dyson, V. [1982] 37444
Ivanov, A.A. [1983] 37599
Jambu-Giraudet, M. [1980] 54167
Jensen, C.U. [1982] 37407
Jensen, C.U. [1984] 39754
Kalantari, I. [1977] 26850
Kalantari, I. [1978] 29262
Kalantari, I. [1979] 74612
Kenzhebaev, S. [1966] 16220
Kharlampovich, O.G. [1983] 39367
Khisamiev, N.G. [1977] 32598
Khisamiev, N.G. [1978] 32599
Khisamiev, N.G. [1979] 32601
Khisamiev, N.G. [1979] 32602
Khisamiev, N.G. [1981] 81571
Khisamiev, N.G. [1983] 40239
Khisamiev, N.G. [1984] 42697
Khisamiev, N.G. [1985] 49409
Kleiman, J.G. [1982] 39318
Kogalovskij, S.R. [1966] 63026
Kozlov, G.T. [1970] 07424
Kreisel, G. [1957] 29369
Kreisel, G. [1958] 07514
Kryazhovskikh, G.V. [1980] 54395
Lachlan, A.H. [1970] 07756
Lavrov, I.A. [1962] 07884
Lin, C. [1981] 35279
Lin, C. [1981] 75552
Lindsay, P.A. [1985] 46372
Lorenzen, P. [1962] 08303
Luo, Libo [1983] 44346
Macintyre, A. [1971] 08464
Macintyre, A. [1972] 08467
Macintyre, A. [1972] 08468

Macintyre, A. [1975] 23079
Macintyre, A. [1981] 90377
Macintyre, A. [1982] 37408
Madison, E.W. [1968] 08499
Madison, E.W. [1968] 08500
Madison, E.W. [1970] 24826
Madison, E.W. [1971] 63572
Makanin, G.S. [1985] 48835
Mal'tsev, A.I. [1960] 08609
Mal'tsev, A.I. [1960] 19257
Mal'tsev, A.I. [1962] 19247
Mal'tsev, A.I. [1962] 41272
Manders, K.L. [1979] 53838
Mann, A. [1982] 39712
Mayoh, B.H. [1967] 08925
Metakides, G. [1975] 17296
Metakides, G. [1977] 23661
Metakides, G. [1978] 33934
Metakides, G. [1979] 55155
Metakides, G. [1980] 55156
Monk, L.G. [1975] 90301
Moschovakis, Y.N. [1966] 21265
Mostowski, T. [1976] 82309
Myasnikov, A.G. [1982] 38582
Neumann, B.H. [1968] 24829
Neumann, B.H. [1973] 24064
Nogina, E.Yu. [1978] 53136
Noskov, G.A. [1982] 39350
Noskov, G.A. [1983] 34170
Noskov, G.A. [1983] 38972
Pappas, P. [1985] 44466
Prestel, A. [1975] 18349
Prestel, A. [1979] 53609
Pride, S.J. [1977] 26231
Rabin, M.O. [1957] 10933
Rabin, M.O. [1960] 33612
Rackoff, C.W. [1974] 24125
Rackoff, C.W. [1976] 14579
Rautenberg, W. [1967] 11046
Remeslennikov, V.N. [1970] 11095
Remeslennikov, V.N. [1983] 39954
Remmel, J.B. [1980] 55232
Remmel, J.B. [1980] 55233
Retzlaff, A.T. [1979] 77762
Robinson, A. [1961] 11249
Robinson, A. [1975] 22933
Robinson, Julia [1959] 11295
Robinson, Julia [1962] 11296
Robinson, Julia [1965] 11298
Robinson, R.M. [1951] 11311
Roche la, P. [1981] 75255
Roman'kov, V.A. [1979] 53388
Romanovskij, N.S. [1980] 77873
Rumely, R. [1980] 55276
Sacerdote, G.S. [1972] 30879
Sacerdote, G.S. [1976] 24067
Schmitt, P.H. [1984] 46136
Schrieber, L. [1985] 47474
Schupp, P.E. [1973] 44402
Seidenberg, A. [1970] 41695
Seidenberg, A. [1975] 41698
Shepherdson, J.C. [1951] 12066
Singer, M.F. [1978] 29244
Slobodskoj, A.M. [1981] 36070
Smith, Rick L. [1981] 36072
Smith, Rick L. [1981] 55685
Soprunov, S.F. [1974] 17402
Suter, G.H. [1973] 12784
Tajtslin, M.A. [1964] 13255
Tajtslin, M.A. [1965] 65641
Tajtslin, M.A. [1966] 13258
Tajtslin, M.A. [1968] 13260
Tajtslin, M.A. [1968] 32029
Tajtslin, M.A. [1970] 28107
Tarski, A. [1931] 16924
Tucker, J.V. [1980] 79494
Umirbaev, U.U. [1984] 42699
Valiev, M.K. [1968] 13814
Vasil'ev, Eh.S. [1973] 13840
Vazhenin, Yu.M. [1974] 79687
Waerden van der, B.L. [1930] 38703
Weispfenning, V. [1976] 14747
Weispfenning, V. [1985] 48460
Wheeler, W.H. [1972] 19104
Wuethrich, H.R. [1976] 43103
Zamyatin, A.P. [1978] 90332
Ziegler, M. [1980] 54050
Ziegler, M. [1982] 35441

C62

Abraham, U. [1984] 40744
Abramson, F.G. [1979] 53788
Adler, A. [1969] 00184
Ajtai, M. [1983] 37538
Apt, K.R. [1972] 00429
Apt, K.R. [1974] 00432
Apt, K.R. [1976] 14789
Apt, K.R. [1978] 29131
Baeten, J. [1984] 45395
Barwise, J. [1970] 00828
Barwise, J. [1971] 00836
Barwise, J. [1974] 00911
Barwise, J. [1975] 23183
Belyakin, N.V. [1970] 29065
Belyakin, N.V. [1983] 41479
Benioff, P.A. [1976] 30647
Cegielski, P. [1982] 34553
Chauvin, A. [1961] 01999
Chauvin, A. [1961] 25473
Clote, P. [1985] 45298
Crossley, J.N. [1974] 21692
David, R. [1982] 35373
David, R. [1985] 46391
Dimitracopoulos, C. [1985] 47569
Ehrenfeucht, A. [1966] 03245
Ellentuck, E. [1981] 72617
Ershov, Yu.L. [1983] 40494
Feferman, S. [1965] 03696
Fenstad, J.E. [1978] 54177
Friedman, H.M. [1970] 14546
Gandy, R.O. [1960] 04774
Gandy, R.O. [1960] 15061
Gandy, R.O. [1961] 22330
Green, J. [1974] 05312
Grilliot, T.J. [1972] 05347
Grzegorczyk, A. [1958] 05398
Grzegorczyk, A. [1962] 05406
Guaspari, D. [1980] 54183
Hajek, P. [1977] 26853
Harrington, L.A. [1975] 05665
Harrington, L.A. [1977] 24274
Hart, J. [1969] 05689
Hirschfeld, J. [1974] 21188
Hirschfeld, J. [1975] 06142
Howard, P.E. [1972] 06260
Jambu-Giraudet, M. [1980] 54167
Jambu-Giraudet, M. [1981] 55535
Jensen, R.B. [1967] 06596
Jensen, R.B. [1970] 22245
Kanovej, V.G. [1974] 17445
Kanovej, V.G. [1979] 54105
Kaufmann, M. [1977] 31400
Kaufmann, M. [1984] 39638
Kaufmann, M. [1984] 41225
Kechris, A.S. [1975] 14831
Kirby, L.A.S. [1981] 34593
Knight, J.F. [1982] 37529
Knight, J.F. [1983] 42733
Knight, J.F. [1984] 33597
Kossak, R. [1983] 34668
Kossak, R. [1984] 41820
Kossak, R. [1984] 44634
Kossak, R. [1985] 42582
Kossak, R. [1985] 47534
Kotlarski, H. [1981] 55248
Kotlarski, H. [1983] 34451
Kotlarski, H. [1984] 45761
Kotlarski, H. [1985] 49911
Kreisel, G. [1961] 07523
Kreisel, G. [1965] 33546
Lachlan, A.H. [1981] 55250
Leeds, S. [1974] 07969
Lipshitz, L. [1978] 31825
Lipton, R.J. [1978] 75578
Loeb, M.H. [1970] 22235
Macintyre, A. [1975] 08474
Macintyre, A. [1982] 37408
Macintyre, A. [1984] 40310
Madison, E.W. [1968] 08499
Madison, E.W. [1968] 08500
Mal'tsev, A.I. [1962] 19247
Marker, D. [1982] 37802
McAloon, K. [1978] 54427
McAloon, K. [1980] 54524
McAloon, K. [1982] 34279
Moschovakis, Y.N. [1971] 09520
Mostowski, Andrzej [1957] 09562
Mostowski, Andrzej [1972] 09588
Nadel, M.E. [1972] 09743
Nadel, M.E. [1974] 16930
Negri, M. [1984] 44715
Nelson, George C. [1973] 17539
Nerode, A. [1966] 09884
Nerode, A. [1980] 76822
Paris, J.B. [1981] 79972
Rabin, M.O. [1958] 10930
Rabin, M.O. [1962] 20907
Richard, D. [1984] 41779
Robinson, A. [1961] 11249
Robinson, Julia [1973] 29547
Schlipf, J.S. [1975] 19103
Schlipf, J.S. [1977] 26851
Schlipf, J.S. [1980] 54283
Schmerl, J.H. [1981] 55178
Scott, D.S. [1960] 11909
Scott, D.S. [1961] 24930
Shilleto, J.R. [1972] 29513
Smorynski, C.A. [1980] 55057
Smorynski, C.A. [1981] 78758
Smorynski, C.A. [1981] 78760
Smorynski, C.A. [1981] 78761

Smorynski, C.A. [1982] 35359
Smorynski, C.A. [1982] 35443
Solovay, R.M. [1985] 46419
Srebrny, M. [1977] 26534
Steel, J.R. [1975] 13142
Steel, J.R. [1978] 29156
Suzuki, Y. [1967] 12790
Suzuki, Y. [1968] 12791
Tanaka, H. [1974] 23068
Tsuboi, A. [1982] 37188
Tsuboi, A. [1983] 47682
Tverskoj, A.A. [1982] 35413
Tverskoj, A.A. [1985] 49425
Vaeaenaenen, J. [1979] 52749
Vetulani, Z. [1984] 44142
Wilkie, A.J. [1980] 79968

C65

Baranskij, V.A. [1970] 00771
Becker, J.A. [1980] 54459
Bondi, I.L. [1973] 01425
Cherlin, G.L. [1980] 54217
Chlebus, B.S. [1980] 56482
Dahn, B.I. [1984] 42268
Dahn, B.I. [1984] 45383
Denef, J. [1984] 38480
Denisov, S.D. [1974] 25962
Dyment, E.Z. [1984] 44722
Emde Boas van, P. [1973] 29008
Ershov, Yu.L. [1968] 03532
Ershov, Yu.L. [1973] 30533
Gurevich, Y. [1979] 54603
Gurevich, Y. [1979] 54604
Gurevich, Y. [1982] 33758
Gurevich, Y. [1983] 33764
Gurevich, Y. [1983] 33765
Gurevich, Y. [1985] 47372
Gurevich, Y. [1985] 48332
Haken, W. [1973] 29825
Hauschild, K. [1971] 27450
Hauschild, K. [1972] 05787
Hay, L. [1977] 30705
Henkin, L. [1959] 22387
Henson, C.W. [1977] 30718
Herre, H. [1981] 34285
Jambu-Giraudet, M. [1980] 54167
Jambu-Giraudet, M. [1981] 55535
Jensen, C.U. [1984] 39754
Kalantari, I. [1979] 53722
Kalantari, I. [1982] 34171
Kalantari, I. [1982] 35452
Kalantari, I. [1983] 33595
Kalantari, I. [1983] 36130
Kalantari, I. [1985] 47472
Kesel'man, D.Ya. [1974] 17588
Kesel'man, D.Ya. [1982] 39437
Kierstead, H.A. [1984] 45164
Ladner, R.E. [1977] 52252
Lerman, M. [1979] 53550
Lerman, M. [1981] 55043
Lerman, M. [1982] 34683
Macintyre, A. [1971] 08464
Manaster, A.B. [1980] 55073
Manaster, A.B. [1980] 55074
Manaster, A.B. [1981] 33275
Manaster, A.B. [1981] 75934
Moschovakis, Y.N. [1964] 09514
Mostowski, T. [1976] 82309
Nelson, George C. [1973] 17539
Nikitin, A.A. [1984] 48426
Nogina, E.Yu. [1966] 49889
Nogina, E.Yu. [1969] 09983
Odifreddi, P. [1983] 34091
Peretyat'kin, M.G. [1973] 25033
Pinus, A.G. [1971] 19598
Pinus, A.G. [1972] 10513
Rautenberg, W. [1962] 11041
Rautenberg, W. [1967] 11046
Rautenberg, W. [1968] 11048
Remmel, J.B. [1981] 77729
Rosenstein, J.G. [1984] 41775
Roy, D.K. [1983] 38265
Roy, D.K. [1985] 47798
Schmerl, J.H. [1980] 56060
Schmerl, J.H. [1981] 55467
Schram, J.M. [1984] 44732
Schwabhaeuser, W. [1983] 40225
Schwarz, S. [1984] 39829
Seese, D.G. [1977] 27132
Seese, D.G. [1978] 52180
Seese, D.G. [1978] 56270
Seese, D.G. [1982] 33532
Shelah, S. [1975] 15024
Singer, M.F. [1978] 29244
Szczerba, L.W. [1965] 12831
Szczerba, L.W. [1979] 79177
Tajmanov, A.D. [1974] 13277
Tarski, A. [1931] 16924
Tucker, J.V. [1980] 79494
Tulipani, S. [1982] 34564
Watnick, R. [1984] 42511
Weese, M. [1972] 14105
Weese, M. [1977] 26551
Wolter, H. [1984] 44686
Ziegler, M. [1976] 24137

C70

Abramson, F.G. [1979] 53788
Barwise, J. [1967] 16732
Barwise, J. [1969] 00825
Barwise, J. [1975] 31017
Barwise, J. [1975] 60316
Barwise, J. [1978] 29280
Barwise, J. [1978] 70698
Cutland, N.J. [1970] 16784
Cutland, N.J. [1972] 02662
Cutland, N.J. [1978] 51893
Ershov, Yu.L. [1983] 40494
Fenstad, J.E. [1978] 54177
Friedman, H.M. [1968] 04643
Friedman, S.D. [1979] 56139
Friedman, S.D. [1981] 33348
Friedman, S.D. [1982] 33350
Fukuyama, M. [1974] 23159
Grant, P.W. [1977] 26439
Green, J. [1974] 05312
Grilliot, T.J. [1974] 23178
Harnik, V. [1980] 56478
Kolaitis, P.G. [1985] 48331
Kreisel, G. [1968] 07540
Kunen, K. [1968] 07631
Makkai, M. [1977] 23657
Makkai, M. [1977] 24202
Miller, Douglas E. [1978] 33521
Nadel, M.E. [1972] 09743
Nadel, M.E. [1974] 16930
Nyberg, A.M. [1976] 14788
Nyberg, A.M. [1977] 28324
Ressayre, J.-P. [1977] 23658
Schlipf, J.S. [1975] 19103
Schlipf, J.S. [1977] 26851
Schlipf, J.S. [1978] 29252
Stark, W.R. [1978] 52964
Stavi, J. [1973] 13133
Vaught, R.L. [1973] 13864

C75

Aczel, P. [1975] 30616
Belyaev, V.Ya. [1979] 32621
Blair, H.A. [1984] 43300
Burgess, J.P. [1975] 03956
Burgess, J.P. [1979] 56715
Campbell, P.J. [1971] 16762
Engeler, E. [1967] 03366
Engeler, E. [1968] 29303
Engeler, E. [1975] 17633
Fitting, M. [1978] 31075
Garland, S.J. [1972] 04799
Gordon, C.E. [1974] 05199
Grilliot, T.J. [1973] 05348
Grilliot, T.J. [1974] 23178
Harnik, V. [1976] 14797
Henson, C.W. [1977] 30718
Hodges, W. [1976] 30711
Knight, J.F. [1983] 42733
Kolaitis, P.G. [1985] 48331
Kossak, R. [1984] 41820
Kossak, R. [1985] 47534
Kreisel, G. [1968] 07540
Kreisel, G. [1975] 27698
Kreisel, G. [1975] 75092
Kunen, K. [1968] 07631
Kuratowski, K. [1937] 07668
Machover, M. [1961] 08457
Macintyre, A. [1972] 08467
Makkai, M. [1973] 20987
Mansfield, R. [1975] 08674
Miller, Douglas E. [1978] 33521
Moschovakis, Y.N. [1972] 09521
Myers, D.L. [1976] 30751
Nepejvoda, L.K. [1977] 55689
Schwichtenberg, H. [1975] 25022
Scott, D.S. [1964] 11916
Vaught, R.L. [1973] 13864
Vaught, R.L. [1973] 13909
Vaught, R.L. [1974] 17478
Wainer, S.S. [1975] 29611
Weispfenning, V. [1976] 14747
Zermelo, E. [1931] 14412

C80

Aczel, P. [1975] 30616
Barwise, J. [1978] 70698
Blass, A.R. [1984] 47716
Blass, A.R. [1984] 47720
Chlebus, B.S. [1980] 56482
Enderton, H.B. [1970] 03352
Fenstad, J.E. [1978] 54177
Grilliot, T.J. [1974] 23178
Hajek, P. [1978] 31131

Hauschild, K. [1981] 73915
Herre, H. [1975] 14283
Herre, H. [1978] 52676
Herre, H. [1981] 34285
Hinman, P.G. [1985] 45303
Ivanov, A.A. [1983] 37599
Kierstead, H.A. [1983] 36120
Kossak, R. [1985] 47534
Krynicki, M. [1977] 33547
Markwald, W. [1954] 08779
Mehlhorn, K. [1974] 63853
Moschovakis, Y.N. [1972] 09521
Mostowski, Andrzej [1956] 09560
Mostowski, Andrzej [1962] 09574
Pinus, A.G. [1976] 50089
Pinus, A.G. [1983] 37103
Pinus, A.G. [1985] 47255
Ressayre, J.-P. [1977] 23658
Roedding, D. [1966] 24963
Rubin, M. [1983] 33531
Seese, D.G. [1972] 11931
Seese, D.G. [1977] 27132
Seese, D.G. [1978] 56270
Seese, D.G. [1982] 33532
Shanin, N.A. [1976] 65058
Shanin, N.A. [1977] 53812
Sobolev, S.K. [1977] 51014
Stillwell, J.C. [1972] 13190
Tharp, L.H. [1974] 13528
Vinner, S. [1975] 13949
Weese, M. [1976] 23814
Weese, M. [1977] 26551
Weese, M. [1981] 33324
Ziegler, M. [1976] 24137

C85

Belyaev, V.Ya. [1979] 32621
Buechi, J.R. [1960] 01690
Buechi, J.R. [1962] 01693
Buechi, J.R. [1965] 01694
Buechi, J.R. [1965] 01697
Buechi, J.R. [1969] 01700
Buechi, J.R. [1983] 33901
Buechi, J.R. [1983] 40515
Cherlin, G.L. [1984] 43903
Compton, K.J. [1984] 47457
Doner, J.E. [1970] 03082
Elgot, C.C. [1966] 03291
Fagin, R. [1974] 18140
Friedrich, U. [1972] 04668
Garland, S.J. [1972] 04799
Garland, S.J. [1974] 24144
Gurevich, Y. [1979] 54603
Gurevich, Y. [1979] 54604
Gurevich, Y. [1982] 33758
Gurevich, Y. [1982] 33759
Gurevich, Y. [1983] 33764
Gurevich, Y. [1983] 33765
Gurevich, Y. [1985] 48332
Heindorf, L. [1984] 39661
Kokorin, A.I. [1978] 55718
Ladner, R.E. [1977] 52252
Lisovik, L.P. [1984] 48409
Litman, A. [1976] 18261
Mostowski, A.Wlodzimierz [1980] 34829
Mostowski, Andrzej [1962] 09574
Muchnik, A.A. [1985] 47256
Muller, D.E. [1985] 48421
Palyutin, E.A. [1971] 64365
Pinus, A.G. [1972] 10513
Pinus, A.G. [1985] 47255
Rabin, M.O. [1968] 29625
Rabin, M.O. [1969] 10942
Rabin, M.O. [1970] 22233
Rabin, M.O. [1971] 14690
Rabin, M.O. [1972] 10943
Rautenberg, W. [1967] 11046
Seese, D.G. [1972] 11931
Seese, D.G. [1977] 27132
Seese, D.G. [1978] 52180
Shelah, S. [1975] 15024
Siefkes, D. [1970] 12145
Siefkes, D. [1970] 27984
Siefkes, D. [1971] 12146
Siefkes, D. [1978] 29136
Turan, G. [1981] 55701

C90

Cherlin, G.L. [1981] 35227
Dosen, K. [1985] 47571
Ellentuck, E. [1977] 27196
Ershov, Yu.L. [1972] 18137
Gabbay, D.M. [1973] 61880
Hajek, P. [1978] 31131
Mostowski, Andrzej [1962] 09574
Smorynski, C.A. [1973] 12535
Sobolev, S.K. [1977] 51014
Troelstra, A.S. [1978] 27952
Vuckovic, V. [1967] 27478
Ziegler, M. [1976] 24137

C95

Tharp, L.H. [1974] 13528
Ziegler, M. [1976] 24137

C96

Mostowski, Andrzej [1979] 82308

C97

Crossley, J.N. [1967] 31684
Crossley, J.N. [1981] 54309
Gandy, R.O. [1977] 16612
Harrington, L.A. [1985] 49810
Lolli, G. [1984] 41493
Mueller, Gert H. [1984] 41750
Sobolev, S.L. [1982] 34492

C98

Barwise, J. [1975] 31017
Barwise, J. [1975] 60316
Barwise, J. [1977] 70117
Bucur, I. [1980] 46245
Burris, S. [1983] 36689
Delon, F. [1985] 47516
Ershov, Yu.L. [1965] 03525
Ershov, Yu.L. [1974] 25764
Ershov, Yu.L. [1980] 72718
Ferrante, J. [1979] 54815
Friedman, H.M. [1975] 04296
Grandjean, E. [1984] 48072
Gurevich, Y. [1985] 48332
Herre, H. [1981] 34285
Kokorin, A.I. [1978] 55718
Kolaitis, P.G. [1985] 48331
Krajewski, S. [1981] 38331
Ladner, R.E. [1980] 75296
Lehmann, G. [1985] 42737
Makanin, G.S. [1985] 48835
Makkai, M. [1977] 24202
Mal'tsev, A.I. [1961] 28706
Malitz, J. [1979] 56167
Marek, W. [1972] 34216
Millar, T.S. [1985] 46374
Mostowski, Andrzej [1965] 09578
Nerode, A. [1985] 46402
Neumann, B.H. [1968] 24829
Prestel, A. [1979] 53609
Rabin, M.O. [1977] 27321
Remeslennikov, V.N. [1983] 39954
Schwabhaeuser, W. [1983] 40225
Shoenfield, J.R. [1967] 22384
Stillwell, J.C. [1977] 52422
Takahashi, S. [1974] 17488
Taylor, W. [1977] 69948

D96

Mostowski, Andrzej [1979] 82308

E05

Baeten, J. [1984] 45395
Blass, A.R. [1985] 45296
Boos, W. [1975] 31627
Burgess, J.P. [1977] 28149
Busch, D.R. [1977] 26482
Carstens, H.G. [1977] 26473
Cleave, J.P. [1968] 02681
Clote, P. [1984] 42432
Ellentuck, E. [1974] 03322
Friedman, S.D. [1980] 54176
Galvin, F. [1973] 04769
Jensen, R.B. [1970] 22245
Kaufmann, M. [1984] 41225
Kechris, A.S. [1981] 74734
Kechris, A.S. [1982] 37240
Kechris, A.S. [1985] 40477
Kossak, R. [1984] 44634
Kranakis, E. [1982] 35456
Kranakis, E. [1983] 38326
Kranakis, E. [1984] 42470
Kranakis, E. [1984] 45390
Kranakis, E. [1985] 47533
Kranakis, E. [1985] 47562
Lefmann, H. [1984] 44423
Louveau, A. [1974] 15025
Louveau, A. [1976] 18266
Magidor, M. [1980] 56487
Manaster, A.B. [1972] 08644
Mansfield, R. [1970] 28913
Masseron, M. [1983] 40760
Mathias, A.R.D. [1972] 17477
Mathias, A.R.D. [1975] 17523
Raisonnier, J. [1984] 44806
Raisonnier, J. [1985] 48248
Seese, D.G. [1982] 33532
Silver, J.H. [1970] 12317
Silver, J.H. [1971] 12319
Silver, J.H. [1971] 19669
Simpson, S.G. [1985] 45316

Solovay, R.M. [1978] 52399
Solovay, R.M. [1978] 52893
Specker, E. [1971] 29960
Stern, J. [1977] 27300
Vaeaenaenen, J. [1979] 52749
Wolfe, P. [1955] 14269
Yates, C.E.M. [1971] 14508

E07

Brook, T. [1977] 80793
Chlebus, B.S. [1980] 56482
Dima, N. [1984] 41381
Emde Boas van, P. [1973] 29008
Emden van, M.H. [1977] 53677
Ershov, Yu.L. [1968] 03532
Eve, J. [1977] 61616
Gurevich, Y. [1979] 54603
Gurevich, Y. [1979] 54604
Gurevich, Y. [1982] 33758
Gurevich, Y. [1983] 33765
Harrington, L.A. [1979] 73861
Hay, L. [1977] 30705
Humberstone, I.L. [1984] 37693
Jambu-Giraudet, M. [1980] 54167
Jambu-Giraudet, M. [1981] 55535
Jongh de, D.H.J. [1977] 26064
Kierstead, H.A. [1981] 36594
Kuratowski, K. [1922] 07652
Kuratowski, K. [1937] 07666
Kuratowski, K. [1937] 07668
Ladner, R.E. [1977] 52252
Lerman, M. [1981] 55043
Lerman, M. [1982] 34683
Levitz, H. [1978] 30732
Manaster, A.B. [1980] 55073
Manaster, A.B. [1980] 55074
Manaster, A.B. [1981] 33275
McBeth, R. [1980] 54978
Moses, M. [1984] 39682
Moszner, Z. [1967] 76611
Muir, A. [1984] 40043
Parikh, R. [1966] 10279
Pinus, A.G. [1971] 19598
Pinus, A.G. [1975] 16662
Roy, D.K. [1983] 38265
Shrejder, Yu.A. [1971] 78575
Weese, M. [1977] 26551
Zermelo, E. [1931] 14412

E10

Baeten, J. [1984] 45395
Bradford, R. [1971] 01574
Buechi, J.R. [1965] 01694
Buechi, J.R. [1965] 01697
Buechi, J.R. [1983] 33901
Buechi, J.R. [1983] 40515
Devlin, K.J. [1975] 16656
Ellentuck, E. [1975] 22911
Ellentuck, E. [1976] 18450
Ellentuck, E. [1978] 51747
Garland, S.J. [1974] 24144
Girard, J.-Y. [1981] 35541
Gurevich, Y. [1983] 33764
Hausdorff, F. [1914] 23280
Hilbert, D. [1926] 45196
Jarnik, V. [1938] 41088
Jongh de, D.H.J. [1977] 26064
Kechris, A.S. [1974] 06989
Kechris, A.S. [1975] 14831
Kruse, A.H. [1960] 07579
Kruse, A.H. [1962] 07582
Levitz, H. [1978] 30732
Litman, A. [1976] 18261
Manaster, A.B. [1968] 08641
Manaster, A.B. [1969] 08642
Sami, R.L. [1984] 42499
Schmidt, Diana [1976] 24318
Schroeder, M.E. [1970] 12984
Shinoda, J. [1984] 42934
Sierpinski, W. [1924] 12158
Srebrny, M. [1977] 26534
Sudan, G. [1927] 12706
Takahashi, Moto-o [1968] 13293
Takeuti, G. [1965] 13329

E15

Adamowicz, Z. [1983] 43740
Adamowicz, Z. [1984] 45775
Addison, J.W. [1957] 29378
Addison, J.W. [1958] 00171
Addison, J.W. [1959] 00173
Addison, J.W. [1962] 00175
Addison, J.W. [1962] 00176
Addison, J.W. [1965] 16215
Addison, J.W. [1968] 00181
Addison, J.W. [1974] 17525
Aleksandrov, P.S. [1916] 38044
Aleksandrov, P.S. [1924] 00266
Alexiewicz, A. [1947] 00267
Amstislavskij, V.I. [1966] 00302
Amstislavskij, V.I. [1966] 00303
Amstislavskij, V.I. [1968] 00305
Amstislavskij, V.I. [1970] 00307
Amstislavskij, V.I. [1970] 00308
Amstislavskij, V.I. [1973] 03825
Amstislavskij, V.I. [1974] 17450
Amstislavskij, V.I. [1975] 60124
Arsenin, V.Ya. [1950] 00496
Arsenin, V.Ya. [1955] 24901
Arsenin, V.Ya. [1977] 80574
Bade, W.G. [1973] 00624
Baire, R. [1905] 05473
Barua, R. [1984] 45772
Becker, H. [1978] 52699
Becker, H. [1980] 54180
Becker, H. [1985] 44736
Blackwell, D. [1967] 01250
Blackwell, D. [1978] 80735
Blass, A.R. [1974] 03908
Blass, A.R. [1984] 40312
Bloom, S.L. [1969] 01320
Bloom, S.L. [1970] 01321
Blue, A.H. [1930] 39319
Boffa, M. [1974] 31022
Braude, E.J. [1971] 01580
Braun, S. [1932] 01585
Braun, S. [1937] 01586
Bressler, D.W. [1964] 01592
Bruckner, A.M. [1969] 01652
Budinas, B.L. [1981] 35767
Budinas, B.L. [1982] 34305
Budinas, B.L. [1983] 47459
Buechi, J.R. [1983] 42601
Bukovsky, L. [1964] 01714
Burgess, J.P. [1975] 03956
Burgess, J.P. [1977] 28149
Burgess, J.P. [1977] 71462
Burgess, J.P. [1978] 28148
Burgess, J.P. [1979] 56715
Burgess, J.P. [1979] 71463
Burgess, J.P. [1982] 37657
Burgess, J.P. [1983] 37856
Burgess, J.P. [1983] 37858
Busch, D.R. [1976] 14769
Busch, D.R. [1976] 18447
Busch, D.R. [1977] 26482
Campbell, P.J. [1971] 16762
Cenzer, D. [1976] 14799
Cenzer, D. [1980] 54969
Cenzer, D. [1982] 37461
Cenzer, D. [1983] 43460
Chikvashvili, R.I. [1975] 21685
Choban, M.M. [1973] 71748
Choquet, G. [1959] 02221
David, R. [1982] 37544
Davies, R.O. [1974] 28188
Davies, R.O. [1977] 28189
Davis, Morton [1964] 26244
Delfino, V. [1978] 54749
Dellacherie, C. [1971] 46943
Dellacherie, C. [1975] 50189
Dellacherie, C. [1977] 51128
Dellacherie, C. [1980] 54044
Demuth, O. [1977] 72287
Dienes, Z.P. [1939] 03009
Ebbinghaus, H.-D. [1976] 50795
Ellentuck, E. [1974] 03322
Engelking, R. [1966] 03375
Feldman, E.D. [1971] 03716
Feldman, E.D. [1972] 03717
Feldman, E.D. [1975] 04234
Fenstad, J.E. [1974] 03747
Filippov, V.P. [1969] 17174
Firmani, B. [1973] 29531
Freiwald, R.C. [1972] 04601
Freiwald, R.C. [1979] 73063
Fremlin, D.H. [1983] 38382
Friedman, H.M. [1970] 03773
Friedman, H.M. [1971] 04649
Friedman, H.M. [1973] 04653
Friedman, H.M. [1973] 21340
Friedman, H.M. [1974] 04658
Galvin, F. [1973] 04769
Ganov, V.A. [1974] 17498
Garland, S.J. [1972] 04799
Grassin, J. [1978] 32137
Grassin, J. [1981] 73595
Griffiths, H.B. [1957] 05337
Griffor, E.R. [1985] 45302
Grigorieff, S. [1977] 50617
Guaspari, D. [1974] 05420
Guaspari, D. [1974] 05421
Guaspari, D. [1975] 17640
Guaspari, D. [1976] 22937
Guaspari, D. [1983] 43128
Guzicki, W. [1973] 30498
Hansell, R.W. [1985] 49732
Harnik, V. [1976] 14797
Harrington, L.A. [1975] 17524
Harrington, L.A. [1975] 21631

Harrington, L.A. [1977] 24274
Harrington, L.A. [1979] 73861
Hausdorff, F. [1914] 23280
Hausdorff, F. [1933] 05798
Hay, L. [1976] 14803
Hay, L. [1982] 34806
Hinman, P.G. [1966] 20914
Hinman, P.G. [1969] 06103
Hinman, P.G. [1973] 06108
Hinman, P.G. [1978] 51378
Hinman, P.G. [1979] 54824
Hrbacek, K. [1978] 52894
Hrbacek, K. [1980] 74290
Hrbacek, K. [1983] 41103
Hurewicz, W. [1930] 06346
Jayne, J.E. [1982] 37743
Kaniewski, J. [1976] 24651
Kanovej, V.G. [1975] 50893
Kanovej, V.G. [1978] 74664
Kanovej, V.G. [1979] 54105
Kanovej, V.G. [1979] 56024
Kanovej, V.G. [1982] 34181
Kanovej, V.G. [1983] 34332
Kanovej, V.G. [1983] 37518
Kantorovich, L. [1930] 39457
Kantorovich, L. [1932] 08214
Kantorovich, L. [1933] 08216
Kastanas, I.G. [1984] 43508
Kechris, A.S. [1972] 06986
Kechris, A.S. [1973] 06988
Kechris, A.S. [1973] 21379
Kechris, A.S. [1974] 06989
Kechris, A.S. [1975] 06990
Kechris, A.S. [1975] 14831
Kechris, A.S. [1975] 30723
Kechris, A.S. [1977] 26243
Kechris, A.S. [1977] 53258
Kechris, A.S. [1978] 30726
Kechris, A.S. [1978] 51832
Kechris, A.S. [1978] 52351
Kechris, A.S. [1978] 52698
Kechris, A.S. [1978] 54897
Kechris, A.S. [1978] 55740
Kechris, A.S. [1978] 56640
Kechris, A.S. [1978] 56714
Kechris, A.S. [1978] 74744
Kechris, A.S. [1979] 34337
Kechris, A.S. [1980] 40463
Kechris, A.S. [1980] 54030
Kechris, A.S. [1981] 34923
Kechris, A.S. [1981] 35283
Kechris, A.S. [1981] 54310
Kechris, A.S. [1981] 74732
Kechris, A.S. [1981] 74733
Kechris, A.S. [1981] 74734
Kechris, A.S. [1982] 37240
Kechris, A.S. [1983] 36754
Kechris, A.S. [1983] 40468
Kechris, A.S. [1985] 40477
Kechris, A.S. [1985] 45614
Keldysh, L.V. [1945] 24914
Keldysh, L.V. [1974] 18221
Kerstan, J. [1961] 07066
Kholshchevnikova, N.N. [1985] 49292
Kiselev, A.A. [1973] 32586
Kiselev, A.A. [1973] 32587
Kiselev, A.A. [1978] 32588
Kline, S.A. [1945] 07237
Kolmogorov, A.N. [1966] 24780
Kondo, M. [1936] 37170
Kondo, M. [1937] 37174
Kondo, M. [1938] 27560
Kondo, M. [1938] 33398
Kondo, M. [1938] 37175
Kondo, M. [1938] 37176
Kondo, M. [1942] 37179
Kondo, M. [1944] 37180
Kondo, M. [1956] 07330
Kondo, M. [1956] 07331
Kondo, M. [1956] 07332
Kondo, M. [1958] 07333
Kondo, M. [1959] 07334
Kondo, M. [1959] 07335
Kondo, M. [1961] 07340
Kondo, M. [1974] 07348
Kozlova, Z.I. [1940] 32604
Kozlova, Z.I. [1962] 07428
Kozlova, Z.I. [1964] 07429
Kozlova, Z.I. [1968] 07430
Kozlova, Z.I. [1969] 07434
Kreisel, G. [1959] 07515
Kruse, A.H. [1969] 22237
Kucera, A. [1985] 45308
Kunugui, K. [1937] 07642
Kunugui, K. [1939] 07643
Kunugui, K. [1940] 07644
Kunugui, K. [1940] 07645
Kuratowski, K. [1922] 07652
Kuratowski, K. [1926] 39828
Kuratowski, K. [1931] 07656
Kuratowski, K. [1931] 07658
Kuratowski, K. [1932] 07659
Kuratowski, K. [1933] 28504
Kuratowski, K. [1934] 07662
Kuratowski, K. [1936] 07664
Kuratowski, K. [1936] 20865
Kuratowski, K. [1937] 07665
Kuratowski, K. [1937] 07666
Kuratowski, K. [1937] 07667
Kuratowski, K. [1937] 07668
Kuratowski, K. [1937] 31806
Kuratowski, K. [1941] 31807
Kuratowski, K. [1948] 07669
Lavrentieff, M. [1925] 41584
Lebesgue, H. [1905] 27561
Lebesgue, H. [1907] 07905
Lindner, R. [1977] 26489
Louveau, A. [1974] 15025
Louveau, A. [1976] 18266
Louveau, A. [1976] 22934
Louveau, A. [1977] 27311
Louveau, A. [1977] 27317
Louveau, A. [1978] 52146
Louveau, A. [1978] 52697
Louveau, A. [1978] 75682
Louveau, A. [1979] 36514
Louveau, A. [1979] 75680
Louveau, A. [1980] 54569
Louveau, A. [1982] 36905
Louveau, A. [1982] 43141
Louveau, A. [1983] 45238
Louveau, A. [1985] 46388
Lunina, M.A. [1977] 75722
Luzin, N.N. [1914] 27672
Luzin, N.N. [1917] 27562
Luzin, N.N. [1918] 38036
Luzin, N.N. [1923] 27564
Luzin, N.N. [1925] 27563
Luzin, N.N. [1925] 41587
Luzin, N.N. [1925] 41588
Luzin, N.N. [1926] 41575
Luzin, N.N. [1927] 38628
Luzin, N.N. [1927] 41559
Luzin, N.N. [1929] 16811
Luzin, N.N. [1929] 39291
Luzin, N.N. [1929] 39294
Luzin, N.N. [1929] 39296
Luzin, N.N. [1930] 08417
Luzin, N.N. [1930] 23405
Luzin, N.N. [1930] 39463
Luzin, N.N. [1930] 39538
Luzin, N.N. [1931] 08418
Luzin, N.N. [1935] 08419
Luzin, N.N. [1935] 28418
Lyapunov, A.A. [1934] 43676
Lyapunov, A.A. [1935] 43678
Lyapunov, A.A. [1936] 08116
Lyapunov, A.A. [1937] 43679
Lyapunov, A.A. [1938] 08117
Lyapunov, A.A. [1939] 24922
Lyapunov, A.A. [1939] 43794
Lyapunov, A.A. [1947] 08119
Lyapunov, A.A. [1947] 08120
Lyapunov, A.A. [1947] 43688
Lyapunov, A.A. [1948] 08121
Lyapunov, A.A. [1948] 43710
Lyapunov, A.A. [1949] 08122
Lyapunov, A.A. [1950] 21021
Lyapunov, A.A. [1950] 21022
Lyapunov, A.A. [1950] 28543
Lyapunov, A.A. [1953] 08125
Lyapunov, A.A. [1953] 08126
Lyapunov, A.A. [1953] 08127
Lyapunov, A.A. [1953] 08128
Lyapunov, A.A. [1957] 08129
Lyapunov, A.A. [1963] 08130
Lyapunov, A.A. [1973] 17231
Lyapunov, A.A. [1973] 17359
Lyubetskij, V.A. [1970] 08210
Lyubetskij, V.A. [1971] 08211
Lyubetskij, V.A. [1976] 54901
Mac Gibbon, B. [1970] 19284
Magidor, M. [1980] 56487
Maitra, A. [1970] 08558
Maitra, A. [1970] 47389
Maitra, A. [1971] 08559
Maitra, A. [1974] 63591
Maitra, A. [1976] 26502
Maitra, A. [1982] 35460
Makkai, M. [1977] 23657
Mansfield, R. [1970] 28913
Mansfield, R. [1971] 08668
Mansfield, R. [1973] 08673
Mansfield, R. [1974] 08675
Mansfield, R. [1975] 08674
Mansfield, R. [1985] 45094
Martin-Loef, P. [1970] 20933
Martin, D.A. [1968] 08796
Martin, D.A. [1969] 08799
Martin, D.A. [1970] 08801
Martin, D.A. [1975] 23023

Martin, D.A. [1977] 27326
Martin, D.A. [1982] 33519
Martin, D.A. [1983] 45240
Martin, D.A. [1985] 46399
Mathias, A.R.D. [1972] 17477
Mathias, A.R.D. [1975] 17523
Mauldin, R.D. [1973] 08906
Mauldin, R.D. [1975] 17569
Maximoff, I. [1940] 08912
Mazurkiewicz, S. [1927] 08935
Menger, K. [1928] 31949
Miller, A.W. [1979] 53140
Miller, A.W. [1983] 35705
Miller, Douglas E. [1978] 33521
Moschovakis, Y.N. [1970] 22217
Moschovakis, Y.N. [1971] 09519
Moschovakis, Y.N. [1973] 26209
Moschovakis, Y.N. [1975] 64058
Moschovakis, Y.N. [1978] 52764
Moschovakis, Y.N. [1980] 76581
Moschovakis, Y.N. [1981] 55369
Moschovakis, Y.N. [1983] 43130
Mostowski, Andrzej [1956] 09560
Mostowski, Andrzej [1962] 33705
Motchane, L. [1956] 09593
Myers, D.L. [1976] 30751
Nikodym, O.M. [1929] 09964
Novikov, P.S. [1935] 10014
Novikov, P.S. [1937] 20900
Novikov, P.S. [1937] 45567
Novikov, P.S. [1939] 10015
Novikov, P.S. [1949] 10019
Novikov, P.S. [1951] 19475
Odifreddi, P. [1977] 23692
Ostaszewski, A.J. [1975] 10191
Paris, J.B. [1972] 19500
Popruzenko, J. [1932] 10656
Raisonnier, J. [1983] 34182
Raisonnier, J. [1984] 44806
Raisonnier, J. [1985] 48248
Rao, B.V. [1969] 16953
Rao, B.V. [1970] 10990
Rao, B.V. [1970] 10991
Rao, B.V. [1978] 29179
Ressayre, J.-P. [1977] 23658
Ribeiro Albuquerque, J. [1944] 00252
Ribeiro Albuquerque, J. [1945] 00253
Ribeiro Albuquerque, J. [1951] 00254
Ribeiro Albuquerque, J. [1952] 00255
Ribeiro Albuquerque, J. [1952] 00256
Ribeiro Albuquerque, J. [1952] 00401
Rogers, C.A. [1968] 11352
Rogers, C.A. [1973] 64880
Romanov, Yu.I. [1969] 11385
Rothberger, F. [1958] 11580
Sacks, G.E. [1969] 14481
Sacks, G.E. [1969] 16822
Sami, R.L. [1984] 42499
Sampei, Y. [1959] 11808
Sampei, Y. [1960] 11809
Sampei, Y. [1961] 11810
Sampei, Y. [1961] 11811
Sampei, Y. [1966] 19651
Sampei, Y. [1968] 11813
Sampei, Y. [1970] 11814
Scarpellini, B. [1965] 11879
Schilling, K. [1984] 44440
Scott, D.S. [1964] 11916
Selivanov, V.L. [1983] 44858
Selivanowski, E. [1927] 41560
Selivanowski, E. [1928] 39107
Shchegol'kov, E.A. [1948] 11891
Shchegol'kov, E.A. [1950] 11892
Shchegol'kov, E.A. [1959] 11894
Shchegol'kov, E.A. [1973] 65096
Shoenfield, J.R. [1961] 21160
Shoenfield, J.R. [1971] 12105
Sierpinski, W. [1918] 38030
Sierpinski, W. [1918] 38038
Sierpinski, W. [1918] 38039
Sierpinski, W. [1918] 38041
Sierpinski, W. [1919] 41759
Sierpinski, W. [1920] 41749
Sierpinski, W. [1924] 12156
Sierpinski, W. [1924] 12158
Sierpinski, W. [1924] 12159
Sierpinski, W. [1924] 12160
Sierpinski, W. [1925] 12161
Sierpinski, W. [1925] 41583
Sierpinski, W. [1925] 41589
Sierpinski, W. [1926] 12162
Sierpinski, W. [1926] 39827
Sierpinski, W. [1926] 41576
Sierpinski, W. [1927] 12164
Sierpinski, W. [1927] 12165
Sierpinski, W. [1927] 12166
Sierpinski, W. [1927] 12168
Sierpinski, W. [1927] 41557
Sierpinski, W. [1927] 41558
Sierpinski, W. [1928] 12169
Sierpinski, W. [1928] 12171
Sierpinski, W. [1928] 12172
Sierpinski, W. [1928] 12173
Sierpinski, W. [1928] 12174
Sierpinski, W. [1928] 39194
Sierpinski, W. [1928] 39825
Sierpinski, W. [1929] 12176
Sierpinski, W. [1929] 12178
Sierpinski, W. [1929] 39815
Sierpinski, W. [1930] 12183
Sierpinski, W. [1930] 12184
Sierpinski, W. [1931] 12185
Sierpinski, W. [1931] 12186
Sierpinski, W. [1931] 12187
Sierpinski, W. [1931] 12188
Sierpinski, W. [1931] 39830
Sierpinski, W. [1932] 12189
Sierpinski, W. [1934] 12195
Sierpinski, W. [1935] 12196
Sierpinski, W. [1935] 12197
Sierpinski, W. [1935] 12198
Sierpinski, W. [1937] 12200
Sierpinski, W. [1938] 12201
Sierpinski, W. [1948] 12225
Sierpinski, W. [1949] 12233
Sierpinski, W. [1949] 12239
Sierpinski, W. [1950] 21089
Sierpinski, W. [1951] 43590
Sierpinski, W. [1963] 12273
Sikorski, R. [1958] 12301
Sikorski, R. [1966] 12315
Silver, J.H. [1970] 12317
Silver, J.H. [1971] 19669
Silver, J.H. [1980] 78616
Simpson, S.G. [1977] 56635
Simpson, S.G. [1982] 38266
Sodnomov, B.S. [1955] 12627
Sodnomov, B.S. [1956] 45667
Solovay, R.M. [1969] 22230
Solovay, R.M. [1978] 52399
Souslin, M. [1917] 38033
Srebrny, M. [1978] 53024
Srebrny, M. [1979] 78903
Staiger, L. [1975] 65959
Steel, J.R. [1980] 54519
Steel, J.R. [1980] 54568
Steel, J.R. [1981] 33283
Steel, J.R. [1981] 35650
Steel, J.R. [1983] 45242
Stern, J. [1973] 13176
Stern, J. [1975] 13180
Stern, J. [1977] 27298
Stern, J. [1977] 27300
Stern, J. [1978] 31515
Stern, J. [1978] 56713
Stern, J. [1979] 36138
Stern, J. [1980] 79002
Suzuki, Y. [1963] 33698
Szpilrajn, E. [1933] 12855
Tanaka, H. [1966] 13250
Tanaka, H. [1967] 13361
Tanaka, H. [1967] 31270
Tanaka, H. [1970] 13362
Tanaka, H. [1970] 13363
Tanaka, H. [1970] 13364
Tanaka, H. [1971] 79236
Tanaka, H. [1972] 13366
Tanaka, H. [1973] 13367
Tang, A. [1981] 46091
Tarski, A. [1931] 16924
Telgarsky, R. [1977] 32018
Uspenskij, V.A. [1985] 47706
Vaught, R.L. [1973] 13864
Vaught, R.L. [1973] 13909
Vaught, R.L. [1974] 17478
Wagner, K. [1977] 26477
Watanabe, H. [1950] 14087
Weiss, T. [1983] 43895
Weitkamp, G. [1982] 37577
Weitkamp, G. [1985] 42291
Welch, P. [1984] 45409
Wesep van, R.A. [1978] 29243
Wesep van, R.A. [1978] 52457
Willard, S. [1971] 14209
Willmott, R. [1969] 14226
Willmott, R. [1971] 14227
Willmott, R. [1975] 79991
Wisniewski, K. [1979] 53128
Wolfe, P. [1955] 14269
Yankov, V.A. [1941] 24916
Yasuda, Y. [1976] 33404
Yasuda, Y. [1981] 33409
Yasuda, Y. [1981] 37193
Zahorski, Z. [1948] 18091
Zalcwasser, Z. [1922] 14343

E20

Banaschewski, B. [1958] 00757
Chimev, K.N. [1984] 49515
Chimev, K.N. [1984] 49516
Emde Boas van, P. [1973] 29008

Misc. Indexes

Jarnik, V. [1938] 41088
Kalmar, L. [1957] 06863
Kruse, A.H. [1969] 22237
Lambert, W.M. [1979] 55239
Leiss, E. [1979] 53757
Lyapunov, A.A. [1939] 43794
Robinson, Julia [1969] 11302
Rogers, C.A. [1973] 64880
Sierpinski, W. [1949] 12239
Wechsung, G. [1973] 42614

E25

Adamowicz, Z. [1983] 43740
Baire, R. [1905] 05473
Bradford, R. [1971] 01574
Ebbinghaus, H.-D. [1976] 50795
Ellentuck, E. [1978] 51747
Feferman, S. [1965] 03696
Friedman, H.M. [1970] 14546
Fuerer, M. [1976] 61859
Jensen, R.B. [1967] 06596
Kanovej, V.G. [1979] 56024
Kiselev, A.A. [1978] 32588
Kruse, A.H. [1960] 07579
Kruse, A.H. [1962] 07582
Kuratowski, K. [1922] 07652
Lebesgue, H. [1907] 07905
Litman, A. [1976] 18261
Loefgren, L. [1966] 27537
Moschovakis, Y.N. [1970] 22217
Platek, R.A. [1969] 10526
Sacks, G.E. [1967] 16821
Sacks, G.E. [1969] 14481
Sacks, G.E. [1969] 16822

E30

Barwise, J. [1974] 00911
Barwise, J. [1975] 31017
Barwise, J. [1975] 60316
Chauvin, A. [1961] 01999
Chauvin, A. [1961] 25473
Collins, G.E. [1970] 02332
Deutsch, M. [1975] 61336
Ershov, Yu.L. [1983] 40494
Feferman, S. [1978] 72806
Friedman, S.D. [1979] 56139
Griffor, E.R. [1984] 44080
Hilbert, D. [1900] 06078
Hilbert, D. [1926] 45196
Menger, K. [1928] 31949
Mostowski, Andrzej [1953] 09553
Rosser, J.B. [1955] 23483
Shinoda, J. [1984] 42934
Trakhtenbrot, B.A. [1956] 19739
Wang, Hao [1965] 14057
Zermelo, E. [1931] 14412

E35

Abraham, U. [1984] 40744
Adamowicz, Z. [1976] 14761
Adamowicz, Z. [1977] 51342
Adamowicz, Z. [1977] 51857
Addison, J.W. [1957] 29378
Barwise, J. [1975] 31017
Becker, J.A. [1980] 54459
Budinas, B.L. [1979] 54375
Budinas, B.L. [1980] 54415
Budinas, B.L. [1980] 55483
Budinas, B.L. [1981] 35767
Budinas, B.L. [1981] 71383
Budinas, B.L. [1982] 34305
Budinas, B.L. [1983] 47459
Bukovsky, L. [1964] 01714
David, R. [1978] 29140
David, R. [1982] 34808
Dawson Jr., J.W. [1973] 02850
Devlin, K.J. [1973] 17243
Devlin, K.J. [1977] 27306
Feferman, S. [1965] 03696
Fraisse, R. [1975] 61773
Fremlin, D.H. [1983] 38382
Friedman, H.M. [1970] 03773
Griffor, E.R. [1984] 44080
Groszek, M.J. [1983] 37641
Gurevich, Y. [1983] 33764
Guzicki, W. [1973] 30498
Harrington, L.A. [1976] 14795
Harrington, L.A. [1977] 24274
Hodges, W. [1976] 30711
Hrbacek, K. [1980] 74290
Jensen, R.B. [1967] 06596
Jensen, R.B. [1970] 22245
Jensen, R.B. [1970] 28527
Jensen, R.B. [1974] 62767
Kanovej, V.G. [1975] 50893
Kanovej, V.G. [1979] 54105
Kanovej, V.G. [1979] 56024
Kanovej, V.G. [1982] 34181
Kanovej, V.G. [1983] 37518
Kechris, A.S. [1985] 45614
Khakhanyan, V.Kh. [1980] 56495
Khakhanyan, V.Kh. [1980] 73754
Khakhanyan, V.Kh. [1981] 36853
Khakhanyan, V.Kh. [1983] 41114
Levy, A. [1970] 20962
Louveau, A. [1976] 22934
Lyubetskij, V.A. [1970] 08210
Lyubetskij, V.A. [1971] 08211
Lyubetskij, V.A. [1976] 54901
Manin, Yu.I. [1977] 51984
Miller, A.W. [1979] 53140
Myers, D.L. [1976] 30751
Novikov, P.S. [1949] 10019
Novikov, P.S. [1951] 19475
Sacks, G.E. [1967] 16821
Sacks, G.E. [1969] 14481
Sacks, G.E. [1969] 16822
Sampei, Y. [1970] 11814
Silver, J.H. [1971] 19669
Skandalis, K. [1977] 26555
Slaman, T.A. [1983] 40710
Smolska-Adamowicz, Z. [1977] 52201
Sodnomov, B.S. [1955] 12627
Sodnomov, B.S. [1956] 45667
Stern, J. [1973] 13176
Stern, J. [1977] 27298
Stern, J. [1977] 27300
Takeuti, G. [1965] 13330
Welch, P. [1985] 42522

E40

Abraham, U. [1984] 40744
Abramson, F.G. [1979] 31649
Adamowicz, Z. [1975] 27206
Adamowicz, Z. [1983] 43740
Adamowicz, Z. [1984] 45774
Adamowicz, Z. [1984] 45775
Ambos-Spies, K. [1984] 45088
Blass, A.R. [1975] 03910
Chong, C.T. [1977] 31058
Chong, C.T. [1979] 53211
David, R. [1982] 35373
David, R. [1982] 37544
David, R. [1985] 46391
Farrington, Patrick [1982] 36949
Feferman, S. [1965] 03696
Fenstad, J.E. [1974] 03747
Friedman, H.M. [1971] 04649
Friedman, H.M. [1973] 21340
Friedman, S.D. [1982] 33350
Friedman, S.D. [1985] 45301
Griffor, E.R. [1985] 49429
Grigorieff, S. [1975] 18177
Hajek, P. [1974] 05528
Hinman, P.G. [1969] 06104
Hodes, H.T. [1980] 56067
Hodes, H.T. [1981] 33271
Jensen, R.B. [1967] 06596
Jockusch Jr., C.G. [1976] 18212
Jockusch Jr., C.G. [1980] 54367
Jockusch Jr., C.G. [1985] 45306
Kanovej, V.G. [1974] 17445
Kanovej, V.G. [1975] 62865
Kanovej, V.G. [1979] 53375
Kechris, A.S. [1978] 52351
Kechris, A.S. [1981] 34923
Kurtz, S.A. [1983] 33598
Maass, W. [1982] 33604
Mansfield, R. [1973] 08673
Mansfield, R. [1974] 08675
Nerode, A. [1985] 45310
Normann, D. [1975] 28312
Odifreddi, P. [1983] 33611
Odifreddi, P. [1983] 38171
Roguski, S. [1974] 11367
Sacks, G.E. [1967] 16821
Sacks, G.E. [1969] 14481
Sacks, G.E. [1969] 16822
Sacks, G.E. [1971] 11750
Sacks, G.E. [1971] 11752
Sampei, Y. [1968] 11813
Selman, A.L. [1972] 11963
Shinoda, J. [1985] 45312
Shore, R.A. [1974] 19679
Shore, R.A. [1982] 33615
Silver, J.H. [1980] 78616
Slaman, T.A. [1985] 47710
Solovay, R.M. [1969] 22230
Solovay, R.M. [1978] 52893
Srebrny, M. [1977] 26534
Stark, W.R. [1978] 52964
Steel, J.R. [1978] 29156
Stern, J. [1973] 13176
Thomason, S.K. [1967] 13585
Thomason, S.K. [1969] 13586
Thomason, S.K. [1970] 13588
Weitkamp, G. [1985] 42291

Yasuda, Y. [1981] 33409
Yasuda, Y. [1981] 37193

E45

Abraham, U. [1984] 40744
Abramson, F.G. [1979] 53788
Adamowicz, Z. [1975] 27206
Adamowicz, Z. [1976] 14761
Adamowicz, Z. [1977] 51342
Adamowicz, Z. [1977] 51857
Adamowicz, Z. [1983] 43740
Adamowicz, Z. [1984] 45774
Addison, J.W. [1957] 29378
Addison, J.W. [1958] 00171
Addison, J.W. [1959] 00173
Apt, K.R. [1974] 00432
Apt, K.R. [1976] 14789
Balcar, B. [1978] 53078
Barwise, J. [1970] 00828
Barwise, J. [1971] 00836
Barwise, J. [1975] 31017
Becker, H. [1978] 52699
Becker, H. [1980] 54180
Blass, A.R. [1975] 03910
Boffa, M. [1974] 31022
Boolos, G. [1968] 01432
Boos, W. [1975] 31627
Budinas, B.L. [1979] 54375
Budinas, B.L. [1980] 55483
Budinas, B.L. [1981] 35767
Budinas, B.L. [1981] 71383
Budinas, B.L. [1982] 34305
Burgess, J.P. [1977] 28149
Burgess, J.P. [1979] 71463
Cenzer, D. [1974] 04278
Chauvin, A. [1962] 02061
Chong, C.T. [1977] 31058
Chong, C.T. [1979] 36886
Chong, C.T. [1979] 54747
Chong, C.T. [1980] 55235
Chong, C.T. [1982] 35280
Cutland, N.J. [1980] 55735
David, R. [1978] 29140
David, R. [1982] 34808
David, R. [1982] 35373
David, R. [1982] 37544
David, R. [1985] 46391
Dawson Jr., J.W. [1973] 02850
Devlin, K.J. [1973] 17243
Devlin, K.J. [1974] 15073
Devlin, K.J. [1975] 16656
Devlin, K.J. [1977] 27306
Farrington, Paddy [1984] 42273
Farrington, Patrick [1982] 36949
Farrington, Patrick [1982] 36950
Farrington, Patrick [1983] 37393
Feferman, S. [1965] 03696
Fraisse, R. [1975] 61773
Friedman, H.M. [1970] 03773
Friedman, H.M. [1971] 04649
Friedman, H.M. [1974] 04658
Friedman, H.M. [1974] 18148
Friedman, S.D. [1981] 33348
Friedman, S.D. [1985] 45301
Friedman, S.D. [1985] 46395
Gostanian, R. [1979] 73542
Gostanian, R. [1979] 73545
Griffor, E.R. [1983] 43126
Griffor, E.R. [1984] 42453
Grigorieff, S. [1975] 18177
Guaspari, D. [1974] 05420
Guaspari, D. [1974] 05421
Guaspari, D. [1980] 54183
Guzicki, W. [1973] 30498
Hajek, P. [1974] 05528
Hajek, P. [1978] 53440
Harrington, L.A. [1975] 05665
Harrington, L.A. [1975] 17524
Harrington, L.A. [1976] 14795
Harrington, L.A. [1977] 26528
Harrington, L.A. [1978] 52767
Hodes, H.T. [1980] 56067
Hodes, H.T. [1981] 33271
Hodes, H.T. [1984] 42462
Hrbacek, K. [1978] 52894
Hrbacek, K. [1980] 74290
Hrbacek, K. [1983] 41103
Jensen, R.B. [1967] 06596
Jensen, R.B. [1970] 22245
Jensen, R.B. [1970] 28527
Jensen, R.B. [1971] 06600
Jensen, R.B. [1974] 62767
Jockusch Jr., C.G. [1976] 18212
Kanovej, V.G. [1974] 17445
Kanovej, V.G. [1975] 62865
Kanovej, V.G. [1979] 54105
Kaufmann, M. [1984] 41225
Kechris, A.S. [1972] 06986
Kechris, A.S. [1973] 21379
Kechris, A.S. [1975] 06990
Kechris, A.S. [1977] 53258
Kechris, A.S. [1978] 30726
Kechris, A.S. [1978] 52698
Kechris, A.S. [1978] 74744
Kechris, A.S. [1981] 35283
Kechris, A.S. [1983] 40468
Kechris, A.S. [1985] 40477
Kechris, A.S. [1985] 45614
Kiselev, A.A. [1973] 32586
Kiselev, A.A. [1973] 32587
Kiselev, A.A. [1978] 32588
Kozlova, Z.I. [1964] 07429
Kranakis, E. [1982] 35456
Kranakis, E. [1982] 35649
Kranakis, E. [1983] 38326
Kranakis, E. [1984] 42470
Kranakis, E. [1984] 45390
Kranakis, E. [1985] 47562
Kreisel, G. [1960] 07519
Leeds, S. [1971] 07967
Lerman, M. [1978] 29152
Levy, A. [1970] 20962
Lucian, M. [1972] 28788
Lukas, J.D. [1974] 08387
Lyubetskij, V.A. [1971] 08211
Maass, W. [1978] 29260
Maass, W. [1978] 31877
Maass, W. [1978] 31878
Machover, M. [1961] 08457
Maitra, A. [1970] 47389
Mansfield, R. [1970] 28913
Mansfield, R. [1973] 08673
Mansfield, R. [1974] 08675
Mansfield, R. [1975] 08674
Martin, D.A. [1969] 08799
Moschovakis, Y.N. [1981] 55369
Mostowski, Andrzej [1956] 09560
Normann, D. [1979] 53435
Novikov, P.S. [1949] 10019
Novikov, P.S. [1951] 19475
Platek, R.A. [1969] 10526
Poss, R.L. [1970] 10678
Putnam, H. [1963] 10837
Richter, W.H. [1975] 23962
Roguski, S. [1974] 11367
Sacks, G.E. [1971] 11750
Sami, R.L. [1984] 42499
Sampei, Y. [1966] 19651
Sampei, Y. [1968] 11813
Sampei, Y. [1970] 11814
Scarpellini, B. [1965] 11879
Schmerl, J.H. [1981] 55178
Shilleto, J.R. [1972] 29513
Shinoda, J. [1984] 42934
Shinoda, J. [1985] 45312
Shoenfield, J.R. [1961] 21160
Shoenfield, J.R. [1971] 12105
Shore, R.A. [1977] 27323
Silver, J.H. [1971] 12319
Silver, J.H. [1971] 19669
Skandalis, K. [1977] 26555
Slaman, T.A. [1985] 45317
Slaman, T.A. [1985] 46378
Slaman, T.A. [1985] 47710
Smolska-Adamowicz, Z. [1977] 52201
Sodnomov, B.S. [1955] 12627
Sodnomov, B.S. [1956] 45667
Solovay, R.M. [1967] 12640
Solovay, R.M. [1969] 22230
Solovay, R.M. [1978] 52399
Solovay, R.M. [1978] 52893
Srebrny, M. [1977] 26534
Srebrny, M. [1978] 53024
Srebrny, M. [1979] 78903
Steel, J.R. [1982] 34562
Stern, J. [1973] 13176
Takahashi, Moto-o [1968] 13293
Takeuti, G. [1965] 13329
Takeuti, G. [1965] 13330
Tanaka, H. [1970] 13362
Tanaka, H. [1970] 13364
Truss, J.K. [1978] 52310
Vetulani, Z. [1984] 44142
Wainer, S.S. [1975] 29611
Wang, Hao [1957] 29384
Welch, P. [1985] 42522

E47

Abramson, F.G. [1979] 53788
Aczel, P. [1974] 04129
Baeten, J. [1984] 45395
Barwise, J. [1971] 00836
Barwise, J. [1975] 31017
Budinas, B.L. [1981] 35767
Budinas, B.L. [1981] 71383
Chang, C.C. [1970] 01977
Chong, C.T. [1979] 54747
Deutsch, M. [1976] 28192
Deutsch, M. [1977] 28193
Devlin, K.J. [1974] 15073
Feldman, E.D. [1975] 04234

Misc. Indexes

Transliteration scheme for Cyrillic

Author names and titles originally in *Cyrillic* have been transliterated into the Roman alphabet using the following scheme. (It is the same as the scheme curently used by Zbl and differs only slightly from that used by MR.)

Cyrillic		Roman
а	А	a
б	Б	b
в	В	v
г	Г	g
д	Д	d
е(ё)	Е(Ё)	e
ж	Ж	zh
з	З	z
и	И	i
й	Й	j
к	К	k

Cyrillic		Roman
л	Л	l
м	М	m
н	Н	n
о	О	o
п	П	p
р	Р	r
с	С	s
т	Т	t
у	У	u
ф	Ф	f
х	Х	kh

Cyrillic		Roman
ц	Ц	ts
ч	Ч	ch
ш	Ш	sh
щ	Щ	shch
ъ	Ъ	”
ы	Ы	y
ь	Ь	’
э	Э	eh
ю	Ю	yu
я	Я	ya